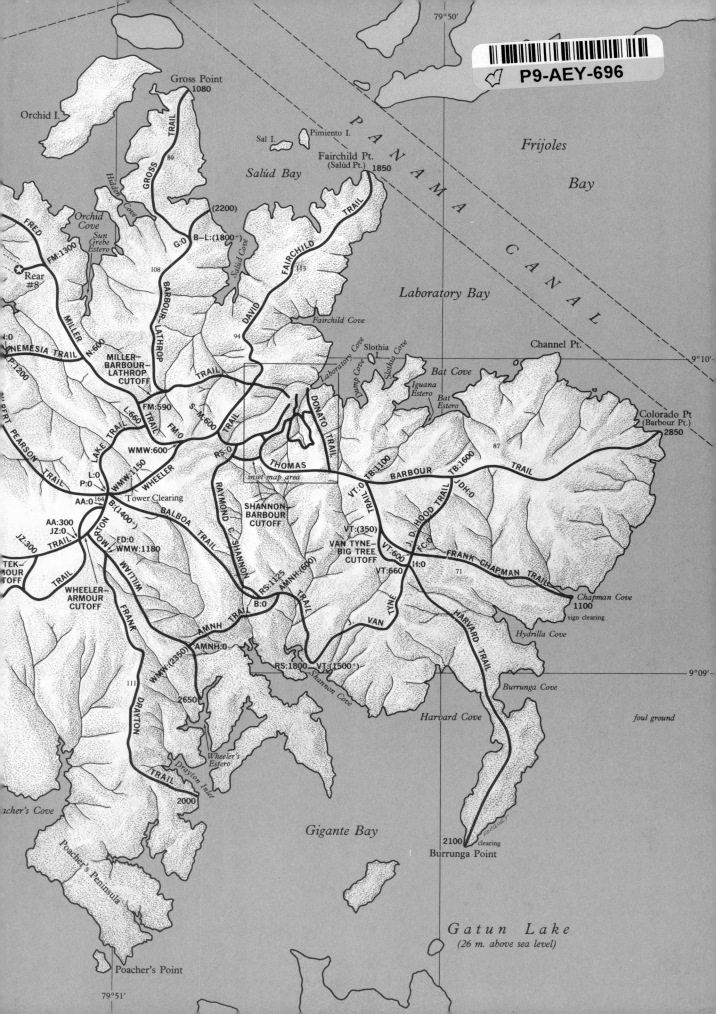

Flora of Barro Colorado Island

Flora of

Barro Colorado Island

THOMAS B. CROAT

STANFORD UNIVERSITY PRESS, STANFORD, CALIFORNIA 1978

The material incorporated in this work was prepared with the support of National Science Foundation Grant No. GB–34502. However, any opinions, findings, conclusions, or recommendations expressed herein are those of the author and do not necessarily reflect the views of NSF.

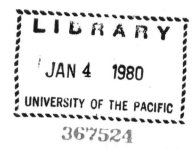

To Patricia,
whose skilled assistance contributed to
every phase of this work

Preface

This revision of the *Flora of Barro Colorado Island* was planned in 1967. The idea was inspired by Dr. Walter H. Lewis, who, in consultation with Dr. Martin Moynihan and other members of the staff of the Smithsonian Tropical Research Institute in Panama, came to feel that the project would be a worthy undertaking. The purpose of the new flora is obvious: because the island has been and will continue to be an area of intense scientific activity, a more complete and up-to-date flora is essential. The only existing description of the vegetation was Paul Standley's much out-of-date and incomplete annotated checklist. The present work is intended primarily for the student of the Barro Colorado Island flora and for others doing scientific work on the island or the isthmus. Barro Colorado Island is one of the few remaining undisturbed and protected areas where sophisticated research on tropical biology can be pursued, and it will become even more important as the alteration or disappearance of natural habitats progresses throughout the tropics.

The preparation of this book has been jointly sponsored by the Missouri Botanical Garden and the Smithsonian Tropical Research Institute. Principal financial support has come from the Missouri Botanical Garden and the National Science Foundation (Grant GB-34502 for research, Grant DEB75–14735 for publication). Special recognition must be given here to Peter H. Raven, who upon his arrival in St. Louis as director of the Missouri Botanical Garden helped me obtain this NSF funding. Substantial support during various phases of the project has come also from the Center for the Study of the Biology of Natural Systems of Washington University, St. Louis.

This work would not have been possible without the excellent herbarium, library, and greenhouses of the Missouri Botanical Garden, where the major portion of the work was done. The quality of the descriptions and ecological commentary was greatly augmented by the many months spent in field work on Barro Colorado Island. I am very grateful to Dr. Moynihan, then director, and the staff of the Smithsonian Tropical Research Institute and to workers at the research station on the island for their assistance. The late Mr. Jay Hayden,

the station's manager during most of the years when I was working there, deserves special praise for his helpfulness. Thanks are also accorded to Dr. A. Standley Rand, staff resident on the island during several of those years, and to Dr. Robert L. Dressler, who helped in numerous ways in his capacity as staff botanist at the Institute. His experience with the island's flora and, especially, his knowledge of the orchids were essential to my work.

The Panama Canal Company assisted the writing of this flora by providing facilities for Summit Herbarium at Summit Garden in the Canal Zone, where I lived and worked from March 1970 to October 1971. I want to thank Mr. Dudley Jones, the Garden's manager, and his staff for being so helpful to me and my family during that time and for providing valuable slat-house space for growing and observing plants. Thanks go also to Roy Sharp and A. I. Bauman of the Community Services Division of the Panama Canal Company, to Walter Lewis of the Missouri Botanical Garden, and to Edwin Tyson of Florida State University for their support in establishing Summit Herbarium. The position at Summit Herbarium provided me with an uninterrupted stay in the Canal Zone that allowed the study of phenological variations in flowering and fruiting.

Dr. Robin Foster's assistance with finding new plants and understanding the ecology of the flora was most valuable. Robin's intense interest and expertise, developed over the course of several years while he was conducting ecological studies on the island, led him to discover many plants new to the island. Because his studies required him to make weekly surveys in the forest, he became very familiar with individual trees. Frequently when I returned with him to climb and collect interesting plants he had found, they proved to be new to the flora. I am particularly grateful to him for help with many ecological matters both during his stay in Panama and later when he returned to posts at Duke University and the University of Chicago.

A number of other ecologists have helped by contributing ecological information and by reviewing portions of the manuscript. Among these I would especially like to thank Herbert Baker, Kamaljit Bawa, Ted Fleming,

Gordon Frankie, Gary Hartshorn, Ray Heithaus, Dan Janzen, Gene Montgomery, Gene Morton, Paul Opler, John Oppenheimer, L. van der Pijl, Stan Rand, Mike Robinson, Christopher Smith, Neal Smith, and Nicholas Smythe.

I am also indebted to the following botanists for their special help with taxonomic problems:

C. K. Allen, Lauraceae
W. R. Anderson, Malpighiaceae
D. A. Austin, Convolvulaceae
V. M. Badillo, Caricaceae
R. C. Barneby, Lauraceae, Menispermaceae
C. C. Berg, Moraceae
L. A. Bernardi, Lauraceae
M. R. Birdsey, Araceae (*Syngonium*)
G. S. Bunting, Araceae
D. G. Burch, Euphorbiaceae (*Chamaesyce*)
W. C. Burger, Moraceae, Piperaceae, Urticaceae
E. L. Core, Cyperaceae (*Scleria*)
R. S. Cowan, Leguminosae
C. L. Cristobal, Sterculiaceae (*Byttneria*)
J. Cuatrecasas, Compositae, Malpighiaceae, Moraceae (*Cecropia*)
G. S. Daniels, Musaceae (*Heliconia*)
W. G. D'Arcy, Solanaceae, Compositae, and all manner of nomenclatural problems
G. Davidse, Gramineae
J. V. A. Dieterle, Cucurbitaceae
R. L. Dressler, Orchidaceae and numerous general identifications
L. H. Durkee, Acanthaceae
J. D. Dwyer, Rubiaceae and many general identifications
T. S. Elias, Leguminosae (*Inga*)
E. Forero, Connaraceae (*Rourea*)
J. A. Fowlie, Orchidaceae (*Chysis*)
A. H. Gentry, Bignoniaceae
W. L. Handlos, Pontederiaceae (*Pontederia*)
G. Harling, Cyclanthaceae
A. Hladik, Araliaceae
R. A. Howard, Polygonaceae (*Coccoloba*)
H. H. Iltis, Capparidaceae
H. S. Irwin, Leguminosae (*Cassia*)
C. E. Jeffrey, Cucurbitaceae
M. C. Johnston, Rhamnaceae (*Colubrina*)
H. Kennedy, Marantaceae
M. Kimnack, Cactaceae (*Epiphyllum*)
R. M. King, Compositae
T. M. Koyama, Cyperaceae
B. A. Krukoff, Leguminosae (*Erythrina*), Loganiaceae (*Strychnos*), Menispermaceae
J. Kuijt, Loranthaceae
D. B. Lellinger, all ferns (special thanks for a supreme effort)
C. L. Lundell, Celastraceae, Myrsinaceae
P. J. M. Maas, Zingiberaceae
B. Maguire, Guttiferae
M. Mathias, Umbelliferae
E. McClintock, Saxifragaceae (*Hydrangea*)
R. McVaugh, Myrtaceae

J. T. Mickel, ferns
H. N. Moldenke, Verbenaceae
H. E. Moore, Jr., Palmae
S. Mori, Lecythidaceae
D. Nicolson, Araceae (especially nomenclature)
J. W. Nowicke, Boraginaceae, Labiatae
R. W. Pohl, Gramineae
D. M. Porter, Burseraceae, Rutaceae
G. T. Prance, Chrysobalanaceae
W. Ramirez B., Moraceae (*Ficus*)
P. H. Raven, Onagraceae
K. E. Roe, Solanaceae (*Solanum*)
K. E. Rogers, Gramineae (*Ichnanthus*)
V. E. Rudd, Leguminosae (especially *Machaerium*)
C. E. Schnell, Melastomataceae (*Conostegia*)
B. G. Schubert, Leguminosae (*Desmodium*)
H. Sleumer, Flacourtiaceae
A. R. Smith, Polypodiaceae (*Thelypteris*)
C. E. Smith, Meliaceae (*Trichilia*)
L. B. Smith, Bromeliaceae
R. R. Smith, Musaceae (*Heliconia*)
T. R. Soderstrom, Gramineae
D. Spellman, Asclepiadaceae
J. A. Steyermark, Rubiaceae
M. T. Stieber, Gramineae (*Ichnanthus*)
P. G. Taylor, Lentibulariaceae
J. H. Thomas, Scrophulariaceae
H. P. Traub, Amaryllidaceae
D. C. Wasshausen, Acanthaceae
G. L. Webster, Euphorbiaceae
F. White, Ebenaceae
H. Wiehler, Gesneriaceae
R. L. Wilbur, Campanulaceae
L. O. Williams, Convolvulaceae, Rubiaceae
J. J. Wurdack, Melastomataceae

Most of the library work for this flora was done at the Missouri Botanical Garden Library, and I am grateful to the librarians, George B. van Schaack and Jim Reed, and especially to Mrs. Carla Lange for her great assistance in finding books and in helping with translations. Dr. W. G. D'Arcy has also been very helpful with problems involving nomenclature and literature sources.

In addition to the herbarium at the Missouri Botanical Garden I have relied heavily on collections at the Gray Herbarium and the New York Botanical Garden and especially the Field Museum and the U.S. National Herbarium. During my numerous visits, the staffs of these institutions were always helpful and I am indebted to them.

Mr. Philip Busey assisted me on the project for the year 1972-73 and Mr. Jim Conrad was my assistant from 1973 until the completion of the project in mid-1974. Their help is gratefully acknowledged. A special debt of thanks is owed to my wife, Patricia, who assisted in almost every facet of the work. I thank her for her invaluable assistance and understanding.

T.B.C.

St. Louis, Missouri
October 1977

Contents

INTRODUCTION

General Climatic Features, 3. Geology and Soil Types, 5.
General Characteristics of the Vegetation, 6. Types of Natural Vegetation, 6.
Habit-and-Habitat Classes, 17. Growth Forms, 17. Floristic Composition, 22.
Sexual Characteristics, 24. Geographical Affinities, 26. Historical and Recent
Changes in the Flora, 28. Phenological Characteristics, 29. History of Panama
and the Canal Zone, 49. History of Botanical Studies, 49. Key Structure, 53.
Family Descriptions, 54. Genus Descriptions, 54. Species Descriptions, 54.
Species Excluded, 61. Maps, 61. Classification and Family Sequence, 61.

THE FLORA

REFERENCE MATERIAL

Flora of Barro Colorado Island

Introduction

Barro Colorado is an island in the Panama Canal Zone lying midway between the Atlantic and Pacific Oceans (9°09′N, 79°51′W). At about 15.6 km² (6 square miles), BCI is the largest island in Gatun Lake, the lake formed between 1911 and 1914 by the damming of the Río Chagres to form the Canal. Gatun Lake, by far the largest expanse of water in the Canal, covers 420 km² (164 square miles) at an elevation of about 25 m (85 feet) above sea level. The ship channel traversing the lake passes along the eastern and northeastern shores of the island, and waves from the passing ships cause substantial erosion along parts of these shores, especially during the dry season, when the trade winds are more forceful.

BCI was set aside as a biological preserve in 1923 and is currently supervised by the Smithsonian Tropical Research Institute, which operates a modern field station at the Laboratory Clearing, on the northeast shore. Aside from this modest little settlement, there is very little that interrupts the dense forest cover: a dozen-odd smaller clearings, some of them used for navigational beacons; a network of trails crossing most of the island; a scattering of tree falls; a few shoreline marshes and silted coves; and an occasional ravine or streambed. And except for a small plateau in the west-central part of the island, the terrain is hilly, the hills dissected by ravines and generally well-drained slopes. The highest point, the Tower Clearing, is 165 m (538 feet) above sea level, 140 m (453 feet) above the level of the lake.

The shoreline, deeply dissected by coves over much of its extent, is altogether about 65 km (40 miles) long, including the shores of its associated islands. The principal island associated with BCI (see the map on the endsheets) is Orchid Island, which is separated from Gross Point by a very narrow passage. Others that are considered part of the biological preserve are Slothia Isle, in Laboratory Bay; Ormosia Isle, near the Front Lighthouse Clearing; and Sal and Pimiento Islands, a pair of small islands northeast of Fairchild Point.

General Climatic Features

Under the Köppen system of climatic classifications, BCI's climate is Am, or Tropical Monsoon Climate.

Annual rainfall on BCI (see Graph 1) ranges from 190 to 360 cm (76 to 143 inches); between 1924 and 1962 it averaged 275 cm (107.3 inches). This compares with an average 328 cm (128 inches) at Colón, on the Atlantic Coast of the Canal Zone, and 177 cm (68 inches) at Balboa, on the Pacific Coast, during the same period. The climate is markedly seasonal, with a sharp dry season usually starting in mid-December and continuing until about the beginning of May. During the dry season, only 18–26 cm (7–10 inches) of rain fall.

The dry season, in fact, sees considerable leaf litter accumulating on the forest floor, and a great deal more light reaches the forest floor than during the rainy season. Most of the steep, rapidly flowing streams dry up completely, but many form small pockets of water that can survive the dry season. What rain does fall during the dry season rapidly disappears, with very little or no runoff. Even after the rains begin, usually in May, less than 2% of the water may leave as runoff. However, by late in the rainy season, in December, nearly 85% of the water falling as rain leaves as runoff (Rubinoff, 1974).

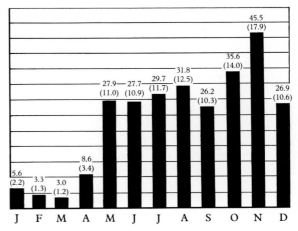

Graph 1. Rainfall on BCI, in cm (and inches). (Source: Panama Canal Company)

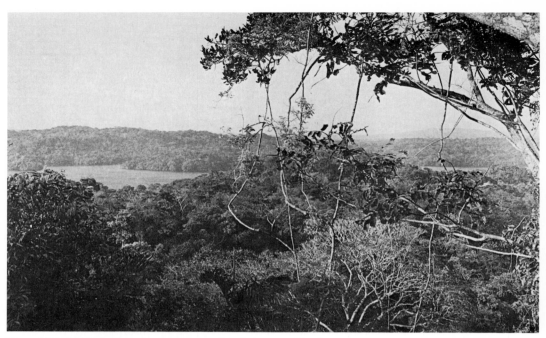

View of Fairchild Point, from the canopy west of the Laboratory Clearing

The dock and laboratory

TABLE 1

Mean relative humidity by month, at 1200 hours, 1973, on BCI

(in percent)

| Month | Mean relative humidity | |
	Clearing	Forest floor
January	78.5%	85.6%
February	72.2	80.6
March	65.9	80.2
April	62.1	76.4
May	63.8	84.0
June	74.5	92.5
July	65.2	92.1
August	66.5	91.8
September	68.0	92.2
October	70.4	92.4
November	76.6	94.2
December	69.3	87.8

SOURCE: Rubinoff, 1974.

In any given month, average relative humidity is appreciably higher in the forest than in such open areas as the Laboratory Clearing. The drop in relative humidity at the onset of the dry season is also less rapid in the forest than in clearings (Rubinoff, 1974). The mean monthly relative humidity for 1973 is presented in Table 1.

Temperature measurements in the forest and in the Laboratory Clearing throughout a single year (1973) are shown in Table 2. The atmospheric temperature may vary from as low as 16.5°C (61.7°F) to as high as 35.5°C (95.9°F), with the lowest temperatures being recorded within the forest during the rainy season and the highest at the Laboratory Clearing in the dry season. With rare exceptions the temperature ranges between 21 and 32°C (70 and 90°F) throughout the year, and the average ambient temperature in the Laboratory Clearing is 27°C (77°F). The seasonal variation in monthly averages is just 2.2°C. In the dry season the range of the monthly averages of diurnal temperature is 9–11°C, and during the rainy season it is 8–9°C. Thus the diurnal temperature variation is greater than the variation between seasons. Temperature on the forest floor is especially constant, the range of the average diurnal temperature being about 6.1°C.

The first three months of the dry season offer the most hours of sunshine, July and August the least (Foster, 1974). During most of the rainy season, and particularly when skies are overcast, conditions are very dark on the forest floor and visibility is poor.

Winds are about twice as forceful at midday in the dry season as at the same time in the rainy season. Measurements of wind velocity for 1973 showed the rainy-season months having an average of about 4.9 km (3 miles) per hour, whereas for the dry season the average was about 9.8 km (6 miles) per hour (Rubinoff, 1974). Wind velocity is always substantially higher at midday than in the morning or evening.

Geology and Soil Types

Much of the present area of Central America was sea floor during early times, and land emerged only during the Oligocene, as a number of islands. In the middle Miocene there was a general uplifting as well as much volcanic activity in the central and western portions of Panama. The final retreat of the isthmus from the sea occurred in Darién during the upper Miocene and Pliocene, although there were later movements of the coastline in the west, in Chiriquí (Torre, 1965). The final uplift could have been in the late Pliocene, "and continuous land connections have existed for three to five million years" (Graham, 1972). Recent data (Emiliana et al., 1972) indicate that the final closing of the isthmus occurred 5.7 million years ago.

The important parent rocks in the Canal Zone are of two kinds—sedimentary rocks, including limestone,

TABLE 2

Atmospheric temperature on BCI, 1973

(degrees C)

| Month | Clearing temperature | | | | | | Forest floor temperature | | | | | |
| | Maximum | | | Minimum | | | Maximum | | | Minimum | | |
	Mean	High	Low	Mean	High	Low	Mean	High	Low	Mean	High	Low
January	31.7	33.0	30.0	23.2	24.5	21.5	28.3	30.0	27.5	22.9	24.5	20.5
February	32.2	33.5	31.0	22.2	24.5	21.0	28.7	29.8	26.5	22.3	24.8	21.2
March	32.5	34.5	29.5	23.2	25.0	21.2	29.4	30.5	29.0	23.2	24.8	21.0
April	33.8	35.5	30.0	22.9	21.5	21.5	30.8	32.5	28.0	22.8	25.0	21.0
May	32.0	34.0	28.5	23.1	25.0	21.0	29.1	30.8	26.0	21.4	24.8	16.8
June	30.1	32.5	26.5	22.6	23.5	21.5	27.1	29.5	25.2	21.8	23.2	16.5
July	30.8	32.0	28.0	22.3	23.5	21.0	27.6	30.0	25.8	20.9	23.2	17.0
August	31.1	33.0	28.5	22.4	24.5	20.0	27.4	29.8	26.0	22.0	24.2	18.5
September	31.5	32.5	29.5	22.3	23.5	21.5	27.4	29.0	26.2	22.0	23.2	20.5
October	30.8	32.0	27.5	22.7	24.0	22.0	27.5	28.8	25.5	22.3	23.2	20.5
November	30.4	31.5	25.5	22.5	23.0	21.5	26.5	29.0	25.0	21.8	23.2	18.0
December	30.2	31.5	29.5	21.8	23.0	21.0	26.1	27.4	25.2	21.5	22.5	20.4
Annual mean and extremes	31.4	35.5	25.5	22.6	25.0	20.0	28.0	32.5	29.0	22.1	25.0	16.5

SOURCE: Rubinoff, 1974.

claystone, shales, and tuffs, and igneous rocks, including basalt, andesite, granodiorite, diorite, metabreccia, and rhyolite (Bennett, 1929).

Bennett (1929) classified BCI soil as Frijoles Clay, which is characterized by its red or deep red color (slightly more brownish in the uppermost 5 cm) and by being well drained, well aerated, slightly acid to alkaline, and friable when moist. Kenoyer (1929) described the soils as "a chocolate brown alluvium underlain by a red subsoil arising from the decomposition of the Bohio sandstones and conglomerates." Soil maps by McDonald (1919) showed all of BCI being made up of Bohio sandstones and conglomerates. According to Woodring (1958), the parent rock is predominantly marine Caimito and non-marine Bohio sedimentary formations (late Oligocene); there is a cap of basalt at the central part of the island and some volcanic rock toward the east.

Knight (1975a) reported finding a second soil type in several areas of the island—a gley, which is an impervious clay subsoil appearing gray mottled with red—at a depth of 59–75 cm, and contrasts it with the Frijoles Clay. According to Foster (1974), the past history of human disturbances appears to have had a more pronounced influence on the soils than the parent material had: in most cases, those areas known to have been disturbed in the past 75 years have developed a hardpan beneath a grayish soil, whereas the less disturbed areas exhibit only a thin brown layer covering a very deep red clay that is almost devoid of humus.

General Characteristics of the Vegetation

The vegetation on Barro Colorado Island is semi-evergreen moist tropical forest (Knight, 1975a). In terms of the classification system devised by Beard (1944; 1955), the oldest forest on the island would appear to be intermediate between Evergreen Seasonal Forest and Semi-Evergreen Seasonal Forest (Bennett, 1963). And in the Holdridge Life-Zone System (Holdridge et al., 1971, discussed below, p. 60), the forest is *tropical moist forest*. The Holdridge system is used in this text as a means of delineating the ecological range of each species within Panama; though not universally accepted, it seems to work for Panama, and it is the only system for which a comprehensive map for Panama has been produced (see pp. 58–59).

Although no major disturbances have taken place on the island since it was set aside as a preserve in 1923, a major portion of the island lay deforested until 1905 (Enders, 1935; Knight, 1975a). This includes much of the eastern third of the island, major promontories on the north side of the island (including Orchid Island), and small portions of the western periphery. According to Enders (1935), most parts of the younger forest (on the east and north sides of the island) may date from as early as 1880, though a few small clearings persisted longer (i.e., the forest that has overtaken them is younger). He reported that the old cocoa plantations on Peña Blanca Peninsula date from 1905, and that much of the Burrunga Peninsula was cleared during 1920–23. One small area

north of the present Laboratory Clearing (formerly a garden for workers) dates from about 1955 (Knight, 1975a). An aerial view of BCI taken in 1927, published by Knight (1975a), shows the approximate distribution of the young and old forest; one published by Hladik and Hladik (1969) shows the forest as it occurs today.

Types of Natural Vegetation

Forest. Most of BCI, then, is covered with forest, perhaps none of it climax forest. Superficially, the forest seems everywhere the same, but there are differences between the young and the old forests. Bennett (1963) described the young forest as having two strata and the old forest as having two or three strata. Though there are clearly differences in the average height of the canopy between the two forest types, I have never been able to distinguish any sharply defined stratification. My impression is one of trees forming a continuous and undifferentiated cover, their crowns fitting together like the pieces of a jigsaw puzzle.

The following are the more common trees, as determined by the density data compiled by Knight (1975a; see his Appendix 1). The list includes species from both young and old forests.

Acalypha diversifolia	*Miconia argentea*
Alseis blackiana	*Mouriri myrtilloides*
Anacardium excelsum	subsp. *parvifolia*
Andira inermis	*Nectandra purpurascens*
Annona hayesii	*Ochroma pyramidale*
Apeiba membranacea	*Oenocarpus panamanus*
A. tibourbou	*Ouratea lucens*
Aspidosperma cruenta	*Platymiscium pinnatum*
Astronium graveolens	*Platypodium elegans*
Bombacopsis quinata	*Poulsenia armata*
Capparis frondosa	*Prioria copaifera*
Casearia guianensis	*Protium panamense*
Cecropia insignis	*Pseudobombax septenatum*
Cordia alliodora	*Pterocarpus officinalis*
C. lasiocalyx	*P. rohrii*
Coussarea curvigemmia	*Quararibea asterolepis*
Desmopsis panamensis	*Quassia amara*
Didymopanax morototoni	*Randia armata*
Dipteryx panamensis	*Rinorea sylvatica*
Erythroxylum multiflorum	*Scheelea zonensis*
Faramea occidentalis	*Simarouba amara*
Gustavia superba	*Sorocea affinis*
Heisteria longipes	*Spondias mombin*
Hirtella americana	*Swartzia simplex*
H. racemosa	*Trattinnickia aspera*
H. triandra	*Trichilia cipo*
Hura crepitans	*Triplaris cumingiana*
Hybanthus prunifolius	*Virola surinamensis*
Luehea seemannii	*Zanthoxylum panamense*
L. speciosa	

The forest on BCI is rich in species. Lang et al. (1971) found that even in the areas where the composition of the forest appears to be most homogeneous, there are as many as 130 species of trees (more than 2.5 cm in diameter at

breast height, both adults and juveniles) in 1.5 hectares (3.7 acres). The species/area curve shows only moderate signs of leveling off in this case. These figures would be much higher if vines, lianas, shrubs, herbs, epiphytes, and hemiepiphytes were included.

Species diversity in the younger forest is about the same as that in the older forest, and is probably increasing very slowly where the forest has reached 50–60 years of age (Knight, 1975a). Knight indicated that the most rapid increase in species diversity occurs during the first 10–15 years of succession, and that after 65 years the increase is slow. So great is the species diversity that, despite several years of collecting on BCI, I discovered a new species for the flora almost every day of collecting, during even the most recent trips. More recently, of 200 collections made by Gene Montgomery in the canopy of the forest, three were species new to the island; and one of these was a species new to Central America. Four species new to the island were collected by Garwood and Foster as late as 1976, and one of these represents a family new to the flora (Humiriaceae). Many of the plants that are being collected for the first time are common on the island. Some are conspicuous but have nevertheless been overlooked: *Celtis schippii*, which was collected only a few years ago for the first time for Panama, is growing at the edge of the Laboratory Clearing; the plant is at least 40 years old and scores of botanists have no doubt walked past it. Many other plants on the island, especially the forest trees, are evidently rare; some are known from only a few individuals, some from only a single plant.

In the young forest, which appears more uniform than the old, the larger trees of the canopy average 18–24 m tall. The understory is usually moderately dense and tangled, and there are few disturbances from tree falls, perhaps because few of the trees are old enough to have been subject to dying or to windfall. Knight (1975a) found the following species of trees only in the young forest: *Annona hayesii, A. spraguei, Casearia guianensis, Didymopanax morototoni, Hymenaea courbaril, Inga mucuna,* and *Stemmadenia grandiflora.* Other species that I have seen only or mostly in areas of young forest include:

Aegiphila panamensis	*Ficus perforata*
Apodanthes caseariae (parasite)	*Heisteria costaricensis*
Bombacopsis quinata	*Nectandra purpurascens*
B. sessilis	*Ocotea pyramidata*
Byrsonima spicata	*Psychotria carthagenensis*
Calycophyllum candidissimum	*Tournefortia cuspidata*
Cryosophila warscewiczii	*Trigonia floribunda*
	Vochysia ferruginea
	Vriesia heliconioides

Knight (1975a) listed *Cordia alliodora, Luehea seemannii, Miconia argentea, Spondias mombin, Scheelea zonensis,* and *Zuelania guidonia* as being more abundant in the young forest.

The canopy trees in the old forest average 22–30 m tall (Bennett, 1963). Species composition is qualitatively similar to that of the young forest, and there is no strong correlation between the age of the forest and its floristic composition (Knight, 1975a, 1975b). Knight has shown that in both the young and the old forest about 15% of the species can be classified as infrequent reproducers— plants incapable of reproducing in the understory environment. The percentage of total species that are frequent reproducers is also about the same (39% in the old forest, 37% in the young forest). In the canopy, it is a different matter: only 7% of the canopy trees of the young forest are classified as frequent reproducers, whereas 73% are infrequent reproducers; by contrast, 48% of the canopy trees in the old forest are frequent reproducers, and only 40% (primarily pioneer species) are infrequent reproducers.

The old forest differs from the young principally in having larger lianas and trees. Several species that are represented in both the old and the young forest are two to five times larger in girth in the old forest than in the young; on this basis, Knight (1975a) has estimated that the old forest is at least twice as old as the young, or as much as 130 years old. But because of the relatively high density of infrequent reproducers, he believes that the vegetation of the old forest is still in a successional stage.

Among the species of trees that Knight found almost exclusively in the old forest are:

Acalypha diversifolia	*Heisteria longipes*
Desmopsis panamensis	*Poulsenia armata*
Erythrina costaricensis var. *panamensis*	*Tovomitopsis nicaraguensis*
	Trattinnickia aspera
Ficus tonduzii	*Virola surinamensis*
Guarea multiflora	*V. sebifera*

He also has the following occurring as large trees only in the old forest:

Andira inermis	*Protium panamense*
Casearia arborea	*P. costaricense*
Guarea glabra	*P. tenuifolium*
Hirtella triandra	subsp. *sessiliflorum*
Hura crepitans	*Quararibea asterolepis*
Inga marginata	*Rheedia acuminata*
Pouteria sapota	*Tetragastris panamensis*
Prioria copaifera	*Trattinnickia aspera*

Other species that I have observed chiefly or only in the old forest include the following:

Epiphytes
Ludovia integrifolia	*Peperomia macrostachya*

Hemiepiphytes
Philodendron fragrantissimum

Herbs
Asplenium delitescens	*Phaeosphaerion*
Calathea micans	*persicariifolium*
Dieffenbachia pittieri	*Pharus parvifolius*
Heliconia irrasa	*Selaginella exaltata*

Vines
Fevillea cordifolia	*Gurania coccinea*

Lianas
Acacia acanthophylla	*Cynanchum recurvum*
Adelobotrys adscendens	*Macfadyena unguis-cati*

Shrubs

Anaxagorea panamensis
Ardisia pellucida
Geonoma cuneata
Hamelia axillaris
Justicia graciliflora
Miconia nervosa
Picramnia latifolia
Psychotria granadensis
Solanum argenteum

Small trees

Bactris coloradonis
Coccoloba acapulcensis
Picramnia latifolia
Xylopia macrantha
Xylosma oligandrum

Trees

Astronium graveolens
Diospyros artanthifolia
Drypetes standleyi
Hampea appendiculata
Inga spectabilis
Laetia thamnia
Nectandra cissiflora
Ocotea oblonga
O. pyramidata
O. skutchii
Pouteria fossicola
Prioria copaifera
Protium costaricense
Psidium anglohondurense
Sloanea terniflora
Symphonia globulifera
Tabernaemontana arborea

Perhaps because of its age and its greater content of mature trees, the old forest appears to have more tree-fall gaps than the young forest does. One result of this difference is the somewhat higher incidence in the older forest of *Cecropia* trees, which establish themselves in the canopy gaps (Knight, 1975a; R. Foster, pers. comm.).

Although climax forest possibly does not exist on BCI, a number of potential climax species can be listed, on the basis of Knight's list of frequently reproducing tree species. The following, each often more than 30 m tall, are listed by Knight as frequent reproducers, and hence are potential climax-canopy species:

Alseis blackiana
Aspidosperma cruenta
Calophyllum longifolium
Guarea multiflora
Heisteria concinna
Licania platypus
Lonchocarpus pentaphyllus
Maquira costaricana
Poulsenia armata
Prioria copaifera
Protium panamense
Pterocarpus rohrii
Quararibea asterolepis
Simarouba amara
Socratea durissima
Tabernaemontana arborea
Tachigalia versicolor
Trichilia cipo
Triplaris cumingiana
Virola sebifera

The following smaller trees, which Knight also lists as frequent reproducers, are potentially a part of the climax flora of the island:

Casearia guianensis
C. sylvestris
Cordia lasiocalyx
Coussarea curvigemmia
Cupania rufescens
C. sylvatica
Desmopsis panamensis
Eugenia nesiotica
Faramea occidentalis
Guarea glabra
Guettarda foliacea
Heisteria longipes
Hirtella americana
H. triandra
Inga fagifolia
I. marginata
Mabea occidentalis
Mouriri myrtilloides
  subsp. *parvifolia*
Oenocarpus panamanus
Picramnia latifolia
Protium tenuifolium
  subsp. *sessiliflorum*
Quassia amara
Randia armata
Rheedia acuminata

Sorocea affinis
Spachea membranacea
Stemmadenia grandiflora
Swartzia simplex

Although more than 70 species of trees reach a height of about 30 m, relatively few grow appreciably taller and occur as emergents above the general level of the canopy. Among these taller trees are:

**Anacardium excelsum* (to 37 m)
Aspidosperma cruenta (35)
A. megalocarpon (40)
Astronium graveolens (35)
Beilschmiedia pendula (40)
Brosimum alicastrum (35)
Calophyllum longifolium (35)
**Cavanillesia platanifolia* (40)
**Cecropia insignis* (40)
Cedrela odorata (40)
Ceiba pentandra (40)
Couratari panamensis (40)
**Dipteryx panamensis* (40)
*Ficus insipida (30)
F. obtusifolia (40)
F. yoponensis (40)
Guarea multiflora (40)
Guatteria dumetorum (35)
**Hyeronima laxiflora* (40)
Myroxylon balsamum var. *pereirae* (40)
Ormosia macrocalyx (40)
Peltogyne purpurea (50)
Pithecellobium macradenium (40)
Prioria copaifera (40)
Pterocarpus rohrii (50)
Simarouba amara (35)
Sterculia apetala (40)
**Tabebuia guayacan* (40)
**Terminalia amazonica* (35)
T. chiriquensis (35)
Tetragastris panamensis (35)
**Trattinnickia aspera* (50)
Schizolobium parahybum (40)
Vantanea occidentalis (40)

Those species marked by an asterisk are listed by Knight (1975a) as infrequent reproducers. Most of the species listed by Knight, then, are large, and some are emergent trees. In addition to those marked with an asterisk, Knight (1975a) lists the following species of large trees as infrequent reproducers:

Apeiba membranacea
A. tibourbou
Bombacopsis quinata
B. sessilis
Byrsonima crassifolia
Casearia arborea
Chrysophyllum cainito
Cordia panamensis
Dendropanax arboreus
Didymopanax morototoni
Enterolobium cyclocarpum
Guapira standleyanum
Hura crepitans
Hymenaea courbaril
Platypodium elegans
Pseudobombax septenatum
Sapium aucuparium
Socratea durissima
Trema micrantha
Turpinia occidentalis
  subsp. *breviflora*
Virola surinamensis
Zuelania guidonia
Zanthoxylum panamense

Clearings. Although man-made clearings constitute much less than 1% of the total area of the island, they contribute 7% to the total flora, representing 197 species. (This total does not include the 53 species of cultivated plants that occur in the Laboratory Clearing.) Most of the natural vegetation occurring in clearings consists of weedy plants, especially of the families Gramineae (37 species), Compositae (27), Leguminosae (17), Solanaceae (13), Cyperaceae (11), Euphorbiaceae (10), Melastomataceae (7), Cucurbitaceae (6), Convolvulaceae (5), Passifloraceae (4), Commelinaceae (3), Phytolaccaceae (3), Rubiaceae (3), and Polypodiaceae (2).

A number of different clearing types occur on the island. These include the Laboratory Clearing, the Tower Clearing (a small clearing at the center of the island), small clearings maintained at the ends of many of the trails, and some clearings along the north and east sides of the island that are maintained for navigational signs for ships in the Panama Canal (of these, the only sizable clearings are those around the Front and Rear #8 Lighthouses).

The large lighthouse clearings are cleared annually, most of the smaller clearings and the Laboratory Clearing more than once per year, and the debris is usually left to decompose. At the time of clearing, the growth is about 2–3 m tall. Regrowth is relatively rapid for those species that survive the cutting back. Since burning is not permitted anywhere on the island, the seeds and roots are not killed, and regrowth follows a similar pattern each year. Herbaceous or suffruticose plants that soon become reestablished include:

Abelmoschus moschatus	Fleischmannia sinclairii
Aeschynomene americana	Heliconia latispatha
var. glandulosa	Hyptis capitata
Blechum brownei	H. mutabilis
Blepharodon mucronatum	Indigofera mucronata
Borreria densiflora	Ipomoea squamosa
B. laevis	Justicia pectoralis
B. latifolia	Lasiacis oaxacensis
B. ocimoides	L. procerrima
Calapogonium mucunioides	Manettia reclinata
Chamaesyce hyssopifolia	Melampodium
C. hypericifolia	divaricatum
Chelonanthus alatus	Melanthera aspera
Chromolaena odorata	Melochia melissifolia
Cissampelos pareira	Melothria pendula
Clibadium surinamense	Mikania micrantha
Clitoria rubiginosa	Mimosa casta
Costus guanaiensis	M. pudica
var. macrostrobilus	Momordica charantia
Cuphea carthagenensis	Oplismenus burmanni
Cyathula prostrata	Panicum fasciculatum
Cyperus diffusus	P. maximum
Desmodium adscendens	P. pilosum
D. canum	Paspalum conjugatum
Diodea ocimifolia	P. virgatum
Dioscorea alata	Phyllanthus urinaria
D. sapindoides	Rolandra fruticosa
Elephantopus mollis	Salvia occidentalis

Setaria paniculifera	Synedrella nodiflora
Sida acuta	Tibouchina longifolia
S. rhombifolia	Tridax procumbens
Spilanthes alba	Verbesina gigantea
Spiracantha cornifolia	Vernonia cinerea

These herbaceous species are accompanied or followed by the more dominant shrubby element of the clearing, which includes:

Cassia fruticosa	Ochroma pyramidale
Cestrum latifolium	Piper auritum
Conostegia speciosa	P. dilatatum
C. xalapensis	Serjania mexicana
Cordia spinescens	Solanum jamaicense
Dioclea guianensis	S. subinerme
Guazuma ulmifolia	Trema micrantha
Hamelia patens	Triumfetta lappula
Lantana camara	Vernonia canescens
Mandevilla villosa	V. patens
Melochia lupulina	

Shorelines and marshes. The flora of the lakeshore contributes the next largest number of herbaceous plants, with 68 species more or less restricted to the shore. Since an additional 197 species are more or less restricted to clearings (excluding cultivated species), these two habitat classes contribute a total of 265 species to the flora. Thus the flora is considerably richer relative to the size of the island than it would be if the area were an undisturbed forest lacking clearings and not surrounded by water. Trees and shrubs may grow very near the edge of the lake where the banks are low, but even where they do there is usually adequate light near the ground for the development of an herbaceous flora along the bank. Most common here are species of Cyperaceae and Gramineae.

There are drastic differences between the shorelines of the northern and eastern sides of the island, on the one hand, and the western and southern shores, on the other. During most of the year, especially during the dry season, the trade winds generate waves that buffet the northern and eastern shores, a condition that is greatly exacerbated by the passing of ships close by the north and east on their way through the Canal. In such areas, and especially on the promontories, the resultant erosion has created high banks with an almost constantly changing shoreline. This pattern creates habitats for the more persistent invading herbs. Among the most common species found on steep eroded banks are the following:

Anthurium brownii	Kohleria tubiflora
Carludovica palmata	Setaria vulpiseta
Chelonanthus alatus	Thelypteris serrata
Gynerium sagittatum	Trichopteris microdonta

The aquatic flora of the wave-swept portion of the shore is poor, often consisting only of Hydrilla. But in the more sheltered coves in the north and east, and along the entire shore of the western and, especially, southern sides of the island, the aquatic flora is very rich, and often blends almost imperceptibly with the remainder of the herbaceous shoreline vegetation. Moreover, hydrarch or

Buttressed trunk of *Ceiba pentandra*, one of the largest trees on the island

The Laboratory Clearing, near the beginning of Snyder-Molino Trail, Kodak House at left center, the Animal House at right front

Ships in the Canal, the channel passing within 120 m (400 ft) of the island

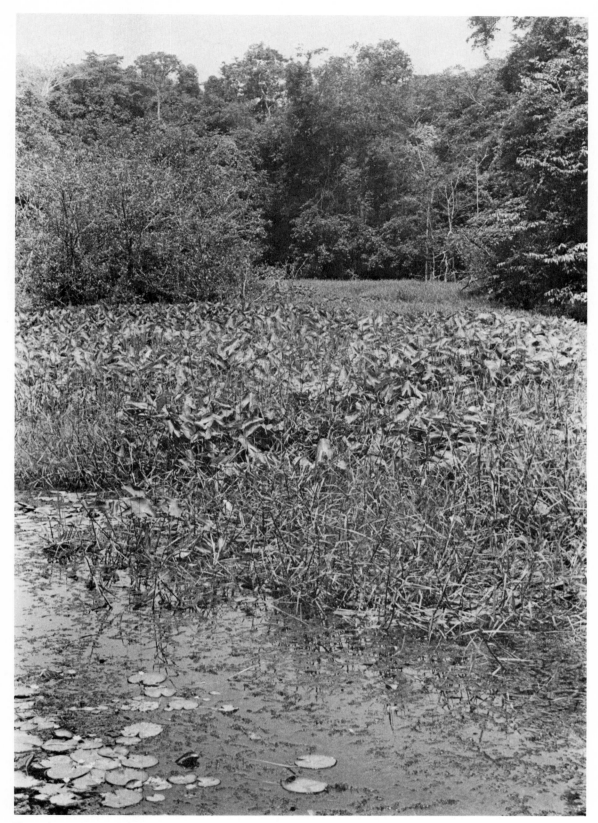

Vegetation on silt deposits in Shannon Cove, on the southern shore

Annona-Acrostichum association, along the western shore

Epiphytes on a floating log: *Nephrolepis biserrata, Hibiscus sororius,* and *Fuirena umbellata*

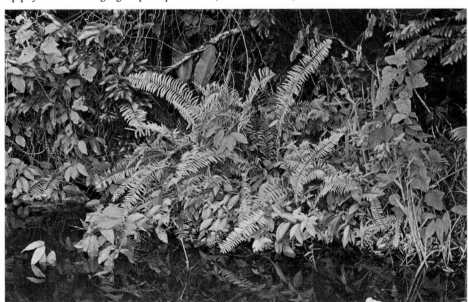

shore-depositing succession is in evidence, and there is often no abrupt end of shore. Similar situations can be found in all of the deeper bays and coves surrounding the island wherever the deposition of silt is great. Many of the coves have silted almost completely shut since the formation of the island. Other large expanses of silt are sometimes exposed during the dry season, when the level of the lake may drop several feet. These are quickly populated with herbaceous plants such as:

Acroceras oryzoides	Panicum grande
Boehmeria cylindrica	P. milliflorum
Cleome parviflora	P. polygonatum
Dennstaedtia cicutaria	P. trichanthum
Gynerium sagittatum	Polygonum punctatum
Hygrophylla guianensis	Scleria mitis
Ludwigia leptocarpa	Thelypteris serrata
L. octovalvis	

The floristic composition of the aquatic community of the island is usually very rich. Certain species are often found in associations that bear special mention. The most consistent of these associations is made up of *Annona glabra*, a small tree occurring near the shore, and *Acrostichum* spp., large aquatic ferns sometimes forming dense stands. These species may be associated with a wide variety of other aquatic species, but especially the following:

Aeschynomene ciliata	Ludwigia leptocarpa
A. sensitiva	L. octovalvis
Andropogon bicornis	Nephrolepis biserrata
A. glomeratus	Passiflora punctata
Begonia patula	Phaseolus trichocarpus
Boehmeria cylindrica	Pontederia rotundifolia
Ceratopteris pteridoides	Rhynchospora corymbosa
Cyperus haspan	Sagittaria lancifolia
C. odoratus	Sarcostemma clausum
Eleocharis caribaea	Stigmaphyllon puberum
E. plicarhachis	Thalia geniculata
Fuirena umbellata	Thelypteris serrata
Habenaria repens	T. totta
Hibiscus sororius	Typha domingensis
Hydrocotyle umbellata	Utricularia foliosa
Isachne polygonoides	U. obtusa
Leersia hexandra	

These species may be found with or separate from the close *Annona-Acrostichum* associations. When not in these associations, the plants often join together to form a floating island of emerged vegetation very rich in species but dominated by *Rhynchospora*, *Sagittaria*, or *Thalia*. Such islands of vegetation, covering water usually 1–2 m deep, may be surrounded by large masses of *Hydrilla verticillata* or, to a lesser extent, by the free-floating aquatics such as *Salvinia radula*, *Eichhornia* spp., and *Pistia stratiotes*. Also present are the rooted but otherwise free-floating plants such as *Nymphaea ampla*, *Limnobium stoloniferum*, *Nymphoides indica*, and *Ludwigia helminthorrhiza*. Thus, the free-floating aquatics, the attached floaters, and the especially abundant *Hydrilla*

verticillata begin the shore-depositing succession by quieting the water and accumulating debris on the lake bottom. *Ceratophyllum demersum*, though less abundant, performs the same function as *Hydrilla*. These two are followed by the emerged aquatic associations and especially by the species-rich floating-island associations. The latter are so floristically variable that no dominant species can be singled out.

Usually nearer the shore, and frequently indistinguishable from the emerged aquatic associations, is an association of shore plants termed the "sedge association" by Kenoyer (1929). This assemblage may be composed of some of the same species found in the floating associations, but the dominant elements include:

Acroceras oryzoides	Panicum grande
Cyperus giganteus	P. mertensii
C. odoratus	Phragmites australis
Fuirena umbellata	Rhynchospora corymbosa
Gynerium sagittatum	Scirpus cubensis
Hymenachne amplexicaulis	

Crinum erubescens, *Hymenachne amplexicaulis*, *Spathiphyllum friedrichsthalii*, and *Montrichardia arborescens* usually occur along the edge of the lakeshore and adjacent to the forest. Here also are such water-tolerant arborescent plants as:

Coccoloba manzanillensis	Pachira aquatica
Cynometra bauhiniifolia	Pithecellobium
Dalbergia brownei	hymeneaefolium
D. monetaria	P. rufescens
Erythrina fusca	Prioria copaifera
Inga sapindoides	Rhabdadenia biflora
Hibiscus bifurcatus	Swartzia simplex

Other species that have been seen mostly along the shore are:

Allophylus psilospermus	Miconia elata
Andira inermis	Nectandra purpurascens
Antirrhoea trichantha	Ochroma pyramidale
Bursera simaruba	Ocotea cernua
Byrsonima crassifolia	Pithecellobium
Cecropia obtusifolia	macradenium
Cochlospermum vitifolium	Quararibea pterocalyx
Coussapoa panamensis	Swartzia panamensis
Eugenia galalonensis	Syzygium jambos
Grias fendleri	Terminalia amazonica
Inga fagifolia	Trichilia hirta
I. mucuna	T. verrucosa
Leucaena multicapitula	Vismia macrophylla
Machaerium kegelii	Xylopia frutescens
Malouetia guatemalensis	

Tree stumps in the lake. A number of tree stumps still remain from the trees that were drowned when the lake waters first rose. These stumps are generally exposed to full sunlight and provide unique epiphytic habitats. At least two hemiepiphytic trees, *Ficus insipida* and *Coussapoa magnifolia*, occur on the tree stumps. The epiphytic herbs on the stumps include:

Tree-stump epiphytes:
Nephrolepis pendula,
Polypodium crassifolium,
Sobralia suaveolens, and
Sarcostemma clausum

Tree stumps in the lake,
offshore between Colorado
Point and the laboratory

Anthurium brownii	Polypodium crassifolium
A. gracile	P. phyllitidis
Catasetum viridiflavum	Sobralia suaveolens
Nephrolepis pendula	Trigonidium egertonianum
Peperomia cordulata	Vittaria lineata

Other nonepiphytic species that commonly occur on tree stumps but are probably always rooted into the debris that accumulates on the trunk are *Eleocharis caribaea, Fuirena umbellata, Paspalidium geminatum,* and *Sarcostemma clausum.*

Trails. There are about 37 km (22 miles) of trails leading to all parts of the island (see p. 61 and the BCI map on the front endsheet and pp. 56–57). Each trail is named, and marked at 100 m intervals by permanent signposts. With this well-marked trail system, BCI becomes a unique place for scientific study, for it is relatively easy to note a position and return later to the same spot.

These trails offer yet another microhabitat, some species of plants having been found principally along their edges. The trails are usually not exposed to full sunlight, but the amount of light reaching the ground along them is considerably greater than that in adjacent areas of the forest. Because the trails range so widely throughout the island and pass through so many different habitats, a large part of the flora of the island can be found within a few meters of the trails. Species that are found principally *along* the trails are the following:

Herbaceous plants

Adiantum humile	O. hirtellus
A. obliquum	Orthoclada laxa
A. petiolatum	Panicum pulchellum
Blechum costaricense	P. pilosum
Calathea panamensis	P. polygonatum
Calyptrocarya glomerulata	Paspalum decumbens
Campelia zanonia	Petiveria alliacea
Costus pulverulentus	Phaeosphaerion
Cyathula prostrata	persicariifolium
Cyperus simplex	Pothomorphe peltata (rare)
Dalechampia tiliifolia	Ruellia metallica
Desmodium axillare	Selaginella arthritica
var. stoloniferum	S. flagellata
D. wydlerianum	S. haematodes
Dimerocostus strobilaceus	S. horizontalis
(rare)	Sida acuta
Diodia denudata	Spathiphyllum
Geophila croatii	phryniifolium
G. repens	Spigelia anthelmia
Gibasis geniculata	S. humboldtiana
Gurania makoyana	Teliostachya alopecuroidea
Ichnanthus brevivaginatus	Thelypteris dentata
I. pallens	T. nicaraguensis
I. tenuis	T. poiteana
Lithachne pauciflora	Xanthosoma
Nautilocalyx panamensis	helleborifolium
Oplismenus burmanni	Xiphidium caeruleum

Shrubs and trees

Acalypha macrostachya	A. pellucida
Ardisia bartlettii	Cephaelis ipecacuanha

Hamelia axillaris	Psychotria carthagenensis
Lycianthes maxonii	P. psychotriifolia
Pavonia dasypetala	P. racemosa
Piper darienense	Vismia billbergiana

Ravines. Ravines offer a unique habitat, because they are moister and darker than any other part of the forest. Streams are never very large, but they flow rapidly, and generally have flowing water throughout the rainy season. The following species are either restricted to or most common in ravines:

Epiphytes and hemiepiphytes

Asplenium auritum	Maxonia apiifolia
A. laetum	Polybotrya villosula
Guzmania lingulata	Sobralia panamensis
var. minor	

Other herbs

Anthurium ochranthum	Hymenocallis pedalis
Asplenium delitescens	Philodendron grandipes
Asplundia alata	Pteris altissima
Bolbitis cladorrhizans	P. grandifolia
Carludovica drudei	Rhodospatha moritziana
Ctenitis protensa	Saccoloma elegans
Cyclanthus bipartitus	Selaginella exaltata
Cyclopeltis semicordata	S. mollis
Danaea nodosa	Tectaria euryloba
Diastema raciferum	Thelypteris balbisii
Dictyoxiphium panamense	T. extensa
Dieffenbachia oerstedii	T. torresiana
Hemidictyum marginatum	Trichomanes diversifrons
Homalomena wendlandii	T. pinnatum

Shrubs and trees

Clidemia	Piper arieianum
purpureo-violacea	P. culebranum
Geonoma cuneata	P. imperiale
G. procumbens	P. pseudo-garagaranum
Leandra dichotoma	P. pubistipulum
Miconia lateriflora	P. viridicaule
Myriocarpa yzabalensis	

Tree ferns

Cnemidaria petiolata	Nephelea cuspidata
Metaxya rostrata	

A few species, such as *Asplenium delitescens* and *Bolbitis cladorrhizans,* may occur directly in the streams on rocks. In a few of the streams there are small waterfalls, but none have developed any plant associations—unlike so many waterfalls and rapid streams elsewhere in Panama where members of the Podostemonaceae might be found. This circumstance is due perhaps to an inadequate amount of light, or more likely to the fact that most streams dry up in the dry season.

Forest floor. In addition to the herbs that grow principally in ravines or along forest trails, a number occur in other parts of the forest. Those that are associated principally with light-gaps created by tree falls include:

Calathea inocephala	C. marantifolia
C. latifolia	Costus allenii

Streambed on Balboa Trail, in the dry season

The forest near Barbour and Lathrop trails

Costus laevis
C. scaber
Heliconia catheta
H. subulata

Ischnosiphon leucophaeus
I. pruinosus
Renealmia alpinia
Scleria pterota

There are other forest-floor herbs not regularly associated with either trails, ravines, light-gaps, or forest edges. These include:

Ferns

Adiantum decoratum
A. fructuosum
A. lunulatum
A. pulverulentum

A. seemannii
Asplenium delitescens
Dictyoxiphium panamense
Diplazium grandifolium

Araceae

Dieffenbachia longispatha
D. pittieri

Dracontium dressleri

Gramineae

Chusquea simpliciflora
Pharus latifolius
P. parvifolius
Rhipidocladum racemiflorum

Streptochaeta sodiroana
S. spicata
Streptogyne americana

Other families

Aechmea magdalenae
Calathea inocephala
Dichorisandra hexandra
Neomarica gracilis

Palmorchis powellii
Peperomia killipi
P. obtusifolia
Renealmia cernua

Aechmea magdalenae is a large conspicuous herb that becomes dominant in some areas of the forest, forming close, virtually impenetrable stands. These are especially in evidence in some of the wet, moderately flat areas along Zetek Trail below the escarpment.

Despite the long lists of herbaceous ground species that do occur in the forest, most of the plants growing in most areas of the forest are not herbaceous. The principal vegetation of the shaded forest floor consists of seedlings of arborescent plants and herbaceous vines. Herbaceous ground plants of any kind are uncommon, except for:

Adiantum lucidum
A. petiolatum
Anthurium ochranthum
Carludovica drudei
Chusquea simpliciflora
Cyclopeltis semicordata
Dieffenbachia longispatha
D. oerstedii
Dictyoxiphium panamense
Diplazium grandifolium

Homalomena wendlandii
Pharus parvifolius
Renealmia cernua
Rhipidocladum
    racemiflorum
Selaginella haematodes
Tectaria incisa
Thelypteris dentata
T. nicaraguensis
T. poiteana

Habit-and-Habitat Classes

Habit diversity, like poor soils and species diversity, is a principal feature of tropical forests, and most habit types are exhibited on BCI. Table 3 lists the number of BCI species in each of a number of habit-and-habitat classes; these classes are employed in the discussions of growth forms that follow and in Table 4 (on geographical affinities, p. 27) and Table 5 (on flowering and fruiting periods, p. 37).

Growth Forms

Arborescent plants. Though shrubs and trees exhibit a continuous range of heights and cannot be easily classified into categories, I have placed them in artificial categories because of obvious ecological differences. For convenience, I have classified as a small tree any species that usually does not become a part of the canopy; usually, these trees are less than 10 m tall. Trees more than 10 m tall are classified as medium-sized or large.

There are 1,316 native or naturalized species of vascular plants on the island. A total of 53 are cultivated and are excluded from further consideration here. Of the native or naturalized species, 481 (36.5%) are arboreal. Of the arboreal species, 34 (7%) are trees that may be taller than 30 m and are known or possible emergents, whereas 177 (37%) are trees 10–30 m tall, 247 (51%) are trees or shrubs less than 10 m tall, 16 (3%) are hemiepiphytic shrubs, and 7 (1.4%) are parasitic shrubs. Of the

TABLE 3

Summary of BCI habit-and-habitat classes

Habit-and-habitat class	Number of species	Percent of native flora (total 1,316)
CRYPTOGAMS		
Epiphytes	41	3.1%
Hemiepiphytes	1	.1
Aquatics	6	.5
Vinelike plants	4	.3
Other terrestrials	47	3.6
Tree ferns	5	.4
TOTAL CRYPTOGAMS	104	7.9
PHANEROGAMS		
Trees more than 10 m tall	211	16.0
Small trees or shrubs (not including plants that are always shrubs)	154	11.7
Shrubs 2(3) m tall	93	7.1
Epiphytic or hemiepiphytic shrubs or trees	16	1.2
Parasitic shrubs	7	.5
TOTAL ARBORESCENT SPECIES	481	36.5
Lianas or climbing woody plants (including ten climbing trees or shrubs)	171	13.0
Vines	83	6.3
Hemiepiphytic or epiphytic vines	11	.8
TOTAL SCANDENT SPECIES	265	20.1
Epiphytic herbs	135	10.3
Primarily aquatic herbs	54	4.1
Primarily clearing herbs	197	15.0
Primarily forest herbs	75	5.7
Parasitic herbs	1	.1
Saprophytic herbs	4	.3
Suffruticose herbs (included in above habitat classes)	(18)	
TOTAL HERBACEOUS SPECIES (not including scandents)	466	35.4
All woody plants (arborescents and lianas)	652	49.5
All herbaceous plants (including scandents)	560	42.6
TOTAL NATIVE PHANEROGAMS	1,212	92.1
Cultivated phanerogams	53	
TOTAL PHANEROGAMS	1,265	
TOTAL VASCULAR PLANTS	1,369	

247 small trees and shrubs, 93 are strictly shrubs (usually less than 3 m tall).

Some of the large trees may extend above the canopy and be classified as possible emergent species (see the list on p. 8), though trees classified as emergents are not always raised above the canopy. (Some individuals flower when they are young and growing at subcanopy levels; others may be found among the emergent trees.)

In general, crown shape is determined by the conditions under which a plant grows. A species with a broad, spreading crown, such as *Enterolobium cyclocarpum,* will usually be found in an area that was once quite open, since such species will not do well as seedlings under closed-canopy conditions. Other species, with narrower crowns, are plants that develop under closed, highly competitive conditions. However, some species that appear to do quite well under crowded conditions, such as *Tachigalia versicolor,* have relatively broad, spreading crowns.

Degree of buttressing is another character often found to be species-diagnostic, though some conditions—such as poor supporting soil, a steep slope, or a stream or lakeshore—tend to accentuate buttressing in most species that develop in this fashion. Buttressing varies considerably both between and within species. Knight (1975a) reported that 22% of the trees less than 10 cm in diameter at breast height in the old forest have buttresses, and that 4% have individual buttresses with an area of more than 0.3 m² (on one face of buttress only). Among the species that usually exhibit buttressing on BCI are:

Acacia glomerosa	*Guatteria dumetorum*
Anacardium excelsum	*Hyeronima laxiflora*
Apeiba membranacea	*Jacaranda copaia*
Aspidosperma cruenta	*Licania platypus*
Beilschmiedia pendula	*Luehea seemannii*
Bombacopsis quinata	*Maquira costaricana*
B. sessilis	*Mosquitoxylon jamaicense*
Calophyllum longifolium	*Ochroma pyramidale*
Castilla elastica	*Ocotea oblonga*
Cedrela odorata	*Pachira aquatica*
Ceiba pentandra	*Poulsenia armata*
Celtis schippii	*Pterocarpus officinalis*
Cespedezia macrophylla	*P. rohrii*
Cordia alliodora	*Quararibea asterolepis*
Couratari panamensis	*Sapium caudatum*
Dipteryx panamensis	*Schizolobium parahybum*
Enterolobium	*Sloanea terniflora*
schomburgkii	*S. zuliinensis*
Ficus bullenei	*Tabernaemontana arborea*
F. costaricana	*Tachigalia versicolor*
F. dugandii	*Terminalia amazonica*
F. insipida	*T. chiriquensis*
F. obtusifolia	*Tetragastris panamensis*
F. tonduzii	*Trattinnickia aspera*
F. trigonata	*Virola surinamensis*
F. yoponensis	*Zanthoxylum panamense*
Guapira standleyanum	

In some of these species, the buttressing may be slight, or it may not always be present on all individuals. Some species typically produce stilt or adventitious roots, which may be effective in providing additional support to the plant. These include the following:

Cecropia spp.	*Protium panamense*
Chamaedorea wendlandii	*Socratea durissima*
Ficus pertusa	*Tovomita longifolia*
Pourouma guianensis	*Trichanthera gigantea*

The trunk height (above ground) of the lowermost branches is also quite variable. In some species, such as *Jacaranda copaia,* the branches may be restricted to the uppermost part of the trunk. Branching is irregular in most cases, but may be markedly regular in the case of *Virola surinamensis,* where the branches are spirally arranged in a more or less systematic manner.

A common feature of many tropical trees, particularly of small or medium-sized trees that are not buttressed, is the production of sucker shoots near the base of the trunk. Sometimes this produces small clumps of trunks, as in the case of *Coccoloba acapulcensis* and *coronata* (Polygonaceae) and *Cupania sylvatica* (Sapindaceae). In other cases, the sucker shoots are not produced until one trunk has become fairly tall, as in *Guettarda foliacea* (Rubiaceae), *Nectandra cissiflora* (Lauraceae), and *Macrocnemum glabrescens* (Rubiaceae). Ecologically, the production of sucker shoots may play an important role, by ensuring a position in the forest for any tree whose major trunk is destroyed by windfall, lightning, or old age. A tree with well-developed sucker shoots attached to a well-developed root system can more quickly fill the void left by its principal trunk than can one that is in the seedling stage at the time the void is created—I have observed sucker shoots that produced a new trunk even where the fall of the main trunk had uprooted much of the root system.

Most of the small trees and shrubs are plants of the forest, though a few occur only in clearings. These include:

Adenaria floribunda	*S. rugosum*
Cordia spinescens	*S. subinerme*
Solanum asperum	*S. umbellatum*
S. jamaicense	*Triumfetta lappula*
S. ochraceo-ferruginum	*Vernonia patens*

As a habit class, small trees and shrubs merge imperceptibly with larger trees, on the one hand, and with suffruticose herbs, on the other. What would normally be an herb in an annually cleared area can become a stout woody shrub in the forest. A good example is *Witheringia solanacea.*

Climbing plants. Climbing plants, including both lianas and herbaceous vines and to some extent much smaller numbers of climbing shrubs and trees, are unique among habit classes in being able to move considerable distances in a relatively short time to reach a source of light. They are thus not so seriously hampered by the low light conditions in the forest as are the herbs.

Lianas are restricted to the forest and are best developed in the oldest forest. Of all growth forms they perhaps best characterize a tropical forest, simply because

they are so infrequent in temperate forests. There are 171 species (13% of the native flora) of lianas, climbing shrubs, and climbing trees on BCI. They are found principally in the following families, which contain 82% of all species of BCI lianas:

Bignoniaceae (24 species)
Leguminosae (22 species)
Sapindaceae (19 species)
Malpighiaceae (13 species)
Apocynaceae (11 species)
Dilleniaceae (8 species)
Vitaceae (6 species)
Smilacaceae (5 species)
Hippocrateaceae (5 species)
Combretaceae (5 species)
Connaraceae (4 species)
Rubiaceae (4 species)
Aristolochiaceae (3 species)
Menispermaceae (3 species)
Loganiaceae (3 species)
Convolvulaceae (3 species)
Verbenaceae (3 species)

Herbaceous vines are fewer in number in the forest than are the lianas. There are 94 herbaceous vines, 11 of which are epiphytic or hemiepiphytic. Vines generally do not grow over the surface of the canopy, as do lianas, but are most often growing *within* the canopy. Among the species of vines that regularly occur in the forest are the following:

Cayaponia granatensis	*Lygodium venustum*
Cissus erosa	*Mendoncia gracilis*
C. microcarpa	*M. littoralis*
C. pseudosicyoides	*Mucuna mutissiana*
C. rhombifolia	*Passiflora nitida*
C. sicyoides	*P. williamsii*
Cynanchum recurvum	*Piper aristolochiifolium*
Dalechampia cissifolia	*Psiguria bignoniacea*
D. dioscoreifolia	*P. warscewiczii*
D. tiliifolia	*Sabicea villosa*
Dioscorea haenkeana	var. *adpressa*
D. macrostachya	*Sicydium coriaceum*
D. polygonoides	*Smilax lanceolata*
D. sapindoides	*S. mollis*
D. urophylla	*S. panamensis*
Fevillea cordifolia	*S. spinosa*
Gurania coccinea	*S. spissa*
G. makoyana	*Wulffia baccata*
Ipomoea phillomega	

Though the distinction between lianas (woody climbers) and vines (herbaceous climbers) is usually readily apparent, it is sometimes difficult to tell if a climbing species is herbaceous or woody. This is especially true of climbers in some of the dicotyledonous families, including the Cucurbitaceae, Aristolochiaceae, Vitaceae, Leguminosae, Passifloraceae, Sapindaceae, Convolvulaceae, Apocynaceae, and Asclepiadaceae. In addition, climbers in two of the monocotyledonous families, the Dioscoreaceae and Smilacaceae, are sometimes woody.

Characteristic of the behavior of lianas is their relatively slow development in girth until they reach the light near the top of the canopy. They then develop larger stems and spread through and especially over the canopy. Also characteristic of both lianas and vines growing on or near the forest floor is their development of long branches that bear reduced leaves or no leaves at all. In either case, the internodes are very much more elongated than comparable parts that have begun to spread through the canopy. Leafless portions are often very long, and in some species they may extend across the forest floor for 100 m or more. Peñalosa (1975) has shown that at least in some cases a very small percentage of all branches reach the top of the canopy. Since most branches specialized for searching are probably operating at a photosynthetic loss, it can be assumed that the canopy portions of the plant are providing food for the searching portions. Since lianas may be very long-lived, a single individual may eventually occupy several spots in the canopy. Relatively few, however, are found to cross from one canopy tree to another. Lianas often survive, after the tree in which they were growing falls, by climbing another tree and becoming reestablished. A history of their success and failure can often be followed by tracing the largest trunk of the plant through the forest. In some cases a fallen liana produces adventitious roots, so that additional parts of the plant are less dependent on the original trunk for water and nutrients.

Lianas and vines have evolved a variety of adaptations that facilitate their climb to the canopy from the forest floor. These adaptations are often specific to a family. Tendrils are used for climbing in the Bignoniaceae, Vitaceae, Sapindaceae, Leguminosae (*Bauhinia*), Rhamnaceae (*Gouania*), and Passifloraceae. Sharp, recurved woody stipules are used in the genus *Machaerium* (Leguminosae). Stout, recurved axillary spines are employed in *Uncaria* (Rubiaceae), *Celtis* (Ulmaceae), *Pisonia* (Nyctaginaceae), and sometimes *Combretum* (Combretaceae) (see the figure on the next page). In *Combretum,* the spines are accompanied by persistent petioles that become woody and help the plant hold its position in the same way the spines do. *Byttneria aculeata* (Sterculiaceae) and *Solanum lanciifolium* (Solanaceae) and some species of *Smilax* (Smilacaceae) bear recurved prickles all along the stem; prickles are also present on the lower blade surface of *Byttneria* and *Solanum lanciifolium*. Such adaptations have the effect of causing a plant to climb by allowing it to work its way upward (but not down again) during movements caused by the wind or by animals.

Lianas of the Hippocrateaceae utilize an unusual method for holding themselves in position. In most members of the family, and perhaps in *Securidaca diversifolia* (Polygalaceae), twining branches grasp branches of other plants with which they come in contact. In some groups, such as *Cissus* (Vitaceae) and *Aristolochia* (Aristolochiaceae), the plants develop a corky periderm, or outer bark, which is less subject to slippage than is a smooth periderm and may thus be helpful to the plant in maintaining its position in the canopy.

The conspicuous swelling of the nodes in *Gnetum*

Understory vegetation along Donato Trail

Two armed forest species: juvenile tree of *Zanthoxylum setulosum* at left, stems of *Combretum decandrum* at right

Liana in the old forest

(Gnetaceae) and in many of the Bignoniaceae probably assists plants in holding their position by preventing them from slipping back. Many lianas and high-climbing herbaceous vines (indeed, about 50% of all scandent plants) have opposite branching, which often prevents the plant from falling by providing a wider area of support.

Vines, as well as plants that are late in becoming woody, such as the Convolvulaceae, Apocynaceae, Piperaceae (*Piper aristolochiifolium*), Rubiaceae (*Manettia* and *Sabicea*), Schizaeaceae, Compositae (*Wulffia*), and Asclepiadaceae, usually climb by twining around the trunks or branches of trees. Vines of the genus *Desmoncus* (Palmae) climb by means of modified leaflets that form stout, recurved spines. *Scleria bracteata* (Cyperaceae), a vine of weedier forest edges, climbs over and into trees by means of retrorse pubescence on its stems and leaves. Juvenile stems of the liana *Prionostemma aspera* bear similar trichomes, as do the young herbaceous stems of many other vines and lianas, such as the Dilleniaceae. Such adaptations are most advantageous higher in the canopy or in open areas where movement by the wind is more effective.

Juvenile stages of a number of species climb the sides of trunks or trees by producing roots that grow into the bark (Araceae; *Hydrangea,* Saxifragaceae) or by using tendrils that are claw-shaped (*Macfadyena unguis-cati,* Bignoniaceae).

A number of species are intermediate in habit in the sense that they may be either arborescent or scandent, depending on the conditions under which they grow. Ten species are scandent or climbing trees or shrubs, rather than lianas. In the canopy or along the lake or at the margins of clearings they may be difficult to separate from lianas, but they generally have a stout, more or less erect, and often unsupported trunk. These include:

Cassia undulata	*S. tenuifolia*
Dalbergia monetaria	*Strychnos brachistantha*
Heterocondylus vitalbis	*Tournefortia angustiflora*
Justicia graciliflora	*T. cuspidata*
Securidaca diversifolia	*Wulffia baccata*

Two other climbing shrubs are hemiepiphytic:

Hydrangea peruviana	*Souroubea sympetala*

Twelve of the lianas may sometimes be climbing shrubs:

Acacia riparia	*C. turczaninowii*
Allamanda cathartica	*Heteropteris laurifolia*
Byttneria aculeata	*Machaerium arboreum*
Chamissoa altissima	*Pisonia aculeata*
Chomelia psilocarpa	*Pouzolzia obliqua*
Connarus panamensis	*Strychnos panamensis*

Herbaceous plants. The herbaceous element of BCI's flora is even more diverse in habit type than is the woody element. Even excluding the many palms that are arborescent in habit and woody in the stems, the herbaceous flora remains more diverse than the woody element. The herbs may be annual, perennial, or suffruticose; they may be erect, sprawling, vinelike, epiphytic, hemiepiphytic,

saprophytic, parasitic, or aquatic; and they include the small palms and other, often giant, monocotyledonous herbs (such as many members of the Musaceae, Zingiberaceae, and Marantaceae) that may compete with the shrubs for light.

Most of the herbaceous ground plants in the flora occur principally in man-made clearings and at the margin of the lake. Of the 560 phanerogamic herbs on BCI (including 94 vines), only 79 (including 4 saprophytes) are restricted to the forest floor. In contrast, 197 herb species (including suffruticose herbs) occur in clearings, and 54 species are aquatic. The smaller number of herbaceous species that inhabit the forest floor is no doubt due to the small amount of light that reaches the ground. Many of these species have broad leaves adapted to absorbing a maximum of light in these relatively dark parts of the forest.

Saprophytic plants, relatively uncommon on BCI, are restricted to the genera *Voyria* (Gentianaceae) and *Thismia* (Burmanniaceae). The orchid genus *Triphora* is also, apparently, to some extent saprophytic.

One herbaceous species, *Apodanthes caseariae* (Rafflesiaceae), survives the relative absence of light near the forest floor by parasitizing the trunks of *Casearia* (Flacourtiaceae). The visible part of the plant consists of a single small flower.

The ecological conditions of the aquatics are unique in the sense that the plants suffer no deficiency of water and rarely a serious deficiency of light. This habit type is discussed in detail in the section on shorelines and marshes (p. 9).

Suffruticose herbs are often difficult to place in a habit class, since, depending on their age, they may either be herbaceous throughout or have a stout woody stem. This pattern of development is especially common in the Malvaceae and Acanthaceae. Suffruticose herbs are common in areas of repeated disturbance, such as clearings and trailsides, since cutting them back seldom causes permanent damage.

Epiphytic and hemiepiphytic plants. Most true epiphytes in BCI's flora are herbaceous (defined as including vines). They are found in the following families: Orchidaceae (82 species), Polypodiaceae (31), Araceae (24), Bromeliaceae (18), Piperaceae (10), Hymenophyllaceae (9), Cactaceae (3), and Begoniaceae, Cyclanthaceae, Gesneriaceae, Lycopodiaceae, and Rubiaceae (1 each). The only woody epiphytes in the flora are *Codonanthe* and *Columnea* (Gesneriaceae).

For the true epiphytes, light does not need to be a limiting factor, since they may be found at any level in the forest, from near the ground to the top of the canopy. Apparently, the advantage of moving higher to receive more light is offset by the increasingly arid conditions at the higher levels and the fewer nutrients available in fallen debris, for the majority of epiphytes are in fact found at lower to intermediate levels. It is interesting to note that the only species known to live at the very top of the canopy in full sunlight is *Aechmea tillandsioides* (Bromeliaceae), which is a tank epiphyte with a built-in

water reservoir, and is usually found growing on ant nests. The nutrients derived from the microfauna in the tank reservoir and from the ant nest apparently compensate for its otherwise nutrient-poor position.

Epiphytes have developed several methods of obtaining an adequate water supply in the relatively acute dry season. In the case of *Trichomanes* (Hymenophyllaceae) and some species of *Polypodium* (Polypodiaceae), plants may be rejuvenated even after serious dessication. Orchids and most *Anthurium* (Araceae) have roots that absorb moisture from the air. The Bromeliaceae bear specialized peltate scales that absorb and hold moisture. Most epiphytes have a thick epidermis that prevents the escape of water, and their thick tissues also act as storage organs for water. In some epiphytes (especially Bromeliaceae and *Anthurium*), the leaves are arranged in rosettes that may catch and hold debris from which the roots may withdraw nutrients and water.

Epiphytes may be found attached on nearly any part of a tree. The larger ones, however, are usually found in the crotches of trunks or branches, owing to the greater support and collection of detritis afforded there. Small epiphytic plants, such as *Codonanthe* and *Columnea*, may attach themselves even to the vertical surfaces of trunks, by means of small roots.

Hemiepiphytes are those that rely on another plant for support but whose roots are in the ground. Most hemiepiphytes are shrubs or small trees, but a few are herbaceous, especially *Monstera, Philodendron,* and *Rhodospatha* (Araceae), as well as *Diodia sarmentosa* (Rubiaceae) and *Polybotrya* and *Maxonia* (Polypodiaceae). Woody hemiepiphytic plants include *Marcgravia* and *Souroubea* (Marcgraviaceae), *Havetiopsis* and *Clusia* (Guttiferae), *Ficus* and *Coussapoa* (Moraceae), *Topobaea* (Melastomataceae), *Oreopanax* (Araliaceae), *Markea* and *Lycianthes* (Solanaceae), *Columnea purpurata* (Gesneriaceae), *Cosmibuena* (Rubiaceae), and *Hydrangea* (Saxifragaceae).

It is not always easy to distinguish true epiphytes from hemiepiphytes. Some groups of herbaceous plants, such as *Philodendron, Rhodospatha,* and *Monstera* in the Araceae, begin life as hemiepiphytes but later may become true epiphytes. Others, such as *Ficus, Clusia,* and *Cosmibuena,* begin as true epiphytes and ultimately send roots to the ground. *Philodendron radiatum* also follows this pattern, though for most *Philodendron* the reverse is true.

The woody hemiepiphytes range in size from the small *Diodia sarmentosa* (Rubiaceae) to *Ficus* and *Coussapoa panamensis* (Moraceae). Most hemiepiphytic shrubs and trees grow high in the canopy and are supported either by a stout branch or by the crotch of a large tree. From this point a stout, usually solitary, stem extends down the side of the supporting trunk to the ground. These species usually have no other obvious adventitious roots that could absorb water. Some species, such as *Clusia* and *Havetiopsis* (Guttiferae) and *Cosmibuena* (Rubiaceae), compensate for this lack by retaining moisture in their very thick, leathery leaves. Still others, such as *Topobaea praecox* (Melastomataceae), are leafless during part of

the dry season, and the plant may be somewhat dormant at this time. Other genera, such as *Coussapoa* (Moraceae) and *Souroubea* (Marcgraviaceae), may produce numerous roots, at least some of which do not reach the ground; these species probably acquire part of their nutrients from the accumulation of falling debris before it reaches the ground.

Ficus is unusual in sending a number of stems to the ground, the stems ultimately coalescing to form a united structure that strangles the host tree.

The hemiepiphytic Araceae attach themselves to tree trunks by means of small rootlets that become fastened to cracks in the tree's bark. Plants such as *Clusia* and some species of *Ficus,* such as *F. obtusifolia,* have more elaborate methods of attachment: in general, seedlings of these species are found on the side of a tree trunk, often growing from a small crack; as the seedling grows and its trunk elongates (usually at a sharp angle to the host trunk), it develops roots that encircle the trunk of the supporting tree as well as its own trunk to bind the two firmly together.

Since hemiepiphytes are usually rooted into the soil, they clearly have an advantage in obtaining water and nutrients that true epiphytes do not share. Nevertheless, the usually small diameter of the stems or trunks must put them at some disadvantage in this respect. The stem of *Markea ulei* (Solanaceae) is particularly small, but this species often compensates by developing a swollen enlargement of the stem that presumably acts as a storage organ during times of stress.

Perhaps the most unusual habit of woody plants, as opposed to herbs, is the parasitic habit. Although there is one parasitic herb in the flora, all other parasitic plants on BCI (seven species) are shrubs or somewhat woody vines of the family Loranthaceae. All seven of these are photosynthetic, but they obtain their water and presumably part of their nutrients from their host. All grow in the canopy and especially along the shore in exposed areas.

Floristic Composition

In his *Flora of Barro Colorado Island* (1933), Standley included a list of the nonvascular plants known from the island. This list, totaling 204 species, included algae in the families Desmidiaceae (34 species), Characeae (1), and Rhodophyceae (1), as well as myxomycete fungi (11), ascomycete fungi (54), basidiomycete fungi (55), imperfect fungi (9), lichens (9), hepatics (11), and mosses (15). Although no thorough modern survey of any of the above groups has been published for BCI, it can be assumed that Standley's list represents only a small portion of the nonvascular plants occurring on the island.

The present work deals only with the vascular flora of BCI, which includes 1,369 taxa. Of these, 104 (8%) are cryptogams, 2 are gymnosperms, 353 (25.8%) are monocotyledons, and 910 (66.5%) are dicotyledons. The average number of native species per genus is 1.9. The largest genera are *Piper* (Piperaceae), 21 species; *Psychotria* (Rubiaceae), 20; *Inga* (Leguminosae), 18; *Ficus* (Mora-

Polypodium phyllitidis and *Nephrolepis pendula,* growing as tree-crotch epiphytes at 20 m (65 ft)

A strangling *Ficus* climbing its host. . .

. . . and the void once filled by the host

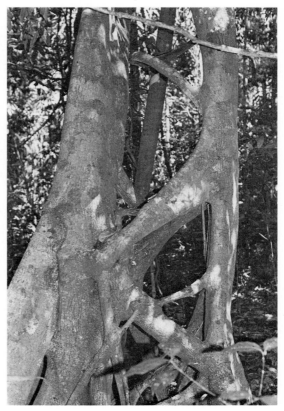

ceae), 17; *Miconia* (Melastomataceae), 14; *Polypodium* (Polypodiaceae), 13; *Philodendron* (Araceae), 13; *Epidendrum* (Orchidaceae), 13; *Anthurium* (Araceae), 12; *Cyperus* (Cyperaceae), 11; *Solanum* (Solanaceae), 11; *Panicum* (Gramineae), 10; *Peperomia* (Piperaceae), 10; and *Trichomanes* (Hymenophyllaceae), 10. A detailed listing of numbers of genera and species per family, the families arranged by order and division, can be found in the section on classification and family sequence (p. 61). A few species of recently introduced cultivated plants are not treated in this work, both because they are considered unimportant and because they cannot be identified with certainty.

A few species that are cultivated elsewhere are treated here as native because they are seemingly naturalized on the island. These include *Chrysophyllum cainito, Mangifera indica, Phoebe mexicana, Syzygium jambos, Theobroma cacao,* and *Vanilla fragrans.* The 53 species that are cultivated on the island are described in this flora, but they have been excluded for the most part from the discussion of seasonal behavior and geographical affinities. The cultivated species, excluding the six that are evidently naturalized, are the following:

Acalypha wilkesiana	*D. trifida*
Amaryllis belladonna	*Ervatamia coronaria*
Anacardium occidentale	*Eugenia uniflora*
Ananas comosus	*Ficus retusa*
Annona muricata	*Garcinia mangostana*
Araucaria excelsa	*Heliconia metallica*
Artocarpus altilis	*H. pogonantha*
Averrhoa carambola	*Hibiscus rosa-sinensis*
Bactris gasipaes	*Ipomoea batatas*
Bambusa arundinacea	*Ixora coccinea*
B. glaucescens	*Jacquinia macrocarpa*
Caesalpinea pulcherrima	*Jatropha curcas*
Cajanus bicolor	*Mammea americana*
Caladium bicolor	*Manihot esculenta*
Calathea villosa	*Musa sapientum*
Capsicum annuum	*Persea americana*
Carica papaya	*Polyscias guilfoylei*
Citrus (6 species)	*Psidium guajava*
Clerodendrum	*Saccharum officinarum*
paniculatum	*Spathodea campanulata*
Cocos nucifera	*Syngonium* sp.
Codiaeum variegatum	*Thunbergia erecta*
Coleus blumei	*Xanthosoma nigrum*
Cordyline fruticosa	*Zingiber officinale*
Dioscorea alata	

Sexual Characteristics

Of the 1,265 phanerogamic plants (including cultivated plants) known for the flora, 304 (24%) have unisexual flowers. Of these, 115 (9%) are dioecious:

Abuta (2 species)
Acalypha macrostachya (also monoecious)
Adelia triloba
Alchornea costaricensis

A. latifolia
Alibertia edulis
Amaioua corymbosa
Apodanthes caseariae
Araucaria excelsa
  (also monoecious; cultivated)
Astronium graveolens
Baccharis trinervis
Bursera simaruba (polygamodioecious)
Carica cauliflora
C. papaya (cultivated)
Catopsis sessiliflora (also monoecious)
Castilla elastica (also monoecious)
Cayaponia granatensis
Cecropia (4 species)
Chamaedorea wendlandiana
Cissampelos (2 species)
Chondrodendron tomentosum
Clusia odorata
Coccoloba coronata
Cordia panamensis
Coussapoa (2 species)
Dioscorea (7 species, 2 cultivated)
Diospyros artanthifolia
Dorstenia contrajerva
Drypetes standleyi
Fevillea cordifolia
Gnetum leyboldii var. *woodsonianum*
Guapira standleyanum
Guarea glabra
G. multiflora
Gurania (3 species)
Gynerium sagittatum
Hampea appendiculata var. *longicalyx*
Havetiopsis flexilis
Hydrilla verticillata
Hyeronima laxiflora
Iresine celosia
Jacaratia spinosa
Limnobium stoloniferum
Maquira costaricana
Margaritaria nobilis
Momordes powellii
Myriocarpa yzabalensis
Neea amplifolia
Ocotea (4 species)
Odontocarya (2 species)
Olmedia aspera
Perebea xanthochyma
Picramnia latifolia
Pisonia aculeata
Pourouma guianensis
Pouteria stipitata
Protium costaricense
P. panamense
P. tenuifolium
Pseudolmedia spuria
Psiguria (2 species)
Randia (2 species)

Rheedia acuminata (polygamodioecious)
R. edulis
Scheelea zonensis (also monoecious)
Sicydium coriaceum
Simarouba amara
Siparuna pauciflora
Smilax (5 species)
Sorocea affinis
Struthanthus orbicularis
Stylogyne standleyi
Tetragastris panamense
Tovomita longifolia
T. stylosa
Trattinnickia aspera
Trichilia (4 species)
Triplaris cumingiana
Trophis racemosa
Urera eggersii
Virola (2 species)
Xylosma (2 species)
Zanthoxylum (4 species)

There are 139 monoecious species in the flora:

Acalypha (4 species, 1 cultivated)
Amaranthus viridis
Araucaria excelsa (cultivated)
Artocarpus altilis (cultivated)
Asplundia alata
Astrocaryum standleyanum
Bactris (5 species, 1 cultivated)
Begonia (3 species)
Boehmeria cylindrica
Brosimum alicastrum
Caladium bicolor (cultivated)
Calyptrocarya glomerulata
Carludovica (2 species)
Castilla elastica (also dioecious)
Catasetum (2 species)
Catopsis sessiliflora (also dioecious)
Cayaponia (2 species)
Cedrela odorata
Celtis iguaneus
Ceratophyllum demersum
·*Chamaesyce* (4 species)
Clibadium surinamense
Coccoloba (4 species)
Cocos nucifera (cultivated)
Codiaeum virgatum (cultivated)
Croton (3 species)
Cyclanthus bipartitus
Dalechampia (3 species)
Desmoncus isthmius
Dieffenbachia (3 species)
Elaeis oleifera
Ficus (17 species, 1 cultivated)
Garcia nutans
Geonoma (3 species)
Homalomena wendlandii
Hura crepitans

Jatropha curcas (cultivated)
Lithachne pauciflora
Ludovia integrifolia
Mabea occidentalis
Mammea americana (cultivated)
Manihot esculenta (cultivated)
Melothria (2 species)
Momordica charantia
Montrichardia arborescens
Musa sapientum (cultivated)
Oenocarpus panamanus
Olyra latifolia
Omphalea diandra
Pharus (2 species)
Philodendron (13 species)
Phoradendron (2 species)
Phyllanthus (3 species)
Pilea microphylla
Pistia stratiotes
Poinsettia heterophylla
Posadaea sphaerocarpa
Poulsenia armata
Pouzolzia obliqua
Sagittaria lancifolia
Sapium (2 species)
Scheelea zonensis (also dioecious)
Scleria (5 species)
Siparuna guianensis
Socratea durissima
Sterculia apetala
Synechanthus warscewiczianus
Syngonium (2 species)
Trema micrantha
Trichospermum mexicanum
Typha domingensis
Xanthosoma (3 species)

The following species, included on both of the preceding lists, may be either monoecious or dioecious:

Acalypha macrostachya
Araucaria excelsa (cultivated)
Castilla elastica
Catopsis sessiliflora
Scheelea zonensis

Finally, there are 54 polygamous species on the island:

Allophylus psilospermus (polygamodioecious)
Baltimora recta
Chaptalia nutans
Cladium jamaicense
Conyza (2 species)
Cupania (4 species)
Eclipta alba
Erechtites hieracifolia
Garcinia mangostana (cultivated)
Heliocarpus popayanensis (gynomonoecious)
Mammea americana (also monoecious, cultivated)
Maytenus schippii
Melampodium divaricatum

Oreopanax capitatus
Paullinia (9 species)
Pluchea odorata
Rhynchospora (3 species)
Schistocarpha oppositifolia
Serjania (9 species)
Synedrella nodiflora
Talisia (2 species)
Tetracera (3 species, androdioecious)
Thinouia myriantha
Tovomitopsis nicaraguensis
Tridax procumbens
Verbesina gigantea
Vismia billbergiana
Vitis tiliifolia
Wedelia trilobata

Geographical Affinities

The BCI flora has substantial geographic affinities with the Central and South American floras. A study of these affinities by Croat and Busey (1975) reviews the geological history of the isthmus and treats the epiphytes, the lianas, and the common tree species. The data presented here embrace all of the species in the flora, except the culti-vated species.

Each species in the flora (with the few exceptions men-tioned later) is assignable to one of five range categories:

Type I   Endemic: restricted to Panama
Type II   Wide endemic: mainly in Panama but extending to Costa Rica and/or northern Colombia
Type III   Central American: mostly from Mexico to Panama
Type IV   South American: mostly from Panama to South America, but also sometimes to Costa Rica if widespread in South America
Type V   Pan-American: mostly from Mexico to South America

Table 4 lists the numbers of BCI species in each of these categories, by major habit-and-habitat classes. Each percentage shown is the percentage of the total habit-and-habitat class assignable to the indicated range category. Since most BCI species that are found also in the West Indies are wide-ranging (Type V) species, it is unneces-sary to treat the West Indian distributions as a separate category. Rather, the number of BCI species occurring also in the West Indies is listed beneath the total for each category (and is *already included* in the upper number). West Indian species are restricted to categories III, IV, and V. For the purpose of this study Trinidad was con-sidered a part of South America and not the West Indies.

A few species —*not reflected in any of the categories*— are known only from the West Indies and Panama, or only from the West Indies plus Costa Rica, Panama, and northern Colombia. Those species known only from Panama and the West Indies are *Habenaria bicornis* (Orchidaceae) and *Polypodium costatum* (Polypodiaceae).

In addition, there are two species—*Tillandsia subulifera* (Bromeliaceae) and *Securidaca tenuifolia* (Polygalaceae) —that occur only in Panama and Trinidad. Three species —*Vriesia ringens* and *V. sanguinolenta* (Bromeliaceae) and *Beilschmiedia pendula* (Lauraceae)—are known from Costa Rica to Colombia as well as from the West Indies. *Hydrilla verticillata* (Hydrocharitaceae) is excluded from consideration; it is known from the Old World, but the range in the New World is poorly known. A number of other Old World species are included if their range in the New World is known. *Sacciolepis striata* (Gramineae), known from the United States, the West Indies, and Panama, is excluded because its range does not fit any of the categories. *Vatairea erythrocarpa* (Leguminosae) is excluded because its determination is doubtful. All of the species considered cultivated are also excluded.

Each of the five range categories is discussed briefly in what follows. In the discussions, climbing shrubs and trees are included with lianas. Because of the limited numbers of species in some of the other groups—the parasitic shrubs, hemiepiphytes, tree ferns, suffruticose herbs, parasitic herbs, and saprophytic herbs—their distribution by category will not be discussed.

Type I: Endemic to Panama. A total of 92 BCI species (7%) are endemic to Panama. There are no endemic aquatic or suffruticose herbs or cryptogams and only three endemic clearing herbs. In each of the remaining habit-and-habitat classes, 6–13% of the species are en-demic to Panama. Thus no class is significantly more endemic than any other.

Type II: Wide endemic. Wide endemics account for 122 BCI species (9%). No aquatic herbs and only two clearing herbs and one suffruticose herb are found in this category. Vascular cryptogams are represented in about the same proportion as seed plants. The remaining classes each constitute from 6 to 15% of this range category. Most significant is the fact that 15% of all shrubs are wide endemics.

Type III: Central American. A total of 180 BCI species (14%) are distributed only in Central America. Of these, 36 (20%) are also found in the West Indies. Very small percentages of aquatic and clearing herbs are found in this category. In each of the remaining habit-and-habitat classes, 7–22% of the species are in this category.

Although there is an insignificant difference between the total numbers of Central American and South Ameri-can species on the island (180 vs. 135), there are signi-ficantly more large trees (those more than 10 m tall) with a Central American distribution than with a South Amer-ican distribution. The same is true of shrubs. Also sig-nificant is the fact that ten large trees of Type III distri-bution occur also in the West Indies, whereas there are only two Type IV species in the West Indies. There are about twice as many Type III arborescent species in the West Indies as Type IV arborescent species, but con-sidering the sample size the numbers are probably not significant.

Terrestrial cryptogams also show a significantly larger number of species with Central American distribution than with South American distribution.

TABLE 4
Geographical affinities of the BCI flora
(lower number is number occurring also in West Indies)

Habit-and-habitat class	Total species		Range category (see text)				
		I	II	III	IV	V	Other
VASCULAR CRYPTOGAMS							
Aquatic cryptogams	6				2	4	
Epiphytic cryptogams	42		2(5%)	5(13%)	7(18%)	27(64%)	1
				2	3	23	
Terrestrial cryptogams	51		5(10%)	11(22%)	2(4%)	33(65%)	
				1		26	
Tree ferns	5		1		1	3	
						2	
TOTAL VASCULAR CRYPTOGAMS	104		8(8%)	16(15%)	12(12%)	67(64%)	1
				3	3	49	
PHANEROGAMS							
Trees more than 10 m tall	211	16(8%)	28(12%)	45(21%)	20(10%)	99(47%)	3
				10	2	44	1
Small trees and shrubs (not including those that are only shrubs)	154	20(13%)	19(12%)	25(17%)	21(13%)	69(45%)	
				5	3	30	
Shrubs (always less than 3 m tall)	93	9(10%)	14(15%)	17(18%)	4(4%)	49(53%)	
				2	1	28	
Hemiepiphytic shrubs or trees	16	1	1	4	3	7	
						2	
Parasitic shrubs	7				3	4	
					3	1	
TOTAL ARBORESCENT PHANEROGAMS	481	46(9%)	62(13%)	91(19%)	51(11%)	228(47%)	3
				22	9	105	1
Lianas (including climbing trees and shrubs)	171	13(8%)	13(8%)	26(15%)	26(15%)	91(53%)	2
				4	1	37	
Vines	83	5(6%)	8(10%)	6(7%)	17(21%)	46(55%)	1
					5	25	
Hemiepiphytic or epiphytic vines	11	1	2	2	1	5	
						2	
TOTAL SCANDENT PHANEROGAMS	265	19(7%)	23(9%)	34(13%)	44(17%)	142(54%)	3
				5	8	60	
Forest herbs	75	7(9%)	7(9%)	8(11%)	3(4%)	49(65%)	1
				1	1	20	1
Aquatic herbs	54			2(4%)	1(2%)	49(91%)	2
				1	1	32	
Clearing herbs	179	3(2%)	2(1%)	7(4%)	5(3%)	162(90%)	
				3	2	127	
Epiphytic herbs	135	16(12%)	18(13%)	20(15%)	16(12%)	62(46%)	3
				1	2	33	
Parasitic and saprophytic herbs	5	1	1	1	1	1	
Suffruticose herbs	18		1	1	2	14	
					1	10	
TOTAL HERBACEOUS PHANEROGAMS	466	27(6%)	29(6%)	39(9%)	28(6%)	337(72%)	6
				6	7	222	1
TOTAL PHANEROGAMS	1,212	92(8%)	114(9%)	164(14%)	123(10%)	707(58%)	12
				33	24	387	2
TOTAL SPECIES	1,316	92(7%)	122(9%)	180(14%)	135(10%)	774(59%)	13
				36	27	436	2

In contrast to the trees, shrubs, and terrestrial cryptogams, there are more herbaceous vines with South American distribution than with Central American distribution (17 (21%) Type IV vs. 6 (7%) Type III).

Type IV: South American. A total of 135 BCI species (10%) have a chiefly South American distribution. Of these, 27 (20%) also occur in the West Indies. There are an insignificant number of aquatic herbs and very small numbers of clearing and forest herbs, terrestrial

cryptogams, and shrubs in this category. In the remaining classes, 10 to 21% of the species have Type IV distributions.

Comparisons between Central and South American distributions are made above, in the discussion of the Type III category. Although the total numbers of Type III and Type IV species in each habit-and-habitat class differ insignificantly, the greater number of species from Central America is somewhat surprising, since the South

American continent must have had a larger number of species to contribute to the isthmian flora. This is perhaps explained by the fact that the same uplift which created the isthmian land bridge in the late Tertiary also elevated the Andes mountains and blocked direct overland migration from a substantial part of the South American continent (Croat & Busey, 1975).

Type V: Pan-American. The largest distribution category for nearly every habit-and-habitat class is Type V, species extending throughout much or all of the tropical regions of North and South America and occasionally the West Indies, as well. Of all BCI species, 774 (59%) are in this category. Of the total Type V species, 436 (56%) are also found in the West Indies.

The habit-and-habitat class with the largest Type V distribution is aquatic herbs, 91% of which are wideranging, including 32 species also found in the West Indies. Close behind are the clearing herbs, 90% of which are Type V, including 127 species in the West Indies. In the remaining habit-and-habitat classes, 46 to 65% of the species are of Type V distribution. Some of these differences may be significant. I would not have expected epiphytic herbs to have one of the lowest percentages of Type V species, since a large proportion of epiphyte seeds are wind-dispersed. But as might be expected, many of the animal-dispersed epiphytic species, such as members of the Araceae, are less widely distributed. It is not surprising that a large proportion of vascular cryptogams are in the Type V category, since these spore-bearing plants are known to be widespread, no doubt owing to the ease with which their minute spores are borne on the wind.

There is little doubt that the species common to Central and South America and the West Indies have reached the West Indies (or migrated from the West Indies) by long-distance dispersal, since there is no evidence that the West Indies were ever connected to the mainland (Darlington, 1957; Graham, 1972). Presumably, long-distance dispersal could explain the migration of these species between Central and South America, as well, though other mechanisms might have been involved.

Of the 1,316 species considered here, only 501 (38%) are found in the West Indies, whereas 954 (72%) are found in Central America (not including endemics in Types I and II), 909 (69%) in South America. This closer relationship to contiguous land masses would seem to be evidence that modern distributions resulted chiefly from overland migrations rather than from long-distance dispersal; for if long-distance dispersal had been more generally important, one would expect a higher percentage of BCI species in the West Indies.

Historical and Recent Changes in the Flora

The flora of BCI has undergone significant modification since the creation of the island in 1914. At that time the island was only a series of hills with rapidly flowing streams. Though these hills included no permanent standing body of water that could support the rich aquatic community which flourishes in the area today, the Río Chagres flowed past on the north and east only a short distance away. And indeed, the channel of the Panama Canal follows roughly the old valley of the Chagres as it passes Barro Colorado Island. By the time Kenoyer made his ecological studies 15 years later (Kenoyer, 1928, 1929), the aquatic vegetation of the shore was much as it is today. His descriptions of the aquatic associations found there differ from mine only in two respects: he listed far fewer species for the associations; and the shoreline now has many fewer emergent tree stumps. Though the advancement of hydrarch succession over half a century has very likely fostered the addition of species to the aquatic associations, Kenoyer's lower species count was probably due primarily to inadequate sampling of the associations.

The presence of large, chiefly water-dispersed trees, such as *Erythrina fusca, Cynometra bauhiniifolia,* and *Pachira aquatica,* which are restricted to the shore, is evidence that such trees were also able, rather early, to invade the newly formed shore. These species are found along the shore in the areas of hydrarch succession but not on the eroded banks on the north and east sides of the island. At the same time, it seems likely that along some parts of the shore there are a number of species persisting that would not normally occur so near water.

Other changes in the flora have been brought on by the silting of the numerous coves and by the resulting formation of sandbars, which have added niches (as discussed on p. 13).

Although less dramatic than the floristic changes along the shore, the changes in the forest have been significant. About half of BCI was used intermittently for agriculture until shortly before its establishment as a preserve (Chapman, 1938). Successional changes were thus very great in the flora, especially in the first few years of the island's existence, and a number of species may have been eliminated through succession. The reduction in the number and size of clearings eliminated a number of plant species and some bird species as well (Willis, 1974). Species of plants that may no longer be present, either as a result of succession or because their weedy habitats were eliminated, include:

Aciotis levyana	*Erechtites hieracifolia*
Alternanthera sessilis	var. *cacalioides*
Amaranthus viridis	*Gomphrena decumbens*
Anthurium flexile	*Hebeclinium*
Bidens pilosa	*macrophyllum*
Casearia corymbosa	*Heliotropium indicum*
Cayaponia glandulosa	*Indigofera mucronata*
C. racemosa	*Iresine celosia*
Centropogon cornutus	*Mandevilla subsagittata*
Columnea purpurata	*Melothria trilobata*
Corchorus siliquosus	*Merremia umbellata*
Crotalaria retusa	*Microtea debilis*
Cyphomandra allophylla	*Pavonia paniculata*
Desmodium cajanifolium	*Piper peracuminatum*
D. distortum	*Posadaea sphaerocarpa*
D. tortuosum	*Rivina humilis*

Schistocarpha oppositifolia   S. rugosum
Siparuna guianensis   Spananthe paniculata
Solanum   Stachytarpheta jamaicense
  ochraceo-ferrugineum

Many of the above are cultivated plants or weeds of cultivated fields that probably disappeared after the large garden area north of the present Laboratory Clearing was allowed to revert to forest. Very likely, many species of crop weeds once present on the island were either missed by Standley (1933) in his listing of species or had already disappeared by 1930. Many weedy plants commonly associated with crops do not persist long once cultivation ceases.

Other very rare or restricted species may also be on the verge of disappearing from the island. These include:

Acalypha arvensis   Elytraria imbricata
Banara guianensis   Lycopodium cernuum
Cleome parviflora   Pavonia dasypetala
Cochlospermum vitifolium   Phytolacca rivinoides
Dioclea guianensis   Pluchea odorata

A number of species whose seedlings do not survive well in the forest are considered by Knight (1975a) to be infrequent reproducers, as discussed and listed on p. 8. Some of these species might be eliminated as succession progresses.

Other species were not seen during the course of my work on the island. They are all believed to be rare, and some may by now have dropped out of the flora. These are:

Abuta panamensis   M. hookeriana
Adiantum lunulatum   Mucuna rostrata
A. seemannii   Ophioglossum reticulatum
Anthephora   Oryza latifolia
  hermaphrodita   Paspalum repens
Asplenium pteropus   Passiflora menispermifolia
Bacopa salzmannii   P. seemannii
Begonia patula   P. williamsii
Bellucia grossularioides   Pitcairnia heterophylla
Blechum serrulatum   Pithecellobium
Ceratopteris pteridoides   barbourianum
Combretum cacoucia   Polygonum acuminatum
Ctenitis sloanei   P. hydropiperoides
Diastema raciferum   Portulaca oleracea
Dicranopteris flexuosa   Prestonia acutifolia
Digitaria ciliaris   Psychotria uliginosa
Dryopteris sordida   Pteris grandifolia
Elaphoglossum hayesii   P. pungens
Fischeria funebris   Rhynchospora micrantha
Gonolobus allenii   Sau.rauia laevigata
Hemidictyum marginatum   Scirpus cubensis
Hyptis brevipes   Securidaca tenuifolia
Ichnanthus tenuis   Spermacoce tenuior
Leptochloa virgata   Stemodia verticillata
Limnobium stoloniferum   Tetrapteris seemannii
Marsdenia crassipes   Thelypteris balbisii
Matalea pinquifolia   Tillandsia fasciculata
Mecardonia procumbens   var. convexispica
Mikania guaco   T. subulifera

Tournefortia maculata   Vriesia gladioliflora
Trichopteris microdonta   V. ringens

Other changes in the flora have been brought about by introduced weeds or pasture grasses. Some, such as Saccharum spontaneum, are believed to have been introduced into Panama recently. Certain other species are believed to be transient clearing weeds that do not persist long but are later reintroduced. The Rear #8 Lighthouse Clearing has been particularly rich in species that at least appear to be transient members of the flora. Each year some additional species are collected there, but some species seen in previous years are often not in evidence—and perhaps the new ones will not be next year. These species, however, are relatively few in number.

Phenological Characteristics

Leaf fall and leaf flushing. Phenological observations of the three classical sorts—flowering, fruiting, and leaf fall—have been made for most species. Observations of leaf fall were based on field notes made on BCI and in adjacent areas of the Canal Zone during the years 1967–74. Observations of flowering and fruiting were made at the same time, and were supplemented by studies of herbarium specimens (Croat, 1975d).

Many plants, particularly the trees and lianas, lose their leaves in the dry season. In many species, the leaves fall at or near the onset of the dry season, but in some, leaf shedding is continuous throughout the dry season. Studies conducted by the U.S. Army Tropic Test Center (1966) at the Albrook Air Force Base test site on the Pacific slope in Panama show that litter fall declines sharply in February and March. Litter accumulation increases until May, then drops sharply. Similar studies by the Smithsonian Environmental Monitoring Program (Rubinoff, 1974) show maximum leaf-litter accumulations in December and January, followed by a rapid decrease in February and further diminishment in April and May. By February, the forest canopy begins to look bare, at least relative to its appearance in the rainy season, and the atmospheric humidity is much lower. Winds, which increase markedly during the dry season, may be felt even at ground level in the depths of the forest.

Leaf litter, which includes falling flowers and fruits and other debris, accumulates to a depth of several inches in some places by the end of the dry season (personal observation). Measurements by Woods and Gallegos on BCI (1970) show that more than 10 metric tons per hectare of litter accumulate during the months June through August.

The beginning of the rainy season brings a rapid increase in the decay of the leaf litter, for the increased soil moisture and atmospheric humidity greatly increase the number of decomposing organisms. The largest part of the leaf litter decomposes within a few weeks of the first rains (I. Healey, pers. comm.). At least in the early stages of the rainy season, some leaf litter may be washed away, for the water currents in the streams can become quite strong—during heavy rainstorms, debris is carried

by water currents along trails even in the flat areas of the forest. The rains, however, serve mostly to compact the litter.

Williams (1941) reported that there is a renewal of litter organisms in May, with the beginning of the rains, and that by the early part of July there is a marked increase in the number of forms present. Fungal organisms as well, which are not common during the dry season, are abundant during the rainy season.

Since many nutrients become available shortly after the onset of the rainy season, it can be assumed that plants are absorbing them at a greater rate during the early weeks of the rainy season, though it is not known how long it takes the plants to assimilate these nutrients. If the assimilation were sufficiently rapid, this influx of nutrients might have some effect on seed germination, leaf maturation, or flower and fruit production. Indeed, emerging seedlings appear to be the most abundant at the end of May and the beginning of June (N. Garwood, pers. comm.), but this may reflect only the increase in soil moisture—that is, it may be unrelated to nutrient availability. Frankie, Baker, and Opler (1974) have shown that for lowland wet forest in Costa Rica, the peak of leaf flushing occurs during the major dry season, especially in February, and a second peak occurs in September, just after the minor dry season. On BCI my general impression is that most flushing of new leaves occurs early in the rainy season. However, random observations on 103 shrubs and trees show that there is no marked difference between the number of species that put on new leaves early in the rainy season and those that put them on in the dry season. Six species show leaf flushing both early in the rainy season and in the dry season, whereas 45 show leaf flushing in the dry season and 42 early in the rainy season. An additional 10 species show leaf flushing both late in the dry season and early in the rainy season, and should be considered as rainy-season leaf flushers. But even if these are included with the rainy-season species, the difference between 45 in the dry and 52 in the rainy is not significant.

Although some species lose and replace their leaves more or less regularly throughout the year, and are never completely leafless, those species that probably contribute most to the accumulation of leaf litter in the dry season are the deciduous species that lose all or nearly all of their leaves for all or part of the dry season; they are the following:

Annona spraguei	*Pseudobombax septenatum*
Bauhinia guianensis	*Pterocarpus officinalis*
Bombacopsis quinata	*Sapium caudatum*
B. sessilis	*Tabebuia guayacan*
Bursera simaruba	*Topobaea praecox*
Cavanillesia platanifolia	*Trichilia hirta*
Cedrela odorata	*Xylophragma*
Ceiba pentandra	*seemannianum*
Cochlospermum vitifolium	*Xylosma chloranthum*
Dalbergia retusa	*Zanthoxylum belizense*
Enterolobium cyclocarpum	*Z. panamense*
Erythrina fusca	*Z. setulosum*
Jacaranda copaia	*Zuelania guidonia*

A few species are leafless during the rainy season. Among these are:

Cordia alliodora	*Ochroma pyramidale*
Erythrina costaricensis	*Triplaris cumingiana*

Many species are leafless for only a short time, usually just prior to flowering; often, the leaves are replaced while the plant is in flower. Among the species in this group are the following:

Anacardium excelsum	*Ormosia coccinea*
Antirrhoea trichantha	var. *subsimplex*
Apeiba membranacea	*Peltogyne purpurea*
A. tibourbou	*Pisonia aculeata*
Casearia corymbosa	*Pithecellobium*
C. guianensis	*macradenium*
Cassia fruticosa	*Platymiscium pinnatum*
Castilla elastica	*Platypodium elegans*
Coccoloba acapulcensis	*Poulsenia armata*
C. manzanillensis	*Psidium anglohondurense*
Combretum decandrum	*Randia armata*
Dendropanax arboreus	*Schizolobium parahybum*
Dipteryx panamensis	*Sloanea terniflora*
Eugenia nesiotica	*Spachea membranacea*
E. oerstedeana	*Spondias mombin*
Genipa americana	*S. radlkoferi*
Hura crepitans	*Sterculia apetala*
Inga fagifolia	*Strychnos panamensis*
Lindackeria laurina	*Tachigalia versicolor*
Lonchocarpus velutinus	*Terminalia amazonica*
Luehea seemannii	*Tetrathylacium johansenii*
L. speciosa	*Trattinnickia aspera*
Machaerium arboreum	*Trichospermum mexicanum*
Malouetia guatemalensis	*Trophis racemosa*
Margaritaria nobilis	*Virola surinamensis*
Omphalea diandra	

Some species lose their leaves more than once per year. These include *Tabebuia rosea* and *Quararibea asterolepis,* which lose their leaves twice a year, and *Ficus* spp., which lose all leaves several times a year. Other species, such as *Beilschmiedia pendula, Byrsonima crassifolia, Jatropha curcas,* and *Guazuma ulmifolia,* replace their leaves gradually, but may at times be almost completely leafless, as well.

Frankie, Baker, and Opler (1974) studied leaf production in a number of species from lowland wet forest in Costa Rica. Many of these same species occur on BCI and may react similarly on BCI, though the seasons are not exactly comparable.

Flowering and fruiting. The data presented in this section are the result of field observations and herbarium studies made between 1967 and 1974, and include observations made during more than three years in Panama and a survey of more than 50,000 herbarium specimens from BCI and adjacent areas.

In these data no attempt has been made to outline the phenology of *individual plants,* though numerous individuals were repeatedly observed. Instead, the data represent what is thought to be the normal phenological variation for *each species*—its historical pattern of flower-

ing. No attempt has been made to include the "broad outlier," especially when the phenology of the species involved is well known. In the better-known species, 95% or more of the flowering or fruiting probably falls within the timespan indicated.

The flowering or fruiting period given for most species is broader than that for any single year; because plants have probably evolved a phenology that is compatible with a particular climatic *condition* (as opposed, say, to calendar month), I have chosen to look at overall phenological patterns rather than what might happen in any one year. The year-to-year variation in phenological pattern for any given species is considerable, and variation can also be great between individuals in a given year, in terms of both timing and duration.

Although other phenological studies have been made (Rovirosa, 1892; Fournier & Salas, 1966; Janzen, 1967b; Smythe, 1970; Foster, 1974; Frankie, Baker & Opler, 1974), this is the only known attempt to define an entire flora in a phenological manner.

Excluded from most aspects of this study were the 53 cultivated species; excluded altogether were the 104 species of vascular cryptogams. Earlier studies (Croat, 1969a) have shown that different habit types exhibit different phenological behavior. In this section and in an earlier version of it (Croat, 1975d), different habitats are also shown to produce different phenological behavior. Graphs of flowering and fruiting activity have been prepared for all major habit-and-habitat classes for the flora:

Herbaceous plants
    All herbaceous species
    Epiphytes and hemiepiphytes
    Vines
    Suffruticose herbs
    Clearing herbs
    Forest herbs (not in light-gaps)
    Herbs of light-gaps and forest edges
    Aquatics
Woody plants
    Trees and shrubs of the forest
        Tall and medium-sized trees (more than 10 m tall)
        Small trees (less than 10 m tall)
        Shrubs (1–2(3) m tall)
    Trees and shrubs of open areas, clearings, etc.
    Lianas
Climbing species vs. arborescent species

In the graphs, the number of species in flower or fruit in any month is recorded, though months for evident deviates were not tallied in cases where the phenology of a species is well known.

Herbaceous plants, as a single class of organisms, are quite diverse in terms of both habit and habitat, and as a result are more finely subdivided here than the woody plants are. In all, there are 560 herbaceous plants in the BCI flora, accounting for 42.6% of the native flora. Of these, there are 94 vines, 135 epiphytes, 330 terrestrial herbs (including 18 suffruticose herbs and four saprophytic herbs), and one herb parasitic on trees. Because

they are inconsequential, saprophytes and parasites are included in the class "forest herbs."

Graphs 2 and 3 show flowering and fruiting curves for all types of herbs studied. I believe that phenological patterns are at least in part determined by fluctuations in climatic conditions. Aquatic herbs and suffruticose herbs are aseasonal, perhaps because they are less subject to the effects of a severe dry season. Aquatics would not be expected to be seasonally cued by availability of pre-

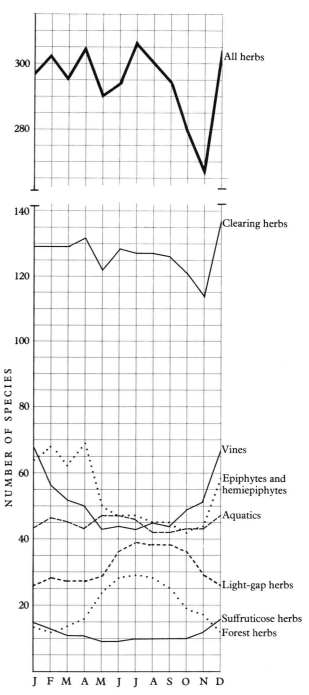

Graph 2. Numbers of herb species in flower, by month and habit-and-habitat class.

cipitate water, but suffruticose herbs, with their well-developed woody root system and underground stems, apparently are also little affected by seasonal changes.

The remaining subclasses of herbs are seasonal. The onset of the dry season, with its reduction of soil moisture and atmospheric humidity, as well as its high insolation, appears to act as a cue to flowering. Flowering times for the different habit-and-habitat classes seem to correlate well with their capacity for withstanding conditions of drought.

Clearing herbs, for example, being the class most exposed to changes in the environment, reach their peak of flowering activity early in the dry season, in December. Flowering then drops off to a relatively steady rate throughout the remainder of the year, except for a slight dip in May and a deeper decline at the end of the rainy season, in November. Because the fruits of most species are small and develop quickly, the fruiting curve closely resembles the flowering curve. Flowering activity in the clearing herbs wanes most at the beginning and end of the rainy season.

Forest herbs, by contrast, reach the peak of their flowering activity early in the rainy season, and their peak of fruiting midway to late in the rainy season. As suggested by Foster (1974), these groups are probably triggered to flower by intense rains following a period of drought. He has shown that typically rainy-season trees will flower in the dry season if a dry period is followed by heavy rains.

For the herbs of forest light-gaps and forest edges, the amount of light received is relatively more stable, and they are protected from excessive insolation. Their flowering and fruiting activity therefore peaks during the rainy season.

Epiphytic herbs do most of their flowering midway to late in the dry season, with small peaks in February and April, which are perhaps a response to the advancing aridity caused by the increasingly leafless canopy of the forest. Most epiphyte fruiting also occurs in the dry season, and the small airborne fruits are dispersed during the same dry season. A smaller peak of fruiting, early in the rainy season in July, consists principally of the animal-dispersed fruits. Of all epiphyte species whose fruiting is restricted to the dry season, 97% produce principally wind-dispersed seeds, the remainder principally animal-dispersed seeds. Rainy-season epiphytes, by contrast, produce wind-dispersed seeds in only 23% of their species, animal-dispersed fruits in 77%. These figures correlate well with the markedly stronger winds of the dry season—to which wind-dispersed seeds or fruits are particularly well adapted. According to Foster (1974), the leaflessness of the canopy may be more important than winds in the dispersal of airborne diaspores, since many are dispersed after the rains are renewed but before the trees have put on new leaves.

Perhaps because herbaceous vines in the forest usually occur in well-lighted areas and are often restricted to exposed surfaces of the canopy, they do not react appreciably differently from those that occur in clearings. I have therefore treated all herbaceous vines as a group. The flowering peak for herbaceous vines is in December

and January, with a second much smaller peak in June. The curve for fruiting in vines, though lacking strong peaks, shows major activity in the dry season. The June peak represents species that appear to be triggered by wet rather than dry conditions.

Of all habit types, the herbs are the least phenologically

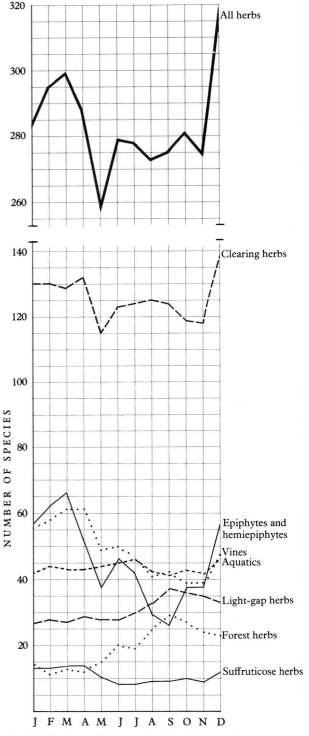

Graph 3. Numbers of herb species in fruit, by month and habit-and-habitat class.

variable. As many as 224 species (40%) flower and fruit most or all of the year.

The graph for all species of herbs (Graph 2) shows a decrease in flowering activity late in the rainy season, but from the low of 267 species flowering in November to the peak of 307 flowering in July is only a 15% increase. Certain categories of herbs do, however, show significant increases in flowering activity. For example, there is an increase of 58% for vines, 65% for epiphytes, and 79% for all forest herbs (excluding vines).

When not restricted to open areas, such as in clearings or along the lakeshore, woody plants tend to be more seasonal than herbaceous plants (Graphs 4 and 5). As a class, the trees and shrubs of open areas are not very seasonal. Three of the forest habit classes—lianas, large and medium-sized trees, and small trees—reach their principal peak of activity in the dry season. Flowering in the lianas is most active from January to March, especially in February, substantially ahead of the flowering peak for large and medium-sized trees. This disparity probably reflects the fact that the bulk of the leaf biomass of lianas is restricted to the surface of the canopy and is thus quickly affected by conditions of drought. The fruiting peak for lianas occurs late in the dry season, in March and April, and is stronger than the flowering peak —a circumstance perhaps due to the fact that many species of lianas produce wind-dispersed fruits, even though they may flower in the rainy season. For example, there are six species of bignoniaceous lianas that flower in the rainy season and fruit in the dry season. Of liana species that fruit only in the dry season, as many as 80% have wind-dispersed seeds, whereas just 22% of the liana species that fruit only in the rainy season have wind-dispersed seeds.

Large and medium-sized trees reach their peak of flowering activity from February to June, especially in March and April, at the end of the dry season. Though triggered by conditions of drought, they do not react so quickly to changes in the environment as the herbs, vines, and lianas do, perhaps owing to the fact that they are much less exposed to the environment than these other classes are. It may be that a great many of the trees flowering in the dry-wet transition period are triggered to flower by the first heavy rains, as suggested by Foster (1974). Fruiting activity in the large and medium-sized trees shows two peaks, one in April and a second, smaller, one in August. The earlier peak is made up for the most part by wind-dispersed species, whereas the second, rainy-season, peak is made up of species that are mostly animal-dispersed.

The flowering peak of March and April for large and medium-sized trees on BCI contrasts rather sharply with the May and July peaks in the wet forest of Costa Rica (Frankie, Baker & Opler, 1974). However, the April and August fruiting peaks on BCI compare well with the May and September peaks in Costa Rica. The flowering and fruiting curves for small trees and shrubs would be somewhat flatter if they also included the class "arborescent in open areas."

Lianas and herbaceous vines are very similar phenologically. If all climbing plants are compared with all

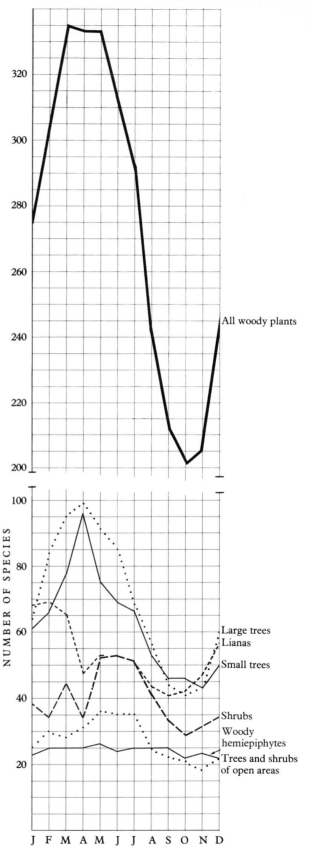

Graph 4. Numbers of woody plant species in flower, by month and habit-and-habitat class.

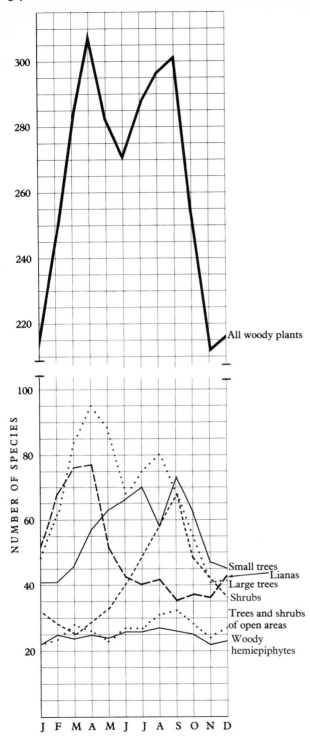

Graph 5. Numbers of woody plant species in fruit, by month and habit-and-habitat class.

arborescent plants (Graph 6), some interesting differences become apparent. As a group, the climbing-plant species reach their peaks of flowering and fruiting earlier than the arboreal plants do. Climbers share the ability to position themselves where they are exposed to light. By the same token, of course, they are subjected to a high degree of exposure when climatic conditions become harsh, as at

the beginning of the dry season. It is, I believe, the onset of the dry season that precipitates flowering in many species, including a great many of the lianas, herbaceous vines, and epiphytes, as well as many trees—though most of the trees flower in the dry-wet transitional period and may be induced to flower prematurely by unseasonally wet conditions. Whether this dry-season phenomenon is in general the result of the drying conditions, photo-periodicity, or otherwise is unknown, but the fact that the flowering period of many species coincides with the dry season is no mere coincidence. The flowering curve for all BCI species considered jointly shows a pronounced peak of activity in the dry season (Graph 7). Fruiting shows two peaks, one in the dry-wet transition period and one in the middle of the rainy season. There is a dearth of activity in both flowering and fruiting during October and November, but acute reactivation in December, with the onset of the dry season.

These conclusions differ significantly from those of Foster (1974), who held that the peak month for overall flowering occurs from one to two months after the start of the rains, i.e., in May or June. My studies for overall flowering show significantly fewer species in flower during May and June than in March and April. Foster's conclusion may derive in part from the fact that he was dealing not with the entire flora but with an area of mostly mature forest containing few of the species that are common in open areas or forest edges. Moreover, his detailed sampling procedures deal principally with fruiting, rather less with flowering.

Notwithstanding the overall flowering peak in the dry season, a number of habit-and-habitat classes and a substantial proportion of the species as a whole are apparently cued to flower sometime after the rainy season begins. Foster (1974) has documented this finding for the BCI *Psychotria* species. Shrubs, forest herbs, and herbs of light-gaps and forest edges all show increased activity in the rainy season. The same can be said for many individual families, especially monocotyledonous herbs such as Marantaceae, Musaceae, Zingiberaceae, and Amaryllidaceae.

It is enlightening to compare the seasonal behavior of these habit-and-habitat classes by examining the number of species in flower in a given month as a percentage of the total number of species in the class (Graphs 8 and 9). Thus, although 96 species of trees more than 10 m tall flower in April, this figure represents only 43% of all such trees, whereas the 43 aquatic herbs that flower in April constitute 75% of all aquatic-herb species. The percent-flowering curves follow the same contours as the absolute flowering curves, but the heights of the curves are substantially different. Even at their peak of flowering activity, the percentage of trees and lianas in flower is smaller than the flowering percentage of any other category. A significantly larger percentage of small trees are flowering or fruiting in every season, peak or low. Several habitat classifications have been combined here for simplification: all small trees and shrubs, including those restricted to open areas, are combined; the curve for all species includes both woody and herbaceous plants, the herbs showing a greater percentage of activity than the woody

and 57% are wind-dispersed. Comparable figures for small trees and shrubs are 35% animal-dispersed fruits and 21% wind-dispersed fruits in the dry season (the others are mechanically dispersed or are not clearly adapted for either animal or wind dispersal) and nearly 100% animal-dispersed fruits in the wet season.

Small trees are those less than 10 m tall, excluding plants that are always shrubs (i.e., plants usually 1–2(3) m tall) and excluding, as well, all small trees and shrubs that are restricted to clearings. Small trees as a class have a strong peak of activity late in the dry season, in April; being understory trees they produce, as one would expect,

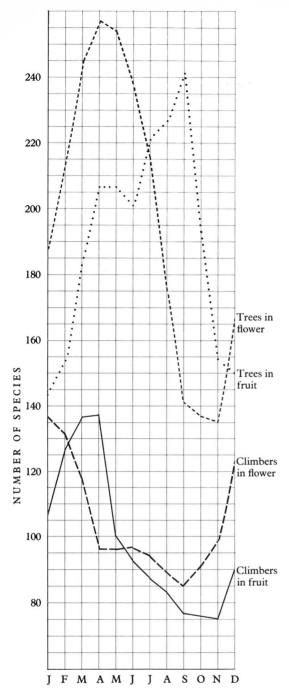

Graph 6. Numbers of species of trees and climbers in flower and fruit, by month.

plants do; and the forest herbs and herbs of forest light-gaps and edges are combined, as well as clearing, epiphytic, and suffruticose herbs.

Comparing the fruiting percentages yields a similar pattern, i.e., the same contour as the absolute fruiting curves and at levels corresponding to the percentages for flowering. Of all large and medium-sized trees (excluding bimodal species) whose fruiting occurs strictly in the rainy season, 85% are animal-dispersed and only 12% are wind-dispersed. Of the comparable species that fruit strictly in the dry season, only 36% are animal-dispersed

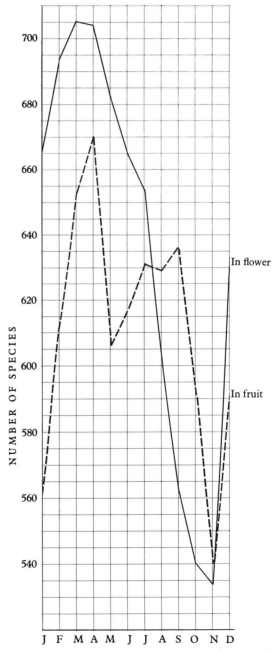

Graph 7. Numbers of vascular plant species in flower and fruit, by month.

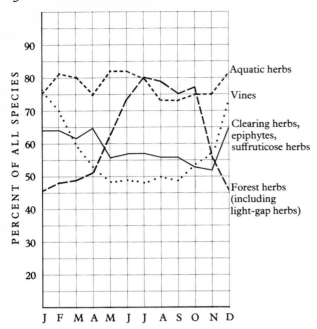

Graph 8. Percentage of species flowering, by month, four groups.

mostly animal-dispersed seeds, for winds are usually not strong enough in the lower levels of the forest, even during the dry season, to disperse successfully most wind-borne seeds or fruits. The fruiting of the small trees, although not as conspicuously peaked as their flowering, occurs mostly in the rainy season, with a small peak in July and another in September.

The flowering activity of shrubs is not very seasonal, but more flowering occurs early in the wet season than at any other time. Shrub fruiting, by contrast, shows a fairly definite peak in September. Finally, trees and shrubs restricted to open areas show no marked seasonal variation as a class.

In contrast to the 15% difference between high and low points for flowering activity in herbs, there is a difference of 67% for all arborescent plants. This finding includes a difference of 134% for large and medium-sized trees, 123% for small trees, 83% for shrubs, and 68% for lianas.

Another useful means of comparing plants of different habits and habitats is to compare the average lengths of the flowering or fruiting seasons. Since these are taken from what I have termed the usual pattern of flowering, they do not represent the length of flowering or fruiting for any individual or even for any species in a given year, but rather the length of time over which a species has been known to be in flower or fruit. Table 5 shows these figures for most of the classes already discussed. The aquatic herbs, which are markedly aseasonal, are shown to flower and fruit more than 9 months per year on the average, whereas such classes as epiphytic herbs, lianas, and trees, which are all seasonal in their flowering behavior, flower and fruit on the average for substantially shorter periods of time.

Despite the fact that many species are distinctly "dry-season species" or "wet-season species," by no means all

of the species that are seasonal are restricted to one of these two periods. Table 6 shows a categorization of seasonality types. It can be seen that at least 346 species are transitional between the two seasons, in either flowering or fruiting behavior. Some of these conclusions can clearly be disregarded, because of year-to-year fluctuations in the onset of the rainy season, but field observations have confirmed that many typically dry-season species, such as *Cochlospermum vitifolium,* may begin flowering in November, the wettest month of the year, or extend into a period that is definitely rainy season. The same is of course true of wet-season species that begin before or end after the rainy season.

A total of 126 species were excluded from consideration here because too little is known about their phenological patterns; a number of these are cultivated species. But all cultivated species for which the phenology is known are included here. Because the bimodal species are so difficult to classify into seasonality types, no attempt has been made to do so, for many of these are also transitional in their flowering or fruiting behavior.

There are 294 species that flower and fruit all year, 164 of which exhibit no discernible peak. For others, there is a peak of activity in the dry or wet season, and in a few cases flowering is bimodal (e.g., with a peak of activity at the beginning of the dry season and another at the beginning of the wet season). In still other cases, the peak of activity is transitional, straddling both the dry and wet seasons.

The most heavily represented seasonality types flower and fruit either in the wet season (176 species) or in the dry season (133 species).

Some species that flower for more than 9 months, and thus overlap the established seasonal-type categories,

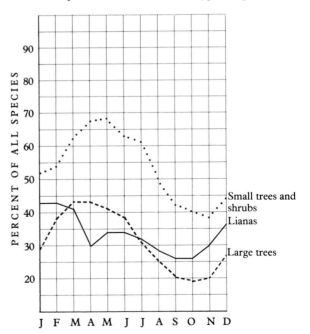

Graph 9. Percentage of species flowering, by month, three more groups.

TABLE 5
Flowering and fruiting extension,
by habit-and-habitat class
(in months)

Habit-and-habitat class	Average flower extension	Average fruit extension
Large trees (to 30 m or more)	4.3	4.4
Medium-sized trees (10–30 m)	3.5	3.3
Small trees or shrubs (less than 10 m)	6.3	6.1
Lianas	4.0	3.8
Vines	6.9	6.5
Epiphytic herbs	4.8	4.0
Clearing herbs	8.6	8.5
Aquatic herbs	9.4	9.1
Forest herbs	7.4	7.2
ALL CLASSES	6.4	6.1

TABLE 6
Flowering and fruiting seasonality types

Seasonality types	Number of species		
	In flower less than 9 months	In flower more than 9 months (peak-period type)	Total
1. Flowers and fruits dry season	133		133
2. Flowers and fruits dry-wet	46		46
3. Flowers and fruits wet season	177		177
4. Flowers and fruits wet-dry	24		24
5. Flowers dry; fruits dry-wet	41	1	42
6. Flowers dry; fruits wet	57	7	64
7. Flowers dry; fruits wet-dry	4		4
8. Flowers dry-wet; fruits wet	90	4	94
9. Flowers dry-wet; fruits wet-dry	12		12
10. Flowers dry-wet; fruits dry	14	1	15
11. Flowers dry; fruits dry 1 year later	7		7
12. Flowers dry-wet; fruits dry-wet 1 year later	7		7
13. Flowers wet; fruits wet-dry	29	1	30
14. Flowers wet; fruits wet 1 year later	3		3
15. Flowers wet; fruits dry	71	7	78
16. Flowers wet; fruits dry-wet	10	2	12
17. Flowers wet-dry; fruits dry	27	4	31
18. Flowers wet-dry; fruits dry-wet	17	2	19
19. Flowers wet-dry; fruits wet	9		9
20. Flowers and fruits bimodally			36
21. Flowers and fruits all year, no peak		164	164
22. Flowers and fruits all year, especially wet		81	81
23. Flowers and fruits all year, especially dry		24	24
24. Flowers and fruits all year, wet-dry peak		9	9
25. Flowers and fruits all year, dry-wet peak		11	11
26. Flowers and fruits all year, bimodal peaks		5	5
27. Seasonality incompletely known			126
28. Flowers dry; fruits all year		1	1
29. Flowers dry-wet; fruits all year		1	1
TOTAL	778	325	1,265

nevertheless exhibit a peak of activity corresponding to a particular seasonal type; these appear in a separate column in Table 6.

From all of these data we can determine whether more species take advantage of the dry season, for flowering or fruiting, than prefer the rainy season. A total of 509 seasonal species (excluding bimodal species) do all or part of their flowering during the dry season; of these, 242 (47%) restrict their flowering to the dry season. A total of 559 seasonal species (excluding bimodal species) do all or part of their flowering in the rainy season; of these, 290 (52%) restrict their flowering to the rainy season.

Comparable figures for the seasonally fruiting species are 462 that set all or part of their fruits in the dry season, 253 (55%) that restrict their fruiting to the dry season, and 542 that set all or part of their fruits in the rainy season (341 (63%) restrict their fruiting to the rainy season).

On the strength of general observations in the field, one might conclude that more species take advantage of the dry season to flower, since it is during this period when many of the more conspicuous species come into bloom. Even the flowering curves indicate that there is a decided preference for flowering in the dry season, since most habit-and-habitat classes reach their peak of activity at that time. However, many trees that attain their peak of flowering late in the dry season and into the dry-wet transitional period may in fact be triggered to flower by the onset of the rainy season. Two other groups, the understory shrubs and forest herbs, definitely reach their peak of flowering activity in the rainy season (Foster, 1974); and both groups are relatively inconspicuous by contrast with the showier dry-season species.

Since the rainy season is substantially longer than the dry season, *more species of plants flower and fruit during the rainy season than during the dry season,* notwithstanding the fact that there is an overall peak of flowering and fruiting activity in the dry season.

The following lists, by family, all of the species assignable to each of the seasonality types* given in Table 6:

1. Flowers and fruits in the dry season

Typhaceae	*Typha domingensis*
Gramineae	*Bothriochloa pertusa*
	Chloris virgata
	Ischaemum indicum
	I. rugosum[1]
	Oplismenus burmanni
	Orthoclada laxa
	Streptogyne americana
Cyperaceae	*Scleria macrophylla*
Bromeliaceae	*Pitcairnia heterophylla*
Dioscoreaceae	*Dioscorea trifida*

*The superscript numbers following some of the species names indicate the following: [1]rarely flowers late wet; [2]rarely fruits early dry; [3]rarely fruits early wet; [4]rarely flowers and fruits early wet; [5]rarely flowers late dry; [6]rarely flowers and fruits early dry; [7]rarely fruits late dry; [8]rarely flowers early wet.

1. Flowers and fruits in the dry season (cont.)

Orchidaceae	*Aspasia principissa*
	Bulbophyllum pachyrrhachis
	Caularthron bilamellatum
	Chysis aurea
	Cochleanthes lipscombiae[1]
	Dichaea panamensis
	Epidendrum imatophyllum[4]
	E. lockhartioides
	E. schlechterianum
	Gongora quinquenervis
	G. tricolor
	Leochilus scriptus[1]
	Lockhartia pittieri
	Maxillaria variabilis
	Mormodes powellii
	Notylia barkeri
	N. pentachne
	Oncidium ampliatum
	O. stipitatum[4]
	Ornithocephalus bicornis[4]
	Polystachya masayensis[1]
	Scaphyglottis graminifolia
	Sobralia panamensis
	Trigonidium egertonianum[4]
Urticaceae	*Myriocarpa yzabalensis*
Polygonaceae	*Coccoloba acapulcensis*
	Triplaris cumingiana
Amaranthaceae	*Iresine angustifolia*
	I. celosia
Phytolaccaceae	*Microtea debilis*
Leguminosae	*Aeschynomene americana* var.
	glandulosa
	Albizia guachapele[3]
	Cajanus bicolor
	Calopogonium caeruleum
	C. mucunioides
	Cassia obtusifolia[1]
	C. reticulata
	Desmodium scorpiurus
	D. triflorum
	Machaerium floribundum
	Mimosa casta
	M. pudica[1]
	Mucuna rostrata
	Rhynchosia pyramidalis
	Teramnus uncinatus
	T. volubilis
Oxalidaceae	*Averrhoa carambola*
Malpighiaceae	*Stigmaphyllon hypargyreum*
	Tetrapteris discolor
Polygalaceae	*Securidaca diversifolia*
Euphorbiaceae	*Adelia triloba*
	Phyllanthus amarus
Anacardiaceae	*Astronium graveolens*
Sapindaceae	*Paullinia fuscescens*
	var. *glabrata*
	Serjania atrolineata[3]
	S. circumvallata
	S. cornigera
	S. decapleuria
	S. mexicana[3]
	S. trachygona
	Thinouia myriantha
Rhamnaceae	*Colubrina glandulosa*
Vitaceae	*Cissus pseudosicyoides*
Elaeocarpaceae	*Sloanea terniflora*
Tiliaceae	*Heliocarpus popayanensis*[3]
	Luehea seemannii[3,1]
	L. speciosa
	Triumfetta lappula
Malvaceae	*Pavonia dasypetala*
	P. paniculata[3]
Bombacaceae	*Bombacopsis quinata*[3]
	B. sessilis[3]
	Cavanillesia platanifolia[3]
	Ceiba pentandra
	Pseudobombax septenatum
Sterculiaceae	*Melochia lupulina*
	M. melissifolia
	Waltheria glomerata
Marcgraviaceae	*Marcgravia nepenthoides*[1]
	Souroubea sympetala[3]
Cochlospermaceae	*Cochlospermum vitifolium*[3]
Violaceae	*Hybanthus prunifolius*[3]
Combretaceae	*Terminalia chiriquensis*[3]
Myrtaceae	*Eugenia galalonensis*
	E. oerstedeana
Gentianaceae	*Schultesia lisianthoides*
Convolvulaceae	*Ipomoea batatas*
	I. squamosa
Labiatae	*Hyptis capitata*
	H. mutabilis
	Salvia occidentalis
Bignoniaceae	*Tabebuia guayacan*[4]
	T. ochracea
Acanthaceae	*Aphelandra sinclairiana*
	Blechum brownei
	B. costaricense
	Elytraria imbricata
	Hygrophila guianensis[1]
	Justicia pectoralis
	Mendoncia gracilis
	Nelsonia brunellodes[4]
	Ruellia metallica
	Teliostachya alopecuroidea
	Trichanthera gigantea
Rubiaceae	*Manettia reclinata*
Compositae	*Ayapana elata*
	Baltimora recta
	Chromolaena odorata[1]
	Elephantopus mollis
	Fleischmannia sinclairii[4]
	Heterocondylus vitalbis[4]
	Koanophyllon wetmorei[1]
	Melanthera aspera
	Mikania guaco
	M. leiostachya

	Neurolaena lobata
	Rolandra fruticosa
	Spiracantha cornifolia
	Synedrella nodiflora
	Verbesina gigantea
	Vernonia canescens
	V. patens

2. Flowers and fruits dry-wet

Alismataceae	*Sagittaria lancifolia*
Gramineae	*Panicum polygonatum*
	Paspalum saccharoides
Cyperaceae	*Calyptrocarya glomerulata*
	Rhynchospora corymbosa
	Scleria mitis
Piperaceae	*Peperomia macrostachya*
	P. obscurifolia
	Piper culebranum
Lacistemaceae	*Lacistema aggregatum*
Moraceae	*Pseudolmedia spuria*
Loranthaceae	*Phoradendron piperoides*
	Struthanthus orbicularis
Amaranthaceae	*Chamissoa altissima*
Chrysobalanaceae	*Hirtella racemosa* (may flower more than once per year)
Leguminosae	*Dalbergia brownei*
Rutaceae	*Citrus reticulata*
Malpighiaceae	*Heteropteris laurifolia*
	Hiraea reclinata
	Mascagnia nervosa
Euphorbiaceae	*Alchornea latifolia*
Sapindaceae	*Paullinia bracteosa*
Malvaceae	*Pavonia rosea*
Guttiferae	*Havetiopsis flexilis*
	Marila laxiflora
Violaceae	*Rinorea sylvatica*
Flacourtiaceae	*Casearia guianensis* var. *guianensis*
	Hasseltia floribunda
	Xylosma oligandrum
Lythraceae	*Cuphea carthagenensis*
Combretaceae	*Combretum cacoucia*
	C. decandrum
	Terminalia amazonica
Myrtaceae	*Eugenia venezuelensis*
Melastomataceae	*Miconia argentea*
	M. elata
Onagraceae	*Ludwigia leptocarpa*
Boraginaceae	*Heliotropium indicum*
	Tournefortia bicolor
	T. cuspidata
Scrophulariaceae	*Bacopa salzmannii*
Rubiaceae	*Bertiera guianensis*
	Diodia denudata
Cucurbitaceae	*Gurania makoyana*
Compositae	*Conyza apurensis*

3. Flowers and fruits wet season

Gramineae	*Andropogon virginicus*
	Lasiacis procerrima
	Panicum mertensii
	P. milleflorum
	Paspalum notatum
	P. plicatulum
	Rhipidocladum racemiflorum
	Saccharum spontaneum
	Setaria paniculifera
	S. vulpiseta
Cyperaceae	*Cyperus giganteus*[2]
	C. rotundus
	C. simplex
	Rhynchospora micrantha
Palmae	*Cryosophila warscewiczii*
Cyclanthaceae	*Carludovica drudei*
	Ludovia integrifolia
Araceae	*Caladium bicolor*
	Dieffenbachia longispatha
	Dracontium dressleri
	Philodendron fragrantissimum
	P. grandipes
	P. inconcinnum
	P. pterotum
	P. radiatum
	P. tripartitum
	Rhodospatha moritziana
	R. wendlandii
	Xanthosoma helleborifolium
	X. nigrum
	X. pilosum
Bromeliaceae	*Aechmea magdalenae*
Commelinaceae	*Campelia zanonia*
	Dichorisandra hexandra
Pontederiaceae	*Eichhornia azurea*
Amaryllidaceae	*Zephyranthes tubispatha*
Musaceae	*Heliconia catheta*
	H. irrasa
	H. latispatha
	H. pogonantha
Zingiberaceae	*Costus allenii*
	C. guianensis var. *macrostrobilus*
	C. laevis
	Renealmia alpinia
	R. cernua[5]
	Zingiber officinale
Marantaceae	*Calathea inocephala*[5]
	C. insignis
	C. latifolia
	C. marantifolia
	C. micans
	C. panamensis
	C. villosa
	Ischnosiphon leucophaeus
	Stromanthe jacquinii
Burmanniaceae	*Thismia panamensis*
Orchidaceae	*Brassia caudata*
	Dichaea trulla
	Encyclia triptera
	Palmorchis powellii

3. Flowers and fruits wet season (cont.)

Piperaceae	*Peperomia mameiana*
Lacistemaceae	*Lozania pittieri*
Ulmaceae	*Celtis iguanaeus*
	Trema micrantha
Moraceae	*Dorstenia contrajerva*
	Trophis racemosa[5]
Urticaceae	*Boehmeria cylindrica*
	Pouzolzia obliqua
	Urera eggersii
Olacaceae	*Heisteria costaricensis*[5]
	H. longipes
Rafflesiaceae	*Apodanthes caseariae*
Polygonaceae	*Coccoloba coronata*
	C. parimensis
Caryophyllaceae	*Drymaria cordata*[6]
Nymphaeaceae	*Nymphaea blanda*
Menispermaceae	*Abuta racemosa*
	Chondrodendron tomentosum
	Odontocarya tamoides var.
	canescens
Lauraceae	*Phoebe mexicana*
Saxifragaceae	*Hydrangea peruviana*
Leguminosae	*Crotalaria vitellina*[6]
	Inga hayesii
	Leucaena multicapitula
	Ormosia coccinea var. *subsimplex*
	Pithecellobium barbourianum
	Pterocarpus officinalis[5]
	P. rohrii
Rutaceae	*Zanthoxylum panamense*
Malpighiaceae	*Byrsonima spicata*
	Hiraea quapara
	Mascagnia hippocrateoides
	Spachea membranacea[5]
	Stigmaphyllon puberum
Polygalaceae	*Polygala paniculata*
Euphorbiaceae	*Acalypha arvensis*
	Croton billbergianus[5]
	Drypetes standleyi
	Margaritaria nobilis
	Poinsettia heterophylla
	Sapium caudatum
Anacardiaceae	*Mosquitoxylum jamaicense*
Sapindaceae	*Cupania cinerea*
	C. latifolia
	Paullinia rugosa
	Serjania pluvialiflorens
Rhamnaceae	*Gouania adenophora*
Vitaceae	*Cissus microcarpa*
	C. rhombifolia
Elaeocarpaceae	*Sloanea zuliinsis*
Bombacaceae	*Quararibea asterolepis*
	Q. pterocalyx
Dilleniaceae	*Doliocarpus major*
	D. olivaceus
	Saurauia laevigata
Flacourtiaceae	*Banara guianensis*
	Casearia arborea

	C. commersoniana
Cactaceae	*Rhipsalis cassytha*
Lecythidaceae	*Gustavia fosteri*
Myrtaceae	*Eugenia coloradensis*
	Myrcia fosteri
	M. gatunensis
	Psidium anglohondurense
Melastomataceae	*Arthrostema alatum*
	Bellucia grossularioides
	Henriettea succosa
	Miconia lateriflora
	M. shattuckii
Onagraceae	*Ludwigia octovalvis*
Araliaceae	*Dendropanax arboreus*
	D. stenodontus[5]
Myrsinaceae	*Ardisia bartlettii*
	Parathesis microcalyx[5]
Sapotaceae	*Pouteria unilocularis*
	P. stipitata
Gentianaceae	*Voyria alba*
	V. tenella
Asclepiadaceae	*Marsdenia crassipes*
Boraginaceae	*Cordia panamensis*
	Tournefortia hirsutissima
Verbenaceae	*Aegiphila cephalophora*
	A. elata
	Vitex cooperi
Solanaceae	*Cestrum latifolium*
	Lycianthes synanthera
	Solanum asperum
	S. jamaicense
Scrophulariaceae	*Lindernia crustacea*
	L. diffusa
Gesneriaceae	*Chrysothemis friedrichsthaliana*
	Diastema raciferum
	Nautilocalyx panamensis
Acanthaceae	*Herpetacanthus panamensis*
Rubiaceae	*Alibertia edulis*
	Amaioua corymbosa
	Cephaelis discolor
	Geophila croatii
	G. repens
	Hamelia axillaris
	Palicourea guianensis[5]
	Psychotria acuminata
	P. brachybotrya
	P. deflexa
	P. micrantha
	P. pubescens
	P. racemosa
	Randia formosa
Cucurbitaceae	*Cayaponia glandulosa*
	Fevillea cordifolia
Compositae	*Clibadium surinamense*
	Conyza bonariensis
	Eleutheranthera ruderalis
	Erechtites hieracifolia var.
	cacalioides
	Fleischmannia microstemon

	Wedelia trilobata[5]
	Wulffia baccata

4. **Flowers and fruits wet-dry**

Gramineae	*Andropogon glomeratus*
	Brachiaria mutica
	Cenchrus brownii
	Ichnanthus pallens
	Lasiacis oaxacensis
	Olyra latifolia
	Panicum grande
	Paspalum microstachyum
	Phragmites australis
	Polytrias amaura
	Schizachyrium microstachyum
Commelinaceae	*Callisia ciliata*
Piperaceae	*Peperomia killipi*
	P. obtusifolia
Leguminosae	*Aeschynomene ciliata*
	Clitoria rubiginosa
Simaroubaceae	*Quassia amara*
Melastomataceae	*Miconia borealis*
	Schwackaea cupheoides
Convolvulaceae	*Ipomoea tiliacea*
Gesneriaceae	*Drymonia serrulata*[6]
Rubiaceae	*Borreria densiflora*
Compositae	*Calea prunifolia*
	Pseudoelephantopus spicatus

5. **Flowers dry; fruits dry-wet**

Cyclanthaceae	*Carludovica palmata*
Araceae	*Anthurium littorale*
Bromeliaceae	*Billbergia macrolepis*
Commelinaceae	*Phaeosphaerion persicariifolium*
Orchidaceae	*Cattleya patinii*
	Epidendrum radicans
	Ionopsis satyrioides
Piperaceae	*Peperomia cordulata*
Moraceae	*Pourouma guianensis*
Chrysobalanaceae	*Licania hypoleuca*
Leguminosae	*Cassia undulata*[1]
	Clitoria javitensis[1]
	Dioclea guianensis[3]
	Enterolobium schomburgkii[3]
	Lonchocarpus velutinus
	Machaerium kegelii
	M. milleflorum
	Schizolobium parahybum
Simaroubaceae	*Picramnia latifolia*
	Simarouba amara var. *typica*
Anacardiaceae	*Anacardium excelsum*
Sapindaceae	*Cupania rufescens*
	C. sylvatica[1]
	Serjania paucidentata
Malvaceae	*Hibiscus bifurcatus*
Dilleniaceae	*Davilla nitida*
	Doliocarpus dentatus
	D. multiflorus
Flacourtiaceae	*Casearia aculeata*

	C. arguta[1]
Cactaceae	*Epiphyllum phyllanthus* var. *columbiense*
Melastomataceae	*Miconia hondurensis*
	M. impetiolaris
Myrsinaceae	*Stylogyne standleyi*
Convolvulaceae	*Merremia umbellata*
Boraginaceae	*Cordia alliodora*
	C. lasiocalyx
Rubiaceae	*Macrocnemum glabrescens*
	Uncaria tomentosa
Cucurbitaceae	*Cayaponia granatensis*
	Melothria trilobata

5a. **Flowers dry; fruits dry-wet (flowers more than 9 months)**

Bombacaceae	*Pachira aquatica*

6. **Flowers dry; fruits wet**

Palmae	*Bactris coloniata*
Araceae	*Anthurium clavigerum*
	A. friedrichsthalii
	A. tetragonum
	Monstera dilacerata
Bromeliaceae	*Aechmea pubescens*
Orchidaceae	*Lockhartia acuta*
Piperaceae	*Piper arieianum*
Moraceae	*Artocarpus altilis*
	Brosimum alicastrum
	Cecropia insignis[7]
Nyctaginaceae	*Pisonia aculeata*[7]
Menispermaceae	*Odontocarya truncata*[7]
Monimiaceae	*Siparuna pauciflora*
Lauraceae	*Beilschmiedia pendula*
	Ocotea skutchii
	Persea americana
Chrysobalanaceae	*Hirtella americana*[8]
Leguminosae	*Acacia melanoceras*
	Bauhinia guianensis
	Hymenaea courbaril
	Inga cocleensis
	I. fagifolia[8]
	I. thibaudiana
	Pithecellobium dinizii
	P. macradenium[7,8]
Rutaceae	*Citrus aurantifolia*
	C. aurantium
	C. sinensis
	Zanthoxylum procerum[8]
	Z. setulosum[7]
Meliaceae	*Trichilia verrucosa*[8]
Anacardiaceae	*Mangifera indica*
Sapindaceae	*Allophylus psilospermus*
	Talisia nervosa
Sterculiaceae	*Herrania purpurea*[7]
Theaceae	*Ternstroemia tepezapote*
Guttiferae	*Garcinia mangostana*
	Rheedia acuminata[7]
	R. edulis

6. Flowers dry; fruits wet (cont.)

Flacourtiaceae	*Laetia procera*
	Zuelania guidonia[7,8]
Passifloraceae	*Passiflora ambigua*
	P. nitida[7,8]
Caricaceae	*Carica cauliflora*
Combretaceae	*Combretum laxum* var. epiphyticum
Melastomataceae	*Miconia prasina*
Theophrastaceae	*Jacquinia macrocarpa*
Loganiaceae	*Strychnos toxifera*
Apocynaceae	*Malouetia guatemalensis*[7]
	Odontadenia puncticulosa
Convolvulaceae	*Maripa panamensis*[7,8]
Solanaceae	*Cestrum megalophyllum*
Bignoniaceae	*Jacaranda copaia*[8]
	Tabebuia rosea
Rubiaceae	*Faramea luteovirens*[8]
Cucurbitaceae	*Gurania coccinea*

6a. Flowers dry; fruits wet (flowers more than 9 months)

Araceae	*Anthurium scandens*
	Syngonium podophyllum
Piperaceae	*Piper grande*
Moraceae	*Cecropia peltata*
Passifloraceae	*Passiflora coriacea*
Solanaceae	*Solanum lanciifolium*
Rubiaceae	*Psychotria limonensis*

7. Flowers dry; fruits wet-dry

Annonaceae	*Desmopsis panamensis*
Leguminosae	*Entada monostachya*
Sapindaceae	*Paullinia baileyi*
Myrtaceae	*Psidium guajava*

8. Flowers dry-wet; fruits wet

Cyperaceae	*Cladium jamaicense*
Palmae	*Bactris barronis*
	B. gasipaes
	Geonoma interrupta
	G. procumbens
Cyclanthaceae	*Asplundia alata*
	Cyclanthus bipartitus
Araceae	*Anthurium ochranthum*
	Homalomena wendlandii
	Philodendron guttiferum
	P. inaequilaterum
	P. nervosum
	P. panamense
	Spathiphyllum phryniifolium
	Syngonium erythrophyllum
Bromeliaceae	*Aechmea setigera*
	A. tillandsioides
Zingiberaceae	*Costus villosissimus*
Orchidaceae	*Ionopsis utricularioides*
	Stellis crescentiicola
Piperaceae	*Peperomia ebingeri*
	P. glabella
	Piper aequale

	P. arboreum
	P. darienense
	P. hispidum
	P. perlasense
	P. pubistipulum
	P. reticulatum
Moraceae	*Castilla elastica*
	Cecropia longipes
	Coussapoa panamensis
	Maquira costaricana
Nyctaginaceae	*Guapira standleyanum*
Annonaceae	*Annona hayesii*
	A. spraguei
Monimiaceae	*Siparuna guianensis*
Lauraceae	*Ocotea cernua*
Leguminosae	*Brownea macrophylla*
	Cynometra bauhiniifolia
	Inga pauciflora
	I. pezizifera
	I. vera subsp. *spuria*
	Lonchocarpus pentaphyllus
	Ormosia panamensis
	Pithecellobium rufescens
Burseraceae	*Protium tenuifolium* var. sessiliflorum
Meliaceae	*Guarea glabra*
	Trichilia montana
Malpighiaceae	*Byrsonima crassifolia*
Euphorbiaceae	*Acalypha diversifolia*
	Alchornea costaricensis
	Jatropha curcas (rarely also bimodal)
	Omphalea diandra
Anacardiaceae	*Spondias mombin*
	S. radlkoferi
Celastraceae	*Maytenus schippii*
Staphyleaceae	*Turpinia occidentalis* subsp. breviflora
Guttiferae	*Tovomita longifolia*
Flacourtiaceae	*Laetia thamnia*
	Tetrathylacium johansenii
	Xylosma chloranthum
Passifloraceae	*Passiflora williamsii*
Lecythidaceae	*Gustavia superba*
Myrtaceae	*Eugenia nesiotica*
Melastomataceae	*Clidemia collina*
	C. purpureo-violacea
	Conostegia bracteata
	Miconia affinis
	M. lacera
	M. lonchophylla
	M. serrulata
	Topobaea praecox
Myrsinaceae	*Ardisia pellucida*
Apocynaceae	*Forsteronia peninsularis*
	F. viridescens
Boraginaceae	*Cordia bicolor*
	Tournefortia maculata
Solanaceae	*Lycianthes maxonii*
Bignoniaceae	*Cydista heterophylla*

Rubiaceae — *Alseis blackiana*
Antirrhoea trichantha
Chimarrhis parviflora
Faramea occidentalis
Hoffmania woodsonii
Psychotria carthagenensis
P. emetica
P. horizontalis
P. pittieri
Randia armata

8a. Flowers dry-wet; fruits wet (flowers more than 9 months)

Amaryllidaceae — *Crinum erubescens*
Leguminosae — *Andira inermis*
Inga quaternata
Euphorbiaceae — *Phyllanthus acuminatus*

9. Flowers dry-wet; fruits wet-dry

Palmae — *Bactris major*
Iridaceae — *Neomarica gracilis*
Orchidaceae — *Campylocentrum micranthum*
Leguminosae — *Myroxylon balsamum* var.
pereirae
Platymiscium pinnatum
Burseraceae — *Bursera simaruba*
Malpighiaceae — *Bunchosia cornifolia*
Euphorbiaceae — *Hura crepitans*
Flacourtiaceae — *Casearia corymbosa*
Lythraceae — *Adenaria floribunda*
Apocynaceae — *Prestonia obovata*
Rubiaceae — *Psychotria furcata*

10. Flowers dry-wet; fruits dry

Bromeliaceae — *Guzmania monostachya*
Tillandsia monadelpha
Orchidaceae — *Epidendrum difforme*
Annonaceae — *Xylopia frutescens*
Leguminosae — *Dalbergia retusa*
Platypodium elegans
Swartzia panamensis
Tachigalia versicolor
Euphorbiaceae — *Garcia nutans*
Apocynaceae — *Odontadenia macrantha*
Bignoniaceae — *Arrabidaea chica*
Ceratophytum tetragonolobum
Macfadyena unguis-cati
Xylophragma seemannianum

10a. Flowers dry-wet; fruits dry (flowers more than 9 months)

Apocynaceae — *Prestonia portobellensis*

11. Flowers dry; fruits dry 1 year later

Bromeliaceae — *Vriesia gladioliflora*
Annonaceae — *Annona glabra*
Leguminosae — *Enterolobium cyclocarpum*
Hippocrateaceae — *Hylenaea praecelsa*
Prionostemma aspera

Sterculiaceae — *Sterculia apetala*
Bignoniaceae — *Pleonotoma variabilis*

12. Flowers dry-wet; fruits dry-wet 1 year later

Palmae — *Scheelea zonensis*
Annonaceae — *Annona muricata*
Leguminosae — *Inga spectabilis*
Hippocrateaceae — *Tontelea richardii*
Apocynaceae — *Lacmellea panamensis*
Tabernaemontana arborea
Rubiaceae — *Tocoyena pittieri*

13. Flowers wet; fruits wet-dry

Palmae — *Geonoma cuneata*
Araceae — *Dieffenbachia pittieri*
Bromeliaceae — *Guzmania lingulata* var. *minor*
Haemodoraceae — *Xiphidium caeruleum*
Zingiberaceae — *Costus pulverulentus*
C. scaber
Marantaceae — *Ischnosiphon pruinosus*
Orchidaceae — *Catasetum bicolor*
Habenaria alata
Piperaceae — *Peperomia ciliolibractea*
Moraceae — *Sorocea affinis*
Annonaceae — *Annona acuminata*
Leguminosae — *Erythrina costaricensis* var.
panamensis
Swartzia simplex var.
grandiflora
Malpighiaceae — *Tetrapteris macrocarpa*
Trigoniaceae — *Trigonia floribunda*
Guttiferae — *Tovomita stylosa*
Vismia macrophylla
Flacourtiaceae — *Casearia sylvestris*
Loganiaceae — *Strychnos panamensis*[5]
Convolvulaceae — *Ipomoea phillomega*
Verbenaceae — *Aegiphila panamensis*
Acanthaceae — *Mendoncia littoralis*
Rubiaceae — *Coussarea curvigemmia*
Isertia haenkeana
Pentagonia macrophylla[5]
Psychotria granadensis
P. uliginosa
Warscewiczia coccinea[5]

13a. Flowers wet; fruits wet-dry (flowers more than 9 months)

Piperaceae — *Piper marginatum*

14. Flowers wet; fruits wet 1 year later

Araceae — *Philodendron hederaceum*
Leguminosae — *Dioclea wilsonii*
Sapotaceae — *Pouteria sapota*

15. Flowers wet; fruits dry

Palmae — *Socratea durissima*
Araceae — *Dieffenbachia oerstedii*
Philodendron sagittifolium
Bromeliaceae — *Catopsis sessiliflora*

15. Flowers wet; fruits dry (cont.)

	Tillandsia anceps
	Vriesia heliconioides
	V. ringens
	V. sanguinolenta
Dioscoreaceae	*Dioscorea sapindoides*
	D. urophylla
Orchidaceae	*Catasetum viridiflavum*
	Dimerandra emarginata
	Epidendrum rigidum
	E. sculptum
	E. strobiliferum
	Liparis elata
	Maxillaria neglecta
	Peristeria elata
	Polystachya foliosa
	Sobralia fragrans
	S. suaveolens
	Trichopilia maculata
	T. subulata
Aristolochiaceae	*Aristolochia gigantea*
Connaraceae	*Cnestidium rufescens*
Leguminosae	*Acacia glomerosa*
	A. hayesii
	A. riparia
	Adenopodia polystachya
	Bauhinia reflexa
	Dipteryx panamensis
	Inga punctata
	Machaerium arboreum[2]
	M. seemannii
	Peltogyne purpurea
Rutaceae	*Zanthoxylum belizense*
Burseraceae	*Trattinnickia aspera*
Meliaceae	*Cedrela odorata*
	Trichilia hirta
Malpighiaceae	*Tetrapteris seemannii*
Sapindaceae	*Paullinia glomerulosa*
	P. pinnata
Tiliaceae	*Apeiba membranacea*
	Corchorus siliquosus
Malvaceae	*Hampea appendiculata* var. *longicalyx*
Sterculiaceae	*Byttneria aculeata*
Dilleniaceae	*Tetracera volubilis*
Guttiferae	*Mammea americana*
Flacourtiaceae	*Lindackeria laurina*
Lythraceae	*Lafoensia punicifolia*
Lecythidaceae	*Couratari panamensis*
Combretaceae	*Combretum laxum* var. *laxum*
Araliaceae	*Didymopanax morototoni*
Sapotaceae	*Chrysophyllum cainito*
Gentianaceae	*Chelonanthus alatus*
Apocynaceae	*Aspidosperma cruenta*
	A. megalocarpon
	Prestonia ipomiifolia
Asclepiadaceae	*Blepharodon mucronatum*
	Matalea pinquifolia
	M. trianae

	M. viridiflora
Bignoniaceae	*Adenocalymma arthropetiolatum*
	Arrabidaea florida
	A. patellifera
	A. verrucosa
	Pithecoctenium crucigerum
	Tynnanthus croatianus
Gesneriaceae	*Kohleria tubiflora*
Rubiaceae	*Cosmibuena skinneri*
	Coutarea hexandra

15a. Flowers wet; fruits dry (flowers more than 9 months)

Orchidaceae	*Maxillaria crassifolia*
Leguminosae	*Swartzia simplex* var. *ochnacea*
Tiliaceae	*Apeiba tibourbou*
Malvaceae	*Sida acuta*
Begoniaceae	*Begonia filipes*
Bignoniaceae	*Anemopaegma chrysoleucum*
	Callichlamys latifolia

16. Flowers wet; fruits dry-wet

Palmae	*Astrocaryum standleyanum*
	Desmoncus isthmius
	Elaeis oleifera
Liliaceae	*Cordyline fruticosa*
Annonaceae	*Xylopia macrantha*
Burseraceae	*Tetragastris panamensis*
Meliaceae	*Guarea multiflora*
Euphorbiaceae	*Dalechampia dioscoreifolia*
Guttiferae	*Calophyllum longifolium*
	Tovomitopsis nicaraguensis

16a. Flowers wet; fruits dry-wet (flowers more than 9 months)

| Piperaceae | *Piper villiramulum* |
| Bignoniaceae | *Martinella obovata* |

17. Flowers wet-dry; fruits dry

Gramineae	*Lasiacis sorghoidea*
Orchidaceae	*Eulophia alta*
	Maxillaria powellii
	M. uncata
Leguminosae	*Desmodium cajanifolium*
	Inga mucuna
	Machaerium microphyllum
	Mucuna mutisiana
Malpighiaceae	*Banisteriopsis cornifolia*
	Hiraea grandifolia
Euphorbiaceae	*Dalechampia tiliifolia*[3]
Sapindaceae	*Paullinia fibrigera*
	P. turbacensis
	Serjania rhombea[3]
Rhamnaceae	*Gouania lupuloides*[3]
Tiliaceae	*Trichospermum mexicanum*
Ochnaceae	*Cespedezia macrophylla*
	Ouratea lucens
Passifloraceae	*Passiflora seemannii*
Begoniaceae	*Begonia guaduensis*
Combretaceae	*Combretum fruticosum*

Apocynaceae	*Prestonia acutifolia*
Bignoniaceae	*Amphilophium paniculatum*
	Arrabidaea candicans
Acanthaceae	*Justicia graciliflora*
Rubiaceae	*Calycophyllum candidissimum*
	Pogonopus speciosus

17a. Flowers wet-dry; fruits dry (flowers more than 9 months)

Commelinaceae	*Gibasis geniculata*
Euphorbiaceae	*Mabea occidentalis*
Apocynaceae	*Mesechites trifida*
Boraginaceae	*Cordia spinescens*

18. Flowers wet-dry; fruits dry-wet

Cyperaceae	*Hypolytrum schraderianum*
	Rhynchospora cephalotes
Dioscoreaceae	*Dioscorea macrostachya*
	(possibly bimodal)
Piperaceae	*Piper carrilloanum*
	P. imperiale
Moraceae	*Perebea xanthochyma*
Aristolochiaceae	*Aristolochia chapmaniana*
Myristicaceae	*Virola surinamensis*
Lauraceae	*Nectandra globosa*
Leguminosae	*Canavalia dictyota*
	Dalbergia monetaria
	Erythrina fusca
Anacardiaceae	*Anacardium occidentale*
Bixaceae	*Bixa orellana*
Passifloraceae	*Passiflora auriculata*
Boraginaceae	*Tournefortia angustiflora*
Cucurbitaceae	*Cayaponia racemosa*

18a. Flowers wet-dry; fruits dry-wet (flowers more than 9 months)

Bombacaceae	*Ochroma pyramidale*
Bignoniaceae	*Pachyptera kerere*

19. Flowers wet-dry; fruits wet

Palmae	*Chamaedorea wendlandiana*
Piperaceae	*Piper cordulatum*
Polygonaceae	*Coccoloba manzanillensis*
Leguminosae	*Dioclea reflexa*
	Inga goldmanii
	I. multijuga
	I. sapindoides
	I. umbellifera
Rubiaceae	*Psychotria marginata*

20. Flowers and fruits bimodally

Gnetaceae	*Gnetum leyboldii* var.
	woodsonianum
Palmae	*Oenocarpus panamanus*
	Synechanthus warscewiczianus
Araceae	*Monstera dubia*
Olacaceae	*Heisteria concinna*
Myristicaceae	*Virola sebifera*
Lauraceae	*Nectandra purpurascens*

Capparidaceae	*Capparis frondosa*
Connaraceae	*Connarus panamensis*
	C. turczaninowii
	Rourea glabra
Leguminosae	*Inga marginata*
	Prioria copaifera
Erythroxylaceae	*Erythroxylum multiflorum*
	E. panamense
Burseraceae	*Protium costaricense*
	P. panamense
Meliaceae	*Trichilia cipo*
Vochysiaceae	*Vochysia ferruginea*
Polygalaceae	*Securidaca tenuifolia*
Euphorbiaceae	*Acalypha macrostachya*
	Hyeronima laxiflora
Hippocrateaceae	*Anthodon panamense*
Sterculiaceae	*Guazuma ulmifolia*
Dilleniaceae	*Tetracera hydrophila*
Guttiferae	*Vismia baccifera*
	V. billbergiana
Araliaceae	*Oreopanax capitatus*
Sapotaceae	*Cynodendron panamense*
Solanaceae	*Cestrum nocturnum*
	Solanum arboreum
	S. argenteum
	S. umbellatum
Rubiaceae	*Cephaelis ipecacuanha*
	Guettarda foliacea
	Psychotria psychotriifolia

21. Flowers and fruits all year, no peak

Gramineae	*Acroceras oryzoides*
	Andropogon leucostachyus
	Bothriochloa intermedia
	Chloris radiata
	Cynodon dactylon
	Hyparrhenia rufa
	Isachne polygonoides
	Oryza latifolia
	Panicum trichoides
	Paspalidium geminatum
	Pharus latifolius
	P. parvifolius
	Schizachyrium brevifolium
	Streptochaeta spicata
Cyperaceae	*Cyperus brevifolius*
	C. diffusus
	C. luzulae
	C. odoratus
	C. sesquiflorus
	C. tenuis
	Eleocharis caribaea
	E. plicarhachis
	Fuirena umbellata
	Rhynchospora nervosa
	Scleria pterota
	S. secans
Araceae	*Anthurium brownii*
	A. gracile

21. Flowers and fruits all year, no peak (cont.)

	Montrichardia arborescens	Passifloraceae	*Passiflora biflora*
	Philodendron scandens		*P. foetida* var. *isthmia*
Commelinaceae	*Commelina erecta*		*P. menispermifolia*
Pontederiaceae	*Pontederia rotundifolia*		*P. punctata*
Smilacaceae	*Smilax mollis*	Caricaceae	*Carica papaya*
	S. panamensis	Begoniaceae	*Begonia patula*
	S. spinosa	Rhizophoraceae	*Cassipourea elliptica*
	S. spissa	Myrtaceae	*Psidium friedrichsthalianum*
Musaceae	*Heliconia mariae*	Melastomataceae	*Aciotis levyana*
Orchidaceae	*Ornithocephalus powellii*		*Adelobotrys adscendens*
	Pleurothallis brighamii		*Clidemia capitellata*
	P. grobyi		*C. dentata*
	Scaphyglottis longicaulis		*C. ocotona*
	Triphora gentianoides		*C. septuplinervia*
Piperaceae	*Pothomorphe peltata*		*Conostegia speciosa*
Moraceae	*Cecropia obtusifolia*		*C. xalapensis*
	Ficus (all species)		*Leandra dichotoma*
	Olmedia aspera		*Miconia nervosa*
Urticaceae	*Pilea microphylla*		*Ossaea quinquenervia*
Loranthaceae	*Oryctanthus alveolatus*		*Tibouchina longifolia*
	O. cordifolius	Onagraceae	*Ludwigia decurrens*
	O. occidentalis		*L. helminthorrhiza*
	Phoradendron quadrangule	Umbelliferae	*Hydrocotyle umbellata*
	Phthirusa pyrifolia	Myrsinaceae	*Ardisia fendleri*
Polygonaceae	*Coccoloba acuminata*	Gentianaceae	*Voyria truncata*
	Polygonum acuminatum	Menyanthaceae	*Nymphoides indica*
	P. hydropiperoides	Apocynaceae	*Allamanda cathartica*
	P. punctatum		*Catharanthus roseus*
Amaranthaceae	*Gomphrena decumbens*		*Ervatamia coronaria*
Phytolaccaceae	*Phytolacca rivinoides*		*Mandevilla subsagittata*
Portulacaceae	*Portulaca oleracea*		*Rhabdadenia biflora*
Menispermaceae	*Cissampelos pareira*		*Stemmadenia grandiflora*
	C. tropaeolifolia		*Thevetia ahouai*
Annonaceae	*Guatteria amplifolia*	Asclepiadaceae	*Asclepias curassavica*
	G. dumetorum		*Sarcostemma clausum*
Capparidaceae	*Cleome parviflora*	Convolvulaceae	*Aniseia martinicensis*
Leguminosae	*Aeschynomene sensitiva*	Verbenaceae	*Lantana camara*
	Crotalaria retusa		*Petrea aspera*
	Desmodium axillare var.		*Stachytarpheta jamaicensis*
	acutifolium	Solanaceae	*Capsicum annuum*
	D. canum		*Cyphomandra hartwegii*
	D. distortum		*Physalis angulata*
	D. tortuosum		*P. pubescens*
Malpighiaceae	*Stigmaphyllon ellipticum*		*Solanum hayesii*
	S. lindenianum		*S. subinerme*
Euphorbiaceae	*Chamaesyce hirta*	Bignoniaceae	*Cydista aequinoctalis*
	C. hypericifolia		*Spathodea campanulata*
	C. hyssopifolia	Gesneriaceae	*Columnea purpurata*
	Dalechampia cissifolia subsp.	Rubiaceae	*Cephaelis tomentosa*
	panamensis		*Diodia ocimifolia*
	Phyllanthus urinaria		*D. sarmentosa*
Hippocrateaceae	*Hippocratea volubilis*		*Ixora coccinea*
Vitaceae	*Cissus sicyoides*	Cucurbitaceae	*Gurania megistantha*
Elaeocarpaceae	*Muntingia calabura*		*Momordica charantia*
Malvaceae	*Hibiscus rosa-sinensis*		*Psiguria bignoniacea*
	H. sororius	Compositae	*Baccharis trinervis*
	Sida rhombifolia		*Chaptalia nutans*
Guttiferae	*Clusia odorata*		*Eclipta alba*
Turneraceae	*Turnera panamensis*		*Emilia sonchifolia*
			Hebeclinium macrophyllum

Melampodium divaricatum
Tridax procumbens
Vernonia cinerea

22. Flowers and fruits all year, especially wet season

Gramineae

Andropogon bicornis
Anthephora hermaphrodita
Axonopus compressus
Digitaria ciliaris
D. horizontalis
Eleusine indica
Gynerium sagittatum
Homolepis aturensis
Leersia hexandra
Leptochloa virgata
Lithachne pauciflora
Panicum fasciculatum
P. pilosum
Paspalum decumbens
P. paniculatum
P. repens
P. virgatum
Rottboellia exaltata
Setaria geniculata

Cyperaceae

Cyperus haspan
Fimbristylis dichotoma
Scleria eggersiana

Araceae

Anthurium acutangulum
A. bakeri
Spathiphyllum friedrichsthalii

Commelinaceae — *Tripogandra serrulata*
Amaryllidaceae — *Hymenocallis pedalis*
Musaceae — *Heliconia vaginalis*
Marantaceae — *Calathea lutea*
Orchidaceae

Epidendrum anceps
Maxillaria alba

Piperaceae — *Piper auritum*
Amaranthaceae

Alternanthera sessilis
Amaranthus viridis

Nyctaginaceae — *Neea amplifolia*
Phytolaccaceae — *Petiveria alliacea*
Chrysobalanaceae — *Hirtella triandra* (flowering 2 or 3 times per year)

Leguminosae

Desmodium adscendens
D. axillare var. *stoloniferum*
Mimosa pigra
Phaseolus peduncularis
P. trichocarpus
Pithecellobium hymeneaefolium
Vigna vexillata

Euphorbiaceae

Chamaesyce thymifolia
Croton hirtus
C. panamensis

Vitaceae — *Cissus erosa*
Malvaceae — *Abelmoschus moschatus*
Guttiferae — *Symphonia globulifera*
Myrtaceae — *Calycolpus warscewiczianus*
Melastomataceae — *Conostegia cinnamomea*
Umbelliferae — *Spananthe paniculata*
Loganiaceae — *Spigelia anthelmia*

S. humboldtiana

Apocynaceae — *Mandevilla villosa*
Convolvulaceae — *Ipomoea quamoclit*
Solanaceae

Cyphomandra allophylla
Solanum ochraceo-ferrugineum
S. rugosum
Witheringia solanacea

Scrophulariaceae — *Scoparia dulcis*
Bignoniaceae — *Phryganocydia corymbosa*
Gesneriaceae

Besleria laxiflora
Codonanthe crassifolia
C. uleana
Columnea billbergiana

Lentibulariaceae — *Utricularia foliosa*
Acanthaceae — *Thunbergia erecta*
Rubiaceae

Borreria latifolia
Chiococca alba
Genipa americana
Posoqueria latifolia
Psychotria brachiata
P. capitata
P. chagrensis
P. grandis
Spermacoce tenuior

Cucurbitaceae

Melothria pendula
Psiguria warscewiczii

Compositae — *Schistocarpha oppositifolia*

23. Flowers and fruits all year, especially dry season

Gramineae

Oplismenus hirtellus
Panicum pulchellum

Cyperaceae

Cyperus densicaespitosus
Scirpus cubensis

Araceae

Anthurium bombacifolium
Stenospermation angustifolium

Musaceae — *Heliconia wagneriana*
Marantaceae — *Thalia geniculata*
Orchidaceae

Maxillaria friedrichsthalii
Psygmorchis pusilla

Piperaceae — *Piper dilatatum*
Amaranthaceae

Alternanthera ficoidea
Cyathula prostrata

Phytolaccaceae — *Rivina humilis*
Leguminosae

Centrosema pubescens
Indigofera mucronata

Dilleniaceae — *Tetracera portobellensis*
Umbelliferae — *Eryngium foetidum*
Solanaceae

Browallia americana
Markea ulei

Scrophulariaceae — *Stemodia verticillata*
Bignoniaceae — *Paragonia pyramidata*
Rubiaceae

Borreria ocimoides
Oldenlandia corymbosa

24. Flowers and fruits all year, especially wet-dry

Gramineae

Hymenachne amplexicaulis
Panicum maximum
P. trichanthum
Sacciolepis striata

Orchidaceae — *Epidendrum nocturnum*

24. Flowers and fruits all year, especially wet-dry (cont.)

	Habenaria repens
Labiatae	*Hyptis brevipes*
Compositae	*Mikania micrantha*
	Spilanthes alba

25. Flowers and fruits all year, especially dry-wet

Gramineae	*Paspalum conjugatum*
	Sporobolus indicus
Araceae	*Anthurium flexile*
Musaceae	*Heliconia metallica*
Zingiberaceae	*Dimerocostus strobilaceus*
Piperaceae	*Peperomia rotundifolia*
Moraceae	*Poulsenia armata*
Leguminosae	*Caesalpinia pulcherrima*
Vitaceae	*Vitis tiliifolia*
Solanaceae	*Solanum antillarum*
Rubiaceae	*Hamelia patens* var. *glabra*

26. Flowers and fruits all year, bimodal peaks

Gramineae	*Ichnanthus tenuis*
Leguminosae	*Cassia fruticosa*
Melastomataceae	*Mouriri myrtilloides* subsp. *parvifolia*
Rubiaceae	*Borreria laevis*
	Sabicea villosa var. *adpressa*

27. Seasonality incompletely known

Araucariaceae	*Araucaria excelsa*
Hydrocharitaceae	*Hydrilla verticillata*
	Limnobium stoloniferum
Gramineae	*Bambusa amplexifolia*
	B. arundinacea
	B. glaucescens
	Chusquea simpliciflora
	Digitaria violascens
	Ichnanthus brevivaginatus
	Saccharum officinarum
Palmae	*Cocos nucifera*
Araceae	*Pistia stratiotes*
	Syngonium sp.
Bromeliaceae	*Ananas comosus*
	Tillandsia bulbosa
	T. fasciculata var. *convexispica*
	T. fasciculata var. *fasciculata*
	T. subulifera
Smilacaceae	*Smilax lanceolata*
Amaryllidaceae	*Amaryllis belladonna*
Dioscoreaceae	*Dioscorea alata*
Musaceae	*Musa sapientum*
Orchidaceae	*Campylocentrum pachyrrhizum*
	Elleanthus longibracteatus
	Encyclia chacaoensis
	E. chimborazoensis
	Epidendrum coronatum
	E. rousseauae
	Notylia albida
	Scaphyglottis prolifera
	Trichocentrum capistratum

Piperaceae	*Piper aristolochiifolium*
	P. peracuminatum
	P. pseudo-garagaranum
	P. viridicaule
Moraceae	*Coussapoa magnifolia*
Ceratophyllaceae	*Ceratophyllum demersum*
Menispermaceae	*Abuta panamensis*
Annonaceae	*Anaxagorea panamensis*
	Crematosperma sp.
	Unonopsis pittieri
Lauraceae	*Nectandra cissiflora*
	N. savannarum
Leguminosae	*Acacia acanthophylla*
	Cymbosema roseum
	Machaerium riparium
	Vatairea erythrocarpa
Humiriaceae	*Vantanea occidentalis*
Rutaceae	*Citrus grandis*
	C. limon
Malpighiaceae	*Hiraea faginea*
Euphorbiaceae	*Codiaeum variegatum*
	Manihot esculenta
	Sapium aucuparium
Sterculiaceae	*Theobroma cacao*
Violaceae	*Rinorea squamata*
Caricaceae	*Jacaratia spinosa*
Myrtaceae	*Eugenia principium*
	E. uniflora
Melastomataceae	*Miconia rufostellulata*
Onagraceae	*Ludwigia torulosa*
Araliaceae	*Polyscias guilfoylei*
Sapotaceae	*Pouteria fossicola*
Loganiaceae	*Strychnos darienensis*
Apocynaceae	*Forsteronia myriantha*
Convolvulaceae	*Operculina codonantha*
Labiatae	*Coleus blumei*
Scrophulariaceae	*Mecardonia procumbens*
Lentibulariaceae	*Utricularia obtusa*
Cucurbitaceae	*Cayaponia denticulata*
	Posadaea sphaerocarpa
	Sicydium coriaceum
Campanulaceae	*Centropogon cornutus*
Compositae	*Clibadium asperum*
	Mikania hookeriana
	M. tonduzii
	Pluchea odorata

27a. Flowers known; fruits unknown

Pontederiaceae	*Eichhornia crassipes*
Dioscoreaceae	*Dioscorea haenkeana*
	D. polygonoides
Orchidaceae	*Coryanthes maculata*
	Epidendrum stangeanum
	Habenaria bicornis
	Lycaste powellii
	Masdevallia livingstoneana
	Maxillaria camaridii
	Pleurothallis trachychlamys
	P. verecunda
	Sievekingia suavis

	Sobralia rolfeana
	Spiranthes lanceolata
	Triphora mexicana
	Vanilla fragrans
	V. pompona
	Xylobium foveatum
Proteaceae	*Roupala montana*
Aristolochiaceae	*Aristolochia pilosa*
Nymphaeaceae	*Nymphaea ampla*
Lauraceae	*Ocotea oblonga*
	O. pyramidata
Chrysobalanaceae	*Licania platypus*
Leguminosae	*Desmodium wydlerianum*
	Inga minutula
	I. ruiziana
	Ormosia macrocalyx
Malpighiaceae	*Malpighia romeroana*
Euphorbiaceae	*Acalypha wilkesiana*
Sapindaceae	*Talisia princeps*
Passifloraceae	*Passiflora vitifolia*
Cactaceae	*Epiphyllum phyllanthus* var. *rubrocoronatum*
Lecythidaceae	*Grias fendleri*
Myrtaceae	*Syzygium jambos*
Ebenaceae	*Diospyros artanthifolia*
Loganiaceae	*Strychnos brachistantha*
Asclepiadaceae	*Cynanchum cubense*
	C. recurvum
	Fischeria funebris
	Gonolobus allenii
Convolvulaceae	*Iseia luxurians*
Verbenaceae	*Clerodendrum paniculatum*
Solanaceae	*Cestrum racemosum*
Bignoniaceae	*Adenocalymma apurense*
	Clytostoma binatum

27b. Flowers unknown; fruits known

Palmae	*Bactris coloradonis*
Ulmaceae	*Celtis schippii*
Sapindaceae	*Paullinia pterocarpa*
Rubiaceae	*Chomelia psilocarpa*

28. Flowers dry; fruits all year

Gramineae	*Streptochaeta sodiroana*

29. Flowers dry-wet; fruits all year

Bignoniaceae	*Stizophyllum riparium*

History of Panama and the Canal Zone

The Spanish first came to Panama in 1501, and the first settlements were established in 1508, at Santa Maria la Antigua de Darién and at Nombre de Dios. During the succeeding decades, the Spanish completed their conquest of the region. In 1718 Panama became part of the viceroyalty of New Granada (Colombia), and the isthmus soon became the principal route for all traffic to and from Peru and environs. Nonetheless, little development took place outside a narrow belt along both coasts in the region of the isthmus.

In 1821 Panama declared independence from Spain and began an unstable alliance with Colombia. Following numerous revolutionary outbreaks, Panama finally split with Colombia in 1903 when the Colombian legislature failed to ratify a treaty with the United States to build a canal across the isthmus. In the same year the United States obtained the Canal Zone through a treaty with Panama and commenced construction of the Panama Canal in 1907. The Canal was finished and in operation by 1914. The opening of the Canal greatly affected the region of the isthmus, and the city of Panamá soon became a major trading and banking center. Much of the region has at times been almost completely deforested, and for a time the Canal Zone administration leased its land and encouraged settlement and farming. Later policy dictated that no new land be leased, and that all existing leases be terminated upon the death of the original lessee. Since then much of the Canal Zone has become reforested, but a number of the original lessees are still living on the land—or their descendents, in any case, have not been evicted. Still other, obviously clandestine, agricultural operations are under way in remote parts of the Canal Zone, especially in the area around Frijoles. The best forest in the Canal Zone besides that on BCI is in Madden Forest (now a reserve) and along the northeast side of the Canal bordering the naval oil pipeline that crosses the isthmus.

Although BCI was created in 1914 by the rising waters of Gatun Lake, no attempt to preserve its forest was made until it was set aside as a permanent reservation by the governor of the Canal Zone on April 17, 1923. Establishment of the preserve was brought about largely through the efforts of Thomas Barbour and James Zetek. The preserve was first operated by the Institute for Tropical Research, and later came under the direction of the Smithsonian Tropical Research Institute, which still manages it.

History of Botanical Studies

Panama and the Canal Zone. The earliest botanical collections in Panama were made about 1700 by James Wallace, who was associated with the old Scottish settlement of New Caledonia, and though they are still extant in London they have not been carefully studied to this day. Serious botanical activity began in Panama about 1825, when J. E. Billberg collected near Portobelo in Colón Province. Other nineteenth-century collectors include B. C. Seemann (1846–49), P. D. Duchassaing (1849–51), A. Fendler (1850), K. Halsted (1850), and Sutton Hayes (1860–63). Early twentieth-century collectors of note in Panama were R. S. Williams (1908), C. W. Powell (1907–27), H. F. Pittier (1911, 1914–15), W. R. Maxon (1911, 1923), E. P. Killip (1917–18, 1922), C. V. Piper (1923), P. C. Standley (1923–25), G. P. Cooper (1927–28), A. M. Chickering (1928), Brother Paul (1934), M. E. Spence Davidson (later Mrs. R. A. Terry; 1938, mainly from Chiriquí), and M. E. Terry and R. A. Terry from Darién (late 1930's).

Until 1930 the collectors who came to Panama visited only a relatively small portion of the country and, with the notable exception of Henri Pittier, most collected in

a relatively restricted area, many of them never leaving the region of the isthmus.

Botanical activity leading to a *Flora of Panama* began in 1934 with the efforts of R. E. Woodson, Jr., and his associates from the Missouri Botanical Garden. Among those who participated in these efforts in the 1930's were R. W. Schery, R. J. Seibert, J. A. Steyermark, P. C. Allen, A. A. Hunter, and Carol Dodge. Particularly notable was Paul C. Allen, who collected more widely in Panama than anyone up to that time. About 10,000 numbers were collected during this prewar era; as a result of this activity, Woodson and Schery initiated the *Flora of Panama* project in 1943. Subsequent collecting in Panama by staff and students of the Missouri Botanical Garden has yielded an estimated 73,000 collections.

During World War II, San José Island in the Bay of Panama was intensively collected by I. M. Johnston, C. O. Erlanson, and others, resulting in publication of a flora, and in 1940–41 H. von Wedel collected extensively in Bocas del Toro. Little other collecting was done in Panama during or after World War II.

Financial support for the *Flora of Panama* project was obtained from the National Science Foundation beginning in 1957 and continues through to the completion of the project, which is expected in 1978. Field work in Panama by the Missouri Botanical Garden was renewed in 1959 by John D. Dwyer, working with Kenton Chambers, William Stern, and John Ebinger (all associated with Yale University). Working independently, Dwyer made three additional expeditions to Panama between 1961 and 1964.

Full-scale reactivation of the field program in Panama began in 1966, under the direction of Walter H. Lewis, with increased financial assistance from the National Science Foundation. Four expeditions were made to Panama under Lewis's direction between 1966 and 1969, accumulating 5,581 collections. Participating in these expeditions were staff and students including John Dwyer, André Robyns, Derek Burch, Tom Croat (principal investigator 1972–77), Marshall Crosby, Duncan Porter (principal investigator 1971–72), Tom Elias, D. F. Austin, Royce L. Oliver, Kenneth R. Robertson, Will H. Blackwell, Jr., Joan W. Nowicke, Bruce Mac-Bryde, John E. Ridgeway, L. H. Durkee, John L. Hawker, Susan E. Verhoek-Williams, Jerry R. Castillon, and Richard K. Baker.

A field station and herbarium were established in Panama in 1969 by the Missouri Botanical Garden and the Panama Canal Company, through the efforts of Walter Lewis, then director of the herbarium at Missouri, and A. I. Baumann, then supervisor of the Community Services Division of the Panama Canal Company. The facility was named Summit Herbarium (SCZ), and it has received support from the National Science Foundation. Others important in the initial organization of the station were Edwin L. Tyson, then with Florida State University in the Canal Zone, and Roy Sharp, then supervisor of Grounds and Maintenance for the Panama Canal Company. The initial collection of approximately 6,000 mounted sheets was made by Tyson and others

under the auspices of the Army Tropic Test Center in the Canal Zone. The field station, with a residence for the curator and a field vehicle, has enabled Missouri to maintain successive collectors in Panama since that time: Tom Croat (1971), Al Gentry (1972), Helen Kennedy (1973), Michael Nee (1974), Scott Mori (1975), and James Folsom (1977), with a repeat by Tom Croat during 1976. Since the establishment of Summit Herbarium, Missouri has made 50,000 collections in Panama.

The herbarium and drying facilities of Summit Herbarium, now located at Ancon in the Canal Zone, are sponsored jointly by the Smithsonian Tropical Research Institute (STRI) and the Missouri Botanical Garden. The drying facilities, now housed in the STRI Tivoli Building, feature about 8,500 watts of electric dryers in a fireproof room and are adequate for a sustained collecting program.

In addition to resident collectors employed by the *Flora of Panama* project, Missouri Botanical Garden has sponsored other expeditions to Panama. In 1972, the National Geographic Society helped sponsor a phytogeographic survey of the Burica Peninsula by Tom Croat. With the assistance of Ron Liesner and Philip Busey, 3,600 collections were made. In 1975, Al Gentry, with support from the National Geographic Society and assistance from Scott Mori, made an expedition to Cerro Tacarcuna on the Colombian border, netting 1,100 collections.

John Dwyer, both independently and with his students, has made about 6,000 collections in Panama on several expeditions since 1964. Among Dwyer's students who have collected in Panama are Joseph Kirkbride, Victoria Hayden, T. S. Elias, B. R. Lallathin, David Spellman, and Richard Wunderlin. Joseph Kirkbride, with James A. Duke, collected about 1,000 numbers in Panama. The most important of these are from an overland expedition from Bocas del Toro on the Caribbean to Chiriquí.

Numerous expeditions to Panama have been made by Tom Croat since 1967 on the Flora of Barro Colorado Island project (also partially NSF funded). Since 1969 the *Flora of Panama* project has funded collecting trips by Walter Lewis, John Dwyer, Duncan Porter, W. G. D'Arcy, and Tom Croat. With support from the *Flora of Panama* project, and with separate NSF funding, W. G. D'Arcy made about 4,000 collections during five visits to Panama.

Other institutions that have been active in Panama include the University of Panama, Duke University, and Florida State University. The University of Panama, principally through the efforts of Mireya D. Correa A. and N. Escobar and their students, has made a significant number of collections in Panama. Robert L. Wilbur, of Duke University, has made eight expeditions to Panama. Students of Wilbur's who have participated in these expeditions include F. Almeda, J. Terri, P. Armond, J. Luteyn, J. Utley, and R. Weaver. Robin Foster, an ecologist trained at Duke, has made numerous collections on Barro Colorado Island and elsewhere in Panama.

Florida State University, through the efforts of R. K. Godfrey, E. L. Tyson, H. Loftin, Sidney McDaniel,

K. Blum, and R. L. Lazor, has also made numerous collections in Panama. Particularly noteworthy are the collections of E. L. Tyson between the years 1962 and 1972. During part of this time he was employed by the Army Tropic Test Center, where he established the collection of approximately 6,000 specimens now incorporated with Summit Herbarium. Noteworthy also are L. R. Holdridge and E. A. Lao, who made collections associated with forestry investigations in Panama during the 1960's, and A. Weston, who made recent collections at high elevations in western Panama.

Most noteworthy among botanists not associated with the above-mentioned institutions are Robert L. Dressler of STRI and Jim Duke. Dressler's wide-ranging and selective collecting has yielded numerous interesting and new species. Duke is best known for his collections from Darién, which were made while he was employed by Battelle Memorial Institute as part of a program investigating a new sea-level canal route.

Barro Colorado Island. Probably the first significant plant collecting on the island was that done by Paul Carpenter Standley, who in the course of 8 days (January 17, 1924, and a week in November 1925) made 800 collections. It was principally on the basis of these collections that Standley (1927) published his first checklist for the island. During the summer of 1927, Leslie A. Kenoyer, of Western State Teachers College in Kalamazoo, Michigan, conducted extensive ecological studies on the island. In the course of his work he made 690 collections, and on the basis of these collections, Kenoyer and Standley published in 1929 the first supplement to the *Flora of Barro Colorado Island*. A second supplement was published by Standley in 1930.

Because of the increase in botanical work on the island during 1931 and 1932, so many collections were accumulated that Standley decided to completely revise the flora. The revision was published in 1933. Collections made during 1931 and 1932 were principally by C. L. Wilson, L. H. and E. Z. Bailey, D. E. Starry, Silvestre Aviles (a local assistant to Bailey), Otis Shattuck, R. H. Wetmore and E. C. Abbe, and R. H. Woodworth and P. A. Vestal. A few collections were also made by C. Ray Carpenter during his studies of monkeys. James Zetek, who together with Thomas Barbour was principally responsible for the establishment of Barro Colorado Island as a preserve, was also the first resident manager. He collected fewer plants than most collectors of his time but his collections are among the most selective. Though Zetek was not a botanist by profession, his correspondence with Standley indicates that he had a keen botanical awareness.

The following list gives the names of all collectors known to have worked on Barro Colorado Island, and when known the number and date of collections:

Abbe, E. C. (see Wetmore and Abbe)
Aviles, S. (few, 1931; see note following list)
Bailey, L. H., and E. Z. Bailey (700, 1931)
Bangham, W. N. (many, late 1920's)
Bartlett, H. H., and T. Lasser (few, 1940)
Brown, W. L. (few)

Busey, P. (few, 1973)
Carleton, M. A. (few)
Carpenter, C. R. (few, 1931–33)
Chardon, C. E. (few)
Chickering, A. M. (few, 1928)
Cook, O. F. (few, early 1920's)
Croat, T. B. (6,614, 1967–75)
D'Arcy, W. G. (few, 1970)
Dare, R. (few, 1972)
Dodge, C. W. (few, 1925)
Dressler, R. L. (few, early 1960's)
Duke, J. A. (few, early 1960's)
Dwyer, J. D. (few, 1961)
Dwyer, Correa, and Pasco (81, 1968)
Ebinger, J. (few, 1960)
Fairchild, G. B. (few, early 1940's)
Faull, J. H. (few)
Folsom, J. (few, 1977)
Foster, R. (1,000, 1969–77)
Garwood, N. (few, 1976–77)
Gentry, A. (few, 1969 and 1971)
Graham, S. (few)
Hayden, M. V. (few, 1965)
Hladik, A. (544, 1967)
Hood, J. D. (few, early 1930's)
Hunnewell, F. W. (few)
Kennedy, H. (few, 1970–72)
Kenoyer, L. A. (680, 1927)
Killip, E. P. (few, 1948)
Knight, D. H. (numerous sterile collections, 1967–68)
Luteyn, J. (few, 1968)
Maxon, W. R. (few, 1923)
McDaniel, S. (few, 1964 and 1972)
Montgomery, G. (300, 1973)
Munch, S. (few, 1973)
Netting, M. G. (few, early 1920's)
Nolla, J. (few)
Oppenheimer, J. (few, 1967–68)
Robyns, A. (few, 1965)
Salvoza, F. M. (few, late 1920's)
Shattuck, O. E. (870, 1930–34)
Standley, P. C. (800, 1924–25)
Starry, D. E. (328, 1931)
Steiner, K. (few, 1976)
Stevens, F. L. (few, 1924)
Stimpson, W. (few, 1967)
Stoutamire, W. P. (few, 1956)
Svenson, H. K. (few)
Terry, M. E., and R. A. Terry (few, late 1930's)
Tyson, E. (few, 1966)
Van Tyne, J. (few, late 1920's)
Weaver, R., and R. Foster (few, 1968)
Wetmore, R. H., and E. C. Abbe (225, 1931–32)
Wetmore, R. H., and R. H. Woodworth (many, late 1920's and early 1930's)
White, P. (few, late 1930)
Wilbur, R. L. (few, 1967)
Wilson, C. L. (158, 1931)
Woodworth, R. H., and P. A. Vestal (450, 1932)
Zetek, J. (few, 1930–43)

Aviles used Shattuck's labels on many of his collections. This has resulted in considerable confusion; the plants collected by Aviles are often cited as Shattuck's collections.

More recent botanical work on the island has been ecologically oriented. Among the projects carried out in the past 10 years are work on bee pollination of orchid flowers, by Robert L. Dressler; on phytosociology, by Dennis H. Knight; on nutrient recycling, by I. N. Healey; and on reproductive potential, by Robin Foster. (Foster was also important as a plant collector, making approximately 1,000 collections on the island.) Zoological work in recent years has dealt principally with studies of animal behavior.

Background of work on the present flora. Standley's final *Flora of Barro Colorado Island,* published in 1933, gives names and brief descriptions of 1,058 species of vascular plants (as well as 201 nonvascular plants). Relatively little collecting was done on the island between the appearance of Standley's *Flora* and the beginning of this work. BCI was even at that time the best-collected and presumably the best-known botanically of any tropical area in the New World. No fewer than 5,000 collections had already been made from an area of about 15.6 km² (6 square miles). Being somewhat of a tropical novice when I began the work, I envisioned a two- or three-year project. I was even told that the flora was so well known that the idea of revising the *Flora* was ill conceived. Knowing what I now know about the richness of lowland tropical rain forests and the very low frequency of many widespread species, I might have argued that such a flora could never be completed.

The project has taken 10 years to complete. My work began at the Missouri Botanical Garden in August 1967, with a two-week trip to the island in December of that year and successive collecting trips in April and May 1968, in September and October 1968, and from January to March 1969. In March 1970, I moved to the Canal Zone to serve as the first curator of Summit Herbarium, and continued my study of the flora of BCI for 18 months. This long, uninterrupted period of study during 1970 and 1971 provided the opportunity to make many of the phenological observations I feel are so important to this work. Many of the descriptions were also written while I was in the field. This is especially important for palms, certain Araceae, and some other plants that cannot, whether because of their ephemeral nature or their size, be easily described from herbarium material or even from fresh specimens carried back to the laboratory on the island or to Summit Garden. In some cases, BCI species that occur elsewhere in the Canal Zone (especially near Summit Garden, where I was living) were collected and used for descriptions. When I left the Canal Zone in 1971, approximately 60% of the species were described, nearly all of them from living material. Some of the species that are the least detrimentally affected by pressing and drying, e.g., grasses and ferns, were left to be described from herbarium material.

The value of describing plants from living material is that many aspects of the flower or fruit that may be lacking in a dried specimen can be recorded—such features as the disposition of the parts before, during, and after anthesis, and the size and shape of the parts before drying. This approach is especially valuable for mature fruits, since collections of them are rarely found in herbaria.

Methods of field work. Barro Colorado Island has been described by many botanists as a difficult place to collect plants. Insects and other pests, such as ticks, are abundant on the island—perhaps because of the abundance of larger animals, which help to sustain the blood-sucking creatures. Other animals, including snakes, are usually no problem, although I was viciously attacked by a collared peccary one morning on the way to the dining hall from my cabin.

The trails, though generally laid out along ridges and thus offering the easiest ascent to the center of the island, are nonetheless often steep and can be hazardous, especially in the rainy season. Since the island is a preserve and no felling of trees is permitted, collecting within the forest is very difficult. Along the trails the forest is more open, but the light is relatively poor. Even with the aid of binoculars it is difficult to determine which plant, of the several overhead, is the one dropping its flowers or fruits. Sometimes, all that can be done is to find the densest area of fallen parts and then, on the basis of the characteristics of the species, choose the tree that appears most suspect. Often, of course, the plant is a liana or hemiepiphyte, generally not visible from subcanopy levels.

To get into the canopy to collect plants I generally used climbing irons (Croat, 1969). In addition, I carried a sheath knife, a machete, a 50-foot cotton rope with a lead weight, binoculars, and often camera equipment. The technique of climbing with climbing irons, once mastered, is not dangerous. I suffered only one serious fall during the BCI work, and that when I too zealously avoided a swarm of wasps. Ants, especially the large stinging *Paraponera,* are sometimes hazardous to climbing, but the smaller biting ants that may cover your arms until they are black are always more of a nuisance. Ants are encountered in nearly every tree, especially the flowering trees.

Climbing the wrong tree is a common problem, but switching to another tree from the canopy is usually relatively easy. The most difficult trees to climb are those that have very slender trunks, and thus offer little lateral stability, and those whose girth is so great that the safety rope will not encircle them. Even these can generally be climbed, however, since the stout lianas or hemiepiphytes that press along the trunk of most large trees can be used for support. Trees with very hard bark or wood are also dangerous, since they cannot be penetrated deeply enough to ensure that the gaffs will not slip out. Trunks covered with lianas or epiphytes can be difficult to climb, as well. It is often necessary in these cases to remove the safety rope to get through the tangle—a safe enough proposition, since it would be difficult to fall out of these tangles even if you wanted to. At times I became so entangled I regretted that I could *not* fall out!

By contrast, collecting along the shore is relatively

easy. Because BCI has so recently come to be an island, the shoreline does not support the exclusively riparian vegetation found in a natural situation. Consequently, some species not normally found along water's edge can still be collected along the shore of the island. Perhaps more important, many shoreline trees represent regrowth of trees that have fallen into the lake, and branching may occur near the ground and over the lake, allowing for easy collecting of many of the same species that in the forest would demand an energetic climb. Lianas are exceedingly difficult to collect in the forest, even from the canopy, because they tend to grow only across the surface of the canopy, and are difficult to reach without venturing far out on a limb. However, along the shore, many liana species come down almost to the water and can be collected easily.

Collecting from a boat can be troublesome, because anything accidentally dropped over the edge may disappear into as much as 20 feet of murky water, even very near the shore. Two cameras, a pair of eyeglasses, binoculars, and a host of other miscellaneous things met their end in this manner during my work on this project. The advantages of the shore—easy access and a multiplicity of species—are somewhat offset by other inconveniences. A boat seldom offers a very stable platform on which to work, especially if the water is choppy, and the larger waves created by passing ships can be even more vexing. On one collecting trip, my boat was struck from the stern by a large wave and immediately sank. Fortunately, I was near the shore. Running onto barely submerged tree stumps is another hazard of shoreline collecting. Still another is the many wasps' nests that hang low over the water; often, I did not spot these until I was hopelessly mired in *Hydrilla* and unable to beat a quick retreat.

The most poorly collected part of the shore lies between the end of Armour Trail and the western tip of Peña Blanca Point. The waters here are choked with tree trunks, and navigation is difficult. The forest that adjoins this area on the western side of the island—the farthest point from the Laboratory Clearing and the boat dock—is also poorly collected, and the area can be counted upon for more species new to the flora.

Key Structure

Since identification is expected to be a major use of this work, keys have been given careful attention. *All statements in the keys apply to BCI specimens,* not necessarily to specimens of the same species collected elsewhere. Where possible, characters for both flowers and fruits, as well as vegetative parts, have been incorporated. The most dependable separating characters are the first in each couplet; no other attempt is made to be consistent about the order of the characters. For the most part, the opposing statements in each couplet are parallel; in some cases, additional characters are added at the end of one half of the couplet, usually on a final dichotomy, to assure the reader that he has the correct plant. In keys extending more than a page, pairs of symbols have been added at the beginning of one or more couplets to facilitate progress through the key.

The keys to families within divisions are not based on traditional characters. I have tried to make these keys easier to use by avoiding technical characters wherever possible. For a key to families based on natural characters, see Hutchinson (1967).

The keys to taxa within families terminate with genera, species, or both. A genus with two or more species on BCI is usually provided with its own key to species, unless the family is small.

The key to major plant groups in the flora is given below (p. 63).

Key to sterile woody plants. Although other sterile keys have been constructed for BCI (Knight, 1970), the one on p. 861 is the first to include all the woody plants. It will thus enable scientists to make surveys of the forest at any time of year. Because the species in the flora are generally quite seasonal, only a small percentage can be collected in fertile condition at a given time. Someone with an interest in studying the vegetation might not be able to manage the many visits to the island that would be required to find all of the species in the sample in flower or fruit. Moreover, a substantial part of the flora will be juvenile in any event. Thus the key to sterile plants will be useful. For several reasons, herbaceous plants are not included: in general, they are much better known and thus relatively easy to identify; they tend to flower and fruit for longer periods; and the incorporation of all the herbaceous plants would have made the key a great deal longer and thus more difficult to construct and more unwieldy to use.

Approximately 700 species are included in the key to sterile woody plants. Excluded from the key are species that are both cultivated and known only from the Laboratory Clearing; species that are usually mostly herbaceous, e.g., *Aristolochia*; and some species that are probably no longer on the island, e.g., *Columnea purpurata*. At the same time, some of the more common coarse herbaceous vines that occur in the forest are included because they will be more frequently encountered in surveys of the forest and would otherwise be difficult to identify.

Needless to say, there are substantial problems attending the construction of a key to sterile plants for so many taxa. The greatest difficulty arises in trying to separate the numerous species having simple alternate leaves. Every attempt has been made to account for intraspecific variation—species that are quite variable appear in the key several times—but in some cases the variation is excessive and the plant may not key out. Moreover, it was impossible to include all juvenile forms in the key, and many of these will not key out. Using the sterile key, approximately 60% of the woody plants on BCI can be identified with little difficulty, the remainder with varying degrees of difficulty. The characters that are generally the most variable are pubescence, leaf shape and size, and the condition of the blade margin (i.e., toothed or entire).

Some plants that will not key out in the sterile key are juvenile forms that do not resemble the adult plants. Others may be species that are new to the flora and have not been found in fertile condition. Any systematic attempt to identify large quantities of sterile material from

the island will probably produce other species new to the flora. Already, many sterile, unidentifiable plants have accumulated that probably represent species new to the flora.

The nature of the key suggests that identifications made with it be thoroughly checked against the text descriptions and compared with herbarium specimens where possible. Where the choices become difficult to make, the troublesome couplets should be marked and the plant keyed out two or more ways, for even if the process yields several possibilities, it is faster and easier to check these in an herbarium than to attempt identification in most other ways. Those who have field-tested the sterile key report that it is difficult to use successfully at first, but that it becomes quite useful after some experience has been gained.

Family Descriptions

Family descriptions apply to genera and species on BCI, except where other data are added parenthetically. Flowers are presumed to have one pistil unless otherwise stated. Information concerning the number of ovules in the ovary and the number of seeds in the fruit is not necessarily repeated; for example, if a fruit is described as being 5-locular with many seeds per locule, it is assumed that the ovary is many-ovulate.

The number of genera and species in a family, as well as the range of the family, have been extracted from standard works such as Willis (1966) and Lawrence (1964) or from recent monographs.

Ecological information of a general nature or pertinent to more than one species is included with the family description. Ecological information of a more detailed nature and pertinent to a particular species — or general information on the only BCI species in a family — follows the species description. The information given is applicable only to those species or genera on BCI, and statements are not necessarily relevant to other species or genera occurring elsewhere. The ecological information given for a genus refers to at least some members of the genus, but not necessarily to all of the species on BCI. Unless otherwise worded, statements concerning possible pollinators or possible dispersal vectors are based on the morphology of the species and what is already known about the pollination and fruit dispersal of related taxa, rather than on actual observation.

When specific information is available concerning the interaction between the plants and animals of BCI, I have included it here. I have frequently mentioned, for example, that white-faced or howler monkeys are known to eat certain fruits (the information is based on the studies of Oppenheimer, Hladik & Hladik, and C. Smith, and citations are given with the text discussions). It can probably be assumed that most of the fruits eaten by these monkeys are also eaten by other fruit-eating animals in the forest, for most fallen fruit that is taken by one of the animals which frequent the forest floor is likely to be taken by many other animals (N. Smythe, pers. comm.). Although little reference is made to them because of the

lack of necessary studies, a number of animals that are seldom referred to, such as lizards (especially iguanas), may be important in the dispersal of many fruits. Most of the references I have made to the fauna have used common names. The following list gives the zoological names of the most commonly mentioned mammals:

Opossum	*Didelphis marsupialis ctensis*
Coati	*Nasua narica panamensis*
Kinkajou	*Potos flavus isthmicus*
Squirrel monkey or marmoset	*Oedipomidas geoffroyi*
Howler monkey	*Alouatta paliata inconsonans*
Night monkey	*Aotus zonalis*
White-faced monkey	*Cebus capucinus imitator*
Tamarin	*Sanguinnus geoffroyi*
Squirrel	*Sciurus gerrardi morulus, Microsciurus alfari venustulus*
Spiny rat	*Proechimys semispinosus panamensis*
Agouti	*Dasyprocta punctata isthmica*
Paca	*Cuniculus paca virgatus*
Collared peccary	*Peccari angulatus bangsi*
White-lipped peccary	*Tayassu peccari spiradens*
Three-toed sloth	*Bradypus griseus*
Two-toed sloth	*Choloepus hoffmanni*

A complete discussion of the mammals of the island is given in Enders (1935). Many publications reporting recent work on the fauna are referred to in the ecological discussions of plant-animal interactions.

Genus Descriptions

To conserve space, formal generic descriptions are not presented, and pertinent information has been incorporated into the family descriptions. For example, "Epiphytic or terrestrial (*Ananas* and *Aechmea*)" says that all the genera are epiphytic except *Ananas* and *Aechmea*. In other cases, morphological or ecological information is given following the genus heading; this is usually done only for large genera, in order to avoid repetition in species descriptions.

Species Descriptions

Synonymy. Synonyms listed herein are restricted to the following: names used by Standley in his revised *Flora of Barro Colorado Island* (1933); names, since changed, that were used by Woodson and Schery in the *Flora of Panama*; and synonyms commonly used for Panamanian species in families that at the time of this writing have not been completed in the *Flora of Panama*.

Common names. The common names reported are those used in Panama or reportedly used in Panama. They have been extracted from herbarium labels and also from the following works: *Flora of Panama* (Woodson & Schery, 1944–   ), *Flora of the Canal Zone* (Standley, 1928), *Flora of Barro Colorado Island* (Standley, 1933), *The Rain Forests of Golfo Dulce* (Allen, 1956), and *Darién Ethnobotanical Dictionary* (Duke, 1968).

Morphology. Species descriptions apply to material

as it occurs on BCI. In some cases, local populations elsewhere may differ in some substantial manner, but such cases are usually rare, and a description will in general be consistent with normal variation throughout the range of a species. Species are usually described in detail; exceptions are sometimes made where (1) too few specimens are available to make a complete description, (2) the species is believed to be no longer present on the island, or (3) the taxon differs only slightly from another and is merely compared with it. Those species described from fresh plants are described in greater detail than those described from herbarium specimens, especially where flowers or mature fruits were collected in the field. Because many descriptions have been made from living material, they differ frequently from those prepared from parts that are dried or secondarily softened before study.

Measurements in parentheses, for example in the span (5)10–15(20), account for extremes in variability, probably less than 5% of the observed individuals of a species, or they are unconfirmed reports for the species from other sources, principally the *Flora of Panama*. In the case of tree height, however, the measurements in parentheses reflect reported heights of the species elsewhere in Panama. The upper height *not* in parentheses is the maximum observed height on BCI. And for purposes of size classification—that is, whether a tree should be designated small, medium-sized, or tall—I considered only those heights recorded on BCI.

Many characters should be presumed unless otherwise stated: stipules absent unless described; blades simple and entire; parts glabrous, unarmed, and smooth. If pubescence is of a generally consistent character throughout the plant, it is usually mentioned only once, near the beginning of the description. Flowers should be assumed to be regular unless described as zygomorphic. Fruits are presumed to be fresh and ripe.

Where practical, terms that are likely to be unfamiliar to the average biologist have been avoided, though a certain amount of technical terminology is unavoidable. No glossary is provided; the reader is referred to Lawrence (1955) and Jackson (1928) for definitions of most botanical terms used here.

The term *domatia* has in general been used rather broadly; often it identifies structures that are sunken or pitlike; but in other contexts it implies mere tufts of pubescence in leaf axils that appear to be forming shelter for insects (though in most cases it has not been confirmed that insects actually inhabit these areas).

Colors used in descriptions are those in common usage. However, for reference purposes, the following list gives the equivalent names and numbers for the most commonly used colors, as given in the *Color Chart of the Royal Horticultural Society* (1966 edition): blue, 100; violet-purple, 77; violet, 82; blue-violet, 93; mulberry (cyclamen-purple), 74; orchid, 84; lavender, 76; thistle (imperial purple), 78; pink, 49; magenta, 66; orange-red, 34; maroon, 60; red-violet, 63; Indian red, 75; brick-red, 180; red, 45; red-orange, 30; salmon, 48; orange, 24; melon, 35; bittersweet (burnt orange), 31.

Specimens cited. The morphological description of a species generally terminates with a citation (in italic type) of the herbarium specimen(s) on which the description is based in part. Usually, only one or two collections are cited, and they are my own collections, unless a better specimen is available or I have never collected the species fertile. The citing of specimens is intended only to help future workers in the event that there is confusion about the taxon in question. Unless otherwise indicated, all collections cited are deposited at the Missouri Botanical Garden (MO). Many are well duplicated, and duplicates are in most cases deposited at Summit Herbarium in the Canal Zone (SCZ), the University of Panama (PMA), the Field Museum (F), the New York Botanical Garden (NY), and Duke University (DUKE). In a few cases no specimens are cited, e.g., those orchids reported by Dressler for which he made no collection, or the species reported by Standley (1933), and likely to be on the island, but for which I could find no specimen.

Abundance on the island. Statements of abundance are based principally on estimates pegged to a scale of 1 to 5, made both by myself and by Robin Foster. Foster's estimates are particularly valuable for the tree flora, with which he is very familiar because of his ecological studies on the island. Measures of abundance are usually: 1, rare; 2, infrequent; 3, occasional; 4, common; 5, abundant.

BCI habitats. Although all of BCI falls within the tropical moist forest life zone of the Holdridge Life-Zone System (see maps on pp. 58–59 and discussion on p. 60), specific habitats are diverse. The major habitats include forest (both old forest and young forest), man-made clearings, and the lakeshore. Within the forest, there are still other specific habitats, such as those along deep ravines, along trails, and in tree-fall areas. The lakeshore itself provides several different types of habitats, such as eroded banks, forest edge (where the forest lies adjacent to the water), clearing edge, and plant-choked marshes.

Phenological statements. The phenological behavior of a species is discussed wherever something can usefully be said. The sections on phenology (p. 29) explain the methods used. Any month listed in parentheses represents less than 5% of the flowering period. Phenological data often provide an additional character for the separation of two species, for even closely related species are often distinct phenologically.

Miscellaneous ecological and taxonomic discussion. Specific ecological data, such as observed visits to the species (by pollinators or feeders), are discussed with the species description. More general commentary is incorporated with the family description; in a few cases, where the comments deal with a single genus, they follow the heading for that genus.

Miscellaneous taxonomic information follows the ecological portion of the species text. Any reference to Standley refers to his 1933 work, unless otherwise indicated.

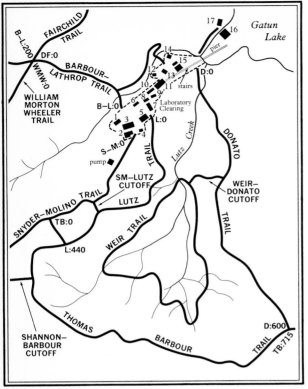

SOURCE: R. W. Rubinoff, *Environmental Monitoring and Baseline Data* (Washington, D.C.: Smithsonian Institution Press, 1974)

1. Tall House
2. Barbour House
3. Main Laboratory
4. Paper House
5. Animal House
6. Kodak House
7. Chapman House
8. Herbarium
9. Office
10. Cook House
11. Dormitory—Dining Hall
12. Smith House
13. ZMA House
14. Boys' House
15. Carpenter Sheds
16. Boat House
17. Toolshed

BARRO COLORADO
ISLAND

Scale in meters

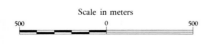

500   0   500

—————— Trail

⌒⌒⌒ Stream

✪ Lighthouse

152 Elevation
Contour interval 6.1 meters
(converted from 20-foot intervals;
rounded to nearest whole number)

AA:300 Trail name initials: meters
from trail beginning
(parentheses indicate approximate
distance)

Adapted from 1927 survey by Panama Canal Company.

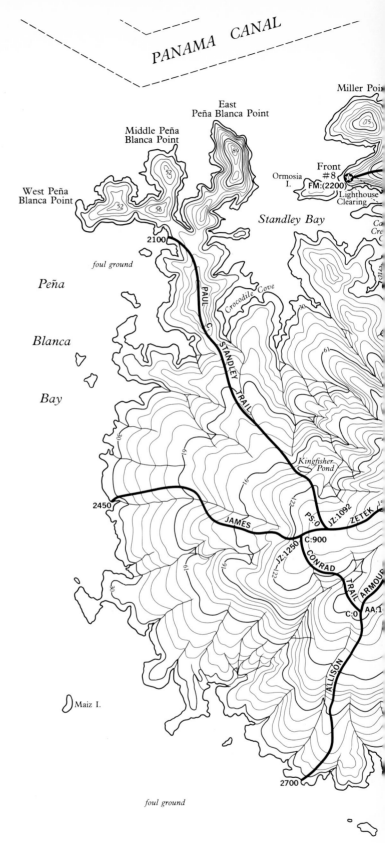

PANAMA CANAL

Miller Poi

East
Peña Blanca Point

Middle Peña
Blanca Point

Front
#8

West Peña
Blanca Point

Ormosia
I.

FM:(2200)

Lighthouse
Clearing

Standley Bay

Ca
Cre

2100

foul ground

Peña

Blanca

Bay

Crocodile Cove

Kingfisher
Pond

2450

JAMES

PS:0

JZ:1092

ZÉTEK

JZ:1250

C:900

CONRAD

TRAIL

ARMOUR

C:0

AA:1

ALLISON

2700

Maiz I.

foul ground

Gatun   Lake
(26 m. above sea level)

Stumpy Bay

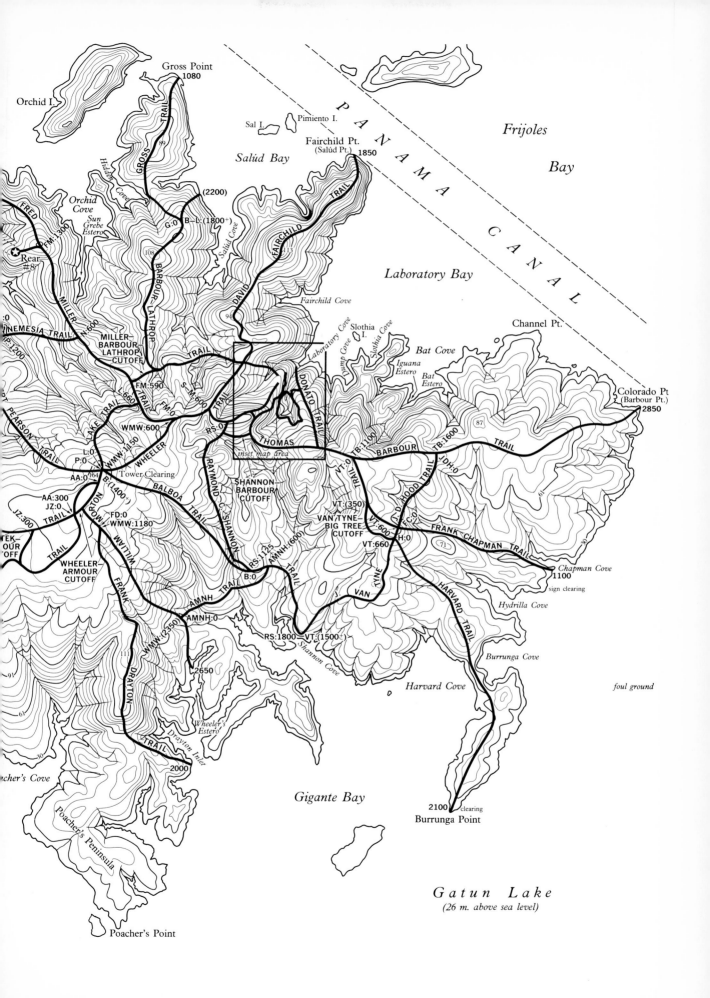

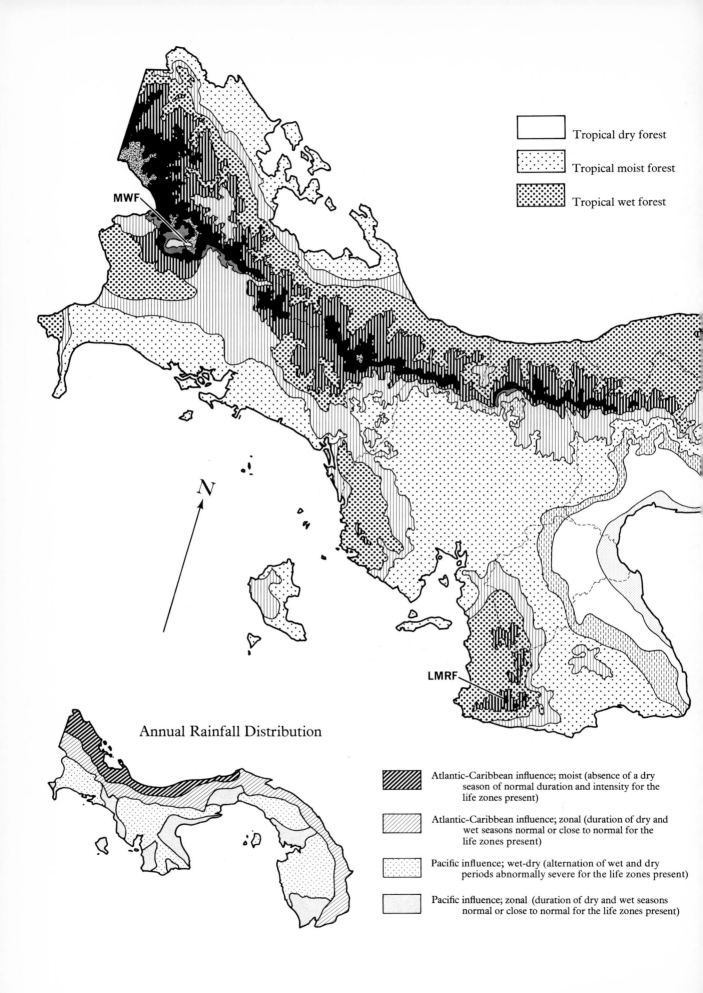

Tropical dry forest

Tropical moist forest

Tropical wet forest

MWF

N

LMRF

Annual Rainfall Distribution

Atlantic-Caribbean influence; moist (absence of a dry
season of normal duration and intensity for the
life zones present)

Atlantic-Caribbean influence; zonal (duration of dry and
wet seasons normal or close to normal for the
life zones present)

Pacific influence; wet-dry (alternation of wet and dry
periods abnormally severe for the life zones present)

Pacific influence; zonal (duration of dry and wet seasons
normal or close to normal for the life zones present)

Holdridge Life Zones for the Republic of Panama

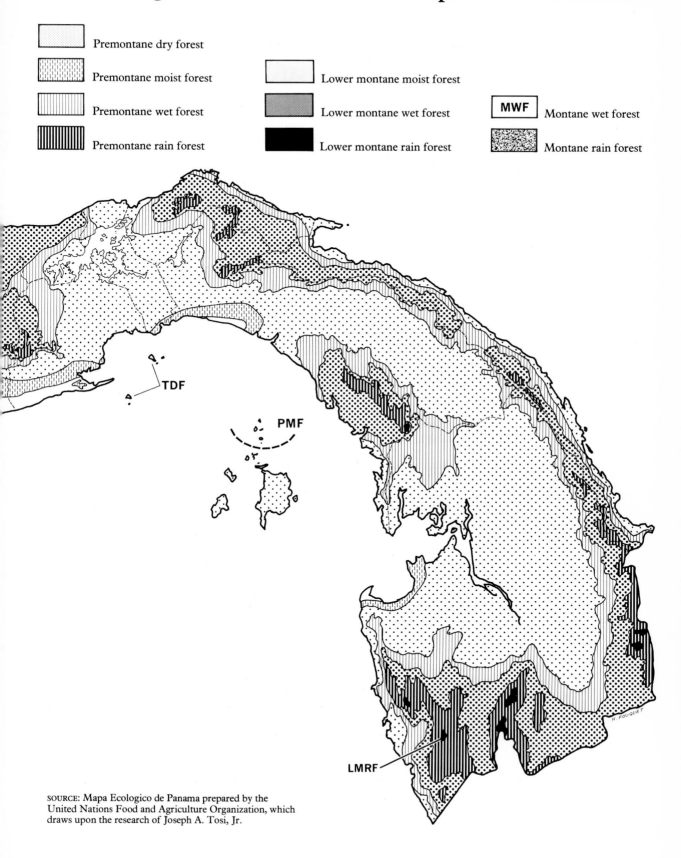

SOURCE: Mapa Ecologico de Panama prepared by the
United Nations Food and Agriculture Organization, which
draws upon the research of Joseph A. Tosi, Jr.

Range and vegetation zones. For the species descriptions, overall range was determined by studies of herbarium specimens and recent monographs. (More details on range are given above, in the discussion of geographical affinities.)

The range of a species within Panama indicates the life zones and provinces in which the species is known to occur. Life zones are based entirely on the Holdridge Life-Zone System (Holdridge et al., 1971; Tosi, 1971). This system of classification is complex, and a thorough understanding requires close study of a detailed explanation. Such an explanation is best supplied in *Life Zone Ecology* by Holdridge (1967), and the life zones are best depicted by his Fig. 1, a two-dimensional diagram representing a three-dimensional set of life zones. The life zones are defined by mean annual values of the three major climatic factors of the environment—heat, precipitation, and moisture (Holdridge, 1967). The amount of moisture available for plant growth in a given site depends not only on the rainfall but also on the amount of moisture lost through evaporation and transpiration. Thus the same life zone may exist from sea level to the continental divide in Panama even though rainfall may be much greater nearer the coast. As the elevation increases, less water is lost through evapotranspiration, and the net amount available to plant populations remains the same. The general progression of life zones pertinent to species occurring on BCI is as follows: premontane dry forest, tropical dry forest, premontane moist forest, tropical moist forest, premontane wet forest, tropical wet forest, premontane rain forest, lower montane moist forest, lower montane wet forest, lower montane rain forest, and montane rain forest. This series of life zones in general progresses from very dry to very wet and, in Panama, also often from lower to higher elevation. Since tropical moist forest is intermediate to the two extremes, a few tropical moist forest species (but only the more ecologically variable) may occur in the extreme wet or dry situations. The average BCI species occurs in tropical moist forests elsewhere in Panama and less frequently in premontane wet forest and tropical wet forest. Since BCI lies in wetter-than-average tropical moist forest, relatively fewer of its species occur in premontane moist forest or tropical dry forest. Relatively few BCI species occur in premontane rain forest, since most such areas are at significantly higher elevations in Panama.

Species that otherwise occur in wetter areas, such as tropical wet forest or premontane rain forest, occur on BCI in comparatively large numbers and do not occur in drier parts of tropical moist forest, such as on the Pacific slope of the isthmus. At the same time, relatively few of the species typically found in drier areas get to BCI.

Life-zone determinations in this work are based principally on geographical information provided by the collector or taken from the *Mapa Ecologico de Panama* prepared by Tosi in 1970 (see pp. 58–59). Additional data are taken from Holdridge & Budowski (1956),

Holdridge (1970), Tosi (1971), and Holdridge et al. (1971). Some of these works are particularly valuable for the larger forest trees, which, although conspicuous elements of the Panamanian flora, are not frequently collected. Provinces are listed in geographical order from west to east on the Atlantic slope and then on the Pacific slope. Locations in the provinces that overlap the continental divide are on the Pacific slope unless otherwise indicated.

In one case I have not followed the Life-Zone map: the northern end of the Navy Pipeline Road northwest of Gamboa in the Canal Zone is actually premontane wet forest, according to L. Holdridge (pers. comm.), though the map indicates that it is tropical moist forest. There are no doubt other errors in the map, but this one is particularly worth noting, since the area is so well collected and contains many species known from BCI.

Considerable criticism (both published and unpublished) has been leveled at the Holdridge Life-Zone System. The arguments against the system are perhaps best voiced by Meyer (1969, pp. 9-12). However, the fact remains that the system in general works quite well. In Panama, those areas indicated as a particular life zone on the map do look alike and have basically the same flora. Until someone produces a more accurate map, the Holdridge Life-Zone map of Panama will be used extensively, and will be quite helpful in understanding the ecological conditions under which a species will grow.

It is not surprising that most BCI species are found in other parts of Panama. In general, if a species is common on BCI it is found in most of the tropical moist forest areas of Panama; the more common (apparent) exceptions, such as absence from San Blas or Chiriquí provinces, are probably due to lack of collecting, and to a lesser extent the same can be said for parts of the Azuero Peninsula and other areas of western Panama. Recent but unstudied collections from tropical moist forest in the vicinity of the Burica Peninsula in lowland Chiriquí Province have produced many new province reports, confirming my belief that the absence of many species from tropical moist forest in Chiriquí is due simply to inadequate collecting.

There is, however, considerable variation within the tropical moist forest life zone, especially if we compare the Pacific slope with the Atlantic slope. Many species are restricted to one or the other slope, depending on ecological requirements.

Photographs. No attempt has been made to illustrate all of the species in the flora. Those photographs selected for publication—all are by the author—were chosen partly on the basis of quality but chiefly for their usefulness in the identification of species. Special consideration was given to those species that had not previously been illustrated elsewhere. Readers who have collected in tropical forests will appreciate the difficulty of obtaining exemplary photographs under conditions of insufficient—or excessive—sunlight.

Species Excluded

A total of 138 names have been excluded from the flora (see p. 911). Most of these are the result of misidentification by Standley at the time of the writing of the *Flora* (1933), but some reflect the different interpretations of later workers. Others are cultivated plants that no longer grow on the island. A few species, such as *Bidens pilosa, Porophyllum ruderale,* and *Lemna cyclostasa,* are species for which no specimens have been found from BCI. I believe that Standley reported some of these names merely because he thought they would be found there.

Maps

The map of the island on the front endsheet is also shown in the text (pp. 56–57). The back endsheet map of Panama shows province boundaries, some cities, and topographical features. The Holdridge Life Zones of Panama (see above), and annual precipitation in Panama, are shown in a series of maps on pages 58–59.

Over the course of the years there have been many modifications and extensions to the island's trail system. Some trails were abandoned while others were added or remade; still others have been diverted so many times by tree falls that they no longer follow the course they originally took.

One of the most significant modifications is the sharp curtailing of Gilbert Pearson Trail, which once began at the west edge of the Laboratory Clearing and ran more or less parallel to Snyder-Molino Trail and William Morton Wheeler Trail, then swung southwest along part of what is now Lake Trail to meet with the present Pearson Trail just west of the Tower Clearing. Other trails have been added to the system as it existed in 1927, and given the names Paul C. Standley, Nemesia, Balboa, Abraham Conrad, American Museum of Natural History (AMNH), Donato, Harvard, and A. Wetmore. Moreover, some of the existing trails have been modified. Frank M. Chapman Trail, which once had its end at Burrunga Cove, was extended south to Burrunga Point and renamed Harvard Trail. A new Chapman Trail was created farther north, to extend more or less east from Van Tyne Trail 660. The end of Fred Miller Trail had been allowed to become very overgrown, stopping at the cutoff to the Rear #8 Lighthouse Clearing (about Miller Trail 1300); it has been reopened recently. The ends of Gross Trail and Fairchild Trail are actually much farther east than the 1927 map indicates. Standley Trail, which once extended to a cove near West Peña Blanca Point, now ends in Peña Blanca Bay. The west (unnamed) fork of what is now J. D. Hood Trail has been abandoned, as has the trail extending between the end of Frank Drayton Trail and Allison-Armour Trail 2200. There are other, more minor, modifications incorporated on the current map, but they need not be mentioned here.

Other physical changes that have been made on the island include the razing of Fuertes House (which began with the assistance of thieves in 1970), Zetek House, Banks House, and the metal tower in what is still called the Tower Clearing, near the center of the island. The houses at the ends of the various trails were built for scientists—largely, I believe, during the time when James Zetek was manager of the island. Zetek was interested in termites, and the houses were variously treated to test the effectiveness of the chemicals against termites. Most were eventually eaten up and became very inhospitable, and were therefore razed. Fuertes House was situated at the end of Pearson Trail next to Fuertes Cove. Banks House was on Burrunga Point, and Zetek House and Drayton House (also called Termite House) were at the ends of the trails so named. Drayton House at this writing still stands, but it is badly infested with termites and will not survive much longer.

The farmers who settled BCI before it became a preserve pursued the typical temporary slash-and-burn techniques that are employed elsewhere in Panama today. Until evicted when the area was set aside as a preserve, they raised maize, manihot, bananas, plantain, pineapples, and the other plants typical of the isthmus.

Many features that were previously unnamed are given names on the endsheet map. Robin Foster and R. K. Enders have been especially helpful in providing names that are known to have been in use. In a normal area (i.e., a stretch of land not given over specifically to scientific study) such tiny bays, coves, and points would go unnamed, but on BCI it will be helpful to other workers to be able to refer to areas of study unequivocally. Egbert and Elizabeth Leigh, Ed Willis, Stan Rand, and Nick Brokaw were of great help in updating the trail system.

Classification and Family Sequence

The classification system used in this book for the vascular cryptogams is that of Scagel et al. (1965); for the phanerogams, that of the traditional Englerian system (De Dalla Torre & Harms, 1963). Genera are alphabetized within families; species within genera. The assignment of genera to families follows the *Flora of Panama,* except in a few cases (such as *Hampea*) that have recently (and persuasively) been transferred to other families.

The family numbers in the list that follows were assigned for convenience; they are used throughout the text and carried in the running heads, as well. The list also gives the number of BCI genera per family and the number of lesser taxa per genus (a species with two BCI varieties, for example, counts as two taxa).

ORDER OF SYSTEMATIC TREATMENT

Major groups and families	Number of genera	Number of lesser taxa
Division Lycopodophyta		
Order Selaginellales		
1. Selaginellaceae	1	6
Order Lycopodiales		
2. Lycopodiaceae	1	2
Division Pterophyta		
Order Marattiales		
3. Marattiaceae	1	1

Order Ophioglossales		
4. Ophioglossaceae	1	1
Order Filicales		
5. Schizaeaceae	2	3
6. Gleicheniaceae	2	3
7. Hymenophyllaceae	2	10
8. Parkeriaceae	1	1
9. Cyatheaceae	4	5
10. Polypodiaceae	27	71
Order Salviniales		
11. Salviniaceae	1	1
TOTAL CRYPTOGAMS	43	104
Division Coniferophyta		
Order Coniferales		
12. Araucariaceae	1	1
Division Gnetophyta		
Order Gnetales		
13. Gnetaceae	1	1
TOTAL GYMNOSPERMS	2	2
Division Anthophyta (=Angiospermae)		
Class Monocotyledoneae		
Order Pandanales		
14. Typhaceae	1	1
Order Alismatales		
15. Alismataceae	1	1
16. Hydrocharitaceae	2	2
Order Graminales		
17. Gramineae (Poaceae)	43	81
18. Cyperaceae	10	28
Order Palmales		
19. Palmae (Arecaceae)	12	18
Order Cyclanthales		
20. Cyclanthaceae	4	5
Order Arales		
21. Araceae	14	46
Order Commelinales		
22. Bromeliaceae	8	20
23. Commelinaceae	7	7
24. Pontederiaceae	2	3
Order Liliales		
25. Liliaceae	1	1
26. Smilacaceae	1	5
27. Haemodoraceae	1	1
28. Amaryllidaceae	4	4
29. Dioscoreaceae	1	7
30. Iridaceae	1	1
Order Scitaminales		
31. Musaceae	2	9
32. Zingiberaceae	4	10
33. Marantaceae	4	12
Order Orchidales		
34. Burmanniaceae	1	1
35. Orchidaceae	45	90
TOTAL MONOCOTS	169	353
Class Dicotyledoneae		
Order Piperales		
36. Piperaceae	3	32
37. Lacistemaceae	2	2
Order Urticales		
38. Ulmaceae	2	3
39. Moraceae	15	35
40. Urticaceae	5	5
Order Proteales		
41. Proteaceae	1	1
Order Santales		
42. Loranthaceae	4	7
43. Olacaceae	1	3
Order Aristolochiales		
44. Aristolochiaceae	1	3
45. Rafflesiaceae	1	1
Order Polygonales		
46. Polygonaceae	3	9
47. Amaranthaceae	6	8
48. Nyctaginaceae	3	3
49. Phytolaccaceae	4	4
50. Portulacaceae	1	1
51. Caryophyllaceae	1	1
Order Ranales		
52. Nymphaeaceae	1	2
53. Ceratophyllaceae	1	1
54. Menispermaceae	4	7
55. Annonaceae	7	13
56. Myristicaceae	1	2
57. Monimiaceae	1	2
58. Lauraceae	5	11
Order Rhoeadales		
59. Capparidaceae	2	2
Order Rosales		
60. Saxifragaceae	1	1
61. Chrysobalanaceae	2	5
62. Connaraceae	3	4
63. Leguminosae (Fabaceae)		
A. Mimosoideae	9	37
B. Caesalpinioideae	11	17
C. Papilionoideae	28	60
Order Geraniales		
64. Oxalidaceae	1	1
65. Humiriaceae	1	1
66. Erythroxylaceae	1	2
67. Rutaceae	2	10
68. Simaroubaceae	3	3
69. Burseraceae	4	6
70. Meliaceae	3	7
71. Malpighiaceae	10	20
72. Trigoniaceae	1	1
73. Vochysiaceae	1	1
74. Polygalaceae	2	3
75. Euphorbiaceae	19	33
Order Sapindales		
76. Anacardiaceae	5	7
77. Celastraceae	1	1
78. Hippocrateaceae	5	5
79. Staphyleaceae	1	1
80. Sapindaceae	6	26
Order Rhamnales		
81. Rhamnaceae	2	3
82. Vitaceae	2	6

Order Malvales		
83. Elaeocarpaceae	2	3
84. Tiliaceae	6	8
85. Malvaceae	5	10
86. Bombacaceae	7	9
87. Sterculiaceae	7	8
Order Parietales		
88. Dilleniaceae	4	9
89. Ochnaceae	2	2
90. Marcgraviaceae	2	2
91. Theaceae	1	1
92. Guttiferae (Clusiaceae)	11	15
93. Bixaceae	1	1
94. Cochlospermaceae	1	1
95. Violaceae	2	3
96. Flacourtiaceae	8	16
97. Turneraceae	1	1
98. Passifloraceae	1	11
99. Caricaceae	2	3
100. Begoniaceae	1	3
Order Opuntiales		
101. Cactaceae	2	3
Order Myrtales		
102. Lythraceae	3	3
103. Lecythidaceae	3	4
104. Rhizophoraceae	1	1
105. Combretaceae	2	7
106. Myrtaceae	5	14
107. Melastomataceae	14	35
108. Onagraceae	1	5
Order Umbellales		
109. Araliaceae	4	5
110. Umbelliferae (Apiaceae)	3	3

Order Primulales		
111. Theophrastaceae	1	1
112. Myrsinaceae	3	5
Order Ebenales		
113. Sapotaceae	2	6
114. Ebenaceae	1	1
Order Gentianales		
115. Loganiaceae	2	6
116. Gentianaceae	3	5
117. Menyanthaceae	1	1
118. Apocynaceae	15	23
119. Asclepiadaceae	8	11
Order Solanales		
120. Convolvulaceae	6	10
121. Boraginaceae	3	11
122. Verbenaceae	6	8
123. Labiatae (Lamiaceae)	3	5
124. Solanaceae	9	25
125. Scrophulariaceae	5	6
126. Bignoniaceae	21	29
127. Gesneriaceae	9	10
128. Lentibulariaceae	1	2
129. Acanthaceae	12	15
Order Rubiales		
130. Rubiaceae	37	67
Order Cucurbitales		
131. Cucurbitaceae	8	15
Order Campanulales		
132. Campanulaceae	1	1
133. Compositae (Asteraceae)	33	42
TOTAL DICOTS	490	910
TOTAL	704	1,369

KEY TO THE VASCULAR PLANTS OF BARRO COLORADO ISLAND

Plants ferns or fernlike, lacking seeds; reproduction by means of spores (vascular cryptogams):
 Leaves scalelike or acicular, minute . . . . LYCOPODOPHYTA, p. 67
 Leaves foliaceous . . . . PTEROPHYTA, p. 71
Plants not fernlike; reproduction by means of true seeds containing embryos (spermatophytes):
 Ovules naked, not included in an ovary of a flower (gymnosperms):
  Xylem vessels present in wood; male cones compound (on BCI represented by a woody liana with opposite leaves, ovate leaf blades, and drupaceous fruits) . . . . GNETOPHYTA (13. GNETACEAE), p. 115
  Xylem vessels lacking in wood; male cones simple (on BCI represented by a cultivated tree with needlelike leaves) . . . . CONIFEROPHYTA (12. ARAUCARIACEAE), p. 113
 Ovule or ovules enclosed by the ovary of a true flower (ANTHOPHYTA/angiosperms):
  Leaves mainly with parallel veins, alternate; stipules lacking; blades often sheathing at base; flower parts usually in 3s or 6s, less frequently in 2s or 4s, never in 5s; stems lacking annular layers or central pith; vascular bundles closed and scattered . . . . MONOCOTYLEDONEAE, p. 117
  Leaves mainly with reticulate veins, alternate or opposite or whorled; stipules present or lacking; blades usually not sheathing at base; flower parts usually in 4s or 5s or in multiples of these; stems usually with annular layers and a central pith; vascular bundles arranged concentrically in stems . . . . DICOTYLEDONEAE, p. 309

The Flora

Lycopodophyta

KEY TO THE LYCOPODOPHYTA

Leaves distichous, the stem with its leaves forming a flat plane; cones ± 4-angled . . . .
. . . . 1. SELAGINELLACEAE
Leaves spiraled, the stem, including its leaves, terete; cones terete . . . . 2. LYCOPODIACEAE

1. SELAGINELLACEAE

Heterosporous annuals or perennials with adventitious roots. Leaves 1-veined, usually dimorphic in 4 ranks, with a minute ligule at base of each leaf. Strobili usually 4-sided, consisting of imbricated megasporophylls at base and microsporophylls at apex; sporangia 1-chambered, containing (in axils of sporophylls) either microspores or megaspores; prothallia very small, unisexual.

Members of the family are distinguished by their minute dimorphic leaves, usually all held in a single plane such that the plant often looks like a single, lacy, highly dissected, compound leaf. The four-sided, cone-like strobili are also a distinguishing character of the Selaginellaceae.

Dispersal may be in part by rain water, which washes away the megacarps (van der Pijl, 1968).

One genus, with about 700 species; tropical, subtropical, and temperate regions.

SELAGINELLA Beauv.

Selaginella arthritica Alston, Arch. Bot. (Forlì)
11:43. 1935
S. conduplicata sensu Spring.

Terrestrial, usually 15–40 cm tall, glabrous; lower parts of stems in adult plants usually creeping, densely rooted, and leafless, the stems becoming erect and sparsely leafy and flexuous near apex, at least the larger stems swollen and markedly articulate, the lateral branches sparse, only the lower few of noticeable length. Median leaves auriculate, appearing affixed well above base, acute at apex, entire; lateral leaves ovate-oblong, acute to blunt at apex, not or weakly auriculate, to ca 5 mm long; lower stem leaves auriculate. Cones 1 mm wide, mostly 5–15 mm long, usually solitary, rarely 2–4 at apex of ultimate branches, sharply square; megaspores few, smooth, light brown. *Croat 8614.*

KEY TO THE SPECIES OF SELAGINELLA

Plants ± erect, usually more than 20 cm tall; main stem not branched from near base:
    Stems reddish at base; leaves minutely and evenly denticulate . . . . . . *S. haematodes* (Kunze) Spring.
    Stems not reddish; leaves entire:
        Plants usually less than 40 cm tall, usually creeping; stems glabrous and markedly articulate,
            usually with few lateral branches (less than 4); plants abundant; fertile parts ca 1 mm
            wide, sharply square . . . . *S. arthritica* Alston
        Plants usually 1 m or more tall, erect, then arching, stems pubescent, not markedly articulate,
            usually with many lateral branches; plants rare; fertile parts ca 2 mm wide, irregular, ±
            terete . . . . *S. exaltata* (Kunze) Spring.
Plants scandent or decumbent, usually less than 20 cm tall; main stem usually branched from near
    base:

Fig. 1. *Selaginella haematodes*

Fig. 2. *Selaginella mollis*

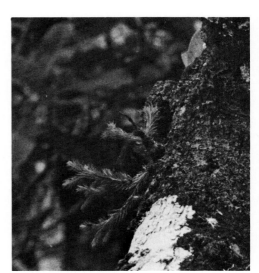

Fig. 3. *Lycopodium dichotomum*

Fig. 4. *Lycopodium dichotomum*

Leaves conspicuously ciliate (at least the base of the lateral leaves):
  Both lateral and stem leaves auriculate, usually appearing affixed well above base; lateral leaves usually more than 2.5 mm long; plants usually long-creeping . . . .
    . . . . *S. horizontalis* (Presl) Spring.
  Leaves not auriculate, clearly basifixed; lateral leaves less than 2 mm long; plants short, not creeping . . . . *S. mollis* A. Braun
Leaves entire or merely short-toothed, not ciliate:
  Stems markedly articulate, never flagelliform at apex of main stem; largest lateral leaves 3.5–5 mm long, entire; median leaves markedly auriculate, appearing affixed well above base . .
    . . . . *S. arthritica* Alston
  Stems not articulate, usually flagelliform at apex; largest lateral leaves usually less than 2 mm long, inconspicuously toothed; median leaves not auriculate, basifixed . . . .
    . . . . *S. flagellata* Spring.

Very abundant throughout the forest, especially along trails.

Juvenile forms often have long-creeping, sparsely leafy stems and much narrower, erect, leafy portions.

Nicaragua to Panama; to 1,700 m elevation. In Panama, known from tropical moist forest in the Canal Zone (BCI only), Bocas del Toro (Shepherd Island), and Darién, and from tropical wet forest in Panamá (Cerro Trinidad).

Selaginella exaltata (Kunze) Spring., Bull. Acad. Roy. Sci. Bruxelles 10:234. 1843

Terrestrial, to 1 m or more tall, erect or frequently arching and scandent; main stem articulate, glabrous or pubescent, sparsely leafy, the smaller stems inarticulate, more closely leafy, usually with short, stiff trichomes. Median leaves asymmetrical, sharply long-acuminate, decurrent at base, entire; lateral leaves ± oblong, sharply acuminate, basifixed, 2–4 mm long, entire; leaves on main and secondary stems sparsely spaced, to 1 cm long, markedly auriculate. Cones terete to squarish at apices of ultimate branches; megaspores prominently white-ruffled. *Shattuck 1158, Croat 16193.*

Rare, in ravines in the old forest.

Standley reported *S. conduplicata* Spring., a synonym of *S. arthritica.* This report may have been based on a misdetermination of *Shattuck 1158,* which is this species.

Panama to Peru. In Panama, known from tropical moist forest in the Canal Zone, San Blas, and Darién, from premontane wet forest in Colón, and from premontane rain forest in Darién.

Selaginella flagellata Spring., Bull. Acad. Roy. Sci. Bruxelles 10:228. 1843

Terrestrial, small and creeping, glabrous; stems moderately weak, not articulate, the lower part leafy, the apical part long-flagellate, sparsely short-branched. Median leaves abruptly and sharply long-acuminate, basifixed and decurrent, not auriculate, the margins inconspicuously toothed and often appearing hyaline; lateral leaves ± ovate-oblong, usually less than 2 mm long, acute to weakly acuminate, not auriculate, the margins inconspicuously toothed. Cones ± inconspicuous, merging with leaves; megaspores minute, white, smooth. *Croat 4098, 6467.*

Infrequent in the forest, usually on trails, sometimes growing on rocks.

Mexico to Bolivia and French Guiana. In Panama, known only from tropical moist forest in the Canal Zone and Panamá.

Selaginella haematodes (Kunze) Spring. in Mart., Fl. Brasil. 1(2):126. 1840

Terrestrial, erect, to 80 cm tall and to 35 cm wide, unbranched in lower half, reddish near base, not articulate, sparsely leafy, regularly branched from near middle, the branches 12–25 cm long, gradually diminished in length toward apex. Leaves uniform, reddish, basifixed, ovate-oblong, usually acute; median leaves gradually long-acuminate, basifixed, minutely toothed; lateral leaves ovate-oblong, acute at apex, basifixed, minutely toothed. Cones numerous, solitary at branch apices, mostly to ca 1 cm long, ± square, the bracts long-acuminate and ± spreading; megaspores minute, white, smooth. *Croat 4111.*

Locally abundant along trails near streams; less frequent elsewhere in the forest.

Easily distinguished by the reddish stem, large size, and triangular blade.

Panama to Bolivia. In Panama, known from tropical moist forest in the Canal Zone, Panamá, and Darien, and from premontane wet and tropical wet forests in Panamá. See Fig. 1.

Selaginella horizontalis (Presl) Spring., Bull. Acad. Roy. Sci. Bruxelles 10:226. 1843
  S. fendleri Baker; *S. sylvatica* Baker

Terrestrial, usually long-creeping and ± regularly short-branched; stems obscurely articulate, lacking flagellate apices, glabrous, prominently rooted along their length, the roots produced from bottom side of stem. Median leaves acuminate, auriculate, and affixed well above base, markedly ciliate at base; lateral leaves ± oblong, 3–4 mm long, acute at apex, auriculate at base, the basal part markedly long-ciliate. Cones short, inconspicuous, solitary at apices of ultimate branches, squarish, the bracts long-tapered, spreading, minutely toothed; megaspores minute, alveolate, light yellow-brown. *Croat 6933, 12670.*

Abundant in the forest, usually along trails; also found in shaded places in clearings.

Costa Rica to Colombia. In Panama, known from tropical moist forest in the Canal Zone and Panamá.

Selaginella mollis A. Braun, Ann. Sci. Nat. Bot., sér. 5, 3:276. 1865

S. schrammii Hieron.

Terrestrial, creeping or appressed, minute, usually less than 10 cm long, glabrous except for long cilia on lower part of leaves; stems not flagellate at apex, leafy throughout at base, short-branched almost throughout, the branches nearly equal in length. Median leaves not much smaller than lateral leaves, ± ovate, abruptly acuminate, short-toothed near apex, ciliate toward base, not auriculate, basifixed; lateral leaves ± oblong, usually acute at apex, not auriculate, basifixed, to ca 2 mm long, prominently long-ciliate at least near base. Cones inconspicuous, constituting a gradual continuation of branch apices, the bracts short-toothed; megaspores minute. *Croat 10759.*

Rare, in the forest, usually on steep shaded banks.

Mexico to Panama and Colombia. In Panama, known from tropical moist forest in the Canal Zone, Bocas del Toro, and Panamá and from premontane wet forest in Panamá.

See Fig. 2.

2. LYCOPODIACEAE

Homosporous perennials; stems and roots dichotomously branched. Leaves spiraled, narrowly linear-lanceolate, 1-veined, usually imbricate. Sporangia adaxial at base of sporophylls, either similar to foliage leaves and at apex of stems or imbricated into distinct strobili; sporangia pouch-shaped, dehiscing vertically across summit; spores numerous, yellowish; prothallia monoecious, mostly saprophytic.

Recognized by their acicular, spirally arranged leaves and frequent dichotomous branching.

As in all cryptogamic families, the plants are dispersed chiefly by wind or water. Some Lycopodiaceae are dispersed by bulbils (Ridley, 1930).

Two genera and about 450 species; tropics to arctic regions.

LYCOPODIUM L.

Lycopodium cernuum L., Sp. Pl. 1103. 1753

Terrestrial, scandent or erect, to ca 1 m tall, creeping, weakly rooted, dichotomously branched; stems mostly less than 4 mm diam, the primary stems sparsely leafy throughout much of their length, the upper parts and secondary stems more densely leafy. Leaves acicular, spiraled, 2–4 mm long. Cones terete, at most apices, often pendent on recurved apices, mostly 4–9 mm long, ca 2 mm diam; sporophylls lanceolate-deltoid, to ca 2 mm long including the attenuate apex, closely imbricated, the margins irregularly lacerate. *Wilson 50, Croat 6448.*

Locally abundant along steep shore banks, especially along Old French Lock Site.

Probably native to Africa; widespread in the New World. In Panama, ecologically variable; known from tropical moist forest in the Canal Zone, Bocas del Toro, San Blas, and Veraguas, from tropical wet forest in Colón and Panamá, from premontane wet forest in Colón, Coclé, and Panamá, from premontane rain forest in Chiriquí and Panamá, and from lower montane rain forest in Chiriquí.

Lycopodium dichotomum Jacq., Enum. Stirp. Plerar. Vindob. 314. 1762

Epiphytic, usually ± erect, 5–20 cm tall; stems densely leafy, obscured, branching dichotomously a few times. Leaves acicular, spiraled, ca 10 mm long and 1 mm broad, moderately thick. Cones lacking; sporangia solitary in upper leaf axils, ± reniform, with a very short stalk in the sinus. *Croat 8029.*

An occasional epiphyte in the forest, usually high in the canopy on branches that are laden with other epiphytes. Individual plants never abundant locally.

United States and Mexico south to Panama, Colombia, Venezuela, and Ecuador. In Panama, known from tropical moist forest in the Canal Zone, Panamá, and Darién, from premontane wet forest in Coclé (El Valle), and from tropical wet forest in Chiriquí and Colón.

See Figs. 3 and 4.

KEY TO THE SPECIES OF LYCOPODIUM

Leaves less than 5 mm long, sparse on stem, most of the stem exposed; sporangia aggregated into distinct cones at branch apices; plants terrestrial, often vinelike .... *L. cernuum* L.
Leaves ca 1 cm long, very dense, obscuring stem; sporangia solitary in upper leaf axils; plants epiphytic in trees, mostly short and erect .... *L. dichotomum* Jacq.

Pterophyta

Genera in this key for which family is not indicated are in the Polypodiaceae, family 10.

Plants free-floating, not attached by roots, with 2 opposite entire leaves floating on the water . . . . .
. . . . 11. SALVINIACEAE (*Salvinia radula* Baker)
Plants attached by roots, not free-floating (aquatic or not):
  Leaves translucent, usually consisting of a single layer of cells; sporangia sessile on a threadlike
    stalk emerging from a bilabiate, tubular, or urceolate indusium, the indusia arising along
    margins of leaflets; plants often minute, usually forming close, ± appressed mats. . . . .
    . . . . 7. HYMENOPHYLLACEAE (in part)
  Leaves and sporangia not as above:
    Plants free-climbing vines, never epiphytic, the stems not appressed to trees:
      Leaves palmately lobed at least at base, not sessile; sporangia borne on protuberances at
        margin of leaflets . . . . 5. SCHIZAEACEAE (*Lygodium*)
      Leaflets pinnatifid, sessile; sporangia borne on lower surface of segments . . . .
        . . . . 6. GLEICHENIACEAE
    Plants not free-climbing vines (if vinelike, the stems appressed to trees):
      ● Leaves simple and not deeply lobed:
        Plants terrestrial; leaves dimorphic:
          Plants less than 10 cm tall; sterile leaf 1 (rarely 2), ovate; fertile leaf 1, slender, nearly
            all sporangial tissue . . . . 4. OPHIOGLOSSACEAE (*Ophioglossum reticulatum* L.)
          Plants more than 20 cm tall; sterile leaves usually 4 or more, narrowly oblong; fertile
            leaves with sporangia restricted to a narrow band along the margin . . . .
            . . . . *Dictyoxiphium panamense* Hook.
        Plants epiphytic; leaves dimorphic or not:
          Sporangia arranged in distinct rounded sori on surface of leaves . . . . . *Polypodium* (in part)
          Sporangia not in distinct rounded sori:
            Sporangia forming an anastomosing pattern over entire underside of leaf; blades
              usually less than 15 cm long, oblanceolate; rhizome usually closely appressed
              . . . . *Anetium citrifolium* (L.) Splitg.
            Sporangia not as above:
              Leaves more than 6 cm wide; sori in narrow lines paralleling the lateral veins . . . .
                . . . . *Asplenium serratum* L.
              Leaves less than 4 cm wide; sori not as above:
                Leaves dimorphic, with sporangia densely covering the reduced fertile leaf; rhi-
                  zome scales conspicuous, reddish-brown, very thin, many times broader
                  than thick; plants frequently sterile . . . . *Elaphoglossum*
                Leaves monomorphic, with sori in rows along margins of leaves; rhizome scales
                  not as above, filiform, not conspicuous:
                  Leaves mostly more than 1 cm wide, widest at middle; sori frequently with
                    short interruptions, more than 1 mm wide . . . .
                    . . . . *Ananthacorus angustifolius* (Sw.) Und. & Max.
                  Leaves less than 5 mm wide, linear; sori usually continuous, less than 1 mm
                    wide . . . . *Vittaria*

● Leaves simple and deeply lobed *or* leaves compound:
  Leaves simple and deeply lobed:
    Leaves ± regularly pinnatifid:
      Leaves monomorphic; sori in round dots on lower surface of leaflets . . . .
      . . . . *Polypodium* (in part)
      Leaves dimorphic; sori immersed in margin . . . .
      . . . . 7. HYMENOPHYLLACEAE (*Trichomanes diversifrons* (Bory) Mett.)
    Leaves not regularly pinnatifid:
      Plants aquatic; leaves dimorphic, the segments of the fertile leaves linear, the sterile
        leaves 2–4-pinnatifid, pentagonal in outline, very thin . . . .
      . . . . 8. PARKERIACEAE (*Ceratopteris pteridoides* (Hook.) Hieron.)
      Plants epiphytic or terrestrial; leaves monomorphic, dichotomously or pinnately
        lobed:
        Sporangia borne in small, pectinate-pinnate, ± reniform clusters from tips of
          teeth on each blade division . . . . 5. SCHIZAEACEAE (*Schizaea elegans* (Vahl) Sw.)
        Sporangia borne over entire blade surface or continuous along margins:
          Plants epiphytic; blades ± dichotomously branched, the lobes mostly ca 1 cm
            wide; leaf tissue extending to base of leaf; sporangia in ± continuous mar-
            ginal rows near apex of lobes . . . .
          . . . . *Dicranoglossum panamense* (Christensen) Lell.
          Plants terrestrial; blades pinnately lobed, the lobes usually more than 5 cm
            wide; leaf tissue not extending to base of leaf; sporangia scattered over
            blade surface . . . . *Tectaria euryloba* (H. Christ) Max.
  Leaves compound:
◆ Leaves 1-pinnate, the leaflets regularly and deeply pinnatifid *or* the leaves more than
    1-pinnate:
    Leaves 1-pinnate, the leaflets regularly and deeply pinnatifid on both sides:
      Lobes of leaflets markedly asymmetrical, especially at base . . . .
      . . . . *Ctenitis protensa* (Afz.) Copel. var. *funesta* (Kunze) Proct.
      Lobes of leaflets ± symmetrical:
        Sori indusiate; upper side of costae with acicular scales . . . . *Thelypteris* (in part)
        Sori exindusiate; upper side of costae lacking acicular scales . . . .
      . . . . *Dryopteris sordida* Max.
    Leaves more than 1-pinnate:
      Leaves dimorphic, the fertile leaves reduced; plants closely appressed, hemiepi-
        phytic climbers, the climbing rhizomes with conspicuous scales:
        Sterile leaves to 3-pinnate, all but the uppermost leaflets divided more than
          halfway to midrib; segments acute at apex, sharply toothed, the lowermost
          often free; rhizome scales fine, ± curly, spreading; sori indusiate, round,
          discrete . . . . *Maxonia apiifolia* (Sw.) Christensen
        Sterile leaves 2-pinnate or 2-pinnate-pinnatifid, most leaflets divided less than
          halfway to midrib; segments nearly rounded at apex, entire to bluntly
          toothed, the lowermost usually not free; rhizome scales coarse, stiff, ap-
          pressed or nearly so; sori exindusiate, apparently continuous along mar-
          gins or over entire surface . . . . *Polybotrya villosula* H. Christ
      Leaves monomorphic; plants terrestrial:
        Plants tree ferns, the adults with a definite trunk (trunk may be short in *Cnemi-*
          daria); leaves armed, at least at base . . . . . . 9. CYATHEACEAE (except *Metaxya*)
        Plants not tree ferns; leaves usually unarmed:
          Lower surface of leaflets white-waxy; sporangia scattered over surface; plants
            growing on exposed banks . . . . *Pityrogramma calomelanos* (L.) Link
          Lower surface of leaflets not white-waxy; sporangia not scattered over sur-
            face; habitats various:
            Leaflets markedly asymmetrical, especially at base, toothed but not equally
              and deeply dissected on both sides (in *Ctenitis,* the leaflets deeply
              lobed but asymmetrical):
            Plants epiphytic; sori exindusiate; leaflets mostly deeply, irregularly
              lobed on both sides . . . .
            . . . . *Ctenitis protensa* Afz. (Copel.) var. *funesta* (Kunze) Proct.
            Plants terrestrial; sori indusiate; leaflets often toothed on 1 side but
             never deeply lobed on both sides . . . . *Adiantum* (in part)
           Leaflets ± symmetrical and regularly pinnatifid:
            Lobes of leaflets entire; plants regularly dichotomously branched, usu-
             ally vinelike; sori round, arranged in a row along midrib of lobe
            . . . . 6. GLEICHENIACEAE

Lobes of leaflets toothed at least near apex; plants not both regularly
    dichotomously branched and vinelike; sori various:
    Leaves not lacy, broadly dissected, the terminal leaflet more than 8
        cm long; sori elongate, along margins of leaflets, with a narrow,
        continuous indusium open on 1 side ... *Pteris* (except *P. grandifolia*)
    Leaves lacy, finely dissected, the terminal leaflet less than 4 cm long;
        sori round:
        Plants tree ferns with a conspicuous trunk; petioles armed with
            spines at least at base . . . . 9. CYATHEACEAE (*Trichopteris*)
        Plants not tree ferns; petioles unarmed:
        Spores borne in cuplike indusia on margin of leaflets, the indusia
            commonly in toothed depressions; leaves 3- or 4-pinnate-
            pinnatifid . . . . *Dennstaedtia cicutaria* (Sw.) Moore
        Spores borne in ± round indusiate or exindusiate sori on surface
            of leaflets, not on margins in cuplike indusia; leaves 2- or
            3-pinnate-pinnatifid, never 4-pinnate-pinnatifid:
            Rachis pubescent, with jointed trichomes on upper side; un-
                derside of leaflets scaly . . . . . . *Ctenitis sloanei* (Poepp.) Mort.
            Rachis glabrous except for dense pubescence along ridge on
                upper side; underside of leaflets hispid, with short-stalked
                glands . . . . *Thelypteris torresiana* (Gaud.) Alston
◆ Leaves 1-pinnate, the leaflets not regularly pinnatifid (i.e., not regularly and deeply
    lobed on both sides):
    Leaves dimorphic, the fertile leaf much reduced:
    Leaflets exactly opposite, caudate-acuminate, ± entire, the veins numerous, close
        and parallel, branched only near base; plants terrestrial . . . .
        . . . . 3. MARATTIACEAE (*Danaea nodosa* (L.) J. Sm.)
    Leaflets not as above; plants epiphytic or terrestrial:
    Plants terrestrial; sterile leaflets shallowly lobed, usually more than 10 cm long,
        the smaller veins anastomosing . . . . . . *Bolbitis cladorrhizans* (Spreng.) Ching
    Plants epiphytic:
    Rachis narrowly winged; leaflets usually less than 6 cm long, usually more
        than 10 pairs of leaflets per leaf . . . . *Lomariopsis vestita* Fourn.
    Rachis not winged; leaflets usually more than 6 cm long, usually less than 6
        pairs of leaflets per leaf . . . . *Bolbitis nicotianifolia* (Sw.) Ching
    Leaves monomorphic:
    Sporangia on margins of leaves:
    Leaflets asymmetrical about midrib . . . . *Adiantum* (in part)
    Leaflets ± symmetrical about midrib:
    Leaflets thin, translucent; sporangia in tubular indusia protruding from leaf-
        let margins . . . . 7. HYMENOPHYLLACEAE (*Trichomanes pinnatum* Hedw.)
    Leaflets and sporangia not as above:
    Sori ± round, interrupted along margins of leaflets . . . .*Nephrolepis*
    Sori continuous along margins of leaflets:
    Rachis dark brown or purplish; leaflets coarsely toothed near apex,
        the lateral veins frequently 1-forked but never anastomosing
        frequently . . . . *Saccoloma elegans* Kaulf.
    Rachis light in color; leaflets ± entire near apex, the lateral veins anas-
        tomosing several times . . . .*Pteris grandifolia* L.
    Sporangia or sori dispersed over surface of blade:
    Plants massive aquatics, growing in dense clusters with the sporangia com-
        pletely covering at least part of all the uppermost leaflets . . . . *Acrostichum*
    Plants not as above, the sporangia arranged in distinct sori:
    ■ Sori ± round:
    Leaflets ± entire (except possibly at apex):
    Leaflets markedly unequal and auriculate on lower proximal edge, over-
        lapping rachis; leaflets generally less than 15 cm long . . . .
        . . . . *Cyclopeltis semicordata* (Sw.) J. Sm.
    Leaflets ± equal at base or at least not auriculate; leaflets generally more
        than 15 cm long:
    Plants epiphytic; sori in rows, each row in an areole between major
        lateral veins; leaflets ± entire at apex . . . . . *Polypodium triseriale* Sw.
    Plants terrestrial; sori not in rows, ± irregularly clustered near mid-
        rib on major lateral veins; leaflets regularly toothed near apex . . .
        . . . . 9. CYATHEACEAE (*Metaxya rostrata* (Willd.) Presl)

Leaflets not entire:
 Leaflets ± regularly toothed and equal in shape . . . .
 . . . . *Thelypteris poiteana* (Bory) Proct.
 Leaflets irregularly lobed or sinuate, at least the lowermost with a larger
  basal lobe . . . . *Tectaria incisa* Cav.
■ Sori linear or at least several times longer than wide (not in dots):
 Sori in 2 continuous lines along midrib of each leaflet . . . . *Blechnum*
 Sori not in 2 continuous lines along leaflet midrib:
  Leaflets more than 20 cm long and 6 cm wide; margins entire; plants
   usually more than 2 m tall . . . . . . *Hemidictyum marginatum* (L.) Presl
  Leaflets usually less than 20 cm long or, if longer, much less than 6 cm
   wide; margins usually toothed; plants usually less than 1.5 m tall:
   Sori short, ± equal, arranged in short rows perpendicular to and be-
    tween lateral veins; plants generally aquatic . . . .
    . . . . *Thelypteris serrata* (Cav.) Alston
   Sori short or long, often markedly unequal in length, arranged in rows
    parallel to lateral veins; plants not aquatic:
    Leaflets mostly less than 6 cm long, frequently blunt or rounded
     and toothed at apex; open side of all indusia facing apex of
     leaflet . . . . *Asplenium* (in part)
    Leaflets usually more than 6 cm long, usually gradually tapered to
     acuminate apex; open face of at least some indusia facing away
     from apex of leaflet (in *Asplenium delitescens,* most facing
     toward apex):
     Leaflets usually more than 2.5 cm wide, obtuse to rounded at
      base on both sides; sori often unequal, the longer ones inter-
      spersed with the shorter . . . . *Diplazium grandifolium* Sw.
     Leaflets usually less than 2.5 cm wide, the base oblique, the
      lower proximal edge usually acute; sori ± equal:
      Terminal leaflet of nearly same shape as lateral leaflets; plants
       epiphytic; rhizome scales clathrate (latticed) . . . .
       . . . . *Asplenium falcinellum* Max.
      Terminal leaflet unlike lateral leaflets, pinnatifid; plants ter-
       restrial; rhizome scales fibrous, not clathrate . . . .
       . . . . *Asplenium delitescens* (Max.) A. R. Smith

3. MARATTIACEAE

Homosporous, terrestrial herbs; stem a dorsiventral rhizome. Leaves usually large, circinate in venation, pinnately compound (in ours); secondary veins on the leaflets
free, closely parallel. Fertile leaves moderately contracted;
sorus a double row of sporangia extending from midrib
to margin, joined to form a synangium, each sporangium opening by a terminal pore; prothallia growing on
ground, green with mycorrhiza.

Recognized by their pinnately compound leaves, the
leaflets opposite, and the secondary veins closely parallel.

Six genera and over 200 species; tropics and subtropics.

DANAEA J. Sm.

Danaea nodosa (L.) J. Sm., Mem. Acad. Roy. Sci.
 (Turin) 5:420. 1793

Terrestrial, usually 1.5–2 m tall; rhizome horizontal.
Leaves 1-pinnate, dimorphic; petioles to ca 1 m long,
sparsely to moderately scaly, prominently ridged above;
leaflets exactly opposite; sterile leaflets usually narrowly
oblong, long-acuminate, obtuse to acute and often slightly
inequilateral at base, 10–40 cm long, 4–5.5 cm wide, the
margin entire to undulate or faintly serrulate near apex,

the midrib sparsely scaly, the scales irregular and deeply
lacerate, the lateral veins 1-forked at base; fertile leaflets
much shorter and narrower, 1.5–2.5 cm broad. Sporangia
in 2 rows, continuous, borne along the veinlets, almost
completely covering the fertile leaflets beneath. *Croat
12301.*

Infrequent, on steep moist creek banks in ravines.

Mexico to Brazil; West Indies. In Panama, known from
tropical moist forest in the Canal Zone, Veraguas, and
Panamá (San José Island), from premontane wet forest
in Panamá (Chimán) and Veraguas, and from tropical
wet forest in Panamá.

See Fig. 5.

4. OPHIOGLOSSACEAE

Homosporous, terrestrial herbs; stems short, lacking
scales. Leaves solitary or few, with dimorphic segments;
blades splitting into a dorsal and a ventral part, the dorsal part sterile, green, with reticulate venation, the ventral
part fertile, arising from dorsal part, spikelike, bearing
a row of large immersed sporangia in each side. Sporangia opening by transverse slits; prothallia subterranean,
saprophytic.

Fig. 5. *Danaea nodosa*

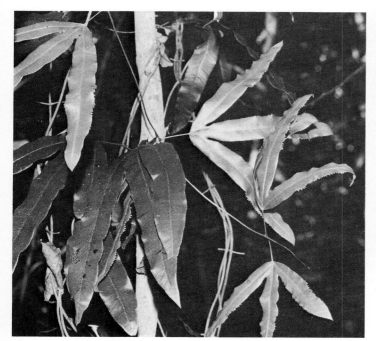

Fig. 6. *Lygodium radiatum*

Fig. 7. *Lygodium venustum*

Recognized by their peculiar leaves, a spikelike fertile segment arising from each green, fan-shaped blade.

Four genera and 70 species; temperate and tropical areas.

OPHIOGLOSSUM L.

Ophioglossum reticulatum L., Sp. Pl. 1063. 1763

Small terrestrial plant, consisting of a single succulent stem bearing 1 sterile leaf (rarely 2) and 1 fertile stalk. Sterile leaf blade arising 2–9 cm above ground, ovate, acute to rounded at apex, cuneate at base, ca 1–4 cm long. Sporangia borne distally on a long, slender, solitary stalk extending 5–20 cm above sterile leaf, the sporangia sessile, thick-walled, lacking annuli. *Dressler 2874.*

Collected once in the Laboratory Clearing. The species has been seen fertile only during the rainy season.

Mexico to Argentina; West Indies; Old World. In Panama, known only from BCI.

5. SCHIZAEACEAE

Homosporous, terrestrial ferns, vines, or herbs; rhizomes creeping or ascending; rachis monopodial, the alternate branches dwarfed, each with 1 pair of leaflets and an abortive bud. Leaves variously palmately or dichotomously lobed; veins usually free. Sporangia biseriate on marginal spikes (each sporangium subtended by an outgrowth serving as an indusium); annuli distal, complete, opening by longitudinal slit; spores tetrahedral; prothallia green, flat.

Distinguished by their vinelike habit (except *Schizaea*), palmately or dichotomously lobed leaves, and marginal, spikelike sori.

Four to six genera and 150 species; mostly tropics, a few species in subtropical or temperate regions.

LYGODIUM Sw.

Lygodium radiatum Prantl, Unters. zur Morph., Schiz. 66. 1881

Slender vine, glabrous throughout; stems slender. Leaves opposite, deeply palmate, with 3–7 lobes (fertile leaves usually 3–5-lobed); petioles slender, mostly 2.5–5.5 cm long; lobes long and narrow, acute to acuminate at apex (rounded and mucronate on some juveniles), 8–25 cm long, 7–25 mm wide, about equal in length on each leaf, serrate. Sporangia borne in small conelike groups exserted along leaf margins, each sporangium solitary, sessile,

subtended by a cup-shaped involucre. *Croat 12803.*

Occasional, in the forest or at the edges of clearings, climbing to a height of 20 m.

Panama and Colombia. In Panama, known from tropical moist forest in the Canal Zone, Bocas del Toro, Colón, San Blas, Panamá, and Darién and from tropical wet forest in Panamá and Colón.

See Fig. 6.

Lygodium venustum Sw., J. Bot. (Schrader) 1801 (2):303. 1803

L. polymorphum (Cav.) H.B.K.

Slender vine; stems, petioles, and upper midrib of leaves densely pubescent. Leaves alternate, sessile or with slender petioles to 7 mm long, unequally palmate, with 3–5 lobes, the central lobe longest, 4–10 cm long, the lateral lobes 1–3 cm long, all lobes ± crenate, moderately to sparsely pubescent. Sporangia sessile, borne in narrow, pubescent, conelike clusters along margins. *Croat 7216.*

Abundant in old clearings; less frequent in the forest along margins of clearings.

Mexico to Peru; West Indies. In Panama, known principally from tropical moist forest in the Canal Zone, Colón, Chiriquí, Herrera, Coclé, Panamá, and Darién; known also from tropical dry forest in Panamá (Taboga Island), from premontane moist forest in Panamá (Saboga Island), and from premontane wet forest in Chiriquí and Panamá (Chimán).

See Fig. 7.

SCHIZAEA J. Sm.

Schizaea elegans (Vahl) Sw., J. Bot. (Schrader) 103. 1801

Erect, terrestrial herb, 20–60 cm tall; rhizome erect, with rufous scales. Petioles 20–40 cm long, darkened at base, sparsely covered with long, flattened or acicular scales, glabrate toward apex; blades dichotomously divided or lobed, the veins repeatedly dichotomous, each leaf segment fan-shaped, narrowly incised at apex, gradually tapered to base. Sporangia sessile on branches of compact, pectinate-pinnate, ± reniform distal segments of blade. *Foster 2795.*

Known from a single collection from the end of Chapman Trail. Elsewhere in Panama the species has been found in fertile condition essentially all year.

Mexico to Brazil and Bolivia; West Indies. In Panama, known from tropical moist forest in the Canal Zone, Herrera, and Panamá and from premontane wet forest in Panamá.

KEY TO THE SPECIES OF SCHIZAEACEAE

Plants erect herbs, less than 60 cm tall; sporangia arranged in pectinate-pinnate, ± reniform structures, only on distal segments of blades . . . . *Schizaea elegans* (Vahl) Sw.
Plants scandent herbs, usually long vines; sporangia arranged along lateral margins of all leaflets:
  Leaflets divided into nearly equal lobes . . . . *Lygodium radiatum* Prantl
  Leaflets with median lobe much longer than lateral lobes . . . . *Lygodium venustum* Sw.

Veins of leaf segments 1-forked; leaflet midribs normally covered with brown scales below; leaf segments 8–16 times longer than wide . . . . *Gleichenia bifida* (Willd.) Spreng.
Veins of leaf segments 2–5-forked; leaflet midribs not covered with brown scales below; leaf segments less than 5 times longer than wide:
    A pair of accessory pinnae at all but ultimate nodes . . . . *Dicranopteris flexuosa* (Schrad.) Und.
    Accessory pinnae lacking, or (rarely) only at lower nodes . . . . *Dicranopteris pectinata* (Willd.) Und.

6. GLEICHENIACEAE

Homosporous, terrestrial herbs, often vinelike; rhizomes long-creeping. Leaves 2-pinnate or more compound (in ours; sometimes pinnate elsewhere), often pseudodichotomous by abortion of terminal bud, usually coriaceous; veins free. Sori exindusiate; sporangia few on lower blade surface, in 2 rows along major divisions, each sporangium with a complete transverse medial annulus opening by a longitudinal slit; prothallia green, flat, costate.

Recognized by their vinelike habit and their dichotomously branched, pinnately compound leaves, the sporangia borne along the segments of the pinnatifid pinnae.

Five or six genera and 130–160 species; tropical, subtropical, and subtemperate areas of the Southern Hemisphere.

DICRANOPTERIS Bernh.

Dicranopteris flexuosa (Schrad.) Und., Bull. Torrey Bot. Club 34:254. 1907

Similar to *D. pectinata,* but with a pair of reduced accessory pinnae at most leaf nodes; the plant glabrous throughout; the veins of the leaflets 2–4-forked, darker, not at all raised. *Kenoyer 4.*

Found in same habitats as *D. pectinata.* Collected once by Kenoyer, and possibly no longer present on the island.

Southern Mexico to Colombia and along the Andes to Brazil and Peru, mainly at lower elevations; West Indies. In Panama, known principally from tropical moist forest in the Canal Zone, Veraguas, Herrera, Panamá, and Darién, but also from premontane wet forest in Panamá (Cerro Campana).

Dicranopteris pectinata (Willd.) Und., Bull. Torrey Bot. Club 34:260. 1907

Free-climbing, vinelike, dichotomously branched herb; rhizome long-creeping, 3–5 mm diam, scabrous from persistent bases of articulate trichomes; primary leaf axes at first erect, eventually arching. Leaflets paired at ultimate nodes, deeply pinnatifid, oblong-lanceolate, mostly 10–25 cm long, 1.5–6 cm wide, sessile, glaucous beneath, glabrate, sometimes with a few brown, stellate scales below; lobes 3–6 mm wide, firm, the veins 3–5-forked, somewhat raised; leaf nodes in lower dichotomies rarely with a pair of reduced accessory pinnae. Sori round, exindusiate, in 2 rows, 1 row on each side of midrib on underside of lobes, each sorus bearing more than 6 sporangia. *Croat 4829.*

Locally abundant on steep, eroding banks on the shore, particularly on the northern shore of Orchid Island and on Gross Point.

For a description of branching in this genus, see Underwood (1907).

Mexico to Brazil, the Guianas, and Bolivia; West Indies. In Panama, known principally from tropical moist forest in the Canal Zone, San Blas, Herrera, Panamá, and Darién; known also from tropical dry forest in Panamá (Taboga Island) and from tropical wet forest in Colón.

GLEICHENIA J. Sm.

Gleichenia bifida (Willd.) Spreng., Syst. Veg. 4:27. 1827

Erect or subscandent, dichotomously branched herb; rhizome creeping, scaly; stems, midribs of leaflets above and below, and lobes of leaflets below ± densely covered with brown arachnoid scales, some scales lanceolate with long-ciliate margins. Leaflets paired at ultimate nodes, deeply pinnatifid, oblong, tapered to long-caudate-acuminate apices, 25–45 cm long, 2–8 cm wide, sessile; lobes sometimes of irregular lengths, mostly 2–4.5 cm long, 2–3 mm wide, moderately thin, the veins 1-forked. Sori round, in 2 rows, 1 row on each side of midrib on underside of lobes, each sorus usually bearing 4 or fewer sporangia. *Croat 11762.*

Locally abundant on steep, eroding banks on the shore, on the north side of the island, often growing with the more abundant *Dicranopteris pectinata* and often preferring the more shaded areas.

Mexico to South America below 2,000 m; West Indies. In Panama, known from tropical moist forest in the Canal Zone and Darién, but probably more common in premontane wet forest in Colón, Chiriquí, Coclé, and Panamá.

7. HYMENOPHYLLACEAE

Homosporous epiphytic or terrestrial herbs; stems usually rhizomes with distichous scales, less commonly erect and radially symmetrical. Leaves usually small; leaflets usually 1 cell thick, lacking stomata, usually coiled in bud; veins usually free. Involucres consisting of urceolate, tubular, or bivalvate marginal extensions of blade (the indusia); sporangia borne on all surfaces of free-ending vein extensions (the receptacles) within the involucre; sporangia opening by a ± longitudinal slit; spores tetrahedral or becoming globose; prothallia filamentous or thallose.

KEY TO THE TAXA OF HYMENOPHYLLACEAE

Involucre valvate, not at all tubular . . . . *Hymenophyllum brevifrons* Kunze
Involucre tubular . . . . *Trichomanes*

Recognized by their lacy, generally transparent leaves and by their tubular, urceolate, or bivalvate marginal indusia.

Three genera and about 650 species; all tropical and temperate regions of the world.

HYMENOPHYLLUM J. Sm.

Hymenophyllum brevifrons Kunze, Bot. Zeitung (Berlin) 5:185. 1847

Epiphyte; rhizome slender, creeping, branched, with numerous, short, filiform, reddish-brown scales, these ultimately deciduous. Leaves simple, deeply and irregularly divided, translucent, glabrous, usually drying brown and curled; petioles ca 1 cm long; blades ± ovate in outline, ca 1 cm or more long; ultimate lobes usually emarginate, never toothed; veins thick, dichotomously branching several times in lobes. Sori usually near apex of blade, marginal, the indusia bivalved throughout (or at least to middle). *Shattuck 1013.*

Probably rare; collected once by Shattuck on Zetek Trail.

An inconspicuous plant, usually growing interspersed with mosses. It could easily be overlooked and is possibly more abundant than meets the eye.

Guatemala to Panama and the Guianas. In Panama, known from tropical moist forest in the Canal Zone and Panamá and from premontane wet forest in Panamá.

TRICHOMANES L.

Many of the smaller, inconspicuous species are easily overlooked and possibly more abundant than they appear. See the genus key for distinguishing characters of *T. curtii* Ros., which has not been identified from BCI but is likely to occur.

Trichomanes diversifrons (Bory) Mett., Sadeb. Nat. Pfl. 1(4):108. 1899

Terrestrial, mostly 20–30 cm tall; rhizome stout, erect; petiole bases closely spaced; rhizome and petioles bearing stiff, dark, threadlike, seemingly jointed scales. Leaves simple, dimorphic; petioles on sterile leaves 3–5 cm long; sterile blades deeply pinnatifid, oblong-lanceolate, tapered to very long filiform apex, 3.5–5 cm broad, glabrous above, with short-jointed filiform scales on underside of veins (particularly midrib); lobes mostly to 5 mm broad, acute at apex, ± uncinate-serrate, the veins anastomosing, the lowermost lobes reduced; sporophylls linear, denticulate, ca 15 cm long and 6 mm wide, on a petiole 9–18 cm long. Sori immersed into margin; sporangia sessile on a filiform, exserted receptacle. *Croat 10813.*

Occasional, on steep creek banks in ravines, usually in dark places.

Throughout American tropics. In Panama, known from tropical moist forest in the Canal Zone, San Blas, Veraguas, and Darién; known also from premontane wet forest in Colón (Santa Rita Ridge), from tropical wet forest in Colón and San Blas, and from premontane rain forest in Darién (Cerro Pirre).

Trichomanes ekmanii Wessels-Boer, Acta Bot. Neerl. 11:319. 1962

Epiphyte; rhizome slender, creeping, scaly, the petiole bases well spaced. Leaves simple; petioles 3–7 mm long, sparsely scaly; blades obovate or oblong to orbicular, rounded to irregularly lobed especially in upper part, rounded or tapered at base, to 2(3) cm long and 1(1.5) cm wide, the midrib distinct, the lateral veins pinnately arranged, and cross-veins lacking; false vein continuous, submarginal. Sori 4–9 per leaf, together near apex or solitary on apical lobes; indusium wholly immersed in margin, lacking lips and not dark-edged, the tube cylindrical and broadly expanded at apex. *Croat 16515.*

Distribution on BCI unknown. Found in wet forests on roots and bases of trees and on fallen logs.

Throughout Central America and northern South America; Greater Antilles. In Panama, known only from tropical moist forest in the Canal Zone.

Trichomanes godmanii Hook. in Baker, J. Linn. Soc., Bot. 9:337. 1866

Tiny epiphyte; rhizome slender, creeping, with fine scales. Leaves simple, glabrous; petioles 4–10 mm long; blades orbicular or obovate, rounded at apex, tapered to obtuse or acute base, 5–20 mm long and wide, entire or slightly lobed, with a distinct costa and pinnate veins in lower blade and flabellate venation above; false veinlets distinct, reticulate; false vein submarginal. Sori 1–9 in upper part of leaf; indusium wholly immersed in margin, lacking lips and not dark-edged, obconic with the mouth expanded, the receptacle long-exserted. *Croat 16202.*

Fairly common; seen densely covering all sides of base of *Scheelea zonensis* (19. Palmae) trunk.

Throughout Central America; Cuba. In Panama, known only from tropical moist forest in the Canal Zone and Panamá.

Trichomanes kapplerianum Sturm in Mart., Fl. Brasil. 1(2):276. 1859
T. hookeri Presl var. *minor* (Jenm.) Domin

Tiny epiphyte; rhizome slender, creeping, scaly. Leaves simple; petioles very short; young blades often orbicular, cordate; mature blades ovate to oblong, irregularly lobed, ± rounded at apex, rounded to narrowly acute at base,

KEY TO THE SPECIES OF TRICHOMANES

Plants terrestrial; leaves often more than 20 cm long:
  Leaves dimorphic, simple, the sterile leaves pinnatifid, the fertile leaves linear; sori immersed in
    margin . . . . *T. diversifrons* (Bory) Mett.
  Leaves ± monomorphic, 1-pinnate; sori along margins of leaflets except near apex . . . .
    . . . . *T. pinnatum* Hedw.
Plants epiphytic; leaves less than 10 cm long:
  All leaves pinnately lobed, most more than 3 cm long, the costa extending to apex of blade:
    Trichomes frequent on blade surface and margins, all trichomes stellate; indusia bell-shaped,
      immersed, the lips not dark-edged . . . . *T. polypodioides* L.
    Trichomes only on margins of blades, the blades otherwise glabrous, the trichomes simple or
      bifid; indusia tubular, partly exserted, the lips dark-edged . . . . *T. krausii* Hook. & Grev.
  At least some leaves not pinnately lobed, most less than 3 cm long, the costa only rarely extend-
    ing to the apex:
    Leaves bearing marginal trichomes but lacking a submarginal false vein; indusia at least partly
      exserted, with 2 distinct dark-edged lips:
      Leaves less than 8 mm long, usually bearing a single terminal sorus . . . .
      . . . . *T. ovale* (Fourn.) Wessels-Boer
      Leaves more than 1 cm long, usually bearing several to many sori:
        Sori often 6 or more per leaf; indusia exserted, not between lobes or in sinuses (species
          not known from BCI but to be expected) . . . . *T. curtii* Ros.
        Sori rarely as many as 5 or 6 per leaf; indusia half immersed, between lobes . . . .
        . . . . *T. punctatum* Poir. subsp. *sphenoides* (Kunze) Wessels-Boer
    Leaves lacking marginal trichomes but bearing a submarginal false vein; indusia wholly im-
      mersed, without lips and not dark-edged:
      Venation reticulate; mature fertile leaves ± orbicular . . . . *T. godmanii* Hook.
      Venation lacking cross-veins; mature fertile leaves elongate:
        Submarginal vein continuous; young leaves ± elongate . . . . *T. ekmanii* Wessels-Boer
        Submarginal vein discontinuous; young leaves often orbicular or cordate . . . .
        . . . . *T. kapplerianum* Sturm

5–30 mm long, 5–15 mm wide, bearing a distinct midrib, pinnate lateral veins in lower part of blade, flabellate venation above; false vein discontinuous, submarginal. Sori usually 3–7, together near apex or on small lobes; indusium wholly immersed in margin, lacking lips and not dark-edged, the mouth expanded; receptacle exserted. *Wilson 87.*

Distribution on BCI not known. Forming mats on tree trunks and moist rocks.

Costa Rica to the Guianas and Amazon basin; Lesser Antilles. In Panama, known only from tropical moist forest on BCI and in Bocas del Toro.

Trichomanes krausii Hook. & Grev., Icon. Fil. t. 149. 1831

Tiny epiphyte; rhizome slender, creeping, scaly. Leaves simple, very thin, somewhat lacy; petioles mostly very short; petioles and lower midribs densely scaly; blades oblong or lanceolate-oblong, irregularly pinnatifid, glabrous except on margins, the segments weakly lobed, 2–4.5(9) cm long, 1–2.5 cm wide; lobes ± linear, irregular, 1–3 mm broad, separated by broad, open sinuses, a large dark stellate scale on a tooth in the sinus, some of the marginal scales simple or bifid; midrib of lobes extending to apex, almost lacking connected lateral veins; false veinlets few, partly parallel to margin. Sori several, solitary in upper lobes; indusium partly immersed, its lips semiorbicular, dark-edged. *Croat 8789.*

Distribution on BCI uncertain. Found on moist, shaded rocks, twigs, and tree trunks to 1,350 m elevation.

Throughout tropical and subtropical America. In Panama, known from tropical moist forest in the Canal Zone and Darién, from premontane wet forest in Panamá, and from tropical wet forest in Chiriquí.

Trichomanes ovale (Fourn.) Wessels-Boer, Acta Bot. Neerl. 11:296. 1962
  T. sphenoides Kunze var. *minor* Ros.

Tiny epiphyte, the blades usually ± appressed. Leaves simple, minute; petioles very short; blades variable, orbicular, ovate, or lanceolate, usually rounded or cordate and broadest at base, 3–8 mm long, 2–6 mm wide, ± entire, with bifid or stellate trichomes on margin; midrib on fertile leaves and some sterile leaves running to apex, the lateral veins pinnate, partly free. Sorus solitary at apex (rarely 2 or 3); indusium partly immersed, at least the lips exserted, the lips dark-edged, small, narrowly winged with stellate scales and with a few rows of brown cells. *Croat 14999.*

Distribution on BCI not known. Found on moist, deeply shaded rocks and tree trunks.

Probably throughout the American tropics, but seldom seen because of its small size. In Panama, known only from tropical moist forest in the Canal Zone and Panamá.

Trichomanes pinnatum Hedw., Fil. Gen. et Sp. t. 4, f. 1. 1799

Terrestrial, usually 25–50 cm tall; rhizome erect, short, bearing brown, threadlike scales. Leaves 1-pinnate,

6–30(40) cm long, often with a long leafless tail at apex, the tail sometimes rooting and producing another plant; petiole and rachis with moderate to sparse, long, thread-like scales; pinnae lanceolate-oblong, blunt to obtuse at apex, obtuse to truncate or subcordate at base (sterile pinnae imbricate at base), 3–6 cm long, 1–1.5 cm wide, minutely and sharply serrate, glabrous except on midrib below, with many lateral, 1–4-forked veins; fertile leaflets somewhat narrower and more widely spaced along rachis. Sori along margin except very near apex; indusium ± tubular, stalked; receptacle exserted, about as long as tube. *Croat 5046.*

Occasional, in ravines on steep banks.

Throughout the American tropics. In Panama, known only from tropical moist forest in the Canal Zone and Panamá and from tropical wet forest in Panamá.

Trichomanes polypodioides L., Sp. Pl. 1098. 1753

Small epiphyte; rhizome slender, creeping, scaly. Leaves simple, very thin, irregularly pinnatifid, the segments lobed, 3–12 cm long, 1.5–2.5 cm wide, oblong to ovate in outline with light-brown, stellate scales on margin and surface; midrib extending to apex with lateral veins at each lobe, these sometimes 1- or 2-forked. Sorus solitary, mostly at apex of lobes or rarely in sinuses; indusium immersed, lacking dark-edged lips but with stellate trichomes on margin, the receptacle long-exserted, often many times longer than involucre. *Croat 11232.*

Distribution on BCI not known. Found on moist tree trunks.

Throughout the American tropics. In Panama, known from tropical moist forest in the Canal Zone (BCI only), Colón, and Panamá, from premontane wet forest in Chiriquí and Panamá, and from tropical wet forest in Panamá (Campo Tres).

Trichomanes punctatum Poir. subsp. **sphenoides**
   (Kunze) Wessels-Boer, Acta Bot. Neerl. 11:301. 1962

Tiny epiphyte; rhizome creeping, slender, densely scaly, the scales dark reddish-brown, threadlike, deciduous. Leaves simple, entire or with short lobes, sometimes split when dry; petioles very short, scaly; blades round-oblong, ovate or obovate, rounded at apex, acute to subcordate at base, 5–20 mm long, 5–15 mm wide, with stellate scales on margin, midrib lacking, the veins very densely flabel-late, the margin entire or with a few ± linear lobes or regularly crenate. Sori few to several, arising from between the lobes; indusium about half immersed, the involucres contracted at neck, the lips circular, narrowly dark-edged, as broad as tube. *Croat 14021.*

Distribution on the island is uncertain.

Very similar to *T. curtii* Ros. (not known from BCI; see the key to the genus).

Throughout Central America and northwestern South America; Greater Antilles. In Panama, known from trop-ical moist forest on BCI and in Bocas del Toro, Colón, Coclé, and Panamá.

8. PARKERIACEAE

Homosporous, aquatic or subaquatic annuals; rhizomes short, ± erect. Leaves alternate, dimorphic, pinnately decompound, the fertile blades erect, repeatedly pinnate-pinnatifid, larger than sterile blades and more finely dissected, the leaflets longer and narrower; veins areolate, included; veinlets absent. Sporangia borne in 1 or 2 sparse rows usually along the margin, globose, ± sessile, thin-walled, protected by continuous revolute margins; an-nulus complete or vestigial; spores triplanate.

Recognized by their aquatic habitat, by their decom-pound, dimorphic leaves, and by sporangia that develop only when the plants are emergent or floating on the sur-face of the water (Schulthorpe, 1967).

One genus, with 2 species; tropics and subtropics.

CERATOPTERIS Brongn.

Ceratopteris pteridoides (Hook.) Hieron., Bot. Jahrb. Syst. 34:561. 1905

Plant floating or partly submerged, usually anchored in soil, the roots borne on the stipe. Leaves very thin, simple, dimorphic, usually glabrous; sterile leaves ± pentagonal, to 25 cm long, about as broad, expanded at about the middle and deeply divided (appearing 2–4-pinnatifid), the ultimate segments mostly 3–10 mm broad, blunt to rounded at apex; leaves of floating plants usually with bulbous stipes; fertile blades to 40 cm long, divided as in sterile blades but with segments more than 2 mm wide, the margins evenly and narrowly revolute. Sporangia large, globose, borne in 1 row along each margin. *Shat-tuck 606.*

Collected once by Shattuck; not collected on BCI in recent years, but probably still occurs there. Seen fertile in July.

Florida, Panama, the Guianas, and Brazil; Greater Antilles. In Panama, known from tropical moist forest in the Canal Zone on the Atlantic slope and from Bocas del Toro.

9. CYATHEACEAE

Mostly arborescent, homosporous; rhizomes usually simple, decumbent or erect, scaly. Leaves borne at apex, articulate or not, 1–4-pinnate, usually very large, lanceo-late-oblong to deltoid-ovate in outline. Sori indusiate or exindusiate, ± globose, borne on veins of underside of blades; sporangia numerous, crowded, radial, in several ranks, the annulus oblique; spores triplanate.

Except for the atypical *Metaxya rostrata*, members of the family on BCI are characterized by being arborescent with large, graceful, compound leaves. They are most readily confused with large, compound-leaved members of the Polypodiaceae (10), but can be distinguished from them by having sporangia with an oblique annulus.

Nine genera and 750 species; tropical regions.

KEY TO THE SPECIES OF CYATHEACEAE

Leaves 1-pinnate . . . . *Metaxya rostrata* (Willd.) Presl
Leaves 2- or 3-pinnate-pinnatifid:
    Sori along margins of pinnatifid lobes of leaflets; leaflets nearly glabrous; lobes divided about
        halfway to midrib, often 8 mm or more broad . . . . *Cnemidaria petiolata* (Hook.) Copel.
    Sori clustered at base of lobes or covering most of the lower surface; leaflets with at least the
        midrib pubescent; lobes divided nearly to base, usually less than 3 mm wide:
        Leaflets pubescent over all of upper surface with sparse, conspicuous, long trichomes; pubes-
            cence of rachis and pinnular rachis conspicuous, erect, 1–3 mm long . . . .
            . . . . *Trichopteris trichiata* (Max.) Tryon
        Leaflets glabrous on upper surface or with few trichomes on midrib of segments; pubescence
            of rachis and pinnular rachis inconspicuous, appressed or very sparse:
            Petiole, rachis, and pinnular rachis armed with spines; leaflets lacking stellate trichomes on
                lower surface; sori exindusiate . . . . *Trichopteris microdonta* (Desv.) Tryon
            Petiole, rachis, and pinnular rachis unarmed; leaflets bearing purplish stellate trichomes on
                lower surface; sori indusiate . . . . *Nephelea cuspidata* (Kunze) Tryon

CNEMIDARIA Presl

Cnemidaria petiolata (Hook.) Copel., Gen. Fil.
    97. 1947
    Hemitelia petiolata Hook.

Small, spiny tree fern, 1.5–3.5 m tall, often fertile when trunk is still very short; trunk to 5 cm diam, conspicuously ribbed and with persistent petiole bases much of its length; leaf scars prominent, the cluster of new leaf crosiers densely scaly. Juvenile plants bearing leaves 1-pinnate, the leaflets with rounded, sharply serrate lobes; mature plants bearing leaves 2-pinnate-pinnatifid, to ca 2 m long, arch-ascending; petioles 50–75 cm long, with lanceolate-linear, brown to reddish-brown scales ca 1 cm long on upper side usually near base, armed throughout but especially on lower side with short spines to ca 5 mm long; rachis pubescent with stiff trichomes on upper surface, sometimes also on upper midrib of leaflets; leaflets usually 5–15 cm long, 1–3 cm wide at base, tapering to acuminate apex, ± regularly lobed approximately halfway to midrib, less so near apex, nearly glabrous; upper leaflets reduced, ultimately confluent. Sori marginal, round, bordering entire leaflet, at least part of the cup-shaped indusium persisting. *Croat 4143.*

    Common in the forest, especially in ravines. May be found in fertile condition throughout the year.

    Panama and Colombia. In Panama, common in some areas of tropical moist forest on the Atlantic slope at least in the Canal Zone, Colón, and San Blas; known also from tropical moist forest in Panamá and tropical wet forest in Colón and along the Darién–San Blas border.
    See Figs. 8 and 9.

METAXYA Presl

Metaxya rostrata (H. & B. ex Willd.) Presl, Tent.
    Pterid. 60. 1836
    Alsophila rostrata Mart.

Terrestrial, lacking a trunk; rhizome with a single siphonostele, densely pubescent with long yellowish tri-

chomes (also a few at base of petiole). Leaves 1-pinnate, to ca 2 m long but arch-ascending, the apex usually drooping; petioles ca 50 cm long; rachis yellowish, dull; leaflets linear-oblong, caudate-acuminate, acute to obtuse and sometimes inequilateral at base, 10–32 cm long, 2–3.5 cm wide, the veins free or forked once usually near midrib, the margin entire except crenate near apex. Sori ± round, on veins in an irregular pattern on either side of midrib. *Croat 9527.*

    Rare, seen only twice; probably restricted to ravines. Apparently fertile throughout the year.

    Easily confused in sterile condition with *Saccoloma elegans* (10. Polypodiaceae), which has marginal sori, but distinguished by having the margins entire (except near apex), the veins rarely forked (except near base), the rachis yellowish and dull, and the rhizome clothed with dense yellow trichomes.

    Guatemala to Peru, Bolivia, and Brazil; West Indies. In Panama, known from tropical moist forest on the Atlantic slope in the Canal Zone, Bocas del Toro, and Colón; known also from tropical moist and premontane wet forests in Panamá.
    See Fig. 10.

NEPHELEA Tryon

Nephelea cuspidata (Kunze) Tryon, Contr. Gray Herb.
    200:40. 1970
    Cyathea punctifera H. Christ

Graceful tree fern, to 7 (15) m tall; trunk to 20 (40) cm diam; trunk and petioles with a few spines to 1 cm long, the spines embedded in a dense layer of irregularly serrate scales and adventitious roots. Leaves 2-pinnate-pinnatifid, to ca 3 m long and 1.5 m or more wide; petiole and rachis atropurpureous, with short, dense, almost granular scales (easily scraped off) especially on underside, the petiole to 50 cm long; pinnular rachis and costules densely pubescent with purplish ascending trichomes; pinnae to ca 1 m long and 30 cm wide; leaflets pinnatifid to near midrib, to 11.5 (15) cm long and to 2.5 (3.5) cm wide, tapered to a

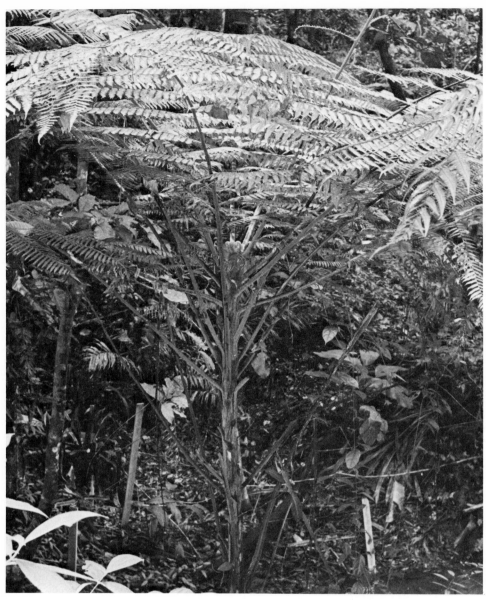

Fig. 8. *Cnemidaria petiolata*

Fig. 9. *Cnemidaria petiolata*

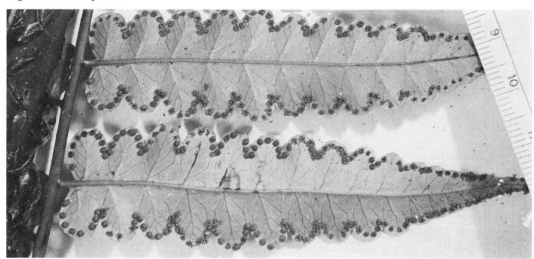

Fig. 10. *Metaxya rostrata*

Fig. 11. *Nephelea cuspidata*

long, bluntly acuminate apex, the lobes 2–3 mm wide, somewhat falcate, entire to minutely crenate; veins mostly 1-forked near base, the major veins sparsely pubescent beneath with purplish, mostly stellate scales. Sori round, usually 4–6, clustered in proximal half of lobes; indusium globose to urn-shaped, dark brown, completely enclosing sporangia, breaking into irregular segments at maturity, deciduous to persistent, scarious. *Croat 6529.*

Rare on the island, restricted to deep ravines. Seasonal behavior undetermined. Seen fertile in April, August, and September.

Nicaragua to Peru, Bolivia, Paraguay, and Brazil. In Panama, known only from tropical moist forest on BCI and in Darién.

See Fig. 11.

TRICHOPTERIS Presl

Trichopteris microdonta (Desv.) Tryon, Contr. Gray Herb. 200:46. 1970
Alsophila microdonta Desv.

Tree fern, 1–5 m tall, slender, the trunk only a few cm diam. Leaves few, 2-pinnate-pinnatifid; petioles to ca 1 m long, lustrous, purplish near base, becoming light brown, glabrate, sparsely scaly near base and bearing numerous distant, narrowly conical spines, these curved and to 1 cm long; scales to ca 1.5 cm long and 2 mm wide, brown, very long-caudate-acuminate; leaves ovate-oblong, abruptly acuminate, to 1.5 m long and 1.2 m broad, the primary rachis brown or yellowish-brown, armed throughout; pinnae petiolate, narrowly oblong, acuminate, 30–60 cm long, 10–25 cm wide; secondary rachis yellowish-strigose above, thinly scurfy-hirtellous below (glabrate in age), distantly aculeate with spines to ca 4 mm long; leaflets ± sessile, ± oblong, long-attenuate, 5–13 cm long, 1.5–3 cm wide, the costa densely yellowish-strigose above, scurfy-hirtellous and with a few minute, mostly caducous scales below, the segments of 19–25 pairs, linear, ± acute, 8–18 mm long, 2–4 (5) mm wide, crenate-serrate, the costules distantly hispid above, thinly scurfy-hirtellous below. Sori in 6–11 pairs, the paraphyses very numerous, equaling sporangia. *Kenoyer 7, Munch s.n.*

Apparently rare, in ravines; more common elsewhere in the Canal Zone.

Mexico to Peru and Brazil, principally at low elevations; West Indies (Isle of Pines). In Panama, known only from tropical moist forest in the Canal Zone.

Trichopteris trichiata (Max.) Tryon, Contr. Gray Herb. 200:44. 1970
Alsophila trichiata Max.

Graceful tree fern, to 3.5 m tall. Leaves 2- or 3-pinnate-pinnatifid on lower parts of large leaves, to 3.5 m long and 2 m wide at the middle; petioles densely pubescent with moderately stiff trichomes, sparsely armed throughout with short spines; rachis with short, soft, villous pubescence interspersed with long, stiff, jointed trichomes; pinnae lanceolate to oblong-lanceolate, 50–90 cm long,

to 30 cm wide, abruptly acuminate and confluent toward apex; leaflets sessile, linear-lanceolate, attenuate-caudate, pinnatifid to beyond the middle, mostly to 9 (15) cm long, to 2 cm wide, both surfaces pubescent especially on veins, the segments 5–10 mm long, 1.5–3 mm wide, usually bidentate at apex. Sori exindusiate, round, on veinlets on proximal two-thirds of lobes, consisting of moderately few sporangia; paraphyses few, scarcely exceeding sporangia. *Croat 11729.*

Apparently rare; collected once on Gross Point Peninsula (*Croat 11729*) and twice elsewhere by Shattuck (*740, 1149*). Probably fertile throughout the year.

The generic name *Trichipteris*, used by some authors, is an orthographic error.

Costa Rica to Ecuador and Venezuela, from near sea level in Costa Rica to ca 1,000 m in Ecuador. In Panama, known from tropical moist forest in the Canal Zone, Panamá, and Darién and from premontane wet forest in the Canal Zone and Panamá.

10. POLYPODIACEAE

Terrestrial or epiphytic (rarely aquatic), homosporous; rhizomes creeping to erect. Leaves pendent to spreading or erect, usually petiolate; blades uniform to strongly dimorphic, simple to multiply pinnate or pinnatifid or further compounded, coiled in bud. Sori various in shape and arrangement on veins on underside of blades, usually in lines or clusters, sometimes over the whole surface; indusia various or lacking, developing from veins or margins of blades; sporangia long-stalked, bearing an incomplete vertical annulus, opening transversely; prothallia green.

About 150 genera and 6,000 species; worldwide. A very diverse family, frequently split into a variety of subfamilies or separate families.

All genera of the Polypodiaceae are keyed out in the key to the Pterophyta (p. 71).

ACROSTICHUM L.

Acrostichum aureum L., Sp. Pl. 1069. 1753
Helecho de manglar

Massive aquatic, to 3 m tall with a stout, scaly rhizome. Leaves 1-pinnate; petioles stout, ribbed, much shorter than the rachis, with large ligulate scales at base; leaflets narrowly oblong, rounded to acute or mucronate at apex, stiff, to 25 cm long and 4 cm wide, revolute and entire, mostly 2–6 cm apart near apex, the veins prominent but glabrous and scarcely if at all raised below. Sporangia covering all or part of underside of uppermost leaflets. *Croat 5288.*

Common in dense stands, chiefly on the south and west shores of the island. No doubt an important element in hydrarch succession along the southern shore, where it forms dense stands in association with *Annona glabra* (55. Annonaceae).

KEY TO THE SPECIES OF ACROSTICHUM
Leaf veins glabrous, scarcely if at all raised . . . . *A. aureum* L.
Leaf veins at least partly pubescent, standing up sharply . . . . *A. danaifolium* Langsd. & Fisch.

Throughout tropical America. In Panama, known from tropical moist forest in the Canal Zone, Bocas del Toro, San Blas, Panamá, and Darién and from tropical wet forest in Colón (Miguel de la Borda).

See fig. on p. 12.

Acrostichum danaifolium Langsd. & Fisch., Icon. Fil. 5, t. 1. 1810

Very similar to *A. aureum*, but the fertile leaves having more of their leaflets fertile; and the leaflets 1–4 cm apart near apex, pubescent at least on the veins, and the veins beneath raised. *Croat 6165*.

Found with *A. aureum* along the shore of the island.

Throughout tropical America. In Panama, known only from BCI and the Pacific slope of the Canal Zone and in Panamá.

ADIANTUM L.

The genus *Adiantum*, distinguished by its oblique leaflets and indusiate marginal sori, provides the dominant terrestrial fern flora on the island.

Adiantum decoratum Max. & Weath., Amer. J. Bot. 19:165. 1932

Terrestrial, to 70 cm tall; rhizome ± erect, with dense linear scales. Leaves 2-pinnate; petiole, rachis, and pinnular rachis with long, conspicuous, ± dense, filiform scales; petioles as long as or longer than blades; leaflets asymmetrical (the midrib nearly submarginal), ± oblong, blunt at apex, truncate and parallel to rachis at base, to 2.5 cm long and 8 mm wide, glabrous, usually glaucous beneath, the sterile margins irregularly, finely serrate. Sori interrupted along upper and distal margins. *Shattuck 282.*

Rare, in the forest.

Mexico to Panama. In Panama, known from tropical moist forest in the Canal Zone, Bocas del Toro, Los Santos, Panamá, and Darién and from premontane wet forest in Panamá (Cerro Azul).

Adiantum fructuosum Spreng., Syst. Veg. 4:113. 1827

Terrestrial, rarely taller than 50(80) cm. Leaves usually several, closely clustered on a short-creeping, scaly rhizome, usually 1-pinnate (often 2-pinnate elsewhere);

KEY TO THE SPECIES OF ADIANTUM
Leaflets borne on slender stalks more than 7 mm long:
  Leaflets ± reniform, rounded at apex, usually less than 4 cm long on stalks less than 2.5 cm long
  . . . . *A. lunulatum* Burm.
  Leaflets ± ovate, acuminate, usually more than 5 cm long on a stalk more than 3 cm long . . . . .
  . . . . *A. seemannii* Hook.
Leaflets not borne on stalks more than 7 mm long:
  Sori continuous along margins:
    Leaflets more than 5 cm long, narrowly triangular; sori continuous along upper and lower
    margins. . . . . *A. lucidum* (Cav.) Sw.
    Leaflets less than 3 cm long, ± falcate; sori continuous on upper margin, sometimes with a
    shorter segment on the distal edge, lacking on lower margin . . . . *A. pulverulentum* L.
  Sori interrupted along margins:
    Rachis bearing long, conspicuously spreading, threadlike scales, clearly visible to the naked
    eye, these interspersed with shorter, fine scales:
      Leaflets pubescent beneath . . . . *A. humile* Kunze
      Leaflets glabrous beneath . . . . *A. decoratum* Max. & Weath.
    Rachis lacking clearly visible scales or bearing scales ± appressed, not spreading, and less
    conspicuous:
      Sterile leaflets unevenly and coarsely serrate on upper and distal margins, the apex usually
      blunt to rounded . . . .*A. fructuosum* Spreng.
      Sterile leaflets evenly and usually finely serrate or biserrate:
        Leaflets not glaucous beneath, the sterile ones usually acute to acuminate . . . .
        . . . . *A. obliquum* Willd.
        Leaflets glaucous beneath (fertile leaflets sometimes pubescent on inner margin), the
        sterile ones usually blunt at apex:
          Leaflets pubescent beneath (often nearly glabrous on juvenile or sterile leaflets), usu-
          ally not tapered much to apex . . . . *A. humile* Kunze
          Leaflets usually glabrous beneath, usually much broader at base than apex . . . .
          . . . . *A. petiolatum* Desv.

petiole and rachis dark brown, with fine, short threadlike scales persisting at least on rachis; leaflets nearly oblong, blunt to acute at apex, 1–3 cm long, glabrate beneath (elsewhere sometimes with filiform scales), midrib very near lower margin, the lower edge ± straight, the proximal edge paralleling the rachis, the upper and distal edges of sterile leaflets unevenly and coarsely serrate. Sori interrupted along upper and distal margins. *Croat 8576.*

Occasional, in the forest.

Distinguished by its small, coarsely serrate leaflets. The species is easily confused with and possibly not separable from *A. tetraphyllum* H. & B. ex Willd., which has a long-creeping rhizome with the petiole bases not closely crowded as in *A. fructuosum.* Moreover, in *A. tetraphyllum* the sterile apices of the leaflets are acute and turned toward the apex of the pinnae, whereas in *A. fructuosum* they are mostly straight and obtuse.

Mexico to Brazil; West Indies. In Panama, known from tropical moist forest in the Canal Zone, San Blas, Panamá, and Darién, from premontane wet forest in Panamá and Coclé, and from tropical wet forest in Colón.

See Fig. 12.

Adiantum humile Kunze, Linnaea 9:80. 1834

A. killipii Max. & Weath.

Terrestrial, usually less than 50 cm tall. Leaves 2-pinnate; petioles shiny and atropurpureous; rachis similar but usually covered with two types of scales—small, fine, ± appressed scales and longer, subulate-filiform scales; leaflets asymmetrical, mostly 1–3 cm long, acute at base, the lateral leaflets ± oblong or sometimes broader at the base with apex usually rounded and the proximal edge nearly paralleling the rachis, the lower surface usually glaucous and usually bearing numerous to few simple trichomes and sometimes filiform scales; sterile leaflets finely and evenly serrate. Sori interrupted along inner, upper, and outer margins, frequently fertile at apex. *Croat 8553.*

Occasional, along trails in the forest.

Belize to Peru, Brazil, and French Guiana; Trinidad. In Panama, known from tropical moist forest on the Atlantic slope of the Canal Zone and in San Blas, and from premontane wet forest in Panamá (Lago Cerro Azul).

See Fig. 13.

Adiantum lucidum (Cav.) Sw., Syn. Fil. 121. 1806

Terrestrial, to 70 cm tall; rhizome moderately short-creeping with slender spreading scales. Leaves usually few, 1-pinnate or rarely 2-pinnate; petiole and rachis dark brown or black, persistently pubescent-scaly (at least on rachis); leaflets asymmetrical, usually 5–10 cm long, ± narrowly triangular, acuminate at apex, markedly inequilateral at base, glabrous or sparsely scaly below; sterile leaflets with margins unevenly and coarsely serrate but the proximal third of lower margin as well as the proximal margin entire; terminal leaflet ± ovate. Sori very

long and continuous on upper and lower margins (except near apex). *Croat 7731.*

Abundant in the forest.

Distinguished by its long pointed leaflets bearing sori along most of both margins. Tryon (1964) reported the species to be 2-pinnate in Peru, a condition that is rare in Panama.

Panama to Peru and the Guianas; Trinidad and Tobago. In Panama, known from tropical moist forest on both slopes of the Canal Zone and in Panamá and Darién, from premontane moist forest in the Canal Zone (Ancón Hill), and from tropical wet forest in Colón.

See Fig. 14.

Adiantum lunulatum Burm., Fl. Ind. 235. 1768

Terrestrial, to 55 cm tall; rhizome erect, with a few dark, lanceolate-linear scales. Leaves 1-pinnate; petiole and rachis dark reddish-brown, shiny; petiolules slender, 0.3–2.5 cm long; leaflets lunate to reniform, lobed on semicircular distal edge, truncate to obtuse on lower edge, to 4 cm long and 1.8 cm wide, glabrous. Sori interrupted along lobes on distal edge. *Shattuck 1058.*

Rare, in the forest.

This species has been confused with the name *Adiantum philippense* L., which should probably be considered a *nomen dubium.*

Mexico to Colombia and Venezuela; West Indies. In Panama, known from tropical moist forest on both slopes of the Canal Zone and in Veraguas; known also from premontane moist forest in Panamá (Panamá City) and from premontane wet forest in Chiriquí, Coclé, and Panamá.

Adiantum obliquum Willd., Sp. Pl. 5:429. 1810

Terrestrial, to 60 cm tall. Leaves mostly 2-pinnate, a few 1-pinnate; petiole atropurpuraceous, sparsely to moderately scaly at least when young, the threadlike, curly scales persisting on rachis; leaflets asymmetrical, mostly 3–6.5 cm long, seldom more than 1.5 cm wide, narrowly tapered, the apex acute to acuminate, the lower edge nearly straight, the proximal edge nearly paralleling the rachis, the underside not glaucous, the sterile margins unevenly and finely biserrate. Sori interrupted along upper and distal margins with the apex usually sterile. *Croat 7815.*

Occasional, along forest trails and no doubt also in other areas of the forest.

Often confused with *A. petiolatum,* but differing in having the leaflets more narrowly triangular and irregularly serrate and in having the apex of the fertile leaflets free of sori and tapered to a sharp point. The leaflets of *A. petiolatum* are glaucous beneath, with the inner edge frequently cordate.

Throughout the West Indies and Central America, as well as from Colombia to Bolivia, Brazil, and the Guianas. In Panama, known from tropical moist forest in the Canal Zone, Bocas del Toro, and Darién and from premontane wet forest in Panamá (Chimán).

Fig. 12. *Adiantum fructuosum*

Fig. 13. *Adiantum humile*

Fig. 14. *Adiantum lucidum*

Fig. 15. *Adiantum petiolatum*

Fig. 16. *Anetium citrifolium*

Fig. 17. *Asplenium delitescens*

Adiantum petiolatum Desv., Ges. Naturf. Freunde
  Berl. Mag. 5:326. 1811

Terrestrial, to 40 cm tall; rhizome with spreading scales,
the petiole bases closely spaced. Leaves 1- or 2-pinnate;
petioles dark, glabrate; leaflets asymmetrical, obscurely
triangular to oblong, rounded to bluntly acuminate at
apex, mostly 2–5 cm long, 1–3.5 cm wide, the proximal
edge frequently semicordate and overlapping the rachis
at base, usually glaucous and glabrous beneath, the sterile margins evenly and finely serrate. Sori interrupted
along upper, lower, and sometimes proximal margins,
frequently extending to very near the apex. *Croat 6948.*

Common on trails in the forest and at the edge of·
clearings.

See the discussion of *A. obliquum* for a comparison of
the two species.

Mexico to Peru, Bolivia, and Brazil; West Indies. In
Panama, known from tropical moist forest in the Canal
Zone (BCI and the Pacific slope), Herrera, Panamá, and
Darién and from tropical dry forest in Panamá (Taboga
Island).

See Fig. 15.

Adiantum pulverulentum L., Sp. Pl. 1096. 1753

Terrestrial, to 1 m tall. Leaves 2-pinnate; petiole and
rachis dark reddish-brown, densely pubescent-scaly, the
scales fine and curly; leaflets asymmetrical, 1.5–3 cm
long, usually of about equal width throughout but gradually curved upward to the sharply toothed, ± acute apex,
the lower edge entire and usually straight, the proximal
edge ± paralleling rachis, glabrate to scaly beneath, the
sterile leaflets and apex of the fertile leaflets coarsely
serrate. Sorus 1 (rarely 2), long, on upper margin, with
occasionally a shorter one on outer margin. *Croat 4340.*

Occasional, in the forest.

Distinguished by the long sorus on the upper edge of
each leaflet.

Mexico to Colombia, Bolivia, Brazil, and the Guianas;
West Indies. In Panama, known from tropical moist
forest in the Canal Zone, Panamá, and Darién, from
premontane wet forest in Colón and Panamá, and from
tropical wet forest in Colón.

Adiantum seemannii Hook., Species filicum
  2(5):81A. 1851

Terrestrial, usually to about 50 cm tall; rhizome short-
creeping, densely scaly. Leaves 1- or 2-pinnate; petiole,
rachis, and petiolules blackish and shiny; leaflets ovate,
acuminate at apex, oblique and obtuse to truncate or
cordate at base, to 9 cm long and 7 cm wide, glabrous
and glaucous beneath, the sterile margins coarsely and
unevenly serrate; petiolules slender, 2–7 cm long. Sori
interrupted along margin. *Aviles 79.*

Uncommon to rare, in the forest.

Not confused with any other species on the island.
Distinguished by its large, ovate, long-petiolate leaflets.

Mexico to Colombia. In Panama, known from tropical
moist forest in the Canal Zone, Bocas del Toro, Pan-

amá, and Darién, from premontane moist forest in Panamá (Panamá City), and from premontane wet forest in
Chiriquí.

ANANTHACORUS Und. & Max.

Ananthacorus angustifolius (Sw.) Und. & Max.,
  Contr. U.S. Natl. Herb. 10:487. 1908

Epiphyte; rhizome moderately short with many fine rootlets densely covered with threadlike, reddish-brown
scales; rhizome scales linear-lanceolate, attenuate, clathrate, iridescent. Leaves simple, entire, sessile, ± linear,
tapering gradually at both ends, mostly 10–30 cm long,
1–1.5 cm wide (the sterile leaves shorter and relatively
broader), ± thick, glabrous to minutely scaly. Sori exindusiate, in continuous or irregularly interrupted lines
very near margins. *Croat 4367.*

Occasional, in the forest, usually on smaller trunks and
branches moderately near the ground.

Throughout much of tropical America. In Panama,
known from tropical moist forest in the Canal Zone,
Bocas del Toro, Veraguas, and Darién, from premontane
wet forest in Chiriquí and Panamá, and from tropical wet
forest in Panamá.

ANETIUM (Kunze) Splitg.

Anetium citrifolium (L.) Splitg., Tijdschr. Natuurl.
  Gesch. Physiol. 7:395. 1840

Epiphyte with a creeping, scaly rhizome; rhizome scales
broadly ovate-attenuate, clathrate, iridescent with short-
ciliate margins; petioles well spaced with many fine rootlets densely covered with threadlike scales. Leaves simple,
entire, glabrous, sessile or nearly so, oblanceolate to obtuse, abruptly acuminate at apex (round on juveniles),
gradually tapered to base, usually less than 15 cm long
and 4 cm wide, thick, the midrib indistinct except at base,
the veins anastomosing. Sporangia scattered chiefly on
veins, forming a reticulum. *Croat 11703.*

Infrequent, in the forest, usually on tree trunks very
near the ground; often on stilt roots of *Scheelea zonensis*
(19. Palmae). Tryon (1964) reported seeing leaves to 100
cm long.

Throughout tropical America. In Panama, known from
tropical moist forest on both slopes in the Canal Zone
and from tropical wet forest in Veraguas (Atlantic slope)
and Colón (Portobelo).

See Fig. 16.

ASPLENIUM L.

The genus occurs in epiphytic habitats similar to those
of *Polypodium*. It can be distinguished by its elongate
indusiate sori along major veins on the underside of
leaves. Most species have oblique leaflets.

KEY TO THE SPECIES OF ASPLENIUM

Leaves simple .... *A. serratum* L.
Leaves 1-pinnate:
  Leaflets usually more than 6 cm long:
    Plants epiphytic; terminal leaflet similar to lateral leaflets, not pinnatifid ..... *A. falcinellum* Max.
    Plants terrestrial or epipatric (on rocks); terminal leaflet unlike lateral leaflets, pinnatifid .....
    .... *A. delitescens* (Max.) A. R. Smith
  Leaflets mostly less than 6 cm long:
    Stipe and lower rachis dark purplish, shiny; proximal half of lower edge of leaflet frequently
      entire .... *A. laetum* Sw.
    Stipe and lower rachis not dark purplish and shiny; proximal half of lower edge of leaflet usu-
      ally toothed:
      Veins mostly 1- or 2-forked; blades not very thin ....
      .... *A. auritum* Sw. var. *auriculatum* (Hook.f.) Mort. & Lell.
      Veins nearly all simple; blades thin .... *A. pteropus* Kaulf.

Asplenium auritum Sw. var. **auriculatum** (Hook.f.) Mort. & Lell., Mem. New York Bot. Gard. 15:19. 1966

Epiphyte, to 35 cm tall. Leaves 1-pinnate, glabrous; leaflets oblong, reduced above and tapered to a narrowly rounded apex, usually with a small auricle on upper side near base, to 3 cm long, less than 1 cm wide, the margins finely and ± irregularly serrate or the proximal one-fourth of the lower edge entire, the veins mostly 1- or 2-forked. Sori several, oblong, 2–5 mm long, along upper edge of lateral veins, sometimes also on auricle, sharply oblique to midrib; indusium ± lunate, thin. *Croat 11293*.

Uncommon, on trees (seen mostly near the ground) or on rocks, within the forest.

Throughout much of tropical America. In Panama, known from tropical moist forest in the Canal Zone, Bocas del Toro, Panamá, and Darién, from premontane wet forest in Chiriquí, Coclé, and Panamá, from tropical wet forest in Colón, Chiriquí, and Darién, and from lower montane wet and lower montane rain forests in Chiriquí.

Asplenium delitescens (Max.) A. R. Smith, Amer. Fern J. 66:120. 1976

Diplazium delitescens Max.

Terrestrial, to 35(43) cm tall; rhizome short-creeping, with minute, dark, fibrous scales, the petiole bases closely spaced. Leaves few, 1-pinnate; petioles 4.5–20 cm long, with a few scales similar to rhizome scales near base; lateral leaflets mostly narrowly lanceolate, gradually long-acuminate, inequilateral at base with the lower proximal edge acute and the upper proximal edge obtuse to truncate and nearly paralleling rachis, 3–12 cm long, 1–2 cm wide, irregularly serrate, sometimes with a small auricle on upper side; veins 3- or 4-forked; terminal leaflet pinnatifid. Sori linear along upper branch of lateral veins, slightly curved, nearer midrib than margin, usually on distal side of veins, much less frequently on both sides; indusia thin, broad, ending abruptly. *Croat 8643*.

Occasional, in the forest, especially older forest.

Its terrestrial habitat, nonclathrate rhizome scales, and pinnatifid terminal leaflet (unlike the lateral ones) distinguish it from *Asplenium falcinellum*, which has a similar leaflet shape. *A. falcinellum*, an epiphyte, has clathrate rhizome scales and a conform terminal leaflet.

Belize to Panama. In Panama, known from tropical moist forest in the Canal Zone (Pacific slope), Panamá, and Darién and from premontane moist forest in Panamá. See Fig. 17.

Asplenium falcinellum Max., Contr. U.S. Natl. Herb. 13:14. 1909

Epiphyte, to 40 cm long, at least sometimes pendent; rhizome scales clathrate. Leaves 1-pinnate, few; petioles mostly 8–10 cm long, ± glabrous; leaflets in 7–12 pairs, narrowly lanceolate, gradually long-acuminate, equally or inequilaterally acute at base (the lower side often more acute than the upper), 6–11 cm long, 9–20 mm wide, entire or irregularly and inconspicuously crenate, the terminal leaflets much like the lateral ones. Sori linear, subequal, on distal side of lateral veins. *Croat 12562*.

Infrequent, along trails in the old forest.

See the discussion of *Asplenium delitescens* for a comparison of the two species.

Guatemala to Panama; near sea level to reportedly 1,200 m. In Panama, known from tropical moist forest in the Canal Zone (on and adjacent to BCI), from premontane wet forest in Veraguas, and from tropical wet forest in Colón (Santa Rita Ridge) and Panamá.

Asplenium laetum Sw., Syn. 79:271. 1806

Epiphyte, usually 25–50 cm tall. Leaves 1-pinnate; petiole and base of rachis atropurpureous and shiny; blades reduced toward apex and tapered to a slender tip; leaflets ± oblong, very unequal at base, the lower edge straight to curved, entire, the proximal margin nearly paralleling rachis, the upper and distal margins serrate, the upper margin infrequently auriculate, 2–6 cm long, 8–15 mm wide, the veins simple or 1-forked. Sori several, linear-oblong, on upper side of lateral veins oblique to midrib, mostly on distal half of leaflets; indusium thin. *Croat 8475*.

Occasional, on rocks in stream beds of larger ravines; locally abundant.

Mexico to Paraguay and Venezuela; the Antilles. In Panama, known from tropical moist forest in the Canal Zone (BCI), Bocas del Toro, and Panamá and from tropical wet forest in Colón.

Asplenium pteropus Kaulf., Enum. Fil. 170. 1824

Epiphyte, to 20 cm tall. Leaves 1-pinnate; petiole and rachis with a narrow strip of foliaceous tissue on either side; leaflets oblong-elliptic, acute at apex, unequal at base with a slight auricle on upper side, 1.5–2.5 cm long, ca 8 mm wide, thin, deeply and ± evenly serrate, the veins nearly all simple. Sori several, oblong, on distal side of lateral veins oblique to midrib, usually in all but basal and apical fourth of leaflets.

Reported by Standley, but no specimens were cited and none has been seen from the island. The species has been seen in Frijoles (*Killip 12151*) and could very easily be overlooked.

Easily distinguished by the thin, deeply serrate leaflets with free veins and by the narrowly margined petiole and rachis.

Guatemala south to Venezuela, Brazil, and Bolivia. In Panama, known from lower montane wet forest (usually above 1,000 m) in Chiriquí and Darién (Cerro Pirre); known also from tropical moist forest on BCI and in San Blas and from premontane wet forest in Panamá (Cerro Campana).

Asplenium serratum L., Sp. Pl. 1079. 1753

Epiphyte, to 70 cm tall; rhizome densely scaly, the scales reddish-brown, threadlike. Leaves simple; petioles to 3 cm long, caniculate; blades oblanceolate, acute at apex, tapered at base, to 70 cm long and 9 cm wide, glabrous except for scales on lower midrib, the scales ± lanceolate, peltate, brown, the midrib raised above, the veins mostly free or branched once. Sori many, linear along distal edge of lateral veins oblique to midrib, of irregular lengths, mostly in midsection of leaf; indusium slender, thin. *Croat 9443.*

Uncommon, on tree trunks in the forest, near the ground or moderately high on the trunks.

Easily distinguished by its simple blade with slender, oblique sori and by its epiphytic habit.

Mexico to Bolivia; the Antilles. In Panama, known from tropical moist forest in the Canal Zone, Bocas del Toro, Panamá, and Darién, from premontane wet forest in Colón and Panamá, and from tropical wet forest in Colón, San Blas, Coclé, and Panamá.

BLECHNUM L.

Blechnum occidentale L., Sp. Pl. ed. 2, 1524. 1763

Terrestrial, 10–100 cm tall; rhizome densely covered with reddish-brown fibrous scales, the lowest part of stipe with lanceolate-linear scales to ca 1 cm long, the scales basi-fixed with a small auricle. Leaves 1-pinnate, slightly dimorphic; petiole and rachis glabrate or granular-scaly, also sometimes with a few larger lanceolate-linear scales, the petiole 1–35 cm long; leaflets sessile, stiff, confluent, very tapered at apex; sterile leaflets lanceolate-elliptic, acuminate, truncate to subcordate at base, to 7 cm long and 2 cm wide, crenulate to usually entire, not articulate; fertile leaflets similar except longer (to 15 cm long) and narrower (to 1.5 cm wide), curved upward and often very weakly auriculate on lower edge near rachis. Sori continuous along either side of the midrib, usually from or near base to near apex, sometimes along the rachis as well and rarely ± perpendicular to the midrib on the auricle; indusium thin, brown, directed toward the midrib. *Croat 6981.*

Locally abundant on steep exposed banks in the Laboratory Clearing and on eroded shores on the north and east sides of the island.

Widespread in New World tropics and subtropics. In Panama, widespread and ecologically variable; known from tropical moist forest in the Canal Zone, Bocas del Toro, Chiriquí, Veraguas, Los Santos, Herrera, Panamá, and Darién, from premontane moist forest in the Canal Zone, from premontane wet forest in Chiriquí, Coclé, and Panamá, from tropical dry forest in Coclé, from tropical wet forest on the Atlantic slope in Veraguas, and from lower montane wet and lower montane rain forests in Chiriquí.

Blechnum serrulatum L. C. Rich., Actes Soc. Hist. Nat. Paris 1:114. 1792

Similar to *B. occidentale*, but having the leaflets serrulate usually throughout (at least near apex), articulate at base, at least the sterile leaflets usually much narrower in relation to their width, seldom more than 1.8 cm wide. *Starry 263.*

Rare or possibly no longer present on BCI; collected once in 1931.

Mexico to Venezuela. In Panama, known only from tropical moist forest in the vicinity of BCI.

BOLBITIS Schott

Bolbitis cladorrhizans (Spreng.) Ching in Christensen, Ind. Suppl. 3:47. 1934
   Leptochilus cladorrhizans (Spreng.) Max.

Terrestrial, to 1 m tall; rhizome stout, erect to short-creeping with thin brown lanceolate-linear scales. Leaves 1-pinnate, dimorphic; petiole and rachis with scattered, thin, lanceolate scales, ridged, light brown; sterile leaflets

---

KEY TO THE SPECIES OF BLECHNUM

Leaflets crenulate to entire, not articulate; plants growing on ± exposed clay banks, locally abundant .... *B. occidentale* L.

Leaflets serrulate at least near apex, articulate; plants growing in moist areas, rare, if present, on BCI .... *B. serrulatum* L. C. Rich.

Fig. 18. *Bolbitis cladorrhizans*

Fig. 19. *Bolbitis nicotianifolia*

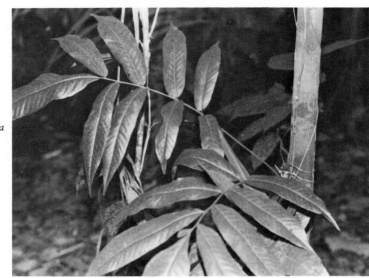

Fig. 20. *Ctenitis protensa* var. *funestra*

KEY TO THE SPECIES OF BOLBITIS

Plants terrestrial, to 1 m tall; leaflets of sterile leaves usually more than 4 cm wide (to 8 cm) . . . . . .
. . . . *B. cladorrhizans* (Spreng.) Ching
Plants creeping, epiphytic; leaflets of sterile leaves less than 3 cm wide . . . .
. . . . *B. nicotianifolia* (Sw.) Ching

---

in 5–7 pairs, ± oblong, long-acuminate, acute to obtuse or rounded and often inequilateral at base, 10–25 cm long and 2–8 cm wide, usually undulate to shallowly lobed and crenate-serrate, confluent toward apex, the veins anastomosing, with scales on midrib beneath like those on petiole; fertile leaflets narrower, oblong to strap-shaped, entire to moderately lobed, 2–20 cm long, 0.5–2.5 cm wide. Sporangia densely covering entire lower surface except midrib. *Croat 5827.*

Common in the forest, usually in creeks or on steep slopes, sometimes growing on stones in rocky stream beds. Seen in fertile condition mostly in the dry season and early in the rainy season.

Nicaragua to Colombia and Venezuela. In Panama, known from tropical moist forest in the Canal Zone and Bocas del Toro, from premontane wet forest in Chiriquí, Coclé, and Darién, and from tropical wet forest in Chiriquí.

See Fig. 18.

Bolbitis nicotianifolia (Sw.) Ching in Christensen, Ind. Suppl. 3:49. 1934
Leptochilus nicotianifolius (Sw.) Christensen

Creeping epiphyte; rhizome closely appressed to trees; scales dark brown, lanceolate-linear, the rhizome scales dense, some deciduous, the petiole scales sparse, mostly near the base. Leaves at first simple, becoming 1-pinnate, ca 1 m long; petioles shorter than blades; sterile leaflets in 1–5 pairs, oblong-elliptic to oblong-oblanceolate, narrowly acuminate, rounded to acute and decurrent at base, 15–40 cm long, 2.5–10 cm wide, entire to irregularly sinuate, the ultimate veins not uniform, the major veins arcuate, parallel to midrib; fertile leaves similar to sterile ones except much reduced, the leaflets 4.5–20 cm long, 0.8–3 cm wide. Sori continuous on underside of blade. *J. H. Faull s.n., Croat 14576.*

Apparently rare.

Honduras south to Colombia, Ecuador, Peru, and the Guianas; West Indies. In Panama, known from tropical moist forest in the Canal Zone and Panamá, from premontane wet forest in Veraguas, and from tropical wet forest in Colón (Río Guanche).

See Fig. 19.

CTENITIS Christensen

Ctenitis protensa (Afz.) Copel. var. **funestra** (Kunze) Proct., Rhodora 63:34. 1961

Epiphyte; rhizome moderately short, the petiole bases closely spaced, covered with light reddish-brown, dense, conspicuous, lanceolate-linear scales, the scales tapered to a threadlike apex, those on petiole similar but shorter. Leaves 1-pinnate-pinnatifid to 2-pinnate, to 25 cm long, tapered to a confluent apex; petioles short; leaflets ± ovate, rounded to acute at apex, merely toothed to deeply pinnatifid (increasingly more deeply dissected toward base), the lobes entire near apex, becoming increasingly toothed toward base, the lowermost lobe at least appearing to be free, itself deeply toothed or even lobed; upper surface and rachis with sessile or stalked glands. Sori minute, rounded, borne on veins below the sinuses, lacking indusium. *Croat 12664.*

Apparently rare, low on tree trunks in the forest.

Panama to Peru, Bolivia, and Brazil; West Indies. In Panama, known from tropical moist forest in the Canal Zone and Panamá and from tropical wet forest in Darién.

See Fig. 20.

Ctenitis sloanei (Poepp.) Mort., Amer. Fern J. 59:66. 1969
Dryopteris ampla (H. & B. ex Willd.) Kunze

Terrestrial, moderately tall. Leaves 2-pinnate-pinnatifid or 3-pinnate; petiole and rachis with deciduous, lanceolate scales especially at nodes; rachis and midrib of leaflets moderately scaly below, densely pubescent above with jointed trichomes; pinnae oblong-lanceolate in outline, ca 40 cm long, tapered to an acuminate apex; secondary pinnae to 10 cm long; leaflets oblong, ca 1.5 cm long, toothed or lobed nearly to middle, blunt at apex, with a few trichomes scattered on lateral veins. Sporangia in round clusters on lateral veins of lobes, in 1 or 2 ± irregular rows along midrib of leaflets. *Wilson 157.*

Collected once on Orchid Island; not seen recently.

Guatemala to Peru and Bolivia; West Indies. In Panama, known from tropical moist forest on BCI and in Panamá, from premontane wet forest in Coclé and Chiriquí, and from tropical wet forest in Colón.

KEY TO THE SPECIES OF CTENITIS

Plants epiphytic, small; leaves 1-pinnate-pinnatifid to barely 2-pinnate-pinnatifid . . . .
. . . . *C. protensa* (Afz.) Copel. var. *funestra* (Kunze) Proct.
Plants terrestrial, moderately tall; leaves 2- or 3-pinnate-pinnatifid . . . . *C. sloanei* (Poepp.) Mort.

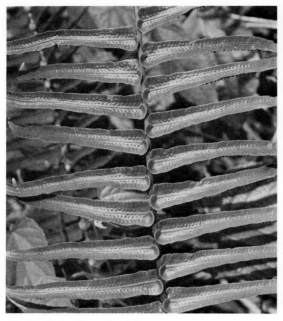

Fig. 21. *Cyclopeltis semicordata*

Fig. 22. *Dicranoglossum panamense*

Fig. 23. *Dictyoxiphium panamense*

CYCLOPELTIS J. Sm.

Cyclopeltis semicordata (Sw.) J. Sm., Bot. Mag. 72,
comp. 36. 1846

Terrestrial, to 1.5 m tall; rhizomes stout, erect, with long,
thin, reddish-brown, lanceolate-linear scales, the scales
becoming less dense along petiole, rachis, and sometimes
midrib of leaflets. Leaves 1-pinnate; petioles much shorter
than rachis, ridged; rachis densely short-pubescent; leaflets oblong-linear, acuminate, inequilateral at base with
the upper proximal edge usually truncate and the lower
proximal edge cordate-auriculate and overlapping rachis,
3–15 cm long, 1–2.2 cm wide, entire or minutely and
sharply serrate near apex. Sori round, scattered in 2 or 3
irregular rows on each side of midrib below. *Croat 4287.*

Common in the forest.

Guatemala to Peru, Bolivia, and Brazil; West Indies. In
Panama, known principally from tropical moist forest in
the Canal Zone, Bocas del Toro, Chiriquí, Veraguas, Los
Santos, Panamá, and Darién, but known also from premontane wet forest in Chiriquí and Panamá and from
tropical wet forest in Colón.

See Fig. 21.

DENNSTAEDTIA Bernh.

Dennstaedtia cicutaria (Sw.) Moore, Ind. Fil.
97. 1857
D. rubiginosa (Kaulf.) Moore

Terrestrial, 1–4 m tall; rhizome stout with dense,
threadlike, reddish-brown scales. Leaves 3- or 4-pinnate-
pinnatifid, finely lacy in appearance; petioles dark brown,
somewhat grooved, glabrate to short-pubescent especially in grooves; rachis more densely pubescent; leaflets
sparsely pubescent on both surfaces, the veins especially
pubescent, the trichomes short and stiff to longer and
multicellular; leaflets mostly to 2.5 cm long, pinnatifid,
the lobes toothed. Sori borne in sinus between lobes of
leaflets; mature indusium saucer- or cup-shaped. *Croat
6406.*

Apparently rare on BCI; collected once on sandbar in
Fuertes Cove.

Mexico to Venezuela, Bolivia, and southern Brazil.
In Panama, known only from tropical moist forest on
BCI and in Panamá and Darién, but probably more wide-
spread than present collections indicate.

DICRANOGLOSSUM J. Sm.

Dicranoglossum panamense (Christensen) Gómez,
Brenesia 8:46. 1976
Eschatogramme panamensis Christensen

Epiphyte, usually 10–40 cm tall; rhizome short-creeping,
densely scaly near petiole bases, the scales ± triangular.
Leaves simple, consisting of several long, irregular lobes;
petioles very short; lobes tapered to acuminate apex, to
1 cm wide, each with a midrib branched from center vein
and no other conspicuous venation, moderately scaly

especially on underside with small, brown, peltate scales,
the scales usually narrowly tapered on one end. Sporangia
in continuous lines along margins, especially on distal
parts of lobes. *Croat 6612.*

Common in the forest.

Reported by Standley as *Eschatogramme furcata* (L.)
Trev., which is from the West Indies and Colombia but
not known from Panama.

Honduras to Ecuador and Peru; West Indies. In Panama, known from tropical moist forest in the Canal Zone,
Panamá, and Darién, from premontane wet forest in
Panamá, and from tropical wet forest in Colón, San Blas,
and Panamá.

See Fig. 22.

DICTYOXIPHIUM Hook.

Dictyoxiphium panamense Hook., Gen. Fil. t. 62. 1840

Terrestrial, to 90 cm tall; rhizome erect with reddish-
brown, lanceolate scales, and with many fibrous rootlets.
Leaves simple, dimorphic; petioles very short or to 2 cm;
sterile blades narrowly oblong, ± entire, narrowly acute
to acuminate at apex, gradually tapered and decurrent at
base, 25–60 cm long, 4–6 cm wide, glabrous except for
a few long lanceolate scales at base of midrib on underside; fertile blades much longer and narrower, tapered
at both ends, 40–90 cm long, 1.5–3 cm wide. Sporangia
in a continuous line along margins, usually beginning
well above base and extending to apex. *Croat 6771.*

Common in some areas of the forest.

Mexico to Colombia (Department of Chocó). In Panama, known from tropical moist forest in the Canal Zone,
Bocas del Toro, San Blas, Veraguas, Panamá, and Darién,
from premontane wet forest in the Canal Zone and Panamá, and from tropical wet forest in Panamá.

See Fig. 23.

DIPLAZIUM Sw.

Diplazium grandifolium Sw., J. Bot. (Schrader)
1800(2):62. 1801

Terrestrial, to ca 1 m tall; rhizome stout, creeping, with
dark scales, the petiole bases closely spaced. Leaves
1-pinnate; petioles dark, to 45 cm long, with dark, deciduous, lanceolate scales near base and short pubescence
throughout; leaflets oblong-lanceolate, acuminate at apex,
obtuse to truncate or rounded and usually inequilateral
at base, to 17 cm long and 4.5 cm wide, reduced and confluent at apex, irregularly toothed, mostly glabrous except
for short pubescence on upper surface of canaliculate
midrib; veins 1–4-forked. Sori of irregular lengths, on 1
or both sides along lateral veins about halfway between
midrib and margin; indusium thin, narrow. *Croat 8329.*

Common in the forest.

Costa Rica to Colombia. In Panama, known from tropical moist forest in the Canal Zone and Panamá, from
premontane wet forest in Chiriquí and Panamá, and from
tropical wet forest in Colón and Panamá.

See Fig. 24.

Fig. 24. *Diplazium grandifolium*

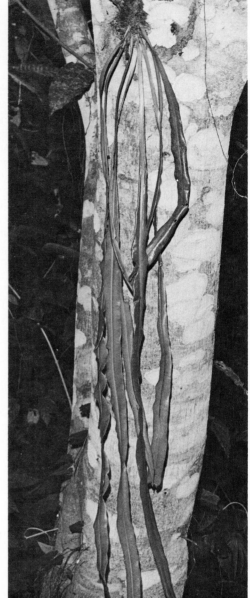

Fig. 25. *Elaphoglossum herminieri*

Fig. 26. *Elaphoglossum sporadolepis*

DRYOPTERIS Adans.

Dryopteris sordida Max., Contr. U.S. Natl. Herb. 24:60. 1922

Terrestrial, to ca 80 cm tall. Leaves 1-pinnate-pinnatifid; petioles dark brown, dull to shiny, somewhat shorter than blades; leaflets sessile, to 15 cm long and 3.5 cm wide, longest at base of leaf, oblong-lanceolate, acuminate at apex, unequal at base, the upper proximal edge obtuse, the lower proximal edge decurrent, glabrous or nearly so, lobed about halfway to midrib, the lobes bluntly toothed near their apex; basal pair of veinlets on each pair of adjacent lobes meeting sinus at or above its base. Sori round, borne on the veins in 2 rows, 1 on either side of the midrib of the lobes. *W. Knight s.n., Kenoyer 34* (*US*).

Not seen in recent years on the island.

This species is similar in appearance to species of *Thelypteris* Sect. *Cyclosorus.*

Known only from Mexico, Guatemala, and Panama. In Panama, known only from tropical moist forest on BCI.

ELAPHOGLOSSUM Schott

Elaphoglossum hayesii (Mett.) Max., Proc. Biol. Soc. Wash. 46:105. 1933

Small epiphyte; rhizomes scaly. Leaves dimorphic, simple; sterile blades linear-oblanceolate, rounded at apex, less than 6 cm long and 1 cm wide, densely and conspicuously scaly, the scales long, lanceolate-linear, light reddish-brown, inconspicuously toothed along their margins; fertile blades spatulate, shorter than sterile leaves. Sporangia covering underside of leaves except for the narrowed base. *Shattuck 593.*

The species was collected once and has not been seen in recent years. Because of its minute size and epiphytic habit, probably high in trees, it could easily be overlooked.

Nicaragua, Costa Rica, and Panama. In Panama, known only from tropical moist forest on BCI and in Panamá.

Elaphoglossum herminieri (Bory & Fée) Moore, Ind. Fil. 16. 1857

Epiphyte, to 1 m long, usually pendent; rhizome with dense, ± linear, thin, light-brown scales to 4.5 cm long and 3 mm broad. Leaves simple and entire, very rarely forked; sterile leaves sessile or with very short petiole,

the blades strap-shaped, tapered to blunt apex, very gradually long-tapered to base, 1.5–3.5 cm wide, entire, sparsely and inconspicuously scaly, the scales light brown, peltate-stellate to branched; fertile leaves reduced. Sporangia densely covering underside of leaves. *Croat 7933.*

Rare, in the forest, growing high in trees. Not seen in fertile condition on BCI.

Guatemala to Ecuador, Venezuela, and the Guianas; West Indies. In Panama, known only from tropical moist forest in the Canal Zone (vicinity of BCI) and from tropical wet forest in Colón.

See Fig. 25.

Elaphoglossum sporadolepis (Kunze) Moore, Ind. Fil. 367. 1857

Epiphyte, 20–40 cm tall; rhizome short-creeping, with dense, reddish-brown, lanceolate-linear scales. Leaves simple, dimorphic; sterile leaves nearly sessile or with a slender petiole to 10 cm long, the blades ± linear, tapered to acuminate apex, long-tapered and decurrent at base, 0.8–2.5 cm wide, entire, sparsely and inconspicuously scaly beneath with minute peltate-stellate scales; fertile leaves narrower, with sessile or minutely stalked glands above. Sporangia densely covering underside of leaves. *Croat 10941.*

Common in the forest, usually high in trees. Collected in fertile condition on BCI only in June. Elsewhere in Panama, fertile collections have also been made in September.

Panama and Venezuela. In Panama, known only from tropical moist forest in the Canal Zone (vicinity of BCI). See Fig. 26.

HEMIDICTYUM Presl

Hemidictyum marginatum (L.) Presl, Tent. Pterid. 111. 1836

Terrestrial, to 2 m tall. Leaves 1-pinnate, glabrous; petiole and rachis canaliculate to ridged; leaflets sessile, oblong, abruptly long-acuminate, inequilateral at base with the upper side ± cuneate and paralleling rachis and the lower side ± truncate, 8–36 cm long, 6–8.5 cm wide, thin, entire, the midvein submarginal, the lateral veins very close, anastomosing in distal third of blade. Sporangia in continuous lines of irregular lengths on distal edge of lateral veins about midway between midrib and margin; indusium thin, very narrow. *Kenoyer 38.*

KEY TO THE SPECIES OF ELAPHOGLOSSUM

Leaves densely and conspicuously scaly, the scales lanceolate-linear, very long . . . .
. . . . *E. hayesii* (Mett.) Max.
Leaves with minute and inconspicuous scales, especially on underside, to almost glabrate:
    Leaf scales stellate to irregularly branched, the branches longer than the width of the united part
        of the scale . . . . *E. herminieri* (Bory & Fée) Moore
    Leaf scales subentire to stellate, the branches, if present, much shorter than the united part of
        the scale . . . . *E. sporadolepis* (Kunze) Moore

Rare, occurring in ravines; not seen in recent years.

Mexico to Peru and Brazil. In Panama, known from tropical moist forest on BCI and in Chiriquí and Darién, as well as from premontane wet forest in Coclé (El Valle) and premontane rain forest in Darién (summit of Cerro Pirre).

LOMARIOPSIS Fée

Lomariopsis vestita Fourn., Bull. Soc. Bot. France 19:250. 1872

Stenochlaena vestita (Fourn.) Und.

Climbing epiphyte; rhizome stout, closely appressed, with dense, long, light-brown, lanceolate-linear scales. Leaves 1-pinnate, to 55 cm long, dimorphic; petiole and lower rachis with scales like those on rhizome; rachis narrowly winged throughout; sterile leaflets usually more than 10 pairs, oblong-lanceolate, usually acute to acuminate at apex, obtuse to truncate at base, 4–8 cm long, 1–2 cm wide, crenate-serrate throughout, glabrous except for sparse, short, gland-tipped trichomes on midrib below, reduced but not confluent at both ends; fertile leaflets linear, mostly 6–8 cm long, ca 2 mm wide. Sporangia densely covering lower surface. *Croat 5316*.

Common in the forest. Fertile during the rainy season (May to September).

Mexico to Panama. In Panama, known from tropical moist forest on BCI and in Bocas del Toro, Veraguas, and Darién, from premontane wet forest in Veraguas, from tropical wet forest in Colón and Panamá, and from lower montane wet forest in Chiriquí.

MAXONIA Christensen

Maxonia apiifolia (Sw.) Christensen, Smithsonian Misc. Collect. 66(9):3. 1916

Hemiepiphytic, closely appressed climber, usually growing from 2–4 m high on tree trunks; rhizome of terrestrial juvenile plants slender, long-creeping, densely scaly, the scales dark reddish-brown, linear-lanceolate, entire, extending only slightly onto petiole bases; rhizome of climbing leaves usually much stouter, 1.5–4.5 cm diam, densely covered with very slender, curly, light reddish-brown scales 1–2 cm or more long, the rootlets long and slender on the side nearest tree, these covered with short, stiff, threadlike scales. Leaves of juvenile plant similar to adult but smaller, to 1 m long; lowermost epiphytic leaves to 50 cm long; adult leaves dimorphic, 3-pinnate-pinnatifid to 4-pinnate; sterile blades to ca 1.5 m long, the petioles ± glabrous except for sparsely scaly base, deeply channeled on upper side, this channel extending throughout upper surface of rachis, pinnular rachis, and tertiary rachis, the canaliculate part densely pubescent except on petiole; pinnae triangular, acuminate; secondary divisions to ca 10 cm long, acuminate and sharply serrate at apex, increasingly more deeply lobed toward base, glabrous except for minute puberulence on tertiary rachis and lower part of midrib of leaflets, the lobes ± oblique, sharply serrate toward apex, the lowermost lobes narrowly obovate, free, or with a minute strip of chlorophyllous tissue along the pinnular rachis. Fertile leaves smaller than sterile leaves, wholly or only partially fertile (if partial, the basal parts fertile); petioles to 25 cm long; blades triangular, 60–80 cm long, borne above sterile leaves; lowermost pinnae to 25 cm long, the leaflets with small, ± rounded lobes each with a prominent free-ending veinlet and with a single sorus on lower surface. Sori round, 1–1.5 mm diam, the indusium round, markedly undulate, attached on one side with an inconspicuous sinus at point of attachment. *Croat 9535*.

Occasional, on trees, especially in deep ravines. All fertile plants were collected from August to October.

This species is distinguished from *Polybotrya villosula* by its three- or four-pinnate leaves and small leaflets (1–3 cm).

Guatemala, Honduras, and Panama; Jamaica and Cuba. In Panama, known only from tropical moist forest on BCI and from premontane wet forest in Panamá (Cerro Campana).

NEPHROLEPIS Schott

Nephrolepis biserrata (Sw.) Schott, Gen. Fil. no. 3. 1834

Epiphyte; rhizome and lower part of petiole with small, thin, lanceolate-linear scales, these affixed subbasally (sometimes moderately scaly throughout petiole and rachis). Leaves 1-pinnate; leaflets lanceolate-linear, long-acuminate on larger leaves, acute to acuminate on smaller ones, obtuse on lower side at base, slightly auriculate on upper side, mostly 7–15 cm long, 1.5–2 cm wide, fibrillose-squamulose and also hirtellous beneath, the margins finely crenate-serrate, the lower leaflets only slightly reduced. Sori round, along margin; indusium orbicular with a very narrow sinus, the sporangia projecting on all sides. *Croat 13106*.

KEY TO THE SPECIES OF NEPHROLEPIS

Indusium ± orbicular, the open side of sorus directed ± toward margin; sporangia projecting on all sides; leaflets commonly acute to acuminate at apex; leaves ± erect . . . . *N. biserrata* (Sw.) Schott

Indusium reniform, the open side of sorus directed toward apex; sporangia projecting on open side; leaflets commonly rounded to blunt at apex; leaves usually long-pendent . . . . *N. pendula* (Raddi) J. Sm.

Locally common, especially on tree stumps and floating logs in deep shady coves in the vicinity of Fuertes Cove.

New and Old World tropics. In Panama, known from tropical moist forest in the Canal Zone, Bocas del Toro, and Panamá and from tropical wet forest in Colón.

See fig. on p. 12.

Nephrolepis pendula (Raddi) J. Sm., J. Bot. (Hooker) 4:197. 1842

Epiphyte, often forming a large mass; rhizome with dense reddish-brown, threadlike scales and also with fewer, thin, lanceolate-linear, irregular scales, the scales affixed subbasally (rarely on petiole). Leaves 1-pinnate, broadly arching and ultimately pendent, essentially glabrous; petioles very short or to ca 30 cm long, the rachis usually 0.7–3 m long; leaflets lanceolate-linear, commonly rounded to acute at apex, rounded to subcordate at base, with upper side slightly auriculate at base, 2–5 cm long, 0.5–1 cm wide, ± glabrous, minutely crenulate. Sori discrete along margin; indusium ± thick, reniform to lunate, directed midway between apex and outer margin. *Croat 5045.*

Abundant in the forest from near the ground to the lower branches of even the largest trees. Common in old leaf bases of *Scheelea zonensis* (19. Palmae) and on dead trees at the margin of the lake. In the dead-tree habitat it generally occurs with a variety of other species, most frequently *Sobralia suaveolens* (35. Orchidaceae), *Polypodium phyllitidis*, and *P. crassifolium.*

Tryon (1964) considered this species to be inseparable from *N. cordifolia* (L.) Presl, which has erect leaves. The plant never approaches that appearance on BCI, however, but rather is always pendent.

Widespread in New and Old World tropics. In Panama, known from tropical moist forest in the Canal Zone, Bocas del Toro, Panamá, and Darién, from premontane wet forest in Chiriquí and Panamá, and from lower montane rain forest in Chiriquí (Cerro Horqueta).

See figs. on pp. 14 and 23.

PITYROGRAMMA Link

Pityrogramma calomelanos (L.) Link, Handb. Gewächse 3:20. 1833

Terrestrial, to 1.3 m tall; rhizome scales linear-lanceolate. Leaves 2-pinnate-pinnatifid (sometimes in part 3-pinnate); petiole and rachis reddish-brown and shiny; pinnae to 20 cm long, tapering to a long, slender, connivant apex, sometimes narrowly decurrent on rachis to next pinna; larger leaflets ± irregularly toothed to pinnatifid, white-waxy beneath, mostly 5–25 mm long. Sporangia dispersed over lower surface. *Croat 6921.*

Locally common in open areas, especially on steep, exposed banks.

Distinguished by the white underside of the leaves.

Most tropical areas of the New World. In Panama, known principally from tropical moist forest in the Canal Zone, Bocas del Toro, San Blas, Veraguas, Los Santos,

Panamá, and Darién; known also from premontane moist forest in Panamá (Panamá City), from premontane wet forest in Colón, Chiriquí, and Panamá, and from tropical wet forest in Chiriquí and Panamá.

See Fig. 27.

POLYBOTRYA H. & B.

Polybotrya villosula H. Christ, Bull. Herb. Boissier, sér. 2, 6:168. 1906

Hemiepiphytic, closely appressed climber growing 2–4 m high on tree trunks; rhizome of terrestrial juvenile plants slender, long-creeping, densely scaly, the scales thick, linear-lanceolate, narrowly acute at apex, dark reddish-brown, 4–8 mm long, not closely imbricated, the surface of the rhizome frequently exposed; rhizome of adult plant stout, 1–3 cm diam, always hemiepiphytic and closely appressed, very densely scaly, the scales ± linear, long-tapered, mostly 1–2 cm long, light reddish-brown, closely imbricated (the rhizome surface not visible). Juvenile leaves much like the adults but usually only 1-pinnate-pinnatifid to 2-pinnate; adult leaves much larger, to ca 1.5 m long, 2-pinnate to 2-pinnate-pinnatifid, ± triangular; petioles about as long as rachis, scaly at base, usually moderately to densely short-pubescent and deeply channeled on upper surface, both channeling and pubescence continuing onto rachis and pinnular rachis, the pubescence of canaliculate part of pinnular rachis and midrib or the leaflets often reddish-brown; pinnae narrowly triangular, the longest pair usually second from base, to 56 cm long, gradually tapered to a slender crenulate apex; leaflets glabrate on upper surface, glabrate below except on midrib or densely short-pilose throughout, ± oblong, rounded to acuminate and crenulate at apex, inequilateral at base, to 12 cm long, becoming confluent toward apex of pinnae, the upper margin of leaflets more prominently toothed or lobed. Fertile leaves smaller, entirely fertile; blades triangular, to 1.2 m long, borne above sterile leaves, 2-pinnate; pinnae triangular at base of leaf, much reduced, linear and simple near apex; leaflets linear, to ca 10 cm long near base of pinnae, ca 2–3 mm wide, glabrous above except for canaliculate midrib. Sori in a broad band along margins of leaflets at maturity appearing to cover entire surface of leaflet. *Croat 10804.*

Common in the forest along Shannon Trail and elsewhere, usually in deep ravines. Plants are usually fertile only during the early rainy season (June and July).

This species has been separated from *P. caudata* Kunze on the grounds of its more dense pubescence. Pubescence is extremely variable, however, even within populations, and further characters are needed to separate the two species.

Belize to Peru, Bolivia, Brazil, and the Guianas; Trinidad. In Panama, known from tropical moist forest in the Canal Zone and Veraguas, from premontane wet forest in Colón and Panamá, and from tropical wet forest in Chiriquí and Panamá.

See Figs. 28 and 29.

Fig. 27. *Pityrogramma calomelanos*

Fig. 28. *Polybotrya villosula*

Fig. 29. *Polybotrya villosula*

POLYPODIUM L.

The genus can be distinguished principally by its epiphytic habit and usually simple, entire, or pinnatifid leaflets with a few rows of discrete, small, round, exindusiate sori. The dominant epiphytic ferns on the island.

Polypodium ciliatum Willd., Sp. Pl. 5:144. 1810

Small, closely appressed epiphyte; rhizome slender, creeping, densely covered with small, thin, lanceolate-linear, peltate scales attenuate to a threadlike apex. Leaves simple, dimorphic, ± evenly spaced on rhizome; petiole very short or to 5 mm long; sterile blades elliptic to lanceolate or rounded, blunt to acute at apex and base, 1–4 cm long, 1–1.5 cm wide, sparsely to densely covered with peltate scales, the scales fimbriate at base and abruptly attenuate to apex; fertile blades oblong-linear, 1–4 cm long and 2–3 mm wide. Sori round, in 2 rows, 1 on each side of midrib, nearly covering blade. *Croat 6618.*

Occasional, in the forest, usually on smaller branches of trees; also occurring on tree branches along the margin of the lake.

Juvenile or aberrant collections of this species may resemble *P. tectum* Kaulf. *Bailey 513* was considered *P. tectum* by Standley.

Costa Rica to Peru, Bolivia, and Amazonian Brazil. In Panama, known only from tropical moist forest in the Canal Zone, Bocas del Toro, Panamá, and Darién.

Polypodium costaricense H. Christ, Bull. Herb. Boissier 4:660. 1896

Epiphyte; rhizome long-creeping, densely and conspicuously scaly, the scales lanceolate, long-acuminate, reddish-brown, affixed subbasally. Leaves simple, very deeply pinnatifid; petioles 1–8 (12) cm long; blades narrowly oblong-lanceolate, tapered to apex, truncate at base, 22–56 cm long and 5–8 (13) cm wide, sparsely short-pubescent but especially on midrib below; lobes 4–5 (9) mm broad, acute to narrowly rounded at apex, reduced toward apex, less so toward base. Sori round, in 2 rows, 1 on either side of midrib of lobes. *Croat 10365, Kenoyer 19.*

Apparently rare on the island; more common above 500 m elevation elsewhere.

Costa Rica and Panama. In Panama, known from tropical moist forest in the Canal Zone (Pacific slope) and Darién, from premontane wet forest in Chiriquí and Coclé (El Valle), and from tropical wet forest in Panamá.

Polypodium costatum Kunze, Linnaea 9:38. 1834

Epiphyte; rhizome short, the petiole bases very close together. Leaves simple, stiff, brittle, glabrous; petioles 2–6 cm long; blades oblong-oblanceolate, long-acuminate, gradually tapered at base, 15–25 cm long, 2.5–4 cm wide, the leafy tissue ending well above the base. Sori round, scattered on lower surface. *Croat 10991, 17384.*

KEY TO THE SPECIES OF POLYPODIUM

Leaves simple, not deeply lobed:
  Sori in 2 rows, 1 on each side of midrib:
    Leaves dimorphic, fertile leaf much narrower . . . . *P. ciliatum* Willd.
    Leaves monomorphic:
      Leaves long-acuminate, more than 10 cm long, prominently scaly below; lateral veins not
        obvious . . . . *P. percussum* Cav.
      Leaves acute to blunt at apex, less than 8 cm long, not prominently scaly below; lateral
        veins obvious . . . . *P. lycopodioides* L.
  Sori in more than 2 rows:
    Sori arranged in distinct rows parallel to lateral veins:
      Sori in a single row between lateral veins; leaf blades thick . . . . *P. crassifolium* L.
      Sori in 2 rows between lateral veins; leaf blades ± thin . . . . *P. phyllitidis* L.
    Sori not in distinct rows:
      Blades with leafy tissue extending to or almost to base of petiole, membranaceous . . . .
      . . . . *P. occultum* H. Christ
      Blades with leafy tissue ending well above base, stiff, not membranaceous . . . .
      . . . . *P. costatum* Kunze
Leaves simple and deeply lobed *or* leaves compound:
  Leaves 1-pinnate . . . . *P. triseriale* Sw.
  Leaves simple and deeply lobed:
    Lobes of blades usually more than 7 mm wide; rhizome scales ± round . . . *P. maritimum* Hieron.
    Lobes of blades mostly less than 7 mm wide; rhizome scales ± lanceolate with a caudate-
      acuminate apex:
      Blades mostly less than 25 cm long:
        Blades densely scaly below . . . . *P. polypodioides* (L.) Watt
        Blades lacking scales below . . . . *P. hygrometricum* Splitg.
      Blades mostly more than 25 cm long:
        Blades densely pubescent near margin on upper surface . . . . *P. pectinatum* L.
        Blades sparsely pubescent to glabrate near margin on upper surface . . . .
        . . . . *P. costaricense* H. Christ

Apparently rare, possibly more abundant and occurring high in trees.

This species can perhaps be confused with *P. crassifolium*, which has a round apex on its leaves.

Panama and the West Indies. In Panama, known from tropical moist forest on both slopes in the Canal Zone and from premontane wet forest in Panamá.

See Fig. 30.

Polypodium crassifolium L., Sp. Pl. 1083. 1753

Epiphyte; rhizome with many fine rootlets, each densely covered with dark, reddish-brown, threadlike scales. Leaves simple, coriaceous, glabrous; blades oblong, round to truncate or often mucronate at apex, tapered at base, the leafy tissue decurrent almost to base of petiole, 13–85 cm long, 3–5 cm wide, covered with a waxlike film when dry. Sori round, large, in single rows between lateral veins in distal third of blade. *Croat 5126.*

Abundant in the forest on larger branches or in the crotches of trees; also frequently found on tree stumps at the margin of the lake. Probably the most common fern on the island.

Mexico to Ecuador, Peru, Brazil, and the Guianas; West Indies. In Panama, known from tropical moist forest in the Canal Zone, Bocas del Toro, Veraguas, Panamá, and Darién, as well as from premontane wet forest in Bocas del Toro, Veraguas, Coclé, and Panamá and from tropical wet forest in Colón (Guásimo).

See Fig. 31 and fig. on p. 14.

Polypodium hygrometricum Splitg., Tijdschr.
Natuurl. Gesch. Physiol. 7:409. 1840
P. truncatulum Ros.

Epiphyte; rhizome long-creeping with small, narrowly triangular, reddish-brown scales. Leaves simple, deeply pinnatifid (nearly to midrib), moderately pubescent throughout with silvery, acicular trichomes; petioles and underside of midribs reddish-brown; blades ± irregularly oblong-elliptic, tapered at both ends, more so at apex, 10–30(50) cm long, 1.5–4(9) cm wide, the lobes 3–5 mm wide, often with basal lobes on mature leaves half or more as long as the longest lobe. Sori round, in 2 rows along margins of lobes. *Croat 6312.*

Uncommon, on tree trunks or rocks, usually near the ground.

Costa Rica to Colombia and Venezuela. In Panama, known only from tropical moist forest on BCI.

Polypodium lycopodioides L., Sp. Pl. 1082. 1753

Small creeping epiphyte; rhizome completely covered with reddish-brown, peltate, lanceolate scales tapered to a threadlike end. Leaves simple, glabrous, nearly sessile; blades elliptic to oblong, round to acute at apex, acute to obtuse at base, 2–8 cm long, 1–2 cm wide; lateral veins dark, prominent, branched many times near margin. Sori round, in 2 rows, 1 on each side of midrib, midway between midrib and margin. *Croat 10160.*

Uncommon, on trees in the forest and along the shore.

Mexico to Peru, Bolivia, and Brazil; West Indies. In Panama, known from tropical moist forest in the Canal Zone, Bocas del Toro, San Blas, Panamá, and Darién and from tropical wet forest in Colón and Coclé.

Polypodium maritimum Hieron., Bot. Jahrb. Syst.
34:527. 1904

Epiphyte; rhizome stout, long-creeping, with appressed, round, peltate scales having irregular margins. Leaves simple, deeply pinnatifid, glabrous; petioles 10–25 cm long; blades triangular, acuminate at apex, truncate at base, 15–60 cm long, 7–15 cm wide, the lobes 10–15 mm broad, obtuse at apex; veins areolate. Sori round, in 2 rows, in areoles along midrib of lobes. *Croat 10363.*

Collected in the Laboratory Clearing. Possibly introduced.

Costa Rica to Colombia, Ecuador, and Venezuela. In Panama, known from tropical moist forest on BCI and in Bocas del Toro, from premontane wet forest in Coclé, and from tropical wet forest in Panamá.

Polypodium occultum H. Christ, Bull. Herb. Boissier,
sér. 2, 5:7. 1905

Epiphyte; rhizome moderately long, with short, lanceolate, reddish-brown, peltate scales tapered to a long acicular apex, the petiole bases closely spaced. Leaves simple, membranaceous; blades narrowly oblanceolate, acuminate, narrowly tapered and decurrent at base, 7–40 cm long, 1.5–4 cm wide, bearing fine erect trichomes usually on both surfaces, the margins minutely irregular. Sori round, scattered over lower surface. *Croat 6281.*

Occasional, in the forest, on mossy faces of tree trunks, frequently associated with *Scheelea zonensis* (19. Palmae).

Costa Rica and Panama. In Panama, known from tropical moist forest in the Canal Zone and Darién and from premontane wet forest in Colón.

Polypodium pectinatum L., Sp. Pl. 1085. 1753

Epiphyte; rhizome long-creeping with linear-triangular, dark reddish-brown scales tapered to an acicular apex. Leaves simple, deeply pinnatifid, with short trichomes throughout; petioles 1–10 cm long; petiole and midrib usually reddish-brown; blades oblong, tapered to a narrow apex, tapered also to base, usually 30–60 cm long, 4–10 cm wide, the lobes 3–8 mm wide, broadest at base, the basal lobes of mature leaves much reduced, less than half as long as longest lobes on blade. Sori round, small, in 2 rows, 1 on either side of midrib of lobes. *Croat 6604.*

Abundant in the forest, usually high in trees on the upper side of larger limbs. Sometimes found on rocky banks along the shore.

Costa Rica to Ecuador, Peru, and the Guianas; West Indies. In Panama, known from tropical moist forest in the Canal Zone (Pacific slope) and Panamá and from premontane wet forest in Coclé (El Valle).

See Fig. 32.

Polypodium percussum Cav., Descr. Pl. 243. 1801

Epiphyte; rhizome creeping, with elongate internodes; rhizome scales peltate, lanceolate or round, reddish-brown; petiole bases well spaced on rhizome. Leaves simple, sparsely covered with round or lanceolate, reddish-brown scales, especially beneath; petioles mostly 1–4 cm long; blades oblong-lanceolate, long-acuminate, tapered to a narrowly acute base, 10–22 cm long, 1.5–3 cm wide, moderately thick, sometimes curled inward and lengthwise during the dry season. Sori round, large, in 2 rows, 1 on each side of midrib of leaf. *Croat 5494.*

Frequent in the forest and on trees along the shore. Sometimes forming dense mats on the upper side of larger tree branches.

Guatemala to Peru, Bolivia, and Brazil. In Panama, ecologically variable; known from tropical moist forest in the Canal Zone, Bocas del Toro, Chiriquí, Veraguas, Los Santos, Panamá, and Darién, from premontane moist forest in the Canal Zone, from premontane wet forest in Bocas del Toro, Veraguas, Coclé, and Panamá, from tropical wet forest in the Canal Zone, Colón, Chiriquí, and Panamá, from lower montane wet forest in Chiriquí, and from premontane rain forest in Darién (Cerro Pirre).

See Fig. 33.

Polypodium phyllitidis L., Sp. Pl. 1083. 1753

Large epiphyte; rhizome stout, moderately long, with many fine rootlets each densely covered with dark reddish-brown, threadlike scales. Leaves simple, large, glabrate; blades oblong-lanceolate, acuminate, tapered to base and narrowly decurrent nearly to base of petiole, 45–100 cm long, 8–12 cm wide, the midrib prominently raised on both surfaces. Sori round, small, in rows along both sides of each lateral vein. *Croat 16532.*

Common in the forest, especially in the crotches of tree trunks close to the ground, but also occurring on smaller limbs and more or less flat surfaces. Less frequent on dead trees at the margin of the lake in full sunlight.

Mexico to Peru, Bolivia, Paraguay, and Brazil; West Indies. In Panama, known from tropical moist forest on the Pacific slope in the Canal Zone, Veraguas, Panamá, and Darién, from premontane wet forest in Panamá, and from tropical wet forest in Chiriquí and Coclé.

See Fig. 34 and fig. on p. 23.

Polypodium polypodioides (L.) Watt, Canad. Naturalist Geol. II, 13:158. 1867

Small epiphyte with creeping rhizome; rhizome scales linear-lanceolate, the margins ciliate; petiole bases well spaced. Leaves simple, deeply pinnatifid, divided nearly to midrib; petioles mostly 2–5 cm long; petioles and underside of blades with dense, lanceolate to round, peltate, reddish-brown scales, their margins scarious and lacerate; blades oblong, thick, narrowest at apex, truncate at base, 3–8 cm long, 1–2.5 cm wide, the lobes ca 2 mm wide, the midrib densely scaly on upper surface. Sori round, in a row along margins of lobes, nearly contiguous. *Croat 5547.*

Infrequent, in the forest, usually growing on larger limbs of trees.

Southern United States to Peru and Argentina. In Panama, ecologically variable; known from tropical moist forest in the Canal Zone, Bocas del Toro, San Blas, Panamá, and Darién, from tropical dry forest in Panamá (Taboga Island), from premontane moist forest in Panamá (Panamá City), from premontane wet forest in Chiriquí, Coclé, and Panamá, from tropical wet forest in Chiriquí and Panamá, and from lower montane moist, wet, and rain forests in Chiriquí.

Polypodium triseriale Sw., J. Bot. (Schrader) 1800(2):26. 1801

Epiphyte; rhizome stout, short-creeping, with many fine, densely scaly rootlets, the rhizome scales ovate-acuminate, dull brown, often caudate-acuminate at apex. Leaves 1-pinnate, glabrous; petioles 8–25 cm long; leaflets strap-shaped, acute to acuminate at apex, obtuse at base, 8–37 cm long, 1.2–3.5 cm wide, the margins entire to minutely sinuate, the veins areolate. Sori round, in 2 or 3 rows beside midrib in areoles. *Croat 8608, Foster 1932.*

Rare, in the forest, apparently growing high in trees. Leaf size is variable.

Standley reported *Shattuck 382,* which is this species, as *P. brasiliense* Poir. It has not been determined whether *P. brasiliense* is a synonym of this species.

North America and Mexico to Peru, Bolivia, and Brazil; West Indies. In Panama, known from tropical moist forest in the Canal Zone, Bocas del Toro, Veraguas, and Darién, from premontane wet forest in Chiriquí, Coclé, and Panamá, and from tropical wet forest in Colón.

PTERIS L.

Recognized by its long, marginal, indusiate sori and its regular, rather oblique pinnae.

Pteris altissima Poir. in Lam., Encycl. Méth. Bot. 5:722. 1804

P. kunzeana J. Agardh

Terrestrial, 0.5–2.5 m tall; rhizome erect to creeping with scales narrowly triangular, thick and dark with a thin, flaky margin. Leaves 2-pinnate, rarely almost 3-pinnate at base, moderately pinnatifid; petioles about as long as blades, light brown, shiny; leaflets oblong-lanceolate, tapered to a narrowly triangular, acuminate, sharply serrate apex, lobed to the middle or beyond, glabrous or minutely appressed-pubescent, especially below, not drying shiny; main midrib awned at middle of each lobe; lobes to ca 1 cm broad, to 1.5 cm on juveniles; sterile margins entire to sharply serrate, sometimes with a broad sinus between terminal lobes; veins areolate with 2 or more areolae along the main midrib between the midribs of lateral lobes. Sori elongate, borne along margin of lobes, including sinus, except near apex of lobes; indusium narrow. *Croat 11285.*

Fig. 30. *Polypodium costatum*

Fig. 31. *Polypodium crassifolium*

Fig. 33. *Polypodium percussum*

Fig. 32.
Polypodium pectinatum

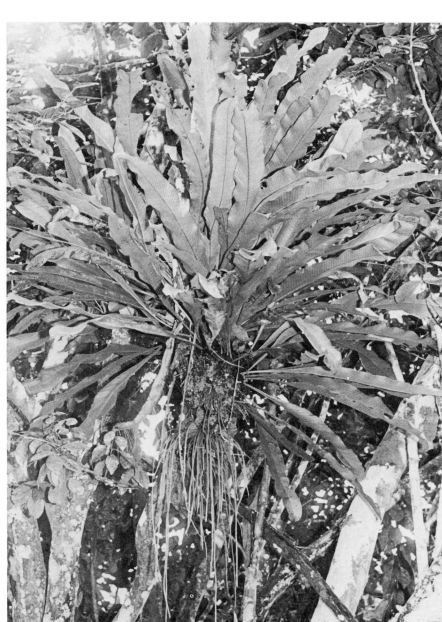

Fig. 34.
Polypodium phyllitidis

KEY TO THE SPECIES OF PTERIS

Leaves 1-pinnate, entire; veins anastomosing in distal half of their length . . . . *P. grandifolia* L.
Leaves 2-pinnate at least at base, pinnatifid; veins free or areolate near midrib:
   Veins free from one another . . . . *P. pungens* Willd.
   Veins areolate (i.e., joining together to form areolae):
      Areole single, along the main midrib between midribs of lobes . . . . *P. propinqua* J. Agardh
      Areoles 2 or 3, along the main midrib between midribs of lobes . . . . *P. altissima* Poir.

Apparently rare, in ravines.

Mexico to Brazil and Bolivia; West Indies. In Panama, known from tropical moist forest on the Pacific slope in the Canal Zone, Chiriquí, Veraguas, and Panamá, from premontane wet forest in Chiriquí, Coclé, and Panamá, from tropical wet forest in Chiriquí and Panamá, and from lower montane moist and wet forests in Chiriquí.

Pteris grandifolia L., Sp. Pl. 1073. 1753

Terrestrial, 1–5 m tall; rhizome stout, to 1.5 cm diam, long-creeping; rhizome scales dense, appressed, light reddish-brown, thin, ± entire, extending onto base of petiole. Leaves 1-pinnate; petioles very stout, about as long as blades, deeply canaliculate; rachis light-colored; leaflets simple, oblong-linear, acuminate, obtuse to acute at base, entire and minutely undulate and hyaline on margin, glabrate to minutely arachnoid-pubescent beneath; veins anastomosing in apical half. Sori borne along much of the margins; indusium narrow. *Kenoyer 50, Shattuck 1169.*

Collected twice on BCI; not seen in recent years. Possibly still present in ravines, although its large size would make it difficult to overlook.

The plant is similar to *Saccoloma elegans,* which has a darker stem and veins seldom branching more than once.

Mexico to Peru; West Indies. In Panama, known only from tropical moist forest on the Atlantic slope of the Canal Zone.

Pteris propinqua J. Agardh, Rec. Spec. Gen. Pterid. 65. 1839

Terrestrial, to 1.5 m tall; otherwise similar to *P. altissima,* except the leaflets having a single areole along the main midrib between midribs of lateral lobes, the blades usually more glossy, at least when dry, the leaf shape not as variable. *Croat 4319.*

Occasional, in the forest. Seen fertile only during the latter part of the rainy season.

The species is doubtfully distinct from *P. altissima;* the areole character separating the two tends to break down.

Mexico to Bolivia and Brazil; Jamaica. In Panama, known from tropical moist forest in the Canal Zone, Bocas del Toro, and Darién and from tropical wet forest in Colón.

See Fig. 35.

Pteris pungens Willd., Sp. Pl. 5:387. 1810

Terrestrial, 0.3–1(2) m tall; rhizome ± erect, stout, scaly, the scales linear, reddish-brown, with irregular margins.

Leaves 2-pinnate at least at base, pinnatifid; petioles scaly at base, about as long as blades, sometimes reddish and armed with minute, blunt spines; leaflets oblong-lanceolate, gradually tapered to a long, narrow, crenate-serrate apex, lobed nearly to main midrib, glabrous; main midrib with two raised margins on upper surface, the margins prominently awned near midrib of each lobe; sterile margins entire to serrate; veins free. Sori borne along margin of lobes; indusium narrow. *Kenoyer 28.*

Not seen on the island in recent years; possibly no longer present.

Mexico to the Guianas and Bolivia; West Indies. In Panama, known from tropical moist forest in the Canal Zone, Bocas del Toro, Panamá, and Darién and from premontane wet forest in Panamá (near Lago Cerro Azul) and Colón (Santa Rita Ridge).

SACCOLOMA Kaulf.

Saccoloma elegans Kaulf., Berlin. Jahrb. Pharm. 1820:51. 1820

Terrestrial, to 1.5 m tall; rhizome stout, erect to decumbent in age, with thick, shiny, dark-brown scales. Leaves 1-pinnate; petioles slightly prickly to touch near base, often scaly; petiole and rachis dark brown or purplish, shiny; leaflets lanceolate-oblong, gradually long-acuminate, obtuse to acute and often unequal at base, 15–30 cm long, 2–4 cm wide; sterile margins crenulate to serrulate; lateral veins frequently 1-forked above base occasionally in distal half of blades. Sori very numerous along margin, contiguous or nearly so, from near base often to very near apex; indusium thin, hemispherical and glabrous. *Croat 8575.*

Infrequent, usually on stream banks in the forest.

Mexico to Bolivia and Brazil; West Indies. In Panama, known from tropical moist forest in the Canal Zone (both slopes), Panamá, and Darién and from premontane wet forest in Chiriquí and Panamá.

See Fig. 36.

TECTARIA Cav.

Tectaria euryloba (H. Christ) Max., Amer. Fern J. 17:6. 1927

Terrestrial, to 90 cm tall; rhizome stout, creeping. Juvenile leaves often nearly entire; adult leaves simple, with a large apical lobe and 2 or 3 pairs of smaller lateral lobes; petioles to 40 cm long with lanceolate-linear scales nearly 1 cm long, most dense near base; blades ovate in outline,

Fig. 35. *Pteris propinqua*

Fig. 36. *Saccoloma elegans*

long-acuminate at apex, cordate at base, to 50 cm long and 35 cm wide, glabrous above, the veins on underside bearing sparse scales similar to those on petiole and appressed trichomes, the smaller veins reticulate; lobes ± oblong-elliptic, long-acuminate, 4.5–8 cm wide, the terminal lobe about twice as wide as lateral lobes (to 15 cm wide). Sori round, small, scattered ± all over lower surface. *Croat 6488.*

Occasional, on stream banks in the forest.

Guatemala to Colombia. In Panama, known from tropical moist forest in the Canal Zone, Bocas del Toro, and Darién and from tropical wet forest in Chiriquí and Panamá.

See Fig. 37.

Tectaria incisa Cav., Descr. Pl. 249. 1802

T. martinicensis (Spreng.) Copel.

Terrestrial, to 90 cm tall; rhizome stout, creeping, the petiole bases very close together. Leaves 1-pinnate; petioles slender, to 45 cm long, densely scaly, the scales fine, short; leaflets ± lanceolate-elliptic, acuminate, variable and often inequilateral at base, mostly 20–30 cm long and 5–10 cm wide, entire to undulate or irregularly lobed, glabrous or septate-pilose or hirsute on both surfaces, the basal pair of leaflets with a large, basiscopic lobe; smaller veins reticulate. Sori round, usually in irregular rows alongside each major lateral vein; indusium round-peltate, pubescent. *Croat 12299.*

Abundant throughout the forest and at the margins of

clearings. The species is probably the most abundant terrestrial fern of the island.

BCI material includes both var. *pilosa* (Fée) Mort., which is pubescent, and var. *incisa*, which is less pubescent and evidently less common.

Costa Rica to Colombia at low elevations. In Panama, known principally from tropical moist forest in the Canal Zone, Bocas del Toro, Colón, San Blas, Chiriquí, Veraguas, Los Santos, Panamá, and Darién, but also from premontane wet forest in Coclé and Panamá, from tropical wet forest in Chiriquí and Panamá, and from lower montane wet forest in Chiriquí.

See Fig. 38.

THELYPTERIS Schmidel

Thelypteris can be distinguished by its terrestrial habit, once-pinnate leaves, and usually toothed to pinnatifid leaflets. Sori are usually round, discrete, and in rows and are sometimes furnished with a peltate indusium.

Thelypteris balbisii (Spreng.) Ching, Bull. Fan Mem. Inst. Biol. 10:250. 1941

T. sprengelii (Kaulf.) O. Kuntze

Terrestrial, to ca 1 m tall; rhizome erect, the scales broad or narrow, pubescent, the rootlets numerous, fine. Leaves 1-pinnate-pinnatifid; petioles glabrous or short-pubescent, very short or to 25 cm long; rachis sparsely pubescent and also with sessile glands; leaflets sessile, the largest to

Fig. 37. *Tectaria euryloba*

Fig. 39. *Thelypteris dentata*

Fig. 38. *Tectaria incisa*

Fig. 40.
Thelypteris nicaraguensis

12(19) cm long and 1.7(2.5) cm wide, abruptly reduced at base and gradually reduced toward apex, dissected to near midrib, sparsely short-pubescent on upper surface, sometimes bearing much longer, ± jointed trichomes on underside chiefly on costule and scattered, orange, shiny, sessile glands on whole surface; lobes 2–3 mm wide, sparsely short-ciliolate. Sori round, moderately close in 2 rows, 1 on either side of midrib of lobes; indusium usually with orange glands on margin. *Croat 6691.*

Apparently rare; collected once on Orchid Island in a forest ravine.

Mexico to Panama; West Indies. In Panama, known from tropical moist forest in the Canal Zone, Bocas del Toro, and Darién, from premontane wet forest in Coclé, and from lower montane wet forest in Chiriquí (Bambito).

Thelypteris dentata (Forssk.) E. St. John, Amer. Fern J. 26:44. 1936
Dryopteris dentata (Forssk.) Christensen

Terrestrial, 55–130 cm tall; rhizome short-creeping. Petioles darkened and often purplish, 15–45 cm long, scaly at base, the scales linear-lanceolate, with dark-brown trichomes on margin; rachis pubescent; leaves 1-pinnate, 40–92 cm long (not including petiole), 14–34 cm wide, tapering evenly toward pinnatifid apex; largest leaflets 7–17 cm long, 1.1–2.7 cm wide, lobed one-half to three-fourths of the way to the base, the lowermost pairs of leaflets ± reduced and auricled at the base, the basal pair of veinlets from adjacent segments broadly united below the sinus with an excurrent vein to the sinus; pubescence of mostly short trichomes less than 0.2 mm long with scattered longer ones. Sori in 2 rows, 1 along each side of midvein of each segment; indusium pubescent. *Croat 6957, 12412.*

Frequent in open places on forest trails and at the edges of clearings; occasional on exposed shore banks.

The species is often confused with *Thelypteris quadrangularis* (Fée) Schelpe var. *quadrangularis.* Determinations of BCI specimens were made by A. R. Smith, who revised *Thelypteris* Sect. *Cyclosorus.*

Probably native to the Old World, but introduced into the United States, West Indies, and Central America. In Panama, known only from tropical moist forest in the Canal Zone.

See Fig. 39.

Thelypteris extensa (Blume) Mort., Amer. Fern J. 49:113. 1959

Terrestrial; rhizome moderately stout, long to short-creeping, scaly at apex, the scales linear-attenuate. Leaves 1-pinnate-pinnatifid; petioles usually short-pubescent, to 80 cm long, dark and scaly at base, light brown above; blades to 80 cm long; rachis and lower costae short-pubescent and with minute, gland-tipped trichomes; leaflets sessile or short-stalked, to 30 cm long and 2.7 cm wide, the longest usually at the base, dissected two-thirds to three-fourths of the way to midrib, the lower surface short-pubescent and with sulfur-yellow, sessile glands (especially on veins), the upper surface, costule, and

margin with longer, stiff trichomes; basal pair of veinlets of each pair of adjacent lobes connivent at the sinus or coming together below the sinus but not actually joining. Sori round, borne on veinlets in a row on either side of the midrib of the lobes; indusium glabrous or sparsely short-pubescent. *Croat 6517.*

Occasional, in forest ravines.

The species is easily distinguished from other *Thelypteris* species on BCI by the sulfur-yellow glands on the lower blade surface.

Native to the Old World: Sri Lanka, southern India, and Burma to southern China, and Malaysia to the Philippines. According to A. R. Smith (1971), it has been introduced and naturalized in Costa Rica, Panama, Colombia, Guyana, and Martinique. In Panama, known only from tropical moist forest in the Canal Zone and San Blas.

Thelypteris nicaraguensis (Fourn.) Mort., Contr. U.S. Natl. Herb. 38:55. 1967
Dryopteris nicaraguensis (Fourn.) Christensen

Terrestrial, 30–100 cm tall; rhizome moderately long-creeping, the petiole bases close together, but only a few leaves borne at any one time. Leaves 1-pinnate-pinnatifid; petioles as long as or much longer than blades; petiole and rachis with moderately dense, short puberulence, the trichomes simple to stalked and minutely tufted at apex; leaflets oblong-elliptic or oblong-lanceolate, pinnatifid, divided one-half to two-thirds of the way to midrib, less so toward apex, ultimately long-acuminate and entire, lobed to near base, the base rounded to cuneate, mostly 10–21 cm long, 2.5–5.5 cm wide, the veins sparsely pubescent on both surfaces, especially below, the trichomes short, stiff, sharply pointed; lobes 5–8 mm wide, curved slightly toward apex of leaflets; 2 or more pairs of connivent veins at the sinus between the lobes. Sori close, round, exindusiate, borne on veins in a row on either side of the midrib of the lobe; sporangia glabrous, interspersed with a few short acicular paraphyses. *Croat 4251.*

Abundant in the forest and at the edges of clearings.

The species is easily confused with *T. tristis* (Kunze) Tryon, which does not occur on BCI.

Guatemala to Panama; usually at less than 300 m altitude. In Panama, known from tropical moist forest in the Canal Zone, Bocas del Toro, San Blas, Panamá, and Darién and from tropical wet forest in Colón.

See Fig. 40.

Thelypteris poiteana (Bory) Proct., Bull. Inst. Jamaica, Sci. Ser. 5:63. 1953
Dryopteris poiteana (Bory) Urban

Terrestrial, to ca 1 m tall; rhizome much like *T. nicaraguensis.* Leaves 1-pinnate; petioles as long as or longer than blades; petiole and especially rachis moderately to densely puberulent, the trichomes mostly stalked-tufted; leaflets sessile, ± oblong, acuminate, rounded to obtuse at base, 11–22 cm long, 3.5–5 cm wide, shallowly lobed throughout except near apex, pubescent on both surfaces (especially underside), the trichomes irregular, sharp and stiff; lower 3 or 4 pairs of veins meniscoid (joining and

forming an excurrent vein that ends before meeting next pair of veins). Sori round, exindusiate, moderately spaced, borne on veins in rows on either side of midrib; sporangia sparsely pubescent. *Croat 6651.*

Common in the forest and at the edges of clearings.

Guatemala to Ecuador below 1,100 m; the Antilles. In Panama, known only from tropical moist forest in the Canal Zone and Bocas del Toro.

Thelypteris serrata (Cav.) Alston, Bull. Misc. Inform. 1932:309. 1932

Terrestrial or aquatic, to 1.3 m tall; rhizome very stout, ± erect, the petiole bases moderately close together, but only a few leaves borne at any one time. Leaves 1-pinnate; petioles as long as or much longer than blades, glabrous to sparsely pubescent, the trichomes simple; rachis more densely pubescent; leaflets narrowly oblong to oblong-lanceolate, acuminate, acute to rounded at base, sessile or with petiole to 1 cm long, mostly (13)20–25(35) cm long and (1.5)3–4 cm wide near middle of blades, only slightly reduced toward base, much decreased upward, the margins uncinate-serrate to entire, glabrate to sparsely pubescent especially on the veins below, the trichomes stiff, straight; veins meniscoid. Sori much longer than broad, curved, exindusiate, the sporangia borne along most opposing pairs of veinlets between the major lateral veins. *Croat 5253.*

Occasional, on eroding banks near the edge of the shore and locally common in floating masses of vegetation along the shores on the south side of the island; elsewhere in the Canal Zone, common on roadbanks.

Widespread, from the United States and Mexico to Argentina and Brazil; West Indies; usually at less than 300 m elevation. In Panama, known from tropical moist forest in the Canal Zone, Bocas del Toro, Chiriquí, and Panamá and from premontane wet forest in Veraguas and Panamá.

Thelypteris torresiana (Gaud.) Alston, Lilloa 30:111. 1960

Terrestrial; rhizome stout, ± erect; petiole bases close and few; rhizome and petiole bases with short, stiff, whitish trichomes as well as reddish-brown lanceolate-linear scales to ca 1 cm long, basifixed and often auriculate, at least 1 surface bearing pubescence like that on the rhizome. Leaves 2-pinnate-pinnatifid (at least appearing so); petioles light in color, as long as or shorter than blades; rachis grooved on upper side and with dense pubescence on the medial ridge, the ridge extending in a narrow line onto midrib of pinnae; pinnae lacy,

sessile, mostly to 12 cm long, oblong-lanceolate and long-acuminate, the leaflets toothed to deeply pinnatifid, divided to very near the base but at least appearing to be free, the upper surface with stiff acicular trichomes on veins, the lower surface bearing similar but longer, jointed trichomes and minute stalked glands. Sori small with moderately few sporangia, borne in a row on either side of midrib of leaflets; indusium small, thin, sparsely long-ciliate. *Croat 7715.*

Apparently rare; collected once along the steep, moist creek bank near Lutz Trail.

Costa Rica and Panama; Hawaii. In Panama, known only from tropical moist forest on BCI and from premontane wet forest in Panamá.

Thelypteris totta (Thunb.) Schelpe, J. S. African Bot. 29:91. 1963

Dryopteris gongylodes (Schkuhr) O. Kuntze; *Nephrodium terminans* Hook.; *T. interrupta* (Willd.) I. Wats.

Terrestrial or aquatic, to 1.8 m tall; rhizome long-creeping, black, the petiole bases widely spaced (2–9 cm apart). Leaves 1-pinnate; petioles stramineous except dark at base, to 1 m long, glabrous or scaly only near base; blades moderately stiff, 30–80 cm long, 14–60 cm wide, gradually tapered to apex; leaflets sessile or short-stalked, the largest 7–30 cm long, 1–2.2 cm wide, lobed mostly one-third to one-half of the way to midrib, the lobes nearly obtuse to rounded at apex, glabrous or scaly on underside, the scales sparse, stramineous, deltoid, the midrib sparsely short-pubescent, the upper surface often also glandular, the glands sessile; 1 or more pairs of veins connivent before reaching the sinus. Sori indusiate, often confluent when mature, in a row on either side of midrib of lobes (forming V-shaped pattern); indusium glabrous or ciliate. *Croat 13970.*

Occasional, in standing water along the shore or on floating islands of vegetation near the shore on the south side of the island.

Pantropical; in the New World in southern Florida, Central and South America, and the West Indies. In Panama, known from tropical moist forest on BCI and from tropical wet forest in Colón (Miguel de la Borda).

VITTARIA J. Sm.

Vittaria graminifolia Kaulf., Enum. Fil. 192. 1824

Superficially identical to *V. lineata,* but the rhizome scales multiseriate to apex (or with a short, uniseriate tip), the cell walls at margin of scale thinner than those in center; the paraphyses stout, distinctly club-shaped,

---

KEY TO THE SPECIES OF VITTARIA

Rhizome scales multiseriate to apex (or with short, uniseriate tip), the cell walls at margin of scale thinner than those at center; paraphyses (structures interspersed with spores) stout, distinctly clavate; spores tetrahedral-globose, trilete . . . . *V. graminifolia* Kaulf.
Rhizome scales with long, uniseriate, filiform tip with cell walls all the same thickness; paraphyses slender, linear to slightly clavate; spores reniform, monolete . . . . *V. lineata* (L.) J. Sm.

reddish-brown or darker, the apical cell enlarged; the spores tetrahedral-globose, trilete. *Croat 4011a.*

Less abundant than *V. lineata,* and doubtfully distinct.

Throughout tropical America. In Panama, known from tropical moist forest on BCI and from premontane wet and lower montane wet forests in Chiriquí.

Vittaria lineata (L.) J. Sm., Mem. Acad. Roy. Sci. (Turin) 5:421. 1793

Pendent epiphyte, 25–75(100) cm long; rhizome with many congested rootlets and scaly, the scales iridescent, lanceolate-linear, 1-costate, with irregular margins and a long uniseriate tip (filiform tip and cell walls all of same thickness), some scales also fine, threadlike, reddish-brown. Leaves simple; petioles scarcely distinguishable from blades; blades linear, 1–2 mm wide, sharp to blunt at apex, glabrous. Sporangia in deep, submarginal grooves, exindusiate; paraphyses slender, linear to slightly club-shaped, tan to light reddish-brown, the apical cells not or not much enlarged; spores reniform, monolete. *Croat 5902.*

Abundant throughout the forest, usually high in trees. Often found with other epiphytic Polypodiaceae.

Throughout tropical America. In Panama, known from tropical moist forest in the Canal Zone, from premontane wet forest in Bocas del Toro and Chiriquí, from tropical wet forest in Colón and Panamá, and from lower montane wet forest in Chiriquí.

11. SALVINIACEAE

Minute, heterosporous, floating aquatics; rhizomes horizontal, freely branched. Leaves straight in vernation, in whorls of 3, of which 2 entire and ± flattened and floating, 1 finely dissected and pendent in water (substituting for a root). Sori borne on the dissected water leaves, subtended by a basifixed indusium; microsporangia numerous, each with 64 microspores, the microspores germinating within the sporangium, the prothallia emerging through its walls as fine tubes, the tubes bearing antheridia apically; megasporangia few, each with a single megaspore, the megaspores producing prothallia bearing archegonia in a manner similar to microsporangia.

Distinguished by their floating aquatic habitat and the two flaplike leaves covered with regularly arranged clusters of trichomes.

Salvinia is often vegetatively dispersed by plants floating away. Any minute part of the stem bearing an axillary bud may produce another plant (Schulthorpe, 1967).

Two genera and 16 species; tropics, subtropics, and warm temperate regions.

SALVINIA Seg.

Salvinia auriculata was reported by Standley (1933), but all material seen has proved to be *S. radula.* It is included in the key, however, because it could easily occur on the island.

Salvinia radula Baker, J. Bot. 24:98. 1886

Very small, floating aquatic, the plants usually less than 4 cm diam; stems short. Stems and fertile leaves bearing slender, jointed trichomes; floating sterile leaves usually flat, usually elliptic to oblong-elliptic or obovate, rounded at apex, subcordate at base, mostly 1–1.5 cm long, the lower surface smooth, with numerous, slender, long, jointed trichomes, the upper surface with ± regular, close rows of short papillae, the rows directed at about a 45° angle from the midrib; papillae bearing usually 4-jointed trichomes (papillae often very reduced and represented only by the trichomes), solitary trichomes sometimes also between the rows. Fertile leaflet much divided, the segments slender, numerous. Sporocarps globular, borne in stalked clusters beneath the leaves, ca 2 mm diam. *Croat 7954.*

Frequent in quiet water at the margin of the lake.

Panama, Colombia, Ecuador, and the Guianas; Cuba. In Panama, known only from tropical moist forest in the Canal Zone and Bocas del Toro.

KEY TO THE SPECIES OF SALVINIA

Floating leaves boat-shaped or conduplicate, usually widest at the broadly cordate base, distinctly broader than long; papillae well developed, 1 mm or more long in center of leaf; leaf tissue usually glabrous between the papillae . . . . *S. auriculata* Aubl.
Floating leaves flat, broadly elliptic to oblong-elliptic to obovate, widest at or above the middle; papillae low or sometimes obsolete; leaf tissue often with single trichomes between the rows of papillae . . . . *S. radula* Baker

---

Coniferophyta

12. ARAUCARIACEAE

Dioecious or monoecious trees, evergreen and resinous; branches regularly whorled. Leaves alternate, awl-shaped, compressed, sharp-pointed. Staminate catkins cylindrical, axillary or terminal; staminate flowers with numerous, spirally arranged stamens, the filaments expanded to form a tough anther-scale; sporangia numerous, free, linear, borne on lower surface of scale; pistillate flowers in terminal heads, the carpels numerous, spirally imbricate, terminally thickened with abrupt sharp apex, the ligule ± adnate to carpel, the ovule 1, immersed in ligule; nucellus free; flowering heads becoming large globular woody cones at maturity, disintegrating.

Two genera and 38 species; tropical and subtropical areas of the Southern Hemisphere (with the exception of Africa).

ARAUCARIA Adr. Juss.

Araucaria excelsa R. Br. in Ait., Hort. Kew ed. 2, 5:412. 1813

Tree, becoming 66 m tall, glabrous; branches ± whorled at regular intervals along the stem. Leaves closely spiraled on numerous approximate branchlets along the stem; juvenile leaves subulate, sharp at apex, flattened or 3-sided, 8–15 mm long, ca 1 mm thick; adult leaves densely imbricate, lanceolate to ovate-triangular, the midrib obscure. Staminate catkins 3.5–5 cm long. Fruiting cones ovoid, 10–12.5 cm long. (Description of flowers and cones taken from Bailey, 1949.) *Croat 44758.*

Cultivated in the Laboratory Clearing. The plant is still juvenile.

Native of Norfolk Island; cultivated throughout the world.

Gnetophyta

13. GNETACEAE

Lianas. Leaves opposite, decussate, petiolate; blades simple, entire, glabrous; venation pinnate; stipules lacking. Flowers unisexual (dioecious), actinomorphic, in strobili, axillary or terminal, simple, or 1- or 2-branched, the whorls of the staminate strobili very closely spaced, subtended by a cup- or saucer-shaped collar of bracts, the whorls of the pistillate strobili separated by conspicuous internodes; perianth tubular, angulate; stamen 1; microsporangia 2, 1-celled; ovule 1, having 2 integuments. Seeds oblong, drupelike.

Though a gymnosperm, the genus *Gnetum* appears in almost every respect like a dicot. It is not, however, to be confused with any other genus.

One genus, with 30 species; humid tropics.

GNETUM L.

Gnetum leyboldii Tul. var. **woodsonianum** Markg., Ann. Missouri Bot. Gard. 52:385. 1965

Dioecious liana, glabrous, highly ramified, contorted, widely spreading (usually in a single large tree), to more than 30 m in canopy; trunk 15–20 cm diam near ground, coarse, unfissured, the nodes enlarged. Leaves opposite, decussate; petioles 5–12 mm long, canaliculate above; blades narrowly to broadly ovate, acute to short-acuminate, acute to rounded at base, 9–15 cm long, 4–10 cm wide, ± thick, the midrib usually ± arched, the margin weakly revolute. Staminate inflorescences axillary and terminal, 1- or 2-branched, the main axis of 1–4 tiers, each 1–4 cm long and terminated by 2 small, opposite, acute, connate bracts subtending a dense cluster of moniliform trichomes, a whorl of 2–9 simple strobili, and the axis of the next higher tier; side branches of 1 or 2 tiers identical to those of the main axis; staminate strobili greenish-brown, 10–18 mm long and 3 mm wide, their stalks slender, 5–10 mm long; staminate flowers numerous, in 6 or 7 whorls, 1–2.5 mm apart (uppermost whorl of a few, usually sterile, pistillate flowers, interspersed in a dense mat of white moniliform trichomes 1.5–1.8 mm long); each whorl of staminate flowers on the strobilus subtended by a collar of circular bracts, these at first cup-shaped, becoming saucer-shaped; staminate perianth obconic, brownish, angulate, 0.8 mm long and 0.3 mm wide; stamen solitary, 1.2 mm long, with 2 unilocular microsporangia. Pistillate inflorescences smaller and more reduced than staminate; strobili simple or 1-branched, the main axis of 3–5 tiers, each terminated by 2 opposite, acute, connate bracts subtending the axis of the next higher tier and a whorl of 5 pistillate flowers embedded in a dense cluster of moniliform trichomes; pistillate flowers brownish-yellow, subglobose, 0.5 mm wide and high, ovule 1 (most soon abort). Fruiting internodes 5–30 mm long, ca 2 mm wide, the nodes swollen; mature seeds oblong, 4.5–5 cm long and ca 2 cm wide, rounded on ends (apex apiculate on drying); external envelope red-orange becoming violet-purple, smooth, leathery, 2 mm thick; inner envelope brown, thin, faintly veined. *Croat 7958*.

Infrequent, in the forest. The species appears to flower twice per season, principally from February to April and at least sometimes again in August or September. The fruits from the dry-season flowering develop to mature size by July or August, whereas the wet-season flowers produce fruits in the early dry season. Foster (pers. comm.) reports that the plants flowering during August and September are not the same individuals as those that flowered in the dry season. However, there are indications that individuals may flower twice per year, since I have observed plants in April with both flowers and fruits. Because it is unlikely that the fruits persisted on the plant all year, they were probably the result of a second flowering. Foster also observed an individual, seen bearing fruits in July and August 1971, with mature-sized green fruits in January 1973. This indicates that if individual plants flower only once a year, i.e., in dry or wet season, they are not necessarily restricted to that single period.

Fruits are probably mammal dispersed. Oppenheimer (1968) reported that white-faced monkeys probably feed on the fruits.

The typical variety of *Gnetum leyboldii* is restricted to the Amazon basin, principally in Brazil, Colombia, and Ecuador. The variety *woodsonianum* extends from Coclé in Panama to northern Colombia. In Panama, it is known principally from tropical moist forest in the Canal Zone, Coclé, and Darién, but also from Cerro Pirre in Darién, which ranges from tropical wet forest to premontane rain and lower montane rain forests.

See Figs. 41 and 42.

Fig. 41. *Gnetum leyboldii* var. *woodsonianum*

Fig. 42. *Gnetum leyboldii* var. *woodsonianum*

ANTHOPHYTA
Monocotyledoneae

KEY TO THE MONOCOTYLEDONEAE

- Plants aquatic:
    Leaf blades narrow, at least 3 times longer than wide, strap-shaped, filiform or grasslike:
        Leaf blades usually ± succulent (at least at base) and basal; flowers ± large and showy, the petals ca 1 cm or more long, white:
            Leaves strap-shaped, of ± equal width throughout, obvious petiolar portion lacking; flowers bisexual, having a prominent perianth tube more than 15 cm long; perianth lobes more than 4 cm long .... 28. AMARYLLIDACEAE (*Crinum erubescens* Ait.)
            Leaves lanceolate to elliptic, the petiole obvious; flowers unisexual, lacking a perianth tube; petals less than 2 cm long .... 15. ALISMATACEAE (*Sagittaria lancifolia* L.)
        Leaf blades not succulent, usually cauline; flowers minute, greenish, usually individually inconspicuous:
            Plants completely submersed; leaves in distinct whorls of 3, less than 4 cm long and 2 mm wide, serrate; flowers unisexual in minute, axillary inflorescences ....
            .... 16. HYDROCHARITACEAE (*Hydrilla verticillata* (L.f.) Royle)
            Plants at least partly held above the surface of the water; leaves not in distinct whorls of 3, more than 4 cm long and 2 mm wide; flowers either bisexual or not, in minute axillary clusters:
                Flowers unisexual, very densely arranged into solid cylindrical inflorescences, the staminate part of the spike separated from the pistillate part by a short sterile section (commonly known as Cattail) .... 14. TYPHACEAE (*Typha domingensis* Pers.)
                Flowers usually bisexual, usually arranged in panicles or ± loose spikes, not in dense cylindrical spikes:
                    Stems usually terete with hollow internodes, never trigonous; leaves 2-ranked, with usually open sheaths; florets enveloped by 2 opposing scales; anthers versatile ...
                    .... 17. GRAMINEAE (in part)
                    Stems usually 3-sided, solid throughout; leaves 3-ranked, usually with closed sheaths; florets subtended by a simple or saccate scale; anthers basifixed ....
                    .... 18. CYPERACEAE (in part)
    Leaf blades broad, less than 3 times longer than broad, never grasslike:
        Plants free-floating on surface of water:
            Leaves more than 20 cm long, prominently petiolate; flowers showy, blue, more than 3.5 cm long .... 24. PONTEDERIACEAE (*Eichhornia*)
            Leaves less than 12 cm long, sessile; flowers inconspicuous, greenish, minute, enclosed within a spathe .... 21. ARACEAE (*Pistia stratiotes* L.)
        Plants not free-floating, rooted in soil:
            Leaf blades lobed at base:
                Plants erect, 1.5–3 m tall; stems spiny; blades with pinnate venation, veins not all closely parallel; flowers unisexual, individually small and inconspicuous ....
                .... 21. ARACEAE (*Montrichardia arborescens* (L.) Schott)
                Plants sprawling over water; stems unarmed; blades with all veins closely parallel; flowers bisexual, showy, blue .... 24. PONTEDERIACEAE (*Pontederia rotundifolia* L.f.)

Leaf blades not cordate at base:
  Plants floating on surface of water; blades ± rounded or oval, less than 7 cm long . . . . . .
    . . . . 16. HYDROCHARITACEAE (*Limnobium stoloniferum* (G. Meyer) Griseb.)
  Plants erect, to more than 1 m tall; blades lanceolate to elliptic, more than 15 cm long . . .
    . . . . 15. ALISMATACEAE (*Sagittaria lancifolia* L.)
● Plants not aquatic:
  Plants lacking chlorophyll (minute terrestrial saprophytes) . . . .
    . . . . 34. BURMANNIACEAE (*Thismia panamensis* (Standl.) Jonk.)
  Plants with chlorophyll:
    Leaves like those of banana plant (large or small but not palmlike), i.e., the lateral veins
      closely parallel, fine and all ± equal, very much less conspicuous than and not parallel
      to the midrib; leaf surface thin, not at all coriaceous:
      Petiole canaliculate on upper side near its apex, lacking a callus (swollen or discolored area);
        bracts boat-shaped and elongate, usually more than 6 cm long, directed at a broad an-
        gle to the main axis of inflorescence, congested and distichous or widely spaced and spi-
        rally arranged; inflorescence always unbranched, sometimes pendent; flowers appear-
        ing ± actinomorphic; stamens 5 or 6 . . . . 31. MUSACEAE
      Petiole terete near its apex, often with a callus; bracts of the inflorescence not elongate or if
        so less than 5 cm long, usually congested and usually spirally arranged with the spike
        ± terete (distichous in *Calathea insignis*) *or* the inflorescence compound; inflorescence
        never pendent; flowers usually asymmetrical; fertile stamen 1, the remainder trans-
        formed into petaloid stamens often more conspicuous than the inner perianth . . . .
        . . . . 33. MARANTACEAE
    Leaves not as above (either not at all banana-like or, if banana-like, with the veins all strictly
      parallel):
      Plants palmlike, the leaves often pinnately or palmately lobed or compound; blade surface
        usually plicate between major veins:
        Perianth segments 4, many, or none; ovules many; inflorescences simple, enclosed in sev-
          eral spathes; staminodia filamentous, usually conspicuous; fruit a fleshy syncarp
          (fruits united; shaped like a large screw in *Cyclanthus*); seeds minute, numerous;
          plants mostly acaulescent or nearly so . . . . 20. CYCLANTHACEAE (in part)
        Perianth segments 6 in 2 whorls; ovules 1–7; inflorescences usually compound (simple in
          Geonoma cuneata and *G. procumbens*), apparently enclosed in a single, often woody
          spathe; staminodia inconspicuous or lacking; fruits usually free from one another;
          seeds usually 1 per fruit (to 3), usually large; plants usually with a trunk (often very
          short in *G. procumbens*) . . . . 19. PALMAE
      Plants not palmlike; blades simple or if lobed the blade surface not plicate between major
        veins:
        ➧ Plants herbaceous vines:
          Plants epiphytic:
            Flowers minute, usually actinomorphic, congested into a dense spike (spadix) sub-
              tended by one or more spathes or spathe scars; fruit a berry or a syncarp of con-
              nate berries:
              Spathes 3, deciduous, leaving scars . . . .
                . . . . 20. CYCLANTHACEAE (*Ludovia integrifolia* (Woods.) Harl.)
              Spathes 1, often persistent, sometimes enveloping spadix after anthesis . . . .
                . . . . 21. ARACEAE (in part)
            Flowers not very minute, usually zygomorphic, not congested into a spike subtended
              by a spathe (in bracteate racemes or panicles); fruit a 3-valved capsule . . . .
              . . . . 35. ORCHIDACEAE (in part)
          Plants not epiphytic:
            Blades having all veins closely parallel; reticulate venation absent; flowers bisexual,
              more than 3 cm diam, blue, showy; plants growing along bodies of water . . . . .
              . . . . 24. PONTEDERIACEAE (*Pontederia rotundifolia* L.f.)
            Blades having 3 or more lateral veins paralleling midrib; reticulate venation conspic-
              uous; flowers unisexual (plants dioecious), less than 1 cm diam, greenish; plants
              usually not growing along bodies of water:
              Ovary superior; flowers in umbelliform clusters, the inflorescences usually not
                long and slender; leaves usually lacking glands; fruit a 1–3-seeded berry . . . .
                . . . . 26. SMILACACEAE
              Ovary inferior; flowers in long spikes, racemes, or narrow, elongate panicles;
                leaves often with plate-shaped glands; fruit a winged capsule . . . .
                . . . . 29. DIOSCOREACEAE (*Dioscorea*)
        ➧ Plants not herbaceous vines:

Plants grasslike or with grasslike leaves conspicuously sheathed at the base, the sheath either closed around the stem or bearing a ligule within at the apex, or both; plants never with equitant leaves (completely folded along midrib):

Flowers individually conspicuous, usually white to blue, not subtended by distichous or imbricate scales; leaves lacking ligules (minute flap of tissue extending above apex of sheath next to stem); fruits bearing 3 or more seeds . . . . 23. COMMELINACEAE (in part)

Flowers individually inconspicuous, usually not colorful, subtended by distichous or imbricate scales; leaves having ligules; fruits bearing 1 seed:

Stems usually terete with hollow internodes, never 3-sided; leaves 2-ranked, with usually open sheaths; florets enveloped by 2 opposing scales; anthers versatile . . . . 17. GRAMINEAE (in part)

Stems usually trigonous, solid throughout; leaves 3-ranked, with usually closed sheaths; florets subtended by a simple or saccate scale; anthers basifixed . . . . 18. CYPERACEAE (in part)

Plants not grasslike, usually lacking a closed sheath; plants with or without equitant leaves:

■ Plants epiphytic:

Flowers minute, individually inconspicuous, congested on a spadix subtended by a spathe; fruits usually fleshy . . . . 21. ARACEAE (in part)

Flowers usually moderately large and showy, not congested into a condensed spike subtended by a spathe; fruits generally capsular:

Leaf blades never plicate or folded, usually closely imbricated at base to form a watertight reservoir; flowers with all petals ± equal; stamens and style not fused together; each fruit a several-seeded berry or a capsule with many conspicuously comose seeds (tufts of silky trichomes on seed) . . . . 22. BROMELIACEAE

Leaf blades usually plicate or folded along the midrib, never closely imbricated to form a watertight reservoir; flowers with 1 petal modified into a lip; stamens and style fused into a column; pollen in pollinia (modified saccate structures); each fruit a capsule with very numerous, minute, inconspicuous, naked seeds . . . . 35. ORCHIDACEAE (in part)

■ Plants not epiphytic:

Flowers usually zygomorphic, with a single prominent lip, the lip usually variously shaped and colored and usually larger than the sepals and petals:

Plants rarely to 1 m tall and if so the leaf blades plicate; leaves often basal; stamens and style fused into a column; lip of flower consisting of a modified petal; pollen contained in pollinia; capsules with very numerous, inconspicuous, naked seeds . . . . 35. ORCHIDACEAE (in part)

Plants usually more than 1 m tall or if less the leaf blades not plicate; leaves either distichous or spirally arranged, never basal; stamens and style not fused into a single structure; lip of flower consisting of modified stamens; pollen in a normal anther borne on a petaloid filament; capsules with moderately large arillate seeds (more than 2 mm long) . . . . 32. ZINGIBERACEAE

Flowers usually actinomorphic or lacking a single prominent lip (1 petal reduced in *Commelina*):

Plants caulescent (with a definite stem):

Shrubs with ± woody stem, ca 2 m or more tall; leaves without a closed sheath . . . . 25. LILIACEAE (*Cordyline fruticosa* (L.) Goepp.)

Herbs usually less than 1 m tall; leaves with a closed sheath . . . . 23. COMMELINACEAE

Plants acaulescent (the leaves arising from root or rhizome):

Flowers numerous, less than 1 cm long, in a broad panicle, white; fruits red berries less than 1 cm long; plants common . . . . 27. HAEMODORACEAE (*Xiphidium caeruleum* Aubl.)

Flowers 1 to several (rarely up to 15), more than 3 cm long, arising from a bracteate scape or laterally from a leaflike peduncle; fruits capsules:

Leaves ensiform (folded along midrib), more than 50 cm long; flowers yellowish with purple markings; seeds red, arillate . . . . : . . . . 30. IRIDACEAE (*Neomarica gracilis* (Herb.) Sprague)

Leaves not folded, less than 50 cm long; flowers white or red-orange; seeds pale green or shiny and black, not arillate . . . . 28. AMARYLLIDACEAE (except *Crinum*)

14. TYPHACEAE

Rhizomatous perennial herbs. Leaves arising from rhizome, having a petiolar sheath; blades simple, linear, entire; venation parallel; stipules absent. Flowers unisexual (monoecious), actinomorphic, in a dense spadix, the staminate flowers borne toward the apex, the pistillate flowers toward the base, each flower subtended by a bractlike spathe; perianth represented by bristles; stamens about 2; filaments basally united; anthers 2-celled, basifixed, dehiscing longitudinally; ovary superior, 1-locular and 1-carpellate; ovule 1, pendulous; style simple, filiform. Fruit a nutlet with a persistent style; seeds with mealy endosperm.

Flowers are wind pollinated (Schulthorpe, 1967). The perianth has bristles that may aid in catching pollen grains.

One genus, with about 15 species; widely distributed.

TYPHA L.

Typha domingensis Pers., Synops. Pl. 2:532. 1807
> *T. angustifolia* L.
> Cattail

Large, monoecious, semiaquatic herb 1–3 m tall; rhizomes long, fleshy. Leaves equitant; blades sword-shaped, acute at apex, ensheathing stem at base, more than 1 m long, 3–10 mm wide. Inflorescences cylindrical, elongate, 1–2.5 cm diam, 50–120 cm long; flowers in dense spikes, the staminate flowers above the pistillate, separated by 3–10 cm of naked peduncle; perianth reduced to bristles; pistillate flowers usually having filiform bractlets with dilated tips; ovary stipitate, 1- or 2-celled. Mature inflorescences becoming fluffy with loosening comose nutlets; mature nutlets ca 1 mm long, attached to a slender stipe ca 5 mm long. *Croat 4671*.

Rare, along the shore. Fertile in the dry season.

This linear-leaved aquatic is confused with no other plant.

The persistent perianth bristles assure wide dispersal in the wind.

Widely distributed in tropics and subtropics throughout the world. In Panama, in marshy areas; known from tropical moist forest in the Canal Zone and Bocas del Toro and from premontane dry forest in Los Santos.

15. ALISMATACEAE

Herbs. Leaves basal, petiolate, simple, with small scales in axils; blades entire, glabrous; venation parallel, converging. Flowers unisexual (in ours), actinomorphic, in axillary whorls; sepals 3, free, imbricate; petals 3, free, imbricate, white; stamens many (in ours), free; anthers 2-celled, extrorse or dehiscing by lateral slits; pistils many; ovaries superior, unilocular; ovules solitary, basal, anatropous or amphitropous; styles and stigmas continuous, acicular. Fruits achenes, the mature ovules lacking endosperm.

Thirteen genera and about 90 species; widely distributed.

SAGITTARIA L.

Sagittaria lancifolia L., Syst. Nat. ed. 10, 1270. 1759

Glabrous, monoecious, aquatic herb, usually 1–1.5 m tall. Leaves basal; petioles fleshy, vaginate in lower third, becoming terete above, usually longer than blades; blades lanceolate to elliptic, gradually tapered to both ends, mostly 25–40 cm long, 4.5–10 cm wide. Inflorescences held above leaves, usually branched; flowers in whorls of 3 at each node, those at the lower nodes usually pistillate, those at the upper nodes staminate; bracts 3 at each node, lanceolate, 5–20 mm long; pedicels 1–3 cm long, ascending; sepals 3, ovate, to 1 cm long; petals 3, obovate, white, to 1.3 cm long, spreading at anthesis; stamens many; filaments free, pubescent in lower half. Fruit a cluster of numerous achenes, the clusters depressed-globose, ca 1 cm wide; achenes ± flattened, oblique, beaked at apex, to 2.5 mm long. *Croat 8446*.

Uncommon, but locally abundant in marshes. Flowers and fruits November to July, principally March to July.

Seeds are probably dispersed by small shorebirds, though many are no doubt spilled locally.

Throughout tropical and subtropical regions of the New World. In Panama, known from marshy areas of tropical moist forest in the Canal Zone, Bocas del Toro, and Darién and from premontane moist forest in the Canal Zone and Panamá.

16. HYDROCHARITACEAE

Submerged, glabrous, aquatic herbs. Leaves whorled (*Hydrilla*) or basal (*Limnobium*); blades simple, entire or serrulate; stipules lacking. Flowers unisexual (dioecious in ours), actinomorphic, the pistillate flowers solitary, the staminate flowers solitary or in umbels arising from axillary spathes; sepals 3, free, imbricate; petals 3, free, imbricate; stamens 3 (*Hydrilla*) or 6–12 (*Limnobium*), free, alternate with the petals; anthers 4-celled, basifixed, dehiscing longitudinally; ovary inferior, 1-locular, 3- or 6-9-carpellate; placentation parietal; ovules many, ana-

KEY TO THE SPECIES OF HYDROCHARITACEAE

Leaves sessile, in whorls of 3–8 along stem; blades 1–2 mm wide and at most 4 cm long, serrulate
.... *Hydrilla verticillata* (L.f.) Royle
Leaves basal with terete petioles 7–15 cm long; blades ± oval, 4–6 cm long, entire ....
.... *Limnobium stoloniferum* (G. Meyer) Griseb.

tropous; styles as many as carpels. Fruits berrylike, bearing 2-6 seeds, the seeds lacking endosperm.

Pistillate flowers come to the surface of the water by the elongating pedicel, and the perianth opens. Sexual parts remain dry as the flower floats on the surface. Staminate flowers break loose from the plant and float to the surface; when the flower opens, the spring-loaded stamens discharge pollen into the air (Ernst-Schwarzenbach, 1945). The pollen must contact the pistillate flowers directly as it falls, because floating pollen cannot reach the stigma of the pistillate flower.

Any part of the plant of *Hydrilla verticillata* is capable of regenerating a new plant (Schulthorpe, 1967). Vegetative reproduction is probably important in the dispersal of this genus, as it is in similar aquatics such as *Elodea* (van der Pijl, 1968).

Fourteen genera and about 100 species; mostly the warmer regions of the world.

HYDRILLA L. C. Rich.

Hydrilla verticillata (L.f.) Royle, Ill. Bot. Himal. 376. 1839

Dioecious, perennial, submerged, freshwater herb; main axis erect, rooting basally, the roots unbranched; stems slender, weak, branched. Leaves sessile, linear to elliptic to obovate, acute at apex, 1–2(4) cm long, 1–2 mm wide, serrulate, translucent, green, having only a midvein; lower leaves alternate or in whorls of 3, the middle and upper leaves in whorls of 3–8, densely clustered and rosette-like at apex of stems; inflorescences axillary, 1-flowered, sessile or subsessile; spathes of 2 connate bracts; staminate spathes solitary, subsessile, globose, in leaf axils, 1.2–1.5 mm long, tearing open when the flower is released; pedicels 1–2 mm long; sepals and petals 3, white, sometimes tinged with red; sepals 1.5–3 mm long and 1 mm wide; petals 2–3 mm long and 0.5 mm wide; stamens 3, having slender, short filaments, the anthers linear. Pistillate spathes ca 5 mm long, sepals 3, white, sometimes with red dots, 1.5–3 mm long and 0.7 mm wide; petals 3, white, 1.5–3 mm long, 0.3–0.5 mm wide, the long hypanthium 1.5–10 cm long; ovary 3–4 mm long; styles 3, 0.8–1 mm long. Fruits elongate, constricted between pairs of seeds, 4–7 mm long, basally surrounded by remnants of the spathe and apically bearing remnants of the hypanthium (1.5–3 cm long); seeds 2–6, in a row, cylindrical, oblong, apiculate on one side, 2–3 mm long, smooth, dark brown. *Croat 7909.*

Extremely abundant around the edges of Gatun Lake. Seasonality not determined.

The habit of *Hydrilla* is much like that of *Elodea,* which has not been found on the island. There is no single known vegetative character that may distinguish the two genera with certainty.

Widely distributed in the Old World: from southern and eastern Europe, Africa, Asia, and Australia. In Panama, known from tropical moist forest in the Canal Zone; these collections are the first from the Americas.

LIMNOBIUM L. C. Rich.

Limnobium stoloniferum (G. Meyer) Griseb., Fl. Brit. W. Ind. 506. 1862

Dioecious, floating or loosely rooted, aquatic herb, stoloniferous. Leaves basal, floating on surface of water and rosulate; petioles terete, fleshy, 7–15 cm long (shorter on juveniles); blades oblong-oval to rounded-oval, 4–6 cm long, 3–3.5 cm wide, obtuse to rounded at apex, acute to truncate at base, entire, spongy-reticulate on lower surface. Flowers unisexual; staminate spathes pedunculate with 2 or 3 long-pedicellate flowers; sepals 3, lanceolate, spreading; petals 3; stamens 6–12; filaments subulate, shorter than the linear anthers. Pistillate spathes of 2 bracts, bearing a short-pedicellate flower; ovary 6–9-carpellate; stigmas as many as carpels, 2-parted. Fruits baccate, many-seeded. *Shattuck 401.*

Rare, around the margin of the lake. Collected once by Shattuck but recently seen also by Dressler (pers. comm.). Seasonality not determined.

Shattuck 401 was reported by Standley (1933) as *Hydrocleys nymphoides.* The species was apparently overlooked by C. den Hartog in the *Flora of Panama* treatment (1973).

Guatemala, Panama, South America; West Indies; doubtless also in other parts of Central America. In Panama, known only from Gatun Lake.

17. GRAMINEAE (POACEAE)

Annual or perennial herbs, rarely woody and shrub- or vinelike (bamboos), sometimes aquatic, rarely armed (*Guadua*); stems terete, often arising from stolons or rhizomes. Leaves alternate, 2-ranked, petiolate; petioles sheathing, ligulate; blades simple, entire; venation parallel. Flowers bisexual or unisexual (monoecious or in *Gynerium* dioecious), more or less zygomorphic, in usually terminal panicles, racemes, or spikes; flowers compounded in the highly modified spikelets consisting generally of 2 bracts, the glumes subtending a rachilla, the rachilla bearing 1 to many sessile flowers, each flower generally subtended by usually 2 bracts, the lemma, and the palea; perianth reduced to 2 or sometimes 3 inconspicuous lodicules; stamens 3 or sometimes 6, free; anthers 2-celled, dehiscing longitudinally; ovary 1, 1-locular, 3-carpellate; ovule 1, usually anatropous; styles 2 or rarely 3, simple; stigmas generally plumose. Fruit a caryopsis (in *Sporobolus* the pericarp free from the seed and subachenial); seeds having starchy endosperm.

Flowers are principally wind pollinated (Faegri & van der Pijl, 1966), though some forest grasses such as *Olyra* and *Lithachne* are visited and perhaps pollinated by insects (Soderstrom & Calderon, 1971). Both of these genera are monoecious and have plumose stigmas. It is believed that the pistillate flowers may receive pollen accidentally from insects that visit the staminate flowers

KEY TO THE TAXA OF GRAMINEAE

Plants bamboolike or at least with one, often woody main culm (*Chusquea* only slightly woody), the branches much smaller than the main culm, often fasciculate or whorled at nodes:

Main culms more than 1 cm wide, often armed; cultivated at the Laboratory Clearing or rare in the forest . . . . *Bambusa*

Main culms less than 1 cm wide, usually much less, unarmed; common in the forest and along the shore:

Blades lacking a prominent tuft of trichomes near the base on lower surface (puberulent throughout below and with longer but not tufted trichomes near base); all veins ± equal; plants usually growing in dense clumps . . . . . *Rhipidocladum racemiflorum* (Steud.) McClure

Blades densely tufted near the base on lower surface (otherwise only minutely scabridulous); midrib more prominent than other veins; plants growing in small clumps or solitary . . . . *Chusquea simpliciflora* Munro

Plants herbaceous (woody with subglobose spikelets in *Lasiacis*), the branches when present not much smaller than the main culm:

Inflorescence a solitary spike or a solitary spikelike raceme or panicle:

Inflorescence with conspicuous, threadlike bristles or awns (longer than the spikelet proper):

Spikelets subtended by straight bristles (more than 1 per spikelet except sometimes in *Setaria vulpiseta*):

Inflorescence a true spike; spikelets sessile, enclosed in burs subtended by a ring of retrorsely barbed bristles . . . . *Cenchrus brownii* R. & S.

Inflorescence a spikelike panicle; spikelets subtended by a few, antrorsely barbed bristles . . . . *Setaria*

Spikelets lacking bristles:

Awns less than 2 cm long, geniculate, then straight to the apex; plants of clearings:

Racemes to 5.5 cm long; spikelets lacking transverse ridges, pubescent throughout; awns less than 8 mm long . . . . *Polytrias amaura* (Miq.) O. Kuntze

Racemes usually more than 6 cm long; spikelets with prominent transverse ridges, densely pubescent only near base; awns ca 15 mm long . . . *Ischaemum rugosum* Salisb.

Awns 2 cm or more long, often prominently curled at apex; plants growing only within the forest:

Blades linear, more than 30 cm long, usually less than 2.2 cm wide . . . . *Streptogyne americana* C. E. Hubb.

Blades mostly narrowly lanceolate-elliptic, less than 30 cm long, more than 2.5 cm wide . . . . *Streptochaeta*

Inflorescence lacking conspicuous, threadlike awns or bristles (the spikelet sometimes narrowed to a point but this seldom as long as the spikelet proper):

Inflorescences less than 3.5 cm long; spikelets obovate, in pairs; plants creeping, often small; growing in clearings . . . . *Paspalum decumbens* Sw.

Inflorescences more than 5 cm long; spikelets not obovate; plants erect:

Inflorescence a spikelike panicle, the branches slender, bearing several pedicellate spikelets; spikelets slender, acuminate, not indurate; plants generally aquatic:

Lemma narrowly acuminate, scabrid on veins . . *Hymenachne amplexicaulis* (Rudge) Nees

Lemma acute to blunt, usually glabrous . . . . *Sacciolepis striata* (L.) Nash

Inflorescence racemose; spikelets indurate; plants not aquatic:

Blades less than 20 cm long; spikelets in groups of 4 on a short stipe, the rachis not cylindrical . . . . *Anthephora hermaphrodita* (L.) O. Kuntze

Blades more than 20 cm long; spikelets paired on a thick, jointed, cylindrical rachis . . . . *Rottboellia exaltata* (L.) L.f.

Inflorescence paniculate or of more than 1 raceme:

● Inflorescence of paired or digitate spikes or racemes or of racemes clustered very near apex; racemes not both densely fuzzy-pubescent and subtended by narrow spathaceous bracts:

Spikelets with conspicuous awns, the awns often geniculate:

Florets on spikes, the spikes several, digitate, not merely closely aggregated at apex; spikelets with 1 fertile floret and a rudiment of several sterile lemmas . . . . *Chloris*

Florets on racemes, the racemes paired or several merely closely aggregated at apex but not digitate:

Racemes merely crowded at apex of inflorescence, not paired; spikelet with a prominent, deep, round pit in the glume . . . . *Bothriochloa pertusa* (L.) A. Camus

Racemes paired; spikelet without depression in the glume:

Racemes usually less than 3 cm long, pubescent throughout with reddish-brown trichomes; inflorescences many, diffuse on plant; peduncles mostly less than 6 cm long . . . . *Hyparrhenia rufa* (Nees) Stapf

Racemes usually more than 3 cm long, nearly glabrous or pubescent with white trichomes; inflorescences few; peduncles mostly more than 6 cm long . . . . *Ischaemum*

Spikelets without awns:
  Spikelets 2 or more at each node of the rachis . . . . *Digitaria*
  Spikelets solitary at each node of the rachis:
    Racemes digitate, 4–7 per inflorescence . . . . *Cynodon dactylon* (L.) Pers.
    Racemes not digitate, paired or congested near apex:
      Spikelets of 3 or more florets each . . . . *Eleusine indica* (L.) Gaertn.
      Spikelets of 1 floret each:
        Spikelets 2 or more times as long as broad, acute at apex; at least some inflorescences with other racemes below the apical pair . . . .
        . . . . *Axonopus compressus* (Sw.) Beauv.
        Spikelets less than 2 times as long as broad, usually rounded or blunt at apex; most inflorescences with a single pair of racemes:
          Spikelets usually less than 2 mm long, with long, fine, white trichomes on margin; plants stoloniferous . . . . *Paspalum conjugatum* Bergius
          Spikelets usually more than 3 mm long, the margins glabrous; plants rhizomatous . . . . *Paspalum notatum* Flugge
• Inflorescence paniculate or racemose with the racemes not clustered near the apex of the peduncle; racemes densely fuzzy-pubescent (the trichomes longer than spikelets) and subtended by narrow spathaceous bracts:
  ◆ Spikelets conspicuously awned or long-acuminate, appearing awned, or subtended by long bristles and/or the inflorescence conspicuously long-pubescent, having an overall fuzzy appearance:
    Leaf blades mostly more than 30 cm long:
      Spikelets subtended by long firm bristles . . . . *Setaria*
      Spikelets not subtended by long firm bristles (sometimes with tufts of silky trichomes):
        Blades more than 1 cm wide; plants usually aquatic or growing near water (at least on BCI):
          Blades more than 1 m long and 4 cm wide; panicles usually more than 1 m long
          . . . . *Gynerium sagittatum* (Aubl.) Beauv.
          Blades usually less than 75 cm long and 3.5 cm wide; panicles usually less than 60 cm long:
            Spikelets not indurate, not awned but the lemma long-acuminate; lemma ± glabrous, spreading at maturity to expose long silky trichomes; frequent on shore . . . . *Phragmites australis* (Cav.) Trin.
            Spikelets indurate, ridged, awned; lemma hispidulous, not spreading at maturity, without long silky trichomes; rare or absent from BCI . . . .
            . . . . *Oryza latifolia* Desv.
        Blades mostly less than 1 cm wide; plants aquatic or terrestrial:
          Spikelets awnless:
            Inflorescence branched once, usually much longer than broad, the branches (racemes) floriferous to the base; plants ca 4 m tall; racemes with tufts of trichomes 1 cm or more long at base of spikelets; racemes not arising from persistent sheaths . . . . *Saccharum*
            Inflorescence branched many times, nearly as broad as long, the branches long, floriferous only at apex; plants usually less than 1.5 m tall; racemes with scattered trichomes usually less than 1 cm long; racemes arising from persistent sheaths . . . . *Andropogon bicornis* L.
          Spikelets awned:
            Racemes not arising from spathes; trichomes at base of spikelets much shorter than spikelets . . . . *Bothriochloa intermedia* (R. Br.) A. Camus
            Racemes arising from (and often partially enveloped at base by) conspicuous narrow spathes; trichomes at base of spikelets usually much longer than the spikelets:
              Racemes scattered in a long loose inflorescence; spathes usually longer than the racemes; ultimate branches glabrous or nearly so just below the spathes . . . . *Andropogon virginicus* L.
              Racemes aggregated in a dense compound inflorescence; spathes often shorter than the racemes and obscured by them; ultimate branches densely villous just below the spathes . . . . *Andropogon glomeratus* (Walt.) B.S.P.
    Leaf blades mostly less than 30 cm long:
    ⊙ Inflorescence less than 6 cm long *or* of few (usually less than 8) widely spaced racemes each less than 4 cm long:
      ■ Spikelets conspicuously awned, the awns many times longer than spikelets:
        Leaf blades mostly less than 5 mm wide, the awns geniculate and twisted . . . .
        . . . . *Schizachyrium brevifolium* Kunth
        Leaf blades mostly more than 5 mm wide, the awns ± straight . . . . *Oplismenus*

■ Spikelets not conspicuously awned (glumes usually long-acuminate):
  Inflorescence with silky pubescence much longer than spikelets; fruits not white
    and shiny . . . . *Andropogon leucostachyus* H.B.K.
  Inflorescence glabrous or inconspicuously pubescent; fruits white and shiny:
    Plants less than 60 cm tall; terminal inflorescence entirely staminate or want-
      ing, the axillary inflorescences each with 1 pistillate spikelet and several
      staminate spikelets about as long as the pistillate and arising from nearly
      the same point; fruits truncate at apex, solitary . . . .
      . . . . *Lithachne pauciflora* (Sw.) Poir.
    Plants 1.5–5 m tall; inflorescences terminal or upper axillary, each branch with
      a single pistillate spikelet and with several much shorter staminate ones on
      branches below; fruits acute at apex . . . . *Olyra latifolia* L.
  ◉ Inflorescence more than 6 cm long, not of widely spaced racemes:
    Plants of the forest; inflorescence not conspicuously pubescent to the naked eye, not
      feathery in appearance:
      Petioles very short; blades lacking cross-veins; spikelets on short branches usually
        less than 3 cm long; glumes very long-acuminate, often 1–2 cm long; fruits
        white, shiny, indurate . . . . *Olyra latifolia* L.
      Petioles usually 1–3 cm long; blades with conspicuous cross-veins; spikelets near
        the ends of slender branches usually more than 6 cm long; glumes merely
        acuminate; fruits inconspicuous . . . . *Orthoclada laxa* (L. C. Rich.) Beauv.
    Plants of clearings or lakeshores; inflorescence conspicuously pubescent, ± feathery:
      Plants usually 2 m or more tall; racemes 15–30 cm long, not emerging from nar-
        row persistent spathes; spikelets very numerous, not awned, borne in 2 rows
        on 1 side of a flattened rachis . . . . *Paspalum saccharoides* Nees
      Plants 1–1.5 m tall; racemes usually less than 10 cm long, emerging from slender,
        persistent spathes; spikelets few, awned, borne on a slender, terete, broadly
        sinuate rachis . . . . . . *Schizachyrium microstachyum* (Desv.) Roseng., Arr. & Izag.
◆ Spikelets neither conspicuously awned nor long-acuminate so as to appear awned; inflorescence not conspicuously long-pubescent and therefore not having an overall fuzzy
  appearance:
  ▲ Inflorescence with primary lateral branches strictly racemose or with slender, ± uni-
    form, raceme-like panicles (the inflorescence thus at least appearing 1-pinnate):
    Uppermost primary lateral branches of inflorescence paired or digitate (arising from
      same point at apex of inflorescence):
      Spikelets solitary at each node of rachis . . . . *Axonopus compressus* (Sw.) Beauv.
      Spikelets paired at each node of rachis . . . . *Digitaria*
    Uppermost primary lateral branches not digitate:
      Spikelets with 3 or 4 florets; lemmas conspicuously flattened, imbricate, ciliate on
        inner margin, sometimes awned . . . . *Leptochloa virgata* (L.) Beauv.
      Spikelets with 1 floret, convex on at least 1 side; lemmas not ciliate or awned:
        Spikelets ± ellipsoid (broadest at or near middle) to obovoid, mostly obtuse to
          rounded at apex (acute at apex in *Paspalum repens*), convex on one side, flat-
          tened on the other (*Brachiaria mutica* may appear convex/flat but is not) . . .
          . . . . *Paspalum*
        Spikelets ± ovoid, acute at apex, convex on both sides:
          Leaf blades ovate-lanceolate, less than 5 cm long; spikelets with 2 or more light,
            pinpoint-size glands on the sterile lemma . . . . *Panicum pulchellum* Raddi
          Leaf blades ± linear, mostly more than 5 cm long; spikelets lacking crateri-
            form glands:
            Rachis or pedicel with hispid trichomes ca 1 mm long:
              Spikelet less than 1.5 mm long; primary lateral branches strictly racemose,
                the rachis with hispid pubescence among spikelets . . . .
                . . . . *Panicum pilosum* Sw.
              Spikelets more than 2 mm long; primary lateral branches appearing race-
                mose but spikelets borne on very short secondary branches, these
                with sparse, long, stiff trichomes . . . . . *Brachiaria mutica* (Forssk.) Stapf
            Rachis and pedicel lacking long-hispid pubescence:
              Rachis with wing broader than width of spikelets . . . *Paspalum repens* Bergius
              Rachis wingless or wing narrower than width of spikelets:
                Spikelets less than 1.5 mm long, usually 2 or more at each node of ra-
                  chis, ± spreading from slender rachis . . . .
                  . . . . *Panicum milleflorum* Hitchc. & Chase
                Spikelets more than 2 mm long, solitary at each node of rachis, closely
                  fitted into winged sinuous rachis . . . . *Paspalidium geminatum* Stapf

▲ Inflorescence with branches compounded 2 or more times, diffuse, open, lacking distinctly racemose branches:
  Blades distinctly long-petiolate above sheath, with prominent cross-veins when dry:
    Spikelets green, uniform (all bisexual), borne near apex of very slender branches; blades mostly less than 17 cm long; fruits enclosed, not emerging, essentially glabrous . . . . *Orthoclada laxa* (L. C. Rich.) Beauv.
    Spikelets brown, unisexual (staminate and much larger pistillate spikelets paired at nodes); blades more than 15 cm long; fruits densely pubescent at apex, emerging . . . . *Pharus*
  Blades lacking petiole above sheath, lacking cross-veins:
    Plants with stout culms, erect, usually more than 1.5 m tall, usually unbranched:
      Blades cordate-clasping at base . . . . *Lasiacis procerrima* (Hack.) Hitchc.
      Blades not cordate-clasping at base:
        Fruits not at all transversely rugose; ligule glabrous or only shallowly fringed at apex, the cilia scarcely longer than the ligule itself; plants aquatic:
          Blades less than 40 cm long and 3 cm wide; spikelets obovate to subglobose, more than 2 mm wide . . . . *Panicum mertensii* Roth
          Blades to 1 m long and 6 cm wide; spikelets oblong-elliptic, less than 1 mm wide . . . . *Panicum grande* Hitchc. & Chase
        Fruits at least weakly transversely rugose; ligule long ciliate from below the apex, the cilia at least twice as long as the ligule itself; plants usually in clearings:
          Blades plicate, more than 3 cm wide . . . . . . *Setaria paniculifera* (Steud.) Fourn.
          Blades not plicate, less than 3 cm wide . . . . *Panicum maximum* Jacq.
    Plants mostly less than 1 m tall or vinelike and clambering:
      Plants with stout, ± woody culms, clambering; spikelets ± round; second glume and lemmas with a tuft of woolly trichomes at apex . . . . *Lasiacis*
      Plants not with stout, ± woody culms, clambering or erect; spikelets round or not, lacking tufts of trichomes:
        Mature spikelets purplish:
          Plants aquatic; spikelets strongly hispid-ciliate, lacking glumes . . . . *Leersia hexandra* Sw.
          Plants not aquatic; spikelets glabrous, not ciliate, having glumes . . . . *Panicum fasciculatum* Sw.
        Mature spikelets green:
          Spikelets with several flowers, more than 6 times as long as wide; lemmas long-ciliate; small, creeping to erect plant . . . . *Ichnanthus*
          Spikelets with 1 fertile flower, usually less than 3 times as long as wide; lemmas not long-ciliate:
            Spikelets more than 3 mm long:
              Spikelets having second glume and sterile lemma laterally compressed and keeled at apex; fruits white, with a small, usually green crest at apex; usually found in moist, open areas . . . . *Acroceras oryzoides* Stapf
              Spikelets not laterally compressed and keeled only at apex; fruits white or tan, lacking a crest:
                Spikelets glabrous on glumes (i.e., on outside), villous near margin of lemmas; fruits inconspicuous, lacking scars or wings near base . . . . *Homolepis aturensis* (H.B.K.) Chase
                Spikelets having glumes minutely scabrid on keel, often throughout, the lemmas glabrous; fruits exposed at maturity, tan, bearing scars or narrow wings near base (a continuation from the rachilla or secondary axis of the inflorescence) . . . . *Ichnanthus*
            Spikelets less than 2 mm long:
            ★ Plants inhabiting marshes, floating islands, ditches, wet thickets, or sandbars at the margin of the lake:
              Leaf blades less than 5 cm long; spikelets obovate, blunt at apex; glumes frequently pubescent; fruits plano-convex, densely pubescent . . . . *Isachne polygonoides* (Lam.) Doell
              Leaf blades mostly more than 5 cm long; spikelets ellipsoid to narrowly ovoid, acute at apex; glumes glabrous; fruits glabrous:
                Spikelets borne on long, slender stalks, 5 mm or more long . . . . *Panicum trichanthum* Nees
                Spikelets borne on short stalks, seldom more than 1–2 mm long . . . . *Panicum polygonatum* Schult.

★ Plants not inhabiting marshes or other wet areas, usually in clearings:
    Leaf blades linear, usually 15 cm or more long, very narrow . . . .
    . . . . *Sporobolus indicus* (L.) R. Br.
    Leaf blades less than 12 cm long:
        Spikelets ca 1 mm long on pedicels 3 mm or more long; blades nar-
            rowly ovoid, less than 7 cm long, to 1.5 cm wide . . . .
            . . . . *Panicum trichoides* Sw.
        Spikelets 1.5–2 mm long on pedicels less than 1.5 mm long; blades
            linear-lanceolate, to 12 cm long and 12 mm wide . . . .
            . . . .*Panicum polygonatum* Schult.

to feed on pollen. *Paspalum virgatum* is probably polli-nated by small noctuid moths, which visit plants in the evening apparently for the sticky, sweet fluid present at the time of flowering (Karr, 1976). In Costa Rica, *Panicum fasciculatum* and *P. laxum* Sw. are heavily visited by the bees *Trigona, Augochloropsis* (Halictidae), and *Caenaugochlora* (Halictidae) (R. Heithaus, pers. comm.).

It could be expected that other species that occur deep in the forest, where wind currents are poor in the dry season, are also insect pollinated. On the other hand *Pharus*, which may occur in dense forest, is believed to be wind pollinated (G. Davidse, pers. comm.). Fewer than 15 grass species are restricted to the forest, of which only the bamboos (including *Rhipidocladum* and *Chusquea*), which rarely flower, and *Pharus, Streptochaeta*, and *Streptogyne* usually occur in dense, unopened forest. The remainder, including *Lithachne, Orthoclada, Olyra, Ichnanthus*, and *Panicum pulchellum*, are generally found in open areas or along trails, where wind pollination may be effective. Except for *Lithachne*, which may be insect pollinated, the species of open areas flower principally in the dry season, when winds are greatest.

Most species occur in clearings. The flowers of *Lasiacis* open between 7:30 and 9:30 A.M., but insect visitors have never been seen (G. Davidse, pers. comm.) so the flowers are apparently wind pollinated.

Grasses are far more diverse in diaspore strategy than in their pollinating agents. Wind plays a principal role (van der Pijl, 1968), especially for those taxa with plumed spikelets or feathery inflorescences such as *Andropogon, Schizachyrium, Gynerium, Saccharum, Paspalum saccharoides, Bothriochloa*, and *Phragmites*. In addition to these, many other species have small disarticulating spikelets or inflorescence parts that are probably also in part wind dispersed.

Small birds disperse the seeds of many species, partic-ularly those with larger or more attractive spikelets, in-cluding *Oryza* and most of the panicoid grasses, as well as species with attractive fruits such as *Lithachne, Olyra*, and especially *Lasiacis*. Investigations by Davidse and Morton (1973) showed that many fruit-eating birds eat the spikelets of *Lasiacis* in great numbers. The glumes contain relatively large amounts of oil, which provides nourishment for the birds, and the caryopsis passes through the bird unharmed. Birds no doubt eat the fruits of a wide variety of grass species, but in general it is uncertain what percentage of the seeds pass through the gizzard unharmed. Ridley (1930) suggested that some seeds may be picked up by birds from the ground as they

look for grit for their gizzard. Enders (1935) reported that iguanas eat *Brachiaria mutica* and *Panicum grande*. Elsewhere many grasses are dispersed by herbivores, which swallow the seeds while grazing and regurgitate or pass them unharmed (Ridley, 1930). *Eleusine indica* and *Cynodon dactylon*, both grasses of clearings, have their seeds carried by ants (Ridley, 1930; Wheeler, 1910).

A number of grasses are adapted for epizoochorous dispersal by appendages on the spikelets. These include *Cenchrus, Pharus, Streptochaeta*, and *Streptogyne*, and probably *Paspalum conjugatum, Oplismenus, Leersia, Oryza, Chloris, Orthoclada*, and most species with feath-ery, disarticulating inflorescences. The awn or awnlike structure of *Oplismenus, Oryza*, and *Chloris* probably serves as much in ensuring disarticulation of the spikelet as in dispersal. Plants of many other species also have prominently geniculate and twisted awns, such as *Hyparrhenia, Ischaemum, Polytrias, Schizachyrium*, and *Bothriochloa*. These probably function in part in epizoochorous dispersal and perhaps also for implantation, such as in the well-known case of *Stipa*.

Those taxa that are restricted to aquatic habitats along the shore probably rely in part on dispersal by water cur-rents. These include *Hymenachne, Isachne, Leersia, Oryza, Brachiaria, Panicum grande, P. mertensii, P. milleflorum, P. polygonatum, P. trichanthum, Paspalidium geminatum, Paspalum repens*, and *Phragmites*.

Unlike most forest species, such as trees with seeds that germinate soon after falling, grasses have seeds that may remain dormant for considerable periods (Corner, 1964).

Some 620–700 genera and 10,000 species; distributed worldwide.

ACROCERAS Stapf

Acroceras oryzoides Stapf in Prain, Fl. Trop. Africa
    9:622. 1920
Panicum zizanoides H.B.K.

Perennial, mostly 0.5–1.5(2) m tall, decumbent-spreading, rooting at lower nodes. Sheaths shorter than the inter-nodes, glabrous or hispid near apex; blades mostly 5–15 cm long, 1–2(3) cm wide, acuminate, cordate-clasping at the base, the margin white and scabrid sometimes with submarginal hispid trichomes. Inflorescence a sparsely branched panicle, mostly terminal, 10–20 cm long; spike-lets paired, narrowly acuminate at apex, rather widely spaced, appressed, 5–6 mm long, glabrous, each pair unequally short-pedicellate; pedicels flattened, scabrid;

first glume keeled, two-thirds as long as spikelet; second glume and sterile lemma equal, laterally compressed and keeled at the apex. Fruits 4–5 mm long, white and shiny, the apex having a green crest. *Croat 7066.*

Common in moist clearings, forming dense stands near the dock and in some areas along the shore.

Most easily distinguished by the crest at the apex of the fruit. The sharply pointed spikelets might function in epizoochorous dispersal.

Mexico to Paraguay; West Indies. Known from tropical moist forest throughout Panama and also occasionally from premontane wet forest in Bocas del Toro (Río Guarumo).

ANDROPOGON L.

Andropogon bicornis L., Sp. Pl. 1046. 1753
  Rabo de chibo, Rabo de venado

Coarse perennial, 1–1.5(2.5) m tall, erect, often growing in large stands of clumps of 4–6 plants each; culms glabrous. At least upper sheaths shorter than the internodes, glabrous, weakly keeled toward apex; blades often 25–60 cm or more long, to 7 mm wide, scabridulous throughout above and on margins and midrib below. Panicle branches numerous, very compound, forming a dense, corymbose, feathery inflorescence; racemes paired, 2–3 cm long, often partly enclosed in narrow inconspicuous spathes; rachis and sterile pedicel conspicuously pilose; spikelets paired, one sessile and perfect, the other pedicellate, staminate, and soon falling; sessile spikelet 2.5–3 mm long, glabrous, awnless; racemes disarticulating below each fertile spikelet. Fruits narrowly oblong, dispersing with the pedicel of the staminate flower. *Croat 14918.*

Common in clearings, often the dominant plant in small navigational-sign clearings along the canal; also commonly encountered in marshes. Seems to flower principally in the rainy season.

Perhaps most easily confused with *Schizachyrium microstachyum,* but that species has flexuous rachises and glabrous blades.

Central Mexico to Panama and south to Venezuela, the Guianas, Bolivia, Paraguay, and eastern Brazil; West Indies. Throughout Panama in tropical moist forest; known also from premontane wet forest in Chiriquí and Coclé and from tropical wet forest in Colón.

Andropogon glomeratus (Walt.) B.S.P., Prel. Cat. N.Y. 67. 1888

Densely tufted, erect perennial, 1–1.5 m high; culms glabrous, to 1 cm thick at modes. Sheaths keeled, often hirsute near apex; blades 30–70 cm or more long, to 1 cm broad, minutely scabridulous on both sides and margin, often with longer trichomes near base. Panicles aggregated, sometimes nearly half the length of the culm; racemes paired, spathes often shorter than racemes and obscured by them; rachis and sterile pedicel villous, the ultimate branches densely villous just below the spathes; spikelets paired, one sessile and perfect, the other pedicellate and staminate (soon falling); sessile spikelet 3–4 mm long with a straight awn 1–1.5 cm long. Fruits narrowly oblong, brown, ca 1.5 mm long. *Woodworth & Vestal 541.*

Occasional, on steep shore banks and in marshes and floating islands. Inflorescences are seen throughout the rainy season into the early dry season.

Similar in habit and appearance to *A. bicornis,* but easily confused with *A. virginicus,* to which it is more closely related.

Southeastern United States and Mexico to Panama and northern South America; West Indies. Probably in tropical moist forest throughout Panama (known from the Canal Zone, Panamá, and San Blas); known also in premontane wet forest areas in Chiriquí.

Andropogon leucostachyus H.B.K., Nov. Gen. & Sp. 1:187. 1816

Small tufted perennial, 25–70 cm tall; culms glabrous. Sheaths narrow, compressed; ligules 1–2 mm long, truncate, minutely erose; blades 5–15 cm long (except on new shoots), 1–3 mm wide, scaberulous. Flowering branches few, sparingly branched, long and slender; racemes paired (sometimes in 3s); peduncles long and slender; spathes long and slender but inconspicuous; rachis and sterile pedicel slender, densely silky-pubescent, the trichomes ca 1 cm long, sessile; spikelets ca 3 mm long, glabrous, awnless; pedicellate spikelets lacking. Fruits narrowly oblong, tan, 2 mm long. *Croat 6916.*

Occasional, in the Laboratory Clearing. Probably flowers and fruits much of the year, but mostly in the early dry and early rainy seasons. It has not been seen in flower in the late rainy season.

KEY TO THE SPECIES OF ANDROPOGON

Spikelets awned:
  Racemes scattered in a long loose inflorescence; spathes usually longer than the racemes; ultimate branches glabrous just below the spathes . . . . *A. virginicus* L.
  Racemes aggregated in a dense compound inflorescence; spathes often shorter than the racemes and obscured by them; ultimate branches densely villous just below the spathes . . . .
  . . . . *A. glomeratus* (Walt.) B.S.P.
Spikelets not awned:
  Plants usually more than 1 m tall; blades much more than 15 cm long; spathes aggregate in a usually dense inflorescence . . . . *A. bicornis* L.
  Plants usually less than 70 cm tall; blades usually less than 15 cm long; spathes not aggregated (the flowering branches few and sparingly branched) . . . . *A. leucostachyus* H.B.K.

Mexico to Argentina; West Indies. In Panama, ecologically variable; known from tropical moist forest in the Canal Zone and Panamá, from premontane wet forest in Colón (Santa Rita Ridge), Chiriquí (Boquete), and Panamá (Cerro Campana), and from lower montane moist forest in Chiriquí (lava flows between Volcán and Cerro Punta).

Andropogon virginicus L., Sp. Pl. 1046. 1753

Densely tufted, erect perennial, 80–150 cm tall; mostly glabrous. Lower leaves in basal cluster; sheaths flattened, keeled, glabrous or hirsute, much shorter than the internodes; blades 15–30 cm or more long (basal leaves longest), to 4 mm wide, the margins obscurely scabridulous. Inflorescences loose and elongate, the flowering branches from the middle to upper nodes; racemes paired, slender, spreading, often partly enclosed in a slender spathe, the spathe exceeding the raceme; rachis flexuous, the ultimate branches glabrous just below the spathes; spikelets paired, one sessile and perfect, 3–4 mm long, the other pedicellate and staminate (soon falling); awn straight, 1–1.5 cm long; first glume acuminate. Fruits narrowly oblong, tan, ca 1 mm long. *Ebinger 590.*

Collected once near the shore in the Drayton House Clearing. Apparently less abundant in Panama than *A. glomeratus.* Seasonal behavior unknown, probably flowers during the rainy season.

Eastern United States and Mexico to Panama; West Indies; Hawaii. In Panama, known from tropical moist forest in the Canal Zone and Panamá (doubtless elsewhere) and from premontane wet forest in Panamá (slopes of Cerro Campana).

ANTHEPHORA Schreb.

Anthephora hermaphrodita (L.) O. Kuntze, Rev. Gen. Pl. 2:759. 1891

Erect annual, to 50 cm high; culms rooting at lower nodes. Leaf sheaths prominently veined, puberulent and papillose-hirsute near apex; blades 5–20 cm long, 3–8 mm wide, puberulent and papillose-hirsute on both surfaces. Racemes 5–10 cm long; spikelets in groups of 4, the first glumes united at the base and forming a false involucre around the rest of the spikelets, the groups short-pedicellate; first glume 5–7 mm long, scabrid on the margins, indurate. Fruits lanceolate, acuminate, tan, ca 3.5 mm long. *Shattuck 418.*

Collected once along the shore; not seen in recent years. Flowers and fruits chiefly in the rainy season.

The spikelets disarticulate from the rachis in groups of four. This unit has no obvious means of dispersal, but since the caryopsis is well protected by the indurate glumes, the spikelets may be bird dispersed.

Throughout tropical America. In Panama, from tropical moist forest in the Canal Zone (both slopes), Colón, and Panamá and from tropical dry forest in Panamá (Taboga Island).

AXONOPUS Beauv.

Axonopus compressus (Sw.) Beauv., Ess. Agrost. 12. 1812
Carpetgrass

Perennial, 20–60 cm tall; stoloniferous, often in dense stands; culms ± erect, flattened, pubescent at nodes. Sheaths flattened, glabrous throughout except at collar and sometimes ciliate on the margins; blades 5–15 cm long, 6–12 mm wide, glabrous below, glabrous to sparsely long-pubescent above, the margins scabrid, usually ciliate at least near base. Inflorescences terminal or upper axillary, with 2–5 racemes 3–10 cm long, the upper 2 conjugate, the others a short distance below; spikelets mostly 2–2.5 mm long, 2 or more times longer than broad, acute at apex, alternate on a triangular, narrowly winged rachis, glabrous or sparsely pubescent; first glume lacking; second glume and sterile lemma acute, exceeding fruit. Fruits to 2 mm long, oblong-elliptic, blunt at apex, whitish, minutely roughened. *Croat 16573.*

Common to locally abundant in the Laboratory Clearing. Flowers all year, especially in the rainy season.

Similar to *Digitaria horizontalis,* but having a single spikelet per node of the inflorescence.

Throughout warmer regions of New and Old Worlds. In Panama, widespread in tropical moist forest; known also in premontane dry forest in Coclé (Aguadulce), tropical dry forest in Panamá (Taboga Island), and lower montane wet forest in Chiriquí (Volcán).

BAMBUSA Schreb.

Bambusa amplexifolia (Presl) Schult.f., Syst. Veg. 7(2):1348. 1830

Bamboo growing in isolated clumps of as many as 10 culms each; culms to ca 8 cm diam and 10 m long, erect to arching, often leaning on other vegetation; culm sheaths triangular when flattened, acute at apex, ca 25 cm long, densely ferruginous-pubescent on outer surface; branches glabrous, long and spreading, heavily armed at nodes with spines mostly 8–25 mm long. Leaves mostly on outer parts of branches, mostly 1–1.5 cm apart on the stem; leaves on juvenile culms 2.5–4.5 cm apart, the blades to 24 cm long and 4.5 cm wide, the sheaths 4–8 cm long; mature sheaths ca 2.5 cm long, glabrous except usually ciliate along one or both margins, the truncate apex with a stout branched trichome on either side 7–13 mm long; petioles 0.5–2 m long; blades oblong-lanceolate, narrowly acute to acuminate at apex, rounded at base, 3–15 cm long, 10–25 mm wide, glabrous, the margins sparsely and inconspicuously scabrid. Inflorescences not seen. *Croat 10101.*

Rare; known only from near the beginning of Armour Trail and along the stream near Chapman Trail 900. Forming small stands in the forest.

Mexico to Colombia and Venezuela. Range in Panama unknown.

KEY TO THE SPECIES OF BAMBUSA

Plants unarmed; main culms usually less than 2 cm wide; leaf blades short-pubescent on lower
  surface, ± pruinose . . . . *B. glaucescens* Munro
Plants armed with spines at least at some nodes; main culms usually more than 6 cm diam; leaf
  blades glabrous, not at all pruinose:
    Plants ca 20 m tall, in large congested clumps, with a dense tangle of many stout, heavily armed,
      leafless branches at base; culms 11–12 cm diam; culm sheaths broad, wrinkled along upper
      margins . . . . *B. arundinacea* Retz.
    Plants 5–10 m (or more) tall, usually solitary or few, lacking a dense tangle of spiny branches at
      base; culms 1–8 cm (or more) diam; culm sheaths ± continuous with the blade, triangular
      when flattened . . . . *B. amplexifolia* (Presl) Schult.f.

Bambusa arundinacea Retz., Obs. Bot. 5:24. 1789.

Erect bamboo to 20(25) m tall; culms olive-green, shiny,
11–12 cm diam, erect to broadly arching, densely rooting
at lower nodes, occurring in congested clumps of ca 4 m
or more in diameter; internodes mostly 10–25 cm long
near the base, the basal ones thick-walled (to ca 1 cm
thick); nodes with a conspicuous leaf-sheath scar and
with many stout branches to 3 cm diam, soon becoming
slender, usually 1- or 2-branched, wide-spreading, forming a tangled mass to 5 m high, heavily armed at the
nodes, the spines to 1.5 cm long; branch apices also spinelike; sheath blades broadly triangular, appressed (reflexed
at upper nodes); branches on main culm to 2 m or more
long, their axes flexuose, the nodes with a prominent ringlike sheath scar and often with stout spines, the spines
often weakly curved, with short internodes; the lower 1.5
cm of the base of each branch encircled by several short
sheathlike bracts, the smaller leafy branches with leaves
clustered mostly toward their apices. Sheaths glabrous
except often short-pubescent along upper margin, truncate to auriculate at apex, often bearing a few, thick,
usually deciduous trichomes; leaf blades linear-lanceolate,
acuminate, (2.5)7–15(20) cm long, (5)8–18(25) mm wide,
glabrous or with a few long trichomes near base on upper
surface, glabrous or puberulent on lower surface. Inflorescences not seen. *Croat 17038.*

Cultivated in the Laboratory Clearing northwest of
the dining hall.

Native to India; introduced occasionally in other tropical areas of the world.

Bambusa glaucescens (Willd.) Sieb. ex Munro, Trans.
  Linn. Soc. London 26:89. 1868

Slender, clumped bamboo, to 5 m tall; culms very numerous, closely aggregated, 1–2 cm diam, erect when young,
leaning in age, appressed-pubescent below internodes,
pruinose when young (including sheath); the reduced
leaves of the main axis tightly appressed to culms, brown,
acuminate at apex, mostly 18–26 cm long (including

sheath), appressed-pubescent outside above ligules;
ligules ca 2 mm high, the margins irregular; branches
clustered at the nodes, of various lengths, to 2 m long,
these in turn branched, the branchlets to ca 1 m long.
Leaves linear, acuminate, acute to rounded at base, mostly
3–18 cm long and 6–16 mm wide, pruinose and short-
pubescent below, ± glabrous above, the margins scabrid,
the midrib weak, the sheaths turning brown in age,
bristled on the upper margin. Inflorescences not seen.
Croat 16563.

Cultivated at the Laboratory Clearing near ZMA
House and the bridge to the animal cages.

Native to Asia, probably China; cultivated throughout
the tropics.

BOTHRIOCHLOA O. Kuntze

Bothriochloa intermedia (R. Br.) A. Camus, Ann. Soc.
  Linn. Lyon n. sér. 76(1930):164. 1931

Perennial, mostly 1–1.5 m tall, erect, usually unbranched;
culms glabrous, striate. Leaves glabrous except for scabrid margins; sheaths ± as long as internodes; blades
25–55 cm long, 3–10 mm wide. Panicles terminal, 6–20
cm long, held above the leaves, the primary axis glabrous,
the branches usually 1- or 2-branched, bearded at their
base; racemes mostly 1.5–4 cm long; rachis and sterile
pedicel villous; spikelets in pairs, one sessile, perfect, ca 3
mm long, the other pedicellate, sterile, ca 2.5 mm long;
sessile spikelet bearing a tuft of trichomes at the base and
a pubescent, reddish-brown, geniculate awn 8–13 mm
long; glumes brown to purplish, the glume of the sessile
spikelet pubescent toward the base. Fruits not seen. *Croat
6973, 16523.*

Frequent in the Laboratory Clearing. Probably in
flower or fruit much of the year.

This species, var. *intermedia,* is introduced on BCI.

India to Australia and Africa; introduced into most
tropical areas. In Panama, known from tropical moist
forest in the Canal Zone, Panamá, and San Blas.

KEY TO THE SPECIES OF BOTHRIOCHLOA

Spikelets awned, bearing a prominent pinhole-like pit in the glume . . . . *B. pertusa* (L.) A. Camus
Spikelets awnless, not with a hole in the glume . . . . *B. intermedia* (R. Br.) A. Camus

Bothriochloa pertusa (L.) A. Camus, Ann. Soc. Linn.
Lyon n. sér. 76(1930):164. 1931

Small annual, to 50 cm tall, usually growing as isolated individuals, usually unbranched, the base sometimes decumbent; culms glabrous, with nodes bearded. Sheaths glabrous or ciliate on margin, shorter than internodes; blades 1.5–10 cm long, to 3 mm wide, conspicuously pubescent on upper and lower surfaces, the trichome bases enlarged. Racemes usually 5 or 6 clustered near apex, 3–6 cm long, held well above leaves; rachis and sterile pedicel prominently long-villous; spikelets in pairs, one sessile and perfect, the other pedicellate and sterile; sessile spikelet awned, 3.5–4 mm long, the glume with a prominent small, round, deep pit above the middle; awns brown, generally twice-geniculate, to ca 2 cm long; pedicellate spikelet awnless, about as long as sessile one. Fruits not seen. *Croat 9206*.

Occasional, in open areas in the Laboratory Clearing during the dry season.

Bothriochloa bladhii (Retz.) S. T. Blake, native to India and Africa, has a pit in the glume similar to that in *B. pertusa,* but it has not been collected on the island.

Panama; West Indies. In Panama, known from tropical moist forest in the Canal Zone and Panamá and from premontane dry forest in Coclé (Río Hato).

BRACHIARIA Griseb.

Brachiaria mutica (Forssk.) Stapf in Prain, Fl. Trop.
Africa 9:526. 1919

Panicum barbinode Trin.; *P. purpurascens* Raddi

Stoloniferous perennial, to 2(6) m long; culms erect or decumbent, rooting and often geniculate at lower nodes, densely villous at nodes. Sheaths mostly longer than the internodes, stiffly papillose-pubescent; blades mostly 15–25 cm long, to 1.5 cm wide, glabrous with finely scabrous margins. Panicles 10–20 cm long, the branches many, raceme-like, mostly 3–5 cm long, densely pubescent at the base, angulate, scabrid; branchlets and pedicels both scabrid and with sparse, long, stiff trichomes; spikelets 2.7–3 mm long, glabrous, often tinged with purple; first glume approximately one-third as long as spikelet; second glume and sterile lemma covering fruit. Fruits pale, longitudinally striate, minutely transversely rugose. *Croat 8679*.

Infrequent but locally common on open, swampy edges of the lake near the dock. Apparently flowers from late in the rainy season to the middle of the dry season (September to March).

This species was reported incorrectly by Standley as *Eriochloa punctata* (L.) Desv. ex Ham.

Throughout tropical and subtropical America. In Panama, from tropical moist forest in the Canal Zone and in premontane moist forest in Panamá (Perlas Islands). See Fig. 43.

CENCHRUS L.

Cenchrus brownii R. & S., Syst. Veg. 2:258. 1817
C. viridis Spreng.
Pega-pega, Cadillo

Erect or decumbent-spreading annual, less than 1 m tall, rooting at lower nodes. Sheaths longer than internodes, flattened, glabrous; blades 10–30 cm long, 5–8(12) mm wide, glabrous or scabrid on upper surface and on margins with longer trichomes at least near the base. Spikes to 10 cm long; spikelets sessile, ca 4 mm long, enclosed in burs, usually 3 in each bur, the burs dense, ca 4 mm wide, subtended by a ring of retrorsely barbed bristles. Fruits not seen. *Shattuck 391*.

Common in open areas in the Canal Zone; perhaps once common also on BCI but possibly no longer occurring there. Flowers throughout the rainy season and into the early dry season.

Southern United States through Central America to Bolivia and Brazil; West Indies; Australia and South Pacific Islands. In Panama, from tropical moist forest all along the Pacific slope and in Bocas del Toro, and from premontane wet forest in Panamá (Perlas Islands).

CHLORIS Sw.

Chloris radiata (L.) Sw., Prodr. Veg. Ind. Occ. 26. 1788

Perennial, to 60 cm tall; culms ± erect, rooting at lower nodes. Sheaths flattened, keeled, longer than the internodes, scabrid to glabrous; blades 4–12 cm long, to ca 5 mm wide, scabrid on upper and lower surfaces. Inflorescences terminal, of subdigitate or closely clustered spikes 4–8 cm long; spikelets to ca 3 mm long, with 1 fertile floret and several reduced sterile lemmas in a club-shaped rudiment; glumes scarious with a stout, scabrid keel; fertile floret 2.5–3 mm long, glabrous except at base and apex below awn, the awn 5–10 mm long, antrorsely scabrid; rudiment minute, almost hidden by fertile floret, its awn 4–6 mm long. Fruits oblong, ca 1 mm long. *Standley 41125*.

Collected once by Standley; not seen recently on the island, but to be expected in clearings. Probably flowers throughout the year.

Throughout most tropical regions of the Hemisphere. From tropical moist forest throughout Panama and from tropical dry forest in Panamá (Taboga Island).

KEY TO THE SPECIES OF CHLORIS

Fertile lemma bearing short, inconspicuous tufts of trichomes, these much shorter than lemma; rudiment slender, acute at apex, often obscured by fertile lemma . . . . *C. radiata* (L.) Sw.
Fertile lemma tufted near apex, the trichomes nearly as long as the lemma; rudiment truncate at apex, not hidden in fertile lemma . . . . *C. virgata* Sw.

Chloris virgata Sw., Fl. Ind. Occ. 203. 1797

Annual, 10–100 cm tall; culms ± erect or decumbent at base. Sheaths flattened, keeled, mostly longer than the internodes, glabrous; blades 4–25 cm long, 3–7 mm wide, mostly glabrous to scaberulous, the margins scabrid. Inflorescences terminal, of several digitate spikes 3–9 cm long; spikelets 3–3.5 mm long, with 1 fertile floret and a rudiment of several sterile lemmas, disarticulating below the fertile lemma; glumes scabrous on the keel, unequal, the second glume with a short awn; fertile floret 3–3.5 mm long, bearded on margins below awn, the awn on the lemma scabrid, to 1 cm long; rudiment stout, 2–2.5 mm long, with an awn 3–7 mm long. Fruits not seen. *Kenoyer 119.*

Collected once by Kenoyer; not seen recently but to be expected in clearings. Apparently flowering principally in the dry season.

A native of Africa, now introduced throughout warmer regions of the world. In Panama, known from tropical moist forest in the Canal Zone.

CHUSQUEA Kunth

Chusquea simpliciflora Munro, Trans. Linn. Soc. London 26:54. 1868
　　Carricillo

Slender, arching or clambering, vinelike plant, to 25 m long; culms solitary or few in a clump, mostly less than 5(8) mm diam, glabrous; branches few to many, 7–30 cm long on sterile plants, the nodes usually ciliate. Sheaths ciliate, the lowermost without blades; blades lanceolate, mostly 5–9 cm long, 8–15 mm broad, nearly glabrous except for long, somewhat tufted trichomes at base on lower surface, the margins and midrib on upper surface scabrous, the lower surface with prominent midrib and usually 3 lateral veins. Fertile shoots rare, 2–8 cm long, their leaves much reduced; panicles very small, of 1–4 spikelets, the spikelets 7–9 mm long; glumes minute; sterile lemmas acuminate; fertile lemmas acute. Fruits not seen. *Shattuck 717, Croat 4358.*

Abundant in the forest, often very abundant locally on the shore. *Shattuck 717* was collected in flower early in the dry season. Elsewhere flowers have been seen in October. Plants probably flower only at intervals of several years.

Might be confused with either *Rhipidocladum racemiflorum* or *Lasiacis divaricata* (L.) Hitchc.; the latter species is not known from BCI.

Guatemala to Panama. In Panama, known from tropi-

cal moist forest in the Canal Zone and its vicinity. See Fig. 44.

CYNODON L. C. Rich.

Cynodon dactylon (L.) Pers., Synops. Pl. 1:85. 1805
　　Bermuda grass

Stoloniferous, widely creeping perennial; rhizomatous; culms slender, 10–40 cm tall, glabrous. Sheaths longer than the internodes, glabrous or sparsely pilose at the apex; blades 2–20 cm long (often less than 5 cm long), 2–4 mm wide, glabrous to scabrous, especially on margins. Inflorescences terminal with several digitate spikes 2–7 cm long, ca 1 mm wide; rachilla prolonged beyond the spikelets; spikelets 1-flowered, 2–3 mm long; glumes nearly equal. Fruits not seen. *Ebinger 246.*

Infrequent in the Laboratory Clearing near the Animal House. Flowering throughout the year.

Under certain environmental conditions the species produces hydrocyanic acid (Blohm, 1962). The toxin affects the nervous and circulatory system of animals by eliminating the normal digestive tract flora, which is necessary for synthesis of the vitamin B complex.

Throughout warmer regions of the world. Probably throughout tropical moist regions of Panama.

DIGITARIA Haller

Digitaria ciliaris (Retz.) Koel., Descr. Gram. 27. 1802
　　D. adscendens Henr.; *D. sanguinalis* sensu auct. non (L.) Scop.

Annual; culms decumbent or geniculate-spreading to erect, often rooting at lower nodes, glabrous. Sheaths glabrous to papillose-hirsute (especially at apex); ligules membranaceous, truncate; blades mostly 4–10 cm long and 4–10 mm wide, minutely scabrous and also sparsely pilose; racemes 2 to several, digitate or in up to 3 whorls; rachis ca 1 mm wide, narrowly winged, its margin scabrous; spikelets paried on rachis, one nearly sessile, the other pedicellate, ca 3 mm long, the first glume minute, the second glume one-half to three-fourths as long as fruit; sterile lemma longer than fruit, prominently veined, the margins often villous. Fruits lanceolate, pale brown, ca 3.5 mm long, bearing several rows of long villous trichomes. *Shattuck 356a.*

Rare weed in clearings; collected once by Shattuck. Flowers throughout the year, principally during rainy season.

Distinguished from *Axonopus compressus*, with which it may be confused, by having two spikelets per node.

KEY TO THE SPECIES OF DIGITARIA

Spikelets ca 3 mm long; rachis ca 1 mm wide . . . . *D. ciliaris* (Retz.) Koel.
Spikelets ca 2 mm long or less; rachis less than 0.7 mm wide:
　　Leaves ± glabrous; spikelets ca 1.3 mm long, lacking scattered, long, white trichomes along
　　　　rachis; second glume equaling fruit . . . . *D. violascens* Link
　　Leaves conspicuously pubescent; spikelets ca 2 mm long with scattered, long, white trichomes
　　　　along rachis; second glume about half as long as fruit . . . . *D. horizontalis* Willd.

Fig. 43. *Brachiaria mutica*

Fig. 44. *Chusquea simpliciflora*

Fig. 45. *Gynerium sagittatum*

This species was treated by Standley (1933) and the *Flora of Panama* as *D. sanguinalis* (L.) Scop. However, *D. sanguinalis* is restricted to temperate regions, whereas this species inhabits tropical and subtropical regions.

Throughout tropics and subtropics of the world, ecologically variable. In Panama, known from most tropical moist forest regions and from premontane wet forest in Chiriquí (Boquete), from tropical wet forest in Colón (Miguel de la Borda), and from premontane dry forest in Coclé (Santa Clara Beach).

Digitaria horizontalis Willd., Enum. Pl. 92. 1809
  Crabgrass

Similar in habit, size, and most other aspects to *D. ciliaris,* except the margins of the leaves lighter in color, scabrous, usually densely pilose; the racemes digitate or more commonly closely grouped on axis; the rachis to 0.5 mm wide, winged, the margins scabrid, but also sometimes bearing a few scattered long trichomes; the spikelets 2 mm long, the first glume minute or lacking, the second glume about half as long as fruit, the sterile lemma glabrous or nearly so. Fruits dark brown, shiny, ca 1.3 mm long. *Croat 8563.*

Frequent weed of clearings. Probably flowers throughout the year, especially in the rainy season.

Most easily distinguished from *D. ciliaris* by its smaller spikelets with a glabrous lemma.

Common in tropical and subtropical regions. In Panama, ecologically variable; from tropical moist forests in Bocas del Toro, Panamá, Darién, and the Canal Zone (doubtless elsewhere), from tropical dry forest in Panamá (Taboga Island), from premontane wet forest in Panamá, and from lower montane wet forest in Chiriquí (Volcán).

Digitaria violascens Link, Hort. Berol. 1:229. 1827

Small perennial, ca 35 cm tall; culms slender, glabrous. Sheaths longer than the internodes, glabrous; blades mostly 8–15 cm long, ca 4 mm wide, glabrous or nearly so. Inflorescences of several, digitate racemes 6–10 cm long; rachis winged, scabrid on margins; spikelets 2 or 3 per node, appressed, ca 1.3 mm long; pedicels short, of variable lengths; first glume absent; second glume and sterile lemma equaling fruit, minutely pubescent at least near margin. Fruits dark brown, shiny, minutely striate. *Croat 16572.*

Collected once in the Laboratory Clearing. That specimen had fruits in early August.

Wide but scattered distribution in the New World; Panama through northern South America; West Indies; occasionally East Asia, from where it may have been introduced. In Panama, from tropical moist forest on BCI and in Colón.

ELEUSINE Gaertn.

Eleusine indica (L.) Gaertn., Fruct. & Sem. Pl. 1:8. 1788
  Goosegrass

Annual, growing in spreading clumps, 20–50(70) cm tall, glabrous except for tufts of trichomes at apex of sheaths and at base of racemes. Sheaths flattened, pubescent near margin at apex, margins hyaline; ligules membranaceous, ca 1 mm long, erose; blades 7–25 cm long, 3–5(8) mm wide, often conduplicate, sometimes sparsely pilose above, scabrous on margins. Inflorescences of 2 to several spikes, the spikes 5–10 cm long, digitate or with 1 or 2 a short distance below; spikelets solitary at each node, each spikelet of few to several flowers, sessile, 2–5 mm long; glumes shorter than the first floret; glumes and lemmas with a scabrid keel, green near the keel, otherwise very light. Fruits dark brown, finely ridged, enclosed in a loose, thin exocarp. *Croat 10305.*

Occasional, in clearings. Flowers throughout the year, principally in the rainy season.

Southern half of United States south to northern Argentina and Uruguay; West Indies; Galápagos Islands; also in the Old World. In Panama, from tropical moist forest in Bocas del Toro, Colón, San Blas, Coclé, Panamá, Darién, and the Canal Zone and from tropical dry forest in Coclé and Panamá.

GYNERIUM Willd. ex Beauv.

Gynerium sagittatum (Aubl.) Beauv., Ess. Agrost. 138, pl. 24, f. 6. 1812
  Uva grass, Wild cane, Cana blanca

Large, dioecious perennial, usually to 3.5 m tall (frequently taller elsewhere); culms glabrous, becoming woody, to 1.5 cm or more diam, frequently branching near the base and sometimes with stout, leafy stolons. Sheaths longer than the internodes, with a long patch of pilose-woolly trichomes on the outside below the blades and also along the margins near the apex; blades to 1 or 2 m long, 4–6 cm wide, forming a fan-shaped summit on a sterile culm. Panicles 1 m or more long, the branches erect to drooping, white turning reddish-brown; pistillate spikelets with glumes very unequal, the first 4–5 mm long, the second 9–12 mm long and scabridulous, the lemma narrow, tapered to a long awn, together with the awn 6–10 mm long, with long spreading-pilose trichomes exceeding the awn; staminate spikelets with glumes ± equal, 2–3 mm long, the lemma glabrous. Fruits narrowly oblong, ca 1 mm long, brown. *Croat 10384.*

Occurring on shore on the steep banks of the north shore and on the few sandy beaches, where large lateral runners may extend considerable distances across the beach front, producing numerous erect branches. Flowers throughout the year, especially in the early rainy season.

Southern Mexico to Paraguay; West Indies. In Panama, known from tropical moist forest in the Canal Zone and vicinity and in Darién.

See Fig. 45.

HOMOLEPIS Chase

Homolepis aturensis (H.B.K.) Chase, Proc. Biol. Soc. Wash. 24:146, f. 12. 1921

Perennial, stoloniferous, ± erect, mostly 20–60 cm tall; culms slender, glabrous. Sheaths mostly shorter than internodes, glabrous but with pubescence on collar and

long-ciliate margins; blades narrowly lanceolate, 4–12 cm long, 1–1.5 cm wide, rounded to subcordate at base, long-pilose above, glabrous to long-pilose below, scabrous and long-ciliate on margins, minutely cross-veined. Panicles 5–10 cm long, narrow, having very slender, few-flowered branches; spikelets ca 7 mm long, acuminate; glumes glabrous, nearly equal, covering sterile lemma and fertile floret, the sterile lemma villous between lateral veins. Fruits indurate, acuminate, white or tan, smooth, shiny. *Croat 5624.*

Common in clearings. Flowers are seen all year, but especially during the rainy season.

Sometimes confused with *Acroceras oryzoides,* but is distinguished by the nearly equal glumes covering the fruit and by the pubescence of the sterile lemma.

Southern Mexico south to Peru, Bolivia, and northern and eastern Brazil. In Panama, ecologically variable; known throughout tropical moist forest and from tropical wet forest in Colón, Panamá, and Bocas del Toro, from premontane rain forest in Panamá (Cerro Jefe), and from lower montane rain forest in Chiriquí (Volcán).

HYMENACHNE Beauv.

Hymenachne amplexicaulis (Rudge) Nees, Agrost. Bras. 276. 1829

Aquatic perennial, mostly 1–2 m tall; culms ± stout, rooting at nodes. Sheaths mostly shorter than internodes, glabrous, sometimes ciliate; blades 15–35 cm long, 1.5–3 cm wide, cordate-clasping at base, glabrous, scabrid on margins and ciliate near the base. Inflorescence a dense, spikelike panicle 8–30(50) cm long, 1–1.5 cm thick; spikelets 3–4 mm long, acuminate; first glume one-fourth to one-half as long as spikelet with scabrous keel; second glume and sterile lemma covering fruit, scabrous on veins, the sterile lemma narrowly acuminate. Fruits oblong-elliptic, ± flattened, dull, light greenish. *Croat 7197.*

Infrequent, usually growing in shallow water at the edge of the lake. Flowers throughout the year, especially in the late rainy and early dry seasons.

The spikelets are probably dispersed in part by water, but the sharply pointed sterile lemma could easily be caught and carried in bird feathers.

Throughout tropics of Western and Eastern hemispheres. In Panama, from tropical moist forest in Coclé, Darién, and the Canal Zone and from tropical dry forest in Coclé (Aguadulce).

HYPARRHENIA N. J. Andersson

Hyparrhenia rufa (Nees) Stapf in Prain, Fl. Trop. Africa 9:304. 1919
Faragua

Erect perennial, 1–2(2.5) m tall; culms glabrous, in large dense clumps. Sheaths glabrous or strigose between veins, especially near margins; blades to ca 35 cm long, 2–8 mm wide, glabrous or scaberulous, usually scabrid on margins. Inflorescences usually open, 20–40 cm or more long, with paired racemes at ends of ultimate branchlets subtended by a narrow spathe; peduncles long, slender, strigose; racemes 2–4 cm long; spikelets 3–4 mm long, densely covered with long reddish-brown trichomes, paired, the pedicellate one awnless; sessile spikelets with geniculate, reddish-brown awn 1.5–2 cm long. Fruits not seen. *Croat 6995.*

Common in the Laboratory Clearing and rarely encountered along exposed areas of the shore. Flowers throughout the year.

This species is the most abundant grass along roadsides in the Canal Zone and one of the most widely used forage grasses in lowland Panama, forming virtually complete stands where grazed and burned.

Veracruz, Mexico, south to Panama, Peru, and Brazil; Cuba, Jamaica; throughout Africa, where it is native. Ecologically wide-ranging throughout Panama.

ICHNANTHUS Beauv.

Ichnanthus brevivaginatus Swall., Phytologia 4:425. 1953

Perennial, ± erect except at decumbent base, rooting at lower nodes, villous especially at nodes. Sheaths much shorter than internodes, prominently ribbed, villous; blades linear-lanceolate to ovate-lanceolate, acuminate, inequilateral and obtuse to rounded at base, 5–10 cm

---

KEY TO THE SPECIES OF ICHNANTHUS

Leaf blades usually less than 3.5 cm long and 1 cm wide, ± linear-lanceolate to short ovate-lanceolate, pubescent with papillose trichomes. . . . . *I. tenuis* (Presl) Hitchc. & Chase
Leaf blades usually more than 5 cm long and 1 cm wide, long linear-lanceolate to obovate-lanceolate, glabrous or pubescent (if pubescent, the trichomes not papillose except at margin of blade just above sheath):
  Inflorescences with 2 or more branches from terminal node of stem; longest panicle 15–26 cm long and 4–5 cm wide, beginning to branch ca 11 cm from base; axillary nodes of stems producing inflorescences; leaf blades obovate . . . . *I. pallens* (Sw.) Benth.
  Inflorescences with 1 branch from terminal node of stem (if 2, neither is long); longest panicle 7–12 cm long and 2–4 cm wide, beginning to branch 2.5–5.5 cm from base; axillary nodes of stems usually lacking visible inflorescences; leaf blades linear-lanceolate
    . . . . *I. brevivaginatus* Swall.

long, 7–20 mm wide, glabrous or with deciduous stiff trichomes on upper surface. Inflorescences with a few short branches from upper node of stem, the longest panicles 7–12 cm long and 2–4 cm wide, beginning to branch 2.5–5.5 cm from base; axillary nodes of stems usually without visible inflorescences; spikelets 3.5–4 mm long, scabrous on keel of first and second glumes; first glume about two-thirds the length of spikelet, acute to acuminate; second glume much longer than fruit, acuminate to ± aristate; sterile lemma slightly longer than fruit, long-ciliate, blunt. Fruits oblong-elliptic, ca 2 mm long, dull white. *Croat 8351, 12868.*

Rare, along trails in the forest, mostly on the western side of the island.

Belize to Panama. In Panama, known from tropical moist forest in the Canal Zone and Panamá and from premontane wet forest in Chiriquí (San Félix).

Ichnanthus pallens (Sw.) Munro ex Benth., Fl. Hongk. 414. 1861

Perennial, decumbent-spreading, rooting at lower nodes, to 1 m or more long, glabrous or sparsely pilose, the nodes not villous. Sheaths much shorter than internodes, glabrous to pilose or papillose-pilose, especially on margins; ligules thin, ca 1 mm long; blades lanceolate to obovate, acuminate, rounded and ± clasping and asymmetrical at base, 5–12 cm long, mostly 5–15 mm wide, scaberulous, finely cross-veined downward. Inflorescences with 2 or more branches from terminal node of stem, the longest panicle 15–26 cm long and 4–5 cm wide, beginning to branch ca 11 cm from base; axillary nodes of stems producing inflorescences; spikelets 3–3.5 mm long, glabrous or sparsely pilose or hirsute; first glume acuminate, one-half to two-thirds as long as spikelet, scabrous on keel; second glume and sterile lemma acuminate, subequal; sterile lemma long-ciliate. Fruits ca 2 mm long, oblong-elliptic, the scars extending downward into very narrow wings. *Croat 4228, 7893.*

Occasional, along trails and at the edges of clearings on the shore. Plants may become locally abundant in open areas of forest trails, especially on slopes. Flowers principally in the late rainy and early dry seasons. Most fruits are gone by the end of the dry season.

Like *I. tenuis,* the species frequently has proliferated spikelets, which give the plants a very different appearance, often more like that of the genus *Poa.*

Throughout tropical America. In Panama, known from tropical moist forest in the Canal Zone, Bocas del Toro, Panamá, and Darién, from tropical dry forest in Panamá (Taboga Island), and from premontane wet forest in Chiriquí.

Ichnanthus tenuis (Presl) Hitchc. & Chase, Contr. U.S. Natl. Herb. 18:334. 1917

Creeping annual; culms forming large loose mats, rooting at lower nodes, freely branching, usually pubescent. Sheaths usually much shorter than internodes, pilose or papillate-pilose with spreading trichomes; blades ±

linear-lanceolate to short ovate-lanceolate, acuminate, 1–3.5(5) cm long, 4–10 mm wide, usually papillate-pilose. Inflorescences of 2 to several spreading racemes, pubescent in axils, the lowermost 1–3 cm long; spikelets 3–4 mm long, appressed, sparsely pilose especially along margins on glumes and sterile lemma; first glume acuminate or attenuate, almost aristate, two-thirds or more as long as spikelet; second glume and sterile lemma ± equal, longer than fruit. Fruits 2–2.5 mm long, oblong-elliptic, the wings reduced to inconspicuous scars. *Ebinger 282.*

Collected once in the small clearing at Fairchild Point. Flowers throughout the year, especially in the early dry season and the middle of the rainy season (November to January and July to September).

Guatemala south to Colombia, Venezuela, the Guianas, and Brazil; Trinidad. In Panama, known from tropical moist forest in the Canal Zone, Bocas del Toro, and Panamá, from premontane moist forest in the Canal Zone and Panamá, and from premontane wet forest in Chiriquí.

ISACHNE R. Br.

Isachne polygonoides (Lam.) Doell in Mart., Fl. Brasil. 2(2):273. 1877

Annual; decumbent-spreading, rooting at lower nodes; culms slender, glabrous, freely branching. Sheaths much shorter than internodes, papillose-hispid and ciliate, prominently veined; blades lanceolate, 2–5 cm long, 5–10(15) mm wide, rounded to subcordate at base, strigose to papillose-hispid especially below, the margins white and scabrid. Panicles diffuse, 5–10 cm long, nearly as wide, the branches slender, glandular-spotted; spikelets subglobose, 1.5–2 mm long; glumes equal, slightly shorter than spikelet, glabrous to sparsely pubescent above middle; florets 2, one staminate and similar in size and texture to the second glume, the other perfect. Fruits ca 1.3 mm long, white, plano-convex, densely pubescent. *Croat 13242.*

Infrequent, in marshes and floating masses of vegetation along the shore. Probably flowers throughout the year.

Costa Rica to Peru and Brazil; West Indies. In Panama, in swamps and marshes throughout a wide range of life zones; known from tropical moist forest in the Canal Zone, Chiriquí, Veraguas, and Panamá (doubtless elsewhere also), from tropical dry forest in Coclé (Aguadulce), from tropical wet forest in Colón (Portobelo), and from premontane wet forest in Chiriquí.

ISCHAEMUM L.

Ischaemum indicum (Houtt.) Merr., J. Arnold Arbor. 19:320. 1938

Erect, stoloniferous plant, mostly to 50 cm tall; culms slender, glabrous except for long tufts at collar. Sheaths slightly shorter than the internodes, long-ciliate, papillose-

pubescent; blades 4–12 cm long, ca 5 mm wide, short-pilose throughout, minutely scabrid on margins. Inflorescences of paired racemes 3–7 cm long; rachis and pedicel pilose-ciliate, disarticulating; spikelets paired, to 5.5 mm long, the long-pedicellate spikelet often not fruitful; glumes rounded at base, narrowly acuminate, pilose above middle, scabrid on margins; lemmas hyaline, the fertile lemma with a geniculate, twisted awn, the awn ca 7 mm long. Fruits not seen. *Croat 7003.*

Locally abundant in the Laboratory Clearing north of the dock and along the stairs to the dining hall. Rather seasonal and conspicuous when in flower, otherwise inconspicuous. Flowering mostly in the dry season, abundantly in the early dry season.

The species resembles *Polytrias amaura,* but is distinguished from it by having only a single spike per peduncle.

Mexico to Ecuador and Brazil; West Indies. Introduced in Panama and currently known only from the vicinity of the Canal Zone.

Ischaemum rugosum Salisb., Icon. Stirp. Rar. 1: pl. I. 1791

Annual; culms ± stout, glabrous except for bearded nodes, freely branching. Sheaths keeled toward apex, glabrous except sometimes near the apex; blades 10–20 cm long, 6–13 mm wide, sparsely pubescent, scabrid on margins. Inflorescences of paired terminal racemes 3–13 cm long, held tightly together when young, often partly enclosed in a sheath; rachis and pedicel thick, long-ciliate; spikelets paired, 3–5 mm long, bearded at base, the pedicellate spikelet often not fruitful; first glume and sessile spikelet strongly transversely ridged, ciliate and acute at apex; lemmas hyaline, the fertile lemma bifid with a long, geniculate, twisted awn to 1.5 cm long between teeth. Fruits ca 1.7 mm long, whitish. *Croat 12941.*

Infrequent in the Laboratory Clearing. Flowering from November to April, principally in the dry season.

Native to the Old World; introduced throughout the tropics. In Panama, from tropical moist forest in the Canal Zone, Bocas del Toro, Coclé, Panamá, and Darién and from tropical dry forest in Los Santos.

LASIACIS (Griseb.) Hitchc.

The genus is related to *Panicum,* but is distinguishable from it by having globose spikelets borne obliquely on pedicels, woolly pubescence at the apex of the spikelet bracts, black coloration of the spikelet, oil production of the glumes and sterile lemma, and very indurate fertile florets. The last three characters are adaptations for bird dispersal of the fruit (Davidse & Morton, 1973).

Lasiacis oaxacensis (Steud.) Hitchc., Proc. Biol. Soc. Wash. 24:145. 1911

Perennial; culms extensively creeping, rooting at nodes, the ultimate parts ± erect, 0.5–2 m tall, branched many times; internodes 2–5 mm thick, mostly herbaceous, solid, rarely partially hollow, glabrous. Sheaths usually glabrous, one or both margins usually ciliate; ligules prominent, 2.5–5 mm long; blades narrowly linear-lanceolate, acuminate, nearly symmetrical at base, (13)17–29 cm long, 1.2–2.4 cm wide, usually scabrous, especially along midrib. Panicles mostly terminal, 16–31 cm long; branches widely spreading, usually naked in lower half to two-thirds; spikelets in pairs or small clusters, mostly 3.8–4.2 mm long; first glume 7–11-veined, about half as long as second glume; second glume and lemmas woolly-pubescent at apex; sterile floret usually bearing a staminate flower; sterile palea as long as fertile floret. Seeds to 3 mm long and 1.6 mm wide. *Croat 7730.*

Common along the margins of larger clearings. Flowers in the latest part of the rainy season and early in the dry season. The fruits develop quickly and may be present during most of the dry season. Davidse and Morton (1973) reported that both *L. oaxacensis* and *L. sorghoidea* are most abundant in the early dry season.

As for *L. sorghoidea,* fruits are dispersed by small birds.

Mexico (Nayarit and Veracruz) south through Central America to Colombia, Venezuela, Ecuador, and Peru; Greater Antilles. In Panama, known from tropical moist forest in the Canal Zone, Bocas del Toro, and Panamá, also known from premontane wet forest and lower montane wet forest in Chiriquí and from premontane moist forest in Panamá (Juan Diaz).

Lasiacis procerrima (Hack.) Hitchc., Proc. Biol. Soc. Wash. 24:145. 1911

Coarse, erect, short-lived perennial or annual, 1–2(4) m tall; culms semiwoody at base, to 1 cm thick, usually unbranched, glabrous, rooting at lower nodes. Sheaths shorter than the internodes near base, closely overlapping near apex, glabrous or puberulent and glaucous, often ciliate near apex; ligules usually 0.5–1.5 cm long; blades linear-lanceolate or lanceolate, cordate-clasping at the base, 15–40 cm long, 2–5 cm wide, glabrous above, glabrate below, scabrous on the margins. Inflorescence a large, diffuse panicle to 1 m long, the branches mostly whorled toward the base, ridged, scabrous on ridges; spikelets broadly obovate, ca 3.5 mm long, sessile to short-pedicellate; first glume one-half to one-third as long as spikelet, lanate-pubescent at tip; second glume and sterile lemma slightly shorter than the fruit, similarly pubescent at tip. Fruits indurate, white and shiny. *Croat 6108.*

Uncommon, on the shore and in the Rear Lighthouse Clearing. Flowers and fruits throughout most of the rainy season, with most flowers appearing in June and July, most fruits by September.

Mexico (Sinaloa and Puebla) south throughout Central America to Colombia, Venezuela, Peru, and western Brazil. In Panama, ecologically variable; known from tropical moist forest all along the Atlantic slope and on the Pacific slope in the Canal Zone and Panamá; known also from tropical dry forest in Panamá (Taboga Island) and from premontane moist forest in Chiriquí and Panamá.

Lasiacis sorghoidea (Desv. ex Ham.) Hitchc. & Chase var. **sorghoidea**, Contr. U.S. Natl. Herb. 18:338. 1917

Carricillo, Millo

Perennial; culms caespitose, 1–10 m long, erect at base, arching and leaning on surrounding vegetation; internodes 5–15 mm thick, hollow, glabrous or with pubescence papillose or reduced to a single line, at least the older ones somewhat woody; nodes glabrous. Sheaths pubescent, rarely glabrate, often papillate-pubescent, especially near apex; ligules inconspicuous, usually to 1.5 mm long; blades mostly elliptic to linear-lanceolate, acuminate at apex, often asymmetrical at base, 8–20 cm long, 1–3.5 cm wide, usually puberulent above and velutinous below, the margins scabrid. Panicles terminal, (5)9–25(35) cm long; branches usually scabrous, at first ascending, spreading when fruiting; spikelets obovate to elliptic, green to purple-black at maturity of fruit, mostly 3.5–4 mm long; first glume 7–11-veined, one-third to one-half as long as second glume; second glume and sterile lemma woolly-pubescent at tip; sterile floret with or without a staminate flower; lemma 9–11-veined; palea one-half as long to same length as fertile floret. Seeds 1.8–2.3 mm long. *Croat 8009.*

Locally abundant along the edges of point clearings and along exposed banks at the margin of the lake; less frequent in the larger clearings. Flowers from the late rainy season to the middle of the dry season, but especially in the early dry season. Most fruits mature from middle to late dry season.

The entire spikelet becomes purple-black at maturity. The caryopsis is carried away with the rest of the spikelet by small birds.

Mexico (Oaxaca and Veracruz) through Central and South America to northern Argentina and southern Brazil; West Indies; usually at less than 1,000 m elevation but ranging to 1,800 m. In Panama, known from tropical moist forest in the Canal Zone, San Blas, and Panamá (doubtless elsewhere also), and from premontane wet forest in Chiriquí.

LEERSIA Sw.

Leersia hexandra Sw., Prodr. Veg. Ind. Occ. 21. 1788

Slender, aquatic perennial, to 1 m tall; culms rooting at lower nodes, erect, retrorsely hirsute at nodes. Sheaths shorter than the internodes, keeled, glabrous to sparsely retrorsely hirsute, ciliate, with a small auricle at apex; ligule to 3 mm long, truncate, fused with auricles; blades mostly 15–30 cm long, 5–12 mm wide, scabrid at least on midrib below. Inflorescence a narrow panicle to 20 cm long, the branches scabrous; spikelets 3 mm long, often purplish, inequilateral in side view, lacking glumes; lemma and palea equal in length, the veins scabrid, the margins hispid-ciliate. Fruits not seen. *Croat 12939.*

Frequent in the marshes at the edge of the lake. Flowers sporadically throughout the year, especially in the rainy season.

The prominent hispid-ciliate margins of the fruits seem to be logically designed for epizoochorous dispersal, but the fruit is probably dispersed by water currents to some extent also.

Virginia to Texas in the United States and south to Ecuador, Argentina, and southern Brazil; West Indies; cosmopolitan in Asia and elsewhere. In Panama, from tropical moist forest in the Canal Zone and Panamá and from tropical dry forest in Coclé.

LEPTOCHLOA Beauv.

Leptochloa virgata (L.) Beauv., Ess. Agrost. 71, 161, 166, pl. 15, f. 1. 1812

Perennial; culms mostly erect, usually about 1 m tall, glabrous. Sheaths as long as or longer than the internodes; blades 10–27 cm long, 4–12 mm wide, minutely scabrid, the margins scabrous. Inflorescence a panicle of slender racemes 5–12 cm long; rachis scabrid; spikelets 3- or 4-flowered, 2–3 mm long; glumes keeled, the keel scabrid, the second glume slightly longer than the first; lemmas flattened, imbricate, sparsely pilose on the margin, sometimes awned. Fruits oblong-elliptic, brown, ca 0.8 mm long. *Stattuck 461.*

Collected once by Shattuck at the end of Miller Trail; not seen recently on the island. Flowers throughout the year, especially during the rainy season.

The small, sharply pointed spikelets are possibly dispersed in part by adhering to passing animals and also in part by wind.

Occurring in most tropical and subtropical areas of the

Western Hemisphere; cosmopolitan in eastern Asia. In Panama, known from tropical moist forest in the Canal Zone, Bocas del Toro, Colón, Panamá, and Darién; known also from premontane moist forest in Los Santos and Panamá and from tropical dry forest in Panamá (Taboga Island).

LITHACHNE Beauv.

Lithachne pauciflora (Sw.) Beauv. ex Poir., Dict. Sci. Nat. 27:60. 1823

Monoecious perennial, to 60 cm tall; culms slender, glabrous, constricted at nodes, somewhat geniculate at the lower nodes. Sheaths except the uppermost shorter than the internodes, ± glabrous, sometimes ciliate, the lower ones bladeless; blades asymmetrical, acuminate, ± rounded to acute on one side of base, obtuse on the other, 5–10 cm long, 1.5–2.5 cm wide, glabrous except on and near the scabrid margins. Inflorescence a small panicle either terminal and staminate or axillary with 1 pistillate spikelet and 2 to several staminate ones; staminate spikelets lacking glumes, 4–6 mm long, soon deciduous; pistillate spikelets lacking first glume; second glume and sterile lemma nearly equal, exceeding fruit, long-acuminate, to 1 cm long, prominently veined, the veins scabrid near the apex. Fruits white, shiny, indurate, triangular, truncate at apex, 4–5 mm long. *Croat 6474.*

Common in the forest, mostly along trails, usually in small clumps. Flowering and fruiting throughout the year, especially in the rainy season.

Easily distinguished by the solitary, white, triangular fruits.

Mexico to northern Argentina; West Indies. In Panama, known from tropical moist forest in the Canal Zone and Panamá and from premontane wet forest in Chiriquí (Boquete).

OLYRA L.

Olyra latifolia L., Syst. Nat. ed. 10, 1261. 1759
O. cordifolia H.B.K.
Carricillo

Monoecious perennial, mostly 2–3 m tall and erect to arching, rarely to 5 m long and clambering; culms becoming woody, to 6 mm diam, often mottled with purple, freely branching at upper nodes. Sheaths glabrous to hispid, usually longer than the internodes, the lowermost short and nearly bladeless; blades lance-linear to lance-oblong, inequilateral, acuminate at apex, acute to rounded at base, very short-petiolate, the lower surface usually glabrous, the margins and midrib on upper surface scabrous, the upper surface papillate or papillate-scabridulous. Panicles terminal or upper-axillary, narrow to pyramidal, 5–17 cm long; branches and peduncles closely scabrid and sparsely hispid, the branches with a single pistillate spikelet at the end, the staminate spikelets scattered along branch below it; staminate spikelet reduced to the awned lemma and palea, deciduous, 4–5 mm long excluding awn, the awn 2–3 mm long; pistillate spikelet lacking first glume; second glume and sterile lemma 1–2 (2.5) cm long, long-acuminate, minutely scabridulous, diverging and exposing fruit at maturity, often purplish along margin. Fruits white, shiny, smooth, indurate, acute at apex, 5–6 mm long. *Croat 6634, 11713.*

Frequent in the forest. Plants flower and fruit principally throughout the rainy season, though they are occasionally fertile during the early dry season.

Plants are extremely variable, especially in size of leaves and inflorescences. The larger plants, which may represent tetraploid races, have previously been separated as *O. cordifolia* H.B.K., but the species is now considered polymorphic.

Throughout most tropical areas of the New World; Africa. In Panama, known from tropical moist forest in the Canal Zone, Veraguas, Panamá, and Darién, from premontane wet forest in Colón, Chiriquí, and Panamá, from tropical wet forest in Colón, and from premontane rain forest in Panamá (summit of Cerro Jefe).

OPLISMENUS Beauv.

Oplismenus burmanni (Retz.) Beauv., Ess. Agrost. 54. 1812
Pajita de ratón

Low, creeping annual, often in dense stands; culms slender, ± decumbent with erect branches, bearing fine soft trichomes, rooting at lower nodes, often purplish. Sheaths usually much shorter than the internodes, long-pubescent especially near the margins, the trichomes longer and sparser above; blades lance-elliptic, acuminate, 2–5 cm long, 5–15 mm wide, sparsely long-pubescent, the margins scabrid. Inflorescence a panicle of 3–6 short, widely spaced racemes to 2 cm long; rachis and spikelet densely villous; spikelets dense, 2–3 mm long; glumes nearly equal, long-awned; awns antrorsely scabrous, 10–14 mm long on first glume, 3–8 mm long on second glume; sterile lemma exceeding fruit, usually awnless. Fruits narrowly elliptic, indurate, 1.5–2 mm long. *Croat 6947.*

Common to locally abundant in the forest on trails and sometimes on stream banks; common in clearings. Flowering chiefly at the beginning of the dry season.

Throughout the tropics of Western and Eastern hemi-

KEY TO THE SPECIES OF OPLISMENUS

Awns antrorsely scabrous; pubescence of rachis obscuring surface, the surface variably hispid and puberulent . . . . *O. burmanni* (Retz.) Beauv.
Awns smooth; pubescence of rachis not obscuring surface, the surface densely and uniformly scabrid . . . . *O. hirtellus* (L.) Beauv.

spheres. In Panama, known from tropical moist forest in the Canal Zone, Bocas del Toro, Panamá, and Darién, from tropical dry forest in Panamá (Taboga Island), from premontane wet forest in Chiriquí, Los Santos, and Panamá, and from lower montane rain forest in Chiriquí.

Oplismenus hirtellus (L.) Beauv., Ess. Agrost. 54. 1812

Freely branching perennial, to 70 cm long; culms decumbent-spreading, rooting at the nodes, glabrous to short-pubescent, the pubescence most dense in slender rows. Sheaths shorter than or ± equal to the internodes, glabrous to hispid, the margins ciliate; blades 3–10.5 cm long, 10–20 mm wide, ± asymmetrical, usually short-pubescent on upper and lower surfaces, the margins scabrid. Inflorescence to 20 cm long, a terminal panicle of 3–7 widely spaced racemes 1.5–4 cm long, the rachis minutely and densely scabrid with pustular hispid trichomes; spikelets 3–4 mm long excluding awns; glumes about equal, two-thirds the length of the sterile lemma, sparsely pubescent, awned; awns glabrous, 5–12 mm long on first glume, 2–4 mm long on second; sterile lemma exceeding fruit, acuminate. Fruits indurate, white, shiny, ca 3 mm long, minutely striate. *Croat 13267.*

Common on trails in the forest and sometimes on rocks in streams. Flowers are most abundant in the early dry season.

Throughout the tropics of the Western Hemisphere. In Panama, known from tropical moist forest in the Canal Zone, Bocas del Toro, Panamá, and Darién, from premontane wet forest in Chiriquí, Panamá, and Darién, and from lower montane wet forest in Chiriquí.

ORTHOCLADA Beauv.

Orthoclada laxa (L. C. Rich.) Beauv., Ess. Agrost. 70, 149, 168. 1812

Perennial, ± erect, 0.5–2 m tall; culms glabrous or nearly so, sometimes geniculate at lower nodes. Sheaths glabrous to short-pubescent, prominently veined; blades 5–20 cm long, 1–3.5 cm wide, the petioles distinct, 2–3 cm long, sparsely hispid above, glabrous to short-pubescent and conspicuously cross-veined below. Panicles diffuse, 15–30 cm long, about as wide, the branches

slender, minutely antrorse-pubescent, the spikelets only near ends; spikelets 6–7 mm long, long-pedicellate, glabrous but with scabrous veins and margins of glumes and lemmas; first glume shorter than second, both shorter than sterile lemma; sterile lemma long-acuminate. Fruits indurate, dark brown, ± glabrous, weakly striate, 2.7–4 mm long. *Croat 7840.*

Frequent in the forest. Flowering mostly in the early dry season.

Spikelets disarticulate immediately below the spikelet and are possibly dispersed by attachment of the long-acuminate sterile lemma to passing animals.

Mexico to Peru and Brazil. In Panama, known from tropical moist forest in the Canal Zone, Bocas del Toro, and Panamá and from tropical wet forest in Colón.

See Fig. 46.

ORYZA L.

Oryza latifolia Desv., Jour. de Bot. (Desv.) 1:77. 1813

Coarse, erect perennial, 1–2 m tall; culms stout. Sheaths longer than the internodes, glabrous, keeled near the apex, the margins scabridulous; blades 15–50 cm long, 1–4 cm wide, sparsely pubescent or scabrous above and below, the margins scabrid. Panicles 25–40 cm long, the branches verticillate, scabrous, with spikelets in upper two-thirds; glumes minute; lemma indurate, 5–6 mm long, stoutly scabrous, minutely striate, awned, the awn 1–2 cm long. *Shattuck 849.*

Collected several times by Shattuck along the shore, but not seen recently. Flowering and fruiting throughout the year.

Belize to Brazil; West Indies; in marshes and along rivers and ditches. In Panama, known from tropical moist forest in the Canal Zone, Bocas del Toro, Colón, Panamá, and Darién.

PANICUM L.

Panicum fasciculatum Sw., Prodr. Veg. Ind. Occ. 22. 1788
Espigadilla, Granadilla

Erect annual, 15–100 cm tall; culms glabrous to hispid. Sheaths hispid, especially on margins; blades 10–25 cm

KEY TO THE SPECIES OF PANICUM

This key includes some common species often confused with *Panicum*.

Spikelets short-pedicellate along one side of the panicle branches forming spikelike or 1-sided racemes:
  Fruits transversely rugose . . . .*Paspalidium geminatum* Stapf
  Fruits smooth, not rugose:
    Blades lanceolate or ovate-lanceolate . . . . *Panicum pulchellum* Raddi
    Blades linear, often elongate, frequently cordate at base:
      Spikelets nearly sessile, densely arranged along 1 side of the rachis:
        Plants usually less than 1 m tall, growing in clearings or on trails; blades less than 20 cm
          long . . . .*Panicum pilosum* Sw.
        Plants usually more than 1 m tall, growing in marshes or moist areas; blades often more
          than 20 cm long . . . . *Panicum milleflorum* Hitchc. & Chase
      Spikelets sessile or pedicellate, in part on short branches along the lower side of the rachis;
        culms with nodes conspicuously bearded . . . .*Panicum polygonatum* Schult.

Fig. 47. *Panicum grande*

Fig. 46. *Orthoclada laxa*

Fig. 48. *Panicum grande*

Spikelets in open or contracted panicles but not on 1-sided racemes:
  Fruits transversely rugose:
    Plants perennial; culms in large coarse clumps, 1–2.5 m tall; spikelets green; cilia of ligule 3–4
      mm long . . . .*Panicum maximum* Jacq.
    Plants annual; culms slender, to 1 m tall; spikes brown to reddish; cilia of ligule ca 1 mm long
      . . . .*Panicum fasciculatum* Sw.
  Fruits not rugose:
    First glume much less than one-fourth as long as the spikelet . . . . *Panicum trichanthum* Nees
    First glume more than one-fourth as long as the spikelet:
      Blades as much as 1 m long and 6 cm wide . . . . *Panicum grande* Hitchc. & Chase
      Blades much smaller:
        Fruits crested at apex . . . .*Acroceras oryzoides* Stapf
        Fruits not crested:
          Panicles 40–60 cm long; branches in verticils . . . . *Panicum mertensii* Roth.
          Panicles not more than 30 cm long; branches not in verticils:
            Inflorescences widely branched; spikelets ca 1 mm long, borne on pedicels many
              times longer than spikelets . . . . *Panicum trichoides* Sw.
            Inflorescences not widely branched; spikelets more than 3 mm long, borne on pedi-
              cels only a few times longer than spikelets . . . . *Ichnanthus pallens* Benth.

---

long, 5–20 mm wide, mostly glabrous but with scabrid margins. Panicles mostly 10–15 cm long, rather dense; branches scabrid; spikelets 2.5–3 mm long, dark purplish-brown, glabrous; first glume one-third to one-half as long as spikelet; second glume and sterile lemma exceeding fruit. Fruits white, 1.5–2 mm long, transversely rugose. *Croat 11612.*

Infrequent in both the Laboratory Clearing and the Rear #8 Lighthouse Clearing. Flowering and fruiting throughout the year, especially during the rainy season.

Throughout most tropical areas of the Western Hemisphere. In Panama, known from tropical moist forest in the Canal Zone, Bocas del Toro, San Blas, Chiriquí, Panamá, and Darién and from tropical wet forest in Colón.

Panicum grande Hitchc. & Chase, Contr. U.S. Natl.
  Herb. 17:529, f. 143. 1915
    Guinea grass

Stoloniferous perennial, 1.5–2 m or more tall; culms hollow, decumbent at the base; nodes with a narrow band of short sericeous trichomes. Sheaths mostly longer than the internodes, glabrous or sparsely hispid at the throat; ligules 2–3 mm long, the upper margin irregular and fringed; larger blades (mid-culm) to 1 m long and 6 cm wide, the margins scabrous. Panicles terminal, 30–60 cm long, to 40 cm wide; axis ± ribbed, the ribs scabrid; branches stiffly ascending, naked at base; spikelets oblong-elliptic, narrowly acute at apex, ca 2.5 mm long, glabrous; first glume to three-fourths the length of the spikelet; second glume as long as the spikelet, ribbed. Fruits 1.3–1.8 mm long, pale, smooth, shiny. *Croat 12700.*

Occasional, along the shore, often forming huge stands during the rainy season on silt deposits in Fuertes Cove. Seasonal behavior uncertain. Probably flowering and fruiting in the rainy season.

Distinguished from other *Panicum* by its large leaves.

Nicaragua to Venezuela; Trinidad. In Panama, known from tropical moist forest in the Canal Zone, Colón, Panamá, and Darién.

See Figs. 47 and 48.

Panicum maximum Jacq., Coll. Not. 1:76. 1786
  Guinea grass

Tall, erect perennial, 1–2.5 m high; culms stout, glabrous but with bearded nodes. Sheaths mostly shorter than the internodes, papillose-hirsute becoming glabrous; ligules truncate and erose with dense cilia behind the ligule 4–5 times as long as the ligule, 3–4 mm long; blades to 70 cm long and 3 cm wide, glabrous to sparsely hispid with scabrous trichomes. Inflorescence a diffuse terminal panicle 20–45 cm long; branches slender, scabrid; spikelets long-pedicellate, 2.7–3.7 mm long, glabrous; first glume one-fourth to one-third as long as spikelet; second glume and sterile lemma nearly equal to each other and exceeding fruit. Fruits ca 2.7 mm long, greenish-white, transversely rugose. *Croat 6001, 11990.*

Common locally at the Laboratory Clearing along the stairs from the dock and in the Lighthouse Clearing. One of the most abundant roadside and pasture grasses in the isthmus.

Introduced throughout the tropics of the Western Hemisphere; native to Africa. In Panama, known from tropical moist forest in the Canal Zone, Veraguas, Los Santos, Herrera, Panamá, and Darién, from tropical dry forest in Panamá, and from premontane dry forest in Coclé.

Panicum mertensii Roth in R. & S., Syst. Veg.
  2:458. 1817
    P. megiston Schult.

Stout, erect perennial, 1–2 m tall; culms glabrous. Sheaths shorter than the internodes, sparsely papillose-pilose; ligules irregular and ciliate at apex, lacking long trichomes behind the ligule; blades 15–40 cm long, 15–30 mm wide, glabrous but with scabrid margins. Inflorescence a terminal panicle 40–60 cm long, the main axis ± stout, the branches slender, whorled at intervals; spikelets obovate to subglobose, pedicellate, 2.7–3.3 mm long, glabrous; first glume about one-fourth as long as spikelet; second glume and sterile lemma covering fruit. Fruits

greenish-white, ca 3 mm long, smooth, obovate. *Croat 11804.*

Infrequent, but locally abundant in quiet marshes. Flowers and fruits in the rainy season, mostly in the early rainy season.

Mexico to Paraguay; Cuba. In Panama, known from tropical moist forest in the Canal Zone and Panamá.

Panicum milleflorum Hitchc. & Chase, Contr. U.S. Natl. Herb. 17:494, f. 70. 1915

Perennial, ± erect, to 2 m tall (usually less), nearly glabrous but with hirsute nodes and papillose-hirsute sheaths; culms rooting at lower nodes. Blades 15–35 cm long, 12–20 mm wide, sometimes sparsely scabrous, the margins scabrous. Panicles 25–45 cm long; branchlets numerous, short, densely flowered, angulate, the angles scabrid, the base of the branchlets with longer trichomes; spikelets secund, ca 1.3 mm long, prominently veined; first glume about half as long as spikelet; second glume ± equaling sterile lemma. Fruits indurate, ca 1 mm long, shiny, white. *Croat 6408.*

Uncommon, occurring in swampy places; collected on the sandbar in Fuertes Cove. Flowering and fruiting in the rainy season.

Belize, and Panama to Brazil. In Panama, known from tropical moist forest in the Canal Zone and Panamá.

Panicum pilosum Sw., Prodr. Veg. Ind. Occ. 22. 1788
Grama de camino

Slender, erect perennial, 25–60 cm tall; culms geniculate and rooting at lower nodes, glabrous. Sheaths longer than the internodes, glabrous except for long cilia on margins especially near apex; ligules lacking; blades 5–22 cm long, 6–13 mm wide, glabrous below, sparsely hispid above, the margins scabrid. Panicles terminal, slender, 5–20 cm long, the branches simple, short densely flowered; rachis both scabrid and long-hispid; spikelets 1.3–1.5 mm long, secund, glabrous; first glume about half as long as spikelet; second glume and sterile lemma covering fruit. Fruits ovoid-ellipsoid ca 1 mm long, white, shiny. *Croat 16551.*

Locally abundant to common in clearings; locally common on trails in the forest. Flowering and fruiting throughout the year, more abundantly in the rainy season.

Mexico to Paraguay and Brazil; West Indies. In Panama, known from tropical moist forest in the Canal Zone, San Blas, Chiriquí, Panamá, and Darién, from tropical dry forest in Coclé and Panamá, from premontane wet forest in Panamá and Darién, and from tropical wet forest in Colón.

Panicum polygonatum Schrad. ex Schult., Mant. 2:256. 1824

Slender perennial; culms decumbent at base, rooting at lower nodes, branching, bearded at the nodes. Sheaths shorter than the internodes, glabrous but with long-ciliate margins; blades 5–12 cm long, 5–12 mm wide, cordate at the base, nearly glabrous but with scabrid margins.

Panicles terminal or axillary, 10–25 cm long, branched many times, the branches angulate and scabrid; spikelets sessile or pedicellate, 1.3–2 mm long, glabrous but with the median veins scabrid near the apex; fertile floret 1; first glume one-fourth to one-half as long as spikelet; second glume and sterile lemma covering fruit. Fruits white, smooth, ca 1.3 mm long. *Croat 5255.*

Infrequent; collected on the sandbar in Fuertes Cove and on Wheeler Trail in moist areas. It may be locally abundant. Flowering and fruiting late in the dry season and throughout the rainy season.

Similar to *P. milleflorum*, which occurs in similar habitat, but distinguished from it by the smaller blades.

Mexico to Paraguay; Trinidad and Jamaica. In Panama, known from tropical moist forest in the Canal Zone, all along the Atlantic slope, and in Panamá and Darién; also known from premontane wet forest in Coclé (El Valle) and from tropical wet forest in Colón.

Panicum pulchellum Raddi, Agrost. Bras. 42. 1823

Perennial; culms very slender, rooting at lower nodes, creeping at base, ultimately ascending, pubescent at least at nodes. Sheaths shorter than the internodes, pilose with long-ciliate margins; blades ovate-lanceolate, somewhat asymmetrical, 1.5–6 cm long, 8–18 mm wide, sparsely pubescent to glabrous above, usually pubescent throughout below, the margins scabrid. Inflorescence a panicle to 15 cm long of short racemes; racemes 0.5–1.5 cm long; rachis and pedicel scabridulous; spikelets to 2 mm long, pubescent; first glume about one-third the length of the spikelet; second glume and sterile lemma covering fruit; sterile lemma with two conspicuous glands near the middle. Fruits ellipsoid, ca 1.2 mm long, pale, shiny. *Croat 9269.*

Uncommon; collected on steep banks and trails in moist areas of the forest. Flowering and fruiting throughout the year, especially during the dry season.

Southern Mexico to Brazil. In Panama, known from tropical moist forest in the Canal Zone, Bocas del Toro, Colón, Los Santos, and Panamá and from premontane wet forest in Chiriquí.

Panicum trichanthum Nees, Agrost. Bras. 210. 1829

Perennial, 1–2 m long; culms clambering, geniculate and rooting at nodes, glabrous. Sheaths shorter than the internodes, glabrous, often ciliate; blades 8–15 cm long, 10–25 mm wide, ± asymmetrical, sparsely pubescent on upper and lower surfaces, usually becoming glabrous, the margins scabrid. Panicles diffuse, 20–30 cm long; branches ± terete, nearly smooth; spikelets 1–1.5 mm long on long slender pedicels, glabrous; fertile floret 1; first glume very small; second glume and sterile lemma exceeding fruit. Fruits ellipsoid, indurate, white, shiny. *Croat 5254.*

Rare, occurring on sandbars in deep coves along the shore. Flowers and fruits throughout the year, but especially in the rainy season.

Mexico to Paraguay; Greater Antilles and Trinidad. In Panama, known from tropical moist forest in the Canal Zone, Bocas del Toro, Los Santos, Panamá, and Darién.

Panicum trichoides Sw., Prodr. Veg. Ind. Occ. 24. 1788

Annual, 20–60 cm tall; culms slender, decumbent at base, sparsely pubescent. Sheaths except the uppermost much shorter than the internodes, papillose-hispid, prominently ciliate; blades ovate-lanceolate, inequilateral at base, 4–7 cm long, 8–15 mm wide, sparsely pubescent on both surfaces, prominently hispid-ciliate. Panicles terminal, 5–20 cm long, the branchlets many, slender; spikelets 1–1.5 mm long on slender pedicels mostly 4–12 mm long, nearly glabrous; first glume one-third to one-half as long as spikelet; second glume and sterile lemma slightly shorter than mature fruit. Fruits white, shiny. *Croat 10259.*

An infrequent weedy grass found in clearings and open areas. Flowers throughout the year.

Mexico to Peru and Brazil; West Indies. In Panama, known from tropical moist forest in the Canal Zone, all along the Atlantic slope, and from Chiriquí, Veraguas, Panamá, and Darién; also known from tropical dry forest in Panamá (Taboga Island) and from premontane wet forest in Coclé and Panamá.

PASPALIDIUM Stapf

Paspalidium geminatum Stapf in Prain, Fl. Trop. Africa 9:583. 1920

Panicum geminatum Forssk.

Perennial, 0.5–1.5 m tall; culms clumped, decumbent and usually succulent at base, glabrous. Sheaths longer than the internodes, glabrous; blades 10–28 cm long, 3–6 mm wide, glabrous below, scaberulous above, the margins scabrid. Inflorescence a panicle 12–30 cm long, of short ascending racemes 2–4 cm long; rachis of racemes sinuous, scabrid, winged; spikelets 2–2.7 cm long, appressed; first glume about one-fourth as long as spikelet; second glume about half as long as spikelet. Fruits oblong-elliptic, white, shiny, weakly transverse-rugose. *Croat 11808.*

Growing in large clumps atop exposed tree stumps in Gigante Bay and off Gross Point. The plants become quite dried during the dry season, when the water recedes well below the clusters. Apparently flowers sporadically throughout the year.

Throughout the tropics. In Panama, known only from tropical moist forest in the Canal Zone.

PASPALUM L.

Paspalum conjugatum Bergius, Acta Helv. Phys.-Math. 7:129, pl. 8. 1762

Stoloniferous perennial, 30–100 cm or more tall; culms sometimes decumbent at base, simple or sparingly branched, glabrous; stolons long, leafy. Sheaths longer than the internodes, pubescent on collar, ciliate; blades 8–17 cm long, 5–15 mm wide, usually sparsely pubescent throughout, the margins scabrous or ciliate. Inflorescences terminal, or 2 (rarely 3) racemes (conjugate or nearly so), 7–16 cm long; rachis winged, the margins scabrid; spikelets ovate, 1.4–2.2 mm long, solitary, short-pedicellate, plano-convex, long-ciliate; first glume and sterile lemma covering fruit. Fruits plano-convex, ovate, 1.4–2.2 mm long. *Croat 5854.*

KEY TO THE SPECIES OF PASPALUM

Inflorescence of solitary racemes (each borne on a separate peduncle arising from a leaf sheath) . . .
. . . . *P. decumbens* Sw.
Inflorescence a panicle or of more than 1 raceme:
  Inflorescence of 2 (rarely 3) subdigitate racemes:
    Spikelets usually less than 2 mm long, with long, fine, white trichomes on margins; plants
      stoloniferous . . . .*P. conjugatum* Bergius
    Spikelets usually more than 3 mm long, the margins glabrous; plants rhizomatous . . . .
    . . . . *P. notatum* Flugge
  Inflorescence a panicle or with racemes not clustered near apex of peduncle:
    Spikelets with long silky trichomes giving the inflorescence a feathery appearance at maturity
    . . . . *P. saccharoides* Nees
    Spikelets not conspicuously pubescent:
      Wing of rachis broader than width of spikelet; plants usually aquatic . . . . *P. repens* Bergius
      Wing of rachis narrower than width of spikelet; plants not aquatic:
        Spikelets mostly more than 2 mm long; fruits light to dark brown:
          Fruits smooth and very shiny; plants usually less than 1 m tall . . . . . .*P. plicatulum* Michx.
          Fruits striate, ± dull; plants usually more than 1 m tall . . . . *P. virgatum* L.
        Spikelets usually less than 1.5 mm long; fruits green:
          Rachis with long hispid trichomes borne sparsely along all of its length; pubescence of
            spikelets patulous, moderately long; plants rare or possibly no longer present on
            BCI . . . . *P. microstachyum* Presl
          Rachis with long hispid trichomes lacking or restricted to near base; pubescence of
            spikelets appressed, ± short; plants common in clearings . . . . *P. paniculatum* L.

Common to locally abundant in clearings. Apparently flowering all year, especially during the late dry season and throughout the rainy season.

Fruits are dispersed by birds and by passing animals, attaching themselves by the long villous trichomes on the margin of the spikelet. Ridley (1930) reported the trichomes to be coated with a viscous substance.

Cosmopolitan tropical weed. In Panama, known from tropical moist forest in the Canal Zone, all along the Atlantic slope, and in Herrera, Panamá, and Darién; known also from premontane wet forest in Chiriquí and Panamá and from tropical wet forest in Colón.

Paspalum decumbens Sw., Prodr. Veg. Ind. Occ. 22. 1788

Small, creeping perennial, usually 10–20(70) cm long; culms decumbent-spreading, rooting at lower nodes, freely branching. Sheaths keeled, the margins usually densely ciliate; blades 2–10(15) cm long, 6–12 mm wide, rounded at base, softly pubescent to nearly glabrous, the margins ciliate and scabrid, 1–8 peduncles arising from upper sheaths. Racemes solitary, 1–3.5 cm long; spikelets paired, plano-convex, short-pedicellate with scabrid pedicels, obovate, 1.3–1.7 mm long, glabrous; first glume ca 0.4 mm long, ciliate at the apex; second glume half as long as spikelet; sterile lemma as long as fruit. Fruits pale, minutely striate. *Croat 8536.*

Infrequent, in open areas of the forest and at the ends of trails along the shore. Apparently flowering and fruiting throughout the year, perhaps more often in the rainy season.

Guatemala to Brazil and Bolivia; West Indies. In Panama, known from tropical moist forest in the Canal Zone, all along the Atlantic slope, in Herrera and Panamá, known also from tropical dry forest in Panamá (Taboga Island), tropical wet forest in Colón, and premontane rain forest in Coclé.

Paspalum microstachyum Presl, Rel. Haenk. 1:215. 1830

Annual, 20–135 cm tall; culms slender, ± erect, geniculate and rooting at lower nodes. Sheaths shorter than the internodes, keeled, glabrous to pubescent; blades 3–30 cm long, 6–20 mm wide, glabrous to pubescent, the margins scabrid. Panicles of 6–35 slender, solitary, or fasciculate racemes 3–8 cm long; rachis narrowly winged, with scattered long trichomes; spikelets paired on slender pedicels, plano-convex, ca 1.5 mm long, short-spreading pubescent; first glume lacking; second glume and sterile lemma covering fruit. Fruits ca 1.5 mm long, pale green, smooth. *Shattuck 465.*

Collected by Shattuck at the end of Miller Trail, probably in the Front Lighthouse Clearing; the species may no longer occur on the island. Flowering and fruiting chiefly at the end of the rainy season.

Guatemala to Brazil. In Panama, known from tropical moist forest in the Canal Zone, Bocas del Toro, Chiriquí, and Panamá and from tropical dry forest in Panamá (Taboga Island).

Paspalum notatum Flugge, Monogr. Pasp. 106. 1810

Rhizomatous perennial, 15–50 cm tall; culms glabrous. Leaves crowded; sheaths usually glabrous, ciliate near apex; blades 4–30 cm long, 3–10 mm wide, glabrous, ciliate near the base. Racemes 2 (rarely 3), subconjugate, 3–12 cm long; spikelets solitary, ovate or obovate, 2.5–3.8 mm long, 2–2.7 mm wide, plano-convex, glabrous; first glume lacking; second glume and sterile lemma covering fruit. Fruits 2.5–3.5 mm long, white, shiny, minutely striate. *Croat 16565.*

Infrequent, in the Laboratory Clearing. Flowering and fruiting chiefly in the early rainy season (June to August).

Mexico to Argentina; West Indies. In Panama, known from premontane wet forest in Chiriquí and Coclé, from tropical moist forest in the Canal Zone, Chiriquí, and Panamá, and from lower montane wet forest in Chiriquí (Volcán).

Paspalum paniculatum L., Syst. Nat. ed. 10, 855. 1759

Perennial, 0.3–1(2) m tall; culms ± stout, clumped, erect, glabrous but with bearded nodes. Sheaths keeled, longer than the internodes, usually papillose-hispid throughout; blades 10–50 cm long, 6–25 mm wide, sparsely to densely hispid on upper and lower surfaces, the margins scabrous, the ligules long-ciliate. Inflorescence 5–30 cm long, a panicle of racemes 4–12 cm long; rachis of racemes bearded at the base; spikelets paired, broadly elliptic, 1.3–1.5 mm long, crowded, plano-convex, softly pubescent; first glume lacking; second glume and sterile lemma barely covering fruit. Fruits smooth, shiny, pale green. *Croat 8489.*

Common in the Laboratory Clearing between the dining hall and the dock as well as in the area north of the dock; may occur in other clearings as well. Apparently flowering throughout the year, but less frequently in the early dry season.

Mexico to Argentina; West Indies; West Africa and Australia. In Panama, known from tropical moist forest in the Canal Zone, Bocas del Toro, San Blas, Chiriquí, Panamá, and Darién and from tropical dry forest in Panamá (Taboga Island).

See Fig. 49.

Paspalum plicatulum Michx., Fl. Bor. Amer. 1:45. 1803

Cabezona

Erect perennial, mostly 50–100 cm tall; culms in small tufts, rarely branched, glabrous. Sheaths longer than the internodes, keeled, glabrous or sparsely pubescent; blades 15–50 cm long, 3–10 mm wide, glabrous or sparsely pilose above, scabrid on and near the margins. Inflorescence a panicle 20–30 cm long of ascending racemes; racemes often tufted at base, 2–10 cm long; spikelets plano-convex, paired, 2–2.8 mm long, on scabrid pedicels, broadly elliptic, glabrous or appressed-pubescent; first glume lacking; second glume and sterile lemma covering fruit, keeled; sterile lemma weakly cross-wrinkled near the margins. Fruits dark brown, smooth, shiny. *Croat 10333.*

Apparently rare, though locally abundant in the Laboratory Clearing. Flowers most abundantly during the rainy season.

Southern United States to Argentina; West Indies. In Panama, known from tropical moist forest in the Canal Zone, Colón, San Blas, Chiriquí, Herrera, Panamá, and Darién, from tropical dry forest in Coclé and Panamá, from premontane wet forest in Colón, Chiriquí, Coclé, and Panamá, from tropical wet forest in Colón, and from lower montane wet forest in Chiriquí (Volcán).

Paspalum repens Bergius, Acta Helv. Phys.-Math. 7:129, pl. 7. 1762

Aquatic perennial; culms spongy, rooting at nodes. Sheaths longer than the internodes, long-pubescent, becoming glabrous, auriculate at apex; blades 10–27 cm long, 10–25 mm wide, scabrid and with fewer long trichomes. Panicles 10–20 cm long, the racemes spreading or reflexed, 3–6 cm long; rachis winged, the wing wider than spikelets, scabrid throughout; spikelets solitary, plano-convex or sometimes convex on both sides, oblong-elliptic to narrowly ovoid, 1.5–2 mm long, glabrous to short-pubescent, whitish; first glume lacking; second glume and sterile lemma exceeding fruit. Fruits 1.4–1.7 mm long, white, smooth. *Shattuck 403.*

Collected by Shattuck apparently at the margin of the lake at or near the end of Thomas Barbour Trail. Since that area of the shoreline has changed drastically over the years, it is unlikely that the same population still exists. The species could still occur on the island, but it has not been seen in recent years. Flowering throughout the year, especially in the rainy season.

United States south to Paraguay. In Panama, known from tropical moist forest in Canal Zone and Panamá.

Paspalum saccharoides Nees in Trin., Gram. Icon. 1, pl. 107. 1828

Perennial, 1–2 m or more long; culms stout, decumbent or creeping at the base, vinelike and clambering or sub-erect, glabrous. Sheaths longer than the internodes, glabrous or sparsely pilose, ciliate with long silky trichomes; blades 15–30 cm long (shorter at the base), 8–15 mm wide, tapering to a long-involute apex, finely pilose above, glabrous below, long-ciliate near base. Inflorescences feathery at maturity, 15–30 cm long, composed of many slender, drooping racemes crowded on a short axis; rachis sinuous, the margins scabrid; spikelets 2–3 mm long, solitary, borne in 2 rows on one side of flattened rachis; first glume lacking; second glume thin, sparsely pubescent, fringed with long silky trichomes 5–8 mm long; sterile lemma slightly shorter than the glume, also thin,

glabrous. Fruits ca 2 mm long, white, smooth. *Croat 11086.*

Occasional, on steep lakeside banks on the north and east sides of the island, especially along Buena Vista Reach near Fairchild Point. The species is usually restricted to steep banks elsewhere in Panama also. Flowering especially at the beginning of the rainy season (April to October).

Seeds are probably wind dispersed.

Costa Rica to Ecuador and Bolivia. In Panama, known from tropical moist forest in the Canal Zone and Panamá, from premontane moist forest in Chiriquí, and from premontane wet forest in Colón, Coclé, and Panamá.

Paspalum virgatum L., Syst. Nat. ed. 10, 855. 1759
  Cabezona

Robust perennial, 1–2 m tall, growing in dense, isolated clumps; culms stout, unbranched, erect. Sheaths longer than the internodes, the lower ones often purplish, the collar and margins glabrous or sparsely hirsute; blades 30–75 cm long, 6–25 mm wide, nearly glabrous but with scabrous margins. Inflorescences 15–30(40) cm long, of few to many thick racemes, the racemes to 15(20) cm long; rachis winged, with a tuft of long trichomes at the base, the margins scabrous; pedicels flattened, the margins scabrid; spikelets obovate, paired, crowded, plano-convex, 2.2–3 mm long, often purplish; first glume lacking; second glume and sterile lemma covering fruit, glabrous or short-appressed pubescent and the margins with longer trichomes especially near the apex. Fruits chestnut-brown, 2–2.7 mm long, minutely striate. *Croat 10244.*

Infrequent, in clearings, especially in the Rear #8 Lighthouse Clearing and on Burrunga Point. Flowering throughout the year, principally in the rainy season.

Texas and Mexico to Brazil; West Indies. In Panama, known from tropical moist forest in the Canal Zone, Bocas del Toro, San Blas, Panamá, and Darién, from premontane wet forest in Colón, Chiriquí, and Panamá, from tropical dry forest in Panamá (Taboga Island), from tropical wet forest in Bocas del Toro and Colón, and from lower montane wet forest in Chiriquí.

PHARUS P. Browne

Pharus latifolius L., Syst. Nat. ed. 10, 1269. 1759
  Pega-pega

Erect, monoecious perennial, 40–80(100) cm tall; culms glabrous. Sheaths longer than the internodes, glabrous; blades oblanceolate to oblong-oblanceolate, acuminate, 10–30 cm long, 2.5–8.5 cm wide, pubescent or glabrous

KEY TO THE SPECIES OF PHARUS

Fruits pubescent only near the tip, curved near apex; blades mostly more than 4 cm wide . . . .
. . . . *P. latifolius* L.
Fruits pubescent nearly to the base, straight; blades less than 4 cm wide . . . . *P. parvifolius* Nash

Fig. 49. *Paspalum paniculatum*

Fig. 50. *Pharus latifolius*

Fig. 51. *Streptochaeta sodiroana*

but with scabrid margins, cross-veined, distinctly petiolate at base, the petioles 2–10 cm long. Panicles 10–30 cm long, open; branches slender, densely pubescent; spikelets appressed, brown, paired, one long-pedicellate and staminate, the other nearly sessile, pistillate, and much larger; staminate spikelets 2.7–4 mm long, minutely pubescent, with 6 stamens, the first glume about half as long as spikelet or less; pistillate spikelets 10–18 mm long; glumes dark brown, minutely pubescent, 3–6 mm shorter than the fruit, the second glume a little longer than the first. Fruits narrowly cylindrical, light brown, densely pubescent on exposed tip with gland-tipped trichomes. *Croat 8592.*

Abundant in all parts of the forest. Flowering and fruiting throughout the year.

The gland-tipped trichomes of the fruit disperse by very effectively attaching to animals.

Belize to Peru and Brazil; West Indies. In Panama, known from tropical moist forest in the Canal Zone, all along the Atlantic slope, and in Chiriquí, Panamá, and Darién and from premontane wet forest in Chiriquí and Panamá.

See Fig. 50.

Pharus parvifolius Nash, Bull. Torrey Bot. Club
   35:301. 1908
   Pega-pega

Monoecious perennial, 60–100 cm or more tall, decumbent at base, rooting at lower nodes, glabrous. Sheaths longer than the internodes, glabrous; blades oblong-lanceolate, acuminate, petiolate, 10–23 cm long, 1.5–3.8 cm wide, cross-veined, glabrous but with minutely scabrid margins. Panicles open, mostly 10–25 cm long; branches slender, pubescent; spikelets appressed, paired, one long-pedicellate and staminate, the other nearly sessile, larger, and pistillate; staminate spikelets 2–3 (4.7) mm long, dark brown, glabrous, the first glume about half as long as spikelet; pistillate spikelets 11–14 mm long; glumes dark brown, about half as long as the fruit. Fruits light brown, slender, densely glandular-pubescent throughout. *Croat 8335.*

Occasional, in the older forest, often locally abundant. Flowering and fruiting throughout the year.

Costa Rica to Brazil; West Indies. In Panama, known from tropical moist forest on BCI and in Darién.

PHRAGMITES Trin.

Phragmites australis (Cav.) Trin. in Steud., Nom. Bot.
   ed. 2, 2:324. 1841
   P. communis Trin.
   Common reed

Coarse perennial, 2–4 (5) m tall, stoloniferous, rhizomatous, usually in large colonies; culms erect, glabrous. Sheaths longer than the internodes, glabrous; blades to 50 cm long, 1–2.5 cm wide, glabrous but with scabrous margins. Panicles 25–55 cm long, plumelike; branches drooping, slender, whorled at base, angulate, scabrous; spikelets 10–20 mm long, bearing several flowers; glumes

unequal, dark brown, the first 4–5 mm long, the second 6–8 mm long; lemmas 10–13 mm long, awned-acuminate; rachillas bearded with long white trichomes nearly as long as the spikelet. Fruits not seen. *Croat 12585.*

Occasional, in shallow water at the edge of the island. The populations on the unprotected north and east sides of the island are often very small, while the populations in the protected areas on the south side are larger. Flowering and fruiting principally in the late rainy and early dry seasons.

Fruits usually disarticulate beneath the long-bearded rachilla, which probably aids in wind dispersal. They are also dispersed to some extent by water currents.

Temperate and tropical areas throughout the world. In Panama, known from tropical moist forest in the Canal Zone, Colón, Panamá, and Darién and from premontane wet forest in Chiriquí (Boquete).

POLYTRIAS Hack.

Polytrias amaura (Buse ex Miq.) O. Kuntze, Rev. Gen.
   Pl. 2:788. 1891
   P. praemorsa (Nees) Hack.
   Java grass

Stoloniferous perennial, to 35 cm tall; culms decumbent-spreading, slender, glabrous with some nodes pubescent. Sheaths glabrous to pilose, longer than the internodes except near apex; blades 2–6 cm long, 1–3 mm wide, pilose on both surfaces, the margins scabrid. Racemes terminal, solitary, 2.5–5.5 cm long, brown; rachis pubescent; spikelets densely brown-pubescent, 2 or 3 per node, 1 or 2 sessile, 1 pedicellate, all perfect, 3–4 mm long excluding awn; first glume truncate and fimbriate at apex; sterile lemma lacking; fertile lemma awned, the awn 6–8 mm long, geniculate, twisted. *Croat 12827.*

Common along the stairs at the Laboratory Clearing. Generally appearing in large flushes in late rainy and early dry seasons.

The genus is supposed to have two sessile spikelets at each node, but specimens from BCI and other areas of Central America (R. Pohl, pers. comm.) have only one sessile spikelet per node.

Introduced from Java into various parts of the Western Hemisphere. In Panama, known only from tropical moist forest in the Canal Zone (BCI and Summit Garden).

RHIPIDOCLADUM McClure

Rhipidocladum racemiflorum (Steud.) McClure,
   Smithsonian Contr. Bot. 9:106. 1973
   Arthrostylidium racemiflorum Steud.
   Carrizo

Slender, arching or clambering, vinelike plant, to 5 m long; culms usually in dense clumps less than 1 m wide with many slender, nodose, and geniculate branches in dense fascicles at broad intervals on culm. Sheaths pubescent, the upper edge extending beyond articulation, with usually 2–4 slender cilia; blades lanceolate, articulate at

base, to 12 cm long, 4–12 mm wide, inconspicuously puberulent on lower surface, the trichomes near the base, glabrous or puberulent on upper surface, sometimes with sparse longer trichomes as well, the veins all about equally prominent, midrib lacking. Flowers very rarely found, in simple racemes 2–8 cm long; spikelets ± crowded, appressed, bearing few flowers; lemmas 6–8 mm long, awned, the awn 1–2 mm long; palea slightly longer than the lemma, the keel pubescent. Fruits not seen. *Croat 8175.*

Abundant in the forest and along the shore. Flowers chiefly in the rainy season, though flowers are rarely seen.

Easily confused with *Chusquea simpliciflora,* but may be distinguished by having blades lacking a distinct midrib. (This flower description is taken from the *Flora of Panama.*)

Southeast Mexico to Colombia. In Panama, known from tropical moist forest in the Canal Zone and Panamá, from tropical wet forest in Panamá (Cerro Azul), and from premontane wet forest in Chiriquí (Boquete).

ROTTBOELLIA L.f.

Rottboellia exaltata (L.) L.f., Nov. Gram. Gen. 40, pl. 1. 1779

Erect perennial, 70–150 cm tall; culms glabrous, with stout roots at lower nodes. Sheaths papillose-hispid; blades 15–45 cm long, 8–16 mm wide, scabrid and sparsely papillose-hispid above, glabrous to scabrid below, the margins scabrous. Racemes terminal or axillary, mostly 5–10 cm long; peduncles scabrous, sheathed; rachis thick, cylindrical, jointed; spikelets appressed, paired, one perfect and sessile, the other staminate and on a thick pedicel; both perfect and staminate spikelets ca 4 mm long, minutely papillate, bifid at apex, the margins scabrid; first glumes indurate; second glume about equal, tightly appressed to rachis. Fruits not seen. *Croat 5853.*

Locally common in the Laboratory Clearing. Flowering and fruiting throughout the year, especially in the rainy season.

Native to the Old World, where it is widespread in tropical Africa, Asia, and Oceania; introduced into Costa Rica, Panama, and the West Indies. In Panama, known only from BCI.

SACCHARUM L.

Saccharum officinarum L., Sp. Pl. 54. 1753
Sugarcane

Erect perennial, to ca 4 m tall; culms to ca 5 cm thick, glabrous; sap sweet. Lower internodes short with overlapping sheaths, the sheaths glabrous or softly pubescent toward the top, densely villous in the throat; ligules truncate, minutely ciliate, ca 5 mm long; blades linear with a prominent midvein, sharply serrate, villous on upper surface at base. Inflorescences very large, of silvery, plumose, panicled racemes, the branches simple, drooping; spikelets all alike, 4–5 mm long, in pairs, one sessile, the other pedicellate, surrounded at the base by long trichomes; rachis disarticulating below spikelets; glumes 1–3-veined, acute to acuminate; sterile lemma similar to glume but hyaline; fertile lemma shorter than glumes, hyaline, awnless, sometimes lacking. Fruits not seen. *Croat 44759.*

Sparsely cultivated by the Panamanian workers in the Laboratory Clearing. Seasonal behavior not determined.

Cultivated throughout the tropics and sporadically in Panama.

Saccharum spontaneum L., Mant. Pl. Altera 183. 1771

Perennial, to ca 3 m tall, sometimes branching; culms stout, glabrous. Sheaths mostly longer than the internodes, glabrous but with villous margins near throat, the old sheaths often persisting on the culms after blades have fallen; blades mostly 25–110 cm long, 5–12 mm wide, glabrous, the margins scabrid. Inflorescences terminal, to ca 25 cm long, bearing very long silky pubescence throughout but especially directly beneath the spikelets; racemes mostly 4–7 cm long; spikelets equal, paired, perfect, 3–4 mm long, one sessile, the other pedicellate; glumes narrowly acute at apex, 1–3-veined, often ± keeled. Fruits not seen. *Croat 11985.*

Collected once along exposed banks on the west side of Peña Blanca Peninsula. Some flowers seen in September. Bor (1960) said that the species flowers in its native range at the end of the rainy season.

The rachilla disarticulates below the spikelets, and the long silky trichomes obviously assist in wind dispersal of the seeds.

Widely distributed in the Old World tropics; introduced into the New World tropics. This is a new report for Panama.

SACCIOLEPIS Nash

Sacciolepis striata (L.) Nash, Bull. Torrey Bot. Club 30:383. 1903

Perennial, 1–2 m tall; culms decumbent-spreading, rooting at lower nodes, glabrous. Sheaths except the uppermost shorter than the internodes, glabrous, the margins ciliate; blades 4–20 cm long, 2–12(15) mm wide, ± glabrous but with scabrid margins. Panicles dense, spikelike, terminal, to 20 cm long; spikelets pedicellate on short

---

KEY TO THE SPECIES OF SACCHARUM

Culms 3–5 cm thick; plants cultivated at the Laboratory Clearing . . . . *S. officinarum* L.
Culms 1–2 cm thick; plants not cultivated . . . . *S. spontaneum* L.

KEY TO THE SPECIES OF SCHIZACHYRIUM

Small, decumbent herbs less than 1 m long; leaf blades less than 4 cm long . . . . . . *S. brevifolium* Kunth
Coarse, erect herbs more than 1 m tall; leaf blades more than 10 cm long . . . .
. . . . *S. microstachyum* (Desv.) Roseng., Arr. & Izag.

branches, 3–4 mm long; first glume ovate, short; second glume and sterile lemma nearly equal to each other, prominently veined, glabrous; second glume inflated-saccate at the base. Fruits ellipsoid, to 1.7 mm long, whitish, indurate. *Bailey & Bailey 387.*

Collected once on the island; probably no longer present. Seasonal behavior not certain; probably fertile in late rainy and early dry season.

Reported by Standley as *S. myuros* (Lam.) R. & S.

Southern United States and Panama; West Indies. In Panama, known only from BCI.

SCHIZACHYRIUM Nees

Schizachyrium brevifolium Nees ex Kunth, Agrost. Bras. 331. 1829

Andropogon brevifolius Sw.

Usually a weak, slender annual, somewhat decumbent at least at base, to 1 m long, ± glabrous but with scabrous margins. Sheaths shorter than the internodes, compressed; blades 1–4 cm long, 1–5 mm wide. Panicle branches slender; racemes solitary, 1–3 cm long, glabrous but with short beard on callus; spikelets paired, one sessile and perfect, the other reduced to an awned rudiment; sessile spikelet 1.5–3 mm long, the awn geniculate, to 8 mm long. Fruits oblong-linear, pale brown, to 1.4 mm long. *Shattuck s.n.*

Collected once on Orchid Island; though the plant has not been seen for some time, it could easily be overlooked and is to be expected. Apparently flowering and fruiting throughout the year.

Fruits primarily dispersed by adhesion to animals and by hydroscopic action of the awn.

Throughout tropics of Western and Eastern hemispheres. Known in Panama from tropical moist forest in the Canal Zone, Herrera, Coclé, Panamá, and Darién.

Schizachyrium microstachyum (Desv.) Roseng., Arr. & Izag., Bol. Fac. Agr. Montev. 103:34. 1968

Coarse perennial, 1–1.5 m tall, often growing in clumps, glabrous. Sheaths mostly shorter than internodes, compressed and keeled; ligules truncate, to 2.7 mm long;

blades 10–30 cm long, 5–10 mm wide, the margins only minutely scaberulous. Inflorescences numerous, large, corymbose, feathery; racemes solitary, 2–3 cm long, the rachis strongly flexuous; rachis and sterile pedicel prominently long-villous; spikelets paired, one sessile and perfect, the other pedicellate, rudimentary, bearing a short awn; sessile spikelets to 4 mm long, awned, the awn ca 1 cm long, geniculate, twisted. Fruits not seen. *Croat 9404.*

Occasional, in clearings, usually in isolated clumps. Flowering and fruiting throughout the rainy season into the early dry season.

Mexico to Brazil; leeward and windward islands of the West Indies. In Panama, known from tropical moist forest on BCI and from premontane wet forest in Chiriquí.

SETARIA Beauv.

Setaria geniculata (Lam.) Beauv., Ess. Agrost. 51:178. 1812

Pajón, Rabo de mono

Perennial, 0.2–1 m tall; culms in dense tufts from short rhizomes, ± erect, branching at the lower nodes, glabrous. Sheaths shorter than the internodes near the base, longer near the apex, glabrous; blades mostly 5–15 cm long, 4–6 mm wide, glabrous to scabrous, often sparsely pilose on the upper surface at the base. Spikelike panicles mostly 2–6 cm long, densely flowered, the axis densely pubescent; bristles 5 or more below each spikelet, usually longer than spikelet; spikelets 2–2.5 mm long, ovoid, glabrous; first glume about one-third as long as spikelet; second glume a little longer than the first; sterile lemma equaling fruit. Fruits white, transversely rugose. *Croat 6811.*

Common in clearings where vegetation is not allowed to grow too tall. Flowering and fruiting throughout the year, especially during the rainy season.

The bristles subtending the spikelets do not appear to aid directly in epizoochorous dispersal, since they remain attached after all fruits have been shed. However, they release the mature disarticulated fruits from the inflorescence upon impact, such as of passing animals, and

KEY TO THE SPECIES OF SETARIA

Leaf blades plicate, more than 4 cm wide . . . . *S. paniculifera* (Steud.) Fourn.
Leaf blades not plicate, less than 3.5 cm wide:
    Plant a small grass less than 1 m tall; inflorescences less than 6 cm long; blades less than 1 cm
        wide . . . . *S. geniculata* (Lam.) Beauv.
    Plant a large grass more than 1 m tall; inflorescences more than 15 cm long; blades more than 2
        cm wide . . . . *S. vulpiseta* (Lam.) R. & S.

perhaps are capable of throwing the fruits out of the inflorescence.

Southern United States to Argentina; West Indies; introduced elsewhere. In Panama, known from tropical moist forest in the Canal Zone, Bocas del Toro, San Blas, Panamá, and Darién, from tropical dry forest in Panamá (Taboga Island), from premontane wet forest in Chiriquí, Coclé, and Darién, and from lower montane wet and rain forests in Chiriquí.

Setaria paniculifera (Steud.) Fourn., Mex. Pl. 2:42. 1886

Pajón, Rabo de mono

Tall perennial, 1.5–3(4) m tall; culms erect, stout, appressed-pubescent at nodes. Sheaths papillose-hispid, more densely near margins; blades plicate, 50–100 cm long, 3–10 cm wide, scabrous. Panicles narrow or loose and open, 40–70 cm long; branches angulate and scabrid; bristles 1 or more below most spikelets, 3–10 times longer than spikelet, antrorsely barbed; spikelets ca 3 mm long, glabrous; first glume about half as long as spikelet; second glume slightly longer than the first; sterile lemma equaling fruit. Fruits lanceolate, finely, transversely rugose. *Croat 6365.*

Forming localized but dense stands in Rear #8 Lighthouse Clearing; also collected along the lakeshore. Flowering and fruiting principally in the rainy season.

Mexico to Colombia and Venezuela; West Indies. In Panama, known from tropical moist forest in the Canal Zone, Bocas del Toro, Panamá, and Darién and from premontane wet forest in Coclé (El Valle).

Setaria vulpiseta (Lam.) R. & S., Syst. Veg. 2:495. 1817

Large perennial, 1–2 m tall; culms in large clumps, erect or geniculate and rooting at lower nodes, glabrous. Sheaths longer than the internodes, glabrous to densely pubescent, especially along margins; blades 25–50 cm long, 2–3.5 cm wide, scabrous especially on underside of veins and on margins. Panicles 15–30 cm long, often spikelike, dense, the axis densely villous; bristles 1 or 2 below each spikelet, 1–2 cm long; spikelets 2–3 mm long, glabrous; first glume about half as long as spikelet; second glume slightly longer than the first; sterile lemma equaling fruit. Fruits transversely rugose, minutely green-crested, white. *Croat 11717.*

Rare; growing on steep, eroded, lakeside banks south of Fairchild Point, and less often in clearings and at the edge of the forest. Flowering and fruiting from June to December, principally from August to October.

Southern Mexico to Argentina; West Indies. In Panama, known from tropical moist forest in the Canal Zone, Bocas del Toro, San Blas, Panamá, and Darién, from

tropical dry forest in Panamá (Taboga Island), and from premontane moist forest in Panamá (Saboga Island).

SPOROBOLUS R. Br.

Sporobolus indicus (L.) R. Br., Prodr. Fl. Nov. Holl. 170. 1810

Slender perennial, 50–100 cm tall; culms erect in small dense clumps, glabrous. Sheaths glabrous, sometimes ciliate; blades 25–60 cm long, 1–4 mm wide, flexuous, the margins scabrid. Panicles 15–30 cm long; branches short, slender, ± ascending, minutely scabrid; spikelets 1.6–1.8 mm long, glabrous, bearing 1 flower; glumes nearly equal, exceeding fruit. Fruits plump, ellipsoid, ca 1 mm long, brown. *Croat 8657.*

Common in the Laboratory Clearing, especially along the stairs to the dock. Flowering and fruiting sporadically throughout the year, especially during the dry season and the early rainy season.

Southeastern United States and Mexico to Colombia and Brazil; West Indies. In Panama, widespread; known from tropical moist forest all along the Atlantic slope and in the Canal Zone, Veraguas, Herrera, Panamá, and Darién; known also from tropical dry forest in Coclé and Panamá, from premontane wet forest in Colón and Chiriquí, from tropical wet forest in Colón, and from lower montane wet forest in Chiriquí.

STREPTOCHAETA Schrad.

Streptochaeta sodiroana Hack., Oesterr. Bot. Z. 40:113. 1890

Erect, coarse perennial, 60–100 cm tall; culms often with stout roots at lower nodes, the leaves sparse below. Sheaths shorter than the internodes, crowded near the apex, the margins densely long-ciliate near the apex; blades oblong-elliptic, asymmetrical, usually acuminate, 18–30 cm long, 6–9.5 cm wide, ± glabrous, cross-veined; petioles short, densely pubescent at apex. Spikes 20–30 cm long, ca 1 cm thick, densely flowered; spikelets appressed, the lemma bearing awns to 10 cm long, antrorsely barbed, the barbs straight below, twisted and curled toward apex, interconnected with other awns; stamens and styles both prominently exserted at anthesis, both white. Inflorescences usually nodding in fruit, the spikelets disarticulating at the base, hanging suspended by their long awns; the indurate bifid palea somewhat divergent at the apex, useful for dispersal. *Croat 8816.*

Common to locally abundant in dense forest. Flowering in the dry season. The fruits present throughout much of the year.

KEY TO THE SPECIES OF STREPTOCHAETA

Blades 6–9.5 cm wide; spikelets many, densely arranged, ca 14 mm long (excluding awns) . . . . *S. sodiroana* Hack.

Blades less than 4.5 cm wide; spikelets few, distant, ca 22 mm long (excluding awns) . . . *S. spicata* Nees

Fruits disperse by attachment of the sharp palea to passing animals. The mass of fruits is held together, however, by the tangled awns and often falls to the ground, where seeds may germinate.

Belize to Panama, Ecuador. In Panama, known from tropical moist forest in the Canal Zone and Panamá.

See Fig. 51.

Streptochaeta spicata Schrad. ex Nees, Agrost. Bras. 537. 1829

Erect perennial, 60–90 cm tall; culms simple or branched few times, minutely pubescent especially near nodes. Sheaths long-ciliate and often auriculate at apex; blades 7–15 cm long, 2.5–4.5 cm wide, acuminate, inequilateral at base, scabrid or glabrous, cross-veined. Spikes 10–14 cm long, with 5–11 rather distant spikelets; axes densely pubescent; spikelets 2–2.5 cm long excluding awn, minutely pubescent, bearing 1 flower; 4 bracts at base of spikelet 2–3 mm long; lemma indurate, tapering into an awn 5–10 cm long, the awn coiled in distal half. Fruits brown, cylindrical, ca 8 mm long, covered by 3 acuminate lodicules. *Croat 11832.*

Occasional, in the forest. Flowering and fruiting throughout the year.

Dispersal is similar to that of *S. sodiroana.*

Guatemala to Ecuador and Brazil; Trinidad. In Panama, known from tropical moist forest in the Canal Zone, Panamá, and Darién and from premontane wet forest in Chiriquí (San Félix) and Panamá (Chimán).

STREPTOGYNE Beauv.

Streptogyne americana C. E. Hubb. in Hook.f., Icon. Pl. 5, pl. 3572. 1956

S. crinata Beauv.

Erect perennial, 30–120 cm tall; culms in dense clumps, glabrous. Sheaths longer than the internodes, pubescent especially near apex, auriculate at apex; blades 30–70 cm long, 1–2.2 cm wide, long-acuminate, glabrous or scaberulous. Racemes solitary, spikelike, 20–40(50) cm long; axes angulate, pubescent; pedicels pubescent, 3–4 mm long; spikelets appressed, 3- or 4-flowered, ca 2.5 cm long; first glume to 5(8) mm long, keeled, the keel scabrid; second glume 12–15 mm long, tapered to an awnlike point, the margins scabrid; lemma 2–2.3 cm long excluding awn, scabrid near apex, the awn ca 2 cm long; stigmas elongate, coiled. Mature fruits suspended from coiled stigmas. *Croat 4343.*

Common in the forest, usually not on trails but sometimes on infrequently used trails. Flowering and fruiting mostly in the early dry season.

Fruits are adapted for epizoochorous dispersal by passing animals.

Mexico to northern Brazil; Trinidad. In Panama, known from tropical moist forest in the Canal Zone and Panamá.

18. CYPERACEAE

Aquatic or terrestrial herbs (*Scleria secans* a scandent vine), usually tufted, rhizomatous or stoloniferous, with 3-sided stems. Leaves mostly crowded and basal with a closed sheath; blades simple, entire or sharply serrulate; venation parallel; stipules lacking. Flowers bisexual or unisexual (monoecious or andromonoecious), in bracteate spikelets, the spikelets solitary or variously compound (often umbellate); perianth reduced to several scales or bristles; stamens 1–3, free, hypogynous; anthers 2-celled, basifixed, dehiscing longitudinally, often having a produced connective; ovary 1, superior, 1-locular, 2- or 3-carpellate, with basal placentation; ovule 1, anatropous; style simple, 2- or 3-branched (depending on the number of carpels). Fruits achenes, sometimes in utricles, with albumen.

Easily mistaken for Gramineae (17), but distinguished by having solid, three-sided stems and three-ranked leaves with usually closed sheaths and basifixed anthers.

KEY TO THE TAXA OF CYPERACEAE

Inflorescences simple and unbranched or if compound the heads ± sessile:
  Spikelets solitary, not subtended by leafy bracts; plants usually aquatic . . . . *Eleocharis*
  Spikelets numerous, in dense heads subtended by leafy bracts; plants rarely aquatic:
    Spikelets ovate, flattened . . . . *Cyperus luzulae* (L.) Retz.
    Spikelets not ovate and flattened:
      Blades usually more than 5 mm wide; heads usually more than 1.5 cm diam . . . .
      . . . . *Rhynchospora cephalotes* (L.) Vahl
      Blades less than 5 mm wide; heads usually less than 1 cm diam:
        Bracts subtending inflorescence usually white, at least at base; spikelets few, 5–10 mm
        long . . . . *Rhynchospora nervosa* (Vahl) Boeck.
        Bracts subtending inflorescence always green (sometimes only 1 or 2); spikelets numerous, less than 4 mm long:
          Spikelets usually ca 3 mm long; scales smooth on keel, blunt at apex with a tiny mucro
          . . . . *Cyperus sesquiflorus* (Torr.) Mattf. & Kuek.
          Spikelets 1.5–2 mm long; scales spinulose-scabrous on keel, subacute at apex with a short cusp:
            Plants perennial, the rhizomes elongate, creeping . . . *Cyperus brevifolius* (Rottb.) Hassk.
            Plants annual, rhizomes lacking . . . . *Cyperus densicaespitosus* Mattf. & Kuek.

Inflorescences compound, the heads not sessile:

Spikelets very numerous, in dense, ± globose heads, the heads few, mostly more than 6 mm diam, radiating on short or long rays at apex of stem:

Heads whitish, borne on very short rays usually less than 1 cm long, the heads lacking leaflike bracts at base; spikelets markedly flattened with numerous scales easily disarticulating; plants not aquatic . . . . *Cyperus luzulae* (L.) Retz.

Heads greenish, on rays more than 2 cm long, subtended by short leaflike bracts; spikelets not markedly flattened, the scales few, not disarticulating; plants aquatic . . . .
. . . . *Scirpus cubensis* Kunth

Spikelets not in dense globose heads radiating from a central point, or if in dense globose heads, the heads less than 6 mm diam:

Achenes white, often shiny, exposed at maturity, ± globose:

Leaves less than 2 mm wide; plants slender annuals; achenes conspicuously transversely rugose . . . . *Rhynchospora micrantha* Vahl

Leaves more than 2 mm wide; plants perennials; achenes not transversely rugose:

Achenes clustered in small glomerules at apex of slender rays . . . .
. . . . *Calyptrocarya glomerulata* (Brongn.) Urban

Achenes solitary, not borne in clusters on slender rays . . . . *Scleria*

Achenes not as above (if exposed at maturity, not white and shiny):

Spikelets, at least in part, solitary on conspicuous peduncles, terete:

Blades less than 4 mm wide; peduncles glabrous . . . . *Fimbristylis dichotoma* (L.) Vahl

Blades more than 1 cm wide; peduncles scabrid . . . . *Hypolytrum schraderianum* Nees

Spikelets sessile, never solitary on conspicuous peduncles, terete or not, *or* if spikelets borne solitary, then not terete:

Plants not aquatic (except sometimes *C. odoratus*); blades less than 1.5 cm wide . . . .
. . . . *Cyperus* (in part)

Plants usually aquatic; blades more than 1.5 cm wide (except *Cladium* with very stiff, spinulose margins):

Spikelets in racemes, at least some racemes on long rays subtended by a whorl of long, leaflike bracts . . . . *Cyperus giganteus* Vahl

Spikelets in umbels or corymbs, not borne on long rays subtended by leaflike bracts:

Leaf blades usually 5- or more-veined; scales of spikelet pubescent . . . .
. . . . *Fuirena umbellata* Rottb.

Leaf blades with only 1 prominent vein; scales of spikelet glabrous or pubescent only on keel:

Bracts subtending ultimate clusters of spikes usually acute at apex, the awns, if present, very short; achenes lacking a tubercle at apex; leaves very stiff, the margins very coarsely spinulose-serrate . . . . *Cladium jamaicense* Crantz

Bracts subtending ultimate clusters of spikes long caudate-acuminate, the awns equaling or longer than remainder of bract; achenes having a cone-shaped tubercle at apex; leaves not very stiff, the margins not coarsely spinulose-serrate . . . . *Rhynchospora corymbosa* (L.) Britt.

Flowers are probably all wind pollinated except for *Rhynchospora nervosa* (Leppik, 1955; Baker, 1963), whose pollination is discussed in the systematic treatment. Other species that need to be investigated for possible insect pollination, because they occur deep in the forest where they cannot rely on wind, are *Cyperus simplex, Hypolytrum schraderianum,* and *Calyptrocarya glomerulata.*

Some taxa, such as *Calyptrocarya* and *Scleria,* with shiny white or colored achenes, are well suited for bird dispersal. Adapted also for dispersal by shore birds and by water are the following principally aquatic genera: *Eleocharis, Scirpus, Fuirena,* and some species of *Cyperus.* Taxa occurring on mud flats, such as *Cyperus,* may have seeds dispersed on the feet of birds. Most members of the family have very small diaspores, which could be eaten easily by small birds. Ducks and other shore birds are very important in the dispersal of aquatic Cyperaceae (Ridley, 1930). The seeds of many species occurring in clearings are probably spilled and scattered by the mechanical action of passing animals. Van der Pijl (1968) said that some Cyperaceae seeds are wind dispersed.

About 70 genera and over 3,500 species; widespread, but particularly numerous in high latitudes, usually in wet places.

CALYPTROCARYA Nees

Calyptrocarya glomerulata (Brongn.) Urban, Symb. Ant. 2:169. 1900

Small perennial, to ca 60 cm tall. Leaves linear, longer than culms, 2–9 mm wide, 3-veined, prominently folded along midrib, glabrous but with minutely scabrid margins. Spikelets minute, in small globular heads less than 4 mm diam on rays of compound or simple axillary corymbs; rays to ca 1.5 cm long; pistillate spikelets terminal, minutely puberulent, bearing 1 flower; style bifid; staminate spikelets lateral, bearing 1–4 flowers; stamen 1. Achenes ovoid, weakly compressed, apiculate, whitish, puberulent, ca 1.3 mm long. *Croat 15249.*

Uncommon; known from a few areas along trails in the forest, collected from Zetek and Balboa trails. Seen fertile April through July.

Mexico to Brazil. In Panama, known from tropical moist forest in the Canal Zone, Bocas del Toro, and Panamá, from premontane moist forest in Panamá, from premontane wet forest in Panamá (Cerro Jefe) and Colón, and from tropical wet forest in Colón.

CLADIUM P. Browne

Cladium jamaicense Crantz, Inst. Rei Herb. 1:362. 1766
Saw grass

Coarse perennial, 1.5–3 m tall; culms ± terete, leafy throughout. Blades to 70 cm long or more, 4–12 mm wide, stiff, the margins and medial rib on lower surface very strongly spinulose-serrulate. Panicles diffuse, to ca 15 cm long, arising from alternate leaf axils; spikelets in clusters of 2–5, terete, ovoid, acute, 4–5 mm long, brown, the uppermost flower perfect; perianth lacking; stamens 2; middle scales of spikelet sometimes with staminate flowers. Achenes ovoid, gradually tapered to a blunt apex, ca 3 mm long, dull, reddish-brown; tubercle lacking.

Reported by Standley to be "common in shallow water at the lake." I have seen no collections, but the plant could be there. Flowers in April and May. The fruits mature during July and August.

This species might be confused with *Rhynchospora corymbosa*, which grows in the same places and has a similar habit; it is possible that Standley confused the two species. *Cladium* can be most easily distinguished from *Rhynchospora corymbosa* by its much coarser leaf margins, its fruits lacking a cone-shaped tubercle at the apex, and its inflorescence lacking antrorsely barbed, attenuate bracteoles.

Throughout tropical and subtropical America. In Panama, known from tropical moist forest on the Atlantic slope in the Canal Zone and from tropical wet forest in Colón (Portobelo).

CYPERUS L.

The styles of *Cyperus* may be cleft two or three times, usually corresponding to whether the achenes are two-sided (lenticular) or three-sided.

Cyperus brevifolius (Rottb.) Endl. ex Hassk., Catal. Hort. Bogor. 24. 1884
Kyllingia brevifolia Rottb.

Very similar to *Cyperus densicaespitosus*, except perennial, the rhizomes usually elongate, and the spikelets slightly smaller. *Croat 12011.*

Occasional, in the Laboratory Clearing. Seasonality uncertain; probably flowers and fruits throughout the year. It has been seen fertile from January to September.

The habit of growth is noteworthy. Since the plant is perennial and develops elongate rhizomes, a whole series

KEY TO THE SPECIES OF CYPERUS

Spikelets clustered in dense, ± globose heads:
  Heads borne on short rays . . . . *C. luzulae* (L.) Retz.
  Heads sessile:
    Scales of spikelets smooth; the headlike spikes often cylindrical, usually 3 in number . . . .
    . . . . *C. sesquiflorus* (Torr.) Mattf. & Kuek.
    Scales of spikelets spinulose-scabrous on keel; the headlike spikes round or ellipsoid, frequently solitary:
      Plants annual, caespitose, lacking elongate rhizomes . . . . *C. densicaespitosus* Mattf. & Kuek.
      Plants perennial, the rhizome elongate, creeping . . . . *C. brevifolius* (Rottb.) Hassk.
Spikelets in open, diffuse inflorescences:
  Ultimate clusters of spikelets forming simple umbels:
    Umbels on rays usually less than 5 cm long; culms long, weak, and flexuous, with most leaves reduced to sheaths; plants growing in swamps . . . . *C. haspan* L.
    Umbels on rays usually 10–20 cm long; culms not weak and flexuous, erect, the leaves not reduced to sheaths; plants not growing in swamps:
      Inflorescences simple; spikelets 1–3 at apex of long, unbranched rays, the rays longer than culms . . . . *C. simplex* H.B.K.
      Inflorescences compound; spikelets several on compound rays, the rays much shorter than culms . . . . *C. diffusus* Vahl
  Ultimate clusters of spikelets racemose:
    Spikelets purple to black . . . . *C. rotundus* L.
    Spikelets green to brown:
      Culms terete, naked; plants aquatic, usually 1.5–2 m tall . . . . *C. giganteus* Vahl
      Culms ± trigonous, leafy at least at base; plants usually terrestrial and less than 1 m tall:
        Spikelets usually more than 1 cm long, sparse along rachis, decurrent, the scale bases becoming corky, the rachilla breaking into 1-fruited joints; inflorescences usually compound . . . . *C. odoratus* L.
        Spikelets less than 1 cm long, very dense on rachis (usually contiguous), the rachilla wings not becoming corky; inflorescences simple . . . . *C. tenuis* Sw.

of plants may be produced in succession. The species thus tends to be more locally gregarious than is *C. densicaespitosus,* which it very closely resembles, except for being somewhat more robust and having creeping rhizomes. Reportedly the most common species of *Cyperus* in Panama, growing in all but the drier areas of the west.

Southern United States to Argentina. In Panama, known from tropical moist forest on both slopes of the Canal Zone and in Bocas del Toro, from premontane wet forest in Chiriquí and Coclé, from tropical wet forest in Colón, and from lower montane wet forest in Chiriquí.

Cyperus densicaespitosus Mattf. & Kuek., Pflanzenr. IV.20 (Heft 101):597. 1936

Kyllingia pumila Michx.

Small annual, 4–30(40) cm tall; culms very slender. Leaves linear, usually shorter than culms, 1–3 mm wide. Bracts subtending inflorescence 3–9 cm long, 2–4 mm wide; heads globose to ovoid, terminal, usually solitary (to 3), sessile, 4–6 mm long; spikelets 1.5–2 mm long; scales membranaceous, pale, keeled, the apex subacute with a short cusp, the keel spinulose-scabrous; stamens 1 or 2; style bifid. Achenes lenticular, elliptic, pale, 1–1.5 mm long. *Croat 5910.*

Occasional, in the Laboratory Clearing. Flowering and fruiting throughout the year, especially in the dry season.

Easily confused with *C. sesquiflorus.* Plants tend to be taller, more slender, and more widely spaced than those of *C. sesquiflorus,* and have spikes tending to be greenish rather than whitish.

Widely distributed in the United States from New York to Ohio and south to Argentina; Africa. In Panama, known from tropical moist forest in the Canal Zone, Panamá, and Darién, from premontane moist forest in the Canal Zone, and from lower montane wet forest in Chiriquí.

Cyperus diffusus Vahl, Enum. Pl. 2:321. 1806

Perennial, 30–60 cm tall; culms slender. Leaves linear, 4–12 mm wide, the underside of midrib and the margins scabrid. Inflorescences subtended by several, long, leaflike bracts, compound; spikelets both at base of inflorescence and on long, spreading rays, the ultimate clusters forming simple umbels, greenish, 10–24-flowered, 5–15(20) mm long; scales green only on keel, with a narrow, incurved tip. Achenes obovate to globose, brown, trigonous with concave sides, smooth, 1–1.5 mm long. *Croat 6912.*

Abundant in clearings, especially near the laboratory. Apparently flowers throughout the year.

Throughout warmer regions of Western and Eastern hemispheres. In Panama, widespread and ecologically variable; known from tropical moist forest all along the Atlantic slope as well as in Chiriquí, Herrera, Panamá, and Darién, from tropical dry forest in Panamá (Taboga Island), from premontane moist forest in the Canal Zone and Panamá, from premontane wet forest in Chiriquí, Coclé, Panamá, and Darién, from tropical wet forest in Panamá and Darién, and from lower montane wet forest in Chiriquí.

Cyperus giganteus Vahl, Enum. Pl. 2:364. 1806

Stout aquatic perennial, 1–2(2.5) m tall; culms terete. Leaves reduced to basal sheaths. Bracts subtending inflorescence leaflike, 10–40 cm long, 0.5–2 cm wide, the margins and the underside of midrib scabrous; inflorescences large, compound; spikes elongate, lax, on spreading rays; spikelets 3–10 mm long, slender, cylindrical, 8–14-flowered; scales straw-colored, green on keel. Achenes minute, 1 mm long, trigonous with 1 face concave, pale yellow or white. *Croat 6168.*

Forming dense stands in a few places along the shore, especially in quiet coves. Flowers in the rainy season, with the fruits persisting into the dry season.

Mexico to Argentina. In Panama, known from tropical moist forest in the Canal Zone, Bocas del Toro, and Panamá, from premontane wet forest in Bocas del Toro, and from tropical wet forest in Colón.

Cyperus haspan L., Sp. Pl. 45. 1753

Slender, glabrous perennial, mostly 50 cm to more than 1 m tall; culms mostly slender, glabrous. Leaves mostly short or reduced to sheaths, sometimes with longer basal leaves usually to 3 mm wide. Bracts subtending inflorescence 2, usually shorter than inflorescence; inflorescences usually compound with several spikelets at apex of simple or branched, umbellate rays; spikelets 5–10(15) mm long, bearing many flowers, flattened, brownish; scales 1.2–1.7 mm long, obtuse and minutely apiculate at apex. Achenes usually white, ca 0.5 mm long, obovoid to ellipsoid, trigonous, with a rough granular surface. *Croat 13241.*

Infrequent, in swamps at the edge of the lake in some areas of the south shore. Flowers and fruits throughout the year, perhaps principally in the rainy season.

Warmer regions of Western and Eastern hemispheres. In Panama, ecologically variable; known from tropical moist forest in the Canal Zone, Veraguas, and Panamá, from tropical dry forest in Coclé, from premontane moist forest in Panamá, from premontane wet forest in Chiriquí, Panamá, and Coclé, from tropical wet forest in Colón and Panamá, and from lower montane wet forest in Chiriquí.

Cyperus luzulae (L.) Retz., Obs. Bot. 4:11. 1786

Perennial, 0.5–1 m tall. Leaves linear, 3–7 mm wide, glabrous or the margins and the underside of midrib scabrous. Bracts subtending inflorescence leaflike, to 30 cm long; spikelets in dense globose heads about 1 cm diam, mostly on short rays, numerous, ovate, flattened, 2–5 mm long, bearing up to 10 or more flowers; scales whitish, membranaceous, boat-shaped, easily disarticulating; stamen 1. Achenes oblong, 1 mm long, trigonous, smooth, light brown. *Croat 5810.*

Common in the Laboratory Clearing. Flowers and fruits throughout the year.

Throughout tropical America. In Panama, known from tropical moist forest all along the Atlantic slope as well as in Veraguas, Los Santos, Herrera, Panamá, and Darién; known also from premontane moist forest in the Canal

Zone and Panamá, from premontane wet forest in Chiriquí, Coclé, Panamá, and Darién, and from tropical wet forest in Coclé (Atlantic slope) and Colón.

Cyperus odoratus L., Sp. Pl. 46. 1753

   C. ferax L.

Annual, usually (10) 50–100 cm tall. Leaves linear, (1.5) 5–14 mm wide. Bracts subtending inflorescence large and leaflike, mostly 15–50 cm long; inflorescences compound (rarely simple), the racemes sessile or on short to long rays (to 20 cm long); spikelets linear, subterete, mostly 7–22 mm long, brownish at maturity; scales 2–3 mm long, striate, ± divergent in age; rachilla strongly winged, the wings becoming corky in age, the spikelets breaking up into single segments. Achenes oblong, 1.5–2.7 mm long, weakly trigonous, dark brown, densely papillose. *Croat 9563, 13239.*

An occasional weed of clearings and marshes. Flowers throughout most of the year.

Well-developed or fruiting specimens are not confused with any other species. Depauperate epiphytic plants on floating logs in the lake may be only 10–25 cm tall and with leaves 1.5–2.5 mm wide. These specimens have a simple inflorescence, but may be distinguished from *C. tenuis* by having corky rachilla wings.

Tropics and subtropics throughout the world. In Panamá, known principally from tropical moist forest in the Canal Zone, Bocas del Toro, San Blas, Los Santos, Panamá, and Darién; known also from tropical dry forest in Coclé, from premontane moist forest in Panamá, from premontane wet forest in Veraguas and Panamá, from tropical wet forest in Colón, and from lower montane wet forest in Chiriquí.

Cyperus rotundus L., Sp. Pl. 45. 1753

   Nut grass, Purple nutsedge, Junco

Slender perennial, 10–60 cm tall. Leaves linear, 2–6 mm wide, usually much shorter than culms. Bracts subtending inflorescence leaflike, short, 1–5 cm long; spikes several, lax, on rays to 6 cm long or very short; spikelets linear, 1–3 cm long, bearing 12–30 flowers; scales purplish to black, keeled, obtuse, Achenes ca 1.5 mm long, obovate-ellipsoid, dark, bluntly trigonous. *Kenoyer 152, Shattuck 357.*

Not seen recently, but to be expected in clearings. Often abundant elsewhere in open areas of the Canal Zone, especially along railroad tracks. Flowers in the rainy season.

Throughout warmer regions of the world. In Panamá, known principally from tropical moist forest in the Canal Zone, Bocas del Toro, Panamá, and Darién; known also from premontane dry forest in Los Santos and from premontane moist forest in the Canal Zone and Panamá.

Cyperus sesquiflorus (Torr.) Mattf. & Kuek., Pflanzenr. IV.20 (Heft 101):591. 1936

   Kyllingia odorata Vahl

Small caespitose perennial, 5–40 cm tall; culms very slender. Leaves linear, shorter or slightly longer than culms, 2–3 mm wide. Bracts subtending inflorescence linear, leaflike, 3–8 cm long; spikes 1–3, terminal, headlike, ovoid to cylindroid, sessile and confluent, 5–12 mm long, the terminal spike tending to be cylindrical, the lateral ones usually ovoid; spikelets 3–3.5 mm long; scales opaque, keeled, the apex blunt with a tiny mucro, the keel smooth; style bifid. Achenes lenticular. *Croat 6458.*

A weedy sedge locally common in the Laboratory Clearing. May be found in flower and fruit throughout most of the year.

The spikes are usually whitish, compared to the greenish color of *C. densicaespitosus*.

Widespread distribution from southern United States to Uruguay; West Indies; tropical areas of Africa and Asia. In Panamá, known from tropical moist forest in the Canal Zone and Darién, from premontane moist forest in Panamá, and from lower montane wet forest in Chiriquí.

Cyperus simplex H.B.K., Nov. Gen. & Sp. 1:207. 1816

Small perennial, 5–35 cm tall; culms slender, glabrous. Leaves linear, 3–6 mm wide, purplish at ground level. Bracts subtending inflorescences leaflike, to 30 cm long; inflorescences simple, with 1–3 sessile spikelets at apex of a ray 10–20 cm long; spikelets 5–15 (20) mm long, many-flowered, compressed; scales with a green keel, acuminate; stamen 1. Achenes subglobose, trigonous, ca 1 mm long, brownish, densely papillose, roughened. *Croat 15157.*

Frequent in the forest. Flowers and fruits principally in the rainy season (April to October).

Southern Mexico to Bolivia and Brazil. In Panamá, known from tropical moist forest in the Canal Zone, Panamá, and Darién and from premontane moist, premontane wet, and tropical wet forests in Panamá.

See Fig. 52.

Cyperus tenuis Sw., Prodr. Veg. Ind. Occ. 20. 1788

   C. caracasanus Kunth

Perennial, the culms 15–60 cm tall, the underground part purplish. Leaves mostly basal, linear, to 4 (5.5) mm wide. Bracts subtending inflorescence long and leaflike; inflorescences of 5–15, simple, loosely cylindrical or flattened spikes; spikes sessile or on short rays, less than 7 cm long including rays; spikelets 5–7 (10) mm long, subterete, 3–8-flowered; scales having green median stripe, the margin broad and thin. Achenes subcylindric-trigonous, slightly curved, beaked at both ends, brownish, minutely papillose, 1.5–1.8 mm long. *Croat 11866.*

Very abundant during part of the year in the Laboratory Clearing. May be found in flower throughout the year.

Mexico to Brazil; West Indies; tropical Africa. In Panamá, known from tropical moist forest in the Canal Zone, San Blas, Panamá, and Darién, from tropical dry forest in Panamá (Taboga Island), from premontane wet forest in Chiriquí and Coclé, and from lower montane wet forest in Chiriquí.

See Fig. 53.

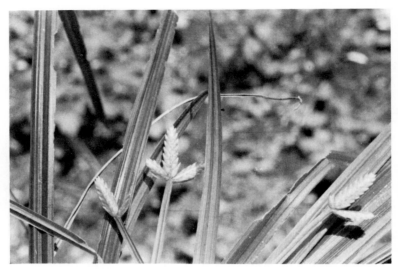

Fig. 52. *Cyperus simplex*

Fig. 53. *Cyperus tenuis*

KEY TO THE SPECIES OF ELEOCHARIS

Spikelets ovoid, 3–5 mm long; mature achenes less than 1 mm long, shiny black, smooth, the tubercle (style base) minute, depressed . . . . *E. caribaea* (Rottb.) S. F. Blake
Spikelets oblong to oblong-linear, 8–25 mm long; mature achenes more than 1 mm long, green to brown, minutely striate, faintly reticulate, the tubercle lanceolate . . . .
. . . . *E. plicarhachis* (Griseb.) Svens.

ELEOCHARIS R. Br.

Eleocharis caribaea (Rottb.) S. F. Blake, Rhodora 20:24. 1918

Glabrous and leafless aquatic, to 45 cm tall, densely tufted with numerous long roots; culms stiff, ca 1 mm wide. Spikelets solitary, at apex of culms, ovoid, obtuse, 3–5 mm long; scales obtuse, yellow-brown, thin, weakly keeled, the lowermost sterile; bristles subtending achene to twice as long as achene; flowers perfect; stamens 1–3; style 2-cleft, exserted. Achenes obovate, weakly compressed, less than 1 mm long, smooth, black, shiny; tubercle (style base) short, depressed. *Ebinger 583.*

Rare; in quiet areas along the south shore of the island in marshes and on tree stumps in the lake. Probably flowers and fruits throughout the year.

Distinguished from *E. plicarhachis* by its much smaller, ovoid spikelets and by its smooth, black, obovate achenes with a depressed tubercle.

Throughout most tropical areas of the world. In Panama, known from tropical moist forest in the Canal Zone and Bocas del Toro, from premontane moist forest in the Canal Zone and Panamá, and from premontane wet and tropical wet forests in Colón.

Eleocharis plicarhachis (Griseb.) Svens., Rhodora 31:158. 1929

Stoloniferous, leafless herb growing on wet soil; culms 20–60 cm long, 2–3 mm thick. Spikelets solitary at apex of culms, oblong to oblong-linear, 8–25 mm long; scales greenish, stiff, striate but central keel lacking; flowers bisexual; stamens 1–3; style 2-cleft. Achenes rounded to obovate, longitudinally striate, green to brown, 1–1.5 mm long; tubercle prominent, lanceolate. *Croat 13230, 13971.*

An occasional component of floating masses of vegetation on the south side of the island. Seen fertile throughout most of the year.

This was mistakenly reported by Standley (1933) as *E. variegata* Presl var. *laxiflora* (Thwaites) Ridley.

Mexico to Paraguay; Cuba. In Panama, known only from tropical moist forest in the Canal Zone and Panamá.

FIMBRISTYLIS Vahl

Fimbristylis dichotoma (L.) Vahl, Enum. Pl. 2:287. 1805

F. annua (Allioni) R. & S.; *F. diphylla* (Retz.) Vahl

Annual or perennial, nearly glabrous, 5–60 cm tall; culms terete, leafy near base. Leaves linear, ca 3 mm wide, mostly arising from base of culm, the margins scabrous. Inflorescences branched, usually loose and open, subtended by few bracts, the bracts longer than inflorescence; spikelets numerous, 5–10 mm long, solitary on peduncles, acute, brownish; scales having a green keel and minute apiculum; flowers perfect; stamens 1–3; style bifid, prominently fimbriate. Achenes obovate, terete, ca 1 mm long, white to brown, with a minute projection at base, sculptured with both vertical grooves and transverse wrinkles. *Croat 8566.*

Common in clearings. Flowers throughout the year, principally in the rainy season.

Low altitudes in temperate and tropical areas of the world. In Panama, known principally from tropical moist forest in the Canal Zone, Bocas del Toro, Colón, Herrera, Panamá, and Darién; known also from tropical dry forest in Coclé and Panamá (Taboga Island), from premontane moist forest in the Canal Zone and Panamá, from premontane wet forest in Chiriquí and Coclé, from tropical wet forest in Colón, and from lower montane wet forest in Chiriquí.

FUIRENA Rottb.

Fuirena umbellata Rottb., Descr. & Icon. 70, pl. 19, fig. 3. 1773

Aquatic herb, 0.5–1.5 m tall; culm often aerenchymatous, several-sided at base, becoming 3-sided above, with creeping rhizomes. Leaves to 40 cm long and 3.5 cm wide, usually minutely scabrid, the veins usually 5 or more. Inflorescences terminal or axillary, branched; spikelets oblong to lanceolate, 4–7(10) mm long; scales scabridulous and sparsely hispidulous, the keel strong, 3-ribbed, extending into an awn, the scale being shed with achene at maturity, the outer bristles filiform, the inner bristles thin, obovate, caudate at apex, enveloping the achene; stamens 3; style trifid. Achenes trigonous, broadly elliptic, light brown, smooth, to 1.3 mm long including their slender apex. *Croat 5340.*

Abundant at the edge of the lake, usually growing in shallow water or on tree stumps. Flowers and fruits throughout the year.

Standley (1933) also reported *Fuirena robusta* Kunth based on *Woodworth & Vestal 464*. This specimen was misidentified and is actually *F. umbellata*.

Throughout the tropics of Western and Eastern hemispheres. In Panama, known from tropical moist forest all along the Atlantic slope and from the Canal Zone and Panamá on the Pacific slope; known also from premontane moist forest in the Canal Zone, from premontane

wet forest in Coclé and Panamá, and from tropical wet forest in Colón and Coclé.

See fig. on p. 12.

HYPOLYTRUM L. C. Rich.

Hypolytrum schraderianum Nees in Mart., Fl. Brasil. 2(1):65, t. 5. 1842

H. nicaraguense Liebm.

Wide-leaved perennial herb, to 1.2 m tall. Leaves longer than culms, 2–3 cm wide, 3-ribbed, the margins scabrous, the outermost leaves basal and bractlike. Inflorescences corymbose-paniculate, mostly 5–8 cm long, often as broad as or broader than long, the lowermost bracts leaf-like, to 40 cm long; branches densely scabrid; spikelets oblong, terete, ca 4 mm long, brown, nearly sessile or peduncles to 5 mm long; the peduncles scabrid; scales rounded at apex; stamens 2; style bifid. Achenes ovoid, ca 2.5 mm long, brown, beaked. *Croat 5932, 8349.*

Occasional in the forest, usually in somewhat open areas. Flowers from the late rainy season to the early dry season. The fruits mature in the late dry and early rainy seasons.

Throughout Central America and south to Brazil. In Panama, known from tropical moist forest in the Canal Zone and Darién and from premontane wet forest in Panamá.

See Fig. 54.

RHYNCHOSPORA Vahl

Rhynchospora cephalotes (L.) Vahl, Enum. Pl. 2:237. 1805

Paja macho de monte

Perennial, usually 75–100 cm tall, with a coarse rhizome. Leaves linear, 5–15 mm wide, the margins scabrous. Bracts subtending inflorescence usually 2 or 3, long, leaf-like; spikelets in a dense, terminal, sessile, ovoid head 1.5–4 cm long; flowers numerous, ca 7 mm long, the upper ones staminate, the lower ones bisexual; scales greenish to light brown with a stout tip; stamens 3; style bifid. Achenes subglobose, weakly flattened, 1.5–2 mm long, light brown, capped by persistent green tubercle and subtended from base by 6 persistent bristles to ca 5 mm long. *Croat 8554.*

Occasional in the forest, but locally common along the shore, usually near water. Flowers from the late rainy season through the dry season. The fruits mature in the

late dry and early rainy seasons.

The species is quite variable elsewhere.

Mexico to Trinidad and Brazil; Jamaica. In Panama, known from tropical dry forest in Herrera, Coclé, Panamá, and Darién, from premontane moist forest in the Canal Zone, from tropical moist forest in the Canal Zone and Colón, from premontane wet forest in Panamá, and from tropical wet forest in Colón and Panamá.

See Fig. 55.

Rhynchospora corymbosa (L.) Britt., Trans. New York Acad. Sci. 11:85. 1892

Coarse herb, to 1.5 m tall, usually aquatic; culms triangular. Leaves linear, to 1 m long, 1–2 cm wide, the midrib scabrous on the underside, the margins scabrous. Dense corymbs arising from each of the upper leaves, to 12 cm long, the bracts caudate-acuminate, the upper flowers staminate, the lower flowers bisexual; spikelets numerous, brown at maturity, 7 mm long, containing a single achene; scales thin, brown, glabrous, the keel green; stamens 3; style bifid. Achenes 2–3 mm long, obovate, brown, with corky conic beak, the bristles exceeding achene. *Croat 13981.*

Uncommon, but locally abundant in some marshes on the south edge of the island. Flowers and fruits principally during the dry season and the early rainy season.

Sometimes the dominant plant in a localized area.

Throughout tropics of Western and Eastern hemispheres. In Panama, known from tropical moist forest in the Canal Zone (Atlantic slope), Bocas del Toro, and Colón, from premontane wet forest in Panamá and Chiriquí, and from tropical wet forest in Colón.

Rhynchospora micrantha Vahl, Enum. Pl. 2:231. 1805

Weak, slender annual, 10–35(50) cm tall, glabrous throughout; culms leafy. Leaves linear, mostly 10–20 cm long, 1–2 mm wide. Bracts slender, scabrid; spikelets in lax, slender corymbs to 1.5 mm long, the upper flowers staminate, the lower flowers bisexual; scales 1-veined, awn-pointed; bristles lacking; stamens 3, style bifid. Achene rounded to obovate, biconvex, white, ca 1 mm long, transversely rugose, the style base slender. *Kenoyer 153.*

Collected once on Orchid Island. Flowers and fruits throughout the rainy season.

The plant grows in similar locations as and looks similar to *Bulbostylis junciformis* (H.B.K.) Lindm., which does not occur on BCI.

KEY TO THE SPECIES OF RHYNCHOSPORA

Inflorescence a single, dense, terminal head:
    Heads less than 1 cm long, usually white; scape bracts usually white at base, less than 15 cm long
        and 5 mm wide; leaves usually less than 5 mm wide . . . . *R. nervosa* (Vahl) Boeck.
    Heads more than 1.5 cm long, greenish; scape bracts green, usually more than 15 cm long and
        5 mm wide; leaves more than 5 mm wide . . . . *R. cephalotes* (L.) Vahl
Inflorescence compound:
    Spikelets less than 2 mm long in widely scattered clusters on slender branches . . . *R. micrantha* Vahl
    Spikelets more than 5 mm long, in a usually congested dense corymb . . . . . . *R. corymbosa* (L.) Britt.

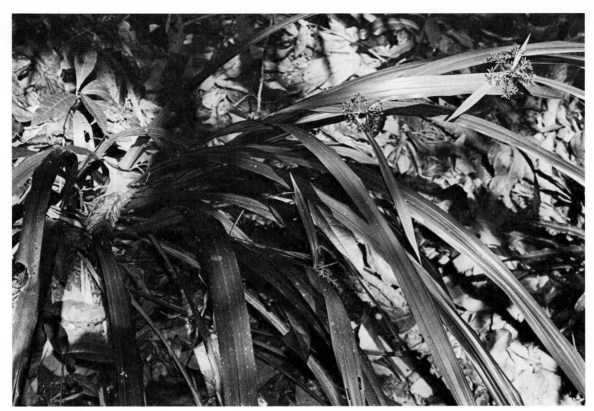

Fig. 54. *Hypolytrum schraderianum*

Fig. 55. *Rhynchospora cephalotes*

Throughout Central America; West Indies; tropical Africa. In Panama, known from tropical moist forest in the Canal Zone, San Blas, and Panamá, from premontane moist forest in the Canal Zone, from premontane wet forest in the Canal Zone and Panamá, and from tropical wet forest in Colón.

Rhynchospora nervosa (Vahl) Boeck. in Vid., Medd. Nat. For. Kjobenh. 1869:43. 1870.

Dichromena ciliata Vahl; *D. radicans* sensu auct.

Junco menudo, Clavo

Weak herb, 10–70 cm tall. Leaves 10–45 cm long, 3–5 mm wide, glabrous or villous on upper surface. Bracts subtending inflorescence 4–6, unequal, 1–15 cm long, usually white near base on upper side; inflorescence a dense head held well above leaves; spikelets 3–15, narrowly ovoid, 5–10 mm long, the flowers bisexual; scales thin, brown to whitish, glabrous or rarely scabrid on keel and margin; stamens 3; style 2-cleft, exserted and white at anthesis. Achenes obovate, weakly flattened, white to brownish, with numerous transverse wrinkles, the style base brown or greenish, ± flattened. *Croat 6808*.

Common in clearings. May be found in flower and fruit throughout the year.

Svenson (1943) in the *Flora of Panama* reported this species to have black or brown achenes, but I have found them to be white or more rarely light brown at maturity. Achenes are exposed along the rachis as the bracts become deciduous, beginning at the base. *Standley 40831*, reported by Svenson as *Dichromena pubera* Vahl, should probably be included here despite the fact that the scales are scabrid on the keel and margins; its achenes are like those of *R. nervosa*, and it does not differ in any other way.

In Central America, flowers open between 9:00 and 11:00 A.M. and are visited by the bees *Trigona* sp., *Bombus mexicanus*, and *Apis mellifera* (Leppik, 1955).

Throughout most tropical areas of the Western Hemisphere. In Panama, known principally from tropical moist forest in the Canal Zone, Bocas del Toro, Colón, Chiriquí, Veraguas, Panamá, and Darién; known also from tropical dry forest in Los Santos, Herrera, Coclé, and Panamá, from premontane dry forest in Los Santos, from premontane moist forest in the Canal Zone and Panamá, from premontane wet forest in Chiriquí and Panamá, and from tropical wet forest in Chiriquí.

SCIRPUS L.

Scirpus cubensis Kunth, Enum. Pl. 2:172. 1837

Stout, glabrous perennial, 30–100 cm tall, usually aquatic. Leaves basal, long and narrow, equaling or exceeding culms and inflorescences, to 1 cm wide except wider at sheathing base, the midrib raised above and usually scabrous, the margins scabrous; leaves drying with numerous minute rectangular reticulations at least near base. Bracts subtending inflorescence leaflike, often more than 60 cm long; inflorescences of dense globose heads, 1–1.5 cm diam on rays 1–10 cm long; spikelets 4–8 mm long; scales acuminate; stamens 2 or 3; style bifid or trifid. Achenes narrowly obovoid, smooth, pale. *Croat 13974*.

Apparently quite rare, though perhaps overlooked in sterile condition; locally abundant in marshes along the shore. Flowers throughout the year, mostly during the dry season.

Throughout most of tropical America. In Panama, known only from tropical moist and premontane moist forests in the Canal Zone.

SCLERIA Bergius

Scleria eggersiana Boeck., Cyp. Nov. 2:41. 1890

Stout herb, 1.5–2 m tall; culms ± smooth or scabrid near apex, often clustered and connected by a ± woody rhizome; ligules broadly to narrowly triangular, to 1 cm long, glabrous or the margins bearing short trichomes. Leaves linear, mostly 1–2.5 cm wide, glabrous or scabrid especially on margins and ribs. Panicles to 35 cm long, sparse, with relatively few ascending branches; spikelets sparse, unisexual; staminate spikelets many-flowered; pistillate spikelets 1-flowered with several empty scales;

KEY TO THE SPECIES OF SCLERIA

Plants climbing and vinelike, often suspended on low shrubs or trees to 3 m or more; leaves 2–7 mm wide; plants generally on lakeshore . . . . *S. secans* (L.) Urban
Plants erect or leaning but never vinelike:
    Plants less than 1 m tall; leaves less than 1 cm wide; inflorescences often purplish; achenes purple or white; hypogynia (cuplike structures beneath fruit) glabrous or weakly ciliate; plants of clearings . . . . *S. pterota* Presl
    Plants 1–3 m tall; leaves more than 1 cm wide; inflorescences not purple; achenes white; hypogynia with upper margin conspicuously ciliate; plants usually growing near water:
        Cilia purple, moderately long . . . . *S. mitis* Bergius
        Cilia brown or tan, usually short or inconspicuous:
            Inflorescences glabrate; achenes to 3 mm long; cilia of hypogynium conspicuous . . . .
            . . . . *S. eggersiana* Boeck.
        Inflorescences conspicuously pubescent; achenes 3.5–6 mm long; cilia of hypogynium inconspicuous . . . . *S. macrophylla* Presl

bracts inconspicuous; hypogynia obscurely 3-lobed, the margins densely ciliate with short brown trichomes (occasionally longer). Achenes white (sometimes brownish), globose, 2.5–3 mm long, often tipped with persistent tubercle. *Croat 11309.*

Occasional, usually in water at the edge of the lake. May be found in fruit throughout much of the year but principally in the rainy season.

Probably more abundant than *S. mitis,* with which it is easily confused; *S. mitis* apparently prefers not to grow directly in water.

Central America to northern South America; West Indies. In Panama, known only from tropical moist forest on the Atlantic slope of the Canal Zone and from tropical wet forest in Colón.

Scleria macrophylla Presl, Rel. Haenk. 1:200. 1828
  S. paludosa (?)Kunth

Stout herb, 1–3 m tall; culms with usually smooth margins; ligules blunt, often pubescent, the veins stout, meeting a collecting vein, the margins thick, glabrous. Leaves linear-lanceolate, 15–60 cm long, 1–7 cm wide, minutely pubescent or glabrous, the margins scabrous. Inflorescences paniculate, terminal or axillary, densely puberulent; spikelets unisexual; staminate spikelets many-flowered; pistillate spikelets with 1 flower and several empty scales; bracts narrow and conspicuous; hypogynia very thick, 3-lobed, the margins minutely ciliate. Achenes rounded, white, shiny, 3.5–6 mm long, often with large, conic, persistent tubercle. *Croat 12815, 13156.*

Uncommon, in marshes on the lakeshore. Probably flowers and fruits toward the beginning of the dry season.

Distinguished by its densely pubescent inflorescence, large achenes, and broad leaves.

Mexico to Brazil and Bolivia. In Panama, known from tropical moist forest in the Canal Zone and Panamá and from premontane moist forest in Panamá.

Scleria mitis Bergius, Kongl. Vetensk Acad. Handl. 26:145. 1765

Stout herb, 1–2 m tall, nearly glabrous; culm margins smooth; ligules narrowly triangulate, glabrous to pubescent, 1–1.5 cm long. Leaves 30–50 cm long, mostly 1.5–2.5 cm wide, somewhat scabrid especially on margins. Panicles narrow, 20–50 cm long, much-branched; spikelets congested, unisexual; staminate spikelets many-flowered; pistillate spikelets with 1 flower and several empty scales; bracts inconspicuous; hypogynia truncate, the rims densely ciliate with long purplish-brown trichomes. Achenes ovoid to ellipsoid, ca 2(4) mm long, smooth and white, often tipped with a persistent, conic tubercle. *Croat 5251, 8010.*

Uncommon; occurring at the edges of clearings, near the shore, and on silt deposits in coves such as the one in Fuertes Cove. Flowers and fruits as early as December, but usually from February to July.

Distinguished by the long, purplish-brown cilia of the hypogynium.

Guatemala to Paraguay; Cuba. In Panama, known from tropical moist forest in the Canal Zone, Bocas del Toro, and Colón, from premontane moist forest in Panamá, and from tropical wet forest in Colón.

Scleria pterota Presl, Isis (Oken) 21:268. 1828
  S. melaleuca Schlecht. & Cham.; *S. pterota* var. *melaleuca* Uitt.

Herb, 30–90 cm tall; culms ± scabrid, the trichomes ascending; ligules triangulate, ca 5 mm long, the veins extending almost to margin; margin thin, ciliate. Leaves to 40 cm long, mostly less than 12 mm wide, ± scabrid. Panicles terminal or axillary, often brown to purplish, the branches short, scabrid, bearing few spikelets; spikelets unisexual; staminate spikelets many-flowered; pistillate spikelets with 1 flower and several empty scales; bracts inconspicuous; hypogynia with 3 broad, rounded, glabrous or weakly ciliate lobes. Achenes globose, 1.5–2.5 cm long, bearing white trichomes in lower half. *Croat 6482.*

Abundant in clearings and open areas of the forest. Flowers and fruits throughout the year.

The BCI material would contain also the variety *melaleuca* Uitt. with black achenes, but the character of fruit color is of doubtful taxonomic value.

Throughout the tropics of the Western Hemisphere. In Panama, known principally from tropical moist forest in the Canal Zone, all along the Atlantic slope, and on the Pacific slope in Chiriquí, Veraguas, Los Santos, Herrera, Panamá, and Darién; known also from tropical dry forest in Panamá, from premontane moist forest in the Canal Zone and Panamá, and from tropical wet forest in Colón.

Scleria secans (L.) Urban, Symb. Ant. 2:169. 1900
  Cortadera

Climbing, vinelike herb, 3–6 m tall; culms with sharp, retrorsely scabrous to hispid trichomes on angles; ligules blunt, usually to 6 mm long, brown and thin above, thick at base, the conspicuous connecting veins generally pilose. Leaves generally less than 30 cm long, 2–7 mm wide, sparsely pubescent with longer trichomes but the margins and underside of ribs retrorsely scabrous. Panicles usually purplish, terminal or axillary; branches sparsely villous; spikelets unisexual; staminate spikelets many-flowered; pistillate spikelets with 1 flower and several empty scales; bracts conspicuous, filiform; hypogynia inconspicuous, glabrous, the margins reflexed, becoming undulate on drying. Achenes usually few, ± ovoid, 2–4 mm long, white, shiny. *Croat 13013.*

Locally quite common along the shore; occasional in the forest and open areas. Flowers and fruits throughout the year.

Of all the plants on the island this is the one most appropriately avoided. It usually grows over low trees at the margin of the lake, and once one is entangled every move worsens the situation. The edges of the culms cut like razors, and someone struggling to get free may pull large masses of the plant down on top of him. White-

faced monkeys have been seen eating the fruits during the latter part of the rainy season when food is scarce.

Mexico to Bolivia and Brazil; West Indies. In Panama, known from tropical moist forest in the Canal Zone, Chiriquí, and Panamá, from tropical dry forest in Panamá (Taboga Island), from premontane wet forest in Colón and Panamá, and from tropical wet forest in Colón.

19. PALMAE
(ARECACEAE)

Unbranched trees, shrubs, or lianas, rarely subacaulescent and appearing herbaceous, often with spiny stems. Leaves alternate, basal or clustered at the apex of the stem, petiolate, the base of the petiole sometimes persistent, usually deciduous leaving a narrow scar ring on the trunk; blades pinnately or palmately compound (sometimes simple and pinnately veined in *Geonoma* or in juveniles); blades and leaflets pleated, the margins usually entire; venation principally parallel, appearing pinnate or palmate; stipules lacking. Inflorescences of spicate or branched interfoliar or intrafoliar spadices; flowers bisexual or unisexual (monoecious or dioecious), actinomorphic, often immersed in a cavity, subtended by 2 envelopes, the primary one usually short and soon falling, the secondary one (the spathe) often persisting until maturation of the fruits; sepals 3, free or united, imbricate; petals 3, usually briefly connate and imbricate, or tepals 6; floral envelopes sometimes accrescent, forming cupule at base of fruit; stamens 6 (many in *Socratea*; 8 or 9 in *Desmoncus*; 3 in *Synechanthus warscewiczianus*), free, in 2 whorls of 3 each; anthers 2-celled, dehiscing longitudinally; ovary superior, 3-locular, 3-carpellate, the carpels united (elsewhere rarely with 3 free carpels); placentation basal or axile (parietal in *Oenocarpus*); ovules 3, 1 per loculus, but usually with only 1 developing, anatropous; styles usually 3, free or fused, or stigmas sessile. Fruits drupes; seeds 1(3), with endosperm.

Because of the characteristic leaves, the Palmae are confused only with the Cyclanthaceae. Leaves of both families are pleated, or the individual pinnae V-shaped. The Palmae may be distinguished from the Cyclanthaceae by generally being caulescent and by having six perianth segments, usually few ovules, and frequently compound inflorescences with a persistent spathe, by lacking conspicuous staminodia, and by having usually large, one-seeded fruits. Palms are unique in being the only monocots tall enough to compete with trees in the forest.

Little is known of the pollination systems of palms. *Astrocaryum* and *Scheelea*, which produce abundant amounts of pollen and are tall enough to be affected by wind currents, are possibly wind pollinated. However, I have seen *Trigona* bees visit the inflorescence of *Scheelea zonensis* in great numbers while it is in flower. The same may be true of *Socratea*, which is one of the tallest species of palm on the island, but it produces relatively small amounts of pollen. *Bactris* flowers are protogynous and all pollinated by small beetles that feed on staminate flowers (Essig, 1971, 1973); no nectar is produced. Studies by Essig (1971) on *Bactris major* in Costa Rica show numerous flower visitors, but he concluded that only nitidulid and curculionid beetles play any role in pollination. *Geonoma* is protandrous. *Asterogyne*, a closely related genus, is pollinated by syrphid flies in Costa Rica (Schmid, 1970). Self-pollination is generally prevented where staminate and pistillate flowers share the same spike by a well-marked protandry.

Fruits are animal dispersed. *Geonoma* as well as possibly *Chamaedorea wendlandiana*, *Synechanthus warscewiczianus*, *Cryosophila warscewiczii*, and *Desmoncus isthmius* are partly bird dispersed. Fallen fruits of *Scheelea* are fed upon by vultures, but it is not known if they are carried away (Foster, pers. comm.). *Astrocaryum standleyanum* is taken by bats (Bonaccorso, 1975). White-faced monkeys eat the mesocarp of *Astrocaryum standleyanum*, *Scheelea zonensis*, *Socratea durissima*, and *Oenocarpus panamanus* without breaking through the tough endocarp. They are also known to swallow whole the fruits of *Desmoncus isthmius* (Hladik & Hladik, 1969; Oppenheimer, 1968). Spider monkeys are reported to eat *Astrocaryum* and *Socratea* fruits (Hladik & Hladik, 1969). *Scheelea zonensis* and *Astrocaryum standleyanum* are taken by monkeys and coatis (Kaufmann, 1962) and by squirrels (Enders, 1935). While fruits of *Astrocaryum*, *Scheelea*, *Socratea*, and *Oenocarpus* may be taken in large part by arboreal frugivores, many find their way to the ground and are further dispersed by rodents and other ground animals. Perhaps most fruits are dispersed by rodents such as spiny rats and agoutis, and by peccaries (Enders, 1935), which are the only groups of animals capable of breaking through the stony endocarp. Agoutis scatter-hoard fruits of *Astrocaryum*, *Bactris*, *Scheelea*, and *Socratea* (N. Smythe, pers. comm.; C. C. Smith, 1975). Smith has found that *Scheelea* often has a second or third ovule develop, and even if the fruit is bitten open by a rodent it is still possible that one seed will remain for germination. On the other hand Janzen (1971) reported that the widespread bruchid beetle *Caryobruchus buscki* Bridw. can be responsible for the destruction of all seeds by eating through carpel walls.

About 236 genera and 2,650 species; widespread, principally in the tropics and subtropics.

KEY TO THE TAXA OF PALMAE

Leaves palmately compound (segments radiating from central point); at least the base of trunk
    armed with branched spines . . . . *Cryosophila warscewiczii* (H. Wendl.) Bartl.
Leaves pinnately compound or simple and pinnately veined (leaflets radiating from 2 sides of a
    central rachis):
    Trunk of stem long and slender, vinelike, erect only as a juvenile; leaflets replaced on end of leaf
        by large recurved spines . . . . *Desmoncus isthmius* Bailey

Trunk erect or, if stemless, leaves without retrorse spines:

Plants armed with long needlelike or flattened spines; leaflet margins usually with tiny sharp setae (sparse on some juvenile *Bactris;* sometimes lacking on *B. gasipaes*):

Trunks less than 7 cm diam; trunkless juvenile leaves with leaflets mostly of same width (if leaflets entire or of different widths, then the cross-veins prominent or the lower surface not densely white-scurfy) .... *Bactris* (in part)

Trunks more than 10 cm diam (trunkless juvenile leaves of *Astrocaryum* with leaflets of greatly different widths or entire, the lower surface densely white-scurfy):

Lower surface of blades covered with a dense whitish, scurfy pubescence able to be scraped off, but the trichomes themselves completely indistinguishable; spathe more than 80 cm long; staminate flowers in part aggregated distally, solitary or in pairs, the subtending bractlets coherent with adjacent bractlets, forming a cupule sometimes as long as the flower .... *Astrocaryum standleyanum* Bailey

Lower surface of blades sparsely pubescent with erect, simple trichomes; spathe less than 70 cm long; staminate flowers not densely aggregated distally but associated with pistillate flowers in triads or irregularly dispersed among triads, the flowers subtended by short, distinct bracteoles .... *Bactris gasipaes* H.B.K.

Plants not armed with long, needlelike or flattened spines; leaflet margins without tiny sharp setae:

Leaf blades not deeply divided............................. *Geonoma cuneata* Spruce

Leaf blades variously pinnate:

Leaflets narrow at base, broadened to 10 cm or more, the lower edge straight, the upper and outer margins irregularly toothed; trunks with large spiny stilt roots .... *Socratea durissima* (Oerst.) H. Wendl.

Leaflets of essentially same width for most of their length, the ends tapering to a point:

Leaflets regularly arranged and of nearly same width throughout:

Leaflets falcate (the ends curved toward apex); lower side of rachis with broad cream-colored band the full length of the leaf; trunks usually with adventitious roots at base .... *Chamaedorea wendlandiana* (Oerst.) Hemsl.

Leaflets straight, the ends not curved toward apex; rachis not with light-colored band along full length of leaf on underside; trunks without adventitious roots:

Plants rare, persisting in old cultivations; lateral costae all equally inconspicuous, the surface flat; trunk more than 20 cm diam .... *Cocos nucifera* L.

Plants common, occurring throughout forest or in swampy areas; lateral costae at least in part raised, the surface sometimes prominently pleated; trunk more than 30 cm diam or less than 12 cm diam:

Rachis of leaf glossy beneath, green when fresh; petioles round at apex, unarmed; trunks less than 12 cm diam, erect, 8–20 m tall .... *Oenocarpus panamanus* Bailey

Rachis of leaf densely furfuraceous beneath, brown; petioles canaliculate at apex, sharply armed; trunks more than 30 cm diam, decumbent at base, to 2 m tall .... *Elaeis oleifera* (H.B.K.) Cort.

Leaflets irregular, either irregularly spaced or of irregular widths or both:

Leaflets inserted on rachis at different angles, some held higher than others; inflorescences massive, ca 2–3 m long; fruits ca 3 cm diam; trees at maturity over 30 cm diam....................................... *Scheelea zonensis* Bailey

Leaflets inserted on rachis at same angle, none raised considerably more than others; inflorescences less than 1.5 m long; fruits less than 1 cm wide:

Leaf scars on trunks ca 1 cm apart; costae of leaflets in outer half of blade (below terminal lobe) decurrent onto rachis and obscuring the central midrib; secondary costae of leaflets scarcely if at all raised on upper surface; trunks seldom more than 1.5 m tall; inflorescences simple, unbranched .... *Geonoma procumbens* Spruce

Leaf scars on trunk mostly 5 cm or more apart; costae of leaflets in outer half of blade (below terminal lobe) not merging with or obscuring central rib of rachis; secondary costae of leaflets much raised on upper surface; trunks usually 2 m tall or more; inflorescences branched:

Petiole and rachis flattened near base of blade on upper surface with the lateral margins sharp; leaflets in about 25 pairs, spaced 1–4 cm apart near apex; leaf scars mostly ca 5 cm apart in middle of trunk; inflorescence branches 2–3 mm thick, branched; fruits 4–6 mm long .... *Geonoma interrupta* (R. & P.) Mart.

Petiole rounded, the rachis prominently ridged just above base of blade; leaflets in 4–6 pairs, spaced up to 10 cm apart near apex; leaf scars to 15 cm apart in middle of trunk; inflorescence branches ca 1 mm thick, unbranched; fruits ca 15 mm long .... *Synechanthus warscewiczianus* H. Wendl.

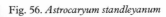

Fig. 56. *Astrocaryum standleyanum*

Fig. 57. *Astrocaryum standleyanum*

Fig. 58.
Bactris barronis

ASTROCARYUM G. Meyer

Astrocaryum standleyanum Bailey, Gentes Herb. 3:88 1933

Black palm, Chunga

Monoecious tree to 15 m tall; juveniles trunkless, the trunk of mature plants to 20 cm dbh, the internodes 30–40 cm long, prominently spiny, the spines slender, 10–15 cm long, flattened, shiny and black or with margins thin and brown, often furfuraceous; spines of petiole, rachis, peduncle, and outer surface of spathe much shorter than of internodes; leaf scars 2–3 cm long. Leaves 2–4 m long; petioles 1 m or more long, canaliculate above; leaflets to 75 cm or more long and 3.5 cm wide (broader on juvenile leaves), with small marginal setae, irregularly spaced and inserted at staggered angles, the apex oblique or cleft, the midrib prominently raised on upper and lower surfaces, the upper surface glabrous, the lower surface covered with a dense whitish scurf; juvenile leaves at first entire, later with few, widely spaced leaflets, these narrow or to 10 cm wide (especially at apex), the petiole at first furfuraceous, its spines to 25 cm long and 8 mm wide. Spathe armed on outer surface, 84–145 cm long, deciduous (rolling inward on drying); spadix somewhat shorter than spathe, ± erect to spreading at anthesis; peduncles terete, armed, equaling rachis; staminate flowers crowded on the outer part of rachilla, solitary or in pairs, falling soon after anthesis; subtending bractlets forming a cupule sometimes as long as flower; stamens 6, exserted; pistillate flowers few, on lower part of rachilla, 15–18 mm long; fruiting spadix pendent. Fruits ± obovoid, orange, 4–4.5 cm long (including the prominent beak), held in a massive pendent cluster, the surface coarsely papillate; nutlets 1–3, with longitudinal black stripes. *Croat 5391, 16626.*

Very common in the younger forest; less abundant in the older forest except on steep ravines. Flowering mostly from May to September. Plants may flower while some fruits are still present on old inflorescences. By the end of the rainy season fruits are one-half to two-thirds full size. They begin to be eaten by the animals when they attain mature size even if still green, beginning in January, with most being removed between March and June.

The species is similar to *A. aculeatum* G. Meyer of northern South America and Trinidad, but that species is described as attaining a height of 25 m with a trunk to 30 cm diam.

Nicaragua to Colombia. In Panama, known from low elevations in wet and very wet regions (Holdridge, 1970); few collections exist but I have seen the species in tropical moist forest in the Canal Zone, Panamá, Colón, and Darién.

See Figs. 56 and 57.

BACTRIS Jacq.

The genus is most easily distinguished by its usually slender (broad in *B. gasipaes*), spiny-ringed trunks and often spiny leaves and spathes. Leaflets usually bear small marginal spicules as well. Flowers have six to twelve anthers and are not sunken in the rachis. Two staminate flowers are usually associated with a single pistillate flower. Floral envelopes often enlarge to form a cupule on the fruit; these may be in two separate series or may be united to form a single series.

Bactris barronis Bailey, Gentes Herb. 3:101. 1933

Small, monoecious tree, to 8 m tall, forming small clusters; trunks to 5 cm diam; internodes prominently spiny. Leaves regularly pinnate but sometimes interrupted (especially on juvenile leaves), usually spineless except on vaginate part of petiole, occasionally with a few spines higher on petiole and rachis; petioles usually more than

KEY TO THE SPECIES OF BACTRIS

Leaflets regularly arranged (or nearly so) on rachis, all directed in about the same plane (sometimes irregular on juvenile plants of *B. barronis,* but these can be recognized by being blackish-green with prominent cross-veins):
 Petioles armed only at the base; rachis usually unarmed; leaflets with prominent cross-veins; rachillae of inflorescence numerous, slender (1–2 mm wide), to 12 cm long; fruits depressed-globose, ca 1 cm diam, prominently short-bristled, held in a close, compact, ± globular cluster, orange at maturity . . . . *B. barronis* Bailey
 Petioles and rachis armed ± throughout; leaflets lacking prominent cross-veins; rachillae of inflorescence usually 9–17, thick (more than 3 mm wide), more than 12 cm long; fruits ± obovate, ca 3 cm diam, not short-bristled, held in a diffuse cluster, purplish at maturity . . . . *B. major* Jacq.
Leaflets irregularly arranged on rachis, directed at different angles (regular only near apex):
 Trunks 10–15 cm diam; spathe 50–70 cm long . . . . *B. gasipaes* H.B.K.
 Trunks less than 5 cm diam; spathe less than 35 cm long:
  Leaflets with prominent cross-veins, the marginal setae conspicuous, mostly 5–10 mm long or longer; fruits orange, the beak obscure to blunt, the cupule with a single ± entire envelope . . . . *B. coloradonis* Bailey
  Leaflets lacking any obvious cross-veins, the marginal setae inconspicuous or absent, mostly 1–3 mm long; fruits not orange, the beak prominent, slender; cupules of 2 series, their margins crenate . . . . *B. coloniata* Bailey

Fig. 59. *Bactris coloniata*

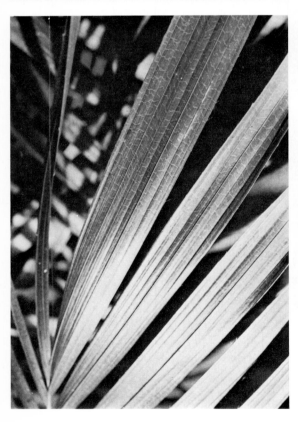

Fig. 60. *Bactris coloradonis*

Fig. 61. *Bactris major*

1 m long, green, terete; spines on petiole and sheath variable in length, mostly 1–9(15) cm long; leaflets 30 or more pairs, 75–100 cm long, 1.5–4.2 cm broad (the uppermost pair often much broader), spaced 2.5–5 cm apart near apex and 7–10 cm apart near base, dull, conspicuously cross-veined between the prominent side ribs, strongly setose marginally, the setae 2–6 mm long; midrib prominently raised on upper surface, the oblique apex 2.5–5 cm long, held ± straight; juvenile leaves blackish-green with prominent cross-veins, at first entire, becoming irregularly pinnate, some leaflets (especially apical pair) much broader than others. Spathe 15–30 cm long, usually persisting in fruit, bearing dense, coarse, black spines usually 1 cm long or less; spadix to 28 cm long, the floriferous part to 15 cm long, rachillae numerous, slender, densely flowered, to 12 cm long, the peduncle very short and stout, tightly curved; flowers cream-colored; staminate flowers not available; pistillate flowers with the calyx cup-shaped, 2 mm wide, 1 mm high, weakly 3-lobed, the corolla tubular, 2.5 mm long, ca 1.5 mm wide, weakly 3-lobed, sparsely appressed-spiculate. Fruits turbinate to pyriform, ca 1 cm diam, usually ± flattened on top with a short beak, green with short bristles, becoming orange and losing most bristles at maturity; mature fruit clusters tight, ± globular, 6–10 cm diam; cupule nearly flat, in 2 unequal series, at least the larger series shallowly lobed; seed 1. *Croat 9272.*

Common in the forest. Flowering in the late dry to early rainy seasons (March to May). Fruits mature in the wet season, June to September.

Distinguished from other *Bactris* by occurring in dense clusters, often associated with juvenile plants, and by having petioles usually spiny only at the base and with very dense, short spadices. Pinnae of adult blades are regular and are inserted at the same angle; in juvenile plants the pinnae may be unequally spaced and stagger-angled.

Known only from Panama in tropical moist forest in the Canal Zone and Panamá.

See Fig. 58.

Bactris coloniata Bailey, Gentes Herb. 3:106. 1933
    Uvito

Small, monoecious tree, 3–4(6) m tall, forming open colonies; plants connected by subterranean stems; trunk 3–4 cm diam; internodes weakly spiny, the spines 1–4 cm long. Leaves 2–3.3 m long, irregularly pinnate, spiny all over or only on base of petiole; petioles terete, to 1.4 m long, furfuraceous on lower side only or all over with only the upper surface near base glabrous, the vaginate base to 40 cm long, spiny throughout its length, the spines solid black to stramineous and black-tipped, 1–7 cm long (mostly 4 cm); rachis ribbed at base, becoming triangulate distally, sparsely spiny, the spines like those of the longer type on petiole; leaflets ca 23 pairs, 30–75 cm long, 2.5–6 cm broad, thin, glossy, lacking prominent side ribs or cross-veins, usually with narrow furfuraceous bands near margins on lower surface, prominently setose at apex, the apex falcate, oblique, very long-caudate and drooping,

the acumen to 6 cm long on upper leaflets, to 14 cm long on lower leaflets; upper leaflets ± regularly arranged, the lower becoming arranged in groups of 4–12, each group 5–20 cm apart, the leaflets of each group inserted at staggered angles, the lowermost pair of pinnae at base of blade often held almost vertical forming nearly a right angle with rachis. Spathe 20–35 cm long, ± caudate at apex, often persisting in fruit, densely setose, the trichomes brown to sienna, at first appressed, later ± erect, the slender, unarmed base nearly as long as broadened part of spathe; spadix with usually 15–20 well-spaced branches, the branches to 25 cm long, 1.5–3.0 mm thick; staminate flowers pedicellate, the pedicel to 1 mm long, slender; sepals 3, very slender, ca 1 mm long; petals 3, accrescent in lower half, acute, ca 4 mm long; pistillate flowers ca 4 mm long, the envelopes striate, the inner ones short and truncate, the outer with 3 acute lobes. Fruits turbinate, 1.5–2.5 cm long, with abrupt, short, slender beak, green to black, sparsely hirtellous, becoming ± glabrous at maturity; cupule double with bluntly scalloped edges; seed 1. *Croat 6772, 8787.*

Abundant in the forest, usually in small or large stands; juveniles often very abundant. Plants are not commonly seen in flower, apparently not flowering every year; numerous individuals flower, but only a small part of the adult population. Flowers in the middle to late dry season, with fruits maturing in the early rainy season, mostly in August and September.

The most abundant and most variable *Bactris* on the island. Included here are *Bailey 505* and *521,* which Bailey considered distinct species.

Known only from Panama in tropical moist forest on the Atlantic slope in the Canal Zone and from premontane wet forest in Panamá.

See Fig. 59.

Bactris coloradonis Bailey, Gentes Herb. 3:104. 1933
    B. coloradensis Burr.

Small, monoecious tree, 5–8 m tall, often growing as single plants; trunk less than 5 cm diam; internodes obscurely spiny, the spines 1–6 cm long, dull, black, broadened at base. Leaves 2.5–3 m long; petioles usually ca 1 m long, round, canaliculate above, slightly rufous beneath, spiny only at base or rarely throughout its length (especially true of juvenile plants), the spines 1–8.5 cm long; rachis usually unarmed, green above, rufous below; leaflets 25 or more pairs, to 70 cm long, 2–4 cm wide, glabrous, gray-green, dull, the midrib very prominent above, scarcely more prominent than strong lateral veins below, the cross-veins very prominent on both surfaces, the upper surface with occasional setae, the lower surface sometimes with rufescent bands, the margins usually sparsely but prominently setose, especially near apex, the setae 5–10 mm long or more; apex of leaflets oblique, 3–4 cm long in the middle and distal parts of leaf, to 12 cm long on proximal leaflets; distal leaflets regularly arranged, becoming clustered in groups of 3 or 4, the groups 2–6 cm apart, but with gaps to 30 cm on either side, the proximal leaflets of each group inserted at a

higher angle and held above the plane of the leaf, the lowermost pair of leaflets at base of blade sometimes inverted with the bottom side up, the apex of leaflet pointing toward apex of blade; juvenile leaves at first entire, becoming pinnate, the outermost pair broad, the leaflets characterized by having extremely long spines on margin and on upper surface. Spathe 25 cm long, densely brown-spiny, usually falling before fruit matures; peduncle short, somewhat flattened, closely curved, densely covered with short, stiff, brown trichomes at least toward apex; fertile part of spadix 15–25 cm long, widely branched, the branches 20–30, to 2 mm thick. Fruits orange, fading on drying, glabrous, ± round to obovoid, 1–2 cm long, obtuse and beakless at apex but drying with a blunt beak; cupule small and shallow, almost entire, the outer series very reduced; seed 1. *Croat 15422.*

Apparently rare on the island, known only from the older forest. Time of flowering unknown. Fruiting individuals have been seen only during June and July.

Easily distinguished by its solitary habit, its irregular leaflets with very prominent cross-veins, and its orange, nearly beakless fruit.

Known only from Panama in tropical moist forest on BCI.

See Fig. 60.

Bactris gasipaes H.B.K., Nov. Gen. & Sp. 1:302. 1816
Gulielma gasipaes (H.B.K.) Bailey; *G. utilis* Oerst.
Pejibaye palm, Pixbá

Monoecious tree, 5–20 m tall; trunk solitary or clustered, 10–15 cm diam, the internodes ca 20 cm long, densely armed with spines to 10 cm long. Leaves irregularly pinnate, to 3 m long; rachis 2 m long or more, sparsely armed; leaflets to 120 pairs, in clusters of 4 each, stagger-angled, 50–60 cm long, ca 3 cm wide, sparsely short-pilose on underside of veins, the margins entire or very sparsely setose, the cross-veins prominent. Inner spathe woody, less than 70 cm long, finely tomentose and armed with spines ca 1 cm long; spadix 30 cm long above branches; peduncle unarmed; rachis short; rachillae 25–30, minutely ferruginous-puberulent, 20–30 cm long, densely flowered, the pistillate flowers interspersed among the staminate ones, often associated with staminate flowers in triads, the flowers subtended by short distinct bracteoles; staminate flowers white, 4–5 mm long, broad and flat at apex, puberulent as on rachillae; petals 3, stout, concave, acute at apex; floral envelope truncate, 8–10 mm long, ca 9 mm wide, the flowers weakly sunken. Fruits orange-red, ovoid, to ca 5 cm long; seed 1. *Croat 11798, 14479.*

Cultivated at the Laboratory Clearing. Flowers in the late dry and early rainy seasons. The fruits mature in the rainy season. The plant at the Laboratory Clearing flowered in mid-May, and though the fruits developed normally for a time, they all fell off without ripening before August.

Similar to *Astrocaryum standleyanum,* but distinguished by the smaller spathe and spadix.

Widely cultivated in Central America and northern South America. In Panama, known from low elevations in very wet regions (Holdridge, 1970); few collections exist, but the species is expected to occur throughout Panama and is perhaps in all provinces.

Bactris major Jacq., Select. Stirp. Am. 280, t. 171, f. 2. 1763
B. balanoidea (Oerst.) H. Wendl.; *B. augustinea* Bailey; *B. superior* Bailey
Mongo lolo, Palma brava

Small, monoecious tree, 3–10 m tall, forming small and dense to large and open thickets (usually small and dense in open areas), sometimes leaning; trunk 4–6 cm diam; leaves often persisting until plants are as much as 5 m tall; internodes 13–26 cm long, prominently spiny, the spines 1–8 cm long, with spineless leaf scars ca 5 cm long; petiole, rachis (especially lower side), and often parts of lower leaf surface ± furfuraceous-scaly. Leaves regularly pinnate, variable, 0.7–4 m long, densely spiny on petiole and most of rachis with spines mostly 1–6.5 cm long, black or brown, the longest near base of petiole to 11 cm long; petioles 1–1.5 m long, the rachis round on lower side (drying sulcate), triangular along most of its length on upper side; pinnae on ± same plane, 30–60 cm long, 0.8–3 cm wide, the sides ± parallel, the lower surface sometimes with furfuraceous bands, the apex oblique, very short or to 10 cm long, the margins prominently dark-setose with small setae 1–3 mm long. Spathe 40–60 cm long, ± woody and persisting in fruit, the broadened part 22–32 cm long, sparsely to densely black-setose with the setae 5–20 mm long, the narrow base of spathe usually lacking setae; spadix with 9–17 stout branches, the peduncle furfuraceous-scaly, the floriferous part 20–30 cm long; flowers cream-colored; staminate flowers sessile, 7–8 mm long, the petals 3, thick, ovate, acuminate, the filaments connate to about the middle; staminodial ring ca 2 mm high; pistillate flowers ca 6 cm long, their calyx accrescent, cuplike, the outer envelope ca 3 mm long at anthesis, the inner envelope ca 1 mm long, soon becoming much larger than the outer ones. Fruits held in a diffuse cluster, ellipsoid to ovoid, to 4 cm long and 3 cm broad, obtuse to rounded at apex with a small point, furfuraceous-scaly, becoming dull purple at maturity; exocarp thick; mesocarp fibrous, sweet; cupule bluntly crenate, the outer series much smaller than the inner; seed 1. *Bailey 162* (type of *B. superior*), *Croat 8567, 10740.*

Common throughout the island; locally very abundant. Usually flowering in the dry season, with the fruits maturing in the rainy season or early dry season. A typical individual flowered in March and lost its fruits in late December. The species may also flower during the middle of the rainy season.

Bactris augustinea Bailey, which represents a more depauperate form of this species, grows in more exposed places, such as along the margin of the lake and on roadsides (e.g., the type locality near Summit Garden, Canal Zone). These plants are much shorter, are colored a lighter green, and have leaves that are much smaller, with leaflets closer together and directed somewhat upward,

compared to those plants occurring in the forest. No other differences have been detected between those plants inhabiting the shore and those in the forest. The specimen of *B. superior* described by Bailey (1933) from BCI is no doubt also *B. major;* it has unusually well-developed leaves and probably represents the other extreme from *B. augustinea*, having been collected within the forest. Otherwise the plant falls well within the limits of variability of *B. major.* Bailey (1933) described his collection *293* as being 15 m tall. I have seen no *Bactris* on BCI as tall as that, but Bailey's photographs, which accompany the type specimen, show a plant that is probably not more than 10 m tall.

Abundant along coasts of Central America and northern South America; Trinidad and Tobago. In Panama, known principally from tropical moist forest on both slopes of the Canal Zone and in Panamá and Darién; known also from tropical dry forest in Los Santos.

See Fig. 61.

CHAMAEDOREA Willd.

Chamaedorea wendlandiana (Oerst.) Hemsl., Biol. Centr.-Amer. Bot. 3:407. 1885

Bodá, Bolá, Caña verde, Nurú, Pacaya

Dioecious shrub or small tree, to 5 (7) m tall, growing as individuals or in open colonies; trunk to 5 cm diam with pronounced leaf scars, usually with adventitious roots at base (rarely higher on trunk). Leaves ca 2.5 m long, glabrous; petioles 40–50 cm long, canaliculate at base, rounded, with cream-colored band on lower surface extending along rachis to tip of blade; blades mostly less than 2 m long; leaflets regular, ca 17 pairs, 40–60 cm long, shortest at both ends of blade, 5–9.8 cm wide, abruptly tapered toward both ends, falcate at distal end; costae several, some prominently raised on upper surface; juvenile leaves entire. Inflorescences 1–3 per plant, in various stages of development, erect until maturity of fruit; spathes to ca 40 cm long, thin, narrow, long-acuminate, soon weathering; spadix branched once, densely flowered, white; rachillae 8–20, strongly divaricate, mostly less than 18 cm long, becoming orange and prominently pitted in fruit; staminate flowers crowded, ca 2 mm long; petals and sepals 3 each, ± equal, ovate; stamens 6, slightly longer than petals; anthers held perpendicular to filament, dehiscing upward; pistillodes about as long as filaments; pistillate flowers similar to staminate flowers but somewhat larger, thicker, and much less crowded, the pistils about as long as petals; stigmas prominent. Fruits oblong, 14–20 mm long, usually light green and shiny but turning black at maturity; exocarp thin; mesocarp fleshy and edible but bitter prior to maturity; seed 1. *Croat 6504, 6506.*

Uncommon, sometimes locally abundant in low areas near streams, such as near Lutz Creek south of the Laboratory Clearing. Flowers mostly late in the rainy season to early in the dry season (September to February), but flowers have also been seen in May. Fruits mature from April to October, but chiefly in the early rainy season (May to July).

The bee *Trigona tataira* (Diptera) was observed collecting pollen. The weevil *Cholus* sp. (Curculionidae, subf. Cholinae) also frequents flowering inflorescences, but does not collect pollen.

Nicaragua, Costa Rica, and Panama. In Panama, known from tropical moist forest in Bocas del Toro, San Blas, both slopes of the Canal Zone, Panamá, and Darién; known also from premontane wet forest in Panamá and Darién.

See Fig. 62.

COCOS L.

Cocos nucifera L., Sp. Pl. 1188. 1753

Coconut, Palma de coco

Monoecious tree, 7–20 (30) m tall; trunk usually more than 20 cm diam, unarmed; leaf scars prominent. Leaves green, 2–6 m long, the rachis and midrib prominently raised, yellowish or light green; petioles including leaf sheath ca 1 m long; leaflets regular, in a single plane, 60–110 cm long, to 5 cm wide, the shorter ones on either end of blade. Spathes ca 1 m long, striate, acuminate; rachillae 30–40, to 40 cm long; staminate flowers to 12 mm long, borne on upper part of interfoliar branching spadices; stamens 6; pistillate flowers globose, ca 3 cm diam, restricted to lower part of spadices; ovary 3-celled. Nuts many, in bunches, obtusely triangular, to 30 cm long, covered by a thick fibrous husk, the interior shell very hard; endosperm solidified into an edible, white substance; interior containing a liquid, whitish endosperm; seed 1. *Croat 7484.*

Rare; cultivated in the Laboratory Clearing and persisting in the forest at old settlement sites. Flowers in February. The fruits require 9 or 10 months to develop.

Pantropical in cultivation. In Panama, most abundant in coastal areas, especially along the Caribbean coast; few collections exist, but probably in cultivation in all provinces.

CRYOSOPHILA Blume

Cryosophila warscewiczii (H. Wendl.) Bartl., Publ. Carnegie Inst. Wash. 461:38. 1935

Acanthorrhiza warscewiczii H. Wendl.; *C. albida* Bartl.; *C. guagara* P. H. Allen

Noli, Palma de escoba

Tree, to 10 m tall; trunks 10–15 cm diam above the enlarged base, much of the trunk bearing spines, at least when young, the spines simple or branched, to 16 (35) cm long, sometimes restricted to base of older plants. Petioles longer than blade, broadly canaliculate; blades round in outline, 1.5–2 m diam or more, palmately veined and irregularly lobed almost to base, lobed completely to base on lower margin, green and glabrous on upper surface, densely whitish-pubescent on lower surface, the trichomes so small and appressed as to appear glaucous

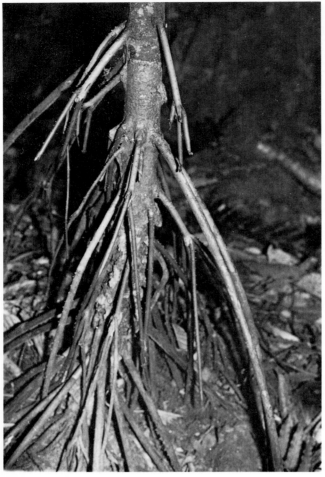

Fig. 62. *Chamaedorea wendlandiana*

Fig. 63. *Cryosophila warscewiczii*

(sometimes arachnoid with brown trichomes on veins); juvenile blades triangular in outline, deeply 2- or 3-lobed (more in age), densely arachnoid-pubescent. Inflorescences interfoliar, nodding, the spathelets (nodifronds) short, acuminate, densely woolly on outside, ultimately deciduous, the outer ones coriaceous, the outermost usually bearing 2 keels, the inner ones longer, each subtending a branch of the spadix, to 15 or 30 cm long, closely enveloping at least lower part of branch; spadix branches at least as long as broad, 30–75 cm long, the main axis stout, the branches very short or to ca 30 cm long, densely flowered; flowers bisexual, 3-parted, glabrous; sepals oblong; petals broadly obovate to rounded, concave, 2.5–4 mm long, the margins thin, the sepals and petals stiff, erect, forming a narrow tube; stamens 6; filaments in pairs, flattened, broad and united at base, slightly longer than petals and cernuous near apex; anthers ca 1 mm long, pendent on outside of corolla just above apex; pistils 3, narrowly ovoid, long-tapered to apex, curved outward just above apex; stigmas held just above.throat of corolla. Fruits round to obovate, mostly 1–2 cm diam, smooth, fleshy and white at maturity; pericarp thin; seed 1, round, 8 mm or more diam. *Croat 17017.*

Common in some areas of the younger forest, especially on the east side of the island. Flowers from May to October in Panama (as late as December elsewhere in Central America), chiefly in August. The fruits develop in 2 to 3 months, mostly between August and December. Populations may flower twice in a year; those seen in flower in October had mature-sized fruits as well.

Recognized by the palmate leaves, the white fleshy fruits, and the branched spines on the trunk, which Bailey (1943) considered to be root thorns. *Cryosophila albida* Bartl. and *C. guagara* P. H. Allen cannot be considered distinct from *C. warscewiczii,* since key characters such as pubescence and fruit size are very variable.

Belize to Panama, no doubt also Colombia. In Panama, known from tropical moist forest on both slopes of the Canal Zone and in Panamá and Darién; known also from premontane wet forest in Coclé (El Valle) and Panamá (Cerro Campana) and from tropical wet forest in Darién.

See Fig. 63.

DESMONCUS Mart.

Desmoncus isthmius Bailey, Gentes Herb. 6:211. 1943
Matamba

Slender, monoecious, widely spreading climber, growing into canopy but usually lower than 8 m; trunk 3–5 cm diam; juveniles usually erect. Leaves usually ca 2 m long; petioles short; rachis with black, flattened spines, often recurved, their bases swollen; base of petiole and sheath with denser, shorter spines, the leaf sheaths extending 3–12 cm above petiole; leaflets alternate, long-lanceolate, acuminate, 12–27 cm long, 2.5–4 cm wide, broad at middle, the margins unarmed, the surfaces glabrous or puberulent, the upper surface sometimes with weakly elevated cross-lines, the midrib pronounced but the side veins

indistinct, the underside often with 1 or more acicular spines; pinnae becoming opposite toward apex of blade, finally replaced by large, opposite, stout, reflexed spines. Spathes to ca 24 cm long, the rachillae 15 or more, slender (less than 2 mm wide), flexuose in fruit, the peduncle and lower part of rachis armed with short prickles (sometimes with pustular bases); flowers either in triads with 1 pistillate between 2 staminate or with staminate flowers solitary near end of rachilla; staminate flowers soon deciduous, ca 8 mm long; calyx short, tridentate; petals 3, ovate, oblique, acuminate, fleshy; stamens usually 8 or 9; filaments fused to petals at base; pistillate flowers with a small annular calyx; corollas much longer than calyx, urceolate, tridentate, with very small, adnate staminodia; pistils ovoid, 3-celled; styles short; stigma trifid. Fruits bright red at maturity, ellipsoid, 1.5–2 cm long, glabrous; exocarp thin; mesocarp fleshy; seed 1, ca twice as long as broad, obscurely 3-sided with a pore on each side and with dark lines radiating from each pore; cupule inconspicuous. *Croat 7759, 11288.*

Frequent in the forest; most abundant in dense thickets and along the shore where vegetation is sufficiently dense to provide support. Flowers in the rainy season. Fruits in the dry season and early rainy season.

Fruits are eaten by white-faced monkeys from April to August (Hladik & Hladik, 1969).

The species is similar to and perhaps inseparable from *D. orthacanthos* Mart. of South America, which was described as a plant forming dense thickets in clearings, with a spathe more than 50 cm long, whereas *D. isthmius* is always in forests and has a much shorter spathe. In addition, the type illustration by Martius (1824) showed the recurved spines to have long filiform appendages. Probably the Panamanian plants will not be found to be distinct from other Central American plants from Mexico south, which are now known as *D. chinanthlensis* Liebm.

Known only from Panama in tropical moist forest in the Canal Zone, Panamá, and Darién and from tropical wet forest in Colón.

ELAEIS Jacq.

Elaeis oleifera (H.B.K.) Cort., Fl. Colomb. 1:203. 1897
Corozo oleifera (H.B.K.) Bailey
Corocito, Corozo, Corozo colorado, American oil palm

Monoecious tree, 4–5 m tall; trunk at first decumbent, the upright part to 2 m high, 30 cm or more thick, bearing old leaf bases; crown usually broader than tree height. Leaves broadly spreading, the tips of the lower leaves often touching the ground; petioles 1–2 m long, 9–12 cm broad near base, broadly canaliculate, the sharp edges bearing spinelike teeth; rachis roughened on underside with brown scales; blades 2–4 m long; leaflets regular, in 60–110 pairs, to ca 1 m long, 4–6 cm wide, the apex oblique, the midrib prominent on upper surface with 2 faint marginal ribs on lower surface. Inflorescences borne among leaf axils; spathes thin, obscure, soon becoming a mass of dilacerating fibers, the outer spathes to 30 cm

Fig. 64. *Elaeis oleifera*

Fig. 65. *Elaeis oleifera*

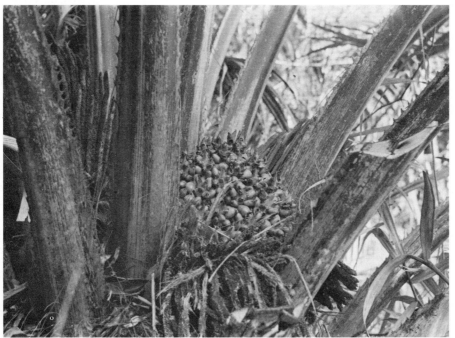

long, the inner to 50 cm long in staminate flowers, to 60 cm long in pistillate; staminate inflorescences soon weathering, the peduncles 20–35 cm long, ca 1 cm thick, the rachillae numerous, 20–25 cm long, ca 1 cm thick; stamens 6, the filaments partly united into a staminal tube ca 3 mm long; pistillate inflorescences 30–40 cm long, the peduncles ca 40 cm long, the rachillae stout, irregular, 4–9 cm long, closely compacted, pointed at apex; flowers white, sunken in rachillae, subtended by 2 or 3 bracts; tepals 6, ca 8 mm long; staminodial rings 6-dentate, ca 2 mm high; styles white, exserted and receptive for about a day, then becoming black; fruiting inflorescences usually broader than high, to 30 cm wide. Fruits orange, of irregular shapes due to mutual pressure, 2–3.5 cm long, often with persistent styles; seeds 1–3. *Croat 5203, 5539.*

Common only at the margin of the lake, but seen also in the seasonally swampy area near Standley Trail 500 and below the escarpment south of Armour Trail 700. Seasonal behavior uncertain. Flowers have been seen only during the early rainy season, but since flowers last for such a short time with the old inflorescences persisting, it cannot be said that flowering does not occur over a longer period of time. The fruits are present all year on some individuals, but most mature during the late dry or early rainy seasons. Individuals may bear more than one fruiting inflorescence in different stages, indicating the possibility that the plant flowers more than once a year.

Standley's plate VI in the *Flora of Barro Colorado Island* (1933), labeled *Corozo oleifera,* is *Scheelea zonensis.*

Bailey (1943) reported that fruits are black at maturity. My observations indicate that the fruits become loose and fall while still orange. However, fruits that are not able to fall free do become black.

Central America to Colombia and the Guianas (perhaps only introduced into northern South America). In Panama, a characteristic element of premontane wet forest (Tosi, 1971); known also from tropical moist forest on BCI and in Darién. Collected from tropical wet forest in Costa Rica (Holdridge et al., 1971).

See Figs. 64 and 65.

GEONOMA Willd.

The genus is distinguished by its slender spadices, which are enveloped at the base of the peduncle by two tubular bracts (spathes) and by its sunken, bilabiate flower pits. Usually there are three unisexual flowers in each pit, one

pistillate between two staminate, though the upper ends of the spadix (or rarely the entire spadix) may be exclusively staminate. Flowers have six tepals in two series. Staminate flowers are white and somewhat exserted, appearing before the pistillate but lasting only a short time. Filaments are united into a tube and anther cells are well separated and spreading. Pistillate flowers are usually purplish with six staminodia and three white stigmas; the stigmas are exserted at anthesis but soon wither.

Geonoma cuneata H. Wendl. ex Spruce, J. Linn. Soc., Bot., 11:104. 1869

G. decurrens H. Wendl. ex Burr.

Small, monoecious shrub. Leaves few, simple, nearly sessile or borne on a narrow trunk 3–4 cm diam and 2 m tall; internodes 1.5–2 cm long; petioles 10–25 cm long; blades deeply 2-lobed at apex (the long acuminate lobes straight or curved inward), 1–1.5 m long, 20–35 cm wide at apex, narrowed toward base, strongly pleated especially near rachis, decurrent on petiole, the margin entire, the veins 3–15 mm apart, variously furfuraceous (at least when young). Spadix simple, held erect or drooping; peduncles 50–70 cm long, glabrous, ensheathed by tubular, densely tomentose spathes at base; rachis 25–40 cm long, 5–9 mm wide, minutely rugose and obscurely appressed-pubescent; flower pits in 7–10, ± oblique series 1–2 mm apart; staminate flowers whitish, exserted; pistillate flowers shorter, embedded. Fruits mostly ellipsoid, 7–8 mm long, somewhat smaller on drying, glabrous, greenish to yellow or pale orange at maturity; exocarp hard; seed 1. *Croat 10825.*

Common in deep woods, particularly in the older forest. Flowering and fruiting data are inconclusive, but most flowering occurs in the rainy season, especially the early rainy season. Fruits develop in both the rainy and the dry seasons. Flowering and fruiting inflorescences may occur on the same plant, indicating either that the fruits require more than a year to develop or, probably more likely, that the plants flower more than once a year.

The only palm in the flora with an entire adult leaf. It may be confused with juveniles of many other palms, which also have entire leaves, but the juvenile leaves of other species are usually much smaller.

Nicaragua to Colombia; elevations below 100 m. In Panama, known from tropical moist forest in the Canal Zone, Bocas del Toro, San Blas, and Panamá and from premontane wet forest in Coclé (El Valle).

See Fig. 66.

KEY TO THE SPECIES OF GEONOMA

Leaf blades simple . . . . *G. cuneata* Spruce
Leaf blades pinnately compound:
    Inflorescences unbranched; plants usually to 1.5 m tall . . . . *G. procumbens* Spruce
    Inflorescences compound, branched many times; plants usually more than 3 m tall . . . .
. . . . *G. interrupta* (R. & P.) Mart.

Fig. 66. *Geonoma cuneata*

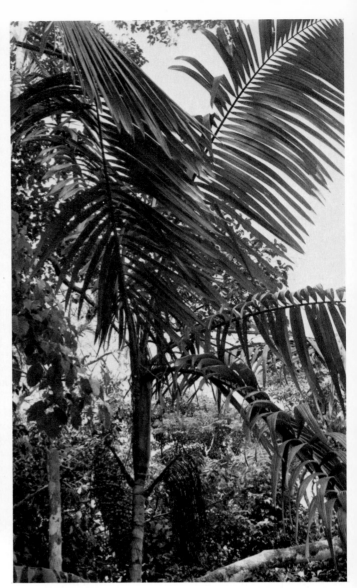

Fig. 68. *Oenocarpus panamanus*

Fig. 67. *Geonoma procumbens*

Geonoma interrupta (R. & P.) Mart., Hist. Nat. Palm. 2:8, t. 7. 1823

G. oxycarpa Mart.; *G. binervia* Oerst.

Small, monoecious tree, to 6 m tall; trunk to 6.5 cm diam; leaf scars prominent, to 8 cm apart. Leaves glabrous, irregularly pinnate, ca 2.5 m long; petioles 40–65 cm long, vaginate at base, flattened above with sharp edges; blades often more than 2 m long; leaflets in 20–30 pairs, abruptly long-acuminate, held in a single plane, 45–60 cm long, 1–9 cm wide (except terminal), spaced 1–10 cm apart, with 1–7 ribs prominent on upper and lower surfaces; rachis flat below, becoming triangulate by middle of blade; juvenile leaves entire. Spadix somewhat maroon, compound, branched many times, 60–75 cm long, often about as broad as long, lightly pubescent; flower pits spirally arranged in 5 series ca 3–6 mm apart; staminate flowers whitish, exserted, ca 4 mm long; pistillate flowers shorter, embedded. Fruits globular-ellipsoid, 4–6 mm long; pericarp slightly fleshy, becoming reddish and finally purple to black at maturity; seed 1. *Croat 7432, 9303.*

Occasional and usually widely dispersed in the forest. Flowering data is inconclusive. Flowers and fruits may be seen in most months of the year, but the flowers appear principally in the dry and early rainy seasons, with the fruits maturing during the rainy season.

Possibly confused with *Synechanthus warscewiczianus*, a plant of similar habit with irregular leaves but with an inflorescence of many slender undivided branches. Some plants possibly are entirely staminate, since the inflorescences wither after flowering. I am using here the broader interpretation of this species used in the *Flora of Surinam* (Wessels-Boer, 1965) and not the narrower interpretation used in the monograph of geonomoid palms (Wessels-Boer, 1968).

Mexico to Panama and northwestern South America; Haiti. In Panama, known from tropical moist forest in the Canal Zone, Bocas del Toro, San Blas, Darién, and Panamá.

Geonoma procumbens H. Wendl. ex Spruce, J. Linn. Soc., Bot. 11:105. 1869

Monoecious shrub, acaulescent or with trunk to 1.5 m tall and 5–6 cm thick; leaf scars 1 cm or less apart. Leaves glabrous, clustered at apex, continuously pinnate but somewhat irregular, to 2 m long; petioles ca 50 cm long or more, vaginate ca 20 cm at base, otherwise terete; rachis becoming sharply ridged above; leaflets in 10–24 pairs, 25–50 cm long, shortest at apex of plant, 2–5 cm wide except terminal pair, the apex long-acuminate and falcate, ± decurrent on rachis at base, the major ribs 1–3, equal, the others smaller; juvenile leaves entire and much smaller but otherwise like adult leaves. Inflorescences about as long as leaves but drooping below them; peduncles enveloped at base by 2 persistent spathes, the inner spathe much longer; spadix simple, maroon, the fertile part to ca 50 cm long, 12–15 mm thick; flower pits in 10–12 almost vertical series and in almost regularly

alternating whorls ca 1 mm apart; rachis minutely rugose and appressed-pubescent; tepals oblong to spatulate, oblique, acute at apex, to 4 mm long, at least the outermost keeled, thickened, concave, and violet-purple at apex; stamens ca 6 mm long, the tube to 4.3 mm long, the free parts spreading. Fruits ovoid-ellipsoid, 9–11 mm long, about half as broad as long, purplish-black at maturity; pericarp hard; seed 1. *Croat 6607, 9790.*

Occasional, preferring the deep shade of areas adjacent to creeks or creek beds. Seasonal behavior uncertain. Flowering is known only in the late dry season and early rainy season. Fruits are seen most of the year, but most mature in the rainy season.

Recognized by the usually short trunk, pinnate leaves, and unbranched spadix. *G. interrupta* has a much longer trunk and a compound inflorescence.

Nicaragua to Colombia. In Panama, known from tropical moist forest in the Canal Zone, Bocas del Toro, Colón, Panamá, and Darién, from premontane wet forest in Colón and Coclé (El Valle), and from tropical wet forest in Panamá and Darién.

See Fig. 67.

OENOCARPUS Mart.

Oenocarpus panamanus Bailey, Gentes Herb. 3:71. 1933

Maquenque

Slender, monoecious tree, 8–20 m tall, 6.5–12 cm dbh, in small clumps of 3–10 individuals, usually with plants of different ages in each clump; trunk smooth, with widely spaced, narrow, discolored leaf scars; crownshaft blackish-green, to 80 cm long. Leaves to 6 m long; petioles short, canaliculate; sheaths very broad, to 75 cm long; rachis near base of blade canaliculate on upper surface, rounded on lower surface, becoming triangulate before apex of blade; leaflets in 30–70 pairs, regular, 2–6 cm apart, 50–88 cm long, the longest at middle of blade, 4.5–6 cm wide, the midrib prominent above, the surface pleated, the apex oblique, its extension 3–5.5 cm long on central pinnae, to 9 cm long on basal ones; juvenile leaves sometimes irregularly pinnate (the youngest with only 2 lobes), glaucous on underside, the petiole with abundant brown persistent fibers at base. Inflorescences borne at base of crownshaft; spathes 2, deciduous at anthesis, the outer spathe of 2 woody valves to ca 30 cm long, the inner spathe a tube that splits on one side, 65–75 cm long; spadix scurfy, 15–25 cm long; peduncles short and stout, ± bulbous at base, the numerous slender flexuose rachillae 40–75 cm long, drooping; flowers unisexual, interspersed on rachillae, usually 1 pistillate flower between 2 staminate flowers but pistillate flowers usually lacking near end of rachillae; some spadices apparently entirely staminate; staminate flowers ca 3 mm long; sepals and petals valvate; stamens 6; filaments short; anthers versatile; pollen dry and powdery; pistillate flowers smaller than staminate flowers, the 3 recurved style lobes often persisting in fruit. Fruits dark blackish-green

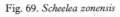
Fig. 69. *Scheelea zonensis*

Fig. 70. *Scheelea zonensis*

Fig. 71. *Socratea durissima*

(drying brown), 2–2.5 cm long, with an abrupt short point, slightly longer than broad, the lower third enveloped by the enlarged, irregularly crenate perianth lobes; seed 1. *Bailey 75* (type), *Croat 8093, 9517.*

Very common on all parts of the island. Flowering and fruiting season undetermined. Apparently flowers more than once a year, at least one wave of flowering occurring in the late rainy to early dry season, mostly November to March. Most fruits apparently mature from the middle of the dry season to early in the rainy season (March to July). Flowers also occur in the early rainy season from June to August, perhaps constituting a second wave of flowering. Plants often bear both flowering and fruiting inflorescences.

Costa Rica and Panama. In Panama, known from moist and wet regions at low elevations (Holdridge, 1970); known from tropical moist forest in the Canal Zone, Panamá, and Darién. Known from premontane wet forest in Costa Rica (Holdridge et al., 1971).

See Fig. 68.

SCHEELEA Karst.

Scheelea zonensis Bailey, Gentes Herb. 3:36. 1933
Palma real, Manaca

Monoecious or dioecious tree, to 30 m or more tall with trunk 30 cm or more diam, the upper part of the trunk very broad with old leaves and inflorescences persisting for several seasons. Blades unarmed, 5–10 m long, often held in a vertical plane; petioles 30–60 cm long, caniculate on upper surface; rachis canaliculate on upper surface at base of blade, becoming sharply triangular toward apex; leaflets very irregular, 1 m or more long, 1–4.5 cm wide, blunt and inequilateral, one side of apex 2.5–3.5 cm longer than the other, the midrib prominent on upper surface, the reticulate veins drying prominent, the pinnae inserted at staggered angles and often in clusters of 2 or 3, the lowermost of each cluster almost erect; juvenile leaves at first entire, the surface much pleated, soon splitting into separate, ± regular pinnae, the uppermost ± united. Spathe to 3 m long, woody, deeply grooved on outside, with a long point, persisting; peduncles massive, usually longer than fertile part of spadix; spadix yellowish-white, over 1 m long, with numerous simple branches usually less than 20 cm long, usually bearing staminate flowers near the end and pistillate flowers closer to peduncle (some trees bear only staminate flowers and others only pistillate flowers); staminate flowers soon falling, their petals subulate, 1.5–2 cm long; stamens 6, often exserted from between petals, less than 5 mm long; filaments short and thin; anthers to ca 4 mm long; pistillate flowers with petals imbricate at base, narrowed toward apex; fruiting inflorescences massive, the rachillae coarse and heavily scarred by old fruit attachments. Fruits ± oblong, brownish-orange, 6 cm or more long, about 3 cm thick, enveloped at base by the large accrescent calyx, the surface minutely longitudinally striate, the beak abrupt, 5–8 mm long; seeds 1–3. *Croat 6100.*

Common throughout the island, especially near the shore and in younger areas of the forest; juvenile plants more abundant than adult. Flowers from April to September with most flowers appearing in the early rainy season. By the end of the rainy season fruits are already of mature size, but green. Many mature by the middle of the dry season, then begin to disappear slowly, and are mostly gone by July. The fruits are usually present and mature at the time of flowering.

Probably most plants bear both staminate and pistillate flowers on the same inflorescence; however, staminate plants (and pistillate, *fide* Bailey, 1943) occur in smaller numbers.

This species does not appear to be different from *S. rostrata* (Oerst.) Burr. from Costa Rica. Standley's plate VI in the *Flora of Barro Colorado Island* (1933), labeled *Corozo oleifera*, is the species; *C. oleifera* is treated here as a synonym of *Elaeis oleifera.*

A variety of small insects visit the huge flowering inflorescences. The heavy fruit crop provides a large portion of the diet of a number of larger animals, including white-faced monkeys and agoutis. The agoutis bury the fruits indiscriminately for later retrieval (Smythe, 1970). Vultures also feed on fallen fruits (R. Foster, pers. comm.).

Costa Rica and Panama; additional field work will probably show the species to range from at least Belize to Colombia. In Panama, common in and characteristic of tropical moist and premontane wet forests (Tosi, 1971); as for most palms, few collections exist, but the species has been seen or collected on both slopes of the Canal Zone and in Colón, Panamá, and Darién.

See Figs. 69 and 70.

SOCRATEA Karst.

Socratea durissima (Oerst.) H. Wendl., Bonplandia 8:103. 1860
Stilt palm, Jira

Monoecious, slender, stilt-rooted tree, 25–30 m tall; wood very hard; trunk to 15 cm diam, ringed; stilt roots 1–3 m high, often replacing the trunk at base, bearing short spines; crownshaft bluish-green, very long and bulging at base. Blades 2 m long or more; petioles rounded on underside; leaflets ± regular, in 8–14 pairs, narrow at base, greatly broadened to deeply toothed at apex (the terminal pair united, truncate), inserted on rachis at an angle (the proximal edge higher on rachis, the distal edge lower on rachis), often splitting in age, the various parts then directed at different angles, the ribs numerous, raised on lower surface; juvenile plants much like adults, the blades entire or with few pinnae. Inflorescences borne below the crownshaft; spathes 2, thin, 40–50 cm long, both deciduous as a single unit; spadix stout with up to 16 stout branches, to 60 cm long; peduncles about as long as or longer than rachis, stout, flattened laterally, drooping in fruit; staminate flowers to 1.2 cm long and 1.5 cm wide, soon falling; sepals 3, minute; petals 3, thick, concave; stamens numerous (ca 50), about as long as petals, the filaments very short; pistillate flower usually borne between 2 staminate flowers, much smaller and

obscured by them. Fruits oblong, 2.5–3 cm long, with a small subapical point, the cupule bowl-shaped, to 1 cm wide; seeds 1–3, to 2.5 cm long, beaked at base and marked with anastomosing lines, enveloped very loosely by thin woody husk at maturity. *Croat 8160, 8638.*

Common throughout the forest; sometimes locally abundant. Seasonal behavior uncertain. Flowering chiefly in the early rainy season (May to August), but sometimes in the dry season as well. Most fruits mature in the middle to late dry season. Juvenile or mature-sized fruits may be found on the ground at most times of the year. Oppenheimer (pers. comm.) reports that fruits are eaten by white-faced monkeys during March and April.

Socratea, Scheelea, and *Desmoncus* are the only palms that are seen regularly in the canopy. *Socratea durissima* is not confused with any other species on BCI, although Bailey (1943) stated that its identification is not positive. However, it has been confused with *S. exorrhiza* (Mart.) H. Wendl., which ranges from Colombia to Brazil, and it is possible the two species are not distinct.

Both spider monkeys and white-faced monkeys eat the thin pulp between the seed and the husk (Oppenheimer, 1968).

Nicaragua, Costa Rica, Panama, and possibly Colombia. In Panama, widely distributed in wet regions at low elevations (Holdridge, 1970) and appearing as a characteristic tree species of tropical moist forest (Holdridge & Budowski, 1956, as *Iriartea exorrhiza* Mart.) and of tropical wet forest (Tosi, 1971); collected from tropical moist forest in the Canal Zone, Panamá, and Darién and from tropical wet forest in Colón and Panamá. Common in premontane wet and premontane rain forests in Costa Rica (Holdridge et al., 1971).

See Fig. 71.

SYNECHANTHUS H. Wendl.

Synechanthus warscewiczianus H. Wendl., Bot. Zeitung (Berlin) 16:145. 1858

S. angustifolius H. Wendl.; *S. ecuadorensis* Burr.; *S. panamensis* H. E. Moore

Palmilla, Bolá

Monoecious tree, to 5(6) m tall; trunk 4–5 cm dbh, sometimes with prop roots at base; internodes 6–8 cm long; leaf scars to 15 cm apart. Leaf blades ca 10, spreading, usually less than 2 m long; petioles becoming rounded or obscurely flattened near base of blade, 50–75 cm long including sheath, sheathing at base, the sheath two-thirds to three-fourths as long as petiole; rachis prominently ridged just beyond base of blade; pinnae few, in 4–6 pairs, irregular in shape, 1.5–2 cm wide, 29–41 cm long, irregularly spaced, to 10 cm apart, the apex long-acuminate, falcate, the terminal pair often irregularly united, the costae many, in part slightly raised on upper surface (sharply so on drying), the larger ones 15–18 mm apart; juvenile leaves entire, bilobed as adults at apex, becoming pinnate with 2 or 3 lobes, often on one side only, with the other side entire. Inflorescences usually below leaves, ± erect in flower, drooping in fruit, to

nearly 1 m long; spathe thin, tightly sheathing peduncle, persisting and becoming weathered; spadix broomlike, on a long slender peduncle to 70 cm long, with many long thin branches mostly at right angles to axis when open; flowers tiny, in biseriate clusters of 4–13, on alternate sides of rachis, only the lowermost of each group pistillate; staminate flowers ca 0.6 mm long, the calyx much less than half as long as petals; stamens 3; filaments elongate, exserted at anthesis; staminodia usually present; pistillate flowers ca 1 mm long, the pistil as long as petals; pistillodes lacking. Fruits oblong, ca 1.5 cm long, light yellow to orange at maturity (drying black); seed 1. *Croat 7010, 11020.*

Occasional, especially in the old forest, preferring shady moist areas near streams. Flowering mostly in the early to middle rainy season (June to September), and fruiting in late rainy to dry season. Less frequently flowering in late dry season, with fruits maturing in the early rainy season.

Recognized by its broomlike inflorescence and irregularly pinnate leaves.

Costa Rica to Colombia and Ecuador; mostly low elevations but to 1,200 m. In Panama, known from tropical moist forest on both slopes in the Canal Zone and in San Blas and Darién, from premontane wet forest in Veraguas (Cerro Tute) and Panamá (Lago Cerro Azul), and from premontane rain forest in Panamá (Cerro Jefe) and Darién (Cerro Pirre).

20. CYCLANTHACEAE

Epiphytic vines or terrestrial, acaulescent or short-stemmed, palmlike herbs arising from a rhizome. Leaves alternate or forming a rosette, petiolate, the petioles dilated at base forming a clasping sheath; blades simple, entire to cleft, often plicate; venation palmate; stipules lacking. Flowers unisexual (monoecious), actinomorphic, the sexes interspersed and closely congested on a simple, axillary spadix subtended by several spathes; perianth cupular or lobed or rudimentary in staminate flowers, many-lobed or rudimentary in pistillate flowers; stamens many; filaments connate at base, epitepalous if a perianth is present; anthers 2-celled, dehiscing longitudinally; staminodia 4 in pistillate flowers, short or flexuously filiform; ovary inferior, 1-locular, 2–4-carpellate; placentation parietal; ovules numerous, anatropous; style none or 1, with 1–4 stigmas. Fruits fleshy syncarps of separate or connate berries; seeds numerous, with copious endosperm.

The Cyclanthaceae may be confused with the Palmae (19), but are distinguished by having conspicuous staminodia, several usually deciduous spathes subtending the simple spadix, and a fleshy syncarpous fruit with numerous small seeds.

Most Cyclanthaceae are visited by small weevils (*Coleoptera*) and other insects. The first day after anthesis the staminodia appear to yield some fragrance, which probably acts as an attractant. Harling (1958) reported that

KEY TO THE TAXA OF CYCLANTHACEAE

Plants vines; leaf blades entire . . . . *Ludovia integrifolia* (Woods.) Harl.
Plants not vines; leaf blades cleft to divided:
   Leaf blades becoming deeply 2-parted, almost divided, the segments usually entire; fruiting
     spadix like a series of concentric rings . . . . *Cyclanthus bipartitus* Poit.
   Leaf blades not deeply 2-parted, either 2-cleft or 4-parted, the segments dentate to deeply split;
     fruiting spadix not a series of concentric rings:
      Leaf blades broadly obovate, bifid halfway at most (the segments often irregularly split); peti-
       oles winged, at most half as long as blade; plants short-stemmed . . . . . *Asplundia alata* Harl.
      Leaf blades rounded in outline, 4-parted; petioles at least half again as long as blade; plants
       acaulescent . . . . *Carludovica*

during this stage the stigmas are receptive and the anthers are unopened, but most species I have observed have the stigmas completely covered by a mass of anthers until later. Moreover, I have seen anthers open and *Trigona* bees collecting pollen at this stage on *Carludovica palmata.* Usually by the following day the staminodia have withered, and Harling reported that the pleasant aroma is replaced by a "sweetish, somewhat sickening" smell but that the weevils are still present on the spadices among the staminodia. He said the weevils are often very destructive but are nevertheless probably the legitimate pollinators of most Cyclanthaceae. *Trigona* bees collecting pollen might also effect pollination, and visits by xylocopid bees were also reported by Harling.

Fruits are mostly bird dispersed, perhaps except for *Cyclanthus,* but they may also be taken by lizards, mammals, and perhaps even ants (van der Pijl, 1968). Fruits of *Carludovica palmata* are eaten by the bat *Micronycteris hirsuta* (Phyllostomidae) (Wilson, 1971) and by the tamarin (Hladik & Hladik, 1969).

According to Harling (1958), a family of 11 genera and 178 species ranging from southern Mexico and Lesser Antilles throughout Central America and the Amazon basin; also in eastern Brazil.

ASPLUNDIA Harl.

Asplundia alata Harl., Acta Horti Berg. 18:223–224. 1958

Large terrestrial herb, ca 2 m tall; stems to 30 cm high and 12 cm thick. Leaves to 1.8 m long, closely spiraled, clustered at apex; petioles about half the length of blade; blades broadly obovate, to ca 1 m long and 60 cm wide near apex, bifid, tapering to narrow base, decurrent onto petiole and continuous with broadly vaginate wing of petiole, becoming very much dissected in age, the sur-

faces pleated, the ribs 3, the lateral ribs branching from the midrib near base, the segment ribs prominently raised on upper surface, with furfuraceous scales scattered on lower surface; juvenile leaves similar to adults in shape, at first entire, splitting in maturity. Inflorescences from leaf axils, often several; peduncles to ca 20 cm long; spathes usually 3 or 4, ± oblong, 6–10 cm long, widely spaced below spadix, the outer one keeled; spadix cylindrical, blunt at apex, 4.5–6 cm long, white in flower, becoming green in fruit; staminate flowers alternating with short, conical pistillate flowers, densely covering spadix, but soon falling free; perianth lobed; pistillate flowers with tepals thick, longer than style at anthesis, light green; staminodia to ca 6 cm long, caducous; stigmas medial, narrow, elongated, dark green. Seeds ca 1.5 mm long and ca 1 mm wide. *Croat 10204, 14088.*

Uncommon, on creek banks and in ravines in the vicinity of the Laboratory Clearing. Flowers from late dry season to the middle of the rainy season, March to August. Mature fruits have been seen from June to September.

Known only from Panama in tropical moist forest in the Canal Zone, San Blas, and Chiriquí and from tropical wet forest in Colón.

CARLUDOVICA R. & P.

Carludovica drudei Mast., Gard. Chron. n.s. 8:714. 1877

Large, acaulescent herb, 2–3 m tall. Petioles to 2.5 m long, terete, vaginate only near base; blades to 1.8 m wide, palmately and unequally 4-lobed, the lobes to 75 cm long, 35–60 cm wide at apex, with 9–16 teeth per lobe, the teeth 5–13 cm long, glabrous and shiny above, dull below; juvenile leaves entire, the apex V-shaped, becoming 4-lobed, the lateral lobes more deeply divided. Peduncles 40–50 cm long; spathes 4, congested immedi-

KEY TO THE SPECIES OF CARLUDOVICA

Lobes of blade toothed less than one-fourth the way to base; blades shiny on upper surface, dull
   on lower surface; stigmas laterally compressed; tepals clearly exceeding stigmas even in flower
   . . . . *C. drudei* Mast.
Lobes of blade toothed deeply, often past middle; blades shiny on both surfaces; stigmas ovate to
   suborbicular; tepals scarcely, if ever, exceeding stigmas . . . . *C. palmata* R. & P.

Fig. 73. *Carludovica drudei*

Fig. 72. *Carludovica drudei,*
inflorescence with staminodia
exserted from pistillate flowers

Fig. 74. *Carludovica drudei,* cross section of spadix

ately below spadix; spadix narrowly cylindrical, 11–12 cm long, 1.5 cm thick in flower, to 22 cm long and 4.5 cm thick in fruit; staminate and pistillate flowers alternating spirally on spadix; staminate flowers in clusters of 4, lacking perianth, the stamens numerous, closely congested, obscuring all of pistillate flower but the staminodium, falling within a few days after anthesis; pistillate flowers sunken into fleshy axis of spadix; tepals 4, 5–6 mm long in flower, distinctly surpassing length of stigmas (to 8 mm long in fruit); staminodium slender, flattened, very long and showy, white, falling soon after anthesis; stigmas 4, laterally compressed; fruiting spadices rupturing at maturity, beginning at apex, to expose bright orange matrix with embedded fruits. Fruits oblong to rounded, ca 10 mm long, 6–8 mm broad; seeds numerous, ± ovoid, ca 2 mm long, flattened. *Croat 10838, 12305.*

Occasional, in the forest, usually along streams, possibly preferring steep banks. Flowers in June. The fruits mature from July to October.

Distinguished by having the leaf lobes toothed much less than halfway to the base.

Lowland forests in Mexico (Chiapas and the Yucatan Peninsula), Costa Rica, Panama, and possibly Colombia. In Panama, known from tropical moist forest on both slopes in the Canal Zone and in Chiriquí and Darién.

See Figs. 72, 73, and 74.

Carludovica palmata R. & P., Syst. Veg. 291. 1798
Panama hat palm, Palmita, Jipijapa, Portorrico, Atadero, Rampira, Iraca, Guachiban, Canagria

Large, glabrous, acaulescent herb, to over 3 m tall. Petioles to 3.5 m long, round; blades palmately lobed, pleated, usually light green in color, to 1 m long and 1.9 m wide, usually with 4 ± irregular lobes to 80 cm long and 85 cm wide at apex, the lobes irregularly toothed to past middle, with 10–16 teeth per lobe. Peduncles 20–45 cm long; spathes usually 3 (4), congested just below spadix, ca 25 cm long, greenish to white, maroon near apex; spadix cylindrical, white in flower, ca 15 cm long and 2.5 cm thick, to 25 cm long and 6 cm thick in fruit; staminate and pistillate flowers alternating spirally as in *C. drudei*; tepals scarcely if at all exceeding stigmas; staminodia 4–6 cm long; stigmas ovate to suborbicular. Fruits orange, ca 1 cm long; seeds many, white, 2 mm long. *Croat 10385.*

Common locally, primarily in clearings and on steep banks of the shoreline, but also in the forest on Orchid Island; common and widespread in the Canal Zone. Flowers mostly in January and February, possibly March. The fruits mature from April to June, rarely later in the rainy season.

Except for leaf differences, tepal length, and shape of stigma the species is very similar to *C. drudei*, which is treated in much greater detail. The leaf segments of *C. palmata* are at least in part toothed more than halfway to base.

Native from Guatemala to Bolivia; introduced in the West Indies. In Panama, known from tropical moist forest in the Canal Zone, Bocas del Toro, Colón, Coclé, Panamá, and Darién.

See Figs. 75 and 76.

CYCLANTHUS Poit.

Cyclanthus bipartitus Poit., Mém. Mus. Hist. Nat. 9:36. 1822
Portorrico, Lengua de buey

Acaulescent, monoecious herb, to 3.5 m tall. Petioles 1.5–2 m or more long; blades to 1.6 m long, almost completely bisected, occasionally trifid, each segment 15–24 cm wide, the base of each lobe oblique, the midrib near upper edge at base, becoming centered by apex. Peduncles 40–90 cm long; spathes 5–7, broadly ovate, 15–25 cm long, the inner cream-colored or pinkish, the outer ± foliaceous; spikes 5–11 cm long when flowering, 1.5–2.5 cm broad, with staminate and pistillate flowers in 10–15 separate alternating whorls, the flowers of each whorl connate; staminate flowers naked; pistillate flowers enclosed by two rims of fleshy tissue, these becoming much enlarged; fruiting syncarps to 6 cm diam, ± pendent. Each segment of syncarp falling off individually, with many seeds in a fleshy matrix; seeds to 1.5 mm long and 1 mm wide. *Croat 5386, 6016.*

Occasional, inhabiting chiefly stream banks and the lakeshore, though numerous juvenile plants are found within the forest away from any streams. Flowers mostly April and May, but also sometimes in June and perhaps later as well. The fruits mature throughout the rainy season, mostly by October.

The soft fleshy segments of the syncarp float and are dispersed in part by water currents and in part by animals or birds.

Guatemala to Peru and Brazil; Lesser Antilles. In Panama, known from tropical moist forest in the Canal Zone, Bocas del Toro, San Blas, Chiriquí, Panamá, and Darién and from tropical wet forest in Panamá (beyond Cerro Jefe) and Colón.

See Fig. 77.

LUDOVIA Brongn.

Ludovia integrifolia (Woods.) Harl., Acta Horti Berg. 18:338. 1958
Carludovica integrifolia Woods.

Nonrosulate epiphytic vine; stems 9–12 mm diam, at first green, becoming brown, forming long, ± tough, brown roots at most nodes; internodes mostly 5–7 cm long. Leaves drying thin, all but the upper 3–8 withering but often persisting; petioles 10–15 cm long, narrowly canaliculate; sheaths thin, extending one-third to almost entire length of petiole; blades narrowly elliptic to oblanceolate, acuminate, cuneate and often ± unequal at base (one side slightly shorter), 13–23 (33) cm long, 4–7 (17) cm wide, weakly bicolorous, the lower surface dull, with 4–7 major veins running almost parallel to the midrib but extending to the blade margin; margins weakly toothed at termination of each veinlet, the midrib raised on both surfaces but most prominent in basal third of blade. Spathes 3, enveloping spadix, acuminate, greenish to white, the outer 6–8 cm long, the inner to 5 cm long, all caducous, breaking loose from base at anthesis and carried

Fig. 75. *Carludovica palmata*

Fig. 76. *Carludovica palmata*

Fig. 78. *Ludovia integrifolia*

Fig. 77. *Cyclanthus bipartitus*

Fig. 79. *Ludovia integrifolia*

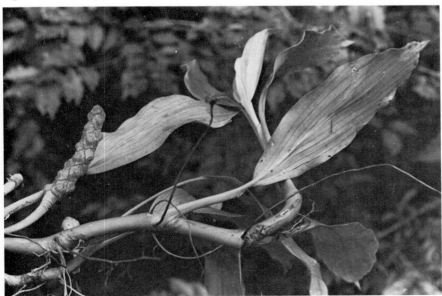

upward by the enlarging spadix; spadix ca 3 cm long at anthesis; staminodia widely dispersed in clusters of 4 (rarely 5), to ca 3 cm long; staminate flowers ca 3.5 mm long, the perianth 20–30-lobed, glandular; pistillate flowers connate with each other, the tepals much reduced, the long, immersed stigmas sessile, longer than broad; fruiting spadices 8.5–9.5 cm long, ca 1.5 cm diam, blunt, becoming orange, fleshy and tasty at maturity, the peduncles with 3 prominent spathe scars; stipes 5–7 mm long; stigmas round, ca 0.5 mm diam. Fruits diamond-shaped, 11–17 mm long, 9–12 mm wide, conical; seeds numerous, ± ovoid, ca 1.5 mm long, embedded in a sweet, sticky, gelatinous matrix at maturity. *Croat 12560.*

Occasional, in the old forest. Reportedly flowers in June. Mature fruits have been recorded in November and December.

Obscurely resembling *Stenospermation* (21. Araceae), but distinguished by having more than one spathe.

Panama to Ecuador on the Pacific slope. In Panama, known from tropical moist forest in the Canal Zone and Darién.

See Figs. 78 and 79.

21. ARACEAE

Epiphytic (two-thirds of BCI species), terrestrial, or rarely aquatic herbs, sometimes with suffruticose parts, occasionally scandent; rarely with milky sap (*Syngonium*), caulescent or acaulescent, glabrous or rarely pubescent or prickly on petioles. Leaves alternate or clustered, petiolate with sheathing bases, simple or divided; blades entire, occasionally perforate; venation parallel or pinnate- or palmate-netted; cataphylls usually present. Flowers bisexual or unisexual and plants monoecious, in terminal or axillary cylindrical spadices, the spadix suspended and sometimes enclosed by an herbaceous spathe; spadix uniform, bearing bisexual flowers or consisting of pistillate and staminate parts, the staminate part above the pistillate and deciduous or rotting away after anthesis; perianth lacking in unisexual flowers (also in bisexual flowers of *Rhodospatha* and *Stenospermation*) or with segments 4 (*Anthurium*), (2)3(4) (*Spathiphyllum*) or 4–8 (*Dracontium*); stamens free or united in unisexual flowers, usually 4, or (2)3(4) (*Spathiphyllum*), 4–6 (*Xanthosoma*), 2–4(6) (*Homalomena*), 3 or 4 (*Syngonium*), and 2–6 (*Philodendron*); anthers 2-celled, dehiscing by apical slits or pores; staminodia sometimes present; ovaries many per spadix, sessile or immersed in spadix (and thus inferior), with 1, 2, or several locules; placentation basal, parietal, axile, or apical; ovules 1 to many; style 1; stigma massive, usually capitate. Fruits berries, free or connate at maturity, usually protruding from tepals in bisexual flowers or exposed by rupturing of the spathe in unisexual flowers; seeds with endosperm.

Araceae are best characterized by the fleshy, cylindrical spadix subtended by a leafy spathe borne on a peduncle. In general, members of the family are not confused with any other, though *Ludovia integrifolia*, a simple-leaved Cyclanthaceae (20), looks much like an aroid.

Flowers of most species with closed spathes are pollinated primarily by large awkward ruteline scarab beetles, which are attracted to the inflorescence by the white spathe and/or the odor. It is not known what they get from their visits. They are often found crawling on the pistillate part of spathate inflorescences even when these are nearly closed. (See the discussion of *Dieffenbachia pittieri*.) In *Anthurium* and *Spathiphyllum* the open spathes also form the attractant organ, but do not trap insects as do closed spathes. The spadix of some species, e.g., *A. ochranthum*, is colored bright yellow or white and may itself attract attention. I have seen male euglossine bees visiting *Anthurium* and *Spathiphyllum* inflorescences in great numbers in the Canal Zone. Dressler (1968a) reported that *Eufriesa pulchra* visits *Anthurium*. Bees are believed to effect pollination of orchids, and it is possible they gather scent compounds from these two genera

---

KEY TO THE TAXA OF ARACEAE

Plants small floating aquatics; inflorescences inconspicuous; leaves spongy, covered with short, few-celled trichomes . . . . *Pistia stratiotes* L.
Plants rooting in soil or epiphytic; inflorescences conspicuous:
    Leaves peltate; blades usually reddish near center and spotted with white, red, or yellow; plants terrestrial; flowers unisexual . . . . *Caladium bicolor* (Ait.) Vent.
    Leaves with petioles basifixed; blades entirely green in color:
● Leaf blades variously divided:
    Plants with one very large, deeply divided leaf arising from a subterranean caudex . . . .
    . . . . *Dracontium dressleri* Croat
    Plants not as above, either epiphytic or with entire leaves, the leaves several to many:
      Leaves pinnatifid, with pinnate divisions in blade at approximately right angles to midrib:
        Blades with basal lobes, the lowermost vein with a prominent branch into each segment; spathe persistent; flowers unisexual; stigmas round . . . .
        . . . . *Philodendron radiatum* Schott
        Blades without basal lobes, the lowermost vein unbranched; spathe deciduous; flowers bisexual; stigmas linear . . . . *Monstera dilacerata* C. Koch
      Leaves variously parted, not pinnatifid, the lobes not primarily at right angles to midrib:
      Blades with 6–18 lobes:
        Plants terrestrial; peduncles arising from the ground; spathe enveloping spadix; spadix of distinct staminate and pistillate parts . . . .
        . . . . *Xanthosoma helleborifolium* (Jacq.) Schott

Plants epiphytic; peduncles not arising from the ground; spathe free; spadix uniform:

Leaf segments lobed, usually more than 50 cm long . . . . *Anthurium clavigerum* Poepp. & Endl.

Leaf segments entire, usually less than 30 cm long . . *Anthurium bombacifolium* Schott

Blades with 3–5 lobes:

Blades round in outline; leaflets of approximately the same size and shape . . . . *Anthurium bombacifolium* Schott

Blades not round in outline; leaflets not of uniform size and shape:

Plants juvenile; caudex short; leaves clustered at apex, commonly directly associated with entire leaves . . . . *Anthurium clavigerum* Poepp. & Endl.

Plants adult; long vines; leaves well spaced, not directly associated with entire leaves:

Leaf veins all parallel, none impressed on upper surface, collecting veins lacking; milky sap lacking; spathe persistent; spadix lacking a sterile staminate part . . . . *Philodendron tripartitum* (Jacq.) Schott

Leaf veins interconnected between major laterals, not parallel, the major veins impressed above, the collecting veins 2 or 3; milky sap common; spathe deciduous; spadix bearing a sterile staminate part above the pistillate part . . . . *Syngonium*

● Leaf blades simple and entire, though often lobed at base:

◆ Leaf blades conspicuously cordate, sagittate, or hastate, lobed at base, the sinus at least one-fourth as long as blade:

Plants epiphytic:

Spathe linear-lanceolate, free; spadix uniform; reticulate venation prominent . . . . *Anthurium brownii* Mast.

Spathe enclosing spadix; spadix divided into staminate and pistillate parts; reticulate venation lacking or not prominent (except *P. hederaceum*, which is also puberulent on petioles) . . . . *Philodendron* (in part)

Plants terrestrial or aquatic, rooted in soil:

Plants acaulescent or with subterranean stem, sometimes pubescent:

Petioles puberulent, armed with short prickles in basal half . . . . *Homalomena wendlandii* Schott

Petioles glabrous or pubescent, but not armed with short prickles . . . . *Xanthosoma*

Plants with definite stem or caudex (the leaves sometimes closely clustered and appearing acaulescent):

Plants commonly growing in standing water, the stem erect, at least 1 m tall, armed with short prickles at least in basal part; spathe deciduous in fruit . . . . *Montrichardia arborescens* (L.) Schott

Plants never in standing water, the stem usually creeping over the ground, short, unarmed; spathe persistent in fruit:

Petioles conspicuously flattened on upper surface; veins all parallel or at least without conspicuous reticulate veins; spathe enveloping spadix; spadix of staminate and pistillate parts . . . . *Philodendron grandipes* Krause

Petioles terete; veins not all parallel, the reticulate veins prominent, forming a collecting vein near margin; spathe free, narrow; spadix uniform with bisexual flowers:

Blades thick; sinus narrow or closed, the lobes often overlapping (at least when flattened out); fruits ± orange; plants growing usually at the margin of the lake on exposed areas (rarely in the forest except as an epiphyte) . . . . *Anthurium brownii* Mast.

Blades thin; sinus always broad, the lobes never overlapping; fruits violet-purple at apex, white at base; plants growing in forest . . . . *Anthurium ochranthum* C. Koch

◆ Blades never conspicuously lobed at base, or if cordate, with basal sinus much less than one-fourth the length of blade:

■ Plants terrestrial (including those in standing water):

Plants acaulescent or nearly so, the caudex nearly subterranean; flowers bisexual (spadix thus uniform); spathe remaining open and persistent . . . . *Spathiphyllum*

Plants caulescent with a thick stout stem, usually short, creeping along the ground then erect; flowers bisexual or unisexual; spathe rather tightly enveloping spadix and persistent or caducous:

Major lateral veins close together, at ca 90° angle from midrib; petioles hard, nodose and geniculate at apex; spathe free, soon deciduous; spadix uniform (i.e., flowers bisexual) . . . . *Rhodospatha moritziana* (Schott) Croat

Major lateral veins not close together, directed decidedly upward; petioles succulent, not nodose or geniculate; spathe persistent, attached to lower part of spadix; spadix with staminate and pistillate parts; sap with odor of oxalic acid . . . . *Dieffenbachia*

■ Plants epiphytic, never terrestrial at maturity except by accident:
  Plants with fertile parts:
    Spadix divided into staminate and pistillate parts; spathe persistent and enveloping
      spadix after opening; plants long epiphytic vines . . . . *Philodendron* (in part)
    Spadix uniform; spathe either deciduous or narrow and free after opening:
      Spathe narrow, free at anthesis, usually persistent; leaf blades with or without col-
        lecting veins near margin; stigmas round or square; plants with or without
        elongate caudices . . . . *Anthurium* (in part)
      Spathe broad or convolute, usually surrounding spadix at anthesis (at least at
        base), soon deciduous; leaf blades lacking collecting veins; stigmas oblong or
        linear; plants with elongate caudices:
        Blades small, less than 7 cm wide, the veins ± parallel to midrib; peduncles re-
          curved at apex before anthesis . . . . *Stenospermation angustifolium* Hemsl.
        Blades very much larger, the veins not parallel to midrib; peduncles always ±
          erect:
          Leaves entire, never perforate or pinnately lobed; primary lateral veins often
            1 cm or less apart; caudex and petioles not finely tuberculate; stigmas
            round or ellipsoid . . . . *Rhodospatha wendlandii* Schott
          Leaves often perforate; primary lateral veins 1.5 cm or more apart; caudex
            and petioles usually finely tuberculate; stigmas linear . . . . *Monstera*
  Plants sterile:
    Blades with ± prominent reticulate veins between major lateral veins:
      Blades perforated or pinnately lobed; plants long slender vines with many short
        roots at most nodes . . . . *Monstera*
      Blades entire, not perforated; plants usually with short caudices . . . .
        . . . . *Anthurium* (in part)
    Blades with all lateral veins parallel or subparallel:
      Lateral veins ± parallel to midrib . . . . *Stenospermation angustifolium* Hemsl.
      Lateral veins never almost parallel to midrib:
        Leaves pinnately lobed . . . . *Monstera dilacerata* C. Koch
        Leaves entire:
          Petioles continuously canaliculate to base of blade . . . .
            . . . . *Rhodospatha wendlandii* Schott
          Petioles winged-vaginate but the sheath always ending short of blade (free
            part of sheath may extend beyond base of blade in *P. guttiferum*) . . . . . .
            . . . . *Philodendron* (in part)

similar to the scent compounds they seek in orchids (Dodson et al., 1969; Hills et al., 1972). A male bee of *Eulaema cingulata* has also been reported to visit inflorescences of *Xanthosoma nigrum* without taking either pollen or nectar (Dodson, 1966). The implication is that bees may be attempting to collect scent compounds from this species as they probably do from *Spathiphyllum* and *Anthurium*. Probably a similar system will be found for other genera with bisexual flowers, such as *Monstera, Rhodospatha,* and *Stenospermation*.

Rhodospatha moritziana has a whitish, syrupy substance covering the spadix just after anthesis that may attract such insects as euglossine bees. Faegri and van der Pijl (1966) reported that these nectarlike substances are secreted by the stigmatic papillae but that the cell contents soon disintegrate. See *Dracontium dressleri* for a discussion of probable fly pollination (Croat, 1975c).

Fruits are endozoochorous, probably predominantly ornithochorous, except perhaps those of *Pistia* and *Spathiphyllum*. This is especially true of *Anthurium,* which has colorful fruits that suddenly emerge from the tepals and may later be suspended slightly from two to four threadlike structures. All of the fruits of the spike may emerge at nearly the same time, as in *A. bakeri* and *A. friedrichsthalii,* or they may be considerably staggered, as in *A. tetragonum*. Once exposed, they are quickly re-

moved. The seeds (usually two to four) are easily forced from the lower end of the berry and are embedded in a clear, jellylike, sticky mesocarp. After eating the sweet mesocarp (no doubt many seeds as well), the bird probably has some seeds stuck to its bill. Whetting its beak on a distant tree, the bird deposits the seeds at an appropriate germination site, where they stick. Bats are also known to take fruits of *Anthurium* (Yazquez-Yanes et al., 1975).

Fruits of *Dracontium* are probably dispersed in a manner similar to *Anthurium*. All genera with unisexual flowers, except *Montrichardia,* have the spathe completely covering at least the lower part of the spadix until the fruits mature. In the case of *Dieffenbachia* and *Philodendron* the pistillate part of the spadix is fused to the spathe, and the fruits are free. At maturity the inflorescence curls, rupturing the spathe and exposing the fruits. Seeds of *Dieffenbachia* seem well suited for bird dispersal, but are probably taken by larger animals as well. In *Philodendron* the fruits are closely aggregated and are probably eaten by monkeys as well as birds. Leaf-cutter ants have been seen carrying the tiny seeds also, and no doubt some are dropped along the way. It is important to point out that, except in the case of *Philodendron radiatum,* seeds germinate on the ground. Later stages of the plant become hemiepiphytic or even truly epiphytic by climbing trees and losing their ground connection.

Caladium and *Xanthosoma* have closely congested fruits that are no doubt dispersed in a manner similar to *Philodendron,* though the spathe seems to fall more by decomposition than by arching of the inflorescence. In *Syngonium* the fruit is syncarpous, and the spathe falls free at maturity of the fruit. Fruits, while not colorful, are fleshy and sweet. The few, relatively large seeds may be swallowed when the tasty mesocarp is eaten. Fruits are probably taken mostly by monkeys but possibly by other animals as well. In *Montrichardia,* the only species having unisexual flowers and lacking a persistent enveloping spathe, the fruits are held in an irregular aggregate much larger than the fruits of any other genus. They are usually water dispersed, but are also reported to be eaten by the Hoatzin bird (*Opisthocoma cristatus*) in South America (Ridley, 1930).

All members of the tribe Monsteroideae, including *Stenospermation, Monstera, Rhodospatha,* and *Spathiphyllum,* have fruits that are apparently unprotected by any structural modification such as a protective spathe. Nicolson (1960) reported that all these genera contain trichosclereids; since in masses these cells may give hardness and mechanical protection (Eames & MacDaniels, 1947), it is possible that they play a part in preventing predation on immature inflorescences.

Monstera and *Stenospermation* have bisexual flowers with uniform inflorescences like *Anthurium,* but the pistils become united and form a syncarpous fruit. At maturity the periphery of the fruit exfoliates in sheets or smaller segments, exposing the soft fleshy locules containing many small seeds. Fruits are probably taken by arboreal frugivores, including birds.

Fruits of *Spathiphyllum* are fully exposed and closely congested like the kernels of a cob of corn, becoming rather loose and easily removed at maturity. They are probably taken by birds and other animals; I have seen large segments of an inflorescence removed as though from a single bite. Bunting (1965) reported that the fruits may be water dispersed. Vegetative reproduction of new plantlets that later split off is a clear possibility for the aquatic *Pistia.*

About 120 genera and 1,800 species; widely distributed but most numerous in the tropics and subtropics.

ANTHURIUM Schott

The genus *Anthurium* appears to be considerably more plastic in its variation in morphological characters than are most genera in the family, with the result that a large number of species have been described more than one time.

Anthurium may be distinguished by having a free spathe that is usually persistent. The spadix is cylindrical and uniform with bisexual flowers having four tepals; the tepals are opposite the four stamens and more or less three-sided at the apex. Fruits have two locules with one or two ovules per locule.

KEY TO THE SPECIES OF ANTHURIUM

Leaves compound:
  Leaflets sessile, undulate or lobed; largest leaflets more than 50 cm long (on adult plants); spadix
    more than 40 cm long .... *A. clavigerum* Poepp. & Endl.
  Leaflets petiolulate and entire; largest leaflets less than 30 cm long; spadix less than 15 cm long
    .... *A. bombacifolium* Schott
Leaves simple:
  Leaves ovate to triangular and cordate, with large basal lobes and a prominent sinus:
    Blades coarse and leathery, drying very stiff, the margins wavy or folded, the sinus narrow or
      closed by overlapping basal lobes (at least on flattening); fruits ± orange; plants usually
      epiphytic except at lake margin .... *A. brownii* Mast.
    Blades thin, the margins never strongly wavy or contorted, the sinus usually open (sometimes
      closed), broad; fruits violet-purple at apex, white at base; plants always terrestrial on BCI
      .... *A. ochranthum* C. Koch
  Leaves without basal lobes or sometimes moderately cordate at base but without large basal
    lobes and a deep sinus:
    Blades without a collecting vein, the primary lateral veins extending apically near the margin
      and closely approaching or merging with it; large epiphyte with obovate to oblanceolate
      leaves and large, ± pendent inflorescences .... *A. tetragonum* Schott*
  Blades with a collecting vein (the primary lateral veins joining before reaching the margin):
    Plants scandent or with elongate, often branched, creeping caudices:
      Blades acute at base, often punctate on lower surface; fruits white at maturity ....
        .... *A. scandens* (Aubl.) Engler
      Blades rounded at base, epunctate; fruits orange or red at maturity .... *A. flexile* Schott
    Plants not scandent, acaulescent or with short unbranched caudices:
    • Plants lacking punctations:
        Blades broadest at or about middle; cataphylls weathering, becoming fibrous, persisting on all but the oldest part of stem; spadix lavender at anthesis; fruits white ...
          .... *A. scandens* (Aubl.) Engler
        Blades broadest usually well above middle; cataphylls not weathering into fibers, ultimately deciduous; spadix not lavender at anthesis; fruits red ....
          .... *A. gracile* (Rudge) Lindl.

- Plants conspicuously punctate, at least on underside of leaf:
  Blades usually more than 10 times longer than wide, of ca equal width throughout;
  plants often pendent or at least the leaves not held stiffly erect; fruits truncate at
  apex, broader than long, pale yellow-orange . . . . *A. friedrichsthalii* Schott
  Blades usually less than 10 times longer than wide, considerably broader in distal half;
  plants not pendent, the leaves held stiffly erect; fruits not as above:
  Blades more than 11 cm wide; peduncles usually more than 30 cm long; blades ±
  thick, broadest at middle, gradually tapered to both ends . . *A. acutangulum* Engler
  Blades usually less than 11 cm wide; peduncles usually less than 30 cm long;
  Blades thin, conspicuously acuminate at apex; collecting vein much more prominent than the lateral veins; spadix less than 11 cm long in flower (to 15 cm in
  fruit); fruits bright red . . . . *A. bakeri* Hook.f.
  Blades thick, obtuse to acute at apex (often with short apiculum); collecting vein
  scarcely if at all more prominent than the lateral veins; spadix usually more
  than 11 cm long in flower; fruits yellowish or orange . . . . *A. littorale* Engler

*Discovered too late for treatment: *A. concolor* Krause, Notizbl. 107:617. 1932. *Croat 8154* from Pearson
Peninsula. Like *A. tetragonum*, but spadix violet-purple and petioles three-ribbed on lower side.

Anthurium acutangulum Engler, Bot. Jahrb. Syst.
25:371. 1898
A. porschianum Krause

Epiphyte; caudex short and thick, usually surrounded by
a dense mass of roots; internodes short; cataphylls thin,
short, weathering into brown fibers. Petioles stout, 9–18
cm long, with a deep narrow channel on upper surface,
geniculate at apex; blades elliptic or oblong-elliptic,
rounded or obtuse at apex and cuspidate-acuminate,
mostly obtuse at base, 20–40 cm long, 11–15 cm wide,
often arching with the tip curved downward, punctate
on underside, ± thick, the major lateral veins in 10–15
pairs, forming a prominent collecting vein near the margin. Inflorescences spreading-pendent; peduncles 30–45
cm long; spathe green (pinkish elsewhere), lanceolate-
acuminate, 9–11 cm long, less than 2 cm wide, withering
in age; spadix sessile or short-stipitate, linear, weakly
tapered to apex, 15–30 cm long, 7–10 mm broad (4–5 mm
when dried); tepals truncate at apex, sharply triangular,
2–3 mm wide. Mature fruits apricot-orange, ca 5 mm
long, 5 mm wide (slightly wider in direction of axis),
drying with small depression at apex. *Croat 4347b, 6740.*

Apparently rare, occurring in the forest and seen only
on small trees within 2–3 m of the ground. Individuals
produce successive inflorescences and may be found in
flower all year, especially during the rainy season. The
fruits probably develop promptly.

Perhaps most easily confused with *A. littorale*, which
has narrower blades, often broadest below the middle,
with reticulate venation usually as prominent on dry
specimens as the lateral veins. Elsewhere in Panama the
species is most easily confused with *A. ramonense* Engler
ex Krause, a species with thick, oblanceolate-elliptic
leaves, thicker petioles, and stouter spadices.

Honduras to Panama; sea level to 900 m. In Panama,
known from tropical moist forest in the Canal Zone and
Bocas del Toro, from premontane wet forest in Bocas
del Toro, and from tropical wet forest in Colón and
Panamá.

See Fig. 80.

Anthurium bakeri Hook.f., Bot. Mag. 32, pl. 6261. 1876
Epiphyte; caudex short and thick, usually less than 10 cm
long and 1.5 cm wide; internodes short, moderately

slender-rooted; cataphylls ± thick, 3–5 cm long, soon
reduced to coarse, brown, persistent fibers. Petioles 3–17
cm long, to 5 mm thick, broadly canaliculate and straight
to moderately geniculate near apex, broadened and shortly
vaginate at base; blades elliptic-lanceolate, broadest at
or about middle, rather abruptly acuminate at apex, acute
at base and weakly decurrent onto margins of petiole,
19–44 cm long, 2.8–9 cm wide, bicolorous, only moderately thick when fresh, thin on drying, held erect largely
by V-shaped construction of blade and by midrib, the
midrib thick, prominently raised on underside, the lower
surface densely punctate; collecting vein 3–8 mm from
margin, prominent on both surfaces, especially the lower,
the lateral veins obscure (moderately prominent on dried
specimens). Inflorescences held ± erect; peduncles slender to moderately stout, 8–29 cm long, shorter than
leaves; spathe green, reflexed, 2–5.5 cm long, 7–20 mm
wide, acute and apiculate at apex, the margin often turned
in, decurrent onto peduncle at base; spadix sessile to
shortly stipitate, bluntly rounded at apex, 2–11 cm long
and 5–7 mm wide in flower, to 15 cm long and 2.5 cm
wide in fruit. Fruits obovate, acute to rounded at apex,
ca 6 mm long, bright red, emerging from tepals and
pendent, suspended on 2 slender white fibers; seeds 2,
oblong, white, ca 3 mm long, somewhat flattened, embedded in a sticky, clear matrix. *Croat 6627, 10882.*

Uncommon, in the forest, usually at low elevations
on reasonably small branches. Probably rare other places
as well, since the plant has seldom been collected despite
its prominently colored inflorescences and frequent flowering. Plants produce a continuing series of inflorescences
throughout the year, especially during the rainy season,
and fruits mature in 3–6 weeks.

Though possibly confused with *A. littorale*, which has
thick leaves lacking a prominent collecting vein, *A. bakeri*
is recognized by its thin, punctate blades, which are
broadest at or about the middle, by the short spadix,
which is scarcely if at all tapered toward the apex, and
by its bright red berries.

Guatemala to Panama (in Costa Rica on the Atlantic
slope as well as from the Osa Peninsula on the Pacific
slope). In Panama, known from tropical moist forest on
BCI and at Summit Garden in the Canal Zone and from
premontane wet forest in Colón (Santa Rita Ridge).

See Fig. 81.

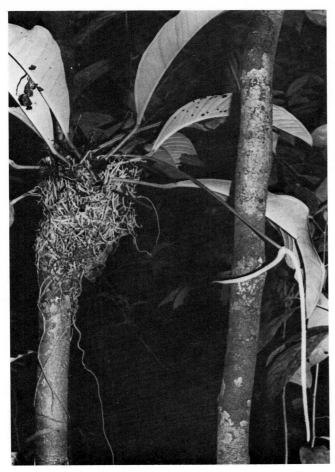

Fig. 80. *Anthurium acutangulum*

Fig. 81. *Anthurium bakeri*

Anthurium bombacifolium Schott, Oesterr. Bot. Z. 8:182. 1858

Creeping epiphyte; caudex long and slender; internodes elongate; cataphylls caducous or weathering to persistent tough fibers. Leaves palmately compound; petioles 24–44 cm long; leaflets 6–11, nearly sessile or on petiole to 5.5 cm long, broadest at about middle, tapered abruptly to cuspidate-acuminate apex and gradually to cuneate-attenuate base, 15–30 cm long, 5–9 cm wide, entire; lateral veins forming 2 collecting veins, 1 marginal, 1 more remote. Peduncles mostly 20–53 cm long; spathe green, 6–10 cm long, 1–2.5 cm wide at base; spadix maroon, short-stipitate or sessile, 3–15 cm long, 7–13 mm broad, attenuate; tepals 6–7 mm long, 4–5 mm wide in fruit. Fruits maroon, 7–8 mm long, 5–6 mm wide; seeds ovoid to oblong, ca 5.3 mm long and 4 mm wide. *Croat 7090, 11700.*

Occasional in the forest in trees, often within 3 m of the ground. May be found in flower or fruit much of the year with most flowering in the dry season. Most fruits develop in the late dry season and in the early rainy season.

Recognized by its thin, palmately compound leaves and elongate caudex.

This species would be considered *A. undatum* Schott from South America if that species is distinct from *A. bombacifolium*, since all the leaflets are subequal and radiate equally from the apex of the petiole. Typical material collected in Honduras, Belize, and Mexico has leaflets pedately arranged, the lateral ones with a distinct auricle on the outer margin.

Mexico to Panama. In Panama, known from tropical moist forest in the Canal Zone, Bocas del Toro, San Blas, Chiriquí, and Darién; known also from premontane moist forest in Panamá, from premontane wet forest in Colón; and from tropical wet forest in Colón and Panamá.

Anthurium brownii Mast., Gard. Chron. n. s. 7:744, f. 139, 140. 1876

Epiphyte, or perhaps secondarily growing terrestrially, often flowering when still rather small; caudex moderately long and densely rooted, 2–5.5 cm thick; cataphylls 5–17 cm long, 2–6 cm broad, soon turning brown, persisting. Leaves coarse and thick, clustered at apex; petioles 25 cm to rarely more than 1 m long, shortly sheathed at base and becoming triangular-canaliculate above, ca 1 cm wide, the geniculum dark, 2 cm long, 1.5 cm wide; blades hastate to trilobate, mostly 27–66 cm long, to 48 cm wide, thick-coriaceous (at least when dried), the upper lobe abruptly acuminate, 10–28 cm broad, the margins broadly undulate, the basal lobes directed outward on young plants, inward on older plants, often overlapping sinus, the sinus to 23 cm deep; major veins above sinus 5–8 pairs, the larger vein of each basal lobe joined by 8 or 9 smaller veins. Peduncles terete, 88–102 cm long, usually longer than petioles; spathe 8–23 cm long, cordate to truncate and 1.5–4.5 cm wide at base, long-acuminate, green, persisting in fruit, often ± twisted; spadix maroon, short-stipitate, 7–28 cm long, long-tapered, 4–15 mm wide in flower, heavy and to 38 cm long and 2.5 cm wide in fruit.

Fruits dense, ± ellipsoid to obovoid, 5–8 mm long, 5 mm wide, orange or orange tinged with red, becoming mostly exserted from tepals and ultimately entirely exserted and dangling on 2 slender filaments; seeds 2, whitish, 4.5–5.3 mm long, 3–3.3 mm wide, embedded in a sticky, transparently whitish matrix. *Croat 8252, 10389.*

Abundant, usually rather high in forest trees or epiphytic on tree roots or old tree stumps at the margin of the lake. Those plants along the lakeshore generally are more massive than those in the forest. Apparently flowers and fruits all year.

Distinguished by the thick, cordate, very undulate blades. Immature plants have been known to flower when the leaves were no longer than 14 cm, scarcely revolute, and almost triangular in shape. Such plants might be thought to be another species.

Costa Rica to Colombia. In Panama, known from tropical moist forest in the Canal Zone, San Blas, and Panamá, from premontane moist forest in the Canal Zone (Ancón), from premontane wet forest in Chiriquí, Coclé, and Panamá, and from tropical wet forest in Colón and Panamá.

See Fig. 82.

Anthurium clavigerum Poepp. & Endl., Nov. Gen. Sp. Pl. 3:84. 1845

A. holtonianum Schott; *A. wendlandii* Schott; *A. panduratum* Mart. non Schott

Large epiphyte; caudex often 1 m or more long, to 5 cm thick, with large, irregular leaf scars. Leaves closely congested near apex of stem, palmately compound, often 1–2 m broad, shorter than petiole; petioles terete, shortly vaginate at base, geniculate at apex; leaflets 7–12, ± sessile, acuminate, the margins sinuate or with 2 to 5 deep lobes, the middle leaflet 50–100 cm or more long, the lateral leaflets becoming progressively shorter; juvenile leaves simple, entire, thick, at first lanceolate, long-caudate, acute at base, later truncate to cordate at base, to 29 cm long and 10 cm wide, becoming palmately compound with 3 or 5 sessile leaflets, the lower pair of leaflets asymmetrical. Inflorescences very stout, usually drooping below leaves; peduncles terete, shorter than petioles; spathe 30–65 cm long, 3–11 cm wide, maroon; spadix 40–80 cm long, 2 cm wide at base tapering to apex, the apex usually withering in fruit; fruiting spadices heavy, to 6 cm wide. Fruits obovate, 6–7 mm long, violet-purple to red-violet, exserted from tepals at maturity; tepals triangular in cross section, thickened and truncate at apex, to 5 mm long at maturity of fruit; seed 1(2), oblong, 3–4 mm long, white, sticky, embedded in a jellylike mesocarp. *Croat 4230, 4654, 8328.*

Common, usually growing at 2–7 m from the ground. Flowering early in the dry season (December) to as late as March (rarely later); juvenile inflorescences may be seen as early as October. The fruits develop from April to October, mostly in June and July. Individuals usually bear no more than two inflorescences, and fruits require about 4 months to develop.

Distinguished from any other plant by its large, pedately compound leaves with lobed leaflets.

Fig. 82. *Anthurium brownii*

Fig. 83. *Anthurium clavigerum*

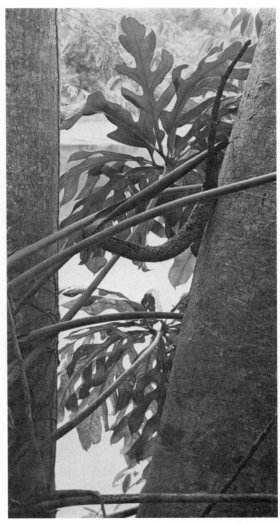

Fig. 84. *Anthurium clavigerum*

Costa Rica to Peru and Brazil; low elevations. In Panama, known from tropical moist forest on both slopes in the Canal Zone and in Darién, from premontane moist forest in Panamá, from premontane wet forest in Chiriquí, and from tropical wet forest in Coclé (Atlantic slope) and Chiriquí.

See Figs. 83 and 84.

Anthurium flexile Schott, Oesterr. Bot. Z. 8:180. 1858

Epiphyte; caudex slender, usually less than 4 mm diam, creeping or scandent, usually with short, fleshy roots at nodes; internodes mostly 2–5 cm apart; cataphylls few, slender, acuminate, to 4 cm long. Petioles slender, 2.5–15 cm long, not obviously geniculate, vaginate only near base or to midlength; blades ovate-elliptic to oblong-oblanceolate, acuminate, narrowly rounded to subcordate at base, 10–20 cm long, 3.5–11 cm wide, thin, epunctate; venation pinnate, the principal lateral veins arising near the base of the blade and uniting with other lateral veins to form a slender collecting vein remote from the margin, often also with an indistinct submarginal collecting vein, the smaller veins distinct on dried specimens. Inflorescences borne from uppermost leaf axils; peduncles very slender, 7–15 cm long (to 20 cm in fruit); spathe lanceolate, acuminate or long-attenuate, rounded or subcordate at base, spreading, 4–8 cm long, 1–1.8 cm wide, pale green; spadix violet-purple at anthesis, sessile to stipitate, 5–9 cm long, of ± equal width throughout but narrowed at apex, less than 5 mm diam at anthesis, to ca 2 cm diam in fruit. Fruits usually developing at base first, ovoid to ellipsoid, often weakly beaked, mostly 8–10 mm long, orange or red. *Starry 60.*

Apparently rare; it has not been seen in recent years. Seasonal behavior uncertain; may be seen in fertile condition most of the year, but with most fruits maturing in the rainy season. Plants produce successive inflorescences over a long period of time, but possibly have two different periods of flowering activity.

Mexico to Colombia. In Panama, known from wetter parts of tropical moist forest in the Canal Zone, Bocas del Toro, Chiriquí, and Darién and from premontane wet forest in Coclé and Darién.

Anthurium friedrichsthalii Schott, Oesterr. Bot. Wochenbl. 5:65. 1855

A. linearifolium Engler; *A. gracile* sensu Standley (1944) non (Rudge) Lindl.

Epiphyte, usually pendent, often loosely attached; caudex usually less than 15 cm long, 1–1.5 cm diam; nodes very close together, moderately rooted, the roots brownish, 3–4 mm wide; leaf scars prominently raised; cataphylls soon reduced to brown fibers, the fibers persistent at upper leaf bases. Leaves crowded; petioles 1.5–14 cm long, mostly less than 10 cm long, ca 2–3 mm wide, moderately to very strongly geniculate at apex, the base enlarged to 1 cm diam, the sheath 12–30 mm long; blades linear, long-acuminate at apex, subacute at base, 12–56 cm long, 0.8–4 (5.8) cm wide (mostly less than 3 cm wide), strongly bicolorous (when fresh), the lower surface light green with numerous brown punctations; midrib promi-

nently raised above, the lateral veins obscure on both surfaces (moderately prominent below when dried), the collecting vein 3–4 mm from margin. Inflorescences pendent, shorter or longer than leaves; peduncles to ca 2.5 mm thick, (9)19–30 cm long (to 40 cm in fruit); spathe linear, 1–5 cm long, 3–5 mm wide, not reflexed; spadix sessile (elsewhere distinctly stipitate), narrowly cylindrical, 3–15 cm long and 4–5 mm wide, to 25 cm long and 2 cm wide in fruit; tepals ca 1.7 mm high and 2 mm wide in flower, the fruiting tepals to 4 mm long and 5 mm wide at base, tapered upward, the apex narrow and truncate. Berries of irregular shape, broader than long, ca 5 mm long, 5–7 mm wide, pale yellow-orange, truncate at apex with a central depression, borne on 4 slender filaments; seeds (1)3–4(5), ± ovate, to 2.7 mm long, greenish or yellowish (darker on one end), embedded in fleshy clear matrix, bearing a slender, mucilaginous, sticky appendage. *Croat 6614, 6980.*

Common in the forest, from near the ground to more than 30 m high, usually pendent from tree branches. Flowers mostly in the dry season (January to April), rarely later in the rainy season. The fruits develop slowly, requiring possibly several months, maturing mostly from June to October.

Recognized by the thick, linear, bicolorous, punctate leaves. Juvenile plants have short, elliptic leaves.

Guatemala to Colombia. In Panama, known from tropical moist forest in the Canal Zone, Bocas del Toro, Coclé, Panamá, and Darién; known also from premontane wet forest in Colón and from tropical wet forest in Colón and Chiriquí.

Anthurium gracile (Rudge) Lindl., Edward's Bot. Reg. 19:1635. 1833

A. scolopendrinum (Ham.) Kunth

Epiphyte, usually erect; caudex 1.5–30 cm long, 7–10 cm wide; internodes moderately close together; roots fleshy, often whitish, 4–6 mm thick; cataphylls 2–9.5 cm long, reddish or brown, the uppermost persisting and remaining intact, ultimately deciduous. Leaves moderately thin, held ± erect; petioles distinctly broadened at base and sheathed to 4.5 cm from base, rounded but with flat rib to 2 mm wide on upper surface, 1–20 cm long, to 3 mm wide, slightly to moderately geniculate and nodose at apex, the geniculum not extending onto blade; blades broadest above middle, gradually acuminate at apex, gradually tapered to long-cuneate base, 11–32 cm long, 0.7–8.5 cm wide, essentially concolorous, glabrous, epunctate; midrib prominent above, the lateral veins and collecting vein indistinct (moderately distinct on drying). Peduncles very slender, ± erect, 21–40 cm long in flower, to 60 cm long in fruit; spathe maroon, 13–25 mm long (to 5 cm in South American material), mostly to 6 mm wide and often joining on opposite side of peduncle, ± pendent; spadix sessile, mostly less than 3 cm long and 4 mm wide in flower, often ± curved, to 10 cm long and 1.5 cm wide in fruit. Berries ± globose, ca 7 mm diam, bright red; seeds 2(4), white, embedded in a very sticky, gelatinous matrix. *Croat 5392, 7957a.*

Occasional, in the forest and on trees in the lake. Ap-

parently lacking any distinct seasonal behavior. Flowers and fruits may be seen at all times of the year.

Distinguished by its thin, oblanceolate, epunctate leaves and persistent, intact cataphylls. The species has been confused with *A. friedrichsthalii*, which has thick, linear, bicolorous, punctate leaves. It may also be confused with *A. bakeri*, which has linear-lanceolate, punctate leaves. The mass of roots is sometimes infested with black stinging ants.

Guatemala to Ecuador and southern Brazil. In Panama, known only from tropical moist forest in the Canal Zone, Bocas del Toro, Panamá, and Darién.

Anthurium littorale Engler, Bot. Jahrb. Syst.
  25:405. 1898

Epiphyte; caudex to 1.5 cm diam, very short or to 30 cm long; leaf scars prominent; internodes short, especially near apex; cataphylls brown, usually nonfibrous, soon deciduous, to 9 cm long and 1 cm wide. Petioles 3–14 cm long (even on a single plant!), conical and to 12 mm thick at base, canaliculate above, weakly to strongly geniculate at apex, the geniculum 1.5–2 cm long, the sheath 1–6 cm long; blades oblong-lanceolate to oblanceolate-oblong, obtuse to acute at apex and often with downturned apiculum, acute at base, the margins continuous with the margins of petiole, the blades 10–38 cm long, 3–11 cm wide, coriaceous, sparsely to moderately punctate on both surfaces, often turned somewhat upward along midrib; midrib often arched, the lateral veins forming 1 or 2 faint collecting veins. Inflorescences 1 or 2 from upper axils; peduncles melon-colored to reddish in flower, 9–25 cm long; spathe whitish, oblong, acuminate, 3–7 cm long, to 1.4 cm wide; spadix (7)9–22 cm long, ca 5 mm wide in flower, melon-colored when young, becoming violet-purple in age; tepals much broader than long, thickened, flat and triangular at apex; ovary depressed-globose, the apex square or rectangular; stigma sessile. Fruits ± globose, 4–5 mm diam, yellowish, orange, or white tinged with orange, the stigma small, buttonlike. *Croat 8821, 13989.*

Frequent at the margin of the lake and in the forest, usually rather high in trees. Flowers primarily from February to April, sometimes as early as January elsewhere in Panama.

The species has been confused with several other undescribed species of *Anthurium* in Panama.

Costa Rica and Panama. In Panama, known from tropical moist forest on BCI and in Bocas del Toro, Colón, San Blas, and Darién, from premontane wet forest in the Canal Zone and Panamá, and from tropical wet forest in Colón, San Blas, and Panamá.

See Fig. 85.

Anthurium ochranthum C. Koch, Ind. Sem. Hort.
  Berol. App. 16. 1853
  A. lapathifolium Schott; *A. triangulum* Engler; *A. baileyi* Standl.

Terrestrial, rarely more than 1.3 m tall; caudex short-creeping over ground, then erect, to 45 cm high (usually shorter), to 3–4 cm thick (rarely thicker); cataphylls short.

Petioles slender, round to canaliculate on upper surface (conspicuously canaliculate on drying), 30–70(90) cm long, the very base vaginate; blades ± ovate-cordate to triangular, abruptly acuminate at apex, cordate at base, the basal lobes to 25 cm long, narrow, directed downward or outward with sinus narrow to open, mostly ca 50 cm long, 33–60 cm long above the sinus, 29–33 cm wide, ± thin, the lower surface much lighter than the upper; major veins above sinus 7–11 pairs, at least the uppermost forming a collecting vein near margin, the basal lobes each with usually 4 veins. Peduncles terete, about as long as petioles; spathe usually less than 15 cm long (but occasionally to 32 cm) and 3.5–5 cm wide, tapered to an attenuate apex, as long as or shorter than spadix, decurrent on peduncle; spadix sessile or on a stipe to 13 mm long, tapering to blunt end, in flower yellow, 13–19 cm long and 10–13 cm wide at base, in fruit violet-purple, to 37 cm long and 2.5 cm wide; ovules ovate, ca 3 mm high; style sessile; sepals yellow, becoming green then violet-purple in fruit, irregular in shape, much thicker and truncate at apex, covering ovule, to 7 mm long in fruit. Fruits obovoid, 7–9 mm long, beaked, violet-purple at apex, white at base, extruded at maturity and suspended on 2 slender white strands; seeds usually 1, greenish, 4–5 mm long, immersed in a fleshy, whitish matrix. *Croat 10219.*

Moderately common in some areas of the forest, especially on steeper slopes. Flowers with successive inflorescences being produced mostly from late March to July, rarely in August and September. The fruits develop within about a month or 6 weeks, maturing mostly from May to September.

The only species of *Anthurium* on BCI that is consistently terrestrial (though *Kenoyer 191* was recorded as epiphytic). In commenting on *A. baileyi* both in his type description and in the *Flora of Panama*, Standley obviously confused sterile plants of *Homalomena wendlandii* with this species. The two often occupy the same habitat, but in *H. wendlandii* the petioles are aculeate and the lower leaf surface is puberulent; *A. ochranthum* never has these characteristics.

Honduras to Panama; probably also Colombia. In Panama, known from tropical moist forest in the Canal Zone, Bocas del Toro, San Blas, Chiriquí, Panamá, and Darién, from premontane moist forest in Panamá, from premontane wet forest in Colón, and from tropical wet forest in Bocas del Toro, Colón, and Panamá.

See Fig. 86.

Anthurium scandens (Aubl.) Engler in Mart., Fl.
  Brasil. 3(2):78. 1878
  A. rigidulum Schott

Creeping epiphyte; caudex elongate, occasionally branching, usually less than 1 m long, with extensive long roots at lower nodes, the internodes usually short but distinct, the leaves not clustered at apex; cataphylls brown, tightly appressed, persistent and often weathering into fine fibers on the leafy part. Petioles usually 2–7.5(10.5) cm long, to two-thirds as long as blade, canaliculate on upper side, nodose at base and apex, geniculate at apex; blades narrowly to broadly elliptic, tapered evenly to both ends,

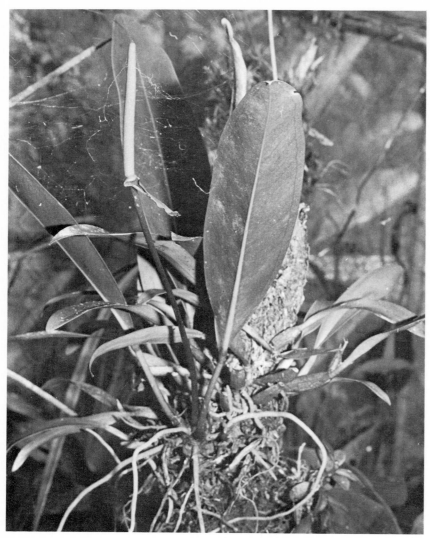

Fig. 85.
Anthurium littorale

Fig. 87. *Anthurium scandens*

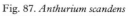

Fig. 86. *Anthurium ochranthum*

Fig. 88. *Anthurium tetragonum*

acute to acuminate and downturned at apex, usually acute at base, 4.5–14 cm long, 2–7 cm wide, subcoriaceous, obscurely to prominently punctate on lower surface; collecting and lateral veins obscure (distinct on dried specimens). Peduncles 2.5–5 cm long, usually exceeding petiole of subtending leaf; spathe lanceolate to oblong-lanceolate, decurrent on peduncle; persisting but becoming dried, (0.4)2–3 cm long, to 12 mm wide at base, acuminate at apex; spadix sessile, usually to 3 cm long and lavender at anthesis, 4–10.5 cm long in fruit, to 17 mm wide; tepals at first lavender becoming whitish, 1 mm high and 2.3 mm broad in flower, to 4 mm high and 6.7 mm broad in fruit; anthers emerging beneath margin of tepal when spadix is quite small. Fruits white, very fleshy and weakly exserted at maturity, ovoid, minutely beaked, ca 7 mm long; seeds usually 5–9, ca 2.5 mm long, ± elliptic, embedded in a thick, clear jelly. *Croat 6292, 14969.*

Infrequent, in the forest, usually high in trees. Though flowers and fruits may be present throughout much of the year, BCI plants flower mostly in the dry season and fruits develop mostly in the rainy season.

Though not so great on BCI, the variability throughout the range of this species is immense. This variability is generally exhibited in the size of plant parts.

Mexico throughout Central America to the Guianas and southern Brazil; Trinidad, Greater Antilles. In Panama, ecologically variable and widespread; known from tropical moist forest in the Canal Zone, Bocas del Toro, Veraguas, Panamá, and Darién, from premontane wet forest in Bocas del Toro, Chiriquí, Coclé, and Panamá, from tropical wet forest in San Blas, Colón, Chiriquí, Coclé (both slopes), Panamá, and Darién, and from premontane rain and lower montane rain forests in Chiriquí.

See Fig. 87.

Anthurium tetragonum Hook. ex Schott, Prodr.
　　Aroid. 1860

Moderate to large, rosulate epiphyte; caudex short, 6–8 cm thick, usually very densely covered with an irregular mass of fleshy roots, the roots often extending above the short petioles. Leaves crowded; petioles 10–20 cm long, 1–2 cm wide, ± flattened on upper surface above geniculum, ± rounded below, becoming carinate on underside after drying, geniculate at base of blade or 5–6 cm below blade, the geniculum about as long as broad; blades broadly obovate to oblanceolate, acute to shortly acuminate at apex, gradually tapered to the usually cuneate base (rarely rounded or subcordate), 60–153 cm long, 25–50 cm wide (rarely smaller when fertile), coriaceous when dried; midrib and veins prominently raised on both surfaces, the major veins mostly 9–14 pairs, arcuate-ascending to margin, not forming a collecting vein. Inflorescences ± pendent below leaves; peduncles terete, 30–64 cm long; spathe green, 15–30 cm long, 2–4 cm wide and decurrent at base, tapered to long-cuspidate point; spadix dull green, to 34 cm long in flower, becoming magenta, to 50 cm long and 3 cm diam at base in fruit, the apex often withered; sepals pale magenta, obovoid, accrescent. Berries ± oblong, 1–1.5 cm long, red, exserted at maturity;

seeds 2, oblong, to 3.3 mm long, flattened, embedded in a clear, jellylike matrix. *Croat 7920, 8512, 10195, 11325.*

Common in the forest, growing from near the ground to ca 20 m high. Quite variable in size, and may be fertile while plants are still fairly small. Flowers mostly in the dry season during February and March. Most fruits mature during the rainy season, July to September. Inflorescences usually begin to develop again by December or earlier.

This species is not confused with any other on the island. The names *A. crassinervium* (Jacq.) Schott, *A. maximum* (Desf.) Engler, and *A. schlechtendalii* Kunth, all reported by Standley for BCI, were apparently based on misdeterminations of *A. tetragonum*. Most determinations were based on the shape of the petiole, a character that is of little value, since drying affects each plant somewhat differently.

Mexico to Panama, probably also Colombia. In Panama, known from tropical moist forest in the Canal Zone, Bocas del Toro, Chiriquí, Panamá, and Darién; known also from premontane moist forest in the Canal Zone (Ancón), from premontane wet forest in Panamá, and from tropical wet forest in Colón and Chiriquí.

See Fig. 88.

CALADIUM Vent.

Caladium bicolor (Ait.) Vent., Descr. Cels. pl. 30. 1801

Terrestrial, acaulescent. Leaves from a depressed-globose rhizome; petioles fleshy, 30–80 cm long, sheathed at base; blades ovate-cordate to ovate-sagittate, acute to acuminate, 10–60 cm long, to 40 cm wide, thin, peltate far above base, glaucescent, usually reddish near center and spotted on upper surface with white, red, or yellow, the basal lobes rounded, directed downward or outward. Inflorescences shorter than leaves; peduncles slender and fleshy; spathe 7–15 cm long, the tube ovoid, greenish, the blade 2–3 times longer than tube, cuspidate, white, weathering away with upper staminate part of spadix after anthesis; sterile staminate part of spadix to 1.5 cm long, ending abruptly at base, its cells longer than broad, continuous with fertile staminate part but discontinuous with pistillate part; pistillate part of spadix ovoid-cylindrical, to 2 cm long; flowers unisexual, naked; staminate flowers with 3–5 connate stamens, the connective flat, angulate marginally; ovary with 2 or 3 cells, several ovules in each locule. Fruits baccate, bearing several to many-seeds; seeds ovoid. *Croat 10184.*

Cultivated at the Laboratory Clearing; seen once on Shannon Trail. Flowers and fruits chiefly throughout the rainy season.

Native to the Amazon region; widely cultivated. In Panama, known from tropical moist forest in the Canal Zone (doubtlessly cultivated elsewhere).

DIEFFENBACHIA Schott

Members of the genus may be distinguished by their naked unisexual flowers, which have four stamens and

KEY TO THE SPECIES OF DIEFFENBACHIA

Stems usually more than 5 cm thick; plants usually well over 1 m tall when fertile; spathes 27–48
cm long; leaf blades more than 40 cm long; both leaf blades and petioles solid green in color
(petioles sometimes with lighter streaks but never splotched) .... *D. longispatha* Engler & Krause
Stems usually less than 3 cm thick; plants usually less than 1 m tall when fertile; spathes usually
less than 25 cm long; leaf blades usually less than 30 cm long; leaf blades often with light
streak on midrib or the petioles speckled:
Leaves narrow, less than 8 cm wide, the base usually acute, sometimes obtuse, never cordate;
midribs and petioles mottled with whitish spots .... *D. pittieri* Engler & Krause
Leaves mostly more than 10 cm wide, often cordate at base; midribs and petioles solid green,
not speckled, the midrib often with a white streak .... *D. oerstedii* Schott

are enclosed after anthesis by the spathe. The stamens
are fused together to form a synandrium with four or
five sides, truncate at the apex; pistillate flowers have four
or five club-shaped staminodia; ovaries have one to three
locules with a single ovule per locule. The sap contains
relatively high amounts of oxalic acid, which gives the
plant a characteristic odor and may burn the skin.

Dieffenbachia longispatha Engler & Krause, Pflanzenr.
IV. 23DC(Heft 64):44. 1915

Terrestrial, 1.2–3.5 m tall; caudex 4–12 cm diam at maturity, at first prostrate, then erect; leaf scars prominent.
Petioles thick and succulent, 27–50 cm long, ± terete
at apex or with faint flat rib on upper surface, prominently sheathed to about middle; blades thick, oblong-
elliptic, short-acuminate, acute to rounded at base, 46–71
cm long, 24–35 cm wide, the edges usually turned upward
near base; midrib flat on upper surface, 1–2 cm wide at
base, prominently raised on lower surface, the major
lateral veins in 2–26 pairs, impressed above, raised below,
the smaller veins indistinct; juvenile blades with acute
base and solid green midrib. Inflorescences usually 1–3;
peduncles 7–25 cm long; spathe green, broadly curved,
long-acuminate, 27–48 cm long, to ca 4 cm broad when
closed, the distal inner surface white when open and 5–6
cm broad; free staminate part of spadix 13–19 cm long,
12–14 mm wide; pistillate part of spadix (except sometimes the uppermost part) fused to spathe; pistillate flowers 10–25, widely spaced, in a single row or scattered,
round or barely bilobed; stigmas 6 mm broad, yellow;
staminodia white, irregular, 2–3 mm long. Fruits 1.5–2
cm diam, often deeply emarginate at both ends appearing
to be a double fruit, yellow to orange; mesocarp ca 2 mm
thick, soft, sweet and tasty at maturity; seeds oblong, 7–8
mm diam, brown, smooth. *Croat 6276, 11321.*

Common throughout the forest, in streams or on sedi-
ment deposits in standing water at the edge of the lake;
usually not highly colonized. Juvenile plants are much
more abundant than adults. Flowers from May to September, although individuals apparently do not flower
every year. The fruits develop from September to December.

The species may not be separable from *D. sequine*
(Jacq.) Schott of Central America and the West Indies,
which has the pistillate flowers closely congested and all
parts much smaller. Until more thorough studies can

be made, it is best to consider *D. longispatha* distinct
from *D. sequine.* Another related but apparently undescribed species occurring in tropical wet forest in Panamá
(Cerro Campana), Coclé (El Valle), and Colón (Portobelo) has flowers very closely congested, but its petiole
is vaginate its entire length. (See *Croat 12710, 13403,
14175.*) *Croat 5709,* a sterile collection from BCI, is
similar and might also be this undescribed taxon, but
it is in some ways intermediate between *D. longispatha*
and *D. oerstedii.* This is the largest of all Central American *Dieffenbachia* species.

Costa Rica, Honduras, and Panama; possibly to Colombia and Venezuela. In Panama, known from tropical
moist forest in the Canal Zone, Bocas del Toro, San Blas,
Chiriquí, and Panamá, from premontane moist forest
in Panamá, from premontane wet forest in Colón, Chiriquí, Coclé, and Panamá, and from tropical wet forest
in Colón.

See Figs. 89, 90, and 91.

Dieffenbachia oerstedii Schott, Oesterr. Bot. Z.
8:179. 1858

Terrestrial, glabrous, usually no more than 1 m tall; stem
2–2.5(3.5) cm diam; caudex long or short, usually creeping over surface of ground; internodes short. Petioles
12–22 cm long, vaginate to near middle or slightly above,
flattened on upper surface at apex; blades oblong-ovate,
abruptly acuminate at apex, cuneate to cordate with the
sides often ± unequal below, 12–30 cm long, 6–15 cm
wide, thin, deep green above, often with light green streak
on midrib in distal half of blade; midrib flat above, 5–8
mm broad, the major veins 7–12 pairs, ± impressed above,
the smaller veins obscurely visible below. Inflorescences
usually 3–5, straight to slightly curved; peduncles 7–12
cm long; spathe green (inner part white at top of blade
when open), 15–23 cm long at anthesis (elongating somewhat after closing of spathe), 2.5–3 cm broad at top when
open, less than 2 cm broad when closed, often becoming
± orange at maturity; free staminate part 7–13 cm long,
6–8 mm broad; pistillate flowers minute, 30–45, closely
spaced; staminodia flattened below, the tips globular.
Fruits ellipsoid, 7–8 mm long, orange to red, the stigmatic area round, ca 2 mm wide. *Croat 5896, 7712.*

Common throughout the forest; often locally abundant
in moist areas. Flowers chiefly in the early rainy season,
May to August. The fruits develop mostly during the

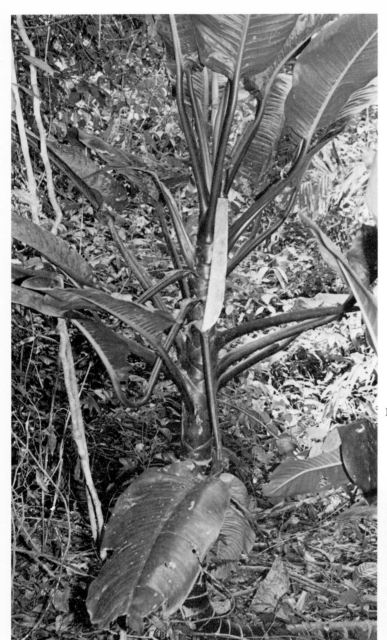

Fig. 89. *Dieffenbachia longispatha*

Fig. 90. *Dieffenbachia longispatha*

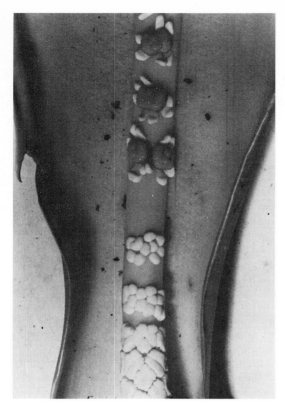

Fig. 91. *Dieffenbachia longispatha*

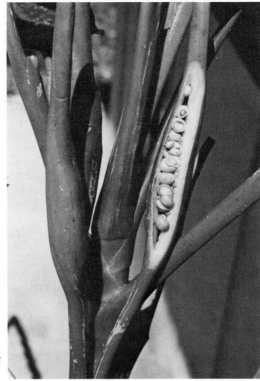

Fig. 93. *Dieffenbachia oerstedii*

Fig. 92. *Dieffenbachia oerstedii*

dry and early rainy seasons of the following year, February to May.

Distinguished by its small size and somewhat cordate leaves, often with a white streak on distal half of midrib.

Guatemala to Panama. In Panama, known chiefly from tropical moist forest in the Canal Zone, Bocas del Toro, Chiriquí, Los Santos, Panamá, and Darién; known also from premontane wet forest in Colón and Panamá.

See Figs. 92 and 93.

Dieffenbachia pittieri Engler & Krause, Pflanzenr. IV.23Dc(Heft 64):42. 1915

Terrestrial, 0.8–1.5 m tall; caudex to 3 cm diam (usually less than 2 cm), weakly rooted, decumbent then erect, covered with thin brown epidermis with reticulate pattern; leaf scars inconspicuous. Petioles 8–12 cm long, dark green with whitish maculations, the spots extending along most of midrib below; sheaths prominent, extending halfway or almost to blade; blades narrowly elliptic, gradually acuminate, often somewhat falcate, cuneate to obtuse (elsewhere subcordate) and often slightly inequilateral at base, 20–30 cm long, 5–8 cm wide; midrib rounded and slightly raised on upper surface, the veins many, obscure above, weak but all more or less equally distinct below. Inflorescences 1–3, 14–20 cm long; peduncles about 4 cm long, somewhat flattened; spathe 12–20 cm long, colored like petioles in basal half, pale green at apex at anthesis, caudate-acuminate, sometimes sharply bent backward below apex; spadix 10–18 cm long, the pistillate and sterile parts united to spathe; staminate part 5–9 cm long, the staminate flowers nearest pistillate ones widely spaced and probably sterile; pistillate flowers usually 20–40, moderately close together, not in obvious rows but scattered, their staminodia white, usually 3, flattened, oblong to clavate, slightly longer than stigmas; stigmas orange, depressed-globose, ca 1.5 mm high; fruiting spadices broadly arched, the spathe mostly deciduous, exposing fruits. Fruits broadly ellipsoid, to ca 1.5 cm long, less than 1 cm wide, orange. *Croat 5759, 6775.*

Occasional, in older forest, possibly most common in the vicinity of Drayton and Armour Trails. Seasonal behavior is uncertain. Believed to flower from May to December and to fruit from October to May.

Easily distinguished by its leaves, which are acute at the base with spotted petioles. BCI populations of the species differ somewhat from others in having slightly narrower leaves, in being acute to obtuse at the base (usually subcordate elsewhere), and in being consistently maculate on the petioles (rare elsewhere). It is possible that additional work will show that the BCI plants, which match others found along the Atlantic slope of Panama, are a distinct species.

When the spathe first opens, the stigmas appear receptive, but the white staminate flowers are so closely compacted that the pollen cannot be shed. Later the staminate flowers become cream-colored and shrink somewhat to expose the thecae of the anthers around the margin. By the time the staminate flowers reach anthesis, the spathe has closed somewhat, all but blocking the opening to the pistillate flowers. This is a general phenomenon for all *Dieffenbachia* (at least those species on BCI) and no doubt helps promote cross-pollination.

Known only from the lowland forest on the Atlantic slope of Panama.

See Figs. 94 and 95.

DRACONTIUM L.

Dracontium dressleri Croat, Selbeyana 1:168. 1975

Terrestrial, consisting of a single large leaf arising from an underground tuber; tuber ovoid, ca 9 cm long and 7 cm wide, the upper part rounded. Leaf 2–4-pinnate to sub-5-pinnate, glabrous; petiole fleshy, to 6.5 cm diam near base, to 3 cm diam at apex, 1–1.8 m long, green blotched with dark brown and pale green in a reptilian pattern, the protuberances irregular, short, especially close together in proximal half; blade thin, with 3 principal divisions, these each leafy mostly in distal two-thirds, the middle section 40–100 cm long, the lateral sections somewhat shorter, the rachises smooth or bearing protuberances, faintly patterned like the petiole; leaflets 3.5–20 cm long, 1–7.5 cm wide, gradually acuminate at apex or bilobed, the lobes acuminate, acute at base and decurrent along the rachis, the decurrent part ± erect on upper side of petiole and irregularly lobed; lateral veins of ultimate segments arcuate-ascending, loop-connected, occasionally interconnected with an equally prominent cross-vein, all major veins impressed on upper surface, raised on lower surface. Inflorescence arising from the apex of the tuber before the leaf is produced; peduncle ca 2 cm long and 1 cm thick, subtended on the outside by a lanceolate sheath, the sheath ca 5 cm long and 2 cm wide, acuminate at apex, with an obscure medial vein, at first violet-purple, soon drying and turning brown; peduncle subtended on the inside by a prophyll ca 2 cm long, this bearing 2 prominent ribs on its outer surface; spathe ± oblong in face view, ca 10 cm long, arcuate in side view, convolute in lower third and to 2.5 cm wide, broadened above and to 4 cm wide, becoming tapered at apex and curving forward to form a portico over the mouth of the spathe, the tip cuspidate-acuminate, the outer surface of spathe dull, violet-purple, tinged with green, bearing 15, ± equidistant veins, the inner surface of spathe shiny, smooth, dark violet-purple except for a weakly transparent area around the spadix, this ca 4 mm wide at any point, the tube of the spathe roomy within; spadix dark violet-purple, short-stipitate, the stipe ca 7 mm long and 6 mm diam, the fertile part ca 2 cm long, 7–9 mm diam, bearing 10 slender appendages at the apex ca 5 mm long; flowers bisexual, closely spiraled in ca 10–12 rows; tepals 4–8 (usually 5 or 6), biseriate, ca 2 mm long, broadened and arched inward at apex, the upper edge truncate, irregularly shaped, mostly triangular to diamond-shaped, ca 1.5–2.5 mm wide, thickened except along the irregular distal margin; stamens usually free, 5–9, to 2.2 mm long, opposite the tepals and emerging slightly above them at anthesis, crowded together by the tepals to form a single locus of pollen; filaments fleshy and swollen toward apex, broader than the anthers, the

Fig. 94. *Dieffenbachia pittieri*

Fig. 95. *Dieffenbachia pittieri*

Fig. 96. *Homalomena wendlandii*

Fig. 97. *Homalomena wendlandii*

inner whorl sometimes fused laterally forming a tube around the pistil, the tube white, streaked with violet-purple; anthers linear-elliptic, ca 4 mm long, vertically dehiscent, the thecae touching at apex, separated at base; pollen light-brown, the grains clinging together in a mass but the mass of pollen not tacky; pistil violet-purple, ca 1.6 mm long; ovary ± oblong, pale-colored, incompletely 2–5-locular (for the genus); ovules solitary; style eccentric, darkly colored, tapered to a slender apex. Fruits unknown, those of the genus surrounded by perianth as in *Anthurium;* seeds rounded to reniform, somewhat compressed. *Dressler s.n.*

Rare, in the forest; collected near the end of Zetek Trail by Dressler. Flowers in the early rainy season before the leaf emerges. The fruits probably also mature in the rainy season. The plant is leafless in the dry season.

The inflorescence presents a typical fly-pollination syndrome (sapromyophily), with a foul aroma, purplish spathe, and easy access to the flowers (Faegri & van der Pijl, 1966). The species also has a transparent window around the spadix at the base of the spathe tube, which is not apparent from the outside but can be seen easily from within if the front of the spathe is closed off. The inner surface of the spathe is very smooth, perhaps making it difficult for an insect visitor to crawl out.

Known from tropical moist forest on BCI and from premontane wet forest in Colón (Achiote).

HOMALOMENA Schott

Homalomena wendlandii Schott, Prodr. Aroid. 308. 1860

Terrestrial, acaulescent. Petioles to 1.3 m long, usually shorter, terete, sheathed in lower third, densely puberulent and armed with short, sharp prickles usually below middle; blades ovate-cordate to ovate-sagittate, abruptly short-acuminate and turned downward at apex, cordate at base, 37–78 cm long, 23–53 cm wide, glabrous above, puberulent below, the basal sinus 13–24 cm deep and 2–10 cm broad at open end with its apex acute; major veins 7–10 pairs above sinus, 3–5 joining in basal lobes, impressed above, markedly raised below. Inflorescences arising from ground; peduncles 5–20 cm long, puberulent; spathe convolute, pale green with white spots, often pure white at apex at anthesis, 19–30 cm long, enveloping spadix, caudate-acuminate at apex, only

slightly bulbous at base, becoming purplish in age (especially base), white within (at least when open); spadix white, 17–26 cm long, the area between staminate and pistillate flowers often pink, its flowers sterile; pistillate part 3.5–7 cm long, to 1.8 cm wide (3 cm in fruit), mostly adnate to spathe; staminate flowers with 2–4(6) stamens; fruiting inflorescences 10 cm long or more and 5 cm wide. Mature fruits obovoid to oblong, to ca 6 mm long, incompletely 2–5-celled, truncate at apex; stigmas persistent, rounded, and depressed with 4 or 5 lobes; seeds many in each cell, ± ellipsoid, to ca 0.7 mm long, longitudinally striate, attached by a long funicle, immersed in a clear, gelatinous matrix. *Croat 14853.*

Very common, preferring steep, moist slopes and creek banks. Flowers from late March to May. Mature fruiting inflorescences have never been seen on BCI but apparently develop soon after flowering.

Recognized by its terrestrial habit, prickly petioles, and puberulent parts (an uncommon feature in the Araceae). The flowering inflorescence has a pungent, sweet aroma resembling anise (also uncommon in the Araceae, which often have rather foul odors). Standley (1944) in the *Flora of Panama* misspelled the genus as *Homalonema.*

The entire inflorescence is often missing above the peduncle (apparently eaten by animals), and this perhaps constitutes the plant's method of dispersal.

Costa Rica and Panama, and southeastern Colombia, along the Río Putomayo (*Schultes & Cabrera 19054, GH*); probably more widespread. In Panama, known from tropical moist forest on the Atlantic slope in the Canal Zone, San Blas, and Chiriquí Provinces, from premontane wet forest in the Canal Zone, Chiriquí, and Panamá, and from tropical wet forest in Colón.

See Figs. 96 and 97.

MONSTERA Adans.

Members of the genus are recognized by the bisexual flowers, which are naked with four stamens, and by the uniform spadix, from which the spathe generally falls free after anthesis. The ovary has two cells, each with two ovules.

Monstera adansonii Schott var. **laniata** (Schott) Mad., Contr. Gray Herb. 207:38. 1977

Epiphytic vine; caudex minutely speckled, smooth, 1–2.5 cm diam, closely appressed to tree or more often free,

KEY TO THE SPECIES OF MONSTERA

Adult blades deeply pinnatifid, not perforate . . . .*M. dilacerata* C. Koch
Adult blades perforate to irregularly pinnatifid (the result of a large perforation reaching the margin), not regularly and deeply pinnatifid:
    Caudex and base of petioles densely and conspicuously tuberculate, the adult caudex often flattened, to 2.5 cm diam; reticulate venation on leaves about as prominent as major lateral veins when dried; spathe coriaceous, rounded at apex . . . . .*M. dubia* (H.B.K.) Engler & Krause
    Caudex and base of petiole usually smooth, the adult caudex terete, less than 1.5 cm diam; reticulate venation on leaves not at all prominulous, much less prominent than major lateral veins; spathe thin, narrowly acuminate at apex . . . .
. . . . *M. adansonii* Schott var. *laniata* (Schott) Mad.

Fig. 98. *Monstera dilacerata*

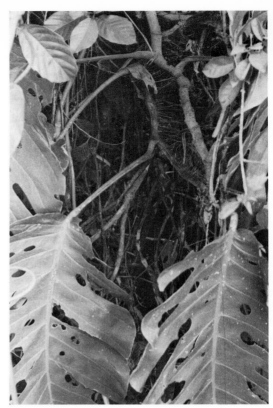

Fig. 100. *Monstera dubia* (see also Figs. 101, 102)

Fig. 99. *Monstera dilacerata*

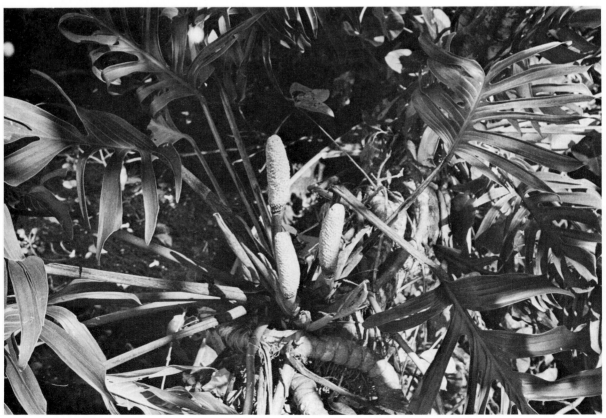

scandent; internodes 6–10 cm long. Petioles minutely speckled, 10–33 cm long, sheathed to near the apex, canaliculate above the sheath, the sheaths to ca 1 cm high on the sides, tapered somewhat toward apex, the sides rounded and free at apex; blades ovate, rarely elliptic, inequilateral, abruptly acuminate, downturned, and weakly inequilateral at apex, obtuse to truncate or subcordate at base, 15–35 cm long, 9–28 cm wide, semiglossy, the lower surface slightly paler; midrib and major lateral veins sunken, slightly paler, the secondary lateral veins visible, dark; preadult leaves much like adult but smaller. Inflorescences solitary or up to 6 per node, each subtended by a bracteole to 12–16 cm long; peduncles flattened somewhat laterally, minutely speckled, 6–9 cm long, to 1 cm diam, to 16 cm long in fruit; spathe creamy-white at anthesis, caducous, acuminate, ca 11 cm long; spadix white at anthesis, soon becoming pale greenish-white, 6–9 cm long, to ca 1 cm wide; fruiting spadices white, to 16 cm long and 2 cm wide. Fruits not coalesced at maturity, to ca 10 mm long and 4 mm wide. *Croat 6225.*

Apparently rare on the island, though common elsewhere in adjacent areas of the Canal Zone; collected once between the dock and Fairchild Point. Flowers and fruits throughout the rainy season, mostly from April to August, less frequently as early as February and as late as October. Fruiting occurs mostly from September to December, less frequently as early as July.

Honduras to Colombia, Venezuela, and the Guianas. In Panama, known principally from tropical moist forest in Bocas del Toro, Colón, Canal Zone, Chiriquí, and Darién, but also from premontane wet forest in Bocas del Toro and from tropical wet forest in Colón and Coclé.

Monstera dilacerata C. Koch, Ind. Sem. Hort. Berol. App. 5. 1855
Ceriman, Anona piña

Epiphytic climber; caudex usually short and thick at maturity (usually less than 1.5 m long), terete with prominent leaf scars; juvenile plants at first terrestrial, often with few leaves. Petioles to 36 cm long, shorter than blade, the sheaths often extending to base of geniculum, the remainder of petiole canaliculate with thin, raised, minutely undulate margins; blades regularly and deeply pinnatifid, rounded to cordate at base, 35–46 cm long, 28–35 cm wide, the apical segment acuminate; lobes 3–9 on each side, 1.5–4.5 cm wide, acute and falcate at apex, dark green and glossy above, paler and drying very darkened below; lateral veins in 1–3 pairs, whitish and raised on underside; juvenile blades thick, ovate-lanceolate, oblique and falcate, acute to acuminate at apex, to about 20 cm long and 9 cm wide, bicolorous (drying black), the major veins in 2–5 pairs; immature plants later forming runners that ascend tree trunks, the leaves then distichously arranged, the blades entire or less frequently perforate to pinnatifid, 7–26 cm long, 7–14 cm wide, the larger with petioles to 18 cm and spreading. Inflorescences several, from upper axils; peduncles 15–23 cm long, subterete with a broad spathe scar at apex; spathe

acute to acuminate at apex, white at anthesis, to 24 cm long, soon deciduous; spadix narrowly oblong, narrowly rounded at apex, mostly 13–16 cm long at anthesis (to 19 cm long in fruit); pistils ca 5 mm long at anthesis (more than 1 cm long in fruit), the sides angulate, the apex truncate, to ca 5 mm diam at anthesis; stigmas linear. Fruits not coalesced at maturity, whitish, to 12 mm long and 5 mm wide. *Croat 7251.*

Juvenile plants common; adults apparently infrequent, occurring at least sometimes within 4 m of the ground on the smooth face of tree trunks in the forest. Flowers in the late rainy season through the dry season (December to March), with the fruits maturing by the early rainy season.

In general the species may be distinguished from other *Monstera* in Panama by having adult blades that are deeply pinnatifid and lacking perforations.

Nicaragua to Venezuela and Peru. In Panama, known from tropical moist forest in the Canal Zone and Darién, from premontane wet forest in Panamá (Cerro Azul), from tropical wet forest in Panamá, and from lower montane rain forest in Chiriquí.

See Figs. 98 and 99.

Monstera dubia (H.B.K.) Engler & Krause, Pflanzenr. IV.23B (Heft 37):117. 1908

Epiphytic vine; caudex mostly 2–2.5 cm thick, usually with many short roots, strongly tuberculate the tuberculae extending onto lower part of petiole; juvenile plants creeping over rocks or up tree trunks, the caudex flattened, to 2.5 cm diam. Petioles stiff and narrow, 20–47 cm long, often twisted at base, broadly canaliculate to near blade; blades mostly ovate-oblong, acute and often somewhat cuspidate at apex, rounded to subcordate at base, 27–83 cm long, 18–52 cm wide, often somewhat falcate and oblique, one side considerably larger, drying ± green, usually with small and large perforations, the latter often reaching margin, the leaf thus irregularly pinnatifid; major veins in 11–14 pairs, the reticulate veins prominent; juvenile leaves at first appressed to their support, ovate-cordate, oblique, shortly acuminate at apex, with basal lobes usually overlapping, 1.5–15 cm long, the petioles very short, flattened; leaves increasing in size with age, becoming more petiolate and distant from tree. Inflorescences several at apex, overtopped by leaves in fruit; peduncles 5–8 cm long, tuberculate, the spathe scar prominent; spathe coriaceous, oblong, rounded on both ends, white to pinkish at anthesis or becoming pale green, 8–12 cm long, 4–8 cm wide when open, remaining open after anthesis, ultimately deciduous, often removed by the growing spadix; spadix 7–11 cm long, 1.5–2.3 cm thick, and white in flower, to 14 cm long and 4–5 cm thick with mature fruit, rupturing to expose seeds; pistils ca 5 mm high in flower (to 1.5 cm in fruit), the sides angulate, the apex truncate, pithy, angulate, 6–7 mm diam, deciduous individually or in sheets (evidence on the ground that plant is in fruit); stigma linear. Seeds broadly oblong, ± flattened, ca 8 mm long and 5 mm wide, surrounded by a sweet, fleshy layer. *Croat 5476, 6225, 6254.*

Fig. 102. *Monstera dubia*

Fig. 101. *Monstera dubia*

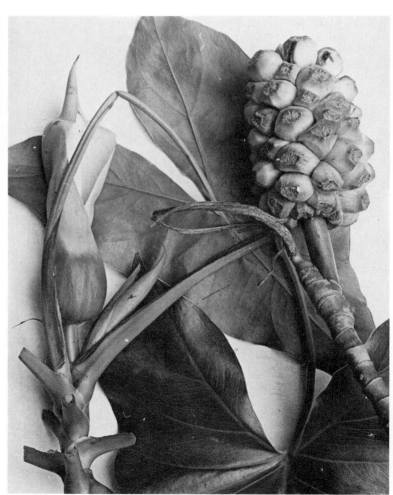

Fig. 103.
Montrichardia arborescens

Juvenile plants are very common and adults common, usually tightly appressed high in trees, but occasionally loosely attached in trees at the margin of the lake. Flowering season uncertain. Since individuals continue to produce inflorescences, a plant may bear a flowering inflorescence at a time when its first inflorescences carry nearly mature fruit. Most flowering, however, appears to begin with the rainy season, and most fruits are mature between August and October. A second and apparently smaller flowering season is in the early dry season, with fruits maturing during March and April.

Distinguished from most *Monstera* in Panama by its warty stems and petioles, its short peduncles, and its short blunt spathe. *Kenoyer 183* was cited in the *Flora of Panama* as *M. deliciosa* Liebm.; the specimen is sterile but is surely *M. dubia*. This species was reported as *M. pertusa* (L.) DeVr. by Standley, but that name cannot be used, since it is a later homonym of *M. pertusa* (Roxb.) Schott, an Indian species.

Southern Costa Rica to Bolivia and western Brazil, and eastward along the Caribbean to Trinidad. In Panama, known from tropical moist forest chiefly on the Atlantic slope, but also on the Pacific slope in Chiriquí.

See Figs. 100, 101, and 102.

MONTRICHARDIA Crueger

Montrichardia arborescens (L.) Schott, Arac. Betreff. 1:4. 1854

Aquatic; caudex 1–3 m high, to 4.5 cm thick near base, mostly 1.5–2 cm thick above, all but occasionally upper internodes armed with short prickles. Petioles 20–45 cm long; sheaths extending to ca middle or beyond, the remainder of the petiole convex below, triangulate above; blades very deeply sagittate, 15–40 cm long or more, often cuspidate at apex, the basal lobes usually acuminate, spreading, often as long as or longer than upper lobe; primary lateral veins in 3 or 4 pairs above sinus, a single basal vein directed into each basal lobe and submarginal at apex of narrow sinus. Peduncles ca one-third as long as spathe or less; spathe ± oblong, 10–18 cm long, convolute basally even at anthesis, opening in upper half with a distinct constriction above spadix, the apex cuspi-date (often twisted forward before anthesis), the tube green at base, the blade white (at least within) at anthesis, the entire spathe ultimately deciduous; spadix mostly 2–4 cm shorter than spathe; staminate part more than 1.5 cm wide, deciduous with spathe, the flowers with 3–6 stamens; lower pistillate part slender, to 4 cm long, usually ca 1 cm wide or less; pistils 1-celled; ovules 1 or 2; stigma sessile, orbicular. Fruit clusters usually broadly oblong, to 11 cm long and 8 cm wide; berries green except for roughened stigmatic area, irregularly developed (some aborted), subglobose and 2.5–3 cm wide when mature, irregularly dehiscent; seeds broad at apex, narrowed below, to 2.5 cm long. *Croat 4961*.

Restricted on BCI to shallow water at the edge of the lake, commonly in somewhat protected areas. Flowers and fruits throughout most of the year.

Distinguished by its aquatic habitat, armed caudices, and large fruits. Seeds float to shore and germinate.

Guatemala and Belize to the Guianas; the Antilles. In Panama, known chiefly from freshwater swamps and river banks all along the Atlantic coast and in Darién.

See Fig. 103.

PHILODENDRON Schott

The genus *Philodendron* can be recognized by an inflorescence that is usually enlarged at the base, with the spadix completely enveloped by the spathe, except for a brief period when the spathe opens to allow entrance of the beetle pollinators. The spadix has unisexual flowers, the short, basal, pistillate part and the longer, upper, staminate part. The ovary has two to several locules with numerous ovules per locule. Stamens number two to six. At maturity the spathe bursts irregularly or at least at the base and falls free to expose the large cluster of closely aggregated fruits. The tannins in the sap of this genus often cause the sap to turn reddish; this is particularly pronounced in *P. sagittifolium* and *panamense*.

Plants of the genus are of three types on BCI: (1) Scandent vines, often hemiephiphytic, especially the juvenile forms: *P. tripartitum, scandens, inconcinnum, hederaceum, guttiferum, nervosum,* and *inaequilaterum.* The internodes are elongate at maturity (when forming an inflorescence)

KEY TO THE SPECIES OF PHILODENDRON

Blades tripartite or pinnatifid:
  Blades tripartite, the center lobe symmetrical, the lateral lobes asymmetrical . . . .
  . . . . *P. tripartitum* (Jacq.) Schott
  Blades pinnatifid, deeply divided . . . . *P. radiatum* Schott
Blades entire:
● Basal lobes well developed and separated by a deep sinus:
  Mature blades usually broadest at about middle, mostly 15–25 cm long (some to 30 cm or, if more, then pubescent on veins); mature leaves widely spaced; caudex 1–2 cm diam near ends, not appressed to its support maturity:
    Blades ± oblong, more than 2 times longer than wide . . . .*P. inconcinnum* Schott
    Blades broadly ovate-cordate, usually less than 1.5 times longer than wide:
      Leaf veins, petioles, and stems shortly pubescent; blades thin, the veins prominently raised on underside . . . .*P. hederaceum* (Jacq.) Schott
      Leaves, petioles, and stems glabrous; blades thick at maturity, the veins not raised on underside . . . . *P. scandens* C. Koch & Sell.

Mature blades broadest below middle, usually more than 35 cm long; mature leaves clustered at ends of thick caudex 3–12 cm diam; caudex tightly appressed to its support:

Blades usually more than 80 cm long, the basal sinus often closed by overlapping lobes; petioles more than 90 cm long and 2.5 cm wide, the upper side flattened with an erect membranaceous wing 4–5 mm high on margin . . . . *P. pterotum* C. Koch & Aug.

Blades usually less than 65 cm long, the basal sinus open; petioles less than 75 cm long and 2 cm wide, rounded or flat on top without membranaceous wing:

Plants terrestrial; major lateral veins in 11–17 pairs, markedly impressed above; petioles usually less than 1 cm wide, flat on upper surface; spathe to 11 cm long . . . . *P. grandipes* Krause

Plants epiphytic or hemiepiphytic, usually at 2 m or more; major lateral veins in usually fewer than 10 pairs, not impressed above; petioles usually more than 1 cm wide, terete or flattened on upper surface; spathe larger, at least at anthesis:

Petioles flat to concave on upper surface; cataphylls red at maturity, persistent as brown fibers; peduncles less than 8 cm long, straight; spathe tube bright red on outside; caudex usually with a slender, ± leafless branch near apex . . . . *P. fragrantissimum* (Hook.) Kunth

Petioles terete; cataphylls reddish or green, caducous; peduncles 6–23 cm long, straight or curved at apex; spathe tube green on outside; caudex lacking slender stem near apex:

Blades thin, the smaller veins readily visible (on both fresh and dry specimens), the midrib and primary lateral veins not noticeably lighter in color than rest of blade; peduncles usually more than 11 cm long (at least at anthesis), often markedly bent just below spathe and prominently white-lineate at apex of peduncle and base of spathe . . . . *P. panamense* Krause

Blades thick, the smaller veins obscure, the midrib and primary lateral veins noticeably lighter in color than rest of blade; peduncles less than 11 cm long, straight, not white-lineate near apex nor on tube of spathe . . . . *P. sagittifolium* Liebm.

● Basal lobes not well developed:

Blades cordate at base, broadest above middle . . . . *P. inconcinnum* Schott

Blades truncate to acute at base, broadest at or below middle:

Petiole sheaths ending far below base of blade (on BCI) . . . . *P. nervosum* (Schult. & Schult.) Kunth

Petiole sheaths extending to or almost to base of blade:

Leaf sheaths with margins only slightly raised, very narrowly rounded at apex, ending decidedly below base of blade; primary lateral veins much more prominent than secondary ones, the secondaries all ± distinguishable . . . . *P. inaequilaterum* Liebm.

Leaf sheaths with margins prominently raised or spreading, usually broadly rounded at apex, ending very near or extending slightly beyond base of blade; primary lateral veins scarcely more prominent than secondary ones, the secondaries usually obscure . . . . *P. guttiferum* Kunth

---

and the degree of plant attachment varies from complete to loose, the plants sometimes hanging pendent from the trees. *Philodendron guttiferum, nervosum,* and *inaequilaterum* lack cataphylls except at the inflorescences; new leaves are protected by the vaginate sheath of the next lower petiole. (2) Epiphytic or hemiepiphytic, thick-stemmed climbers: *P. radiatum, sagittifolium, fragrantissimum, panamense,* and *pterotum.* The caudex is usually short (less than a few meters) and 4 cm or more in diameter, with adult leaves closely clustered near the apex of the caudex. (3) Terrestrial: *P. grandipes* is always terrestrial on BCI and everywhere else I have seen it, though Standley reported it as an epiphyte.

Philodendron fragrantissimum (Hook.) Kunth, Enum. Pl. 3:49. 1841

Thick-stemmed, closely appressed, epiphytic or hemi-epiphytic climber at maturity, usually forming a slender, leafless, dangling stem at apex; caudex 3–6 cm diam; cataphylls 11–25 cm long, reddish at maturity, persisting

among leaf bases as brown weathered fibers; cut parts with red sap (especially spathe). Adult leaves clustered near apex; petioles 32–73 cm long, flat to slightly concave on upper surface with rigid, marginal rib; blades triangular, short-acuminate at apex, cordate at base, 33–52 cm long, 26–39 cm wide, the short broad lobes forming a sinus 4–10 cm deep and widely broadened to 8–16 cm; major veins in 6–9 pairs, not very prominent, the smaller veins all distinct; juvenile leaves narrowly ovate, acute to rounded or truncate at base, sheathed half or more of their length, spaced 3 cm apart or less, becoming slightly cordate with the petioles only slightly sheathed, the internodes longer. Inflorescences several to numerous, clustered amid weathered cataphylls; peduncles short, 4–8 cm long; spathe 14–19 cm long, the blade white to greenish, abruptly long-acuminate, the tube red, to 4.5 cm broad; spadix 13–16 cm long, the pistillate part 4–5 cm long, red with white flowers, ultimately with abundant red sap when cut. Fruits broadly oblong, 7–10 mm long, truncate at apex; seeds many, narrowly oblong, ca 1.5 mm long. *Croat 9410, 11007.*

Fig. 105. *Philodendron fragrantissimum*

Fig. 104.
Philodendron fragrantissimum

Fig. 106. *Philodendron fragrantissimum*

Fig. 107. *Philodendron grandipes*

Fig. 108. *Philodendron grandipes*

Fig. 109. *Philodendron guttiferum*

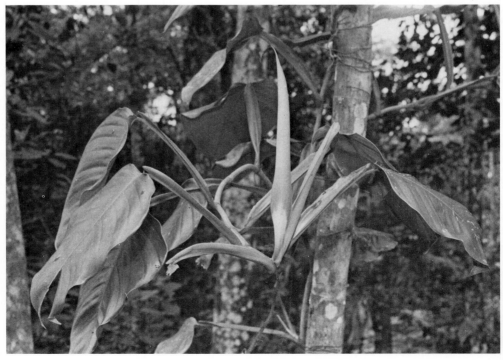

Fairly common in some areas of the island, particularly the older forest. Flowers chiefly in the early rainy season (June to August), less frequently in the late dry season. The fruits develop in October or later.

This species is most easily confused with *P. sagittifolium* or *P. panamense*. It differs from both in having the petioles flattened on the upper surface, in bearing persistent cataphylls, and in forming a slender, usually unbranched stem at the apex, which usually dangles back to the ground, creeps across the forest floor, and climbs another tree. This method of vegetative reproduction appears to be absent in the other thick-stemmed, appressed species on BCI. *P. fragrantissimum* is similar also to *P. clementis* (C. Wright) Griseb. of Cuba, which may prove to be synonymous. A sterile collection from Nicaragua (*Molino 15043*) may also be this species.

Panama to Amazonian Peru, Venezuela, and the Guianas. In Panama, known from tropical moist forest in the Canal Zone and Bocas del Toro, from premontane wet forest in Colón and Panamá, and from tropical wet forest in Colón and Panamá.

See Figs. 104, 105, and 106.

Philodendron grandipes Krause, Pflanzenr. IV.23Db(Heft 60):48. 1913

Terrestrial; stems 2–3 cm thick, usually less than 1 m long, creeping along the ground, leafy the last 15–30 cm, the leaves mostly clustered toward apex; internodes very short; cataphylls 12–25 cm long, green to pinkish, 2-ribbed, persisting as brown fibers. Petioles 20–70 cm long (mostly more than 40 cm), flattened on upper side, with the lateral margins sharply raised; mature blades ovate-cordate, acute and falcate-cuspidate at apex with the tip downturned, 20–43 cm long, 13–30 cm wide, bicolorous, shiny above, the basal sinus 3–11 cm deep, broader than deep on small leaves, usually 8–11 cm deep and closed by overlapping basal lobes on larger leaves; major veins in 11–17 pairs above sinus, much impressed above, dark and prominently raised below, the smaller parallel veins distinct; juvenile blades narrowly elliptic to ovate, acute to rounded at base, soon truncate to cordate, usually less than 15 cm long. Peduncles 8–11 cm long, white-lineate; spathe green inside and out (except limb white at anthesis), enveloping spadix, 9–11 cm long, 2.5–3.8 cm wide at base, ca 1.5 cm at apex (when closed), sometimes with long-acuminate tip; spadix very short-stipitate, 7–8 cm long, the pistillate part 2 cm long, ca 1 cm wide, to ca 2 cm wide in fruit. Fruits obovoid, truncate at apex, the stigma rounded, with 3–5 minute punctations. *Croat 11886.*

Apparently rare, restricted to steep creek banks. Flowers in the early rainy season (May to August), rarely earlier or later (April to October). Most of the fruits mature from August to October.

Distinguished from other species in Panama by being terrestrial and by having the upper petiole surface flat.

Known only from Panama, from tropical moist forest on the Atlantic slope of the Canal Zone and in Bocas del Toro and San Blas, from premontane wet forest in Colón

and Coclé, and from tropical wet forest in Colón, Chiriquí, and Coclé.

See Figs. 107 and 108.

Philodendron guttiferum Kunth, Enum. Pl. 3:51. 1841
P. rigidifolium Krause

Hemiepiphytic vine, frequently branched; stems gray to brown, 1 cm or less diam, only loosely attached to trees, frequently pendent. Petioles 8–18 cm long, the sheath broad, extending almost to or slightly beyond base of blade, the sides of sheath 5–10 mm high on mature leaves, one side clearly exceeding the other; leaf blades ovate to ovate-elliptic, those of climbing stems mostly 15–26 cm long and 9–15 cm wide, spaced 2–16 cm apart, those of terrestrial creeping stems smaller, mostly less than 14 cm long and 7 cm wide, the dangling stems often with leafless portions 2–4 m long, the internodes to 30 cm long; major veins of all leaves 5–8 per side, scarcely more prominent than the obscure smaller veins. Peduncles stout, 1–3 cm long; spathe green or yellow-green, red-lineolate inside base of tube, 8–26 cm long, spadix sessile, 10–23 cm long, the staminate part white, the pistillate part green, to 5.5 cm long. Fruits white, to ca 6 mm long and 3 mm wide. *Croat 10912, 11776.*

Abundant. Flowers chiefly from the late dry season to the early rainy season (February to September) but mostly during the early rainy season (June to July). The fruits apparently develop within the rainy season of the same year.

Distinguished by its scandent habit and broadly winged petiole. Cut parts are usually quite aromatic. Most easily confused with *P. inaequilaterum*, which has a similar habit but has the petioles winged nearly to the leaf base.

Southern Mexico to Panama, French Guiana, and Peru. In Panama, ecologically variable; known from tropical moist forest in the Canal Zone, Bocas del Toro, and Darién, from premontane wet forest in Colón, Chiriquí, and Panamá, and from premontane rain and lower montane rain forests in Chiriquí.

See Fig. 109.

Philodendron hederaceum (Jacq.) Schott, Wiener Z. Kunst 3:780. 1829
P. jacquinii Schott; *P. erlansonii* Johnston

Hemiepiphytic vine; stems, petioles, and veins of lower leaf surface shortly setose-pubescent; internodes 6–15 cm apart and usually ca 1.5 cm diam (to 3.5 cm); periderm paper-thin, sometimes peeling from larger stems. Petioles to 40 cm long, round in cross section, those subtending the inflorescence winged-vaginate; blades ovate, acute to short-acuminate and mucronate at apex, cordate at base with narrow sinus, 14–46 cm long, 11–35 cm wide, papyraceous; veins in 4–6 pairs, prominently impressed above and raised below; juvenile leaves similar to adult, though narrower and with the petioles vaginate-winged almost half their length. Peduncles glabrous, to 14 cm long in fruit; spathe glabrous, 13–19 cm long, oblique and inflated-bulbous at base (much of the space empty

in flower), to 7 cm wide in fruit, green outside, red inside, the tube green or white inside, falling free in fruit; spadix sessile, ca 8 cm long, the staminate part to 12 cm long, the pistillate part to 3 cm long and 3.5 cm diam; styles to 5 mm long; stigmas 3. Immature fruits greenish, irregular, filling entire cavity; mature fruiting spadix naked with fruit cluster 8–9 cm long and to 5 cm wide; mature fruits pale orange, ca 1 cm long, 5 mm wide; seeds usually 4–6 per fruit, ovoid, white, ca 4 mm long, only moderately sticky. *Croat 9259.*

Apparently rare on BCI, though common elsewhere in the Canal Zone. Flowers in the middle of the rainy season (July to September). The fruits mature by early in the rainy season of the following year (May to August). Inflorescences hang on leafless caudices during most of the dry season.

Distinguished by the cordate blade, the puberulence on the leaf veins, petioles, and caudices, and by the bulbous-based spathe and long-styled ovaries. The species was for many years confused nomenclaturally with *P. scandens* C. Koch & Sell.; for a discussion of this, see papers by Dugand (1945) and Bunting (1963).

Mexico (Veracruz) to Colombia, Venezuela, and the Guianas. In Panama, known from tropical moist forest in the Canal Zone, Herrera, Panamá, and Darién, from tropical dry forest in Los Santos, from premontane moist forest in Los Santos, and from premontane wet forest in Panamá.

See Fig. 110.

Philodendron inaequilaterum Liebm., Vidensk. Meddel. 16. 1850

P. coerulescens Engler

Hemiepiphytic vine, slender, branched many times, the main axis loosely attached; branches widespread, some very long and almost leafless, usually less than 1.5 cm diam, usually green when fresh and almost the same color as petioles (drying brown, sharply contrasting with the green petioles); internodes 5–22 cm (short near ends of branches). Petioles 5–24 cm long, canaliculate, narrowly sheathed almost to base of blade, the sheath margins usually 5 mm high, held erect (those subtending the branches of inflorescences larger); blades broadly ovate or elliptic-ovate, frequently oblique, abruptly acuminate at apex, rounded or truncate at base (slightly subcordate on largest leaves), usually 15–30 cm long but to 40 cm long and 20 cm wide on climbing, appressed stems, thin, usually drying dark; major veins 9–13 per side (to 20 on largest blades), upstanding and more prominent than the minor veins, the minor veins obscure but distinguishable on fresh leaves; juvenile leaves like adults except usually narrower in relation to length. Inflorescences few; peduncles short and stout, 2–4 cm long; spathe usually whitish or green and minutely white-lineate, 12–20 cm long, cuspidate, often recurved below apex at anthesis, usually not reclosing tightly after anthesis, the slender staminate part of spadix remaining protruded; spadix with short stipe, the staminate part narrowly tapered,

broadly curved forward, long-persistent after anthesis, the pistillate part 2.5–6 cm long, the fruiting spadix to 9 cm long and 2.5 cm wide. Fruits orange, angulate, to 4 mm long and 2 mm wide; seeds 6–20, usually ca 14, narrowly oblong, to 1 mm wide, immersed in a clear, sweet, watery matrix at maturity. *Croat 5831, 15570.*

Abundant in the forest, sometimes climbing to the top of the canopy; the most abundant *Philodendron* on BCI. Flowers mostly from April to May, occasionally as early as March or as late as September. The fruits mature mostly during June and July.

This species is most easily confused with *P. guttiferum*, a species of similar habit, but is distinguished by having a narrow sheath ending well below the base of the blade; *P. guttiferum* has a broad sheath ending nearly at or beyond the base of the blade. Blades of *P. inaequilaterum* may also be confused with those of *Rhodospatha wendlandii*, but they lack the prominently geniculate petioles toward the apex of the plant that are so characteristic of *R. wendlandii*.

Mexico, Panama, Colombia, Venezuela, Ecuador, and Peru, and no doubt elsewhere in Central America. In Panama, known from tropical moist forest in the Canal Zone, Panamá, and Darién, from premontane wet forest in Coclé and Chiriquí, and from premontane rain forest in Darién.

Philodendron inconcinnum Schott, Syn. Aroid. 81. 1856

Narrow-stemmed, hemiepiphytic vine, loosely attached at maturity; stems brown at maturity; internodes 4–15 cm long (closer at apex), about 1–2 cm thick, usually with 1–3 roots at nodes, the roots 20 cm or more long; periderm crisply and longitudinally folded when dried, often peeling off; cataphylls pale green, 6–12 cm long, with 1 or 2 faint ribs. Petioles 7–15 cm long, sheathed 1.5–2.5 cm at base (those subtending branches or inflorescences more completely sheathed); blades oblong or oblong-ovate, acute to rounded and cuspidate at apex, shallowly cordate at base, 16–30 cm long, 7–10 cm wide; the lobes rounded at base, the sinus generally shallow, square at apex; major veins in 4–6 pairs, impressed above, scarcely more prominent than smaller veins below, the secondary veins somewhat obscure; juvenile blades constricted above petiole, the basal lobes hastate. Peduncles 5–11 cm long; spathe 10–18 cm long, only slightly broader at base, green outside, the inside pale green but reddish at base of tube; spadix subsessile, only slightly shorter than spathe, the pistillate part 3–4 cm long in flower, 8 cm or more in fruit. Fruits white, to ca 6 mm long and 3.5 mm wide. *Croat 5064.*

Apparently uncommon, seen only a few times along the shore of Orchid Island. Known to flower at the beginning of the rainy season (late April to July). The fruits develop within a few months, probably mostly in July and August.

Distinguished from all other species by its small, oblong, shallowly cordate blade. The species was confused

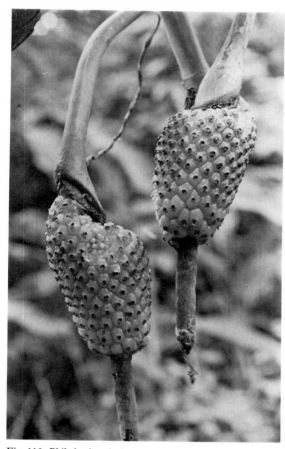

Fig. 110. *Philodendron hederaceum*

Fig. 112. *Philodendron panamense*

Fig. 111. *Philodendron nervosum*

with *P. wendlandii* Schott by Standley and was called by that name in the *Flora of Panama*. However, *P. wendlandii* has rosulate leaves and thick stems with much larger, more or less oblanceolate-obovate leaves; it also occurs in Panama.

Costa Rica, Panama, and Venezuela. In Panama, known from tropical moist forest in the Canal Zone, Bocas del Toro, and Darién and from premontane wet forest in Colón.

Philodendron nervosum (Schult. & Schult.) Kunth, Enum. Pl. 3:51. 1841
P. karstenianum Schott

Narrow-stemmed, hemiepiphytic climber, with mature stems tightly appressed or more frequently scandent; stems green when fresh (becoming tan and longitudinally wrinkled on drying, contrasting sharply with green petioles); internodes 3–10 cm, becoming even shorter at flowering apex; cataphylls lacking except beneath inflorescences. Petioles 11–34 cm long, broadly sheathed at least half of their length, the sheath 5–10 mm high, the remainder of the petiole flattened or canaliculate on upper surface; blades mostly ovate to oblong-elliptic, frequently somewhat oblique, rounded to acute and abruptly long-acuminate at apex, broadly rounded to short-cuneate at base, 17–37(46) cm long, 8–16(22) cm wide (much smaller on trailing stems); major veins mostly in 6–11 pairs, scarcely more prominent than the smaller veins, the secondary veins obscure (at least when fresh); juvenile leaves similar to mature leaves but smaller. Inflorescences few, emerging from leaf sheaths of uppermost nodes; prophylls green, to 12 cm long, with 2 narrow, longitudinal ribs on outer surface; peduncles about 10 cm long at maturity; spathe 12–16 cm long, green, caudate-acuminate at apex, scarcely widened at base; spadix stipitate or nearly sessile, slightly shorter than spathe, the pistillate part 2.5–4 cm long, the pistils slender; stigma doughnut-shaped in fruit. Fruits white, to 7 mm long and 4 mm wide. *Croat 6500.*

Occasional, in the forest. Flowers principally in the dry season and early rainy season (February to May), but also collected in flower in the late rainy season.

Recognized by its elongate, obscurely veined leaves with petioles winged much of their length. Plants on Cerro Campana (Panamá Province) often have petioles winged all the way to the blade.

Guatemala to Venezuela, Ecuador, and Peru. In Panama, known from tropical moist forest in the Canal Zone and Bocas del Toro, from premontane wet forest in the Canal Zone, Coclé, and Panamá, and from tropical wet forest in Colón, Chiriquí, Coclé, and Panamá.
See Fig. 111.

Philodendron panamense Krause, Pflanzenr. IV.23Db(Heft 60):65. 1913

Large-stemmed, epiphytic or hemiepiphytic climber, growing at all heights; stems of medium length, 4–6 cm diam; internodes to 15 cm long, becoming much shorter near apex; cataphylls deciduous. Petioles 60–70 cm long, solid green, minutely grooved, ± terete with narrow flattened rib on upper surface, scarcely vaginate at base; mature blades broadly triangulate, short-acuminate, thin, rounded at apex, deeply cordate at base, 40–65 cm long, 30–47 cm wide, the lobes broadly rounded, the sinus 10–18 cm deep and 6–11 cm broad; major veins in 7–9 pairs above sinus, the veins into basal lobes united, the smaller veins easily distinguished; juvenile leaves at first acute to rounded at base, broadest at middle, soon becoming weakly to strongly cordate, the petioles terete, winged, broadly sheathed half to three-fourths their length. Inflorescences few, long-pedunculate, the peduncles 7–23 cm long (mostly more than 15 cm), slightly to moderately bent just below spathe; peduncle and spathe both strongly white-lineate; spathe 11–18 cm long, pale green on inside, the outside with blade whitish and tube green; spadix sessile, slightly shorter than spathe, the pistillate part to 6.5 cm long. Berries white, ca 6 mm long; seeds many, white, narrowly cylindrical, 1.3 mm long, sticky. *Croat 10264, 10867.*

Common in the forest. One of the more abundant species, growing closely appressed to trees at 2–12 m from the ground, often with several large caudices in the same tree. Flowering from late March to July. The fruits develop mostly from June to August.

Most easily confused with *P. fragrantissimum*, which has much-flattened petioles, and with *P. sagittifolium*, which has much thicker, more narrowly triangulate leaves and dark-green petioles with light-green specks.

Honduras, Costa Rica, Panama, and Ecuador, no doubt in Nicaragua as well. In Panama, known from tropical moist forest on both slopes of the Canal Zone, from premontane wet and tropical wet forests in Panamá and from lower montane wet forest in Chiriquí.
See Fig. 112.

Philodendron pterotum C. Koch & Aug., Ind. Sem. Hort. Berol. App. 6. 1854

Large-stemmed, epiphytic or hemiepiphytic climber; stems usually short, 5–12 cm diam. Leaves closely clustered near apex; petioles 90–120 cm long, rounded below, flattened above (the basal 15–20 cm vaginate), the upper margins with a narrow, erect, wavy, membranaceous wing 4–5 mm high on margin; blades ovate, short-acuminate, cordate at base, 65–130 cm long, 46–100 cm wide, the lobes sometimes overlapping, the sinus narrow and acute at apex; major veins in about 10 pairs above sinus, the veins extending into basal lobes about 7, the smaller veins obscure; juvenile stems loosely attached to tree, the internodes long, the leaves broadly ovate, glistening on upper surface, the youngest rounded at base, becoming increasingly cordate with size, the smaller veins easily distinguished, the petioles flattened, 2-edged. Inflorescences in leaf axils, 6–22 or more near apex of single stem; peduncles 6–12 cm long, white-lineate; spathe 20–25 cm long, long-acuminate at apex, the tube purple outside, red within at base, the limb green outside, white within; spadix sessile, ca 20 cm long, the pistillate part 5 cm long, to 6 cm long in fruit. Fruits obovoid, to ca 1 cm long,

with a buttonlike style bearing 5 or 6 minute depressions. *Croat 6581, 6640.*

Occasional, in the forest on trees at 2–8 m. Flowers on BCI from May to July. The fruits mature usually before the end of the rainy season in December. At Summit Garden in the Canal Zone populations begin to flower in the early dry season, continuing until the middle of the rainy season, with the fruits mostly gone by the end of the rainy season.

Distinguished by its very large leaves with a flattened petiole bearing thin marginal ribs.

Costa Rica and Panama. In Panama, known from tropical moist forest in the Canal Zone (BCI and Summit Garden) and Chiriquí and from tropical wet forest in Colón.

See Figs. 113 and 114.

Philodendron radiatum Schott, Oesterr. Bot. Wochenbl. 3:378. 1853

P. augustinum C. Koch; *P. polytomum* Schott

Thick-stemmed, epiphytic or hemiepiphytic climber; stems closely attached, 5–12 cm thick; cataphylls green to pale pink, linear-lanceolate, to ca 30 cm long, weathering and persistent; sap sticky, clear. Leaves clustered near end of stem; petioles to 1 m or more, often ± dark-speckled, the sheath usually 8–10 cm long or less; blades broadly ovate-cordate in outline, 35–80 cm long and 30–65 cm wide or larger, often ± dark-speckled along major veins, pinnately dissected once or twice nearly to midrib into 5–10 linear-lanceolate segments on each side, the segments 2–4 cm wide, acuminate, thin except for very stout midrib on lower surface, the secondary veins obscure, closely parallel, the uppermost segments undivided, the middle segments becoming pinnately lobed or lobulate, the basal segments shorter and coherent, their major veins joining to form a stout trunk vein; juvenile leaves ovate-oblong, cordate and subentire, becoming shallowly or deeply incised-lobed. Inflorescences several (usually 4 or 5 or more), closely congested in upper axils; peduncles 5–9 cm long; spathe 20–25 cm long, apiculate at apex, the tube green, the blade whitish, often with sparse dark punctations; spadix slightly shorter than spathe, the pistillate part 6–8 cm long; style round; fruiting spadices exposed by rupture of the deciduous spathes, ca 5 cm diam. Fruits whitish, many-seeded; seeds oblong, sticky. *Croat 6060, 6124, 7178.*

Abundant in the forest, usually very high in trees on stout limbs or in the crotches of branches, often nearer the ground along the shore. Seasonal behavior uncertain. Probably flowers from the middle to late rainy season, with the fruits maturing during the rainy season of the following year.

Juvenile plants are wholly epiphytic, eventually sending long roots to the ground. The roots are generally 5–20 mm thick, mostly unbranched, and densely warty throughout their length, with aborted rootlets. When the end nears the ground, it branches profusely and roots. Thus, plants become hemiepiphytic in age, though such roots probably provide only a part of the plant's nutrients.

Blades of the species are variable (Bunting, 1965), and while the type description and drawing of *P. radiatum* do not closely match most of our material, it is probable that this taxon is represented in Central America by a single, very variable species. Should separate entities be segregated, our material matches most closely *P. augustinum* C. Koch and *P. polytomum* Schott, which are clearly synonymous with each ohter.

Seeds of this species have been observed being carried by leaf-cutter ants; however, since ants rarely ascend another tree in their return to their nest, this method of dispersal is of dubious value.

Range uncertain; Mexico to Panama and no doubt in South America as well. In Panama, apparently restricted to the Atlantic slope, where it is known from tropical moist forest in the Canal Zone and Bocas del Toro and from premontane wet forest in Colón.

See Fig. 115.

Philodendron sagittifolium Liebm., Vidensk. Meddel. 17. 1850

Closely appressed, epiphytic or hemiepiphytic climber; stems 4–6 cm diam; internodes to 15 cm long, but 1–3 cm long near apex; nodes often with thick, very long, brownish roots, the roots with thin periderm, longitudinally folded on drying, often peeling off; periderm of caudex similar to that of roots but lighter in color on drying and with numerous horizontal cracks; cataphylls ca 30 cm long, dark red at maturity (sometimes green), with 2 prominent ribs on outer surface, deciduous. Petioles 30–60 cm long, round in cross section, dark green with light-green specks, with an obscure, flattened sheath about 6 cm long at base; blades narrowly triangular-ovate, short-acuminate at apex, cordate at base, about 35–65 cm long, usually narrower than 40 cm, thick, with broadly revolute margins, the basal sinus 8–13 cm long, 3–6 cm wide, obtuse to acute at apex; major veins light green on upper surface, prominent against dark-green leaf color, in 7–9 pairs above sinus, the veins into the basal lobes united, the smaller veins obscurely visible on lower surface; juvenile leaves thick, oblanceolate, acuminate at apex, acute to rounded at the narrow base, sheathed three-fourths the length of petiole, the blades becoming elliptic, then increasingly broadened and cordate to nearly hastate at base. Inflorescences few to several; peduncles 6–11 cm long; spathe 12–22 cm long, the blade reddish (at least on margin) on outside, white (at anthesis) or yellowish-green within, the tube usually greenish on outside, usually red within at base; spadix slightly shorter than spathe, the pistillate part 5–8.5 cm long, to 10.5 cm long in fruit. Fruits ca 7 mm long at maturity; stigma round, weakly raised, with 10–12 minute, round depressions around periphery; seeds numerous, narrowly ellipsoid, ca 1.5 mm long, weakly striate longitudinally when dried, with a weak constriction near one end. *Croat 5052, 6334, 10901.*

Occasional, on trees at usually less than 4 m above the ground; juvenile plants may be epiphytic on rocks. Apparently flowering later than most *Philodendron* on BCI,

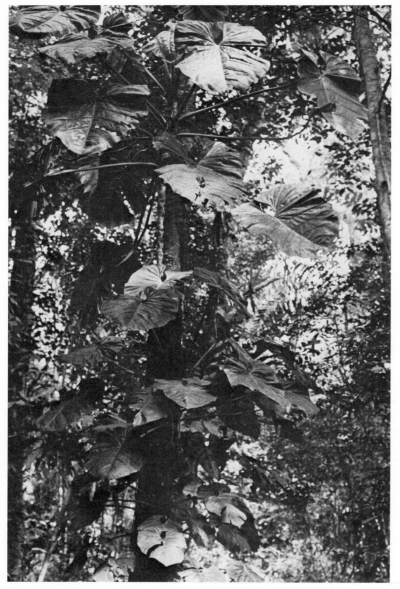

Fig. 113. *Philodendron pterotum*

Fig. 114. *Philodendron pterotum*

Fig. 115. *Philodendron radiatum*

Fig. 117.
Philodendron sagittifolium

Fig. 116. *Philodendron sagittifolium*

often not until August or September. At Summit Garden in the Canal Zone plants may begin to develop inflorescences by the early dry season and flower from March through July. The fruits probably do not mature before the middle of the dry season.

Most easily distinguished by its elongate blade, which is longer in relation to its width than any other large-leaved ovate-cordate *Philodendron;* distinguished also by its large red leaf sheath and by its thick leaves with prominent whitish midrib and veins on the upper surface, and speckled, round petiole.

Mexico to Colombia, possibly throughout northern South America as well. In Panama, common all along the Atlantic slope, and known at higher elevations on the Pacific slope; collected from tropical moist forest in the Canal Zone and San Blas (Puerto Obaldia), from premontane wet forest in Panamá (Cerro Azul and Cerro Campana), and from tropical wet forest in Colón (Miguel de la Borda).

See Figs. 116 and 117.

Philodendron scandens C. Koch & Sell., Ind. Sem. Hort. Berol. App. 14. 1853
   P. oxycardium Schott; *P. harlowii* Johnston

Narrow-stemmed vine, glabrous, occasionally branching, widely spreading and dangling from trees at maturity; stems green, with clusters of brown roots to 10 cm long and prominent cataphyll scars; cataphylls ca 10 cm long, pale green, caducous. Petioles 9–25 cm long or more, to 1 cm broad at base, terete, without invagination at base or invaginate to about middle; blades broadly ovate-cordate, 15–36 cm long, 10–27 cm wide, thick, the lobes directed outward or inward and overlapped, the sinus usually 3–7 cm deep, usually deeper than broad; primary veins in 4 or 5 pairs, the veins extending into the basal lobes united; juvenile plants usually hemiepiphytic and closely appressed, the blades ovate, caudate-acuminate, the upper surface dark green (sometimes reddish-green), glistening with minute, close papillations, the lower surface somewhat maroon. Peduncles 8–10 cm long; spathe green, cuspidate at apex, only slightly broader at base, 14–16 cm long, red inside especially at base of tube; spadix short-stipitate, the pistillate part ca 6 cm long; stigmas with thin margins. Fruits with numerous seeds, the seeds oblong-ellipsoid, slightly constricted at one end, ca 1 mm long. *Croat 7129, 10383.*

Common, especially along the shore, where it may festoon trees or hang long-pendent from tree branches. Probably flowers and fruits all year.

Not confused with any other species on the island. The Panamanian material of this species has been assigned to subsp. *scandens* f. *scandens* by Bunting (1968).

Throughout Mexico and Central America and much of tropical lowland South America; West Indies. In Panama, known from tropical moist forest in the Canal Zone, Bocas del Toro, Panamá, and Darién, from premontane moist forest in Panamá (Tocumen), from premontane wet forest in Coclé, and from tropical wet forest in Colón, San Blas, and Chiriquí.

See Fig. 118.

Philodendron tripartitum (Jacq.) Schott, Melet. 1:19. 1832

Branched, narrow-stemmed vine; stems brown except near apex; internodes 3–12 cm long; nodes 1–1.5 cm thick, drying with a ± loose, paper-thin periderm; cataphylls elongate-lanceolate, 18–33 cm long, greenish, thin, un-ribbed, caducous; sap, especially in the inflorescence, very aromatic. Petioles mostly 20–60 cm long; blades 3-parted, thin, the segments oblong-lanceolate, 15–30 cm long, to 13 cm wide, the lateral segments conspicuously oblique; veins close, subequal, ascending, even the smaller veins conspicuous; juvenile plants hemiepiphytic, the terminal lobe of blade elliptic to oblanceolate, the basal lobes small, slender and spreading, the lower blade surface green. Peduncles 3–13 cm long; spathe 14–21 cm long (only slightly broadened in lower half), the tube greenish, white on upper margins of blade and inside when open, maroon inside base in fruit; spadix sessile or borne on short stipe, the pistillate part 3–11 cm long, green at anthesis, the staminate part white. Fruits white, ca 4 mm long, irregularly angulate, 2–2.7 mm wide; seeds usually 6–8, cylindrical, ca 1.5 mm long, sticky. *Croat 10741.*

Occasional, climbing over trees. Flowers to some extent throughout the rainy season but especially during the early part (May to July); elsewhere flowers rarely in the dry season. The fruits develop chiefly in the middle to late rainy season.

Because of similar shape, juvenile leaves of this species are confused with juvenile leaves of *Syngonium erythrophyllum,* in which the underside of the blade is violet-purple and the venation is markedly different.

Mexico to Colombia, Venezuela, and Brazil; Jamaica. In Panama, known from tropical moist forest in the Canal Zone and Bocas del Toro, from premontane moist forest in Panamá, and from tropical wet forest in Coclé.

PISTIA L.

Pistia stratiotes L., Sp. Pl. 963. 1753
   Water lettuce, Lechuga de agua

Floating aquatic, nearly acaulescent; caudex sometimes producing stolons with new rosettes of leaves at apex. Leaves forming dense rosette; petioles very short; blades ± obovate, rounded or emarginate at apex, cuneate at base, mostly 5–17 cm long, 2–7 cm wide, thick and spongy, pubescent on both surfaces with short, few-celled trichomes; veins in 5–15 pairs, parallel, prominent beneath. Inflorescences small and inconspicuous, subsessile, borne among leaves; spathe white, constricted at middle, the lateral margins connate to middle; spadix shorter than spathe and adnate to it for two-thirds of its length, with a single naked pistillate flower and 2–8 naked staminate flowers arranged in verticils; stamens 2, connate to form a synandrium; ovary obliquely attached to spadix, 1-celled; ovules numerous with basal placentation. Seeds cylindrical, minute. *Croat 8431.*

Occasional, at the lake margin, usually in quiet waters. Flowering period unknown.

Throughout the tropics of the world. Probably throughout Panama at lower elevations; known from tropical moist forest on both slopes in the Canal Zone and in Panamá and Darién.

RHODOSPATHA Poepp.

Anepsias, previously considered a distinct genus, is included with *Rhodospatha*. See the discussion following *Rhodospatha moritziana*.

Rhodospatha moritziana (Schott), comb. nov.
Anepsias moritzianus Schott, Gen. Aroid. pl. 73. 1858

Terrestrial; caudex thick, mostly or entirely creeping over the ground, usually less than 1 m long, to 5 cm diam below leaves, tapering to a narrow end entering the soil, secured by smaller roots along its length. Leaves closely spaced, imbricate; petioles minutely speckled with light green, on adult plants usually 30–50 cm long, vaginate-winged about three-fourths its length, one side higher, to 2.5 cm wide at base, ± nodose and geniculate just below blade; blades broadest in middle and tapering evenly to both ends, short-acuminate at apex, cuneate to obtuse at base, usually 50–76 cm long, 20–42 cm wide, glabrous or minutely papillate, dark green and glossy above, light brownish-green below, densely covered with minute reddish-brown spots, midrib sunken and light green above, raised below; major lateral veins in 24–34 pairs, somewhat impressed above, 10–25 mm apart. Inflorescences solitary in leaf axils; peduncles to 40 cm long; spathe boat-shaped, white, 15–26 cm long, with cuspidate apiculum ca 2.5 cm long, decurrent on and tightly enveloping spadix, to 12 cm broad when open, 5–6 cm deep, soon deciduous, the margins revolute; spadix uniform, salmon-pink, slightly shorter than spathe, densely many-flowered, short-stipitate, cylindrical, blunt on end, 16–20 mm wide, becoming covered with a sticky thick solution shortly after anthesis; flowers perfect, naked; stamens 4; anthers narrowly pointed; ovary ca 3 mm long, the sides angulate, 1-locular but with a partial division suggesting 2 locules, each division containing 1 placenta with numerous ovules; style as broad as ovary; stigma linear. Berries small (mature berries not seen), prismatic, truncate; seeds minute (probably less than 0.5 mm), elbow-shaped, very numerous, embedded in a sticky matrix. *Croat 11275, 12297.*

Uncommon; locally abundant, occurring on steep, moist creek banks in the vicinity of the laboratory. Flowers principally throughout the middle to late rainy season

(August to November). Fruiting inflorescences have never been seen, but fruits probably develop in a short time.

Possibly confused with *R. wendlandii*, which is, however, always epiphytic as an adult. Simmonds's report (1950) of this species from Trinidad was possibly an error, because he described it as a high-climbing epiphyte, a habit I have never encountered among the exclusively terrestrial Panamanian specimens. Standley reported the species from Chiriquí (Burica Peninsula) on the basis of *Woodson & Schery 929*, but that specimen may be of another species of *Rhodospatha*.

Traditionally *Anepsias* has been separated from *Rhodospatha* by having two to six cells in the ovary, compared to two in *Rhodospatha*. The ovules of those *Rhodospatha* investigated on BCI were indeed bilocular, but all of the *Anepsias* specimens investigated were found to have an incomplete septum and thus were unilocular. Therefore, it appears likely that the divisions of the ovary are variable and cannot be used as a character of generic separation; the genus *Anepsias* is here reduced to synonymy under *Rhodospatha*.

Costa Rica to Colombia and Venezuela. In Panama, probably restricted to the Atlantic slope and known from tropical moist forest in the Canal Zone and San Blas and from tropical wet forest in Colón.

See Figs. 119 and 120.

Rhodospatha wendlandii Schott, J. Bot. 2:52. 1864
R. forgetii N. E. Brown

Epiphytic climber; caudex usually unbranched, 1–2(7) m long, to 4 cm thick, tightly appressed to its support; juvenile plants terrestrial, the stems creeping, rooting at nodes. Leaves closely spaced near apex of stem, well spaced below; petioles 25–72 cm long, canaliculate on upper surface to base of blade, vaginate-winged most of length (the wing weathering away), prominently swollen and geniculate below blade; blades oblong-lanceolate, cuspidate-acuminate at apex, usually rounded to truncate at base; midrib much raised and narrow below, the veins branching from midrib at nearly 90°, the major lateral veins 6–20 mm apart; leaves of juveniles gray-green, ovate to elliptic, acuminate, rounded to acute at base, mostly 4–10 cm long; juvenile leaves on climbing stems lanceolate-elliptic, dark green, mostly 18–35 cm long, acuminate at apex, acute to narrowly acute at base. Flowering inflorescences from upper axils; peduncles 12–23 cm long; spathe 20–44 cm long, 9–18 cm wide when expanded, white to somewhat pinkish, deciduous; spadix rose-pink, short-stipitate (the stipe to 2 cm long), slightly shorter than spathe, cylindrical, blunt at apex, ca 1.5 cm

KEY TO THE SPECIES OF RHODOSPATHA

Plants epiphytic at maturity; blades drying blackened, rounded to truncate at base, the major lateral veins often less than 10 mm apart . . . . *R. wendlandii* Schott
Plants terrestrial at maturity; blades drying bronze-colored beneath (the result of numerous minute reddish-brown spots), attenuate at base, the major lateral veins usually more than 10 mm apart
. . . . *R. moritziana* (Schott) Croat

Fig. 118. *Philodendron scandens*

Fig. 119. *Rhodospatha moritziana*

Fig. 120. *Rhodospatha moritziana*

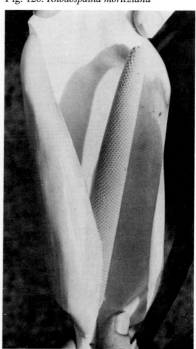

wide, sometimes recurved in fruit; flowers mostly perfect, naked; stamens 4, ca 3 mm long, tightly compressed in small cavity between adjoining pistils; ovary 3–4 mm long, the stigma round or ellipsoid; fruiting spadices usually moderately short, ca 3 cm wide. Berries less than 1 cm long, obovoid, prismatic and truncate, 2-celled, the septum broad; ovules numerous, elbow-shaped, sticking together in a gelantinous mass; seeds round, flattened, ca 1 mm long. *Croat 11406.*

Very common in trees, to 10 m high; even more abundant as juvenile plants, occasionally carpeting the forest floor. The adults usually lose their connection with the ground. Flowers mostly throughout the rainy season, from July to December. The fruits probably develop within about 1 month after flowering. Fruiting inflorescences are seldom seen, and mature fruiting spadices seldom last long. They are apparently removed by arboreal animals, including to some extent birds.

The species is possibly confused with *Philodendron inaequilaterum* with its large oblong leaves, but *Rhodospatha* may also be distinguished by having the uppermost petioles very strongly geniculate, with the blade hanging downward. Reported by Standley in the *Flora of Panama* as *R. forgetii* N. E. Brown, but the Panamanian material matches very well the type of *R. wendlandii.* The inflorescence on the type specimen is smaller than those in Panama, but size is quite variable even in central Panama, where the spathe varies from 20 to 40 cm in length. It is also likely that *R. costaricensis* Engler & Krause, *R. nervosa* Lundell, and *R. roseospadix* Mat. are synonymous with *R. wendlandii.*

It is uncertain how the plants prevent predation on the fruit, since the fruits are much like those of *Philodendron* at maturity but do not have the benefit of the protective spathe before maturity.

Costa Rica and Panama, probably also along all of the lowland Atlantic slope of Mexico and Central America to Colombia. In Panama, known from tropical moist forest on the Atlantic slope in the Canal Zone, from premontane wet forest in Coclé, and from tropical wet forest in Coclé and Darién.

See Fig. 121.

SPATHIPHYLLUM Schott

The genus *Spathiphyllum* is distinguished by its terrestrial habit, its uniform inflorescence of bisexual flowers, and its persistent spathe. Flowers are bisexual with three (rarely two or four) tepals and stamens equal in number to the tepals.

Spathiphyllum friedrichsthalii Schott, Aroid. 2, pl. 4. 1853

Acaulescent plants, robust, often more than 1 m tall, usually growing in large clones in shallow water. Petioles 30–60 cm long, sheathed to about middle or more, otherwise terete, weakly geniculate at apex; blades narrowly elliptic, narrowed to both ends, gradually long-acuminate to abruptly short-acuminate, attenuate to acute and somewhat decurrent at base, 28–70 cm long, 7–22 cm wide; primary lateral veins many, the smaller veins ± distinct. Inflorescences about as high as leaves; spathe boat-shaped, ± elliptic, oblique, cuspidate and somewhat asymmetrical at apex, long-decurrent at base, (10)13–32 cm long, (4)5–11 cm wide, white in flower except for green midrib, green in fruit; spadix cylindrical, blunt, sessile or short-stipitate, white, 3–7 cm long, less than 2 cm wide in flower; tepals ca 2.3 mm long and 1.7 mm wide; pistils 3-locular, greatly exceeding perianth, to 6 mm long in flower, to more than 1 cm long in fruit, the apex green; ovules (3)5–8 per locule; fruiting spadices green, somewhat longer but chiefly broader, to 3 cm wide (including protruding styles). Fruits obovoid; seeds usually 2–11 per fruit, warty and very irregular, brown at maturity, to 4 mm long. *Croat 11766, 11802.*

Locally common in secluded coves in shallow water. Flowering plants may be seen throughout most of the year, especially during the rainy season.

Distinguished from *S. phryniifolium* by its aquatic habit and white spathe. The species is exceedingly variable in the size of parts, possibly reflecting the age of the clone. It is confused with *S. blandum* Schott in parts of Central America.

The sweet aroma of the flowers is very characteristic and can be detected from a considerable distance. In flower the inflorescences are often visited by *Trigona* bees.

Nicaragua to Colombia on the Atlantic slope and along the Pacific slope from lower Panama to southern Colombia. In Panama, known from tropical moist forest on both slopes of the Canal Zone and in Bocas del Toro, Colón, and Darién, from premontane wet forest in Colón, Chiriquí, and Coclé, from tropical wet forest in Bocas del Toro, Colón, and Darién, and from lower montane wet forest in Chiriquí.

See Fig. 122.

Spathiphyllum phryniifolium Schott, Oesterr. Bot. Wochenbl. 7:59. 1857
S. zetekianum Standl.

Acaulescent herb, quite variable, to more than 1 m tall. Petioles 9–100 cm long, the sheath to middle or, less

KEY TO THE SPECIES OF SPATHIPHYLLUM

Plants usually growing in water at the lake margin; spathe white at anthesis; leaf blade and spathe usually acute to narrowly acute at base; pistils with 5–8 ovules per locule . . . . *S. friedrichsthalii* Schott
Plants growing in moist areas in the forest; spathe usually green at anthesis; leaf blade and spathe usually obtuse to rounded at base (acute on smaller plants); pistils with 1 or 2 ovules per locule . . . . *S. phryniifolium* Schott

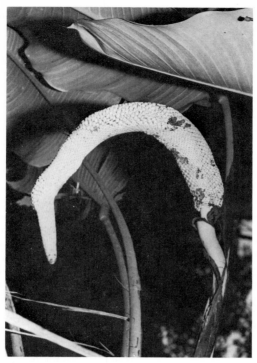

Fig. 121. *Rhodospatha wendlandii*

Fig. 122. *Spathiphyllum friedrichsthalii*

Fig. 123. *Spathiphyllum phryniifolium*

frequently, to just beneath geniculum; blades lanceolate to oblong-elliptic, gradually to abruptly acuminate, obtuse to subtruncate or acute on smaller plants at base, 12–60 cm long, 4–22 cm wide; veins many and close together, the major ones impressed above, raised below. Inflorescences held well above leaves; spathe boat-shaped, about shape of leaf blades, attenuate at apex, oblique and rounded to acute at base, green at anthesis, 7–26 (33) cm long; spadix cylindrical, white at anthesis, 2–8 cm long, 8–15 mm wide, to 13 cm long and 15 mm wide in fruit, borne on stipe ca 1 cm long; perianth segments truncate, their outer margins thin; pistil (2) 3-locular; style conic, greatly exceeding perianth, persisting in fruit; ovules 1 or 2 per locule. Fruits obovoid; seeds usually irregular, smooth or foveolate. *Croat 10930, 17050.*

Occasional, in the forest along trails or streams. Flowers mostly from the middle of the dry season to the middle of the rainy season (March to September), especially in the early rainy season. The fruits develop mostly within 2 months and mature chiefly in the rainy season.

The species has been confused with *S. floribundum* N. E. Brown and *S. patinii* (Hogg) N. E. Brown, both Colombian species; *S. patinii* is known only from cultivation. Further work needs to be done on the species throughout the range. Bunting (1960) wrote that plants from the Perlas Islands (Panamá Province) and Guanacaste most clearly match the type.

Costa Rica (Guanacaste) and Panama. In Panama, known from tropical moist forest in the Canal Zone, Bocas del Toro, San Blas, Veraguas, Los Santos, and Panamá, from premontane wet forest in Coclé and Panamá, from tropical wet forest in Colón, and from lower montane rain forest in Chiriquí.

See Fig. 123.

STENOSPERMATION Schott

Stenospermation angustifolium Hemsl., Biol. Centr.-
   Amer. Bot. 3:425. 1885

Epiphytic vine; stems 4–8 mm diam, weakly rooting at lower nodes; internodes mostly 1–5 cm long. Petioles 1.5–8.5 cm long, the sheath thick, extending almost to blade, the upper edge free; blades lanceolate to narrowly elliptic, sharply acuminate, cuneate to rounded and often unequal at base, 4–18 cm long, 1–7 cm wide, entire, the lower leaves deciduous; midrib distinguishable almost to apex, all lateral veins equally indistinct, ± parallel to

midrib. Peduncles 6–8 cm long, frequently recurved near apex before anthesis; spathe acuminate, 4–6 cm long, soon deciduous; spadix 2–4 cm long in flower, ca 4 mm diam, the stipe obsolete or to 3.3 mm long; flowers bisexual, naked; stamens 4; pistils ca 4 mm diam, 2-locular; style raised, cylindrical; fruiting spadices white, fleshy. Seeds obovate, to 1.8 mm long, tan, with many vertical greenish lines, constricted slightly below base. *Croat 14038.*

Occasional, growing high in trees. Flowers and fruits throughout the year, perhaps chiefly in the dry season.

Seeds are probably dispersed by birds.

Costa Rica and Panama. In Panama, known from tropical moist forest in the Canal Zone, Bocas del Toro, Chiriquí, and Veraguas, from premontane wet forest in Colón, Coclé, and Panamá, from tropical wet forest in Chiriquí, and from premontane rain forest in Panamá.

SYNGONIUM Schott

The genus *Syngonium* is distinguished by its epiphytic habit, syncarpous fruits, and unisexual flowers borne on separate parts of the spadix and separated by sterile staminate flowers. Its leaves have two or three collecting veins, and tertiary veins regularly interconnect the major laterals. The genus is closest to *Xanthosoma.*

Syngonium sp.

Herbaceous vine, probably becoming hemiepiphytic in its natural habitat; stems slender, densely short-puberulent, the caudices puberulent, the underside of leaf veins and the petioles papillate-puberulent, the plant otherwise glabrous. Petioles to 17 cm long, sheathed one-fourth to one-half their length, the lower side ribbed, the ribs close together, weakly raised, papillate-puberulent; blades thin, tripartite, the lobes distinct, the terminal lobe ± elliptic, acuminate on both ends, to 17 cm long and 5.5 cm wide, the lateral lobes distinctly inequilateral, the outer half rounded to cordate, the inner edge diminishing before base, the slender base to ca 1 cm long; reticulate veins distinct, at least below, the midrib green or slightly achlorophyllous; juvenile blades ovate to ovate-elliptic, acuminate, cordate or hastate at base, solid green or ± achlorophyllous along midrib. Fertile parts unknown. *Croat 17016.*

Cultivated in the Laboratory Clearing.

KEY TO THE SPECIES OF SYNGONIUM

Petiole and veins of lower blade surface papillate-puberulent; plants cultivated in the Laboratory
   Clearing . . . . *S.* sp.
Petiole and blades glabrous; plants occurring in the forest or at the lake margin:
   Adult leaf blades 5-lobed or 3-lobed with conspicuous auricles, thus appearing 5-lobed; juvenile
      blades with terminal and basal lobes narrowly acuminate, never violet-purple beneath . . . .
      . . . . *S. podophyllum* Schott
   Adult leaf blades 3-lobed; juvenile blades acute or very narrowly rounded and apiculate at apex,
      the basal lobes usually very blunt, violet-purple beneath when young . . . .
      . . . . *S. erythrophyllum* Bunt.

The species is similar to *S. mauroanum* Birds. ex Bunt. of northwestern Panama, but can be distinguished by the puberulent caudices and the papillate-puberulent petioles and underside of veins.

The BCI plants perhaps represent a new species, but since they are never fertile it is not possible to provide the species with a name. The plants may be some aberrant cultivar. No similar plants have been found elsewhere in Panama.

Syngonium erythrophyllum Birds. ex Bunt., Baileya 14:17. 1966

Hemiepiphytic vine; caudex usually less than 1 cm diam, covered with thin, minutely papillate, flaky periderm, branched, usually well attached high on tree trunks; sap milky; juvenile plants at first terrestrial, frequently as single plants, some branching with slender, leafy, somewhat scandent shoots, later climbing trees. Petioles 11–20 cm long, vaginate-winged to near apex; blades deeply tripartite, dark green above, paler beneath, acuminate to rounded at apex, ending abruptly with minute apiculum, to 21 cm long and 8.5 cm wide, the lobes distinct, the median lobe elliptic to lanceolate-elliptic, the lateral lobes asymmetrical, lanceolate-oblong, to 16 cm long and 6 cm wide, blunt at apex, oblique at base, cuneate on upper side, the lower side broader and rounded or subcordate; major lateral veins usually in 3 pairs, the smaller veins distinct below; juvenile blades simple, lanceolate-ovate and cordate, the lobes rounded and short, the apex acute to rounded, the blades 3–9 cm long, blackish-green above, at first green below, soon becoming violet-purple, the major lateral veins mostly near base, the intercostals weaker, anastomosing; juvenile blades on climbing stems larger, entire, hastate to trilobate, dark green above, deep violet-purple beneath, 9–17 cm long, becoming indistinctly 3-lobed. Inflorescences axillary, commonly 2 or 3; peduncles 4–11 cm long, recurved in fruit; spathe as long as or longer than spadix, the tube green, 4–6 cm long, the blade white, 6–11 cm long, ca 5.5 cm wide, turning brown, withering or deciduous; spadix to 9.5 cm long, the fertile staminate part 6–7 cm long, the pistillate part 1.7 cm long; pollen white, in arachnoid clusters. Fruits ovoid, usually ca 3.5 cm long, soft at maturity, the fruiting blade of the spathe opening when fruit matures; seeds white, ± oblong, rounded on one side, apiculate and angled on the other, 6–7 mm long. *Croat 6253, 14955.*

Juvenile plants are very abundant in the forest, both creeping on the ground and climbing on trees; adults, not commonly seen, are usually high in trees in the forest and occasionally along the shore. Appears to flower in the dry season and the early rainy season (March to July). The fruits mature in the rainy season (mostly May to September).

Leaf coloration of juvenile plants distinguishes the species from juveniles of *S. podophyllum*. Though resembling juvenile leaves of *Syngonium* in shape and habit, those of *Philodendron tripartitum* can be distinguished by having many, closely spaced, parallel veins; the smaller veins of *Syngonium* are always reticulate.

Known only from Panama in tropical moist forest on BCI and in Chiriquí and from premontane wet forest in Chiriquí and Panamá.

Syngonium podophyllum Schott, Bot. Zeitung (Berlin) 9:85. 1851

Hemiepiphytic climbing vine; caudex ca 1 cm diam, branching; adult plants with milky sap; juvenile plants at first terrestrial as single individuals or more often with trailing stems, rooting at nodes. Petioles to 50 cm long, sheathed one-half to two-thirds their length (the sheath free above), rounded above sheath; blades usually 3-lobed, dark green above, pale below, 12–27 cm long, the lobes connected or free, the lowermost variously auriculate below, the blades sometimes becoming 5-lobed, the ear free, the middle segment of blade ± ovate to elliptic, to 32 cm long and 16 cm wide; major lateral veins in 3–5 pairs, impressed above, raised below, the collecting veins 2 or 3, irregular, the smaller veins all distinct; juvenile leaves simple, cordate, 7–14 cm long, becoming sagittate, acuminate, the terminal lobe somewhat constricted and with margin broadly undulate at base, the basal lobes usually ± triangulate, acuminate, directed toward stem or prominently outward. Inflorescences axillary, mostly 4–9 per axil; peduncles usually less than 9 cm long at anthesis (to 13 cm) and usually ± twisted and pendent in fruit; spathe 9–11 cm long, the blade 6–7 cm long, ca 5 cm wide and white when open, soon deciduous; spadix short-stipitate, ca 8 cm long, the staminate part 4.5–6 cm long; stigmas sessile, white, raised and rounded in flower, flattened and brown in fruit, the tube eventually turning yellow to yellow-orange at maturity and opening to expose the brown compound fruit. Fruits ovoid, ca 4 cm long, 2.5 cm wide, the outer surface scurfy; seeds numerous, ovoid, 7–8 mm long, 5–6 mm wide, brown, enveloped in a grayish tissue, embedded in a sweet, soft, white pulp. *Croat 16218.*

Abundant in the forest, commonly rather high in trees. The species may be seen in flower from November to August and perhaps all year, but chiefly throughout the dry season. Individuals flower over a long period, with inflorescences appearing sequentially, and thus bear fruits in successive stages. The fruits mature mostly in the rainy season, beginning to mature shortly after the season starts.

Range is uncertain, but probably throughout Central America. In Panama, known from tropical moist forest in the Canal Zone, Bocas del Toro, Chiriquí, Veraguas, Panamá, and Darién, and from tropical wet forest in Bocas del Toro and Panamá.

See Figs. 124 and 125.

XANTHOSOMA Schott

The genus *Xanthosoma* is distinguished by its terrestrial habit and unisexual flowers borne on separate parts of the spadix and separated by sterile staminate flowers. It is in many respects like the epiphytic genus *Syngonium*, but lacks the syncarpous fruit and milky sap (may have milky sap elsewhere). The flowers are naked, with four to

KEY TO THE SPECIES OF XANTHOSOMA

Leaf blades deeply pedately lobed . . . . *X. helleborifolium* (Jacq.) Schott
Leaf blades entire:
 Underside of blades glabrous, usually pruinose . . . . *X. nigrum* (Vell.) Stellf.
 Underside of blades conspicuously pubescent and not pruinose . . . . *X. pilosum* C. Koch & Aug.

six stamens; the ovaries have two to four locules, and each locule has few to many ovules.

Xanthosoma helleborifolium (Jacq.) Schott, Oesterr. Bot. Z. 15:33. 1865

Terrestrial, occasionally more than 1 m tall, glabrous; caudex tuberous, the leaves and peduncles arising from the ground. Petioles thick and succulent, the lower part broadly vaginate-winged; petioles and parts of leaf rachis with characteristic textured pattern of maroon and white but predominantly maroon; blades reniform in outline, deeply dissected, the central lobe to 30 cm long and 7 cm wide; rachis branching, curving to both sides; leaflets 5–18, thin, sessile and diminishing markedly in size toward either end, oblong or lanceolate, acuminate, cuneate at base; veins prominent, the collecting vein within 5 mm of margin. Peduncles 40 cm long or more, textured like petiole and rachis; spathe to 19 cm long, enveloping spadix, the blade white or greenish at anthesis, soon withering, the tube green, persisting, broad in fruit; spadix white, to 17 cm long, with a broad short stipe, the staminate part ca 10 cm long, the sterile staminate part constricted apically, much broadened basally, the pistillate part ca 2.5 cm long. Fruits not seen. *Croat 10892, 11483.*

A common species of clearings and very open trails in the forest. The plant dies back after the beginning of the dry season and reappears shortly after the rains begin, flowering mostly from May to September, with inflorescences repeatedly produced. The fruits develop within about 1 month.

El Salvador to the Guianas and Amazonian Peru; the Antilles. In Panama, known from tropical moist forest in the Canal Zone and in San Blas and Panamá and from premontane wet forest in Coclé (El Valle).

See Fig. 126.

Xanthosoma nigrum (Vell.) Stellf., Tribuna Farm. 12:20. 1944

X. violaceum Schott

Acaulescent, plant robust, glabrous, with large tuberous rhizome, often more than 1 m tall. Petioles 30–85 cm long, fleshy, rounded except for vaginate basal part; blades sagittate-ovate, short-acuminate at apex, 20–70 cm long, 15–45 cm wide, pruinose beneath at least when young, the basal lobes ± triangular, obtuse, directed outward, to 36 cm long, the sinus open, acute; costae 3; midrib plus large veins extending almost to tip of either lobe, the major lateral veins in ca 6 pairs above sinus, all veins merging with a prominent collecting vein near the margin. Peduncles to 30 cm long; spathe to 21 cm long, green on basal bulbous part, white apically, to 8 cm broad when open; spadix to 17 cm long, emitting sweet odor when open, the fertile staminate part white, 10 cm long, the sterile staminate part bulbous at base, 4.5 cm long, the pistillate part yellowish, to 2.5 cm long. Fruits not seen.

The species has never been collected on the island, but has been seen cultivated near the dock in the Laboratory Clearing. Flowering and fruiting during much of the rainy season in the Canal Zone.

A native of unknown parts of tropical America; cultivated and naturalized throughout much of tropical America, Africa, and Asia. In Panama, known from tropical moist forest in the Canal Zone and Panamá and from tropical dry forest in Panamá (Taboga Island).

Xanthosoma pilosum C. Koch & Aug., Ind. Sem. Hort. Berol. App. 2. 1855

Primrose malanga, Badu, Coro, Otó

Essentially acaulescent plant, most parts densely short-villous (upper leaf surface only moderately so); roots shallow, with many fine smaller roots, to 3 cm broad; cataphylls to about 20 cm long, prominently 2-ribbed. Petioles fleshy, 15–40 cm long, vaginate-winged to near middle or beyond, then terete; blades mostly cordate-ovate, cuspidate-acuminate at apex, thin, sometimes mottled with white spots, the basal sinus open to closed with the basal lobes often overlapping, the medial vein of lobes marginal at apex of sinus; major veins impressed above, 4 or 5 pairs above sinus, the collecting vein prominent. Peduncles usually shorter than leaves, to 33 cm long; spathe to 17 cm long, the tube green, the blade white, ca 10 cm long, to 4.5 cm wide when open, usually purplish within near mouth of tube; spadix 13 cm long, the fertile staminate part 5–7 cm long, the sterile staminate part purple-violet, the pistillate part ca 3 cm long; fruiting inflorescences nearly globular. Fruits ± obovoid, to ca 7 mm diam, whitish, the blade opening or weathering away by the time the fruits mature; seeds many, irregular, ± elliptic, white, ca 1 mm long. *Croat 11256.*

Apparently restricted to open areas in the vicinity of the laboratory, often locally abundant. Plants usually disappear during the dry season and reappear with the onset of the rains. Flowers chiefly from May to October, the fruits developing within about 1 month.

Standley reported both *X. pilosum* and *X. mexicanum* Liebm. in the *Flora of Panama*. Engler reported *X. mexicanum* only from Mexico and *X. pilosum* from Panama and Costa Rica. In my opinion Panamanian material is not separable into two species and the two will prove, in all probability, to be synonymous. If so, the name *X. mexicanum* would have precedence.

Costa Rica to Colombia. In Panama, ecologically variable; known from tropical moist forest in the Canal Zone,

Fig. 124. *Syngonium podophyllum*

Fig. 125. *Syngonium podophyllum*

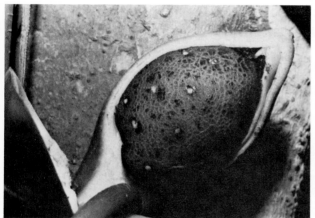

Fig. 126. *Xanthosoma helleborifolium*

Fig. 127. *Xanthosoma pilosum*

Chiriquí, Veraguas, Panamá, and Darién, from tropical dry forest in Panamá, from premontane moist forest in Los Santos and Panamá, from premontane wet forest in Chiriquí, Coclé, and Panamá, and from premontane rain forest in Darién.

See Fig. 127.

22. BROMELIACEAE

Epiphytic or rarely terrestrial (*Aechmea, Ananas*) herbs. Leaves spirally arranged in a basal rosette, often forming a watertight tank; blades acicular to ligulate, simple, entire or spiny-serrate, lacking hairlike trichomes but often with peltate scales; venation parallel; stipules lacking. Flowers bisexual or functionally unisexual (dioecious), actinomorphic, in terminal, usually conspicuously bracteate panicles, racemes, or spikes; sepals 3, free or connate; petals 3, free or connate, showy; stamens 6, free or connate, epipetalous or not; anthers 2-celled, versatile, dehiscing introrsely by vertical slits; ovary superior (partly superior in *Pitcairnia*) or inferior, 3-locular, 3-carpellate; placentation axile; ovules numerous, anatropous; style 1; stigmas 3. Fruits berries (*Aechmea, Ananas*) or capsules; seeds naked, winged or plumose, with mealy endosperm.

Bromeliaceae are not confused with members of any other family and may be recognized by their rosulate, acicular or sword-shaped leaves and usually colorfully bracteated inflorescences.

The flowers are usually colorful and often produce copious nectar. Nectaries are usually well protected by floral bracts, and the plants require specialized pollinators with adaptations for reaching the nectar. Though flowers may be green, yellow, white, or blue, the floral and scape bracts of the larger-flowered species are usually red and attract hummingbirds. Some, especially *Vriesia*, which has a more open throat, may also be visited by long-tongued bees.

Diaspore strategy has diverged in two directions in the family, those with superior ovaries having capsular fruits with plumose, wind-dispersed seeds, and those with inferior ovaries having fleshy fruits dispersed by animals, chiefly birds. Among the latter, *Aechmea magdalenae* is eaten by coatis (Kaufmann, 1962), and *A. tillandsioides*, often associated with ant nests, is partly dispersed by ants, I believe.

About 45 genera and over 1,000 species; almost exclusively in the tropics and subtropics of the New World.

AECHMEA R. & P.

Aechmea magdalenae (André) André ex Baker, Handb. Bromel. 65. 1889
Pita, Pingwing

Terrestrial. Leaves in a somewhat spreading rosette, sessile or from a short, stout trunk, linear, acuminate, to 2.5 m long and 5–10 cm wide, pale-lepidote between veins below, the midrib broadly sunken, the margins

---

KEY TO THE TAXA OF BROMELIACEAE

Plants terrestrial:
    Plants cultivated, seldom more than 1 m tall; fruits usually more than 15 cm long . . . .
    . . . . *Ananas comosus* (L.) Merr.
    Plants native, most more than 1.5 m tall; fruits 5–6 cm long . . . .*Aechmea magdalenae* (André) Baker
Plants epiphytic:
    Blades at least in part with margins toothed:
        Blades dimorphic, the outer ones reduced to spinose-serrate, brownish spines, the largest
            blades less than 1.5 cm wide; ovaries superior or only partly inferior; flowering plants
            less than 25 cm tall . . . . *Pitcairnia heterophylla* (Lindl.) Beer
        Blades all ± alike, the largest more than 2 cm wide; ovaries inferior:
            Leaves blotched with silver, usually less than 3 cm wide, often very long and pendent; in-
                florescence simple, pendulous; scape bracts not red . . . . . *Billbergia macrolepis* L. B. Smith
            Leaves solid green, often more than 3 cm wide, usually erect (occasionally pendent in *A.*
                tillandsioides); inflorescence branched, often erect; scape bracts red . . . . .*Aechmea* (in part)
    Blades with margins entire:
        Leaves broad below but quickly tapered to a very long, narrow tip; most of blade very narrow,
            less than 1 cm wide throughout much of its length; petals lacking scales within . . . .
        . . . . *Tillandsia*
        Leaves with apex blunt or abruptly acuminate; most of blade not very narrow in relation to
            base; petals each bearing 2 scales on inner surface:
            Leaves usually less than 20 cm long; scape and floral bracts less than 1 cm long, shorter than
                sepals; scape bracts shorter than internodes . . . . *Catopsis sessiliflora* (R. & P.) Mez
            Leaves usually more than 20 cm long; scape and floral bracts usually much more than 1 cm
                long, longer than sepals; scape bracts usually longer than internodes:
                Flowers polystichous, the flowers and bracts arranged in a whorl or spiral, not in 1 or 2
                    planes . . . . *Guzmania*
                Flowers distichous or secund, the bracts and flowers arranged along either 1 or 2 sides of
                  scape, not in spirals or whorls . . . . *Vriesia*

Fig. 128. *Aechmea magdalenae*

Fig. 129. *Aechmea pubescens*

KEY TO THE SPECIES OF AECHMEA

Plants terrestrial. . . . . *A. magdalenae* (André) Baker
Plants epiphytic:
    Scape bracts below inflorescence entire; leaves dull (especially beneath), their appressed scales
        conspicuous . . . .*A. pubescens* Baker
    Scape bracts below inflorescence prominently toothed; leaves ± shiny, their appressed scales
        inconspicuous:
        Branches of inflorescence bearing dark, needlelike spines; floral bracts forming tubular sheaths
            around flowers . . . . *A. setigera* Schult.
        Branches of inflorescence lacking spines; floral bracts boat-shaped, merely subtending flower
        . . . .*A. tillandsioides* (Mart.) Baker var. *kienastii* (Mez) L. B. Smith

sparsely armed with stout, uncinate spines, the spines near the apex antrorse, those toward middle and base retrorse. Inflorescences simple or compound, on stout terete stalk to ca 4 cm diam, if compound, the heads closely clustered, subequal; heads sessile, hemispherical or globose, rarely more elongate, 10–15 cm diam; floral bracts red, recurved, spinose-serrate, to 6.5 cm long, cinereous-lepidote beneath; flowers sessile, yellow, to 5 cm long, few exserted at a time; calyx lobes sharply pointed, persisting in fruit; petals acute, to 4 cm long, bearing 2 minute truncate scales well above base; ovary inferior. Berries elliptic to ovate, 5–6 cm long, to ca 2 cm diam, fleshy, irregular, angulate, yellow, becoming orange and soft at maturity; mesocarp orange, fibrous, sweet and very tasty; seeds 6–12 or more, very irregular, ca 6 mm long, dark brown, shiny, weakly striate. *Croat 6473, 11341.*

Abundant throughout the forest, often forming dense stands, as along Zetek Trail near 1300. Probably the most objectionable plant on the island because of its fierce spines. Apparently flowering and fruiting chiefly in the rainy season.

Fruits are both colorful and very tasty and are frequently torn open, apparently by coatis or other animals.

Mexico to Ecuador. In Panama, apparently restricted to tropical moist forest on the Atlantic slope and in Darién.

See Fig. 128.

Aechmea pubescens Baker, J. Bot. 17:135. 1879

Epiphyte, to about 1 m tall (usually less than 60 cm tall). Leaves forming a rosette, ligulate, strongly narrowed below middle, acuminate or acute at apex, moderately lepidote to glabrous above, dull and densely lepidote below, to 1.2 m long and 2–4 cm wide, drying moderately thin except for somewhat thickened midrib, the margins prominently armed with straight to recurved spines, the basal sheath unarmed, often purplish at least within. Scapes ± equaling leaves, white-lanate, soon glabrous; scape bracts acuminate, well spaced, bright red, those subtending the branches of the inflorescence spreading; inflorescence usually bipinnate (the lower branches sometimes divided), 10–40 cm long (usually short on BCI), most parts at least at first white-woolly-pubescent; spikes narrow, 8- to 16-flowered; floral bracts ovate, acuminate, 10–13 mm long; flowers sessile; sepals asymmetrical;

petals obtuse, ca 1 cm long; ovary inferior. Berries ca 1 cm long and 6 mm wide, fleshy, dark bluish-green, contrasting sharply with the lighter green floral bracts; seeds many, narrowly oblong, to 3.5 mm long, white, immersed in a sweet, white, watery matrix, the funiculus slender, sticky. *Croat 5826, 10905.*

Occasional, in the forest, much less abundant than *A. tillandsioides* and preferring lower areas out of full sunlight. Flowers principally from January to March. The fruits mature mostly from May to July.

Honduras to Colombia. In Panama, known chiefly from tropical moist forest in the Canal Zone, Bocas del Toro, Colón, Panamá, and Darién; known also from tropical wet forest in Colón and Coclé and from premontane moist forest in Panamá (Juan Diaz).

See Fig. 129.

Aechmea setigera Mart. ex Schult. in R. & S., Syst. Veg. 7(2):1273. 1830

Epiphyte, usually to 1.5 m tall. Leaves ligulate, acute to rounded with a triangular apiculum, to 1 m long, 3.5–7 cm wide, glabrous above, densely lepidote below, the margins with black spines to 1 cm long, the spines antrorse at apex of blade, otherwise spreading, the sheath of blade suborbicular, to 10 cm long, brown at base. Scapes usually arched in flower; scape bracts linear-lanceolate, acuminate, to 20 cm long, bright red; inflorescences densely flowered, bipinnate, cylindrical, often more than 1 m long, the spikes with 2–4 fertile flowers, the uppermost reduced; floral bracts forming a tubular sheath around calyx, the outermost bracts with a slender dark spine at apex to 2.5 cm long; sepals coriaceous, tapered to apex, ca 2 cm long, tightly enveloping lower part of corolla, persistent in fruit; petals pale yellow or greenish-yellow, to 3.5 cm long, spreading above the sepals, each subtended within by 2 fimbriate scales; stamens exserted; style with 3 branches; stigmas oblique, twisted together in bud, their margins and apex pubescent; ovary inferior; nectar copious, accumulating in flower tube. Berries fleshy, 8-seeded; seeds narrowly ovoid, ca 5 mm long and 2 mm wide, reddish. *Croat 8033.*

Occasional, in the forest, usually high in trees. Flowers from February to May. The fruits mature in the rainy season.

A variety of creatures, including ants, lives at the base

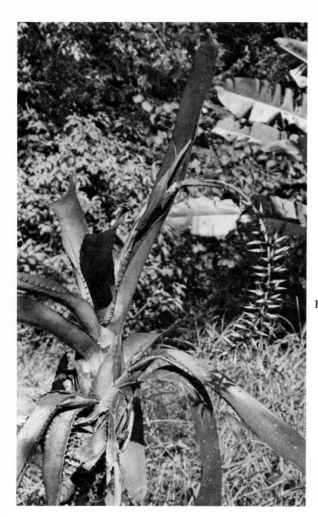

Fig. 130. *Aechmea setigera*

Fig. 131. *Aechmea tillandsioides* var. *kienastii*

of the leaves. The fruits are no doubt chiefly dispersed by birds though they are apparently neither colorful nor very well exposed at maturity.

Panama, Colombia, the Guianas, and Amazonian Brazil. In Panama, known from tropical moist forest in the Canal Zone, Colón, Panamá, and Darién, most abundantly in wettest areas such as the Atlantic coast.

See Fig. 130.

Aechmea tillandsioides (Mart.) Baker var. **kienastii** (E. Morr. ex Mez) L. B. Smith, Caldasia 1(5):5. 1942

Epiphyte. Leaf blades linear, sharply tapered and acuminate at apex, 50–120 cm long, slightly to greatly exceeding inflorescence, 1–7 cm wide, minutely appressed-lepidote, the margins with dark spinelike teeth to 4 mm long, those near apex antrorse, those near base retrorse. Scapes slender or to ca 1 cm thick, white-woolly-pubescent at first; scape bracts widely spaced, lanceolate, 3–10 cm long, bright red, sharply toothed like the leaves; inflorescences digitately or pinnately compound to simple, densely white-woolly when young; primary bracts like scape bracts, suberect, longer or shorter than spikes; spikes spreading to erect, oblong, to 12 cm long and 2 cm wide, distichously 6–30-flowered; rachis square, the wings angled, adnate to base of floral bracts; floral bracts red, imbricate at anthesis, spreading and persistent in fruit, broadly elliptic, acute, mucronulate, 7–15 mm long; sepals asymmetrical, elliptic, mucronulate, 7–10 mm long, ± free; petals acute, mucronulate, 13–16 mm long, bearing 2 fimbriate scales; ovary inferior. Berries oblong-ovoid with a prominent beak, to 2.5 cm long, at first red at apex, white at base, becoming blue overall, the exocarp thin; seeds ca 20–40, ± oblong, to 3.4 mm long, brown, minutely longitudinally striate, surrounded by a sweet watery matrix, each seed bearing a slender, sticky, mucilaginous appendage on one end to 1 cm long. *Croat 6655, 8267, 13806.*

Common in the tops of trees in the full sunlight; less common on exposed branches over the edge of the lake. Inflorescences usually begin to open in the early dry season, but individual plants may flower over a long period, perhaps for the full flowering season. Flowering from at least January to August. Mature fruits have been seen from May to December.

Fruits are no doubt chiefly dispersed by birds. The seeds ooze from the base when the fruit is pressed in the midsection, in the manner of *Anthurium* (21. Araceae). Part of the seeds may stick to the bird's beak, owing to the slender mucilaginous funicles.

The species *A. tillandsioides* ranges from Mexico throughout Central America and much of South America. The variety *kienastii* is known from Mexico, Central America, Colombia, and the Amazon basin. The typical variety is in Colombia, the Amazon basin, and the Guianas. In Panama, apparently restricted to the Atlantic slope; known from tropical moist forest in the Canal Zone and Bocas del Toro and from tropical moist forest, premontane wet forest, and tropical wet forest in Colón.

See Fig. 131.

ANANAS Mill.

Ananas comosus (L.) Merr., Interpr. Rumph. Amb. 133. 1917

Pineapple, Piña

Terrestrial, to ca 1 m tall. Leaves forming a basal rosette; blades sword-shaped, heavily armed on margin. Inflorescences globular or ellipsoid, headlike, borne on a stout stem much shorter than leaves; flowers bisexual, sterile, sessile, spirally arranged and sunken into dense, swollen, pulpy rachis; calyx very short; petals with 2 ligules at base; ovary inferior; style filiform, 3-branched. Fruits syncarps formed of the spiny-toothed floral bracts, the aborted ovaries, and the thickened, fleshy, sweet, edible rachis.

Cultivated at the Laboratory Clearing.

Native to Brazil; widely cultivated in the tropics.

BILLBERGIA Thunb.

Billbergia macrolepis L. B. Smith, Contr. Gray Herb. 114:3, pl. I, f. 6. 1936

Slender epiphyte, often pendent. Leaves linear, acuminate, usually 1–1.8 m long, 1–4 cm wide, densely pale-lepidote, the lower surface marked with small to large, whitish spots, the margins sparsely spinose-serrate below middle, the spines 2–3 mm long, mostly antrorse, sometimes retrorse. Scapes pendent, less than 5 mm thick; scape bracts imbricate near base, more widely spaced higher on scape, lanceolate, acuminate, to 2.8 cm long, membranaceous; inflorescences simple, cylindrical, to 40 cm long, densely covered with a meallike pubescence of tiny trichomes in minute clusters; floral bracts spreading-oblong, the lower ones exceeding flower, the upper ones ovate and shorter than ovary; flowers sessile, suberect, broadly acute and apiculate, to 10 mm long, borne on an epigynous tube ca 3 mm long; petals linear, acute, to 4 cm long and 4 mm wide, bronze-green, spirally recurved at anthesis (*fide* Smith (1944) in *Flora of Panama*), bearing 2 scales at base; ovary inferior, subglobose, ca 15 mm long, coarsely sulcate, densely covered with pubescence similar to that of rachis. Fruits 1.5–1.8 mm long, fleshy; seeds numerous. *Croat 9255.*

Infrequent in the forest, usually moderately high in the canopy, rarely near the ground. Flowers in the dry season. The fruits probably mature in the late dry and early rainy seasons.

Fruits are probably dispersed by mammals or birds.

Costa Rica and Panama. In Panama, known only from tropical moist forest in the Canal Zone and Panamá.

CATOPSIS Griseb.

Catopsis sessiliflora (R. & P.) Mez in DC., Monogr. Phan. 9:625. 1896

C. sessiliflora var. *dioica* L. B. Smith

Dioecious (rarely monoecious) epiphyte; glabrous. Leaves ligulate, arching, divergent, rounded or obtuse and apiculate at apex, to 22 cm long, 1.2–2.5 cm wide, obscurely lepidote, moderately thin, the outermost markedly re-

duced, acute at apex, becoming bractlike, often strongly recurved. Scapes generally much longer than leaves, erect, slender; scape bracts broadly elliptic, apiculate, erect, green, to ca 1 cm long, much shorter than internodes; staminate inflorescences widely spaced, bipinnate, pyramidal, the axis flexuous to geniculate; branches ascending, bearing many widely spaced flowers; rachis very slender; floral bracts ovate, 3–4 mm long, equaling or shorter than sepals; flowers spreading or subspreading; sepals asymmetrical, to 4.5 mm long; petals elliptic, obtuse, 4–6 mm long, greenish-yellow; stamens 6, in 2 unequal whorls, the filaments flattened; anthers slightly longer than broad; ovary ovoid, ca 1.5 mm long; style short; stigmas 3, slender, nonfunctional. Pistillate inflorescences usually simple, sometimes branched, 2.5–11 cm long; floral bracts broadly ovate, obtuse, much shorter than sepals, 3–7 mm long; sepals asymmetrical, suborbicular, 7–8 mm long, the apical edge discolored in age; petals free, lance-ovate, only slightly exserted, not appendaged, white; style very short; ovary superior, broadly ovate. Capsules ovoid, short-beaked, 10–13 mm long; seeds ca 1.5 mm long, slender, with an apical coma, the coma to 3.5 cm long, pale brown, folded 2 or 3 times in the capsule. *Croat 5230, 8263, Shattuck 604.*

Occasional, in the forest; perhaps at one time more abundant on the island. Seasonal behavior uncertain, apparently flowers mostly in the rainy season. Fruits mature in the dry season.

Because of the dimorphic nature of the inflorescences, both sexes of this species have been described as distinct species and were regarded as separate taxa in the *Flora of Panama* (Smith, 1944). Dr. Smith (pers. comm.) concurs in the reduction of the variety *dioica*, the name given the staminate form of the plant.

Southern Mexico to southern Brazil; West Indies. In Panama, known only from tropical moist forest in the Canal Zone and Bocas del Toro.

GUZMANIA R. & P.

Guzmania lingulata (L.) Mez var. **minor** (Mez) L. B. Smith & Pittend., Phytologia 7:105. 1960

G. minor Mez

Glabrous epiphyte. Leaves thin, ascending then spreading, the plant thus often broader than long; sheaths ovate; blades ligulate, acute or acuminate, caudate at apex,

mostly 20–35 cm long and 1–2.5 cm wide, the inner leaves held well above scape. Inflorescences simple, flat-topped and cup-shaped, few-flowered, 15–25 cm long, the outer scape bracts leaflike, the inner ones orange-red, grading to red-orange tipped with yellow-orange, the innermost ones yellow-orange tipped with white, all becoming green in fruit; floral bracts white tinged with yellow, ca 4.5 cm long, scarcely longer than flowers; sepals slender, white; petals linear, ± fused for most of their length, ca 3.5 cm long, yellow tipped with white; stamens 6, about as long as petals; filaments ± flattened, adnate to petals much of their length; anthers ca 6.5 mm long; pollen white, sticky, amassing in large clusters; ovary superior. Capsules ± 3-sided, ca 3 cm long and 5 mm wide, brown, acute at apex, smooth or rugose, the 3 valves splitting at maturity; seeds numerous, 1.5–2 mm long, the seminiferous areas brown, 3–4 mm long, bearing a folded tuft (coma) of brownish trichomes from base, the trichomes fused together at the fold in the middle, the distal part fused to the seminiferous area of the seed. *Croat 10897.*

Common to locally abundant in the forest, especially in moist ravines, usually growing rather close to the ground on small trees or branches. Flowers mostly in the early rainy season, less often from late in the dry season to the middle of the rainy season. The fruits mature from the latter half of the rainy season through the dry season.

The seeds of *G. lingulata* differ in their construction from those of *G. monostachya*. Although both species have the coma refolded at the middle, the distal end in *G. lingulata* is folded to the seed body, whereas it is not in *G. monostachya*. Upon drying, the coma of *G. lingulata* opens to form an airy and globular mass.

Belize to Panama, Colombia, Venezuela, the Guianas, Brazil, and Bolivia; West Indies. In Panama, known from wetter regions of tropical moist forest in the Canal Zone, Bocas del Toro, and Darién, as well as from premontane wet forest in Bocas del Toro, Coclé, Panamá, and Darién and from tropical wet forest in Coclé (La Mesa), Panamá (Cerro Jefe), and Darién (Cerro Pirre).

See Figs. 132 and 133.

Guzmania monostachya (L.) Rusby ex Mez in DC., Monogr. Phan. 9:905. 1896

Epiphyte, to 44 cm high, often with several plants in dense clusters. Leaves ligulate, acute to acuminate, 15–41

Fig. 132. *Guzmania lingulata* var. *minor*

Fig. 133. *Guzmania lingulata* var. *minor*

Fig. 134. *Guzmania monostachya*

cm long, 2–3 cm wide except at the broadly ovate base, glabrous at least in age. Scapes about as long as leaves; scape bracts imbricate, ovate, acuminate, the lower ones green and leaflike, the upper ones increasingly striped with purple, ultimately merging imperceptibly with floral bracts; inflorescences simple, oblong, the lower floral bracts to 4 cm long, those intermediate shorter and becoming tinged with dark violet, the uppermost bright red to orange-red, slightly exserted, all floral bracts soon fading and withering; flowers spirally arranged in upper half of scape; sepals lanceolate, ca 1.5 cm long, brown, indurate; petals white, ca 3 cm long, connate to near apex, rounded and concave at apex, sharply constricted at apex of sepals; stamens 6, included (held well below apex of tube); filaments flattened, adnate to petals, broader than anthers; anthers narrowly tapered to both ends, ca 5 mm long; ovary superior; style shorter than anthers, 3-branched, the branches ca 6 mm long. Capsules to 3.5 cm long, narrowly pointed at apex; valves rough and brown outside, shiny and black within; seeds 1.5–2 cm long, the seminiferous area ca 3.5 mm long, dark reddish-brown, the coma fused midway and reflexed back to seed, the refolded part not fused to the seminiferous area. *Croat 6715, 15058.*

Infrequently encountered in the forest, possibly more common at higher levels of the forest. Flowers in the late dry and early rainy seasons, mostly June and July. The fruits mature during the following dry season, mostly from December to March.

Southern Florida, and Nicaragua to Panama, Colombia, Venezuela, Ecuador, Peru, and Bolivia; West Indies. In Panama, known from tropical moist forest in the Canal Zone, Bocas del Toro, Colón, and Panamá and from premontane wet forest in Bocas del Toro and Panamá.

See Fig. 134.

PITCAIRNIA L'Hér.

Pitcairnia heterophylla (Lindl.) Beer, Bromel. 68. 1857

Epiphyte, the flowering plants 8–12(20) cm tall. Leaves of flowering plants numerous, closely imbricate; blades dimorphic, the outermost reduced to awnlike, spinose-serrate spines, blackened at least at base, the bases ovate to rounded, sheathed, the inner leaves usually to 20 cm long and 15 mm wide, sometimes to 70 cm long and narrower, shorter than inflorescence, becoming decreasingly narrowed and spinose-serrate, softly woolly-pubescent, the innermost glabrous; vegetative stems like those of flowering plants but the inner leaves 50–100 cm long. Scapes scarcely exceeding leaves; scape bracts ovate, acuminate to spinose (especially the lowermost) at apex; inflorescences 3–12-flowered, simple, capitate or subspicate; floral bracts like upper scape bracts, entire, shorter than sepals; pedicels ca 3 mm long; flowers erect; sepals narrowly triangular, ca 3 cm long; petals linear, white (sometimes red), to 5.5 cm long, bearing a saccate, retuse scale well above base; stamens 6, slightly shorter than petals; anthers linear, ca 9 mm long; ovary superior for most of its length. Capsules narrowly ovoid, acute, shorter than sepals; seeds brown, caudate on both ends,

the tails white, flattened, twisted, ca 3 mm long, both directed at 45° angle to the seed, parallel to each other. *Aviles 61.*

Collected once by Aviles, but not seen in recent years. The plant is an inconspicuous one, not common at such low elevations. Flowers and fruits during the dry season.

The seeds are probably wind dispersed though not well adapted for it. The capsules do not open wide, but the tails of the seeds are no doubt hygroscopically active, which could cause them to be slowly loosened from the capsule.

Southern Mexico to Venezuela and Ecuador; probably more prevalent in the Canal Zone before original deforestation. In Panama, mostly at elevations from 500–1,500 m in premontane wet, tropical wet, and premontane rain forests in Chiriquí, Coclé, and Panamá (Cerro Campana).

TILLANDSIA L.

Tillandsia anceps Lodd., Bot. Cab. 8, pl. 771. 1823

Cogollos

Epiphyte. Leaves many, equaling or exceeding inflorescences in length, very narrowly triangular, acuminate, 15–40 cm long, 7–12 mm wide, densely and minutely appressed-lepidote, recurving; sheaths triangular-ovate, with purple stripes. Scapes erect, very short; scape bracts imbricate, mostly ovate and acute; inflorescences simple, elliptic, strongly flattened, 10–15 cm long, to 5.5 cm wide; floral bracts densely imbricate, boat-shaped, acute, to 4 cm long, as many as 20 of them fertile, each subtending a flower; flowers appearing one at a time, soon withering; sepals narrowly lanceolate, acute, ca 3 cm long, keeled; petals more than twice as long as sepals, the claw linear, white, the blade lanceolate-elliptic, blue, recurved at anthesis; stamens deeply included, exceeding the style; ovary superior. Capsules cylindrical, to 2.5 cm long, rugose outside; seeds ca 1.7 cm long, comose, the seminiferous area ca 3 mm long, the coma fused and refolded at middle. *Croat 8503, Shattuck 560.*

Apparently rare, though the number of earlier collections indicates possibly greater abundance. Flowers in the middle to late rainy season. The fruits mature in the dry season.

Guatemala to Colombia, Venezuela, and the Guianas; Trinidad. In Panama, known only from tropical moist forest in the Canal Zone and Darién.

Tillandsia bulbosa Hook., Exot. Fl. pl. 173. 1826

Epiphyte; usually growing in dense clusters. Leaves often exceeding inflorescences in length, covered with fine, appressed-cinerous scales; sheaths orbicular, 2–5 cm long, inflated and sometimes housing ants, abruptly contracted into blade; blades involute-subulate, acuminate, to 30 cm long, 2–7 mm wide, contorted and spreading. Scapes erect; scape bracts often exceeding inflorescence; inflorescences simple to subdigitate, distichous, red or green; spikes spreading, lanceolate, acute, flattened, 2–5 cm long, 2–8-flowered; floral bracts erect, imbricate, ovate, acute,

KEY TO THE SPECIES OF TILLANDSIA

Spikes more than 5 cm long *and* floral bracts conspicuously imbricate:
    Spikes ± elliptic, always solitary; scape usually inconspicuous, often hidden by leaves, less than
        10 cm long; sheath of leaf often with purplish stripes; stamens included . . . . . . *T. anceps* Lodd.
    Spikes ± oblong, frequently digitate (sometimes simple); scape usually conspicuous, more than
        10 cm long; sheath of leaf not with purple stripes; stamens excluded:
        Spikes markedly flattened . . . . *T. fasciculata* Sw. var. *fasciculata*
        Spikes only slightly flattened . . . . *T. fasciculata* Sw. var. *convexispica* Mez
Spikes less than 5 cm long *or* floral bracts not conspicuously imbricate:
    Leaf sheath orbicular and inflated; blades widely spreading, contorted; spikes 2- to 8-flowered . .
        . . . . *T. bulbosa* Hook.
    Leaf sheath not orbicular and inflated; blades ± erect, not contorted; spikes with less than 7 or
        more than 10 flowers:
        Leaves thick, the sheaths about half as long as blades; outer leaves greatly reduced; spikes 4- to
            6-flowered; plants rare or absent . . . . *T. subulifera* Mez
        Leaves thin, the sheaths less than one-fourth the length of blade; outer leaves (except for some
            withered ones) not greatly reduced; spikes with more than 10 flowers when fully open;
            plants common . . . . *T. monadelpha* (E. Morr.) Baker

ca 15 mm long, exceeding sepals, keeled, densely lepidote; flowers sessile; sepals oblong, apiculate, ca 13 mm long; petals violet-blue, 3–4 cm long; stamens exserted; ovary superior. Capsules narrowly cylindrical, 3–4 cm long, the valves 3, roughened outside, brown and shiny inside; seeds ca 2.5 cm long, comose, the coma white, very fine, fused and refolded midway. *Croat 6098, 7048.*

Occasional, in the forest, usually rather high in trees. Flowering time uncertain. Apparently flowering most of the year, but most mature fruits seen in the dry season.

Southern Mexico to Colombia, Venezuela, the Guianas, and Brazil; West Indies. In Panama, known principally from tropical moist forest in the Canal Zone, Bocas del Toro, San Blas, and Darién, but known also from tropical wet forest in Colón (Guásimo).

See Fig. 135.

Tillandsia fasciculata Sw. var. **fasciculata**, Prodr. Veg.
    Ind. Occ. 56. 1788

Epiphyte, 20–100 cm tall (usually less than 60 cm), with 1 to several stems. Leaves rosulate, 2–3 cm broad above the ovate sheath, narrowed to 1 cm or less for most of its length, often red to purple near base, sometimes also apically, gradually narrowed to a sharp point, mostly less than 40 cm long, finely lepidote, the lower ones spreading, the upper ones ± erect, narrower, and merging with scape bracts. Scape bracts gradually tapered to acicular apex; inflorescences simple or digitate; spikes flattened, reddish- to greenish-yellow, mostly 12–18 cm long, to 3.5 cm broad; floral bracts acute, strongly keeled near apex, coriaceous; sepals linear-lanceolate, ca 3.5 cm long, shorter than bracts; petals ca 7 cm long, slender, violet above middle; stamens 6, exserted, violet above middle, 3 of them long and equaling style, the other 3 somewhat shorter; style to ca 8 cm long with the 3 bristly stigmas somewhat twisted together; ovary superior. Capsules 3.5–4 cm long, the 3 valves twisting and spreading widely at maturity, one remaining in the bract, the inner valve surface shiny and black; seeds very numerous, ca 2 cm long, comose, the seminiferous area slender, 2–3 mm

long, brown, the coma weakly fused and strongly folded back near middle, the lower half of the strands held together in 3 or more clusters at apex, the outer half of the strands of each cluster spreading widely. *Croat 6171, 9523.*

Uncommon; seen only along the southeast shore, but no doubt occurring on upper branches of trees in the forest as well. Seasonal behavior uncertain. Possibly flowering and fruiting throughout the year, but most flowering collections have been made in January and July at the beginning of the dry and rainy seasons, respectively. Mature fruits have been seen chiefly in the dry season.

Mexico to Colombia; Florida, West Indies, Trinidad, and the Guianas. In Panama, known from drier parts of tropical moist forest in the Canal Zone and Darién.

See Fig. 136.

Tillandsia fasciculata Sw. var. **convexispica** Mez in
    DC., Monogr. Phan. 9:683. 1896

Like the typical variety, except the spikes to 20 cm long, only slightly flattened. *Chickering 63.*

Collected once on BCI, but not seen in recent years. Seasonality unknown, but probably like that of the typical variety.

Mexico, Guatemala, Belize, and Panama; Jamaica. In Panama, known only from tropical moist forest on BCI.

Tillandsia monadelpha (E. Morr.) Baker, J. Bot.
    25:281. 1887

Epiphyte. Leaves very narrowly triangular, gradually tapered to the ovate sheath, mostly 10–30 cm long, ca 1 cm wide, thin, often purplish in age. Scapes erect, slender, 20–33 cm long, shorter or longer than leaves; scape bracts lanceolate-elliptic; inflorescences simple, oblong, distichous, 4–10 cm long, ca 22-flowered, compressed; rachis flexuous; floral bracts ovate, acute, ca 17 mm long, equaling sepals, at first erect, soon spreading; flowers sessile, ca 3 cm long; sepals equal, short-connate, lanceolate-elliptic, keeled; petals white, rarely seen, the

Fig. 135. *Tillandsia bulbosa*

Fig. 136. *Tillandsia fasciculata* var. *fasciculata*

Fig. 137. *Vriesia sanguinolenta*

blade ± oblong, spreading at anthesis, soon withering; stamens deeply included, exceeding style; ovary superior. Capsules narrowly cylindrical, 4–7 cm long, the 3 valves dark brown and shiny inside, spreading widely and twisting to release the comose seeds; seeds 2.5–3 cm long, the seminiferous part ca 3 mm long, the coma fused and refolded midway. *Croat 8233.*

Common in the forest; epiphytic at various levels. Flowers chiefly in the late dry and early rainy seasons. Buds appear during the early part of the dry season, but open flowers are rarely seen. The fruits mature chiefly in the dry season of the following year, but some may open during the rainy season as they are of mature size long before the end of the rainy season.

Guatemala to Colombia, Ecuador, and the Guianas; Trinidad. In Panama, ecologically variable; known from tropical moist forest in the Canal Zone, Panamá, and Darién and from premontane wet and tropical wet forests in Colón (Santa Rita Ridge) and Panamá (Cerro Campana and Cerro Jefe).

Tillandsia subulifera Mez, Feddes Repert. 16:74. 1919

Epiphyte, 15–19 cm tall. Leaves few, stiffly erect, thick, sheathed about half their length, narrowly tapered and folded together toward apex, appressed-lepidote, sometimes with faint white cross-bands, the inner blades to 18 cm long and ca 5 mm wide above sheath, the outer ones greatly reduced. Scapes erect, mostly concealed by leaves; scape bracts imbricate; inflorescences simple, oblong, distichous, 5–7 cm long, 4–6-flowered; axis geniculate, mostly exposed; floral bracts erect, elliptic, broadly acute, ca 2 cm long, shorter than sepals, incurved and ± keeled at apex; flowers short-pedicellate; sepals free, elliptic, narrowly obtuse, ca 2 cm long; petals tubular-erect, ca 3 cm long, yellowish; stamens exserted; ovary superior. Capsules narrowly cylindrical, ca 6 cm long. *Chickering 62, Shattuck 1166.*

Collected twice on the island, but not seen in recent years; rare if still present. Seasonality unknown.

Panama and Trinidad (further collecting will no doubt show a wider range). In Panama, known only from tropical moist forest in the Canal Zone.

VRIESIA Lindl.

Vriesia gladioliflora (Wendl.) Ant., Wiener Ill. Gart.-Zeitung 5:97. 1880

Epiphyte, but apparently also terrestrial (perhaps accidentally), to 1 m tall. Leaves ligulate, broadly acute or obtuse and apiculate at apex, ca 60 cm long, 6–8 cm wide, purplish when young, becoming deep green, glabrous above, obscurely punctulate-lepidote beneath. Scapes erect, very stout; scape bracts imbricate, elliptic, abruptly acute; inflorescences simple, of many dense flowers, subcylindrical, 20–40 cm long, ca 5 cm wide; floral bracts distichous, coriaceous, erect, imbricate, broadly ovate, obtuse, 4.5–5.5 cm long, equaling or exceeding sepals, green, purplish toward apex; pedicels very short, stout; sepals broadly elliptic, obtuse at apex; petals ligulate, greenish-white, with suborbicular blade, 4–7 cm long, bearing 2 subincised scales at base; stamens included; ovary superior. Capsules to 3.5 cm long, 3-valved, brown, each valve consisting of an outer oblong part and an inner, ± ellipsoid part, the inner black and shiny within; seeds ca 2 cm long, comose, the seminiferous area ca 4 mm long and brown, the coma refolded at its middle, the distal part not fused to the seminiferous area. *Shattuck 524.*

Collected once on the island, but not seen in recent years; apparently rare elsewhere in Panama. Flowering time uncertain, probably mostly in the late rainy season. The fruits mature in the dry season.

Mexico, Guatemala, Belize, Costa Rica, Panama, and Colombia (probably from Mexico to Colombia). In Panama, known only from tropical moist forest in the Canal Zone.

Vriesia heliconioides (H.B.K.) Hook. ex Walp., Ann. Bot. Syst. 3:623. 1852

Epiphyte, rarely more than 40 cm tall (mostly to 30 cm) including inflorescence. Leaves ligulate, acuminate at apex, mostly 12–30 cm long and 3 cm wide, subglabrous, often recurled. Scapes erect, exceeding leaves when fully mature; scape bracts imbricate, the lower ones foliaceous,

KEY TO THE SPECIES OF VRIESIA

Flowers secund (borne along 1 side of inflorescence):
  Floral bracts acuminate . . . . *V. ringens* (Griseb.) Harms
  Floral bracts obtuse to abruptly acute at apex:
    Floral bracts smooth when dried, not closely imbricate, directed to one side with the flowers; much of the rachis exposed; uncommon but locally abundant . . . . *V. sanguinolenta* Cogn. & Marchal
    Floral bracts rugulose when dried, closely imbricate, not directed to one side with the flowers; most of the rachis hidden; plants rare or no longer present on the island . . . . *V. gladioliflora* (Wendl.) Ant.
Flowers distichous (borne along 2 sides of inflorescence):
  Inflorescence long and narrow, subcylindrical at anthesis; floral bracts not red; petals with suborbicular blade . . . . *V. gladioliflora* (Wendl.) Ant.
  Inflorescence flattened, not at all cylindrical; floral bracts red with white margins; petals long and slender . . . . *V. heliconioides* (H.B.K.) Walp.

the upper ones reddish; inflorescences flattened; floral bracts very colorful, distichous, ovate, broadly spreading (much as for *Heliconia*, 31. Musaceae), cuspidate and turned upward at apex, red except white on the upper margins; flowers white, generally appearing one at a time; sepals rigid, keeled, acuminate, to 3.5 cm long, green sometimes tinged with red; corollas to 6.5 cm long, curved outward, fused to calyx near base, bearing 2 slender scales on inner surface above base; stamens 6, included; anthers dark, ca 3.3 cm long; pollen oblong, white, tacky; style ± equaling stamens, the 3 branches ca 4 mm long; ovary superior. Capsules to ca 6 cm long and 7 mm wide, narrowly acuminate at apex, fully exposed by weathered bracts at maturity, the valves spreading but not markedly twisting, black and shiny inside; seeds ca 3 cm long, the seminiferous area brown, the coma white, refolded, the folded area obscurely fused and the distal part free. *Croat 11130.*

Locally abundant in the vicinity of the coves on the eastern side of Burrunga Peninsula; not known elsewhere. Flowers from late May to August. The fruits mature during the dry season of the following year.

Guatemala to Bolivia and southwestern Brazil. In Panama, known only from tropical moist forest in the Canal Zone, Bocas del Toro, Panamá, and Darién.

Vriesia ringens (Griseb.) Harms, Notizbl. Bot. Gart. Berlin-Dahlem 10:801. 1929

Epiphyte, variable in size, to ca 1 m tall. Leaves ligulate, acute to acuminate, mostly 30–90 cm long, 3–6 cm wide, obscurely punctate-lepidote on underside. Scapes stout, erect, held above leaves; scape bracts lanceolate-elliptic, pale green, closely imbricate; inflorescences laxly compound, rarely simple, to 50 cm long; branches suberect, bearing few secund flowers and several imbricate sterile bracts at base; floral bracts broadly ovate, acuminate, 3–6.5 cm long, enfolding flowers and exceeding sepals of at least the lowest flowers, green or brownish, weakly keeled toward apex; sepals elliptic, acuminate, 2.5–3.5 cm long, ca 1.3 cm wide, subcoriaceous; petals white or yellowish, coiling and recurved, bearing 2 spatulate, acute scales at base; stamens exserted; ovary superior. Capsules oblong, ca 4 cm long, black; seeds nearly 2 cm long, basally comose, the seminiferous area brown, ca 3 mm long, the coma refolded and fused at middle, the distal part free. *Shattuck 337.*

Collected once on the island, but not seen in recent years. Flowers in the rainy season. The fruits mature in the dry season.

Costa Rica, Panama, and Colombia; West Indies. In Panama, known from tropical moist forest in the Canal Zone, Colón, and Darién and from tropical wet forest in Colón (Guásimo).

Vriesia sanguinolenta Cogn. & Marchal, Pl. Ornem. 2:52. 1874

Epiphyte, to ca 1 m tall. Leaves ligulate, mostly gradually acuminate, sometimes rounded and long-apiculate, to 1 m or more long, (4.5)8–10 cm wide, obscurely punctate-lepidote. Scapes erect, much longer than leaves, mostly 1.2–2 m long; scape bracts closely imbricate, long-acuminate, coriaceous; inflorescences simple or having few branches, 25–100 cm long; branches suberect, bearing 11 to 19 secund flowers; rachis to 1 cm thick, strongly angled on drying; floral bracts broad, elliptic to suborbicular, abruptly acute, to 5 cm long, directed to one side like the flowers; pedicels stout, ca 1 cm long; sepals ± ovate, mostly obtuse, 2–4.5 cm long, rigid; petals greenish-white, bearing 2 scales at base; ovary superior. Capsules oblong-ellipsoid, pointed at apex, 3.5–6 cm long, dark brown and shiny, tightly enveloped by persistent sepals; seeds 2–2.5 cm long, comose, the seminiferous area ca 4 mm long, brown, the coma white, fused midway and refolded, the distal end free to spread. *Croat 12879.*

Uncommon, but locally abundant in some areas along the shore, usually in trees fairly low over the water; probably also occurring in upper branches of canopy trees. Seasonal behavior uncertain. Flowering probably occurs in the rainy season on BCI; in upland regions of Chiriquí flowers have been seen in February. The fruits mature during the dry season.

Costa Rica, Panama, Colombia; Cuba and Jamaica. In Panama, common along the Atlantic slope in tropical moist and tropical wet forests, but also abundant at higher elevations in lower montane wet forest in Chiriquí.

See Fig. 137.

23. COMMELINACEAE

Erect to sprawling, succulent herbs; stems with enlarged, often rooting nodes. Leaves alternate; petioles sheathing; blades sessile or not, simple, entire. Flowers withering rapidly, bisexual, actinomorphic or zygomorphic, solitary or on simple or paniculate, often umbelliform, helicoid cymes; sepals 3, free, imbricate, equal or 1 much reduced; petals 3, showy, free, equal or the anterior reduced; stamens 3 (*Callisia*), or 6 with 3 frequently longer, showy, and sterile; anthers 2-celled, the cells parallel or divergent, dehiscing longitudinally or by a terminal pore (*Dichorisandra*); ovary superior, 3-locular, 3-carpellate; placentation axile, the ovules 1–6, orthotropous; style 1; stigma 1, capitate or simple. Fruits loculicidal capsules; seeds 1 to several, with copious, mealy endosperm.

Members of the family are most easily recognized by being herbaceous and by having sheathing stipules and usually small, blue, white, or pink, ephemeral flowers sometimes subtended by spathaceous bracts.

Flowers are mostly open and are probably pollinated by small insects. In species with six stamens, the three longer, sterile stamens bear colored trichomes, an adaptation probably acting as an attractant. Many are zygomorphic, with obvious adaptations toward a specialized pollinator.

Several taxa (*Phaeosphaerion, Campelia, Dichorisandra*) have colorful fruits well suited for bird dispersal. The remainder have tiny seeds in small, thin-shelled capsules. Many seeds are no doubt merely spilled.

About 45 genera and 550 species; warm regions of the world.

KEY TO THE SPECIES OF COMMELINACEAE

Ultimate cymes subtended by conspicuous foliaceous bracts, large compared to the cyme:
  Spathe united at base; flowers usually blue or lavender .... *Commelina erecta* L.
  Spathe open to base; flowers white:
    Spathe solitary; inflorescence essentially sessile; capsules white at maturity, pearllike, with a
      thin, hard, outer shell; leaf blades ending abruptly at petiole, less than 3.5 cm wide .....
      .... *Phaeosphaerion persicariifolium* (DC.) C. B. Clarke
    Spathes paired; inflorescences on long, often branched peduncles; fruits purple, fleshy; leaf
      blades tapering onto petiole, often more than 4 cm wide ..... *Campelia zanonia* (L.) H.B.K.
Ultimate cymes with inconspicuous bracts, small compared to inflorescence:
  Flowers blue; sepals more than 7 mm long; plants usually more than 1 m tall, growing in the
    forest.................................... *Dichorisandra hexandra* (Aubl.) Standl.
  Flowers white to pale lavender; sepals less than 4 mm long; plants less than 1 m tall, growing in
    the forest or in clearings:
    Leaf sheaths pilose all over; leaf margins pilose .... *Gibasis geniculata* (Jacq.) Rohw.
    Leaf sheaths pubescent only on margins; leaf margins scabrid:
      Stamens 3, equal; sepals and pedicels eglandular; inflorescences usually with 3 or fewer
        umbelliform clusters per axil, on long peduncles; flowers pink; leaves glabrous, equi-
        lateral at base .... *Callisia ciliata* H.B.K.
      Stamens 6, unequal, the 3 longer ones sterile and bearded; sepals and pedicels often glan-
        dular; inflorescences often with more than 3 umbelliform clusters per axil; flowers
        usually white; leaves usually pubescent, at least on underside, inequilateral at base
        with one side rounded, the other obtuse .... *Tripogandra serrulata* (Vahl) Handl.

CALLISIA Loefl.

Callisia ciliata H.B.K., Nov. Gen. & Sp. 1:261. 1816

Decumbent to erect herb, usually less than 30 cm long. Leaves sheathed and sessile, the sheaths glabrous but with pilose margins, sometimes red-striate; blades oblong-lanceolate to lanceolate, acuminate, obtuse to acute at base, 1.8–8 cm long, 5–17 mm wide, glabrous except the margins scabrid or (rarely) the midrib pubescent on upper surface. Flowers in pedunculate, terminal or sub-terminal, umbelliform clusters, the clusters 1–3(4) per axil, often solitary; peduncles mostly 1.5–3.5 cm long, hispidulous; pedicels and sepals glabrous, eglandular; pedicels ca 3.5 mm long; sepals 3, oblong-ovate, 3.5–4 mm long, bluntly acute at apex, thin, prominently keeled especially at apex; petals 3, pink, ± oblong, ± equaling sepals; stamens 3, ± equaling petals; filaments thin, flattened, broader at base; anther 7–8 mm long, widely separated by an expanded connective, the thecae unequal; ovary ovoid; style simple, ca 0.7 mm long, persisting in fruit. Capsules subglobose, 3-sulcate, 2.5 mm long; seeds 6, 2 per carpel, ca 1 mm long, irregularly trigonous, white, foveolate with a depression on the side opposite the point of attachment. *Croat 6081*.

Infrequent, in the Laboratory Clearing. Flowers and fruits throughout the rainy season and in the early dry season, mostly from August to January.

The species may be confused with *Tripogandra serrulata*, but is distinguished by having flowers with pink petals and three stamens.

The capsule possibly never dehisces; it may be dispersed by birds because of the fleshy, colorful calyx enclosing it at maturity.

Panama and Colombia. In Panama, known from tropical moist forest in the Canal Zone and Panamá and from premontane moist forest in Panamá.

CAMPELIA L. C. Rich.

Campelia zanonia (L.) H.B.K., Nov. Gen. & Sp. 1:264. 1816

Coyontura

Somewhat succulent herb, mostly 1.5 m tall or less; stems mostly 1 cm thick except at enlarged nodes, sometimes with few, widely arching branches. Leaves sessile or obscurely petiolate, usually broadest in middle, gradually tapered to both ends, narrowly acuminate at apex, mostly 25–35 cm long and 4–8 cm wide, widely spaced at base of plant, becoming closely spaced near apex, usually glabrous but the margins pilose near base. Inflorescences lateral, 8–30 cm long, often extending above apex of stem, simple or branched, bracteate at base of branches; peduncles usually branched, glabrous or velutinous, slender, bearing 2 or 4 paired foliaceous bracts at apex, the bracts 1.5–5 cm long; flowers umbellate above bracts; sepals 3, somewhat zygomorphic, fleshy, becoming enlarged, crosier-shaped, and purple, to 9 mm long, enclosing fruit at maturity; petals 3, white, broadly acute or rounded, ca 6 mm long, spreading at anthesis; stamens 6, weakly exserted, 8–9 mm long, often clustered to one side of the flower opposite the style; style bent to one side near apex, sometimes merged with anthers. Capsules fleshy, white to purplish, to 2.3 mm long, enclosed within an attractive, fruitlike calyx, 3-carpellate but usually forming 1 or 2 seeds; seeds black, minutely reticulate, ellipsoid and ± flattened on one side. *Croat 8610*.

Uncommon, throughout the forest, though locally very abundant in the swampy area extending northward from Armour Trail 900. Flowers and fruits throughout the rainy season. The fruits are especially abundant near the end of the rainy season.

The capsules possibly never dehisce. Because they are wholly enveloped by the fleshy colorful calyx at maturity,

they are probably dispersed intact by the birds that feed on them.

Mexico to Bolivia and Brazil; Greater Antilles. In Panama, known from tropical moist forest in the Canal Zone, Bocas del Toro, Panamá, Darién, from premontane moist forest in Coclé, from premontane wet forest in Coclé and Panamá, and from lower montane wet forest in Chiriquí.

COMMELINA L.

Commelina erecta L., Sp. Pl. 41. 1753
  C. elegans H.B.K.
  Cadillo

Trailing or suberect herb; stems branching, glabrous or sparsely pubescent with maroon lines. Leaves mostly widely spaced, sessile or nearly so, the sheaths 8–20 mm long, usually pubescent; blades ± elliptic to lanceolate, gradually acuminate at apex, cuneate to rounded at base, mostly 3–14 cm long, 2–3.5 cm wide, moderately pubescent with longer trichomes above, densely short-pubescent below. Inflorescences terminal or subterminal; spathes ovate, truncate and united on one side, usually puberulent and with longer trichomes at base, usually filled with a watery or gelatinous fluid; peduncles short, each with several short-pedicellate flowers; flowers usually emerging one at a time, protruding above spathe at anthesis, later withdrawing, white to blue or very pale lavender; sepals 3, transparent, one much reduced, the others obovate; petals 3, the anterior one reduced, the posterior ones prominently clawed, the blade mostly 6 mm long and 10 mm wide, often overlapping; stamens 6, 3 reduced, the other 3 unequal in size, the shortest 1 medial with an enlarged anther, the lateral 2 ± equal; style curved, longer than lateral stamens, the stigma held just in front of the anthers of the lateral stamens. Capsules fleshy, dehiscent, 2-valved, to 4 mm long. *Croat 11697.*

The species is common in clearings, where it may form dense stands. Flowers and fruits throughout the year.

Throughout the American tropics and subtropics. In Panama, known from tropical moist forest in the Canal Zone, San Blas, and Panamá, from tropical dry forest in Coclé and Panamá, and from premontane moist forest in the Canal Zone.

DICHORISANDRA Mikan

Dichorisandra hexandra (Aubl.) Standl. in Standl. & Cald., Lista Prelim. Pl. El Salvador 48. 1925

Plant erect or clambering, 1–3 m tall (usually 1.5 m and erect); stems slender, glabrous to minutely puberulent, the lower part with conspicuous green-and-white striations. Leaves sessile or short-petiolate, the sheaths ca 2 cm long, often maroon and somewhat ciliate at apex; blades ovate to ovate-elliptic, abruptly long-acuminate, tapering to base, often ± oblique at base, mostly 10–20 cm long, 2.5–5 cm wide, thin, glabrous to indefinitely puberulous beneath, the margins ± undulate. Inflorescences terminal, paniculate, 3–15 cm long, the branches

subtended by narrow bracts 0.5–3 cm long; flowers few to many; pedicels to 5 mm long; sepals 3, ca 12 mm long; petals 3, blue or white with bluish edges, lanceolate, 7–17 mm long; stamens 6, all fertile, the filaments fused to petals at base, the anthers blue, 4–8 mm long. Capsules obovoid, 1–1.5 cm long; seeds 3–5, orange-arillate. *Croat 11666.*

Frequent in the forest. Flowers in the rainy season (July to November), with the fruits maturing from September to December.

Seeds are probably dispersed by birds.

Guatemala to Brazil and Paraguay. In Panama, known from tropical moist forest in the Canal Zone, Bocas del Toro, San Blas, Chiriquí, Los Santos, Panamá, and Darién; known also from premontane moist forest in the Canal Zone and Panamá and from premontane wet forest in Colón, Chiriquí, Coclé, and Panamá.

GIBASIS Raf.

Gibasis geniculata (Jacq.) Rohw., Farinosae Veg. El. Salv. 143. 1956
  Tradescantia geniculata Jacq.; *Aneilema geniculata* (Jacq.) Woods.

Sparsely villous herb, to ca 60 cm tall; stems ± succulent, creeping at base, rooting at lower nodes, with a single row of trichomes on internodes. Leaves ± sessile; sheaths pilose all over; blades ovate to ovate-lanceolate, 3–7 cm long, sparsely pilose on both surfaces. Inflorescences terminal, slender, to 7 cm long, dichotomously branched; flowers 3-parted; petals ovate, to 3 mm long, white, broader than sepals; sepals ovate-lanceolate, to 3 mm long and 1 mm wide; stamens 6, somewhat shorter than petals, interspersed with and slightly exceeding numerous moniliform trichomes; style ± equaling stamens; stigma minutely bristly. Capsules dry, 3-valved, 3-loculed, to 2 mm long, each locule with 1 or 2 seeds; seeds 1–1.5 mm long, each bearing a prominent funicular scar. *Croat 12799.*

Rare, in isolated, partially open areas along forest trails; often very abundant locally. Flowering and fruiting to some extent throughout the year, but mostly in the late rainy season and in the dry season.

Capsules do not appear to be always dehiscent; some may merely weather open. Seeds are possibly carried away by small insects.

Mexico to Bolivia and Brazil; West Indies. In Panama, known from tropical moist forest in the Canal Zone, Bocas del Toro, Colón, and Darién; doubtlessly elsewhere in tropical moist forest.

PHAEOSPHAERION Hassk.

Phaeosphaerion persicariifolium (DC.) C. B. Clarke in DC., Monogr. Phan. 3:137. 1881

Erect or ± reclining herb, 30–60 cm tall; stems usually unbranched, rooting at nodes, glabrous or sparsely pubescent. Leaf sheaths with ferruginous trichomes; blades elliptic-lanceolate, narrowly acuminate, inequilateral at

base (one side acute, the other rounded), 6–12 cm long, 1.5–3.5 cm wide, pilose on both surfaces, the upper margins long-ciliate. Inflorescences on short, subterminal branches; flowers scorpioid, pedicellate, usually in clusters of 3 or 4, subtended by a broadly ovate spathe 2–3 cm long; sepals 3, free, hyaline; petals 3, white, the anterior petal reduced; stamens 6, the upper 3 sterile with sagittate anthers, the lower 3 fertile. Capsules ovoid, ca 5 mm long, pearly white, the exocarp thin, fragile; seeds 4–6, very small, black. *Croat 14070.*

Rare, along trails in the older forest. Probably flowers in the early dry season, with the fruits maturing later in the dry season or in the early rainy season (February to June).

The shiny, white, fragile fruits are probably attractive to birds, but apparently would not provide anything of food value. Crushing the thin-shelled capsule (possibly even while in flight), birds would scatter the tiny seeds.

Guatemala to Peru. In Panama, known from tropical moist forest in the Canal Zone, Bocas del Toro, Panamá, and Darién, from tropical dry forest in Coclé, and from premontane moist forest in the Canal Zone.

TRIPOGANDRA Raf.

Tripogandra serrulata (Vahl) Handl., Baileya 17:33. 1970

Decumbent to erect herb, rooting at nodes, usually 30–100 cm long, often with red striations on stems and sheaths. Leaves sheathed and sessile, the sheaths glabrous except the margins pilose; blades oblong-lanceolate, acuminate, oblique at base with one side ± rounded to subcordate, the other obtuse, 3–14 cm long, 7–25 mm wide, usually pubescent especially below (on BCI; sometimes glabrous elsewhere), the margins scabrid. Flowers in pedunculate, terminal or subterminal, umbelliform clusters, few to numerous in axils; peduncles 7–30(50) mm long (mostly 10–15 mm long), glabrous to hispidulous; pedicels very short or to 6 mm long, often glandular; sepals 3, ± glabrous, often glandular, broadly oblong, boat-shaped, 3.5–4.5 mm long; petals 3, white or pale lavender, usually equaling sepals; stamens 6, 3 long and sterile with anthers yellow, 3 short and fertile with anthers white; filaments of the sterile stamens with long, jointed, gland-tipped trichomes especially near apex; pistil glabrous; style short; stigma simple, brush-like, ± equaling the short functional anthers in height. Capsules subglobose, ca 2 mm long, 3-carpellate; seeds 2 per carpel, irregularly trigonous, ca 1 mm long, gray, foveolate with a depression

on the side opposite the point of attachment. *Croat 9186.*

Common in the Laboratory Clearing. Flowers and fruits throughout the year, especially in the rainy season. The flowers open in midmorning and close in midafternoon, apparently regardless of weather conditions.

Central Mexico to Peru, Venezuela, and the Guianas; West Indies. In Panama, known from tropical moist forest in the Canal Zone, Bocas del Toro, San Blas, Panamá, and Darién, from premontane wet forest in Chiriquí (Boquete) and Coclé (El Valle), and from tropical wet forest in Colón (Miguel de la Borda).

24. PONTEDERIACEAE

Aquatic, succulent herbs. Leaves opposite or whorled; petioles sheathing, sometimes inflated; blades simple, entire; venation parallel. Flowers bisexual, somewhat zygomorphic, in axillary spikes subtended by spathaceous leaf sheaths; perianth 6-parted, obscurely 2-seriate, bilabiate, the lobes basally connate, bluish, hyaline; stamens 6, obscurely 2-seriate; anthers 2-celled, introrse, dehiscing by vertical slits; ovary superior, 1- or 3-locular; 3-carpellate; placentation axile when 3-locular, parietal when 1-locular; ovules many, anatropous; style 1; stigma 3-lobed, 6-lobed, or moplike and capitate. Fruits achenes, utricles, or many-seeded, 3-celled capsules; seeds with copious mealy endosperm.

Pontederiaceae are distinguished by aquatic habitat and clusters of large, pale blue, somewhat zygomorphic flowers.

The type of heterostyly varies from place to place in *Eichhornia crassipes* (Schulthorpe, 1967) and *Pontederia rotundifolia.* Lowden (1973) found that populations of *P. rotundifolia* often had only one of the three style forms and suspected that many such populations are the result of a single introduction of seed followed by clonal establishment and a certain amount of self-pollination. H. Baker (pers. comm.) reports that such populations usually have flowers with medium-length styles or sometimes long styles, but never short styles. Absence or paucity of one of the style types in different areas suggests that cross-pollination is only rarely achieved (Schulthorpe, 1967).

The utricles are buoyant, and dispersal is believed to be by water currents (Schultz, 1942; Lowden, 1973). Vegetative reproduction is important to population establishment in *Eichhornia*, which produces new stems from the rhizome, and in *Pontederia* (Lowden, 1973). Van der Pijl (1968) reported that in other areas ducks may be active in the dispersal of *Pontederia*. The spinulose peri-

KEY TO THE SPECIES OF PONTEDERIACEAE

Leaves deeply sagittate-cordate; perianth usually less than 2.5 cm long . . . . *Pontederia rotundifolia* L.f.
Leaves not cordate at base; perianth usually more than 3.5 cm long:
    Petioles inflated (especially shorter ones); stems short, condensed, with long roots from base;
        perianth lobes entire . . . . *Eichhornia crassipes* (Mart.) Solms
    Petioles not inflated; stems elongate, rooting from nodes; perianth lobes erose . . . .
    . . . . *Eichhornia azurea* (Sw.) Kunth

gone bases are ideally suited for epizoochorous dispersal, as suggested by Lowden (1973).

Flowers are probably bee pollinated. Lovell (1920) reported that *Halictoides novae-angeliae* feeds on and pollinates only *Pontederia cordata* L. Flowers usually open soon after sunrise, but on cloudy days opening may be delayed (Agharkar & Benerji, 1930).

Seven genera and about 30 species; mostly in the tropics and subtropics but extending into temperate America.

EICHHORNIA Kunth

Eichhornia azurea (Sw.) Kunth, Enum. Pl. 4:129. 1843
Water hyacinth

Aquatic, usually rooted in soil, sometimes free-floating; stems elongate, with long, pendent roots at lower nodes. Petioles to 30 cm long, not markedly inflated; blades round to obovate, rounded or mucronate at apex, truncate to obtuse at base, to 15 cm long, lacking midvein. Spikes many-flowered, axillary, with a subensheathing spathe at base; flowers 6-parted; perianth ± funnelform, 3.5–5.5 cm long, purplish-blue with a yellow spot on the upper, expanded perianth lobe, the tube 2–3 cm long, glandular-pubescent, the lobes erose-margined, decurrent; stamens 6, unequal, the 3 shorter ones included; filaments adnate to tube, glandular-pubescent; style ca 2 cm long, glabrous; stigma red, capitate. Capsules with seeds 1–1.6 mm long, less than 1 mm wide. *Shattuck 409*.

Rare, along the shore. Flowers and fruits throughout the rainy season.

Mexico to Argentina; the Antilles. In Panama, known from tropical moist forest in the Canal Zone and its vicinity and in Bocas del Toro.

Eichhornia crassipes (Mart.) Solms in DC., Monogr. Phan. 4:527. 1883
Water hyacinth, Lechuga de agua

Aquatic, free-floating; stems short and condensed with many long roots at the base. Petioles 2–30 cm long, very inflated in short-petiolate leaves; blades ± round, short-acuminate to rounded at apex, obtuse to truncate at base, 1–6(8) cm long, lacking midvein. Spikes many-flowered, axillary, with a subensheathing spathe at base; perianth 6-parted, ± funnelform, 4–6 cm long, lavender, the upper lobe darker at base and with a yellow spot at center, the tube 1–1.5 cm long, the lobes not erose; stamens 6, unequal, the shorter 3 slightly exserted, the longer 3 well exserted; filaments adnate to tube, glandular-pubescent; ovary ca 8 mm long, 3 mm wide, gradually tapered to apex; ovules minute; style elongate; stigma capitate, mop-like. Capsules with seeds ca 1 mm long, ca 0.5 mm wide. *Croat 7905*.

Rare, around the edge of the island. Flowers from August to March, mostly in the late rainy season and especially in the early dry season.

The species is heterostylous, though a single population may have only one type of style. Too few observations have been made of the *Eichhornia* around the lakeshore

of BCI to ascertain if more than a single style type exists there.

After the fruits have been floating for a day or so, water absorption causes them to split longitudinally from the pressure of the swelling mucilage (Parija, 1934).

United States to Paraguay; the Antilles. In Panama, known from tropical moist forest in the Canal Zone, Bocas del Toro, and Chiriquí.

PONTEDERIA L.

Pontederia rotundifolia L.f., Suppl. Pl. Syst. Veg. 192. 1781
Pickerelweed

Succulent aquatic; stems to 3 m long, prostrate or floating, finally ascending, rooting at lower nodes. Leaves emergent; petioles 17–54 cm long (those of the floral leaf shoot as short as 5 cm long); petiole sheaths with ligule to 7.5 cm long (those in axil of floral shoot to 24 cm long); blades ovate-cordate, reniform or sagittate, 15–34 cm long, about as wide as long, the sinus open or narrow, the veins closely spaced, radiating from apex of sinus but arcuate-ascending near margin. Spikes axillary, many-flowered; peduncles 6–25 cm long, spathate beneath flowers; flowers bilabiate, tristylous, congested at apex of the pilose rachis; perianth blue, 1.5–2.2 cm long, 6-lobed to near base, sparsely pubescent, the hardened bases spinulose-ridged, the largest lobe grooved with a bilobed yellow spot; stamens 6, unequal, in 2 groups of 3 each; anthers introrse, blue; filaments adnate to tube basally, glandular-pubescent near apex; style 1, of 3 possible lengths: (1) short with medium and long stamens; (2) medium with short and long stamens; or (3) long with short and medium stamens; stigma trilobate, each lobe bifid. Fruits ovoid, indehiscent utricles; seed 1, ovoid, 6–7 mm long. *Croat 7065*.

Locally abundant near the dock and rare elsewhere, in marshy areas especially on the south side of the island. Flowers and fruits throughout the year.

Small black bees have been seen visiting the flowers.

Guatemala to Paraguay. In Panama, known only from tropical moist forest in the Canal Zone (especially the Atlantic slope) and in Darién.

25. LILIACEAE

Shrubs. Leaves alternate, spiraled, petiolate; petioles sheathing; blades simple, entire; venation parallel-pinnate. Flowers bisexual, actinomorphic, in terminal, bracteate panicles or racemes; tepals 6, subequal, basally connate; stamens 6, epitepalous, opposite tepals; anthers 2-celled; ovary superior, 3-locular, 3-carpellate; placentation axile; ovules many, anatropous; style 1, simple; stigma small, capitate. Fruits many-seeded berries; seeds with copious endosperm.

Approximately 240–250 genera and 3,500–3,700 species; worldwide but less commonly tropical. The Smilacaceae (26), which are sometimes considered to be

in this family, are not included here. Liliaceae are well represented in Panama (for a tropical area), but only one cultivated species occurs on BCI.

CORDYLINE Commers. ex Adr. Juss.

Cordyline fruticosa (L.) A. Chev. ex Goepp., Nov. Actorum Acad. Caes. Leop.-Carol. Nat. Cur. 25:53. 1855

Taetsia fruticosa (L.) Merr.

Shrub, to 2.5 m tall; stems sparingly branched. Leaves clustered at ends of stems, narrowly elliptic to linear, acuminate, tapered to base and decurrent on petiole, 15–35 cm long, 5–9 cm wide, glabrous, often reddish along margins and midrib. Flowers white, sessile, alternate on branches of open panicles to 30 cm long, each flower subtended by 3 small bracts; perianth tubular, 6-lobed about halfway to base, perpendicular to rachis; stamens 6, included; style slender, included. Berries ± globose, 3–5 mm diam, red at maturity. *Croat 6149.*

Cultivated in the Laboratory Clearing. Probably flowers chiefly in the rainy season (particularly May, July, and September). The fruits mature mostly in the dry season and the early rainy season.

Pollination system unknown. Fruits are well suited for bird dispersal.

Cultivated in tropical America. In Panama, known from tropical moist forest on BCI and in Colón; no doubt cultivated elsewhere also.

26. SMILACACEAE

Rhizomatous, tendriled vines; stems often prickly, woody at the base. Leaves alternate; petioles sheathing; blades simple, palmately veined or pliveined; stipules lacking. Flowers unisexual (dioecious), actinomorphic, in axillary umbels, solitary or in a short-branched series of umbels; tepals 6, free; stamens 6, free; anthers incompletely 2-celled, introrse; ovary superior, 3-locular, 3-carpellate; placentation axile; ovules 1 or 2 per locule, pendulous; styles 3. Fruits 1–3-seeded berries; seeds with hard endosperm.

Closely allied to the Liliaceae (25). These dioecious, mostly herbaceous vines generally have pliveined leaves and umbels of six-parted flowers.

The flowers have the bowl-shaped blossom typical of insect pollination.

Fruits are endozoochorous, dispersed probably by birds (Ridley, 1930) and, since the vines offer little support, by skillful climbers such as monkeys.

Four genera (only 1 in the New World) and about 300 species; widely distributed in tropical and temperate regions.

SMILAX L.

Smilax lanceolata L., Sp. Pl. pl. 1031. 1753

S. domingensis Willd.

Essentially similar to *S. spinosa* except the plant never armed except on older stems; the flowers larger, the tepals more than 4 mm long. *Croat 14955.*

Rare, in the forest.

Croat 14955 may be *S. spinosa* with an unusually large flower. *Shattuck 150* is a sterile plant—probably *S. spinosa,* since it has spines on the young stem. No other collections from BCI can be assigned to this species with certainty. Since fruiting specimens of *S. lanceolata* and *S. spinosa* are not distinguishable, the validity of this species seems in doubt.

Mexico to Panama; West Indies. In Panama, known from tropical moist forest in the Canal Zone, Bocas del Toro, Colón, Veraguas, Herrera, Panamá, and Darién; known also from premontane wet forest in Bocas del Toro and Panamá and from tropical wet forest in Bocas del Toro, Coclé, and Panamá.

Smilax mollis H. & B. ex Willd., Sp. Pl. 4:785. 1806

Vine, somewhat woody at base; stems terete, unarmed, persistently short-pilose or subtomentose. Petioles 1–3 cm long, densely pubescent; blades ovate to ovate-oblong, acute or apiculate at apex, cordate at base, 8–18 cm long, 3–9 cm wide, 7-veined at base, densely pubescent when young, persistently pubescent on underside of veins. Flowers greenish-white, in solitary axillary umbels (rarely in a short, branched series of umbels); peduncles, pedicels, and receptacles densely pubescent, the tepals glabrous except for tuft at apex; staminate umbels with peduncles 1–4 cm long; pedicels 3–4 mm long, inserted on a globose receptacle; tepals 6, narrowly oblong, ca 4 mm long, ca 1 mm wide; filaments 6, 2–3.5 mm long;

KEY TO THE SPECIES OF SMILAX

Plants pubescent, conspicuously so at least on stems, petioles and receptacles . . . . *S. mollis* Willd.
Plants glabrous:
    Peduncles shorter than subtending petioles:
        Tepals less than 2.8 mm long; often spiny on younger stems . . . . *S. spinosa* P. Mill.
        Tepals more than 4 mm long; never spiny except on oldest stems . . . . *S. lanceolata* L.
    Peduncles longer than subtending petioles:
        Staminate flowers sessile, staminate umbels always solitary; pistillate flowers with 3 staminodia; secondary veins subparallel . . . . *S. spissa* Killip & Mort.
        Staminate flowers pedicellate, staminate umbels often in branched series; pistillate flowers with 6 staminodia; secondary veins reticulate . . . . *S. panamensis* Morong

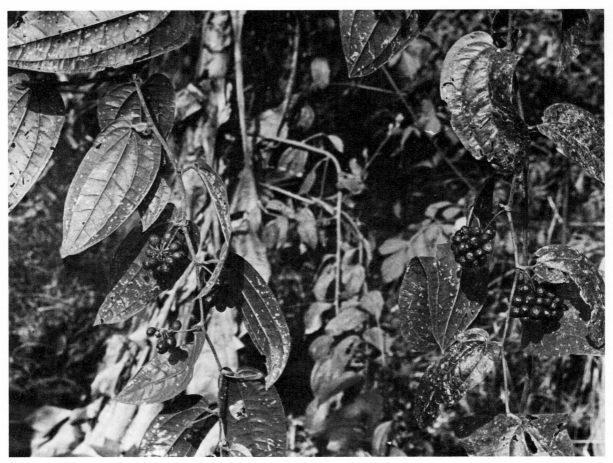

Fig. 138. *Smilax mollis*

Fig. 139. *Smilax spissa*, staminate flowers

anthers ca 1 mm long. Pistillate umbels with peduncles 1–3 cm long, usually longer than the subtending petiole; pedicels 3–5 mm long; perianth segments 6, lanceolate-oblong, ca 3 mm long and 1 mm wide; staminodia 3; styles 3. Fruit clusters ca 5 cm diam; berries globose, 5–10 mm diam, green to red and finally purplish-black at maturity; seeds 1–3, embedded in a fleshy matrix. *Croat 6462.*

Occasional, in the forest. Flowers and fruits throughout the year.

Southern Mexico to Panama. In Panama, known from tropical moist forest in the Canal Zone, Colón, San Blas, Panamá, and Darién, from premontane wet forest in Chiriquí (Boquete), and from premontane rain forest in Darién (Cerro Pirre).

See Fig. 138.

Smilax panamensis Morong, Bull. Torrey Bot. Club 21:441. 1894
  Zarza

Vine, climbing into the canopy, somewhat woody at base, glabrous; stems terete, armed when juvenile with broad-based spines to 2 cm long. Petioles 1–3 cm long; blades lanceolate or lanceolate-oblong to ovate or ovate-oblong, short-acuminate, obtuse to truncate or rounded at base, 10–23 cm long, 3–12 cm wide, 5–7-veined at base, all other veins reticulate. Flowers 6-parted, greenish-yellow or greenish-white, the umbels solitary and axillary or sometimes in a short, branched series; peduncles flattened, longer than subtending petioles; staminate umbels with peduncles 6–20 mm long; pedicels 4–8 mm long; tepals lanceolate, 4–6 mm long, ca 1.5 mm wide; filaments 6, 1–2 mm long; anthers linear, 2–2.5 mm long. Pistillate umbels with peduncles 1–2.5 cm long; pedicels 5–15 mm long; staminodia 6; styles 3. Berries globose, 7–12 mm diam, green to red and finally black, 1–3 seeded. *Croat 6475.*

Rare, in the forest. Flowers and fruits throughout the year.

Guatemala to Panama. In Panama, known from tropical moist forest in the Canal Zone and Darién and from premontane wet and lower montane rain forests in Chiriquí.

Smilax spinosa P. Mill., Gard. Dict. ed. 8, no. 8. 1768
  Zarza

Somewhat woody vine, glabrous, usually armed with sparse, stout, recurved prickles 5–6 mm long, especially on older stems; young stems ± angulate. Petioles 5–20 mm long; blades ovate-oblong or ovate to lanceolate, acute or abruptly acuminate and apiculate at apex, acute to obtuse or cordate at base, 6.5–14 cm long, 2.5–9 cm wide, palmately 5–7-veined at base, occasionally aculeate on petioles and underside of major veins. Flowers 6-parted, cream-colored or greenish, in axillary umbels; peduncles 1–9 mm long, much shorter than subtending petioles; pedicels 4–13 mm long, interspersed with minute bracteoles on the globose receptacle; staminate umbels solitary or in a branched series of 3–5 umbels; tepals

ovate-oblong, 2–2.8 mm long, 1–1.4 mm wide; stamens ca 1.5 mm long; filaments longer or shorter than anthers. Pistillate umbels solitary; tepals oblong-lanceolate, 1.5–2.8 mm long; staminodia 3 or 6, ca 1 mm long; styles 3. Berries subglobose, green to dull red or black at maturity, 5–10 mm long; seeds 1–3, embedded in a fleshy matrix. *Croat 4305, 14955.*

Rare, in the forest. Apparently flowers and fruits throughout the year.

Mexico to Panama; West Indies. Range within Panama uncertain due to confusion with *S. lanceolata;* known from tropical moist forest in the Canal Zone, Bocas del Toro, Veraguas, Los Santos, and Panamá and from premontane moist forest in the Canal Zone. Specimens that are probably this species are seen also from tropical moist forest in Darién and from premontane moist, tropical dry, and premontane rain (Cerro Jefe) forests in Panamá.

Smilax spissa Killip & Mort., Publ. Carnegie Inst. Wash. 461:273. 1936

Vine, ± glabrous overall, the older stems somewhat woody; stems terete, unarmed. Petioles 5–10(20) mm long, sheathed at base, with a pair of tendrils at apex of sheath; blades oblong or ovate-oblong, acuminate, rounded to obtuse at base, 6–18 cm long, 7–8 cm wide, 3-veined at base with an additional inconspicuous pair of submarginal veins, the secondary veins subparallel, connecting the primary veins. Flowers 6-parted, greenish-white, sessile, congested on globose, bracteate receptacle, in solitary, axillary umbels; peduncles longer than subtending petioles, the fruiting peduncles 1.5–2.5 cm long; staminate umbels on peduncles 2–4 cm long, always solitary; tepals unequal, the outer ovate-oblong, 3–4 mm long, ca 1.5 mm wide, the inner narrowly oval, 2–3 mm long, ca 1 mm wide; filaments 1–1.5 mm long; anthers linear, 1.5–2 mm long. Pistillate flowers with pedicels 1–2 mm long (ca 5 mm long in fruit); tepals 6, unequal, the outer narrowly ovate, 3–3.5 mm long, ca 1.8 mm wide, the inner slightly shorter, ca 1 mm wide; ovary ca 2 mm long, subglobose; staminodia 3, minute, oblong, ca 0.5 mm long and ca 0.2 mm wide; styles 3, ca 0.5 mm long, flattened, nearly as broad as long. Berries ± globose, ca 1.5 cm diam, green turning red and finally black, 1–3-seeded. *Croat 8772, 15260, Zetek 5587.*

Rare, in the forest. Flowers and fruits throughout the year.

Costa Rica and Panama. In Panama, known only from tropical moist forest in the Canal Zone.

See Fig. 139.

27. HAEMODORACEAE

Perennial herbs; rhizomes ± elongate, horizontal. Leaves alternate, sheathing the stem at the base; blades simple; venation parallel. Flowers bisexual, actinomorphic, in terminal panicles or scorpioid cymes; tepals 6, free, showy; stamens 3, free, opposite the inner tepals; anthers 2-celled, longitudinally dehiscent; ovary subinferior, 3-locular, 3-carpellate; placentation axile; ovules many;

style simple. Fruits many-seeded berries; seeds with copious endosperm.

Easily confused with the Liliaceae (25).

Some 22 genera with about 120 species; Southern Hemisphere and tropical and North America.

XIPHIDIUM Aubl.

Xiphidium caeruleum Aubl., Hist. Pl. Guiane Fr. 1:33, t. 11. 1775

Palma, Palma del norte, Palmita

Perennial herb, 30–80 cm tall; rhizome creeping then erect, weakly rooted. Leaves linear, equitant, 20–50 cm long, overlapping in two ranks, the lateral margins free only near base, fused toward apex, minutely serrulate near apex. Panicles 7–35 cm long, 3–16 cm wide, the branches few to many, circinately coiled in bud, held ± horizontal at maturity, the flowers on upper side; flowers white, distinctly pedicellate; tepals 6, narrowly ovate, 4–9 mm long, spreading at anthesis, long-persistent; stamens 3, ± erect; filaments flattened, ca 2 mm long; ovary globose; style simple, ca 4 mm long, slightly longer than ovary, becoming ± curved to one side. Berries dull red at maturity, ca 5 mm long; seeds numerous, red, ± rounded, irregularly dented, densely papillate, less than 1 mm diam. *Croat 11785.*

Common along trails in the forest, along the lake margin, and in clearings, often locally abundant. Flowers throughout the rainy season, especially in July and August. The fruits mature in the late rainy and early dry seasons.

This conspicuous plant appears flattened, with equitant, irislike leaves.

Flowers are open (dish-shaped) and seem well suited for pollination by small, unspecialized insects. The fruit, though appearing to be a capsule with three distinct fleshy valves, is apparently never dehiscent, even though the fruit easily separates into three parts. The whole top of the fruit is removed, generally, and at least part of the seeds are removed, apparently by some animal.

Mexico to Bolivia and Brazil; introduced into the West Indies. In Panama, known from tropical moist forest in the Canal Zone, Bocas del Toro, Panamá, and Darién, from premontane moist forest in the Canal Zone and Panamá, and from premontane wet forest in Chiriquí.

28. AMARYLLIDACEAE

Erect or aquatic, glabrous herbs arising from bulbs. Leaves basal, simple; blades linear, succulent; venation parallel; stipules lacking. Flowers 1 to many, bisexual, actinomorphic, in umbels on leafless stalks; perianth tubular, 6-lobed, showy; stamens 6, attached to the perianth, united at base of filaments, 2-celled, versatile, dehiscing longitudinally; ovary inferior, 3-locular; placentation axile; ovules several, anatropous; style 1, slender; stigmas 3, or 1 and trifid. Fruits loculicidal, tardily dehiscent capsules; seeds with fleshy endosperm.

Recognized by their showy, lilylike flowers; often confused with both the Liliaceae (25) and the Iridaceae (30) in other areas, but this should cause no problem on BCI.

Flowers on *Crinum* and *Hymenocallis* are white with very long, slender, tubular corollas and are probably pollinated by hawk moths, though the flowers are open during the day and are no doubt visited by a variety of pollen feeders. Most smaller insects could not effect pollination, however, since the stamens and style are widely separated.

Fruits are probably water dispersed. Van der Pijl (1968) reported that some species of *Hymenocallis* have viviparous seeds that burst through the pericarp and may be deposited on the ground by the reclining shaft of the inflorescence. Seeds of *Crinum* are buoyant because of the lightness of the albumen (Ridley, 1930).

About 90 genera and more than 1,000 species; widespread in warm temperate regions, less numerous in the tropics.

AMARYLLIS L.

Amaryllis belladonna L., Sp. Pl. 293. 1753

Hippeastrum puniceum Urban

Bulbous perennial; bulb globose, producing runners or stolons. Leaves few, basal, strap-shaped, not developing

---

KEY TO THE TAXA OF AMARYLLIDACEAE

Leaves ca 1 cm wide or less; flowers solitary; perianth lobes 3–4.5 cm long, white; capsule 3-valved, ca 0.7 cm long and 1 cm wide; seeds 2 per carpel, flattened, shiny, black . . . . *Zephyranthes tubispatha* Herb.
Leaves more than 2.5 cm wide; flowers 2–15; perianth lobes more than 7 cm long, orange-red or white; fruits not as above:
    Flowers orange-red, 2–4 per inflorescence; cultivated in the Laboratory Clearing . . . . *Amaryllis belladonna* L.
    Flowers white, usually 4–12 or more per inflorescence; plants cultivated or aquatic along lake margin or in wet areas in forest:
        Leaves less than 50 cm long; filaments united into a prominent, exserted, white, staminal tube, green toward apex; usually in wet areas in forest . . . . *Hymenocallis pedalis* Herb.
        Leaves usually 1–2 m long; stamens distinct, not connected by a membrane into an exserted tube, usually purple toward apex . . . . *Crinum erubescens* Ait.

fully until after flowering, bluntly acute at apex, gradually tapered to sheathing base, 30–70 cm long, 1–5 cm wide. Scapes hollow, as long as or longer than leaves, with a pair of lanceolate bracts to 6 cm long at apex; flowers 2–4; pedicels to 7 cm long; perianth funnelform, showy, red-orange, greenish within near base, to 11 cm long and wide, the tube 2–3 cm long, the lobes 6–8 cm long, to 3.5 cm wide, the inner ones narrower; stamens 6, to 10 cm long, shorter than tepals, prominently arched toward center of perianth; filaments orange above middle; anthers ca 7 mm long; style ca 10 cm long; stigma exceeding stamens, capitate but obscurely 3-lobed. Fruits not seen. *Croat 14052.*

Cultivated at the Laboratory Clearing.

Mexico to South America; West Indies; Old World tropics. In Panama, cultivated at various places.

CRINUM L.

Crinum erubescens Ait., Hort. Kew 1:413. 1789
Lirio

Large aquatic herb; bulb ovoid, 7–10 cm diam. Leaves basal, strap-shaped, mostly acute at apex, 1–2 m long and to 8.5 cm wide, thick and succulent, broadly channeled. Flowering scapes stout, usually somewhat shorter than leaves; involucral bracts 7–11 cm long, pointed; flowers 4–12 per inflorescence, sessile; perianth salverform, 15–25 cm long, white, usually tinged with maroon, often persisting in fruit, the tube slender, green, the lobes 6, lanceolate, 7–11 cm long; stamens 6, long-exserted, to 7 cm long or more, white at base, maroon above; anthers versatile, more than 1 cm long; style maroon, exceeding stamens; stigma weakly trilobate. Capsules fleshy, irregularly rounded, to 7 cm long; seeds 2–4, naked, of very irregular size and shape, mostly ca 5 cm long. *Croat 11803.*

Locally abundant, usually in protected areas. Flowering plants are apparently restricted to the shoreline. The species may be found in flower and fruit during any part of the year and shows no marked seasonality, though perhaps the greatest flowering activity occurs during the dry season and the early rainy season. The same population may have both flowering and fruiting inflorescences simultaneously.

The buoyant seeds are usually washed ashore, where they germinate, but they have been reported to germinate while still afloat.

Widespread, often cultivated, in tropical America. In Panama, known from tropical moist forest in the Canal Zone, Bocas del Toro and Darién and from tropical wet forest in Bocas del Toro.

See Fig. 140.

HYMENOCALLIS Salisb.

Hymenocallis pedalis Herb., App. 44. 1821

Glabrous herb; bulb subcylindrical, to 6 cm diam including leaf bases. Leaves basal, numerous, ascending-spreading, tightly ensheathing one another at base, ± succulent, deciduous, the basal part persisting; blades oblong-oblanceolate, acuminate and ± flat near apex, gradually tapered toward the canaliculate, subpetiolar base, 20–45 cm long, 3–7 cm wide. Inflorescences to 15-flowered, usually held well above leaves; scapes to 45 cm long, compressed, glaucous, arising from middle of rosette of leaves, 2.5–3 cm by 1–1.5 cm in cross section; involucral bracts narrowly triangular, 4–10 cm long; flowers several, salverform, white, to 31 cm long, sweetly aromatic at anthesis, the tube to 18 cm long and 5 mm wide, green and glaucous except near apex beneath lobes; perianth lobes 6, lance-linear, to 12 cm long and 7 mm wide, spreading, gradually tapered to apex; stamens 6, widely exserted; filaments 6–7 cm long (free portion), green near apex, white below middle and fused into an exserted funnelform tube; anthers ca 1.5 cm long; pollen orange, tacky; style simple, green, slightly longer than stamens, at first erect, finally deflected to one side; stigma weakly 3-lobed. Fruits ovoid with the corolla persistent, the carpel wall thin; ovules 2, ovoid or almond-shaped, enlarging to ca 3 cm long; seeds buoyant, the seed coat soft, green, leathery. *Croat 17085.*

Rare, in the forest, possibly persisting from previous cultivation, usually in moist places but not in standing water. Though the plants have been seen elsewhere in flower during most of the year, they flower principally during the rainy season. An individual plant transferred to the greenhouses at the Missouri Botanical Garden flowered for about 2 weeks during late October and early November. Although no pollination took place, ovaries developed somewhat for 2–3 weeks and then shriveled. One year later the same plant flowered during the second and third weeks of October. A total of seven flowers were produced, one or two per day; flowers opened in the evening and lasted most of the following day before withering. At anthesis flowers are sweetly aromatic. Artificial pollination between sexual parts of the same flower and between other simultaneously opened flowers resulted in good fruit set. Within 2 weeks of pollination ovules had burst through the wall of the thin carpel. The fruits enlarged until they reached ca 3 cm in length, in about 1 month. The seeds then became loosened. The first fruits fell by the second week of November, the last by the end of November. By this time the inflorescence had drooped until it was lying on the ground.

Crinum erubescens is similar, but its leaves are much larger and ensiform, the colored part of its filaments and the style are maroon rather than green, and the prominent staminal tube of *Hymenocallis* is lacking. The species is confused with *H. littoralis* Salisb., which was reported by Traub (1962) from Mexico, Colombia, and the Guianas. *H. pedalis* is distinguished from *H. littoralis* by having the perianth lobes free from the staminal tube, whereas in *H. littoralis* the perianth lobes are adnate to the base of the staminal tube.

Mature fruits remain green but are very buoyant, suggesting that they probably are loosened by stream currents and carried away for dispersal.

Belize (*fide Yuncker et al. 8843*), Panama to Brazil.

Fig. 140. *Crinum erubescens*

Fig. 141. *Hymenocallis pedalis*

Fig. 142. *Hymenocallis pedalis,*
naked ovules bursting from ovary wall—
those at right appearing like fruits

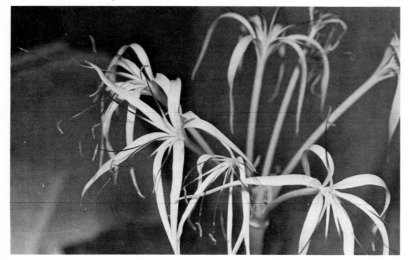

In Panama, known principally from tropical moist forest on the Atlantic slope, but also from tropical moist forest in the Perlas Islands (Panamá).

See Figs. 141 and 142.

ZEPHYRANTHES Herb.

Zephyranthes tubispatha Herb., App. 7:96. 1821

Glabrous, acaulescent herb, 10–50 cm tall. Leaves arising from an underground bulb, this globose, to ca 3 cm diam; blades linear, narrowly acute at apex, to 50 cm long and 3–9 mm wide, moderately thick, canaliculate and often purplish near ground level. Scapes terete, hollow, ± equaling leaves, bearing 1 flower, the single cylindric spathe with a tube to 2 cm long, narrowly attenuate on one side to a deeply cleft tip; pedicel slender, to 5 cm long; flower white; perianth tube less than 5 mm long, the lobes obovate or oblanceolate, acute to bluntly acuminate at apex, 3–4.5 cm long, 6–16 mm wide; filaments 1.5–2.5 cm long, inserted at base of tube; anthers yellow, linear, attached below middle; style slender, 2.5–3.7 cm long, trifid; stigma lobes linear, to 2.5 mm long, narrowly tapered and weakly recurved at apex, the stigmatic surface on upper edge, chiefly in outer two-thirds of lobes. Capsules 3-valved, ca 8 mm long and 10 mm wide, yellowish or brown; seeds 6, 2 per carpel, ± ovoid, ca 6–7 mm long, somewhat flattened, shiny, black. *Croat 14889, 17365.*

Growing in the Laboratory Clearing, apparently not the result of cultivation. Flowers in the early rainy season (May to August). The fruits develop quickly, probably within one month of flowering.

Nothing is known of the dispersal of the seeds. Since they are shiny and black, they might be picked by birds, or, just as likely, they may merely spill out of the capsule.

Southern United States, Guatemala, Panama, Peru, and Argentina; Jamaica; no doubt more widespread. In Panama, known only from tropical moist forest on the Atlantic slope in Bocas del Toro, the Canal Zone, and San Blas (Puerto Obaldía).

29. DIOSCOREACEAE

Twining herbaceous vines arising from tubers. Leaves alternate or partly opposite, petiolate; blades mostly simple, sometimes 3–5-lobed, the margins entire; venation palmate; stipules lacking. Flowers unisexual (dioecious), actinomorphic, in axillary, bracteate fascicles, panicles, or spikes, the inflorescences often several per node; perianth 6-lobed, greenish, persistent; stamens 6, epipetalous, in 2 series, the outer fertile, the inner sometimes reduced to staminodia; anthers 2-celled, introrse or extrorse, basifixed, dehiscing longitudinally; ovary inferior, 3-locular, 3-carpellate; placentation axile; ovules 2 per cell, anatropous; styles 3. Fruits winged or angled, 3-valved, loculicidal capsules; seeds winged, disk-shaped or samaroid, with endosperm.

In Panama the family consists only of *Dioscorea*. The leaves lack parallel veins and thus are not typically monocotyledonous.

Flowers of *Dioscorea* are usually inconspicuous and often poorly known. They are open, with easy access to the nectaries.

The usually winged fruits are much more conspicuous than the flowers. The wing flutters when the wind is blowing, dislodging and dispersing the seeds. Most *Dioscorea* vines flower in the late rainy and early dry seasons, with the fruits maturing in the dry season. Though *D. urophylla* flowers in the rainy season, its seeds are also dispersed in the dry season.

Five, six, or more genera and 600–750 species; widely distributed.

DIOSCOREA L.

Dioscorea is recognized by its scandent habit, by its tiny, dioecious, six-parted, greenish flowers arranged in spikes or racemes, and by its three-winged or three-angled capsules with winged seeds. Among the diagnostic characters of many species is the presence or absence of plate-shaped glands and pellucid lineations. The Smilacaceae (26) are also twining with small green, dioecious flowers, but differ in having the flowers arranged in umbels and the fruits fleshy.

Dioscorea alata L., Sp. Pl. 2:1033. 1753
Ñame de agua, Ñame, Yam, Winged yam

Dioecious, twining vine; stems stout, 4-winged, twining to the right, the wing extending onto petiole. Leaves opposite; petioles 5–14 cm long; blades ovate-cordate, acuminate, with a broad to narrow basal sinus, 10–21 cm long, 6–14 cm wide; glands small, dark, scattered; veins (7)9–11(13), palmate. Staminate inflorescences in narrow axillary panicles 20–30 cm long; pistillate inflorescences stout simple spikes, solitary in axils, to 25 cm long; rachis glabrous; flowers yellow; tepals 6, very thick-

KEY TO THE SPECIES OF DIOSCOREA

Stems and petioles winged:
  Leaf blades with apical lobe deeply 3–5-lobed; rachis pubescent .... *D. trifida* L.f.
  Leaf blades with apical lobe not lobed; rachis glabrous .... *D. alata* L.
Stems and petioles not winged:
  Leaves more than 20 cm long and 12 cm wide, sparsely pubescent only on secondary and tertiary veins of lower surface; inflorescence paniculate; staminate flowers solitary; fruits more than 5 cm wide .... *D. haenkeana* Presl

Leaves less than 20 cm long, less than 12 cm wide, glabrous or with pubescence not restricted to
    secondary and tertiary veins; fruits less than 3 cm wide:
    Blades not pellucid-lineolate; rachis pubescent or with minutely scabrid ribs:
        Rachis and flowers densely pubescent; fertile stamens 6; lower side of blade with veins
            smooth, bearing dark plate-shaped glands near base . . . . *D. sapindoides* Presl
        Rachis merely minutely scabrid on ribs; flowers glabrous; fertile stamens 3; lower side of
            blade with veins minutely scabridulous (with magnification), eglandular . . . .
        . . . . *D. polygonoides* Willd.
    Blades pellucid-lineolate; rachis glabrous and smooth:
        Lower surface of blade eglandular; rachis minutely scabridulous; flowers with 3 stamens
            and 3 bifid staminodia . . . . *D. polygonoides* Willd.
        Lower surface of blade with glands either numerous and all over (best seen under $10\times$
            magnification) or large, whitish, and crateriform near axils of lateral veins below;
            stamens 6:
            Glands of lower surface minute, all over; staminate flowers solitary, sessile; pistillate
                flowers with 6 minute staminodia; fruits oblong, usually tuberculate, acute to obtuse
                at apex, rounded to subcordate at base . . . . *D. urophylla* Hemsl.
            Glands of lower surface crateriform, whitish, restricted to area along veins especially near
                base; staminate flowers on short, stalked fascicles (sometimes appearing solitary as
                only one flower opens at a time); pistillate flowers lacking staminodia; fruits usually
                obovate, smooth, truncate to emarginate at apex, abruptly tapered at base . . . .
        . . . . *D. macrostachya* Benth.

---

ened, ca 3 mm long; style 3, the branches bifid. Capsules 3-winged, suborbicular or broader than long, smooth; seeds winged. *Croat 11616.*

Occasional in the Laboratory and Lighthouse Clearings, apparently persisting from cultivation.

Stems and tubers contain the alkaloid dioscorine, which may paralyze the central nervous system (Blohm, 1962).

Native to southern Asia; introduced generally throughout the tropics.

Dioscorea haenkeana Presl, Rel. Haenk. 1:135. 1827

Dioecious twining vine, climbing into canopy; stems terete, smooth except for some very fine longitudinal ribs (at least when dried). Leaves alternate; petioles 8–12 cm long, ± terete; blades ovate, acuminate, cordate at base (sinus ca 4–6 cm deep), 16–30 cm long, 11–20 cm wide, upper surface glabrous, rugose in age, the lower surface with sparse, inconspicuous, dark, plate-shaped glands; veins 9–11, conspicuous, palmate, the major veins arcuate-ascending, impressed above on older leaves, the secondary veins nearly perpendicular to primary ones, the tertiary reticulations less conspicuous, all veins minutely papillate, the secondary and tertiary veins sparsely pilose. Panicles axillary, about as long as leaves; staminate flowers solitary at nodes, sessile or with a pedicel 1–2 mm long; perianth ca 1.5–2 mm long; bracts sharply acuminate, ca 1.5 mm long; pistillate flowers not seen. Capsules ca 3.5 cm long and 6 cm wide, the inner surface of the valves brown, shiny; seeds disk-shaped, ca 2 cm or more in diam, the seminiferous areas at the center. *Croat 11880a, Montgomery 15.*

Collected on Drayton and Wheeler Trails and near the Laboratory Clearing from the top of the canopy. Flowers have been seen in August and September. Old fruits have been found on the ground in late August, however, which indicates that plants may flower as early as July.

Panama and Peru; this is the first report from Central America.

Dioscorea macrostachya Benth., Pl. Hartweg. 73. 1839

Dioecious twining vine, glabrous; stems twisting to the right. Leaves alternate; petioles slender, 2–7 cm long; blades ovate-cordate, narrowly acuminate, with a broad or narrow sinus at base, 4–12 cm long, 2–9.5 cm wide, usually obscurely pellucid-lineate, with a few, scattered, rounded, crateriform, whitish glands on lower surface especially in axils of veins; veins usually 9, palmate. Flowers unisexual; staminate inflorescences usually solitary in leaf axils, 4–12(30) cm long, simple or branched; flowers solitary, borne on short-stalked fascicles, usually with a single open flower and 1 or more buds; peduncles 0.5–3 mm long, usually with a slender bract about midway; perianth segments violet-purple except sometimes greenish on margins, broadly ovate, to 1 mm long, spreading at anthesis; stamens 6, all fertile, borne on the hypanthium at the center of the flower; anthers about as broad as long, dehiscing extrorsely or upward; filaments stout, shorter than the width of the anther; pistillode stout, slightly exceeding stamens. Pistillate inflorescences usually solitary and unbranched, commonly somewhat longer than staminate ones, usually to 20 cm long; pistillate flowers solitary, ± sessile, to 5 mm long at anthesis, bearing 1 or 2 lanceolate bracts at base; perianth lobes as in staminate flowers; staminodia lacking; styles 3, short and stout, each with a pair of short, divergent branches; ovary sharply 3-angled, soon curved upward along the pendent rachis. Capsules 3-winged, 2–3 cm long, tan, smooth, the valves obovate to broadly oblong, truncate to emarginate at apex, abruptly tapered and short-stipitate at base; seeds flat, ± oblong, 8–12 mm long, 5–8 mm wide, winged all around, brown, one side usually straight, the seminiferous area to 5 mm long and 3.5 mm wide. *Shattuck 680.*

Apparently rare on the island, though moderately common in surrounding areas of the Canal Zone. Flowers throughout most of the year, especially in the very late rainy season and in the dry season (November to March). The species may flower twice per year, since there seems to be a second burst of flowering in the early rainy season. Most fruits probably mature during the late dry and early rainy seasons.

Although Standley reported *D. macrostachya*, the specimen he cited (*Shattuck 582*) was *D. polygonoides*.

Mexico to Panama. In Panama, known principally from tropical moist, premontane wet, and tropical wet forests on the Atlantic slope, but also from premontane wet forest at higher elevations on the Pacific slope in Panamá and from tropical moist forest in Veraguas and Darién; apparently rare in premontane moist forest in the Canal Zone on the Pacific slope.

Dioscorea polygonoides H. & B. ex Willd., Sp. Pl. 4:795. 1806

Dioecious twining vine; stems, petioles, and inflorescences weakly ribbed, the ribs minutely scabrid; stems twisting to the left. Leaves alternate; petioles 1.5–6 cm long; blades ovate-orbicular to ovate, narrowly acuminate, sometimes falcate, deeply to broadly cordate at base, mostly 4–11 cm long, 2.5–8.5 cm broad, obscurely pellucid-lineolate, eglandular; veins (9)11, palmate, the cauline veins 3, the veins of lower surface often minutely scabrid (especially after drying); juvenile plants often with a single ovate-cordate leaf arising from an underground stem, the petiole to 1.3 cm long, the blade often mottled with gray; leaves of the juvenile climbing stems larger than on adults, to 22 cm long and 13.5 cm wide, the petiole to 8 cm long. Inflorescences axillary; rachises minutely scabrid; flowers to 2 mm broad; staminate spikes 2–4 in leaf axils, 10–30 cm long, usually unbranched, sparsley flowered; flowers greenish, sessile, in few-flowered fascicles, the perianth segments ovate, 1–1.3 mm long, united in lower half, each with a ± prominent medial vein; stamens 3, inserted on perianth; anthers extrorse, separated by a broad connective; staminodia 3, bifid at apex, about as long as stamens; ovary rudiment conspicuous. Pistillate spikes solitary or paired, unbranched, less than 15 cm long, the flowers sessile; stamens lacking; ovary ca 3 mm long; styles 3, minute, conical; stigmas bifid. Capsules 3-winged, obovate to oval, 2.5–3 cm long, to ca 2 cm wide, glabrous; seeds winged all around. *Croat 12877, Shattuck 582*.

Occasional, in the forest and at the margins of clearings; probably about as abundant as *D. urophylla* with which it is confused. Seasonal behavior uncertain. Apparently flowers in the early dry season (December to February); the fruiting season is not known.

Standley cited *Shattuck 582*, which is this species, as *D. macrostachya*. Morton in the *Flora of Panama* (1945) reported that the species lacks pellucid lineations. Though they are not as prominent as in *D. urophylla*, pellucid lineations are usually present, at least after the specimens are dried.

Mexico to Peru and Brazil; West Indies. In Panama, known only from tropical moist forest in the Canal Zone and in adjacent Panamá.

Dioscorea sapindoides Presl, Rel. Haenk. 1:33. 1830

Dioecious twining vine, extending into trees in the forest; stems slender, terete, usually sparsely and inconspicuously pubescent, twisting to the right. Leaves alternate, thin; petioles 3–5(8) cm long, pubescent, flattened above, minutely ribbed marginally, somewhat angulate on lower side, both ends thickened; blades ovate-oblong, abruptly acuminate, cordate at base, (4)9–15(19) cm long, 5.5–11 cm wide, glabrous, not pellucid-lineate, bearing dark, scattered, platelike glands on lower surface especially at base in vein axils; veins 9(11) (cauline veins 3), palmate, impressed above, raised below (intervenous areas sometimes becoming achlorophyllous in age); juvenile plants with leaves mottled green above, violet-purple below, the basal lobes sometimes ± truncate. Inflorescences axillary; rachises, pedicels, perianth segments, and ovaries short-pilose; staminate inflorescences 1–3 per axil, the rachis angulate, 5–25(30) cm long, usually simple; flowers in numerous, short, several-flowered racemes, the pedicels short, each subtended by a narrowly triangulate bract; perianth to 2 mm long, 6-lobed ca two-thirds of the way to base, glabrous within; stamens 6, less than 1 mm long, included; filaments connate into a tube. Pistillate inflorescences unbranched, solitary or paired, 6–30 cm long, the flowers 5–6 mm long, the perianth to 2 mm long, lobed to about middle, glabrous within; staminodia 6, minute; styles 3, ca 0.7 mm long, stout; stigmas bifid. Capsules 3-winged, oblong, obtuse to rounded at apex, rounded to subcordate at base, to 2 cm long and 1.1 cm wide, glabrous or sparsely short-pilose along median at maturity; seeds oblong, samaroid, 10–13 mm long, 4–5 mm wide. *Croat 12804, 12837*.

Rare; known only from the vicinity of Fairchild Point and the Lighthouse Clearing. Flowers in the late rainy and early dry seasons (October to December). The fruits probably mature in the dry season.

Mexico, Costa Rica, and Panama; no doubt in other parts of Central America as well. In Panama, ecologically variable; known chiefly from premontane moist forest in the Canal Zone and Panamá, but also from tropical moist forest in the Canal Zone and Panamá and from premontane wet forest in Panamá (on the summit of Cerro Campana).

Dioscorea trifida L.f., Suppl. Pl. Syst. Veg. 427. 1781
Ñame, Yampi, Yam

Dioecious twining vine, nearly glabrous; stems conspicuously 4-winged on the angles. Leaves alternate; petioles to 15 cm long; blades cordate, 3–5-lobed, 6–23 cm long, as broad as long, glabrous to minutely puberulous on upper surface and on the veins below, pellucid-lineolate, not obviously glandular, the lobes acute to acuminate, the terminal lobe with 3 veins; veins 9–13, palmate. Staminate spikes axillary, 2–5 per axil, their flowers greenish,

solitary, short-pedicellate, the ovary pubescent; tepals 6, oblong-lanceolate, to 2.7 mm long; stamens 6, all fertile, ca 1.3 mm long, attached to tepals; anthers introrse; pistillate flowers not seen. Capsules 3-winged, ca 2.7 cm long and 1.7 cm wide (*fide* Morton (1945) in *Flora of Panama*), puberulent. *Croat 7242.*

Cultivated at the Laboratory Clearing. Seasonal behavior uncertain. Probably flowering and fruiting in the dry season.

Native to South America; cultivated widely, at least from Guatemala to Peru and Brazil, and also in the West Indies. In Panama, known from tropical moist forest in the Canal Zone and Panamá and from premontane moist forest in Panamá.

Dioscorea urophylla Hemsl., Biol. Centr.-Amer. Bot. 3:361. 1884

Bejuco de saina

Dioecious twining vine, glabrous; stems usually twisting to the left, with short broad-based spines at the base of larger leaves. Leaves alternate to opposite; petioles 3.5–9.5 cm long (to 14 cm long on juvenile plants); blades ovate, narrowly long-acuminate, truncate to weakly cordate at base, 5–14 cm long, 3–8.5 (14) cm wide, thin, conspicuously pellucid-lineolate and glandular-dotted on lower surface (glands sparse to lacking on upper surface); veins 7–9, palmate. Staminate spikes 1–3 per axil, to ca 50 cm long (usually less than 10 cm), unbranched or branched (especially when long); flowers greenish-yellow, solitary, sessile; tepals oblong, 2–3 mm long, acute at apex, connate at base, spreading at anthesis; stamens 6 (in 2 whorls of 3 each), ca 1.3 mm long, inserted on and held at about same level as tepals; anthers introrse; pollen golden, tacky; pistillode much shorter than filaments, trilobate. Pistillate spikes 1 per axil, 10–20 cm long; flowers sessile; tepals 6, 2–3 mm long; stamens 6 (probably nonfunctional but with some pollen), adnate to lower part of tepal, ca 0.7 mm long; ovary glabrous, ca 3.5 mm long; style single, 1.3–2 mm long, conical; stigmas bilobed, flattened. Capsules 3-winged, oblong-elliptic, acute or apiculate at apex, ± rounded below, 2.3–3 cm long, 1.3–1.6 cm wide, 3-valved, the valves coriaceous, tuberculate (especially medially); seeds samaroid, 2–2.5 cm long (including wing), 6–8 mm wide, brown. *Croat 6704, Foster 1177.*

Occasional, in the forest. Flowers principally in the rainy season (June to September, occasionally as early as April). The fruits mature during the following dry season (December to April).

Capsules of this species are almost identical to those of *D. cymosula* Hemsl., which can be distinguished by having the persisting flower and pedicel densely pilose, whereas *D. urophylla* is wholly glabrous. The species is most easily confused with *D. polygonoides* and *D. macrostachya.*

Panama and Peru; probably in intervening regions as well. In Panama, known from drier regions of tropical moist forest in the Canal Zone (both slopes) and Herrera, Panamá, and Darién; known also from premontane moist forest in the Canal Zone and Panamá and from premontane wet forest in Colón (Santa Rita Ridge).

See Fig. 143.

30. IRIDACEAE

Perennial, rhizomatous herbs. Leaves alternate, basal, sessile; blades simple, entire; venation parallel; stipules lacking. Flowers soon withering, bisexual, actinomorphic, borne in a few-flowered, bracteate cluster arising from below apex of leaflike peduncle; sepals 3, free, showy; petals 3, free, showy; stamens 3; filaments free; anthers 2-celled, extrorse, basifixed, dehiscing by longitudinal slits, coherent to the style; ovary inferior, 3-locular, 3-carpellate; placentation axile; ovules numerous, anatropous; style 1, 3-branched; stigmas transverse, basal. Fruits 3-valved, loculicidal capsules; seeds many, arillate, with copious endosperm.

About 60–70 genera and 800–1,500 species; widely distributed. (Adams (1972) reported the higher number and Willis (1966) the lower number.)

NEOMARICA Sprague

Neomarica gracilis (Herb.) Sprague, Bull. Misc. Inform. 1928:280. 1928

Marica gracilis Herb.

Rhizomatous herb, 50–100 cm tall, glabrous. Leaves basal in a 2-ranked, fan-shaped cluster; blades ensiform, long-acuminate, tapered to base, 30–70 cm long, 1–2 cm wide, the midrib conspicuously raised. Flowering scapes leaflike, the peduncle appearing lateral and overtopped by the terminal leaflike spathe 30–40 cm long; peduncles 1–3 cm long, ensheathed at base with spathes 1–3 cm long; flowers 2–5, 3–5 cm long, soon withering, the tube obsolete; sepals 3, obovate, spreading, white with yellow and brown to purple markings at base; petals 3, much smaller, oblanceolate, spreading to about the middle, then prominently arched upward and inward toward center of flower, prominently reflexed at apex, yellowish with prominent purple transverse markings, especially toward apex; sexual parts not available for study. Capsules oblong, 2–3 cm long, green; seeds many, red, irregular, ca 5 mm long. *Croat 4130.*

Rare, in the forest. Flowers from January to June, principally in March and April. The fruits mature from June to January, principally from August to December.

Flowers are rather specialized and are open for only one day (Bailey, 1949). They seem suited to bee pollination. Flowering stems fall over and root at the tips (*fide* collection label of *Standley 31341*). The fruits have arillate seeds.

Mexico to Brazil. In Panama, known from tropical moist forest in the Canal Zone and Colón and from premontane wet forest in Coclé (El Valle).

31. MUSACEAE

Stout, unbranched, rhizomatous herbs; stems very short or formed by closely imbricated, sheathing petioles. Leaves alternate, distichous (*Heliconia*) or spiral (*Musa*); blades simple, entire; venation pinnate and at about a 90° angle to midrib. Flowers bisexual (*Heliconia*) or unisexual (monoecious in *Musa*), weakly zygomorphic, closely congested and arranged colaterally in racemes

KEY TO THE TAXA OF MUSACEAE

Leaves and flowers spirally arranged; flowers unisexual by abortion, the staminate flowers at the apex of the inflorescence soon falling; calyx tubular; fruit an elongate, indehiscent berry (the banana); plants cultivated at the Laboratory Clearing .... *Musa sapientum* L.

Leaves and flowers distichously arranged (i.e., in 2 ranks); flowers bisexual; calyx not tubular (of 3 sepals, free or partially united to corolla); fruits tardily dehiscent capsules, rounded to oblong, less than 2 cm long .... *Heliconia*

(*Musa*) or of many-flowered cincinni subtended or enveloped by conspicuous, colorful, distichous bracts (*Heliconia*); perianth of 6, ± united segments in 2 series, the inner showy; stamens 6, including 1 sterile staminodium, often petaloid; anthers 2-celled, dehiscing longitudinally; ovary inferior, 3-locular, 3-carpellate; placentation axile; ovules many and anatropous (*Musa*) or solitary and basal (*Heliconia*); style 1; stigma lobed or 3. Fruits fleshy, usually blue capsules dehiscing into 3 cocci (*Heliconia*) or a long berry (*Musa*); seeds with endosperm.

Members of the family are easily confused with the Marantaceae (33) in juvenile or sterile stages, since that family may also have similar, broad, banana-like leaves. The Musaceae have a canaliculate petiole, however, whereas in the Marantaceae the petioles have a hardened callus (round in cross section) on the upper part and are occasionally geniculate.

Flowers are narrowly tubular and produce abundant nectar. Those of *Heliconia* are generally white, green, or yellow, but attract hummingbirds because of their colorful red bracts (Chapman, 1931; Stiles, 1975).

The often blue, fleshy capsules of *Heliconia* are frequently exserted from the bracts by an elongating, often white pedicel. Despite the fact that fruits appear to be capsular, they are taken before dehiscing. The fruits are dispersed by birds. Stiles (1975) recorded 27 birds (chiefly manakins) that disperse these fruits in Costa Rica; the most important were *Manacus candei, Pipra mentalis,* and *Pipro morpha oleagina* (pers. comm.). Aggregated species such as *H. latispatha* are pollinated by territorial hummingbirds (usually not hermits), whereas isolated forest species are pollinated by wandering, nonterritorial hummingbirds (usually hermits) (Linhart, 1973; Stiles, 1975). Nectar production for *Heliconia* increases during the morning and usually levels off by midday, whereas visits by hummingbirds decrease rapidly as the morning progresses (Stiles, 1975).

Five genera and 150 species; widespread in the tropics.

HELICONIA L.   (Platanillo)

Heliconia catheta R. R. Smith, Phytologia 30:65–66. 1975

Caulescent rhizomatous herb, 3–5 m tall, of musaceous habit; stem flattened, ca 5 cm diam by 3 cm or less; stem, petiole, and midrib pruinose. Petioles 50–100 cm long, stout; blades oblong, abruptly acuminate at apex, obtuse-truncate to cordate at base with one lobe longer than the other, 50–135 (175) cm long, (15) 20–36 cm wide, glabrous; veins 4–6 mm apart. Inflorescence pendent from uppermost leaves, hanging 70–100 cm long from stout

KEY TO THE SPECIES OF HELICONIA

All bracts of the inflorescence, except perhaps the lowermost, overlapping to some extent (on fully opened inflorescences):

Bracts overlapping for most of their length, entirely red, mostly blunt or short-acuminate at apex .... *H. mariae* Hook.f.

Bracts overlapping only at their base, usually red but with greenish or yellowish margins, long-acuminate at apex .... *H. wagneriana* O. G. Petersen

Bracts of the inflorescence relatively distant, not overlapping except before fully opened:

Inflorescences pendent:

Bracts with free margins and apex both yellow or yellow-green, the length of the upper bract margins broadly concave; middle bracts more than 10 cm long ...... *H. catheta* R. R. Smith

Bracts red throughout, the length of the upper margins ± straight to convex; middle bracts less than 10 cm long .... *H. pogonantha* Cuf.

Inflorescences erect:

Bracts densely villous .... *H. irrasa* R. R. Smith

Bracts essentially glabrous:

Bracts spirally arranged, broadened at base, concealing flower pedicels ... *H. latispatha* Benth.

Bracts not spirally arranged, narrow at base, exposing flower pedicels:

Bracts red, the lower ones often leafy; flowers yellow; lower portion of ovary and fruit yellow; leaf blades green on underside; plants usually of shoreline and clearings ... . *H. vaginalis* Benth.

Bracts green, the lower ones usually not leafy; flowers pink or red; lower portion of ovary and fruit white; leaf blades maroon on underside; plants cultivated .... *H. metallica* Hook.

Fig. 143. *Dioscorea urophylla*

Fig. 144. *Heliconia irrasa*

Fig. 145. *Heliconia latispatha*

erect peduncle more than 30 cm long and to 1.5 cm diam; most parts minutely pubescent; peduncle and rachis bright red, the rachis strongly flexuous; bracts (10)12–20, spirally arranged, red except for broad yellow band on distal and free margins, usually pubescent only near base, 2–5.5 cm high at their base, tapered to a point, the free margin broadly concave along its length; basal bracts (those nearest peduncle) to 30 cm long (often green), the remaining bracts decreasing in length, 8–10 cm long at apex of inflorescences, sometimes deciduous, the lowermost internodes 5–6 cm long; pedicels ca 1 cm long (to 2.5 cm in fruit); flowers yellow, 6–10 per bract; perianth 4–6 cm long with tips protruding from the bract. Fruits at first yellow, turning blue, ca 2 cm long and less than 1.5 cm wide, puberulent, protruding through bracts at maturity; seeds 3, oblong, black, warty. *Croat 6387.*

Occasional, in clearings, in open areas along trails, and especially along the shore. Flowers from June to September. The fruits mature from July to October. Both flowers and fruits may be present on the same inflorescence.

Distinguished by the pruinose stems and petioles and by the pendent inflorescences with widely spaced, red and yellow bracts. Standley (1933) and Woodson and Schery (1945a) confused this species with *H. platystachya* Baker, a South American species, which has fewer and shorter branch bracts with smaller flowers and occurs at higher elevations.

Known only from Panama in tropical moist forest in the Canal Zone, Colón, Chiriquí, Panamá, and Darién.

Heliconia irrasa Lane ex R. R. Smith, Phytologia 30:68–70. 1975

Caulescent, rhizomatous herb, 1.5–2(3) m tall, of musaceous habit; stem slender, to ca 5 cm diam. Petioles to ca 50 cm long, woolly-pubescent around sheath; blades oblong to oblong-ovate, acuminate, obtuse to rounded at base, 30–100 cm long, 7–30 cm wide; major veins 3–7 mm apart. Inflorescence erect, or nodding in fruit, densely villous-woolly throughout; peduncle 15–30 cm long; rachis moderately flexuous; bracts usually 6–8, the lowermost often leafy, the others spirally arranged, 6–17 cm long, long-acuminate, usually curling on ends, broadest near base, densely villous, 2–6 cm high, orange to yellowish-brown, the free margin red and undulate; internodes of lower bracts 2–3 cm long; flowers 5–6 cm long; perianth yellow, 4–5.2 cm long. Fruits broadly triangular, ca 1 cm long, ca 6 per bract, greenish to yellow, becoming purple. *Croat 12194.*

Occasional, in moist shady areas of the older forest. Flowers primarily from June to September, sometimes as early as April. The fruits mature mostly from September to December.

Distinguished by the erect inflorescence with yellowish, densely pubescent bracts that are often curled on the ends.

Known only from Panama from tropical moist forest in the Canal Zone, Colón, Panamá, and Darién, from premontane moist forest in Coclé and Panamá, and from tropical wet forest in Colón and Coclé (La Mesa).

See Fig. 144.

Heliconia latispatha Benth., Bot. Voy. Sulphur 170. 1846

Platanillo, Guacamaya

Caulescent, rhizomatous herb, 1.5–3 m tall; stem flattened, 4–5 cm wide by 2–2.5 cm thick. Petioles often somewhat pruinose, 30–40 cm long; blades broadly oblong, rounded and cuspidate at apex, rounded to truncate at base, 50–100 cm long, 15–30 cm wide; veins 4–10 cm apart, branching from midrib at ca 90° angle. Inflorescence glabrous, erect; peduncle 15–30 cm long; rachis strongly flexuous, usually greenish; bracts broad at base, long-acuminate, spirally arranged, the lower 1–3 bearing leaves (occasionally with blades to 1 m long), the leafless ones 4–25 cm long, 2–3.5 cm high at base, red or orange except the lower proximal margin yellowish-green; flowers 4–5 cm long, orange to yellow with green margins, lying ± flat in bract and mostly concealed by it. Fruits purple at maturity, broadly oblong, weakly trigonous, 8–10 mm long, emerging from bracts, the pedicels remaining concealed; seeds 3, ca 7 mm long. *Croat 6368.*

Very common in the Lighthouse and Laboratory Clearings and occasional on the shore; a ubiquitous species along roadsides and the railroad right-of-way in the Canal Zone. Flowers throughout the rainy season, rarely during the dry season. Stiles (1975) reported that the species flowers mostly from May to August in Costa Rica. The fruits develop within about 1 month and are abundant on some plants during the middle to late rainy season.

Recognized by its erect, glabrous inflorescence with much-spiraled, broad, red bracts often leafy at the base.

Mexico to Colombia, Venezuela, and Ecuador; possibly elsewhere in South America. In Panama, known from tropical moist forest on both slopes in the Canal Zone and in probably all provinces.

See Fig. 145.

Heliconia mariae Hook.f., J. Proc. Linn. Soc., Bot. 7:69. 1864

Beef-steak heliconia

Caulescent, rhizomatous herb, 3–6 m tall; stem flattened, to 12 cm broad and 4–6 cm thick. Petioles 50–100 cm long above the sheath; blades elliptic, blunt and cuspidate at apex, slightly lobed to rounded at base, to 2 m or more long and 60 cm wide (sometimes smaller on lower part of plant); veins 8–20 cm apart. Inflorescence glabrous or minutely pubescent, pendent on stout peduncle 50 cm or more long, with about 40 cm of peduncle exposed, the fertile part 20–80 cm long, 10–13 cm wide, tapering only slightly to apex; bracts 20–70, red, distichous, all but the lowermost closely imbricated and 3.5–8 cm long, broadest in the middle, acute to acuminate at apex; flowers only slightly protruding from bracts, white at base, reddish at apex, 2.5–4 cm long; ovary white, becoming lavender to blue. Fruits oblong, ca 1.5 cm wide and 9 mm wide, 3-angled, bright blue, exserted on a fleshy white peduncle, 3-seeded; seeds oblong, 8–10 cm long, 3–4 mm wide, covered with a fleshy white matrix. *Croat 8689.*

Known only from the vicinity of the Laboratory Clearing, usually at the edge of the forest and along Creek #8.

Fig. 146. *Heliconia mariae*

Fig. 147. *Heliconia pogonantha*

Probably flowering and fruiting throughout the year, but new inflorescences begin to open in the middle to late rainy season. Some flowers and mature fruits are generally present on the same inflorescence.

The only species of *Heliconia* on the island with a pendent inflorescence and imbricated bracts.

Belize to Panama and northern South America. In Panama, known from tropical moist forest in the Canal Zone (Atlantic slope), Bocas del Toro, Colón, and Darién.
See Fig. 146.

Heliconia metallica Planch. & Lind. ex Hook., Bot. Mag. 88, t. 5315. 1862

Caulescent, rhizomatous herb, 1–3 m tall; stem to ca 5 cm diam. Leaves much like *H. vaginalis* but very dark green above, maroon below, 20–120 cm long and 9–25 cm wide. Inflorescence glabrous, erect; peduncle to 50 cm long; bracts 5–7, green, glabrous or long-tomentose, 2.5–8 cm long, 1.5 cm or less deep; rachis ± straight; flowers reddish-pink, 4–5 cm long; all flowers parts extending above bracts; pedicel and lower half of ovary white, the upper rim becoming green in fruit. Fruits truncate, ca 1 cm long, green at apex, white at base, becoming blue at maturity; seeds ± pyramidal, ca 6 mm long and wide, gray-brown, rough. *Croat 16562.*

Cultivated in the Laboratory Clearing south of the Animal House near the edge of the forest. Introduced to the island from El Valle. Probably flowering all year, but especially in the dry and early rainy seasons.

Distinguished by the greenish bracts, reddish-pink flowers, and purplish lower blade surfaces.

Panama and Colombia, possibly elsewhere in South America. In Panama, known from tropical moist forest in the Canal Zone and Darién, from premontane wet forest in Coclé (El Valle) and Panamá (Cerro Campana), and from tropical wet forest in Coclé (La Mesa) and Darién.

Heliconia pogonantha Cuf., Arch. Bot. Sist. 9:191. 1933
H. pendula Wawra

Caulescent, rhizomatous herb, 3–5(6) m tall. Petioles ca 1 m long; blades broadly oblong, acute to acuminate, rounded at base and sometimes inequilateral, 1.5–2 m long and ca 50 cm wide, sometimes glaucous on underside. Inflorescence pendulous, red, 0.5–1(2) m long, inconspicuously puberulent; rachis flexuous; bracts broadly ovate, 6–18 cm long, red, strongly reflexed at anthesis; flowers 8–30, yellow, ca 4.5 cm long. Fruits pedicellate, ca 1 cm long, bluish. *Croat 12422.*

Cultivated at the Laboratory Clearing; introduced to the island by Dressler from the Caribbean slope. The inflorescence begins to emerge at the beginning of the rainy season, requiring nearly 1 month to become fully extended. The flowers emerge from the bracts before the inflorescence is fully extended and may still be present into the dry season of the following year. Stiles (1975) reported the species to flower all year in Costa Rica. The fruits develop within about a month.

Although possibly confused with *H. longa* (Griggs) Winkl., this species has branch bracts that are fleshy, not coriaceous, and only slightly longer than high. *H. longa* has coriaceous branch bracts that are long and slender (the longer ones four or more times longer than broad).

Costa Rica and Panama. In Panama, known from tropical wet forest along the Caribbean slope and in the Canal Zone.
See Fig. 147.

Heliconia vaginalis Benth., Bot. Voy. Sulphur 171. 1846

Caulescent, rhizomatous herb, 1–3 m tall, glabrous except the midrib of younger leaves minutely furfuraceous. Petioles often marked with maroon and sheathing stem to base of leaves; blades elliptic-oblong, long-acuminate at apex, obtuse to rounded below, mostly 30–100 cm long, 12–20 cm broad (the lowermost often tiny), dark green above, lighter below, the minute cross-veins visible. Inflorescence glabrous, erect; peduncle at least 3 cm long; bracts distichous, 6–9, red, 12–18 mm high at base, the lower ones often leafy and to 40 cm long, the uppermost 3–4 cm long; rachis somewhat flexuous; flowers 4–5.5 cm long, yellow except the tip green; pedicels and lower part of ovary yellow, the upper edge of ovary dark green, protruding above bract, prominently so in fruit. Fruits conspicuously truncate, ca 1 cm high, about as broad as long; immature fruit yellow with apex and peduncle green, mature fruits bluish-purple; seeds minute, brown. *Croat 11137.*

Occasional; particularly common on the shoreline, also common in clearings, rare in the forest. Perhaps the most common *Heliconia* on the island. Flowering and fruiting to some extent throughout the year, especially in the rainy season. Most plants have many fruits in various stages of development between the late rainy and early dry seasons.

Distinguished by the narrow, solid red, leafy bracts and yellow flowers.

Mexico to Ecuador. In Panama, known from tropical moist forest in the Canal Zone, Bocas del Toro, Colón, San Blas, Panamá, and Darién, from premontane wet forest in Colón (Río Providencia) and Panamá (Cerro Campana), and from tropical wet forest in Coclé (La Mesa).

Heliconia wagneriana O. G. Petersen in Mart., Fl. Brasil. 3(3):12. 1890

Caulescent, rhizomatous herb, to 6 m tall; stem flattened. Petioles mostly 40–60 cm long above vagination, the younger ones often with white waxy covering; blades oblong, short-acuminate at apex, broadly rounded to obtuse at base, 90–140 cm long, 20–30 cm broad. Inflorescence erect; peduncle stout, 15–50 cm long; bracts 6–12, usually pale red with greenish or yellowish margins, to 5 cm high at base, mostly 7–21 cm long, gradually long-acuminate, uniformly ascending, the lowermost often leafy; flowers 4–6 cm long, the tube white basally, greenish apically, curved toward apex; fertile stamens 5, exserted, equaling style. Fruits oblong, ca 1.5 cm long, blue, protruding on elongating, tubular, white pedicel at maturity. *Croat 6970.*

Known only from the vicinity of the Laboratory Clear-

ing. Flowers and fruits throughout the year, especially in the dry season. Stiles (1975) reported the species to flower only during the dry season in Costa Rica.

Distinguished by the moderately short, erect, stout inflorescence with closely imbricating bracts. Although described in *Flora Brasiliensis* by Petersen, it is listed there as only from Panama.

Probably pollinated by hummingbirds.

Honduras to Panama; Trinidad. In Panama, known from tropical moist forest in the Canal Zone, Colón, San Blas, Chiriquí and Darién, from premontane wet forest in Coclé, and from tropical wet forest in Bocas del Toro (Río Guarumo).

MUSA L.

Musa sapientum L., Syst. Nat. ed. 10, 1303. 1759
Banana

Caulescent, rhizomatous herb, to 6 m tall; stem round, to 15 cm thick, tapering upward. Leaves spiraled, the stout successive sheaths forming the trunk; blades broadly elliptic, to more than 2 m long, both the petiole and the lower surface of blade white-waxy. Inflorescence a terminal spike, emerging from leaf sheaths, drooping to pendent; flowers unisexual, in flat clusters beneath broad, spirally arranged, purplish bracts, the distal clusters on the rachis functionally staminate, the proximal clusters functionally pistillate; flowers of both sexes with the calyx tubular, the corolla of 1 petal; staminate flowers and bracts deciduous; stamens usually 6, 1 a staminodium. Fruit a fleshy elongate berry.

Cultivated in the Laboratory Clearing.

Juvenile plants are similar to *Heliconia,* but the species may be distinguished by its round stem and petioles, which are continuous with the margin of the blade.

Native to India, but now widely cultivated in the tropics.

32. ZINGIBERACEAE

Perennial rhizomatous or tuberous herbs, usually aromatic. Leaves alternate, spiraled, or distichous (*Renealmia*), sessile to petiolate; petioles sheathed, ligulate; venation closely spaced, parallel-pinnate. Flowers bisexual, zygomorphic, in terminal or scapose, bracteate, spiciform inflorescences; calyx trilobate; petals 3, unequal, showy, the posterior lobe largest, or corolla trilobate; fertile stamen 1; sterile stamens 2, united (opposite the fertile anther) into a petaloid labellum (the labellum small in *Renealmia*); anther 2-celled, dehiscing longitudinally; ovary inferior, 3-locular, 3-carpellate (2-locular and 2-carpellate in *Dimerocostus*); placentation axile; ovules many, anatropous; style 1; stigma capitate or bilamellate. Fruits usually 3-valved, loculicidal capsules (irregularly dehiscent in *Zingiber;* 2-valved and tardily dehiscent in *Dimerocostus*); seeds many, arillate, with copious hard endosperm.

Zingiberaceae are distinguished by their spirally or distichously arranged leaves with nearly parallel veins and by their closely bracteate terminal inflorescences with usually showy flowers bearing a single petaloid stamen. They may also be distinguished by their usually capsular fruits with many small arillate seeds.

Hummingbirds pollinate the small red or red-orange tubular flowers of *Costus pulverulentus* and *C. scaber.* They possibly pollinate *Renealmia cernua* and *R. alpinia,* which have short, tubular, reddish-orange flowers. Other species of *Costus,* including *C. allenii, C. guanaiensis,* and *C. laevis,* are pollinated by both male and female bees of the genera *Euglossa, Eulaema,* and *Euplusia* (Maas, 1972; Dodson, 1966). Both male and female bees of *Euglossa ignita, Eulaema meriana, E. speciosa, E. polychroma,* and *Euplusia surinamensis* visit flowers of *Costus villosissimus* (Dodson, 1966).

The fruits are eaten by animals. The soft, white, generally indehiscent capsules of *Costus* are exposed against the generally red inner bract surface as the bracts reflex at maturity. The seeds are shiny, black, and bear a lacy, sticky aril ideally suited to sticking to the beaks of birds. The fruits are eaten by other animals as well. Monkeys have been observed eating fruits of *Renealmia cernua* (Oppenheimer, 1968). Inflorescences of *Renealmia* are produced near the ground, perhaps more because of the pollination system than because of the dispersal strategy (van der Pijl, 1968).

About 50 genera with 1,500 species; tropics and subtropics.

KEY TO THE TAXA OF ZINGIBERACEAE

Leaves distichous (2-ranked); flowers less than 3 cm long;
  Corolla yellow-green, the lip trilobate; leaves less than 2.5 cm wide; inflorescences spikes bearing closely spaced, green bracts .... *Zingiber officinale* Rosc.
  Corolla yellow or red, the lip not trilobate; leaves mostly more than 2.5 cm wide; inflorescences not as above .... *Renealmia*
Leaves spirally arranged; flowers more than 3 cm long:
  Flowers white, very large; calyx 3–3.5 cm long, becoming brown and extending well above the bracts (at least in fruit); ovary bilocular ....
  .... *Dimerocostus strobilaceus* O. Kuntze subsp. *strobilaceus*
  Flowers variously colored but never entirely white; calyx less than 2 cm long, never exserted, even in fruit; ovary trilocular .... *Costus*

KEY TO THE SPECIES OF COSTUS

All bracts with foliaceous appendages; plants growing in clearings or open areas:
  Flowers yellow; plants densely ferruginous-villous-hirsute, the trichomes on the bracts to 2 mm
    long . . . . *C. villosissimus* Jacq.
  Flowers white, often striped with red in the labellum; plants puberulus, the trichomes to 1 mm
    long . . . . *C. guanaiensis* Rusby var. *macrostrobilus* (K. Schum.) Maas
Bracts lacking appendages or only the lower ones appendaged; plants usually growing in the forest:
  Plants rarely more than 1 m tall; inflorescences fusiform (sharp-pointed at apex); margins of
    bracts dilacerating into fibers; plants common; flowers red, tubular, 50–70 mm long . . . . . .
    . . . . *C. pulverulentus* Presl
  Plants usually 1.5–3.5 m tall; inflorescences ± cylindrical; margins of bracts usually not dilac-
    erating into fibers; plants infrequent:
    Plants densely brownish-pilose; labellum yellowish striped with red or reddish striped with
      yellow; bracts green . . . .*C. allenii* Maas
    Plants glabrous or nearly so:
      Inflorescences 1–3.5(4.5) cm wide; flowers tubular, 3–4 cm long; labellum inconspicuous,
        exceeding corolla at most a few mm, less than 20 mm wide; midrib on upper side of
        leaves densely strigillose; bracts red to orange-red . . . . *C. scaber* R. & P.
      Inflorescences 3–7(9) cm wide; flowers 5–9 cm long; labellum conspicuous, greatly exceed-
        ing corolla, more than 50 mm wide; midrib on upper side of leaves glabrous; bracts
        green to yellowish-green in the exposed part . . . . *C. laevis* R. & P.

COSTUS L.

The genus is distinguished by the unbranched stems and the spirally arranged leaves bearing a prominent sheath, which encircles the stem. In some species the stem itself may be spiraled. Flowers are compacted into a conelike terminal spike (in ours) and appear singly or few at a time over a long period. The fleshy capsules are usually white and contrast sharply with the usually red interior surface of the bracts as the bracts fold out in age to expose the fruits.

Costus allenii Maas, Fl. Neotr. Monogr. no. 8, Zingiberaceae, Costoideae 61. 1972

Plants to 2(3.5) m tall, often in moderately large populations, brownish-pilose on most parts, especially on
sheaths, petioles, and lower midrib of leaves; stem ca
1.5 cm diam. Leaves nearly sessile, oblanceolate-elliptic,
caudate-acuminate, tapered to obtuse base, mostly 25–40
cm long, 9–16 cm wide. Inflorescences ± ovoid, 4.5–
6.5(11) cm long, 3.5–4.5(7) cm wide; bracts rounded,
3–4 cm long and wide, green, the covered portion red,
the upper ones inconspicuously pubescent, the lowermost
conspicuously villous and with reduced leaves, the upper
margin thin, glabrous, and reddish, the callus green;
bracteole 2–2.5 mm long; calyx 8–10 mm long, the lobes
2–4 mm long; corolla white, ca 5 cm long, the lobes obovate, ca 3.5 cm long, 1.5–2.0 cm wide; labellum 5.5–7 cm
long, pale red striped with yellowish-white or pale yellow
striped with red at apex, yellowish-white at base, unequally trilobate, the central lobe heavily tinged with
yellow at base; petals 3, lanceolate-elliptic, obtuse to apiculate, yellowish-white, ± transparent, 1.5–5 cm long;
stamen 3–4.5 cm long, reddish and reflexed at apex;
style borne between the 2 thecae, its broad stigma held

just above the thecae. Fruiting inflorescence and fruits
similar to *C. laevis;* capsules ellipsoid, 8–20 mm long,
glabrous or puberulous at apex; seeds black. *Croat 16198.*

Infrequent, but sometimes locally abundant along
streams or in marshy areas. Apparently flowering and
fruiting during the rainy season.

Most easily confused with *C. laevis* because of the
similar color of the labellum. It is distinguished, however,
by being densely and softly pubescent, while *C. laevis*
is nearly glabrous; *C. laevis* is also a much larger, stouter
plant. Maas (1972) suggested this may be a hybrid of
C. villosissimus and *C. laevis.*

Pollinated by large bees, as are *C. laevis* and *C. guanaiensis.*

Panama, Venezuela, Colombia, and Ecuador. In Panama, known from tropical moist forest in the Canal Zone
and from premontane wet forest in Colón, Coclé (El
Valle), and Panamá (Cerro Jefe).

Costus guanaiensis Rusby var. **macrostrobilus** (K.
  Schum.) Maas, Fl. Neotr. Monogr. no. 8, Zingibera-
  ceae, Costoideae 52. 1972
  C. friedrichsenii sensu Woods. non O. G. Petersen

Plants moderately stout, usually densely puberulent,
2–4(6) m tall; stem straight, usually 2–5 cm diam, the
lower 1.5–2 m leafless. Leaves narrowly ovate to narrowly
obovate, gradually acuminate, narrow and cuneate to
rounded at base, all but the uppermost 30–65 cm long,
7–17 cm wide, shortening at apex and merging inconspicuously with leaflike bracts, puberulent on both surfaces, the trichomes soft, to 1 mm long. Inflorescence
broadly ovoid; bracts green (red at base within), all with
foliaceous tips, 1.5–6 cm long or more, the outermost
pubescent like the leaves, those on inner whorls glabrous;
bracteole 2.5–4 cm long, red; calyx to 16(22) mm long,

Fig. 148. *Costus guanaiensis* var. *macrostrobilus*

Fig. 149. *Costus laevis*

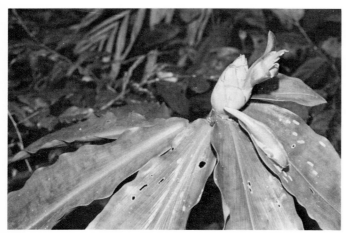

Fig. 150. *Costus pulverulentus*

Fig. 151. *Costus pulverulentus,* fruiting inflorescence

red, with 3 acute lobes; corolla white, 7–8.5(10) cm long, 3–3.5 cm wide, the lobes 3, almost free; stamen 5–8 cm long; labellum 7.5–11 cm long, white with red to pink lines, sometimes white-pubescent within on either side of stamen, its central lobe fimbriate, usually marked with yellow. Capsules 1–3.5 cm long; seeds irregularly 3-sided to round, 2–3 mm long, black, enveloped in a white lacerate aril. *Croat 6388.*

Apparently uncommon; collected at Rear #8 Lighthouse Clearing. Flowers in the rainy season, chiefly from June to September. The fruits are shed usually by the end of the rainy season.

Distinguished by having foliaceous tips on all bracts.

Guatemala and Belize to Guyana, Brazil, and Peru. In Panama, known from tropical moist forest in the Canal Zone, Bocas del Toro, Panamá, and Darién, from premontane moist forest in the Canal Zone and Panamá, from premontane wet forest in Chiriquí, and from tropical wet forest in Panamá.

See Fig. 148.

Costus laevis R. & P., Syst. Veg. 3. 1798

Stout, essentially glabrous plant, 1–3.5(6) m tall; stem 1.5–3 cm diam, the lower part leafless. Petioles lacking or to 3 cm long; blades oblanceolate, gradually long-acuminate, cuneate at base, mostly 44–53 cm long, 10–16 cm wide; basal and apical leaves much smaller, the lowermost reduced to bulbous projections. Inflorescence ovoid to oblong, somewhat narrowed at both ends, 5–10 cm long (to 25 cm long in fruit), 3–7(9) cm wide; bracts glabrate to appressed-pubescent, red, lacking leaflike appendages or with only the lowermost bearing leaflike appendages, the others rounded to broadly oblong and green at apex; bracteole 2–3 cm long; calyx 8–13(20) mm long, red or white, trilobate; corolla 5–9 cm long, white or faintly red and yellowish at base; labellum 6–9.5 cm long, 4–5.5 cm wide at apex, unequally trilobate, the middle lobe curved forward and lacerate, its color variable, usually red to maroon with white lines or white to yellowish with red markings, the center yellowish within; stamen red and recurved at apex, 3.5–5 cm long, the margin minutely toothed. Capsules white, ellipsoid, 1–2.5 cm long, sharply contrasting with inner surface of opened bracts at maturity; seeds subglobose to irregularly 3-sided, 2–3 mm long, black, enveloped in a fleshy lacerate aril. *Croat 15586.*

Occasional, in the forest, especially in open areas of tree falls and along streams. Flowers in the rainy season, chiefly from June to September. The fruits are usually shed by late in the rainy season or in the earliest part of the dry season (December).

Probably pollinated by euglossine bees. At maturity the fruiting inflorescence may be as much as 15 cm broad when the bracts are spread open. This species is the tallest *Costus* on the island. It is often nearly leafless for the first 2 m.

Guatemala south to western South America as far as Bolivia. In Panama, known from tropical moist forest in the Canal Zone, Bocas del Toro, San Blas, Panamá,

and Darién, from premontane wet forest in Bocas del Toro, Coclé, and Panamá, and from tropical wet forest in Colón.

See Fig. 149.

Costus pulverulentus Presl, Rel. Haenk. 1:41. 1830

C. sanguineus Donn. Sm.

Plant usually about 1 m or less tall; stem commonly spiraled, less than 1 cm diam, often reddish at least near base. Leaves moderately few and well spaced (except the uppermost), narrowly obovate, abruptly acuminate at apex, narrowed to broadly-cuneate at base, 12–26(30) cm long, 4–9.5(12) cm wide, essentially glabrous or puberulent on underside, the sheath-ligule extended above the leaf base, its upper side fringed with dilacerating fibers. Inflorescence usually 3–7 cm long, ellipsoid to sharply pointed; bracts 2.5–4.5 cm long, red to orange-red (rarely greenish), the margins dilacerating into fibers, the innermost usually glabrous on outer surface except at margin; bracteole ca 2 cm long, red; flowers tubular, red, 5–7 cm long, ca 1 cm wide; calyx red, shallowly trilobate, 7–10 mm long, persistent in fruit; corolla lobes ± oblong, 1–1.5 cm wide, flared to 4 cm wide above, the center lobe barely exceeding stamen; labellum shorter than stamen, 5-lobed, the 3 central lobes narrow and toothlike. Capsules ellipsoid, 1–2.5 cm long; seeds many, black, 2–3 mm long, irregularly rounded to irregularly 3-angled, covered with a shredded white aril. *Croat 9222.*

Abundant in the forest, particularly along trails, but also in old tree-fall areas. Flowering mostly in the middle to late rainy season, especially from July to October. The fruits are shed by the late rainy season or early dry season. A second but distinctly smaller flush of flowering occurs in the dry season (March and April), with the fruits being shed at the beginning of the rainy season.

Maas (1972) indicated considerable variability in the height of this species, 0.5 to 2.5 m and rarely to 3.5 m tall. On BCI it is rarely more than 1 m tall. The species may be confused only with *C. scaber.*

Pollinated by hummingbirds.

Mexico to Colombia, Venezuela, and Ecuador. In Panama, known from tropical moist forest in the Canal Zone, Bocas del Toro, Chiriquí (Burica Peninsula), Los Santos, Panamá, and Darién and from premontane wet forest in Coclé (El Valle).

See Figs. 150 and 151.

Costus scaber R. & P., Syst. Veg. 2, pl. 3. 1798

Slender herb, mostly 1.5–2(3) m tall, usually in clusters of 3–8 plants; stem usually less than 1 cm thick, sometimes spiraled at apex; stems, lower leaf surfaces, and inflorescence bracts usually inconspicuously puberulent. Leaves ± sessile, mostly narrowly elliptic-obovate, acuminate at apex, narrowed to rounded or subcordate at base, all but the uppermost 10–25(30) cm long and 3–6(11) cm wide; midrib on upperside densely strigillose. Inflorescence 5–12 cm long, 2.5–4.5 cm wide, oblong, bluntly rounded at apex; bracts red to red-orange, obtuse

at apex, 2–3.5 cm long and wide, usually puberulent, the margin thin, glabrous but opaque, only the covered part dilacerating into fibers, the callus prominent, green or yellowish; bracteole 1–1.5 cm long, reddish; flowers orange-red, 3–4 cm long, ca 1 cm wide; calyx reddish, 3–7 mm long, shallowly trilobate, the lobes obtuse; corolla lobes oblong-obovate, thin, ca 1 cm wide, red; labellum red or red-orange, oblong-obovate (when spread out), thick, slightly exceeding corolla lobes, the margins curved upward, partially enclosing stamen; stamen about as long as labellum, usually yellow at apex; style fit in slot between the 2 thecae of anther, obcordate and thickened apically; stigma rounded at apex, consisting of 2 flaps, the margins ciliate. Capsules ellipsoid to subglobose, 7–12 mm long, glabrous to densely puberulous at apex; seeds black. *Croat 14868.*

Occasional, in the old forest, usually in old tree-fall areas or along trails; often in large local stands. Flowers in the rainy season, chiefly in June and July, but some flowering continues throughout the rainy season. The fruits are shed mostly in the late rainy and early dry seasons.

Pollinated by hummingbirds.

Mexico to the Guianas, Brazil and Bolivia; West Indies. In Panama, known from tropical moist forest in the Canal Zone, Bocas del Toro (Isla Colón), Colón (Isla Grande), Panamá, and Darién and from premontane wet forest in Chiriquí, Coclé, and Panamá.

Costus villosissimus Jacq., Fragm. Bot. 55, pl. 80. 1809
  C. hirsutus Presl
  Cañagria, Caña de mico

Plant 2–3(6) m tall, conspicuously ferruginous villous-hirsute especially on blades, upper part of sheaths, and tips of inflorescence bracts. Petioles very short; blades oblong-elliptic, acuminate, cuneate to inconspicuously subcordate at base, 15–30(62) cm long, 5–13(14) cm wide. Inflorescence 5–10 cm long (to 20 cm in fruit); bracts green, 5–8 cm long, the lowermost longer, leaflike, the tips spreading or recurved, acute, foliaceous, unexposed parts of bracts red on both surfaces at least in fruit; bracteole 1.5–3 cm long; flowers with all parts yellow (rarely with white corolla); corolla lobes narrow, 6–7.5 cm long, to 3.7 cm wide; labellum 8–10 cm long; anther to 8 cm long, the thecae well below apex. Capsules white, ellipsoid, 15–25 mm long, minutely puberulous at apex; seeds black. *Croat 11722.*

Very common along roadsides in parts of the Canal Zone adjacent to BCI and probably abundant on the island at one time; seen recently only a few times along the shore and in sign clearings, but to be expected sporadically in larger clearings. Beginning to flower in late March or April, reaching a peak of flowering in June or July, and continuing until October. The fruits mature mostly in the middle to late rainy season.

Pollinated by large bees.

Costa Rica to Colombia, Venezuela, the Guianas, Ecuador, and Peru; West Indies (Jamaica and Lesser Antilles). In Panama, known from tropical moist forest in the Canal Zone, Colón, San Blas, Herrera, Panamá, and Darién, from tropical dry forest in Panamá (Taboga Island), from premontane moist forest in the Canal Zone (Ancón Hill), and from premontane wet forest in Chiriquí, Coclé, and Panamá.

DIMEROCOSTUS O. Kuntze

Dimerocostus strobilaceus O. Kuntze subsp. **strobilaceus,** Rev. Gen. 2:687. 1891
  D. uniflorus (Poepp. ex O. G. Petersen) K. Schum.
  Cañagria

Stout herb, 3–4(6) m tall; stem 1–4 cm diam, often leaning and somewhat spiraled, bearing leaves only near apex. Leaves sessile or on obscure petiole, closely spiraled, oblong-oblanceolate, narrowly acuminate, gradually narrowed to base, 20–50 cm long, 6–10 cm wide, glabrous above, minutely sericeous or glabrous below, the midrib prominently impressed above, raised below; leaf sheaths densely pubescent, ciliate on upper edge. Inflorescence spiciform, ± spirally contorted, 15–30 cm long, 5–9 cm broad; bracts numerous, closely sheathing, 2–3 cm long, usually with a linear callus near apex; bracteole surrounding each flower, bilobed, winged laterally, exceeding bracts; calyx green becoming brown in fruit, coriaceous, sericeous, 3–3.5 cm long at anthesis, 2 of 3 lobes with linear callus; flowers white, usually only 1 open at a time; corolla lobes 3, narrowly oblong, 5.5–7(8) cm long; labellum 8–11 cm long and broad, the center of the tube opposite the stamen yellow; stamen solitary, 3–4 cm long; ovary sericeous, ca 1 cm long at anthesis, to 4 cm in fruit. Fruits tardily dehiscent, 2-celled capsules, green above, whitish below; seeds numerous, closely packed in irregular rows, shiny and black, ca 4 mm long, oblong, angulate, flat above, with a rounded apical depression. *Croat 5100.*

Occasional, in the forest, usually in open places along trails; locally common along the shore. The flowers and fruits are seen all year, though perhaps with heaviest flowering late in the dry season and during the rainy season. Individual plants often have both flowers and fruits in various stages, but occasionally plants are seen with fruits and without flowers or buds.

A conspicuous species, distinguished from *Costus* by its stiff calyces, which turn brown and persist in fruit. The fruits eventually weather and dehisce.

Honduras to Ecuador and Peru. In Panama, known from tropical moist forest in the Canal Zone, Bocas del Toro, Panamá, and Darién and from tropical wet forest in Colón.

RENEALMIA L.f.

Renealmia alpinia (Rottb.) Maas, Acta Bot. Neerl. 24:474. 1975
  R. exaltata L.f.

Large herb, 2.5–4 m tall. Leaves distichous, oblong-elliptic, acute or weakly acuminate at apex, decurrent at

KEY TO THE SPECIES OF RENEALMIA

Inflorescences arising directly from the rhizome (scapose), paniculate (at least not conelike), red; leaves often more than 10 cm wide.............................. *R. alpinia* (Rottb.) Maas
Inflorescences terminal at apex of stem, conelike, orange; leaves usually less than 10 cm wide..... .... *R. cernua* (Sw.) Macbr.

Two other species of *Renealmia* that could be expected on BCI are *R. occidentalis* (Sw.) Sweet and *R. mexicana* Ulotr. ex O. G. Petersen.

---

base onto obscure petiole, to 84 cm long and 19 cm wide (mostly less), the sheath and ligule slightly longer than base of petiole. Inflorescence scapose, arising from rhizome near base of stems, 2–45 cm tall, the base enveloped with imbricate bracts, these 1–10 cm long and deciduous; flowers red, to 2.7 cm long, weakly pubescent, arranged in short panicles to ca 5 cm long; calyx thin, trilobate in upper third, the lobes narrowly acute; corolla trilobate, the lobes rounded, ± cucullate; labellum scarcely exceeding corolla, pale yellow, with 4 short lobes at apex, the inner pair of lobes often acute, the outer pair round; stamen solitary, thick, shorter than corolla; apex of style much thickened. Capsules 3-locular, ± ellipsoid, to 3.5 cm long and 2 cm wide, red, drying with many closely spaced, prominent, longitudinal striations; valves fleshy, 3 mm thick; seeds many, white, 3–4 mm long, bearing a long, sticky, arillate appendage at base. *Croat 6483.*

Infrequent, in the forest, usually occurring in tree-fall areas, not persisting long after being completely shaded. Seedlings of the species are more abundant. Flowers and fruits within the rainy season. The flowers bloom over a long period, chiefly from July to September. Mature fruits may be present on an inflorescence that is still producing flowers.

Belize to Brazil; the Antilles. In Panama, known from tropical moist forest in the Canal Zone and Darién, from premontane wet forest in Chiriquí and Coclé, and from tropical wet forest in Colón.

See Fig. 152.

Renealmia cernua (Sw.) Macbr., Field Mus. Nat. Hist., Bot. Ser. 11:14. 1931

Aromatic herb, 1.5–2.5(3) m tall. Leaves distichous; petioles sheathed, the sheath patterned with small rectangular areas, often extending beyond base of blade; blades oblong to oblong-elliptic, gradually to abruptly acuminate, 4–40 cm long, 2–12 cm wide; veins closely spaced and prominent on underside. Inflorescence sessile or short-pedunculate at apex of stem, 4–15 cm long, conelike, ovoid to oblong; bracts usually orange, persistent, boat-shaped, 2–3 cm long; calyx acutely trilobate, to 1.5 cm long; corolla tubular, trilobate, orange to yellow, to ca 2.5 cm long when mature, slightly exceeding bracts; labellum orange, trilobate, only slightly longer than corolla; style thick at apex. Capsules ovoid to ellipsoid, ± equaling length of fruiting calyx, prominently striate with whitish lines, topped by the persistent green calyx exserted well above the bracts. *Croat 11882.*

Common in all parts of the forest. Flowers from the

late dry season through the rainy season, chiefly from June to September. The fruits mature mostly in the middle to late rainy season, from August to December. Juvenile inflorescences are common by the beginning of the dry season in December.

The fruits are reported as fleshy, irregularly dehiscent capsules, but I have never seen them dehisce. They are probably dispersed by birds.

Guatemala to Peru. In Panama, known from tropical moist forest in the Canal Zone, Bocas del Toro, Colón, San Blas, Chiriquí, Panamá, and Darién, from premontane wet forest in Colón, Chiriquí, Coclé, Panamá, and Darién, from tropical wet forest in Colón and Darién, from premontane rain forest in Coclé, and from lower montane rain forest in Chiriquí.

See Fig. 153.

ZINGIBER Boehm.

Zingiber officinale Rosc., Trans. Linn. Soc. London 8:348. 1807

Common ginger

Herb, ca 1 m tall; rhizome tuberous, aromatic. Leaves distichous, sessile, lanceolate to linear-lanceolate, gradually narrowed to both ends, sheathing at base, 15–30 cm long, 1.5–2.5 cm wide, thin. Inflorescences spicate, usually arising from the root (sometimes from stems), shorter than the stems; peduncles bracteate, their bracts merging with the bracts of the spike; spikes 4.5–7 cm long, 1.5–2.5 cm broad; floral bracts ovate, ca 2.5 cm long, green, the margin scarious; corolla yellow-green, cylindrical, the tube enlarged at apex, the lobes lanceolate, ca 2 cm long, longer than lips; lip deflexed, purple with yellow spots; fertile stamen solitary, the filament short, the connective usually produced into a long spur exceeding the lip. Capsules 3-valved, rupturing irregularly. *Croat 6349.*

Cultivated at Rear #8 Lighthouse Clearing. Flowers apparently in the rainy season.

Probably native to the Pacific islands, but cultivated throughout tropical regions of the world. In Panama, collections have been made only from BCI.

33. MARANTACEAE

Perennial, caulescent or acaulescent herbs, rarely aquatic; roots often tuberous. Leaves alternate, distichous or basal; petioles pulvinate at apex, sheathing the stem at base; blades simple, entire; venation pinnate. Inflorescences

Fig. 152. *Renealmia alpinia*

Fig. 153. *Renealmia cernua*

KEY TO THE TAXA OF MARANTACEAE

Plants stemless, the leaves arising from the ground .... *Calathea* (in part)
Plants with a stem, at least part of the leaves arising from the stem:
  Apex of blade oblique, the tip markedly offset from center; pulvinus with annular ring at base . .
    .... *Ischnosiphon pruinosus* (Reg.) O. G. Petersen
  Apex of blade not markedly oblique, the tip not offset from center; pulvinus without annular
    ring at base:
    Blades pruinose beneath (covered with white waxy layer; on older specimens, the waxy layer
      is no longer visible):
      Leaves often more than 50 cm long and 25 cm wide, obtuse or rounded at apex; spikes 3–5
        cm broad, the bracts nearly as broad as high .... *Calathea lutea* (Aubl.) Schult.
      Leaves seldom more than 35 cm long and 20 cm broad, acuminate to caudate at apex; spikes
        less than 1 cm broad, the bracts much higher than broad ....
        .... *Ischnosiphon leucophaeus* (Poepp. & Endl.) Koern.
    Blades not pruinose beneath (sometimes weakly pruinose in *Thalia geniculata*):
      Spikes markedly flattened and more than 4.5 cm broad ...... *Calathea insignis* O. G. Petersen
      Spikes ± cylindrical or somewhat flattened and less than 1 cm wide:
        Plants aquatic or at least occurring at edge of shoreline marshes; flowers purple, in open
          paniculate inflorescences; uncommon .... *Thalia geniculata* L.
        Plants never aquatic, generally within forest or in clearings:
          Plants with numerous leaves forming a close spiral at the apex of a stout, canelike stem;
            most blades less than 12 cm broad; bracts orange, in open paniculate inflores-
            cences, deciduous .... *Stromanthe jacquinii* (R. & S.) Kenn. & Nic.
          Plants never with leaves forming a close spiral at the apex of a stout stem; blades usu-
            ally more than 12 cm broad; bracts persistent in capitate inflorescences:
            Flowers cream-colored; lower surface of blade softly pilose; major veins scarcely
              raised (not pleated), midrib pubescent in furrow above; sheath glabrate ....
              .... *Calathea marantifolia* Standl.
            Flowers purple; lower leaf surface glabrous; major lateral veins raised, leaf pleated;
              midrib glabrous above; sheath densely pilose ... *Calathea latifolia* (Link) Klotzsch

bracteate panicles or racemes, often capitate, usually terminal, sometimes arising directly from the rhizome; bracts usually congested, spirally arranged; spikes more or less terete; flowers bisexual, zygomorphic; sepals 3, free; corolla deeply 3-lobed; stamens petaloid, the androecium basically of 2 whorls, the outer whorl usually with 1–3 staminodia, the inner whorl with 1 fertile stamen and 1 or 2 staminodia; anther 1-celled, lateral, dehiscing by a vertical slit; ovary inferior, 3-locular (usually only 1 locule fertile), 3-carpellate; placentation axile; ovules 1 per locule, seemingly basal; style 1, simple, (may be twisted, lobed, etc.); stigma terminal. Fruits loculicidal capsules; seeds 1–3, often arillate, with copious endosperm.

Members of the family share with the Cannaceae and Musaceae (31) usually broad leaves, with the closely parallel, fine lateral veins much less prominent than the midrib. They may be distinguished by the pulvinate, terete petiole, the very asymmetrical flowers with much modified, often petaloid staminodia, and the short, spirally arranged bracts (or, if distichously arranged, the compound inflorescence).

The flowers of most Marantaceae are bee pollinated (Kennedy, 1973). After alighting on the flower, the bee commences to feed, thus triggering a small flap on the cucullatum that acts as a release for the springlike style, which bears a loose mass of pollen just below the stigma. The pollen is deposited into a depression in the style before anthesis. The style springs forward, striking the bee on the "chin," first with the stigma (picking up pollen deposited there during a previous visit) and then with the pollen load. In some species, such as *Calathea marantifolia* and *C. latifolia,* the flower buds do not open until forced open by a bee. *Calathea insignis* is visited by both sexes of *Euglossa ignita, E. dodsoni, E. gorgonensis,* and *E. hansoni,* as well as by *Eulaema cingulata* and *E. polychroma.* All visits are to collect pollen (Dodson, 1966). On BCI most species of Marantaceae are pollinated by *Euglossa imperialis,* but sometimes also by *Eulaema cingulata* (H. Kennedy, pers. comm.). I have seen large tabinid flies visiting *Calathea inocephala. Calathea panamensis* has flowers that are cleistogamous (H. Kennedy, pers. comm.). *Stromanthe jacquinii* is visited by hummingbirds (H. Kennedy, pers. comm.).

The fruits are capsular, and some usually develop to maturity while the plant is still flowering. Seeds usually emerge from the capsule by swelling of the pedicel and become exposed as the bracts are reflexed or weather away; they are probably bird dispersed.

About 30 genera with 350–400 species; mostly in damp, shady, tropical or subtropical habitats.

CALATHEA G. Meyer

Calathea inocephala (O. Kuntze) Kenn. & Nic., Ann. Missouri Bot. Gard. 62:501. 1975

   C. barbillana Cuf.; *Phyllodes inocephalum* O. Kuntze

Acaulescent herb, to more than 2 m tall. Petioles ca 1.5 m long, round, vaginate on lower part, the pulvinus 10–16

KEY TO THE SPECIES OF CALATHEA

Plants stemless, the leaves arising from the ground:
  Plants more than 1 m tall; leaf blades glabrous beneath, the midrib above glabrous, often with lighter green band running the length of the blade between midrib and margin when still juvenile; adult blades 25–45 cm broad; inflorescences globose, ca 6 cm wide . . . . *C. inocephala* (O. Kuntze) Kenn. & Nic.
  Plants less than 1 m tall; leaf blades minutely to conspicuously pubescent beneath; midrib above pubescent; blades mostly less than 15 cm broad; inflorescences usually not globose, usually less than 4 cm wide:
    Blades pubescent above, marked with several brown or dark green spots along the blade between the midrib and the margins; bracts of inflorescence 2-ranked; plants cultivated in the Laboratory Clearing . . . . *C. villosa* (Lodd.) Lindl.
    Blades glabrous above except midrib; leaf uniformly green above; bracts of inflorescence spirally arranged:
      Blades green beneath; leaf sheaths wide, spreading, subglabrous, extending to base of pulvinus; inflorescence sessile, more than 4 cm long; flowers yellow . . . *C. panamensis* Standl.
      Blades often maroon beneath; leaf sheaths puberulent, narrow, ending well below the pulvinus; inflorescence long-pedunculate (7 cm or more), held as high as or higher than leaves; flowers white . . . . *C. micans* (Math.) Koern.
Plants with a stem, at least part of the leaves arising from the stem:
  Blades pruinose beneath . . . . *C. lutea* (Aubl.) Schult.
  Blades not pruinose beneath:
    Spikes markedly flattened and more than 4.5 cm broad . . . . *C. insignis* O. G. Petersen
    Spikes ± cylindrical or somewhat flattened and less than 1 cm wide:
      Flowers cream-colored; lower blade surface softly pilose; major veins scarcely raised (not pleated), midrib pubescent in furrow above; sheath glabrate . . . . . . *C. marantifolia* Standl.
      Flowers purple; lower leaf surface glabrous; major lateral veins raised, leaf pleated; midrib glabrous above; sheath densely pilose . . . . *C. latifolia* (Link) Klotzsch

cm long, darker green in color; blades very broadly elliptic, blunt at apex, truncate to rounded at base, 45–100 cm long, 25–45 cm wide, ± concolorous, the very margin often reddish; veins about 10–15 mm apart, branching from midrib at 45° angle; blades of juvenile plants occasionally with streaked, light green bands midway between midrib and margin. Inflorescences ± globose, 6–12 cm long, ca 6 cm wide; peduncles 50–80(150) cm long; bracts many, ca 5 cm long, becoming very weathered; flowers densely aggregated, pale orangish-yellow or cream-colored, 5 cm long, soon perishing. Capsules orange, pedicellate, strongly 3-sided, less than 2 cm long and about as broad; seeds 1–3, irregularly oblong-ovoid, ca 1 cm long, 7–8 mm wide, violet-blue, with an irregular white aril at base. *Croat 16576.*

Common in the forest, at least as a juvenile plant. Flowers from mid-April to September, mostly in June after the rainy season has begun. The fruits develop to maturity by August and may constitute most of the head by late in the flowering season; the last fruits persist until about mid-January.

Easily recognized by its short broad leaves and globular inflorescences, which often age and weather while the plant is still flowering. Sterile plants may be confused with *C. lutea,* but the blades are never pruinose on the underside as in that species. Unlike most species of *Calathea,* the fruits are a conspicuous feature of the plant.

Mexico to Peru. In Panama, known from tropical moist forest in the Canal Zone, Bocas del Toro, Chiriquí, Coclé, Panamá, and Darién and from premontane wet forest in Chiriquí.

Calathea insignis O. G. Petersen in Mart., Fl. Brasil. 3(3):124. 1890

Caulescent, moderately stout herb, 2–3 m tall; stems ± swollen at base of petiole. Petioles long, often scabrid and brown near base; blades ovate to ovate-elliptic, short-acuminate, rounded or obtuse at base, usually more than 1.5 times longer than broad, 35–70 cm broad, dark green on upper surface with pale green midrib, glabrous except the midrib appressed-pubescent on upper surface. Inflorescences terminal on stem; spikes 1–3, laterally flattened, broadly oblong, 15–40 cm long; peduncles to 25 cm long, 4.5–6 cm broad; bracts greenish-yellow, 2-ranked, perpendicular to the rachis, broadly reniform, glabrate, the upper edge rounded; flowers pale yellow, 2.5–3.5 cm long. Fruits similar to those of *C. latifolia. Aviles 85b.*

Though long known to be present on Orchid Island, a population of the species was discovered by Foster near Barbour Trail 1500 only recently. Flowers principally in the rainy season (June to November), but also in March and April at higher elevations in Panama. The fruits develop soon and are usually already of mature size while the plant is still in flower.

The species prefers a moist habitat, often occurring in coastal and estuarine flood plains elsewhere in Panama; it does not grow well in deep forest. It is an invading species.

Coastal areas from Mexico to Colombia and Peru. In Panama, known from tropical moist forest in the Canal Zone, Bocas del Toro, San Blas, Panamá, and Darién, from premontane wet forest in Bocas del Toro, Colón,

and Coclé, and from tropical wet forest in Bocas del Toro and Colón.

Calathea latifolia (Willd. ex Link) Klotzsch in R. Schomb., Reisen Brit.-Guiana 3:918. 1848

C. allouia var. violacea sensu Woods. non Lindl.

Bijao, Faldita morada, Sal, Sweet corn root

Caulescent, 1–2 m tall; roots bearing large, edible tuber-like storage organs. Petioles with short-pilose sheath, the pulvinus to 7.5 cm long, somewhat brownish; blades ovate to oblong-ovate, obtuse to short-acuminate, rounded to subtruncate at base, ca 70 cm long and 30 cm wide, bicolorous, frequently with light purple bands between midrib and margin, the surface glabrous, pleated; midrib glabrous to minutely puberulent, the lateral veins sparsely puberulent, raised. Spikes solitary, ovoid to cylindrical, 6–14 cm long, 3–5.5 cm wide; peduncle 7–30 cm long; bracts reniform, spiral, closely imbricated, to 2.5 cm high and 4 cm broad, green and maroon, the apex ± rounded; flowers not opening spontaneously, borne in pairs with as many as 13 pairs subtended by a single bract; petals unequal, obovate-elliptic, purple, ca 2 cm long; calyx to 3.5 cm high, whitish at base, purple above or with tips tinged with pale violet, persistent in fruit, turning darker purple, the lobes unequal, free; staminodia white to cream-colored, shorter than corolla, sometimes tinged with purple at apex. Capsules obovoid, ca 1 cm long; seeds usually 3, trigonous, rugose, ca 5 mm long, bearing a basal aril ca half as long as seed. *Croat 4271, 11989.*

Adult plants uncommon in the forest, more common in clearings; juvenile plants sometimes common. Flowers throughout the rainy season (July to December). The fruits develop to mature size in about 2 months and are usually present in the same inflorescence throughout much of the flowering season.

Plants of this species may be distinguished by the light purple bands that extend over much of the blade midway between the margin and the midrib, chiefly on the lower surface. The other distinguishing feature is the short-pilose pubescence on the sheaths, which is usually visible to the naked eye. In contrast, the sheaths of *C. marantifolia* vary from glabrous to rather densely pubescent; the trichomes are generally appressed, and always very inconspicuous, scarcely or not at all visible to the naked eye. Juvenile plants can be distinguished from other species by having the sheath villous at least at its apex. The cream-colored flower form of *C. latifolia,* which occurs elsewhere in the Canal Zone, lacks the purple bands on the leaf. Woodson and Schery in the *Flora of Panama* (1945b) mistakenly reported this species as a variety of *C. allouia* (Aubl.) Lindl. (*C. allouia* var. *violacea*). This name is based on a misinterpretation of *Calathea violacea* Lindl., which is restricted to Brazil. The plant is fairly shade tolerant.

On BCI *C. latifolia* is pollinated by the bee *Euglossa imperialis* (elsewhere by *Eulaema cingulata* and *Eulaema nigrita*) (H. Kennedy, pers. comm.).

Panama, Colombia (Meta Valley near the Macarena), and Venezuela (drier coastal areas); Trinidad. In Panama, known from tropical moist forest in the Canal Zone,

Panamá, and Darién, from premontane moist forest in the Canal Zone and Panamá, from premontane wet forest in Colón, Chiriquí, Coclé, and Panamá, and from tropical wet forest in Colón.

See Fig. 154.

Calathea lutea (Aubl.) Schult., Mant. 1:8. 1822

Hoja blanca

Caulescent herb, 2–3 m tall, mostly glabrous. Petioles 1.2–2 m long, the sheath less than 2 m long, geniculate at petiole, the pulvinus to 17 cm long, brownish; blades broadly elliptic to almost rounded, obtuse or abruptly acuminate at apex, obtuse to truncate and abruptly decurrent at base, mostly 50–100 cm long (to 150 cm) and 25–60 cm wide, prominently whitish-pruinose on lower surface, the margins held ± erect. Inflorescences emerging from sheath of subtending leaf; peduncles 6–17 cm long, unbranched; spikes oblong, ± flattened, 10–30 cm long, 3–5 cm broad; bracts yellowish when young, reddish or bronze in age, 2-ranked, imbricate, nearly orbicular, 3.5–5 cm long, often lobed at apex with mucronate point; staminodia usually pale yellow, occasionally whitish; petals 4–4.5 cm long. Fruits capsules; seeds orange, with a brilliant orange aril. *Croat 6992.*

Almost never in the forest except in tree-fall areas; occasional on creek beds and shoreline soil deposits. Flowers and fruits throughout the year, but activity is greatest during the early rainy season. The fruits develop quickly.

Distinguished by its leaves, which are white beneath, and by its oblong, somewhat rounded spikes. This is the most aggressive species of the genus, according to H. Kennedy (pers. comm.).

The species has been seen pollinated by euglossine bees and is robbed of nectar by hummingbirds.

Coastal areas from Mexico to Peru and Brazil; principally in coastal marshes and disturbed areas along both coasts. In Panama, known from tropical moist forest in the Canal Zone, Bocas del Toro, Colón, Chiriquí, Los Santos, and Darién, from tropical dry forest in Coclé, from premontane wet forest in Chiriquí and Panamá (Cerro Campana), and from tropical wet forest in Colón (near Portobelo).

See Fig. 155.

Calathea marantifolia Standl., J. Wash. Acad. Sci. 17:250. 1927

C. allouia (Aubl.) Lindl. sensu Woods. (1942) non Aubl.;
C. lagunae Woods.

Caulescent herb, 1–2 m tall. Blades oblong to oblong-elliptic, abruptly short-acuminate, obtuse to rounded at base, to 45 cm long and 25 cm wide, the edges turned upward, the upper surface glabrous, the midrib with a line of hirsute pubescence, the lower surface dull, short-pilose, the pubescence diminishing but extending onto petiole and sheath, the sheath almost glabrate. Spikes broadly oblong, at apex of leafy stem, solitary; peduncles to 21 cm long; bracts minutely pubescent, very broad, scarcely 1 cm high, persistent; flowers cream, 4–5 cm long, not opening spontaneously; sepals ca 2.5 cm long,

Fig. 154. *Calathea latifolia*

Fig. 155. *Calathea lutea*

Fig. 156.
Calathea panamensis

white to yellowish, persisting in fruit. Fruits obovate; seeds 3, slate-gray with white aril. *Croat 12222.*

Occasional in the forest and rare in clearings. Flowers mostly from June to October; may flower much later in tropical wet forests. The fruits soon mature to adult size and, according to H. Kennedy (pers. comm.), are dispersed in about 2 months.

Juvenile plants may be recognized by the soft short-pilose pubescence on the leaf surfaces. This species has been confused with *C. macrosepala* K. Schum., which occurs in Veraguas and Los Santos Provinces.

Guatemala to Ecuador. In Panama, known from tropical moist forest in the Canal Zone, Bocas del Toro, San Blas, Chiriquí, Coclé, Panamá, and Darién and from tropical wet forest in Colón (Río Guanche).

Calathea micans (Math.) Koern., Cat. 126. 1862

C. microcephala (Poepp. & Endl.) Koern.

Acaulescent, to 30 cm tall. Petioles often equaling blades, short-pilose; leaf sheaths puberulent, narrow, ending well below pulvinus; blades ovate to oblong-elliptic, acute to short-acuminate, acute to rounded at base, 6–15 cm long, 2.5–8 cm wide, glabrous above except the midrib, short-pilose and usually purple below. Spikes small, long-pedunculate, equal to or longer than leaves, 1.5–2.5 cm long; bracts few, lanceolate or ovate-lanceolate, spirally arranged; flowers white or the labellum tinged purple. Fruits not studied. *Croat 6638.*

Common locally, but apparently restricted to certain areas of the older forest. Flowers throughout the year in Panama, but in regions such as BCI with a prominent dry season plants die back altogether during the dry season and reappear shortly after the rainy season begins.

Guatemala to Peru and Brazil. In Panama, known from tropical moist forest in the Canal Zone, Bocas del Toro, and Darién, from premontane wet forest in Colón and Coclé, and from tropical wet forest in Veraguas (Atlantic slope) and Coclé.

Calathea panamensis Rowl. ex Standl., J. Wash. Acad. Sci. 15:4. 1925

Acaulescent, to 65 cm tall; roots bearing edible tubers. Petioles broadly winged to base of pulvinus, somewhat shorter than blade; blades oblong or obovate-elliptic, shortly and abruptly acuminate to blunt at apex, obtuse or rounded at base, 25–38 cm long, 6–15 cm wide, often ± asymmetrical, dark green and glabrous above except the midrib, paler and moderately pubescent with soft trichomes below. Inflorescences terminal, arising among leaf sheaths, sessile or very short-pedunculate; bracts mostly 3–7 cm long, lanceolate, spirally arranged; flowers yellow, cleistogamous, ca 5 cm long, exserted ca 1.5 cm, the tube very slender, ca 3.5 cm long; corolla lobes ± unequal, to 17 mm long, 5–9 mm wide, elliptic and boat-shaped; staminodia petaloid, irregular, shorter than petals, united with filament; anther ca 3.2 mm long; style stiff, curved. Capsules ca 1 cm long; seeds 3, brown, ca 5 mm long, the aril with 2 lateral, pointed appendages. *Croat 11786.*

Uncommon in the forest, but locally abundant in a few places along trails, notably on Hood, Van Tyne, and Wheeler trails. The plant dies back during the middle of the dry season and reappears the following rainy season, about May. Flowers mostly from June to October. The fruits apparently develop quickly, but their time of dispersal is unknown.

Recognized by its small size and its yellow flowers inconspicuously situated below the leaves. The flowers are cleistogamous, and reproduction is by means of autogamy.

Costa Rica (Guanacaste) and Panama. In Panama, known from seasonally dry parts of tropical moist forest in the Canal Zone, Panamá and Darién and from premontane moist forest in Panamá. BCI has about the upper limit of rainfall for the species.

See Fig. 156.

Calathea villosa (Lodd.) Lindl., Edward's Bot. Reg. 31, pl. 14. 1845

Moderately pubescent, acaulescent herb, to 80 cm tall; roots with tubers. Petioles 1.5–6 cm long; blades oblong-elliptic to obovate, short-acuminate, obtuse at base, 15–60 cm long, 8–15 cm wide, pubescent on upper surface, with 5–7 dark-green markings equally spaced on blade between margin and midrib. Inflorescences on leafless scape usually overtopping leaves; bracts 4–7, 2-ranked, not imbricate, 2–3 cm long; flowers yellow, autogamous, 3.5–4 cm long. Fruits not studied. *Croat 10913.*

Cultivated at the Laboratory Clearing. Flowers during the rainy season, beginning about July.

Costa Rica to Brazil. In Panama, known from seasonally dry parts of tropical moist forest and from premontane moist forest in the Canal Zone and Panama; known also from premontane wet forest in Panamá (Cerro Campana).

ISCHNOSIPHON Koern.

Ischnosiphon leucophaeus (Poepp. & Endl.) Koern., Bull. Soc. Imp. Naturalistes Moscou 35(1):91. 1862

Caulescent herb, 1–2 m tall. Leaves chiefly basal; petioles to 65 cm long; blades broadly oval, gradually to abruptly acuminate, 15–33 cm long, 12–20 cm wide, glabrous, pruinose on underside. Spikes usually in clusters of 2–6, sessile, narrowly cylindrical, 10–15 cm long, to 1 cm broad; bracts oblong, narrow and 2.5–3 cm high, somewhat pruinose; flowers white, 3–4 cm long. Fruits not studied. *Croat 4360.*

Uncommon, in the forest. Flowering and fruiting throughout the rainy season (May to December).

Recognized by its very narrow, cylindrical spikes.

Panama to Brazil. In Panama, known from tropical moist forest in the Canal Zone, Bocas del Toro, San Blas, Panamá, and Darién and from tropical dry forest in Panama (Taboga Island).

Ischnosiphon pruinosus (Reg.) O. G. Petersen, Bot. Tidsskr. 18:264, pl. 18. 1892

Pleiostachya pruinosa (Reg.) K. Schum.

Caulescent herb, 2–3 m tall. Petioles 39–84 cm long, at least those bearing inflorescences in their axils vaginate

Fig. 157. *Ischnosiphon pruinosus*

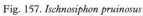

Fig. 158. *Thalia geniculata*

Fig. 159. *Thismia panamensis*

more than three-fourths their length, ± pubescent up to the callus, glabrous above, the pulvinus with a raised annular ring 5–7 cm below the blade; blades elliptic-oblong, 25–85 cm long, 12–28 cm wide, with one side 4–8 cm narrower, very oblique and falcate-cuspidate at apex, obtuse at base, the lower edges often held erect, the midrib and the underside of blade maroon (sometimes green or with only a maroon band along one margin). Inflorescences 2 or 3 pedunculate clusters per leaf axil; peduncles 14–43 cm long, the branches flattened, each subtended by a bract 5–14 cm long; spikes flattened, green, 9–19 cm long, to 2 cm broad, at first pruinose; flowers exserted from tightly appressed bracts, 4–5 cm long, white, the outer staminodium to ca 1 cm wide, violet-purple, the cucullatum pale yellow, the staminodium callosum tinged with violet-purple at apex. Fruits 1-seeded capsules ca 13 mm long, apparently falling out of the weathered and drooping inflorescence; seeds 3-sided, 7–8 mm long. *Croat 4116.*

Frequent in the forest, often in open places such as tree-fall areas. Flowers principally from July to October; fruiting inflorescences persist into the early dry season before weathering away.

Pollinated by the bee *Euglossa imperialis.*

Belize to Panama. In Panama, known from tropical moist forest on both slopes of the Canal Zone and in Bocas del Toro, Chiriquí (Burica Peninsula), Panamá, and Darién; known also from premontane moist forest in Veraguas and Panamá and from premontane wet forest in Chiriquí.

See Fig. 157.

STROMANTHE Sond.

Stromanthe jacquinii (R. & S.) Kenn. & Nic., Ann. Missouri Bot. Gard. 62:501–2. 1975

S. lutea (Jacq.) Eichl. non Aubl.

Platanillo de montana

Caulescent rhizomatous herb, 1–2 m tall. Leaves 2-ranked, atop a leafless stem 30–185 cm tall; petioles mostly 19–37 cm long, the sheath narrow, extending more than halfway to blade, often to base of pulvinus, the pulvinus 15–25 mm long; blades oblong-elliptic, abruptly short-acuminate, broadly obtuse to rounded at base, mostly 26–46 cm long, 9–18 cm wide (smaller on juveniles), faintly pleated, all veins ± equal. Inflorescence usually solitary, arising from a leaf sheath, held above the leaves; peduncle 50–75 cm long; spikes diffusely and paniculately compound, the rachis closely flexuous; bracts orange, mostly broadly ovate, 1–2.5 cm long, soon deciduous; flowers ca 1 cm long, pedicellate, in pairs on a common stalk, subtended by a bracteole; sepals 3, free, ca 7 mm long, slightly shorter than petals; petals united at base, white with red markings; staminodia petaloid, slightly shorter than petals. Fruits ca 1 cm long; seed 1, black, on a short white aril. *Croat 11815.*

Uncommon in the forest and at the edge of the forest along the lake. Flowers mostly throughout the rainy season (June to December), principally in September; rarely flowering earlier (in the dry season) elsewhere in Panama at higher elevations. The fruits develop rapidly and are present on the same inflorescence with the flowers, soon shedding to expose the rachis.

Costa Rica to Venezuela. In Panama, known principally from tropical moist forest in the Canal Zone, Colón, San Blas, Panamá, and Darién; known also from premontane wet forest in the Canal Zone, Colón, and Panamá, from tropical wet forest in Colón, Panamá, and Darién, and from premontane rain forest in Darién (Cerro Pirre).

THALIA L.

Thalia geniculata L., Sp. Pl. 1193. 1753

Swamp lily, Platanillo

Caulescent, perennial herb, usually 2–3 m tall. Leaves both basal and cauline; petioles usually much longer than blade; blades broadest at base, gradually tapering, ovate to ovate-lanceolate, gradually or abruptly short-acuminate at apex, obtuse or rounded at base, 20–75 cm long, 5–30 cm wide, glabrous, sometimes ± pruinose on underside. Inflorescences widely branched, very diffuse panicles, many-flowered; peduncles often very long, exceeding leaves; rachis of spikes sharply flexuous, the internodes 5–10 mm long; bracts 1–2.5 cm long, usually bluish, caducous; flowers in pairs subtended by a pair of unequal bracts, green or bluish, caducous, the longest 1.5–2.5 cm long, sessile; sepals 3, oblong to obovate, violet at least medially, rounded at apex; labellum obovate, clawed, ca 2 cm long, the blade lavender, to 1.2 cm wide; petals very irregular, violet; ovary obovate, ca 2.5 mm long; style spirally twisted, spring-loaded, bearing two slender appendages. Fruits nutlike, indehiscent, 1-seeded. *Croat 5455, Shattuck 971.*

Rare; restricted to undisturbed marshy areas along the shore. Flowers throughout the year, but especially in the dry season. The fruits develop quickly and often share the same inflorescence, as in *Stromanthe jacquinii.*

Florida and Mexico to Argentina; Greater Antilles. In Panama, known principally from tropical moist forest on the Pacific slope in the Canal Zone, Chiriquí, Panamá, and Darién; known also from tropical dry forest in Herrera and Panamá, from premontane moist forest in Panamá, and from premontane wet forest in Chiriquí.

See Fig. 158.

34. BURMANNIACEAE

Saprophytic, colorless herbs arising from a tuberous underground part. Leaves lacking. Flowers bisexual, zygomorphic, solitary, terminal, bracteate; perianth urceolate-campanulate, 6-lobed; stamens 6, attached to the perianth tube, alternating with small, triangular appendages; filaments attached to the appendages; anthers sessile, 2-celled, dehiscing transversely; ovary inferior, 1-locular, 3-carpellate; placentas 3, parietal, stalked;

ovules many; style 1, thick-filiform, trifid, bearing terminal stigmas. Fruit a fleshy capsule; seeds very numerous, with little endosperm.

The single tiny saprophyte on BCI is not confused with any other species.

Eleven genera with about 100 species; tropics and subtropics.

THISMIA Griff.

Thismia panamensis (Standl.) Jonk., Monogr. Burm. 234. 1938
Ophiomeris panamensis Standl.

Small, white to flesh-colored, saprophytic herb, 4–12 cm tall, arising from a tuberous root, leafless; stem slender, 1-flowered. Flower urceolate-campanulate, subtended by 4 bracts ca 4 mm long; perianth 6-lobed, to 12 mm long, zygomorphic, persisting in fruit, the bulged part with 3 slits, the outer lobes short and reflexed, the inner lobes spreading, tapered to filiform purplish appendages 3–4 cm long, the throat 3–4 mm diam, surrounded by a 6-lobed rim; stamens 6, to 5.5 mm long, pendent below throat, emarginate at apex, each bearing a pair of small lateral appendages, the thecae borne on their outer surface (against inner wall of perianth); ovary inferior; style to 3.5 mm long, 3-branched, the branches densely pubescent and longer than the unbranched part, the stigmatic part terminal. Capsule fleshy, to ca 3 mm long; seeds very numerous, minute. *Croat 10848.*

Frequent in the forest. Flowering and fruiting from June to September; not obvious at other times of the year.

The flower is specialized and probably pollinated by small insects. The small fleshy fruit may be dispersed by small birds, or perhaps the minute seeds are carried by insects. Van der Pijl (1968) suggested earthworms as possible dispersers of seeds in the family.

Known only from Panama, from tropical moist forest on BCI and from tropical wet forest in Colón (near Portobelo).

See Fig. 159.

35. ORCHIDACEAE

Terrestrial or epiphytic, perennial herbs (*Vanilla* scandent), very rarely saprophytic, usually with pseudobulbs. Leaves alternate, distichous or spiraled; blades simple, entire, fleshy, obscurely veined, rarely the leaves lacking and the roots photosynthetic. Flowers bisexual (rarely unisexual and monoecious (*Catasetum*) or dioecious (*Mormodes, Catasetum*), zygomorphic, in bracteate, generally simple racemes (panicles in *Oncidium*); sepals 3, free, somtimes showy; petals 3, the central petal (the lip)

KEY TO THE TAXA OF ORCHIDACEAE

- Plants terrestrial:
  Leaf blades more than 30 cm long; sepals more than 6 mm wide:
     Flowers white, ± globose, very fleshy; sepals often about as wide as long; fruits more than 1.5 cm wide; leaves mostly more than 6 cm wide .... *Peristeria elata* Hook.
     Flowers greenish, not globose, not very fleshy; sepals much longer than wide; fruits less than 1.5 cm diam; leaves mostly less than 6 cm wide .... *Eulophia alta* (L.) Fawc. & Rendle
  Leaf blades less than 30 cm long; sepals usually less than 6 mm wide:
     Leaves very reduced on stem *or* leaves in a basal rosette, often not present at time of flowering:
        Flowers more than 1.5 cm long; scape bracts 2–2.5 cm long; both stem and scape bracts pubescent .... *Spiranthes lanceolata* (Aubl.) León
        Flowers less than 1 cm long; scape bracts and stem not as above; both stem and scape bracts usually glabrous:
           Plants to 15 cm tall, usually leafless at time of flowering, ± saprophytic; lip 7–8 mm long .... *Triphora gentianoides* (Sw.) Ames & Schlechter
           Plants 15–40 cm tall, usually bearing leaves at time of flowering, not saprophytic; lip 4–5.5 mm long .... *Liparis elata* Lindl.
     Leaves borne along stem, always present at time of flowering (not to be confused with scape bracts), either usually more than 5 cm long or, if less, ± ovate:
        Plants usually less than 15 cm tall; leaves narrowly to broadly ovate, less than 2.5 cm long; flowers 1–3, terminal .... *Triphora mexicana* (S. Wats.) Schlechter
        Plants 20–100 cm or more tall; leaves narrowly elliptic to almost linear, usually more than 4 cm long; flowers usually much more numerous or not terminal:
           Inflorescence of axillary racemes or panicles at usually leafless nodes, much shorter than leaves; flowers lacking a spur; leaves conspicuously petiolate .... *Palmorchis powellii* (Ames) Schweinf. & Corr.
           Inflorescence of terminal racemes held above leaves; leaves not conspicuously petiolate:
              Flowers green or green and white; lip less than 12 mm long, bearing a conspicuous spur .... *Habenaria*
              Flowers brick-red or orange; lip more than 12 mm long, lacking a spur .... *Epidendrum radicans* Lindl.

● Plants epiphytic, normally growing in trees or on rocks (fallen epiphytes may thrive in exposed positions, but this will cause little confusion on BCI):
  Plants fleshy-leaved vines . . . . *Vanilla*
  Plants not vines:
    Plants lacking both pseudobulbs and leaves, the gray-green roots functioning as photosynthetic organs . . . . *Campylocentrum pachyrrhizum* (Reichb.f.) Rolfe
    Plants with pseudobulbs and/or normal leaves:
      Inflorescence terminal on stem or pseudobulb or restricted to upper leaf axils (may appear upper-axillary in *Maxillaria* but then inflorescence arising from base of pseudobulb):
        Leaves plicate (with several folds longitudinally):
          Inflorescence normally with only 1 or 2 flowers open at a time; flowers more than 2 cm long . . . . *Sobralia*
          Inflorescence many-flowered; flowers less than 1 cm long:
            Leaf blades ± elliptic, thin, rosulate at apex of short pseudobulb; pseudobulbs inconspicuous, usually enveloped by leaves . . . . *Liparis elata* Lindl.
            Leaf blades lanceolate, not thin, scattered along stem, the lower deciduous; pseudobulb lacking . . . . *Elleanthus longibracteatus* (Griseb.) Fawc.
        Leaves conduplicate (folded once along midrib):
          Leaves solitary on each stem, the stems not thickened:
            Flowers less than 3 mm long; leaves prominently striate longitudinally when dry . . . . *Stelis crescentiicola* Schlechter
            Flowers more than 4 mm long; leaves not prominently striate when dry . . . *Pleurothallis*
          Leaves 2 to several on each stem, *or* the stems definitely thickened:
            Lip adnate to the column for the length of the column; stems usually slender with several or many leaves . . . . *Epidendrum* (except *E. rousseauae* Schlechter)
            Lip at least partly free from the column:
              Pseudobulbs cigar-shaped, hollow; flowers white or pink, cleistogamous . . . . *Caularthron bilamellatum* (Reichb.f.) Schult.
              Pseudobulbs various and solid *or* pseudobulbs lacking:
                Leaves or leaf scars usually numerous, scattered on stem:
                  Leaves equitant (laterally flattened, V-shaped); flowers white or yellow; column wings lateral, not surpassing anther; stem not thickened . . . *Lockhartia*
                  Leaves normal; flowers rose-colored, serial on a condensed raceme; column wings surpassing anther; stem thickened . . . . *Dimerandra emarginata* (G. Meyer) Hoehne
                Leaves few (usually 1–3), terminal on stem or pseudobulb (*Scaphyglottis* forms a series of superimposed pseudobulbs, but each bears terminal leaves):
                  Pseudobulbs club-shaped, stalked; leaves 2 or 3, about 3 times longer than broad; flowers orchid-lavender . . . . *Cattleya patinii* Cogn.
                  Pseudobulbs ovoid or ellipsoid; leaves more than 3 times longer than broad; flower color various:
                    Flowers usually more than 1 cm long (to 8 mm in *E. triptera*); lip entire, cordate, concave; flowers white with purplish lines or spots on lip . . . . *Encyclia*
                    Flowers less than 7 mm long; lip not as above:
                      Stems with a single ovoid pseudobulb; flowers in a raceme or panicle, yellow or greenish-yellow; lip with mealy powder . . . . *Polystachya*
                      Stems with several swollen, superimposed segments (slender pseudobulbs); flowers clustered at tips of segments; lip not with mealy powder . . . . *Scaphyglottis*
      Inflorescence lateral on stem, usually from base of stem or pseudobulb:
      ◆ Leaves plicate (*Xylobium foveatum* sometimes appearing conduplicate):
        ■ Leaves several, scattered along pseudobulb:
          Pseudobulbs cigar-shaped, pendent, stalked; flowers fleshy, with a trilobate lip and 8 pollinia . . . . *Chysis aurea* Lindl.
          Pseudobulbs oblong or cigar-shaped, erect, not stalked; flowers usually unisexual, the staminate flowers with "sensitive" columns, ejecting the 2 pollinia violently when triggered:
            Lip ± saccate in both staminate and pistillate flowers; staminate column pointed, usually with 2 slender "antennae" beneath; inflorescence from base of pseudobulb; old pseudobulbs spiny above after leaves have fallen . . . . *Catasetum*
            Lip not saccate; column lacking antennae, pointed, with a slender apical bristle, connivent with lip in staminate flowers; inflorescence usually from middle of pseudobulb, the flowers asymmetrical, both column and lip twisted, the staminate and pistillate flowers usually similar; old pseudobulbs not spiny . . . . *Mormodes powellii* Schlechter

■ Leaves few, terminal on pseudobulb:
  Peduncles each with a single flower (but several peduncles produced at once) . . . . . .
  . . . . *Lycaste powellii* Schlechter
  Peduncles with several or many flowers:
    Each pseudobulb with a single leaf; flowers small (12–20 mm diam), the lip
      concave . . . . *Sievekingia suavis* Reichb.f.
    Each pseudobulb with 2 or 3 leaves:
      Inflorescence erect; lip trilobate, fleshy, simple; flowers cream-colored . . . .
      . . . . *Xylobium foveatum* (Lindl.) Nich.
      Inflorescence pendent; lip complex; flowers yellow to brown with red or purple
        markings:
        Flowers large (ca 10 cm diam), 2–4; lip forming a cup containing a clear
          liquid produced by glands at base of lip . . . . . . *Coryanthes maculata* Hook.
        Flowers small (3–4 cm diam), numerous; lip not cuplike . . . . *Gongora*
◆ Leaves conduplicate:
  Plants lacking pseudobulbs (often present in *Ionopsis,* but minute and ensheathed by
    leaves), the leaves thin or laterally flattened:
    Stems longer than leaves:
      Leaves laterally flattened; flowers yellow or white . . . . *Lockhartia*
      Leaves normal:
        Inflorescence of many white, spurred flowers (one-sided and toothbrushlike) . .
        . . . . *Campylocentrum micranthum* (Lindl.) Maury
        Inflorescence of single flowers . . . . *Dichaea*
    Stems much shorter than leaves:
      Leaves normal (± linear in *Ionopsis satyrioides*):
        Inflorescence of several to many flowers . . . .*Ionopsis*
        Inflorescence of 1 flower:
          Leaves less than 12 cm long; flowers less than 12 mm long . . . .
          . . . . *Masdevallia livingstoneana* Reichb.f.
          Leaves more than 12 cm long; flowers more than 30 mm long . . . .
          . . . . *Cochleanthes lipscombiae* (Rolfe) Garay
      Leaves laterally flattened; plants fanlike:
        Inflorescence of few to many white or green flowers . . . . *Ornithocephalus*
        Inflorescence with proportionately large yellow flowers produced serially . . . . .
        . . . . *Psygmorchis pusilla* (L.) Dods. & Dressl.
  Plants with pseudobulbs *or* the leaves very thick and fleshy:
    Flowers always 1 per inflorescence (though many 1-flowered peduncles may be pro-
      duced at one time):
      Sepals much larger than petals or lip, connivent to about middle, abruptly re-
        flexed; peduncle erect, much taller than pseudobulbs; flowers brownish-
        purple . . . . *Trigonidium egertonianum* Lindl.
      Sepals and petals subequal, the sepals not abruptly reflexed at middle; inflores-
        cence usually shorter than pseudobulbs or subequal:
        Column with a fringed hood at apex; lip trumpet-shaped; pseudobulbs 1-leaved,
          with spotted sheaths . . . . *Trichopilia maculata* Reichb.f.
        Column lacking apical hood (or if hooded the margin entire); lip never trumpet-
          shaped; pseudobulbs with sheaths not spotted . . . . *Maxillaria*
    Flowers usually few, several, or many, produced serially on the same peduncle:
      Lip united with column for length of column; flowers green . . . .
      . . . . *Epidendrum rousseauae* Schlechter
      Lip at least partly free from column:
        Pseudobulbs more or less 4-angled, usually widely separated on a creeping rhi-
          zome; rachis of inflorescence thick, fleshy; small flowers sessile on the
          rachis, the flower diam less than that of the rachis . . . .
          . . . . *Bulbophyllum pachyrrhachis* (Reichb.f.) Griseb.
        Plants not with this combination of features; rachis never thick and fleshy:
          Flowers very flat, yellow and brown; lip with a fleshy, lumpy callus; column
            with lateral, fanlike wings; inflorescence a many-flowered panicle, much
            surpassing leaves (inflorescence equal to or shorter than leaves in *Oncid-
            ium stipitatum* with long, terete leaves) . . . . *Oncidium*
          Plants not with this combination of features:
            ▲ Pseudobulbs with 2 or 3 terminal leaves, the leaves markedly flattened:
              Lip adnate to basal half of column and then abruptly diverging; pseudo-
                bulbs stalked; sepals and petals acute, but not long-acuminate;
                plants common on tree trunks . . . . *Aspasia principissa* Reichb.f.
              Lip free from column; pseudobulbs not stalked; sepals and petals long-
                acuminate, spidery; plants rare . . . . *Brassia caudata* (L.) Lindl.

▲ Pseudobulbs with only 1 terminal leaf *or* the terminal leaf aborted:
   Anther parallel with axis of column; inflorescence many-flowered, often
      pendent; lip clawed; column without a fringed hood over anther;
      flowers white or green . . . . *Notylia*
   Anther terminal and caplike on column:
      Base of lip saccate, retrorse; pseudobulbs very reduced . . . .
         . . . . *Trichocentrum capistratum* Linden & Reichb.f.
      Base of lip not saccate and retrorse; pseudobulbs usually conspicuous
         (scarcely wider than leaf in *Trichopilia subulata*):
         Pseudobulbs narrowly cylindric; apex of column with a fringed
            hood; flowers white; lip more or less enclosing column or
            fringed . . . . *Trichopilia subulata* (Sw.) Reichb.f.
         Pseudobulbs compressed, ellipsoid; apex of column lacking fringed
            hood; lip neither enclosing column nor fringed . . . .
            . . . . *Leochilus scriptus* (Scheidw.) Reichb.f.

---

usually larger; 1 or more sepals sometimes forming a nectariferous spur; stamens forming a column with the stigma either 1 and terminal or 2 and lateral; anthers 2-celled, introrse; ovary inferior, 1-locular, 3-carpellate; placentation parietal; ovules numerous, anatropous; stylar portion of the column stout, with 3 sessile stigmas or stigmatic lobes. Fruits 3-valved, longitudinally dehiscent capsules; seeds many, lacking endosperm.

Orchidaceae are characterized by the unusual, zygomorphic flowers of three sepals and three petals, one of which is very different from the rest, and by the stamen(s) united to the style. Other distinguishing features include the usually epiphytic habit, the swollen, pseudobulbous stems, and the usually ribbed, capsular fruits with many tiny seeds.

Orchids are more highly modified and specialized for insect pollination than any other family (Baker, 1963). Species-specific attraction of pollinators is characteristic of the more highly evolved species of orchids (Dodson et al., 1969; Dressler, 1968a). There are often no other barriers to cross-pollination of many genera of orchids, and artificial hybrids are common (Proctor & Yeo, 1973). Most orchids are adapted to bee pollination. Genera of BCI orchids with known bee pollinators include *Aspasia, Bulbophyllum, Catasetum, Cochleanthes, Coryanthes, Dichaea, Encyclia, Epidendrum, Eulophia, Gongora, Lockhartia, Lycaste, Maxillaria, Mormodes, Notylia, Oncidium, Ornithocephalus, Peristeria, Sievekingia, Sobralia, Spiranthes, Trichocentrum, Trigonidium, Vanilla,* and *Xylobium.* Because so many of the species on BCI have known pollinators, a record of these specific pollinators will be found with the respective species. All reports of pollination are taken from Dodson (1965b) or Dressler (1968a) or are known through personal communication from R. Dressler. Probably the most important pollinators of orchids on BCI are bees of the genera *Euglossa, Eulaema,* and *Euplusia* (tribe Euglossinae). Bees visit flowers to collect pollen and/or nectar, but in general pollen is collected by the female bees. Male euglossine bees often collect floral fragrances from flowers and may as a result be important in pollination; moreover, they may be particularly important in long-range dispersal of pollinia, since they do not remain affiliated with the brood and are wide-ranging (Williams & Dodson, 1972). Many of the floral fragrances have been isolated as pure chemi-

cals and are found to attract certain bees even when dissociated from the flowers that produce them (Hills, Williams & Dodson, 1968).

Birds are probably responsible for pollination of orange-flowered species of *Elleanthus.* Elsewhere birds have been reported visiting *Masdevallia,* but not the BCI species. Flies are also reported as pollinators of *Bulbophyllum, Masdevallia, Pleurothallis,* and *Liparis.*

The genus *Habenaria* is pollinated by moths, though at least one species elsewhere is visited by butterflies. *Epidendrum difforme* is pollinated by the moth *Amastus acona.*

Wasps, which pollinate *Leochilus, Brassia,* and *Encyclia,* are in general much less effective pollinators than bees. The lip of the orchid forms a landing platform for bees and wasps. The basal part of the lip leads the pollinator down the front of the column. Upon backing out of this "tunnel," the pollinator usually gets some stigmatic liquid on its back. A little farther on it contacts the pollinia, which sticks to its back. In more advanced orchids other mechanisms, including springlike traps, are employed in depositing the pollinia on the insect (Dodson et al., 1969).

Ants associated with some orchid species probably do not pollinate, but may be useful to the plant in preventing predation by phytophagous insects.

Most species have seeds that are wind dispersed, but *Vanilla* has fleshy fruits dispersed by animals (Ridley, 1930; van der Pijl, 1968). In most species the capsule splits by three or six lateral slits, but remains attached at the apex. Wind passing through the fruits carries the seeds away.

At least 600 genera and 20,000 species; worldwide but with the greatest diversity in the tropics.

ASPASIA Lindl.

Aspasia principissa Reichb.f., Bot. Zeitung (Berlin) 10:367. 1852
   A. epidendroides var. *principissa* (Reichb.f.) P. H. Allen

Epiphyte; pseudobulbs oblong-elliptic, flattened, to 16 cm long and 4.5 cm wide, 2-leaved, stipitate, with leaflike imbricating bracts at base. Petioles conduplicate; blades lanceolate to ligular, 10–41 cm long, 1.7–5.5 cm wide.

Fig. 160. *Aspasia principissa*

Fig. 162. *Brassia caudata*

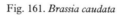

Fig. 161. *Brassia caudata*

Racemes 1 or 2, erect, few-flowered, from base of pseudo-bulb; sepals and lateral petals lanceolate to oblong, acute at apex, pale green striped with brown or purple, to 4 cm long, the petals slightly broader than sepals; lip ± fiddle-shaped, to 2.5 cm wide, white to pale yellow with pink or lavender streaks, wavy on margin, clawed at base, the claw fused to column at base then diverging abruptly; column to 2.5 cm long. Fruits narrowly fusiform, ribbed, to 8 cm long, usually bearing the persistent sepals. *Croat 8496.*

Common in the forest, sometimes occurring on tree trunks near the ground. Flowers principally in the dry season (late December to mid-May), mostly from mid-January to mid-February.

Pollinated by the bees *Eulaema* and *Exaerte* (*fide* Dressler, pers. comm.).

Guatemala to Panama. Widespread in tropical moist forests throughout Panama.

See Fig. 160.

BRASSIA R. Br.

Brassia caudata (L.) Lindl., Edward's Bot. Reg. 10, t. 832. 1824

Epiphyte; pseudobulbs linear to oblong-elliptic, ca 14 cm long and 3.5 cm wide, flattened with sharp edges, 2(3)-leaved, with 4–6 usually papery bracts at base. Petioles conduplicate; leaves ligular to elliptic-oblong, 16–27 cm long, 2.5–8 cm wide, symmetrical to strongly asymmetrical at apex. Inflorescences 1 or 2, unbranched, arching, 6–12-flowered, to 35 cm long, from lateral base of pseudobulb; sepals caudate-acuminate and spreading, light yellow-green, the longer lateral ones 12–30 cm long; petals similar to sepals but much shorter; lip oblong-lanceolate, acuminate at apex, free from column; sepals, petals, and lip similarly mottled with violet-purple near base, the basal part of lip with 2 short, erect teeth in front of 2 erect orange-tipped lamina. Fruits oblong-elliptic, to 7 cm long and 1.3 cm wide, ± acuminate on both ends, 5-ribbed. *Croat 11780, 14628.*

Apparently rare. Most plants in Panama were seen in flower in the early rainy season (May to August). The fruits may develop to mature size by August, but the time of dehiscence is unknown.

Widespread; Florida and Mexico to Venezuela, Brazil, and Bolivia; Greater Antilles. In Panama, known from tropical moist forest in the Canal Zone, Chiriquí, Los Santos, Panamá, and Darién.

See Figs. 161 and 162.

BULBOPHYLLUM Thouars

Bulbophyllum pachyrrhachis (Reichb.f.) Griseb., Fl. Brit. W. Ind. 613. 1864

Epiphyte, 10–45 cm tall; pseudobulbs short, subconical, strongly 4-angulate, well spaced along creeping rhizome. Leaves 2, linear-lanceolate, conduplicate, 7–20 cm long, to 2.4 cm wide. Inflorescences from base of pseudobulb, erect or arching, 10–45 cm long; rachis thick and fleshy; flowers many, small and inconspicuous, green-yellow spotted with purple, sessile in pits of rachis, subtended by ovate bracts; sepals ovate, 4–5 mm long, the dorsal sepal free, arching over the column, the lateral ones connate at base and adnate to base of column; petals ca one-third as long as sepals, oblong-elliptic; lip entire, fleshy. Fruits ovoid, ca 1 cm long and 6 mm wide, bearing many closely-spaced ribs. *Shattuck 777.*

Not seen in recent years on BCI. Elsewhere flowers and fruits principally in the dry season (December to May). Populations of plants have been seen in March with individuals bearing either flowers or full-sized fruits, indicating that the species may flower more than once a year.

Mexico to Panama; Greater Antilles, Trinidad. In Panama, known from tropical moist forest in the Canal Zone, Veraguas, and Panamá.

CAMPYLOCENTRUM Benth.

Campylocentrum micranthum (Lindl.) Maury, J. Bot. (Morot) 3:273. 1889
C. panamense Ames

Epiphyte; pseudobulbs lacking; stems unbranched, to 35 cm long. Leaves alternate, conduplicate, distichous, 4–9 cm long, 1.2–2 cm wide, articulate at base, ultimately deciduous below, alternating along stem with long, thickened, whitish roots. Inflorescences short, recurved, densely flowered racemes from base of roots; flowers small, secund and distichously arranged on scape, white or greenish; sepals to 4.5 mm long, ± linear; petals to 4 mm long; lip trilobate, to 4.5 mm long, produced at base into a spur ca 4 mm long. Fruits ± oblong, somewhat curved, longitudinally grooved, 6–9 mm long. *Croat 4623.*

Fairly common in the forest and around clearings, preferring sunlight. Plants flower for a moderately long period of time and may bear fruits on lower inflorescence branches while still in flower, in the late dry and early rainy seasons, mostly from April to June. Peak fruiting

KEY TO THE SPECIES OF CAMPYLOCENTRUM

Leaves regular, borne along slender stems . . . . *C. micranthum* (Lindl.) Maury
Leaves lacking, the roots gray-green, chlorophyllous . . . . *C. pachyrrhizum* (Reichb.f.) Rolfe

Other species of *Campylocentrum*, e.g., *C. poeppigii* Rolfe, may also appear on the island (R. Dressler, pers. comm.).

season is unknown, but plants with developed fruits appear to be most common during the rainy season and early dry season.

R. Dressler (pers. comm.) believes that *C. panamense* Ames, treated here as a synonym, may be a distinct species occurring in the forest, while *C. micranthum* is the plant found in clearings.

Mexico to Trinidad, Guyana, Brazil, and Peru; Greater Antilles. In Panama, reported from tropical moist forest in the Canal Zone, Bocas del Toro, San Blas, Panamá, and Darién.

Campylocentrum pachyrrhizum (Reichb.f.) Rolfe, Orchid Rev. 11:246. 1903

Leafless epiphyte; roots chlorophyllous, flattened, thin, 20–30 cm long, ca 2 mm wide. Racemes short, usually 2–3 cm long, densely flowered; floral bracts prominent, dark brown, the margins erose, persisting on fruiting inflorescences; flowers to 1.5 mm long; sepals very pale cream, subequal, to 4.5 mm long and 1.5 mm wide, acute at apex, the ventral margin of lateral sepals strongly inrolled; petals to 4 mm long and 1 mm wide, white, the ventral margin inrolled in lower half; lip white, spurred, shorter than petals, narrowly acute, the lower edges inrolled, the spur nearly as long as the blade, extending well below the sepals. Fruits ovate-oblong, prominently ridged, ca 6 mm long. *Shattuck 844.*

Rare; collected once by Shattuck at Gross Point. His collection, made in late March, bears mature fruits.

Florida, West Indies, Trinidad, French Guiana, Venezuela, and Panama (no doubt in Colombia as well). In Panama, only from the Canal Zone (BCI and Summit Garden).

See Fig. 163.

CATASETUM L. C. Rich. ex Kunth

Catasetum bicolor Klotzsch, Allg. Gartenzeitung 22:337. 1854

Monoecious epiphyte; pseudobulbs subconic or cylindrical, 4–9.5 cm long, 2.5–4 cm wide, enveloped by the imbricating, persistent leaf bases. Leaves plicate, elliptic-lanceolate, acute or acuminate, deciduous at the end of the growing season. Flowers unisexual, dimorphic, the staminate and pistillate flowers on separate racemes from base of pseudobulb, greenish-brown with tinges of maroon, the petals darker than the sepals; staminate inflorescences arched or pendent, to 15 cm long, the sepals and petals subequal, ± lanceolate, to 3 cm long, the lip short, white to pale yellow with reddish-brown spots, promi-

nently saccate, bearing 3 narrow lobes at apex and 2 additional slender lobes near base; pistillate racemes ± rare, few-flowered, 8–10 cm long, erect, the sepals coriaceous, elliptic-lanceolate, acute, the petals similar to sepals, the lip fleshy, yellowish-green, forming a deep pocket, the apex broadly triangular. Fruits not seen. *Zetek s.n.* (collected in 1942).

Collected only once on the island. Flowers throughout the rainy season and in the early dry season (June to January).

The species is vegetatively identical to *C. warczewitzii* Lindl. & Paxt., a species thus far unknown on BCI.

Pollinated by *Euglossa cordata* (Dodson, 1967a; Dressler, 1968a), by *E. cyanaspis* (Dressler, 1968a), and by *E. variabilis* (R. Dressler, pers. comm.).

Panama to Venezuela and Brazil. Apparently a wide-ranging species ecologically; on the basis of few collections, it appears that the species ranges from tropical moist forest at elevations of 24 m, such as around Gatun Lake, to tropical wet forest and premontane rain forest at elevations of more than 1,000 m. In Panama, known from Colón (Cerro Santa Rita) and Chiriquí (Volcan Chiriquí).

Catasetum viridiflavum Hook., Bot. Mag. 69, t. 4017. 1843

Monoecious or dioecious epiphyte; pseudobulbs large, fusiform, to 25 cm long, enveloped by leaf bases, the older leafless ones persisting. Leaves 6–12-plicate, elliptic-lanceolate, acuminate, 20–48 cm long, 3.5–12 cm wide. Flowers unisexual, dimorphic, on separate, stout racemes from base of pseudobulb; staminate inflorescences erect or arching, 25–70 cm long; staminate flowers 2–12, lacking aroma, to 5 cm long, pale green, aging yellow, the sepals oblong to oblanceolate, acuminate to cuspidate, to 5 cm long, the petals ovate, acuminate, to 4.5 cm long, the lip very firm, subglobose, to 3.5 cm long, green but inside of distal margin yellow (the color fading through to outer surface), the lateral margin ciliate, weakly spurred at base, the column beaked and exserted from lip, the lower edge with 2 antennae, one curved laterally, the other extending out into the spur depression, the anther beaked, ca 1.5 cm long, the pollinia ovoid, ca 6 mm long, ejected from column with considerable force; pistillate racemes infrequent, stouter than staminate racemes; pistillate flowers 2–4, similar to staminate flowers but the sepals and petals fleshy and smaller (to 2.5 cm long), the lip rounded at base, 3–4 cm long, persisting in fruit. Fruits fusiform, heavy and fleshy, to 9 cm long, bluish-green, with 5 broad ribs. *Croat 5515, 5546, 11298.*

Common on tree stumps in the lake; occasionally high in trees in the forest. Flowers from the late dry season

---

KEY TO THE SPECIES OF CATASETUM

Lip of staminate flowers 1 cm or less, with 5 elongate lobes (3 apical, 2 basal) . . . . . *C. bicolor* Klotzsch
Lip of staminate flowers 2 cm or more, lacking elongate lobes . . . . *C. viridiflavum* Hook.

Fig. 163. *Campylocentrum pachyrrhizum*

Fig. 164. *Catasetum viridiflavum*

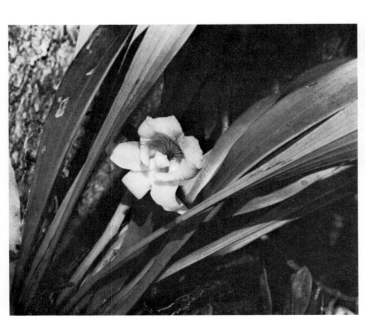

Fig. 165. *Cochleanthes lipscombiae*

(April) to the late rainy season (November), mostly from May to September. The fruits develop to full size by November, probably dehiscing chiefly during the dry season.

Recognized by the large pseudobulb and the plicate leaves, as well as by the thick waxy flowers and large fruits.

Pollinated by *Eulaema cingulata* (Dodson, 1965a) in the early morning hours (Dressler, 1968; Hills, Williams & Dodson, 1972).

Allen (1949) reported that the pollinia are ejected with any disturbance of the anther or antennae. My investigations of three different flowers show that only the applicator is released from its position in the column by forceful movement of the base of the antennae. The anther can easily be removed without releasing the translator arm from its position, and the distal parts of the antennae can be moved violently without reaction. However, any movement of the sticky white applicator after its release from the column causes the translator arm to release violently. These flowers were possibly immature.

Known only from Panama in tropical moist forest in the Canal Zone, Colón, Veraguas (Coiba Island), and Panamá and from premontane wet forest in Coclé (hills south of El Valle).

See Fig. 164.

CATTLEYA Lindl.

Cattleya patinii Cogn., Dict. Icon. des Orch. t. 25. 1900
  C. skinneri Batem. var. *autumnalis* P. H. Allen

Epiphyte (or elsewhere semiterrestrial), to 40 cm tall; pseudobulbs fusiform to cylindrical, usually narrowed toward base, 15–30 cm long, usually 2-leaved from apex, enveloped by leafy bracts at base. Leaves conduplicate, elliptic to oblong-lanceolate, obtuse, coriaceous, 10–15 cm long, 3.5–5.5 cm wide. Inflorescences simple, terminal, few-flowered, erect racemes subtended by spathaceous bracts; flowers large and showy, both the sepals and petals 3–5 cm long, orchid-lavender, the petals broader; lip sessile, very deep orchid-lavender, folded along the edges forming a tube around the column 3–4.5 cm long. Fruits 5–7 cm long, narrowly tapered on both ends, not prominently ridged.

Rare; R. Dressler has seen the species on BCI, but no collections have been made. Flowers in the dry season, mostly from January to March. The fruits develop during the dry season and shed their seeds by early in the rainy season. Costa Rican and Guatemalan collections bearing the name *C. skinneri* Batem. show flowers from January to March.

Recognized by its large, terminal, showy, purplish flowers, the lip large, tube-shaped, and darker than the petals and sepals.

Costa Rica to Colombia and Venezuela; Trinidad. In Panama, range uncertain; known from tropical moist forest in the Canal Zone, Los Santos, and Darién and from premontane moist forest in Panamá (around Bejuco).

CAULARTHRON Raf.

Caularthron bilamellatum (Reichb.f.) Schult., Bot. Mus. Leafl. 18:42. 1958
  Diacrium bilamellatum (Reichb.f.) Hemsl.

Epiphyte, 20–65 cm tall; stems pseudobulbous, fusiform to cylindrical, to 16 cm long, hollow and often inhabited by ants, enveloped by close, short bracts, bearing 1–4 strap-shaped leaves at apex. Leaves conduplicate, 6–20 cm long, to 2.5 cm wide. Inflorescences terminal, few-flowered, long-pedunculate racemes; peduncles 5–45 cm long, bearing numerous short sheaths; flowers white or pink, often cleistogamous, pedicellate, 1 or few open at a time; sepals 11–16 mm long, slightly longer and narrower than the petals, thickened toward apex, equaling lateral petals; lip free from the column, with 2 horns on upper side near base. Fruits elliptic, ribbed capsules, 2.5–3 cm long at maturity. *Croat 7762, 8709.*

Occasional over trees at the margin of the lake, and infrequent within the forest. Flowering chiefly in February and March (rarely into the early rainy season). The fruits develop to full size by March, and seeds are usually shed by late in the dry season.

Mexico to Panama and across northern South America to Trinidad. In Panama, a lowland, principally coastal species known from tropical moist forest in the Canal Zone, Bocas del Toro, Colón, and Chiriquí, Panamá, and Darién and from tropical wet forest in Colón.

CHYSIS Lindl.

Chysis aurea Lindl., Edward's Bot. Reg. 23, t. 1937. 1837

Epiphyte; stems pseudobulbous, cylindrical, fusiform or club-shaped, branched, often pendent, 5–50 cm long, 1–2 cm wide. Leaves plicate, lanceolate, acute or acuminate, 6–40 cm long, to 5 cm wide, moderately thin, ± undulate, the persistent imbricating bases enclosing stems. Racemes short, usually solitary from axils of new leaf growth; floral bracts pale green, the veins light violet-purple; flowers 3–7, 2.5–4 cm long; sepals fleshy, yellow at apex, yellow-cream at base, 2–3 cm long; petals less fleshy than sepals, about as long as lateral sepals, creamy white with red veins and spotting in apical half; lip yellow or white, trilobate, the middle lobe sometimes purple with paler markings; pollinia 8. Fruits narrowly ellipsoid, 5 cm long and 2.5 cm wide, ribs broad.

Apparently rare on BCI and throughout Panama. Reported by R. Dressler (pers. comm.) for the island, but no collections seen. Seasonal behavior uncertain. Probably flowers and fruits principally during the dry season.

Though still somewhat uncertain, Fowlie (1971) believed Panamanian material belongs to *C. aurea*, of Venezuela and Colombia, and that it is distinct, therefore, from *C. maculata* (Hook.) Fowlie, a species ranging from Costa Rica north to Nicaragua and perhaps Honduras.

Panama, Colombia, and Venezuela. In Panama, known

only from tropical moist forest in the Canal Zone (BCI and the Pipeline Road north of Gamboa).

COCHLEANTHES Raf.

Cochleanthes lipscombiae (Rolfe) Garay, Orquideologia 4:152. 1969

Caespitose epiphyte; pseudobulbs lacking. Petioles conduplicate, 6–7 cm long, occasionally persisting; leaves moderately thin, elliptic-oblanceolate, acuminate to acute, mostly 12–30 cm long, tapering to petiole, the lowermost bractlike. Inflorescences 1-flowered, arching or somewhat pendulous, bracteate scapes 6–15 cm long, from axils of lowermost bracts; flowers large, with clovelike aroma; sepals white, 3–3.5 cm long, the laterals directed downward, the dorsal erect and recurved; lateral petals similar to dorsal sepal, directed outward; lip to 4.5 cm long and 3.5 cm wide, white tinged with violet on margin and with violet lines on face within, the lateral margins incurved around column; column white, shorter than lateral lobe of lip, 12–18 mm high. Fruits not seen. *Croat 12711.*

Apparently rare, occurring in the forest. Seasonality uncertain. It has been reported to flower at the end of the dry season (*Powell 11*), but has been collected in flower on BCI in late November. An individual plant transplanted to Summit Garden flowered twice, a few weeks apart, early in the dry season (January and February).

Allen (1949) reported that the flowers are fragrant during the morning. They are pollinated by the bee *Eulaema meriana* (R. Dressler, pers. comm.).

Known only from Panama, principally from tropical moist forest in the Canal Zone, but also from premontane wet forest in Coclé (El Valle).

See Fig. 165.

CORYANTHES Hook.

Coryanthes maculata Hook., Bot. Mag. 58, t. 3102. 1831

Epiphyte; pseudobulbs subcylindrical, strongly ridged and tapering to apex, 6.5–15 cm long, 2-leaved at apex. Leaves lanceolate, 30–60 cm long, 4–10 cm wide, plicate, strongly veined. Racemes 30–60 cm long, pendent from base of pseudobulbs; flowers 2–4, ca 10 cm diam, showy, of variable color and complex shape; sepals and petals membranaceous, usually similarly colored, clear yellow to pale brown with a few purplish spots or pale purple or reddish-brown; lip very fleshy, complexly 4-parted, waxy, yellow, forming a cup containing a clear liquid produced by glands at the base of the lip. Fruits not seen.

Apparently rare; reported by R. Dressler (pers. comm.) for the island, but no specimens have been collected.

Herbarium collections give evidence that the species flowers during spring (dry-season) months.

According to Allen (1949) the plants are usually found high in trees but often also on ant nests in association with *Epidendrum imatophyllum*. The unique pollination of this species was described by P. H. Allen in the *Flora of Panama* (Ann. Missouri Bot. Gard. 36:66–67. 1949). The bee *Euglossa cordata* was observed visiting this species in Trinidad (Dodson, 1965b) and in Panama (Dressler, 1968a).

Costa Rica to Venezuela, the Guianas, Brazil, and Peru; Trinidad. In Panama, known from tropical moist forest in the Canal Zone and adjacent Panamá.

DICHAEA Lindl.

Dichaea panamensis Lindl., Gen. & Sp. Orch. Pl. 209. 1833

Caespitose epiphyte; pseudobulbs lacking, stem unbranched, 4–18 cm long. Leaves conduplicate, 2-ranked, narrowly linear-lanceolate, acute or acuminate, sheathing at base, 1–2 (4) cm long, to ca 5 mm wide. Flowers numerous, solitary from leaf axils, on slender pedicels to 1.5 cm long, white spotted with lavender (sometimes completely lavender elsewhere), to 1.5 cm long; sepals acute to acuminate, the lateral ones falcate; petals shorter and broader than sepals; lip sagittate, recurved, the lateral lobes folded over the column; column short; pollinia held beneath a purple, caplike flap. Fruits narrowly obovate, ca 1 cm long, weakly ribbed. *Croat 8099.*

Abundant in the forest, often growing in dense mat, occasionally in association with *Masdevallia livingstoneana* and species of *Pleurothallis*. Flowers throughout the dry season, especially in February and March, rarely during the dry season elswhere in Panama.

Easily recognized by its small size and distichous leaves, most of which subtend a small flower.

Probably pollinated by the bee *Euglossa cordata* (Dressler, 1968a).

Mexico to Venezuela and Peru, probably also in Ecuador; according to Ospina (1958) it ranges along the entire Pacific coast of Colombia. In Panama, ecologically wide-ranging; known from tropical moist forest in the Canal Zone, Bocas del Toro, and Panamá, from premontane wet forest in Coclé (El Valle), and from tropical wet forest along the Caribbean slope in Veraguas and Colón.

Dichaea trulla Reichb.f., Beitr. Orch. Central Amer. 104. 1866

D. powellii Schlechter

Erect or ± pendent, epiphytic herb; pseudobulbs lacking; stem unbranched, 15–45 cm long, enveloped by condup-

KEY TO THE SPECIES OF DICHAEA

Leaves 1–2 (4) cm long . . . . *D. panamensis* Lindl.
Leaves 6–13 cm long . . . . *D. trulla* Reichb.f.

Fig. 166. *Dimerandra emarginata*

Fig. 167. *Encyclia chacaoensis*

licate leaf bases, the terminal portion bearing 6–12 leaves. Leaves distichously imbricating; blades spreading, linear-ligular, acute to acuminate, 6–13 cm long, to 6 mm wide, articulated at base, eventually deciduous. Flowers few, solitary from leaf axils; pedicel with a cucullate bract at apex; sepals and petals subequal, ± lance-elliptic, pale yellow-green, to 1.3 cm long and 4 mm wide, ± acute; lip purplish or green marked with purple, the claw at base broad and ligular, the blade somewhat expanded, trilobate, folded into an open-sided tube, the lateral lobes retrorse; column shorter than lip. Capsules glabrous, shiny, smooth, to 1.6 cm long. *Shattuck 1012.*

Apparently rare, in the forest. Known to flower in the early rainy season (May to August).

Easily distinguished from *D. panamensis* by having flowers many times shorter than the leaves.

Honduras to Venezuela, Guyana, and Brazil. In Panama, known from tropical moist forest in the Canal Zone and from premontane wet forest in Colón and Panamá; no doubt occurring in tropical moist forest in Darién since it has been collected in Colombia (Chocó, near the Panamanian border).

DIMERANDRA Schlechter

Dimerandra emarginata (G. Meyer) Hoehne, Bol. Agric. (São Paulo) 34:618. 1934
Epidendrum stenopetalum Hook.

Caespitose epiphyte, 15–40 cm long; pseudobulbs lacking; stems leafy, slender, with prominent longitudinal grooves, often somewhat flexuous; internodes usually prominently ribbed, often swollen. Leaves conduplicate, linear to oblong, 2.5–11(14) cm long, 0.4–1.5 cm wide, unequally, shallowly bilobed at apex. Flowers rose-colored, in 1 to few, short, few-flowered racemes; sepals and petals nearly alike, 10–13 mm long, acute or acuminate; lip ± obovate, 11–13 mm long. Capsules ellipsoid, ca 2.5 cm long, prominently ribbed. *Croat 4682, Shattuck 221.*

Common in the forest, usually rather high in trees. Flowers in the rainy season (August to December). The fruits reach full size by January and dehisce at least by April; some fruits persist without dehiscing well into the rainy season.

Mexico to Central America, Venezuela, Trinidad, the Guianas, Brazil, and Ecuador. In Panama, known from tropical moist forest along both slopes of the Canal Zone and in Bocas del Toro, Panamá, Herrera (Las Minas), and Darién.

See Fig. 166.

ELLEANTHUS Presl

Elleanthus longibracteatus (Lindl. ex Griseb.) Fawc., Fl. Pl. Jam. 38. 1893
E. trilobatus Ames & Schweinf.

Epiphytic (or elsewhere terrestrial); pseudobulbs lacking; stems slender and reedlike, to 120 cm tall. Leaves ± equidistant on stem, somewhat lanceolate, acuminate, to 18 cm long, plicate, becoming bractlike near apex; basal leaves deciduous. Inflorescences terminal, dense, many-flowered, bracteate, congested racemes to 9 cm long; lower bracts to 2.5 cm long, the upper ones reduced; flowers white or yellowish, ca 8 mm long; dorsal sepal apiculate, the lateral sepals somewhat oblique, acute; petals linear-oblong, obtuse, to 7 mm long; lip trilobulate, 6–7.5 mm broad, fimbriate-lacerate in outer half, the base with a pouchlike enlargement. Fruits elliptic-oblong, 1–2 cm long. *Shattuck 201.*

Apparently rare, collected once on Pearson Trail. Seasonal behavior not determined. The species has been collected in flower in the Canal Zone in May.

Costa Rica to Colombia. In Panama, known principally from tropical wet forest in Colón (Portobelo) and Panamá (Cerro Campana); known also from tropical moist forest in the Canal Zone and Panamá.

ENCYCLIA Hook.

Encyclia chacaoensis (Reichb.f.) Dressl., Phytologia 21:436. 1971
Epidendrum ionophlebium Reichb.f.

Repent or caespitose epiphyte, to 40 cm tall; pseudobulbs usually narrowly ovoid, 3–9 cm long. Leaves 2(3), borne at apex of pseudobulb, ligulate or elliptic, obtuse at apex, 10–35 cm long, 1.5–4 cm broad. Inflorescences terminal, to 15 cm long, bearing few to several flowers; flowers greenish to white; sepals 1.4–3.5 cm long; petals somewhat shorter, elliptic-oblanceolate to oblanceolate, acute or acuminate; lip to 2.2 cm long, ovate to orbicular, obtuse to apiculate at apex, clawed at base, colored as the petals but with purplish stripes. Fruits winged at maturity. *Shattuck 799.*

Not seen on the island in recent years. Flowering collections from Panama were all made between January and July.

Mexico to Venezuela. In Panama, known from tropical moist forest in the Canal Zone, Veraguas, and Los Santos and from premontane wet forest in Chiriquí.

See Fig. 167.

KEY TO THE SPECIES OF ENCYCLIA

Pseudobulbs bearing 1 leaf . . . . *E. chimborazoensis* (Schlechter) Dressl.
Pseudobulbs bearing 2 or more leaves:
    Flowers less than 1 cm long, the lip ± trilobate; leaves less than 15 cm long and 2 cm wide . . . .
    . . . . *E. triptera* (Brongn.) Dressl. & Poll.
    Flowers more than 2 cm long, the lip not trilobate; leaves usually more than 15 cm long and 2
      cm wide . . . . *E. chacaoensis* (Reichb.f.) Dressl.

Encyclia chimborazoensis (Schlechter) Dressl., Phytologia 21:440. 1971

Repent or caespitose epiphyte, to ca 40 cm high; pseudobulbs narrowly oblong, somewhat flattened, stipitate at base, 5–9 cm long, to 1.3 cm wide and 8 mm thick, borne on a flexuous stem. Leaves solitary from apex of pseudobulb, conduplicate, ligulate, acute to obtuse, (10)14–20(30) cm long, (1.5)2–3(5) cm wide, coriaceous. Inflorescences terminal, few-flowered, to 15 cm long, ensheathed at base, the sheath to 2.5 cm long; flowers with sweet, intense aroma, subtended by a narrowly acute bract; sepals narrowly lanceolate, acuminate, to 3 cm long; petals oblanceolate, acuminate, to 1.7 cm long; both sepals and petals greenish-white to yellowish in age, dotted near base with violet-purple; lip broadly ovate, to 1.8 cm long and 1.1 cm wide, narrowly long-acuminate, striped throughout with violet-purple, the lateral margins upturned. Fruits not seen. *Croat 15573, Shattuck 551.*

Rare, in the forest. Flowering usually November to January (R. Dressler, pers. comm.).

Central Panama through western Colombia to Ecuador. In Panama, known from tropical moist forest on BCI and from premontane wet forest in Coclé (El Valle).

Encyclia triptera (Brongn.) Dressl. & Poll., Phytologia 21:438. 1971

E. pygmaea (Hook.) Dressl.; *Epidendrum pygmaeum* Hook.

Small, repent epiphyte, the rhizome creeping; pseudobulbs cylindrical to fusiform, 2–10 cm long, scattered along the rhizome, 2-leaved at apex. Leaves narrowly elliptic to oblong-oval, obtuse to acute, to 15 cm long and 2 cm broad. Inflorescences short, sessile in upper leaf axils; bearing few to several flowers; flowers inconspicuous; sepals and petals pale green; sepals to 12 mm long; petals linear, acute, to 8 mm long and 1 mm broad; lip ± trilobate, white with 1–3 purplish spots or streaks on the middle lobe, ± rounded and clawed, 3–8 mm long. Fruits not seen. *Dressler s.n. (STRI).*

Rare. The few Panamanian collections indicate that the species flowers in the rainy season.

Florida and Mexico to Brazil; Trinidad, Greater Antilles. In Panama, known from tropical moist forest in the Canal Zone and Darién and from premontane wet forest in Chiriquí and Coclé (El Valle).

EPIDENDRUM L.

Epidendrum anceps Jacq., Select. Stirp. Am. 224, t. 138. 1763

Epiphyte, 15–100 cm tall; pseudobulbs lacking; stem unbranched, usually somewhat 2-edged. Petioles articulate below blade; leaves conduplicate, ligulate to elliptic-lanceolate, acute or obtuse, 4–25 cm long, 1–5 cm wide, largest on the middle of the stem. Inflorescences simple or branched, terminal, racemose or subcapitate; peduncles very long, 10–45 cm long, longer than rachis, 2-edged, covered with scarious sheaths; flowers greenish- or brownish-yellow; sepals 5–10 mm long and 2–4 mm wide; petals slightly shorter and to 1.5 mm broad, sometimes reduced, threadlike, and nearly as long as sepals; lip somewhat cordate, the claw adnate to the column and somewhat enclosing it at apex. Fruits not seen. *Croat 9793.*

Uncommon, usually occurring rather high in trees; abundance is perhaps underestimated, since its flowers are not at all conspicuous. Flowering may occur throughout the year with the same plant producing several inflorescences; most flowers have been seen from April through November. The fruits develop rather quickly and are not often seen.

A very variable species.

Throughout most tropical and subtropical regions of the Western Hemisphere. In Panama, known chiefly from tropical moist forest in the Canal Zone, Bocas del Toro, and Panamá; known also from premontane wet forest in Chiriquí and Panamá (Cerro Campana).

Epidendrum coronatum R. & P., Syst. Veg. 242. 1798

E. moyobambae Kranzl.; *E. subpatens* Schlechter

Caespitose, pendent or ascending epiphyte, ca 80 cm long; pseudobulbs lacking; stem unbranched. Leaves elliptic-lanceolate to elliptic-ovate, acute, 6–16 cm long, 1.5–4 cm wide, coriaceous. Racemes terminal and pendent, to 40 cm long, long-pedunculate, the rachis longer than the peduncle, bearing few to many flowers; flowers greenish to yellowish or whitish; sepals mostly oblanceolate, 17–25 mm long, 5–9 mm wide; petals slightly smaller than sepals, of similar shape; lip broader than long, trilobate, the claw adnate to the column. Fruits not seen. *Croat 5462.*

Apparently rare, occurring in trees in the forest. Seasonal behavior uncertain. Collected once in flower in May.

Mexico to Venezuela, Brazil, Peru, and Ecuador; Trinidad. In Panama, known only from tropical moist forest in the Canal Zone and Veraguas.

Epidendrum difforme Jacq., Enum. Syst. Pl. Ins. Carib. 29. 1760

Caespitose or repent epiphyte, to 50 cm tall; pseudobulbs lacking; stems straight to flexuous, covered with leaf sheaths. Leaves usually ± oblong, blunt at apex and often emarginate, mostly 4–10(12) cm long and 0.7–3.5 cm wide. Inflorescences short, terminal, subumbellate racemes; flowers green or yellowish (especially in age), long-pedicellate, few to several; sepals 12–35 mm long; petals 7–30 mm long; both petals and sepals narrow; lip waxy, reniform, broader than long, 12–30 mm broad, the claw adnate to the column. Fruits ellipsoid, ca 3.5 cm long and 2 cm diam. *Croat 10106.*

Occasional, sometimes locally abundant; a variable species usually high in trees. Flowering to some extent all year, mostly in the early rainy season, from April to August, but especially in July and August. Mature dehiscing fruits are seen mostly in the dry season.

Pollinated by the moth *Amastus acona* in Ecuador (Dodson & Frymire, 1961).

KEY TO THE SPECIES OF EPIDENDRUM

Encyclia is included in this key because of possible confusion of the two genera.

Stems with true pseudobulbs:
  Pseudobulbs bearing 1 leaf:
    Inflorescences borne from near base of pseudobulb; leaf blades not linear, less than 10 cm long . . . . *Epidendrum rousseauae* Schlechter
    Inflorescences borne at apex of pseudobulb:
      Leaves linear, 4–12 cm long; flowers minute . . . . *Epidendrum stangeanum* Reichb.f.
      Leaves ligulate, more than 10 cm long; flowers more than 2 cm long . . . .
      . . . . *Encyclia chimborazoensis* (Schlechter) Dressl.
  Pseudobulbs bearing 2 or more leaves:
    Flowers less than 1 cm long, the lip ± trilobate; leaves less than 15 cm long and 2 cm wide
    . . . . *Encyclia triptera* (Brongn.) Dressl. & Poll.
    Flowers more than 2 cm long, the lip not trilobate; leaves usually more than 15 cm long and 2 cm wide . . . . *Encyclia chacaoensis* (Reichb.f.) Dressl.
Stems lacking true pseudobulbs (although the stems sometimes thickened):
  Leaves equitant or semiequitant (folded along the midrib); inflorescences composed of 1 to several, ± sessile flowers in the upper leaf axils:
    Plants forming dense mats, their stems appressed to tree, not more than 8 cm tall; sepals more more than 8 mm long . . . . *Epidendrum schlechterianum* Ames
    Plants erect, often more than 8 cm tall; sepals less than 8 mm long . . . .
    . . . . *Epidendrum lockhartioides* Schlechter
  Leaves not equitant; inflorescences racemose (in *E. sculptum,* 2 terminal flowers):
    Inflorescences long-pedunculate;
      Flowers reddish or purplish:
        Plants with long, adventitious roots near base; lip of the flowers markedly trilobate, the middle lobe smaller than the lateral lobes; plants rare . . . .
        . . . . *Epidendrum radicans* Pav. ex Lindl.
        Plants lacking long adventitious roots; lip of the flowers obscurely trilobate, the middle lobe larger than the lateral lobes; plants common . . . . . *Epidendrum imatophyllum* Lindl.
      Flowers greenish to yellow or white:
        Sepals less than 1 cm long; peduncle much longer than rachis of inflorescence; flowers brownish- to greenish-yellow . . . .*Epidendrum anceps* Jacq.
        Sepals more than 1.5 cm long; peduncle shorter than rachis of inflorescence; flowers greenish- to yellowish-white . . . . *Epidendrum coronatum* R. & P.
    Inflorescences not long-pedunculate:
      Inflorescences of 2 terminal flowers subtended by spathaceous bracts . . . .
      . . . . *Epidendrum sculptum* Reichb.f.
      Inflorescences a raceme of more than 2 flowers:
        Flowers white; sepals and petals more than 3.5 cm long, filiform to linear; plants common
        . . . . *Epidendrum nocturnum* Jacq.
        Flowers greenish to yellowish; sepals and petals less than 3.5 cm long:
          Flowers in a short, many-flowered, subumbellate raceme; sepals and petals usually more than 1 cm long . . . . *Epidendrum difforme* Jacq.
          Flowers in a strict raceme dispersed along rachis; sepals and petals less than 1 cm long:
            Sepals and petals less than 5 mm long; pedicel and ovary less than 1 cm long . . . . . .
            . . . . *Epidendrum strobiliferum* Reichb.f.
            Sepals and petals more than 5 mm long; pedicel and ovary more than 1.5 cm long
            . . . . *Epidendrum rigidum* Jacq.

Throughout the New World tropics. In Panama, known from tropical moist forest in the Canal Zone, Veraguas (Coiba Island), Darién, and Panamá, from premontane wet forest in Chiriquí and Coclé, and from lower montane wet forest in Chiriquí.

Epidendrum imatophyllum Lindl., Gen. & Sp. Orch. Pl. 106. 1831

Erect epiphyte, usually 75–100 cm tall; pseudobulbs lacking, stems slender, somewhat weak. Leaves ± ligulate, mostly to 12(20) cm long and 1.5(3) cm wide, acute or obtuse at apex, articulate at apex of sheath. Flowers orchid to lavender or violet-purple (the lip more intense), to ca 3 cm wide, in corymbiform terminal racemes, subtended by linear-lanceolate bracts; peduncles 6–20 cm long; sepals and lateral petals subequal, 1–2 cm long, acuminate, slender at base; lip obscurely trilobate (the middle lobe largest), shorter than lateral petals, clawed at base, the claw adnate to the column, the outer margins lacerate-dentate; column with 2 mamillate calluses at apex, these continuous with lamellate calluses on lip.

Fig. 168. *Epidendrum imatophyllum*

Fig. 170. *Lycaste powellii*

Fig. 169. *Epidendrum nocturnum*

Fruits ellipsoid, ca 4 cm long, narrowly tapered to both ends, the ribs ± prominent. *Croat 8257.*

Occasional, in the canopy of the forest or on exposed branches along the shore. Flowering from January to July, chiefly in the dry season (February to April). The fruits develop quickly and may be present on a flowering raceme.

Often associated with ant nests and with *Aechmea tillandsioides* var. *kienastii* (22. Bromeliaceae). Similar to *E. radicans,* but lacking the adventitious roots on the stem characteristic of that species.

Mexico to Panama, Venezuela, the Guianas, Brazil, Peru, and Ecuador; Trinidad. In Panama, known from low elevations in tropical moist forest in the Canal Zone, Los Santos, Panamá, and Darién.

See Fig. 168.

Epidendrum lockhartioides Schlechter, Feddes Repert. Beih. 19:39. 1923

Densely caespitose epiphyte, to 15(25) cm tall; pseudobulbs lacking. Leaves laterally compressed, acute at apex, 1–3.5 cm long, the narrow base subequitant, ensheathing stem, the upper leaves crowded. Flowers green to brownish-yellow, solitary in axils of each of several uppermost leaves; sepals ± lanceolate, cucullate, 6–8 mm long, to 3 mm broad, keeled; petals linear, slightly arcuate, obtuse at apex, 4–7 mm long; lip clawed, the claw adnate to the column, the blade subrounded, 4–5 mm broad, often apiculate, the midvein thickened. Fruits not seen.

No specimens seen from BCI, but reported to occur by R. Dressler. Flowers elswhere in Panama at the beginning of the dry season (December to February).

Costa Rica and Panama. In Panama, known primarily from premontane wet forest in Coclé (El Valle) and Panamá (Cerro Azul and the Río Boqueron above Madden Lake); known also from tropical moist forest in the Canal Zone and from tropical wet forest in Panamá (Cerro Jefe).

Epidendrum nocturnum Jacq., Enum. Syst. Pl. Ins. Carib. 29. 1760

Caespitose epiphyte, 20–60(100) cm tall; pseudobulbs lacking; stem canelike, unbranched, often covered with leaf sheaths or becoming naked. Leaves mostly ligulate to elliptic-oblong, acute to obtuse at apex, 8–16 cm long, 1–3.5(6.5) cm wide, ± fleshy. Racemes terminal, few-flowered; peduncles short; flowers large, white, only 1 opening at a time; sepals alike, linear, 3.5–7(9) cm long, to 5 mm broad; petals smaller than sepals, 3–5(8) cm long, linear; lip prominently trilobate, to 6.5 cm long, clawed, the claw adnate to the column, the lateral lobes of lip acute, 1–4 cm long, the middle lobe subfiliform, 2–4 cm long, usually much longer than the lateral lobes. Fruits ellipsoid, usually to 5 cm long, narrowly tapered to apex, the old flower usually persisting, the ribs with prominent margins. *Croat 12619.*

Frequent in the forest trees and on branches along the shore. Flowering throughout most of the year, especially in the late rainy season and early dry season (November to January). Mature-sized fruits are common in February.

Throughout most tropical regions of the Western Hemisphere; also in Sierra Leone, Africa. In Panama, most common in tropical moist forest in the Canal Zone and Bocas del Toro; known also from premontane wet forest in Coclé (El Valle), Colón (Santa Rita Ridge), and Chiriquí.

See Fig. 169.

Epidendrum radicans Pav. ex Lindl., Gen. & Sp. Orch. Pl. 104. 1831

Terrestrial; pseudobulbs lacking; stems short or to 1 m, erect, pendent or sprawling, simple or branched, usually with long whitish roots opposite some of the leaf bases. Leaves ligulate to ovate, blunt at apex, 1.5–8(12) cm long, 0.6–2(3.5) cm wide, coriaceous. Inflorescences terminal, the racemes lax, subumbellate or paniculate, borne on long bracteate scapes; flowers variable in size and color, from red to white but usually brick-red or orange (Colombian and Venezuelan forms violet), 12–22 mm long, the lobes acute; lip suborbicular-cordate, trilobate (the middle lobe smallest), 7–17 mm long, the claw adnate to the column. Fruits oblong-elliptic, narrowly tapered at base. *Shattuck 768, Woodworth & Vestal 703.*

Not seen on the island recently; collected twice in 1932 on floating islands of vegetation along the shore. Elsewhere preferring open, sunny areas. Flowers throughout the dry season (January to April), rarely during the rainy season. The fruits develop quickly, and mature-sized fruits may be found on the same inflorescence with flowers.

The species is similar to *E. ibaguense* H.B.K., which is also terrestrial but roots only at the base; *E. radicans* is terrestrial and has roots forming all along the stem. The species may also be confused with *E. decipiens* Lindl., which is an epiphyte that has roots only near the base.

This species was reported to be visited by swallow-tailed skippers, *Papilio* sp., and also by the hummingbird *Amazalia* sp., believed by Dodson to be the legitimate pollinator (1962).

Mexico and Central America to southwestern Colombia. In Panama, known principally from premontane wet forest and tropical wet forest at higher elevations in Chiriquí, Coclé (El Valle), and Panamá (Cerro Campana); known also from tropical moist forest in the Canal Zone and Bocas del Toro.

Epidendrum rigidum Jacq., Enum. Syst. Pl. Ins. Carib. 29. 1760

Small epiphyte; pseudobulbs lacking; stem repent, unbranched, covered with amplexicaul leaf sheaths. Leaves linear to oblong-elliptic, obtuse to rounded at apex, emarginate to bilobed, 3–13 cm long, 0.5–2 cm wide, thick. Racemes terminal, 5–15 cm long, several-flowered, short-pedunculate; bracts to 1.5 cm long; flowers yellow to greenish; sepals oblong to narrowly ovate, mostly 5–9 mm long, 2–4 mm wide, the lateral ones sometimes obovate; petals ± obovate, oblique, narrower but about as long as sepals; lip cordate-orbicular, 4–6 mm long, about as broad as long, clawed, the claw adnate to the column, the

lateral margins revolute; ovary and pedicel together more than 1.5 cm long. Fruits oblong-ellipsoid to ellipsoid, 1.5–2.5(3) cm long, the old flower persisting at the apex. *Shattuck 345, 649.*

Rare; collected recently by R. Dressler. Elsewhere in Panama flowers from April to December in the middle to late rainy season, especially from September to December. Mature-sized fruits have been seen throughout the dry season.

Throughout most tropical and subtropical regions of the Western Hemisphere. In Panama, know principally from tropical moist forest in the Canal Zone, Bocas del Toro, Colón, Panamá, and Darién; known also from tropical wet forest at higher elevations in Panamá (Cerro Brewster).

Epidendrum rousseauae Schlechter, Beih. Bot. Centralbl. 36, abt. 2:407. 1918
E. laterale Rolfe

Small epiphyte, sometimes trailing; pseudobulbs slender, to 3.5 cm long, 1-leaved; stems simple or branched. Leaves ligulate to elliptic-oblong, 5–10 cm long, 1.5–3 cm wide, obtuse to somewhat rounded and mucronate at apex, thick. Racemes arising from base of pseudobulb, several-flowered, to 12 cm long; bracts small; flowers greenish; sepals narrowly lanceolate, acute, fleshy, ca 1.2 cm long; petals as long as sepals, linear; lip trilobate, ca 7 mm long, deeply cordate at base, clawed, the claw adnate to the column for the length of the column. Fruits not seen. *Shattuck 347.*

Not seen recently on the island but to be expected. Seasonal behavior not determined. *Shattuck 347* had fruits nearly mature in November.

Known only from Panama, principally from premontane wet forest in Chiriquí (Monte Lirio), Coclé (El Valle), and Panamá (headwaters of Río Corso); known also from tropical moist forest in the Canal Zone and adjacent Panamá.

Epidendrum schlechterianum Ames, Sched. Orch. 7:9, fig. 1. 1924

Densely caespitose, dwarf epiphyte, to 8 cm tall; pseudobulbs lacking; stems covered with overlapping leaf bases. Leaves ± equitant, linear-oblong, acute, 1–3 cm long, to 1 cm wide, fleshy. Inflorescences of 1 to few sessile flowers in terminal leaf axils; flowers pale greenish-brown tinged with pink; sepals usually ± lanceolate, 0.8–2 cm long, 2–4.5 mm wide, the lateral ones slightly shorter than dorsal one and connate at their base with the claw of the lip; petals similar to sepals except smaller; lip long-clawed, trilobate, 9–15 mm long, nearly as broad as long, the claw adnate to the column, the lateral lobes small, the terminal lobe obovate-cuneate, bilobed or deeply bifid. Fruits ± globular, ca 1 cm long, prominently 3-angled. *Croat 8199.*

Occasional, in the forest, usually on the upper surface of larger branches high in the canopy. Flowers in Panama principally during the early dry season (December to February). The fruits develop quickly; ones of mature size are seen frequently in February.

Mexico to Panama, Venezuela, Surinam, Brazil, and Peru; Jamaica. In Panama, known only from tropical moist forest in the vicinity of the Canal Zone and in Veraguas (Coiba Island). R. Dressler reports that he would expect it in most of the drier forests of Panama (pers. comm.).

Epidendrum sculptum Reichb.f., Bonplandia 2:89. 1854

Pendent or repent epiphyte, 10–50 cm long; pseudobulbs lacking; stems simple or branched, densely leaved, the old petiole bases persisting. Leaves oblong to lanceolate-oblong, obtuse to rounded and emarginate at apex, 2–6 cm long, 1–1.8 cm wide, thick. Inflorescences of 1–3 (usually 2) terminal flowers subtended by spathaceous bracts; flowers green to greenish-yellow, reportedly with a fetid odor; sepals oblong-lanceolate, 1–1.5 cm long, ca 3 mm wide; petals similar to sepals; lip about as long as sepals, clawed, lanceolate-ovate, acute, trilobate, the lateral lobes rounded and about half as long as the lanceolate middle lobe, the claw adnate to the column. Fruits ± ellipsoid, remaining enveloped by spathaceous bracts. *Shattuck 558.*

Not seen in recent years on the island. In Panama, flowers principally from the middle to late rainy season (September to November). The fruits are of mature size by the early dry season.

Known from Panama and the Guianas. In Panama, known principally from tropical moist forest in the Canal Zone and Panamá; known also from premontane wet forest in Colón (Río Indio de Fato, *Pittier 4265*).

Epidendrum stangeanum Reichb.f., Gard. Chron. n.s. 15:462. 1881

Small, repent or caespitose epiphyte, to 20 cm long, the rhizome slender, creeping; pseudobulbs 1.5–5 cm long, to 5 mm diam, 1-leaved. Leaf linear, often inrolled at margins, 4–12 cm long, to 3 mm wide. Raceme in leaf axil, several-flowered, shorter than the subtending leaf; flowers small, tan; sepals lanceolate, 5–7 mm long, 1.5–3 mm wide; lip clawed, deltoid to ovate-lanceolate, apiculate, ca 5 mm long, the claw adnate to the column. Fruits narrowly ellipsoid, ca 12 mm long and 4.5 mm wide, 6-ribbed, 3 stouter ribs alternate with the other 3. *Shattuck 454.*

Not seen in recent years on the island. On the basis of a few collections, it appears that this species flowers only in the late rainy season (October to December).

Known from Costa Rica and Panama. In Panama, known only from tropical moist forest in the Canal Zone.

Epidendrum strobiliferum Reichb.f., Ned. Kruidk. Arch. 4:333. 1859

Small, repent or caespitose epiphyte, to 20 cm long; pseudobulbs lacking; stems simple or branched, covered with dense, amplexicaul leaf sheaths. Leaves ligulate to linear-lanceolate, acute to obtuse and usually bilobed at apex, 0.8–4.5 cm long, 2–10 mm wide, thick. Racemes short, few-flowered, terminal; bracts to 8 mm long, cucullate, chartaceous; flowers small, yellowish or white; sepals

± lanceolate, 3.5–5 mm long, 1.2–2 mm wide, the dorsal one narrower; petals linear-oblanceolate, smaller than sepals; lip ovate-cordate, acute, to 3.5 mm long and 3 mm wide, adnate to base of the column. Fruits ellipsoid, ca 1 cm long, the old flowers persisting at the apex. *Shattuck 550.*

Not seen on the island in recent years. On the basis of few collections, it appears that the species flowers exclusively in the rainy season (August to December), especially from August to September, with the fruits maturing in the early dry season.

Florida and Mexico through Central America to Panama, Venezuela, the Guianas, Brazil, and Peru; Trinidad, West Indies. In Panama, known from tropical moist forest in the Canal Zone and Panamá.

EULOPHIA R. Br. ex Lindl.

Eulophia alta (L.) Fawc. & Rendle, Fl. Jam. 1:112, t. 22, figs. 4–8. 1910

Glabrous, rhizomatous, terrestrial herb, to 130 cm tall. Petioles tightly ensheathing stem; leaves appearing sessile, linear-lanceolate, gradually tapered to apex, to 80 cm or more long, 5.5–6 cm wide, the midrib prominently canaliculate in basal half of blade, whitish and prominently ridged on underside, the ridges diminishing by apical fourth of blade. Inflorescences long, stout, erect racemes 70–100 cm long, terminating leafless scapes produced directly from the rhizome; flowers olive or greenish-tan; sepals free, lanceolate to oblanceolate, acute, to 25 mm long and 8 mm wide; petals erect, oblong, obtuse, ± equal to sepals; lip trilobate, shaded with rose or purple, the middle lobe with numerous denticulate longitudinal veins. Fruits 3.5–4 cm long, ellipsoid, with 6 prominent ribs, the pedicel ca 1.5 cm long. *Croat 12809.*

Rare; known only from Rear #8 Lighthouse Clearing, where it usually persists after each cutting. Flowers during the middle to late rainy season and the earliest part of the dry season. Mature-sized fruits have been seen in the early dry season (December). Seeds are probably dispersed during the dry season.

Florida and Mexico to Peru and Brazil; West Indies; West Africa. In Panama, known from tropical moist forest in the Canal Zone, from premontane wet forest in Coclé (El Valle), and from tropical wet forest in Colón.

GONGORA R. & P.

Gongora quinquenervis R. & P., Syst. Veg. 227. 1798
  G. maculata Lindl.

Epiphyte; pseudobulbs ovoid, 4.5–6.5 cm long, to 3.5 cm wide, dark brown, strongly ridged, enveloped at base by fibrous, imbricating bracts, bearing 2 or 3 terminal leaves. Leaves ± lance-elliptic, usually acuminate, plicate, 25–50 cm long, 5–12 cm wide, moderately thin. Racemes ± pendulous from base of pseudobulb, 30–60 cm long; flowers numerous (to 30), well spaced, grotesque, strongly fragrant, with all parts yellow and spotted or banded with reddish-brown, mostly membranaceous; pedicels slender, 2–3.5 cm long; dorsal sepal lanceolate, acuminate, to 1.8 cm long, free and erect, inserted about midway on the column, the lateral sepals strongly reflexed, obliquely ovate, acuminate, to 2.5 cm long, inserted on the column foot; petals slender, acuminate, to 8 mm long, inserted on column and fused at base to lateral sepals; lip fleshy, clawed, complexly bipartite, 1.5–2.5 cm long, with 2 slender, lateral, ± recurved projections from base and with 2 slender, erect, lateral antennae nearer the apex; column erect, ± arcuate, 1.5–2 cm long. Fruits not seen. *Dressler 2858.*

Apparently rare on BCI. Reportedly flowering irregularly more than once a year (*Powell 76*). Flowers principally in the dry season, but may flower in the early rainy season (to July) as well. The fruits apparently develop quickly and disperse their seeds chiefly in the dry season.

BCI plants are pollinated by the bee *Euglossa tridentata* Mourre, but have also attracted *E. cordata, E. townsendii, E. hemichlora,* and *E. cyanaspis* (Dressler, 1968a; Baker, 1963).

Mexico to Brazil and Peru. In Panama, known from tropical moist forest in the Canal Zone, Bocas del Toro, and Panamá, from premontane wet forest in Coclé (El Valle). Reported also from Colón (R. Dressler, pers. comm.).

Gongora tricolor (Lindl.) Reichb.f., Bonplandia 2:93. 1854
  G. maculata Lindl. var. *tricolor* Lindl.

Epiphyte, similar to *G. quinquenervis* but the flowers more richly colored; the sepals rich yellow to orange, usually sparsely blotched with dark red or dark purple; the petals greenish-yellow, usually spotted with dark red; the lip very fleshy, its base (the hypochile) broader than in *G. quinquenervis,* the basal horns of the lip hemispheric and fleshy (thicker than those of *G. quinquenervis*).

Apparently rare on BCI; reported by Dressler, but no specimens seen from the island. More common than *G. quinquenervis* (*fide* R. Dressler, pers. comm.). Flowering pattern is apparently the same as that of *G. quinquenervis.* Dressler (1968a) reported it to flower during March and April.

Pollinated consistently by the bee *Euglossa ganura.* Visits are made in the morning until 11:00 A.M. (Dressler, 1966).

KEY TO THE SPECIES OF GONGORA

Hypochile when seen from above relatively narrow, the base with lateral ligular projections . . . . *G. quinquenervis* R. & P.
Hypochile when seen from above relatively broad, the base auriculate or with short, lateral, fleshy, hemispheric horns . . . . *G. tricolor* (Lindl.) Reichb.f.

KEY TO THE SPECIES OF HABENARIA

Petals entire; lip entire . . . . *H. alata* Hook.f.
Petals bifid; lip trilobate:
    Spur at base of lip less than 1.5 cm long; plants found in swampy places or on floating debris
    . . . . *H. repens* Nutt.
    Spur at base of lip more than 3 cm long; plants not found in swampy or aquatic habitats . . . .
    . . . .*H. bicornis* Lindl.

Known from Panama and Peru. In Panama, apparently with the same range as *G. quinquenervis.* Dressler (1966) reported it to be common on the Atlantic slope in central Panama.

HABENARIA Willd.

Habenaria alata Hook.f., Exot. Fl. t. 169. 1826

Terrestrial, erect, 20–70 cm tall; pseudobulbs lacking. Leaves scattered along slender stems, linear-lanceolate, 6–14 cm long, 1–2 cm wide, much reduced at base and apex of stems. Inflorescences terminal, subracemose, 4–20 cm long; flowers dense, pale green; pedicels very short; sepals to 1 cm long, the dorsal sepal ovate to suborbicular, concave, to 7 mm wide, broader than the laterals, the lateral sepals ± lanceolate, acute to acuminate; petals ± entire, slightly smaller than the sepals, lanceolate, ± acute; lip ± lanceolate, 6–8 mm long, ca 2 mm broad; spur at base of lip recurved, subclavate, to 1.3 cm long. Fruits sharply ribbed to winged, ca 17 mm long and 6 mm wide, the lower fourth hidden by the persistent bract. *Kenoyer 249.*

Not seen in recent years on BCI. Flowers throughout the rainy season. The fruits are common in the late wet and early dry seasons.

Mexico to Colombia, Venezuela, and Bolivia; West Indies. In Panama, ecologically wide ranging; known from tropical moist forest in the Canal Zone and from lower montane wet forest in Chiriquí (Volcán).

Habenaria bicornis Lindl., Gen. & Sp. Orch. Pl. 309. 1835

Terrestrial, erect, to 65 cm tall; pseudobulbs lacking. Leaves scattered along slender stems, linear-lanceolate, 5–25 cm long. Racemes terminal, many-flowered, ca 8 cm long, bracteate; flowers green with white petals; sepals 6–10 mm long, ca 5 mm wide; petals bifid, the lobes unequal, to 9 mm long; lip trilobate, 7–12 mm long, the middle lobe shortest; spur at base of lip recurved, 4–5 cm long. Fruits narrowly ellipsoid, ca 2.5 cm long and 6 mm wide, prominently 6-ribbed.

No specimens seen for the island, but reported to occur

by R. Dressler. Flowers in July, according to the label on *Powell 315*; flowering collections have also been made in September.

Panama and the West Indies. In Panama, known from tropical moist forest in the Canal Zone and Panamá.

Habenaria repens Nutt., Gen. N. Amer. Pl. 2:190. 1818

Terrestrial, ± aquatic, erect, to 1 m tall; pseudobulbs lacking. Leaves scattered along slender stems, ± linear-elliptic, to 20 cm long, reduced toward apex. Racemes terminal on stems, densely flowered, bracteate, 10–15 cm long; flowers light green; sepals 4–8 mm long; petals 4–9 mm long, bipartite, the lobes subequal; lip tripartite, 5–10 mm long, the middle lobe slightly shorter; spur at base of lip recurved, to 15 mm long. Fruits much like *H. alata* but perhaps more rounded at apex. *Bailey & Bailey 663.*

Not seen on the island in recent years; found on floating debris and in swampy places. Possibly flowering all year, but mostly in the late rainy and early dry seasons. Both flowering and fruiting individuals have been collected from the same area during February.

Scattered from southeastern United States to Brazil; Trinidad, Greater Antilles. In Panama, known only from tropical moist forest on BCI.

IONOPSIS Kunth

Ionopsis satyrioides (Sw.) Reichb.f., Ann. Bot. Syst. 6:683. 1863

Small epiphyte; pseudobulbs minute or lacking. Leaves ± linear, from a very short stem, narrowly acute at apex, mostly 3–8 cm long, 1–4 mm wide, about half as thick as wide, sulcate along one edge. Inflorescences arising from among the leaves, simple or sparsely branched, equaling or exceeding length of leaves, 3–10 cm long; peduncles slender, with several sheathing bracts 3–5 mm long; flowers ca 6 mm long; sepals brownish-cream to rose-purple, united for ca one-third their length, acute and ± recurved at apex, the midvein ± thickened on outside; petals white with purplish markings on veins, thinner than sepals, to 5.5 mm long and 1.8 mm wide; lip

KEY TO THE SPECIES OF IONOPSIS

Leaves less than 5 mm wide, almost as thick as wide; inflorescences less than 10 cm long . . . .
. . . . *I. satyrioides* (Sw.) Reichb.f.
Leaves 5–15 wide, considerably wider than thick; inflorescences 15–60 cm long . . . .
. . . . *I. utricularioides* (Sw.) Lindl.

± oblong, truncate, to 5.5 mm long and 2.5 mm wide, white with 2 purplish lines medially, with a yellow, bifurcate callus near base just above the short, terete claw. Fruits oblong-elliptic, the body ca 1.5 cm long and 1 mm wide, narrowly tapered at both ends, the basal stipe to 7 mm long. *Dressler s.n.*

Rare. Flowers mostly in March and April. Mature fruits have been seen in April.

Belize, Guatemala, Costa Rica, and Panama (possibly other parts of Central America as well) across northern South America to Venezuela and the Guianas; West Indies. In Panama, known only from tropical moist forest in the Canal Zone and Chiriquí (Concepción).

Ionopsis utricularioides (Sw.) Lindl., Coll. Bot. t. 39-A. 1825

Epiphyte; pseudobulbs minute or lacking; stems very short. Leaves linear-lanceolate, acute, 4–16 cm long, 5–15 mm wide, keeled on lower surface, ensheathing stem at base. Inflorescences paniculate, from leaf axils; scapes slender, erect to arching, 15–60 cm long; flowers several, mostly 6–7 mm long, diffuse, lavender or violet to white; sepals subequal, ± oblong-lanceolate, acute, the dorsal sepal free, concave, the laterals connate at the base and produced into a short, broad sac; petals subequal to dorsal sepal, elliptic-ovate; lip twice as long as sepals, deeply emarginate and bilobed, contracted at base into a fleshy claw, the claw adnate to the base of the short, stout column; anther terminal, caplike. Fruits ellipsoid, 2–2.5 cm long, moderately ribbed and thin-walled. *Croat 4679.*

Occasional, in the Laboratory Clearing. Flowers in the late dry season (March to May, especially April). The fruits mature in the rainy season (June to October).

Florida and Mexico along the Caribbean coast and along the eastern side of the Andes to Brazil and Paraguay; West Indies; Galápagos Islands. In Panama, known from tropical moist forest in the Canal Zone, Panamá, and Darién.

LEOCHILUS Knowl. & Westc.

Leochilus scriptus (Scheidw.) Reichb.f., Xenia Orch. 1:15, t. 6. 1854

Caespitose, erect epiphyte, 10–20 cm tall; pseudobulbs compressed, ellipsoid, 2–5 cm long, enveloped at base by several imbricating bracts. Leaves usually solitary from apex of pseudobulb, ligulate, shortly bifid at the apex, 6–14 cm long, 1–2.8 cm wide, short-petiolate. Racemes 1 or 2 from basal leaf axils, ± equaling leaves; flowers 1.5–2 cm diam, greenish-yellow with purplish-brown markings especially near center of each segment; sepals subequal, spreading, 8–12 mm long, 2.5–5 mm wide, the dorsal sepal concave, keeled; petals similar to dorsal sepal, lanceolate, acute; lip obovate-oblong, truncate, to 12 mm long and 7 mm wide, keeled on lower surface, thickened at base and adnate to the base of the column; column semiterete with 2 extended ligular arms below the stigma; anther terminal and caplike. Fruits narrowly ellipsoid, to

ca 4 cm long, the body 2–2.2 cm long, 6-ribbed, 3 ribs obscure, the 3 alternate ribs very flattened. *Dressler 2900.*

Rare. Flowers in Panama chiefly in the late rainy and early dry seasons (November to February). The fruits probably develop quickly; mature fruits have been seen in February.

Guatemala to Panama. In Panama, known from tropical moist forest on both slopes of the Canal Zone and in Veraguas and Panamá.

LIPARIS L. C. Rich.

Liparis elata Lindl., Edward's Bot. Reg. 14, t. 1775. 1828

Terrestrial where humus abundant, or epiphytic, 15–40 cm tall; pseudobulbs inconspicuous, usually enveloped by leaves; stems short, becoming fleshy, sheathed with broad petiole bases. Leaves 3 or 4, rosulate, at apex of pseudobulb; blades ± elliptic, 5–30 cm long, 2.5–11 cm wide, plicate. Racemes terminal on stems, equaling or exceeding leaves; pedicels slender, 5–10 mm long; flowers light greenish-yellow with purplish-brown veins; sepals unequal, oblong to ovate, obtuse, 4–7 mm long, the lateral sepals arcuate; petals linear to linear-oblanceolate, obtuse, arcuate, smaller than sepals; lip ± obovate, truncate or emarginate and purplish-brown at apex, 4–5.5 mm long, with 2 tubercles at base. Fruits oblong-elliptic, ca 1 cm long. *Dressler 2864.*

Rare. Flowers mostly in the early rainy season. Fruits are seen throughout the dry season. The fruiting inflorescence sometimes persists on the old leafless stem until the next flowering.

Florida and Mexico to Peru and Brazil; West Indies. In Panama, known chiefly from premontane wet forest at middle elevations such as in Panamá (Cerro Campana); known also from tropical moist forest in the Canal Zone and Panamá.

LOCKHARTIA Hook.

Lockhartia acuta (Lindl.) Reichb.f., Bot. Zeitung (Berlin) 10:767. 1852

L. pallida Reichb.f.

Caespitose epiphyte; pseudobulbs lacking; stem flattened, unbranched, covered with equitant, distichously imbricating leaves. Blades acute, short, to 3 cm long. Flowers small, yellowish or white with a yellow lip, in 1–3 divaricately branched panicles to 8 cm long at or near apex; pedicels slender, subtended by small subcordate bracts; sepals and lateral petals subequal, ovate, ± rounded at apex, strongly concave, to ca 4 mm long; lip 4.7–6 mm long, somewhat expanded and bilobed at apex, fused to the short column near the base; column with lateral wings, the wings not surpassing apex; pollinia 2, pale yellow, teardrop-shaped. Fruits ± obovate, 6–10 mm long, pruinose, smooth, splitting into 3 valves. *Croat 8056.*

Common high in trees in the forest. Flowers in the dry season (December to April), especially in the early dry season, with the fruits maturing in the early rainy season.

Inflorescences may appear successively with both flowering and fruiting inflorescences present on the same plant.

Panama across northern South America to Trinidad.
In Panama, known from tropical moist forest in the Canal
Zone and Panamá, apparently more frequently on the
Pacific slope.

Lockhartia pittieri Schlechter, Feddes Repert.
12:216. 1913

Caespitose epiphyte; pseudobulbs lacking; stem flattened,
unbranched, obscured by equitant, distichously imbricating leaves. Blades acute, 2–3.5 cm long. Inflorescences
short, solitary, 1-flowered scapes from upper leaf axils;
flowers yellow with an orange callus on inner surface of
lip near base; sepals elliptic-lanceolate, acute, 4–5 mm
long, with apiculate tips; petals broader than sepals,
obtuse to acute; lip 7–8 mm long, ± convex and deeply
emarginate, the margins raised, thickened, and papillose
at apex with an erect, central spur. Fruits not seen.

No specimens seen for BCI, but reported by R. Dressler. Apparently flowers in the early dry season (*Powell
372*).

Belize to Panama. In Panama, known from tropical
moist forest in the Canal Zone and Darién.

LYCASTE Lindl.

Lycaste powellii Schlechter, Feddes Repert. Beih.
17:65. 1922

Epiphyte (or elsewhere pseudoterrestrial, growing in
sandy places, *fide Powell 15*); pseudobulbs elliptic-ovoid,
laterally compressed, 3–6 cm long, smooth to ± ridged,
bracteate at base, the uppermost bracts leaflike. Leaves
2 or 3, terminal on pseudobulb, ± lanceolate, acute to
acuminate, 20–35 cm long, 3–3.5 cm wide, ultimately deciduous, ± plicate. Inflorescences 1–4, erect, 1-flowered,
5–15 cm long from base of pseudobulb before leaves
have fallen; peduncles slender, with several sheathing
bracts along their length and a solitary flower at apex;
flowers relatively large, very fragrant; sepals usually pale
green, 2.5–3 cm long, the apices strongly recurved, the
laterals adnate to base of column; petals about equal to
sepals, usually white with rose markings; lip trilobate,
to 2.5 cm long, nearly as wide, colored like petals, the
lateral lobes erect, the middle lobe broadly obtuse and
somewhat reflexed; column 8–9 mm long, the undersur-
face conspicuously pubescent. Fruits not seen. *Dressler
2868*.

Rare, growing in wooded ravines (Allen (1949) in the
Flora of Panama). Flowers in July.

The species is closely related to *L. brevispatha* Klotzsch
ex Lindl., which is reported from higher elevations.

Known only from Panama, principally from premontane and tropical wet forests in Coclé (El Valle) and
Panamá (Cerro Campana); known less frequently from
tropical moist forest such as on BCI.

See Fig. 170.

MASDEVALLIA R. & P.

Masdevallia livingstoneana Reichb.f., Gard. Chron.
ser. 2, 2:322. 1874

Caespitose epiphyte, 6–13 cm tall; pseudobulbs lacking.
Leaves solitary on very short secondary stems; blades
oblanceolate, 5–11 cm long. Inflorescences shorter to
somewhat longer than leaves; flowers to 12 mm long,
pale yellow or white with the throat violet-purple, solitary on peduncles from base of leaves, the fragrance
strong, sweet; sepals 1.5–2 mm long, connate into a tube
for half their length, the free part of dorsal sepal thickened, ligulate, to 12 mm long and 3 mm broad, arcuate-
spreading, the lateral sepals similar to the dorsal sepal but
slightly wider; petals and lip enclosed within calyx tube;
petals 4–5 mm long with a single longitudinal callus from
below middle to apex; lip ± oblong, 4–6 mm long. Fruits
not seen. *Croat 9120*.

Occasional, in the forest usually high in trees on
branches. Flowers from February to April.

Reported only from Panama; possibly also in Costa
Rica. In Panama, known from tropical moist forest and
premontane wet forest in the Canal Zone, Colón, and
Panamá.

MAXILLARIA R. & P.

Maxillaria alba (Hook.) Lindl., Gen. & Sp. Orch. Pl.
143. 1832

Epiphyte with elongate, rhizomatous stems enveloped by
closely imbricating, persistent bracts; pseudobulbs inserted at an acute angle to stem, ± overlapping, narrowly
elliptic, flattened and 2-edged, 4–5 cm long, 1-leaved.
Leaves ligulate, acute to bilobed at apex, conduplicate,
25–40 cm long, 1–2 cm wide. Flowers white to cream
except for yellow on lip, solitary from axils of bracts of
new growth, several open at once; pedicels 3–4 cm long,
longer than bracts; sepals subequal, free, spreading, 2–2.5
cm long, to ca 5 mm wide, the laterals adnate to base of
column; petals ca equal to sepals; lip obscurely trilobate,
concave, slightly curved, yellow or white with yellow

KEY TO THE SPECIES OF MAXILLARIA

Leaves not borne on conspicuously thickened secondary stems (all but the leaves of the small, inconspicuous pseudobulb are actually foliaceous bracts); plants caespitose (lacking an elongate
rhizome) . . . . *M. crassifolia* (Lindl.) Reichb.f.
Leaves borne on conspicuously thickened secondary stems (i.e., pseudobulbs):
Plants caespitose (growing in a flattened tuft); pseudobulbs all ± erect, among weathered bracts;
scapes 1–4, 1-flowered, arising from among the pseudobulbs, 4–13 cm long, the bracts continuous along its length, mostly more than 1.5 cm long, scarcely if at all overlapping . . . . . .
. . . . *M. powellii* Schlechter
Plants caulescent, erect, scandent, or pendent; pseudobulbs inserted at an oblique angle along an
elongate, sometimes branching stem:
Leaves very numerous, 1 per pseudobulb, mostly less than 4 cm long and 5 mm wide, ± tri-
angular in cross section . . . . *M. uncata* Lindl.
Leaves much larger, not as above:
Flowers ca 6 mm long, produced in dense clusters at base of pseudobulb on new growth,
each flower enveloped by 2 glumaceous bracts; stem bracts closely spaced, ca 1.5 cm
long or less; fruits globular, ca 5 mm long . . . . *M. neglecta* (Schlechter) L. O. Wms.
Flowers usually much larger, mostly more than 1 cm long, not densely clustered or enveloped by bracts; stem bracts usually longer than 1.5 cm (short in *M. friedrichsthalii*):
Scapes closely bracteate throughout their length, usually several from base of mature
pseudobulb; stem bracts mostly less than 1.5 cm long; leaves usually 2 or 3 (some-
times 4) per pseudobulb . . . . *M. friedrichsthalii* Reichb.f.
Scapes not bracteate (at least not toward apex), solitary from base of mature pseudobulb
and/or from bract axils on flush of new growth; stem bracts mostly more than 1.5
cm long; leaves 1 or 2 per pseudobulb:
Leaves usually 2 per pseudobulb; bracts of new growth unequally bilobed at apex;
pseudobulbs often widely scattered along stem; flowers usually more than 2.5 cm
long . . . . *M. camaridii* Reichb.f.
Leaves usually 1 per pseudobulb; bracts of new growth acute at apex; pseudobulbs
usually closely positioned; flowers usually less than 2.5 cm long:
Leaves usually less than 20 cm long; flowers usually less than 1.2 cm long . . . .
. . . . *M. variabilis* Lindl.
Leaves usually more than 25 cm long; flowers usually more than 2 cm long . . . .
. . . . *M. alba* (Hook.) Lindl.

lobes, 1.2–1.5 cm long, ca 5 mm wide, contracted at base and articulated with foot of column. Fruit not seen. *Dressler 2951* (F).

Collected once in the forest. Flowers mostly in the rainy season, but flowering collections have also been made during the dry season.

Allen (1949) reported that flowers are produced more or less simultaneously, in contrast to most other species of *Maxillaria*, which produce flowers in succession.

Guatemala to Brazil; West Indies. In Panama, known from tropical moist forest on both slopes in the Canal Zone and in Panamá and Darién; known also from pre-montane wet forest in Colón (Santa Rita Ridge) and Chiriquí (Boquete).

Maxillaria camaridii Reichb.f., Hamburger Garten-Blumenzeitung 19:547. 1863
Camaridium ochroleucum Lindl.

Epiphyte with long, cylindrical, pendent stems covered with sheathing bracts; pseudobulbs inserted at an angle on the stem, 3–7 cm long, flattened and 2-edged. Leaves (1)2 at apex of pseudobulb, ± linear, shallowly and unequally bilobed at apex, 10–30 cm long, conduplicate. Flowers solitary on short peduncles, produced in succes-sive pairs from bract axils on new growth, fragrant; sepals

white, subequal, 2.5–3.5 cm long, the laterals adnate at base to base of column; petals white, nearly equaling sepals, widely spreading; lip white outside, yellow inside with purplish-brown, transverse streaks, trilobate, 10–12 mm long and wide, contracted at base and articulated with foot of column. Fruits not seen. *Shattuck 341, 346, 348.*

Not seen recently on BCI. Collected by Shattuck dur-ing November, but reportedly flowering in the dry season as well.

Allen (1949) in the *Flora of Panama* reported that each plant flowers three or four times during a season.

Guatemala to Panama and northeastern South America in the Guianas and Trinidad. In Panama, known from tropical moist forest in the Canal Zone and Panamá and from premontane wet forest west of the Canal Zone (El Valle in Coclé and Cerro Campana in Panamá).

Maxillaria crassifolia (Lindl.) Reichb.f., Bonplandia 2:16. 1854

Caespitose, erect epiphyte, 20–40 cm tall; pseudobulbs not obvious. Leaves linear-lanceolate, 1–3.8 cm wide, all but the leaf of the small, inconspicuous pseudobulb actually foliaceous bracts. Flowers solitary on very short scapes from upper bract axils; sepals subequal, spreading,

yellowish, lanceolate, acute, 1.5–2 cm long, the laterals adnate at base to foot of column; petals yellow, slightly smaller than sepals; lip obscurely trilobate, yellow with red spots to dark red, 12–14 mm long, concave, the lateral margins erect, contracted at base and articulated with foot of column. Fruits oblong, 2–3 cm long, ca 5 mm wide, closely ribbed.

No specimens seen for BCI, but reported by R. Dressler. Flowering collections have been made throughout much of the year, but in Panama mostly flowers in the rainy season with the fruits maturing in the early dry season.

Florida and Mexico to Brazil; Greater Antilles. In Panama, known principally from tropical moist forest in the Canal Zone, Bocas del Toro, Panamá, and Darién; reported also from an unknown locality in Chiriquí at 1,200 m elevation.

Maxillaria friedrichsthalii Reichb.f., Bot. Zeitung (Berlin) 10:858. 1852

Erect or pendent epiphyte; stems slender, covered by closely imbricating bracts; upper pair of bracts often foliaceous; pseudobulbs oblong-elliptic, rugose when dry, 1.5–5 cm long, inserted at an angle on the stem. Leaves 2 or 3 (sometimes 1 or 4) from apex of pseudobulb, ligulate, 4–18 cm long, to 1.5 cm wide, conduplicate and short-petiolate at base. Flowers solitary on 1 to several scapes from base of mature pseudobulb; scapes 2–4 cm long, bracteate; sepals and petals usually curved in the same direction; sepals subequal, linear-lanceolate, 1.5–3 cm long, greenish-yellow or greenish-lavender, the laterals adnate to base of column; petals 1–2.5 cm long, greenish-yellow, somewhat smaller than sepals; lip entire, linear-oblanceolate, 1–2.5 cm long, 3–6 mm wide, pale yellow-green, geniculate and ± S-shaped in side view, the lateral margins ± erect, dark maroon and thickened at apex, articulated with foot of column. Fruits ± oblong, narrowly tapered at apex, 2.5–3 cm long, ca 6 mm wide, bluntly 3-sided, slightly pruinose, the 3 ribs ca 1.5 mm wide, very thin-winged marginally, the old flower persisting. *Shattuck 453.*

Collected once on the island, but not seen in recent years. Flowers mostly in the early dry season, rarely in the rainy season. Apparently mature-sized fruits were seen undehisced in late July.

Grows clinging to the sides of trees in partial shade (*fide Powell 136*).

Mexico to Panama. In Panama, known chiefly from premontane moist forest on the Pacific slope in Chiriquí, Coclé, and Panamá at intermediate elevations; known also from tropical moist forest in the Canal Zone and Panamá.

Maxillaria neglecta (Schlechter) L. O. Wms., Ann. Missouri Bot. Gard. 29:348. 1942

Ornithidium anceps Reichb.f. non *M. anceps* A. & S.

Erect or pendulous epiphyte; stems frequently branching, covered by closely imbricating bracts; pseudobulbs variable, ligulate to suborbicular, 1.5–4.5 cm long, inserted at an acute angle on the stem, with 2 long bracts enveloping base. Leaves solitary from apex of pseudobulb, ligulate, 6–21 cm long, 1–2.5 cm wide, conduplicate and short-petiolate at base. Inflorescences many-flowered, produced in ± dense clusters from base of pseudobulbs; flowers small, solitary, usually enveloped in 2 glumaceous bracts borne on short scapes; sepals concave, ca 6 mm long, not spreading, yellow or white, the laterals broadly rhombic, oblique at apex, adnate at base to column; petals slightly shorter than sepals, colored similarly; lip yellow, geniculate in profile, ca 6 mm long, trilobate near apex, the lateral lobes erect. Fruits globose, ca 5 mm long, rugose, with 3 broad flat ridges. *Shattuck 544.*

Collected once on BCI; common elsewhere in Panama. Flowers throughout the rainy season. The fruits develop to full size by the early dry season, no doubt dehiscing during the dry season.

The vegetative habit is exceedingly variable, especially the shape and size of the pseudobulb.

Belize to Panama. In Panama, ecologically widespread; known primarily from higher elevations in premontane wet forest in Chiriquí and Panamá and from tropical wet forest in Colón and Los Santos; known also from tropical moist forest in the Canal Zone; possibly occurring in premontane rain forest and lower montane rain forest in Chiriquí.

Maxillaria powellii Schlechter, Feddes Repert. Beih. 17:70. 1922

Caespitose epiphyte; pseudobulbs elliptic-ovoid, 2–4 cm long, close, compressed, enclosed at base by papery bracts later weathering into fibers. Leaves solitary from apex of pseudobulb, ligulate, acute to bifid at apex, 15–40 cm long, 2–3 cm wide, conduplicate and petiolate at base. Inflorescences of 1–4 slender, 1-flowered, bracteate scapes 4–13 cm long from base of pseudobulb; bracts continuous along length of scape, mostly 1.5–2 cm long, scarcely if at all overlapping; sepals free, spreading, subequal, yellow or tan, 1.5–2 cm long, ± obtuse at apex, the laterals adnate at base to foot of column; petals 1.5–1.7 cm long, yellow or tan, obliquely ligulate, held ± erect on either side of the column; lip conspicuously trilobate at apex, ca 1.5 cm long, yellow or tan except the middle lobe sometimes reddish-brown, the lobes curved downward above the column. Fruits not seen. *Croat 12526a.*

Rare, in the forest. Flowers in the late rainy season and the early dry season (October to January), especially in the late rainy season.

According to Allen (1949) in the *Flora of Panama,* the species is doubtfully distinct from *M. ringens* Reichb.f.

Known only from Panama chiefly from premontane wet forest at intermediate elevations in Coclé and Panamá (west of the Canal Zone); known also from tropical moist forest in the vicinity of the Canal Zone.

Maxillaria uncata Lindl., Edward's Bot. Reg. 23, sub. t. 1986. 1837

Creeping or pendulous epiphyte; stems with many minute pseudobulbs ca 1 cm long enveloped at base by brown sheaths. Each pseudobulb with a single, thick,

sharp leaf, ± V-shaped in cross-section, mostly 2–4(8) cm long and 3–5(10) mm wide. Flowers usually solitary, 15–18 mm long, translucent with maroon or red stripes; dorsal sepal short, 7–9 mm long, equaling petals in length, the lateral sepals ca 1 cm long, oblique at base and adnate to foot of column to form a projection; the lip slender, extending from gibbous base, ± equaling the lateral sepals, the sides folded inward; column flat inside, rounded outside, bearing 4 small yellow pollinia at its apex, the apical hood entire, its front edge easily removed as a narrow rectangular strip bearing the pollinia with it. Capsules ca 1 cm long, with 3 boat-shaped valves, persisting after seeds are shed. *Croat 6744.*

Common in the forest, usually on tree branches high in the canopy. Flowers throughout the rainy season and the earliest part of the dry season, especially in the rainy season. The fruits mature quickly, but most are apparently mature during the dry season.

Mexico to Peru, Brazil, Venezuela, and the Guianas. In Panama, common from tropical moist forest in the Canal Zone, Colón, Veraguas, and Darién, from premontane wet forest on both slopes in the Canal Zone, Coclé, and Panamá, and from tropical wet forest in Coclé and Panamá.

Maxillaria variabilis Batem. ex Lindl., Edward's Bot. Reg. 23, sub. t. 1986. 1837

Erect or pendulous epiphyte; stems simple, bracteate, 5–30 cm long; pseudobulbs elliptic-oblong, 1.5–6 cm long, inserted at an angle on the stem. Leaves solitary from apex of pseudobulb, ligulate, shallowly bilobed at apex, 3–20 cm long, to 1.5 cm wide, conduplicate and petiolate at base. Flowers usually solitary on short scapes from axils of bracts of new growth or from base of pseudobulb, to 1.2 cm long, colored white, yellow marked with red, or entirely dark red; sepals subequal, ± lanceolate, 6–12 mm long, spreading, the laterals adnate at base to column; petals similar but slightly shorter than sepals; lip entire or obscurely trilobate, 6–12 mm long, half as wide, the lateral margins erect, the base articulated with foot of column. Fruits not seen.

No specimens seen from BCI, but reported by Dressler. Flowers principally in the dry season (especially early in the dry season), infrequently in the rainy season.

A vegetatively variable species. Reported by Ospina (1958) to be generally terrestrial in Colombia.

Mexico to Colombia, the Guianas. In Panama, ecologically wide ranging; known principally from premontane wet forest and lower montane wet forest in Chiriquí;

known also from tropical moist forest in the Canal Zone and Panamá.

MORMODES Lindl.

Mormodes powellii Schlechter, Feddes Repert. Beih. 17:55. 1922

Dioecious epiphyte, often on rotting wood; pseudobulbs elongate, ± cylindrical, tapering gradually toward apex. Leaves several, scattered along stem, plicate, deciduous, the bases persistent, imbricating. Racemes arching, ca 30 cm long, produced from near base to middle of pseudobulb (often leafless when in flower); staminate and pistillate flowers similar; sepals and petals subequal, spreading, linear-lanceolate, acuminate, to 5 cm long, greenish to yellowish-brown to cream-colored; lip elliptic- to rhombic-ovate, with a short claw, the lateral margins often strongly recurved; column twisted to one side, pointed, with a slender apical bristle. Fruits not seen. *Croat 7794.*

Seen only in the epiphyte house, but reported for the island by Dressler. Flowers in the dry season, from December to March.

Reported by R. Dressler to be distinct from *M. colossus* Reichb.f., with which it was included in synonymy in the *Flora of Panama* (Allen, 1949). *Mormodes colossus* occurs in western Panama and Costa Rica and can be distinguished from *M. powellii* by having a much wider lip.

The species is pollinated by the bee *Euglossa tridentata* (Dressler, 1968a).

Known only from Panama, occurring in tropical moist forest in the Canal Zone and Colón and in premontane wet forest in Coclé (El Valle).

NOTYLIA Lindl.

All species of the genus are apparently pollinated by euglossine bees (Dressler, 1968a).

Notylia albida Klotzsch in Otto & Dietr., Allg. Gartenzeitung 19:281. 1851

N. panamensis Ames

Epiphyte; pseudobulbs ca 2 cm long and 1 cm wide, partially enveloped in papery imbricating bracts, the bracts soon weathering away. Petioles conduplicate, very short; blades solitary at apex of pseudobulb, oblong-lanceolate, obtuse and unequally bilobed at apex, to 15 cm long and 3.5 cm wide, coriaceous. Racemes pendulous, from base of current pseudobulb, densely flowered,

KEY TO THE SPECIES OF NOTYLIA

Column distinctly papillose . . . . *N. pentachne* Reichb.f.
Column glabrous:
    Dorsal sepal broadly elliptic-oblanceolate, obtuse, ca 5 mm wide, distinctly broader than the lip
    . . . . *N. albida* Klotzsch
    Dorsal sepal linear-lanceolate, subacute, 2 mm wide or less, distinctly narrower than the lip . . . .
    . . . . *N. barkeri* Lindl.

to 20 cm long; flowers white; pedicels slender, subtended by scarious, narrowly triangular, acuminate bracts; sepals subequal, the dorsal sepal broadly elliptic-oblanceolate, obtuse, concave, ca 7 mm long, 5 mm wide, the lateral sepals fused, the united segments linear-lanceolate, to 8 mm long, 2 mm wide; petals elliptic-lanceolate, acute, equaling sepals; lip ca 6 mm long, clawed, obliquely inserted at base of column, the blade sagittate, ca 3 mm wide, with a short-keeled callus; column slender, terete, ca 3 mm long. Fruits not seen. *Dressler 2908* (*US*).

Apparently rare. The Dressler collection was flowering in June.

The species is pollinated by the bee *Euglossa hemichlora* (Dressler, 1968a).

Panama and Colombia. In Panama, known from tropical moist forest in the Canal Zone, Colón (Achiote), and Darién.

Notylia barkeri Lindl., Edward's Bot. Reg. n.s. 1, Misc. 90. 1838

Small epiphyte, often with long, pendent roots; pseudobulbs oblong, to 2.5 cm long and 1 cm wide, enveloped at base by several imbricating bracts, the uppermost bracts foliaceous. Leaves solitary from apex of pseudobulb, ligulate, 8–18 cm long, 1.3–4 cm wide, conduplicate at base. Racemes 1 or 2, arching, 8–30 cm long (usually less than 15 cm long), from base of pseudobulb, many-flowered; flowers white; pedicels filiform, 2–6(10) mm long, subtended by minute bracts; sepals subequal, linear-lanceolate, ± spreading, the dorsal sepal free, 3–6 mm long, 1–2 mm wide, the laterals connate partway, 3–5 mm long, 1–2 mm wide together; petals 2.5–5 mm long, 1–1.5 mm wide; lip clawed and continuous with base of column, 3–5 mm long, 1–2 mm wide, the basal part with a distinct keeled callus; column terete, glabrous, 2–3 mm long. Fruits narrowly ellipsoid, ca 1.5 cm long, with low flat ribs.

No specimens seen for BCI, but reported by R. Dressler. Apparently flowers in the late dry season (March and April).

Flowers are pollinated by the bee *Euglossa* (Dressler, 1968a).

Mexico to Panama; apparently widespread ecologically, occurring from near sea level to 1,200 m. In Panama, known from tropical moist forest in the Canal Zone, from premontane moist forest in the Canal Zone (Balboa), and from premontane wet forest in Chiriquí and Veraguas.

Notylia pentachne Reichb.f., Bonplandia 2:90. 1854

Epiphyte, usually less than 15(25) cm high; pseudobulbs oblong, to 2.5 cm long, with foliaceous bracts at base, 1-leaved. Leaves oblong, to 20 cm long, 2–5 cm wide, tapered to short, conduplicate petiole. Flowers many, in 1 or 2 erect to pendulous racemes to 35 cm long (usually less than 25 cm long), from base of pseudobulb; pedicels slender, mostly 6–10 mm long; sepals green, 8–10 mm long, slender, recurved at apex, the lateral sepals connate for more than half their length; petals 7–8 mm long, slender, obliquely lanceolate, acuminate, whitish with a green or orange spot above base; lip white, narrowly deltoid above, with a basal claw; column slender, terete, densely papillose, 3–5 mm long; anther light green, ca 2.3 mm long; pollinia yellow on a long white appendage, sticky at apex. Fruits ± ellipsoid, to 4 cm long including beak, the beak tapered, prominent. *Croat 4745, 5708.*

Occasional, in the forest, usually rather high in trees. Flowers in the dry season, mostly from March to late May. The fruits develop quickly, and an old inflorescence may bear fruits during the late dry season while another inflorescence bears flowers.

Pollinated by the bee *Eulaema cingulata* (Dressler, 1968a).

Known only from Panama, principally from tropical moist forest in the Canal Zone, Veraguas (Santa Fé), and Darién.

ONCIDIUM Sw.

Oncidium ampliatum Lindl., Gen. & Sp. Orch. Pl. 202:1833

Epiphyte; pseudobulbs ovoid to orbicular, flattened or sometimes angular, 3–12 cm long, 3–9 cm wide, often purple-spotted, sometimes with large weathered bracts at base. Leaves 1–3 at apex of pseudobulb, narrowly oblanceolate to ligular, obtuse to subacute, 8–40 cm long, 3–12 cm wide, conduplicate and short-petiolate at base. Panicles 1 or 2, produced laterally from base of pseudobulb, much longer than leaves; flowers very flat, ca 2.5 cm diam; sepals subequal, 6–10 mm long, 3.5–5 mm wide, brownish-yellow, broader than sepals, 7–11 mm long, 5–9 mm wide, clawed at base, the blades suborbicular; lip trilobate, 1.5–2 cm long, the lateral lobes small, the center lobe deeply emarginate, 1.5–3 cm wide, the base of the lip with a fleshy, lumpy callus colored white with red spots, contracted into a short claw adnate to base of column; column 3–4 mm long, 3-winged at apex. Fruits not seen. *Croat 6746, Woodworth & Vestal 529, 706.*

Infrequent, in the forest and on stumps at the margin of the lake. According to Allen (1949), once common in the valley of the Río Chagres. Flowers usually in the early dry season (December to March), especially in February and March.

KEY TO THE SPECIES OF ONCIDIUM

Pseudobulbs conspicuous, compressed .... *O. ampliatum* Lindl.
Pseudobulbs lacking, or rudimentary and inconspicuous .... *O. stipitatum* Lindl.

Easily recognized by its more or less disk-shaped, often spotted pseudobulbs.

Guatemala to Venezuela, Trinidad, Peru, and Ecuador. In Panama, known principally from tropical moist forest in the Canal Zone; known also from tropical wet forest in Veraguas (on the Pacific slope near Bahía Honda).

Oncidium stipitatum Lindl. in Benth., Bot. Voy. Sulphur 172. 1846

Caespitose epiphyte; pseudobulbs lacking or 5–10 mm long, usually ensheathed in papery bracts, bearing a single leaf. Leaves terete, longitudinally grooved, 24–70 cm long, to 1 cm wide, becoming pendent and usually reddish-brown in age. Panicles solitary, many-flowered, from base of pseudobulb, ± equaling leaves; flowers variable in size, to 2.5 cm long; sepals and lateral petals subequal, clawed at base, yellow heavily marked with reddish brown; lip trilobate, fiddle-shaped, clawed, yellow, to 2 cm long, nearly as broad as long, the lateral lobes obovate-spatulate, with a short claw, separated from the broad bilobed middle lobe by a narrow isthmus; base of lip with a fleshy, toothed callus; column with lateral, fanlike wing. Fruits ± ellipsoid, tapered toward apex, to 2.5 cm long, with flat ridges. *Croat 8396.*

Occasional, along the margin of the lake, usually not in full sun. Flowers from January to April, especially in February and March. The fruits develop quickly and are dispersed usually by May (or as late as July).

Pollinated by a bee of the genus *Centris* (Dodson, 1965a).

Known only from Panama, principally in tropical moist forest in central Canal Zone, Colón, and Panamá; known also from premontane wet forest in Panamá (Cerro Campana).

See Fig. 171.

ORNITHOCEPHALUS Hook.

Ornithocephalus bicornis Lindl. in Benth., Bot. Voy. Sulphur 172. 1846

Small epiphyte; pseudobulbs lacking; stems very short, hidden by leaves. Leaves 5–12 cm long, flattened laterally, arranged in a suborbicular fan; blades oblong-lanceolate, obliquely acute, articulated to the conduplicate, imbricating, persistent base, the base to 3 cm long, obliquely acute at apex, resembling a pseudobulb. Racemes 1–17, from basal leaf axils, many-flowered, slender, equaling or exceeding leaf length; rachis densely erect-pubescent with short, simple or branched trichomes; flowers ca 4 mm diam, greenish; sepals suborbicular, concave, obtuse,

glandular-puberulent outside, keeled medially, often tinged with yellow; petals as large as sepals, obovate, ca 2 mm diam, concave, thin, white toward apex, green at base, with a prominent ciliate keel; lip green, slender, strongly incurving, to 5.2 mm long and 1.2 mm wide, the base with a cushionlike callus and two papillate horns, the apex acute, white, held just above the ornate column; pollinia with a slender stipe extending down the column; stigma basal on column. Fruits ellipsoid, ca 5 mm long, densely pubescent with simple or branched trichomes. *Croat 8243.*

Occasional, in the forest. Flowers mostly in the dry season (December to April) or as late as July, with fruits developing quickly and commonly maturing in May.

Pollinated by the bee *Paratetrapedia calcarata* (Dodson, 1967a).

Mexico to Panama, Colombia, Venezuela, and Ecuador. In Panama, known from tropical moist forest in central Canal Zone, Veraguas, Panamá, and Darién.

Ornithocephalus powellii Schlechter, Feddes Repert. Beih. 17:88. 1922

Small, fan-shaped epiphyte; pseudobulbs lacking; stems short, hidden by leaves. Leaves grayish-green, few, obliquely ligulate, acute, often apiculate, 3–7 cm long, to 1.2 cm broad, articulate with a conduplicate base to 3 cm long and often oblique at apex. Racemes 1 or 2 from basal leaf axils, to 10 cm long, few-flowered; rachis laterally compressed, narrowly winged, subglabrous to puberulent with bracts to 7 mm long; flowers green, to ca 1.5 cm diam; sepals oblong-ovate, ca 4 mm long, keeled, the margins ciliate to serrulate; petals obovate-flabellate, to ca 6 mm long, often broader than long, round at apex, ciliate to serrulate; lip fiddle-shaped, to ca 1 cm long, the apex serrulate; disk with a fleshy, bilobed callus. Fruits not seen. *Croat 5487.*

Rare. Flowering records are scattered; apparently flowers intermittently in both the dry and rainy seasons.

Known only from Panama, from tropical moist forest in the Canal Zone and Panamá.

PALMORCHIS Rodr.

Palmorchis powellii (Ames) Schweinf. & Corr., Bot. Mus. Leafl. 8:119. 1940
Rolfea powellii Ames

Terrestrial, often caespitose, to 60 cm tall; stems slender, reedlike. Leaves few, scattered along stem, moderately thin, plicate; petioles to 8 cm long; blades ± elliptic, 15–30 cm long, 3–15 cm wide. Inflorescences usually

axillary, often borne at leafless nodes, racemose or paniculate, to 7 cm long, bracteate at base of flowers; bracts ovate, acute, 5–6 mm long; flowers white, 1.5–2 cm long; sepals and petals ± spatulate, obtuse, 3 mm wide; lip somewhat trilobate, about as long as but broader than sepals and petals, adnate to base of column; column ca 12 mm long. Fruits narrowly oblong, to 5 cm long, ca 5 mm wide, weakly ribbed. *Foster 2362.*

Rare on the island; seen recently in the older forest. Apparently flowers in the early rainy season (July and August). The fruits develop quickly, and the seeds may be dispersed in the rainy season.

The related *P. trilobulata* L. O. Wms. is found in tropical moist forest and in premontane wet forest in Coclé (El Valle) and Panamá (Chimán).

Apparently restricted to Panama, in tropical moist forest on BCI and in Panamá.

PERISTERIA Hook.

Peristeria elata Hook., Bot. Mag. 58, t. 3116. 1831
Holy Ghost, Dove orchid

Terrestrial, to 1.3 m tall; pseudobulbs ovoid, 4–12 cm long, 4–8 cm wide, enveloped at base by foliaceous bracts. Leaves 3–5 from apex of pseudobulb, deciduous; blades 30–100 cm long, 6–14 cm wide, plicate. Racemes solitary from base of pseudobulb, 80–130 cm tall, erect, few-flowered; flowers subglobose, fleshy, white, fragrant, opening 2–4 at a time from the base first; pedicels to 4 cm long; sepals subequal, 2.5–3 cm long, 2–3 cm wide, the laterals connate at base; petals 2–2.5 cm long, 1.5–2 cm wide; lip with lateral wings spotted red, articulated at base with the broad claw, thickened at base, with an apical crest; column ca 1 cm long. Fruits ellipsoid, 4–5.5 cm long, to 2 cm wide. *Shattuck 854.*

A plant of open areas, probably eliminated by succession. Flowers in the rainy season. Fruits in the dry season.

The national flower of the Republic of Panama, the species is highly favored by orchid growers, but has disappeared in some areas.

Pollinated by the bee *Euplusia concava* and also visited by *Euglossa* sp. (Dressler, 1968a).

Costa Rica to Colombia and Venezuela. In Panama, known from premontane moist forest and tropical moist forest in Panamá and from premontane wet forest at low to medium elevations in Colón, Coclé, and Panamá.

PLEUROTHALLIS R. Br.

Pleurothallis brighamii S. Wats., Proc. Amer. Acad. Arts 23:285. 1888
P. acrisepala Ames & Schweinf.

Caespitose epiphyte; pseudobulbs lacking; stems very short, each bearing a single leaf. Leaves ± oblanceolate, obtuse or acute, 1.5–9 cm long, the veins obscure except the midrib, the base ensheathed in slender, scarious bracts. Flowers 1 to few in terminal fascicles borne on slender peduncles usually above the leaves; peduncles usually persisting; sepals acute, maroon, fused at base, 7–9 mm long, the lateral sepals united to the middle or beyond; petals 2–3.5 mm long, solid maroon, the apex pointed; lip narrow, 2–4 mm long, 1–2 mm wide, lighter and papillate at apex, the margin ciliate, auriculate on both sides. Fruits ellipsoid, to 1 cm long, green or olive, tinged or striped with purple. *Croat 9787.*

Abundant in the forest, perhaps the most abundant orchid on the island; growing high or low in trees. Apparently flowers repeatedly throughout much of the year.

Guatemala to Panama. In Panama, known from tropical moist forest on both slopes of the Canal Zone and in Bocas del Toro, Colón, Panamá, and Darién.

Pleurothallis grobyi Batem. ex Lindl., Edward's Bot. Reg. 21, t. 1797. 1835
P. marginata Lindl.

Densely caespitose epiphyte; pseudobulbs lacking; stems very short, bearing a single leaf. Leaves oblanceolate to rarely suborbicular, obtuse to rounded and usually minutely emarginate at apex, attenuate to petiole, 1.5–7 cm long, 2–8 mm wide, drying bicolorous, the margin dark, thickened. Flowers pale yellow-green, few to several, in slender weak racemes 5–12 cm long, usually much exceeding leaf; both peduncles and petioles surrounded by dried sheaths; dorsal sepal 4–6(9) mm long, the lateral sepals weakly fused laterally, somewhat longer than dorsal sepal; petals erect, much shorter, hidden within sepals, marked with violet-purple; lip oblong, similar to lateral petals but longer; column white, equaling or exceeding petals, winged, the wings extending above anther in a point. Fruits not seen. *Croat 10800.*

Common high in trees on shaded, moss-laden branches. Apparently flowering repeatedly throughout the year.

KEY TO THE SPECIES OF PLEUROTHALLIS

Stems caespitose (tufted); leaves obtuse or at most acute at apex:
  Inflorescences fascicles borne at apex of long peduncles, of 1 to several flowers; peduncles often
    persisting; leaves 7 or 8 times longer than wide, ± concolorous . . . . *P. brighamii* S. Wats.
  Inflorescences fragile racemes; peduncles not persisting; leaves usually less than 6 times longer
    than broad, bicolorous (at least when dried) . . . . *P. grobyi* Lindl.
Stems elongate, erect or repent; leaves ± sharply pointed at apex:
  Inflorescences short, few-flowered fascicles, much shorter than the leaves; secondary stems to 6
    cm long; leaves to 7 cm long . . . . *P. trachychlamys* Schlechter
  Inflorescences few-flowered racemes, ca half as long as leaves; secondary stems to 19 cm long;
    leaves 6–16 cm long . . . . *P. verecunda* Schlechter

Fig. 171. *Oncidium stipitatum*

Fig. 172.
Pleurothallis verecunda

Common throughout the New World tropics. In Panama, known principally from tropical moist forest on both slopes of the Canal Zone and in Panamá; known also from tropical wet forest in Colón (Río Indio and Portobelo).

Pleurothallis trachychlamys Schlechter, Feddes Repert. Beih. 17:23. 1922

Small repent epiphyte, to ca 11 cm high; pseudobulbs lacking; rhizome concealed in coarse sheaths, the sheaths bearing ± hispid trichomes; secondary stems to 6 cm long, covered with a scurfy sheath, each stem bearing a solitary leaf. Leaves narrowly elliptic to lanceolate, narrowly acute at both ends, 4.5–7 cm long, 0.4–1 cm wide. Flowers white, very short-lived, appearing in succession in short fascicles near the base of the upper leaf surface; sepals oblong-lanceolate, acute, 4–5 mm long, the lateral sepals united at base; petals lanceolate, to 4 mm long; lip oblong-oval, obtuse, to 2 mm long, with 2 erect lateral lobes near the middle and a small callus at the base, the apex recurved; column two-thirds as long as lip with the lateral margins raised. Fruits not seen.

No specimens seen for BCI, but reported to occur by R. Dressler. Known to flower in November.

Costa Rica to Venezuela and possibly Peru. In Panama, known from tropical moist forest in the Canal Zone and Panamá.

Pleurothallis verecunda Schlechter, Feddes Repert. Beih. 17:24. 1922

Erect or repent epiphyte, to ca 20 cm high; pseudobulbs lacking; secondary stems slender, 1-leaved, to 19 cm long, enveloped tightly at base by a scarious sheath, arising from a creeping, densely rooting rhizome. Leaves lanceolate-ligulate, acute to acuminate, narrowly acute at base, 6–16 cm long, 1–2.5 cm wide, thick; blade, secondary stem, and pedicels often violet-purple. Flowers few, in terminal racemes about half as long as leaves, yellow-green, all parts lightly or heavily marked with violet-purple especially near base; sepals narrowly pointed, 5.5–9 mm long, with a thick medial rib, persisting in fruit; petals oblong-oblanceolate, much shorter and enclosed within sepals, exceeding length of column; lip thick, ± oblong, 2.5–3.5 mm long, with 2 short erect teeth laterally below the middle. Fruits oblong, somewhat oblique, broadened toward apex, 2.5–3.5 cm long. *Croat 16200*.

Infrequent, in the forest and in *Citrus* (66. Rutaceae) trees at the Laboratory Clearing. Flowers principally in the rainy season but also in the late dry season (April to October). The fruits have been seen in January but also in mid-June.

Costa Rica and Panama, In Panama, known from tropical moist forest in the Canal Zone and tropical wet forest in Colón (Río Indio).

See Fig. 172.

POLYSTACHYA Hook.

Polystachya foliosa (Lindl.) Reichb.f., Ann. Bot. Syst. 6:640. 1863

P. cerea Lindl; *P. minor* Fawc. & Rendle

Caespitose epiphyte, 10–30 (60) cm tall; stems short; pseudobulbs ovoid, enclosed in the leaf sheaths, with 2–5 leaves from apex. Leaves 3–15 (25) cm long, 0.5–3 cm wide. Panicles terminal, 5–15 (30) cm long; peduncles stout, with short racemose branches; rachis glabrous to sparsely puberulent; flowers yellow, maturing at base of raceme first, each subtended by a persistent bract; sepals 3–4.5 mm long, 1.5–3 mm wide; petals much smaller; lip 2.5–3.5 mm long, 2–3 mm wide, trilobate, the middle lobe largest, farinose (covered with a mealy powder); ovary glabrous. Fruits ellipsoid, ca 1 cm long, ridged. *Croat 6707*.

Frequent in the forest and on tree branches over the edge of the lake. Flowers chiefly in September and October. The fruits mature chiefly in the dry season (January to March).

Mexico through tropical South America; West Indies. In Panama, known from tropical moist forest chiefly on the Pacific slope of the Canal Zone and in Panamá and from premontane wet forest in Chiriquí (Lino) and Coclé.

Polystachya masayensis Reichb.f., Bonplandia 3:217. 1855

Tiny epiphyte, usually 3–10 cm tall; stems very short; pseudobulb inconspicuous, ovoid, usually enclosed in leaf sheaths, bearing 1–5 leaves at apex. Leaves mostly 2.5–8 cm long, 4–8 mm wide, acute to rounded at apex. Inflorescences terminal racemes, usually 1–2.5 cm long, shorter than the leaves; peduncles enclosed in leaf sheaths; rachis densely puberulent; flowers greenish-yellow, ca 3 mm long, each subtended by a persistent bract; dorsal sepal somewhat smaller than the laterals; petals slender; lip somewhat constricted at center, farinose, the apex acute; ovary puberulent. Fruits ellipsoid, ca 7 mm long, puberulent. *Croat 7370*.

Apparently uncommon, usually growing high in trees in the forest. Flowers from late November to March, perhaps over a longer span.

Mexico, Nicaragua, Costa Rica, and Panama. In Panama, known from tropical moist forest in the vicinity of the Canal Zone and from premontane wet forest in Colón, Chiriquí, and Coclé.

KEY TO THE SPECIES OF POLYSTACHYA

Ovary, fruit, and rachis usually glabrous; inflorescences branched many times in large plants; plants mostly more than 10 cm tall . . . . *P. foliosa* (Lindl.) Reichb.f.
Ovary, fruit, and rachis usually pubescent (± puberulent); inflorescences usually unbranched, 1–2 cm long; plants usually less than 10 cm tall . . . . *P. masayensis* Reichb.f.

PSYGMORCHIS Dods. & Dressl.

Psygmorchis pusilla (L.) Dods. & Dressl., Phytologia 24:288. 1972
Oncidium pusillum (L.) Reichb.f.

Dwarf epiphyte, to 8 cm tall; pseudobulbs lacking. Leaves ± dense, spreading like a fan, equitant, ± linear, 2–6 cm long, to 1 cm wide. Inflorescences 1–6 from base of leaves, about equaling leaves, consisting of long scapes, the apices with several acute, strongly compressed, imbricating sheaths; flowers produced in succession from axils of sheaths; flowers 2–2.5 cm long; sepals free, spreading, bright yellow, keeled and apiculate, the dorsal sepal ca 5 mm long, nearly as wide, the lateral sepals 4–5 mm long, 1–1.5 mm wide, hidden by lateral lobes of lip; petals to 8 mm long and 4 mm wide, bright yellow streaked with reddish-brown, the margins undulate; lip trilobate, to 2 cm long and broad, bright yellow, the lateral lobes ± orbicular, the middle lobe deeply emarginate, the base with an elaborate subquadrate process; column ca 3 mm long, with prominent, denticulate, lateral wings. Fruits ellipsoid, ca 1.5 cm long, with slender ridges. *Croat 4157.*

Rare on BCI, more common elsewhere in Panama. Flowers during most of the year, but principally in the dry season. Known to flower more than once per season since plants often bear both flowers and fruits on separate inflorescences.

Often found in large colonies (especially on *Citrus*, 66. Rutaceae, trees) covering many of the branches.

Mexico to Venezuela and south to Brazil and Bolivia; Trinidad. In Panama, known from tropical moist forest on both slopes of the Canal Zone and in Bocas del Toro, Panamá, and Darién.

SCAPHYGLOTTIS Poepp. & Endl.

Scaphyglottis graminifolia (R. & P.) Poepp. & Endl., Now. Gen. Sp. Pl. 1:59. 1836
S. behrii (Reichb.f.) Benth. & Hook. ex Hemsl.

Caespitose epiphyte, to 45 cm tall; stems densely clustered and slender, consisting of several superimposed, weakly swollen, finely ridged segments (the lower segments most slender), occasionally branched, the old leafless stems persisting. Leaves usually 2, near apex of each segment, usually only the uppermost persisting, linear, 5–25 cm long, 1.5–4.5 mm broad, obscurely emarginate at apex. Inflorescences single flowers or more commonly few-flowered fascicles or abbreviated, few-flowered racemes, borne at apex of stems; flowers white, 3.5–4.5 mm long; sepals 3–4.5 mm long, 1–2 mm wide; petals as long as sepals, 0.5–1 mm wide; lip 3.5–5 mm long, 2–3.5 mm wide, entire or obscurely trilobate; column narrowly winged. Fruits oblong-elliptic, ca 1 cm long (including the long narrowly tapered base), ca 2 mm wide. *Croat 8079.*

Common in the forest, usually high in trees. Flowers in the early dry season (December to March), especially in January and February. The fruits mature in the middle to late dry season.

Confused with *S. longicaulis,* which is usually a much smaller plant bearing mostly solitary flowers.

Guatemala to Venezuela, Brazil, Peru, and Ecuador. In Panama, no doubt widespread; known from tropical moist forest in the Canal Zone and adjacent Panamá, from premontane wet forest in Chiriquí, and from tropical wet forest in Veraguas (Bahía Honda).

Scaphyglottis longicaulis S. Wats., Proc. Amer. Acad. Arts 23:286. 1888
S. unguiculata Schlechter

Epiphyte, to 30 cm tall; stems densely clustered, usually short, slender, less than 4 mm diam, consisting of a few swollen, superimposed segments, occasionally somewhat maroon, the old leafless stems persisting; stem sheaths (or their scars) restricted to base. Leaves usually 2 near apex of each segment, usually only the uppermost persisting, linear, 4–16 cm long, 2–7 mm wide, obscurely emarginate at apex. Flowers white, 5–7 mm long, usually solitary at apex of the short upper stem segment; sepals 5–7 mm long, 1–2 mm wide, the laterals united at base; petals 5–7 mm long, ca 1 mm wide; lip prominently trilobate, 5–7 mm long, 3–4 mm wide, acute, usually tinged with maroon; column winged laterally, tinged with purple especially on pollinia. Fruits narrowly ellipsoid, ca 7 mm long. *Croat 10098.*

Common in the forest, usually high in trees. Apparently flowers irregularly throughout the year, possibly with two flowering seasons; flowering has been observed in March, April June, September, November, and December.

KEY TO THE SPECIES OF SCAPHYGLOTTIS

Leaves linear-lanceolate, relatively short and broad (usually no more than 8 times longer than broad), mostly less than 6 cm long . . . . *S. prolifera* Cogn.
Leaves linear, many times longer than broad, most more than 6 cm long (if less only 1–2 mm wide):
    Lip trilobate, the lobes of ± equal length; flowers usually 1 per stem (at least only 1 open); stems usually less than 10 cm long, bearing only 1 or 2 sheaths, the sheaths usually near the base, often loose and spreading . . . . *S. longicaulis* S. Wats.
    Lip entire or obscurely trilobate (if lobed, the lateral lobes shorter than the middle lobe); flowers usually several in fascicles or racemes; stems often more than 10 cm long, bearing numerous, closely fitting sheaths (or with numerous sheath scars) . . . . *S. graminifolia* (R. & P.) Poepp. & Endl.

Fig. 173. *Scaphyglottis longicaulis*

Fig. 174. *Sobralia suaveolens*

Most easily confused with *S. graminifolia*. See the genus key for distinctions.

Guatemala to Colombia. In Panama, known from tropical moist forest in the Canal Zone, Colón, San Blas, Panamá, and Darién.

See Fig. 173.

Scaphyglottis prolifera Cogn. in Mart., Fl. Brasil. 3(5):15. 1898

S. cuneata Schlechter

Small caespitose epiphyte, 7–25 cm tall; stems slender with slender, fusiform segments superimposed along their length. Leaves usually 2 at apex of each segment, all but the lowermost persisting, linear-lanceolate, 1–6 cm long, 1–7 mm wide, emarginate at apex. Flowers white, ± sessile, solitary or few in fascicles at apex of stem segments; sepals 3.5–5 mm long, 1–2.5 mm wide; petals as long as lateral sepals, 0.5–0.6 mm wide; lip entire or trilobate, 4–6 mm long, 2.5–4 mm wide. Fruits not seen. *Shattuck 549.*

Uncommon, high in trees. Flowering season undetermined. Seen flowering in both the late rainy season (October) and the dry season (December and February).

Guatemala to Brazil and Bolivia; West Indies. In Panama, known from tropical moist forest in the Canal Zone and Panamá, from premontane wet forest in Coclé (El Valle), from tropical wet forest in Panamá (Cerro Campana), and from premontane rain forest in Coclé Cerro Pilón) and Panamá (Cerro Jefe).

SIEVEKINGIA Reichb.f.

Sievekingia suavis Reichb.f., Beitr. Syst. Pfl. 3. 1871

Dwarf epiphyte; pseudobulbs ovate, 1.5–3 cm long, with 2 or 3 papery bracts at base. Leaves solitary at apex of pseudobulb, elliptic-lanceolate, 8–25 cm long, 1.5–3 cm wide, plicate. Racemes short, few-flowered, pendent, from base of pseudobulb; sepals subequal, 10–17 mm long, 4–8 mm wide, pale yellow; petals orange, 10–15 mm long, 3–6 mm wide; lip concave, orange, with purplish spots, 8–11 mm long, nearly as wide; column 6–8 mm long, green with broad orange wings. Fruits not seen.

Not seen on BCI, but Dressler has collected the pollinia from bees on the island and is certain that the species

occurs there. Seasonal behavior uncertain. Costa Rican plants are known to flower in March.

Pollinated by *Euglossa dodsoni* bees (Dressler, 1968a).

Costa Rica and Panama. In Panama, known from tropical moist forest on BCI and reported from Colón and the Atlantic slope of central Panama.

SOBRALIA R. & P.

Sobralia fragrans Lindl., Gard. Chron. 598. 1853

Small epiphyte (rarely terrestrial), 15–25(45) cm tall; pseudobulbs lacking; stems reedlike but flattened, usually with only 1 leaf (to 3). Leaves narrowly elliptic to oblong-elliptic, narrowly acute to acuminate at apex, (6)9–12(23) cm long, 1.5–3.5 cm wide, plicate, articulated at base with a conduplicate, persistent, sheathing base. Flowers solitary at apex of stem; sepals ± equal, linear to oblong-lanceolate, acute or apiculate apically, 3–4 cm long, 4–7 mm wide, white to cream or pale lavender (at least externally), connate at base; petals ± equaling sepals, white or cream with a yellow throat; lip oval to obovate, 3–3.5 cm long, 1.5–2 cm wide, white with raised yellow ridges, the central ridge terminating in ragged, hairlike projections, the base of the lip with several thickened veins and 2 callus thickenings; column 16–18 mm long. Fruits slender, to ca 6 cm long and 5 mm wide, closely ribbed.

Reported by Dressler; one specimen was seen from the island and transferred to the epiphyte house, but no collection was made. The species flowers throughout the rainy season (June to November), especially in July and August. Fruiting plants are common in the dry season.

Guatemala to Colombia (Chocó) and Venezuela. In Panama, known from tropical moist forest in the Canal Zone, Panamá, and Darién, from premontane wet forest in Chiriquí and Veraguas, and from premontane rain forest in Panamá (Cerro Jefe).

Sobralia panamensis Schlechter, Feddes Repert. Beih. 17:11. 1922

Epiphyte, to 1.5 m tall; stems reedlike, terete, simple or branching, lepidote or furfuraceous. Leaves spaced along stem, elliptic to broadly elliptic-lanceolate, acuminate, obtuse at base, 5–22 cm long, 2–7.5 cm wide, plicate; leaf sheaths pubescent, closely clasping stem. Flowers solitary

---

KEY TO THE SPECIES OF SOBRALIA

Plants usually less than 50 cm tall; most leaves less than 4.5 cm wide; sepals less than 4 cm long:
   Flowers borne from bract at apex of stem, the distance to the next lower leaf at least 7 cm; stems markedly flattened, often with a single leaf; sepals 3–4 cm long . . . . *S. fragrans* Lindl.
   Flowers borne in axil of uppermost leaf, the distance to the next lower leaf usually less than 5 cm; stems ± terete, usually bearing 3 or more leaves; sepals less than 2.8 cm long . . . .
   . . . . *S. suaveolens* Reichb.f.
Plants usually more than 75 cm tall; most leaves at least 5 cm wide; sepals more than 4.5 cm long:
   Stems and leaf sheaths minutely furfuraceous or scurfy; flowers lavender to red-violet, enclosed at base by narrowly acute bracts; fruits to ca 2 cm wide . . . . *S. panamensis* Schlechter
   Stems and leaf sheaths glabrous; flowers yellow, borne between the uppermost leaf and a leaflike bract; fruits less than 1 cm wide . . . . *S. rolfeana* Schlechter

at apex of stems, lavender to red-violet or whitish; sepals ± equal, ligulate to lanceolate, 4.5–6 cm long, 12–16 mm wide, the laterals ± oblique; petals slightly shorter than sepals, broader, to 2 cm wide, obtuse or acute; both sepals and petals weakly recurved near apex; lip colored as petals but much darker, 4–6 cm long, 3–4 cm wide, forming a tube around the column, flaring abruptly at apex, undulate on the margin, with 2 small calluses at base; column to 3 cm long. Fruits elongate, ca 12 cm long and 2 cm wide, the ribs stout and widely spaced. *Croat 4314.*

Occasional, in the forest, usually on tree trunks in deep shade above 3 meters. Flowering season uncertain. Seen in flower in December and March, and both flowers and mature-sized fruits have been seen in February.

The species can easily be distinguished by its scurfy stems. It is similar in other aspects to *S. rolfeana,* which has yellow flowers and lacks the scurfy stems. *S. panamensis* appears to be most similar to *S. decora* Batem., a species ranging from Mexico to Panama, and is perhaps not separable from it.

Known only from Panama, in tropical moist forest in the Canal Zone, Coclé, and Panamá.

Sobralia rolfeana Schlechter, Feddes Repert. Beih. 17:12. 1922

Epiphyte, 75–100 cm tall; stem reedlike, terete, unbranched, glabrous. Leaves spaced along stem mostly above the middle, elliptic to oblanceolate-elliptic, acute to very abruptly acuminate and downturned at apex, obtuse at base, 7–21 cm long, 2.5–7.5 cm wide, glabrous and smooth, plicate, the sheath often somewhat flattened. Flowers solitary at apex of stem, creamy yellow with a lemon-yellow throat, arising from between a reduced leaf and a much smaller, leaflike bract; sepals 5.5–7.5 cm long, 8–16 mm broad, ligulate to elliptic or elliptic-lanceolate, acute, the laterals ± oblique; petals 5.5–6.5 cm long, 0.8–1.4 cm wide, ligulate to oblong-ligulate, obtuse or acute; lip obovate, retuse, 5.5–7 cm long, 2.5–4.7 cm wide, lacerate-dentate at apex, obscurely bilamellate at base; column 2–3 cm long. Fruits to 12 cm long and 7 mm wide, closely ribbed, light brown. *Croat 6259.*

Apparently rare, in the forest. Flowering more than once during the rainy season; flowers have been seen in August and September borne on an inflorescence with a fruit of mature size.

Collections of the species from Cerro Campana are often much shorter (to as little as 20 cm tall) than those found on BCI and in other lowland areas.

Known only from Panama, principally from tropical moist forest in the Canal Zone and Bocas del Toro (probably also in Darién); known also from premontane wet forest in Panamá (Cerro Campana).

Sobralia suaveolens Reichb.f., Gard. Chron. n.s. 9:622. 1878

Epiphyte, 20–50 cm tall; stems reedlike, few-leaved. Leaves narrowly elliptic, 5–19 cm long, 2–4.5 cm wide, plicate; leaf sheaths glabrous, not verrucose. Flowers 2 or 3 from axil of terminal leaf; sepals and petals ± equal, 2–2.5 cm long, 4–5 mm wide, white to pale yellow; lip

as long as petals, darker yellow, 1–1.5 cm wide, trilobate, the longitudinal lamellae several, becoming lacerated at apex and branching into 2 groups at base; column about 8 mm long. Fruits slender, 4–7 cm long, ca 4 mm wide, closely ribbed, the stout ribs forming a cagelike structure persisting long after seeds are dispersed. *Croat 6788, 8253, 12639.*

Occasional, in the forest; common on tree stumps at the edge of the lake and often forming the dominant element of the tree-stump vegetation. Flowers chiefly from October to December. The fruits attain full size by February and March and probably dehisce in the late dry season.

Known only from Panama, from tropical moist forest in the Canal Zone and Panamá.

See Fig. 174 and fig. on p. 14.

SPIRANTHES L. C. Rich.

Spiranthes lanceolata (Aubl.) León, Contr. Ocas. Mus. Hist. Nat. Colegio "De La Salle," 8:358. 1946
    S. orchiodes (Sw.) A. Rich.

Terrestrial, 20–70 cm tall; stem unbranched, stiffly erect, slender, pubescent, leafless at flowering, sheathed in thin bracts. Leaves few, basal, appearing after flowering, elliptic to elliptic-lanceolate, 15–21 cm long, 2.5–4 cm wide, thin; scape bracts lanceolate, to 1.5 cm long. Spikes terminal, many-flowered; sepals unequal, connivent or spreading at tip, yellow-green, often flushed with pink or white on inner surface, pubescent on outer surface; dorsal sepal 13–21 mm long, 3–6 mm wide, the lateral sepals 18–27 mm long, 3–4.5 mm wide; petals salmon-pink, 12–15 mm long; lip lanceolate, clawed, 15–23 mm long, 5.5–9 mm wide, the lateral margins erect, the margins of the claw fleshy and pubescent. Fruits not seen. *D'Arcy 4292.*

Collected once in Rear #8 Lighthouse Clearing. Flowers in the late dry season and the early rainy season (April to May, rarely as late as August).

Stenorrhynchus sp., as reported by Standley from a collection by Kenoyer, was apparently this species, although no such collection annotated by him has been located.

Scattered throughout tropical America. In Panama, known principally from tropical moist forest in the Canal Zone, Chiriquí, and Panamá (the Chiriquí collection near Remedios could have been made in premontane wet forest).

STELIS Sw.

Stelis crescentiicola Schlechter, Feddes Repert. 16:442. 1920

Small, caespitose epiphyte; thickened pseudobulbs lacking; stems short, slender, bearing a single leaf. Blades oblanceolate, mostly rounded or emarginate at apex, attenuate at base, 3.5–10 cm long, 0.6–1.5 cm wide, glabrous, coriaceous (prominently striate longitudinally when dried). Inflorescences densely flowered, terminal

racemes 10–22 cm long; flowers pale violet-purple, persisting in fruit, minute (to 2.7 mm long); sepals ± ovate, acute or obtuse, to 1.3 mm long, papillate to minutely glandular; petals minute, much smaller than sepals, broader than long, held to either side of the minute trilobate lip. Fruits obliquely 3-sided, ± ellipsoid, 4–6 mm long, green. *Croat 14938.*

Apparently uncommon, though because of its habit in trees it is often overlooked. Flowers in the late dry season (April) and the early rainy season (May). The fruits develop by June.

In habit the plant is similar to *Pleurothallis* and may occur in the same places as *P. brighamii* and *P. grobyi.*

Costa Rica and Panama. In Panama, known from tropical moist forest in the Canal Zone, Panamá, and Darién and from premontane wet forest and tropical wet forest in Colón (Santa Rita Ridge and Portobelo, respectively).

TRICHOCENTRUM Poepp. & Endl.

Trichocentrum capistratum Linden & Reichb.f., Gard. Chron. 1257. 1871
    T. panamense Rolfe

Dwarf epiphyte, 3–12 cm tall; pseudobulbs minute. Leaves very fleshy, solitary at apex of pseudobulbs, narrowly elliptic-lanceolate, acute, 2.5–10 cm long, 0.8–2.5 cm wide. Scapes pendent, arising from base of leaf, 1–4 cm long, elongating and successively producing solitary flowers; sepals and petals subequal, spreading, 10–12 mm long, 3–4 mm wide, pale yellowish-green; lip 10–12 mm long, ca 6 mm wide, white with purplish blotch at base, adnate at base to base of column and expanded into an obscurely 4-lobed spur; column about 5 mm long. Fruits not seen.

One specimen seen in the epiphyte house, the specimen having been collected by Dressler on the island. Seasonal behavior unknown.

The species is variable in the size and shape of its leaves and the length of its inflorescence.

Pollinated by the bee *Euglossa cordata* (Dodson, 1967a).

Costa Rica to Colombia and Venezuela. In Panama, known from tropical moist forest in the Canal Zone (both slopes) and from premontane wet forest in Coclé (Río Antón, 500–600 m) and Panamá (Cerro Campana).

TRICHOPILIA Lindl.

Trichopilia maculata Reichb.f., Bonplandia 3:215. 1855

Small epiphyte, 7–20 cm tall; pseudobulbs oblong-elliptic, strongly flattened, 2–5 cm long, densely clustered and inserted at an acute angle on a short rhizome, envel-

oped at base by several densely dark-spotted, imbricate, papery bracts. Leaves solitary at apex of pseudobulb, ± oblong-elliptic, 5–12 cm long, 2–3.5 cm wide. Scapes solitary, 1-flowered, to 6 cm long, from base of pseudobulb; sepals and petals lanceolate, acuminate, free and spreading, subequal, 3–4 cm long, ca 5 mm wide, greenish-yellow; lip forming a tube around the column, flared at apex, ca 3.5 cm long and 1.7 cm wide, white to pale yellow with orange-red streaks, contracted and adnate at base to base of column; column slender, with a tripartite, minutely denticulate hood at apex covering the anther. Fruits not seen. *Shattuck 555.*

Not seen on BCI in recent years. Apparently flowering more than once during the rainy season (September to December). Moderately well-developed fruits have been seen in early December on a flowering plant; full-sized fruits have been seen in January.

Known only from Panama, from tropical moist forest in central Canal Zone and Panamá.

Trichopilia subulata (Sw.) Reichb.f., Flora 48:278. 1865

Caespitose epiphyte, 5–25 cm tall; pseudobulbs cylindrical, inconspicuous, 1–2.5 cm long, less than 5 mm wide, enclosed in papery deciduous sheaths. Leaves leathery, solitary on pseudobulb, linear to subterete, 5–22 cm long, less than 1 cm wide. Racemes ± erect or pendulous, 2.5–6 cm long, from base of pseudobulb; flowers 5 or 6 on slender pedicels subtended by spathaceous bracts; sepals and petals subequal, linear-lanceolate, acuminate, 18–23 mm long, 2.5–4 mm wide, translucent white to pale yellow; lip ± obovate, concave, 15–20 mm long, nearly as wide, white spotted with rose-purple, the margin denticulate to prominently lacerate, adnate at base to base of column; column 7–9 mm long with an entire, denticulate to fimbriate hood at apex covering the anthers; anthers terminal and caplike. Fruits narrowly-ellipsoid, ca 2 cm long. *Croat 8500.*

Rare, in the forest. Probably flowers in the rainy season. Mature-sized fruits have been seen in March.

Known from Panama to Peru; Trinidad and Cuba. In Panama, known principally from premontane wet and tropical wet forests in Coclé (El Valle) and Panamá (Cerro Campana); known also from tropical moist forest in the vicinity of the Canal Zone.

TRIGONIDIUM Lindl.

Trigonidium egertonianum Batem. ex Lindl., Edward's Bot. Reg. n.s. 1, misc. 73. 1838

Epiphyte; pseudobulbs ± ovoid, compressed and ridged, densely clustered, ca 5 cm long and 3.5 cm wide, sub-

KEY TO THE SPECIES OF TRICHOPILIA

Pseudobulbs elliptic-oblong, strongly flattened; leaves subcoriaceous, elliptic-lanceolate, 2–3.5 cm wide; lip tubular . . . . *T. maculata* Reichb.f.
Pseudobulbs subcylindrical; leaves fleshy, narrowly linear to semiterete, 10 mm or less wide; lip not tubular . . . . *T. subulata* (Sw.) Reichb.f.

Fig. 175. *Trigonidium egertonianum*

Fig. 176.
Triphora gentianoides

tended by imbricating bracts; the bracts weathering into fibers. Leaves 2, from apex of pseudobulb, linear-lanceolate, 20–60 cm long, 1–3 cm wide, conduplicate at base. Inflorescences 1 to several, erect bracteated scapes bearing 1 flower, ± the same height as leaves, arising from base of pseudobulb; sepals free, mostly acute, 3–4 cm long, faintly brown- to purple-striped, usually tan to yellow, the dorsal sepal erect, the laterals recurved at apices; petals much smaller, more prominently striate, purple and with a glandlike thickening near apex; lip thick, trilobate, recurved at apex, about a third as long as petals, the sides closely fit around the smaller column; column caniculate on inner surface, the margins purple; anther solitary near apex; pollinia orange. Fruits narrowly ellipsoid, ca 5 cm long. *Croat 8414.*

Abundant on tree stumps at the margin of the lake. Flowers from December to July, chiefly in the dry season months of January to March. Some fruits develop to mature size by late February and have been found unopened as late as early April.

Mexico to Colombia, Venezuela, and Ecuador. In Panama, known from tropical moist forest in the Canal Zone, Bocas del Toro, Colón, Veraguas, and Panamá; known also from tropical wet forest in Veraguas (Bahía Honda) and Colón (Cerro Santa Rita, 400 m).

See Fig. 175.

TRIPHORA Nutt.

Triphora gentianoides (Sw.) Ames & Schlechter in Ames, Orchid. 7:5. 1922
Pogonia cubensis Reichb.f.; *T. cubensis* (Reichb.f.) Ames

Apparently saprophytic, terrestrial herb, to 15 cm tall; stem usually solitary, slender, arising from one end of a tuber. Leaves few, lanceolate to suborbicular, reduced to sheathing bracts above, 1–2 cm long, 0.5–1 cm wide. Flowers several, solitary from upper axils, forming corymbiform racemes; petals and sepals 7–8 mm long, 1–1.5 mm wide, obtuse to acute, green, the sepals tinged with maroon; lip lanceolate-obovate, clawed, trilobate, 7–8 mm long, ca 3 mm wide, white with light green on 3 crested keels, the middle lobe subrotund. Fruits not seen. *Kenoyer 250.*

Rare. Flowering season unknown; the plant seems to appear at irregular intervals.

Reported by Dunsterville and Garay (1965) to have cleistogamous flowers.

Florida and Mexico, Guatemala, Panama, Venezuela, and Ecuador; the Antilles; probably in the intervening areas also. In Panama, known only from tropical moist forest on BCI.

See Fig. 176.

Triphora mexicana (S. Wats.) Schlechter, Feddes Repert. 17:139. 1921
Pogonia mexicana S. Wats.

Terrestrial, to ca 15 cm tall; stem single, slender, ± fleshy, arising from a small tuber. Leaves few, alternate, narrowly to broadly ovate, obtuse to abruptly short-acuminate, obtuse to rounded at base, 8–20 mm long, 5–17 mm wide, thin, 5-veined, the uppermost and lowermost usually much reduced. Pedicels to 12 mm long; flowers 1–3, borne in axils of upper leaves or leaflike bracts; sepals and petals white, narrowly acute at apex; dorsal sepal narrowly lanceolate, to 2 cm long, the lateral sepals linear to narrowly elliptic, subfalcate, to 1.7 cm long; petals similar to lateral sepals; lip ± obovate, trilobate, slightly shorter than lateral sepals, white, with 3 green medial lines, the terminal lobe rounded, marked along the margin with red-violet, the lateral lobes turned upward along the column, infused submarginally with red-violet; column free, thickened somewhat laterally, slightly shorter than lateral lobes of lip; anther red-violet. Fruits not seen. *Croat 15567.*

Rare; seen only once in the forest on BCI. Known to flower in July.

According to R. Dressler, the species may be somewhat saprophytic and, as a consequence, may go for long periods without producing a vegetative shoot.

Mexico, Guatemala, and Panama. In Panama, known only from tropical moist forest in the Canal Zone.

VANILLA P. Mill.

Vanilla fragrans (Salisb.) Ames, Sched. Orch. 7:36. 1924
V. planifolia Andr.
Vanilla

Scandent epiphytic vine; stems thick, often branched, 10–15 mm long. Leaves oblong-elliptic to oblong-oblanceolate, usually acuminate, 8–20(26) cm long, 2–4.5(8) cm wide, fleshy. Racemes axillary (rarely terminal), 5–7 cm long, bracteate at nodes; bracts oblong, usually obtuse, to 1 cm long, persistent at nodes; pedicels 2–4 cm long; flowers white; sepals linear to oblanceolate, 4–7 cm long, 1–1.5 cm wide; petals similar to sepals but smaller; lip obscurely trilobate, curved around sides of column, 4–5 cm long, 1.5–3 cm wide, with a tuft of trichomes at base and streaked with longitudinal, verrucose lines, the middle lobe ± recurved at apex; column ca 3 cm long, pubescent on outer surface. Capsules terete, linear, very aromatic, to 25 cm long, 8–15 mm diam, pendent. *Croat 4528.*

Occasional, in the forest, often climbing to 10 m or more in the canopy. Flowering in March and April; the

---

KEY TO THE SPECIES OF TRIPHORA

Leaves usually lanceolate; sepals and petals 7–8 mm long . . . . *T. gentianoides* (Sw.) Ames & Schlechter
Leaves narrowly or broadly ovate; sepals and petals 10–20 mm long . . . .
. . . . *T. mexicana* (S. Wats.) Schlechter

Lip with verrucose lines or papillae, somewhat trilobate, not visibly longer than sepals and petals
.... *V. fragrans* (Salisb.) Ames
Lip without verrucose lines or papillae, not at all trilobate, visibly longer than sepals and petals . . .
.... *V. pompona* Schiede

plant rarely flowers and when it does the flowers perish so quickly that they are seldom seen. The fruit set appears to be exceedingly poor also since the fruits are not seen any more frequently than the flowers.

Flowers are pollinated by the bee *Melipona beechii* (Dodson, 1967a).

Mexico to Panama; cultivated throughout the tropics of the world (Boriquet, 1954). In Panama, known from tropical moist forest in the Canal Zone, San Blas, and Panamá.

Vanilla pompona Schiede, Linnaea 4:573. 1829
Vanilla

Epiphytic vine; stems thick, fleshy. Leaves variable in shape, mostly ± oblong-elliptic, often oblique, acute or obtuse at apex, acute to rounded at base, 10–30 cm long, 3.5–14 cm wide. Racemes axillary, to 16 cm long, several-flowered; floral bracts lanceolate to ovate, concave, ca 1 cm long, persistent; petals and sepals greenish-white; sepals ± linear-oblanceolate, obtuse or acute, 7–8.5 cm long, ca 1 cm wide; petals similar to sepals but somewhat smaller and with a midrib outside; lip slightly longer than petals, clawed, to 4 cm wide, enveloping the column and adnate to it, mostly orange-yellow, the outer margin crenulate; column ca 6 cm long. Fruits fusiform capsules to 1.5 cm long, 1.5–2 cm diam. *Wilson 119.*

Not seen in recent years on BCI. Seasonal behavior undetermined. The few flowering collections seen were made in March and April.

Southeastern Mexico to Colombia, Venezuela, the Guianas, Brazil, Paraguay, and Ecuador; Trinidad; probably occurring as an escaped cultivar throughout part of this range. In Panama, known only from tropical moist forest in the Canal Zone.

XYLOBIUM Lindl.

Xylobium foveatum (Lindl.) Nich., Dict. Gard. 4:225. 1887

Epiphyte; pseudobulbs ovoid or subconic, mostly about 8 cm long and 3 cm wide, tapered to apex, ± smooth, enveloped in fibrous imbricating sheaths. Leaves 2 or 3 at apex of pseudobulb, lanceolate to oblanceolate, acute or acuminate, plicate (often weakly so), tapering to a conduplicate petiole, to 50 cm long (including petiole) and 7 cm wide. Inflorescences erect, bracteated, many-flowered racemes from base of pseudobulb, 12–30 cm long; flowers creamy-white, the lip with reddish veins, pedicellate, fragrant, ca 1.5 cm long; sepals subequal, free, the dorsal sepal free, the laterals produced below the pedicel and adnate to foot of column; petals ± equaling dorsal sepal; lip trilobate, the lateral margins twined upward; apex of column with a hooded cavity bearing the anther on its inner margin. Fruits not seen. *Aviles 32.*

Rare; collected only once. Flowering from November to March, principally in the dry season.

Mexico to Peru, the Guianas; Jamaica. In Panama, known from tropical moist forest in the Canal Zone and Panamá. Ospina (1958) reported that the species usually occurs in Colombia at 1,200–1,500 m elevation!

ANTHOPHYTA
Dicotyledoneae

KEY TO THE DICOTYLEDONEAE

All families in Key 5 are also keyed out in Keys 6–9. Key 5 is included so that decisions based on floral characters of minute flowers can be avoided.

Plants aquatic . . . . KEY 1
Plants not aquatic:
    Leaves compound:
        Leaves pinnately or bipinnately compound with more than 3 leaflets . . . . KEY 2
        Leaves palmately compound or with only 2 or 3 leaflets . . . . KEY 3
    Leaves simple or lacking:
        Plants lacking leaves and/or chlorophyll . . . . KEY 4
        Plants with both leaves and chlorophyll:
            Flowers individually inconspicuous, minute, usually crowded, the perianth often lacking
                or undifferentiated into calyx and corolla . . . . KEY 5
            Flowers as above or not:
                Leaves opposite, ternate, or quaternate:
                    Ovary inferior . . . . KEY 6
                    Ovary superior . . . . KEY 7
                Leaves alternate, spiraled or whorled:
                    Ovary inferior . . . . KEY 8
                    Ovary superior . . . . KEY 9

KEY 1: PLANTS AQUATIC

Petals 5 (rarely 4), yellow or white with yellow spots at base, caducous; stamens 10, in 2 series; ovary inferior; capsules cylindrical . . . . 108. ONAGRACEAE (*Ludwigia*, in part)
Petals not as above, either not yellow or white with yellow spots or not 5; stamens not 10 in 2 series; ovary inferior or superior; fruits not capsular or, if capsular, then not cylindrical:
• Blades or blade segments narrow, at least 3 times longer than wide, ligulate, filiform or grasslike:
    Leaves simple, alternate or basal, not submerged:
        Leaves succulent, basal . . . . 28. AMARYLLIDACEAE (*Crinum erubescens* Ait.)*
        Leaves not succulent, alternate . . . . 17. GRAMINEAE (in part)*
    Leaves dissected or verticillate, primarily submerged:
        Blades simple . . . . 16. HYDROCHARITACEAE (*Hydrilla verticillata* (L.f.) Royle)*
        Blades dissected or forked (also inconspicuous in Lentibulariaceae):
            Plants free-floating, at or near surface . . . . 128. LENTIBULARIACEAE
            Plants rooted . . . . 53. CERATOPHYLLACEAE (*Ceratophyllum demersum* L.)

*Not a dicotyledon; included here because of superficial similarity.

● Blades broad, not dissected, less than 3 times as long as wide:

Blades ± orbicular, often floating on surface, conspicuously palmately veined below:

Largest blades more than 20 cm long, deeply dentate . . . . 52. NYMPHAEACEAE

Largest blades less than 20 cm long, generally entire to undulate:

Blades more than 8 cm long, purplish beneath; flowers bisexual; petals fimbriate . . . .
. . . . 117. MENYANTHACEAE (*Nymphoides indica* L.)

Blades less than 7.5 cm long, not purplish beneath; flowers unisexual or bisexual; petals not fimbriate:

Flowers unisexual, 1–3, subtended by 1 or 2 spathes . . . .
. . . . 16. HYDROCHARITACEAE (*Limnobium stoloniferum* (G. Meyer) Griseb.)*

Flowers bisexual, borne solitary in leaf axils or in dense, long-pedunculate umbels:

Flowers regular, white, numerous, in dense, long-pedunculate umbels; leaves with the petioles many times longer than blades, the blades peltate . . . .
. . . . 110. UMBELLIFERAE (*Hydrocotyle umbellata* L.)

Flowers bilabiate, blue, solitary in leaf axils; leaves sessile, not peltate . . . .
. . . . 125. SCROPHULARIACEAE (*Bacopa salzmannii* Edw.)

Blades not orbicular, not floating on surface, not conspicuously veined below:

Blades entire:

Blades more than 20 cm long; petioles not inflated . . . .
. . . . 15. ALISMATACEAE (*Sagittaria lancifolia* L.)*

Blades less than 20 cm long; petioles often inflated . . . . 24. PONTEDERIACEAE

Blades serrate:

Blades peltate . . . . 110. UMBELLIFERAE (*Hydrocotyle umbellata* L.)

Blades basifixed:

Blades ± equilateral at base; flowers minute, clustered in axils . . . .
. . . . 40. URTICACEAE (*Boehmeria cylindrica* (L.) Sw.)

Blades inequilateral at base; flowers showy, in terminal inflorescences . . . .
. . . . 100. BEGONIACEAE (*Begonia patula* Haw.)

KEY 2: BLADES PINNATELY OR BIPINNATELY COMPOUND WITH
MORE THAN THREE LEAFLETS

Leaves opposite:

Petals and sepals 5, free; stamens 5; pistil 1, with 3 free carpels . . . .
. . . . 79. STAPHYLEACEAE (*Turpinia occidentalis* subsp. *breviflora* Croat)

Petals and sepals united; stamens 4 or 10; pistil of 1 or 2 united carpels:

Stamens 10, united into a staminal tube which is open on one side; fruits indehiscent,
1-seeded . . . . 63C. PAPILIONOIDEAE (*Platymiscium pinnatum* (Jacq.) Dug.)

Stamens 4, borne on corolla tube, the filaments not united; fruits dehiscent, many-seeded . . . .
. . . . 126. BIGNONIACEAE (in part)

Leaves alternate:

Flowers with the petals lacking, the calyx 6-lobed in 2 series; stamens 3; blades simple, the branchlets only appearing as pinnately compound leaves, the branchlets deciduous . . . .
. . . . 75. EUPHORBIACEAE (*Phyllanthus*)

Flowers with the calyx and corolla each 4- or 5-lobed or 4- or 5-parted (segments 3 in some Burseraceae, sometimes 6 in Connaraceae; *Prioria* (Leguminosae) with 5 petaloid sepals and 2 calyx-like bracts also included here); stamens 4 to many; leaves pinnately or bipinnately compound:

◆ Style 1, the stigma simple or obscurely lobed:

Calyx with the sepals united, variously lobed (or obsolete in some *Mimosa*):

Ovary 1-carpellate, 1-locular; fruits usually ± elongate, 2-valved capsules, rarely drupes or samaras; flowers actinomorphic, densely congested in heads or umbels or spikes, with usually many (as few as 4 in *Mimosa*) exserted stamens giving the inflorescences a fluffy appearance (Mimosoideae) or flowers weakly to strongly zygomorphic (Caesalpinioideae, Papilionoideae) . . . . 63. LEGUMINOSAE (in part)

Ovary with 2 or more carpels and locules (*Trattinnickia* 1-seeded by abortion); fruits usually capsular, fleshy or with more than 2 valves; flowers actinomorphic, the stamens to twice as many as petals, but flowers not congested with the stamens exserted giving the inflorescence a fluffy appearance:

Stamens united into a tube or attached to the gynophore . . . . 70. MELIACEAE

Stamens free, in 2 whorls . . . . 69. BURSERACEAE

*Not a dicotyledon; included here because of superficial similarity.

Calyx with the sepals free or nearly so:

 Flowers with the corolla sympetalous or markedly zygomorphic; seeds not arillate . . . . .
. . . . 63. Leguminosae (in part)

 Flowers actinomorphic and with the petals free (weakly zygomorphic in *Paullinia* and
Serjania); seeds arillate or not:

  Petals bearing petaloid scales on inner surface; stamens 8; stipules present on lianas
. . . . 80. Sapindaceae (in part)

  Petals lacking scales; stamens 5 or 10(12) or more in 2 or more series; stipules lacking:

   Plants lianas . . . . 62. Connaraceae (*Connarus*)

   Plants trees or shrubs:

    Rachis conspicuously winged; leaves imparipinnate . . . .
. . . . 68. Simaroubaceae (*Quassia amara* L.)

    Rachis not winged; leaves paripinnate . . . . 63B. Caesalpinioideae (in part)

◂ Styles 2 or more or parted 2 or more times:

 Plants lianas lacking tendrils; flowers lacking a disk; seeds arillate . . 62. Connaraceae (in part)

 Plants shrubs, trees, or tendriled lianas; flowers with a disk (except Oxalidaceae); seeds arillate or not:

  Ovary inferior; leaflets conspicuously serrate, usually white-margined; plants cultivated
. . . . 109. Araliaceae (*Polyscias guilfoylei* (Cogn. & Marchal) Bailey)

  Ovary superior; leaflets not white-margined; plants cultivated or not:

   Leaves glandular-punctate; plants frequently armed; ovary lobed and elevated on a
disk; sap often aromatic; fruits of usually several follicles with a shiny seed . . . . .
. . . . 67. Rutaceae (*Zanthoxylum*)

   Leaves not punctate; plants unarmed; ovary usually not lobed and not elevated on a
disk; sap usually not distinctively aromatic; fruits various:

    Petals bearing petaloid scales on inner surface; stamens 8; plants tendriled lianas or
trees; fruits capsular, the seeds arillate . . . . 80. Sapindaceae (in part)

    Petals not appendaged; stamens 5, 10, or more; plants not scandent; fruits not cap-
sular, the seeds not arillate:

     Plants cultivated trees in the Laboratory Clearing; stamens 5 with 5 more reduced
to staminodia; ovules many per locule . . . .
. . . . 65. Oxalidaceae (*Averrhoa carambola* L.)

     Plants not cultivated; stamens 5 or 10, equal or unequal, but never 5 and 5; ovules
1 or 2(3) per locule:

      Flowers in leaf-opposed spikes (*Picramnia*) or flowers greenish, in terminal
panicles (*Simarouba*) . . . . 68. Simaroubaceae

      Flowers white (except *Astronium* with the flowers greenish-yellow, the styles
3, the stamens 5, and the drupes carried in winglike, accrescent sepals), in
terminal panicles . . . . 76. Anacardiaceae (in part)

KEY 3: BLADES PALMATELY COMPOUND OR WITH ONLY
TWO OR THREE LEAFLETS

Leaves opposite:

 Flowers blue; blades with simple trichomes (at least when young); fruits ± globose berries;
seeds few, not winged . . . . 122. Verbenaceae (*Vitex cooperi* Standl.)

 Flowers yellow, pinkish, or whitish; blades glabrous or with stellate trichomes; fruits elongated
capsules; seeds usually many, usually winged . . . . 126. Bignoniaceae (in part)

Leaves alternate:

 Ovary inferior:

  Plants large trees; flowers in compound umbels; leaflets densely rufous-pubescent beneath;
styles 2; seeds 2 . . . . 109. Araliaceae (*Didymopanax morototoni* (Aubl.) Dec. & Planch.)

  Plants vines; flowers solitary or in cymose panicles; leaflets not rufous-pubescent; style 1;
seeds usually many . . . . 131. Cucurbitaceae (in part)

 Ovary superior (sometimes subinferior in *Ceiba*):

 ■ Most leaves with more than 3 leaflets:

  Plants herbs to 60 cm tall . . . . 59. Capparidaceae (*Cleome parviflora* H.B.K. subsp. *parviflora*)

  Plants shrubs or trees:

   Plants shrubs or small trees; flowers and fruits borne on trunk . . . .
. . . . 87. Sterculiaceae (*Herrania purpurea* (Pitt.) R. E. Schult.)

   Plants large trees; flowers borne on branches:

    Flowers usually large, with many exserted anthers (5 in *Ceiba*); fruits capsules, the
seeds embedded in kapok (except *Pachira*) . . . . 86. Bombacaceae (in part)

    Flowers with 8 stamens; fruits berries, the seeds enveloped in a fleshy matrix . . . .
. . . . 99. Caricaceae (*Jacaratia spinosa* (Aubl.) A. DC.)

■ Most leaves with 2 or 3 leaflets:
 Stipules lacking:
  Plants herbs to 60 cm tall; stamens 6; fruits fusiform siliques borne on a gynophore . . . . .
   . . . . 59. CAPPARIDACEAE (*Cleome parviflora* H.B.K. subsp. *parviflora*)
  Plants shrubs, lianas, or trees; stamens 8–12; fruits not as above:
   Leaves with 2 leaflets; flowers in small conelike inflorescences . . . .
    . . . . 63B. CAESALPINIOIDEAE (*Cynometra bauhiniifolia* Benth.)
   Leaves with 3 or more leaflets:
    Sepals and petals 4; plants shrubs or small trees, never scandent; fruits of 1(2) obo-
     void, fleshy cocci; petals with scales appendaged on inner surface . . . .
     . . . . 80. SAPINDACEAE (*Allophylus psilospermus* Radlk.)
    Sepals and petals 5; plants usually lianas; fruits leathery follicles with a single,
     shiny, arillate seed; petals not appendaged . . . .
     . . . . 62. CONNARACEAE (*Connarus panamensis* Griseb.)
 Stipules present:
  Styles 3:
   Inflorescences subtended by and partially enclosed by a pair of large bracts; flowers
    apetalous, distinctly unisexual; stamens many . . . . 75. EUPHORBIACEAE (*Dalechampia*)
   Inflorescences not partially enclosed in large bracts; flowers at least appearing bisexual,
    the petals 4; stamens usually 8 . . . . 80. SAPINDACEAE (in part)
  Style 1, simple:
   Inflorescences leaf-opposed; flowers 4-parted; stamens 4 . . . . 82. VITACEAE (*Cissus,* in part)
   Inflorescences axillary or terminal; flowers 5-parted; stamens 5 or more . . . .
    . . . . 63. LEGUMINOSAE (in part)

KEY 4: PLANTS LACKING LEAVES AND/OR CHLOROPHYLL

Plants growing on the ground, saprophytic:
 Flowers zygomorphic; perianth 6-parted, appendaged; stamens 6 . . . .
  . . . . 34. BURMANNIACEAE (*Thismia panamensis* (Standl.) Jonk.)*
 Flowers actinomorphic; perianth 5-parted, not appendaged; stamens 5 . . . .
  . . . . 116. GENTIANACEAE (*Voyria*)
Plants growing on trees, epiphytic or parasitic:
 Plants parasitic, consisting only of flowers less than 1 cm diam . . . .
  . . . . 45. RAFFLESIACEAE (*Apodanthes caseariae* Poit.)
 Plants epiphytic, consisting of green, angular or flattened stems and large flowers . . 101. CACTACEAE

KEY 5: BLADES SIMPLE; FLOWERS INCONSPICUOUS, MINUTE, USUALLY CROWDED; PERIANTH OFTEN LACKING OR UNDIFFERENTIATED INTO CALYX AND COROLLA

All families in this key are also keyed out in Keys 6–9. This key is included so
that decisions based on floral characters of minute flowers can be avoided.

Leaves opposite:
 Flowers obscured by bracts *or* flowers in dense heads subtended by a series of bracts:
  Flowers in dense heads subtended by a series of greenish bracts; corolla tubular at least at
   base; seeds usually with bristles or scales borne at apex . . . . 133. COMPOSITAE (in part)
  Flowers individually hidden by whitish scarious bracts; corolla lacking; seeds shiny, black,
   lacking appendages . . . . 47. AMARANTHACEAE (in part)
 Flowers not obscured by bracts and not in dense heads subtended by a series of bracts:
  Plants parasitic shrubs or vines with thick leathery leaves . . . . 42. LORANTHACEAE
  Plants herbs, the leaves not thick and leathery:
   Flowers with a perianth, usually of 3–5 tepals, with 3–5 stamens or with a simple style;
    leaves often opposite on stem but alternate on branches (opposing leaves unequal in
    Pilea) . . . . 40. URTICACEAE (in part)
   Flowers lacking a perianth, consisting of a single stamen or 3 styles, subtended by a con-
    spicuous stalked gland; leaves all opposite, or alternate at base and opposite above . . . .
    . . . . 75. EUPHORBIACEAE (in part)
Leaves alternate:
◻ Stipules lacking:

*Not a dicotyledon; included here because of superficial similarity.

Plants scandent; blades palmately veined:
  Flowers bisexual, congested in pencil-like spikes . . . . 36. PIPERACEAE (in part)
  Flowers unisexual, in panicles or racemes, often fasciculate . . . . 54. MENISPERMACEAE
Plants not scandent; most blades not palmately veined:
  Flowers in pencil-like spikes; plants epiphytic herbs (*Peperomia*), terrestrial herbs (*Pothomorphe*), or shrubs or trees with leaf-opposed spikes (*Piper*) . . . . . 36. PIPERACEAE (in part)
  Flowers not in pencil-like spikes:
    Plants forest trees with red sap . . . . 56. MYRISTICACEAE
    Plants shrubs or herbs (if treelike, never in forest), the sap not red:
      Flowers with tubular corollas, in dense heads surrounded by series of bracts . . . .
      . . . . 133. COMPOSITAE (in part)
      Flowers with free sepals and corolla lacking, hidden by scarious white bracts . . . .
      . . . . 47. AMARANTHACEAE (in part)
□ Stipules present:
  Plants small herbs (less than 20 cm tall), with fleshy leaves and yellow flowers crowded in the
    axils . . . . 50. PORTULACACEAE (*Portulaca oleracea* L.)
  Plants large herbs, shrubs, vines, or trees:
    Petioles with conspicuous glands at apex . . . . 75. EUPHORBIACEAE (in part)
    Petioles lacking glands:
      Blades lobed . . . . 39. MORACEAE (in part)
      Blades not lobed:
        Blades palmately veined at base, the strong basal vein extending ca half the length of
          the blade or more:
          Style 1; fruits green achenes less than 4 mm long, usually hidden in bracts . . . .
          . . . . 40. URTICACEAE (in part)
          Styles 2; fruits drupes either green and more than 1 cm long or red at maturity . . . . .
          . . . . 38. ULMACEAE
        Blades not palmately veined at base or weakly so, the basal veins not extending half-
          way into blade:
          Flowers not in spikelike racemes; trees usually with milky or colored, viscid sap;
            stipules often large and amplexicaul . . . . 39. MORACEAE (in part)
          Flowers in spikelike racemes; sap milky or not; stipules lateral, not encircling stems:
            Styles 2 or bifid; stamens 4; fruits drupes, the seeds not arillate; sap milky . . . .
            . . . . 39. MORACEAE (in part)
            Styles 3; stamen 1; fruits fleshy 3-valved capsules, the seeds arillate; sap not milky
            . . . . 37. LACISTEMACEAE

KEY 6: LEAVES SIMPLE, OPPOSITE (OR TERNATE OR QUATERNATE); OVARY INFERIOR

Flowers with the corolla tubular or the perianth undifferentiated into calyx and corolla:
  Stipules interpetiolar; calyx and corolla usually 4- or 5-lobed, usually regular . . . . . 130. RUBIACEAE
  Stipules lacking or, if present, not continuous between petioles; perianth undifferentiated into
    typical calyx and corolla or bilabiate and zygomorphic:
    Corolla bilabiate, zygomorphic; stems juicy; plants epiphytic or loosely rooted in soil . . . .
    . . . . 127. GESNERIACEAE (in part)
    Corolla not bilabiate or lacking; stems not juicy; plants terrestrial or parasitic (Loranthaceae):
      Flowers with tubular corollas in dense bracteate heads, the calyx reduced to bristles or
        scales . . . . 133. COMPOSITAE (in part)
      Flowers lacking a corolla or with free, undifferentiated tepals, not in dense bracteate heads:
        Plants parasitic shrubs or vines; tepals free; blades thick, leathery . . . . . . 42. LORANTHACEAE
        Plants terrestrial shrubs, trees, climbing shrubs, or lianas; calyx cupular, 5-lobed, the
          corolla lacking; blades not thick; ovary appearing inferior . . . . 48. NYCTAGINACEAE
Flowers with the petals free, the calyx present and ± typical:
  Plants herbs . . . . 110. UMBELLIFERAE (in part)
  Plants not herbs:
    ▲ Plants lianas or hemiepiphytic shrubs, often somewhat scandent:
      Plants lianas; fruits winged or prominently angled, more than 10 mm long . . . .
      . . . . 105. COMBRETACEAE (*Combretum*)
      Plants hemiepiphytic shrubs; fruits small, winged capsules to 2 mm long or fleshy berries:
        Fruits fleshy, many-seeded berries to 10 mm long; style 1; inflorescences with all flowers
          alike . . . . 107. MELASTOMATACEAE (*Topobaea praecox* Gleason)
        Fruits small, winged capsules to 2 mm long; inflorescences with some neuter flowers
          consisting of showy calycine lobes, the bisexual flowers with 2 styles . . . .
          . . . . 60. SAXIFRAGACEAE (*Hydrangea peruviana* Moric.)

▲ Plants trees or shrubs, never hemiepiphytic or scandent:
   Stamens equal in number to petals and ± hidden by petals; leaves with large glands at base
      below . . . . .81. RHAMNACEAE (*Colubrina glandulosa* Perk.)
   Stamens more than petals; leaves lacking large basal glands:
      Stamens many; most parts pellucid-punctate . . . . 106. MYRTACEAE
      Stamens twice as many as petals; plants not pellucid-punctate:
         Leaves pliveined or nearly so (except *Mouriri*, which is not epiphytic); flowers all
            bisexual; plants terrestrial (except *Topobaea*); fruits berries . . . .
            . . . . 107. MELASTOMATACEAE (in part)
         Leaves not pliveined; inflorescences with neuter flowers consisting only of enlarged
            calycine lobes; plants epiphytic; fruits small capsules . . . .
            . . . . .60. SAXIFRAGACEAE (*Hydrangea peruviana* Moric.)

KEY 7: LEAVES SIMPLE, OPPOSITE (OR TERNATE OR QUATERNATE);
OVARY SUPERIOR

Plants dioecious lianas with opposite, decussate leaves and swollen nodes; flowers with naked
   ovules, in bracteate strobili; seeds oblong, drupelike, reddish; leaves broadly ovate, coria-
   ceous, glabrous . . . . 13. GNETACEAE (*Gnetum leyboldii* Tul. var. *woodsonianum* Markg.)*
Plants not as above:
★ Petals united at least at base; perianth differentiated into calyx and corolla:
   Flowers zygomorphic, the corolla bilabiate (except *Trichanthera*, Acanthaceae); stamens 4 and
      didynamous or 2:
      Ovary 4-celled (2-carpellate, 2-loculate) with 1 ovule per cell; stems usually square; plants
         often aromatic; herbs of open areas . . . . 123. LABIATAE
      Ovary 1- or 2-celled (except *Clerodendrum*, Verbenaceae, with red flowers, cultivated at the
         Laboratory Clearing); stems usually not square; plants not distinctly aromatic; herbs
         or shrubs in open areas or forest:
         Ovary 1-locular with more than 4 ovules; stems juicy; plants epiphytic or loosely rooted;
            corolla often gibbous at base . . . . 127. GESNERIACEAE
         Ovary not 1-celled or if so with 1 or 2 ovules:
            Ovules borne on hooklike funicle, flattened at maturity and forcibly ejected by funicle;
               inflorescences usually prominently bracteate; leaves with cystoliths (except *Men-
               doncia*, with 1 cell and 2 ovules) . . . . 129. ACANTHACEAE
            Ovules sessile or on short, not hooked funicles, ± round at maturity, not ejected; in-
               florescences not bracteate; leaves lacking cystoliths:
               Plants small herbs; ovules many per locule . . . . 125. SCROPHULARIACEAE
               Plants trees, shrubs, lianas, or large herbs; ovules 1 per locule . . . .
                  . . . . 122. VERBENACEAE (in part)
   Flowers regular (weakly zygomorphic in *Lantana*, Verbenaceae); stamens usually alternate
      with the petals or corolla lobes or stamens 5 (except *Chelonanthus*, with green flowers):
      Sap milky; pistil apocarpous, with 1 style and 2 separate, unicarpellate, unilocular ovaries,
         each with few to many ovules (except *Allamanda cathartica*, Apocynaceae, which has
         quaternate leaves, large yellow flowers, and spiny fruits); plants frequently herbaceous
         vines:
         Corolla twisted in bud; stamens and style held within corolla tube, blocking entrance;
            anther sacs normal . . . . 118. APOCYNACEAE (in part)
         Corolla not twisted in bud; stamens and style joined in a column at the center of the
            flaring corolla tube; anther sacs formed into a pollinium, connected in pairs . . . .
            . . . . 119. ASCLEPIADACEAE
      Sap not milky; pistil simple; style 1:
         Flowers distinctly unisexual, with one of the sexes greatly reduced; calyx-like structure
            beneath perianth of 1–3 free parts (these actually bracts, not true sepals) . . . .
            . . . . 48. NYCTAGINACEAE
         Flowers usually bisexual or at least not distinctly unisexual; calyx typical or at least not
            of 1–3 free bractlike structures:
            Stipules interpetiolar or with stipular lines or scars; plants lianas with pliveined leaves
               or plants herbs . . . . .115. LOGANIACEAE
            Stipules lacking; plants herbs, lianas, shrubs, or trees, the leaves never pliveined:
               Leaf pairs usually very unequal; corolla usually plicate . . . . . 124. SOLANACEAE (in part)
               Leaf pairs ± equal; corolla not plicate:
                  Ovary 2-carpellate, 1-locular, with many ovules per locule; fruits 2-valved cap-
                     sules with many, minute, wind-dispersed seeds . . . 116. GENTIANACEAE (in part)
                  Ovary 2–4-carpellate and 2–4-locular, with 1 ovule per locule; fruits of several
                     nutlets or drupes of several pyrenes . . . . 122. VERBENACEAE (in part)

*Not a dicotyledon; included here because of superficial similarity.

★ Petals free or lacking *or* perianth not differentiated into calyx and corolla:
    Petals lacking *or* perianth undifferentiated into calyx and corolla:
        Plants herbs:
            Perianth lacking; inflorescences cyathia . . . . 75. EUPHORBIACEAE (in part)
            Perianth present; inflorescences lacking glands:
                Flowers not obscured by scarious bracts; cystoliths linear or punctiform in blades;
                    blades less than 5 mm long or strongly palmately veined at base . . . .
                     . . . . 40. URTICACEAE (in part)
                Flowers obscured by scarious bracts; cystoliths lacking; blades more than 1 cm long,
                    not palmately veined at base . . . . 47. AMARANTHACEAE (in part)
        Plants trees, shrubs, or lianas:
            Leaf blades with numerous, closely parallel veins; tertiary veins lacking; sap yellow;
                plants large trees . . . . 92. GUTTIFERAE (*Calophyllum longifolium* Willd.)
            Leaf blades lacking numerous, closely parallel veins; tertiary veins present; sap not yel-
                low; plants shrubs or small trees:
                Inflorescences terminal, corymbose; calyx 5-lobed, the corolla lacking; stamens 6–10;
                    fruits achenes enclosed in the accrescent calyx; seed 1, not arillate; sap not aro-
                    matic . . . . 48. NYCTAGINACEAE
                Inflorescences axillary or cauliflorous cymes; tepals 4–8, borne on a hypanthium;
                    stamens many; fruits drupelike aggregates enclosed in the hypanthium; seeds
                    several, arillate; sap aromatic . . . . 57. MONIMIACEAE
    Petals free, the calyx present:
        Plants trees or shrubs:
            Styles or stigmas usually 3 or more (2 in *Spachea*, Malpighiaceae):
                Stamens 5, hidden by petals; leaves with large basal glands below . . . .
                     . . . . 81. RHAMNACEAE (*Colubrina glandulosa* Perk.)
                Stamens 4, 8, 10, or more:
                    Calyx lobes with pairs of conspicuous glands; sap not colored . . . .
                     . . . . 71. MALPIGHIACEAE (in part)
                  Calyx lacking glands; sap yellow, orange, or red . . . . 92. GUTTIFERAE
            Style and stigma 1:
                Stamens as many as petals *or* stamen 1 with 2 staminodia; flowers frequently zygo-
                  morphic:
                    Petals 3, orange; plants large trees; seeds winged; stamens not appendaged . . . .
                     . . . . 73. VOCHYSIACEAE (*Vochysia ferruginea* Mart.)
                    Petals 5, white; plants shrubs or small trees; seeds not winged; stamens appendaged
                     . . . . 95. VIOLACEAE (*Rinorea*)
                Stamens more numerous than petals; flowers actinomorphic:
                  Flowers with calyx and corolla fused into a hypanthium at base, petals crumpled in
                    bud; ovary 1- or 2-locular, 2-carpellate, with many ovules per locule; seeds not
                    arillate . . . . 102. LYTHRACEAE (except *Cuphea*)
                Flowers lacking a hypanthium; petals not crumpled; ovary 3-locular, 3-carpellate,
                  with 2 ovules per locule; seeds arillate . . . .
                     . . . . 104. RHIZOPHORACEAE (*Cassipourea elliptica* (Sw.) Poir.)
        Plants herbs, vines, or lianas:
            Plants herbs:
                 Leaves pliveined or nearly so . . . . 107. MELASTOMATACEAE (in part)
                 Leaves pinnately or palmately veined:
                    Plants tiny prostrate herbs; blades reniform to ovate; petals bifid; styles 3 . . . .
                     . . . . 51. CARYOPHYLLACEAE (*Drymaria cordata* (L.) R. & S.)
                  Plants erect herbs; blades lanceolate to oblanceolate; petals entire; style 1, simple:
                    Flowers with ± zygomorphic, tubular hypanthium and 6 violet petals; stamens
                    11 or 12 . . . . 102. LYTHRACEAE (*Cuphea carthagenensis* (Jacq.) Macbr.)
                  Flowers lacking a hypanthium and with 4 white petals; stamens 4 and didyn-
                    amous . . . . 125. SCROPHULARIACEAE (*Scoparia dulcis* L.)
            Plants vines or lianas:
                 Leaves pliveined . . . . 107. MELASTOMATACEAE (*Adelobotrys adscendens* (Sw.) Tr.)
                 Leaves not pliveined:
                  Flowers zygomorphic and white; stamens united into a cleft tube; fruits single cap-
                    sules . . . . 72. TRIGONIACEAE (*Trigonia floribunda* Oerst.)
                  Flowers actinomorphic or yellow or red and zygomorphic; stamens not united into
                  a cleft tube; fruits not capsular or fruits of 3 capsular mericarps:
                    Sepals with conspicuous glands; stamens 10 (sometimes with 4 reduced); disk
                    lacking; styles or stigmas 3; fruits of 2 separable cocci, schizocarps of 3 sam-
                    aras, or drupes of 1–3 pyrenes . . . . 71. MALPIGHIACEAE (in part)
                  Sepals eglandular; stamens 3, strap-shaped, inserted on a prominent disk; style
                    and stigma 1; fruits usually of 3 capsular mericarps (drupes in *Tontelea*) . . . .
                     . . . . 78. HIPPOCRATEACEAE

KEY 8: BLADES SIMPLE, ALTERNATE, SPIRALED OR WHORLED; OVARY INFERIOR

Perianth segments undifferentiated *or* flowers with calyx and/or corolla lacking *or* flowers very zygomorphic or in bracteate heads:

 Plants vines or lianas:

  Flowers solitary or in short racemes; flowers zygomorphic, modified into a large, complex, fly-trapping structure . . . . 44. ARISTOLOCHIACEAE

  Flowers in bracteate heads; flowers actinomorphic, tubular . . . . 133. COMPOSITAE (in part)

 Plants not vines or lianas:

  Stipules present:

   Plants herbs or suffrutices; sap clear, not viscid; blades conspicuously inequilateral at base; tepals of staminate flowers 2, or 4 in unequal pairs, those of pistillate flowers 4 or 5 . . . . 100. BEGONIACEAE

   Plants trees; sap usually colored, viscid; blades ± equilateral; tepals not as above . . . . 39. MORACEAE (in part)

  Stipules lacking:

   Perianth parts free; flowers in a corymb . . . . 103. LECYTHIDACEAE (*Gustavia*)

   Perianth tubular; flowers solitary or in heads or spikes:

    Flowers solitary; pedicels more than 3.5 cm long . . . . 132. CAMPANULACEAE (*Centropogon cornutus* (L.) Druce)

    Flowers in heads or spikes; pedicels much shorter:

     Flowers in dense bracteate heads; corolla present; calyx reduced to bristles or scales; stamens 5 . . . . 133. COMPOSITAE (in part)

     Flowers in long spikes; corolla lacking; calyx tubular, 5-lobed; stamens 8 or 10 . . . . 105. COMBRETACEAE (*Terminalia*)

Perianth differentiated into a ± typical calyx and corolla:

 Plants herbs or suffrutescent herbs, not vines:

  Flowers solitary in leaf axils:

   Flowers with the corolla tubular, zygomorphic, 6–8 cm long, red or pink . . . . 132. CAMPANULACEAE (*Centropogon cornutus* (L.) Druce)

   Flowers of free petals, actinomorphic, less than 2 cm long, yellow . . . . 108. ONAGRACEAE

  Flowers not solitary:

   Flowers congested in leaf axils; stipules present; blades succulent . . . . 50. PORTULACACEAE

   Flowers in spicate heads or umbels; stipules lacking; blades not succulent . . . . 110. UMBELLIFERAE (*Hydrocotyle umbellata* L.)

 Plants vines, lianas, or trees:

  Plants vines or lianas:

   Flowers unisexual, with the corolla tubular; fruits many-seeded pepos . . . . 131. CUCURBITACEAE (in part)

   Flowers bisexual, with the petals free; fruits winged:

    Tendrils present; spines lacking; stamens 4 or 5; stipules present; fruits with the 3 mericarps 2-winged . . . . 81. RHAMNACEAE (*Gouania*)

    Tendrils lacking; spines on stems; stamens 8–10; stipules lacking; fruits 5-winged . . . . 105. COMBRETACEAE (*Combretum*, in part)

  Plants shrubs or trees:

   Flowers more than 10 mm long:

    Flowers solitary; style exceeding staminal column . . . . 86. BOMBACACEAE (*Quararibea*)

    Flowers in clusters on trunks or in corymbs or panicles; style very short . . . . 103. LECYTHIDACEAE (in part)

   Flowers less than 5 mm long:

    Blades with large glands at base below; flowers in axillary thyrses 1–5 cm long; fruits explosively dehiscent capsules . . . . 81. RHAMNACEAE (*Colubrina glandulosa* Perk.)

    Blades eglandular; flowers in terminal panicles with ultimate divisions in umbels or heads; fruits berries . . . . 109. ARALIACEAE (in part)

KEY 9: BLADES SIMPLE, ALTERNATE, SPIRALED OR WHORLED; OVARY SUPERIOR

● Perianth undifferentiated into calyx and corolla *or* flowers apetalous:

 ◆ Stipules or tendrils present:

  ■ Blades markedly palmately veined, the veins extending nearly midway or further from base into the blade:

   ○ Blades lobed:

    ◇ Blades large, usually more than 20 cm long:

     Blades peltate . . . . 39. MORACEAE (*Cecropia*)

     Blades basifixed:

Flowers more than 10 mm long, greenish with violet-purple markings; petioles not
   hollow; stipules axillary . . . . . 87. STERCULIACEAE (*Sterculia apetala* (Jacq.) Karst.)
Flowers less than 5 mm long, greenish; petioles hollow; stipules amplexicaul . . . . . .
   . . . . 39. MORACEAE (*Pourouma aspera* Trec.)
◇ Blades less than 20 cm long:
   Leaves conspicuously pubescent:
      Plants shrubs; most parts stellate-pubescent; flowers bisexual; stamens 5 or 15
         (rarely 10). . . . . 84. TILIACEAE (*Triumfetta lappula* L.)
      Plants vines; trichomes not stellate; flowers unisexual (plants monoecious); stamens
         more than 20 . . . . 75. EUPHORBIACEAE (*Dalechampia*)
   Leaves essentially glabrous:
      Blades peltate, usually trilobate, not glaucous below; plants vines; sap not colored;
         flowers bisexual . . . . .98. PASSIFLORACEAE (*Passiflora coriacea* Adr. Juss.)
      Blades basifixed, usually with more than 3 lobes, glaucous below; plants herbs or
         shrubs, not climbing; sap milky; flowers unisexual (plants monoecious). . . . . . .
         . . . . 75. EUPHORBIACEAE (*Manihot esculenta* Crantz)
   ○ Blades not lobed:
      Stipules amplexicaul, large and conspicuous; sap thick, usually colored . . . .
         . . . . 39. MORACEAE (in part)
      Stipules not amplexicaul, usually inconspicuous or reduced; sap usually not thick and
         colored (violet-purple in *Omphalea*, Euphorbiaceae):
         Stems usually conspicuously swollen at nodes; flowers naked, closely aggregated in
            firm, peltate-bracteate pencil-like spikes . . . . 36. PIPERACEAE (*Piper*)*
         Stems not swollen at nodes; flowers with the perianth usually present, variously ar-
            ranged but not as above:
            Styles 3; stamens usually more than or fewer than perianth lobes; fruits 3-valved,
               explosively dehiscent capsules (except *Omphalea*) . . . .75. EUPHORBIACEAE (in part)
            Styles 2 or 1; stamens equal in number to and opposite perianth lobes (perianth
               lacking in staminate *Myriocarpa*, Urticaceae); fruits indehiscent:
               Styles 2; fruits drupes, either green and more than 1 cm long or red at maturity
                  . . . . 38. ULMACEAE
               Style 1; fruits achenes, green and less than 4 mm long, hidden in bracts . . . .
                  . . . . 40. URTICACEAE (in part)
■ Blades not palmately veined:
   Pistillate inflorescences of numerous flowers congested into dense, capitate, bracteate clus-
      ters developing into fleshy aggregate fruits (*Ficus* flowers contained within fleshy
      receptacles). . . . . 39. MORACEAE (in part)
   Pistillate inflorescences not as above *or* flowers bisexual:
      Pistil with 1 simple style and 1 ovule (2 ovules in *Licania*, Chrysobalanaceae); fruits
         1-seeded:
         Plants herbs; stamens 8; blades inconspicuously pubescent . . . .
            . . . . 49. PHYTOLACCACEAE (*Petiveria alliacea* L.)
         Plants shrubs; stamens 3 or 4; blades glabrous above, densely white-arachnoid-
            pubescent below or with linear cystoliths visible on both surfaces:
            Flowers 4-parted, in long slender spikes to 60 cm long; leaves nearly glabrous be-
               neath with conspicuous linear cystoliths radiating from areoles . . . .
               . . . . .40. URTICACEAE (*Myriocarpa yzabalensis* (Donn. Sm.) Killip)
            Flowers 5-parted, in terminal panicles 10–20 cm long; leaves densely white-
               arachnoid-pubescent beneath . . 61. CHRYSOBALANACEAE (*Licania hypoleuca* Benth.)
      Pistil with 2 or more styles or with 2 or more ovules; fruits various:
         Stipules ocreate (encircling stem and forming sheath around stem); styles 3 and mostly
            free (2 in *Polygonum acuminatum*); stamens 6–9; fruits usually shiny, trigonous
            or lenticular achenes . . . . 46. POLYGONACEAE
         Stipules not ocreate; styles or stigmas 1–3, connate or free; stamens 1 to many; fruits
            not as above:
            Stems usually conspicuously swollen at nodes; flowers naked, closely aggregated in
               firm, peltate-bracteate, pencil-like spikes . . . . 36. PIPERACEAE (*Piper*)
            Stems not swollen at nodes; flowers with perianth usually present, variously ar-
               ranged but not as above:
               ▲ Styles 3, free or connate *or* stigma trifid:
                  Stamen 1; flowers bisexual, lacking staminodia . . . . 37. LACISTEMACEAE
                  Stamens 2 or more; flowers bisexual or unisexual:
                     Blades pellucid-punctate; staminodia conspicuous, alternating with stamens;
                        flowers bisexual . . . . 96. FLACOURTIACEAE (*Casearia*)
                     Blades not pellucid-punctate; staminodia lacking; flowers unisexual (mo-
                        noecious or dioecious) . . . . .75. EUPHORBIACEAE (in part)

*Included here because of stipule-like structures.

▲ Styles 1 or 2:

Sap often milky; styles 2 or conspicuously bilobed or style 1 and agglutinated to stamens:

Flowers unisexual; corolla never contorted in bud; styles 2 or conspicuously bilobed; plants trees or shrubs .... 39. MORACEAE (in part)

Flowers bisexual; corolla contorted in bud; style 1, agglutinated to stamens (except *Ervatamia*); plants lianas (*Ervatamia* a cultivated shrub) ......
.... 118. APOCYNACEAE (in part)

Sap clear; style 1 and simple:

Blades usually pellucid-punctate or with gland-tipped teeth; fruits drupes or unarmed capsules; seeds 1 or many .... 96. FLACOURTIACEAE (in part)

Blades not pellucid-punctate, not with glandular teeth; fruits spiny capsules with 1 or 2 seeds *or* fruits drupes:

Fruits drupaceous, fleshy when mature, the outside smooth, never armed; seed 1 .... 61. CHRYSOBALANACEAE (*Licania hypoleuca* Benth.)

Fruits capsular, hard at maturity (usually woody), the outside spiny; seeds 1 or 2 .... 83. ELAEOCARPACEAE (*Sloanea*)

◆ Stipules and tendrils lacking:

Perianth lacking; spikes leaf-opposed, solitary, pencil-like; plants mostly epiphytic herbs or shrubs with swollen leaf nodes .... 36. PIPERACEAE

Perianth present (except *Poinsettia*, Euphorbiaceae); spikes not pencil-like, not leaf-opposed or solitary; plants not epiphytic, usually lacking swollen leaf nodes:

Filaments completely united into a tube; sap in trunk and branches reddish; fruits 2-valved, the seed arillate .... 56. MYRISTICACEAE

Filaments free or partly free; sap not reddish; fruits and seeds various:

Carpels simple or compound and united; styles 3 (free or connate) or conspicuously bifid or trifid:

Flowers unisexual; pistillate flowers with 1 petal; fruits small red drupes less than 7 mm long .... 54. MENISPERMACEAE (*Cissampelos*)

Flowers unisexual or bisexual; perianth lobes 2 or more; fruits shiny achenes or explosively dehiscent capsules (rarely fleshy and more than 8 cm broad):

Stamens and perianth lobes 3 or 5, the stamens alternate with lobes; fruits shiny achenes .... 47. AMARANTHACEAE (in part)

Stamens and perianth lobes unequal or 2, 4, or 8 of each; fruits usually explosively dehiscent capsules, rarely drupes .... 75. EUPHORBIACEAE (in part)

Carpels free, each with 1 style, *or* carpels simple or compound and united; styles 1, 2, or many:

Gynoecium of 1–6, ± free carpels; plants vines or lianas; blades palmately veined; petals sometimes present but minute; sepals in whorls ....
.... 54. MENISPERMACEAE (in part)

Gynoecium not of ± free carpels; plants of variable habit; blades variously veined; petals lacking or indistinguishable from sepals; perianth not in whorls:

Perianth parts 6; stamens in 3 series of 3 each, at least the outer series alternating with conspicuous glands; fruits 1-seeded, usually subtended by a conspicuous cupule (accrescent, persistent calyx tube) .... 58. LAURACEAE

Perianth parts 3–5; stamens not in series, lacking glands; seeds 1 to many, fruits not subtended by cupule:

Plants herbs or vines; flowers bisexual:

Flowers hidden by scarious bracts; style 1 .... 47. AMARANTHACEAE (in part)

Flowers not hidden by bracts; styles 1, 2, or many .. 49. PHYTOLACCACEAE (in part)

Plants trees or shrubs; flowers bisexual or unisexual:

Stamens many; blades usually glandular-toothed or pellucid-punctate ....
.... 96. FLACOURTIACEAE (in part)

Stamens 10 or fewer; blades not with gland-tipped teeth or pellucid punctations:

Stamens 4, opposite 4 free sepals; flowers more than 6 mm long, bisexual; juvenile leaves compound ...... 41. PROTEACEAE (*Roupala montana* Aubl.)

Stamens 3–10, not opposite lobes of calyx; plants dioecious, the flowers less than 4 mm long; juvenile leaves simple .... 56. MYRISTICACEAE

● Perianth differentiated into calyx and corolla:

★ Corolla fused into a tube at least at base, actinomorphic (flowers of *Browallia*, Solanaceae, zygomorphic, and those of Marcgraviaceae appearing zygomorphic owing to nectaries and calyx):

Stamens as many as and alternate with petals *or* stamens fewer than petals; flowers bisexual:

Flowers markedly zygomorphic, with a prominent, external, club-shaped nectary more than 1.5 cm long .... 90. MARCGRAVIACEAE (*Souroubea sympetala* Gilg.)

Flowers not zygomorphic, lacking prominent external nectaries:

✿ Plants lianas or vines:

Plants armed on stems and blades *or* plants hemiepiphytic .... 124. SOLANACEAE (in part)

Plants unarmed, not hemiepiphytic:
　Corolla more than 2 cm long, pleated vertically, usually unlobed . . . .
　　. . . . . 120. CONVOLVULACEAE
　Corolla less than 1 cm long, not pleated, usually lobed ca one-third its length . . . . . .
　　. . . . . 121. BORAGINACEAE (*Tournefortia*)
◐ Plants shrubs, trees, or herbs:
　Corolla twisted in bud; sap milky; pistil with carpels free . . . . 118. APOCYNACEAE (in part)
　Corolla not twisted in bud; sap not milky; pistil with the carpels united:
　　Ovary with many ovules . . . . 124. SOLANACEAE (in part)
　　Ovary with 2 or 4 ovules . . . . 121. BORAGINACEAE (in part)
Stamens equal in number to and opposite the corolla lobes *or* stamens more numerous than
　corolla lobes *or* plants dioecious:
　Plants vines, lianas, or scandent hemiepiphytic shrubs:
　　Flowers with conspicuous nectaries; plants hemiepiphytic shrubs; fruits berries with
　　　many seeds . . . . 90. MARCGRAVIACEAE (*Marcgravia nepenthoides* Seem.)
　　Flowers lacking nectaries; plants vines, not hemiepiphytic; fruits drupaceous . . . .
　　　. . . . . 54. MENISPERMACEAE (*Cissampelos*)
　Plants not scandent:
　　Plants unbranched trees; leaves with petioles more than 30 cm long; blades more than 20
　　　cm long, palmately lobed . . . . 99. CARICACEAE (*Carica*)
　　Plants normally branched trees or shrubs; leaves with petioles less than 15 cm long;
　　　blades usually less than 20 cm long, unlobed (except *Jatropha*):
　　　Flowers with many stamens and a single, simple style; capsules irregularly dehiscent,
　　　　with bright red seeds . . . . 91. THEACEAE (*Ternstroemia tepezapote* Schlecht. & Cham.)
　　　Flowers with stamens equal to or double the number of corolla lobes *or* flowers with
　　　　many stamens and more than 1 style:
　　　　Stamens twice as many as corolla lobes, in 2 series . . . . 43. OLACACEAE
　　　　Stamens as many as corolla lobes *or* stamens many:
　　　　　Stamens as many as corolla lobes; style 1:
　　　　　　Flowers orange, more than 1 cm diam; blades coriaceous, stiff, with a sharp
　　　　　　　apiculum . . . . 111. THEOPHRASTACEAE (*Jacquinia macrocarpa* Cav.)
　　　　　　Flowers not orange, less than 1 cm diam; blades not stiff, lacking a sharp apicu-
　　　　　　　lum:
　　　　　　　Blades pellucid-punctate or opaque-punctate or opaque-lineate; sap not
　　　　　　　　milky; corolla deeply lobed to near base; stamens with the free part of
　　　　　　　　filaments evident; plants mostly shrubs . . . . 112. MYRSINACEAE
　　　　　　　Blades not punctate; sap milky; corolla shallowly lobed; stamens with the
　　　　　　　　anthers sessile and borne at apex; plants mostly large trees more than 10
　　　　　　　　m tall . . . . 113. SAPOTACEAE
　　　　　More stamens than corolla lobes; styles several:
　　　　　　Blades palmately veined and usually palmately lobed; plants monoecious;
　　　　　　　corolla 5-lobed; fruits capsules . . . . . . 75. EUPHORBIACEAE (*Jatropha curcas* L.)
　　　　　　Blades pinnately veined and unlobed; plants dioecious or flowers bisexual;
　　　　　　　corolla 4-lobed or mostly 6-lobed; fruits berries:
　　　　　　　Corolla 4-lobed; blades with stellate trichomes; berries white at maturity . . .
　　　　　　　　. . . . . 88. DILLENIACEAE (*Saurauia laevigata* Tr. & Planch.)
　　　　　　　Corolla mostly 6-lobed; blades with simple trichomes; berries not white at
　　　　　　　　maturity . . . . 114. EBENACEAE (*Diospyros artanthifolia* Mart.)
★ Corolla of free petals (Papilionoideae sometimes with keel petals united), zygomorphic or actino-
　morphic:
◉ Stamens as many as petals:
　Plants scandent and unarmed:
　　Petals large, more than 1 cm long, showy; flowers solitary or paired in axils; tendrils
　　　axillary; fruits berries with many seeds . . . . 98. PASSIFLORACEAE (in part)
　　Petals small, less than 5 mm long:
　　　Flowers bisexual, in cymose inflorescences; petals easily visible; tendrils leaf-opposed;
　　　　fruits with 1 or 2 seeds . . . . 82. VITACEAE (in part)
　　　Flowers unisexual, in racemose inflorescences; petals minute, hidden; tendrils lacking;
　　　　gynoecium of usually 6 free carpels, developing into 6 drupes (or fewer by abor-
　　　　tion) . . . . 54. MENISPERMACEAE (in part)
　Plants not scandent (except *Byttneria*, Sterculiaceae, which is conspicuously armed):
　　Blades palmately veined at base:
　　　Flowers zygomorphic, 3–4 cm long; stipules small, triangular, persistent, white; sta-
　　　　mens appendaged . . . . 95. VIOLACEAE (*Hybanthus prunifolius* (Schult.) Schulze)
　　　Flowers actinomorphic, less than 1 cm long; stipules usually caducous:
　　　　Blades with prominent glands at base; flowers with a conspicuous disk; plants ±
　　　　　glabrous, with simple pubescence on young parts . . . .
　　　　　. . . . . 81. RHAMNACEAE (*Colubrina glandulosa* Perk.)

Blades eglandular; flowers lacking a disk; plants usually with stellate or branched
trichomes . . . . 87. STERCULIACEAE (in part)
Blades not palmately veined:
Plants with blades and sepals usually pellucid-punctate . . . . 112. MYRSINACEAE
Plants not pellucid-punctate:
Flowers solitary or few, in condensed axillary inflorescences less than 3 cm long:
Petals bright orange, more than 20 mm long; calyx of 5 free sepals; blade margins
glandular-crenate . . . . 97. TURNERACEAE (*Turnera panamensis* Urban)
Petals cream-colored, less than 5 mm long; calyx weakly lobed; blades entire . . . .
. . . . 77. CELASTRACEAE (*Maytenus schippii* Lund.)
Flowers many, in racemes or panicles more than 5 cm long:
Flowers white, in large terminal panicles more than 20 cm long . . . .
. . . . 76. ANACARDIACEAE (*Mangifera indica* L.)
Flowers lavender, in slender racemes usually less than 20 cm long . . . .
. . . . 61. CHRYSOBALANACEAE (*Hirtella racemosa* Lam.)
◉ Stamens not equal in number to petals:
Stamens twice as many as petals:
Stamens 10 or 12, ± equal, in 2 series of 5 or 6 each:
Stipules conspicuous; blades with a brownish band of minute trichomes along mid-
rib beneath; petals with ligulate appendage inside; calyx not accrescent in fruit
. . . . 66. ERYTHROXYLACEAE
Stipules lacking; blades lacking discolored band; petals not appendaged; calyx accres-
cent, red and showy in fruit . . . . 43. OLACACEAE
Stamens 6 or 10, equal or unequal or united, not in 2 series of 5 or 6 each:
Plants large trees or shrubs, in the forest; stamens 10, equal or with 1 or 4 longer and
fertile:
Plants large forest trees; stamens unequal; pedicels expanding into fleshy hypocarps
in fruit; fruits solitary . . . . 76. ANACARDIACEAE (*Anacardium*)
Plants shrubs; stamens equal; pedicels not expanding; fruits of 5 elliptic drupelets
. . . . 89. OCHNACEAE (*Ouratea lucens* (H.B.K.) Engler)
Plants herbs, scandent shrubs, vines, or lianas (*Dalbergia brownei*, Leguminosae, sometimes an erect shrub, but always occurring along the shore); stamens 6 or 10:
Petals 3; flowers actinomorphic; stamens 6; fruits drupaceous . . . .
. . . . 54. MENISPERMACEAE (except *Cissampelos*)
Petals 5; flowers zygomorphic; stamens 10; fruits legumes with 1–4 seeds . . . .
. . . . 63. LEGUMINOSAE (in part)
Stamens not twice as many as petals:
Stamens conspicuously united in 1 or more clusters:
Plants herbs, vines, lianas, or scandent shrubs; flowers zygomorphic (except staminate
flowers of *Cissampelos*); petals 1, 3, or 4; stamens 4 or 8:
Flowers bisexual, zygomorphic, the petals 3; stamens 8, connate most of their length
. . . . 74. POLYGALACEAE
Flowers unisexual, the pistillate flowers with 1 petal and sepal; stamens 4, the an-
thers sessile on filament tube . . . . 54. MENISPERMACEAE (*Cissampelos*)
Plants shrubs, trees, or suffruticose herbs, not scandent; flowers actinomorphic; petals
4 or 5; stamens 15 to many, variously united into 1 or 5 clusters:
Petals with bifid, ligulate appendage ca 5 mm long (petal ca 4 mm long excluding
appendage); blades very unequal at base; fruits nearly globose capsules with
short, ± pyramidal tubercles . . . . . . 87. STERCULIACEAE (*Guazuma ulmifolia* Lam.)
Petals lacking apical appendages; blades equilateral or only slightly unequal at base;
fruits not as above:
Anthers 2-celled, dehiscent by 2 pores; capsules transversely elliptic (*Apeiba*) or
seeds winged (*Luehea*) . . . . 84. TILIACEAE (in part)
Anthers 1-celled, dehiscent by 1 pore or slit:
Blades usually toothed (sometimes also lobed); fruits dehiscent capsules or
schizocarps separating into mericarps . . . . 85. MALVACEAE
Blades not toothed (sometimes lobed); fruits capsules with seeds enveloped in
kapok *or* fruits broadly 5-winged, to 15 cm long and 8 cm wide *or* fruits
drupaceous . . . . 86. BOMBACACEAE (in part)
Stamens apparently free (weakly connate at base in *Vantanea*, Humiriaceae):
Petal 1, more than 3 cm long and wider than long; stamens of 2 sizes . . . . 63B.
CAESALPINIOIDEAE (*Swartzia simplex* (Sw.) Spreng. var. *ochnacea* (A. DC.) Cowan)
Petals 2 or more:
□ Blades deeply lobed; calyx with 2 outer and 3 inner sepals; fruits capsules with
wind-dispersed seeds in a cottony mass . . . .
. . . . 94. COCHLOSPERMACEAE (*Cochlospermum vitifolium* (Willd.) Spreng.)

□ Blades not lobed or shallowly so; flowers and fruits not as above:
  Plants lianas . . . . 88. DILLENIACEAE (in part)
  Plants not scandent, herbs, shrubs, or trees:
    Blades palmately veined at base:
      Blades very unequal at base, one side cordate and ± overlapping stem . . . . .
        . . . . 83. ELAEOCARPACEAE (*Muntingia calabura* L.)
      Blades ± equilateral:
        Pedicels with 5 conspicuous glands below the calyx . . . .
          . . . . 93. BIXACEAE (*Bixa orellana* L.)
        Pedicels eglandular:
          Calyx tubular:
            Calyx barely lobed; style 1, slightly tripartite . . . . 85. MALVACEAE
              (*Hampea appendiculata* (J. D. Sm.) Standl. var. *longicalyx* Fryx.)
            Calyx lobed to middle; styles 3, bifid . . . . 75. EUPHORBIACEAE (in part)
          Calyx with the sepals free or nearly so:
            Sepals 4 or 5; plants usually stellate-pubescent; fruits mostly cap-
              sular; ovary with 2 to many cells; placentation axile . . . .
              . . . . 84. TILIACEAE (in part)
            Sepals 3 (except 4 in *Hasseltia*); plants usually lacking stellate pubes-
              cence, often pellucid-punctate (except *Lindackeria*); fruits bac-
              cate (except *Lindackeria*); ovary with 1 cell; placentation parietal
              . . . . 96. FLACOURTIACEAE (in part)
    Blades not palmately veined:
      Blades pellucid-punctate; petioles usually winged . . . . . . 67. RUTACEAE (*Citrus*)
      Blades not pellucid-punctate; petioles not winged:
        Petals 3, 4, 6, or more (5 or 6 in staminate flowers, lacking in pistillate
          flowers of *Codiaeum*, Euphorbiaceae):
          Petals 4, thin; petioles variable in length on the same plant, 1–20 cm
            long; fruits purple schizocarps . . . .
            . . . . 59. CAPPARIDACEAE (*Capparis frondosa* Jacq.)
          Petals 3, 6, or more, thin or fleshy; petioles ± uniform; fruits monocar-
            pic berries or fleshy aggregates or capsules:
            Petals (5)6–13, not fleshy; fruits capsular . . . . 75. EUPHORBIACEAE (in part)
            Petals 3 or 6, usually fleshy; fruits monocarpic berries or fleshy ag-
              gregates; sepals 3, regular . . . . 55. ANNONACEAE
        Petals 5:
          Flowers bright yellow; leaves to 1 m long; fruits capsules . . . .
            . . . . 89. OCHNACEAE (*Cespedezia macrophylla* Seem.)
          Flowers greenish-white or white; fruits drupes:
            Stamens to ca 80; drupes ca 3.5 cm long, the endocarps hard, ovoid-
              ellipsoid, with 5 broad ribs alternating with 5 oblong valves . . . .
              . . . . .65. HUMIRIACEAE (*Vantanea occidentalis* Cuatr.)
            Stamens 3–15; fruits not as above . . . . . . 61. CHRYSOBALANACEAE (in-part)

36. PIPERACEAE

Small trees, shrubs, or epiphytic or terrestrial herbs, rarely vines, often fleshy, usually with swollen nodes. Leaves alternate; petioles often sheathing the stem, those at flowering nodes also with stipule-like prophylls; blades simple, entire, basifixed or sometimes peltate; venation pinnate or palmate; stipules adnate to petiole when present. Flowers bisexual, in dense, axillary, leaf-opposed, or rarely terminal spikes (compound in *Pothomorphe*); flowers subtended by a peltate bract; perianth lacking; stamens 2–5; anthers 2-celled, dehiscing longitudinally; pistal 1; ovary superior, 1-locular, 2–5-carpellate; placentation basal; ovule 1, orthotropous; styles 1–4; stigma lobes 1–4. Fruits tiny drupes; seed with endosperm.

Distinguished by the slender spikes with the flowers usually so closely congested as to be continuous. In addition, *Piper* may be distinguished by the woody stems with swollen nodes and *Peperomia* by the epiphytic or loosely terrestrial habit and the fleshy stems.

Flowers are apparently protandrous, and the stamens are exserted above the bracts. *Trigona* bees are important pollinators of *Piper* (R. Dressler, pers. comm.). In Costa Rica, *Piper* species are often visited by *Trigona* and *Exomalopsis* bees and by *Strangalia* (Cerambycidae), galerucine, and galemud beetles (Chrysomelidae) (Heithaus, 1973).

The tiny fruits of *Piper* are probably endozoochorous through small birds (Ridley, 1930) and bats (Bonaccorso, 1975; Wilson, 1971; Heithaus, Fleming & Opler, 1975). Seeds of several species of *Piper*, including *P. auritum* and *P. hispidum*, were gathered by the bat *Artibeus jamaicensis* in Mexico (Yazquez-Yanes et al., 1975). Fruits of *Peperomia* are dispersed perhaps in part in the same manner, but some species have very tiny, sticky fruits that are probably dispersed on the beaks and feathers of birds also. *Peperomia macrostachya* is often associated with ant nests and it may be that its seeds are dispersed in part by ants.

Ten to twelve genera with more than 2,000 species;

mostly in the tropics. Although the family is represented in the Old World also, it is especially well developed in Latin America.

PEPEROMIA R. & P.

Plants of the genus are distinguished by being epiphytic or weakly terrestrial herbs with usually fleshy stems and leaves. Flowers are sessile and perfect with two stamens. Each flower is subtended by a triangular or rhombic bract, which usually hides the flower and ovary until maturity of the fruit. The minute, one-seeded, drupelike fruits are well exserted from the rachis at maturity; they are sticky and probably epizoochorous.

Peperomia ciliolibractea C. DC., Candollea 1:360, 383. 1923

Stoloniferous herb, epiphytic (or terrestrial where there is considerable debris); stems usually decumbent but soon ascending, thick, rooting profusely at lower nodes; stems, petioles, and peduncles densely puberulent. Leaves deciduous below, crowded toward apex; petioles 2–5(8) cm long; blades oblong-ovate to subobovate, shortly sharp-acuminate, rounded to subcordate or acute at base, 5–14 cm long, 2–7 cm wide, puberulent on both surfaces, especially on veins below, 9–11-pliveined mostly in basal half, drying membranaceous with moderately strong glandular dots below, the margins ciliate. Spikes 10–18 cm long, from upper leaf axils, ca 4 mm wide in fruit; bracts round, reddish, glandular-dotted, sparsely short-ciliate. Fruits subglobose, ca 1 mm diam, the lower two-thirds of each fruit yellowish, the upper part reddish with glandular dots, the apex pointed and oblique; stigma essentially apical. *Croat 6309, 6848.*

Infrequent, in the forest, usually near the ground. Flowering chiefly in the middle of the rainy season (September to October). Fruiting chiefly in the late rainy season and the earliest part of the dry season (October to January).

Known only from Panama, from tropical moist forest around Gatun Lake and in eastern Panamá.

Peperomia cordulata C. DC., J. Bot. 4:137. 1866

Glabrous, pendent, epiphytic herb, to ca 75 cm long; stems and leaves thick and succulent; internodes 2–6 cm long. Leaves peltate within ca 1 cm of the margin; petioles stout, 1–5 cm long; blades broadly ovate to round-ovate, acute to acuminate at apex, rounded at base, 3–11 cm long, 2.5–8.5 cm wide, drying thin but firm, palmately veined, the veins obscure. Spikes 12–19 cm long, on short, bracteate, leaf-opposed branches; peduncles 1–2.5 cm long, the subtending bract lanceolate; floral bracts round. Fruits subglobose, ca 1 mm diam, only slightly exserted; stigma central on an oblique, bluntly tongue-shaped style. *Croat 7770, 8386.*

Infrequent, at the margin of the lake, generally low over the lake surface. Beginning to flower mostly with the onset of the dry season. The fruits develop by the middle to late dry season and the early rainy season.

Commonly associated with *Trigona* bee nests, often on tree stumps in the lake.

Known only from Panama, from tropical moist forest in the Canal Zone (around Madden Lake) and in Colón. See Fig. 177.

Peperomia ebingeri Yunck., Ann. Missouri Bot. Gard. 53:263. 1966

Creeping epiphyte; stems weak, puberulent. Petioles 1–3 mm long, usually perpendicular to blade; blades ± orbicular, to 11 mm long, usually broader than long, puberulent to glabrous but minutely ciliate, the upper surface dark green, minutely and densely papillate with the few veins lighter green, the lower surface light green, smooth. Spikes less than 1 cm long; peduncles 3–8 mm long;

Leaf blades seldom if ever orbicular, mostly more than 1.5 cm long:
  All leaf blades rounded to emarginate at apex; fruits ellipsoid, tapered to a slender hooked
    beak at apex . . . . *P. obtusifolia* (L.) A. Dietr.
  Leaf blades seldom, if ever, rounded at apex:
    Blades often subcordate at base; petioles mostly more than 2.5 cm long; plants stolonif-
      erous, stiffly ascending; stems usually short and unbranched:
      Blades pinnately veined; stems, petioles, and lower blade surfaces densely puberulent;
        rachis of inflorescence not densely papillate . . . . *P. ciliolibractea* C. DC.
      Blades palmately veined; stems, petioles, and lower blade surfaces glabrous or nearly
        so; rachis of inflorescence densely papillate . . . . *P. killipi* Trel.
    Blades seldom if ever subcordate at base; petioles less than 2.5 cm long; plants usually
      sprawling, erect to pendent; stems often long and branched:
      Fruits cylindrical, more than 1.5 mm long; style tongue-shaped, flat; leaves mostly more
        than 4 cm long; plants frequently long-pendent . . . . *P. macrostachya* (Vahl) A. Dietr.
      Fruits subglobose to broadly ovate; style mammilliform (not flattened); leaves generally
        less than 4 cm long; plants sprawling to erect, never long-pendent:
      Plants with prominent dark glandular dots (when dried) on all parts but especially
        on underside of blade; stems pubescent only along 2 lines below petiole, if at all
        . . . . *P. glabella* (Sw.) A. Dietr.
      Plants lacking prominent, dark, glandular punctations; stems densely pubescent all
        over . . . . *P. obscurifolia* C. DC.

---

bracts round, thick, usually puberulent on upper surface in apical half. Fruits subglobose, less than 0.5 mm diam; stigma a small brown tuft of trichomes. *Croat 10802, 11456, Ebinger 165* (type).

Fairly common throughout the forest, often on trunks of trees near the ground. Flowering and fruiting during the dry season and the early rainy season, with the fruits occurring chiefly in the early rainy season. Some populations remain sterile during much of the rainy season.

The species is distinguished by its small round leaves, minutely puberulent bracts, and short inflorescences. It is similar to *P. rotundifolia*, but never with such a scandent habit as that species.

Guatemala, Costa Rica, and Panama. In Panama, known only from tropical moist forest on BCI.

Peperomia glabella (Sw.) A. Dietr., Sp. Pl. 1:156. 1831
  P. conjungens Trel.; *P. glabella* var. *nervulosa* (C. DC.) Yunck.

Epiphyte, nearly glabrous, weak-stemmed, sprawling or erect, to 25 cm high; punctations obscure or lacking when fresh but prominent red to black on most parts when dry; stems often somewhat reddish, flexuous near apex. Petioles 1–11 mm long, often ciliate, the pubescence usually continuing in 2 lines on stem; blades ovate-elliptic to lanceolate-elliptic, acuminate and often downturned at apex, acute at base, (1.5)2.5–5(8) cm long, (0.5)1–2(3) cm wide, thick, weakly 3-veined. Spikes to 15 cm long, solitary or few at apex of stem or from upper axils; peduncles to 1 cm long when fresh; bracts round; anthers 2, exposed between bract and style. Fruits globose-ovoid, ca 1 mm diam, sticky, covered (except style) with glands, exserted on stalk at maturity, the stalk conical, about as long as fruit; style persistent, oblique, tongue-shaped; stigma subapical. *Croat 5478, 8625*.

Abundant in the forest at most levels. Flowering mostly in the dry season and the earliest part of the rainy season; flowering plants sometimes are seen in the very late rainy season. The fruits develop rapidly, but most mature

during the early rainy season. Burger (1971) reported the species to flower and fruit all year in Costa Rica.

About as common as *P. obscurifolia* and confused with that species, but may be distinguished by its nearly glabrous stem and dark punctations when dry. The variety *nervulosa* (C. DC.) Yunck. is not considered distinct because leaf shape and size are quite variable. Burger (1971) reported the species to range from sea level to 2,400 m.

Guatemala to Colombia, Venezuela, the Guianas, and Brazil; West Indies. In Panama, known from tropical moist forest in the Canal Zone, Panamá, and Darién and from premontane wet forest in Colón and Chiriquí.

Peperomia killipi Trel., Bot. Gaz. (Crawfordsville) 73:143. 1922

Erect, short-stemmed epiphyte, 15–30 cm tall (including spikes); stems subrhizomatous, decumbent then ascending, rooting at lower nodes, glabrous. Leaves mostly clustered near apex; petioles 2.5–9 cm long, usually as long as or longer than blade, the stout base persisting on stem; blades broadly ovate and cordate, acute or bluntly acuminate at apex, 3–7 cm long, 1.5–4.5 cm wide, bicolorous when fresh, inconspicuously appressed-pubescent above, glabrous beneath, palmately 7-veined. Spikes often several, from upper axils, extending well above leaves, 5–15 cm long, to ca 2.5 mm wide (1.5 mm dry); peduncles to 7 cm long, glabrous; rachis conspicuously papillate; bracts round, minute. Fruits ovoid, ca 0.5 mm diam, black; stigma subapical on an oblique, bluntly tongue-shaped style. *Croat 12857, Shattuck 596*.

Rare; seen once on Wheeler Trail growing on rotting wood. Flowering and fruiting mostly in the late rainy season (October to December), perhaps to the earliest part of the dry season (January).

The species is similar to *P. pseudo-dependens* C. DC. and *P. lignescens* C. DC. of Costa Rica. See Burger (1971) for a discussion of the differences.

Known only from Panama, from tropical moist forest

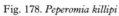

Fig. 178. *Peperomia killipi*

Fig. 177. *Peperomia cordulata*

Fig. 179. *Peperomia macrostachya*

in the Canal Zone, Panamá, and Darién and from premontane wet forest in Chiriquí.

See Fig. 178.

Peperomia macrostachya (Vahl) A. Dietr. in L., Sp.
Pl. ed. 6, 1:149. 1831
P. caudulilimba C. DC. var. *longependula* C. DC.; *P. gatunensis* C. DC.; *P. caudulilimba* C. DC. var. *cylindribacca* (C. DC.) Yunck.

Epiphytic herb, variable in habit, seldom erect, commonly pendent or creeping vertically or horizontally over larger tree branches, to 1.5 m long; larger plants often associated with ant nests; stems, petioles, and peduncles glabrous to puberulent or villous; stems 4–7 mm diam. Leaves variable; petioles 1–27 mm long (short near apex of stem, longer below); blades ovate to ovate-elliptic or lance-elliptic, gradually or abruptly acuminate to rarely acute at apex (often downturned), obtuse to acute at base, 3.5–15 cm long, 0.8–5.5 cm wide, usually ± succulent (drying papyraceous to moderately thick), the veins 2–5, obscure, from near base, the midrib sunken above, the margin sometimes minutely ciliate. Spikes usually 1–3 at apex of stems, mostly 10–15 cm long in flower, less than 2 mm wide, to 22 cm long in fruit; peduncles 1–2.5 cm long, sometimes violet-purple; bracts round. Fruits subcylindrical, narrowed toward apex, 2–3 mm long (as little as 1.5 mm on drying), usually flesh-colored, covered with numerous, sticky globules; style oblique, tongue-shaped; stigma central on style. *Croat 13805*.

Frequent in the forest, especially the older forest, generally high in trees but below the canopy. Flowering chiefly in the dry season and the early rainy season. The fruits develop rapidly, but most mature in the late dry and early rainy seasons; most fruits are gone by the middle of the rainy season.

No doubt the most variable *Peperomia* on BCI. The habit differences indicate that more than one species is included, but no characters are sufficiently uniform to separate the material. The species is similar to and probably not separable from *P. portobellensis* Beurl., described from Portobelo (Colón Province). Standley (1933) misspelled *P. caudulilimba* as *P. cordulilimba*.

Sometimes associated with *Aechmea tillandsioides* var. *kienastii* (22. Bromeliaceae) on the same ant nest.

Mexico to South America; sea level to 1,800 m. In Panama, known from tropical moist forest in the Canal Zone, Bocas del Toro, Panamá, and Darién, from premontane moist forest in Coclé, and from premontane wet forest in Colón and Chiriquí.

See Fig. 179.

Peperomia mameiana C. DC. ex Schroed., Candollea
3:128. 1926

Rhizomatous, glabrous, epiphytic herb; stems and leaves succulent; stems to 1 cm or more thick when fresh, rooting at nodes; internodes 1–3 cm long. Petioles 5–10 cm long, stout, reddish at least at base; blades elliptic, lanceolate-elliptic to oblanceolate, acute to bluntly acuminate at apex, gradually tapered at base, 15–27 cm long, 4–10 cm wide, the veins on each side 5–7, pinnate, the midrib stout below, the margin weakly revolute. Inflorescences terminal or axillary, 20–40 cm long, paniculate, the major axes of the inflorescences and peduncles reddish; peduncles 0.6–5 cm long; spikes 6–15 cm long, ca 3 mm thick, white at anthesis; bracts round; stigma centrally disposed on style. Fruits ellipsoid, ca 1 mm long. *Croat 9015, Kenoyer 303*.

Apparently rare on BCI; usually occurring on rocks or trees near the ground. Flowers chiefly from the middle to late rainy season (August to December). The fruits apparently develop soon after flowering.

Costa Rica and Panama; at elevations less than 800 m. In Panama, known from tropical moist forest in the Canal Zone (vicinity of Gatun Lake), Bocas del Toro, and eastern Panamá. Collected from tropical wet forest on the Osa Peninsula of Costa Rica.

See Fig. 180.

Peperomia obscurifolia C. DC., Candollea 1:357. 1923
P. baileyae Trel.; *P. chrysleri* Yunck.

Epiphyte; stems creeping or erect, to ca 25 cm high, flexuous near apex; internodes sometimes reddish; stems, petioles, peduncles, and sometimes midribs of lower surfaces sparsely to densely pubescent with straight, usually appressed trichomes. Petioles 1–10 mm long; blades elliptic to lance-elliptic, acute or rarely acuminate at apex, acute to cuneate at base, 1–5.5 (6.5) cm long, 0.6–2.5 cm wide, bicolorous, the veins 3–5, the basal 1 cm coalesced with midrib on larger leaves. Spikes 5–16 cm long, ca 2 mm thick; peduncles 1.5–2.5 cm long; bracts round, ca 0.8 mm wide. Fruits globose, sticky, ca 0.5 mm diam, brown, exserted at maturity; style oblique, tongue-shaped, as broad as fruit body; stigma punctiform on center of style. *Croat 9792, 10178*.

Abundant in the forest, most frequently on lower vegetation or on large rocks. Flowers chiefly in the late dry season and the early rainy season. The fruits develop quickly. Both flowering and fruiting spikes may be present on the same plant.

Similar to *P. glabella*, but differing by having pubescence throughout on the stems and by lacking black punctations when dry (obscure but light punctations may appear). Leaf blades are extremely variable. Some blades are so thick as to be lenticular in cross section, while others on the same plant, usually higher on the stem, are thin. Some plants of the species lack the very thick leaves.

Known only from central Panama, from tropical moist forest in the Canal Zone, from premontane moist forest in Colón and Coclé, and from premontane wet forest in Coclé.

See Fig. 181.

Peperomia obtusifolia (L.) A. Dietr. in L., Sp. Pl. ed. 6,
1:154. 1831

Fleshy, stoloniferous herb, epiphytic or terrestrial, nearly glabrous; stems thick, decumbent-ascending, to 15 cm or more high, rooting at lower nodes. Petioles 0.5–4 cm long;

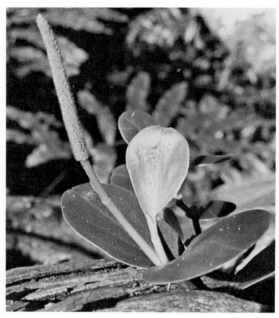

Fig. 180. *Peperomia mameiana*

Fig. 182. *Peperomia obtusifolia*

Fig. 181. *Peperomia obscurifolia*

blades ± obovate, obtuse to emarginate (usually rounded) at apex, cuneate at base, 4–12 cm long, 2.5–6 cm wide, thick, sometimes weakly pubescent on lower surface, pliveined mostly below middle, the veins obscure. Spikes densely flowered, solitary or paired on a 1-bracted, terminal, leaf-opposed, or axillary stalk, 5–15 cm long, to 5 cm wide in fruit; peduncles and stalks weakly puberulent, together 3–5 cm long; floral bracts round. Fruits ellipsoid, ca 1 mm long, tapered at apex into a slender, terminally hooked beak; stigma on upper side at base of beak. *Croat 6610.*

Rare, in the forest. On the basis of the few observations and herbarium collections of the species, it appears that flowering and fruiting occur in the middle to late rainy season, perhaps extending into the early dry season.

When terrestrial, the plant is usually rooted in soil composed of a considerable amount of debris.

Throughout tropical regions of the Western Hemisphere; tropical Africa and Madagascar; usually between sea level and 1,200 m elevation. In Panama, known from tropical moist forest in the Canal Zone, Veraguas, and Panamá and from tropical wet forest in Coclé.

See Fig. 182.

Peperomia rotundifolia (L.) H.B.K., Nov. Gen. & Sp. 1:65. 1816
Poleo

Small, creeping, epiphytic herb; stems scarcely 1 mm wide, glabrous or pubescent. Petioles to 5 mm long, basal, usually perpendicular to blade; blades orbicular, to 12 mm diam, fleshy, bicolorous, often sparsely crisp-pubescent above. Spikes to 2 cm long, little more than 1 mm wide, often on short lateral branches; peduncles short; bracts round-peltate, glabrous. Fruits globose-ovoid, less than 1 mm long; stigma subapical. *Croat 9446.*

Common in the forest, usually creeping or pendent from slender branches and within 6 m of the ground. Flowering chiefly in the dry season and the early rainy season, though it has been seen in flower in the middle of the rainy season; perhaps less seasonal than most *Peperomia.*

The species is apparently closely related to *P. ebingeri.* See that species for discussion.

Throughout the tropics of the Western Hemisphere; from sea level to 2,200 m elevation. In Panama, known from tropical moist forest in the Canal Zone, Bocas del Toro, Los Santos, Panamá, and Darién and from premontane wet forest in Colón and Chiriquí.

PIPER L.
(Gusanillo, Hinojo, Cordoncillo, Barbasco, Canotillo)

The genus is distinguished by its swollen nodes and leaf-opposed, cylindrical spikes with minute, usually sessile, perfect flowers. Each flower has two to five stamens (usually four) and is subtended by a round to triangular, usually ciliolate, peltate bract; styles are present or absent, the stigmas two, three, or four. Fruits are minute (usually 1–3 mm long), one-seeded, and drupelike; they are usually exserted from the bracts at maturity.

A noteworthy feature of the genus is the development of a modified prophyll in those leaf axils that form an inflorescence. Nodes that do not form inflorescences protect the new growth by a sheathing leaf base. The prophyll may be confused with a stipule (Burger, 1972).

Piper aequale Vahl, Eclog. Amer. 1:4, pl. 3. 1796

Glabrous shrub, 1–3 m tall; branches widely spreading. Petioles 5–25 mm long (shortest on ovate leaves); blades lanceolate-ovate to ovate, gradually to abruptly long-acuminate, subinequilaterally obtuse to rounded at base, 10–27 cm long, 4.5–9 cm wide, the major lateral veins usually 4 or 5 pairs, chiefly in basal half; prophylls prominent, slender and pointed. Spikes erect in flower and fruit, 8–10 cm long, 2–4 mm wide in fruit; peduncles equaling or slightly shorter than petiole; bracts inconspicuously fringed. Fruits green, glabrous, bluntly 3-sided, papillate, the small round stigmatic area weakly depressed; stigmas 3, undifferentiated. *Croat 12213.*

Very common in the forest. Juvenile inflorescences appear with new leaf formation during the middle to late rainy season, with the flowers occurring chiefly during the dry season and in the earliest part of the next rainy season. The fruits begin to develop in the rainy season; most fruits mature in the middle to late rainy season and are still present when the new spikes begin appearing.

The species normally has gradually long-acuminate leaves that are broadest well below the middle of the blade. It is in this condition that the leaves of the species differ from the otherwise similar leaves of *P. darienense* and *P. perlasense,* which also have lateral veins extending into the upper three-fourths of the blade. Some specimens differ, however, in having ovate-elliptic leaves that are broadest at about the middle and have shorter petioles and peduncles. Though they are included here with *P. aequale,* these collections (*Croat 8719, 10096, 16577*) may prove to represent another species. Standley (1933)

KEY TO THE SPECIES OF PIPER

Leaves palmately veined or parallel-veined from the lowermost 5 mm:
  Plant a twining vine . . . . *P. aristolochiifolium* (Trel.) Yunck.
  Plants shrubs or trees:
    Shrubs or trees to 8 m tall in the forest; leaves usually more than 15 cm wide, the margins glabrous; fruits papillate, with a prominent, glabrous, apical disk . . . . *P. reticulatum* L.
    Shrubs to 2 m in clearings; leaves usually less than 15 cm wide, without prominent punctations, the margins pubescent; fruits lacking any evident apical disk . . . . *P. marginatum* Jacq.

Leaves pinnately veined:

  Major lateral veins arising along at least three-fourths of blade:

    Leaf bases strongly unequal, one side usually 5–15 mm shorter . . . . *P. arboreum* Aubl.

    Leaf bases ± equal or sometimes oblique with one side only slightly shorter (to 5 mm):

      Petioles strongly vaginate-winged to blade; leaves bronzed beneath when dry, bicolorous fresh; veins prominently loop-connected . . . . *P. cordulatum* C. DC.

      Petioles usually not vaginate-winged (weakly so in *P. grande* and *P. arieianum*); leaves usually greenish beneath:

        Bracts cupulate, glabrous; fruits free on rachis, exserted, very short-styled; plants usually less than 1 m tall; leaves without glandular dots . . . . *P. darienense* C. DC.

        Bracts ± peltate; fruits congested on rachis; plants usually more than 1 m tall:

          Blades mostly more than 8 cm wide, ± glabrous beneath except for narrow band between margin and collecting vein . . . . *P. grande* Vahl

          Blades mostly less than 6 cm wide:

            Leaves glandular-dotted on underside . . . . *P. arieianum* C. DC.

            Leaves not obviously glandular-dotted . . . . *P. perlasense* Yunck.

  Most lateral veins arising in basal half of blade (to basal two-thirds):

    Leaves mostly 20–25 cm long or more, 10–20 cm wide or more:

      Leaves scabrous . . . . *P. peracuminatum* C. DC.

      Leaves not scabrous:

        Leaf margins densely ciliolate, drying thin . . . . *P. auritum* H.B.K.

        Leaf margins glabrous or essentially so:

          Young twigs, nodes, and/or petioles and midribs below ± fleshy-warty (particularly younger petioles) . . . . *P. imperiale* (Miq.) C. DC.

          Stems glabrous, not fleshy-warty:

            Leaves minutely puberulent on all major veins, gray when dry; fruits short-papillate; bracts glabrous . . . . *P. carrilloanum* C. DC.

            Leaves either glabrate or with sparse loose pubescence (particularly between margin and collecting vein); fruits not papillate; bracts ciliate with a prominent papilla . . . . *P. grande* Vahl

  Leaves mostly less than 20 cm long and 10 cm wide or, if larger, not lobed at base:

    Leaves at least somewhat scabrous or markedly asperous above:

      Pubescence of stems conspicuously long, the trichomes of smaller stems often longer than the width of the stem on which they are borne; plants usually less than 1.5 m tall, uncommon, apparently restricted to ravines in the forest:

        Leaves dark glandular-dotted, smooth to only slightly scabrous; petioles to ca 5 mm long . . . . *P. pseudo-garagaranum* Trel.

        Leaves not glandular-dotted, markedly scabrous above; petioles to 5 mm long at apex of plant, grading to 2.5 cm long nearer base . . . . *P. viridicaule* Trel.

      Pubescence of stems not conspicuously long, the trichomes of smaller stems not longer than the width of the stem on which they are borne; plants common in the forest or occurring in clearings only (those in forest more than 2 m tall):

        Upper blade surface not markedly asperous; trichomes of lower veins ± appressed-ascending; plants found in clearings, usually 1.5–2 m tall . . . *P. dilatatum* L. C. Rich.

        Upper blade surface markedly asperous; trichomes of lower veins erect, at least not appressed-ascending:

          Lower blade surface with prominent reddish-brown glandular dots (visible when fresh or dried); intervenous areas with many long trichomes; plants found in clearings, usually less than 2 m tall . . . . *P. villiramulum* C. DC.

          Lower blade surface without prominent glandular dots (sometimes obscure black dots visible when dried); intervenous areas glabrous or with very few small trichomes (elsewhere in Panama with longer trichomes in intervenous areas); blades drying dark gray; plants found in the forest, usually 3–6 m tall . . . . *P. hispidum* Sw.

    Leaves not noticeably scabrous:

      Spikes short, usually less than 2 cm long and 5 mm wide at anthesis, to 3 cm long in fruit; stems and petioles hirsute; plants usually about 1.5 m or less tall . . . . *P. pubistipulum* C. DC.

      Spikes 6 cm long or longer at anthesis; otherwise not as above:

        ● Stems glabrous; leaves nearly glabrous:

          Leaves totally glabrous below . . . . *P. aequale* Vahl

          Leaves either puberulent on veins below or with short sparse trichomes between margin and collecting vein:

            Leaves puberulent beneath on major veins below, gray when dried; fruits short-papillate; floral bracts minute, lacking a dome-shaped apex, glabrous . . . . *P. carrilloanum* C. DC.

Leaves mostly glabrous below except for band of coarse short trichomes between margin and collecting vein; fruits not papillate; bracts large, dome-shaped, ciliate . . . . *P. grande* Vahl
- Stems and leaves pubescent:
  Leaves (especially older ones) smooth and ± glabrous above . . . .
  . . . . *P. culebranum* Schroed.
  Leaves pubescent above, at least on midrib:
  Upper leaf surface glabrate or with shorter scabrous trichomes over surface, the longer pubescence restricted to veins . . . . *P. dilatatum* L. C. Rich.
  Upper leaf surface sparsely long-villous and glabrescent, the pubescence not restricted as above . . . . *P. pseudo-garagaranum* Trel.

---

treated *P. aequale* as *P. frostii* Trel., a name that was never published.

Honduras to northern South America; West Indies; sea level to 2,000 m. In Panama, known from tropical moist forest in the Canal Zone, Bocas del Toro, Panamá, and Darién and from premontane wet forest in Colón, Coclé, and Darién.

See Fig. 183.

Piper arboreum Aubl., Hist. Pl. Guiane Fr. 1:23. 1775

   P. subnudispicum Trel.

Glabrous shrub, 2–7 m. Petioles 2–20 mm long, those at branching nodes vaginate-winged almost to blade; blades lanceolate-elliptic, gradually acuminate to short-acuminate, acute to subcordate and markedly unequal at base (one side 5–15 mm shorter), 14–26 cm long, 3.5–10(12) cm wide, very dark green and shiny above, duller beneath, veined throughout, the veins all prominent, with 7–10 pairs of major lateral veins. Spikes erect when juvenile, usually pendent in fruit, 5–14 cm long in fruit, and to 5 mm diam; peduncles 5–17 mm long; bracts sharply triangular (appearing round in bud), densely ciliate marginally, often more conspicuously so on upper margin. Fruits compressed laterally and much longer in the direction of the rachis, truncate but depressed at the style, sparsely puberulent to papillate-puberulent; stigmas 3, short, deltoid, persistent. *Croat 6685, 12605.*

Infrequent, in the forest. Probably flowers mostly in the dry season and the early rainy season, with most fruits maturing in the early rainy season. Burger (1971) reported the species to flower and fruit from November to May in Costa Rica.

Distinguished by having the leaves veined throughout and inequilateral at base. BCI specimens show no sign of the verrucose-warty stems reported by Yuncker (1950).

Guatemala to northern South America; West Indies; sea level to 1,500 m. In Panama, known from tropical moist forest in the Canal Zone, Darién, and Panamá, but more common from premontane wet forest in Panamá and Chiriquí, from tropical wet forest in Colón, and from premontane rain forest in Chiriquí.

Piper arieianum C. DC., Anales Inst. Fís.-Geogr. Nac. Costa Rica 9:166. 1897

   P. acutissimum Trel.
   Cordoncillo

Glabrous shrub, usually 1–1.5 m tall. Petioles 5–10 mm long near apex of plant, to 3.3 cm long below; blades lanceolate-elliptic, sharply and narrowly acuminate at apex, mostly subequilaterally acute at base, somewhat decurrent, 10–19 cm long, 2.5–6 cm wide, the major lateral veins 6–10 pairs, extending to apex, often weakly loop-connected, glabrous or sparsely pubescent beneath, drying with glandular dots. Spikes 3–9 cm long; peduncles very slender, 7–20 mm long, glabrate to puberulent; bracts peltate, triangular, sparsely short-ciliate. Fruits bluntly trigonous, ± granular-papillate, truncate and concave at apex; stigmas 3, short, strap-shaped, often persisting in fruit. *Croat 14574.*

Occasional, apparently preferring moist areas of gulleys. Seasonal behavior uncertain. Inflorescences develop during the late rainy season and flower mostly in the dry season. The fruits probably mature in the early rainy season.

Shorter plants may be confused with *P. darienense*, but have longer petioles and short-ciliolate bracts. The species is perhaps most easily confused with *P. perlasense*, but that species lacks the dark glandular dots and has leaf blades more prominently inequilateral with one side obtuse or rounded. It may also be confused with *P. cordulatum* but differs in having prominent reticulate veins on fresh leaves and glandular dots on dried leaves.

Nicaragua to Colombia; Trinidad; usually less than 500 m elevation. In Panama, known from tropical moist forest in the Canal Zone, Herrera, Panamá, and Darién and from premontane wet forest in Panamá (Chimán).

Piper aristolochiifolium (Trel.) Yunck., Ann. Missouri Bot. Gard. 37:18. 1950

Twining vine; stems and petioles densely and inconspicuously short-puberulent; internodes elongate, to 15 cm or more long. Petioles 3.5–4.5 cm long, invaginate; blades ovate-cordate, abruptly acuminate, 9–17 cm long, 6–13 cm wide, drying dark, ± glabrous above, glabrous to sparsely puberulent on veins below, the veins 5–7, palmate, the reticulate veins prominulous. Inflorescences unknown. *Croat 12929, 13245a.*

Apparently rare in the forest, though surely more common than it would appear; it has been collected only in the forest in the vicinity of the Laboratory Clearing.

The collections are sterile but closely match the type, which was also sterile. The plant was described as a shrub by Yuncker in the *Flora of Panama,* but the BCI plants are clearly vinelike.

Known only from the Canal Zone (BCI and the Frijoles type collection).

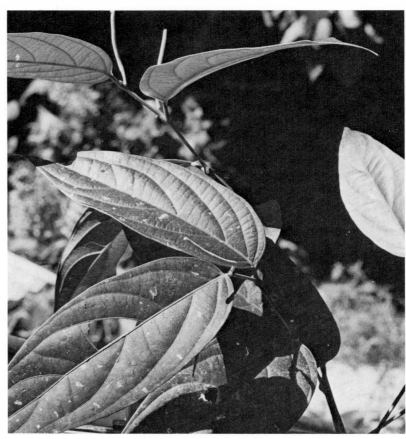

Fig. 183. *Piper aequale*

Fig. 184. *Piper auritum*

Piper auritum H.B.K., Nov. Gen. & Sp. 1:54. 1816
Cowfoot, Cordoncillo, Santa María

Tree, 2.5–8 m tall, the trunk to ca 10 cm diam; wood soft; cut parts sweetly aromatic; stems ± glabrous. Petioles 4–12 cm long, conspicuously winged-vaginate much of their length; blades ± ovate, acute to short-acuminate at apex, cordate at base with very unequal lobes and one side 1–2 cm shorter on petiole, mostly 20–40 cm long, 12–27 cm wide, pinnately veined above the sinus, sparsely short-pubescent above, especially on veins, densely so below, densely white-ciliolate. Spikes stout, 10–27 cm long, to 8 mm or more thick in fruit; peduncles 2–8 cm long; bracts ± round and densely ciliate with long white trichomes. Fruits small, obpyramidal-trigonous; stigmas 3, sessile. *Croat 6976.*

Uncommon; occurring sporadically, usually at the edge of clearings. Apparently not markedly seasonal in flowering behavior, but perhaps with more flowering activity in the rainy season. Plants tend to have more fruiting spikes in the dry season.

Mexico to Colombia; West Indies; sea level to 1,200 m. In Panama, known from tropical moist forest all along the Atlantic slope and in the Canal Zone, Chiriquí, Panamá, and Darién; known also from tropical dry forest in Coclé, from premontane wet forest in Colón, Chiriquí, Coclé, and Panamá, and from tropical wet forest in Colón. See Fig. 184.

Piper carrilloanum C. DC., Bull. Soc. Roy. Bot. Belgique sér. 2, 30(1):209. 1891
P. paulownifolium C. DC.

Shrub, 2–4 m tall; glabrous but with the veins of lower leaf surface puberulent. Petioles usually 2–5(21) cm long, winged-vaginate about halfway on some larger leaves; blades subequilaterally ovate, abruptly acuminate at apex, rounded at base, the lower ones subcordate, becoming cordate on larger lowermost leaves, usually 14–21 cm long, 6–12 cm wide (to 38 cm long and 30 cm wide on largest leaves), the pairs of major lateral veins usually 4–6 (more on large cordate leaves), chiefly in lower half, often pinkish on fresh leaves, the blade drying more or less grayish. Spikes 6–20 (25) cm long, 2–4 mm thick, scarcely thicker in fruit than in flower; peduncles 5–20 mm long; bracts small, ± triangular, glabrous. Fruits short-papillate; stigmas 3, sessile, recurved. *Croat 6319.*

Abundant in the forest. Flowering spikes are present during the late rainy season and throughout most of the dry season. The fruits mature in the late dry season and throughout the rainy season (to December).

May be confused with *P. grande,* but distinguishable by the dense, short puberulence on all the veins; *P. grande* is glabrous except along the margin of the blade.

Hladik and Hladik (1969) reported that fruits of this species are eaten by the tamarin (*Sanguinus geoffroyi*) during January and February. Probably only juvenile fruits or flowers were eaten, since the fruits do not mature until later than this.

Nicaragua to Colombia and Ecuador; sea level to 1,500 m. In Panama, known from tropical moist forest in the Canal Zone, Bocas del Toro, Panamá, and Darién and from premontane wet forest in Colón and Chiriquí. See Fig. 185.

Piper cordulatum C. DC., J. Bot. 4:217. 1866

Glabrous shrub, usually less than 2 m tall. Petioles 1–2 cm long, prominently vaginate-winged throughout; blades moderately thick, lanceolate or lanceolate-oblong, sharply acuminate, rounded to acute at base (one side often slightly shorter), 10–22 cm long, 3–6.5 cm wide, markedly bicolorous especially when fresh, the lower surface pale green becoming brown on drying, the veins obscure above, only moderately prominent below, forming collecting vein. Spikes apiculate, green, to 8.5 cm long and 6 mm wide and pendent in fruit; peduncles nearly 1 cm long in fruit; bracts triangular and cupulate beneath. Fruits ovoid, glabrous, the apex depressed at center; stigmas 3(4), short, strap-shaped. *Croat 7307, 11537.*

Abundant in the forest. Though both flowering and fruiting plants may be found throughout the year, young inflorescences generally begin to appear near the middle of the rainy season (chiefly in September and October). The first flowering inflorescences appear in the latter part of the rainy season and in the early dry season, as early as November but chiefly from December to February. The fruits develop as early as March, chiefly from May to August, with some persisting until late in the rainy season.

Easily distinguished by its narrow, bicolorous leaves and sparsely fruited, pendent spikes. The fruits are so large that there are fewer than ten around the circumference of the spike.

Known only from Panama, from tropical moist forest in the Canal Zone and Panamá and from premontane wet forest in Panamá (Cerro Azul).

Piper culebranum C. DC. ex Schroed., Candollea 3:136. 1926

Shrub or small tree, usually 1.5–4 m tall; stems sometimes warty with lenticels; all parts except upper surface of older leaves sparsely to moderately villous (at least when young), the trichomes long, mostly erect and straight (crisp and crinkled when dry). Petioles 4–22 mm long; blades elliptic to elliptic-oblanceolate or elliptic-obovate, abruptly long-acuminate, unequal at base, one side 1–5 mm shorter and acute to obtuse, the longer side obtuse to subcordate, the blades 12–20 cm long, 4.5–8(9.5) cm wide, the upper side smooth, sparsely villous when young, becoming smooth, glabrate, and often ± rugulose in age (drying gray-green), the lower surface villous throughout but especially on the veins, the major lateral veins usually in 4 or 5 pairs arising mostly in basal half of blade. Spikes 7–9.5 cm long, pendent when juvenile, erect at anthesis, ca 3 mm thick at maturity; peduncles 8–15 mm long; bracts very densely fringed, obscuring apex and overlapping in flower, the cilia longer than width of bract proper. Fruits subcylindrical, truncate, ± depressed medially, glabrous; styles

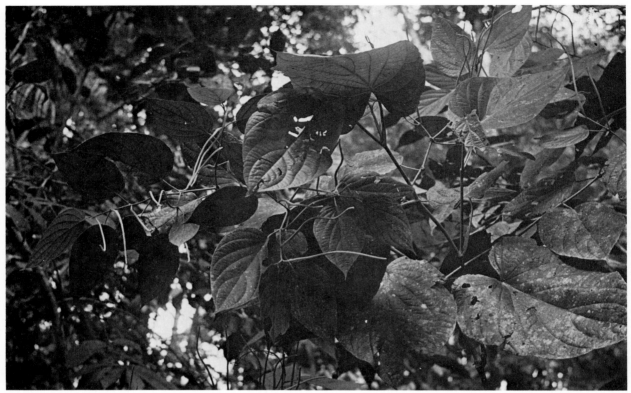

Fig. 185. *Piper carrilloanum*

Fig. 186. *Piper culebranum*

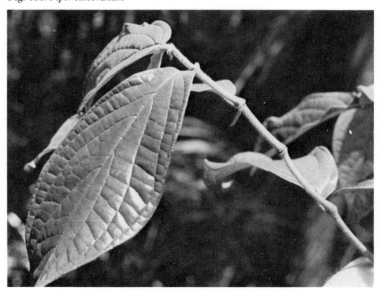

usually 3, short and thick, deciduous. *Croat 5179, 7733*.

Occasional, in the forest, apparently preferring ravines, most commonly a tree 3–3.5 m tall. Flowers in the dry season and the early rainy season. The fruits begin to mature by the late dry season, but are dispersed chiefly in the rainy season. Plants may go for a considerable portion of the rainy season in totally sterile condition, putting on the first juvenile spikes as early as September.

Because of similar height and leaves of the same size and shape, *P. culebranum* may be easily confused with *P. hispidum*, which has short scabrous pubescence and leaves with glandular dots. In contrast, *P. culebranum* has long-villous pubescence (crisp-villous when dry), inflorescence bracts with long, dense cilia, and no glandular dots on the leaves.

Piper culebranum was treated as a synonym of *P. colonense* C. DC. by Burger (1971) in *Flora Costaricensis*. He believes, however, that he may have overclumped (pers. comm.) and that *P. culebranum* should be separate from *P. colonense*.

Costa Rica and Panama. In Panama, known from tropical moist forest in the Canal Zone and Panamá and from tropical wet forest in Panamá (Cerro Jefe).

See Fig. 186.

Piper darienense C. DC. in DC., Prodr. 16(1):374. 1869
P. laxispicum Trel.
Duerme boca

Shrub, mostly less than 60 cm tall (occasionally to 120 cm), glabrous, mostly unbranched and leafless in basal one-half to three-fourths. Petioles usually less than 6 mm long (to 13 mm); blades mostly oblong to lanceolate-elliptic, long-acuminate, subequilaterally acute to rounded at base, 10–19 cm long, 2–7 cm wide, the lateral veins in 6–9 pairs, loop-connected, extending to the apex. Spikes 1.5–5 cm long, white at anthesis, irregular, to 8 mm thick in fruit; peduncles ca 5 mm long; stamens 4 per flower; pistil conical, much longer than stamens; stigmas 3 or 4; bracts cupulate, glabrous, usually partly enclosing the lowermost anther of each flower, much smaller than fruits. Fruits mostly exserted, rounded, to 3 mm long including style, weakly papillate-puberulent, weakly ribbed, becoming tetragonous on drying; seeds brown, with 4 ribs. *Croat 6550, 8582*.

Common in the forest. Flowers chiefly in the dry season and the earliest part of the rainy season. The fruits mature mostly in the middle to late rainy season, by the time the next season's spikes are beginning to develop.

Leaf shape is similar to that of *P. aequale*, and the blade is especially similar to that of *P. perlasense*.

Nicaragua to northern Colombia; usually sea level to 200 m. In Panama, known from tropical moist forest in the Canal Zone, Bocas del Toro, San Blas, Chiriquí, Panamá, and Darién.

Piper dilatatum L. C. Rich., Actes Soc. Hist. Nat. Paris 1:105. 1792
P. leptocladum C. DC.; *P. diazanum* Trel.

Shrub, to 3 m tall, usually 1.5–2 m; upper internodes short, sparsely crisp-pubescent. Petioles villous, 5–10 mm long; blades subrhombic-elliptic or subovate, gradually acuminate, often unequal and obtuse to subcordate at base, mostly 10–19 cm long, 3.5–7 cm wide, usually drying thin, puberulent to scabrous above (trichomes longer on veins), often glabrate or the trichomes restricted to veins in age, crisp or appressed-villous below, especially on veins, the glandular dots obscure when fresh, inconspicuous when dry, the lateral veins in 4 or 5 pairs, arising in basal half of blade; prophylls lunate. Spikes whitish in flower, 6–9 cm long in fruit, ca 3 mm wide; bracts densely white-ciliate, contiguous. Fruits brown or black, obpyramidal, 3-sided and truncate at apex with sparse white pubescence; styles usually 3, slender, short, deciduous. *Croat 12263*.

Occasional in clearings, open areas in the forest, and exposed areas on the lakeshore, particularly abundant in the Laboratory Clearing. Lacking strong seasonal flowering behavior, it can be seen with flowers or fruits most of the year. A flush of flowering occurs with the beginning of the dry season; flowering then continues through most of the rainy season. Most fruits mature at the end of the rainy season and in the early dry season.

This species was mistaken for *P. pseudo-cativalense* Trel. by Standley in his *Flora of Barro Colorado Island*. It may be confused with both *P. hispidum* and *P. culebranum*, but can be distinguished by having short and appressed trichomes on the lower leaf surface, a leaf scar, and a lunate prophyll. The other species are usually much taller and are restricted to the forest.

Mexico throughout continental tropical America; Lesser Antilles and Trinidad; sea level to 200 m. In Panama, known from tropical moist forest in the Canal Zone, Los Santos, and Panamá, from tropical dry forest in Coclé, and from tropical wet forest in Colón.

Piper grande Vahl, Eclog. Amer. 2:3, pl. 11. 1798
P. pseudo-variabile Trel.

Shrub or small tree, 1.5–5 m tall, nearly glabrous. Petioles 1–3(7) cm long and winged-vaginate on larger leaves; blades ovate, somewhat inequilateral, abruptly acuminate, acute or subacute at base of upper leaves, obtuse to subcordate or occasionally cordate at base of the larger lower leaves, 16–31 cm long, 8–23 cm wide, glabrous or with scattered trichomes on the veins but usually with a narrow band of coarse trichomes between margin and collecting vein, the major lateral veins usually in 4–6 pairs, mostly in the basal half. Spikes to 16 cm long, 4–6 mm wide; peduncles 5–10 mm long; bracts large, round, with a prominent dome-shaped apex, the margin densely and obscurely short-ciliate (may be hidden), the stalk of the bract villous. Fruits cushion-shaped, very bluntly trigonous, borne on a slender, ± flattened stalk at maturity, obscured by bracts until maturity; styles 3, moderately long, slender, their thick bases persisting. *Croat 4591, 6501*.

One of the most abundant species of *Piper* on the island, usually in the forest. Though spikes in various stages of development may be present throughout most of the year, most flowering occurs during the dry season when plants may have only flowering spikes; some

flowering continues through most or all of the rainy season. The fruits mature chiefly during the rainy season, mostly from June to December, but some plants still have fruits until just before flowering resumes in January.

Easily confused with *P. carrilloanum,* but that species is densely puberulent on the veins. Yuncker (1950) was perhaps referring to the persistent style bases when he called the stigmas sessile.

Nicaragua to northern South America; usually less than 800 m elevation. In Panama, known from tropical moist forest in the Canal Zone, Colón, Veraguas, Coclé, Panamá, and Darién, from premontane moist forest in the Canal Zone, and from premontane wet forest in Colón, Chiriquí, Coclé, and Panamá.

Piper hispidum Sw., Prodr. Veg. Ind. Occ. 15. 1788

Shrub, usually 3–6 m tall; stems inconspicuously hispid or glabrate in age basally, the trichomes moderately short, rigid, mostly directed outward or ascending, often more dense about the nodes, of varying lengths. Petioles usually 3–8 mm long; blades mostly elliptic to elliptic-ovate, acuminate, obliquely inequilateral, one side 2–5 mm shorter at base and mostly acute, the longer side usually obtuse to rounded at base, the blades mostly 8–20 cm long, 3.5–9 cm wide, drying very dark, pustular-scabrous above with longer stiff trichomes on veins, the lower surface usually with obscure glandular dots and with long straight trichomes mostly restricted to veins (remaining straight on drying), the intervenous areas glabrous or with shorter, less conspicuous trichomes (elsewhere in Panama with many long trichomes), the major lateral veins in 4–6 pairs, mostly in basal half. Spikes mucronate, mostly 9–12 cm long in fruit, ca 3 cm wide; peduncles usually 10 mm long; bracts triangular to round, sparsely short-ciliolate, the cilia shorter than width of bract. Fruits glandular-dotted, obovoid, ± truncate and usually puberulent at apex; stigmas 3, minute, sessile. *Croat 10191.*

Occasional, in the forest. Flowering mostly in the dry season and the early rainy season. Fruiting in the rainy season. Plants are usually completely sterile during the late rainy season.

The species is confused with *P. dilatatum* and *P. culebranum,* which are pubescent and have leaves of similar shape. Yuncker (1950) recognized two varieties of the species on BCI in addition to the typical variety. The variety *trachyderma* (Trel.) Yunck. differs in having the short stout trichomes on the upper internodes curved upward and more or less appressed. Considering the extreme variability in species of *Piper,* the character is hardly worthy of recognition.

Range is uncertain, owing to the confusion in the taxonomy of this complex taxon. The species is described from the West Indies and probably occurs throughout the tropics of the Western Hemisphere. In Panama, known from tropical moist forest in the Canal Zone, Bocas del Toro, San Blas, Panamá, and Darién, from premontane wet forest in Chiriquí and Coclé, and from lower montane wet forest in Chiriquí. Burger (1971) reported the species to be absent below 500 m elevation on the Pacific slope in Costa Rica.

See Fig. 187.

Piper imperiale (Miq.) C. DC. in DC., Prodr. 16(1):339. 1869

Shrub or small tree, 2–5 m tall; upper internodes stout, with warty excrescences especially on nodes, these extending onto petiole and midrib. Petioles usually 4–8 cm long, vaginate nearly to blade; blades broadly ovate-cordate, acute or short-acuminate, unequally lobed at base, mostly 30–50 cm long, 20–29 cm wide, glabrate or minutely puberulent above, glabrate to appressed-puberulent below, short-crisp-villous along veins. Spikes 30–50 cm long or more, to 1 cm thick; peduncles 4–6 cm long or more; bracts ± triangular, the apex prominently raised, densely ciliate, the cilia long, irregular, often denser on apical side; stamens 4, conspicuous. Fruits obovoid to oblong, glabrous; styles 3, slender, soon deciduous, the stout, raised, stylar base persisting in fruit. *Croat 12195.*

Apparently rare, known only from the stream north of the Laboratory Clearing, probably occurring in ravines elsewhere. Spikes have been seen beginning to develop in the middle of the rainy season, flowering usually in the late rainy season or the early dry season. The fruits probably require several months to develop; mature fruits have been collected in other areas in March and June.

Easily distinguished by its cordate leaves and the warty excrescences on the stems and petioles.

Costa Rica and Panama; sea level to 2,000 m, but usually 1,000–2,000 m. In Panama, known from tropical moist forest in the Canal Zone, San Blas, and Darién, from tropical dry forest in Panamá (Taboga Island), from premontane wet forest in Coclé, and from tropical wet forest in Colón, Panamá, and Darién.

Piper marginatum Jacq., Icon. Pl. Rar. 2:2, pl. 215. 1786

P. san-joseanum C. DC.

Shrub or small tree, usually ca 2(3) m tall, ± glabrous; branches often black when dry. Petioles to 6 cm long, vaginate-winged to blade; blades thin, round-ovate, sharply acuminate, usually cordate at base, 8–20 cm long, glabrous except at margin, drying with pellucid dots, the veins 9–13, palmate. Spikes usually 10–16 cm long, 2–4 mm thick in fruit; peduncles ca 1 cm long; bracts rounded, densely ciliate with long white trichomes (± contiguous in spiral arrangements). Fruits obpyramidal, smooth, depressed-truncate at apex; stigmas 3, linear, deciduous. *Croat 4582, 6908.*

Occasional, in clearings. Showing little seasonal flowering behavior, which is no doubt beneficial to its weedy, invading nature. Most flowering commences sometime after the first rains (usually in July) and continues into the earliest part of the dry season, occasionally to the end of the dry season. The fruits develop quickly, most maturing in the late rainy and early dry seasons.

Cut parts are strongly and pleasantly aromatic.

Guatemala to Ecuador and Brazil; West Indies; usually sea level to 1,200 m. In Panama, known from tropical moist forest all along the Pacific slope and on the Atlantic slope in the Canal Zone and San Blas, from tropical dry forest in Herrera, Coclé, and Panamá, from premontane moist forest in the Canal Zone and Veraguas, from pre-

Fig. 187. *Piper hispidum*

Fig. 188. *Piper marginatum*

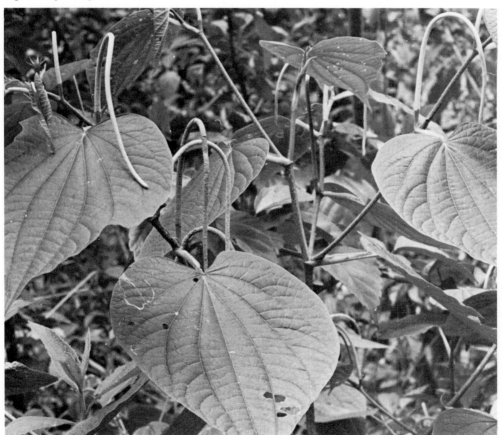

montane wet forest in Colón, Veraguas, Los Santos, Coclé, and Panamá, and from tropical wet forest in Colón and Darién.

See Fig. 188.

Piper peracuminatum C. DC., Smithsonian Misc. Collect. 71(6):9. 1920

Shrub or small tree, 2–3 m tall; stems with hirsute trichomes of varying lengths; stems and lower leaf surface with stiff, erect trichomes of different lengths as well as longer flexuous trichomes. Petioles usually 5–10 mm long, vaginate-winged; blades elliptic-obovate, abruptly acuminate, cordate at base, inequilateral (the longer side 2–4 mm shorter at petiole and covering petiole), 16–27 cm long, 8–14 cm wide, scabrous above, the major lateral veins usually in 6 or 7 pairs, mostly in basal half. Spikes 10–13 cm long, ca 5 mm wide; peduncles about 2 cm long; bracts with short, stiff cilia. Fruits oblong; stigmas 3, recurved.

Rare or no longer present on the island. The species was reported by Standley, but no specimens were cited nor have any specimens been located. No plant resembling this species has been seen recently on BCI, though the species does occur in the isthmus on the Atlantic slope and could be on BCI. Seasonal behavior uncertain; fertile plants have been seen in March in other parts of Panama.

Costa Rica and Panama on the lowland Caribbean coastal plain. In Panama, known only from tropical moist forest in the Canal Zone (around Gatun Lake) and in Colón.

Piper perlasense Yunck., Ann. Missouri Bot. Gard. 37:67. 1950

Glabrous shrub, 2–3 m tall; branches slender, usually widely branched. Petioles 4–10 mm long (usually ca 5 mm); blades mostly lanceolate, gradually long-acuminate, somewhat falcate, subequilateral at base, one side slightly shorter and acute, the longer side obtuse to rounded, the blades 6–16 cm long, 2–4.7 cm wide, the major lateral veins in 6–8 pairs, to all but perhaps apical fourth of blade, strongly ascending. Spikes erect, 5–8 cm long, 2–3 mm wide in fruit; peduncles ca 5 mm long; bracts rhombic or rounded, fringed. Fruits glabrous, obtusely 3-sided, slightly beaked; stigma single, round, sessile. *Croat 8069, Tyson 4203.*

Occasional, in the forest. Flowering mostly in the dry and early rainy seasons. The fruits mature in the rainy season, mostly from June to September.

Leaves are similar in shape to *P. aequale,* but may be distinguished by having the major lateral veins extending into at least three-fourths of the blade. The leaves of this species resemble those of *P. darienense,* but that species is distinguished by having cupulate, glabrous floral bracts.

Known only from Panama, from tropical moist forest in the Canal Zone and Panamá (San Jose Island).

Piper pseudo-garagaranum Trel., Contr. U.S. Natl. Herb. 26:28. 1927

Shrub, 1–2 m tall; flowering internodes short and slender, villous, the trichomes appearing jointed or multicellular,

directed outward, the stem glabrate downward. Petioles usually 2–4 mm long; blades thin, lanceolate-elliptic, gradually very long-acuminate, inequilateral at base, one side 2–4 mm shorter and obtuse, the longer side rounded to subcordate, the blades 9–15 cm long, 3–5 cm wide, sparsely long-villous on both sides, particularly on veins beneath, the trichomes occasionally more than 4 mm long, both surfaces but especially the lower with obscure to prominent reddish glandular dots, the major lateral veins in the basal half in 4 pairs. Spikes to 7.5 cm long at maturity; peduncles less than 1 cm long, with red glandular dots. Fruits 3-sided, truncate at apex, red-glandular; stigma sessile, unlobed or obscurely 2–3-lobed. *Croat 10829, Standley 41164.*

Apparently rare; it has been collected in Creek #7 on Shannon Trail. Seasonal behavior is uncertain.

Distinguished by its narrow, long-pointed leaves with long, villous trichomes and glandular dots. Similar to and possibly not distinct from *P. viridicaule.*

Known only from Panama, from tropical moist forest in the Canal Zone, Chiriquí, and Darién.

Piper pubistipulum C. DC., Smithsonian Misc. Collect. 71(6):5. 1920
P. pubistipulum C. DC. var. *estylosum* Trel.

Shrub, 1–1.5 m tall; internodes short; stems and petioles densely, retrorsely hirsute, often matted with debris. Petioles 3–12 mm long, deeply vaginate at base; blades oblong-lanceolate to narrowly ovate (drying firm, grayish), gradually acuminate, narrowed and inequilateral at base (one side 2–4 mm shorter), 10–20 cm long, 3.5–7 cm wide, dark green and glabrous above (except sometimes at base of midrib), duller and hirsute below especially on the raised veins, the lateral veins in lower half usually 4 pairs, the intercostals prominent and raised. Spikes very small in flower, mucronate, to 3 cm long and 6 mm wide at maturity; peduncles 5–9 mm long, hirsute; bracts cupulate, triangular. Fruits 4-angled, irregularly ridged; styles 2, prominent, deciduous. *Croat 8502.*

Rare; found in ravines and low areas in the forest. Flowering chiefly in the dry season and the early rainy season. Fruiting from the early to middle rainy season. Juvenile spikes may begin to develop before the previous year's fruits have shed.

Fruiting inflorescences are sometimes quite irregular, owing to unequal development of fruits. The species is easily recognized by its short stature, its dingy, densely pubescent stem, and its short spikes.

Known only from central Panama, from tropical moist forest in the Canal Zone and Panamá and from premontane wet forest in Panamá (Cerro Campana).

See Fig. 189.

Piper reticulatum L., Sp. Pl. 29. 1753
P. smilacifolium H.B.K.
Canotillo

Shrub or tree, to 8 m tall; ± glabrous. Petioles mostly 1–2 cm long, channeled above; blades ovate-elliptic to broadly ovate, abruptly acuminate and usually downturned at apex with arched midrib, subequilateral and

Fig. 189. *Piper pubistipulum*

Fig. 190. *Piper reticulatum*

acute to rounded or truncate below (rarely cordate except often so on juveniles), 10–32 cm long, 5–27 cm wide (juvenile blades to 36 cm long and 32 cm wide), the lower surface often sparsely puberulent, the veins 5–9, palmate, prominently raised. Spikes erect, as little as 4 cm long at anthesis, ca 2–2.5 mm wide, to 16 cm long in fruit and 4–6 mm thick; peduncles slender, 1–2 cm long; bracts small, their stalk and margin bearded. Fruits obovoid, papillate, with a broad, smooth apical disk; styles 3 or 4, short and thick. *Croat 14651.*

Occasional, in the forest. Flowering mostly from the middle of the dry season into the rainy season. The fruits develop throughout the rainy season, perhaps chiefly in the early rainy season, particularly in June and July.

Because of its peltate venation it could be confused with *P. marginatum,* a species with smaller, thinner, more cordate leaves occurring in clearings and at the margins of the forest. The species is distinguished from all others on BCI by the prominent apical disk on its fruit. Standley listed this taxon as *P. smilacifolium* C. DC., which is not a valid name.

Nicaragua to northern South America; West Indies; sea level to 700 m. In Panama, known from tropical moist forest in the Canal Zone, Bocas del Toro, Colón, Chiriquí, Panamá, and Darién.

See Fig. 190.

Piper villiramulum C. DC., Smithsonian Misc. Collect. 71(6):11. 1920

Shrub, 2–3 m tall; stems villous (crisp-villous when dry), glandular-pubescent and sometimes gland-dotted. Petioles 5–10 mm long; blades ovate-elliptic, sharply long-acuminate, unequal at base with one side 1–6 mm shorter and acute to obtuse (the longer side rounded to subcordate), 8–17 cm long, 3–7 cm wide, densely scabrous above, interspersed with longer straight trichomes, the surface rugulose in age, densely villous below, prominently dotted with reddish-brown glands, the major lateral veins chiefly in basal half, in 5 or 6 pairs. Spikes to 9 cm long and 4 mm diam; peduncles 5–10 mm long; bracts small, round, densely ciliolate with short white trichomes especially on upper edge. Fruits dark brown, obovoid, ± flattened and puberulent with white glands at apex; stigma small, sessile. *Croat 9400, 12812.*

Uncommon, in clearings. Flowering and fruiting to some extent throughout the year. Flowering is initiated mostly in the late rainy season, with the fruits maturing mostly in the dry and early rainy seasons.

Distinguished by the scabrous leaves, often rugose in age, bearing prominent, brown, glandular dots.

Nicaragua to Panama; sea level to 1,000 m. In Panama, known from tropical moist forest in the Canal Zone, Bocas del Toro, San Blas, Los Santos, Panamá, and Darién, from tropical dry forest in Los Santos, from premontane wet forest in Colón, and from tropical wet forest in Colón, Panamá, and Darién.

Piper viridicaule Trel., Contr. U.S. Natl. Herb. 26:32. 1927

Slender shrub, usually 1–1.5 (2) m tall; stems slender; plants pubescent, the stems, petioles, and midrib below with conspicuous trichomes to ca 2 mm or more long and appearing jointed. Petioles short near apex of plant, 4–5 mm long, grading to 2.5 cm long near base; blades lanceolate-elliptic to ovate-elliptic, sharply long-acuminate, moderately to prominently inequilateral and rounded to subcordate at base, 9–12 cm long, 3–7 cm wide, thin, lacking glandular dots, with long slender trichomes on the lower surface and with both long slender and scabrous trichomes on upper surface, the major lateral veins in 3–5 pairs, in basal half of blade. Spikes unknown. *Croat 11282.*

Rare, occurring in steep ravines along Shannon Trail in the same habitat as *P. pseudo-garagaranum.* Seasonal behavior not determined.

The species is very similar to *P. pseudo-garagaranum,* and additional material may prove them inseparable. However, *P. viridicaule* differs in its longer petioles, its lack of glandular dots, and its prominently scabrous upper blade surface.

Known only from BCI.

POTHOMORPHE Miq.

Pothomorphe peltata (L.) Miq., Comm. Phyt. 37. 1840
Piper peltatum L.
Santa María, Hinojo

Herb, usually to 1.5 m tall, glabrous except for sparse to dense short white trichomes on veins of blade, especially below. Petioles 9–22 cm long, the sheaths prominent; blades peltate ca one-third from base, suborbicular, short-acuminate, cordate at base, to 28 cm diam, whitish below with minute, dense, glandular dots, the major veins usually 9–15, palmate. Spikes several on an axillary stalk to 10 cm long; peduncles to 15 mm long; bracts triangular to round, prominently white-ciliate. Fruits small, obpyramidal, sharply 3-sided, truncate; stigmas 3. *Croat 4051.*

Abundant locally in the Laboratory Clearing and occasionally growing along trails. An exceedingly weedy plant showing little seasonal variation in flowering.

Throughout the tropics of the world. In Panama, usually below 1,000 m; known from tropical moist forest in the Canal Zone, Bocas del Toro, San Blas, Panamá, and Darién, from tropical wet forest in Colón, and from premontane rain forest in Chiriquí (Boquete), Colón, and Coclé (El Valle).

37. LACISTEMACEAE

Shrubs or trees. Leaves alternate, petiolate; blades simple, entire or serrate; venation pinnate; stipules present, caducous. Flowers bisexual, apetalous, ± zygomorphic, in axillary, bracteate spikes or racemes; perianth 4-lobed, the lobes subequal; disk present, prominent, hypogynous; stamen 1; anther 2-celled; thecae separated by a forked connective, dehiscing longitudinally; ovary superior, 1-locular, 3-carpellate; placentation parietal; ovules 1 or 2 per placenta, anatropous; styles 3 or stigma trifid. Fruit a 3-valved, loculicidal capsule (berrylike at first); seeds 3 or by abortion 1, arillate, with endosperm.

Lacistemaceae are trees or shrubs distinguished by

KEY TO THE SPECIES OF LACISTEMACEAE

Inflorescences less than 3 cm long; capsules more than 5 mm diam; axils of lower lateral veins not tufted . . . . *Lacistema aggregatum* (Berg) Rusby
Inflorescences more than 4 cm long; capsules less than 4 mm diam; axils of lower lateral veins usually conspicuously tufted (axils glabrous elsewhere) . . . . *Lozania pittieri* (S. F. Blake) L. B. Smith

their reduced flowers bearing a single stamen and by their small, fleshy, capsular fruits with arillate seeds. The family will be treated by H. Sleumer (pers. comm.) as Flacourtiaceae in his revision of that family.

Flowers are minute and clustered, each with a single anther. Pollination system is unknown.

The fruits display arillate seeds, and though dispersed principally by birds, they are also taken by monkeys (Oppenheimer, 1968).

Two genera and about 27 species; tropical America.

LACISTEMA Sw.

Lacistema aggregatum (Berg) Rusby, Bull. New York Bot. Gard. 4:447. 1907

Tree, usually 6–20 m tall, glabrate or with sparse puberulence on young stems and underside of leaf. Petioles 5–10(20) mm long; blades lanceolate, oblong-elliptic or oblanceolate-acuminate, acute to rounded at base, 10–16 cm long, 4.5–7 cm wide, entire or remotely crenate, the veins drying wrinkled, usually barbate in the axils of lateral veins on the lower surface, often pubescent all along midrib at least when young; stipules to 1 cm long, caducous. Flowers greenish, sessile, in narrow, cylindrical, bracteated spikes, the spikes 1–3 cm long, 4–12 per axil; bracts cupulate, broader than long, one subtending each flower; perianth segments 4, ca 0.5 mm long, unequal, erose; ovary and stamen centrally situated on a broad fleshy disk, the disk subtended by free bracteoles (usually 3) exceeding width of disk; stamen solitary, ca 2 mm long, exceeding pistil, the connective bifurcate; styles 3, short. Capsules ovoid, ca 1 cm long, red, short-stalked, splitting ± irregularly into 2 or 3 valves; seed 1, ca 7 mm long, surrounded by a fleshy, bitter, white aril. *Croat 5691, 8402.*

Common in the forest, especially in the young forest. Flowers in the dry and early rainy seasons, from January to May (rarely to July), usually in the latter half of the dry season. The fruits mature from April to June (sometimes August).

The inner wall of one of the valves of the capsule becomes free and folds along the median. After being forced out, the seed is suspended on a slender white fiber from near the apex of the inner wall. Though the seeds are no doubt principally dispersed by birds, Oppenheimer (1968) reported that white-faced monkeys eat the white aril associated with the seed.

Mexico to Colombia, Venezuela, the Guianas, and Peru; Trinidad, Greater Antilles. In Panama, known from tropical moist forest in the Canal Zone, Bocas del Toro, Colón, Veraguas, Los Santos, Herrera, and Darién, from premontane wet forest in Coclé and Panamá, and

from tropical wet forest in Colón (Guásimo). Reported from premontane rain forest in Costa Rica (Holdridge et al., 1971).

See Figs. 191 and 192.

LOZANIA Mutis ex Cald.

Lozania pittieri (S. F. Blake) L. B. Smith, Phytologia 1(3):138. 1935

   L. pedicellata (Standl.) L. B. Smith

Shrub or small tree, to 8 m tall; stems and twigs brittle; younger stems, petioles, and axes of inflorescences strigose. Petioles 3–8 mm long; blades elliptic to obovate-elliptic, acuminate, acute at base, 8–14 cm long, 2.5–4.5 cm wide, entire to obscurely toothed, glabrous above, the veins below weakly strigose and often pilose as well, especially in axils of lateral veins; stipules 1.5–2 mm long, caducous, the scar at most half-encircling stem. Flowers green, ca 1.5 mm diam, in solitary, axillary racemes to 9 cm long; axis and bracts pubescent; bracts minute; pedicels ca 1.5 mm long; perianth oblique, 4-lobed, the lobes ± rounded, concave, the uppermost at first erect and ± enclosing the pistil, later spreading to expose the pistil, the others ± spreading, subtending the fleshy reniform disk; stamen 1, to 1 mm long, at first held near the upper perianth lobe with the anthers unopened, the filament later elongating, spreading away from the pistil, the anthers opening; anthers longitudinally dehiscent, broader than long, the thecae distinct, directed upward; ovary minute, pubescent; styles 3, recurved (appearing as 3 hooks). Capsules globose, ca 4 mm diam, sparsely puberulent; seeds 3, minute, with a red aril. *Foster 2364.*

Apparently rare, but at least locally common. Collected by R. Foster on Balboa Trail 10 (*Foster 2364*) and by O. Shattuck at Gross Point (*Shattuck 972*). Flowering on BCI mostly in July and August. The fruits mature in August and September (mostly in August). Rarely flowering elsewhere in Panama in March, with the fruits maturing in April. Allen (1956) reported the species to flower in December on the Osa Peninsula in Costa Rica.

Costa Rica to northernmost Venezuela. In Panama, known from tropical moist forest in the Canal Zone, Bocas del Toro, and San Blas, from premontane wet forest in Colón and Panamá, and from tropical wet forest in Colón and Veraguas.

38. ULMACEAE

Trees and shrubs, sometimes scandent and armed. Leaves alternate, petiolate; blades simple, generally serrate, palmately veined at base; stipules present, caducous. Flowers

Fig. 191. *Lacistema aggregatum*

Fig. 192. *Lacistema aggregatum*

Fig. 193. *Trema micrantha*

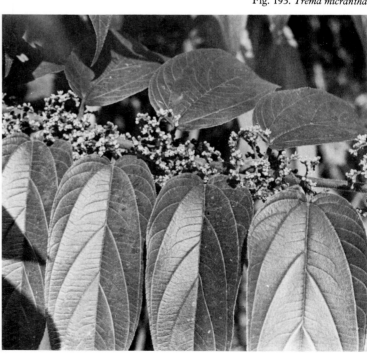

KEY TO THE SPECIES OF ULMACEAE

Plants lianas or climbing shrubs, the stems armed . . . . *Celtis iguanaeus* (Jacq.) Sarg.
Plants unarmed trees:
    Leaf blades ± glabrous except for midrib, not cordate at base; fruits more than 1 cm long . . . . .
    . . . . *Celtis schippii* Standl.
    Leaf blades pubescent, asperous especially above, cordate at base; fruits less than 5 mm long . . .
    . . . . *Trema micrantha* (L.) Blume

unisexual or bisexual (monoecious or polygamomonoe-cious), apetalous, sometimes obscurely zygomorphic, solitary or in axillary cymes; sepals 5, basally connate; stamens 5, free, opposite the sepals; anthers 2-celled, dehiscing longitudinally; ovary superior, 1-locular, 2-carpellate; placentation apical; ovule 1, anatropous, pendulous; styles 2, basally fused (in *Celtis,* the stigmas bifurcate). Fruits drupes; seed lacking endosperm.

Members of the family resemble both the Moraceae (39), from which they differ by lacking milky or viscid sap and by having usually inequilateral leaf bases, and the Urticaceae (40), from which they differ by having a two-carpellate ovary and two styles. Some Urticaceae also have unequal leaf bases.

Pollination system unknown. Wind pollination is common in the order Urticales (Faegri & van der Pijl, 1966).

Fruits are probably all animal dispersed. *Trema micrantha* has small red fruits and is probably bird dispersed. Van der Pijl (1968) reported that the seeds of *Celtis iguanaeus* are eaten by iguanas.

Sixteen genera and about 300 species; widely distributed.

CELTIS L.

Celtis iguanaeus (Jacq.) Sarg., Silv. North Amer.
    7:64. 1895
    Hackberry, Cagalara

Andromonoecious (with staminate and bisexual flowers on the same plant) liana or climbing shrub, 10–25 m long, the trunk usually less than 10 cm diam, at least the smaller branches armed with recurved spines. Petioles 5–10 mm long; blades ± ovate, acute to acuminate at apex, obtuse to subcordate, 3-veined, and sometimes inequilateral at base, 3–11 cm long, 1.5–4.5 cm wide, sparsely pubescent to glabrous, serrate, the veins on underside prominently raised. Inflorescences cymose, axillary, the smaller branches and often pedicels and calyces minutely puberulent; flowers usually 5-parted, pedicellate, 1.5–2 mm long, green or greenish-yellow; sepals ± oblong, the margins scarious, ± ciliate; petals lacking; staminate flowers with a rudimentary pistil ca 1.5 mm long, the stamens opposite the sepals, the filaments at first curved inward, ca 1.7 mm long, the anthers attached apically near the base of the pistillode, becoming free, the filament then straightening rapidly to fling pollen from the anthers; pollen powdery; bisexual flowers similar to staminate flowers, the ovary 1-locular. Drupes ovoid, ca 1 cm long, ± 2-edged, yellow, orange, or red. *Croat 14649.*

Occasional, in the forest, climbing high into the canopy. Flowers in the early rainy season. The fruits mature in the middle to late rainy season.

Mexico to Argentina; the Antilles. In Panama, known from tropical moist forest in the Canal Zone, Panamá, and Darién.

Celtis schippii Trel. ex Standl., Field Mus. Nat. Hist.,
    Bot. Ser. 12:409. 1936

Tree to 30 m tall, to 60 cm dbh, often buttressed 1.5 m; bark thin, planar, unfissured or with minute cracks; inner bark thick, with brown streaks; sparsely short-pubescent on young stems and leaves, ± glabrous in age except for midrib above and below. Petioles 5–9 mm long; blades ovate-elliptic to elliptic or ovate, acuminate, obtuse and slightly inequilateral at base, ± decurrent onto petiole, 6–14 cm long and 3.5–7 cm wide (to 24 cm long and 10 cm wide on juveniles), 3-veined at base, the larger lateral veins with inconspicuous cavelike domatia in axils beneath (especially the apical axils) with 2–4 pairs of major laterals above the basal pair. Flowers axillary, solitary; pedicels to 7 mm long in fruit; sepals 5, suborbicular, to 1 mm long, persistent, ciliate. Drupes narrowly ovoid, with a pungent odor, green at maturity, to 1.8 cm long, glabrous; pericarp leathery, ca 1 mm thick; seed with a brown testa, ca 1 cm long. *Croat 16212.*

Rare; known from a few individuals in the old forest. Flowering unknown. Fruits in July; elsewhere in Central America mature-sized fruits have been seen in March and September.

Belize, Costa Rica, and Panama. In Panama, known only from tropical moist forest on BCI.

TREMA Lour.

Trema micrantha (L.) Blume, Mus. Bot. Lugduno-
    Batavum 2:58. 1856
    T. integerrima (Beurl.) Standl.
    Capulín macho, Capulín, Jordancillo

Monoecious tree, 2–20 m tall, the trunk to ca 16 cm dbh; bark light brown, thin, prominently lenticellate; inner bark moderately thick, tan; pubescent all over with erect trichomes, the branchlets becoming glabrate; sap ± pungent, somewhat foul-smelling. Petioles 5–10(18) mm long; blades narrowly lanceolate, long-acuminate, cordate and often inequilateral at base, 5–15 cm long, 1.5–5 cm wide, asperous especially above, serrate to subentire; stipules lanceolate, minute. Cymes axillary, to 1.5 cm long; flowers minute, 5-parted, green, greenish-yellow, or whitish; sepals to 1.3 mm long, minutely pubescent

outside; petals lacking; anthers of the staminate flowers contained within the boat-shaped tepal in bud, the tension of the elongating filament released when the anther slips from the sepal and springs upward, the thecae directed upward; pollen white, powdery; pistillode columnar; pistillate flowers similar to staminate flowers, the ovary globose, with 2 bifid styles, the stylar bases persisting in fruit. Fruits ovoid or subglobose, 3–4 mm long, red. *Croat 6242.*

Occasional at the edges of clearings, rare in the forest. Flowers and fruits principally in the rainy season.

Superficially similar to *Pouzolzia obliqua* (40. Urticaceae), which has four-parted flowers and an achene with a single long style.

In Ecuador and some other areas, *Trema integerrima* (Beurl.) Standl. is recognized as a distinct species, being a larger tree occurring in forests rather than in disturbed areas. The differences on BCI seem insignificant, though two collections, *Aviles 58a* and *Hayden 4,* have the short erect pubescence and the entire thin leaves attributed to *T. integerrima.* Both are considered here to be only forms of *T. micrantha.*

Widely distributed in the tropics and subtropics of the Western Hemisphere. In Panama, ecologically variable and widespread in cutover areas; known from tropical moist forest throughout Panama, from tropical dry forest in Coclé, from premontane moist forest in the Canal Zone, from premontane wet forest in Chiriquí, Coclé, and Panamá, from tropical wet forest in Colón and Veraguas, and from premontane rain and lower montane wet forests in Chiriquí.

See Fig. 193.

39. MORACEAE

Trees, shrubs, or herbs (*Dorstenia*), sometimes buttressed, often epiphytic or hemiepiphytic (ultimately freestanding in usually strangling *Ficus*); sap viscid, usually milky, brown, black, or merely turbid; stems sometimes hollow. Leaves alternate, usually basifixed (peltate in *Cecropia*); blades usually simple and pinnately veined (deeply palmately lobed or divided and palmately veined in *Cecropia* and *Pourouma,*), entire to serrate; stipules present, usually conspicuous and ensheathing bud, leaving a scar encircling the stem. Flowers unisexual (dioecious or monoecious), actinomorphic, in cymes, spathaceous catkins, or hollow receptacles; perianth obsolete, tubular or (2)4(5)-lobed; stamens 1, 2, or 4 (numerous in *Castilla*), free; anthers 2-celled, versatile, dehiscing longitudinally (circumscissile in *Brosimum*); ovary 1, superior to inferior, 1-locular, 1- or 2-carpellate; stigmas usually 2 (1 in *Ficus, Coussapoa,* and *Cecropia*); ovule 1, anatropous and pendulous from near the apex of the locule or campylotropous (basal and erect in *Cecropia, Coussapoa,* and *Pourouma*). Fruits diverse, basically drupes or achenes (1-seeded drupes in *Trophis, Pourouma, Sorocea,* and *Maquira;* false drupes in *Brosimum,* consisting of a single

pistillate flower surrounded by a receptacle and all old staminate flowers; fleshy syncarps in *Artocarpus, Castilla, Poulsenia,* and *Coussapoa;* achenes closely aggregated into fleshy, catkinlike structures in *Cecropia* or encased in the receptacle, completely in *Ficus,* partially in *Dorstenia*).

Members of the family are most easily distinguished by the latex, by the paired stipules often forming a cap over the bud and leaving a cylindrical scar, and by the unisexual flowers, the pistillate ones having only two styles.

Most species of Moraceae, including *Brosimum, Coussapoa, Poulsenia,* and *Cecropia,* have numerous aggregated flowers with exserted anthers ideally suited to pollen-feeding insects. *Olmedia* and *Trophis* are possibly wind pollinated. In both cases, stamens are spring-loaded. When the stamens are released from the tepals, the pollen is catapulted from the anthers. Also possibly wind pollinated are *Cecropia, Artocarpus, Brosimum, Castilla, Coussapoa, Pourouma, Poulsenia,* and *Maquira.* Bawa and Opler (1975) reported that *Cecropia* offers no nectar, which further indicates that it may be wind pollinated. *Ficus* pollination is treated at length in the genus discussion.

The seeds of all species of Moraceae are animal dispersed. Most species, especially *Cecropia* (Leck, 1972), *Coussapoa, Olmedia* (Chapman, 1931), *Maquira, Sorocea,* and *Trophis,* are dispersed by birds. All species, however, are also dispersed by other animals, particularly arboreal frugivores. *Olmedia aspera, Maquira costaricana,* all *Cecropia,* and nearly all *Ficus* species were reported to be taken by white-faced monkeys (Oppenheimer, 1968). Most species of *Cecropia* and *Ficus* and *Brosimum alicastrum* (Hladik & Hladik, 1969) are also taken by spider and howler monkeys. *Cecropia longipes* and *Olmedia aspera* are taken by tamarins (Hladik & Hladik, 1969). *Cecropia peltata* is eaten by coatis (Kaufmann, 1962). *Ficus* is an important part of the diet of bats. In addition to *Piper* (36. Piperaceae) and *Cecropia,* bats depend heavily on *Ficus* for food (E. Tyson, pers. comm.). Of the 23 species of plants found in bat detritus of *Artibeus jamaicensis* Leach in Mexico (Yazquez-Yanes et al., 1975), 28 percent were Moraceae, including seeds of *Pseudolmedia, Cecropia obtusifolia, Ficus insipida, F. obtusifolia, Poulsenia armata,* and *Brosimum alicastrum.* Bonaccorso (1975) also reported *Brosimum* and *Poulsenia* to be dispersed by bats. Fruits of *Cecropia peltata* are also bat dispersed (Goodwin & Greenhall, 1961). R. Foster (pers. comm.) believes bats are the principal dispersing agent for *Cecropia* and green-fruited figs. *Cecropia obtusifolia* was the most well-represented plant species in studies of bat detritus samples made in Mexico (Yazquez-Yanes et al., 1975).

Birds may take small colored figs. Since most species of *Ficus* must germinate high in the canopy of the forest and develop as true epiphytes, fruits that fall to the ground and are further dispersed by forest floor inhabitants are wasted. Enders (1935) reported that opossums, squirrels, spiny rats, and collared peccaries eat figs.

Despite their close association with *Cecropia,* ants of

KEY TO THE TAXA OF MORACEAE

Plants herbs; inflorescences cup-shaped . . . . *Dorstenia contrajerva* L.
Plants trees or shrubs; inflorescences various:
　Flowers borne on the inner surface of a ± globose, hollow receptacle, its apex a small opening
　　closed by scales; stems solid; stipules fully amplexicaul, usually conspicuous . . . . *Ficus*
　Flowers not borne on inner surface of a ± closed receptacle; stems often hollow; stipules fully
　　amplexicaul or lateral:
　　Leaves conspicuously lobed:
　　　Leaves pinnately lobed; sap white . . . . *Artocarpus altilis* (Park.) Fosb.
　　　Leaves palmately lobed; sap brown or black:
　　　　Flowers in spikes; leaves peltate . . . . *Cecropia*
　　　　Flowers in cymes; leaves basifixed . . . . *Pourouma guianensis* Aubl.
　　Leaves not lobed:
　　　Plants armed; staminate flowers in dense globose heads 1–2.3 cm diam; pistillate flowers
　　　　more than 5 mm long, 5–9 in ovoid heads . . . . *Poulsenia armata* (Miq.) Standl.
　　　Plants unarmed; flowers various:
　　　　Flowers in spikes or racemes:
　　　　　Staminate spikes densely flowered; pistillate flowers and fruits sessile; leaves asperous
　　　　　. . . . . *Trophis racemosa* (L.) Urban
　　　　　Staminate spikes loosely flowered; pistillate flowers and fruits pedicellate; leaves
　　　　　　smooth . . . . *Sorocea affinis* Hemsl.
　　　　Flowers solitary or in globose or discoid heads:
　　　　　Leaves ovate, spirally arranged; petioles more than 3 cm long; staminate inflorescences
　　　　　　dichotomously branched 2–4 times with several ± globose heads ca 5 mm diam:
　　　　　　Leaves usually cordate at base, the major lateral veins in 6–8 pairs above base; pis-
　　　　　　　tillate inflorescences (heads) 2–4-lobed . . . . *Coussapoa magnifolia* Trec.
　　　　　　Leaves usually obtuse to truncate at base, the major lateral veins in 10–20 pairs
　　　　　　　above base; pistillate inflorescences (heads) not lobed . . . .
　　　　　　　. . . . . *Coussapoa panamensis* Pitt.
　　　　　Leaves not ovate, distichous; petioles less than 2 cm long; staminate inflorescences not
　　　　　　compound:
　　　　　　Inflorescences bisexual, globular, ca 5–10 mm diam with many staminate flowers
　　　　　　　and 1 or 2 pistillate flowers at center of cluster . . . .
　　　　　　　. . . . . *Brosimum alicastrum* Sw. subsp. *bolivarense* (Pitt.) C. C. Berg
　　　　　　Inflorescences unisexual:
　　　　　　　Staminate inflorescences fan-shaped; pistillate flowers in discoid heads; perianth
　　　　　　　　lacking; fruits syncarps (all carpels united); blades strongly cordate, dentate
　　　　　　　　. . . . . *Castilla elastica* Sessé
　　　　　　　Staminate inflorescences not fan-shaped; pistillate flowers solitary or in discoid
　　　　　　　　heads; perianth distinct; fruits solitary or the carpels ± free; blades cuneate
　　　　　　　　to narrowly rounded at base, mostly entire (undulate-serrate in *Perebea*):
　　　　　　　　Pistillate flowers dense in discoid heads:
　　　　　　　　　Leaves and stems glabrous or the stems inconspicuously puberulent; blades
　　　　　　　　　　entire; fruiting heads to 5 cm diam; fruits to 2 cm long . . . .
　　　　　　　　　　. . . . . *Maquira costaricana* (Standl.) C. C. Berg
　　　　　　　　　Leaves (at least on midrib) and stems conspicuously hispidulous; blades
　　　　　　　　　　undulate-serrate toward apex; fruiting heads ca 2 cm diam; fruits less
　　　　　　　　　　than 1 cm long . . . . *Perebea xanthochyma* Karst.
　　　　　　　　Pistillate flowers solitary, small, bracteate at base:
　　　　　　　　　Leaves asperous; staminate flowers with a distinct perianth and 4 stamens;
　　　　　　　　　　ovary superior; leaves remotely undulate-serrate at apex . . . .
　　　　　　　　　　. . . . . *Olmedia aspera* R. & P.
　　　　　　　　　Leaves smooth; staminate flowers with vestigial perianth, the heads unorgan-
　　　　　　　　　　ized; ovary inferior to semi-inferior; leaves entire . . . .
　　　　　　　　　　. . . . . *Pseudolmedia spuria* (Sw.) Griseb.

the genus *Azteca* do not function in pollination of flowers or dispersal of seeds (Wheeler, 1910). These ants are beneficial to the plant, however, in warding off predators such as leaf-cutter ants (*Atta*) (Janzen, 1969).

Reportedly 53 (Willis, 1966) to 73 (Lawrence, 1964) genera and 1,000–1,500 species. The family is worldwide in distribution, principally in tropical or subtropical areas with comparatively few species in temperate areas.

ARTOCARPUS Forst. & Forst.f.

Artocarpus altilis (Park.) Fosb., J. Wash. Acad. Sci. 31:95. 1941.

A. communis Forst. & Forst.f.

Breadfruit

Monoecious tree, usually less than 12(20) m tall; trunk lenticellate; outer bark thin; inner bark thick, granular, with copious milky sap; wood yellowish. Petioles stout, (2)4–7 cm long; blades ovate in outline, deeply pinnately lobed, acuminate at apices of lobes, the blades acute at base, 30–80(100) cm long, 15–40(65) cm wide, thick and leathery, scabrous, usually pubescent below and on veins above; stipules fully amplexicaul, to 15 cm long, conspicuously appressed-pubescent, deciduous. Peduncles stout, 5–8 cm long; staminate spikes club-shaped, 20–40 cm long, ca 3 cm diam at apex, narrower at base, the staminate flowers dense; perianth 2- or 4-lobed, ca 3 mm long; stamen solitary; anther exserted, oblong, bilobed. Pistillate flowers in globular or oblong spikes; perianth tubular, embedded in the fleshy rachis; style exserted; stigmas entire, 2- or 3-lobed; fruiting pedicels to 7.5 cm long, 1.5–2 cm wide. Syncarp ellipsoid or rounded, 10–30 cm long, green and soft at maturity, the surface asperous with numerous round or isodiametrical segments or echinate; seeds many. *Croat 10121.*

Cultivated in the Laboratory Clearing. Flowers in the late dry season. The fruits mature in the middle to late rainy season.

Native to the South Pacific islands; cultivated throughout the tropics of the world and in various places in Panama.

BROSIMUM Sw.

Brosimum alicastrum Sw. subsp. **bolivarense** (Pitt.) C. C. Berg, Acta Bot. Neerl. 19:326. 1970.

B. bernadetteae Woods.; *Helicostylis latifolia* Pitt.

Berba

Monoecious tree, 3–35 m tall; bark thin, with prominent leaf scars and irregular horizontal raised lines; inner bark smooth, thick, tan; sap forming milky droplets. Petioles stout, 4–14 mm long; blades elliptic-obovate to elliptic, acuminate to mucronate at apex, broadly cuneate to rounded at base, 6–15(20) cm long, 3–6.5(8) cm wide, glabrous, coriaceous, the major lateral veins raised below, with a conspicuous submarginal collecting vein and prominulous reticulate veins; stipules nearly encircling stem, 5–9(15) mm long, deciduous. Flowers dense, in globular clusters 4–9 mm diam, completely concealed before anthesis by short-stipitate, round, peltate bracts; peduncles obsolete or to 5 mm long; perianth obsolete; staminate flowers many, the stamen solitary, the anther circular, eccentrically peltate, ca 1 mm diam, dehiscing by 2 basal valves; pistillate flowers 1 or 2 at center of clusters, the stigmas deeply 2-lobed, the lobes exserted 4–7 mm before staminate flowers open, spreading, subu-

late. Fruit a false drupe, ± globose, 1–1.5(2) cm diam, with minute round protuberances, with an apical depression; seed 1. *Croat 10306, 11647.*

Common in the forest. Seasonal behavior uncertain. Apparently flowers from November to May, mostly during the dry season. The fruits mature from May to October and are eaten when ripe by monkeys (Hladik & Hladik, 1969).

Carpenter (1934) reported that fruits of this species are second only to *Ficus* as food for most animals of the forest. Bats play a principal role in their dispersal (R. Foster, pers. comm.). The outer shell of the fruit is often thrown to the ground.

The subspecies *bolivarense* is distinguished from the subspecies *alicastrum* by having anthers with free thecae; anthers of the subspecies *alicastrum* are peltate with the thecae fused.

The typical subspecies ranges from Mexico to Costa Rica and the West Indies. The subspecies *bolivarense* ranges from Costa Rica through the Andes to Guyana and Brazil (Acre Territory). In Panama, known from tropical moist forest in the Canal Zone, Chiriquí, Panamá, and Darién. Reported from tropical wet forest in Costa Rica (Holdridge et al., 1971).

CASTILLA Sessé

Castilla elastica Sessé in Cerv., Gaz. Lit. Mexico, Suppl. 7. 1794

C. panamensis Cook

Rubber tree, Mastate blanco, Caucho, Hule, Ule

Monoecious or dioecious tree, 10–30 m tall, with low buttresses; young parts densely yellowish-pubescent; stems hollow; sap white. Leaves pendulous, deciduous; petioles to 1 cm long; stipules fully amplexicaul, 3–9 cm long, densely golden, appressed-pubescent, deciduous; blades oblong to oblong-obovate, acuminate at apex, cordate at base, 20–45(55) cm long, 8–18(25) cm wide, minutely ciliate-denticulate, asperous especially below with short, spreading, often golden trichomes. Flowers inserted on large, flattened, unisexual receptacles covered with imbricate bracts, the receptacles axillary or at defoliated nodes, involucrate; primary staminate inflorescences (2)4(6) per axil, fan-shaped, conduplicate, to 15 mm long and 25 mm wide; peduncles 3–10 mm long; bracts in 10–12 series; perianth absent; stamens numerous, scattered among the bractlets; filaments to 3.5 mm long; anthers oval, ca 1 mm long; complemental staminate inflorescences (accompanying pistillate inflorescences) usually 2, similar to primary ones or funnel-shaped to cup-shaped. Pistillate receptacles usually solitary, thickly discoid, 1–2 cm diam, 1–2 cm thick, nearly sessile; involucral bracts in 5–10 series; flowers mostly 15–30; perianth tubular, (1.5)2–3 mm long, shallowly 4(5)-lobed, fleshy, short-velutinous, accrescent; ovary subinferior; styles dimorphic, to 1.5 mm long with stigmas 3–6 mm long or 2–3 mm long with stigmas less than 3 mm long; stigmas 2, rarely 3. Fruits thick, discoid syncarps 2.5–4.5

cm wide; seeds many, ca 1 cm long and enclosed in the orange, fleshy, accrescent perianth. *Croat 5335.*

Occasional in the forest on the western side of the island. Probably flowers in the late dry and early rainy seasons (April and May). Berg (1972) reported that the species flowers all year. The fruits are mature in May and June. The leaves are lost in the dry season.

Widespread on both coasts from Mexico to Panama and along the western coasts of Colombia and Ecuador. In Panama, known from tropical moist forest in the Canal Zone, Bocas del Toro, Panamá, and Darién, from premontane moist forest in the Canal Zone and Panamá, from premontane wet forest in Coclé (El Valle), and from tropical wet forest in Colón (Guásimo).

CECROPIA L.   (Guarumo, Trompy, Trumpet tree)

Cecropia is distinguished by its slender, pendent, unisexual catkins; staminate catkins fall soon after anthesis. Flowers have a minute tubular perianth, the staminate with two stamens, the pistillate with a barely exserted, pencil-shaped style. Each flower produces a tiny achene. Leaves are peltate and spirally arranged on stout, hollow branches, which frequently house myrmecophilous ants. The leaves are a favorite food of sloths. The genus is characteristic of secondary areas and usually has soft wood and black sap.

Cecropia insignis Liebm., K. Danske Vidensk. Selsk. Skr. Naturvidensk. Math. Afd., ser. 5, 318. 1851
C. eximia Cuatr.
Guarumo blanco

Dioecious tree, to about 40 m high and 70 cm dbh, broadly branched; trunk with prominent lenticels and often stilt-rooted; outer bark thin; inner bark reddish. Petioles stout, arachnoid-villous and longitudinally striate, the basal pulvinus brown; stipules commonly red when falling, mostly 17–26(40) cm long and 5–8 cm broad, densely hirtellous on outer surface, with 2 prominent ridges; mature leaves prominently clustered at ends of branches, up to 90 cm wide, divided more than three-fourths of the way to center, the lobes 7–9, acute to obtuse at apex, conspicuously narrowed at base, to 22 cm broad, the margins held conspicuously upward, the upper surface smooth and shiny, glabrous except for sparse arachnoid trichomes deciduous in age, the lower surface paler and minutely canescent beneath (at least between veins). Staminate spathes oblong, 10–12(16) cm long, 5–6.5 cm wide, rounded at apex with long apiculate tip, the spadices held erect in clusters of usually 6 or 7, 12–15 cm long, 8–10 mm wide, whitish or green, the basal stipes

KEY TO THE SPECIES OF CECROPIA

Leaves smooth above, remaining intact after falling to ground; basal stipe of spadices glabrous, purplish, pruinose (covered with thin, waxy layer); pistillate spadices yellow (except when very young) . . . . *C. insignis* Liebm.
Leaves scabrous or rough on upper surface, usually rolling up after falling to ground; basal stipe of spadices not pruinose, variously pubescent; pistillate spadices white or greenish:
  Leaves often divided more than three-fourths of the way to base, usually flat or only slightly folded; staminate spadices usually in clusters of less than 10; pistillate spadices 25–30 cm long when fully expanded . . . . *C. obtusifolia* Bertol.
  Leaves usually divided about halfway to base or less, the surface of the blade much folded, not able to be flattened; staminate spadices in clusters of 12–60; pistillate spadices less than 12 cm long when fully expanded:
    Pistillate spadices in clusters of 4–6, the common peduncle not more than twice as long as spadix; basal pulvinus of petiole with uniform trichomes, the velvetlike layer of trichomes not interspersed with longer white trichomes . . . . *C. peltata* L.
    Pistillate spadices in clusters of 6–12, the common peduncle 5–8 times as long as spadix; basal pulvinus of petiole with the brown velvetlike layer of trichomes interspersed with dense, longer, white trichomes . . . . *C. longipes* Pitt.

KEY TO THE SPECIES OF CECROPIA (ON THE BASIS OF STAMINATE SPADICES)

The following key is provided because staminate inflorescences fall soon after anthesis and are often found on the ground.

Staminate spadices less than 10 per peduncle:
  Spadices 4 mm thick; basal stipes ± green, puberulent, to 10 mm long; leaves scabridulous above, flat to somewhat pleated . . . . *C. obtusifolia* Bertol.
  Spadices ca 1 cm thick; basal stipes glabrous, purple, pruinose, 12–16 mm long; leaves smooth above, much pleated . . . . *C. insignis* Liebm.
Staminate spadices 12 or more per peduncle:
  Spadices usually less than 6 cm long . . . . *C. peltata* L.
  Spadices usually more than 6 cm long . . . . *C. longipes* Pitt.

Fig. 194. *Cecropia insignis*

Fig. 195. *Cecropia longipes*

purplish, broad, 12–16 mm long, pruinose, the common peduncles 5–10 cm long; pistillate spathes as those of staminate, the spadices in clusters of 4–7, pale yellow, 8–10 cm long and ca 8 mm thick at anthesis, becoming bright yellow in fruit and to 15 cm long and 1.3 cm thick, the basal stipes thick, pruinose, ca 1 cm long, the common peduncles 8–13 cm long, gray-hirtellous. Achenes ovate-elliptic, somewhat flattened, 1.7–2.3 mm long, ca 1 mm wide, conspicuously muricate, reddish-brown. *Croat 7023 ♂.*

Common in the forest, even in the older forest. Many individuals grow in the vicinity of the Laboratory Clearing. Flowers in the dry season (December to April). The fruits mature from April to August with a peak in July.

Easily distinguished by the large, smooth, very deeply divided blades and the spadices with pruinose stipes.

Nicaragua to Colombia. In Panama, known from tropical moist forest in the Canal Zone, Bocas del Toro, and Los Santos and from tropical wet forest in Colón (Salúd). Reported from premontane wet and rain forests in Costa Rica (Holdridge et al., 1971).

See Fig. 194.

Cecropia longipes Pitt., Contr. U.S. Natl. Herb. 18:227. 1917

Guarumo poludo

Dioecious tree, to 15 m tall; branches stout, 4–5 cm diam at apex. Petioles gray-hirtellous, 30–75 cm long, the brown velutinous trichomes interspersed with longer white trichomes; stipules ca 7–9 cm long; mature leaves to ca 90 cm wide, usually divided less than halfway to center, dark green and scabridulous above, paler and subarachnoid-puberulent below; lobes 9–13, broadly rounded at apex, not narrowed at base. Staminate spathes carrot-shaped before anthesis, pendent, reddish-brown, densely hirsute, 9–12 cm long, the spadices in clusters of 16–50, to 13 cm long and 2–3 mm diam, the stipes 1–2 cm long, gray-hirtellous, the common peduncles 7–12 cm long, densely hirsute; perianth tubular; stamens 2. Pistillate spathes reddish, 2-ribbed, the spadices in clusters of 6–12, 8–9 cm long and 5 mm diam at anthesis, 10–12 cm long and ca 1 cm diam in fruit, the stipes hirtellous, to 1 cm long, the common peduncles greatly accrescent in fruit, 50–80 cm long, moderately covered with sharp and stiff, somewhat urticating trichomes. Achenes tan, ellipsoid to narrowly ovate, smooth when fresh, ca 2.3 mm long, to 1.3 mm wide. *Croat 15248a.*

Apparently rare; known only from the vicinity of the Laboratory Clearing. In the Canal Zone, the plant is uncommon, growing as isolated plants in open areas. Both of the pistillate plants growing in the Laboratory Clearing set an abundance of fruit. Flowers at the beginning of the rainy season (April to June). The fruits mature from July to September (sometimes November).

Known only from Panama, from tropical moist forest on both slopes of the Canal Zone and in Darién and from premontane moist forest on the Pacific slope of the Canal Zone (Fort Kobbe).

See Fig. 195.

Cecropia obtusifolia Bertol., Fl. Guat. 39. 1840

C. mexicana Hemsl.; *C. panamensis* Hemsl.

Guarumo, Trumpet tree

Dioecious tree, mostly 5–10 m tall; trunk moderately slender, the young branches stout, ferruginous-hirtellous to glabrate. Petioles densely short-pubescent; stipules 7–11 cm long; mature leaves usually divided more than halfway to center, scabridulous and minutely arachnoid above, paler and minutely cinereous below; lobes usually 9–15, rounded to shortly acuminate at apex, somewhat to moderately narrowed at base, the lower free margin held ± flat to moderately erect, the leaves thus ± pleated, the area surrounding the petiole flat to 3–4 cm in all directions from center; veins, petioles, and spathes often ± maroon. Staminate spathes (11)12–20 cm long, ca 2 cm wide at anthesis, gradually acuminate, the spadices usually in clusters of 3–9, (10)12–18 cm long, 3–4 mm thick, subsessile or with puberulent stipes to 1 cm long, the common peduncle 8–12 cm long; pistillate spathes 16–20 cm long at anthesis, usually arachnoid on outside, villous inside, the spadices in clusters of 2–4, 17–30 cm long, ca 5 mm diam at anthesis, 6–9 mm diam and fleshy in fruit, the common peduncle 8–20(24) cm long, usually rough with short erect trichomes. Achenes ovate to broadly oblong, somewhat flattened, the edges acute, 3.3–3.7 mm long, to 2.7 mm broad, tan to white. *Croat 11716, 11800.*

Very common along the edge of the lake, particularly the northern and the eroded eastern shorelines; occasional in the forest, especially in tree-fall areas, some trees being found in the older forest. Flowers and fruits throughout the year.

Most easily confused with *C. peltata,* which has leaves conspicuously pleated to the center with shorter, broader lobes and shorter spadices.

Fruits are taken by spider monkeys from April to August (Hladik & Hladik, 1969).

Southern Mexico to Panama, Ecuador, and probably Colombia. In Panama, a wide-ranging and ecologically variable species and a characteristic tree of tropical moist forest (Holdridge & Budowski, 1956); known from tropical moist forest in the Canal Zone, Bocas del Toro, Veraguas, Panamá, and Darién (no doubt elsewhere also), from premontane wet forest in Panamá, from tropical wet forest in Colón and Darién, and from premontane rain forest in Panamá (summit of Cerro Jefe).

See Fig. 196.

Cecropia peltata L., Syst. Nat. ed. 10, 1286. 1759

C. arachnoidea Pitt.

Guarumo, Trumpet tree

Dioecious tree, 6–20 m tall. Petioles densely hirtellous and arachnoid, the basal pulvinus a uniform mass of dense trichomes; stipules 6–9 cm long; leaves lobed less than three-fourths of the way to middle, usually about halfway; lobes 9–11, obtuse to rounded at apex, scarcely or not at all contracted at base, the lateral margins held prominently upward and prominently pleated between

lobes to almost center of leaf, scabridulous above, paler and densely arachnoid-villous to glabrate below, the veins green. Staminate spathes usually 4–7 cm long, softly arachnoid outside, glabrous inside, the spadices in clusters of 12–46, 2.5–6 cm long and 3–4 mm diam, the stipes short, minutely hirtellous to glabrate, the common peduncle puberulent, 3.5–13 cm long, minutely hirtellous; perianth tubular; stamens two. Pistillate spathes mostly less than 7 cm long but sometimes to 10 cm at anthesis, broadly conic, mucronate, the spadices in clusters of 4–6, 3–5 cm long and ca 5 mm wide at anthesis, to 15 cm long and 1 cm wide in fruit, subsessile or with an inconspicuous sparsely pubescent stipe, the common peduncle 4.5–12(17) cm long, hirtellous. Achenes oblong, acute at apex, ca 1.7 mm long, to 1 cm wide, muricate, greenish. *Croat 11749, 11831.*

Common along the shore and in the old tree-fall areas of the older forest. Flowering seems to begin in the dry season and continue nearly all year. The fruits mature all year, possibly with a peak in the early rainy season.

Closely related to and perhaps inseparable from *C. obtusa* Trec., which ranges from the Guianas to Peru, Brazil, and Paraguay.

Mexico to Colombia, Venezuela, and the Guianas; West Indies. In Panama, characteristic of tropical moist and premontane moist forests (Tosi, 1971); known from tropical moist forest in the Canal Zone, Bocas del Toro, Panamá, and Darién and from tropical wet forest in Colón, Veraguas, and Darién. Reported from premontane wet and premontane rain forests in Costa Rica (Holdridge et al., 1971).

See Fig. 197.

COUSSAPOA Aubl.

Coussapoa magnifolia Trec., Ann. Sci. Nat. Bot., sér. 3, 8:98. 1847

   C. nymphiifolia Standl.

Dioecious trees, to 10 m tall, usually hemiepiphytic; branches hollow; young parts minutely ferruginous-tomentulose, glabrate in age. Leaves spiral; petioles 5–15 cm long; blades broadly oval, rounded at apex, usually cordate at base, 15–30(36) cm long, 10–25 cm wide, somewhat asperous on both surfaces, entire or indefinitely sinuate on margin; venation palmate at base, the major lateral veins in 6–8 pairs above base; stipules fully amplexicaul, 3–5 cm long. Inflorescences upper-axillary, shorter than subtending petioles; staminate heads many, globular, 4–10 mm diam, the inflorescences dichotomously branched 3 or 4 times, the flowers with 4 free tepals and 2 fused stamens; pistillate heads 3–5 cm broad, obscurely compound or lobed with 2 to 4 lobes, the flowers with a short style, the stigma barely exserted. Syncarps loosely coherent, quite fleshy. *Croat 5125, 15064.*

Rare; known from areas along the shore of the bay south of Orchid Island, usually epiphytic on submerged trees in the lake. Possibly also occurring in the canopy of the forest. Probably begins to flower in the early dry season and continues for much of the year.

Panama to Peru and Brazil. In Panama, known only from tropical moist forest on the Atlantic slope of the Canal Zone.

Coussapoa panamensis Pitt., Contr. U.S. Natl. Herb. 18:226. 1917

Dioecious tree, usually to 13 m tall, usually hemiepiphytic, attached to host at least when young by circumferal grasping roots; young stems and petioles reddish-brown, scurfy, glabrate, or puberulent; sap viscid, clear, reddish or yellowish. Leaves spiral; petioles 3–7 cm long; stipules fully amplexicaul, 1–7 cm long, usually pubescent; blades ovate, obtuse and often apiculate at apex, subcordate to rounded or truncate at base, 10–20 cm long, 6.5–12 cm wide, glabrous above, whitish below with dense, appressed, arachnoid trichomes, the margins sinuate to crenate; venation palmate at base, the major lateral veins in 10–20 pairs above base. Inflorescences of pedunculate globular heads ca 5 mm broad, the staminate inflorescences dichotomously compounded 2–4 times, equaling or shorter than subtending petiole, the flowers with 4 free tepals and 2 fused stamens; pistillate inflorescences simple, usually 2 per node, the peduncles 4–6 cm long, the heads ca 1 cm diam at anthesis, the style very short, the stigma barely exserted, the peduncle and ovary minutely pubescent. Syncarps ± globose, fleshy at maturity, orange, ca 2 cm diam. *Croat 5698, 7839.*

Occasional, along the shore and in the forest. Flowers mostly in the dry season (February to June), with the fruits maturing in late rainy season.

Southern Mexico to Panama. In Panama, known from tropical moist forest in the Canal Zone, Bocas del Toro, and Panamá and from premontane wet and tropical wet forests in Colón.

See Fig. 198.

DORSTENIA L.

Dorstenia contrajerva L., Sp. Pl. 121. 1753

Herb, usually less than 30 cm tall; stems short, the plants often appearing acaulescent; leaves and stems juicy. Leaves crowded, spirally arranged; petioles 7–25 cm long; blades extremely variable in shape and size, commonly deeply pinnatifid, basifixed, 8–24 cm long, 9–20 cm wide, scabridulous or inconspicuously puberulent, the lobes acuminate, often contracted toward base, the leaf base truncate to cordate. Peduncles usually longer than petioles, to 36 cm long; receptacles yellowish, centrally peltate, variously radiate to ± quadrangular, 1.5–3.5 cm diam, accrescent and to 5 cm diam in fruit. Seeds ca 1 mm long, ± rounded, weakly tuberculate. *Foster 2791.*

Collected once at Shannon Creek. Plants are fertile from May to August.

Southern Mexico to Peru and the Guianas; the Antilles. In Panama, known from tropical dry forest in Panamá (Taboga Island), from tropical moist forest in the Canal Zone, Bocas del Toro, and Darién, and from premontane wet forest in Coclé.

Fig. 196. *Cecropia obtusifolia*

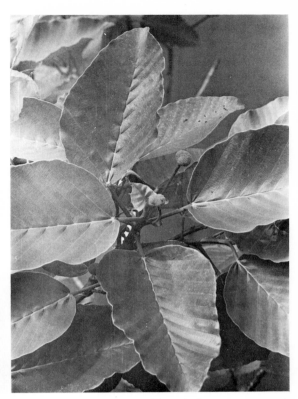

Fig. 198. *Coussapoa panamensis*

Fig. 197. *Cecropia peltata*

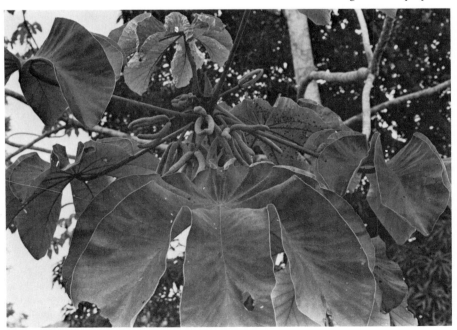

FICUS L.   (Higo, Higueron)

The genus is recognized by its fruitlike inflorescence (the fig), by its usually viscid milky sap, and by its prominent amplexicaul stipules. The minute, unisexual, naked flowers of both sexes are interspersed on the inner surface of the receptacle, which becomes fleshy when the fig is mature. Pistillate flowers consist of both functional pistillate flowers, which are generally sessile and mature into viable achenes, and sterile pistillate flowers (gall flowers), which are generally stalked and function as incubators for the larvae of the wasp that pollinates the flower.

Ficus is pollinated by tiny, specialized wasps. *Ficus* flowers are unisexual, borne on the inner surface of a globose receptacle, with the receptacle having an apical pore closed by a series of interlocking bracts. Pistillate flowers are of two kinds, long-styled and short-styled. Flowers are pollinated by usually host-specific female agaonid wasps (Ramirez B., 1970), which enter the fig through the ostiole (carrying pollen from another tree) when the ostiolar scales loosen and the pistillate flowers are ready for pollination. All pistillate flowers in a receptacle mature synchronously. Wasps probably pollinate both long- and short-styled pistillate flowers, then lay eggs in the short-styled flowers and later die within the fig. The long-styled flowers that are pollinated develop normally and produce seeds, while the short-styled flowers that also received agaonid eggs become "gall" flowers, each nourishing a single wasp larva. The period of time between fertilization of the pistillate flowers and maturation of the staminate flowers, when the wasps reach the adult stage, is called the interfloral phase. Maturation of staminate flowers occurs synchronously with the emergence of the adult wasps from the gall flowers. The interfloral phase varies greatly from species to species.

Copulation takes place before the female wasps leave their galls. Once they leave the galls they go directly to the anthers, collect pollen, then leave through holes in the fig wall made by the males. The females fly to a tree of the same species that is ready for pollination. Fig development on any tree is usually highly synchronized so that pollination of all figs of a single tree occurs on the same day. There is also usually a more or less synchronous ripening of figs on a single tree. This occurs only after the fig wasps have emerged from the fig and ensures that they are not eaten by frugivores. The activity of different individuals of a species is not synchronized, however, and the species may be found in all stages at any one time. Recent studies by A. S. Rand (pers. comm.) on *Ficus yoponensis* and *F. insipida* indicate that although there is no synchronization of flowering between different individuals of a species, an individual tree may show some degree of fidelity to a particular pattern of flowering on consecutive years. Rand has found that these two species have one period of flowering sometime between November and January and may have a second or third period of flowering later in the year. Because no mention can be made of flowering and fruiting behavior in a seasonal sense, the length of the interfloral phase will be given when it is known.

The two New World subgenera of *Ficus*, which both occur on BCI, are morphologically and ecologically distinct:

Pharmacosyce: Free-growing trees, the result of a seed that germinated on the ground; figs solitary in axils with three or three-lobed basal bracts; flowers pollinated by *Tetrapus* wasps; stamens two; anthers dehisce and pollen comes out naturally from thecal sacs; ovary pale, unspotted; tepals of pistillate flowers narrowly deltoid or lanceolate; seeds usually dry, not sticky.

KEY TO THE SPECIES OF FICUS

Leaves ovate-cordate . . . .*F. nymphiifolia* P. Mill.
Leaves not ovate-cordate:
    Young stems bearing dense, reddish-brown, spreading pubescence; leaves usually asperous
        above; figs conspicuously pubescent:
        Figs cylindrical (conspicuously longer than broad); leaves puberulent or hispidulous below
            with straight trichomes; ostioles not wide-tubular . . . . *F. popenoei* Standl.
        Figs globose; leaves softly villous below; ostioles wide-tubular . . . .*F. bullenei* I. M. Johnston
    Young stems not densely rufous-pubescent; leaves smooth above; figs glabrous or pubescent,
        usually not conspicuously pubescent:
• Figs solitary in leaf axils; trees not strangling; trunk terete, often buttressed:
        Petioles with thin, scurfy, brown periderm; stipules less than 2.5 cm long; leaves often very
            minutely puberulent below; figs asperous . . . . *F. maxima* P. Mill.
        Petioles not scurfy; stipules mostly more than 2.5 cm long; leaves glabrous on both surfaces:
            Major lateral veins in less than 15 pairs, rather distant; collecting vein below as prominent as lateral veins or nearly so; trees usually less than 15 m tall; figs usually sessile,
               depressed-globose . . . .*F. tonduzii* Standl.
            Major lateral veins in 15–30 pairs, close; collecting vein below usually more obscure
               than lateral veins; trees usually 25 m or more tall:
               Figs to 1.8 cm diam, the slender, tubular ostiole to 3 mm long; major lateral veins
                  less than 4 mm apart . . . . *F. yoponensis* Desv.
               Figs to 4 cm diam, the ostiole only slightly raised; major lateral veins mostly more
                  than 5 mm apart . . . . *F. insipida* Willd.

• Figs paired in axils (sometimes with one aborted); trees usually strangling; trunks often contorted, buttressed or not:
    Figs on a distinct peduncle:
        Leaves less than 5 cm wide:
            Figs less than 7 mm diam; ostioles mammillate; leaf blades rounded to bluntly acuminate at apex . . . . *F. perforata* L.
            Figs more than 7 mm diam; ostioles sunken; leaf blades long-acuminate . . . *F. pertusa* L.f.
        Larger leaves more than 5 cm wide:
            Figs more than 15 mm diam, reddish to purplish when ripe; stipules usually pubescent with long appressed trichomes; figs few, usually restricted to leafy axils . . . .
            . . . . *F. trigonata* L.
            Figs less than 15 mm diam, greenish or yellowish when ripe (except purplish in *F. citrifolia*); stipules glabrous or minutely puberulent; figs numerous, often at leafless axils (except *F. dugandii*):
                Small epiphytic trees; trunks gray; sap clear; leaf blades acuminate, coriaceous; figs ± globose, the ostiole flat, not prominent; peduncles 4–15 mm long . . . .
                . . . . *F. citrifolia* P. Mill.
                Large buttressed trees; trunks brownish; sap milky; leaf blades thin, long-acuminate; figs turbinate, the ostiole mammillate; peduncles less than 4 mm long . . . .
                . . . . *F. dugandii* Standl.
    Figs sessile or nearly so (*F. obtusifolia* rarely with peduncles to 8 mm long, but usually at least appearing sessile):
        Bracteoles at base of figs 3; leaf margins conspicuously white, matching midrib when dry; plants cultivated at the Laboratory Clearing . . . .*F. retusa* L.
        Bracteoles at base of figs 2; leaf margins not conspicuously white; plants not cultivated:
            Figs less than 8 mm diam; leaves 3-veined near base (sometimes 5-veined), the major lateral veins in fewer than 5 pairs, long-appressed-pubescent below . . . .
            . . . . *F. colubrinae* Standl.
            Figs more than 10 mm diam; leaves not 3-veined near base, the major lateral veins in more than 5 pairs, glabrous or pubescent:
                Figs conspicuously velvety pubescent, maturing green; leaves and figs densely clustered at apex of stem; bracts more than 1 cm long . . . . *F. obtusifolia* H.B.K.
                Figs not velvety pubescent, maturing reddish or green with red, purple, or brown spots; leaves and figs not clustered at apex of stem; bracts less than 1 cm long:
                    Leaves distinctly acuminate; stipules glabrous or puberulent:
                        Figs with purplish vertical stripes, the ostioles purplish; small trees usually less than 10 m tall; petioles less than 3.5 cm long . . . . *F. paraensis* (Miq.) Miq.
                        Figs not marked with purple; usually large trees more than 10 m tall; petioles often more than 4 cm long . . . . *F. dugandii* Standl.
                    Leaves rounded to obscurely acuminate; stipules with long trichomes when young:
                        Figs less than 14 mm wide, whitish or yellowish-green, becoming reddish to red, the ostiole not prominently raised (the surface often forming a few, small, radial ridges on drying); stipules subpersistent. . . . .
                        . . . . *F. costaricana* (Liebm.) Miq.
                        Figs more than 14 mm wide (when mature), usually green with reddish spots or entirely violet-purple, the ostiole raised and buttonlike (the surface not drying with radial ridges); stipules deciduous . . . . *F. trigonata* L.

Urostigma: Epiphytic or hemiepiphytic and usually with the strangling habit (i.e., roots coalescing around host tree), the result of a seed that germinated on another tree; figs paired in axils, their basal bracts two or two-lobed; flowers pollinated by *Blastophaga* subg. *Pegoscapus* wasps; stamen one; anthers dehisce but pollen does not come out naturally; ovary with a red spot at base of style; tepals of all flowers hooded; seeds usually sticky, thus sticking to trees where they are deposited by bats, which eat them, and birds, often no doubt as they clean their beaks. See Fig. 199 and fig. on p. 23.

I am indebted to William Ramirez B. for the information given here concerning length of the interfloral phase, fruit dispersal agents, and ecology.

Ficus bullenei I. M. Johnston, Sargentia 8:113. 1949

Tree, to 21 m tall, usually a strangler; trunk to 60 cm dbh, grayish or reddish, with many close pustular lenticels, buttressed at base, some buttresses stiltlike (perhaps only by erosion); young stems and petioles densely reddish-brown spreading-pubescent; sap copious, milky. Petioles 1–3(4) cm long; stipules triangular, 1–2 cm long, deciduous, the pubescence densely appressed, reddish-brown; blades broadly elliptic to obovate, short-acuminate, rounded to subcordate and sometimes inequilateral at base, 7.5–26 cm long, 4.5–13 cm wide, entire to ± erose, asperous above, the pubescence dense, reddish-brown (lighter than stems and petioles), spreading, especially

Fig. 199. *Ficus,* showing the void once occupied by the host

Fig. 200. *Ficus bullenei*

Fig. 201. *Ficus costaricana*

below and on major veins above; juvenile leaves with coarsely crenate margins. Figs paired, globose, to 1.7 cm diam, subsessile, bearing conspicuous, ± spreading pubescence, greenish at maturity; ostioles surrounded by a wide, tubular, erect to spreading ring to 3 mm high; basal bracts 2, ovate, 4–5 mm long, bearing dense reddish-brown pubescence. *Croat 5557*.

Rare in the forest; occasional on shore. Interfloral phase is 37 days.

Costa Rica and Panama. In Panama, known from tropical moist forest in the Canal Zone and Panamá (San José Island).

See Fig. 200.

Ficus citrifolia P. Mill., Gard. Dict. ed. 8, *Ficus* no. 10. 1768

Tree, to 12(20) m tall, originally epiphytic, usually rather short and forming a solid trunk in exposed areas (such as along the shore); bark grayish; twigs minutely puberulent, smooth except for small lenticels; sap copious, watery. Leaves glabrous; petioles slender, 5–13 cm long; stipules narrowly triangular, 1–3 cm long, glabrous, deciduous; blades ovate, elliptic, oblong, or rarely obovate, acuminate, mostly obtuse to rounded or subcordate at base, 10–25 cm long, 4–9 cm wide, more or less coriaceous, folded along midrib, the veins in 4–12 pairs. Figs ± globose, 8–14 mm diam at maturity, borne in pairs (one frequently falling) in dense clusters among and below leaves, green with many pale green dots or becoming violet-red; peduncles 4–15 mm long, glabrous; ostioles flat; basal bracts 2, usually bluntly rounded, 2–3 mm long. *Croat 7891*.

Common, at least along the shore; rare or possibly absent from the older forest.

Occasionally visited by bats.

Florida to Paraguay; West Indies. In Panama, known from tropical moist forest in the Canal Zone and Panamá (San José Island) and from premontane dry forest in Coclé (Santa Clara Beach).

Ficus colubrinae Standl., Contr. U.S. Natl. Herb. 20:16. 1917

Hemiepiphytic tree, usually less than 6 m tall, probably never a strangler, the trunk to 20 cm diam; stems sparsely long-pubescent to glabrate in age; sap copious, milky. Petioles 5–25 mm long, often conspicuously long-appressed-pubescent; stipules 5–20 mm long, ovate, with long stiff trichomes, sometimes glabrate in age; blades obovate to (less frequently) ovate, abruptly short-acuminate, acute to rounded at base, 6.5–10.5 cm long, 3–7 cm wide, glabrous or long-pubescent especially on veins below, the veins at base 3(5), palmate, the major lateral veins in usually 2 pairs above base (sometimes 3 or 4). Figs paired, globose, 5–8 mm diam, glabrous, sessile, red at maturity; ostioles slightly raised; basal bracts 2, semicircular, to 2 mm long, with long stiff trichomes. *Croat 6822*.

Collected mostly along the shore; occasional in the forest.

Ficus colubrinae was considered synonymous with

F. hartwegii (Miq.) Miq. by DeWolf (1960) in the *Flora of Panama*. W. C. Burger (pers. comm.) believes the two species are distinct, with *F. colubrinae* found only from sea level to 800 m and *F. hartwegii* from 900 m to 1,600 m.

Guatemala and Belize to central Panama; sea level to 800 m. In Panama, known from tropical moist forest on the Atlantic slope of the Canal Zone and in Bocas del Toro, Colón, and Darién, from tropical wet forest in Colón, and from premontane wet forest in Coclé (El Valle).

Ficus costaricana (Liebm.) Miq., Ann. Mus. Bot. Lugduno-Batavum 3:298. 1867

Tree of conspicuously strangling habit, to 30 m tall, glabrous except the stipules and floral bracts densely pubescent and the young stems, petioles, and leaf blades below or midribs above sometimes with long trichomes; bark often reddish-brown; sap milky. Petioles 1–4 cm long; stipules lanceolate, 1.5–3 cm long, densely pubescent (possibly glabrate in age), subpersistent; blades obovate or less frequently elliptic, rounded or obtuse at apex (acuminate on juveniles), rounded to subcordate at base, 8–20(32) cm long, 4.5–11(13) cm wide, the veins in 6–10 pairs, usually 3(5)-veined at base. Figs paired, depressed-globose, 12–14 mm diam, borne among leaves, glabrous, sessile, yellowish-green to pale yellow becoming reddish with darker red spots, finally entirely red at maturity; basal bracts 2, usually with long appressed trichomes, to 1 cm long. *Croat 9505*.

Common in the forest, occasional along the shoreline; a large tree occurs at Zetek Trail 310. The species is one of the most well-developed stranglers on the island.

The fruits are eaten by birds. Ants often form permanent trails on the branches to patrol the figs against predators. Escaping female wasps jump downward soon after emerging from the ostiole of the fig to prevent being eaten by the ants (W. Ramirez B., pers. comm.).

Guatemala to Panama. In Panama, known from tropical moist forest on BCI and in Chiriquí (Davíd) and from premontane wet forest in Chiriquí (Caldera). Reported from tropical dry forest in Costa Rica (Holdridge et al., 1971).

See Fig. 201.

Ficus dugandii Standl., Trop. Woods 32:20. 1932
F. turbinata Pitt. non Willd.

Hemiepiphytic tree, to 30 m tall, apparently not a strangler, the trunk ca 1.5 m dbh, buttressed; stems and leaves glabrous; bark pale reddish-brown, lenticellate; sap copious, milky. Petioles 2.5–7 cm long; stipules minutely puberulent to glabrate, ca 7 mm long; blades ovate-elliptic, acuminate, obtuse to rounded or truncate at base, 9–18 cm long, 5–8 cm wide, moderately thin, the major veins in 10–13 pairs, often appearing 3-veined at base, the reticulate veins of lower surface very distinct. Figs paired, turbinate, to 14 mm diam, puberulent, green with brownish spots at maturity; peduncles obsolete or to 4 mm long; ostioles raised; basal bracts 2, semicircular, puberulent, to 3 mm long. *Croat 12863*.

Adults rare, seen north of Zetek Trail 1300 and on

Barbour Trail. Seedlings more common, seen both along the shore and in the forest, always epiphytic and near the ground, such as on rotting fallen trees.

The species was considered to be a synonym of *F. citrifolia* by DeWolf (1960) in the *Flora of Panama,* but the two have been shown to be distinct by Ramirez B. (1970), who has made detailed studies of their pollination systems. Though they are morphologically similar and often difficult to distinguish in the herbarium, they are much easier to distinguish in the field. *F. dugandii* frequently grows along river banks in the Canal Zone, where it is a large tree with the solid trunk indicative of plants that grow from the ground up (or, in the case of fig trees, as epiphytes from near the ground) rather than from the treetops down as in the case of stranglers.

Costa Rica to Colombia and Venezuela. In Panama, known only from tropical moist forest in the Canal Zone (BCI and Gamboa).

Ficus insipida Willd., Sp. Pl. 4:1143. 1806

Free-growing tree, 30(40) m tall and ca 70 cm dbh, prominently buttressed; bark light gray, ± smooth; essentially glabrous but the fig puberulent; sap milky, copious. Petioles 2–5.5 cm long; stipules linear, 1–12 cm long, glabrous, deciduous; blades ± elliptic, obtuse to short-acuminate at apex, obtuse to rounded at base, 10–20 cm long, 5–9 cm wide (juvenile leaves to 40 cm long and 16 cm wide), the veins in 15–30 pairs, 2–4 veins arising sharply at base. Figs solitary, to 4 cm diam, borne among leaves, glabrous or minutely pubescent, yellowish-green at maturity with lighter spots; peduncles mostly 5–15 mm long; ostioles mammillate; basal bracts 3, to 3 mm long, semicircular, glabrous or pubescent. *Croat 8220.*

Abundant at least in younger areas of the forest; seedlings of this species have been seen growing in marshes on the south side of the island. One of the most abundant figs on the island. Interfloral phase is 15 days.

Seedlings require a tree fall or clearing to have sufficient light to grow. This species has been confused with *F. glabrata* H.B.K., which always occurs in riparian habitats. *F. crassiuscula* Warb. was considered a synonym by DeWolf in the *Flora of Panama,* but this species has been shown to be distinct by Burger (1974), who said it occurs only at elevations greater than 1,100 m in Costa Rica and Panama, whereas *F. insipida* occurs only at elevations less than 500 m. Ramirez B. (1970) also reported that the two species have different pollinators.

Southern Mexico to Brazil. In Panama, from tropical moist forest in the Canal Zone, Panamá, and Darién.

Ficus maxima P. Mill., Gard. Dict. ed. 8, *Ficus* no. 6. 1768

Free-growing tree, to 20(30) m tall, to 1 m dbh; outer bark thin, light brown, coarse but planar; inner bark thick, light, marbled; smaller twigs and petioles with a scurfy brown periderm; sap clear or turbid, copious. Petioles 1–3(4) cm long; stipules lanceolate, 1–2.5 cm long, usually glabrous; blades obovate to elliptic, usually abruptly short-acuminate, cuneate and 3-veined at base,

ending abruptly on petiole, 9–16.5(25) cm long, 5–8(11) cm wide, stiff, asperous and very inconspicuously pubescent beneath or glabrous. Figs solitary, globose, 2–3 mm diam, reddish inside before maturity, green outside at maturity, the surface ± punctate and asperous with stiff, short trichomes; peduncles 5–10(25) mm long, minutely pubescent; ostioles inconspicuous, ± flat; basal bracts 3, short, rounded. *Croat 8363.*

Occasional, in the forest. Interfloral phase is 30 days.

The species is not very conspicuous because the trunk is not light-colored, as are many other figs, nor is it conspicuously buttressed.

Southern Mexico to the Amazon basin; Cuba, Jamaica. In Panama, known from tropical moist forest in the Canal Zone, Los Santos, Panamá, and Darién and from premontane moist forest in the Canal Zone.

See Fig. 202.

Ficus nymphiifolia P. Mill., Gard. Dict. ed. 8, *Ficus* no. 9. 1768

Tree, usually less than 15 m tall, of inconspicuously strangling habit; stems minutely puberulent near apex, otherwise glabrous; sap milky. Petioles to 14 cm long, commonly more than three-fourths as long as blade, occasionally longer; stipules narrowly triangular, mostly 12–30(40) mm long; blades ovate, abruptly acuminate and downturned at apex, deeply cordate at base, mostly 9–20 cm long, 7–15 cm wide, palmately veined at base, the major lateral veins in 4–6 pairs above base; juvenile leaves reddish when young. Figs globose, 15–28 mm diam, minutely puberulent, somewhat glaucous, borne in pairs among leaves (one often aborting), green at maturity, with a faint sweet odor; ostioles flat or weakly raised; basal bracts 2, connate, usually rounded (sometimes acute) at apex. *Croat 12708.*

Rare; seen both in the forest and along the shore. Often epiphytic on palms. It has been seen epiphytic at 30 m in the canopy on *Sapium caudatum* (75. Euphorbiaceae). Interfloral phase is 20 days.

Figs of this species (and apparently most others that are green at maturity) have a pleasant odor and are eaten by bats.

Costa Rica to the Guianas and Brazil. In Panama, known from tropical moist forest in the Canal Zone and Darién and from tropical wet forest in Colón (Salúd). Reported from premontane wet forest in Costa Rica (Holdridge et al., 1971).

Ficus obtusifolia H.B.K., Nov. Gen. & Sp. 2:49. 1817, non Roxb. 1814

Tree, to 40 m tall, usually a conspicuous strangler (juveniles especially), usually glabrous; trunk often unbuttressed; sap milky, copious. Leaves clustered near ends of stems; petioles 2–4 cm long; stipules 2–3 cm long, ovate, caducous; blades obovate or oblanceolate, rounded to obtuse at apex, cuneate at base, 15–25 cm long, 6–12 cm wide, often drying grayish above, the veins in 8–11 pairs. Figs paired, globose, 1.5–3 cm diam, green, borne among leaves, glabrous or minutely velvety pubescent;

Fig. 202. *Ficus maxima*

Fig. 203. *Ficus paraensis*

peduncles obsolete or to 8 mm long; ostioles usually slightly raised; basal bracts 2, to 2 cm long, semicircular, minutely pubescent, sometimes split. *Croat 8276.*

Common; the most abundant and conspicuous strangler on the island. Interfloral phase is 45 days.

Visited by bats but generally not visited by birds. Juvenile plants are usually conspicuously strapped to the host tree with their roots enveloping the tree's trunk.

This species was reported by Standley as *F. involuta* (Liebm.) Miq., which was considered a synonym of *F. obtusifolia* by DeWolf (1960) in the *Flora of Panama* and by Burger (1975) in his *Flora Costaricensis.* However, W. Ramirez B. (pers. comm.) considers *F. obtusifolia* a distinct species from *F. involuta,* which does not occur on BCI.

Mexico to Peru. In Panama, known from tropical moist forest in the Canal Zone and San Blas and from premontane wet forest in Panamá (Cerro Campana).

Ficus paraensis (Miq.) Miq., Ann. Mus. Bot. Lugduno-Batavum 3:298. 1867
    F. panamensis Standl.

Hemiepiphytic tree, to 10(20) m tall, apparently not a strangler; branches conspicuously roughened with lenticels and pedicel scars; sap copious, often not very milky. Petioles 5–35 mm long; stipules 1.5–2.5(3.5) cm long, glabrous or minutely puberulent; blades oblong to oblong-oblanceolate, abruptly acuminate and markedly downturned at apex; rounded to subcordate at base, 9–18 cm long, 4–8 cm wide, glabrous or minutely puberulent, the major lateral veins in 9–18 pairs, the reticulate veins prominent. Figs paired, globose, 10–18 mm diam, borne among leaves, green with maroon streaks becoming yellow with violet-purple streaks; peduncles short and stout or lacking; ostioles violet-purple, raised; basal bracts 2, ovate, puberulent, often somewhat maroon. *Croat 12680.*

Frequent, especially on the shore. Interfloral phase is 60 days.

Ant nests are nearly always on the branches of this species, often with aerial roots in them.

Southern Mexico to the Guianas, Peru, and Brazil. In Panama, known from tropical moist forest in the Canal Zone, Panamá, and Darién and from premontane wet forest in Colón (Nombre de Dios).

See Fig. 203.

Ficus perforata L., Pl. Surin. 17. 1775
    F. oerstediana (Miq.) Miq.

Hemiepiphytic tree, to 15(30) m tall, apparently not a strangler, the trunk slender; stems glabrous in age, often pubescent when young; sap milky. Petioles 5–10 mm long, sometimes pubescent; stipules 5–12 mm long, ovate, glabrous or minutely puberulent; blades obovate, obtuse to very shortly blunt-acuminate at apex, cuneate to narrowly rounded at base, 4–9 cm long, 2–4.5 cm wide, the major lateral veins not raised below, faint above. Figs paired, globose, to 7 mm diam, borne among leaves, reddish at maturity; peduncles to 12 mm long, minutely

puberulent or glabrous; ostioles mammillate; basal bracts 2, to 2 mm long, ovate. *Croat 6693.*

Possibly restricted to the shoreline or in the younger forest. Interfloral phase is 30 days.

A tree planted at the north of the library (*Croat 9009*) is an introduced species from the Old World (*F. retusa* L.), which can easily be confused with *F. perforata,* especially when sterile.

Guatemala and Belize to Colombia; Greater Antilles. In Panama, known from tropical moist forest in the Canal Zone, Veraguas, and Darién and from premontane wet forest in Panamá (Cerro Campana).

Ficus pertusa L.f., Suppl. Pl. Syst. Veg. 442. 1781

Hemiepiphytic tree, 8–25 m or more tall; trunk smooth, sometimes with somewhat stilted roots; branches spreading, the youngest (including petioles and stipules) often puberulent. Petioles usually 5–15(30) mm long; stipules 5–13 mm long, slender, sharp-pointed; blades elliptic to oblong-lanceolate, abruptly long-acuminate, cuneate to rounded at base, mostly 3.5–7 cm long, 2–3.5 cm wide, pellucid-punctate when fresh, glabrous or with puberulence on midrib near base. Figs paired (one usually aborting), globose, usually 10–18 mm diam, puberulent, green with lighter flecks becoming greenish-yellow at maturity; peduncles 2–7 mm long; puberulent; ostioles prominently sunken; basal bracts 2, semicircular, to 4 mm long. *Croat 12274.*

Apparently rare; seen along the shortcut of Lutz Trail behind the Animal House and on the shore of Fuertes Cove.

Distinguishing characters are small leaves and figs usually greater than 8 mm diam. The species may be confused with two other small-leaved species, *Ficus perforata,* which has fruits less than 8 mm diam, and *F. retusa,* which has sessile fruits.

Mexico to Paraguay; West Indies. In Panama, known from tropical moist forest on BCI and in Bocas del Toro and Darién.

See Fig. 204.

Ficus popenoei Standl., Publ. Field Columbian Mus., Bot. Ser. 4:301. 1929

A conspicuous strangler, occasionally forming a solid but gnarled trunk, to 17(25) m tall; young stems with dense, rufous, spreading pubescence, the stems glabrate in age; sap milky. Petioles 1–2 cm long, rufous-pubescent; stipules 6–15 mm long, densely rufous-pubescent; blades obovate to broadly elliptic, rounded at apex, rounded or subcordate at base, 9–21 cm long, 5.5–10.5 cm wide, tawny-pubescent below and on veins above, asperous. Figs paired, cylindrical, about twice as long as broad, to 2.5 cm long and 1.5 cm wide, densely tawny-pubescent, green at maturity; peduncles 2–5 mm long, pubescent; ostioles flat or slightly raised; basal bracts 2, ovate, to 5 mm long. *Croat 11768.*

Occasional. A tree near Zetek Trail 600 has a trunk ca 1 m diam. Interfloral phase is 43 days.

Fruits dispersed by bats.

Belize to Colombia. In Panama, known from tropical moist forest in the Canal Zone (BCI) and Panamá (San José Island) and from premontane wet forest in Coclé (El Valle).

Ficus retusa L., Mant. Pl. 129. 1767

F. nitida Thunb.

Glabrous tree, 8 m tall, the trunk ca 30 cm dbh; bark on twigs whitish, peeling; sap milky. Petioles 1–2 cm long, drying reddish-brown; stipules lanceolate, ca 5 mm long, drying reddish-brown; blades obovate to elliptic, acuminate, acute at base, 7–10 cm long, 2.5–4 cm wide, minutely pellucid-punctate, the margin thickened, whitish (at least when dry). Figs paired, sessile, globose, 6–10 mm diam; ostioles ± flat; basal bracts 3, ± ovate, ca 2 mm long. *Croat 10116.*

Planted at the north edge of the Laboratory Clearing.

Ficus retusa can be confused with *F. perforata*, which has similar leaves but which also has figs borne on a distinct peduncle, whereas those of *F. retusa* are sessile.

Native to lowland tropics of Southeast Asia.

Reported by Standley in *Flora of the Panama Canal Zone* and planted in Balboa and Panamá City. Known from Darién in areas where cultivars are not expected (*Duke & Bristan 413, Stern et al. 785*).

Ficus tonduzii Standl., Contr. U.S. Natl. Herb. 20:8. 1917

Free-growing tree, 3–10(20) m tall, usually rather small; trunk buttressed, to 50 cm dbh; bark planar, brown, with many round, raised lenticels; sap milky, copious. Petioles to 5 cm long; stipules ovate, 1–3(6) cm long; blades mostly broadly elliptic, rounded or acuminate at apex, obtuse to rounded or truncate at base, 6–19(30) cm long, 5–12(17) cm wide, glabrous, stiff, the major lateral veins in 7–12 pairs, branching from midrib at ca 90° angle, loop-connected, the reticulate veins distinct. Figs solitary, depressed-globose, 2.5–3.5 cm diam, glabrous or minutely scabrid, depressed at apex, borne among leaves, green with lighter spots; peduncles obsolete or very short; ostioles flat or beaked; basal bracts 3, to 6 mm broad, ± rounded at apex. *Croat 8621.*

Occasional, chiefly in the older forest. Many of the trees in the forest are relatively small, mostly less than 10 m tall; at least some produce fruit. Elsewhere in Panama, at least on the Atlantic slope, the species is often riparian and taller than is commonly encountered on BCI (usually to 15 m). Interfloral phase is 31 days.

Easily distinguished by the leaves.

The fruits are distributed by bats.

Honduras to Ecuador. In Panama, known from tropical moist forest in the Canal Zone and Chiriquí and from premontane wet forest in the Canal Zone.

Ficus trigonata L., Pl. Surin. 17. 1775

Commonly a strangler but often a large tree, to 30(40) m tall, ca 4 m wide at base, 80–150 cm wide above buttresses; stems ± glabrous; bark brownish; buttressing roots to 3 m in open areas or along the shore; sap copious, milky. Petioles 2–5 cm long; stipules 1.5–3 cm long with long silky trichomes, deciduous; blades ± elliptic to ovate-elliptic, infrequently obovate, abruptly short-acuminate (sometimes rounded) at apex (the acumen sharp or blunt), obtuse, rounded, or subcordate at base, mostly 13–25 cm long and 7–10 cm wide, glabrous, the midrib above white, 3–5-veined at base, the major veins in 5–11 pairs, the smallest reticulate veins distinct. Figs paired, globose, 1.5–3 cm diam, puberulous, sparsely covered with round reddish dots; peduncles obsolete or to 7 mm long; ostioles raised and buttonlike, flat on top, ca 5 mm wide; basal bracts 2, rounded, 1–5 mm long, 4–7 mm broad, becoming brown, bilobed. *Croat 12719.*

Uncommon, in the forest and along the shore. Interfloral phase is 27 days.

Distinguished by the fig with red spots and buttonlike ostiole. Most easily confused with *F. costaricana;* see the genus key for distinguishing characters.

Figs are distributed by bats.

Southern Mexico to Colombia on the Caribbean slope; Greater Antilles. In Panama, known from tropical moist forest on both slopes in the Canal Zone and in Panamá (San José Island) and Darién; known also from premontane dry forest (Penonomé) and premontane wet forest (El Valle) in Coclé.

Ficus yoponensis Desv., Ann. Sci. Nat. Bot., sér. 2, 18:310. 1842

Free-growing tree, to 40(50) m tall; trunk buttressed, ca 1 m diam, light gray, somewhat smooth; stems and leaves glabrous; outer bark planar with weak fissures and rows of lenticels; inner bark tan, in inverted V-shaped bands; sap copious, milky. Petioles 1–2.5 cm long; stipules linear, 3–5 cm long, caducous; blades elliptic to oblong-elliptic, acuminate, acute to obtuse at base, 6–11 cm long and 2.5–4 cm wide (juveniles with blades to 28 cm long and 5 cm wide), the major lateral veins in 15–30 pairs, mostly less than 4 mm apart, forming obscure collecting vein. Figs solitary, globose, to 1.8 cm diam, green at maturity, often purplish at apex, mottled with irregular, lighter green, weakly pustular areas; peduncles 3–11 mm long; ostioles narrowly tubular, to 3 mm long; basal bracts 3, free, ca 1 mm long. *Croat 15060.*

Common, especially in the younger forest. Seedlings require considerable light to survive, which explains their greater abundance in the younger forest.

Figs are distributed by bats.

Mexico (Chiapas) to Colombia and Venezuela; sea level to 1,600 but usually at 500 to 1,200 m. In Panama, known from tropical moist forest on BCI and from premontane wet forest in Chiriquí (Boquete).

MAQUIRA Aubl.

Maquira costaricana (Standl.) C. C. Berg, Acta Bot. Neerl. 18:463. 1969

Dioecious tree, ca 20 m tall, the trunk mostly 10–20 cm diam, with narrow buttresses ca 1 m high, glabrous but with puberulent stems; bark thin, brown; sap in trunk

Fig. 204. *Ficus pertusa*

Fig. 205. *Olmedia aspera*

copious, light brown, viscid, at first not obvious, becoming milky. Petioles to 1.5 cm, stout, callous; stipules paired, lateral, ovate, ca 5 mm long, caducous; blades oblong-elliptic, abruptly long-acuminate, obtuse to slightly rounded at base, 11–20 cm long, 4–8 cm wide, with a conspicuous collecting vein. Staminate inflorescences discoid, 1–3 per axil, 5–10 mm diam; peduncles 2–8 mm long; involucral bracts in ca 5 series; flowers free; perianth to 1 mm long, 4-lobed, puberulent, the lobes obtuse; stamens 4, to ca 1.5 mm long. Pistillate involucrate heads solitary in axils, discoid, 7–10 mm long and 10–17 mm wide at anthesis, subsessile or pedunculate; bracts ovate to rounded, acute to obtuse at apex, in 3–6 series, to 2.5 mm wide, minutely pubescent, their margins scarious, ciliate; perianth 2–2.5 mm long, fleshy, tubular, truncate or obscurely 4-lobed, minutely pubescent outside, tightly enclosing and eventually adnate to ovary; stigmas sessile, bilobed, the lobes flat, strap-shaped, densely puberulent, persisting in fruit. Fruiting heads to 5 cm diam; fruits oblong-elliptic to obovate, to 2 cm long and 1.8 cm wide, with a depression around the style at apex (often with minute ridges when immature), nearly glabrous, red at maturity; exocarp thin; mesocarp sweet, fleshy, to 3 mm thick; seed obovate, smooth, brown, ca 1.3 cm long. *Croat 9786, 15248.*

Occasional, but locally common. Flowers from January to May, with the fruits maturing mostly from May to July.

The fruits are eaten by toucans (label on *Duke 12280*).

Nicaragua to Peru; to 850 m elevation. In Panama, known from tropical moist forest in the Canal Zone, Panamá, and Darién.

OLMEDIA R. & P.

Olmedia aspera R. & P., Syst. Veg. 257. 1798

Dioecious shrub or tree, to ca 6(20) m tall; younger stems, petioles, and peduncles with coarse, erect or appressed trichomes; stems eventually glabrate or puberulent; sap whitish, watery, becoming clear with white clots. Leaves alternate; petioles 5–7(15) mm long; stipules fully amplexicaul, small; blades variable, mostly oblanceolate-elliptic, subcaudate-acuminate at apex, remotely serrate-undulate toward apex, cuneate at base, mostly 10–25 cm long, 2.5–9 cm wide, often somewhat falcate, glabrate and weakly asperous above, asperous below with short inconspicuous trichomes. Staminate flowers numerous, in involucrate discoid heads to ca 1 cm diam; perianth green, apiculate in bud, 4-lobed, the lobes acuminate, scabrid outside; stamens 4, inflexed in bud, springing out violently upon the bursting open of the thin perianth lobes, becoming strongly reflexed; pollen white, powdery. Pistillate inflorescences small, axillary, solitary or clustered, ovoid, consisting of a solitary flower surrounded by involucrate bracts, the bracts orange within at maturity, opening broadly to expose fruit; ovary superior; ovule and style pubescent at apex; stigmas with 2 filiform exserted lobes. Fruits sweet, orange, false (?) drupes ca 8 mm diam;

seed rounded, ca 5 mm diam. *Croat 5073, 10168.*

Common in both the young and the old forests. Flowers and fruits throughout the year.

The flowers are visited by small bees.

Costa Rica through the Andes to Bolivia. In Panama, known from tropical moist forest in the Canal Zone, Bocas del Toro, Colón, Chiriquí, Los Santos, Panamá, and Darién; known also from premontane wet forest in Coclé and Panamá.

See Fig. 205.

PEREBEA Aubl.

Perebea xanthochyma Karst., Fl. Columb. 2:23, t. 112. 1861

Dioecious or monoecious tree, to 10(35) m tall, usually to ca 15 cm dbh; branches slender, slightly flexuous, conspicuously but sparsely hispidulous (except densely so when young), sometimes becoming glabrate, the trichomes usually acropetal; sap yellowish, turning brown or reddish. Leaves alternate, distichous; petioles 3–5 mm long; petioles and midrib of blades pubescent like the stems; stipules fully amplexicaul, lanceolate, 5–10 mm long, densely hispidulous; blades oblong-elliptic to narrowly obovate, abruptly and narrowly long-acuminate to subcuspidate at apex, obtuse to rounded at base, (6)16–28(48) cm long, (2)4–10(20) cm wide, glabrate on upper surface except on midrib, glabrate to sparsely hispidulous on smaller veins below, densely hispidulous on midrib below, the lateral veins in 7–23 pairs, the margins entire to ± coarsely undulate-serrate toward apex. Inflorescences axillary, solitary or clustered, involucrate, discoid; staminate inflorescences 3–6(10) mm diam, the peduncles 1–3(6) mm long, the bracts ca 25–60, deltoid to ovate, acute at apex, in 4–8 series, the flowers 10 or more; perianth to 1.1 mm long; tepals 4, free, cucullate, obtuse, yellowish-puberulent; stamens 4, free, nearly included; anthers broadly oval. Pistillate inflorescences solitary or accompanied by staminate ones, 4–15 mm diam, subsessile to pedunculate, the peduncles slender, to 3 mm long, the bracts 20–90, deltoid to ovate, in 4–10 series, the flowers numerous, all fertile; perianth entire to 4-lobed, to ca 2 mm long, hispidulous, accrescent and ± pulpy in fruit but essentially free; ovary superior to sub-inferior; style central; stigma lobes short, broad. Fruiting heads subsessile, 1–2 cm diam; fruits ovoid, orange, hispidulous, to ca 9 mm long, together forming a weakly united syncarp. *Knight 3502, 1510.*

Apparently rare; known only from the forest near Standley Trail 1000 and along AMNH Trail. Seasonality uncertain. Flowers principally from August to April. The fruits mature mostly from January to June.

Costa Rica to Peru. In Panama, restricted to the Atlantic Coast; known from wetter areas of tropical moist forest in the Canal Zone and Bocas del Toro, from premontane wet forest in the Canal Zone (Pipeline Road) and Bocas del Toro (Punta Peña, vicinity of Chiriquicito), and from tropical wet forest in Colón (Portobelo) and Panamá.

POULSENIA Eggers

Poulsenia armata (Miq.) Standl., Trop. Woods
33:4. 1933
Cucuá, Cocuá, Maragua, Mastate, Namaqua

Monoecious tree, to 27 m tall and 90 cm dbh, with low buttresses; outer bark thick; stems, petioles, stipules, and lower midrib of leaf armed with short sharp spines 1–3 mm long; sap copious, yellowish-brown. Leaves alternate; petioles stout, 1–4 cm long, armed; stipules amplexicaul, 1.5–3 cm long, armed; blades variable, obliquely ovate to oblong-elliptic, obtuse to short-acuminate at apex, obtuse to rounded and inequilateral at base, 10–40(70) cm long, 6–25(35) cm wide, usually glabrous in age, coriaceous. Staminate inflorescences globose, 1–2.3 cm diam, bearing many dense flowers; tepals 4, barely united; stamens 4, slightly exserted. Pistillate inflorescences ovoid, 1–1.5 cm diam, yellowish with (3)5–9 flowers; perianth 4-dentate; stigmas deeply 2-lobed, exserted. Fruiting heads subglobose or ovoid, to 2.5 cm long, the individual carpels conic, irregularly angulate, sharply pointed at apex, the involucral bracts sharply pointed; seeds 1 per carpel, ovoid, shiny, brown, to 4 mm long. *Croat 8533, 9280.*

Abundant throughout the forest. Flowers and fruits throughout the year, especially during the dry and early rainy seasons. The leaves, which are somewhat resistant to decay, are usually lost and replaced during the dry season.

Red spider monkeys have been seen eating the syncarp.

Mexico to Bolivia. In Panama, known from tropical moist forest in the Canal Zone, San Blas, and Darién, from premontane wet forest in the Canal Zone, and from tropical wet forest in Colón and Darién.

See Fig. 206.

POUROUMA Aubl.

Pourouma guianensis Aubl., Hist. Pl. Guiane Fr.
2:892, t. 341. 1775
Mangabé, Guarumo macho, Viranjo

Dioecious tree, 10–25(30) m tall; trunk light brown, with irregular, horizontal rings; branchlets stout, hollow, defoliate except near apex; young parts bearing dense reddish-brown pubescence with minute branched trichomes (appearing papillate) and usually also golden, pilose, deciduous pubescence especially around stipule scars and on stipules; bark soft; sap clear or light yellow in trunk, turbid or milky usually turning black in younger parts. Leaves spiral, crowded at apex of branchlets; petioles mostly 10–25 cm long, terete, hollow, longitudinally striate or grooved, with minute reddish-brown trichomes interspersed with longer, ± hispid pubescence; stipules amplexicaul, 7–15 cm long, caducous, the pubescence dense, minute, reddish-brown, often also golden-pilose; blades ovate in outline, entire to irregularly sinuate with (3)5(7) deep lobes, usually abruptly short-acuminate at apex, deeply cordate at base, mostly 20–45(50) cm long and 20–50 cm wide, asperous above with minute

trichomes on surface and longer hispid pubescence on midribs of lobes, the pubescence below dense, white, arachnoid-tomentose interspersed with sparse hispid bristles especially on veins, the veins parallel, raised and darker than surface below, the parallel tertiary veins connecting the major lateral veins prominulous below. Inflorescences upper-axillary, repeatedly cymose; inflorescence branches, pedicels, and pistillate flowers reddish-brown granular-pubescent; flowers unisexual, sessile; staminate flowers sessile or borne on a thick pedicel; perianth deeply 3- or 4-lobed, ca 1 mm long; stamens 3 or 4; anthers ovate, yellowish. Pistillate flowers borne on a stout pedicel about as long and as broad as the flower; perianth ovoid-tubular, 3.5–4 mm long, with a small opening at apex; stigmas peltate-discoid, papillate, exserted. Drupes ovoid, 1–1.5 cm long, bearing minute reddish-brown pubescence, capped by the persistent, discoid stigma, purplish-black and juicy at maturity. *Croat 8097, 8100.*

Mature plants uncommon; seedlings common in the forest. Mature plants are locally common along Conrad Trail. Flowers in the early dry season. The fruits mature in the late dry season.

Belize to Colombia, Venezuela, and the Guianas. In Panama, a characteristic tree species in tropical moist forest (Tosi, 1971), known from BCI and in Darién; known also from premontane wet forest in the Canal Zone (Pipeline Road) and from tropical wet forest in Darién. Reported from premontane rain forest in Costa Rica (Holdridge et al., 1971).

See Fig. 207.

PSEUDOLMEDIA Trec.

Pseudolmedia spuria (Sw.) Griseb., Fl. Brit. W. Ind.
152. 1859
Caciqui, Cucuá

Dioecious tree, 8–20(30) m tall; branchlets slender, weakly flexuous, soon glabrous; sap white. Leaves alternate, distichous; petioles 3–10 mm long; stipules very narrowly lanceolate, long-acuminate, fully amplexicaul, mostly 5–7 mm long, rarely to 2 cm; blades ± oblong-elliptic, acuminate to caudate-acuminate at apex, acute to obtuse at base, 9–15(17) cm long, 3–5(6) cm wide, glabrous or nearly so. Flowers unisexual; staminate inflorescences in discoid, usually paired, sessile or subsessile heads 8–10 mm diam, the bracts broadly obtuse, minutely puberulent, in 2–4 series, the bracteoles interspersed with flowers; perianth vestigial; stamen solitary, to 3 mm long; anthers ± oblong, basifixed, apiculate. Pistillate flowers ovoid, solitary, sessile, involucrate, ca 2 mm diam, the bracts in 3–6 series, imbricate, softly puberulent; perianth tubular, 2–2.5 mm long, minutely 4-lobed; ovary inferior to semi-inferior; style to 1.5 mm long; stigmas 2, 4–6 mm long, widely exserted. Fruits false drupes, globose to cylindroid, 9–14 mm long, red; mesocarp thin. *Croat 11935.*

Rare; seen only once. The BCI collection is sterile, but almost certainly represents this species. Descriptions of

Fig. 206. *Poulsenia armata*

Fig. 207. *Pourouma guianensis*

the flowers and fruits were based on the *Flora of Panama* (Woodson & Schery, 1960) and on herbarium material from Central America. Flowers elsewhere from January to May, with the fruits maturing from February to June.

Mexico (Chiapas) to Panama; Greater Antilles; sea level to 900 m. In Panama, known from tropical moist forest in the Canal Zone, Bocas del Toro, and Darién (Duke, 1968).

SOROCEA St.-Hil.

Sorocea affinis Hemsl., Biol. Centr.-Amer. Bot. 3:150. 1883
Cauchillo

Dioecious tree, usually less than 6 m tall (to 15 m), glabrous but the leaves occasionally puberulent; branchlets grayish, sometimes lenticellate; sap milky. Leaves alternate; petioles 5–10 mm long; stipules lateral, ovate, paired, 2–5 mm long, caducous; blades oblong to oblong-obovate, caudate-acuminate at apex, narrowing to acute or rounded base, 7–16(18) cm long, 2.5–5.5(7.2) cm wide, entire or rarely bluntly serrate, the major lateral veins conspicuous below, loop-connected. Racemes puberulent, axillary; staminate racemes soon falling, 2–8 cm long, cream-colored, puberulent, the flowers numerous, bowl-shaped, ca 2 mm long, equaling length of pedicel, not dense, interspersed with minute peltate bracts; tepals 4, fleshy, broadly oval, concave; stamens 4, opposite and as long as tepals; filaments arched. Pistillate racemes usually 1–8 cm long in flower (to 6 cm in fruit), the flowers to 3 mm long, numerous, not dense, interspersed with many small bracts; pedicels pink and elongating to 6–10 mm in fruit; perianth depressed-globose, thick, ca 1.5 mm long and to 2.3 mm wide, the upper margin held tightly around the style; style bifid, with short, stout, flattened lobes spreading and recurving above throat of perianth; ovary ovoid, weakly pubescent. Drupes subglobose, ca 8 mm diam, green with an orange or red tip, becoming bright red and finally purple at maturity; seed 1. *Croat 4106, 5765.*

Frequent in the forest. Flowers sometimes starting in April, but mostly June to November, with the fruits maturing from September to December (sometimes February).

Guatemala to Panama. In Panama, known principally from tropical moist forest in the Canal Zone, Bocas del Toro, Colón, Veraguas, Panamá, and Darién; known also from tropical dry forest in Panamá (Taboga Island), from premontane moist forest in Panamá, from premontane wet forest in Coclé (El Valle), and from tropical wet forest in Colón.

TROPHIS P. Browne

Trophis racemosa (L.) Urban, Symb. Ant. 4:195. 1905
Breadnut, Gallote, Lechosa, Morillo, Ojocho macho, Ramón

Dioecious tree, to 17 m tall, to 25 cm dbh; trunk with broad, horizontal leaf scars; outer bark thin, sometimes reddish; inner bark thick, granular, tan; glabrous all over or sparsely puberulent on young stems, petioles, leaf veins below, and fruit; sap thin, copious, milky. Leaves alternate; petioles to 1(1.6) cm long; stipules lanceolate, lateral, minute, subpersistent; blades mostly oblong-elliptic to obovate, abruptly long-acuminate, rounded to obtuse at base, 7–15(23) cm long, 2.5–6(10) cm wide, entire or minutely serrate near apex, slightly asperous above and below. Staminate spikes pendent, 4–6.5 cm long, ca 5–6 mm wide, very densely flowered; tepals 4, broadly oval, to ca 1.7 mm long, densely and minutely pubescent; stamens 4, 2–3 mm long, strongly inflexed in bud; filament elongating, the anther springing free to fling the powdery white pollen into the air; pistillode ca 0.5 mm long. Pistillate spikes densely grayish-velutinous, usually less than 2 cm long, the flowers 4–15 per spike, sessile, ovoid, or conic; perianth minutely 4-lobed at the narrowly opened apex, 2–4.5 mm long; style branches 2, 2–5.5 mm long, long-exserted. Drupes sessile, ovoid to globular, ca 1 cm diam, densely and minutely velutinous, red becoming purple at maturity. *Croat 15247, Shattuck 1164.*

Occasional in both the young and the old forests. Seasonal behavior uncertain. Flowers principally from June to August, sometimes as early as April or as late as December, with the fruits maturing from October to December. Allen (1956) reported that the species flowers in February on the Osa Peninsula in Costa Rica.

Mexico to Peru and Brazil; Greater Antilles. In Panama, known from tropical moist forest in the Canal Zone, Bocas del Toro, Chiriquí, Veraguas, Panamá, and Darién and from premontane moist forest in the Canal Zone.

40. URTICACEAE

Trees, shrubs, or herbs (succulent in *Pilea*), sometimes scandent or epiphytic. Leaves alternate or opposite, petiolate; blades simple, serrate to entire (*Pilea*), sometimes palmately veined at the base; stipules present. Flowers unisexual (monoecious or dioecious), actinomorphic, in generally axillary, bracteate cymes or glomerules, sometimes spicate; tepals 3–5, biseriate, undifferentiated (or perianth lacking); stamens 3–5, opposite the tepals, free, springing out elastically at anthesis to throw pollen; anthers 2-celled, dehiscing longitudinally; ovary superior, 1-locular, 1-carpellate; placentation apical; ovule 1, falsely orthotropous; style and stigma 1; stigma penicillate. Fruits achenes; seed solitary, with endosperm.

Members of the family are distinguished by the frequently stinging trichomes, the minute greenish flowers, and the ovary with one carpel and one style.

Flowers are wind pollinated (Faegri & van der Pijl, 1966). The dry powdery pollen is catapulted by the elastic movement of the stamens at anthesis. G. Frankie (pers. comm.) reports that in Costa Rica stingless bees may aid in the opening of staminate flowers of *Myriocarpa* by their persistent probing of unopened flowers.

Dispersal of the minute achenes may be by birds, wind, or water. *Boehmeria cylindrica* and *Myriocarpa yzabalensis* usually grow in swampy areas or along water-

KEY TO THE SPECIES OF URTICACEAE

Leaves opposite, at least along main stem:
  Leaves less than 1 cm long, one leaf of a pair ca half the size of the other; blades entire; plants often minute, always less than 30 cm tall . . . . *Pilea microphylla* (L.) Liebm.
  Leaves more than 2 cm long, pairs ± equal; blades serrate-dentate; plants more than 50 cm tall
  . . . . *Boehmeria cylindrica* (L.) Sw.
Leaves alternate:
  Flowers in long slender spikes; leaves often more than 8 cm wide . . . .
  . . . . *Myriocarpa yzabalensis* (Donn. Sm.) Killip
  Flowers not in slender spikes; leaves less than 8 cm wide:
    Petioles less than 1 cm long; blades unequal at base . . . . *Pouzolzia obliqua* (Poepp.) Wedd.
    Petioles 1–7 cm long; blades ± equal at base . . . . *Urera eggersii* Hieron.

---

courses, and their seeds may be water dispersed. *Pilea microphylla* somehow finds itself in every concrete crack in shady areas; possibly its minute seeds are carried around by ants or rain wash. Van der Pijl (1968) reported that in some species of *Pilea* the staminodia eject the entire fruit.

Some 49 genera with about 2,000 species; widespread but mostly in the tropics.

BOEHMERIA Jacq.

Boehmeria cylindrica (L.) Sw., Prodr. Veg. Ind. Occ. 34. 1788

Monoecious herb, sometimes epiphytic, erect, to 1.5 m high. Leaves opposite on stem, alternate on any branches; petioles 5–25 mm long, sparsely pubescent; blades narrowly ovate or lanceolate, acuminate, rounded to subcordate at base, 4.5–11 (18) cm long, 2.5–4 (7) cm wide, serrate-dentate, glabrous or with inconspicuous pubescence on lower surface of veins. Flowers minute, pinkish, in discrete, unisexual, globular clusters along axillary spikes 2–8 cm long and often leafy near the tip; staminate perianth 4-lobed; pistillate perianth 4-toothed, ca 2 mm long, nearly sessile. Achenes ± round, compressed, ca 1 mm diam, brown. *Croat 12691.*

Uncommon, usually occurring in swampy areas along the shore. Flowers and fruits in the rainy season.

Canada south to southeastern Brazil; West Indies. In Panama, known only from tropical moist forest in the Canal Zone and Bocas del Toro.

See Fig. 208.

MYRIOCARPA Benth.

Myriocarpa yzabalensis (Donn. Sm.) Killip, Proc. Biol. Soc. Wash. 40:29. 1927
  Cow itch

Dioecious shrub or small tree, 3–8 (10) m tall; stems ribbed below petioles, with harsh pubescence. Leaves alternate; petioles one-fourth to as long as blade; blades ovate to ovate-elliptic, abruptly acuminate, mostly rounded at base, 15–40 cm long, 10–25 cm broad, punctate, with linear cystoliths conspicuous when dry, glabrous

above, sparsely pubescent on veins below. Staminate spikes to 10 cm long, erect or ± pendent, bracteate, the flowers to 1.2 mm diam; tepals 4; stamens spring flowers open and throw pollen. Pistillate spikes 30–60 (90) cm long, slender, pendent; perianth lacking; style oblique on short stalk; stigma brushlike, straight and divergent at anthesis, later curling. Achenes scabrid, ca 1 mm long, soon falling. *Croat 4196, 8048.*

Frequent along creek banks in the younger forest. Flowers in the early dry season. The fruits mature in the middle to late dry season.

The achenes are light enough to be blown and probably are carried downstream by water currents. Their scabrid trichomes no doubt play a role in dispersal.

Guatemala to Panama. In Panama, known from tropical moist forest in the Canal Zone, Bocas del Toro, Colón, Los Santos, Panamá and Darién, from premontane wet forest in Panamá (Cerro Campana), and from tropical wet forest in Darién.

PILEA Lindl.

Pilea microphylla (L.) Liebm., K. Danske Vidensk. Selsk. Skr. Naturvidensk. Math. Afd., ser. 5, 2:296. 1851

Monoecious (rarely dioecious), glabrous, succulent herb, often minute but to 30 cm high. Leaves opposite, one of each pair smaller; petioles to 2 mm long; blades obovate, mostly obtuse at apex, decurrent at base, the larger to 1 cm long and 5 mm wide, the smaller to ca 3 mm long and 1.5 mm wide, with linear cystoliths. Flowers white, in minute, globular, axillary, ± sessile clusters; staminate flowers mostly 4-parted; pistillate flowers usually 3-parted, the middle segment larger than lateral ones. Achenes ovate, ca 0.5 mm long, brown. *Croat 12966.*

Locally abundant in shady or moist places in the Laboratory Clearing, especially on sidewalks, concrete, or rock walls; rare elsewhere. Flowers and fruits throughout the year.

Throughout the American tropics. In Panama, known from tropical moist forest in the Canal Zone, Bocas del Toro, Colón, San Blas, and Panamá, from premontane wet forest in Coclé and Panamá, and from tropical wet forest in Colón (Portobelo).

Fig. 208. *Boehmeria cylindrica*

Fig. 209. *Pouzolzia obliqua*

Fig. 210. *Urera eggersii*

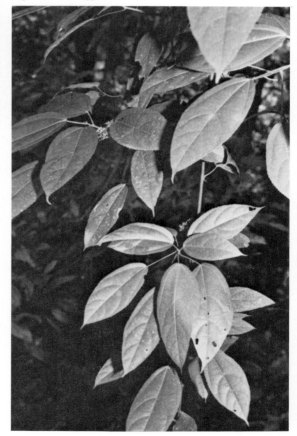

POUZOLZIA Gaud.

Pouzolzia obliqua (Poepp.) Wedd., Arch. Mus. Hist. Nat. 9:405. 1857

Monoecious liana or climbing shrub, to 2 (5) m tall; most parts densely hirsute. Leaves alternate; petioles less than 1 cm long; stipules paired, lanceolate, 3–10 mm long, persistent; blades oblong-lanceolate to oblong, acuminate, rounded to subcordate and inequilateral at base, 2–12 (15) cm long, 1.5–4.5 (5) cm wide, scabrous above with dense punctiform cystoliths. Flowers minute, white or greenish, in small, globular, androgynous (bisexual with staminate flowers at apex) or unisexual clusters; staminate flowers 4-parted; pistillate perianth tubular, 2- or 4-toothed, strongly veined. Achenes ovoid-ellipsoid, acute at apex, 2–4 mm diam, light brown and shiny. *Croat 11773.*

Occasional, along the shore. Flowers and fruits during the rainy season.

Guatemala to Venezuela and Peru. In Panama, known from tropical moist forest in the Canal Zone, Bocas del Toro, San Blas, Coclé, and Panamá.

See Fig. 209.

URERA Gaud.

Urera eggersii Hieron., Bot. Jahrb. 20, Beibl. 49:3. 1895
 Ortiga, Palo ortigo

Dioecious liana or vine, sometimes pendent from trees (elsewhere in Panama sometimes a small tree); young stems densely hirsute, glabrous in age. Leaves alternate; petioles 1–7 cm long, sparsely pilose; stipules acute, 1.5–2 mm long, caducous; blades ovate-oblong to obovate-oblong (rarely ovate elsewhere), acuminate, obtuse to subcordate at base, 9–17 cm long, 3.5–7.5 cm wide, short-hirsute on veins below or glabrate, entire to obscurely sinuate or crenate-dentate, the linear cystoliths of the upper surface mostly radiating from scattered, pellucid-punctations, those of the lower surface aligned mostly parallel to secondary and tertiary veins. Inflorescences dichotomously branched cymes, 1–5 cm long, about as long as wide; peduncles to 2 cm long, usually less than 1 cm long; peduncles and branches of inflorescence puberulent. Flowers unisexual, minute, white or greenish, inconspicuous; staminate flowers 4- or 5-parted, sessile, in glomerules; pistillate flowers 4-parted, usually solitary, short-pedicellate, the style very short, the stigma ± capitate, densely pubescent. Achenes ± globular, to ca 3 mm diam, orange, partially covered by fleshy, enlarged perianth. *Croat 12179.*

Uncommon, in the forest. Flowers in the late dry and early rainy seasons, from April to September, especially in July and August. The fruits mature mostly from September to December, but sometimes in the early dry season.

Possibly several species are involved. In the *Flora of Panama* (Killip, 1960), this species was treated as *U. elata* Griseb., which is endemic to the West Indies. *Urera eggersii* is extremely variable throughout its range; it is described as being a shrub or tree in other regions. The description here reflects mostly material collected on BCI. Plants from BCI differ from the type specimen from Ecuador in drying darker, in having much less pubescence, and in having conspicuous cystoliths. It is possible that the BCI plants represent a new species, but the problem warrants much more field work before a decision can be made. *Urera boliviensis* Herz. and *U. killipii* Standl. & Steyerm. may also be synonymous with *U. eggersii.*

Costa Rica to Colombia, Ecuador, and Bolivia. In Panama, known from tropical moist forest in the Canal Zone, Bocas del Toro, San Blas, Panamá, and Darién, from premontane wet forest in Chiriquí, Veraguas, and Coclé, and from tropical wet forest in Colón, Los Santos, and Darién.

See Fig. 210.

41. PROTEACEAE

Trees with aromatic wood. Leaves alternate, petiolate; blades heteromorphic, the juvenile ones pinnate, the adult ones simple with pinnate venation; stipules lacking. Flowers bisexual, apetalous, actinomorphic, in terminal or axillary, bracteate racemes; sepals 4, free, petaloid; disk present; stamens 4, opposite the sepals; anthers 2-celled, longitudinally dehiscent; ovary 1, superior, 1-locular, 1-carpellate; placentation pendulous; ovules 2; style 1; stigma 1, bulbous. Fruits 2-valved follicles; seeds 2, winged.

The family is represented on BCI only by the genus *Roupala,* which is distinguished by its dimorphic leaves and by the spikelike racemes of flowers with four tepals and a slender, exserted, clublike style.

Inflorescences are brush-type pollination units in the sense of Faegri and van der Pijl (1966). Seeds are winged and apparently wind dispersed.

Some 62 genera and 1,400 species; mainly in the drier regions of the Southern Hemisphere.

ROUPALA Aubl.

Roupala montana Aubl., Hist. Pl. Guiane Fr. 1:83. 1775
 R. darienensis Pitt.

Tree, to 20 m tall; wood aromatic when cut; young stems ferruginous-strigillose. Leaves alternate, heteromorphic; juvenile leaves pinnately compound, imparipinnate, the leaflets to ca 19, coarsely serrate; adult leaves simple; petioles 1–6 cm long; blades ovate, mostly acute or acuminate at apex, obtuse to cuneate at base, 5–12 cm long, 2–9 cm wide, subcoriaceous to coriaceous, grayish-green, mostly entire to undulate, rarely serrate, glabrous but the midvein above ferruginous-strigillose. Flowers cream-colored, on multi-flowered, terminal or axillary racemes; rachis 9–18 cm long, ferruginous-strigillose; pedicels ca 3 mm long, strigillose; sepals 4, linear-oblanceolate, ca 8 mm long, widely reflexed, strigillose outside, gla-

brous inside; petals lacking; stamens 4, inserted about midway on sepals, strap-shaped, ca 3 mm long, nearly sessile; disk of 4 hypogynous glands alternate with sepals; ovary densely strigose; style and stigma linear-oblanceolate, together ca 5.5 mm long, glabrous. Follicles flattened, elliptic to obovate, 2.5–4 cm long, ca 1.5 cm wide, glabrous, often obscurely spurred at base; seeds 2, oval, winged, ca 1.5 mm long and 8 mm wide. *Croat 14647.*

Seen only once at the junction of Snyder-Molino and Barbour trails. Flowers January to May. The fruits reported in June.

Mexico to Peru, Bolivia, and Brazil. In Panama, widespread and ecologically variable; known from tropical moist forest in the Canal Zone, Veraguas, Herrera, Coclé, Panamá, and Darién, from tropical dry forest in Coclé and Panamá, from premontane moist forest in the Canal Zone and Panamá, from premontane wet forest in Chiriquí, Coclé, and Panamá, and from tropical wet and lower montane rain forests in Darién.

42. LORANTHACEAE

Glabrous, parasitic shrubs or vines; nodes usually swollen, the internodes with scalelike cataphylls. Leaves opposite, sessile or petiolate; blades simple, entire, leathery; venation generally obscure (may be palmately veined at base); stipules lacking. Flowers bisexual or unisexual (dioecious or monoecious), actinomorphic, borne on a cupular receptacle (the calyculus) in axillary spikes of few to many flowers; perianth in 1 or 2 similar series, the outer series truncate or 4- or 6-parted, the inner series (lacking in ours) 6-parted; stamens equal in number to perianth lobes and borne on them; anthers 2-celled, generally introrse, dehiscing longitudinally, often with a connective extending above the thecae; ovary inferior, formed by the receptacle, 1-locular (obscurely so), presumably 3- or 4-carpellate; placental area central, basal; ovule 1; style 1, simple; stigma capitate or lacking in staminate flowers. Fruits berries; seeds few, with endosperm.

This family of parasitic shrubs with brittle stems is not confused with any other. Plants are particularly conspicuous in the dry season if their host happens to be a deciduous tree. All species are photosynthetic and attached firmly by modified roots (haustoria), which connect directly to the vascular system of the host. No systematic study has been made with regard to host preference but, if any exists, it is not apparent. The *Citrus* species (67. Rutaceae) of the Laboratory Clearing abound with a number of species of Loranthaceae, especially *Oryctanthus* and *Phthirusa pyrifolia*, but this is perhaps due to the fact that these trees are roosts for a multitude of birds. Sexual parts of the opposite sex for species with unisexual flowers may be absent altogether as in *Phoradendron* or merely reduced and sterile as in *Struthanthus*.

The minute greenish flowers are pollinated by small insects (Kuijt, 1969).

The sticky-seeded berries are typically bird dispersed (Ridley, 1930; Kuijt, 1969). Seeds are frequently seen germinating on tree branches where birds have roosted. Van der Pijl (1968) reported that some Loranthaceae have seeds with a built-in water reserve, which allows them to germinate on dry branches in the absence of water.

About 30–40 genera with 1,000–1,400 species; mainly in the tropics.

KEY TO THE SPECIES OF LORANTHACEAE

Rachis of the spike with depressions in which the flowers and fruits are attached:
  Spikes of several joints with bracteoles beneath each joint; stems with paired scales (cataphylls) on some of the internodes, at least on the lowermost internode of each branch; tepals 3:
    Scales present at the base of each internode; stems terete; leaves more than 2.5 cm wide, acuminate . . . . *Phoradendron piperoides* (H.B.K.) Trel.
    Scales present only on the lowermost internode of each branch; stems 4-angled when young; leaves less than 2.5 cm wide, not acuminate . . . . *Phoradendron quadrangule* (H.B.K.) Krug & Urban
  Spikes lacking joints or bracteoles; stems lacking paired scales; tepals 6:
    Leaves more or less cuneate; fruits attached perpendicularly to and entirely exposed from rachis; rachis glabrous, less than 2 mm wide . . . . *Oryctanthus occidentalis* (L.) Eichl.
    Leaves rounded or cordate at base; fruits attached antrorsely to and partly sunken into cavities of rachis; rachis scaly, more than 2 mm wide:
      Leaves petiolate, rounded at base; stems terete . . . . *Oryctanthus alveolatus* (H.B.K.) Kuijt
      Leaves usually sessile, cordate at base; stems usually sharply 2(4)-angled . . . . *Oryctanthus cordifolius* (Presl) Urban
Rachis lacking depressions:
  Plants herbaceous vines, usually much more than 1 m long; leaves orbicular, less than 5 cm long; flowers more than 5 mm long; fruits ca 10 mm long . . . . *Struthanthus orbicularis* (H.B.K.) Blume
  Plants shrubs, to ca 1 m long; leaves ovate-elliptic, more than 5 cm long; flowers less than 2 mm long; fruits ca 5 mm long . . . . *Phthirusa pyrifolia* (H.B.K.) Eichl.

ORYCTANTHUS Eichl.

Oryctanthus alveolatus (H.B.K.) Kuijt, Bot. Jahrb. Syst. 95(4):504. 1976

Mato palo

Erect, parasitic shrub; stems terete, to 70 cm long; periderm rufous-scaly but glabrescent. Leaves glabrous; petioles short, to 5 mm long; blades ± ovate, rounded at apex and base, slightly decurrent on petiole, not at all clasping stem, 4–9 cm long, 3–5.5 (rarely 9) cm wide, coriaceous, drying brownish; venation palmate, the midvein not reaching apex, the principal vein usually branching. Spikes axillary, usually 2 per axil, 2–3.5 cm long, sessile or on rufous peduncles to 5 mm long; rachis becoming more than 3 mm wide in fruit, reddish, scaly; flowers sunken into pockets in rachis, oblique to the rachis; margins of bracteoles united with rachis, appearing indistinct from rachis; flowers minute, bisexual, greenish; tepals 6; stamens 6, the filaments fused to tepals; anthers strongly dimorphic, the 2 inner pollen sacs much smaller than the outer ones; style 1. Fruits ovoid-oblong, ca 5 mm long, greenish, sometimes with yellowish base, the apex exceeded by the margin of the persistent, oily calyculus; seed 1, obdeltoid, ca 2 mm long, drying black. *Croat 6566.*

Frequent along the shore and at the edges of clearings. Flowers and fruits throughout the year.

This species was confused with *O. spicatus* (Jacq.) Eichl. in the *Flora of Panama* by Rizzini (1960) and by Standley in the *Flora of Barro Colorado Island* (1933). *O. spicatus* ranges from Guatemala to Venezuela, Brazil, and Peru. It has not been reported from Panama, but is to be expected, since it occurs in Colombia (Chocó).

Costa Rica to Colombia, Venezuela, the Guianas, Brazil, and Bolivia; Trinidad. In Panama, known from tropical moist forest in the Canal Zone, Herrera, Panamá, and Darién, from tropical dry forest in Coclé, from premontane moist forest in the Canal Zone and Panamá, and from premontane wet forest in Chiriquí.

Oryctanthus cordifolius (Presl) Urban, Bot. Jahrb. Syst. 24:30. 1898

Erect, parasitic shrub; stems prominently 2(4)-angled, especially when young. Leaves sessile, thick, ovate, obtusely acute to rounded at apex, rounded to cordate at base, 7–12 cm long, 4–8 cm wide, the lateral veins from near the base 7–9, obscure, visible when dried. Spikes axillary, solitary, 1–8 cm long; peduncles 5–15 mm long; bracteoles with margins free; flowers and buds red, sunken obliquely into rachis in 4 distinct rows; tepals 6, ca 2 mm long. Fruits ellipsoid to obovoid, 5 mm long, 2.5 mm wide, perpendicular to axis, 1-seeded, purplish-black when mature (Kuijt, 1976). *Shattuck 425.*

The Shattuck collection has not been located but it is assumed to have been properly identified; the species has been collected elsewhere in the Canal Zone. Apparently flowers and fruits all year.

The species is similar to *O. occidentalis* except for its usually two-sided stem and cordate leaf blades.

Mexico to Caribbean and Pacific Colombia, Guyana

(based on a single collection). In Panama, ecologically variable; known from tropical moist forest in the Canal Zone, Bocas del Toro, Colón and Panamá, from tropical dry forest in Coclé (Penonomé), from premontane moist forest in Panamá (Punta Paitilla), and from premontane wet forest in Coclé (El Valle).

Oryctanthus occidentalis (L.) Eichl. in Mart., Fl. Brasil. 5(2):89. 1868

Erect, parasitic shrub; stems terete, usually less than 60 cm long; periderm rufous, scaly. Leaves glabrous; petioles lacking or to 5 mm long; blades ovate (may be lanceolate when young), acute to rounded at apex, ± obtuse at base and decurrent on petiole but not clasping stem, 7–11 cm long, 2–7 cm wide, coriaceous, shiny, green (bronze when young), drying chestnut and green, the veins purple. Spikes axillary, usually 2 or 3 per axil, conelike, to 3 cm long; peduncles rufous, 5–10 mm long; rachis to 2 mm wide (rarely to 3 mm), green, glabrous; flowers bisexual, sunken in rachis, perpendicular to the rachis and arranged in 4 rows; margins of the bracteoles distinct from the rachis; tepals 6, dimorphic, green, acute at apex, ca 1.2 mm long; stamens 6, each fused to lower part of tepal; filaments red; anthers dimorphic, both types at or above middle of tepal, each with 4 locules or with 2 locules, or both these types alternating, then the 4-celled anthers lower than the 2-celled anthers, the connective extending above thecae; style simple; stigma capitate, held at about the level of the anthers; nectariferous disk surrounding style at base (Kuijt, 1976). Fruits cylindrical, to 4 mm long and 2 mm wide, yellow- and green-striped, drying dark; seed 1, ellipsoid, ca 1 mm long, brown. *Croat 6443.*

Common, often in exposed, sunny areas, especially along the shore. Even individual plants may flower throughout the year.

The fruits are taken by a variety of birds, including pigeons, manakins, flycatchers, tanagers, and finches (Leck, 1972).

Costa Rica to Ecuador, Venezuela, the Guianas, Brazil, and Peru; Jamaica. In Panama, known from tropical moist forest in the Canal Zone, Bocas del Toro, Colón, Chiriquí, Panamá, and Darién and from premontane wet forest in Bocas del Toro, Coclé, and Panamá.

See Fig. 211.

PHORADENDRON Nutt.

Phoradendron piperoides (H.B.K.) Trel., Monogr. Phoradendron 145. 1916

Monoecious, parasitic shrub; stems terete, branched, green when fresh (drying rugose with a yellowish cast), a pair of cataphylls on almost every internode, the lower internodes usually with 2–5 pairs of cataphylls. Petioles less than 5 mm long; blades obliquely ovate, bluntly acuminate, ± cuneate at base, 6–13 cm long, 2.5–5.5 cm wide, coriaceous, green when fresh, drying rugose and brownish to almost black if young, glabrous, often marginally crisped, the veins lacking or obscurely pinnate on underside. Inflorescences of axillary, jointed spikes

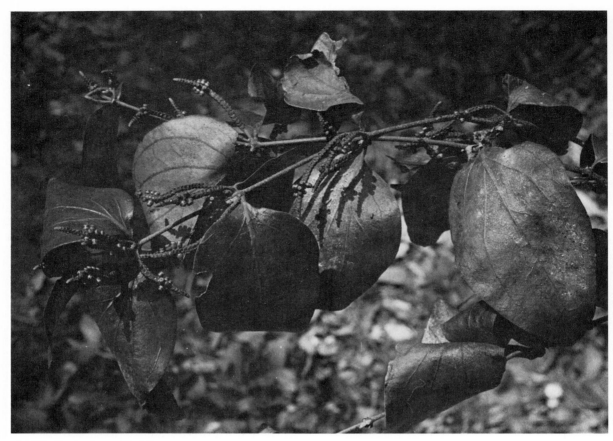

Fig. 211. *Oryctanthus occidentalis*

Fig. 212. *Phoradendron piperoides*

2.5–4.5 cm long, the spikes 4–6(8) per node, the joints 4–6 per spike, each subtended by two small, perfoliate bracts to 2 mm long; peduncles obscure or lacking; flowers unisexual, yellow, generally 12–20 per joint, 4-ranked, minute; tepals 3. Fruits attached from depressions in the rachis, ellipsoid, ca 4 mm long when mature, pale yellow to orange-brown. *Croat 8408.*

Occasional, in open areas and along the shore. Flowers and fruits from January to August.

The proper name for this species cannot be *P. piperoides* (H.B.K.) Trel. J. Kuijt (pers. comm.), who has studied the type of *Loranthus piperoides* H.B.K., reports that it is not the same species as that going by the name *P. piperoides*. Until someone can study the types of other old synonyms, however, the name will be used.

Mexico to Argentina; West Indies. In Panama, known from tropical moist forest in the Canal Zone, Bocas del Toro, Coclé, Panamá, and Darién, from premontane wet forest in Bocas del Toro and Chiriquí, and from premontane rain forest in Chiriquí.

See Fig. 212.

Phoradendron quadrangule (H.B.K.) Krug & Urban, Bot. Jahrb. Syst. 24:35. 1897

P. seibertii Rizz.; *P. venezuelense* Trel.

Mahoho, Mongreo

Monoecious, parasitic shrub, ± glabrous, only the inflorescence obtaining a yellowish cast when dry; stems square at least when young, a pair of cataphylls on only the lower internode of each branch. Petioles ca 2 mm long; blades elliptic-oblong, somewhat asymmetrical, mostly rounded at apex, sometimes mucronulate, attenuate to the base, 3–6 cm long, 0.7–2.5 cm wide, coriaceous, drying green to brown, rugulose, glabrous, the veins 3 and parallel or lacking. Spikes axillary, jointed, 1 or 2 per node, 1.5–4 cm long, the joints 3 or 4 per spike, each subtended by 2 minute bracts to 15 mm long; peduncles 3–10 mm long; flowers yellowish-green, generally 10–24 per joint, minute; tepals 3. Fruits attached from depressions in the rachis, globose, ca 3 mm long, yellow. *Croat 7398.*

Occasional, in the Laboratory Clearing. Flowers and fruits throughout the year.

Panama, Colombia, and Venezuela; West Indies. In Panama, known from tropical moist forest in the Canal Zone, Bocas del Toro, Colón, Veraguas, Panamá, and Darién, from tropical dry forest in Herrera and Coclé, and from premontane moist forest in the Canal Zone, Los Santos, and Panamá.

See Fig. 213.

PHTHIRUSA Mart.

Phthirusa pyrifolia (H.B.K.) Eichl. in Mart., Fl. Brasil. 5(2):36. 1868

Parasitic shrub, sometimes to about 1 m long; branches flattened, 2-edged (especially the younger ones), the scurfy margins extending onto midrib of blades. Leaves glabrous; petioles 3–10 mm long; blades ovate-elliptic, obtuse to rounded at apex, rounded at base, 5–10 cm long, 3–5 cm wide, coriaceous, the veins often visible. Spikes 1 or 2 per axil, to 15 cm long; flowers bisexual, 4–6-parted, maroon, 1–2 mm long, in sessile ternations, these widely spaced along the furfuraceous, lenticellate spikes; calyculus truncate; tepals valvate, oblong, acute; stamens shorter than tepals; filaments thick, fused to lower half of tepal; anthers about as broad as long, held at the level of the style; ovary red; stigma sessile; nectary prominent, the nectar rather copious. Fruits oblong, ca 5 mm long, becoming yellow or orange at maturity, drying gray to brown; exocarp thin but leathery; mesocarp sticky, white; seed 1. *Croat 6812, 7919.*

Common in trees of the Laboratory Clearing; frequent elsewhere. Flowers and fruits throughout the year.

The fruits are a favorite of small birds.

Belize to Peru, Bolivia, and Brazil. In Panama known from tropical moist forest in the Canal Zone, Bocas del Toro, Colón, and Veraguas, from premontane moist forest in Panamá, and from premontane wet forest in Chiriquí, Panamá, and Darién.

STRUTHANTHUS Mart.

Struthanthus orbicularis (H.B.K.) Blume in R. & S., Syst. Veg. 7(2):1731. 1830

Mata rama, God bush

Dioecious, somewhat woody, glabrous, parasitic vine; stems terete with thick adventitious roots at the nodes. Petioles ca 1 cm long, sometimes twining, if so their blades reduced; blades suborbicular, rounded at apex, obtuse at base, 3–5 cm long, 2.5–4 cm wide, coriaceous, the veins generally apparent. Flowers with strong, sweet aroma, greenish or greenish-yellow, fairly conspicuous, 5–7 mm long, in sets of 3 on slender spikelike inflorescences to 15 cm long, the ternations ± sessile; tepals 6, slender, free, acute, spreading above middle; stamens of staminate flowers fused to lower half of tepals, exserted at anthesis (above recurving tepals); filaments thickened just below anther (providing some protection to nectary), 3 filaments somewhat longer than the others; anthers introrse, about as broad as long; stigma held midway between upper and lower sets of anthers; nectary small, the nectar not copious. Fruits ellipsoid, ca 1 cm long, faintly orange to rust-red or purple, with copious milky latex; seed solitary. *Croat 12620, White 121.*

Uncommon, on trees near the edge of the lake in the vicinity of Colorado Point and Gigante Bay. Flowers and fruits from January to September.

Mexico to Peru and western Brazil. In Panama, widespread and ecologically variable; known from tropical moist forest in the Canal Zone, Bocas del Toro, San Blas, Veraguas, Panamá, and Darién, from premontane dry forest in Herrera and Coclé, from tropical dry forest in Panamá (Taboga Island), from premontane moist forest in the Canal Zone, Los Santos, and Panamá, from premontane wet forest in Chiriquí and Coclé, and from premontane rain forest in Darién (Cerro Pirre).

43. OLACACEAE

Glabrous trees or shrubs. Leaves alternate, petiolate; blades simple, entire; venation pinnate; stipules lacking. Flowers bisexual, actinomorphic, in axillary, cymose clusters; calyx 5(6)-lobed, accrescent and colorful in fruit; petals 5(6), briefly connate at very base, valvate; stamens 10(12), in two whorls, free; anthers 2-celled, longitudinally dehiscent; ovary superior, 3-locular, 3-carpellate; placentation apical; ovules solitary in each locule, pendulous; style 1, short; stigma 1, trilobate. Fruits drupes borne on the showy, accrescent, red calyx; seed solitary, with copious endosperm.

Flowers of *Heisteria*, the only genus of the family on BCI, are very small and inconspicuous. Distinguishing features of the flower include the superior ovary, the stamens numbering twice the petals, and the freely pendulous ovules. The colorful accrescent calyx of the fruiting plant is the most distinctive feature of the genus.

The system for pollination of the minute greenish flowers is unknown.

The fruits are clearly suited for bird dispersal. *Heisteria* fruits are dispersed by pigeons in Martinique (Ridley, 1930).

Some 27 genera with about 250 species; tropics.

HEISTERIA Jacq.

Heisteria concinna Standl., Publ. Field Columbian Mus., Bot. Ser. 8:137. 1930

Naranjillo, Ajicillo, Chorola

Glabrous tree, to 20 m tall, to ca 30 cm dbh; outer bark thin, minutely fissured (often bumpy below), peeling off easily after slash; inner bark ± thin, granular, its outer surface reddish with irregular green strips; sap with a foul, pungent odor. Leaves shiny; petioles 9–16 mm long; blades ovate, acuminate, obtuse at base, 10–15(23) cm long, 3.5–8(14.5) cm wide, coriaceous, the primary lateral veins not conspicuous, scarcely lighter than the surrounding tissue. Flowers greenish-white, ca 2 mm long; pedicels ca 5 mm long, clustered in axillary fascicles, 8–12 mm long in fruit; calyx shallow, accrescent in fruit, the lobes 5, acute; petals 5, valvate, floccose within, boat-shaped; stamens 10; anthers 0.2–0.3 mm long and wide; ovary oblate-spheroid. Fruiting calyces conspicuously 5-lobed, shorter than drupe, ca 2 cm broad, red; drupes ± globose (or broadly ellipsoid), 10–15 mm diam, white at maturity. *Croat 12559, Foster 1491.*

Common in the forest. Flowers more than once per season, apparently mostly in the late rainy season (No-vember and probably earlier as well), with the fruits developing from January to April. A second flowering may occur in the early rainy season, but it is apparently smaller; the fruits of the second flowering have been seen mature in October and November.

Costa Rica and Panama. In Panama, known from tropical moist forest in the Canal Zone, Los Santos, Herrera, and Darién.

Heisteria costaricensis Donn. Sm., Bot. Gaz. (Crawfordsville) 19:254. 1894

Slender glabrous shrub, to 2.5 m tall; young stems weakly flexuous, the internodes with 2 ribs. Petioles 5–11 mm long; blades linear-lanceolate, tapered to apex, rounded at base, 14–25 cm long, (1)2.5–5.5 cm wide, chartaceous, the primary lateral veins many and variable in number. Flowers greenish-white; pedicels clustered in axillary fascicles, ca 1 mm long, 6–12 mm long and jointed in fruit; calyx accrescent in fruit, the lobes 5, shallow; petals 5, valvate; stamens 10, free; anthers to 0.5 mm long. Fruiting calyces becoming bright red, longer than drupe, crateriform, to 2 cm diam, shallowly 5-lobed, spreading; drupes ovoid, ca 8 mm long, vertically ribbed, black with a single white seed. *Croat 4190, 11131a.*

Occasional, in the younger forest, particularly east and south of the laboratory. Flowers mostly from April to July. The fruits mature principally from July to December.

The leaves of this species are quite variable, the broadest leaves of some specimens approaching those of *H. macrophylla* Oerst. *Standley 40877* was originally identified as *H. macrophylla* for the *Flora of Panama* (Nevling, 1961) and apparently formed the basis for Nevling's report of this species for BCI.

Costa Rica and Panama. In Panama, growing mostly on the Atlantic slope at low and middle elevations; known from tropical moist forest in the Canal Zone and San Blas, from premontane wet forest in the Canal Zone and Panamá, and from tropical wet forest in Colón and Darién. Reported from premontane rain forest in Costa Rica (Holdridge et al., 1971).

See Fig. 214.

Heisteria longipes Standl., J. Wash. Acad. Sci. 17:8. 1927

Naranjillo colorado, Coloradito

Glabrous shrub or small tree, to 10(20) m tall; inner bark red; wood pinkish, hard; younger stems olive-green, the internodes weakly ribbed. Petioles 5–15 mm long, cana-

KEY TO THE SPECIES OF HEISTERIA

Leaves linear-lanceolate, more than 3 times as long as wide .... *H. costaricensis* Donn. Sm.
Leaves more or less ovate, less than 3 times as long as wide:
    Leaves shiny, coriaceous, the primary lateral veins not conspicuous; fruiting calyces conspicuously 5-lobed; drupes white .... *H. concinna* Standl.
    Leaves not shiny, subcoriaceous, the primary lateral veins conspicuous below; fruiting calyces at most shallowly 5-lobed; drupes black .... *H. longipes* Standl.

Fig. 213. *Phoradendron quadrangule*

Fig. 214. *Heisteria costaricensis*

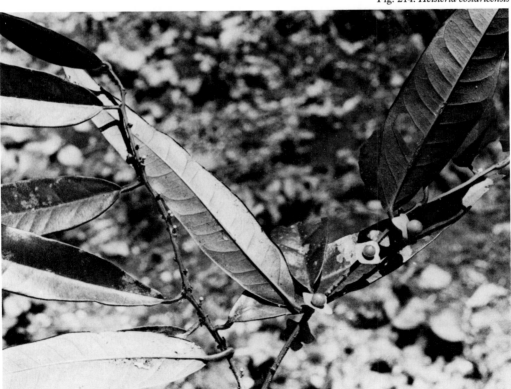

liculate; blades ovate, elliptic or obovate, acute to acuminate at apex, acute at base, 3.5–18 cm long, 2–7.5 cm wide, subcoriaceous, the major lateral veins in 6–9 pairs, conspicuous below, distinctly lighter than the surrounding tissue. Flowers greenish, ca 2.5 mm long; pedicels ca 5 mm long, clustered in axillary fascicles; calyx shallow, accrescent in fruit, the lobes 5, acute; petals 5, greenish, valvate, boat-shaped; stamens 10, shorter than petals, included; filaments strap-shaped, narrowed just below anther; anthers ca 0.3 mm long and wide; ovary ovate. Fuiting calyces about as long as drupe, ca 2 cm broad, red, shallowly 5-lobed or undulate; drupes ellipsoid, ca 1 cm long, black. *Croat 4532, 6052.*

Common in the forest, especially the old forest. Flowers principally from April to July (especially in May and June). The fruits develop to mature size by July, but mostly in August and September.

Costa Rica to Colombia. In Panama, widespread; most common at low to middle elevations; known from tropical moist forest in the Canal Zone, Colón, Chiriquí, and Darién, from premontane wet forest in Veraguas, Coclé, Panamá, and Darién, and from tropical wet forest in Colón and Coclé.

44. ARISTOLOCHIACEAE

Lianas, the younger stems herbaceous. Leaves alternate, petiolate, simple; blades entire, somewhat pubescent; venation pinnate (may be subpinnate at base); stipules lacking; pseudostipules lacking (in ours). Flowers protogynous, bisexual, apetalous, zygomorphic, solitary in the leaf axils or on short racemes; calyx tubular, consisting of 3 united sepals, inflated at the base (the inflated part called the utricle), expanded at the limb, showy, the tube forming an annulus where it meets the limb and also forming an inverted narrowed structure on the lower end (the syrinx), which extends into the utricle; petals lacking; styles, stigmas, and stamens united in a crownlike gynostegium; stamens 6, adnate to the style; anthers 4-celled, extrorse, dehiscing longitudinally; ovary inferior, 6-loculate; placentation axile; ovules several to many per placenta, anatropous; styles 6, forming subcapitate, stigmatic lobes. Fruits capsules; dehiscence acropetal, septifragal; seeds numerous, flat, with copious endosperm.

Distinguished by their unusual flowers and fruits. The complex flower is a greatly modified calyx. For a descriptive account of flower morphology refer to Pfeifer (1966).

Pollinated by insects, usually flies (Diptera), which are attracted to the flowers by a carrionlike odor and the intense colors (usually purplish). The corolla tube has rigid retrorse trichomes on the first day after opening when the stigmas are receptive (Petch, 1924; Corner, 1964). The pollinators are prevented from leaving the flower by these trichomes, so they find their way to the utricle. The following day the anthers dehisce, shedding pollen on the insect, and the retrorse trichomes wilt, allowing the insect to escape carrying the pollen.

For a description of typical seed dispersal, see the discussion of *Aristolochia chapmaniana.*

Ten genera and about 450 species; primarily in the tropics but extending to most temperate regions.

ARISTOLOCHIA L.

Aristolochia chapmaniana Standl., Contr. Arnold Arbor. 5:60. 1933

Twining liana, sparsely hispidulous; trunk corky and deeply fissured near ground, to ca 5 cm diam. Leaves alternate; petioles 8–22 mm long; blades oblong-spatulate, acute to obtuse at apex, deeply cordate at base, 10–20 cm long, 3–6 cm wide, glabrate above, inconspicuously short-pubescent below. Flowers solitary, axillary, usually slightly arched, dark purple-brown; calyx expanded into a narrow ellipsoid balloon ca 6 cm long, then constricted into a tube ca 3.5 cm long, the tube split on one side to form an oblong-lanceolate limb to 7 cm long; corolla lacking; stamens 6, the anthers sessile and adnate to style; styles 6, connate; stigmas capitate. Capsules ovate-cylindrical, prominently 6-ribbed, 10–12 cm long, ca 5 cm wide, on a long stipe, dehiscing from apex to base, the 6 valves spreading widely to disperse seeds but remaining attached on both ends, the inner valves diverging through outer valves at apex and attached to outer valves by a ladderlike series of fibers, the fibers acting to slow dispersal of seeds; seeds very numerous, flattened, stacked in 6 vertical rows, the seminiferous part cordate, ca 5 mm diam, bearing 2 unequal pairs of lateral wings ca 2 cm wide, the sides of the wings ± rounded, the longer pair ca 1.2 cm long, the shorter pair ca 0.8 cm long. *Croat 5958, Shattuck 413* (type).

Occasional, in the forest and along the shore. Flowers from at least November to January. The fruits, with some seeds, reported from April to September. Capsules may hang on all year.

Central America and northern South America. In Panama, known from tropical moist forest in the Canal

---

KEY TO THE SPECIES OF ARISTOLOCHIA

Plants conspicuously pilose on most parts; limb of calyx less than 3 cm long; capsules ovoid, ca 6 cm long . . . . *A. pilosa* H.B.K.
Plants not pilose; limb of calyx more than 3 cm long; capsules ± cylindrical:
    Blades deeply and narrowly cordate at base, scarcely if at all broader at base; calyx limb narrow, ca 7 cm long . . . . *A. chapmaniana* Standl.
    Blades truncate or shallowly and broadly cordate at base, conspicuously ovate; calyx limb 10–20 cm long . . . . *A. gigantea* Mart. & Zucc.

Fig. 215. *Aristolochia chapmaniana*

Fig. 217. *Aristolochia pilosa*

Fig. 216. *Aristolochia chapmaniana*

Zone and Panamá and from premontane wet forest in Coclé.

See Figs. 215 and 216.

Aristolochia gigantea Mart. & Zucc., Nov. Gen. Sp. Pl. 1:75, t. 48. 1824
A. sylvicola Standl.

Zaragosa

Twining liana, glabrous but the leaf blade below arachnoid-pubescent. Leaves alternate; petioles slender, ca 4–6 cm long; blades ovate, acuminate, truncate to shallowly cordate at base, 10–15(20) cm long, 7–11(12) cm wide, grayish below with arachnoid trichomes. Flowers in short racemes of few flowers, white, densely speckled with pink and maroon; calyx expanded into a ± campanulate balloon to 8 cm long, constricted into a short, reflexed tube opening into a large limb to 20 cm long; corolla lacking; stamens 6, the anthers sessile and adnate to style; styles 6, ± connate; stamens capitate. Capsules cylindrical, to 13 cm long and ca 3 cm wide, dehiscing from base, bearing many seeds; seeds 7–8 mm long, flattened, elliptic, the margins bordering a prominent medial rib, very revolute, the dehisced valves ca 1.5 cm wide. *Croat 11260, Shattuck 640.*

Rare, in the forest. Some flowers have been seen in September and December. Maturity time of the fruits is uncertain, probably in the dry season.

Known only from Panama, in tropical moist forest in the Canal Zone and Bocas del Toro and from tropical wet forest in Darién.

Aristolochia pilosa H.B.K., Nov. Gen. & Sp. 2:146, t. 113. 1817
A. costaricensis (Klotzsch) Duch.

Slender, twining liana, conspicuously pilose on most parts. Leaves alternate; petioles slender, to 7 cm long; blades ovate, acute at apex, deeply cordate at base, 6–25 cm long, 5–15 cm wide, glabrescent above. Flowers solitary in axils, speckled white and brownish-purple; calyx expanded into a short ellipsoid balloon to 3 cm long, then constricted obliquely into a campanulate tube ca 2.5 cm long, finally opening into a narrow, fringed limb 1.5–2.5 cm long; corolla lacking; stamens 6, the anthers sessile and adnate to style; styles 6, connate; stigmas capitate. Capsules ovoid, ca 6 cm long and 4 cm wide, dehiscing from base, bearing many seeds; seeds disk-shaped, ca 1 cm diam. *Croat 9011.*

Occasional, around the Laboratory Clearing. Flowers mostly from January to May, sometimes later in the rainy season also. The fruits have not been seen on the island.

Central America and northern South America. In Panama, known from tropical moist forest in the Canal Zone, Bocas del Toro, Panamá, and Darién.

See Fig. 217.

45. RAFFLESIACEAE

Parasitic herbs lacking chlorophyll, growing (on BCI) on the wood of *Casearia* spp. (96. Flacourtiaceae). Leaves lacking. Flowers unisexual, apetalous, actinomorphic, solitary, subtended by 2 series of bracts, the lower of 2 bracts, the upper of 4; sepals 4, free, imbricate, subequal, caducous; stamens indefinite, usually many, in 2 series, epigynous; ovary subinferior, 1-locular, 4-carpellate; placentation parietal; ovules many; style 1; stigma apparently an annulus around the style apex. Fruits berries; seeds many, with endosperm.

Apodanthes may be difficult to recognize as a flowering plant because of its leafless habit. It may appear to be simply a cauliflorous flower of the host plant.

About 8 genera and 50 species; tropics, primarily in the Old World.

APODANTHES Poit.

Apodanthes caseariae Poit., Ann. Sci. Nat. (Paris) 3:422. 1824
A. flacourtiae Karst.; *A. panamensis* Vatt., nomen nudum

Small, leafless, parasitic herb, emerging from the bark on trunk and larger branches of *Casearia* (96. Flacourtiaceae). Flowers solitary; pistillate flowers white to cream, sitting in a cuplike receptacle formed in the outer bark of the host plant, the receptacle 1–2 mm high and 2–3 mm wide; basal bracts 2, ± orbicular, concave, thick, 2–2.7 mm long; upper bracts 4, narrowly rounded at apex, to 5 mm long, fused to lower one-half or one-third of the ovary and pressed inward against the ovary; calyx segments 4, obovate, imbricate, white, slightly inequilateral, spreading at about the middle, narrowed at the base and borne in a minute depression near the apex of the ovary; petals lacking; ovary ovoid, orange, ca 3.5 mm long; ovules numerous, white, minute, borne on several weak invaginations in the outer walls of the unilocular ovary; style thick and stout, to ca 2 mm long, broader at apex; stigma minutely papillate; staminate flowers (Gentry, 1973a) "with a short central stylar column terminating in a hemispheric stigma-like cap; stamens inserted around the stylar column below the cap in 2 series of ca 19 each, subsessile. Fruit a berry." *Croat 17021.*

Infrequent; known from several locations in the young forest. Perhaps more frequently encountered on *Casearia aculeata* than on *C. guianensis*. Apparently not noticeable the rest of the year, plants have been seen only during August and September.

Plants are inconspicuous and easily could be missed, but the species is not confused with any other. Vattimo (1971) used the name *A. panamensis* for Panamanian populations, but the name was invalidly published. It is listed as a synonym here only to avoid future misunderstandings about the name.

Flowers were seen visited by a small, yellowish-brown bee. R. Foster (pers. comm.) also reports that the flowers are visited by small butterflies and even mosquitos.

Fruits are seemingly adapted to bird dispersal, but if so it is uncertain how the seeds find themselves on only the two species of *Casearia* on which they occur—surely birds aren't so particular where they perch! This unexplained fidelity of host and parasite indicates a more

complex system of fruit dispersal. A host-dependent phytophagous insect might be suspected.

Range is difficult to determine because of the inconspicuous habit of the plant. Probably ranging from Belize to Brazil. In Panama, known from tropical moist forest in the Canal Zone and from premontane wet forest in Coclé (El Valle).

See Fig. 218.

46. POLYGONACEAE

Trees, shrubs, lianas, or succulent-stemmed herbs (stems hollow in *Triplaris*). Leaves alternate, petiolate; blades simple, entire; venation pinnate; stipules present as sheathing ocreae. Flowers bisexual or unisexual (functionally monoecious in *Coccoloba*, dioecious in *Triplaris*), actinomorphic, on terminal spikes, racemes, or panicles in ocreolate fascicles of 1 to few flowers; tepals 4–6, free, in 1 series (in *Triplaris*, 6, connate, in 2 series), accrescent in fruit, becoming winged or bladdery; stamens 6–9, in basically 2 series, free or adnate; anthers 2-celled, at least the outer introrse, dehiscing longitudinally; ovary superior, 1-locular, 2–4-carpellate, subtended by an annular, nectar-secreting disk; placentation seemingly basal; ovule 1, orthotropous; styles and stigmas 2 or 3. Fruits achenes (enveloped by the fleshy perianth lobes and drupelike in *Coccoloba*); seed with mealy endosperm.

Polygonaceae are distinguished by their ocreate stipules.

Pollination systems are unknown. Some members of the family have dimorphic heterostyly (Faegri & van der Pijl, 1966).

The seeds of *Triplaris cumingiana* are wind dispersed, but those of probably most Polygonaceae are principally bird dispersed. *Coccoloba* generally has a thin fleshy mesocarp and a well-protected seed, but in the case of *C. acuminata* the shiny dark seed is displayed against the greatly accrescent, fleshy, white sepals. The achenes of *Polygonum* are probably similarly displayed, but since most occur in aquatic habitats they are perhaps in part water dispersed as well. Van der Pijl (1968) suggested that seeds of *Polygonum* may be dispersed on the feet of shore birds. Despite their close association with *Triplaris*, ants of the genus *Pseudomyrmex* do not function in pollination or fruit dispersal (Wheeler, 1910). They do benefit the plant, however, by warding off predators such as leaf-cutter ants (*Atta*). The sting of the genus *Pseudomyrmex* is severe and no doubt provides *Triplaris* with considerable protection.

About 40 genera with some 800 species; mostly of North Temperate distribution.

COCCOLOBA P. Browne

Coccoloba acapulcensis Standl., Proc. Biol. Soc. Wash. 33:66. 1920

Monoecious tree, 4–8 m tall, glabrous but with the lower midrib of leaf minutely hirtellous; trunk slender, sometimes forming adventitious roots. Leaves borne individually at apex of short shoots to ca 1 cm long; blades ovate, narrowly very long-acuminate at apex, rounded at base, 4–10 cm long, 2–6 cm wide, the midrib raised above; major lateral veins in 3 or 4 pairs. Flowers brownish-purple, borne in racemes to 2 cm long from near apex of short shoot before leaves appear; rachis hirtellous; bracts ± lanceolate, ca 0.5 mm long; pedicels ca 1 mm long; staminate flowers borne individually along rachis (usually less than 12 per inflorescence); tepals usually 5, oblong-obovate, rounded at apex, ca 1.5 mm long, ± spreading; stamens usually 9, to 4.5 mm long; pistillate flowers not seen. Fruits narrowly ovoid, red, to 9 mm long and 6 mm wide (dried), often moderately well developed before leaves appear. *Croat 8339, 14991.*

Apparently rare on the island; collected on Standley Trail 1340 and north of Zetek Trail on the escarpment in the old forest. Flowers in early April before the leaves appear. The fruits usually develop by late April, and the trees have leaves by the time the fruit fully matures (see *Gentry 5163*, Madden Lake).

Midrib raised above, usually glabrous or inconspicuously puberulent (never with conspicuous long trichomes):
  Leaves less than 10 cm long, ovate, narrowly very long-acuminate at apex, rounded at base, borne on a short shoot about 1 cm long; inflorescences ca 1–2 cm long . . . .
    . . . . *C. acapulcensis* Standl.
  Leaves usually more than 10 cm long, not shaped as above; inflorescences much longer:
    Leaves lanceolate to ovate-lanceolate, gradually acuminate, often less than 5 cm wide, with prominent axillary tufts on lateral veins of lower surface; inflorescences spicate; fruiting perianths red or white; plant usually a shrub . . . . *C. acuminata* H.B.K.
    Leaves not as above, often more than 5 cm wide, generally lacking axillary tufts below; inflorescences racemose; fruiting perianths never red or white; plants usually a tree or vine:
      Plant usually a vine; blades often subrounded, broadly rounded at base, usually blunt to rounded at apex, the midrib often with some short puberulence above; lepidote scales on blades reddish-brown . . . . *C. parimensis* Benth.
      Plant a tree; blades acuminate at apex, not subrounded; midribs usually glabrous above; lepidote scales usually pale, easily seen under magnification . . . . *C. coronata* Jacq.

Panamanian material, though identified as *C. acapulcensis* by R. A. Howard, differs radically from specimens elsewhere in Central America. However, the species is quite variable.

Mexico to Panama. In Panama, known from tropical moist forest in the Canal Zone (BCI and near Madden Lake) and from tropical wet forest in adjacent Colón (Río Piña–Río Media divide). The species tends to grow only in regions where there are extensive outcrops of limestone.

See Fig. 219.

Coccoloba acuminata H.B.K., Nov. Gen. & Sp. 2:176. 1817

Slender, functionally monoecious shrub or tree, 2–8 m tall; stems glabrous to puberulent, with sparse round lenticels; ocreae to 1 cm long. Petioles from base of ocrea, 4–15 mm long; blades mostly oblong-lanceolate or oblong-elliptic, acuminate, cuneate to nearly rounded at base, mostly 6–18 (22) cm long, 2–5 (8) cm wide, glabrous above, glabrous to weakly pubescent on midrib and veins below, the vein axils tufted, the midrib arched. Inflorescences terminal, spicate, slender, mostly 15–25 cm long; rachis ridged, swollen at each flower cluster; staminate flowers red, ca 2 mm long, appearing successively from the apex of a very short, stout, bracteated stalk; perianth lobes 5, ovate, imbricate, 6–8 mm long, fleshy, accrescent, reddish turning white at maturity; stamens 8 (9), included, the filaments fused into a tube ca half their length; anthers blocking entrance to nonfunctional ovary. Pistillate flowers usually borne singly at each nodule of rachis; ovary 3-sided; styles 3, short, held somewhat below anthers; nectar stored within staminal tube. Fruits subglobose, 6–8 mm wide, the achene strongly 3-sided, shiny black or brown, enveloped by 3 of the 5 perianth lobes and contrasting sharply with them in color, protruding at maturity. *Croat 15101.*

Common along some areas of the shore. Flowers and fruits throughout the year.

Panama to Ecuador, Peru, and Brazil. In Panama, known from tropical moist forest in the Canal Zone, Panamá, and Darién and from tropical wet forest in Darién.

Coccoloba coronata Jacq., Enum. Syst. Pl. Ins. Carib. 19. 1760

C. obovata H.B.K.

Peña blanca, Papaturra blanca

Apparently dioecious tree, 4–13 m tall, the trunks mostly less than 15 cm diam and clustered, usually producing sucker shoots from near base, the internodes often hollow; branches glabrous, the younger parts with lepidote scales (striate when dry); bark smooth. Petioles ± terete, 1–2.5 cm long, inserted above base of stipule (about midway on stipule); blades ± ovate to obovate, abruptly to gradually acuminate at apex, acute to obtuse or subcordate and sometimes inequilateral at base, 8–24 cm long, 3–9.5 cm wide, glabrous but with both surfaces bearing ± prominent lepidote scales; midrib raised above; juvenile leaves red to wine-colored, to 28 cm long and 14 cm wide or larger. Inflorescences slender, solitary, erect, on short lateral branches; flowers numerous, white, ca 2 mm long, subtended by brown ocreolae; staminate inflorescences to 25 cm long; stamens usually 8, exserted. Pistillate inflorescences 16–20 cm long, the flowers sessile or short-pedicellate; staminodia usually 8, short; styles 3; stigmas spatulate, flattened. Fruits sessile or short-stalked, ovate, ± terete in cross section when fresh (prominently striate when dry), to 12 mm long and 8 mm wide. *Croat 11962.*

Occasional, in the forest. Usually flowering from late June to August, less frequently in September. The fruits mature from September to December.

The hollow stems may be inhabited by small ants.

Guatemala to Colombia and Venezuela; Trinidad, southern Windward Islands. In Panama, known principally from tropical moist forest in the Canal Zone, San Blas, Chiriquí, Veraguas, Panamá, and Darién; known also from tropical dry forest in Herrera (Pesé) and from premontane wet forest in Chiriquí, Coclé, and Panamá.

See Fig. 220.

Coccoloba manzanillensis Beurl., Prim. Fl. Portobello in K. Vetensk. Acad. Handl. 142. 1854 (1856)

C. nematostachya (Griseb.) Lindau

Hueso

Monoecious tree, usually less than 13 m tall; branchlets densely appressed-pubescent; stems at first grayish-

Fig. 218. *Apodanthes caseariae*

Fig. 219. *Coccoloba acapulcensis*

Fig. 220. *Coccoloba coronata*

Fig. 221. *Coccoloba manzanillensis*

Fig. 222. *Coccoloba manzanillensis*

pubescent, soon becoming glabrous, gray, conspicuously lenticellate (striate when dried); ocreae divided to middle or below, to 4.5(9) cm long. Petioles 1–2 cm long (to 4.5 cm on juveniles), inserted at base of ocrea, puberulent (also with longer trichomes near base); blades oblong-elliptic to obovate, short-acuminate (often downturned) to rounded at apex, narrowing below middle, narrowly cordate or rounded at base, mostly 9–28 cm long, 4.5–19 cm wide, young, thin, and often reddish when in flower, mature, thick, and bullate in fruit, ± glabrous above but with the veins densely brown-hirsute (trichomes sometimes deciduous in age), the veins sunken above, conspicuously raised and puberulent to hirsute below. Inflorescences spikelike racemes to 40 cm long, pendent from the ends of the main stem or from short lateral branches near apex; flowers greenish, ovate, ca 1.5 mm long, functionally unisexual; staminate flowers in clusters of 3 or 4; tepals 4 or 5, green; stamens 6–8, long-exserted, spreading; pollen white, moderately tacky; pistillode with 2 or 3 styles. Pistillate flowers densely congested on racemes, borne singly at each nodule. Fruits globose (beaked and striate on drying), ca 6 mm diam, reddish to violet-purple at first, fleshy, sweet, and purple when mature; achenes ovoid, beaked, dark brown, ca 4 mm long, drying trigonous. *Croat 9558, 15071.*

Adult plants common along the shore but uncommon within the forest; juveniles are abundant within the forest, especially in the younger forest. Juveniles are typically unbranched with hollow, densely pubescent stems and leaves that are very large (usually 35–50 cm long and 18–27 cm wide), short-petiolate, elliptic to obovate or oblanceolate, with pilose, deeply divided stipules to 11 cm long. Plants lose their leaves shortly before flowering and generally flower as new leaves are emerging. Flowers from late rainy season into the dry season. The fruits are mature in the rainy season, mostly from June to September.

Known only from Panama, from tropical moist forest in the Canal Zone, Panamá, and Darién and from tropical wet forest in Colón and Panamá.

See Figs. 221 and 222.

Coccoloba parimensis Benth., London J. Bot. 4:626. 1845

  C. leptostachya Standl.

Monoecious liana, extending into top of canopy (juvenile sometimes appearing to be suberect shrub; plant reportedly a tree elsewhere); stems minutely puberulent, becoming glabrous (striate when dry), often ± flattened. Petioles mostly 2–3 cm long, (to 5 cm on juveniles) weakly canaliculate above; stipules ocreate, 2.5–5 cm long, subpersistent; blades broadly ovate to elliptic or obovate, rounded to acuminate at apex, rounded to subcordate and sometimes inequilateral at base, mostly (5)10–25 cm long and (3)5–16 cm wide (juveniles to 33 cm long and 22 cm wide), both surfaces with minute, reddish-brown, lepidote scales, the midrib raised, ± glabrous to minutely puberulent above, all veins conspicuously raised below, usually loop-connected. Inflorescences axillary or terminal on short, lateral, leafy branches, usually solitary; all parts puberulent; flowers greenish, ca 2 mm long, each surrounded by a thin brown ocreola; staminate inflorescences to 16 cm long, the flowers clustered in distinct, somewhat spiral whorls; stamens slightly exserted; pistil rudimentary. Pistillate inflorescences usually less than 10 cm long, the flowers solitary at each nodule; stamens rudimentary; pistil 2 mm long; styles 3. Fruiting pedicels usually 3–4 mm long; fruits ovoid to globose, to 1 cm long, black, with a thin, fleshy mesocarp; seed 1, ovoid, ca 7 mm long, brown. *Croat 4777, 15161.*

Frequent in the forest. Flowers at the beginning of the rainy season in May and June. The fruits usually mature by September and November, sometimes as early as August.

Less common than *C. manzanillensis.*

Known from Panama, Colombia, Peru, and Brazil. In Panama, known only from tropical moist forest in the Canal Zone and Panamá.

POLYGONUM L.

Polygonum acuminatum H.B.K., Nov. Gen. & Sp. 2:178. 1817

Robust herb, to 3 m tall (usually less), often growing in water; moderately to densely strigose on most parts. Leaves sessile or subsessile; blades linear-lanceolate, long-acuminate at apex, abruptly decurrent at base, 10–30 cm long, 1–3.5 cm wide, strigose-ciliate, inconspicuously pellucid-punctate; ocreae 2–3 cm long, apically ciliate with long strigose pubescence 1–1.5 cm long. Inflorescences terminal, spikelike racemes or panicles of few branches; flowers continuous on rachis, fasciculate, whitish, subtended by ciliate, imbricate ocreolae 2–3.5 mm long; pedicels exceeding ocreolae ca 1 mm, conspicu-

---

KEY TO THE SPECIES OF POLYGONUM

Leaves conspicuously strigose all over; ocreae with long apical cilia to 1.5 cm long . . . .
. . . . *P. acuminatum* H.B.K.
Leaves glabrous or pubescent on veins only; ocreae with apical cilia lacking or less than 1 cm long:
  Leaves, ocreae, and tepals conspicuously dark-punctate; midribs of leaves subglabrous . . . .
  . . . . *P. punctatum* S. Elliott
  Leaves, ocreae, and tepals inconspicuously pellucid-punctate; midribs of leaves strigose . . . .
  . . . . *P. hydropiperoides* Michx.

ously articulate at apex; tepals 4 (5), ovate, 2.5–3.5 mm long; stamens usually 6, exserted, ca 3.5 mm long; ovary lenticular; styles 2, 2–3 mm long. Achenes lenticular, 2–2.5 mm long, 1.5–2 mm wide, brown to black, beaked. *White 142.*

Rare, in marshy habitats. Seasonal behavior uncertain. Probably flowers and fruits throughout the year.

Mexico to Argentina; West Indies. In Panama, known only from tropical moist forest in the Canal Zone.

Polygonum hydropiperoides Michx., Fl. Bor. Amer. 1:239. 1803

Slender herb, to ca 1 m tall, often ± reclining, subglabrous. Petioles usually less than 5 mm long; blades linear-lanceolate, long-acuminate at apex, acute and decurrent on petiole at base, 4–15 cm long, 0.5–1.5 cm wide, inconspicuously pellucid-punctate, usually strigose on midrib and veins below, minutely ciliate; ocreae 1–3 cm long, with strigose apical cilia to 8 mm long. Inflorescences terminal, spikelike racemes or panicles of few branches; flowers interrupted along rachis, fasciculate, light purplish to greenish, subtended by ciliate ocreolae 2–3 mm long; pedicels exceeding ocreolae ca 1 mm, articulate at apex; tepals usually 5, ovate, 2–3 mm long; stamens usually 9, ca 1.5 mm long; ovary trigonous; styles 3, ca 1 mm long. Achenes trigonous, 2–3 mm long, brown to black, inconspicuously beaked. *Shattuck 840.*

Collected once by Shattuck at Gigante Bay. Apparently flowering and fruiting throughout the year.

Canada to South America. In Panama, known only from BCI.

Polygonum punctatum S. Elliott, Bot. S. Carolina & Georgia 1:455. 1817

Chilillo, Chili de perro

Slender herb, to 70 cm tall, nearly glabrous. Leaves, ocreae, and tepals conspicuously dark-punctate; petioles less than 5 mm long; blades linear-lanceolate, acuminate at apex, narrowly decurrent on petiole at base, 2–10 cm long, 0.5–2 cm wide, glabrous but minutely ciliate and sometimes with the midrib near base sparsely strigose; ocreae 5–15 mm long, with strigose cilia at apex usually 7–10 mm long. Inflorescences terminal, spikelike racemes or panicles of few branches; flowers interrupted along rachis, fasciculate, white or greenish, subtended by ciliate ocreolae 2–3 mm long; pedicels exceeding ocreolae 1–2 mm, articulate at apex; tepals 5, ovate, 3–4 mm long; stamens usually 8, 1.5–2 mm long; ovary trigonous; styles 3, ca 1 mm long. Achenes trigonous, 2.5–4 mm long, brown to black, somewhat beaked. *Croat 5247.*

Rare, in marshy places, especially sandbars. Flowers throughout the year, possibly with a peak in the late rainy and early dry seasons.

Canada to Argentina; West Indies. In Panama, known from tropical moist forest in the Canal Zone and Panamá, from tropical dry forest in Coclé, and from premontane wet forest in Chiriquí and Coclé.

TRIPLARIS Loefl.

Triplaris cumingiana Fisch. & C. Meyer, Mém. Acad. Imp. Sci. Saint-Pétersbourg, Sér. 6, 6:149. 1840

Palo santo, Guayabo hormiguero, Vara santa

Dioecious tree, usually 10–20 m tall; trunk smooth, 12–30 cm dbh; bark light brown, thin, peeling off; stems hollow. Petioles very short or to 2 cm long, canaliculate; blades mostly oblong-elliptic, acuminate, obtuse at base, 15–30 cm long, 4–12 cm wide, glabrous but with the veins below strigose (especially midrib). Inflorescences from upper axils; staminate inflorescences of spikes to 35 cm long and 1.5 cm wide; flowers subsessile, usually in pairs, emerging one at a time from densely pubescent, spathaceous ocreolae; perianth greenish, in one series of 3 linear and 3 narrowly triangular tepals 3–4 mm long, connate for about half their length; stamens 9, exserted; anthers introrse, versatile. Pistillate inflorescences of racemes to ca 20 cm long (to 30 cm in fruit); pedicels 2–9 mm long; calyx sericeous, ca 1 cm long at anthesis (greatly accrescent and becoming red in age), the lobes narrowly triangular to linear, 2–3 times the length of the tube and spreading at anthesis; petals ± linear, exceeding tube, fused to base of tube; styles 3, the inner surface stigmatic in upper two-thirds. Achenes sharply trigonous (the surfaces ± flat), 8–12 mm long, shiny, brown, persistent within and dispersed by the enlarged calyx, the calyx to 6 cm long, pubescent, 3-winged, the wings pinkish, spreading, 3.5–4.5 cm long, 6–7 mm wide. *Croat 4633, 8165.*

Usually locally common; otherwise only occasional in both the young and old forests. Plants may begin to flower when as little as 11 m tall and 12 cm dbh. Flowers from February to April, chiefly in March. The fruits begin to mature by February and are dispersed chiefly in March and April but also in May. Plants lose their leaves in July and August.

Stems are inhabited by very aggressive ants (*Pseudomyrmex triplaridis* Forel.), whose sting is quite severe.

Costa Rica to Ecuador; cultivated in the West Indies and elsewhere. In Panama, known from tropical moist forest in the Canal Zone, San Blas (Permé), Los Santos, Panamá, and Darién and from tropical wet forest in Panamá (Cerro Campana).

See Fig. 223.

47. AMARANTHACEAE

Erect or decumbent herbs, vines, or clambering shrubs. Leaves alternate or opposite, petiolate, simple; blades entire to somewhat irregular; venation pinnate; stipules lacking. Flowers bisexual or unisexual, apetalous, in bracteate panicles, spikes, or axillary glomerules; sepals 5 (or by reduction 3), free, overlapping, scarious; stamens 5 (or by reduction 3), alternate with petals, united below into a tube, filamentous; staminodia present or absent; anthers 2- or 4-celled, introrse, with dorsal-median at-

KEY TO THE SPECIES OF AMARANTHACEAE

Leaves alternate:
    Sepals 5; flowers bisexual; leaf blades usually more than 7 cm long; plants occasional . . . .
    . . . . *Chamissoa altissima* (Jacq.) H.B.K.
    Sepals 3; flowers monoecious or polygamomonoecious; leaf blades less than 6 cm long; plants
    rare or no longer present on the island . . . . *Amaranthus viridis* L.
Leaves opposite:
    Inflorescences ± globose, not more than 2 cm long:
        Mature leaves pilose; stigmas 2 . . . . *Gomphrena decumbens* Jacq.
        Mature leaves at most sparsely villous; stigma 1:
            Sepals longer than utricle, 3–5-veined . . . . *Alternanthera ficoidea* (L.) R. Br.
            Sepals shorter than utricle, 1-veined . . . . *Alternanthera sessilis* (L.) R. Br.
    Inflorescences not globose, usually more than 2 cm long:
        Inflorescence a simple spike, the flowers in glomerules, becoming reflexed and hooked . . . . . .
        . . . . *Cyathula prostrata* (L.) Blume
        Inflorescence a panicle of spikes, the flowers not reflexed and not associated with hooks:
            Leaf blades near base of plant ovate . . . . *Iresine celosia* L.
            Leaf blades near base of plant lanceolate to linear . . . . *Iresine angustifolia* Euphr.

tachment, dehiscing by longitudinal slits; ovary superior, unilocular; ovule solitary (in ours), basal, campylotropous; style single, conspicuously trifid; stigma capitate or long-bifurcate or trifurcate. Fruits indehiscent or circumscissily dehiscent utricles (1-seeded, with a loose pericarp); seed usually with a shiny testa and abundant endosperm.

Distinguished by their small, densely bracteate, usually greenish inflorescences with inconspicuous apetalous flowers and small shiny seeds.

The flowers are probably wind pollinated or self-pollinated (H. Baker, pers. comm.).

Seeds are dispersed chiefly by small birds. Ridley (1930) reported that seeds of some genera, including *Gomphrena*, are possibly wind dispersed by means of the plumelike pubescence of the achene or the glumaceous flower and broad persistent bracts. Some seeds are eaten by browsing animals and passed unharmed (Ridley, 1930), and some are harvested by ants (Wheeler, 1910).

About 50–65 genera with 500–850 species; mostly in the tropics of Africa and America.

ALTERNANTHERA Forssk.

Alternanthera ficoidea (L.) R. Br., Prodr. 1:417. 1810

Decumbent, sprawling perennial herb; stems branching, to ca 1 m long; stems velutinous above and on nodes below, the trichomes antrorsely hispidulous. Leaves opposite; petioles 2–10 mm long; blades broadly ovate to elliptic or obovate, acute and mucronulate at apex, cuneate at base, to 6 cm long and 3 cm wide, villous when young, becoming sparsely villous to glabrate. Inflorescences of sessile, whitish, axillary, ovoid or globose tufts to 1 cm long; bracts and bracteoles ± equal, ± ovate, to 3 mm long, acuminate; flowers bisexual; sepals 5, similar to bracts but the outer 3 broader, 3-veined, 3–5 mm long; stamens 5, united below, exceeded by pseudostaminodia;

style 2–3 times longer than the single, capitate stigma. Utricles indehiscent, suborbicular, membranaceous, to 1.5 mm long, shorter than sepals; seeds reddish-brown, to 1.2 mm long. *Croat 9240.*

Abundant in the Laboratory Clearing, usually growing over and supported by other low vegetation. Flowers and fruits all year though probably with a peak of activity in the dry season.

BCI plants apparently intergrade with *A. halimifolia* (Lam.) Standl., as they show characters intermediate with that species, especially the moderate to dense, plumose pubescence.

Florida and central Mexico south to Paraguay; West Indies. In Panama, an occasional weed growing in clearings and along streams; known from tropical moist forest in the Canal Zone, Bocas del Toro, Chiriquí, Veraguas, Panamá, and Darién, from premontane moist forest in the Canal Zone, from premontane wet forest in Coclé, and from tropical wet forest in Colón.

See Fig. 224.

Alternanthera sessilis (L.) R. Br., Prodr. 1:417. 1810
Sanguinaria

Similar to *A. ficoidea* except the petioles 1–5 mm long; the blades to 4 cm long and 1.5 cm wide; the inflorescences of axillary tufts to 4 mm long; the bracts and bracteoles 1 mm long or less; the sepals 5, 1-veined, to 1.5 mm long; the utricles 1.5–2 mm long, slightly exceeding the sepals. *Standley 40948.*

Perhaps no longer present on the island, but reported by Standley to have been frequent. Probably flowers all year, but especially during the rainy season.

Honduras to Brazil; West Indies. In Panama, the few collections are widely scattered and the plant occasionally extends to aquatic habitats; known from tropical moist forest in the Canal Zone, Bocas del Toro, Los Santos, Panamá, and Darién.

Fig. 223.
Triplaris cumingiana

Fig. 224.
Alternanthera ficoidea

Fig. 225. *Chamissoa altissima*

AMARANTHUS L.

Amaranthus viridis L., Sp. Pl. ed. 2, 1405. 1763
A. gracilis Desf.
Bledo, Calalú

Polygamomonoecious or monoecious annual herb, to ca 1 m tall, glabrous; stems branched. Leaves alternate; petioles 0.5–4 cm long; blades ± deltoid, emarginate to rounded and mucronate at apex, truncate to subacute at base, 1.5–6 cm long, 1–4 cm wide. Flowers in closely congested, ± cylindrical thyrses ca 5 mm wide, arranged in racemes; bracts and bracteoles ca 1 mm long; sepals 3, rounded and mucronate at apex, 1–1.5 mm long; stamens 3, equaling sepals; ovary compressed-globose; style 1, minute; stigmas 3, longer than style. Utricles indehiscent, strongly rugose; seed 1, cochleate-orbicular, ca 1 mm broad, dull, dark. *Kenoyer 348.*

Collected once and not seen recently, but it could be expected in clearings. Flowers and fruits most of the year, especially in the rainy season.

Presumably native to the Old World tropics, but introduced into the New World tropics and subtropics. In Panama, known only from tropical moist forest in the Canal Zone.

CHAMISSOA Kunth

Chamissoa altissima (Jacq.) H.B.K., Nov. Gen. & Sp. 2:197. 1817

Suffruticose vine or clambering shrub, glabrous but with the fertile parts stellate-villous. Leaves alternate; petioles 0.5–3.5 cm long; blades lanceolate to ovate, broadly acuminate and mucronate at apex, ± obtuse at base, 7–15 cm long, 3–6 cm wide, villous when young, becoming glabrous. Inflorescences of glomerules arranged in axillary spikes to 10 cm long or in terminal or upper-axillary, often leafy panicles to 20 cm long; flowers bisexual, subtended by membranaceous-margined bracts to 1.5 mm long; sepals 5, subequal, ovate, acute, 5-veined, the central vein more conspicuous; stamens 5; filaments to 3 mm long, the lower third united; ovary ovate, crowned by a narrow circular ring; style to 1 mm long; stigmas 2, less than half the length of style. Utricles ovate, thin-walled, 4–5 mm long, exserted from bracts at maturity, weakly crowned at apex, circumscissily dehiscent; seed rounded and flattened, ca 2 mm diam, shiny, round, enveloped in an aril, the aril foamy, glistening, transparent, 2-valved, extruding from lower valve and becoming somewhat enlarged. *Croat 7082.*

Occasional in clearings. Flowers principally throughout the dry season, from December to April, but sometimes as late as June and elsewhere in Panama as early as September.

Central Mexico to northern Argentina; West Indies. In Panama, known from tropical moist forest in the Canal Zone, Bocas del Toro, Colón, San Blas, Los Santos, and Darién, from tropical dry forest in Los Santos, Coclé, and Panamá, from premontane wet forest in Coclé, and from tropical wet forest on the Atlantic slope in Coclé.

See Fig. 225.

CYATHULA Blume

Cyathula prostrata (L.) Blume, Bijdr. Ned. Ind. 549. 1826
Cadillo

Erect or decumbent herb, to 1 m tall, usually rooting at lower nodes, sparsely puberulent to strigose on most parts. Leaves opposite; petioles to 2 cm long; blades ovate to obovate, often rhombate, acuminate, acute at base, 3–10 cm long, 1.5–5.5 cm wide. Flowers bisexual, in glomerules (often in 3s) to 5 mm long in axillary or terminal, deflexed spikes to 20 cm long, the spikes interrupted along more than half their length when mature, the glomerules at maturity with ca 20 flowers modified into hooks to 1.5 mm long; sepals 5, subequal, lanceolate, ca 2 mm long, pubescent, the veins to 3; stamens 5, united basally into a short tube projected into 5 pseudostaminodia; pseudostaminodia alternate with stamens, regularly bifurcate when young, trifurcate when mature; style 1; stigma capitate, slightly longer than stamens. Utricles ellipsoid, conspicuously operculate, irregularly circumscissily dehiscent, the cap chartaceous, the remainder hyaline; seed ovate, ca 1 mm long, tan to reddish-brown. *Croat 4072.*

Very abundant on slopes in the Laboratory Clearing; rare on forest trails. Flowers and fruits throughout the year, probably peaking in the dry season.

Widely scattered in Central America and northern South America; West Indies; common and widespread throughout the Old World tropics, where it is probably native. In Panama, known from tropical moist forest in the Canal Zone, Bocas del Toro, and Darién, from premontane wet forest in Coclé, and from tropical wet forest in Colón.

GOMPHRENA L.

Gomphrena decumbens Jacq., Hort. Schoenbr. 4:41. 1804
G. dispersa Standl.

Decumbent herb, to 50 cm long, rooting at nodes and often forming dense mats, pilose on most parts. Leaves opposite; petioles obsolete or to 5 mm long; blades oblong-obovate to oblong-ovate, acute and mucronate at apex, rounded to attenuate at base, 2–5(10) cm long, 0.5–1.5(5) cm wide. Inflorescences of terminal or axillary, ± globose, pedunculate heads ca 1–1.5 cm diam; peduncles very short; flowers bisexual, subtended by whitish bracts, the largest bract to 6 mm long; sepals 5, to 5 mm long, mostly woolly-pubescent from base, becoming indurate; stamens 5, sessile, alternate with bilobed pseudostaminodia; stigmas 2; style persistent. Utricles to 2 mm long, indehiscent; seed reddish-brown, ca 1.5 cm diam. *Shattuck 633.*

Collected once on the island; rare or no longer present but to be expected. Flowers essentially all year elsewhere in Panama.

Florida and Texas to northern Argentina; West Indies. In Panama, known from tropical moist forest in the Canal Zone, Bocas del Toro, San Blas, Chiriquí, Veraguas,

Panamá, and Darién and from premontane moist forest in the Canal Zone.

IRESINE P. Browne

Iresine angustifolia Euphr., Beskr. St. Barth. 165. 1795

Erect or reclining suffruticose herb, 1 m tall; stems with several light ridges; young parts strigilose. Leaves opposite; petioles 0.5–1.5 cm long; blades lanceolate to linear-lanceolate in lower part of plant, linear in upper flowering part, acuminate and mucronate at apex, acute to attenuate at base, 5–9 cm long, 1.5–3 cm wide. Inflorescences spicate, ovoid to linear, usually ca 1 cm long and 5 mm wide, pedunculate, alternate, subtended by linear leaves, in panicles merging with vegetative part of plant, the reproductive part to 50 cm long or more; flowers bisexual, subtended by transparent, mucronate bracts to 1 mm long; sepals 5, acute, 1–1.5 mm long, pilose and subtended by dense woolly pubescence; stamens 5; stigmas 2, not much longer than style. Utricles fragile, indehiscent, capped, falling with sepals; seed lens-shaped, ca 0.7 mm diam, reddish-brown to dark brown. *Croat 8710.*

Infrequent, in the Laboratory Clearing. Flowers and fruits mostly in the dry season (February to May).

Baja California and Mexico south to Brazil; West Indies. In Panama, ecologically variable; known from tropical moist forest in the Canal Zone and Darién, from premontane dry forest and tropical dry forest in Coclé, from premontane moist forest in the Canal Zone, from premontane wet forest in Colón and Panamá, and from tropical wet forest in Colón.

Iresine celosia L., Syst. Nat., ed. 10, 1291. 1759

Dioecious herb, clambering or erect, to 3 m long, glabrate but with the young parts sometimes densely puberulent, particularly lower leaf surface. Leaves opposite; petioles 0.5–3(6) cm long; blades ovate, acuminate and mucronate at apex, rounded to cuneate at base, 1.5–9(15) cm long, 0.5–4(7) cm wide. Flowers in panicles of spikes, the spikes ± sessile, filiform when young, becoming linear, to 1 cm long and 3 mm wide; bracts and sepals membranaceous, pinkish-green when fresh; sepals 5, acute at apex, 0.5–1 mm long, pilose and subtended by dense woolly tufts in the pistillate flowers; stamens 5; stigmas 2, much longer than style. Utricles fragile, indehiscent, capped, falling with sepals; seed lens-shaped, ca 5 mm diam, varying from reddish-brown to dark brown. *Shattuck 738.*

Not seen on the island in recent years but to be expected in clearings. Flowers and fruits principally from January to May, sometimes as early as December or as late as July.

Florida, Texas, and Mexico south to northern Argentina; West Indies. In Panama, ecologically variable; known principally from tropical moist forest in the Canal Zone, Bocas del Toro, Veraguas, Los Santos, Coclé, Panamá, and Darién, also from tropical dry forest in Herrera and Panamá, from premontane wet forest in Chiriquí, Coclé, and Panamá, and from premontane rain, lower montane wet, and lower montane rain forests in Chiriquí.

48. NYCTAGINACEAE

Trees, shrubs, or lianas (if lianas, with axillary spines); young branchlets often ferruginous-pubescent. Leaves opposite or nearly so, petiolate; blades simple, entire; venation pinnate; stipules lacking. Flowers unisexual (dioecious), apetalous, in terminal, bracteate, corymbose clusters; calyx 5-lobed, petaloid; stamens 6–10; filaments united near the base; anthers 2-celled, dorsifixed, dehiscing longitudinally; ovary superior, 1-locular, 1-carpellate; placentation basal; ovule solitary, anatropous or campylotropous; style 1, slender; stigma divided and appearing brushlike. Fruits achenes, enclosed in the persistent, modified calyx (i.e., the anthocarp); seed 1, with endosperm.

Members of the family are distinguished by the simple, often subopposite leaves, which usually dry darkened, by the distinctive anthocarpic fruit, and by the unisexual apetalous flowers, which usually have an undulate perianth limb and a brushlike or much-divided style.

Pollination systems are unknown.

Fruits of *Neea* and *Guapira* are endozoochorous. Those of *Neea* are chiefly bird dispersed but are also taken by monkeys. Oppenheimer (1968) reported that white-faced monkeys usually remove the pulp and discard the seed. Fruits of *Guapira* are probably dispersed in part by birds, but white-faced monkeys eat them as well. *Pisonia aculeata* has sticky fruits that appear suited for

---

KEY TO THE SPECIES OF NYCTAGINACEAE

Plants lianas with stout axillary spines . . . . *Pisonia aculeata* L.
Plants shrubs or trees, unarmed:
    Shrubs usually less than 2 m tall (to 7 m); staminate flowers tubular-urceolate, the stamens included; pistillate flowers with ovary sessile, the style included; anthocarps reddish to violet-purple at maturity, usually less than 10 mm long, the surface drying smooth . . . .
    . . . . *Neea amplifolia* Donn. Sm.
    Trees to 20 m or more tall; staminate flowers campanulate, the stamens ca half-exserted; pistillate flowers with ovary stipitate, the style exserted; anthocarps dark blue-violet at maturity, mostly more than 10 mm long, drying with prominent longitudinal grooves . . . .
    . . . . *Guapira standleyanum* Woods.

epizoochory by becoming attached to the fur of animals or to the feathers or beaks of birds.

About 30 genera with 300 species; mostly in the American subtropics and tropics.

GUAPIRA Aubl.

Guapira standleyanum Woods., Ann. Missouri Bot. Gard. 48:404. 1961

Dioecious tree, to 20(35) m tall and 70 cm dbh, sometimes buttressed, often ribbed near base; outer bark very thin, flaking in minute pieces; inner bark tan; sap moderately strong and somewhat foul-smelling; glabrous but with the branchlets and inflorescences densely ferruginous-tomentose and the petioles and midribs puberulent. Leaves opposite, crowded; petioles 1–3 cm long, variable on same stem; blades variable, ovate to elliptic to obovate, acute to blunt or long-acuminate at apex with tip downturned, acute to rounded at base, 5.5–20 cm long, 2.5–9 cm wide, often inequilateral, broadly undulate, drying gray-brown. Inflorescences terminal, corymbose-thyrsiform, 4–13 cm long, all exposed parts minutely ferruginous-pubescent; flowering peduncles to 8 cm long; flowers unisexual, greenish-yellow, 3–7 mm long, densely tomentose; perianth campanulate, flared abruptly above base, the rim undulate, not flared; staminate flowers campanulate; stamens 7–10, about half-exserted; filaments united near base. Pistillate flowers with the perianth accrescent and persistent in fruit; ovary stipitate; style exserted, whitish at anthesis; stigma with many divisions. Anthocarps oblong-ellipsoid, 12–16 mm long, minutely ferruginous-pubescent, becoming dark blue-violet at maturity, longitudinally striate when dry; seed 1, whitish, ribbed, oblong-ellipsoid, to 13 mm long; mesocarp fleshy, purple, to 2 mm thick. *Croat 5555, 5704.*

Frequent in the forest. Flowers from February to June. The fruits mature mostly from May to July.

More field work is necessary on this species before the name can be certain. *Guapira standleyanum* does not differ appreciably from the type of *G. costaricana* Standl. from the Nicoya Peninsula of Guanacaste Province in Costa Rica. Nor does it differ greatly from *G. itzana* Lund. from Guatemala or from the following species of *Torrubia*, which are also to be considered *Guapira: T. uberrima* Standl. (Colombia), *T. rusbyana* (Heim.) Standl. (Venezuela), and *T. myrtiflora* Standl. (Peru). Probably the species is much more widespread, but currently it should be considered to be endemic to Panama.

Differing considerably from *Guapira standleyanum*, however, are plants going by the name *Guapira costaricana* Standl. from premontane wet forest at higher elevations such as El Valle in Coclé and Loma Prieta in Los Santos (e.g., *Lewis et al. 2199*). This is possibly a new species.

Known only from Panama, where it is ecologically variable; known from tropical dry forest in Panamá, from tropical moist forest in the Canal Zone and Darién, and from probably premontane wet forest in Colón (Santa Rita Ridge).

See Fig. 226.

NEEA R. & P.

Neea amplifolia Donn. Sm., Bot. Gaz. (Crawfordsville) 61:386. 1916
  N. pittieri Standl.

Dioecious shrub, usually less than 2 m tall (to 7 m); younger branches densely ferruginous-pubescent, becoming glabrate in age. Leaves opposite or nearly so; petioles 0.5–5 cm long, canaliculate, often somewhat reddish; blades ± elliptic, ovate to obovate-elliptic, abruptly to gradually long-acuminate, obtuse to attenuate at base, 7–36 cm long, 3–15 cm wide, entire, glabrous above, glabrate to puberulent below. Inflorescences terminal, obscurely dichasial thyrses, essentially glabrous to densely ferruginous-pubescent, 4–15 cm long, bearing few to many flowers; staminate flowers narrowly tubular-urceolate, 5–10 mm long, about 2.5 mm wide, the perianth with 5 short lobes, the limb weakly spreading; stamens 8, to 5.5 mm long, included, attached near base of tube; filaments of different lengths, united into a short tube around the sessile pistillode. Pistillate flowers similar to staminate flowers but with a prominent constriction about one-third of the way down the perianth tube, thickened within at point of constriction; style and staminodia (usually 9) fitting tightly through the constriction, included, later exposed when the upper third of perianth above the constriction withers and falls; staminodia held just above constriction; ovary narrowly elliptic, sessile, at first loosely enveloped by perianth, by maturity completely filling it; style slender; stigma with few divisions. Anthocarp elliptic-oblong, ca 1 cm long, at first reddish, becoming violet-purple at maturity, the persistent perianth fleshy and sweet; seed solitary, somewhat shorter than fruit. *Croat 4213, 5626.*

Frequent in the forest. Flowers throughout the year, most commonly from March to September. The fruits are most common from June to December.

Standley (1933) also reported *N. psychotrioides* Donn. Sm., which is a Costa Rican species earlier confused with *N. laetevirens* Standl. of the Atlantic slope of Panama. *N. laetevirens* probably does not occur on BCI. Most of the material from BCI assigned the name *N. psychotrioides* is *N. amplifolia*. A single sterile collection, *Shattuck 121*, is in doubt; it does not appear to be typical of *N. amplifolia*. However, it can be stated with confidence that a second species of *Neea* does not now occur on the island. Leaves of the species are consistently host to various cryptogamic epiphytes, including mosses and lichens.

Costa Rica and Panama. In Panama, known principally from tropical moist forest in the Canal Zone, all along the Atlantic slope, and in Darién; known also from premontane wet forest in Colón, Chiriquí, and Coclé and from tropical wet forest in Colón.

See Fig. 227.

PISONIA L.

Pisonia aculeata L., Sp. Pl. 1026. 1753

Dioecious, climbing shrub or liana, growing into canopy, its climbing aided by stout, recurved, axillary spines on

Fig. 226. *Guapira standleyanum*

Fig. 228. *Pisonia aculeata*

Fig. 227. *Neea amplifolia*

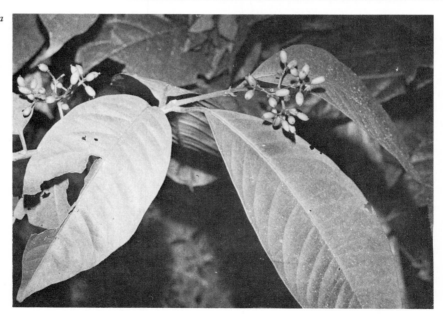

smaller stems, nearly glabrous to sparsely villous, especially on younger parts, underside of leaves, and axes of inflorescences; trunk to 12 cm diam, unarmed; branching divaricate. Leaves opposite to subopposite, the pairs often markedly unequal; petioles 0.5–3 cm long; blades variable, mostly obovate to obovate-elliptic, acute to short-acuminate at apex, rounded to acute at base, 2–10 cm long, 1.5–5 cm wide, usually glabrate above except midrib. Inflorescences usually terminal on condensed short shoots; flowers greenish-yellow, numerous, in dense, ± globular, short-pedunculate clusters to 4 cm diam; pedicels short, bearing few bracts; pedicels and perianth densely and coarsely short-pubescent; staminate flowers campanulate, ca 3 mm long, the limb 5–7 mm diam, the lobes 5, acute; stamens 6–8 (often 7), unequal, widely exserted, to 8 mm long; anthers as broad as or broader than long, longitudinally dehiscent, dorsifixed at base; pistillode with a slender style and brushlike stigma, usually held to one side and above rim. Pistillate flowers tubular, ca 2.5 mm long; style short-exserted; stigma brushlike; staminodia reduced. Fruiting inflorescences usually much expanded; anthocarps club-shaped and 5-sided, to 1.5 cm long, densely short-pubescent, bearing a longitudinal row of prominent stalked glands on the angles, the glands sticky. *Croat 5390, 8313.*

Occasional, in the canopy of the forest, sometimes hanging down over the edge of the lake. Plants may lose their leaves just before flowering. Flowering usually occurs from January to April, with the fruits maturing as early as February.

The sticky, stipitate glands on the fruit presumably function in distribution by adhering to the feathers of birds.

Widely distributed in the American tropics. In Panama, known from tropical moist forest in the Canal Zone and Panamá, from premontane moist forest in Panamá, and from premontane wet forest in Coclé.

See Fig. 228.

49. PHYTOLACCACEAE

Herbs or weak shrubs, sometimes with a garlic odor. Leaves alternate, petiolate; blades simple, entire; venation pinnate; stipules minute or lacking. Flowers bisexual, actinomorphic, in terminal or extra-axillary racemes, bracteate; tepals 4 or 5, briefly united; stamens 4 to many, free, inserted on the hypogynous disk; anthers 2-celled, dorsifixed, dehiscing longitudinally; ovary superior, 1-carpellate (*Rivina* and *Petiveria*), 2-carpellate (*Microtea*), or many-carpellate (*Phytolacca*); locules the same number as carpels; ovules 1 per locule, on axile placentas; styles the same number as carpels. Fruits many-celled berries (*Phytolacca*), drupes (*Rivina*, *Microtea*), or uncinate achenes (*Petriveria*); seeds with much endosperm.

The species of Phytolaccaceae on BCI are individually distinct, but there is relatively little that typifies them as a family.

Pollination systems are unknown.

The fruits are chiefly ornithochorous in *Phytolacca rivinoides* and *Rivina humilis* and epizoochorous in *Microtea debilis* and *Petiveria alliacea*. In *Petiveria* fruits are attached by the hooked apices of the achene. In *Microtea debilis* the fruits are attached to animals by the numerous viscid protuberances on the exocarp. Seeds of *Rivina humilis* have been found in the gut of ducks (Ridley, 1930).

Seventeen genera with over 120 species; widespread but mostly in the American subtropics and tropics.

MICROTEA Sw.

Microtea debilis Sw., Prodr. Veg. Ind. Occ. 53. 1788

Decumbent herb, 15–50 cm long; stems sharply angled. Petioles 5–20 mm long, winged nearly to base; blades variable in shape, commonly ovate, elliptic, or rhomboid, acute at apex, long-attenuate at base, 1–4 cm long, 1–2.8 cm wide, glabrous. Racemes terminal or upper-axillary, 1.5–4 cm long; bracts lanceolate, ca 1 mm long, very thin, persistent; pedicels ca 1 mm long; tepals 5, lanceolate, less than 1 mm long, white; stamens 5, ca 0.5 mm long; ovary globose, ca 0.5 mm diam; stigmas 2. Drupes ± globose, 1–1.5 mm diam, with ± viscid tubercles often united into a honeycomb pattern; seed black, closely filling the capsule.

Reported by Standley (1933) for BCI, but no specimens have been seen. The species occurs in weedy areas of the Canal Zone and may appear from time to time on the island. Flowers and fruits principally in the dry season.

Guatemala to Peru and Brazil; West Indies. In Panama, known from tropical moist forest in the Canal Zone,

---

KEY TO THE SPECIES OF PHYTOLACCACEAE

Racemes less than 15 cm long:
  Tepals 4, more than 2 mm long; fruits red drupes more than 4 mm diam; leaves usually more
    than 4 cm long . . . . *Rivina humilis* L.
  Tepals 5, less than 1 mm long; fruits achenes less than 2 mm diam; leaves less than 4 cm long
    . . . . *Microtea debilis* Sw.
Racemes more than 15 cm long:
  Flowers subsessile; tepals 4; fruits linear achenes less than 1 cm long, bearing 4 uncinate green
    appendages at apex . . . . *Petiveria alliacea* L.
  Flowers with pedicels more than 5 mm long; tepals 5; fruits globose berries 2–6 mm diam, black
    and juicy . . . . *Phytolacca rivinoides* Kunth & Bouché

Bocas del Toro, Panamá, and Darién and from tropical dry forest in Panamá (Taboga Island).

PETIVERIA L.

Petiveria alliacea L., Sp. Pl. 342. 1753
Garlic weed, Anamú

Herb, to ca 1 m tall, woody at base in age; cut parts with a foul aroma; inconspicuously appressed-pubescent on stems, axes of inflorescences, and leaves, especially on veins. Petioles 5–20 cm long; blades narrowly elliptic to obovate, acuminate or acute at apex, acute at base, 5–16 cm long, 2–6 cm wide. Flowers white, in sparsely flowered, terminal or upper-axillary racemes 15–40 cm long; pedicels ca 1 mm long; tepals 4, white or greenish-white (sometimes pinkish), spreading, united at base, 3–5 mm long, erect, green and persistent in fruit; stamens 8, fully exposed at anthesis, to ca 3 mm long; filaments unequal; anthers linear, dorsifixed; ovary densely pubescent, bearing 4 uncinate green appendages at apex, the hooks becoming enlarged and prominent in fruit (probably aiding in epizoochorous dispersal); stigma finely dissected, subapical. Achenes linear, bilobed at apex, to 8 mm long, each lobe bearing hooks to 4 mm long. *Croat 17051.*

Uncommon, in the forest beyond the Tower Clearing. Flowers and fruits throughout the year, principally in the rainy season.

Southern United States to Argentina; West Indies. In Panama, known from tropical moist forest in the Canal Zone, Bocas del Toro, Chiriquí, Panamá, and Darién, from tropical dry forest in Los Santos, Herrera, and Panamá, from premontane moist forest in the Canal Zone, and from premontane wet forest in Chiriquí and Coclé.

PHYTOLACCA L.

Phytolacca rivinoides Kunth & Bouché, Ind. Sem. Hort. Berol. 15. 1848
Pokeberry, Jaboncillo, Calalú

Robust, glabrous herb, often woody at base, usually 1–2 m tall, sometimes treelike and to as much as 4 m tall; branches angulate. Petioles 1–4.5 cm long; blades elliptic, oblong-elliptic, or lanceolate, acuminate and often mucronate at apex, decurrent at base, 9–17 cm long, 4.5–7 cm wide. Racemes terminal or upper-axillary, 15–50(70) cm long; rachis and pedicels often reddish; pedicels 7–13 mm long; basal bracts 1.5–2 mm long; bracteoles 2, ca 0.5 mm long; tepals 5, ovate, ca 2 mm long and 1.5 mm wide, white to reddish; stamens 9–17(22), 2–3 mm long, in 2 whorls on a hypogynous disk; ovary globose, 10–16-carpellate, united; styles connivent at base, free above. Berries ± globose, 2–6 mm diam, purple-black and juicy at maturity, ribbed when dry. *Hladik 110, Shattuck 1150.*

The species was once common on the island, judging from collections made, but now is apparently rare along the shore in clearings. Probably flowers and fruits throughout the year.

Mexico to Bolivia; West Indies. In Panama, known from tropical moist forest in the Canal Zone, Bocas del Toro, San Blas, Los Santos, Panamá, and Darién, from premontane wet forest in Colón, Chiriquí, and Panamá, and from premontane rain forest in Coclé and Darién.

RIVINA L.

Rivina humilis L., Sp. Pl. 121. 1753

Herb, sometimes woody at base, usually to ca 1 m tall; stems ribbed. Petioles 1–6 cm long; blades mostly ovate, acuminate, rounded to truncate at base, 2.5–12 cm long, 1.5–6 cm wide, glabrous to finely pubescent. Racemes terminal or axillary, 6–15 cm long; pedicels to 3 mm in flower, to 8 mm long in fruit; tepals 4, spatulate, 2–3.5 mm long, greenish in bud, white or pinkish at anthesis; stamens 4, alternate with tepals and inserted at base of perianth, 1–2 mm long; anthers held at level of stigma; ovary ellipsoid to obovoid, 1-carpellate; style ca 0.5 mm long; stigma capitate. Drupes red, broadly ovoid, 4–6 mm long, 3.5–5 mm broad; mesocarp juicy; seed ca 2.5 mm wide, round, weakly flattened, the pubescence dense, stiff.

Reported by Standley (1933), but no specimens have been seen. A weedy species in the Canal Zone, it can be expected on BCI. Flowers and fruits throughout the year, principally in the dry season.

Southern United States to Argentina; West Indies (Cayman Islands); Africa, Asia, and Australia. In Panama, known from tropical moist forest in the Canal Zone, Bocas del Toro, Colón, San Blas, Panamá, and Darién, from tropical dry and premontane wet forests in Coclé, and from tropical wet forest in Chiriquí.

50. PORTULACACEAE

Succulent, glabrous, prostrate herbs. Leaves alternate, petiolate, fleshy, succulent; blades simple, entire; venation obscure; stipules minute. Flowers bisexual, actinomorphic, crowded in the axils; sepals 2, briefly united, subequal; petals 4, briefly united, yellow; stamens 6–15, inserted at base of petals; anthers 2-celled, longitudinally dehiscent; ovary half-inferior, 1-locular, 2- or 3-carpellate; placentation basal; ovules many, campylotropous; styles and stigmas 4–6. Fruits circumscissile capsules; seeds many, with copious endosperm.

Members of the family can usually be distinguished by having fleshy leaves and flowers with two sepals. *Portulaca oleracea*, the only species on BCI, is native to northern Africa and distributed throughout the world. For centuries it was a medicinal plant and pot herb. (Ridley, 1930).

Pollination system is unknown.

Seeds are apparently dispersed by the wind or merely spilled locally. Elsewhere, ants of the genus *Pheidole* harvest seeds (Wheeler, 1910).

Nineteen genera with more than 500 species; widely distributed in the subtropics and tropics.

PORTULACA L.

Portulaca oleracea L., Sp. Pl. 445. 1753

Pusley, Verdolaga

Glabrous herb, usually less than 20 cm long, prostrate and radially spreading. Petioles 1–8 mm long; stipules minute, fimbriate; blades elliptic to obovoid, obtuse at apex, cuneate at base, 1–3 cm long, 0.5–1 cm wide. Inflorescences near apex of stems; flowers sessile, yellow; sepals 2, ovate, 3–4.5 mm long, connate below, unequal, ± persistent; petals 4, 3–4.5 mm long; stamens 6–15; anthers globose; style 4–6-lobed. Capsules ovoid to fusiform, ca 2.5 mm diam; seeds minute.

Reported by Standley as "a rare weed in open places." No specimens from the island have been seen, but the species should be expected there. Elsewhere in Panama apparently flowers and fruits throughout the year.

A weed in temperate and tropical regions of the New World; native to the Old World tropics. In Panama, known from disturbed areas of tropical moist forest in the Canal Zone, Bocas del Toro, Colón, Los Santos, Coclé, Panamá, and Darién; known also from premontane wet forest in Chiriquí (Boquete).

51. CARYOPHYLLACEAE

Annual herbs. Leaves opposite, petiolate; blades simple, entire, palmately veined at base; stipules present. Flowers bisexual, actinomorphic, in bracteate, terminal, axillary cymes of 5 to many flowers (rarely 1); sepals 5, free; petals 5, free, bifid, white; stamens 2–5, slightly connate at base; anthers 2-celled, versatile, dehiscing longitudinally; ovary superior; placentation free central; ovules numerous, campylotropous; styles 3, united at base. Fruits longitudinally dehiscent capsules; seeds usually several to many; endosperm hard.

Caryophyllaceae are of relatively little consequence in tropical regions. They can be distinguished by a combination of opposite leaves, connate leaf bases, free central placentation, and unilocular capsules.

About 80 genera with over 2,000 species; mostly in North Temperate regions.

DRYMARIA Willd.

Drymaria cordata (L.) Willd. ex R. & S., Syst. Veg. 5:406. 1819

Weak herb; stems prostrate and spreading or erect, 10–30 cm long, rooting at lower nodes, branching along length of stem. Petioles wing-margined, 2.5–5 mm long, connected by stipules; stipules scarious, laciniate, to 2.5 mm long; blades reniform to ovate, rounded to acute at apex, subcordate to rounded at base, 5–25 mm long, 5–30 mm wide, palmately veined, crenulate. Inflorescences of terminal or axillary dichasial cymes of few to many flowers or rarely the flowers solitary in axils; pedicels canescent-glandular, to 5 mm long; sepals 5, elliptic, scarious-

margined, ca 3 mm long, 3(5)-veined; petals 5, white, deeply bifid and Y-shaped, the rounded sinus between lobes ca 2 mm deep, the lobes linear; stamens 2 or 3 (to 5); filaments flattened, 2–2.5 mm long; anthers suborbicular, ca 0.2 mm long; ovary ellipsoid, ca 2.5 mm long; styles 3, free nearly to bases. Capsules ovoid, 1.5–2.5 mm long, the 3 valves entire; seeds 1–12, 1–1.5 mm broad, dark, reddish-brown, tuberculate. *Croat 4163.*

Frequent in clearings. Flowering and fruiting from May to January.

Pollination system is unknown. Seeds are probably spilled locally.

Pantropical. In Panama, known from tropical moist forest in the Canal Zone, Bocas del Toro, and Panamá, from premontane wet forest in Coclé, from tropical wet forest in Panamá, and from lower montane wet forest in Chiriquí.

52. NYMPHAEACEAE

Aquatic, rhizomatous, perennial herbs, the leaves and flowers floating. Leaves alternate or basal, long-petiolate; blades simple, entire or more often serrate, with a deep basal sinus; venation palmate; stipules lacking. Flowers bisexual, actinomorphic, often fragrant; sepals 4–8, mostly free; petals several to many, white; stamens many; anthers 2-celled, introrse, dehiscing longitudinally; pistils many, immersed in the receptacle; ovary thus semi-inferior, 1-locular; placentation parietal; ovules many per ovary, anatropous; style 1; stigmas many, radiate. Fruits baccate, ripening under water and dehiscing irregularly; seeds many, arillate, operculate, with little endosperm.

These large floating-leaved aquatics are confused only with Menyanthaceae (117), which have a similar habit but have five-parted flowers in umbels and are easily separated.

Nymphaea ampla, with flowers opening by day, are visited by pollen feeders; *Trigona* bees visit them in large numbers. Pollination of *Nymphaea blanda*, whose flowers open at night, has never been observed, but it should be by a completely different group of pollinators. At least some species of *Nymphaea* are trap flowers. In greenhouse studies *N. citrina* Peter had a definite pistillate and staminate phase (Faegri & van der Pijl, 1966). During the pistillate phase, the stigma secretes nectar. Later the stamens of the inner whorls close over the stigma and thus the flowers present only pollen. The flowers are thus considered protogynous. It is not known if the same or similar activities are carried out by *Nymphaea* species on BCI. According to Schulthorpe (1967) most Nymphaeaceae are nectarless and are pollinated by beetles, small flies, and "sweat" bees, which "crawl indiscriminately over flowers." He reported that *N. ampla* L. is mostly odorless and homogamous (stamens and style developed at the same time) and that the species is largely self-pollinated.

Seeds are probably mostly water dispersed, though those of *N. blanda*, which is often found in isolated and

seasonal swamps, must also be carried around by birds or other animals. Van der Pijl (1968) reported that in other areas species of *Nymphaea* are sometimes distributed by ducks. He also reported that the seeds of *N. alba* L. are heavier than water but float because of a transformed arilloid.

Eight genera with about 80 species; cosmopolitan.

NYMPHAEA L.

Nymphaea ampla (Salisb.) DC., Reg. Veg. Syst. Nat. 2:54. 1821
Waterlily

Aquatic, glabrous herb. Leaves with very long petioles, arising from a rhizome rooted in soil; blades floating on water, suborbicular, 10–45 cm long, green above, usually purple-mottled below, the veins raised on lower surface, the margin prominently toothed, the sinus deep. Flowers to ca 15 cm diam at anthesis, long-pedunculate, opening diurnally, held somewhat above surface of water (buds below surface); sepals 4–8, greenish with purple striations, 6–10 cm long; petals white, slightly shorter and more numerous than sepals, blunt at apex; stamens numerous, the outer progressively longer, the connective produced above, the innermost stamens usually folded over the bowl-shaped surface of carpels; stigmas slender, radiating from center, 5–7 mm long. Fruits 2–3 cm broad, maturing beneath water, irregularly dehiscent; seeds many, subglobose, ca 1 mm diam, strigillose, indurate. *Croat 5715.*

Uncommon, in quiet waters on the south edge of the island. Flowers seen from February to July on BCI; elsewhere in Central America no seasonal variation is obvious from specimens. Fruit maturity time uncertain.

Development of the flower primordia is very slow and may require three to four years for flowers to develop fully (Cutter, 1957).

The bee *Trigona cupira* have been seen in great numbers collecting pollen from this species.

Throughout tropical and subtropical regions of the New World. In Panama, known from wet places in tropical moist forest in the Canal Zone and adjacent Panamá.

See Fig. 229.

Nymphaea blanda G. Meyer, Prim. Fl. Esseq. 201. 1818

Rhizomatous, aquatic herb, nearly glabrous, floating. Petioles 15–50 cm long; blades suborbicular, blunt to rounded at apex, subcordate (the sinus 1–5 cm deep), 5–15 cm long, 3–10 cm wide, entire, green below. Flowers

long-pedunculate, floating on water, opening at night, mostly 7–9 cm broad; sepals 4, lanceolate-ovate to acute, 3–5 cm long, green outside, greenish-white inside; petals to 14, lanceolate-oblong, acute at apex, white, the outermost often tinged with green at apex, about as long as sepals, the innermost much shorter; stamens numerous; flattened, 1–2.5 cm long, arched inward, the outermost thicker and longer, gradually reduced in length inward; anthers introrse, comprising about half of stamen, the connective rounded at apex; carpels ca 20, connate laterally, the entire unit to 1.2 cm diam. Fruits 1.5–2.5 cm broad, 1–2 cm long, maturing under water, pulpy, irregularly dehiscent; seeds many, globose, operculate, ca 1.5 mm long, strigillose. *Croat 12232.*

Rare; known to occur intermittently in swamps. The species has been collected in the swampy area between Armour Trail 900 and Zetek Trail 300. Flowers in the rainy season. The plants disappear during the dry season.

Guatemala to northeastern South America. In Panama, known from wet areas of tropical moist forest in the Canal Zone, Coclé, Panamá, and Darién.

53. CERATOPHYLLACEAE

Submerged, perennial aquatics. Leaves whorled, sessile; blades linear, dichotomously divided; stipules lacking. Flowers unisexual (monoecious), apetalous, actinomorphic, solitary at the nodes; sepals 6–15, basally connate; stamens 10–20, spirally arranged on a flat receptacle; anthers subsessile, 2-celled, dehiscing longitudinally, with a colored connective; ovary superior, 1-locular, 1-carpellate; placentation parietal; ovule 1, pendulous, anatropous; style 1, slender; stigma undifferentiated. Fruits nuts; seed lacking endosperm.

The family is represented by a single species on BCI, which may easily be recognized by its aquatic habit and whorled, dichotomously branched leaves.

Flowers are pollinated under the surface of the water (hyphydrophily of Faegri & van der Pijl, 1966). After the staminate flowers open, anthers float to the surface by means of their expanded connective, where they dehisce. The pollen eventually sinks and comes in contact with the filiform style of the axillary pistillate flowers (Schulthorpe, 1967).

The fruits are dispersed by water currents or aquatic birds. Plants may rely more on vegetative reproduction than on sexual reproduction. According to Ridley (1930), the widespread distribution of *Ceratophyllum demersum* is due to epizoochory by birds of its minute spiny fruits.

One genus, with 4 or more species; cosmopolitan.

KEY TO THE SPECIES OF NYMPHAEA

Leaves often more than 20 cm long, usually purplish beneath, toothed on margin, the veins coarse, prominently raised on lower surface .... *N. ampla* (Salisb.) DC.
Leaves less than 15 cm long, usually green beneath, entire, the veins fine, not raised on lower surface .... *N. blanda* G. Meyer

CERATOPHYLLUM L.

Ceratophyllum demersum L., Sp. Pl. 992. 1753

Monoecious, submerged, aquatic herb, rootless, to about 1 m long, dark green, unbranched or branched, with a single lateral branch produced at a node. Leaves whorled, 6–12 per node, 1–1.5 cm long, dichotomously dissected once or twice, the divisions mostly filiform, minutely denticulate, the whitish teeth on a broad base of green tissue. Flowers rare, unisexual, sessile, solitary in axil of one leaf of a whorl, ca 1 mm long; staminate flowers with sepals 10–15, basally connate; corolla lacking; stamens 10–20, spirally arranged on a flat receptacle; filaments very short; anthers linear-oblong, with a thickened, produced connective. Pistillate flowers with perianth similar to staminate flowers; ovary gradually tapered to slender style; pistil longer than sepals; stigma 1, undifferentiated from style. Nuts also rare, axillary, tuberculate, with 2 spines at base, ca 4 mm long, tipped by the long persistent style. *Croat 5738.*

Occasional, in coves around the island, occurring with but much less common than *Hydrilla verticillata* (L.f.) Royle (16. Hydrocharitaceae). Seasonal behavior unknown.

This distinctive aquatic is not confused with any other in Panama.

Cosmopolitan distribution. In Panama, it occurs probably in all lakes or slow-moving, freshwater habitats.

54. MENISPERMACEAE

Lianas or vines. Leaves alternate, petiolate; blades simple, entire or crenate (may be slightly lobed in *Odontocarya*); venation palmate or blades only palmiveined at base; stipules lacking. Flowers unisexual (dioecious), generally actinomorphic, in axillary, bracteate, often fasciculate panicles or racemes; sepals and petals 3 or 6 (1 or 4 in *Cissampelos;* petals lacking in *Abuta*), the sepals free, in 1 or 2 series; petals often connate; stamens 6 (or 4 in *Cissampelos*), opposite the petals; anthers 2- or 4-celled, the thecae often separated by a connective, dehiscing longitudinally; gynoecium of 1–6 free pistils; ovary superior, 1-locular; placentation parietal; ovules 2, aborting to 1, anatropous; style 1, very short, or the stigma sessile; stigma lobed or incised. Fruits drupaceous; seed usually lacking endosperm.

These uncommon vines or lianas are distinguished by unisexual and usually three-parted flowers, often bearing a double whorl of sepals, leaves palmately veined (at the base), and curved seeds.

Pollination systems are unknown. Plants are both dioecious and not very common.

Species with known fruits (and certainly *Cissampelos* species) are probably eaten by birds. *Cissampelos* are also eaten by white-faced monkeys (Oppenheimer, 1968).

About 65–70 genera and 350–400 species; mostly in the tropics.

ABUTA Aubl.

Abuta panamensis (Standl.) Kruk. & Barn., Mem. New York Bot. Gard. 20:22. 1970
Hyperbaena panamensis Standl.

Dioecious liana to 35 m high or tree to 6 m tall; trunk not known, probably flattened; stems terete, the older stems lenticellate. Leaves alternate; petioles (1)4–6(9) cm long, usually curved and enlarged near apex; blades variable, elliptic to oblong or ovate, acute and apiculate or short-acuminate at apex, obtuse to rounded or truncate at base,

---

KEY TO THE SPECIES OF MENISPERMACEAE

Blades peltate; staminate flowers with 4 sepals and 4 stamens, the petals connate; pistillate flowers
    with 1 sepal and 1 petal:
    Inflorescence bracts to 1.5 cm long and wide; carpels glabrous; drupes 5–7 mm long; leaves and
        stems sparsely pubescent, the trichomes long and spreading . . . . *Cissampelos tropaeolifolia* DC.
    Inflorescence bracts usually less than 1 cm long; carpels densely pubescent; drupes 4–5 mm
        long; leaves and stems moderately to densely pubescent, the trichomes long and sericeous
        or short and tomentose . . . . *Cissampelos pareira* L.
Blades basifixed; staminate flowers with 6–16 sepals and 6 stamens; petals various; pistillate flowers
    with 6 sepals and 6 petals or petals lacking:
    Blades densely white-woolly beneath . . . . *Chondrodendron tomentosum* R. & P.
    Blades not densely white-woolly beneath:
        Plants herbaceous vines; blades broadly ovate, usually conspicuously cordate, with domatia
            or glands at base; sepals 6; petals 6:
            Blades short-pubescent at least on veins below; drupes ca 1 cm long . . . .
                    . . . . *Odontocarya tamoides* (DC.) Miers var. *canescens* (Miers) Barn.
            Blades glabrous; drupes ca 2 cm long . . . . *Odontocarya truncata* Standl.
        Plants woody lianas; leaf blades mostly elliptic, rarely subcordate, lacking domatia or glands
            at base; sepals 6 in 2 series; petals lacking:
            Petioles mostly more than 4 cm long; staminate inflorescences more than 10 cm long; flow-
              ers obviously pedicellate . . . . *Abuta panamensis* (Standl.) Kruk. & Barn.
            Petioles usually less than 3 cm long; staminate inflorescences less than 10 cm long; flowers
              sessile or nearly so . . . . *Abuta racemosa* (Thunb.) Tr. & Planch.

Fig. 229. *Nymphaea ampla*

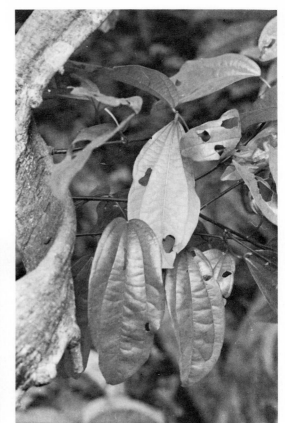

Fig. 230. *Abuta racemosa*

Fig. 231. *Chondrodendron tomentosum*

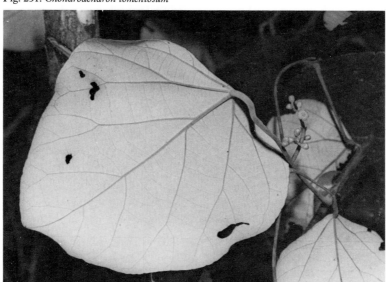

6–24 cm long, 3–14 cm wide, glabrous above, sparsely puberulent or glabrous below; veins at base 3–5. Staminate inflorescences racemose, to 10–15 cm long, the branches reduced to short-stalked, 3-flowered cymes; branches and pedicels appressed-pubescent; pedicels to 3 mm long; outer sepals 3, narrowly ovate, ca 1 mm long, the inner sepals 3, broadly ovate to obovate, 2–2.8 mm long, fleshy, granular-puberulent on inside, ± strigose on outside; petals lacking; stamens 6, 1.5–2.2 mm long; filaments somewhat flattened, pubescent at least on one margin, connate into a column toward the base; anthers ca 0.2 mm long, dehiscent longitudinally; pistillode lacking. Pistillate inflorescences racemose, axillary, pedunculate or subsessile, 7–12 cm long, with few flowers; sepals and petals similar to staminate flowers; staminodia 6, 2–2.3 mm long, short-pilose in basal half, glabrous above; ovary densely pubescent; styles 3. Fruiting peduncles stout, ca 2 mm long; fruits of 3 drupes, the drupes oblong, ca 2.5 cm long and 1.3 cm diam, yellow-orange, densely short-pubescent, sessile. *Bangham 484, Bailey 101, Standley 28417.*

A number of collections exist, but the plant has not been seen recently on the island. Seasonality not determined. Flowers mostly from July to September elsewhere in Central America, with the fruits maturing from March to May. Some flowers have also been seen in November and February.

Mexico to Panama. In Panama, known from tropical moist forest in the Canal Zone and Veraguas.

Abuta racemosa (Thunb.) Tr. & Planch., Ann. Sci. Nat. Bot., sér. 4, 17:48. 1862

Dioecious liana, growing into canopy; trunk flattened, 5–20 cm diam; stems puberulent when young, becoming glabrous, the apices often attenuated and twining. Leaves alternate; petioles 0.7–4.5 cm long, usually curved and enlarged near apex; blades oblong-elliptic to oblanceolate, acuminate to bluntly acuminate, rounded, truncate, or subcordate at base, 5–15.5 cm long, 2–8.5 cm wide, stiff, glabrous but with sparse appressed pubescence below and stiff erect trichomes on major veins above, the margins entire to obscurely crenulate; veins at base 3–5, the midrib arched. Staminate inflorescences narrowly pseudoracemose, 5–10 cm long, the branchlets reduced to short-stalked, 3-flowered cymules; pedicels to 1.7 mm long; pedicels and sepals with dense, erect or appressed puberulence; outer sepals 3, narrowly triangular or lanceolate, ciliate, ca 1.3 mm long, the inner sepals 3, ovate, thick, valvate, held closely together in basal half, somewhat spreading above the middle, acute at apex, ± glabrous to granular-puberulent on inside, with appressed grayish pubescence on outside; petals lacking; stamens 6, included, slightly more than 1 mm long, in 2 series, the inner 3 somewhat larger with the filaments connate into a column and the anthers swollen, the thecae apical, widely separated by a swollen connective, the outer 3 with the filaments free and the anthers minute. Pistillate inflorescences similar to staminate but apparently shorter; pistil 3-carpellate; ovary densely sericeous, curved; styles

apical and possibly fused to each other in flower but free at least in juvenile fruit, curved outward, pointed at apex. Fruits 3-parted, the carpels drupaceous, oblong-obovate, orange to black at maturity, 2–2.3 cm long, 1.1–1.2 cm diam, tomentulose when young, glabrate; seeds 1 per carpel, each surrounded by a fleshy white mesocarp. *Croat 15004.*

Occasionally seen, though likely to be common in the canopy; the flowers are small and difficult to see. Most collections have been made in tree-fall areas. Flowers in the early rainy season (May to July), with the fruits probably maturing from August to October.

Panama, Colombia, and Bolivia. In Panama, known only from tropical moist forest on BCI and in Darién. See Fig. 230.

CHONDRODENDRON R. & P.

Chondrodendron tomentosum R. & P., Syst. Veg. 261. 1798

C. hypoleucum Standl.

Large dioecious canopy liana; trunk near the ground to ca 10 cm diam; stems terete, puberulent to tomentose (striate when dry). Petioles 4–14 cm long, puberulent to tomentose; blades ovate to suborbicular, blunt to apiculate (rarely emarginate) at apex, truncate to cordate at base, 10–20 cm long, 9–18 cm wide, subcoriaceous, glabrous above, white-woolly below; veins at base 5, palmate, with 2–4 additional pairs of major lateral veins above. Staminate inflorescences paniculate; panicles fascicled in leaf axils, often on older stems, to 10 cm long, the branches short; all axes densely tomentose; flowers greenish-white; sepals 9–16, the outermost reduced, tomentose or ciliolate, the inner 6–8 much larger, to 3 mm long, oblong-obovate to obovate, recurved, glabrate or inconspicuously pubescent, ciliolate; petals usually 6, minute and obscured by sepals, less than 1 mm long; stamens 6, ca 1.5 mm long; anthers ± horizontal, affixed subbasally, the connective produced beyond the thecae and usually directed inward. Pistillate inflorescences and flowers similar to staminate but with gynoecium of 6 pistils, the pubescence dense, short-appressed; style 1 per carpel, simple, ca 0.5 mm long, slightly curved. Drupes oblong-ovoid, ca 1.5 mm long, narrowed at base into a short stipe, tomentose to glabrous in age. *Croat 13801.*

Uncommon. Known to flower in late July and early August.

Panama to Peru and Bolivia. In Panama, known from tropical moist forest on BCI and in Darién. See Fig. 231.

CISSAMPELOS L.

Cissampelos pareira L., Sp. Pl. 1031. 1753

Bejuco de cerca, Alcotán

Generally a slender twining vine; stems, petioles, and lower surface of leaves loosely pubescent with long and

sericeous or short and puberulent trichomes, the upper leaf surface glabrate or sparsely pubescent. Leaves obscurely peltate; petioles 3.5–7 cm long, slender; blades broadly ovate, usually obtuse and mucronate at apex, truncate to cordate at base, 2–12 cm long and wide; venation palmate. Staminate inflorescences lacy and branched many times, usually axillary, often at leafless nodes; bracts less than 1 cm long; flowers greenish, ca 3 mm diam; sepals 4, much longer than the cone-shaped corolla; stamens united, the apex flaring; anthers 4. Pistillate flowers fasciculate along axillary branches, the branches often several at a node, the fascicles each subtended by a small, ovate or reniform, leaflike bract; sepals and petals each 1 per flower, caducous, borne at the gibbous base of the carpel; carpel 1, densely pubescent; stigma deeply trilobate. Drupes suborbicular, ± flattened, red, 4–5 mm long, pubescent; seed brown, flattened, verrucose, 3–4 mm long, the embryo horseshoe-shaped. *Croat 4397, 11997.*

Generally abundant in clearings; rare in the forest. Flowers and fruits throughout the year.

Throughout tropics of the world. In Panama, known from tropical moist forest in the Canal Zone, Bocas del Toro, Veraguas, Los Santos, Panamá, and Darién, from premontane moist forest in the Canal Zone and Panamá, from premontane wet forest in Chiriquí, Los Santos, Coclé, and Panamá, and from tropical wet forest in Coclé and Darién.

Cissampelos tropaeolifolia DC., Reg. Veg. Syst. Nat.
  1:532. 1818

Twining vine; stems, leaves, and inflorescence branches usually with long whitish trichomes. Leaves peltate; petioles slender, 3–7 cm long; blades ovate to suborbicular, rounded to obtuse or acuminate and mucronate at apex, truncate to cordate at base, 5–11 cm long and wide, glabrous to sparsely long-pubescent above, paler and usually long-pubescent or puberulent below; venation palmate. Staminate inflorescences of fasciculate dichasia in leaf axils or on short axillary branches in the axils of reduced leaves or bracts, the bracts to 1.5 cm long and 2.5 cm wide; flowers minute; sepals 4, 1–1.5 mm long, ca 1 mm wide, usually pubescent on outside; corolla greenish, campanulate, 0.5–1 mm diam, glabrous; stamens with filaments united into a tube; anthers 4, borne on filament tube, glabrous. Pistillate flowers fasciculate in the axils of bracts on secondary branches; sepal 1, ovate to obovate, ca 1 mm long, glabrous or puberulent; petal 1, greenish, suborbicular, ca 0.5 mm long and to 1 mm wide; carpel 1, gibbous, sessile, glabrous; stigma trilobate. Drupes obovoid, flattened, 5–7 mm long, 4–5 mm wide, red, pubescent; seed verrucose, the embryo horseshoe-shaped. *Croat 9265.*

Rare in the forest. Flowers and fruits throughout the year.

Southern Mexico to northern South America. In Panama, known from tropical moist forest in the Canal Zone, Bocas del Toro, San Blas, Chiriquí, and Darién, from

premontane wet forest in the Canal Zone and Chiriquí, and from tropical wet forest in Darién.

ODONTOCARYA Miers

Odontocarya tamoides (DC.) Miers var. **canescens**
  (Miers) Barn., Mem. New York Bot. Gard.
  20:89. 1970
  O. paupera (Griseb.) Diels

Slender vine; older stems lenticellate. Leaves alternate; petioles slender, 3–8.5 cm long; blades ovate to subhastate, obtuse to acuminate and mucronulate at apex, usually shallowly cordate at base (acute on juveniles), 3–10 cm long, 3–8.5 cm wide, glabrous above, with conspicuous glandular areas at base of blades near petiole, pubescent below with short, spreading trichomes especially on veins; venation at base palmate. Inflorescences axillary or supra-axillary, solitary; rachis glabrous or puberulent; staminate inflorescences pseudoracemose, to 22 cm long, with flowers in fascicles of 2–6, occasionally with several short branches; pedicels 3–6 mm long; sepals 6, membranaceous, the outer 3 ± ovate, to 1.5 mm long and 1 mm wide, the inner 3 ± oblanceolate, 2–4 mm long, 1.5–2.5 mm wide; petals 6, the outer 3 greenish, somewhat fleshy, bowl-shaped, 2–2.5 mm long, 1.5–2 mm wide, the inner 3 similar but smaller; stamens 6, 1–2 mm long, the filaments connate about half their length; anthers 6, ca 0.5 mm long. Pistillate inflorescences strictly racemose, with 6–12 loosely spaced flowers; pedicels to 1.4 cm in fruit; sepals and petals as in staminate flowers; staminodia 6; ovary glabrous. Fruits of 1–3 drupes, ellipsoid, ca 1 cm long and 7 mm diam, reddish-orange, drying blackish. *Croat 5631, 5640.*

Rare in clearings and occasional in the canopy of the forest. Seasonal behavior uncertain. Known to flower and fruit from May to October.

Southern Mexico to Colombia, Venezuela, the Guianas, and northeastern Brazil; Lesser Antilles. In Panama, known from tropical moist forest in the Canal Zone, Panamá, and Darién and from tropical dry forest in Panamá.

Odontocarya truncata Standl., J. Arnold Arbor.
  11:121. 1930

Glabrous, slender vine. Leaves alternate; petioles 4.5–7.5 cm long; blades ovate, acuminate, truncate to shallowly cordate at base, 8–13 cm long, 5–9 cm wide, with sunken, discolored domatia in axils on lower surface; venation palmate at base. Inflorescences racemose, axillary (sometimes at leafless nodes), solitary; sepals of staminate flowers 6, membranaceous, the outer 3 ovate, ca 0.5 mm long and wide, the inner 3 obovate, concave, ca 2 mm long and 1.5 mm wide; petals of staminate flowers 6, narrowly oblanceolate, ca 1.5 mm long and 0.5 mm wide; stamens 6, ca 1.3 mm long; filaments ca 1 mm long, the outer 3 connate to the middle, the inner 3 connate their entire length. Pistillate infructescences 10–20 cm long; fruiting pedicels 1–1.5 cm long; pistillate flowers unknown. Fruits

of 1–3 drupes, these oblong-ellipsoid, ca 2 cm long and 1–2 cm diam, orange-yellow, drying blackish, with a small ridge around the margin. *Croat 4945*.

Rare, at the edge of clearings. Seasonal behavior uncertain. Known to flower in January and fruit from April to September.

Costa Rica to northern Colombia. In Panama, known from tropical moist forest in the Canal Zone (on and adjacent to BCI).

55. ANNONACEAE

Trees or shrubs. Leaves alternate, petiolate, simple, entire; venation pinnate; stipules lacking. Flowers bisexual, actinomorphic, solitary or several in a cluster; sepals 3, free or connate, valvate; petals 6, in 2 series (the outer may be reduced or lacking), free or connate, imbricate or valvate; stamens numerous, free; anthers 4-celled, introrse, dehiscing longitudinally, the connective expanded above the anther; gynoecium of numerous, separate pistils; ovary superior, 1-locular; placentation parietal (may seem basal); ovules 1 to several, anatropous; style 1 and short or lacking; stigma 1, simple. Fruits monocarpic berries or fleshy aggregates (*Annona*); seeds with copious, ruminate endosperm.

Annonaceae are confused with no other family when in flower or fruit. The flowers are usually large and greenish, with thick petals and many stamens and ovules. The fruits are either fleshy aggregates or composed of several monocarps. The ruminate endosperm of the seeds is also diagnostic of the family.

The open flowers provide no access problems. They also offer little or no nectar and are probably pollinated by a wide variety of insects, especially pollen feeders with chewing mouth parts such as beetles (Corner, 1964).

KEY TO THE SPECIES OF ANNONACEAE

Inflorescences in leaf axils:
  Most blades more than 20 cm long:
    Petals imbricate; fruits ellipsoid, ca 1 cm long, pointed at both ends ....
    .... *Guatteria amplifolia* Tr. & Planch.
    Petals valvate; fruits globose, ca 1.5–2 cm diam .... *Unonopsis pittieri* Saff.
  All blades less than 20 cm long:
    Blades less than 6 cm long, lanceolate; flowers nearly sessile; branches conspicuously arcuate-
      ascending .... *Xylopia frutescens* Aubl.
    Blades more than 6 cm long, ± elliptic; flowers pedicellate; branches not arcuate:
      Sepals united to above middle; monocarps reniform, several-seeded, 3 or more cm long in
        dense globular clusters .... *Xylopia macrantha* Tr. & Planch.
      Sepals free; monocarps not as above:
        Outer petals imbricate, densely pubescent; monocarps ellipsoid, indehiscent, to 2.5 cm
          long, with 1 seed .... *Guatteria dumetorum* Fries
        Outer petals valvate, glabrate; monocarps golf-club-shaped, dehiscent, more than 3 cm
          long, with 2 orange seeds .... *Anaxagorea panamensis* Standl.
Inflorescences not axillary, usually ± opposite leaves or internodal:
  Petals 6, ± equal in size, narrowly lanceolate, more than 3 times longer than broad; fruits of
    several monocarps, stipitate, ± cylindrical, red, with several disk-shaped seeds ....
    .... *Desmopsis panamensis* (Rob.) Saff.
  Petals with the 3 outer much larger than the 3 inner (or 3 inner petals rudimentary), the outer
    usually ovate (except *A. hayesii*), less than 3 times longer than broad; fruits with mericarps
    united into a single many-seeded mass:
    Petals 6:
      Outer petals connate ca 5 mm above base with a long narrow acumen; fruits less than 8 cm
        long, densely pubescent, lacking spines .... *Annona hayesii* Standl.
      Outer petals free, valvate, ovate; fruits more than 8 cm long, glabrous or spiny:
        Plants cultivated at Laboratory Clearing; fruits spiny .... *Annona muricata* L.
        Plants restricted to swamps at edge of lake; fruits smooth, glabrous ...... *Annona glabra* L.
    Petals with the 3 inner lacking or rudimentary:
      Flowers less than 2.5 cm long and 2.5 cm wide (petal size assumed from size of other species
        and size of calyx); fruits of many 1-seeded, stipitate monocarps .... *Crematosperma* sp.
      Flowers either longer or wider than 2.5 cm; fruits not of 1-seeded, stipitate mericarps, either
        globose or ± ellipsoidal aggregates (1 aggregate fruit per flower):
        Leaves glabrous; blades mostly less than 9 cm long; petals less than 15 mm long; fruits
          to 3 cm wide with very short protuberances, explosive at maturity ....
          .... *Annona acuminata* Saff.
        Leaves densely tomentose below; blades more than 10 cm long; petals more than 18 mm
          long; fruits to 6 cm wide, shaggy with long-attenuate protuberances, not explosive at
          maturity .... *Annona spraguei* Saff.

Fig. 232. *Anaxagorea panamensis*

Fig. 233.
Anaxagorea panamensis

Fig. 234.
Annona glabra

Guatteria, Annona, and Xylopia are pollinated by small beetles of the families Nitulidae and Curculionidae (Gottsberger, 1970). Gottsberger reported that the flowers are protogynous, with anthesis lasting a long time, and usually somewhat closed at the end of anthesis to prevent the entry of larger beetles, which would completely destroy the flowers. Beetles are apparently attracted to the flowers by a strong foul aroma (Faegri & van der Pijl, 1966); Gottsberger (1970) likened the smell to rotting fruit.

The fruits are dispersed chiefly by birds (Ridley, 1930) and mammals, except Anaxagorea, which has mechanically dispersed fruits. Bats are also listed as fruit dispersers by Ridley (1930). Seeds of Guatteria and Xylopia are bird dispersed (Gottsberger, 1970). Annona muricata is eaten by the bat Artibeus jamaicensis (Phyllostomidae) in Trinidad (Goodwin & Greenhall, 1961) and by a woodpecker in Jamaica (Ridley, 1930). Fruits of Annona acuminata are eaten by coati (Kaufmann, 1962). Several species of Annona, as well as Desmopsis panamensis, are eaten by the white-faced monkey (Oppenheimer, 1968); the seeds and pericarp of Desmopsis were often spit out. Annona spraguei is also eaten by spider monkeys, and fruits of Guatteria dumetorum and G. amplifolia are eaten by howler monkeys (Carpenter, 1934).

About 100–120 genera and 2,000 or more species; mostly in the tropics.

ANAXAGOREA St.-Hil.

Anaxagorea panamensis Standl., J. Wash. Acad. Sci. 15:101. 1925

Shrub, to 3 m tall; parts puberulent when young, glabrous in age. Petioles 5–9 mm long; blades oblong-elliptic to lanceolate-oblong, acuminate, acute to obtuse at base, 10–16 cm long, 2.5–5 cm wide. Flowers pale yellow, solitary in axils; pedicels 1.5–2.5 cm long, with a minute bracteole near apex; sepals 3, ovate, ca 8 mm long; petals valvate, the outer 3 narrowly oblong, ca 2 cm long, the inner 3 ovate, ca 1 cm long. Monocarps obliquely club-shaped, 3–4 cm long, apiculate at apex; seeds 2 per carpel, ± elliptic, flat, shiny, orange, forcibly expelled by contraction of walls of monocarp (fide R. Foster). Croat 11158, 12569.

Rare; known from a single area in the old forest north of Zetek Trail 600. Flowers in July and January. Mature fruits were seen from November to January.

Known only from Panama, from tropical moist forest in the Canal Zone (on and adjacent to BCI) and Panamá.

See Figs. 232 and 233.

ANNONA L.

Annona acuminata Saff., Contr. U.S. Natl. Herb. 16:274, pl. 97. 1913
Camaron

Small tree, (3)4–5(7) m tall; young stems, petioles, and lower surface of blades with sparse appressed-ferruginous pubescence, soon glabrous. Petioles to 6 mm long; blades lanceolate to oblong-elliptic or oblanceolate, acuminate, acute and decurrent at base, 4–11.5 cm long, 1.5–3.5 cm wide. Flowers solitary, on stems opposite leaves; pedicels to 1.6 cm long, bearing a small bracteole near base; flower buds depressed-globose; calyx ca 5 mm long, saucer-shaped with 3 minute lobes, with appressed-ferruginous pubescence; petals 3, valvate, white, green in bud, ovate-rounded, thick, obtuse at apex, 10–15 mm long; stamens 2.5 mm long, numerous. Fruits fleshy aggregates, green, 2–3 cm wide, 1.8–2.3 cm long, rounded above, flat below, the surface covered with prominent projections 1–2 mm high, subtended by the thick, flattened, 3-sided calyx, bursting open irregularly from apex at maturity to expose bright orange interior; seeds 1 per carpel, broadest at apex, gradually tapered to base, weakly angulate, ca 1 cm long, tan. Croat 4379, 11757.

Frequent in the forest. Flowers from May through August. The fruits mature from November to March, sometimes beginning in October.

Known only from Panama, from tropical moist forest in the Canal Zone and Panamá and from premontane moist forest in Panamá.

Annona glabra L., Sp. Pl. 537. 1753
Pond apple, Anón de puerco, Anón

Glabrous tree, to 10 m tall, usually to about 6 m and 15(25) cm dbh; trunk swollen below water level. Petioles canaliculate, brownish above, 1–2 cm long; blades ovate-elliptic to oblong-elliptic, acute at apex, rounded and ± decurrent at base, 5–16 cm long, 3–8 cm wide. Flowers solitary, supra-axillary (rarely opposite leaves); pedicels 1.5–3 cm long; sepals 3, short, minutely apiculate; petals valvate, thick, ovate, white, the outer 3 larger, red inside at base and often spotted with red, to 3 cm long, the inner 3 somewhat shorter, often red nearly all over the inside; stamens numerous, 3–4 mm long, the connective thickened, papillose; stigmas many, ca 1 mm long. Fruits fleshy aggregates, ovoid, rounded at apex, to 15 cm long and 9 cm wide, green, densely speckled, glabrous; seeds numerous, 1 per carpel. Croat 5019, 7791.

Frequent in and restricted to shoreline marshes, chiefly on the south and west sides of the island, often in association with Acrostichum (10. Polypodiaceae). Flowers mainly in the dry season (February to June). The fruits are probably mature from December to May, but hang on the tree all year.

Mexico to Ecuador and Brazil; West Indies, western Africa. In Panama, known from tropical moist forest in the Canal Zone, Bocas del Toro, San Blas, Panamá, and Darién and from tropical wet forest in Colón.

See Fig. 234 and fig. on p. 12.

Annona hayesii Saff. ex Standl., J. Wash. Acad. Sci. 15:102. 1925

Slender tree or shrub, to 4 m tall, less than 8 cm dbh; stems puberulent when young, glabrate in age. Leaves puberulent when young, glabrous except on veins in age; petioles less than 1 cm long; blades obovate to oblong-elliptic, abruptly acuminate, rounded to obtuse at base,

10–23 cm long, 5–9 cm wide. Flowers greenish, 1 to few, opposite leaves or internodal; pedicels to 1.5 cm long; sepals 3, ovate, to ca 4 mm long, acuminate; petals valvate, the outer 3 to 3.2 cm long, connate to 5 mm at base, abruptly narrowed ca 1 cm above base, dark violet-purple inside at base, the inner three 5–7 mm long, connate with outer petals at base; stamens numerous, 1–1.5 mm long. Fruits fleshy aggregates, oblong-ellipsoid, 4–6 cm long, ca 3 cm diam, densely pubescent, smooth or with slightly raised anastomosing lines, these forming irregular isodiametric cells; seeds numerous, 1 per carpel. *Croat 11428, 14879.*

Occasional to locally common in the forest. Numerous trees grow beyond Snyder-Molino Trail 300, at Shannon Trail 0, and along Miller and Lake trails. Flowers from April to July. The fruits mature probably from July to September (sometimes to October).

Known only from Panama, from tropical moist forest in the Canal Zone, Panamá and Darién and from premontane moist forest in the Canal Zone.

Annona muricata L., Sp. Pl. 536. 1753

Soursop, Guanábana

Tree, to 7 m tall; young stems and sometimes petioles ferruginous-tomentose. Petioles ca 5 mm long; blades oblong-elliptic to oblong-obovate, acuminate, obtuse to rounded at base, 6–12 cm long, 2.5–5 cm wide, glabrous. Flowers solitary along stems; pedicels 2–3 cm long; sepals 3, ovate, less than 5 mm long; petals valvate, the outer 3 ovate, free, thick, 2–3 cm long, the inner 3 thinner and smaller than the outer; stamens numerous. Fruits fleshy aggregates, ovoid-ellipsoid, to 20(25) cm long and 10(12) cm diam, with white juicy pulp, dark green, bearing stout fleshy spines; seeds numerous, 1 per carpel. *Croat 5176, 5898.*

Cultivated at the Laboratory Clearing near the dock. Flowers from January to August. The fruits, which hang on all year, are known to mature in July and August, but probably in the dry season also.

Cultivated throughout tropical America; introduced into West Africa. In Panama, collections exist from tropical moist forest in the Canal Zone and San Blas, but the plant is cultivated in many places.

Annona spraguei Saff., Contr. U.S. Natl. Herb. 16:270. 1913

Chirimoya, Negrito

Tree, to 15 m tall; outer bark thin, with broad shallow fissures in young trees; inner bark thin; branchlets tomentose when young, glabrous in age; sap at first with pungent odor, soon fading. Leaves deciduous, short-pilose all over, more sparsely above; petioles 5–15 mm long; blades oblong-lanceolate to narrowly elliptic, acuminate, rounded to obtuse at base, 15–30 cm long, 6–11 cm wide. Flowers solitary, arising between nodes, globose in bud, on pedicels 1–2 cm long; sepals 3, connate at base, the lobes deltoid, to 1.3 cm long; petals 3, valvate, to 6 mm thick, broadly ovate, acute at apex, to 3 cm long, moderately pubescent outside with a purple spot inside near

base; stamens numerous, ca 4 mm long. Fruits fleshy aggregates, ± globose, to 6 cm diam, green, shaggy with dense, long-attenuate protuberances; seeds many, 1 per carpel, oblong, to 1 cm long. *Croat 5070, 6750.*

Frequent in the forest, especially in the young forest. Leaves are lost in the middle of the dry season. The flowers often appear at the same time as the new leaves, usually from February to June. The fruits are probably mature from June to October.

Known only from Panama, from tropical moist forest in the Canal Zone, Panamá, and Darién, from premontane wet forest in the Canal Zone, and from tropical wet forest in Colón.

See Fig. 235.

CREMATOSPERMA R. E. Fries

Crematosperma sp.

Tree, to ca 15 m; stems weakly appressed-pubescent when young, soon glabrous, lenticellate. Leaves glabrous, firm; petioles 2–4 mm long, canaliculate on upper surface; blades oblong-lanceolate, narrowly acute to acuminate at apex, acute at base, 8.5–15 cm long, 1.7–3.5 cm wide, green and shiny on upper surface, paler below, weakly glaucous; major lateral veins 10–14 pairs, weakly loop-connected 2–5 mm from margin. Mature flowers not known; flower buds suborbicular, the sepals broadly ovoid-rounded, ciliate. Fruiting inflorescences interaxillary; fruiting pedicels ca 9 mm long, 2.5 mm thick, articulate near the middle, bearing a bract above the articulation, the bract deltoid, ca 1 mm long, ciliate, the old sepals broadly triangular, ca 4 mm long; monocarps numerous, oblong-ellipsoid, 1.5–1.8 mm long, obtuse on both ends, red at maturity; seed 1, only slightly smaller than monocarp. *Garwood & Foster 442.*

Apparently rare, in the old forest; collected once recently on Wheeler Trail.

The species may be new to science, but owing to the difficulty of the group I am reluctant to describe it from a fruiting collection.

Known only from Panama on BCI.

DESMOPSIS Saff.

Desmopsis panamensis (Rob.) Saff., Bull. Torrey Bot. Club 43:185, pl. 7. 1916

Tree, usually 4–6(8) m tall; young branches densely ferruginous-pubescent. Petioles to 6 mm long; blades ± elliptic to elliptic-lanceolate, acuminate, obtuse to acute at base, 8–22 cm long, 3–10 cm wide, ± glabrous above except on midrib, brownish-puberulent beneath, especially on veins. Flowers 1 or 2, opposite leaves, with short brownish pubescence; peduncles short with an ovate-cordate bract 4–16 mm long near apex; pedicels to 7 cm long; sepals 3, small, triangular; petals 6, ± equal, free, at first green, becoming greenish-yellow at maturity, narrowly triangular, to 3 cm long, valvate or slightly imbricate at apex, the margins revolute, the apex curled inward; anthers many, sessile, to 1.7 mm long; carpels

Fig. 235. *Annona spraguei*

Fig. 236. *Desmopsis panamensis*

Fig. 237. *Guatteria amplifolia*

Fig. 238. *Guatteria amplifolia*

Fig. 239. *Unonopsis pittieri*

Fig. 241. *Xylopia frutescens*

Fig. 240. *Xylopia frutescens*

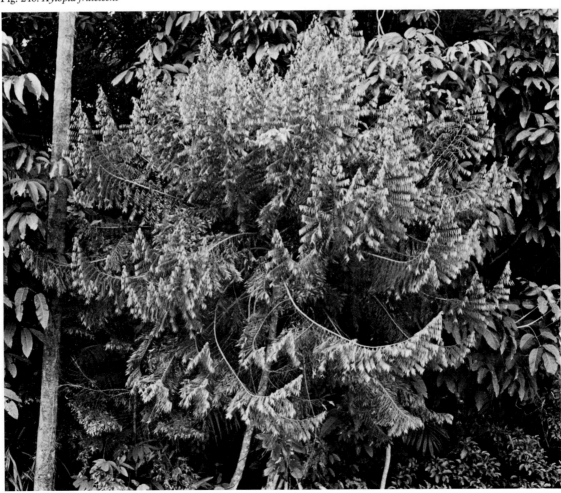

numerous, the styles held tightly together and exceeding anthers; ovules 2–8 per carpel. Monocarps densely short-pubescent, cylindrical, rounded on both ends, longer than broad, to 25 mm long, becoming soft and brick red at maturity on stipes to 1 cm long; seeds usually 4–6, disk-shaped. *Croat 7361, 8793.*

Abundant in the forest, especially in the old forest. Flowers mostly during the dry season (January to April), rarely during the rainy season. Fruit maturity time uncertain, but the fruits are eaten by white-faced monkeys from October to May (J. Oppenheimer, pers. comm.).

Known only from Panama, from tropical moist forest in the Canal Zone, Bocas del Toro, and Darién.

See Fig. 236.

GUATTERIA R. & P.

Guatteria amplifolia Tr. & Planch., Ann. Sci. Nat. Bot., sér. 4, 17:35. 1862

Tree, to 8 m tall; branches often long and arching; branchlets glabrous. Petioles short, ± curved, thick; blades moderately thick, elliptic to oblong-elliptic, abruptly or gradually acuminate, rounded or emarginate at base, mostly 20–35 cm long and 7–13 cm wide, glabrous but with inconspicuous pubescence on veins below; lateral veins impressed above, raised below, loop-connected, some reticulate veins prominent. Flowers usually solitary in axils, covered with dense short pubescence; pedicels ca 1 cm long, articulate, with 1 to several small bracts below the articulation; calyx lobes 3, acute, divided nearly to base, ca 1 cm long, often persisting in fruit; petals 6, imbricate, ± equal, triangular-oblong, ca 2 cm long, to 14 mm wide, green or yellowish; stamens very numerous, ca 2 mm long, forming a dense, round, cushion-shaped mass around style; style somewhat longer than stamens, its many stigmas held tightly together, ca 1 mm long. Monocarps ellipsoid, short-pubescent, ca 1 cm long, purple, on red stipes 10–15 mm long, 1-seeded, usually on older branches. *Croat 12216, 12693a.*

Occasional in the young forest; abundant on Orchid Island (R. Foster, pers. comm.). The flowers are seen throughout the year, less commonly from April through June. Fruit maturity time uncertain.

Mexico to Colombia. In Panama, known from tropical moist forest in the Canal Zone, Bocas del Toro, San Blas, and Panamá, from premontane moist forest in Panamá, and from premontane wet forest in the Canal Zone and Panamá. Reported from tropical wet and premontane rain forests in Costa Rica (Holdridge et al., 1971).

See Figs. 237 and 238.

Guatteria dumetorum Fries, Kongl. Svenska Veten-skapsakad. Handl., ser. 3, 24(10):12. 1948

Tree, 20–35 m tall, to 80 cm dbh, sparsely buttressed; outer bark coarse, dark, flaky; inner bark tan, unmarked; younger branchlets appressed-pubescent, becoming glabrate; sap with sweet aroma. Petioles 6–10 mm long, pubescent; blades narrowly elliptic, acuminate, acute at base and decurrent onto petiole, 6.5–15 cm long, 2.5–5

cm wide, glabrate above, sparsely appressed-pubescent below, the surfaces smooth when fresh, minutely warty when dry. Flowers greenish-yellow, 1 or 2 in leaf axils; pedicels 2–2.5 cm long, sparsely to moderately pubescent, articulate ca 5 mm above base; sepals and petals densely sericeous to tomentose; sepals 3, valvate, 3–4 mm long, recurved; petals 6, imbricate, ± equal, rounded at apex, ca 1 cm long; staminal cluster ca 6 mm wide, somewhat 3-sided, burnt orange to tan; anthers ca 1 mm long. Monocarps 6–30, ellipsoid, acute at both ends, 1–2.5 cm long, purplish, 1-seeded, on a stipe to 12 mm long. *Croat 7738, 14040.*

Frequent in the old forest. Flowers throughout the year, mainly from February through August, with some individuals flowering more than once a year. Time of fruit maturity not known.

Known only from Panama, from tropical moist forest on BCI and in Darién and from premontane wet forest in the Canal Zone (Pipeline Road) and Colón.

UNONOPSIS Fries

Unonopsis pittieri Saff., J. Wash. Acad. Sci. 15:102. 1925
Yava blanca

Slender tree, to 20 m tall, with pyramidal crown; stems dark, often lenticellate; parts minutely sericeous when young, glabrate in age. Petioles swollen, stout, ca 6 mm long; blades elliptic to obovate-elliptic, obtuse to acuminate at apex, obtuse to rounded at base, 20–30 cm long, 6–9 cm wide. Flowers solitary in axils on leafless branches; pedicels slender, 2–4 cm long, articulate near middle with small bracteoles above and below the articulation; sepals 3, valvate, 1.5–3 mm long, triangular, connate; petals 6, valvate, sericeous, broadly ovate, thick, the outer three 8–9(20) mm long, the inner 3 thicker, 5–6(15) mm long; stamens ca 1.5 mm long. Monocarps 11 or 12, globose, 1.5–2 cm diam, brown or black at maturity, 1-seeded, on a stipe 1–1.5 cm long. *Croat 9577.*

Infrequent, in the forest, especially the old forest. Seasonal behavior uncertain; probably flowering and fruiting intermittently all year. The flowers have been seen in February, July, and August. Full-sized fruits have been seen throughout the year.

Costa Rica and Panama. In Panama, known from tropical moist forest in the Canal Zone and Bocas del Toro, from premontane wet forest in Colón and Panamá, and from tropical wet forest in Coclé.

See Fig. 239.

XYLOPIA L.

Xylopia frutescens Aubl., Hist. Pl. Guiane Fr. 1:602, t. 292. 1775
Malagueto hembra

Slender tree, to 19(25) m tall; branches clustered toward apex of tree, arcuate, ascending; stems when young pilose, in age dark, glabrate, and lenticellate. Petioles pilose, 3–5 mm long; blades lanceolate, acuminate, acute to ob-

Fig. 242. *Virola sebifera*

Fig. 243. *Virola sebifera*

tuse at base, 3–6 cm long, 8–10(15) mm wide, stiff, glabrous above, sericeous below. Flowers ca 1 cm long, in short axillary clusters of 1–5 each, sweetly aromatic; pedicels very short, bracteate; sepals 3, ovate, to 2.3 mm long, connate at base, sericeous outside, glabrous inside; petals 6, valvate, greenish-white and ± erect at anthesis, oblong, blunt at apex, the outer 3 to 11 mm long and 2.5 mm wide, sericeous outside, the inner 3 slightly shorter and much narrower; stamens greenish, ca 1.3 mm long; style triangular-conic, to 5.3 mm long, white, fleshy, soon deciduous; stigma simple. Monocarps irregular, ca 1.5 times longer than wide, to 1.2 cm long and 7 mm wide, rounded at both ends, orange to red, (1)2(3)-seeded, the short stipe off-center; seeds ovoid, black, ca 6 mm long. *Croat 4868, 8418.*

Occasional to locally common along the margin of the lake and in the young forest. Flowers mainly from April to June. The fruits are mature mostly in the dry season, from January to April (sometimes from November).

Guatemala to southern Brazil. In Panama, ecologically variable; a typical component of tropical dry forest (Holdridge & Budowski, 1956) and tropical moist forest (Tosi, 1971), but known also from premontane dry forest in Coclé, from premontane moist forest in Panamá, from premontane wet forest in the Canal Zone and Panamá, and from tropical wet forest in Coclé.

See Figs. 240 and 241.

Xylopia macrantha Tr. & Planch., Ann. Sci. Nat. Bot., sér. 4, 17:38. 1862
Corobá, Rayado

Small tree, to 10 m tall and 10 cm dbh; stems sericeous-villous when young, glabrous and lenticellate in age. Petioles ca 5 mm long, sericeous to glabrate; blades oblong-elliptic, acuminate, rounded at base, 6–14 cm long, 2–5 cm wide, glabrous above, sericeous below when young, becoming glabrate in age. Flowers solitary at sometimes defoliated leaf axils; pedicels stout, sericeous, to 9 mm long, bracteate; sepals 3, connate to above middle, 1–1.5 cm long, sericeous; petals 6, valvate, thick, the outer 3 oblong, 2–2.5 cm long, ± tomentose, the inner 3 quadrangular-prismatic, 1–1.5 cm long, ca 3 mm thick. Monocarps 10–27, reniform to oblong, 3–4.5 cm long, puberulent, dehiscing along one side and opening out flat to expose seeds, the valves thick and fleshy, red-orange within; seeds several. *Croat 6236.*

Occasional, in the old forest and along the shore on Burrunga Point. Flowers from May to October. The fruits probably mature from February through May.

Panama, Colombia, and Venezuela. In Panama, known from tropical moist forest in the Canal Zone and from tropical wet forest in Colón and Panamá.

56. MYRISTICACEAE

Trees or shrubs with red sap. Leaves alternate, petiolate; blades simple, entire, sometimes stellate-pubescent; venation pinnate; stipules lacking. Flowers unisexual (dioecious), actinomorphic, in axillary panicles, subspicate racemes, or fascicles; perianth 3- or 4-lobed; stamens 3–8, equal or double the number of lobes; filaments partly or completely connate in a column; anthers 2-celled, extrorse, dehiscing longitudinally; ovary superior, 1-locular, 1-carpellate; placentation parietal (seemingly basal); ovule 1, anatropous; style 1; stigma 1. Fruits 2-valved, dehiscent drupes; seed 1, arillate, with ruminate endosperm.

Recognized by the tiny, unisexual flowers with fused stamens, the red sap in the trunk and branches, and the two-valved fruits each with a large arillate seed having ruminate endosperm.

Pollination systems are unknown.

The endozoochorous fruits are taken principally by arboreal frugivores, though many are found on the ground and may be dispersed by rodents and other animals. The aril around the seed of *Virola sebifera* is eaten by white-faced, spider, and howler monkeys (Hladik & Hladik, 1969; Carpenter, 1934). Oppenheimer (1968) reported that white-faced monkeys sometimes swallow the seed. Birds, including motmots, toucans, and trogons, are also fond of *Virola sebifera* (Chapman, 1931).

Eighteen genera and about 300 species; tropics.

VIROLA Aubl.

Virola sebifera Aubl., Hist. Pl. Guiane Fr. 2:904. 1775
V. panamensis (Hemsl.) Warb.; *V. warburgii* Pitt.
Velario colorado, Copidijo, Wild nutmeg, Bogamani, Fruta dorada, Gorgoran, Malaguela de montaña, Tabegua, Mancha

Dioecious tree, to 30(40) m tall, to 30 cm dbh; outer bark hard, dark, minutely fissured vertically (flaking off when slashed); inner bark moderately thick, with ± viscid, acrid sap forming near the wood; branches often clustered in distinct whorls on the trunk; younger branches, underside of leaves, inflorescences and fruits densely pubescent with reddish-brown stellate trichomes; sap reddish, with a sweet aroma. Petioles 8–25 mm long, stout, terete; blades mostly ± oblong to obovate-oblong, long-acuminate, cordate to truncate or rounded at base, mostly 20–40 cm long, 6–15 cm wide, glabrous above. Staminate flowers in much-branched supra-axillary panicles 6–12 cm long; pedicels 1–4 mm long; perianth 1.5–3 mm long, 3- or 4-lobed about one-third of its length; stamens 3(4), equaling number of lobes, fused into a column. Pistillate flowers solitary or clustered in racemes 3–7 cm long; stigma sessile, obscurely bilobed. Capsules globose to

KEY TO THE SPECIES OF VIROLA

Older leaves densely pubescent below with stalked trichomes; fruits 10–30 per inflorescence, dark reddish-brown, usually less than 2 cm wide . . . . *V. sebifera* Aubl.
Older leaves glabrate, the trichomes (if any) sessile; fruits 3–8 per inflorescence, light orange, usually 3–3.5 cm wide . . . . *V. surinamensis* (Rol.) Warb.

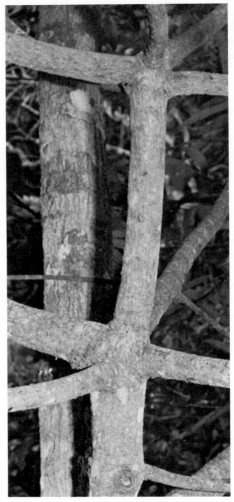

Fig. 244. *Virola surinamensis*

Fig. 245. *Virola surinamensis*

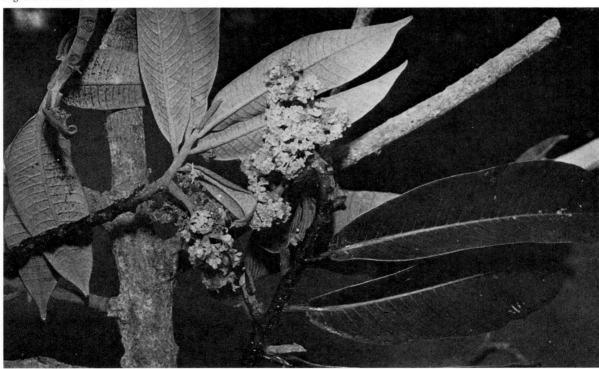

ellipsoid, ca 3 cm long, dark reddish-brown, dehiscing by 2 thick woody valves to expose a single seed; valves bearing scars on the inside made by the aril; seed ca 2 cm long, ellipsoid, the endosperm markedly ruminate, the aril red, irregularly laciniate. *Croat 4275, 4539.*

Frequent, especially in the old forest. Apparently flowering twice a year, from December to April, especially in January and February, and again from June to August. Mature-sized fruits are seen nearly all year, but the length of the fruit maturation period is unknown.

Nicaragua to Peru, Bolivia, and Brazil. In Panama, known from tropical moist forest in the Canal Zone, Bocas del Toro, Chiriquí, Panamá, and Darién, from premontane wet forest in Panamá, and from tropical wet forest in Colón and Darién. Reported from premontane rain forest in Costa Rica (Holdridge et al., 1971).

See Figs. 242 and 243.

Virola surinamensis (Rol.) Warb., Nova Acta Acad. Caes. Leop.-Carol. German. Nat. Cur. 68:208. 1897
V. nobilis A. C. Smith

Dioecious tree, to 30 m or more tall, to ca 60 cm dbh, often moderately buttressed; outer bark coarse, hard, shallowly fissured, reddish-brown; inner bark tan, reddish on its outer surface; branches often spiraled or clustered, extending nearly horizontally; parts when young bearing ferruginous, sessile, stellate pubescence, glabrate in age; sap red, lacking distinctive odor. Petioles canaliculate, 5–10 mm long; blades oblong, acuminate, rounded to acute at base, 9–16 cm long, 1.5–4.5 cm wide, coriaceous; major lateral veins in 20–30 pairs. All parts of inflorescences densely short-pubescent, the trichomes mostly stellate; pedicels ca 1.5 mm long; perianth ca 2 mm long, 3- or 4-lobed usually to middle or beyond, the lobes thick, acute to rounded at apex, spreading at anthesis; staminate flowers in fascicles on panicles to 4 cm long; anthers mostly (2)3(6), connate to apex. Pistillate flowers in clusters of 3 to many, in racemes to 5 cm long; ovary 1-carpellate, ± ovate; stigma sessile, 2-cleft. Capsules ovoid-ellipsoid, thick-walled, light orange, 3–3.5 cm long, bearing dense, short, stellate pubescence; valves 2, woody, ca 5 mm thick, splitting widely at maturity; seed 1, ellipsoid, ca 2 cm long, the aril deeply laciniate, red at maturity (white until just before maturity), fleshy, tasty but becoming bitter soon after being chewed. *Croat 7488, 8090.*

Common in the forest, especially in the older forest; probably less abundant than *V. sebifera.* Flowers from June to March, especially November to February. The fruits mature from April to August (sometimes from February).

Costa Rica and Panama, the Guianas, and Brazil; Lesser Antilles. In Panama, known from tropical moist forest on BCI and in adjacent parts of the Canal Zone and from premontane wet forest in Panamá (Cerro Azul). Reported from tropical wet forest in Costa Rica (Holdridge et al., 1971).

See Figs. 244 and 245.

57. MONIMIACEAE

Strongly scented shrubs or small trees. Leaves opposite, petiolate; blades simple, generally entire, sometimes stellate-pubescent; venation pinnate; stipules lacking. Flowers unisexual (monoecious or dioecious), actinomorphic, in axillary or cauliflorous cymes; tepals 4–8, connate, borne on a hypanthium; stamens many; anthers 2-celled, basifixed, dehiscing longitudinally; pistils many, 1-carpellate; ovaries superior, 1-locular; placentation parietal; ovule 1 per locule, anatropous; styles simple, basally connate. Fruits drupelike aggregates, enclosed in the hypanthium, opening irregularly; seeds arillate, several, with copious oily endosperm.

The family is represented on BCI only by *Siparuna,* which can be distinguished by the very aromatic sap, by the greenish dioecious flowers having a conspicuous hypanthium and no obvious tepals, and by the aggregate, berrylike fruits, which rupture irregularly at maturity to expose the colorful interior.

Pollination system is unknown.

The hypanthium dehisces like a fruit, and the arillate seedlike fruitlets are probably bird dispersed. Though the fleshy hypanthium might be eaten by larger animals, it is not tasty.

Twenty genera and 150 species (Willis, 1966) or 32 genera and 350 species (Lawrence, 1964); chiefly in the Southern Hemisphere tropics, especially Madagascar, Polynesia, and Australia.

SIPARUNA Aubl.

Siparuna guianensis Aubl., Hist. Pl. Guiane Fr. 2:865. 1775
Hierba de pasmo, Pasmo

Monoecious shrubs or small trees, mostly to 5 m tall but occasionally taller; stems reddish, stellate-puberulent particularly when young, malodorous. Leaves opposite; petioles 5–8 mm long; blades elliptic-oblong, acuminate, round to obtuse at base, 8–18 cm long, 4–8 cm wide, densely and minutely pellucid-glandular, sparsely strigillose and stellate-puberulent above, sparsely stellate-

KEY TO THE SPECIES OF SIPARUNA

Blades sparsely stellate-puberulent below, not soft to the touch; petioles less than 1 cm long . . . . . .
. . . . *S. guianensis* Aubl.
Blades densely stellate-pubescent below, soft to the touch; petioles more than 1 cm long . . . .
. . . . *S. pauciflora* (Beurl.) A. DC.

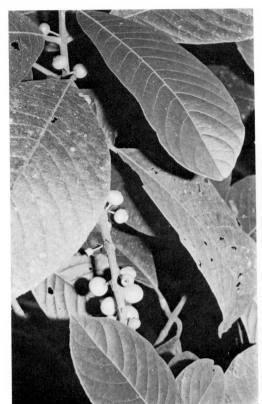

Fig. 246. *Siparuna pauciflora*

Fig. 247. *Siparuna pauciflora*

Fig. 248. *Beilschmiedia pendula*

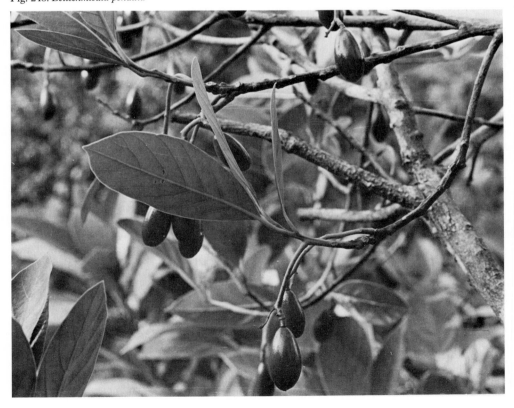

puberulent below, densely yellowish-puberulent when young; major lateral veins in 7–9 pairs. Flowers greenish, in axillary cymes, the cymes 1 or 2 on each side of stem, to 1.5 cm long (to 3 cm long in fruit), mostly unisexual, with 5–15 flowers, the puberulence yellowish, stellate; flowers to 2 mm long on pedicels to 3 mm long; tepals 4–6, deltoid; staminate flowers with 10–14 stamens; pistillate flowers with basally connate styles. Fruits several per cyme, drupaceous, aggregate, enclosed in the hypanthium, ± globose, to 1.5 cm diam, reddish-yellow when mature, bursting irregularly; seeds about 6, tuberculate, grayish. *White 250.*

Not collected in recent years on the island; several earlier collections were made, but the species may have been replaced in the course of succession in recent years. Elsewhere in the Canal Zone, the plant is occasional in disturbed areas. Flowers in April and May. The fruits mature from August to October. What is referred to as seeds within the hypanthium are the fruits of individual flowers, but it seems best to think of them as seeds of a single fruit.

Costa Rica to Colombia, Peru, and Brazil. In Panama, known from low to middle elevations in all regions of tropical moist forest (except in Bocas del Toro) and from premontane wet forest in Chiriquí (Tolé) and Panamá (Lago Cerro Azul).

Siparuna pauciflora (Beurl.) A. DC. in DC., Prodr.
16(2):696. 1868
Limoncillo, Pasmo, Pasmo tetano

Dioecious shrub or small tree, usually 4–6(10) m tall, shortly stellate-pubescent all over, densely so except on upper leaf surfaces; upper branches quadrangular or flattened; sap aromatic. Leaves opposite; petioles 1.5–4 cm long; blades broadly elliptic-obovate (may be narrowly elliptic elsewhere), generally short-acuminate to acute at apex, ± obtuse at base, 15–32 cm long, 7–17 cm wide, thin, obscurely pellucid-punctate, strigillose above but with densely stellate-pubescent veins, densely stellate-pubescent below; lateral veins in 12–15 pairs. Flowers usually many, in cymes congested in axils or borne on stems below leaves; pedicels 2–10 mm long; tepals 4–8, connate to form a broadly flattened annulus with a small central orifice, pale orange at anthesis; staminate inflorescences often short-pedunculate; stamens usually 12–30, ca 1 mm long, in several series, the filaments flattened laterally; pistillate inflorescences ± ses-

sile, the styles many, short, densely papillate, exserted through the central orifice. Fruits drupaceous, aggregate, enclosed in the hypanthium, globose, with a crateriform depression at apex, 1.5–2 cm diam, light green at maturity, bursting irregularly to display usually fewer than 10 seeds; seeds ± oblong, verrucose, enveloped at least on one side by a thin, fleshy, rose-colored aril. *Croat 6284, 13216.*

Frequent in the forest. Flowers regularly from January to April. The fruits may reach full size by June, but do not mature until September or October.

Costa Rica to Peru; mostly at low elevations. In Panama, widespread and common throughout tropical moist forests; known also from premontane wet forests in Panamá (Cerro Azul) and Chiriquí (Tolé) and from tropical wet forest in Coclé (above El Valle), Panamá (slopes of Cerro Jefe), and Darién.

See Figs. 246 and 247.

58. LAURACEAE

Trees or shrubs, usually with weakly aromatic sap. Leaves alternate, petiolate; blades simple, entire or undulate; venation pinnate; stipules lacking. Flowers bisexual, or unisexual and dioecious (in *Ocotea*), actinomorphic, in terminal or axillary panicles, often clustered; perianth undifferentiated, arising from a hypanthium, deeply 6-lobed, the lobes unequal or subequal, generally white, often fragrant, the tubular part persisting in fruit; stamens free, epipetalous, in 3 series of 3 each, variously modified, the outer and middle series of 3 each subtended by and alternating with glands, the inner series usually bearing glands and alternating with staminodia (these ideally a fourth series of stamens), their filaments usually longer than those of the outer series, the glands usually conspicuous; anthers 2- or 4-celled (if 4, usually superimposed, but arclike in *Nectandra*), basifixed, dehiscing by flaplike valves opening upward, the outer 2 series introrse, the inner series extrorse, usually longer than the style; ovary superior, 1-locular, seemingly 1-carpellate; placentation parietal; ovule 1, anatropous, pendulous; style 1; stigma simple. Fruits drupes, often surrounded at the base by the cupular, persistent perianth tube; seed 1, lacking endosperm.

Certainly one of the most poorly known families of tropical America. Individuals of the family are among the principal components of the most poorly known lowland

KEY TO THE TAXA OF LAURACEAE
(ON THE BASIS OF TRADITIONAL SEXUAL CHARACTERS)

Anthers 2-celled; fruits lacking cupule; lower leaf surface with dense, minute, very appressed, whitish trichomes and prominulous, very closely spaced reticulate veins, the major lateral veins essentially glabrous . . . . *Beilschmiedia pendula* (Sw.) Hemsl.
Anthers 4-celled; fruits with or without cupule; lower leaf surface glabrous, with erect trichomes, or if with dense appressed trichomes then lacking closely spaced prominulous reticulate veins with glabrous lateral veins:
Fruits very large, usually 10–12 cm long at maturity, lacking a cupule; staminodia large, cordate, stipitate (also true of *Phoebe*); lower leaf surface bearing conspicuous, erect pubescence . . .
. . . . *Persea americana* Mill.

Fruits small, less than 6 cm long, borne in a cupule; staminodia inconspicuous and sessile or
lacking; lower leaf surface glabrous or nearly so or bearing ± appressed pubescence (±
erect but ascending in *Nectandra cissiflora*):
Blades 3-veined, the lateral veins prominently pliveined at base, extending to near apex; cupule of fruit conspicuously lobed . . . . *Phoebe mexicana* Meisn.
Blades not 3-veined, not pliveined, the lateral veins not extending to near apex; cupule of fruit
truncate:
Perianth lobes reflexed or spreading, usually fleshy; outer series of anthers fan-shaped, the
4 thecae arranged in a gradual arc; flowers bisexual . . . . *Nectandra*
Perianth lobes erect, often thin; outer series of anthers broadly oblong, the 4 thecae arranged in 2 distinct or only slightly overlapping planes; flowers unisexual . . . . *Ocotea*

KEY TO THE TAXA OF LAURACEAE
(CHIEFLY ON THE BASIS OF STERILE CHARACTERS)

Leaf blades definitely pubescent on lower surface:
Pubescence of lower blade surface erect or at least not closely appressed:
Blades usually broadly rounded at apex or with an abrupt, short acumen, the upper surface
glabrous; flowers ca 10 mm wide; fruits more than 8 cm long . . . . *Persea americana* Mill.
Blades usually gradually long-acuminate at apex, the upper surface moderately pubescent at
least on midrib; flowers less than 7 mm wide; fruits less than 6 cm long . . . .
. . . . *Nectandra cissiflora* Nees
Pubescence of lower blade surface mostly appressed:
Axils of lower blade surface bearing distinct domatia (pits or tufts of trichomes):
Axillary domatia of elongated slitlike pits; blades usually broadest above the midlle . . . . . .
. . . . *Ocotea oblonga* (Meisn.) Mez
Axillary domatia of ± rounded tufts of villous pubescence, never of slitlike pits:
Flowers 10 mm or more wide, the perianth lobes broadly spreading; fruits ca 1 cm long
. . . . *Nectandra globosa* (Aubl.) Mez
Flowers less than 10 mm wide, the perianth lobes ± erect; fruits 4–6 cm long . . . .
. . . . *Ocotea skutchii* C. K. Allen
Axils of lower blade surface usually lacking domatia . . . . *Beilschmiedia pendula* (Sw.) Hemsl.
Leaf blades essentially glabrous (sometimes with axillary tufts of trichomes):
Blades 3-veined or with the lower pairs of lateral veins longer than the middle and upper pairs:
Blades 3-veined, markedly pliveined at base, the lateral veins extending nearly to apex; calyx
lobes persistent in fruit . . . . *Phoebe mexicana* Meisn.
Blades with several pairs of lateral veins, not pliveined, the lateral veins not extending to near
apex; calyx lobes deciduous, forming a truncate cup at maturity:
Perianth lobes erect, ca 1 mm long, glabrous; peduncles usually less than 1.5 cm long;
blades usually prominently arched along midrib . . . . *Ocotea cernua* (Nees) Mez
Perianth lobes spreading, ca 1.5–2 mm long, densely pubescent; peduncles usually more
than 2 cm long; blades usually not prominently arched along midrib . . . .
. . . . *Nectandra savannarum* (Standl. & Steyerm.) C. K. Allen
Blades not 3-veined, the lower pairs of lateral veins shorter than the middle and upper pairs:
Blades bearing domatia in axils of at least lowermost lateral veins:
Blades widest at or below middle, markedly arched along midrib; reticulate venation of
upper surface prominulous; flowers bisexual; perianth lobes spreading; fruits rounded,
to 1.5 cm long . . . . *Nectandra purpurascens* (R. & P.) Mez
Blades widest at or above middle, not arched along midrib; reticulate venation of upper
surface not prominulous; flowers unisexual; perianth lobes erect; fruits ± oblong or
ellipsoid, 4–6 cm long . . . . *Ocotea skutchii* C. K. Allen
Blades lacking axillary domatia . . . . *Ocotea pyramidata* Brandegee

forest regions. They are often widely dispersed and difficult to distinguish and collect. Nearly three times as many
Lauraceae are now known from BCI as when Standley
wrote his flora; this increase is probably greater than for
any other group, and no doubt even more species will
still be found.

Lauraceae are easily recognized by their unusual primitive flowers, with stamens generally in two or three series
alternating with the glands, and especially by the flaplike
valves over the thecae of the anthers. The fruits are
equally distinctive, frequently being subtended by or set

in an often colorful cupule. The plants are often large
trees, usually with pleasant-smelling sap. The leaves
have lateral veins that usually extend downward along
the side of the midrib before finally merging with it.
Many other tropical species have this "lauraceous vein-
ing" but none have the characteristic as frequently or
conspicuously as do the Lauraceae.

The flowers are primitive and open. Pollination systems are unknown.

Smaller-fruited species, such as *Nectandra globosa, N.
purpurascens, Phoebe mexicana*, and *Ocotea cernua*, are at

least in part bird dispersed, whereas the larger-fruited species are probably dispersed chiefly by arboreal mammals. Both *Ocotea skutchii* and *Beilschmiedia pendula* fruits have been found on the forest floor with the thin exocarp scraped off, apparently the work of monkeys or birds. Large birds often regurgitate seeds after they have removed the outer layer. This may explain how many Lauraceae fruits, suited to bird dispersal but with seeds poorly protected, can survive being eaten by birds.

Some 45 genera with 2,000–2,500 species; mostly in the subtropics and tropics of Asia and America.

BEILSCHMIEDIA Nees

Beilschmiedia pendula (Sw.) Hemsl., Biol. Centr.-
  Amer. Bot. 3:70. 1882
  Hufelandia pendula (Sw.) Nees

Tree, 13–40 m tall, to 75 cm dbh; buttresses ± lacking or to 1 m; outer bark reddish-brown and scarcely fissured on younger trees, becoming loosened in squarish patches in age, the older parts becoming marked with shallow depressions; inner bark granular with darker lines in tangential section; stems with dense, short, brown, appressed pubescence, the younger stems ribbed; sap sweet, with pleasant aroma. Petioles 8–15(25) mm long, canaliculate; blades elliptic to oblong-elliptic, abruptly short-acuminate (rarely with a long acumen), cuneate to attenuate at base, decurrent onto petiole, mostly 10–20 cm long, 5–9 cm wide, ± glabrous above, the pubescence below short, whitish, appressed; reticulate veins prominulous, very closely spaced. Panicles terminal and from axils of older leaves, 10–15 cm long; flowers bisexual, greenish-yellow, ca 3 mm long; hypanthium lobes 6, ovate, thin at margin, pubescent inside and out; stamens 9, 2-celled, the connective produced and ± fleshy at apex, the outer series of 6 somewhat shorter than the inner 3, 1.3–1.7 mm long, pubescent and ciliate, subtended by and alternating with 6 fleshy yellow glands; filaments stout, fused to perianth lobes, the inner series of 3 stamens alternating with shorter ovate staminodia; ovary glabrous; style simple, shorter than inner stamens. Fruits lacking a cupule, oblong or ellipsoid, shiny, to 5.5 cm long and 2 cm diam, purple-brown, with a single embryo and 2 large yellow-green cotyledons surrounded by a thin green pericarp ca 1.5 mm thick. *Croat 12928, 14063.*

Common in the old forest. Flowers in the early dry season (December and January). The fruits attain full size by March, but ripen during May and June. Leaves are replaced slowly at about the time of flowering.

Not confused with any other species.

Wilson (1971) reported that the fruits are taken by the bat *Micronycteris hirsuta*. Monkeys eat the thin outer skin and the scant mesocarp, discarding the remainder. Seeds germinate soon after falling.

Costa Rica and Panama; West Indies. In Panama, known from tropical moist forest in the Canal Zone and from tropical wet forest in Los Santos (Coabal). Possibly also in Ecuador (*Little 6294*).

See Fig. 248.

NECTANDRA Rol. ex Rottb.

The genus *Nectandra* is easily confused with *Ocotea*. On BCI, *Nectandra* may be distinguished by having bisexual flowers with the perianth lobes fleshy to membranaceous and spreading to reflexed. The outer series of anthers is ovate and more or less flattened, with their four thecae arranged in an arc on the inner surface. The inner series of anthers is more or less quadrangular, the four cells arranged in two overlapping planes, the upper ones laterally dehiscent, the lower ones extrorsely dehiscent.

Nectandra cissiflora Nees, Syst. Laur. 296. 1836

Tree, to 27 m tall and 30 cm dbh; trunk weakly involute throughout much of its length, with warty protuberances (probably deciduous branch bases); outer bark unfissured, thin, obscurely lenticellate; inner bark thin, granular; young stems densely rufous-tomentose, with a flat-topped rib extending below each petiole, these persisting on older stems; sap with faint sweet aroma. Petioles 1–2 cm long, densely tomentose; blades oblong-elliptic to oblanceolate, gradually long-acuminate and often twisted at apex, attenuate to acute and decurrent at base, 10–24 cm long, 4–8 cm wide, moderately pubescent above at least on midrib (the trichomes more dense on veins, brownish, straight, leaning forward), tomentose on veins below (the trichomes not completely appressed), otherwise the surface below moderately appressed-pubescent, the margins ± revolute; midrib arched. Panicles from upper axils, to 20 cm long, with a slender axis, branched mostly in the upper two-thirds; axes, peduncles, and pedicels

KEY TO THE SPECIES OF NECTANDRA

Lower leaf surface distinctly pubescent:
  Pubescence of major veins of lower leaf surface mostly closely appressed; flowers usually more
    than 1 cm wide . . . . *N. globosa* (Aubl.) Mez
  Pubescence of major veins of lower leaf surface mostly erect, never closely appressed; flowers
    usually less than 7 mm wide . . . . *N. cissiflora* Nees
Lower leaf surface essentially glabrous:
  Blades usually with fewer than 6 pairs of lateral veins, the lowermost pairs much longer than
    those at middle of blade . . . . *N. savannarum* (Standl. & Steyerm.) C. K. Allen
  Blades usually with more than 7 pairs of lateral veins, the lowermost pairs usually much shorter
    than those at middle of blade . . . . *N. purpurascens* (R. & P.) Mez

densely short-pubescent; flowers bisexual, mostly clustered at ends of branches; pedicels 1–3 mm long in flower, 5–12 mm long in fruit, densely grayish-puberulent; flowers cream-colored, ca 5 mm wide; perianth lobes sessile, subequal, spreading at anthesis, ovate-triangular, acute at apex, weakly grayish-pubescent outside, granular-puberulent inside; stamens 9, the 6 outer ones opposite perianth lobes, sessile, the outer side weakly pubescent, the subtending glands fused into a ring around the ovary, about half as high as the outer stamens, the 3 inner stamens apparently fused to the glandular ring; ovary ovoid; style short; stigma simple. Fruits fleshy, 1-seeded, round, to 1.3 cm diam, the cupule green, obconical, 9–13 mm long and ca 1 cm wide at apex, glabrous to weakly pubescent. *Croat 14989.*

Rare; known above the escarpment north of Zetek Trail and on the slope northwest of the laboratory. Saplings are seen in the old forest. Seasonal behavior uncertain; probably fruits in the early rainy season.

The plant north of Zetek Trail has sucker shoots near the base with densely rufous-tomentose stems.

Panama to Peru and Brazil. In Panama, known only from tropical moist forest on BCI.

Nectandra globosa (Aubl.) Mez, Jahrb. Königl. Bot. Gart. Berlin 5:415. 1889
Sigua, Sweetwood

Tree, usually 3–10(15) m tall; wood soft, light in weight; branchlets and inflorescences minutely and densely pubescent. Petioles stout, to 1.5 cm long; blades elliptic to oblong-elliptic, acuminate and often ± twisted to one side, rounded to acute at base, 10–32 cm long, 3–9 cm wide, ± glabrous (sparsely and obscurely pubescent) and shiny above, dull below and densely and minutely appressed-pubescent, often with tufts in axils. Panicles axillary, usually at end of branches, mostly 7–20 cm long; flowers bisexual, white, with sweet odor, to 13 mm wide; perianth lobes 6, acute to obtuse, broadly spreading, fleshy, pubescent outside, papillose inside; stamens almost sessile, about as broad as long, topped by a broad, white, papillose connective about half their height, the outer 6 subtended by and alternate with fleshy rounded glands, the inner 3 held above the stigma, affixed to a thick disk, alternating with staminodia, the disk about as high as ovary; ovary glabrous, round; style about as long as ovary; stigma globular, probably receptive before pollen is released. Fruits ellipsoid, ca 1 cm long, black at maturity. *Croat 4888, 7847.*

Frequent along the shore, at least in some areas of the northern part of the island and on Orchid Island; occasional elsewhere along the shore. Flowers from late November to early April, chiefly from December to March. The fruits mature mostly during April and May.

Distinguished from all other Lauraceae on BCI by the large flowers. *Nectandra globosa* is uniformly and persistently pubescent on the lower leaf surface. It can be confused with *N. glabrescens* Benth., which occurs along the Atlantic slope of Panama and in South America to the Guianas, Bolivia, and Brazil (Planalto). *N. glabrescens* has leaves much less pubescent, at least in age, and flowers

principally from July through October. It also differs from *N. globosa* in having a depressed-globose ovary with a very short style.

Mexico to Panama, the Guianas, and Peru; West Indies. In Panama, known from tropical moist forest in the Canal Zone, Bocas del Toro, Panamá, and Darién. Reported from premontane rain forest in Costa Rica (Holdridge et al., 1971).

Nectandra purpurascens (R. & P.) Mez, Jahrb. Königl. Bot. Gart. Berlin 5:443. 1889
N. latifolia (H.B.K.) Mez
Sigua blanca, Sigua negra

Small tree, to 17 m tall and 27 cm dbh, glabrous except young parts with short appressed trichomes. Petioles 1–1.5 cm long, slightly canaliculate; blades ovate to elliptic or lanceolate-elliptic, long-acuminate and sometimes falcate at apex, acute to rounded at base, 6–20 cm long, 3–6 cm wide, usually bearing weak tufts of trichomes (domatia) in lower vein axils on underside of leaves; midrib arched, the reticulate venation conspicuous on both surfaces. Panicles 6–12 cm long, terminal or in upper axils; flowers bisexual, white, ca 5 mm wide, with sweet aroma, in usually subcorymbiform panicles; perianth lobes 6, fleshy, spreading, densely short-pubescent, the inner 3 usually narrower; stamens exserted, the outer 6 with short filaments and the anthers ± reniform, the glands white or orange, subtending and alternating with outer stamens (often larger than anthers), the 3 inner stamens closely surrounding and ± equaling the length of the style, alternating with 3 shorter staminodia; ovary glabrous. Fruits fleshy, 1-seeded, ± rounded, green and speckled, becoming black at maturity, to 1.5 cm diam, the pedicel and cupule conical, bright red, together ca 1 cm long. *Croat 9556.*

Frequent along the shore, especially on the north side of the island; occasional in the younger forest. Some flowers may be seen in all months of the year, but flowering occurs principally in two waves, one at the beginning of the dry season (December to March) and one at the beginning of the rainy season (May to July). The fruits usually develop within a few months, but some have been seen as late as October. Individual plants flowering in March were seen again with both flowers and fruits in the rainy season.

Standley mistakenly reported this species as *N. glabrescens* Benth., which has affinities with *N. globosa* and occurs in Panama only along the Atlantic slope.

Nicaragua to Panama, south to Ecuador, Peru, and Brazil; at low elevations. In Panama, known principally from tropical moist forest in the Canal Zone, but also from tropical wet forest in Colón.

See Fig. 249.

Nectandra savannarum (Standl. & Steyerm.) C. K. Allen, J. Arnold Arbor. 26:382. 1945

Tree, 7 m tall; stems obscurely and densely strigillose. Petioles pubescent, 5–9 mm long, canaliculate; blades narrowly ovate to lanceolate-elliptic, gradually acuminate

to caudate-acuminate, acute to attenuate at base, 5–17 cm long, 1.5–5 cm wide, moderately thin, essentially glabrous above, nearly glabrous below but with inconspicuous tufts in axils, the margins weakly undulate; lateral veins usually in 4–6 pairs, the reticulate venation inconspicuous especially above. Inflorescences 2–10 cm long, slender; peduncles 1.5–4.5 cm long, the branches short; pedicels ca 3 mm long; inflorescence branches, pedicels, and hypanthia tomentulose especially the terminal parts; flowers bisexual, 5–6 mm wide; perianth lobes 6, spreading, ca 1.5–2 mm long, glandular-tomentose to villous inside and out, hispid outside toward base; outer stamens 6, obovate, ca 0.5 mm long, the short filaments and outer surface of anthers pubescent, the inner stamens 3, spatulate, ca 0.7 mm long, the filaments almost equaling anthers, pubescent, the anthers pubescent to near apex; staminodia club-shaped, shorter than and alternating with inner stamens; glands subglobose, glabrous, ca 0.5 mm high; ovary ovoid, ca 1 mm long, glabrate, tapered abruptly to style; style equaling inner stamens. Fruits subglobose, apiculate, the cupule coral-orange, to 4 mm long, ca 8 mm diam and 3–4 mm deep. *Foster 960.*

Rare; collected once along the shore on the east side of Peña Blanca Peninsula southwest of Orchid Island. Seasonal behavior uncertain. *Foster 960* had flowers in early June; elsewhere in Central America flowering specimens have been collected principally in the dry season.

Description of the fruits, which are incompletely known, is based on Allen (1945).

The BCI material differs only slightly from specimens from other areas in Central America that usually have leaves more coriaceous and frequently much more pubescent on the stems and underside of leaves, rarely also on the midrib and veins above.

Guatemala, Belize, Honduras, and Panama; no doubt in areas between Honduras and Panama. In Panama, known only from tropical moist forest on BCI.

OCOTEA Aubl.

The genus *Ocotea* is easily confused with *Nectandra*. On BCI, *Ocotea* can be distinguished by having unisexual flowers (dioecious) with the perianth lobes usually thin

and erect at anthesis. The outer series of anthers is more or less fan-shaped with the four thecae arranged in two distinct horizontal planes on the inner surface. The inner series of anthers is more or less quadrangular with the four thecae arranged in two distinct planes. The two upper thecae are laterally dehiscent, and the two lower extrorsely dehiscent.

Ocotea cernua (Nees) Mez, Jahrb. Königl. Bot. Gart. Berlin 5:377. 1889
O. caudata (Nees) Mez
Sigua, Encibe

Dioecious tree, to 12 m tall, glabrous but sometimes with sparse pubescence on young stems, petioles, and midribs below and dense pubescence at tips of stems; stems weakly ribbed below petioles. Petioles 8–20 mm long; blades elliptic to oblong-elliptic, usually long-acuminate, acute to obtuse at base, 4.5–18 cm long, 2–7 cm wide, the margins minutely undulate; midrib arched and raised, the veins in 3–5 pairs. Flowers cream-white, unisexual, with aroma of fresh peaches, 3–4 mm wide, in solitary axillary panicles usually 3–7 cm long; peduncles to ca 1 cm long; pedicels 1–4 mm long; perianth lobes 6, slightly unequal, ovate, blunt to acute, slightly more than 1 mm long, erect at anthesis, glabrous outside, hispidulous inside near base, borne on a campanulate hyphanthium; stamens of staminate flowers in 2 series, the anthers of the outer 6 sessile, ca 0.8 mm long, opposite and fused to base of perianth lobes, subtended by and alternating with glands, the glands broader than long, the inner stamens 3, ca 1 mm long, hispidulous on inner surface, bearing 2 small yellowish glands on outer side at base, the filaments short, the anthers oblong; pistillode cylindrical, ca 0.5 mm long; pistillate flowers similar to staminate flowers but with smaller, apparently nonfunctional stamens (about half as high as pistil); ovary and style together to 1.5 mm long, the style short; stigma conspicuous, tripartite. Fruits ellipsoid, apiculate, to 1.5 cm long and 1 cm wide, black, the lower third tightly set into cupule, the cupule to 6 mm long and 11 mm wide, brownish. *Croat 15559 ♂, 8116 ♀, Shattuck 1140.*

Uncommon, along the shore, especially on the north side of the island and in the young forest. Flowers from

---

KEY TO THE SPECIES OF OCOTEA

Leaves definitely pubescent:
  Leaf blades with slitlike pits in axils of major lateral veins . . . .*O. oblonga* (Meisn.) Mez
  Leaf blades with ± rounded villous tufts in axils of major lateral veins . . . . . *O. skutchii* C. K. Allen
Leaves glabrous or essentially so:
  Leaf blades bearing tufts of trichomes in axils of major lateral veins . . . . *O. skutchii* C. K. Allen
  Leaf blades lacking axillary tufts:
    Leaves with fewer than 5 pairs of lateral veins, the blades usually less than 5 cm wide, usually markedly arched along the midrib and caudate-acuminate at apex; midrib usually raised; major lateral veins not whitish on upper surface; perianth lobes ca 1 mm long . . . .
    . . . . *O. cernua* (Nees) Mez
  Leaves with more than 7 pairs of lateral veins, the blades usually more than 5 cm wide, usually not markedly arched along midrib, merely acuminate, not caudate-acuminate; midrib usually sunken or flat; major lateral veins whitish on upper surface; perianth lobes 2–2.7 mm long . . . . *O. pyramidata* Brandegee

Fig. 249. *Nectandra purpurascens*

Fig. 250. *Ocotea skutchii*

Fig. 251. *Phoebe mexicana*

February to March (sometimes to May and rarely as late as July). The fruits mature mostly in August and September.

The species is not separable from *O. caudata* (Nees) Mez, which ranges widely in South America. The character used by Mez, i.e., degree of prominence of reticulate veins, is not adequate for separation of the South American material.

Southern Mexico to Panama, Guyana, Peru, and Brazil (Mato Grosso); Windward Islands. In Panama, known from tropical moist forest in the Canal Zone, Bocas del Toro, Chiriquí, and Darién. Known from premontane wet forest in Costa Rica (Holdridge et al, 1971).

Ocotea oblonga (Meisn.) Mez, Jahrb. Königl. Bot.
Gart. Berlin. 5:366. 1889
Sigua

Dioecious tree, to 30 m tall, to 45 cm dbh, conspicuously buttressed; bark coarse, conspicuously lenticellate, becoming moderately fissured on exposed parts above; branches mostly from near apex, widely spreading; young stems densely ferruginous-tomentose, with 2 ribs below the petioles; sap with ± pungent odor. Leaves drying blackened; petioles mostly 1–2 cm long, densely appressed-puberulent; blades narrowly elliptic to broadly oblanceolate, acute to acuminate with the acumen broadly rounded, narrowly acute to attenuate at base, 10–22 cm long, 3–6 cm wide, glossy and glabrous above except along the prominently raised, squarish midrib, densely and obscurely appressed-pubescent below, with slitlike pit domatia near the axils, the domatia usually puberulent within, directed at a sharp angle to the lateral veins; major lateral veins in about 7 pairs, the reticulate veins obscure. Panicles axillary or terminal, to 13 cm long; flowers, pedicels, and inflorescence branches densely short-pubescent; flowers unisexual; staminate flowers not available; pistillate flowers to 1.7 mm long, about as wide as long, greenish-white at anthesis; perianth lobes 6, erect to spreading, ovate, ca 1.5 mm long, densely papillose-puberulent on inside; outer series of stamens 6, ca 0.7 mm long, the filaments short, weakly pubescent on outside near base, the glands ± orbicular and flattened, the inner stamens 3, the filaments about half as long as anthers, pubescent; anthers ± quadrangular, pubescent inside on basal half; ovary ovoid, ± glabrous; style short; stigma conspicuous, held just above the inner stamens. Fruits ellipsoid, to 14 mm long and 8 mm wide, the basal cupule subflattened, ca 5 mm wide. *Aviles 9, Croat 16201, 16515.*

Rare; a few individuals known in the old forest south of the big trees on Armour Trail. The flowers have been seen on BCI during July and August.

Determination of the BCI plants was made by C. K. Allen. Staminate flowers remain unknown.

Characters of the fruits are taken from the type description.

Bolivian specimens bearing this name may be the same, but the leaves lack the slitlike domatia in the axils of the lateral veins.

Panama and the Guianas, but no doubt more widespread in South America. In Panama, known only from tropical moist forest on BCI.

Ocotea pyramidata Blake ex Brandegee, Univ. Calif.
Publ. Bot. 7:326. 1920
Sigua

Dioecious tree, to ca 27 m tall and ca 50 cm dbh; outer bark unfissured, with numerous raised lenticels; inner bark inconspicuous; branchlets minutely and densely puberulent, becoming glabrous, conspicuously flattened, with a rib extending downward from each petiole. Petioles 1–3 cm long, canaliculate above; blades elliptic to lanceolate-elliptic, acuminate, somewhat falcate and downturned apically, acute at base, 7–20 cm long, 2.5–6.5 cm wide, glabrous or minutely appressed-pubescent on lower surface especially on veins; midrib arched, usually sunken above, at least apical part of midrib and the major lateral veins whitish, the major lateral veins in 7–10 pairs, the reticulate venation prominulous on both sides, especially below, and whitish above. Panicles axillary or subterminal, to 7 cm long, usually in upper axils (rarely from older branches); flowers unisexual; staminate flowers greenish-white, ca 5 mm long; pedicels ca 1.5 mm long; perianth lobes 6, thin, blunt, ca 2.7 mm long and 2 mm wide, puberulent inside near base; stamens included, the outer series of 6 opposite perianth lobes, 2–2.5 mm long, subtended by and alternating with glands, the filaments ± flattened, about equaling anthers, fused toward base to perianth lobes, the anthers ovate, the glands ca 0.7 mm wide, cushion-shaped, glabrous, the 3 inner stamens slightly shorter than the outer series, surrounded by the glands, their anthers narrowly ovate, held closely around style, shedding pollen after lengthening and overtopping stigma; pistillode slender, glabrous, merging imperceptibly with style; stigma held at or just above the level of the anthers of inner stamens; pistillate flowers not available. Fruits unknown. *Croat 12805.*

Rare; known only from Fairchild Trail 1675, on the shore south of Colorado Point, and in the young forest above the escarpment. All the plants seen were staminate. Flowers at the beginning of the dry season.

The BCI plants are a perfect match for the type (*Purpus 8456*) from Zacuapan, Veracruz, Mexico. The species probably also occurs all along the Atlantic slope of Central America.

The three individual plants known to me from BCI are very distant from one another. Though there are undoubtedly more individuals on the island, their infrequency and the relatively short flowering season make pollination a difficult and interesting phenomenon.

Known only from BCI and Veracruz, Mexico.

Ocotea skutchii C. K. Allen, J. Arnold Arbor.
26:352. 1945
O. williamsii P. H. Allen

Dioecious tree, 15–30 m tall; branchlets slender, ± terete, grayish-sericeous when young, glabrate in age; sap with sweet aroma. Petioles short and obscure or to 1.5 cm long; blades oblong-obovate to oblong-elliptic, moderately

acuminate to long-acuminate (the acumen blunt), narrowly acute to attenuate at base, 7–15 cm long, 3–6 cm wide, weakly coriaceous, the upper surface glabrous, the lower surface inconspicuously and ± densely appressed-pubescent, in age glabrous but with well-developed villous axillary tufts; midrib flat or slightly raised above, the lateral veins mostly in 7–9 pairs, mostly branching from midrib at ca 30° angle, the reticulate venation prominulous below. Panicles upper-axillary or subterminal, 6–15 cm long, branched several times, the branches grayish-puberulent, the trichomes becoming appressed on the upper parts of the branchlets; pedicels and hypanthium lobes gray-sericeous; pedicels ca 2 mm long; flowers unisexual, ca 3 mm long, whitish; perianth 6-lobed nearly to base, deciduous from the hypanthium usually as a unit, the lobes ± equal, imbricate, ovate-elliptic, moderately thick, narrowly rounded at apex, papillate-puberulent within; outer stamens 6, suborbicular, somewhat flattened, ca 1 mm long, the filaments very short, merging almost imperceptibly with anthers, papillate-puberulent on outside and along medial part of connective on inside, the glands globose, about half as high as stamens, glabrous except near base, the inner stamens 3, to ca 1.5 mm long, the filaments papillate-pubescent, the anthers as long as filaments, narrowly ovoid, papillate-pubescent on outside at base; pistil ca 1.5 mm long; ovary ovoid, glabrous, about twice as long as style; stigma ± triangular. Fruits ± ellipsoid to narrowly ovoid, 4–6 cm long, 2–2.5 cm wide, minutely apiculate at apex, at first dark green with minute speckles of lighter green, becoming black at maturity; exocarp thin, black; mesocarp less than 2 mm thick, the 2 large cotyledons red; cupule 1.5–2.5 cm long, the upper part saucer-shaped, 1.1–1.4 cm wide at apex, 2–4 mm deep, green to brown or reddish-brown at maturity of fruit. *Croat 8150, 9780, 14846.*

Occasional or rare, in the older forest. Flowers in the late dry season (April). The fruits mature in the early rainy season (late May to July). One collection tentatively determined as this species (*Shattuck 535*) has juvenile fruits in early December. Allen (1956) reported the species (as *O. williamsii* P. H. Allen) to flower in May in Costa Rica.

As the leaves of this species age, they become increasingly darkened on drying and the appressed pubescence of the lower leaf surface almost disappears. In addition, axillary domatia become better developed on mature leaves. Fruits of the species fall at about the same time as those of *Beilschmiedia pendula*, which are the same size and shape, but *B. pendula* differs in having yellowish-green cotyledons rather than the red ones of *O. skutchii*.

Costa Rica and Panama. In Panama, known only from tropical moist forest on BCI.

See Fig. 250.

PERSEA P. Mill.

Persea americana P. Mill., Gard. Dict. ed. 8. 1768
Aguacate, Avocado, Avocado pear

Tree, to 15(20) m tall. Leaves falling shortly before or during time of flowering, soon replaced; petioles 1.5–5 cm long; blades ovate to obovate-oblong, abruptly acuminate to rounded at apex, usually obtuse at base, mostly 10–30 cm long and 6–16 cm wide, glabrous above, softly pubescent below; veins all prominent. Panicles axillary or subterminal; flowers fragrant, bisexual, greenish-white or greenish-yellow, ca 1 cm wide, weakly puberulent all over, 6–7 mm long; perianth lobes 6, acute, spreading at anthesis; stamens 9, ± erect; filaments distinct, puberulent, the outer 6 ca 2.5–3.5 mm long, the inner 3 longer, ± equaling the slender style; anthers with 4 thecae in 2 planes; staminodia prominent, yellow, short-stipitate; ovary ovoid, pubescent; style short; stigma discoid. Fruits ovoid or pear-shaped drupes, usually 10–12 cm long; exocarp thin, leathery; mesocarp thick; seed ovoid, to 5 cm long. *Croat 4162, 7491.*

Cultivated at the Laboratory Clearing and rarely encountered in the forest, presumably persisting from old settlement sites. Flowers mostly in the early dry season. The fruits mature in the early rainy season.

Probably native to Mexico; widely cultivated in Panama and elsewhere in the tropics and subtropics. Reported from premontane wet, tropical wet, premontane rain, lower montane wet, and lower montane rain forests in Costa Rica (Holdridge et al., 1971).

PHOEBE Nees

Phoebe mexicana Meisn. in DC., Prodr. 15(1):31. 1864
Sigua blanca

Tree, to 25 m tall and 30 cm dbh; outer bark planar, minutely lenticellate; inner bark tan, granular, moderately thin; glabrous but with fine, minute trichomes on younger stems, petioles, axes of inflorescences, and pedicels; smaller stems angulate with prominent ribs extending below petioles; sap with a faint sweet odor. Petioles stout, broadly canaliculate, to 2.5 cm long; blades ± elliptic to oblong-elliptic, rounded or bluntly to narrowly acuminate at apex, often falcate and downcurved, obtuse to acute or attenuate at base, 9–29 cm long, 4–14.5 cm wide, conspicuously 3-veined from above base, pliveined at base, coriaceous, the lower axils often weakly tufted with trichomes. Flowers greenish-white, bisexual, 2–3 mm long, in numerous axillary or subterminal racemose panicles 12–20 cm long; pedicels 3–5 mm long; perianth lobes 6, unequal (the outer 3 shorter), ovate, obtuse to rounded at apex, glabrous or weakly pubescent outside, ± sericeous inside, erect at anthesis, persistent in fruit; stamens 9, subequal, usually longer than outer perianth lobes, shorter than inner ones; filaments longer than anthers, the outer 6 fused to perianth lobes in basal half, ca 1.7 mm long; anthers with 4 thecae in 2 definite planes, the inner 3 anthers extrorse, to 2 mm long or more, adjacent to and equaling or longer than style, alternating with 3 cordate stipitate staminodia; glands longer than and alternating with filaments between the outer and inner whorl of stamens; ovary glabrous, ± equaling length of style; style ± simple. Fruits obovoid to ellipsoid, 1–1.5 cm long and 7–10 mm wide, dark green with light green spots, the cupule campanulate, with the enlarged perianth lobes ca 2 mm long, glabrous. *Croat 14819.*

Occasional, along the shore, uncommon in the old forest, and rare in the young forest. Flowers mostly in

May and June. The fruits mature by July and August.

Monographic work will probably show that this species is much more wide ranging. The species is variable and should probably include *P. elongata* (Vahl) Nees of the West Indies, *P. costaricana* Mez & Pitt. of Costa Rica and western Panama, and *P. cinnamomifolia* (H.B.K.) Nees of South America from Colombia and Venezuela to Peru.

Mexico to Panama, Colombia, and Venezuela. In Panama, occasional in tropical moist forest in the Canal Zone, Panamá, and Darién, especially on the Atlantic slope, or at medium elevations of the Pacific slope such as in Panamá (lower slopes of Cerro Campana).

See Fig. 251.

59. CAPPARIDACEAE

Small trees, shrubs, or annual herbs often with spines at nodes. Leaves alternate, sessile or petiolate; blades simple or palmately compound, entire; venation pinnate; stipules lacking. Flowers bisexual, actinomorphic, in terminal or upper-axillary, bracteate racemes of few flowers; sepals 4, free; petals 4, free; receptacle elongated into a discoid androgynophore; stamens 6 or many; anthers 4-celled, dorsifixed near base, dehiscing longitudinally; ovary superior, 1-locular, 2-carpellate; placentas 2, parietal; ovules few to many, campylotropous; style 1 or the stigma sessile. Fruits dry or fleshy siliques (2 valves falling away from a central frame on which the seeds are attached), dehiscing regularly or irregularly lengthwise; ovules numerous, lacking endosperm.

Recognized by their slender gynophore, which bears the ovary.

Corner (1964) implied that the family is principally bird pollinated, but characteristics of bat-pollinated flowers are present. Some species of *Cleome* (*C. anomala* H.B.K.) are known to be bat pollinated (Vogel, 1958). *Capparis frondosa* is animal dispersed. Its purplish fruit with whitish seeds suggest possible bat dispersal. At least some species of *Cleome* have eliasomes associated with their seeds and are ant dispersed; this is true of *C. pilosa* Benth. and *C. aculeata* L. and is possibly true of *C. parviflora* as well (H. Iltis, pers. comm.).

Some 46 genera and 700–800 species; mostly in the drier parts of the subtropics and tropics.

CAPPARIS L.

Capparis frondosa Jacq., Enum. Syst. Pl. Ins. Carib. 24. 1760

C. baducca sensu auct. non Rheed. ex L.

Shrub or small tree, to 5 m tall. Leaves simple; petioles quite variable in length, very short or to 20 cm long (variable even on the same branch), pulvinate at both ends; blades oblong-lanceolate to oblong-oblanceolate, mostly acuminate, acute to obtuse at base, 4.5–22 cm long and 1.8–9 cm wide (variable in size even on the same branch); midrib raised above. Flowers in short racemes of few flowers in the upper leaf axils; pedicels 1–3 cm long (2–4 cm long in fruit); sepals 4, bluntly triangular, ca 1.5 mm long, open in bud, each with a minute disk gland at base within; petals 4, ovate, ca 1 cm long, obtuse at apex, white; stamens ca 100, 1.5 cm long; pistil oblong, ca 3 mm long, on a gynophore ca 1 cm long (ca 3 cm long in fruit). Fruits oblong, fleshy, tardily dehiscent siliques to 9 cm long, becoming maroon when mature, ca 1.5 cm wide, dehiscing irregularly lengthwise to expose seeds, the seeds ca 20, sticky, ± spherical, ca 8 mm diam. *Croat 5861, 7802.*

Common throughout the forest. Seasonality uncertain. Flowers in January and February and from July to September. White-faced monkeys eat the fruits from November to February (Oppenheimer, 1968), but fruits are also seen from March to May.

The species is distinguished by a combination of many-staminate flowers, petioles of greatly differing lengths, and purple siliques borne on a gynophore.

Southern Mexico to Peru and Brazil; West Indies. In Panama, ecologically variable; known from tropical moist forest on both slopes in the Canal Zone and in Los Santos, Panamá, and Darién, from tropical dry forest in Panamá (Taboga Island), from premontane wet forest in Panamá, and from premontane rain forest in Darién (Cerro Pirre).

CLEOME L.

Cleome parviflora H.B.K. subsp. **parviflora**, Nov. Gen. & Sp. 5:83. 1821

C. houstoni sensu Standl. (1933) non R. Br.; *C. panamensis* Standl.

Weak, annual herb, to 60 cm tall; stems at most sparsely pubescent, the nodes with retrorse spines to 2 mm long. Leaves palmately compound (or simple near base of plant); petioles 3–6 mm long; leaflets 3(5), ± lanceolate, acuminate, 3–6 cm long, strigillose and becoming glabrous above, occasionally with minute spines on midrib below. Inflorescences terminal racemes of few flowers, mostly less than 15 cm long in flower but to 30 cm long; pedicels 1–1.5 cm long, very slender, subtended by foliaceous bracts; sepals lanceolate, (2)3–5 mm long; petals pale to deep pink, greenish-purple, green, or white, 3–6 mm long, with a claw about one-third as long as blade; stamens 6; gynophore 1–9(11) mm long. Fruits narrowly fusiform siliques, 2–8 cm long and ca 3 mm wide, dehiscing regularly lengthwise; seeds numerous, reniform, 1.5–2.3 mm long. *Croat 6403.*

KEY TO THE SPECIES OF CAPPARIDACEAE

Plants shrubs or small trees; leaves simple . . . . *Capparis frondosa* Jacq.
Plants herbs; leaves palmately compound . . . . *Cleome parviflora* H.B.K. subsp. *parviflora*

Infrequent; known only from sandbars in coves around the edge of the island. Seasonal behavior undetermined. Flowers and fruits at least from May to September.

The species can be recognized by a combination of palmately compound leaves, four clawed petals much exceeded by the stamens, and siliques borne on a gynophore. The species was mistakenly reported by Standley (1933) as *Cleome houstoni* R. Br., a species endemic to Cuba.

Mexico to the Guianas (Surinam), Ecuador, northern Peru, and Amazonian Brazil; usually at elevations of 10 to 250 m, usually in marshes. In Panama, known from tropical moist forest in the Canal Zone on the Atlantic slope and in Darién.

60. SAXIFRAGACEAE

Woody climbers. Leaves opposite, petiolate; blades simple; venation pinnate; stipules lacking. Flowers bisexual or neuter (sterile), actinomorphic, in generally terminal, bracteate cymes; calyx 4-toothed; petals 4, free, showy; stamens 8, free; anthers 2-celled, dehiscing longitudinally; ovary inferior, of 2(3) locules and carpels; placentation axile; ovules many, anatropous; styles 2; stigmas capitate; neuter flowers consisting only of short, broadly expanded, deeply divided calyces. Fruits septicidally dehiscent capsules; seeds many, with abundant endosperm.

The family is represented on the island only by *Hydrangea*, which is distinguished by being a climbing shrub, somewhat hemiepiphytic, with showy, sterile flowers and less conspicuous fertile flowers in the same inflorescence.

Hydrangea is probably pollinated by insects in the orders Hymenoptera, Diptera, Coleoptera, and Lepidoptera (McClintock, 1957).

The tiny seeds are possibly wind dispersed.

About 80 genera and 1,200 species; primarily in temperate North America.

HYDRANGEA L.

Hydrangea peruviana Moric. in DC., Prodr. 4:14. 1830
Climbing liana-like shrub, usually tightly fastened to supporting tree; older stems with a flaky brown periderm; younger stems and inflorescences with scalelike stellate pubescence. Leaves subcoriaceous; petioles 1–2 cm long; blades oval to elliptic, round to acute at apex, obtuse at base, 6–15(23) cm long, 3–7(14) cm wide, bicolorous, both surfaces sparsely covered with brown stellate scales, those of the upper surface often ± sunken, the margins ± entire; midrib usually arched, the reticulate venation visible only on lower surface. Inflorescences at or near

apex, open and spreading; branches 4–7 cm long, at first enclosed in a large bud 1.5–2 cm broad of several spathelike bracts; sterile flowers at first white, becoming pale green, to 1.5 cm diam, of 4 rounded sepals, the pedicels mostly 1.5 cm long; fertile flowers maroon, short-pedicellate; calyx obscurely 4-toothed; petals 4, ca 1.7 mm long, caducous; stamens 8, inconspicuous, to 0.7 mm long, shorter than style; styles 2, spatulate; stigmas marginal, 2 mm long at maturity. Capsules ca 2 mm long, splitting open at apex between styles; seeds less than 1 mm long, linear, very numerous. *Croat 11850.*

Rare, occurring high in the canopy. Juvenile plants with their small, more or less ovate leaves and densely rooted, closely appressed stems are usually common climbing trees in the vicinity of the adult plants. Flowers principally from July to September (elsewhere often flowering in January). The fruits develop quickly and are usually present on all but the youngest inflorescences.

H. oerstediana Briq., reported in the *Flora of Panama* (McClintock, 1950), may be merely a form of *H. peruviana.*

Costa Rica to Peru, possibly as far north as Mexico. In Panama, known principally at high to middle elevations from lower montane wet forest in Chiriquí, from tropical wet forest in Coclé, and from premontane wet forest in the Canal Zone (Pipeline Road), Chiriquí, and Coclé; known much less frequently from tropical moist forest at lower elevations in the Canal Zone and Panamá.

61. CHRYSOBALANACEAE

Trees or shrubs; sap sometimes colored. Leaves alternate, petiolate; blades simple, entire; venation pinnate; stipules present. Flowers bisexual, ± actinomorphic, in terminal or axillary panicles or racemes; receptacles present; calyx 5-lobed, the lobes imbricate; petals 5 (rarely lacking), imbricate, free, inserted on the margin of the disk at the top of the hypanthium; stamens 3–15, free, sometimes unilateral; anthers 2-locular, longitudinally dehiscent; ovary superior, with 1 fertile carpel (and 2 aborted carpels), 1-locular; placentation basal; ovules 2; style filiform, basally attached at one side of ovary; stigma truncate or slightly lobed. Fruits drupaceous; seed exalbuminous.

Distinguished by having their styles attached basally to the ovary. Though no other clearly defined morphological features distinguish them, all the BCI species are distinctive and are not easily confused with members of any other family.

The flowers of the family are open. Pollinators are unknown.

The fruits of *Hirtella* are dark-colored and fleshy. They are possibly dispersed chiefly by bats, but are also

KEY TO THE GENERA OF CHRYSOBALANACEAE

Stamens more than 8 mm long; inflorescences narrow racemose panicles or racemes; leaf blades pubescent but never arachnoid-tomentose . . . . *Hirtella*
Stamens less than 5 mm long; inflorescences ± pyramidal panicles; leaf blades white arachnoid-tomentose on lower surface or completely glabrous . . . . *Licania*

taken by monkeys. *H. triandra* is taken by the white-faced monkey, which eats the fleshy part and spits out the seed (Oppenheimer, 1968). *Licania platypus* is also probably bat dispersed (R. Foster, pers. comm.). A *Licania* species was among those whose seeds were found in bat detritus of *Artibeus jamaicensis* in Mexico.

Seventeen genera and about 420 species; lowland tropics of Western and Eastern hemispheres.

HIRTELLA L.

Hirtella americana L., Sp. Pl. 34. 1753

Pigeon plum

Tree, to 20 m tall, to 30 cm dbh; young branchlets and petioles densely rufous-tomentulose. Petioles stout, very short; stipules subulate, paired, velutinous, 5–9 mm long, subpersistent; blades oblong to oblong-elliptic, short-acuminate, rounded at base, 7–15 cm long, 2.5–6.5 cm wide, glabrous to sparsely pubescent above, velutinous below especially on veins, the trichomes denser on veins, with a few round glands near base below. Inflorescences narrow terminal panicles 10–20 cm long, densely pubescent except on petals and inner flower parts; branches with rounded or oval, gland-tipped bracts, the glands sessile or short-stipitate; pedicels 1–2 mm long; calyx lobes 5, oval, rounded at apex, often tinged with purple; petals 5, oval, ca 4 mm long, white to reddish, spreading at anthesis; stamens 3, ca 1 cm long, mounted on a whitish disk (2 aborted stamens sometimes visible as subulate trichomes); filaments white, straight and divergent at anthesis; style erect, ± equaling ovary, violet-purple near apex, villous near base. Drupes ellipsoid, 1.5–2 cm long, sparsely pubescent to glabrous, black, shiny; pericarp thin, fleshy. *Croat 4866, 7751.*

Occasional, in the forest, especially the younger forest. Flowers from February to May. The fruits probably mature in the early to middle rainy season.

Southern Mexico to northern Colombia and Venezuela. In Panama, known from tropical moist forest in the Canal Zone, Veraguas, Herrera, Panamá, and Darién and from tropical dry forest in Coclé (Penonomé).

Hirtella racemosa Lam., Encycl. Méth. Bot. 3:133. 1789

H. americana sensu Aubl. non L.

Small tree or arching shrub, 1–5 m tall, usually less than 10 cm dbh; young branches usually moderately pubescent. Petioles 1–3 mm long, densely hispid to glabrous; stipules paired, linear, 2–5 mm long, persistent; blades elliptic to oblong, bluntly acuminate at apex, obtuse to subcordate at base, 4–12(19) cm long, 1.5–5(7) cm wide, sparsely appressed-pubescent on midrib above and below and sometimes on surface below. Racemes axillary or terminal, 5–15(29) cm long; rachis and pedicels puberulent; bracts and bracteoles 1–3 mm long, narrowly triangular, appressed-pubescent, usually with 1 or more round, sessile glands near base; pedicels slender, (1.5)4–11 mm long, perpendicular to rachis; calyx lobes 5, ovate, 2–3 mm long, sparsely puberulent; petals 5, obovate to elliptic, 3–5 mm long, lavender; stamens 5–7; filaments 1–2 cm long, lavender; ovary densely pilose-tomentose; style to 1.5 cm long. Drupes oblong-obovoid, to 1.5 cm long, ca 6 mm diam, purplish-brown turning black, sparsely pubescent. *Croat 9518.*

Frequent in the forest and along the shore. Flowers from November to June, principally in the dry season. The species may flower more than once a year. The fruits probably mature in a few months, mostly in the late dry and early rainy seasons.

Panama through Colombia, Venezuela, and the Guianas to Peru, northern Bolivia, and Amazonian Brazil. In Panama, known from tropical moist forest in the Canal Zone, Colón, Los Santos, Panamá, and Darién and from tropical dry forest in Coclé (Penonomé) and Panamá (Taboga Island). Reported from premontane moist, premontane wet, and premontane rain forests in Costa Rica (Holdridge et al., 1971).

See Fig. 252.

Hirtella triandra Sw., Prodr. Veg. Ind. Occ. 51. 1788

Camaroncillo, Carapoto, Chicharron, Conejo, Wild pigeon plum

Tree or shrub, to 20 m; trunk to 23 cm dbh; outer bark thin, fissured, weakly flaky; inner bark reddish-brown, granular; wood light brown, hard, heavy; youngest stems pubescent; sap inconspicuous. Petioles to 5 mm long, moderately pubescent; stipules paired, subulate, to 5 mm long, subpersistent; blades elliptic-oblong, acuminate at apex, acute to rounded or subcordate at base, 4–12(14.5) cm long, 1.5–5 cm wide, glabrous above except on midrib, sparsely pubescent below, the trichomes denser on veins. Flowers in terminal racemose panicles to 8 cm long; rachis, branches, and pedicels tomentulose; pedicels slender, 2–3 mm long (appearing longer since flowers are often solitary at apices of branches); bracts and bracteoles narrowly lanceolate, 1–3.5 mm long, pubescent, lack-

---

KEY TO THE SPECIES OF HIRTELLA

Inflorescences strict racemes (pedicels ± perpendicular to axis); stamens 5–7; flowers lavender . . . . *H. racemosa* Lam.
Inflorescences short-branched panicles; stamens 3; flowers white:
    Branches of inflorescences and younger stems very densely golden-brown tomentose; panicles usually more than 10 cm long, many times longer than broad; bracteoles of inflorescence bearing conspicuous glands; drupes black, shiny, sparsely pubescent . . . . *H. americana* L.
    Branches of inflorescences and younger stems not very densely golden-brown tomentose; panicles less than 8 cm long, often about as broad as long; bracteoles of inflorescence lacking conspicuous glands; drupes brownish-purple, densely golden-brown tomentose . . . . *H. triandra* Sw.

Fig. 252. *Hirtella racemosa*

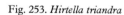

Fig. 253. *Hirtella triandra*

ing glands; calyx lobes 5, ovate, 3–4 mm long, rounded at apex, reflexed at anthesis, puberulent on both sides; petals 5, broadly elliptic, ca 5 mm long, rounded at both ends, white; stamens 3, long-exserted, arising from side of flower; filaments purplish above, 1–2 cm long, fused into a low ring arising from rim of hypanthium (remnants of a 4th and a 5th stamen persisting as sharp lobes on this ring); style ± equaling stamens and opposite them at anthesis, pilose and white below middle, purple above; ovary pilose-tomentose; stigma small, capitate. Drupes ovoid-oblong to rounded, ca 2.3 cm long, densely short-pubescent; exocarp thin, brownish-purple; mesocarp fleshy, sweet, tasty; seed ± ovate, ca 1.7 cm long, with irregular longitudinal grooves, abruptly narrowed at base. *Croat 7171, 10969.*

Common in the forest, mostly in the old forest. Flowers sporadically throughout the year, principally from November to May. Individuals flower more than once a year, possibly as many as three times. The fruits mature throughout the year; white-faced monkeys eat the fruits in June and July (Oppenheimer, 1968).

Unusual growths were found on *Croat 6010*, which consisted of long-stipitate, obovate, greenish-white, fruitlike structures covered with short erect trichomes. The structures were regularly disposed on the stems and were at first believed to be fruits. Dr. Ghillean Prance (New York Botanical Garden), who made sections of the structures, has confirmed that they are not fruits. They are possibly due to gall insects.

Central Mexico to northern and western South America, Bolivia, and Brazil; West Indies. In Panama, known from tropical moist forest all along the Atlantic slope and in Chiriquí, Panamá, and Darién and from premontane wet forest in Colón.

See Fig. 253.

LICANIA Aubl.

Licania hypoleuca Benth., Bot. Voy. Sulphur 91, t. 32. 1844

Garapata

Tree, to 15 m tall; trunk ca 20 cm dbh; outer bark unfissured, minutely roughened, thin; inner bark reddish-brown, thin, hard; branches enlarged at base; young branchlets and petioles puberulent; sap with faint, pungent odor. Petioles 0.5–1 cm long; stipules minute, paired, subpersistent; blades ovate, acuminate at apex, obtuse to rounded at base, 5.5–10 cm long, 2.5–5 cm wide, glabrous above, whitish below with dense, minute, arachnoid trichomes. Flowers minute, in terminal or upper-axillary panicles 10–20 cm long, the branches and pedicels densely and minutely puberulent; hypanthium campanulate or turbinate, to 2 mm long; calyx lobes 5, minute,

ovate, acute; petals lacking; stamens 3, less than 1 mm long; anthers about half as long as filaments; ovary white-hirsute; style to 2 mm long, adjacent to the 2 stamenless calyx lobes. Fruits pyriform to obovate, to 2.5 cm long and 2 cm diam, at first pink to red, becoming white and soft at maturity, usually with a minute depression at apex; mesocarp thick, fleshy, sweet, somewhat pithy; seed obovate, ca 1 cm diam. *Croat 14648.*

Infrequent, in both the younger and older forests. Probably flowers in the early dry season. The fruits are mature mostly from April to June.

Southern Mexico to Colombia, Venezuela, the Guianas, and Amazonian Brazil. In Panama, known from tropical moist forest on BCI and in Darién and from premontane wet forest in Panamá (Cerro Azul).

Licania platypus (Hemsl.) Fritsch, Ann. K.K. Naturhist. Hofmus. 4:53. 1889

Sangre, Wild pear, Zapote

Tree, 10–30(50) m tall; trunk to 75 cm dbh (sometimes buttressed to 1–2.5 m elsewhere); branches and leaves glabrous; branchlets often reddish; sap red in age. Petioles ca 1 cm long; stipules ovate, ca 2.5 mm long, stiff, adnate to petiole at base, persistent; blades mostly narrowly oblong-elliptic, shortly acuminate at apex, acute to rounded at base, 10–20(30) cm long, 3–6(8) cm wide, lustrous above, pale and ± glaucous below, with small round flattened or sunken glands below, especially near margin. Panicles terminal or upper-axillary, 10–25(35) cm long, the branches gray-tomentose, flattened at base; flowers minute, white, fragrant, sessile or very short-pedicellate; hypanthium, sepals, and edges of petals gray-tomentose; hypanthium ± turbinate, ca 1.5–2 mm diam; sepals 5, triangular, ca 1 mm long and wide, spreading; petals 5, obovate, 2–3 mm long; stamens 15, glabrous; filaments ca 3 mm long, attached separately to disk; style 5–6 mm long. Drupes variable in size, reported to 20 cm long and 14 cm diam, green turning brown; mesocarp granular, yellow, juicy, sweet; seed usually 1, ovate-oblong, flattened, to 5 cm or more long. *Croat 8695, 11851.*

Frequent in the forest, especially the old forest. Flowers principally in the dry season, especially from February to April. Time of fruit maturation is uncertain, since fruit size is so variable, but fruits believed to be mature were falling in June and late August. Some fruits probably persist on the tree much longer.

At maturity, the mesocarp smells much like fresh pumpkin.

Southern Mexico (both coasts) south to Colombia; reported also from the valley of the Magdalena (Jimenez S., 1970). In Panama, known from tropical moist forest in the Canal Zone, Chiriquí, Panamá, and Darién; Allen (unpublished) reported the species to be very common

KEY TO THE SPECIES OF LICANIA

Stamens 15; petals 5; leaves glabrous; sap red; fruit to 20 cm long; branchlets reddish . . . . *L. platypus* (Hemsl.) Fritsch
Stamens 3; petals lacking; leaves densely whitish-arachnoid below; sap not colored; fruits to 2.5 cm long; branchlets not reddish . . . . *L. hypoleuca* Benth.

Fig. 256. *Connarus panamensis*

Fig. 254. *Licania platypus*

Fig. 255. *Licania platypus*

on the dry Pacific coast, and Johnston (1949) reported that it appears to prefer growing on well-drained terraces in ravines.

See Figs. 254 and 255.

62. CONNARACEAE

Lianas or climbing shrubs. Leaves alternate, petiolate, imparipinnate; leaflets entire; T-shaped trichomes may be present; venation pinnate; stipules lacking. Flowers bisexual, actinomorphic, often glandular and fragrant, in axillary or terminal panicles; sepals 5, free or connate at base, imbricate or valvate, persistent in fruit; petals 5(6), free or connate at base, imbricate, white; stamens 10(12), in 2 series, briefly united at base, the shorter series opposite petals; anthers 2-celled, dehiscing longitudinally; pistils 1 (*Connarus*) or 5 (*Cnestidium, Rourea*); ovaries superior, 1-locular, 1–5-carpellate; placentation subbasal from the inner angle; ovules 2, orthotropous or anatropous; style 1 per carpel; stigma simple, capitate. Fruits dehiscent, sessile or stalked follicles (in 5-pistillate flowers only 1 develops) with a single seed; seed black, arillate, the aril yellow or orange; endosperm lacking.

Members of this family may be easily mistaken for Leguminosae (63) because of their imparipinnate leaves, swollen pulvini on petiolules, and legume-like follicles. Connaraceae are distinguished from the legumes by their arillate seed.

Flowers are small, white, and open, especially in *Rourea* and *Cnestidium,* and appear well suited for pollination by small insects, especially bees.

Seeds are black and shiny and are enveloped at the base by an aril of contrasting color, usually yellow, orange, or red. They are bird dispersed.

Some 24 genera with over 300 species; pantropical.

CNESTIDIUM Planch.

Cnestidium rufescens Planch., Linnaea 23:440. 1850

Liana, usually slender; younger branches, petioles, rachises, lower midribs, axes of inflorescences, pedicels, and calyces densely ferruginous-pubescent. Leaves imparipinnate, 10–25 cm long; leaf scars prominent; petioles 2–6 cm long; petiolules to 3 mm long; leaflets 7–11, alternate to opposite, obovate-oblong to elliptic, acute to bluntly acuminate, obtuse to rounded at base, persistently ferruginous-pubescent, especially below, the terminal leaflet 3.5–12 cm long, 1.5–6 cm wide, the others somewhat smaller. Flowers 5–6 mm diam, numerous, slightly aromatic, in upper-axillary panicles 6–15 cm long; pedicels very short; sepals 5, oblong-ovate, acute to blunt, to 1.5(3) mm long, slightly accrescent in fruit; petals 5, white, oblong, acute to rounded at apex, to 3.5 mm long, spreading to recurved; stamens 10, in alternate cycles of 2 lengths, united briefly at base, the longest to 2.5 mm; ovary and base of styles densely pubescent with stiff white trichomes; ovary of usually 5 free carpels; styles to 2.5 mm long, erect; stigmas simple. Follicles sessile, solitary, densely ferruginous-tomentose, ca 1.5 cm long; seed 1, oblong, black, shiny, ca 1 cm long, the basal aril orange. *Croat 7878, 16701.*

Infrequent, along the shore and at the margin of the forest in clearings. Flowers from July to September. The fruits mature by the early dry season (December to March).

Southern Mexico to Colombia; Cuba. Common in most of Panama at lower elevations; known principally from tropical moist forest in the Canal Zone, Bocas del Toro, Colón, Herrera, Panamá, and Darién; known also from premontane moist forest in the Canal Zone and Panamá, from tropical dry forest in Coclé and Panamá (Taboga Island), and from premontane wet forest in Chiriquí and Panamá.

CONNARUS L.

Connarus panamensis Griseb., Bonplandia 6:6. 1858

Liana or climbing shrub, erect when juvenile, growing into lower part of canopy; trunk less than 2 cm dbh, with soft, thin, flaking bark; sap red with faint sweet aroma. Leaves weakly pubescent to glabrate; petioles 5–14 cm long, swollen and wrinkled at base; petiolules enlarged, wrinkled; leaflets 3(5), elliptic, acuminate at apex,

KEY TO THE SPECIES OF CONNARACEAE

Stems, leaves, and follicles densely ferruginous-pubescent . . . . *Cnestidium rufescens* Planch.
Stems, leaves, and follicles weakly pubescent to glabrous in age:
 Flowers with 5 carpels; follicles sessile, ± terete, less than 8 mm diam; inflorescences never ferruginous-tomentose; leaflets ± appressed-pubescent on veins below . . . . *Rourea glabra* H.B.K.
 Flowers with 1 carpel; follicles stipitate, somewhat compressed, more than 12 mm diam; at least younger parts of inflorescence densely ferruginous-tomentose; leaflets glabrous or nearly so:
  Leaflets mostly 3, never 7, the margins often erose (as though eaten); flowers sessile, in dense clusters along branches; axes of inflorescences and young fruits bearing very dense, pile-like tomentum of erect simple trichomes; inflorescence branches usually stout, moderately short, and stiffly spreading (sometimes slender and dangling) . . . . *Connarus panamensis* Griseb.
  Leaflets 3 or 7, the margins usually smooth; flowers pedicellate, in small clusters or solitary along branches of inflorescence; axes of inflorescences and young fruits bearing mostly appressed pubescence of obscurely T-shaped trichomes (sometimes with erect trichomes on younger parts but these soon deciduous); inflorescence branches slender, usually long and dangling . . . . *Connarus turczaninowii* Tr.

rounded at base with petiolule often attached slightly above base, 10–25 cm long, 3–10 cm wide (juveniles to 35 cm long and 18 cm wide). Inflorescences spicate-paniculate, terminal, densely ferruginous-tomentulose, the pubescence eventually deciduous, the axis simple or branched; flowers sessile or subsessile, sweetly aromatic; sepals 5, ovate, 1.5–2.3 mm long, often about as broad as long; petals 5, white, to 3 mm long, often with red glandular dots; stamens 10, in 2 whorls, the inner whorl much shorter, the longer filaments to 2.3 mm long, often with stipitate glands near apex; carpel 1; ovary erect-pubescent; style 1, short, conic, off-center in fruit. Follicles compressed-obovoid, 1.5–2.5 cm long, red-orange, the stalk off-center, ca 5 mm long, the valves nearly glabrous at maturity (any remaining pubescence, usually on stipe, dense and erect), dehiscing along one side to expose a single seed; seed shiny, black, arillate, ca 1.5 cm long, the aril thin, yellow, to about midway on seed, its margins ± fimbriate. *Croat 7986, 11100.*

Abundant along the shore; less common in the canopy of the forest, especially the young forest. Seasonal behavior uncertain. Some flowers and fruits have been seen throughout much of the year, but individuals flower at least twice a year, mostly at the beginning of the dry season, with their fruits developing within 3 months. Another and possibly even larger surge of flowering occurs during the early rainy season (July to September), but with other plants flowering even later in the rainy season. The latter part of the dry season and the earliest part of the rainy season (April to June) appear to show the least flowering activity.

Costa Rica to Colombia. In Panama, known principally from tropical moist forest in the Canal Zone, Veraguas, Herrera, Panamá, and Darién; known also from tropical dry forest in Coclé and Panamá (Taboga Island), from premontane moist forest in the Canal Zone and Panamá, and from premontane wet forest in Chiriquí.

See Fig. 256.

Connarus turczaninowii Tr., Ann. Sci. Nat. Bot., sér. 5, 16:364. 1872

Liana or climbing shrub, ± glabrous except inflorescences; trunk often stout, growing into lower part of canopy. Leaves compound; petioles pulvinate at base; petiolules stout, callous; leaflets 3 or 7, opposite or subopposite (rarely alternate), lanceolate-elliptic to oblanceolate, acuminate and often downturned at apex, acute to rounded at base, 4–15 cm long, 2–7 cm wide, stiff. Inflorescences terminal, usually branched only once, the branches slender, drooping, 8–45 cm long; branches, pedicels, and calyces rufous-tomentose at least when young, the simple erect trichomes usually soon deciduous, leaving mostly appressed, T-shaped trichomes, these also eventually deciduous; flowers 5-parted, 4–5 mm long, with a definite but short pedicel and a strong, sweet aroma; sepals to 2.5 mm long, usually much longer than broad; petals white, ± oblong, about twice as long as sepals; stamens 10, in 2 whorls, the inner 5 shorter, the longer 5 nearly equaling petals; filaments and connective with gland-tipped trichomes; carpel 1; ovary narrowly

ovoid, with dense, ± appressed pubescence (the trichomes obscurely T-shaped); style 1, short. Fruits nearly identical to those of *C. panamensis* but with any remaining pubescence (often persisting on stipe) short, appressed, often whitish, and T-shaped. *Croat 8397, 12599.*

Common along the lakeshore; less common in the canopy of the forest. Seasonal behavior uncertain. Some flowers have been seen throughout most of the year; individuals clearly flower more than once a year, but perhaps no more than twice a year. Mature fruits are frequently found on plants at the time of their next flowering.

The flowers are visited by the bee *Trigona tataira.*

Known only from Panama, from tropical moist forest in the Canal Zone, Colón, and Darién.

ROUREA Aubl.

Rourea glabra H.B.K., Nov. Gen. & Sp. 7:41. 1824
R. adenophora S. F. Blake
Mata negro

Liana. Leaves imparipinnate; petioles 2–6 cm long, terete with swollen basal pulvinus; petiolules pulvinate, 2–4 mm long; leaflets 3–7(9), alternate or subopposite, ovate-elliptic to oblong-elliptic, usually acuminate at apex, rounded to obtuse at base, 3–14 cm long, 1.5–6 cm wide, glabrous above except on midrib, glabrate to puberulent below especially on veins; midrib often arched. Panicles 5–15 cm long, from upper axils; branches and pedicels usually puberulent; flowers white, fragrant, ovoid in bud; calyx to ca 3.5 mm long at anthesis, larger and much thickened in fruit, glabrate or puberulent, with or without minute glands, the 5 lobes imbricate, divided to beyond middle; petals 5(6), ± oblong and rounded at apex, strongly reflexed at anthesis, usually 5–6 mm long, soon falling; stamens 10(12), alternating long and short, the longer to 4.5 mm long, the shorter to 3 mm long; filaments fused in a ring at base; anthers about as broad as long, opening in bud, the thecae opening broadly; pollen removed soon after anthesis; ovary of 5(6) carpels, pubescent; styles ca 4 mm long, pubescent on base. Follicles 1–1.5 cm long, cylindrical, slightly curved, orange-red to bittersweet at maturity (at least at apex), splitting open on one side to expose a single seed; seed shiny, black, slightly shorter than follicle, the aril yellow to orange, ca 5 mm long, about one-third as long as seed. *Croat 17030.*

Occasional, along the shore and in advanced clearings; also in the canopy, but usually on trunks below the top of the canopy. Flowers twice a year, once at the beginning of the dry season (December to February) and again from July to September. The fruits usually develop within 3 months, but some may persist until the next flowering.

In my opinion *R. adenophora* S. F. Blake is not a good species. It is supposed to differ from *R. glabra* by having the calyces and pedicels glandular. Presence or absence of pubescence on the upper midrib has also been used as a character to separate these taxa, but my investigations show the pubescence to be variable even on the same plant, often deciduous on some leaves and present on others.

Seeds are reportedly poisonous (Blohm, 1962).

Southern Mexico to the Guianas and Brazil; West Indies. In Panama, known principally from tropical moist forest in the Canal Zone, San Blas, Veraguas, Los Santos, Panamá, and Darién; known also from tropical dry forest in Coclé and Panamá (Taboga Island), from premontane moist forest in the Canal Zone and Panamá, and from premontane wet forest in Colón and Panamá.

animals, most endozoochorous diaspores of the family are bird dispersed. A smaller number are dispersed by larger animals. Large-seeded legumes are hoarded by agoutis and spiny rats (N. Smythe, pers. comm.). The next most important type of dispersal for the family is anemochory. Fewer are hydrochorous, epizoochorous, or autochorous.

About 600 genera and 12,000–14,000 species; worldwide.

63. LEGUMINOSAE (FABACEAE)

Trees, shrubs, or herbs, sometimes scandent, tendriled, and/or armed. Leaves alternate (opposite in *Platymiscium*), petiolate; blades simple, pinnate or bipinnate, entire or subentire; venation pinnate (palmate in *Bauhinia* and the blades bilobed); stipules usually present. Inflorescences terminal or axillary, paniculate, racemose, spicate, or sometimes globose heads; flowers bisexual, actinomorphic or zygomorphic; calyx 5-lobed, the lobes subequal, rarely free, rarely obsolete; petals 5, equal or unequal, free, or the 2 anterior petals united, usually showy; stamens 10 to many, free, monadelphous or diadelphous; anthers 2-celled, dehiscing longitudinally; ovary superior, 1-locular, 1-carpellate; placentation parietal; ovules 2 to many, in 2 rows on the ventral suture, anatropous; style 1, simple; stigma 1. Fruits legumes, loments, follicles, or drupes; seeds generally lacking endosperm.

An exceedingly diverse, large, and important tropical family most easily confused with the Connaraceae (62), which differ in having follicles with arillate seeds.

Because of the nature of the leguminous fruit, which generally offers little in the way of a meal for larger

63A. MIMOSOIDEAE

Leaves bipinnate (merely pinnate in *Inga* and *Pithecellobium rufescens*); stipules present. Flowers actinomorphic, in cylindrical spikes or globose heads; floral bracteoles present in *Mimosa;* calyx cupular, rarely obsolete, the lobes imbricate or valvate; corolla lobes usually valvate; stamens many, often polyadelphous, equal, considerably longer than corolla.

Members of the subfamily Mimosoideae are most easily distinguished by their bipinnate leaves (except *Inga* and *Pithecellobium rufescens,* which usually have conspicuous glands at the apex of their petioles) and by their actinomorphic flowers with numerous stamens (at least twice the number of petals).

The flowers of the Mimosoideae are open with exserted sexual parts. In most cases the pollination unit includes the entire inflorescence; in all cases the pollen is easily accessible. Flowers may be pollinated by unspecialized pollen feeders, which crawl over the surface of the inflorescence, or by fluttering or hovering insects or birds. In general, the larger flowers or inflorescences, which provide nectar in quantity, are probably visited by bats or birds. *Inga vera* is known to be pollinated by several

Leaves bipinnate:
  Inflorescences various, not globular or subglobular; flowers arising from an elongated rachis more than 2 cm long; leaflets at least in part 1 cm or more wide:
    Pinnae bifoliolate; stems with paired, recurved, nodal thorns; corollas ca 10 mm long; legumes reddish, linear-moniliform, less than 3 cm wide .... *Pithecellobium hymeneaefolium* (H. & B.) Benth.
    Pinnae with 3 or more pairs of leaflets; stems unarmed; corollas less than 5 mm long; legumes more than 5 cm wide:
      Inflorescences terminal racemes of spikes; legumes less than 7 cm wide; seeds ellipsoid, less than 1.5 cm long, dispersed with transverse segments of legume .... *Adenopodia polystachya* (L.) J. Dixon
      Inflorescences in supra-axillary spikes; legumes more than 8 cm wide; seeds disk-shaped, 5–6 cm diam, not dispersed with transverse segments of legume .... *Entada monostachya* DC.
  Inflorescences globular or with very brief floral rachises less than 5 mm long:
    Leaflets more than 1 cm wide:
      Peduncles more than 3 cm long; flowers pedicellate; petioles bearing large cupular gland 1 cm long at apex; legumes curled:
        Pedicels less than 2.5 mm long; gland between basal pair of pinnae ca 1 cm long, the others at successively higher nodes small; mature fruits curved into nearly complete circle .... *Pithecellobium macradenium* Pitt.
        Pedicels more than 8 mm long; gland between all pairs of pinnae minute; mature fruits oblong-linear, never curved into circle .... *Albizia guachapele* (Kunth) Dug.
      Peduncles less than 2 cm long; flowers sessile; petioles bearing small gland at apex:
        Inflorescences paniculate with many globose heads ca 1 cm wide; individual flowers tiny, difficult to distinguish without magnification; pinnae usually in (2)4 pairs .... *Leucaena multicapitula* Schery
        Inflorescences of several spikes issuing from nodes along stem; individual flowers clearly visible, the corolla ca 5 mm long; pinnae usually in 2(3) pairs .... *Pithecellobium dinizii* Ducke
    Leaflets less than 1 cm wide:
      Plants conspicuously armed:
        Petiole and rachis eglandular; stamens fewer than 10; legumes armed .... *Mimosa*
        Petiole or rachis with small glands; stamens more than 10; legumes unarmed .... *Acacia*
      Plants unarmed:
        Flowers on pedicels usually more than 1 cm long; legumes flat, straight, more than 15 cm long .... *Albizia guachapele* (Kunth) Dug.
        Flowers sessile or on pedicels less than 3 mm long, in small globular heads (less than 2 cm diam); legumes curled:
          Leaflets in fewer than 15 pairs per pinna; leaves with petiole, rachis, and lower leaflet surface tawny-tomentose; legumes less than 2 cm wide, reddish inside .... *Pithecellobium barbourianum* Standl.
          Leaflets usually in more than 15 pairs per pinna; leaves not tawny-tomentose; legumes more than 2 cm wide, not reddish inside .... *Enterolobium*

---

bat species (Heithaus, Fleming & Opler, 1975).

Sphingid moths have been seen visiting the flowers in the early morning and early evening in Costa Rica (G. Frankie, pers. comm.). R. Heithaus (pers. comm.) also suggests hummingbirds and large trap line bees as pollinators.

Evidence exists of specialization in the pollen arrangement in the Mimosoideae. Different taxa form polyads with varying numbers of pollen grains per polyad (Guinet, 1969; Sorsa, 1969). The number of pollen grains per polyad usually correlates very well with the number of ovules per carpel (T. Elias, pers. comm.). This probably indicates a specialization in pollinators, and clusters of pollen are probably carried by nectar feeders rather than by the relatively unspecialized pollen feeders.

The largest dispersal category of Mimosoideae is endozoochory, especially mammalian. Examples are *Enterolobium cyclocarpum*, *E. schomburgkii*, and possibly all

species of *Inga*. Oppenheimer (1968) reported that fruits of perhaps all species of *Inga* are eaten by the white-faced monkey. Some *Inga* may be water dispersed as well, especially those occurring only along the lakeshore. Quite likely these species of *Inga*, as well as similar shoreline species of *Pithecellobium*, are dispersed in part by fish or reptiles as is suggested by van der Pijl (1968).

Birds are probably the chief agents of dispersal for seeds of all *Pithecellobium* but especially for those of *P. barbourianum* and *P. rufescens*, which have well-developed arilloids clearly suited to bird dispersal. Seeds of some species of *Acacia*, such as *A. hayesii* and *A. riparia*, as well as those of *Leucaena multicapitula*, may also be bird dispersed, because the fruits open at maturity and the dark seeds are then displayed against the light inner valve surface.

Seeds of all *Mimosa* on BCI are chiefly epizoochorous. They are perhaps also wind dispersed, since they are

much flattened and break up into small segments. Water may also play a part in the dispersal of *M. casta* and *M. pigra*, since these generally occur near water; water currents play an important role in the dispersal of *Entada monostachya*. Van der Pijl (1968) suggested that *Albizia*, which lacks arilloids, is frequently wind and water dispersed; the seeds themselves sink but are dispersed while enclosed in the buoyant pod (Ridley, 1930).

ACACIA P. Mill.

Acacia acanthophylla (Britt. & Rose) Standl., Publ. Field Mus. Nat. Hist., Bot. Ser. 18:488. 1937

Lianas climbing into canopy; trunk and branches sulcate, generally glabrous, the ribs (3)4 (to several), pubescent, armed with prickles, the prickles numerous, 3–5 mm long, broad-based, recurved. Leaves bipinnate with 4–9 pairs of pinnae; rachis and petiole glabrous to puberulent, usually armed and sometimes with many slender stalked glands ca 0.5 mm long, the glands sometimes at nodes on rachis; leaflets in 8–25 pairs per pinna, oblong-elliptic, oblique, acute to ± rounded and apiculate at apex, truncate at base and attached at a corner, 8–14 mm long, 1.5–4 mm wide, the margins entire or sparsely ciliate. Flowers in short, paniculate spikes; calyx and corolla glabrous. Legumes tan, 15–20 cm long, ca 3.5 cm wide, flat. *Croat 15568*.

Collected once in the old forest above the escarpment. Seasonal behavior unknown. Immature fruits have been seen in December.

The flower and fruit description given here is taken from the *Flora of Guatemala* (Standley & Steyermark, 1946). BCI vegetative material differs from the descriptions of other authors by having larger leaflets (to 14 mm long and 4 mm wide, compared to 10 mm long and 2 mm wide).

Mexico, Guatemala, Honduras, Costa Rica, and Panama. In Panama, known only from tropical moist forest on BCI.

Acacia glomerosa Benth., Hooker's J. Bot. Kew Gard. Misc. 1:521. 1842

Moderate to large tree, often buttressed, unarmed or armed with recurved prickles on stems and rachises; branchlets becoming glabrous in age. Leaves bipinnate with 6–8 pairs of pinnae; petioles ca 4–5 cm long, glandular near the base; leaflets many (ca 30 pairs per pinna), oblong to subfalcate, apiculate at apex, truncate at base, 8–12 mm long, ca 2 mm wide, puberulent, the underside paler; midvein submarginal. Flowers cream or white, fragrant, in terminal panicles, the heads small, dense, globose, to 12 mm diam, bearing 12–20 flowers; calyx ca 1 mm long, strigillose; corolla ca 2 mm long, deeply 5-lobed; stamens numerous, showy, white, to 7 mm long. Legumes narrowly oblong, flattened, 10–20 cm long, ca 3 cm wide, becoming glabrous; seeds few. *Aviles 10b*.

Collected once; not seen in recent years. Flowers mostly from September to January, especially from September to November. The fruits mature mostly from January to May.

The species is perhaps conspecific with *A. polyphylla* DC., which ranges from Colombia to Brazil. *Aviles 10b* differs from most material of *A. glomerosa* in that both surfaces of the leaflets are more densely pubescent with appressed trichomes. The inflorescence is also somewhat more diffuse than the typical *A. glomerosa*, and the pedicels are more slender.

Mexico to southern Brazil. In Panama, collected only from tropical moist forest on BCI, but reported by Holdridge et al. (1971) from moist and wet areas at low elevations.

Acacia hayesii Benth., Trans. Linn. Soc. London 30:524. 1875

Large canopy liana; trunk to 12 cm diam near the ground, with prominent raised horizontal lenticels; all but the smallest woody parts glabrous, ± 5-sided, the angles with prominent retrorse prickles. Leaves 30–60 cm long, bipinnate with 9–14 pairs of pinnae, each pinna 6–24 cm

---

KEY TO THE SPECIES OF ACACIA

Plants armed with large, paired, stipular spines resembling bull horns; flowers yellow; fruits subterete . . . . *A. melanoceras* Beurl.
Plants armed with recurved prickles or unarmed; flowers usually white; fruits flat:
    Plants large buttressed trees . . . . *A. glomerosa* Benth.
    Plants lianas or small, vinelike or arching shrubs:
        Leaflets 3–7 mm long (usually 4–5 mm), glabrous or with the lower edge nearest the rachis conspicuously pubescent; fruits less than 15 cm long and less than 2.5 cm wide, conspicuously rufous-pubescent; flowers in globular heads . . . . *A. riparia* H.B.K.
        Leaflets 6–14 mm long (usually 8–12 mm), glabrous or pubescent, but the trichomes never restricted to lower inner edge; fruits more than 15 cm long and more than 2.5 cm wide, tan, inconspicuously pubescent; flowers in short, oblong spikes:
            Branches predominantly quadrangular, sulcate; petioles usually displaying more than 2 large sessile glands; rachis usually armed with recurved prickles; interfoliar glands on rachis usually present . . . . *A. acanthophylla* (Britt. & Rose) Standl.
            Branches predominantly 5-angled to several-angled; petioles with 1(2) conspicuous sessile glands; rachis usually not armed; interfoliar glands on rachis usually lacking . . . . *A. hayesii* Benth.

long; petiole and rachis puberulent; petioles bearing 1 or 2 round or oblong, sessile glands; rachis canaliculate above, often aculeate below, with glands at the nodes; leaflets in 20–30 pairs per pinna, oblong, blunt at apex, truncate and very inequilateral at base, mostly 9–14 mm long, ca 2.5 mm wide, glabrate to pubescent especially below, the trichomes appressed or erect. Inflorescences large terminal panicles of pedunculate spikes; peduncles 1–2 cm long, puberulent; spikes 10–12 mm long, dense; flowers glabrous, subsessile, greenish; calyx cupulate, ca 2 mm long, as broad as long; corolla cylindric-campanulate, ca 4 mm long; stamens numerous, 7–8 mm long. Legumes thin, linear to oblong, 15–24 cm long, 2.5–3.5 cm wide, softly puberulent, acute or acuminate apically, rounded at base, the margins markedly raised, the surface broadly undulate; seeds ± disk-shaped, 7–8 mm diam, dark, displayed against the light inner valve surface, borne on a slender funiculus to 13 mm long. *Croat 6202.*

Common within the forest, generally in the canopy, but the leafy branches may occasionally be seen at lower levels where the vine has fallen. Elsewhere this species becomes a tree (*fide Flora of Panama*), but on BCI its habit is always decidedly vinelike. Flowers in late October and November. The fruits develop to more or less mature size by late January and are dispersed throughout the rest of the dry season and early rainy season (to early May). Plants lose their leaves during the dry season.

The flowering description is based on the *Flora of Panama* (Woodson & Schery, 1950).

Honduras, Costa Rica, and Panama. In Panama, known only from tropical moist forest in the Canal Zone, eastern Panamá, and Darién.

Acacia melanoceras Beurl., Kongl. Svenska Vetenskap-sakad. Handl. 1854:123. 1856

Shrub or small tree, to 6 m tall; branchlets and rachis puberulent, otherwise glabrous; branchlets armed with large spines, the spines 3–4.5 cm long, hornlike, paired, hollow, black, stipular. Leaves bipinnate with 15–26 pairs of pinnae; petioles 1–3 cm long, with several raised glands on upper side near base; rachis 12–25 cm long, usually with a single raised gland between pairs of pinnae; leaflets in 15–30 pairs per pinna, oblong, rounded to obtuse at apex, truncate and inequilateral at base, to 7 mm long, 1–2 mm wide. Inflorescences terminal, raceme-like groupings of pedunculate heads; peduncles short, subtended by a 3- or 4-parted involucre; heads globular, densely flowered; floral bracts peltate, ± equaling flowers; flowers minute, yellow; calyx cupulate, ca 1 mm long, obscurely lobed; corolla funnelform, somewhat longer than calyx, puberulent apically; stamens numerous, 2–3 mm long. Legumes linear-oblong, to 11 cm long and 1.5 cm wide, ± compressed-subterete, short-beaked, longitudinally striate, glabrous, tardily dehiscent. *Croat 6667.*

Occasional, in the forest, especially on the west side of the island. Seasonal behavior uncertain. Probably flowering in February and March. Mature fruits have been seen in May.

Stipular spines often house ants that bite fiercely and remove any vegetation contacting the plant. The plant rewards the ants with sugars from the petiolar glands and with protein from the small beltian bodies along the margins of the leaflets (Janzen, 1967a). According to D. Janzen (pers. comm.), seeds merely spill out of the pods at maturity but may be dispersed further from the ground. Unless later regurgitated, seeds would probably not survive the passage through a bird.

Costa Rica and Panama. In Panama, known from tropical moist forest in the Canal Zone and from tropical wet forest in Colón (Portobelo).

Acacia riparia H.B.K., Nov. Gen. & Sp. 6:276. 1824

Vine or scandent shrub, to 5 m tall; branchlets ± angulate, the angles of the branchlets and sometimes the rachis armed with recurved prickles; petiole and rachis above puberulent and leaf margins ciliate, otherwise glabrous. Leaves bipinnate with 8–15 pairs of pinnae, each pinna 2.5–4(5) cm long; petioles 1–3 cm long, with 2 oblong glands on upper side; rachis 5–8 cm long, with an oblong gland at base of each pinna mostly in upper portions; leaflets in 20–40 pairs per pinna, oblong, rounded at apex, rounded to truncate at base, 4–7 mm long, ca 1 mm wide, glabrous but ciliate and sometimes with tufts in inside axil at base of leaf; midrib subcentral. Inflorescences terminal or upper-axillary; peduncles 1–2 cm long; flowers ± glabrous, sessile, clustered in white globular heads to ca 2 cm wide; rachis less than 5 mm long; calyx 1–2 mm long, acutely lobed; corolla tubular, 3–4 mm long, acutely lobed. Legumes flat, thin, mostly 10–15 cm long, 2–2.5 cm wide, tomentulose, stipitate, with a sharp beak 5 mm long at apex of fruit; seeds disk-shaped, ca 6 mm long, borne on a slender funiculus, the funiculus attached laterally to the seed, with a sharp bend near its middle. *Croat 8000, Foster 1392.*

Uncommon, on the shore on the eastern and southern sides of the island. Flowers in the rainy season. The fruits are of mature size by January and February, but probably are not dispersed until late in the dry season (April).

Panama to Ecuador, Bolivia, and southern Brazil. In Panama, known from tropical moist forest on BCI and from tropical dry forest in Panamá (Taboga Island).

See Fig. 257.

ADENOPODIA Presl

Adenopodia polystachya (L.) J. Dixon, comb. nov.
Mimosa polystachya L., Sp. Pl. 520. 1753; *Entada polystachya* (L.) DC., Prodr. 2:425. 1825; Mém. Leg. 422, 434, t. 61, 62. 1826

Liana; trunk to 15 cm dbh; stems striate; pinnular rachises and inflorescences puberulent, otherwise ± glabrous. Leaves bipinnate with 2–5 (mostly 3 or 4) pairs of pinnae; petiole and rachis lacking glands; leaflets in 5–7 pairs per pinna, oblong, rounded at both ends, oblique at base, 1.5–4 cm long, 0.5–1.8 cm wide. Inflorescences terminal racemes to 25 cm long, of many slender spikes to 10 cm

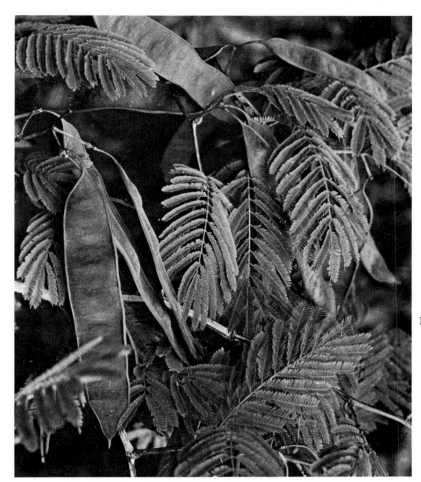

Fig. 257. *Acacia riparia*

Fig. 258. *Albizia guachapele*

Fig. 259. *Entada monostachya*

Fig. 260. *Enterolobium cyclocarpum*

long; flowers small, sessile or short-pedicellate, mostly white, but reportedly also reddish, with a foul odor; calyx cupulate, shallowly lobed to subentire, ca 1 mm long; corolla of 5 petals, the petals ± free, 2.5–3 cm long, acute; stamens 10, somewhat exserted, ca 4 mm long. Legumes oblong, 15–30(40) cm long, 5–7.5 cm wide, flat, thin, curved; exocarp thin, peeling away at maturity, the valves then breaking into narrow, wind-dispersed, transverse segments, each carrying a small seed; seeds ellipsoid, 1–1.4 cm long, shiny, brown. *Foster 731, Starry 318.*

Occasional, in the canopy of the forest; seldom seen, except for its fallen rectangular fruit segments. Flowers from July to September. Mature fruits were seen in December.

The transfer of *Entada polystachya* (L.) DC. to *Adenopodia* was convincingly argued by John Dixon (unpublished Ph.D. dissertation, University of Southern Illinois, Carbondale, 1965). *Adenopodia* and *Entada* as represented on BCI are two very different taxa both morphologically and ecologically.

Mexico to central South America; West Indies. In Panama, known from tropical moist forest in the Canal Zone, Bocas del Toro, Chiriquí, and Panamá and from tropical dry forest in Herrera (Pesé).

ALBIZIA Durazz.

Albizia guachapele (Kunth) Dug., Phytologia 13:389. 1966
Albizia longepedata (Pitt.) Britt. & Rose

Deciduous tree, to 15 m tall, tomentose to sparsely tomentose all over. Leaves bipinnate with 3–5 pairs of pinnae; petioles 3–6 cm long with a round sessile gland near the middle of upper surface; stipules acicular, caducous; rachis 7–15 cm long, usually with a sessile gland at insertion of each pair of pinnae; pinnular rachis 8–12 cm long, with similar glands at insertion of each pair of leaflets; leaflets in 3–7 pairs per pinna, variable, ovate to obovate, asymmetrical, rounded or emarginate at apex, obtuse and inequilateral at base, 1.5–3(4) cm long, 1–2(2.5) cm wide. Inflorescences 1–3, subterminal in axils; peduncles 3–7(9) cm long; pedicels 8–20 mm long; bracts linear, caducous, small; flowers in dense umbels, dimorphic, the central flower large and sessile; calyx 5–7 mm long (10 mm long in central flower), the lobes acute, ca 1 mm long; corolla green, 8–11 mm long (20 mm long in central flower), the lobes acute, ca 2 mm long, glabrous inside; stamens long-exserted, ca 2 cm long, white, the staminal tube included. Legumes oblong-linear, flat, 15–20 cm long, 2–3.5 cm wide, markedly short-pubescent, with a narrow marginal rib, tardily dehiscent. *Croat 8707.*

Apparently rare, known only from the cove between Slothia Island and Colorado Point. Flowers from December to March, in the early dry season. The fruits mature from January to May. Leaves are lost in the early dry season.

Guatemala to northern South America. In Panama, known from low elevations in moist, monsoon areas (Holdridge, 1970).
See Fig. 258.

ENTADA Adans.

Entada monostachya DC., Prodr. 2:425. 1825
E. gigas (L.) Fawc. & Rendle

Tendriled liana; trunk to 40 cm diam near the ground; larger stems with densely and minutely fissured, rusty-brown outer bark; smaller stems green, shiny, almost glabrous. Leaves bipinnate with usually 2 pairs of pinnae, the pinnae opposite, each 6–10 cm long; rachis 6–12 cm long, ending with a simple tendril to 15 cm long; petiolules short, with prominent basal pulvinus; leaflets in (3)5(6) pairs per pinna, asymmetrically oblong, blunt to emarginate at apex, unequal at base, mostly 2.5–5(7) cm long, 1.2–3 cm wide, glossy above, dull below; midrib puberulent. Inflorescences densely flowered, slender, spikelike racemes inserted shortly above leaf axils, to 25 cm long; flowers green, with a strong and sweet but disagreeable odor, maturing from base of rachis upward; pedicels short; calyx cupulate, about 1 mm long, the lobes minute; petals elliptic, ca 3 mm long and 1 mm wide, the margins sometimes scarious; stamens 10, ca 6 mm long; filaments white, the connective bearing a minute, stalked, papillate food body at apex. Legumes very large, 30–120 cm long, 10–13 cm wide, forming a broad spiral, the margins raised; seeds button-shaped, 5–6 cm diam, to 2 cm thick, dark and shiny at maturity. *Croat 4957, 7869.*

Occasional, in the canopy of the forest and pendent from the trees on the lakeshore. Seasonal behavior uncertain. Flowers from November to May. The fruits reach mature size by the late rainy season, but it is not known when they are dispersed, perhaps during the dry season.

Recognized by the huge fruits, which require about a year to mature. Seeds either are removed from the stout valves by animals, which tear holes in the valves, or fall from the weathered valves. The species is common in riparian situations, and water is known to play an important role in seed dispersal. Seeds are apparently carried great distances by rivers and may often be found in great abundance on ocean beaches.

Central America and tropical South America; West Indies; West Africa. In Panama, known from tropical moist forest in the Canal Zone, Bocas del Toro, Panamá, and Darién.
See Fig. 259.

ENTEROLOBIUM Mart.

Enterolobium cyclocarpum (Jacq.) Griseb., Fl. Brit. W. Ind. 226. 1860
Coratú, Corotú, Curutú, Ear tree, Genisero, Guanacaste, Jarina

Tree, to 30 m tall and 1.5(3) m dbh, branched from near the ground, the crown widely spreading; outer bark with

KEY TO THE SPECIES OF ENTEROLOBIUM

Leaves with 4–15 pairs of pinnae and 15–30 pairs of leaflets per pinna; stems, petioles, and rachises usually glabrate, not ferruginous .... *E. cyclocarpum* (Jacq.) Griseb.
Leaves usually with more than 15 pairs of pinnae and more than 60 pair of leaflets per pinna; young stems, petioles, and rachises ferruginous-tomentulose .... *E. schomburgkii* Benth.

rough fissures; wood reddish-brown; branchlets usually glabrous in age. Leaves sometimes several at a node, bipinnate with 4–15 pairs of opposite pinnae, puberulent; petioles 4–8 cm long, with a round gland near middle of upper surface; rachis and pinnular rachis with glands at insertions near apex; leaflets in 15–30 pairs per pinna, ± oblong, decidedly inequilateral, acute at apex, rounded at base, 8–15 mm long, 2–4 mm wide, sometimes glabrate. Inflorescences axillary in groups of 1–3, white, puberulent to glabrate throughout, 2–4(6) cm long; heads globular, ca 2 cm wide; bracts minute; flowers sessile, green; calyx 2–3 mm long, the lobes to 0.5 mm long, subacute; corolla 5–6 mm long, the lobes to 1.5 mm long, subacute; stamens to 12 mm long; filaments united in basal half; anthers white; style ± equaling stamens. Legumes reniform, flat, glabrous, 4–6 cm wide, curved into a nearly complete or overlapping circle to 11 cm diam; seeds ovoid-compressed, 1.5–2 cm long, to 1 cm wide, smooth, brown, bearing a lighter submarginal ring on either flattened surface. *Croat 7365, 8708.*

Uncommon, persisting as large trees in areas of previous cultivation. Flowers principally in the dry season (January to May). The fruits mature one year later in the dry season. The tree is leafless for a short time before flowering and produces new leaves at about the same time as the flowers. Neal Smith (pers. comm.) reports that an individual of *E. cyclocarpum* flowered in January in three consecutive years.

Tropical Mexico to northern South America; introduced into the West Indies and West Africa. In Panama, known from tropical moist forest in the Canal Zone, Bocas del Toro, Chiriquí, and Darién; reported also from premontane moist forest in Panamá (Tosi, 1971).

See Fig. 260.

Enterolobium schomburgkii Benth., Trans. Linn. Soc.
    London 30:599. 1875
    Mimosa wilsonii Standl.
    Jarina

Tree, to 25 m tall; trunk weakly buttressed, to 45 cm dbh; outer bark unfissured, rough, hard, thin; inner bark reddish-brown, granular; young branches, petioles, ra-

chises, and pinnular rachises rufous-tomentose; sap with an indistinct odor. Leaves bipinnate with 10–25 pairs of opposite pinnae; petioles 1.5–3 cm long, swollen basally, bearing a small, sessile gland ca 0.5 mm wide just above the pulvinus; stipules minute, caducous; rachis to 25 cm long, ridged along the center of the upper surface, with a similar gland at each node; pinnular rachis 4–10 cm long with 1 to several glands at terminal nodes; leaflets in 50–70 pairs per pinna, linear-oblong, less than 5 mm long and 0.8 mm wide, glabrous above, rufous-puberulent below, the midrib eccentric. Inflorescences rufous-tomentulose, axillary, in groups of 3–8; peduncles 1–4 cm long; heads short-spicate, appearing globular with the axis ca 2.5 mm long; bracts miniscule; flowers sessile to subsessile, green with white anthers; calyx turbinate, ca 2.5 mm long, the teeth 5, obtuse to acute, ca 1 mm long; corolla 2.5–5 mm long, the 5 segments acute, densely sericeous outside, glabrous inside; stamens 10, to 10 mm long, united at base into a short tube (if the tube extends nearly the entire length of the stamens, the stamens reduced to staminodia). Legumes reniform, 2–2.5 cm wide, curved into an overlapping circle to 5 cm diam, flat, reddish-brown, glabrous; seeds ellipsoid- to triangular-compressed, 6–8 mm long, smooth, brown, ringed on the largest circumference with a lighter area. *Croat 14986, Wilson 130, Woodworth & Vestal 688.*

Rare; a large tree grows at Miller Trail 875. Flowers in the dry season (February to April). The fruits mature from April to May (or later).

Mexico and Guatemala to Brazil. In Panama, scattered and possibly cultivated; known from tropical moist forest in the Canal Zone and Panamá, from premontane wet forest in Chiriquí, and from tropical wet forest in Panamá.

INGA Scop.

Small to medium-sized trees. Leaves alternate, paripinnate; stipules usually minute; rachis with interfoliar glands, the glands filiform, conic, or patelliform; leaflets opposite, often asymmetrical, the basal leaflets reduced in size. Inflorescences axillary or terminal, racemose or spicate, with a peduncular and floriferous part; flowers

KEY TO THE SPECIES OF INGA

Rachis not winged:
    Bracts large, obscuring calyces; terminal leaflets more than 20 cm long ... *I. spectabilis* (Vahl) Willd.
    Bracts not obscuring calyces; terminal leaflets less than 20 cm long:
        Inflorescences capituliform or umbelliform (rachis of inflorescence very short); flowers on pedicels 3–8 mm long; legumes short, curved, about half as thick as broad, densely ferruginous; leaflets in 3 or 4 pairs .... *I. quaternata* Poepp.
    Inflorescences spicate or racemose; flowers sessile or on pedicels less than 1 mm long:

All leaves with fewer than 4 pairs of leaflets:
  Leaflets obovate; calyx less than 2 mm long; legumes with rounded sides at maturity . . . .
    . . . . *I. fagifolia* (L.) Benth.
  Leaflets elliptic to lanceolate; calyx more than 2 mm long; legumes flat at maturity . . . . .
    . . . . *I. punctata* Willd.
Most leaves with more than 3 pairs of leaflets:
  Calyx usually less than 2 mm long; leaves with fewer than 6 pairs of leaflets; young stems
    glabrous or puberulent . . . . *I. pezizifera* Benth.
  Calyx more than 3 mm long; usually some leaves with more than 5 pairs of leaflets;
    young stems mostly ferruginous-pubescent:
    Upper leaflets usually narrowly long-obovate; corollas less than 9 mm long . . . .
      . . . . *I. ruiziana* G. Don
    Upper leaflets usually elliptic or lanceolate; corollas more than 10 mm long:
      Calyx more than 7 mm long; leaflets in 7–10(13) pairs . . . . *I. multijuga* Benth.
      Calyx less than 7 mm long; leaflets in 4–8 pairs:
        Lower leaflets usually asymmetrical; legumes flat; leaf rachis often with a linear
          appendage ca 4 mm long at apex; corollas often more than 15 mm long . . . . .
          . . . . *I. thibaudiana* DC.
        Lower leaflets ± symmetrical; legumes subterete; leaf rachis without appendage
          at apex; corollas less than 15 mm long . . . . *I. cocleensis* Pitt.
Rachis winged:
  Calyx at anthesis less than 6 mm long:
    Leaflets densely pubescent below; legumes subterete . . . . *I. minutula* (Schery) Elias
    Leaflets glabrous or sparsely pubescent below; legumes flat:
      Inflorescences umbelliform; flowers and fruits on long slender pedicels . . . .
        . . . . *I. umbellifera* (Vahl) Steud.
      Inflorescences racemose; flowers and fruits almost sessile:
        Most leaves with more than 3 pairs of leaflets; floral rachis shorter than peduncle, less
          than 3 cm long; legumes at least 2.5 cm wide at maturity . . . . *I. pezizifera* Benth.
        Most leaves with 2 pairs of leaflets; floral rachis longer than peduncle, more than 4 cm
          long; legumes less than 2 cm wide at maturity:
          Leaflets obovate, in 2 or 3 pairs . . . . *I. fagifolia* (L.) Benth.
          Leaflets elliptic, in 2 pairs . . . . *I. marginata* Willd.
  Calyx at anthesis more than 6 mm long:
    Stems bearing dense reddish-brown pubescence (in age); terminal leaflets more than 7 cm
      wide; legumes usually 20 cm or more long, bearing dense reddish-brown pubescence:
      Corollas slender, less than 5 mm wide, sparsely sericeous; leaflets lacking a gland on midrib
        . . . . *I. mucuna* Walp. & Duch.
      Corollas thick, more than 8 mm wide, densely sericeous; leaflets with a gland on midrib
        within 2 cm of base (sometimes obscured by pubescence) . . . . *I. goldmanii* Pitt.
    Stems not bearing dense reddish-brown pubescence, at least not in age:
      Bracts large, obscuring calyx; terminal leaflets more than 9 cm wide . . . .
        . . . . *I. spectabilis* (Vahl) Willd.
      Bracts small, the calyx visible; terminal leaflets frequently less than 9 cm wide:
        Corollas more than 20 mm long; petioles usually more than 20 mm long; legumes tetra-
          gonal, ± glabrous (at least in age) . . . . *I. sapindoides* Willd.
        Corollas less than 20 mm long; petioles usually less than 20 mm long; legumes flat or
          terete, usually densely pubescent:
          Some leaves with more than 4 pairs of leaflets; legumes terete . . . .
            . . . . *I. vera* Willd. subsp. *spuria* (Willd.) J. León
          All leaves with fewer than 5 pairs of leaflets (rarely 5 in *I. hayesii*); legumes terete or
            flat:
            Calyx sparsely pilose to glabrate, not bearing dense reddish-brown pubescence;
              marginal ridge of legume narrow, the fruit remaining ± flat until nearly fully
              mature; trichomes of fruit sparse enough to be individually distinguished with-
              out difficulty . . . . *I. hayesii* Benth.
            Calyx densely pubescent, the young branches, petioles, and rachises bearing dense
              reddish-brown pubescence; marginal rib of legume broad and rounded, giving
              the fruit an overall terete appearance; trichomes of fruit so dense and short as
              to be not easily distinguished individually . . . . *I. pauciflora* Walp. & Duch.

---

usually many, with sweet aroma, subtended by bracts; calyx usually tubular, 5-toothed, usually somewhat irregularly lobed, sometimes splitting; corolla gamopetalous, tubular to tubular-funnelform, usually greenish, pubescent outside, glabrous inside; stamens numerous, fused into a tube at the base, the tube length variable relative to length of free filaments, even between flowers on same branch; filaments white at anthesis; style slender,

± equaling anthers; stigma discoid. Legumes linear; seeds surrounded by a sweet, pulpy, white mesocarp.

Inga flowers are visited by *Eulaema* bees in Ecuador (Dodson & Frymire, 1961). In Panama members of the genus are commonly called Guava or Ice cream beans.

Inga cocleensis Pitt., Contr. U.S. Natl. Herb.
18:211. 1916
Nacaspiro, Cuje

Tree, to 13 m tall, ± tomentose throughout, the trichomes sparser on upper leaflet surface, the pubescence reddish-brown except for surfaces of leaflet and corolla. Rachis terete; petiolules 1–2 mm long; leaflets in 4–8 pairs per pinna, ± elliptic, sometimes inequilateral, acute at apex, obtuse to rounded at base, the terminal pair 7–11 cm long and 2–4 cm wide. Spikes solitary or in groups of 2 or 3; peduncles mostly 2–3 cm long; rachis mostly 2–4 cm long; bracts nearly obsolete, less than 1 mm long; flowers sessile; calyx tubular, 4–6 mm long, the lobes acute, ca 1 mm long; corolla 10–15 mm long, the lobes acute, ca 2 mm long; stamens and style ca 2.5 cm long. Legumes subterete, to 35 cm long and 2.5 cm diam. *Croat 7941.*

Infrequent or rare, along the shore and in the forest. Flowers in the dry season, with the fruits maturing in several months.

Guatemala to Panama. In Panama, known from tropical moist forest in the Canal Zone and Coclé and from premontane wet forest in Panamá.

Inga fagifolia (L.) Willd. ex Benth., Trans. Linn. Soc.
London 30:607. 1875
I. laurina (Sw.) Willd.
Guavo, Sweet pea, Guama

Tree, to 8(23) m; calyx and axis of inflorescences puberulent, otherwise ± glabrous. Leaves with (1)2 or 3 pairs of leaflets; petioles terete to margined; rachis terete to narrowly winged; leaflets obovate to elliptic, retuse to obtuse or bluntly short-acuminate, acute and slightly inequilateral at base, the terminal pair 6–14 cm long, 3–6 cm wide. Spikes usually 1–5 in axils of terminal leaves; peduncles mostly 1–3 cm long; rachis mostly 4–9 cm long; bracts nearly obsolete, less than 1 mm long; flowers sessile; calyx tubular, 1–2 mm long, the teeth shallow, acute; corolla funnelform, 3–6 mm long, the lobes acute, ca 1 mm long, somewhat pubescent at apex; staminal tube included to long-exserted; stamens and style to ca 12 mm long. Legumes flat and ribbed to bulged in the middle with raised margins at maturity, mostly 7–15 cm long, 1.5–2.5 cm wide. *Croat 5537, 6119.*

Common at least on the shore and in the young forest. Flowers mostly from January to May. The fruits mature mainly from August to November.

Mexico to Bolivia, Brazil, and Paraguay; Trinidad. In Panama, known from tropical moist forest in the Canal Zone, Veraguas, Herrera, Panamá, and Darién, from tropical dry forest in Coclé (Penonomé), and from premontane wet forest in Panamá (Cerro Azul).

Inga goldmanii Pitt., Contr. U.S. Natl. Herb.
18:198. 1916
Guavo de mono

Tree, to 20 m tall, densely ferruginous-hirsute all over to sparsely so or glabrate on leaflet surface and sericeous on corolla. Rachis conspicuously winged; interfoliar glands stalked, to ± 3 mm tall; leaflets in 3–5 pairs, ovate to elliptic, acute to caudate-acuminate, rounded to slightly cordate and asymmetrical at base, sometimes ± bullate between veins, the terminal pair 11–24 cm long, 6–13 cm wide, usually with a gland on midrib above within 2 cm of base (sometimes obscured by pubescence); juvenile leaves less densely pubescent and to 80 cm long with the terminal leaflets to 42 cm long and 19 cm wide. Spikes axillary, usually solitary; peduncles 3–9 cm long, stout; rachis 10–20 cm long; bracts to 6 mm long; flowers sessile; calyx campanulate, 12–20 mm long, the lobes acute, 3–7 mm long; corolla campanulate, sericeous, 20–28 mm long, more than 8 mm wide; stamens and style to ca 50 mm; style sometimes persistent on growing legume. Legumes flat or twisted, 10–25 cm long, 4–6 cm wide, to 1.5 cm thick, densely ferruginous-hirsute; valves splitting open (usually from bottom) at maturity to expose seeds; seeds oblong, to 1.8 cm long, not colorful. *Croat 8300, 13107.*

Occasional, in the forest. Flowers mostly from November to February. The fruits probably mature in the rainy season.

Similar to *I. mucuna,* but distinguished by the short corolla, more reddish-brown fruits, and the gland on the midrib of the leaflet above.

Costa Rica and Panama. In Panama, known from tropical moist forest in the Canal Zone, all along the Atlantic slope, and in Veraguas, Los Santos, and Panamá. See Fig. 261.

Inga hayesii Benth., Trans. Linn. Soc. London
30:617. 1875
Guavo, Guamo

Tree, to 15 m tall; branchlets ferruginous-pilose when young, glabrate in age. Leaves sparsely pilose on axis and veins; petioles 5–40 mm long; rachis conspicuously winged; leaflets in 2–4(5) pairs, ovate-elliptic to obovate, acuminate to broadly acute, acute to rounded at base, usually asymmetrical, the terminal pair 7–14 cm long, 3–6 cm wide. Spikes 1 to several per axil; peduncles very short or to 1.5 cm long, strigose; rachis strigose, 1.5–2.5 cm long; flowers sessile; bracts ca 3 mm long; calyx 7–10 mm long, puberulent but with pilose base, longitudinally striate, the lobes acute to subacute, 1–2 mm long; corolla 10–19 mm long, densely white-sericeous, the lobes narrow, to ca 5 mm long; stamens and style to ca 45 mm long. Legumes flat until nearly fully mature, 4–15 cm long, 1.5–2.5 cm wide, strigose, usually with low, thin, transverse wrinkles. *Zetek 3497.*

Probably at one time more abundant than at present; the species was collected several times by Zetek when there were more disturbed areas on the island. Juvenile

plants have been seen in the forest, and the species is fairly common in the Canal Zone along roadsides. Seasonal behavior uncertain. Flowers principally from May to July, infrequently as early as January, with the fruits probably maturing in the rainy season.

Easily confused with *I. pauciflora*, except in fruit, and sometimes with *I. vera*.

Known only from Panama, from tropical moist forest on the Pacific slope in the Canal Zone, Panamá, and Darién and from premontane wet forest in Panamá (Cerro Azul).

Inga marginata Willd., Sp. Pl. 4:1015. 1806

Guavo de mono, Sweetwood

Tree, to 20 m tall; leaflet midribs above and axes and calyces of inflorescences puberulent, otherwise glabrous. Rachis and petiole winged, sometimes narrowly so; leaflets in 2 pairs, ± elliptic, asymmetrical in lower half, acuminate, ± acute at base, the terminal pair 4.5–12 cm long, 2–4.5 cm wide. Spikes axillary, in groups of 1–4; peduncles usually less than 1 cm long; rachis 4–11 cm long; bracts linear, ca 2 mm long; flowers usually sessile; calyx ca 1 mm long, shallowly lobed; corolla 2–4 mm long, the lobes acute, ca 1 mm long; stamens and style to ca 10 mm long. Legumes flat, the seminiferous areas bulged, 8–14 cm long, 1–2 cm wide. *Croat 8185.*

Occasional, in the forest, perhaps more abundant along the shore. Apparently flowers twice a year, principally in February and March, but also in September and October. The fruits appear to develop quickly, probably maturing 1 to 3 months after flowering.

Similar to *I. fagifolia* and *I. pezizifera*, but separated by always having four leaflets and a winged rachis.

Costa Rica to Bolivia and Brazil. In Panama, known from tropical moist forest in the Canal Zone, Bocas del Toro, Colón, Chiriquí, and Darién and from premontane wet forest in Coclé (El Valle).

Inga minutula (Schery) Elias, Phytologia 14:211–12. 1967

I. edulis Mart.; *I. oerstediana* Benth.

Guavo, Guava, Ice cream bean

Tree, 4–20 m tall, densely tomentose on most parts, sparsely so on upper surface of leaflets; stems with prominent ribs extending below petioles. Petioles terete, to 7.5 cm long; rachis winged, the interfoliar glands unusually large for the genus, 3–5 mm wide; leaflets in 3–6 pairs, oblong to lanceolate-elliptic, acuminate, rounded at base, the terminal pair 8–25 cm long, 3–11 cm wide. Spikes axillary in groups of 1–7; peduncle, floral rachis, and calyx cinereous-tomentose; rachis flattened, angulate; bracts ca 4 mm long and ca 2 mm wide; flowers sessile; calyx 3–5 mm long, irregularly lobed, the sinuses 1–4 mm deep; corolla 7–11 mm long, sericeous, the lobes 1–4 mm long; stamens and style to ca 30 mm long. Legumes ± terete, many-ribbed, 15–40(120) cm long, 1.5–2.5 cm diam. *Croat 11721, 11966.*

Occasional, in the forest, perhaps more abundant on the shore. Flowers from July to November, principally in

August and September. Fruit maturity time not determined. Mature fruits in September in Costa Rica (Frankie, Baker, & Opler, 1974).

Mexico to South America; West Indies. In Panama, known from most areas of tropical moist forest and from tropical wet forest in Coclé. Reported from premontane wet forest in Costa Rica (Holdridge et al., 1971).

See Fig. 262.

Inga mucuna Walp. & Duch. in Walp., Ann. Bot. Syst. 2:456. 1851–52

Tree, to 10(20) m tall; branchlets densely ferruginous-velutinous. Leaves densely rufous to golden-tomentose on axis and veins, sparsely so on leaf surface; rachis widely winged; petioles terete; leaflets in 3 or 4 pairs, ovate to broadly elliptic, acute to acuminate, obtuse to rounded and asymmetrical at base, the terminal pair 13–21 cm long, 6–12 cm wide. Spikes axillary, usually solitary or paired, the axis densely brownish-tomentose; bracts to ca 8 mm long; flowers sessile; calyx 15–21 mm long, glabrous except on tips of teeth, conspicuously longitudinally striate, the lobes to ca 2 mm long, blunt, bearing conspicuous tuft of trichomes at apex; corolla 35–55 mm long, white-sericeous, the lobes to ca 6 mm long; stamens and style to ca 80 mm long. Legumes flat, densely ferruginous-hirsute, 25–45 cm long, 4–6 cm wide, often twisted, the margins raised, rounded. *Croat 6858, 12959.*

Uncommon, in the young forest and along the shore. Flowers from August to February. The fruits are most common in the dry season.

Similar to *I. goldmanii*, but having a long slender corolla, fruit of a lighter reddish-brown color, and no gland near the base of midrib on upper leaflet surface.

Known only from Panama, from tropical moist forest in the Canal Zone, San Blas, Panamá, and Darién.

See Fig. 263.

Inga multijuga Benth., Trans. Linn. Soc. London 30:615. 1875

Tree, to 20 m tall and 40 cm dbh; inner bark reddish, granular; branchlets densely ferruginous-tomentose when young, glabrate in age; lenticels prominent. Petiole and rachis terete, tomentose; leaflets in (5)7–10(13) pairs, narrowly ovate to oblong-elliptic or oblong, acute to acuminate, obtuse to rounded at base, glabrous above except on midrib, tomentose below on veins, sparsely so on surface, the terminal pair 7–18 cm long, 2–4.3 cm wide. Spikes terminal or axillary, usually solitary or paired, the axis and calyx densely ferruginous-tomentose; bracts subulate, ± 1 mm long; flowers sessile; calyx 7–10 mm long, densely villous, the lobes to ca 1 mm long; corolla 2–3 cm long, sericeous, the lobes narrow, to ca 4 mm long; stamens and style to ca 40 mm long. Legumes flat, 20–30 cm long, to 4.5 cm broad, tomentose when young, glabrate in age, remaining flat, often markedly curved, the margins stout, raised. *Croat 8282, Foster 1418.*

Infrequent, along the shore. Flowers from September

Fig. 262. *Inga minutula*

Fig. 261. *Inga goldmanii*

Fig. 263. *Inga mucuna*

to April, principally at the beginning of the dry season. The fruits mature in the early to middle rainy season.

Distinguished from *I. thibaudiana* by normally having nine or ten pairs of leaflets on some leaves and by having larger fruits, which are glabrate in age.

Honduras to Panama. In Panama, known from tropical moist forest in the Canal Zone, San Blas, Chiriquí, Veraguas, Panamá, and Darién, from premontane wet forest in the Canal Zone and Panamá, from tropical wet forest in Veraguas, and from premontane rain forest in Panamá (summit of Cerro Jefe).

Inga pauciflora Walp. & Duch., Linnaea 23:746. 1850

Tree, to 10 m tall; branchlets densely ferruginous-tomentose when young, glabrate in age. Leaves velutinous especially on axes and veins, sparsely so on upper surface; petioles very short, terete; rachis cuneate-winged; leaflets in 3 or 4 pairs, ovate to elliptic, acute to short-acuminate, obtuse to rounded at base, asymmetrical, the terminal pair 7–16 cm long, 3–7 cm wide, the basal pair often much reduced. Spikes axillary or terminal, solitary or in pairs; peduncle, rachis, and calyx densely ferruginous-tomentose; peduncles 1.5–2.5 cm long; rachis 1.5–3.5 cm long; bracts to 3 mm long; flowers sessile; calyx 8–15 mm long, the lobes to ca 3 mm long; corolla 13–18 mm long, sericeous, the lobes ca 3 mm long; stamens and style to ca 50 mm long. Legumes subterete, with broad ribs, densely ferruginous-pubescent, mostly 4–20 cm long, 1–2 cm diam, often curled, yellowish-brown at maturity. *Croat 6105, 13967.*

Common along the shore. Flowers from March to July. The fruits probably mature from July to October.

Easily confused with *I. hayesii* and *I. vera*. It may be distinguished from *I. hayesii* by its more or less terete fruit and more densely pubescent calyx, and from *I. vera* by having fewer than five pairs of leaflets.

Known only from Panama, from tropical moist forest in the Canal Zone, Veraguas, Panamá, and Darién and from premontane wet forest in Panamá (Cerro Campana). See Fig. 264.

Inga pezizifera Benth. in Hook., London J. Bot. 4:587. 1845

Tree, to 27 m tall and 40 cm dbh, glabrous to puberulent throughout; trunk ± involuted with many prominent horizontal wrinkles; outer bark thin; inner bark reddish; young stems very prominently lenticellate. Petioles margined; rachis margined to narrowly winged; leaflets in 2–5 (usually 4) pairs, narrowly ovate to lanceolate-elliptic or oblanceolate, acuminate, rounded to acute at base, slightly asymmetrical, the terminal pair 8–22 cm long, 2.5–8 cm wide. Spikes axillary in groups of 1–6; peduncles 1.5–3.5 cm long; rachis shorter than peduncle, very densely flowered, 0.5–3 cm long; bracts ca 5 mm long; flowers sessile; calyx 1–2(3) mm long, the lobes ca 0.5 mm long, blunt; corolla 4–8 mm long, the lobes blunt to acute, ca 1 mm long; stamens and style ca 1 cm long. Legumes flat, 10–22 cm long, (2)2.5–3 cm wide, ± glabrous, ± straight, the marginal ribs prominent, the

seminiferous areas individually raised, the axes of the inflorescences persisting with warty bases of fallen fruits. *Croat 14881.*

Common in the forest, especially the young forest. Flowers most commonly in May and June, beginning as early as December. The fruits mature in the rainy season.

Separated from *I. ruiziana* by the lanceolate-elliptic leaflets, which are always in less than six pairs, and by the shorter calyx, usually less than 2 mm long.

Panama to Brazil. In Panama, known from tropical moist forest in the Canal Zone and from premontane wet forest in Panamá (Cerro Jefe).

Inga punctata Willd., Sp. Pl. 4:1016. 1806

I. leptoloba Schlecht.

Guavita cansa-boca, Bribri, Guava, Guava del mono

Tree, to 15 m tall; branchlets sparsely strigose when young, glabrate in age, minutely lenticellate, minutely ribbed below petioles (at least when dried). Leaves strigose throughout, especially on axis and midrib of leaflets; petiole and rachis terete to narrowly margined; leaflets in 2 or 3 pairs, lanceolate to elliptic, long-acuminate, acute to rounded at base, the terminal pair 9–18 cm long, 3.5–7.5 cm wide. Spikes axillary, in groups of 1–7; peduncle and rachis strigose; peduncles 2–4 cm long; rachis much shorter, 1–2.5 cm long; bracts subulate, to ± 2 mm long; calyx puberulent, 3–5 mm long, the lobes ± 1 mm long; corolla moderately sericeous, 5–9 mm long, the lobes to 2 mm long; stamens and style to ca 2 cm long. Legumes flat, to 16 cm long and 2 cm wide, minutely puberulent, the margins raised. *Croat 7418.*

Common in the forest. Flowers throughout the rainy season (May to December, rarely to January). One individual on BCI flowered for a week in early June in two consecutive years. Mature fruits seen in late dry season.

Fruits are similar to those of *I. pezizifera*, but that species lacks persistent warty fruit bases.

Mexico to Peru; West Indies. In Panama, known from tropical moist forest in the Canal Zone, Bocas del Toro, San Blas, Veraguas, Panamá, and Darién, from premontane wet forest in Colon, Chiriquí, Coclé, and Panamá, from tropical wet forest in Panamá and Darién and from lower montane wet forest in Chiriquí. Reported from premontane rain forest in Costa Rica (Holdridge et al., 1971).

Inga quaternata Poepp. in Poepp. & Endl., Nov. Gen. Sp. Pl. 3:79. 1845

I. roussoviana Pitt.

Guavito, Cansa boca, Bribri

Tree, to 15 m tall; branchlets densely ferruginous-tomentose to glabrous-lenticellate. Petiole and rachis ± terete, tomentose; leaflets in 3 or 4 pairs, oblanceolate to oblong-obovate, acuminate to obtuse at apex, acute to obtuse at base, puberulent especially on veins, the terminal pair 7–19 cm long, 3–7.5 cm wide. Racemes axillary, solitary (2 to 3 elsewhere); peduncle and floral rachis densely ferruginous-tomentose; peduncles 0.5–2.5(3.5) cm long; rachis very short, to 5 mm long, giving the inflorescence a capituliform appearance; bracts ca 1 mm

Fig. 264. *Inga pauciflora*

Fig. 265. *Inga ruiziana*

Fig. 266. *Inga umbellifera*

long; pedicel and calyx villous; pedicel 3–8 mm long; calyx 3–6 mm long, the lobes to 1 mm long; corolla sericeous, 6–11 mm long, the lobes to 1.5 mm long; stamens and style to 15 mm long. Legumes flat, 3–9(18) cm long, 1.5–3 cm wide, 1–1.5 cm thick, curved, ferruginous-tomentose when young to glabrate in age, ribbed on margins. *Croat 7699, Foster 1013.*

Common, at least on the shore. The flowers often appear on denuded branchlets or on branchlets with new leaves. Flowers possibly throughout the year, mostly in the late rainy and early dry seasons. Mature-size fruits have been seen from June to September.

According to the label on *Foster 1013,* the white pulp of the green fruits is eaten by the white-faced monkeys.

Mexico to Panama and possibly Colombia. In Panama, known from tropical moist forest in the Canal Zone, Bocas del Toro, and San Blas, from tropical dry forest in Coclé (Penonomé), and from premontane wet forest in Chiriquí (San Félix). Reported from tropical wet forest in Costa Rica (Holdridge et al., 1971).

Inga ruiziana G. Don, Gen. Hist. Dichl. Pl. 2:391. 1832
   I. confusa Britt. & Rose
   Bribri

Tree, to 9(25) m tall; branchlets lenticellate, minutely ferruginous-pubescent, sometimes ribbed below petioles. Petiole and rachis unwinged, minutely margined, minutely pubescent; leaflets in (4)6–7(8) pairs, obovate-oblong to obovate, short-acuminate, obtuse to rounded at base, glabrate but with puberulent veins, the terminal pair 15–25(34) cm long, 4.5–8(12) cm wide. Spikes axillary (appearing paniculate on new growth) in groups of 1–4; peduncle and rachis tomentose; peduncles 1–2.5(4.5) cm long; rachis 1–1.5 cm long; flowers dense, very short-pedicellate; bracts ca 1 mm long; pedicel and calyx sparsely pubescent; calyx 3–5 mm long, the lobes ca 0.5 mm long; corolla minutely, sparsely pubescent, 7–9 mm long, the lobes ca 1.5 mm long; stamens and style to 4 cm long. Legumes flat, minutely puberulent, curved slightly, to 16 cm long and 4 cm wide, the marginal ribs prominent. *Croat 8436.*

Infrequent or rare; collected along the shore of Gigante Bay and the shore of Bat Cove. Seasonal behavior uncertain. Flowers from August to April, possibly with individuals flowering twice a year. Fruit maturity period not determined.

Nicaragua to Peru and Brazil. In Panama, known from tropical moist forest in the Canal Zone, Bocas del Toro, Colón, Panamá, and Darién.

See Fig. 265.

Inga sapindoides Willd., Sp. Pl. 4:1012. 1806
   I. panamensis Seem.; *I. pittieri* Micheli
   Bribri, Guavo

Tree, to 15 m tall; branchlets ferruginous-puberulent when young, glabrate and prominently lenticellate in age. Petioles ribbed, 2–7 cm long; rachis winged; petiole and rachis sparsely ferruginous-pubescent; leaflets in 2–5 pairs, obovate to elliptic, acute to acuminate, obtuse to rounded at base, densely tomentose to puberulent

especially on midrib and veins below, the terminal pair 11–26 cm long, 4.5–12 cm wide. Spikes usually axillary in groups of 1–3; peduncles usually sparsely tomentose, 1–5(6) cm long; floral rachis much shorter, 0.5–2(3) cm long; bracts to 12 mm long; calyx and corolla sparsely sericeous; calyx 6–13(17) mm long, the lobes to 2 mm long; corolla 18–30 mm long, the lobes to 3 mm long; stamens and styles 4–5.5 cm long. Legumes tetragonal with prominently ribbed corners, 11–30 cm long, 1.5–3.5 cm wide, ca 1.5 cm thick, glabrate to densely pubescent. *Croat 5025, 7345.*

Common, at least along the shore. Seasonal behavior uncertain. Flowers from August to May, especially from November to February. Most fruits mature in the rainy season.

An extremely variable species. Often planted for shade in coffee plantations.

Mexico to Panama; West Indies. In Panama, known from tropical moist forest in the Canal Zone, Bocas del Toro, San Blas, Coclé, Panamá, and Darién and from premontane wet forest in Chiriquí (San Félix).

Inga spectabilis (Vahl) Willd., Sp. Pl. 4:1017. 1806
   Bribri, Guavo, Monkey tambrin, Guava de castilla, Guava real

Tree, to 22 m tall and 34 cm dbh; outer bark smooth, light brown; inner bark pale reddish-brown or tan; branchlets 4-angled, ± lenticellate, glabrate in age. Leaves glabrate in age above except on veins, sparsely short-pilose below, more densely on veins; petiole and rachis narrowly winged to only margined; leaflets in 2 pairs (rarely 3), sessile, broadly elliptic to obovate, rounded to acute or mucronate at apex, obtuse to subcordate and inequilateral at base, the terminal pair 17–28 cm long, 9–17 cm wide, the basal pair much reduced. Flowers in capituliform spikes, the spikes axillary or terminal, in groups of 1–6, with dense bracts; peduncles tomentose, 3–8 cm long; rachis 1–3 cm long; bracts large, to 14 mm long and 11 mm wide, obscuring calyx, tomentose, ovate, deciduous as flowers open; calyx tomentose, 7–9 mm long, the lobes 2.5 mm long; corolla short-sericeous, 15–22 cm long, the lobes to 3 mm long; stamens and style to ca 25 mm long. Legumes flat, 30–70 cm long, 5–8 cm wide, 1.5–3 cm thick, glabrous, the margins not elevated. *Croat 10480, Shattuck 925.*

Uncommon, possibly restricted to the older forest. Flowers from March to August with a peak in June and July. The fruits probably mature from February to May of the following year.

Mexico and Costa Rica to Colombia and Venezuela. In Panama, known from tropical moist forest in the Canal Zone, Bocas del Toro, San Blas, Chiriquí, Herrera, Coclé, Panamá, and Darién, from tropical dry forest in Panamá (Taboga Island), from premontane moist forest in Los Santos (Pocrí), from premontane wet forest in Chiriquí (Tolé), and from tropical wet forest in Colón.

Inga thibaudiana DC., Mém. Leg. 12:439. 1826

Tree, to 8 m tall; branchlets, petioles, rachises, leaflets, below, and midribs above densely short-pubescent; lenti-

cels ± prominent, small. Petiole and rachis not winged; leaflets in 4–8 pairs, variable in shape, asymmetrical, mostly narrowly ovate to elliptic, acute to long-acuminate, obtuse to rounded at base, puberulent above except on midrib, the terminal pair 7–10(15) cm long, 2.5–4(6) cm wide. Spikes axillary or terminal, in groups of 1–4; peduncles 3–4 cm long, densely tomentose; bracts 1 mm long; calyx 3–5 mm long, the lobes acute, ca 0.5 mm long, tomentose; corolla 14–20 mm long, sericeous-villous, the lobes to 1.5 mm long. Legumes flat, 6–20 cm long, 1.5–2.5 cm wide, bearing dense short, ferruginous-tomentose pubescence, the marginal ribs prominent. *Foster 703, Shattuck 1122.*

Rare; collected along the shore of Gigante Bay and Burrunga Point. Seasonal behavior uncertain. Flowers from January to May. The fruits were mature in the rainy season.

Confused with *I. multijuga,* which often has nine or ten pairs of leaflets.

Belize to Panama; Trinidad. In Panama, known from tropical moist forest in the Canal Zone and Darién and from tropical wet forest in Colón (Salúd) and Darién.

Inga umbellifera (Vahl) Steud., Bot. Nom. Phan. 431. 1821
    I. gracilipes Standl.

Tree, to 10 m tall; young branches and inflorescences puberulent, otherwise glabrous. Petioles unwinged or winged near apex; rachis winged; leaflets in 2 or 3 pairs, elliptic to oblong-elliptic, acuminate, obtuse to rounded at base, the terminal pair 11–17 cm long, 4–6.5 cm wide. Racemes umbelliform, axillary and solitary (appearing paniculate on new growth); peduncles 2–4(5) cm long; rachis very short, 3–4 mm long; bracts subulate, ca 4 mm long; flowers long-pedicellate; pedicels slender, 8–12(15) mm long; calyx 3–5 mm long, the lobes acute, ca 0.5 mm long; corolla 9–14(17) mm long, the lobes acute, to ca 2 mm long; stamens and style to ca 20 mm long. Legumes flat, 6–18 cm long, 1.5–3.5 cm wide, straight or curved, glabrate in age, the marginal ribs prominent. *Croat 7877, 11889.*

Common, at least on the shore. Flowers from August to May, possibly twice a year. Fruit maturity period uncertain; mature fruits have been seen in the early rainy season.

Inga quaternata, the only other species in the genus with long-pedicellate flowers, is distinguished by its unwinged rachis.

Panama to Peru and Brazil. In Panama, known from tropical moist forest in the Canal Zone and Darién and from premontane wet forest in Colón and Panamá.

See Fig. 266.

Inga vera Willd. subsp. **spuria** (Willd.) J. León, Ann. Missouri Bot. Gard. 53:339. 1966
    Guava, Coralillo

Tree, to 12(20) m tall, ca 30 cm dbh; branchlets sometimes ribbed below petioles, tomentose when young, glabrate in age, lenticellate. Petiole and rachis tomentose; petioles winged or not, 8–40 mm long; rachis winged;

leaflets in 4–8 pairs (rarely 9 or 10), oblong-elliptic to lanceolate, acuminate, obtuse to rounded at base, tomentose to sparsely pubescent, more densely on veins, the terminal pair 4.5–12(17) cm long, 2.5–5.5 cm wide. Spikes axillary or terminal, in groups of 1 to many; peduncles 3–5(7) cm long, tomentose; rachis 1–3(5) cm long, tomentose; bracts ca 3 mm long; calyx densely villous, 11–19 mm long, sometimes striate or minutely ribbed, the lobes acute, ca 3 mm long; corolla lanate, 15–20 mm long, the lobes acute, ca 4 mm long; stamens and style to 80 mm long. Legumes terete, densely tomentose, usually less than 15 cm long (to 30 cm), to 2 cm wide, usually curved, the base blunt, the apex pointed, the margins prominently raised, the lateral margins rounded and narrowly ribbed, the intervening area of the valve minutely transversely ribbed. *Zetek 3464.*

Though several old collections were made by Zetek, the species has been collected only once in recent years, along the shore of Gigante Bay. Flowers throughout the dry season (January to June), rarely in the rainy season; an individual plant may flower twice a year. Synchronous bursts of flowering have been observed in April. The fruits may be of mature size by June, with seed dispersal in the rainy season.

Mexico to northern South America; Greater Antilles. In Panama, known from tropical moist forest in the Canal Zone, Bocas del Toro, Veraguas, Los Santos, Herrera, Coclé, Panamá, and Darién, from tropical dry forest in Coclé, from premontane moist forest in Los Santos, and from premontane wet forest in Chiriquí (Boquete).

LEUCAENA Benth.

Leucaena multicapitula Schery, Ann. Missouri Bot. Gard. 37:302. 1950

Tree, to 20 m tall; branchlets glabrate and lenticellate in age. Leaves bipinnate with 2–4 pairs of opposite pinnae, with minute appressed pubescence above and below; petioles 3–5 cm long with a small, sessile, conical gland near apex; rachis 5–9 cm long, usually eglandular; pinnular rachis to 9 cm long with a gland at the apex; leaflets in 3–5 pairs per pinna, ovate to elliptic, acute at apex, acute to rounded at base, 1.5–5.5 cm long, 1–2.5 cm wide. Inflorescences terminal and upper-axillary, compounded 3 times, the heads orbicular, ca 1 cm wide, in simple or compound fascicles, the branches 1–5 per axil, to 20 cm long; peduncles ca 1 cm long, tomentose; flowers green, sessile, each subtended by a peltate bract as long as calyx; calyx ca 1.1 mm long; petals to 3.3 mm long, linear-oblong, weakly adnate to calyx; stamens 10, free, exserted ca 3 mm; style 1–2 mm longer than stamens. Legumes linear, flat, 7–14 cm long, 2.5–4.5 cm wide, thin, glabrous; seeds ovate, flat, ca 8 mm long, dark brown with prominent margins. *Croat 11735.*

Uncommon; collected on the shore of Miller Peninsula south of Orchid Island and at Gross Point. Flowers from June to August (sometimes as early as May) in central Panama. The fruits mature from July to October.

Albizia adinocephala (Donn. Sm.) Britt. & Rose, as reported by Standley (1933), is clearly this species. *Bailey*

281 has no more than ten stamens per flower, and the upper sides of the leaflets are not conspicuously reticulate-veined, as is *A. adinocephala.*

The species has been confused with *L. trichodes* (Jacq.) Benth., a closely related South American species, and is distinguished from it by having its inflorescence more highly branched.

The dark-colored seeds may either be displayed against the opened light-colored interior of the valves or fall free as the valves open.

Known only from Panama, from tropical moist forest in the Canal Zone, Coclé, Panamá, and Darién.

See Fig. 267.

MIMOSA L.

Mimosa casta L., Sp. Pl. 518. 1753

Vine, rarely to 5.5 m long, ± glabrous but sometimes with puberulent lower surface of leaflets, armed throughout with stout, recurved thorns. Leaves bipinnate with 1 pair of pinnae; petioles to 10 cm long, eglandular; stipules lanceolate, paired, to ca 5 mm long, striate; leaflets in 3–6 pairs per pinna, inequilateral, elliptic to lanceolate-oblong, acute at apex, oblique and rounded at base, 1–3 cm long, 7–10 mm wide, setose-ciliate on margins and sometimes on surface below; venation at base palmate. Flowers minute, in globular, white, pedunculate, axillary clusters ca 1–1.5 cm diam, the outermost in each head staminate; floral bracteoles linear-lanceolate, 1–2 mm long, pectinate-ciliate at base; calyx obsolete; corolla greenish, 4-lobed, ca 2.5 mm long; stamens (4)5, 5–8 mm long; style ± equaling stamens. Legumes usually 10–20 in a dense, usually globular cluster, flat, 3–4.5 cm long, ca 1 cm wide, usually 3–6-seeded, each segment distinct, the margins ± sinuate with laterally directed setae. *Croat 12821, 13157.*

Rare; seen only on the shore at the edge of Rear #8 Lighthouse Clearing. Locally abundant in the Canal Zone along roadsides, especially near water. Flowers and fruits from December to January.

Panama and the northern coast of South America to Brazil; Jamaica and Lesser Antilles. In Panama, known only from tropical moist forest in the Canal Zone.

See Fig. 268.

Mimosa pigra L., Cent. Pl. 1:13. 1755

Bashful plant, Zarza, Dormilón

Usually an erect shrub, less than 2 m tall; branchlets setose-hispid and armed with stout, recurved thorns.

Leaves bipinnate with 8–15 pairs of pinnae; stipules ovate, to 8 mm long, appressed-pubescent, subpersistent; petiole and rachis hispid with trichomes ca 2 mm long, eglandular, armed, the spines stout to slender, straight or recurved, to 4 mm long; petioles less than 1 cm long; rachis 10–20 cm long; pinnae opposite, 3–6 cm long, hispid; leaflets in 20–30 pairs per pinna, oblong, acute at apex, rounded to truncate at base, ca 8 mm long and 1 mm wide, appressed-pubescent especially below, the margins setose-ciliate; veins few, longitudinal. Flowers in dense, globular, white, pedunculate heads in terminal axils; peduncles hispid, 3–5 cm long; floral bracteoles linear, 2–3 mm long, pectinate above middle; calyx 1–2 mm long, cleft ca two-thirds its length, the lobes setaceous, glabrous; corolla 2–4 mm long, glabrous except short-hispid near apex, 4-lobed for nearly half its length; stamens 8, 4–5 mm long; style about as long as stamens. Legumes in dense clusters of 10–15, linear, 4–8 cm long, 1–1.5 cm wide, densely setose-hispid, flat, stipitate, with 7–20 segments, the margins persistent, the rectangular, 1-seeded segments 2.5–4 mm wide, falling free individually. *Croat 6868.*

Common in clearings at the margin of the lake, apparently preferring moist habitats. Flowers and fruits throughout the year, especially from May to August.

Pantropical. In Panama, ecologically variable; known from tropical moist forest in the Canal Zone, Bocas del Toro, San Blas, Veraguas, Coclé, Panamá, and Darién, from tropical dry and premontane moist forests in Los Santos, from premontane wet forest in Chiriquí, and from tropical wet forest in Colón and Darién.

Mimosa pudica L., Sp. Pl. 518. 1753

Sensitive plant, Dormidera, Ciérrate, Cierra tus puertas, Shameface, Shameweed

Prostrate herb, usually less than 1 m tall, but occasionally to 2 m or more when undisturbed; stems weakly ribbed, sparsely setose and armed with sharp spines. Leaves bipinnate with (1)2 closely spaced pairs of pinnae, the pinnae mostly 3.5–6.5 cm long; leaflets in 10–20 pairs per pinna, sensitive, folding upon disturbance, oblong, 3–12 mm long, 1–2 mm wide, glabrous, ciliate. Flowers dense, globular, pink clusters (congested racemes) on peduncles 1–2.5 cm long; floral bracteoles linear-lanceolate or linear-oblanceolate, setose; rachis elongating in fruit; calyx nearly obsolete; corolla funnelform, 1–2 mm long; stamens 4, ca 7 mm long; style sharp at apex, ± equaling stamens. Legumes brown, oblong, mostly 3- or 4-seeded, equally notched on both sides, glabrous but with multidirectional setae on lateral margins. *Croat 12712, 12813.*

Very abundant in the Laboratory Clearing; rare else-

KEY TO THE SPECIES OF MIMOSA

One pair of pinnae per leaf; leaflets usually more than 15 mm long; 3–6 pairs of leaflets per pinna
. . . . *M. casta* L.
Two or more pairs of pinnae per leaf; leaflets less than 15 mm long; more than 10 pairs of leaflets per pinna:
    Pinnae per leaf in 2 pairs; flowers pinkish; legumes setose-aculeate only on margin . . . . *M. pudica* L.
    Pinnae per leaf in 8–15 pairs; flowers white; legumes densely setose-hispid throughout . . *M. pigra* L.

Fig. 268. *Mimosa casta*

Fig. 267. *Leucaena multicapitula*

Fig. 269. *Pithecellobium barbourianum*

Fig. 270. *Pithecellobium hymeneaefolium*

where in clearings. Flowers and fruits principally from November to April.

Pantropical; native to northern South America. Throughout Panama in all life zones where disturbed habitats exist.

PITHECELLOBIUM Mart.

Pithecellobium barbourianum Standl., Contr. Arnold Arbor. 5:74, pl. 11. 1933

Unarmed tree, to 12 m tall and 25 cm dbh; outer bark thin, coarse, cracked; inner bark tan, finely granular; branchlets, leaf axes, and lower surface of leaves densely villous with reddish-brown trichomes; sap with weak, pungent odor. Leaves bipinnate, to 20 cm long, with 7–11 pairs of pinnae; petioles with a gland near apex; rachis and pinnular rachis with glands at nodes; leaflets in 9–16 pairs per pinna, oblong, oblique, ca 10 mm long, 4–5 mm wide, thick, glabrous above except on midrib, pubescent below, the margins revolute. Flowers minute, sessile, in globular, long-pedunculate, axillary heads; peduncles ca 6 cm long; flowering rachis ca 4 mm long; calyx funnelform, ca 3 mm long, golden-brown, hirsutulous, shallowly 5- or 6-dentate; corolla funnelform, ca 4 mm long, the outside bearing dense, golden-brown, hirsute pubescence; stamens ca 13, the filaments united basally. Legumes prominently coiled and somewhat constricted between each seed, to ca 11 cm long and 1.4 cm wide, brown, densely short-villous, twisting open irregularly at maturity to expose black seeds on the red-orange inner surface of the valves. *Shattuck 237.*

Represented from the island only by the type collection made at Zetek Trail 2400. Flowers in October. Some fruits known from August to December.

Known only from Panama, from tropical moist forest in the Canal Zone (BCI and adjacent areas around Gatun Lake).

See Fig. 269.

Pithecellobium dinizii Ducke, Arch. Jard. Bot. Rio de Janeiro 3:66. 1922
P. umbriflorum Ducke

Tree, to 8 m tall; young stems, leaf axes, and inflorescence branches tomentose. Leaves bipinnate with 2 or 3 pairs of opposite pinnae; petioles 4–6 cm long with a small gland near the base; stipules paired, oblong, acute, to 1 cm long,

persistent; rachis slightly longer with a similar gland at each node; pinnular rachis 10–15 cm long with a gland at all but the lowermost nodes; leaflets in 4–8(9) pairs per pinna, opposite, sessile, asymmetrical, oblong to elliptic, acuminate often ending in a slender dark cusp, acute to rounded at base, 2–9 cm long, 1–3 cm wide (to 6 cm long and 2 cm wide in South America), gradually reduced in size toward base, sparsely to densely pubescent on midrib. Flowers dense on densely clustered, axillary spikes to 2 cm long, usually at defoliated nodes; peduncles short; floral rachis swollen beneath flower, forming a small shelf; bracts ca 1.5 mm long; calyx to 4 mm long, irregularly lobed, usually deeply divided on one side, sparsely pubescent and ciliate; corolla white, to 5 mm long, sparsely pubescent, ± regularly 5-lobed; staminal tube exserted; anthers and apex of filaments red-violet; anthers broader than long, attached basally, the thecae directed upward. Legumes ca 8 cm long and 1 cm wide, bearing dense, golden-brown, tomentose pubescence, flattened with a narrow marginal ridge. *Croat 17024, 17027, Foster 1676.*

Known only from the shore along Chapman Cove. Flowers in the dry season. The fruits mature in the early rainy season.

Mature specimens on BCI differ from mature South American specimens by having larger leaflets, with the lower edge often obtuse to acute, compared to the somewhat auriculate edge in the South American material.

Panama, Colombia (*Cuatrecasas 16075*, Valle on the Pacific coast), Peru (Loreto), and Brazil (*Krukoff 1302*, "Amazonas, Calama on Río Madeira"). In Panama, known only from BCI.

Pithecellobium hymeneaefolium (H. & B.) Benth. in Hook., London J. Bot. 3:198. 1844
P. panamense Walp. & Duch.

Liana or climbing shrub, to 7 m long, minutely puberulent throughout but with the leaves and older stems glabrous to glabrate; stems minutely lenticellate, armed, the thorns paired, recurved, to 1 cm long at most nodes, persisting on older stems. Leaves bipinnate with 1 pair of pinnae; petiole and pinnular rachis each with a sessile cupular gland at apex; petioles 2–4 cm long, canaliculate; pinnae bifoliolate; pinnular rachis 1–2 cm long; leaflets very asymmetrical, ovate to ± elliptic, bluntly acute to rounded at apex, oblique at base with one side rounded, 4.5–12 cm long, 3–6 cm wide, sometimes barbate on

---

KEY TO THE SPECIES OF PITHECELLOBIUM

Leaves pinnate . . . . *P. rufescens* (Benth.) Pitt.
Leaves bipinnate:
    Leaflets in 1 pair per pinna . . . . *P. hymeneaefolium* (H. & B.) Benth.
    Leaflets in 4 or more pairs per pinna:
        Leaflets less than 1 cm wide; leaves with more than 6 pairs of pinnae . . . . *P. barbourianum* Standl.
        Leaflets more than 1 cm wide; leaves with less than 6 pairs of pinnae:
            Peduncles more than 5 cm long; petioles bearing large cupular gland at apex; legumes curled; pinnae in 3–5 pairs per leaf . . . . *P. macradenium* Pitt.
            Peduncles less than 5 cm long; petioles bearing small gland at apex; legumes flat; pinnae in 2 or 3 pairs per leaf . . . . *P. dinizii* Ducke

narrow side (*fide Flora of Panama*, Woodson & Schery, 1950). Flowers showy in long-pedunculate, axillary or terminal spikes, greenish with anthers white; peduncles to 4.5(6) cm long; rachis ca 7 cm long; calyx cupular, ca 2.5 mm long; corolla funnelform, ca 10 mm long, the lobes acute, ca 2 mm long; stamens 25–40, united at base, the staminal tube long-exserted, 2–4 cm long. Legumes 5–10 cm long, 1.5–2 cm wide, curved, densely tomentose, becoming bright red and twisting open at maturity; inner valve surface bright orange-red; seeds ovoid, black, ca 1.5 cm long and 1.2 cm wide, somewhat compressed, shiny, dangling from a very irregular pale red-orange funiculus and enveloped on most of one side by a finely dissected aril of the same color. *Croat 8176, Foster 1030.*

Occasional, along the shore on north side of the island and on Orchid Island. Flowers throughout the year, especially during the early rainy season (May to August). The fruits mature mostly from June to September.

Panama to Venezuela. In Panama, known from tropical moist forest in the Canal Zone, Los Santos, Panamá, and Darién, from tropical dry forest in Los Santos and Panamá (Taboga Island), and from premontane wet forest in Panamá.

See Fig. 270.

Pithecellobium macradenium Pitt., Contr. U.S. Natl. Herb. 20:465. 1922

Tree, to 30(40) m tall, to 1 m dbh; young branchlets with brownish short pubescence, becoming glabrous and sometimes lenticellate in age. Leaves bipinnate with 3–5 pairs of pinnae, the axes minutely pubescent; petioles 4–7 cm long; stipules acicular, caducous; rachis 8–16 cm long, bearing a large cupular gland to 1 cm long between pairs of basal pinnae and a smaller gland at node of successive pair; pinnular rachis to 15 cm long, with small glands at nodes; leaflets in 7–12 pairs per pinna, asymmetrical, ovate to oblong, obtuse to rounded at apex, usually obtuse at base, 2–5 cm long, 1–2 cm wide, minutely appressed-puberulent with longer trichomes on midrib. Flowers in condensed umbels or short spikes; peduncles 5–12 cm long, ferruginous-tomentose; floral rachis ca 5 mm long; pedicels 1–2.5 mm long; flowers pale green, 9–12 mm long excluding stamens, with minute, brown, appressed trichomes outside; calyx 5–6 mm long, the lobes irregular, ca 2 mm long, sometimes ± cleft on one side; corolla usually 5(6)-lobed ca one-third its length, fused to the staminal tube near base; stamens many, ca 3 cm long, the staminal tube nearly as long as corolla. Legumes linear, short-stipitate, curved into a nearly complete circle ca 6 cm diam, to 12 cm long and 2–2.5 cm wide, very thick at seminiferous areas, the valves splitting apart and twisting to display seeds; seeds flattened, ovoid, ca 8 mm long, on alternate segments of the valve. *Croat 5205.*

Uncommon; known only from a few places along the north shore and in the forest on the north side of the island. Flowers principally from February to May, with the fruits maturing from April to September, chiefly in the rainy season.

Costa Rica and Panama. In Panama, reported by Hol-

dridge (1970) from wet regions at low elevations; collected only from tropical moist forest on the Atlantic slope in the Canal Zone. Reported from tropical wet forest in Costa Rica (Holdridge et al., 1971).

See Figs. 271 and 272.

Pithecellobium rufescens (Benth.) Pitt., Contr. U.S. Natl. Herb. 18:181. 1916

Coralillo, Flor de indio, Harino, Jarino

Shrub or tree, to 7(12) m tall; branchlets, leaf axes, and inflorescences ferruginous-tomentulose; branchlets lenticellate. Leaves pinnate, usually with 3–5 pairs of pinnae; petioles very short; stipules acicular, caducous; rachis to 15 cm long with small gland at each node; leaflets variable, oblong to broadly elliptic, rounded to short-acuminate, rounded to obtuse at base, 5–13 cm long, 3.5–10 cm wide, weakly pubescent to glabrous above except on midrib, sparsely pubescent below, more densely on midrib. Flowers white, sessile, in globular, pedunculate, subterminal heads ca 2 cm wide; peduncles 1–2 cm long, ferruginous-tomentose; floral bracts linear, to 4 mm long, giving the head a burrlike appearance before anthesis; calyx 1–2 mm long, shallowly lobed, pubescent at apex of lobes; corolla 6–8 mm long, the lobes ca 1 mm long, pubescent at apex. Legumes linear-moniliform, curled, to 15 cm long, reddish, splitting and twisting at maturity to expose black, ellipsoid or orbicular seeds ca 7–10 mm long. *Croat 5355, 5925.*

Common along the shore on the north side of the island. Seedlings have been seen in the forest at Wheeler Trail 1900. Flowers mostly in the dry season (January to June). The fruits mature mostly in August and September (less often as early as February or as late as December).

Costa Rica, Panama, and Colombia. In Panama, known from tropical moist forest in the Canal Zone, Colón, San Blas, Los Santos, Herrera, Panamá, and Darién, from premontane moist forest in the Canal Zone, from premontane wet forest in Colón, Coclé, and Darién, and from premontane rain forest in Panamá (summit of Cerro Jefe).

See Fig. 273.

63B. CAESALPINIOIDEAE

Leaves simple or pinnate; stipels lacking. Flowers zygomorphic, in racemes or panicles, ebracteate; calyx sometimes 4-parted by union of 2 lobes, the lobes usually imbricate; petals 5 with 4 petals imbricate and the adaxial petal (upper) held within the bud, or petal 1 (*Swartzia*) or lacking (*Prioria*); stamens 10, or numerous, sometimes monadelphous.

Members of the subfamily Caesalpinioideae are most easily distinguished by their open, weakly zygomorphic flowers and ten free stamens. Zygomorphy is usually due to having only one petal or to having one petal unlike the others; this petal (the adaxial one) is held within the bud.

Flowers of most species are insect pollinated. Bees are known to be regular visitors of some species of *Cassia* and *Bauhinia*, which have both pollinating and feeding

KEY TO THE TAXA OF CAESALPINIOIDEAE

Leaves simple *or* leaflets 2 and plants vines:
  Leaflets 2 or blades deeply cleft at apex . . . . *Bauhinia*
  Leaves simple, the blades not cleft at apex . . . .
    . . . . *Swartzia simplex* (Sw.) Spreng. var. *ochnacea* (A. DC.) Cowan
At least some leaves compound *or* leaflets 2 and plant not a vine:
  Leaves bipinnate:
    Leaves with 12 or more pairs of leaflets per pinna; flowers yellow; plants large forest trees . . .
      . . . . *Schizolobium parahybum* (Vell.) S. F. Blake
    Leaves with 9–11 pairs of leaflets per pinna; flowers red-orange; plant a cultivated shrub in
      the Laboratory Clearing . . . . *Caesalpinia pulcherrima* (L.) Sw.
  Leaves pinnate:
    Leaves with 2 or 4 leaflets:
      All leaves bifoliolate:
        Inflorescences terminal; petals more than 10 mm long; sides of leaflets not coming to-
          gether on petiolule at base, the gap 2–7 mm; blades not emarginate . . . .
          . . . . *Hymenaea courbaril* L.
        Inflorescences terminal or axillary; petals less than 5 mm long; fruits less than 5 cm long;
          sides of leaflets coming together on petiolule; blades sometimes emarginate:
          Leaflets emarginate; dominant axis of inflorescence less than 1 cm long . . . .
            . . . . *Cynometra bauhiniifolia* Benth.
          Leaflets acuminate; dominant axis of inflorescence more than 10 cm long . . . .
            . . . . *Peltogyne purpurea* Pitt.
      Most leaves 4-foliolate:
        Leaves lacking gland between lower pair of leaflets; flowers very small, greenish-white;
          legumes suborbicular, concave or flat on one side, convex on the other . . . .
          . . . . *Prioria copaifera* Griseb.
        Leaves bearing prominent gland on rachis between lower pair of leaflets; flowers more
          than 1 cm long, yellow; legumes linear-cylindrical . . . . *Cassia* (in part)
    Leaves not with 2 or 4 leaflets:
      Leaves imparipinnate, with 3 or 5 leaflets:
        Leaves 3-foliolate or simple; inflorescences less than 10 cm long, bearing few flowers;
          legumes subterete, less than 5 cm long . . . .
          . . . . *Swartzia simplex* (Sw.) Spreng. var. *grandiflora* (Raddi) Cowan
        Leaves 3- or 5-foliolate; inflorescences more than 20 cm long, bearing many flowers;
          legumes broad, flattened, more than 15 cm long . . . . *Swartzia panamensis* Benth.
      Leaves paripinnate:
        Leaflets 4 or 6; plants rare or absent from the island . . . . *Cassia obtusifolia* L.
        Leaflets more than 8:
          Flowers orange, in large, dense, usually globular clusters; leaflets in 5–7 pairs; legumes
            densely brownish-tomentose . . . . *Brownea macrophylla* Linden
          Flowers yellow, in elongate racemes or spikes; leaflets in 7–14 pairs; legumes not
            densely brownish-tomentose:
            Plants small trees or shrubs of swampy areas; flowers markedly pedicellate; leaflets
              in 9–14 pairs; fruits oblong-linear, to 2 cm wide; seeds many . . . .
              . . . . *Cassia reticulata* Willd.
            Plants large forest trees; flowers ± sessile; leaflets in 7–9 pairs; fruits flat, elliptic,
              more than 3 cm wide; seed 1 . . . . *Tachigalia versicolor* Standl. & L. O. Wms.

anthers (Faegri & van der Pijl, 1966). While the bee (usually *Xylocopa* in *Cassia*) "milks" the feeding anthers, it vibrates its wings, raising a cloud of pollen from the pollinating anthers. Some grains land on the insect's back despite the fact that it sits atop the androecium. Some species of *Bauhinia* (not BCI species) and *Hymenaea courbaril* are known to be visited by several species of bat (Heithaus, Fleming & Opler, 1975).

No category of seed dispersal predominates. A number of species are endozoochorous, including *Cassia* spp., *Swartzia simplex,* and possibly also *Peltogyne purpurea;* these are probably taken mostly by birds. Van der Pijl (1968) reported the fruits of *Swartzia prouacensis* (Aubl.)

Amsh. to be bat dispersed, but they are more long-dangling than those of *Swartzia simplex.* Fruits of *Cynometra retusa* Britt. & Rose are taken by the bat *Artibeus jamaicensis* in Mexico (Yazquez-Yanes et al., 1975), and *Cassia undulata* is dispersed by bats (Bonaccorso, 1975). Howard Irwin (pers. comm.) has seen the pods of *Cassia fruticosa* and *C. undulata* being worked by mammals and lizards. The fruits of *Cynometra bauhiniifolia* and *Hymenaea courbaril* are taken by mammalian frugivores, but may also be dispersed from the ground by rodents. *Hymenaea* is reported to be dispersed by rodents in Central America and Mexico (J. Langenheim, pers. comm.). *Prioria copaifera* is sometimes eaten by white-faced mon-

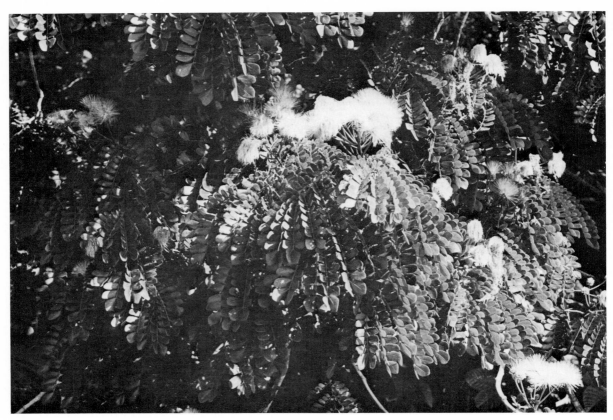

Fig. 271. *Pithecellobium macradenium*

Fig. 272. *Pithecellobium macradenium*

Fig. 273. *Pithecellobium rufescens*

Fig. 274. *Bauhinia guianensis*

keys (Hladik & Hladik, 1969) and also by peccaries (Enders, 1935) and agoutis (N. Smythe, pers. comm.) when food is scarce.

Wind-dispersed species include *Schizolobium parahybum* and *Tachigalia versicolor*. Water plays a role in the dispersal of the seeds of *Prioria copaifera* and possibly also of *Swartzia panamensis* and *Cassia reticulata*. Ridley (1930) reported the seeds of *Cynometra bauhiniifolia* and *Hymenaea courbaril* to be buoyant and possibly water dispersed. Pods of *Hymenaea* can survive long periods of time in both fresh and salt water, but quickly decompose when attacked by soil microorganisms (J. Langenheim, pers. comm.).

Bauhinia reflexa and *B. guianensis* have autochorous, disk-shaped seeds, which are thrown long distances by the elastically dehiscent valves.

BAUHINIA L.

Bauhinia guianensis Aubl., Hist. Pl. Guiane Fr. 1:377. 1775

B. excisa (Griseb.) Hemsl.; *B. manca* Standl.

Bejuco de cadena, Bejuco de mono

Tendriled liana; trunk to 20 cm diam near the ground; pubescence when present appressed, 0.3 mm long; older stems flattened and ± regularly folded, winged on both sides, the old stem scars thus borne medially; younger stems terete, appressed-pubescent; tendrils simple, watchspring-like. Leaves simple; petioles 2.5–3.5 cm long, swollen at both ends; blades ovate, usually cleft about one-third their length at apex on mature plants or to the base on younger plants, cordate at base, 6–10 cm long, 4–10.5 cm wide, glabrous above, densely pubescent below, the pubescence appressed-ferruginous, often not conspicuous. Racemes terminal or axillary, to 15 cm long; pubescence on all parts ± densely appressed-ferruginous; lower flowers opening first, the ovoid flower buds above the flowers conspicuous, longitudinally striate, 4–7 mm long; bracts 1–3, subulate, ca 1 mm long; flowers sweetly aromatic, subsessile; calyx ca 7 mm long and expanding to 13 mm wide in flower, the lobes obtuse, ca 2 mm long; petals 5, white, obovate, 4 of them ca 2 cm long and 1 cm wide, puberulent inside, the fifth petal narrower and shorter, keeled, recurved at apex; stamens 10, of various lengths, some to 8 mm long; anthers held within the keeled petal both above and below the stigma; style stout, its stigma protruding into the keeled petal. Legumes flat, oblong, ca 8 cm long and 2.5 cm wide, glabrous, apiculate at apex on one side; seeds 2–4, disk-shaped, 0.5 cm diam. *Croat 10931, 13810.*

Occasional, in the forest. Flowers February and March (sometimes to April). The fruits apparently mature in the early rainy season, but have been seen only in June. The species loses its leaves in the early dry season.

Because this is an exceedingly variable species, it has been placed under many different names throughout its range.

Southern Mexico to Bolivia and southern Brazil; Trinidad. In Panama, known from tropical moist forest in the Canal Zone, Colón, San Blas, Veraguas, Panamá, and Darién and from premontane rain forest in Coclé.

See Fig. 274.

Bauhinia reflexa Schery, Ann. Missouri Bot. Gard. 38:317. 1951

Tendriled liana; pubescence brown-hirsute on most parts, especially branchlets, petioles, pedicels, and veins of leaf below, the trichomes ca 1 mm long; older stems flattened, irregularly twisted and folded, bearing a single wing much thicker than the major axis that bears the old stem scars. Leaves simple; stipules 5–10 mm long, foliaceous, curved, oblique, caducous; petioles 2–4 cm long (to 9 cm on juveniles), swollen at both ends; blades broadly ovate to orbicular, cleft about one-third their length when mature, cordate at base, 4–12 cm long, about as broad as long, glabrous above, bearing an apicule to ca 3 mm long at base of cleft; basal veins 11–13; juvenile leaves to 18 cm long, often cleft to base. Racemes terminal or axillary, to ca 15 cm long; flowers with 1–3 linear bracts ca 1 cm long; pedicels ca 1 cm long; calyx cup-shaped, conspicuously ca 15-ribbed, 6–8 mm long, expanding to ca 10 mm wide, the lobes slender, reflexed to spreading, ca 6 mm long and 1 mm wide; petals 5, obovate and clawed, 2–2.5 cm long, pale lavender to pink, glabrous within except at base, with dense, ± appressed brown trichomes outside ca 1 mm long, 4 of the petals spreading, ca 6 mm wide, the fifth ca 4.5 mm wide and turned sharply inward toward the lower petals, usually reflexed above the middle, the base thickened and curled inward along the margins, partly enclosing the stamens; stamens 10, unequal, the longest to 1 cm, the shortest ± equaling the length of the style; filaments thick; anthers ca 1.3 mm long, dehiscing toward the reduced petal, forming an oblique surface appressed against the reduced petal; style to ca 7 mm long, nearly surrounded by the anthers; stigma oblique; ovary elongate, coarsely pubescent. Legumes linear-oblong, flat, 8–12 cm long, 2.8–3.5 cm wide, glabrous or sparsely long-pubescent, apiculate at apex on one side, explosively dehiscent; seeds usually 5 or 6 or fewer by abortion, disk-shaped, dark brown, shiny, 0.5–2 cm diam. *Croat 6152, 8126.*

Occasional, along the edge of the lake; no doubt in the canopy of the forest as well. Flowers from May to September. Mature fruits have been seen from June to November and also in February.

KEY TO THE SPECIES OF BAUHINIA

Young stems, petioles, and rachises of inflorescences hirsute, the trichomes ca 1 mm long . . . . *B. reflexa* Schery

Parts glabrous or pubescent, the trichomes appressed, ca 0.3 mm long . . . . *B. guianensis* Aubl.

Panama and northwestern Colombia. In Panama, known from tropical moist forest in the Canal Zone, San Blas, and Darién.

BROWNEA Jacq.

Brownea macrophylla Linden, Cat. no. 18:11. 1863

Cuchillito, Ariza, Palo de la cruz, Rosa del monte, Cacique negro

Small tree, to 7(10) m tall; young branchlets, petioles, rachises, and petiolules tomentulose. Leaves pinnate; petioles swollen, ca 1 cm long (to 10 cm long on non-flowering stems); petiolules swollen, to 5 mm long; leaflets in 5–7 pairs, narrowly elliptic to elliptic-lanceolate or elliptic-oblanceolate, 10–25 cm long, 2.5–6 cm wide, long-acuminate, asymmetrical at base and acute to subcordate, scattered-puberulent below, glaucescent, bearing a single globose gland ca 1 mm diam on the underside of leaflet at base. Inflorescences axillary or nearly terminal, densely flowered, capitate to short-racemose; peduncles, pedicels, and the outside of the bracteolar sheaths tomentose; pedicels ca 5 mm long; bracteolar sheaths bilobed, 3–4 cm long, ± enclosing calyx; sepals usually 4, somewhat petaloid, obovoid, ca 3 cm long and 0.6–3 cm wide; petals orange, 5, obovate and clawed, 3–3.5 cm long and 1–1.5 cm wide, rounded, oblique at base, the narrow claw ca 1 cm long; stamens 10–15(20), 8 cm long, united at base into a tube 2.5 cm long, the tube open on the upper side, wrapping only three-fourths of the way around the pistil; pistil narrow, 10 cm long; ovary tomentose. Legumes flat, slightly turgid, oblong, peaked, to 20 cm long and 4.5 cm wide, brownish-tomentulose when immature, with heavy ridges along both edges; seeds flat, rectangular, 3.5 cm long, 2.5 cm wide, black. *Croat 7200, 14660.*

Rare; known only from one area along the shore of Bat Cove, and possibly a remnant from cultivation. Flowers from November to May, principally in the late dry season. The fruits mature mostly in the early rainy season.

Panama and northern South America. In Panama, known only from tropical moist forest in the Canal Zone, Panamá, and Darién.

CAESALPINIA L.

Caesalpinia pulcherrima (L.) Sw., Obs. Bot. 166. 1791

Gallito, Barbados pride, Bird of paradise flower, Flower bence

Glabrous shrub or small tree; older stems bearing sharp prickles to 1 cm long. Leaves bipinnate, with 7–9 pairs of opposite pinnae; petioles to 5(8) cm long; rachis and pinnular rachis bearing minute, acicular, glandlike structure at each node; petiolules ca 5 mm long; leaflets usually in 9–11 pairs per pinna, opposite or subopposite, oblong or obovate, rounded or emarginate at apex, obtuse to rounded at base, slightly inequilateral, ca 1.5 cm long and 7 mm wide (to 2.3 cm long and 10 mm wide). Racemes subcorymbose, terminal or subterminal; bracts caducous, subulate, to 3 mm long; pedicels 2–6 cm long; calyx tube 3–4 mm long, 5-lobed, the lobes obovate, to 1.5 cm long; petals bright orange, 5, obovate, clawed, spreading, ca 2 cm long, 1 reduced, ca 5 mm wide; stamens 10, ca 4 cm long, unequal, the filaments orange, villous at the base; style orange, extended about 2 cm beyond stamens. Legumes flat, 9–12 cm long, broader above the middle, beaked at apex, obliquely acute at both ends, pendent, elastically dehiscent; seeds about 6, ovate, ca 8 mm long, brown. *Croat 4856.*

Cultivated at the Laboratory Clearing. Probably flowers and fruits throughout the year, especially in the late dry season to the middle of the rainy season.

Pantropical in cultivation; probably native to tropical Asia. In Panama, known from tropical moist forest in the Canal Zone and Panamá and from tropical dry forest in Los Santos and Panamá.

CASSIA L.

Cassia fruticosa P. Mill., Gard. Dict. ed. 8, no. 10. 1768

Shrub or small tree, mostly to 5 m tall (occasionally to 10 m), puberulent all over, especially on lower leaflet surface. Leaves paripinnate; stipules linear, caducous; petioles 2–4 cm long; rachis 1–4 cm long, bearing stipule-like, subconic gland between pairs of leaflets, the terminal gland often missing; petiolules to 6 mm long, stout; leaflets 4, elliptic, sometimes inequilateral, acuminate, obtuse to rounded at base, 6–24 cm long, 2–9 cm wide; juvenile blades often pruinose beneath. Inflorescences terminal or subterminal, paniculate to racemose; bracts brown, subulate, ca 3 mm long, caducous; pedicels to 4 cm long; calyx with a disklike base, the sepals rounded, imbricate, ca 12 mm long; petals 5, mostly obovate-suborbicular, 2.5–3 cm long, with a slender claw at base ca 1 mm long, yellow to orange-yellow, spreading at anthesis; stamens 10, the 3 uppermost aborted, minute, the 4 fertile stamens curved, the 3 fertile stamens nearest the style curved to sigmoid, often with the pores reduced to 1 at apex; anthers ca 7 mm long, with 2 apical pores; style densely pubescent, emerging from one side of flower and broadly curved;

KEY TO THE SPECIES OF CASSIA

Leaves sometimes with more than 4 leaflets:
    Leaflets 4 or 6 . . . . *C. obtusifolia* L.
    Leaflets 18–28 . . . . *C. reticulata* Willd.
Leaves all with 4 leaflets:
    Petals less than 2 cm long; floral bracts and stipules persistent; leaflets mostly less than 3.5 cm
        wide; glands on rachis between both pairs of leaflets . . . . *C. undulata* Benth.
    Petals more than 2 cm long; floral bracts and stipules caducous; leaflets mostly more than 3.5 cm
        wide; gland on rachis between apical pair of leaflets usually missing . . . . . . *C. fruticosa* P. Mill.

Fig. 275. *Cassia reticulata*

Fig. 276. *Cassia undulata*

stigma deeply cupular, held above the anthers. Legumes linear-cylindrical, straight or slightly curved, 20–30 cm long, ca 1 cm diam at maturity, splitting dorsally; seeds 60–80, transverse, black. *Croat 6012, Hladik 234.*

Sparse in the forest, along the lakeshore, and at the edges of clearings; elsewhere often common. Flowers and fruits apparently throughout the year, especially in the dry season and in the middle of the rainy season, from August to September, with little flowering activity at the beginning of the rainy season. Fruit maturity time not determined. Leaves are replaced in the dry season.

The variety *gatunensis* (Britt.) Schery intergrades with the typical variety and probably does not merit recognition. Both forms can be seen on BCI.

Mexico to Brazil. In Panama, known from tropical moist forest in the Canal Zone, Bocas del Toro, Veraguas, Colón, San Blas, and Darién, from premontane dry forest in Los Santos, from premontane wet forest in Chiriquí, Coclé, and Panamá, and from premontane rain forest in Panamá.

Cassia obtusifolia L., Sp. Pl. 377. 1753

Senna, Dormidera

Slender erect herb or suffrutex, to 1.2 m tall. Leaves paripinnate; stipules linear, to 1.5 cm long; rachis bearing a gland between the leaflets of basal pair; leaflets in 2 or 3 pairs, obovate, inequilateral, obtuse and mucronate at apex, rounded to cuneate at base, 1.5–2 cm long, 1–1.5 cm wide, ciliolate, glabrous above, pubescent and glaucescent below. Flowers solitary or paired at the nodes, oblique on pedicels; pedicels ca 11 mm long; sepals ovate to oblong, to 8 mm long, ciliate; petals 5, yellow, obovate, to 11 mm long and 5 mm wide; fertile stamens 7, of unequal sizes, dehiscent by a single terminal pore; anthers 2–4 mm long; filaments 1–2.5 mm long; ovary pubescent. Legumes linear, arcuate, to 17 cm long, ca 4 mm wide, subglabrous, somewhat quadrangular; seeds rhomboid, ca 4 mm long and 2.2 mm wide, dark brown, shiny.

No specimens have been seen, but to be expected in clearings. Flowers and fruits mostly in the dry season (November to April). The fruits persist until about May.

Reported by Standley for BCI as *C. tora* L., an Old World species.

Cosmopolitan in the tropics and subtropics; native to India. In Panama, known from tropical moist forest in the Canal Zone, Bocas del Toro, Colón, Veraguas, Herrera, Panamá, and Darién.

Cassia reticulata Willd., Enum. Pl. 1:443. 1809

Laureño, Wild senna

Tree, usually 2–4(6) m tall; branchlets puberulent, pithy. Leaves paripinnate, mostly 20–50 cm long; petioles with prominent basal pulvinus; leaflets in 9–14 pairs, oblong, acute and downturned at apex, obtuse-rounded at base, 4–13 cm long, 1.5–5.5 cm wide, usually densely pubescent on both surfaces, especially below. Racemes terminal and upper-axillary; flowers subtended in bud by a yellow, ovate, caducous bract ca 2 cm long; pedicels 2–5 mm

long; sepals elliptic, to 13 mm long; petals 5, yellow, conspicuously dark-veined, rounded at apex and turning inward at anthesis, obtuse at base, ca 16 mm long, with a slender claw ca 1 mm long; stamens 10, the outer 2 large, fertile, with the anthers prominently curved, ca 11 mm long, with 2 apical pores and lateral slits (sometimes not opening), each loculus almost filled with a juicy matrix, the 4 medial stamens much smaller, 4 stamens aborted; pistil puberulent, held between the 2 large stamens; style slender, recurved, the stigmatic surface sunken. Legumes flattened, to 15 cm long and 2 cm wide, marginally ribbed, breaking into many, 1-seeded, linear parts at maturity. *Wetmore & Abbe 166.*

Occasional in marshy areas near Frijoles and elsewhere in the isthmus, but not seen recently on BCI. On the basis of old collections, it was once common on the island, but apparently prefers more disturbed, swampy areas than now exist. Flowers mostly in the dry season; Allen (1956) reported the species to flower from late August to February. The fruits persist for a long time. Time of dehiscence is not known.

Mexico to Bolivia and Brazil. In Panama, known from tropical moist forest in the Canal Zone, Bocas del Toro, Colón, Los Santos, Herrera, Panamá, and Darién and from tropical wet forest in Coclé.

See Fig. 275.

Cassia undulata Benth. in Hook., J. Bot. (Hooker) 2:76. 1840

Slender arching shrub or liana, 2–5 m high; older stems glabrous, terete or angulate, lenticellate; younger stems and petioles sparsely to densely pubescent with spreading trichomes. Leaves paripinnate, set on a prominent woody base, mostly less than 15 cm long; stipules conspicuously falcate, subpersistent; petiole and rachis ribbed above; rachis bearing a raised gland between leaflets of each pair; petiolules short; leaflets 4, lanceolate-subfalcate, acuminate, unequal at base, to 11 cm long and 4 cm wide, ± glabrous above, dull and with appressed trichomes below, ciliate. Inflorescences racemose-paniculate, terminal or subterminal, puberulent; bracts ovate-lanceolate, ca 8 mm long, mucronate, persistent; pedicels ca 20 mm long; sepals elliptic to obovate, 7–8 mm long; petals 5, yellow, rounded, cuneate at base, to 15 mm long; anthers conspicuous, 4 large (ca 7 mm long) and 3 small (ca 4 mm long), blunt, with usually 2 terminal pores; style long, pubescent, emerging from one side of flower, recurved over anthers; stigma deeply cup-shaped, the outer margin minutely fringed. Legumes linear, irregularly cylindrical, 10–20 cm long, ca 1 cm thick, usually splitting open on one side; seeds many, smooth, brown, ± elliptical, flattened, to 5 mm long, stuck together laterally by a sweet, sticky, reddish-black, tarlike substance. *Croat 4827.*

Abundant along the shore. Flowers from November to April, mostly during the dry season. The fruits mature from April to June, possibly later also.

Standley confused this species, which he named *C. hayesiana* (Britt. & Rose) Standl., with *Cassia maxonii* (Britt. & Rose) Schery.

The flower unfolds to some extent well in advance of being functional, the petals remaining erect, the anthers closed, and the style recurved with its apex between the stamens. Later (sequence unknown) the anthers become functional, the style uncurls somewhat (the stigma still protruding downward), and the stigmatic cup begins to secrete nectar. Pollen is apparently deposited on the stigma by insects seeking the nectar.

Southern Mexico to northern South America; Trinidad. In Panama, known from tropical moist forest in the Canal Zone and Veraguas and from premontane wet forest in the Canal Zone and Panamá.

See Fig. 276.

CYNOMETRA L.

Cynometra bauhiniifolia Benth. in Hook., J. Bot. (Hooker) 2:99. 1840

Tree, to 15 m tall, ca 40 cm dbh; outer bark thin, sandpapery, unfissured, dark brown; inner bark thick, hard, reddish-brown with irregular, white, radial lines, the sap sweet, not aromatic; stems conspicuously lenticellate. Leaves bifoliolate, stiff; petioles to 5 mm long; leaflets very inequilateral, emarginate at apex, acute at base, 3–6 cm long, 1–2.8 cm wide, glabrous or sparsely pubescent on midrib. Fascicles short-pedunculate, 5–15 mm long, 1 to several in leaf axils, conspicuous and conelike in bud; bracts prominently veined, 1–2.3 mm long; pedicels crisp-villous, 4–8 mm long; sepals 4, ca 3 mm long, membranaceous, deciduous; petals 5, white, to 3.6 mm long, unequal, inserted on a disklike receptacle; stamens exserted. Legumes oblong to subrotund, 3.5–4.5 cm diam, minutely and densely tomentose, with a hard exocarp to 2 mm thick; seed 1. *Foster 1663.*

Known only from the shore of Chapman Cove. Seasonal behavior uncertain. Flowers have been seen in April and July, and fruits in July and September.

Guatemala to Argentina. In Panama, known only from tropical moist forest in the Canal Zone and Panamá and from tropical wet forest in Colón.

HYMENAEA L.

Hymenaea courbaril L., Sp. Pl. 1192. 1753
West Indian locust, Algarrobo, Courbaril, Cuapinol, Quapinol

Tree, mostly to 20(30) m tall and 50(200) cm dbh; outer bark brown, closely lenticellate, bitter tasting, pale orange beneath surface, glabrous; wood reddish-brown, hard. Leaves bifoliolate; petioles 1–2 cm long, rugose when dry; leaflets narrowly oblong to elliptic-lanceolate, asymmetrical, short-acuminate, unequally rounded at base, 4–10 cm long, 2–5 cm wide, coriaceous, punctate, the midrib conspicuous below. Inflorescences terminal, subcorymbose, to ca 8(12) cm long, the branches puberulent, the parts jointed and articulate; flowers white or purplish, soon falling, probably opening at night; bracts caducous; pedicels thick, ca 7 mm long; calyx tube ca 8 mm long, 4-lobed, the lobes ovate to oblong, expanding

to ca 15 mm long, coriaceous, densely tomentose inside, easily caducous; petals 5, white, sometimes tinged with purple, rounded, 1.5–2 cm long, ca 9 mm wide, clawed below, the claw ca 1.5 mm long; stamens 10, alternately short and long, the long ones to 2 cm long; style attached laterally at apex of ovary, directed somewhat to one side of the flower; stigma held above the lower anthers and at some distance from the divergent longer set. Legumes oblong, flattened, to 17 cm long and 6.5 cm wide, turgid, hard, reddish-brown; seeds (2)4–6, embedded in sticky pulp. *Croat 10209.*

Rare, in the young forest. Flowers during the dry season and the early rainy season (December to May). The fruits mature chiefly during the rainy season, especially late in the rainy season.

The flowers are believed to be bat pollinated.

Mexico to Peru, Paraguay, and Brazil; West Indies. In Panama, ecologically variable; known from tropical moist forest in the Canal Zone, Veraguas, Herrera, Panamá, and Darién, from premontane dry forest in Herrera, from tropical dry forest in Coclé, from premontane moist forest in the Canal Zone and Panamá, and from premontane wet forest in Chiriquí, Coclé, and Darién. Tosi (1971) listed this species as characteristic of tropical dry, premontane moist, and tropical moist forests in Panama. Reported also from tropical wet forest in Costa Rica (Holdridge et al., 1971).

PELTOGYNE J. Vogel

Peltogyne purpurea Pitt., J. Wash. Acad. Sci. 5:471. 1915
Nazareno, Morado, Purple heart

Tall tree, to 50 m tall and 1 m dbh; wood dark purple, hard; branchlets slender. Leaves bifoliolate; petioles to 2 cm long; petiolules 3–4 mm long; leaflets lanceolate-elliptic, subfalcate, long-acuminate, obliquely rounded or obtuse at base, 5–7 cm long, 2–3 cm wide, the veins reticulate. Inflorescences racemose, terminal or subterminal, bearing few to many flowers; branches of the inflorescence 5–7 cm long, the branches, pedicels, and calyces tomentose; pedicels to 2 mm long, nodose in middle; sepals ca 2.5 mm long, obtuse; petals 5, ca 3 mm long, obovate; stamens ca 3 mm long; ovary tomentose; style slender, maroon, ca 2 mm long. Legumes broadly obovate, flattened, ca 3 cm long and 1.6 cm wide, glabrous, mucronulate at apex, somewhat arcuate above, rounded below, on pedicels to 8 mm long; seminiferous area indistinct; seed 1, ca 2 cm long, obliquely ovate, depressed, persistent on the dehisced fruit, hanging by the funicle.

Reported by Standley for the island, but no collection was cited and none has been seen. The only material of this species from the Canal Zone that I have seen was growing at Summit Garden. It is possible that the species is not native to central Panama.

Allen (1956) reported that in Costa Rica individuals flower several times from August to December at two-week intervals, with flowers lasting about three days and the fruits maturing in February. The leaves of these plants were briefly deciduous in the early dry season.

Costa Rica and Panama. In Panama, known from tropical moist forest in the Canal Zone and reported to be common in Darién in tropical moist and tropical wet forests (Duke, 1968; Allen, unpubl.; Lamb, 1953). Tosi (1971) listed the species as characteristic of premontane wet forest in Panama. Holdridge (1970) reported it from tropical moist and tropical wet forests on well-drained soils.

PRIORIA Griseb.

Prioria copaifera Griseb., Fl. Brit. W. Ind. 215. 1860

Amansa mujer, Cativo, Cantivo

Tree, to 40 m tall; trunk to 75(150) cm dbh, unfissured, coarse, roughly lenticellate; inner bark light brown becoming darker in time, the sap usually black, sweet, with faint aroma; stems lenticellate, puberulent. Leaves compound, glabrous; petiole and rachis lenticellate; petioles 1–3 cm long; rachis longer, to 4 cm; petiolules callous-rugose, 5–10 mm long; leaflets (2)4, ± elliptic, acuminate, rounded at base, 4–16 cm long, 2.5–8 cm wide, inequilateral. Inflorescences to ca 30 cm long, branched, the spikes to 10 cm long, unbranched, bearing many flowers; flowers greenish-white, sessile, ca 4 mm diam, subtended by 2 bracts, forming a cuplike structure ca 1 mm deep; sepals petaloid, spreading, rounded, concave, ca 2.5 mm long, obscurely pellucid-punctate, ciliate; petals lacking; stamens 10, ca 5 mm long, exserted and weakly spreading; filaments villous below the middle; pistil to ca 3 mm long; ovary villous, gradually tapered to a simple style. Legumes suborbicular, to 10 cm long and 7 cm wide, concave or flattened on one side, convex on the other. *Croat 6860.*

Common along the shore on the southern side of the island (e.g., on the shore of Gigante Bay) and in the old forest along Armour Trail; apparently rare or absent elsewhere. Elsewhere in Panama the species usually occurs along the margins of rivers and swamps. Seasonal behavior is uncertain. Flowering is possibly bimodal, with a major period of flowering in September and October (possibly in response to the slight decrease in rainfall in September) and another period of flowering in the dry and earliest rainy season (December to May). The fruits mature in about 6 months.

White-faced monkeys eat the fruits in April and May (Hladik & Hladik, 1969), and Standley (1928) found peccaries to be fond of them. Seeds are very buoyant and are often found floating in the water at the edge of the lake. According to Chapman (1938), seeds are not eaten by animals.

Nicaragua to Colombia; Jamaica. In Panama, common in tropical moist forest on the Atlantic slope in the Canal Zone, Bocas del Toro, San Blas, and Darién, but also common in tropical wet forest in Darién, where it forms nearly pure stands in flooded areas (Holdridge & Budowski, 1956; Holdridge, 1970; Lamb, 1953). Tosi (1971) listed the species as characteristic of tropical wet forest in Panama, but also common in tropical moist forest.

See Figs. 277 and 278.

SCHIZOLOBIUM J. Vogel

Schizolobium parahybum (Vell.) S. F. Blake, Contr. U.S. Natl. Herb. 20:240. 1919

Indio, Tinecú

Tree, 25–30(40) m tall, 30(100) cm dbh, with narrow, low buttresses; outer bark hard, planar, with minute fissures and fine lenticels; inner bark thin, granular, whitish, the sap sweet-tasting; branches glabrous to sparsely puberulent; juvenile plants often with very resinous stems. Leaves bipinnate, very large, to 130 cm long, with ca 8 pairs of opposite pinnae; petioles ca 30 cm long; rachis flattened or sulcate above; petiolules ca 1 mm long; leaflets mostly in 12–22 pairs per pinna, ± oblong, rounded to emarginate at apex, often with a minute apiculum, rounded at base, ± equilateral, 1.3–3 cm long, ca 0.9 cm wide, appressed-pubescent especially below, the lateral veins obscure. Inflorescences axillary or terminal racemes or panicles of several racemes; bracts lanceolate, ca 2 mm long; pedicels to 1 cm long; calyx deeply lobed, ferruginous-tomentose, the tube turbinate, 2–3 mm long, the lobes ovate-elliptic, 3–7 mm long; petals 5, yellow, obovate-spatulate, to 2 cm long, 4–6 mm wide; stamens 10, ± equaling petals; ovary subfalcate, hispid-tomentose. Legumes oblanceolate, somewhat asymmetrical, to 12 cm long, 3–5 cm wide, the venation conspicuously reticulate; seed 1, round, flat, ca 8 mm diam, shiny, brown, borne in a thin endocarp, the endocarp shaped like the fruit with a one-sided wing. *Croat 12968, Zetek 6018.*

Rare, in the forest near the Laboratory Clearing. Flowers in the early dry season (December to February); the principal flowering period may be followed by a smaller one. The fruits mature from the middle of the dry to the early rainy seasons and require about a year to develop. Plants lose their leaves just prior to flowering and replace them after flowering.

Juvenile plants can be recognized by their large leaves and very viscid stems.

Caribbean coast of southern Mexico to southern Brazil. In Panama, collected from tropical moist forest on the Atlantic slope of the Canal Zone and in Panamá and Darién; reported also from tropical wet forest in Costa Rica (Holdridge et al., 1971). Holdridge (1970) reported the species to be common in secondary growth in moist and wet areas in Panama.

SWARTZIA Schreb.

Swartzia panamensis Benth. in Mart., Fl. Brasil. 15(2):38. 1870

Cutarro, Malvecino

Tree, 4–30 m tall; wood hard, dark brown. Leaves imparipinnate; petioles 4–7 cm long; petiolules 4–6 mm long; leaflets 3 or 5, ovate-lanceolate to elliptic-lanceolate, acuminate, rounded to acute at base, 7.5–17(22) cm long, 4–7(8.5) cm wide, glabrous above, shortly patulous-pubescent on veins below, the surface often with short, appressed trichomes, the lowermost blades smaller and

Fig. 277. *Prioria copaifera*

Fig. 278. *Prioria copaifera*

KEY TO THE TAXA OF SWARTZIA

Leaves simple . . . . *S. simplex* (Sw.) Spreng. var. *ochnacea* (A. DC.) Cowan
Leaves with 3 or 5 leaflets:
    Most leaves 3-foliolate (never 5-foliolate); inflorescences less than 10 cm long, bearing few
        flowers; legumes subterete, less than 5 cm long . . . .
        . . . . *S. simplex* (Sw.) Spreng. var. *grandiflora* (Raddi) Cowan
    Leaves 3- or 5-foliolate; inflorescences more than 20 cm long, bearing many flowers; legumes
        broad, flattened, more than 15 cm long . . . . *S. panamensis* Benth.

± ovate. Inflorescences axillary, from leaf scars on old branches, pendent, 40–60 cm long, racemose, minutely strigillose on most exterior parts; bracts minute; pedicels 12–20(30) mm long; buds many, globose, to nearly 1 cm diam, bearing a leathery covering at anthesis, splitting into (3)4 or 5(6) irregular parts; petal 1, pale yellow, clawed, the claw 5–8 mm long, the blade ± rounded, 2–3 cm long; stamens unequal, the larger (7)10–12 with filaments 8–12 mm long and anthers ca 4 mm long, the smaller ones numerous, their anthers ca 2 mm long; ovary glabrous. Legumes oblong, flat, 19–26 cm long excluding the persistent style to 2 cm long, 8–10 cm wide, minutely white-lenticellate; seeds few (usually less than 4), ± rounded in outline, flattened, to ca 6 cm diam, ± fleshy. *Croat 8202.*

Rare; seen along the shore of Burrunga Peninsula as small but mature individuals. Seasonal behavior uncertain. In Panama the species is believed to flower from December to July, especially in February and April. The fruits require about a year to develop and are commonly mature or nearly so at the time of the next season's flowering. Plants may have both flowers and mature fruits simultaneously. Allen (1956) reported that the species apparently flowers and fruits at irregular intervals throughout the year on the Osa Peninsula in Costa Rica and that it produces a new flush of leaves in December.

Easily distinguished by its long pendulous raceme of flowers, each with one yellow petal, and by its large oblong fruits.

Honduras (Lancetilla Valley) to Panama. In Panama, known from tropical moist forest in the Canal Zone, Colón, San Blas, Panamá, and Darién.

See Fig. 279.

Swartzia simplex (Sw.) Spreng. var. **grandiflora** (Raddi) Cowan, Flora Neotropica 1:172. 1968
Naranjita

Tree, to 15 m tall, ca 25 cm dbh; outer bark thin, light brown; inner bark thin, ± granular, the sap with ± strong, pungent odor. Differing from var. *ochnacea* by having most of the leaves trifoliolate, the lateral pair smaller than the terminal leaflet; juvenile leaves often with 5 leaflets, the rachis narrowly winged. *Croat 14965.*

Occasional, apparently more common within the forest than on the shore, generally on slopes or in ravines; much less abundant than var. *ochnacea*. Flowers abundantly from June to September, possibly year-round. Mature fruits have been seen from July to March.

Ranges throughout most of the areas where var. *och-*

nacea is found, but known also from northwestern Venezuela, western Peru, and Bolivia. In Panama, known from tropical moist forest in the Canal Zone, San Blas, Panamá, and Darién and from premontane wet forest in Panamá.

Swartzia simplex (Sw.) Spreng. var. **ochnacea** (A. DC.) Cowan, Flora Neotropica 1:178. 1968
Naranjita

Tree, usually to 15 m, glabrous or minutely strigillose and glabrate all over. Leaves simple; stipules minute, subpersistent; petioles ca 5(35) mm long, with a produced wing at the apex on the upper surface; blades elliptic to oblong-elliptic, bluntly acuminate, acute to rounded at base, quite variable in size, (4)8–18(24) cm long, (2.5)3–7.5(8) cm wide, the midrib often arched, the sides folded somewhat upward along the midrib. Racemes short, upper-axillary or terminal, mostly 4–10 cm long; pedicels ca 4 mm long, elongating to 10 mm; buds few, round or ellipsoid to obovoid (looking like small fruits), 7–12 mm diam, bearing a leathery covering splitting at anthesis into 3–6 irregular sepals, the sepals recurving and soon deciduous; petal 1, ± rounded-cordate, broader than long, palmately veined and clawed at base, to 3.5 cm long and 4.5 cm wide, pale yellow; stamens many, of 2 sizes, the smaller more numerous, to 1.5 cm long, the larger 5–15, to 2 cm long; filaments and gynoecium both curved toward the petal; ovary and style to 3 cm long, the ovary glabrous. Legumes orange, 2.5–5 cm long, ovoid to oblong, borne on a prominent stipe, apiculate at apex, the valves curling inward along their margins at maturity to expel seeds; seeds 1 or 2 (sometimes 4), irregular in shape, smooth, black, shiny, borne on a long funiculus, enveloped at base by prominent, white, bitter-tasting aril. *Croat 4097, 10847.*

Frequent in the forest; locally and sporadically abundant along the shore. Flowers along the shore during much of the year, especially during the dry season and most abundantly by May; flowering begins in the forest usually in June. Most fruits seem to mature in January and February (sometimes to May).

Western Mexico (Nayarit) to western Guatemala and on both coasts of Honduras, Costa Rica, and Panama, south to both coasts of Colombia and to eastern lowlands of Peru and Ecuador; known also from the east-central coast of Brazil. In Panama, known from tropical moist forest in the Canal Zone, Bocas del Toro, Chiriquí (Burica Peninsula), Panamá, and Darién and from premontane wet forest in Panamá.

See Fig. 280.

Fig. 279. *Swartzia panamensis*

Fig. 280. *Swartzia simplex*
var. *ochnacea*

Fig. 281. *Tachigalia versicolor*

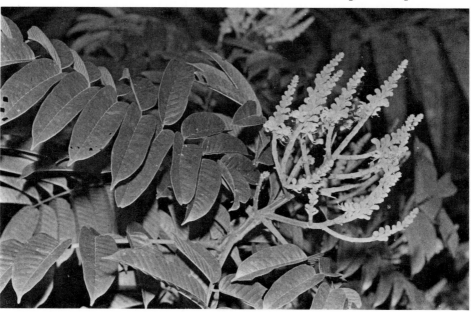

TACHIGALIA Aubl.

Tachigalia versicolor Standl. & L. O. Wms., Ceiba 3:27. 1952

Tree, to 30 m tall, ca 75 cm dbh, often buttressed to 10 m; branches widely spreading, the crown nearly as broad as the tree is tall; outer bark thin, reddish-brown (especially on buttresses), minutely fissured, bearing irregular small lenticels; inner bark thin, granular, reddish-brown, the sap sweet-tasting, with a faint pleasant aroma. Leaves pinnate; petioles solid, angled, 5–9 cm long (much longer on juvenile plants); petiolules pulvinate, ca 4 mm long; leaflets in 7–9 pairs, oblong, short-acuminate, obliquely rounded to obtuse at base, the 1 or 2 pairs nearest the stem somewhat shorter than the rest, the longest 9–14 cm long, 3–4.5 cm wide, minutely puberulent below, subcoriaceous, the primary lateral veins in 13–15 pairs. Inflorescences terminal panicles of spikes, rufous-pubescent, the spikes ca 12 cm long; flowers pink in bud, pale yellow when open, each flower subtended by a subulate, caducous bract ca 3 mm long; calyx tube oblique, ca 5 mm long, the lobes 5, ca 5 mm long; petals 5, spatulate, ca 6 mm long, lanate-pubescent; stamens 10, curved, 7–12 mm long, orange, lanate in basal fourth, particularly on the inner side; pistil ca 6 mm long, very densely orange-lanate. Legumes flat, forming a single, elliptic, fibrous wing, 11–15 cm long, 4–5 cm wide, cordate to rounded at point of attachment; seed 1, flat, rectangular, ca 2 cm long and 1.3 cm wide, olive-brown. *Croat 9575, 12568.*

Common in the forest. Flowers from March to July, sometimes from January. The fruits may develop to mature size by August, but they remain green until the following dry season. Most trees drop old leaves and produce new ones in August and September, sometimes briefly remaining leafless. The few trees that flower drop their leaves at the start of the dry season. The large flat green fruits give the tree a leafy appearance well into the dry season before the exocarp peels off and the seeds are dispersed by the wind. All plants that were observed to flower on BCI and elsewhere have died following flowering, i.e., after fruiting no new leaves emerged. Studies by R. Foster (pers. comm.) both on BCI and elsewhere have conclusively proved that the species is monocarpic.

Costa Rica to northern Colombia. In Panama, known from tropical moist forest in the Canal Zone and Panamá, from premontane wet forest in the northeast part of the Canal Zone, Colón, and Darién, and from tropical wet forest in Colón.

See Fig. 281.

63C. PAPILIONOIDEAE

Leaves usually trifoliolate or pinnate, rarely unifoliolate (simple in *Crotalaria retusa, Dalbergia brownei*); stipules and stipels usually present. Flowers zygomorphic, in racemes or panicles, ebracteate; calyx 5-parted, the carinal tooth (posterior tooth) usually longest, anterior teeth often united; petals imbricate; stamens 10, monadelphous (all stamens united) or more frequently diadelphous (all but one stamen united).

Members of the subfamily Papilionoideae are most easily distinguished by their very zygomorphic flowers with the adaxial petal (banner) held outside the bud and by their usually united stamens.

Flowers of the subfamily are typically bee pollinated, but some, such as those of *Erythrina costaricensis* var. *panamensis*, are hummingbird pollinated. *Mucuna* species

---

KEY TO THE TAXA OF PAPILIONOIDEAE

- Leaves all trifoliolate:
  - Plants trees or shrubs (soft-wooded in *Cajanus*), not at all scandent:
    - Plants trees in the forest or along the shore; flowers bright red to pale orange, ca 5 cm long or more; leaflets mostly more than 5 cm wide .... *Erythrina*
    - Plants shrubs or trees in clearings, soft-wooded, often cultivated; flowers yellowish at least in part, less than 3 cm long; leaflets less than 3.5 cm wide:
      - Flowers completely yellow; fruits terete, ca 2 cm long, lacking purple stripes; leaves lacking glands below .... *Crotalaria vitellina* J. Ker
      - Flowers yellow, often marked with red, purple, or brown; fruits flattened, more than 5 cm long, with dark stripes; leaves with microscopic globular glands below .... *Cajanus bicolor* DC.
  - Plants herbs, vines, or lianas (stems suffruticose in *Desmodium canum* and *D. cajanifolium*):
    - Plants lianas, at least the older parts definitely woody:
      - Calyx 5–20 mm long; flowers purplish to pink:
        - Free stamen entirely free; upper lobe of calyx minutely bifid; plants semiherbaceous .... *Cymbosema roseum* Benth.
        - Free stamen united apically to staminal column at anthesis; upper lobe of calyx not minutely bifid; plants woody .... *Dioclea*
      - Calyx at anthesis less than 5 mm long *or* more than 2 cm long; flowers pink, white, or greenish-yellow:
        - Calyx less than 5 mm long; corolla greenish-yellow, less than 6 mm long; legumes less than 3 cm long, constricted between seeds, the seeds usually 2; younger parts with globular orange glands .... *Rhynchosia pyramidalis* (Lam.) Urban

Calyx more than 20 mm long; corolla more than 4 cm long; legumes more than 7 cm long, the seeds several to many; plant parts lacking glands:

Flowers yellowish on long pendent racemes; legumes bearing stiff, irritating trichomes .... *Mucuna rostrata* Benth.

Flowers pinkish, the racemes not pendent; legumes softly brown-villous .... *Clitoria javitensis* H.B.K.

◆ Plants herbs or herbaceous vines, rarely woody at base:

Corolla more than 1.5 cm long:

Stems and petioles bearing conspicuous, spreading, brownish trichomes 1.5–2 mm long (in *Clitoria rubiginosa*, the trichomes 1 mm long, the keel white with violet medial lines):

Calyx more than 1.5 cm long; standard white with violet medial lines, often more than 3 cm long, the keel petals straight; legumes glabrescent, with submarginal ridges; apex of terminal leaflet often rounded or blunt, bearing short apiculum or tuft of trichomes .... *Clitoria rubiginosa* Adr. Juss.

Calyx less than 1.5 cm long; standard white to bluish becoming yellow during the day, usually less than 3 cm long, the keel petals twisted; legumes conspicuously hirsute, not ridged; apex of terminal leaflet usually sharply acute .... *Vigna vexillata* (L.) A. Rich.

Stems and petioles lacking long, brown, spreading trichomes 1.5–2 mm long:

Flowers flesh-colored or yellowish; calyx very densely pubescent, the trichomes totally obscuring surface, of 2 sizes (the longer often sparse); legumes densely covered with irritating trichomes; inflorescences pendent, the peduncles usually more than 15 cm long (sometimes less than 15 cm long in *M. rostrata*) .... *Mucuna*

Flowers reddish to blue; calyx glabrous or not as densely pubescent as above; legumes not densely covered with irritating trichomes; inflorescences erect, the peduncles usually less than 15 cm long:

Flowers marked with blue; legumes less than 8 mm wide, with both lateral margins raised; flowers subtended by conspicuous, ovate, striate bracts .... *Centrosema pubescens* Benth.

Flowers pinkish to reddish; legumes more than 15 mm wide, not with *both* lateral margins raised; flowers subtended by inconspicuous bracts:

Flowers red (lavender when dry), to 3.5 cm long; calyx pubescent becoming glabrous, with both long-appressed and short erect trichomes; legumes less than 5 cm long, densely pubescent with appressed acicular trichomes, lacking submarginal ridges .... *Cymbosema roseum* Benth.

Flowers lavender, to 2.5 cm long; calyx ± glabrous; legumes more than 10 cm long, ± glabrous, the submarginal ridges prominent .. *Canavalia dictyota* Piper

Corolla less than 1.5 cm long:

Stems, petioles, peduncles, and legumes conspicuously long-pilose, the trichomes more than 1.5 mm long:

Flowers yellow; calyx shallowly lobed; stipules oblong, with a long basal lobe .... *Phaseolus trichocarpus* C. Wright

Flowers blue; calyx narrowly lobed to two-thirds its length; stipules acute, lacking a basal appendage .... *Calopogonium mucunioides* Desv.

Stems, petioles, peduncles, and legumes not conspicuously long-pilose or the trichomes less than 1.5 mm long:

Corolla less than 7 mm long, bluish or tinged with blue:

Flowers in terminal inflorescences or, if axillary, in elongate racemes .... *Desmodium* (in part)

Flowers in small axillary clusters or much-reduced axillary racemes:

Leaflets less than 1 cm long, obovate, emarginate or truncate at apex .... *Desmodium triflorum* (L.) DC.

Leaflets more than 1 cm long, ± lanceolate, acute to rounded at apex ..... *Teramnus*

Corolla more than 8 mm long, white, yellow, or blue (*Calopogonium*):

Flowers blue; keel petals not coiled; legumes usually more than 4 cm long and 6 mm wide; secondary lateral veins of leaflets prominulous, perpendicular to and often extending between the pairs of primary lateral veins .... *Calopogonium caeruleum* (Benth.) Sauv.

Flowers white to yellowish (sometimes lavender); keel petals coiled or not; legumes usually less than 4 cm long *or* less than 4 mm wide; secondary lateral veins of leaflets reticulate:

Plants erect herbs or suffrutices; terminal leaflets acute at base .... *Crotalaria vitellina* J. Ker

Plants vines; terminal leaflets rounded to truncate at base .... *Phaseolus*

- Leaves mostly not trifoliolate:
    - Plants herbaceous or essentially so:
        - Leaflets mostly less than 5 mm wide .... *Aeschynomene*
        - Leaflets or leaves more than 5 mm wide:
            - Leaves simple; calyx more than 10 mm long; flowers yellow .... *Crotalaria retusa* L.
            - Leaves imparipinnate, the leaflets 5 or 7; calyx less than 5 mm long; flowers red ....
            .... *Indigofera mucronata* DC.
    - Plants woody:
        - Leaves simple; flowers white, ca 1 cm long; legumes flat, ± oblong, the seeds 1 or 2 ....
        .... *Dalbergia brownei* (Jacq.) Urban
        - Leaves compound:
            - Plants lianas or climbing shrubs:
                - Fruits flattened, orbicular, ca 4 cm long; flowers white; leaflets 1–5, abruptly acuminate, the acumen to 1.8 cm long; plants growing along the shore .... *Dalbergia monetaria* L.f.
                - Fruits samaroid, with a conspicuous, usually curved wing; flowers white to purple; leaves not as above; plants growing in the canopy of the forest or along the shore ....
                .... *Machaerium*
            - Plants trees, never scandent:
                - Leaves opposite; flowers golden-orange, precocious; fruits oblong-elliptic, 1-seeded, with a broad marginal wing .... *Platymiscium pinnatum* (Jacq.) Dug.
                - Leaves alternate:
                    - Rachises of leaves conspicuously winged; flowers ± red-violet; fruits large, 1-seeded, oblong drupes .... *Dipteryx panamensis* (Pitt.) Rec. & Mell.
                    - Rachises of leaves not winged:
                        - Leaflets rounded to emarginate at apex; fruits samaroid, the seminiferous area distal; leaflets mostly more than 4 cm long; plants usually more than 10 m tall ....
                        .... *Platypodium elegans* J. Vogel
                        - Leaflets acute to acuminate at apex (sometimes slightly emarginate in *Dalbergia retusa*):
                            - Fruit drupaceous, 1-seeded, green, ± globose, 3–5 cm long; stipels usually persistent at bases of leaflets; flowers purple, ca 1 cm long ....
                            .... *Andira inermis* (W. Wright) H.B.K.
                            - Fruits not drupaceous, flattened, often winged (except *Ormosia*):
                                - Fruits not winged *or* the wing lateral and the seminiferous area at apex:
                                    - Fruits prominently winged, indehiscent, the seminiferous area at apex; flowers whitish or yellowish; petals to ca 1 cm long; major lateral veins of leaflets mostly less than 5 mm apart, scarcely more prominent than reticulate veins...
                                    .... *Myroxylon balsamum* (L.) Harms var. *pereirae* (Royle) Harms
                                    - Fruits not winged, flattened, dehiscent with colored seeds; flowers purple; petals more than 1 cm long; major lateral veins of leaflets usually more than 5 mm apart, much more prominent than reticulate veins ..... *Ormosia*
                                - Fruits winged, the wing surrounding the seminiferous area:
                                    - Fruits about as long as broad:
                                        - Fruits thin, indehiscent, with wing more than 1 cm broad; flowers pale orange-yellow or violet-purple; sap red .... *Pterocarpus*
                                        - Fruits thick except at margin, dehiscent, with wing very narrow, the seeds 1–4, red; flowers lilac; sap not colored .... *Ormosia panamensis* Seem.
                                    - Fruits much longer than broad:
                                        - Fruits samaroid, the wing tissue all apical ...... *Vatairea erythrocarpa* Ducke
                                        - Fruits not samaroid, the wing tissue surrounding the seminiferous area:
                                            - Fruits glabrous, conspicuously stipitate, the stipe ca 1 cm long; flowers white................................. *Dalbergia retusa* Hemsl.
                                            - Fruits pubescent to glabrate, not conspicuously stipitate; flowers purple or greenish to cream-colored .... *Lonchocarpus*

are bird or bat pollinated (Faegri & van der Pijl, 1966; Baker, 1970; Proctor & Yeo, 1973); *Mucuna mutisiana* is reported to be bat pollinated (Vogel, 1958). In the typical papilionoid flower the banner acts as the chief attractive part of the flower. The keel petals generally enclose and protect the stamens and pistil. These parts sometimes project from the keel, but are generally hidden until the pollinator (typically a bee) enters the flower.

Bees usually alight on the keel of wing petals and probe the flower for nectar. The style is often equipped with antrorse trichomes (e.g., *Vigna*) and, acting like a piston, pushes out a load of pollen with each successive visit. Pollen is deposited on the lower side of the insect.

In *Centrosema, Canavalia,* and *Clitoria,* the flower is turned upside down (Faegri & van der Pijl, 1966). The bee alights on the banner but, while probing for nectar

in the same manner, receives pollen on its back. In *Phaseolus,* which has an erect flower, the keel, pistil, and stamens are spirally wound. In some cases, notably species of *Desmodium,* the filaments are held under tension until released by a visitor, at which time all their pollen is shed at once in an explosive manner. *Mucuna* also has an explosive mechanism for the release of pollen (Baker, 1970; Proctor & Yeo, 1973).

Papilionoid fruits show a different seed development than do the Mimosoideae and Caesalpinioideae, with more extended dormancy and greater development of the pericarp, which assists in dispersal (van der Pijl, 1968).

Species with small colored seeds well suited for bird dispersal include *Erythrina costaricensis* var. *panamensis, Rhynchosia pyramidalis,* and all species of *Ormosia.* These are mimetic, looking like arillate seeds and no doubt dispersed by forest-dwelling birds. Other species that occur in clearings also have fruits suited for bird dispersal; these usually have dark, shiny seeds displayed against the generally paler inner valve surface. Moderately large seeds or those borne in pods too short to twist at dehiscence are usually displayed on untwisted (or only slightly twisted) valves, for example, *Calopogonium caeruleum, C. mucunioides,* and *Dioclea guianensis.* Other species with smaller seeds have the valves of the legume markedly twisted in such a way as to display a single seed within each twist of the valve; this modification is an adaptation that makes the seeds more conspicuous by exposing alternately the usually dark outer surface and the light-colored inner surface bearing the seed. Among such probably bird-dispersed species with twisted valves are *Phaseolus peduncularis, Teramnus uncinatus, T. volubilis, Vigna vexillata,* and *Centrosema pubescens.*

Other endozoochorous Papilionoideae include *Dipteryx panamensis,* the pod of which is eaten by most frugivores (including bats), as well as most species of *Dioclea* and *Andira inermis.* Kinkajous reportedly eat fruits of *Dipteryx* at night (Chapman, 1931), and several animals eat them by day including agoutis, peccaries, coatis, and monkeys (Carpenter, 1934; Hladik & Hladik, 1969; Kaufmann, 1962; Enders, 1935; Bonaccorso, 1975). Partially eaten fruits or seeds are dropped to the ground by monkeys. Both *Andira* and *Mucuna* have been suggested as genera with bat-dispersed fruits. The bat *Artibeus jamaicensis* (Phyllostomidae) takes fruits of *Andira inermis* in Trinidad (Goodwin & Greenhall, 1961). D. Janzen (pers. comm.) suggests that forest-floor rodents move seeds of *Mucuna.*

Species of *Desmodium* are epizoochorous with flattened fruits, which break up into segments covered with hooked trichomes. *Aeschynomene americana* and *A. ciliata* are at least partly epizoochorous.

Autochorous species include *Clitoria javitensis* and *Canavalia dictyota,* the pods of which have thick, woody, elastically dehiscent valves, and apparently *Indigofera mucronata,* whose thin valves become twisted from the apex to the base. With each turn of the spiral valve a few of the tiny seeds are pushed out; they are small enough to be windblown a short distance.

Other species with wind dispersed seeds include *Dalbergia retusa, Myroxylon balsamum, Platymiscium pinnatum, Platypodium elegans,* and all species of *Pterocarpus, Machaerium,* and *Lonchocarpus,* but none of these are exceptionally good fliers.

Species that have buoyant seeds and are at least in part hydrochorous include *Mucuna mutisiana, Dalbergia brownei, D. monetaria, Dioclea reflexa* (and no doubt others of this genus), *Canavalia dictyota,* and *Erythrina fusca.* Seeds of both *Dioclea* and *Andira inermis* are reported to be river dispersed (Ridley, 1930). The flat seeds of *Dalbergia brownei* float nicely, owing to surface tension, but once sunk they will not resurface; they are probably wind dispersed to a lesser extent. *Aeschynomene ciliata* and especially *A. sensitiva* are also hydrochorous. Seeds of *Vigna* float (Ridley, 1930). Guppy (1912) reported *Dioclea reflexa* to be sea dispersed. Rain wash may be important in the dispersal of *Desmodium triflorum* and *Crotalaria* (Ridley, 1930).

The inflated pod of *Crotalaria* is not adapted for wind or water dispersal, since it dehisces before breaking free from the plant, and seeds often fall free from the funiculus even before dehiscence. Probably the pod snaps open rapidly and throws the seeds.

AESCHYNOMENE L.

Aeschynomene americana L. var. **glandulosa** (Poir.) Rudd, Contr. U.S. Natl. Herb. 32:26. 1955
Pega-pega

Erect or somewhat reclining herb, 1–3 m tall; stems and inflorescences sparsely hirsute. Leaves pinnate, 2.5–6.5 cm long, short-petiolate; stipules narrowly lanceolate, to ca 12 mm long, with a narrowly lanceolate appendage ca 8 mm long at base pointing antrorsely; leaflets folding together top to top upon dessication, in mostly 12–30 pairs, oblong, obliquely apiculate, oblique at base, 6–9 mm long and less than 2 mm wide, ± scarious, the costae 3, parallel, conspicuous below. Inflorescences axillary, racemose, to 5 cm long; bracts conspicuous, foliaceous,

KEY TO THE SPECIES OF AESCHYNOMENE

Leaflets with 2 to several costae .... *A. americana* L. var. *glandulosa* (Poir.) Rudd
Leaflets with 1 costa:
    Plants with stems, petioles, and rachises conspicuously hispid, sometimes glandular, the trichomes to 4 mm long .... *A. ciliata* J. Vogel
    Plants strigose to glabrous, any trichomes much shorter than 3 mm .... *A. sensitiva* Sw.

broadly lanceolate, to 3 mm long and 1.5 mm wide, the veins 5–15, conspicuous, parallel, the margins setaceous; pedicels to 10 mm long; flowers few, oblique on pedicels, to 9 mm long; calyx bilobed, hispid, to 5.5 mm long, subtended by 2 bracts nearly equaling calyx lobes; standard 8 mm long, rounded, pale orange with a yellow spot near its base, often marked with dark red as well, held erect at anthesis; wings slightly shorter than standard, the keel fused only on outer edge, loosely holding stamens and style; stamens in 2 clusters of 5 each, equaling length of style, shedding pollen before opening. Loments stipitate, 3–9-lobed, to 26 mm long and 2.5 mm wide, densely pubescent, the lobes ca 3 mm diam, the upper margin of loment ± entire, the lower margin deeply constricted between each seed; seeds 2–3 mm long, 1.5–2 mm wide. *Croat 13154.*

Occasional, in the Lighthouse Clearing. Flowers and fruits mostly from December to February.

The plants are often more or less reclining in age, perhaps dispersing fruits by lateral displacement. The plant has been seen visited briefly by small black bees (possibly *Trigona*) in rapid succession. The fruits do not stick to clothing (at least before maturity), although they have many trichomes.

Florida and southern Mexico to Argentina; West Indies; introduced into West Africa. In Panama, known from tropical moist forest in the Canal Zone, Bocas del Toro, Veraguas, Herrera, and Panamá, from premontane dry forest in Herrera, from premontane moist forest in Panamá, and from premontane wet forest in Chiriquí.

Aeschynomene ciliata J. Vogel, Linnaea 12:84. 1838
   A. hispida sensu Standl. non Sw.

Herb, 1–2.5 m tall; stems and rachises hispid, the trichomes yellow, glandular, 2–4 mm long. Leaves pinnate, 10–15 cm long; stipules 1–2 cm long, acute; petioles 8–10 mm long; leaflets in 15–20 pairs, short-petiolulate, oblong, rounded to retuse at apex, subcordate and slightly asymmetrical at base, 10–15(30) mm long, 3–5(8) mm wide, glabrous, lighter below, the midvein central. Cymes terminal, to ca 4 cm long; bracts foliaceous, ovate, subacuminate, ca 4 mm long and 2 mm wide, setaceous on margins; flowers few, yellow; calyx bilabiate, the upper lip trifid, the lower lip bifid, ca 6 mm long, subtended by 2 bracts; corolla reflexed, the standard 8–10 mm long, the claw 2 mm long, the blade orbicular, 5–8 mm diam, noticeably serrulate-ciliate; stamens ca 8 mm long. Loments mostly 8–10-articulate on stipes 5–10 mm long, smooth, hispid, the margins entire; articles ca 4 mm long and 6 mm wide; seeds 3–4 mm long and ca 2 mm wide. *Foster 1403.*

Rare; collected once at the margin of the lake near the dock. Apparently flowers and fruits principally from November to January.

Southern Mexico to Ecuador and Amazonian Brazil; Jamaica. In Panama, known only from tropical moist forest in the Canal Zone and the immediate vicinity, usually near water.

Aeschynomene sensitiva Sw., Prodr. Veg. Ind. Occ. 107. 1788

Erect, slender-stemmed herb, to 2(4) m tall, glabrous to densely hispidulous appearing glabrous, drying slightly dark. Leaves pinnate; stipules narrowly subulate, 5–20 mm long, caducous; petiole and rachis hispidulous; rachis to ca 5 cm long; leaflets in 5–20 pairs, oblong, rounded and mucronate at apex, obliquely rounded to subcordate at base, 4–6 mm long, 1–2 mm wide, glabrous, the midvein central, obscure. Panicles generally upper-axillary; peduncles and pedicels hispidulous; flowers few, yellow; bracts ovate, 1–2 mm long, ciliolate; calyx ca 4 mm long, the lips at most obscurely dentate; petals ca 7 mm long, obscurely ciliate; stamens ca 6 mm long. Loments drying dark, mostly 7- or 8-articulate, ca 5 cm long, on stipes ca 4 mm long, only slightly notched between articles, sparingly pubescent, appearing glabrous, ± smooth, the upper margin essentially entire, the lower margin crenate; articles ca 5 mm long and wide; seeds obliquely reniform, ca 3 mm long and 2 mm wide, black. *Croat 4199, 13232.*

Growing in water at the shore near the dock. Probably flowers and fruits year-round.

Southern Mexico to central South America as far as Paraguay and Argentina; West Indies. In Panama, known from tropical moist forest in the Canal Zone, Bocas del Toro, Herrera, Coclé, Panamá, and Darién; known also from tropical dry forest in Coclé and premontane wet and tropical wet forests in Colón.

ANDIRA Adr. Juss.

Andira inermis (W. Wright) H.B.K., Nov. Gen. & Sp. 6:385. 1824
   Geoffroea inermis W. Wright
   Cabbage bark, Cocú, Arenillo, Pilon, Quira, Almendro del río

Tree, to 10(15) m tall, 30–40 cm dbh; bark with disagreeable odor; stems glabrous. Leaves alternate, imparipinnate; stipules prominent especially on juvenile plants, persistent, slender, to 1 cm long; leaflets in (5)7–15 pairs, opposite to subopposite along the rachis, oblong, acuminate, rounded at base, mostly 4–9.5(13) cm long, 1.5–4.5(5) cm wide, glabrate to minutely puberulent on midrib. Panicles terminal, 15–30 cm long; branches, pedicels, and calyces brown-puberulent; flowers red-violet, subsessile, ca 1 cm long, with sweet aroma; calyx campanulate, the teeth short, ± equal, ca 3 mm long; standard darker inside in a broad submarginal band, white at center below; wings slightly exceeding length of keel, the keel petals held together loosely by overlapping on outer edge; stamens diadelphous, the staminal sheath open, glabrous; filaments alternating long and short, curved inward somewhat; style sharply curved inward, glabrous; stigma small, brushlike, longer than the anthers. Drupes ligneous, oval, to ca 5 cm long, the exocarp green, woody, thick; seed 1, oval, slightly shorter than drupe. *Croat 5342, 12272.*

Common along the margin of the lake and occasional in the young forest. Flowers and fruits may be seen throughout the year. Flowers mostly in the middle to late dry and early rainy seasons (February to May), with most fruits maturing in the middle to late rainy season, especially in September and October. A second flush of flowering may occur in October, with the fruits maturing in January and February.

The fruit and bark of this species are poisonous (Allen, 1956; Blohm, 1962). The species is an obligate outcrosser, since it is self-incompatible (Bawa, 1974).

Southern Florida and Mexico south to Bolivia and the Amazon basin; West Indies; introduced to West Africa. In Panama, known from tropical moist forest in the Canal Zone, Bocas del Toro, Veraguas, Herrera, Panamá, and Darién; known also from premontane moist forest in the Canal Zone and from premontane wet forest in Chiriquí, Coclé, and Darién. Characteristic of tropical dry, tropical moist, and premontane moist forests in Panama (Tosi, 1971).

CAJANUS DC.

Cajanus bicolor DC., Cat. Hort. Bot. Monsp. 85. 1813

Pigeon pea, Guandú, Frijol de palo

Shrub, woody at base, to 3 m tall; stems minutely sericeous, 3-ridged from the nodes. Leaves trifoliolate; stipules lanceolate, 2.5–3 mm long; petioles 1–4 cm long; rachis 5–15 mm long; leaflets narrowly elliptic, acute or slightly acuminate at apex, cuneate at base, softly puberulent and grayish-green above, glandular-dotted, tomentose, and pale below, the terminal leaflet 5–10 cm long, 2–3.5 cm wide, the veins at base 3. Racemes terminal or upper-axillary, 2–6 cm long, the flowers often paired at the nodes; pedicels 5–10 mm long; calyx bilabiate, to 8 mm long, lobed half its length, the lobes acute to acuminate, the upper 2 ± united along their common side; petals to 2 cm long, yellow, the standard often red to maroon outside. Legumes oblong, 5–10 cm long, 1–1.5 cm wide, puberulent, strongly beaked, yellowish-green with brown-striped mottling, with oblique constrictions between the seeds on the side of the legume, the calyx persisting at base; seeds 2–5, flattened, ca 7 mm diam, gray, the hilum prominent. *Croat 7238.*

Cultivated in the Laboratory Clearing. Flowers and fruits primarily from January to March, but sometimes flowering much earlier, even in the late rainy season. The fruits develop in about 1 month.

Probably native to the East Indies; introduced sporadically in the New World tropics. In Panama, known from tropical moist forest in the Canal Zone, Bocas del Toro, Coclé, Panamá, and Darién.

CALOPOGONIUM Desv.

Calopogonium caeruleum (Benth.) Sauv., Ann. Acad. Havana 5:337. 1869

Vine, to 5 m long, softly orange-tomentose all over. Leaves trifoliolate; stipules inconspicuous, caducous; petioles 3–10 cm long; rachis lacking or to 1.5 cm long; petiolules slightly pulvinate, articulate, ca 5 mm long; leaflets rhombic-ovate, rounded to acuminate at apex, obtuse to slightly subcordate at base, 5–11 cm long, 4–8 cm wide, the veins at base 3, the secondary lateral veins prominulous, perpendicular to primary lateral veins and often extending between pairs of primary laterals. Racemes axillary, 2–40 cm long; flowers in fascicles arising from tubercles 2–8 mm long; bracts subulate, ca 2 mm long; pedicels 1–2 mm long; calyx ca 4 mm long, densely hispid, bilabiate, lobed ca half its length, the upper lip shallowly bifid, the lower lip with 3 linear teeth; petals blue, to 10.5 mm long; standard emarginate at apex, flat and erect, marked at center with green; wings and keel slightly shorter than standard, the wings adhering weakly to sides of keel, the keel fused apically, not enclosing style or stamens; stamens diadelphous, ca 4 mm long; filaments free above from about the middle; style straight, antrorsely hispidulous, almost equaling standard, much longer than anthers; nectar copious, storing in calyx. Legumes narrowly oblong, flattened, acuminate, 3.5–6 cm long, 6–8 mm wide; seeds 4–8, ovoid, ca 5 mm long, reddish-brown. *Hladik 136.*

Rare; collected only once. Occurring commonly along roadsides and at the edges of clearings in other areas of the Canal Zone. Flowers mostly in January and February (sometimes to March), with the fruits maturing in March and April.

Southern Mexico to northern South America; West Indies. In Panama, known principally from tropical moist forest in the Canal Zone, Bocas del Toro, Chiriquí, Veraguas, Herrera, Panamá, and Darién; known also from premontane moist forest in Herrera and premontane wet forest in Chiriquí.

Calopogonium mucunioides Desv., Ann. Sci. Nat. Bot., sér. 1, 9:423. 1826

Twining vine, to 2 m long, densely hispid, the trichomes to 2 mm long, many, especially on fruit and stem, pustular at base. Leaves trifoliolate; stipules acute, to 5 mm long,

---

KEY TO THE SPECIES OF CALOPOGONIUM

Stems, petioles, calyces, and fruits densely reddish-brown-tomentose, the trichomes not individually visible; inflorescences usually more than 20 cm long; fruits usually more than 3.5 cm long
. . . . *C. caeruleum* (Benth.) Sauv.
Stems, petioles, calyces, and fruits sparsely long-hispid, the trichomes to ca 2 mm long; inflorescences less than 15 cm long; fruits usually less than 3.5 cm long . . . . *C. mucunioides* Desv.

striate; petioles 3–10 cm long; petiolules of terminal leaflet ca 1 cm long, of lateral leaflets ca 2 mm long; terminal leaflet ± ovate, acute to rounded at apex, rounded at base, 4.5–13 cm long, 3–9 cm wide, the lateral leaflets inequilateral, the veins at base 3. Racemes axillary, to 15 cm long; pedicels to 3 mm long; bracts linear, to 5 mm long, striate; flowers blue, inconspicuous, 8–10 mm long, clustered; calyx slender, densely hirsute, ca 6 mm long, lobed ca two-thirds its length, the lobes subequal, subulate; standard 6–8 mm long, ± spatulate, emarginate; wings white, slender, ± equaling standard, weakly fused to the shorter and inconspicuous keel, the keel petals ± free; stamens diadelphous, loosely held, ca 4 mm long, straight; filaments united in basal third, shedding pollen in bud; style equaling anthers, straight; stigma globular, somewhat eccentric, sticky. Legumes narrowly oblong, flattened, briefly acuminate, 2.5–3.5 cm long, ca 5 mm wide, densely brown-hispid, often occurring in dense clusters; seeds 5–9, ovoid, ca 5 mm long, light brown. *Croat 6925, 13169a.*

Occasional, in clearings. Flowers mostly in December and January. The fruits mature from December to March.

Southern Mexico to northern South America; West Indies; introduced into the Old World tropics. In Panama, known from tropical moist forest in the Canal Zone, Los Santos, Herrera, and Panamá; known also from tropical dry forest in Coclé, from premontane moist forest in the Canal Zone and Panamá, and from premontane wet forest in Coclé.

CANAVALIA DC.

Canavalia dictyota Piper, Contr. U.S. Natl. Herb. 20:574. 1925

Twining herbaceous vine. Leaves trifoliolate; petioles 3–9 cm long, densely pubescent; rachis 1.5–4.5 cm long; leaflets lanceolate-elliptic, rounded to acute at apex, obtuse at base, 9–19 cm long, 4–12 cm wide, glabrous above, sparsely strigillose below. Flowers lavender, in axillary racemes; peduncles ca 10 cm long; rachises to 20 cm long; pedicels to 2 mm long, elongating to 1 cm and thickening greatly in fruit; calyx bilabiate, glabrous, to 13 mm long and 7 mm wide, with the 2 superior teeth ± rounded and to 5 mm long, the lower 3 teeth acute and to 2 mm long; standard pale lavender, to 2.5 cm long, the margins revolute, strengthening petal; wings and keel white with lavender tips, somewhat shorter than standard, the keel petals fused only briefly below their summit; stamens diadelphous, to 2 cm long, arching into keel petals; filaments united about four-fifths of their length; style ± equaling the stamens, apparently receptive when pollen is shed; nectaries 2, on either side of the ovary. Legumes to 20 cm long, to 3.5 cm wide, glabrous, brown, the upper margin broad with prominent submarginal ridges and an apical beak; valves at maturity twisting open violently and throwing the seeds; seeds 6–9, vertically oriented, to 2 cm long, 1.5 cm wide, and 1.2 cm thick, brown with dark brown and black markings, buoyant. *Croat 4794.*

Occasional, in open areas, especially along the shore near the dock. Flowers from November to February (rarely later). The fruits ripen mostly from February to June, with most mature perhaps by April.

May be confused with *C. brasiliensis* Mart. ex Benth., which differs chiefly in having the leaves more broadly ovate and the seeds not buoyant.

The nectaries of the flower can be approached by the pollinator only at the open side of the staminal cluster. Bees land on the standard, which is strengthened by its revolute margin and supported beneath by the two large calyx lobes. In forcing its way in, the bee pushes the keel upward and rubs the stationary stigma and anthers with its back.

Panama to the northern Amazon basin and eastern Brazil; West Indies. In Panama, known only from tropical moist forest in the Canal Zone, San Blas, and Panamá.

CENTROSEMA (DC.) Benth.

Centrosema pubescens Benth., Comm. Leg. Gen. 55. 1837
Campanilla, Caracucha

Slender vine. Leaves trifoliolate; stipules ovate, ca 2 mm long; petioles 1.5–4 cm long; petiolules to 15 mm long on terminal leaflet, ca 2 mm long on lateral leaflets; leaflets lanceolate to oblong, acute to acuminate, rounded at base, villous above, velutinous below, the terminal leaflet 5–13 cm long, 2.5–5.5 cm wide, the lateral leaflets somewhat reduced, the veins prominent on both surfaces. Racemes axillary; peduncles 8–11 cm long; rachises to 2 cm long; bracts ovate, acute, striate, to 8 mm long, caducous; calyx bifid, shallowly lobed, ca 10 mm long, pubescent, becoming glabrate in age except for carinal lobe, the carinal lobe arcuate, slender, ca 5 mm long; standard ± rounded to emarginate, to 3.5 cm long and 4 cm wide, sparsely pubescent and usually white outside, blue to orchid or violet-purple inside, prominently marked with purple and white along median line, often yellow at base, clawed at base; keel and wings white or tipped with violet, ca 3 cm long, the keel at first weakly sealed on both sides except at base; stamens 10, diadelphous, enclosed within keel, the staminal column forming an arched open tube below, fan-shaped at apex; style to ca 3.5 cm long, held within the open staminal tube, flat, truncate and puberulent at apex, longer than anthers, the outer margin bristled. Legumes 10–14 cm long, ca 7 mm wide, long-acuminate at apex, sparsely appressed-pubescent, the lateral margins raised; seeds as many as 25, black, round, flattened, ca 3 mm diam. *Croat 4398, 6986.*

An occasional weed in clearings. Flowers and fruits throughout most of the year, especially in the dry season, but principally in December and January.

The style extends from the keel beyond the anthers and presses against the wing and keel petals. Both of its flat surfaces are covered with a thick, pasty layer of pollen.

Standley (1933) also reported *Centrosema virginianum* (L.) Benth., but that species does not appear to be distinct from *C. pubescens* Benth. in Panama. It is reported to

range from Mexico to Panama, but possibly does not come as far south as Panama.

Mexico to tropical South America; West Indies; introduced in the Old World tropics. In Panama, known principally from tropical moist forest in the Canal Zone, Bocas del Toro, San Blas, Chiriquí, Veraguas, Los Santos, Herrera, Panamá, and Darién; known also from premontane dry forest in Herrera, from tropical dry forest in Coclé and Panamá, from premontane wet forest in the Canal Zone, Coclé, and Panamá, and from tropical wet forest in Darién.

CLITORIA L.

Clitoria javitensis H.B.K., J. Linn. Soc., Bot. 2:42. 1858
   C. portobellensis Beurl.; *C. arborescens* sensu auct. non H.B.K.
   Peronil

Liana, becoming shrubby elsewhere in open areas, lacking tendrils; younger stems reddish-brown, to 12 mm diam. Leaves trifoliolate; stipules lanceolate, ca 7 mm long, pubescent, persistent; petioles 1–15 cm long, reddish-villous; rachis of terminal leaflet to 30 mm long; petiolules of lateral leaflets ca 4 mm long, pulvinate and articulate at apex; leaflets elliptic to obovate, abruptly short-acuminate, rounded at base, 6–22 cm long and 3.5–11 cm wide, dark above, paler below, glabrous above, ± pubescent below at least on the midrib. Racemes axillary, to 30 cm long; bracts and bracteoles ovate, acute, ca 7 mm long, striate; pedicels to 4 mm long; calyx to 3.5 cm long and 1 cm wide, the lobes sharply and abruptly acuminate, the carinal tooth to 8 mm long, the others somewhat shorter; standard white with divergent red lines at center, often infused throughout with pink, rounded at apex, the sides folded over other petals, to ca 8 cm long; wings white, fused to keel above, the keel white, the 2 sides free except at ± twisted apex, enclosing style and stamens, the wings and keel each to ca 6.5 cm long; stamens of 2 lengths, the free stamen and alternate stamens of the staminal tube short, the others slightly longer, the longest to ca 5.5 cm long, considerably exceeded by the densely pubescent style. Legumes on stipes 3 cm long, linear, flattened, sometimes broader toward the apex, to 27 cm long and 2.5 cm wide, abruptly long-acuminate (the acumen to 4 cm long), softly brown-villous; valves twisting when dry at maturity and capable of straightening or twisting repeatedly with changes in humidity; seeds 7–11, ca 12 mm long, 10 mm wide, and 5 mm thick, black. *Croat 4853, 8497.*

Common in the forest, growing to the top of the canopy, and at the edges of clearings. Flowers from October to May, mostly early in the dry season, in January and February. The fruits mature from January to May, mostly in April and May.

Pollen is shed in the bud. The style and anthers are released violently by the pollinator.

This species has been confused with *C. arborescens* Ait. from the Lesser Antilles and the Guianas and also with *C. glaberrima* Pitt., which generally occurs in drier regions of Panama than does *C. javitensis.* It is possibly conspecific with *C. leptostachya* Benth.

Panama to the Guianas, northern Brazil, Peru, and Ecuador. In Panama, known principally from tropical moist forest in the Canal Zone, Colón, Chiriquí, Veraguas, Panamá, and Darién; known also from tropical dry forest in Panamá (Taboga Island) and from premontane wet forest in Chiriquí.

Clitoria rubiginosa Adr. Juss. in Pers., Synops. Pl.
   2:303. 1807

Vine; stems, peduncles, calyces, and underside of leaves moderately brown-pilose. Leaves trifoliolate; stipules ovate, acute, ca 4 mm long; petioles 1.5–4.5 cm long; rachis of terminal leaflet subtending the stipels 1–2 cm long; petiolules pulvinate and articulate; leaflets lance-elliptic to oblong, rounded at base, 2.5–7.5 cm long, 1.5–4 cm wide, glabrous above, villous and paler below, the terminal leaflet rounded-acute to retuse at apex, often bearing short apiculum or tuft of trichomes. Panicles axillary, bearing few flowers; peduncles to 15 cm long; rachises 2–3 cm long; calyx 2–3 cm long, acutely lobed ca one-third its length, the lobes subequal; standard white, drying yellow, tinged with violet along median line below apex, to 4 cm long, enclosing other petals and surpassing them by ca 2 cm; keel straight; stamens ca 2.5 cm long, arcuate. Legumes stipitate, somewhat flattened, linear-oblong, 3–5.5 cm long, with 2 ribs ca 4 mm below the upper edge, glabrous; seeds to ca 10, globose, ca 3 mm diam. *Woodworth & Vestal 693.*

This species occurs sporadically in the Canal Zone, but has not been seen or collected in recent years on the island. The plant could be expected to occur in the Lighthouse Clearing. Flowers at least from September to March. The fruits develop quickly and are usually present at the same time as the flowers.

Southern Mexico to tropical South America; West Indies; introduced in West Africa. In Panama, known from tropical moist forest in the Canal Zone, Chiriquí, Coclé, and Panamá, from premontane moist forest in the Canal Zone, and from premontane wet forest in Chiriquí.

KEY TO THE SPECIES OF CLITORIA

Plants lianas, common in the forest; flowers pinkish, more than 5 cm long . . . . *C. javitensis* H.B.K.
Plants vines, in clearings, rare or absent; flowers mostly white, less than 5 cm long . . . .
. . . . *C. rubiginosa* Adr. Juss.

KEY TO THE SPECIES OF CROTALARIA

Leaves simple .... *C. retusa* L.
Leaves trifoliolate .... *C. vitellina* J. Ker

CROTALARIA L.

Crotalaria retusa L., Sp. Pl. 715. 1753

Gallito, Frijolillo

Erect herb, usually less than 1 m tall; stems ribbed, appressed-pubescent. Leaves simple, nearly sessile; blades obovate, rounded to emarginate at apex, tapered gradually to base, 3–10 cm long, 1.5–2.5 cm wide, glabrous above, appressed-pubescent below, generally bluish-green and lighter below. Racemes terminal, ca 20 cm long; pedicels ca 1 cm long, spreading to reflexed; calyx strigillose, ca 1.5 cm long, lobed ca two-thirds its length, the lobes subequal, acute, those subtending the standard broader; standard ca 2.5 cm long and 3 cm wide, yellow with dark nectar guidelines near base inside and with a broad maroon to brown spot outside; wings ca 2 cm long, ca 1.2 cm wide, covering keel at apex, the keel deeply saccate, the common margins fused but the free portion folded near apex to form a funnel, the lower margins of the keel densely pubescent; stamens 10, fused in basal half into an open-sided tube with alternating long and short anthers, the longer anthers ca 3 mm long, borne on the shorter filaments, shedding pollen in bud, the shorter anthers probably sterile; style strongly bent at about the height of the sterile anthers, the stigma flat-tipped, ciliolate, the trichomes distal, antrorse; pollen about the consistency of cake frosting, squeezed out by stigma and trichomes. Pods oblong, inflated, glabrate, ca 3.5 cm long, brown to black; seeds ca 10, reniform, flat, ca 2 mm diam, becoming black at maturity. *Croat 8659.*

Rare, in clearings; collected once in recent years in the Laboratory Clearing, where it had possibly been planted. The species was once abundant on the island. With the reduction in the number of clearings, it has probably disappeared, though it might easily reoccur in one of the two large clearings. Flowers and fruits throughout the year.

A cosmopolitan tropical weed. In Panama, known from tropical moist forest in the Canal Zone, Bocas del Toro, Colón, Coclé, Panamá, and Darién, from premontane dry forest in Los Santos, and from premontane wet and tropical wet forests in Colón.

Crotalaria vitellina J. Ker in Lindl., Edward's Bot.
Reg. 6, pl. 447. 1820

Zapatito del Obispo

Erect herb or suffrutex, 1–1.5 m tall; stems sericeous. Leaves trifoliolate; stipules linear-lanceolate, to ca 4 mm long, caducous; petioles to 5 cm long, sericeous; leaflets lanceolate-elliptic, apiculate, cuneate at base, 5–7.5 cm long and 1.8–2.5 cm wide, glabrous above, hirtellous and glaucous below, the secondary veins reticulate, the terminal leaflet acute at base. Racemes terminal or axillary, to 16 cm long, the flowers bracteate, oblique on pedicels; bracts subulate, ca 3 mm long; pedicels 5(6) mm long; calyx ca 7 mm long, lobed three-fourths of its length, hirtellous, the lobes slenderly lanceolate, to 5 mm long; petals yellow, tinged with purple, with thin longitudinal ribs; standard to 1 cm long; wings to 8 mm long with a claw ca 2 mm long, the keel united along upper common margins, to 12 mm long; style linear, strigose, curving into sac formed by keel, then following bottom surface of keel forward. Legumes terete, ca 2 cm long and 7 mm wide, the curved remnant of the style often still attached, ca 8 mm long, the trichomes less than 0.2 mm long; seeds ca 10. *Aviles 13.*

Not seen on the island in recent years. Flowers from July to January. The fruits develop within about 1 month.

Mexico to Brazil. In Panama, known from tropical moist forest in the Canal Zone, Veraguas, Herrera, and Panamá, from premontane dry forest in Herrera (Chitré), from premontane moist forest in the Canal Zone and Panamá, and from premontane wet forest in Chiriquí and Coclé.

CYMBOSEMA Benth.

Cymbosema roseum Benth. in Hook., J. Bot. (Hooker)
2:61. 1840

Semiherbaceous liana climbing to the top of trees, soft-villous all over, generally with tendrils. Leaves trifoliolate; stipules acuminate, ca 2 mm long; petioles 3–7(12) cm long; leaflets broadly lanceolate to elliptic, broadly rounded to acute and sometimes retuse at apex, rounded-truncate at base, 6–12 cm long, 3–6 cm wide, darker above than below, the veins finely reticulate. Racemes terminal; peduncles to 25 cm long; rachises to ca 10 cm long; flowers bright red turning lavender on drying; pedicels ca 2 mm long; calyx ca 1 cm long, lobed ca one-fourth its length, bilabiate, the upper lip with 2 minute teeth, the lower with 3 acute teeth, gibbous at base; petals subequal, to ca 3.5 cm long; standard oblong-ovate, ca 1 cm wide; wings ca 5 mm wide, the keel flaring open; stamens 10, diadelphous, barely exceeding the petals. Legumes oblong-elliptic, 4.5 cm long and 2 cm wide, beaked, densely appressed-pubescent, the trichomes acicular, the beak attenuate, to 2 cm long; seeds 3–6. *Croat 8304.*

Apparently rare, seen only along the shore. Some flowers have been seen in October and February, and mature fruits in February and July.

May be confused with *Dioclea guianensis,* but *Cymbo-*

sema is distinguished by having a free stamen associated with the banner, long petals, and fruits with a falcate stigma even at maturity.

Mexico (Chiapas), Costa Rica, Panama, Colombia, and Peru. In Panama, known only from tropical moist forest in the Canal Zone and Panamá.

DALBERGIA L.f.

Dalbergia brownei (Jacq.) Urban, Symb. Ant. 4:295. 1905

Bejuco frijolillo, Sisa guidup

Shrub, small tree, or small liana, to 5 m tall; young stems and petioles rufous-pubescent. Leaves simple; petioles to 1.5 cm long, pulvinate and often articulate at both ends, less often basally; blades ovate, obtuse to acute at apex, rounded to subcordate at base, 5–8 cm long, 3–5 cm wide, inconspicuously strigillose when young, possibly glabrous in age, shiny above, pale below. Panicles short, axillary or subterminal; flowers fragrant, white; pedicels 2–5 mm long, pubescent; calyx ca 4–5 mm long, bilabiate, lobed about half its length, the upper lobe shallowly bifid, the lowermost tooth elongated and subulate, the calyx subtended by bracts, the bracts 2, opposite, obtuse, ca 1 mm long; petals ca 10 mm long, obovate, the keel petals united; stamens 10, diadelphous, ca 5 mm long, the staminal column open basally. Legumes usually 1 per inflorescence, 1- or 2-seeded, oblong, mostly 1.5–2 cm long (to 3 cm long if 2-seeded), 1 cm wide, and ca 3 mm thick, glabrous, speckled; seeds flattened, brown, to ca 7 mm diam. *Croat 7903, Foster 955.*

Occasional, along the shore, usually partly in water. Flowers mostly from February to May. The fruits mature from April to August.

Florida and southern Mexico to Colombia and Venezuela; West Indies. In Panama, known from tropical moist forest in the Canal Zone, Bocas del Toro, San Blas, Panamá, and Darién, from tropical dry forest in Panamá, and from premontane moist forest in the Canal Zone and Los Santos.

Dalbergia monetaria L.f., Suppl. 317. 1781

Liana or climbing shrub, to 13 m tall, forming dense masses along shore; stems minutely fissured and bearing prominent round lenticels. Leaves compound; stipules caducous; petioles 1–2 cm long; rachis puberulent when young, 2–6 cm long, often zigzag with the leaflets alternate; leaflets (1)2–5, ovate to elliptic, abruptly long-acuminate (the acumen to 1.8 cm long), rounded or obtuse at base, 5.5–12 cm long, 3–6 cm wide, glabrous and shiny above except for pubescent midrib, glabrous and pale below, the reticulate veins prominulous but not raised. Racemes axillary; rachises to ca 2 cm long, ± strigose; pedicels slender, 2–3 mm long; calyx inflated, campanulate, to 2 mm long and wide, ± irregularly lobed ca one-third its length, the teeth acute and subequal; petals 5–7 mm long, white, clawed ca half their length, the wings truncate to subcordate and oblique at base; stamens diadelphous, 4–5 mm long, the column split into 2 fascicles; ovary stipitate, the stipe slender, about as long as ovary; style bent sharply forward. Legumes ± orbicular, flattened, becoming bowl-shaped in age, 3.5–4.5 cm long, 2–3 mm thick, glabrous, borne on stipes 5–7 mm long, one margin sometimes raised, the surface often with raised reticulations; seeds brown, discoid, ca 1.5 cm diam, the margin thin. *Croat 13960, Foster 1349.*

Distribution unknown, but locally abundant along the northern shore of Gigante Bay. Flowers in the late rainy season and early dry season, especially in early dry season. The fruits mature in the late dry season and early rainy season (to July).

Mexico along the Atlantic slope of Central America to northern South America; West Indies. In Panama, known from tropical moist forest in the Canal Zone and probably in Bocas del Toro; known also from premontane and tropical wet forests in Colón.

Dalbergia retusa Hemsl., Diag. Pl. Mex. 8. 1878, non Baillon 1884

Cocobola, Rosewood

Tree, to 20 m tall, ca 30 cm dbh; outer bark coarse, becoming fissured and loosened in age; wood dark, hard, heavy. Leaves alternate, pinnate; stipules caducous, ovate-elliptic, to 2 cm long, persistent on juvenile plants; petioles to 6 cm long, puberulent; rachis 10–20 cm long; leaflets 7–15, alternate on rachis, oblong to ovate-oblong, obtuse and often slightly emarginate at apex, obtuse to rounded at base, 6–10 cm long, 2.5–3.5 cm wide, appressed-pubescent below when young, becoming glabrate in age. Panicles axillary or terminal, appearing just before or at same time as new leaves; branches, pedicels, and calyces sparsely to densely pubescent, the trichomes short, brown, appressed; pedicels 3–4 mm long; calyx 5–7 mm long, lobed to one-third its length, weakly bilabiate, the teeth acute; petals white, 1–1.4(2) cm long; standard emarginate, clawed; keel petals fused at apex; stamens fused into a single tube with an open slit apically, markedly curved at apex, exceeded by the style. Legumes ± oblong, flat, 6–8 cm long, ca 2 cm wide, borne on stipes ca 1 cm long, glaucous, glabrous, mostly 1-seeded (if 2-seeded, the fruit to 13 cm long), the winged area of

KEY TO THE SPECIES OF DALBERGIA

Leaves simple; plants usually growing in water around edge of lake . . . . *D. brownei* (Jacq.) Urban
Leaves with 2–15 leaflets; plants usually not growing in water:
  Plants trees; leaflets 7–15, ± oblong; fruits oblong . . . . *D. retusa* Hemsl.
  Plants lianas or climbing shrubs; leaflets 2–5, ovate to elliptic; fruits orbicular . . . . *D. monetaria* L.f.

equal thickness throughout; seeds discoid, to ca 1 cm diam, medial on fruit body. *Croat 5320.*

A single tree, which undoubtedly was planted, grows near the laboratory. Herbarium collections from central Panama have shown flowers throughout much of the year, but the tree on BCI loses its leaves in the dry season and then flowers as the new leaves are emerging from February to July, usually in March and April. The species may flower only every other year (Frankie, Baker, & Opler, 1974). The fruits develop to full size within 2 months, but persist until the following dry season.

Mexico, Nicaragua, Costa Rica, and Panama. In Panama, known from tropical moist forest in the Canal Zone, Panamá, and Darién and from premontane moist forest in Colón and Panamá.

See Fig. 282.

DESMODIUM Desv.

The genus consists of herbs or suffruticose shrubs with alternate, stipulate, trifoliolate leaves. The flowers are racemosely arranged in pairs. The style is longer than the stamens and is enclosed with the stamens within the keel under tension. The stamens and style are held in position until triggered by a flower visitor, when they spring forward and throw pollen.

The genus is distinguished by the flattened loment of one-seeded, indehiscent sections, usually bearing uncinate trichomes.

Desmodium adscendens (Sw.) DC., Prodr. 2:332. 1825
   Beggarlice, Pega-pega

Perennial, often rooting at the nodes, to 50 cm long; stems slender, white-pilose, the trichomes rubbing off easily. Leaves trifoliolate; stipules long-attenuate, ca 7 mm long; petioles to ca 12 mm long; stipels subulate, to 4 mm long; petiolule of terminal leaflet ca 5 mm long, the petiolules of lateral leaflets ca 2 mm long; leaflets ± elliptic, ± rounded at both ends, often emarginate at apex, 1.5–4 cm long, 1–2 cm wide, ± glabrous above, appressed-pilose and pale below. Racemes terminal, slender, 5–20 cm long; peduncles 2–3.5 cm long; pedicels ascending, 7–10 mm long; flowers 2 or 3 per node of inflorescence, blue to pinkish-blue; calyx 2–4 mm long, bilabiate, lobed three-fourths of its length, the teeth slender, subequal; corolla 4–5 mm long, inconspicuous, withering quickly. Loments

KEY TO THE SPECIES OF DESMODIUM

Leaflets often emarginate at apex:
    All leaflets less than 1 cm long, rarely rounded at apex; upper margin of fruits nearly entire . . . .
        . . . . *D. triflorum* (L.) DC.
    Most leaflets more than 1 cm long, often rounded at apex; upper margin of fruits nearly entire
        or as deeply incised as lower:
        Segments of fruit less than 2 times longer than broad . . . . *D. adscendens* (Sw.) DC.
        Segments of fruit ca 3 times longer than broad . . . . *D. scorpiurus* (Sw.) Desv.
Leaflets not emarginate at apex:
    Fruits with 1 or 2 segments; leaves acute to acuminate at apex:
        Fruits on stipes 1–2 mm long; flowers borne singly . . . . *D. wydlerianum* Urban
        Fruits on stipes more than 4 mm long; flowers on paired pedicels:
            Stems and petioles conspicuously villous, the trichomes 0.5–1 mm long . . . .
                . . . . *D. axillare* (Sw.) DC. var. *acutifolium* (O. Kuntze) Urban
            Stems and petioles minutely puberulent, the trichomes usually 0.1 mm long . . . .
                . . . . *D. axillare* (Sw.) DC. var. *stoloniferum* (Poir.) Schub.
    Fruits with mostly 3 to many segments; leaves rounded to acute at apex:
        Fruits deeply lobed on both margins; most terminal leaflets at least 4 cm long; flowers usually
            not single at nodes of inflorescence:
            Segments of fruit 5–6 mm long; most petioles below inflorescence less than 1.5 cm long . . .
                . . . . *D. cajanifolium* (H.B.K.) DC.
            Segments of fruit 2.5–5 mm long; most petioles below inflorescence more than 1.5 cm long:
                Pedicels mostly 1–1.6 cm long; fruiting nodes on rachis (6)10–30 mm apart; segments of
                    fruit 3–3.5(4) mm long, 2.6–3.5 mm wide . . . . *D. tortuosum* (Sw.) DC.
                Pedicels usually less than 1 cm long; fruiting nodes on rachis (4)6–10 mm apart; seg-
                    ments of fruit smaller, 1.5–2.5 mm long, 1.5–2 mm wide . . . .
                        . . . . *D. distortum* (Aubl.) J. F. Macbr.
        Fruits deeply lobed only on lower margin; leaflets less than 4 cm long *or* flowers borne singly
            at nodes of inflorescence:
            Flowers usually borne singly at nodes of inflorescence; terminal leaflets acute at apex, at
                least 3 times as long as petioles; plants often shrublike . . . .
                    . . . . *D. canum* (J. F. Gmel.) Schinz & Thell.
            Flowers (not fruits) in 2s or 3s at nodes of inflorescence; terminal leaflets obtuse to rounded
                at apex, less than 3 times as long as petioles; plants usually sprawling or prostrate . . . .
                    . . . . *D. adscendens* (Sw.) DC.

to 20 mm long and ca 3 mm wide, densely uncinate-puberulent, the upper margin entire, the lower margin lobed to about the middle between seeds; articles 2–4(5), 4–5 mm long. *Croat 11846.*

Common in clearings. Flowers and fruits all year, perhaps mainly from May to September.

Mexico to South America; West Indies; tropical Africa. In Panama, widespread and ecologically variable; known from tropical moist forest in the Canal Zone, Bocas del Toro, San Blas, Los Santos, and Darién, from tropical wet forest in Colón, Panamá, and Darién, from premontane wet forest in Coclé, and from premontane rain forest in Chiriquí and Coclé.

Desmodium axillare (Sw.) DC., Prodr. 2:333. 1825

Repent perennial; stems creeping over ground, usually rooting at nodes. Leaves trifoliolate; stipules 5–10 mm long, connate about one-third their length; petioles 4–7 cm long, puberulent, the pulvinus at base terete; rachis of terminal leaflet 4–10 mm long; leaflets ovate, ± inequilateral, 4–8 cm long, 3–6 cm wide, glabrous above, white-pilose below, the reticulate veins often conspicuously raised and sometimes close together below. Racemes axillary, erect, elongate; peduncles 15–20 cm long; axis elongating to ca 20 cm; bracts ovate, acuminate, ca 4 mm long, strigillose, concave at base; pedicels paired, ca 5 mm long (to 15 mm in fruit); flowers pink to lavender, spring-loaded, ca 5 mm long, often paired; calyx 1.5–2 mm long, acutely lobed to half the length, the lobes 4, ± equal, deltoid, puberulent to hispid, purplish; corolla ca 4.5 mm long, pink to rose or reddish-violet; standard emarginate at apex; stamens ca 4 mm long, ± equal, the staminal column open basally, spring-loaded, forcibly emerging from closed keel on disturbance. Loments at right angles on stipes 5–6 mm long, each with the upper margin entire, the lower margin deeply lobed; segments usually 2, more or less reniform, 8–13 mm long.

Desmodium axillare (Sw.) DC. var. **acutifolium** (O. Kuntze) Urban, Symb. Ant. 4:292. 1905

Stems conspicuously villous, the trichomes 0.5–1 mm long. Leaflets strigillose above, thickly pilose below. Loments 10–15 mm long, densely white-tomentose; segments 2. *Croat 7063.*

Occasional, in clearings. Probably flowers and fruits year-round.

Belize to South America; Jamaica. In Panama, known only from tropical moist forest in the Canal Zone, Colón, Panamá, and Darién.

Desmodium axillare (Sw.) DC. var. **stoloniferum** (L. C. Rich. ex Poir.) Schub., J. Arnold Arbor. 44:289. 1963

Stems minutely puberulent, the trichomes ca 0.1 mm long. Leaflets glabrous above, densely pilose below. Loments 20–25 mm long, yellowish-hispid-uncinate, on stipes ca 7 mm long; segments 2. *Croat 6630.*

Frequent along trails in the forest. Possibly flowers

and fruits year-round, but more abundantly in the early to middle rainy season.

Central America; Greater Antilles. In Panama, known only from tropical moist forest in the Canal Zone.

Desmodium cajanifolium (H.B.K.) DC., Prodr. 2:331. 1825

Pega-pega

Stout erect suffrutex, usually 1–3 m tall; stems finely pubescent to pilose, to 2 cm thick at base. Leaves trifoliolate; stipules 3–7 mm long, often persistent; petioles 5–10(30) mm long, sericeous; rachis of terminal leaflet 3–10 mm long; leaflets lanceolate to ovate, acute to rounded at apex, obtuse to rounded and inequilateral at base, 4.5–7 cm long, 2–3 cm wide, puberulent above, densely soft-pilose and paler below, the reticulate veins conspicuously raised below. Inflorescences terminal, racemose-paniculate, some lower racemes axillary, to ca 40 cm long and 25(30) cm wide; floral bracts deciduous, subulate, ca 1 mm long; pedicels ca 3 mm long (to 5 mm in fruit); calyx 2.5–3 mm long, bilabiate, acutely 4-lobed to middle, hispidulous; corolla 4–6 mm long; standard violet (turning blue), marked inside on either side and above the base of the medial groove with a greenish spot; wings the same color as the standard on their upper margin (nearest stamens), the keel usually white; stamens diadelphous, 5–6 mm long, shedding pollen in bud, only loosely held by keel. Loments subsessile, lobed from both margins but more deeply so from below, the isthmus between joints very narrow, sparsely puberulent; segments 4–6, each 5–6 mm long, chartaceous, the veins reticulate. *Shattuck 561.*

Possibly no longer occurring on the island, though abundant at Frijoles and to be expected in clearings. In central Panama, flowers and fruits from November to April (sometimes as early as October). Since fruits are probably wind dispersed, most probably mature during the dry season.

Southern Mexico to South America. In Panama, known from tropical moist forest in the Canal Zone, Veraguas, and Panamá.

Desmodium canum (J. F. Gmel.) Schinz & Thell., Mem. Soc. Neuchat. Sci. Nat. 5:371. 1914

D. frutescens (Jacq.) Schindl.

Pega-pega, Pegadera

Perennial suffrutex, becoming over 1 m tall, puberulent, the trichomes sparse on upper leaflet surface, often uncinate especially on inflorescence. Leaves trifoliolate; stipules ovate, attenuate, often connate, persistent, 6–10 mm long, ciliate, becoming brown and striate; petioles 5–15 mm long; stipels acicular, paired, 3–4 mm long, subpersistent; leaflets ovate to elliptic, acute to obtuse at apex, rounded to slightly subcordate at base, 2–7 cm long, 1–4 cm wide. Racemes terminal or subterminal; pedicels 5–8 mm long; flowers pale blue, ca 4 mm long, borne singly at nodes of inflorescence; calyx ca 4 mm long, somewhat reddish, the vexillar lobes joined to near the apex; standard rounded, ca 3 mm wide, with 2 green

spots near base; wings shorter than keel, the keel petals at first joined near apex; stamens 3–4 mm long, bent toward standard at apex, of 2 lengths, the free stamen and alternate stamens of tube shorter; style longer than the stamens, enclosed under tension with stamens within keel, released by pollinator, springing forward toward standard and throwing pollen; ovary densely villous, the hairs minute. Loments narrow, sparsely pubescent, the upper margin entire, the lower margin deeply lobed; segments usually 3–7. *Croat 6928.*

Occasional in clearings and rare in disturbed areas on the shore. Flowers and fruits throughout the year.

The pubescence on the fruit apparently does not attach to passing mammals as in the other species of *Desmodium.* The fruits are probably wind dispersed, the height of the plant allowing the seeds to be borne for some distance by the wind.

Throughout the New World tropics; tropical Africa. In Panama, widespread and ecologically variable; known from tropical moist forest in the Canal Zone, Bocas del Toro, Colón, San Blas, Chiriquí, Panamá, and Darién, from premontane dry forest in Herrera and Coclé, from premontane wet forest in the Canal Zone, Colón, Coclé, and Panamá, and from tropical wet forest in Darién.

Desmodium distortum (Aubl.) J. F. Macbr., Publ. Field Columbian Mus., Bot. Ser. 8:101. 1930
D. asperum (Poir.) Desv.

Erect perennial to 2 m tall; stems uncinate-hirtellous. Leaves trifoliolate; stipules 10–15 mm long, clasping; petioles 2–5 cm long; stipels noticeable; leaflets ± elliptic-ovate, rounded at apex and base, hispidulous or hirsute-puberulent on both surfaces, the terminal leaflet 4–8 cm long and 1–3.5 cm wide, the lateral leaflets smaller, the reticulate veins prominulous below. Panicles of racemes large, terminal, the lower racemes axillary; bracts inconspicuous, subulate, ca 3 mm long; pedicels ca 7 mm long (to 12 mm in fruit); flowers small, purplish; calyx 4-lobed, ca 3 mm long; standard ca 5 mm long. Loments briefly stipitate, lobed equally from both margins, uncinate-puberulent; segments 5 or 6, orbicular, ca 2.5 mm diam. *Starry 307.*

Collected only once at the edge of the Laboratory Clearing; not seen in recent years. Apparently flowers and fruits throughout the year.

Mexico to South America; West Indies. In Panama, ecologically variable, known from tropical moist forest in the Canal Zone and Darién, from tropical dry forest in Herrera and Panamá, from premontane moist forest in the Canal Zone, from premontane wet forest in Chiriquí and Coclé, and from tropical wet forest in Darién.

Desmodium scorpiurus (Sw.) Desv., J. Bot. Agric. 1:122. 1813

Sprawling, weak-stemmed herb, often from a stout woody root; stems and leaves villous to hispidulous. Leaves trifoliolate; stipules often cordate at lower edge, ca 4 mm long; petioles 1–5 cm long; leaflets rounded, elliptic or

oblong, usually emarginate to rounded at apex, obtuse to rounded at base, sparsely strigose above, more densely so below, the terminal leaflet 0.8–3.5 cm long, 0.8–2.5 cm wide, the lateral leaflets slightly smaller. Racemes terminal, slender; pedicels 5–10 mm long; flowers few, bluish or whitish, 3–4 mm long; calyx deeply divided, ca 3 mm long; stamens didynamous, shorter than style; filaments fused to near apex; style sharply turned inward. Loments turgid, mostly 2.5–3 cm long and 1.5 mm wide, lobed along both margins, densely uncinate-puberulent, very sticky; segments 5–7, separating easily, oblong, ca 5 mm long. *Croat 6923, 15580.*

Occasional in open areas of the Laboratory Clearing. Flowers and fruits mainly from January to March (sometimes from December to April).

This is a variable species in terms of vegetative characters. Though superficially similar to *D. adscendens,* it may be distinguished by its generally uncinate pubescence, its broad, ovate-cordate stipules, and (when the fruit is mature) its very distinctive, oblong articles.

Mexico to Venezuela, the Guianas, Brazil, and Bolivia; West Indies; naturalized in parts of tropical East Indies and West Africa. In Panama, known from tropical moist forest in the Canal Zone, Veraguas, and Panamá, from tropical dry forest in Los Santos, and from premontane wet forest in Chiriquí.

Desmodium tortuosum (Sw.) DC., Prodr. 2:332. 1825
D. purpureum (P. Mill.) Fawc. & Rendle

Erect or straggling herb, to 2 m high; stems, petioles, and inflorescences puberulent, the trichomes dense, uncinate, glandular. Leaves trifoliolate; stipules striate, attenuate, to 1(1.4) cm long; petioles 1–2(5) cm long; leaflets lanceolate to ovate, obtuse to acute at apex, cuneate to obtuse at base, uncinate-puberulent and pilose on both surfaces, the terminal leaflet 3–6.5(11) cm long, 1.5–2.5(4.5) cm wide, the lateral leaflets somewhat smaller, the reticulate venation prominent. Racemes or panicles terminal or subterminal; pedicels 5–16 mm long, ascending; calyx 2–3 mm long, lobed to three-fourths of its length, the lobes linear, attenuate; corolla white or pink, fading light blue or mauve, 3–4 mm long; stamens didynamous, 4–5 mm long, the staminal column arched downward, the free stamen arched upward; style slender, to 2 mm long, exceeding the stamens, upon maturation of the fruit acquiring a contortion at the base and thus held ± persistently and obliquely on fruit. Fruiting rachis with nodes (6)10–30 mm apart; loments stipitate, lobed equally on each side; segments (2)3–6(7), each 3–4 mm long, 2.5–3.5 mm wide, uncinate-pubescent. *Standley 40950.*

Seen only once on the island, possibly no longer occurring there. A plant of weedy open areas. Apparently flowers and fruits all year.

Distinguished from other species by the prominently reticulate venation, the large persistent stipules, and the long, ascending pedicels.

Throughout American tropics and subtropics; introduced in most other tropical areas of the world. In Pan-

Fig. 282. *Dalbergia retusa*

Fig. 283. *Dioclea reflexa*

Fig. 284. *Dioclea wilsonii*

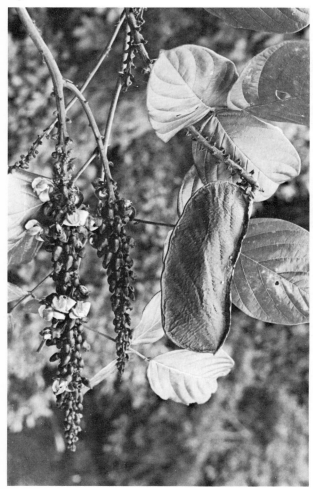

ama, known primarily from tropical moist forest in the Canal Zone, Bocas del Toro, Colón, and Darién; known also from tropical dry forest in Panamá (Taboga Island) and from premontane moist forest in the Canal Zone (Balboa).

Desmodium triflorum (L.) DC., Prodr. 2:334. 1825

Pega-pega

Prostrate, rooting, often matting herb; stems slender, usually pilose, to 30 cm long. Leaves trifoliolate; stipules ovate, ca 2.5 mm long; petioles ca 4 mm long; leaflets broadly obovate, emarginate or truncate at apex, obtuse at base, glabrous above, moderately to densely sericeous below, the terminal leaflet 7–10 mm long, 6–10 mm wide, the lateral leaflets smaller. Flowers bright lavender, small, in axillary clusters of 1–3; pedicels 4–7 mm long; calyx deeply dentate, 4–5 mm long, pilose; corolla little longer than the calyx. Loments not stipitate, curved upward, entire along upper margin, undulate on lower margin, uncinate-puberulent, the trichomes retrorse, reticulate; segments 4–6, ca 3 mm long. *Hladik 89.*

Collected once in the Laboratory Clearing. Flowers and fruits probably only in the dry season (December to April).

Pantropical. In Panama, known from tropical moist forest in the Canal Zone (Atlantic slope), Bocas del Toro, and Colón and from tropical dry forest in Panamá (Taboga Island).

Desmodium wydlerianum Urban, Symb. Ant. 2:302. 1900

Prostrate or ascending perennial herb, rooting at nodes; stems uncinate-puberulent. Leaves trifoliolate; stipules to 7 mm long, caducous; petioles 3–6 cm long; leaflets ovate to ovate-deltoid, acute or gradually acuminate at apex, ± truncate at base, sparsely strigillose on both surfaces, thinly and palely pubescent below, the terminal leaflet 5–8 cm long, the lateral leaflets smaller. Racemes axillary, elongate; pedicels solitary, slender, ca 1.5 cm long; calyx ca 3.5 mm long; corolla blue or pink, ca 4.5 mm long. Loments briefly stipitate, downcurved, entire on upper margin, cleft from the lower margin, uncinate-

pubescent; segments 2, each 8–10 mm long. *Croat 14958.*

Rare, in the forest along trails. Flowers at least in May and June.

Guatemala to northwestern South America. In Panama, known only from tropical moist forest on BCI and in Darién.

DIOCLEA Kunth

Dioclea guianensis Benth., Comm. Leg. Gen. 70. 1837

Haba de monte

Liana; stems short-pilose. Leaves trifoliolate; stipules subulate, to 1.5 cm long, subpersistent; petioles usually 3–5 cm long (to 12 cm on juvenile leaves), the pulvinus terete, basal; rachis usually ca 2 cm long (to 4.5 cm on juvenile leaves); stipels acicular, 5–15 mm long; petiolules 5–10 mm long, swollen; leaflets broadly elliptic, acuminate, rounded to somewhat cordate at base, 7–11 cm long, 4–7 cm wide, sparsely appressed-pilose all over or glabrate above except on midrib (young leaves sericeous above). Racemes long-stalked, to 50 cm long; rachises studded with tuberculate knobs on which the flowers develop; pedicels 2–4 mm long at anthesis, enlarging in fruit; calyx 10–12 mm long, lobed, the lobes 4, acuminate, spreading; standard, wings, and tip of keel violet; standard rounded, 16–20 mm wide, emarginate, reflexed, with a small yellow guide spot at base in center; wings ca 17 mm long, erect, joined to keel near base, the keel weakly fused apically; stamens and style curved upward near apex, equal in length, the free stamen and alternate stamens of the staminal tube somewhat shorter than the others; anthers and style fully exposed to a pollinator approaching the nectary from the base of the standard; pollen tacky, adhering to anther after dehiscing; nectary prominently raised. Legumes densely pubescent (less so at maturity), 7–10 cm long, ca 14 mm wide, the upper margin with sharp lateral ribs; segments 8–10. *Croat 4841.*

Infrequent, along the shore and in clearings, especially the Lighthouse Clearing. Probably more abundant in earlier years, since it grows in weedier areas in other parts of the Canal Zone. Flowers mostly throughout the dry season (December to April). The fruits mature mostly in

KEY TO THE SPECIES OF DIOCLEA

Stipules basifixed, lacking a produced lobe; fruits less than 1.5 cm wide; stamens with the anthers uniform, all fertile; plants of secondary areas, usually at the edges of clearings . . . . *D. guianensis* Benth.
Stipules not appearing basifixed, with a produced basal lobe; fruits more than 1.5 cm wide; stamens with 5 fertile and 5 sterile anthers; plants usually in the canopy of the forest:
 Legumes indehiscent, the lower edge (the narrowest edge not associated with the beak at the distal end) knifelike or at least not deeply sulcate along its length; inflorescence bracts and young fruits dark ferruginous-pubescent (at least on drying), the bracts linear, erect . . . . *D. wilsonii* Standl.
 Legumes dehiscent, the lower edge not knife-edged, deeply sulcate along its length; inflorescence bracts and young fruits yellowish-brown-pubescent, the bracts linear to linear-lanceolate, usually reflexed . . . . *D. reflexa* Hook.f.

the late dry and early rainy seasons (March to May).

Panama to northern South America. In Panama, known from tropical moist forest in the Canal Zone, Colón, Coclé, Panamá, and Darién, from tropical dry forest in Coclé, from premontane moist forest in the Canal Zone, and from premontane wet forest in Chiriquí.

Dioclea reflexa Hook.f. in Hook., Nig. Fl. 306. 1849

Twining liana; stems to 5 cm diam; vegetative branchlets strigillose. Leaves trifoliolate; stipules subulate, to 2 cm long, with a produced basal lobe; petioles 4–7 cm long; leaflets elliptic-obovate, briefly acuminate, rounded-truncate at base, 6–14 cm long, 4–9 cm wide, glabrous above, sparsely strigillose below, thinly coriaceous, shiny. Racemes to 30 cm long, rufous-tomentose; pedicels at anthesis 5–7 mm long, enlarging in fruit; calyx densely pubescent, the trichomes appressed, brown, the subtending bracts linear to narrowly lanceolate, usually reflexed, the carinal tooth curved inward; standard violet to purple, to 1.5 cm long, recurved, rounded and emarginate, broader than long, the upper margin violet to orchid, lobed and clawed at base; stamens diadelphous, weakly connate to form a tube, the tube curved forward, the free stamen and 4 additional alternate stamens of the tube sterile, the 5 fertile stamens longer; style recurved near apex in bud, then straightening; stigma round, at height of sterile anthers, later at height of fertile anthers as the style lengthens. Legumes oblong, somewhat turgid, 12–18 cm long, 5–6 cm wide, at most only slightly concave along the upper margin, constricted somewhat in thickness between the seeds, sulcate along the lower margin, rufous-tomentose when young, the upper suture with a broad costa on either side; seeds 2 or 3, orbicular, ca 3 cm diam, dark brown. *Croat 5017, 8390.*

Occasional, along the shore and in the canopy of the forest. Flowers from September to April, mostly from December to March. The fruits are full size from January to May, but their time of maturity is not known, possibly in the rainy season.

The stigma is at first recurved near the apex in bud, at which time the five fertile anthers are undehisced. Before the petals open, the stigma straightens and is disposed among the sterile anthers below the fertile anthers, which then shed their pollen.

Pantropical. In Panama, known from tropical moist forest in the Canal Zone, Bocas del Toro, San Blas, Coclé, Panamá, and Darién.

See Fig. 283.

Dioclea wilsonii Standl., Publ. Field Columbian Mus., Bot. Ser. 4:310. 1929

D. violacea sensu auct. non Mart. ex Benth.

Liana; young branchlets and petioles long-pilose. Leaves trifoliolate; stipules with a produced basal lobe; petioles 4–9 cm long; leaflets elliptic, rounded or sometimes very abruptly acuminate at apex, rounded, truncate, or subcordate at base, 7–15 cm long, 4–13 cm wide, becoming glabrous above, strigillose below. Racemes to 45 cm long; inflorescences and bracts dark ferruginous-pubescent;

rachises studded with tuberculate knobs on which the flowers develop; bracts linear, erect; pedicels 4–6 mm long at anthesis; flowers purple; calyx to 1.2 cm long; standard to 3 cm long, longer than wide; stamens with 5 anthers fertile, 5 sterile. Legumes completely indehiscent, 10–14 cm long, to 5.5 cm wide and 1–1.7 cm thick, purplish-pilose (dark ferruginous-pubescent when young), only slightly concave along the upper margin, not at all constricted between the seeds, obliquely sulcate, not sulcate along the lower margin; seeds plump, 3 or 4, ca 3 cm diam, brown with black mottling. *Croat 6726.*

Occasional, along the shore; probably less common than *D. reflexa*. Flowers mostly in the rainy season, especially in October. The fruits are nearly full size by November or December and seem to ripen by May.

Seeds of this species cannot easily be distinguished from those of *D. reflexa*. The species has been confused with *D. violacea* Mart. ex Benth., which R. Maxwell considers to be equal to *D. paraguariensis* Hassl. (pers. comm.).

Belize to Panama, the Guianas, and Brazil; Trinidad and Granada. In Panama, known only from tropical moist forest in the Canal Zone, Bocas del Toro, Herrera, Panamá (Perlas Islands), and Darién.

See Fig. 284.

DIPTERYX Willd.

Dipteryx panamensis (Pitt.) Rec. & Mell, Timbers Trop. Am. 303. 1924

Oleiocarpon panamense (Pitt.) Dwyer

Almendro, Almendro corozo, Ebo, Tonka bean

Tree, to 40 m tall; trunk more than 1 m dbh, usually weakly buttressed; outer bark light brown, thick, hard, the surface granular and loose; inner bark tan (becoming brown in time), the cells isodiametric. Leaves alternate, paripinnate; stipules conspicuous, linear, to ca 15 cm long, 1 cm wide, caducous; petioles flat, canaliculate at base, 6–15 cm long; petiole and rachis revolute-winged; rachis 15–30 cm long, continuing past the terminal pair of leaflets into a conspicuous flange 2–7 cm long; petiolules pulvinate, 4–6 mm long, subtended by acute stipels ca 4 mm long; leaflets 10–16, alternate on rachis or opposite, elliptic to oblong, acute to retuse at apex, ± inequilaterally obtuse at base, 7–22 cm long, 2.5–7.5 cm wide. Panicles terminal, to 40 cm long; pedicels 3–5 mm long, expanded at apex; flowers conspicuous, red-violet, ca 2 cm long; calyx with 2 lobes, these much expanded, obliquely oblong, spreading laterally, to 1.7(2) cm long, the carinal and lateral teeth very reduced; calyx orchid-colored, sparsely punctate and puberulent; standard to 1.5 cm long and 1.3 cm wide, emarginate, green at base medially, surrounded by a dark band of red-violet, the claw short and strongly recurved; stamens to 1.4 cm long, subequal; filaments fused in basal two-thirds, directed upward near apex; style a little longer than stamens, directed sharply upward; ovary glabrous. Drupes borne on thick stipes ca 8 mm long, plump, oblong, ca 6 cm long and 3 cm wide, at first gray-green-pubescent, be-

coming brown-pubescent, at maturity filled with an oily, fragrant liquid, the liquid crystalizing when the fruit dries; seed 1, flattened, ca 5 cm long and 1.5 cm wide. *Croat 6084, Foster 1326, Hladik 441.*

Common to locally abundant in most areas of the forest. Flowers from May to August, generally in July. The fruits develop to mature size by the late rainy season, with most maturing from January to April. Leaves are deciduous during part of the dry season.

Recognized by the light granular bark.

Bees are very active visitors when the tree is in full flower. The fruits are taken by white-faced, spider, and howler monkeys as early as November (Hladik & Hladik, 1969). White-faced monkeys eat the fruits mostly from January to April (Oppenheimer, 1968), removing a part of the mesocarp surrounding the seed before discarding the remainder. Fruits are also taken by bats, rodents, and coatis (R. Foster, pers. comm.).

Costa Rica to Colombia. In Panama, known from tropical moist forest in the Canal Zone, Bocas del Toro, Colón, Panamá, and Darién. The species is a characteristic tree in tropical moist forest in Panama (Holdridge & Budowski, 1956).

ERYTHRINA L.

Erythrina costaricensis Micheli var. **panamensis** (Standl.), comb. nov.

E. panamensis Standl., J. Wash. Acad. Sci. 17:10. 1927.
Holotype: *Pittier 2656 (US)*
Palo santo, Pito, Collins machete

Shrub or small tree, usually 3–7 m tall; trunk and branches armed with short, stout, corky prickles; branches few, stout. Leaves trifoliolate; stipules linear-lanceolate, ca 1 cm long, caducous; petioles 17–34 cm long, often with prickles; petiolules 6–12 mm long, articulate at base; leaflets broadly ovate, acuminate, rounded or truncate at base, thin, ± glabrous above, closely matted with long appressed trichomes below, the terminal leaflet to 28 cm long and 19 cm broad, the lateral leaflets somewhat smaller, inequilateral at base. Racemes congested at ends of leafless branches, soon falling; pedicels 2–3 mm long; flowers red; calyx bilobed, ca 2.5 cm long; standard to 9 cm long, ca 1 cm wide, conduplicate, shaped like the blade of a machete; other petals much reduced, shorter than calyx; stamens 10, 9 united more than halfway; anthers held at 5 levels in a 1-2-4-2-1 arrangement, the lowermost being on the free filament; style densely pubescent except near apex, at first shorter than most an-

thers, longer than anthers by the time the flower falls. Legumes reddish, densely pubescent, 15–25 cm long, long-pointed at apex, deeply constricted between seeds with the isthmus 2–20 mm long, splitting open and twisting slightly at maturity to expose seeds; seeds several, bright red, ca 8 mm long. *Croat 6560, 12710.*

Common in the forest, especially the old forest. Flowers appear before the leaves from September to December (especially in September and October), rarely later. The fruits mature mostly during November and December and have usually shed all their seeds by March. The leaves fall around September, growing out again during the early dry season.

The banner of the flower is weakly sealed along its outer margins. When the flower first opens, the style is held below all but the lowermost anther. At this time even the shorter anthers are shedding pollen. The stigmatic surface of the style is somewhat cupular and is filled with a jellylike plug that collects much pollen. The anthers continue shedding pollen until the style has grown beyond the longest stamen. At this time, when the flower falls, the stigma is flattened, without the jellylike substance. It has been presumed that hummingbirds pollinate the plant, but it is difficult to see how the plant prevents self-pollination—possibly the jellylike substance prevents self-pollination.

Although Krukoff (1939) originally distinguished both *E. costaricensis* and *E. panamensis* Standl., he has in recent years clumped them. Though admitting extreme variability in the taxa, I feel that many of the Panamanian materials of this species, including BCI plants, are at least subspecifically distinct. The BCI plants have larger, thinner leaves and fruits that are usually so constricted between the seeds that the seed-filled segments are broader than long and scarcely more than 1 cm apart. Flowers are bright red and appear while the plant is leafless; Costa Rican specimens are often reported as orange-flowered and flower when the plant has leaves. They also have smaller, thicker leaves and fruits that are not nearly so constricted between the seeds, so that the seed-filled segments are usually as long as to much longer than broad. Costa Rican plants are often reported to be large trees, whereas the BCI plants are never more than 7 m tall.

Panama to Colombia. In Panama, known from tropical moist forest in the Canal Zone, Bocas del Toro, Chiriquí, Panamá, and Darién, from tropical wet forest in Colón, Coclé, and Darién, and from premontane rain forest in Darién.

See Figs. 285 and 286.

KEY TO THE SPECIES OF ERYTHRINA

Plants small trees in the forest; corolla bright red, the standard narrowly tubular, more than 6 cm long, the other petals much reduced; legumes deeply constricted between seeds; trunk bearing small, conical spines . . . . *E. costaricensis* Micheli var. *panamensis* (Standl.) Croat
Plants large trees on the shore; corolla pale orange, the standard reflexed, not tubular, ca 5 cm long; legumes only slightly constricted between seeds; trunk bearing broad corky spines . . . .
. . . . *E. fusca* Lour.

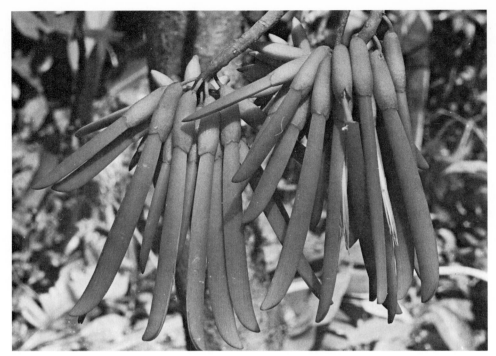

Fig. 286. *Erythrina costaricensis* var. *panamensis*

Fig. 285. *Erythrina costaricensis*
var. *panamensis*

Fig. 287. *Erythrina fusca*

Erythrina fusca Lour., Fl. Cochinch. 427. 1790

E. glauca Willd.

Palo santo, Gallito, Bois immortelle, Palo bobo, Pito, Immortal

Tree, mostly 10–20 m tall; trunk ca 1 m dbh; outer bark grayish, coarse, sparsely covered with broad corky prickles; wood white to yellowish, moderately soft; branches glabrous, sparsely armed with short prickles; petiolar bases persistent, prominent; on juveniles, the prickles larger and extending onto petioles, rachises, and midribs. Leaves trifoliolate; stipules caducous; petioles 8–18 cm long; rachis 4–8 cm long; both petiole and rachis with 2 glands at their apices; leaflets ± ovate, ± rounded at both ends or acute at apex, glabrous above, with a dense mat of white appressed trichomes below, the terminal leaflet 8–14 cm long and 7–12 cm wide, the lateral leaflets smaller. Flowers thick, mostly 3 per node, in large, terminal, somewhat pendent racemes; pedicels stout, turned away from apex, ca 2 cm long; flowers showy, pale orange; calyx spathelike; standard spatulate, ca 5 cm long, reflexed; keel open at apex, ca 2.5 cm long, the wings somewhat shorter, greenish below, bright orange above; stamens diadelphous, green, gradually arched, about halfway exserted; filaments of 3 lengths in 5-4-1 arrangement, the free stamen one of the shorter 5; anthers held in an open pattern over the entrance to the nectaries through the open end of the keel; style bent sharply away from anthers just below apex; stigma usually between the shorter 2 sets of anthers in length; nectar copious. Legumes ca 19 cm long and 2 cm wide, densely brown-tomentose, pointed at apex, weakly ribbed on margins; seeds several, ellipsoid, dark brown, ca 12 mm long, possibly expelled forcibly. *Croat 8203.*

Rare, on the shore. Elsewhere in the Canal Zone common to locally abundant, generally near bodies of water. The leaves are deciduous shortly after flowering. Flowers from November to March, usually in February. The fruits mature from February to May, mostly in May.

The plant is probably hummingbird pollinated. The curare-like alkaloids erthraline, erythramine, and erythratine have been obtained from this species (Blohm, 1962).

Guatemala throughout the Amazon basin; West Indies; widespread in the Old World tropics. In Panama, known only from tropical moist forest in the Canal Zone, Bocas del Toro, Coclé, Panamá, and Darién. The species sometimes forms pure stands in freshwater marshes (Holdridge, 1970).

See Fig. 287.

INDIGOFERA L.

Indigofera mucronata Spreng. ex DC., Prodr. 2:227. 1825

Añil

Perennial herb, erect or usually decumbent, sometimes greatly elongate and ± scandent, most parts densely to sparsely strigose; stems slender, branched, angulate, to ca 1 m long or sometimes much longer. Leaves imparipinnate; stipules subulate, ca 3 mm long; petioles 1.5–2 cm long; rachis 1.5–4 cm long; leaflets 5–7, ovate to elliptic, ± rounded at both ends, apiculate at apex, 1.5–2.5(4) cm long, 0.8–1.5(2.5) cm wide, paler below. Racemes axillary, lax, often greatly elongate in fruit, to 20 cm long, mostly longer than the leaves; pedicels to 2 mm long, reflexed in fruit; bracteoles subulate, ca 1 mm long; calyx 2–3 mm long, weakly bilobed, deeply toothed, the teeth attenuate; corolla 5–6 mm long, shades of red; stamens diadelphous; style longer than stamens. Legumes linear, 2.5–3.5(4) cm long, ca 1.5 mm wide, with a persistent cusp at apex ca 2 mm long, reflexed; seeds 10–15, ca 2 mm long and 0.8 mm wide, dark brown or maroon. *Shattuck 562.*

Collected once on the island, but possibly no longer there. It could be expected, however, in the larger clearings and along the shore. Elsewhere the species may be found in flower throughout most of the year, but most flowering and fruiting occurs during the dry season (December to March).

Mexico to Argentina. In Panama, known from tropical moist forest in the Canal Zone, Bocas del Toro, Los Santos, Herrera, Panamá, and Darién, from premontane moist forest in the Canal Zone, and from premontane wet forest in Chiriquí and Panamá.

LONCHOCARPUS Kunth

Lonchocarpus pentaphyllus (Poir.) DC., Prodr. 2:259. 1825

Gallito

Tree, to 30 m tall; trunk to 50 cm dbh; outer bark roughened in age; inner bark granular, often with fine radial reddish lines, the outer margin wavy, the sap with a weakly pungent odor; stems minutely rufous-strigillose, becoming prominently lenticellate in age. Leaves alternate, imparipinnate, to 35 cm long; stipules inconspicuous, caducous; petioles 5–15 cm long; petiolules 5–7 mm

---

KEY TO THE SPECIES OF LONCHOCARPUS

Pedicels golden-sericeous; calyx drying golden-brown; veins on underside of leaflets not prominulous, the trichomes scarcely longer than height of veinlets; fruits glabrous to touch . . . .
. . . . *L. pentaphyllus* (Poir.) DC.

Pedicels white-villous; calyx drying blackish; veins on underside of leaflets prominulous, the trichomes usually several times longer than height of veinlets; fruits conspicuously pubescent to touch . . . . *L. velutinus* Seem.

long; leaflets 5–9(13), ovate to ± elliptic-oblong or oblong-lanceolate, acute or abruptly acuminate at apex, rounded to obtuse at base, 6–14(24) cm long, mostly 4–7 cm wide, ± glabrous above, densely but minutely strigillose below, the trichomes short, appressed. Racemes mostly 8–14 cm long, solitary in axils or several on short, leafless, lateral branches; peduncles and all exposed parts of flower densely but minutely golden-pubescent; flowers usually purple, rarely tan or cream-colored, 7–9 mm long, often paired on short, secondary peduncles; calyx cupular, drying golden; standard somewhat spreading at anthesis, with a green center inside; wings and keel erect; stamens monadelphous, the tube straight, the anterior stamen free at base, the tube flared to either side providing an opening to the nectary; pistil pubescent; style curved sharply inward; stigma remote from anthers; nectar copious. Legumes flattened, elliptic, 5–8 cm long, 1.5–2.5 cm wide, acute to rounded at both ends, apiculate, glabrate, tan, the margin ca 1 mm wide; seeds 1 or 2, central. *Croat 4804, 5227.*

Occasional along the shore; rare in the forest. Flowers in March and April or later in the rainy season. The fruits mature mostly from May to October.

The stems are generally hollow and infested with ants.

Central and South America; West Indies. In Panama, known from tropical moist forest in the Canal Zone, Colón, Veraguas, and Panamá and from tropical wet forest in Colón.

Lonchocarpus velutinus Benth. ex Seem., Bot. Voy. Herald 111. 1853

Gallote

Tree, to 18 m tall; trunk to 25(30) cm dbh; branchlets strigillose. Leaves alternate, imparipinnate, to 30 cm long; stipules inconspicuous, caducous; petioles 2.5–10 cm

long; petiolules ca 6 mm long; leaflets 5–9, ± elliptic-obovate, mostly rounded-acuminate at apex, obtuse at base, 6–13 cm long, 3–7 cm wide, minutely puberulent appearing glabrous above, velutinous below, the trichomes not appressed, pilelike, the veins on the lower surface very conspicuous, anastomosing. Panicles of racemes axillary or terminal, 6–12 cm long; peduncles and all exposed parts of the flower densely pubescent, the trichomes short, appressed, whitish; flowers reddish-purple, ca 6 mm long, solitary or on secondary peduncles of 2 or 3 flowers; calyx flattened-cupular, drying black; petals conspicuously pubescent; standard somewhat spreading at anthesis; wings and keel erect. Legumes flattened, elliptic, at least 5–6 cm long, 1.5–2 cm wide, mostly rounded at apex, acute at base, apiculate, tan, the margins generally ca 1 mm wide, the pubescence dense, minute, golden, strigillose, soft to the touch and with a slight sheen; seed 1, central. *Croat 5121, 7772.*

Occasional, along the shore and in the forest. Flowers in February and March. The fruits mature from March through May. Most of the leaves fall around January, and the new leaves are immature or generally absent when flowering begins.

The stems are generally not hollow.

Central America to Colombia. In Panama, known from tropical moist forest in the Canal Zone, Chiriquí, Veraguas, and Panamá and from premontane moist forest in the Canal Zone.

See Fig. 288.

MACHAERIUM Pers.

The genus is easily distinguished by its stout, recurved, modified stipules, which carry the shrublike lianas into the canopy. The fruits are one-seeded samaras.

KEY TO THE SPECIES OF MACHAERIUM

Blades of leaflets rounded, obtuse, or rarely acute at apex, never acuminate, mostly less than 1 cm wide:
  Leaflets mostly more than 30, the blades mostly less than 12 mm long . . . .
    . . . . *M. microphyllum* (E. Meyer) Standl.
  Leaflets usually fewer than 30, the blades mostly more than 15 mm long:
    Plants rare; secondary veins conspicuously straight and parallel, usually 16–24 per cm; fruits ± glabrous in age . . . . *M. riparium* Brandegee
    Plants common; secondary veins not conspicuously straight and parallel, usually 8–12 per cm; fruits pubescent . . . . *M. milleflorum* Pitt.
Blades of most leaflets acuminate, mostly more than 1 cm wide:
  Largest blades mostly less than 2.5 cm wide, often rufous-pilose on petiole and underside of midrib . . . . *M. seemannii* Seem.
  Largest blades mostly more than 3 cm wide, glabrous or pubescent on petiole and underside of midrib but not rufous-pilose:
    Leaflets 5–7; midrib glabrous, the stems lacking long, stiff setae . . . . *M. arboreum* (Jacq.) J. Vogel
    Leaflets 9–17; midrib pubescent or the stems bearing conspicuous, stiff setae:
      Leaflets 11–17, the underside glabrous, the major lateral veins scarcely more prominent than the reticulate veins; stems bearing conspicuous slender setae and thick, recurving, stipular spines; flowers to 1.5 cm long . . . . *M. kegelii* Meisn.
      Leaflets 9–11, the underside strigillose, the major lateral veins much more prominent than the reticulate veins; stems lacking slender setae but sometimes bearing thick stipular spines; flowers less than 10 mm long . . . . *M. floribundum* Benth.

Machaerium arboreum (Jacq.) J. Vogel, Linnaea 11:182. 1837

Liana or erect to climbing shrub, to 8 m tall, glabrous except on inflorescences; spinescent stipules lacking. Leaves compound; rachis 4–14 cm long; petiolules 2–6 mm long, wrinkled; leaflets 5–7, alternate on rachis, ovate to elliptic, gradually to abruptly acuminate, often long-acuminate, usually obtuse to rounded at base, less frequently acute or slightly cordate at base, 2.5–11 cm long, 1.5–6 cm wide, drying dark brown, often arched along the midrib, the sides folded ± upward, the veins prominulous on lower surface; juvenile leaves much larger, more abruptly and conspicuously long-acuminate than adult leaves, to 17 cm long and 8 cm wide. Panicles 3–9 cm long, several per axil in upper axils, the branches densely ferruginous-tomentose; bracteoles broadly reniform, prominently striate, ciliate, glabrous to inconspicuously pubescent; flowers white, sessile; calyx ca 4.5 mm long, prominently striate, glabrate to inconspicuously pubescent, more so on the lobes, the lobes subequal, the lateral lobe ± deltoid, the carinal tooth much narrower; standard to 9 mm long and 13 mm wide; keel and wings only slightly shorter, the outer surface of all petals conspicuously and densely golden-brown sericeous. Fruits samaroid, 6–8 cm long, 1.5–2 cm wide, becoming glabrous in age, borne on narrow stipes ca 7 mm long, the upper margin thick, ± straight, the lower margin thin, curved; seminiferous area brown, the wing yellow-brown with prominent, dark, reticulate veins. *Croat 7951, Foster 1301.*

Frequent along the shore and in the canopy; juvenile plants have been found in the forest. Probably the most common *Machaerium* in the canopy. Flowers mostly in the late rainy season and the earliest part of the dry season, especially from October to December. The fruits are fully developed by February and are dispersed in the dry season. Leaves are replaced at the end of the dry season.

The species has been confused with *M. darienense* Pitt., which is now considered to be synonymous with *M. striatum* J. R. Johnston and does not occur on BCI.

Mexico to Colombia and Venezuela. In Panama, known only from tropical moist forest in the Canal Zone and Darién.

Machaerium floribundum Benth., J. Proc. Linn. Soc., Bot. 4, Suppl. 68. 1860
M. woodworthii Standl.

Liana; branches with coarse, short, spinescent stipules; branchlets ± glabrous, bearing thick, spinescent stipules 5–10 mm long; outer bark smooth and flaking. Leaves compound; petiole and rachis usually inconspicuously pubescent; petiolules wrinkled, 2–6 mm long; leaflets 9–11, alternate or opposite on rachis, ovate, oblong, or obovate, obtuse to abruptly acuminate, obtuse to rounded at base, 3–10 cm long, 1.5–4.5 cm wide (juveniles to 19 cm long and 8.5 cm wide), glabrous on upper surface, minutely appressed-pubescent on lower surface, the major lateral veins 10–14, much more conspicuous than secondary veins and merging imperceptibly with the enlarged, discolored leaflet margin. Panicles axillary, 4–7 cm long, the branches densely golden-puberulent; bracteoles ca 1.3 mm long; flowers mostly sessile, white with a purple throat; calyx urceolate, 3–4 mm long, the lobes very short, obtuse; standard glabrous, ca 6.5 mm long, abruptly flexed near the middle; keel and wings to ca 5 mm long; stamens monadelphous, the filaments about half the length of the sheath; ovary pubescent, the trichomes dense, brown, appressed. Fruits unknown. *Croat 10093, 17033, Woodworth & Vestal 422* (type of *M. woodworthii* Standl.).

Infrequent, in the forest and along the shore. Seasonal behavior uncertain. Flowers at least in February. The fruits probably mature in the same dry season.

Mexico to Venezuela and Peru. In Panama, known only from tropical moist forest on BCI.

Machaerium kegelii Meisn., Linnaea 11:194. 1837
M. pachyphyllum Pitt.

Liana or arching and scandent tree, to ca 15 m tall; trunk to ca 12 cm diam, the upper part straight, rarely branched, usually arching or pendent from trees; old stems armed with recurved, spiny, woody stipules and conspicuous, stiff, long setae. Leaves compound, clustered near stem apex; stipules triangular, to 2 cm long and 1 cm wide, pubescent, paired, becoming woody in age; petioles to 8 cm long, the pulvinus conspicuous, terete, sparsely pubescent; petiolules stout, ca 5 mm long; leaflets 11–17, mostly oblong, long-acuminate, rounded to truncate or obtuse at base, 4–17 cm long, 2–6 cm wide, usually glabrous or reddish-villous on midrib below, coriaceous, the reticulate veins prominulous, about as conspicuous as the major lateral veins, the margins conspicuously enlarged and discolored. Panicles axillary or terminal, to 90 cm long, usually less than 50 cm long, on short, lateral, usually leafy branches on the terminal 3–4 m of the stem; rachises, peduncles, pedicels, and calyces bearing dense velvety-brown pubescence; flowers to 1.5 cm long; calyx subtended by 2 ovate bracts appressed to the anterior lobes, the carinal tooth acute, exceeding lateral lobes; standard spreading, rufous-pubescent outside, glabrous and lavender inside at least near apex, the center greenish, fringed with violet-purple; wings and keel usually white, weakly held together, the keel petals fused along their outer margin; free stamen united to the staminal tube, all stamens ± equal, the apical fourth free, strongly turned inward, the free part alternating with short erect appendages; style exceeding the anthers; pistil densely appressed-pilose. Samaras oblong, ca 8 cm long and 2 cm wide, distally winged; seminiferous area narrower, ca 2 cm long, rufous-villous, the trichomes interspersed with much stouter trichomes that have dark swollen bases. *Croat 5346, 13984.*

Occasional, in the forest and along the shore. Flowering infrequently, usually in the early dry season (February and March). Immature fruits have been seen in early April; fruits probably mature in April and May. Leaves are deciduous just before flowering.

Easily distinguished by its larger inflorescences and flowers and by the conspicuous setae on the stems. This

Fig. 288. *Lonchocarpus velutinus*

Fig. 290. *Machaerium kegelii*

Fig. 289. *Machaerium kegelii*

species was reported by Standley (1933) and by Johnston (1949) as *M. marginatum* Standl. Although the paratype of that species is referable to *M. kegelii*, the holotype of *M. marginatum* is referable to *M. microphyllum.*

The pollinator is apparently a strong bee, since the flower is strong and the keel is not easily moved.

Costa Rica and Panama. In Panama, known from tropical moist forest in the Canal Zone (both slopes) and Panamá (San José Island), from tropical dry forest in Coclé (Penonomé), and premontane moist forest in Panamá (Panamá City).

See Figs. 289 and 290.

Machaerium microphyllum (E. Meyer) Standl., J. Wash. Acad. Sci. 15:459. 1925

M. isadelphum (E. Meyer) Amsh.; *M. marginatum* Standl.

Slender liana (elsewhere sometimes a tree), armed at the nodes with pairs of stout recurved spines; outer bark usually smooth. Leaves compound, 8–15 cm long; stipules sharply triangular, becoming brown in age; rachis ferruginous-pubescent; leaflets 40–60, subsessile, oblong, rounded on both ends, 6–12(23) mm long, 2–6 mm wide, glabrous or pubescent, folding tightly under water stress. Panicles terminal, rarely axillary, 10–30 cm long with 1–3 short, densely rufous-pubescent racemes at each node, each raceme subtended by a recurved spine; flowers purple or light purple, usually less than 10 mm long; calyx maroon, 4–5 mm long, tubular, subtended by 2 rounded bracts; standard usually whitish outside, lavender inside, recurved, rounded and emarginate; wings and keel recurved, the keel petals fused only in apical third; ovary pubescent. Samaras borne on slender stipes, 5 cm long and 1 cm wide, strongly curved above the flat seminiferous area. *Croat 12704, 13969.*

Common on the shore and in the canopy of the forest. Flowers from October to January. The fruits mature from December to April.

The leafless stems can be recognized on the forest floor by the pairs of large, straight or slightly recurved spines, which are stout, sharp, and 1–2 cm long.

Panama to northern South America. In Panama, known from premontane dry forest in the Canal Zone and Panamá, from tropical dry forest in Herrera, and from tropical moist forest in the Canal Zone, Colón, San Blas, Chiriquí, Veraguas, Panamá, and Darién.

Machaerium milleflorum Pitt., Contr. U.S. Natl. Herb. 20:119. 1918

M. purpurascens Pitt.

Vine or liana in the canopy; trunk to 16 cm diam, unarmed but often with prominent horizontal ridges (old spine bases), the slash with a fairly strong foul odor, ultimately producing sparse red sap; small spiny branches extended over the forest floor for great distances, then climbing into the canopy by means of stipules, the stipules stout, paired, recurved, spinescent, to 6 mm long; at least the younger stems pubescent. Leaves compound, usually 3–19 cm long; petioles 1–2 cm long, densely golden-pilose marginally; leaflets 10–70, oblong, rounded

to acute at apex, rounded at base, 0.7–2.5(4) cm long, (3)5–10(20) mm wide, bicolorous, ± glabrous or sparsely pubescent above, soft- and often appressed-pubescent at least on midrib below; juvenile leaves similar to adult leaves in shape but to 9.5 cm long and 3 cm wide, sparsely to densely pubescent. Panicles densely floriferous, 20–70 cm long, the branches densely ferruginous-tomentose; calyx violet-purple, ca 3.3 mm long, glabrous, the teeth mostly acute; corolla violet-purple; standard 4–5 mm long, densely appressed-pubescent outside, emarginate at apex, strongly reflexed at anthesis; wings shorter than standard, held erect, the keel petals fused in the distal half; stamens fused in 2 groups of 5 each, turned inward in apical third, slightly shorter than the style; style similarly bent. Samaras reddish, in dense purplish clusters, stipitate for ca 4 mm, 3.5–5.5 cm long, ca 1 cm wide, narrowed above the seminiferous area; seminiferous area ca 1 cm long, more densely golden-villous than the wing. *Croat 8371, 13813.*

Common along the shore, but probably flowering in the forest also since juvenile plants have been seen there. Flowering generally during February, rarely as late as March or April. Fruits are dispersed from March to May.

Johnston (1949) reported that the species has its active period of vegetative growth in the early dry season just before flowering.

Costa Rica to Colombia. In Panama, known from tropical moist forest on both slopes of the Canal Zone and in Colón, Panamá, and Darién; known also from tropical dry forest in Coclé and Panamá and from premontane moist forest in the Canal Zone and Panamá.

Machaerium riparium Brandegee, Univ. Calif. Publ. Bot. 6:500. 1919

Suberect shrub or usually a liana, to ca 12 m long; branchlets densely ferruginous to glabrate; stipular spines retrorsely curved, 5–7 mm long, the bases to 3 mm wide. Leaves compound, 10–20 cm long; petioles ca 1 cm long; leaflets in 10–15(20) pairs, ± oblong, rounded to rarely retuse at apex, mucronate, rounded at base, 1–4 cm long, 0.8–2 cm wide, strigose to glabrous, paler below than above, the secondary veins many, fine, parallel. Panicles terminal, densely flowered; bracts acute, to 6 mm long; pedicels ca 2 mm long; calyx cup-shaped, 3–4 mm long, puberulent on and near the shallow lobes, drying dark; corolla purple to pale violet, the standard 7–8 mm long; exposed parts of petals sericeous. Legumes stipitate for ca 3 mm, samaroid, 3.5–4.5(5.8) cm long and 0.8–1.4(1.8) cm wide, glabrous in age, rounded at apex; seminiferous area 1.6–2 cm long, the wing venation conspicuously reticulate. *Montgomery 128, Woodworth & Vestal 327.*

Apparently rare on the island; known from two collections. Seasonal behavior unknown.

The species is similar to *M. milleflorum,* but may be distinguished by the fine, parallel, secondary veins of the leaflets.

Mexico, Belize, and Panama. In Panama, known only from BCI.

Fig. 291. *Mucuna mutisiana*

Fig. 292. *Mucuna mutisiana,*
fruits only (the leaves are
of *Tetracera* sp., 88. Dilleniaceae)

Fig. 293.
Ormosia coccinea
var. *subsimplex*

Machaerium seemannii Benth. ex Seem., Bot. Voy. Herald 110. 1853

Liana or less often a climbing shrub, to 4 m or more long; mature branchlets glabrous, lacking spinescent stipules; younger plants with some branches leafless and with well-developed spinescent stipules. Leaves compound; rachis 5–9 (18) cm long; rachis, petiolule, and underside of blade sparsely ferruginous-pilose; petiolules 1–3 mm long; leaflets 6–13, oblong-lanceolate, ovate or narrowly elliptic, gradually acuminate and often falcate at apex, obtuse to rounded at base (rarely acute), 1.5–7.5 cm long, 0.7–2.7 cm wide, thinly coriaceous. Panicles 1–5 cm long, congested in the upper axils; branches densely ferruginous-tomentose; bracteoles reniform, densely appressed-pubescent; flowers purple, sessile; calyx 3–4 mm long, densely appressed-pubescent, the lobes inconspicuous; standard to 8 mm long; keel and wings only slightly shorter, the outer parts of all petals golden-brown-sericeous. Samaras borne on narrow stipes (to 7 mm long), 6–9.5 cm long, 1.8–2.5 cm wide, very densely golden-puberulent all over, the carinal margin conspicuously thickened to 1.5 mm thick, straight to curved. *Croat 6049, 7983.*

Frequent along the shore and in the forest to the top of the canopy; juvenile plants have been seen along trails. Flowering mostly during September and October but sometimes much later and in the early dry season. The fruits mature by January or February and are dispersed in the dry season (to April).

Guatemala to Colombia. In Panama, known principally from tropical moist forest in the Canal Zone (principally on the Pacific slope) and in Colón, Chiriquí (Puerto Armuelles), Panamá, and Darién; known also from premontane wet forest in Chiriquí (Boquete) and Coclé and from tropical wet forest in Colón (Salúd).

MUCUNA Adans.

Mucuna mutisiana (H.B.K.) DC., Prodr. 2:406. 1825
Ojo de venado

Large herbaceous vine, climbing to 15 m high in trees, somewhat twining. Leaves trifoliolate, usually drying black; stipules linear-lanceolate, to 5 mm long; petioles 4–9 cm long; petiolules dark, slightly pulvinate, articulate, subtended by stipels at base; leaflets ovate to elliptic, acuminate, oblique at base, 7–16 cm long, 4–7 cm wide, often drying dark, the veins at base 3 or 4. Inflorescences densely tomentose, in many, umbellate clusters of usually 3 flowers each at the end of pendent peduncles to 2 (5) m long, each umbel subtended by deciduous ovate bracts, their nodose bases persisting on the peduncle; pedicels spreading, to 5 cm long, curved down at apex, the flow-ers pendent; calyx cup-shaped, 11–13 mm long, dark-ferruginous, the trichomes numerous, slender, irritating, ca 2 mm long, the lobes inconspicuous except for the carinal tooth; petals inconspicuously colored, violet to flesh-colored; standard rounded, emarginate at apex, to 4 cm long; keel greenish, the wings ca 5 cm long and 1.5 cm wide, twisted, clawed and markedly auriculate at base; stamens 10, 9 fused into a tube except near apex, the free stamen and 4 alternate stamens of the staminal tube bearing shorter anthers, these anthers perpendicular to filament, cordate at base, attached medially, with one end bearing a conspicuous loose tuft of brown trichomes, the other anthers erect, glabrous; anthers shedding pollen in bud; style longer than anthers in bud, pushing through point of keel before keel is fully opened. Legumes oblong, to 13 cm long and 6 cm wide, constricted between seeds, densely covered with sharp, reddish-brown trichomes and numerous, narrow, transverse ridges to 1.5 cm high; seeds usually 1 or 2, round, flattened, 2.5–4 cm wide, dark brown, with a conspicuous flat scar three-fourths of the way around their circumference. *Croat 7426, 12611.*

Uncommon, hanging over the edge of the lake along the shore, perhaps also in the forest. More abundant in the Canal Zone along roads. Flowers from October to January, especially in November. The fruits mature from January to April, but seeds may hang on the pods until June. Plants have been seen putting on new leaves in August.

The nectary is functional prior to the opening of the flower. Bats, as pollinators (R. Foster, pers. comm.), probably gain entrance by pushing the banner back and forcing open the auricles of the wings. The wing petals are fused to the keel petals at the base and consequently are pulled apart from the base to the apex. The style and staminal tube then spring forward and strike the pollinator's abdomen.

Specimens of this species (*Wetmore & Abbe 53* and *Shattuck 53*) were misidentified as *M. bracteata* Dwyer, which may not be a distinct species. *M mutisiana* has also been confused with *M. urens* (L.) Medic., which also may not be specifically distinct.

Panama to Venezuela. In Panama, known principally from tropical moist forest in the Canal Zone, San Blas, Panamá (San José Island), and Darién; known also from premontane moist forest in the Canal Zone.

See Figs. 291 and 292.

Mucuna rostrata Benth. in Mart., Fl. Brasil. 15 (1):171, pl. 47. 1859
Pica-pica

Liana; branchlets glabrous, drying dark. Leaves trifoliolate; petioles 5–10 cm long, pulvinate about 1 cm at base;

KEY TO THE SPECIES OF MUCUNA

Flowers flesh-colored, less than 5 cm long; peduncles to 2 m or more long; legumes with conspicuous, raised, transverse lamellae . . . . *M. mutisiana* (H.B.K.) DC.
Flowers yellowish, usually more than 5 cm long; peduncles less than 1 m long; legumes lacking ridges . . . . *M. rostrata* Benth.

leaflets rhombic-ovate, abruptly acuminate, broadly rounded at base, 10–14 cm long, 6–9 cm wide, the veins at base 3–5, sparsely strigose below. Racemes pendent; peduncles 8–18 cm long; flowers yellow; calyx 5-lobed ca half its length, 2–2.5 cm long, densely pubescent, often also with many, appressed, slender, irritating trichomes; corolla 6–8(9) cm long; standard orbicular, 4–5 cm long, coriaceous, glabrous, the claw 5–7 mm long; wings 7–8(9) cm long, ca 16 mm wide, basally auriculate, the keel as long as the wings; stamens diadelphous. Legumes oblong, at least 7 cm long and 2–3.5 cm wide, curved, beaked, bearing erect stinging trichomes, lacking ridges or crests; seeds several, orbicular, ca 2–5 cm diam, somewhat compressed, black. *Shattuck 489.*

Collected once by Shattuck, but not seen in recent years and possibly no longer present on the island. Flowers in the early dry season, usually in December but as late as March. The fruits probably mature late in the dry season.

Belize to the Amazon basin. In Panama, known only from tropical moist forest on BCI and in Darién.

MYROXYLON L.f.

Myroxylon balsamum (L.) Harms var. **pereirae** (Royle) Harms, Notizbl. Königl. Bot. Gart. Berlin 5:95. 1908
Balsam, Balsam of Peru

Tree, to 35(40) m tall and ca 1 m dbh; trunk straight; branches ascending; outer bark smooth, bearing abundant lenticels; heartwood reddish-brown; sapwood pale, the freshly cut bark emitting a pungent, obnoxious odor. Leaves alternate, generally imparipinnate, fragrant when crushed; petioles 1–4 cm long; rachis and petiolules pubescent, terete; rachis 5–15 cm long; leaflets 5–10, alternate on rachis, lanceolate to elliptic, acute to acuminate, obtuse at base, 3–11 cm long, 1.8–4 cm wide. Racemes axillary, to 20 cm long, closely cinereous-tomentose; pedicels 1–1.5 cm long; flowers oblique on pedicel; calyx campanulate, 3.5–4.5(6) mm long, with obvious, fine ribs, the lobes ca 1.5 mm long; standard orbicular, ca 9 mm diam, cordate basally, the claw ca 1 mm long; wings elliptic to narrowly spatulate, ca 1 cm long, ca 4 mm wide, their claws 2 mm long, the keel subelliptic, ca 8.5 mm long and 3 mm wide, its claw ca 1.5 mm long; ovary sparsely villous, borne on stipe 2 mm long; style subulate. Samaras narrowly obovate, to 11 cm long, glabrous, on

stipes to 1 cm long; wing to 8 cm long, 1–2 cm wide, the seminiferous area turgid, at the apex of the fruit, obliquely oblong, 2–3 cm wide, ca 1 cm thick, extending as a low, longitudinal ridge into the pedicel; appearance of fruit much like that of *Platypodium elegans. Ebinger 215.*

Apparently rare; collected once on the island. Flowers from January to June. Fruits from September to March.

Southern Mexico to the Amazon basin of Brazil and Peru; planted as a street tree in Ceylon and in tropical botanical gardens. In Panama, known only from tropical moist forest in the Canal Zone, Panamá, and Darién. Characteristic of tropical moist forest in Panama (Tosi, 1971). Holdridge (1970) reported it for low and medium elevations in moist and wet areas with a dry season.

ORMOSIA G. Jackson

Ormosia coccinea (Aubl.) G. Jackson var. **subsimplex** (Spruce ex Benth.) Rudd, Contr. U.S. Natl. Herb. 32:328–29. 1955
Alcornoque, Pernillo de monte

Tree, to at least 25 m; trunk commonly 60(100) cm dbh; outer bark thin, hard, with a ± reticulate pattern of very shallow fissures, flaking easily; inner bark granular, its outer surface reddish with large whitish spots; stems weakly flexuous, with ribs extending downward below each petiole; stems, petioles, and midribs densely pubescent, the trichomes matted, short, brown. Leaves alternate, imparipinnate, 25–40 cm long; petioles 6–10 cm long, the basal pulvinus prominent; petiolules ca 6 mm long; leaflets usually 9, oblong, abruptly acuminate and downturned at apex, obtuse to slightly cordate at base, 7–18 cm long, 4–7.5 cm wide, coriaceous, glabrous above, sparsely appressed-pubescent and somewhat viscid below, more densely so on veins, the margins turned somewhat upward and wavy in age; secondary veins in 10–16 pairs, prominently raised on lower surface. Inflorescences cymose-paniculate, terminal, to 30 cm long; flowers purple; calyx tomentulose, 7–9 mm long, 5-lobed to half its length; corolla 1–1.5 cm long, rose-purple. Legumes flattened, oblong to obovate, 3.5–5 cm long, 2–2.5 cm wide, glabrous, abruptly narrowed at apex, the valves reddish-brown at maturity, opening to expose seeds; seed 1 (rarely 2), red and black, somewhat flattened, ca 1 cm long. *Croat 5120.*

Occasional, in the forest and along the shore. Flowers from June to August. The fruits probably mature in 4–6

KEY TO THE SPECIES OF ORMOSIA

Underside of leaflet midribs conspicuously brown-tomentose; leaf bases rounded to slightly cordate; seeds bicolorous, red and black; pods obliquely obovate to elliptic, lacking an alate margin . . . . *O. coccinea* (Aubl.) G. Jackson var. *subsimplex* (Benth.) Rudd
Underside of leaflet midribs glabrous or essentially so, yellow-green:
  Leaflets acuminate, very inconspicuously golden-pubescent below; mature legumes suborbicular, light brown, ± winged along margins . . . . *O. panamensis* Seem.
  Leaflets very abruptly acuminate to rounded at apex, glabrous on both surfaces; mature legumes much longer than wide, dark, with a thick marginal suture and a submarginal ridge . . . . . .
  . . . . *O. macrocalyx* Ducke

weeks, but may persist with the colorful seeds exposed until about flowering time of the next year. Plants are deciduous before flowering.

Panama to Brazil. In Panama, known only from tropical moist forest on BCI.

See Fig. 293.

Ormosia macrocalyx Ducke, Arch. Jard. Bot. Rio de Janeiro 3:137. 1922

Tree, to 30(40) m tall, ca 50 cm dbh; outer bark planar, ca 1 cm thick, minutely fissured, the intervening areas with thin, smooth, brown strips; inner bark granular, tan; branchlets minutely strigillose. Leaves alternate, imparipinnate, 15–40 cm long; petioles 4.5–9 cm long, conspicuously swollen at base, glabrous to minutely pubescent; petiolules pulvinate, 5–8 mm long; leaflets usually 9, elliptic-oblong to ovate or obovate, abruptly acuminate to rounded at apex, acute at base, 5.5–15 cm long, 3.5–10 cm wide, ± glabrous; secondary veins in 4 or 5 pairs. Inflorescences terminal, paniculate, to 30 cm long; flowers purplish; calyx 8–15 mm long, grayish-pubescent, 5-lobed; petals ca 2 cm long; standard minutely biappendiculate at base. Legumes ± circular to oblong, flattened, to ca 10 cm long and ca 3 cm wide, the lateral margins straight or weakly constricted between the seeds, the free margin with a submarginal ridge, the surface glabrous, dull, depressed between seeds; seeds 1–4, solid red, ± ovoid, ca 1 cm long. *Croat 13217.*

Apparently rare; known only from the old forest south of Armour Trail above the escarpment and from near the tower at Barbour Trail 700. Probably flowers in the middle of the rainy season. The fruits persist most of the year.

Southern Mexico to Brazil; at elevations under 100 m. In Panama, known from tropical moist forest in the Canal Zone, Bocas del Toro, and Veraguas and from premontane wet forest in Chiriquí (Progreso).

Ormosia panamensis Benth. ex Seem., Bot. Voy. Herald 111. 1853

Alcornoque, Sur espino

Tree, to 8(30) m tall; young stems golden-sericeous. Leaves alternate, compound, 9–17 cm long; petioles 5–8 cm long, pulvinate at base; petiolules pulvinate, ca 5 mm long, golden-sericeous; leaflets 5–9, oblong-lanceolate, gradually long-acuminate at apex, obtuse at base, 10–14 cm long, 3–4.5 cm wide, inconspicuously golden-puberulent below, the secondary veins in 10–16 pairs, ± straight, 5–20 mm apart. Flowers lilac; calyx densely fulvous-sericeous, 8–11 mm long, 5-lobed; corolla nearly 2 cm long. Legumes suborbicular when mature, flattened,

ca 4 cm diam, not constricted between seeds, the valves thick with an alate margin 5–15 mm wide; seeds 1–4, separated by septa, ca 1.5 cm long, dark red. *Foster 1128.*

Collected only once along the shore of a cove by Peña Blanca Point; seen also south of Armour Trail 1200 by R. Foster. Seasonal behavior uncertain. Some flowers have been seen from March to June, with the fruits maturing from August to December. Trees have been seen in July with both full-sized fruits and very juvenile fruits, indicating that individuals may flower more than once per year.

Guatemala, Costa Rica, and Panama. In Panama, known from tropical moist forest on BCI and in Bocas del Toro and Chiriquí and from Cerro Pajita in Coclé, which is possibly tropical wet forest.

PHASEOLUS L.

Phaseolus peduncularis H.B.K., Nov. Gen . & Sp. 6:447. 1824

Slender, herbaceous vine; stems twining or prostrate, and puberulent, pilose, or glabrate. Leaves trifoliolate; stipules lanceolate, to 6 mm long; petioles 2–7 cm long; rachis 5–10 mm long; leaflets ovate-deltoid, acuminate or acute, truncate at base, thin, sparsely strigillose, the terminal leaflet 2.5–8.5 cm long, 2–6 cm wide, the lateral leaflets smaller, the basal veins 3. Inflorescences fasciculate-racemose, often crowded at apex of peduncle; peduncles equaling or surpassing leaves; rachises 2–5 cm long; pedicels to 2 mm long; flowers pale violet, green in bud; bracteoles shorter than calyx; calyx ca 3 mm long, closely subtended by acute bracts to one-third the length of the calyx; petals white to pale blue on outer margins; standard 1–1.5 cm long, about as broad as long; keel ciliate, closely enveloping style and stamens, turning sideways after anthesis and spreading the outer margins of the standard apart; stamens diadelphous; filaments slender, free in apical half; style bearing sharp erect trichomes near apex. Legumes 4–5 cm long, to 3 mm wide, acuminate at apex, strigose, with a minute marginal rib on both sides of each valve, the valves twisting at maturity; seeds subcylindrical, greenish-white with purple mottling, to 3 mm long, leaning out from the dehisced valves. *Croat 5256, 7032.*

Common in clearings, usually growing among other weedy plants. Flowering and fruiting year-round with a peak of flowering during the middle of the rainy season (August to October). The fruits develop quickly.

Normal movements of the flower parts do not appear to force much pollen from the open end of the carinal tube.

Guatemala to northern South America; West Indies.

KEY TO THE SPECIES OF PHASEOLUS

Stipules with prominent basal lobe; bracteoles longer than calyx; petals yellow; legumes 6–7 mm wide, pilose . . . . *P. trichocarpus* C. Wright
Stipules lacking basal lobe; bracteoles shorter than calyx; petals white or pale lavender; legumes ca 3 mm wide, strigose. . . . . *P. peduncularis* H.B.K.

In Panama, known from tropical moist forest in the Canal Zone, Bocas del Toro, San Blas, Coclé, Panamá, and Darién and from tropical wet forest in Darién.

Phaseolus trichocarpus C. Wright in Sauv., Anales Acad. Ci. Méd. Habana 5:337. 1868

Slender, twining, herbaceous vine; younger stems orange-pilose, the trichomes often retrorse. Leaves trifoliolate; stipules persistent, lanceolate, to 9 mm long, appendaged at base, the appendage to 4 mm long, usually oblong and blunt, sometimes divided and narrowly acute; petioles 2–5 cm long, orange-pilose; rachis 5–10 mm long; leaflets oblong-lanceolate, acute at apex, rounded-truncate at base, 3.5–12 cm long, 0.8–3 cm wide, ciliate, essentially glabrous above but the veins strigose, strigose below. Racemes bearing 1–3 pairs of opposite pedicels per inflorescence, each raceme subtended by a peduncle 8–14 cm long; pedicels becoming spiraled, to 5 mm long; base of the flowering rachis (just above the lowermost pair of flowers) with conspicuous glandular area to 3 mm long; flowers few, subtended by 2 narrow bracts to 5 mm long, these longer than calyx; calyx ca 2.5 mm long, lobed, the lobes 5, shallow, subequal; corolla yellow, to 1 cm long and wide; standard cordate, strongly cucullate, 6 mm long and 10 mm wide; wings broadly obovate, 7 mm long and 6 mm wide, the keel spiraled; stamens diadelphous; anthers 3.2 mm long, 1.5 mm wide, wider at base; ovary pubescent; style 8 mm long; stigma recurved, ca 7.5 mm long, pubescent along one edge. Legumes oblong, 2.8–3.2 cm long, 6–7 mm wide, only slightly inflated, short-beaked, turning black at maturity, long-yellow-pilose; seeds 7–9, ovoid, to 3 mm long and 2 mm wide, black with a prominent white hilum. *Croat 5717, 8250.*

Known from the shoreline on the south and west sides of the island. Probably flowers and fruits throughout the year, chiefly from May to December. The fruits apparently develop quickly.

This species has been mistaken for *Vigna unguiculata* (L.) Walp., which also has appendaged stipules. The *Vigna* is distinguished by having leaves at most lanceolate, not oblong, and pods 10 cm or more long, not pilose, with lighter seeds and a dark hilum.

Aviles 909, identified by Standley as *Vigna repens* (L.) O. Kuntze (= *V. luteola* (Jacq.) Benth.), is an aberrant specimen of this species. It differs only in having the produced part of the stipule bilobed and acute.

Panama to the Guianas; Greater Antilles. In Panama, known only from tropical moist forest in the Canal Zone, Chiriquí, Panamá, and Darién.

PLATYMISCIUM J. Vogel

Platymiscium pinnatum (Jacq.) Dug., Contr. Hist. Nat. Colomb. 1:11. 1938

P. polystachyum Benth. ex Seem.

Quirá, Swamp kaway, Sangrillo, Panama redwood

Tree, to 30 m tall and 40 cm dbh; outer bark light brown, fissured, ± flaky, soft, ringed (like *Tabebuia guayacan*, 25. Bignoniaceae); wood reddish-brown, hard. Leaves opposite, imparipinnate; stipules interpetiolar, lanceolate, to 1.2 cm long, deciduous, leaving distinct scars; leaflets in 4–7 pairs, opposite on the rachis, ovate to elliptic, acuminate, obtuse to rounded at base, 5–11 (22) cm long, 2.5–4.5(15) cm wide, glabrous, entire and sometimes undulate. Racemes or panicles 9–17 cm long, terminal, subtending new leaves or in lower leafless axils; pedicels 3–6 mm long, glabrous; bracteoles ovate; flowers 1–1.5 cm long, often paired at nodes, yellow-orange tinged with purplish-brown toward center, especially on standard; calyx narrowly campanulate, glabrous to warty-puberulent, with short, acute teeth, the lateral and carinal teeth somewhat shorter than the vexillar teeth; standard orbicular, emarginate at apex; keel petals fused at apex and on outer edge, ciliate below point of fusion; stamens ca 1 cm long, in 2 fascicles of 5 each, loosely fused to form an open tube; anthers reniform, about as broad as long; ovary and style glabrous, the style bent sharply inward, much longer than anthers. Fruits samaroid, oblong-elliptic, obtuse to rounded at apex, acute at base, flat, 5–11 cm long, to 3.5 cm wide, thin, on stipes ca 1 cm long, the winged portion of same thickness throughout; seminiferous area medial, to 3.5 cm long, 1 cm wide; seed 1. *Croat 5218, 5652, Standley 40945.*

Occasional, in the forest. Leaves fall shortly before the flowers appear, and new leaves are produced with the flowers from March to May (rarely earlier). The species may flower only every other year. The fruits are dispersed one year later in the dry season (R. Foster, pers. comm.).

From a distance the flowers and new leaves appear somewhat brown.

Guatemala to Panama on the Pacific slope and along the Atlantic slope from Colombia to Trinidad. In Panama, known from tropical moist forest on both slopes in the Canal Zone and on the Pacific slope in Chiriquí and from the Azuero Peninsula to Darién. Characteristic in Panama of tropical dry forest (Holdridge & Budowski, 1956) and tropical moist forest (Tosi, 1971). Reported from tropical wet forest in Costa Rica (Holdridge et al., 1971).

PLATYPODIUM J. Vogel

Platypodium elegans J. Vogel, Linnaea 11:420. 1837

P. maxonianum Pitt.

Carcuera, Costilla, Tigre, Canalua

Tree, to 30 m or more, usually 75–100 cm dbh; trunk with conspicuous, deep, irregular, longitudinal invaginations in age; bark soft, minutely fissured, the sap becoming reddish-brown in time, with a foul odor at least in younger parts. Leaves alternate, imparipinnate, to 25 cm long; petiole, rachis, petiolule, and underside of blade puberulent; leaflets 10–20, oblong, rounded to emarginate at apex, obtuse to rounded at base, 2.5–7.5 cm long, 1–3 cm wide, glabrous above. Inflorescences lax racemes from upper axils; pedicels 8–12 mm long, bracteate near apex; bracts oblong, ca 2 mm long, persistent; calyx ca 4 mm long, turbinate, sparsely appressed-pubescent, the carinal tooth acute, equaling lateral teeth, the vexillar teeth ±

united; corolla yellow-orange; petals clawed; standard reniform, ca 1.8 cm long; keel petals fused only near apex; stamens ca 1 cm long, 2 stamens free, the others in clusters of 4 each; stigma about as long as anthers. Samaras obliquely oblong, to 13 cm long and 3 cm wide, on stipes ca 2 cm long; seminiferous area distal. *Croat 5411.*

Common. Flowers synchronously every two years from April to June (sometimes as early as March), apparently initiated by the beginning of the rainy season. The fruits may be full size by June but mature over a long period, ripening by December or January. Leaves are replaced in the dry season before the time of flowering. A few individuals are out of phase (R. Foster, pers. comm.).

Panama to Venezuela, and in southeastern Brazil and Paraguay. In Panama, known from tropical moist forest in the Canal Zone, Chiriquí, Panamá, and Darién.

PTEROCARPUS Jacq.

Pterocarpus officinalis Jacq., Select. Stirp. Am. 283, f. 92. 1763
Bloodwood, Dragon blood tree, Sangre de drago, Chuella, Suela

Tree, to 20(30) m tall; trunk to 30(90) cm dbh, often widely buttressed; wood yellow or whitish, the sap copious, red; trichomes on younger parts and inflorescence axes minute, appressed-ferruginous, otherwise glabrous. Leaves alternate; petioles 2.5–11 cm long; petiolules 4–6 mm long; leaflets 5–9, oblong to ovate-oblong, acuminate, rounded to truncate at base, 3.5–17 cm long, 2.5–7.5 cm wide, the terminal leaflet largest. Panicles terminal or upper-axillary, to ca 25 cm long; calyces and pedicels reddish-brown; pedicels to 4 mm long; calyx ca 6 mm long, markedly toothed, glabrous except on the upper margin; standard obovate, ca 1 cm long, rounded at apex, yellow-orange, with a red, inverted V-shaped area in the middle, the lateral margins folding back along these lines; wings shorter than the standard, exceeding keel, the keel petals weakly connate on outer upper margins; stamens weakly connate, often free above the middle; ovary glabrous or bearing a few trichomes on one edge. Fruits indehiscent, stipitate for ca 3 mm, suborbicular, ± flattened; wing asymmetrical, not quite oblong on 1-seeded fruits, to ca 5 cm long on 2-seeded ones, dark brown or black when mature, glabrous; seminiferous area ca 1 cm thick, the margins thick and hard, distinctly veined, corky; seeds 1 or 2, to ca 3 cm long. *Knight 1043.*

Rare. Flowers in central Panama from April to October, but mostly in June and July. The fruits mature from August to October (sometimes to December). Leaves fall in the early dry season.

Although the Knight collection cited here is sterile, it is almost certainly this species. Knight reported it as common, but most of the specimens he had were *P. rohrii*. *Bangham 502*, reported as this species in the *Flora of Panama*, is really *P. rohrii* also. R. Foster (pers. comm.) has seen fruits of *P. officinalis* along the shore. The description given here is based on material from nearby areas of the Canal Zone.

On the Atlantic slope from Belize to the mouth of the Amazon and on the Pacific slope from Panama to Ecuador; West Indies. In Panama, known from tropical moist forest in the Canal Zone, Bocas del Toro, Panamá, and Darién and from premontane wet forest in Colón. Typical of tropical moist forest in Panama (Tosi, 1971) and occasionally forming pure stands next to mangroves (Holdridge & Budowski, 1956).

Pterocarpus rohrii Vahl, Symb. Bot. 2:79. 1791
P. hayesii Hemsl.
Bloodwood, Sangre de gallo

Tree, to 50 m tall; trunk 30–100 cm dbh, buttressed; outer bark whitish, thin, often loose; inner bark moderately thick, with numerous red streaks exuding minute sap droplets, the streaks forming concentric rings especially near wood, the sap copious, red; branchlets glabrous to densely pubescent. Leaves alternate; stipules caducous, linear to ± falcate, 3–10 mm long; petioles 2.5–4.5 cm long, pulvinate at base; rachis glabrous to densely ferruginous-pubescent, 3–12(20) cm long; petiolules 3–6(9) mm long; leaflets (3)5–9(12), ovate to elliptic, bluntly acuminate, acute to rounded or rarely slightly cordate at base, 3.5–10(13) cm long, 2–4(5.5) cm wide, glabrous above, usually with short, appressed pubescence below, the midrib sometimes arched with the sides folded somewhat upward. Racemes usually axillary, to 12 cm long, shorter than leaves; calyces and pedicels densely pubescent, the trichomes short, golden-brown, appressed; pedicels 5–7(10) mm long; flowers pinkish-orange, ca 1.5 cm long; calyx to 7(11) mm long, oblique, narrowly campanulate, the teeth blunt, 1–2 mm long, the vexillar teeth larger; standard ca 1.5 cm long, emarginate, violet-purple at the center, clawed, the claw ca 4.5 mm long; keel and wings slightly shorter than standard, both marked with violet-purple, the keel petals weakly connate on their outer edge; stamens diadelphous, those of the tube weakly connate; pollen golden; ovary appressed-pilose, uniformly oblong; style longer than stamens. Fruits round to oval,

---

KEY TO THE SPECIES OF PTEROCARPUS

Inflorescences paniculate, branched many times; fruits stipitate, much thickened medially, the wing narrow; underside of leaves glabrous at least in age; leaves and branchlets usually drying blackened . . . . *P. officinalis* Jacq.
Inflorescences racemose or, rarely, branched only at base; fruits ± sessile, scarcely thickened medially, the wing broad; underside of leaves bearing sparse, ± appressed, short, inconspicuous, brownish trichomes; leaves and branchlets drying green . . . . *P. rohrii* Vahl

flat and thin, mostly 5.5–7 cm long and 5–6 cm wide, usually unequal at the base, minutely pubescent, yellowish- to rusty-brown, indehiscent; seminiferous area central, ca 3 cm long, slightly raised, the margins very thin, reticulate, often minutely wrinkled; seed 1. *Croat 14866, 16623, 16627.*

Frequent in the forest, possibly less abundant on the western side of the island. Flowers every other year (R. Foster, pers. comm), usually in May and June, less frequently earlier in the dry season. The fruits mature from August to November.

Southern Mexico to the upper reaches of the Amazon, as far south as Bolivia, and in the lowlands along the Atlantic as far south as southern Brazil. In Panama, known from tropical moist forest in the Canal Zone, Bocas del Toro, Panamá, and Darién.

RHYNCHOSIA Lour.

Rhynchosia pyramidalis (Lam.) Urban, Feddes Repert. 15:318. 1918
R. phaseoloides DC.

Liana; older stems flattened and several cm broad, the larger with 3 ribs below each petiole, the lateral ribs becoming prominent and corky in age, the smaller stems puberulent; most vegetative parts bearing raised, orange glands. Leaves trifoliolate, to 15 cm long; petioles 2.5–5 cm long, swollen at base, hispidulous; rachis with a raised margin above; leaflets ovate to rhombic-ovate, acuminate at apex, obtuse to rounded at base, mostly 4–16 cm long, 3.5–11.5 cm wide, glabrous to sparsely pubescent especially on veins below, the lateral leaflets inequilateral, the veins at base palmate. Racemes apical, to ca 15 cm long; flowers dense, yellowish-green, short-pedicellate; calyx to 2 mm long, puberulent and glandular-dotted; standard obovate, clawed, ca 6 mm long, purple-striate, puberulent and glandular-dotted outside; wings narrow, puberulent outside near apex, fused to keel near base, the keel petals fused in apical half, weakly enclosing stamens; stamens diadelphous; style bearing antrorse trichomes near apex; stigma borne among the anthers. Legumes ca 1.5 cm long and 1 cm wide, constricted between the seeds, puberulent; seeds 2, black and red. *Croat 13817.*

Occasional, in all parts of the forest; possibly extending into the canopy. Though some flowers have been seen along the edge of the lake, they are rarely seen in the forest. Flowers from January to March. The fruits develop soon, but seeds may persist throughout the rainy season.

Baja California, Sonora, and southern Mexico south to Peru, and the Guianas to eastern Brazil and Argentina;

West Indies. In Panama, known from tropical moist forest in the Canal Zone, Bocas del Toro, Colón, Coclé, and Darién.

TERAMNUS P. Browne

Teramnus uncinatus (L.) Sw., Prodr. Veg. Ind. Occ. 105. 1788
Glycine uncinata (L.) J. F. Macbr.

Vine; most parts sericeous. Leaves trifoliolate; stipules ovate, ca 3 mm long; petioles 3–6 cm long; petiolules pulvinate, articulate; leaflets ovate to oblong-lanceolate, 5–8 cm long, acute to obtuse at apex, rounded or slightly subcordate at base, green and densely strigose above, pale and densely sericeous below. Racemes axillary, slender, bearing few flowers; calyx pilose, 5–6 mm long, the 5 lobes narrow; corolla purplish or whitish, only slightly exceeding the calyx. Legumes flattened, 4–7 cm long, 3–4 mm wide, densely long-villous, beaked; seeds numerous, oblong, ca 3 mm long, shiny, brown. *Croat 13151.*

Collected at the edge of the forest, in shade near the Animal House in the Laboratory Clearing. Flowers and fruits in the early dry season, from December to February.

Southern Mexico to South America; West Indies; introduced into tropical Africa. In Panama, known from tropical moist forest in the Canal Zone, Los Santos, Panamá, and Darién, from premontane wet forest in Chiriquí, and from premontane rain forest in Los Santos.

Teramnus volubilis Sw., Prodr. Veg. Ind. Occ. 105. 1788

Slender twining vine; stems pubescent to glabrate. Leaves trifoliolate; stipules ovate, ca 2 mm long; petioles 2–4 cm long; petiolules minutely pulvinate; leaflets lanceolate to oblong-elliptic, narrowly rounded at apex, rounded at base, glabrous above, sparsely strigose below, the terminal leaflet 2–6 cm long, 1.5–2 cm wide, the lateral leaflets smaller. Flowers solitary to few, axillary or racemose; calyx 3.5–4 mm long, 4-lobed by fusion of the 2 upper lobes; corolla ca 5 mm long; standard ± spatulate, emarginate and recurved near apex at anthesis, white or tinged with violet; wings pale violet, held erect, attached to keel near apex, the keel white, much shorter than wings, fused only near apex; stamens monadelphous, alternate; anthers small. Legumes 2–3(7) cm long, ca 2 mm wide, somewhat flattened, with an obliquely descending beak, appressed-pubescent; seeds oblong, ca 2 mm long, olive-green. *Croat 13222.*

Apparently rare; collected once in the Laboratory

KEY TO THE SPECIES OF TERAMNUS

Leaflets densely sericeous below; calyx 5-lobed, about as long as corolla; fruits densely long-villous . . . . *T. uncinatus* (L.) Sw.
Leaflets sparsely strigose below; calyx 4-lobed, much shorter than corolla; fruits appressed-pubescent . . . . *T. volubilis* Sw.

Clearing. Apparently flowers and fruits mostly in the dry season.

Costa Rica to northern South America; Trinidad and Jamaica. In Panama, known only from tropical moist forest on BCI and in Darién.

VATAIREA Aubl.

Vatairea erythrocarpa Ducke, Arch. Jard. Bot. Rio de Janeiro 5:139, 192, pl. 12, f. 25. 1930

Tree, to ca 27 m tall, ca 30 cm dbh; outer bark with many shallow fissures; inner bark smooth, light-colored, the sap with a faint, sweet aroma; young stems puberulent, soon almost glabrous, lenticellate. Leaves alternate, imparipinnate, 25–40 cm long; stipules narrowly deltoid, ca 1.5 mm long; petioles 5–11 cm long; petioles, rachises, and petiolules puberulent; petiolules 4–5 mm long; leaflets 11–15, alternate to opposite, narrowly ovate to oblong-elliptic, bluntly acuminate at apex and downturned, rounded to truncate at base, (3)5.5–11 cm long, (2.5)3–5.5 cm wide, glabrous or weakly appressed-pubescent on midrib of lower surface, drying subcoriaceous, the margins somewhat thickened and turned down; major veins in 6–10 pairs, anastomosing and somewhat loop-connected 5–10 mm from the margin, the reticulate veins prominulous on dried specimens. Inflorescences terminal; flowers not seen. Fruits samaroid, ca 8.5 cm long, 1-seeded, yellowish-green, glabrous except sparsely appressed-pubescent on ventral margin of seminiferous area, the seminiferous area ca 3 cm long, oval, bearing a conspicuous, almost medial ridge, this ridge continuous with the dorsal margin of the wing; wing obovate, mucronate at apex, thin, ca 5.5 cm long, the dorsal margin thickened. *Croat 16624, Folsom 3509, Foster s.n.*

Apparently rare; known only from the beginning of Snyder-Molino Trail and at the junction of Armour and Conrad trails. R. Foster collected fruits in the late dry season.

The leaves and twigs of this species are indistinguishable from those of *Pterocarpus officinalis,* making sterile determination impossible.

Sterile collections by Folsom and Croat are presumably leaves of this species, which represents a new genus for the island's flora. The determination is doubtful, because the species has been known only from Brazil. Because determination is in doubt, it is not considered in the section of the introduction on geographic affinities.

VIGNA Savi

Vigna vexillata (L.) A. Rich. in Sagra, Hist. Cuba 10:191. 1845

Twining herbaceous vine; stems and petioles conspicuously brownish-pilose. Leaves trifoliolate; stipules narrowly lanceolate, 4–7 mm long, sagittate at base, the lobes to 1 mm long; petioles 1.5–4 cm long; rachis 5–15 mm long; leaflets generally oblong-lanceolate, usually narrowly acute at apex, obtuse to truncate at base, 2.5–11 cm long, 1.5–5 cm wide, ± strigillose above and below, the lateral leaflets oblique. Racemes bearing few flowers; peduncles 5–18 cm long, with stout, brown, retrorse trichomes near apex; pedicels opposite, in 1 or 2 pairs; calyx to 12 mm long, the lobes ± equal, the lowermost lobe longest; petals white to blue when opening in the morning with yellow spots on either side of the standard at base, becoming pale yellow through the course of the day, then falling; standard to 2.5 cm long and 3.7 cm wide, emarginate at apex; wings free, ca 1.5 cm wide, the keel petals free to about the middle, enveloping the stamens and very much twisted to one side; style protruding from the tube formed by the outer part of the keel petals, ca 8 mm longer than stamens, the trichomes fine, retrorse, villous, near the apex. Legumes linear, narrowly acute at apex, 7–10 cm long, ca 5 mm wide, dark brown, pilose; seeds many, oblong, to 5 mm long, dark brown, with a white hilum. *Croat 5625.*

Infrequent, in small clearings and along the shore. Apparently flowers throughout the year, more abundantly during the rainy season. Though flowers are produced every day, few fruits are set—those seen were mature in the early dry season.

Standley (1933) reported *Vigna repens* (L.) O. Kuntze on the basis of a misidentified collection of *Phaseolus trichocarpus.*

Pantropical. In Panama, known from tropical moist forest in the Canal Zone, Bocas del Toro, Panamá, and Darién and from premontane wet forest in Bocas del Toro.

See Fig. 294.

64. OXALIDACEAE

Trees. Leaves alternate, imparipinnate, petiolate; leaflets entire; venation pinnate; stipules lacking. Flowers bisexual, actinomorphic, in axillary, cymose panicles; calyx deeply 5-lobed, imbricate; petals 5; stamens 10, in 2 series, 5 fertile, 5 sterile; anthers 2-celled, versatile, dehiscing longitudinally; ovary superior, 5-locular, 5-carpellate; placentation axile; ovules many, anatropous; styles 5; stigmas capitate. Fruits fragrant, deeply ridged berries; seeds many, with little fleshy endosperm.

The family is represented on BCI by a single cultivated species. Corner (1964) reported that *Averrhoa* is possibly bat dispersed.

Seven genera and about 950 species; subtropics and tropics.

AVERRHOA L.

Averrhoa carambola L., Sp. Pl. 428. 1753
Carambola

Tree, to 10 m tall; twigs reddish and generally tomentose. Leaves imparipinnate; rachis 10–16 cm long, ferruginous-tomentose; petiolules ca 2 mm long, swollen, tomentose;

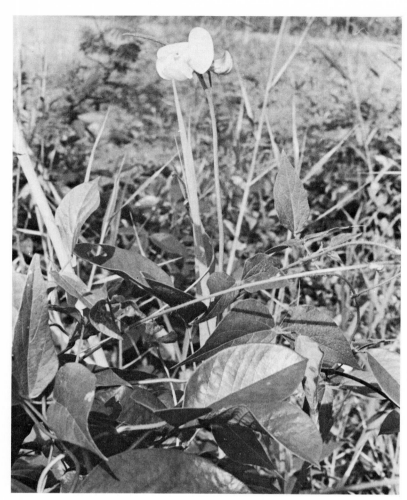

Fig. 294. *Vigna vexillata*

Fig. 295. *Erythroxylum panamense*

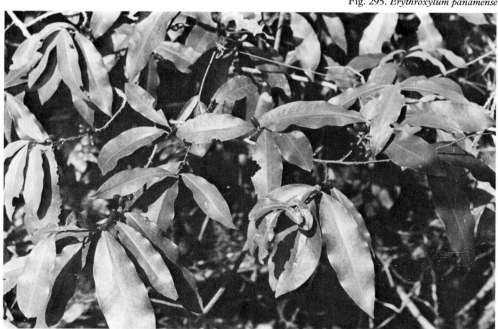

leaflets 7–11 (15), opposite or alternate, the lowermost leaflets smallest, the terminal leaflets largest, ovate, 6–9 cm long, 3–4 cm wide, acuminate, rounded at base, mostly glabrous but strigillose on and near veins below and near margins, dark green above, light green below and near veins above. Inflorescences reddish, axillary, at first growing near ends of branches, later overtopped by new leaf growth and occurring on older wood behind leaves; flowers 5-parted, on articulate pedicels; calyx ca 3 mm long, deeply 5-lobed, reddish, on a tube ca 0.5 mm long; corolla ca 5 mm long, rotate, the petals ± obovate, maroon, paler on margins, puberulent inside, weakly connate just above base; stamens 10, in 2 series, the outer 5 fertile, short, alternating with even shorter staminodia, the longer set of 5 sterile; filaments pubescent; anthers held horizontally, the pollen shed in bud; style simple, short, deeply encircled by sterile stamens. Berries ovoid or ellipsoid, 8–14 cm long, usually with 5 deep ribs, yellow; mesocarp sour tasting. *Croat 4558, 7036.*

Cultivated at the laboratory. Flowers from December to January. The fruits mature in April. It may flower and fruit at other times as well.

Native to Malaya, but incidentally cultivated throughout the tropics and presumably naturalized in Central America.

65. HUMIRIACEAE

Large trees; wood reddish, hard. Leaves alternate, simple, petiolate, usually more or less coriaceous; blades entire; venation pinnate; stipules small, deciduous. Inflorescences paniculate; pedicels short, articulate; bracts and bracteoles deciduous, small; flowers with sepals more or less united into a cupular calyx; petals free, thick; stamens many, in 3 or 4 rows, connate basally in a tube surrounding ovary; anthers ovate-lanceolate, the thecae 2, bilocular, attached at the lower side, the connective thick; disk cupular; ovary superior, sessile, 5-locular; ovules 2 per cell, anatropous; integuments 2; style 1, as long as or longer than stamens; stigma somewhat thickened, 5-lobed. Fruits drupes, the surface smooth; endocarp woody, with 5 linear-oblong valves extending most of the length of the fruit; seeds 1 or 2, oblong.

Humiriaceae in Panama can be recognized by being large trees with unbranched trunks, rather smooth bark, reddish wood, and a dense rounded crown with dark green, coriaceous leaves and by having drupaceous fruits with woody, usually very hard endocarps, 1 or 2 seeds, and many resin-filled cavities.

Eight genera and 50 species; American tropics and western Africa.

VANTANEA Aubl.

Vantanea occidentalis Cuatr., Trop. Woods 96:40. 1950

Tree, to 40 m; branchlets glabrous. Leaves alternate, coriaceous; petioles 1–10 mm long, flattened toward apex, canaliculate and thickened at base; blades elliptic to obovate-elliptic, obtuse to bluntly acuminate at apex, the

tip downturned and emarginate, cuneate at base, 6–16 cm long, 3–8.5 cm broad. Panicles terminal; branchlets dichotomously articulate, subcoriaceous, puberulent; bracts caducous; flowers (*fide* Cuatrecasas, 1961) many; pedicels thick, puberulent, 1–2.5 mm long; sepals orbicular, glabrous or nearly so, 6–7 mm long, bearing a single gland outside; petals oblong, obtuse at apex, ca 9 mm long and 3 mm wide, glabrous, greenish-white outside, white inside; stamens many (ca 80); filaments unequal, ca 8 mm long, linear, white, glabrous, united at base; anthers small, ovate-lanceolate, the connective thick at base, acute at apex; disk cupular, 1 mm high, minutely denticulate; ovary 3–4 mm high, hirsute; style 5 mm long. Drupes ovoid-ellipsoid, narrowed at both ends, ca 3.5 cm long and 1.2–1.7 cm thick; endocarp woody, 2.2–3.3 cm long, 1.2–1.7 cm thick, ellipsoid-ovoid, acute at apex, tapering to a rounded base, the surface moderately smooth, with 5 broad ribs, the ribs alternating with 5 oblong valves, the valves 1.8–2.1 cm long, 3–4 mm wide (only 1 of them removable); seed 1, oblong, ca 2 cm long. *Garwood & Foster 400.*

Apparently rare; discovered recently between Wheeler Trail 2400 and 2500. Seasonality uncertain. Flowering has occurred in August. Mature-sized fruits have been collected in February, and the old endocarps may be seen on the ground year-round.

Central Panama to western Colombia. In Panama, most common in premontane wet forest and tropical wet forest (Gentry, 1975); known from premontane wet forest in the Canal Zone (Pipeline Road) and Veraguas (Cerro Tute). The BCI collection is the first report from tropical moist forest.

66. ERYTHROXYLACEAE

Trees or shrubs. Leaves alternate, petiolate, simple, entire, glabrous or nearly so; venation pinnate; stipules present and sometimes greatly expanded. Flowers bisexual, actinomorphic, terminal or axillary, solitary or in fascicles or clusters; calyx campanulate, 5-lobed; petals 5, free, appendaged at base; stamens 10, in 2 series, united at base; anthers 2-celled, basifixed, dehiscing longitudinally; ovary superior, 3-locular (only 1 fertile); placentation axile; ovule 1, anatropous, pendulous; styles 3; stigmas capitate. Fruits drupaceous, prominently sulcate on drying; seed with fleshy endosperm.

The family is represented in the New World only by *Erythroxylum*, which can be distinguished by having ten stamens in two unequal series, appendages at the base of the petals, and clavate or capitate stigmas. Members of the family often have prominent stipules, a discolored band along the lower midrib, and an ellipsoid fruit that dries markedly sulcate and usually has the old stamens persisting around the base. Bawa and Opler (1975) reported *Erythroxylum lucidum* var. *costaricense* O. E. Schultz to be dioecious. Dioecy should be looked for in other species as well.

Four genera and about 250 species; primarily in the neotropics, but also in Africa, chiefly Madagascar.

ERYTHROXYLUM P. Browne

Erythroxylum multiflorum Lund., Amer. Midl.
Naturalist 29:474. 1943

Glabrous shrub or tree, to 6 m tall; stems marked with
scars of deciduous stipules. Stipules lanceolate, 1–6 cm
long, brown, overlapping at apex, subpersistent; petioles
1–2 cm long, canaliculate; blades oblong-elliptic, acute or
short-acuminate and downturned at apex, acute at base
and often decurrent, 10–24 cm long, 3.5–10 cm wide (to
34 cm long and 15 cm wide on seedlings), ± leathery,
bicolorous, glossy above, lighter and dull beneath, the
margins revolute especially near base; reticulate veins
distinct. Flowers white, in dense axillary or terminal
clusters, interspersed with brown striate bracts 5–8 mm
long; pedicels 5–10 mm long, 5-angulate, the edges
winged; calyx lobed, persisting in fruit, the lobes 5, ovate,
ca 5–7 mm long; petals 5, ovate, smaller than calyx lobes,
bearing a petaloid ligule on the inner surface; stamens 10,
united basally into a tube with 10 small teeth at apex;
styles 3, persistent, slightly exserted; ovary 3-celled.
Drupes ± ovoid-ellipsoid, ca 1 cm long, fleshy, the thick,
corky pedicels persisting among bracts after fruits have
fallen. *Croat 8265, Foster 1331.*

Occasional, in the forest; seen also along the shore.
Seasonal behavior uncertain. Some flowers have been
seen from May to September and again in February and
March. The fruits are probably mature in about 1 month.

Recognized by the large, brown, persistent stipules.
This species was considered *Erythroxylum amplum*
Benth. by Standley (1933), but *E. amplum* occurs only
in South America. *E. multiflorum* belongs to a difficult
species complex, of which *E. macrocalyx* Cav. is the
earliest name. All of these species are characterized by
having conspicuous clusters of stipules associated usually
with the younger branches.

Known only from lowland central Panama, from tropical moist forest in the Canal Zone, and from tropical wet
forest in Colón.

Erythroxylum panamense Turcz., Bull. Soc. Imp.
Naturalistes Moscou 36(1):581. 1863

Tree or shrub, to 3 m; young stems lenticellate, reddish-
brown. Leaves very sparsely pubescent throughout; stipules deltoid, ca 3 mm long; petioles 3–7 mm long, drying
brown, canaliculate; blades ± elliptic, acute at apex and
base, 4.5–11 cm long, 2–4 cm wide, with discolored area
ca 5 mm wide on both sides of midrib below. Flowers
solitary or fasciculate on stems, 5-parted; pedicels 1–2
mm long, with well-marked, rounded angles; calyx deeply
lobed, the lobes valvate, acute, ca 1.3 mm long; petals

greenish, 4 mm long, rounded at apex, keeled medially,
with a white, bifid, petaloid ligule inside to 2.7 mm long
(the petals appearing to be borne on the ligule); stamens
10, to 4 mm long, fused into a tube 1.3 mm long at base,
5 borne on rim of tube, 5 from somewhat within tube
near apex (free part of filaments ± irregular); ovary
oblong, glabrous; styles 3, united to about middle, then
curved outward; stigmas capitate; nectar apparently
stored above appendages on petals. Drupes ellipsoid, red,
7–10 mm long; exocarp thin; mesocarp fleshy; seed white,
drying ribbed. *Croat 8662, 11127.*

Frequent in the forest. Some flowers have been seen
from February to May and also in September and October. The fruits mature within 1 month.

Endemic to Panama; known from lower elevations
in tropical moist forest on the Atlantic slope of the Canal
Zone and in Colón, San Blas, Los Santos, Panamá, and
Darién, from premontane wet forest in Los Santos (Loma
Prieta), and from tropical wet forest in Colón (Río
Guanche) and Darién (Cerro Pirre).

See Fig. 295.

67. RUTACEAE

Trees, usually armed; leaves, flowers, and fruits usually
punctate-glandular, the glands with aromatic oils. Leaves
alternate, simple (*Citrus*) or pinnate (*Zanthoxylum*),
petiolate; petioles often winged (in *Citrus*, the wing
representing reduced leaflets of a compound leaf); blades
pellucid-punctate, occasionally stellate-pubescent; venation pinnate; stipules lacking. Flowers fragrant, bisexual
and few or solitary in axils (*Citrus*) or dioecious and
paniculate (*Zanthoxylum*); sepals 4 or 5, imbricate, basally
connate; petals 4 or 5, imbricate, free; stamens as many as
petals and in one whorl (*Zanthoxylum*) or 2 to several
times as many as petals (*Citrus*); filaments weakly fused;
anthers 2-celled, introrse, dehiscing longitudinally; disk
present; ovary superior, 4- or 5-locular (8–15 in *Citrus*),
4- or 5-carpellate, often lobed; placentation axile; ovules
1 to several per locule; styles 1 (*Citrus*) or 1 per carpel
(*Zanthoxylum*), connate. Fruits berries with a leathery
rind (the hesperidium; *Citrus*) or of several follicles
(*Zanthoxylum*); seeds with or without endosperm.

Members of the family are distinguished by the frequent spines, the translucent pellucid dots on the leaves,
the lobed ovary elevated on the disk, and the frequently
aromatic sap.

The flowers of *Zanthoxylum* are open. Their pollination system is unknown, but they are probably insect
pollinated. I have seen flowers of *Citrus* visited by small
bees and hummingbirds.

KEY TO THE GENERA OF RUTACEAE

Flowers more than 10 mm long; stamens numerous; fruits large, fleshy, with many seeds; leaves
    appearing simple; plants cultivated trees in the Laboratory Clearing . . . . *Citrus*
Flowers less than 5 mm long; stamens 4 or 5; fruits of 1–5 follicles each dehiscing on one side with
    a single shiny seed; leaves pinnately compound; plants large forest trees . . . . *Zanthoxylum*

Seeds of most *Zanthoxylum* are well suited for bird dispersal, but those of *Z. procerum* are apparently autochorous, being released by the hydroscopic contractions of the inner wall of the follicle. The same inner wall is used in other species to project the shiny black seed out of the follicle, where it may be displayed on the pendent funicle. Other animals, such as monkeys, may also play an important role in dispersal of *Zanthoxylum*. Howler and white-faced monkeys are reported to take the seeds (Oppenheimer, 1968; Hladik & Hladik, 1969). *Citrus* species are eaten by spider monkeys (Hladik & Hladik, 1969) and no doubt by other mammals as well.

About 150 genera and over 1,500 species; widely distributed in warm regions, most numerous in South Africa and Australia.

CITRUS L.

Citrus aurantifolia (Christman) Swingle, J. Wash.
    Acad. Sci. 3:465. 1913
    Lime, Limón, Limón verde

Shrub or tree, 2–8 m tall, glabrous, usually with numerous axillary spines 5–17 mm long. Leaves simple; petioles oblanceolate-winged, 5–28 mm long, 2–9 mm wide, articulated at apex; blades ovate to ± elliptic, acute to bluntly acuminate at apex (the acumen sometimes emarginate), obtuse to rounded at base, 4–13 cm long, 2–6 cm wide, crenate. Flowers solitary to few, in short corymbose cymes, (4)5-parted; pedicels 3–5 mm long; calyx shallowly 4- or 5-lobed; petals white, oblong-lanceolate,

11–14 mm long, 3–5 mm wide; stamens 20–25; filaments 4–6 mm long, connate into small clusters; anthers 2–3 mm long; disk ca 1 mm high; ovary obovoid, 2–3 mm long; style 2–3 mm long, soon deciduous; stigma subglobose. Fruits ellipsoid at maturity, 3.5–6 cm diam, mammillate at apex, green to yellow, sour, the core solid. *Croat 9190, 14945.*

Cultivated at the Laboratory Clearing; reported by Standley (1933) to be naturalized at some places on the island. Flowers mostly during the dry season. The fruits mature mainly in the rainy season.

Native apparently to East Pakistan, Assam, Burma, Thailand, and Malaysia; now widely cultivated in the tropics and subtropics and throughout Panama.

Citrus aurantium L., Sp. Pl. 782. 1753
    Sour orange, Seville orange

Tree, 3–6 m tall, unarmed or the few axillary spines to 5 mm long. Leaves simple, glabrous; petioles broadly winged at apex, 14–27 mm long, usually 10–15 mm wide, narrowed to an almost wingless base, articulated at apex; blades broadly elliptic to obovate, bluntly acuminate and ± emarginate at apex, cuneate at base, 5–12 cm long, 3–7.5 cm wide, crenate. Flowers 5-parted, solitary in axils; pedicels puberulent, 9–11 mm long in fruit; calyx puberulent, the lobes ca 2 mm long; petals white. Fruits globose, 4–6 cm diam, depressed or flattened at apex, orange, sour, the core hollow. *Croat 9184, 14944.*

Cultivated at the laboratory but uncommon. Flowers in the dry season. The fruits mature in the rainy season.

KEY TO THE SPECIES OF CITRUS

Petioles broadly winged, the wings oblanceolate to apically broadened, 2–15 mm wide:
    Fruits 10 cm or more diam; petioles and lower surface of blades sparsely pubescent (at least on
        veins) . . . . *C. grandis* (L.) Osbeck
    Fruits less than 8 cm diam; petioles and lower surface of blades glabrous:
        Petiole wings oblanceolate, 2–9 mm wide; branchlets usually with numerous axillary spines
            5–17 mm long; mature fruits ellipsoid, mammillate apically, green to yellow, the core
            solid . . . . *C. aurantifolia* (Christman) Swingle
        Petiole wings apically broadened, mostly 10–15 mm wide; branchlets usually spineless, or the
            few spines 5 mm long or less; mature fruits globose, depressed or flattened apically,
            orange to red, the core hollow . . . . *C. aurantium* L.
Petioles narrowly winged or narrowly margined, the wings usually 2 mm wide or less (rarely wider
    in *C. sinensis*):
    Branchlets usually spineless; mature fruits depressed-globose, green to orange or reddish-
        orange, the peel easily separating from the segments . . . . *C. reticulata* Blanco
    Branchlets usually with at least a few axillary spines; mature fruits globose or ovoid, the peel not
        easily separating from the segments:
        Petioles winged; axillary spines few, 5 mm long or less; mature fruits globose, flattened or
            depressed apically, yellow-green to orange, sweet . . . . *C. sinensis* (L.) Osbeck
        Petioles margined; axillary spines usually many, to 27 mm long; mature fruits ovoid, mammil-
            late apically, light yellow, sour . . . . *C. limon* (L.) Burm.f.

Native apparently to East Pakistan, Assam, Burma, and adjacent southwestern China; now widely cultivated in the tropics and subtropics and throughout Panama.

Citrus grandis (L.) Osbeck, Dagbok Ofwer Ostin. Resa 98. 1757
Shaddock, Pummelo

Tree, 3–10 m tall, unarmed or bearing axillary spines; twigs pubescent. Leaves simple; petioles broadly obcordate-winged, articulate at apex; blades oval to broadly ovate, rounded to retuse at apex (bluntly pointed on young shoots), 8–20 cm long, 6–15 cm wide, sparsely pubescent at least on veins below. Flowers 5-parted, solitary or usually clustered in axils, 2–2.5 cm long, very fragrant; petals white, broad; stamens 20–25; ovary globose; style columnar; stigma large, capitate. Fruits globose or depressed-globose, 10–15 cm diam, smooth, pale yellow, sour; segments ca 12. *Croat 4210.*

Cultivated in the Laboratory Clearing between the dock and the boathouse, now engulfed in secondary vegetation. Seasonal behavior unknown.

Native to southeastern Asia and India; now widely cultivated in the tropics.

Citrus limon (L.) Burm.f., Fl. Ind. 173. 1768
Lemon

Tree, to 6 m tall, glabrous, heavily armed with axillary spines 5–27 mm long. Leaves simple; petioles narrowly margined, 5–7 mm long, ca 2 mm wide, articulate at apex; blades elliptic to oval, rounded to bluntly acuminate at apex, widely cuneate at base, 4–11 cm long, 1.5–6 cm wide, crenate. Flowers 5-parted, solitary or clustered in axils, 8–17 mm long; petals pinkish outside, white inside; stamens 20 or more; ovary tapered at apex. Fruits oblong to ovoid, 5–10 cm long, mammillate at apex, light yellow, somewhat roughened, sour; segments 8–10. *Croat 10767.*

Occasional; cultivated at the Laboratory Clearing. Seasonal behavior not determined.

Apparently native to northeastern India, Bangladesh, northern Burma, and southern China; now widely cultivated in the tropics and subtropics and throughout Panama.

Citrus reticulata Blanco, Fl. Filip. 610. 1837
Tangerine, Mandarin orange

Tree, to 8 m tall, glabrous, unarmed; stems prominently ribbed below petioles. Leaves simple; petioles narrowly winged near apex, 6–14 mm long, ca 2 mm wide, articulate at apex; blades ovate to elliptic, blunt to retuse at apex, obtuse to acute at base, 3–8.5 cm long, 1.5–4.5 cm wide, densely punctate, ± stiff, bicolorous, subentire to crenate. Flowers strongly aromatic, 5-parted, solitary (or paired) in axils; calyx short, thick, lobed to about middle, with prominent submarginal bumps, the margin thin, often minutely ciliolate; petals oblong-obovate, acute to rounded at apex, white, spreading at anthesis, 11–14 mm long, ca 3 mm wide, with sparse, large, glandular dots; stamens ca 20, ± unequal, 8–10 mm long, the filaments weakly fused laterally; anthers oblong, the connective prolonged into a greenish knob; pollen

yellow-golden, tacky, adhering in large clusters; ovary depressed-globose, glabrous; disk prominent; style shorter than stamens; stigma globose. Fruits depressed-globose, 3–3.5 cm diam, orange at maturity, sweet, the rind easily separating. *Croat 14579.*

Common; cultivated in the Laboratory Clearing. Flowers throughout the dry season and the early rainy season. The fruits mature in the late dry season and in the rainy season.

Flowers are visited by hummingbirds.

Apparently native to Indochina; now widely cultivated in the tropics and subtropics and throughout Panama.

Citrus sinensis (L.) Osbeck, Reise Ostind. China 250. 1765
Sweet orange, Orange, Naranja

Shrub or tree, to 8 m tall; axillary spines few, 2–5 mm long. Leaves simple; petioles usually with a narrow, oblanceolate wing 6–18 mm long and 1–2(5) mm wide, articulate at apex; blades ± elliptic, acute or obtuse and retuse at apex, cuneate at base, 3.5–9 cm long, 1.5–4.5 cm wide, crenulate. Flowers solitary in axils, 5-parted; pedicels 7–12 mm long in fruit; calyx to 6 mm diam, the lobes acute, ciliate, to 2 mm long; petals white. Fruits globose, 4–12 cm diam, yellow-green to orange, sweet, the rind not separating easily from segments. *Croat 10765.*

Occasional; cultivated in the Laboratory Clearing. Flowers in the dry season. The fruits mature in the rainy season.

Apparently native to Bangladesh, northern Burma, southeastern China, and Indochina; now widely cultivated in the tropics and subtropics and throughout Panama.

ZANTHOXYLUM L.

Zanthoxylum belizense Lund., Contr. Univ. Michigan Herb. 6:35. 1941
Arcabú, Tachuelo

Dioecious tree, 13–30 m tall, to almost 1 m dbh; trunk of younger trees armed, the prickles large, corky, horizontally flattened, to ca 5 cm wide and 1 cm thick (somewhat rounded and numerous on juvenile plants), generally deciduous on older trees, the scar often visible; branches and branchlets with ribs extending downward from petioles and with occasional small conical prickles; outer bark thin, brown, sparsely stellate-pubescent. Leaves pinnate, generally imparipinnate or the terminal pair bearing the scar of an aborted terminal leaflet (occasionally with 1 leaflet merely appearing terminal), 16–67 cm long; petioles mostly 5–10 cm long, sparsely stellate-pubescent; leaflets 8–20(26), subopposite to alternate, oblong-elliptic to oblong, abruptly acuminate, acute to obtuse at base, 4.5–16(21) cm long, 1.5–5.5(7) cm wide, sessile or obscurely petiolulate, pellucid-punctate, ± entire and revolute on margin, dark green, shiny and sparsely stellate-pubescent above, duller and densely pubescent below, the trichomes stellate, mostly sessile; juvenile leaves as much as 1.5 m long, the leaflets 22 cm long and 7.5 cm wide. Panicles terminal and upper-

KEY TO THE SPECIES OF ZANTHOXYLUM

Leaves densely pubescent on lower surface:
    Trichomes stellate; corky prickles on trunk flattened, horizontally oriented . . . . . . *Z. belizense* Lund.
    Trichomes unbranched; corky prickles on trunk rounded in outline, somewhat vertically oriented . . . . *Z. setulosum* P. Wils.
Leaves glabrous or minutely pubescent on lower surface:
    Petals 3; follicle 1, the valves deciduous; leaflets equilateral and ± decurrent at base, markedly pellucid-punctate on margin, otherwise opaque, both surfaces densely lepidote, the underside glabrous, lacking prickles . . . . *Z. procerum* Donn. Sm.
    Petals 5; follicles 3 or 4 (rarely 2), the valves persistent; leaflets inequilateral at base, not decurrent, pellucid-punctate all over, lacking lepidote scales, the underside often minutely puberulent, often bearing 1 or 2 long prickles . . . . *Z. panamense* P. Wils.

axillary, 15–33 cm long, widely branched, the branches and pedicels sparsely stellate-pubescent; branchlets scaly, the scales deltoid, ciliate; pedicels to 1.5 mm long; calyx triangular, ciliate; flowers unisexual, greenish-white, 5-lobed, to 3.7 mm wide; petals ± elliptic, acute, imbricate, 1.4–3.3 mm long; stamens 5, alternate, broadly exserted, to 4 mm long in staminate flowers, shorter and sterile in pistillate flowers; ovary 5-lobed, pubescent; style short; stigma simple. Fruits of 1 or 2 globose follicles, 3.5–5 mm long, punctate-verrucose; seeds dark brown, shiny, somewhat shorter than follicle. *Croat 12497.*

Occasional in the forest, though sometimes locally abundant in older forest. Flowers from August to October; individual plants may flower for at least a month. The fruits mature from January to March. Leaves fall off in the dry season, but the new ones all grow out before flowering begins.

Southern Mexico to Panama and possibly Colombia. In Panama, known from tropical moist forest in the Canal Zone, Colón, and San Blas, from premontane wet forest in Colón (Santa Rita Ridge), and from tropical wet forest in Colón (Icacal).

See Fig. 296.

Zanthoxylum panamense P. Wils., Contr. U.S. Natl.
    Herb. 20:479. 1922
    Arcabú, Acabú, Alcabú, Prickly holly, Prickly yellow, Lagarto

Medium to large, dioecious tree, to 28 m tall; trunk to 75 cm dbh, buttressed to ca 1.5 m, armed, the prickles conical, corky, the base to 3.5 cm long, the apex somewhat flattened laterally, deciduous from older trunks at base, persistent above and on branches (especially the smaller ones); outer bark thin, not deeply fissured, often very roughened; inner bark thick, tan, granular, flaking upon slash; twigs puberulent, the trichomes appressed or uncinate; sap usually bitter, not noticeably aromatic. Leaves alternate, compound, imparipinnate or rarely paripinnate, mostly 15–60 cm long; petioles minutely puberulent, mostly 3–9 cm long, the basal pulvinus pronounced; rachis often canaliculate above, broadly so below leaflets, glabrous to minutely puberulent, sometimes armed; petiolules 3–5(7) mm long; leaflets 10–20, opposite to subopposite, sometimes alternate especially basally, mostly ± oblong-elliptic, abruptly acuminate (the acumen sharp or blunt), rarely rounded at apex,

rounded or cuneate and usually inequilateral at base (except terminal leaflet), 4–19 cm long, 1.5–7 cm wide, entire or obscurely crenulate, glabrous above or inconspicuously puberulent especially on midrib, glabrous or with inconspicuous, short, appressed trichomes below, occasionally bearing 1 or more, long, sharp prickles on midrib beneath, both surfaces shiny, with small and numerous pellucid dots (sometimes obscure before drying), sometimes with a few much larger, plate-shaped glands on both surfaces. Panicles terminal, 20–30 cm long (shorter on pistillate plants), highly branched, the branches densely hispidulous; pedicels ca 1 mm long; flowers unisexual, white or greenish, 5-parted; calyx puberulent, the lobes short, acute to rounded; petals ± elliptic, acute at apex, 1.5–3 mm long, to 1 mm wide; staminate flowers with stamens 5, exserted, the filaments ca 3 mm long, alternating with lobes of pistillode, the anthers oblong, the pistillode minutely pubescent; pistillate flowers lacking stamens, the ovary broadly obovoid, 3- or 4-lobed, to 1.5 mm long, glabrous and glandular-dotted, the styles (2)3, the stigmas broadly discoid, nearly sessile, round or obtusely 3-sided, more than three-fourths as broad as ovary. Fruits of usually 3 or 4 brown follicles, each 6–8 mm long, puberulent, dehiscing from an apical, medial suture, the valves persistent; seeds 1 per follicle, shiny, black or dark brown, 3–5 mm long, suspended from capsule on a tough fiber. *Croat 6249* and *12574* (large-leaved form), *14885* and *16589a* (small-leaved form).

Occasional, especially in the young forest. Flowers from April to October, mostly from June to September; plants may flower only once every two years. The fruits mature mostly from June to December. Plants have been seen replacing their leaves over a short span of time in the late dry season.

Probably the most variable *Zanthoxylum* on the island, especially in terms of leaf shape. Some BCI plants have consistently much smaller leaflets than others, and additional morphological and phenological study may show them to be distinct species. Plants of the small-leaved form generally flower and fruit ahead of those with larger leaves, although the seasons overlap. This is further evidence that two taxa may be involved in what is being called *Z. panamense*. The small-leaved plant corresponds with the type.

This species and *Z. setulosum* are different from *Z. procerum* in their manner of dehiscence. The fruits of

Fig. 296. *Zanthoxylum belizense*

Fig. 297. *Zanthoxylum panamense*

Fig. 299. *Zanthoxylum setulosum*

Fig. 298. *Zanthoxylum procerum*

Z. procerum expel their seeds from the follicle by the elastic movements of the inner wall of the carpel, whereas *Z. panamense* is thought to be bird dispersed because of the shiny black seeds displayed outside the valves on a slender fiber.

The prickles on the trunk may be hollow and are often inhabited by ants.

Costa Rica and Panama. In Panama, known from tropical moist forest in the Canal Zone, Bocas del Toro, Colón, Chiriquí, Panamá, and Darién.

See Fig. 297.

Zanthoxylum procerum Donn. Sm., Bot. Gaz. (Crawfordsville) 23:4. 1897

Alcabú, Ikor, Lagarto

Dioecious tree, 8–17(20) m tall; trunk to 18 cm dbh, armed, the prickles corky, conical, with the sharp apex usually slightly off-center, the sides with weak vertical grooves; branches sparsely armed with short prickles; outer bark with lenticels arranged in irregular vertical streaks; inner bark tan; twigs soon glabrous. Leaves alternate, imparipinnate or paripinnate, 12–45 cm long (to 60 cm on juveniles), glabrous; petioles 1–10(14) cm long, the pulvinus moderately small to much expanded; rachis canaliculate above, especially beneath insertion of each leaflet; petiolules 3–7 mm long; leaflets (2)4–18, unarmed, mostly opposite, sometimes alternate especially basally, broadest in middle, narrowed gradually to either end, gradually acuminate at apex, cuneate to attenuate with both sides ± equally decurrent at base, 4.5–14.5 cm long, 1.5–5 cm wide (to 25 cm long and 8 cm wide on juveniles), the margins usually minutely crenate and conspicuously pellucid-punctate, the surfaces ± shiny and glabrous, with numerous, minute, lepidote scales on both surfaces, lacking pellucid dots except near margin, often somewhat viscid when dried. Panicles terminal, 10–20 cm long, the branches many, sparsely puberulent, densely lenticellate and punctate, with markedly constricted articulations at the base at maturity; flowers unisexual, with sweet aroma, ca 2 mm long, 3-parted; calyx lobes triangular; petals broadly oblong, rounded at apex; staminate flowers with the stamens 5, well exserted, 2–3 mm long, the pistillode minute, conical, the style short; pistillate flowers unknown. Fruits of 1 brown follicle, 3–3.3 mm diam, globose, densely covered with lenticels and minute glandular projections, on a short stalk ca 7 mm long, the valves 2, persistent, translucent, hydroscopic, dehiscing at maturity, folding together laterally to expel seed; seed 1, rounded, ca 3 mm diam, black with brown reticulations. *Croat 11670, 12189.*

Frequent in the forest. Flowers from February to May (sometimes from January). The fruits mature from June to November, mostly in August and September. Allen (1956) reported this species to flower in August and September in the Golfo Dulce area of Costa Rica, although he was possibly dealing with a different species.

Distinguished by the small, glabrous leaves, which are acute at the base, and the three-parted flower. Cut parts of the plant have a very strong odor similar to that of the citron fruit, and the flowers have a sweet odor. The valves of the fruit are capable of expanding and contracting repeatedly during alternate periods of high and low humidity.

Southern Mexico to Panama. In Panama, known from tropical moist forest in the Canal Zone, Panamá, and Darién and from premontane wet forest in Chiriquí (Progreso) and Panamá (Cerro Campana). Reported from tropical wet forest in Costa Rica (Holdridge et al., 1971).

See Fig. 298.

Zanthoxylum setulosum P. Wils., Contr. U.S. Natl. Herb. 20:480. 1922

Prickly yellow, Arcabú, Acabú, Alcabú, Tachuelo

Dioecious tree, (3)6–12(20) m tall; trunk 10–15(20) cm dbh, armed, the prickles many, large, corky, oval or rounded in basal outline, the apex rounded with a sharp point set somewhat off-center; outer bark ± smooth, light brown, with many raised lenticels; inner bark tan; wood pale yellow, the cambial layer fluted. Leaves alternate, pinnate, 15–50 cm long, conspicuously pubescent (particularly below), the trichomes erect, simple; petioles terete; rachis margined above, sometimes bearing small prickles; leaflets 15–27, subopposite, ± sessile, ovate to oblong or oblong-elliptic, acuminate, acute to rounded at base, usually inequilateral at base, 2.5–10(15) cm long, 1.5–3.5 cm wide, entire or crenulate, obscurely pellucid-punctate, sometimes glabrate in age above, rarely with a few prickles on midrib below. Panicles terminal or upper-axillary, ± congested, to 21 cm long, the branches short, sparsely puberulent; pedicels 2–3.7 mm long, glabrous or crisp-pubescent; flowers unisexual, 5-parted, 2.7 mm long; calyx ca 1 mm long, usually glabrous, rounded at apex; petals elliptic, boat-shaped, spreading, white to greenish-white, the veins prominent, 2–2.3 mm long, rounded at apex, glabrous or inconspicuously pubescent inside; stamens of staminate flowers 5, included or exserted and spreading, 1.7–2.7 mm long, the anthers about as long as or much shorter than filaments, the pistillode small, 1–4-lobed, the styles 1–4; pistillate flowers not seen. Fruits of (2)4(5) follicles, tan or brown, glabrous, the valves persistent, muricate, the inner valve whitish, curling from the base to force seed from follicle; seeds 1 per follicle, ± globose or ovoid, black, shiny, 2.5–3.3 mm long, suspended on a strong slender fiber. *Croat 5430.*

Frequent in the forest. Flowers in the late dry season (March and April). The fruits develop soon but may persist for a long time, maturing from April to October (sometimes to December). Leaves are shed in the dry season, and the trees may be bare for more than a month.

Ants may inhabit old spines and hollow parts of some stems.

Panama and possibly Costa Rica. In Panama, known principally from tropical moist forest on the Pacific slope in the Canal Zone, Panamá, and Darién; known also from tropical dry forest in Los Santos, from premontane moist forest in Coclé (La Pintada), and from premontane wet forest on Coclé (El Copé).

See Fig. 299 and fig. on p. 20.

KEY TO THE SPECIES OF SIMAROUBACEAE

Flowers large (more than 2 cm long), bright pink, bisexual; fruits on a broad, red receptacle, the
 drupes 3–5, black, ovoid; rachis and petiole conspicuously winged . . . . *Quassia amara* L.
Flowers small (less than 1 cm long), greenish, white, or cream, unisexual; fruits not as above; rachis
 and petiole not winged:
  Leaflets mostly more than 7; inflorescences usually branched, open . . . .
  . . . . *Simarouba amara* Aubl. var. *typica* Cronq.
  Leaflets 5–7; inflorescences unbranched . . . . *Picramnia latifolia* Tul.

68. SIMAROUBACEAE

Trees or shrubs, with bitter sap. Leaves alternate, pinnate, petiolate; leaflets entire; venation pinnate; stipules lacking. Flowers actinomorphic, in leaf-opposed racemes or spikes (*Picramnia, Quassia*) or in terminal panicles (*Simarouba*), bisexual (*Quassia*) or unisexual (dioecious in *Simarouba* and *Picramnia*); calyx 4- or 5-lobed or 4- or 5-parted, imbricate; petals 5, imbricate; stamens 5, opposite the petals, or 10 (*Quassia, Simarouba*), inserted on a toral disk; anthers 2-celled, dehiscing longitudinally; ovary superior, mostly 5-locular and 5-carpellate (2- or 3-carpellate in *Picramnia*, the carpels weakly united in *Quassia* and *Simarouba*); placentation axile; ovules 2 per locule (1 in *Picramnia*); styles 2, or solitary and 2- or 5-lobed. Fruits berries (*Picramnia*) or several drupes (*Quassia, Simarouba*); seeds lacking endosperm.

The family is related to the Rutaceae (67), but lacks the pellucid dots on the leaves. The BCI species have little in common and are all individually unique and easily distinguished.

The flowers of *Quassia amara* are principally hummingbird pollinated. *Simarouba* and *Picramnia* are probably best suited for insect pollination.

The fruits of all species are mostly bird dispersed, but no doubt attract some attention from climbing frugivores as well. The bright red receptacles are taken by white-faced monkeys (Oppenheimer, 1968), and fruits of *Simarouba amara* are eaten by howler monkeys (Hladik & Hladik, 1969). Fruits of *Quassia amara* are taken by birds (Duke, 1968).

About 30 genera and 200 species; mostly in the tropics.

PICRAMNIA Sw.

Picramnia latifolia Tul., Ann. Sci. Nat. Bot., sér. 3, 7:258. 1847
 Canjura

Dioecious shrub or small tree, to 7(10) m tall; bark and wood with bitter sap. Leaves imparipinnate; petiolules ca 5 mm long, swollen throughout; leaflets mostly 5 or 7, alternate, elliptic to elliptic-ovate, acuminate, acute to rounded at base, 6–15 cm long, 3.5–6.5 cm wide, glabrous. Spikes opposite leaves, the staminate ones densely flowered, to 5 cm long, the pistillate ones long, pendent, to 46 cm long, usually on leafless stems; rachis and calyx scabridulous; calyx bowl-shaped, (4)5-lobed to about middle; staminate flowers with calyx ca 1.5 mm long; petals (4)5, white, cream, or greenish-white, linear-lanceolate, ca 2.5 mm long, glabrous; stamens (4)5, long-

exserted; anthers orange. Pistillate flowers numerous; calyx lobes and petals ovate, 1.5–2 mm long; ovary densely appressed-pubescent; style short; stigmas 2, stout, recurled, persisting in fruit. Fruits broadly oblong, to 1 cm long, orange-red; seeds 1 or 2, red outside. *Croat 5263*.

Frequent in some areas of both young and old forests and along the shore in shady areas. Flowers in the early dry season (January to March), with the fruits maturing in the late dry or early rainy seasons (April to June).

Costa Rica to Colombia. In Panama, a typical component of tropical dry forest (Holdridge & Budowski, 1956); known also from tropical moist forest in the Canal Zone, Bocas del Toro, Los Santos, and Darién.

QUASSIA L.

Quassia amara L., Sp. Pl. ed. 2, 553. 1763
 Cruceta, Guavito, Guavito amargo, Guavo amargo, Hombre grande, Puesilde, Quassia, Bitterwood

Shrub or small tree, to 4(8) m tall, glabrous; sap bitter. Leaves imparipinnate; rachis and petiole winged; leaflets 3 or 5, elliptic to oblanceolate, abruptly acuminate, gradually tapered to base, 5–16 cm long, 3–6.5 cm wide, sessile, bicolorous. Flowers 5-parted, in elongate, terminal racemes (rarely paniculate basally); branches of inflorescences and pedicels pinkish; pedicels 1–4 cm long; sepals minute, free; corolla 2.5–4.5 cm long at anthesis, the slender petals glabrous, bright pink outside, white inside, soon falling; stamens 10, exserted at anthesis; filaments curiously hooked, flattened and bearded near base; anthers yellow, longitudinally dehiscent; styles connate, equal to or longer than stamens; stigma 1, simple or slightly lobed. Drupes 4 or 5 (rarely 2 or 3), black, ovoid, 1–1.5 cm long, on a broad red receptacle; seed suspended from apex. *Croat 4038, 4754*.

Common in the forest, especially in some areas. Flowers principally in the late rainy and early dry seasons (August to March). The fruits mature within about 2 months, mostly from December to February.

Distinguished by the distinctive leaf and quinine-like sap in all parts of the plant. Prior to the opening of the flower, a small hole 2–3 mm diam may be produced near the base of the corolla. The organism that makes the hole is not known, but hummingbirds have been seen using the holes to remove nectar.

The fruits are eaten by white-faced monkeys (Hladik & Hladik, 1969).

Southern Mexico to northern South America. In Panama, known principally from tropical moist forest in the

Canal Zone, Bocas del Toro, Colón, Chiriquí, Veraguas, Coclé, Panamá, and Darién; known also from tropical dry forest in Panamá (Taboga Island) and from tropical wet forest in Panamá and Darién.

SIMAROUBA Aubl.

Simarouba amara Aubl. var. **typica** Cronq., Bull. Torrey Bot. Club 71:229. 1944

Aceituno

Dioecious tree, 5–35 m tall, to 70 cm dbh, glabrous except for inflorescence; outer bark hard, with many minute lenticels and small, ± vertical cracks; inner bark and wood with bitter sap; older stems minutely fissured. Leaves pinnate; petioles 4–8 cm long; leaflets (5)9–13(21), oblong-elliptic, rounded or abruptly short-acuminate, acute to obtuse at base, 4–12(15) cm long, 2–4(6) cm wide, dark green and shiny above, lighter green below; lateral veins inconspicuous. Flowers unisexual, campanulate, green, in terminal panicles ca 30 cm long; peduncles and pedicels often minutely puberulent; pedicels 5–15 mm long; calyx shallowly bowl-shaped, 5-lobed, the lobes acute, minutely ciliate; petals 5, ovate to elliptic, beaked inside at apex; staminate flowers 4–5 mm long; stamens 10, equaling lobes, in 2 whorls, the inner whorl to ca 4 mm long, the outer whorl slightly shorter; filaments subulate, the appendages densely tomentose; gynoecium rudimentary, cushion-shaped, weakly 5-lobed, the trichomes tufted, forming a ring around outer edge; pistillate flowers 3–3.5 mm long; pistil ± oblong, deeply 5-lobed, 5-locular; style ca 1 mm long, stout; stigmas 5, slender; carpels 5, soon becoming free (usually not more than 3 surviving until maturity); staminodia much shorter than pistil, the filaments villous near the middle. Fruits of 3–5 drupes to 17 mm long, with a medial ridge, green becoming red-orange or black at maturity; seeds ellipsoid, to ca 14 mm long. *Croat 8441, 9508.*

Occasional, in the forest; locally common, at least in the vicinity of Zetek Trail 250. Flowers in the late dry season (March and April). The fruits mature within 2 months.

The fruits are taken by many birds, including flycatchers, motmots, thrushes, and chachalacas (Duke, 1968).

Guatemala to Panama and south to Brazil; Lesser Antilles. In Panama, known from tropical moist forest on BCI and from premontane wet forest in Panamá (Cerro Campana). Reported from tropical wet and premontane rain forests in Costa Rica (Holdridge et al., 1971).

See Figs. 300 and 301.

69. BURSERACEAE

Trees, often with aromatic sap. Leaves alternate, pinnate, petiolate; leaflets entire; venation pinnate; stipules lacking. Flowers usually functionally unisexual (dioecious) but similar in appearance, actinomorphic, minute, in axillary or rarely terminal, cymose panicles or racemes; calyx cupulate, 3–5-lobed; petals 3–5, free or rarely connate into a lobed tube; disk present; stamens usually 2 times the number of corolla lobes, in 2 whorls; anthers 2-celled, introrse, dehiscing longitudinally; ovary superior, 2–5-locular, 2–5-carpellate; placentation axile; ovules 2 per locule, anatropous; style 1; stigma 1, simple or in some cases 2–5-lobed. Fruits drupaceous, each a tardily dehiscent capsule or a drupe with 1–5 pyrenes (*Trattin-*

KEY TO THE SPECIES OF BURSERACEAE

Petals united to form a tube at base:
    Leaves conspicuously asperous, scabridulous on both surfaces; flowers 3-parted, reddish; fruits ovoid, ca 1 cm diam, indehiscent, the seeds (1)2 . . . . *Trattinnickia aspera* (Standl.) Swart
    Leaves not asperous, sparingly pubescent; flowers 4- or 5-parted, greenish-yellow; fruits broadly turbinate to depressed-globose, more than 2.5 cm diam, the valves deciduous, the seeds usually 3 or more . . . . *Tetragastris panamensis* (Engler) O. Kuntze
Petals free:
    Leaves deciduous, the flowers appearing with new leaves; fruits ± 3-sided, the seed 1, white, 3-sided, indurate, remaining attached at base after valves fall; bark on trunk and larger branches papery, reddish-brown, often peeling . . . . *Bursera simaruba* (L.) Sarg.
    Leaves not deciduous, the flowers concurrent with mature leaves; fruits or seeds not 3-sided, the seeds covered with a fleshy white aril, falling from fruit at maturity; bark on trunk and branches not papery, reddish-brown, not peeling:
        Leaves and inflorescences glabrous; fruits glabrous; mature fruits usually obtuse to acute at both ends . . . . *Protium panamense* (Rose) I. M. Johnston
        Leaves and inflorescences pubescent (at least inflorescence branches; leaves glabrescent in *P. tenuifolium* subsp. *sessiliflorum*); fruits pubescent (often scantily so at maturity); mature fruits obtuse to rounded at base:
            Flowers 4-parted, pedicellate; petioles, rachises, and petiolules densely brownish-hirtellous, the leaf blades (at least below) sparsely hirtellous; fruits ± ovoid, mostly obtuse at base and at apex, the seed 1 . . . . *Protium costaricense* (Rose) Engler
            Flowers 5-parted, sessile or nearly so; petioles, rachises, and petiolules not densely brownish-hirtellous, the leaf blades glabrous except for minute papillae; fruits usually rounded at both ends, 2–5-lobed, the seeds frequently 2–5 . . . . *Protium tenuifolium* Engler subsp. *sessiliflorum* (Rose) Porter

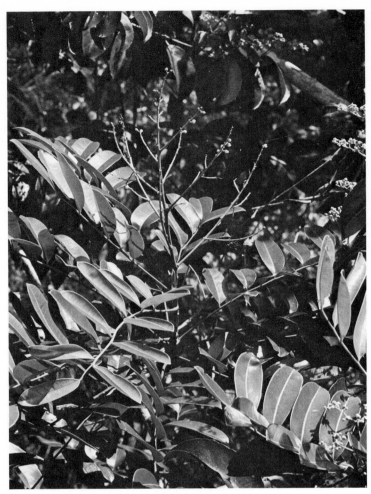

Fig. 300. *Simarouba amara* var. *typica*

Fig. 301. *Simarouba amara* var. *typica*

nickia); seeds 1 (rarely 2) per pyrene, lacking endosperm.

Members of the family may be confused with Anacardiaceae (76) and Meliaceae (70), both of which have alternate, similarly compound leaves with leaflets frequently inequilateral at the base. The BCI species of Burseraceae can be distinguished from these other families by the presence of resin ducts in the bark, by the free stamens and single short style, and by the absence of pellucid dots. The sap of most species is resinous and has a characteristic aroma that is faintly turpentine-like but pleasant.

The pollination system is unknown, but the flowers are best suited to insect pollination.

The fruits are probably chiefly bird dispersed. Those of *Protium* and *Tetragastris* have white arillate seeds, which are displayed against the often bright-red inner carpel wall after one of the valves falls free. Seeds are often pendent, ideally suited to bird dispersal. The pyrenes or pits are very hard and probably pass through the digestive tract unharmed. The fruits are also taken by white-faced and spider monkeys (Oppenheimer, 1968), which eat the fleshy aril. Fruits of *Tetragastris panamensis* are taken by howler monkeys (Carpenter, 1934).

About 20 genera and 600 species; in the tropics.

BURSERA Jacq. ex L.

Bursera simaruba (L.) Sarg., Gard. & Forest 3:260. 1890

Naked Indian, Almácigo, Carate, Huechichi, Indio desnudo

Dioecious or polygamodioecious tree, 5–25 m tall, to ca 40(100) cm dbh; bark coppery red, shiny, thin, peeling to expose green layer beneath; sap, at least of fruit, very aromatic. Leaves pinnate, deciduous, clustered at apex of branches, densely woolly when young to glabrate except on veins below in age; petioles to 14 cm long; leaflets 5–7(9), ovate-elliptic to lanceolate-elliptic (terminal ones usually obovate), long-acuminate, inequilateral and obtuse to rounded at base, 4.5–14.5 cm long, 2.5–8 cm wide, entire. Flowers 3–5-parted, functionally unisexual, in axillary raceme-like panicles (sometimes appearing terminal before leaves appear), appearing ± with new leaves; calyx bowl-shaped, shallowly 5-lobed, the lobes acute to blunt; petals greenish-white, narrowly ovate, acute, cucullate, 2–3 mm long, spreading at anthesis, later recurved; stamens twice the number of and shorter than petals, those opposite the petals usually ± spreading, the alternate ones erect; pollen golden-yellow, covering all sides of anther at anthesis; pistillode ovoid, white, glabrous, scarcely longer than weakly lobed and undulate disk. Capsules drupaceous, ellipsoid, obtusely 3-sided, maturing reddish-brown, ca 1 cm long, a single valve falling free at maturity to expose the pyrene, followed by both remaining valves falling off as a unit, leaving the pyrene attached at base; pyrene 1(2), 1-seeded, 3-angled, bony, lenticular-ovoid, white, ca 7 mm long. *Croat 5325.*

Occasional, but locally common, especially along the shore on the northwest side of the island and on Orchid Island. Flowers in the late dry and early rainy seasons (March to the middle of June), just before or during the onset of new leaves. Mature fruits may be seen through-

out much of the year, but most fruits mature during the late rainy or early dry seasons of the following year, generally after the plant has lost its leaves. Most trees are bare by February, beginning to put on new leaves and flowering by March or April. Allen (1956) reported that the trees are leafless throughout most of the dry season in Costa Rica.

Easily recognized by the reddish-brown, papery bark.

Southern Florida and northeastern Mexico to Colombia, Venezuela, and the Guianas; West Indies. In Panama, a characteristic species in tropical moist forest (Tosi, 1971), known all along the Pacific slope; known also from premontane dry forest in Coclé, from tropical dry forest in Los Santos, Coclé, and Panamá, and from premontane moist forest in the Canal Zone and Panamá. Reported from tropical wet forest in Costa Rica (Holdridge et al., 1971).

PROTIUM Burm.f.

Protium costaricense (Rose) Engler, Nat. Pfl. ed. 2, 19a:414. 1931

P. salvozae Standl.

Functionally dioecious tree, to 11(25) m tall, the trunk usually less than 20 cm dbh; outer bark minutely roughened; inner bark reddish, forming minute, cloudy, viscid droplets; sap lacking typical burseraceous aroma; stems, petioles, rachises, inflorescence branches, and veins of leaflet pubescent, the trichomes stiff, erect, brownish, of varying lengths. Leaves pinnate; petioles 4–6.5 cm long, flattened on upper surface; leaflets (3)5–7(9), mostly oblong-elliptic, abruptly acuminate, obtuse to acute at base, slightly inequilateral, 6–17 cm long, 2.5–8 cm wide. Flowers cream, 4-parted, in axillary panicles 1–13.5 cm long, often branched from near base; pedicels 1.5–3 mm long, puberulent; calyx bowl-shaped, broader than long, 1–2 mm long, shallowly lobed, the lobes acute to rounded; petals ca 3 mm long, narrowly ovate, acute and weakly spreading at apex; stamens 8, arising from beneath a prominent disk, the disk weakly lobed, half as high as ovary in pistillate flowers (2 mm tall), as high as ovary in staminate flowers; filaments inflexed; anthers held directly over style; pollen white, tacky; ovary ovoid, appressed-pubescent, much shorter than stamens; style ± sessile; stigma 4-lobed. Capsules drupaceous, ± ovoid, 1.5–2 cm long, violet-purple to brownish, sparsely lenticellate, obtuse at both ends, the valves 2(5), one falling free; pyrene 1, tan, 1-seeded, ca 1 cm long, falling free from funiculus (this ca 5 mm from apex) and suspended on a narrow band of tissue, enveloped in a white, sweet, fleshy aril displayed against the bright red inner valve surface. *Croat 8262, 14926.*

Rare, in the older forest along Zetek and Drayton trails. Apparently flowers twice per season, once in February and early March, with the fruits maturing in March and April, and again in May and June, with the fruits developing mostly in August and September.

Known from Costa Rica and Panama. In Panama, known only from tropical moist and premontane wet forests in the Canal Zone on the Atlantic slope; no doubt occurring on the Atlantic slope of western Panama also.

Fig. 302.
Protium panamense

Fig. 303. *Tetragastris panamensis*

Fig. 304. *Trattinnickia aspera*

Protium panamense (Rose) I. M. Johnston, Contr.
Gray Herb. 70:72. 1924
Copá

Functionally unisexual tree, usually less than 12 m tall
(to 40 m elsewhere), ± stilt-rooted at base, glabrous;
outer bark thin, smooth, the leaf scars on younger trees
ca 5 cm broad; inner bark pinkish; sap with sweet strong
odor typical of family. Leaves pinnate; petioles 6–13 cm
long, flattened on upper surface, swollen at base; petiolules 1–4(8) cm long, swollen at both ends; leaflets
3–7(9), mostly ovate-lanceolate to oblong, bluntly short-
acuminate, obtuse to rounded at base, 13–35 cm long,
6–9(16) cm wide, coriaceous. Inflorescences paniculate
(rarely racemose), axillary or terminal, sometimes cauliflorous, mostly to 15 cm long; flowers greenish-white
to greenish-yellow, 4(5)-parted, with a strong, sweet
aroma; pedicels to ca 4 mm long; calyx ± truncate or
shallowly lobed, very short; petals 3 mm long, valvate,
± spreading at anthesis, papillose-puberulent on margins,
the apex acute; stamens 8(10), 2 mm long; filaments
broadened below, recessed somewhat in fleshy yellowish
disk; pistil conical, short; stigma simple, almost sessile,
persistent in fruit. Capsules drupaceous, red at maturity,
often in a large congested cluster, ellipsoid to ovoid,
apiculate, shortly stipitate, to 2.8 cm long, the valves 2
or 4, unequal, the smaller falling free at maturity; pyrene
1(4), to ca 1.5 cm long, bearing 1 or 2 seeds, greenish,
covered with a thick, white, fleshy, sweet mesocarp ca 3
mm thick, attached to larger valve near apex, becoming
pendent, displayed against red inner valve surface. *Croat
4834, 11110.*

Abundant, especially in the younger forest and along
the shore; rare in the older forest. Flowers usually twice
per year (sometimes three times per year), once in the
early dry season (January to February, rarely as late as
April), once in the early rainy season (usually July and
August), and rarely again in the late rainy season. Fruit
maturation is considerably more staggered, and mature
fruits have been seen from February to August, often at
the same time as flowering.

Though flowers are functionally unisexual, there are
few conspicuous differences between staminate and
pistillate flowers or inflorescences.

Known only from Panama, probably extending into
both Costa Rica and Colombia. In Panama, known only
on the Atlantic slope from tropical moist forest in the
Canal Zone, Bocas del Toro, and San Blas, from tropical
wet forest in Colón and Coclé, and from premontane wet
forest in the Canal Zone and Colón.

See Fig. 302.

Protium tenuifolium Engler subsp. **sessiliflorum**
(Rose) Porter, Ann. Missouri Bot. Gard. 56:475. 1969
P. sessiliflorum (Rose) Standl.; *P. neglectum* var. *panamense*
Swart; *P. neglectum* var. *sessiliflorum* (Rose) Swart
Animé, Chutras, Comida de mono

Functionally unisexual tree, to 18 m tall, to 27 cm dbh;
outer bark thin, flaking in firm sheets, lenticellate; inner
bark moderately thick, reddish; sap slowly forming viscid,
cloudy droplets after slash, with strong, sweet aroma;
stems glabrous, prominently brown-lenticellate. Leaves
imparipinnate; petioles flattened above with sharp margins; petiolules thickened at both ends; leaflets 5–9(13),
mostly oblong to oblong-elliptic, acuminate with blunt
acumen, acute to rounded at base, often ± inequilateral,
10–20(27) cm long, 5–9(12) cm wide, glabrous except
for minute papillae. Panicles axillary, near stem apex, the
branches ferruginous; staminate inflorescences somewhat
larger than pistillate (to 24 cm long); flowers 5-parted,
sessile, 3.5–5 mm long; calyx to 2 mm long, cupulate,
5-lobed, minutely pubescent, the lobes acute; petals
lanceolate, yellow-green, thickened medially on outside,
ca 3 mm long, erect to somewhat spreading at anthesis;
stamens 10, alternately long and short, the longest to 2
mm long, the shortest to 1.3 mm long, scarcely longer
than the prominent, pubescent disk; ovary ovoid, densely
pubescent; style short, the ovary and style together ca
2.7 mm long; stigma 5-lobed, persisting in fruit. Capsules
drupaceous, bearing 1–5 pyrenes, ovoid to depressed-
globose, with 4 or 5 lobes (rarely 2- or 3-lobed), mostly
2–3 cm diam, red at maturity, usually rounded at both
ends, the carpels 4 or 5, each dehiscing by a single valve,
the valve falling free to expose the white mesocarp surrounding pyrene; pyrenes black, 1 per carpel, 1-seeded,
ovoid in face view, irregular in side view. *Croat 11109,
14822.*

Common along the shore; occasional in the forest.
Flowers chiefly in April and May. The fruits mature
chiefly in the middle of the rainy season (August to
October).

Small bees, possibly *Trigona*, have been seen visiting
the flowers. The fruits are edible.

Costa Rica and Panama. In Panama, known only from
tropical moist forest in the Canal Zone, Chiriquí, Panamá, and Darién. Reported from tropical wet forest in
Costa Rica (Holdridge et al., 1971).

TETRAGASTRIS Gaertn.

Tetragastris panamensis (Engler) O. Kuntze, Rev.
Gen. Pl. 1:107. 1891
Animé

Dioecious tree, to 35 m tall and 60 cm dbh, scarcely
buttressed; outer bark rough, unfissured, flaky. Leaves
imparipinnate; petioles flattened on upper surface, nearly
glabrous; leaflets 7–9(11), ± oblong-elliptic, acuminate,
acute to nearly rounded at base (lateral leaflets often
inequilateral), 5–20 cm long, 2–7 cm wide, nearly glabrous, stiff, the midrib arched. Panicles axillary (or terminal by abortion of stem apex), loosely branched,
2–15 cm long, with most exposed parts ± appressed-
puberulent; flowers (4)5-parted; pedicels ca 1 mm long;
calyx cup-shaped, shallowly lobed, to 1.5 mm long;
corolla yellowish-green, 3–5 mm long, lobed ca one-
third its length, the lobes acute, thick, erect; stamens
(8)10, included, abortive in pistillate flowers; disk annular, (8)10-lobed; ovary ovoid, (4)5-lobed and (4)5-locular,
sparsely pubescent; style short, pyramidal; stigma 5-lobed.
Capsules drupaceous, purplish or reddish to brown,
broadly turbinate to depressed-globose, round in cross

section, to 2.5 cm long and 3.5 cm wide, usually with all 5 carpels developing, rarely 3 or 4, the thick wedge-shaped valves red within, falling free at maturity; mesocarp spongy, sweet; pyrenes 1 per carpel, 1-seeded, ca 1.5 cm long (including aril), attached subapically and situated between slender, red, platelike partitions of the main body, covered with a fleshy, sweet, white aril ca 2 mm thick. *Croat 6823, 11195.*

Common in the forest. Flowers chiefly in June and July (rarely as late as August and September); individuals of the species may flower every two years. The fruits mature mainly from March to the middle of May (rarely to July).

On BCI, fruits are generally found on the ground, though usually only the star-shaped main axis and loose valves remain. The seeds are quickly taken by large birds and mammals.

Belize to Peru and Brazil. In Panama, ecologically variable; characteristic of tropical moist forest (Tosi, 1971), known in the Canal Zone, San Blas, Veraguas, Panamá, and Darién; known also from tropical dry forest in Coclé, from premontane moist forest in the Canal Zone and Panamá, from premontane wet forest in Panamá and Darién, and from tropical wet forest in Colón. Reported from premontane rain forest in Costa Rica (Holdridge et al., 1971).

See Fig. 303.

TRATTINNICKIA Willd.

Trattinnickia aspera (Standl.) Swart, Recueil Trav. Bot. Néerl. 39:426. 1942
Protium asperum Standl.
Caraño

Functionally dioecious tree, 25–50 m tall, to ca 75 cm dbh, weakly buttressed, the base becoming very roughened and warty; outer bark hard, dark brown; inner bark granular, tan; sap with faint, pleasant aroma (not characteristically burseraceous). Leaves deciduous, imparipinnate; petioles flat above, with a raised marginal rib; rachis triangulate on upper surface, rounded below; petiole and rachis shortly pubescent; leaflets 7–11 (to 19 on juveniles), oblong-ovate to oblong-elliptic, acuminate, cuneate to cordate at base (sometimes inequilateral on lateral leaflets), 10–20(30) cm long, 4–6(12) cm wide, asperous and scabridulous on both surfaces, sparsely hispidulous below especially on veins, stiff, the margins entire and ± undulate; lateral veins prominent below. Flowers 3-parted, functionally unisexual, in terminal panicles to 29 cm long, usually in dense glomerules, the branches sharply angulate, densely floriferous, hispidulous; pedicels to 4 mm long at anthesis, broadened apically, flattened; calyx sericeous inside, ca 2.7 mm long, caducous; corolla ± urceolate, ca 5 mm long, dull red (tinged with green in pistillate flowers), trilobate, the lobes acute, divided ca halfway to base in staminate flowers and one-fourth to three-fourths the way to base in pistillate flowers; stamens 6 (rarely 10), almost sessile (abortive in pistillate flowers); filaments strap-shaped; anthers oblong, ca 2 mm long, introrse; ovary ± ovoid, glabrous, 2–2.7 mm long at

anthesis (reduced in staminate flowers); styles 2, short and thick. Fruits drupaceous, ovoid, ca 1 cm long, blunt or rounded at apex, smooth (drying ± wrinkled and acute at apex), indehiscent, violet-purple at maturity; pyrenes (1)2, each 1-seeded. *Croat 11667, 11881, 13932.*

Frequent in the old forest. Flowers in late July and August. The fruits mature during the dry season from January to April. Trees lose all their leaves during the early rainy season, but are renewed soon.

Easily distinguished from all other species by its sandpapery, compound leaves with the petiole flattened on the upper surface. Because of the differences in stamen size and condition of the ovary, the sex of any tree can be determined in the field by fallen flowers. Staminate flowers usually fall soon after the pollen has been shed, in most cases with the corolla still attached.

Fruits are probably consumed whole by large birds and mammals, since there is relatively little mesocarp.

Known only from Panama; characteristic of tropical wet forest, principally on the Atlantic slope (Tosi, 1971); known also from tropical moist forest in the Canal Zone, Bocas del Toro, and Panamá and from premontane wet forest in the Canal Zone (Pipeline Road).

See Fig. 304.

70. MELIACEAE

Functionally dioecious or monoecious (*Cedrela*) trees, the wood often scented. Leaves alternate, petiolate, pinnately compound; leaflets entire; venation pinnate; stipules lacking. Flowers appearing bisexual (see discussion below), actinomorphic, in terminal or axillary, cymose panicles bearing few to many flowers; perianth mostly 5-parted; calyx variously lobed; petals free or attached to the staminal tube; disk present; stamens as many or twice as many as the petals; filaments variously united into a tube; anthers 2-celled, longitudinally dehiscent; ovary superior, 2–5-locular and 2–5-carpellate; placentation axile; ovules 1, 2, or many per locule, pendulous, anatropous; style 1; stigma capitate or discoid. Fruits loculicidally or septicidally dehiscent capsules, with 1, 2, or many (*Cedrela*) seeds per cell; seeds often winged, often arillate, with or without fleshy endosperm.

Members of the family may be confused with Burseraceae (69) and Anacardiaceae (76). They can be distinguished by having the stamens united into a tube or adnate to the gynophore and by having discoid or capitate stigmas. In addition, leaves are sometimes paripinnate (*Guarea*), and the petioles are usually markedly flattened on the upper side.

Styles (1972) believed *Guarea, Cedrela,* and *Trichilia* to be functionally dioecious. Flowers of the two sexes are only slightly modified, with the stamens and pistil reduced in the respective unisexual flowers. I have called dioecious only those species that I am confident are dioecious, on the basis of field observations on BCI. Some species certainly appear to have bisexual flowers. For example, all trees of *Trichilia cipo* observed set fruit after flowering. For most species not enough observations were made to be certain of their breeding behavior.

KEY TO THE SPECIES OF MELIACEAE

Leaflets mostly in more than 6 pairs (10–22 leaflets):
    Leaflets equilateral at base; leaves to 2 m long; flowers ca 1 cm long; fruits ± globose, to 3.5 cm diam; seeds with a red aril . . . . *Guarea multiflora* Adr. Juss.
    Leaflets inequilateral at base:
        Flowers ca 3 mm long; capsules subglobose, 3-valved, ca 1 cm long; seeds arillate; leaflets thin, usually less than 4.5 cm wide, lacking axillary domatia; bark not coarsely fissured . . . . *Trichilia hirta* L.
        Flowers 6–10 mm long; capsules oblong-ellipsoid, 5-valved, to 4.5 cm long; seeds winged; leaflets thick, often more than 4.5 cm wide, bearing pocketlike axillary domatia; bark coarsely fissured . . . . *Cedrela odorata* L.
Leaflets in 6 or fewer pairs:
    Flowers 6–10 mm long; anthers borne on inner surface of staminal tube; disk columnar or obsolete; fruits 4- or 5-valved, not warty, usually conspicuously covered with raised lenticels . . . . *Guarea glabra* Vahl
    Flowers less than 4 mm long; anthers borne at apex of staminal tube; disk ring-shaped; fruits mostly 3-valved, somewhat to markedly warty, lacking lenticels:
        Filaments united only in basal half; terminal leaflet larger than lateral leaflets, 20–32 cm long, 6.5–16 cm wide; inflorescences axillary, usually less than 8 cm long . . . . *Trichilia montana* H.B.K.
        Filaments united to form a staminal tube; terminal leaflet usually ± equaling lateral leaflets, less than 20 cm long and 8 cm wide; inflorescences subterminal, often more than 8 cm long:
            Leaflets heteromorphic, 3–5 full-sized and 1 or 2 pairs abortive; midrib flat; staminal tube pubescent; fruits round, green, only slightly bumpy . . . . *Trichilia verrucosa* C. DC.
            Leaflets regular, 5–9, all full-sized; midrib sharply raised; staminal tube glabrous; fruits ellipsoid to oblong, orange, conspicuously warty . . . . *Trichilia cipo* (Adr. Juss.) C. DC.

Flowers are somewhat specialized and nectar is usually present. The anthers provide little pollen, and the pollination system is unknown.

Seeds of *Cedrela odorata* are wind dispersed. *Guarea* and *Trichilia* are probably for the most part bird dispersed, but fruits of *Trichilia cipo* are taken by white-faced, spider, and howler monkeys (Oppenheimer, 1968; Hladik & Hladik, 1969).

Fifty genera and about 1,000–1,400 species; mostly in the tropics.

CEDRELA Scop.

Cedrela odorata L., Syst. Nat. ed. 10, 940. 1759
    Cedro, Spanish cedar, Cigar-box cedar

Monoecious tree, to 40 m tall; trunk reddish-brown especially near base, grayish above, deeply fissured, sometimes buttressed to 2 m; outer bark hard, persistent; sweet odor of cedar from slash. Leaves alternate, glabrous, to 60 cm long; leaflets in 5–11 pairs, opposite or subopposite, ovate-elliptic, acuminate, asymmetrical and acute to rounded at base, 6–17 cm long, 3–5.5 cm wide, the axils of major veins with pocketlike domatia on lower surface. Panicles pyramidal, open, terminal; branchlets generally at right angles to axis; bracts caducous; inflorescence branches and pedicels glabrous to minutely puberulent; pedicels to 2 mm long (to 5 mm and thickened in fruit); flowers functionally unisexual; calyx cupulate, 1.5–3 mm long, usually split on one side, ± irregularly 5-dentate, usually glabrous; petals 5, oblong-elliptic, 6–9 mm long, fused to gynophore medially, puberulent

outside, white; stamens 5, slightly shorter than petals; filaments free above, ± fleshy, uniform; anthers introrse, ca 1 mm long; style slightly longer than stamens; stigma green, thick, capitate, almost as long as petals. Capsules oblong-ellipsoid, to 4.5 cm long, green with light brown scurfy patches, becoming brown, dehiscing from apex, 5-valved, the valves thin; seeds many, samaroid, ca 3 cm long, with membranaceous wing, suspended from the seminiferous apex on the central column, the column thick, corky, 5-ribbed. *Croat 10332.*

Rare; known from only a few places in the forest—one tree is at Barbour Trail 1400. Flowers usually in the early rainy season. The fruits mature in the dry season. Trees are deciduous in the early dry season.

Northern Mexico south to Peru, Bolivia, and Brazil; West Indies; introduced into the Old World. In Panama, known from tropical moist forest on BCI and in Bocas del Toro and Darién and from premontane wet forest in Chiriquí; known also from tropical dry forest in Panamá (Tosi, 1971). The closely related *C. angustifolia* Sessé and Moc. is more widespread in tropical moist and tropical dry forests along the Pacific slope.

See Fig. 305.

GUAREA Allem. ex L.

Guarea glabra Vahl, Eclog. Amer. 3:8. 1807
    Cedro macho

Dioecious tree, 3–15 m tall, the trunk to 16 cm dbh; outer bark soft, flaky, easily scraped off; stems roughened with fine lenticels; young petioles and leaves ± sericeous,

Fig. 305. *Cedrela odorata*

Fig. 306. *Guarea multiflora*

Fig. 307. *Guarea multiflora*

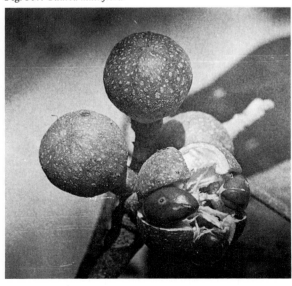

KEY TO THE SPECIES OF GUAREA

Leaflets in usually fewer than 6 pairs, less than 7 cm wide . . . . *G. glabra* Vahl
Leaflets in usually more than 6 pairs, frequently more than 7 cm wide . . . . *G. multiflora* Adr. Juss.

becoming glabrate. Leaves paripinnate, to 54 cm long; petioles flat and margined above, swollen at base; petiolules 5–10 mm long, swollen; leaflets 6–12, ovate-elliptic to oblong, acuminate and downturned at apex, acute to obtuse at base, 7–25 cm long, 4–7 cm wide, glabrous above, glabrate below. Inflorescences racemose, axillary or on older leafless branches, to 23 cm long; flowers with sweet aroma, functionally unisexual, well spaced, ca 1 cm long, solitary along rachis or clustered on very short branches; calyx cupular, 2–4 mm long, irregularly lobed, the lobes often fewer than petals, puberulent outside; petals white, 4 or 5(6), to ca 10 mm long, usually puberulent outside, spreading or recurved near middle at anthesis; staminal tube cylindrical, white, 6–8 mm long; anthers 8 or 10(12), borne inside near rim; ovary ± glabrous to sericeous, borne on a gynophore broadened just below ovary; style exserted above staminal tube; stigma capitate. Capsules subglobose to pyriform, to 2.5 cm diam, usually borne on a short stipe, reddish, variously mottled with lenticels in age, dehiscent by 4 or 5 woody valves; seeds 1 or 2 per carpel, covered with a red aril, irregular. *Croat 8493, 8814.*

Frequent in the forest. Seasonal behavior uncertain. Flowers at least from December to July, especially from March to June. The fruits probably mature mostly from April to September, especially in the rainy season.

A variable species, it has been identified by C. E. Smith under three other names as well, including *G. guidonia* (L.) Sleumer, *G. kunthiana* Adr. Juss., and *G. tonduzii* C. DC. It can be reported with certainty that *G. tonduzii* does not occur on BCI and furthermore that there is but a single species in this complex thus far collected on the island. The pubescence of the ovary, used in distinguishing *G. glabra* and *G. guidonia*, is at best a poor character.

A large caterpillar with stinging hairs has been seen eating the leaves.

Mexico to Colombia, Venezuela, and Ecuador; West Indies. In Panama, known from tropical moist forest in the Canal Zone, Bocas del Toro, San Blas, Panamá, and Darién and from tropical wet forest in Colón and Chiriquí; probably occurring in most areas of tropical moist and tropical wet forests.

Guarea multiflora Adr. Juss., Mém. Mus. Hist. Nat. 19:284. 1830

Dioecious tree, 5–40 m tall; bark flaking. Leaves paripinnate, mostly 30–200 cm long; petioles flattened above; rachis and petiole puberulent to tomentose; petiolules 3–7 mm long, ± glabrous above, glabrous to minutely scabrid below, becoming glabrous in age; leaflets 8–34, elliptic to oblong-elliptic, acute to acuminate or sometimes rounded at apex, acute to rounded at base, 7–30 cm long, 4.5–12 cm wide. Flowers in slender, sparsely branched, axillary panicles to 35 cm long; all exposed parts tomentose; pedicels short; calyx cupular, to 3 mm long, glabrous inside, the lobes blunt to acute, mostly 4 or 5(6); petals 4 or 5(6), oblong, blunt at apex, spreading at anthesis, ca 1 cm long, staminal tube slightly shorter than petals, ± glabrous; anthers 8 or 10(12), mounted within tube below rim, ca 1.3 mm long; ovary usually 5-sulcate, mounted on a stout gynophore, the ovary and style sericeous; stigma discoid, to 1.5 mm broad, held just above apex of staminal tube. Capsules ± pyriform, 3–3.5 cm diam, reddish to reddish-brown, scurfy with numerous lenticels and persistent short pubescence, dehiscing by 4–6 woody valves to expose seeds; seeds 1(2) per carpel, ellipsoid, to 2 cm or more long, covered with a thin red aril. *Croat 8443, 11300.*

Rare, in the forest. Flowers from June to October. The fruits mature mostly from February to June.

Woodworth & Vestal 747, cited by Standley as *G. guarea* P. Wils. (= *G. guidonia* (L.) Sleumer), is *G. multiflora*, since he said it has 8–20 leaflets. C. E. Smith (1965) reported the specimen in the *Flora of Panama* as *G. guidonia*, but said the species has no more than 14 leaflets.

Southern Mexico to Bolivia. In Panama, known from tropical moist forest in the Canal Zone, Bocas del Toro, San Blas, Panamá, and Darién.

See Figs. 306 and 307.

TRICHILIA P. Browne

Trichilia cipo (Adr. Juss.) C. DC. in Mart., Fl. Brasil. 11(1):214. 1878

Trichilia tuberculata C. DC.

Alfaje, Alfajeo colorado, Camfine, Fosforito

Tree, to 30 m tall and 40 cm dbh (often much shorter along the shore); outer bark hard, closely fissured; inner bark red; younger stems, leaf rachises, and axes of inflorescences densely strigose and puberulent (in Panama) or sparsely strigose, sparsely puberulent, or nearly glabrous (often in South America). Leaves alternate, imparipinnate, mostly 15–30 cm long; petioles flat on upper surface with marginal ribs; leaflets apparently never

KEY TO THE SPECIES OF TRICHILIA

Capsules very densely and conspicuously pubescent; staminal tube entire or the anthers merely alternating with apicula:
    Leaflets more than 9, the uppermost less than 15 cm long and 4.5 cm wide . . . . *T. hirta* L.
    Leaflets fewer than 7, the uppermost more than 15 cm long and 6.5 cm wide . . . *T. montana* H.B.K.

Capsules glabrous or only sparsely and inconspicuously pubescent; staminal tube divided for at least one-third its length:
    Capsules ± globose, green, only slightly bumpy, not conspicuously warty; staminal tube pubescent; leaflets usually 5 or fewer, usually with 1 or 2 pairs of abortive leaflets at base, the midrib not raised . . . . *T. verrucosa* C. DC.
    Capsules oblong or ellipsoid, orange, conspicuously warty; staminal tube glabrous; leaflets usually more than 5, lacking abortive leaflets, the midrib prominently raised . . . .
. . . . *T. cipo* (Adr. Juss.) C. DC.

---

dimorphic, usually 5–9 (rarely 3 or 10), usually alternate, ± elliptic, acuminate, obtuse-attenuate at base, 4–21 cm long, 1.5–8 cm wide, usually glabrous or minutely puberulent and/or strigose on midrib on both surfaces, rarely strigose below; midrib sharply raised. Panicles axillary (often appearing terminal), mostly 8–20 cm long; flowers white or greenish-yellow, 4–6-parted, as wide as long, globular in bud, the trichomes (at least on calyx) erect or appressed; calyx saucer-shaped, the lobes acute or blunt; petals valvate, ovate, to ca 2 mm long, glabrous or sparsely appressed-pubescent outside, glabrous inside, the margins papillose-puberulent; staminal tube two-thirds to three-fourths as long as petals, broader than long, glabrous outside and in but sometimes pubescent inside near apex, the lobes narrowly acute; anthers twice as many as petals, very narrowly ovoid, ca 0.5 mm long, alternating with and usually longer than lobes of staminal tube; ovary broadly ovoid, densely appressed-pubescent with white trichomes all over; style usually glabrous. Capsules ellipsoid, 11–18 mm long, orange at maturity, prominently tuberculate, the 3 (4) valves folding back to expose shiny red aril; seeds 1 or 2, smooth, ellipsoid, ca 1 cm long. *Croat 12271, 12500.*

Abundant as seedlings; common as adults in the forest, especially in the old forest. Flowering is bimodal, but principally from April to June, especially in June; a second but much smaller flowering period may occur from September to November. It is not known whether the second flowering occurs each year or only sporadically. The fruits from the first flowering mature mostly during September and October, especially in October; mature fruits from the second flowering have been seen in June.

Guatemala to the Guianas and Bolivia. In Panama, known from tropical moist forest in the Canal Zone, principally on the Atlantic slope, and in Bocas del Toro, Colón, San Blas, Chiriquí (Progreso), Panamá, and Darién; known also from tropical wet forest in Colón (Salúd).

Trichilia hirta L., Syst. Nat. ed. 10, 2:1020. 1759
    Conejo colorado, Huesito, Souca

Tree, to 10 m tall; young stems, petioles, rachises, veins of blades below, and inflorescence branches sparsely hirsute. Leaves pinnate, imparipinnate or paripinnate, to 30 (45) cm long; petioles 3–5 (9) cm long, pulvinate at base, the pulvinus ± flat above; leaflets (9)13–21, opposite or subopposite near apex, sometimes becoming alternate toward base, ± elliptic, long-acuminate, acute to rounded and slightly inequilateral at base, 5–9 (14) cm long, 1.5–3 (4.4) cm wide, sparsely ciliate. Panicles

3–15 (18) cm long, solitary or paired (when paired, 1 usually very short) in upper axils; lateral branches short; peduncles and basal part of axis often ± flattened; pedicels to 1 mm long, with minute narrowly acute bracts at base; flowers 5-parted, 2–3 mm long; calyx saucer-shaped, the lobes short, acute; petals oblong, acute at apex, white, densely papillate (especially on margins); stamens 10, free except near base, to 2.3 mm long, included; filaments flattened, contiguous, forming a loose staminal tube, stiff-villous on margins and within above middle, the lateral margins weakly prolonged at apex; anthers ca 0.7 mm long, narrowly ovoid, introrsely dehiscent, sparsely villous all over; ovary orange, depressed, ca 1.2 mm wide, about half as high, glabrous to densely pubescent; style thick, ca 2.3 mm long, swollen slightly below apex, villous, especially near base; stigma simple, depressed; nectar stored within stamens at base of ovary. Capsules subglobose, obtuse and apiculate at apex, ca 1 cm long, brown to violet-purple at maturity, densely and minutely papillate-puberulent and sparsely hirtellous at least when young, 3-valved, the open capsule bowl-shaped with a prominent narrow midrib inside; seeds (2) 3 (4), ovoid, ca 6 mm long, covered with a fleshy orange aril. *Croat 15099.*

Apparently rare; collected on the west edge of Bat Cove. Flowers mostly from May to July. The fruits mature during the dry season of the following year, at least in some cases while the trees are bare.

Smith (1965) in the *Flora of Panama* stated that the inflorescences are racemose. He uses this character in the key to separate this species from *T. tomentosa* H.B.K. However, most of the Panamanian materials I have seen have paniculate inflorescences. Reported to be dioecious (Bawa & Opler, 1975).

Mexico to Brazil; West Indies. In Panama, reported by Smith (1965) to be a species of fence rows, forest margins, and forests; known from tropical moist forest in the Canal Zone, Chiriquí, Panamá, and Darién.

Trichilia montana H.B.K., Nov. Gen. & Sp.
    7:226. 1825

Tree, to 10 (25) m tall, slender; outer bark light brown, very smooth, with minute whitish lenticels; inner bark and wood whitish, the sap with sweet aroma; branches low-arching; most parts glabrous but with leaf axes, young stems, and inflorescences puberulent. Leaves imparipinnate, to 45 cm long; petioles mostly 4–8 cm long; leaflets (3) 5 (7), ± elliptic, acuminate, acute to obtuse at base, the terminal leaflet 20–32 cm long, 6.5–16 cm wide, the lower leaflets reduced in size. Flowers 3–4 mm long, ± sessile, white, cream or greenish, 4–6-parted,

in axillary panicles to 8 cm long (usually to 4 cm long, sometimes reduced to appear fasciculate); calyx ca 1 mm long, the lobes acute; petals ± oblong; stamens 8–12, to 3.5 mm long; filaments united into a tube in basal half, the tube orange, fleshy within, the free part villous, broader than anther; anthers villous; ovary bearing straight erect trichomes; style short. Capsules obovate, ca 1 cm long, yellowish or reddish, densely velutinous, sometimes transversely striate or bearing dense, flat, sharp protuberances ca 1 mm long, the valves 3 or 4; seeds 1 or 2, ca 6 mm long, black, partly covered by a red-orange aril. *Croat 14847, Foster 1123.*

Rare; seen in the vicinity of the large *Ceiba pentandra* trees (86. Bombacaceae) between Armour Trail 700 and Zetek Trail 400. Flowers from April to August (sometimes from March), with the fruits maturing from June to September (sometimes from April). Plants from the Osa Peninsula in Costa Rica flowered in October, with the fruits maturing in January and February.

Mexico to Brazil. In Panama, known from tropical moist forest in the Canal Zone, Bocas del Toro, Chiriquí, and Darién and from tropical wet forest in Darién. Reported from premontane wet forest in Costa Rica (Holdridge et al., 1971).

Trichilia verrucosa C. DC., Monogr. Phan. 1:695. 1878

Tree, 5–20 m tall; outer bark minutely fissured, thin (bearing short, raised, horizontal lines on inner surface); inner bark reddish-brown on the outer edge, bearing impressed short horizontal lines, the sap with sweet, distinctive aroma; young twigs, rachises, axes of inflorescences and of leaves moderately short-appressed-pubescent, glabrous in age. Leaves alternate, pinnate, mostly 8–25 cm long, essentially glabrous; petioles flat on upper surface with marginal ribs; leaflets usually alternate, often dimorphic, those of full size usually 3–5, often with 1 or 2 pairs of minute, aborted leaflets at base, rarely with 7 full-sized leaflets, the blades ± elliptic, acuminate, acute to obtuse at base, 6–16 cm long, 2.5–6(7) cm wide; major lateral veins mostly 8–12, scarcely raised

below, joining before the margin, the midrib usually flat above. Panicles solitary in upper axils, 5–17 cm long; flowers greenish, mostly 5-parted; calyx bowl-shaped, shallowly lobed, 1–2 mm long, glabrate; petals ovate, 2–3 mm long, usually glabrous but with minute papillae on margin; staminal tube one-half to two-thirds as long as petals, much broader than long, moderately to sparsely pubescent outside at least toward base (often glabrous outside in South America), sparsely to densely pubescent inside; anthers twice the number of petals, slender, slightly tapered toward apex, ca 1 mm long, dehiscing along lateral margins, alternating with and ± equaling acicular lobes of staminal tube; ovary broadly ovoid, 1–2 mm wide, with 2 or 3(4) locules, the trichomes dense, appressed, white on basal half; style ± equaling ovary at anthesis; stigma capitate, inconspicuously lobed, held about at height of anthers. Fruits ± globose, green at maturity, weakly verrucose, 2–3 cm diam, ± glabrous, dehiscing by 3 valves; seeds 1 or 2, covered with a shiny red aril. *Croat 5354, 6035.*

Rare; collected only along the shore between Bat Cove and the dock. Flowers from March to May. The fruits mature from August to September.

The species was considered synonymous with *T. cipo* by Smith (1965) in the *Flora of Panama*, but has been shown to be distinct from that species (Croat, 1975a).

Panama to the Guianas, Brazil, Bolivia, and Peru; Trinidad. In Panama, known from tropical moist forest on both slopes in the Canal Zone, Colón, Los Santos, Panamá, and Darién.

71. MALPIGHIACEAE

Trees, shrubs, or scandent lianas (may be only slightly woody in *Stigmaphyllon*). Leaves opposite, petiolate; petioles often glandular; blades simple, entire or irregularly toothed when juvenile; venation pinnate; stipules present; T-shaped trichomes frequently present. Flowers bisexual, actinomorphic or zygomorphic principally by modification of one petal, in axillary panicles, pseudo-

---

KEY TO THE TAXA OF MALPIGHIACEAE

Plants trees or shrubs, not scandent; fruits drupes, cocci, or capsules (except *Heteropteris*):
  Inflorescences much-branched, usually pyramidal panicles; fruits winged . . . .
  . . . . *Heteropteris laurifolia* (L.) Adr. Juss.
  Inflorescences racemose or pseudocorymbose; fruits not winged:
    Leaves densely and conspicuously soft-tomentose below . . . . *Byrsonima crassifolia* (L.) H.B.K.
    Leaves not tomentose below:
      Flowers yellow:
        Leaves glaucous and with appressed, white trichomes below, the younger parts not furfuraceous; flowers mostly 8–16; fruits 1.5–2.5 cm long, orange to red, the pyrenes usually 2 . . . . *Bunchosia cornifolia* H.B.K.
        Leaves not glaucous, ± glabrous in age, the younger parts densely furfuraceous; flowers many, usually more than 16; fruits ca 1 cm diam, green to yellow-orange, the pyrenes 3 . . . . *Byrsonima spicata* (Cav.) H.B.K.
      Flowers pink, red, or whitish:
        Inflorescences pendent, to 20 cm long, the fertile part more than 5 cm long; fruits cocci of 2 carpels, ca 5 mm long . . . . *Spachea membranacea* Cuatr.
        Inflorescences erect or nearly so, to ca 8 cm long, the fertile part less than 4 cm long; fruits drupelike with 3 pyrenes, to 10 mm long . . . . *Malpighia romeroana* Cuatr.

Plants scandent; fruits of 3 winged samaras:
  Wings of fruit lateral, each samara with 2 or 4 wings (sometimes united at apex into a nearly complete circle):
    Lateral wings of samara deeply 2-lobed, the lobes narrow, elongate (contiguous or separate at the base); inflorescences mostly terminal panicles . . . . *Tetrapteris*
    Lateral wings of samara not lobate, the lobes often as broad as or broader than long; inflorescences various, mostly axillary:
      Pedicels appearing articulate, with 2 bracteoles at the node or articulation; stipules small, at base of petioles; inflorescences panicles . . . . *Mascagnia*
      Pedicels sessile; stipules usually conspicuous, borne somewhere above base of petiole; inflorescences umbellate . . . . *Hiraea*
  Wings of fruit dorsal, each samara with essentially one wing (lateral wings when present much smaller):
    Inflorescences umbels of few flowers or series of widely separated umbels; stamens with 2 or more reduced and sterile; petioles bearing sessile glands at apex . . . . *Stigmaphyllon*
    Inflorescences panicles of many flowers, not appearing distinctly umbellate or as a series of umbels; stamens all fertile:
      Inflorescence branches reddish-brown; petioles eglandular; samaras lacking reduced lateral wings or crests . . . . *Heteropteris laurifolia* (L.) Adr. Juss.
      Inflorescence branches not reddish-brown; petioles usually bearing 2 sessile glands near apex; samaras with reduced lateral wings or crests . . . . *Banisteriopsis cornifolia* Small

racemes, or umbels of many flowers; sepals 5, free, often with conspicuous glands; petals 5, showy, free, clawed, fringed or toothed, alternate to the sepals; stamens (8)10, equal or unequal, sometimes with 4 reduced to staminodia; anthers 2-celled, introrse, dehiscing longitudinally, often with a thickened connective; ovary superior, mostly 3-locular, 3-carpellate; placentation axile; ovules solitary in each locule, pendulous, semianatropous; styles 3, free (rarely connate or with only 1 developed), sometimes forming a hood over the stamens; stigmas 1 per style or 3 when styles united, entire or minutely lobed. Fruits usually schizocarps of three samaras, sometimes drupes with 1–3 pyrenes (*Bunchosia, Byrsonima, Malpighia*), or sometimes berries of 2 separable cocci (*Spachea*).

Distinguished by the usually T-shaped (medicentric) trichomes, glandular calyx, and fringed petals. Most species are lianas.

Many species are markedly zygomorphic. Flowers are insect pollinated, and I have seen them visited by small bees. Vogel (1958) reported the family to be pollinated by anthophorid bees, usually *Centris*.

Fruits are mostly wind dispersed. A few species, including *Spachea membranacea, Bunchosia cornifolia, Malpighia romeroana*, and species of *Byrsonima*, are endozoochorous. *Malpighia romeroana* might be bat dispersed, since a similar species, *M. glabra* L., is eaten by bats in Trinidad (Goodwin & Greenhall, 1961).

About 60 genera and 850 species; mostly in the American tropics.

BANISTERIOPSIS C. B. Robinson ex Small

Banisteriopsis cornifolia (H.B.K.) C. B. Robinson ex Small, N. Amer. Fl. 25:132. 1910
  B. cornifolia (H.B.K.) Spreng.

Liana; young stems, petioles, and blades below densely to sparsely pubescent, the trichomes appressed, T-shaped. Petioles 5–10 mm long, bearing 1 or 2 sessile glands along each side near apex; blades ovate to elliptic, acumi-

nate (the acumen often downturned), rounded at base, 9–16(18) cm long, 4.5–7 cm wide, glabrous and somewhat shiny and rugulose above. Panicles axillary or terminal, solitary or several per axil, 4–15 cm long, branched many times; peduncles, pedicels, and sepals densely pubescent, the trichomes short, appressed; peduncles to 3.5 cm long; pedicels sessile, to 1 cm long (to 1.5 cm long in fruit); sepals ovate, 2–3 mm long, bearing large exterior glands; petals clawed, 4–7 mm long, yellow, glabrous, caducous; stamens and styles glabrous; stamens unequal; anthers ± oblong, glandular; styles 3, distinct, ± equaling stamens. Fruits of 1–3 samaras, each samara with a single dorsal wing thickened along the inner side; seminiferous area ca 8 mm long, densely pubescent, raised, with a small wing or ridge on both sides; dorsal wing 2.5–3.5 cm long, ca 1 cm wide, bearing short appressed pubescence especially near seed and a small protrusion on inner margin just above seed. *Foster 1478.*

Rare, in the forest, usually growing in the canopy. Probably flowers in late rainy and early dry seasons, with the fruits maturing in the dry season.

Southern Mexico to Peru. In Panama, known from tropical moist forest on BCI and from premontane wet forest in Panamá (Cerro Campana).

BUNCHOSIA L. C. Rich. ex Adr. Juss.

Bunchosia cornifolia H.B.K., Nov. Gen. & Sp. 5:154. 1822

Shrub or tree, to 6 m tall, pubescent especially on the younger and lower parts and on the leaf blades below, the trichomes appressed, T-shaped; stems becoming glabrous in age, the outer bark very thin, grayish, the lenticels and outer bark easily scraped off. Petioles to 1 cm long; blades ± elliptic to oblong-elliptic, acuminate, obtuse to acute at base and decurrent-revolute onto petiole, 6–15(35) cm long, 3–6(14) cm wide, the upper surface soon glabrous, the lower surface persistently pubescent and also glaucous, with scattered glands especially

near base, the margin often somewhat undulate; dried leaves making a characteristic oily deposit on drying papers. Pseudoracemes axillary, 5–8 cm long; flowers mostly 8–16, 1–1.5 cm broad, 5-parted; sepals oblong-elliptic, rounded at apex, each lobe inflexed, bearing 2 conspicuous, large glands; petals yellow, clawed, rounded to oblong, spreading, the margin irregular, 1 or more petals markedly concave, 1 petal somewhat erect, its margin entire; stamens 8, subequal, in a staggered whorl, 5 held in close proximity to style, the longest ca 2.7 mm long; filaments thickened at base, weakly fused; anthers dehiscing inward, the connective somewhat thickened apically; styles 3, to 1.5 mm long; stigmas 3, short, fleshy, held just above anthers; ovary weakly pubescent. Drupes ovoid, orange turning red at maturity, 1.5–2.5 cm long, the style bases persistent; exocarp thin; mesocarp pasty, eventually sweet and tasty, ca 2–3 mm thick; pyrenes 2(3), ovoid-hemispheroid, to 17 mm long, with a thin reticulate covering. *Croat 11107.*

Known from a few areas along the northern shore. Elsewhere in the Canal Zone the plant is common. Flowers from December to July, principally from April to July. The fruits mature from June to January, sometimes on flowering plants.

Fruits are probably dispersed chiefly by birds and other arboreal frugivores.

Southern Mexico to Colombia and Ecuador. In Panama, known from tropical moist forest in the Canal Zone, Bocas del Toro, Panamá, and Darién and from tropical wet forest in Darién. Reported from premontane rain forest in Costa Rica (Holdridge et al., 1971).

BYRSONIMA L. C. Rich. ex Adr. Juss.

Byrsonima crassifolia (L.) H.B.K., Nov. Gen. & Sp.
5:149. 1822
Nance, Nance blanco, Nance colorado

Tree, 4–13 m tall, to 30 cm dbh; bark fissured and lenticellate; wood dull reddish-brown, hard, heavy; younger parts densely downy-tomentose (sparsely on upper leaf surface and becoming glabrate); stems bearing prominent leaf scars. Petioles ca 1 cm long, stout; blades obovate to elliptic or ovate, acuminate, narrowed to an acute or obtuse base, 7–14 cm long, 3–8 cm wide, densely pubescent below becoming glabrate except on midrib in age; midrib ± arched. Pseudoracemes terminal, usually solitary, to 20 cm long; pedicels to 1.5 cm long; flowers many, yellow, becoming red-orange in age; sepals each bearing 2 conspicuous glands, blunt, recurved, glabrous inside; petals clawed, 10–13 mm long, the blade orbicular, often concave, ± equaling length of claw, the margin irregular, 1 petal often smaller and held somewhat erect, the others spreading to reflexed; stamens 10, 4–5 mm long, interspersed with long straight trichomes; anthers introrse, equaling length of styles, the thecae prominently raised,

shedding pollen in bud, the connective thickened; ovary usually pubescent; styles 3, distinct, slender, longer than stamens, persisting on young fruits. Drupes ± globose, 1–1.5 cm diam, green turning yellow to reddish; pyrenes 1–3. *Croat 6068, 8702.*

Locally common along the shore; infrequent in the forest, usually near old settlement sites. Flowers from November to July, principally from March to June. Each tree flowers for about 6 weeks. The fruits mature principally in August and September. Leaves turn old and reddish, falling in the dry season and gradually growing out again in March just before the greatest flush of flowering. Allen (1956) reported that the species had fruits during April and May in Costa Rica, but the fruits were probably not ripe.

Byrsonima crassifolia may be confused with *B. cumingiana* Adr. Juss., which ranges from Nicaragua to Colombia. *B. cumingiana* is distinguished by having a glabrous ovary and leaves that are more thickly coriaceous and rugose above.

The fruits are probably dispersed by large birds and by mammals.

Veracruz, Mexico, south to Brazil and Paraguay; West Indies. In Panama, widespread and ecologically variable; known from tropical moist forest in the Canal Zone, Bocas del Toro, Veraguas, Herrera, Coclé, Panamá, and Darién, from premontane dry forest in Coclé, from tropical dry forest in Los Santos and Panamá, from premontane moist forest in the Canal Zone, from premontane wet forest in Chiriquí, and from tropical wet forest in Colón. Tosi (1971) listed this species as characteristic of tropical dry and tropical moist forests in Panama.

Byrsonima spicata (Cav.) H.B.K., Nov. Gen. & Sp.
5:147. 1822
B. coriacea (Sw.) Kunth
Nance, Nancillo

Tree, to 22 m tall, the trunk to 40 cm dbh; outer bark thin, light brown, with many minute fissures; inner bark brownish-red, moderately thick, the sap lacking odor; young parts ferruginous, appressed-pubescent. Stipules narrowly triangular, ca 3 mm long; petioles to 8 mm long; blades oblong-elliptic, acuminate, acute and decurrent at base, 5–10(14) cm long, 1.5–3(4.5) cm wide, glabrate or sparsely pubescent below and on midrib above in age. Pseudoracemes terminal, 6–10 cm long; pedicels 4–8 mm long; flowers many, ca 1.5 cm diam, yellow; sepals to 2 mm long, lobed, glandular, the glands 10, oblong-obovate, yellow, the lobes curved outward at apex; petals subequal, ± orbicular, concave, clawed, 4–6 mm diam, subentire, 1 petal flat and somewhat more erect than the others; stamens 10, in 2 series, nearly erect, 3–3.5 mm long; filaments pubescent, nearly as long as anthers; anthers introrse, the connective swollen, somewhat prolonged at apex; styles 3, narrowly pointed, distinct, slightly longer

KEY TO THE SPECIES OF BYRSONIMA

Leaves softly and conspicuously tomentose below . . . . *B. crassifolia* (L.) H.B.K.
Leaves glabrate to sparsely pubescent below . . . . *B. spicata* (Cav.) H.B.K.

than stamens, persisting in fruit. Drupes ± globose, ca 10 mm diam, glabrous, green to yellow or yellow-orange at maturity; pyrenes 3. *Croat 11133.*

Infrequent; known from the younger forest on Colorado Peninsula. Flowers from June to August, with the fruits maturing from August to October.

The fruits are probably dispersed by birds or other arboreal frugivores.

Costa Rica to Colombia and Venezuela; West Indies. In Panama, known from tropical moist forest in the Canal Zone and Darién and from tropical wet forest in Colón (Salúd) and Darién.

HETEROPTERIS Kunth

Heteropteris laurifolia (L.) Adr. Juss., Ann. Sci. Nat. Bot., sér. 2, 13:276. 1840

Liana or climbing shrub (rarely elsewhere a small tree or shrub), 1–8 m tall; young stems and petioles pubescent, the trichomes dense, short, appressed. Petioles to 1.2 cm long; blades oblong-elliptic, acuminate, acute to obtuse at base, 10–17 cm long, 3.5–6(7.5) cm wide, glabrate, sometimes bearing small glands near margin at base. Panicles solitary, terminal or upper-axillary, to 12 cm long; inflorescence branches, pedicels, and sepals bearing dense, ferruginous, T-shaped trichomes; sepals lanceolate to oblong, recurved at apex, 2.5–4 mm long, eglandular or bearing glands covering the basal half; petals yellow, clawed, 4–5 mm long, glabrous; stamens 10, 3–4 mm long, ± equal, enlarged and fused at base; anthers ca 1 mm long; styles 3, distinct, truncate at apex, ± equaling stamens; stigmas ± equal. Schizocarps of 3 samaras, each samara with a well-developed dorsal wing, ferruginous-tomentose especially on the seminiferous area; wings 2–3 cm long, with a small protrusion on the apical end near the seed. *Croat 5122, 9550.*

Known only from several areas along the shore. Flowers from January to May, with the fruits maturing from March to June.

Mexico to Peru, Brazil, and the Guianas; Greater Antilles. In Panama, known from tropical moist forest in the Canal Zone and Panamá from tropical wet forest in Colón (Santa Rita Ridge); collected in 1839 in what was probably tropical wet forest on the Pacific slope of Veraguas.

HIRAEA Jacq.

Hiraea faginea (Sw.) Niedenzu, Gen. Hiraea 16. 1906

Liana; younger stems, axes of inflorescences, sepals, and lower leaf surfaces densely pubescent, the trichomes appressed, T-shaped, those of the stem very short, those of the lower leaf surface light brown. Petioles 4–6 mm long; stipules paired, subulate, to ca 4 mm long, borne near apex on upper surface of petiole; blades ± oblong-elliptic, acuminate, narrowly rounded to subcordate at base, 9–16 cm long, 3.5–6 cm wide, glabrous above, the trichomes of the lower surface contiguous. Umbels small, axillary; peduncles short, appearing solitary, branched, mostly shorter than pedicels; pedicels to 1 cm long; sepals ovate, all but 1 usually bearing conspicuous glands; petals clawed, yellow, orbicular, ca 6 mm wide, their margins lacerate, the claw to 3.3 mm long, the petal opposite the eglandular sepal somewhat erect, smaller than the other petals (to ca 4.5 mm wide), its margin with glandular teeth; stamens 10, ± equal, glabrous; styles 3, pubescent at base, 1 erect, the 2 opposite the erect petal arched-spreading. Schizocarps of 3 samaras (sometimes 1 or 2 by abortion), each samara with 2 large lateral wings; seminiferous area ± globular, ca 5 mm long, densely appressed-pubescent; wings shaped like butterfly wings, ca 2.5 cm wide and 3 cm long, thin, moderately appressed-pubescent especially near seed. *Foster 1105.*

Rare, along the shore. Some flowers were seen in January, June, and July, and mature fruits in August.

Nicaragua to Panama, Venezuela, probably also Colombia. In Panama, known from tropical moist forest on BCI and in Darién. Standley (1928) in the *Flora of the Canal Zone* says the species is frequent on the Atlantic slope.

Hiraea grandifolia* Standl. & L. O. Wms., Ceiba 3:116. 1952

Large, woody canopy vine, the trunk 10 cm or more broad near base; bark brown, densely lenticellate; all but the older stems very densely ferruginous-tomentose. Petioles densely furfuraceous, 1.5–3 cm long; stipules paired, prominent, subulate, at about the middle of peti-

*José Cuatrecasas finds *H. grandifolia* from Costa Rica to be distinct; the Panamanian species will be published as *H. croatii* Cuatr. in the *Flora of Panama.*

KEY TO THE SPECIES OF HIRAEA

Lower leaf surface not very densely and conspicuously pubescent except on midrib or when very young; leaves usually acute or rounded at apex and less than 20 cm long .... *H. reclinata* Jacq.
Lower leaf surface densely and conspicuously pubescent; leaves acuminate at apex or more than 20 cm long:
   Umbels markedly pedunculate, the peduncle more than 1.5 cm long ... *H. quapara* (Aubl.) Sprague
   Umbels sessile or the peduncle less than 1 cm long:
      Leaves rounded to cuspidate or very short-acuminate at apex, the larger ones mostly more than 10 cm wide, pubescent on veins above; stems very densely and conspicuously ferruginous-pubescent .... *H. grandifolia* Standl. & L. O. Wms.
      Leaves ± narrowly acuminate, the larger ones mostly less than 7 cm wide, glabrous above; stems not as above .... *H. faginea* (Sw.) Niedenzu

ole; blades obovate, rounded and cuspidate to short-acuminate at apex, tapered to a narrow, cuneate, rounded or subcordate base, 12–30(67) cm long, 6–14(28) cm wide, glabrous except on veins above, densely pubescent below with branched trichomes; major lateral veins in 9–14 pairs, impressed above, raised below. Umbels axillary; peduncles short or to 8 mm long, 2–4 per axil; peduncles, pedicels, and calyces densely tomentose; pedicels 1–2 cm long; flowers 5-parted; sepals blunt, to 3 mm long, 4 of these bearing 2 red-orange, shiny, ovate glands to 1.5 mm long; petals yellow, clawed, the margins fimbriate, the petal opposite the eglandular sepal somewhat erect and marked with orange, its margins fimbriate, gland-tipped; stamens 10, of 2 lengths, shorter than styles and curling after anthers have fallen; styles 3, interspersed with dense straight trichomes, 2 twisted to either side of the erect petal; stigmas subterminal. Schizocarps of 3 samaras; wings lateral, ± reniform to semicircular, thin, 1–3 cm long, 2–5.5 cm wide, sparsely to moderately pubescent, the body densely pubescent with stiff straight trichomes. *Croat 12697.*

Frequent in the forest. Flowers from November to February. The fruits mature from February to April.

Seedlings often consist only of a long leafless stem.

Costa Rica and Panama. In Panama, known from tropical moist forest on BCI and in Colón.

See Fig. 308.

Hiraea quapara (Aubl.) Spraque, J. Bot. 62:22. 1924
  H. smilacina Standl.

Liana; outer bark of older stems thin, reddish-brown, becoming fissured; most parts except upper leaf surface densely pubescent with T-shaped trichomes, those of the stems, petioles, peduncles, pedicels, and calyces appressed-pubescent. Petioles 1–2 cm long, with 2 sessile glands at apex; stipules 2, subulate, to 2 mm long, borne at about the middle of petiole; blades sublanceolate-elliptic to elliptic, gradually long-acuminate, ovate to subcordate at base, mostly 15–25 cm long, 6–11 cm wide, glabrous above except on midrib, softly pubescent all over the underside with stalked T-shaped trichomes. Umbels solitary, axillary, with small bracteoles at apex of peduncles; peduncles 2.5–4 cm long; pedicels 13–20 mm long; flowers (6)35–60 per peduncle, 10–12 mm diam; sepals triangular-ovate, 4 of these sometimes bearing 2 large glands; petals yellow, ovate-suborbicular, subcordate at base, 4–5 mm diam, the margin sinuate, fimbriate on all but the inner small petal; stamens 10, ± equal; ovary hirsute; styles 3, 1 erect, 2 curved, the apices narrowly spatulate-truncate with the outer angle acute or shortly apiculate. Schizocarps of 3 samaras, softly hirsute especially on seed; wings lateral, ± reniform, 0.5–2 cm long, 2–3 cm wide, thin. *Croat 6842, 11931.*

Occasional, in the canopy of the forest. Flowers in July and August. The fruits mature in October and November.

Costa Rica (possibly Belize) to Colombia and the Guianas. In Panama, known only from tropical moist forest on BCI.

Hiraea reclinata Jacq., Select. Stirp. Am. 137, t. 174, f. 42. 1763
  H. obovata (H.B.K.) Niedenzu

Liana, ranging widely in forest over low vegetation or ± pendent from larger trees; younger parts, including peduncles, pedicels, and calyces, densely pubescent with T-shaped trichomes. Petioles usually ca 5 mm long, bearing paired sessile glands near apex; stipules subulate, borne about midway on petiole, slender, 3–5 mm long; blades elliptic to obovate, acute or rounded at apex (sometimes acuminate especially when young), tapered to a subcordate base, 5–15(22) cm long, 2.5–8.5 cm wide, glabrate to sparsely pubescent below; veins below densely appressed-pubescent. Umbels small; peduncles nearly obsolete or to 1.5 cm long; pedicels 9–25 mm long; sepals triangular to ovate, 4 usually glandular, the glands prominent, ± stalked, greenish-yellow, round to ellipsoid, ca 1.5 mm long; petals yellow, often marked with orange near base, to ca 7 mm long, clawed, the blade orbicular, the petal opposite the eglandular sepal with a fimbriate margin; stamens 10, of irregular sizes, some directed inward beneath styles; ovary lobed, the lobes densely pubescent; styles 3, ± flattened, curved, to 4.5 mm long; stigmas green, minute, lateral, directed inward. Schizocarps of 3 samaras; seminiferous area sparsely setaceous; wings lateral, orbicular to reniform, 1.2–1.6 cm long, 2–3 cm wide, thin, with sparse, appressed, T-shaped trichomes and prominent reticulate veins. *Croat 5614, 14028.*

Common in the forest. Found at all levels to over 30 m, but rarely flowering at the lower levels. Flowers from January to May. The fruits mature from February to June.

Mexico, Guatemala (probably throughout Central America), and Panama to Venezuela and Brazil. In Panama, known from tropical moist forest on BCI and in Colón, Panamá, and Darién.

MALPIGHIA L.

Malpighia romeroana Cuatr., Webbia 13:561. 1958

Shrub, to 2.5 m tall; stems and leaves glabrate. Petioles 2–4 mm long; blades elliptic, long-acuminate, acute at base, 10–16 cm long, 3.5–6 cm wide, sometimes bearing 4 glands at base ca midway between margin and midrib. Pseudocorymbs solitary in leaf axils, 3–8 cm long, bearing many flowers; inflorescence branches, pedicels, and calyces with sparse T-shaped trichomes; peduncles 1.5–3 cm long, bracteate about midway and at nodes; pedicels 5–8 mm long; sepals 5, ovate-elliptic, bearing 6 large, oblong-elliptic glands; petals pinkish outside, white inside, clawed, glabrous, to 5 mm long; stamens 10, unequal; styles 3, free, 1 reduced. Fruits drupaceous, ovoid, to 1 cm long and wide, red, with 3 crestlike appendages, drying ± 6-ridged; pyrenes 3. *Croat 6716, Foster 1590.*

Occasional. Flowers principally from August to October (sometimes to February). Time of fruit maturity is not known.

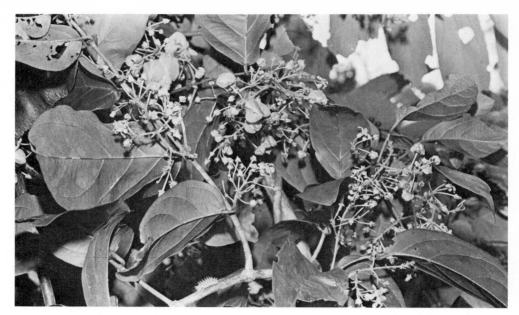

Fig. 309. *Mascagnia nervosa*

Fig. 308. *Hiraea grandifolia*

Fig. 310. *Stigmaphyllon ellipticum*

KEY TO THE SPECIES OF MASCAGNIA

Petioles bearing 2 large glands at apex; flowers yellow to orange.............................
.... *M. hippocrateoides* (Tr. & Planch.) Niedenzu
Petioles eglandular (pair of glands at base of blade); flowers pinkish .... *M. nervosa* Niedenzu

Panama and Colombia. In Panama, known from tropical moist forest in the Canal Zone and Darién.

MASCAGNIA Bertero

Mascagnia hippocrateoides (Tr. & Planch.) Niedenzu, Arb. Bot. Inst. Lyc. Braunsberg 3:24. 1908
Hiraea hippocrateoides Tr. & Planch.

Liana, growing into canopy to ca 30 m; stems glabrate to glabrous in age. Petioles 7–15 mm long, canaliculate, biglandular near apex, the glands 2–3 mm broad; blades ovate to ovate-elliptic, acuminate, rounded at base, 5.5–20 cm long, 3–10 cm wide, glabrous. Panicles axillary, to 25 cm long, branched many times; flowering peduncles 5–12 mm long, bearing 2 small bracteoles and a large sessile gland at apex; bracteoles paired at nodes of inflorescence, 2–12 mm long, the larger ones near the base often toothed, 1 of each pair bearing a large gland; inflorescence branches and pedicels moderately pubescent with short T-shaped trichomes (somewhat ferruginous); pedicels pedunculate (appearing articulate just below flower), 2–4 mm long, enlarged at apex; sepals five, 1.5–3 mm long, oblique on pedicel, glabrous, curved inward against stamens, the lateral margins turned outward, the glands 9, mostly 2 per sepal, ± irregular, greenish, to 1.5 mm long; petals 5, clawed, obovate, 4 of these yellow, rounded at apex, undulate and rounded to truncate at base, 8–10 mm long, entire, the fifth petal orange (at least on margin), minutely undulate and fimbriate on margin, 11–13 mm long, to 7 mm wide, usually somewhat erect and held above the entire petals; stamens 10, spreading, unequal, 3 directed toward the orange petal between 2 of the styles, the longest to 4 mm long; anthers nearly as broad as long; ovary 3-parted, each part with 3 ribs, densely pubescent; styles 3, somewhat spreading, to ca 4 mm long, the upper edge prolonged into a point on the outer edge opposite the stigmatic surface. Schizocarps of 3 samaras, sparsely appressed-pubescent with short T-shaped trichomes; dorsal wings small, to ca 5 mm long, subglabrous, the lateral wings fan-shaped, 1.8–2.5 cm long, 1.7–3.3 cm wide, subglabrous. *Croat 14867, Foster 880, 2321.*

Occasional, in the forest. Flowers in May and June. The fruits probably develop to full size in a month.

Panama, Colombia, and Ecuador. In Panama, known only from tropical moist forest on BCI.

Mascagnia nervosa Niedenzu, Arb. Bot. Inst. Lyc. Braunsberg 3:12. 1908

Liana or scandent shrub, to 3 m or more long; stems bearing sparse T-shaped trichomes when young, glabrate in age. Petioles 8–15 mm long, sparsely pubescent, eglandular; blades ovate-elliptic, acuminate, obtuse to rounded at base, 7–16 cm long, 4.5–9 cm wide, glabrous above,

glabrous below or moderately to densely pubescent on veins, especially midrib, the trichomes T-shaped, the glands 2, sessile, at base beside midrib below. Panicles axillary, divaricately branched, bracteolate at nodes; inflorescence branches, pedicels, and sepals short-pubescent; branches ending in a corymb of 4–12 flowers; peduncles ca 2 mm long, thicker than pedicel, minutely bracteolate at apex; pedicels pedunculate, 8–18 mm long; calyx bearing usually 8 large, green glands; sepals 5, ovate-oblong, 1.5–2 mm long, glabrous to sparsely pubescent; petals pink, clawed, ± equal, 4–4.5 mm long, keeled, entire; stamens 10, unequal (2 much larger than the others), ca 2 mm long; anthers oblong, introrse, yellowish, the connective conspicuously swollen on the outer side; ovary densely whitish-hirsute; styles 3, exceeding stamens, with a narrow lateral arm at apex. Schizocarps of 3 samaras; seminiferous area moderately appressed-pubescent; dorsal wings reduced, the lateral wings thin, ± semicircular, ca 3 cm long and 1.5 cm wide, often not divided at apex. *Croat 11681.*

Rare; growing in the canopy and along the shore. Flowers sporadically, mostly in the rainy season (March to October), with the fruits maturing from April to October.

Panama, Colombia, and Venezuela; Trinidad. In Panama, known from tropical moist forest on BCI and in adjacent areas of the Canal Zone and Panamá.

See Fig. 309.

SPACHEA Adr. Juss.

Spachea membranacea Cuatr., Webbia 13:548–49. 1958

Slender tree, to 10 m tall; trunk sap often somewhat milky; young stems, petioles, midribs of leaves, and inflorescence branches strigose, the trichomes ferruginous, T-shaped. Petioles 6–9 mm long; stipules minute, ovate, axillary; blades ovate-elliptic to oblanceolate, long-acuminate, rounded at base, 4–18 cm long, 1.5–6 cm broad, glabrate except on midrib above and below, bearing several round glands, usually 1 on each side of midrib below and others sometimes scattered on both surfaces, the margin entire, revolute in age. Racemes long, pendent, axillary, 10–20 cm long, the axis and peduncles sparsely ferruginous-pubescent with ± appressed, T-shaped trichomes; peduncles to ca 5 mm long, bearing 1 flower, 2 small bracteoles midway, and 5 inconspicuous glandular appendages at apex; pedicels appearing articulate, 5–10 mm long, pink; flower to 9 mm diam; sepals acute to rounded and reflexed at apex, ciliate, biglandular, the glands prominent, white, to 3 mm long, narrowed and curved outward at apex; petals magenta, obovate, ca 4 mm long, ciliate, spreading; stamens 10, ca 2 mm long,

shorter than sepals, ± equal; pistil bipartite, each part with a slender style; styles longer than ovary; stigmas ± divergent, held at about the level of anthers, persisting in fruit. Cocci fleshy, with a thin fleshy red covering at maturity; the carpels 2, separable, indehiscent, ovoid, 3.5–5 mm long and ca 2 mm wide; seeds 1 per carpel, hemispheroid, tan, smooth, ca 4 mm long. *Croat 6397, Foster 873.*

Frequent in the young forest. Flowers in the early rainy season (April to June). The fruits mature from June to August, though old fruits have been seen until February or March.

Fruits are probably dispersed chiefly by birds.

The species has been confused with *S. elegans* (G. Meyer) Adr. Juss., a species from the Lesser Antilles, the Guianas, and the Amazon basin, and it closely resembles *S. herbert-smithii* (Rusby) Cuatr. from Colombia. Both those species, however, have a prominent stalked gland on each one-flowered peduncle.

Colombia and Panama. In Panama, known only from tropical moist forest on BCI and in Darién.

STIGMAPHYLLON Adr. Juss.

Stigmaphyllon ellipticum (H.B.K.) Adr. Juss., Ann. Sci. Nat. Bot., sér. 2, 13:290. 1840

Liana, essentially glabrous except for appressed T-shaped trichomes on some younger parts; stems reddish-brown. Petioles 1–3 cm long, bearing a pair of apical glands; blades ovate-elliptic to oblong-ovate, usually abruptly acuminate and often downturned at apex, obtuse to rounded at base, mostly 6–13 cm long, 2.5–6.5 cm wide, glaucous below. Pseudoumbels axillary, bearing few flowers; peduncles sparsely pubescent, 1.5–3.5 cm long, sometimes with sparsely pubescent bracteoles at apex; pedicels mostly 1–1.5(2) cm long, glabrate; flowers to 2.5 cm wide; sepals rounded at apex, all but 1 bearing 2 conspicuous glands; petals 12–14 mm long, yellow, rounded, clawed at base, the margins fimbriate, the petals on either side of eglandular sepal somewhat to markedly cupular and usually fitted together after anthesis (cupped over the fertile parts); stamens 10, unequal, 6 long and equaling style, 3 of these situated beneath or adjacent to stigmas, producing much pollen, 3 long stamens sterile or producing little pollen, 2 short fertile stamens on either side of eglandular sepal, the other 2 short stamens on the opposite side of the flower sterile; anthers whitish, globular, fleshy, the connective produced above; pollen tan, sticking together loosely; styles 3, the stigmas green, the 1 opposite the eglandular sepal spatulate with an active gland near apex, the other 2 stigmas adjacent, broadly expanded, with a slit on the inner side, this oozing pollen from the anther below. Fruits usually of 3 samaras fused at base, reddish at maturity, especially the outer margin and apex, pubescent especially on seeds; dorsal wing 2–3 cm long, ca 1 cm wide, with a small crest or lobe on inner margin above seed, the lateral wings reduced, variously lobed or sometimes divided into 2–4 narrow appendages. *Croat 6438, 7901.*

Common in the forest, along the shore, and in clearings, growing over vegetation or along the ground. Flowers and fruits throughout the year.

Veracruz, Mexico, to Colombia and Ecuador; West Indies. In Panama, ecologically variable; known from tropical moist forest in the Canal Zone, Herrera, Panamá, and Darién, from tropical dry forest in Panamá, from premontane wet forest in Panamá and Chiriquí, and from tropical wet forest in Colón.

See Fig. 310.

Stigmaphyllon hypargyreum Tr. & Planch., Ann. Sci. Nat. Bot., sér. 4, 18:318. 1862

Vine; branches, lower leaf surfaces, peduncles, pedicels, and sepals densely sericeous. Petioles 1–7 cm long, bearing 2 sessile, lateral glands at apex; blades broadly ovate, acuminate, rounded to truncate or subcordate at base, 8–13(15) cm long, 5–9(10) cm wide, glabrous above except on midrib, densely sericeous below, mostly eglandular, entire or irregularly crenate. Pseudoumbels terminal and axillary, simple or branched, pedunculate, of 10–16 flowers; peduncles 2–4 mm long; pedicels 5–8 mm long; sepals turned inward above, 4 bearing basal glands 2 mm long, 1 eglandular; petals rounded, clawed, at least the outer margin fimbriate, the 2 petals on either side of the eglandular sepal larger (to 9 mm diam), concave, yellow or with orange on center basally, the other 3 petals

KEY TO THE SPECIES OF STIGMAPHYLLON

Lower blade surface densely whitish-sericeous, the trichomes contiguous, the surface not visible
. . . . *S. hypargyreum* Tr. & Planch.
Lower blade surface not densely sericeous:
  Lower leaf surface glaucous, the veins of adult leaves glabrous to sparsely pubescent; blades
    mostly ending abruptly at petiole; petals mostly more than 10 mm wide . . . .
    . . . . *S. ellipticum* (H.B.K.) Adr. Juss.
  Lower leaf surface not glaucous, the veins of adult leaves densely pubescent; blades mostly
    ending gradually at petiole; petals mostly less than 8 mm wide:
    Larger leaf blades broadly ovate, palmately veined at base, the trichomes of lower blade sur-
      face short (less than 1 mm long); petals yellow or tinged with orange on one petal; sam-
      aras constricted at base, broader at apex . . . . *S. lindenianum* Adr. Juss.
    Larger leaf blades ± elliptic, not palmately veined at base, the trichomes of lower blade sur-
      face ca 2 mm long; petals yellow with violet-purple tips; samaras broad at base, tapering
      to apex . . . . *S. puberum* (L. C. Rich.) Adr. Juss.

reduced, orange to red, the middle one often held higher, ± erect, its claw thicker; stamens 10, 6 long, fertile; anthers of long stamens covered (at least in part) by flap-like apices of styles, the connective enlarged, introrsely dehiscent; pollen tacky; styles 3, the stigmas expanded, the stigma of the style opposite the eglandular sepal concave. Schizocarps of 3 samaras; samaras 2–4 cm long, pubescent on seed and along thickened inner margin of wing; wing dorsal, 3–5 mm wide at base, enlarged to 1 cm wide at apex; seeds ± tuberculate. *Croat 7726, 13476.*

Infrequent, along the shore. Flowers in January and February. The fruits mature from February to April.

Trigona tataira bees visit the flowers (pers. obs.).

Panama and Colombia. In Panama, known from tropical moist forest in the Canal Zone and Los Santos.

Stigmaphyllon lindenianum Adr. Juss., Arch. Mus. Hist. Nat. 3:362. 1843

Liana, moderately pubescent all over, the trichomes small, ± appressed, T-shaped, sparse between veins on upper leaf surfaces. Petioles 1–5 cm long, bearing 2 sessile, lateral glands at apex; blades ovate, obtuse to broadly acuminate, cordate or truncate to abruptly acute at base, 5–16 cm long, 4–11 cm wide, entire to variably and irregularly toothed (deeply lobed when juvenile), bicolorous, palmately veined at base, drying dark. Pseudoumbels several, often 1 sessile at each inflorescence node and 2 pedunculate at ultimate branches; peduncles 2–5 mm long, bearing 2 small bracteoles at apex; pedicels appearing articulate, very short or to 6 mm long; sepals 5, acute to acuminate, appressed-pubescent, held inward at apex, 4 bearing 2 thick glands ca 2 mm long; petals 5, yellow, caducous, rounded at apex, narrowed to a claw at base, ca 8 mm long, the margin ± revolute and irregularly toothed, the petal opposite the eglandular sepal with a thickened claw, the blade sometimes marked with orange at base, ± erect, the petals on either side of eglandular sepal with the lower margin upturned and cupulate; stamens 10, 2 or 3 elongating into hoods of styles, 4 reduced and sterile; anthers dehiscing inward, the connective much thickened; styles 3, to 4 mm long; stigmas broadened and hooded, ± divergent. Fruits of 3 samaras, often becoming reddish, appressed-pubescent especially on seeds; wing dorsal, 3–4 cm long, ca 5 mm wide at base, broadened to ca 9 mm near apex. *Croat 14083.*

Occasional, at the margin of clearings and in the canopy to a height of more than 30 m. Flowers and fruits throughout the year.

Mexico to Venezuela. In Panama, known from premontane moist forest in the Canal Zone, from tropical moist forest in the Canal Zone, Bocas del Toro, San Blas, Chiriquí, Los Santos, Panamá, and Darién, from premontane wet forest in the Canal Zone and Panamá, and from tropical wet forest in Colón, Chiriquí, and Darién.

See Fig. 311.

Stigmaphyllon puberum (L. C. Rich.) Adr. Juss., Ann. Sci. Nat. Bot., sér. 2, 13:289. 1840

Vine, moderately to densely appressed-pubescent with T-shaped trichomes on most parts. Petioles 1–4 cm long,

bearing 2 sessile, lateral glands at apex; blades ovate to elliptic, long-acuminate, obtuse to rounded at base, 8–17 cm long, 3–10 cm wide, glabrous above except on midrib, moderately pubescent all over the underside. Pseudoumbels axillary, serial, to 12 cm long, divaricately branched, usually 1 sessile at each node; peduncles 2–4 mm long, minutely bracteolate at apex; pedicels appearing articulate, 2–5 mm long; inflorescence branches, peduncles, pedicels, and calyces densely pubescent; sepals to 4 mm long, ovate-oblong, rounded at apex, held against stamens, 4 bearing 2 broadly oblong glands 2–2.3 mm long, the fifth sepal sometimes with much-reduced glands, the glands with a thin, sticky covering; petals yellow, ± rounded, clawed, to 9 mm broad, the margins fimbriate, the 3 opposite the eglandular sepal spreading, usually tinged with purple-violet, the 2 petals on either side of the eglandular sepal ± erect, strongly concave, yellow or purple-violet at center outside; stamens 10, unequal, 4 reduced and sterile; styles 3; stigmas thin and broadly spreading, 1 bilateral. Schizocarps of 3 samaras; samaras pubescent especially on seed, dorsally winged; wing wider at base, 2–3 cm long, to 1.2 cm wide at base, ca 5 mm wide near apex. *Croat 11303.*

Rare; seen on the shore of Burrunga Point. The flowers were seen in June and July.

Guatemala to Peru, Brazil, and the Guianas; West Indies. In Panama, known from tropical moist forest in the Canal Zone and Darién.

TETRAPTERIS Cav.

Tetrapteris discolor (G. Meyer) DC., Prodr. 1:587. 1824

Liana or scandent shrub; young branches densely whitish-sericeous with appressed, T-shaped trichomes, glabrous in age. Petioles 3–15 mm long, canaliculate, eglandular; blades ovate-lanceolate to ovate, acuminate, cuneate to obtuse at base, 6–16 cm long, 2.5–7 cm wide, ± thin, glabrous or nearly so at least in age. Panicles terminal and upper-axillary, with numerous pairs of reduced, ± rounded, leaflike bracts, the branches sericeous; flowers 9–12 mm diam, in umbels; floral bracts oval or elliptic, 5–30 mm long, glabrate or sparsely appressed-pubescent especially along midrib; peduncles 2–5 mm long; pedicels slender, 2–5 mm long; sepals ± ovate, obtuse, glabrous but sparsely ciliate, the glands oval, 2.5–3 mm long; petals yellow, ± elliptic, rounded at apex, sagittate at base, 4–5.5 mm long, entire; stamens 10; anthers 1.2 mm long; styles 3, ca 2 mm long. Samaras 3, sparsely pubescent, the trichomes white, appressed, more dense on seminiferous area; wings moderately thin, the superior wings obovate-oblong to oblong, 1.2–2 cm long, 6–10 mm wide, the inferior wings ovate, 6–8 mm long, 4–5 mm wide, the dorsal crest 2–3 mm high, the appendages between the dorsal crest and the wings 1–5, irregularly linear or ovate, longer or shorter than dorsal crest. *Wetmore & Abbe 186, Wilson 149.*

Known only from several old collections made from along the shore on the east side of the island. Although these were made at a time when that area was much more

Fig. 311. *Stigmaphyllon lindenianum*

Fig. 312. *Tetrapteris macrocarpa*

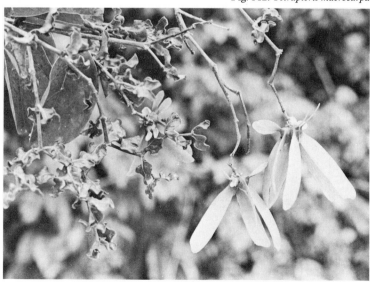

KEY TO THE SPECIES OF TETRAPTERIS

Fruits bearing several appendages between the lateral wings and the dorsal crest, the largest wing less than 2 cm long; flowers less than 12 mm wide; inflorescence bracts glabrate or sparsely pubescent:

Samaras with several appendages on the seminiferous area between the dorsal and lateral wings (these distinct from the base and spreading); leaf blades gradually long-acuminate, 2.5–3 times longer than broad; stems densely pubescent, the pubescence persisting in age . . . .
. . . . *T. discolor* (G. Meyer) DC.
Samaras lacking appendages between the dorsal and lateral wings; leaf blades acute to shortly acuminate, usually 2–2.5 times longer than broad; stems sparsely pubescent when young, the pubescence usually not persisting . . . . *T. seemannii* Tr. & Planch.
Fruits lacking appendages between the lateral wings and the dorsal crest, the largest wing more than 3 cm long; flowers more than 13 mm wide; inflorescence bracts densely sericeous . . . . . .
. . . . *T. macrocarpa* I. M. Johnston

---

disturbed, the species probably still occurs on the island. Flowers and fruits in the early dry season.

Mexico to Bolivia; the Antilles. In Panama, known from tropical moist forest in the Canal Zone, Bocas del Toro, Colón, and Panamá and from premontane wet forest in Coclé (El Valle).

Tetrapteris macrocarpa I. M. Johnston, Sargentia 8:172. 1949

Liana, climbing into canopy; stems and axes of inflorescences densely pubescent, the trichomes whitish, appressed or stalked, T-shaped. Petioles canaliculate, 5–20 mm long; blades ± ovate to ovate-elliptic, acute to short-acuminate, rounded to subcordate at base, 4–15 cm long, 2.3–10 cm long, moderately thick, pubescent when young, glabrate in age, the trichomes dense, ± appressed. Panicles terminal or upper-axillary, with numerous, ± rounded, leaflike bracts; flowers 13–17 mm broad, in umbels; floral bracts pinkish, ovate to rounded, densely pubescent all over, 10–15 mm long; peduncles 2–4 mm long; pedicels 2–5 mm long; sepals ovate, blunt at apex, sericeous, the glands 2.5–4 mm long; petals yellow, prominently clawed, the blade ca 7 mm wide, ± rounded but weakly sagittate at base, the margin irregular; stamens 10; anthers ca 1.5 mm long; styles 3, ca 2 mm long. Samaras 3, persistently sericeous, the trichomes moderately dense all over; wings moderately thick, the superior wings ± oblong, rounded to ± truncate on the end, 3–5.5 cm long, 10–14 mm wide, not at all confluent with the inferior wing, the inferior wings spatulate, 6–15 mm long, 3–6 mm wide, the dorsal crest 2–3 mm high. *Croat 12703, Foster 1913.*

Occasional, along the shore and in the canopy of the forest. Flowers from July to September in the early rainy season. The fruits mature to full size from October to December and are dispersed in the late rainy and early dry seasons.

Standley and Steyermark erroneously called this taxon *T. acapulcensis* in the *Flora of Guatemala* (1946b). According to Cuatrecasas (1958) that species was erroneously reported as being described from Mexico; the species was actually described from Colombia.

Mexico to Panama. In Panama, known from tropical moist forest in the Canal Zone, Bocas del Toro, Colón, Veraguas, Panamá, and Darién and from premontane wet forest in the Canal Zone, Chiriquí, and Panamá. See Fig. 312.

Tetrapteris seemannii Tr. & Planch., Prodr. Ann. Sci. Nat., sér. 4, 18:335. 1862

T. alloicarpha Rusby; *Banisteria ferruginea* Seem.

Liana or scandent shrub; stems slender, sparsely pubescent with T-shaped trichomes when young, soon glabrous, bearing interpetiolar ridges at least when young. Leaves thin; petioles 5–11 mm long, glabrous; blades elliptic to oblong-elliptic or ovate-elliptic, short-acuminate to acute at apex (the acumen downturned), mostly acute to obtuse but sometimes rounded at base, 5.5–13(16) cm long, 2–7 cm wide, glabrous above, glabrous below except the midrib at least sometimes sericeous when young, both surfaces with scattered glands less than 0.5 mm diam; major lateral veins 4–7 per side. Panicles terminal or upper-axillary, usually exceeding the leaves; branches, peduncles, and pedicels sericeous, soon glabrate; bracts inconspicuous, oval to elliptic, appressed-pubescent especially near base, 5–20 mm long; peduncles 2–5 mm long; pedicels slender, 2–5 mm long; sepals narrowly ovate, obtuse, 2.5–3 mm long, glabrous but ciliate; calycine glands thick, shiny, 2–2.5 mm long; petals clawed, yellow, probably turning orange in age, glabrous, the blade obovate to oblong-elliptic, sagittate at base, 4–5.5 mm long, 2.6–4 mm wide, the margin entire, the medial vein thickened toward the base, the claw thick, 1.5–2 mm long; anthers elliptic, glabrous, 1–1.2 mm long; styles held ca 1.5 mm above the sepals, slightly longer than stamens. Samaras stramineous to brown, glabrous, the wings thin but rigid, the veins often conspicuous, the superior wing narrowly spatulate, 15–30 mm long, 6–10 mm wide, the inferior wing ovate to obovate or rarely spatulate-oblong, 6–9 mm long, 4–6 mm wide, the adjacent margins of the superior and inferior wings confluent at base, the dorsal wing semiorbicular, rigid, 2–4 mm high; appendages between dorsal and lateral wings usually lacking; seed glabrous, ca 6 mm diam. *Aviles 17b.*

Collected once by Aviles; not seen in recent years, but the species could easily be overlooked. Flowers in Sep-

tember and October (rarely as late as November). The fruits develop to mature size by late November and are dispersed in the dry season beginning in December and are usually lost by late March.

Tetrapteris seemannii is easily confused with *T. discolor*, especially in flower. *T. discolor* tends to be more densely pubescent, with more pubescence persisting on the stem in age. Its leaves tend to be gradually long-acuminate and 2.5–3 times longer than wide. *T. seemannii*, on the other hand, has blades merely acute to short-acuminate at apex and usually 2–2.5 times longer than wide. Its stems bear less pubescence on the younger parts and are soon glabrate.

Panama and Colombia. In Panama, known from tropical moist forest in the Canal Zone (both slopes), Chiriquí, Los Santos, Panamá, and Darién and from premontane moist forest in Panamá.

72. TRIGONIACEAE

Lianas. Leaves opposite, petiolate; blades simple, entire; venation pinnate; stipules present, interpetiolar. Panicles terminal, thyrsoid; flowers bisexual, zygomorphic; sepals 3 or 4, basally connate, unequal; petals 5, free, white, very unequal, the posterior one largest and spurred; stamens 10, with 4 usually sterile, the filaments basally united into a tube, the tube cleft opposite the spurred petal; anthers 2-locular, introrse, longitudinally dehiscent; glandular disk present; ovary superior, 3-locular, 3-carpellate; placentation axile; ovules several per locule; style 1; stigma truncate. Fruits 3-valved, septicidal capsules; seeds several per locule, woolly, with endosperm.

The family is represented on BCI only by *Trigonia*. This liana can be recognized by its zygomorphic flowers with the stamens only partly fertile and united into a cleft tube.

Flowers are zygomorphic and seem well suited to pollination by small bees.

Seeds are wind dispersed.

About 4 genera and 40 species; tropical South America, Madagascar, and Malaya.

TRIGONIA Aubl.

Trigonia floribunda Oerst., Vidensk. Meddel. 38. 1856

Liana, growing into canopy, the ultimate branches usually pendent; young stems and inflorescence branches densely grayish-tomentose to glabrate in age; stems angulate to terete in age. Leaves moderately pubescent to glabrate; petioles 1–1.5(2) cm long; stipules interpetiolar, caducous, leaving conspicuous scar; blades obovate-oblong to oblong-elliptic, acute to short-acuminate, obtuse to rounded at base, 9–15 cm long, 4.5–7.5 cm wide. Panicles terminal, thyrsoid, to 23 cm long; pedicels to 2 mm long; flowers ca 7 mm diam; sepals 3, irregular, 2.5–4 mm long, recurved, acute to blunt, variously imbricate, glabrous inside, densely tomentose outside, one or both margins woolly; petals 5, white, unequal, 3–4 mm long, rounded at apex, the lateral petals narrower, obovate-spatulate,

± erect, the anterior petals folded medially and concave inside, weakly enclosing the cluster of anthers, the posterior petal spreading, ± rounded at apex, yellow to brown medially, the base produced into a ± globose spur, densely bearded at its apex inside, enclosing a glandular-crenate projection of the disk (appearing to be continuous with staminal tube); fertile stamens 6; filaments 10, unequal, connate more than half their length, forming an open tube around style; ovary and style together ca 3 mm long; ovary ovoid, densely white-villous; style slightly longer than stamens; stigma ± oblique. Capsules ca 2 cm long, 1.5 cm broad, bifid at apex, dark, rugose, glabrous, the valves coriaceous; seeds oval, 6–9 in each locule, pubescent, the trichomes long, yellowish, silky. *Croat 16581*.

Rare; known only from a few locations in the younger forest, though doubtlessly more abundant. Because of its growth habit high in the canopy and its inconspicuous white flowers it is seldom seen. Flowers in the middle of the rainy season. The fruits are probably mature in the late rainy and early dry seasons.

Mexico to Colombia. In Panama, known from tropical moist forest in the Canal Zone and Darién and from tropical wet forest in Darién.

73. VOCHYSIACEAE

Trees. Leaves opposite, petiolate; blades simple, subentire; venation pinnate; stipules present. Thyrses terminal, axillary; flowers bisexual, zygomorphic; calyx 5-lobed, 1 of the lobes spurred; petals 3, free, unequal, showy; fertile stamen 1; anther 2-celled, introrse, dehiscing longitudinally; staminodia 2; ovary superior, 3-locular, 3-carpellate; placentation axile; ovules 2 per locule, anatropous; style 1; stigma lateral. Fruits 3-valved, loculicidal capsules; seeds 3, winged, exalbuminous.

The family is represented in Panama only by *Vochysia*. It is distinguished by having opposite leaves and many brownish-orange, zygomorphic flowers with five sepals (one spurred) and three petals.

The flowers are visited by bees and butterflies (R. Dressler, pers. comm.).

Seeds are wind dispersed. Plants tend to be very underdispersed, with most of the plants in large clumps, indicating that the seeds are not carried very far and/or have few predators.

Six genera and about 200 species; tropical America and West Africa.

VOCHYSIA Aubl. mut. Poir.

Vochysia ferruginea Mart. in Mart. & Zucc., Nov. Gen. Sp. Pl. 1:151, t. 92. 1824
   Yemeri mayo, Yemeri macho, Yemeri wood, Flor de mayo, Pegle, Mocri, Mecri, Palo malin

Tree, 6–25 m tall (taller elsewhere); trunk usually to 30 cm dbh; bark smooth; branches widely spreading, the crown somewhat flattened; branchlets, rachises, and lower leaf surfaces softly ferruginous-tomentulose. Leaves

short-petiolate; stipules minute; blades elliptic-oblong, long-acuminate, acute to obtuse at base, glabrous and shiny above except on midrib, 6–13 cm long, 2–4.5 cm wide, the margin revolute. Cincinni terminal and axillary, cylindrical, ca 15 cm long; flowers 1–5, zygomorphic, yellow-orange; calyx lobes 5, unequal, 1 equaling petals and spurred at base, the spur recurved; petals 3, oblong-spatulate, to 1 cm long; stamen 1, opposite the anterior petal, slightly shorter than petals; anther boat-shaped; staminodia 2; style equaling anther, sunken in face of anther in bud; stigma lateral. Capsules 3-angular, ± oblong-obovoid, 3-celled, obtuse to subretuse at apex, 1.7–2.5 cm long, 5–8 mm wide, slightly verrucose, gray-green; seeds to 2.3 cm long, with a unilateral wing, the limb tomentose. *Croat 5428, Foster 1112.*

Occasional; locally abundant in young areas of the forest, especially north of Barbour Trail, on Gross Peninsula, and on Orchid Island. Flowers from late March to early July, mostly from April to June, with individual trees flowering for as long as 2 months. A second, much smaller flowering may occur in September and October. The fruits mature from August to October.

Nicaragua to Peru and Brazil. In Panama, known from tropical moist forest in the Canal Zone and Panamá and from premontane wet forest in Colón and Panamá. Reported from tropical wet and premontane rain forests in Costa Rica (Holdridge et al., 1971).

74. POLYGALACEAE

Scandent shrubs, short lianas, or annual herbs. Leaves alternate, petiolate; blades simple, entire or undulate; venation pinnate (vein 1 in *Polygala*); stipules lacking. Racemes terminal or axillary; flowers bisexual, zygomorphic, subtended by 1 bract and 2 bracteoles; sepals 5, 2 enlarged; petals 3, the lower one keel-shaped; stamens 8, monadelphous; anthers 1-celled, dehiscing by terminal pores; ovary superior, 2-locular (1-locular by abortion in *Securidaca*), 2-carpellate; placentation axile; ovules 1 per locule, pendulous, anatropous; style 1; stigmas 2 or bilobed. Fruits samaras (*Securidaca*) or 2-valved, loculicidally dehiscent capsules; seeds arillate in *Polygala*, with soft endosperm.

Though the flower parts are not analogous, the flowers resemble those of Leguminosae (63). Members of the family may be distinguished by having simple leaves (relatively few Leguminosae have simple leaves) and a bilocular ovary.

Flowers are well suited for bee pollination, possibly by very small bees in the case of *Polygala paniculata*.

Seeds of *Securidaca* are wind dispersed. Those of *Polygala paniculata* are possibly just spilled locally, but Ridley (1930) reported the presence of elaiosomes on the seeds of *Polygala* and thus suggested ant dispersal.

Twelve genera and about 800 species; widely distributed.

POLYGALA L.

Polygala paniculata L., Syst. Nat. ed. 10, 1154. 1759

Annual herb, to 50 cm tall, usually branched many times; stems densely and minutely stalked-glandular. Leaves short-petiolate; blades mostly linear-lanceolate, acute, to 3.2 cm long and 3.5 mm wide; vein 1. Racemes long, slender, terminal or upper-axillary, 3–7 mm long; flowers red-violet, pedicellate, 2–2.5 mm long; sepals 5, the outer 3 minute, green or tinged with white or red-violet, the inner 2 equaling petals, red-violet, oblong, rounded at apex; petals 3, the upper 2 white, ± connate laterally and enfolding the lower petal, the lower petal keel-shaped, crested, enclosing stamens and style, the crest 6–10-lobed, the lobes often tinged with red-violet; stamens 8, arranged in a fan-shaped pattern within the lower petal; filaments connate nearly to apex; ovary obovoid to orbicular, 0.6–0.8 mm long; style curved, flattened, recessed among stamens; stigmas 2, widely spaced, on either end of the staminal cluster. Capsules ± oblong, 1.8–2.6 mm long, 2-valved; seeds oblong, to 1.8 mm long, black, densely erect-pubescent, arillate at base, the aril slender, bilobed, more than one-third as long as seed. *Croat 11988.*

Seasonally frequent in clearings. Flowers and fruits principally in the rainy season.

Pantropic weed. Throughout Panama in weedy situations in tropical dry, tropical moist, tropical wet, and premontane wet forests; known from the Canal Zone, Bocas del Toro, Colón, San Blas, Chiriquí, Herrera, Coclé, Panamá, and Darién.

See Fig. 313.

SECURIDACA L.

Securidaca diversifolia (L.) S. F. Blake in Standl., Contr. U.S. Natl. Herb. 23:594. 1923

Bejuco amarrar, Bejuco mulato, Elsota

Scandent shrub or liana, usually to 4 m but sometimes also in canopy; strigose all over but the upper leaf surface glabrate; stems bearing raised round glands at the

---

KEY TO THE SPECIES OF POLYGALACEAE

Plants herbs less than 50 cm tall; leaves less than 5 mm wide .... *Polygala paniculata* L.
Plants lianas or scandent shrubs; leaves more than 10 mm wide:
    Leaves and stems usually strigose or glabrate, usually drying green; petals less than 7 mm long;
        samaras with a small triangular lobe or tooth above seed partly free from wing; flowering
        principally from February to April .... *Securidaca diversifolia* (L.) S. F. Blake
    Leaves and stems softly and densely pilose; petals more than 8 mm long; samaras with a small
        stipitate gland above on margin of wing; flowering principally in May and June ....
        .... *Securidaca tenuifolia* Chodat

Fig. 313. *Polygala paniculata*

Fig. 314. *Securidaca diversifolia*

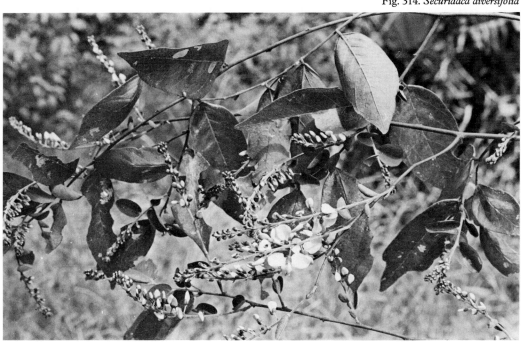

base of each petiole. Petioles 5–10 mm long; blades ovate-elliptic, bluntly acuminate at apex, rounded to obtuse at base, 6–11 cm long, 3–4.5 cm wide (much reduced and oval on flowering branches). Racemes terminal or axillary, to 11 cm long; pedicels ca 5 mm long; flowers 1.5–2 cm broad; enlarged sepals 2, petaloid, lavender, ± rounded and clawed, the other 3 ovate, to ca 3 mm long; petals 3, red-violet, the upper petals subspatulate, 5–7 mm long, the lower 1 keeled, 7.5–10 mm long, enclosing sexual parts, the apex with a folded, fimbriate crest; stamens 8, united basally into a thick sheath, the apical one-third free, tapered, the inside pubescent; anthers much shorter than stigma; style curved inward; stigmas bilobed. Samaras 4–6 cm long; wing ± oblanceolate, 3–5 cm long, 1–1.5 cm wide; seed apical, 5–8 mm long, margined above, the margin extended above into a small triangular lobe partly free from rest of wing. *Croat 8660.*

Occasional, along the edge of the lake, in older clearings, or at the edges of clearings; rarely high in the forest. Flowers in the dry season, usually from February to April (rarely in May in the early rainy season). Most fruits are mature in the late dry season.

Stems sometimes twine in a manner characteristic of the Hippocrateaceae (78).

Mexico to Peru; Lesser Antilles. In Panama, known from tropical dry forest in Coclé and Panamá, from premontane moist forest in Los Santos and Panamá, from tropical moist forest in the Canal Zone, Bocas del Toro, Panamá, and Darién, and from premontane wet forest in Coclé and Panamá.

See Fig. 314.

Securidaca tenuifolia Chodat, Bull. Herb. Boissier 3:545. 1895

Scandent shrub or liana; stems bearing raised, round, stipular glands at the junction of each petiole; most parts softly short-pilose. Petioles 2–4 mm long; blades oblong-ovate to ovate, obtuse to rounded and sometimes emarginate at apex, rounded to truncate at base, 3.5–9.5 cm long, 2–3.5(4) cm wide, sparsely pilose or glabrate above, densely velvety-pilose below. Racemes terminal, 3–3.5 cm long; pedicels 5–8 mm long; flowers magenta, to 1 cm long; outer sepals 2, enlarged, 3.5–5 mm long, the other 3 oval, 2.5–4 mm long; petals 3, the upper 2 obovate or obovate-spatulate, 8–11 mm long, the lower 1 keeled, enclosing sexual parts, folded, with a minute fimbriate crest at apex less than 1 mm long; stamens 8, united at base into a sheath, pubescent at base; anthers much shorter than stigma; style 7.5–11 mm long; stigma bilobed. Samaras 4.5–7 cm long, densely short-pilose at least on seminiferous part; wing obovate-oblong, 4–6 cm

long, 1–2 cm wide, with a small stipitate gland on margin near seed; seed apical, 5–8 mm long. *Woodworth & Vestal 505.*

Not seen in recent years; rare or no longer present on the island. The species apparently flowers more than once per year, mostly in May and June (rarely early July), with the fruits maturing in July and August, and again in September, with the fruits maturing in January.

Woodworth & Vestal 505 was collected in flower in February. It is intermediate in several ways between *S. tenuifolia* and *S. diversifolia* and may represent only an aberrant collection of the latter species.

Panama; Trinidad. In Panama, known from tropical moist forest in the Canal Zone, Veraguas, Panamá, and Darién and from premontane wet forest in Colón (Santa Rita Ridge).

75. EUPHORBIACEAE

Trees, shrubs, lianas, vines, or annual or perennial, erect or repent herbs, with or without spines, often with milky sap. Leaves alternate or less commonly opposite, petiolate; blades simple or palmately compound or lobed, entire to serrate, often stellate-pubescent; venation pinnate or occasionally palmate at base; stipules present (may be reduced). Flowers unisexual (monoecious or dioecious), actinomorphic, the inflorescences various, usually axillary, often specialized into condensed partial inflorescences (cyathia); calyx 2–11-lobed or lacking; petals 4–13 and free or commonly lacking; stamens 1 to many, in 1 or 2 series, often united by their filaments; anthers 2-celled, dehiscing longitudinally; intrastaminal disk often present; ovary superior, (2)3-locular, 2- or 3-carpellate (4-locular and 4-carpellate in *Phyllanthus*, 4- or 5- in *Margaritaria*, 10- in *Hura*); placentation axile; ovules 1(2) per locule, anatropous; styles 2–5, free or connate, simple or lobed; stigmas discoid (sometimes compound). Fruits usually explosive, septicidally dehiscent capsules or schizocarps, often with mericarps attached apically to columella, or fruits drupaceous (*Drypetes*, *Hyeronima*) or pyrenes; seeds 1 or 2 per locule or 1 per fruit by abortion; endosperm usually copious, rarely lacking.

Together with the Flacourtiaceae (96), one of the most morphologically diverse families. Euphorbiaceae are most easily recognized by their unisexual flowers and generally three-valved woody capsules with elastically dehiscent cocci separating from a persistent columella.

Most species are probably insect pollinated, although a few species are adapted for wind pollination (*Acalypha* and perhaps also *Adelia triloba*). Most flowers are unisexual but otherwise not very specialized. The tacky

KEY TO THE TAXA OF EUPHORBIACEAE

- Leaves conspicuously lobed or compound:
  Plants vines . . . . *Dalechampia* (in part)
  Plants erect herbs, shrubs, or trees:
      Leaves deeply lobed, glaucous beneath; capsules ribbed . . . . *Manihot esculenta* Crantz
      Leaves shallowly lobed, not glaucous beneath; capsules not ribbed . . . . *Jatropha curcas* L.

● Leaves simple and not lobed:
　　Plants herbs:
　　　　Leaves all less than 1.5 cm long:
　　　　　　Leaves opposite; plants repent, forming mats; stipules toothed or conspicuously ciliolate; capsules ca 1 mm diam . . . . *Chamaesyce thymifolia* (L.) Millsp.
　　　　　　Leaves alternate; plants erect or ascending; stipules entire; capsules ca 2 mm diam . . . . *Phyllanthus* (in part)
　　　　Leaves sometimes more than 1.5 cm long:
　　　　　　Leaves opposite:
　　　　　　　　Leaves more than 5 cm long . . . . *Poinsettia heterophylla* (L.) Klotzsch & Garcke
　　　　　　　　Leaves less than 3.5 cm long . . . . *Chamaesyce* (in part)
　　　　　　Leaves alternate:
　　　　　　　　Leaves palmately veined at base; stems and petioles conspicuously pubescent:
　　　　　　　　　　Petioles bearing stalked glands at apex; stems with acicular, stiffly spreading, stellate trichomes with one branch larger and 2 mm or more long . . . . . . *Croton hirtus* L'Hér.
　　　　　　　　　　Petioles lacking glands; stems with simple trichomes . . . . *Acalypha arvensis* Poepp.
　　　　　　　　Leaves not palmately veined at base; stems and petioles not conspicuously pubescent; petioles eglandular:
　　　　　　　　　　Leaves less than 4.5 cm long; inflorescences axillary; capsules on slender pedicels more than 1 cm long:
　　　　　　　　　　　　Leaves linear to oblong-ovate, obtuse to mucronate at apex, 2–6(9) mm wide; capsules with pedicels less than 5 mm long . . . . *Phyllanthus urinaria* L.
　　　　　　　　　　　　Leaves ovate to elliptic, acuminate at apex, 10–25 mm wide; capsules with pedicels 5–12 mm long . . . . *Phyllanthus acuminatus* Vahl
　　　　　　　　　　Leaves more than 5 cm long; inflorescences terminal; capsules nearly sessile . . . . *Poinsettia heterophylla* (L.) Klotzsch & Garcke
　　Plants vines, shrubs, or trees:
　　　　Leaves pinnately veined at base:
　　　　　　Petioles bearing conspicuous stalked glands at apex . . . . *Sapium*
　　　　　　Petioles lacking glands at apex:
　　　　　　　　Flowers borne in fascicles or short racemes less than 1 cm long:
　　　　　　　　　　Inflorescences terminal (becoming overtopped by branches in fruit); pedicels more than 1 cm long; fruits 3-valved capsules more than 3 cm wide . . . *Garcia nutans* Vahl
　　　　　　　　　　Inflorescences axillary; pedicels less than 5 mm long; fruits drupes ca 1.5 cm wide or capsules ca 1 cm wide with 5 or 6 cocci:
　　　　　　　　　　　　Ovary densely pubescent, the stigmas discoid to weakly lobed; fruits broadly ovoid, indehiscent; stipules lacking; leaf acumen thickened at apex, hyaline or discolored . . . . *Drypetes standleyi* Webster
　　　　　　　　　　　　Ovary glabrous, the stigmas free, on frequently bifid to tripartite styles; fruits depressed-globose, irregularly dehiscent; stipules subpersistent; leaf acumen not thickened . . . . *Margaritaria nobilis* L.f.
　　　　　　　　Flowers borne in panicles or racemes more than 5 cm long:
　　　　　　　　　　Plant densely covered with lepidote scales; leaf blades more than 5 cm wide; plants dioecious, the pistillate inflorescences producing many fruits; fruits ca 3 mm diam, fleshy . . . . *Hyeronima laxiflora* (Tul.) Müll. Arg.
　　　　　　　　　　Plants glabrous, not bearing lepidote scales; leaf blades less than 5 cm wide; plants monoecious, each inflorescence producing few fruits; fruits more than 1 cm diam (*Codiaeum* on BCI not setting fruit):
　　　　　　　　　　　　Leaves strap-shaped, often variegated; plants cultivated shrubs in the Laboratory Clearing . . . . *Codiaeum variegatum* (L.) Blume
　　　　　　　　　　　　Leaves oblong, not variegated; plants common in the forest . . . . *Mabea occidentalis* Benth.
　　　　Leaves conspicuously palmately veined at base:
　　　　◆ Petioles bearing conspicuous, usually stalked glands at apex:
　　　　　　Plants vines or lianas:
　　　　　　　　Plants herbaceous vines; glands at apex of petiole stalked, at least in part acicular; flowers borne in conspicuous, foliaceous bracts; fruits capsules less than 1.5 cm diam . . . . *Dalechampia dioscoreifolia* Poepp.
　　　　　　　　Plants lianas; glands at apex of petiole sessile, disk-shaped; flowers not borne in foliaceous bracts; fruits fleshy, more than 6 cm diam . . . . *Omphalea diandra* L.
　　　　　　Plants trees:
　　　　　　　　Plants large trees with spiny trunks; leaf blades glabrous or bearing simple trichomes on lower surface, usually conspicuously toothed . . . . *Hura crepitans* L.
　　　　　　　　Plants small to medium-sized trees, unarmed; leaf blades stellate-pubescent, ± entire . . . . *Croton* (in part)

↝Petioles lacking glands:
  Leaves ovate-cordate:
    Plants cultivated shrubs in Laboratory Clearing; leaves usually reddish; plants ±
      glabrous . . . . *Acalypha wilkesiana* Müll. Arg.
    Plants trees in forest; leaves not reddish; plants conspicuously and densely pubescent
      . . . . *Acalypha macrostachya* Jacq.
  Leaves not ovate-cordate, mostly ± elliptic and at most subcordate at base:
    Leaves with glands at base of blade on lower surface; trichomes stellate (may be
      branched from near base); capsules 2-valved, with 2 persistent, simple styles usu-
      ally 7–20 mm long . . . . *Alchornea*
    Leaves eglandular; trichomes simple or lacking; capsules trilobate, lacking a conspic-
      uous style:
      Plants dioecious; flowers clustered in axils; capsules more than 7 mm diam, borne
        on a pedicel more than 1.5 cm long; stipules inconspicuous . . . .
        . . . . *Adelia triloba* (Müll. Arg.) Hemsl.
      Plants monoecious; flowers in dense catkinlike spikes; capsules ca 3 mm diam, ses-
        sile, borne at base of spike; stipules lanceolate, persistent, 4–8 mm long . . . . . .
        . . . . *Acalypha diversifolia* Jacq.

pollen of *Hura crepitans* oozes forth in great abundance, but the pollination system is unknown. The extrafloral nectaries of *Chamaesyce* and *Poinsettia* provide their only attractants for insects and birds. G. Webster (pers. comm.) reports that most Euphorbiaceae are pollinated by *Diptera* (midges) and that *Dalechampia* is pollinated by carpenter bees.

Seeds of most species are mechanically dispersed by the elastically dehiscent cocci that break free from the persistent columella. This is a very effective means of dispersal. *Hura crepitans* has been known to throw seeds 14 m (van der Pijl, 1968). A few taxa have fruits that are variously colored and mostly bird dispersed. They may be indehiscent (*Hyeronima*) or dehiscent with a colorful seed (*Sapium, Alchornea*). *Margaritaria nobilis* has mimetic fruits, not fleshy, but brightly colored, apparently functioning to deceive fruit-eating birds (van der Pijl, 1968). *Drypetes* and possibly also *Omphalea* are mammal dispersed. Fruits of *Sapium* are taken by white-faced monkeys (Oppenheimer, 1968). Both white-faced and howler monkeys take fruits of *Hyeronima* (Oppenheimer, 1968; Carpenter, 1934). *Acalypha* have beetle-like seeds which may be taken by carnivorous ants. Fruits of *Mabea occidentalis*, while basically mechanically dispersed, are colored and are also taken by white-faced monkeys (Oppenheimer, 1968) before the pericarp becomes fully woody. It is not known if the seeds are viable after passing through the monkeys. Hydrochory may be important in *Omphalea diandra* (Ridley, 1930). Seeds of *Croton* and *Euphorbia* (*Chamaesyce*) are harvested by ants in the southeastern United States (Wheeler, 1910). This is possibly also true of *Chamaesyce* in the tropics.

More than 300 genera and 7,000 species; widely distributed, but mainly in the tropics.

ACALYPHA L.

Acalypha arvensis Poepp. in Poepp. & Endl., Nov. Gen. Sp. Pl. 3:21. 1841

Monoecious herb, 20–50 cm tall; stems and petioles moderately to densely pubescent, at least the smaller trichomes conspicuously recurved. Leaves alternate, simple; stipules lanceolate, to ca 1 cm long; petioles mostly 1–2.5 cm long; blades ovate, acute at apex, cuneate at base, mostly 3–6 cm long and 1.5–3 cm wide, strigose-hirsute above, glabrous below except on veins, the margins crenate-serrate; veins palmate at base. Spikes axillary; bisexual spikes ellipsoid to short-cylindrical, the peduncles 6–12 mm long, the pistillate part to 2 cm long and 1.5 cm wide, the staminate part usually a terminal projection 4–7 mm long, bearing 5–10 nodes of flowers and sometimes a bractless pistillate flower. Pistillate flowers solitary, sessile; floral bracts 4–8 mm long, with usually 5 lobes, the medial lobes mostly 2.5–5 mm long; ovary apically 3-keeled, hispidulous, eglandular; styles 1.7–4 mm long, with 3–5 distal branches. Staminate spikes uninterrupted, ca 5–20 mm long, the peduncles 3–25 mm long. Capsules sharply keeled, hispid to hispidulous; seeds 1–3, ellipsoid, to 1.3 mm long, grayish. *Croat 17748.*

KEY TO THE SPECIES OF ACALYPHA

Plants herbs, less than 50 cm tall . . . . *A. arvensis* Poepp.
Plants shrubs or trees, more than 1 m tall:
  Stems and petioles conspicuously pilose . . . . *A. macrostachya* Jacq.
  Stems and petioles ± glabrous:
    Leaves ovate-cordate, usually reddish; plants cultivated shrubs in the Laboratory Clearing . . .
      . . . . *A. wilkesiana* Müll. Arg.
    Leaves ± elliptic, not reddish; plants abundant in the forest . . . . *A. diversifolia* Jacq.

Fig. 315. *Acalypha diversifolia*

Fig. 316. *Acalypha macrostachya*

Fig. 317. *Adelia triloba*

Collected once in the clearing north of the dock. Flowers and fruits principally in the rainy season.

Throughout the New World tropics. In Panama, known from tropical moist forest in the Canal Zone, Bocas del Toro, Chiriquí, Veraguas, Panamá, and Darién, from premontane wet forest in Bocas del Toro, Chiriquí, and Coclé, and from lower montane rain forest in Chiriquí (Cerro Horqueta).

Acalphya diversifolia Jacq., Hort. Schoenbr. 2:63, t. 244. 1797

Monoecious shrub or small tree, to 6 m tall; stems sparsely pubescent. Leaves alternate, simple; stipules lanceolate, 4–8 mm long, persistent; petioles 1.5–3 cm long; blades ± elliptic, long-acuminate, narrowed and obtuse or subcordate at base, 5–15(22) cm long, 1.5–8(10) cm wide, sparsely pubescent (more densely on veins), serrate. Spikes densely flowered, axillary, 3–8 cm long; flowers minute, greenish-white, chiefly staminate; petals and disk lacking; pistillate flowers, when present, few in number and restricted to base of spike, sessile, 2 or 3 per bract; bracts ovate, to 4.5 mm broad, toothed, larger in fruit; staminate calyx 4-lobed, somewhat raised on pedicel at anthesis; ovary and capsule hispidulous and muricate; styles 3, free, pubescent, divided at apex into many branches, persisting in fruit. Capsules trilobate, to 3 mm diam, sparsely pubescent, explosively dehiscent; seeds 3, 1 per carpel, ± ovoid, 1–1.2 mm long, black, densely punctate. *Croat 8544.*

Abundant in the old forest. On BCI, flowers mostly from March to May (sometimes from February). The fruits are seen from May to July. Elsewhere, such as in the drier areas of the Azuero Peninsula and in Chiriquí, flowering collections have been made in August and September.

This species, perhaps more than any other, is subject to attacks by gall-forming insects. Galls of three distinctly different forms have been found. One form often looks so similar to a fruit that it was at first considered to be one. Specimens bearing all three types of galls are deposited at the Missouri Botanical Garden.

Southern Mexico to Peru, Bolivia, and Brazil. Throughout Panama; occurring on both slopes in tropical moist forest at mostly lower elevations, though known also from premontane wet forest in Bocas del Toro, Coclé, and Panamá and from tropical wet forest in Coclé and Panamá.

See Fig. 315.

Acalypha macrostachya Jacq., Hort. Schoenbr. 2:63, t. 245. 1797

Monoecious or dioecious shrub or small tree, 3–5(8) m tall; stems often clustered, densely pubescent. Leaves alternate, simple; stipules paired, lanceolate, long-acuminate, pilose, caducous leaving a scar; petioles 12–25 cm long, pilose; blades ± ovate, long-acuminate, truncate to cordate at base, 10–21 cm long, 7–15 cm wide, pilose especially on veins, crenate-serrate. Spikes axillary, usually unisexual, usually a few pistillate spikes alternating with many staminate ones on apical part of stem, the staminate spikes usually 14–20 cm long, the pistillate spikes to 22 cm long in flower (to ca 30 cm long in fruit); petals and disk lacking; staminate flowers 4-lobed, ca 1 mm or less wide; filaments fleshy, the thecae divergent; pollen ± tacky; pistillate flowers solitary, subtended by foliaceous bracts, the bracts 2.5–5 mm long (to 8 mm in fruit), their margins deeply toothed; styles 3, ca 1 cm long, red, slender, branched many times. Capsules trilobate, to 4.3 mm diam, hispid, explosively dehiscent; seeds 3, 1 per carpel, ellipsoid, ca 2 mm long. *Croat 8602.*

Occasional, in the forest, generally in areas of disturbance along trails. Flowers principally from February to May, less frequently elsewhere in Panama during August and September. The fruits develop within about 6 weeks, the majority being dispersed during the late dry and early rainy seasons.

Although the stigmas are ideally suited to catching wind-dispersed pollen, the pollen is tacky and does not appear to be blown from the staminate flowers. However, the entire staminate flower is light and easily loosened and may be wind borne. Though capsules are explosively dispersed, the large bracts may act as receptacles for the dispersed seeds and thus serve as a wind ballast.

Southern Mexico to Peru, Bolivia, and Brazil. In Panama, ranging about the same as *A. diversifolia*, but less abundant and preferring weedier habitats.

See Fig. 316.

Acalypha wilkesiana Müll. Arg. in DC., Prodr. 15(2):817. 1866

Copperleaf, Jacob's coat

Monoecious shrub, to 2.5 m tall; stems sparsely pubescent in age, the young parts densely pubescent. Leaves alternate, simple; petioles to 7 cm long, with a broad band of trichomes above near apex; blades ovate-cordate, abruptly acuminate, 10–25 cm long, 8–15 cm wide, dentate with gland-tipped teeth, maroon beneath, sparsely pubescent on veins; veins at base 5, palmate. Spikes axillary, 6–20 cm long, 3–8 mm wide, reddish, the pistillate bracts 5–6 mm long, divided almost to base, purple-violet; pistil to 1 mm long, 1.2 mm wide; styles 3, free, purple-violet, to 5 mm long, each with numerous filiform segments. Fruits not seen. *Croat 8658.*

Cultivated in the Laboratory Clearing. Flowers in the dry season.

BCI plants would be best considered var. *macageana* W. Miller.

Native to the Pacific Islands (Standley, 1928); cultivated throughout the world.

ADELIA L.

Adelia triloba (Müll. Arg.) Hemsl., Biol. Centr.-Amer. Bot. 3:130. 1883

Dioecious shrub or small tree, usually to 5(15) m tall; trunk slender, to 20 cm dbh; outer bark peeling; branchlets with thin, grayish periderm, sparsely puberulent, rarely short and tapered to a sharp spine at apex. Leaves

alternate, simple; stipules inconspicuous, lanceolate, less than 1 mm long; petioles 4–9 mm long; blades elliptic to obovate, cuspidate-acuminate, tapering to base and usually subcordate, 6–12 cm long, 2–6 cm wide, glabrous but with axillary tufts below and puberulence on major veins. Inflorescences glomerulate; flowers small, cream to white, apetalous, the disk fleshy; staminate flowers pedicellate, the pedicels 4–7 mm long, puberulent; calyx lobes 4 or 5, elliptic, reflexed, ca 2.5 mm long; stamens mostly 13–16 but to more than 20, forming a ± globular mass; filaments connate into a short column terminated by a pistillode; pistillate flowers on glabrate pedicels 2–7 cm long in fruit; calyx lobes 6, ± lanceolate, 3–6.5 mm long; ovary densely pubescent; styles free or connate basally, spreading, fimbriate-lacerate. Capsules prominently trilobate, 7–11 mm diam, about two-thirds as long as broad, pubescent, the valves woody, dehiscing explosively into 6 parts, leaving a persistent columella; seeds 3, globose, smooth, black and shiny, glaucous. *Croat 9347, 13160.*

Frequent along the shore and in the young forest. Flowers from December to February. Fruits from late January to early April.

Known only from Costa Rica and Panama. In Panama, known from tropical moist forest in the Canal Zone, Bocas del Toro, Panamá, and Darién. Reported from premontane wet forest in Costa Rica (Holdridge et al., 1971).

See Fig. 317.

ALCHORNEA Sw.

Alchornea costaricensis Pax & Hoffm., Pflanzenr. IV. 147. VII (Heft 63):235. 1914

Dioecious tree, 5–15 m tall, usually less than 30 cm dbh; trunk often forming sucker shoots near base. Leaves alternate, simple; stipules inconspicuous, less than 0.5 mm long; petioles 1.5–3.5 (9) cm long, short-pubescent; blades elliptic or ovate-elliptic, long-acuminate, obtuse at base, 8–18 cm long, 3–6.5 cm wide, pinnately veined, 3-veined at base, minutely and sparsely stellate-pubescent below and on veins above, becoming glabrate, remotely crenate-dentate, with 2 foliar glands at base of blade. Spikes stellate-pubescent, the staminate axes unbranched, mostly 2–18 cm long, the pistillate spikes 2–8 (10) cm long; staminate flowers apetalous, subsessile, to 1.5 mm diam, clustered with 3 or 4 per bract; calyx lobes 3 or 4; stamens 7 or 8, in 2 series; filaments ± equaling anthers, confluent with disk; pistillate flowers apetalous, solitary, densely pubescent; calyx 4-lobed to about midway; ovary

ellipsoid, ca 1.5 mm long, densely pubescent; styles 2, more than twice length of ovary, usually persisting in fruit, simple, ± divergent, 7–10 mm long; stigmatic surface glabrous. Capsules subglobose, green to brown, 2-valved, 5–7 mm long, 6.5–8.5 mm wide, the valves falling free at maturity to expose bright red seeds; seeds 2, 1 per carpel, ellipsoid, 4–6 mm long, displayed on the central columella, flattened on 1 side, the testa brown, irregularly tuberculate, covered with a bright red, thin, pulpy layer. *Croat 10914, 14546, 14619.*

Frequent in the forest. Flowers from March to June, with the fruits maturing mainly from May to July. According to R. Foster (pers. comm.), individuals may flower twice in quick succession following the start of the rainy season. Flowers of the second wave are produced at the time the first fruits ripen.

Costa Rica and Panama. In Panama, known from tropical moist forest in the Canal Zone, Bocas del Toro, Panamá, and Darién.

See Fig. 318.

Alchornea latifolia Sw., Prodr. Veg. Ind. Occ. 98. 1788

Dioecious tree, to 15 (22) m tall, ca 45 cm dbh; stems and leaves glabrate. Leaves alternate, simple; stipules minute, caducous; petioles 2.5–5 cm long; blades ovate to elliptic, acuminate, obtuse to rounded at base, 8–18 cm long and 4–8.5 cm wide (larger on juveniles), the trichomes tufted, stellate, branched from base in vein axils below, the glands 2–4 at base in vein axils and scattered on both surfaces, the margins entire or toothed, the teeth minute, remote, gland-tipped; venation pinnate, 3-veined at base. Spikes axillary, stellate-pubescent, the flowers apetalous, the staminate axes 5–15 (30) cm long, branched many times, the lateral branches 1–3 cm long, the pistillate spikes simple or branched few times near the base, 5–20 cm long; staminate flowers subsessile; calyx lobes 2–4, to 1.5 mm long; stamens 8; pistillate flowers on stout pedicels 1–1.5 mm long; ovary stellate-pubescent; styles 2, simple, 1–2 cm long. Capsules to 11 mm long and wide, bilocular, dark purplish-brown; seeds 2, tuberculate, red, ca 6 mm long and nearly as wide. *Croat 4871, 5024.*

Rare, in the forest; apparently more abundant elsewhere in Panama on the Atlantic slope. Seasonal behavior uncertain. Flowers mainly in April and May. The fruits mature mostly from April to June. In Puerto Rico it is reported to flower all year (Little & Wadsworth, 1964).

Mexico to Panama, introduced into southern Florida; Greater Antilles. In Panama, known from tropical moist forest in the Canal Zone and Bocas del Toro, from pre-

---

KEY TO THE SPECIES OF ALCHORNEA

Staminate spikes unbranched; leaf axils not tufted on underside; leaves usually drying green . . . . *A. costaricensis* Pax & Hoffm.
Staminate spikes compound; leaf axils conspicuously tufted on underside; leaves usually drying brown . . . . *A. latifolia* Sw.

Fig. 318. *Alchornea costaricensis*

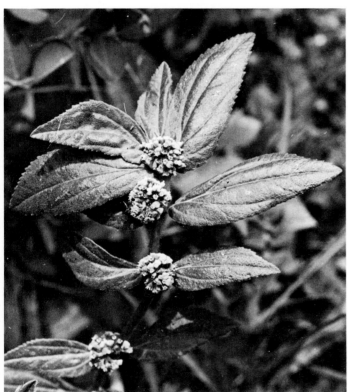

Fig. 319. *Chamaesyce hirta*

Fig. 320. *Chamaesyce hyssopifolia*

montane wet forest in Panamá (Cerro Campana) and Colón (Santa Rita Ridge), and from premontane rain forest in Panamá (summit of Cerro Jefe).

CHAMAESYCE S. F. Gray

Chamaesyce hirta (L.) Millsp., Publ. Field Columbian Mus., Bot. Ser. 2:303. 1909

Euphorbia hirta L.

Hierba de pollo, Saca teta, Golondrina

Monoecious annual herb, erect or decumbent at base, pubescent, some trichomes short and ± uncinate, some longer and straight. Leaves opposite, simple; stipules minute, paired, acicular; petioles 2–4 mm long; blades ± elliptic, asymmetrical, obtuse to acute at apex, slightly cordate and oblique at base, 2–3 (5.5) cm long, 10–15 (25) mm wide, serrate. Flowers naked, in cyathia, the glomerules terminal and axillary, leafless, strigose (often reddish), to ca 1 cm wide; disk bearing 4 glands and minute, white, petaloid appendages; staminate flowers with a single red anther; pistillate flowers solitary, medial in the cyathia; styles 3, deeply bifid. Capsules ovoid, to 1.7 mm long and wide, strigose, held to one side of the cyathia toward maturity by bending of the pedicel, explosively dehiscent, the seeds being thrown several meters; seeds 3, 1 per carpel, ovoid, less than 1 mm long, angulate, reddish-brown. *Croat 8508.*

Abundant in clearings. Flowers and fruits throughout the year. The most abundant species of *Chamaesyce* on the island.

Pantropic weed. In Panama, ecologically variable; known from tropical moist forest in the Canal Zone, all along the Atlantic slope, and in Chiriquí, Herrera, Panamá, and Darién, from tropical dry forest in Los Santos, Herrera, Coclé, and Panamá, from premontane moist forest in the Canal Zone and Panamá, from premontane wet forest in Bocas del Toro, and from tropical wet forest in Colón.

See Fig. 319.

Chamaesyce hypericifolia (L.) Millsp., Publ. Field Columbian Mus., Bot. Ser. 2:302. 1909

Euphorbia hypericifolia L.

Hierba de pollo

Monoecious, glabrous herb, usually erect, to 60 cm tall; stems somewhat woody at base, ca 3 mm diam, reddish.

Leaves opposite, simple; petioles 2–4 mm long; stipules interpetiolar, subpersistent, triangular or sometimes forked especially at apex of plant; blades ovate-elliptic to obovate-elliptic, obtuse to acute at apex, rounded and inequilateral at base, 1.5–3 (3.5) cm long, 7–15 mm wide, serrate. Flowers naked, the cyathia glabrous, the glomerules loose, short-stalked, axillary or terminal, leafless or bearing a few inconspicuous reduced leaves; disk with glands suborbicular, their appendages white or pink; stamen 1; styles 3, bifid. Capsules subspherical, to 1.6 mm diam (1.2 mm when dry); seeds reddish-brown, less than 1 mm long, obtusely 3-angulate. *Croat 9201.*

Occasional in clearings, less abundant than *C. hirta* or *C. hyssopifolia;* perhaps preferring taller-grass areas than *C. hyssopifolia.* Found fertile throughout the year.

Throughout the New World tropics and subtropics. In Panama, known principally from tropical moist forest in the Canal Zone, all along the Atlantic slope, and in Panamá and Darién; known also from premontane moist forest in Panamá, from premontane wet forest in Panamá, and from tropical wet forest in Colón.

Chamaesyce hyssopifolia (L.) Small, J. New York Bot. Gard. 3:429. 1905

Euphorbia brasiliensis Lam.

Golondrina

Monoecious herb, to 50 (70) cm tall, essentially glabrous all over; stems straw-colored, reddish, or brown. Leaves opposite, simple; stipules minute, interpetiolar; petioles 1–2 mm long; blades variable, obovate to ovate, obtuse at apex, rounded to slightly cordate and oblique at base, 1–3 cm long, 6–10 mm wide, serrate. Flowers naked, the cyathia glabrous, the glomerules loose, short-stalked, terminal or axillary, usually noticeably leafed; disk with glands elliptic or suborbicular, their appendages white or pink; stamen 1; styles 3, bifid. Capsules oblong-ovoid, glabrous, to 1.8 mm long, 1.7–2 mm wide, sometimes broader than long; seeds ovoid, ca 1 mm long, reddish-brown, angulate. *Croat 16526.*

Common in the Laboratory Clearing. Flowers and fruits throughout the year.

Standley (1933) erroneously reported this as *Euphorbia brasiliensis* L.

Throughout tropical America; adventive in the Old World. In Panama, ecologically variable; known from tropical moist forest throughout Panama and from premontane dry forest in Los Santos, from tropical dry

KEY TO THE SPECIES OF CHAMAESYCE

Capsules and stems pubescent, at least when young:
    Leaves less than 5 mm wide; flowers borne on leafy lateral branches, the clusters less than 4 mm wide; plants usually growing prostrate . . . . *C. thymifolia* (L.) Millsp.
    Leaves more than 5 mm wide; flowers borne in leafless glomerules much more than 4 mm wide; plants growing erect . . . . *C. hirta* (L.) Millsp.
Capsules and stems glabrous:
    Flowers in almost leafless glomerules; capsules subspherical, less than 1.6 mm wide . . . .
    . . . . *C. hypericifolia* (L.) Millsp.
    Flowers in leafy dichasia; capsules oblong-ovoid, more than 1.7 mm wide . . . .
    . . . . *C. hyssopifolia* (L.) Small

forest in Coclé and Panamá, from premontane moist forest in Panamá, from premontane wet forest in Bocas del Toro, Colón, Chiriquí, and Panamá, and from tropical wet forest in Colón.

See Fig. 320.

Chamaesyce thymifolia (L.) Millsp., Publ. Field
   Columbian Mus., Bot. Ser. 2:412. 1916

Monoecious annual herb, commonly with a thick prostrate root, usually forming dense mats; stems usually villous above and glabrous below, often reddish-brown; sap ± milky. Leaves opposite, simple; stipules distinct, linear, toothed; blades oblong-elliptic, rounded to acute at apex, rounded to subcordate and inequilateral at base, 4–10 mm long, 2–5 mm wide, glabrous above, glabrous to sparsely pubescent below, inconspicuously serrate. Flowers naked, the cyathia 1 to several in axils of lateral branches, the glomerules less than 4 mm wide; disk with glands broadly elliptic to rounded, reddish; stamen 1; styles 3, bifid. Capsules long-ovoid, appressed-pubescent, to 1.3 mm long, not fully exserted but splitting cyathium at maturity; seeds conical-ovoid, to 1 mm long, 4-angled, tan. *Croat 16527.*

Collected once on the mossy sides of a concrete drainage ditch near the laboratory. Flowers and fruits throughout the year, especially during the rainy season.

Throughout the tropics of the New World. In Panama, known throughout from tropical moist forest, as well as from tropical dry forest in Coclé, from premontane moist forest in the Canal Zone, and from tropical wet forest in Colón.

CODIAEUM Adr. Juss.

Codiaeum variegatum (L.) Blume, Bijdr. Ned. Ind.
   606. 1826
   Croton

Monoecious, small, glabrous tree, to 5 m tall. Leaves alternate, simple; petioles 1–2 cm long; stipules ca 2 cm long, linear; blades strap-shaped, acute or apiculate at apex, acute and decurrent at base, 25–30 cm long, ca 1 cm wide (to ca 5 cm wide elsewhere), reddish or mottled with various shades of green, the margins entire and revolute. Racemes upper-axillary, 15–20 cm long, solitary, subtended by a broad leaflike bract ca 2.5 cm long; staminate flowers clustered under the bracts; sepals 5 (rarely 3 or 6), oblong-elliptic, ca 3 mm long; petals 5 or 6; stamens 15–30 or more, inserted on receptacle, the filaments free; pistillate flowers white, solitary, on articulated pedicels 1–2 cm long; petals lacking; calyx 5-lobed; disk of 5–15 glands; ovary 3-celled, each cell with a single ovule; styles 3, distinct. Capsules globose, dehiscing into 2-valved cocci; seeds 2 per coccus (*fide* Bailey, 1949). *Croat 7037.*

Cultivated in the Laboratory Clearing. It has been reported in flower from January to August. No fruits are set on BCI.

Native to the Pacific Islands (Standley, 1928); widely cultivated. In Panama, known only from tropical dry forest on BCI and in Panamá (Taboga Island).

CROTON L.

Croton billbergianus Müll. Arg., Linnaea 34:98. 1865
   Sangrillo, Baquero

Monoecious shrub or tree, 3–10 m tall, densely stellate-pubescent all over except on upper leaf surface; sap copious, ± sticky, yellowish, fragrant. Leaves alternate, simple; stipules subulate-lanceolate, 5–7 mm long; petioles to 6 cm long, with 2 or more sessile, round, yellow glands at apex on lower side; blades usually ovate, acuminate, usually cordate at base, 7–25 cm long, 5–18 cm wide, entire; basal veins 6–9, the cauline veins in 6–10 pairs. Racemes terminal (paniculate by aggregation), 6–15(20) cm long, bisexual or staminate; pedicels 3–5 mm long; calyx cupulate, 4–5 mm long, greenish-white at anthesis, glabrous inside, lobed to ca middle, the lobes 5, broadly acute and somewhat spreading; receptacles villous; staminate flowers terminal on (or full length of) racemes; petals 5, ± elliptic, ca 3.5 mm long, villous; stamens 14–16, well exserted, erect or spreading, to 4.5 mm long; filaments villous; pistillate flowers 1–6 on basal part of raceme, developing before staminate flowers; calyx 5-lobed; petals lacking; ovary broader than high, subtended by a 5-parted, ringlike disk; styles 3, connate into a short column, each style bifid 2 or 3 times, the ends exserted as much as 4 mm and spreading, white, persisting in fruit, the ovary and basal parts of style densely stellate. Capsules obovoid to depressed-globose, to 1 cm diam, stellate-tomentose, green, the columella 4–4.5 mm long; seeds 3, obovoid, 4–5 mm long, 3–4 mm broad. *Croat 14848, 16546.*

Frequent in the forest, mostly the older forest; also along the shore. Flowers from April to August, especially in April and May. The fruits mature from June to October, especially in August and September.

Known only from Panama, from tropical moist forest in the Canal Zone, Bocas del Toro, Colón, Panamá, and Darién and from tropical wet forest in Colón and Panamá.

KEY TO THE SPECIES OF CROTON

Leaves sometimes opposite; plants herbaceous . . . . *C. hirtus* L'Hér.
Leaves alternate; plants arborescent:
    Glands at apex of petiole sessile, only on lower surface; inflorescences mostly less than 15 cm
        long . . . . *C. billbergianus* Müll. Arg.
    Glands at apex of petiole stalked, on upper and lower surfaces; inflorescences mostly more than
        20 cm long . . . . *C. panamensis* (Klotzsch) Müll. Arg.

Croton hirtus L'Hér., Stirp. Nov. 17, t. 9. 1785

Monoecious herb, to 40 cm tall; stems retrorsely stellate-hispid. Leaves alternate, sometimes opposite at branch nodes, simple; stipules subulate, 5–8 mm long; petioles to 3.5 cm long, bearing 2 stalked glands at apex, densely stellate-pubescent; blades ovate to elliptic, acute to obtuse at apex, broadly cuneate at base, 2–6 cm long, 1.5–4 cm wide, one side sometimes shorter, sparsely stellate-pubescent all over, irregularly crenate, basal veins 5, the cauline veins in 3–6 pairs. Inflorescences terminal, bisexual, ± sessile, to 2.5 cm long, subtended by reduced leaves; staminate flowers white, terminal, 5-parted, ca 3 mm wide; calyx lobed to about middle; petals oblong, equaling calyx lobes; stamens 10 or 11, exserted, in 2 whorls of 5 each plus a central stamen; filaments glabrous; pistillate flowers subsessile; calyx lobes usually 4, slender, unequal; disk 5-angled; petals 5, ellipsoid, to 0.3 mm long; styles 3, free to connate at base, bipartite to bifid. Capsules explosive, subglobose, ca 3.7 mm diam, stellate-hispidulous; seeds compressed, dark brown, smooth, ca 3 mm long, 2 mm wide. *Croat 5855.*

Infrequent in the Laboratory Clearing. Flowers and fruits throughout the year, especially in the early rainy season.

Widespread in disturbed areas of tropical America; West Africa; Malaya. In Panama, known principally from tropical moist forest in the Canal Zone, Colón, Herrera, Coclé, and Panamá; known also from premontane dry forest in Los Santos, from premontane moist forest in the Canal Zone, and from premontane wet forest in Chiriquí.

Croton panamensis (Klotzsch) Müll. Arg. in DC., Prodr. 15(2):546. 1866

Sangaree, Sangrillo

Monoecious tree, to 20 m tall (rarely to 30 m); most parts densely stellate-pubescent; sap becoming red in time. Leaves alternate, simple; stipules linear, to 2 cm long; petioles 8–20 cm long, with a row of stalked orange glands at apex; blades ovate, long-acuminate, cordate at base, 12–25(30) cm long, 10–23 cm wide, softly whitish-pubescent below, entire to denticulate; basal veins 5–7, the cauline veins in 6–10 pairs. Racemes terminal, ± pendent, (15)20–40 cm long; flowers pedicellate, 5-parted; staminate flowers usually distal, to 7 mm diam; sepals acute; petals obovate, both petals and receptacle densely villous; stamens 13–20, exserted and broadly spreading, to ca 4.5 mm long; anthers dehiscing along lateral margins; pollen tacky; pistillate flowers extending to ca one-third of the raceme; calyx deeply lobed, the lobes acute,

persistent; petals usually filiform, inconspicuous; styles 3, bifid or bipartite. Capsules subglobose, ca 6.5 mm diam; seeds 3–4 mm long, 2.6–2.8 mm wide, brown, obscurely costate-roughened. *Croat 16529a.*

Though abundant in other parts of the Canal Zone, the plant is now rare on BCI, having been collected only three times along the shore. Flowers to some extent throughout the year, but the principal flowering season is the early rainy season (June to August, especially July), with the fruits maturing usually by September. A second burst of flowering occurs in the late wet and early dry seasons (October to January), with the fruits developing from April to July. Individual plants may flower twice, since plants in flower in July have had mature fruits also. Allen (1956) reported that the species flowers only from September to November in Costa Rica, so perhaps there are different phenological races of the species.

Allen (1956) reported that the species may form small, almost pure stands along forest margins.

Pollination system is unknown: the pollen does not seem to be wind borne; staminate flowers do not appear to have nectaries, though a faint aroma is present; and, since the pollen is not very copious, it would not appear to be visited by pollen collectors.

Mexico to Colombia. In Panama, known principally from tropical moist forest in the Canal Zone, Bocas del Toro, Veraguas, Herrera, Panamá, and Darién; known also from premontane wet forest in Chiriquí, Coclé, and Panamá and from lower montane rain forest in Chiriquí.

DALECHAMPIA L.

Dalechampia cissifolia Poepp. subsp. **panamensis** (Pax & Hoffm.) Webster, Ann. Missouri Bot. Gard. 54:193. 1967

D. panamensis Pax & Hoffm.

Monoecious vine, sparsely short-pubescent all over, more densely on petioles and pedicels. Leaves alternate, compound; stipules paired, linear, 5–10 mm long, persistent; petioles 1.5–6 cm long, with 2 small acicular stipels at apex; leaflets 3, ± narrowly elliptic, acute to acuminate, acute to rounded at base, 4–12 cm long, 1–3 cm wide, the lateral leaflets markedly asymmetrical, all margins with minute, apiculate teeth. Inflorescences bisexual, axillary; peduncles 1–2(2.5) cm long; flowers subtended by a pair of involucral bracts, the bracts trilobate, greenish, ca 1.5–2.5 cm long, pubescent on veins and margins; staminate flowers 8 or 9, in a nearly sessile cymule; pedicels articulate, 2–3 mm long; calyx lobes 4 or 5; disk and

---

KEY TO THE SPECIES OF DALECHAMPIA

Leaves simple and unlobed . . . . *D. dioscoreifolia* Poepp.
Leaves trifoliolate or trilobate:
    Leaves trifoliolate; involucral bracts green at anthesis . . . .
    . . . . *D. cissifolia* Poepp. subsp. *panamensis* (Pax & Hoffm.) Webster
    Leaves trilobate; involucral bracts cream-colored at anthesis . . . . *D. tiliifolia* Lam.

petals lacking; stamens ca 25, the staminal column minute; pistillate flowers 3, subtended by 2 bractlets ca 3 mm long; calyx lobes 3–11, unequal; disk and petals lacking; ovary densely hispidulous; styles 3, connate, 5–6 mm long; stigma 1, compound, concave, 0.5–1 mm wide. Capsules 6.5–8.5 mm diam, with 3 cocci 5–7 mm long; seeds nearly globose, ca 3–3.5 mm diam, mottled. *Croat 7449*.

Known only from the Laboratory Clearing in the shade at the edge of the forest. Apparently flowering and fruiting throughout the year, possibly flowering twice.

Guatemala to Peru. In Panama, known from tropical moist forest in the Canal Zone, Colón, and Panamá and from premontane wet forest in Panamá (Cerro Campana).

Dalechampia dioscoreifolia Poepp. in Poepp. & Endl., Nov. Gen. Sp. Pl. 3:20. 1841

Monoecious vine; stems, petioles, and blades pubescent (sparsely so on blade above between veins), the trichomes fine, short, white. Leaves alternate, simple; stipules linear-lanceolate, to 6 mm long, paired, persistent; petioles 2–11 cm long, with 2–4 glands at apex, 2 of them to 3 mm long, acicular; blades ovate, short-acuminate, cordate at base, 4–15 cm long, often as broad as or broader than long, subentire to obscurely glandular-toothed; basal veins 5, the cauline veins in 1–3 pairs. Inflorescences bisexual, axillary, short-pedunculate; flowers subtended by a pair of involucral bracts, the bracts ovate-cordate, caudate-acuminate, reddish particularly on veins, the margins lacerate-toothed; staminate flowers with 4 calyx lobes; disk and petals lacking; stamens 20–30; pistillate flowers with the sepals 6–11, deeply pectinate-lobed; disk and petals lacking; ovary densely hispidulous; styles 3, connate, 3–5.5 mm long; stigma 1, disk-shaped, 2–3.5 mm wide. Capsules deeply trilobate, ca 12 mm diam; seeds 3, ca 5 mm long, ± cube-shaped, dark-mottled. *Croat 14456*.

Uncommon or rare, seen infrequently in lower levels of the canopy; two old collections are from Wheeler Trail. Probably more abundant than would appear, owing to its inconspicuous manner of growth. Flowering mostly throughout the rainy season, but often as early as April. The fruits mature mostly in the dry season and early rainy season.

Costa Rica to Peru. In Panama, known from tropical moist forest in the Canal Zone, Bocas del Toro, Colón, Panamá, and Darién and from premontane wet forest in Panamá.

Dalechampia tiliifolia Lam., Encycl. Méth. Bot. 2:257. 1786

Monoecious vine; all parts densely short-pubescent. Leaves alternate, simple; stipules 3–5 mm long, caducous; petioles 2.5–14 cm long; blades variable, entire or trilobate to about middle, mostly 10–16 cm long, usually broader than long, the margins serrulate to entire. Inflorescences borne within large involucral bracts on short axillary shoots; involucral bracts ovate, exceeding inflorescence, prominently veined, 3-dentate at apex, white at anthesis,

becoming green; both the staminate and pistillate cymules contained within same inflorescence, the old staminate cymule persisting; staminate flowers with (3)4–6 calyx lobes; disk and petals lacking; stamens 30–45; pistillate flowers with 9–12 fimbriate calyx lobes; ovary densely hispidulous; styles 3, connate, ca 1 cm long; stigma 1, compound, peltate, concave, 1.5–2.5 mm wide. Capsules 3 per cymule, ca 1 cm diam, densely hispid, the trichomes irritating; seeds ca 4 mm long, mottled. *Wetmore & Abbe 156*.

An abundant plant in disturbed areas in the vicinity of the Canal Zone, but it has not been seen recently on the island. Flowers from October to February. The fruits mature mostly from February to April (sometimes to July).

Honduras to Brazil. In Panama, known principally from tropical moist forest in the Canal Zone, Bocas del Toro, Colón, Veraguas, Los Santos, Herrera, Coclé, Panamá, and Darién; known also from tropical dry forest in Panamá (Taboga Island), from premontane moist forest in the Canal Zone and Panamá, and from premontane wet forest in Coclé and Panamá.

DRYPETES Vahl

Drypetes standleyi Webster, Madroño 24:65. 1977

Dioecious tree, ca 30 m tall and 45 cm dbh, glabrous all over; outer bark thin, unfissured, prominently lenticellate, the lenticels round to flattened horizontally; inner bark tan; sap at first with a faint, sweet aroma. Leaves alternate, simple, estipulate; petioles less than 1 cm long; blades variable, usually oblong-elliptic to elliptic, acuminate, obtuse and decurrent at base, 8–12 cm long, 4–6 cm wide, the acumen thickened, hyaline or discolored. Flowers green, sessile, apetalous, in axillary fascicles; pedicels and sepals minutely puberulent; sepals 4, obovate, concave, rounded at apex, 2.5–4 mm long, glabrous or appressed-pubescent inside, ciliate; staminate flowers numerous; pedicels slender, to 1 cm long; stamens 8, in 2, ± spreading series, 2.5–3 mm long, arising from margin of a flat cross-shaped pistillode; anthers oblong, to 1.5 cm long, dehiscing laterally; pistillate flowers 1–5 per axil; pedicel thick, to 4 mm long; ovary ovoid, densely gray-sericeous, borne on a flat pubescent disk; stigmas discoid, sessile, green, ca 1.5 mm broad, persisting in fruit. Drupes broadly ovoid, to 2 cm long and 1.5 cm wide, densely pubescent, the trichomes ± appressed; seed 1. *Croat 14849, 16516*.

Rare; known only from three individuals above the escarpment in the old forest. Seasonal behavior uncertain. Flowers at least in May and June. The fruits probably mature in July and August. Elsewhere the fruits have been observed to be cream-colored or yellowish at maturity (G. Webster, pers. comm.).

Panama and Venezuela. In Panama, known from tropical moist forest in the Canal Zone, from tropical wet forest in Colón (Santa Rita Ridge), and from lower montane wet forest in Veraguas (Cerro Tute).

See Fig. 321.

Fig. 321. *Drypetes standleyi*

Fig. 322. *Hyeronima laxiflora*

Fig. 323. *Hyeronima laxiflora*

GARCIA Vahl

Garcia nutans Vahl in Rohr, Skr. Naturhist-Selsk. 2:217, t. 9. 1792

Monoecious shrub or tree, to 15 m; latex scanty; twigs tomentose, soon glabrous. Leaves alternate, simple, moderately thick, estipulate; petioles variable, 1–9 cm long, pulvinate at apex; blades ± elliptic, abruptly acuminate to obtuse at apex, obtuse to rounded at base, 8–19(25) cm long, 2.5–8 cm wide, glabrous above, pilose to glabrous below, stellate-tomentose on midrib when young; basal pair of lateral veins arising from midrib below the pulvinus. Flowers unisexual, the cymes terminal, pedunculate, reduced, of 1 or 2 pistillate and several staminate flowers; pedicels 1–3.5 cm long, pubescent; calyx rupturing into 2 or 3, valvate segments; petals 6–13, slender, 10–17 mm long, sericeous, longer than calyx, pinkish to dark red; staminate flowers with calyx persistent; disk with numerous segments; stamens numerous (60–150); filaments free; anthers apiculate; pistillate flowers with calyx deciduous; disk lobed; ovary densely sericeous; styles 3, thick, 1.5–2 mm long, reflexed, bifid. Capsules trilobate, 2–2.5 cm long, 3–4 cm diam, sericeous; seeds 3, globose, to 17 mm diam. *Knight 1090.*

Rare; collected only once. The plant was sterile but there is no doubt of its determination. Seasonal behavior uncertain. The fruits apparently mature in the dry season. Costa Rican collections show flowers mostly in the late dry season and early rainy season.

Mexico to Colombia; Greater Antilles. Apparently rare in Panama, perhaps only from drier parts of tropical moist forest in the Canal Zone and Los Santos.

HURA L.

Hura crepitans L., Sp. Pl. 1008. 1753
Sandbox, Javillo, Coquillo macho, Nuno, Nune, Tronador, White cedar

Large monoecious tree, to 25 m tall; trunk to 1(2) m dbh, bearing hard conical spines; branches glabrous; second-year growth with a thin, smooth, brown periderm. Leaves alternate, simple, deciduous; petioles to 20 cm long, with 2 glands at apex; blades ovate, abruptly acuminate, cordate at base, 11–25 cm long, 7–15 cm wide, glabrous above, sparsely long-pubescent to glabrous below, crenate, the teeth often gland-tipped. Receptacle fleshy, conical, 2–4.5 cm long; flowers lacking petals and disk; staminate peduncles to 10 cm long, the flowers red, emerging from the ruptured tissue of a large, conical, fleshy receptacle, to 4 mm high and 3 mm diam, maturing acropetally (toward apex), numbering well over 100, as many as 80 functional at one time; calyx cupulate; pollen yellow, oozing simultaneously from many pores around circumference; pistillate flowers solitary in upper axils; pedicels 2–5 cm long, woody in fruit; calyx cupulate, 3–5 mm long, truncate or shallowly 5-lobed; styles 3, connate into a slender column 2–5 cm long, terminated by a fleshy disk to 1 cm diam with radiating tips to 1 cm long; ovary with ca 15 carpels, each with 1 ovule. Cap-

sules oblate, 5–8(10) cm diam, 2.5–5 cm long, woody, dehiscing explosively into many concentric cocci; seeds many, disk-shaped, 1.5–2 cm diam. *Croat 4907, 5791.*

Common in both the young and old forests. Flowers from April to December. The fruits mature during the late rainy season and in the dry season. Leaves are lost and replaced during the dry season.

Sap and seeds are poisonous to humans and irritating to the skin (Blohm, 1962). Fruits dehisce with a loud report. They are eaten by macaws and monkeys (Allen, 1956).

Costa Rica to Peru and Brazil; West Indies; introduced into California, Florida, the Bahamas, and the Old World tropics (Allen, 1956). In Panama, known principally from and a characteristic component of tropical moist forest (Tosi, 1971) in the Canal Zone, Bocas del Toro, Chiriquí, Veraguas, Los Santos, Panamá, and Darién; known also from tropical dry forest in Los Santos, from premontane moist forest in Panamá, and from premontane wet forest in Chiriquí.

HYERONIMA Allem.

Hyeronima laxiflora (Tul.) Müll. Arg., Linnaea 34:67. 1865
Bully tree, Palo chancho, Pilon, Zapatero, Platano, Pantano

Dioecious tree, to 40 m tall; trunk usually less than 1 m dbh, buttressed, the buttresses sometimes continuous with exposed lateral roots extending over ground surface as much as 30 m from tree; bark reddish-brown and often weathered (appearing chewed up) at base; all younger parts densely lepidote. Leaves alternate, simple; stipules lanceolate, to 1.5 cm long, deciduous; petioles 3–10 cm long; blades broadly elliptic, abruptly acuminate, obtuse to rounded at base, 7–23(30) cm long, 4–12(19) cm wide; juvenile blades 30 cm long and 16 cm wide, ovate-cordate, inflated and rounded at base, usually inhabited by ants, the stipules enlarged. Panicles upper-axillary, the branches densely lepidote, the staminate panicles to 17 cm long, the pistillate ones to 10 cm long and 15 cm broad; main axis with bracts 5–15 mm long, the lateral axis with minute bracts; flowers apetalous, greenish-white; staminate flowers on minute pedicels; calyx cupuliform, shallowly 3- or 4-lobed, less than 1 mm high; disk nearly equaling calyx; stamens 4; filaments to 1 mm long, free; anthers with an enlarged connective with 2 anther sacs pendent and divergent at anthesis; pistillate flowers on stout pedicels 1–2 mm long; calyx cupuliform, shallowly 3- or 4-lobed, to 0.8 mm high; disk smaller than calyx; ovary ovoid, to 1 mm high; styles 3, obsolete; stigmas minute, bifid. Drupes green turning red then purple-black at maturity, subglobose, ca 3 mm diam, sweet; seed 1, ellipsoid, ca 2 mm long, brown. *Croat 8403, 14964.*

Common in the forest, especially the old forest. Flowers mostly from March to June, also sporadically in the late rainy season. The fruits mature mainly from March to July, also sporadically in the late rainy season.

The fruits are eaten by white-faced monkeys.

Panama to the Guianas. In Panama, known from tropical moist forest in the Canal Zone, Bocas del Toro, Panamá, and Darién, from premontane wet forest in Chiriquí, and from tropical wet forest in Colón.

See Figs. 322 and 323.

JATROPHA L.

Jatropha curcas L., Sp. Pl. 1006. 1753

Coquillo, Physic nut, Piñon, Arbol santo

Monoecious shrub or small tree, usually 2–5(6) m tall, glabrous but with puberulence on young leaves and inflorescence branches; stems thick, the leaf scars close and pronounced; sap red in time. Leaves alternate, simple; stipules minute, acicular, caducous; petioles 7–15 cm long; blades ovate, unlobed or with 5–7 shallow lobes, acuminate, cordate, 10–25 cm long, 9–15 cm wide; veins palmate at base. Flowers green, 5-parted, in short terminal dichasia becoming 10–25 cm long; staminate flowers ca 7 mm diam; pedicels 1–5 mm long, puberulous; calyx deeply lobed, the lobes rounded at apex and recurved; petals obovate to oblong, rounded and recurved at apex, to 6 mm long, densely villous inside, weakly imbricate and united below middle; disk segments free, ellipsoid, white, ca 1 mm long; nectar copious, enclosed by petals; stamens 8–10, to 6 mm long, in a close erect cluster; anthers oblong, to 1.7 mm long, extrorse; pollen pale yellow; pistillate flowers on pedicels 5–9 mm long (to 13 mm in fruit); calyx lobes ± oblong-lanceolate, 7–9 mm long; petals as in staminate flowers; ovary glabrous; styles 3, slender, 1.5 mm long, connate in basal half, dilated into massive stigmas. Capsules ± globose, smooth, turning yellow, 2.5–3 cm long, at first fleshy but eventually drying and dehiscent; seeds 3, 1.5–2.2 cm long, black, broadly oblong. *Croat 10223, 12606.*

Cultivated in the Laboratory Clearing. The plant loses leaves for a short time during the dry season, then flowers shortly after the new leaves grow out at the beginning of the rainy season (April to August). These fruits mature mostly from June to September. The plant rarely flowers again at the beginning of the dry season, with the fruits maturing in the late dry season.

Seeds are purgative and may cause death if eaten in quantity (Standley, 1928; Blohm, 1962).

Probably native to northern Central America, but widely cultivated throughout the tropics. In Panama, known principally from tropical moist forest in the Canal Zone, Bocas del Toro, Colón, Panamá, and Darién, but also known frequently from premontane dry forest in Los Santos and Herrera and from tropical dry forest in Los Santos, Coclé, and Panamá.

MABEA Aubl.

Mabea occidentalis Benth., Hooker's J. Bot. Kew Gard. Misc. 6:364. 1854

Monoecious shrub or small tree, usually 4–6(10) m tall, glabrous all over except on inflorescences; sap whitish, not copious. Leaves alternate, simple; stipules paired,

linear, ca 5 mm long, caducous; petioles to 8 mm long; blades oblong, long-acuminate at apex, obtuse to rounded at base, 6–13 cm long, 2.5–4.5 cm wide, the margins subentire to crenulate and often undulate. Inflorescences pendent from leaf axils, ± granular-puberulent on axes and pedicels, to 12 cm long; staminate flowers in groups of 3, the peduncles stout, short, bearing a conspicuous biglandular bract at base; pedicels slender, articulate above base, ca 1 cm long; flowers globular, maroon, 1.5–2 mm wide; calyx of 3–5 ovate lobes; petals and disk lacking; anthers many; pistillate flowers solitary at basal nodes, 3–7 per inflorescence; pedicels 1–1.5 cm long, subtended by a mostly eglandular bract; sepals 6, maroon, bifid at apex, subequal; petals and disk lacking; ovary 3-carpellate, each carpel with 1 ovule; styles 3, connate into a slender column more than 1 cm long, free and simple near apex, densely whitish-pubescent. Capsules subglobose, ca 1.5 cm diam, 3-valved, explosively dehiscent, the valves in turn splitting into 2 parts; seeds 1 per carpel, ca 8 mm long, suspended by a fleshy funicle at apex. *Croat 7026, 7303.*

Frequent in the forest, especially the young forest. Flowers sporadically throughout the year, mostly from October to April. Most fruits mature during the dry season. Individuals are often found in the dry season with new flowering inflorescences that produce fruits in the rainy season.

Mexico to Brazil. In Panama, known from tropical moist forest in the Canal Zone, Colón, Veraguas, Panamá, and Darién, from tropical dry forest in Panamá (Taboga Island), from premontane moist forest in Panamá, from premontane wet forest in Colón, and from premontane rain forest in Panamá (summit of Cerro Jefe).

MANIHOT P. Mill.

Manihot esculenta Crantz, Inst. Rei Herb. 1:167. 1766

Yuca, Cassava

Glabrous, monoecious shrub, 1–3 m tall; stems with milky sap. Leaves alternate, simple; stipules lanceolate, acuminate, 7–8 mm long, deciduous; petioles to 12(17) cm long; blades deeply lobed, acuminate, 8–12(17) cm long, 1–3.5(5) cm wide, glaucous below, the lobes 3–5(7), narrowly elliptic. Panicles axillary and terminal, to 10 cm long in flower, longer in fruit; staminate flowers racemose along main axis; pedicels 4–6 mm long; calyx campanulate, 3–7 mm long, yellowish-green, 5-lobed, puberulent on outside; petals lacking; disk ca 2 mm wide; stamens 10; filaments to 12 mm long, unequal; anthers ca 2 mm long; pistillode lacking; pistillate flowers several along basal branches; pedicels slender and 7–12 mm in flower, stout and to 17 mm in fruit; calyx 5-lobed to near middle, to 10 mm long; petals lacking; disk ca 3 mm wide; ovary 6-winged or 6-ribbed; styles three, ca 2 mm long, dilated and divided into capitate tips. Capsules subglobose, 1.5–2 cm long, 6-ribbed; seeds 3, compressed, ca 9 mm long and 6 mm wide, smooth, with a darker beak at apex. *Croat 7248.*

Cultivated in the Laboratory Clearing. Seasonal behavior uncertain. Flowers at least from October to Jan-

uary, with mature fruits known from January and July.

Plant tissues contain hydrocyanic acid (Blohm, 1962).

Apparently native to South America (Brazil), but widely cultivated in the tropics. In Panama, known from tropical moist forest in the Canal Zone, Bocas del Toro, San Blas, and Darién.

MARGARITARIA L.f.

Margaritaria nobilis L.f., Suppl. Pl. Syst. Veg. 428. 1781

Phyllanthus nobilis (L.f.) Müll. Arg.

Dioecious shrub or tree, to 15 m tall, to 20 cm dbh; outer bark thin, flaky; inner bark thin, rose-colored, with faint longitudinal grains; sap at first with a sweet aroma; stems glabrous; branchlets conspicuously warty-lenticellate. Leaves alternate, simple, glabrous or sometimes bearing short trichomes on petioles and on veins below; stipules paired, triangular, 1–2 mm long, subpersistent; petioles 3–5 mm long; blades ovate-elliptic to elliptic, acuminate, obtuse to rounded and decurrent at base, 7–12(19) cm long, 3–7.5 cm wide, moderately thin; smaller reticulate veins clearly visible. Inflorescences axillary, bearing few flowers; flowers unisexual, green, 4-parted, apetalous; disk round to minutely 4-lobed, annular, 2.5–3 mm wide, the lobes opposite calyx lobes; staminate flowers green, 5–6 mm wide; pedicels 2.5–5 mm long; calyx lobes somewhat unequal, ovate to suborbicular, spreading, 1–2 mm long; stamens 4 and opposite the calyx lobes; filaments to 2 mm long; anthers minute, slightly longer than broad, extrorse; pistillate flowers solitary or clustered or in short racemes; pedicels stout, ca 3 mm long, becoming 5–14 mm long in fruit; calyx lobes 1–2 mm long, about as broad as long, rounded at apex; disk thin, round or minutely 4-lobed; ovary (3)4- or 5-carpellate, ovoid, weakly angled, glabrous or minutely puberulent, together with style ca 2.5 mm long; styles 1 per carpel, entire or bifid or tripartite. Capsules with 4–6 cocci, depressed-globose, ca 1 cm diam, dehiscing irregularly; seeds paired in each coccus. *Croat 6104, 14634, 14656, 14915.*

Occasional in the forest, especially the young forest. Flowers mostly from May to July, rarely later in the rainy season. The fruits mature mostly from July to October. Leaves fall late in the dry season, growing out again at the time of flowering.

Mexico to Peru and Brazil; West Indies. In Panama, known from tropical moist forest in the Canal Zone, Bocas del Toro, Colón, San Blas, Los Santos, Panamá, and Darién, from premontane wet forest in Chiriquí and Panamá, and from tropical wet forest in Colón.

OMPHALEA L.

Omphalea diandra L., Syst. Nat. ed. 10, 2:1264. 1759

Monoecious liana, to more than 30 m high in canopy, sparsely pubescent to glabrate all over except on inflorescence; trunk to 7 cm dbh; outer bark unfissured; inner bark tan with raised white areas on outer surface, the sap violet-purple, often forming droplets between inner

bark and wood. Leaves alternate, simple; stipules minute; petioles to 11 cm long, with 2 large round glands at apex; blades ovate to broadly elliptic, bluntly mucronate at apex (rarely abruptly acuminate), rounded to truncate or slightly cordate at base, 8–21 cm long, 5–15 cm wide, entire; veins at base 3–5, the major lateral veins in 2 or 3 pairs. Panicles 10–50 cm long, terminal or upperaxillary, bracteate; bracts biglandular, the basal 2–3 mm long, the apical 2–3.5 cm long and very narrow; branches and bracts densely pubescent; flowers apetalous, green or yellowish, clustered in cymules in 3 possible arrangements: with all staminate flowers, with a central pistillate flower and several staminate flowers, or with a solitary pistillate flower at apex of branches; staminate flowers globular; pedicels ca 2 mm long; calyx lobes 4, biseriate, 1.5–2.5 mm long and wide, glabrate and ciliate; stamens 2 (rarely 3); anthers ca 0.7 mm long; pistillate flowers on pedicels 1–2 mm long; calyx lobes 4, ovate, ca 2 mm long and wide, pubescent to glabrate; styles 3, connate, the column stout, densely pubescent, 2–2.5 mm long. Fruits ± round, 8–12 cm diam, fleshy but ultimately dehiscent into 3 woody cocci (*fide Flora of Panama*); seeds 1 per coccus, compressed-globose, 4–4.5 cm diam, brown or black, slightly rugose. *Croat 5236.*

Uncommonly encountered in the forest and at the margin of the lake. Since the plant is a high-canopy vine with inconspicuous flowers, it might be more common than collections and observations indicate. Flowers mostly throughout the dry season and in the early rainy season. Nearly mature-sized fruits are seen late in the dry season, probably dehiscing in the early rainy season. Leaves fall for at least a short time in the middle of the dry season.

The fruits are reported to be capsular, but I have never seen them open. They are possibly mammal dispersed or, as some have suggested, in part water dispersed, since the plant perhaps grows more frequently near water. Larvae of the moth *Urena fulgens* (Urenidae) are leaf-miners apparently restricted to this species (Neal Smith, pers. comm.).

Honduras to Peru and Brazil; West Indies. In Panama, known from tropical moist forest in the Canal Zone, Bocas del Toro, Veraguas, Panamá, and Darién and from tropical wet forest in Chiriquí.

See Fig. 324.

PHYLLANTHUS L.

Phyllanthus acuminatus Vahl, Symb. Bot. 2:95. 1791

P. conami Sw.

Jobitillo

Glabrous, soft-wooded, monoecious herb or shrub, usually less than 3(8) m tall, sparsely branched, the branches usually bipinnatiform (the earliest simply pinnate), the primary axis to 65 cm long, margined below branchlets, the ultimate branches 8–20, each 5–25 cm long, with 7–20 leaves. Leaves alternate, simple; stipules triangular, ca 1 mm long, blunt; petioles 1.5–3 mm long; blades ovate to elliptic, gradually acuminate, acute to obtuse at base, 2–4.5 cm long, 1–2.5 cm wide. Flowers apetalous,

Fig. 324. *Omphalea diandra*

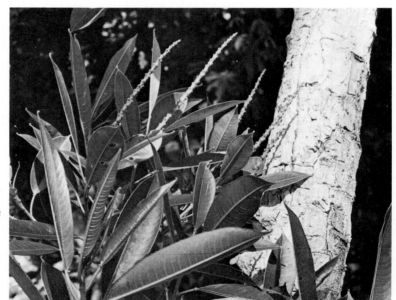

Fig. 325. *Sapium caudatum*

KEY TO THE SPECIES OF PHYLLANTHUS

Leaves more than 1.5 cm long, gradually acuminate . . . . *P. acuminatus* Vahl
Leaves less than 1.5 cm long, rounded or mucronate at apex:
    Leaves rounded at apex, glabrous; seeds not transversely ribbed, not pitted . . . *P. amarus* H. Schum.
    Leaves mucronate at apex, bearing very short trichomes near margins below; seeds transversely
        ribbed, often pitted on sides . . . .*P. urinaria* L.

small, in bisexual cymules in axils of ultimate branches; staminate flowers 5–20 per cymule; pedicels to 5 mm long; calyx lobes 6, in 2 unequal series, the inner rounded, to 1.2 mm long; disk prominent, trilobate; stamens 3; filaments connate into a short column; anthers fused at base, forming a 3-sided structure, dehiscing laterally; pistillate flowers 1 per cymule; pedicels 5–12 mm long; calyx lobes 6, in 2 series, the inner broadly ovate, to 1.7 mm long; disk cup-shaped, trilobate; ovary trilobate; styles 3, flattened, spreading, bifid at apex. Capsules 4.5–5 mm diam, prominently trilobate, explosively dehiscent; seeds plano-convex, ca 2.5 mm long, smooth, reddish-brown. *Croat 10240.*

Occasional, usually in tree falls or clearings; sometimes locally common along the shore. Flowers throughout the year, but principally from February to August. The fruits are mature mostly from July to September. Plants may have mature fruits while still in flower.

Mexico to northern Argentina; West Indies. In Panama, known principally from tropical moist forest in the Canal Zone, Bocas del Toro, Colón, San Blas, Chiriquí, Veraguas, Los Santos, Panamá, and Darién; known also from tropical dry forest in Coclé, from premontane moist forest in the Canal Zone, from premontane wet forest in Chiriquí, Coclé, and Panamá, and from tropical wet forest in Colón.

Phyllanthus amarus H. Schum., Beskr. Guiana Pl.
    421. 1827
    P. niruri L. var. *amarus* (H. Schum. & Thonn.) Leandri

Monoecious herb, 10–50 cm tall, glabrous; branchlets deciduous, 4–10 cm long, very slender. Leaves alternate, simple, thin; stipules lanceolate, ca 1 mm long; petioles to 0.5 mm long; blades elliptic-oblong, ± rounded at both ends, 5–11 mm long, 3–6 mm wide. Flowers minute, apetalous, greenish, 5-parted, usually 2 in each axil, the lowermost axils with staminate flowers, the others with 1 staminate and 1 pistillate flower; staminate flowers on pedicels 0.5–1.3 mm long; calyx lobes ± ovate, ca 0.5 mm long; stamens (2) 3; filaments connate into a column ca 0.3 mm long; pistillate flowers on pedicels to 1–2 mm long in fruit; calyx lobes obovate-oblong, to 1 mm long; disk flat, 5-lobed; ovary 3-carpellate, 2 ovules per locule; styles 3, free, ca 0.2 mm long, bifid at apex. Capsules globular, with 3 cocci, ca 2 mm diam, dehiscent, leaving persistent columella; seeds 2 per locule, trigonous, ca 1 mm long. *Shattuck 398.*

Collected once by Shattuck, but not seen in recent years on the island; to be expected as a weed in clearings. Seasonal behavior not determined. The few specimens

seen had flowers and fruits in the dry season.

Throughout the tropics. In Panama, known from tropical moist forest in the Canal Zone (both slopes), Chiriquí, and Panamá.

Phyllanthus urinaria L., Sp. Pl. 982. 1753

Monoecious herb, to 80 cm tall; branchlets deciduous, 5–10 cm long; stems and petioles glabrous. Leaves alternate, simple; stipules lanceolate, to 1.5 mm long, paired, the pairs unequal; petioles 1 mm or less long; blades linear to oblong-obovate, obtuse to mucronate at apex, rounded at base, 8–25 mm long, 2–6 (9) mm wide, glabrous but with hispidulous margins below, entire. Flowers minute, axillary, greenish-white, apetalous; staminate flowers 5–7, in cymules in upper axils of branchlets; calyx lobes 6, ca 0.5 mm long; stamens 3; filaments connate their entire length; pistillate flowers solitary in lower branchlet axils; pedicels ca 0.5 mm long; calyx lobes 6, less than 1 mm long; ovary papillate; styles 3, connate into a column, bifid at apex. Capsules globular, ca 2 mm diam, green, often reddish-tinged, nearly sessile, dehiscing into 3 cocci, leaving a persistent columella; seeds 1 per coccus, ca 1 mm long, transversely ribbed, often pitted on sides. *Croat 16528.*

Common in clearings. Flowers and fruits at various times throughout the year.

Native to tropical Asia; introduced throughout tropical America. In Panama, known from tropical moist forest in the Canal Zone, Bocas del Toro, Colón, San Blas, Chiriquí, and Panamá, from premontane moist forest in the Canal Zone, from premontane wet forest in Coclé and Panamá, and from tropical wet forest in Colón and Coclé.

POINSETTIA R. Grah.

Poinsettia heterophylla (L.) Klotzsch & Garcke,
    Monatsber. Königl. Preuss. Akad. Wiss. Berlin
    1859:253. 1859
    Euphorbia heterophylla L.

Monoecious herb, to 120 cm tall, with milky latex; stems ± glabrous. Leaves alternate at base of plant, opposite above, petiolate; stipules lacking or minute; blades mostly elliptic, 5–11 cm long, 2–5 cm wide, glabrous or sparsely pubescent. Floral bracts leaflike, green or sometimes purple-spotted; cyathia glabrous, terminal, usually bearing 1 stipitate, tubelike gland, several staminate flowers, and 1 pedicellate pistillate flower, this developing well in advance of the staminate

flowers; flowers naked; staminate flowers with 1 stamen; pistillate flowers with the styles 3, joined at base, each bifid most of its length. Capsules dehiscent, 3-celled, to 2.5 mm long; seeds 1 per cell, ovoid, sculptured. *Croat 4070, 5952.*

Occasional in clearings, seasonally abundant. Flowers and fruits principally in the rainy season.

Throughout tropical America. In Panama, known from tropical moist forest in the Canal Zone, Bocas del Toro, Chiriquí, Veraguas, and Panamá, from tropical dry forest in Herrera and Coclé, from premontane moist forest in the Canal Zone, and from premontane wet forest in Bocas del Toro and Chiriquí.

SAPIUM P. Browne

Sapium aucuparium Jacq., Enum. Syst. Pl. Ins. Carib. 31. 1760
S. jamaicense Sw.
Wild fig, Olivo, Nipa

Glabrous, monoecious tree, to 20 m tall; bark minutely, shallowly fissured. Leaves alternate, simple, membranaceous; stipules obsolete; petioles 2–4 cm long, with 2 stipitate glands at apex; blades elliptic to obovate, acuminate (the acumen sometimes cucullate), obtuse to rounded at base, 9–12 cm long, 3–5 cm wide, subentire to finely serrate. Spikes in clusters of 3–5 (rarely fewer), to 10 cm long, bisexual; flowers with corolla and disk lacking; staminate flowers in groups of 2 or 3(5), the subtending bracts deltoid, to 3 mm long, bearing 2 large glands at base, the apical part caducous; calyx to ca 1 mm long, 2-lipped; stamens 2; pistillate flowers 8–15, solitary at basal nodes of spike, the basal bracts like staminate bracts but with smaller glands; calyx cupular, bilobed, the lobes unequal, free almost to base; ovary orbicular; styles 3, simple, caducous, to 2 mm long, connate for about half their length, the apical half uncinate. Capsules ovoid, to 8 mm long; seeds usually 3, flattened-ovoid, the outer seed coat red. *Kenoyer 661.*

Collected once on the island, and possibly still occurs there. Seasonal behavior not certain. Flowers mostly in March, but sometimes also in the rainy season during September and October. The fruits mature probably mostly from May to July, but full-sized fruits have also been seen in October.

Mexico to northern South America; Cuba, Jamaica. In Panama, known principally from tropical moist forest along the Atlantic slopes in the Canal Zone, Bocas del Toro, and San Blas; known also from tropical moist forest in Los Santos and from premontane wet forest in Chiriquí.

Sapium caudatum Pitt., Contr. U.S. Natl. Herb. 20:127. 1918

Monoecious tree, to 30 m tall and 90 cm dbh, glabrous all over. Trunk weakly buttressed; younger trunks and branches bearing short, stout, branched spines; outer bark thin, hard, in ± rectangular patches, easily peeling, with horizontal scars 6–20 cm long staggered on trunk; inner bark very granular; sap copious, milky, especially in branches. Leaves alternate, simple, somewhat coriaceous; stipules ca 1 mm long, triangular, subpersistent; petioles 2–5.5 cm long, with 2 stipitate glands near apex; blades oblong-elliptic to obovate, with a hooded acumen, obtuse at base, (9)13–18(27) cm long, 3–5(8) cm wide, the margins revolute, irregularly and minutely toothed; juvenile leaves thinner, the teeth much longer and gland-tipped. Spikes solitary, terminal or in uppermost axils, bisexual, to 25 cm long; corolla and disk lacking; staminate flowers globular to pyriform, to 1.3 mm long, in groups of 7–14, inserted on axis above a minute bract, the bract bearing a pair of flattened glands ca 3 mm long, the flowers and glands violet-purple; calyx 4-lobed, ca 0.5 mm long, the pairs of lobes unequal; stamens 2, included; anthers extrorse; pollen orange, tacky; pistillate flowers 5–10, solitary at basal nodes, the bract as in staminate flowers but usually larger; calyx (2)3–5-lobed, sometimes obscurely so; ovary orbicular; styles 3, simple, ca 2 mm long, united only at base, in part deciduous. Capsules ovoid, to 1 cm long, short-stipitate, with 6 longitudinal grooves, splitting into 3 segments; seeds 1 per segment, compressed-ovoid, ca 6 mm long, minutely warty, covered most of its length at maturity with a bright red, thin, pulpy layer. *Croat 14998, Foster 1787.*

Occasional in the forest. Flowers from late May to July. The fruits mature from July to September. Leaves fall in the early dry season, usually in February, and grow back toward the end of the dry season or in the early rainy season.

Range uncertain, possibly restricted to Panama where it is known principally from tropical moist forest in the Canal Zone, Veraguas, Panamá, and Darién and also from lower montane wet forest in Chiriquí (near Cerro Punta).

See Fig. 325.

76. ANACARDIACEAE

Trees. Leaves alternate, petiolate, simple (*Anacardium* and *Mangifera*) or pinnately compound; blades entire or undulate, glabrous or pubescent; venation pinnate; stipules lacking. Plants dioecious or polygamodioecious (*Astronium*, *Spondias*); flowers bisexual or unisexual,

---

KEY TO THE SPECIES OF SAPIUM

Spikes in clusters of 3–5 (rarely fewer); staminate flowers 2–5 per node .... *S. aucuparium* Jacq.
Spikes solitary; staminate flowers 7–14 per node .... *S. caudatum* Pitt.

KEY TO THE SPECIES OF ANACARDIACEAE

Leaves simple:
  Blades acute or acuminate at apex .... *Mangifera indica* L.
  Blades mostly rounded or emarginate at apex:
    Mature leaves mostly more than 15 cm long; trees more than 20 m tall; stamens 10, 4 longer
      than the rest .... *Anacardium excelsum* (Bertero & Balb.) Skeels
    Mature leaves mostly less than 15 cm long; trees less than 10 m tall; stamens 10, 1 much
      longer than the rest .... *Anacardium occidentale* L.
Leaves compound:
  Stamens 10; fruits more than 2.5 cm long and 1.5 cm wide; leaves usually sharply acuminate:
    Pubescent parts minutely puberulent (trichomes short and straight); pedicels and calyces
      usually pubescent; bark coarsely fissured; fruits orange at maturity; endocarp ovoid .....
      .... *Spondias mombin* L.
    Pubescent parts short-villous (trichomes seldom straight, if so not very short); pedicels and
      calyces usually glabrous; bark not coarsely fissured (with paper-thin strips of periderm);
      fruits green at maturity; endocarp oblong .... *Spondias radlkoferi* Donn. Sm.
  Stamens 5; fruits less than 1.5 cm long and 0.6 cm wide; leaves not acuminate or bluntly
    acuminate:
    Flowers pedicellate, unisexual; styles 3, separate to ovary; pistillode lacking in staminate
      flowers; fruits greenish, with winglike, accrescent sepals .... *Astronium graveolens* Jacq.
    Flowers sessile, bisexual; styles tripartite only at apex; pistil present in staminate flowers;
      fruits reddish, lacking accrescent sepals .... *Mosquitoxylum jamaicense* Krug & Urban

actinomorphic, numerous, in terminal panicles; sepals 5, free, imbricate in bud; petals 5, free, imbricate or subvalvate in bud; intrastaminal disk present or absent (*Anacardium*); stamens 5, or (8)10 with 4 or 1 longer fertile stamens; filaments united basally; anthers 2-celled, introrse, basifixed, dehiscing longitudinally; ovary superior, 1-locular (sometimes 3-); placentation basal; ovule 1, anatropous; styles 3 (3–5 in *Spondias*), or style 1 and tripartite. Fruits drupes or nutlike seeds (*Anacardium*) with little or no endosperm.

Confused with Burseraceae (69) and Meliaceae (70), and distinguished from these families by the combination of an intrastaminal disk and a drupaceous or nutlike fruit.

Flowers are more or less open with easily accessible nectaries. They are probably pollinated by a wide variety of small bees and other small insects. Though bats have been found bearing pollen of *Anacardium* and *Mangifera indica*, it is presumed that they collected it accidentally while visiting the fruits (Heithaus, Fleming & Opler, 1975).

Diaspores are carried chiefly by mammals and, to a lesser extent, by birds, though rodents play an important role in the dispersal of *Spondias* (Smythe, 1970; Croat, 1974c). Bats are known to disperse fruits of *Anacardium excelsum* and *Spondias* (Heithaus, Fleming & Opler, 1975; Bonaccorso, 1975; Yazquez-Yanes et al., 1975). Fruits of *Mangifera* are eaten by white-faced and spider monkeys (Hladik & Hladik, 1969) and by coatis (Kaufmann, 1962). *Anacardium excelsum* is taken by coatis (Kaufmann, 1962) and also by white-faced and howler monkeys, which eat only the fleshy receptacle and discard the seeds. All of the larger monkeys and the coati eat the mesocarp of *Spondias mombin* (Hladik & Hladik, 1969). *Astronium graveolens* is wind dispersed by the winged calyx.

About 60–70 genera and 600 species; mainly in the subtropics or tropics, but widely distributed.

ANACARDIUM L.

Anacardium excelsum (Bertero & Balb.) Skeels, U.S.D.A. Bur. Pl. Industr. Bull. 242:36. 1912
Espavé, Wild cashew

Tree, 15–37 m tall; trunk often to 160 cm dbh, somewhat buttressed at base; outer bark coarse, deeply fissured, sometimes flaking loose in large patches, even younger bark (on trees 60 cm diam) with many narrow vertical fissures; inner bark pale pinkish-orange, forming minute close droplets of rust-colored sap soon after slash. Leaves simple, alternate, glabrous; petioles 1–2 cm long, pulvinate at base; blades long-obovate, rounded and sometimes emarginate at apex, tapering, obtuse and decurrent at base, 15–31 cm long, 6.5–10.5 cm wide, broadly undulate; veins lighter than surface. Panicles terminal, sparsely to densely pubescent all over, 15–35 cm long; pedicels 1–6 mm long; flowers bisexual, 5-parted, with strong, sweet, clovelike aroma; calyx lobes ovate, ± fleshy, densely ferruginous-pubescent; petals oblong-linear, ca 6 mm long, ± adnate to staminal tube, recurved at anthesis, cream-colored or green, ferruginous-pubescent; stamens usually 10, 4 much longer, exserted to 4 mm; filaments villous nearly full length; ovary minute; style 1, narrow, simple, equaling length of ovary. Nuts reniform, 3–4 cm long, glabrous, green at maturity, borne on a twisted or recurled green hypocarp ca 3 cm long and 5 mm wide. *Croat 7757, 8518.*

Abundant in the forest and along the lakeshore. Flowers for 6–8 weeks from February to April (sometimes to May). The fruits mature mostly from March

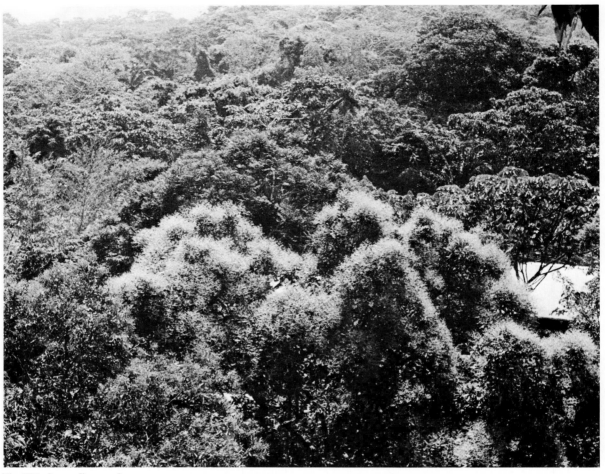

Fig. 326. *Anacardium excelsum*

Fig. 327. *Anacardium excelsum*

to May. Leaves fall in the early dry season and grow in again within 2 or 3 days or sometimes even before the old ones fall.

Nuts contain anacardic acid and a caustic oil called cardol. They are poisonous before they are roasted (Blohm, 1962). Fruits are eaten by white-faced and howler monkeys (Hladik & Hladik, 1969) and by the bat *Micronycteris hirsuta* (Wilson, 1971).

Costa Rica to Ecuador and Venezuela. In Panama, a characteristic tree species in tropical moist forest (Tosi, 1971; Holdridge & Budowski, 1956); known also from tropical dry forest in Coclé and from premontane moist forest in Veraguas and Panamá (Panamá City). Reported from tropical wet forest in Costa Rica (Holdridge et al., 1971).

See Figs. 326 and 327.

Anacardium occidentale L., Sp. Pl. 383. 1753

Cashew, Marañón

Small cultivated tree, usually to 8 m tall and 25–30 cm dbh; branches roughened somewhat with prominent lenticels. Leaves simple, alternate; petioles 1–1.5 cm long; blades obovate, rounded to emarginate at apex, mostly acute at base, 6–16 cm long, 4.5–8 cm wide, glabrous; midrib and major lateral veins impressed and paler above, raised and paler below, the reticulate veins obscure. Panicles terminal; flowers bisexual, pedicellate, 5-parted, fragrant, puberulent on exposed parts outside; calyx lobes sharp; petals reflexed at about middle, sharply lobed, 1–2 mm wide, white to reddish near middle at anthesis, otherwise greenish; stamens 10, 1 much longer, exserted, the shorter ones held about midway in tube formed by recurved petals, often of different lengths, sometimes sterile; anthers occasionally with one theca held higher than the other; filaments fused near base; style 1, minute; nectar collecting both within ring of petals and between petals and sepals. Nuts reniform, 2–3.5 cm long, violet-purple to brown, borne at apex of an obovate, bright red, edible, fleshy hypocarp (the accrescent pedicel); mesocarp thin; seed 1. *Croat 6917.*

Cultivated at the Laboratory Clearing. Flowers from November to May (sometimes from October). The fruits mature from April to July (sometimes from December).

Native to the American tropics, but cultivated throughout the tropics. In Panama, known principally from tropical moist forest in the Canal Zone (both slopes), Chiriquí, Veraguas, Los Santos, Herrera, Panamá, and Darién; known also from tropical dry forest in Panamá (Taboga Island), from premontane wet forest in Colón, Chiriquí, and Panamá, and from premontane moist forest in Panamá.

ASTRONIUM Jacq.

Astronium graveolens Jacq., Enum. Syst. Pl. Ins. Carib. 33. 1760

Gonzolo alves, Zorro, Tigrillo

Dioecious tree, to 35 m tall and 80(100) cm dbh; outer bark lenticellate, peeling off in small, ± round patches leaving irregular shallow depressions; inner bark yellowish, granular, thick. Leaves imparipinnate, alternate; rachis and petiole 20–35 cm long; leaflets (2)9–15, mostly opposite; petiolules to 6 mm long; blades lanceolate-oblong to elliptic, bluntly acuminate, acute to rounded and sometimes inequilateral at base, 4–14 cm long, 2–7 cm wide, glabrous or pubescent especially on veins below. Panicles terminal, to 25 cm long; pedicels 1–3 mm long (much longer in fruit), articulate; flowers unisexual, ca 2 mm long, 5-parted, yellowish-green; sepals ovate to elliptic, minute, enlarging to 1.5 cm and persisting as fruit matures; petals to 3 mm long, ± elliptic; staminate flowers with 5 stamens to 3 mm long; pistillode lacking; pistillate flowers with the styles 3, ca 1 mm long, persistent or deciduous; stigmas small, disciform; ovary ovoid-oblong, with a single subapical ovule; staminodia reduced. Drupes narrowly oblong, 1–1.5 cm long, subterete. *Croat 13492.*

A few individuals are known in the old forest; seedlings, however, are rather common. Plants lose their leaves for a short time just before flowering, and new leaves appear with the flowers in the late dry season. The fruits develop quickly and are dispersed in the late dry and early rainy seasons. In Mexico, the fruits mature from April to June (Pennington & Sarukhan, 1968).

Southern Mexico to Bolivia. In Panama, a typical component of tropical dry and tropical moist forests (Tosi, 1971); known also from premontane moist forest in Panamá and from premontane wet forest in Chiriquí.

MANGIFERA L.

Mangifera indica L., Sp. Pl. 200.1753

Mango

Tree to 40 m tall, to 1 m dbh; sap milky; stems and leaves glabrous. Leaves simple, alternate, dense at apex of branchlets, more widely spaced below; petioles 1–6 cm long; blades mostly oblong, acute to long-acuminate, cuneate at base, 9–35 cm long, 2–7 cm wide, often undulate. Panicles terminal, 20–50 cm long; flowers bisexual, 5-parted; sepals ovate, to 3 mm long; petals greenish-white, ± oblong, to 5 mm long; stamens 5, 1 large and fertile, the other 4 reduced to staminodia; ovary obliquely subglobose or the style eccentric, slender, about equal to fertile stamen. Drupes oblong or semireniform, 10–20(30) cm long, greenish-yellow or orange at maturity; mesocarp thick, juicy, yellow or orange; seed oval, flattened. *Croat 5435, 7107.*

Cultivated at the Laboratory Clearing and persisting in several places in the younger forest and at the margin of the lake. Flowers in the dry season. The fruits are mature from April to July.

Touching these plants, especially the fruit, causes some people to be poisoned in the same manner in which they are from poison ivy (Blohm, 1962).

Native to Asia; cultivated throughout the tropics and throughout Panama. In Panama, known from tropical moist forest in the Canal Zone, from tropical dry forest in Panamá (Taboga Island), from premontane moist

Fig. 328. *Mosquitoxylum jamaicense*

Fig. 330. *Spondias mombin*

Fig. 329. *Spondias mombin*

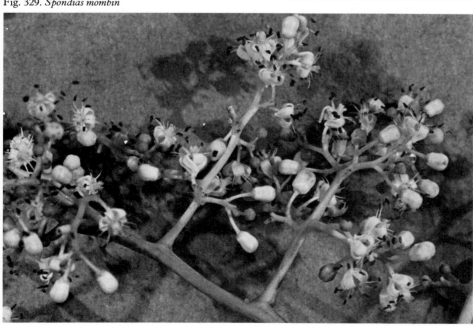

forest in Los Santos, and from premontane wet forest in Chiriquí.

MOSQUITOXYLUM Krug & Urban

Mosquitoxylum jamaicense Krug & Urban, Notizbl. Königl. Bot. Gart. Berlin 1:79. 1895

Mosquito wood, Jobillo, Carbonero

Tree to 30 m high; trunk to 60 cm dbh, weakly buttressed; outer bark weakly fissured; inner bark thin, reddish, forming white sap droplets when cut; wood white. Leaves imparipinnate, alternate, 10–30 cm long, clustered toward apex of branches; leaflets 5–17, usually opposite, short-petiolulate; blades mostly oblong, variable and emarginate to acuminate at apex, cuneate and unequal at base, 2.5–10 cm long, 1.5–3.5 cm wide, usually pubescent on both surfaces especially on midrib. Panicles 7–27 cm long, hirtellous; flowers minute, 5-parted, sessile, bisexual, each subtended by 3, persistent, deltoid bracts ca 1 mm long; sepals ovate, ca 1 mm long, ciliate; petals ca 1 mm long; stamens 5, ca 1 mm long, inserted at margin of disk; disk cupular, 5-lobed, each lobe secondarily lobed; style 1, persistent, tripartite at apex. Drupes 6–9 mm long, red, shiny, glabrous, laterally oblique and somewhat flattened; mesocarp thin, bitter but edible; seed 1, its funiculus subbasal. *Croat 12565.*

Uncommon, in the forest. Seasonal behavior uncertain. Flowers from June to August, but probably later in the rainy season as well. The fruits mature from July to January, but mostly from August to December. The January fruiting record is from higher elevations in Panama in tropical wet forest.

The tree may be infested with moderately large, biting ants when in fruit. The fruits are probably dispersed by birds.

Mexico to Panama; West Indies. In Panama, known from tropical moist forest in the Canal Zone and Daríen, from premontane wet forest in the Canal Zone, Colón, and Panamá, and from tropical wet forest in Coclé.

See Fig. 328.

SPONDIAS L.

Spondias mombin L., Sp. Pl. 371. 1753

Hog plum, Jobo, Wild plum

Tree, mostly 10–30 m tall, to 60 cm dbh; outer bark gray, deeply and coarsely fissured, the raised segments hard, rough, the inner margin irregular; inner bark variously colored, usually with triangular patches of red or tangerine alternating with white; wood soft, white; sap clear, at least never forming viscid droplets; at least the youngest branches puberulent. Leaves imparipinnate, alternate, to 60 cm long (to 70 cm on juveniles); petiole and rachis usually finely puberulent; leaflets mostly (3)9–17, opposite or subopposite; petiolules 6–9(14) mm long; blades oblong to ovate, usually acuminate, acute to rounded and asymmetrical at base, 3–20 cm long and 1.5–7 cm wide, usually ± glabrous but with puberulence on midribs and major veins above and

below, minutely revolute with a prominent submarginal vein; midrib of larger leaflets arched, the reticulate veins prominulous. Panicles terminal, to 60 cm long; branches, peduncles, pedicels, and calyces usually puberulent; pedicels 1–5 mm long, usually articulate near base; flowers 5–7 mm wide, 5-parted, bisexual or rarely pistillate, globular to obovoid in bud; calyx shallow, the lobes short, triangular, sharply acute, usually minutely puberulent, ciliate; petals white, acute and inflexed-apiculate at apex, somewhat reflexed at anthesis; stamens 10, exserted, 1.5–3 mm long, alternating with fleshy, undulate segments of disk; disk fleshy, undulate-lobed, to ca 1 mm wide, the width of one side less than width of clump of styles; styles usually 4 or 5 (rarely 3), much shorter than stamens at anthesis; stigmas linear, on dorsal surface of style near apex. Fruits oblong to obovoid, 2.5–3 cm long, yellow to orange at maturity; mesocarp to 6 mm thick, fleshy, sweet and tasty; endocarps obovoid, 2–2.5 cm long, hard, covered by a tough, coarse, fibrous matrix. *Croat 10751, 14090.*

Frequent in the forest. Flowering principally from March to June (rarely earlier), but most abundantly in April and May. The fruits are mature from July to October, mostly in August and September. Leaves fall during the early part of the dry season, beginning in December and January, and grow in again before flowering commences.

The species is reportedly dioecious in Mexico (Pennington & Sarukhan, 1968) and monoecious in Costa Rica (Bawa & Opler, 1975). In Panama, flowers are apparently mostly bisexual with some pistillate flowers also.

Throughout tropical America; introduced in tropical Africa and East Indies. In Panama, a typical component of tropical moist forest (Tosi, 1971), known principally from the Canal Zone, Bocas del Toro, Colón, Panamá, and Daríen; known also from tropical dry forest in Coclé, from premontane moist forest in Panamá (Farfan Beach), and from premontane wet forest in Chiriquí (Progreso).

See Figs. 329 and 330.

Spondias radlkoferi Donn. Sm., Bot. Gaz. (Crawfordsville) 16: 194. 1891

Tree, to 30 m tall and to 75 cm dbh; outer bark not deeply fissured, the surface with thin, narrow strips of periderm; inner bark similar to *S. mombin* except producing whitish, viscid droplets within a short time after being cut; younger branches glabrate to sparsely crisp-villous to densely villous becoming glabrate. Leaves imparipinnate, alternate, to 54 cm long, usually sparsely crisp-villous on petiole, rachis, upper midrib, and underside of leaflets, especially on younger leaves; leaflets mostly 7–19; blades ovate to oblong-elliptic or oblong, abruptly long-acuminate, acute to subcordate and markedly inequilateral at base, 2.5–16 cm long and 1.8–6 cm wide, ± revolute, usually ciliate, submarginal vein usually lacking. Panicles terminal and upper-axillary, to 55 cm long; axes and rarely pedicels sparsely to densely crisp-villous; pedicels glabrous or less often pubescent,

articulate usually 0.5–2.5 mm below calyx (the articulation sometimes obscured by bracteoles); flowers 5-parted, usually bisexual, rarely pistillate, the first open ones usually appearing with new leaves, the buds usually ± pyriform; calyx cupulate, the lobes thick, prominent, rounded to blunt-triangular, ca 1 mm long, usually glabrous all over; petals ± oblong-elliptic, acute and inflexed-apiculate at apex, 2.3–4.3 mm long, white or greenish-white, recurved at anthesis, the veins 3 (including marginal vein); stamens 10, 1.7–2 mm long, in 2 series, exserted at anthesis; disk to 2.3 mm wide, fleshy, undulate-lobed, the width of one side more than width of clump of styles; ovary subglobose, pubescent; styles usually 3 or 4 (rarely 5), usually free and shorter than stamens at anthesis in bisexual flowers; stigmatic surfaces linear, on dorsal surface near apex; pistillate flowers rare; styles to ca 2 mm long, ca twice as long as stamens, united below middle; stigmatic surface ovate, turned inward. Fruits 3–3.5 cm long, minutely pubescent when immature, oblong to obovate and green at maturity; mesocarp thin, green, with a ± unripened flavor, faintly sweet to acidic; endocarp oblong, nearly as long as fruit, hard, covered by a tough, coarse, fibrous matrix. *Croat 11682.*

Frequent in the forest. Flowers 4–6 weeks later than *S. mombin* on the island, but since the flowering periods overlap, both species may be seen flowering at the same time. Flowers mostly from April to June, especially in May and June. The fruits mature from September to December, especially in October and November. Leaves fall shortly before flowering, growing in again at the time of flowering.

The species has long been confused with *S. mombin* (Croat, 1974c).

Southern Mexico (Veracruz, Chiapas, Campeche) through Central America into Colombia and Venezuela. In Panama, known from tropical moist forest in the Canal Zone, Bocas del Toro, and Panamá (El Llano), and from premontane wet forest in Chiriquí.

77. CELASTRACEAE

Trees. Leaves alternate, petiolate; blades simple, entire; venation pinnate; stipules present. Flowers polygamous, functionally unisexual, actinomorphic, solitary or in condensed racemes; calyx weakly 5-lobed; petals 4 or 5, imbricate; disk present; stamens 4 or 5, alternate with petals, arising from below the disk; anthers 2-celled, introrse, dehiscing longitudinally; ovary superior, 2–4-locular, 2–4-carpellate, immersed in the disk; placentation axile; ovules 2 per placenta, anatropous; stigma sessile, 2–4-lobed. Fruits loculicidal capsules; seeds usually 2, arillate, with endosperm.

Celastraceae are most easily confused with Hippocrateaceae (78), which are only lianas on BCI. The family is represented on BCI by a single species. It may be recognized by its two-valved capsule with arillate seeds.

About 80 genera and over 1,000 species; widely distributed.

MAYTENUS Mol.

Maytenus schippii Lund., Phytologia 1:305. 1939

Glabrous tree, 7–10 m tall. Stipules ± deltoid, less than 1 mm long, ± entire; petioles 5–10 mm long, narrowly winged; blades elliptic to oblong-elliptic, abruptly acuminate, attenuate and decurrent at base, 6–12 cm long, 3–5.5 cm wide, moderately thick, drying gray-green, the acumen obtuse, the margins entire and weakly revolute. Racemes very short, condensed; pedicels 2–6 mm long, articulate at base; flowers functionally unisexual, solitary, appearing before or with new leaves; calyx ca 3 mm wide, weakly 5-lobed, the lobes irregular; petals 4 or 5, ovate, rounded at apex, ca 2 mm long and 1.6 mm wide, imbricate in bud, spreading at anthesis, cream to yellowish; stamens 4 or 5, inserted below the broad, flattened disk, ca 1.5 mm long; filaments broadest and ± flattened at base; anthers introrse; ovary embedded in disk and confluent with it, 2–4-locular; stigma ± sessile, 2–4-lobed. Capsules usually few, often apparently paired, obovate, ca 1.5 cm long, smooth, orange; valves usually 2, coriaceous, opening broadly to expose seeds; seeds usually 2, arillate, the aril white, about as long as seed. *Croat 6135, Hayden 144.*

Rare, along the shore of Bat Cove and Channel Point. Flowers in the dry and early rainy seasons. The fruits mature from July to September or later.

The pollination system is unknown. The arillate seeds are well suited for bird dispersal.

The *Flora of Panama* treatment by Edwin and Hou (1975) included *Hayden 144* under *M. guyanensis* Klotzsch & Schomb. However, that collection differs in no important way from other BCI material included under *M. schippii*, and I believe it was collected from the same tree as *Croat 6135. Stern et al. 863* from Darién is also *M. schippii.*

Mexico to Panama. In Panama, known from tropical moist forest on BCI and in Darién and from tropical wet forest in Colón (Salúd).

78. HIPPOCRATEACEAE

Lianas, occasionally with colored sap. Leaves opposite, petiolate; blades simple, entire; venation pinnate; stipules when present small, interpetiolar. Cymes terminal or axillary, bracteolate; flowers bisexual, actinomorphic, occasionally aromatic; sepals 5, free, imbricate; petals 5, free, imbricate or subvalvate (*Hippocratea*) or valvate (*Prionostemma*); stamens 3, usually inserted within the disk; anthers 2-celled, dehiscing by a transverse slit; ovary superior, 3-locular, 3-carpellate; placentation axile; ovules 2 to many per cell, anatropous; style 1, simple; stigma 3-sided. Fruits of 3 capsular mericarps with many seeds per mericarp and the seeds winged, or the fruits drupaceous (*Tontelea*); endosperm lacking.

Closely allied to the Celastraceae (77). Members of the family are most easily recognized by being lianas with usually small greenish flowers, by the three, usually

KEY TO THE SPECIES OF HIPPOCRATEACEAE

Fruits drupes, more than 10 cm long with a thick pericarp; seeds wingless, embedded in a fleshy matrix; flowers yellow-orange, 5–7 mm diam; leaves more than 7 cm wide, the petiole more than 1 cm long . . . . *Tontelea richardii* (Peyr.) A. C. Smith
Fruits 3 capsular mericarps; seeds winged; flowers greenish, 2–15 mm diam; leaves less than 7 cm wide, or if wider on a petiole less than 1 cm long:
   Mericarps connate, forming a shallow, trilobate bowl; petals narrowly ovate, serrate, ca 4 mm long . . . .*Anthodon panamense* A. C. Smith
   Mericarps attached to receptacle separately; petals not as above:
      Mericarps 1 cm or more thick; flowers fragile, ca 2 mm diam; seeds irregularly oblong, less than 5 per mericarp, probably water-dispersed, the wing smaller than the seminiferous part . . . . *Hylenaea praecelsa* (Miers) A. C. Smith
      Mericarps ± flat, less than 5 mm thick; flowers not falling so easily, more than 5 mm diam; seeds samaroid, more than 4 per mericarp, wind-dispersed, the wing larger than the seminiferous part:
         Leaves asperous; flowers more than 9 mm diam; mericarps ca 1.5 times longer than broad, somewhat swollen; sap red . . . . *Prionostemma aspera* (Lam.) Miers
         Leaves smooth; flowers less than 9 mm diam; mericarps ca 2 times longer than broad, flat; sap clear . . . . *Hippocratea volubilis* L.

fused stamens mounted on a conspicuous disk, and by the three-parted fruits with winged seeds. On BCI only *Tontelea* has simple fruits.

The flowers are greenish and mostly open. Pollination systems are unknown. The fruits of *Anthodon panamense*, *Hippocratea volubilis*, and *Prionostemma aspera* have winged, more or less samaroid, wind-dispersed seeds. Seeds of *Hylenaea*, though also released from a capsular mericarp, are fewer in number, fleshy, and probably mostly water dispersed. Seeds of *Tontelea* are dispersed by agoutis and probably other terrestrial animals as well.

Eighteen genera; tropics of both hemispheres.

ANTHODON R. & P.

Anthodon panamense A. C. Smith, Brittonia 3:422, f. 8 (a–f). 1940
   Hippocratea malpighiifolia sensu Standl. non Rudge
   Bejuco de estrella

Glabrous liana; stems terete (often faintly quadrangular on drying). Petioles 5–8 mm long; stipules lacking or inconspicuous and early deciduous; blades elliptic to oblong-elliptic, obtuse to acuminate at apex, obtuse to acute or rounded at base, 5–13.5 cm long, 2.5–5.7 cm wide. Inflorescences cymose, axillary or terminal, dichotomously branched, to 6 cm long; flowers greenish-yellow to olive-green, 10–12 mm diam, 5-parted; sepals semi-orbicular, ca 1 mm long, broader than long; petals ± ovate-lanceolate, 4–5 mm long, blunt at apex, serrate; disk prominent; stamens 3, ca 1.5 mm long, held closely against the 3-sided stigma; filaments flattened, broadened at base; anthers broadly reniform, broader than long, dehiscing by a transverse slit; ovary ovoid; style short, slightly shorter than stamens. Fruits trilobate, shallowly bowl-shaped, to 15 cm wide, pendent at maturity; mericarps 3, conspicuously flattened, rounded to emarginate at apex, connate 3–5 cm at their bases, dehiscing along a median suture to allow seeds to slowly drift out, the margins thin; seeds 10–14 per mericarp, thin-winged,

ca 2 cm long and 2.5 cm wide including wings. *Croat 11734, 12622.*

Occasional, in the canopy of the forest. Flowers most abundantly from July through August, but also from December to April. Fruit maturity time uncertain.

Known only from Panama, from tropical moist forest on the Atlantic slope of the Canal Zone and in Bocas del Toro, Colón, and Darién.

See Figs. 331 and 332.

HIPPOCRATEA L.

Hippocratea volubilis L., Sp. Pl. 1191. 1753

Liana; branches usually opposite, the older branches minutely fissured, the younger branches glabrous to puberulent. Leaves opposite; stipules deltoid, ca 2 mm long, widely separated on stem; petioles short, canaliculate; blades usually elliptic to oblong-elliptic, bluntly cuspidate to acuminate and often downturned at apex, rounded to acute at base, mostly 4–11 (17) cm long, 2–6 (8) cm wide, glabrous, the margins subentire to crenate, continuous with edges of petiole. Inflorescences axillary and terminal, cymose-paniculate with dichotomous branching, mostly 6–15 cm long, most parts minutely tomentulose; flowers greenish-white or yellowish, with a faint sweet aroma, 5-parted, 6–8 mm diam, short-pedicellate, bracteolate; sepals rounded, minute; petals 3–4 mm long, glabrous at base inside, completely pubescent at apex or bearing an inverted Y-shaped fringe of trichomes; disk prominent; stamens 3, erect at anthesis, hiding style, later folding outward to expose style; filaments flattened; style ca 1 mm long. Mericarps 3, flattened, oblong or elliptic, 4–6 (8) cm long, mostly to 3 cm broad, rounded or deeply emarginate at apex; seeds samaroid, mostly 3.5–5 cm long; seminiferous area apical, darker, 13–25 mm long. *Croat 5040, 7275.*

Very abundant in the canopy and over the water at the edge of the lake. Flowers most of the year except in the early rainy season. Fruit maturity time uncertain.

Fig. 331. *Anthodon panamense*

Fig. 332. *Anthodon panamense*

Fig. 333. *Hippocratea volubilis*

Southern Florida and Mexico to northern Argentina. In Panama, known from tropical moist forest in the Canal Zone, Bocas del Toro, Panamá, and Darién and from premontane wet forest in Colón.

See Fig. 333.

HYLENAEA Miers

Hylenaea praecelsa (Miers) A. C. Smith, Brittonia 3:410. 1940

Salacia praecelsa (Miers) Standl.

Colmillo de puerco, Gorotillo

Nearly glabrous liana; outer bark gray, thin, becoming fissured and peeling to expose reddish-brown inner bark; sap sometimes red in older wood. Stipules lacking or minute and early deciduous; petioles short, to 8 mm long; blades ± elliptic, acuminate, rounded or subcordate at base, mostly 6–26(33) cm long, 2–11 cm wide, somewhat coriaceous, the sides ± upturned, broadly undulate. Inflorescences axillary, to 15 cm long, branched many times, fragile; pedicels ca 1.5 mm long, slender; flowers greenish, to 2 mm diam, 5-parted; sepals minute; petals spreading, ca 1 mm long, papillose-puberulent; stamens 3, minute, included. Mericarps 3, divergent, capsular, longer than broad, rounded or emarginate at apex, 7.5–10 cm long, 5–6 cm wide, 1–2 cm thick; exocarp light brown, ± roughened; seeds usually 4 per mericarp, 4–7.5 cm long, 1–1.5 cm wide, fleshy, buoyant, irregularly angulate. *Croat 7677.*

Occasional in the canopy; more common at the edge of the lake. Flowers in the dry season, with the fruits maturing 9–12 months later.

Flowering inflorescences disappear quickly. The fruits are quite distinctive and are probably principally water dispersed.

Known from Panama and Colombia (Chocó). In Panama, known from tropical moist forest on the Atlantic slope of the Canal Zone and in San Blas and Panamá (San José Island).

PRIONOSTEMMA Miers

Prionostemma aspera (Lam.) Miers, Trans. Linn. Soc. London 28:355. 1872

Hippocratea malpighiifolia Rudge

Liana climbing into canopy; stems retrorsely asperous when young, glabrous in age; sap red. Leaves apparently estipulate; petioles 5–10 mm long; blades ovate-oblong to oblong-elliptic or oblong, short-acuminate, rounded to subcordate at base, 4.5–15 cm long, 2–7 cm wide, ± scabrous and asperous above and below. Inflorescences axillary, 3.5–10 cm long; inflorescence branches and petals minutely puberulent; flowers yellowish-green, 9–15 mm diam, pedicellate, 5-parted; sepals ± rounded, fimbriate; petals broadly spatulate, clawed, rounded at apex, the margin minutely toothed; disk prominent, to ca 8 mm broad; stamens 3, to 3.3 mm long; filaments strap-shaped, broadened below, at first erect, later twisting and holding the anther close to disk; style conic,

minutely puberulent; stigma simple. Mericarps 3, divergent, capsular, to 7 cm long and 4.5 cm wide, less than 5 mm thick, truncate, obtuse or slightly emarginate at apex, obtuse at base; seeds 4–8 per mericarp, winged; seminiferous area darker, to 14 mm long and 7 mm broad; wing elliptic or obovate, to 6 cm long and 3 cm broad. *Croat 7675.*

Frequent in the canopy of the forest. Flowers in February and March. The fruits possibly mature 1 year later.

Easily recognized by the moderately large green flowers, which fall to the ground. Juvenile plants of this species have two noteworthy features: the stems exude copious red sap when cut; and the smaller branches are very asperous with many, very short, stout-based trichomes, which no doubt assist the plants in climbing. Like most other species of Hippocrateaceae, climbing is assisted at least in later stages of growth by the twining of lateral branches.

Panama to Bolivia. In Panama, known only from tropical moist forest on BCI and in Darién.

See Fig. 334.

TONTELEA Aubl.

Tontelea richardii (Peyr.) A. C. Smith, Brittonia 3:478. 1940

Bejuco de canjura

Glabrous liana climbing into canopy; stems terete. Stipules deltoid, 1–2 mm long, often leaving conspicuous, usually unconnected scars; petioles 12–17 mm long; blades ovate-oblong to elliptic-oblong, abruptly acuminate to mucronate at apex, rounded at base, 12–24 cm long, 7–11 cm wide, coriaceous. Inflorescences terminal or in uppermost axils, 3–12 cm long; flowers 5-parted, sweetly aromatic, 5–7 mm diam; calyx thick, ca 3.3 mm diam, deeply lobed, the lobes ± orbicular, imbricate; petals spreading, obovate, to 2.7 mm long, rounded at apex, thick, yellow-orange becoming burnt-orange in age; disk fleshy, bowl-shaped; stamens 3, arising from between disk and ovary, erect, to ca 1 mm long; filaments flattened, broader toward base; anthers obcordate, about as broad as long, extrorse; ovary 3-sided; styles 3, very short, united into equilateral triangle. Fruits heavy drupes to 17 cm long and 12 cm broad; exocarp brown and scurfy outside, 1–1.5 mm thick; locules 3; seeds 2 per locule, large, irregular, to 6 cm wide, with abundant white endosperm, covered with a thick scurfy layer of tissue. *Croat 7932, 15069.*

Common in the forest, usually high in the canopy; occasional along the shore and at lower levels. Flowers from December to June. The fruits mature about 1 year later.

The large cannonball-like fruits often fall to the ground and break open or are opened by rodents. Some fruits are buried by agoutis, which dig them up late in the rainy season when food is scarce. Some germinate where they are buried.

Panama and the Guianas. In Panama, reported only from tropical moist forest on BCI.

See Fig. 335.

Fig. 334. *Prionostemma aspera*

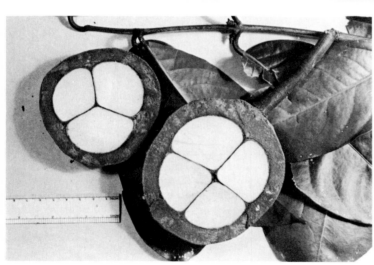

Fig. 335. *Tontelea richardii*

Fig. 336. *Turpinia occidentalis*
subsp. *breviflora*

79. STAPHYLEACEAE

Trees or shrubs. Leaves opposite, petiolate, pinnately compound; leaflets serrate; venation pinnate; stipules present. Panicles terminal; flowers bisexual and actinomorphic; sepals 5, free; petals 5, free; stamens 5, arising from between the lobes of the conspicuous disk, alternate with the petals; anthers 2-celled, dehiscing longitudinally; ovary superior, 3-locular, 3-carpellate; placentation axile; ovules few, anatropous; styles 3. Fruits berries; seeds with fleshy endosperm.

Sometimes confused with Caprifoliaceae but with no other family on BCI. Distinguished by the opposite, pinnately compound leaves with serrate-margined leaflets and by the small white flowers with pistils of three free carpels.

Pollination system is unknown.

Fruits are endozoochorous, apparently by arboreal frugivores. Many are found on the ground as well and are probably further dispersed by other animals. *Turpinia pinnata* Schlechter, with a very similar fruit, is taken by bats in Mexico (Yazquez-Yanes et al., 1975).

Five to seven genera and about 60 species; mostly from North America.

TURPINIA Vent.

Turpinia occidentalis G. Don subsp. **breviflora** Croat, Ann. Missouri Bot. Gard. 63:397. 1976

T. paniculata sensu auct. non Vent.

Tree, to 18 m tall and 30 cm dbh, ± glabrous all over; outer bark with many, small, closely spaced, vertical fissures; inner bark brown with white markings; sap without odor. Leaves opposite; petioles 4–6(11) cm long; petiolules less than 1 cm long on lateral leaflets, longer on terminal leaflet; leaflets 3–9, elliptic or ovate-elliptic, acuminate, obtuse to rounded at base, 6–13 cm long, 2.5–5 cm wide, sharply to obscurely serrate. Panicles terminal, branched many times, to 30 cm long; pedicels to 1.5 mm long; flowers 5-parted, fragrant; sepals ± irregular, concave, at least 1 somewhat longer than others, to 2.7 mm long, rounded at apex, persisting in fruit; petals white, obovate, rounded above, 2.3–2.7 mm long; stamens 5, as long as and alternating with petals from between the lobes of the fluted disk; anthers ovate, attached at center, the thecae directed upward; pistil of 3 free carpels; styles connate at anthesis, later becoming free; stigmas united, at about the level of anthers. Berries yellow, subglobose to obovate, 3-locular, to 2 cm diam, with 3 radial grooves at apex; seeds several per locule, irregularly ovate, 4–5 mm long, smooth, orange-brown. *Croat 15048, 17048*.

Occasional, in the forest. Flowers principally from April to June. The fruits mature from July to September.

The subspecies *occidentalis* occurs in Panama also but usually at elevations above 1,000 m. It is distinguished by having flowers more than 3.5 mm long.

Southern Mexico to Colombia; West Indies; most abundant in lower middle America and Panama, from sea level to 850 m. In Panama, known from tropical moist forest in the Canal Zone and Darién and from premontane wet forest in Veraguas, Coclé, and Panamá. See Fig. 336.

80. SAPINDACEAE

Trees or shrubs (*Allophylus, Cupania, Talisia*) or watch-spring-tendriled lianas (*Paullinia, Serjania, Thinouia*), the climbing species generally with milky sap and complex wood structure. Leaves alternate, petiolate, pinnate or compound-pinnate; leaflets sometimes lobed, entire or serrate; venation pinnate (palmately veined at base in *Thinouia*); stipules present only in climbing species. Cymose inflorescences or thyrses terminal or axillary, bracteate, sometimes racemose; flowers unisexual or

---

KEY TO THE TAXA OF SAPINDACEAE

- Leaves more than pinnate, at least on basal pair of pinnae:
    Leaves with only the basal pair of pinnae ternate, the remainder pinnate . . . .
    . . . . *Paullinia glomerulosa* Radlk.
    Leaves strictly biternate (3 sets of 3 leaflets) *or* leaves with more than 3 sets of 3 leaflets *or* leaves
        more than bipinnate:
        Leaves more than bipinnate *or* with more than 3 sets of 3 leaflets:
            Leaflets less than 2 cm wide, glabrate; petioles unwinged . . . . *Serjania trachygona* Radlk.
            Leaflets more than 2 cm wide, minutely hispid with the vein axils densely tomentose; peti-
                oles and rachises conspicuously winged . . . . *Serjania mexicana* (L.) Willd.
        Leaves strictly biternate (3 sets of 3 leaflets):
            Lower blade surface bearing tufts of trichomes in vein axils, otherwise glabrate:
                Rachis winged; young stems prominently lenticellate; flowers in congested, unbranched,
                    ± cylindrical inflorescences, the flowers ± contiguous; fruits suborbicular, 3-
                    angled, red, woody capsules . . . . *Paullinia fuscescens* H.B.K. var. *glabrata* Croat
                Rachis not winged; young stems not lenticellate; flowers diffuse, often in branched ter-
                    minal racemose panicles; fruits ovate-cordate or ovate-elliptic, 3-winged schizo-
                    carps:
                    Cells of fruit and ovaries glabrous; stems with simple wood . . . .
                        . . . . *Serjania pluvialiflorens* Croat
                    Cells of fruit and ovary puberulent; stems with composite wood, the large central core
                        surrounded by 10 regular peripheral bundles . . . . *Serjania decapleuria* Croat
            Lower blade surface lacking axillary tufts . . . . *Serjania* (in part)

● Leaves all pinnate:
   Leaflets always 3:
      Plants shrubs or trees lacking tendrils; cocci 1 or 2, red, obovoid, less than 1 cm long . . . . .
      . . . . *Allophylus psilospermus* Radlk.
      Plants tendriled lianas:
         Leaflets bearing axillary tufts of trichomes on underside; stems not ribbed or striate,
            lenticellate:
            Leaflets palmately veined at base; stamens long-exserted; petals less than 1 mm long;
               fruits schizocarps of 3 samaras, more than 5 cm long . . . .
               . . . . *Thinouia myriantha* Tr. & Planch.
            Leaflets not palmately veined; stamens ± included; petals ca 5 mm long; fruits 3-
               sided, thick-walled capsules, less than 2.5 cm long . . . . . *Paullinia turbacensis* H.B.K.
         Leaflets lacking axillary tufts of trichomes on underside; stems striate or ribbed, not len-
            ticellate . . . . *Serjania circumvallata* Radlk.
   Leaflets usually more than 3:
      Leaflets 10 or more . . . . *Talisia*
      Leaflets 3–9:
         Petioles or rachises winged . . . . *Paullinia* (in part)
         Petioles and rachises unwinged:
            Plants lianas; leaflets opposite, the apical 3 attached at the same point on rachis, the
              base of terminal leaflet different from base of lateral leaflets:
            Plants essentially glabrous . . . . *Paullinia pterocarpa* Tr. & Planch.
            Plants conspicuously brownish-hirsute on most parts . . . . *Paullinia rugosa* Radlk.
            Plants shrubs or trees; leaflets alternate or only subopposite, the apical 3 not attached
              at the same point on rachis, the terminal leaflet as others . . . . *Cupania*

---

sometimes bisexual (polygamous); sepals 4 or 5, free or briefly connate, the lobes imbricate; petals 4 or 5, free, clawed, often with petaloid appendages (scales); extra-staminal nectariferous disk present, sometimes glandular; stamens 8 (sometimes 5, 7, 9, or 10), free (united in *Serjania*); anthers 2-celled, versatile or basifixed, introrse, longitudinally dehiscent; ovary superior, (2)3-locular, (2)3-carpellate; placentation axile; ovules 1 or 2 per locule; style simple or trifid. Fruits samaroid mericarps (*Serjania, Thinouia*) or capsules (*Cupania, Paullinia*), rarely indehiscent cocci (*Allophylus*) or indurate berries (*Talisia*); seeds sometimes arillate (*Cupania, Paullinia*), lacking endosperm.

Sapindaceae are distinguished by the pinnate or compound-pinnate leaves, by the watchspring tendrils on the lianas, by the small, usually polygamodioecious flowers with scale- or gland-appendaged petals, and by the usually schizocarpous and samaroid or capsular, arillate fruits with shiny black seeds.

Flowers are generally pollinated by small bees. *Trigona* are frequently seen visiting flowers of *Serjania* and *Paullinia*. In Costa Rica, *Serjania* is visited by *Augochloropsis* (Halictidae), *Paratetrapedia* (Anthophoridae), *Trigona*, and *Melipona*. Also in Costa Rica, *Allophylus* is visited by *Trigona, Exomalopsis* (Anthophoridae), and *Phthiria* (Bombyliidae). D. Janzen (pers. comm.) reports that in Costa Rica and Mexico, however, *Serjania* and *Paullinia* are pollinated by *Ptiloglossa* very early in the morning, and that the rest of the bee visitors are robbing pollen and nectar. Heithaus (1973) reported small chrysomelid beetles to visit *Allophylus* in Costa Rica.

Serjania and *Thinouia* are wind dispersed. *Paullinia* and *Cupania* are principally bird dispersed, although Oppenheimer (1968) reported a number of species, including *Cupania rufescens, C. sylvatica*, and *Paullinia turbacensis*, to be taken by white-faced monkeys, and seeds

of *Cupania rufescens* have been eaten by coatis (Kaufmann, 1962). *Allophylus* is perhaps equally dispersed by birds and other animals such as the white-faced monkey (Oppenheimer, 1968). Many seeds fall to the ground and may be further dispersed. *Talisia* is probably mammal dispersed.

Some 150 genera and about 2,000 species; subtropics and tropics.

ALLOPHYLUS L.

Allophylus psilospermus Radlk., Sitzungsber. Math.-Phys. Cl. Königl. Bayer. Akad. Wiss. München 20:230. 1890
A. panamensis Radlk.

Slender, polygamous tree, usually 5–10(20) m tall; stems terete, glabrous except when young. Leaves trifoliolate; petioles somewhat flattened laterally, 2–5(7) cm long, strigose; leaflets ± elliptic, subsessile, acuminate at apex, acute and sometimes inequilateral at base, 2.5–26 cm long, 1.5–7.5 cm wide, remotely serrate, ± glabrous but with sparsely pubescent veins and sometimes tufts of trichomes in axils of veins below. Thyrses puberulent, simple or paniculate, chiefly axillary; flowers 4-parted, campanulate, cream or greenish-white, ca 2 mm wide, pedicellate; sepals ciliolate, orbicular to obovate; petals obovate, 1–2 mm long, the scales united with petals to near apex, densely villous, each subtended by a yellowish basal gland; staminate flowers with the stamens 8, well exserted, to ca 2.3 mm long, clustered on one side of the flower; filaments villous at least near base; pistillode obscure; pistillate flowers with the stamens less than 1 mm long, included; ovary 2-celled, minutely hispidulous; style borne between 2 cells of ovary, 1.5–2 mm long, persisting in fruit but very oblique, held at base of fruit;

stigmas 2, recurved, densely papillate. Fruits of 2 obovoid cocci (1 often aborting), 6–9 mm long, yellow to orange, finally red and glabrate at maturity with a thin fleshy pericarp covering the seed; seed 2–3 mm diam. *Croat 5862, 10754.*

Frequent in the forest, especially the young forest; possibly most abundant along the shore. Flowers in the dry season (January to May), with the fruits maturing in the early rainy season (May to August).

The species is variable throughout Panama. BCI plants are generally glabrate on the surface of the leaves and sparsely pubescent on the veins, sometimes with axillary tufts of trichomes in the vein axils. The closely related *A. occidentalis* Sw. (not on BCI) is puberulent to villous over the entire leaf surface.

Mexico to Panama; West Indies. In Panama, known principally from tropical moist forest in the Canal Zone, Chiriquí, Veraguas, Los Santos, Panamá, and Darién; known also from tropical dry forest in Los Santos and Coclé, from premontane moist forest in Los Santos (Punta Mala), from premontane wet forest in Chiriquí (Tolé), and from tropical wet forest in Coclé (El Valle). Standley (1928) reported the species to be occasional on the Atlantic slope of the Canal Zone. Reported from premontane rain forest in Costa Rica (Holdridge et al., 1971).

CUPANIA L.

The scale of the petals is usually fused to the margin of the petal in *Cupania* so that the petal and the scale form a small pocket when pulled apart at the apex. All species are polygamous.

Cupania cinerea Poepp. in Poepp. & Endl., Nov. Gen. Sp. Pl. 3:38. 1843

    C. costaricensis Radlk.

    Gorgojo blanco, Gorgojo, Gorgojero

Tree, to ca 8(10) m tall; stems terete, conspicuously lenticellate, often flexuous, glabrous but with young stems tomentose. Leaves pinnate; petioles 1–6 cm long, tomentose to glabrate; petiolules ca 5 mm long; leaflets

3–7, obovate, rounded to truncate or emarginate at apex, acute at base, 6–17 cm long, 3.5–7 cm wide, densely white-tomentose below, glabrate above, serrate-dentate. Panicles dense, terminal or subterminal, racemose; sepals and petals 5; sepals ovate, tomentose, greenish; petals ± obovate, white, to 2 mm long, villous, the blades fused to the scales laterally (deeply divided in middle); stamens 8, inserted on inner edge of disk; filaments villous below middle; disk fleshy, ca 0.7 mm high, tomentose; staminate flowers with the stamens exserted, 2.5–3 mm long, the ovary abortive, pubescent, lacking style; bisexual flowers with the stamens 1.5–2 mm long, only slightly exserted, the ovary ovoid, obtusely 3-angulate, with a stout style about as long as ovary, the style and ovary together ca 3 mm long; stigmas 3, recurved; ovary, style, and stigmas tomentose. Capsules obovate, ± rounded to bilobed, short-stipitate, ca 1.5 cm long and 1 cm wide, pale greenish-tomentose or brown-tomentose outside, woolly inside; seeds oblong, more than 1 cm long, shiny black, the lower half covered with an orange aril. *Croat 6696, 11767.*

Occasional, on the northern side of the island along the shore. Flowers in the early rainy season (June and July). The fruits mature in September and October.

The species is closest to *C. latifolia.*

Costa Rica to Colombia, Venezuela, Peru, and Bolivia; West Indies. In Panama, known from tropical moist forest in the Canal Zone, Bocas del Toro, Panamá, and Darién and from tropical wet forest in Colón.

See Fig. 337.

Cupania latifolia H.B.K., Nov. Gen. & Sp. 5:126. 1821

    C. papillosa Radlk.

Tree, 4–20 m tall; trunk 5–25 cm dbh, often somewhat twisted; outer bark smooth, bearing fine, granular lenticels which wipe off easily; inner bark thin, tan; sap with faint, pleasant aroma; stems densely lenticellate, the younger parts (including petioles and branches of inflorescence) densely brown-tomentose. Leaves pinnate (juvenile plants often with some large simple leaves); petioles to 6 cm long (15 cm long on juvenile leaves);

KEY TO THE SPECIES OF CUPANIA

Leaf blades ± glabrous below or the pubescence inconspicuous and usually sparse (except possibly on veins) (*C. latifolia* is sometimes moderately pubescent but the trichomes are not contiguous or long):
    Leaflets crenate to wavy, obtuse to truncate at apex; inflorescences usually branched many times, often 15 cm or more long; fruits ± round, brown, to 2.5 cm diam (when fresh) . . . .*C. latifolia* H.B.K.
    Leaflets entire, acute to acuminate at apex; inflorescence usually unbranched, usually less than 10 cm long; fruits depressed-globose, markedly 3–6-lobed, orange, usually less than 2 cm diam . . . . *C. sylvatica* Seem.
Leaf blades conspicuously pubescent below, the trichomes either long or very close:
    Younger stems and branches densely brown-hirsute; blades hirsute below, the trichomes not contiguous; flowers crowded on inflorescence; capsules with narrow wings . . . .*C. rufescens* Tr. & Planch.
    Young stems and branches glabrous or very short-pubescent; blades white-tomentose below, the trichomes contiguous; flowers not crowded on inflorescence; capsules not winged . . . .*C. cinerea* Poepp.

Fig. 337. *Cupania cinerea*

Fig. 338.
Cupania latifolia

Fig. 339. *Cupania rufescens*

Fig. 340. *Cupania sylvatica*

leaflets 3–11, oblong-elliptic to obovate, obtuse to truncate and sometimes emarginate at apex, obtuse to acute and inequilateral at base, 8–20 cm long and 2–6.5 cm wide (to 35 cm long and 15 cm wide on juveniles), densely short-pubescent to glabrate on veins above, usually papillate and glabrous to sparsely pubescent and with inconspicuous stalked glands below, the margin obscurely crenate to wavy. Inflorescences 15 cm or more long, paniculate, the panicles axillary or subterminal, to ca 2.5 cm long; flowers white, 5-parted, ca 2–2.7 mm long; sepals ovate to oblong, tomentose; petals villous, obovate, rounded or emarginate at apex, the margin fringed, the scales fused to margins of petals, deeply divided in middle; disk thick, bowl-shaped, weakly lobed, densely velutinous except on inner margin; stamens 8, villous below middle; staminate flowers with the stamens to 3.5 mm long, exserted, the pistillode with 3 minute styles; bisexual flowers with the stamens ca 1.5 mm long; ovary ovoid, tomentose, gradually tapered to a stout style; style and stigmas pubescent, together about equaling the ovary; stigmas 3, ca 1.5 mm long, divergent. Capsules subglobose to trilobate, short-stipitate, to ca 2.5 cm diam at maturity, greatly shrinking upon drying and becoming more markedly trilobate; valves thick, woody, densely dark-brown-tomentose outside, woolly inside; seeds black, shiny, ca 1 cm long, enveloped at base with an orange-yellow aril. *Croat 11076, 11981.*

Occasional, in the forest and at least near the lake margin in the vicinity of the laboratory cove and on the northern side of the island. Flowers in June and July. The fruits mature in September and October.

Like most other *Cupania*, this species is variable throughout its range, especially in the type and degree of pubescence. It merges almost imperceptibly at times with a number of other species, including *C. americana* L. (West Indies and northern South America), *C. scrobiculata* L. C. Rich., and *C. oblongifolia* Mart. (Brazil). *C. papillosa* Radlk., segregated by Radlkofer on the basis of the dense papillations and glandular trichomes on the lower surface, is only an extreme form of this species.

On BCI this species is closest to *C. cinerea* and may hybridize with it. *Croat 11981* bears pubescence intermediate between the two species.

Panama to Colombia, Venezuela, Ecuador, Peru, and Amazonian Brazil. In Panama, known from tropical moist forest in the Canal Zone, Colón, and Panamá and from tropical wet forest in Colón (Portobelo).

See Fig. 338.

Cupania rufescens Tr. & Planch., Ann. Sci. Nat. Bot., sér. 4, 28:374. 1862
   C. fulvida Tr. & Planch.
   Candelillo

Tree, to 15 m tall; trunk to 25 cm dbh, sulcate at least when young; outer bark with shallow, horizontal and vertical fissures, flaking off to expose reddish inner bark; wood cream-colored, with sweet odor; younger stems densely ferruginous-hirsute, obscurely 5-ribbed. Leaves pinnate; rachis conspicuously hirsute; leaflets 3–7(9),

obovate-oblong, mostly rounded at apex (sometimes obtuse or acute), often inequilateral at base, the larger 7–22(33) cm long, 3.5–10 cm wide, glabrous above but with densely pubescent midrib, conspicuously hirsute below especially on veins, the margins entire to usually denticulate (serrate on juvenile plants); all veins prominent, the major laterals impressed above; simple leaves on juvenile plants to 45 cm long and 15 cm wide. Panicles stout, densely floriferous, upper-axillary, ± equaling leaves, densely hirsute; flowers white, 5-parted; calyx densely pubescent, 2.8–3.3 mm long, equaling corolla, regular; petals obovate, pubescent, the scales fused along the margin at base; disk orange, prominent, nearly glabrous; stamens 8, the filaments villous on basal three-fourths; staminate flowers opening before bisexual flowers and mostly deciduous when bisexual flowers open, with the stamens to 3.5 mm long, exserted; anthers ca 1 mm long, attached to filament at middle, the thecae divergent in lower half; ovary and style very densely pubescent, the trichomes stiff, straight, usually exceeding stigmas; ovary narrowly ovoid, strongly 3-angulate; style and stigmas less than 2 mm long; stigmas 3, ± erect; bisexual flowers with the stamens 2–2.5 mm long, not exserted; anthers smaller than in staminate flowers, otherwise similar; ovary ovoid, ca 2 mm high, the pubescence as in staminate flowers but with style and stigmas not exceeded by trichomes; style ± equaling ovary; stigmas 3, recurved. Capsules in dense clusters, burnt-orange to Indian red, sharply 3-sided, 1.5–2 cm broad, nearly as long as broad, densely pubescent outside, less so inside, dehiscing broadly along the angles, each angle narrowly winged; seeds 3, obovoid, ca 1 cm long, black, shiny, partly enveloped by a yellow to greenish aril. *Croat 14627.*

Frequent in the forest, especially the young forest. Seedlings are often abundant; they are more deeply toothed and more densely hirsute and often look quite unlike adult plants. Flowers in the early to middle dry season, usually in February and March. The fruits mature in late dry and early rainy seasons, from late April to June. Dehisced fruit valves hang on the tree all year.

Seeds are probably bird dispersed.

Mexico to Colombia, Venezuela, the Guianas, and Brazil. In Panama, known from tropical moist forest in the Canal Zone, Bocas del Toro, Los Santos, Panamá, and Darién, from tropical wet forest in Darién, and from premontane rain forest in Chiriquí and Veraguas.

See Fig. 339.

Cupania sylvatica Seem., Bot. Voy. Herald 93. 1853
   C. seemannii Tr. & Planch.; *Talisia svensonii* Standl.

Tree, to 8 m tall; trunk to 16 cm dbh, often in clumps of 2 or more of various sizes, sometimes with sucker shoots at base; bark smooth; younger branches and inflorescences densely rufous-tomentose, otherwise glabrous to inconspicuously puberulent. Leaves pinnate, to 45 cm long; petiolules 3–15 mm long, pulvinate at base; leaflets 3–7, ± elliptic, acuminate, sometimes weakly falcate, acute to obtuse at base (sometimes inequilateral), 4–30 cm long,

1.7–12 cm wide, the axils of veins sometimes weakly tufted below. Inflorescences axillary, 1–3 per axil, to 12 cm long; pedicels very short, subtended by tiny bracts; flowers white, 5-parted, 4–5 mm wide, with a faint sweet aroma, in short clusters of 3 each, most parts villous; calyx cupular, divided to near base, the lobes ovate, erect; petals obovate to rhombic, rounded to acute, to 2.7 mm long, the scales nearly as long as or longer than petal, ± emarginate, fused to petal along both margins and forming a pocket; ovary ± rounded; stigmas 3 in bisexual flowers, lacking in staminate flowers. Capsules few or many in dense globose clusters, 3–6-lobed, broader than long, to 2 cm wide, densely tomentose, bright red-orange to red at maturity at least at apex; valves 3, densely silky-pubescent inside; seeds 2 or 3, ovoid or subglobose, ca 1 cm long, 8–10 cm wide, shiny, black, at least the apex exserted from the open capsules, the aril fleshy, white, ± acidy but tasty. *Croat 5080, 13481.*

Common at least in some areas of the island. Flowers from the late rainy season throughout the dry season (November to May), with the fruits maturing mostly from March through June.

Panama and northern Colombia. In Panama, known from tropical moist forest in the Canal Zone and Darién, from tropical dry forest in Panamá (Taboga Island), and from premontane wet forest in Panamá.

See Fig. 340.

PAULLINIA L.

The genus is easily confused with *Serjania* when sterile or in flower. Usually leaves are once-pinnate and thus easily separable from *Serjania*, which generally has biternate leaves. However, both genera may have trifoliolate leaves or leaves that are more than once-pinnate. In *Paullinia*, generally only the basal part of the blade is more than once-pinnate and the apical part is merely pinnate; in *Serjania*, the blades, if more than once-pinnate, are usually more than once-pinnate farther toward the apex as well.

Inflorescences of *Paullinia* tend to be narrow and spikelike, while those of Serjania are more frequently paniculate. Flowers are like those of *Serjania* (see the discussion following that genus). Fruits are definitive for separation of the two genera. In *Paullinia* the fruit is generally an unwinged (or obscurely winged), often woody capsule. In *Serjania*, the fruit is tricoccal, with each coccus samaroid. In *Paullinia*, seeds are arillate and animal dispersed. In *Serjania*, each individual samara of the three-winged schizocarp is wind dispersed.

All *Paullinia* species are apparently polygamous.

Paullinia baileyi Standl., Contr. Arnold Arbor. 5:95, pl. 14. 1933

Woody liana; stems 6-ribbed or grooved, long-hispid, with milky sap; tendrils axillary, forked, bracteate at fork, the arms watchspring-like; pubescence brown-pilose all over but with upper leaf surface between veins glabrous. Leaves pinnate; stipules paired, lanceolate, to 3 cm long, ciliate; petioles 1–12 cm long; rachis winged; leaflets 5, ± elliptic or elliptic-oblong, acuminate, acute at base (lower pair often rounded or subcordate), 2.5–22 cm long, 1.5–11 cm wide, remotely dentate. Thyrses short, glomerulate, in leaf axils or borne on tendrils; flowers white, ca 5 mm long; sepals 5, oblong, glabrous; petals 4, oblong to obovate, acute, the scales ca three-fourths as long as petals, their crests yellow, slender, pointed, the scales of the anterior petals held together by villous trichomes, the glands of the anterior petals slender, erect, flattened; stamens 8; filaments ± glabrous, fused into a tube at base; ovary 3-angled, glabrous; styles 3, longer than

KEY TO THE SPECIES OF PAULLINIA

Leaves more than pinnate:
  Leaves with basal pair of pinnae ternate, otherwise pinnate, the leaflets usually 11 . . . . *P. glomerulosa* Radlk.
  Leaves strictly biternate (3 sets of 3 leaflets) . . . . *P. fuscescens* H.B.K. var. *glabrata* Croat
Leaves pinnate:
  Leaflets 3 . . . . *P. turbacensis* H.B.K.
  Leaflets usually more than 3:
    Petiole and rachis not winged:
      Plants essentially glabrous . . . . *P. pterocarpa* Tr. & Planch.
      Plants conspicuously brownish-hirsute on most parts . . . .*P. rugosa* Radlk.
    Petiole or rachis winged:
      Petioles not winged, densely long-hispid; leaflets and stems densely hirsute . . . . *P. baileyi* Standl.
      Petioles winged, not densely long-hispid; leaflets glabrous but possibly with tufted axils:
        Stems terete; petioles glabrous or minutely and obscurely puberulent; leaflets usually entire; pubescence of capsules dense, short, rufous, tomentose . . . . . . *P. fibrigera* Radlk.
        Stems ribbed or striate; petioles not puberulent; leaflets usually toothed; pubescence of capsules not dense, short, rufous, tomentose:
          Leaflets lacking tufted axils below; stipules often persistent, more than 2 cm long; capsules ellipsoid, more than 1.5 cm wide . . . . *P. bracteosa* Radlk.
          Leaflets tufted in axils below; stipules less than 2 cm long; capsules pyriform, less than 1 cm wide . . . . *P. pinnata* L.

staminal tube. Capsules reddish, suborbicular, 1–1.5 cm long, 3-celled, 3-winged, glabrous; seeds 1 or 2, oblong-obovate, dark, shiny, covered at base with a white aril. *Croat 4000a, 8723.*

Occasional, in the forest. Flowers mostly in the late dry season (March and April). Most fruits mature in the late rainy season, some in the early dry season.

The flowers and fruits usually occur on leafless stems near the ground.

Known only from Panama. In Panama, known from tropical moist forest in the Canal Zone and Bocas del Toro and from premontane rain forest in Colón and Panamá (Cerro Jefe).

Paullinia bracteosa Radlk., Bull. Herb. Boissier, sér. 2, 5:321. 1905

Liana; young stems 5-sulcate, the older stems winged, the wings 3, broad, corky; sap milky; stems, petioles, and rachises bearing sparse long pubescence. Leaves pinnate; stipules paired, lanceolate, 2–4 cm long, to 1.5 cm broad, minutely pubescent, often persistent, ciliate; petioles, rachises, and blades ciliate; petioles to 22 cm long, winged; rachis winged; leaflets 5, ± elliptic, acuminate, acute at base, 8–30 cm long, 4–12 cm wide, glabrous above but with pubescent midrib and principal veins, glabrate to hispidulous all over below, crenate above the middle. Thyrses dense, arranged ± densely on strong, slender, densely rufous, bracteate, axillary inflorescences; flowers white, 4–7 mm long; sepals 5, unequal, suborbicular, concave, tomentulose, ciliolate (the largest to 5.3 mm long); petals 4, equaling or exceeding sepals, glabrous, the scales two-thirds to three-fourths as long as petals (± enclosed within concavity of petal), the margin and reflexed appendage markedly bearded, the crest prominent, orange, the glands of the anterior petals whitish, erect, flattened laterally, broader than high, pubescent especially at base; stamens 7 or 8; filaments densely villous, to 3 mm long, held closely together around pistil; anthers dehiscing laterally; ovary bearded at base of staminate styles; styles 3, in staminate flowers short and reduced (held well below the anthers), in bisexual flowers 3-branched to below the middle, to 2.3 mm long, held well above the anthers, the unbranched part pubescent. Capsules ellipsoid to turbinate, 3–5 cm long, rounded to subcordate and apiculate at apex, tapered to base, terete to broadly 3-sided, striate, puberulent to glabrous; seed solitary, to 2 cm long. *Croat 4814, 5363.*

Occasional, on the shore and in the forest; seedlings common. Flowers from the early dry to early rainy seasons (December to August), with the fruits maturing mostly from February to August.

The fruits are perhaps tardily dehiscent—an opened capsule is rarely seen.

Costa Rica (Atlantic slope and the Osa Peninsula) south to Venezuela and the western and south-central parts of the Amazon basin from the Río Madeira in Brazil to Peru and Bolivia. In Panama, known from tropical moist forest in the Canal Zone, Bocas del Toro, San Blas, Chiriquí, Los Santos, Panamá, and Darién.

See Fig. 341.

Paullinia fibrigera Radlk., Smithsonian Misc. Collect. 61(24):2. 1914

Liana; stems terete except when very young, minutely puberulent and lenticellate; sap milky, forming droplets near periphery of stems. Leaves pinnate, glabrous but with puberulent petioles, rachises, and midribs on upper sides; stipules ovate, small; petioles to 10 cm long; petiole and rachis winged; leaflets 5, oblong-elliptic, obtuse to acuminate at apex, obtuse at base, 7–23 cm long, 3–9 cm wide, punctate on both surfaces (especially below), bearing axillary tufts below, entire, sometimes with a single tooth on each side. Inflorescences slender, usually unbranched, upper-axillary, usually to 12 cm long (to 25 cm long in fruit); flowers white, to ca 3 mm long, solitary or in few-flowered thyrses; sepals 5, tomentulose, unequal (2 reduced), 1 concave and partially enclosing the anterior petals; petals 4, glabrous, the scales shorter, their reflexed appendages villous, their crests yellow, bilobed, the lobes slender, the anterior petals subtended by glands, the glands large, orange, erect, flattened, puberulent outside, glabrous inside; stamens 8, shorter than petals; filaments villous, weakly fused at base; ovary ovoid, terete, densely tomentose; styles short, stout. Capsules globose to broadly clavate, beaked at apex, gradually narrowed to stipitate base, densely rufous-tomentose, to 2 cm long and 1.5 cm wide; valves thin; seed 1, ellipsoid, laterally compressed, ca 1.2 cm long. *Croat 6731, 12841.*

Common along the shore and in the canopy of the forest. Flowers in the late rainy season through the early dry season (October to February). The fruits mature in the dry season.

Probably the most common *Paullinia* in the canopy. The species is most easily confused with *P. pinnata*, but can be distinguished by having a terete stem and a densely rufous-tomentose fruit. Vegetation dries dark grayish in contrast to the green of *P. pinnata*.

Known only from Panama, from tropical moist forest in the Canal Zone, Chiriquí, and Darién and from tropical wet forest in Panamá (Cerro Jefe).

See Fig. 342.

Paullinia fuscescens H.B.K. var. glabrata Croat, Ann. Missouri Bot. Gard. 63:480. 1976
Hierba de alacran

Tendriled liana; stems terete, inconspicuously pubescent and bearing many small lenticels; tendrils borne on a short branch, usually bifid. Leaves biternate, to 12 cm long; petioles with a raised pubescent margin on upper surface; rachis narrowly winged, the wing sometimes widest in apical half; blades ± elliptic or oblanceolate, bluntly acute to acuminate (rarely rounded to emarginate) at apex, attenuate to base, sparsely crenate, glabrous above except on sharply raised midrib, glabrate below but with tufts of trichomes in axils, the terminal leaflets 3–9 cm long, 3.3(5) cm wide. Inflorescences terminal or axillary, borne often on the same short branches as the tendrils, to 10 cm long and 1 cm wide; pedicels short, 1.5–2.5 mm long at anthesis, puberulent, articulate about middle or above; flowers with a faint, very sweet aroma,

Fig. 341. *Paullinia bracteosa*

Fig. 342. *Paullinia fibrigera*

Fig. 343. *Paullinia glomerulosa*

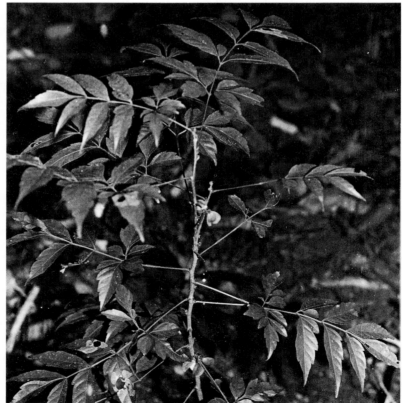

ca 2.5–3.5 mm long; sepals sparsely to moderately appressed-pubescent, boat-shaped, rounded at apex; petals white, the glands of all 4 petals ovoid, yellow to orange, the scales of the anterior petals to 1.5 mm long, about as broad as long, the reflexed appendage densely bearded, the scale crest slender, 0.7 mm long, often as long as reflexed appendage; stamens 8, the staminate flowers with the short stamens to 1.5 mm long, the longer stamens to 2.5 mm long, the pistillate flowers with the stamens ca 1.5 mm long, about as long as ovary; anthers narrowly ellipsoid; pollen golden-brown, tacky; pistil small, well concealed by the densely pubescent filaments; stigmas 3, short, 0.7 mm long. Capsules suborbicular to broadly obovate, to 2 cm long, red, sharply 3-angled, glabrate to sparsely pubescent, short-stipitate, ± beaked at apex, the medial crest sharply raised, the wings 7–8 mm wide; seeds 1–3, oblong, shiny, black, the lower half enclosed in a white aril. *Croat 13235, 13809.*

Occasional, in the canopy and at the margin of the lake. Flowers throughout the dry season. The fruits develop within about 1 month.

It is difficult to distinguish flowering or sterile collections from *Paullinia costaricensis* Radlk., which occurs elsewhere in Panama and Central America.

Mexico to Panama along the Atlantic slope. In Panama, known from tropical moist forest in the Canal Zone, from tropical wet forest in Veraguas, and from an undetermined life zone in Chiriquí (Cerro Vaca).

Paullinia glomerulosa Radlk., Monogr. Paull. 257. 1895

Liana; stems terete, glabrous; tendrils forked. Leaves subbipinnate (lower set of pinnae biternate), pubescent on upper sides of petioles, rachises, and midribs and on lower midrib and major veins, otherwise glabrous; stipules linear-lanceolate, to 1 cm long; petioles 1–2.5 cm long; petiole and rachis narrowly winged, wider at apex; leaflets usually 11, lanceolate, acuminate, acute at base, subsessile, 1.5–10 cm long, 1–3 cm wide, both surfaces often bearing granular punctations, ciliate, serrate-dentate in apical half to entire; foliage of young plants gray-green. Thyrses in leaf axils or on leafless stems often near the ground, aggregated, glomerulate; flowers ca 3 mm long, white; pedicels ca 4 mm long; sepals 5, glabrous, orbicular to obovate and petaloid, to 3 mm long, spreading; petals 4, narrowly obovate, sparsely granular-puberulent, the scales broader than petals, ca half as long, glabrous but with villous margin, the scales of the anterior petals with simple orange crests, the reflexed appendage densely pubescent on margin only, the anterior glands glabrous, ca 1 mm long, ca twice as long as broad; stamens 8, the 3 anterior ones only slightly reduced; filaments glabrous, somewhat flattened; styles 3, short. Capsules suborbicular, sessile, glabrate, 1–1.5 cm long, reddish, with 3 narrow wings to 4 mm wide; valves splitting between the wings; seed solitary, black, shiny, ca 7 mm long, sparsely and softly pubescent with short trichomes, subtended by a white, fleshy aril. *Croat 7997.*

Occasional, in the forest; juveniles often seen as an erect, shrublike plant. Flowers in the middle of the rainy

season (June to October). The fruits mature in the dry season.

Mexico, Panama, and Venezuela; West Indies. In Panama, known from tropical moist forest on both slopes of the Canal Zone and in Bocas del Toro, Panamá, and Darién, and from premontane moist forest in the Canal Zone (Farfan Beach), and from premontane wet forest in Panamá (Cerro Campana).

See Fig. 343.

Paullinia pinnata L., Sp. Pl. 366. 1753
Barbasco

Liana; older stems 3-ribbed, the younger ones mostly 6-ribbed, glabrous to puberulent; tendrils forked. Leaves pinnate, strigillose on upper sides of petioles, rachises, and midribs; stipules lanceolate-sericeous, 18 mm long, deciduous; petioles to 8 cm long; petiole and rachis winged, ciliate; leaflets 5, ± elliptic, short-acuminate, obtuse at base, 5–11 cm long, 2.5–4.5 cm wide, bearing granular punctations on both surfaces and tufted axils below, remotely crenate. Thyrses solitary, 5–25 cm long, axillary or borne at fork of tendrils, spikelike; flowers white or yellowish, ca 3 mm wide, wider than long; sepals 5, ± equal, tomentulose; petals 4, each fused with the scales to form single, broadly clavate structures, these structures densely villous, ca 1.5 mm long with a pocket at the apex, regular, alternating with sepals; stamens 8, exserted, ca 2.5 mm long; filaments flattened, villous below middle; pistil short; staminate flowers with the styles not obvious. Capsules broadly clavate, 2–4 cm long, rounded to truncate and abruptly acuminate at apex, gradually tapered to long-stipitate base, red, round to bluntly 3-sided in cross section; seeds 1–3, ellipsoid, 1.5–3 cm long, black, shiny, partly enclosed by a white aril. *Croat 11111, 12845.*

Uncommon; seldom seen now, though many older collections exist. Flowers principally in the middle of the rainy season (June to August). The fruits mature in the early dry season (December to February). Rarely flowers in the dry season, with the fruits maturing in the middle of the rainy season.

The species seems to intergrade with *P. clavigera* Schlechter in Central America.

Seeds are reportedly poisonous (Blohm, 1962; Standley, 1928).

Throughout tropical America and in tropical central Africa. In Panama, known from tropical moist forest in the Canal Zone, Bocas del Toro, San Blas, Panamá, and Darién and from premontane wet forest in Panamá (Cerro Azul).

See Fig. 344.

Paullinia pterocarpa Tr. & Planch., Ann. Sci. Nat. Bot., sér. 4, 18:356. 1862

Glabrous liana; stems terete or 3-ribbed. Leaves pinnate; stipules lanceolate-linear; petioles unwinged, ribbed above, 5–9 cm long; rachis unwinged; leaflets 5, opposite, ovate to elliptic, long-acuminate, 6–22 cm long, 3.5–8 cm wide, entire to remotely coarse-crenate, the terminal

Fig. 344. *Paullinia pinnata*

Fig. 345. *Paullinia rugosa*

Fig. 346. *Paullinia turbacensis*

leaflet attenuate at base, the lateral leaflets obtuse at base. Thyrses densely clustered, glomerulate, axillary (rarely solitary); flowers white, ca 4.5 mm long; sepals 5, glabrous, irregular, ciliate; petals 4, obovate, often asymmetrical, about twice as long as their scales, the anterior scales ciliolate, about as broad as long, their inner margins extending laterally and overlapping, the reflexed appendage as broad as long, ciliolate, the glands on anterior petals ovoid, hispidulous, on lateral petals lacking; stamens 8; filaments flattened, glabrous; ovary pubescent near apex; staminate flowers with 3 sessile stigmas; bisexual flowers with the style 3-branched at middle. Fruits broadly 3-winged, red, 3-celled (only 1 functional), ca 2 cm long, glabrous; valves striate on outside, tomentose inside; seed solitary, ca 9 mm long, ovoid, black, the lower half white-arillate. *Aviles 46, Croat 14850.*

Rare; collected once in flower by Aviles and once sterile recently. Elsewhere the fruits mature in the middle of the rainy season (August to October).

Distinguished by being glabrous and having five leaflets, with the rachises and petioles unwinged.

Costa Rica to Peru. In Panama, known from tropical moist forest on BCI and in Bocas del Toro, Chiriquí, Coclé, Panamá, and Darién, from premontane wet forest in Panamá (Cerro Campana), and from tropical wet forest in Colón, Coclé, and Panamá.

Paullinia rugosa Benth. ex Radlk., Monogr. Serj. 75. 1875
P. fimbriata Radlk.

Liana; stems 5-ribbed, obscurely so in age, ca 2–2.5 cm diam, with milky sap; stems, tendrils, petioles, and rachises bearing dense, reddish-brown, hirsute pubescence; tendrils stout, forked. Leaves pinnate; stipules semiorbicular, sericeous, to 1 cm diam, the margins fimbriate; petioles 5–18 cm long, terete; leaflets 5, opposite, broadly elliptic, acuminate, attenuate to rounded and sometimes inequilateral at base, 12–35 cm long, 7–17 cm wide, hirtellous above on veins, hirsute below especially on veins, bearing apiculate teeth at ends of lateral veins. Thyrses short, closely congested on slender bracteate spikelike inflorescences, simple and borne in axils or on tendrils or compound and terminal; flowers white, ca 3.5 mm long; sepals 5, unequal, tomentulose, orbicular or ovate and acute; petals 4, obovate, the scales with the lateral margin and the distal edge of appendage bearded, the crest orange, weakly lobed, the glands about as broad as long, sparsely pubescent, weakly concave on outside; stamens 8; filaments densely pubescent; ovary and outer surface of styles densely pubescent. Capsules bright red-orange, suborbicular, to 1.5 cm diam, ± round in cross section, densely tomentose; seed solitary, subglobular,

ca 1 cm diam, black, covered by a thin white aril. *Croat 6036, 11783.*

Adult plants frequent in the forest; seedlings common and recognizable by the large orbicular fimbriate stipules. Flowers in the late rainy season (August to November). The fruits mature within 1 month.

Costa Rica to Peru and Brazil. In Panama, known from tropical moist forest in the Canal Zone and Colón and from premontane wet forest in the Canal Zone (Pipeline Road), Veraguas, and Panamá (Cerro Jefe).

See Fig. 345.

Paullinia turbacensis H.B.K., Nov. Gen. & Sp. 5:114. 1821
P. wetmorei Standl.

Tendriled liana; stems terete, ± glabrous, bearing conspicuous brown lenticels; sap at periphery milky, not copious. Leaves trifoliolate; petioles to 14 cm long, with marginal ribs above, glabrate; leaflets broadly elliptic, acuminate or acute at apex, attenuate to rounded at base, 5–15 cm long, 3–12 cm wide (to 23 cm long and 11 cm wide as juveniles), glabrate but with axillary tufts below, subentire to remotely crenate-dentate. Thyrses small, helicoid, to 7 cm long, solitary or clustered at nodes, often on leafless stems near the ground but occurring to 5 m high; flowers white, ca 5 mm long; petals with the scales ca two-thirds as long as petals, yellow-crested, the anterior scales bilobed, their appendages pubescent, the glands white, acute, densely pubescent; staminal cluster equaling petals; filaments densely pubescent; nectar abundant, stored near the large glands; ovary glabrous; styles pubescent. Capsules elliptic to obovate, 2 cm long, red, short-pubescent, the 3 boat-shaped valves broadly spreading to expose seeds; seeds 1–3, black, shiny, less than 1 cm long, covered on the lower half by a fleshy white aril, dangling on a thin funiculus when fully displayed. *Croat 7213, 7817.*

Frequent in the forest. Flowers in the early dry season (November to February). The fruits mature from February to April.

Mexico, Guatemala, Panama, and Colombia. In Panama, known from tropical moist forest on both slopes of the Canal Zone and in Colón, San Blas, Chiriquí, Panamá, and Darién and from premontane wet forest in Panamá (Cerro Jefe) and Darién.

See Fig. 346.

SERJANIA C. J. Schum.

Serjania is easily confused with *Paullinia.* (See the discussion of that genus for distinguishing features of vegetation and fruits.) Flowers of both genera have five

KEY TO THE SPECIES OF SERJANIA

Leaves trifoliolate (ternate) or bipinnately compound:
  Leaves trifoliolate; stems and leaves glabrous . . . . *S. circumvallata* Radlk.
  Leaves bipinnately compound:
    Leaves at most 3-pinnate; leaflets more than 3 cm wide . . . . *S. mexicana* (L.) Willd.
    Leaves often 4- or 5-pinnate; leaflets less than 2 cm wide . . . . *S. trachygona* Radlk.

Leaves biternate (3 sets of 3 each):
  Rachis winged:
    Seminiferous area (cell) of fruit glabrous; larger stems armed with short spines; wood simple
    . . . . *S. mexicana* (L.) Willd.
    Seminiferous area of fruit conspicuously pubescent; stems unarmed; wood simple or
      composite:
      Lower surface of leaflets conspicuously pubescent above, densely hirtellous to short-pilose
        below; leaflets, flowers, and fruits lacking dark lines . . . . *S. rhombea* Radlk.
      Lower surface of leaflets glabrous; leaflets, flowers, and fruits sometimes with conspicuous
        dark lines:
        Cells of fruit whitish-hirtellous; lower leaflet surface, flowers, and fruits with conspic-
          uous dark black lines; peripheral vascular bundles usually flattened, irregular, often
          more than 1 mm wide . . . . *S. atrolineata* Sauv. & Wright
        Cells of fruit brown-hispid; lower leaflet surface, flowers, and fruits lacking conspicuous
          black lines; peripheral vascular bundles usually terete, regular, usually less than 1
          mm wide . . . . *S. paucidentata* DC.
  Rachis unwinged:
    Cells of fruit and ovaries glabrous or essentially so . . . . *S. pluvialiflorens* Croat
    Cells of fruit or ovaries conspicuously pubescent:
      Lower leaflet surface densely and conspicuously pubescent, the trichomes usually reddish-
        brown; flowers ca 7 mm long; fruits subrectangular-oblong, the cells conspicuously
        brown-hispid, the trichomes ca 2 mm long, each cell bearing a sharp horn . . . .
        . . . . *S. cornigera* Turcz.
      Lower leaflet surface glabrous or only very sparsely pubescent, any trichomes not reddish-
        brown; flowers less than 5 mm long; fruits ovate-cordate, the cells not brown-hispid,
        not at all horned:
        Apical leaflet rachis conspicuously margined; lower leaflet surface with conspicuous
          blackish lines visible to naked eye; stems with 3 flattened vascular bundles; fruits
          constricted below the cell . . . . *S. atrolineata* Sauv. & Wright
        Apical leaflet rachis not conspicuously margined; lower leaflet surface lacking conspic-
          uous blackish lines visible to the naked eye; stems usually with 10 weak ribs and 10
          vascular bundles; fruits not at all constricted below the cell . . . . . . *S. decapleuria* Croat

---

unequal sepals and four petals. Each petal bears a petal-oid appendage (called a scale) on its inner surface, usually one-half to three-fourths as long as the petal. The scales of the anterior petals are larger than those of the lateral petals and are usually held closely together. They are provided with reflexed appendages, which are generally densely villous at least on their common margin. The appendages of the scales are usually coherent by means of their intertwining trichomes. The upper part of the scale, referred to as the crest, is usually colored yellow and is variously shaped. At least the anterior petals are borne on or subtended by conspicuous, white or yellow disk glands, which apparently function as nectaries. Stamens are arranged in a tight whorl surrounding the pistil, and the entire structure is set to one side of the flower opposite the anterior petals. The three stamens adjacent to the pair of anterior petals are usually much shorter than the other five. The pistil of the staminate flowers is much reduced and has three obscure style branches; the pistil of the bisexual flowers is much larger and has three prominent styles.

Plants in the genus are usually polygamous, but possibly dioecious in *S. cornigera*.

Serjania atrolineata Suav. & Wright, Fl. Cubana 24. 1873

Tendriled liana, glabrate in age (juveniles with pubescent midribs); stems ribbed; tendrils bifid. Leaves biternate; stipules minute, ovate; petioles, rachises, and petiolules ribbed to narrowly winged; leaflets ± elliptic to lanceolate, obtuse to acuminate, acute at base, 4–8.5 cm long, 1.5–3 cm wide, usually dentate above middle, ciliate when juvenile, with distinct, irregular black lines on lower surface (sometimes inconspicuous when fresh). Thyrses small, either racemose and solitary in leaf axils and on tendrils or paniculate, open, and terminal; flowers white, ca 3 mm long; sepals densely tomentose outside; petals each subtended by a broad green gland, the scales of the anterior petals yellow, bilobed at apex, held in front of and slightly below the cluster of stamens; stamens 7–9; filaments pubescent; ovary tomentose. Fruits ovate-cordate, ca 2.5 cm long, constricted below the cell, the cells blackened and whitish-hirtellous on inner margin, conspicuously marked with irregular black lines. *Croat 4673, 7685.*

Frequent along the shore and no doubt in the canopy of the forest as well. Flowers throughout the dry season. The fruits mature within about 1 month.

Mexico to northern South America; West Indies. In Panama, known from tropical moist forest in the Canal Zone and San Blas and all along the Pacific slope, from tropical dry forest in Panamá, from premontane moist forest in the Canal Zone and Panamá, and from premontane wet forest in Colón.

Serjania circumvallata Radlk., Monogr. Serj. 345. 1875

Glabrous, tendriled liana; younger stems striate, the older stems grooved, flattened to 3-angled. Leaves trifoliolate and pinnate; petioles 2–7 cm long, with marginal ribs

above; petiolules very short or to 13 mm long; leaflets lanceolate-elliptic, ovate or ± elliptic, acuminate, attenuate to rounded at base, entire to obtusely crenate above the middle, the teeth usually glandular, the terminal leaflet usually 10–14 cm long, 6–8 cm wide, the lateral leaflets somewhat smaller. Thyrses solitary in leaf axils or in terminal paniculate racemes, the ultimate branches densely bracteate and cincinnal; branches, pedicels, and calyces minutely tomentose; flowers white, ca 3 mm long; sepals 5, unequal, oblong to obovate, the 2 opposite the staminal cluster enlarged, obovate, petaloid; petals 4, obovate, glabrous but with villous margins, glandular inside near middle, to 1.7 mm long, all borne atop large glands, the anterior petals with an obscure, densely villous appendage medially and with a yellow crest, lobed to about midway and extending above the petals, the scales of the lateral petals lacking an obvious appendage; stamens 8; filaments villous to near the apex, the longest to 2.3 mm long, exserted; all anthers facing anterior petals; ovary minute, glabrous, 3-sided, glandular; styles 3, equaling ovary. Fruits ovate, subcordate, 4.5–5 cm long, less than 3.5 cm wide, glabrous, the cells ca 7 mm long, rugose, the veins of the wings prominulous. *Croat 12668, Wetmore & Abbe 195, Woodworth & Vestal 638.*

Rare; collected near Fuertes House and near the Redwood House at the end of Armour Trail and on Drayton Trail. Flowers and fruits in the dry season.

Distinguished by the glabrous, trifoliolate leaves with blunt, gland-tipped teeth.

At present known from Costa Rica to Colombia, but possibly more widespread in South America where a number of other similarly trifoliolate species occur. In Panama, known only from tropical moist forest on BCI and in Bocas del Toro.

Serjania cornigera Turcz., Bull. Soc. Imp. Naturalistes Moscou 32:267. 1859

Tendriled liana, densely rufous-pubescent except sparsely so on upper surface of leaflets; stems 5-ribbed. Leaves biternate; petioles mostly 3–6.5 cm long, ribbed above; rachis margined, unwinged; leaflets ± elliptic, blunt to acuminate, abruptly tapered to base (especially lateral leaflets), remotely serrate-dentate, the terminal leaflet mostly 7–13 cm long, 3.5–5.5 cm wide, the lateral leaflets gradually smaller. Thyrses paniculate, terminal or upper axillary, the branches to 15 cm long; pedicels to 1 cm long; flowers white, ca 7 mm long, unisexual; sepals 5, tomentulose; petals 4, oblong-obovate, ciliolate, the 2 anterior petals subtended by a rounded gland, minutely pubescent on lower margins, depressed on outer face, white in staminate flowers, purple at apex in bisexual flowers, the scales of the anterior petals to two-thirds the length of the petals, the crest yellow, thin, emarginate, the appendage villous, slender, pendent to just above glands (scales of lateral petals lacking appendages); staminate flowers with the 5 longer stamens to 6.5 mm long, the shorter ones to 4 mm long; filaments sparsely villous; style nonfunctional; pistillate flowers with the stamens 8, nonfunctional, to 3.5 mm long; ovary pubescent; style

3–4 mm long; stigmas 3, ca 1 mm long. Fruits subrectangular-oblong, ca 4 cm long, the cells rufous-pubescent, subapical, hirtellous and setose with a hornlike projection, the wings hirtellous. *Croat 7823, 12714, 12715.*

Occasional. Flowers in the earliest part of the dry season, with the fruits maturing in the middle to late dry season.

This species is possibly dioecious and is unusual for its large flowers.

Honduras to Panama. In Panama, known from tropical moist forest in the Canal Zone, Colón, and Daríen.

See Fig. 347.

Serjania decapleuria Croat, Ann. Missouri Bot. Gard. 63:515. 1976

Tendriled liana; stems weakly 10-costate; branchlets bearing short crisp pubescence, the branchlets, petioles, rachises, and petiolules minutely appressed-puberulent; tendrils forked near apex. Leaves biternate; petioles 2–10 cm long; rachis unwinged; leaflets sessile to short-petiolulate (juveniles often on longer petiolules), oblong-ovate to elliptic, acute to acuminate at apex, acute to attenuate at base, entire to dentate with irregular black lines (at least on drying), glabrous but with strigillose midrib and main veins above, the basal axils sometimes bearing very slight tufts of trichomes below, the terminal leaflet 6–12 cm long, 2.5–5 cm wide. Thyrses in solitary and axillary racemes or in terminal paniculate racemes; flowers white, ca 4 mm long; sepals tomentulose; petals obovate, the scales of the anterior petals three-fourths as long as petals, the crest as broad as or broader than high, the upper margin thin, the appendages about as wide as long, bearded, fused laterally, the anterior glands broader than high, the lateral glands about half as broad as anterior glands; stamens 8; filaments ± flattened, villous; ovary and style pubescent; style 3-branched; stigmas apical. Fruits ovate, subglabrous, 3.5–4 cm long, the cells puberulent, the wings sometimes reddish, subglabrate. *Croat 7704.*

Occasional, along the shore. Flowers and fruits throughout the dry season.

Costa Rica to Colombia. In Panama, known only from tropical moist forest in the Canal Zone.

Serjania mexicana (L.) Willd., Sp. Pl. 2:465. 1799
S. nesites I. M. Johnston
Barbasco

Tendriled liana; trunk to 7 cm diam, involuted, twisted, warty, lacking milky sap (at least sometimes); stems with milky sap, glabrous to villous especially when young, (3)5-ribbed (the ribs on larger stems in turn 2-ribbed), often sparsely armed with short prickles, especially larger stems; tendrils bifid, axillary. Leaves biternate to bipinnate or tripinnate, 10–40 cm long, often much reduced on inflorescence; stipules linear, paired, ca 1 cm long (on juveniles); petioles with marginal ribs above; rachis winged; leaflets 9–26, ovate to elliptic, acute to bluntly acuminate, rounded to attenuate at base, 2–8 cm long,

Fig. 347. *Serjania cornigera*

Fig. 348. *Serjania mexicana,* juvenile stem and leaves

Fig. 349. *Serjania paucidentata*

1.5–5 cm wide, ± glabrous to sparsely pubescent especially on veins, the margins sinuate-dentate near apex (conspicuously dentate on juveniles). Thyrses in solitary and axillary racemes or on tendrils or in terminal or axillary racemose panicles; flowers white, sweetly aromatic, ca 4 mm long; pedicel and calyx densely pubescent; sepals elliptic, ca 2.5 mm long, reflexed or spreading at anthesis; petals spatulate to obovate, 2.3–3.5 mm long, the glands large, ovoid, orange, glabrous, subtending petals, the scales of the anterior petals orbicular, nearly three-fourths as long as petals, their appendages slender, attached laterally to appendage of neighboring scale by villous pubescence, pendent nearly to base of scale, the crest yellow, hammer-shaped, the lateral petals borne on large glands; stamens 8; filaments flattened, sparsely villous; ovary 3-sided, glabrous. Fruits ovate-cordate, 1.7–2.7 cm long, glabrous, the cells with raised veins, the wing sometimes not constricted above seed. *Croat 13959, 14611.*

Common both as an adult plant and as a seedling in the forest. Flowers from February to April (sometimes from January). The fruits mature from March to May.

This is vegetatively the most variable species of *Serjania* on the island.

Mexico to Colombia and Venezuela. In Panama, known from tropical moist forest in the Canal Zone, Bocas del Toro, Chiriquí, Veraguas, Los Santos, Herrera, Panamá, and Daríen, from premontane wet forest in Chiriquí (Boquete) and Veraguas, and from tropical wet forest in Coclé and Panamá.

See Fig. 348.

Serjania paucidentata DC., Prodr. 1:603. 1824
  Paullinia protracta Steud.

Liana; stems glabrous, 6-ribbed but with 3 more prominent, sulcate when young, glabrous; wood composite, the vascular bundles usually 3, small, terete, peripheral. Leaves biternate, glabrous, 8–15(26) cm long; petioles 1.5–4(8) cm long, canaliculate on upper surface; rachis narrowly winged; leaflets narrowly elliptic to oblanceolate, acute to acuminate and often deeply incised on both sides beneath acumen, attenuate at base, 3–13 cm long, 1.8–5 cm wide, thick, mostly entire except for a few crenate, often glandular teeth near apex, the acumen blunt with a gland-tipped apiculum. Inflorescences terminal or usually axillary and borne on tendrils, slender, to ca 20 cm long; pedicels densely tomentose, to ca 1.5 mm long; rachis weakly tomentose; flowers white, to 4.5 mm long; sepals ovate to obovate, densely tomentose outside, glabrous inside, to 2 mm long; petals narrowly obovate, glabrous, to 4.3 mm long, the anterior scales ca three-fourths as long as petals, stiffly pubescent on margins, the appendage deflexed, slender, extending down to about the middle of the scale, densely stiff-pubescent throughout, the crest slightly bilobed, orange, equaling the length of appendage; disk glands 2 or 4, semicircular, nearly glabrous; stamens villous, flattened; staminate flowers with the stamens to 4 mm long, the pistil minute, 3-sided, hispid at apex; bisexual flowers not seen. Fruits narrowly

ovate-cordate, weakly constricted above the cells, 2.3–2.8 cm long, sparsely hispidulous on wing, densely brown-hispid on cell. *Croat 7887.*

Infrequent; collected several times along the shore on the east side of the island. Flowers from February to April. The fruits are distributed mostly in the dry season, but may persist until as late as June.

Serjania paucidentata is closest to *S. mexicana,* but can be distinguished by having larger flowers, hispid fruits, unarmed stems, and composite wood.

Mexico to the Guianas, Brazil and Peru; Trinidad. In Panama, known only from tropical moist forest on BCI.

See Fig. 349.

Serjania pluvialiflorens Croat, Ann. Missouri Bot.
  Gard. 63:522. 1976

Tendriled liana; stems terete, glabrate to sparsely puberulent. Leaves biternate; petiole and rachis ± angled and slightly ribbed, usually pubescent at least on upper side; petioles to 6 cm long; leaflets elliptic to ovate, ± sessile, obtuse to acuminate, acute at base, 3–9 cm long, 2–4 cm broad, crenate above middle (the teeth usually glandular), the surface pellucid-punctate, glabrous but sometimes with short-pubescent veins and usually with axillary tufts on lower surface. Thyrses on solitary racemes borne on tendrils and in leaf axils or densely congested in terminal paniculate racemes; pedicels to 1.5 mm long; flowers white, to ca 4.5 mm long; sepals orbicular to obovate, glabrate to tomentulose outside, pubescent or glabrous inside; petals obovate, the scales of the anterior petals two-thirds the length of petals, subtended by a triangular to oblong, ± glabrous gland, the crest yellow, glabrous, the appendage villous, the lateral petals with or without subtending glands, their scales lacking crest; stamens 8; filaments flattened, villous; ovary glabrous or villous near apex and on dorsal side of stigmas. Fruits ovate-cordate, glabrous, the cells deeply rugose, 2–3 cm long, the wing reddish. *Croat 12421.*

Rare. Flowers and fruits in the middle to late rainy season (September to November).

Known only from Panama, from tropical moist forest in the Canal Zone and Bocas del Toro.

Serjania rhombea Radlk., Monogr. Serj. 324. 1875

Tendriled liana; stems 6-ribbed, brown-hirtellous especially on ribs. Leaves biternate, mostly to 15 cm long; rachis narrowly winged; leaflets ovate to rhomboid, obtuse to acuminate at apex, attenuate at base, coarsely and obtusely toothed above middle, softly pubescent (especially below and on raised midrib above), the terminal leaflet mostly 5–9.5 cm long, 2.5–6 cm broad, the lateral leaflets smaller. Thyrses in racemes 5–15 cm long and borne in axils and on tendrils, or in terminal racemose panicles; inflorescence branches softly short-pubescent; flowers white, ca 3 mm long; sepals ovate, short-tomentose; petals narrowly obovate, borne atop large greenish glands, spreading broadly at anthesis, the scales of the anterior petals with the margin villous and turned

inward, the appendage short, villous, the crest broader than long, bifid, recurved toward outside; staminal cluster spreading, leaning away from axis, otherwise typical; filaments sparsely villous, all ± recurved at apex; ovary glabrous. Fruits cordate-ovate, 1.7–2.2 cm long, the cells sparsely to densely (when young) villous, the wing ± glabrous or subvillous, often reddish at maturity. *Wetmore & Woodworth 83.*

Perhaps the most abundant species in central Panama, particularly in disturbed areas, and to be expected on BCI. A sterile collection is the only record of the species from the island (*Wetmore & Woodworth 83*). In the Canal Zone, flowers in the late rainy and dry seasons (October to April), with the fruits maturing throughout the dry season into the early rainy season (January to May).

The flower is unlike any other of the genus. Being spread open with the colored crown held to the outside, it presents a different type of pollinating unit.

Mexico to Colombia, Venezuela, and Ecuador. In Panama, known from tropical moist forest in the Canal Zone, Colón, San Blas, Panamá, Los Santos, and Darién, from tropical dry forest in Coclé, from premontane moist forest in the Canal Zone and Panamá, and from premontane wet forest in Coclé and Panamá.

Serjania trachygona Radlk., Monogr. Serj. 327. 1875

S. deltoidea Radlk.

Tendriled liana; younger stems merely striate or obscurely 3–6-lobed, the older stems prominently 3-ribbed, each rib consisting of 2 small strands loosely attached to a much larger central core; sap milky; tendrils forked; stems, petioles, and rachises puberulent to hirsute. Leaves 2- or 3(5)-pinnate in 4–6 sets; petioles ribbed; rachis narrowly winged; leaflets small, acute to acuminate, acute to attenuate at base, 1–3 cm long, 0.7–2 cm wide, the terminal leaflet rhomboid, often trilobate, the lateral leaflets ovate-elliptic, crenate, sessile, glabrate to hirsute (especially on midrib); juvenile leaves usually with more leaflets than adults. Thyrses racemose and axillary or in terminal racemose panicles; flowers white, 2–3 mm long, short-pedicellate; petals ca 1.7 mm long, obovate, the anterior petals borne on the outer face of large glands, their scales with yellow, prominently bilobed crests, the appendages densely villous all over, united as a single unit, pendent to apex of glands, the lateral petals borne atop glands, their crests slender, entire, usually white; staminal cluster leaning away from the axis; filaments villous; ovary glabrous. Fruits ovate-cordate, 1.5–2 cm long, as broad as or broader than long, weakly viscid (at least on the cell when dried), the cells sparsely hirsute with raised veins, the wing glabrous or sparsely hirsute on inner margin, reddish at maturity. *Croat 7867, 13123.*

Occasional, in the forest. Flowers in the early dry season (December to February). The fruits mature in the late dry season (February to April).

The species is variable throughout its range in size of flowers, leaflet shape, and degree of compounding of the leaves.

Panama, Peru, and Bolivia, and probably more widespread in South America. In Panama, known only from tropical moist forest in the Canal Zone, Colón, and Panamá.

See Fig. 350.

TALISIA Aubl.

Talisia nervosa Radlk., Smithsonian Misc. Collect. 61(24):4. 1914

Mamón de monte

Polygamous shrub or small tree, usually less than 5 m tall; trunk to 6 cm dbh, unbranched (unless previously damaged). Leaves pinnately compound, clustered near apex, often 1 m or more long, lacking reduced leaflets at apex; petioles ca 25 cm long, terete; petiolules swollen, 5–10 mm long; leaflets usually in 5–8 pairs, oblong-elliptic, acute at apex and base, 20–45 cm long, 6.5–13 cm wide, glabrous above, glabrous to puberulent below. Thyrses small, arranged in panicles to 70 cm long, widely branched, the major branches ribbed; branches, pedicels, and calyces puberulent to tomentulose; pedicels short, to 2 mm long, articulate below calyx; calyx bowl-shaped, to ca 2 mm long, 5-lobed to about middle, the lobes indurate, acute, ciliate; petals 5, oblong, white, 3–5 mm long, acute to blunt at apex, spreading above calyx, glabrous, the scale ± exceeding petal, sericeous, tufted and slightly spreading at apex; disk raised, 5-angulate, the points alternating with petals; stamens 5 or 8, equaling scales, 3 often reduced or aborted; filaments weakly pubescent, shorter than anthers; anthers oblong, 1–4 mm long, the connective beaked at apex; staminate flowers with the ovary densely pubescent, less than 1 mm long; style lacking; stigmas 3, minute, hidden by the pubescence of ovary; bisexual flowers with the ovary ovoid; style nearly glabrous, about as long as ovary; stigmas capitate, held at about the level of anthers. Fruits ± ellipsoid to globose, brown, sharply apiculate at apex, sparsely pubescent, minutely lenticellate, usually 2–3.5 cm long, with a thick (2–3 mm) woody pericarp; seeds ellipsoid, flat on side if more than 1, 1.5–2.5 cm long, embedded in a firm or jellylike, whitish to orange mesocarp. *Croat 8236, 14921.*

Occasional in both the young and old forests; abundant along Snyder-Molino Trail 400–700. Flowers in the dry season (December to April), with the fruits maturing in

---

KEY TO THE SPECIES OF TALISIA

Calyx less than 3 mm long; leaflets in fewer than 9 pairs, more than 7 cm wide; reduced leaves lacking; filaments villous . . . . *T. nervosa* Radlk.
Calyx more than 3 mm long; leaflets in more than 9 pairs, the largest less than 7 cm wide; reduced leaves often present at apex of stem; filaments glabrous . . . . *T. princeps* Oliv.

Fig. 350. *Serjania trachygona*

Fig. 351. *Talisia princeps*

Fig. 352. *Talisia princeps*

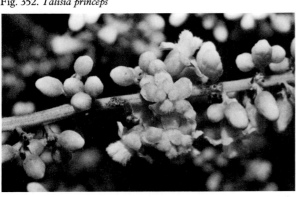

the middle to late rainy season, chiefly from July to October.

The genus *Talisia* is said to be polygamodioecious, i.e., functionally dioecious but with a few bisexual flowers or flowers of the opposite sex. However, at least some specimens of this species are polygamous, i.e., with large numbers of both bisexual and staminate flowers.

Costa Rica to Colombia. In Panama, known from tropical moist forest in the Canal Zone, Bocas del Toro, Veraguas (Coiba Island), Panamá, and Darién and from tropical wet forest in Colón (Santa Rita Ridge) and Darién.

Talisia princeps Oliv., Hooker's Icon. Pl. 18, t. 1769. 1888

Polygamous tree to 6 m tall; trunk slender, to ca 3 cm dbh, prominently ribbed. Leaves pinnately compound, to 1 m long or more, sparsely to densely puberulous except on blade surface; petiolules pulvinate; leaflets in 6–14 pairs, opposite or alternate, oblong to oblong-obovate, abruptly acuminate, obtuse to acute at base, slightly inequilateral at base, 7–28 cm long, 2.5–6 cm wide, ± entire; apical leaves (called Kataphylls by Uittien, 1937) often reduced or aborted, 20–30 cm long, the petiole and rachis flattened, acicular, the leaflets narrowly oblong to linear, brown, dry, achlorophyllous, less than 2 cm long, with fascicles of still further reduced leaves in their axils like adult leaves. Thyrses arranged in large terminal panicles to 70 cm long, the lower branches to ca 40 cm long, the branches ribbed; bracts linear-lanceolate, 8–10 mm long; branches, bracts, pedicels, and calyces puberulent; calyx coriaceous, ca 4 mm long, deeply 5-lobed, the lobes ovate, ciliate, their margins thin; petals 5, white, regular, narrowly obovate, ca 7 mm long, rounded at apex, spreading and concave in upper half, sericeous near base outside, glabrous inside, the scales fused to lower third of petal inside, the free part erect, very densely white-velutinous, ± equaling petals; stamens 8; filaments glabrous, longer than anthers; anthers ca 1.5 mm long, the connective beaked at apex; disk prominently 5-lobed, the lobes densely villous at apex; staminate flowers with the stamens ca 5 mm long, the 5 outer ones each subtending a large pubescent lobe of the disk, alternating with the petals, the 3 inner ones surrounding the pistil; ovary minute, less than 1 mm long, densely sericeous, the trichomes obscuring the 3 triangular stigmatic lobes; bisexual flowers with the stamens ca 3 mm long; ovary, style, and stigmas together to 6 mm long; ovary ovoid, 2.5–3 mm long, gradually tapered into style; style stout, thick, ca 1.5 mm long; stigmas 3, short, triangular, usually obscured by pubescence of style. Fruits (*fide* Radlkofer) to 2.5 cm long and 1.5 cm diam, acuminate, tomentulose, 1-celled, 1-seeded. *Croat 8820, 12494, Zetek 3570.*

Occasional, in the forest. Flowers in October and November. Time of fruit maturation not known.

Some plants observed in flower on BCI did not set any fruit. Possibly the plants are functionally dioecious rather than polygamous, as is the general rule in the family.

Except for the pubescence of the leaves and petals, the species is very close to *T. megaphylla* Sagot from the Guianas and Amazonian Brazil. The species has other seemingly close relatives in South America, including *T. stricta* (Tr. & Planch.) Radlk. (Ecuador), *T. cupularis* Radlk. (Brazil), *T. hemidasya* Radlk. (Surinam), and *T. tiricensis* Steyerm. & Maguire (Venezuela).

Panama and Venezuela. In Panama, known from tropical moist forest in the Canal Zone, Bocas del Toro, and Darién, from premontane wet forest in Darién, and from tropical wet forest in Panamá (Serrania de Majé).

See Figs. 351 and 352.

THINOUIA Tr. & Planch.

Thinouia myriantha Tr. & Planch., Ann. Sci. Nat. Bot., sér. 4, 369. 1862

T. tomocarpa Standl.

Polygamous, tendriled canopy liana; trunks to 12 cm diam, involuted; younger stems terete, pubescent, lenticellate, the trichomes minute, appressed, ferruginous; tendrils watchspring-like. Leaves trifoliolate; petioles 2–4 cm long; terminal petiolule mostly 2–2.5 cm long, the lateral ones ca half as long; leaflets mostly ovate to ovate-elliptic, acute to acuminate, obtuse to rounded (or rarely subcordate) at base, 6–12 (15) cm long, 2.5–9 cm wide, sparsely pubescent when young, ± glabrous in age except on midrib above and in vein axils below, the midrib arched, the sides folded somewhat upward along midrib, the margins undulate, entire, or crenate near apex, the teeth blunt, usually glandular; veins at base 3. Cincinni in umbellate clusters 3–4 cm diam, borne on short axillary or terminal branches, the branches to 7 cm long, often bearing tendrils; branches, peduncles, pedicels, and calyces appressed-puberulous; pedicels mostly 2.5–3.5 mm long; flowers ± actinomorphic, minute, greenish; calyx bowl-shaped, lobed, the lobes 5, uniform, acute, 0.7–1 mm long; petals 5, uniform, obovate, ca 0.4 mm long, each bearing from inside the base a deeply bifid, pubescent scale ca 1 mm long; disk prominent, ca 1.7 mm broad, glabrous; stamens 8, actinomorphic, in a tight cluster around pistil at center of flower, 3 erect and held closely together, 5 somewhat divergent; filaments villous in basal half, attached to anthers midway; anthers about as broad as long, the thecae usually separate below middle; pollen tacky; staminate flowers with the stamens ca 2.7 mm long; pistil minute; styles lacking; bisexual flowers with the stamens less than 2 mm long; pistil stipitate, to 3 mm long, 3-angled, weakly to densely appressed-pubescent; style short; stigmas 3, ca 1 mm long, pubescent, recurved. Fruits narrowly ovate-cordate, to 6.5 cm long, brown, glabrous, borne on a slender stalk 5 mm long, the samaras 3, to 1.9 mm wide, the cells lunate, submarginal, ca 1.5 cm long, the veins prominulous. *Croat 13803.*

Rare; known from several places on Armour and Zetek trails. Flowers and fruits in the dry season. On BCI, plants that flowered in late February had mature fruits during April and early May.

The species is variable throughout its range in pubes-

cence, leaf shape, petal size, and particularly the shape and degree of pubescence of the flower petal scales. Herbarium specimens with the name *T. marginata* Tr. & Planch. are also this species, but the name was apparently never published. Some Peruvian specimens identified as *T. repens* Radlk. are also this species. The type (from Paraguay) was not seen.

Probably along the entire Atlantic slope of Central America from Belize and Guatemala to Colombia, Venezuela, the Guianas, and Peru; in Central America known presently only from Belize and Panama. In Panama, known from tropical moist forest in the Canal Zone and Darién and from tropical wet forest in Darién.

81. RHAMNACEAE

Trees or tendriled lianas. Leaves alternate or subopposite, petiolate; blades simple, entire to serrulate; venation pinnate, somewhat palmate at base; stipules present, minute. Racemes terminal or axillary, cymose (may be glomerate); flowers bisexual, actinomorphic, many; calyx cupular, (4)5-lobed, the lobes valvate; petals (4)5, free, alternate with calyx lobes; stamens (4)5, opposite the petals, arising between the lobes of the disk; disk prominent, sometimes nectariferous; anthers 2-celled, longitudinally dehiscent; ovary inferior (*Gouania*) or semi-inferior (*Colubrina*), 3-locular, 3-carpellate; placentation basal; ovules 1 per locule; style trilobate. Fruits schizocarps of 3, 2-winged mesocarps (*Gouania*) or explosively dehiscent 3-valved capsules (*Colubrina*), with little or no endosperm.

Rhamnaceae may be recognized by their open, bowl-shaped flowers with stamens borne opposite the usually prominently concave petals.

Pollination system is unknown.

Seeds are wind dispersed (*Gouania*) or autochorous by means of explosively dehiscent capsules (*Colubrina*). Seeds of *Colubrina* are buoyant because of an air space between the cotyledons (Ridley, 1930) and may be further dispersed by water currents.

About 58 genera and 900 species; widespread but mostly in the tropics and warmer temperate regions.

COLUBRINA L. C. Rich. ex Brongn.

Colubrina glandulosa Perk., Bot. Jahrb. Syst. 45:465. 1911

C. rufa (Vell.) Reiss. var. *glandulosa* (Perk.) M. C. Johnston
Spanish elm

Tree, (3)6–20(40) m tall, to ca 50 cm dbh, ± glabrous but with ferruginous, ± antrorse puberulence on younger parts including peduncles, pedicels, and cupules; bark moderately coarse in age. Leaves opposite or nearly so; stipules subulate, 3–10 mm long, caducous; petioles 0.5–18 cm long; blades ± ovate-elliptic, acute to short-acuminate at apex, rounded to very shallowly cordate at base, 7–15(25) cm long, 2.5–10 cm wide, the lower surface dull, often inconspicuously pubescent on veins and with some moderately large submarginal glands, the upper surface shiny, the margin weakly revolute, markedly so near petiole. Thyrses axillary, 1–5 cm long; peduncles 1–7 mm long; pedicels 1–4 mm long, 3–12 mm long in fruit; flowers 10–50, 5-parted; petals, sepals, and stamens borne on apex of floral cup, the cup 2.5–3 mm wide; sepals yellowish, ± triangular, ca 1 mm long, with a prominent medial ridge inside; petals ± oblong, ca 1 mm long; stamens opposite and slightly longer than petals; anthers less than 0.5 mm long; disk prominent, filling floral cup at anthesis and hiding the 3-celled ovary; style much shorter than stamens, tripartite one-fourth to two-thirds its length. Capsules explosively dehiscent, ± globose, 6–8 mm long, dark brown or black, glabrous, enveloped only slightly at base by the cup and disk at maturity; endocarp hard, separating into 3 dry endocarpids; seeds 1–3, ± obovate, 4–5 mm long, dark brown, shiny. *Croat 7349, Wetmore & Abbe 165.*

Apparently rare, collected only along the shore but no doubt growing in the forest as well. Flowers in the earliest part of the dry season, from December to January. The fruits mature from February to April.

Panama to Colombia, Venezuela, the Guianas, Brazil, and Peru. In Panama, known only from tropical moist forest in the Canal Zone, Los Santos (south of Macaracas), and Panamá.

GOUANIA Jacq.

Gouania adenophora Pilg., Notizblatt 6:314. 1915

Liana, climbing into canopy of forest; younger stems densely puberulent, the older stems glabrous; outer bark grayish. Petioles 1–3 cm long, canaliculate, weakly pubescent to glabrous; blades ovate to ovate-elliptic, abruptly acuminate and downturned, rounded to subcordate at base, 4–14 cm long, 2.5–9 cm wide, glabrous above, sparsely and inconspicuously pubescent below especially on the major veins; major lateral veins in 4–6 pairs, arcuate-ascending, the 2 basal pairs arising at the base of leaf, the tertiary veins more or less conspicuous, extending almost straight between primary lateral veins. Cymes umbelloid, less than 1 cm long, arranged in terminal racemes 15–25 cm long; peduncles bracteate, the bracts deltoid, acute, pubescent; flowers bisexual, densely aggregated, numerous; pedicels ca 1 mm long, densely

KEY TO THE SPECIES OF RHAMNACEAE

Plants trees, lacking tendrils; leaves subopposite; fruits not winged, of thick-walled, explosively
    dehiscent capsules . . . . *Colubrina glandulosa* Perk.
Plants tendriled lianas; leaves alternate; fruits winged:
    Wings of fruit 11–12 mm high and 7 mm wide . . . . *Gouania adenophora* Pilg.
    Wings of fruit 5–6 mm high and 5 mm wide . . . . *Gouania lupuloides* (L.) Urban

villous; calyx bowl-shaped, densely villous especially toward base, the lobes deltoid, ca 1 mm long; petals obovate, cupulate, with the margins partially enclosing the subtending stamen, rounded at apex, somewhat clawed at base; stamens 5, borne on the margin of the disk with the petals, ca 1 mm long; disk 5-sided, the inner margin adjacent to the ovary with 3 erect lobes opposite the stigmas; pistil ca 0.3 mm long, depressed in disk; stigmas 3, about as long as style. Schizocarps of 3, 2-winged mericarps, each 8–11 mm wide, glabrous except near apex, the body 7–9 mm long; wings reniform, 8–12 mm long; seeds narrowly ellipsoid, flattened, 4 mm long. *Foster s.n.*

Apparently rare, though reported by R. Foster (pers. comm.) to drop fruits in the old forest near Armour Trail 700 and Drayton Trail 100. Flowering in July and August. Fruiting in September and October.

Determination of the species is doubtful, because the plant most closely matches a Peruvian plant. Perhaps it is a new species, but because of the polymorphic nature of *Gouania* I am reluctant to describe it as new.

Peru and Panama. In Panama, known from tropical moist forest in the Canal Zone (BCI), from premontane wet forest in the Canal Zone (Pipeline Road) and Panamá (El Llano–Cartí Road), and from tropical wet forest in Colón (Río Guanche) and Panamá (El Llano–Cartí Road).

Gouania lupuloides (L.) Urban, Symb. Ant. 4:378. 1910
  G. polygama (Jacq.) Urban
  Jaboncillo

Tendriled liana; stems striate, densely pubescent; tendrils watchspring-like, terminating short branches, becoming woody in age. Leaves alternate, short-petiolate; blades ± ovate, acuminate to acute or cuspidate at apex, obtuse to subcordate at base, 5–11 cm long, 2–6.5 cm wide, sparsely to densely pubescent, the pubescence variable; veins at base 3. Racemes terminal or upper-axillary, spikelike, 5–15 cm long; flowers mostly bisexual, sometimes pistillate or staminate by reduction, sessile or subsessile, ca 3 mm wide, in few-flowered glomerules, opening 1 to few at a time in each glomerule; calyx cupular, white or greenish, acutely 4- or 5-lobed, pubescent on outer surface, the lobes ca 1 mm long; petals minute, equaling calyx lobes, partially enclosing stamens; stamens (4)5, mounted on rim and alternate with lobes of the disk; disk prominent, cupular; anthers emerging above petals; styles 3, short at anthesis, later elongating and exceeding height of disk lobes, the tips becoming recurved. Schizocarps of 3, 2-winged mericarps, the central part sparsely to densely pubescent, 3–4 mm long; wings rounded, 5–6 mm high and ca 5 mm broad, glabrous to densely reddish-brown-pubescent, splitting medially at maturity; mericarps 3, consisting of one-third of the central axis and one-half of each of 2 wings. *Croat 5744, 7075.*

Abundant at the edges of clearings and occasional in the forest canopy. Flowers from November to March, principally in the early dry season; flowering is rare in March and even more rare during the rainy season. The fruits develop to mature size as early as January, but are dispersed from February to May (rarely June or later), especially in March and April.

The cuplike disk may become partly filled with a sweet watery nectar, especially while the anthers are shedding pollen. The nectar appears to be absent when the style is receptive.

The species is represented on BCI by two distinct races. The less common of the two is characterized by dense reddish-brown pubescence all over but especially on the stems, the lower leaf surfaces, and the seminiferous areas of the fruit. Examples of the more pubescent form include *Croat 7274, 7984, 12699, 12739, 13482, Shattuck 290, 444, 523,* and *Woodworth & Vestal 326.* Although the difference between the two races is striking on BCI, variation of *G. lupuloides* throughout Panama is so great that they cannot be recognized at any higher taxonomic level. Apparently identical collections made in Mexico and Brazil have been identified as *G. tomentosa* Jacq. and *G. mollis* (L.) Urban. Further monographic work with *Gouania* may well prove that these two BCI variants should be considered distinct at the varietal or even the specific level.

Mexico to northern South America; West Indies. In Panama, ecologically wide-ranging, probably occurring in all areas of tropical moist forest; known also from tropical dry forest in Coclé (near Antón), Herrera, and Los Santos, from premontane wet forest in Chiriquí (Boquete), and from premontane rain forest (south of Volcán) in Chiriquí.

82. VITACEAE

Lianas and vines, with leaf-opposed tendrils and often with swollen nodes. Leaves alternate (lower ones sometimes opposite), petiolate; blades simple and trilobate or trifoliolate, serrate; venation palmate at least at base; stipules interpetiolar. Inflorescences cymose, appearing corymbose, umbellate or paniculate; flowers primarily bisexual (bisexual and staminate in *Vitis*), actinomorphic, 4-parted (*Cissus*) or 5-parted (*Vitis*); calyx shallowly toothed or lobed; petals valvate, free (apically fused in *Vitis*); stamens inserted at base of disk, of the same number as petals and opposite them; anthers 2-celled, introrse, longitudinally dehiscent; ovary superior, 2-locular, 2-carpellate; placentation axile; ovules 2, anatropous; style 1; stigma discoid (may be slightly bilobed). Fruits berries; seeds 1 or 2, with copious endosperm.

Members of the family are recognized by their climbing habit, the apparently terminal buds developing into lateral tendrils, the leaf-opposed, branched, cymose inflorescences, and the valvate petals opposite the stamens.

The small open flowers are probably insect pollinated. In Costa Rica, *Cissus* is visited by the vespid wasps *Polistes* and *Stelopolybia* (Heithaus, 1973).

The fruits are well suited for bird dispersal and those of *Cissus sicyoides* are much sought by birds (Duke, 1968).

Twelve genera and 700 species; primarily in the tropics but extending to temperate regions.

KEY TO THE SPECIES OF VITACEAE

Leaves simple:
  Flowers 5-parted, in panicles; petals fused at apex; blades on lower surface ± densely floccose-tomentose .... *Vitis tiliifolia* R. & S.
  Flowers 4-parted, in cymes branched many times; petals free; blades glabrous or velutinous only on veins below:
    Plants glabrous or the trichomes villous and simple; pedicels always glabrous; blades ovate-oblong, drying green, the larger truncate to only slightly cordate (not ovate-cordate) .... *Cissus sicyoides* L.
    Plants pubescent at least on veins of lower leaf surface and on inflorescences with minute, close, puberulent trichomes and also usually with appressed T-shaped trichomes; pedicels pubescent; blades ± ovate, drying blackened, the larger broadly ovate-cordate .... *Cissus pseudosicyoides* Croat
Leaves trifoliolate:
  Plants densely pubescent, some trichomes gland-tipped; terminal leaflets ± rhombic; flowers pale yellow to yellowish-green .... *Cissus rhombifolia* Vahl
  Plants glabrous or sparsely pubescent on surface with simple trichomes; terminal leaflets ± elliptic; flowers red to orange:
    Mature peduncles less than 2.5 cm long; leaves often velutinous only on veins below, with tufted axils; stems not winged; fruits obovoid, more than 6 mm long .... *Cissus microcarpa* Vahl
    Mature peduncles more than 5 cm long; leaves glabrous or sparsely pilose below, lacking tufted axils; stems often winged; fruits orbicular, less than 6 mm diam .... *Cissus erosa* L. C. Rich.

CISSUS L.

Cissus erosa L. C. Rich., Actes Soc. Hist. Nat. Paris 1:106. 1792

C. salutaris Kunth ex H.B.K.

Vine with simple tendrils, glabrous or with scattered pilose pubescence throughout; stems angled or winged on margins, usually maroon at nodes and speckled with maroon all over (not obvious on dried specimens), the older stems woody. Leaves trifoliolate; stipules ovate, subpersistent, minute (to 5 mm long); petioles 2–6 cm long, maroon at base and apex, angled and often winged on margins; leaflets crenate-serrate, the terminal one ± elliptic, obtuse to bluntly acuminate at apex, cuneate at base, 4.5–16 cm long, 1.5–8 cm wide, on a short petiolule, the lateral leaflets smaller than terminal leaflet, inequilateral, ovate to ovate-elliptic, acute to obtuse at apex, rounded to subcuneate at base, 3.5–12 cm long, 2–6 cm wide. Cymes corymbiform, opposing leaves, congested in pseudoumbels; peduncles tetragonal, 5–12 cm long; flowers 4-parted, red; pedicels 2–5 mm long, villous; calyx cupular, the lobes short or obscure; petals valvate, acute at apex, cucullate within, often falling free at anthesis; stamens 4, set between notches in the thick disk, erect at anthesis, later recurved, somewhat reflexed; anthers open in bud; filaments and connective red; style short; stigma simple. Fruits orbicular, to 6 mm diam, purple-black at maturity; seeds usually 2. *Croat 12943.*

Occasional, on the shore and at the edge of clearings; possibly also in the forest canopy. One of the most common species of *Cissus* on the island. Flowers throughout the year, especially in the early rainy season. The fruits develop rapidly.

Mexico to Colombia, Venezuela, the Guianas, Peru,

and Bolivia; West Indies. In Panama, known from tropical moist forest in the Canal Zone, Colón, San Blas, Veraguas, Herrera, and Panamá, from premontane moist forest in the Canal Zone and Los Santos, from premontane wet forest in Colón, Coclé, and Panamá, and from tropical wet forest in Colón.

Cissus microcarpa Vahl, Eclog. Amer. 1:16. 1796

Tendriled climbing vine, becoming woody; stems to 7 cm diam, usually angulate, sometimes bearing 2-sided, narrow, peltate scales. Leaves trifoliolate; stipules (on youngest stems) broadly ovate, to 4 mm long; petioles 2–4.5(6) cm long; leaflets velutinous especially on veins below with appressed, long, flattened, T-shaped trichomes on veins, the major veins crisped-villous at least near tufted axils, the margins obscurely mucronate-serrate, the terminal leaflet broadly elliptic to ovate-elliptic, acuminate, cuneate at base, 4–10 cm long, 1.5–5.5 cm wide, the lateral leaflets somewhat smaller and inequilateral. Cymes corymbiform, umbellate, 3–8 cm long; peduncles 6–18 mm long; inflorescence branches reddish, moderately pubescent, the pedicels and branches of the inflorescence with long, flattened, T-shaped trichomes; pedicels to 4 mm long; flowers 4-parted, red; calyx cupular, nearly truncate, minute; corolla ca 1.5 mm long, the lobes ovate, valvate; stamens 4; pistil 1–1.5 mm long. Fruits obovoid, 7–9 mm long, green to orange at maturity; seed usually 1, pyriform, to 7 mm long. *Croat 6396, 11890.*

Frequent along the margin of the lake; found also in trees in the forest to 10 m high. Flowers and fruits throughout the rainy season.

Mexico to Brazil; Greater Antilles. In Panama, known from tropical moist forest all along the Atlantic slope and

Fig. 353. *Cissus microcarpa*

Fig. 354. *Cissus pseudosicyoides*

in the Canal Zone, Veraguas, Herrera, and Panamá and from premontane wet forest in the Canal Zone, Coclé, and Panamá.

See Fig. 353.

Cissus pseudosicyoides Croat, Ann. Missouri Bot. Gard. 60:564. 1973

Tendriled herbaceous vine, probably ultimately arising from a woody stem; at least smaller stems, petioles, and veins of leaf blades (especially below) densely and inconspicuously puberulent; the same parts but also the axes of the inflorescences, pedicels, and leaf surfaces often sparsely pubescent with flattened, ± appressed, T-shaped trichomes; stems of juvenile parts often white-speckled. Leaves simple, thin, usually drying dark, dimorphic; larger leaves borne below the inflorescences, on petioles mostly 7–11 cm long, ovate-cordate, as broad or nearly as broad as long, 9–15 cm long and 9–12 cm wide, the lateral veins above sinus in 3–6 pairs, extending into apiculate teeth along margins of blade, a single strong trunk vein extending into each basal lobe, the sinus about as deep as broad; smaller leaves higher on stem and opposite inflorescences, on petioles mostly 2–8 cm long, usually narrowly ovate, truncate to obtuse or acute at base (rarely cordate), mostly 3–10 cm long and 2–8 cm wide, otherwise like larger leaves. Cymes terminal or opposite upper leaves, small, congested, branched, umbelliform, 1–4 cm long, about as broad as long; peduncles mostly 2–10 mm long at anthesis (somewhat longer in fruit), densely appressed-pubescent with T-shaped trichomes, densely bracteate at apex, the bracts minute with margins glabrous or very inconspicuously ciliate; pedicels terete, 1.5–3.5 mm long, sparsely pubescent, the trichomes as on peduncles but usually smaller; calyx spreading, ± bowl-shaped, inconspicuously 4-lobed, narrower than buds, nearly glabrous; buds ovoid, 1.5–2 mm long, drying with ridges along margins of petals; petals 4, free, broadly oblong, obtuse and cucullate at apex inside, usually white or cream (rarely red); stamens opposite petals; filaments to ca 1 mm long, equaling or longer than anthers; anthers nearly as broad as long, dehiscing laterally; stigma simple, to ca 1.5 mm long. Fruits ± globose, to 6 mm diam, apparently green at maturity; exocarp and mesocarp thin; seed 1, round, only slightly smaller than dimensions of fruit. *Croat 7017* (type).

Occasional, along the shore and at the edge of the Laboratory Clearing. Flowers at the beginning of the dry season in December and January on BCI (rarely elsewhere as late as March) or in the rainy season (late July to October); individual plants may flower for 1 month or more. The fruits develop promptly, are usually present with flowers, and are usually gone by March.

The fruits are probably dispersed by small to medium-sized birds. It is not known whether the fruits become brightly colored. Observations on BCI indicate that the fruits are probably removed before turning color. *Cissus pseudosicyoides* has been confused with *C. sicyoides* (Croat, 1973), but can be distinguished most easily by its pubes-

cent pedicels, dimorphic leaves, and T-shaped trichomes on the midrib of the blade.

Costa Rica (Guanacaste Province) to northern Colombia. In Panama, widespread in lowland areas; known principally from the Pacific slope in drier areas of tropical moist forest, but also from premontane dry forest in Panamá (Juan Diáz) and from premontane wet forest in Panamá (Chimán).

See Fig. 354.

Cissus rhombifolia Vahl, Eclog. Amer. 1:11. 1796
Batilla

Vine, the older parts woody; most parts conspicuously pubescent with both short gland-tipped trichomes and longer eglandular trichomes, the longer ones often red on those surfaces of stems and petioles, upper veins, and margins of leaves exposed to the sun; stems 4- or 5-angulate to terete, swollen at nodes; tendrils simple, sometimes reddish. Leaves trifoliolate; stipules narrowly lanceolate to ovate, 3–12 mm long, subpersistent; petioles 3–8 cm long, ± angulate; leaflets serrate, the terminal leaflet rhombic, acuminate, cuneate at base, 7–13 cm long, 3–6.5 cm wide, the lateral leaflets inequilateral, acute to acuminate, rounded on one side, acute on the other at base; veins impressed above, raised below. Cymes corymbiform, congested in pseudoumbels; peduncles 1–3 cm long; pedicels 3–8 mm long; flowers 4-parted; calyx bowl-shaped, broadest at base, pubescent, to 1.3 mm long, the teeth obscure; petals pale yellow to greenish-yellow, ± oblong-ovate, acute, to 2.3 mm long, spreading to recurved at anthesis, caducous; stamens 4, erect, ca 1.8 mm long, weakly adnate to base of petal; disk prominently 4-lobed; style to 2.7 mm long, simple. Fruits obovoid, 6–10 mm long, black at maturity; seeds 1 or 2.

Occasional, in disturbed areas in other parts of the Canal Zone and to be expected on BCI, but no recent collections have been made there. The species was reported by Standley (1933) but some older collections identified as this species have proved to be *C. microcarpa*, including *Shattuck 252* cited in the *Flora of Panama* (Elias, 1968). Flowers and fruits throughout the rainy season.

Mexico to Colombia, Venezuela, Peru, and Bolivia; West Indies. In Panama, ecologically variable; known from tropical moist forest in the Canal Zone, Bocas del Toro, San Blas, Chiriquí, Veraguas, Los Santos, Panamá, and Darién, from tropical dry forest in Herrera and Coclé, from premontane moist forest in the Canal Zone and Panamá, and from premontane wet forest in Bocas del Toro, Chiriquí, Coclé, and Panamá.

Cissus sicyoides L., Syst. Nat. ed. 10, 897. 1759
Rockrope, Bejuco loco, Uru cimarrona

Tendriled vine, the herbaceous stems terete, usually arising ultimately from a woody stem; most parts except pedicels and calyces villous; tendrils usually simple. Leaves simple; petioles mostly 1.5–5.5 cm long; blades all ± of same shape, narrowly ovate to oblong-ovate,

Fig. 355. *Vitis tiliifolia*

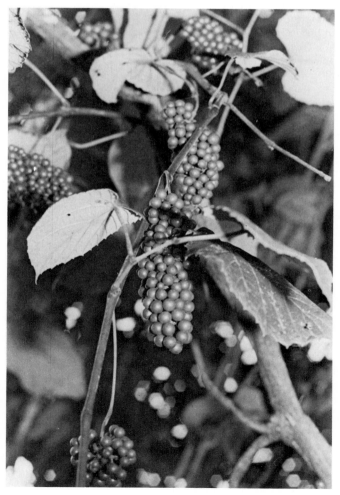

Fig. 356. *Vitis tiliifolia*

acuminate to acute at apex, mostly obtuse to truncate at base, sometimes subcordate to rarely cordate, the lower surface usually villous especially on veins, rarely nearly glabrous (frequently glabrous elsewhere), the upper surface usually glabrous except on veins; veins arising from base usually few to several, the lateral veins in 4 or 5 pairs above the basal ones, entering the apiculate teeth on margins of blade. Cymes terminal or opposite upper leaves, branched, umbelliform, as broad as or broader than long; peduncles 1–5 cm long at anthesis, to 6.5 cm long in fruit; bracts of inflorescence usually ciliate; pedicels glabrous, 2–3 mm long; flowers 2–3 mm long, greenish-white to white or pale yellow (often red elsewhere in bright exposed areas); calyx spreading, usually with 4 small lobes, glabrous, usually wider than the unopened corolla; petals 4, oblong, free, falling soon after opening; stamens 4, shorter than and opposite petals, arising from between the lobes of the disk; disk prominent, 4-lobed, persisting in fruit; style to 1 mm long. Fruits obovoid, to 6 mm long, green becoming red then black at maturity; seed usually 1. *Croat 4581a.*

Occasional, in and around the edge of the Laboratory Clearing. Flowers principally in the dry season, but some flowers may be seen all year. Individuals probably flower several times per year; certainly they flower more than once, since flowering plants frequently bear juvenile or mature fruit from an earlier flowering. The fruits probably mature within 1 or 2 months.

Southern United States throughout Central America and much of South America; West Indies. In Panama, known from tropical moist forest in the Canal Zone, all along the Atlantic slope, and in Chiriquí, Los Santos, Herrera, Panamá, and Darién; known also from tropical dry forest in Los Santos, Herrera, Coclé, and Panamá, from premontane moist forest in the Canal Zone and Panamá, and from premontane wet forest in Panamá.

VITIS L.

Vitis tiliifolia H. & B. ex R. & S., Syst. Veg. 5:320. 1819
Grape, Uva, Bejuco de agua

Polygamous, tendriled liana, the youngest parts herbaceous; stems angulate to terete; stems, peduncles, petioles, and lower leaf surfaces, and veins above densely floccose-tomentose at least when young; tendrils forked. Leaves simple; petioles 2–8 cm long; blades ovate-cordate, acuminate, 8–14 cm long, 6.5–12 cm wide, shallowly trilobate, serrate. Panicles leaf-opposed, to 20 cm long; peduncles 4–10(15) cm long; flowers bisexual or functionally staminate, greenish, ca 2 mm long; calyx bowl-shaped, the margin undulate; corolla to 1.7 mm long, the petals 5, ± oblong, fused at their apex and caducous as a unit; stamens 5, soon spreading, those of bisexual flowers somewhat shorter than those of staminate flowers; disk prominent, more than half as high as calyx; ovary weakly 5-lobed at base, ovoid; style short. Fruits ± globose, to 8 mm diam, becoming brown and sour but tasty at maturity; seeds 2. *Croat 10238, 11138.*

Frequent in the canopy, at the edges of clearings, and along the margin of the lake. Flowers and fruits throughout the year, principally during the dry and early rainy seasons.

Mexico to Colombia; West Indies. In Panama, known from tropical moist forest in the Canal Zone, all along the Atlantic slope, and in Panamá and Darién, from premontane moist forest in the Canal Zone, and from premontane wet forest in Chiriquí, Coclé, and Darién.

See Figs. 355 and 356.

83. ELAEOCARPACEAE

Trees or shrubs (*Muntingia*) frequently with pubescent stems. Leaves alternate (or subopposite), petiolate, simple, undulate to serrate, pubescent; venation pinnate throughout or palmate at base; stipules present. Cymes axillary, compound; flowers bisexual, actinomorphic, solitary, few or many; sepals 4–6, united into a flat plate below the receptacle; petals 5, free (*Muntingia*) or lacking (*Sloanea*); stamens many, free, arising from a disk; anthers 2-celled, dehiscing by terminal pores; ovary superior, 4- or 5-locular, 4- or 5-carpellate; placentation axile; ovules 2 per locule, anatropous, pendulous; style 1, simple, 4-angled, or stigmas radiate. Fruits loculicidal, 4-valved capsules with 1 or 2 arillate seeds or berries with many seeds (*Muntingia*); endosperm abundant.

Elaeocarpaceae are separated from the Tiliaceae (84) chiefly on the basis of the mucilaginous canals and ducts.

Muntingia and *Sloanea* seemingly have little in common in superficial appearance of flower and fruit structure. The flowers of both are open and seemingly unspecialized.

Both the bright-red, and many-seeded berries of *Muntingia* and the capsular fruits of *Sloanea* with one or two arillate seeds are probably chiefly bird dispersed. *Muntingia calabura* is known to be bat dispersed also (Heithaus, Fleming & Opler, 1975; Yazquez-Yanes et al.,

---

KEY TO THE SPECIES OF ELAEOCARPACEAE

Blades inequilateral, palmately veined at base, densely stellate-pubescent between veins below; flowers with white petals; fruit baccate, naked . . . . *Muntingia calabura* L.
Blades equilateral, pinnately veined throughout, glabrescent between veins below; flowers apetalous, not white; fruit capsular, densely covered with spines or trichomes:
    Blades more than 15 cm long; fruits with spines more than 2 cm long . . . . *Sloanea zuliinsis* Pitt.
    Blades less than 15 cm long; fruits covered with clinging trichomes at most 4 mm long . . . .
    . . . . *Sloanea terniflora* (DC.) Standl.

1975). White-faced monkeys may eat the pulp around the seed (Oppenheimer, 1968).

Twelve genera and 350 species; tropics and subtropics.

MUNTINGIA L.

Muntingia calabura L., Sp. Pl. 509. 1753
Majaguillo, Pacito, Periquito

Shrub or small tree, rarely to 10 m tall and ca 15 cm dbh; stems hirsute; bark black. Leaves alternate; stipules linear, 3–5 mm long, hirsute, sometimes paired on one side of petiole; petioles ca 5 mm long, densely hirsute; blades inequilateral, ± elliptical, acute to gradually acuminate, strongly inequilateral at base, obtuse on one side, ± cordate on the other, 4–10(11) cm long, 2–3.5 cm wide, with sparse, sessile, stellate pubescence on upper surface and dense, soft, whitish, stellate pubescence on lower surface, serrate; palmate veins at base 3–5. Inflorescences supra-axillary, usually 1-flowered; pedicels 1–2 cm long; flowers 5-parted, with a disagreeable odor, 1.5–2 cm diam; sepals lanceolate, very long-acuminate, to 7 mm long, pubescent on both surfaces, softly so inside; petals fugacious, white to creamy white, rounded at apex, 7–9 mm long, the outer margin undulate; stamens numerous, plainly visible, 4–5 mm long, erect to spreading, borne on outer margin of the disk; disk narrow, the rim velutinous; ovary ovoid, glabrous; style short, thick; stigma radiate, 5-lobed. Berries ovate, baccate, bright red, sweet, rugose when dried, subtended by the receptacle trichomes and the filaments; seeds many, ovoid, ca 0.5 mm long. *C. L. Wilson 152.*

Probably once widespread on the island when there were more clearings, but possibly no longer existing there. Flowering and fruiting throughout the year elsewhere.

Widely distributed in tropical America in secondary areas; introduced elsewhere in the world. In Panama, known chiefly from tropical moist forest in the Canal Zone (Pacific slope), Colón, Chiriquí, Los Santos, Herrera, Panamá, and Darién; known also from premontane moist forest in Panamá.

SLOANEA L.

Sloanea terniflora (Moc. & Sessé ex DC.) Standl., Trop. Woods 79:10. 1944
Terciopelo

Tree, to 30 m tall; trunk buttressed; bark smooth, black; branchlets grayish, but generally hidden by a dense golden puberulence when young. Leaves alternate to subopposite; petioles 5–15 mm long, densely golden-puberulent when young; blades elliptical to obovate, rounded to obtuse at apex, ± cuneate-obtuse and equilateral at base, 6–8(15) cm long, 3–5(7.5) cm wide, entire to sinuate, becoming glabrous above, lighter and glabrous below except on veins; midrib and major lateral veins prominent. Inflorescence 3–7 cm long, umbellately 1–3 flowered, axillary or terminal; peduncles 1.5–5 cm long;

pedicels to 1.5 cm long. Peduncles and pedicels reddish, glabrate in age; flowers apetalous, maroon; sepals 4, valvate, ovate, ca 7 mm long, the margins densely puberulent; stamens many, ca 4 mm long, pale yellow; filaments scarcely 1 mm long; anthers opening by an apical pore, the connective produced into a fugacious awn ca 1 mm long; ovary 2–2.5 cm long, tufted-puberulent; style glabrous above, 4-lobed; stigmas 1 per lobe. Capsules ellipsoid, to 2.5 cm long; valves 4, 3–4 mm thick, densely pubescent, the trichomes easily detached, reddish, antrorsely barbed, ca 3 mm long; seeds 1 or 2, ellipsoid, ca 1 cm long and 5 mm wide, nearly enveloped in an aril, the aril firmly attached in basal fourth of the seed. *Croat 9778.*

Frequent in the old forest as canopy trees. Flowers in the dry season (February to April). The fruits mature by the late dry season, though old valves may persist long after the seeds are removed, even until the time of the next flowering. K. Bawa (pers. comm.) reports that the species flowers for only 8–10 days in Costa Rica.

Trichomes of the fruit, which are irritating and cling to the skin, are no doubt beneficial in preventing the fruits from being eaten before maturity. At maturity the arillate seeds are displayed on the open inner valve surface and are probably removed by birds.

Central Mexico to Peru, Bolivia, and Brazil. In Panama, known from tropical moist forest in the Canal Zone (from BCI to the Pacific coast), Chiriquí, Veraguas, Los Santos, Herrera, Coclé, and Panamá.

Sloanea zuliinsis Pitt., Bol. Com. Ind. Venezuela 4(34):31. 1923
S. microcephala Standl.

Tree, usually 5–18 m tall; trunk with narrow buttresses; bark minutely lenticellate, with many horizontal wrinkles; twigs minutely brown-puberulent. Leaves alternate; stipules lanceolate, 12–30 mm long; petioles 5–12(21) cm long, subpuberulent; blades mostly elliptical, ± acute at apex, subacute to rounded at base, 20–40(60) cm long, 8–15(25) cm wide, subcoriaceous, becoming glabrate except on veins below, sinuate; midrib and veins prominently raised below. Cymes compound, the parts minutely puberulent; peduncles 12–26 cm long; pedicels 5–12 mm long; peduncles and pedicels subtended by bracts; flowers apetalous, yellowish, 2–3 mm long; sepals united into a flat, 4–6-lobed plate, beneath the receptacle; stamens many, to 2 mm long, congested to form a tight, subglobose mass to 4 mm across; ovary to 1.5 mm long; style to 1.5 cm long, tapered from base, sometimes 4-angled, glabrous above, the apex obtuse, entire. Capsules ellipsoid, 1.5–2.5 cm long; valves 4, 2–4 mm thick, densely covered with spines, the spines very slender, 2–3 cm long, tapered, yellow-green to reddish, minutely and antrorsely puberulent, not easily removed; seeds 1 or 2, ca 1.5 cm long and 7 mm wide, almost completely covered by aril, the aril unequally 6-lobed, firmly attached to basal third of seed, the three large lobes irregularly laciniate. *Foster 1316.*

Occasional, in the forest, most commonly seen on

Drayton Trail. Flowers principally in September and October. The fruits are mature from October to December.

Sterile collections from Chiriquí, possibly this species, occur at altitudes to 1,900 m. These trees may be 27 m tall.

Pacific coast of Costa Rica to Venezuela (southwest side of Lake Maracaibo). In Panama, known from tropical moist forest in the Canal Zone and Darién.

84. TILIACEAE

Trees or shrubs (sometimes suffrutescent in *Corchorus*), usually with stellate pubescence. Leaves alternate, petiolate; blades simple, sometimes lobed, the margins variously serrate; veins palmate at base; stipules present. Cymes or panicles terminal, axillary or leaf-opposed; flowers bisexual or unisexual (monoecious in *Trichospermum*; gynodioecious in *Heliocarpus*), sometimes apetalous (*Triumfetta*), actinomorphic, solitary; sepals 4 or 5, free, valvate; petals 4 or 5, free, imbricate, showy, sometimes reduced or lacking; stamens 5 (*Triumfetta*) to many (in bisexual flowers), free or rarely connate into a short tube; anthers 2-celled, introrse or extrorse, basifixed or versatile, dehiscing longitudinally; pistil 1; ovary superior, sometimes on a gynophore, 2- or 5-locular and 2- or 5-carpellate; placentation axile; ovules many per locule, anatropous; style 1; stigma capitate to bilobed. Fruits 2- or 5-valved loculicidal capsules, often spiny; seeds albuminate, winged or unwinged.

Members of the family are most easily confused with the Malvaceae (85) and the Sterculiaceae (87), principally because the leaves are palmately veined at base. They are distinguished by the nearly distinct stamens, the two-celled anthers, and the typically cymose inflorescences.

Flowers of all species are probably insect pollinated. Because of the usual abundance of generally exserted stamens and exserted style, it is suspected that the flowers may be pollinated by indiscriminate pollen collectors. In Costa Rica, the bee *Eulaema polychroma* "buzzes" pollen from the anthers of *Apeiba*. *Luehea* in Costa Rica is nocturnally pollinated by beetles and perhaps by moths (Heithaus, 1973).

Diaspores are diverse. Seeds of *Triumfetta lappula* are epizoochorous. *Apeiba membranacea* and *A. tibourbou* are endozoochorous, at least those of *A. membranacea* being dispersed chiefly by monkeys but also by coatis (Kaufmann, 1962) and parrots (Chapman, 1931) (see that species for discussion). Enders (1935) reported that peccaries also eat the fruits in times of food scarcity. Several species, including *Trichospermum mexicanum*, *Heliocarpus popayanensis*, and species of *Luehea*, are wind dispersed. *Corchorus siliquosus* has small seeds that are probably shaken slowly from the linear capsule and blown away by the wind.

KEY TO THE SPECIES OF TILIACEAE

Plants shrubs less than 3 m tall; flowers small, less than 6 mm long, yellow; fruits small burrs less than 1 cm diam *or* linear capsules to 6 cm long:
  Petals lacking; fruits small burrs less than 1 cm diam; leaf blades broadly ovate, usually 3–5-lobed; plants common . . . . *Triumfetta lappula* L.
  Petals 5; fruits unarmed cylindrical capsules to 6 cm long and 2 mm wide; leaf blades ± elliptic, not lobed; plants probably no longer occurring on the island . . . . *Corchorus siliquosus* L.
Plants trees more than 5 m tall; flowers more than 1 cm long *or* not yellow; fruits variable:
  Petals magenta; fruits flattened, obovate, cordate and apiculate at apex, densely pubescent; leaves oblong-lanceolate, very long-acuminate, sometimes with white tufts of trichomes in axils of basal veins on lower surface of blade . . . . *Trichospermum mexicanum* (DC.) Baill.
  Petals white or yellow; fruits and leaves not as above:
    Flowers 4-parted, greenish-white, less than 5 mm long; inflorescences often more than 12 cm long; capsules minute, less than 5 mm long, prominently bristled on margin; leaves broadly ovate, often shallowly trilobate . . . . *Heliocarpus popayanensis* H.B.K.
    Flowers usually 5-parted, yellow or white, more than 1 cm long; inflorescences usually less than 12 cm long (except *Luehea seemannii* longer); capsules more than 2 cm long:
      Bracteoles 9, conspicuous, valvate, immediately subtending flowers; petals glandular-thickened at base inside; fruits distinctly capsular, longer than broad, unarmed:
        Flowers ca 1.5 cm long; capsules deeply 5-sulcate, to 2.5 cm long . . . . *Luehea seemannii* Tr. & Planch.
        Flowers ca 3 cm long; capsules 5-angled, not sulcate, more than 2.5 cm long . . . . *Luehea speciosa* Willd.
      Bracteoles few, imbricated and distant from base of calyx; petals eglandular; fruits indehiscent or only tardily so, broader than long, armed:
        Leaves with tufts of brown trichomes in vein axils below, subentire; stems, petioles, and inflorescence branches not pilose; capsules like sea urchins, bearing short, stiff, conic spines . . . . *Apeiba membranacea* Benth.
        Leaves lacking tufts in vein axils, serrate; stems, petioles, and inflorescence branches conspicuously pilose; capsules bearing long, flexible, stout spines . . . . *Apeiba tibourbou* Aubl.

Fig. 357. *Apeiba membranacea*

Fig. 358. *Heliocarpus popayanensis*

Fig. 358. *Heliocarpus popayanensis*

About 50 genera and about 600 species; mainly in the tropics.

APEIBA Aubl.

Apeiba membranacea Spruce ex Benth., J. Proc. Linn. Soc., Bot. 5:61, Suppl. 2. 1861

A. aspera Aubl.

Peinecillo, Monkey comb, Cortezo

Tree, 10–30 m tall; trunk to 75 cm dbh, often weakly buttressed and ribbed above buttresses; outer bark thin, flaky, often minutely fissured with small lenticels in vertical rows; inner bark with narrow, radial, V-shaped wedges, these with large pores exuding clear, viscid droplets, the sap with sweet aroma; wood white, soft; stems bearing ± ferruginous, stellate pubescence when young, glabrous in age. Stipules ovate, to 5 mm long, deciduous; petioles slightly swollen at apex, 1.5–3 cm long, minutely stellate-pubescent; blades oblong-elliptic to obovate-elliptic, acute to acuminate, rounded to subcordate at base, 8–25 cm long, 3–10.5 cm wide, sparsely and minutely fimbrillate-lepidote below with tufts of brown trichomes in vein axils, ± entire; palmate veins at base 3. Panicles open, to 8 cm long, opposite leaves; pubescence of inflorescence branches, pedicels, and calyces short, rufous, stellate, tomentose; pedicels 1–2 cm long; flowers few, 5-parted; sepals lanceolate, 1–2.5 cm long, 3–6 mm wide; petals yellow, spatulate or obovate, 1.5–2 cm long, 7–13 mm wide, glabrous; stamens numerous, 3.5–5.5 mm long, long-pilose at base; style ca 12 mm long; stigma shortly denticulate. Fruits shaped like flattened sea urchins, to 6 cm diam and 1.5 cm thick, densely covered with sharp conic spines; seeds numerous, light brown, irregular, to 4 mm long. *Croat 5213, 7281.*

Abundant in some areas of the forest and common all over the island. Flowers mainly in the rainy season (May to December) with flower buds appearing as early as March. The fruits develop in about 9 months and are seen from November to May. White-faced monkeys eat the larvae infesting the fruits from January to May (Oppenheimer, 1968). Leaves fall in the dry season and grow in again soon.

The fruits are perhaps tardily dehiscent, but probably all fruits with viable seeds are opened by monkeys. One surface is removed and the fruits are picked out. Oppenheimer (1968) reported seeing the monkeys pick out grubs that frequently infect part of the seeds while discarding the viable, uninfected seeds. Macaws have been seen eating the fruits on BCI, and Chapman (1929) reported that the fruits are eaten by the Amazona parrot.

Mexico, Costa Rica, Panama, and western South America to Bolivia. In Panama, a characteristic component of tropical dry forest (Holdridge & Budowski, 1956) and tropical moist forest (Tosi, 1971); known from tropical moist forest in the Canal Zone, Bocas del Toro, Colón, Panamá, and Darién, from premontane wet forest in Colón and Coclé, and from tropical wet forest in Colón and Darién. Reported from premontane rain forest in Costa Rica (Holdridge et al., 1971).

See Fig. 357.

Apeiba tibourbou Aubl., Hist. Pl. Guiane Fr. 1:538, t. 213. 1775

Monkey comb, Cortezo, Cortés, Peinecillo (Chiriquí), Peine de mico, Fruto de piojo

Tree, 6–20(30) m tall, ca 25 cm dbh; outer bark thin, weakly fissured; inner bark white to tan with granular areas interspersed with round pockets containing thick sap; pubescence ferruginous, both long simple trichomes and shorter stellate trichomes on stems, petioles, lower midribs, axes of inflorescences, and calyces. Stipules triangular, to 2 cm long, subpersistent; petioles 1–3 cm long, long-hispid; blades ± oblong-elliptic, acuminate, subcordate at base, 10–30(33) cm long, 6–12(15) cm wide, sparsely stellate-puberulent above with long simple trichomes on midrib, densely stellate-arachnoid below, serrate; palmate veins at base 3–7. Panicles to 11 cm long, opposite leaves; pedicels to 1.8 cm long; flowers few, 4- or 5-parted; sepals ± lanceolate, 1.5–2.2 cm long, spreading at anthesis, thick, densely long-pubescent outside, glabrous inside; petals glabrous, spatulate to narrowly obovate, to 1.6 cm long, yellow or less commonly white, spreading at anthesis; stamens numerous, yellow, ca 7 mm long, the outermost irregularly united into a tube, sometimes sterile; anthers to ca 4 mm long, usually sparsely pilose; ovary globose, densely pubescent; style 8–12 mm long, ± equaling petals; stigma shortly denticulate. Capsules depressed-globose to globose, to 8 cm diam (including bristles), densely covered with bristles, the bristles long, flexible, stout, to 1.5 cm long; seeds numerous, depressed-globose, ca 2.5 mm diam. *Croat 4003a, 8170.*

Frequent in the forest. Flowers mainly in the rainy season (May to December), although a few flowers are seen all year. The fruits probably develop in 6–9 months and are most abundant in the dry season. Leaves fall in the dry season and new leaves are seen in the early rainy season.

Throughout tropical America. In Panama, common in secondary growth and typical of tropical moist forest (Tosi, 1971); known from tropical moist forest in the Canal Zone, Colón, San Blas, Veraguas, Herrera, Chiriquí, Panamá, and Darién, from tropical dry forest in Coclé and Panamá, from premontane moist forest in the Canal Zone, Veraguas, and Panamá, from premontane wet forest in Chiriquí, Panamá, and Darién, and from tropical wet forest in Colón and Darién.

CORCHORUS L.

Corchorus siliquosus L., Sp. Pl. 529. 1753

C. orinocensis sensu Standl.

Broomweed, Escobilla

Small shrub, usually less than 1(2) m tall; stems with 1 or 2 vertical bands of short, simple, erect trichomes. Reduced leaves in axils often 1 to several; stipules paired, persistent, bristle-like, to 2.5 mm long; petioles 5–25 mm long, short-pubescent on upper side; blades ovate to elliptic, acute to acuminate, rounded to subcordate at base, 2–7 cm long, 1–3.5 cm wide, nearly glabrous, crenate-serrate; palmate veins at base 3. Pedicels 4–8

mm long; flowers 1–3, in axils or opposite leaves, 5-parted; sepals linear-lanceolate, ca 8 mm long, glabrous; petals obovate, ca 6 mm long, glabrous, yellow; stamens numerous, free, ca 5 mm long; style ca 2.5 mm long; stigma subbilobed, densely papillate. Capsules linear, erect, 4–6 cm long, ca 2 mm diam, minutely puberulent; seeds numerous, ca 1 mm long. *Aviles 12.*

Probably once abundant when the island was weedier; not seen in recent years, but to be expected in the larger clearings. Seasonal behavior uncertain. Elsewhere in Panama flowering most abundantly in the rainy season, with the fruits seen during the dry season.

Native to East Asia; now common as a weed throughout the American tropics. In Panama, known from tropical moist forest all along the Atlantic slope and in the Canal Zone, Panamá, and Darién, from tropical dry forest in Panamá (Taboga Island), from premontane moist forest in the Canal Zone and Panamá, and from premontane wet forest in Colón, Coclé, and Panamá.

HELIOCARPUS L.

Heliocarpus popayanensis H.B.K., Nov. Gen. & Sp. 5:341. 1823

Majaguillo, Majagua

Gynomonoecious tree (with pistillate and bisexual flowers on the same plant), 6–10(30) m tall; branches bearing dense, ferruginous, stellate pubescence when young, glabrous in age. Stipules lanceolate, paired, to 1 cm long, caducous; petioles stellate-pubescent, 4–8 cm long; blades broadly ovate, acuminate (often with 3 shallow acuminate lobes), truncate to cordate at base, 8–25 cm long, 4–20 cm wide, finely serrate, the pubescence sparse, simple, stellate above, densely stellate below; palmate veins at base 5–7. Cymes small, along panicles to 25 cm long, usually terminal, bearing both pistillate and bisexual flowers (the bisexual ones possibly functionally staminate); flowers several, 4-parted, pedicellate (those of the bisexual flowers somewhat longer); sepals narrowly lanceolate, acute, tomentulose outside, glabrous inside; bisexual flowers with petals 4, narrowly spatulate, white, shorter than sepals; stamens many, 3–4 mm long; anthers dehiscing upward; pollen ± tacky; ovary minute; style bifid; pistillate flowers with petals reduced or lacking; staminodia numerous, minute; style to 2 mm long, bifid about one-third its length; stigma lobes usually obscurely 2- or 3-lobed. Capsules ellipsoid, flattened, 2-valved, ca 4 mm long, the pubescence short, stellate, the bristles long, plumose, in 2 rows along the margin; seeds usually 2, mostly compressed-ovoid, 1.7–2.5 mm long. *Croat 14051.*

Common at the margin of the forest around the Laboratory Clearing. Flowers in the dry season (December to March). The fruits mature quickly, mostly from March to June, with flowers and fruits often on the same tree.

Capsules probably dehisce on the tree, but instead of releasing the seeds, each valve may be carried away separately with the seed still attached.

Southern Mexico to northern Argentina. In Panama,

common in secondary growth; known from tropical moist forest in the Canal Zone, Colón, and Darién, from premontane wet forest in Coclé, from tropical wet forest in Chiriquí and Darién, and from premontane rain, lower montane wet, and lower montane rain forests in Chiriquí.

See Fig. 358.

LUEHEA Willd.

Luehea seemannii Tr. & Planch., Ann. Sci. Nat. Bot., sér. 4, 17:348. 1862

Guácimo, Guácimo molenillo

Tree, usually 15–30 m tall; trunk to 125 cm dbh, the buttresses 1–3 m high, to 4 m wide at base, usually continuous with ribs on trunk; outer bark thin, peeling easily, with prominent, round, evenly distributed lenticels; inner bark with lighter streaks oozing thick, clear, sometimes sweet sap; stems somewhat flexuous, especially near apex; branchlets, petioles, and inflorescence branches bearing dense, ferruginous, stellate, tomentose pubescence. Petioles ca 1 cm long, thick; blades mostly oblong-elliptic, somewhat asymmetrical (especially at base), acuminate at apex, rounded to subcordate at base, 5–40 cm long, 2–15 cm wide, densely brown-arachnoid below, very sparsely so and shiny above, the margins irregularly serrate especially above middle. Thyrses terminal or upper-axillary, 2–8(16) cm long; flowers 5-parted, enclosed by bracteoles in bud, to 2.5 cm wide when open, the bracteoles 9, densely pubescent, deciduous; sepals free, spreading, ca 12 mm long, oblong-lanceolate, densely pubescent outside; petals spatulate, white or yellow, ± equaling sepals, pubescent near base; stamens many, ca 9 mm long, in 5 clusters each united at the base, slightly shorter than style; style thick, capitate, ca 5 mm long; stigma obscurely 5-lobed. Capsules ± elliptic to obovoid, deeply 5-grooved, to 2.5 cm long, densely brown-tomentulose; seeds many, 6–10 mm long, winged. *Croat 7327, 7996.*

Common in the young forest. Flowers in the late rainy and early dry seasons. The fruits mature late in the dry season and early in the rainy season (March to July). Leaves are gradually lost in the dry season and are renewed sometime from April to July.

Central America to Panama; reported by Jimenez S. (1970) from Colombia (Magdalena River Valley). In Panama, a characteristic tree species in tropical moist forest (Holdridge & Budowski, 1956; Tosi, 1971); known from tropical moist forest throughout Panama, from premontane moist forest in the Canal Zone and Panamá, and from premontane rain forest in Panamá (summit of Cerro Jefe). Reported from tropical dry, premontane wet, and tropical wet forests in Costa Rica (Holdridge et al., 1971).

Luehea speciosa Willd., Ges. Naturf. Freunde Berlin Neue Schriften 3:410, t. 5. 1801

Guácimo, Guácimo molenillo

Tree, 5–20(25) m tall, to 70 cm dbh, the pubescence stellate, ferruginous, dense except on upper leaf surface, the older stems ± glabrous. Leaves deciduous; stipules

ovate-lanceolate, 10–20 mm long, subpersistent; petioles 8–12 mm long; blades broadly elliptic to ovate, abruptly acuminate at apex, truncate, rounded or subcordate and sometimes inequilateral at base, 8–24 cm long, 6–12 cm wide, serrate. Panicles open, terminal or upper-axillary, to 12 cm long; pedicels 1–3.5 cm long; bracteoles 9 or 10, deciduous, lanceolate-linear, to 2 cm long and 4 mm wide, stellate-pubescent on both sides; flowers 5-parted; sepals lanceolate-oblong, 3–4 cm long, to 9 mm wide, glabrous inside; petals light yellow, obovate, to 3.5 cm long and 1.5 cm wide, glabrous but with villous base inside, the margin irregular; stamens many, in 5 or 10 clusters, densely white-hirsute at base, 1–2 cm long; style 1.5–2.5 cm long; stigma capitate. Capsules oblong-obovate, to 4.5 cm long and 1.5 cm wide, obtusely 5-angled, rounded at apex, woody, ferruginous-tomentose; seeds very numerous, winged, flattened, oblique, ca 10 mm long and 4 mm wide, overlapping in 2 rows in each of 5 carpels, the valves opening slightly at maturity. *Croat 4222, 7871.*

Occasional in the young forest; more common elsewhere in the Canal Zone in disturbed areas and along roads. The species flowers in the early dry season while old leaves are still on the tree. The fruits mature in the late dry season, but the valves persist on the tree long after the seeds have been shed. Leaves probably fall late in the dry season and grow in again in the early rainy season.

Mexico to Colombia and Brazil (as far south as Saõ Paulo *fide* Rizzini, 1971); Cuba. In Panama, known from tropical moist forest in the Canal Zone, Colón, San Blas, Herrera, Panamá, and Darién, from tropical dry forest in Coclé and Panamá (Taboga Island), from premontane moist forest in the Canal Zone and Panamá, from premontane wet forest in Coclé and Panamá, and from tropical wet forest in Panamá and Darién.

See Fig. 359.

TRICHOSPERMUM Blume

Trichospermum mexicanum (DC.) Baill., Hist. Pl. 4:179. 1872
Belotia panamensis Pitt.

Monoecious tree, to 15(22) m tall; young branches, petioles, and pedicels densely stellate-pubescent. Stipules ca 5 mm long, lanceolate, caducous; petioles 1.5–2.5 cm long; blades oblong-lanceolate to oblong-elliptic, long-acuminate, obtuse to rounded at base, 12–22 cm long, 3.5–9 cm wide, sparsely puberulent above with simple forked trichomes, moderately stellate-pubescent below, sometimes with tufts of trichomes in basal vein axils, entire or serrulate; palmate veins at base 3. Panicles axillary or terminal, bearing flowers of a single sex or of both sexes; pedicels ca 6 mm long with small, caducous bracteoles about 3 mm below base of sepals; flowers many, (4)5-parted; sepals pink, to 1.5 cm long, acute and hooded, densely stellate-tomentose outside; petals magenta, blunt at apex, shorter and narrower than sepals, pubescent outside, ± glabrous inside except around basal gland; stamens numerous; staminate flowers with stamens ca 1

cm long; filaments united into a broad, undulate, pubescent ring below base; style lacking; stigma obscurely lobed; pistillate flowers opening before staminate flowers; stamens apparently nonfunctional, to 3.3 mm long; style 1–4.3 mm long; stigma bilobed, the lobes divided many times. Capsules depressed-obovate, emarginate and apiculate at apex, to 2 cm long and 2.5 cm wide, densely stellate-pubescent; seeds many, broadly ellipsoid, to 2.5 mm long, long-ciliate on margins. *Croat 12844.*

Common in the forest. Flowers from November to January (sometimes to March). The fruits mature mostly from February to April.

Staminate flowers appear to greatly outnumber pistillate flowers on any tree, but pistillate flowers are more abundant on some branches than others. Staminate flowers continue to open after most or all of the pistillate flowers of an inflorescence have been fertilized and after some have developed mature-sized fruits.

Southern Mexico to Ecuador along the Pacific slope. In Panama, common in secondary growth; known from tropical moist forest in the Canal Zone, San Blas, Panamá, and Darién, from premontane moist forest in Veraguas and Panamá, from premontane wet forest in Chiriquí, Coclé, and Panamá, and from tropical wet forest in Colón and Darién.

See Fig. 360.

TRIUMFETTA L.

Triumfetta lappula L., Sp. Pl. 444. 1753
Cadillo, Cepa de caballo, Abrojo

Shrub, usually 1–2.5 m tall; most parts ± densely stellate-pubescent. Petioles mostly 2–8 cm long; blades ovate, 3–5-lobed, acuminate at apex, obtuse to rounded or subcordate at base, 5–12 cm long, 4–10 cm wide, irregularly serrate, the serrations often glandular; palmate veins at base 3–5. Cymes condensed, upper-axillary; sepals reddish-brown, narrowly oblong, about 5 mm long, pubescent outside, with a short apiculum near apex, recurved after anthesis; petals lacking; stamens 5 or 15, (rarely 10), yellow, ca 3 mm long, alternately long and short, the longer ones about the height of the style; anthers not shedding pollen in bud; style shorter than stamens in bud; stigmas 2(3), slender, open, later elongating to height of tallest stamens, mostly or completely closed when anthers shed pollen; ovary with uncinate trichomes. Capsules ellipsoid, to 1 cm long (including spines), covered with long, uncinate spines; seeds 2 per cell, pyriform, ca 2 mm long. *Croat 7476, 7778.*

Common in the Rear #8 Lighthouse Clearing. Flowers early in the dry season (December to February). The fruits mature mostly from February to April.

Throughout tropical America. In Panama, widespread and ecologically variable; known from tropical moist forest throughout the country, as well as from tropical dry forest in Panamá and Coclé, from premontane moist forest in the Canal Zone, Los Santos, and Panamá, from premontane wet forest in Colón, Coclé, and Panamá, and from tropical wet forest in Darién.

See Fig. 361.

Fig. 359.
Luehea speciosa

Fig. 360. *Trichospermum*
mexicanum

Fig. 361. *Triumfetta lappula*

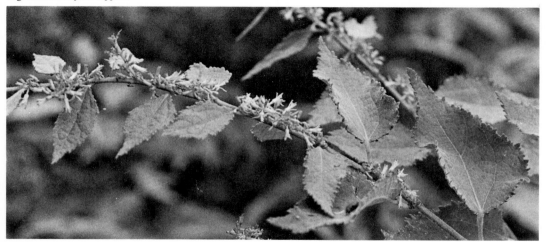

85. MALVACEAE

Trees, shrubs, or erect or sprawling herbs, often with mucilaginous sap, usually stellate-pubescent. Leaves alternate, petiolate; blades simple, usually palmately lobed, entire or serrate; venation palmate; stipules present. Flowers bisexual (dioecious in *Hampea*), solitary and axillary or few and corymbose, usually subtended by a crown of conspicuous epicalyx bracts; calyx 5-lobed; petals 5, free, showy; stamens many, in 1 or 2 whorls, united by their filaments into a column; anthers 1-celled, reniform, dehiscing longitudinally; ovary superior, the locules 2 to many, the carpels the same number as locules; placentation axile; ovules 1 to many per locule; style 1, branched, the branches equal to or twice as many as carpels; stigmas capitate to discoid. Fruits loculicidally dehiscent capsules or mericarps; seeds 1 or 2 per locule or many per locule (*Abelmoschus* and *Hibiscus*), sometimes arillate (*Hampea*), with oily endosperm.

Malvaceae are distinguished by their flowers, with a prominent staminal column and epicalyx bracts, and by their usually capsular fruits.

The flowers are generally open. They produce copious pollen and at least some species produce nectar. *Hibiscus rosa-sinensis* at the Laboratory Clearing is visited by hummingbirds, and male *Euplusia surinamensis* bees have been observed collecting nectar. Papilionid butterflies are its normal pollinators in Asia (H. Baker, pers. comm.). Most species of Malvaceae are probably pollinated by pollen-feeding insects. Flowers of *Sida rhombifolia* are reportedly self-pollinated in Java (van der Pijl, 1930), but they are also visited by the cotton-stainer beetle *Dysdercus cingularis* (van der Pijl, 1930) and by numerous bees (Doctors van Leewen, 1938).

Some 75–85 genera and 1,500 species; all temperate and tropical regions.

ABELMOSCHUS Medic.

Abelmoschus moschatus Medic., Malv.-Fam. 46. 1787
Musk okra, Wild okra

Herb or suffrutex, to 3 (4) m tall, sparsely to densely hirsute, the trichomes of stem retrorse and spreading. Stipules linear, to 8 mm long; petioles to 15 cm long; blades nearly circular in outline, palmately lobed, 10–20 cm long, 12–24 cm wide, cordate or subsagittate at base, the lobes 3–5, acuminate, narrow, deep, irregularly serrate-dentate. Flowers solitary in upper axils; pedicels 3–6 cm long in flower, to 10 cm in fruit; epicalyx with 8–10 linear bracteoles to 17 mm long, ca 2 mm wide, persistent; calyx spathaceous, 2–4 cm long, 5-dentate, splitting laterally at anthesis; petals 5, obovate, to 8 cm long, yellow with a dark purple spot at base, asymmetrical; staminal tube ca one-third as long as corolla; style slightly longer than staminal tube. Capsules ovoid, 5–7 cm diam, acuminate, densely pubescent; seeds many, globose-reniform, ca 3.5 mm long, prominently brown-striate. *Croat 6376*.

KEY TO THE SPECIES OF MALVACEAE

Leaves deeply cordate at base, broadly ovate to ± circular in outline:
  Leaves mostly palmately lobed to beyond middle:
    Flowers yellow with a purple spot at base; stems and petioles with long trichomes . . . .
    . . . . *Abelmoschus moschatus* Medic.
    Flowers red to magenta; stems and petioles with short, pustular-based prickles . . . .
    . . . . *Hibiscus bifurcatus* Cav.
  Leaves shallowly lobed or not lobed:
    Epicalyx bracteoles broadened at apex into ± reniform blade, shorter than calyx . . . .
    . . . . *Hibiscus sororius* L.f.
    Epicalyx bracteoles lance-linear, acuminate at apex, exceeding calyx . . . .
    . . . . *Pavonia dasypetala* Turcz.
Leaves not cordate or very shallowly so:
  Epicalyx bracteoles exceeding calyx:
    Leaves obovate; petals white; herbs usually less than 30 cm tall . . . . *Pavonia rosea* Schlechter
    Leaves ovate; petals yellow; herbs more than 1 m tall . . . . *Pavonia paniculata* Cav.
  Epicalyx bracteoles shorter than calyx or lacking:
    Plants trees, usually more than 10 m high; blades entire, broadly ovate . . . .
    . . . . *Hampea appendiculata* (Donn. Sm.) Standl. var. *longicalyx* Fryx.
    Plants herbs or shrubs, less than 4 m tall; blades toothed:
      Flowers red; staminal tube long-exserted; epicalyx bracts lance-linear; plants usually more
        than 2 m tall, cultivated . . . . *Hibiscus rosa-sinensis* L.
      Flowers white to yellow-orange; staminal tube not exserted; epicalyx bracts lacking; plants
        usually less than 2 m tall:
        Leaves ± ovate-elliptic, moderately stellate-pubescent below, the trichomes not obscur-
          ing surface; stipules linear, usually to 15 mm long; pedicels to 12 mm long . . . .
          . . . . *Sida acuta* Burm.f.
        Leaves ± rhomboid, densely stellate-pubescent below, the trichomes obscuring surface;
          stipules subulate, usually less than 5 mm long; pedicels to 3.5 cm long . . . .
          . . . . *Sida rhombifolia* L.

Fig. 362. *Hampea appendiculata* var. *longicalyx*

Fig. 363. *Hibiscus bifurcatus*

Collected once in Rear #8 Lighthouse Clearing. Flowers and fruits elsewhere throughout much of the year, especially during the rainy season.

Native to southeast Asia; cultivated in the tropics throughout the world. In Panama, known principally from tropical moist forest in the Canal Zone, Bocas del Toro, Colón, Chiriquí, Herrera, Panamá, and Darién; known also from premontane wet forest in Colón.

HAMPEA Schlechter

Hampea appendiculata (Donn. Sm.) Standl. var. **longicalyx** Fryx., Brittonia 21:391–92. 1969

Dioecious tree, 10–17 m tall, to 40 cm dbh, ± densely stellate-tomentose (sparsely so on upper leaf surface); bark smooth, the wood creamy gray, very soft; stems and petioles reddish-brown. Stipules caducous, less than 2 mm long; petioles 3–7(11) cm long, one-half to one-third length of blade; blades broadly ovate to elliptic, acuminate to long-acuminate at apex, subcordate or obtuse at base, 9–21 cm long, 6–16 cm wide, often with minute, round, erect auricles over petiole at base, entire; palmate veins at base 5 (rarely 7), all veins prominently raised below, with conspicuous glands below especially at lower axils. Involucral bracteoles 3, subulate, 1–3 mm long, inserted at base of calyx, deciduous; flowers unisexual, axillary, in groups of 1–10, ca 2.5 cm diam; pedicels 1–3(5) cm long; epicalyx bracteoles lanceolate, ca 1–5 mm long; calyx truncate or irregularly 3- or 4-lobed to near middle, 8–15 mm long, glabrous inside; petals 5, oblong-elliptic to obovate, fused at base to the staminal tube, white to yellow, 15–25 mm long, glabrous inside, spreading at anthesis; stamens numerous, connate into staminal tube; staminate flowers with the stamens of irregular lengths, 5–11 mm long within a cluster; anthers oblong, 1–1.5 mm long; pistil lacking; pistillate flowers with the stamens much reduced, nonfunctional; style filiform, ca 1 cm long. Capsules obovate to elliptic, rounded at apex, densely stellate-tomentose, 2–3 cm long, the valves 3, spreading widely at maturity to expose seeds; seeds 1 or 2 per locule, ± ovoid, ca 1 cm long, black and shiny, covered on one side by a fleshy, white aril. *Croat 12495.*

Uncommon, in the older forest. Flowers in the rainy season (July to November, especially in September and October). The fruits are mature in the dry season, principally in February and March.

The capsule is reportedly explosively dehiscent, but it is doubtful that this serves to disperse the seeds completely since they are well adapted to bird dispersal. The typical variety of *H. appendiculata* differs in having longer stipules (3–9 mm long), smaller flowers (less than 1.5 cm diam), and a shorter calyx (5–6 mm long).

The species is known from Honduras, Costa Rica, and Panama. The variety *longicalyx* is known only from Panama, where it is ecologically variable, occurring in tropical moist forest on BCI but also in premontane moist, tropical wet, and premontane rain forests in Coclé, Panamá, and Darién.

See Fig. 362.

HIBISCUS L.

Hibiscus bifurcatus Cav., Mon. Cl. Diss. Dec. 146, t. 51(1). 1787
Algodoncito

Sprawling suffrutex or shrub, to 4 m tall; stems and petioles sparsely stellate-pubescent (often in lines); stems, petioles, pedicels, and major veins with retrorse pustular-based prickles. Stipules linear-subulate; petioles 2–10.5 cm long; blades ovate in outline, lobed mostly to beyond middle, deeply cordate at base, 9–16 cm long, sparsely pubescent with both simple and stellate trichomes, the lobes 3–5, mostly narrowly ovate, acuminate, the margins dentate-serrate. Flowers solitary in upper axils; pedicels 2–7 cm long; epicalyx bracteoles 9–13, linear, unequally bifurcate at apex, 1.5–2.5 cm long, pubescent; calyx 1–2 cm long, lobed to about middle, pubescent, accrescent in fruit, the lobes acuminate, 3-veined, the midvein glandular, the gland about midway, secreting sweet nectar; petals 5, obovate, 5–7(9) cm long, rounded at apex, magenta or rose; staminal tube dark violet, sparsely covered with anthers throughout, half to fully as long as corolla; style branches 5, held above staminal tube; stigmas hirtellous. Capsules ca 2 cm long, slightly shorter than accrescent calyx, broadly ovoid, mucronulate, sericeous; seeds many, irregular, 3–4 mm long, minutely papillate. *Croat 4254.*

Rare; possibly restricted to marshes or wet areas at the edge of the lake. Apparently flowers principally in the dry season, especially in the early dry season, with most fruits maturing late in the dry season and in the early rainy season.

Mexico to the Guianas and Brazil; West Indies. In Panama, known from tropical moist forest in the Canal Zone and Panamá and perhaps more abundantly from tropical wet forest in Colón.

See Fig. 363.

Hibiscus rosa-sinensis L., Sp. Pl. 694. 1753
Chinese hibiscus, Papo, Tapo, Papo de la reina

Shrub, to 3(7) m tall; young leaves and stems with dense, simple or branched trichomes, glabrate in age. Stipules lanceolate-linear, 6–8 mm long, paired; petioles 1.5–4 cm long; blades ovate, acuminate, obtuse to cuneate at base, 8–15 cm long, 5–9 cm wide, coarsely serrate. Flowers solitary in axils; pedicel, epicalyx, and calyx sparsely to densely stellate-pubescent; epicalyx bracteoles 5–7, narrow, to 1 cm long; calyx tubular-campanulate, 2–3 cm long, lobed ca one-fourth its length; petals 5, red, obovate, 7–9 cm long; staminal tube long-exserted, to 12 cm long, 5-lobed at apex; filaments to 1 cm long, attached in upper third of tube; style exceeding staminal tube ca 1 cm, the branches pubescent. Fruits not seen. *Croat 5467.*

Cultivated at the laboratory and also present in the Tower Clearing. Flowers throughout the year.

Probably native to tropical Asia; cultivated throughout the tropics of the world. In Panama, known from disturbed areas in tropical moist forest in the Canal Zone, Bocas del Toro, San Blas, and Panamá and from premontane moist forest in Panamá.

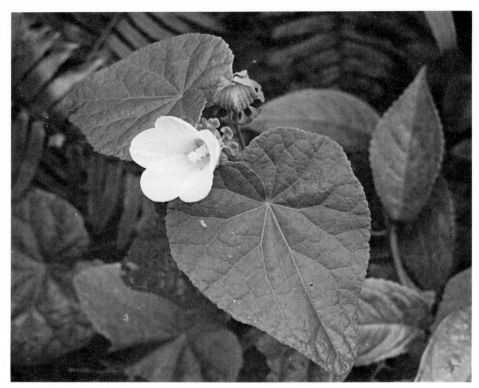

Fig. 364. *Hibiscus sororius*

Fig. 365. *Sida acuta*

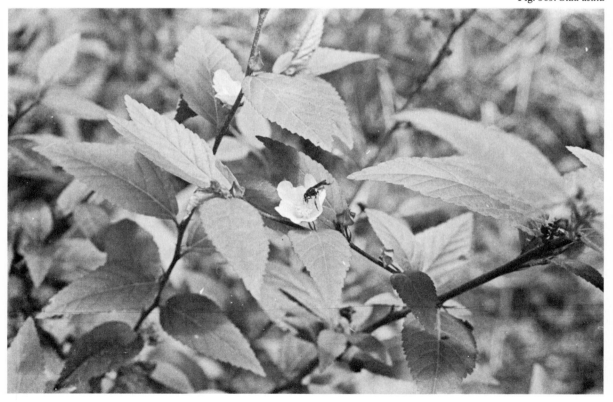

Hibiscus sororius L.f., Suppl. Pl. Syst. Veg. 311. 1781

H. sororius L.f. forma *albiflorus* Standl.

Herb or suffrutex, to 2 m tall, densely stellate-pubescent but sparsely on upper blade surface. Stipules minute; petioles 3–9(11) cm long; blades ovate (to irregularly angular), obtuse at apex, deeply cordate at base, 5–16 cm long, 5–13 cm wide, irregularly crenate; palmate veins at base 5–7. Flowers solitary in upper axils, long-pedicellate; epicalyx bracteoles markedly broadened at apex into a ± reniform blade; calyx 5-lobed to middle or beyond, 2–3 cm long, 5–7-veined, persisting in fruit; petals 5, obovate, 5–7.5 cm long, usually white, often magenta in bud; staminal tube antheriferous throughout, about half as long as corolla; pollen sticky, clinging together in chainlike masses; styles 5, occasionally splitting staminal tube, recurving and coming into direct contact with anthers; stigmas minutely fimbrillate, slightly exceeding staminal tube; nectar accumulating between calyx and petals. Capsules oblong-elliptic, ca 2.5 cm long, contained within larger, accrescent calyx, densely hispid, the valves 5, opening at maturity; seeds subglobose-reniform, less than 2 mm long, with glandular droplets, buoyant. *Croat 13231.*

Common in marshy areas on the south side of the island. Usually a component of the floating masses of vegetation (i.e., *Annona-Acrostichum* formations). Flowers and fruits throughout the year.

Throughout the tropics of Central and South America; Cuba. In Panama, known from tropical moist forest in the Canal Zone, Bocas del Toro, Panamá, and Darién, usually in swamps.

See Fig. 364 and fig. on p. 12.

PAVONIA Cav.

Pavonia dasypetala Turcz., Bull. Soc. Imp. Naturalistes Moscou 31(1):189. 1858

Shrub or small tree, to 5 m tall; pubescence dense, stellate, often ± viscid especially on inflorescences. Stipules narrowly ovate, to 20 mm long, 2–4 mm wide; petioles 5–15 cm long; blades broadly ovate, acuminate, cordate at base, 12–23 cm long, 10–20 cm wide, sometimes shallowly trilobate, irregularly serrate-dentate. Flowers solitary in axils or paniculate and terminal; pedicels 1–2.5 cm long, articulate near apex; epicalyx bracteoles 12–16, lanceolate-linear, 1.5–2.5 cm long; calyx cupular, hidden by epicalyx, 2–4 mm long; petals 5, rose, narrowly obovate, 3–5 cm long; staminal tube shorter than corolla, lobed, antheriferous on upper part; filaments ca 6 mm long; style exceeding staminal tube, the branches 10, to 7 mm long. Mericarps 5, obovoid, to 6 mm long and 3 mm wide, ± 3-sided, black; seeds ca 4.5 mm long, black, surrounded by a whitish aril. *Croat 4330, Foster 1455.*

Rare, collected twice in recent years on Zetek Trail; probably more abundant in previous years. Elsewhere in the Canal Zone the species usually occurs along roadsides or in open areas. Flowers in the early dry season. The fruits mature principally from January to March.

Costa Rica to Colombia and Venezuela. In Panama, common in premontane wet forest in the Canal Zone and Panamá; known also from tropical moist forest in the Canal Zone and Panamá and from tropical wet forest in Panamá (Cerro Jefe).

Pavonia paniculata Cav., Mon. Cl. Diss. Dec. 135, t. 46(2). 1787

Pape

Herb or shrub, to 3 m tall; trichomes of stems, petioles, and inflorescence branches both long and short, mostly simple, sometimes gland-tipped. Stipules lanceolate, 1–1.5 cm long, persistent, paired; petioles 3–10 cm long; blades ovate to shallowly trilobate, acute to acuminate at apex, truncate to shallowly cordate at base, 5–12 cm long, 4–10 cm wide, crenate-serrate, sparsely pubescent above, moderately stellate-pubescent below; palmate veins 7–9. Flowers solitary in axils, becoming paniculiform or subcorymbose at ends of branches by reduction of leaves; pedicels very short or to 3 cm long; epicalyx hirsute, with 6–12 linear bracteoles 1–2 cm long; calyx cupular, ca 5 mm long, 5-lobed to near middle, long-ciliate; petals 5, yellow, obovate, 1–1.5 cm long, spreading at anthesis, glabrous except inside at base; staminal tube ca half as long as petals, antheriferous throughout; filaments ca 5 mm long; style exceeding staminal tube, the branches 10. Mericarps 5, obovoid, ± 3-sided, 3–4 mm long, minutely tuberculate and appressed-puberulent; seeds trigonous-reniform, ca 2.5 mm long. *Shattuck 731.*

Apparently rare or absent, collected twice in earlier years but not seen recently. The species generally grows in weedy areas along roadsides in the Canal Zone. Flowers from November to March, especially in December and January. Most fruits are mature in February and March.

Mexico to Argentina; Greater Antilles. In Panama, known from tropical moist forest in the Canal Zone, Bocas del Toro, and Panamá, from premontane moist forest in the Canal Zone and Panamá (Saboga Island), and from premontane wet forest in Chiriquí (Boquete), Coclé (El Valle), and Panamá.

Pavonia rosea Schlechter, Linnaea 11:355. 1837

Suffruticose herb, usually less than 30 cm tall; stem usually unbranched; most parts with stellate or less frequently hirsute pubescence but the upper blade surface often glabrous. Stipules paired, linear, 6–8 mm long, persistent; petioles 5–15 mm long; blades obovate, acute to acuminate at apex, tapered to a slightly cordate base, 9–18 cm long, 3–8 cm wide, irregularly dentate. Inflorescences usually terminal, short; flowers few, to ca 1.3 cm diam; epicalyx bracteoles 7–11, linear, united at base, to ca 1 cm long; calyx 3–5 mm long, shorter than epicalyx, lobed to middle, the lobes triangular; petals white, ± obovate, 7–10 mm long, stellate-puberulent outside, spreading at anthesis; staminal tube bearing anthers just below apex; style branches 10, slightly longer than staminal tube, pinkish; stigmas capitate. Fruits to 8 mm long; mericarps 5, prominently 3-ribbed, spiny, the spines

5–7 mm long, 3 per mericarp, retrorsely barbed; seeds trigonous, 4–5 mm long, glabrous. *Croat 9254, 10942.*

Common along trails in the older forest, often forming continuous stands below the escarpment on Zetek Trail; common to rare elsewhere. Flowers from March to November, especially in the early rainy season. The fruits develop rather quickly, and mature fruits may share the same inflorescence with flowers.

Mexico to Colombia and Brazil; Greater Antilles. In Panama, known principally from tropical moist forest in the Canal Zone, Bocas del Toro, Panamá, and Darién; known also from tropical dry forest in Coclé (Penonomé), from premontane wet forest in Bocas del Toro (Río Guarumo) and Coclé (El Valle), and from premontane and lower montane rain forests in Chiriquí.

SIDA L.

Sida acuta Burm.f., Fl. Ind. 147. 1768

Escobilla

Suffrutex, to 1(1.8) m tall, sparsely pubescent with simple and stellate trichomes. Leaves distichous; stipules lanceolate-linear, 1–1.5 cm long, ca 1 mm wide, persistent; petioles less than 1 cm long; blades narrowly ovate, acute at apex, obtuse to rounded at base, 4.5–10 cm long, 2–3.5 cm wide, serrate-dentate; palmate veins 3. Flowers 5-parted, solitary or in small pseudoumbels, axillary, 18–25 mm broad; pedicels to 12 mm long; calyx campanulate, to 1 cm long, foliaceous, lobed to near middle, moderately accrescent; petals obovate, to 1 cm long and 8 mm broad, yellow or white, oblique; staminal tube ca 4 mm long; filaments ca 2 mm long, with gland-tipped trichomes; style 4–5.5 mm long. Mericarps 5, 7, or 9 (rarely to 12), 3-sided, with 2 ± parallel beaks, 2.5–4 mm long, often densely stellate-pubescent; seeds trigonous, ca 2 mm long. *Croat 10249.*

Locally common in clearings. Flowers and fruits throughout the year, but apparently with the heaviest concentration of flowering in the rainy season with the fruits maturing mostly in the dry season.

Flowers of *S. acuta* reportedly open between 8 A.M. and 1 P.M. (van der Pijl, 1930).

Though beaks of the mericarps are not recurved to assist in epizoochorous dispersal as in *S. rhombifolia*, the closely parallel, usually densely stellate pubescence of the beaks serves nicely to attach the mericarps to passing animals.

Throughout the tropics of Western and Eastern hemispheres. In Panama, abundant and widespread in tropical moist forest; known also from premontane moist forest in Darién (Punta Patiño), from premontane wet forest in Coclé, Panamá, and Darién, and from tropical wet forest in Colón.

See Fig. 365.

Sida rhombifolia L., Sp. Pl. 684. 1753

Escobilla, Hierba de puerco

Herb or suffrutex, to 1.5 m tall; stems usually branched many times, minutely stellate-pubescent. Leaves minutely stellate-pubescent, especially below; stipules subulate, to 5 mm long; petioles 3–5 mm long; blades variable in shape and size, usually ± rhombic, acute to obtuse at apex, obtuse to cuneate at base, 1.5–7.5(8.5) cm long, 0.8–2.5(3.5) cm wide, serrate; palmate veins 3(5). Flowers 5-parted, ca 1 cm wide, solitary in axils, becoming corymbose at apex of stem; pedicels to 3.5 cm long, articulate usually above middle; calyx campanulate, 5-angulate, ca 6 mm long, the lobes deltoid, acuminate; petals ovate, oblique, clawed, emarginate at apex, mostly 7–9 mm long, yellow-orange; staminal tube ca 3.5(6) mm long; styles 4(8), connate to half their length. Mericarps usually 8–12, 3-sided, with a single beak, to 5.5 mm long (including beak); seeds bluntly 3-sided, ca 2 mm long. *Croat 6926.*

Locally common in clearings. Flowers and fruits throughout the year.

The single beak of the mericarp is often recurved and sparsely stellate-pubescent, both adaptations for epizoochorous dispersal.

Throughout tropics and subtropics of the world. In Panama, ecologically widespread, occurring in all provinces and in life zones ranging from tropical moist forest through tropical wet forest.

86. BOMBACACEAE

Trees, stellate-pubescent on some parts and sometimes armed. Leaves alternate, petiolate; blades simple or palmately compound, sometimes lobed, the margins entire; venation pinnate or palmate; stipules present. Flowers bisexual, actinomorphic, bracteolate, solitary or cymose, terminal or axillary; calyx cupular or campanulate, truncate or lobed; petals 5, imbricate, showy; stamens 5 (*Ceiba*) to many, the filaments united into a tube or the entire stamens forming a column (*Ochroma, Quararibea*); anthers 1–3 per stamen, 1- or 2-celled, often horseshoe-shaped, straightening at anthesis, dehiscing longitudinally; ovary superior (subinferior in *Quararibea*), 5-locular (2-locular in *Quararibea*), 2–5-carpellate; placentation axile; ovules 2 to many per locule, anatropous; style with the stigma usually 5-lobed. Fruits usually loculicidal capsules opening by 3–5 valves, the seeds with little or no endosperm and enveloped by woolly mass of trichomes; less frequently fruits pulpy and indehiscent (*Quararibea*) or capsular, the seeds large, not enveloped in a woolly mass of trichomes (*Cavanillesia, Pachira*).

Most members of the family are recognized by their powder-puff-like stamens and large capsular fruits with kapok enveloping the seeds. The genus *Quararibea* is atypical in both respects.

The flowers with large powder-puff-like clusters of stamens (*Bombacopsis, Pachira, Pseudobombax*) are pollinated most effectively by birds and bats. *Bombacopsis quinata, Pseudobombax septenatum, Ceiba pentandra, Ochroma pyramidale,* and *Quararibea asterolepis* are known to be pollinated by bats (Baker & Harris, 1959; Heithaus, Opler & Baker, 1974; Heithaus, Fleming & Opler, 1975; Vogel, 1958; Bonaccorso, 1975). *Ochroma*

KEY TO THE SPECIES OF BOMBACACEAE

Leaves present and simple:
  Leaves ovate-cordate, usually lobed:
    Flowers solitary, more than 10 cm long, not precocious; capsules not winged, more than 6
      times longer than wide; leaves shallowly sublobate . . . . . *Ochroma pyramidale* (Lam.) Urban
    Flowers not solitary (in terminal cymes), less than 3 cm long, precocious; capsules winged,
      about as wide as long; leaves deeply lobed (usually on ground at time of flowering) . . . . .
      . . . . *Cavanillesia platanifolia* (H. & B.) H.B.K.
  Leaves ± elliptic, not cordate or lobed:
    Calyx with 10, narrow, longitudinal wings; fruits ca 4 cm diam; plants usually less than 12 m
      tall . . . . *Quararibea pterocalyx* Hemsl.
    Calyx lacking wings; fruits ca 2 cm wide; plants more than 20 m tall . . . .
      . . . . *Quararibea asterolepis* Pitt.
Leaves lacking or palmately compound:
  Leaves always present with flowers; leaf blades acuminate; flowers more than 16 cm long and
    anthers more than 3 mm long; stamens scarlet toward apex; fruits with relatively few large
    seeds (more than 2 cm long); trees growing usually only in wet areas . . . *Pachira aquatica* Aubl.
  Leaves usually lacking at time of flowering; leaf blades acuminate or not; flowers less than 16
    cm long *or* anthers less than 3 mm long; filaments whitish toward apex; fruits with many
    small seeds in kapok; trees not restricted to wet areas:
    Flowers less than 5 cm long; fruits to more than 4 times longer than broad *or* fruits winged:
      Anthers numerous, the staminal tube with numerous deep divisions; fruits broadly winged;
        twigs lacking noticeable scars; trunk unarmed, only slightly buttressed, with promi-
        nent raised rings every meter or so; leaves simple but lobed . . . .
        . . . . *Cavanillesia platanifolia* (H. & B.) H.B.K.
      Anthers 10–15, borne on 5 divisions of staminal tube; fruits wingless; twigs with definite
        stipule scars; trunk usually armed, markedly buttressed; leaves palmately compound . .
        . . . . *Ceiba pentandra* (L.) Gaertn.
    Flowers more than 6 cm long; fruit wingless, less than 3 times longer than broad:
      Fruits ellipsoid, more than 11 cm long, striped green and purplish-brown; leaflets inarticu-
        late; trunk with green stripes; twigs whitish-pruinose; pedicels usually more than 1.8
        cm long . . . . *Pseudobombax septenatum* (Jacq.) Dug.
      Fruits oblong, less than 8 cm long, not striped; leaflets articulate; trunk not striped; twigs
        not whitish-pruinose; pedicels usually less than 1.8 cm long:
        Trunk and branches armed; leaflets ± acuminate at apex . . . .
        . . . . *Bombacopsis quinata* (Jacq.) Dug.
        Trunk and branches unarmed; leaflets rounded or emarginate at apex . . . .
        . . . . *Bombacopsis sessilis* (Benth.) Pitt.

pyramidale is also visited by white-faced and squirrel monkeys (Oppenheimer, 1968; Enders, 1935) as well as kinkajous, marmosets, opossums, parrots, and oropendulas (Chapman, 1938) (see the discussion under that species). Faegri and van der Pijl (1966) reported that movements of the *Ceiba pentandra* flower in the wind may cause self-pollination. Baker (1973) reported that it is freely visited by hummingbirds in Mexico.

Bombacopsis, Ceiba, Ochroma, and *Pseudobombax* have small seeds that are enveloped within a mass of kapok-like fibers and are wind dispersed, as is the winged seed of *Cavanillesia.* The mass of fibers easily supports the seed in water. *Quararibea* and *Pachira* are mammal or water dispersed. *Quararibea asterolepis* fruits are taken by white-faced monkeys (Oppenheimer, pers. comm.), howler, spider, and night monkeys. Fallen fruits are further dispersed by opossums (Hladik & Hladik, 1969) and possibly also by reptiles. *Quararibea* fruits are taken by the bat *Artibeus jamaicensis* Leach in Mexico (Yazquez-Yanes et al., 1975).

About 25 genera and some 150 species; mostly in the American tropics.

BOMBACOPSIS Pitt.

Bombacopsis quinata (Jacq.) Dug., Contr. Hist. Nat. Colomb. 1:2. 1938
  B. fendleri (Seem.) Pitt.
  Spiny cedar, Cedro espinoso

Deciduous tree, to 30(40) m tall, to 1 m dbh; trunk armed, often broadly buttressed at base; branches armed; wood reddish-brown, light, soft. Leaves alternate, digi-tately compound, glabrous; stipules lanceolate, caducous; petioles to 12 cm long, canaliculate above; petiolules very short, to 8 mm long; leaflets 5(7), oblong-obovate, acute to acuminate, acute and tapered at base, decurrent on articulate petiolule, 4–17 cm long, 2–8 cm wide, entire. Cymes growing from younger wood; flowers 1 to few, precocious, 7–11 cm long; pedicels 0.4–0.8 mm long, the bracteoles at apex 3, subtending the generally 5-glandular receptacle; calyx ± campanulate, undulate to flat-topped, to 1 cm high and 1 cm wide; both calyx and pedicel generally reddish-brown (at least when dry) and shortly tufted-puberulent, appearing velvety; petals 5, imbricate,

Fig. 366. *Bombacopsis quinata*

Fig. 367. *Bombacopsis quinata*

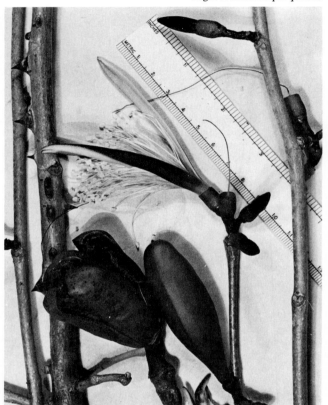

Fig. 368.
Bombacopsis sessilis

to 10 cm long and 8 mm wide, lighter than calyx, tan, puberulent and ca 14-veined inside, bicolorous in 2 stripes and appearing stellate-pubescent outside, tan on the side protected by next petal and light reddish-brown on exposed side; stamens ca 150, the free portion ca 6 cm long and united basally into a column ca 2 cm long; anthers horseshoe-shaped, dehiscing by unfolding and becoming ± straight, more than 2 mm long; style less than 10 cm long; stigma 5-lobed. Capsules oblong, ca 8 cm long, the valves 5, cinnamon, mucronulate, opening to expose pale yellowish kapok and seeds; seeds several, small, less than 5 mm long. *Croat 8380, 8705.*

Occasional, in the forest, principally in the younger forest. Flowers in February and March. The fruits mature and fall in March or later. Leaves fall during the dry season, growing in again in April and persisting until December or January.

Nicaragua to Colombia and Venezuela. In Panama, a characteristic component of tropical dry forest (Holdridge & Budowski, 1956) and of tropical and premontane moist forests (Tosi, 1971); known from tropical moist forest in the Canal Zone, Panamá, and Darién, from tropical dry forest in Coclé, from premontane moist forest in Panamá, and from premontane wet forest in Panamá (Chimán).

See Figs. 366 and 367.

Bombacopsis sessilis (Benth.) Pitt., Contr. U.S. Natl. Herb. 18:162. 1916
Ceibo, Yuca de monte, Ceibo nuno

Deciduous tree, 15–30 m tall, mostly 20–65 cm dbh; trunk unarmed, with greenish bark, with narrow plank buttresses to 1.5 m high; outer bark thin, grayish, with many vertical rows of lenticels, flaking; inner bark tan, thick, with irregular dark spots near periphery; branches at ± regular intervals, self-pruning, the lowermost branches deciduous; branchlets roughened with lenticels and old petiolar bases. Leaves palmately compound, glabrous; stipules lanceolate; petioles ± equaling leaflets, enlarged at both ends; blades (4)5–7(9), ± narrowly obovate, emarginate at apex, 4–18 cm long, 1.5–6.5 cm wide. Flowers showy, white to pink, terminal on short lateral branches, solitary or as many as 5 on a branch, appearing before or after leaves fall, 13–24 cm long; pedicels 5–18(32) mm long; calyx tubular, truncate or weakly lobed, about 1 cm long and 1 cm wide; petals 5, stellate outside, imbricate in bud, drying greenish, generally not markedly striped, and darker than *B. quinata;* stamens ca 250, fused into a column ca 8 cm long, the free part of filaments ca 6 cm long; anthers horseshoe-shaped, dehiscing by straightening, less than 2 mm long, shedding some pollen in bud; style bright red; stigma of 5 sharp lobes, pubescent inside. Capsules oblong, somewhat more elongate than *B. quinata,* the valves 5, woody, yellow-brown when fresh, opening to expose pale, often brownish kapok and seeds; seeds numerous, medium-sized, to 1 cm. *Croat 8654, 12976.*

Frequent in the forest, especially in the younger forest. Flowers from December to February. The fruits mature from March, persisting to October. Trees are bare for a short time in the dry season.

Costa Rica and Panama; cultivated in Cuba and Jamaica. In Panama, known from tropical moist forest in the Canal Zone, Veraguas, and Panamá, from tropical dry forest in Coclé and Panamá, from premontane moist forest in the Canal Zone, Veraguas, and Panamá, from premontane wet forest in Chiriquí and Panamá, and from tropical wet forest in Colón.

See Fig. 368.

CAVANILLESIA R. & P.

Cavanillesia platanifolia (H. & B.) H.B.K., Nov. Gen. & Sp. 5:306. 1823
Cuipo, Quipo, Bongo, Hamati

Deciduous tree, to 40 m tall, small-crowned; trunk usually 1–1.5 m diam, pale, often somewhat bulging ca 2 m above base, markedly ringed at intervals of ca 1.5 m, the surface planar but roughened with lenticels in age; inner bark granular. Leaves simple, clustered at apex of branches; stipules ovate, 5 mm long; petioles 10–25 cm long; blades ovate, palmately lobed (young leaves entire), the lobes 3–7, acuminate, cordate at base, to 30 cm long and wide, glabrate above, densely puberulent below. Cymes contracted, axillary; pedicels ca 1 cm long, surmounted by 3 fugacious bracteoles; flowers red, precocious, 2–3 cm long; calyx bell-shaped, irregularly 5-lobed, ca 1.5 cm long; calyx and pedicel shortly ferruginous-tomentulose; petals 5, strap-shaped, to 2.5 cm long and ca 5 mm wide, red, ferruginous-tomentulose toward obtuse apex; stamens many, reddish, to 2 cm long, united into a column about one-third their length; anthers ca 1 mm wide, broader than long, dehiscing upward; ovary 4- or 5-celled, the ovules 2 per cell; style straight, exceeding stamens by 5–10 mm; stigma 5-forked. Capsules hard, narrow, linear, to 12 cm long, with 4 or 5 conspicuous wings; wings semicircular, membranous and markedly veined, each to 15 cm long and 8 cm wide, red and green when mature; seeds 4 or 5. *Croat 8348.*

Occasional, in the forest. Flowers in March and April. The fruits mature in April and May. Leaves fall in December or January and reappear in May and June.

The fruits, despite their large wings, are poor fliers, and most land within a few hundred feet of the parent plant.

Nicaragua to Peru. In Panama, a typical component of tropical moist forest (Tosi, 1971) and described as the most abundant tree in upland Darién (Lamb, 1953); known from tropical moist forest in the Canal Zone, Los Santos, Panamá, and Darién, mostly on limestone soils (R. Foster, pers. comm.).

See Fig. 369.

CEIBA P. Mill.

Ceiba pentandra (L.) Gaertn., Fruct. & Sem. Pl. 2:244, t. 133. 1791
Silkcotton tree, Ceiba, Cotton tree, Bongo

Tree, to 40 m tall; trunk ca 1.5 (2.5) m dbh, armed at least when young, with large, curving plank buttresses

Fig. 369. *Cavanillesia platanifolia*

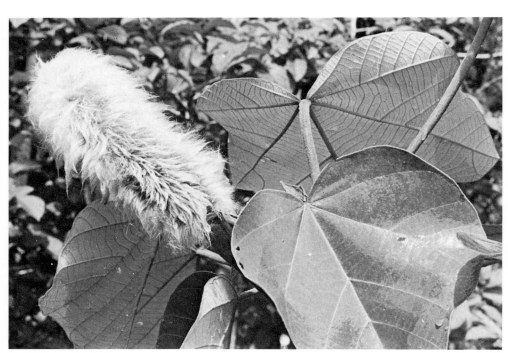

Fig. 370.
Ochroma pyramidale

to 10 m high and 10 m wide at base; outer bark grayish-brown with horizontal lenticular lines. Leaves clustered at tips of branchlets, palmately compound, glabrous; stipules ovate, to 6 mm long, caducous; petioles 5–23 cm long; leaflets 5–9, oblong-lanceolate, acuminate, narrowed and obtuse to acute at base, 10–21 cm long, 2–4.5 cm wide, ± entire. Flowers precocious, in clusters at tips of branchlets, ca 3 cm long; pedicels ca 2.5 cm long; calyx bell-shaped, ca 1 cm long and wide, with 4 or 5 small lobes, glabrous, drying dark; petals 5, oblong, ca 3 cm long and 1 cm wide, light yellow; stamens 5, united at base into a column; anthers 2 or 3 per stamen, linear, 1-celled, spiraled; style slightly surpassing stamens, surmounted by a capitate stigma. Capsules elongate-elliptic to obovoid, 10–26 cm long, to 4 cm diam, the valves 5, greenish, opening to expose copious grayish kapok and seeds; seeds numerous, round, ca 5 mm long. *Croat 14071.*

Frequent in the forest. Flowering usually from December through February, but individuals never flower 2 years in succession and sometimes only once in 5 years. Mature fruits are seen from January to March. Allen (1956) reported flowers in Costa Rica in February, with the fruits maturing in late March. Leaves fall in the late rainy season to the early dry season. In reproductive years individuals may be leafless much longer than during sterile years (R. Foster, pers. comm.).

In open areas the crown of the tree may be wider than the tree is tall.

Pantropical. In Panama, a typical component of tropical dry and tropical moist forests (Holdridge & Budowski, 1956; Tosi, 1971), but growing in many climates and places (Holdridge, 1970); known from tropical moist forest in the Canal Zone, Bocas del Toro, Chiriquí, and Panamá.

See fig. on p. 10.

OCHROMA Sw.

Ochroma pyramidale (Cav. ex Lam.) Urban, Feddes Repert. Beih. 5:123. 1920
  O. lagopus Sw.; *O. limonensis* Rowlee
  Balsa, Balso, Lana, Cotton tree

Tree, usually less than 12 m tall but to 30 m where persisting in the older forest, 30–180 cm dbh, sometimes buttressed; wood very lightweight; branchlets, pedicels, lower blade surfaces, petioles, and exposed parts of calyces with a dense layer of brown, stellate, tufted trichomes. Leaves simple, clustered at ends of branchlets; stipules ovate, to 1.5 cm long; petioles thick, 3–40 cm long; blades ovate, acuminate, cordate, 9–40 cm long, 8–35 cm wide, entire to shallowly 3–5-lobed, glabrous above in age, pale below; palmate veins at base 7(9). Flowers not precocious, leathery, with a pleasant aroma when young; pedicels to 12.5 cm long and 1.5 cm broad; calyx lobes stiffly coriaceous, unequal in shape, to 5.5 cm long, sericeous-villous inside; petals obovate-spatulate, to 15 cm long and 5 cm wide, whitish; staminal column 10–12 cm long, the antheriferous part 5–6 cm long, enlarged at anthesis; pollen with a foul odor, at least in age,

evenly distributed over column surface; anthers sessile, longitudinally dehiscent, the theca 1; style completely enveloped by stamens in bud, about as long as stamens at anthesis; stigma stout, spirally 5-sulcate. Capsules long, narrow, to 25 cm long and 2.5 cm diam, black and glabrous outside, densely lanate inside, the valves 5; seeds many, small, ca 5 mm long. *Croat 4637, 7214.*

Common along the margin of the lake or in disturbed areas, rare in the older forest where it is a large tree. Flowers from August to May, mostly from November to March. The fruits mature from February to August. Leaves fall around June and are replaced in August.

The flowers open at night, sometimes persisting to the following day. At anthesis the anthers, which are somewhat spirally arranged around the style, are unfurled and spread broadly. By the following day they are again closely twisted around the stigma. White-faced monkeys have been seen in the daytime poking their faces into the flower, possibly searching for insects, and their faces become liberally covered with pollen. They probably do not regularly effect pollination because the stigma is then covered, but the flowers may be visited by nocturnal mammals in the same way. Birds often cut holes near the base to obtain nectar. Seeds are principally wind dispersed but probably also water dispersed, made buoyant by the brownish kapok fibers within which they are enveloped.

Native to the New World; pantropical in cultivation. In Panama, known from tropical moist forest in the Canal Zone, Bocas del Toro, Colón, Veraguas, Los Santos, Panamá, and Darién, from tropical dry forest in Panamá (Taboga Island), from premontane moist forest in the Canal Zone and Panamá, from premontane wet forest in Chiriquí, and from tropical wet forest in Veraguas (Atlantic slope).

See Fig. 370.

PACHIRA Aubl.

Pachira aquatica Aubl., Hist. Pl. Guiane Fr. 2:726, t. 291, 292. 1775
  Provision tree, Sapote longo

Evergreen tree, to 23 m high and 70 cm dbh, often buttressed; outer bark hard, planar, thin, with weak distant vertical fissures; inner bark thick, reddish, marbled with white. Leaves palmately compound, glabrous; stipules ovate, ca 1 cm long; petioles to 24 cm long, often ribbed, swollen at both ends; leaflets 5–7(9), oblong-obovate to elliptic, caudate-acuminate to apiculate at apex, tapered to an acute base and decurrent on petiolule, 5–29 cm long, 3–15 cm wide, whitish-lepidote especially below. Flowers sweetly aromatic, usually solitary in upper axils; pedicels stout, 1–5.5 cm long; calyx ± tubular, truncate, the lobes obscure; petals 5, valvate, linear, greenish-white to brown, 17–34 cm long, ca 1.5 cm wide, curled outward at anthesis, stellate-puberulent outside, glabrous to villous inside; stamens many, scarlet in apical third, white basally, erect to spreading, slightly shorter than petals, variously united in small clusters basally to middle, the clusters finally uniting with staminal column; anthers

Fig. 372. *Pseudobombax septenatum*

Fig. 373.
Pseudobombax septenatum

Fig. 371. *Pachira aquatica*

horseshoe-shaped, dehiscing by straightening; ovary broadly ovoid, ca 1 cm long; style colored like stamens but several cm longer; stigma of 5 tiny lobes. Capsules reddish-brown, elliptic, oblong-elliptic, or subglobose, shallowly 5-sulcate, mostly to 20(30) cm long and 10(12) cm wide, the valves 5, densely ferruginous outside, appressed-silky-pubescent within; seeds usually 2 or 3 per carpel, irregularly angulate, mostly 3–4.5 cm long at maturity, brown, buoyant, embedded in solid, white, fleshy mesocarp. *Croat 5613.*

Known only from the edge of the lake. Flowers to some extent all year, mostly from February to April (sometimes from December to July). Most fruits mature from March to August. New leaves appear around May.

The buoyant seeds are no doubt dispersed by water currents.

Southern Mexico to northern Peru and Brazil; cultivated in the West Indies. In Panama, a characteristic species in tropical moist and tropical wet forest areas (Tosi, 1971), forming small, pure stands in Darién (Holdridge & Budowski, 1956) and usually growing beside fresh water; known from tropical moist forest in the Canal Zone, Bocas del Toro, Colón, Panamá, and Darién, from premontane wet forest in Panamá, and from tropical wet forest in Colón. Reported from premontane rain forest in Costa Rica (Holdridge et al., 1971).

See Fig. 371.

PSEUDOBOMBAX Dug.

Pseudobombax septenatum (Jacq.) Dug., Caldasia
2:65. 1943
Bombax barrigon (Seem.) Dec.
Barrigon, Ceibo barrigon

Deciduous tree, to 25 m tall; trunk to 140 cm dbh; outer bark thin, unarmed, usually with smooth, green, striped areas interspersed with corky areas; inner bark thick, mottled reddish and white; branchlets whitish-pruinose, generally stout and more than 1 cm diam at apex. Leaves palmately compound, glabrous; stipules broadly ovate, to 12 mm long, caducous; petioles 10–56(68) cm long; leaflets subsessile or on petiolules to 1 cm long, inarticulate, obovate, short-acuminate, narrowed to an acute base and decurrent on petiolule, 7–29 cm long, 4–14 cm wide, sometimes sparsely lepidote below, entire. Inflorescences cymose, toward ends of branchlets; flowers 1–3, precocious or not, 7–10 cm long; pedicels 2–4 cm long; receptacle glandular below; calyx cup-shaped, truncate, 1 cm long and 2–3 cm wide; petals 5, leathery, ± white, ± linear, 7–9 cm long and ca 2 cm wide, curling below calyx; stamens more than 1,000, whitish, in a dense powder puff atop calyx, to 9 cm wide and 6 cm high, in 5 bifurcate clusters; column about 1 cm high; anthers horseshoe-shaped, dehiscing by straightening; style about as long as or somewhat surpassing stamens. Capsules ± ellipsoid, green, irregularly and longitudinally striped with purplish-brown, to 18 cm long, about half as broad, filled with grayish kapok; seeds numerous, ca 5 mm long. *Croat 8191.*

Common throughout the forest. Flowers mostly from January to March. The fruits mature from February to April. Leaves fall in January or February and grow in again in May.

Nicaragua to Brazil and Peru. In Panama, a typical component of tropical moist forest in the Canal Zone, Panamá, and Darién (Tosi, 1971); known also from tropical dry forest in Panamá (Taboga Island), from premontane moist forest in the Canal Zone and Panamá, and from tropical wet forest in Colón.

See Figs. 372 and 373.

QUARARIBEA Aubl.

Quararibea asterolepis Pitt., Feddes Repert.
13:316. 1914
Molenillo, Garrocho, Panula, Cinco dedos, Guayabillo

Tree, to 30 m tall; trunk to 75 cm dbh, buttressed to 2 m; outer bark very thin, peeling, the sap with characteristic spicy aroma; branches drooping. Leaves simple; stipules lanceolate, 5 mm long, caducous; petioles mostly 1–2 cm long, densely fimbriate-lepidote; blades ± elliptic, obtuse to acuminate, obtuse to rounded at base, 8–15(25) cm long, 2.5–8(10) cm wide, glabrous but with dense layer of stellate scales below, sparser above, entire; venation pinnate. Flowers solitary, opposite leaves, mostly on young branchlets but sometimes nearly cauliflorous, ca 3 cm long; pedicels 1–1.5 cm long, bracteolate; calyx ± conical, to 1.5 cm long and 7 mm wide in flower, the lobes 4, ca 3 mm long, wingless; both calyx and pedicel green, densely fimbriate-lepidote; petals 5, oblong, to 2.5 cm long and 5 mm wide, whitish, with short stellate trichomes; stamens joined into a column for most of their length, the column ca 2 mm long, enlarged at apex, lobed, the lobes 5, ca 2 cm long at apex, not obviously separate; anthers 8 per column lobe, sessile, extrorse, longitudinally dehiscent; style surpassing staminal column by about 1 mm, flaring out slightly at summit, appearing several-lobed due to extension of ribs found on style but actually bilobed. Nuts ca 2 cm diam, ± ovate, buff-colored, subtended by and at least one-third enclosed by enlarged calyx, densely short-lepidote, with prominent mammillate apex 5 mm long; seeds more than 1 cm long, 1 (rarely 2) fertile. *Croat 5934, 6719.*

Common to locally abundant. Flowers in June and July. The fruits mature from August to November. Leaves fall around February.

The aroma of *Quararibea* trees can be detected from some distance. Even herbarium specimens retain the aroma for years.

Costa Rica and Panama. In Panama, known from tropical moist forest in the Canal Zone, Bocas del Toro, Chiriquí, Panamá, and Darién and from premontane wet forest in Coclé.

See Figs. 374 and 375.

Quararibea pterocalyx Hemsl., Diag. Pl. Mex. 4. 1878
Wild palm

Tree, usually less than 10 m tall; twigs with a strong spicy odor. Leaves simple; stipules lanceolate, to 5 mm long,

Fig. 374. *Quararibea asterolepis*

Fig. 375. *Quararibea asterolepis*

Fig. 376. *Quararibea pterocalyx*

caducous; petioles densely tomentulose, less so in age, 1.5–2.5 cm long; blades broadly elliptic, rounded to acuminate at apex, rounded at base, 13–30 cm long, 7–13 cm wide, glabrous above or tomentulose on veins, densely tomentulose below with whitish stellate trichomes, entire, venation pinnate. Flowers solitary, mostly opposite leaves on young branchlets, 12 cm long; pedicels 1–1.5 cm long, bracteolate; calyx tubular, to 4 cm long and ca 1 cm wide when in flower, the lobes 3, to 4 mm long; both calyx and pedicel generally yellowish, short-tomentulose, with 10 longitudinal wings at least 1 mm high; petals 5, linear, at least 10 cm long when fully expanded and 1–1.5 cm wide, white, with trichomes in scattered groups; staminal column 8–13 cm long, enlarged at its apex, with numerous anthers; anthers subsessile, extrorse, longitudinally dehiscent; style about as long as staminal column, slightly enlarged at apex. Nuts subglobose, ca 4 cm diam, greenish, subtended by and at least two-thirds enclosed by enlarged calyx, tomentulose, with a prominent mammillate apex, generally 2-celled; seeds 1 per cell, more than 2 cm long. *Croat 6828, Foster 1031.*

Occasional along the shore and in the forest on Pearson Peninsula. Flowers from May through July. The fruits mature in September and October. Leaves may be replaced twice per year (R. Foster, pers. comm.).

Fruits are dispersed chiefly by mammals. They are buoyant and may be dispersed by water as well.

Known only from Panama, from tropical moist forest in the Canal Zone, Bocas del Toro, Panamá, and Darién, from premontane moist forest in Panamá, and from premontane wet forest in Darién.

See Fig. 376.

87. STERCULIACEAE

Trees or shrubs (somewhat herbaceous in *Melochia*, scandent with armed stems in *Byttneria*), usually stellate-pubescent. Leaves alternate, petiolate; blades simple and entire or 3–5-lobed (palmately compound in *Herrania*); venation basally palmate; stipules present. Flowers bisexual (or unisexual and functionally monoecious in *Sterculia*), essentially actinomorphic, in axillary umbels and cymes; calyx mostly 5-lobed (trilobate or sepals 3 in *Guazuma, Herrania*); petals 5, free, showy, sometimes appendaged (lacking in *Sterculia*); stamens 5 or 15, united basally; anthers 2-celled, dehiscing longitudinally; ovary superior, 4- or 5-locular and 4- or 5-carpellate; placentation axile; ovules 2 to several per locule, anatropous; styles 5, or style 1 and simple or 2–5-lobed. Fruits leathery or woody berries, or the carpels separating as dehiscent dry cocci; seeds with copious endosperm.

The herbaceous species are often confused with the Malvaceae (85) because the flowers are similar and the leaves are palmately veined at the base and usually toothed. The Sterculiaceae may be distinguished by the monadelphous stamens of two cells, differing from those of the Malvaceae, which have a single cell. In addition, several genera have conspicuously appendaged petals.

Most species are probably insect pollinated, but *Sterculia apetala* and *Herrania purpurea*, with easily accessible purplish flowers, are possibly fly pollinated. G. Frankie (pers. comm.) says flies are common visitors to *Sterculia apetala* flowers in Guanacaste, Costa Rica, and he has seen staphylinid beetles in flowers of *Herrania purpurea* in La Selva, Costa Rica.

Many species have seeds that are clearly eaten by animals, including *Theobroma cacao, Herrania purpurea, Guazuma ulmifolia,* and *Sterculia apetala.* White-faced monkeys eat *S. apetala,* sometimes swallowing the seed but usually eating off the two outer layers of tissue surrounding the seed (Oppenheimer, 1968). Van der Pijl (1968) reported that rodents disperse the seeds of *Theobroma cacao* by eating only the fleshy sarcotesta pulp around the seed, leaving the seed unharmed. Bats also disperse *Guazuma* fruits (Bonaccorso, 1975).

About 70 genera and 1,000 species; mostly tropical.

KEY TO THE SPECIES OF STERCULIACEAE

Plants coarsely armed, vinelike . . . . *Byttneria aculeata* Jacq.
Plants unarmed:
  Leaves digitately compound, the leaflets 5, borne on long petiolules . . . .
  . . . . *Herrania purpurea* (Pitt.) R. E. Schult.
  Leaves not digitately compound:
    Leaves deeply 3–5-lobed; flowers large, ca 2 cm wide . . . . *Sterculia apetala* (Jacq.) Karst.
    Leaves unlobed:
      Blades more than 15 cm long, the petioles pulvinate at both ends . . . . *Theobroma cacao* L.
      Blades less than 15 cm long, the petioles not pulvinate:
        Stems hollow:
          Calyx more than 3 mm long, becoming 8 mm long, hiding the fruit from the side;
          bracts ca 2 mm long . . . . *Melochia lupulina* Sw.
          Calyx less than 3 mm long, inconspicuous behind bracts 4–5 mm long . . . .
          . . . . *Melochia melissifolia* Benth.
        Stems not hollow:
          Blades becoming oblique at base; inflorescences loose, essentially all flowers visible
          . . . . *Guazuma ulmifolia* Lam.
          Blades not oblique at base; inflorescences compact, some of the flowers partly hidden
          . . . . *Waltheria glomerata* Presl

BYTTNERIA Loefl.

Byttneria aculeata Jacq., Select. Stirp. Am. 76. 1763
Espino hueco, Zarza, Rabo de iguana, Rangay

Woody vine or scandent shrub; older trunk often breaking into 5 distinct, terete stems; stems 5-angulate and hollow; stems, petioles, and often midribs below with small, sharp, recurved prickles, the stems and petioles sometimes short-pilose on one side. Stipules triangular, 1–2 mm long, caducous; petioles 0.5–2 cm long (to 5 cm on juveniles); blades variable, usually lanceolate to elliptic, acuminate, obtuse to rounded, truncate, or subcordate, usually 4–11 cm long and 1.5–4.5 cm wide (juveniles to 19 cm long), entire to remotely serrate, glabrous or sparsely short-pilose above, often blotched with silver; palmate veins at base 3–5. Umbels axillary; peduncles and pedicels sparsely pubescent, slender; peduncles mostly 5–10 mm long; pedicels variable on a single inflorescence, 1–10 mm long; flowers few, 5-parted, white, yellow or greenish; calyx 5-lobed to near base, to 4 mm long, glabrous; petals linear-spatulate, ca 4 mm long but only 1 mm wide, yellowish; stamens 5, the staminal tube minute; anthers opposite petals; staminodia 5, alternate with petals; pistil minute; style very short; stigma inconspicuously 5-lobed (breaking into 5 cocci). Capsules globular, 1.5–2.5 cm diam (including spines), 5-lobed; seeds 5, reniform, ca 4 mm long, 2 mm wide, falling to display the 5-lobed receptacle. *Croat 6767, Wetmore & Abbe 8.*

Occasional, in the forest. Seedlings, which are recognized by having leaf blades with a light medial discoloration, are often locally abundant in clearings. Flowers mostly in October and November. The fruits mature from December to February.

Mexico to Bolivia; introduced into Polynesia. In Panama, known from tropical moist forest in the Canal Zone, Bocas del Toro, Veraguas, Panamá, and Darién, from tropical dry forest in Coclé and Panamá, and from premontane moist forest in the Canal Zone.

See Fig. 377.

GUAZUMA P. Mill.

Guazuma ulmifolia Lam., Encycl. Méth. Bot. 3:52. 1789
Cabeza de negrito, Guácimo, Guácimo de ternero, Bastard cedar, West Indian elm

Tree, generally 6–25 m tall; trunk to 60 cm dbh; stems, petioles, lower leaf surfaces, and axes of inflorescences densely stellate-tomentose. Stipules minute, caducous; petioles 5–20 mm long; blades lanceolate, acuminate, rounded to cordate and oblique at base, 6–16 cm long, 2–6 cm wide, glabrate or more often stellate-puberulent on both surfaces, the margins irregularly and finely toothed; palmate veins at base 3–7. Flowers in short, loose, axillary, thyrsiform clusters; pedicels 3–6 mm long; sepals 3, yellow, unequal, stellate-puberulent outside, recurved at anthesis; petals 5, obovate, ca 2.5 mm high, yellowish-green, hooded and bifid at apex, fused to free part of filament, the claw red with several red lines ex-

tending up petals, the appendage of the petal deeply bifid, ca 4 mm long; stamens 15, in 5 groups of 3 each, opposite petals, the staminal tube minute, red; pistil minute; staminodia 5, triangular, alternate with petals. Fruits globose to oval, 2–4 cm long, indehiscent or opening partway along 4 or 5, ± regular, longitudinal fissures, covered with many stout brown tubercles, the tubercles separating at maturity to expose the irregularly porate white surface; seeds numerous, irregularly ovoid, 2.5–3.5 mm long, borne among the fibrous inner parts, apparently falling through the holes in the exocarp, the testa maculate, covered by a membrane, the membrane very thin, mucilaginous and very sticky when wet. *Croat 6780, 8484.*

Uncommon, in the forest and at the edge of the Laboratory Clearing. Flowers mostly from March to May and from September to November. The fruits have been seen from March to May and in September.

Though fruits are never completely dehiscent, some seeds may fall through the fissures in the fruit that often form even while the fruits are still on the tree. The fruit has a sweet pericarp and is dispersed principally by bats (R. Foster, pers. comm.).

Mexico to Paraguay; West Indies; introduced into Asia and western tropical Africa. In Panama, widespread and ecologically variable, common in secondary areas and typical of tropical moist forest in Panama (Tosi, 1971); known from tropical moist forest throughout the country and also from premontane dry forest in Los Santos and Coclé, from tropical dry forest in Coclé and Panamá, from premontane moist forest in the Canal Zone and Panamá, and from premontane wet forest in Coclé and Panamá.

HERRANIA Goudot

Herrania purpurea (Pitt.) R. E. Schult., Caldasia 2:333. 1944
Theobroma purpurea Pitt.
Cacao cimarrón, Chocolatillo, Wild cacao

Shrub or small tree, to 5.5 (10) m tall; young stems and petioles densely stellate-pubescent. Leaves digitately compound, deciduous, to 80 cm long; stipules linear, to 5 cm long, caducous; petioles stout, to 40 cm long; petiolules densely pubescent, 3–5 mm long; leaflets 5, narrowly obovate, acuminate, cuneate at base, stellate-pubescent on veins above and below and sparsely on surface below, entire or obscurely crenate, the terminal leaflet to 53 cm long and 19 cm wide, the lateral leaflets somewhat smaller; lateral veins in 10–15 pairs, usually ending in minute teeth on margins. Flowers borne on trunk, globular in bud, all parts violet-purple to maroon, to 1.5 cm wide; calyx deeply trilobate, densely stellate-pubescent, the lobes broadly ovate, longer than petals or staminodia; petals 5, thick and deeply cucullate, papillate, the apex recurved, bearing a slender appendage ca 1 cm long, the inner surface with stout ridges; stamens (10)15, in 5 groups of (2)3 each, the staminal tube short, subtended by a densely pubescent disk; filaments short and stout, (2)3 held within the cucullate petal; style ca 1.5 mm long, tubular, the upper edge sharply 5-lobed, glabrous; ovary

10-ribbed, densely stellate-pubescent with stinging trichomes; staminodia papillate, obovate, recurved over and obscuring petals (except for the erect appendage). Fruits orange at maturity, to 7.5 cm long and 4 cm wide, the ribs and pubescence of ovary persisting; seeds many, to about 1.5 cm long, each surrounded by a pulpy, white, sweet mesocarp. *Croat 6791, 9281.*

Occasional in both the young and old forests, locally common. Flowers mostly from December to February. The fruits mature mostly in the early rainy season (April to May), but have been seen in August and September.

The flower structure is strange. While the style is sunken between the staminodia, it is quite accessible. The anthers are concealed so well, however, that it would be interesting to see what organism effects pollination.

Costa Rica and Panama. In Panama, a characteristic tree species of tropical moist forest (Holdridge & Budowski, 1956); known from tropical moist forest in the Canal Zone, Bocas del Toro, San Blas, and Darién, from premontane wet forest in Chiriquí, and from tropical wet forest in Colón and Darién.

See Fig. 378.

MELOCHIA L.

Melochia lupulina Sw., Prodr. Veg. Ind. Occ. 97. 1788

Shrub, herbaceous when young, mostly erect, to 2 m high; stems hollow; stems and petioles villous. Leaves simple; stipules lanceolate-linear, to 4 mm long, caducous; petioles slender, 1.5–5 cm long; blades ovate, acuminate, rounded to cordate at base, 3.5–8.5(10) cm long, 2.5–6.5(7) cm wide, softly velutinous especially below, crenate-serrate; palmate veins at base 5–7. Cymes small, dense, axillary; peduncles, pedicels, bracts, and calyces stellate-puberulent; peduncles short, ca 3 mm long; pedicels slender, 2–4 mm long; bracts linear, ca 2 mm long; calyx campanulate, ca 4 mm long, enlarging to ca 8 mm, 5-lobed to near middle, glabrous inside; petals 5, clawed, ca 4.5 cm long, white; stamens 5, opposite petals; staminal tube and style heteromorphous; styles 5, free, 1.7–2.8 mm long; staminodium 1.6–2.5 mm long. Fruits capsules, mostly hidden by enlarged calyx, ca 2 mm long and to 3 mm diam, pale, opening by 5 slits to expose seed; seed 1, ca 1 mm diam. *Croat 7471.*

Uncommon along the shore on exposed banks, occasional in Rear #8 Lighthouse Clearing. Flowers mainly from December to April. The fruits mature mostly from January to April.

Scattered from Mexico to Peru; Jamaica. In Panama, known from tropical moist forest in the Canal Zone, Bocas del Toro, Chiriquí, Coclé, Panamá, and Darién, from premontane moist forest in the Canal Zone, and from premontane wet forest in Chiriquí (Boquete) and Coclé.

Melochia melissifolia Benth., J. Bot. (Hooker) 4:129. 1842

Herb or suffrutex, to 1 m tall; stems hollow, with both fine stellate trichomes and simple, red, gland-tipped trichomes. Leaves simple, hispidulous; stipules linear; blades ± lanceolate-ovate, acute, ± rounded at base, to 3(4.5) cm long and 2(2.5) cm wide, serrate-crenate. Glomerules axillary, subsessile; bracts subtending flowers prominent, linear, to 5 mm long; calyx persistent, ca 2 mm long, with 5 acute lobes; petals 5, pinkish; ± obovate, ca 2(3.4) mm long, slender at base; stamens 5, united into tube at base, equaling styles; styles 5, free to the base; ovary sericeous. Capsules subglobose, to 3 mm diam, ca 2 mm high, hispidulous, splitting into 4 or 5 cocci, each coccus eventually splitting into 2 valves; seeds 1 or 2 per coccus, shaped like orange wedges, ca 1.5 mm long, black, the surface weakly ribbed. *Croat 7458.*

Locally abundant in the Rear #8 Lighthouse Clearing along the path. Flowers and fruits during the dry season, especially in December and January.

Fruit dispersal method is uncertain. Seeds are perhaps thrown to some extent when capsules open. They are neither viscid nor armed.

Costa Rica to the Guianas; Cuba; tropical Africa. In Panama, known from tropical moist forest in the Canal Zone, Colón, and Panamá, from tropical dry forest in Coclé, and from premontane moist forest in Panamá.

STERCULIA L.

Sterculia apetala (Jacq.) Karst., Fl. Columb. 2:35, pl. 118. 1861

Panama, Panama wood

Functionally monoecious tree, to 40 m tall, to 1 m dbh; outer bark hard, thin, light brown, sandpapery, the lenticels many, closely spaced, raised; inner bark thick, tan, granular; sap with strong, pungent odor. Leaves deciduous; stipules ovate, 5–8 mm long, axillary, caducous; petioles to 25 cm long, densely to sparsely stellate-pubescent; blades deeply and palmately 3–5-lobed, rounded to acute at apex of lobes, cordate at base, to 35 cm long and 45 cm wide, glabrous above except on veins especially near base, softly stellate-arachnoid below to glabrate in age; basal veins usually 5–7, each lobe with 3–7 pinnate veins. Panicles or racemes axillary or subterminal; flowers bisexual or functionally staminate, predominately staminate, with a strong spicy odor; calyx bowl-shaped, greenish, ca 2 cm broad and 1 cm deep, flexible, coriaceous, densely covered outside with short, violet-purple, stellate trichomes, striate and glabrate inside, the lobes acute, recurved; petals lacking; gynophore of staminate flowers hook-shaped, shorter than rim, sparsely glandular-puberulent, also hispidulous at apex; stamens 15, sessile on a raised disk at apex of gynophore; anthers extrorse, dehiscing in bud; pollen ± tacky; pistil rudimentary, sunken deeply at base of staminal disk, produced above into a slender style, gynophore of pistillate flowers similar and shorter, the staminal disk reduced, the nonfunctional stamens much shorter than the pistil; pistil stellate-tomentulose, 5-carpellate, 5-sulcate, the carpels separating at maturity; style solitary, stout, bent to one side; stigma capitate. Follicles obovoid, to 8 cm long and 5 cm wide, short-tomentose outside, with dense,

Fig. 377. *Byttneria aculeata*

Fig. 378.
Herrania purpurea

Fig. 379. *Sterculia apetala*

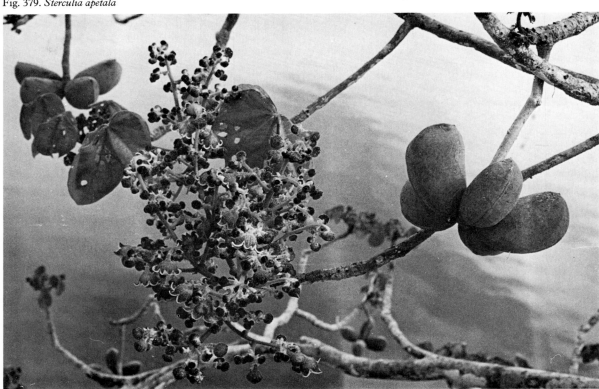

erect, orangish, stinging trichomes inside, opening very widely along the ventral suture, the periderm woody, 5–7 mm thick; seeds 2–4, oblong-ellipsoid, ca 2 cm long, 1.2 cm diam, covered with trichomes similar to inside of pod, then with a thin, tough, chestnut-brown layer, and then with a thin white layer. *Croat 13244.*

Frequent in the forest. The plant loses its leaves at the beginning of the dry season. After a short span of leaflessness, new leaves are produced. Flowering occurs usually every 2 years and the flowers appear shortly before or shortly after the onset of new leaves, usually in February. The fruits mature about 1 year later, mostly from January to April. Elsewhere in Panama the species may flower twice a year, in November or December and again in March.

All fruits on an individual tree dehisce within about 1 month. Those observed by J. Derr (pers. comm.) in Guanacaste, Costa Rica, fell to the ground within 1 week. The seeds are heavily parasitized by cotton stainer bugs (*Dysdercus bimaculatus*), and any seeds falling beneath the tree are killed by these bugs. Pods that are removed some distance by monkeys, agoutis, or other animals usually have some uneaten seeds discarded because the frugivore becomes distracted by the irritating trichomes lining the pod (Janzen, 1972). In Guanacaste fruit pods are opened by parrots and orange-chinned parakeets and by magpie jays which take the seeds after the pods have opened (J. Derr, pers. comm.).

Mexico to Peru and Brazil; West Indies. In Panama, a typical component of tropical dry forest (Holdridge & Budowski, 1956) and of tropical moist forest (Tosi, 1971); known from tropical moist forest in the Canal Zone, San Blas, Panamá, and Darién, from tropical dry forest in Coclé, from premontane moist forest in the Canal Zone and Panamá, and from premontane wet forest in Chiriquí. Reported from tropical wet forest in Costa Rica (Holdridge et al., 1971).

See Fig. 379.

THEOBROMA L.

Theobroma cacao L., Sp. Pl. 782. 1753

Cacao

Tree, usually 4–8(12) m tall, usually branched from near base and often widely spreading; trunk and larger branches with the persistent bases of the inflorescences; young branchlets and petioles pubescent. Leaves alternate; stipules subulate, deciduous; petioles variable in length, with 2 pulvini; blades ± oblong, acuminate, rounded or obtuse at base, 15–50 cm long, 4–15 cm wide, sparsely stellate-tomentose above and below when young or on veins above, the lower surface a lighter green than the upper; palmate veins at base 3–5. Cymes minute, borne on small tubercles on branches below leaves and on trunk; peduncles and pedicels pubescent; flowers 5-parted, appearing fasciculate; sepals ± lanceolate, 5–8 mm long in flower, acuminate, variously colored; petals white, 3–4 mm long, hooded, the hood enclosing apex of stamens, with 3 maroon veins inside, the lateral veins stronger, the blade of the petal ± obovate, usually

obliquely mucronate, the claw S-shaped; fertile filaments 5, alternating with staminodia, reflexed, to 2 mm long; anthers 2 per filament (thus the thecae 4); styles appearing single (actually 5 coherent), surpassing fertile filaments but not staminodia; staminodia 5, erect, narrow and acute, ca 0.5 mm long, ± maroon. Fruits usually oblong-ellipsoid, 10–25 cm long, yellow to brown at maturity; exocarp thick, fleshy, 10-ribbed, densely stellate-pubescent; seeds 20–40 in a sweet pulp, roundish in cross section, 2–4 cm long and 1.2–2 cm wide, brown. *Croat 8131, 12311.*

Infrequent, in the forest. Some flowers have been seen in February and September. Time of fruit maturity is not known.

Native to the New World tropics; cultivated in the Old World tropics. In Panama, known from tropical moist forest in the Canal Zone, Bocas del Toro, Herrera, Panamá, and Darién and from premontane moist forest in Panamá.

WALTHERIA L.

Waltheria glomerata Presl, Rel. Haenk. 2:152. 1835

Slender shrub, 1.5–2 m high; most parts stellate-pubescent. Leaves alternate; stipules lanceolate, ca 9 mm long; petioles 5–10 mm long; blades ± elliptic, acute at apex, obtuse to subcordate at base, 4–13 cm long, 1.5–8 cm wide, irregularly serrate; reticulate veins obvious. Glomerules dense, axillary, globose; peduncles short; flowers 5-parted, sessile, subtended by several linear-obovate bracts to 6 mm long; calyx tubular, to 6 mm long, the lobes triangular-oval, ca 2 mm long, tomentulose; petals narrowly spatulate, 6 mm long, creamy white to somewhat yellowish; stamens 5, united into a tube for only 1 mm; filaments free for ca 4 mm; anthers at about the same height as calyx and corolla; style ca 3 mm long, simple. Fruits capsules, 2-carpellate, ovate, apiculate, 1-celled, 1 mm long and 0.5 mm wide, densely plumose-papillate (the trichomes to 1 mm long); seeds 2, ca 0.5 mm long, attached to septa at base of ovary. *Croat 13477.*

Rare, known only from the shore on the north edge of Orchid Island, but to be expected in clearings and other disturbed areas. Flowers mostly from December to February. Mature fruits have been seen in March.

Mexico to Panama. In Panama, widespread and ecologically variable; known from tropical moist forest in the Canal Zone (both slopes), Veraguas, Los Santos, Herrera, Panamá, and Darién, from tropical dry forest in Coclé, Herrera, and Panamá, from premontane moist forest in the Canal Zone and Panamá, from premontane wet forest in Chiriquí, Coclé, and Panamá, and from tropical wet forest in Panamá.

88. DILLENIACEAE

Lianas or rarely scandent shrubs (*Tetracera*) or trees (*Saurauia*); bark usually thin, flaking. Leaves alternate, petiolate; blades simple, entire or serrate, the pubescence stellate (*Saurauia*), simple (*Doliocarpus* and *Davilla*),

KEY TO THE TAXA OF DILLENIACEAE

Plants trees, rare; petals connate at base, falling as a unit with stamens; fruits baccate; seeds many, small, reticulate, lacking an aril . . . . *Saurauia laevigata* Tr. & Planch.

Plants lianas; petals free, caducous; fruits dehiscent (fruitlike structure in *Davilla*); seeds few, moderately large, smooth, arillate:

Leaves markedly asperous, pustular and stellate-pubescent above; flowers unisexual; fruits pyriform follicles, gradually tapered to a beaked apex, the pericarp indurate, not at all fleshy; aril thin, lacerate, split on one side, red (usually drying pale yellow) . . . . *Tetracera*

Leaves smooth to scabridulous, not markedly pustular and stellate-pubescent above; flowers bisexual; fruits berrylike, the apex rounded (often with a slender persistent style), the pericarp fleshy at maturity or fruits indehiscent and enclosed in 2 cuplike sepals; aril fleshy, entire, white and almost or entirely enveloping seed:

Sepals unequal, 2 enlarging, becoming orange and fitting together like petri dishes to enclose ovule and persisting stamens (and ultimately the fruit); entire fruitlike structure usually less than 6 mm diam . . . . *Davilla nitida* (Vahl) Kub.

Sepals nearly equal, not becoming orange, not enclosing ovule and stamens; fruits more than 6 mm diam . . . . *Doliocarpus*

or with both simple and stellate trichomes arising from circular pustules; venation pinnate, the secondary veins nearly parallel; stipules lacking. Panicles or glomerules terminal, axillary, or cauliflorous; flowers bisexual or unisexual (androdioecious in *Tetracera*, i.e., flowers either bisexual or staminate), many, actinomorphic, fragrant; sepals imbricate, usually 5 (2-6 in *Doliocarpus*, usually 4 in *Saurauia*); petals (1)2-5(6), free or connate, imbricate, white, caducous (except in *Saurauia*); stamens numerous, exserted; anthers 2-celled, with parallel or basally divergent thecae, dehiscing lengthwise or by apical pores, the connective expanded; pistils 1-5; ovaries superior, 1-locular, 1-carpellate; placentation parietal or ventral; ovules usually 2 (2-12 in *Tetracera*, indefinite in *Saurauia*); styles as many as pistils. Fruits follicular or baccate (or indehiscent); seeds 1-4 per carpel, arillate (except *Saurauia*), with copious fleshy endosperm.

The flowers are of the brush type with stamens and styles exserted. They are probably pollinated by pollen-feeding insects.

The fruits are best suited for bird dispersal. *Tetracera* and *Doliocarpus* usually have shiny black seeds enveloped in part by a red or white aril. *Doliocarpus* fruits are also eaten by white-faced, howler, and spider monkeys (Hladik & Hladik, 1969). Oppenheimer (1968) reported that the fruits of *D. major* may make up a major portion of the diet of white-faced monkeys during the early part of October. The fruits are eaten whole and may pass through the digestive tract intact. It is very likely that the fruits of *D. olivaceus*, which also matures when fruits are scarce, are taken by monkeys as well. It is interesting to note that both of these species of *Doliocarpus* have larger fruits, no doubt more attractive to larger animals, than do *D. dentatus* or *D. multiflorus*, which fruit principally in the dry season or early rainy season when food is not scarce.

Eleven genera and over 300 species; mostly in the subtropics and tropics.

Most BCI species can be distinguished by being lianas bearing flowers with numerous stamens and fruits with arillate seeds. Many have leaves that are conspicuously asperous.

DAVILLA Vand.

Davilla nitida (Vahl) Kub., Mitt. Bot. Staatssamml. München 7:95. 1971

D. multiflora (DC.) St.-Hil.

Liana; outer bark loose, thin, brown, often peeling. Petioles short, narrowly-winged; blades elliptic to obovate, acute to short-acuminate or rounded at apex with a short downturned apiculum, obtuse to attenuate and decurrent at base, 6–10 cm long, 2.5–6 cm wide, glabrate to minutely scabrid with longer trichomes on midrib above, the surface punctate with persistent bases of scabrid trichomes, glabrous or scabridulous below except sparsely hirsute on veins, the midrib often arched. Panicles terminal or upper-axillary; rachis often extended well beyond floriferous part; flowers many, pedicellate, fragrant; sepals 5, unequal (2 enlarged), ± maroon in bud, orange after anthesis, soon closing (perhaps only after pollination) to enclose persistent stamens and developing ovary; petals usually 5, obovate, ca 5 mm long, spreading to reflexed at anthesis, caducous; stamens numerous, ca 5 mm long, bright yellow, dehiscing in bud; carpel 1, globose; style bent in bud; stigma discoid, minutely papillate, held slightly above anthers soon after anthesis. Fruits small, rounded, black, minutely white-arillate, held within 2 orange dishlike sepals until maturity when the sepals open, indehiscent and dispersed by birds; seed 1. *Croat 14610.*

Abundant on trees on the shore, and frequent in the canopy of the forest within the island. Flowers in the early dry season, mostly from January to March; the flowers last for a very short time. The fruits may appear mature from February to August. Infected fruits may persist most of the year.

Standley mistakenly made BCI reports for both *Davilla rugosa* Poir., which is a distinct species from South America, and *D. kunthii* St.-Hil., which is a synonym of *D. aspera* (Aubl.) Benoist.

Pollen appears to be removed rather quickly from the anthers. Since it is somewhat tacky it is unlikely that

considerable amounts blow away. The flowers are ideally
suited to pollen feeders and are possibly pollinated by
them. The species has been seen visited by the bee *Trigona cupira.*

The fruits fall or are taken by birds mostly in the early
rainy season, beginning usually in May. The fruit crop
is often very highly infested by Curculionidae beetles,
which eat the seeds and then escape as soon as the sepals
open.

Southern Mexico to northern Brazil; the Antilles. In
Panama, known principally from tropical moist forest in
the Canal Zone, Colón, Veraguas, Coclé, and Panamá,
but known also from premontane wet forest in Chiriquí.

DOLIOCARPUS Rol.

Doliocarpus dentatus (Aubl.) Standl., J. Wash. Acad.
Sci. 15:286. 1925

Liana; younger parts appressed-pubescent with long
slender trichomes especially on stems, petioles, and major
veins of leaves, glabrate in age. Petioles 5–15 mm long,
narrowly winged; blades ± elliptic to obovate or oblanceolate, acuminate, acute to obtuse at base, 7–20 cm
long, 3–7 cm wide. Glomerules dense, at usually leafless
axils; pedicels 5–10 mm long in flower, to 25 mm long in
fruit; flowers 5–8 mm broad; sepals 3–5, obovate, to 5
mm long, slightly pubescent outside, glabrous to slightly
pubescent inside; petals 5, white, 4–5 mm long; stamens
numerous, sometimes persisting; ovary 1, glabrous; style
to 1.5 mm long; stigma peltate, often persisting. Fruits
globose, 6–9 mm diam, glabrous, red, dehiscing regularly
along median suture, the valves remaining intact; seeds
2, ca 5 mm long, enveloped except at apex by the fleshy
white aril. *Croat 5144, 9516.*

Frequent in the forest and at the edge of the lake; sometimes locally common at the top of the canopy. Flowers
throughout the dry season, from December to April, but
particularly in February. The fruits mature in central
Panama from March to May (possibly later), chiefly in
April. At least some individuals may flower more than
once per season, with individuals flowering in March
or April already bearing fruits from an earlier flowering.

The fruits were reported to be indehiscent by Hunter
in the *Flora of Panama* (1965). They are smaller and more
numerous than for any other species of *Doliocarpus.*

Mexico to Paraguay; Cuba, Trinidad, Tobago. In
Panama, known from tropical moist forest in the Canal
Zone, Panamá, and Darién, from tropical dry forest in
Coclé, from premontane moist forest in the Canal Zone
and Panamá, and from premontane wet forest in Coclé.

Doliocarpus major J. F. Gmel. in L., Syst. Nat. ed.
13, 2:805. 1791

Liana; bark thin and flaking; older stems often prominently angulate and winged, the younger stems ± flexuous, usually angulate. Leaves coriaceous; blades elliptic
to oblong-elliptic, abruptly acuminate and often downturned at apex, obtuse and decurrent at base onto markedly canaliculate petiole, 5–18 cm long, 2–8 cm wide,
± glabrous but with the midrib often pubescent, prominently punctate on drying, entire to sparsely serrate, the
midrib often arched; juvenile leaves gradually acuminate,
prominently serrate, to 32 cm long and 13 cm wide.
Flowers 2–3 cm broad, in axillary glomerules or less
frequently in 2-flowered cincinni; pedicels 5–12 mm long
(10–20 mm long in fruit); sepals 4 or 5, pubescent, to
7 mm long, inconspicuously sericeous inside, persisting
in fruit; petals 2 or 3, white or reddish in age, 1–1.5 cm
long; stamens numerous, exserted; ovary 1, pubescent;
style 6–8 mm long. Fruits red, subglobose, 10–13 mm
diam, densely puberulent, dehiscing regularly along
median suture, the style 1, often subapical; seeds 2, ca 8
mm long, shiny, black, surrounded except at apex by a
fleshy, white, sweet aril, the aril very astringent before
maturity. *Croat 6833, 12954.*

Common along the shore and high in trees in the forest.
Flowering mostly during June and July, rarely as late
as September or as early as May. The fruits mature in
the rainy season, mostly from July to November, perhaps
chiefly in October, rarely as late as the early dry season.
At least some individuals flower twice per season, and
they may bear mature fruits when they flower the second
time. Leaves usually fall before flowering occurs, dropping mostly at one time and growing in again soon.

This species was mistakenly reported by Standley as
D. brevipedicellatus Garcke, a species which occurs in
the Canal Zone but not on BCI and which is distinguished from *D. major* by having a glabrous ovary.

Nicaragua to Panama, and in South America from
Venezuela to the Guianas, Amazonian Brazil, and Peru.

KEY TO THE SPECIES OF DOLIOCARPUS

Ovary and fruit pubescent; style 4–8 mm long, sometimes exceeding sepals; sepals sericeous inside:
  Leaves punctate; pedicels 5–12 mm long; fruits usually less than 1.5 cm diam, the 2 valves re-
    maining intact and convex after dehiscence . . . . *D. major* J. F. Gmel.
  Leaves not punctate; pedicels 2–4 mm long; fruits usually more than 1.5 cm diam, the valves
    splitting and spreading widely after dehiscence . . . . *D. olivaceus* Standl.
Ovary and fruit glabrous to slightly pubescent; style 1–2 mm long, not exceeding sepals; sepals
  glabrous to slightly pubescent inside:
  Inflorescences glomerulate (flowers all arising from a single area) . . . . *D. dentatus* (Aubl.) Standl.
  Inflorescences in fascicles of few-flowered cincinni (at least some flowers arising from a branched
    axis) . . . . *D. multiflorus* Standl.

Fig. 380.
Doliocarpus major

Fig. 381. *Doliocarpus olivaceus*

Fig. 382.
Doliocarpus olivaceus

In Panama, known principally from tropical moist forest in the Canal Zone (both slopes), all along the Atlantic slope, and in Darién; known also from premontane wet forest in Colón and Darién.

See Fig. 380.

Doliocarpus multiflorus Standl., J. Wash. Acad. Sci. 15:285. 1925

Liana; all but the youngest stems with thin, flaky bark. Petioles 1–2.5 cm long, canaliculate; blades obovate to elliptic, abruptly acuminate, mostly acute at base, 9–25 cm long, 3.5–10 cm wide, glabrous above, at first appressed-pubescent on veins below, the axils often tufted, glabrate in age. Cincinni axillary or borne on leaf-less stems, fasciculate, the axes pubescent; flowers few, pedicellate, 7–9 mm broad; sepals 5, ± rounded and concave, glabrous to slightly pubescent on both surfaces; petals 3 or 4, white, 3–5 mm long, soon falling; stamens many, persistent, ca 5 mm long; ovary 1, glabrous to slightly pubescent; style 1–2 mm long; stigma peltate. Fruits globose, ca 1 cm diam, red or purplish-red, often with sparse, ± appressed trichomes, splitting regularly into 2 valves; seeds 2. *Croat 9507, 13488.*

Apparently uncommon in flower; seen both along the shore and in the vicinity of the Laboratory Clearing where sterile plants are frequent in the canopy. Seasonal behavior poorly known. Flowers from January to March (sometimes to April). The fruits probably mature mostly in April and May.

Sometimes confused with specimens of *D. major,* but may be distinguished by lacking punctate leaves and having glabrous fruits. This species was reported by Standley as *D. multiflorus* Standl., but that name was considered synonymous with *D. guianensis* (Aubl.) Gilg. by Hunter in the *Flora of Panama* (1965). Kubitzki (1971) considered *D. multiflorus* and *D. guianensis* as distinct species, however, with *D. guianensis* restricted to Venezuela and the Guianas. *D. guianensis* is distinguished by having a pubescent ovary.

Belize to Panama; Cuba. In Panama, known only from tropical moist forest in the Canal Zone (Atlantic slope) and Colón, though very likely to be found on the Pacific slope in Darién.

Doliocarpus olivaceus Sprague & L. O. Wms. ex Standl., Contr. U.S. Natl. Herb. 27:265. 1928

Liana; stems and leaves glabrate. Petioles 1–3 cm long, often narrowly winged, slightly enlarged at base; blades oblong-elliptic to oblong-obovate, acuminate, acute to obtuse at base, 8–22 cm long, 4–9 cm wide, subentire to obscurely dentate. Fascicles small, axillary, usually on defoliated stems; pedicels 4–10 mm long; sepals 3 or 4 (sometimes to 6), irregular, mostly rounded or ovate, concave, glabrous outside, appressed-pubescent inside; petals 2 or 3 (sometimes to 5), obovate, 10–15 mm long, 6–10 mm wide, pale yellow, soon falling; stamens 35–60, to 7 mm long, deciduous; anthers ca 1 mm long; ovary subglobose, densely pubescent with stiff straight tri-

chomes; style glabrous, to 7 mm long. Fruits globose, 1.5–2 cm diam, red at maturity, densely puberulent and with sparser hispid trichomes, dehiscing into 2 valves at maturity, the valves fleshy, flattening out, splitting irregularly at about the middle and spreading widely; seeds 2, mostly enveloped by a fleshy white aril. *Croat 6820, 14873.*

Occasional in the forest and at the edge of the lake; common in the canopy at least in the vicinity of the Laboratory Clearing. More abundant in the old forest than any other *Doliocarpus.* Flowers chiefly in May and June. The fruits mature from August to December, especially in September and October.

The manner of dehiscence of the fruit is probably unique among *Doliocarpus* in Panama. The fruits are larger than those of any other species of the genus in Panama.

Known only from Panama, from tropical moist forest in the Canal Zone and Panamá and from premontane moist forest in Panamá.

See Figs. 381 and 382.

SAURAUIA Willd.

Saurauia laevigata Tr. & Planch., Ann. Sci. Nat. Bot., sér. 4, 18:267. 1862
   S. zetekiana Standl.

Tree, 5–30 m tall; stems with prominent leaf scars; younger parts and axes of inflorescences sparsely to densely stellate-pubescent. Petioles 1–3 (9) cm long; blades ± elliptic to obovate, acute to abruptly acuminate, obtuse at base, 6–22 cm long, 2–10 cm wide, glabrous but with sparse, scalelike, stellate trichomes often on veins, the margins serrate to serrulate, the midrib usually arched, the lateral margins held somewhat erect. Inflorescences thyrsiform, from upper axils, to 20 cm long; pedicels 1–4 (6) mm long; flowers white, sweetly aromatic, usually 4-parted, rarely 3- or 5-parted; sepals ± rounded, ciliate; petals ± oblong, ca 5 mm long, rounded to emarginate at apex, spreading at anthesis, connate at base; stamens 20–30, to 4 mm long, weakly fused to base of petals and borne among villous tufts; anthers yellow, to 1.7 mm long, the thecae divergent, dehiscing chiefly at apex; ovary ± obovoid, the locules equaling the petals, ca 1.3 mm long; styles usually 4, ca 1 mm long. Berries globose, ca 1 cm diam, usually (3)4(5)-sulcate, glabrous, white and fleshy at maturity; seeds many, ca 1 mm long, markedly reticulate to alveolate. *Bangham 578, Salvoza 998.*

Apparently rare if still present on the island; not seen in recent years. Uncommon in adjacent areas of the Canal Zone. Flowers in July and August (elsewhere in Panama rarely as early as May or as late as September). The fruits mature by September or October.

Southern Mexico to Colombia. In Panama, mainly in upland areas but, according to Hunter in the *Flora of Panama* (1965), one of the few species capable of spreading across lowland barriers. In Panama, known from

tropical moist forest in the Canal Zone, from premontane wet forest in Coclé and Panamá, and from tropical wet forest in Coclé.

TETRACERA L.

Tetracera hydrophila Tr. & Planch., Ann. Sci. Nat. Bot., sér. 4, 17:20. 1862

T. ovalifolia sensu auct. non DC.

Androdioecious liana; trichomes of younger parts dense and stellate, interspersed with sparse and simple ones, the leaves and stems usually glabrate in age. Petioles 1.5–3.5 cm long, winged mostly in apical half; blades ± ovate to elliptic, rounded to abruptly short-acuminate at apex, rounded to obtuse at base, 4–17 cm long, 4–9 cm wide (to 28 cm long and 13 cm wide on juvenile leaves), chiefly stellate-pubescent except on veins, pustular, entire or becoming serrulate toward apex; lateral veins in fewer than 15 pairs. Flowers unisexual, 1–1.5 cm wide, pedicellate; sepals 5, obovate, to 5 mm long (longer in fruit), glabrous or slightly pubescent inside; petals 4 or 5, white, obovate, 7–9 mm long; stamens very numerous, 5 mm long, somewhat reduced and nonfunctional in pistillate flowers; carpels 2–5, pyriform; styles 1 per carpel, 1–2 mm long, held above stamens in pistillate flowers, reduced in staminate flowers; stigma simple. Follicles 2–5, glabrous, tawny, 6–8 mm long (excluding style), dehiscing along inner side; seeds 1–4, ± reniform, ca 4 mm long, shiny, black, enveloped by a lacerate red aril. *Croat 6575.*

Uncommon; collected only along the shore and along Lutz Creek. Seasonal behavior uncertain. On BCI no flowers have been collected. All collections in central Panama, however, indicate that the species may flower twice a year, once in the dry season, with the fruits maturing from July to October, and once in the late rainy season (October and November), with the fruits maturing in the middle of the dry season.

Although Standley reported this taxon as *T. sessiliflora* Tr. and Planch., a synonym of *T. portobellensis*, his description leaves no doubt that he was referring to *T. hydrophila*. *T. hydrophila* was mistakenly reported by Hunter in the *Flora of Panama* (1965) as *T. ovalifolia* DC., a synonym of *T. costata* subsp. *rotundifolia* (J. E. Smith) Kub., known only from the Guianas.

Belize to Panama, Colombia, and Ecuador. In Panama,

probably occurring in all lowland areas; known from tropical moist forest in the Canal Zone, Bocas del Toro, and Panamá (San José Island) and from premontane wet forest in Bocas del Toro.

Tetracera portobellensis Beurl., Kongl. Vetensk. Acad. Handl. 113. 1854

T. sessiliflora Tr. & Planch.

Androdioecious liana or scandent shrub; young stems, petioles, and axes of inflorescences with moderate to dense stellate pubescence interspersed with simple trichomes. Petioles 3–10 mm long, winged full length; blades elliptic to obovate, rounded to abruptly acuminate (rarely short-acuminate), acute to obtuse and decurrent on petiole at base, 5–18 cm long, 3–8 cm wide, the pubescence minute, pustular, stellate on both surfaces, the simple trichomes mostly restricted to veins; lateral veins in 15–25 pairs. Panicles spikelike; flowers unisexual, usually subsessile, 4–5 mm broad, closely congested; sepals 5; petals (2) 3 (4), pale yellow to greenish-white, rounded at apex, soon falling, the margin folded inward; stamens numerous, to ca 5 mm long, reduced in pistillate flowers; pistils reduced in staminate flowers; style usually directed to one side; stigma ± spatulate-truncate. Follicle 1, glabrous, shiny, dark brown, 5–7 mm long (excluding prominent beak), dehiscing on one side from style downward to expose seed; seeds 1–4, shiny, black, 3–4 mm long, enveloped at least in part by a deeply lacerate red aril (this becoming pale yellow on drying). *Croat 9375, 13997.*

Frequent along the shore and in the forest. Flowering may be induced throughout much of the year, but the chief flowering period is the early dry season (December to March, especially January and February). Flowering appears to be induced by drought. Severing a vine in the proper stage of bud will soon cause the flowers to open. The fruits mature mostly from March to May (sometimes from February).

This species was mistakenly reported by Standley (1933) as *T. oblongata* DC. Although Standley reported *T. sessiliflora* Tr. & Planch., a synonym of *T. portobellensis*, his description leaves no doubt that he was referring to *T. hydrophila* Tr. & Planch.

Mexico to Colombia. In Panama, growing primarily in the wet lowland areas of the Atlantic slope; known from tropical moist forest in the Canal Zone (Atlantic slope)

KEY TO THE SPECIES OF TETRACERA

Blades with usually more than 15 pairs of major lateral veins; follicle 1; petioles usually less than 1 cm long; flowers sessile or nearly so, in very dense glomerules . . . . *T. portobellensis* Beurl.
Blades with usually fewer than 15 pairs of major lateral veins; follicles 2–5; petioles usually longer than 1 cm; flowers usually pedicellate, the glomerules not very densely congested:
    Blades usually rounded to truncate at base, the upper surface usually moderately smooth; petioles winged mostly only to about middle; sepals sparsely pubescent or glabrous inside . . . . *T. hydrophila* Tr. & Planch.
    Blades usually obtuse to attenuate at base, the upper surface coarsely asperous; petioles winged usually to base; sepals densely pubescent inside . . . . *T. volubilis* L.

and Bocas del Toro, from tropical dry forest in Darién (Garachiné), and from premontane wet forest in Colón. See Fig. 383.

Tetracera volubilis L., Sp. Pl. 533. 1753

Chumico, Pasmo de sol

Androdioecious liana; bark on older stems thin and flaky; most parts with both stellate and simple trichomes. Petioles 5–20 mm long, winged to base; blades obovate to ± elliptic, rounded and abruptly acuminate to obtuse at apex, obtuse to attenuate at base, decurrent on petiole, 6–20 cm long, 3–9.5 cm wide, entire to serrate, the simple trichomes sparse except on veins of lower leaf surface, the pubescence of upper surface often deciduous except for pustular bases; lateral veins in 10–15(20) pairs. Flowers unisexual, 5–10 mm wide, pedicellate, rarely subsessile; sepals 5, densely pubescent inside; petals 3–5, obovate, to 5 mm long, white; stamens very numerous, about equaling petals, reduced in pistillate flowers; carpels 4 or 5, pyriform, usually pubescent with stiff white trichomes near apex, reduced in staminate flowers; styles 1 per carpel, ca 1.5 mm long. Follicles 4 or 5, glabrous to apically barbed, brown at maturity, 6–8 mm long, dehiscing along inner side; aril red (pale yellow on drying), at first enclosing seed; seeds 1–4, shiny, 3–4 mm long. *Croat 6569, 6802.*

Frequent along the shore and within the forest, sometimes at the top of the canopy. Flowers usually from August to November, especially in September and October, rarely as early as June. The fruits mature chiefly from January to March.

Mexico to Colombia, Venezuela, Amazonian Brazil, and Peru; Cuba and Jamaica. In Panama, known principally from tropical moist forest in the Canal Zone, Panamá, and Darién, but also from tropical dry forest in Coclé and from premontane moist forest in Panamá.

89. OCHNACEAE

Glabrous trees or shrubs. Leaves alternate, petiolate, sometimes clustered at apex of stems; blades simple, undulate or serrate; venation pinnate; stipules present. Panicles racemose, terminal; flowers bisexual, actinomorphic (or zygomorphic by disposition of sexual parts in *Cespedezia*), solitary or fasciculate; sepals 5, shortly united at base; petals 5, free, yellow; stamens 10 or many; anthers 2-celled, basifixed, dehiscing by oblique terminal pores; pistil 1; ovary superior, borne on conspicuous toral disk or gynophore, deeply lobed (5-locular and 5-carpellate with axile placentation in *Ouratea,* 1-locular and 5-carpellate with intrusive parietal placentation in *Cespedezia*); style 1; stigma undifferentiated or 5-radiate. Fruits septicidally dehiscent capsules with many winged seeds (*Cespedezia*) or 5 monocarpic drupelets of 1 seed on the swollen torus; seeds with or without endosperm.

The two BCI species have little in common superficially but are individually unique and not confused with any other species.

Flowers are well suited for insect pollination. Those of *Cespedezia* are probably pollinated by large bees.

Drupes of *Ouratea lucens* are well suited for ornithochory, though they may be dispersed in part by animals other than birds. Seeds of *Cespedezia macrophylla* are wind dispersed.

About 25–40 genera with 400–600 species; subtropics and tropics.

CESPEDEZIA Goudot

Cespedezia macrophylla Seem., Bot. Voy. Herald 97. 1853

Membrillo, Membrillo macho, John Crow wood

Tree, 5–20(25) m tall, to 50(70) dbh, often buttressed, with aerial roots. Leaves clustered at apex of branches; stipules densely imbricate, to 7 cm long, ca 1 cm wide, becoming weathered and subligneous; petioles to 1 cm long; blades linear-obovate, usually obtuse at apex, cuneate at base, 20–100 cm long, 15–25 cm wide, glabrous, coriaceous, the margins uneven; lateral veins prominent. Panicles terminal, large, exceeding uppermost leaves; flowers 5-parted, somewhat zygomorphic; sepals ca 5 mm long, thick, divergent at anthesis; petals yellow, obovate, concave, 2–3 cm long; stamens ca 80; filaments ca 1 cm long; anthers linear, ca 1 cm long, clustered opposite pistil; pistil stipitate, held to one side of flower, the stipe ca 6 mm long; ovary ca 1 cm long and 5 mm wide; style minute. Capsules narrowly ellipsoid, to 4(6.5) cm long and 1 cm diam, splitting longitudinally into 5 narrow segments; seeds many, small, with 2 long, narrow, lateral wings, the seminiferous area ca 2 mm long and 1 mm wide, the wings to 1 cm long, ca 1 mm wide at seed, tapering to a narrow point. *Croat 8163.*

Occasional, on the shore. Flowers principally in the late rainy season to the middle of the dry season (August to March). The fruits probably mature principally in the dry season.

The flowering plant is impressive for its huge yellow panicles.

Nicaragua to Colombia. In Panama, characteristic of

KEY TO THE SPECIES OF OCHNACEAE

Plants large trees; leaves to 1 m long, clustered at apex of branches; panicles to 1 m long; fruits
capsular, with many small winged seeds . . . . *Cespedezia macrophylla* Seem.
Plants shrubs or small trees; leaves less than 25 cm long, not clustered; panicles racemose, less than
20 cm long; fruits with 5 sessile drupelets attached to a red receptacle . . . .
. . . . *Ouratea lucens* (H.B.K.) Engler

Fig. 383. *Tetracera portobellensis*

Fig. 385.
Souroubea sympetala

Fig. 384. *Marcgravia nepenthoides*

tropical wet forest (Tosi, 1971); known also from tropical moist forest all along the Atlantic slope and from the Canal Zone, Panamá, and Darién, from tropical dry forest in Coclé (Penonomé), and from premontane wet forest in Coclé (El Valle).

OURATEA Aubl.

Ouratea lucens (H.B.K.) Engler in Mart., Fl. Brasil. 12(2):350. 1896
O. nitida (Sw.) Engler; *O. wrightii* (Van Tiegh.) Riley
Wild pigeon plum

Glabrous shrub or small tree, to 8 m tall. Stipules lanceolate, 4–6 mm long; petioles 5–12 mm long; blades variable, lanceolate to oblong-elliptic to rarely obovate, acuminate, nearly rounded to cuneate at base, 5–22 cm long, 2–7.5 cm wide, the margins serrate to serrulate; veins not raised except midrib below. Inflorescences racemose, terminal or upper-axillary, 2–14 cm long, bracteate especially at base, the bracts lanceolate, to 5 mm long; pedicels 5–10 mm long, articulate at base; flowers 5-parted in bud; sepals 5–7(9) mm long, ovate; petals obovate, to 9 mm long, soon falling; stamens 10, connivent; anthers subsessile, subulate, to 5 mm long; style ca 6 mm long, subpersistent. Fruits of 5 elliptic drupelets, black at maturity, to 10 mm long, sessile on a receptacle, the receptacle prominent, ± rounded or club-shaped, red; seeds 1 per drupelet. *Croat 8727.*

Frequent in the forest. Flowers from the late rainy season to the middle of the dry season (October to March). The fruits mature primarily in the dry season (December to April), though the receptacle may well persist until July.

Mexico to Panama. In Panama, ecologically variable; Bocas del Toro, Los Santos, Panamá, and Darién, from premontane wet forest in Chiriquí, Panamá, and Darién, from tropical wet forest in Darién, and from premontane rain forest in Panamá and Darién.

90. MARCGRAVIACEAE

Hemiepiphytic lianas. Leaves alternate, petiolate; blades simple, entire, glabrous, coriaceous, with inconspicuous hypophyllous glands; venation pinnate; stipules lacking. Racemes or umbels terminal, bracteate, with prominent nectaries; flowers bisexual, actinomorphic, often fragrant; sepals 4 or 5, free; corolla 5-lobed; stamens 5 (*Souroubea*) or many (*Marcgravia*); anthers 2-celled, dehiscing longitudinally; ovary superior, 5-locular (*Souroubea*) or about 10-locular (*Marcgravia*); ovules numerous; stigma sessile, 5-radiate. Fruits leathery, fleshy capsules; seeds many, with little or no endosperm.

Marcgraviaceae are recognized by being liana-like epiphytes with prominent nectaries subtending the flowers.

Marcgravia nepenthoides is probably pollinated by bats (Vogel, 1958). At least some species of *Marcgravia* are also pollinated by hummingbirds (Proctor & Yeo, 1973). The pollination system of *Souroubea sympetala* is unknown, but because of the red coloration of the inflorescence, I suspect hummingbirds.

Four or five genera and 100–125 species; New World tropics.

MARCGRAVIA L.

Marcgravia nepenthoides Seem., J. Bot. 8:245. 1870
Glabrous hemiepiphytic shrub or liana; trunk usually less than 6 cm diam, usually ascending a single tree, the ultimate branches slender, long-pendent, the smaller branches somewhat angulate. Petioles less than 1 cm long; blades obovate-oblong to elliptic-oblong, acuminate, rounded at base, 10–23(35) cm long, 4–7(9) cm wide, with a row of minute glands along margins below. Inflorescences terminal, umbelliform, pendent; pedicels spreading, to 3.5 cm long (to 4.5 cm in fruit), turned downward just before the apex; flowers 25–30, radially disposed from a ± globose rachis; bracteoles and sepals ± suborbicular; sepals 4, persistent in fruit, to 4 mm long and 6.5 mm wide, the inner 2 smaller; petals connate into a deciduous cap 8–10 mm long; stamens 25–30, pendent; filaments linear, flattened, 3–8 mm long; anthers ca 2 mm long; ovary subglobose, ca 3 mm diam, ca 10-locular; style short; nectaries 5 or 6, dark violet-purple, dipper-shaped, 6–9 cm long, the cup to 3.5 cm long and to 2.8 cm wide, somewhat constricted at apex, the stalk to 5.5 cm long. Capsules depressed-globose, apiculate, to 1.2 cm diam, brown; exocarp thin; mesocarp fleshy, red; seeds numerous, ca 1 mm long. *Croat 7033.*

Occasional, in the forest; juvenile plants ascending trees are more common. Flowers and fruits most abundantly in the dry season, although some flowers are seen as early as September. Plants are generally not found in the crown of a tree but rather on the trunk below the crown.

On the Atlantic slope from Belize to Panama. In Panama, known from tropical moist forest in the Canal Zone and Bocas del Toro.

See Fig. 384.

KEY TO THE SPECIES OF MARCGRAVIACEAE

Inflorescences umbelliform, with 25–30 radially disposed flowers subtended by 5 or 6 dipper-shaped nectaries; capsules less than 1.5 cm diam . . . . *Marcgravia nepenthoides* Seem.
Inflorescences racemose, the continuation of an unbranched stem; capsules more than 2 cm diam . . . . *Souroubea sympetala* Gilg.

SOUROUBEA Aubl.

Souroubea sympetala Gilg., Bot. Jahrb. Syst. 25 Beibl. 60:32. 1898

S. guianensis Aubl.

Hemiepiphytic, vinelike shrub, usually fastened to other vegetation with numerous, long, aerial roots; stems with papery-brown outer bark, often rooting at nodes. Petioles to 12 mm long; blades mostly obovate, rounded to obtuse at apex, acute at base, 8–14(16) cm long, 3.5–6(8) cm wide, thick, with 1 or 2 rows of minute glands within 1 cm of margins below. Inflorescences terminal, mostly 15–25 cm long, the continuation of an unbranched stem; pedicels 1–4 cm long, reduced near apex, densely and minutely ferruginous-pubescent; sepals 5, rounded, 2–3 mm long, very thick, imbricate, persisting; corolla greenish-yellow, to 14 mm wide, divided to basal third, broadly spreading or reflexed, rounded at apex, the lobes 5; stamens 5, alternate; anthers held above rim, 2–3 mm long, reddish; filaments about as long as thick; stigma sessile, 5-lobed, becoming soft and appearing to be receptive only after corolla has fallen, the lobes rupturing irregularly in radial pattern from center, becoming sticky; nectaries subtending flowers club-shaped, markedly auriculate, straddling pedicel, the auricles red, flattened, to ca 18 mm long. Capsules depressed-globose, 2–3 cm broad, the 5 thick valves deciduous at maturity, exposing a bright red-orange, fleshy mesocarp; seeds many, reddish-brown, sausage-shaped, 4–5 mm long, partly exposed from mesocarp. *Croat 7272, 7950.*

Occasional, on the shore and within the forest where the branches may be found in the top of the crown. Flowers from November to July, mostly from December to March. Fruit maturity time is uncertain.

In some cases the flowers never open; the apical part of the tube is removed as a calyptra before the lobes unfold.

Scattered from Belize to Peru. In Panama, known from tropical moist forest in the Canal Zone, Bocas del Toro, Colón, and Panamá and from tropical wet forest in Colón.

See Fig. 385.

91. THEACEAE

Glabrous shrubs or trees. Leaves alternate, spirally disposed, petiolate; blades simple, ± entire; venation pinnate, obscure; stipules lacking. Flowers bisexual, actinomorphic, solitary in axils, subtended by 2 sepaloid bracteoles; sepals 5, free, unequal, imbricate; petals 5, briefly connate at base, whitish; stamens many, free, in 2 series; anthers 2-celled, dehiscing longitudinally; ovary superior, 2-locular, probably 4–6-carpellate; placentation axile; ovules 4 or 5 per cell, anatropous; style 1; stigma punctiform. Fruits irregularly dehiscent capsules; seeds few, with scanty endosperm.

Members of the family are distinguished by the several whorls of numerous stamens adnate to the petals, the unequal sepals, and the irregularly dehiscent fruit with bright red seeds.

Flowers have numerous exserted stamens and an exserted style. Their pollination system is unknown.

The bright red seeds are probably dispersed by birds.

Thirty genera and 500 species; principally in the Old World tropics, but with many in the New World tropics and some in temperate areas.

TERNSTROEMIA Mutis ex L.f.

Ternstroemia tepezapote Schlechter & Cham., Linnaea 6:420. 1831

T. seemannii Tr. & Planch.

Manglillo, Manzanilla de sabana

Glabrous shrub or tree, 2–15(20) m tall, to 30 cm dbh; bark reddish-brown, thin. Leaves alternate, clustered at apex of branches, thick; petioles less than 1 cm long; blades oblong-elliptic to oblanceolate, bluntly acuminate, obtuse to rounded at base, 5–12 cm long, 2.5–5 cm wide; veins except midrib indistinct. Flowers 5-parted, solitary in axils, pleasantly aromatic (anise-like), on pedicels 2–5 cm long; calyx subtended by 2 sepaloid bracteoles to 4 mm long; sepals unequal, ± broadly ovate, ca 1 cm long, white to greenish-white; petals white, yellow at apex, ± equaling length of sepals, connate at base; stamens many, included; filaments united and adnate to corolla at base; anthers with a prolonged connective at apex; ovary conical, 2-celled; ovules 4 or 5 per cell; style slender, pointed. Fruits broadly ovoid, 1–2 cm wide, cream-colored at maturity, often with tinges of red, the outer wall thick, bursting irregularly at maturity; seeds few, ± ovoid-cylindrical, to ca 10 mm long and 5 mm wide at base, with a thin, bright-red, pulpy covering. *Croat 14963.*

Rare; collected along the shore of Orchid Island, along the northern shore of BCI, and in the forest on Zetek Trail. Flowers in the late dry season, in March and April; sometimes flowering a second time from late September to November. The fruits mature from July to September.

Mexico to Panama. In Panama, known from tropical moist forest in the Canal Zone, Bocas del Toro, Veraguas, Herrera, Panamá, and Darién, from tropical dry forest in Coclé, from premontane wet forest in Colón, Chiriquí, Coclé, and Panamá, and from tropical wet forest in Panamá and Darién.

See Fig. 386.

92. GUTTIFERAE (CLUSIACEAE)

Terrestrial or hemiepiphytic trees or shrubs, usually glabrous, with milky or colored sap. Leaves opposite, petiolate; blades simple, entire, usually coriaceous; venation pinnate, the lateral veins parallel; stipules lacking. Panicles, racemes, or umbellate cymes (*Symphonia*), terminal or axillary, bracteate; flowers mostly unisexual (dioecious or polygamodioecious), sometimes bisexual (*Symphonia, Marila, Vismia, Tovomitopsis nicaraguensis*), mostly fragrant, actinomorphic; sepals 2–5, free or basally connate, mostly subequal; petals 4–6 and free or lacking;

KEY TO THE TAXA OF GUTTIFERAE

Leaves rounded or emarginate at apex, usually obovate (except *Calophyllum*):
  Lateral veins sharply ascending at ca 45° angle or less to midrib; trees epiphytic; flowers pink to greenish:
    Flowers and fruits 5–9-parted; petals white to pinkish, more than 2 cm long; fruits more than 2.5 cm broad . . . . *Clusia odorata* Seem.
    Flowers and fruits 4-parted; petals green, less than 1 cm long; fruits less than 1.5 cm broad . . . . *Havetiopsis flexilis* Planch. & Tr.
  Lateral veins nearly perpendicular to midrib, branching at 60° angle or more; trees not epiphytic; flowers yellow or white:
    Leaves usually more than twice as long as broad, lacking crossveins between major lateral veins; pedicels less than 1 cm long; inflorescences usually more than 3 cm long; flowers yellow, less than 1 cm long; seed 1; trees common in the forest . . . . *Calophyllum longifolium* Willd.
    Leaves usually less than twice as long as broad, with prominent crossveins between major lateral veins; pedicels more than 1 cm long; inflorescences less than 3 cm long; flowers white; seeds 2–4; trees cultivated in the Laboratory Clearing . . . . *Mammea americana* L.
Leaves acute to acuminate at apex:
  Plants bearing dense, brown, stellate trichomes especially on the lower leaf surfaces and axes of inflorescences, the leaf blades pellucid- or opaque-punctate; petals densely woolly inside . . . . *Vismia*
  Plants glabrous, punctate or not; petals not woolly-pubescent inside:
    Flowers solitary or fasciculate:
      Bisexual flowers solitary or paired; style sessile, the conspicuous, broad, 5–7-lobed stigma persisting in fruit; tree cultivated in the Laboratory Clearing . . . . *Garcinia mangostana* L.
      Flowers fasciculate; style not sessile or at least not 5–7-lobed and persisting in fruit:
        Flowers red, more than 1 cm diam at anthesis; inflorescences terminal; leaves less than 8 cm long; tree usually more than 20 m tall . . . . *Symphonia globulifera* L.f.
        Flowers white, less than 1 cm diam at anthesis; inflorescences axillary; at least some leaves more than 8 cm long; tree usually less than 10 m tall . . . . *Rheedia*
    Flowers paniculate or racemose:
      Flowers in pendent racemes; fruits linear capsules, with many minute comose seeds; secondary lateral veins closely parallel, perpendicular to primary lateral veins and connecting them . . . . *Marila laxiflora* Rusby
      Flowers in panicles; fruits globular or obovate, the seeds not as above; secondary lateral veins not at all parallel, not perpendicular to or connecting primary lateral veins:
        Flower buds globular and less than 5 mm diam; fruits globular, the 5 styles reduced to black dots at the apex . . . . *Tovomitopsis nicaraguensis* Planch. & Tr.
        Flower buds oblong or globular and more than 5 mm diam; fruits obovate, the styles long, conspicuous, persistent . . . . *Tovomita*

stamens many (4 in *Havetiopsis*), sometimes fascicled, sometimes inserted on the disk or reduced to staminodia; anthers 2-celled, introrse, dehiscing longitudinally; ovary superior, 1- to many-locular, 3- to many-carpellate; placentation axile; ovules 1 to many per locule, anatropous; styles as many as the carpels, united or free; stigmas as many as carpels, often sessile. Fruits usually fleshy capsules, sometimes drupaceous or baccate, with several valves; seeds sometimes arillate, lacking endosperm.

Guttiferae are most easily recognized by the usually thick leaves frequently with closely parallel veins, the colored sap (usually yellow, orange, or red), and the frequently unisexual flowers.

Flowers are relatively primitive and open. *Symphonia globulifera* is pollinated by hummingbirds. In Costa Rica, *Clusia* flowers are visited by *Eulaema* (D. Janzen, pers. comm.). See additional comments under *Clusia* and *Symphonia*.

The fruits of several native species, including *Sym-* *phonia globulifera, Calophyllum longifolium,* and both species of *Rheedia,* are chiefly mammal dispersed. Fruits of *Rheedia acuminata* and *R. edulis* are taken by white-faced monkeys (Oppenheimer, 1968; Hladik & Hladik, 1969), which generally swallow the seeds. *Rheedia edulis* fruits are also taken by the bat *Artibeus jamaicensis* Leach in Mexico (Yazquez-Yanes et al., 1975). Species that have capsules with small arillate seeds or colorful fruit interiors, such as *Tovomitopsis* and species of *Tovomita* and *Clusia,* are probably chiefly bird dispersed. However, howler monkeys are reported to eat the fruits of *Clusia odorata* (Carpenter, 1934). *Vismia,* with indehiscent, baccate, many-seeded fruits, is probably also chiefly bird dispersed. *Marila* has wind-dispersed seeds. Fruits of *Calophyllum* are hydrochorous (Ridley, 1930) and are also taken by bats (Bonaccorso, 1975; Yazquez-Yanes et al., 1975).

About 40–45 genera and 1,000 species; mostly in the subtropics and tropics.

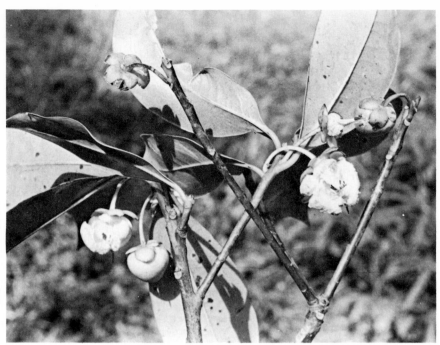

Fig. 386.
Ternstroemia tepezapote

Fig. 387. *Clusia odorata,*
roots encircling supporting tree

CALOPHYLLUM L.

Calophyllum longifolium Willd., Ges. Naturf.
Freunde Berl. Mag. 5:80. 1811
María, Calaba

Polygamous tree, to 35 m high; trunk ca 1 m dbh, buttressed; outer bark with short fissures, thick, irregular on the inner edge; inner bark reddish with white mottling, forming sap droplets; sap yellow, viscid, with the odor of fresh pumpkin; wood white, soft, lightweight; plant glabrous except for dense, minute, ferruginous trichomes on young stems and inflorescences. Petioles 2–4 cm long, stout; blades oblong, rounded to emarginate at apex, mostly obtuse at base, 11–20 cm long, 4.5–9.5 cm wide (to 30 cm long and 8.5 cm wide on juveniles), coriaceous; lateral veins very fine and parallel, secondary lateral veins lacking. Racemes short, to 4.5 cm long; inflorescence branches, petals, and sepals densely ferruginous-tomentose; pedicels less than 5 mm long; buds globose; sepals 4, ± orbicular to broadly elliptic, concave, ca 6–8 mm long, yellow, markedly unequal in width on a single flower, the outermost wider, about half as wide as long; petals lacking; stamens 10–12; filaments ca 2 mm long; anthers ca 1.8 mm long; ovary ovoid, ca 4.5 mm long; style ca 2 mm long; stigma capitate, much broader than the style. Fruits round, green, ca 3 cm diam; exocarp thin, green; mesocarp ± corky, thick; endocarp thin, shell-like, containing a single large seed. *Croat 8532, Foster 1480.*

Frequent in the forest, especially in the younger forest. Flowers throughout the rainy season, principally in late rainy season (October to November, rarely as late as January). The fruits mature principally in the late dry and early rainy seasons (March to July). J. Oppenheimer (pers. comm.) reports that white-faced monkeys eat the fruits from May to August.

Panama, Colombia, Surinam, Brazil, and Peru. In Panama, known from tropical moist forest in the Canal Zone, Panamá, and Darién and from tropical wet forest in Colón and Darién.

CLUSIA L.

Clusia odorata Seem., Bot. Voy. Herald 89. 1853
Copey, Cope, Matapalo, Tar gum tree

Dioecious, hemiepiphytic tree, to 10 m tall; trunk to 15 cm dbh; branches spiraled, stiff and spreading, densely rooting at the nodes, the roots often extending back to the host tree and enveloping it for support; stems glabrous with prominent leaf scars; sap pale yellow to bright orange (in time), not copious. Petioles to 2 cm long; blades obovate, rounded at apex (sometimes acute to acuminate on juvenile leaves), acute to obtuse at base, 5–15 cm long, 3–10 cm wide, very thick, dark green and glossy above, dull below, coriaceous; veins indistinct, drying with many fine veins. Panicles stout, bracteated, divaricately branched, terminal; buds depressed-globose; pedicels stout, to 1 cm long; flowers thick, 5-parted, spreading at anthesis, to 4.5 cm diam, white to pink, the lobes often marked with maroon or magenta at base

or apex; sepals 2, concave, unequal, persisting in fruit; petals 4–10, broadly oblong, 2–2.5 cm long; androecium of staminate flowers cushion-shaped, ca 1 cm wide and 5 mm high with numerous anthers; pollen white, granular, compressed in round or oval pouches ca 1 mm diam, embedded just beneath the sticky surface of the disk; pistillate flowers similar to staminate but the gynoecium divided into (4)9(15) segments, the petals falling soon after anthesis, the sepals closing. Capsules subglobose, 2.5–3 cm long, dehiscing into 5–9 parts at maturity; valves thick, ± spreading to expose the colorful mass of seeds; seeds numerous, reddish, curved-oblong, ca 5 mm long, embedded in a deeply divided, bright-orange aril. *Croat 14957.*

Frequent epiphyte in the forest or on exposed shorelines; epiphytic on rocks or on trees near the edge of the water. Plants are usually supported on tree trunks or large branches but their branches may grow into the top of the canopy. Apparently flowers and fruits sporadically throughout the year.

Where plants occur along the shore, the branches usually root at the nodes with the roots hanging in the water. The pollen sacs of the staminate flowers burst under the slightest pressure, and abundant pollen oozes out (sometimes explosively). Though the pollen is not at all tacky, the surface of the androecium is covered with an abundance of yellowish, tasteless, resinlike substance, which no doubt allows the pollinator to pick up pollen (perhaps on its feet). Several species of small bees, including *Trigona* spp., have been seen visiting the flowers.

Nicaragua to Panama, probably to Colombia. In Panama, known from tropical moist forest in the Canal Zone, Colón, San Blas, Veraguas, Herrera, Panamá, and Darién, from premontane wet forest in Coclé (El Valle) and Panamá (Cerro Campana), and from tropical wet forest in Coclé (La Mesa) and Panamá.

See Fig. 387.

GARCINIA L.

Garcinia mangostana L., Sp. Pl. 443. 1753
Mangosteen

Small, glabrous, polygamous tree, to 10 m tall; sap yellow. Petioles ca 2 cm long, with a swollen appendage above the articulation at the base; blades oblong-elliptic, acuminate and downturned at apex, obtuse to rounded at base, 15–22 cm long, 6.5–10.5 cm wide, bicolorous, darker above, coriaceous. Bisexual flowers solitary or paired, terminal; staminate flowers in 3–9-flowered, terminal fascicles; pedicels 1.5–2.5 cm long, stout; sepals 4, concave, coriaceous, persistent, often reddish inside; petals 4–6, fleshy, white to reddish, 1–1.7 cm long; bisexual flowers with many stamens ca 5 mm long (much shorter than the pistil); pistil bearing a broad, flat, 4–7-lobed stigma; stigma enlarged and persistent in fruit. Fruits ± depressed-globose, to 5 cm long and 6.5 cm wide, pale green to purplish at maturity; exocarp thick, with copious yellow sap; mesocarp sweet, tasty; seeds 5–7, ellipsoid, to 2.5 cm long. *Croat 4561, 5800.*

Cultivated at the Laboratory Clearing. Flowers in

March and April, with the fruits maturing in May and June.

Native to the Malay region; cultivated in tropical areas, widely so in the Canal Zone.

HAVETIOPSIS Planch. & Tr.

Havetiopsis flexilis Spruce ex Planch. & Tr., Ann. Sci. Nat. Bot., sér. 4, 14:246. 1860

Dioecious, hemiepiphytic tree, usually less than 4 m tall, growing in upper levels of canopy or on stumps along the lake shore, glabrous. Petioles 5–15 mm long; blades obovate, rounded or emarginate at apex, acute at base, 6–10 cm long, 4–6 cm wide, coriaceous. Inflorescences divaricately branched, terminal, short, shorter than the subtending leaves; axes flattened or quadrangulate, each branch subtended by a minute bract; pedicels short, the bracts apical, opposite, connate at base; flowers unisexual; buds globular or depressed-globular; sepals 4, concave, round, to ca 5 mm long, the margins ± scarious, the outermost sepals smaller, united at base; petals 4, round, ca 5 mm long, greenish marked with violet-purple, very thick especially the inner pair; stamens 4, ca 2 mm long; filaments very thick basally, about as long as the anthers, narrowed below the anther; anthers introrse; pollen white, powdery; pistillate flowers similar to staminate ones but with the stamens to 2.3 mm long, nonfunctional, the anthers much shorter than the filaments; sepals and petals persisting in fruit; styles 4, short, persistent in fruit. Capsules subglobular, to 1.2 cm diam, often obscurely 4-sided and weakly 4-lobed at apex, 4-valved, 4-locular, green or tinged with purple at maturity, splitting open to display seeds; seeds many, orange, oblong, ca 3.5 mm long, enveloped in a red-orange matrix. *Croat 14884, 16195.*

Frequent in the forest and along the shore. Flowers in the dry season (February to May), with the fruits maturing mostly in the rainy season (April to October).

Panama to Colombia, Venezuela, Brazil, and Peru. In Panama, known only from tropical moist forest on BCI. This is the first report for the genus in Central America.

See Fig. 388.

MAMMEA L.

Mammea americana L., Sp. Pl. 512. 1753

Mamey de cartagena

Monoecious to polygamous tree, to 20 m tall; trunk to 50 cm dbh; bark ± smooth; sap pale yellow. Petioles to ca 1 cm long; blades oval to obovate, rounded or emarginate at apex, rounded to obtuse at base, 8–16 cm long, 5–10 cm wide, thick, with pellucid glandular dots; major lateral veins departing midrib at nearly 90°, the cross-veins numerous and about as prominent as lateral veins. Flowers fragrant, axillary, solitary or few, clustered on short stout stalks; pedicels 1–1.5 cm long; buds globose; sepals 2, nearly round, concave, ca 1.5 cm long; petals white, 4–6 (usually 6), obovate, spreading, 2–2.5 cm long; staminate flowers with a cluster of yellow stamens ca 2

cm diam, ca 1 cm long; pistillate flowers with the pistil 2–4-celled; style short, bilobed; stigmas ± reniform. Fruits ± globose, apiculate, to ca 12 cm diam, brownish; exocarp thick; mesocarp red to yellow, with white sap; seeds 2–4, oblong, ca two-thirds the length of the fruit, reddish-brown. *Croat 5787.*

Cultivated in the Laboratory Clearing. Flowers in the rainy season, especially the early rainy season. The fruits require most of the year to develop and probably mature during the dry season of the following year.

Native to the West Indies; spread by cultivation to most parts of the New World tropics; known also from the Old World tropics. In Panama, cultivated.

MARILA Sw.

Marila laxiflora Rusby, Mem. Torrey Bot. Club 6:9. 1896

Small tree, to 20 m tall, glabrous except for inconspicuous pubescence on young stems, petioles, and inflorescence branches. Petioles 6–15 mm long; blades oblong-elliptic, acuminate, obtuse to rounded at base, 11–25 cm long, 5–9 cm wide, often with conspicuous pellucid dots in areoles below; midrib and major lateral veins prominently raised below, the major lateral veins connected by numerous parallel, sinuate, tertiary veins mostly at ± right angles to major laterals, the reticulate veins below anastomosing. Racemes axillary, 15–20 cm long, pendent, sparsely flowered; flowers green, fragrant, mostly bisexual; pedicels stout, ca 8 mm long, ± perpendicular to rachis; sepals 4 or 5, ovate, 7–9 mm long; petals 4 or 5, ± oblong or elliptic, about as long as calyx; stamens numerous, yellowish-brown, forming a ± globular mass; filaments nearly free; stigma 1, ± sessile, subentire, sticky. Capsules ± linear, 4–7 cm long, less than 5 mm diam, splitting longitudinally into 2 or 4 segments; seeds many, dark, ellipsoid, ca 0.6 mm long, comose at both ends, the trichomes light brown, 2–3 mm long. *Croat 10806.*

Collected twice near the Laboratory Clearing, not seen otherwise. Flowers from March to August, chiefly in the rainy season (July to August). Mature fruits known from April and November.

The genus is unusual among the Guttiferae of Panama in having wind-dispersed seeds.

Known only from Panama, from wetter regions of tropical moist forest in the Canal Zone, Coclé, Panamá, and Darién, from premontane wet forest in the Canal Zone (Pipeline Road), Colón (Achiote), and Panamá (Cerro Campana), and from premontane rain forest in Panamá (Cerro Jefe).

RHEEDIA L.

Rheedia acuminata (R. & P.) Planch. & Tr., Ann. Sci. Nat. Bot., sér. 4, 14:314. 1860

R. madruno (H.B.K.) Planch. & Tr.

Madroño, Fruta de mono, Machari, Satro, Cero

Glabrous, polygamodioecious tree, to 9(20) m tall, less than 15 cm dbh; outer bark very thin; inner bark reddish;

KEY TO THE SPECIES OF RHEEDIA

Pedicels often more than 2.5 cm long; fruits usually ovoid, to 5 cm long, with dense protuberances
.... *R. acuminata* (R. & P.) Planch. & Tr.
Pedicels usually less than 2.5 cm long; fruits globular, less than 3 cm diam, smooth ....
.... *R. edulis* (Seem.) Planch. & Tr.

wood hard; sap yellow, in branches and trunk. Petioles 1–2.5 cm long, somewhat swollen at base with a short appendage on the inner side above the articulation; blades elliptic, acuminate, mostly acute at base, 7.5–22 cm long, 2.5–8 cm wide, the midrib often arched; lateral veins and submarginal collecting vein visible on both surfaces (prominulous above when dry). Fascicles axillary, sessile, often at leafless nodes; pedicels slender, usually 2.5–3.5 cm long (rarely shorter); sepals 2, rounded at apex, ± united at base, ca 3 mm long; petals 4, nearly orbicular, to 7 mm long, creamy-white, strongly reflexed at anthesis; bisexual flowers with fewer stamens than the staminate flowers; style short; stigma discoid, as broad as or broader than ovary. Fruits ovoid to globular, often weakly flattened, to 5 cm long and 4 cm wide, yellow, densely covered with puberulent, flattened protuberances to 3 mm long; exocarp thick, moderately hard; mesocarp thin, sweet; seeds usually 2, longer than broad. *Croat 8240, 13847.*

Occasional, in the forest. Flowers in the dry season (rarely in the late rainy season). The fruits mature mostly from April to August.

Monkeys are fond of the fruits. Mature fruits are usually all removed in a very short time.

Mexico to Peru. In Panama, known from wetter parts of tropical moist forest in the Canal Zone, Chiriquí, Panamá, and Darién and from premontane wet and tropical wet forests in Colón, Panamá, and Darién.

Rheedia edulis (Seem.) Planch. & Tr., Ann. Sci. Nat. Bot., sér. 4, 14:310. 1860

Sastra, Cero, Chaparrón

Glabrous, polygamodioecious tree, to 10(30) m tall; trunk usually less than 15 cm dbh, with sparse yellow sap; outer bark thin; inner bark reddish; younger parts with white or pale yellowish sap. Petioles 1–2.5 cm long, somewhat swollen at base with a short appendage on the inner side above the articulation; blades narrowly elliptic, acute to weakly acuminate, ± attenuate at base, 12–20(25) cm long, 4–7(10) cm wide, thick; lateral veins and collecting vein scarcely visible, not prominulous above when dry. Fascicles dense, globular, sessile, ca 4 cm wide, usually borne on leafless stems; pedicels 7–20 mm long; flowers small, aromatic, numerous; sepals 2, concave, rounded at apex, ± united at base; petals 4, ovate to orbicular, rounded at apex, white, 5–7 mm long, strongly reflexed at anthesis; staminate flowers with the stamens free, 16–30 (usually nearer 16), ca 3 mm long, erect to somewhat spreading, borne around a fleshy disk; anthers minute, about as broad as long, the thecae mostly directed upward; pistillode broad, cushion-shaped; bisexual flowers with the stamens 6–12, ca 2 mm long,

alternating with the lobes of a fleshy disk; ovary ovoid, ca 2.7 mm long; style very short; stigma discoid, ca three-fourths as broad as ovary. Fruits subglobular to ovoid, 2–2.8 cm long, to 2.3 cm wide, glabrous, at first green and densely whitish-punctate, becoming pale yellow to orange at maturity; exocarp thin, ca 2 mm thick, leathery; mesocarp thick, fleshy, white, tasty, sticking to seeds; seeds 1 or 2, oblong, to 2 cm long and 1 cm wide, with yellow latex, covered with a thin, brown, ± fibrous layer. *Croat 13921, 14454.*

Frequent in the forest. Flowers from January to April, especially in February and March (rarely earlier or later elsewhere in Panama). The fruits mature principally from May to July, but some persist much later into the rainy season.

Sterile individuals are difficult to distinguish from *R. acuminata.*

Mexico, Panama, and Peru. In Panama, usually lower than 900 m elevation; known principally from wetter parts of tropical moist forest in the Canal Zone, Bocas del Toro, Chiriquí, Veraguas, Los Santos, Panamá, and Darién, but also from premontane wet and tropical wet forests in Los Santos (Loma Prieta and along the Serrania de Cañazas).

See Fig. 389.

SYMPHONIA L.f.

Symphonia globulifera L.f., Suppl. Pl. Syst. Veg. 302. 1781

Barillo, Bogum, Cerillo, Cero, Sambo gum

Glabrous tree, usually to 30 m or more tall; trunk 50–120 cm diam, usually bearing numerous, large, adventitious roots near the base; outer bark thick, brown, slightly fissured; inner bark thick, much lighter, forming sap droplets when cut; sap yellow, copious. Petioles less than 1 cm long, canaliculate; blades narrowly elliptic, acuminate, acute to obtuse at base, 5–8 cm long, 2–3 cm wide; veins numerous, 1–2 mm apart. Inflorescences of large, globular, terminal, umbelliform fascicles to ca 6 cm diam, borne among the leaves; flowers numerous, bisexual, red; pedicels 5–15 mm long (to 25 mm in fruit), drying angulate; sepals 5, imbricate, unequal, suborbicular, coriaceous, usually broader than long, to 5 mm long, 5–8 mm wide; corolla depressed-globose, ca 15 mm wide and 8 mm high; petals strongly concave and imbricate, their apical margins directed inward and contacting the androecium; disk cupular, at the base of the staminal tube; staminal tube to 15 mm long and 4.5 mm wide, markedly infolded at about the middle, incised to ca 5 mm at apex between the anthers, the lobes 5, arcuate-spreading, each bearing 2–4 linear anthers on the outer

Fig. 388. *Havetiopsis flexilis*

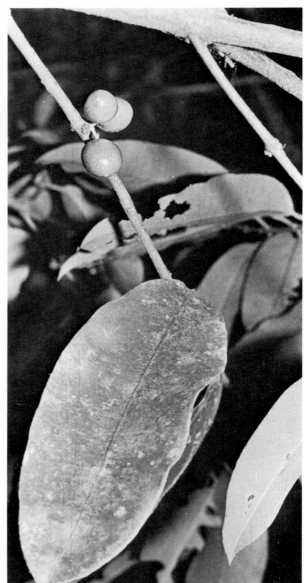

Fig. 389. *Rheedia edulis*

Fig. 390. *Symphonia globulifera*

surface well below the apex; anthers to 2.7 mm long; ovary 5-locular; ovules 2–8 per locule; style to ca 6 mm long, frequently persisting in fruit, the lobes 4, stout, spreading, inserted between the lobes of the staminal tube. Fruits ± round, 2.5–3.5(4) cm diam; exocarp thin, leathery; mesocarp fibrous, 6–9 mm thick, yellow-green or brown; seeds usually 1 (to 3), 2–2.5 cm long, ellipsoid to subglobose, red, with yellow sap forming in droplets when cut, marked externally by irregular grooves. *Croat 9517.*

Frequent in the older forest. Flowers throughout the year but with a peak from June to August; individual trees may remain in flower for 2 months. The fruits mature in 1 or 2 months, so trees may still be bearing flowers when their first fruits mature.

The flowers are visited frequently by hummingbirds on BCI.

Belize south along the Atlantic slope to Panama and tropical South America; West Indies; tropical Africa; appearing to like water and growing in great abundance in swamp forests. In Panama, a characteristic species of tropical moist and tropical wet forest (Tosi, 1971), growing from sea level to 300 m; known from wetter parts of tropical moist forest in the Canal Zone, Bocas del Toro, Colón, San Blas, and Darién, from premontane wet forest in Chiriquí, Coclé, Colón, and Panamá, and from tropical wet forest in Colón, Coclé, and Panamá. Reported from premontane rain forest in Costa Rica (Holdridge et al., 1971).

See Fig. 390.

TOVOMITA Aubl.

Tovomita longifolia (L. C. Rich.) Hochr., Ann. Conserv. Jard. Bot. Genève 21:66. 1919

Glabrous, dioecious tree, to 12 m tall; trunk to ca 20 cm dbh, the base often with weakly developed stilt-roots; branches slender, often drooping; sap yellow. Petioles 2–2.5(5) cm long; blades oblong-obovate, acute to bluntly acuminate at apex (rarely rounded), acute at base, (10)12–35 cm long, (4.5)7.5–13 cm wide; lateral veins in 6–12 pairs. Panicles terminal, short-pedunculate, to 10 cm long, the branches divaricate, often ± flattened; flowers unisexual, many, ca 3 cm broad, arranged in dichasia, the buds globose; sepals 4, the outer pair smaller, strongly concave, reflexed at anthesis; petals 5, elliptic to oblong or obovate, to 13 mm long, white or yellowish; staminate flowers with numerous stamens in a cluster 15 mm wide and ca 5 mm long, free or somewhat connate; filaments fleshy, broader than the tapered anther, the innermost

curved outward; pistillode glabrous, conical, with 5 short styles; pistillate flowers fewer, somewhat smaller; ovary 4- or 5-locular; ovules 1(2) per locule, erect; style very short; stigmas 5, cuneate-obovate; staminodia many, ca 2.5 mm long. Fruits obovoid, ca 4 cm long (including stipe), rounded at apex, with 5 persistent styles to 5 mm long, the stigmas discoid, ca 2.5 mm diam; staminodia persistent. *Croat 14452, Dare 1116.*

Rare, in the forest, known only from the vicinity of the Laboratory Clearing. Throughout its range the species has been collected in flower nearly all year but especially in the late dry or early rainy seasons. On BCI flowers have been seen in late April and in May. Most fruits were mature in the rainy season.

Panama to the Guianas and Brazil. In Panama, known from tropical moist forest on BCI, from tropical wet forest in Colón, from premontane wet or tropical wet forest in Panamá (Río Tuquesa), and from premontane rain forest in Panamá (Cerro Jefe).

See Fig. 391.

Tovomita stylosa Hemsl., Biol. Centr.-Amer. Bot. 1:88. 1879

Glabrous, dioecious shrub or small tree, to 7 m tall; sap yellow, more abundant in twigs than in trunk. Petioles 1–3 cm long, canaliculate; blades broadly elliptic to obovate, acute to abruptly acuminate and ± downturned at apex, cuneate to obtuse at base, 3–15 cm long, 2–6.5 cm wide, entire; lateral veins in 3–7 pairs, ± impressed above. Panicles terminal, bracteate, less than 4 cm long; bracts acute, mostly 1–2.3 mm long; branches flattened; pedicels 4–13 mm long; flowers unisexual, the buds oblong; staminate flowers cylindrical in bud; calyx bilobed, spathaceous, splitting on one side and recurving to expose other parts, ca 1 cm long, greenish; petals 4, ± equaling calyx, white, free, to 2 mm wide, hooked inward at apex; stamens many, 5–7.5 mm long; filaments white; anthers minute, longitudinally dehiscent; pistillode to 1 mm high, with 4 sessile stigmas; pistillate flowers similar to staminate ones; ovary glabrous; styles 4, only slightly exceeding staminodia, persisting in fruit; staminodia numerous, ca 7 mm long. Fruits ± obovate, 1.5–2 cm long, 4-lobed (at least when dry), splitting to expose red interior when ripe, 4-carpellate, each carpel with one seed; exocarp usually with scurfy, lenticellate areas; seeds oblong, ca 7 mm long. *Croat 11164, Oppenheimer 1521.*

Occasional, in the forest. Flowering in the rainy season (June to September), rarely earlier during the dry season elsewhere. Some fruits develop to full size by August but most mature in the dry season.

KEY TO THE SPECIES OF TOVOMITA

Leaves less than 15 cm long; flower buds oblong; stamens fewer than 20; fruits 1.5–2 cm long, 4-carpellate, the surface with scurfy areas; styles 4, slender . . . . *T. stylosa* Hemsl.
Leaves more than 15 cm long; flower buds globular, often more than 1 cm diam; stamens 20 or more; fruits ca 4 cm long, 5-carpellate, the surface smooth; styles 5, stout . . . . *T. longifolia* (L. C. Rich.) Hochr.

Fig. 392. *Tovomita stylosa*

Fig. 391.
Tovomita longifolia

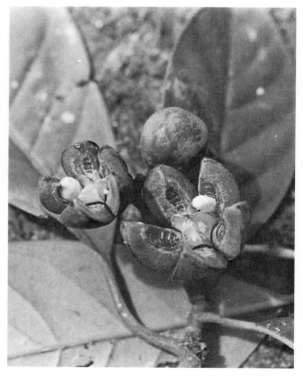

Fig. 393. *Tovomitopsis nicaraguensis*

Known from Panama and Colombia (Chocó). In Panama, known from tropical moist forest in the Canal Zone and Darién and from premontane wet and tropical wet forests in Colón, Panamá, and Darién.

See Fig. 392.

TOVOMITOPSIS Planch. & Tr.

Tovomitopsis nicaraguensis Oerst. ex Planch. & Tr., Ann. Sci. Nat. Bot., sér. 4, 14:266. 1860

Tovomita nicaraguensis (Oerst.) L. O. Wms.; *Chrysochlamys eclipes* L. O. Wms.

Polygamous or dioecious shrub or small tree, 3–11 m tall, to 24 cm dbh, essentially glabrous; sap yellow, not copious. Petioles to 3 cm long; blades elliptic, acuminate, acute at base, mostly 12–22 cm long, 4.5–11 cm wide, moderately thick, drying brown to rusty, especially below; lateral veins in 6–11 pairs. Panicles terminal, to 16 cm long, the branches divaricate, each subtended by a small bract; pedicels stout, ca 5 mm long; flowers bisexual, ca 5 mm long, arranged in dichasia; sepals 5, reddish, broadly rounded, concave, very unequal, the outermost reduced, the inner few with the margin thin, ± erose; petals 5, yellow, ovate, ca 4 mm long, ± concave, thick medially, thin along the margin; staminate flowers not seen (possibly lacking); bisexual flowers with the stamens to ca 20, ca 1.5 mm long, of irregular sizes, longer and shorter ones generally alternating; filaments flattened and fused to at least the middle; pistil ca 2 mm long; stigmas 5, pitlike, subapical. Fruits red, round to obovoid, ca 2 cm long, with 5 black dots at the apex (style bases), dehiscing into 5 parts at maturity to expose seeds; valves thick, 3–4 cm long, tasting bitter even at maturity; seeds 1 per carpel, ± oblong, black, ca 1 cm long, mounted between the wings of the 5-lobed central column, the aril white, not sweet but edible. *Croat 4651, 11479, 12663.*

Common in the old forest. Flowers from July to September. The fruits mature in late dry and early rainy seasons (March to June).

It is uncertain if the species is dioecious or, as is often the case in the Guttiferae, polygamous.

Nicaragua to Panama. In Panama, known from tropical moist forest in the Canal Zone, Bocas del Toro, and Darién and from premontane wet forest in the Canal Zone (Pipeline Road) and Panamá (Cerro Jefe).

See Fig. 393.

VISMIA Vand.*

Vismia baccifera (L.) Tr. & Planch., Ann. Sci. Nat. Bot., sér. 4, 18:300. 1862

Achiote tigre, Sangre de perro, Sangrillo, Pinta-mozo

Shrub or small tree, to 5 m tall; older stems with reddish-brown flaky bark; younger stems, petioles, lower blade surfaces, sepals, and axes of inflorescences densely stellate-tomentose with usually brown, sessile trichomes; sap yellow to bright orange, drying red. Petioles 1–2.5 cm long; blades extremely variable, mostly ovate to elliptic, acute to acuminate, obtuse to rounded or truncate at base, 7–17 cm long, 2.5–10 cm wide, green above, yellowish-brown to dark brown below (depending on extent of ferruginous pubescence), reddish-pellucid-punctate. Panicles terminal, mostly to 8 cm long; pedicels ca 5 mm long; flowers bisexual, ca 1 cm long, 5-parted, without conspicuous aroma; sepals thick, 7–9 mm long, acute, valvate, closing to protect ovary after flowering, spreading in fruit, at least some with one or more margins thin and glabrous; petals obovate, round at apex, to ca 1.3 cm long, yellowish to transparent with vertical orange streaks, glabrous outside, densely woolly inside; staminal columns 5, exceeding styles, less than 7 mm long, alternating with short woolly staminodia; anthers many, directed outward against woolly inner surface of petals; pistil ovoid, becoming 5-lobed, glabrous, sometimes punctate; styles 5, diverging outward between staminal columns, persisting in fruit; stigmas capitate; nectar copious, stored chiefly in the calyx. Berries ovoid, ca 1.5 cm long, firm, green and fleshy at maturity, the styles, calyx, and weathered petals persisting; seeds many, cylindrical, straight or curved, 2–2.7 mm long, in several vertical stacks, faintly reticulate. *Croat 6695.*

Abundant in older clearings; common along the edge of the lake. Apparently flowers twice per year. The principal flowering season is during the early dry season,

---

KEY TO THE SPECIES OF VISMIA

Leaf blades shallowly cordate at base, with red opaque punctations; petals white with purple specks or streaks, recurved at anthesis .... *V. macrophylla* Kunth
Leaf blades acute to rounded at base, with black opaque or reddish pellucid punctations; petals green to pale orange:
  Leaves broadest at the base, usually more than 6 cm wide, sometimes ferruginous-tomentose; sepals generally not reflexed in fruit; flowers bisexual; petals yellowish to transparent with orange streaks .... *V. baccifera* (L.) Tr. & Planch.
  Leaves broadest near the middle, usually less than 6 cm wide; sepals strongly recurved in fruit; flowers both bisexual and pistillate in the same inflorescence; petals greenish ....
  .... *V. billbergiana* Beurl.

---

*After this treatment went to press, Norman K. B. Robson, British Museum of Natural History, submitted his treatment of *Vismia* for publication in the *Flora of Panama*. Two of the species he recognized from the BCI material I have treated as *V. baccifera*: *V. latisepala* N. Robson (Panama and Colombia), represented by *Croat 4910, 6110, 6416, 6695, 10736, 11089,* and *11286,* and *V. panamensis* Duch. & Walp. (Nicaragua to Panama), represented by *Croat 4614, 4619, 4956, 5523, 5636,* and *8664.*

with the first fruits maturing by late April. A second wave of flowering occurs during the early rainy season, mostly in July, with scattered flowering later in the rainy season; the fruits are apparently all mature before the end of the rainy season.

The species is quite variable in terms of both leaf shape and indument. Standley (1933) reported the species under three names, *V. dealbata* H.B.K., *V. ferruginea* H.B.K., and *V. guianensis* (Aubl.) Pers. *V. guianensis* does not occur on BCI. The white- and brown-leaved forms were treated by Ewan (1962) in his treatment of South American *Vismia* as the subsp. *dealbata* (H.B.K.) Ewan and subsp. *ferruginea* (H.B.K.) Ewan, respectively.

Small green halictid bees visit flowers of this species. The bee generally alights atop the corolla, pushes its way into the flower between the petals and staminal clusters, and generally disappears from sight. Leaving the flower, the bee may back out or turn around and come out frontwards.

Mexico to Colombia, Venezuela, the Guianas, and Brazil. In Panama, ecologically variable; known from tropical moist forest in the Canal Zone, Veraguas, Panamá, and Darién, from tropical dry forest in Coclé, from premontane moist forest in the Canal Zone and Panamá, from premontane wet forest in Panamá, from tropical wet forest in Colón and Panamá, and from premontane rain forest in Chiriquí and Panamá.

See Fig. 394.

Vismia billbergiana Beurl., Handl. Svenska Vet. Akad., sér. 4, 2:117. 1856
V. viridiflora Tr. & Planch.

Polygamous shrub or slender tree, to 6 m tall; sap yellow or orange; stems, peduncles, pedicels, exposed parts of calyces, and lower leaf surfaces with an inconspicuous, thin, usually dense layer of stellate trichomes. Petioles 6–13 mm long; blades narrowly ovate-elliptic, acuminate, obtuse to rounded at base, 7–13 cm long, 2.5–5 cm wide, with black opaque punctations especially on lower surface. Panicles terminal or upper-axillary, to 10 cm long; flowers 5-parted, with both bisexual and pistillate flowers on the same inflorescence; bisexual flowers ca 1 cm long; sepals to 7 mm long, glabrate inside, persisting and strongly reflexed in fruit; petals obovate, green and glabrous outside, densely white-woolly inside, ca 1 cm long; stamens numerous, included; staminal columns 5, opposite the petals, densely woolly, alternating with staminodia; ovary ± glabrous, 5-ribbed; styles 5, recurved, protruding outward between staminal columns; stigma capitate; nectar copious around the base of the ovary; staminodia 5, prominent, densely pubescent, less than 2 mm long, sometimes yellow; pistillate flowers with the sepals 5–6 mm long, 1 sepal with a villous line of trichomes inside; petals lacking; staminodia, ovary, and styles as in bisexual flowers. Berries ovoid, 1–1.5 cm long, green. *Croat 5089.*

Uncommon in the forest along trails, occasional at some points along the shoreline. Flowers principally in the dry season, especially in January and February, with most fruits maturing in April and May. Individuals often flower again in April or May when they have mature fruits, and a few flowers are also seen in September and October.

The species differs from *V. baccifera* and *V. macrophylla* in preferring to grow along the edges of shady areas and not in open areas.

Costa Rica to Colombia. In Panama, known from premontane moist forest in Panamá, from tropical moist forest in the Canal Zone, Bocas del Toro, San Blas, Coclé, and Panamá, from premontane moist forest in the Canal Zone, Veraguas, and Coclé, and from tropical wet forest in Colón, Veraguas, Panamá, and Darién.

See Fig. 395.

Vismia macrophylla Kunth in H.B.K., Nov. Gen. & Sp. 5:184. 1822
V. angusta Miq.; *V. latifolia* sensu Reich. in Mart. non (Aubl.) Choisy
Sangrillo, Pinta-mozo

Tree, 4–12 m tall, usually less than 15 cm dbh; bark usually loose and shaggy; sap copious, orange, drying red; stems, petioles, and axes of inflorescences densely stellate-ferruginous. Petioles 1.5–2.5 cm long; blades narrowly ovate-oblong, acuminate, rounded to subcordate at base, 17–30 cm long, 5.5–9 cm wide, the upper surface glabrous except for midrib, the lower surface, especially the veins, densely stellate-ferruginous and with reddish opaque punctations. Panicles terminal, ca 6 cm long; peduncles to 6 cm long, bearing a pair of leaflike bracts at the apex; major branches usually opposite; pedicels short, less than 2 mm long, thick; flowers bisexual, 5-parted, aromatic, ca 12 mm long; calyx to ca 7 mm long, ovate, broadly acute at apex, densely stellate-ferruginous, thick except for the thin scarious margins on one or both sides of 3 sepals; petals, inner surface of sepals, upper edge of ovary, and styles with conspicuous violet-purple punctations or streaks; petals white with purple spots, oblong, to 13 mm long and 3 mm wide, the apex blunt to acute and recurved from about the middle, densely matted-villous inside; staminal columns 5, flattened and glabrous at base, densely villous above; stamens many, of various lengths, directed outward against the villous surface of petals or becoming exserted; ovary ovoid, glabrous; styles 5, exserted, to 6.5 mm long, united near the base, persistent in fruit; stigma thick, bilobed; staminodia orange-red, ca 2 mm long, glabrous below, villous above, alternating with petals. Berries globose to ovoid, ca 1.5 cm long, green or olive-brown; seeds numerous, cylindrical, straight or curved, reddish-brown, faintly reticulate, ca 3 mm long. *Croat 5322, 6434.*

Occasional, along the shore on the north side of the island. Flowers early in the rainy season, mostly from May to August. The fruits develop to mature size before the end of the rainy season, but some persist through the dry season.

The species is popular with insects in the Canal Zone when it is in flower, and wasps, bees, and butterflies of several species visit it. Small bees of the family Anthophoridae protrude their heads into the flower and are possibly a legitimate pollinator. Wilson (1971) reported

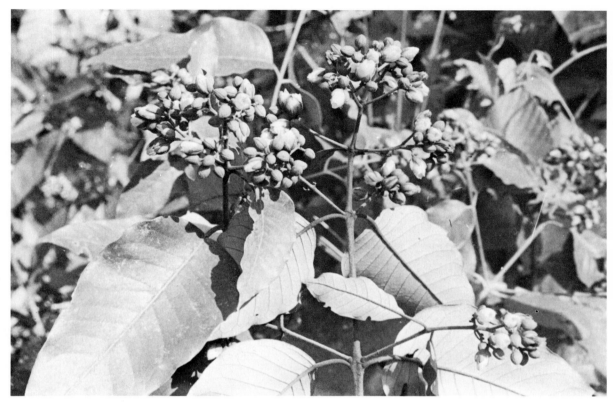

Fig. 394. *Vismia baccifera*

Fig. 395. *Vismia billbergiana*

that fruits are eaten by the bat *Micronycteris hirsuta.* The fruits are probably removed principally by birds.

Belize south to the Guianas, Brazil, Bolivia, and Peru. In Panama, known from tropical moist forest in the Canal Zone, Bocas del Toro, San Blas, Chiriquí, and Darién, from tropical dry forest in Panamá (Taboga Island), from premontane wet forest in Veraguas and Panamá, and from tropical wet forest in Colón and Veraguas.

93. BIXACEAE

Shrubs or small trees, with red-orange sap. Leaves alternate, petiolate; blades simple, entire; venation pinnate, palmate at the base; stipules present. Panicles terminal; flowers bisexual, actinomorphic; sepals 5, free, imbricate; petals 5, free, imbricate, showy; stamens numerous, free, on a thick receptacle; anthers 2-celled, dehiscing by 2 apical pores; ovary superior, 1-locular, 2-carpellate; placentation parietal; ovules numerous, anatropous; style 1, simple; stigma shortly bilobed. Fruits usually spiny, loculicidal capsules; seeds numerous, with fleshy testa, with endosperm.

Bixa can be recognized by having numerous stamens and a spiny reddish capsule.

One genus; native in tropical America, now widely distributed in the tropics.

BIXA L.

Bixa orellana L., Sp. Pl. 512. 1753
   Achote, Achiote, Anatto

Shrub or small tree, 1.5–8(15) m tall; trunk to 10(15) cm dbh; branchlets reddish-brown. Stipules narrowly ovate, ca 1 cm long, caducous; petioles slender, 2–10(14) cm long, slightly enlarged at apex; blades ovate or ovate-lanceolate, acuminate, rounded to truncate or cordate at base, 5–27 cm long, 3–18 cm wide, glabrous or sparsely lepidote above, densely lepidote below, the scales minute, reddish-brown; palmate veins at base 5. Panicles terminal, variable in size, mostly 5–15 cm long; inflorescence branches, pedicels, and calyces densely reddish-brown-lepidote outside; pedicels to 1 cm long, with 5 conspicuous glands below calyx; sepals 5, orbicular, ca 1 cm diam, concave; petals 5, obovate, 2–3 cm long, 1–2 cm wide, white (rarely pinkish); stamens many, 1–1.5 cm long; filaments slender, white, glabrous; anthers yellow, horseshoe-shaped, 1–1.5 mm long; ovary globose to pyriform, densely to sparsely bristly; style 1–1.5 cm long. Capsules variable in size, shape, and indument throughout the species' range, usually ± globose to ovoid, 1.5–4.5 cm long, brown to red, with sparse to dense, short, flexible spines; seeds many, ± irregular, red, ca 5 mm long, the testa fleshy. *Croat 7875, 12294.*

Occasional, along the shore. Flowers from September to February (sometimes from August). The fruits mature from January to June, with old dehisced capsules hanging on as late as September.

In Costa Rica bees are the most common visitors and presumed pollinators, especially *Eulaema* (D. Janzen, pers. comm.), *Xylocopa, Trigona,* and *Melipona* (Heithaus, 1973). In Ecuador the flowers are visited by female *Eulaema cingulata, E. polychroma, E. homichlora,* and *E. meriana* for pollen (Dodson & Frymire, 1961).

The spines on the fruits are not functional in epizoochorous dispersal, but probably prevent the fruits from being eaten prematurely.

A red condiment and cosmetic is obtained from the seeds (Duke, 1968).

Throughout the tropics and subtropics of the world. In Panama, from tropical moist forest in all provinces, from premontane moist forest in Los Santos and Panamá, from premontane wet forest in Panamá (Chimán), and from tropical wet forest in Colón.

94. COCHLOSPERMACEAE

Trees or shrubs. Leaves alternate, petiolate; blades simple, palmately lobed; venation at base palmate; stipules present. Panicles terminal; flowers bisexual, slightly zygomorphic; sepals 4 or 5, free, unequal, imbricate; petals 4 or 5, free, imbricate, showy; stamens numerous, unequal, free; anthers 2-celled, basifixed or dorsifixed, dehiscing by terminal porelike slits; ovary superior, 1-locular, 3–5-carpellate; placentation parietal, often intruding; ovules numerous, anatropous; style 1; stigma simple, minute, dentate. Fruits loculicidal capsules; seeds densely long-lanate, with oily endosperm.

Three genera and 255 species; tropics.

COCHLOSPERMUM Kunth

Cochlospermum vitifolium (Willd.) Spreng., Syst. Veg. 2:596. 1825
   Brazilian rose, Poro-poro

Tree, 3–12 m tall, mostly to 10 cm dbh (sometimes to 70 cm dbh elsewhere); wood very soft; branches few; branchlets densely pubescent, becoming sparsely so in age. Stipules subulate, small, caducous; petioles to 30 cm long; blades usually palmately 5-lobed, cordate at base, mostly 12–25 cm wide, glabrous above, puberulent below, the lobes acute to acuminate, the margins usually crenate. Flowers yellow, pedicellate, the inflorescence branches recurved near apex; sepals 4 or 5, unequal, the 2 outer sepals ovate to oblong-obovate, usually rounded at apex, 12–18 mm long, 7–9 mm wide, the 3 inner sepals mostly obovate to subrotund, rounded at apex, 20–22 mm long, 16–20 mm wide; petals 4 or 5, obovate, 5–6 cm long, often emarginate at apex; stamens yellow, numerous, the outer ones longer, curved inward near apex; anthers slender, somewhat curved; pollen dehiscing from a single apical pore; style longer than stamens, at first ± erect, later recurved and about equaling height of anthers, the apical part somewhat hooked; stigma simple. Capsules ± obovoid, 5-valved, to 8 cm long; seeds reniform, 4–5 mm long, bearing many cottonlike fibers. *Croat 7690.*

Probably once common, now rare; known only from

a few places along the shore, especially the north edge of Orchid Island. Flowers throughout the dry season (December to April). The fruits mature in late dry and early rainy seasons (March to July). Leaves fall throughout most of the dry season.

The species can be recognized at once by the large yellow flower with numerous stamens and by the palmately lobed leaf. A large pollinator, such as a bat, bird, or large bee, would be the most effective for this large flower with protruding style. Primary pollinators in Costa Rica are large bees, principally in the families Anthophoridae and Xylocopidae (G. Frankie, pers. comm.). Capsules open broadly, but the wind-dispersed seeds leave only a few at a time, the rest being held by the outer capsule valves, which curve sharply inward and hold the cottony mass of seeds.

Mexico to northern South America. In Panama, a characteristic component of tropical dry forest (Holdridge & Budowski, 1956) and a common invader (Holdridge, 1970); known from tropical moist forest in the Canal Zone, Herrera, Panamá, and Darién, from premontane moist forest in the Canal Zone, Los Santos, and Coclé, and from tropical wet forest in Panamá (Cerro Campana); cultivated in San Blas (Duke, 1968).

See Fig. 396.

95. VIOLACEAE

Trees or shrubs. Leaves alternate (*Hybanthus*) or opposite (*Rinorea*), petiolate; blades simple, entire to serrate; venation pinnate, frequently palmate at base; stipules present. Flowers bisexual, actinomorphic (*Rinorea*) or zygomorphic (*Hybanthus*), solitary in the axils (*Hybanthus*) or in few-flowered terminal racemes (*Rinorea*); sepals 5, free, imbricate; petals 5, equal (unequal in *Hybanthus*), ± white; stamens 5, free; anthers 2-celled, with a broadly expanded, scalelike connective, dehiscing longitudinally; ovary superior, 1-locular, 3-carpellate; placentation parietal; ovules 1 or 3 per placenta; style 1; stigma simple. Fruits 3-valved, loculicidally and elastically dehiscent capsules; seeds few, with fleshy endosperm.

Members of the family may be distinguished on BCI by having appendaged stamens and three-sided capsules with a few globose seeds.

Flowers are insect pollinated, those of *Hybanthus* probably by bees.

Seeds are probably bird dispersed. Oppenheimer (1968) reported that, while white-faced monkeys eat immature seeds of *Hybanthus prunifolius*, they do not take mature seeds. Mature capsules of *Hybanthus* become constricted and forcibly expel the seeds (R. Foster, pers. comm.).

Some 22 genera and 900 species; widely distributed.

HYBANTHUS Jacq.

Hybanthus prunifolius (Schult.) Schulze, Notizbl. Bot. Gart. Berlin-Dahlem 12:114. 1934

H. anomalus (H.B.K.) Melch.

Shrub or small tree, 1–6 m tall, sparsely pubescent on lower leaf surfaces and midribs above, densely so on young stems, pedicels, and bracts. Leaves alternate; stipules triangular, to 3 mm long, whitish, subpersistent; petioles to 7 mm long; blades elliptic, acuminate, acute at base, 5–15 cm long, 2.5–6 cm wide, remotely to prominently serrate-crenate. Flowers solitary in axils, puberulent; pedicels ca 1.5 cm long, articulate above middle, bracteolate at base; sepals 5, triangular-ovate, acute, ca 9 mm long, persistent; petals 5, white, unequal, the anterior petal 3.5–4 cm long, 1.5–2.5 cm wide, spatulate, with the median lobe green to pale yellow below middle, puberulent inside near base, the remaining petals ± oblong, the lateral ones ca 1.5 cm long, longer than the posterior ones; stamens 5; anthers ca 3 cm long, the connective appendages brown, much exceeding anthers, weakly fused laterally to form nectariferous appendages at bases of anthers; filaments very short; style ca 5 mm long, curved near apex toward anterior petal, persistent; stigma terminal, held well above staminal appendages. Capsules ellipsoid, 3-sided, acute, 1–2 cm long; seeds ovoid, 4–5 mm long, tan. *Croat 8900.*

Often very abundant in the forest, more common in the old forest; one of the most abundant species on the island. Flowers in sporadic bursts throughout the dry season, rarely in the middle of the rainy season, usually within 2 weeks after a heavy rain following a period of drought (R. Foster, pers. comm.).

Recognized by the persistent whitish stipules and bracteoles. On one occasion many of the trees were stripped of their leaves by a small caterpillar, leaving the understory vegetation very open.

Costa Rica to Colombia and Venezuela. In Panama, known from tropical moist forest in the Canal Zone, San Blas, Coclé, Panamá, and Darién, from tropical wet forest in Colón and Darién, and from premontane rain forest in Darién. Reported from premontane wet forest in Costa Rica (Holdridge et al., 1971).

KEY TO THE SPECIES OF VIOLACEAE

Leaves alternate; flowers solitary . . . . *Hybanthus prunifolius* (Schult.) Schulze
Leaves opposite; flowers in short racemes:
    Petals slightly longer than sepals, the sepals 3–4.5 mm long; bracteoles exceeding pedicels; leaf
        blades usually inequilateral at base and rounded to subcordate . . . .
         . . . . *Rinorea sylvatica* (Seem.) O. Kuntze
    Petals more than twice as long as sepals, the sepals 1–1.5 mm long; bracteoles shorter than pedi-
        cels; leaf blades usually equilateral and obtuse to acute at base . . . *Rinorea squamata* S. F. Blake

Fig. 396. *Cochlospermum vitifolium*

Fig. 397. *Rinorea sylvatica*

RINOREA Aubl.

Rinorea squamata S. F. Blake, Contr. U.S. Natl. Herb.
20:516. 1924

Guayacillo, Molenillo

Shrub or small tree, 2–5(12) m tall, 5–10(20) cm dbh, minutely ferruginous-puberulent and sparsely hirsute on young stems, petioles, leaf veins (especially below), inflorescence branches, pedicels, bracteoles, and calyces. Leaves opposite; stipules small, deciduous; petioles 3–8 mm long; blades mostly obovate-elliptic, caudate-acuminate, obtuse to acute and usually ± equilateral at base, 6–13 cm long, 2.5–5 cm wide, subentire to serrulate. Racemes terminal, to 6 cm long; pedicels to 2.5 mm long; bracteoles broadly ovate, less than 1 mm long; sepals 5, ovate, ± equal, 1–1.5 mm long; petals 5, ± equal, oblong, 4–4.5 mm long, white or yellowish, minutely ciliolate; stamens 5, 2.5–3 mm long; filaments free, ca 1 mm long, with a dorsal appendage nearly as long as and attached to basal half of filaments; anthers ca 1 mm long, the connective expanded, brownish; style ca 2.5 mm long, glabrous. Capsules ovoid, ca 2 cm long, trilobate, hirsute-hispid; valves thick; seeds few, ca 6 mm long. *Croat 8771.*

Occasional, in the forest. Seasonal behavior uncertain. Flowers at least in October and March. The fruits are known from June and July.

Costa Rica and Panama. In Panama, known from tropical moist forest in the Canal Zone, Bocas del Toro, Panamá, and Darién and from tropical wet forest in Darién.

Rinorea sylvatica (Seem.) O. Kuntze, Rev. Gen. Pl.
1:42. 1891

Shrub or small tree, mostly 2.5–3.5 m tall; young branches, petioles, and leaf veins below puberulent and hirsute. Leaves opposite; stipules triangular, to 6 mm long, deciduous; petioles 3–8 mm long; blades obovate to elliptic, acuminate, usually inequilateral and rounded to subcordate at base, 9–16 cm long, 4–6.5 cm wide, subentire to crenulate. Racemes terminal, to 4 cm long; bracteoles ovate, 2–3 mm long, longer than pedicels; rachises, pedicels, sepals, and midribs of petals appressed-hirtellous; pedicels to 2 mm long; flowers few, aromatic;

sepals 5, unequal, acuminate, the longest nearly equaling petals; petals 5, cream-colored, to 5 mm long, acute and spreading at apex; stamens 5, to 4 mm long, included; filaments free, to 1 mm long, stout, lacking appendages; anthers sagittate, ca 1.5 mm long, the connective expanded, burnt orange; ovary bearing appressed rufous pubescence; style simple, ca 2.5 cm long, included, pubescent on basal half. Capsules ellipsoid, markedly trilobate, ca 2.5 cm long, splitting widely at maturity; valves puberulent, reticulate; seeds 3 per valve, ca 4.5 mm long, densely puberulent. *Croat 8463, 8481.*

Common at least in some areas of the forest. Flowers and fruits principally from December to May, rarely in the middle of the rainy season. Flower buds are preformed at the end of the rainy season, and the flowers open rapidly after a heavy rain following a period of drought (R. Foster, pers. comm.).

Costa Rica to Colombia. In Panama, known from tropical moist forest in the Canal Zone, Bocas del Toro, Panamá, and Darién.

See Fig. 397.

96. FLACOURTIACEAE

Trees or shrubs, the branchlets sometimes spiny. Leaves alternate, petiolate; blades simple, entire or serrate, often pellucid-punctate; venation pinnate (palmate at base in *Hasseltia*); stipules present or absent. Inflorescences terminal, axillary, or leaf-opposed, cymose, fasciculate or spicate, bracteate; flowers bisexual or unisexual (dioecious), actinomorphic; calyx 4- or 5-lobed, the lobes united at base (3 sepals in *Lindackeria*, 4 in *Hasseltia*); petals 4 or 5 (6–12 in *Lindackeria*), alternate with sepals, or frequently lacking; stamens (6)10 to many, alternating with equal number of staminodia, or staminodia lacking; anthers 2-celled, introrse, dehiscing longitudinally; intrastaminal disk often present; ovary superior, sometimes stalked, 1-locular, 3- or 4-carpellate; placentation parietal; ovules many; style 1, simple or 2–5-branched, or styles 3; stigma simple (sometimes lobed), globular. Fruits berries with few to many seeds or 3-valved capsules with arillate seeds and abundant endosperm.

A morphologically very diverse family. Most species have serrate leaf margins, and many have pellucid dots

KEY TO THE TAXA OF FLACOURTIACEAE

Inflorescences fasciculate and axillary; petals lacking:
  Flowers unisexual; leaves lacking pellucid punctations, lacking stipules; plants with branches
      often bearing spines . . . . *Xylosma*
  Flowers bisexual; leaves pellucid-punctate (except *Casearia commersoniana*), stipulate; plants
      unarmed (except *C. aculeata*):
    Leaves densely pubescent, especially petioles; mature fruits more than 5 cm diam; flowers
        precocious, in axils of terminal leaflets only . . . . *Zuelania guidonia* (Sw.) Britt. & Millsp.
    Leaves not densely pubescent; mature fruits less than 5 cm diam; flowers not precocious (or
        precocious and axillary all along branches in *C. aculeata*):
      Stamens 6–12(15), alternating with staminodial appendages; capsules usually less than 2
          cm diam (to 2.5 cm diam in *C. arguta*); younger parts of plant often pubescent . . . . . .
          . . . . *Casearia*
      Stamens 10 or more, lacking appendages; capsules usually more than 2 cm diam (1.5–2 cm
          in *Laetia procera*); younger parts of plant mostly glabrous . . . . *Laetia*

Inflorescences not fasciculate, terminal or upper-axillary; petals lacking or not:
Leaf blades entire; petioles more than 5 cm long; petals more numerous than sepals; fruits bearing prickles . . . . *Lindackeria laurina* Presl
Leaf blades usually toothed; petioles usually less than 4 cm long; petals lacking or as many as sepals; fruits lacking prickles:
Petals present:
Leaves with 3 palmate veins at base, ± glabrous, with 2 large glands at base of blade above . . . . *Hasseltia floribunda* H.B.K.
Leaves pinnately veined throughout, conspicuously pubescent below, with 1 or 2 stalked glands at apex of petiole below . . . . *Banara guianensis* Aubl.
Petals lacking:
Flowers many, less than 2 mm broad, sessile, in congested spikes . . . . *Tetrathylacium johansenii* Standl.
Flowers relatively few, more than 8 mm broad, pedicellate, in open inflorescences:
Staminodia alternating with stamens; inflorescences cymose-corymbiform; leaf blades rounded to subcordate at base, equilateral . . . . *Casearia corymbosa* H.B.K.
Staminodia lacking; inflorescences dichasial; blades oblique at base, acute to obtuse . . . . *Laetia thamnia* L.

or glandular teeth. All except *Banara, Hasseltia, Tetrathylacium,* and *Xylosma* have fleshy capsules with several to many seeds embedded in an orange or red matrix.

Flowers are pollinated by many kinds of small insects (Heithaus, 1973).

Seeds of species with capsular arillate seeds, such as *Casearia, Laetia, Lindackeria,* and *Zuelania,* are probably dispersed chiefly by birds, as are those with small colorful berries, such as *Hasseltia floribunda* and *Xylosma.* Fruits of *H. floribunda* are also taken by the white-faced monkey (Hladik & Hladik, 1969). Those of *Tetrathylacium johansenii* are bat dispersed (Bonaccorso, 1975).

Some 84–93 genera and about 1,000 species; tropics and subtropics.

BANARA Aubl.

Banara guianensis Aubl., Hist. Pl. Guiane Fr. 1:548, pl. 217. 1775

Shrub or small tree, to 6 m tall; stems with short appressed pubescence when young, glabrate in age. Stipules triangular, minute, persistent; petioles 3–15 mm long, densely appressed-pubescent, with 1 or 2 stalked glands near apex on lower surface; blades oblong-ovate to oblong-elliptic, acuminate, usually rounded to truncate at base, 7–16 cm long, 3–7 cm wide, the pubescence short, appressed, sparse above, the margins serrate, the teeth gland-tipped. Panicles narrow, to 12 cm long, terminal or opposite leaves; inflorescence branches, sepals, and petals densely whitish-tomentose; peduncles to 5 cm long; pedicels ca 5–9 mm long; flowers 4-parted; sepals ovate, 4–6 mm long, persistent in fruit; petals narrowly ovate, greenish-yellow, alternating with sepals, subpersistent in fruit; stamens many; style simple. Berries globose or depressed-globose, ca 8 mm diam, glabrous, green to blackish, with a long stout apiculum at apex; seeds many, tiny. *Foster 1175.*

Rare, in the forest. Flowers from May to July (rarely as early as January elsewhere). The fruits are mature principally from July to September.

The fruits are probably chiefly bird dispersed.

Costa Rica to northern South America. In Panama, known from tropical moist forest in the Canal Zone, Bocas del Toro, Veraguas, Coclé, and Panamá, from tropical dry forest in Coclé, from premontane wet forest in Chiriquí, Coclé, and Panamá, from tropical wet forest in Colón and Coclé, and from premontane rain forest in Coclé.

CASEARIA Jacq.

The genus is recognized by its conspicuously pellucid-punctate leaves with small stipules and its bisexual, apetalous flowers. The flowers usually grow in bracteate fasciculate inflorescences in leaf axils and have four to six sepals and six to fifteen stamens, which alternate with staminodia.

Casearia aculeata Jacq., Enum. Pl. Carib. 21. 1760
C. guianensis (Aubl.) Urban var. *rafflesioides* Croat

Tree, ca 5 m tall; trunk ca 7 cm dbh, with simple spines near base; branches and branchlets glabrate to densely strigillose or puberulent; branches arching, wide-spreading, often with straight, stout, sharp branch-spines. Stipules deltoid to narrowly triangular, ca 2 mm long; petioles 2–8 mm long; blades ovate to elliptic or obovate or oblanceolate, acuminate to acute or rounded at apex, obtuse to attenuate and decurrent at base, 2.5–11 cm long, 1.3–4 cm wide, glabrous or with sparse, ± appressed pubescence, weakly or not at all pellucid-punctate, the margins crenate-serrate, the teeth obscure to sharp, glandular and incurled. Fascicles sparse, sessile, axillary; bracteoles connate, sparsely pubescent, whitish to translucent, very thin, the outermost ± triangular, very short, the inner ones oval, rounded at apex; pedicels 1.5–3 mm long, slender, to 3 mm wide when dried, sparsely villous, articulate at about middle, usually longer than bracteoles; flowers greenish-white, ca 7 mm diam; sepals 4–6, 3–3.5(5.5) mm long, strigillose outside, blunt to rounded at apex, spreading at anthesis; corolla lacking; stamens 7 or 8(9), ca 4 mm long, erect at anthesis, fused into a ring at base, alternating with densely villous stami-

KEY TO THE SPECIES OF CASEARIA

Stigmas 3:

    Style simple; blades densely pellucid-punctate; calyx 2–3 mm long . . . . *C. sylvestris* Sw.

    Style with 3 branches; blades lacking pellucid-punctations; calyx 3–5 mm long . . . .

    . . . . *C. commersoniana* Camb.

Stigma 1:

    Inflorescences cymose-corymbiform; capsules to 1.5 cm long . . . . *C. corymbosa* H.B.K.

    Inflorescences fasciculate:

        Fascicles distinctly short-pedunculate; blades oblong, finely toothed . . . .

        . . . . *C. arborea* (L. C. Rich.) Urban

        Fascicles sessile or nearly so:

            Leaf blades coarsely serrate; sepals 5–7 mm long, gradually tapered to a ± sharp-pointed apex, united at base into a tube 1–1.5 mm long; leaves present at time of flowering, falling at time of fruiting; fruits to 2.5 cm diam; plants rare or no longer present on the island . . . . *C. arguta* H.B.K.

            Leaf blades entire or shallowly toothed; sepals 3–6 mm long, ± oblong, usually rounded at apex, not markedly united at base; leaves present or absent at flowering time, present at time of fruiting; fruits less than 1 cm wide; plants occasional to common on BCI:

                Leaves elliptic (rarely obovate), mature at time of flowering, drying green on flowering collections, to 9.5 cm long and 4.5 cm wide (usually to 7.5 cm long and 2.5 cm wide); stipules deltoid to narrowly triangular, 1.5–2 mm long; branchlets frequently with stout, sharp branch-spines (branches ending in a sharp spine); pedicels articulate usually at about middle . . . . *C. aculeata* Jacq.

                Leaves obovate, usually very young or even lacking at time of flowering, drying blackish on flowering collections, the blades of at least the larger leaves more than 9.5 cm long and 4.5 cm wide; stipules narrowly triangular to subulate, 2–5 mm long; branchlets lacking branch-spines; pedicels articulate usually well below middle . . . . *C. guianensis* (Aubl.) Urban

nodia, the staminal tube glabrous outside, pubescent inside; filaments ± glabrous; anthers 1 mm long, introrse, equaling height of style; pollen yellowish, ± tacky; ovary sparsely villous, narrowly ovate; style short; stigma globular, viscid, short-puberulent. Capsules 3-valved, round to ellipsoid, to ca 1 cm long, pale green to white, often marked with purple; valves maroon inside, marked with prominent white spots; seeds of irregular shapes, ca 4 mm long, enveloped in a pale orange aril. *Croat 11777, 13268, 14057.*

Occasional, in the forest. Flowers in March and April. The fruits mature from April to June.

Distinguished from *C. guianensis* by having elliptic leaves present at the time of flowering and by having sharp branch-spines. Other differences are discussed elsewhere (Croat, 1975b).

H. Sleumer considers *Casearia stjohnii* I. M. Johnston synonymous (pers. comm.); I consider it distinct (Croat, 1975b), though probably more closely related to *C. aculeata* than (as I had originally thought) to *C. guianensis*. *Casearia stjohnii* flowers later than *C. aculeata* (June and July), with the fruits maturing during August and September. It also has thicker, longer inflorescence bracts and much stouter, more densely pubescent pedicels.

Mexico to Colombia, Venezuela, Peru, and Brazil. In Panama, known principally from wetter parts of tropical moist forest in the Canal Zone, Panamá, and Darién.

Casearia arborea (L. C. Rich.) Urban, Symb. Ant. 4:421. 1910

Tree, to 20 m tall and 25 cm dbh; branches puberulent when young. Petioles to 5 mm long; blades oblong-elliptic, often weakly inequilateral, long-acuminate, acute to rounded at base, 6–15 cm long, 2–6 cm wide, conspicuously pellucid-punctate or pellucid-lineolate, finely crenulate-serrate; veins puberulent. Fascicles dense, stalked; peduncles to 4 mm long; pedicels to 4 mm long, articulated near middle; pedicels and calyces densely short-pubescent; flowers white to greenish; calyx 4–4.5 mm long, the lobes 5, united at base, persistent in fruit; corolla lacking; stamens 10, slightly shorter than calyx lobes; filaments unequal; staminodia shorter than shortest anthers, cupular, villous at apex; ovary pubescent at apex; style short, simple; stigma globular. Capsules subglobose, apiculate, 4–5 mm long, pilose at apex, 3-valved, splitting at maturity; seeds 1–6, pyriform, brown, punctate; aril orange, fimbriate-lacerate. *Croat 11779, 16210.*

Occasional, in the forest. Flowers mostly in June and July (rarely as early as March). The fruits mature from July to September (sometimes to January).

Belize to the Guianas, Brazil, Peru, and Bolivia; Greater Antilles. In Panama, known from tropical moist forest in the Canal Zone, from premontane wet forest in the Canal Zone and Colón, and from premontane rain forest in Panamá.

Casearia arguta H.B.K., Nov. Gen. & Sp. 5:364. 1823

    Pica lengua, Raspa lengua

Shrub or small tree, 2–10(12) m tall; trunk to 12 cm diam; branchlets often densely ferruginous-pubescent, becoming glabrous and lenticellate. Leaves deciduous; petioles usually less than 5 mm long; blades ± oblong-elliptic, acuminate, ± acute at base, 9–17 cm long, 2.5–6 cm wide, usually pellucid-punctate, pubescent on veins on

both surfaces (especially below), the margins coarsely serrate with gland-tipped teeth. Fascicles dense, sessile, axillary, ca 2 cm broad; pedicels somewhat shorter than flowers, articulated near base; flowers greenish-white, with a moderately sweet aroma, 4–5 mm long; calyx 5-lobed, united near base, spreading at anthesis; corolla lacking; stamens usually 10, stiffly erect, alternating with very pubescent staminodia, the 5 alternating with calyx lobes somewhat shorter than the others; anthers with apical bristles; style held above anthers in bud; stigma round, minutely bristled, the surface with an abundant sticky fluid; ovary conspicuously pubescent, the pubescence merging with that of staminodia. Capsules ± globose, to 2.5 cm diam, yellow at maturity, nearly glabrous; seeds numerous, irregular, to 8 mm long, embedded in a sweet, juicy, orange matrix.

Common in the Canal Zone and to be expected on the island, but not seen in recent years. It was reported by Standley for the island, but no collections have been found. Flowers in the late rainy and early dry seasons; according to Allen (1956), it flowers several times at short intervals. The fruits mature in the late dry and early rainy seasons. Plants lose their leaves in the late dry season, while still bearing fruit, and remain leafless for a short time.

The flowers are visited by *Trigona* bees.

Throughout the tropics of the Western Hemisphere. In Panama, known principally from tropical moist forest in the Canal Zone, Bocas del Toro, Herrera, Veraguas, Panamá, and Darién; known also from tropical dry forest in Herrera and Coclé and from premontane wet forest in Chiriquí and Coclé.

Casearia commersoniana Camb. in St.-Hil., Adr. Juss. & Camb., Fl. Bras. Merid. 2:235. 1830

C. javitensis sensu auct. non H.B.K.

Small tree, usually less than 10(15) m tall; trunk 10–20 cm dbh; bark smooth; branches ± glabrous. Petioles to 5 mm long; blades ± narrowly elliptic, long-acuminate, acute to obtuse at base, 7–25 cm long, 2.5–9 cm wide, glabrous, lacking pellucid punctations, coarsely serrate to crenate or nearly entire. Fascicles axillary, to 2.5 cm diam; pedicels and sepals pubescent; pedicels to 1 cm long, articulated below middle; flowers few to many, white or greenish; sepals 4 or 5, 3–4(5) mm long; corolla lacking; stamens 9–15, longer than sepals; ovary sparsely to rather densely pubescent; style with 3 short branches; stigmas capitate. Capsules subglobose, ca 1 cm diam, apiculate, green turning red or brown; seeds 1 or 2, subglobose, 4–5 mm long. *Croat 6389, 6561.*

Occasional, on the shore or along trails. Flowers mostly from June to September, with the fruits maturing from August to October.

Southern Mexico to northern South America. In Panama, known principally from tropical moist forest in the Canal Zone, all along the Atlantic slope, and in Chiriquí, Panamá, and Darién; known also from premontane wet forest in Chiriquí and Panamá and from tropical wet forest in Colón and Panamá.

See Fig. 398.

Casearia corymbosa H.B.K., Nov. Gen. & Sp. 5:366. 1823

Comida de loro, Mamar, Mako, Carano

Deciduous shrub or small tree, 1.5–7.5 m tall. Petioles 3–15 mm long; blades ± elliptic to oblong-obovate, variable but usually blunt-acuminate at apex, narrowed to an obtuse to subcordate base, 6–16 cm long, 3.5–5.5 cm wide, conspicuously pellucid-punctate and pellucid-lineate, glabrous above, glabrous to puberulent below, minutely serrulate to crenulate-serrate. Corymbs axillary, cymose; pedicels to 6 mm long, articulated below middle; flowers white or greenish; calyx ca 5 mm long, 5-lobed; corolla lacking; stamens 8, ca 3 mm long; filaments villous; anthers introrse; staminodia densely pubescent, one-third to one-half as long as stamens; ovary sparsely pilose; stigma ± globose, held at level of anthers. Capsules globose to ellipsoid, with usually 3 longitudinal ridges, to 1.5 cm long, glabrous, orange to red at maturity; seeds 1–3, ± ovoid, to 7 mm long, covered with resinous glands, with an incomplete red aril.

Reported by Standley (1933); the species could not easily be confused by him with any other, but I have not seen any collections from the island. Elsewhere in Panama, flowers commonly in the late dry season and early rainy season (April and May). The fruits are mature in the late rainy and early dry seasons. Leaves fall in the dry season and grow in again at the time of flowering.

The species has been confused with *Casearia nitida* (L.) Jacq. in the *Flora of Panama* (Robyns, 1968) and elsewhere. As indicated by Adams (1972), *Casearia nitida* occurs only in the West Indies and is distinct from *C. corymbosa.*

Mexico to Colombia. In Panama, ecologically variable; most commonly found in tropical moist forest in the Canal Zone, along the Atlantic slope, and in Veraguas, Panamá, and Darién; known also from premontane dry forest in the Canal Zone, from tropical dry forest in Los Santos and Panamá, and from premontane wet forest in Colón (Nombre de Dios).

Casearia guianensis (Aubl.) Urban, Symb. Ant. 3:322. 1902

Palo de la cruz

Shrub or small tree, (1.5)3–6(10) m tall; trunk to 12 cm dbh; stems glabrous to puberulent. Leaves deciduous; stipules subulate, 2–5 mm long, densely pubescent, caducous; petioles 2–10 mm long; blades mostly obovate, abruptly acuminate, acute and ± decurrent at base, 5–18 cm long, 2–8.5 cm wide, glabrous or puberulent on veins especially below, pellucid-punctate and pellucid-lineate, ± entire to shallowly crenate-serrate, the teeth often sharp. Fascicles dense, in old leaf axils; pedicels and calyx puberulent; pedicels 2.7–4.7 mm long at anthesis, articulated shortly above base, elongating to 8 mm in fruit; flowers white, with a faint sweet aroma, precocious; calyx deeply 4–6-lobed, 4–6 mm long, the lobes spreading; corolla lacking; stamens (7)8, ± glabrous, 2.5–3.5 mm long, alternating and uniting with much shorter, densely villous staminodia; ovary villous; stigma globular,

Fig. 398. *Casearia commersoniana*

Fig. 399. *Hasseltia floribunda*

held ± at level of anthers. Capsules ellipsoid, 8–14 mm long, green often becoming violet-purple at least at apex, splitting into 3 parts; valves red inside; seeds several, irregular, brown, ca 3 mm long, enveloped in an orange aril. *Croat 8391, 8747.*

Common along the shore and at the margins of clearings; occasional in the young forest. Flowers commonly from March to May. The fruits mature from March through June. Leaves fall before flowering, and new leaves begin to develop near the time of flowering in the late dry season.

The species is similar to *Casearia aculeata.* See that species for a discussion.

Costa Rica to Venezuela, the Guianas, and Brazil (Pará). In Panama, known only from tropical moist forest in the Canal Zone, Bocas del Toro, Veraguas, Herrera, Panamá, and Darién.

Casearia sylvestris Sw., Fl. Ind. Occ. 752. 1798
Corta lengua

Shrub or small tree, 2–10 m tall; trunk to 12 cm dbh; branchlets puberulent when young, sometimes on one side only. Stipules minute, caducous; petioles 3–10 mm long, canaliculate; blades ± oblong-elliptic, long-acuminate and downturned at apex, ± acute to obtuse at base and sometimes inequilateral, mostly 2–14 cm long, 2.5–5 cm wide, glabrous, conspicuously pellucid-punctate, the margins entire to obscurely toothed, undulate. Flowers greenish-white or purple, 2–3 mm long, in dense axillary clusters; pedicels to 5 mm long, articulate near middle, the basal part persisting; calyx lobes 5, broadly ovate, divided to near base; corolla lacking; stamens 10; filaments united to calyx tube, unequal, the longer ones opposite calyx lobes; anthers broader than long, of 2 lengths, held horizontal, dehiscing downward; staminodia broad, densely pubescent, alternating with and shorter than stamens; ovary ± glabrous; style at level between long and short anthers; stigmas 3, globular, ± sessile. Capsules ± globose, 3–4 mm long, apiculate, pubescent at apex, surrounded by persistent calyx, reddish or purple at maturity, 3-valved; seeds 2–6, ca 2 mm long, with an incomplete red aril. *Croat 4977, 6442.*

Frequent in the forest, especially the young forest, and along the shore. Flowers principally from August to December with sporadic flowering in the dry season. The fruits mature in about 1 month.

Throughout tropical America. In Panama, known principally from tropical moist forest in the Canal Zone, all along the Atlantic slope, and in Chiriquí, Veraguas, Coclé, Panamá, and Darién; known also from premontane wet forest in Chiriquí and Panamá.

HASSELTIA Kunth

Hasseltia floribunda H.B.K., Nov. Gen. & Sp. 7:232, pl. 651. 1825
Raspa lengua, Parimontón

Small tree, 4–20 m tall; trunk to 30 cm dbh, often forming stilt-roots when leaning; outer bark smooth, thin; inner bark tan, granular, streaked with harder dark areas; young stems, petioles, and midribs above sparsely pubescent. Petioles 1–4 cm long; blades ± elliptic, acuminate, obtuse at base and with 2 green glands on upper surface, 7–22 cm long, 4–11 cm wide, coarsely serrate; palmate veins at base 3, with a second pair of lateral veins arising in apical half to two-thirds of blade. Umbels divaricately branched, often broader than long; inflorescence branches, pedicels, and flowers puberulent; flowers 4-parted; sepals acute at apex; petals white, narrowly spatulate, to 2.7 mm long, equaling sepals; both petals and sepals recurved at anthesis; stamens numerous, 3–4 mm long; ovary ± stalked, nearly glabrous; style to 2 mm long; stigma simple. Berries subglobose, to 8 mm long, red turning dark violet-purple at maturity; seeds 1 or 2, to 5 mm long. *Croat 7273.*

Common in the forest, especially the young forest. Flowers mostly in the dry season, but as late as June. Most fruits mature in the late dry and early rainy seasons. Plants may develop a second inflorescence from the base of the infructescence at about the time the fruits are maturing.

Fruits are eaten by white-faced monkeys in June (Hladik & Hladik, 1969).

Honduras to northern South America. In Panama, known from tropical moist forest in the Canal Zone, Bocas del Toro, Colón, Panamá, and Darién and from tropical wet forest in Colón (Portobelo). Reported from premontane wet and premontane rain forests in Costa Rica (Holdridge et al., 1971).

See Fig. 399.

LAETIA Loefl. ex L.

Laetia procera (Poepp. & Endl.) Eichl. in Mart., Fl. Brasil. 13(1):453. 1871
Casearia belizensis Standl.

Tall tree, probably exceeding 30 m; branches and petioles usually glaucous. Leaves distichous; stipules minute, caducous; petioles 4–15 mm long; blades elliptic-oblong, abruptly short-acuminate, rounded to minutely cordate at base, usually 7–22 cm long, 3.5–5 cm wide, minutely appressed-denticulate, somewhat pellucid-punctate, paler

---

KEY TO THE SPECIES OF LAETIA

Leaves usually more than 14 cm long, rounded to subcordate at base; fruits glabrous . . . .
. . . . *L. procera* (Poepp. & Endl.) Eichl.
Leaves less than 12 cm long, acute at base; fruits bearing dense, short, brown, tomentulose
pubescence . . . . *L. thamnia* L.

below than above. Umbelliform fascicles of 4–25(30) flowers, arising from leaf axils or from points ca 5 mm above axils or taking place of leaf along stem, arising on basal two-thirds of branch; pedicels 5–12 mm long; sepals somewhat petaloid, obtuse-reflexing, 2–4 mm long; corolla lacking; stamens 12–20; filaments ca 2 mm long; anthers ca 1 mm long, oblong, style simple, 1–2 mm long, persisting on fruit. Capsules berrylike, 1.5–2 cm thick, glabrous, subglobose; seeds 10–20, pitted, ca 3 mm long. *Knight 1301.*

Apparently rare; restricted to the old forest. Seasonal behavior not certain. Flowers mostly from April to July and fruits in August and September elsewhere.

Belize to Panama and the Guianas to Brazil; West Indies. In Panama, known only from tropical moist forest in the Canal Zone. Reported from premontane wet and tropical wet forests in Costa Rica (Holdridge et al., 1971).

Laetia thamnia L., Pl. Jam. Pugil. 31. 1759

Conejo

Glabrous tree, to 10 m tall; branches lenticellate, terete. Stipules minute, caducous; petioles canaliculate, ca 1 cm long; blades ± elliptic, acuminate, acute and inequilateral at base, 5–12 cm long, 2–3.5 cm wide, pellucid-punctate and glabrous, crenulate. Flowers white to greenish-white, in axillary dichasia to ca 8 cm long, usually only a few open at a time; bracts minute; pedicels puberulent, 9–13 mm long, articulated near base; sepals 4, caducous, ± elliptic or boat-shaped, rounded at apex, to 9 mm long; corolla and staminodia lacking; stamens many, to 8 mm long; filaments often puberulent; ovary ovoid; style short; stigma simple. Capsules ± globose, to 4 cm diam, with dense, short, brown, tomentulose pubescence; exocarp thick; valves 3, thick, red inside when dehisced; seeds irregular, rounded, to 6 mm long, embedded in an orange mesocarp. *Croat 13977.*

Occasional, in some areas of the old forest and along the margin of the lake on the south side of the island. Flowers mostly from March to May, with the fruits maturing from June to September. Frankie, Baker, and Opler (1974) reported that the species may flower more than once a year. Leaves probably fall during the dry season.

Southern Mexico to northern South America; West Indies. In Panama, known from tropical moist forest in the Canal Zone, Veraguas (Coiba Island), and Panamá and from tropical wet forest in Colón (Miguel de la Borda).

LINDACKERIA Presl

Lindackeria laurina Presl, Rel. Haenk. 2:89, pl. 65. 1835

Oncoba laurina (Presl) Warb.

Guavo cimarron, Carbonero, Uvre, Chopo cucullo, Amarillo carbonero

Tree, 3–8(15) m tall, 5–35 cm dbh; wood yellow; stems weakly striate, brown or with a thin gray flaky crust, the leaf scars 3-sided, raised. Petioles 5–11 cm long, canaliculate, somewhat swollen at apex; blades ± oblong-elliptic, long-acuminate, obtuse to rounded at base, 10–30 cm long, 5–11 cm wide, glabrous, the midrib raised above. Panicles axillary and terminal, to ca 20 cm long; pedicels to 1.5 cm long; flowers ca 1.2 cm diam, conspicuous and sweetly scented; sepals 3, reflexed at anthesis, caducous; petals 6–12, white, blunt at apex; stamens numerous, unequal, exserted; anthers slightly bifid at apex; style 4–6 mm long; stigma shortly trilobate. Capsules 1–1.5 cm diam, covered with stiff bristles, dehiscing into 3 or 4 parts to expose seed; seed usually solitary, smooth, broadly ovate, ca 7 mm long, flattened on one side, gray-coated, with a shiny, bright red, lateral aril. *Croat 6097, 7753.*

Locally abundant along some areas of the shore, especially on the southern side of the island; frequent in the forest. Flowers principally in the rainy season from May to October, rarely as early as April or as late as December. The fruits mature in the early dry season (January to March). Many leaves fall in late June.

Southern Mexico to northern South America. In Panama, known from tropical moist forest in the Canal Zone, Chiriquí, Panamá, and Darién, from premontane wet forest in Chiriquí and Panamá, and from tropical wet forest in Colón.

TETRATHYLACIUM Poepp. & Endl.

Tetrathylacium johansenii Standl., J. Wash. Acad. Sci. 15:479. 1925

Tree, 5–30 m tall; trunk ca 55 cm dbh; bark thin, brown, with many small vertical fissures; branchlets glabrous to puberulent. Stipules foliaceous, ± oblong, to 17 mm long; petioles to 7 mm long; blades ± oblong, acuminate, usually broadest above middle, gradually narrowed to rounded or subcordate base, mostly 12–20(25) cm long, 3–6(9) cm wide, glabrous except sometimes on midrib below, with depressions in axils of larger veins below, entire to usually obscurely toothed, the teeth gland-tipped. Inflorescences paniculate-spicate, to 8 cm long; flowers bisexual, white, very numerous, sessile, angulate, congested on secondary rachises to 4.5 cm long; calyx ca 2 mm long, with 4 minute lobes; corolla lacking; stamens 4, at first included with the anthers directed inward, becoming exserted with the anthers turned outward and folding over apex of sepals; anthers as broad as long; style short; stigma simple, held just below and adjacent to anthers. Fruits baccate, globose to obovoid, to 2.5 cm long, glabrous at maturity; seeds numerous, ovoid, to 2 mm long. *Croat 10152.*

Uncommon to rare. The tree usually loses its leaves in the early rainy season and immediately puts on new ones. Flower buds appear with the new leaves, and most flowers are seen from April through June. The fruits mature quickly, mostly from May to August. Mature fruits have been collected in Colón (Santa Rita Ridge) in January.

The fruits are probably dispersed chiefly by mammals.

Fig. 400. *Tetrathylacium johansenii*

Fig. 401. *Xylosma chloranthum*

Fig. 402.
Xylosma oligandrum

Costa Rica to Colombia. In Panama, known from tropical moist forest in the Canal Zone, San Blas, and Darién and from premontane wet forest in Colón (Santa Rita Ridge).

See Fig. 400.

XYLOSMA Forst.f.

Xylosma chloranthum Donn. Sm., Bot. Gaz. (Crawfordsville) 57:415–16. 1914

Dioecious shrub or tree, to 5 m tall, glabrous except for puberulence on young stems; trunk usually with large branched spines; stems prominently lenticellate, unarmed or armed with straight stiff axillary spines to 2.5 cm long (perhaps only on juveniles). Petioles 3–5 mm long; blades narrowly ovate to lanceolate-elliptic, long-acuminate, acute to rounded at base, 8–19 cm long, 2–6 cm wide, obtusely glandular-crenate, shiny on both surfaces, lacking pellucid-punctations; reticulate veins prominulous on both sides. Flowers fasciculate, yellowish, usually growing in leafless axils; pedicels 3–7 mm long, puberulent, articulate in basal third of staminate flowers and near middle of fruiting pedicel; sepals 4 or 5, very broadly ovate to nearly rounded, 1.5–2 mm long, ferruginous-puberulent and glandular near apex, ± ciliate near apex; corolla lacking; stamens 22–29 (35), to 6 mm long, glabrous; disk prominently undulate-lobed; style (3) 4 (5)-lobed, the lobes short and spreading. Berries ellipsoid, ca 6 mm long; seeds few, ovoid, to 4.5 mm long. *Croat 10087, 11617, Foster 785.*

Infrequent, in the forest. Flowers during the dry season and especially in the early rainy season, usually in May while plants are still devoid of leaves. The fruits develop within a short time, often even before the leaves have been replaced, usually in May and June.

The taxonomy of the genus *Xylosma* is among the most seriously in need of work of all Central American plant genera. The names used here thus will remain doubtful until a complete revision of the group can be made. This taxon has been confused with *X. excelsa* Standl. & L. O. Wms., which differs by being much taller (13–25 m), by being polygamous, and by bearing leaves at flowering time. *X. chloranthum* has also been confused with *X. intermedia* (Seem.) Tr. & Planch., which has polygamous flowers and has been transferred to the genus *Eichlerodendron.*

Belize, Guatemala, Costa Rica, and Panama, probably throughout Central America. In Panama, known from tropical moist forest in the Canal Zone and Panamá and

from premontane wet forest in Colón (Santa Rita Ridge), Chiriquí (near Boquete), and Panamá (Cerro Azul). Reported from premontane rain forest in Costa Rica (Holdridge et al., 1971).

See Fig. 401.

Xylosma oligandrum Donn. Sm., Bot. Gaz. (Crawfordsville) 23:235. 1897

X. sylvicola Standl.

Dioecious tree, to (2) 5 (10) m tall, often branching near ground; trunk spineless (elsewhere with branched spines); larger branches sparsely armed with a few simple spines to 1 cm long; stems glabrous. Petioles ca 5 mm long, glabrous to puberulent; blades elliptic, acuminate, acute to obtuse at base, 8–16 cm long, 3.5–7.5 cm wide, glabrous except sometimes puberulent on midrib below, the margins with minute, gland-tipped teeth. Inflorescences very short racemes usually less than 3.3 cm long, often appearing glomerulate; pedicels 5–8 mm long, not articulate, puberulent, subtended by triangular to oblong bracts to 2 mm long; flowers unisexual; calyx deeply 4-lobed, to 1.3 mm long, the lobes ovate, rounded at apex, minutely ciliate, often weakly keeled, glabrous on outside, weakly pubescent inside; corolla lacking; stamens 8–16, to 2.7 mm long, glabrous; anthers as broad as long; disk of staminate flowers 9–12-lobed, the lobes rounded, irregular, the disk of pistillate flowers entire or bipartite; ovary ellipsoid to obovoid, ca 1.5 mm long; style with 2 short flattened branches, each with 2 rounded lobes. Berries ellipsoid to obovoid, 8–12 mm long, to 9 mm wide, becoming orange, then bright red, finally violet-purple at maturity; exocarp thin; mesocarp fleshy; seeds 1–3, ovoid, ca 5 mm long. *Croat 14640, 14642, Knight 1201.*

Occasional in some areas of the older forest; not seen elsewhere on the island. Flowers from March to July, mostly in the early rainy season. Mature fruits have been seen from April to September.

Knight 1201 has both staminate- and pistillate-flowered twigs. If the two branches were taken from the same plant, the species is not always dioecious. Field observations show only the dioecious condition, however. In contrast to *X. chloranthum,* this species flowers and fruits without losing leaves. The plants of this species from Coclé (El Valle) differ in having smaller, sometimes thicker leaves 3–10 cm long and 1.5–4.5 cm wide. Calyx lobes also average slightly larger and are more pubescent on the inner surface. The species corresponds to *Xylosma* species #2 in the *Flora of Panama* (Robyns, 1968).

Mexico (Chiapas) to Panama. In Panama, known from

KEY TO THE SPECIES OF XYLOSMA

Leaves lanceolate-elliptic, usually more than 2.5 times as long as broad, with conspicuous crenate teeth; branched spines on trunk of mature plant; pedicels articulate; sepals pubescent only at glandular apex; stamens 22 or more . . . . *X. chloranthum* Donn. Sm.

Leaves ± elliptic, usually less than 2.5 times as long as broad, minutely toothed; branched spines lacking; pedicels not articulate; sepals pubescent inside, not glandular at apex; stamens 16 or less . . . . *X. oligandrum* Donn. Sm.

Fig. 404. *Zuelania guidonia*

Fig. 403. *Zuelania guidonia*

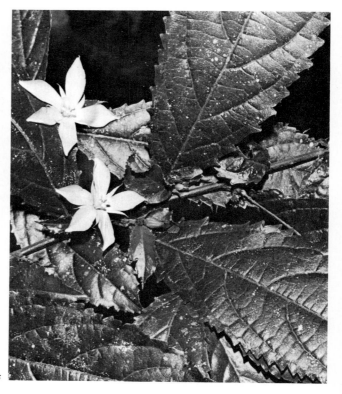

Fig. 405. *Turnera panamensis*

tropical moist forest on BCI and from premontane wet forest in Coclé (El Valle).

See Fig. 402.

ZUELANIA A. Rich.

Zuelania guidonia (Sw.) Britt. & Millsp., Bahama Fl. 285. 1920

Cagajón, Caranon

Tree, 10–25 m tall, to 50 cm dbh; branches dark, conspicuously covered with light lenticels (and raised leaf bases on older wood). Leaves deciduous, often clustered near apex of branchlets; stipules lanceolate, 3–6 mm long, densely pubescent; petioles 5–20 mm long, densely brown-hirsute; blades ± oblong-elliptic, short-acuminate, rounded to subcordate at base, 6–16(25) cm long, 2–6(9) cm wide, densely pubescent below, sparsely so above, pellucid-punctate, obscurely serrulate. Inflorescences of globose fascicles in uppermost axils of leafless branches; flowers precocious, ca 1.5 cm broad, white or yellow, with a faint, rather foul aroma; pedicels and sepals conspicuously pubescent; pedicels to 18 mm long, articulate a few mm above base; sepals 5, 6–7 mm long, ± rounded at apex; corolla lacking; stamens numerous, exserted, ca 4 mm long, interspersed with staminodia about half as long; both stamens and staminodia united into a flat platelike base; anthers introrse; ovary densely short-tomentose; style short; stigma as broad as ovary. Capsules fleshy, depressed-globose, somewhat 3-sided, rounded at apex, subcordate at base, mostly 6–8 cm diam, usually dark brown at maturity, softly pubescent; valves thick, exposing bright orange matrix and seeds; seeds many, obovoid, ca 4 mm long, angular, white. *Croat 4962, 13995.*

Frequent in the forest, especially the young forest. Leaves fall sometime before flowering. Flowers mostly from February to May, before or during the onset of new leaves. Leaves are often not full size before the fruits reach maturity from April to June.

Southern Mexico to Panama; West Indies, Trinidad. In Panama, known from tropical moist forest in the Canal Zone, Panamá, and Darién and from tropical dry forest in Coclé (Penonomé). Reported from premontane moist forest in Costa Rica (Holdridge et al., 1971).

See Figs. 403 and 404.

97. TURNERACEAE

Shrub. Leaves alternate, petiolate; blades simple, serrate; venation pinnate; stipules present. Flowers bisexual, actinomorphic, solitary in axils, often on short branchlets, heterostylous (in BCI species); sepals 5, free, imbricate; petals 5, free, clawed, showy; stamens 5, free, attached to hypanthium; anthers 2-celled, dehiscing longitudinally; ovary superior, 1-locular, 3-carpellate; placentation parietal; ovules many, anatropous; styles 3; stigmas penicillate. Fruits 3-valved, loculicidal capsules; seeds many, arillate, with endosperm.

Turnera is distinguished by the leaves with glandular-crenate margins, the heterostylous orange flowers with a hypanthium and fimbriate-tipped stigmas, and the arillate seeds.

Flowers are insect pollinated (Doctors van Leewen, 1938). *Turnera* is pollinated by small bees in Guanacaste, Costa Rica (D. Janzen, pers. comm.).

The arillate seeds are probably ornithochorous.

Seven or 8 genera and about 120 species; American and African subtropics and tropics.

TURNERA L.

Turnera panamensis Urban, Jahrb. Königl. Bot. Gart. Berlin 2:92. 1883

Shrub, to 4 m tall; branches appressed-puberulent. Stipules subulate, to 4 mm long, subpersistent; petioles to 5 mm long, caniculate above; blades ± elliptic, acuminate, acute and decurrent at base, mostly 10–15(18) cm long and 3.5–5(6) cm wide, glabrate above, puberulent below, glandular-crenate; veins prominent; flowers 5-parted, solitary, on short, bracteate, axillary branches, heterostylous; sepals ribbed, long-acuminate, to 1.5 cm long; petals orange, 2–3.5 cm long, blunt at apex, inserted near throat of hypanthium, widely spreading at anthesis, caducous; stamens 5; filaments either to 6 mm or to 12 mm long, their respective styles to 8 mm or to 5 mm long; stigmas numerous. Capsules 3-valved, ovoid, 6–13 mm long, minutely tuberculate, brown at maturity; seeds numerous, fewer than 10 per valve, ± oblong-curved, puberulent, striate, subtended by an aril, the aril collar-shaped, displayed on valves of open capsule. *Croat 7335.*

Frequent in the forest, especially in the old forest, and along the shore. Flowers and fruits sporadically throughout the year. Plants flower in a distinct synchronous wave if a period of drought is followed by a heavy rain (R. Foster, pers. comm.).

Distinguished by the orange flowers and glandular-crenate leaves.

Known only from Panama, from tropical moist forest in the Canal Zone and Panamá.

See Fig. 405.

98. PASSIFLORACEAE

Vines or lianas, climbing by means of simple, axillary tendrils. Leaves alternate; petioles usually bearing prominent, often stalked glands; blades simple, usually 2- or 3-lobed, the margins entire or serrate; venation palmate, palmate at base, or rarely strictly pinnate; stipules present. Flowers bisexual, actinomorphic, rarely apetalous, solitary or paired (rarely in racemes), axillary; peduncles jointed (often mistaken for a pedicel), bearing 3, often setaceous bracts; sepals 5, free, horned below the tip, often showy; petals 5, free, showy (lacking in *P. coriacea*); corona (petaloid structure between petals and stamens) of several series; corona and perianth parts attached to hypanthium; operculum (ring of tissue within corona)

membranous, sometimes plicate; stamens 5, attached to androgynophore (the fused staminate and pistillate parts); anthers 2-celled, versatile, dehiscing longitudinally; ovary superior, at apex of androgynophore, 1-locular, 3-carpellate; placentation parietal; ovules many, anatropous; styles 3 or stigmas sessile; stigmas 3, capitate. Fruits berries; seeds many, with fleshy endosperm, embedded in usually clear, fleshy, sweet mesocarp.

The family is represented on the island only by the genus *Passiflora*. The genus is recognized by being tendriled vines with glandular petioles, by the showy, solitary or paired flowers with a modified corona and an androgynophore (complex structure containing both staminate and pistillate parts), and by the fleshy, many-seeded, usually tasty fruits.

The flowers are very specialized. Most probably follow the behavior of *Passiflora foetida* and *P. vitifolia* as reported by Janzen (1968). After the flower opens, the styles become deflected, which promotes outcrossing. During first visits to the flower the pollinating organism receives only pollen, but later visits bring it in contact with the stigmas. *P. vitifolia*, which is hummingbird pollinated, has strong stigmal deflection 30–180 minutes after opening. Janzen (1968) reported that the flowers open asynchronously between 5:30 and 7:30 A.M. and are repeatedly visited. Possibly all blue- and white-flowered species are bee pollinated (G. Frankie, pers. comm.), though the possibility of self-pollination (Lewis, 1966) cannot be overlooked. *P. foetida* is reported by Janzen to open synchronously in the early morning hours and is visited almost immediately by bees of the genus *Ptiloglossa* (Colletidae); stigmal deflections take place within 30 minutes and all visiting activity ends in about 45 minutes. Flowers of *Passiflora* last but a single day and most wither well before the day is over.

Fruits are endozoochorous. Most seeds of the small, thin-walled fruits are probably dispersed chiefly by birds, whereas the larger-fruited species, such as *P. ambigua*, *P. menispermifolia*, *P. nitida*, *P. seemannii*, *P. vitifolia*, and *P. williamsii*, are probably dispersed chiefly by arboreal frugivores. Peccaries and white-tailed deer eat fruits of *P. vitifolia* (N. Smythe, pers. comm.). Many of these species have rinds so thick and leathery that only the larger birds would attempt to peck them open. Once fruits are opened by monkeys and other animals, birds might be instrumental in the dispersal of any seeds left, but the fleshy, sweet, translucent substance covering the seeds of nearly all species of *Passiflora* is very tasty, and opened fruits are seldom found with any seeds remaining. I suspect that animals spit out the seeds after sucking on them for a while because the seeds become very bitter or astringent.

Twelve genera and 600 species; subtropics and tropics.

PASSIFLORA L.

Passiflora ambigua Hemsl., Bot. Mag. 128, pl. 7822. 1902

Coarse vine, essentially glabrous, with simple tendrils in leaf axils. Stipules filiform, inconspicuous, not persistent; petioles 1.5–3 cm long, bearing 2 thick glands 0.5 to 1 cm above base; blades ovate to oblong-elliptic, acuminate, obtuse to subcordate at base, 10–20 cm long, 5–8(9) cm wide, subcoriaceous. Flowers solitary, adjacent to tendrils; peduncles ca 3 cm long (to 7 cm long in fruit), surmounted by 2 involucral bracts, the bracts free, trifid at apex, ca 3 cm long or longer, subfoliaceous; sepals 5, narrowly oblong, 4–6 cm long, maroon, with a slender appendage 2–4 mm long at apex; petals 5, narrowly

KEY TO THE SPECIES OF PASSIFLORA

Stipules subreniform, 1.5 cm long and 1 cm wide . . . . *P. menispermifolia* H.B.K.
Stipules mostly linear, to 2 mm wide:
  Blades peltate . . . . *P. coriacea* Adr. Juss.
  Blades basifixed:
    Blades 2- or 3-lobed at least to middle:
      Blades deeply 3-lobed, cordate at base; flowers and fruits more than 3 cm across:
        Palmate leaf veins 5; flowers deep scarlet, at least 10 cm diam; plants common . . . .
        . . . . *P. vitifolia* H.B.K.
        Palmate leaf veins 3; flowers white with pink, at most 7 cm diam; plants rare . . . .
        . . . . *P. williamsii* Killip
      Blades broadly 2-lobed, rounded at base; flowers and fruits less than 3 cm across:
        Blades more than 3 cm long along midvein; flowers paired . . . . *P. biflora* Lam.
        Blades less than 2.5 cm long along midvein; flowers generally solitary . . . . *P. punctata* L.
  Blades at most shallowly lobed (juvenile leaves of *P. seemannii* lobed):
    Blades shallowly 3-lobed or with the lobes reduced to angles:
      Petioles bearing 2 large auriculate glands near the base; flowers less than 3 cm diam, greenish-white . . . . *P. auriculata* H.B.K.
      Petioles eglandular; flowers more than 3 cm diam, white . . . . *P. foetida* L. var. *isthmia* Killip
    Blades not lobed:
      Blades deeply cordate and palmately veined . . . . *P. seemannii* Griseb.
      Blades at most rounded at base, not palmately veined:
        Blades at least serrulate; sepals lacking appendages . . . . *P. nitida* H.B.K.
        Blades entire; sepals with slender appendage at apex . . . . *P. ambigua* Hemsl.

lanceolate, 3–4 cm long, white outside, purplish inside; stamens 5; filaments ca half as long as corolla, purplish. Fruits ovoid to globose, to 15 cm long, greenish-yellow with whitish spots; exocarp 1–2 cm thick; seeds many, ca 8 mm long, with many small pits. *Croat 11858, Zetek 3480, s.n.*

Apparently rare; collected only once recently. Flowers from December to April. Mature fruits have been collected in the early rainy season, from May to July.

Southern Mexico to Colombia (Chocó). In Panama, known from tropical moist forest in the Canal Zone (around Gatun Lake) and Bocas del Toro, from premontane wet forest in Coclé and Panamá, and from tropical wet forest in Colón and Panamá; no doubt in Darién as well since it has been found in adjacent Colombia.

Passiflora auriculata H.B.K., Nov. Gen. & Sp. 2:131. 1817

Vine, with simple axillary tendrils. Stipules bristle-like, to 1 cm long; petioles 1.5–2.5 cm long, biglandular near base, the glands large, ear-shaped, to ca 3 mm long; blades ± ovate, shallowly trilobate (generally with the lateral lobes reduced to angles), 7–15 cm long, 4–10 cm wide, glabrous to puberulent, with numerous brownish glands ca 1 mm diam. Flowers paired in leaf axils; peduncles ca 1 cm long in flower (ca 1.5 cm long in fruit), inconspicuously jointed above middle, the bracts inconspicuous; sepals 5, oblong-lanceolate, ca 1 cm long, foliaceous; petals 5, linear, shorter than sepals, whitish; filaments of corona about as long as sepals; stamens 5, united to ca 4 mm at base. Berries globose, 1–2 cm diam, yellow-green, becoming black when mature, scattered-puberulent; seeds many, ca 3.5 mm long, reticulate, dark. *Croat 5617.*

Occasional along the shore and to be expected in clearings. Flowers from October to March (sometimes from August), with the fruits maturing from March to July.

Nicaragua to Brazil and Bolivia. In Panama, known principally from premontane wet forest in Colón, Coclé, Panamá, and Darién; known also from tropical moist forest in the Canal Zone (Atlantic slope) and Bocas del Toro and from tropical wet forest in Panamá and Darién.

Passiflora biflora Lam., Encycl. Méth. Bot. 3:36. 1789
Guate-guate, Camacarlata, Calzoncillo

Vine, woody at base, essentially glabrous; stems striate. Stipules minutely setose, with 4 or 5 pairs of round glands essentially only in the central part, the first pair at the very base; petioles eglandular, to 2 cm long; blades broadly bilobed (occasionally with a small central lobe as well), V-shaped, obtuse to rounded at base, 3–5(10) cm long in the center, 4–9 cm wide, distinctly lighter below, glabrescent, with a slender apiculum at apex of midvein and each lobe; palmate veins 3. Flowers paired in axils, 3–4 cm diam; peduncles to 2 cm long, jointed at or below the middle, with 3 slender bracts at articulation; hypanthium ca 3 mm deep, depressed at point of attachment to pedicel; sepals 5, lanceolate, 1.5–2 cm long, hooded, creamy white; petals 5, lanceolate, 1–1.5 cm long, creamy white; corona filaments in 2 series, the outer of

oblong filaments ca 7 mm long, yellow, the inner of narrowly linear filaments ca 5 mm long, reddish; operculum closely plicate, 3 mm high; gynophore ca 5 mm long; stamens 5, fused into tube around style; anthers dorsifixed, held below stigma, spreading and dehiscing extrorsely; ovary ± tomentulose. Berries globose, 1.5–1.8 cm diam, yellow-green and mottled, becoming purple when mature; seeds many, to 3 mm long, reticulate, becoming dark. *Croat 5416, 5707.*

Occasional, along the shore over low vegetation in *Annona-Acrostichum* associations and in clearings. Flowers and fruits throughout the year.

Mexico to Colombia and Venezuela. In Panama, widespread and ecologically variable; known from tropical moist forest in all provinces, from tropical dry forest in Coclé and Panamá, from premontane moist forest in the Canal Zone and Panamá, from premontane wet forest in Chiriquí, Veraguas, and Panamá, from tropical wet forest in Colón and Darién, and from premontane rain forest in Chiriquí.

Passiflora coriacea Adr. Juss., Ann. Mus. Natl. Hist. Nat. 6:109. 1805

Vine, sometimes trailing on the ground, ± glabrous. Stipules narrowly linear, ca 5 mm long, persistent; petioles 1–4 cm long, bearing a pair of ear-shaped glands in apical half; blades peltate (the petiole attached 3–13 mm from base of blade), trilobate, broadly rounded to cordate at base, 3–7 cm long, 6–25 cm wide, the central lobe occasionally suppressed to a mucro, the lateral lobes nearly horizontal, abruptly acuminate; palmate veins 5. Flowers generally paired, axillary but also racemose, the racemes opposite the leaves; peduncles jointed above the middle, the bracts inconspicuous and deciduous; sepals 5, lanceolate, ca 1.5 cm long, greenish-white to cream-colored, apetalous; corona in 2 series, the outer of filaments to 10 mm long, the inner of a few glandular filaments at most 5 mm long; operculum not generally plicate, ca 2 mm high; gynophore ca 10 mm long; ovary glabrous. Berries globose, 1–2 cm diam, the exocarp generally fleshy, glaucous, green and perhaps speckled when immature, maturing dark blue; seeds many, ca 5 mm long, with about a dozen contiguous pits on each side, drying dark. *Croat 5414a, 6732.*

Occasional, on the shore and in the young forest. Flowers throughout the year, principally in the dry season, with most fruits maturing from May to July.

Southern Mexico to the Guianas, Bolivia, and Peru. In Panama, known from tropical moist forest in the central Canal Zone and from premontane wet forest in Coclé and Panamá.

Passiflora foetida L. var. **isthmia** Killip, Publ. Field Mus. Nat. Hist., Bot. Ser. 19:497. 1938
P. hispida DC.

Vine, often creeping on the ground, yellow-pubescent on all parts, the trichomes ca 2 mm long, sometimes gland-tipped. Stipules about half encircling stem, deeply cleft with linear lobes, bearing glandular trichomes; petioles

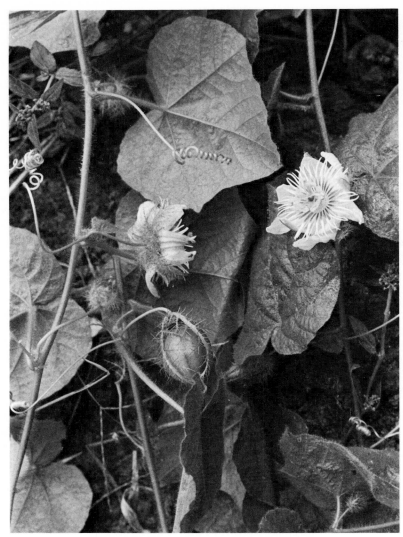

Fig. 406. *Passiflora foetida* var. *isthmia*

Fig. 407. *Passiflora nitida*

2–8 cm long, eglandular but with glandular trichomes; blades variable, deltoid in ours, shallowly trilobate (the lobes abruptly acuminate), cordate at base, 5–8 cm long and 5–8 cm wide; palmate veins 5–7. Flowers strongly scented, solitary in leaf axils; peduncles 3–6 cm long, jointed directly beneath the flower, the bracts foliaceous, 2–3.5 cm long, 2- or 3-pinnatifid, the ultimate segments filiform, gland-tipped; sepals 5, ± lanceolate, ca 2 cm long, white; petals 5, similar to sepals but slightly shorter; filaments of corona in several series, the longer to 1.5 cm long, purple below, the shorter ca 2 mm long; operculum erect, ca 1 mm long, denticulate. Berries inflated, ovoid, yellowish, ca 2.5 cm long, minutely apiculate at apex, subcordate at base, usually obscurely 3-sulcate and with 3 indistinct lines of dehiscence (apparently nonfunctional?); exocarp very thin; seeds numerous, ovoid-flattened, ca 5 mm long and 2.5 mm wide, slightly sculptured, black, shiny, trilobate at base, bearing at apex a slender funicle ca twice length of seed. *Croat 4641, 7400.*

Occasional, in clearings. Flowers and fruits throughout the year. In Costa Rica the species begins to flower in June (beginning of the rainy season) and flowers into the following dry season (Janzen, 1968).

The species is a cosmopolitan weed native throughout the American tropics; the variety *isthmia* extends from Panama along the Pacific coast to Ecuador. In Panama, widespread in weedy areas; known from tropical moist forest in the Canal Zone, all along the Atlantic slope, and in Panamá and Darién, from tropical dry forest in Panamá, from premontane wet forest in Bocas del Toro and Chiriquí, and from tropical wet forest in Colón and Darién.

See Fig. 406.

Passiflora menispermifolia H.B.K., Nov. Gen. & Sp. 2:137. 1817

Vine; stems densely hispid-hirsute. Stipules conspicuous, subreniform, ca 1.5 cm long and 1 cm wide, cuspidate at one end, usually glandular-denticulate; petioles 1.5–5(9) cm long, bearing 4–8 glands; blades suborbicular, 8–14 cm long, 8–13 cm wide, angulately trilobate, cordate at base, sometimes glandular-denticulate, definitely lighter and pilose below, densely pilose above, the middle lobes much larger, to 8 cm wide; palmate veins 5–7. Flowers solitary; peduncles 2–3 cm long, jointed directly beneath the flower, the bracts narrowly lanceolate, ca 10 mm long, at most 3 mm wide; sepals 5, lanceolate-oblong, 2–3 cm long, greenish-white, awned at apex, the awn little, off-white; petals 5, oblong, 2.5–3.5 cm long, light violet; filaments of corona in several series, the outermost to 2.5 cm long, glandular; operculum membranous, erect, divided above into numerous filaments ca 5 mm long; stigmas 3, ca 13 mm long. Berries narrowly ovoid, 6 cm long or more, ca 2 cm diam, glabrous; seeds ca 5 mm long, coarsely reticulate (fruit description from *Flora of Panama*, Woodson & Schery, 1958). *Aviles 21b, Woodworth & Vestal 634.*

Not seen on the island in recent years. Seasonal behavior uncertain. Probably flowering and fruiting throughout the year.

Nicaragua to northwestern South America. In Panama, known from tropical moist forest in the Canal Zone, Bocas del Toro, Colón, Veraguas, and Panamá and from premontane wet forest in Chiriquí (Boquete).

Passiflora nitida H.B.K., Nov. Gen. & Sp. 2:130. 1817

Coarse vine, glabrous. Stipules linear; petioles 1.5–2 cm long, biglandular above the middle; blades ovate-oblong, acuminate, rounded to subcordate at base, 11–19 cm long, 6–11 cm wide, subcoriaceous, serrulate. Flowers solitary; peduncles 2–3.5 cm long, jointed directly beneath the flower, bracteate, the bracts 2, conspicuous, ovate-elliptic, ca 3.5 cm long and 2.5 cm wide, rounded at apex and base; sepals 5, oblong-elliptic, to 5 cm long, greenish outside, white inside; petals 6, similar, white; filaments of corona about 4 cm long, white with pink and blue; stamens 5, purple; pistil pinkish. Berries obovoid, blunt at apex, red-orange to orange-red or magenta, densely punctate with lighter dots (these often run together), 6–8 cm long, 4.5–6 cm wide, glabrous, the withered flower often persisting; pericarp 1–1.5 cm thick, spongy; seeds obovate, ca 7 mm long, punctate, black. *Shattuck 675, Zetek 3620.*

Occasional, high in the forest canopy. Flowers in Panama from December to May. Mature fruits have been seen throughout the year, mostly from April to September.

The Panamanian collections of this species have generally greater dimensions than those from South America. Also, the petiolar glands are slightly below the apex (about one-third of the way down the petiole), instead of at the apex as on South American species.

Panama to the Guianas, northern Brazil, and Peru. In Panama, known from tropical moist forest on BCI, from premontane wet forest in the Canal Zone, Colón, and Panamá, and from premontane rain forest in Darién (summit of Cerro Pirre).

See Fig. 407.

Passiflora punctata L., Sp. Pl. 957. 1753
P. misera H.B.K.

Slender vine, essentially glabrous; stems striate. Stipules usually persistent, narrowly subulate, ca 2 mm long; leaf blades of 2, ± oblong lobes directed at nearly 180° angle to each other (rarely with a small central lobe), truncate or subcordate at base, 1–2.5 cm long, 5–13 cm broad, the lobes rounded and mucronulate, the lower surface glaucescent with 3–5 pairs of round glands usually along midribs especially in axils of lateral veins; palmate veins 3. Flowers generally solitary in leaf axils; peduncles 3–9 cm long, obscurely jointed immediately below to 1 cm below the flowers, the bracts inconspicuous, setaceous, ca 1 mm long; calyx tube campanulate, ca 4 mm deep, the lobes 5, lanceolate, ca 1.5 cm long, light yellowish-green; petals 5, oblong, ca 1 cm long, greenish-white; filaments of corona in 2 series, the outer to 10 mm long, mostly

pale purple, the inner ca 5 mm long, purplish, narrower; operculum plicate, ca 3 mm high, incurved, purplish; gynophore slender, to 8 mm long, purple; ovary narrowly ellipsoid, to 4 mm long, purple, capitellate. Berries globose, ca 1.5 cm diam, deep blue; seeds ovate, ca 3 mm long, transversely ridged, the ridges rugulose. *Croat 8273, 8278.*

Uncommon, along the shore. Flowers and fruits throughout the year.

The distinction is obscure between this species and *P. misera* H.B.K., which is therefore not recognized here.

The petals, sepals, and outer series of corona filaments reflexed somewhat at anthesis. Anthesis occurs in the early morning, and plants collected at 10:30 A.M. usually have all flowers closed. Nectar is probably enclosed within the operculum, within which many holes are seen.

Panama to Argentina. In Panama, known only from tropical moist forest around Gatun Lake.

Passiflora seemannii Griseb., Bonplandia 6:7. 1858
　　Guate-guate

Stout vine, glabrous; stems to at least 5 mm diam, striate. Stipules somewhat persistent, narrowly linear, 10–15 mm long; petioles 3–10 cm long, biglandular near apex; blades broadly ovate-cordate, rounded and mucronulate to apiculate at apex, 5–10 cm long, 6–11 cm wide, the basal sinus ca 3 cm deep, the lower surface glaucous, eglandular, the margins subentire to serrulate; palmate veins 7–9; juvenile leaves trilobate. Flowers solitary in leaf axils, ca 8 cm diam; peduncles 6–9 cm long (to 10 cm long when in fruit), jointed directly beneath flowers, the bracts 3, fused, 2.5–5 cm long, the lobes broadly ovate, white, purple-tinged; hypanthium ca 2 cm deep; sepals 5, ovate-lanceolate, 3.5–4 cm long, appendaged, greenish-white, sometimes tinged with violet; petals 5, oblong-lanceolate, ca 3.5 cm long, purple, the entire perianth strongly reflexed when open; corona filaments in 2 series, the inner erect, incurved at apex, to 2.5 cm long, banded with purple and white, the outer ca 1 cm long; staminal filaments fused except at apex; anthers shedding pollen after beginning of fruit development; operculum 2 mm high, denticulate; gynophore ca 2 cm long in flower; ovary ovoid; styles 3; stigmas broad, yellow. Berries broadly oblong to ovoid, at least 4–5 cm long and 3.5 cm broad, green, weakly pruinose, densely speckled with light green spots, the exocarp thick, white inside; seeds many, stalked, orbicular-ovate, ca 3 mm long, punctate. *Shattuck 692, Wetmore & Abbe 147.*

Uncommon along the shore. Flowers mostly from October to March, sometimes from August. The fruits mature from January to March.

Native to Panama and Colombia; cultivated elsewhere. In Panama, known principally from tropical moist forest in the Canal Zone, all along the Atlantic slope, and in Panamá and Darién; known also from premontane wet forest in Chiriquí and Panamá and from tropical wet forest in Colón and Panamá.

See Fig. 408.

Passiflora vitifolia H.B.K., Nov. Gen. & Sp. 2:138. 1817
　　Guate-guate, Pasionaria, Granadilla, Granadillo de monte

Liana; stems to at least 1.5 cm diam, ± densely ferruginous-pubescent on all parts. Stipules deciduous, subsetaceous, ca 5 mm long; petioles 2–6 cm long, inconspicuously glandular toward base; blades deeply trilobate, ± deeply cordate, 7–14 cm long, 9–14 cm wide, the lobes acute to acuminate at apex, the central lobe larger, the upper surface sparsely strigillose except on tomentulose veins, the lower surface softly ferruginous-pubescent, the glands minute, 2 to several, around base of each sinus; palmate veins 5. Flowers solitary in axils, often borne on young leafless branches, 10–15 cm wide when open; peduncles 3–5.5 cm long (to 6.5 cm long in fruit), the bracts 3, free, lanceolate, 2.5–3 cm long, glabrescent, glandular-serrate, generally with a larger (1 mm across) disk-shaped pair of glands toward the base; sepals 5, narrowly lanceolate, 6–8 cm long, 1–2 cm wide, with a long slender appendage at apex; petals 5, linear-lanceolate, 4–6 cm long; both sepals and petals deep scarlet to magenta, reflexed when open; corona filaments in 3 series, the outer longest, to 2 cm long, erect, bright red or bright yellow; anthers 5, green, held perpendicular on gynophore; operculum deflexed, to 1 cm long, fimbriate; gynophore ca 1 cm long, pale red; ovary densely tomentulose. Berries ovoid, 5–8 cm diam, puberulent, greenish-yellow with darker stripes and bands of lighter splotches, very fragrant; seeds ± flattened, ca 6 mm long and 5 mm wide, reticulate with numerous small punctations, light-colored. *Croat 4766, 8319.*

Frequent in the forest. Flowers from December to May, elsewhere in Panama also from July to November. Fruit maturity time not determined.

Native to lowland forests from Nicaragua to Venezuela and Peru; cultivated in the West Indies and elsewhere. In Panama, known from tropical moist forest in the Canal Zone, Bocas del Toro, San Blas, Panamá, and Darién, from premontane wet forest in Colón, Coclé, and Panamá, and from tropical wet forest in Colón, Coclé, Panamá, and Darién.

Passiflora williamsii Killip, J. Wash. Acad. Sci. 12:262. 1922

Coarse vine, ± densely puberulent on all parts. Stipules 6–7 mm long; petioles 4–6 cm long, bearing a pair of prominent glands 1–2 cm from base and sometimes a second pair above them; blades trilobate to about middle, truncate to generally deeply cordate at base, 9–13 cm long and wide, the lobes acuminate, the central lobe stronger and contracted toward the base, the upper surface glabrescent except on strigillose veins, the lower surface minutely crisp-villous, sparsely so in age, distinctly biglandular at base of each sinus, the glands ca 1 mm long, the margins subentire to serrulate; palmate veins at base 3. Flowers solitary in axils, 6–7 cm diam; peduncles to 2 cm long in flower, the bracts 3, united at base, ca 2.5 cm long, tomentulose; sepals 5, oblong, 2.5–3.5 cm long, with a slender appendage at apex; petals 5, oblong-spatulate,

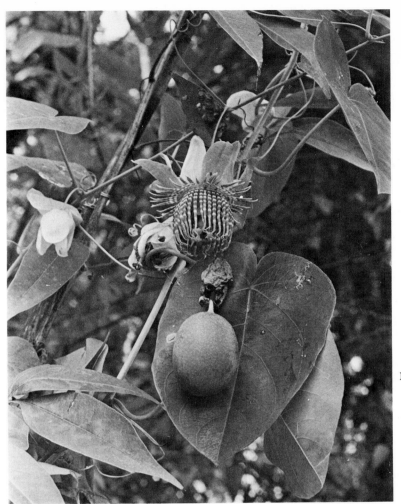

Fig. 408. *Passiflora seemannii*

Fig. 409. *Carica cauliflora*

ca 2 cm long, white inside, spotted with dark pink; corona with the outer filaments shorter, 6–7 mm long, the middle series 2–2.5 cm long; operculum 2 mm long, entire; ovary densely white-tomentose. Berries globose, puberulent, to 5 cm long, green, lacking markings. *Zetek 4352.*

Collected only once on the island and not seen in recent years. Flowers elsewhere in Panama from February to June, and fruits from July.

Known only from Panama, from tropical moist forest in the Canal Zone and Los Santos (Loma Prieta) and from premontane rain forest in Darién.

99. CARICACEAE

Dioecious trees and shrubs with milky sap; trunk branched or not, herbaceous to soft-wooded. Leaves alternate, petiolate, clustered at ends of stems; blades palmately compound (*Jacaratia*) or simple and palmately lobed with palmate veins at base (*Carica*); stipules lacking. Corymbs, cymes, or racemes upper-axillary; flowers actinomorphic, few (pistillate) or many (staminate); calyx small, 5-lobed; corolla tubular and 5-lobed (staminate) or 5-lobed to near base (pistillate), showy; stamens 8 or 10, in 2 series; anthers 2-celled, dehiscing longitudinally; ovary superior, unilocular, 5-carpellate; placentation parietal, often intruding; ovules numerous; stigmas 5, free, linear, simple or branched. Fruits large berries; seeds many, with fleshy endosperm.

These pachycaulous, soft-wooded trees, which may look like overgrown herbs, are not confused with any other plants on the island.

The natural method of pollination is unknown. A number of different birds and insects, including hummingbirds, butterflies, *Trigona* bees, hawkmoths, and noctuid moths, have been seen taking nectar from staminate flowers (Traub et al., 1942; Allan, 1963). In contrast, pistillate flowers have no nectar and are less scented, but they have been observed being visited by moths and mosquitoes (Baker, 1973). Parthenocarpy has also been reported for *Carica* (Badillo, 1971), at least for *C. papaya*.

Although there is no *Carica papaya* growing wild on the island, the tasty fruits of *C. cauliflora* become soft enough to be pecked open by birds and would be attractive to other animals also. The bats *Carollia perspicillata* and *Artibeus jamaicensis* (Phyllostomidae) eat fruits of *C. papaya* in Trinidad (Goodwin & Greenhall, 1961). The cultivated *C. papaya* on BCI attracts opossums and prehensile-tailed porcupines (N. Smythe, pers. comm.).

Four genera and 31 species (Badillo, 1971), mostly of the genus *Carica*; mostly in the neotropics (one genus with two species in Africa).

CARICA L.

Carica cauliflora Jacq., Hort. Schoenbr. 3:33, t. 311. 1798

Unbranched dioecious tree, 3–6 m tall, to 10 cm dbh, mostly glabrous. Leaves in a dense terminal crown; petioles mostly more than 50 cm long, hollow, with slightly milky sap; blades ovate, shallowly 5-lobed when mature, cordate, 30–70 cm long, the lobes acute at apex, often broadly incised; palmate veins at base 5. Staminate inflorescences dichotomously branched, ca 30 cm long; pedicels very short; flowers many, ca 3 cm long; calyx deeply 5-lobed, ca 1.5 mm long, green; corolla light yellow, 5-lobed, the lobes spreading, linear-oblong, 12–20 mm long; anthers 10, nearly sessile, linear-spatulate, to 4 mm long. Pistillate inflorescences contracted, ca 7-flowered, 2.5–5 cm long; calyx lobes ca 2 mm long; corolla 5-lobed to base, ca 2.5 cm long, white or yellowish, the apices reddish; ovary glabrous; style ca 3 mm long; stigmas 5, linear, ca 1 cm long. Fruits maturing well below leaves, sometimes near base of trunk, borne on thick warty protuberances, the fruiting pedicels to 5 cm long and 8 mm diam; berries obovate, ca 8 cm long and 6 cm wide, minutely apiculate, becoming orange; exocarp to 7 mm thick, forming whitish droplets when cut; mesocarp sweet, pithy, white to translucent, with aroma of fresh grapes; seeds many, ovate, 10–12 mm long, the ridges 5 or 6, deep, corky. *Foster 1394.*

Rare; collected once by Aviles and once by Foster near the shore between the ends of Zetek and Armour trails. Flowers in the dry season (February and March). Some fruits have been seen in June and November.

Southern Mexico to Colombia and Venezuela; Trinidad. In Panama, known only from tropical moist forest in the Canal Zone, Veraguas, and Darién.

See Fig. 409.

Carica papaya L., Sp. Pl. 1466. 1753

Papaya

Unbranched tree, mostly to 6 m tall, essentially glabrous, dioecious or rarely polygamous or monoecious. Leaves in a dense terminal crown; petioles to 70 cm long; blades ovate, deeply 5–9-lobed, cordate at base, the lobes broadly pinnatifid when mature, 20–40 cm long, obscurely acuminate; palmate veins at base 7–9. Staminate inflorescences axillary; flowers many, short-pedicellate, ca 2 cm long; sepals 5, ca 1 mm long, green; corolla cream-yellow, the tube slender, the limb 5-lobed, the lobes oblong; stamens in 2 series, ca 2.5 mm long. Pistillate inflorescences of 1–3 flowers on pedicels to 2.5 cm long; calyx

KEY TO THE SPECIES OF CARICACEAE

Leaves compound; trees more than 20 m tall . . . . *Jacaratia spinosa* (Aubl.) A. DC.
Leaves simple, lobed; trees less than 8 m tall, soft-wooded:
  Leaves shallowly lobed, the palmate veins at base 5 . . . . *Carica cauliflora* Jacq.
  Leaves deeply lobed, the palmate veins at base 7–9 . . . . *Carica papaya* L.

lobes ca 1 mm long, green; corolla 5-lobed to base, 5–7 cm long, cream-yellow; style to 1 cm long; stigmas 5, usually with 1 or 2 short branches. Fruits maturing within crown of leaves; berries variable. *Croat 5907.*

Cultivated at the Laboratory Clearing. Flowers sporadically year-round.

Sex changes in the flowers are reported under different ecological conditions (Percival, 1965).

Native to the American tropics; now widespread in cultivation. In Panama, known from tropical moist forest in the Canal Zone, Bocas del Toro, Veraguas, Panamá, and Darién, from premontane moist forest in the Canal Zone, from premontane wet forest in Panamá (Chimán), and from tropical wet forest in Colón (Portobelo).

JACARATIA C. DC.

Jacaratia spinosa (Aubl.) A. DC., Prodr. 15(1):419. 1864

Dioecious tree, 20–40 m tall; trunk 0.8–1 m dbh, armed or unarmed, branched near apex, the branches slender, short, ascending, armed with short, stout, conical spines at least near their ends. Leaves palmately compound; petioles 10–25 cm long; leaflets 5–12, lanceolate, acuminate at apex, narrowly attenuated at base to an obscure petiolule 6–18 cm long, 2–6 cm wide, dark green above, glaucous below. Staminate inflorescences pedunculate, laxly cymose, 7–9 cm long; peduncles 6–12 cm long; bracts minute; pedicels 3–4 mm long; flowers many, tubular; calyx tube 1 mm long; corolla greenish-white, the tube 10–14 mm long, 3–4 mm wide, the lobes ovate-rounded, 7–8 mm long; stamens 8, in 2 unequal series, united in a short tube; filaments pilose; pistillode filiform, to 11 mm long. Pistillate flowers usually solitary, the other flowers not developing; peduncles 8–10 cm long; bracts 11–13 mm long; calyx tube 1–1.5 mm long; corolla divided almost to base, the tube 2 mm long, the lobes fleshy, 2–3.3 cm long, 7 mm wide; ovary narrowly ovoid, sub-pentagonal, 11–20 mm long; styles lacking; stigmas 5, narrowly linear, 7–8 mm long. Fruits ± ellipsoid, sometimes ovoid, obtuse or sometimes apiculate at apex, obtuse at base, to 12 cm long and 3.5 cm diam, pendent, yellow or orangish; seeds many, ovoid. *Foster 2816.*

Collected once on Balboa Trail. Seasonal behavior uncertain. Some flowers have been seen in March and mature fruits in September.

Nicaragua and the Guianas to northern Argentina, probably from Colombia and Venezuela. In Panama, known only from tropical moist forest on BCI and from tropical wet forest in Colón.

100. BEGONIACEAE

Fleshy, monoecious, annual or perennial, erect or scandent, terrestrial, epiphytic, or semiaquatic herbs. Leaves alternate, petiolate; blades simple, serrate, mostly asymmetrical at base; venation pinnate and usually also palmate at base; stipules present. Cymes terminal or axillary, bracteate; flowers few, unisexual, usually slightly zygomorphic; staminate flowers with 2 or 4 tepals, when 4, the outer pair valvate, the inner pair imbricate; stamens numerous in many whorls, free or basally connate, inserted on the receptacle; anthers 2-celled, basifixed, longitudinally dehiscent; pistillate flowers with 4 or 5 separate, undifferentiated, imbricate, showy tepals; ovary inferior, 3-locular; placentation axile; ovules numerous, anatropous; styles 3, bifid; stigmas spiraled, strongly papillose on all sides. Fruits unequally 3-winged, loculicidal capsules; seeds many, minute, lacking endosperm.

In Panama the family consists of only *Begonia*, which can be recognized by its succulent stems, asymmetrical leaves, unisexual flowers with four unequal tepals, and asymmetrical winged fruits.

Pollination system is unknown.

Though the fruits are winged, they are not themselves dispersed. At maturity the walls of the fruit split open along the bottom on all three sides, allowing the seeds to pass through slowly. Seeds are minute, dispersed at least in part by the fluttering of the fruit in the wind. The wing causes the fruit to flutter, ensuring that most seeds fall out only when there are air currents to carry them away. Seeds are also partly dispersed by rain wash (Ridley, 1930).

Five genera and over 800 species, mostly of *Begonia;* widespread in the subtropics and tropics.

BEGONIA L.

Begonia filipes Benth., Bot. Voy. Sulphur 101. 1845

Ala de angel, Hierba de agua

Weak, monoecious, annual herb, 15–60 cm tall, usually epiphytic; stems juicy, glabrous, reddish, weakly ridged. Petioles to 5 cm long; stipules oblong-ovate, 6–12 mm long, translucent, the apex very slender; blades very asymmetrical, semiovate, acuminate, rounded at base, 3–10 cm long, 1–5 cm wide, appressed-pilose above, glabrous below, the margins crenate-dentate, ciliate. Cymes axillary or terminal; flowers white, few; bracts with deeply lacerate margins; staminate tepals 2, ovate, to 3.5 mm long, their margins sealed before anthesis; stamens to 18, on a short column; pistillate tepals 4(5),

---

KEY TO THE SPECIES OF BEGONIA

Leaves cordate, not strongly asymmetrical; herbs growing in marshy habitats . . . . *B. patula* Haw.
Leaves not cordate, strongly asymmetrical; herbs not growing in marshy habitats:
    Plants suffrutescent perennials; all tepals more than 5 mm long; largest wing on fruit more than
        1.2 cm wide . . . . *B. guaduensis* H.B.K.
    Plants herbaceous annuals; pistillate tepals less than 5 mm long; wings of fruit less than 1.2 cm
        wide . . . . *B. filipes* Benth.

ca 2 mm long; styles 3, bipartite; stigmas spiraled, receptive before opening of staminate tepals. Capsules glabrous, 3-winged, the wings very unequal, the largest to 1 cm wide; seeds ± round, minute, very numerous. *Croat 4318.*

Occasional, in clearings and in somewhat open areas of the forest, loosely rooted in soil or more commonly epiphytic on mossy tree trunks; possibly locally abundant. Flowers and fruits all year, but flowering begins principally in the late rainy season. Most fruits mature in the early dry season. Leaves sometimes fall when plants have mature fruits.

Costa Rica to Colombia. In Panama, known from tropical moist forest in the Canal Zone, Colón, Chiriquí (San Félix), Panamá, and Darién.

Begonia guaduensis H.B.K., Nov. Gen. & Sp. 7:178. 1825
    B. serratifolia C. DC.

Monoecious, erect or scandent, perennial herb, to 1(2) m long; stems woody at base, glabrous. Stipules oblong-ovate, acuminate, ca 8 mm long; petioles 2–20 mm long; blades strongly asymmetrical, lanceolate to elliptic-lanceolate, somewhat acuminate, inequilateral at base, one side cuneate, the other rounded, 5–9 cm long, 1.5–4 cm wide, membranaceous, glabrous, the margins doubly crenate-serrate, ciliate. Cymes dichotomously branched; flowers pinkish-white; bracts often deciduous, ovate, ca 5 mm long; staminate tepals 4, unequal, the outer 2 ovate, 8–15 mm long, the inner 2 elliptic-lanceolate, about half as long as outer; stamens numerous; pistillate tepals 5, subequal, elliptic, to 10 mm long; styles 3, bipartite; stigmas branched and finally spiraled. Capsules elliptic-oblanceolate, 3-winged, the wings very unequal, the largest triangular, ascending, 1.5 cm wide; seeds very numerous, elliptic, 0.3 mm long. *Croat 7374.*

Rare, in the forest. Flowers as early as September but usually not until December. The fruits develop quickly, some of mature size usually occurring with flowers on the same inflorescence. Plants are often leafless or nearly so when the plant is in full fruit.

Panama to Colombia and Venezuela. In Panama, known from tropical moist forest in the Canal Zone,

Chiriquí (San Félix), Los Santos (Tonosí), and Darién and from premontane wet forest in Veraguas (west of Soná).

Begonia patula Haw., J. Wash. Acad. Sci. 40:245. 1950
    B. ciliibracteola C. DC.; *B. fischeri* Schrank var. *tovarensis* (Klotzsch) Irmsch.

Erect, monoecious, perennial herb, sometimes aquatic, to 1.2 m tall, fleshy; stems red, sparsely brown-pilose to glabrous. Stipules ovate-oblong, to 1 cm long, caducous; petioles 5–45 mm long; blades weakly asymmetrical, ± broadly ovate, acute or rounded at apex, cordate at base, 2–9 cm long, about as broad as or broader than long, glabrous above, mostly brown-pilose below, the margins crenate-serrate, ciliate; veins palmate at base. Inflorescences cymes of few flowers; bracts persistent, ovate, 2–4 mm long; staminate tepals 4, the outer 2 orbicular and ca 8 mm long, the inner 2 smaller, narrowly obovate; stamens numerous; pistillate tepals 5, obovate, 3–6 mm long; styles 3, bipartite, stigmas spiraled. Capsules 10–15 mm long, the 3 wings decurrent, very unequal, the largest ascending, often hooked, to 23 mm wide; seeds minute, fusiform. *Bailey & Bailey 644, Kenoyer 460.*

At least at one time a component of the floating masses of vegetation that border the shores, but not seen in recent years. Probably flowers and fruits throughout the year.

Mexico to Colombia, Venezuela, Peru, and Bolivia; Cuba. In Panama, known from marshy areas in tropical moist forest in the Canal Zone and Panamá and from premontane wet forest in Coclé (El Valle).

101. CACTACEAE

Epiphytic succulent herbs or shrubs, climbing by roots; stems chlorophyllous, terete to angled or flattened, with watery sap; areoles generally naked. Leaves lacking. Flowers bisexual, actinomorphic, solitary, sessile on the side of the stem; tepals 5 to many, petaloid, weakly differentiated, connate in a hypanthium; stamens twice as many as the tepals or numerous, spirally attached to hypanthium; anthers 2-celled, dehiscing longitudinally; ovary superior, 1-locular, 3- to many-carpellate; placentas

KEY TO THE TAXA OF CACTACEAE

Stems terete; flowers minute, white; fruits round . . . . *Rhipsalis cassytha* Gaertn.
Stems terete only at base, mostly angulate or prominently flattened:
  Flowers 7.5–9(11) cm long; filaments white to yellowish . . . .
  . . . . *Epiphyllum phyllanthus* (L.) Haw. var. *columbiense* (Weber) Kimn.
  Flowers 24–29 cm long; filaments white or orange to reddish:
    Filaments orange; tepals 4–5 cm long; fruits 4.5–7.5 cm long . . . .
    . . . . *Epiphyllum phyllanthus* (L.) Haw. var. *rubrocoronatum* Kimn.
    Filaments white; tepals mostly 2–3 cm long (rarely to 4); fruits 7–9 cm long . . . .
    . . . . *Epiphyllum phyllanthus* (L.) Haw. var. *phyllanthus*

Epiphyllum phyllanthus var. *phyllanthus*, which occurs in the Canal Zone, may occur on the island. Some specimens (*Aviles 103b*) appear closest to that variety, but since they are sterile it is impossible to be certain.

few to several, parietal; ovules numerous, anatropous; style 1; stigmas as many as carpels, radiating. Fruits fleshy berries; seeds few to many, the endosperm viscid (*Rhipsalis*) or lacking.

The generally epiphytic, leafless aspect of the plants makes them hard to confuse with any other family.

Although nothing is known of the pollination of *Rhipsalis cassytha*, it can be concluded with little doubt that *Epiphyllum phyllanthus* is pollinated by hawkmoths (Porsch, 1939). Kimnach (1964) reported that flowers of *Epiphyllum* open at night and have a strong fragrance exuded during the night. Both of these features are strongly correlated with hawkmoth or other moth pollination. Moreover, because the nectar source is at the base of the very long, slender floral tube, the hawkmoth is probably the only pollinator capable of reaching it.

The fruits have tiny seeds embedded in a sweet sticky matrix. The larger fleshy fruits of *Epiphyllum* have been observed being pecked open by birds. There is no reason, however, to believe that they are not also dispersed by other arboreal animals. The small sticky seeds, which must germinate on tree branches, are most ideally suited to bird dispersal.

Genera mostly ill defined and of uncertain number, species about 1,800; almost all native in subtropical and tropical America.

EPIPHYLLUM Haw.

Epiphyllum phyllanthus (L.) Haw. var. **columbiense** (Weber) Kimn., Cact. Succ. J. (Los Angeles) 36:114. 1964

Leafless epiphyte, glabrous; stems green, usually flexible and pendent, 3-sided in basal half, flattened and to 3 cm wide distally, the margins ± regularly and obtusely serrate-lobed. Flowers borne along margins of ultimate stems, 7.5–9(11) cm long, the limb 4.5–6 cm wide; perianth tube white, 5.5–9.5 cm long, bracteolate, the bracteoles slender, appressed at base, recurved at apex; tepals many, the outer 1.5–3 cm long, the inner 2–2.5 cm long; stamens many, cream-colored to yellow; filaments ca 7 mm long; anthers ca 1 mm long; style 6.5–8 cm long; stigma lobes ca 5 mm long. Berries green, 4–4.5 cm long, to 2.5 cm wide; seeds numerous, ca 2.5 mm long, the surface with closely spaced pits. *Croat 8256, Shattuck 603.*

Frequent, high in the canopy of the forest. Seasonal behavior not determined.

Costa Rica to Colombia and Ecuador. In Panama, known only from tropical moist forest in the Canal Zone and Chiriquí (Puerto Armuelles).

Epiphyllum phyllanthus (L.) Haw. var. **rubrocoronatum** Kimn., Cact. Succ. J. (Los Angeles) 36:110. 1964

Leafless epiphyte, glabrous, often occurring in association with ant nests; stems green, usually stiff and pendent, terete at base, becoming 3-sided and mostly less than 2.5 cm broad, at apex 2-edged and to 7(9) cm wide and less than 5 mm thick, the margins regularly and obtusely crenate-lobed. Flowers borne along margins of ultimate stems, 24–29 cm long, the limb 9–11 cm wide; perianth tube usually pinkish, 21–26 cm long, slenderly bracteolate; tepals many, slender, 4–5 cm long, the outer pinkish, the inner white and broader than the outer; stamens many, orange; filaments exserted 2–3 cm; anthers ca 3 mm long; style exserted 2.5–4 cm above throat; stigma lobes 9–13, 5–6 mm long, both style and stigma orange. Berries angulate, 4.5–7.5 cm long, to 3 cm wide, magenta, the bracteoles persisting; seeds numerous, ca 3 mm long. *Croat 14059.*

Apparently uncommon, though possibly common since it usually grows high in the canopy at 10–30 m, sometimes on branches in the crotch of a tree but generally on top of the canopy in association with *Aechmea tillandsioides* var. *kienastii* (22. Bromeliaceae). Seasonal behavior uncertain. Some flowers have been seen in the dry season. Some fruits have been seen in the late dry and early rainy seasons.

Panama, Colombia, and Ecuador. In Panama, known only from tropical moist forest in the Canal Zone.

RHIPSALIS Gaertn.

Rhipsalis cassytha Gaertn., Fruct. & Sem. Pl. 1:137. 1788

Mistletoe cactus

Slender, leafless epiphyte, 1–2(9) m long; stems green, succulent, terete, pendent, sparsely branched, dichotomous or in false whorls, ca 5 mm thick, the areoles very inconspicuous, glabrous to minutely puberulent. Flowers lateral, solitary, sessile, greenish-white; tepals ca 2 mm long, the segments few (sometimes as few as 5), ± free, usually spreading; stamens shorter than and about twice as numerous as tepals; filaments inserted on margin of hypanthium; style longer than tepals; stigmas 3 or 4. Berries fleshy, depressed-globose, to 9 mm long and 6 mm wide; exocarp thin; seeds several, to 1.3 mm long, about twice as long as broad, usually curved, dark brown, embedded in a sweet, sticky, watery matrix. *Croat 6742.*

Occasional, high in the canopy. Seasonal behavior not determined.

The fruits are probably dispersed by birds. Removing the fruit usually squeezes the seeds from the lower end, an action which probably frequently results in seeds' sticking to the outer edge of a bird's beak. Later, when the bird cleans its beak or feeds on bark insects, the seeds are deposited where they can germinate.

Throughout tropical America; Ceylon, tropical Africa. In Panama, known only from tropical moist forest in the Canal Zone and Bocas del Toro.

102. LYTHRACEAE

Trees, shrubs, or herbs. Leaves opposite, petiolate; blades simple, entire, sometimes punctate; venation pinnate; stipules present, minute. Flowers with a hypanthium, bisexual, actinomorphic (slightly zygomorphic in *Cuphea*), sometimes glandular, solitary and terminal (in

KEY TO THE SPECIES OF LYTHRACEAE

Flowers large, fleshy, parted 12 or more times, white to yellow; fruits woody capsules more than
    4 cm diam with winged seeds; trees usually more than 5 m tall . . . . *Lafoensia punicifolia* DC.
Flowers small, 3–6-parted, white to magenta or violet; fruits small capsules less than 1 cm diam
    with wingless seeds; trees or herbs less than 2.5 m tall:
    Plants with small orange punctations on leaves, flowers, and fruits (black when dry), shrubs or
        small trees more than 1 m tall; petals white . . . . *Adenaria floribunda* H.B.K.
    Plants lacking punctations, herbs less than 50 cm tall; petals violet to reddish-violet or violet-
        purple . . . . *Cuphea carthagenensis* (Jacq.) Macbr.

Lafoensia) or in axillary clusters; calyx 4–6- or many-lobed (*Lafoensia*); petals showy, free, of the same number as and alternate with calyx lobes, arising from upper inner surface of hypanthium, crumpled in bud; stamens twice as many as petals, in 2 series, usually unequal; anthers 2-celled, introrse, dorsifixed, dehiscing longitudinally; ovary superior, 1- or 2-locular, 2-carpellate; placentation axile; ovules many, anatropous; style 1; stigma capitate. Fruits capsules, variously dehiscent, sometimes explosively so; seeds many, sometimes winged (*Lafoensia*), lacking endosperm.

A small but diverse family. The BCI species seemingly have little in common and are apparently diverse in both their pollination and diaspore strategies. Their only common features are the hypanthium, the superior ovary, and the crumpled petals. All are distinct species not confused with any other.

The flowers of both *Cuphea* and *Adenaria* are heterostylous and seem suited to insect pollination. *Cuphea*, which is zygomorphic, is probably pollinated by small bees. H. Baker (pers. comm.) reports some *Cuphea* and *Adenaria* to be self-pollinated, however. *Lafoensia punicifolia* is bat pollinated (Vogel, 1958; Baker, 1973). I have seen petals falling from open flowers in the early morning, which also suggests nocturnal pollination.

The fruits of *Adenaria floribunda*, though several-seeded, are apparently indehiscent. The thin exocarp shatters easily, usually causing the cluster of fruits to break up as well. I suspect the fruits are taken by birds, which scatter some of the seeds and eat others. Diaspores of *Cuphea* are autochorous; those of *Lafoensia* are wind dispersed.

Some 22 genera and 500 species; in all regions, but most numerous in the American tropics.

ADENARIA Kunth

Adenaria floribunda H.B.K., Nov. Gen. & Sp.
    6:188. 1824
    Fruta de pavo

Shrub, to 2.5 m tall; stems pubescent and glandular, the younger ones often tetragonal and purplish-red. Petioles short, ca 5 mm long; blades lanceolate, acuminate, acute to obtuse at base, mostly 5–13 cm long and 2–5 cm wide, sparsely glandular on both sides (drying as black dots), glabrous to inconspicuously but densely puberulent above, puberulent below, the trichomes larger below and chiefly on veins. Flowers 4- or 5-parted, many, tristylous,

in axillary short-pedunculate clusters; peduncles 1–7 mm long; pedicels 5–7 mm long; hypanthium campanulate, 2–4 mm long; calyx lobes deltoid; petals obovate, white, to ca 2.5 mm long in the short- and long-styled flowers, to 4 mm long in the intermediate-styled flowers; stamens 8 or 10, the short-styled flowers with the stamens 3–4.5 mm long, the style to 1 mm long, the intermediate-styled flowers with stamens 4.5–5.5 mm long, the style to 1.5 mm long, the long-styled flowers with stamens to 3 mm long, the style to 3 mm long; ovary ovoid, short-stipitate; ovary and style pubescent; outside of hypanthium, calyx, petals, ovary, and style with orange glands (drying black). Fruits thin-walled, indehiscent capsules, ovoid or globose, bearing the persistent style, becoming red at apex, ca 4–5 mm long; seeds many, clustered in a globular mass, ca 1 mm long, broadest at apex, tapered to narrow base. *Croat 8325, 14942.*

Occasional, in clearings. Flowers from March to November. The fruits probably mature within 3 or 4 months.

Mexico to Argentina. In Panama, known from tropical moist forest in the Canal Zone, Colón, Panamá, and Darién and from premontane moist forest in the Canal Zone and Panamá.

CUPHEA P. Browne

Cuphea carthagenensis (Jacq.) Macbr., Publ. Field
    Columbian Mus., Bot. Ser. 8:124. 1930

Herb, to 50 cm tall, mostly erect; stems brownish, with long, straight, glandular trichomes interspersed with short curved trichomes, subwoody at base in age. Stipules minute; petioles less than 5 mm long; blades lanceolate to elliptic or ovate, acute at apex, acute to attenuate at base, 1–5 cm long, 0.5–2 cm wide, strigose and scabrous below, bearing sparse, long, straight trichomes above, the margins bearing short trichomes. Flowers 6-parted, generally solitary in reduced interpetiolar cymes, the leaves on the inflorescence much reduced near apex; peduncles and pedicels short; hypanthium ± swollen on one side and weakly gibbous at base, to 7 mm long, prominently ribbed, with long trichomes on ribs, enclosing fruit at maturity; calyx lobes short, deltoid; petals spatulate, violet to reddish-violet or violet-purple, to 3 mm long, borne on upper margin of hypanthium, spreading; stamens (11)12, unequal, attached about midway on hypanthium, included; anthers violet, minute, introrse; ovary thin-walled; style to 1 mm long; stigma broader than style. Fruits capsules, bursting at maturity under

tension of the placenta, the placenta stout, reflexing and flinging seeds from the ovary; seeds 3–7, orbicular to elliptic, somewhat flattened, 1–2.5 mm long, with a narrow margin. *Croat 11991.*

Occasional, in the Lighthouse Clearing. Flowers from April to September.

United States, Central America, and South America. In Panama, widespread and ecologically variable; known from tropical moist forest in the Canal Zone, Bocas del Toro, Colón, Herrera, and Panamá, from tropical dry forest in Coclé and Panamá (Taboga Island), from premontane moist forest in the Canal Zone and Panamá, from premontane wet forest in Chiriquí, Coclé, and Panamá, and from lower montane wet forest and lower montane rain forest in Chiriquí.

LAFOENSIA Vand.

Lafoensia punicifolia DC., Mém. Soc. Phys. Genève 3(2):86, t. 1. 1828
  Amarillo, Amarillo fruto, Amarillo de fruto, Amarillo papito, Pino amarillo

Glabrous tree, usually 10–15(27) m tall; trunk to 1 m dbh; outer bark coarse, fissured, brown, thick; inner bark fine, thick, reddish, blending into lemon-yellow wood; sap lacking odor. Leaves deciduous; petioles 4–7 mm long; blades oblong-elliptic, acuminate, obtuse at base, 3–9 cm long, 1–3 cm wide, the acumen often broad at apex, the lower surface with a small pore at apex of midrib. Flowers 12–16-parted, solitary but clustered at apex in leafy pseudoracemes; pedicels 2–3 cm long, stout; bracteoles 2, small, deciduous, at base of flower; hypanthium campanulate, 2–4 cm long, nearly as wide, lobed ca one-third its length, the lobes acuminate; petals yellow or greenish-white, attached ca 5 mm below opening of hypanthium, obovate, 2.5–3 cm long; stamens many, long-exserted; filaments to 9 cm long, becoming reddish, united at base into a fleshy perigynous cup continuous

with the ovary; ovary turbinate, ca 1 cm long; style to 10 cm long. Capsules woody, ellipsoid, 4.5–6.5 cm long, brown, pointed at apex; pericarp 3–5 mm thick; seeds numerous, oblong, 2–4 cm long, winged. *Croat 7948.*

Common along the shore; occasional in the forest. Flowers from September to December. The fruits mature in the dry season (January to April).

This genus was reported by Nevling in the *Flora of Panama* (1958) to have indehiscent fruits, but the fruits of *L. punicifolia* are irregularly dehiscent. After the exocarp of the fruit falls free, the mass of winged seeds are exposed and blow away, usually a layer at a time.

Scattered from Mexico to Bolivia. In Panama, reported by Holdridge (1970) from low to middle elevations in moist areas and by Tosi (1971) as a characteristic component of tropical moist forest; known from tropical moist forest on BCI and in Panamá and Darién and from premontane moist forest in Panamá.

See Figs. 410 and 411.

103. LECYTHIDACEAE

Glabrous trees, occasionally buttressed. Leaves alternate, often clustered at ends of branches, petiolate; blades simple, entire or serrate; venation pinnate; stipules lacking. Corymbs or panicles subterminal, axillary or cauliflorous, bracteate; flowers large, bisexual, actinomorphic, occasionally fragrant; calyx generally 6-lobed, the fused portion forming a hypanthium (sometimes obscure in *Gustavia*); petals 4–12, equal or unequal; stamens many, in several series, often arising at the base from an androphore (occasionally only on one side of androphore in *Couratari*); anthers 2-celled, basifixed, dehiscing longitudinally; ovary inferior, 2–6-locular; placentation axile or pendulous; ovules 1 to several per locule, anatropous; style simple; stigma of several radiate slits. Fruits circumscissilely dehiscent capsules or indehiscent and baccate with a leathery exocarp (1-seeded berries in *Grias*,

---

KEY TO THE SPECIES OF LECYTHIDACEAE

Leaves less than 18 cm long; capsules woody, ± cylindrical pyxidia to 12 cm long . . . . *Couratari panamensis* Standl.
Leaves to 1 m or more long; fruits not as above, fleshy pyxidia or berries:
  Flowers small; petals 4(5), to 2 cm long; fruits ellipsoid, the seed 1 . . . . *Grias fendleri* Seem.
  Flowers large; petals 6–12, to 7 cm long; fruits subglobose, the seeds several:
    Flowers always cauline (borne below the leaves); fruits 7–10 cm wide . . . . *Gustavia superba* (H.B.K.) Berg
    Flowers always or mostly subterminal (borne above the leaves):
      Leaf blades 73–100 cm long, 10–21 cm wide, with 45–54 pairs of lateral veins; flowers always subterminal; floral bracts 80–105 mm long, 30–35 mm wide; bracteoles 45–65 mm long, 34–48 mm wide; calyx lobes ovate to oblong, 29–55 mm long, 24–55 mm wide; fruits with an orange endocarp . . . . *Gustavia grandibracteata* Croat & Mori
      Leaf blades 30–72 cm long, 4.5–10 cm wide, with 22–31 pairs of lateral veins; flowers mostly subterminal, infrequently cauline; floral bracts 8–12 mm long, 5–7 mm wide; bracteoles 7–11 mm long, 5–8 mm wide; calyx lobes broadly triangular, 4 mm long, 14 mm wide; fruits with a creamy white endocarp . . . . *Gustavia fosteri* Mori

Gustavia grandibracteata will probably be found on the island, although it has not been seen yet. It is common around swamps in the Canal Zone.

Fig. 411. *Lafoensia punicifolia*

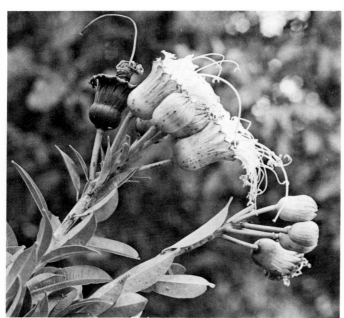

Fig. 410.
Lafoensia punicifolia

Fig. 412. *Couratari panamensis*

1- to few-seeded pyxidia in *Gustavia*); seeds 1 to several per locule, sometimes winged (*Couratari*), lacking endosperm.

Members of the family are distinguished by their unusual flowers with numerous petals, many stamens (sometimes on one side of the androphore and folded over), and an inferior ovary with a sessile, simple stigma.

The flowers are best suited to bee pollination. (See the discussion under *Gustavia superba*.) They do not last more than 1 day whether they are pollinated or not (Prance, 1976).

The fruits of *Grias* and *Gustavia* are dispersed by arboreal frugivores such as white-faced and howler monkeys (Oppenheimer, 1968; Hladik & Hladik, 1969; Carpenter, 1934), though agoutis, spiny rats, and peccaries are important in transporting seeds (Enders, 1935; N. Smythe, pers. comm.). Generally the orange mesocarp is eaten, leaving the seeds unharmed—perhaps they are toxic, as Mori (1970) showed for those of *Lecythis*. *Couratari panamensis* has wind-dispersed seeds.

Either 24 genera and about 450 species (Adams, 1972) or 15 genera and 325 species (Willis, 1966); mostly in wet tropical regions.

COURATARI Aubl.

Couratari panamensis Standl., Publ. Field Columbian Mus., Bot. Ser. 4:239. 1929

Coco, Coco de mono, Coquito, Congolo, Carapelo

Tree, to 40 m tall, to 75 cm or more dbh; buttresses 1–2 m high; outer bark reddish-brown, hard, moderately coarse, with weak vertical fissures; inner bark pale; wood reddish-brown, hard; stems minutely ferruginous-hirtellous, at least when young, angulate on drying. Petioles 5–15 mm long, glabrate; blades oblong to oblong-elliptic, acute to short-acuminate, obtuse to rounded at base, 5.5–18 cm long, 2–8(10) cm wide, hirtellous to glabrate on midrib and veins, the margins irregular. Panicles terminal, bearing few to many flowers; flowers with a faint, somewhat unpleasant aroma, precocious or occurring with new leaves, sometimes on a few leafless branches while other branches fully leaved; pedicels 1.5–2.5 cm long; calyx deeply 6-lobed, the lobes imbricate, ± ovate, obtuse; petals 6, lavender to rose, to ca 2.5 cm long, rounded at apex, often minutely fimbriate, of unequal widths, 2 petals ± spatulate, usually broader, to 1.3 cm wide, the others narrowly oblong, to 8 mm wide; androphore produced into a prominent hood spirally involute then reversely revolute, the whole structure subglobular, slightly broader than high, to 1.7 cm wide, ± magenta and white, the basal part of the androphore shaped like a soap dish, whitish inside, magenta on margins; stamens mostly 13–18, arranged around the stigmatic orifice, ca 1 mm long; anthers rounded, flattened, about as long as filament; ovary 3-celled; stylar orifice ca 1.5 mm diam; stigma ± sessile, minutely 3-radiate. Capsules woody, narrowly cylindrical-campanulate pyxidia, slightly asymmetrical, brown, conspicuously lenticellate, 7–12 cm long, 3–4 cm diam, pendent at maturity; cap and the strong central placenta falling free to allow dispersal of seeds through orifice; seeds several per locule, prominently winged, thin, ca 7 cm long. *Croat 8061, 16507.*

Rare; only a few individuals known on the island. Flowers from July through August, but an individual may flower only once every two years. Allen (1956) reported flowering in late September on the Osa Peninsula in Costa Rica. The fruits mature in February and March.

Wind is probably important in dislodging the seeds from the dangling, pendent fruits.

Costa Rica and Panama. In Panama, known principally from tropical moist forest in the Canal Zone (Atlantic slope), Bocas del Toro, and Panamá; known also from premontane wet forest in the Canal Zone (Pipeline Road). Reported from tropical wet forest in Costa Rica (Holdridge et al., 1971).

See Fig. 412.

GRIAS L.

Grias fendleri Seem., Bot. Voy. Herald 126. 1854

Jaguey, Membrillo, Membrillo macho, Sapo

Glabrous tree, to 7(25) m tall; trunk smooth; wood hard; branches stout, thick. Leaves alternate, clustered at apices of branches, sessile; blades oblanceolate, acuminate at apex, gradually attenuate from above middle to a decurrent subpetiolar base, to 1 m long, 25(40) cm wide. Umbelliform clusters cauliflorus, on both trunk and branches below leaves, bearing few flowers; flowers sweetly aromatic; hypanthium and calyx ca 8 mm long; calyx often splitting irregularly in age; petals 4(5), ovate, white, to 17 mm long, to 11 mm wide, sometimes unequal, the innermost reduced or more markedly concave than the others; stamens numerous, the outermost longer, to 9 mm long; filaments fleshy, often turned inward near apex, the thecae 2 per stamen, loosely attached; ovary bright red at apex; stigma sessile, consisting of (3)4 equidistant radial slits, apparently secreting a viscous fluid. Berries ellipsoid, indehiscent, becoming ribbed in age, to 5 cm long (only immature fruits seen); seed 1. *Croat 13961.*

Known from a single area along the shore of Gigante Bay. Flowers asynchronously from February to May. Time of fruit maturity is uncertain.

Costa Rica (Osa Peninsula) and Panama. In Panama, known from tropical moist forest in the Canal Zone (Atlantic slope), Bocas del Toro, and Chiriquí and from tropical wet forest in Colón. Tosi (1971) reported this genus to be characteristic of tropical wet forest in Panama. Reported from premontane wet forest in Costa Rica (Holdridge et al., 1971).

See Fig. 413.

GUSTAVIA L.

Gustavia fosteri Mori, Brittonia 30:340. 1978

Small tree, to 4 m tall; leaf-bearing branches 5–8 mm diam, the internodes 4–21 cm long. Leaves in 1–5 clusters at apices of branches, 9–16 per cluster; petioles 0.5–5 cm long, 3–5 mm thick at blade juncture; blades oblanceolate to narrowly oblanceolate, attenuate to acuminate

Fig. 413. *Grias fendleri*

Fig. 414. *Gustavia superba*

Fig. 415. *Gustavia superba*

at apex, tapered from middle to an acute base, 30–72 cm long, 4.5–10 cm wide, glabrous, chartaceous, the margins serrate apically, entire basally; lateral veins in 22–31 pairs, the intramarginal vein well developed. Inflorescences subterminal, infrequently cauline, with up to 13 flowers; rachis 5–6.5 cm long, 9–12 mm wide at base; pedicels 6–7.5 cm long; basal bract 1, triangular to narrowly triangular, 8–12 mm long, 5–7 mm wide, the bracteoles 2, ovate to narrowly ovate, cucullate, 7–11 mm long, 5–8 mm wide, inserted slightly above the middle; flowers 10–14 cm diam; calyx lobes 4, broadly triangular, 4 mm long, 14 mm wide; petals 8, narrowly obovate to oblanceolate, 5.5–6.5 cm long, 2–2.5 cm wide, light pink throughout; androphore connate at base, light pink, 11–12 mm high; outermost filaments white at the base, dark pink above, 18–19 mm long; anthers yellow, 2.5–3.5 mm long; ovary puberulent, 5-locular, white-pubescent at apex; ovules 20–30 per locule; style conical, 3 mm long; stigma with 5(6) lobes. Pyxidia globose, truncate at apex, 5 cm long, 5 cm wide, obscurely 4-costate, the calyx lobes persistent, the operculum nearly as wide as the fruit; funiculi white, 5 mm long, 4–5 mm wide; seeds several, cream-colored, irregularly 3-sided in cross section, 3–3.5 cm long, 2.2–2.4 cm diam, surrounded by a creamy white endocarp. *Foster 2790* (holotype, *MO*; isotype, *F*).

Apparently rare; collected once by R. Foster near the junction of Snyder-Molino and Wheeler trails. The species has been observed in flower in June and July, and with mature fruits in September and October.

The closest relative of *G. fosteri* is *G. angustifolia* Benth. of the dry coastal forests of Ecuador. The principal difference between these two species is the glabrous lower leaf surface of *G. fosteri*, in contrast to the pubescent surface of *G. angustifolia*.

The species is easily confused with the much more abundant *G. superba*, but is distinguished by having its leaves grouped in several verticils at the branch ends instead of scattered along the ends of the branches, by its mostly terminal instead of cauline inflorescences, by its calyx of four broadly triangular lobes (the calyx in *G. superba* is entire), and by its petals being light pink throughout instead of white with flushes of pink. Also the fruit has a white instead of orange mesocarp at maturity, and the leaves of *G. fosteri* tend to be smaller than leaves of *G. superba*.

Known only from BCI.

Gustavia superba (H.B.K.) Berg, Linnaea 27:444. 1856
    Membrillo, Membrillo hembra

Trees, mostly 6–10(15) m tall; trunks usually less than 20 cm dbh, branching infrequently; wood moderately hard. Leaves alternate, crowded at ends of stout branches; petioles very short or to 11 cm long on juveniles; blades elliptic-oblanceolate, abruptly acuminate, narrowly cuneate and more or less decurrent onto petiole, mostly to 1 m or more long and 25 cm wide, glabrous, coarsely toothed. Corymb short, beneath leaves; pedicels stout, to ca 8 cm long, minutely puberulent; flowers mostly to 12 cm diam; hypanthium turbinate, to 2.5 cm wide; calyx obscure; petals 6–12, unequal, ± spatulate, oblong, to 7 cm long and 3.5 cm wide, white to cream or spotted with lavender, especially on outside; stamens many, ca 4 cm long, fused one-third their length into a yellowish ring; filaments fleshy, arched inward, often tinged with lavender above, constricted just below anther; stigma short-conical. Pyxidia depressed-globose, 7–10 cm broad, to ca 8 cm long, indehiscent, baccate, with a prominent raised ring at apex, green or pale yellow to orange at maturity with irregular brown lenticels and minute longitudinal ribs, often with a foul odor at maturity; seeds few, fleshy, irregularly shaped, to ca 6 cm long. *Croat 5095, 8812.*

Abundant in the forest, most abundant in the younger forest. Flowers from March to June (sometimes from January). The fruits mature from June to August, sometimes from April and especially in July; they are eaten by white-faced monkeys as early as April (Hladik & Hladik, 1969).

At anthesis the flowers have a sweet aroma, but they become foul-smelling like the fruit in age. Euglossine-like bees have been observed making rapid and repeated visits to the flower. Either the bee enters the staminal tube or it ruffles rapidly through the stamens as if to shake pollen loose. Usually the bee enters the staminal tube and in every case entry was from the lower edge, which is V-shaped, owing to the inflexed stamens. The upper edge of the V-shaped opening through which the bee enters is directly over the stigma. Mori & Kallunki (1976) reported *Gustavia superba* to be self-compatible and observed the species being visited by various bees, including a species of *Melipona* (Apidae), various species of *Trigona* (Apidae), a species of *Xylocopa* (Xylocopidae), and a species of Halictidae. R. Foster (pers. comm.) suggests that bats may be more important in pollination of *Gustavia* and bees may be only incidental pollinators.

The species is easily confused with *G. grandibracteata* Croat & Mori, which is to be expected on the island. The key for this family gives the distinguishing characteristics. Woodson and Schery in the *Flora of Panama* (1958a) reported the range of *G. superba* to be as far south as Ecuador; they were confusing the species with *G. angustifolia* Benth., which occurs there.

Costa Rica to Colombia. In Panama, known from tropical moist forest in the Canal Zone (Atlantic slope), Panamá, and Darién, from tropical dry forest in Panamá (Taboga Island), from premontane moist forest in the Canal Zone and Panamá, and from premontane wet forest in Coclé.

See Figs. 414 and 415.

104. RHIZOPHORACEAE

Shrubs or trees. Leaves opposite, petiolate; blades simple, dentate; venation pinnate; stipules present, interpetiolar. Fascicles bracteate, upper-axillary; flowers bisexual, actinomorphic, mostly numerous; calyx campanulate, 4- or 5-lobed, the lobes valvate; petals 4 or 5, free, white;

stamens many, united basally by the filaments, inserted on a disk; anthers 4-celled, introrse, dehiscing longitudinally; ovary superior, 3-locular, 3-carpellate; placentation axile; ovules 2 per locule, pendulous, anatropous; style 1; stigma capitate. Fruits 3-valved, tardily dehiscent, septicidal capsules; seeds about 3, arillate, usually with endosperm.

Only two genera occur in Panama. *Rhizophora*, a genus of mangroves, does not occur on BCI. *Cassipourea* is distinguished by its opposite leaves, its interpetiolar stipules, and its spatulate, laciniate petals. It is surely insect pollinated. Fruits are adapted for dispersal by birds.

Sixteen genera and 120 or more species; tropics.

CASSIPOUREA Aubl.

Cassipourea elliptica (Sw.) Poir., Dict. Suppl. 2:131. 1811
 C. podantha Standl.
 Huesito, Limoncillo, Goat wood

Shrub or tree, to 13(17) m tall; stems nearly glabrous, ± roughened with lenticels. Stipules interpetiolar, 4–5 mm long, caducous; petioles 3–10 mm long; blades ovate to elliptic, acute to acuminate at apex, cuneate at base, 5–16 cm long, 2–6.5 cm wide, glabrous, often weakly toothed in apical half. Fascicles upper-axillary; flowers 4- or 5-parted, usually numerous; pedicels 2–5 mm long, articulate beneath flower; calyx campanulate, valvate in bud, thick, to 5 mm long, glabrous outside, sericeous inside, exuding a yellowish sap when cut, persistent and reddish in fruit; petals white, spatulate, thin and laciniate, usually pilose, to ca 7 mm long; stamens 15–25, included, ca 7 mm long; filaments slender, often united at base; anthers to ca 1 mm long, 4-celled (owing to a partition in the thecae); ovary depressed-globose; style to 6.5 mm long, persistent in fruit; style and apex of ovary sericeous; stigma capitate. Capsules elliptic to obovoid, ca 1 cm long, yellowish at maturity, fleshy, glabrous except for few trichomes at apex, splitting into 3 parts to expose seeds; seeds 2(4), brown, covered with a white, much-folded aril. *Croat 5371, 6132.*

Frequent along the shore on the north side of the island; uncommon in the young forest. Flowers and fruits throughout the year, especially from April to August. Allen (1956) reported the species to be common in mangrove swamps and to flower in May. Frankie, Baker, and Opler (1974) reported the species to be bimodal in its flowering, with some flowers during January and February and some during July and August in Costa Rica; the fruits there mature from July to October.

Allen (1956) reported the tree to be as tall as 30 m. The flowers have abundant pollen and seem well suited to pollination by pollen-collecting insects.

Seeds are shiny and arillate and are probably bird dispersed.

Guatemala to Panama; West Indies. In Panama, known from tropical moist forest in the Canal Zone, Bocas del Toro, Panamá, and Darién, from premontane wet forest in the Canal Zone and Panamá, from tropical wet forest in Colón, Panamá, and Darién, and from premontane rain forest in Darién (Cerro Pirre).

105. COMBRETACEAE

Trees, shrubs, or lianas, sometimes bearing spines. Leaves alternate or opposite, petiolate; blades simple, entire; venation pinnate; stipules lacking. Inflorescences spikes, panicles, or simple or paniculate racemes, axillary, bracteate; flowers bisexual, actinomorphic or slightly zygomorphic; calyx campanulate or cupulate, forming a hypanthium, 4- or 5-lobed; petals 4 or 5 or lacking (*Terminalia*), attached to the calyx; stamens 8–10, free, in 2 series, inserted on the calyx tube; intrastaminal disk present; anthers 2-celled, versatile, dehiscing longitudinally; ovary very small and inconspicuous when immature, inferior, 1-locular, apparently 1-carpellate; placentation apical; ovules usually 2, sometimes 4–6, anatropous, all suspended from a single funiculus; style 1, slender; stigma simple, capitate or undifferentiated. Fruits indehiscent pseudocarps, with 2–5 wings or ridges; seed single, lacking endosperm.

Combretaceae are distinguished by the usually small, white, open flowers congested into spikes or racemes with exserted stamens (double the number of perianth lobes), by having all ovules suspended from the apex by a slender funiculus, and by the usually winged fruits.

The flowers are mostly small, white, open, and bowl- or cup-shaped with stamens and style long and protruding. They are usually very numerous and fit the brush-type pollination syndrome of van der Pijl (1966). Small amounts of easily accessible nectar have usually been observed. I have seen flowers of *Combretum cacoucia* and *C. fruticosum* visited by hummingbirds.

The winged fruits of all species except *Combretum cacoucia* and *C. laxum* var. *epiphyticum* are wind dispersed and generally good fliers. *Combretum cacoucia* and *C. laxum* var. *epiphyticum* have fruits that are very buoyant and no doubt chiefly water dispersed (Croat, 1974b).

Eighteen genera and about 500 species; subtropics and tropics.

KEY TO THE GENERA OF COMBRETACEAE

Plants lianas; petals present; fruits 4- or 5-winged or -ridged, the wings or ridges all equal . . . . *Combretum*
Plants trees; petals lacking; fruits with 2 prominent wings and 1–3 shallow ridges or vestigial wings . . . . *Terminalia*

COMBRETUM Loefl.

Combretum cacoucia Exell in Sandw., Bull. Misc. Inform. 1931:469. 1931

Liana; branchlets rufous-pilose. Leaves opposite; petioles 5–10 mm long, articulate above base; blades broadly ovate to oblong-elliptic, mostly acuminate at apex, subcordate at base, 8–20 cm long, 4–8 cm wide, shiny when fresh, at most short-pilose on veins below, chartaceous. Racemes to 35 cm long; flowers bright red, slightly zygomorphic, each subtended by a lanceolate, caducous bract to 15 mm long; pedicels ca 5 mm long; lower hypanthium (ovary) ca 5 mm long, the upper hypanthium campanulate, to 2 cm long; pedicels, hypanthia, and calyx lobes densely ferruginous-tomentose; calyx lobes 5, triangular, to 5 mm long; petals 5, ovate to elliptic, to 1 cm long, mucronate, affixed at apex of hypanthium; stamens 10, attached at 2 different heights on calyx tube, 1.5–2.5 cm long; ovary topped by a strigose disk; style ca 3 cm long, barely exceeding stamens. Fruits narrowly ellipsoid, longitudinally 5-ridged, 5–6 cm long, dry. *Wilson 116.*

Collected once on the island. Probably rare this far south of the Caribbean coast, the plant is much more common in regions near the Caribbean. Flowers in the dry season (January to May). Some mature fruits have been seen in April.

Hummingbirds, no doubt the natural pollinator, have been observed visiting these slender, bright-red, tubular flowers with abundant nectar.

Belize to Brazil. In Panama, principally near the Caribbean coast; known from tropical moist forest in the Canal Zone, Bocas del Toro, Colón, and San Blas, from premontane wet forest in the Canal Zone and Colón, and from tropical wet forest in Colón.

Combretum decandrum Jacq., Enum. Syst. Pl. Ins. Carib. 19. 1760

Liana; trunk to 12 cm diam near ground; outer bark coarse, flaky, thin; inner bark weakly and irregularly ringed; branches bearing stout, straight or recurved spines to 4.5 cm long (mostly less than 1.5 cm long on upper stems); juvenile stems slender, leafless, chlorophyllous, bearing slender spines becoming somewhat recurved in age. Leaves alternate or subopposite, deciduous; petioles 5–15 mm long, articulate 1–5 mm above base, the lower part accrescent, persistent; blades narrowly ovate to oblong-elliptic, acuminate, rounded or subcordate and often inequilateral at base, 5–15 cm long, 3–7 cm wide, mostly pellucid-punctate, glabrous or pubescent on veins below, the axils usually with tufts of trichomes, the midrib commonly arched. Panicles large, terminal and upper-axillary, becoming axillary by new growth; flowers white, subsessile or short-pedicellate, 5-parted, actinomorphic, usually 6–7 mm wide; calyx and hypanthium together ca 4 mm long, the upper hypanthium and calyx campanulate to cup-shaped, glabrous, the lower hypanthium glabrous or weakly pubescent with erect trichomes; calyx lobes triangular, ca 0.5 mm long; petals 5, obovate-elliptic, 2–3 mm long, pubescent; stamens 10, exserted; filaments ca 4 mm long; disk present; style ca 4 mm long. Fruits ovoid to orbicular, 1–1.4 cm long, acute to rounded and weakly emarginate at apex, acute at base, broadly 5-winged, glabrous, brown at maturity. *Croat 5545, 14818.*

Often seen as juvenile or adult leafless plants on the forest floor, but rarely encountered with leaves except high in the canopy or along the shore, where it is common. Accrescent petiole bases serve the same function as spines, i.e., assisting in climbing by preventing the branches from slipping. Flowers from March to May, rarely later in the rainy season. The fruits mature from April to July.

Fallen fruits of this species may be easily distinguished from those of *C. laxum* and *C. fruticosum* by being smaller and having five wings.

Mexico to Colombia. In Panama, known from tropical moist forest in the Canal Zone and Los Santos and from premontane moist forest in Darién.

See Figs. 416, 417, and fig. on p. 20.

Fig. 416. *Combretum decandrum*

Fig. 417. *Combretum decandrum*

Combretum fruticosum (Loefl.) Stuntz, U.S.D.A.
Bur. Pl. Industr. Invent. Seeds 31:86. 1914

C. farinosum H.B.K.

Chupachupa

Liana; periderm of older stems sometimes peeling off as fibers; leaves, rachises, and calyces bearing somewhat sunken scales (especially lower leaf surfaces). Leaves opposite; petioles 7–10 mm long; blades ± elliptic, acuminate and mostly falcate at apex, cuneate or rounded at base, 6–12 cm long, 2–6 cm wide, yellowish below owing to golden scales. Panicles terminal or upper-axillary; flowers 4-parted, actinomorphic, pellucid-punctate, yellowish-green or yellow, sessile or subsessile, densely arranged along branches, the branches 1-sided, spikelike, mostly 9–12 cm long; hypanthium (ovary and calyx tube together) ca 1 cm long, the upper part campanulate, golden-scaly; calyx lobes short, acute, exceeding petals; petals minute, yellow; stamens 8, yellowish, 15–20 mm long, subtended by the pilose, funnel-shaped disk; stamens and style folded in bud; anthers shedding some pollen in bud; style unfolding first, equaling or exceeding anthers after anthesis. Fruits more or less orbicular, rounded to emarginate at apex, obtuse to subcordate at base, generally 12–18 mm long, 13–25 mm wide, the wings 4, flexible, to 9 mm wide, reddish-brown to purplish, sparsely covered with dark brown scales. *Croat 7947, Hladik 39.*

Occasional, on the shore and in the canopy of the forest. Flowers principally from November to January. The fruits mature mostly from January to April.

The flowers produce copious nectar and are frequented by hummingbirds, other small birds, and insects. Loose fallen fruits can be distinguished by being four-winged and conspicuously lepidote.

Mexico to Argentina. In Panama, ecologically variable; known from tropical moist forest in the Canal Zone, Veraguas, Herrera, Panamá, and Darién, from premontane dry forest in Panamá, from tropical dry forest in Herrera, Coclé, and Panamá, from premontane moist forest in the Canal Zone, Los Santos, and Panamá, and from premontane wet forest in Veraguas and Panamá.

See Fig. 418.

Combretum laxum Jacq. var. **laxum,** Enum. Syst.
Pl. Ins. Carib. 19. 1760

Liana. Leaves opposite or subopposite; petioles 2–7 mm long, ± glabrous; blades lanceolate to oblong-elliptic or ovate-elliptic, acuminate at apex, obtuse to rounded and inconspicuously subcordate at base (the sinus 1–2 mm deep), glabrous, often obscurely punctate, prominently arched along midrib, drying dark. Panicles terminal or upper-axillary; flowers sessile, actinomorphic, white or yellowish, very fragrant, 4-parted; lower hypanthium ovate-oblong, densely dark-strigose in basal two-thirds, sparsely so above, the upper hypanthium cup-shaped, ca 1.5 mm long including lobes, sparsely strigose; calyx lobes broadly triangular; petals rounded, ca 1 mm wide, white, spreading, ± clawed at base; stamens 8, ca 4 mm long, exserted; style to 4 mm long. Fruits ovoid to suborbicular, emarginate at apex, cordate at base, 1.5–2 cm long, 1.5–1.7 cm wide, yellowish-brown, winged, the wings 4, to 7 mm wide, the body of the fruit to 4 mm wide. *Croat 8405, Foster 1327.*

Frequent in the forest and along the shore, usually high in the canopy. Flowers mostly in October and November, but the flowers sometimes persist until February. The fruits mature from January to March.

Loose fallen fruits can be distinguished by having four wings and by being glabrous or at least not conspicuously lepidote as in *C. fruticosum.*

Mexico to northern Argentina. In Panama, known from tropical moist forest in the Canal Zone, Panamá, and Darién.

See Fig. 419.

Combretum laxum Jacq. var. **epiphyticum** (Pitt.)
Croat, Phytologia 28:189. 1974

C. epiphyticum Pitt.

Shrub or low sprawling liana, usually to 8 m tall. Leaves opposite; petioles 2–7 mm long, densely ferruginous-tomentose; blades ovate-elliptic to oblong-elliptic to obovate, gradually to abruptly acuminate at apex, rounded at base and inconspicuously subcordate (the sinus 2–4 mm deep), 6–16 cm long, 2.5–7 cm wide, the upper surface sparsely short-pilose throughout, the trichomes somewhat denser on midrib and deciduous in age except along midrib, the lower surface similarly pubescent but the trichomes denser, persisting in age; lateral veins in 10–14 pairs, the secondary lateral veins usually continuous between primary lateral veins. Panicles terminal or upper-axillary; inflorescence branches opposite or in whorls of 3, the lower ones subtended by small leaves, the upper ones by lanceolate, caducous bracts ca 4 mm long; rachises, peduncles, and ovary very densely ferruginous-tomentose; flowers sessile, actinomorphic, closely aggregated, subtended by a subulate, caducous bracteole to 1.5 mm long; lower hypanthium ovoid, ca 1 mm long, the upper hypanthium cup-shaped, 1.2–1.5 mm long including the calyx lobes; calyx lobes 4, triangular; petals 4, broadly obovate, ca 1 mm diam, white, glabrous; stamens 8, ca 4 mm long, glabrous; anthers reddish-brown, broader than long, ca 0.5 mm broad; disk thin, dark brown, adnate to calyx, glabrous; style glabrous, subulate. Fruits ovoid in outline, acute at apex, obtuse at base, 1.7–2.2 cm long, prominently ridged longitudinally, the ridges 4, sharp, winglike, extending less than halfway to center of fruit. *Croat 5090.*

Apparently rare; known only from the shore of Peña Blanca Peninsula. Flowers in March and April. The fruits mature in August and September.

The taxon was considered conspecific with *C. laxum* by Exell in the *Flora of Panama* (1958), but recent studies (Croat, 1974b) show this taxon to differ substantially from *C. laxum*, especially in phenology and fruit morphology (see key).

Known only from Panama, from tropical moist forest in the Canal Zone (Atlantic slope) and from tropical wet forest in Veraguas (Bahía Honda). Apparently always associated with bodies of water.

See Fig. 420.

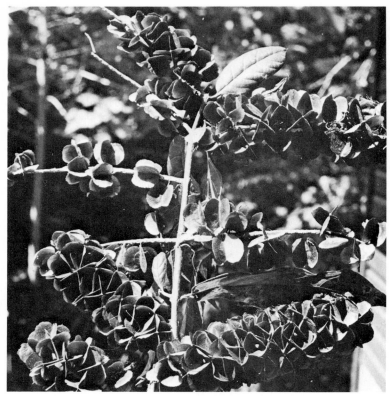

Fig. 418.
Combretum fruticosum

Fig. 419. *Combretum laxum*
var. *laxum*

Fig. 420. *Combretum laxum* var. *epiphyticum*

Fig. 421. *Terminalia amazonica*

Fig. 422. *Terminalia amazonica*

KEY TO THE SPECIES OF TERMINALIA

Stamens less than 4 mm long; wings of fruit less than 7 mm wide; leaves mostly subcoriaceous . . .
. . . . *T. amazonica* (J. F. Gmel.) Exell
Stamens more than 4 mm long; wings of fruit more than 10 mm wide; leaves mostly membranous
at maturity . . . . *T. chiriquensis* Pitt.

TERMINALIA L.

Terminalia amazonica (J. F. Gmel.) Exell in Pulle, Fl. Surinam 3:173. 1935

T. hayesii Pitt.

Amarillo, Amarillo real, Guayabo de montana, Nargusta

Tree, to 35 m tall, often with sucker shoots; trunk buttressed to 2 m high, to 1 m or more diam above buttresses; outer bark brown, soft, with shallow vertical fissures, flaking; inner bark reddish or tan, hard, ringed; branches sympodial; young parts rufous-pubescent. Leaves alternate, clustered at ends of branchlets; petioles rufous-pubescent; blades obovate, obtuse to abruptly acuminate at apex, cuneate at base, 6–9 cm long, 2–3.5 cm wide, mostly subcoriaceous, sparsely pubescent, becoming glabrate except for tufts of trichomes in axils below. Spikes slender, pubescent, axillary, 6–9 cm long; flowers 5-parted, greenish-white; receptacle slender at base, broadened and shallowly cupular above; calyx lobes deltoid; petals lacking; disk densely pilose; stamens 10, exserted, to 3.5 mm long; style usually shorter than stamens; stigma simple. Fruits winged, the wings 2, to 5 mm long and 7 mm wide, green turning brown. *Croat 5606, 7858.*

Common tree in the forest, sometimes in water at the edge of the lake; seedlings have been seen in marshes. Flowers and fruits from February to May (sometimes to July). Trees are leafless for a short time, with the flowers appearing before and during the onset of new leaf growth.

Pennington and Sarukhan (1968) reported this species to be 75 m tall and 3 m dbh in southern Mexico.

Mexico to Guyana, Brazil, and Peru; Trinidad. In Panama, a characteristic component of tropical moist forest (Tosi, 1971); known from tropical moist forest in the Canal Zone, Bocas del Toro, Colón, Panamá, and Darién, from premontane moist forest in the Canal Zone, from premontane wet forest in Panamá, and from tropical wet forest in Colón.

See Figs. 421 and 422.

Terminalia chiriquensis Pitt., Contr. U.S. Natl. Herb. 18:238. 1917

Guayabo de monte

Tree, to 35 m tall; trunk to 90 cm dbh; buttresses many, to 1.5 m high, occasionally with many protruding knots; outer bark loose, falling away from trunk in large pieces; inner bark at first pale lemon-yellow, soon becoming tan; stems rufous-pubescent when young, soon glabrous and fissured. Leaves alternate; petioles 3–10 mm long; blades ± narrowly elliptic, acuminate, acute and sometimes falcate and inequilateral at base, mostly 5–15 cm long, 3–6 cm wide, mostly membranous, glabrous, minutely

and densely pellucid-punctate, arched along the midrib; reticulate veins prominent. Inflorescences terminal, to 15 cm long; flowers yellowish-green, sessile; upper receptacle cup-shaped, ca 2.5 mm long; calyx lobes triangular, reflexed, ca 1 mm long, pubescent outside, lanate inside; petals lacking; stamens 8, exserted, to 7 mm long. Fruits broadly 2-winged, with a single medial ridge on one side, to 2.5 cm long and 3.8 cm wide, emarginate at apex with the tissue sharply twisted to one side. *Croat 13922.*

Uncommon, in the forest. Flowers mostly in the early dry season, especially in January. The fruits mature quickly and are dispersed mostly by the beginning of the rainy season. Leaves drop in June.

Possibly more than one species exists. R. Foster (pers. comm.) has detected striking differences in fruits from different plants of this species. In one type the seeds are about half as long as broad, with the medial ridge straight. In the other type, the seed is 1.5 times as long as broad, with the medial groove twisted into a weak S from one end to the other. No intermediate forms have been seen and nothing is known of the significance of the differences. According to the label on *Skutch 3990* (Costa Rica), the flowers are protogynous. The typical *T. chiriquensis* has a smooth white trunk.

Allen (1956) reported this species as *T. lucens* Hoffm., but that species was published as *T. lucens* Hoffm. ex Mart. and is from Brazil.

Mexico to Panama. In Panama, known from tropical moist forest in the Canal Zone, Bocas del Toro, Chiriquí, and Los Santos.

106. MYRTACEAE

Trees or occasionally shrubs; most parts punctate. Leaves sometimes aromatic, opposite, petiolate; blades simple, entire to undulate; venation pinnate; stipules lacking. Flowers sometimes solitary but mostly in terminal or axillary, cymose spikes or racemes, sometimes panicles, the flowers mostly sweetly aromatic, bisexual, actinomorphic, subtended by a pair of bracteoles; calyx 4- or 5-parted; petals 4 or 5, imbricate, mostly white; stamens many; anthers 2-celled, introrse, dorsifixed, dehiscing longitudinally; ovary inferior, 1-locular, 3-carpellate; placentation parietal; ovules 2 to many per placenta; style and stigma 1. Fruits berries; seeds usually 1, sometimes many (*Calycolpus, Psidium*), with little or no endosperm.

Members of the family are easily distinguished by the opposite, glandular-punctate leaves, the usually small flowers with many stamens and caducous petals, and the persistent glandular sepals at the apex of the fruit.

The flowers are of the brush type, with many exserted stamens, an exserted style, and nectar. In Costa Rica

Eugenia oerstedeana attracts the bees *Anlacoscelis* (Anthophoridae) and *Trigona* (Heithaus, 1973).

The fruits are endozoochorous. Those of *Myrcia*, *Calycolpus*, and most species of *Eugenia* are small, colorful, and fleshy, suggesting bird dispersal. Ridley (1930) reported that *Eugenia* is dispersed principally by birds, fruit bats, and monkeys. Several species of *Eugenia* on BCI are eaten by arboreal frugivores, especially *E. coloradensis* and *E. nesiotica*. Bats take the fruits of *E. nesiotica* (Wilson, 1971), *Psidium guajava* (Goodwin & Greenhall, 1961), and *Syzygium jambos*. Oppenheimer (1968) reported that white-faced monkeys usually remove only the pulp of *Eugenia nesiotica* and spit the seed out. *Psidium* species and *Syzygium jambos* are probably taken by larger animals also. *Psidium guajava* is reportedly eaten by spider monkeys (Hladik & Hladik, 1969) and by coatis (Kaufmann, 1962).

About 60 genera and nearly 3,000 species; chiefly in the tropics of the Southern Hemisphere.

CALYCOLPUS Berg

Calycolpus warscewiczianus Berg, Linnaea
  27:382. 1856
  Guayabillo

Glabrous tree, to 7 m tall. Petioles 2–3 mm long; blades ovate to oblong-elliptic, acuminate, obtuse to rounded at base, mostly 4–9 cm long, 1.5–3 cm wide, pellucid-punctate. Flowers usually solitary (to 3) in axils, sometimes pseudoterminal and in clusters of 5 or fewer,

sweetly aromatic, 5-parted, to 4 cm diam; pedicels 1–5 cm long, with a pair of minute, caducous bracteoles below calyx; calyx lobed almost to base, the lobes oblong, to 1 cm long, with a faint midrib, persisting in fruit; petals obovate-oblong, ca 1.5 cm long, white or white tinged with pink (especially on outside), spreading at anthesis; stamens many; style simple; ovules numerous. Berries subglobose, ca 1 cm long; seeds many, ca 3 mm long, tan. *Croat 7682*.

Occasional, along the shore. Flowers and fruits principally during the dry season (December to April), but also during the rainy season.

Wilson (1971) reported that the fruits are eaten by the bat *Micronycteris hirsuta*.

Costa Rica and Panama. In Panama, most commonly from tropical moist forest in the Canal Zone, Panamá, and Coclé, but known also from premontane dry forest in Coclé (Penonomé) and from premontane wet forest in Panamá (Cerro Campana).

See Fig. 423.

EUGENIA L.

Eugenia coloradensis Standl., Trop. Woods
  52:27. 1937
  E. melanosticta Standl.
  Guayabito de monte

Tree, to 15 m tall, glabrous except for sparse pubescence on midrib of leaves above and all over young leaves. Petioles 5–10 mm long; blades mostly oblong-elliptic,

● Flowers fasciculate or peduncles very short:
    Pedicels more than 1 cm long in flower; leaves glabrous in age, ovate, less than 6 cm long
      (more than half as broad); fruits ribbed, sometimes more than 2.5 cm diam; plants rare
      or absent . . . . *E. uniflora* L.
    Pedicels less than 1 cm long in flower; leaves usually pubescent at least below, not broadly
      ovate; fruits not ribbed, less than 1 cm diam:
      Leaves appressed-pubescent on both surfaces; fruits ca 8 mm diam at maturity . . . .
      . . . . *E. galalonensis* (Griseb.) Krug & Urban
      Leaves not appressed-pubescent; fruits less than 8 mm diam:
        Leaf blades less than 6 cm long, usually glabrous above except on midrib . . . .
        . . . . *E. principium* McVaugh
        Leaf blades more than 6 cm long, both surfaces covered with ferruginous soft pubescence
        . . . . *E. venezuelensis* Berg

---

acute to bluntly acuminate, rounded to obtuse at base, 6–13(17) cm long, 2.5–6(7.5) cm wide, pellucid-punctate. Racemes axillary or terminal, of few flowers, to 7 cm long, sometimes gathered into a panicle at apex of branches; pedicels 5–10 mm long; flowers pellucid-punctate, 4-parted; sepals unequal, suborbicular or longer than broad, to 2.5 mm long, persistent and somewhat accrescent in fruit; petals oblong to obovate, 5–7 mm long, white, concave, ciliate; stamens numerous, the innermost much shorter than style, gradually increasing in length outwardly, the outermost as long as style; style ca 5 mm long; stigma simple. Berries ovoid-globose, to 2.5 cm long and 1.5 cm wide, violet-purple to dark violet at maturity; seed 1. *Croat 15001, Foster 1185.*

Infrequent, in the forest and along the shore. Some flowers were seen in bud in June. The fruits have been seen mostly from July to October.

Costa Rica and Panama. In Panama, known only from tropical moist forest in the Canal Zone and Darién.

Eugenia galalonensis (Griseb.) Krug & Urban, Bot. Jahrb. Syst. 19:641. 1895

Tree, to 6(15) m tall; young stems, petioles, and leaves minutely appressed-pubescent to glabrate in age. Petioles 4–7(10) mm long; blades ± elliptic, bluntly to narrowly acuminate, mostly acute at base, 5.5–10 cm long, 3–4.5 cm wide, pellucid-punctate; reticulate veins obscure. Racemes 2 or 3, superposed, axillary, less than 1.5 cm long; pedicels 2–7 mm long; bracts ovate-triangular, less than 1 mm long; bracteoles broadly ovate, connate, forming an involucre to 1.5 mm wide and 0.5 mm high; flowers 4-parted, the buds 2.5–3.5 mm long; hypanthium campanulate, ca 1 mm long, appressed-pubescent; calyx lobes rounded, in unequal pairs; petals greenish-white; stamens ca 50, to 5 mm long; disk pubescent around base of style; style 5–6.5 mm long; ovary bilocular; ovules 8–10 per locule; stigma simple. Fruits globose, 6–8 mm diam, glabrate, green turning yellow then red and finally maroon at maturity; seed 1, globose, 4–5 mm diam, black. *Croat 8280.*

Seen once on the shore near Drayton House, but no flowers have been seen on the island. In Central America flowers from December to May; BCI plants probably flower in December and January. Full-sized fruits have been seen on BCI in February.

The flower description given here is based on the *Flora of Guatemala* (McVaugh, 1963), where the fruits are described as obovoid or short-ellipsoid.

Guatemala to Colombia (Santa Marta). In Panama, known only from tropical moist forest on BCI.

Eugenia nesiotica Standl., Field Mus. Nat. Hist., Bot. Ser. 17:203. 1937
Guayabillo

Tree, 10–15 m tall; bark flaking; young stems, petioles, and midribs and margins of leaves short-rufous-pubescent, glabrate in age. Leaves deciduous; petioles 2–6 mm long; blades ovate to elliptic, bluntly acuminate, obtuse to acute at base, 4–5 cm long, 1–2 cm wide, pellucid-punctate; reticulate veins prominulous. Fascicles axillary; pedicels slender, 1–2 cm long, short-pubescent; flowers precocious, 4-parted; sepals ovate, 2–3 mm long, reflexed at anthesis, puberulent, often persisting in fruit; petals orbicular, 3–5 mm long, white, glabrous or ciliate; stamens many; style to 5 mm long; stigma simple. Berries globose, 1.5–3 cm diam, red at maturity; exocarp thin, punctate; mesocarp thick, fleshy, sweet; seed 1, round, 6–10 mm diam. *Croat 10314, Foster 776.*

Frequent in the forest, especially in the young forest. Flowers in the late dry or early rainy seasons after the leaves fall. The fruits mature mostly in late May and June (sometimes in July). Leaves grow in again after flowering.

The species is distinguished from other species of *Eugenia* on the island by its prominulous reticulate veins. *Starry 109,* cited by Standley (1933) as *Eugenia* sp., is this species.

Wilson (1971) reported that fruits are eaten by the bat *Micronycteris hirsuta.*

Known only from Panama, from tropical moist and premontane moist forests in the Canal Zone.

See Fig. 424.

Eugenia oerstedeana Berg, Linnaea 27:285. 1856
Sequarra

Tree, mostly 8–15 m tall, to 20 cm dbh; bark smooth, light brown, soft; youngest stems brown, inconspicuously puberulent. Petioles 3–5 mm long; blades ± elliptic, long-acuminate, acute or obtuse at base, 4–8 cm long,

Fig. 423. *Calycolpus warscewiczianus*

Fig. 424. *Eugenia nesiotica*

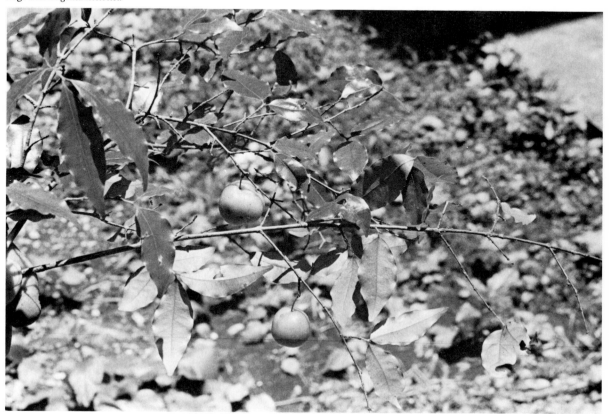

2–3.5 cm wide, glabrous except for puberulent midrib above, pellucid-punctate, the margins slightly undulate. Racemes axillary, of few flowers; inflorescence branches minutely puberulent; pedicels mostly 5–12 mm long; flowers white, 4-parted; sepals and petals glandular-dotted; sepals ovate, 1–1.5 mm long, persisting in fruit, reflexed at anthesis; petals white, broadly ovate, 2–3 mm long; stamens many; style simple. Berries globose to ellipsoid, to 9 mm long, bright red at maturity; seed 1, ± reniform, 4–5 mm long. *Croat 7486, 7761.*

Occasional, in the forest. Flowers in January and February. The fruits probably mature in February and March. At least some plants lose all their leaves in mid-February, with the new leaves growing in again soon.

Mexico, Costa Rica, and Panama. In Panama, known from tropical moist forest in the Canal Zone, Chiriquí, Panamá, and Darién.

Eugenia principium McVaugh, Fieldiana, Bot. 29:451. 1963

Shrub or small tree, to 3 m tall; young stems, petioles, midribs, pedicels, and hypanthia bearing dense, short, erect trichomes. Petioles 1–2 mm long; blades ovate or elliptic, bluntly acuminate, mostly obtuse at base, 3.5–6 cm long, 1.5–2.5 cm wide, pellucid-punctate, ± thick, the margins revolute and ciliate. Racemes or umbels abbreviated, nearly glabrous; pedicels slender, to 1 cm long, bearing 2 bracts beneath the hypanthium; flowers 4-parted; sepals ovate, ca 2 mm long, spreading at anthesis, glandular-dotted, persisting in fruit; petals obovate, ca 5 mm long, pellucid-dotted, ciliate; stamens numerous, ca 5 mm long; style simple. Berries globose, 4–5 mm diam, at first green, turning orange and finally black, glandular-dotted; seed 1, globose, ca 4 mm diam. *Croat 5457.*

Rare, seen only on the shore north of the Front #8 Lighthouse Clearing. Seasonal behavior not determined.

Mexico and Panama, probably elsewhere in Central America. In Panama, known from tropical moist forest in the Canal Zone (Atlantic slope), Panamá, and Darién.

Eugenia uniflora L., Sp. Pl. 470. 1753
Surinam cherry

Shrub or small tree, to 10 m tall; stems glabrate in age. Petioles ca 2 mm long; blades ovate, short-acuminate, rounded to subcordate at base, 2–6 cm long, 1.5–3 cm wide, glabrous in age, pellucid-punctate. Flowers 4-parted, solitary or in fascicles, axillary; pedicels slender, to 2 cm long; sepals ovate, to 4 mm long, ciliate; petals white, obovate, to 12 mm long; stamens many; style 1.

Berries depressed-globose, to 3 cm diam, ribbed, red at maturity; seed 1. *Shattuck 71.*

Collected once, possibly from a cultivated plant, since the species is reportedly cultivated and has not been seen in recent years on the island. Seasonal behavior not determined.

Native to tropical America; cultivated throughout the tropics and subtropics. In Panama, known only from this collection.

Eugenia venezuelensis Berg, Linnaea 27:188. 1854
E. origanoides Berg; *E. banghamii* Standl.

Shrub or small tree, to 6 m tall; stems densely spreading-pubescent, less so in age. Petioles stout, ca 3 mm long, densely pubescent; blades ± elliptic, acuminate, obtuse to rounded at base, 4–12 cm long, 1.5–4.5 cm wide, pellucid-punctate, the surfaces sparsely pubescent, the midrib densely so, the margins revolute and ciliate. Glomerules dense, axillary, among leaves or at defoliated nodes, less than 1 cm long, densely pubescent; pedicels obsolete or nearly so; flowers 4-parted; sepals ovate, 1–2 mm long, spreading at anthesis, ciliate, with ± inconspicuous glandular dots, persisting in fruit; petals white, obovate, 3–4 mm long, glabrous; anthers many, long-exserted; style long-exserted, more than 1 cm long. Berries globose, 4–7 mm diam, at first green turning red and finally black at maturity, glabrous, glandular-dotted; seed 1, globose, 4–5 mm diam. *Croat 5556, 6131.*

Occasional, along the shore. Flowers mostly from February to August. The fruits develop in about 1 month.

Mexico to Colombia and Venezuela. In Panama, known from tropical moist forest in the Canal Zone, San Blas, Coclé, Panamá, and Darién, from premontane dry forest in Coclé (near Penonomé), and from premontane wet forest in Panamá (Cerro Azul).

MYRCIA DC. ex Guillem.

Myrcia fosteri Croat, Ann. Missouri Bot. Gard. 61:886. 1974

Slender tree, 3–8 m tall; trunk to ca 7 cm dbh. Petioles 2–6 mm long, bearing sparse to dense, ± appressed pubescence of short brownish trichomes; blades lanceolate to elliptic, abruptly caudate-acuminate, acute or rarely obtuse at base, 3–8(10) cm long, 1–3(3.5) cm wide, inconspicuously pubescent all over, more densely so on midrib above and below, glabrous above in age except on midrib, pellucid-punctate, thin, the acumen often as much as one-fourth the length of the blade, the margins entire, ± revolute especially near base; second-

KEY TO THE SPECIES OF MYRCIA

Largest leaf blades less than 10 cm long and 3.5 cm wide; flower buds 1.5–2 mm long; fruits globose to depressed-globose, less than 5 mm long .... *M. fosteri* Croat
Largest leaf blades more than 10 cm long and 3.5 cm wide; flower buds 2–3 mm long; fruits ellipsoid to obovoid, to 1 cm long .... *M. gatunensis* Standl.

ary veins not prominulous below when dry. Racemes or panicles axillary, 1–2.5 cm long; pedicels 1–1.5 mm long; flowers 5-parted; calyx to ca 2.5 mm wide, the lobes short-triangular to rounded; axes of inflorescences, pedicels, and hypanthia bearing dense, erect to more commonly appressed pubescence; petals orbicular, ca 1.7 mm long, pellucid-punctate, white, soon falling; stamens numerous, to 3.5 mm long; ovary and basal half of style short-villous, together to 5.5 mm long; style simple. Berries globose to depressed-globose, to ca 5 mm long, 6–8 mm diam, green turning blue-gray and finally blue-black at maturity; seed 1, ± globose, ca 4 mm diam, smooth, the seed coat thin, brown. *Croat 15147* (type), *Foster 2371*.

Occasional in the younger forest; locally abundant along Miller Trail 100–300 and at Wheeler Trail 700. Flowers in late June and in July. The fruits mature in September and October.

Known only from Panama, from tropical moist forest on BCI and in Darién (*Tyson et al. 4756, Stern et al. 719*).

Myrcia gatunensis Standl., Publ. Field Columbian Mus., Bot. Ser. 4:154. 1929
Pimiento

Tree, 3–12 m tall; stems and inflorescences minutely pubescent. Petioles 2–6 mm long, bearing sparse to dense, appressed to erect pubescence; blades lanceolate-oblong or elliptical, gradually acuminate, acute to obtuse or rarely rounded at base, 5–16 cm long, 1.5–5 cm wide, glabrous and ± shiny above except on midrib, glabrous to inconspicuously puberulent below, usually glabrous in age, pellucid-punctate, usually moderately thick, the acumen usually less than one-sixth the length of the blade, the margins entire, weakly revolute; veins of lower surface ± prominulous when dried. Panicles axillary or terminal, 1.5–9 cm long; pedicels to ca 2 cm long; flowers 5-parted; receptacle produced into a densely silvery-puberulent, button-shaped structure above calyx; calyx to 4 mm wide, the lobes ca 1.5 mm long, ± rounded, pellucid-punctate; pedicel and calyx densely pubescent with short silvery trichomes, those of the calyx very dense and mostly appressed; petals orbicular, 1.5–2 mm long, white, pellucid-punctate, soon falling; stamens numerous; style pubescent in basal half. Berries ellipsoid to obovoid, prominently pellucid-punctate, to ca 1 cm long and 8 mm wide, usually considerably longer than broad, green becoming greenish-yellow and finally

blue-black at maturity; seed 1. *Croat 6133, Foster 1120.*

Frequent along the shore on the eastern side of the island, apparently preferring forest edges near water. Flowering principally from June to August. Flowering collections believed to be this species were made in Bocas del Toro in February. The fruits mature mostly in August and September, sometimes as early as June.

There is considerable doubt that *M. gatunensis* Standl. is the oldest name for this taxon. Material of what I believe to be the same species has been identified by McVaugh and others as *M. fallax* (L. C. Rich.) DC. and *M. splendens* (Sw.) DC. In calculating the range for *M. gatunensis* I have limited myself to material that I am confident is the same species as the BCI material. The species is probably more wide-ranging, however, and it is certainly not restricted to BCI as is indicated in the *Flora of Panama* (Amshoff, 1958).

Known only from Panama, but probably ranging into Costa Rica and Colombia. In Panama, known from tropical moist forest in the Canal Zone, Bocas del Toro, and Darién and from tropical wet forest in Colón (Salúd).

PSIDIUM L.

Psidium anglohondurense (Lund.) McVaugh, Wrightia 2(3):123. 1961
Eugenia schippii Standl.

Glabrous tree, 10–20 m tall, to 40 cm dbh; stems terete, grayish, the nodes somewhat to conspicuously swollen, the branching mostly dichotomous. Petioles 1–5 mm long, ± terete or flattened on upper surface, sometimes roughened; blades oblong-elliptic, gradually long-acuminate (the acumen sometimes downturned), broadly to narrowly acute to very weakly attenuate at base, 7–16 (20) cm long, 2–5 cm wide, markedly bicolorous, both surfaces dull, the upper surface with sunken midrib, the lower surface and midrib prominently punctate, the margins weakly revolute; lateral veins 2–6 mm apart, obscure, with a collecting vein 2–3 mm from margin, the reticulate veins very obscure; new flush of leaves ± maroon. Flowers few to several at leafless, swollen nodes; pedicels 2–5 mm long, the bracts solitary or paired, borne apically, ± round or ovate to ± acute, pellucid-punctate, ciliate, usually closely appressed to the hypanthium; flower buds globose to obovoid, ca 7 mm long; calyx completely closed in bud, splitting ± irregularly, 2–5-lobed, the lobes rounded to acute at apex, 3–3.5 mm long, persisting

KEY TO THE SPECIES OF PSIDIUM
Plants decidedly pubescent on most parts; lateral veins of leaves in 12 pairs or more and prominently raised; plants cultivated at the Laboratory Clearing .... *P. guajava* L.
Plants ± glabrous; lateral veins of leaves in 8 pairs or fewer or not at all raised; plants of the forest:
Twigs 4-sided; fruits usually more than 3.5 cm diam; seeds many; blades with the lateral veins in 8 pairs or fewer, prominulous .... *P. friedrichsthalianum* (Berg) Niedenzu
Twigs terete; fruits usually less than 3.5 cm diam; seeds 2; blades with the lateral veins in 12 pairs or more, obscure .... *P. anglohondurense* (Lund.) McVaugh

Fig. 425. *Psidium
friedrichsthalianum*

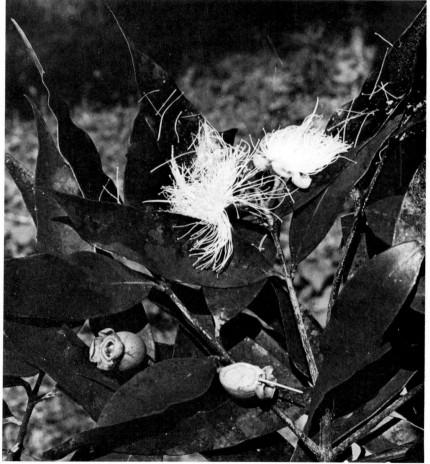

Fig. 426. *Syzygium jambos*

in fruit; petals 5, ± rounded, concave, inconspicuously ciliate, white, ca 5 mm long; stamens numerous; stamens and style ± equaling petals; style glabrous, narrowly tapered to apex; ovules numerous, flattened and reniform, less than 1 mm diam. Berries globose to depressed-globose, 2–3.5 cm diam, green tinged with red (probably also becoming completely red) when mature; seeds usually 2, horseshoe-shaped, ca 2 cm long and 3 mm wide. *Croat 16213.*

Apparently rare, in the old forest; known from the vicinity of Zetek Trail 300–400, where seedlings are common. Mature fruits have been seen on BCI in July. Both flowers and immature fruits are seen elsewhere in Panama in August. In Belize the flowers are seen in May and mature fruits in August and November. Owing to its unusual C-shaped embryo, which lacks a hard bony testa, it is possible that the species does not even belong in the genus *Psidium* (R. McVaugh, pers. comm.). Panamanian material differs from that of plants from Belize in having shorter pedicels and calyx lobes that are essentially glabrous inside rather than strongly pubescent.

Fruits on *Lao & Holdridge 194* (Salúd, Colón Province) were heavily infested with gall-forming insects, which caused the fruits to be somewhat lobed.

Guatemala, Belize, and Panama. In Panama, known from tropical moist forest on BCI and from tropical wet forest in Colón (Salúd).

Psidium friedrichsthalianum (Berg) Niedenzu in Engler & Prantl, Nat. Pfl. 3(7):69. 1893

Guayabo de agua, Wild guavo

Small tree, nearly glabrous, to 9(12) m tall; bark thin, reddish-brown, peeling in thin strips (similar to *P. guajava*); wood hard, fine-grained; branchlets with 4 thin narrow ribs, appearing square; young stems and petioles sometimes bearing fine, sparse pubescence. Leaves opposite or nearly so; petioles 3–6 mm long; blades oblong or elliptic, acuminate, obtuse to rounded at base, 4–12 cm long, 2.5–5 cm wide; major lateral veins in ca 8 pairs, faint above and prominulous below. Cymes axillary, of 1–3 flowers; peduncles 1–2.5 cm long; calyx closed and globose in bud with a short apiculum, 2- or 3-lobed when open, sometimes persisting in fruit, stiff, concave inside, ca 1 cm long; petals 5, orbicular, 1–1.5 cm long, white; stamens numerous; anthers oblong, dorsifixed, dehiscing longitudinally; ovary glabrous; stigma peltate. Berries ± globose, 3–6 cm diam, green to yellowish at maturity, white inside, with a tart, acidlike taste; seeds many, irregularly obovoid or reniform, 6–8 mm long, tan. *Croat 16584.*

Apparently uncommon as an adult, known only from the forest near Wheeler Trail 1300; seedlings common in some areas of the forest above the escarpment south of the Big Trees on Armour Trail. Probably flowering and fruiting all year.

Mexico to Panama. In Panama, known from tropical moist forest in the Canal Zone, Bocas del Toro, Chiriquí, and Panamá.

See Fig. 425.

Psidium guajava L., Sp. Pl. 470. 1753

Guava, Guayaha, Guayava peluda

Shrub or small tree, to 5(10) m tall, to 25 cm dbh; stems glandular-dotted; young stems, petioles, lower leaf surfaces, and midribs above densely short-pubescent. Petioles ca 5 mm long; blades oblong-elliptic, obtuse or apiculate at apex, obtuse to rounded at base, 7–12 cm long, 3.5–6 cm wide, glabrate in age, pellucid-punctate; lateral veins in 12 pairs or more, prominently raised. Flowers solitary in axils, 5-parted; pedicels 1–2 cm long; pedicel and outside of calyx short-pubescent; calyx lobes triangular, to 1 cm long, sericeous inside; petals ovate, white, 1.5–2 cm long, sparsely pubescent outside, glabrous inside; stamens numerous; style simple. Berries globose or pear-shaped, to 6 cm diam, green turning yellow-pink; seeds many. *Croat 5299, 8385.*

Cultivated in the Laboratory Clearing; growing spontaneously on the shore at the end of Peña Blanca Peninsula and on Orchid and Slothia islands. Flowers in the dry season. The fruits mature in the late rainy season and the early dry season.

Immature fruits are eaten by white-faced and spider monkeys (Hladik & Hladik, 1969; Oppenheimer, 1968).

Cultivated throughout the tropics; common in secondary growth. In Panama, known from tropical moist forest in the Canal Zone, Bocas del Toro, San Blas, Panamá, and Darién, from premontane dry forest in Coclé (Penonomé), and from premontane wet forest in Coclé (El Valle) and Panamá (Cerro Jefe).

SYZYGIUM Gaertn.

Syzygium jambos (L.) Alston in Trimen, Handb. Fl. Ceylon, Suppl. 6:115. 1931

Eugenia jambos L.

Pamarosa, Rose apple

Glabrous trees, to 15 m tall, ca 30 cm dbh. Petioles 5–10 mm long; blades oblong-lanceolate, acuminate, acute to obtuse at base, 12–22 cm long, 3–5 cm wide, pellucid-punctate. Flowers 1–5 at ends of branchlets, 4-parted, glabrous; pedicels ca 1.5 cm long; receptacle 1–1.5 cm long; sepals semicircular, recurved at anthesis, 5–7 mm long, persistent in fruit; petals white, 1–2 cm long, ovate, pellucid-punctate; stamens many; stamens and style long-exserted. Berries subglobose to pear-shaped, 2–3 cm diam, green to pinkish or yellowish at maturity; seeds 2 to several. *Croat 7413.*

Infrequent, along some areas of the shore. The species is apparently naturalized but may be the result of former habitation. Flowers mainly in the dry season (rarely November to as late as May). Fruit maturity time has not been determined.

Native to the East Indies; cultivated and escaped throughout tropical America. In Panama, reported from low and middle elevations in moist and wet areas, often in pastures (Holdridge, 1970); known from tropical moist forest in the Canal Zone, Bocas del Toro, Colón, and Panamá.

See Fig. 426.

107. MELASTOMATACEAE

Shrubs and small trees, sometimes herbs, rarely vines (*Adelobotrys*), rarely hemiepiphytic (*Topobaea*); stems occasionally winged or angled. Leaves opposite, petiolate (sessile in *Mouriri*); blades simple, entire or serrate; principal veins palmate at base, usually pliveined or rarely (in *Mouriri*) only the midrib present; stipules lacking. Flowers bisexual, actinomorphic (the androecium and style may be slightly zygomorphic); perianth 4–7(8)-parted; calyx lobed, arising from hypanthium, sometimes with a small exterior tooth on the hypanthium at each calyx lobe; petals free, usually clawed, showy; stamens about twice as many as petals, free; anthers 2- or 4-celled, opening by terminal pores, the connective prolonged at base into a prominent appendage; ovary superior or inferior, mostly 3–5-locular, 3–5-carpellate; placentation axile; ovules numerous, anatropous; style 1, simple; stigma ± capitate. Fruits loculicidal capsules if from a superior ovary, ± enclosed in the persistent hypanthium, or berries if from an inferior ovary; seeds exalbuminous, usually numerous (1 in *Mouriri*).

Members of the family (except *Mouriri myrtilloides*) can be distinguished by the pliveined leaves and by the usually fleshy fruits with many, small, usually wedge-shaped seeds. The anthers are usually tapered to the apex and have terminal pores.

The flowers are probably pollinated by bees, especially of the genus *Trigona*. I have observed large bees (probably *Eulaema*) drumming pollen from *Topobaea praecox* (see that species description for details). In Costa Rica *Miconia argentea* is primarily visited by *Melipona beechii* (G. Frankie, pers. comm.). *Schwackaea cupheoides* in Costa Rica is visited by the halictid bees *Caenaugochlora* and *Augochlora*.

The fruits are chiefly endozoochorous. Those that are chiefly bird dispersed include *Leandra dichotoma*, *Ossaea quinquenervia*, and *Mouriri myrtilloides*, as well as species of *Conostegia*, *Clidemia*, and *Miconia*. Fruits of *Topobaea praecox* and possibly *Henriettea succosa* are also taken by birds. Fruits of *Bellucia grossularioides*, though possibly dispersed by mammals or larger birds, become soft at maturity and could be pecked open by smaller birds. Oppenheimer (1968) reported that mature fruits of *Miconia argentea* are eaten by white-faced monkeys. They are also taken by coatis (Kaufmann, 1962). Hladik and Hladik (1969) reported that fruits of *Mouriri myrtilloides* are eaten by white-faced, spider, and howler monkeys nearly all year.

Seeds of several species are wind dispersed, including *Schwackaea cupheoides*, *Tibouchina longifolia*, *Arthrostema alatum*, and *Adelobotrys adscendens*.

About 200–240 genera and 3,000–4,500 species; throughout the tropics, but most abundant in South America.

ACIOTIS D. Don

Aciotis levyana Cogn. in Mart., Fl. Brasil. 14(3):460. 1885

Herb, to 1 m tall, pubescent, the trichomes long, simple, chiefly gland-tipped toward the apex of blade and on petioles and stems; stems square, the corners usually winged and ciliolate. Petioles 1–4 cm long, often winged laterally; blades ± ovate to oblong-ovate, acuminate, obtuse to subcordate at base, mostly 6–13 cm long, 3.5–6.5 cm wide; veins 5–7. Cymes terminal, divaricately branched, the branches usually square and weakly winged; flowers 4-parted; hypanthium globose, fleshy, ± inflated, both the hypanthium and calyx lobes hispid; calyx lobes minute, curved outward; petals to 4.3 mm long, soon falling, white or pinkish; stamens 8, erect, mounted on rim of hypanthium, to 3.5 mm long, shorter than sepals at anthesis, then elongating; ovary superior; style slightly longer, straight. Capsules globose, thin-walled, dehiscing irregularly; seeds cochleate (shaped like a snail shell), brown, somewhat foveolate, less than 1 mm long. *Bailey & Bailey 581*.

Collected once on the island and, though abundant elsewhere in Panama, not seen recently on the island. The species could be expected to occur from time to time in

---

KEY TO THE TAXA OF MELASTOMATACEAE

Fruits loculicidal capsules (not obvious in *Aciotis levyana*); ovary superior, free from the hypanthium; stems usually either winged or more or less angulate (not winged or angulate in *Tibouchina* or *Adelobotrys*):
  Stems conspicuously winged:
    Stem wings markedly ciliolate with long simple or glandular trichomes; capsules round . . . . .
      . . . . *Aciotis levyana* Cogn.
    Stem wings glabrous (stems may be sparsely pubescent with glandular trichomes); capsules campanulate . . . . *Arthrostema alatum* Tr.
  Stems terete or square but lacking conspicuous wings:
    Plants climbing vines in the forest; calyx tube widely spreading . . . .
      . . . . *Adelobotrys adscendens* (Sw.) Tr.
    Plants erect herbs in open areas; calyx tube not widely spreading:
      Stems glabrous or with pubescence restricted chiefly to angles; trichomes of upper blade surface free to base; flowers solitary, the petals usually pinkish; capsules oblong . . . . . .
        . . . . *Schwackaea cupheoides* (Benth.) Durand
      Stems pubescent all over; trichomes of upper leaf surface adnate to the epidermis for a part of their length; flowers in cymes, the petals white (rarely pinkish); capsules ovate to rounded . . . . *Tibouchina longifolia* (Vahl) Baill.

Fruits berries; ovary wholly or partly inferior; stems terete:
    Leaves with a single vein (the midrib), the blades sessile, to 8.5 cm long; anthers with 2 short
        longitudinal slits; berries with 1 seed . . . .
        . . . . *Mouriri myrtilloides* (Sw.) Poir. subsp. *parvifolia* (Benth.) Morley
    Leaves with 3 to several veins; blades petiolate, usually more than 7 cm long; anthers opening
        by terminal or subterminal pores; berries with many seeds:
        Flowers each subtended by 2 pairs of decussate bracts on base of hypanthium; petals thin,
            rose-purple, ca 1.5 cm long; anthers connate in a ring, with 2 pores; plants usually hemi-
            epiphytic . . . . *Topobaea praecox* Gleason
        Flowers not individually subtended by decussate bracts; petals white (to pink) or much
            smaller than 1.5 cm; anthers free, usually with a single pore (2 in *Bellucia*); plants
            shrubs, trees, or vines, not hemiepiphytic:
            Plants vines; calyx plate-shaped, widely spreading, the teeth minute or obscure; petals ca 1
                cm long, rounded at apex; blades thick, rounded to cordate at base . . . .
                . . . . *Adelobotrys adscendens* (Sw.) Tr.
            Plants trees or shrubs; calyx not plate-shaped or widely spreading, often conspicuously
              toothed:
              Inflorescences lateral or axillary (rarely terminal in *Clidemia capitellata*):
                Hypanthium and calyx lobes together more than 1.2 cm long:
                  Leaf blades glabrous on veins below (except sometimes along margins of veins);
                      stems glabrous or inconspicuously puberulent; hypanthium more than 2 cm
                      wide . . . . *Bellucia grossularioides* (L.) Tr.
                  Leaf blades appressed-pubescent on veins below; stems conspicuously appressed-
                      pubescent; hypanthium less than 1 cm wide . . . . *Henriettea succosa* (Aubl.) DC.
                Hypanthium and calyx lobes together less than 1 cm long:
                    Leaf blades obtuse to rounded or cordate at base, not decurrent on petiole . . . .
                    . . . . *Clidemia* (in part)
                  Leaf blades attenuate at base and decurrent on petiole:
                    Inflorescences usually at leafless nodes, 2 cm long or less; major lateral veins ±
                      evenly spaced in basal fourth of blade . . . . *Clidemia septuplinervia* Cogn.
                  Inflorescences usually at current leaf nodes, 2–5 cm long; major lateral veins
                    principally from a single point near base of blade . . . .
                  . . . . *Ossaea quinquenervia* (P. Mill.) Cogn.
            Inflorescences terminal:
               Petals acute or acuminate; panicles long and slender, often reddish; pubescence of
                  inflorescences and stems simple, long, spreading or retrorse; leaves conspicuously
                  pubescent . . . . *Leandra dichotoma* (D. Don) Cogn.
               Petals obtuse to rounded (acute in *Conostegia cinnamomea* but its stems not hirsute):
                  Calyx calyptrate in bud (i.e., its lobes forming a united cap over petals, not opening
                    in the regular manner but becoming severed from the flower at its base and
                    falling free) . . . . *Conostegia*
                Calyx open in bud, the lobes present at anthesis (often small):
                  Plants with 3 types of pubescence on younger stems and petioles (stellate-tomen-
                    tose, glandular-hirsute, and simple-hirsute); petals 8–10 mm long . . . .
                    . . . . *Clidemia octona* (Bonpl.) L. O. Wms.
                  Plants never with 3 types of pubescence at one time; petals less than 5 mm long
                    . . . . *Miconia*

---

the larger clearings. Flowers and fruits throughout the year.

Guatemala to Ecuador. In Panama, known from tropical moist forest in the Canal Zone and Bocas del Toro, from premontane wet forest in Coclé and Panamá, and from tropical wet forest on the Atlantic slope in Veraguas, Colón, and Coclé.

ADELOBOTRYS DC.

Adelobotrys adscendens (Sw.) Tr., J. Bot. 5:210. 1867

Liana or scandent shrub climbing to several meters; stems terete to ribbed, becoming glabrate in age, sometimes with fine roots along their length; younger parts, petioles, and inflorescences densely brown-strigose. Petioles 6–30 mm long; blades ovate to subrotund, acuminate, rounded to cordate at base, 4.5–15 cm long, 3–7 cm wide, thick, entire or denticulate, glabrous or strigose above, strigose below; veins usually 5. Panicles 10–40 cm long, of umbels of 2–6 flowers; pedicels 2–5 mm long; flowers 5-parted; hypanthium ca 5 mm long, higher than ovary; calyx tube widely spreading, persisting and to 6 mm wide in fruit, the teeth minute or obscure; petals obovate, white to pinkish, ca 1 cm long, rounded at apex, spreading; stamens 10, about as long as petals; filaments flattened, sometimes blue; anthers linear, deflexed, 6–10 mm long, the connective elaborate, with a short part directed upward and a longer, dorsal, forked appendage directed basally ± parallel to the thecae; ovary superior; style shorter than stamens, becoming recurved near apex. Fruiting hypanthia conspicuously ribbed at maturity, to 8 mm long, the

outer wall of fruit and the calyx eventually weathering away, the thin inner wall held in place by the stout ribs, the ribs then attached only at base; seeds very numerous, obcylindrical, ca 1 mm long, wind dispersed. *Foster 2090.*

Rare; collected once in the old forest. Probably flowers and fruits throughout the year.

The fruits are not obviously capsular until maturity. Seeds are slowly dispersed as if by a saltshaker.

Mexico to Peru (Loreto), Bolivia, and Brazil; Jamaica. In Panama, known from tropical moist forest on BCI and in Coclé and Darién, from premontane wet forest in Chiriquí, Coclé, and Panamá, and from tropical wet forest in Colón, Coclé, Panamá, and Darién.

ARTHROSTEMA Pav. ex D. Don

Arthrostema alatum Tr., Trans. Linn. Soc. London 28:35. 1871

Herb, to 60 cm tall; stems succulent, square, the edges narrowly winged; trichomes sparse, red, gland-tipped, on stems and inflorescences, including hypanthia and petals. Petioles to 3.5 cm long, winged; blades ovate, acute to acuminate, truncate to subcordate at base, 3–7 cm long, 1.5–4.5 cm wide, setose above, glabrous below, serrulate and ciliate; veins 5. Cymes terminal, divaricately branched, usually with a solitary flower between branches; flowers 4-parted, ± sessile; hypanthium campanulate, to 4 mm long; calyx tube obscurely lobed, accrescent in fruit; petals pink to lavender, ca 3 mm long; stamens 8, ± equal; anthers with the connective prolonged somewhat at base; filaments united to calyx tube; ovary superior; style straight, about as long as stamens. Capsules campanulate, ca 1 cm long, 4-valved, loosening at base and acting as a wind ballast; seeds numerous, minute, cochleate, with grooves and minute tubercles on the rounded side. *Croat 6150.*

Uncommon, in the Laboratory Clearing. Flowers from August to October. Mature fruits seen in October.

Southern Mexico to Venezuela. In Panama, known from tropical moist forest in the Canal Zone and Panamá and from premontane moist forest in Panamá.

BELLUCIA Neck. ex Raf.

Bellucia grossularioides (L.) Tr., Trans. Linn. Soc. London 28:141. 1871

Small tree, to 5(10) m tall, glabrous except sometimes for puberulence on stems and lower leaf surfaces. Petioles to 6 cm, stout; blades broadly elliptic, obtuse to short-acuminate, obtuse to rounded at base, 15–35 cm long, 10–20 cm wide, entire, pliveined, the major veins 3–5, soon becoming glabrous. Cymes axillary, frequently on leafless branches and on trunk; pedicels 1–2.4 cm long; flowers with an intense, sweet aroma at anthesis, ca 6 cm wide; hypanthium hemispherical; calyx 1.5–2 cm long, hyaline, dehiscing irregularly at anthesis and falling free from the hypanthium; petals 5 or 6 (to 8), oblong, thick, white or pink, to 3 cm long, broader and often very oblique at apex, spreading at anthesis; stamens 10–16,

to 1.8 mm long; filaments about as long as anthers; anthers obovate, turned inward, yellowish, with 2 small terminal pores, both the thecae and connective swollen; ovary inferior; style straight, stout, at first medially positioned, soon pushed to one side of flower; stigma large, to 4 mm wide, capitate, held well above stamens. Berries ± globose, to ca 3 cm diam, pale green at maturity; seeds tan, numerous, ca 0.7 mm long, about twice as long as broad. *Shattuck 1119.*

Collected by O. Shattuck on Gigante Bay and possibly still occurring there, although the species is undoubtedly more abundant elsewhere. Flowers during July and August; mature fruits have been seen in October.

The species has been confused with *B. axinanthera* Tr., which is more abundant and apparently more ecologically variable as well, occurring in Panamanian life zones ranging from premontane moist to premontane rain forests. *B. costaricensis* Cogn., reported by Gleason (1958) in the *Flora of Panama*, is a synonym of *B. axinanthera.* The calyx of *B. axinanthera* differs from that of *B. grossularioides* by being regularly dehiscent, with lobes that are thick, regularly acute, and persistent in fruit. The calyx lobes of *B. grossularioides* are hyaline, and their dehiscence is irregular and often circumscissile.

Southern Mexico to Amazonian Brazil and Peru. In Panama, known from tropical moist forest in the Canal Zone, Bocas del Toro, San Blas, and Darién.

CLIDEMIA D. Don

Clidemia is distinguished by having axillary inflorescences and petals rounded at the apex. The ovary is inferior. Most species are conspicuously pubescent.

Clidemia capitellata (Bonpl.) D. Don ex DC., Prodr. 3:159. 1828
    C. neglecta D. Don

Shrub, 1–3 m tall; stems, inflorescences, and lower leaf surfaces densely pubescent with both stalked-stellate and simple trichomes, the simple trichomes often glandular-tipped, usually longer than the stellate ones. Petioles 1–5 cm long; blades ovate to ovate-oblong, acuminate, obtuse to rounded at base, 6–15(20) cm long, 3–7(9) cm wide, minutely toothed, pubescent above, the trichomes long, simple, set on bulbous projections of surface appearing as depressions on lower surface; veins 5–7. Flowers sessile, 5-parted, clustered and well spaced on a strong central axis and at the ends of short branches; hypanthium and calyx tube together 3–4 mm long; calyx lobes 1–1.5 mm long, the exterior teeth slender, ca 1 mm longer than calyx lobes; petals white, 4.5–5 mm long, spreading at anthesis and later closing; stamens folded in bud, all directed to one side at anthesis, the terminal pore of the anthers held somewhat beneath style; style erect or held to one side; stigma held just above the ± closed petals after stamens shrivel. Berries subglobose, purple, densely pubescent, ca 6 mm diam; seeds minute, numerous. *Croat 11770.*

Though the species is common in the Canal Zone in disturbed areas, it is uncommon on the island and has

KEY TO THE SPECIES OF CLIDEMIA

Flowers 4-parted, the flowers or at least the fruits often occurring on a leafless stem below leaves:
  Leaf blades usually attenuate at base and decurrent on petiole, not cordate . . . .
  . . . . *C. septuplinervia* Cogn.
  Leaf blades cordate or subcordate at base . . . . *C. purpureo-violacea* Cogn.
Flowers 5- to 8-parted:
  Pubescences of simple trichomes only:
    Plants with stems, blades, and inflorescence branches setose, the trichomes to 1 cm long . . . . .
    . . . . *C. collina* Gleason
    Plants ± hirsute all over, the trichomes ca 1 mm long or less . . . . *C. dentata* DC.
  Pubescence (at least on stems) of 3 types: stellate, simple, and simple gland-tipped:
    Trichomes of upper blade surface often on conspicuous projections appearing as depressions
      from the underside; simple trichomes only slightly or a few times longer than the stellate
      ones; petals 4–5 mm long . . . . *C. capitellata* (Bonpl.) DC.
    Trichomes of upper blade surface distributed evenly, the blades lacking projections; simple
      trichomes usually many times longer than the stellate ones; petals 8–13 mm long . . . .
      . . . . *C. octona* (Bonpl.) L. O. Wms.

been collected only at Gross Point. Flowers and fruits throughout the year.

Recognized by its unusual inflorescences and pustular leaves.

Southern Mexico to Colombia; West Indies. In Panama, known principally from tropical moist forest in the Canal Zone, Bocas del Toro, Chiriquí, Veraguas, Panamá, and Darién Provinces; known also from premontane moist forest in the Canal Zone, from premontane wet forest in Bocas del Toro, Colón, Chiriquí, Coclé, and Panamá, and from tropical wet forest in Darién and on the Atlantic slope in Veraguas, Colón, and Coclé.

See Fig. 427.

Clidemia collina Gleason, Phytologia 3:359. 1950

Slender shrub, to 2 m tall; stems, petioles, major veins of lower blade surfaces, and inflorescences conspicuously setose, the trichomes of stems mostly to 1 cm long, often deciduous in age. Petioles 1–3 cm long, the basal part densely setose and usually encrusted with ant detritus, the apical part with swollen domatia inhabited by small brown ants; blades ± elliptic, gradually acuminate, narrowly rounded at base, 15–32 cm long, 5.5–11.5 cm wide, very sparsely setose above, more densely so on major veins below, conspicuously ciliate, pliveined, the veins 7. Flowers 5-parted, in densely setose, branched, axillary clusters; pedicels pink; hypanthium and calyx tube campanulate, ca 3 mm long, the upper part with setae 6–10 mm long; calyx lobes blunt, 1–2 mm long, the exterior teeth slender, setose, about as long as or longer than sepals; petals white, oblong-obovate, rounded at apex, 4–6 mm long; stamens straight, yellow, to 5.5 mm long; anthers to 3 mm long with a single terminal pore; style only slightly exceeding stamens. Fruiting hypanthia unknown. *Foster 2355.*

Apparently rare on BCI, collected once south of Zetek Trail 800. Seasonal behavior not determined. Flowers at least in April and May, with some fruits maturing in August.

Known only from Panama, from tropical moist forest on BCI and in Bocas del Toro and from premontane wet forest in Coclé (El Valle).

Clidemia dentata D. Don ex DC., Prodr. 3:158. 1828

Shrub, 1–4 m tall, densely pubescent with long, stiff, erect or recurved trichomes. Leaves paired, one often smaller than the other; petioles to 1.5 cm long; blades narrowly ovate-elliptic, acuminate, acute to rounded and usually unequal at base, 6–20 cm long, 2.5–8 cm wide, entire or minutely serrate, pliveined, the veins 5–7. Inflorescences short, paniculate or racemose, bearing few flowers; hypanthium and calyx bell-shaped, 3–4 mm long; calyx truncate, the exterior teeth subulate, to 4 mm long; petals 5, white, turning brown on drying, oblong, 5–7 mm long; stamens to 4 mm long; filaments arcuate below anthers; style slightly shorter than petals; stigma simple. Berries subglobose, to 1 cm diam, green turning blue-black at maturity; seeds minute, numerous. *Croat 11940.*

Uncommon, within the forest and at the edges of clearings. Flowers and fruits throughout the year.

Southern Mexico to Bolivia and Brazil. In Panama, known from tropical moist forest in the Canal Zone, Bocas del Toro, San Blas, Coclé, and Darién, from premontane moist forest in the Canal Zone, from premontane wet forest in Bocas del Toro and Coclé, and from tropical wet forest in Colón, Coclé, and Darién.

See Fig. 428.

Clidemia octona (Bonpl.) L. O. Wms., Fieldiana, Bot. 29:558. 1963
Heterotrichum octonum (Bonpl.) DC.

Shrub, (1)2–3(5) m tall; stems, petioles, peduncles, and hypanthia with both stellate and simple trichomes, the simple ones in part very long, in part short and glandular. Leaf pairs sometimes unequal; petioles 2–10 cm long; blades ovate, acuminate, cordate at base, 8–20 cm long, 6–12 cm wide, chiefly with long simple trichomes above and short stellate ones below, denticulate; veins 7. Panicles upper-axillary and subterminal; flowers 7- or 8-parted; hypanthium and calyx lobes together to 5 mm long, the exterior teeth slender, much longer than calyx lobes, ca 1.5 mm long; petals obovate, 8–13 mm long, white, broadly spreading; stamens folded in bud, erect and radially symmetrical at anthesis; anthers ca 6 mm long, gradually tapered to apex with a single termi-

Fig. 427.
Clidemia capitellata

Fig. 428.
Clidemia dentata

Fig. 429.
Clidemia octona

nal pore; style stout, erect in bud, elongating after anthesis, held to one side well above anthers, to ca 1 cm long. Berries purple at maturity, 1–1.5 cm diam, the pubescence of the hypanthium persisting; seeds minute, numerous. *Croat 11848.*

Occasional, at the edge of the forest. Flowers and fruits throughout the year.

Southern Mexico to Peru and Brazil; Cuba, Jamaica. In Panama, known from premontane moist forest in the Canal Zone and Veraguas, from tropical moist forest in the Canal Zone, Bocas del Toro, San Blas, Veraguas, Los Santos, Herrera, Panamá, and Darién, from premontane wet forest in the Canal Zone, Chiriquí, and Panamá, and from tropical wet forest in Panamá and Darién.

See Fig. 429.

Clidemia purpureo-violacea Cogn., Bull. Soc. Roy. Bot. Belgique sér. 2, 30(1):263. 1891

Slender shrub, 1–5 m tall, often arching; stems glabrous in age. Petioles 1–8 cm long, bearing minute, furfuraceous, stellate pubescence; blades broadly ovate, acuminate, rounded to cordate at base, 12–22 cm long, 3–14 cm wide, minutely stellate-pubescent on veins below and above near base, minutely denticulate and ciliate, ± maroon below; veins 5–7. Inflorescences axillary, small, paniculate, to ca 1.5 cm long, mostly below leaves; branches of inflorescence inconspicuously stellate-pubescent; flowers 4-parted, pale red, ca 4 mm diam; hypanthium ± globose, 2–3 mm long, minutely pubescent, the trichomes borne on minute papillae, often deciduous; exterior teeth ending in deflexed setae ca 0.5 mm long; calyx lobes 0.5–0.7 mm long, markedly recurved with a submarginal projection forming the crown; petals oblong, rounded at apex, ca 1.5 mm long; stamens erect, ca 3 mm long; style ca 2 mm long, becoming directed to one side. Berries to 5 mm long, ± globose, at first red-violet, turning purple-black at maturity; seeds minute, numerous. *Croat 14551.*

Rare, known only from deep ravines in the vicinity of Creek #8. Flowers and fruits from April to December. Most fruits mature during the rainy season.

Costa Rica to Colombia. In Panama, known principally from tropical moist forest in the Canal Zone, Bocas del Toro, San Blas, and Darién; known also from premontane wet forest in Panamá and from tropical wet forest in Darién.

Clidemia septuplinervia Cogn. in Mart., Fl. Brasil. 14(4):506. 1888

Shrub, 1.5–3(7) m tall; stems, petioles, lower leaf surfaces, axes of inflorescences, and hypanthia bearing very minute, reddish-brown, granular puberulence. Leaf pairs sometimes unequal; petioles to 6 cm long; blades ovate to elliptic, acuminate, attenuate (rarely rounded) and sometimes inequilateral at base, decurrent onto petiole, 11–24 cm long, 5–12 cm wide, glabrous above, sparsely ciliate, subentire, pliveined, the veins 5–7. Cymes axillary, minute, ca 1 cm long, usually occurring on leafless stems; hypanthium ovoid or oblong, to 3 mm long; calyx lobes 4, rounded, ca 1 mm long, reddish, the exterior teeth slender, 3–4 times as long as sepals with a few long bristles; petals 4, pink or white. Berries globose, 3–5 mm diam, green then red turning blue at maturity, minutely puberulent; seeds minute, numerous. *Foster 1096.*

Collected once west of Armour Trail 100–200. Flowers and fruits throughout the year.

Belize to Colombia, Peru, and Brazil; Jamaica. In Panama, known from tropical moist forest in the Canal Zone, Bocas del Toro, San Blas, and Darién, from tropical wet forest in Veraguas (Atlantic slope), Colón, Panamá, and Darién, and from premontane rain forest in Darién.

CONOSTEGIA D. Don

Conostegia is distinguished by having terminal inflorescences, petals obtuse or rounded at apex (acute in *C. cinnamomea*), and flower buds with a calyptrate calyx (i.e., the entire calyx deciduous as a unit leaving an irregular scar). The ovary is inferior.

Conostegia bracteata Tr., J. Bot. 5:209. 1867

Shrub, to 3.5 m tall, conspicuously hirsute except on petals and inner flower parts, the trichomes long, simple, sometimes slightly branching, especially on main veins and below inflorescences. Petioles to 2 cm long; blades narrowly elliptic to oblanceolate, acuminate, acute to obtuse at base, 9.5–20 cm long, 4–8 cm wide, obscurely serrulate; veins 3. Panicles to ca 10 cm long; flowers in small glomerules subtended by oblong-ovate bracts to 8 mm long; hypanthium 2.5–3 mm long at anthesis; petals (5)6(7), obovate, rounded to truncate at apex,

KEY TO THE SPECIES OF CONOSTEGIA

Leaf blades glabrous above or nearly so at least at maturity; trichomes of lower surface sessile, stellate (minute in *C. cinnamomea*):
    Lower leaf surface completely concealed by a fine close mat of brownish stellate trichomes .... *C. xalapensis* (Bonpl.) DC.
    Lower leaf surface not concealed by trichomes, sparsely pubescent at maturity (often superficially glabrous) .... *C. cinnamomea* (Beurl.) Wurdack
Leaf blades conspicuously pubescent above, the trichomes chiefly or entirely simple:
    Trichomes of lower blade surface all simple (rarely a few on principal veins branched but not stellate) .... *C. bracteata* Tr.
    Trichomes of lower blade surface often both simple and stellate, the stellate trichomes mostly stalked .... *C. speciosa* Naud.

Fig. 430. *Conostegia bracteata*

Fig. 431.
Conostegia cinnamomea

Fig. 432. *Henriettea succosa*

white, 7–8 mm long; bracts and petals burnt-orange when dry; stamens 12–18, 5–6 mm long; filaments geniculate below anthers; anthers with terminal pores all directed toward style in vicinity of stigma; style straight. Berries round, hirsute, black and shiny, 8–12 mm wide; seeds minute, numerous. *Croat 11079.*

Occasional, in the forest. Flowers from March to July (rarely later). The fruits mature from August to November.

Nicaragua to Panama. In Panama, known from tropical moist forest in the Canal Zone, Bocas del Toro, Colón, Panamá, and Darién, from premontane wet forest in Coclé and Panamá, from tropical wet forest in Panamá and Darién, and from premontane rain forest in Darién.

See Fig. 430.

Conostegia cinnamomea (Beurl.) Wurdack, Phytologia 38:287. 1978

C. micromeris Standl.

Shrub, usually 1.5–3.5 m tall; younger parts furfuraceous with stellate trichomes, soon glabrate. Petioles 2–5(10) mm long; blades broadest at about middle, gradually tapered to both ends, acuminate, mostly 6–22 cm long, 2.5–8.5 cm wide, bicolorous, glabrate above, sparsely stellate-pubescent below especially on veins, pliveined, the margins entire or minutely wavy (sinuate), the veins 5–7. Terminal compound dichasia 3–6 cm long, divaricately branched, the branches opposite, each subtended by a pair of narrow bracts; flowers sessile or short-pedunculate, minutely furfuraceous, the buds 3–4.7 cm long, distinctly apiculate, short-pedicellate; hypanthium 2–3 mm long at anthesis; petals (4)5, white to pink, 4–4.7 mm long, acute at apex, recurved, soon falling; stamens (8)10, erect at anthesis; filaments flattened, slightly longer than the anthers, markedly curved; anthers with terminal pores; style ca 5 mm long. Berries depressed-globose, to 6 mm diam, green turning light blue then deep purple at maturity; seeds minute, numerous. *Croat 6250.*

Frequent in the forest. Flowers and fruits throughout the year, principally from July to September.

Recognized by its small, inconspicuous inflorescences with beaked buds. The flowers fall off soon, and the plant seldom appears to be conspicuously in flower.

Nicaragua to western Colombia. In Panama, known principally from tropical moist forest in the Canal Zone, Colón, Panamá, and Darién; known also from premontane wet forest in Coclé and from tropical wet and premontane rain forests in Darién.

See Fig. 431.

Conostegia speciosa Naud., Ann. Sci. Nat. Bot., sér. 3, 16:109. 1851

Shrub, seldom more than 2 m tall, very densely pubescent on all external surfaces, the trichomes of stems and inflorescences stalked-stellate. Petioles 1–4 cm long; blades ± ovate, acuminate, obtuse to rounded at base, 10–27 cm long, 5.5–13 cm wide, rugulose and mostly simple-pubescent above, with both stalked-stellate and simple

trichomes below, minutely serrate, pliveined, the veins 5–7, raised below. Panicles congested, ± maroon, terminal; flowers ca 1.5 cm diam, opening few at a time, often on same inflorescence with fruits, the buds violet, ovoid; petals 6 or 7, white to pink, ± rounded; stamens 12 or 14, directed to one side, geniculate below anther; anthers all with apices clustered around stigma; style short and thick, strongly bent to one side near apex; stigma green. Berries densely pubescent, dark purple, ca 1.5 cm diam; seeds minute, numerous. *Croat 6340.*

Common in clearings; occasional along the margin of the lake. Flowers and fruits throughout the year.

Foster 1134 and *1211,* collected along the shore on Peña Blanca Point, represent an extreme variation of *C. speciosa* or possibly a distinct variety of it. It has been suggested by C. Schnell (pers. comm.) that they are hybrids between *C. speciosa* Naud. and *C. subcrustulata* (Beurl.) Tr. In general appearance they look like *C. speciosa,* but they have only sessile stellate trichomes on the lower leaf surface as in *C. subcrustulata.*

Costa Rica to northern South America; Jamaica. In Panama, known principally from tropical moist forest in the Canal Zone, Bocas del Toro, San Blas, Veraguas, Panamá, and Darién; known also from tropical dry forest in Panamá (Taboga Island), from premontane moist forest in the Canal Zone and Panamá, from premontane wet forest in Coclé, and from tropical wet forest in Colón and Coclé.

Conostegia xalapensis (Bonpl.) D. Don ex DC., Prodr. 3:175. 1828

Canallito

Shrub or small tree, to 10 m tall (often only 3–4 m on BCI); all parts except interior parts of flowers and upper blade surfaces densely pubescent, the trichomes ± sessile, brown, stellate. Petioles 1–2.5 cm long; blades ± lanceolate, acuminate, obtuse to cuneate at base, 5–25 cm long, 1.7–8 cm wide, glabrate above in age, usually serrate, pliveined, the veins 3–5. Inflorescences terminal, paniculate; flower buds pyriform; petals 5, glabrous, white (drying yellow), obovate, ca 5 mm long; stamens 10; filaments bent sharply below anthers; anthers with terminal pores, all directed to one side of the flower (as is the stigma), the connective thickened on upper side; style bent to one side, thickened at apex; stigma papillate. Berries densely stellate-pubescent, green to purple at maturity, to 8 mm diam; seeds minute, numerous. *Croat 12281.*

Uncommon, collected only along the shore on the north side of the island. Flowers and fruits throughout the year.

Southern Mexico to northern South America; Cuba. In Panama, ecologically variable; known from tropical moist forest in the Canal Zone, Bocas del Toro, San Blas, and Darién, from premontane wet forest in Chiriquí, Coclé, and Panamá, from tropical wet forest in Veraguas (Atlantic slope), Colón, Panamá, and Darién, and from premontane rain, lower montane moist, and lower montane wet forests in Chiriquí.

HENRIETTEA DC.

Henriettea succosa (Aubl.) DC., Prodr. 3:178. 1828

Tree, to 10 m tall; young stems and petioles densely strigose. Petioles 1–2 cm long, stout; blades elliptic to obovate, short-acuminate, obtuse at base, 11–22 cm long, 6–10 cm wide, ciliate, pliveined, the upper surface scabrous especially on veins, the lower surface strigose especially on veins, the trichomes densely matted, stout, acropetal, densely branched near base, the veins 3–5. Flowers 5-parted, few, short-pedicellate, in small axillary clusters on usually leafless stems; hypanthium and calyx together 1.2–1.5 cm long at anthesis, densely and coarsely strigose; calyx prolonged, the lobes acute, the exterior teeth minute or concealed; petals obovate, clawed, white or pinkish, with a conspicuous medial vein, pubescent near apex and outside along vein, to 1.2 cm long, the margins thin, often minutely lacerate; stamens and style erect at one side, the style then curving to one side, the stamens to the other; stamens 10, purple, ca 1.5 cm long; anthers ca 8 mm long, subulate, slightly S-shaped, ± equaling length of filaments, with a terminal pore, the connective lobed at base; ovary inferior; style enlarged at apex. Berries remaining enclosed in the green hypanthium, ± globose, to 1 cm diam, turning purple; seeds minute, numerous. *Croat 6592.*

Rare, collected recently only along the shore of Gross Peninsula. Apparently flowers throughout the rainy season, rarely in the late dry season. Fruit maturity time undetermined.

Central America to the Guianas and eastern Brazil; Trinidad. In Panama, known from tropical moist forest in the Canal Zone, Bocas del Toro, and Panamá, from premontane moist forest in Panamá, from premontane wet forest in Colón and Panamá, and from tropical wet and premontane rain forests in Panamá.

See Fig. 432.

LEANDRA Raddi

Leandra dichotoma (D. Don) Cogn. in Mart., Fl. Brasil. 14(4):200. 1886

Shrub, 1–3 m tall, densely hirsute except on petals and inner flower parts, the trichomes on lower leaf surfaces of various lengths, those of stems and axes of inflorescences retrorse at least at base. Petioles 2–9 cm long; blades ovate to ovate-elliptic, long-acuminate, obtuse to rounded at base, 11–25 cm long, 5.5–15 cm wide, ciliate and denticulate; veins 5–7. Panicles slender, usually terminal, reddish, usually 10–20 cm long; flowers small, usually 5-parted; hypanthium often with gland-tipped trichomes, together with calyx ca 3 mm long, the lobes minute, the exterior teeth subulate, longer than lobes; petals acute, 2.5–3 mm long, white to pinkish; stamens usually twice the number of petals; ovary inferior; style longer than stamens, erect or ± curved to one side. Berries round, to 6 mm diam, purple, hirsute; seeds minute, numerous. *Croat 12507.*

Common, especially along streams. Apparently flowers and fruits throughout the year.

Belize and Guatemala to Panama and Bolivia. In Panama, known from tropical moist forest in the Canal Zone, all along the Atlantic slope, and in Veraguas and Darién on the Pacific slope, from premontane wet forest in Colón, Coclé, and Panamá, from tropical wet forest in Colón, Coclé, Panamá, and Darién, and from premontane rain forest in Bocas del Toro and Panamá.

See Fig. 433.

MICONIA R. & P.

Miconia is distinguished mostly by the terminal inflorescences, the petals rounded at the apex, and the open calyx buds. It has an inferior ovary. The fleshy fruits have numerous minute seeds.

Miconia affinis DC., Prodr. 3:187. 1828

M. microcarpa DC.; *M. beurlingii* Tr.

Shrub or small tree, usually 3–7 m tall; younger parts of stems and petioles, panicles, and hypanthia stellate-tomentose. Petioles 1–2 cm long; blades ± oblong-elliptic, acuminate, acute to obtuse at base, 9–28 cm long, 3.5–12 cm wide, glabrate to sparsely appressed-stellate on both surfaces (at least on veins below), the midrib sometimes arched; veins from the base 3–5. Panicles terminal, to 15 cm long, almost as broad; hypanthium 1.5–2.5 mm long; calyx lobes 5, rounded to triangular, the exterior teeth often nearly as long, slender; petals 5, white, to 2.7 mm long, spreading; stamens 10; style straight, erect, at first much shorter than stamens, elongating to ca 6 mm and ± equaling pores of stamens; stamens ± erect; filaments curved, geniculate below anthers. Axes of fruiting inflorescences reddish; berries depressed-globose, green becoming whitish, turning blue and finally purple-black at maturity, 3 mm long, 6 mm wide; seeds numerous, minute. *Croat 11443.*

Occasional, in the forest. Flowers mostly in the dry season (January to June), with the fruits maturing mainly from June to October. Plants may flower more than once per year.

The species has been confused with *M. hyperprasina* Naud., which is known only from southern Mexico and Guatemala.

Elsewhere in Panama, *Trigona* bees have been seen collecting pollen from this species.

Mexico to Panama, Colombia, and the Amazon basin in Peru, Brazil, Venezuela, and French Guiana; Trinidad. In Panama, known from tropical moist forest in the Canal Zone, Bocas del Toro, Panamá, and Darién, from premontane wet forest in Colón, Coclé, and Panamá, and from premontane rain forest in Panamá and Darién.

Miconia argentea (Sw.) DC., Prodr. 3:182. 1828

Cainillo, Cainillo de cerro, Dos caras, Friega platos, Gorgojo, Gorgojillo, Mancha-mancha, Oreja de mula, Palo negro, Papelillo

Tree, to 15(20) m tall, canescent except on upper leaf surfaces, the trichomes dense, fine, stellate; trunk to 50 cm dbh, with coarse shaggy bark in age; younger stems flattened, the edges twisted at the nodes. Leaves de-

KEY TO THE SPECIES OF MICONIA

Lower blade surface *conspicuously* pubescent:
  Trichomes simple (sometimes bristled, branched, or rarely stellate in *M. shattuckii*):
    Upper leaf surface very sparsely pubescent with simple trichomes, the lower surface with
      both simple and bristled trichomes (rarely stellate) .... *M. shattuckii* Standl.
    Both leaf surfaces rather conspicuously pubescent, the trichomes simple, not branched or
      barbed along their length:
      Blades to 6 cm wide; pubescence on upper stems, inflorescence, and petioles of purplish
        trichomes to 10 mm long .... *M. lacera* (Bonpl.) Naud.
      Blades more than 6 cm wide; pubescence densely strigose, the trichomes less than 5 mm
        long.. *M. nervosa* (Smith) Tr.
  Trichomes stellate:
    Leaves sessile; trichomes of lower leaf surface not contiguous ....
    .... *M. impetiolaris* (Sw.) DC. var. *impetiolaris*
    Leaves petiolate:
      Blades very sparsely pubescent, with simple trichomes above and sessile stellate trichomes
        below, denser on veins .... *M. rufostellulata* Pitt.
      Blades ± glabrous above, the lower surface with stellate trichomes contiguous throughout:
        Hypanthium much longer than broad, together with the calyx to 5 mm long; calyx flared,
          broader than hypanthium at anthesis; petals more than 5 mm long, densely stellate-
          pubescent outside; fruits more than 8 mm diam at maturity, densely stellate-
          pubescent; leaf blades usually 2 times longer than broad ..... *M. serrulata* (DC.) Naud.
        Hypanthium not much longer than broad, together with the calyx ca 2 mm long; calyx
          not flared, not obviously broader than hypanthium at anthesis; petals to 3 mm long,
          glabrous; mature fruits usually less than 6 mm diam (rarely to 8 mm in *M. argentea*),
          not densely stellate-pubescent; leaf blades usually ca 1.5 times longer than broad:
          Blades whitish below, the veins scarcely if at all darker than the surface; branchlet
            pubescence pale, stellate, the trichomes ± strictly sessile; trees often large, the
            trunks frequently more than 15 cm dbh, the bark then usually shaggy ....
            .... *M. argentea* (Sw.) DC.
          Blades tan to brown below, the veins considerably darker than the surface; branchlet
            pubescence dark brown, the trichomes at least in part ± stalked and not strictly
            stellate; trees small, the trunks usually less than 10 cm dbh, the bark deeply fis-
            sured but not shaggy .... *M. elata* (Sw.) DC.
Lower blade surface glabrous or essentially so (at least at maturity):
  Plants wholly glabrous (*M. lateriflora* sometimes with minute, appressed, simple trichomes):
    Blade margins often obscurely crenate, sparsely ciliate; hypanthium oblong, its lobes to 2 mm
      long.. *M. lateriflora* Cogn.
    Blade margins entire, not ciliate; hypanthium narrowly bell-shaped or round, its lobes very
      short or obsolete:
      Blades broadest below the middle, often rounded or obtuse at base, drying dark, usually
        less than 4.5 cm wide; anthers linear, not tapering to apex .... *M. borealis* Gleason
      Blades broadest at middle, usually acute to obtuse at base, drying green, bicolorous, mostly
        more than 4.5 cm wide; anthers subulate, tapering to apex .... *M. hondurensis* Donn. Sm.
  Plants with younger parts usually densely stellate-pubescent, some trichomes persisting on
    smaller stems in age:
    Upper blade surface sparsely pubescent with simple trichomes, the lower surface often ma-
      roon; plants slender arching shrubs .... *M. rufostellulata* Pitt.
    Upper blade surface glabrous or very sparsely and inconspicuously pubescent with stellate
      trichomes:
      Blades ending abruptly on petiole, the lateral veins departing midrib at or near base:
        Blades ± ovate or lanceolate, rounded to slightly subcordate at base, broadest well below
          the middle, usually less than 4.5 cm wide; flowering hypanthium less than 1.5 mm
          long; fruits to 3 mm diam .... *M. borealis* Gleason
        Blades elliptic to oblong-elliptic, acute to obtuse at base, broadest at about the middle,
          usually more than 4.5 cm wide; flowering hypanthium 2–3 mm long; fruits to 6 mm
          diam .... *M. affinis* DC.
      Blades ± decurrent on petiole, the lateral veins usually departing midrib somewhat above
        the base:
        Blades usually undulate-denticulate, drying green, abruptly narrowed at the base, the
          decurrent portion broad, often irregular and undulate, ending abruptly very near
          base of petiole .... *M. prasina* (Sw.) DC.
        Blades usually entire or nearly so, drying blackened, gradually narrowed at the base, the
          decurrent portion narrow, gradually diminishing, almost imperceptibly, well above
          base of petiole .... *M. lonchophylla* Naud.

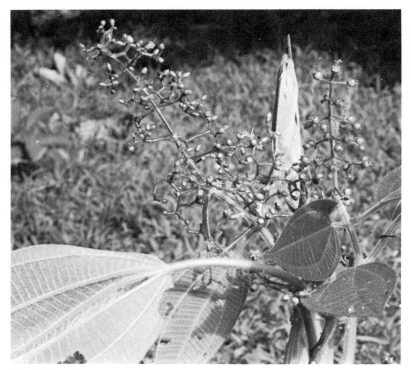

Fig. 433. *Leandra dichotoma*

Fig. 434. *Miconia argentea*

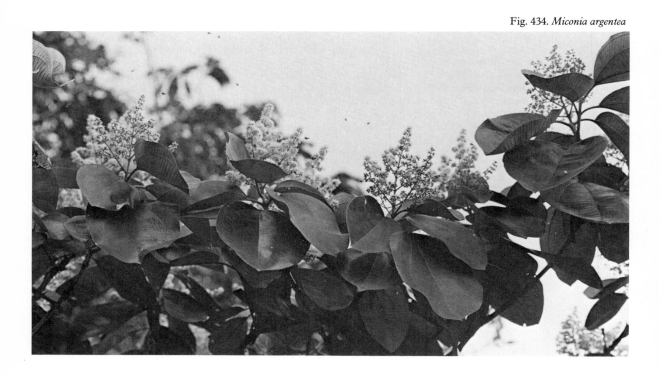

cussate; petioles 2–6 cm long; blades broadly elliptic, abruptly short-acuminate to acute at apex, rounded at base, 8–30 cm long, 6–18 cm wide, entire or denticulate, veins from base 5; (juveniles with the leaf blades ± ovate, to 38 cm long and 27 cm wide, the upper surface bearing sparse simple trichomes, the margins more denticulate, ciliate). Panicles terminal, to 25 cm long, widely branched; flowers many, sessile, 5-parted; hypanthium and calyx together ca 2 mm long, the calyx with obscure lobes; petals inequilateral, white, 2–3 mm long; stamens 10, erect at anthesis, of 2 sizes, the longer series somewhat longer than style, the shorter series slightly shorter than style, the connective of larger set prolonged at base into 2 small lateral lobes; anthers with a terminal pore as broad as anther; style to 4 mm long; stigma capitate. Berries globose, 4–8 mm diam, blue-purple at maturity, stellate-pubescent; seeds numerous, minute. *Croat 13104.*

Common in the forest, especially the younger forest; seedlings common. Flowers principally in the dry season (December to May), with the fruits maturing from January to June. J. Oppenheimer (pers. comm.) says fruits are eaten from March to May by white-faced monkeys. Trees are leafless during the dry season.

E. Morton (pers. comm.) reports that elsewhere in the Canal Zone fruits are eaten while still green by a variety of birds.

Southern Mexico to Panama. In Panama, ecologically variable; known from tropical moist forest in the Canal Zone, Colón, San Blas, Chiriquí, Veraguas, Los Santos, Panamá, and Darién, from premontane dry forest in Los Santos, from tropical dry forest in Coclé, from premontane wet forest in Chiriquí, Coclé, and Panamá, and from tropical wet forest in Colón and Panamá.

See Fig. 434.

Miconia borealis Gleason, Bull. Torrey Bot. Club 55:118. 1928

Shrub or small tree, to 5 m tall; younger stems finely stellate-pubescent, glabrous and often reddish in age. Petioles 5–15 mm long; blades narrowly ovate to lanceolate-oblong, long-acuminate, rounded to slightly cordate at base, 6–12(15) cm long, 2.5–4.5 cm wide, entire, sparsely stellate-pubescent becoming glabrate, drying dark; veins 3 (rarely 5). Panicles terminal and upper-axillary, to 10 cm long; flowers 5-parted; hypanthium 1–1.5 mm long, obscurely lobed, ± glabrous, lacking exterior teeth; petals white, recurved, ca 2 mm long; stamens folded in bud, held erect and exceeding style at anthesis, the style then lengthening to equal height of anthers (held in close proximity); filaments flat, recurving over edge of flower and withdrawing anthers from vicinity of style; anthers white, the apical pore as wide as anther. Inflorescences purplish in fruit; berries blue-black, ca 3 mm diam; seeds numerous, minute, white, translucent. *Foster 1345.*

Probably common on the island at one time, now rare, persisting only on the shore. The plant is abundant in the Canal Zone along Gaillard Highway between Gamboa and the Pacific coast. Flowers June to February, with the fruits maturing from August to May.

The style may not be receptive until the pollen has been removed from the anthers.

Southern Mexico to Panama; Cuba. In Panama, known from tropical moist forest in all provinces, from tropical dry forest in Panamá (Taboga Island), from premontane moist forest in the Canal Zone and Panamá, from premontane wet forest in Colón, Chiriquí, Veraguas, Coclé, and Panamá, and from tropical wet forest in Colón, Coclé, Panamá, and Darién.

Miconia elata (Sw.) DC., Prodr. 3:182. 1828

Tree, 3.5–11(13) m tall; trunk usually 6–15 cm dbh; bark with moderately close, deep vertical fissures, but not flaking; young stems, axes of inflorescences, petioles, and underside of veins densely rufous-pubescent, the trichomes pinnately branched, often somewhat stalked. Petioles to 9 cm long; blades broadly elliptic, obtuse to abruptly short-acuminate, obtuse to rounded at base, 9–35 cm long, 6–18 cm wide, denticulate, the surface below canescent and completely covered with trichomes, these sessile, stellate, in part with a brown center; veins 5, dark. Panicles large, terminal, 18–22 cm long, branched many times, the branches ± flattened or angulate; flowers ± sessile; hypanthium and calyx together 1.5–2 mm long, campanulate, the teeth obscure or lacking; petals 5, white, obovate, to 1.5 mm long. Berries round to depressed-globose, to 6 mm diam, white to lavender-blue, probably becoming purple-black at full maturity; seeds numerous, minute. *Croat 5693, 15070.*

Known only from shoreline north of the dock and south of Fairchild Point, but undoubtedly more abundant. Seasonal behavior uncertain. Flowers at least in January, February, and May. Mature fruits have been seen in April, May, and October.

Easily confused with *M. argentea*, which is similar in most respects except pubescence, trunk bark, and stature.

Mexico to Panama and Venezuela; West Indies. In Panama, known only from tropical moist forest in the Canal Zone and Panamá and from premontane wet forest in the Canal Zone.

Miconia hondurensis Donn. Sm., Bot. Gaz. (Crawfordsville) 40:3. 1905

Tree, to 12 m tall; glabrous except on youngest parts; trunk to 17 cm dbh; bark soft, vertically fissured, flaky. Petioles 1–4 cm long; blades elliptic, acuminate, acute to obtuse at base, mostly 9–20 cm long, 4.5–8 cm wide, moderately bicolorous, pliveined, the veins 3, the lateral veins departing midrib very near base. Panicles terminal, mostly less than 15 cm long; pedicels short, often minutely pubescent together with basal part of hypanthium; flowers 8–10 mm long; hypanthium and calyx together 3–4 mm long; calyx flared, weakly 5-lobed; petals 5, white, oblong, to ca 5 mm long, spreading at anthesis; stamens 10 (often slightly unequal), the longest ± equaling petals; filaments geniculate; anthers tapered to apex, directed to one side, the connective lobed at base; style ca 12 mm long, broadly curved at anthesis; stigma curved in same direction as anthers. Fruiting inflorescences

violet-purple; berries round, 7–8 mm diam, black, glabrate; seeds lunate, ca 2.7 mm long. *Croat 13800, 14467.*

Uncommon, in the forest. Flowers in the middle of the dry season (February and March), with the fruits maturing in April and May.

Belize to Panama. In Panama, known from tropical moist forest in the Canal Zone and Darién and from premontane wet forest in the Canal Zone.

Miconia impetiolaris (Sw.) D. Don ex DC. var. **impetiolaris,** Prodr. 3:183. 1828

Oreja de mula, Dos caras

Shrub or tree, ca 3(5) m tall, bearing conspicuous, ferruginous-stellate pubescence except on older stems, upper leaf surfaces, petals, and inner flower parts. Leaves sessile; blades obovate-elliptic to oblong-elliptic, gradually to abruptly acuminate, cordate at base, 12–45 cm long, 7–15 cm wide, soon glabrous above, entire or denticulate, pliveined, the surface visible between the trichomes below, the veins 3–5. Panicles sessile, to 27 cm long, almost as wide as long; flowers white; hypanthium truncate, 2.5–3 mm long; petals 5, white, reflexed, rounded at apex; stamens 10, somewhat spreading, in 2 series, the longer series with anthers held just above stigma; anthers curved, the connective expanded and toothed at base of anther; style straight, erect, ca 6 mm long. Berries globose, to 9 mm diam, sparsely stellate-puberulent, green then bright red becoming black at maturity; seeds numerous, minute. *Croat 8052.*

Occasional in some areas of the forest, especially near the Tower Clearing and near the end of Barbour Trail; rare or absent elsewhere. Few observations have been made of the variety *impetiolaris,* but BCI individuals flower in the middle of the dry season (February and March), with the fruits maturing in the late dry season to early rainy season (April and May).

The species is represented on the island by the typical variety. The variety *panduriformis,* well represented elsewhere in the Canal Zone, flowers synchronously at least twice a year. The first synchronous flowering, possibly the largest, occurs in late February or March. Later flowerings may involve other individuals or the same inflorescence, making it possible for a single inflorescence to have both mature fruits and a new batch of flowers.

Southern Mexico to Panama; West Indies. In Panama, known principally from tropical moist forest in the Canal Zone, Bocas del Toro, Chiriquí, Veraguas, Panamá, and Darién; known also from premontane moist forest in the Canal Zone.

Miconia lacera (Bonpl.) Naud., Ann. Sci. Nat. Bot., sér. 3, 16:152. 1851

Shrub, to 2(4) m tall; stems, petioles, and axes of inflorescences with many violet-purple to red-violet trichomes mostly 5–10 mm long. Petioles 1–3.5 cm long; blades ± ovate-lanceolate to narrowly elliptic, acuminate, obtuse to rounded or slightly subcordate at base, 5–15 cm long, 2–6.5 cm wide, more or less pilose on both sides especially on veins, entire to serrulate and ciliate; veins from base 3–5. Panicles very congested in flower, branched

or unbranched, to 4 cm long, the trichomes at first whitish becoming colored in fruit; flowers ca 6 mm wide, secund, in a staggered row; hypanthium usually glabrous; calyx spreading, its lobes deeply lacerate and markedly ciliate, the exterior teeth equaling calyx lobes; petals 5, white to pinkish, 3–4 mm long, often emarginate at apex; stamens and style folded in bud; stamens red-violet, ± directed to one side; anthers 1.7 mm long, their terminal pores held in close proximity to style; style directed like the stamens. Berries subglobose, 6–7 mm diam, black, shiny; seeds numerous, minute, brown. *Croat 6126.*

Occasional in clearings and infrequent within the forest. Elsewhere in the Canal Zone it is frequent in open areas. Plants flower more than once per year and may bear both flowers and fruits on different branches. Flowers from March to August. The fruits mature principally from June to October but are seen all year.

The plant apparently does not produce nectar and is probably visited by small pollen feeders. The anthers do not forcibly expel the pollen, but the grains come out readily if the stamens are pulled back and then released.

Mexico to Peru; West Indies, Trinidad. In Panama, widespread and ecologically variable; known from tropical dry forest in Panamá (Taboga Island), from premontane moist forest in the Canal Zone, from tropical moist forest in the Canal Zone, Bocas del Toro, San Blas, Chiriquí, Herrera, Coclé, Panamá, and Darién, from premontane wet forest in the Canal Zone, Colón, Coclé, and Panamá, from tropical wet forest in Veraguas (Atlantic slope), Colón, Coclé, Panamá, and Darién, and from premontane rain forest in Panamá.

See Fig. 435.

Miconia lateriflora Cogn., Bol. Mus. Paraense Hist. Nat. 5:255. 1909

Ossaea disparilis Standl.; *O. disparilis* Standl. var. *adenophora* Standl.

Shrub, to 3 m tall, ± glabrous. Petioles 1–3(5) cm long; blades ± elliptic, narrowly long-acuminate, attenuate to obtuse at base, 7–20(25) cm long, 3–9(12) cm wide, entire to crenulate and ciliate, the cilia continuous with the lateral veins; veins 3. Panicles terminal, loose and open, sparsely branched, the branches often red in fruit; flowers few, usually 4-parted, sessile, in terminal glomerules; inflorescence branches often sparsely stellate-pubescent; hypanthium tubular, 3.5–4 mm long, often sparsely stellate-pubescent, sometimes also glandular-pilose; calyx nearly obsolete, the external teeth narrowly triangular, 1.5–2 mm long; petals white, obovate (sometimes obcordate), 1.6–1.8 mm long; stamens to ca 5.3 mm long; anthers subulate, 3–3.5 mm long, the connective minutely bilobed at base; style erect, as long as stamens. Fruiting hypanthia ovoid, ca 2 cm long, strongly 8-ribbed, blue; seeds numerous, minute. *Croat 10822.*

Apparently rare on the island, seen on Shannon Trail in a steep ravine and at Pearson Trail 400. Flowers principally in June and July, with the fruits maturing probably in September and October.

Belize to northern Brazil; Trinidad. In Panama, known from tropical moist forest in the Canal Zone, Bocas del Toro, and Darién, from premontane wet forest in Colón

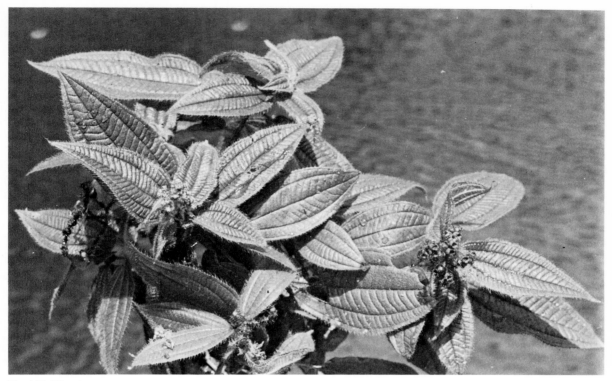

Fig. 435. *Miconia lacera*

Fig. 436. *Miconia nervosa*

Fig. 437.
Miconia prasina

and Panamá, and from tropical wet forest in Bocas del Toro, Panamá, and Darién.

Miconia lonchophylla Naud., Ann. Sci. Nat. Bot., sér. 3, 16:176. 1851

Shrub or small tree, to 6 m tall; young stems, petioles, and axes of inflorescences moderately to densely pubescent with small, sessile, stellate trichomes. Petioles 6–25 mm long; blades narrowly elliptic, acuminate, attenuate and decurrent on petiole at base, 9–29 cm long, 3–8 cm wide, pliveined, drying blackened, sparsely stellate-pubescent below, usually soon becoming glabrate above, the veins sometimes pubescent, the margins entire or rarely sinuate to denticulate, the midrib often arched; major veins 3, usually arising well above base, with a weak submarginal vein. Panicles terminal, to ca 15 cm long; hypanthium and calyx ca 2 mm long, sparsely stellate-pubescent, the calyx teeth thin, obtuse to rounded, the external teeth thick, minute; petals 3, oblong, 1.8–2.7 mm long, often markedly oblique at apex, white; stamens 10; anthers 1.5–2 mm long, oblong-linear, the connective shortly prolonged at base, the two terminal pores about as broad as thecae; style about as long as stamens; stigma at first cupuliform, becoming capitate. Fruiting inflorescences with red axes; berries depressed-globose, 5–6 mm diam, at first red, becoming black at maturity; seeds numerous, minute. *Croat 14609, 15052.*

Adult plants rare, seen only on Orchid Island and on the eastern side of Gross Peninsula; juveniles slightly more common. Known from few observations. Flowers at least from the late dry to early rainy seasons (April to May), with the fruits maturing shortly afterwards (May to July).

Costa Rica to Colombia and Venezuela. In Panama, known from tropical moist forest in the Canal Zone, Colón, and Darién and from premontane wet forest in Panamá. Reported from premontane rain forest in Costa Rica (Holdridge et al., 1971).

Miconia nervosa (Smith) Tr., Trans. Linn. Soc. London 28:111. 1871

Shrub, 2–3 m tall, densely soft-pubescent, the trichomes on petals and inner flower parts simple, ± stiff, those on stems, petioles, and axes of inflorescences appressed-ascending. Leaf pairs often unequal; petioles 5–20 mm long; blades elliptic, acuminate at apex, acute to attenuate at base, 7–28 cm long, 3.5–12 cm wide, pliveined, ciliate, sparsely pubescent on upper surface, softly pubescent to subsericeous on lower surface, the pubescence much denser on veins especially above, the veins 5–7; juveniles with the blades purplish below. Panicles narrow, branched or unbranched, 5–15 cm long, usually less than 3 cm wide, the main axis reddish; flowers 5-parted, sessile; hypanthium 2.5–4 mm long, appressed-pubescent; petals strongly reflexed at anthesis, white, rounded at apex; stamens isomorphic; anthers tapered toward apex, the connective inconspicuously prolonged at base; ovary pubescent; style ca 6 mm long; stigma truncate. Berries depressed-globose, to 12 mm wide, pale orange-red becoming gray-blue to purple at maturity, fleshy, densely

pubescent; seeds brown, ± sticky, narrowly wedge-shaped, ca 0.7 mm long. *Croat 12872.*

Occasional in the older forest; uncommon elsewhere. Flowers and fruits throughout the year.

Southern Mexico to Bolivia. In Panama, known from tropical moist forest in the Canal Zone, Bocas del Toro, Veraguas, and Darién, from premontane wet forest in Colón, Chiriquí, and Panamá, from tropical wet forest in Veraguas (Atlantic slope), Colón, and Darién, and from premontane rain forest in Darién.

See Fig. 436.

Miconia prasina (Sw.) DC., Prodr. 3:188. 1828

Shrub or tree, usually 2–8 m tall; stems, petioles, axes of inflorescences, and hypanthia bearing ± dense, ferruginous, stellate pubescence (especially stems and petioles). Petioles very short; blades elliptic to narrowly elliptic, acuminate (the acumen often somewhat twisted), acute at base and decurrent onto petiole, 12–26 cm long, 4.5–14 cm wide, entire or more commonly undulate-denticulate, glabrate or sparsely stellate-pubescent above (at least in age), pubescent with sparse, sessile, stellate trichomes below at least on veins, pliveined, the veins 5, sometimes reddish. Panicles terminal, 8–23 cm long; hypanthium 2–3 mm long, reddish; calyx lobes 5, obtuse to rounded or obscure, the submarginal teeth small and inconspicuous or lacking; petals 5, obovate, to 3 mm long, white, densely papillate-puberulent in distal three-fourths, strongly recurved at anthesis; stamens 10, at first erect; filaments curved, geniculate below anther; anthers to 3 mm long, tapered to apex, the terminal pore held in vicinity of style, later recurved over sides of hypanthium; style erect, straight, to ca 5.7 mm long. Berries depressed-globose, ca 6–9 mm diam, to 5 mm long when fully mature, lavender-blue, becoming purple at maturity, sparsely pubescent; seeds numerous, minute. *Croat 8555.*

Collected once on Gross Peninsula. Populations usually flower twice during the dry season, once near the onset of the dry season and again in about the middle of the dry season. At the second flowering the plants may also bear fruits from the earlier flowering period. The second flowering is the heavier, with most fruits maturing in the early rainy season, chiefly in July and August.

Southern Mexico to Bolivia and Paraguay; West Indies. In Panama, known from tropical moist forest in the Canal Zone, Bocas del Toro, Veraguas, Herrera, Panamá, and Darién, from premontane moist forest in the Canal Zone (Ancón), from premontane wet forest in Colón and Chiriquí, from tropical wet forest in Colón, Coclé, Panamá, and Darién, and from premontane rain forest in Panamá.

See Fig. 437.

Miconia rufostellulata Pitt., J. Wash. Acad. Sci. 13:390. 1923

Shrub, to 2 m tall, often vinelike; stems, petioles, inflorescences, and underside of veins sparsely to densely, minutely stellate-pubescent; stems glabrous in age. Petioles 4–12 mm long; blades lanceolate, acuminate, rounded to subcordate at base, 4–14 cm long, 1.5–3.5 cm wide, setose above, maroon below, entire to denticulate, ciliate,

the cilia terminating the small veins; veins 3, from base or pliveined from very near base. Panicles or racemes small, terminal; bracts subulate; flowers 4-parted; hypanthium cup-shaped, 1.5–2 mm long; calyx to 2.2 mm long, the lobes triangular, persisting in fruit, the exterior teeth subulate and divergent at apex, often longer than sepals; petals oblong-obovate, white, to 2.5 mm long; stamens dimorphic; anthers 2–2.5 mm long, the connective of the larger pair dilated; stigma capitate. Berries globose, 4–5 mm diam, 10-ribbed, sparsely stellate-pubescent; mature fruits not seen. *Croat 15106.*

Rare, along Barbour and Chapman trails. In Darién, flowers in March and fruits in June and July.

Characterized by the reddish lower leaf surface. Description of the flower parts is based on the type.

Known only from the type collection from tropical moist forest in the Canal Zone and Darién and from premontane wet forest in the Canal Zone (Pipeline Road).

Miconia serrulata (DC.) Naud., Ann. Sci. Nat. Bot., sér. 3, 16:118. 1851

Shrub or tree, 3–10 m tall; stems, axes of inflorescences, petioles, and veins of lower leaf surfaces bearing dense, brown, stellate pubescence, the trichomes of stems, inflorescences, and petioles dark brown. Petioles 3–9 cm long; blades ± elliptic, acuminate, narrowly rounded or subcordate at base, 15–33 cm long, 5–14(21) cm wide, glabrous above, densely matted with stellate trichomes below, denticulate; veins 5–7, from base. Panicles large, terminal, 10–15 cm long, congested toward ends of branches; flowering hypanthium oblong, much longer than broad, the hypanthium and calyx together ca 5 mm long, very densely stellate-canescent; calyx tube 1–2.5 mm long, markedly flared, the lobes obtuse to rounded, the exterior teeth lacking; petals 5, white, obovate, 5–9 mm long, rounded to irregular at apex, densely stellate-pubescent outside, glabrous inside, often densely ciliate with simple trichomes; stamens 10, isomorphic; anthers purplish, slightly curved, tapered toward apex, ca 4.5(5.8) mm long, the connective pubescent with glandular trichomes, prolonged into 2 short lobes at base; style to 9 mm long, thickened at apex. Berries ca 1 cm diam at maturity, densely stellate-pubescent; seeds numerous, minute. *Croat 12504.*

Collected near Zetek Trail 500 and 1015. Flowering in the late dry season (and probably early rainy season). The fruits develop chiefly in the late rainy season (August to November).

Mexico to Peru, Bolivia, and southeast Brazil; West Indies. In Panama, known from tropical moist forest in the Canal Zone, Bocas del Toro, and Darién, from tropical wet forest in Colón and Darién, and from premontane rain forest in Darién. Reported from premontane wet forest in Costa Rica (Holdridge et al., 1971).

Miconia shattuckii Standl., Contr. Arnold Arbor. 5:119, pl. 16. 1933

Shrub or small tree, usually 3–3.5 m tall; branches, petioles, and inflorescences densely to sparsely tomentose, the trichomes with a single irregular axis and numerous,

short, sharp branches. Petioles stout, 1–5 cm long; blades broadly ovate-elliptic, acuminate, rounded to shallowly cordate at base, 20–35 cm long, 12–20 cm wide, dark green and glabrous to sparsely villous above, more densely pubescent below, with both straight and branched trichomes, the branched ones usually restricted to veins, entire or the margins entire or somewhat irregular; veins 5–7, from base (1 weak and submarginal), impressed. Inflorescences terminal, reddish, paniculate, to 12 cm long; pedicels 1.7–2.3 mm long, minutely bracteate; flowers white; hypanthium and calyx pinkish, pubescent with both simple and stellate trichomes, together 2.3–3 mm long; sepals 5, acute, the external teeth minute, often green; petals 5, oblong-obovate, 4.2–5.2 mm long, to 2.4 mm wide, rounded at apex; stamens slightly shorter than corolla; filament geniculate below anther; anthers linear-attenuate, to 2 mm long, each with 1 apical pore; style ca 4 mm long. Berries depressed-globose, 7–10 mm diam, pink when young, becoming violet then black at maturity; seeds numerous, minute. *Croat 12190, 14552.*

Rare; known from a single locality on a creek bank on Shannon Trail. The type was collected earlier on Snyder-Molino Trail. Flowers and fruits in the rainy season. Apparently flowers twice per year, at the beginning of the rainy season and toward the middle of the rainy season. The fruits mature in about 1 month.

Colombia (Antioquia) and Panama. In Panama, known from tropical moist forest on BCI and from lower montane rain forest in Darién.

MOURIRI Aubl.

Mouriri myrtilloides (Sw.) Poir. subsp. **parvifolia** (Benth.) Morley, Brittonia 23:422. 1971

M. parvifolia Benth.

Guayabillo, Arracheche, Solacra, Cierito, Kenna

Shrub, usually 2–4(15) m tall, essentially glabrous, divaricately branched; younger stems ribbed, the ribs 4, thin, sharp, deciduous on older wood. Leaves sessile; blades lanceolate or ovate-lanceolate, acute to acuminate, rounded to subcordate and inequilateral at base, 3–8.5 cm long, 2–3 cm wide, pellucid-punctate, the margins undulate; vein 1 (the midrib). Fascicles axillary, bearing few flowers; pedicels with opposite-decussate bracteoles; flowers 5-parted; hypanthium and calyx together 5–6 mm long, campanulate, the lobes narrowly triangular, to 3 mm long, persisting in fruit; petals white, acuminate, 4–6 mm long, soon falling; stamens 10, exserted, shorter than style; anthers ca 2 mm long, with 2 subterminal, short, porelike slits, the connective darker, thickened toward base, bearing a dorsal gland prolonged at base; ovary inferior; style elongating to more than twice the length of the flower. Berries round, 7–13 mm diam, with persistent hypanthia, green turning deep purple at maturity, indehiscent or splitting open irregularly at maturity, ephemerally sweet; exocarp thin; mesocarp purplish, 2–3 mm thick; seed 1, subglobose, to 7 mm broad, with minute longitudinal ridges. *Croat 11790.*

Common in the forest. Flowers sporadically throughout the year, mostly in August and September but also

in the middle of the dry season. The fruits probably develop in about a month.

Leaves are similar to *Eugenia nesiotica* (106. Myrtaceae), a tree that flowers before putting on new leaves. Though Melastomataceae are said to lack stipules, this species usually bears interpetiolar structures that are at least stipule-like.

Mexico to Colombia, Ecuador, Peru, Bolivia, and western Brazil. In Panama, known from tropical moist forest in the Canal Zone, Bocas del Toro, Chiriquí, Los Santos, Panamá, and Darién, from tropical dry forest in Coclé, from premontane wet forest in Coclé and Panamá, and from tropical wet forest in Coclé.

OSSAEA DC.

Ossaea quinquenervia (P. Mill.) Cogn. in A. DC., Monogr. Phan. 7:1064. 1891

O. diversifolia (Bonpl.) Cogn.

Fruta de pava

Shrub, to 1(1.5) m tall; young stems, petioles, and inflorescences bearing dense, minute, brown, stellate pubescence. Petioles 1–3(5) cm long; blades ovate to broadly elliptic, acute to short-acuminate, abruptly attenuate and decurrent nearly to base of petiole, 6–22 cm long, 3.5–11.5 cm wide, glabrate to sparsely pilose above, densely stellate on veins below, denticulate and ciliate with reddish pilose trichomes, pliveined, the veins 5. Cymes loosely branched, paniculate, axillary, 2–5 cm long; flowers 5-parted; hypanthium pink, subglobose to campanulate, glandular-puberulent, sometimes sparsely setose (especially elsewhere); hypanthium and calyx tube together ca 1 mm long; calyx tube prominent, the lobes less than 1 mm long and widely spreading, the lobes and exterior teeth subulate; petals narrowly triangular to ovate, minute, ca 1.2 mm long, white, rounded to acute at apex, with a subapical tooth on outside, spreading at anthesis; stamens 10, erect, to 3.3 mm long, the connective extending somewhat beyond filament; anthers with apical pores; ovary inferior; style erect, to 5.7 mm long. Berries red, turning purple-black, subglobose, to ca 4 mm diam; seeds minute, numerous. *Croat 11429.*

Locally common at the edges of clearings; less frequent in the forest. Flowers and fruits throughout the year.

Costa Rica to northern South America. In Panama, known principally from tropical moist forest in the Canal Zone, all along the Atlantic slope, and in Chiriquí, Los Santos, Panamá, and Darién on the Pacific slope; known also from premontane moist forest in Panamá (Panamá City), from premontane wet forest in Panamá, and from tropical wet forest in Colón, Panamá, and Darién.

SCHWACKAEA Cogn.

Schwackaea cupheoides (Benth.) Cogn. ex Durand, Index Gen. Phan. 132. 1888

Branched herb, 7–75 cm tall; stems square, reddish, glabrous to pubescent on nodes and along angles. Petioles very short, to 8 mm long; blades ovate to narrowly elliptic, acute to acuminate, acute at base, 1–6 cm long, 0.3–2 cm wide, entire, sparsely pubescent, the trichomes above longer and sparser than below; veins 3. Flowers 4-parted, sessile, solitary in leaf axils; hypanthium ca 4 mm long at anthesis (excluding teeth), strigose, the lobes slender, ca 2 mm long, strigose, ciliate; petals pink (rarely white), ca 5 mm long; stamens 8, dimorphic, the outer 4 lavender, their connective longer than the anther, at right angles to filament, the inner 4 stamens yellow, with the anthers much reduced; ovary superior; style ± equaling height of anthers. Capsules oblong, ca 1 cm long, ribbed, the ribs 8, corky, tuberculate, setose; seeds cochleate, minute, numerous. *Croat 12279.*

Occasional, in clearings and on exposed lakeshore banks. Flowers principally from August to January. The fruits probably mature from September to February.

Capsules probably split open from the apex. The minute globular seeds pass slowly through the passages formed between the long cilia of the persistent sepals, probably only while the wind is blowing.

Southern Mexico to Panama and Colombia. In Panama, known from tropical moist forest in the Canal Zone, San Blas, Veraguas, Los Santos, and Panamá, from premontane wet forest in Chiriquí, Coclé, and Panamá, and from tropical wet forest in Colón.

See Fig. 438.

TIBOUCHINA Aubl.

Tibouchina longifolia (Vahl) Baill., Adansonia 12:74. 1877

Herb, to 1.5 m tall, becoming woody in age, appressed-pilose. Petioles 5–15 mm long; blades narrowly lanceolate, acute to acuminate, acute to obtuse at base, 4–11 cm long, 1–3.5 cm wide, entire, the trichomes above adnate to the leaf for part of their length; veins 5–7. Cymes terminating stems and branches; flowers 5-parted, often many; hypanthium and calyx densely setose, the lobes slender, 3–4 mm long; petals white (rarely pinkish), obovate, 6–8 mm long, ciliate; stamens 10, folded at anthesis, later unfolding to 5 mm long; anthers yellow, ca 3 mm long, tapered to apex, with a single terminal pore (eventually equaling or exceeding stigma), curved, lobulate at base; ovary superior; style erect; stigma simple. Capsules subglobose, ca 3 mm long; seeds many, cochleate, tuberculate. *Croat 4611.*

Usually locally abundant in clearings. Flowers and fruits throughout the year.

The capsules are somewhat nodding at maturity and are irregularly dehiscent by slender cracks near the apex. The minute seeds pass through when the capsules are shaken in the wind.

Southern Mexico to western Brazil and Bolivia; West Indies. In Panama, known principally from tropical moist forest in the Canal Zone, Bocas del Toro, Chiriquí, Los Santos, Herrera, Panamá, and Darién; known also from premontane moist forest in Panamá, from premontane wet forest in Chiriquí, Coclé, and Panamá, from tropical wet forest in Coclé (Atlantic slope) and Panamá, and from lower montane rain forest in Chiriquí.

TOPOBAEA Aubl.

Topobaea praecox Gleason, Phytologia 3:355. 1950

Hemiepiphytic shrub, to 4 m tall, often perched in crotches of trees, not commonly found in top of crown, the trunk to the ground slender, 5–6 cm diam, not twining; stems terete, glabrous; branches arching, pendent, to 4 m long, the youngest stellate-furfuraceous, the internodes short on flowering branches. Leaves immature at flowering time, clustered; petioles 1–5 cm long, stellate-pubescent; blades ovate to oval, acuminate, obtuse to subcordate at base, 3–16 cm long, 1.5–11 cm wide, glabrous above, weakly stellate-pubescent below especially on veins, entire, pliveined or basally veined, the veins 5, the midrib arched. Fascicles axillary, of usually 3 or 4 (to 5) flowers, usually on leafless nodes; pedicels less than 1 cm long; bracts in 2 pairs, at base of each flower, decussate, distinct to the base, broadly rounded, emarginate and recurved at apex, 7–8 mm long, glabrous to sparsely and minutely stellate-pubescent; flowers ca 3 cm wide, 6-parted; hypanthium and calyx to 1 cm long; calyx lobes rounded to obtuse, curved inward in fruit, ± glabrous, tinged with red; petals magenta, obovate, sometimes oblique, 1.5(2.2) cm long, thick at base; stamens 12, to 2.3 cm long; filaments flexed below anthers; anthers to 1.2 cm long, coherent laterally except at apex, horizontal and directed to one side of flower with the style, with 2 subapical pores; ovary inferior; style to 1.6 cm long, held just in front of anthers; stigma simple. Fruiting hypanthia and calyces to ca 1.2 cm long, bowl-shaped; berries contained within persistent sepals; seeds sticky, numerous, dark brown, oblong, to 1.3 mm long, embedded within a fleshy white matrix. *Croat 14953.*

Occasional in the forest high in trees, rarely lower on the lakeshore; probably always somewhat epiphytic. Flowers from March to June, with the fruits probably maturing from June to September. Leaves fall off during the late dry season and are replaced in the early rainy season.

A large bee, possibly *Eulaema mexicana,* was observed drumming pollen out of flowers in an upside-down position. The process takes less than 10 seconds per flower.

Allen (unpublished manuscript) said that a related species, *T. regeliana* Cogn., is at first a strangler, eventually developing a trunk of its own.

Costa Rica and Panama. In Panama, known from tropical moist forest in the Canal Zone, San Blas, and Darién and from premontane wet and tropical wet forests in Coclé, Panamá, and Darién.

108. ONAGRACEAE

Aquatic herbs or erect or prostrate terrestrial herbs; stems sometimes winged. Leaves alternate, sessile or petiolate; blades simple, entire; venation pinnate; stipules lacking. Flowers bisexual, actinomorphic, solitary, axillary, bracteolate; calyx 4–6-lobed, forming a hypanthium, the lobes valvate; petals the same number as calyx lobes, clawed, yellow or occasionally white with a yellow spot, free, soon falling; stamens twice the number of calyx lobes, in 2 series; anthers 2-celled, dehiscing longitudinally; ovary inferior, generally 4-locular, 4-carpellate; placentation axile; ovules anatropous, many on each placenta; style simple; stigma capitate. Fruits loculicidally dehiscent capsules; seeds many, lacking endosperm, the raphe (scar of the ovular stalk) often diagnostic.

The family is represented on BCI only by *Ludwigia.*

Distinguished by the usually four-parted flowers, with nearly rounded, clawed, yellow petals and an inferior, multiovulate ovary terminated by a hypanthium and persistent sepals.

Flowers of some species are pollinated at least in part by stingless bees, halictid bees, and other small bees. Pollen grains of the Onagraceae are held together by viscid elastic threads. Bees specialized on this family have pollen baskets made up of long unbranched bristles (Proctor & Yeo, 1973). Some species are self-pollinated (*Ludwigia decurrens*) or mostly self-pollinated (*L. octovalvis*) (P. Raven, pers. comm.).

Plants grow principally in aquatic situations, and the minute, usually buoyant seeds are probably dispersed partly by wind and partly by water.

Some 21 genera and about 650 species; primarily in temperate and subtropical regions.

LUDWIGIA L.

Ludwigia decurrens Walt., Fl. Carol. 89. 1788
Jussiaea decurrens (Walt.) DC.

Erect, ± glabrous herb, usually to 1 m tall, branched; stems usually 4-winged. Blades subsessile, lanceolate to linear, attenuate to apex, ± acute at base, mostly 5–12 cm long, 1–3.5 cm wide, prominent lateral veins in 11–16 pairs, conspicuously loop-connected. Flowers solitary in upper axils, 4-parted; sepals lanceolate-ovate, 7–10 mm long; petals yellow, 8–12 mm long; stamens 8, 2–3 mm long; style ca 2 mm long; stigma globose, ca 2 mm thick. Capsules clavate-obpyramidal, (8)12–20 mm long, 4-angled or 4-winged, the corners dark-ribbed, weather-

---

KEY TO THE SPECIES OF LUDWIGIA

Leaves obovate to suborbicular, the petioles conspicuous, one-fourth to fully as long as blade; plants floating or prostrate on soil, rare . . . . *L. helminthorrhiza* (Mart.) Hara

Leaves lanceolate to linear, the petioles lacking or short, much less than one-fourth as long as blade; plants often aquatic, but always erect:

Flowers 4-parted; seeds multiseriate in each locule of the capsule, not enclosed in a corky persistent endocarp:

Capsules club-shaped, conspicuously broadened toward apex, 8–20 mm long; plants rare or no longer present on the island . . . . *L. decurrens* Walt.

Capsules ± cylindrical, 2–5 cm long; plants common . . . . *L. octovalvis* (Jacq.) Raven

Flowers 5- or 6-parted; seeds uniseriate or biseriate in each locule, each seed surrounded by its corky persistent endocarp:
Plants conspicuously pubescent; capsules ± cylindrical, 2–4.5 cm long; seeds in 1 series in each locule, free within a horseshoe-shaped endocarp; plants common . . . . *L. leptocarpa* (Nutt.) Hara
Plants glabrous or essentially so (the stems and fruits drying viscid); capsules ± oblong, less than 1 cm long; seeds in 2 series in each locule, entirely enclosed within endocarp; plants possibly no longer present . . . . *L. torulosa* (Arn.) Hara

ing into fibers; seeds subcylindrical, less than 1 mm long, free, multiseriate in each locule, the raphe ca one-fifth the width of the body. *Bailey & Bailey 384.*

Possibly no longer occurring on the island; last collected by the Baileys along the shore. Probably flowers and fruits throughout the year.

Southern United States to northern Argentina. In Panama, known from tropical moist forest in the Canal Zone and Bocas del Toro and from tropical dry forest in Coclé.

Ludwigia helminthorrhiza (Mart.) Hara, J. Jap. Bot. 28:292. 1953
Jussiaea natans H. & B.

Aquatic herb, floating on water or prostrate on mud, ± glabrous; stem conspicuously rooting at nodes, seldom branching. Petioles 1–3 cm long; blades obovate to suborbicular, blunt at apex, abruptly narrowed to petiole, 2–7 cm long, 1.5–4.5 cm wide; lateral veins conspicuous, with a loop-connecting vein. Flowers solitary in axils, usually 5-parted; sepals lanceolate-ovate, ca 5 mm long; petals white with basal yellow spot, oblong-obovate, 8–14 mm long, shortly clawed; stamens 10, unequal, 3–5 mm long; anthers 1–2 mm long; style white, 4–7 mm long; stigma green, slightly lobed, ca 1.5 mm wide. Capsules subcylindrical, 2–3 cm long, ca 3 mm thick; endocarp hard, shining, ca 1.5 mm long; seeds contained in and adnate to the endocarp, uniseriate in each locule. *Shattuck 1132.*

Apparently once common at the edge of the lake and no doubt still occurring in marshy areas on the southern and western edges of the island. Apparently flowers and fruits throughout the year.

Southern Mexico to Peru and Paraguay. In Panama, known from tropical moist forest in the Canal Zone, Panamá, and Darién.

Ludwigia leptocarpa (Nutt.) Hara, J. Jap. Bot. 28:292. 1953
Jussiaea leptocarpa Nutt.

Herb usually about 1 m tall or shrub to 2.5 m, rooting at lower nodes when in water; stems angulate-winged from below petiole, sparsely white-pilose and puberulent. Blades sessile or on petioles to 10 mm long, lanceolate, usually acute or acuminate at apex, gradually tapered to base, 4–8(14) cm long, 0.7–2.8 cm wide, usually puberulent on both surfaces, often with longer trichomes on veins. Flowers solitary in axils, 5- or 6-parted; pedicels to ca 2 cm long; sepals lanceolate, acuminate, 5–8 mm

long; petals yellow, obovate, rounded at apex, usually 8–10 mm long (rarely smaller); stamens 10 or 12, in 2 series, the shorter series opposite petals, the longer ones alternate, about as high as stigma; anthers extrorse; style stout, 2–4 mm long; stigma capitate (becoming fleshy); nectaries in U-shaped depressions around base of outer filaments, protected by a close series of arching trichomes from either side; nectar copious. Capsules subcylindrical, striate, 2.5–4.5 cm long, to 3 mm thick, long-pubescent; seeds uniseriate in each locule, ca 1 mm long, surrounded by the horseshoe-shaped corky endocarp. *Croat 11299, 13972.*

Frequent in swampy or moist areas on the shore; often a component of floating masses of vegetation. Flowers and fruits from January to July.

Throughout the tropics and subtropics of the New World. In Panama, known from tropical moist forest in the Canal Zone and Bocas del Toro, from tropical dry forest in Coclé, and from premontane wet forest in Chiriquí.

See Fig. 439.

Ludwigia octovalvis (Jacq.) Raven, Kew Bull. 15:476. 1962
Jussiaea suffruticosa L.

Aquatic herb, usually to 1.5 m tall; stems usually branched and weakly or rather densely pubescent. Leaves variable, sessile or short-petiolate, lanceolate or oblong to linear, acute to acuminate at apex, acute to attenuate at base, the larger leaves 6–15 cm long, 0.5–3.5 cm wide. Flowers solitary in axils, 4-parted; sepals ovate or lanceolate, 3–15 mm long; petals obovate, yellow, rounded or emarginate at apex, to 1.5 cm long; stamens 8, in 2 series, the outer 4 opposite petals, the inner 4 around style, longer than style in bud; anthers extrorse; style 1.5–3.5 mm long; stigma ± globular, held just above anthers at anthesis. Capsules ± cylindrical, 2–5 cm long, striate with 8 darker ribs, weathering at maturity, splitting apart longitudinally; seeds multiseriate in each locule, free, minute, with the raphe enlarged and as large as body of seed, the seeds dispersed by wind and water. *Croat 6863, 15245.*

Common along the shore, usually in standing water, occasionally as an epiphyte on floating debris. Flowers and fruits principally in the rainy season.

This species as treated here includes both *Jussiaea suffruticosa* L. var. *octofila* (DC.) Munz and var. *ligustrifolia* (H.B.K.) Griseb. as treated by Munz in the *Flora of Panama* (1959).

Throughout the tropics of the world. In Panama, ecologically variable; known from tropical moist forest

Fig. 439. *Ludwigia leptocarpa*

Fig. 438. *Schwackaea cupheoides*

Fig. 440.
Ludwigia octovalvis

in all areas, from tropical dry forest in Coclé, from premontane wet forest in Chiriquí, from tropical wet forest in Colón, Coclé, and Darién, and from lower montane wet forest in Chiriquí.

See Fig. 440.

Ludwigia torulosa (Arn.) Hara, J. Jap. Bot. 28:294. 1953

Erect glabrous herb; stems reddish-brown and viscid (at least on drying). Petioles less than 1 cm long; blades narrowly lanceolate, long and gradually tapered to apex, acute at base, 8–13 cm long, 0.5–1.7 cm wide. Flowers solitary in axils, usually 5-parted. Fruits ± oblong, constricted in the middle, ± viscid (at least on drying), less than 1 cm long; seeds many, biseriate in each locule, surrounded by endocarp. *Shattuck 1161.*

Collected once by Shattuck in August 1934 at Zetek House. This collection was not considered by either Standley or Munz in the *Flora of Panama.* Seasonality not known.

Belize, Panama, the Guianas, Brazil, and Bolivia; Cuba, Santo Domingo. In Panama, known only from BCI.

109. ARALIACEAE

Terrestrial or epiphytic shrubs or trees. Leaves alternate, petiolate (the petioles often variable in length), simple to pinnately or palmately compound; blades entire, serrate, or lobed, sometimes with T-shaped or branched trichomes; venation pinnate; stipules present, sometimes ligulate. Panicles terminal, compound, the ultimate divisions heads or umbels; flowers bisexual or polygamous (*Oreopanax*), greenish; calyx inconspicuous, with about 5 teeth; petals 5, valvate, arising from the epigynous disk; stamens 5–10, free, alternate with the petals; anthers 2-celled, dorsifixed, dehiscing longitudinally; ovary inferior, 2–10-locular; placentation axile; ovules solitary in each locule, pendulous, anatropous; styles 2–10, free or connate; stigmas terminal or on inner surfaces. Fruits berries; seeds as many as carpels, with copious endosperm.

Members of the family are confused with only the Umbelliferae (110) and are most easily recognized by being arborescent and by having panicles with the flowers arranged in small umbels. Except for *Didymopanax* (with a usually laterally compressed fruit), fruits are also characteristic, being more or less globose with a prominent ring around the persisting styles.

The flowers are very open, with dorsifixed versatile anthers that turn inside out, making the pollen easily accessible. Nectar, if present, is also accessible. Flowers are probably pollinated by a wide range of small insects. Hladik (1970) reported that they are visited primarily by bees of the genera *Trigona* and *Melipona*.

The fruits are probably dispersed chiefly by birds, but may be taken by white-faced monkeys (Enders, 1935; Hladik & Hladik, 1969) and perhaps other animals. Fruits of *Didymopanax morototoni* are taken by a variety of birds including guans, manakins, toucans, honeycreepers, pigeons, and woodpeckers (Chapman, 1938). Fruits of *Dendropanax arboreus* are taken by the bat *Artibeus jamaicensis* Leach in Mexico (Yazquez-Yanes et al., 1975).

About 65 genera and about 800 species; mostly in the tropics.

DENDROPANAX Dec. & Planch.

Dendropanax arboreus (L.) Dec. & Planch., Rev. Hort. sér. 4, 3:107. 1854
Gilbertia arborea (L.) Marchal
Vaquero

Tree, 10–20(25) m tall, 25–40(70) cm dbh, glabrous; bark with prominent lenticels, not deeply fissured; wood soft; ultimate branches whorled from apex of stout stems; stems brittle. Leaves simple, alternate; petioles 1–8 cm long; blades variable, mostly elliptic, also ovate-elliptic and obovate-elliptic, obtuse to acuminate and downturned at apex, attenuate to rounded at base, 5.5–15 cm long, 2–8 cm wide, usually entire, rarely with remote apiculate teeth. Racemes terminal, 6–12 cm long, of 3–20 pedunculate umbels, frequently with many umbels along rachis below terminal whorl of umbels; primary peduncle

---

KEY TO THE SPECIES OF ARALIACEAE

Leaves compound:
    Leaves pinnately 3–9-foliolate; blades sharply serrate; plants cultivated in the Laboratory
        Clearing . . . . *Polyscias guilfoylei* (Bull) Bailey
    Leaves palmately 7–12 foliolate; blades entire; plants frequent in clearings and along the shore
    . . . . *Didymopanax morototoni* (Aubl.) Dec. & Planch.
Leaves simple:
    Flowers sessile or nearly so in dense heads, the heads staminate or bisexual; inflorescence
        branches usually stellate-pubescent . . . . *Oreopanax capitatus* (Jacq.) Dec. & Planch.
    Flowers pedicellate in umbels, bisexual; inflorescence branches glabrous:
        Leaves usually ± elliptic, infrequently toothed; calyx truncate to weakly lobed; petals more
            than 2 mm long; fruit ca 10 mm diam; seeds rugose, notched at apex . . . .
        . . . . *Dendropanax arboreus* (L.) Dec. & Planch.
        Leaves usually ± oblong, usually with remote, minute teeth; calyx with 5 deltoid teeth; petals
          usually less than 1.5 mm long; fruits ca 6 mm diam; seeds appearing smooth, grooved
          along rounded outer margin . . . . *Dendropanax stenodontus* (Standl.) A. C. Smith

and rachis 1.5–6 cm long, bracteate at nodes, the floriferous peduncles 1–5 cm long, often bracteate, the bracts
free, ovate, ca 1 cm long; pedicels 3–9 mm long; flowers
(3)10–30(50) per umbel, greenish-white, ca 6 mm diam,
5–7-parted; calyx ± bowl-shaped, truncate and undulate
to remotely 5-lobed; petals 5, valvate, (1.3)2–2.5 mm
long, acute and cucullate at apex, spreading at anthesis,
later reflexed; stamens 5(7), alternate with and longer
than petals, to 4 mm long, erect at anthesis, later spreading; anthers ca 1 mm long; styles 5(7), connate at base,
the connate part flat or slightly conical, free and held
tightly together at apex, later spreading, persistent in
fruit, ca 0.5 mm long. Berries globose, ca 1 cm diam,
smooth, drying deeply sulcate between seeds with ±
sharp ridges, black-purple at maturity, with a ringlike
scar around persistent styles; seeds 5(7), ± semicircular,
flattened, with a sharp margin, ca 7 mm long, tan, notched
at apex on inner margin, minutely muricate. *Croat 11751,
17047.*

Frequent in the forest. Flowers from June to December, principally in July and August. The fruits mature
in about 1 month. Individuals may bear mature fruits
as well as flowers on separate inflorescences. Leaves are
lost and quickly replaced in the dry season.

Mexico to Colombia, Venezuela, Ecuador, Peru, and
Bolivia; West Indies. In Panama, known from tropical
moist forest in the Canal Zone, Bocas del Toro, Veraguas,
Los Santos, and Darién Provinces, from premontane wet
forest in the Canal Zone, Chiriquí, Coclé, Panamá, and
Darién, and from lower montane wet forest in Chiriquí.
Reported from premontane moist, tropical wet, and premontane rain forests in Costa Rica (Holdridge et al.,
1971).

See Figs. 441 and 442.

Dendropanax stenodontus (Standl.) A. C. Smith,
  Trop. Woods 66:3. 1941

Shrub or small tree, 1–2.5(4) m tall, glabrous; stems
striate when dry. Leaves simple, alternate; petioles 1–8
cm long; blades variable, frequently oblong, also elliptic,
narrowly elliptic, oval, oblong-ovate or oblong-obovate,
usually abruptly long-acuminate at apex, attenuate to
rounded at base, 6–25 cm long, 2–10 cm wide, usually
with sparse, minute, apiculate teeth. Racemes terminal,
2–8 cm long, of 3–15 pedunculate umbels, usually with
few umbels on rachis below terminal whorl of umbels;
primary peduncle and rachis 4–7 cm long, bracteate at
nodes, the floriferous peduncles 1–5 cm long, bracteate
near middle, the bracts free, ovate, ca 0.5 mm long; pedicels 2–9 mm long; flowers 4–35 per umbel, greenish-
white to greenish-purple, ca 4.7 mm diam at anthesis,
usually drying black to purple; calyx ± bowl-shaped,
lobed, 1–1.8 mm long (mostly less than 1.5 mm), acute
and cucullate at apex, spreading at anthesis, later reflexed,
the lobes 5, deltoid to apiculate; petals narrowly ovate,
3–3.5 mm long; stamens 5, alternate with and as long as
or slightly longer than petals, erect to spreading; anthers
ca 0.7 mm long, the thecae directed upward at anthesis;
styles 5–9, connate, ca 1 mm long, conical at base, free at
apex and held tightly together at anthesis, later spreading,

persistent in fruit. Berries globose, smooth, ca 6 mm long,
drying slightly sulcate between seeds; seeds 5, shaped like
orange segments, ca 4.3 mm long, tan, deeply grooved on
one side along rounded outer margin. *Croat 5409, 11427.*

Occasional, in somewhat disturbed areas at the edges
of clearings or on shore; one individual known in the
older forest. Flowers from April to August, mostly in
July and August, rarely earlier or later. The fruits mature
from May to September, mostly in July and August.
Individuals may flower two or more times per season
and can be found with both flowering and fruiting inflorescences.

Mexico (Chiapas) to Panama. In Panama, known principally from tropical moist forest on the Atlantic slope
of the Canal Zone, in adjacent parts of Colón, and in San
Blas; known also from premontane wet forest in Colón
(near María Chiquita).

See Fig. 443.

DIDYMOPANAX Dec. & Planch.

Didymopanax morototoni (Aubl.) Dec. & Planch.,
  Rev. Hort. sér. 4, 3:109. 1854
  Mangabe, Gargorán, Pavo, Pava

Tree, to 25(30) m tall; trunk to 40(70) cm dbh; outer
bark planar, thin, ± mottled, marked with vertical rows
of minute lenticels, these rubbing off easily; inner bark
moderately thick, coarse, mottled tan and white; sap with
sweet, strong odor; leaves and inflorescences clustered at
apex. Leaves alternate, palmately compound; petioles to
1 m long with a basal ligule ca 1 cm long; petiolules 3–14
cm long; leaflets 7–12, oblong or oblong-oblanceolate,
acuminate, obtuse to rounded or subcordate at base, 8–45
cm long, 3–19 cm wide, entire, glabrous above, densely
and softly ferruginous-pubescent below with T-shaped
trichomes; juvenile leaves much thinner, the lower surface
with numerous, small, appressed, T-shaped trichomes,
the upper surface with long simple trichomes much like
those on the conspicuously ciliolate margins. Panicles
compound, terminal, 15–50 cm long, with racemose
umbels, the pubescence appressed, grayish, the branches
subtended by small bracts; flowers bisexual, 5-parted,
7–15 per umbel; calyx cupuliform, 1–1.5 mm long; petals
oblong, 1.5–2.5 mm long, glabrous inside; stamens
1.5–2.5 mm long; styles 2, free, ca 1 mm long, erect,
spreading in fruit. Berries transversely oblong, flattened,
cordate at both ends, ca 5 mm long and 8–10 mm wide;
seeds 2. *Croat 4394, 6853; 14584* (juvenile leaves).

Frequent in clearings and along the shore. Flowers
from July to December, but mostly in the late rainy
season. Allen (1956) reported that the flowers are present
from November to February on the Osa Peninsula in
Costa Rica. The fruits are common from January to May.

The species is frequently visited by small birds during
the fruiting period.

Mexico to Brazil and Argentina; West Indies. In Panama, characteristic of tropical dry and tropical moist
forests (Holdridge & Budowski, 1956; Tosi, 1971); known
also from premontane moist and premontane wet forests
in Panamá and from tropical wet forest in Colón.

Fig. 441. *Dendropanax arboreus*

Fig. 442. *Dendropanax arboreus*

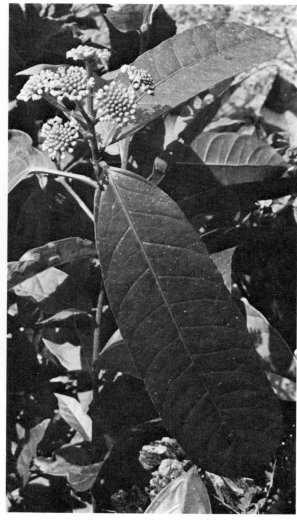

Fig. 443.
Dendropanax stenodontus

OREOPANAX Dec. & Planch.

Oreopanax capitatus (Jacq.) Dec. & Planch., Rev.
Hort. sér. 4, 3:108. 1854

Polygamous shrub or tree, to 15 m, sometimes hemiepi-phytic and growing near top of canopy, glabrous except for inflorescences. Leaves simple, whorled near apex of branches; petioles 5–25 cm long, terete; blades ovate, elliptic or obovate, 10–30(35) cm long, 4–16(25) cm wide, acute to long-acuminate at apex, acute to rounded at base, the larger ones conspicuously undulate. Panicles terminal, 15–20(30) cm long, the heads dense, racemose, staminate or bisexual, the inflorescence branches with branched trichomes; bracts at branching nodes small; staminate heads10–25-flowered, the flowers with 1 or 2 styles; bisexual heads 5–12-flowered, the flowers with 5–10 styles, the styles to ca 1 mm long; stamens to 3 mm long; both types of heads with calyx ca 1 mm long; petals 5, white, oblong, 2–2.5 mm long. Berries subglobose, to 8 mm diam; seeds few. *Croat 11861.*

Frequent in the forest. Seasonality uncertain. Flowers and fruits in the early dry season (January and February), but mostly during the early to the middle of the rainy season (June to October), especially in July and August.

Though generally found to be a small to medium-sized tree growing in the normal manner, at least one plant was seen growing as a hemiepiphyte on another tree at ca 30 m.

Mexico to South America; West Indies. In Panama, known from tropical moist forest in the Canal Zone, Bocas del Toro, and Darién, from premontane wet forest in Chiriquí and Coclé, and from lower montane wet and lower montane rain forests in Chiriquí.

POLYSCIAS Forst. & Forst.f.

Polyscias guilfoylei (Bull) Bailey, Rhodora 18:153. 1916
Nothopanax guilfoylei (Bull) Merr.
Wild coffee, Coffee tree

Glabrous shrub, 5–7 m tall but usually pruned; stems with elongate lenticels. Leaves imparipinnate; petiole, rachis, and petiolules often purplish and marked with light green streaks; leaflets 3–9, variable, orbicular to ovate-elliptic, obtuse to short-acuminate at apex, acute to rounded at base, attenuate to petiole, mostly 5–12 cm long, 3–7 cm wide, sharply serrate, the margins some-times white. Panicles large, open, the umbels long-

pedunculate; flowers small, 4- or 5-parted, pedicellate; ovary 5–8-locular. Fruits not seen. *Croat 14549.*

Cultivated in the Laboratory Clearing and usually pruned. Seasonal behavior not determined.

The variety *laciniata* Bailey, with deeply incised margins, is cultivated at Summit Garden (Canal Zone).

Native to Polynesia; cultivated in the American tropics. In Panama, known from tropical moist forest in the Canal Zone and Bocas del Toro.

110. UMBELLIFERAE (APIACEAE)

Glabrous perennial herbs (*Spananthe* annual), aquatic (*Hydrocotyle*) or terrestrial, sometimes strong-scented (*Eryngium*). Leaves opposite, alternate, or basal, petiolate; blades simple, usually serrate, spiny, sometimes tripartite in *Eryngium,* peltate in *Hydrocotyle;* venation generally palmate (obscure in *Eryngium*); stipules lacking (*Spananthe* with bristly cilia). Umbels or heads terminal and axillary; flowers bisexual, actinomorphic; calyx 5-lobed (lacking in *Hydrocotyle*); petals 5, free, white; stamens 5, free, alternating with the petals, inserted on an epigynous disk; ovary inferior, 2-locular, 2-carpellate; placentation axile; ovules 1 per locule, anatropous, pendulous; styles 2, basally swollen and forming a depressed stylopodium (lacking in *Eryngium*); stigmatic tips swollen, ± globular. Fruits schizocarps of 2 mericarps; seeds 1 per mericarp, with abundant endosperm.

Umbelliferae are characterized by the usually sheathing petioles, the usually umbellate inflorescences, the bicarpellate, bilocular, inferior ovary, and the schizocarpic fruit.

The flowers of most species are self-compatible, but protandrous, and are pollinated by insects, especially Diptera and unspecialized Hymenoptera (Bell, 1971). Self-pollination is believed to be effected by thrips, which are found in great abundance on the inflorescences.

The fruits do not have any apparent means of dispersal. *Hydrocotyle umbellata* must rely on water currents or perhaps water fowl to disperse its seeds because of its consistent aquatic habitat around the margin of the lake. Leaves of *Eryngium foetidum* are used as a condiment in Panama, and the species has no doubt been dispersed by man.

About 275 genera and 3,000 species; primarily in the North Temperate regions.

KEY TO THE SPECIES OF UMBELLIFERAE

Plants aquatic; leaves orbicular, peltate .... *Hydrocotyle umbellata* L.
Plants not aquatic; leaves not orbicular, not peltate:
    Plants with basal rosette of linear-oblanceolate leaves; flowers sessile, in densely bracteate heads
    .... *Eryngium foetidum* L.
    Plants lacking basal leaves; flowers pedicellate, in long pedunculate umbels ....
    .... *Spananthe paniculata* Jacq.

ERYNGIUM L.

Eryngium foetidum L., Sp. Pl. 232. 1753

Culantro, Culantro coyote, Fitweed, Spiritweed

Glabrous herb from a stout taproot, to 60 cm tall. Leaves dimorphic; blades of basal rosette linear-oblanceolate, rounded at apex, cuneate at base, 10–16 cm long, 2.5–4 cm wide, serrate; upper blades opposite, sessile, oblanceolate, often tripartite, apiculate at apex, acute to obtuse at base, 2–3 cm long, 0.5–1.5 cm wide, spinulose-serrate. Flower heads dense, bracteate, cylindrical, ca 1 cm long, subtended by 5 or 6 leaflike bracts greatly exceeding the heads; flowers 5-parted, white, minute, sessile, congested, subtended by a bracteole; bracteoles narrow, exceeding the fruit; sepals widely separated, ovate, mucronate at apex, ± equaling petals; petals ca 0.5 mm long, refolded inward; stamens nearly twice as long as styles, shed with corolla after anthesis; styles 2, ca 1 mm long, somewhat spreading, exserted earlier than stamens. Fruits ± globose, ca 2 mm long, conspicuously muricate. *Croat 8672.*

Common in the Laboratory Clearing. Flowers and fruits throughout the year, but the flowers are initiated principally in the dry season.

The plant has a foul aroma in all parts. Crushed leaves are savored by the natives as a condiment in foods.

Throughout the tropics of the New World; introduced into tropical Africa. In Panama, growing in clearings and weedy areas; known from tropical moist forest in the Canal Zone, all along the Atlantic slope, and in Panamá and Darién, from tropical dry forest in Coclé and Panamá, from premontane moist forest in the Canal Zone, and from premontane wet forest in Chiriquí, Coclé, and Panamá.

HYDROCOTYLE L.

Hydrocotyle umbellata L., Sp. Pl. 234. 1753

Glabrous herb, usually aquatic, creeping, densely rooting at nodes. Leaves single at each node; petioles 5–20(40) cm long, fleshy; blades peltate, orbicular, 2–7.5 cm diam, crenate or crenately lobed. Umbels simple, axillary; peduncles long, erect, ± equaling leaves; flowers 5-parted, several to many on rays 2–25 mm long; sepals lacking; petals 5, white, ca 1 mm long, spreading at anthesis, becoming recurved; stamens spreading, less than half as long as or equaling petals; styles 2, spreading, or directed toward one another; stigmas ± globular. Fruits ellipsoid, 2-carpellate (rarely 3-carpellate), 1–2 mm long, 2–3 mm broad. *Croat 13233.*

Uncommon, restricted to swampy areas on the south side of the island at the lake's edge; a component of the floating marshes. Flowering and fruiting year-round.

Throughout warmer regions of the New World; tropical southern Africa. In Panama, known from aquatic situations in tropical moist forest in the Canal Zone, Bocas del Toro, Colón, Chiriquí, and Panamá, from tropical dry forest in Coclé, and from premontane wet forest in Chiriquí and Coclé.

SPANANTHE Jacq.

Spananthe paniculata Jacq., Coll. 3:247. 1789

Herb, 20–150 cm tall; stems hollow. Leaves opposite; stipules minute, bristly-ciliate; petioles 1–8 cm long, bearing a tuft of fine trichomes at apex; blades deltoid-ovate or ovate, acuminate, truncate to cordate at base, 2.5–9(14) cm long, 2–6(14) cm wide, dentate-crenate, sparsely setose-pubescent on veins, paler below. Umbels simple, axillary; peduncles 1–14 cm long, finely pubescent especially near apex; pedicels slender, 7–10 mm long; flowers 5-parted, minute, white; calyx ca 1 mm long, with apiculate teeth; petals obovate, ca 2 mm long; stamens to 1.5 mm long; anthers ca 0.4 mm long, 0.5 mm wide; styles 2, to 1 mm long; stigma simple. Fruits ovoid, 2–4 mm long, ca 2 mm wide. *Aviles 18.*

Collected once on the island; probably no longer occurring there. Seasonal behavior not determined. Flowers and fruits most abundantly in the rainy season.

Throughout warmer regions of the New World. In Panama, known only from tropical moist forest in the Canal Zone.

111. THEOPHRASTACEAE

Small trees. Leaves alternate, ± sessile; blades simple, entire; venation pinnate; stipules lacking. Racemes terminal, bearing several flowers; flowers aromatic, bisexual, actinomorphic; sepals 5, free; corolla tubular, 5-lobed, the lobes imbricate, alternating with 5 petaloid staminodia; stamens 5, epipetalous, opposite the corolla lobes; filaments united at the base and adnate to the corolla tube; anthers 2-celled, extrorse, longitudinally dehiscent; ovary superior, 1-locular, 5-carpellate; placentation free-central; ovules many, anatropous; style 1; stigma subsessile, capitate. Fruits irregularly dehiscent capsules; seeds few, immersed in pulp, with fleshy endosperm.

The family is represented on BCI only by *Jacquinia macrocarpa*, which can be distinguished by its sharply apiculate leaf blades and its coriaceous, bright-orange flowers. In Costa Rica hummingbirds visit the flowers on rare occasions (D. Janzen, pers. comm.). Fruits are probably animal dispersed.

Some 32 genera and about 1,000 species; subtropics and tropics.

JACQUINIA L.

Jacquinia macrocarpa Cav., Icon. Pl. 5:55, t. 483. 1799

J. aurantiaca Ait.

Small tree. Leaves very stiff, elliptic to obovate, acute to rounded at apex with a sharp apiculum, acute at base, 6–10 cm long, 2–3.5 cm wide, ± sessile, glabrous except on upper midrib. Racemes corymbiform; flowers pleasantly and strongly aromatic; sepals coriaceous, persisting in fruit, the lobes rounded, imbricate, the margins thin; corolla bright orange, coriaceous, the lobes 5, blunt,

alternating with 5 petaloid staminodia; stamens 5, extrorse, held closely together around the style (the pollen may be shed in bud), after several days folding back against tube to expose the much shorter stigma; ovary unilocular; ovules many, on a stalked basal placenta; stigma capitate, covered with a viscid liquid. Fruits oblong-elliptic, apiculate, to 7 cm long, yellowish; exocarp thick, woody; seeds several, dark brown, hard, shiny. *Croat 12885.*

Cultivated at the Laboratory Clearing. Flowers sporadically throughout the year, mostly in the dry season and the early rainy season. The species is not deciduous on BCI (R. Foster, pers. comm.). In Costa Rica, plants lose their leaves within 8 weeks of the first rains and are usually completely leafless by the first week of July, remaining leafless through the rest of the rainy season until late November or December. Flower buds are present within 1 month after new leaf production, and nearly all flowers are open within 2 months. The fruits begin to mature in the middle of the rainy season.

The species has been called *J. aurantiaca* Ait. in Panama; it is not known for certain that *J. aurantiaca* is synonymous with *J. macrocarpa.*

Mexico south to Panama and possibly as far as Ecuador. In Panama, restricted to the Pacific coast at low elevations, frequently found growing with *Hippomane mancinella* (75. Euphorbiaceae; not on BCI); known from premontane dry forest in Los Santos and Herrera, from tropical dry forest in Coclé and Panamá, and from tropical moist forest in the Canal Zone.

112. MYRSINACEAE

Shrubs or sometimes trees; most parts pellucid- or opaque-punctate. Leaves alternate, petiolate; blades simple, entire or sometimes serrulate, sometimes with stellate pubescence; venation pinnate; stipules lacking. Inflorescences terminal or sometimes axillary, ± paniculate, sometimes bracteate; flowers bisexual or sometimes unisexual (*Stylogyne* dioecious), actinomorphic; calyx deeply (4)5-lobed; corolla (4)5-lobed to near base; stamens of the same number as and opposite corolla lobes,

epipetalous, distinct; anthers 2-celled, dorsifixed, introrse, dehiscing longitudinally or by apical pores; ovary superior, 1-locular, (4)5-carpellate; placentation axile; ovules 3 to many; style short; stigma simple. Fruits drupes with fleshy exocarp and stony endocarp; seed with copious endosperm.

Members of the family can be recognized usually by the alternate leaves, by the characteristic pellucid or opaque dots on the leaves and most floral and fruit parts, and by the globose, one-seeded fruits.

Pollination systems are unknown.

The fruits are probably eaten by birds (Ridley, 1930).

Some 32 genera and about 1,000 species; subtropics and tropics.

ARDISIA Sw.

Ardisia bartlettii Lund., Contr. Univ. Michigan Herb. 7:37. 1942

Slender shrub or small tree, to 4 m tall, glabrous. Petioles mostly 5–10 mm long; blades elliptic, weakly or markedly acuminate, acute to attenuate and decurrent at base, mostly 6–15 cm long, 2.5–4.5 cm wide, entire or inconspicuously toothed, densely pellucid-lineolate, the midrib somewhat raised on both surfaces, the lateral veins obscure (at least when fresh). Racemes terminal, corymbose or subumbellate, lavender to red-violet throughout, short, mostly less than 5 cm long; pedicels slender, 4–13 mm long; flowers 5-parted, to 14 mm wide, pellucid-punctate; calyx lobes 5, densely punctate, ca 2 mm long, becoming green, persisting in fruit; corolla ca 7 mm long in bud, ca 12 mm wide when open, the lobes 5, ovate, spreading to recurved, free to near base; stamens 4–5 mm long; filaments short, stout, united with corolla at base; anthers yellow, pointed, 3–4 mm long, dehiscing by apical pores; style simple, slender, slightly exceeding stamens, shorter than petals, at first erect, becoming bent to one side. Drupes subglobose, fleshy, at first ± maroon, becoming purple-black at maturity, 6–8 mm diam; seed globose, ca 5 mm diam. *Croat 11271.*

Frequent in the forest. Flowers from May to September (sometimes from March), with the fruits maturing

KEY TO THE SPECIES OF MYRSINACEAE

Inflorescences axillary; flowers unisexual; corolla white . . . . *Stylogyne standleyi* Lund.
Inflorescences usually terminal; flowers bisexual; corolla reddish (white in *Ardisia fendleri*):
  Leaf blades bearing wide bands of minute brownish trichomes on underside along midrib; inflorescence branches, pedicels, and calyces finely reddish-tomentose . . . .
  . . . . *Parathesis microcalyx* Donn. Sm.
  Leaf blades lacking bands of brownish trichomes; inflorescences glabrous or lepidote:
    Leaf blades often more than 20 cm long, the margins pectinate-dentate with subulate teeth
    . . . .*Ardisia pellucida* Oerst.
    Leaf blades less than 20 cm long, entire:
      Inflorescences 10–20 cm long; flowers white; leaf blades more than 4.5 cm wide . . . .
      . . . . *Ardisia fendleri* Lund.
      Inflorescences ca 5 cm long; flowers purplish; leaf blades less than 4.5 cm wide . . . .
      . . . . *Ardisia bartlettii* Lund.

Fig. 444. *Ardisia fendleri*

Fig. 445. *Parathesis microcalyx*

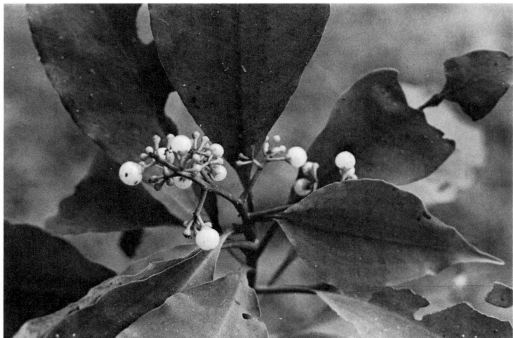

from June to October (sometimes to December).

Known only from Panama, from tropical moist forest in the Canal Zone, San Blas, and Darién.

Ardisia fendleri Lund., Wrightia 4:45. 1968

Shrub or small tree 2–8 m tall, ± glabrous. Petioles 5–10 mm long, marginate; blades elliptic or oblong-elliptic, acute to shortly acuminate, obtuse at base, 10–20 cm long, 4.5–9.5 cm wide, sparsely appressed-furfuraceous below when young, glabrous in age, with dense, opaque, orange punctations. Inflorescences terminal and axillary, 5–15 cm long, paniculate, the ultimate flower clusters subcorymbose; inflorescence branches and pedicels red; pedicels slender, 3–7 mm long; flowers 5-parted; sepals ovate, 1–1.5 mm long, punctate, persistent in fruit; corolla 3–6 mm long, white, lobed nearly to base, the lobes ± elliptic, 2–3 mm wide, lineate-punctate; stamens ca 4 mm long; anthers ca twice as long as filaments, dehiscing by apical pores; style simple, ca 5 mm long. Drupes globose, 6–7 mm diam, fleshy, green turning red then purple and finally black at maturity; seed globose, hard. *Croat 5560.*

Uncommon. Probably flowers and fruits throughout the year.

Known only from Panama, from tropical moist forest on the Atlantic slope of the Canal Zone and from premontane wet forest in Colón and Coclé.

See Fig. 444.

Ardisia pellucida Oerst., Vidensk. Meddel. 1861:130, t. 2. 1861

A. myriodonta Standl.

Shrub, usually less than 60 cm tall (to much taller elsewhere); stems, petioles, lower leaf surfaces, and exposed parts of inflorescences sparsely granular-puberulent. Petioles 1–1.5(2.5) cm long; blades ± elliptic to oblanceolate, acuminate, cuneate at base, 18–32 cm long, 6–10.5 cm wide, decurrent on petiole as marginal ribs, glabrous and shiny above, dull and often purplish below (with minute red-orange punctations when held to light), pectinate-dentate, the teeth subulate. Panicles small, terminal, violet, usually less than 4 (to 20) cm long, about as broad as long; pedicels 4–11 mm long, slender; calyx deeply lobed, 2–3 mm long, persistent, the lobes 4 or 5, ovate, acuminate, recurved, with orange pellucid punctations near apex; corolla 4–6 mm long, 4- or 5-lobed to near base, the lobes glandular-punctate near apex, violet-purple outside, light violet inside; stamens 4 or 5, 1.8–2(4) mm long, somewhat exserted; filaments fused to corolla at base; anthers yellow, equaling filaments, dehiscing laterally, beginning at apex; style simple, slender, 2–4(5) mm long. Drupes depressed-globose, ca 6 mm diam, purple-black at maturity; seed depressed-globose, 4–5 mm diam, brown. *Croat 11156, 14473.*

Occasional, in the old forest above the escarpment. Flowers from March to June (sometimes to August). The fruits mature from May to September.

Lundell (1971) reported that the plant may reach a height of 7 m.

Mexico to Colombia. In Panama, known from tropical moist forest in the Canal Zone, Panamá, and Darién and from premontane wet forest in Darién.

PARATHESIS Hook.

Parathesis microcalyx Donn. Sm., Bot. Gaz. (Crawfordsville) 48:295. 1909

Shrub, 1–2 m tall; older stems with a smooth, light-brown, corky periderm, the younger stems, lower surface of petioles, underside of midribs, and all exposed parts of inflorescences densely covered with brown, scalelike, stellate trichomes. Petioles 1–2.5 cm long, the upper surface glabrous and concave; blades elliptic to oblanceolate, acuminate, attenuate and decurrent at base, 6–14 cm long, 2.5–4.5 cm wide, densely pellucid-punctate, glabrous except for a narrow band of sparse stellate trichomes on either side of midrib below, thin, entire, the margins somewhat undulate. Panicles congested, terminal, subcorymbiform, 3–7 cm long; pedicels 2–4 mm long; flowers 5-parted, to 5 mm broad; inflorescence branches, pedicels, and calyces finely reddish-tomentose; calyx ca 1.3 mm long, lobed about midway, the lobes triangular, black-punctate, persisting in fruit; corolla magenta, pellucid-lineate, lobed to near base, the lobes slender, papillate-puberulent inside, ca 4 mm long, recurved at anthesis; stamens ca 2 mm long; filaments flattened, broadened somewhat and fused to corolla at base; anthers yellow, equaling filaments in length, dehiscing by longitudinal slits; ovary ovoid, furfuraceous at apex; style simple, slender, to 8 mm long. Drupes white to greenish, ± globose, 4–6 mm diam, smooth, shiny, sparsely stellate-pubescent (densely in juveniles); seed white, shiny, of same shape as fruit, a bit smaller. *Croat 9576, 12416.*

Abundant locally in the swampy area north of Zetek 300; unknown elsewhere on the island. Flowers from April to August, with the fruits maturing from July to September.

Nicaragua to Panama. In Panama, known from tropical moist forest in the Canal Zone and Darién and from tropical wet forest in Darién.

See Fig. 445.

STYLOGYNE A. DC.

Stylogyne standleyi Lund., Wrightia 3:110. 1964

Glabrous, dioecious shrub or small tree, usually to ca 3(7) m tall. Petioles ca 1 cm long or less; blades oblong-elliptic to lanceolate-oblong, gradually acuminate (the acumen often twisted), obtuse and decurrent on petiole at base, 12–30 cm long, 6–11 cm wide, minutely pellucid-punctate, thick. Panicles axillary, to 7.5 cm long, sessile or subsessile, pale, glabrous; pedicels 2–4 mm long; flowers 5-parted, white; staminate flowers ca 5.5 mm long; sepals 5, ovate, ca 1.4 mm long, 5-lobed to near base, asymmetrical; stamens to 6 mm long, exserted; filaments to 5 mm long; anthers ca 1.3 mm long, dehiscing

by introrse slits; ovary abortive; style more than 1 mm long. Pistillate flowers with the sepals 5, thick, ovate, densely orange-punctate, 1–1.4 mm long, imbricate, closing to protect young ovary after corolla falls; corolla ca 4.5 cm long, 5-lobed to near base, revolute, asymmetrical, orange-punctate; stamens ca 3 mm long, included; filaments slender, ± equaling anthers; anthers ca 1.3 mm long; ovary ovoid, glabrous; style 1.5–3 mm long, simple. Fruiting inflorescences with peduncles and pedicels red; drupes subglobose, 7–9 mm diam, violet to blue-violet or black at maturity; exocarp thin; mesocarp fleshy, sweet; seed globose, ca 5 mm diam. *Croat 4878, Foster 1457.*

Occasional to locally common in the forest. Flowers in the early dry season (December to February). The fruits mature from January to September, primarily from March to June.

Costa Rica to Colombia and Venezuela. In Panama, known from tropical moist forest in the Canal Zone, Bocas del Toro, Panamá, and Darién and from premontane moist forest in the Canal Zone and Panamá.

113. SAPOTACEAE

Trees with white sap. Leaves alternate, petiolate; blades simple, entire; venation pinnate; stipules lacking; T-shaped trichomes frequently present. Flower clusters or fascicles in axils or at leafless nodes; flowers bisexual or unisexual (*Pouteria stipitata* dioecious), actinomorphic; sepals 4–6 (8–12 in *Pouteria sapota* and *P. fossicola*), free or basally connate; corolla ± tubular, 4- or 5-lobed; stamens 4–6, inserted just below the corolla lobes, in *Pouteria* alternating with an equal number of staminodia, the staminodia alternate with the corolla lobes; anthers 2-celled, basifixed, dehiscing longitudinally; ovary superior, 4- or 5-locular and 4- or 5-carpellate; placentation variously axile; ovules 1 per locule, anatropous; style simple, obscurely 4- or 5-lobed. Fruits berries; seeds 1 or several (*Pouteria stipitata*), lacking endosperm (*Pouteria*) or with copious oily endosperm (*Chrysophyllum, Cynodendron*).

With the Lauraceae (58), one of the most difficult tropical families, probably with many undescribed species. Sapotaceae can most easily be distinguished by the milky sap and by the small gamopetalous flowers with the stamens borne at the apex of the short tube and usually alternating with staminodia (*Pouteria*).

The flowers are best suited to insect pollination.

The fruits are eaten by animals, principally arboreal mammals, or by birds, as in the case of the smaller-fruited species or those with moderately thin exocarps such as *Cynodendron panamense*. *Chrysophyllum* and *Cynodendron* are eaten by most frugivorous mammals (Carpenter, 1934; Hladik & Hladik, 1969; Goodwin & Greenhall, 1961; N. Smythe, pers. comm.). *Pouteria fossicola* has large fruits that have been found intact long distances from the tree, presumably rolling after falling to the ground.

About 35–75 poorly defined genera and about 800 species; tropics and subtropics.

KEY TO THE SPECIES OF SAPOTACEAE

Secondary lateral veins ± paralleling primary laterals; staminodia lacking:
  Blades coppery-brown-sericeous on lower surface; corolla lobes equaling or slightly exceeding the tube; stigma 7–12-lobed; fruits usually 5–10 cm diam at maturity . . . . *Chrysophyllum cainito* L.
  Blades glabrate or with relatively few whitish trichomes on lower surface; corolla lobes less than half as long as tube; stigma 5-lobed; fruits usually less than 3 cm diam . . . . *Cynodendron panamense* (Pitt.) Aubreville
Secondary laterals not paralleling primary laterals (often perpendicular to them); staminodia present (appearing to be lobes of corolla, the corolla thus appearing to have large and small lobes alternately disposed):
  Leaves mostly more than 16 cm long, usually more than 6 cm wide; sepals 8–12 or more; fruits more than 7 cm long at maturity:
    Primary lateral veins in 20–50 pairs; sepals often emarginate or more deeply lobed at apex; fruits ellipsoid, ovoid, or subglobose, brown, mealy-roughened on outer surface . . . . *Pouteria sapota* (Jacq.) H. E. Moore & Stearn
    Primary lateral veins in 15–25 pairs; sepals entire or scarcely emarginate; fruits obovoid, yellowish-green, the surface smooth and shiny (drying with wrinkles) or lenticellate . . . . *Pouteria fossicola* Cronq.
  Leaves less than 16 cm long, less than 6 cm wide; sepals 4–6; fruits less than 4 cm long at maturity:
    Leaf blades usually broadest well beyond middle; corolla 4-lobed, usually less than one-third of way to base, the lobes alternating with staminodia; staminodia similar to corolla lobes but very different from stamens; fruits more than 2.5 cm broad at maturity, yellowish . . . . *Pouteria stipitata* Cronq.
    Leaf blades broadest near middle; corolla usually 6-lobed almost to base of tube; staminodia very different from corolla lobes, similar to stamens; fruits less than 2 cm broad at maturity, purplish . . . . *Pouteria unilocularis* (Donn. Sm.) Baehni

CHRYSOPHYLLUM L.

Chrysophyllum cainito L., Sp. Pl. 192. 1753
Caimito, Star apple

Tree, to 20(25) m tall, ca 70 cm dbh; bark with white latex; wood pinkish, hard, heavy; branchlets and petioles densely brown-sericeous. Petioles 9–16 mm long; blades elliptic, acute or more often abruptly acuminate at apex, obtuse at base, 6.5–12 cm long, 3.5–6 cm wide, glabrous above except on midrib, densely brown-sericeous below, the midrib sunken above; major veins in 15–30 pairs, scarcely more prominent than the secondary lateral veins, forming an obscure collecting vein near margin. Fascicles axillary; pedicels ca 1 cm long; flowers cream-colored, numerous; sepals 5, ca 1.5 mm long; pedicels and outside of sepals densely brown-sericeous; corolla mostly 5-lobed about halfway, ca 4 mm long, sericeous outside; stamens equaling, opposite, and attached to corolla lobes, included; style to 0.5 mm long; stigma with 7–12 marginal lobes. Berries purple, subglobose, 5–10 cm diam; seeds several, oblique-obovate, flattened, to 2.5 cm long, the testa hard, lustrous. *Croat 4630, Foster 1221.*

Rare in the old forest; known also from the Laboratory Clearing and from the shore of Bat Cove. Flowers from July to September. The fruits mature during the dry season of the following year.

Probably native to the West Indies; commonly cultivated and naturalized in the lowland tropics from Mexico to northern South America; introduced into western Africa. In Panama, introduced throughout tropical moist forest and known from tropical dry forest in Panamá (Taboga Island), from premontane moist forest in Panamá, and from premontane wet forest in Chiriquí, Veraguas, Coclé, and Panamá.

CYNODENDRON Baehni

Cynodendron panamense (Pitt.) Aubreville, Mem. New York Bot. Gard. 23:215. 1972
Chrysophyllum panamense Pitt.
Cafecillo, Camito

Tree, to 20(25) m tall, to ca 40 cm dbh; outer bark moderately thin, fissured, soft; inner bark reddish, coarsely fibrous; sap white, usually not copious. Petioles 8–25 mm long; blades ± elliptic, acuminate, acute to obtuse at base, 10–25(33) cm long, 5.5–11(14) cm wide, glabrous above, glabrate to obscurely appressed-pubescent below with short T-shaped trichomes; lateral veins in 10–20 pairs, forming a loose collecting vein near the margin. Fascicles axillary; pedicels to 7 mm long; flowers 5-parted, ca 6 mm long, several to numerous; calyx ca 2 mm long, deeply lobed, the lobes ovate to suborbicular, imbricate; pedicels, calyx, and corolla pubescent with appressed T-shaped trichomes; corolla urceolate, to 5.5 mm long, the lobes short, deltoid to broadly ovate, ± spreading; stamens equaling and opposite the lobes, attached at base of lobes, ca 0.5 mm long; filaments fused to corolla except at deltoid apex; anthers deltoid-sagittate; pollen scant;

ovary sericeous; style 1–1.5 mm long; stigma 5-lobed. Fruits oblate-spheroid to ovoid, usually depressed at apex and also around pedicel, ca 2 cm diam, purplish-brown; exocarp thin, with some milky sap; mesocarp whitish, fleshy, sweet, tasty; seeds oblique-obovate, flattened, 1–2.5 cm long, brown, the hilum lateral, extending nearly the length of seed. *Croat 7825.*

Frequent in the forest. The species flowers twice per year. The principal flowering period is from June to October, especially in July and August, with the fruits maturing in the late dry season, March and April. A secondary and much smaller flowering period is from January to April, mostly in March and April, with the fruits maturing in the rainy season.

Costa Rica and Panama. In Panama, apparently more common on the Atlantic slope; known principally from tropical moist forest in the Canal Zone, Bocas del Toro, Panamá, and Darién, but also from tropical wet forest in Colón (Salúd). Reported from premontane wet forest in Costa Rica (Holdridge et al., 1971).

POUTERIA Aubl.

Pouteria fossicola Cronq., Lloydia 9:289. 1946
Grias megacarpa Dwyer

Tree, to 12 m tall, usually less than 12 cm dbh; bark thin, brown, flaky; sap milky; branchlets glabrous, with prominent leaf scars. Leaves clustered at apex of branchlets; petioles 1.5–4.5 cm long; blades obovate, usually acuminate, tapered to an obtuse or rounded base, 12–30 cm long, 6–13 cm wide, finely white-strigillose on major veins below and sometimes sparsely so on lower or upper surfaces, drying thin; major veins in 15–25 pairs, the secondary lateral veins mostly perpendicular to the primary lateral veins. Flowers densely clustered on stems below leaves; pedicels ca 5 mm long; sepals ca 8, sericeous except on margins, increasing in size centripetally, to 6 mm long; other flower parts unknown. Fruits obovoid, to 10 cm long and 7.5 cm wide, fleshy and yellow-green, the surface smooth and shiny with a few irregular lenticels; seed 1, obovoid, to 6.5 cm long, fleshy, the seed-coat firm, yellowish. *Croat 16636.*

Rare, known from the young forest north-northwest of the Tower Clearing; some fruits have been seen in the old forest south of Armour Trail 700 and on Orchid Island. Flowering season unknown. Mature fruits have been seen in May and July.

Known only from Panama, from tropical moist forest on BCI and from tropical wet forest in Colón (Salúd).

Pouteria sapota (Jacq.) H. E. Moore & Stearn, Taxon 16:383. 1967
Calocarpum mammosum (L.) Pierre; *P. mammosum* (L.) Cronq.
Mamey, Mamey de tierra

Tree, to 30(40) m tall, to 60 cm dbh; wood buff to reddish, hard and heavy; sap milky; branchlets brown-tomentose. Leaves alternate, clustered near ends of

branches; petioles 1–4.5 cm long; blades oblanceolate to narrowly obovate, usually shortly acuminate, narrowly acute at base, 10–40 cm long, 4–14 cm wide, glabrous above, pubescent below, becoming glabrate in age, the trichomes sparse to dense, brownish, appressed or shortly stalked, T-shaped; primary lateral veins in 20–50 pairs, the secondary laterals perpendicular to and connecting the primary laterals, more prominent than the closely spaced reticulate veins. Flowers subsessile, clustered at leafless nodes; sepals 8–12, spirally imbricate, ± orbicular, often emarginate or more deeply bilobed at apex, 2–6 mm long (the innermost longest), densely sericeous with T-shaped trichomes except on the thin, glabrous margin; corolla ± cylindrical, 6–10 mm long, 4- or 5-lobed to about the middle or less; stamens attached at top of tube; filaments 2–3.5 mm long; staminodia linear-lanceolate, 2–3 mm long; ovary densely ascending-sericeous, 5-locular; style 3.5–7 mm long, pubescent to about the middle; stigma 5-lobed. Fruits fleshy, ellipsoid, ovoid, or subglobose, sandpapery outside, 8–20 cm long, brown; mesocarp usually reddish, often milky; seed 1, ellipsoid, 5–6 cm or more long, brown and shiny, with a broad dull scar extending the length of seed. *Knight 1513.*

Collected once on the island. Elsewhere flowers from May to August, with the large fruits maturing almost a year later.

Mexico to northern South America; possibly native only in Mexico and along the Atlantic slope of Central America. In Panama, known from tropical moist forest on BCI and in San Blas, Chiriquí, Herrera, Coclé, and Panamá and from tropical dry forest in Panamá (Taboga Island).

Pouteria stipitata Cronq., Lloydia 9:265. 1946

Dioecious(?) tree, to 25(35) m tall, to 25 cm dbh; outer bark thin, brown, fissured vertically, often flaking off to expose areas of inner bark; inner bark white to tan, lighter than the wood; wood tan, hard; sap milky, sometimes with a strong, foul, pungent odor; stems appressed-pubescent to glabrate in age. Petioles appressed-pubescent, 1–2.5 cm long; blades oblanceolate to oblong-elliptic, acuminate, obtuse to cuneate at base, 9–16 cm long, 2–4.5(5.5) cm wide, glabrous in age except for appressed trichomes at base of midrib below; major lateral veins usually 5–7 on each side, the secondary veins irregularly anastomosing (only faintly prominulous when dried). Flower clusters dense, often at leafless nodes; pedicels 2–5 mm long; flowers greenish-white; pedicels and calyces densely pubescent, the trichomes short, appressed, T-shaped, less conspicuous on calyx; sepals 4(5), ovate to obovate, ca 2.7 mm long, thin, rounded at apex, ciliate, the inner pair thinner, its margins ± scarious; corolla tubular, 2.5–4 mm long, lobed, the lobes 4, large, alternating with 4 smaller staminodial lobes, all lobes ciliate; stamens 4, mounted on basal third of corolla tube, alternating with corolla lobes, sometimes directed inward and perpendicular to the corolla, the apices forming a cross; filaments ca twice as long as anthers, narrowed below anthers;

anthers narrowly deltoid, to 1.7 mm long, opposite corolla lobes and shorter than rim; ovary 4-locular, ovoid, very densely hirsute, the trichomes white, ascending, ca twice as long as ovary; style simple, ca twice as long as ovary; stigma simple, held at about the level of apex of corolla tube. Fruits obovoid, ellipsoid or obliquely ellipsoid, to 3.5 cm long and 3 cm diam, yellow, densely velutinous and with longer, stalked, T-shaped trichomes when juvenile, yellow and often glabrate at maturity; seeds 1–4, ellipsoid, ca 1.5 cm long. *Croat 6116, 10293.*

Frequent throughout the forest. Flowers in May and June, with the fruits maturing in August and September.

The species is probably dioecious, since many trees that flowered set no fruit while others set abundant fruit.

Costa Rica and Panama. In Panama, known from tropical moist forest on BCI and in Panamá (San José Island). Reported from premontane rain forest in Costa Rica (Holdridge et al., 1971).

Pouteria unilocularis (Donn. Sm.) Baehni, Candollea 9:273. 1942

Tree, to 25 m tall, mostly 20–50 cm dbh; outer bark thin, brown, fissured and often flaky (at least near base), fine-grained; inner bark tan to light brown, moderately thick, fine-grained; wood white; sap milky, with a foul, pungent odor. Petioles 5–10 mm long, canaliculate; blades oblong-elliptic, narrowly acuminate, acute to attenuate at base, 6–15 cm long, 2.5–5.5 cm wide, ± glabrous, stiff, usually folded along arched midrib, the reticulate venation prominulous. Flower clusters dense, of usually 15–25 flowers, often at leafless nodes; pedicels 4–5 mm long; flowers greenish-white, ca 2 mm long; sepals 4–6, thick, ovate to rounded, blunt at apex, 1.5–2 mm long; pedicels and sepals ± equally ferruginous-pubescent with minute T-shaped appressed trichomes, the sepals ± patulose-pubescent inside; corolla usually 6-lobed, ca 4.5 mm wide, 1.7–2 mm long, lobed nearly to the thickened base, the lobes equal, rounded at apex, ciliate (otherwise glabrous); stamens 5 or 6, shorter than and borne opposite the corolla lobes; filaments fused to corolla tube most of their length; anthers narrowly ovate, less than 1 mm long, pubescent; staminodia 5 or 6, broadly ovate, nearly sessile, borne at about the level of the anthers and alternating with the petals, pubescent at least marginally with trichomes like cilia of corolla lobes; ovary depressed-globose, densely woolly at base, glabrous apically; style to 1.3 mm long, narrowly conical, ± equaling length of style; stigma simple. Fruits ellipsoid, usually weakly oblique, 2.5–2.8 cm long, 1.5–1.9 cm diam, violet-purple or green heavily tinged with violet-purple at maturity; exocarp yellowish-green inside, hard, to 4 mm thick, with slight milky sap at periphery; seed 1, ellipsoid, ca 1.8 cm long, somewhat compressed, brown, surrounded by a thin, very tasty, sweet, gray mesocarp (the taste soon disappearing), the hilum extending nearly the full length of seed, more than 3 mm wide. *Croat 14874, 17052.*

Occasional, in the forest. Flowers in the early rainy season (June). The fruits mature in August and September.

I suspect that flowers of the species are unisexual, since the flowers investigated showed no sign of pollen.

Known from Guatemala, Belize, Costa Rica, and Panama; no doubt also occurring in intervening areas. In Panama, known from BCI and from tropical wet forest in Panamá (El Llano–Cartí Road).

See Fig. 446.

114. EBENACEAE

Trees. Leaves alternate, petiolate; blades simple, entire; venation pinnate; stipules lacking. Racemes short, axillary; flowers unisexual (dioecious), actinomorphic; calyx campanulate, 6-lobed, persistent and accrescent in fruit; corolla deeply lobed, the lobes 6, imbricate, white; stamens many; anthers 2-celled, introrse, dehiscing longitudinally; ovary superior, with several locules and carpels; placentation axile; ovules 2 per locule, anatropous; styles several, simple. Fruits berries; seeds many, with copious hard endosperm.

The family is represented in Panama only by *Diospyros*, which is distinguished by its unisexual, 6-parted flowers. Staminate flowers have numerous stamens with prolonged connectives. Flowers of *Diospyros* are probably insect pollinated. Fleshy fruits are no doubt animal dispersed.

Two genera and about 450 species; subtropics and tropics of Western and Eastern hemispheres.

DIOSPYROS L.

Diospyros artanthifolia Mart., Fl. Brasil. 7:7. 1856

Slender dioecious tree, ca 4–15 m tall, to ca 20 cm dbh, unbranched for much of its length; smaller branches long and drooping; most parts moderately to densely hirsute (especially dense on young parts), the upper blade surfaces, petals, and inner flower parts glabrous. Petioles 5 mm long; blades lanceolate to oblong-lanceolate, narrowly acuminate, obtuse to rounded at base, 6–14 cm long, 2.5–4 cm wide (larger in South America), glabrous above except on midrib. Staminate flowers in short axillary cymes, usually only one flower opening at a time; pubescence of branches, pedicels, and calyces crisp-villous; buds narrowly ovate; calyx campanulate, to 3.7 mm long, appressed-pubescent inside except near base, the lobes 6, irregular, triangular, extending one-third to one-half the way to base; corolla white, ca 1 cm long, lobed to near base, the lobes 6, somewhat imbricate, thick, 4 mm wide, with a longitudinal line of sparse pubescence somewhat off-center outside, this diminishing at about middle, the apex asymmetrical and recurved at anthesis; stamens ca 35–40, to 2.7 mm long; filaments less than 11 mm long, narrowed at apex, pubescent along inner side, the trichomes stiff, straight, translucent, extending entire length of connective; anthers apiculate at apex, the connective prolonged one-third to one-half the length of the thecae. Pistillate flowers in ca 5-flowered axillary cymes, subsessile or with pedicels to 5 mm long; corolla and calyx like staminate flowers; staminodia ca 6; ovary ovoid-conical, ca 4 mm wide and long; styles 4 or 5, ca 4 mm long, divergent, united only at base; locules 8 or 10; ovules 1 per locule. Berries subglobose, 8-celled, hispidulous, to 4 cm diam; seeds 8 or 10 or fewer by abortion, ca 20 mm long and 9 mm wide, ca 6 mm thick, black, the endosperm smooth; fruiting calyx scarcely accrescent. (Description of pistillate flowers and fruits taken from White, 1978.)

Rare, in the old forest. Seen in flower in May.

The flowers are open and unspecialized, but because of the rarity of the species and its dioecious condition, it might be expected to have some specialized pollinators.

The fruits are probably mammal dispersed.

Panama to Brazil. In Panama, known from tropical moist forest in the Canal Zone and Darién, from premontane wet forest in the Canal Zone, and from tropical wet forest in Colón.

115. LOGANIACEAE

Scandent shrubs, lianas, suffrutices, or annual herbs; stems often tendriled or spiny. Leaves opposite, petiolate; blades simple, entire (sometimes undulate); venation

---

KEY TO THE SPECIES OF LOGANIACEAE

Leaves pinnately veined:
  Spikes usually solitary (rarely 2); calyx lobes glabrous, ± equal; capsules smooth, the persistent
    base usually rounded on the ends . . . . *Spigelia humboldtiana* Cham. & Schlechter
  Spikes often 2–5; calyx lobes and keel scabrid on margins, often unequal; capsules muricate
    externally, the persistent base pointed on the ends . . . . *Spigelia anthelmia* L.
Leaves pliveined:
  Inflorescences axillary; leaf axils of major lateral veins below tufted with rufous trichomes;
    fruits yellow, ca 4 cm diam . . . . *Strychnos darienensis* Seem.
  Inflorescences terminal; leaf axils not tufted *or* the trichomes not rufous:
    Corolla tube not exceeding calyx; leaf axils tufted with short, white trichomes; larger stems
      armed with stout spines . . . . *Strychnos brachistantha* Standl.
    Corolla tube much longer than calyx; leaf axils not tufted:
      Plants with stems, leaf surfaces, inflorescence branches, and corolla conspicuously rufous-
        hirsute; fruits gray-green to bluish-green, 4–7 cm diam . . . . *Strychnos toxifera* Benth.
      Plants not rufous-hirsute, the leaf pubescence principally on veins below, the corolla gla-
        brous outside; fruits green becoming orange, 6–9 cm diam . . . *Strychnos panamensis* Seem.

pinnate (*Strychnos* pliveined); stipules interpetiolar (sometimes reduced to an interpetiolar line in *Strychnos*). Inflorescences various, in bracteolate, modified dichasia; flowers bisexual, actinomorphic; calyx 4- or 5-lobed; corolla ± funnelform, 4- or 5-lobed; stamens as many as and alternate with corolla lobes, epipetalous; anthers 2-celled, introrse, dehiscing longitudinally; ovary superior, 2-locular, 2-carpellate; placentation axile; ovules many, amphitropous or anatropous; style 1; stigma ± capitate. Fruits explosive, septicidal capsules (*Spigelia*) or hard-shelled berries; seeds many, with fleshy or bony endosperm.

Members of the family can be distinguished by a combination of opposite leaves with interpetiolar stipules or stipular lines and a superior, bilocular ovary.

The flowers are tubular and often very slender (especially *Strychnos*) and are probably pollinated by butterflies or moths. Some *Spigelia* are self-pollinated (H. Baker, pers. comm.).

Seeds of *Spigelia* are mechanically dispersed by means of explosively dehiscent capsules. The seeds of *Strychnos* are endozoochorous. They are embedded in a fleshy sweet matrix and are probably taken by arboreal frugivores large enough to break open the thick, usually hardened exocarp. Spider monkeys are very fond of *Strychnos* on BCI. Agoutis eat the partly eaten fruit dropped by the monkeys (N. Smythe, pers. comm.).

About 30 genera and 800 species; subtropics and tropics.

SPIGELIA L.

Spigelia anthelmia L., Sp. Pl. 149. 1753

   S. multispica Steud.

   Worm grass

Annual herb, small or to more than 1 m tall, nearly glabrous, simple or branched few times, leafless or nearly so except on ultimate segments. Leaves opposite-decussate and petiolate on stems, appearing whorled and usually sessile or nearly so at apex; stipules broadly deltoid, interpetiolar, ca 5 mm long; blades variable but mostly lanceolate-oblong, acute to acuminate, acute at base, 3–18 cm long, 1–6.5 cm wide, scabridulous above and on veins below. Spikes (1)2–5, terminal, simple or infrequently branched, 3–18 cm long; pedicels obsolete or to 1 mm long; flowers closely aggregated near apex, in all stages and eventually continuous with fruits on basal part of spike; sepals 5, linear-lanceolate, ca 3 mm long, thickened at base, keeled, the margins and keel scabrid; corolla narrowly funnelform, 6–15 mm long, white with a red-violet medial line below each lobe, strongly pleated, the lobes 5, acute, spreading; stamens 5, included; filaments fused to tube in basal half, markedly thickened near point of attachment, the free part arched inward; anthers yellow; style slightly exserted above throat, the apical half swollen, pubescent near apex. Capsules explosive, ca 4 mm long and 6 mm wide, 2-carpellate, sulcate medially, bilobed, compressed at right angles to septum, conspicuously muricate except at base, the persistent base pointed at either end; seeds brown, irregular, to 2

mm long, conspicuously tuberculate, mostly 4–6 per carpel. *Croat 7145.*

Uncommon in weedy areas of the Laboratory Clearing and much less frequent on forest trails; once locally abundant in the vicinity of the old orchidarium north of Kodak House. Flowers throughout the year, principally in the rainy season. The fruits mature quickly.

This species contains the toxic, volatile alkaloid spigeline (Blohm, 1962). When the capsule bursts open, the capsule walls become vertically compressed, and the wall and the seeds are usually thrown a distance of more than 1 meter.

Throughout American tropics; tropical Africa; Indonesia. In Panama, known principally from tropical moist forest all along the Atlantic slope and from Los Santos to Darién on the Pacific slope; known also from tropical dry forest in Los Santos and Panamá, from premontane moist forest in the Canal Zone, and from tropical wet forest in Colón, Panamá, and Darién.

Spigelia humboldtiana Cham. & Schlechter, Linnaea 1:200. 1826

Herb or suffrutex, usually less than 50 cm tall; stems often purplish, with 2 pubescent ribs beneath each stipule. Leaves opposite-decussate, those at apex crowded or whorled, connected by an interpetiolar line or sheath; stipules weblike between petioles, rounded or truncate at apex; petioles obsolete or to 1.5 cm long; blades ± lanceolate-ovate, acute to acuminate, obtuse to rounded and decurrent on petiole at base, 3–8(10) cm long, 2–3.5(4.5) cm wide, glabrous or with short stiff trichomes especially above. Flowers in 1 or 2 usually simple spikes; calyx to 4 mm long, glabrous, the lobes 5, slender, ± equal; corolla white, 1.5–2 cm long, slender, funnelform, abruptly flared above, the lobes 5, ± ovate, acute, often marginally tinged with purple; stamens 5, included, the free parts beyond flare of corolla tube curved inward; anthers held tightly against style, introrse; style densely pubescent at and above level of anthers, the short trichomes removing pollen from thecae as style elongates; nectar rather sparse, stored at base of tube around ovary. Capsules 4–6 mm broad, smooth externally, the persistent base rounded at both ends; seeds brownish, ovoid, to ca 1.5 mm long, densely muricate. *Croat 15572.*

Frequent along trails or tree-fall areas of the forest and in clearings in shady areas. Flowers throughout the year, principally in the rainy season (May to September). The fruits develop quickly.

Mexico to Argentina. In Panama, known from tropical moist forest in the Canal Zone, Bocas del Toro, San Blas, Chiriquí, Coclé, and Darién, from premontane wet forest in Chiriquí, and from premontane rain forest in Darién.

STRYCHNOS L.

Strychnos brachistantha Standl., Field Mus. Nat. Hist., Bot. Ser. 12:412. 1936

Scandent shrub, often to 30 m long; larger stems armed with stout spines (straight on juveniles, recurved on

Fig. 446.
Pouteria unilocularis

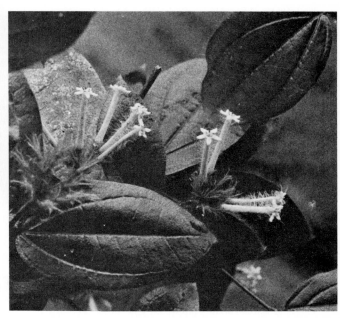

Fig. 448.
Strychnos toxifera

Fig. 447. *Strychnos panamensis*

adults); smaller stems sometimes square; branchlets puberulent with minute trichomes, glabrous in age; tendrils forked, the arms coiled like watchsprings. Petioles ca 3 mm long, puberulent with minute trichomes; blades lanceolate to narrowly elliptic, very long-acuminate, obtuse to rounded at base, 2.5–14 cm long, 1–4 cm wide, glabrous except for puberulent midrib below and tufted axils (with short white trichomes) of central pair of veins, entire, minutely and conspicuously undulate on margins, pliveined, the veins 3(5), the secondary lateral veins ± conspicuous in apical two-thirds of blade. Paniculate cymes terminal, usually hemispheric and bearing many flowers; calyx lobes (4)5, lanceolate to narrowly ovate, acute to acuminate, 1–2 mm long; corolla ca 2 cm long, (4)5-lobed almost to base, the tube ca 0.6 mm long, the lobes oblong, obtuse, densely villous inside, ca twice as long as tube; anthers 5, ca 0.6 mm long, glabrous. Fruits globose, 6–7 cm diam, orange-yellow at maturity; exocarp ca 6 mm thick, smooth; seeds 5–11, obliquely pyramidal, 12–17 mm long, the mesocarp yellowish, tasty. *Croat 14475.*

Apparently rare in the forest; collected once on Drayton Trail. Seasonal behavior uncertain. Flowers in June and July. Old fruits have been seen in September.

Mexico south along the Atlantic slope of Belize, Guatemala, and Nicaragua to Panama. In Panama, known only from tropical moist forest in the Canal Zone.

Strychnos darienensis Seem., Bot. Voy. Herald 166. 1854

Liana; stems divaricately branched, minutely puberulent to glabrate in age; tendrils stout, curled. Petioles and midribs sparsely pubescent; petioles mostly 2–7 mm long; blades usually narrowly elliptic or lanceolate, acuminate, rounded to acute at base, 6–17 cm long, 3–6 cm wide, ± thick, with brown trichomes on lower surface in axils of central vein pair, pliveined, the veins 3 or 5. Thyrses short, axillary, to 3 cm long; flowers 4(5)-parted; inflorescence branches and calyx lobes to 2 mm long; corolla yellowish, bearded inside tube, 2–3 mm long; filaments attached to corolla tube; style long-exserted, to 7 mm long, glabrous. Fruits globose or ovoid, ca 4 cm diam, yellow at maturity; seeds few, irregularly oblong, to 2 cm long and 1.5 cm wide. *Croat 15079, Foster 1465.*

Rare, in the forest. On BCI some flowers have been seen in December.

Nicaragua and Costa Rica south to the Guianas, Brazil, Ecuador, and Peru. In Panama, known from tropical moist forest in the Canal Zone, Veraguas, and Darién and from tropical dry forest in Panamá (Taboga Island).

Strychnos panamensis Seem., Bot. Voy. Herald 166. 1854

Canjura, Fruta de murciélago

Liana or climbing shrub; tendrils short-coiled, pubescent, becoming woody; stems sparsely long-pubescent to glabrous in age. Petioles 1–5(15) mm long, sparsely pubescent especially when young; blades lanceolate to elliptic,

usually acuminate, acute to rounded (rarely cordate) at base, 4.5–12 cm long, 1.5–4(5) cm wide, sometimes sparsely pubescent at base of midrib below, pliveined, the veins 3(5). Cymes pedunculate, terminal; pedicels obsolete or to 3.5 mm long; flowers (4)5-parted; sepals ± lanceolate, to 4 mm long, ciliate; corolla salverform, 8–24 mm long, the tube greenish-white, very minutely papillate outside, with a broad band of moniliform trichomes on apical third inside, the lobes white, slender, to 4.7 mm long, recurved at anthesis, densely papillate inside; stamens exserted to 3 mm above the rim; anthers ca 8 mm long, attached subbasally; style exserted about twice as far as stamens, expanded somewhat and flat at apex. Fruits globose, 1 to several on a stout woody stalk, at first blue-green, turning orange and usually 6–9 cm diam at maturity; seeds many, 2–3 cm long, irregular, embedded in a fleshy, sweet, orange pulp. *Croat 10229, 12595.*

Common along the shore; apparently less abundant in the forest. Usually flowering from April to August, chiefly from May to July, rarely late in the rainy season or in the early dry season. Full-sized fruits are seen by the late rainy season. They turn orange in the dry season and are removed usually before flowering begins again. Plants usually put on new leaves shortly before flowering.

Pacific slope of tropical Mexico to northeastern Venezuela and northern Colombia. In Panama, known from tropical moist forest in the Canal Zone, Chiriquí, Coclé, Panamá, and Darién, from tropical dry forest in Panamá (Taboga Island), and from premontane rain forest in Panamá (summit of Cerro Jefe).

See Fig. 447.

Strychnos toxifera Schomb. ex Benth., J. Bot. (Hooker) 3:240. 1841

Urari

Liana with long reddish trichomes on most parts; oldest stems glabrous; stems divaricately branched; tendrils coiled. Petioles to 6 mm long; blades ovate to elliptic, acuminate, rounded to subcordate at base, 6–20 cm long, 3–8 cm wide, pliveined, the veins 3(5). Cymes terminal, densely reddish-pilose; pedicels less than 5 mm long; flowers 5-parted; sepals lanceolate-linear, to 7 mm long; corolla white or yellow, salverform, the tube to 1.5 cm long, the lobes ca 2 mm long, densely bearded inside; anthers barely exposed at apex of corolla, ca 1 cm long; filaments attached to corolla tube; ovary subglobose, ca 1 mm diam; style exserted ca 2 mm; stigma capitate. Fruits globose, 4–7 cm diam, gray-green turning bluish-green; seeds usually 10–15, ca 2.5 cm diam. *Croat 6766, 7278.*

Occasional, in the forest. Flowers in the early dry season. The fruits mature in the rainy season.

This species is the source of toxiferin, one of the most potent curare alkaloids (Blohm, 1962).

Costa Rica to Ecuador and Amazonian Brazil. In Panama, known from tropical moist forest in the Canal Zone, San Blas, and Darién.

See Fig. 448.

KEY TO THE TAXA OF GENTIANACEAE

Plants saprophytic, lacking chlorophyll . . . . *Voyria*
Plants not saprophytic; leaves green:
    Herbs small, less than 1 m tall; leaves usually less than 4 cm wide; flowers reddish, 4-parted, less
       than 2 cm long; capsules less than 1 cm long . . . . *Schultesia lisianthoides* (Griseb.) Hemsl.
    Herbs large, coarse, more than 1 m tall; leaves usually more than 5 cm wide; flowers green,
       5-parted, 2–3 cm long; capsules usually more than 1 cm long . . . .
      . . . . *Chelonanthus alatus* (Aubl.) Pulle

116. GENTIANACEAE

Perennial or annual, erect, glabrous herbs (saprophytic and achlorophyllous in *Voyria*). Leaves opposite (sometimes crowded to base), sessile or short-petiolate (scalelike in *Voyria*); blades simple, entire; venation subpinnate (sometimes a few principal veins diverging from near base); stipules lacking. Flowers terminal or axillary, bisexual, mostly actinomorphic (somewhat zygomorphic in *Chelonanthus*), solitary or in simple or compound dichasia; calyx 4- or 5-lobed or 4- or 5-parted; corolla funnelform, 4- or 5-lobed, showy; stamens the same number as and alternate with corolla lobes, epipetalous; anthers borne on filaments or sessile (*Voyria*), 2-celled, introrse, dorsifixed, dehiscing longitudinally; ovary superior, 1-locular (or 2-locular by intrusion of the placentae), 2-carpellate; placentation parietal; ovules numerous, anatropous; style simple; stigma capitate or bilobed. Fruits septicidal, 2-valved capsules; seeds numerous (sometimes winged in *Voyria*), with copious endosperm.

Members of the family on BCI have little in common superficially but are all very distinctive.

Chelonanthus is probably pollinated by moderately large bees. Other pollination systems are not known.

Chelonanthus seeds are minute and are probably wind dispersed. *Voyria* also has minute seeds, but grows on the forest floor where wind dispersal is less likely. Possibly seeds are carried away by insects, birds, or isopods or are transported by water. Water currents on the forest floor capable of transporting such tiny seeds are not uncommon in the rainy season when the fruits mature.

About 70–80 genera and 900–1,000 species; worldwide.

CHELONANTHUS Gilg.

Chelonanthus alatus (Aubl.) Pulle, Enum. Pl. Surinam 376. 1906
    Arbol de mal casada

Herb, to 1–1.5(5) m; stems square, glabrous, the corners with thin ribs. Leaves opposite, sessile; blades broadly ovate, acute to short-acuminate at apex, attenuate at base, 4–18 cm long, 3–10 cm wide, thin. Inflorescences terminal, simple or compound dichasia, appearing racemose; flowers pedicellate, usually borne on one side of branches, subtended by ovate bracts 2–3 mm long; calyx ca 6 mm long, coriaceous, the lobes 5, blunt, imbricate, keeled; corolla campanulate, pale green, ± fleshy, weakly 2-lipped, 2–3 cm long, the lobes 5, short, acute but

reflexed so as to appear truncate; stamens 5, held tightly against the lower lip; filaments with a marked bend about midway, thick and flattened at base, fused to basal part of corolla tube; anthers shedding pollen before anthesis, sometimes adhering to each other and to style before anthesis; style broadly bilobed, scarcely exceeding stamens, appressed to anthers. Capsules ± oblong, brown at maturity, 1–1.5 cm long, the valves thick, held between calyx and strong persistent style base; seeds very numerous, minute, somewhat sticky, square or rectangular. *Croat 7258.*

Occasional in clearings, especially the Rear #8 Lighthouse Clearing; infrequent along the shore. Flowers principally in the rainy season, with the fruits maturing during the dry season.

The flowers are probably pollinated by medium-sized bees.

Capsule valves are hygroscopic, arching in dry conditions and allowing wind to pass through to carry away the tiny seeds and straightening again under wet conditions to close the gap.

Mexico to Brazil. In Panama, known principally from premontane wet forest in Colón, Chiriquí, Coclé, and Panamá; known also from tropical moist forest in the Canal Zone and Darién, from tropical wet forest in Darién, and from premontane rain forest in Chiriquí.

See Fig. 449.

SCHULTESIA Mart.

Schultesia lisianthoides (Griseb.) Benth. & Hook. ex Hemsl., Biol. Centr.-Amer. Bot. 2:348. 1882
    Sulfatillo

Erect herb, 7–90 cm tall, ± glabrous, unbranched or with few, strongly ascending branches mostly from near apex. Leaves opposite, sessile; blades mostly ovate, acute to short-acuminate at apex, attenuate to amplexicaul at base, 1.5–8(10) cm long, 0.5–4(4.5) cm wide, thin. Inflorescences terminal or upper-axillary, simple or compound dichasia, bearing few flowers; pedicels 1–10 mm long, subtended by bracts, the bracts paired, narrowly ovate, persistent, to 1.5(2) cm long; flowers 4-parted, lavender, 1–1.7 cm long; calyx lobed to near base, ca 6 mm long, the lobes ± lanceolate, strongly keeled; calyx and corolla withering but persisting at apex of fruit; corolla funnelform, shallowly 4-lobed, 1–1.7 cm long; stamens 4–5 mm long; style simple, 2–3 mm long. Capsules fusiform, 6–9 mm long, 2–3 mm diam; valves 2,

Fig. 451. *Nymphoides indica*

Fig. 449. *Chelonanthus alatus*

Fig. 450. *Voyria alba* at left, *V. tenella* at right

thick, held together at apex by withered corolla; seeds minute, irregularly rounded. *Croat 7783.*

Rare, in clearings along the shore. Flowers and fruits throughout the dry season (December to April).

Seeds are probably wind dispersed in the manner of *Chelonanthus alatus.*

Southern Mexico to northern South America. In Panama, known from tropical moist forest in the Canal Zone, Colón, Veraguas, Los Santos, Herrera, Panamá, and Darién, from tropical dry forest in Los Santos and Coclé, from premontane moist forest in Panamá, and from premontane wet forest in Chiriquí and Panamá.

VOYRIA Aubl.

Voyria alba (Standl.) L. O. Wms., Fieldiana, Bot. 31:413. 1968
Leiphaimos albus Standl.

Saprophytic herb, white, glabrous, 6–13 (20) cm tall, usually unbranched; stems to 1 mm thick; scales opposite, united, to 4 mm long, borne at the nodes. Flowers 5-parted, ca 1 cm long, in terminal dichasia of few flowers; calyx lobed about midway, ca 3.5 mm long, the lobes very slender; corolla narrowly tubular, the lobes narrowly triangular, to ca 3 mm long, spreading; stamens 5, extrorse, included; filaments connate to tube most of their length; anthers ca 2 mm long, united laterally in a ring surrounding style, the connective narrowed into slender appendages at the base; pollen white, moderately tacky; ovary elongate, slightly broader toward base, 5–6 mm long, gradually narrowed to the style; stigma discoid, ca 0.7 mm broad, much broader than style, slightly exserted from throat at anthesis. Capsules 6–8 mm long; seeds very numerous and minute. *Croat 11180.*

Rare, though sometimes locally abundant in moist areas of the forest, often in rich leaf litter. Flowers and fruits throughout the rainy season. Plants disappear in the dry season.

Panama and Colombia. In Panama, known from tropical moist forest in the Canal Zone, Colón, San Blas, and Panamá and from premontane wet forest in Colón.
See Fig. 450.

Voyria tenella Hook., Bot. Misc. 1:47, pl. 25, f. B. 1829
Leiphaimos simplex (Griseb.) Standl.

Saprophytic herb, to 20 cm tall, unbranched, white except for corolla lobes and base of tube, crosier-shaped when young; stems ca 1 mm thick; scales few, ovate, to 5 mm

long. Flowers pendent, erect at anthesis, solitary, terminal, 5-parted, to 1.7 cm long; calyx hyaline, 2.5–4 mm long; corolla tubular, to 2 cm long, the lobes spreading, to 5 mm long and 1.8 mm wide, violet to blue-violet, rounded at apex, the tube swollen and pale orange in vicinity of ovary, the apical part of throat with ascending short pubescence, with an orange ring at apex just below corolla lobes; anthers 5, ± sessile, ca 1 mm long, extrorse, blunt at base, connate into a ring, the connective thin, broader toward apex; ovary short-stipitate, narrowly oblong-ellipsoid, to 6.7 mm long (including stipe), narrowed to short style, glandular at base on either side, the glands 2, stalked, to 4.7 mm long, somewhat sunken into ovary; style and stigma together ca 2.7 mm long; stigma discoid and dome-shaped, held just above stamens and ca 1.7 mm below throat of tube. Capsules ellipsoid; seeds very numerous, winged, to 1 mm long. *Croat 6662.*

Infrequent to locally common in the forest, usually in rich, moist areas. Flowers and fruits throughout the rainy season. Plants disappear in the dry season.

Costa Rica to northern South America. In Panama, known from tropical moist forest in the Canal Zone, Bocas del Toro, Colón, San Blas, Panamá, and Darién and from premontane wet forest in Colón and Panamá.
See Fig. 450.

Voyria truncata (Standl.) Standl. & Steyerm., Publ. Field Mus. Nat. Hist., Bot. Ser. 23:78. 1944

Saprophytic herb, to 15 (20) cm tall, ± glabrous, unbranched or branched few times; stems enclosed by overlapping cauline scales at base; scales dull red, opposite, connate more than half their length, 3–5 mm long. Flowers solitary, 5-parted, 3–5.5 cm long; calyx ca 5 mm long, campanulate; corolla pale yellow to lavender, the narrow tube 3–4 cm long, the lobes spreading, ovate, 7–15 mm long; stamens included, 1–3 mm long; gynoecium to 3.6 cm long; ovary sessile, oblong, truncate apically, to 8–9 mm long, glabrous, eglandular; style filiform, to 2.7 mm long; stigma capitate-peltate, the margins sinuate. Capsules to 1.5 cm long, ca 5 mm diam; seeds wingless, numerous. *Croat 4296a.*

Rare, in rich, dark, moist forest. Seasonal behavior not determined. The few specimens, collected throughout much of the year, indicate that the species may not have a seasonal flowering behavior.

Guatemala, Nicaragua, and Panama. In Panama, known from tropical moist forest in the Canal Zone, Bocas del Toro, and San Blas and from premontane wet forest in Coclé (near El Valle) and Panamá (near Cerro Campana).

KEY TO THE SPECIES OF VOYRIA

Flowers few to many; flowers and plants white . . . . *V. alba* (Standl.) L. O. Wms.
Flowers solitary; flowers or cauline scales usually colored:
    Cauline scales many, crowded and overlapping at base of plant, enclosing stem; corolla usually
        more than 3 cm long . . . . *V. truncata* (Standl.) Standl. & Steyerm.
    Cauline scales few, inconspicuous; corolla usually less than 2 cm long . . . . *V. tenella* Hook.

117. MENYANTHACEAE

Aquatic herbs. Leaves solitary, alternate, petiolate; blades simple, entire to repand; venation palmate at base; stipules lacking. Umbels terminal; flowers bisexual, actinomorphic; calyx 5-parted; corolla deeply 5-lobed, showy, the lobes induplicate-valvate, alternating with sepals; stamens 5, epipetalous; anthers 2-celled, versatile, dehiscing longitudinally; ovary superior, 1-locular, 2-carpellate; placentation parietal; ovules many, anatropous; style simple; stigma 2-lipped. Fruits capsules, indehiscent or irregularly dehiscent; seeds many, with copious endosperm.

The flowers of *Nymphoides indica* are probably pollinated by small, diurnal bees or flies during the few hours they are open. The *Nymphoides indica* complex is polymorphic in the spectral qualities of its flowers under ultraviolet light. The yellow-flowered races of the West Indies have more contrasting patterns (ultraviolet light is absorbed by the center and reflected by the outer edges) than does the white-flowered race, which occurs in Panama (Ornduff & Mosquin, 1969). After flowering, the pedicels become deflexed and the fruits mature under water (R. Ornduff, pers. comm.).

After capsules ripen and dehisce, seeds surface and are water dispersed. They ultimately sink to the bottom and germinate. Some of the floating seeds are possibly eaten by aquatic birds and thus find their way to the bottom sooner than others.

Five genera and 40 species of swamp or aquatic plants; worldwide.

NYMPHOIDES Seguier

Nymphoides indica (L.) O. Kuntze, Rev. Gen. Pl. 2:429. 1891

N. humboldtianum (H.B.K.) O. Kuntze

Glabrous, succulent, aquatic herb from submerged rhizomes; stems petiole-like, fleshy, to 1.5 m long. Leaves solitary at apices of stems; blades ± orbicular, obtuse to rounded at apex, cordate at base, mostly 6–18 cm diam, usually violet-purple to purple on underside, thick. Fascicles dense, umbelliform, borne at nodes; pedicels to 7 cm long, fleshy; calyx lobes 5, lanceolate, acute and cupped inward at apex, ca 6 mm long; both pedicels and sepals glandular-punctate; corolla white, 1–1.8 cm long, 5-lobed for two-thirds its length, the lobes acute, spreading at anthesis, densely pilose inside, the trichomes antrorsely barbed; stamens 5, erect, 5–7 mm long; filaments fused to tube, alternating with tufts of trichomes near summit of tube; ovary narrowly conical, glabrous; style slender; stigma lobes flattened, the margins thickened and undulate. Capsules ellipsoid, 4–6 mm long, tardily dehiscent, eventually rupturing irregularly; seeds 10–18, orbicular, ca 1 mm diam. *Croat 8215.*

Occasional, in quiet lagoons at the edge of the lake, principally on the south side of the island. Probably flowers sporadically through the year.

The small seeds are probably dispersed by water currents.

Throughout the tropics of the world. In Panama, known principally from tropical moist forest in the Canal Zone and Panamá, but also from premontane wet forest in Coclé (El Valle) and from tropical wet forest in Colón (Miguel de la Borda).

See Fig. 451.

118. APOCYNACEAE

Trees, shrubs, lianas, vines, or herbs; sap milky. Leaves alternate, opposite, or whorled, simple; blades entire; venation pinnate; stipules sometimes present, minute; glands frequently in the leaf axils. Flowers bisexual, actinomorphic, 2 to many, in dichasial cymes, thyrses, corymbs or racemes; sepals 5, equal or unequal, sometimes with squamellae inside at base, basally connate, imbricate; corolla tubular, 5-lobed, showy, contorted in bud; stamens 5, attached part of their length to the corolla and alternate with lobes; anthers 2-celled, introrse, dehiscing longitudinally, free or adnate to stigma; gynoecium usually of 2 distinct, unicarpellate, unilocular ovaries, the placentation parietal; less frequently (*Lacmellea*, *Allamanda*) the gynoecium consisting of a syncarpous, 2-carpellate, 2-locular ovary, the placentation axile; style 1 per pistil (fused together on apocarpous species, later breaking apart with development of the fruit); stigma 1, usually massive and complex, usually pentagonal above, umbrella-shaped below. Fruits usually apocarpous, 2 follicles with many winged seeds, or rarely drupelike berries with few seeds (*Thevetia*, *Lacmellea*) or capsules (*Allamanda*).

Members of the family are very diverse in habit, but can be distinguished by a combination of milky sap and showy flowers contorted in bud. The flowers are somewhat twisted after opening also. All of the vines and the genus *Malouetia* may also be distinguished by their connivent anthers, which are glued to the stigma. Sterile vines are confused with Asclepiadaceae (119).

The flowers are mostly specialized with ample, well-protected nectar. The nearly closed throat of most species indicates they must be pollinated by long-tongued bees or possibly butterflies or hawkmoths. Others are perhaps pollinated by moths (see the discussion of *Odontadenia puncticulosa*). The flowers of *O. macrantha* and of *Stemmadenia grandiflora,* because of the larger, more open corolla, probably have different kinds of pollinators. Bees of the genus *Eulaema* were seen pollinating flowers of *Prestonia* in Ecuador (Dodson & Frymire, 1961) and of *Stemmadenia* in Costa Rica (D. Janzen, pers. comm.). R. Dressler (pers. comm.) reports that *Thevetia ahouai* and *Prestonia* are pollinated by euglossine bees. *Thevetia* is also visited by hummingbirds (H. Baker, pers. comm.). In Veracruz, Mexico, *Tabernaemontana* is pollinated by butterflies (D. Janzen, pers. comm.).

Most seeds are wind dispersed; those of *Lacmellea*

KEY TO THE TAXA OF APOCYNACEAE

Leaves alternate, ternate, or quaternate (i.e., more than 2 at a node):
  Leaves quaternate; flowers large, yellow; fruits less than 8 cm long, spiny .... *Allamanda cathartica* L.
  Leaves alternate; flowers minute and greenish or flowers whitish with corolla tube less than 4 cm long; fruits not spiny:
    Flowers whitish, more than 2 cm long; fruits red leathery drupes with 2–4 seeds .... *Thevetia ahouai* (L.) A. DC.
    Flowers minute and yellowish to greenish-yellow; fruits large woody follicles with disk-shaped, winged seeds:
      Leaves with major lateral veins 15–20, easily visible, 4–9 mm apart; calyx 4–5 mm long; corolla ca 10 mm diam; fruits usually less than 10 cm long; seeds to ca 7 cm diam .... *Aspidosperma megalocarpon* Müll. Arg.
      Leaves with major veins 30–40, moderately obscure, 2–6 mm apart; calyx 2–3 mm long; corolla ca 6 mm diam; fruits usually more than 11 cm long; seeds 9–10 cm diam ....*Aspidosperma cruenta* Woods.
Leaves opposite:
  Plants trees, herbs, or shrubs; fruits not long and slender (fusiform in *Malouetia guatemalensis*):
    Plants small herbs growing in open areas, rare; follicles slender, terete, less than 3 cm long ... . *Catharanthus roseus* (L.) G. Don
    Plants trees or shrubs:
      Corolla tube less than 5 mm long, the lobes slender, narrowly tapered to sharp apex; follicles fusiform, 10–15 cm long, more than 5 times longer than broad .... *Malouetia guatemalensis* (Müll. Arg.) Standl.
      Corolla tube more than 6 mm long, the lobes broad or blunt at apex; fruits less than 8 cm long, less than 2 times longer than broad:
        Calyx lobes unequal (the larger more than 1 cm long) *or* corolla tube less than 1 cm long at maturity; fruits ± reniform follicles, with many seeds, the seeds embedded in an orange matrix:
          Flowers pale orange, more than 4.5 cm wide; sepals unequal, the larger more than 1 cm long; fruits smooth outside, acuminate or pointed at apex............... .... *Stemmadenia grandiflora* (Jacq.) Miers
          Flowers white, less than 2.5 cm wide; sepals equal, less than 3 mm long; fruits scurfy on outside, rounded at apex .... *Tabernaemontana arborea* Rose
        Calyx lobes ± equal and less than 1 cm long *and* corolla tube more than 1 cm long at maturity; fruits drupaceous, with few seeds (less than 5):
          Corolla double (i.e., ± 10-lobed); fruits apparently not set; plants cultivated shrubs in the Laboratory Clearing .... *Ervatamia coronaria* (Jacq.) Stapf
          Corolla single, 5-lobed; fruits yellow-orange, ± ellipsoid, with a persistent style; plants tall forest trees .... *Lacmellea panamensis* (Woods.) Markg.
  Plants lianas or herbaceous vines; fruits slender follicles more than 5 cm long (oblong-ellipsoid in *Odontadenia macrantha*):
    Flowers less than 1 cm long; leaves frequently with pitlike axillary domatia below.... *Forsteronia*
    Flowers more than 1 cm long; leaves lacking pitlike axillary domatia below:
      Leaves with glands at or just below base of midrib or scattered along midrib on upper surface:
        Flowers greenish-white or greenish-yellow; inflorescences compound; calyx lobes more than 2 mm long; corolla tube less than 2.5 cm long; leaves ± obtuse to rounded at base, the glands at base of midrib only .... *Mesechites trifida* (Jacq.) Müll. Arg.
        Flowers yellow with reddish throat; inflorescences simple racemes; calyx lobes less than 2 mm long; corolla tube more than 2.5 cm long; leaves hastate or cordate at base, the glands sparse, along midrib .... *Mandevilla*
      Leaves lacking glands:
        Inflorescences usually of 2 (3) flowers; flowers white, with yellow throat; calyx lobes lacking squamellae inside .... *Rhabdadenia biflora* (Jacq.) Müll. Arg.
        Inflorescences of several to many flowers; flowers greenish, yellow, or orange (*Prestonia portobellensis* with white corolla lobes marked with orchid lines); calyx lobes bearing squamellae inside:
          Corolla more than 5 cm long; fruits either oblong-ellipsoid and more than 3 cm wide *or* fruits cylindrical-fusiform and plants not densely rufescent .... *Odontadenia*
          Corolla less than 4 cm long; fruits long and slender (fusiform in *P. ipomiifolia*, the plants densely rufescent) .... *Prestonia*

and *Thevetia* are probably mammal dispersed, and those of *Stemmadenia* and *Tabernaemontana* are probably dispersed by both birds and mammals. White-faced monkeys remove the pulpy mesocarp of *Thevetia ahouai* and *Stemmadenia grandiflora*, then spit out the seeds (Oppenheimer, 1968). *Lacmellea panamensis* is eaten by howler monkeys and tamarins (Carpenter, 1934; Enders, 1935). In tropical dry forest in Costa Rica, the fruits of *Stemmadenia donnell-smithii* (Rose) Woods. are taken by several bird species (McDiarmid, Ricklefs & Foster, 1977). Larger birds could easily feed on fruits of *Lacmellea* and *Thevetia* as well.

About 180–300 genera and about 1,300–2,000 species; all regions.

ALLAMANDA L.

Allamanda cathartica L., Mant. Pl. Altera 214. 1771

Liana or arching shrub, to 9 m tall; younger stems sparsely pubescent, becoming glabrous in age. Leaves quaternate, sometimes opposite or alternate apically, subsessile or borne on petioles to 5 mm long; blades obovate to oblanceolate, shortly and abruptly acuminate, obtuse to acute and decurrent on petiole at base, 8–16 cm long, 3–6 cm wide, glabrous or sparsely pubescent below especially on veins, with a submarginal collecting vein. Flowers few, in subterminal cymes; pedicels 2–7 mm long; bracteoles minute; calyx lobes 5, irregular, ± oblong-linear, acute, to 14 mm long; corolla yellow, 8–12 cm wide, the tube to 3.5 cm long, the throat broadened and as long as tube, the lobes 2.5–3.5 cm long, ± rounded, imbricate; stamens situated just above tube; filaments fused to tube, retrorsely barbellate near apex; anthers sharply triangular, free but with their apices held tightly together, the ring of anthers covered above by long, straight trichomes pointing inward from wall of corolla; ovary at first smooth, becoming spiny; style slender, its apex broadly expanded, plungerlike, situated inside the ring of anthers; stigmas 2, short. Capsules ± ellipsoid to obovoid, spiny, 5–8 cm long excluding spines, green at maturity, the spines 5–20 mm long; seeds numerous, disk-shaped, 2–2.5 cm diam, concentrically winged. *Croat 6573.*

Uncommon, on the shore. Plants may be found in flower throughout the year.

Distinguished by its large yellow flowers, quaternate leaves, and spiny capsules. All parts of the plant are reportedly poisonous (Blohm, 1962).

The ovary is surrounded by an active nectary. The broad cylindrical part of the style is also covered with a thick substance, which may attract the pollinator. In the plant's native habitat, the pollinator is unknown, but the organism would need a tongue 3.5–4 cm long to reach the nectar at the base of the constricted tube. Presumably the pollinator's head is thrust into the upper flared part of the corolla. Fruits are occasionally produced in central Panama, so some pollinator must be effective.

The winged seeds are no doubt to some extent wind dispersed, but they are not very good fliers.

Native to northeastern South America, but widely cultivated and escaped in the tropics of the Western and Eastern hemispheres. In Panama, known principally from tropical moist forest in the Canal Zone, Bocas del Toro, Colón, San Blas, Panamá, and Darién; known also from premontane wet forest in Chiriquí.

ASPIDOSPERMA Mart. & Zucc.

Aspidosperma cruenta Woods., Amer. J. Bot. 22:684. 1935

Alcarreto

Tree, to 35 m tall; trunk to 1 m dbh; outer bark coarse, lenticellate, thin; inner bark thick, granular, light brown, the sap yellow to red, more copious in branches than in trunk, sometimes whitish in juveniles, faintly aromatic; stems densely pubescent when young, glabrous in age, ± ribbed. Leaves alternate; petioles 2–4 cm long, glabrate; blades oblong-elliptic, acute at apex, acute to obtuse at base, 7–12(26) cm long, 2.5–4.5(7) cm wide, shiny above, dull and inconspicuously pubescent and sometimes pruinose below, the margins revolute. Thyrses terminal or upper-axillary, corymbose; pedicels 1–2 mm long; calyx 5-lobed, the lobes ca 2 mm long, densely pubescent; corolla tubular-salverform, greenish-yellow, glabrous except on tube inside, the tube 5-ribbed, ca 4 mm long, the lobes lanceolate, ca 1.5 mm long; stamens 5, attached near middle of tube; anthers free, to 1 mm long; filaments about as long as anthers; ovary compressed; stigma 1, capitate. Follicles paired, woody, flattened, shaped like the head of a golf club, to 16 cm long and 10 cm wide, bearing a thick rib medially, brown-tomentose; seeds numerous, flat, 8–10 cm diam, with a thin concentric wing. *Croat 8159.*

Uncommon, found on Pearson Peninsula and at Fairchild Trail 500. Flowers from May to July. The fruits are shed mostly in March and April. Individuals do not appear to flower during the year in which fruits are shed —perhaps they flower only in alternate years (R. Foster, pers. comm.).

Seedlings have both red and yellow sap, the red in the center of the stem. The species has been confused with *A. megalocarpon* Müll. Arg., but bears little resemblance to that species (Gomez Pampa, 1966).

Mexico to Colombia, Venezuela, and the Guianas. In Panama, a typical component of tropical moist forest (Tosi, 1971), known in the Canal Zone and Panamá; known also from premontane wet forest in Colón (Santa Rita Ridge) and from tropical wet forest in Colón.

Aspidosperma megalocarpon Müll. Arg., Linnaea 30:400. 1860

Pelmax, Bayalté, Huichichi, Volador, Ballester

Tree, 20–40 m tall, the trunk conspicuously fluted or involute, to 20–80 cm diam; outer bark gray, relatively smooth, becoming weakly fissured in age; inner bark pale yellow, granular, the sap white, copious; stems and petioles, especially stems, densely mealy-granular-pubescent when young, soon glabrous; stems soon conspicuously

lenticellate. Leaves alternate; petioles 6–15 mm long, weakly canaliculate; blades oblong-elliptic to narrowly elliptic, gradually acuminate at apex, acute to cuneate at base, (3.7)6–14 cm long, 2.2–7.5 cm wide, green above, pale and weakly glaucous below (drying grayish), glabrous or inconspicuously pubescent along midrib, the margins weakly recurved. Panicles axillary, 2–3 cm long, pubescent; pedicels 2–3 mm long; flowers fragrant, ca 1 cm diam; calyx grayish-green, tubular, 4–5 mm long, the lobes rounded, 1–1.5 mm long, pubescent on outside; corolla pale yellow, lobed to about the middle, the tube slender, pubescent inside, the lobes oblong-elliptic, asymmetrical; stamens 5, inserted at throat of tube; ovary ca 1 mm long, glabrous; styles 2, glabrous, united, ca 1.5 mm long; stigma simple. Fruits brown, paired, reniform to suborbicular, flattened, 5.5–12 cm long, densely and obscurely lenticellate, densely mealy-pubescent, becoming glabrous in age; seeds several, tan, orbicular, papery thin, to 7 cm diam, the funicle slender, attached near center of seed, to ca 2.5 cm long. *Garwood & Foster 443.*

Rare in the old forest; collected recently on AMNH Trail. Apparently flowers in the rainy season in Panama — it has been seen in bud in March. Fruiting occurs in the following dry season, in February and March. In Mexico the species is reported to flower from April to September and to fruit from August to February (Pennington & Sarukhan, 1968).

Aspidosperma megalocarpon was confused by Standley (1928) with *A. cruenta,* which is more common on BCI.

Mexico to Colombia (Chocó). In Panama, known from low elevations in tropical moist forest in the Canal Zone, San Blas (Puerto Obaldía), Chiriquí (Burica Peninsula), and Darién.

CATHARANTHUS G. Don

Catharanthus roseus (L.) G. Don, Gen. Hist. Dichl. Pl. 4:95. 1837(1838)

Lochnera rosea (L.) Reichb.

Herb, to 60 cm tall, usually woody at base, sparsely pubescent on leaves and young stems; sap milky. Leaves opposite; petioles 3–10 mm long, with axillary glands; blades obovate to spatulate, mucronulate, acute to attenuate at base, mostly 3.5–6 cm long, 0.7–3 cm wide, slightly succulent. Flowers 1–4, clustered near apex, on pedicels 1–3 mm long; calyx lobes 5, linear-lanceolate,

3–7 mm long, sparsely pubescent; corolla salverform, white to pinkish, the tube 2–2.5 cm long, pubescent inside at throat, the lobes 5, ± obovate, apiculate, 1.5–2.5 cm long; stamens included, attached near apex of tube; anthers free, ca 2 mm long; style 1; stigma 1, massive, pentagonal-umbraculiform. Follicles paired, slender, terete, 1.5–2.5 cm long, longitudinally ridged, with stiff, short trichomes; seeds not seen. *Croat 9389, Ebinger 204.*

Rare, in clearings. Flowers and fruits throughout the year.

Native to Madagascar, but cultivated and escaped throughout the tropics. In Panama, known from tropical moist forest in the Canal Zone and Herrera.

ERVATAMIA Stapf

Ervatamia coronaria (Jacq.) Stapf, Fl. Trop. Africa 4(1):127. 1904

Glabrous ornamental shrub, to 1.5 m tall; sap milky. Leaves opposite; stipules small, glandular, axillary; petioles short; blades elliptic, acuminate, acute to attenuate and decurrent on petiole at base, to 16 cm long and 6 cm wide, glossy. Cymes terminal or axillary; pedicels 1.3–2.5 cm long; flowers ca 4 cm wide, few; calyx lobes ovate, 2–4 mm long; corolla white, double, the tube to 2.5 cm long and greenish, the lobes rounded at apex; stamens in 2 series, the outer attached to the corolla, with anthers free, the inner forming a tube, with anthers reduced and sterile; style 1; stigmas 2. Fruits not seen. *Croat 4629.*

Cultivated in the Laboratory Clearing. Some flowers have been seen at various times throughout the year.

The plant apparently never sets fruit on BCI.

Native to India, but cultivated and escaped throughout the tropics. In Panama, cultivated in tropical moist forest in the Canal Zone and Panamá.

FORSTERONIA G. Meyer

Forsteronia myriantha Donn. Sm., Bot. Gaz. (Crawfordsville) 27:435. 1899

Liana; stems minutely puberulent to papillate when young, becoming glabrate and conspicuously lenticellate. Leaves opposite; stipules inconspicuous; petioles slender, 1–6 mm long, canaliculate, puberulent; blades narrowly elliptic to oval, acute to acuminate at apex, broadly acute to obtuse or rounded at base, 4–12 cm long, 2–5.4 cm

KEY TO THE SPECIES OF FORSTERONIA

Leaf blades usually conspicuously pubescent on lower surface, at least along midrib, usually lacking axillary pitlike domatia (or domatia inconspicuous); reticulate venation of lower surface obscure or not closely spaced (veins more than 1 mm apart) . . . . *F. myriantha* Donn. Sm.
Leaf blades usually glabrous on lower surface, with conspicuous pitlike axillary domatia; reticulate venation very closely spaced and easily visible (veins less than 1 mm apart):
Leaf blades to 9 cm long and 3.5 cm wide; midrib lacking glands near base . . . .
. . . . *F. peninsularis* Woods.
Leaf blades usually more than 10 cm long and 3.5 cm wide; midrib with 2 triangular raised glands near base . . . . *F. viridescens* S. F. Blake

wide, the upper surface glabrous but with a usually puberulent midrib, the midrib bearing 2 acute glands near petiole, the lower surface glabrous, the midrib and major lateral veins villous and puberulent, the vein axils sometimes with inconspicuous pitlike domatia; reticulate veins not raised but easily visible. Inflorescences terminal, ± hemispherical; pedicels to 1 mm long; bracteoles narrowly acute, to 3 mm long; bracteoles, rachises, and calyces densely puberulent-papillate to puberulent; flowers white, to 5.5 mm wide; calyx lobes ovate, broadly acute to obtuse, ca 1 mm long, the squamellae inside minute, many; corolla lobed to about the middle, to ca 4 mm long, glabrous to very minutely papillate on outside, villosulous inside on tube, the lobes oblong-ovate, spreading at anthesis; stamens to ca 3 mm long, adhering to style; anthers ca 1.5 mm long, glabrous, exserted at anthesis; ovary ovoid, apocarpous, minutely papillate-puberulent, ca 3 mm long; stigma 1. Fruits unknown. *Foster 4107.*

Apparently rare, in the forest. Reportedly flowers only in February in Panama; flowers elsewhere in April and May.

Guatemala to Panama. In Panama, known only from tropical moist forest in the Canal Zone.

Forsteronia peninsularis Woods., Ann. Missouri Bot. Gard. 22:215. 1935

Liana, ± glabrous; stems slender, usually less than 1 cm diam; outer bark thin, brown; sap thick, grayish-white. Leaves opposite; stipules inconspicuous; petioles 4–5 mm long, canaliculate; blades narrowly elliptic, acuminate and downturned at apex, acute at base, 2.5–8.5 cm long, 1–3.3 cm wide, glabrous or nearly so, often with pitlike axillary domatia below, the margins revolute; major lateral veins 4–6, obscure. Inflorescences terminal, ± flat-topped, dichasial cymes, bearing several to many flowers, the branches minutely puberulent, each subtended by an acute bract to 1.3 mm long; pedicels ca 1.5 mm long; flowers 5-parted, ca 7 mm long (excluding pedicel); calyx 2.3 mm long, thickest at base, the lobes acute, thickened medially, imbricate, inconspicuously ciliolate, bearing minute squamellae at base inside, these shorter than nectaries; corolla pale yellow, to 8 mm wide, minutely and inconspicuously puberulent except at base, the tube 2 mm long, as broad as long, with a densely villous ring at apex inside, the lobes spreading, to ca 3.7 mm long, acute at apex, ciliolate; stamens ca 4.7 mm long, exserted at anthesis; filaments fused to tube near base, weakly connate at least below anthers; anthers connate, ca 2.3 mm long, sagittate at base; ovary pubescent; stigma 1, ± conical, held tightly inside anthers, sticky; nectaries ovate, ± orange, ca 0.5 mm long, pubescent near apical margin, the nectar stored in corolla tube. Follicles paired, to 15 cm long, 3–4 mm diam, terete, glabrous; seeds oblong-linear, concave on one side, ca 7 mm long, pubescent, with a tuft of trichomes at apex. *Croat 14000, Foster 950.*

Apparently uncommon, in the forest. Flowers from the middle of the dry season to the early rainy season.

Individuals flower for at least a month, with some fruits of nearly mature size on flowering plants. Mature fruiting season unknown.

Belize, Guatemala, Honduras, and Panama, probably ranging all along the Atlantic slope of Central America. In Panama, known from tropical moist forest on BCI and from premontane wet forest in Colón (Santa Rita Ridge) and Los Santos.

Forsteronia viridescens S. F. Blake, Contr. Gray Herb. 52:80. 1917

Liana; stems ± reddish-brown, minutely lenticellate, ferruginous-pubescent when young; sap milky. Leaves opposite; stipules inconspicuous; petioles less than 5 mm long; blades ± oblong-elliptic, acute to bluntly short-acuminate, obtuse to acute at base, 7–15.5 cm long, 3–6.5 cm wide, moderately thick, ± glabrous, with glands at base of midrib above, often with large or small pitlike domatia in or near axils below. Thyrses axillary or terminal, 4–8 cm long, ± ferruginous-pubescent on all exposed parts; pedicels very short or to 1.5 mm long; bracteoles triangular, to 1.5 mm long; calyx lobes triangular, 1–2.3 mm long, bearing 2 squamellae inside; corolla subcampanulate, cream or white, the tube ca 1 mm long, the lobes ca 2 mm long; stamens ca 1.5 mm long, half-exserted; filaments free; anthers connate, ca 1 mm long; ovary papillate; stigma fusiform, acute, sticky and held tightly inside anthers. Follicles commonly solitary (the result of 1 aborting), slender, mostly 25–50 cm long, to 7 mm wide, bearing dense, short, ferruginous-tomentose pubescence; seeds slender, 1–1.8 cm long, with a tuft of brownish trichomes 2–4.5 cm long at apex. *Foster 950, Weaver & Foster 1614.*

Apparently rare, in the forest. Flowers from March to May in Panama, with the fruits developing to maturity in September and October. The species may flower twice per year, since it flowers chiefly during September in Belize.

Belize, Guatemala, and Panama, probably ranging along the Atlantic slope of Central America. In Panama, known only from tropical moist forest on BCI and in Panamá (San José Island), from premontane wet forest in the Canal Zone (Pipeline Road) and Panamá (Cerro Campana), and from tropical wet forest in Colón.

LACMELLEA Karst.

Lacmellea panamensis (Woods.) Markg., Notizbl. Bot. Gart. Berlin-Dahlem 15:622. 1941
L. edulis sensu auct. non Karst.

Tree, to 20 m tall, ± glabrous, sometimes armed with short stout spines; outer bark coarse, thin, hard; inner bark thick, tan; sap milky, abundant in trunk, branches, and fruits. Leaves opposite; petioles to 1 cm long; blades oblong-elliptic, acute to long-acuminate, obtuse and decurrent on petiole at base, 5–13 cm long, 2–4 cm wide. Flowers congested in axillary cymes; pedicels 2–4 mm long; calyx lobes ± rounded, 1–1.5 mm long, ciliate;

corolla narrowly tubular, white, drying burnt-orange, the tube to 3 cm long, puberulent inside, somewhat inflated just above base and at insertion of anthers near apex of tube, the limb to 1.5 cm diam, the lobes spreading, rounded to acute; anthers free, ca 4 mm long; style shorter than anthers at anthesis; stigma 1, cylindrical, papillose. Fruits broadly ellipsoid to obovoid, 2–3 cm long, yellow-orange, with a short persistent style to 3 mm long; exocarp ± leathery; mesocarp fleshy, sweet and tasty, with copious milky sap; seeds 1–4, usually 2, ellipsoid, usually flattened on one side, ca 1.5 cm long. *Croat 5373.*

Occasional in the forest, most abundant on the slope toward the tower; not seen in higher levels of the old forest. Flowers from March to August, mostly from April to June. The fruits mature from February to June, mostly from March to May of the following year. Juvenile fruits are common in the late rainy season. Leaves of this species are a favorite of sloths and as much as one-fifth of the foliage is sometimes cropped (E. Montgomery, pers. comm.).

Belize, Costa Rica, and Panama, probably ranging along the Atlantic slope of Central America. In Panama, a characteristic species in tropical wet forest (Tosi, 1971); known also from tropical moist forest in the Canal Zone and Darién.

MALOUETIA A. DC.

Malouetia guatemalensis (Müll. Arg.) Standl., J. Wash. Acad. Sci. 15:459. 1925

Mostly glabrous tree, to 15 m tall; sap milky. Leaves opposite; petioles ca 1 cm long; blades elliptic to somewhat ovate, acuminate, obtuse and decurrent on petiole and sometimes inequilateral at base, 8–22 cm long, 2.5–10 cm wide. Inflorescences axillary or occasionally terminal, usually branched, the clusters dense, umbelliform, the branches short, stout, woody; peduncles short, with short bracts; pedicels 3–5 mm long; flowers 5-parted; calyx 3–4 mm long, the lobes acute, stout, glabrate or puberulent; corolla white, ca 12 mm long in bud, minutely puberulent outside and on lobes inside, lobed to slightly beyond middle, the lobes slender, spreading at anthesis, the limb to 2 cm wide, the tube constricted just above sepals, 4–5 mm long; stamens included; anthers ca 1.8 mm long, connate, glued to stigma; ovary rounded above, minutely puberulent; style ca 1.5 mm long, club-shaped; stigma 1, ± cupulate; nectary conspicuous, deeply lobed, more than half as high as ovary, the lobes usually 5. Follicles paired, fusiform, to 15 cm long and 2.3 cm wide, glabrous; seeds compressed, ca 2 cm long. *Croat 5667, 13486.*

Occasional, at least along the margin of the lake. Flowers mostly from late January to the middle of April (elsewhere on the Atlantic slope of Panama, some flowers are also reported from September to November). Fruits in the late dry and early rainy seasons (April to October). The plant loses its leaves just before flowering and may begin flowering before new leaves are of full size; it is probably not leafless for a long period, putting on new leaves at once.

Throughout Central America. In Panama, known from tropical moist forest in the Canal Zone (Atlantic slope), Bocas del Toro, San Blas, and Darién and from premontane wet and tropical wet forests in Colón.

See Fig. 452.

MANDEVILLA Lindl.

Mandevilla subsagittata (R. & P.) Woods., Ann. Missouri Bot. Gard. 19:69. 1932

Vine; stems sparsely pubescent. Leaves opposite; petioles 2–7 mm long, sparsely pubescent; blades elliptic to oblong-elliptic, acute to acuminate, narrowly cordate at base, 3–8 cm long, 1–3.5 cm wide, glabrous or pubescent especially along veins and margins, sparsely glandular along midrib. Racemes axillary, 3–6(9) cm long; pedicels 3–5 mm long; bracts linear-lanceolate, 1–5 mm long, scarious; calyx lobes 5, narrowly triangular, 1–1.5 mm long, with a single squamella inside; corolla salverform, yellow with a red throat, the tube 2.5–3 cm long, curved and somewhat gibbous, the lobes 5, spreading, to ca 1.5 cm long; stamens included, attached near throat; anthers ca 1.5 mm long, connate, glued to stigma; nectaries 5, sometimes fewer and united, ca half as high as ovary; stigma 1, pentagonal-umbraculiform. Follicles paired, slender, 8–12 cm long, usually swollen at ± equal intervals; seeds oblong, ca 5 mm long, densely pubescent with a tuft of brownish trichomes ca 1 cm long at apex, the seminiferous area held within swollen areas of follicle. *Brown 191, White 132.*

Apparently rare; collected only twice on the island. Seasonal behavior uncertain.

Mexico to northern South America. In Panama, known from tropical moist forest in the Canal Zone and Panamá and from premontane wet forest in Chiriquí.

Mandevilla villosa (Miers) Woods., Ann. Missouri Bot. Gard. 19:70. 1932

Vine or slender liana; stems sparsely to densely pubescent. Leaves opposite; petioles pubescent, 1–2 cm long; blades elliptic to obovate, acute to acuminate, truncate

KEY TO THE SPECIES OF MANDEVILLA

Bracts of inflorescences slender, scarious, 1–5 mm long; calyx lobes ca 1 mm long . . . . *M. subsagittata* (R. & P.) Woods.
Bracts of inflorescences foliaceous, 10–30 mm long; calyx lobes 1.5–3.3 mm long . . . . *M. villosa* (Miers) Woods.

Fig. 452.
Malouetia guatemalensis

Fig. 453. *Mandevilla villosa*

Fig. 454. *Mesechites trifida*

to subcordate at base, 3.5–8 cm long, 1.5–4 cm wide, glabrous or sparsely pubescent above, glabrous to densely pubescent below, sparsely glandular along midrib. Cymes short, axillary; pedicels short, each subtended by a bract, the bracts lanceolate-caudate, foliaceous, ciliolate, 1–3 cm long; flowers 5-parted; calyx 1.5–3.3 mm long, turbinate, constricted below the short, triangular lobes; corolla pale yellow, to 3 cm long, the lobes spreading, to 2.3 cm long and 1 cm wide, reddish at base and rim of tube; stamens united to tube; anthers connate in a ring just below rim of corolla, 4 mm long, the free part of filament and the upper half of tube densely pubescent with stiff, white trichomes; style glabrous; stigma 1, pentagonal-umbraculiform, contained inside anthers; nectaries growing together, nearly as high as ovary. Follicles slender, ± moniliform, terete, ca 10–15 cm long, sparsely pubescent; seeds awl-shaped, ca 8 mm long. *Croat 10243, 11614.*

Occasional in clearings, especially the Lighthouse Clearing. Apparently flowers throughout much of the year, but most specimens have been collected in the rainy season, from May to December. Flowering plants may be found with mature fruits from the previous flowering, indicating that individual plants probably flower more than once per year.

Southern Mexico to Venezuela. In Panama, known from tropical moist forest in the Canal Zone, Bocas del Toro, Panamá, and Darién and from premontane wet forest in Coclé.

See Fig. 453.

MESECHITES Müll. Arg.

Mesechites trifida (Jacq.) Müll. Arg. in Mart., Fl. Brasil. 6(1):151. 1860

Liana, usually ± herbaceous, mostly glabrous; sap milky; stem with a prominent interpetiolar ridge. Leaves opposite; petioles 5–15 mm long (to 25 mm), usually with 2 pairs of subconical glands at apex, the lower pair sometimes fused into 1 large gland; blades elliptic to oblong-elliptic or ovate, acute to short-acuminate or mucronate at apex, obtuse to rounded at base, 6–11 cm long, 2–4(5.5) cm wide. Racemes short, branched or unbranched, cymose; pedicels to ca 1 cm long; flowers 5-parted, few open at any time; calyx lobes ca 3 mm long, blunt; corolla tube 1.5–2.5 cm long, constricted about midway, green apically, greenish basally, the limb spreading, ca 2.5 cm wide, the lobes green at base, white on margins and at apex; stamens included, 4.7 mm long; filaments fused to

tube, densely pubescent near apex; anthers ca 4 mm long, connate and glued to stigma, the connective pilose; ovary sparsely and minutely muricate; style 1; stigma 1, fusiform-umbraculiform; nectaries separate or growing together; nectar copious. Follicles paired, slender, 15–40 cm long, terete, less than 5 mm diam; seeds many, linear, ca 1 cm long, densely pubescent, with a tuft of trichomes ca 2.5 cm long at apex. *Croat 6698, 7243.*

Occasional along the margins of the forest, no doubt occurring in the canopy as well. Flowers throughout the year, principally in the late rainy to early dry seasons. Fruit maturity time uncertain.

Southern Mexico to northern South America. In Panama, known principally from tropical moist forest in the Canal Zone, Bocas del Toro, San Blas, Veraguas, and Panamá; known also from tropical dry forest in Panamá, from premontane moist forest in the Canal Zone and Panamá, and from premontane wet forest in Coclé and Panamá.

See Fig. 454.

ODONTADENIA Benth.

Odontadenia macrantha (R. & S.) Markg., Fl. Surinam 4(1):461. 1937

O. grandiflora Miq.; *O. hoffmannseggiana* (Steud.) Woods.

Liana, ± glabrous; sap milky, copious in stems; stems hollow, with an interpetiolar ridge. Leaves opposite, sometimes alternate near apex; petioles 1–2 cm long, sometimes purplish; blades ± elliptic, usually acuminate, obtuse to acute at base, 7–23(36) cm long, 4–10(17) cm wide, somewhat folded along midrib, the midrib ± arched, impressed above with a distinct central rib in the groove. Thyrses axillary or terminal, irregularly branched, glabrous; peduncle 4–9 cm long; pedicels 1–2(3) cm long; bracts minute; flowers ca 7 mm long; calyx lobes ovate, to 6 mm long, each with 1 or 2 squamellae inside; corolla funnelform, ca 6 cm long, orange-yellow, the lobes ca 3 cm long, spreading, tinged with red-orange inside near throat, the tube orange, bulbous, 5-lobed, ca 1 cm long, markedly constricted above stamens, the throat ca 3 cm long; stamens near base of tube; filaments and connective of the anthers densely short-pubescent; anthers sagittate, connate, glued to stigma; style 1; stigma 1, fusiform. Follicles paired, obovoid when young, becoming oblong-ellipsoid at maturity, to 22 cm long and 6 cm wide, rounded at base, narrowly rounded and weakly oblique at apex, glabrous, green tinged with brown; seeds many, the seminiferous portion

---

KEY TO THE SPECIES OF ODONTADENIA

Corolla yellow-orange; fruits more than 3 cm wide, glabrous; leaves not bicolorous, the midrib canaliculate above, with a raised center ridge, conspicuous reticulate veins lacking . . . . *O. macrantha* (R. & S.) Markg.

Corolla greenish; fruits less than 2 cm wide, densely pubescent; leaves bicolorous, the midrib canaliculate above but lacking a raised center ridge, the reticulate veins conspicuous . . . . *O. puncticulosa* (L. C. Rich.) Pulle

Fig. 455. *Odontadenia macrantha*

Fig. 456. *Odontadenia macrantha*

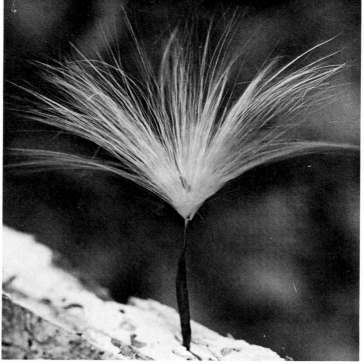

Fig. 457.
Odontadenia macrantha

terete, brown, ca 4 cm long, narrowed to a slender stalk and bearing a plume of silky white trichomes ca 7 cm long at apex. *Croat 8315, 16597.*

Common in the forest and along the lakeshore. Flowers throughout the dry season and in the early rainy season, with the fruits maturing during the dry season of the following year.

These widely dispersed, buoyant seeds are commonly found throughout the forest during the dry season. The trichomes of the plume are spreading to erect, with the seed usually suspended vertically. In *O. puncticulosa,* the trichomes of the plume are spread in all directions, but chiefly arched downward and below the seed. As a result, the alignment of the seed during flight is far less dependable than in *O. macrantha.*

Costa Rica south through tropical South America; Trinidad. In Panama, known from tropical moist forest in the Canal Zone, San Blas, and Panamá, from premontane dry forest in Herrera, and from premontane wet forest in Panamá.

See Figs. 455, 456, and 457.

Odontadenia puncticulosa (L. C. Rich.) Pulle, Enum. Pl. Surinam 383. 1906

Liana; sap milky, copious; stems glabrous, hollow and lenticellate in age, with a prominent interpetiolar ridge. Leaves opposite, glabrous; stipules 2, broadly ovate, softly puberulent, 2–3 mm long; petioles 1–2 cm long, stout; blades ± elliptic to ovate, shortly and bluntly acuminate, obtuse to subcordate at base on larger leaves, 12–24 cm long, 6–12.5 cm wide, bicolorous, the midrib above pale green, impressed, lacking a central rib; reticulate veins prominulous and very closely spaced below. Thyrses axillary or terminal; axes of inflorescences, pedicels, and calyces densely puberulent; pedicels to 3 cm long; flowers 5-parted; calyx lobes narrowly ovate, 4–9 mm long, each with several squamellae inside; corolla tubular, 5–6.5 cm long, the tube pale green, minutely puberulent, constricted ca 1 cm above base, the lobes white, spreading at anthesis, ca 1 cm long but with one side extended laterally 2–2.5 cm; stamens included, attached at base of corolla throat, ca 2 cm long; filaments and anthers pubescent; anthers 5–6 mm long, connate, glued to the fusiform stigma; stigma 1; nectaries growing together, covering ovary. Follicles paired, widely divergent, cylindrical, 11–16 cm long, 1.5–2 cm diam, densely pubescent with short erect trichomes; pericarp with a thin leathery outer layer and a thicker woody layer; seeds many, the seminiferous portion cylindrical to angulate or flattened, 13–20 mm long, obscurely striate, brown, narrowed only a few millimeters at apex; plume at apex light brown, to 6 cm long. *Croat 8633, Knight 6988.*

Occasional, in the forest, probably less common than *O. macrantha.* Individuals flower intermittently throughout the dry season. Some fruits have been seen dispersing seeds in August.

Because flowers are a pale greenish color, with a very long corolla tube, and fall profusely in the early morning, I suspect that the plant is pollinated by hawkmoths.

Costa Rica, Panama, the Guianas, and Brazil. In Panama, known from tropical moist forest in the Canal Zone, from premontane wet forest in Colón and Panamá, and from tropical wet forest in Colón.

See Figs. 458, 459, and 460.

PRESTONIA R. Br.

Prestonia acutifolia (Benth.) K. Schum. in Engler & Prantl, Nat. Pfl. 4(2):188. 1895

Vine or slender liana; stems with a prominent interpetiolar ridge, the younger stems sparsely puberulent, the older stems lenticellate. Leaves opposite; petioles 0.3–1 cm long, sometimes bearing triangular glands; blades elliptic, acute to acuminate, acute to obtuse at base, 5–10 cm long, 1.5–3.5 cm wide, glabrous above, glabrous to weakly short-pubescent below. Racemes simple, axillary, most parts inconspicuously short-puberulent; pedicels 3–8 mm long; bracteoles minute; flowers 5-parted; calyx lobes triangular, 1.5–2.3 mm long, with a single lacerate squamella inside; corolla salverform, greenish-yellow, with 5 epistaminal appendages deeply included and minute, the tube ca 1.5 cm long, the lobes obliquely obovate, 5–6 mm long; stamens barely exserted; anthers ca 4 mm long, glabrous, connate and glued to stigma; stigma 1; nectaries joined at base, about as high as ovary. Follicles not seen, no doubt paired, terete, and slender as in other *Prestonia. Croat 4401, Woodworth & Vestal 498(F).*

Collected recently, only at the Lighthouse Clearing. Seasonal behavior uncertain. Elsewhere in Panama, flowers from November to March.

Panama to South America. In Panama, known from tropical moist forest in the Canal Zone and Panamá, from premontane moist forest in the Canal Zone, and from premontane rain forest in Darién (Cerro Pirre).

KEY TO THE SPECIES OF PRESTONIA

Leaves less than 3.5 cm wide; calyx lobes less than 3 mm long ...... *P. acutifolia* (Benth.) K. Schum.
Leaves mostly more than 5 cm wide; calyx lobes more than 10 mm long:
  Plants conspicuously brown-pilose on most parts, rare .... *P. ipomiifolia* A. DC.
  Plants essentially glabrous or puberulent:
    Corolla lobes yellow or yellow-green; nectaries thick and fleshy, shorter than ovary; stems
      often conspicuously covered with large corky lenticels .... *P. obovata* Standl.
    Corolla lobes white with orchid markings; nectaries thin, ± translucent, longer than ovary;
      stems lacking conspicuous, corky lenticels .... *P. portobellensis* (Beurl.) Woods.

Fig. 458. *Odontadenia puncticulosa*

Fig. 459. *Odontadenia puncticulosa*

Fig. 460.
Odontadenia puncticulosa

Prestonia ipomiifolia A. DC., Prodr. 8:429. 1844

Liana, conspicuously brown-pilose on most parts; younger stems herbaceous. Leaves opposite; petioles 0.5–7 cm long (usually less than 2 cm long), sometimes with conspicuous glands at base; blades elliptic to ovate, acuminate, rounded to subcordate at base, 10–20(30) cm long, 4–8(14) cm wide. Racemes congested at leaf nodes; peduncles ca 1.5–2.5(5) cm long; floral rachis 2.5–3.5 cm long; bracts lanceolate, 1–1.5 cm long; pedicels 3–13 mm long; flowers 5-parted, dense; calyx lobed nearly to base, the lobes lanceolate, 8–15 mm long, glabrate inside, with a single triangular squamella ca 2 mm long inside each lobe; corolla yellow, salverform, to 3 cm long, lobed about one-third its length; stamens slightly exserted from throat, the epistaminal appendages prominent, about as long as orifice; anthers 5–6 mm long, glabrous, connate and glued to stigma; stigma 1; nectaries growing together, higher than ovary. Follicles paired, widely diverging, fusiform, ca 10 cm long, to 2 cm diam, densely rufous-pubescent. *Croat 12411, Shattuck 706.*

Rare, in the forest and along the shore. Flowers in the rainy season (May to November), with the fruits maturing in the dry season (December to February).

Panama, Colombia, and the Guianas. In Panama, ecologically variable; known from tropical moist forest in the Canal Zone, Los Santos, Panamá, and Darién, from tropical dry forest in Panamá (Taboga Island), from premontane moist forest in the Canal Zone (Ancón), and from premontane wet forest in Coclé and Panamá.

Prestonia obovata Standl., J. Wash. Acad. Sci. 15:459. 1925

Liana, essentially glabrous; sap milky; younger stems often with corky lenticels; older stems becoming totally covered with corky tubercles. Leaves opposite; petioles 1–3 cm long, swollen, with several sharp glands in axils; blades mostly obovate, acute to rounded and cuspidate at apex, rounded to obtuse at base, 14–28 cm long, 9–18.5 cm wide, coriaceous, glabrous to minutely puberulent; elsewhere venation sometimes prominent below, with a submarginal collecting vein. Corymbs axillary, long-pedunculate; peduncles and pedicels densely puberulent to glabrous; pedicels to 4 cm long; flowers 5-parted; calyx lobes triangular, 10–15(20) mm long, pubescent at least near base, with a single squamella inside each lobe; corolla salverform, yellow, 2.5–3.5 cm long, puberulent outside, with 5 white epistaminal appendages held above the prominently raised rim, the rim white to pink, the tube ca 2 cm long, pubescent near anthers; stamens included (just below rim); anthers ca 4 mm long, connate and glued to stigma; stigma 1; nectaries thick and fleshy, shorter than ovary. Follicles paired, 25–35 cm long, often covered with corky tubercles like those of stem; seeds ca 2 mm long, with tufted trichomes at apex, the trichomes 3.5–4 cm long. *Aviles 38.*

Rare, in the forest. Flowers during the dry season and in the early rainy season (January to September, especially in July). The fruits mature in the late rainy to early dry seasons (October to January), especially in the dry season.

Known only from Panama, from tropical moist forest in the Canal Zone, Chiriquí, and Darién, from premontane wet forest in Colón and Panamá, and from tropical wet forest in Panamá and Darién.

Prestonia portobellensis (Beurl.) Woods., Ann. Missouri Bot. Gard. 18:553. 1931

Liana; stems with a prominent interpetiolar ridge, at first coarsely pubescent, minutely lenticellate; sap milky. Leaves opposite, glabrous; petioles ca 1 cm long, glandular in axils; blades elliptic to oblong-elliptic or obovate, very abruptly acuminate and falcate at apex, obtuse to rounded at base, 9–20(30) cm long, 2.5–10(15) cm wide. Corymbs axillary, pedunculate; peduncles mostly 1–6 cm long; pedicels 1–2.5 cm long, usually puberulent; flowers 5-parted; calyx lobes 10–15 mm long, ovate to narrowly triangular, each bearing a deltoid squamella inside at base; corolla salverform, constricted just below lobes, the tube greenish, to 18 mm long, pubescent inside at level of stamen attachment, bearing 5 slender appendages above stamens, the tube with a raised yellow rim, the lobes obliquely obovate, white with orchid lines near margin, to 14 mm long, spreading at anthesis; stamens half-exserted; anthers 4–5 mm long, connate and glued to stigma; nectaries thick at base, growing together, the margins thin, toothed, higher than ovary. Follicles paired, often united at apex, 20–40 cm long and 6 mm wide, glabrous; seeds to 14 mm long, ± fusiform, flattened, glabrous but bearing an apical tuft of trichomes 2–3 cm long. *Croat 13497.*

Common along margins of the forest, possibly occurring in the canopy as well. May be found in flower during most of the year, principally in the dry season and early rainy season. The fruits probably mature mostly in the dry season.

Southern Mexico to Panama. In Panama, ecologically variable; known from tropical moist forest in the Canal Zone, Bocas del Toro, San Blas, Los Santos, Herrera, Panamá, and Darién, from tropical dry forest in Panamá (Taboga Island), from premontane moist forest in Panamá, from premontane wet forest in Colón, Chiriquí, Coclé, and Panamá, and from tropical wet forest in Colón, Coclé, and Panamá.

See Fig. 461.

RHABDADENIA Müll. Arg.

Rhabdadenia biflora (Jacq.) Müll. Arg. in Mart., Fl. Brasil. 6(1):175. 1860

Mangrove vine, Clavelito

Glabrous liana; sap milky. Leaves opposite; petioles 1–1.5 cm long; blades oblong to oblong-elliptic, acuminate to mucronate, obtuse to acute and decurrent on petiole at base, 4–10 cm long, 1–5 cm wide. Dichasia axillary or subterminal, usually of 1–3 flowers; peduncles 2–8 cm long; pedicels mostly to 1.5 cm long; flowers 5-parted; calyx to 1 cm long, lacking squamellae; corolla white, 5.5–7 cm long, the tube 1.5–2 cm long, retrorsely bearded at point of staminal attachment and below, the

Fig. 461. *Prestonia portobellensis*

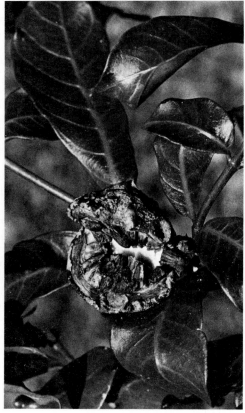

Fig. 462. *Stemmadenia grandiflora,*
dehisced fruit
with exposed seeds

Fig. 463.
Stemmadenia grandiflora

throat conical, pale yellow basally, 2–3 cm long, the lobes somewhat spreading; stamens included, attached at apex of tube proper; anthers ca 4 mm long, connate and glued to stigma, densely bearded apically; nectaries separate or slightly connate. Follicles slender, 10–14 cm long; seeds acicular, 2–3 cm long, with a tuft of long trichomes at apex. *Croat 6574, 8296.*

Occasional, in marshes on the south side of the island. The flowers are seen more or less throughout the year.

Southern Florida through Central America to northern South America; West Indies. In Panama, known principally from tropical moist forest in the Canal Zone, Bocas del Toro, Colón, San Blas, Chiriquí, Panamá, and Darién; known also from premontane moist forest in Panamá and from tropical wet forest in Colón.

STEMMADENIA Benth.

Stemmadenia grandiflora (Jacq.) Miers, Apoc. S. Amer. 75. 1878

Huevo de gato, Lechugo, Venenillo

Glabrous shrub or tree, to 7 m tall; sap milky, copious in all parts; stems grayish-brown, with a prominent interpetiolar ridge. Leaves opposite, the pairs sometimes unequal; petioles 3–10 mm long, (leaving a conspicuous scar); blades mostly oblanceolate, abruptly acuminate (sometimes falcate), acute to obtuse at base, 3–11 (16) cm long, 1.5–4 (6) cm wide, bicolorous. Cymes of few flowers; peduncles short; pedicels to 1 cm long, with a single small bracteole attached ca midway on pedicel; flowers 5-parted; calyx lobes 1.5–3 cm long, irregular, 2 narrow and flat, 3 folded with 1 or more margins reflected; corolla pale orange, the tube 2.5–3.5 cm long, the lobes spreading, ca 5 cm wide, to 2.5–3.5 cm long, much contorted; stamens included, attached about 2 cm above base of tube, held in a tight circle by thick pubescent ribs of corolla wall; anthers ca 4 mm long, free; style equaling stamens; stigmas 2, minute. Follicles reniform, usually paired, ca 5 cm long, smooth on the outside, the apex acuminate, the very thick valves splitting open along upper side to expose fleshy orange mesocarp; seeds many, oblong to rounded, 5–10 mm long, papillate and deeply grooved. *Croat 4223, 11979.*

Frequent in the younger forest, especially along the shore. Flowers and fruits throughout the year.

Seeds are individually borne on flaps of the mesocarp (or the funiculus is much expanded). The flaps probably flip the seeds out of the fruit to some extent.

Mexico to northern South America. In Panama, ecologically variable; known from tropical moist forest in the Canal Zone, Bocas del Toro, San Blas, Chiriquí, Veraguas, Los Santos, Herrera, Panamá, and Darién, from premontane dry forest in Los Santos, from tropical dry forest in Coclé and Panamá, from premontane moist forest in the Canal Zone and Panamá, from premontane wet forest in Coclé, and from tropical wet forest in Colón, Panamá, and Darién.

See Figs. 462 and 463.

TABERNAEMONTANA L.

Tabernaemontana arborea Rose in Donn. Sm., Bot. Gaz. (Crawfordsville) 18:206. 1893

Wild orange

Tree, to 20 m tall and 55 cm dbh, weakly buttressed, sometimes involuted, usually glabrous; bark with small vertical rows of lenticels; wood hard, heavy; sap milky, copious. Leaves opposite; petioles 2–11 mm long, with a prominent interpetiolar ridge (leaving a prominent scar), glandular in axils; blades mostly elliptic to obovate, short-acuminate, obtuse to acute at base, 5–18 cm long, 2.5–6 cm wide, dotted with brown glands below. Corymbs terminal; pedicels to 1 cm long; bracteoles 1.5–2 mm long; flowers fragrant, many, 5-parted; calyx lobes ovate, 1.5–3 mm long, each with several slender squamellae inside at base; corolla white, the limb ca 2 cm wide, the tube to 9 mm long, swollen in basal half, pubescent inside above the point of staminal attachment; stamens included, attached just above base of tube; anthers ca 3.2 mm long, free; style and stigma 1, 5-lobed, ca 2 mm long; nectaries lacking. Follicles 2, reniform, to 7.3 cm long and 5 cm wide; exocarp roughened, splitting at maturity to expose orange matrix and brown seeds; seeds oblong, many, to 13 mm long, longitudinally striate. *Croat 5242, 5419.*

Common in and mostly restricted to the older forest. Flowers mainly from March to June. Flowering time and intensity may vary from year to year. R. Foster (pers. comm.) reports that in one year trees flowered synchronously and heavily over a few weeks, but that in the next year there was intermittent flowering over several months. Time of fruit maturity is uncertain, possibly during the dry season (January to March). The fruits are eaten by white-faced monkeys in June (J. Oppenheimer, pers. comm.).

Guatemala to Panama. In Panama, known from tropical moist forest in the Canal Zone, Bocas del Toro, San Blas, Panamá, and Darién, from premontane wet forest in Panamá, and from tropical wet and premontane rain forests in Darién (Cerro Pirre).

See Fig. 464.

THEVETIA L.

Thevetia ahouai (L.) A. DC., Prodr. 8:345. 1844

T. nitida (H.B.K.) A. DC.

Huevo de tigre, Huevo de gato, Cojon de gato, Lavaperro

Shrub or small tree, usually to 6 m or less tall, ± glabrous; sap milky, copious. Leaves alternate; petioles very short or to ca 1 cm long, glandular in axils; blades usually oblanceolate, acute to acuminate, tapered to an acute base, 14–25 (30) cm long, 5–7 (8) cm wide, bicolorous, drying blackened. Cymes mostly terminal, irregular; pedicels to 3 cm long; flowers 5-parted; calyx lobes ovate, 5–7 mm long, spreading, with numerous slender squamellae inside at base; corolla cream, pale yellow or white, the tube

Fig. 464. *Tabernaemontana arborea*

Fig. 465. *Thevetia ahouai*

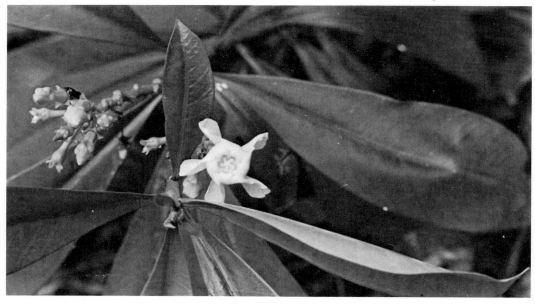

2.5–4 cm long, densely pubescent near stamens, closed near apex by 5 triangular appendages held opposite and just above stamens, the lobes spreading, twisted, to 2.5 cm long, rounded at apex; stamens included, attached near apex of tube; anthers ca 1.8 mm long, free; style with an expanded apex, the lower margin revolute; stigmas 2, thick. Fruits bright red, fleshy, broader than long, ca 3 cm long, 5 cm wide, and 3.5 cm thick; seeds 2–4, ± ovoid, rounded on one side, depressed on the other with a markedly raised margin. *Croat 5032, 8359.*

Frequent in the forest, especially the younger forest. The flowers and mature fruits have been seen throughout the year.

Southern Mexico to northern South America. In Panama, known from tropical moist forest in all regions, from tropical dry forest in Panamá (Taboga Island), from premontane moist forest in the Canal Zone and Panamá, from premontane wet forest in the Canal Zone, Panamá, and Darién, and from tropical wet forest in Colón.

See Fig. 465.

119. ASCLEPIADACEAE

Lianas, vines, or erect herbs with milky sap. Leaves opposite or ternate, simple; blades entire; venation pinnate; stipules lacking. Racemes or umbels terminal or axillary; flowers bisexual, actinomorphic, few to many; sepals 5, basally connate, usually imbricate; corolla 5-lobed, the lobes valvate, showy (in *Asclepias* a colorful corollar corona is present and associated with hornlike appendages); stigma and stamens united into a 5-lobed gynandrium; stamens 5; anthers 2-celled; pollen agglutinated and borne in specialized saclike pollinia, these connected to a uniting gland and to each other by translator arms, the pollinia each separately set into slits in the gynostegium; pistil 1; ovaries 2, superior or nearly so, unilocular; placentation ventral or parietal; ovules anatropous and pendulous; each ovary bearing a style but sharing a single, common, 5-lobed, often much-enlarged stigma. Fruits double follicles (commonly 1 aborted), dehiscing adaxially; seeds silky-comose, with little endosperm.

Members of the family are most easily confused with

KEY TO THE SPECIES OF ASCLEPIADACEAE

Plants erect herbs, to 1 m or more tall . . . .*Asclepias curassavica* L.
Plants vines, herbaceous or somewhat woody:
  Stems glabrous or nearly so:
    Corolla lobes glabrous (except for cilia on margin); petioles to 1.5 cm long . . . .
    . . . . *Marsdenia crassipes* Hemsl.
    Corolla lobes conspicuously pubescent on at least part of one or both surfaces:
      Leaf blades ovate to suborbicular, cordate at base, 3.5–12 cm wide; flowers to 6 mm diam
      . . . . *Cynanchum cubense* (A. Rich.) Woods.
      Leaf blades linear to oblong, usually not cordate at base, less than 3 cm wide; flowers more than 10 mm diam:
        Petioles less than 8 mm long; stems lacking interpetiolar ridges and glands . . . .
        . . . . *Sarcostemma clausum* (Jacq.) R. & S.
        Petioles more than 10 mm long on mature leaves; stems with prominent interpetiolar ridge and glands . . . . *Blepharodon mucronatum* (Schlecht.) Dec.
  Stems conspicuously pubescent:
    Vegetative parts with only 1 type of trichome, the trichomes whitish (sometimes deciduous):
      Leaf blades cordate at base; flowers with petals ca 1.5 cm long . . . . . . *Gonolobus allenii* Woods.
      Leaf blades obtuse to rounded at base, not cordate; flowers minute, less than 5 mm diam
      . . . . *Cynanchum recurvum* (Rusby) Spellm.
    Vegetative parts with both long, multicellular, usually brownish trichomes and shorter, glandular or puberulent pubescence:
      Petals light yellow to whitish or greenish-white, not reticulate-veined, with the margins strongly crisped; sepals ± equaling petals; gynandrium with large, inflated, dorsal appendages on anthers . . . . *Fischeria funebris* (Donn. Sm.) S. F. Blake
      Petals green-bronze to orange, often reticulate-veined, with the margins not crisped; sepals much shorter than petals; gynandrium lacking dorsal appendages on anthers:
        Corolla suburceolate, the tube somewhat constricted toward apex, often more than 10 mm long; leaves usually more than 12 cm long . . . . *Matalea trianae* (Trin.) Spellm.
        Corolla rotate, the tube usually not constricted toward apex, usually 5 mm long or less; leaves usually less than 12 cm long:
          Corolla usually reflexed; petals less than 6 mm long (usually 3–4 mm); blades usually broadest below middle, the basal sinus usually narrow and 10 mm or more deep
          . . . . *Matalea pinguifolia* (Standl.) Woods.
          Corolla spreading; petals more than 6 mm long (usually 12–14 mm); blades often broadest above middle, the sinus usually open and less than 5 mm deep . . . .
          . . . . *Matalea viridiflora* (G. Meyer) Woods.

members of the Apocynaceae (118), which share the same habit and have milky sap. In a few genera, such as *Prestonia* in the Apocynaceae, the resemblance of the Asclepiadaceae is very close. Distinguishing characters of the Asclepiadaceae are the pollen in pollinia and the corona derived from the filaments of the stamens. When a corona is present in the Apocynaceae, it is derived from the corolla.

Flowers are insect pollinated, probably by small bees. *Asclepias* is pollinated by butterflies (Heithaus, 1973). I have also seen *Sarcostemma clausum* visited by butterflies. Insects visit the flower in such a way that the sticky gland bearing the pollinia attaches itself to the insect's legs and is carred away to be placed on the stigma of the next plant.

Seeds are exclusively wind dispersed and are remarkably well adapted to remaining airborne with slight amounts of wind.

Some 130–250 genera and about 2,000 species; predominantly in the tropics.

ASCLEPIAS L.   (Milkweed)

Asclepias curassavica L., Sp. Pl. 215. 1753
  Niño muerto, Pasorín, Yuquillo, Mata

Erect herb, sometimes more than 1 m tall; stems ± glabrous below, pubescent above; sap milky. Leaves opposite or ternate; petioles 1–2 cm long; blades mostly lanceolate, gradually tapered to both ends, 5–18 cm long, 1–3.2(4) cm wide, glabrous or weakly pubescent. Umbels several, terminal; pedicels crisp-pubescent, 1–2 cm long; calyx green, hidden by reflexed petals, 5-lobed, 2–3 mm long; corolla usually orange-red, 6–8 mm long; corona segments (hoods) bright yellow, broadly ovate, 4–5 mm long, shorter than narrow inner horns of corona; nectar accumulating at base of corona or filling corona almost to brim. Follicles held erect, fusiform, 8–10 cm long and 1–1.5 cm wide, glabrous; seeds brown, ovate, flat, 5–6 mm long, topped by comose tuft ca 2.5 cm long. *Croat 6374.*

Occasional, in clearings. Flowers and fruits throughout the year. Individual plants flower more than once a year.

Plants contain the toxic glucoside asclepidadin (Blohm, 1962). Flowers are probably self-compatible (D. Spellman, pers. comm.).

Southern Florida and Mexico to Argentina; West Indies. A cosmopolitan weed in the tropics, introduced to the Old World as an ornamental. In Panama, known from most areas of tropical moist forest, from tropical dry forest in Herrera, from premontane moist forest in the Canal Zone, Los Santos, and Panamá, from premontane wet forest in Chiriquí, Coclé, and Panamá, and from tropical wet forest in Coclé (Atlantic slope).

BLEPHARODON Dec.

Blepharodon mucronatum (Schlecht.) Dec. in DC., Prodr. 8:603. 1844

Slender vine with copious milky sap, glabrous except on corolla and sometimes upper midrib of leaf; stems with nodes somewhat enlarged, with prominent interpetiolar ridge bearing several (usually 3 or 4) glands. Leaves opposite; petioles 1–1.5 cm long; blades ± oblong to elliptic, acuminate and downturned at apex, rounded or obtuse at base, 3–7.5 cm long, 1–2.5(3) cm wide, bicolorous. Cymes axillary, pedunculate, umbellate, shorter than leaves; peduncles usually shorter than pedicels; pedicels 1–2.5 cm long, slender; flowers 1–1.4 cm wide, 5-parted; calyx lobes oblong-ovate, ca 1.5 mm long, obtuse, glabrous, pale-margined; corolla greenish-white, spreading, deeply lobed, the lobes lanceolate-oblong to triangular, 5–6 mm long, obtuse, densely short-villous inside especially along margin, glabrous outside; gynostegium ca 3 mm long, the corona with 5 semivesicular sacs affixed to staminal tube; anthers terminating in an inflexed membrane; pollinia pendulous, ovoid. Follicles glabrous, fusiform, 7–9 cm long, ca 2 cm wide at widest part, rounded at base, broadest below middle, gradually tapered to narrow, blunt tip; seeds ca 5 mm long, narrowly spatulate, one side prominently papillate, the apical tuft of trichomes ca 4 cm long. *Croat 6375, 13159.*

Seen in Rear #8 Lighthouse Clearing. Flowers from August to October. The fruits mature in the early dry season (December to March).

Very similar to *Sarcostemma clausum,* but almost totally glabrous, with interpetiolar ridges and glands, and with fruits rounded at the base and tapered at the apex.

Mexico to South America. In Panama, known from tropical moist forest in the Canal Zone, Veraguas, Panamá, and Darién, from premontane moist forest in the Canal Zone and Panamá, and from premontane wet forest in Colón, Coclé, and Panamá.

See Figs. 466 and 467.

CYNANCHUM L.

Cynanchum cubense (A. Rich.) Woods., Ann. Missouri Bot. Gard. 28:213. 1941

Large herbaceous vine, climbing to ca 8 m in forest; cut parts with abundant milky sap, this drying chalklike; stems glabrous. Leaves opposite; petioles 4.5–8(12) cm long; blades ovate to suborbicular, acuminate and downturned at apex, cordate at base, 7–15 cm long, 3.5–12 cm wide (juvenile blades to 30 cm long, 20 cm wide), glabrous, with a series of platelike glands at base above, the sides of the blade held somewhat upward. Inflorescences axillary, racemose, becoming panicles of racemes, to 35 cm long, the branches puberulent; pedicels ca 5 mm long; flowers greenish, 5-parted; calyx lobes glabrous, 1.5–3 mm long; petals ± oblong, ca 3 mm long, flexed at anthesis, the apical half with crisp white trichomes on upper surface; corona shallowly 5-lobed, the lobes ca 6 mm long; gynostegium broadened and ca 2.7 mm wide at apex; pollinia pendulous. Follicles ovate-oblong, terete, acuminate, smooth, 15–21 cm long, 7–10 cm diam; seeds bearing a coma of silky trichomes. *Croat 11175.*

Rare, in the forest. Flowers in July and August. Fruit maturation time unknown.

Cuba, Panama, Colombia, and perhaps southern Mexico. In Panama, known from tropical moist forest in the Canal Zone, Coclé, and Panamá.

Fig. 466. *Blepharodon mucronatum*

Fig. 467. *Blepharodon mucronatum*

Fig. 468.
Matalea viridiflora

Cynanchum recurvum (Rusby) Spellm., Phytologia
25:438. 1973

Slender liana; stems less than 1 cm wide; copious milky
sap in all parts; stems, petioles, and leaf blades above
weakly strigose. Leaves opposite; petioles 5–9 mm long;
blades oblong-elliptic or oblong-ovate, acute to obtuse
and often apiculate at apex, obtuse to rounded at base,
2.5–5.8 cm long, 1.5–2.7 cm wide, weakly bicolorous,
with a pair of glands at base, weakly strigose above, the
pubescence below sparse, appressed, short; reticulation
indistinct, most major lateral veins faint. Inflorescences
axillary panicles of umbels, to 15 cm long; pedicels 2–4
mm long; flowers 5-parted, rotate-campanulate, yellow-
green, 3–4 mm diam, in well-spaced fascicles of usually
5–10 flowers; sepals ovate, rounded or obtuse at apex,
1–1.3 mm long, with a single, small, slender squamella
alternating with each sepal; corolla to 9 mm long, lobed to
middle or beyond, densely short-pubescent inside except
just below corona, the lobes spreading; corona less than
1 mm high, lobed below pollinia, the lobes short, opposite corolla lobes; gynostegium short, about as long as
corolla tube. Follicles unknown, but probably ± fusiform, long-acuminate at apex, 4–5 cm long. *Croat 15000.*

Rare; collected once above the escarpment north of
Zetek Trail 200. The plant has been collected in flower
only in June.

Costa Rica and Panama to Brazil (mostly Brazil). In
Panama, known only from tropical moist forest on BCI.

FISCHERIA DC.

Fischeria funebris (Donn. Sm.) S. F. Blake, J. Wash.
Acad. Sci. 14:293. 1924
  F. martiana var. *funebebris* Donn. Sm.

Suffrutescent vine; sap milky; stems and leaves sparsely
to densely spreading-hirsute and sparsely puberulent,
the hirsute trichomes septate. Leaves opposite; petioles
1–3.5(7) cm long; blades ovate-elliptic to elliptic, abruptly
short-acuminate at apex, cordate at base, 7–15(22) cm
long, 3.5–7(15) cm wide, the trichomes often more dense
and longer on veins especially below. Umbels axillary;
peduncles, pedicels, and calyces sparsely to densely hirsute; peduncles 2.5–10(22) cm long; pedicels 1–3.5 cm
long; flowers white, sometimes tinged with pink, the
buds ± globular; calyx lobes 5, lanceolate-linear, long-
attenuate, 7–15 mm long; corolla lobes 5, ± ovate, 5.5–8
mm long, undulate and densely ciliate, densely hirtellous;
corona ± entire, 1.5–2 mm long; the column smooth,
1.5–2 mm high; gynostegium 2.5–3 mm high; stamens
with the inflated portions suborbicular in surface outline, the staminal membranes usually triangular and
covering most of the stigma head; ovaries usually sparsely
pubescent. Follicles ± fusiform, broader near base, ca
17 cm long and 5 cm diam, blunt at apex; periderm thick;
seeds plumed at apex with long white trichomes. *Bailey
& Bailey 286, Shattuck 816, Foster s.n.*

Collected recently by R. Foster. Seasonal behavior not
determined. Flowers at least from March to June.

Southern Costa Rica to Colombia. In Panama, known

from tropical moist forest in the Canal Zone, San Blas,
Panamá, and Darién and from premontane wet forest
in Coclé (El Valle).

GONOLOBUS Michx.

Gonolobus allenii Woods., Ann. Missouri Bot. Gard.
27:333, f. 1. 1940

Vine; stems and leaves sparsely pilose. Leaves opposite;
petioles slender, 2–4 cm long; blades ovate-oblong, acuminate, deeply cordate at base, 5–10 cm long, 2–4 cm
wide, with 2 or 3 minute, fingerlike glands at base of
midrib. Flowers 2–4 in axillary, pedunculate clusters,
5-parted, 2–3 cm diam; peduncles 1–2 cm long; pedicels
2–3 cm long; calyx lobes linear-lanceolate, ca 7(10) cm
long, sparsely pilose outside; corolla greenish, lobed to
below middle, the lobes ovate-lanceolate, 1–2 cm long;
outer corona fleshy, the margin bearded. Follicles unknown. *Wetmore & Abbe 202.*

Collected once on BCI, but not seen in recent years.
This specimen had flowers in January.

Known only from Panama. The type is from premontane wet forest in Coclé (El Valle); the species is known
also from tropical moist forest on BCI and from tropical
wet forest in Panamá.

MARSDENIA R. Br.

Marsdenia crassipes Hemsl., Biol. Centr.-Amer. Bot.
2:337. 1882

Slender glabrous liana; stems terete. Leaves opposite;
petioles 1–1.5 cm long; blades ovate to oblong-elliptic,
abruptly short-acuminate, usually rounded at base, 8–15
cm long, 5–7 cm wide. Inflorescences axillary, subumbellate, bearing few flowers; peduncles 1–2 mm long; pedicels ca 1 mm long; flowers 5-parted, greenish-yellow;
calyx lobes ± ovate, ca 2 mm long, ciliolate; corolla
lobes ovate-oblong, ca 2 mm long, spreading, ciliolate,
with a narrow, whitish margin; corona segments ca 2.5
mm long. Follicles narrowly ellipsoid, ca 13 cm long and
3.5 cm wide; exocarp thick; seeds many, with tuft of long
white trichomes at apex. *Shattuck 341(?), Garwood 103.*

Rare. Seasonal behavior undetermined. Flowers at least
in June, July, and October, with the fruits maturing in
June and August.

Shattuck 341 is a sterile collection probably referable
to this species.

Known only from Panama, from tropical moist forest
in the Canal Zone (central region), San Blas, and Panamá.

MATALEA Aubl.

Matalea pinquifolia (Standl.) Woods., Ann. Missouri
Bot. Gard. 28:235. 1941
  Vincetoxicum pinquifolium Standl.

Herbaceous or suffrutescent vine; stems densely puberulent and hirsute, the hirsute trichomes fewer, straight,
jointed, to ca 3 mm long, often falling away in age. Leaves

opposite, glabrous or with short puberulence; petioles 2–6 cm long; blades ovate-cordate, acuminate, 6–10 cm long, 3–7 cm wide. Inflorescences brown-puberulent, axillary, subumbellate; pedicels to 3 cm long, longer than peduncles; calyx lobes 5, sharp, puberulent outside, glabrous inside, to 3 mm long; corolla lobes 5, reflexed, somewhat orange, reticulate-veined, puberulent outside, long-pilose inside along one margin and apex, 4–6 mm long; corona slightly shorter than gynostegium; gynostegium purple. Follicles ± lanceolate, long-tapered at apex, ± rounded at base, 10–13 cm long, 2–4.5 cm wide, pilose and long-tuberculate-muricate; seeds many, brown, flat, with long white trichomes tufted at apex. *Shattuck 612.*

Rare. Flowers mostly from October to December, but also in June. Mature fruits have been seen only in April.

Costa Rica to Colombia, Ecuador, Venezuela, the Guianas, and eastern Brazil; Trinidad. In Panama, known from tropical moist forest in the Canal Zone and Panamá, from premontane moist forest in Panamá, and from lower montane wet forest in Chiriquí.

Matalea trianae (Dec. ex Trin.) Spellm., Ann. Missouri Bot. Gard. 62:140. 1975

Suffruticose vine; stems, petioles, midribs of leaf blades, peduncles, and pedicels both densely puberulent and brown-hispid, the hispid trichomes multicellular, to 1.5 mm long. Leaves opposite; petioles 1–3(7) cm long; blades broadly ovate to suborbicular, acute to obtuse and long-cuspidate at apex, subcordate at base, (9)12–18 cm long, (4.2)5–14 cm wide, chartaceous, densely hirsute on both surfaces; major lateral veins in ca 6 pairs. Inflorescences umbelliform, bearing few to several flowers; peduncles 1–2 cm long; pedicels 7–12 mm long; flowers 2–3 cm long, somewhat fetid; calyx green, lobed to the base, the lobes lanceolate to ovate-lanceolate, acute at apex, 10–12 mm long, 5–8 mm wide, puberulent on outside, glabrous inside, ciliolate; corolla pale yellowish-green becoming brownish-green, salverform, to ca 4 cm diam, the tube 9–12 mm long, glabrous, constricted at the throat inside, the lobes ovate to suborbicular, 9–12 mm long, reticulate-veined, sparsely hispidulous on outside near apex, hispidulous-papillate on inside; corona of 5 distinct fleshy segments, the apices nearly truncate and equaling corolla tube or slightly exserted; gynostegium ca 6 mm long. Follicles attenuate-ovoid, ca 11 cm long, 3 cm diam, conspicuously 7-winged the entire length of fruit (based on Colombian specimen). *Steiner 81.*

Apparently rare; growing at the edge of the Laboratory Clearing, apparently a recent introduction to the island. Flowers principally in the rainy season, from June to December, especially from July to September. Mature fruits are known only from the dry season, from January to March.

Northern Costa Rica along the Pacific slope into western Colombia and into the Atlantic lowlands of Venezuela. In Panama, ecologically variable; known from tropical dry forest in Coclé and Panamá, from premontane moist forest in Panamá, and from drier parts of tropical moist forest in the Canal Zone and Panamá.

Matalea viridiflora (G. Meyer) Woods., Ann. Missouri Bot. Gard. 28:235. 1941

Vine; sap milky; vegetative parts, especially stems, with strigose to hirsute, multicellular, usually brownish trichomes and also with shorter glandular trichomes. Leaves opposite; petioles 1–4 cm long; blades ovate to oblong-ovate, acuminate, narrowly cordate to subcordate at base, 2.5–11 cm long, 1–5 cm wide, bicolorous, sparsely pubescent on both surfaces, the sinus generally open and shallow, mostly less than 5 mm deep. Cymes axillary, umbelliform; peduncles less than 2 cm long; pedicels mostly 1–2 cm long; flowers 5-parted, few, 2–2.5 cm wide; calyx deeply lobed, the lobes linear-lanceolate, ciliate, 4–5 mm long; corolla spreading, yellowish-green, 12–14 mm long, deeply lobed, glabrous on outside, puberulent inside especially near base, with black reticulate veins, bearing a mammillate process near gynandrium opposite sinus of lobes, the lobes lanceolate-ovate, acute; gynostegium rounded to obtusely 5-angulate, short. Follicles narrowly ovoid, ca 2 cm diam, abruptly tapered toward apex, narrowly 5-ridged, glabrous; seeds narrowly ovate, flattened, ca 5 mm long, brown, minutely papillate and ± viscid, with a sculpture depression on one side. *Croat 6102, 12882, Shattuck 727.*

Rare, in the Laboratory Clearing. Flowers mostly from August to December. Mature fruits have been seen in the middle of January and in February.

Costa Rica to Colombia, Ecuador, Venezuela, the Guianas, and eastern Brazil; Trinidad. In Panama, known from tropical moist forest in the Canal Zone, Bocas del Toro, Chiriquí, Panamá, and Darién, from premontane moist forest in the Canal Zone, and from premontane wet forest in Chiriquí and Panamá.

See Fig. 468.

SARCOSTEMMA R. Br.

Sarcostemma clausum (Jacq.) R. & S., Syst. Veg. 6:114. 1820

Funestrum clausum (Jacq.) Schlechter; *F. seibertii* Woods.

Chiefly herbaceous vine, branched many times, glabrous or sparsely pubescent with loose white trichomes easily wiped off; all cut parts with copious milky sap. Leaves opposite; petioles 5–7(9) mm long; blades linear to elliptic-oblong, acuminate, rounded or weakly cordate at base, 3–7 cm long, 0.5–1.5(3) cm wide. Inflorescences of stalked, axillary umbels, mostly bearing to 30 flowers; peduncles 6–9 cm long; pedicels 12–20 mm long; flowers 5-parted, white, 1–2 cm long (ca 2 cm wide when open); calyx lobes 5, short, ca 3 mm long, sharp-pointed; petals white or pinkish outside, 6–7 mm long, imbricate, ciliate, densely puberulent inside; hoods 5, white, encircling broad gynostegium; pollinia cylindrical, pale yellow, ca 1.2 mm long; style weakly bifurcate, slightly exceeding hoods; nectaries cuplike at base of and alternating with hoods, the nectar may fill the space between hoods (over the pollinia) by capillary action. Follicles 5–9 cm long, ca 1.2 cm thick; seeds numerous, flat, brown, less than 6 mm long, bearing a terminal tuft of fine trichomes 2–3 cm long. *Croat 11928.*

Occasional, on the shore, particularly on the south side of the island; often represented in *Annona-Acrostichum* formations. Flowers throughout the year. The fruits are rarely seen.

Throughout American tropics and subtropics. In Panama, known from tropical moist forest in the Canal Zone, Bocas del Toro, Herrera, Coclé, and Panamá and from premontane wet forest in Coclé.

See fig. on p. 14.

120. CONVOLVULACEAE

Terrestrial (sometimes temporarily growing in water), herbaceous vines or lianas (*Maripa*). Leaves alternate, petiolate; blades simple, entire or pinnately parted (*Ipomoea quamoclit* with the lobes filiform); venation pinnate or palmate; stipules lacking. Flowers bisexual, actinomorphic, solitary or in axillary, bracteate umbels or dichasia of few to many flowers; sepals 5, free, often unequal, imbricate, persistent in fruit; corolla funnelform or salverform, entire or 5-lobed, showy, pleated vertically; stamens 5, attached to base of corolla and alternate with lobes, unequal; anthers 2-celled, introrse or extrorse, dorsifixed, dehiscing longitudinally; nectariferous intrastaminal disk often present; ovary superior, 2- or 3-locular, 2–4-carpellate; placentation axile; ovules 1 or 2 per locule, anatropous; style 1; stigma subglobose or lobed, the lobes 2, hemispherical, minutely ridged, papillose-puberulent; nectary a fleshy ring around the ovary. Fruits loculicidal capsules or indehiscent; seeds usually 4, with hard, cartilaginous endosperm.

Convolvulaceae are recognized by being vines without tendrils and with large, showy, funnelform flowers with a capitate to globular stigma and stamens of usually unequal lengths. Fruits are more or less globular, have several seeds, and are contained within the persistent calyces.

Most species are bee pollinated. Most of the *Ipomoea* are visited by matinal bees. *Melitoma euglossoides* (Anthophoridae) is oligolectic to *Ipomoea*. In the Guanacaste area of Costa Rica *Aniseia martinicensis* was observed being visited principally by halictid bees (*Dialictus* and *Augochlora*) and by *Ancyloscelis armata*. The same genus also is visited by *Melitoma euglossoides* (Schlising, 1972). *Merremia*, which does not flower until late morning, as well as *Aniseia*, *Iseia*, and *Maripa*, are probably also bee pollinated. Some species of South American *Maripa* are visited by *Trigona* spp.

Ipomoea quamoclit has a slender tubular red flower. According to R. Cruden (pers. comm.), plants produce abundant nectar between 6:00 and 9:00 A.M., during which time they are visited by hummingbirds. Since the plants are self-fertile, the birds may merely cause self-pollination and only occasional outcrossing. The flowers are visited later in the day by a variety of butterflies. Most plants in cultivation are self-compatible and may set seed without apparent pollination. (I am indebted to D. Austin for much of this information.)

Few species have obvious means of dispersal. *Maripa panamensis* is endozoochorous, probably mostly by monkeys (Oppenheimer, 1968) and parrots. *Aniseia martinicensis*, which grows around the margin of the lake, is probably water dispersed. *Iseia* may also be in part water dispersed (D. Austin, pers. comm.), but it often grows away from bodies of water. According to Austin (pers. comm.) a number of species are probably dependent on man to a great extent for their dispersal, including *Merremia umbellata* and *Ipomoea quamoclit*. Seeds of both of these species are lightweight and might also be wind dispersed.

About 50–55 genera with 1,500–1,650 species; generally distributed.

ANISEIA Choisy

Aniseia martinicensis (Jacq.) Choisy, Mém. Soc. Phys. Genève 8:66. 1839

Herbaceous vine; stems sparsely pubescent with fine appressed trichomes, directed toward base of plant. Petioles 1–2.5 cm long; blades oblong-elliptic, acute and

KEY TO THE TAXA OF CONVOLVULACEAE

Leaves cordate at base:
  Flowers yellow .... *Merremia umbellata* (L.) Hallier f.
  Flowers not yellow, reddish-blue or white:
    Flowers solitary on long thickened pedicels; fruits transversely dehiscent capsules ....
    .... *Operculina codonantha* (Benth.) Hallier f.
    Flowers 5 or more in cymes; fruits longitudinally dehiscent or indehiscent .... *Ipomoea*
Leaves not cordate at base:
  Leaves pinnately parted into linear lobes; flowers red .... *Ipomoea quamoclit* L.
  Leaves entire:
    Flowers lavender; fruits with a prominent beak; flowers often more than 5 per inflorescence
    .... *Maripa panamensis* Hemsl.
    Flowers white; fruits lacking beak; flowers 1–4 per inflorescence:
      Sepals subequal; corolla 3–4 cm long; leaves rounded or emarginate at apex; fruits indehiscent, less than 1.5 cm diam .... *Iseia luxurians* (Moric.) O'Don.
      Sepals with 3 enlarged, winglike; corolla 2–3 cm long; leaves acute to obtuse and apiculate at apex; fruits dehiscent, ca 2 cm long ....
      .... *Aniseia martinicensis* (Jacq.) Choisy

Fig. 469. *Aniseia martinicensis*

Fig. 470. *Aniseia martinicensis*

apiculate at apex, acute at base, 5–10 cm long, 1.8–3.5 cm wide, glabrous above, appressed-pubescent below. Flowers axillary, mostly solitary or in dichasia of few flowers; calyx ca 1.5 cm long, with the inner 3 sepals winglike, broadly ovate, the outer 2 shorter, sharply pointed; corolla campanulate, pleated, 2–3 cm long, ca 2.5 cm wide when open, white, the triangular areas exposed in bud densely pubescent, the margin with 5 short apiculae; stamens equaling style, little more than half the length of the corolla; filaments pubescent in basal half and united with tube most of their length; pollen shed when stigma apparently receptive; ovary subglobose, ca 5 mm long; style ca 9 mm long. Capsules ovoid, ca 2 cm long, 4-valved; valves scarcely exceeding the persistent sepals, brown outside, silvery and shiny inside; seeds 4, ca 6 mm long, round in outline, ± wedge-shaped in cross section, brown, bearing short stellate trichomes, at least one margin fringed with a row of brown scales. *Croat 8308.*

Occasional, in marshy areas along the shore, particularly in protected areas. Flowers and fruits throughout the year. Mature fruits may develop while the plant is still flowering.

Seeds float. Their fringe of scales possibly aids in their establishment in a place suitable for germination. Because the dark seeds stand out against the silvery inner valve surfaces, they may be distributed by birds also, though it seems no part is edible unless the exocarp is crushed.

Schlising (1972) said the style is shorter than the stamens.

Southern Florida and Mexico to Colombia, Venezuela, and Peru; the Antilles. In Panama, known from tropical moist forest in the Canal Zone, Los Santos, and Panamá and from tropical dry forest in Herrera.

See Figs. 469 and 470.

IPOMOEA L.

Ipomoea batatas (L.) Poir. in Lam., Encycl. Méth.
Bot. 6:14. 1804
I. triloba sensu auct. non L.
Camote, Sweet potato

Vine; stems and petioles usually very sparsely long-pubescent. Petioles 3–12 cm long, square; blades ovate,

long-acuminate at apex, deeply cordate and palmately veined at base, 2.5–12(20) cm long, 2.5–10(18) cm wide, entire, dentate or deeply lobed, glabrous except the veins very sparsely pubescent. Cymes axillary, bearing several flowers (7–15); peduncles 4–9 cm long; pedicels bracteate, 1–6 mm long; sepals 5, oblong, 8–10 mm long, medially ribbed, narrowly long-acuminate, sometimes with long trichomes on outside, usually ciliolate; corolla to ca 3.5 cm long (longer elsewhere), the tube violet outside, red-violet inside, darker near base, glabrous except for white-tipped, glandular trichomes near base inside (also on base of filaments), the limb white, broadly spreading at anthesis (as all *Ipomoea* species); stamens 5, of unequal lengths, 1.3–2.5 cm long; ovary densely hispid; style ca 2 mm long; stigmas obscurely bilobed. Capsules ovoid, ca 5 mm long, splitting into 3 valves, glabrous or hispid on all but basal third; seeds 4, globose, ca 3 mm long, dark brown, glabrous. *Croat 4563, 7250.*

Occasional, in the Laboratory Clearing. Flowers and fruits from October to July, especially in the dry season.

The species is variable and elsewhere may have lobed leaves.

Probably native to Mexico; now cultivated throughout the tropics. In Panama, ecologically variable; known from tropical moist forest in the Canal Zone, Colón, San Blas, Panamá, and Darién, from premontane moist forest in Colón, Chiriquí, and Panamá, and from premontane wet forest in Chiriquí, Coclé, and Panamá.

Ipomoea phillomega (Vell.) House, Ann. New York
Acad. Sci. 18:246. 1908

Suffruticose liana. Petioles 2–9 cm long; blades broadly ovate-cordate, acuminate, 7–20 cm long, 5–20 cm wide, glabrate to moderately pubescent especially on veins below, maroon beneath when young. Flowers in long-pedunculate, dense, axillary clusters; peduncles to 21 cm long, stout; pedicels to 3 cm long, slender; sepals unequal, suborbicular to broadly ovate, 15–18 mm long, rounded at apex or emarginate, ciliate, the outermost 2 larger, enveloping the inner 3; corolla funnelform, red-violet, 4.5–5.5(8) cm long, glabrous, the tube with sides parallel, 1.1–1.5 cm wide, contracted at base below stamens, abruptly spreading at apex, the limb thin, to ca 4 cm wide, rolling inward upon slightest desiccation; stamens to 1.2 cm long, borne on thick, knoblike, short-

---

KEY TO THE SPECIES OF IPOMOEA

Blades pinnately divided into linear lobes; corolla bright red, less than 5 mm wide .... *I. quamoclit* L.
Blades not pinnately divided into linear lobes; corolla not bright red, mostly funnelform:
  Sepals acuminate at apex:
    Sepals ciliate; corolla usually ca 3.5 cm long; stems bearing sparse, long, deciduous trichomes
    .... *I. batatas* (L.) Poir.
    Sepals glabrous or very sparsely ciliate; corolla usually 4.5–5.5 cm long; stems glabrous .....
    .... *I. tiliacea* (Willd.) Choisy
  Sepals rounded at apex:
    Sepals glabrous, scarious on margins with a submarginal apiculum; leaves glabrous below,
      the cordate lobes often sagittate .... *I. squamosa* Choisy
    Sepals ciliate, not scarious; leaves pubescent below, the cordate lobes rounded ....
    .... *I. phillomega* (Vell.) House

glandular projections, these alternating with similar but smaller projections in a ring immediately below stamens; filaments unequal, red-violet; anthers oblong, basifixed and versatile, 5–7 mm long; style ca 2 cm long, weakly bifid at apex. Capsules globose, partially enclosed in persistent sepals, 1–1.5 cm diam; seeds 4, 3-sided, bearing long dense trichomes at base, with an "eye" at apex on inside. *Croat 16687.*

Occasional, in the forest. Flowers from July to September on BCI. The fruits may mature by late August and be present on the same inflorescence as flowers. Plants may flower continuously for at least 2 months. Austin (1975) reported that the species flowers from June through February elsewhere and perhaps flowers year-round.

Guatemala to Guyana and Peru; West Indies. In Panama, known from tropical moist forest in the Canal Zone, Bocas del Toro, Colón, Panamá, and Darién and from premontane wet and tropical wet forests in Veraguas (Atlantic slope), Colón, and Panamá.

See Fig. 471.

Ipomoea quamoclit L., Sp. Pl. 159. 1753

Cundeamor

Slender, glabrous vine. Petioles to 5 cm long; blades pinnately divided, to 9 cm long and 6 cm wide, the lobes linear, ca 0.5 mm wide. Flowers solitary or clustered 2–5 per cyme on long axillary peduncles to 9 cm long; sepals 5, unequal, 4–8 mm long, 3-ribbed, with mucronate tips; corolla red, 2.5–3 cm long, the tube less than 4 mm wide, the lobes 5, acute; stamens 5, unequal, the longest nearly equaling lobes; filaments pubescent in basal third, fused to tube near base; style equaling longest stamen, coiled and much longer than stamens in bud; stigma globular. Capsules ovoid, 6–8 mm diam; seeds 4, black, with patches of short trichomes. *Croat 4153.*

A rare weedy plant occurring in clearings. Flowers to some extent throughout the year, but especially in the rainy season and early dry season.

Native to tropical America, now widespread in the tropics. In Panama, known from tropical moist forest in the Canal Zone, Bocas del Toro, San Blas, Chiriquí, and Panamá and from premontane moist forest in the Canal Zone and Panamá.

Ipomoea squamosa Choisy in DC., Prodr. 9:376. 1845

I. morelli Duch. & Walp.; *I. vestallii* Standl.

Batatilla

Herbaceous vine, glabrate except for short pubescence on petioles and veins on upper surfaces of young leaves. Petioles to 5 cm long; blades ovate-cordate, gradually acuminate, cordate at base, 5–13 cm long, 2.5–7 cm wide, the lobes ± sagittate. Flowers in axillary, corymbiform clusters; peduncles to 9 cm long, stout; pedicels less than 1 cm long; sepals glabrous, suborbicular, rounded at apex, the 2 outer ones 3–5 mm long, the 3 inner ones 5–10 mm long, the margin scarious, forming a submarginal apiculum; corolla funnelform, 5–8.5 cm long, lavender outside; dark red-violet inside; stamens of irregular lengths, 2–3 cm long at anthesis; filaments fused to lower 1 cm of tube and glandular-villous near base of free part;

anthers introrse, to 4.5 mm long, shedding pollen in bud; style ca 2 cm long, held above anthers in bud but well below anthers at anthesis; stigma white, bilobed; nectary ringlike; nectar copious. Fruits not seen. *Croat 7463.*

Infrequent, in clearings. Flowers in the dry season (December to March).

Mexico to Peru. In Panama, known from tropical moist forest in the Canal Zone and Panamá and from premontane moist forest in Panamá.

See Fig. 472.

Ipomoea tiliacea (Willd.) Choisy in DC., Prodr. 9:375. 1845

Vine, glabrous except sometimes with long trichomes along midribs of blades. Petioles 3–9 cm long; blades ovate-cordate, acuminate, 6–11(15) cm long, 5–8 cm wide. Flowers in cymose, axillary, pedunculate clusters; peduncles stout, to 6 cm long; pedicels 0.5–1.5 cm long; sepals 5, ovate, ca 1 cm long, acuminate, glabrous (or 1 pubescent), 1 sepal often curling around ovary after flower falls; corolla 4.5–5.5(6) cm long, lavender to whitish outside, red-violet on tube inside, the rim broadly flaring, the lobes 5, minute, apiculate; stamens 5, all of different lengths, 1.5–2.5 cm long; filaments pubescent, fused to tube near base; anthers opening in bud; style slightly exceeding stamens; stigma lobed, the lobes 2, hemispherical. Capsules ± globose, 6–10 mm diam, splitting into 4 valves; exocarp thin; seeds 4, dark brown, glabrous. *Croat 6869, 7002.*

Apparently uncommon; collected at Burrunga Point at the edge of the clearing. Flowers from July to February. The fruits probably mature in 1 or 2 months. Old capsules may hang on for a long time.

When the level of the lake is high, the stems may be standing in water.

Tropics of Western and Eastern hemispheres. In Panama, restricted to the Atlantic slope; known from tropical moist forest all along the Atlantic slope and from tropical wet forest in Veraguas, Coclé, and Colón.

ISEIA O'Don.

Iseia luxurians (Moric.) O'Don., Bol. Soc. Argent. Bot. 5:77. 1953

Liana, growing over the canopy, the younger parts herbaceous; stems inconspicuously appressed-pubescent to glabrate. Petioles 1–2.5 cm long; blades ± oblong to ovate, rounded to emarginate and apiculate at apex, rounded to subcordate at base, mostly 4–12 cm long, 2.5–9 cm wide, glabrate or inconspicuously appressed-pubescent on lower surface (especially on younger leaves), the older leaves undulate. Cymes axillary; flowers numerous, 1–4 on each branch, 4–5 cm long at anthesis, densely appressed-puberulent; sepals 5, elliptic, 7–12 mm long, densely to inconspicuously pubescent, rounded and minutely toothed at apex; corolla white, funnelform, 3–4 cm long, broadened above calyx and deeply pleated most of its length, glabrous inside, the outer exposed ribs with long, brown trichomes; stamens 5, unequal, 10–14 mm long; filaments fused in basal half, the fused

Fig. 471. *Ipomoea phillomega*

Fig. 472.
Ipomoea squamosa

part with short, white, glandular-tipped trichomes; anthers to 3.3 mm long, sparsely covered with globular glands, the apex becoming ± recurved; pollen shed in bud; style to 1.5 cm long; stigma globose, weakly bilobed, apparently receptive, held well above anthers. Fruits subglobose, 9–14 mm diam, black, indehiscent, ± glabrous; seeds 4, 4–6 mm long, pubescent on margins. *Croat 12798.*

Rare, seen only in the vicinity of Fairchild Point; locally very abundant. Flowers from November to January. Fruit maturity time not known.

Insects occasionally bore small holes in the thick base of the flower to reach the nectaries.

Honduras to Argentina; Trinidad. In Panama, known from tropical moist forest in the Canal Zone, Panamá, and Darién and from tropical dry forest in Herrera.

MARIPA Aubl.

Maripa panamensis Hemsl., Biol. Centr.-Amer. Bot. 2:382. 1882

Canopy liana; older stems generally glabrous, slender and contorted, becoming much flattened, the younger stems often weakly fissured vertically. Leaves glabrous, alternate; petioles 0.5–2.5 cm long; blades oblong-elliptic, acute to short-acuminate or rounded and often emarginate at apex, obtuse to rounded at base, 4–22 cm long, 2.5–11 cm wide, stiff, glabrous with minute scales above, folded along midrib, the midrib arched. Panicles terminal, thyrsiform; inflorescence branches, pedicels, and calyces ferruginous-pubescent; bracts triangular, caducous; pedicels 3–8 mm long; sepals 5, rounded apically, to ca 1 cm long, usually tinged with lavender, closing tightly after corolla falls and persisting in fruit; corolla funnelform, 3–5 cm long, lavender, densely sericeous outside on exposed parts and tomentose on pleats, glabrous inside except at point of attachment with filaments; stamens 5, ca 2.3 cm long, included; filaments fused to tube ca 1 cm from base; ovary oblong, acute at apex; style ca 3 times length of ovary; ovary and style together ca 1.5 cm long. Capsules ellipsoid to depressed-globose, to 3.5 cm long (including beak) and 2.8 cm diam, glabrous, smooth, sulcate when dry, light green at maturity, with a prominent beaked apex, dehiscing by 2 leathery valves; seeds (1)2(4), oblong-elliptic, flattened on one side, to ca 2 cm long and 12 mm wide, covered with a syrupy substance, this thin, brown, difficult to remove, at first sweet. *Croat 6158, 10157.*

Abundant in the top of the canopy, often occurring over the margin of the lake; rarely flowering near the ground in the forest. Flowers from December to May. The fruits mature from April to September.

The species sometimes produces abnormally small flowers, as little as 1.5 cm long, that open normally (*Croat 34357*). These plants are otherwise identical to normal plants, and the smaller-flower forms are not believed to be another species.

Fruits are commonly torn open by animals before they can dehisce.

Panama to Colombia and Venezuela. In Panama, known from tropical moist forest in the Canal Zone (Atlantic slope), Colón, and San Blas and from tropical moist, premontane wet, and tropical wet forests in Panamá and Darién.

MERREMIA Dennst. ex Endl.

Merremia umbellata (L.) Hallier f. in Engler, Bot. Jahrb. Syst. 16:552. 1893
Ipomoea polyanthes R. & S.
Batatilla amarilla

Vine; stems and leaves densely pubescent when young, sparsely so in age. Petioles 1–8 cm long; blades ovate, acuminate at apex, cordate at base, 4.5–14 cm long, 2.5–8 cm wide. Umbels of several to many flowers; peduncles to 9 cm long, pubescent; pedicels slender, 1–2.5 cm long; sepals 5, suborbicular to oblong, rounded at apex, to 1 cm long; corolla campanulate, 3–3.5 cm long, yellow; stamens 5; filaments ± equal; anthers spiraled after dehiscence; style 1; stigma globose. Capsules ± globular, ca 8 mm diam, longitudinally dehiscent by 4–6 valves; exocarp thin; seeds 4–6, dark brown, densely short-pubescent. *Croat 7792.*

Probably common on the island at one time; now occasional in lighthouse clearings. The plant is common in the more open areas at Frijoles. Flowers to some extent throughout the year, but especially in the dry season. The fruits persist for several months. Old inflorescences are seen throughout the year.

Widespread in the tropics. In Panama, ecologically variable; known from tropical moist forest in the Canal Zone, Bocas del Toro, Colón, San Blas, Chiriquí, Herrera, Panamá, and Darién, from tropical dry forest in Herrera and Panamá, from premontane moist forest in the Canal Zone and Panamá, from premontane wet forest in Coclé, and from tropical wet forest in Colón and Coclé.

OPERCULINA S. Manso

Operculina codonantha (Benth.) Hallier f. in Engler, Bot. Jahrb. Syst. 16:550. 1893

Liana or herbaceous vine, glabrous. Petioles 2–3 cm long; blades narrowly triangular, acuminate at apex, cordate at base, 8–15 cm long, 4–8 cm wide. Flowers axillary, solitary; pedicels greatly elongate, much thickened toward apex; sepals 5, broadly elliptic, 2.5–3 cm long, obtuse at apex, enlarged and thickened in fruit; corolla broadly campanulate, to 9 cm long, white or lavender; stamens and style included. Capsules depressed-globose, 3–3.5 cm diam, dehiscent at or above middle by a circumcissile epicarp, subtended by the enlarged sepals; seeds 4, black, glabrous. *Wetmore & Abbe 246.*

Collected once on the island, but not seen in recent years. Seasonal behavior not determined.

Panama, Ecuador, and Peru. In Panama, known only from two old collections from tropical moist forest in the Canal Zone (BCI and Frijoles).

121. BORAGINACEAE

Trees or climbing shrubs, lianas, vines, or erect herbs. Leaves alternate or rarely subopposite, petiolate; blades simple, entire or serrate; venation pinnate; stipules lacking. Cymes terminal or axillary, scorpioid, spicate or open, bearing several to many flowers, sometimes arranged in panicles; flowers bisexual (or functionally unisexual), actinomorphic; sepals 5, free or weakly connate, imbricate, or calyx merely 5-lobed or toothed; corolla tubular, 5-lobed or toothed; stamens 5, epipetalous, alternate with corolla lobes, sometimes connate apically; filaments partly fused to tube; anthers 2-celled, introrse, basifixed or basally dorsifixed, dehiscing longitudinally; ovary superior, 2-locular but becoming falsely 4-locular, 2-carpellate; placentation axile; ovules 2 per carpel, anatropous; style 1, simple, with the stigma peltate or conical (*Heliotropium, Tournefortia*) or the style twice bifid with the stigma capitate or clavate (*Cordia*). Fruits drupes (*Cordia*) or fleshy and separating at maturity into 2–4 nutlets; seeds generally lacking endosperm.

Members of the family are most easily recognized by their usually conspicuous, helicoid, cymose inflorescences. Some exhibit dimorphic heterostyly (Faegri & van der Pijl, 1966). All Boraginaceae have small, white, tubular or narrowly campanulate corollas, many of which must be pollinated by long-tongued bees or by butterflies. The calyces generally allow good protection of the nectaries. A small beetle was seen in the flower of *Cordia spinescens*. I have seen flowering plants of *Cordia panamensis* visited by butterflies, wasps, and several beetles, including curculionids.

While *Cordia alliodora* has wind-dispersed seeds, those of *Heliotropium indicum, Cordia spinescens,* and all species of *Tournefortia* are probably chiefly bird dispersed. *Cordia bicolor, C. lasiocalyx,* and *C. panamensis* are dispersed by both mammals and birds. *Cordia bicolor* and *C. lasiocalyx* are reportedly swallowed by white-faced monkeys without being bitten into: the seed coat gives off a strongly astringent substance when broken; and defecated seeds germinate (Oppenheimer, 1968). *Cordia bicolor* is eaten by the bat *Artibeus jamaicensis* in Trinidad (Goodwin & Greenhall, 1961).

Some 100 genera and about 2,000 species; of cosmopolitan distribution.

CORDIA L.

Cordia alliodora (R. & P.) Cham., Linnaea 8:121. 1833
Laurel, Laurel blanco, Laurel negro, Capa

Tree, to 25 m tall and 40(90) cm dbh; buttresses weak; bark smooth to coarse, thin, usually coarsely lenticellate, sometimes deeply fissured; inner bark and wood white instantly after slashing, turning tan within seconds and brown within a few minutes; sap with sweet aroma; younger stems and branches, leaves, calyces, pedicels, and axes of inflorescences sparsely to densely stellate-pubescent; nodes of stems often swollen and inhabited by ants. Leaves alternate; petioles to 3.5 cm long; blades ± elliptic, acuminate, usually obtuse at base, usually 7–18 cm long, 3–8 cm wide. Cymes irregular, loosely spreading, terminal, the primary branches to ca 20 cm long; flowers bisexual, short-pedicellate, ca 12 mm wide, with a sweet, moderately strong aroma; calyx tubular, ca 5 mm long, 10-ribbed, with 5 small teeth; corolla tube

Fig. 473. *Cordia alliodora*

Fig. 474.
Cordia lasiocalyx

mostly enclosed by calyx, the lobes 5, ± spreading at anthesis; stamens 5, slightly shorter than lobes; filaments fused to tube, the free part curved inward and pubescent so as to block entrance to nectary; style exserted from tube but held well below anthers, its branches 4, directed laterally and forming an H-shaped structure above mouth of tube; nectar stored within base of corolla tube. Fruits 1-seeded nuts, cylindrical, ca 6 mm long, persisting within corolla, the corolla soon turning brown and also persisting. *Croat 8104*.

Common in most parts of the forest, particularly abundant in the younger forest. Flowers throughout the dry season, chiefly in February and March. The fruits develop quickly and are dispersed mostly in April and May. Plants begin to lose their leaves at the beginning of the rainy season (May) after most fruits have fallen, remain leafless for 1 or 2 months, and produce new leaves by August or September.

The species is unusual in that it loses its leaves during the early part of the rainy season when most species are actively vegetating. This phenomenon has as yet been unexplained. Despite this apparent handicap, the species is nevertheless a successful competitor. The fruits are dispersed when the dried, buoyant flower is released and blown away.

Throughout the American tropics. In Panama, a common invader (Holdridge, 1970) and characteristic of premontane moist, tropical moist, and premontane wet forests (Tosi, 1971); known from tropical moist forest in the Canal Zone, all along the Atlantic slope, and from Los Santos to Darién on the Pacific slope and from tropical wet forest in Colón (west of the Canal Zone).

See Fig. 473.

Cordia bicolor DC., Prodr. 9:485. 1845

Tree, usually 8–14(20) m tall; younger parts appressed-pubescent. Leaves alternate; petioles usually 4–8 mm long; blades elliptic to ovate, acuminate to acute at apex, obtuse to rounded at base, 10–18 cm long, 5–8 cm wide, ± entire, darker and moderately appressed-pubescent above, paler below with slender appressed trichomes on veins, the trichomes obscuring the surface. Panicles open, with flowers in small cymes; flowers bisexual; calyx tubular-campanulate, ca 4 mm long, with 5 small, ± unequal teeth, persistent in fruit; corolla white, salverform, 4.5–6 cm long, 5-lobed ca one-third its length; stamens 5, exserted; filaments ca 3 mm long; anthers ca 1.5 mm long; ovary ovoid; style lobed, the lobes 4, exserted, clavate. Drupes ovoid, ca 1 cm long, attached at base, dull green turning yellow, with dense, short, appressed trichomes; seeds ovoid, ca 8 mm long. *Croat 5630*.

Frequent in the forest. Flowers from February to May, especially in February and March, with the fruits maturing from May to July. Different branches of the tree may not flower synchronously.

Sporadic in Central America and northeastern South America; southernmost Lesser Antilles. In Panama, known from tropical moist forest in the Canal Zone, Panamá, and Darién and from premontane wet forest in Panamá (Altos de Río Pacora, ca 800 m elevation).

Cordia lasiocalyx Pitt., Contr. U.S. Natl. Herb. 18:251. 1917

Shrub or small tree, to 4 m tall, glabrous except for sparse, short trichomes on lower leaf surfaces, often divaricately branched with a leaf between the branches. Leaves alternate; petioles 4–8(10) mm long; blades oblong-elliptic to obovate, abruptly caudate-acuminate, acute to obtuse and decurrent at base, 6–14 cm long, 2.5–5.5 cm wide, entire. Panicles axillary or terminal, open, to ca 6 cm long; flowers nearly sessile, perfect; calyx cupulate, 2.5–3 mm long, the lobes 3, deltoid, ca 2 mm long; corolla funnelform, 8–10 mm long, white, 5-lobed ca half its length; stamens 5, exserted; filaments 3–3.5 mm long; anthers ca 1.4 mm long; ovary ovoid; style slender; stigmas 4, clavate. Drupes irregularly ovoid, ca 5 mm long and 9–15 mm wide, white, attached at its side; exocarp thin; mesocarp thin, gelatinous, sticky, sweet; seed ca 1 cm long. *Croat 8572*.

Frequent in the forest; juvenile plants more common than adults. Flowers mostly from February to April, especially in March, with the fruits maturing from late April to June.

Known only from Panama, from tropical moist forest in the Canal Zone and Darién and from tropical dry forest in Darién (Garachiné).

See Fig. 474.

Cordia panamensis Riley, Bull. Misc. Inform. 1927:135. 1927

Shrub or slender tree, to 7(15) m tall, functionally dioecious (see discussion below), strigose nearly throughout and hirsute on stems and axes of inflorescences. Leaves alternate or subopposite, dimorphic; petioles 5–15 mm long; larger blades ovate to ovate-elliptic, acute at apex, ± rounded at base, 16–30 cm long, 7–15 cm wide, ± entire, asperous, the midrib arched; smaller blades subopposite to the larger ones and usually less than half their length, ovate-orbicular. Inflorescences large, usually terminal, the cymes paniculate; flowers white, sessile, functionally unisexual; calyx 3.3–4 mm long, 5-dentate, pubescent; corolla ca 4 mm long; staminate flowers with corolla campanulate, lobed more than one-third its length, the lobes 5, recurved; stamens 5, 4–5 mm long, exserted ca 2.3 mm above rim; filaments united to basal two-thirds of tube, pubescent above point of attachment; ovary and style together to 3 mm long; ovary depressed-globose, orange; style narrowly conical, included, branched, each branch bifid near apex; nectar abundant; pistillate flowers with corolla salverform; stamens reduced, included, less than 2 mm long; ovary ovoid-oblong; style well developed, exserted, branched, each branch bifid near apex, the lobes clavate and recurved. Drupes subglobose to ovate, ca 1 cm diam, white at maturity; seeds tan, subglobose, ca 6 mm long. *Croat 10390, 11967*.

Occasional, in the forest. Flowers in early rainy season (May to July). The fruits mature rarely as early as June, mostly in September and October, rarely as late as November.

According to Nowicke in the *Flora of Panama* (1969) and Bawa and Opler (1975), the species is functionally

dioecious. However, a tree on the slopes of Cerro Campana (*Croat 14667*), apparently staminate, was probably the same tree from which fruit was collected a year earlier.

El Salvador to Panama and probably northern South America; reported as a doubtful species from the Valley of the Magdalena River by Jimenez S. (1970). In Panama, wide-ranging; known from all tropical moist forest areas except in Bocas del Toro; known also from tropical dry forest in Los Santos (Las Tablas) and from tropical wet forest in Colón (west of the Canal Zone).

Cordia spinescens L., Mant. Pl. Altera 206. 1771
C. ferruginea (Lam.) R. & S.

Arching shrub or small tree, to 5 m tall. Leaves alternate; petioles 1–2.5 cm long; blades ovate to elliptic, acute to acuminate, rounded to obtuse at base, 8–16 cm long, 4–9 cm wide, usually irregularly dentate, sparsely strigose above, lighter and softly pubescent below. Inflorescences branched or unbranched, axillary; peduncle partly united to petiole; flowers sessile, in spikes; calyx cupular, ca 2–3 mm long, its lobes 5, curving inward to protect ovary after flower falls, enlarging and becoming red in fruit; corolla white, 3–4 mm long, glabrous outside, with gland-tipped trichomes inside, the lobes 5, short, toothed; stamens 5, slightly exserted at anthesis; filaments fused to tube near base, the tube with gland-tipped trichomes below point of fusion with stamens; anthers shedding at least some pollen in bud; style shorter than corolla at anthesis, branched twice; stigmas 4, folded and held just above anthers in bud, apparently receptive. Drupes ovoid, 4–5 mm long, bright red; seed with prominent reticulations, covered with a thin, fleshy mericarp. *Croat 12811.*

Occasional along the shoreline, abundant in Rear #8 Lighthouse Clearing. Flowers and fruits throughout the year, but most flowering occurs in the middle of the rainy season to the early dry season (August to February). Mature fruits are most abundant in the dry season. Since the plant may flower over a long period of time, the same plant may bear flowers and mature fruits.

The plant is probably pollinated by pollen feeders, such as the small beetle seen visiting some open flowers. The fruits are often transformed into galls by insects.

Mexico to Bolivia. In Panama, ecologically widespread; known principally from all regions of tropical moist forest except in the Azuero Peninsula and from tropical wet forest on the Atlantic slope in Colón, Coclé, and Veraguas, but also from a variety of life zones at higher elevations on the Pacific slope, including premontane wet forest in Chiriquí and Coclé and premontane rain, lower montane wet, and lower montane rain forests in Chiriquí.

HELIOTROPIUM L.

Heliotropium indicum L., Sp. Pl. 130. 1753
Flor de Alacran

Coarse herb, to 1.5 m tall; stems conspicuously pubescent with short and longer trichomes. Petioles 1–8 cm long, often ± winged; blades ± ovate, acute at apex, obtuse to rounded and long-attenuate at base, 4–14 cm long, 2–7 cm wide, sparsely pubescent on both sides, the margins irregularly sinuate. Spikes narrow, usually terminal, to 28 cm long; flowers sessile; sepals 5, subulate, 1.5–2 mm long; corolla salverform, blue, the tube 3–4 mm long, the lobes 5, ca 1 mm long; stamens ± sessile; anthers less than 1 mm long; ovary weakly 4-lobed; style stout, simple; stigma conical. Fruits ± ovoid, separating into 4 nutlets at maturity, the nutlets 2–3 mm long, angulate; seeds 1 or 2 per nutlet. *Starry 249.*

Apparently no longer present on the island, though probably common when the island was more weedy; to be expected in the larger clearings and possibly also on exposed areas along the shore. Elsewhere in Panama, flowers and fruits from February to July. The fruits develop quickly. An inflorescence usually contains flowers at the apex and mature fruits at the base.

Widely distributed in the American tropics and Panama. In Panama, ecologically diverse; known from most areas of tropical moist forest as well as from tropical dry forest in Panamá (Taboga Island) and Los Santos (Pocrí) and from tropical wet forest on the Atlantic slope.

TOURNEFORTIA L.

Tournefortia angustiflora R. & P., Fl. Peruv. 2:25, pl. 151. 1799

Scandent shrub, to 3 m tall; most parts sparsely pubescent. Petioles very short or to 4 cm long; blades ovate-elliptic to lanceolate-elliptic, acute to acuminate, obtuse to rounded at base, 7–23 cm long, 2.5–10 cm wide, ± entire. Inflorescences of terminal, scorpioid cymes, the branches spikelike, to 8 cm long; flowers sessile or short-pedicellate, 5-parted; sepals slightly connate at base, ovate, 1–1.5 mm long; corolla white to yellow, the tube 10–12 mm long, the lobes ca 2 mm long; stamens ± sessile, borne midway on corolla tube; filaments fused in lower half; anthers ca 2 mm long; ovary ovoid; stigma sessile. Fruits globose, 4–5 mm diam, white, separating into 2–4 nutlets at maturity; nutlets brown, ca 2.5 mm long. *Croat 9100.*

Infrequent, at the edges of clearings; seen at the Laboratory Clearing. Flowers from October to May, chiefly

---

KEY TO THE SPECIES OF TOURNEFORTIA
Stems conspicuously pubescent with spreading trichomes, the trichomes 3 mm long or more:
    Trichomes on stem 4–6 mm long; sepals 5–7(9) mm long, the trichomes of 2 sizes . . . .
    . . . . *T. cuspidata* H.B.K.
    Trichomes on stem to 4 mm long; sepals 2–4 mm long, the trichomes uniform, short . . . .
    . . . . *T. hirsutissima* L.

Stems not conspicuously pubescent with spreading trichomes, the trichomes if present usually less than 1 mm long:
    Corolla tube more than 10 mm long, nearly glabrous . . . . *T. angustiflora* R. & P.
    Corolla tube less than 7 mm long, conspicuously pubescent:
        Stems and leaves glabrous; corolla lobes more than 1.5 mm long; plants frequent along the
            shore . . . . *T. bicolor* Sw.
        Stems and leaves conspicuously pubescent; corolla lobes less than 1 mm long; plants possibly
            no longer present on the island . . . . *T. maculata* Jacq.

from December to April. The fruits mature within about a month and are usually present on the flowering plants after about January, the last being seen about August.

Honduras to Peru. In Panama, ecologically diverse; known from all areas of tropical moist forest except in Chiriquí, from premontane wet forest in Coclé (El Valle), from tropical wet forest in Coclé (Boca de Toabré), and from premontane rain forest in Coclé (Cerro Pilón). See Fig. 475.

Tournefortia bicolor Sw., Prodr. Veg. Ind. Occ. 40. 1788

Liana or arching shrub. Petioles 0.5–2 cm long; blades elliptic to oblong-elliptic, acute and falcate at apex, obtuse at base, 6–14 cm long, 3–7.5 cm wide, ± glabrous, entire, often drying blackish. Cymes dense, scorpioid; flowers 5-parted, sessile; sepals slightly connate at base, subulate, 1.5–2 mm long, glabrous; corolla white, the tube ca 5 mm long, the lobes 1.5–2 mm long; stamens nearly sessile, attached ca one-third of the length of corolla tube; anthers ca 1.5 mm long; ovary ovoid; stigma 1, sessile, conical. Fruits globose, 5–6 mm diam, white at maturity, containing 4 nutlets; nutlets brown, ca 4 mm long. *Croat 9560.*

Frequent along the shore. Flowers from December to July, very rarely in the middle of the rainy season, mostly during February and March. The fruits mature within a month and are most common in March and April.

Throughout American tropics. In Panama, known from most areas of tropical moist forest, as well as from tropical dry forest in Panamá (Taboga Island), from premontane moist forest in Los Santos (Punta Mala), from premontane wet forest in Chiriquí (Boquete) and Coclé (El Valle), from tropical wet forest in Colón (Miguel de la Borda), and from premontane rain forest in Colón (Santa Rita Ridge) and Panamá (Cerro Jefe). See Fig. 476.

Tournefortia cuspidata H.B.K., Nov. Gen. & Sp. 3:83. ed. fol. 1818
    T. obscura DC.

Vinelike shrub, sometimes growing into canopy at 10 m or more; stems, petioles, and inflorescences densely pubescent, the long trichomes erect, brown, hirsute, 4–6 mm long, the short ones gray, puberulent, less than 0.2 mm long. Leaves short-petiolate; blades usually ± ovate, acuminate, often falcate and downturned at apex, obtuse to rounded at base, 10–19 cm long, 5–8 cm wide, pubescent on both surfaces especially on veins beneath. Cymes dense, mostly terminal, helicoid; flowers 5-parted, sessile;

sepals narrowly pointed, usually 5–7 mm long; corolla tubular, to 1 cm long, the tube greenish, bearing long appressed trichomes, tapered to flaring rim, the limb white, ca 5 mm wide; stamens fused to tube just above somewhat bulbous base; anthers shedding pollen in bud; ovary ovoid, glabrous; style simple, very short, somewhat globular, held well below anthers; lateral rim of stigma with a ± sticky secretion. Fruits ovoid to obovoid, 3–4 mm long, fleshy, held within persistent sepals, separating into 4 nutlets at maturity; nutlets irregularly ellipsoid, brown, ca 2.5 mm long. *Croat 13114.*

Occasional on the shore, generally climbing to less than 5 m; uncommon in younger areas of the forest, climbing into the canopy to 10 m or more. Flowers in Panama in all months of the year, but predominantly from January to July. The fruits develop quickly and may be found on flowering inflorescences, mostly maturing in the late dry and early rainy seasons.

Central America and northern South America; West Indies. In Panama, known from tropical moist forest in the Canal Zone, Bocas del Toro, Colón, Panamá, and Darién and from tropical wet forest in Colón (Miguel de la Borda). See Fig. 477.

Tournefortia hirsutissima L., Sp. Pl. 140. 1753

Vine or liana, pubescent, the trichomes stiff, 1–4 mm long. Petioles 0.5–1.8 cm long; blades ovate to elliptic, acute to acuminate, rounded to broadly acute at base, 6–17 cm long, 3–8 cm wide, thin. Cymes axillary or terminal, scorpioid; flowers 5-parted, sessile, aromatic; sepals to 3.5 mm long, valvate, narrowly pointed; corolla tubular, white, sericeous outside, glabrous inside, the tube to 9 mm long, the limb to 7 mm broad, the lobes acute to obtuse and apiculate, pleated along middle; stamens dark, sessile, mounted nearly midway on tube; ovary ovoid; stigma conical, sessile, ca three-fourths as broad as ovary. Fruits ± globose, to 7 mm long, white, fleshy, weakly pubescent, containing 4 nutlets ca 2.5 mm long. *Croat 6246, 11718.*

Common, at least along the shore; probably climbing into lower parts of the canopy. Flowers from April to October, especially in June and July. The fruits develop quickly and are usually present on the flowering inflorescences.

The flowers are visited by butterflies (*Heliconius*) and small bees.

Florida and throughout Mexico and Central America to Panama, Venezuela, Peru, and Bolivia. In Panama, known principally from tropical moist forest in the Canal

Fig. 475. *Tournefortia angustiflora*

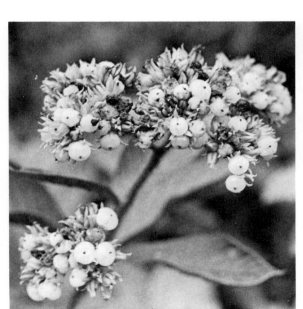

Fig. 476. *Tournefortia bicolor*

Fig. 477.
Tournefortia cuspidata

Zone, Bocas del Toro, Panamá, and Darién; known also from premontane wet forest in Chiriquí (Boquete and Río Chiriquí Viejo).

Tournefortia maculata Jacq., Enum. Syst. Pl. Ins. Carib. 14. 1760

  T. peruviana Poir.

Liana or climbing shrub, to 5 m tall; young parts pubescent. Petioles 1–2 cm long; blades ovate to elliptic, acute to acuminate, obtuse at base, 5–14 cm long, 2.5–5 cm wide, glabrous to moderately pubescent, often drying dark. Cymes arranged in irregular panicles, the branches to ca 5 cm long; pedicels very short or to 2 mm in fruit; sepals 5, free, ± ovate, 1.5–2 mm long, sparsely pubescent; corolla white to yellow, the tube to 6 mm long, the lobes 5, subulate, less than 1 mm long; stamens 5, sessile, attached midway on tube; anthers ca 1 mm long, connate apically; ovary ovoid; style simple, 2–3 mm long; stigma conical. Fruits distinctly lobed, 4–5 mm diam, orange to yellow and separating into 4 nutlets at maturity; nutlets ca 3.5 mm long. *Bailey & Bailey 237 (F)*.

Collected only once on the island and not seen in recent years. Elsewhere in Panama, flowers principally from April to July, with the fruits developing quickly and appearing most abundantly in July.

Mexico to Peru and Brazil; West Indies. In Panama, ecologically variable; known from tropical moist forest in the Canal Zone, Veraguas (Coiba Island and Santiago), and Panamá, from premontane dry forest in Panamá (Taboga Island), and from premontane wet and lower montane wet forests in Darién (Cerro Pirre). It appears likely that the species will be found to occur in tropical wet forest as well.

122. VERBENACEAE

Trees, shrubs, lianas, or herbs; stems and twigs sometimes quadrangular. Leaves opposite, petiolate; blades simple, entire or serrate, sometimes lobed (*Clerodendrum*), sometimes glandular beneath; venation pinnate; stipules lacking. Panicles, racemes, spikes, cymes, or heads terminal or axillary; flowers bisexual, actinomorphic or zygomorphic; calyx 4- or 5-lobed; corolla tubular, 4- or 5-lobed, often showy; stamens 4(5), didynamous or equal; anthers 2-celled, dehiscing longitudinally; ovary superior, 2–4-carpellate, the locules as many as carpels or twice as many (by false septation); placentation axile; ovules 1 per apparent locule, anatropous; style 1; stigma simple or bilobed. Fruits of nutlets or dupes of several pyrenes; seeds lacking endosperm.

Verbenaceae are easily confused with the Labiatae (123), but are distinguished by the undivided ovary, the terminal style, and inflorescences that are not verticillate.

Flowers are probably all insect pollinated (Chapman, 1970). *Lantana camara* is visited by several species of butterfly, as species of *Aegiphila* probably are (Chapman, 1970). *Stachytarpheta* in Costa Rica is heavily visited by skippers (Hesperiidae) and by *Eulaema, Euglossa,* and large anthophorid bees (Heithaus, 1973).

Lantana camara is chiefly bird dispersed (Leck, 1972).

KEY TO THE SPECIES OF VERBENACEAE

Leaves compound (opposite and each trifoliolate) . . . . *Vitex cooperi* Standl.
Leaves simple:
  Leaves toothed or lobed:
    Leaves broadly cordate at base, semiorbicular with 3–7 short-acuminate lobes; inflorescences large terminal panicles . . . . *Clerodendrum paniculatum* L.
    Leaves not cordate, ovate, finely serrate; inflorescences not paniculiform:
      Flowers in small, long-pedunculate heads; plants common weeds . . . . *Lantana camara* L.
      Flowers scattered along long slender spikes; plants rare, probably no longer on the island . . . . *Stachytarpheta jamaicensis* (L.) Vahl
  Leaves entire:
    Flowers with calyx and corolla violet; calyx lobes exceeding corolla; inflorescences racemes; leaves stiff, asperous below . . . . *Petrea aspera* Turcz.
    Flowers white, yellow, or greenish; calyx shorter than corolla; inflorescence not racemose; leaves not stiff and asperous:
      Trichomes of plants (stems, inflorescence branches, and calyces) dense, appressed, white, ca 1 mm long; inflorescences capituliform; fruits in virtually contiguous clusters at maturity . . . . *Aegiphila cephalophora* Standl.
      Trichomes lacking or less than 1 mm long; inflorescences ± open panicles; fruits not contiguous at maturity:
        Calyx lobed to one-third its length, in fruit splitting to the base; inflorescence branches and young stems terete to slightly angulate; fruits pointed at apex; leaves usually less than 2.5 times as long as broad, usually more than 7 cm wide, often softly villous below . . . . *Aegiphila elata* Sw.
        Calyx entire to minutely lobed, in fruit ± entire; inflorescence branches and young stems usually 4-sulcate; fruits truncate to depressed at apex; leaves usually more than 2.5 times as long as broad, less than 7 cm broad, strigillose, puberulent below . . . . *Aegiphila panamensis* Moldenke

Most species of *Aegiphila* are probably bird dispersed and probably also dispersed to some extent by other animals. *Petrea aspera* and possibly also *Stachytarpheta* are wind dispersed.

About 100 genera and some 2,600 species; primarily in the subtropics and tropics in the Southern Hemisphere.

AEGIPHILA Jacq.

Flowers of *Aegiphila* were reported by Moldenke (1934) as being either male-predominant or female-predominant. I am considering these to be merely short-styled and long-styled forms, respectively.

Aegiphila cephalophora Standl., Publ. Field Columbian Mus., Bot. Ser. 4:156. 1929

Liana, becoming arborescent, to 10 m long; trunk 6–8 cm diam near ground; stems, petioles, and inflorescence branches bearing dense, long, appressed pubescence. Petioles 0.5–1.1 cm long; blades elliptic to ovate-elliptic, acuminate, obtuse to rounded at base, 10–16 cm long, 5–6.5 cm wide, sparsely appressed-pubescent above but more densely so on midrib, the trichomes below dense, acropetal, appressed only on midrib. Cymes terminal or axillary, in dense capitate heads; inflorescence branches and calyces with dense long trichomes; flowers sessile or subsessile; calyx ca 3.5 mm long, enlarging to ca 5 mm in fruit, lobed in apical third, glabrous inside, the lobes 4, obtuse to rounded, persisting in fruit; corolla white, lobed, 6–12 mm long, the tube 4–7 mm long, the lobes 4, ovate, spreading; stamens 4, in long-styled forms to 1 mm long, in short-styled forms to 10 mm long; filaments fused to basal two-thirds of tube; anthers oblong, ca 0.8 mm long; style with the stigma bifid, in long-styled forms the style and stigma each to 8 mm long, in short-styled forms the style and stigma each ca 1.5 mm long. Fruits globose to ellipsoid, to 9 mm diam, in dense, ± globose clusters, green becoming orange at maturity, the surface granular; seeds (2)3 or 4, irregular, to 5 mm long. *Croat 12543, 16511.*

Common; normally growing high in the canopy, but occasionally flowering near the ground in tree-fall areas. Flowers from late June to September (sometimes to December). The fruits mature from September to December.

Apparently endemic to the Canal Zone in tropical moist forest.

Aegiphila elata Sw., Prodr. Veg. Ind. Occ. 31. 1788
Bejuco de peine mico, Guairo santo, Guaro

Liana, becoming shrubby, to 6 m high; stems, petioles, inflorescence branches, and calyces moderately short-pubescent. Petioles 1–2 cm long; blades elliptic to ovate-elliptic, acuminate, obtuse to rounded and often somewhat inequilateral at base, 10–20 cm long, 5–10 cm wide, short-villous below especially on veins, puberulent on veins above, with glandular punctations near midrib especially near base. Panicles axillary or terminal, often thyrsoid; pedicels to 5 mm long; calyx bell-shaped, ca

4 mm long, lobed ca one-third its length, in fruit glabrate, persistent, enlarging to ca 12 mm long, the lobes 4(5), sometimes irregular, ovate, with a short acumen; corolla greenish to cream-colored, 8–9 mm long, the lobes 4(5), ovate, acute to blunt at apex, glabrous outside, puberulent inside on tube and on basal part of filaments; stamens 4, long-exserted, 10–12 mm long; anthers oblong, ca 1.3 mm long; pollen white, tacky; style ca 2.5 mm long, slender; stigma bifid, ca 2 mm long. Fruits ± globose, to 1 cm diam, mostly enclosed by the weathering calyx, bright orange at maturity, glabrous; seeds 2–4, to 6 mm long. *Croat 14639, Foster 1240.*

Infrequent, in the forest. Some flowers have been seen in May and old fruits in September.

A few plants showed the calyx to be merely split and essentially bilobed, with one of the lobes bearing two minute teeth.

Mexico to Colombia, Venezuela, and the Guianas; West Indies. In Panama, known from tropical moist forest in the Canal Zone, Panamá, and Darién and from tropical wet forest in Colón (Santa Rita Ridge).

Aegiphila panamensis Moldenke, Trop. Woods 25:14. 1931

Shrub or tree, to 15 m tall, to 15 cm dbh; young stems and inflorescence branches usually 4-sulcate; stems, leaves, and exposed parts of inflorescences glabrate to densely pubescent, the trichomes very short, stiff, erect or appressed. Petioles to 1.2 cm long; blades elliptic to narrowly ovate-elliptic, acuminate, acute to obtuse at base, 8–19 cm long, 3–7 cm wide. Panicles axillary and terminal, to 13 cm long, the cymes open; flowers numerous; calyx and pedicel like an inverted cone; pedicels ca 4 mm long; calyx ± entire, ca 2 mm long, enlarging to ca 4 mm; corolla cream, white, or yellowish, puberulent outside, 7–9 mm long, tubular below, soon falling, the lobes 4, ca 3 mm long, globular in bud; stamens 4, in long-styled flowers to 1.5 mm long, in short-styled flowers to 10 mm long; filaments coiled in bud, pubescent near point of fusion with tube; long-styled forms with the style long-exserted, to 10 mm long, the stigma bifid, ca 3.5 mm long, the short-styled forms with the style inserted, the style and stigma each to ca 2 mm long. Fruits drupaceous, ± oblong, truncate and ± depressed at apex, to 12 mm long and 9 mm wide, encircled at base by the accrescent calyx, green turning orange at maturity; seeds 4, white, only slightly shorter than the fruit. *Croat 4094, 6749, 6780.*

Occasional, as a shrub in clearings and open areas or as a tree in the younger forest. Flowers from July to December (sometimes from April). The fruits mature from October to January.

This species is similar to *A. martinicensis* Jacq.

Southern Mexico, Costa Rica, and Panama. In Panama, known from tropical moist forest in the Canal Zone, Bocas del Toro, San Blas, Panamá, and Darién, from premontane moist forest in the Canal Zone and Panamá, from premontane wet forest in Chiriquí, and from tropical wet forest in Coclé.

See Fig. 478.

Fig. 478. *Aegiphila panamensis*

Fig. 479.
Petrea aspera

CLERODENDRUM L.

Clerodendrum paniculatum L., Mant. Pl. 90. 1767

Scarlet glorybower, Danger flower, Red pagoda flower

Herb or shrub, to 3 m tall; stems angulate, glabrous except for rings of long trichomes at leaf nodes. Petioles to 30 cm long; blades broadly ovate to suborbicular, acuminate and lobed, cordate and palmiveined at base, 9–25 cm long, 12–30 cm wide, with conspicuous glands 3–8 mm apart on margins, the lobes 5 (sometimes 3 or 7), shallow, acuminate, the lower surface punctate, with many small, whitish, peltate scales. Inflorescences terminal panicles ca 15 cm high and ca 20 cm wide, with reduced leaves at branching nodes; inflorescence branches rose-red, minutely puberulent; pedicels to 12 mm long; flowers 5-parted; calyx campanulate, ca 13 mm long, the tube orange, the lobes obtuse, ca 3 mm long, glandular before bud opens, red; corolla red-orange, the tube slender, 1.5–2 cm long, the lobes paler, spreading, ca 6 mm long; stamens 4, orange-red, exserted to ca 3.5 cm above throat, curved; style exserted ca two-thirds as far as stamens, minutely bilobed at apex, violet-purple. Drupes small, green, ± enclosed in persistent calyx. *Croat 7000.*

Cultivated in the Laboratory Clearing. Flowers principally in the rainy season. No fruits have been seen on BCI.

Native to Asia; cultivated in Europe and in the American tropics and subtropics. In Panama, known only from tropical moist forest in the Canal Zone.

LANTANA L.

Lantana camara L., Sp. Pl. 627. 1753

Pasorín, Bandera español, Camara, Cinco negritos, Chichiquelite

Herb, usually less than 1.5 m tall, sometimes woody, hirtellous on most parts, sparsely so on upper leaf surfaces. Petioles 4–20 mm long; blades ovate, acute to acuminate, rounded to obtuse at base, then abruptly attenuate to half the length of petioles, 4–11 cm long, 2–5 cm wide, minutely crenate-serrate, scabrous or scaberulous above. Flower heads 1.5–2 cm diam; peduncles to 8 cm long; flowers sessile, bracteate, puberulent outside, the outermost of each head orange-red to burnt-orange, the innermost orange, the bracts acute, ca 3 mm long; calyx ca 2.5 mm long, obscurely 4-lobed, ciliate; corolla tubular, ca 10 mm long, 4-lobed, the tube somewhat expanded and bent above the middle, the limb ca 4 mm wide; stamens 4, fused to tube at 2 levels about midway down tube, the free part of filaments short; anthers often shedding pollen in bud, their thecae distinct; style held somewhat below the lowest pair of anthers; stigma attached laterally, becoming fleshy after anthers have lost their pollen, sometimes secreting a nectarlike substance visited by tiny insects; nectary small, but the nectar often filling base of corolla tube. Drupes round, several to many in a tight cluster, grayish-blue, to 5 mm diam; exocarp thin; mesocarp sweet, tasty; seeds obovate, to 4 mm long. *Croat 6004.*

Abundant in weedier areas of clearings. Flowers and fruits sporadically throughout the year.

Hummingbirds (*Damophila julie* and *Amazilia tzacatl*) visit the flowers for nectar (Leck, 1972).

The fruits are taken by a variety of birds, including manakins, flycatchers, honeycreepers, and tanagers (Leck, 1972).

Pantropical. In Panama, known from tropical moist forest in the Canal Zone, San Blas, Herrera, and Darién, from premontane moist forest in Veraguas and Los Santos, and from premontane wet forest in Coclé.

PETREA L.

Petrea aspera Turcz., Bull. Soc. Imp. Naturalistes Moscou 36(3):211. 1863

Viuda, Flor de mayo, Flor de la cruz

Liana, growing into canopy, appearing to be an arching shrub when juvenile; young parts sparsely short-pubescent, glabrate in age; stems sometimes 4-angled. Petioles to 1.5 cm long; blades elliptic, often broadly so, rounded to acuminate at apex, acute to rounded at base, 7–21 cm long, 3.5–11 cm wide, usually stiff, asperous below. Racemes long, ± pendent, upper-axillary, to ca 30 cm long; pedicels ca 11 cm long; flowers 5-parted; calyx light violet, the tube membranous, ridged, ca 6 mm long, the lobes 5, thin, membranous, narrowly obovate, 2–2.5 cm long, a crownlike projection occurring at summit of tube and at base of lobes, ca 3 mm high, acutely 5-lobed to near base, often drying brown, persisting and acting as dispersal mechanism for fruit; corolla violet, salverform, ca 1.5 cm long, lobed to middle, somewhat zygomorphic, puberulent on both surfaces, the tube villous inside near apex, 1 lobe with a white spot near its base; stamens 4, fused to basal two-thirds of tube; filaments villous, ca 2 mm long; anthers held closely together on one side of the tube at its rim; pollen somewhat sticky, adhering to anthers after being shed; style bent slightly at apex in the direction of the white-spotted corolla lobe; stigma held just below anthers. Fruits of 1 or 2 tiny nutlets completely enclosed by the persistent calyx; nutlets ovoid, ca 1.5 mm long, glabrous. *Croat 9432.*

Abundant in the canopy and at the edge of the forest over the lake; one plant grows as an epiphyte from a large ant nest (*Croat 9432*). Flowering and fruiting throughout the year, often in synchronous waves throughout the forest.

Petrea volubilis L. was reported by Moldenke (1973) in the *Flora of Panama* for BCI based on *Aviles 14* and *Shattuck 412*. These specimens do not differ from the typical *P. aspera*, however, and I believe that only a single species of *Petrea* occurs on the island.

Widespread in tropical America from northern Mexico to southern Brazil; Cuba, West Indies; widely cultivated. In Panama, known from tropical moist forest in the Canal Zone, San Blas, Veraguas, Los Santos, Panamá, and Darién.

See Fig. 479.

STACHYTARPHETA Vahl

Stachytarpheta jamaicensis (L.) Vahl, Enum. Pl.
1:206. 1804

Berbena, Blue porterweed, Gervao, Jamaica, Verbena,
Vervain, Rinchao, Verbena azul, Brazilian tea, Cola de millo

Herb, to ca 1 m tall, sparingly pubescent to glabrate,
sometimes slightly woody basally. Petioles absent or to
1 cm long; blades oblong to oval, obtuse to acute at apex,
cuneate at base, decurrent onto petiole, to ca 11 cm long
and to ca 4.5 cm wide, scaberulous above, glabrous below
but with sparse jointed trichomes on primary veins, the
margins serrate and ciliate-scabrous. Spikes terminal,
to 20(50) cm long, terete, 3–4(5) mm wide; flowers at
first erect, later embedded in the thickened rachis; bracts
subtending pits in which flowers are immersed, acute,
to 5 mm long and ca 1–1.5 mm wide, the margins scarious; calyx tubular, ca 5 mm long, shallowly 4-toothed on
the side facing away from stem, notched to 1 mm on side
next to stem; corolla salverform, shades of blue, violet,
and purple, ca 10 mm long, the tube slightly curved,
the limb ca 8 mm wide; perfect stamens 2, inserted above
middle of corolla tube, included; filaments very short;
thecae divergent; staminodia 2, posterior, minute; style
ca 6 mm long, filiform; stigma terminal, persistent on
fruit after corolla falls. Fruits oblong-linear, emerging
from rachis pits at maturity, dry, splitting into cocci,
the cocci 2, long, hard, narrow; seeds 1 per coccus, linear.
Woodworth & Vestal 739.

Rare. Flowering and fruiting throughout the year.

Alabama and Mexico to Ecuador and Brazil; West
Indies; introduced into parts of tropical Africa, Asia,
Australia, and Oceania. In Panama, known from tropical
moist forest in the Canal Zone, Colón, San Blas, Chiriquí, Los Santos, Panamá, and Darién, from premontane
wet forest in Chiriquí and Panamá, and from tropical
wet forest in Bocas del Toro and Colón.

VITEX L.

Vitex cooperi Standl., Trop. Woods 16:26, 29, 32,
nomen nudum 1928; Publ. Field Columbian Mus.,
Bot. Ser. 4:256. 1929

Tree, to 26 m tall; trunk buttressed and fluted, to 75 cm
dbh; outer bark thin, light-colored; branchlets acutely
tetragonal to terete or compressed, densely puberulent
when young. Leaves trifoliolate, opposite; petioles 1.3–9.5
cm long, finely appressed-puberulent; petiolules 1–8
mm long, the central one longest, flattened, margined;
leaflets unequal with the lateral pair smaller and narrower, the central leaflet rounded and acuminate at apex,
acute to acuminate at base, 4–22 cm long, 2.5–10.5 cm
wide, entire, shiny and glabrous above, finely puberulent
on veins beneath when young, becoming glabrate in age.
Inflorescences axillary, solitary, 3.5–14 cm long, 4–6
cm wide, dichotomously branched, finely puberulent
throughout, bearing many flowers; peduncles conspicuously flattened, 4.5–7.5 cm long; pedicels 1–3 mm long;
bracts ca 1.5 cm long and 4 mm wide; flowers not seen,
reportedly blue and lavender. Fruits not seen.

Rare, known only from the vicinity of the Laboratory
Clearing; collected sterile once by R. Foster. Elsewhere
in Panama, flowers during June and July, with the fruits
maturing during July and August.

Guatemala to Panama; sea level to 600 m. In Panama,
known from tropical moist forest in the Canal Zone and
Darién and from premontane wet forest in Chiriquí
(Progreso); reportedly fairly common on the Atlantic
watershed around Gatun Lake (*Fisher 1*), and no doubt
more widespread and common than collections indicate.

123. LABIATAE (LAMIACEAE)

Annual or perennial herbs, erect or vinelike, pubescent,
usually aromatic; stems usually quadrangular. Leaves
opposite, sessile or petiolate; blades simple, serrate; venation pinnate; stipules lacking. Cymes bracteate, capituliform or verticillate, contracted, axillary or terminal;
flowers bisexual, zygomorphic; calyx tubular or campanulate, 5-lobed; corolla bilabiate, 5-lobed; stamens 2, or
4 and didynamous, epipetalous, often forcefully ejected
from flower; anthers 2-celled, the thecae often divergent, introrse, dehiscing longitudinally; ovary superior,
4-lobed, 2-locular, 2-carpellate; placentation basal; ovules
4, anatropous; style 1, arising from depression between
lobes; stigma bilobed. Fruits of 4 nutlets (in *Salvia* usually only 1 maturing); seeds 1 per nutlet, with little or
no endosperm.

KEY TO THE SPECIES OF LABIATAE

Inflorescences capituliform; flowers white; stamens 4:
    Calyx tubular, more than 5 mm long; nutlets ca 1 mm long . . . . *Hyptis capitata* Jacq.
    Calyx campanulate, less than 4 mm long; nutlets less than 1 mm long . . . . *Hyptis brevipes* Poit.
Inflorescences verticillate; flowers blue, lavender, or violet; stamens 2 or 4:
    Herbs small, cultivated, once planted at the laboratory, probably no longer present on the island;
        leaves often variegated . . . . *Coleus blumei* Benth.
    Herbs small or large, not cultivated:
        Flowers pedicellate; stamens 2; fruiting calyx with short blunt teeth, conspicuously covered
          with large gland-tipped trichomes . . . . *Salvia occidentalis* Sw.
        Flowers sessile; stamens 4; fruiting calyx with long slender teeth, lacking gland-tipped tri-
          chomes but with sessile glands . . . . *Hyptis mutabilis* (A. Rich.) Briq.

Recognized by the opposite leaves, squarish stems, and pleasant aroma. Members of the family are closely related to the Boraginaceae (121) and Verbenaceae (122), but are generally not confused with either of these families on BCI.

All BCI species have typical gullet flowers, which are scentless and generally bee pollinated (Faegri & van der Pijl, 1966). The flower is two-lipped, the lower lip serving as a landing platform. The genus *Hyptis* probably has the same pollination syndrome as that of *Eriope* studied by Harley (1971) in Brazil, since most *Hyptis* have the same morphology and triggered release of stamens. *Eriope* flowers open early in the day and are visited by tiny bees. The stamens then spring out, the style later elongates past the stamens, and the stamens become deflexed with the lower lip. The flower usually falls before the day is over.

Nutlets are small and usually numerous. *Hyptis* nutlets exude mucilage copiously when wetted (H. Baker, pers. comm.). They are probably in part dispersed by small birds of clearings and in part merely spilled and scattered by passing animals or the wind. The burlike pseudoheads of *Hyptis brevipes* and *H. capitata* are ballasts set loose by passing animals (van der Pijl, 1968). Ridley (1930) suggested that *H. brevipes* might be epizoochorous by means of its long-pubescent sepaloid points.

About 180–200 genera and 3,200–3,500 species; cosmopolitan but concentrated in the Mediterranean region.

COLEUS Lour.

Coleus blumei Benth., Lab. Gen. et Sp. 56. 1832
Coleus, Pompolluda, Chontadua, Jacob's coat

Sprawling herb, to 70 cm long; younger stems and petioles densely pubescent. Petioles 1–4 cm long; blades ovate, acute at apex, obtuse to truncate at base, 2–15 cm long, 1.5–10 cm wide, sparsely pubescent above and on veins below, resin-dotted below, crenate, sometimes variegated. Inflorescences terminal, branched, 15–30 cm long; flowers in cymes, purplish; pedicels 1–2 mm long; calyx ca 2 mm long, resin-dotted, unevenly lobed and toothed; corolla ca 1 cm long, the tube recurved, the upper lip bilobed, the lower lip entire, enlarged, boat-shaped; stamens 4; style bifid near apex. Nutlets 4, ca 1 mm diam. *Shattuck 158.*

Formerly cultivated at the laboratory; not seen in recent years, but so commonly cultivated in Panama it is included here. Seasonality uncertain. Possibly flowers and fruits year-round.

Native to the East Indies; cultivated in many places. In Panama, known from BCI and Coclé (El Valle), but no doubt cultivated elsewhere.

HYPTIS Jacq.

Hyptis brevipes Poit., Ann. Mus. Natl. Hist. Nat. 7:465. 1806

Herb, to 60 cm tall, pubescent throughout. Petioles obsolete or to 1 cm long; blades lanceolate-elliptic or ovate, acute at apex, attenuate at base, to 5 cm long and 2 cm wide, irregularly or doubly serrate, resin-dotted. Inflorescences capitulate, ca 1 cm diam; peduncles to 1 cm long; bracts awl-shaped, ciliate, to 6 mm long; flowers sessile, tufted at base; calyx campanulate, to 3.5 mm long, with 5 spinulose teeth ca 2 mm long; corolla bilabiate, 5-lobed, ca 3 mm long, white; stamens 4, weakly exserted; style bifid at apex. Nutlets 4, oblong-ovoid, ca 0.6 mm long. *Shattuck 479.*

Collected once by O. Shattuck on Wheeler Trail; possibly no longer occurring on the island. Seasonal behavior uncertain. Elsewhere flowers and fruits to some extent all year, but possibly with most flowering from November to January.

Native to Brazil, but now a weed throughout tropical America, Asia, and Polynesia. In Panama, known from tropical moist forest in the Canal Zone and Panamá and from tropical wet forest in Colón.

Hyptis capitata Jacq., Coll. 1:102. 1787
Suspiro de monte

Herb, to 2 m tall, sparsely to moderately pubescent throughout. Petioles slender, 5–20 mm long; blades ovate to rhombic, acute at apex, attenuate at base, 5–12 cm long, 2–6 cm wide, irregularly serrate. Flower heads at first hemispheroid, later globular; flowers numerous, many open at a time in each head; calyx scabrid, hispid at base outside, with a pubescent line inside above nutlets, the lobes 5, slender, at first equaling tube, the tube much longer in fruit; corolla white, ca 4 mm long, slightly exceeding calyx, bilabiate, the middle lobe of upper lip hooded, loosely enclosing the style and 2 longest stamens, folding backward following anthesis; anthers red, closed at anthesis; filaments curling over the edge of upper lip in time; stigma held midway between long and short pairs of stamens, its 2 lobes open at anthesis. Nutlets usually 4, ovoid, ca 1 mm long, smooth. *Croat 4406, 8683.*

Common in clearings. Flowers mostly in the early dry season, with the fruits maturing late in the dry season.

This species differs considerably from *H. mutabilis* in its pollination behavior. Those individuals of *H. mutabilis* observed shed the pollen in the bud, and the stamens were thrown forward violently after being opened forcibly (while the stigma was still unopened). In *H. capitata* only two stamens are held within the upper lip and then only very loosely. Their flowers are probably not forced open and the stamens, when released, do not spring forward violently. Both protandrous and protogynous forms were observed in a single population of *H. capitata*, though the anthers in some plants were shed in bud, with the style remaining unopened until most of the pollen was shed.

Mexico to Colombia, Venezuela, Ecuador, and Peru; Asia and Polynesia. In Panama, widespread; known from tropical moist forest in the Canal Zone, Bocas del Toro, San Blas, Veraguas, Los Santos, Panamá, and Darién, from premontane moist forest in Los Santos and Panamá, and from premontane wet forest in Chiriquí, Coclé, and Panamá.

Hyptis mutabilis (A. Rich.) Briq., Bull. Herb. Boissier 4:788. 1896

Herb, 2–3(3.5) m tall, aromatic, moderately pubescent on most parts. Petioles obsolete or to 2 cm long; blades ovate or rhombic, acute to acuminate, cuneate to truncate at base, 3–6 cm long, 1.5–3.5 cm wide, serrate or doubly serrate. Panicles verticillate, the verticils each with 3–10 flowers; flowers sessile, emerging 1 at a time in each verticil; calyx toothed, the teeth 5, slender, alternating with erect trichomes in flower (once the flower has fallen the trichomes bend inward to close opening), enlarging and enclosing nutlets in fruit; corolla violet, pubescent, ca 4 mm long, bilabiate, the upper lip bilobed, with a white spot below rim, the lower lip trilobate, the center lobe hooded and marginally fringed, enclosing style and stamens; stamens 4, as long as lobes, fused to tube below lower lip, pubescent distally; anthers withering in age; style at first equaling stamens, later elongating; stigma bilobed, closed when first released, later opening. Nutlets usually 3 or 4, ca 2 mm long, contained within the expanded calyx. *Croat 7461.*

Locally abundant in the Lighthouse Clearing, usually a dominant plant at certain times of the year. Flowers very early in the dry season, with the fruits maturing in the late dry season.

Pollination is effected when an insect (presumably a tiny bee) attempts to force its way into the corolla tube. This releases the stamens and style, which are under tension because of being pushed forward by a flap of tissue at the base of the hooded lobe. See the discussion under *Hyptis capitata* for a comparison of the two species.

A common weed throughout tropical and subtropical America; introduced into the Old World tropics. In Panama, growing in disturbed areas in a variety of life zones; known from tropical moist forest in the Canal Zone, Veraguas, Herrera, Coclé, Panamá, and Darién, from tropical dry forest in Los Santos and Herrera, from premontane moist forest in Los Santos, and from premontane wet forest in Chiriquí and Coclé.

SALVIA L.

Salvia occidentalis Sw., Prodr. Veg. Ind. Occ. 14. 1788

Erect or sprawling herb, usually 30–40 cm long, occasionally vinelike and very long, sparsely pubescent throughout. Petioles 5–20 mm long; blades ovate to triangular, acute at apex, obtuse to attenuate or truncate at base, 1.5–6 cm long, 1–3 cm wide, crenate-serrate. Panicles verticillate, the flowers 6–10 in a verticil; calyx tubular, prominently veined, with capitate glandular trichomes persisting in fruit; corolla ca 2.5 mm long, the tube white, the lower lobe weakly trilobate, blue with white stripes; stamens 2, affixed near apex of tube; filaments with hook-shaped appendages on inner side, these acting as levers to force stamens inward when pressed upon by an insect entering the tube; style held slightly above anthers and adjacent to upper lip; stigma bilobed, the lobes flared. Nutlets smooth, 0.5–1.5 mm long, usually only 1 maturing. *Croat 6943.*

An occasional weedy plant of clearings. Flowers in the early dry season, with the fruits maturing in the late dry season.

The fruits are dispersed by means of the sticky trichomes on the persisting calyx.

Throughout the American tropics. In Panama, growing in disturbed areas in a variety of life zones; known from tropical moist forest in the Canal Zone, Herrera, Coclé, and Panamá, from tropical dry forest in Los Santos, Herrera, and Panamá, from premontane moist forest in Los Santos, and from premontane wet forest in Coclé (El Valle).

124. SOLANACEAE

Trees, shrubs, vines, or herbs, terrestrial or hemiepiphytic (*Markea* and *Lycianthes*); stems often prickly; stellate pubescence often present. Leaves alternate or subopposite, petiolate; blades simple, entire or lobed; venation pinnate; stipules lacking. Flowers bisexual, actinomorphic (zygomorphic in *Browallia*), generally solitary or in cymose axillary clusters; calyx 5-lobed; corolla (4)5-lobed; stamens (4)5 or 4 and didynamous (*Browallia*), epipetalous; anthers 2-celled, dehiscing longitudinally or poricidally (*Solanum*); ovary superior, 1- or 2-locular, 2-carpellate; placentation axile; ovules many (few in *Cestrum*); style 1; stigma minutely bilobed. Fruits generally berries with many seeds, rarely septicidally dehiscent capsules (*Browallia*); seeds with fleshy endosperm.

Members of the family are most easily distinguished by their actinomorphic, gamopetalous, usually plicate corollas and berries of many seeds. They are closely related to the Scrophulariaceae (125), which generally have zygomorphic flowers.

All species are probably insect pollinated. *Solanum* and *Cyphomandra* (G. Frankie, pers. comm.), with apical pores, are mostly pollinated by drumming, pollen-collecting bees. I have seen *Solanum subinerme* visited in this way by *Melipona*, and G. Frankie (pers. comm.) has observed xylocopids drumming flowers of *S. tridynamum* Dun. (*S. amazonicum* Ker.) in Mexico. *Cestrum* species have flowers that open late in the afternoon and are mostly moth pollinated. The flowers fall before mid-morning of the following day.

The fruits are endozoochorous, except perhaps in the case of *Browallia americana*. *Capsicum* has been widely dispersed by man. Most species are dispersed at least in part by birds (Ridley, 1930), but probably especially those of *Cestrum, Lycianthes, Physalis,* and *Witheringia solanacea*, which have fruits so small as to be unattractive to larger animals. Smaller fruits may be dispersed in their entirety by birds. Because species of Solanaceae have fairly thin-walled fruits, birds probably pick open the larger fruits and take the seeds. On the other hand, spiny rats eat fruits of *Physalis* (Hladik & Hladik, 1969), and white-faced monkeys are reported to eat fruits of *Cestrum* (Oppenheimer, 1968). The fleshy part of the *Cestrum* fruit is sweet, but the seed is bitter and is swallowed whole. Bats may be important in the distribution of fruits

KEY TO THE TAXA OF SOLANACEAE

Plants herbs to ca 1 m tall:
- Flowers blue, zygomorphic; fruits capsular . . . . *Browallia americana* L.
- Flowers not blue, actinomorphic; fruits berries:
  - Flowers pale yellow or white (often with a dark "eye" at base), solitary in leaf axils; fruits globose berries surrounded by the inflated calyx . . . . *Physalis*
  - Flowers greenish or yellow-green, lacking a dark throat, not solitary in leaf axils; fruits not as above:
    - Leaf blades ± alike, not deeply lobed; flowers in dense axillary clusters; fruits ca 8 mm diam, orange-red; plants occasional in clearings . . . . *Witheringia solanacea* L'Hér.
    - Leaf blades dimorphic, some entire, others deeply lobed; flowers few in pendent axillary racemes; fruits ca 2 cm diam with green stripes when young; plants probably not on island . . . . *Cyphomandra allophylla* (Miers) Hemsl.
Plants trees or shrubs more than 1 m tall:
- Corolla narrowly tubular, the tube much longer than lobes; fruits obovoid or ellipsoid, less than 9 mm diam, purple or black at maturity; calyx lobes acute . . . . *Cestrum*
- Corolla not narrowly tubular, the tube usually as short as or much shorter than lobes (tube longer than lobes in *Markea* but corolla campanulate):
  - Corolla lobes divided nearly to base:
    - Flowers in very long, pendent cymes in age; petals greenish; anthers to 10 mm long, the connective thickened dorsally; fruits 3–5 cm diam . . . . *Cyphomandra hartwegii* (Miers) Dun.
    - Flowers not in long, pendent cymes; petals white or lavender; anthers with the connective not thickened:
      - Calyx truncate, sometimes with submarginal teeth reduced to bumps; plants unarmed shrubs . . . . *Lycianthes*
      - Calyx 5-lobed (with 5 minute, marginal teeth and appearing truncate in *S. subinerme*); plants shrubs, vines, or herbs, often prickly . . . . *Solanum*
  - Corolla lobes not divided to near base:
    - Calyx distinctly lobed; plants hemiepiphytic; leaves ± coriaceous, glabrous beneath; fruits ovoid, yellowish, ca 1.5 cm long; plants in forest . . . . *Markea ulei* (Damm.) Cuatr.
    - Calyx truncate; plants not hemiepiphytic; leaves thin, pubescent beneath; fruits globose or elongate, red at maturity; plants common in clearings:
      - Flowers solitary in axils; anthers bluish, not apiculate; fruits narrowly ovoid, more than 3 cm long; pubescence of leaves ± restricted to vein axils below . . . . *Capsicum annuum* L. var. *annuum*
      - Flowers clustered in axils; anthers not bluish, apiculate; fruits globular, ca 8 mm diam; pubescence of leaves not restricted to vein axils below . . . . *Witheringia solanacea* L'Hér.

as well, especially the long-pendent fruits of *Cyphomandra* and the larger arborescent or lianous *Solanum* species such as *S. hayesii* and *S. lanciifolium. Markea ulei* is taken by bats (Bonaccorso, 1975). Heithaus, Fleming and Opler (1975) reported species of Solanaceae to be taken by several species of bat. A species of *Solanum* is taken by the bat *Artibeus jamaicensis* Leach in Mexico (Yazquez-Yanes et al., 1975).

About 80 genera and over 3,000 species; concentrated in the New World.

BROWALLIA L.

Browallia americana L., Sp. Pl. 631. 1753

Chavelita de monte

Herb, usually less than 1 m tall, puberulent to sparsely villous. Leaves alternate, simple; petioles to 2 cm long; blades ovate, acute to acuminate, acute at base, 1.5–7 cm long, 1–4 cm wide, entire. Flowers 5-parted, short-pedicellate, solitary in axils, 1.5–2 cm long; calyx 6–8(10) mm long, narrowly campanulate, striate, pubescent; corolla tube greenish, slender, pubescent, the limb lavender to blue, the throat 1–1.5 cm wide, minutely puberulent apically, with a prominent white or green spot near throat (nectar guide); stamens 4, the upper pair situated in throat near apex opposite nectar guide, the connective broadened, violet-purple, pubescent, their anthers dehiscing into cup-shaped cavities on one side of the much-thickened, elaborate style apex, the lower pair of stamens held slightly below the upper pair, their filaments hooked at apex; ovary 1–1.5 mm long, appressed-pubescent at apex; stigma bilobed. Fruits 4-valved, ellipsoid capsules contained within the accrescent calyx; seeds irregularly round, to 1.5 mm long, muricate. *Croat 10258.*

Uncommon; found in shady places at the Laboratory Clearing. Probably flowering and fruiting all year, especially in the dry season.

American tropics. In Panama, ecologically variable; most common at middle and upper elevations, occurring in most life zones in highland Chiriquí and mountainous regions of central Panama; less frequently from tropical dry forest in Panamá (Taboga Island) and from tropical moist forest in the Canal Zone, Panamá, and Darién.

CAPSICUM L.

Capsicum annuum L. var. annuum, Sp. Pl.
188–89. 1753

C. frutescens sensu Standl.

Red pepper, Chile, Aji, Aji picante

Shrub, less than 2 m tall. Leaves alternate or subopposite with a large leaf opposed by a smaller one; petioles to 3 cm long, densely pubescent; blades ovate to elliptic, falcate-acuminate, attenuate at base, 6–11(15) cm long, 2.5–5 cm wide, puberulent above and on veins below, tufted in veins axils below, ciliate. Flowers 5-parted, solitary (rarely paired) in axils; pedicels to more than 2 cm long; calyx truncate, ca 3 mm long, nearly glabrous, the teeth 5, blunt, submarginal, accrescent and persisting in fruit; corolla greenish-white, divided one-half to one-third of the way to base, to ca 17 mm diam, reflexed at anthesis, glabrous on outside, papillate-puberulent inside and on margins, the lobes acute; anthers bluish; filaments violet-purple near apex, fused to tube basally; style slightly exceeding anthers. Berries variable, often ovoid, ca 4.5 cm long, bright red at maturity. *Croat 9007.*

Cultivated at the laboratory. Flowering and fruiting all year, apparently with most fruits maturing during the rainy season.

Probably native to tropical America; cultivated throughout the world and throughout Panama.

CESTRUM L.

Cestrum latifolium Lam., Illustr. 2:5, no. 2275. 1793

Shrub or small tree, to 3.5(12) m tall; stems, especially on younger parts, petioles, and leaves, especially on veins below, sparsely to densely pubescent with weak multicellular trichomes. Leaves alternate; petioles 1–2(4) cm long; blades ovate to elliptic, acuminate, acute to obtuse at base, 7–16(25) cm long, 3.5–8.5 cm wide. Racemes or panicles short, congested, axillary, the branches and calyces pubescent; flowers 5-parted, to 3.5 cm long; calyx 1–3 mm long, the lobes acute; corolla narrowly tubular, 13–18(20) mm long, whitish-yellow or greenish, the lobes 5(6), narrowly acute, 2–2.5 mm long, pubescent outside along margins, glabrous inside; stamens 5, included; filaments fused to tube most of their length; anthers about as broad as long, held just below rim; stigma truncate; style held just above anthers. Berries

obovoid, pink to black at maturity, 5–9 mm long; seeds several. *Croat 11996.*

Uncommon, known from Rear #8 Lighthouse Clearing and from the forest near the Laboratory Clearing. Flowers from April to September, but mostly in the early rainy season, from May to July. The fruits mature from July to November, mostly from August to October.

Nicaragua to the Guianas and Brazil; the Antilles. In Panama, known from tropical moist forest in the Canal Zone, San Blas, Veraguas, Panamá, and Darién, from tropical dry forest in Los Santos, from premontane moist forest in the Canal Zone, from premontane wet forest in Panamá, and from tropical wet forest in Chiriquí.

Cestrum megalophyllum Dun. in DC., Prodr.
13(1):638. 1852

C. baenitzii Ling.

Shrub or small tree, to 8(10) m tall; outer bark thin; young stems green, sparsely crisp-pubescent, glabrous in age. Leaves alternate; petioles to 1.3(2.5) cm long, glabrous; blades narrowly elliptic to obovate-elliptic, acuminate, gradually tapered to an acute or rounded base, 12–25(35) cm long, 3.5–8(12) cm wide, ± glabrous. Racemes short, axillary, fasciculate; peduncles crisp-pubescent, bracteate; flowers 5-parted, usually many, often on leafless stems; calyx cupular, 2–3.5(4) mm long; corolla salverform, ca 1.5 cm long, the tube greenish, constricted just below lobes, the lobes white, spreading at anthesis, 4–5 mm long; stamens included; filaments fused to tube in basal two-thirds, pubescent near their point of fusion, turned inward below anthers; style slightly longer than stamens; stigma capitate, held just above anthers; nectary inconspicuous at base of ovary. Berries ellipsoid, 6–10 mm long, green becoming whitish then light violet and finally dark violet when mature; mesocarp thin, fleshy, white; seeds 1 or 2, minutely papillate. *Croat 14568.*

Occasional, in the forest. Flowering from November to June, mostly in the dry season, from February to April. The fruits mature in the early rainy season, from May to August.

Guatemala to Venezuela. In Panama, known from tropical moist forest in the Canal Zone, Bocas del Toro, Los Santos, and Darién, from premontane wet forest in Colón and Chiriquí, from tropical wet forest in Colón, and from lower montane wet forest in Chiriquí.

See Fig. 480.

KEY TO THE SPECIES OF CESTRUM

Mature leaves pubescent beneath . . . . *C. latifolium* Lam.
Mature leaves glabrous beneath:
    Largest leaves more than 15 cm long, gradually tapered to a usually acute base, often broadest
        above middle . . . . *C. megalophyllum* Dun.
    Largest leaves less than 15 cm long, not gradually tapered to base, broadest at or below middle:
        Filaments conspicuously dentate near base; leaf blades usually acute at base . . . . *C. nocturnum* L.
        Filaments not dentate at base; leaf blades usually obtuse to rounded at base . . . .
        . . . . *C. racemosum* R. & P.

Cestrum nocturnum L., Sp. Pl. 191. 1753

Dama de noche

Shrub or small tree, to 5 m tall. Leaves alternate; petioles ca 1 cm long; blades ovate, acute to acuminate, obtuse to rounded at base, to 11 cm long, to 5.5 cm wide, minutely puberulent when young, glabrous in age. Panicles axillary or terminal, congested; peduncles with scattered leaf-like bracts; pedicels to 4 mm long, with minute bracteoles; flowers 5-parted; calyx campanulate, puberulent, 2 mm long; corolla greenish to yellowish-white, 14–17 mm long, glabrous outside; stamens equal; filaments ca 3 mm long, dentate near point of insertion; stigma slightly exserted. Berries ellipsoid, to 10 mm long, black at maturity; seeds not studied. *Bangham 429.*

Collected once on the shore. Seasonal behavior uncertain. According to our records, flowers mostly in the dry season, but also in August.

Native to the Antilles, but cultivated and escaped in many other places in tropical America. In Panama, known from tropical moist forest in the Canal Zone and Bocas del Toro and from tropical wet forest in Colón and Chiriquí.

Cestrum racemosum R. & P., Fl. Peruv. 2:29, pl. 154. 1799

C. panamense Standl.; *C. racemosum* R. & P. var. *panamense* (Standl.) Franc.

Yedi

Tree, to 12 m tall. Leaves alternate; petioles ca 1–2 cm long, glabrate; blades lanceolate-elliptic, long-acuminate, mostly obtuse to rounded at base, 7–15 cm long, 2.5–5 cm wide, ± glabrous. Cymes branched, axillary or terminal, 3–9 cm long, the branches crisp-villous; flowers 5-parted; calyx 2.3–3.3 mm long, the lobes acute, ciliate; corolla to 15 mm long, slender, cream or greenish-white, the lobes villous inside; stamens included; filaments fused to tube most of their length; style flat on end, held slightly above anthers; nectaries golden-yellow. Berries obovoid or ellipsoid, ca 6 mm long, black, with thickened pedicels ca 2 mm long; seeds usually 3, wedge-shaped (like orange segments), densely papillate, ca 4 mm long, embedded in a juicy white matrix, very bitter before maturity, sweet at maturity. *Croat 8899.*

Uncommon, known from the edge of the Laboratory Clearing. Flowering from December to May, mostly in the late dry season and the early rainy season (April to May), rarely as late as September. Frequently flowering more than once per season. The fruits mature mostly from February to December.

Belize to Brazil. In Panama, known from tropical moist forest in the Canal Zone, Bocas del Toro, Panamá, and Darién, from premontane wet forest in Chiriquí and

Coclé, and from tropical wet forest in Colón and Coclé. Reported from premontane rain forest in Costa Rica (Holdridge et al., 1971).

CYPHOMANDRA Mart. ex Sendt.

Cyphomandra allophylla (Miers) Hemsl., Biol. Centr.-Amer. Bot. 2:417. 1882

Solanum allophyllum (Miers) Standl.

Hierba de gallinazo, Hierba gallota

Erect herb, to 1 m tall, glabrate. Leaves 2 or 3 subopposite at a node (often with 1 leaf lobed and the other 2 entire), dimorphic; petioles 1.5–5 cm long, narrowly winged; blades broadly ovate, acuminate, rounded at base, entire or 3–5-lobed, the entire blades 5–6 cm long and 3–4 cm wide, the lobed blades 9–11 cm long and 6–10 cm wide. Racemes short, at the dichotomies of stems; peduncles 1–2 cm long; pedicels ca 1 cm long, broadening at apex; flowers 1–6; calyx ca 2 mm long; corolla white or yellowish, 10–12 mm long, lobed about halfway; filaments ca 1 mm long, attached to dorsal surface of a membranous ring inserted near base of corolla tube; anthers lanceolate, ca 6 mm long, with 2 terminal pores; ovary elongate, conical; stigma punctiform. Berries ellipsoid, ca 2 cm long, white at maturity, sometimes striped with green when young; seeds yellow, flattened, ca 1.5 mm long. *Aviles 105, Zetek 5040.*

Collected twice; not seen in recent years, but it could reoccur in clearings. Flowering and fruiting mostly in the rainy season, but probably to some extent throughout the year.

Known only from the Pacific slope of Panama, from tropical moist forest in the Canal Zone, Panamá, and Darién and from tropical dry forest in the Canal Zone.

Cyphomandra hartwegii (Miers) Dun. in DC., Prodr. 13(1):401. 1852

C. heterophylla Donn. Sm., nomen nudum; *C. costaricensis* Donn. Sm.; *C. mollicella* Standl.

Monca prieto

Soft-wooded shrub or tree, to 5 m tall; most parts sparsely to densely puberulent. Leaves alternate or subtended by a smaller leaf; petioles mostly 3–6(10) cm long; blades ovate, acuminate and often downturned at apex, unequally obtuse to cordate at base, 6–25(30) cm long, 5–13 cm wide, the upper surface sparsely puberulent to glabrate, some of the trichomes usually glandular-tipped; juvenile leaves often deeply lobed, to 45 cm long and 25 cm wide. Cymes upper-axillary, indeterminate, to 50 cm or more long before flowers cease being produced, pendent; flowers 5-parted, closely spaced on peduncle; pedicels to 3 cm long in flower, elongating and

KEY TO THE SPECIES OF CYPHOMANDRA

Plants herbs to 1 m tall; leaves 2 or 3 per node, subopposite . . . . *C. allophylla* (Miers) Hemsl.
Plants shrubs or trees to 5 m tall; leaves alternate or opposite with 1 leaf of each pair much smaller
. . . . *C. hartwegii* (Miers) Dun.

Fig. 481.
Cyphomandra hartwegii

Fig. 480. *Cestrum megalophyllum*

Fig. 482. *Markea ulei*

Fig. 483. *Markea ulei*

thickening at apex in fruit; calyx ca 5 mm long, the lobes short, acute to obtuse; corolla green, 1.5–3.5 cm long, the lobes long-tapered, free to calyx; filaments thick, short, fused to corolla tube; anthers to 10 mm long, papillate, the connective swollen, purple, the pores terminal; stigma held just above anthers in bud, elongating after anthesis. Berries globose to ellipsoid, 3–5 cm long, green with lighter stripes, yellow at maturity; seeds numerous, disk-shaped, reniform to round, to 6 mm long. *Croat 9016.*

Uncommon, appearing in tree-fall areas of the forest and at the edges of clearings. The plant flowers unceasingly for long periods, with individual plants flowering for as much as a year or more.

Visited by male *Eulaema bombiformis* bees (*fide* label on *Dressler 3070*).

Honduras to Brazil and Bolivia. In Panama, known from tropical moist forest in the Canal Zone, Bocas del Toro, and Panamá, from premontane wet forest in Chiriquí and Panamá, from tropical wet forest in Colón, and from premontane rain forest in Colón, Coclé, and Panamá.

See Fig. 481.

LYCIANTHES (Dun.) Hassl.

Lycianthes maxonii Standl., J. Wash. Acad. Sci. 17:14. 1927

L. maxonii Standl. var. *appendiculata* Standl.

Slender shrub, to 1.5 m tall; younger stems and petioles puberulent; nodes somewhat swollen. Leaves alternate; petioles to 5 mm long; blades mostly oblong-obovate, acuminate, cuneate to obtuse at base, 7–15 cm long, 2–5 cm wide, glabrous, entire, minutely undulate; major veins few, arcuate-ascending. Fascicles very short, axillary, usually with only 1–3(6) flowers or fruits at any time; pedicels 1–2.5 cm long in flower, to ca 3.5 cm long in fruit; flowers 5-parted; calyx truncate, with 5 short, wart-like protuberances below rim; corolla violet, 5–10 mm long, divided to near base, the lobes spreading at anthesis; stamens yellow, shorter than corolla, held together at apex in a ring; filaments united to tube in basal half, very short; anthers 4 mm long, with 2 terminal pores; ovary ovoid, glabrous; style straight, held well above anthers, nearly equaling length of corolla. Berries globose to obovoid, to 1.2 cm long, orange becoming red at maturity; seeds ca 4 mm long, longer than broad. *Croat 6307.*

Frequent, especially in the old forest. Flowers from March to December, mostly from April to September. The fruits develop within about a month, maturing from April to January, mostly from July to October. Plants usually have flowers and fruits simultaneously.

Usually growing about a meter tall, but becoming nearly twice as tall in areas of the older forest.

Nicaragua to Panama. In Panama, known from tropical moist forest in the Canal Zone, Panamá, and Darién, from premontane wet forest in Panamá, from tropical wet forest in Colón, from lower montane wet forest in Chiriquí, and from premontane rain forest in Panamá (summit of Cerro Jefe).

Lycianthes synanthera (Sendt.) Bitter, Abh. Naturwiss. Vereine Bremen 23(2):499–500. 1919

Shrub or tree, usually hemiepiphytic, ca 2(10) m tall, ± glabrous except for tufted axils on lower leaf surfaces. Leaves alternate or subopposite with the pairs unequal; petioles 5–30 mm long; blades ± elliptic, acuminate, acute to obtuse at base, 5–20 cm long, 3–8.5 cm wide. Flowers solitary or in fascicles of few flowers; pedicels ca 1.5 mm long; calyx cyathiform, nearly truncate, 2–3 mm long, becoming woody and 5–6 mm long in fruit; corolla 5-lobed to near base, ca 1 cm long, purplish; stamens 5, ± equal; anthers 6–7 m long, yellow, united into an ellipsoid column ca 3 mm diam; style exserted. Berries globose, ca 7 mm diam, fleshy; seeds discoid and margined, ca 1.5 mm long, foveate, yellow. *Croat 11899.*

Rare, known only from the area east of Wheeler Trail 1600. Flowers from May to September. The fruits mature from June to October.

Mexico to Panama. In Panama, known from tropical moist forest in the Canal Zone, Bocas del Toro, Colón, Veraguas, and Panamá, from premontane wet forest in Chiriquí, from tropical wet forest in Coclé and Panamá, and from premontane rain forest in Coclé.

MARKEA L. C. Rich.

Markea ulei (Damm.) Cuatr., Feddes Repert. 61:78. 1958

M. panamensis Standl.

Hemiepiphytic shrub, often associated with ant nests; smaller stems fleshy, sometimes arising from a large, swollen, tuberous stem perched in the crotch of a tree, with only a slender root trailing to the ground; stems and leaves glabrous. Leaves alternate or subopposite with the pairs equal or unequal; petioles less than 1 cm long; blades obovate to broadly elliptic, acuminate, acute to rounded at base, 6–20 cm long, 3–8.5 cm wide, coriaceous. Panicles short, axillary, cymose; peduncles stout, to ca 7 cm long; pedicels 10–14 mm long; flowers 5-parted, ca 1.3 cm long; calyx deeply divided, the lobes oblong-ovate, curved inward, to ca 7 mm long, persisting in fruit; corolla campanulate, to 14 mm long, greenish-yellow, densely pubescent outside with very short glandular

---

KEY TO THE SPECIES OF LYCIANTHES

Plants hemiepiphytic; corolla more than 1 cm long . . . . *L. synanthera* (Sendt.) Bitter
Plants terrestrial; corolla less than 1 cm long . . . . *L. maxonii* Standl.

trichomes, some tinged with purple, the lobes broadly rounded, imbricate, 3–4 mm long; stamens included; filaments fused to tube two-thirds their length; anthers yellow, ca 4 mm long, with longitudinal dehiscence from apex, the connective purplish; ovary yellow; style included, nearly equaling lobes, club-shaped at apex. Berries ovoid, glabrous, smooth, yellow becoming white and fleshy at maturity, 1–1.5 cm long; seeds light brown, ca 2.7 mm long, curved, thicker at one end. *Croat 8911.*

Occasional, occurring from within a few feet above the ground to more than 30 m high. Flowers and fruits throughout the year. Flowering may be heaviest during the middle of the dry season. The fruits probably develop in about a month.

Central Panama to Peru. In Panamá, known from tropical moist forest in the Canal Zone, Panamá, and Darién and from tropical wet and premontane rain forests in Colón.

See Figs. 482 and 483.

PHYSALIS L.

Physalis angulata L., Sp. Pl. 183. 1753

Herb, generally 25–100 cm tall, glabrous or sparsely and inconspicuously pubescent with antrorse appressed trichomes on stems, leaves, pedicels, calyces, and especially major veins of lower leaf surfaces. Leaves alternate; petioles mostly 1–8 cm long; blades ± ovate to ovate-lanceolate, acute to acuminate, acute to rounded at base, mostly 3–14 cm long, 1.5–10 cm wide, sparsely and irregularly toothed to entire. Flowers 5-parted, solitary in leaf axils; pedicels slender, 6–12 mm long (longer in fruit); flowering calyx 4–7 mm long, divided to near middle, the lobes sharply acute; corolla pale yellow to white, ± campanulate, 6–12 mm long, not dark-dotted inside; anthers blue, (1)2–2.5 mm long, considerably shorter than filaments; ovary ovoid, ca 1 mm long; style 3 mm long. Fruiting calyces inflated, ovoid, 2–3.5 cm long and ca 2 cm wide, 10-angled or ribbed, greenish, glabrous to inconspicuously pubescent (especially on lobes); berries globose, ca 1 cm diam, yellow; seeds numerous, disk-shaped, ca 1 mm long. *Ebinger 594.*

Rare, occurring in clearings. Flowering and fruiting throughout the year.

The only collection known from the island is a mixed collection by Ebinger (*594*), which also contains a specimen of *P. pubescens.* The species is similar to *P. pubescens,* but distinguished by being glabrate or sparsely short-pubescent, by its smooth fruit, and by the lack of a dark eye in the corolla.

Manitoba to southern Argentina; much of the Old World. In Panama, known from tropical moist forest in the Canal Zone, Bocas del Toro, Herrera, Panamá, and Darién, from tropical dry forest in Panamá (Taboga Island), and from tropical wet forest in Colón.

Physalis pubescens L., Sp. Pl. 183. 1753

Herb, less than 1 m tall, generally densely pubescent on stems, leaves, pedicels, and calyces with long multicellular trichomes. Leaves alternate; petioles mostly 1–7 cm long; blades ovate, acute to acuminate, acute to subcordate and often unequal at base, mostly 2.5–9 cm long, 1.5–6 cm wide, entire or lobed to bluntly toothed. Flowers 5-parted, solitary in leaf axils; pedicels slender, mostly 3–6 mm long, to ca 1 cm long in fruit; flowering calyx 4–10 mm long, divided to middle or beyond, the lobes sharply acute; corolla campanulate, pale yellow, to 1 cm long, with a brown to black or purple spot at base inside and pubescence below the spots; anthers blue, 2–3 mm long, about as long as filaments; ovary ovoid, ca 1.5 mm long; style 4 mm long; stigma capitate. Fruiting calyces inflated, ovoid, to 3 cm long and 2 cm wide, sharply 5-angled with 10 ribs, greenish; berries globose, 1–1.8 cm diam, yellow; seeds numerous, disk-shaped, reticulate, ca 1 mm wide. *Shattuck 45.*

Uncommon, occurring in clearings. Flowering and fruiting throughout the year. Individual plants flower over a long period and most have both flowers and fruits present.

The fruits may be dispersed by animals or wind—the inflated calyx enables the wind to carry them for short distances.

Northeastern United States to Chile and Argentina; warmer regions of the Old World. In Panama, ecologically variable; known from tropical moist forest in the Canal Zone, Bocas del Toro, Panamá, and Darién, from tropical dry and premontane moist forests in Coclé, from premontane wet forest in Panamá, and from tropical wet forest in Los Santos.

SOLANUM L.

Solanum antillarum O. E. Schulz in Urban, Symb. Ant. 6:164–66. 1909–10

S. parcebarbatum Bitter

Hoja hedionda

Small, almost glabrous tree, 1.5–3 m tall. Leaves opposite in unequal pairs or alternate; petioles obscure or to 1.5 cm long; blades elliptic, gradually acuminate, attenuate and decurrent at base, 3.5–15 cm long, 2–7.5 cm wide, the axils of the lower surface with crisp-villous trichomes. Cymes short, axillary; flowers white, 5-parted,

---

KEY TO THE SPECIES OF PHYSALIS

Most parts conspicuously pubescent, the trichomes usually crisp, multicellular; fruits sharply 5-angulate; corolla usually with a dark "eye" at base .... *P. pubescens* L.
Most parts glabrate or shortly and inconspicuously pubescent, the trichomes not multicellular; fruits not sharply angulate; corolla lacking dark spots .... *P. angulata* L.

KEY TO THE SPECIES OF SOLANUM

Plants usually armed with prickles (usually only on trunk in *S. hayesii*); anthers elongate, attenuate
    at apex:
    Calyx with 5 minute, marginal teeth . . . . *S. subinerme* Jacq.
    Calyx deeply 5-lobed:
       Leaves subsessile, the petioles less than 7 mm long . . . . *S. jamaicense* P. Mill.
       Leaves petiolate, the petioles usually more than 1 cm long:
          Calyx armed; plants scandent . . . . *S. lanciifolium* Jacq.
          Calyx usually unarmed; plants erect:
             Stems unarmed; leaves glabrate above . . . . *S. hayesii* Fern.
             Stems prickly; leaves densely stalked-stellate above . . . *S. ochraceo-ferrugineum* (Dun.) Fern.
Plants unarmed; anthers elliptic or cylindrical, obtuse at apex (except in *S. hayesii*):
    Leaves glabrous or bearing simple trichomes:
       Vein axils tufted below . . . . *S. antillarum* O. E. Schulz
       Vein axils not tufted below . . . . *S. arboreum* Dun.
    Leaves bearing stellate or branched trichomes:
       Most leaf blades more than 8 cm wide, the upper surface nearly glabrous, the lower surface
          densely and softly stellate-pubescent; plants stout trees usually more than 6 m tall . . . . . .
          . . . . *S. hayesii* Fern.
       Most leaf blades less than 8 cm wide, the pubescence various but seldom as above; plants
          shrubs or small trees usually less than 3 m tall (*S. rugosum* rarely to 5 m):
          Flowers in lateral corymbs of many flowers; leaves obtuse or rounded at base, minutely
             and densely white-tomentose below . . . . *S. argenteum* Poir.
          Flowers in large terminal cymes; leaves acute to attenuate at base:
             Leaves mostly narrowly oblong-elliptic and less than 5 cm wide; trichomes on lower
                blade surface stalked; calyx lobed halfway to base . . . . *S. asperum* L. C. Rich.
             Leaves mostly elliptic or ovate and more than 5 cm wide; trichomes on lower blade sur-
                face stalked or sessile; calyx variously lobed:
                Trichomes on stems and inflorescences mostly long-stalked; calyx lobed much more
                   than halfway to base . . . . *S. umbellatum* P. Mill.
                Trichomes on stems and inflorescences mostly sessile; calyx lobed halfway to base . . .
                   . . . . *S. rugosum* Dun.

---

pedicellate; peduncles, pedicels, and calyces enlarged in fruit; calyx with 5 short thick lobes; corolla white, 8–9 mm broad at anthesis, deeply lobed, the lobes acute with a median rib; stamens nearly sessile, to 2 mm long; anthers yellow, rounded on dorsal side, the 2 terminal pores directed inward; style to twice as long as stamens. Berries globose, ca 1 cm diam, purple to brownish at maturity; seeds disk-shaped, ca 2.5 mm diam. *Aviles 72, D'Arcy 3988, Woodworth & Vestal 555.*

Occasional, occurring at the edge of the forest along the margin of the lake. Flowering and fruiting to some extent all year but mostly in the dry season and early rainy season.

Recognized by its nearly glabrous leaves with tufted axils below.

Mexico to Venezuela; the Antilles. In Panama, known from tropical moist forest in the Canal Zone, Veraguas, and Panamá, from premontane wet forest in Chiriquí and Coclé, from tropical wet forest in Coclé, and from premontane rain forest in Coclé.

Solanum arboreum H. & B. ex Dun., Sol. Syn.
    20. 1816
    S. kenoyeri Standl.

Shrub, 1–2.5(8) m tall, unarmed and ± glabrous. Leaves alternate or subopposite in unequal pairs; petioles less than 1 cm long; blades ± elliptic, long-acuminate, cuneate at base, the larger leaves 20–30 cm long, 8–12 cm

wide, the smaller leaves ca 4 cm long and 3 cm wide. Cymes very short, congested, axillary; pedicels to 1 cm long (lenticellate and to 13 mm long in fruit); flowers 5-parted; calyx usually less than half as long as corolla, lobed to middle or beyond, the lobes ovate, acute; corolla white, to 4(8) mm long, lobed to near base, the lobes ovate, acute, ribbed medially; stamens to 3.5 mm long, orange; anthers nearly sessile, to 3 mm long, the terminal pores directed inward; style elongating to ca 7 mm long, at first erect, becoming directed to one side. Berries ± globose, ca 1 cm diam; seeds many, more or less ellipsoid, ca 3 mm long, 2 mm wide, black. *Croat 11151.*

Rare, in the forest. Flowers and fruits throughout the year, but the flowering is heaviest in April and May and again in August. Most fruits mature in June and July and later in October. Flowering plants frequently bear mature fruits.

Mexico to Peru. In Panama, known from tropical moist forest in the Canal Zone, Bocas del Toro, San Blas, Los Santos, Panamá, and Darién.

See Fig. 484.

Solanum argenteum Poir. in Lam., Encycl. Méth.
    Bot., Suppl. 3:755. 1814
    S. lepidotum Dun.; *S. salviifolium* Standl. non Lam.

Slender shrub, to 2.5(7) m tall; stems (at least when young), inflorescences, and lower leaf surfaces densely pubescent, the trichomes sessile, short-stalked, white,

Fig. 485. *Solanum hayesii*

Fig. 484. *Solanum arboreum*

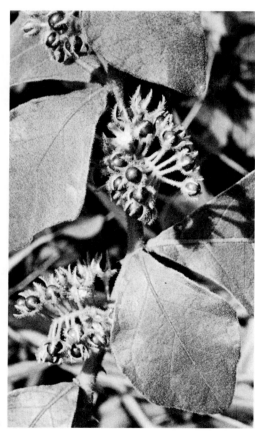

Fig. 486. *Solanum jamaicense*

stellate. Leaves usually alternate, sometimes subtended by a much smaller subopposite leaf; petioles ca 0.5(1) cm long; blades narrowly ovate, acuminate, rounded and often subequal at base, 7–14(20) cm long, 3–7 cm wide, dark green above, white and sparsely stellate-pubescent below (more densely so on midrib). Corymbs upper-axillary, bearing many flowers; flowers 5-parted; calyx weakly lobed, enlarged in fruit, glabrous inside, the lobes obtuse; corolla white, 5–7 cm long, divided ca three-fourths of the way to base, the tube glabrous; style somewhat shorter than lobes, sparsely stellate-pubescent at base; stamens oblong; filaments short; anthers ca 2 mm long, their terminal pores directed inward. Berries globose, 6–8 mm diam, black, glabrate. *Foster 1711.*

Rare, collected north of Zetek Trail 400 in the old forest and once along the shore (*Wetmore & Abbe 200*). Seasonal behavior uncertain. Apparently flowers and fruits twice per year, once in the dry season (January to April) and once in the rainy season (July and August).

Mexico to Brazil. In Panama, known from tropical moist forest in the Canal Zone, Bocas del Toro, and Darién, from premontane wet forest in Panamá, from tropical wet forest in Coclé and Panamá, from premontane rain forest in Panamá (summit of Cerro Jefe), and from lower montane wet forest in Chiriquí.

Solanum asperum L. C. Rich., Actes Soc. Hist. Nat. Paris 1:107. 1792

Shrub or small tree, to 3.5(7) m tall; pubescence brown, stellate, dense except on upper leaf surfaces. Leaves alternate; petioles less than 1 cm long; blades narrowly oblong-elliptic, acuminate, attenuate at base and decurrent on petiole, 8–21 cm long, 1.5–6 cm wide. Inflorescences pseudoterminal, congested, the branches terminating in cymes of many flowers; flowers 5-parted, 10–15 mm wide, densely stellate-tomentose; calyx lobed to about middle, the lobes acute, enlarging and persisting in fruit; corolla white, lobed more than midway or to near base, ca 8 mm long, ± glabrous inside; stamens less than 5 mm long; filaments stout, fused at base to corolla; anthers oblong, blunt, with 2 apical pores and longitudinal slits; ovary at first conspicuously pubescent; style stellate-tomentose, emerging from bud well above anthers, ca 5 mm long. Berries globose, ca 1 cm diam, sparsely puberulent; seeds numerous, ovoid, flattened, ca 1.5 mm long. *Croat 12839.*

Uncommon, occurring in disturbed areas. Flowering mostly throughout the rainy season, with the fruits developing rather soon and persisting into the dry season.

Easily confused with *S. umbellatum,* which tends to have slightly wider leaf blades (to 8 cm) and a more deeply lobed calyx (one-half to three-fourths of the way to base).

Belize to Brazil. In Panama, known from tropical moist forest in the Canal Zone and Panamá, from premontane wet forest in the Canal Zone, Colón, and Panamá, from tropical wet forest in Colón, and from premontane rain forest in Coclé, Panamá, and Darién.

Solanum hayesii Fern., Proc. Amer. Acad. Arts 35:560. 1900

Shrub or tree, to 12 m tall; trunk usually armed with prickles; bark smooth; wood soft; stems densely matted with sessile stellate trichomes, usually unarmed except on juveniles. Leaves alternate; petioles 1.5–4.5(6) cm long; blades ovate, abruptly acuminate, obtuse to rounded and inequilateral at base, 11–29 cm long, 6–17 cm wide, shiny and glabrous to sparsely stellate-pubescent above (especially on younger leaves), very densely pubescent below with stalked stellate trichomes; juvenile plants with the petioles armed, the leaves lobed, to 42 cm long and 30 cm wide. Inflorescences of usually once-branched, indeterminate racemes on stems below upper leaves, to 6 cm long (longer in fruit); flowers 5-parted, white, densely stellate-tomentose, to 2 cm wide; calyx 3–4 mm long, the lobes ± truncate and mucronate at apex; corolla deeply lobed, ca 1 cm long; stamens yellow; filaments thick, fused to tube; anthers 3–4.5 mm long, with both terminal pores and longitudinal slits; pollen shed while style short; style thick, to 2 mm long, elongating to 7 mm. Berries globose, to 13 mm diam, yellow at maturity; seeds many, ± ovoid, ca 3 mm long. *Croat 7706.*

Frequent in the forest; rare along the shore. Flowers throughout much of the year; known to flower from September to May, mostly from November to May. The fruits apparently require several months to mature, but mature fruits may be found on trees that are still flowering during the dry season.

Trees in the species bear flowers all of one type, either short-styled or long-styled. The short styles are 2 mm long, held well beneath the anthers. The long styles first equal the height of the anthers, then lengthen to 7 mm, well above the anthers. Pollen is ready to shed at anthesis, and the stigma is apparently receptive at the same time.

Nicaragua to Peru. In Panama, known from tropical moist forest in the Canal Zone and Bocas del Toro, from premontane wet forest in Colón, Coclé, and Panamá, from tropical wet forest in Colón and Coclé, and from premontane rain forest in Panamá.

See Fig. 485.

Solanum jamaicense P. Mill., Gard. Dict. ed. 8. 1768
Friega plata

Shrub, to 2(3) m tall, densely stellate-pubescent on all exposed parts, the trichomes chiefly stalked; stems and lower midribs armed. Leaves alternate or subopposite in unequal pairs, sessile or nearly so; blades irregular, ± rhomboid, acute at apex, cuneate to attenuate at base, the larger leaves 12–21 cm long, 7–15 cm wide, often 3–5-lobed, the smaller leaves 4–11 cm long, 2–6 cm wide, usually rhomboid. Racemes short, congested, axillary, subumbellate; pedicels to ca 2 cm long; flowers 5-parted; calyx divided to near base, the lobes narrowly triangular, to 5 mm long, densely pubescent with long stalked trichomes; corolla white, 6–9(11) mm long, lobed to near

base, the lobes narrowly triangular, ± glabrous inside; stamens ca 4.7 mm long; filaments less than 1 mm long; anthers elongate, attenuate at apex; style 6–7 mm long; ovary and style sparsely pubescent. Berries globose, ca 6 mm diam, orange to red at maturity; seeds oblong, orange, flattened, 1.5–2 mm long. *Croat 11987.*

Common in Rear #8 Lighthouse Clearing. Flowering from June to October, especially in July. The fruits mature from August to December, a good many of them in October.

Florida to Panama and parts of South America; the Antilles. In Panama, ecologically variable; known from tropical moist forest in the Canal Zone, San Blas, Veraguas, Herrera, Panamá, and Darién, from premontane moist forest in the Canal Zone, from premontane wet forest in Panamá, from tropical wet forest in Colón and Panamá, and from premontane rain forest in Panamá.

See Fig. 486.

Solanum lanciifolium Jacq., Coll. 2:286. 1788

S. scabrum Vahl non P. Mill.; *S. donnell-smithii* Coult.

Araño gato

Vine or liana, often growing into canopy; stems, petioles, and veins of leaves below armed with recurved spines and stellate-tipped bristles. Leaves alternate or subopposite in slightly unequal pairs; petioles to 4 cm long, stellate-pubescent; blades elliptic, acuminate, acute to rounded at base, 9–25 cm long, 4–8 cm wide, entire to pinnately lobed, the pubescence of both surfaces sparse, short-stalked, sessile, stellate; juveniles with the leaves to 13 cm broad. Racemes axillary, to 7 cm long; peduncles, pedicels, and calyces stellate-pubescent; pedicels to ca 12 mm long; flowers 5-parted; calyx 1–2 mm long, spiny, the lobes abruptly pointed, broadly spreading to recurved, bearing both stellate trichomes and larger simple spines, ± warty in fruit; corolla lobes lanceolate, violet (owing to dense violet trichomes) and densely stellate-tomentose outside, white and ± glabrous inside, ca 15 mm long and 3 mm wide, spreading, the tips curved up and in; anthers yellow, ± sessile, to 13 mm long, tapered to apex, with terminal pores; style about half as long as anthers in bud, longer than anthers at anthesis. Berries globose, 1.3–3 cm diam at maturity, sparsely stellate-pubescent to glabrate, orange or red; seeds yellow, flattened, ca 2.5 mm long. *Croat 14914.*

Seedlings common in the forest; adult plants, usually climbing into the lower parts of the canopy, occasional. Flowers and fruits throughout the year, but the flowers are especially abundant in April.

Mexico to Brazil; West Indies. In Panama, widespread and ecologically variable; known from tropical moist forest in all areas, from premontane dry forest in Los Santos, from tropical dry forest in Los Santos, Coclé, and Panamá, from premontane moist forest in Los Santos, from premontane wet forest in Colón and Chiriquí, from tropical wet forest in Colón and Coclé, from premontane rain forest in Coclé, and from lower montane rain forest in Chiriquí.

Solanum ochraceo-ferrugineum (Dun.) Fern., Proc. Amer. Acad. Arts 35:560. 1900

S. isthmicum Bitter

Shrub or small tree, to 3.5 m tall, densely stellate-pubescent; stems armed with sharp recurved prickles. Leaves alternate or subopposite with the pairs unequal; petioles 1–6 cm long; blades broadly ovate, acuminate, obtuse to truncate and often unequal at base, 9–20 cm long, 5–14 cm wide, subentire to shallowly lobed. Racemes branched, axillary; pedicels 1–1.8 cm long; flowers 5-parted, to 3 cm broad; calyx to 1 cm long, lobed to about middle, the lobes acuminate; corolla white, lobed to beyond middle, 3–5 cm diam and spreading at anthesis, stellate outside, glabrous inside; stamens ca 1 cm long; filaments short; anthers yellow with terminal pores; style somewhat longer than anthers. Berries globose to ovoid, 1–1.5 cm long, yellow at maturity; seeds yellow, compressed, ca 2.5 mm long. *Wetmore & Abbe 176, Woodworth & Vestal 625.*

Rare in clearings; probably more abundant when the vegetation was younger. Flowering to some extent throughout the year, but especially at the beginning of the rainy season in May. Plants apparently may flower more than once per year, because mature fruits have been found on flowering plants and the fruits no doubt require less than a year to mature.

Mexico to South America. In Panama, widespread and ecologically variable; known from tropical moist forest in all areas, from premontane dry forest in Coclé, from tropical dry forest in Coclé, from premontane moist forest in Los Santos, from premontane wet forest in Chiriquí, from tropical wet forest in Panamá, and from lower montane rain forest in Chiriquí.

Solanum rugosum Dun. in DC., Prodr. 13(1):108. 1852

Shrub or small tree, to 5 m tall; stems bearing dense, sessile, stellate trichomes. Leaves alternate or subopposite in unequal pairs; petioles 5–30 mm long; blades broadly lanceolate to elliptic, acuminate, acute to attenuate at base, 8–25 cm long, 2.5–7 cm wide, stellate-pubescent above and below. Cymes dense, long-stalked; peduncles to 20 cm long, branched near apex, stellate-pubescent; pedicels 2–9 mm long; flowers 5-parted; calyx 3–5 mm long; corolla white, ca 1.5 cm broad; stamens ca 3 mm long; ovary broadly ovoid, ca 1.5 mm long; style 3 mm long. Berries ca 1 cm wide, glabrate, yellow at maturity; seeds ca 2 mm long. *Aviles 64, Shattuck 475.*

Several collections exist from previous years, but the plant has not been seen recently on the island. Flowering and fruiting all year, with the flowers especially abundant in the first half of the rainy season.

Belize to Peru and Brazil; the Antilles. In Panama, growing most abundantly in higher and wetter areas; known from tropical moist forest in the Canal Zone, Bocas del Toro, Panamá, and Darién, from premontane wet forest in Colón, Coclé, and Panamá, from tropical wet forest in Colón and Coclé, and from premontane rain forest in Panamá.

Fig. 487. *Solanum umbellatum*

Fig. 488. *Witheringia solanacea*

Fig. 489. *Witheringia solanacea*

Solanum subinerme Jacq., Enum. Syst. Pl. Ins. Carib. 15. 1760

Shrub, to 1.5(4) m tall, densely to moderately stellate-pubescent; stems armed with stout recurved spines. Leaves alternate or subopposite in unequal pairs; petioles to 2.5 cm long; blades broadly elliptic, short-acuminate, obtuse to rounded and unequal at base, 5.5–12.5(15) cm long, 3.5–7.5 cm wide. Racemes upper-axillary, unbranched; pedicels 6–15 mm long; flowers 5-parted; calyx 2.3–3 mm long, with 5 narrow teeth; corolla violet, 1.5–2 cm long, lobed nearly to base, glabrous to sparsely pubescent inside, spreading at anthesis; stamens ca 1–2 cm long; anthers to 1.5 cm long, yellow, attenuate at apex, with terminal pores; filaments very short; ovary and basal part of style stellate-pubescent; style 5–15 mm long, white with a green tip, exceeding anthers and curved to one side. Berries globose, 1–1.5 cm diam, red or orange; seeds many, brown, flattened, 3–3.5 mm long, 2.5–2.8 mm wide, ca 1.5 mm thick, weakly alveolate. *Croat 7479.*

Occasional, in Rear #8 Lighthouse Clearing. Common elsewhere in the Canal Zone. Flowering and fruiting nearly all year, but with fewer flowers during the late dry and early rainy seasons.

Heterostylous forms may exist within the species (short-styled: *Croat 7479, Tyson & Blum 2000*; long-styled: *Croat 7149, Zetek 5043*).

Mexico to South America; the Antilles. In Panama, known from tropical moist forest in the Canal Zone, San Blas, and Panamá, from premontane moist forest in the Canal Zone, from premontane wet forest in the Canal Zone and Panamá, from tropical wet forest in Panamá, and from premontane rain forest in Coclé and Panamá.

Solanum umbellatum P. Mill., Gard. Dict. ed. 8, no. 27. 1768

Shrub, to 7 m tall (usually less than 2.5 m), unarmed, densely stellate-pubescent, the trichomes of stems and inflorescences coarse and long-stalked. Leaves alternate or subopposite in unequal pairs; petioles to 2.5 cm long; blades narrowly elliptic, long-acuminate, cuneate and decurrent on petiole at base, 8–25 cm long, 2–8 cm wide. Inflorescences pseudoterminal, branched many times, the branches terminating in helicoid cymes of many flowers; pedicels 3–6 mm long; flowers 5-parted, 1–1.5 cm wide, densely stellate-tomentose; calyx 2–5 mm long, lobed one-half to three-quarters its length, stipitate-glandular; corolla 1–1.5 cm long, white, lobed to about middle, ± glabrous inside; stamens ca 4 mm long; anthers cylindrical, obtuse at apex, with terminal pores directed inward; ovary and style ± glabrous; style to ca 5 mm long. Berries round, ca 1 cm diam, glabrous, yellow at maturity; seeds many, brown, irregularly ellipsoid to obovoid, ca 1.5 mm long, 1.2 mm wide, flattened, ca 0.5 mm thick, the surface alveolate. *Croat 5106.*

Rare, found at the edge of the Laboratory Clearing. Flowering mostly in the first half of the rainy season, with some flowering at the beginning of the dry season as well. The fruits develop quickly and may be found with flowers.

Mexico to Colombia; Greater Antilles. In Panama,

known from tropical moist forest in the Canal Zone, Bocas del Toro, Chiriquí, Veraguas, Los Santos, Herrera, Panamá, and Darién, from tropical dry forest in Panamá (Taboga Island), from premontane moist forest in the Canal Zone, from premontane wet forest in Chiriquí and Panamá, from tropical wet forest in Colón and Coclé, from premontane rain forest in Chiriquí and Panamá, and from lower montane rain forest in Chiriquí.

See Fig. 487.

WITHERINGIA L'Hér.

Witheringia solanacea L'Hér., Sert. Angl. 33, pl. 1. 1788
Capsicum macrophyllum (Dun.) Standl.

Suffrutescent herb or shrub, to 3(4) m tall; stems, petioles, and midribs often purplish when young, nearly glabrous or sparsely pubescent with short, stiff, uncinate or appressed trichomes. Leaves alternate or subopposite in unequal pairs; petioles mostly 1–5 cm long, flat and marginally ridged above, the ridges extending along stem to next node; blades ovate to elliptic-oblong, acuminate, obtuse to subcordate and sometimes somewhat inequilateral at base, mostly 8–20 cm long, 3.5–11 cm wide, glabrous or sparsely scabrous and moderately shiny above, duller below, the trichomes sparse or restricted to veins. Fascicles dense, axillary; flowers minutely pubescent, greenish-yellow; calyx truncate, sparsely pubescent, to 2 mm long; corolla 4- or 5-lobed, 6–9 mm long, divided about halfway to base, the lobes spreading; stamens 4 or 5, broad, ca 5.3 mm long; filaments adnate to tube basally, densely villous in basal half; anthers acute at apex; ovary ovoid, glabrous, ca 1 mm long; style 4 mm long. Berries depressed-globose, ca 8 mm diam, orange-red; seeds many, ca 1.3 mm long, densely alveolate-ridged. *Croat 12199.*

Occasional, in the Laboratory Clearing; rarely encountered in the forest as a shrub. Flowering and fruiting all year, but especially during the rainy season. Often occurring with both flowers and mature fruits.

Distinguished by the dense axillary clusters of greenish-yellow flowers and bright red fruits.

Mexico to Brazil; the Antilles. In Panama, known from tropical moist forest in the Canal Zone, Bocas del Toro, San Blas, Chiriquí, Los Santos, Panamá, and Darién, from premontane wet forest in Chiriquí, Coclé, and Panamá, from tropical wet forest in Colón, Coclé, and Panamá, from premontane rain forest in Chiriquí, Panamá, and Darién, and from lower montane wet forest in Chiriquí.

See Figs. 488 and 489.

125. SCROPHULARIACEAE

Annual herbs, sometimes suffruticose, sometimes aquatic and succulent. Leaves opposite, obscurely petiolate; blades simple, entire to serrate, often glandular-punctate; venation pinnate and usually palmate at base, or strictly palmate; stipules lacking. Flowers bisexual, solitary in the

KEY TO THE SPECIES OF SCROPHULARIACEAE

Corolla rotate, actinomorphic, densely hirsute inside at base, the lobes equal, all spreading, much longer than the tube; plants often 30–60 cm or more tall . . . . *Scoparia dulcis* L.
Corolla tubular, zygomorphic (bilabiate), not densely hirsute inside at base, the lobes unequal, partly erect, partly spreading, shorter than the tube; plants usually reclining, rarely more than 15 cm tall (except *Stemodia verticillata,* which is usually erect):
    Sepals free, markedly unequal:
        Stems densely villous (visible to naked eye); at least one sepal cordate at base . . . .
        . . . . *Bacopa salzmannii* (Benth.) Edw.
        Stems glabrous; sepals not cordate at base . . . . *Mecardonia procumbens* (P. Mill.) Small
    Sepals fused at base, mostly equal or subequal, not cordate at base:
        Leaf blades conspicuously dark-punctate beneath; corolla more than 12 mm long; plants often erect and more than 30 cm tall . . . . *Stemodia verticillata* (P. Mill.) Hassl.
        Leaf blades not punctate beneath; corolla less than 7 mm long; plants usually reclining or at least less than 15 cm tall:
        Flowers sessile; capsules more than 6 mm long, acute at apex and greatly exceeding sepals
        . . . . *Lindernia diffusa* (L.) Wettst.
        Flowers long-pedicellate; capsules rounded or truncate at apex, shorter than or about as long as sepals . . . . *Lindernia crustacea* (L.) F. W. Müll.

leaf axils, sometimes in a terminal raceme, zygomorphic or actinomorphic (*Scoparia*); calyx 4- or 5-lobed, the lobes imbricate, ± equal or very unequal; corolla 4- or 5-lobed, rotate and 4-lobed near the base (*Scoparia*) or bilabiate, the upper lip entire or 2-lobed, the lower lip usually spreading, 3-lobed; stamens 4, free or didynamous and fused to tube much of their length; anthers 2-celled, introrse, dehiscing longitudinally; ovary superior, 2-locular, 2-carpellate; placentation axile; ovules many, anatropous; style 1, apically dilated; stigma usually bilamellate. Fruits 2-valved septicidal capsules; seeds many, with fleshy endosperm.

Scrophulariaceae are sometimes confused with the Solanaceae (124), Bignoniaceae (126), and Gesneriaceae (127). They are usually distinguished by being small herbs (on BCI) with small, usually zygomorphic, non-plicate corollas, bilocular ovaries, and small bivalved capsules.

Flowers are insect pollinated. BCI species are typically scentless (Percival, 1965) and well adapted for pollination by small bees, except perhaps *Scoparia dulcis,* which is not zygomorphic.

Seeds do not appear to be well adapted to wide dispersal and are probably spilled locally. Most species grow in open, weedy areas. Elsewhere most Scrophulariaceae are dispersed by browsing animals, which eat the seeds along with the vegetation and pass the seeds unharmed (Ridley, 1930).

About 220 genera and 3,000 species; cosmopolitan.

BACOPA Aubl.

Bacopa salzmannii (Benth.) Wettst. ex Edw., Bol. Commiss. Geogr. Estado São Paulo 13:176, 181. 1897
    B. violacea (Pennell) Standl.

Small aquatic or terrestrial herb; stems simple or branched, thick, succulent, densely brownish-villous. Leaves sessile; blades rounded-ovate, rounded or very obtuse at apex, cordate and clasping at base, usually 8–15

mm long, punctate, glabrous above, glabrous to villous on midrib below; veins several, palmate. Flowers solitary in axils; pedicels usually longer than leaves; calyx 5-parted, the lobes imbricate, the posterior lobe larger, cordate, very obtuse, 4–5 mm long, long-ciliate, the lateral lobes linear-lanceolate; corolla bilabiate, blue or white, 8 mm long, the upper lip 2-lobed, the lower lip 3-lobed; stamens 4, didynamous, included, held against upper lip. Capsules ± globose, bisulcate; seeds small, numerous. *Shattuck s.n.* (July 15, 1934).

Collected once by Shattuck; possibly still occurring along the shore, but probably not persisting at such low elevations. Seasonal behavior uncertain. Specimens have been collected mostly in the dry season and early rainy season with flowers and fruits.

Southern Mexico to Brazil; from sea level to 1,500 m elevation. In Panama, known from tropical moist forest in the Canal Zone and Panamá.

LINDERNIA Allioni

Lindernia crustacea (L.) F. W. Müll., Syst. Census Austr. 1:97. 1882
    Torenia crustacea (L.) Cham. & Schlecht.

Weak, annual herb, creeping to ascending, 5–15 cm high, inconspicuously scaberulous; stems of older plants branched below many times, rooting at lower nodes, sometimes flushed with purple. Petioles to 4 mm long; blades variable in shape, ± ovate, obtuse to acute at apex, subcordate to rounded at base and decurrent to base of petiole, 12 mm long and 12 mm wide or smaller, crenate to serrate, ciliolate; veins pinnate. Flowers solitary in the axils, the terminal flowers often in an inflorescence of 3; pedicels to 12 mm; calyx ± bilabiate, 5-parted, 3–4 mm long; corolla bilabiate, ± 5 mm long, violet to white, the lower lip spreading, 3-lobed; stamens 4, in 2 pairs, held next to upper lip; filaments of one stamen pair longer; thecae divergent; staminodia knotty, arising from near the bottom of the free part of the filament; style to

25 mm long, persistent until later stages of fruit; stigmas 2. Capsules ovoid, to 4 mm long, septicidally dehiscent; seeds numerous, tan, ca 0.3 mm long, the surface granular. *Croat 6579.*

A common plant of open areas of the Laboratory Clearing; no doubt in other clearings as well. Flowers and fruits from May to October.

Throughout the tropics and subtropics. In Panama, known from tropical moist forest in the Canal Zone, San Blas, and Darién and from tropical wet forest in Colón.

Lindernia diffusa (L.) Wettst. in Engler & Prantl, Nat. Pfl. 4(3b):79. 1891

Weak, puberulent, annual herb, creeping to ascending, to ca 12 cm high; stems weakly branching on older plants, rooting at nodes, often flushed with purple. Petioles to 3 mm long, much shorter in plants exposed to sun; blades ovate to orbicular, obtuse to acute at apex, obtuse to cuneate at base, to 2 cm long and 1.5 cm wide, crenate to dentate; veins pinnate. Flowers solitary in axils or clustered at ends of stems, sessile to subsessile; calyx 5-parted, regular or bilabiate, the lobes slender, deeply toothed to about middle, keeled at base, scaberulous on keel and usually on margins; corolla ca 6 mm long, falling soon after anthesis, bilabiate, the upper lip straight, entire, the lower lip with 3 recurved lobes, blue-violet or with the 3 lower lobes white and the upper lobe blue-violet, the tube often yellow below the lower lip; stamens 4, in 2 pairs, held next to upper lip; filaments fused to tube basally, the free part arched inward; anthers of a pair fused together forming a cross-shaped pattern, the filaments of the upper pair with a prominent staminodium arising from the base of the free parts; style 2.5 mm long, positioned beneath the upper lip, remaining on fruit until later stages; stigmas 2. Capsules narrowly ovoid, acute at apex, scaberulous, 6–8 mm long, 2-valved; seeds numerous, tan, ± ovoid, ca 0.6 mm long, the surface granular. *Croat 11845.*

Locally abundant; known only from open areas of the Laboratory Clearing. Flowers and fruits throughout the rainy season.

Throughout the tropics; possibly native to the Old World tropics. In Panama, known from tropical moist forest in the Canal Zone.

MECARDONIA R. & P.

Mecardonia procumbens (P. Mill.) Small, Fl. S.E. U.S. 1065, 1338. 1903

Bacopa procumbens (P. Mill.) Greenm.

Glabrous, ephemeral herb, creeping or ascending, sometimes rooting at the lower nodes; branches mostly from the base, to 15 cm long, slender, flexuous; stems sometimes strongly angled, sometimes glandular, the glands dark brown, multicellular, peltate; roots mostly short and fibrous. Petioles very short or indistinct; blades ovate, obtuse at apex, cuneate at base, mostly 10–15 mm long, 8–10 mm wide, glandular-punctate on lower surface, the margins crenate in distal two-thirds of blade; major

lateral veins usually 2 or 3 per side. Flowers solitary, seldom geminate, at first terminal, soon axillary, ca 8 mm long; pedicels slender, ca 3 mm long in flower, longer in fruit, the bracts basal, 2–4 mm long, linear, mostly entire; sepals free, the outer 3 ovate, 7 mm long and 2–3 mm wide, accrescent in fruit (becoming 9 mm long and 5 mm wide), acute at apex, obtuse or rounded at base, mostly eglandular, the inner 2 linear, slightly shorter than the outer ones; corolla tubular, exserted, yellow with purple lines, 5–8 mm long, bilabiate, the lobes rounded, sometimes irregularly crenulate, the upper lip reflexed, emarginate, the lower lip 3-lobed, scarcely reflexed, the throat bearded below the upper lip with clavate, unicellular trichomes; stamens 4, ca 3.5 mm long, held against one side of the tube beneath upper lip (a rudimentary staminodium sometimes present); filaments glabrous, inserted at unequal heights in the basal half of the tube, free ca 2 mm; thecae 2 per anther, oblong, held apart on expanded connectives, the connectives consisting of a discoid or ellipsoid expansion at the filament apex and one slender arm to each theca; ovary narrowly ovoid, longitudinally sulcate; style 1.5 mm long, slightly curved, apically flattened; stigma a linear crest on the style. Capsules narrowly ovoid, ca 4 mm long, the walls stramineous, dehiscent loculicidally and septicidally from the apex, enclosed in accrescent calyx, the placenta enlarged, linear, persistent on the withered capsule; seeds numerous, ovoid, 0.3 mm long, longitudinally ridged with a reddish-brown reticulum. *White 149.*

The species is apparently rare and has not been collected since 1938, though it could easily be overlooked.

Southern United States through Mexico and Central and South America to the West Indies and Argentina. In Panama, ecologically variable; known primarily from the drier parts of tropical moist forest in the Canal Zone, San Blas, Chiriquí, Los Santos, Herrera, Panamá, and Darién, but also from premontane dry forest and tropical dry forest in Los Santos, from premontane moist forest in the Canal Zone and Panamá, and from premontane wet forest in Chiriquí and Coclé.

SCOPARIA L.

Scoparia dulcis L., Sp. Pl. 116. 1753

Escoba dulce, Escobilla amarga, Sweet broom

Herb, very small or to 1 m tall, branched, essentially glabrous; stems 6-ribbed. Petioles obscure; blades oblanceolate, acute at apex, narrowly tapered to base, to 3.5 cm long and 1.2 cm wide, densely punctate, crenate; major veins usually 2 or 3 on each side. Flowers solitary or clustered in leaf axils, 4-parted, 4–5 mm wide; pedicels slender, 4–6 mm long; sepals ovate, rounded at apex, 3-veined, ciliate, ca 1.3 mm long; corolla deeply lobed, to 2.7 mm long, densely pilose inside at base, the lobes rounded at apex, white or tinged with lavender; stamens 4, exserted, ca 2 mm long; anthers as broad as long, dehiscing upward; ovary ovoid, glabrous; style to 2.7 mm long, simple, longer than ovary. Capsules ovoid, 2-valved; seeds numerous, minute, irregularly ovoid, reticulate. *Croat 4160.*

Occasional in clearings. Probably flowering and fruiting throughout the year, especially from April to December.

Throughout the tropics. In Panama, known principally from tropical moist forest in all provinces; also known from premontane wet forest in Colón and Coclé and from tropical dry forest in Panamá.

STEMODIA L.

Stemodia verticillata (P. Mill.) Hassl., Contr. Fl. Chaco (Trab. Mus. Farmac. Fac. Ciencias Méd. Buenos Aires) 21:110. 1909

S. parviflora Ait.

Herb, suberect to reclining, to 75 cm long; stems slender, densely viscid-villous. Petioles slender, 2–10 mm long; blades broadly ovate, elliptic-ovate or rhombic-ovate, obtuse to acute at apex, rounded to truncate at base and attenuate-decurrent onto petiole, 0.5–3.5 cm long, 0.4–2.5 cm wide, crenate, sparsely villous on both surfaces, markedly dotted with dark glands beneath. Flowers usually solitary in axils; pedicels and calyces sparsely to densely villous; pedicels very slender, mostly 1–3.5 cm long; sepals 5-parted, subequal, 7–9 mm long, linear-subulate; corolla bilabiate, 12–17 mm long, sparsely short-villous outside, white with purplish streaks in throat, the lower lip spreading, 3-lobed; stamens 4, in 2 pairs, included, held against upper lip; thecae disjunct, stipitate; style dilated at apex. Capsules ovoid; valves 4; seeds ca 5 mm long, flattened. *Standley 41004.*

In clearings; collected by Standley but not seen recently, although it is to be expected. Probably flowers and fruits throughout the year, especially in the dry season.

Mexico to Argentina; West Indies. In Panama, known from tropical dry forest in Panamá, from premontane moist forest in the Canal Zone, and from tropical moist forest in the Canal Zone, Bocas del Toro, San Blas, Los Santos, and Darién.

126. BIGNONIACEAE

Trees or more commonly tendriled lianas; stems sometimes angled; axillary buds with the outer scales often pseudostipular and sometimes foliaceous; interpetiolar glands often present on stem. Leaves opposite, bipinnate or palmately compound, petiolate; blades mostly pinnate and 2- or 3-foliolate, the terminal leaflet often replaced by a simple or forked tendril; leaflets entire or serrate, often glandular; venation mostly pinnate; stipules lacking. Racemes, cymes, thyrses, or rarely panicles terminal or axillary, usually bracteolate, of few or many flowers; flowers bisexual, zygomorphic; calyx cupular or campanulate, truncate to 5-lobed; corolla tubular, bilabiate, generally 5-lobed, the lobes imbricate; stamens 4, usually didynamous, attached to the corolla tube; staminodium usually present; anthers 2-celled, the 2 thecae usually widely divergent; ovary 1, superior, 1- or 2-locular; placentation of 4 vertical ridges, parietal in 1-locular ovaries, axile in 2-locular ovaries; ovules numerous, anatropous; hypogynous disk present; style slender; stigma 2-lobed. Fruits 2-valved, septicidal or loculicidal capsules; seeds numerous, winged or rarely the wing almost lacking, without endosperm.

Members of the family are characterized by flowers with a tubular, showy, bilabiate corolla, four stamens, and anthers with two widely divergent thecae, and by capsular fruits with usually prominently winged seeds.

Pollination is mostly by insects, especially bees, though *Martinella obovata* is pollinated by hummingbirds.

KEY TO THE TAXA OF BIGNONIACEAE

Plants trees, lacking tendrils:
  Leaves palmately compound, the leaflets 3–7 . . . . *Tabebuia*
  Leaves pinnately or bipinnately compound:
    Leaves pinnately compound; flowers orange; tree cultivated in the Laboratory Clearing . . . . .
    . . . . *Spathodea campanulata* Beauv.
    Leaves bipinnately compound, the leaflets 3–25 per pinna; flowers lavender-blue; trees common in the forest . . . . *Jacaranda copaia* (Aubl.) D. Don
Plants tendriled lianas:
● Flowers predominantly white or yellow (sometimes marked with purple in *Amphilophium*; for fruiting specimens, see the following key to fruits):
◆ Young stems square with sharp angles or with 6–8 prominent angles or ribs; tendrils trifid:
  Leaves 2- or 3-ternate; young stems square; flowers whitish-yellow; tendrils arising at nodes, not replacing terminal leaflets; capsules linear . . . .
  . . . . *Pleonotoma variabilis* (Jacq.) Miers
Leaves with 2 or 3 leaflets; young stems with 6–8 angles or ribs; flowers white; tendrils replacing terminal leaflets; capsules oblong to elliptical:
  Corolla 3–4 cm long, sometimes marked with purple; blades whitish below with stellate trichomes in axils; fruits with smooth surface, usually less than 16 cm long; pseudostipules small, caducous . . . . *Amphilophium paniculatum* (L.) H.B.K.
  Corolla 4–6 cm long, white turning yellowish in age; blades not whitish below, with plate-shaped glands in axils; fruits with spiny surface, more than 16 cm long; pseudostipules 4 per node, linear-oblong, to 9 mm long . . . .
  . . . . *Pithecoctenium crucigerum* (L.) A. Gentry

◆ Stems terete or inconspicuously angled; tendrils trifid or not:
  Interpetiolar glandular fields conspicuous; tendrils trifid:
    Flowers white throughout; pseudostipules vertically 3-seriate, to 7 mm long; fruits oblong, acute to acuminate at both ends, mostly 12–25 cm long, 3–3.5 cm wide; seeds thick, wingless . . . .*Pachyptera kerere* (Aubl.) Sandw.
    Flowers yellow-orange, pale yellow, or white with a yellow throat; pseudostipules not serially arranged in 2 sets of 3 pairs; fruits blunt on one end or less than 2 cm wide:
      Flowers white or pale yellow; leaflets with cavelike axillary domatia below; pseudostipules 3-lobed; fruits 3–4 cm wide, blunt or subcordate on lower end . . . .
      . . . . *Ceratophytum tetragonolobum* (Jacq.) Sprague & Sandw.
      Flowers bright yellow-orange; leaflets lacking domatia in lower axils; pseudostipules simple, inconspicuous; fruits less than 2 cm wide, tapered to lower end . . . .
      . . . . *Macfadyena unguis-cati* (L.) A. Gentry
  Interpetiolar glands lacking; tendrils simple or trifid:
    Flowers yellow; tendrils simple:
      Calyx fleshy, completely enveloping corolla in bud (looking much like a fruit), more than 2.5 cm long; capsules compressed-ellipsoid, 6–11.5 cm wide, the valves very woody, more than 5 mm thick . . . . *Callichlamys latifolia* (L. C. Rich.) K. Schum.
      Calyx not fleshy, not completely enveloping corolla in bud, less than 1 cm long; capsules oblong or ellipsoid, less than 6 cm wide, the valves usually less than 5 mm thick:
        Flowers not subtended by bracts; calyx truncate, much broader than corolla tube; pseudostipules foliaceous, obovate; capsules ellipsoid, 3–5 cm wide . . . .
        . . . . *Anemopaegma chrysoleucum* (H.B.K.) Sandw.
        Flowers subtended by conspicuous bracts; calyx with at least minute teeth and usually bilabiately split, not much broader than corolla tube; pseudostipules not foliaceous; capsules oblong, less than 3 cm wide (those of *A. arthropetiolatum* not known) . . . .*Adenocalymma*
    Flowers white or cream; tendrils simple or trifid:
      Flowers less than 2 cm long; leaflets lacking conspicuous glands; fruits linear, with prominent, raised lateral margins . . . . *Tynnanthus croatianus* A. Gentry
      Flowers more than 4 cm long; leaflets with glands at least at axils below; fruits not as above:
        Flowers cream, sparingly yellowish-glandular-lepidote on outside; leaflets with conspicuous glands throughout; fruits linear, gradually attenuate to apex . . . . .
        . . . . *Stizophyllum riparium* (H.B.K.) Sandw.
        Flowers white or lavender, with prominent nectar guides from lobes to deep within corolla; leaflets with dense clusters of glands in axils; fruits linear, rounded at end, shiny on surface . . . . *Cydista aequinoctalis* (L.) Miers
● Flowers not white or yellow (shades of purple, blue, pink):
  Corolla usually less than 4.5 cm long; tendrils simple:
    Calyx broadly flared, not tubular; corolla pubescent on lobes, the tube glabrous; pseudostipules foliaceous, falcate, to 1 cm long, caducous; leaflet midrib and veins densely villous below, the axils tufted . . . . *Arrabidaea patellifera* (Schlecht.) Sandw.
    Calyx tubular or cupular, not flared at apex; corolla densely pubescent throughout; pseudostipules usually inconspicuous; leaflet midrib glabrous or short-pubescent, the axils glabrous or minutely tufted:
      Flowers precocious; corolla densely stellate-pubescent; mature leaves stellate at least on midrib; fruits oblong, 9–20 cm long, ca 5 cm wide, smooth; seeds ± round with transparent wings . . . . *Xylophragma seemannianum* (O. Kuntze) Sandw.
      Flowers and fruits not as above:
        Blades densely puberulent below, whitish, the trichomes contiguous; fruits often rough with coarse, irregular, raised lenticels . . . . . . *Arrabidaea candicans* (L. C. Rich.) DC.
        Blades not densely puberulent below; fruits not as above:
          Pseudostipules lanceolate, sharp-pointed, 2–7 mm long; blades drying gray or blackish . . . . *Paragonia pyramidata* (L. C. Rich.) Bur.
          Pseudostipules inconspicuous or lacking:
            Corolla less than 1.8 cm long; veins of leaflet puberulent below, the surface conspicuously lepidote; blades drying green-brown . . . . *Arrabidaea florida* DC.
            Corolla more than 2 cm long; veins of leaflet glabrate below, the surface minutely and sparsely lepidote; blades and fruits drying reddish . . . .
            . . . . *Arrabidaea chica* (H. & B.) Verl.
  Corolla usually more than 4.5 cm long; tendrils simple or branched:
  ▲ Calyx more than 12 mm long:
    Flowers dark purple; leaflets cordate to truncate at base, puberulent below on midrib and lateral veins, lacking tufted axils; tendrils trifid, persistent; fruits more than 60 cm long . . . . *Martinella obovata* (H.B.K.) Bur. & K. Schum.

Flowers not dark purple; leaflets obtuse or rounded at base, ± glabrous below, with or without tufted axils; tendrils simple or caducous; fruits less than 50 cm long:
   Calyx spathaceous; corolla frequently more than 7 cm long; leaves glabrous; petioles glabrous; fruits smooth . . . . *Phryganocydia corymbosa* (Vent.) K. Schum.
   Calyx ± equally lobed at apex; corolla less than 7 cm long; leaves glabrate with axillary tufts below; petioles short-pubescent; fruits very rough, short-tuberculate
   . . . . *Arrabidaea verrucosa* (Standl.) A. Gentry
▲ Calyx less than 10 mm long:
   Flowers precocious (appearing when leaves are very young):
      Flowers glandular-lepidote; leaves simple or less often 2- or 3-foliolate, lacking trichomes; capsules long, linear; seeds with opaque wings . . . . *Cydista heterophylla* Seib.
      Flowers densely stellate-pubescent; leaves 3-foliolate, or 2-foliolate with a terminal tendril, with stellate trichomes at least on veins below; capsules oblong, to 18(25) cm long; seeds thick, opaque, the wings transparent . . . .
      . . . . *Xylophragma seemannianum* (O. Kuntze) Sandw.
   Flowers not precocious:
      Flowers lepidote throughout, the nectar guides conspicuous, extending from lobes to near base on 3-lobed side; vein axils of leaflets with fields of contiguous glands below; fruits linear, dark, shiny, smooth . . . . *Cydista aequinoctalis* (L.) Miers
      Flowers not glandular-lepidote throughout, with or without prominent nectar guides; vein axils of leaflets with none or few glands below; fruits linear and rough or ellipsoid to globose:
         Corolla densely pubescent outside with moniliform trichomes, ± contiguous; tube constricted within at base, the constriction pubescent; apex of petiole with conspicuous plate-shaped glands; fruits linear, bumpy . . . .
         . . . . *Paragonia pyramidata* (L. C. Rich.) Bur.
         Corolla sparsely puberulent and lepidote, the tube not constricted, with nectar guides to near base; apex of petiole lacking conspicuous glands; fruits ellipsoid to globose . . . . *Clytostoma binatum* (Thunb.) Sandw.

KEY TO THE BIGNONIACEAE
(ON THE BASIS OF FRUIT CHARACTERS)

■ Capsules not linear:
   Valves spiny-tuberculate:
      Capsules ca 5 cm long, nearly as broad . . . . *Clytostoma binatum* (Thunb.) Sandw.
      Capsules to 32 cm long and 8 cm broad . . . . *Pithecoctenium crucigerum* (L.) A. Gentry
   Valves not spiny-tuberculate:
      Capsules to about twice as long as broad; plants trees or lianas:
         Plants trees; capsules compressed-oblong, rounded to truncate on both ends, sessile, drying dark brown to black; seeds flat, small-bodied with a thin, orbicular wing; wing hyaline-membranaceous with radial, brownish striations, clearly demarcated from seed body
         . . . . *Jacaranda copaia* (Aubl.) D. Don
         Plants lianas; capsules elliptic, compressed, acute at apex, stipitate ca 1 cm at base, drying yellow-brown; seeds woody, somewhat flattened; wing essentially absent . . . .
         . . . . *Anemopaegma chrysoleucum* (H.B.K.) Sandw.
      Capsules more than 2.5 times as long as broad; plants lianas:
         Capsules oblong:
            Capsules 2–2.6 cm thick; seeds (including wing) more than 3 times wider than long, the wing not clearly demarcated from seed body . . . .
            . . . . *Ceratophytum tetragonolobum* (Jacq.) Sprague & Sandw.
            Capsules less than 1.5 cm thick; seeds (including wing) ± orbicular or less than 3 times wider than long, the wing absent or clearly demarcated from seed body:
               Valves with a prominent medial ridge; seeds lacking a wing . . . .
               . . . . *Pachyptera kerere* (Aubl.) Sandw.
               Valves lacking a medial ridge; seeds with transparent wing . . . .
               . . . . *Xylophragma seemannianum* (O. Kuntze) Sandw.
         Capsules oblong-elliptic:
            Capsules more than 20 cm long and more than 6 cm wide . . . .
            . . . . *Callichlamys latifolia* (L. C. Rich.) K. Schum.
            Capsules less than 16 cm long and less than 5 cm wide:
               Capsules blunt on both ends, the surface usually roughened and densely lepidote, the valves with a medial groove or ridge . . . . *Amphilophium paniculatum* (L.) H.B.K.
               Capsules pointed at apex, stipitate at base, the surface ± smooth, not densely lepidote, the valves lacking a medial groove or ridge . . . .
               . . . . *Anemopaegma chrysoleucum* (H.B.K.) Sandw.

■ Capsules linear, less than 3 cm wide and many times longer than broad:
  Capsules ± terete in cross section; plants trees or lianas:
    Capsules bumpy with coarse lenticels or tuberculate:
      Plants trees; capsules with irregular tubercles; seeds with transparent wings . . . .
      . . . . *Tabebuia guayacan* (Seem.) Hemsl.
      Plants lianas:
        Capsules less than 1.5 cm wide, the ends usually acute, the surface densely and minutely
          muricate, the bumps discrete and visible to the naked eye; seeds less than 1.5 cm
          long and less than 4.5 cm wide . . . . *Paragonia pyramidata* (L. C. Rich.) Bur.
        Capsules more than 2.5 cm wide at maturity, the ends both rounded, the surface irregular
          and densely covered with lenticels but not densely and minutely muricate; seeds
          more than 1.7 cm long and more than 5.1 cm wide . . . .
          . . . . *Adenocalymma apurense* (H.B.K.) Sandw.
    Capsules lacking bumps or tubercles:
      Capsules densely pubescent, the trichomes golden-brown, deciduous in patches . . . .
      . . . . *Tabebuia ochracea* (Cham.) Standl. var. *neochrysantha* A. Gentry
      Capsules lacking trichomes . . . . *Tabebuia rosea* (Bertol.) DC.
  Capsules flattened in cross section; plants lianas:
    Valves with midvein distinctly raised; seeds with hyaline wings sharply demarcated from
      body . . . . *Arrabidaea*
    Valves with midvein not raised (except *Macfadyena* slightly raised) or indistinct; seeds with
      opaque wings not sharply demarcated from body (except *Tynnanthus*):
      Valves conspicuously warty or with sharp tubercles:
        Valves with the surface like coarse sandpaper, the bumps as broad as long; valves less
          than 1.4 cm wide; seeds less than 1.5 cm long . . *Paragonia pyramidata* (L. C. Rich.) Bur.
        Valves with the surface coarsely tuberculate, the bumps conical and several times longer
          than broad; valves more than 1.3 cm wide (to 2.5 cm); seeds more than 1.5 cm long
          . . . . *Arrabidaea verrucosa* (Standl.) A. Gentry
      Valves not conspicuously bumpy or tuberculate, mostly smooth or only irregularly bumpy:
        Valves each with 2 raised longitudinal ridges (ca halfway between middle and margin) or
          the valves with lateral margins raised and the corners sharply angled:
          Fruits with raised lateral margins and square corners, without submarginal ridges, less
            than 1.2 cm wide; seeds 1.7–2.9 cm wide, with the wings sharply demarcated
            from body . . . . *Tynnanthus croatianus* A. Gentry
          Fruits lacking raised lateral margins and square corners, with 2 submarginal longi-
            tudinal ridges, usually more than 1.5 cm wide; seeds mostly 2.6–5.3 cm wide,
            with the wings not sharply demarcated from body . . . . *Cydista heterophylla* Seib.
        Valves lacking longitudinal or marginal ridges and square corners:
          Valves puberulent to very short-villous, at least along margins, thin:
            Fruits less than 8 mm wide, not much flattened; valves densely villous throughout,
              the trichomes short, crisped, not straight . . *Stizophyllum riparium* (H.B.K.) Sandw.
            Fruits more than 1 cm wide, flattened; valves puberulent near margins, the tri-
              chomes straight, erect . . . . *Martinella obovata* (H.B.K.) Bur. & K. Schum.
          Valves lacking trichomes:
            Fruits obtuse to rounded at apex:
              Seeds with the wings gradually more transparent toward margin; valves tending
                to be black, merely bending up on the sides, not curling lengthwise . . . .
                . . . . *Cydista aequinoctalis* (L.) Miers
              Seeds with the wings evenly opaque with a distinct hyaline margin; valves more
                gray, curling lengthwise after falling . . . .
                . . . . *Phryganocydia corymbosa* (Vent.) K. Schum.
            Fruits acute to acuminate at apex:
              Fruits less than 30 cm long and more than 2 cm wide; valves thick, irregularly
               warty; seeds with the wings opaque nearly throughout . . . .
                . . . . *Pleonotoma variabilis* (Jacq.) Miers
            Fruits more than 50 cm long and less than 2 cm wide; valves thin, ± smooth;
              seeds with the wings hyaline at tip . . . . . . *Macfadyena unguis-cati* (L.) A. Gentry

Spathodea campanulata is pollinated by perching birds in the Old World, where it is native (probably by Icteridae in the New World). Species pollinated mostly by euglossine bees include *Anemopaegma chrysoleucum*, *Arrabidaea verrucosa*, *Callichlamys latifolia*, *Cydista aequinoctalis*, *Pachyptera kerere*, *Phryganocydia corym-* *bosa*, and *Stizophyllum riparium*. Species pollinated mostly by euglossine or anthophorid bees include *Arrabidaea candicans*, *A. patellifera*, and *Xylophragma seemannianum*. Species pollinated mostly by anthophorid bees include *Tabebuia guayacan*, *T. rosea*, and possibly *Macfadyena unguis-cati*. *Tynnanthus croatianus* and *Arrabi-*

daea florida, both with small flowers, are pollinated by skippers (Hesperiidae) and small bees. *Ceratophytum tetragonolobum, Paragonia pyramidata*, and *Pithecoctenium crucigerum* are probably pollinated in part by *Xylocopa* bees and perhaps in part by anthophorid bees. The flowers of *Amphilophium paniculatum* remain closed until forced open by large bees.

The seeds of most species are wind dispersed. *Anemopaegma chrysoleucum, Clytostoma binatum*, and *Pachyptera kerere* have water-dispersed seeds. *Macfadyena unguis-cati* and *Adenocalymma arthropetiolatum* are probably at least partly water dispersed.

Gentry (1974) described five types of phenological classes for the Bignoniaceae. Type 1, associated with bat pollination, is not found on BCI. Type 2, "steady state" pattern, in which a few flowers are produced per day over a long time period, is represented by *Stizophyllum riparium*. Type 3, "cornucopia" pattern, with large numbers of flowers over several weeks, is the most common type among species on BCI. Type 4, "big bang" pattern, represents those species that flower profusely for a short time, such as *Tabebuia guayacan*. Type 5, "multiple bang" pattern, is similar to Type 4, but represents species that flower profusely several times per year; it is best represented on BCI by *Phryganocydia corymbosa*. [Data on pollination, dispersal, and phenology are derived partially from Gentry (1972; 1974).]

About 120 genera and 650–800 species; mostly in tropical America, but also in the tropics and subtropics of the Old World.

ADENOCALYMMA Mart. ex Meisn.

Adenocalymma apurense (H.B.K.) Sandw., Lilloa 3:461. 1938

A. inundatum Mart. ex DC.

Liana; trunk to 10 cm diam, often 4-sulcate; bark smooth, gray; stems with 4 phloem arms in cross section; twigs subterete, glabrous to slightly lepidote, usually drying black with whitish lenticels, the nodes lacking glandular fields; tendrils simple; pseudostipules small, pointed, ovate. Leaves 2- or 3-foliolate; petioles 3.2–6.5 cm long; petiolules 0.9–4 cm long; leaflets ovate to elliptic-ovate, acuminate, rounded to subcordate at base, 4.5–17 cm long, 2.5–8 cm wide, weakly lepidote on both surfaces, drying dark gray or olive, the margins wavy, with a narrow, conspicuous, cartilaginous border. Racemes axillary or terminal; flowers each subtended by a deciduous bract ca 1 cm long; rachises, pedicels, and calyces puberulent;

calyx cupular, 5–8 mm long, bilabiately split for 1–2 mm, usually with plate-shaped glands; corolla bright yellow, tubular-funnelform, 3–7 cm long, puberulent outside, glandular-lepidote inside on lobes; stamens included, the longest pair ca 1.8 cm long; anthers slightly divaricate, 2–3 mm long; staminodium 4–5 mm long; pistil 2.4–2.5 cm long; ovary cylindrical, 3–3.5 cm long, lepidote. Capsules oblong, 10–27 cm long, 2.5–3 cm wide, 1.4–2.2 cm thick, rounded at both ends, not compressed; valves woody and somewhat thickened, gray, with numerous, raised, tannish lenticels, the midrib obscure; seeds 1.7–2.1 cm long, 5.1–7.6 cm wide, the body thick, the wings thin, brown. *Foster 1116 (DUKE)*.

Apparently rare; collected once on Peña Blanca Point along the shore. Flowers elsewhere mostly during the late dry and early rainy seasons (April to August); collected in flower also in February.

The fruits are possibly water dispersed, since plants are usually found along streams or lakes (Gentry, 1972).

Mexico to the Guianas. Gentry (1973b) reported the species to be characteristic of tropical dry forest and edaphically dry regions of tropical moist forest, but none of the specimens he cited are from tropical dry forest according to the Holdridge Life-Zone map of Panama. In Panama, apparently preferring drier regions along streams; known from tropical moist forest in the Canal Zone, Chiriquí, Los Santos, Panamá, and Darién and from premontane moist forest in Darién (Punta Patiño).

Adenocalymma arthropetiolatum A. Gentry, Ann. Missouri Bot. Gard. 60:789. 1973

Liana; trunk to 3 cm diam; bark smooth, greenish; stems with 4 phloem arms in cross section; twigs subterete to tetragonal, ± glabrous, drying tan to greenish, the nodes lacking glandular fields; tendrils simple; pseudostipules small, 4-scaled, usually narrow. Leaves 2- or 3-foliolate; petioles 1–6.5 cm long, puberulent; petiolules 0.5–4 cm long; leaflets narrowly ovate, acute to acuminate, rounded to shortly subcordate at base, 5–17 cm long, 2–7.5 cm wide, glabrous to puberulent along main veins; reticulate veins prominulous. Racemes terminal or axillary; flowers with a faint musky odor (*fide* Gentry), each subtended by a deciduous bract; rachises, pedicels, and calyces puberulent; calyx cupular, 5–15 mm long, 5-toothed but ± bilabiate, usually with submarginal, sunken glands; corolla bright yellow, 4–8 cm long, puberulent outside and on lobes inside, the tube slender, the apical part campanulate, the 2 upper lobes erect, extending horizontally in front of the tube, the other 3 variously reflexed; sta-

---

KEY TO THE SPECIES OF ADENOCALYMMA

Leaflets ovate, drying dark gray or olive, with a whitish, cartilaginous margin, usually lepidote; twigs usually drying dark with pale lenticels; corolla tubular-funnelform; anthers with the thecae only slightly divergent, the connective apiculate . . . . *A. apurense* (H.B.K.) Sandw.
Leaflets narrowly ovate, drying green, lacking a whitish margin, usually not lepidote; twigs drying tan; corolla tubular-campanulate; anthers with the thecae very divergent, the connective not extended . . . . *A. arthropetiolatum* A. Gentry

mens slightly exserted, the pairs nearly equal, the longer 3.5–3.9 cm long; staminodium 12–21 mm long; pistil 5–6 cm long; ovary narrowly cylindrical, 4–5 mm long, 1.5–2 mm wide, puberulent. Capsules unknown (those of the genus oblong-linear, thick, flattened, woody). *Standley 40912 (US).*

Apparently rare; collected once. Flowers during the late rainy season in October and November.

Known only from Panama, along rivers in tropical moist forest and premontane wet forest in the Canal Zone and Panamá.

AMPHILOPHIUM Kunth

Amphilophium paniculatum (L.) H.B.K., Nov. Gen. & Sp. 3:116. 1819

Liana; branchlets, petioles, and rachises angled, hexagonal, lepidote, pubescent with stellate and simple trichomes; stems with conspicuous interpetiolar ridges; tendrils trifid; pseudostipules sickle-shaped, to 5 mm long, densely lepidote and stellate-pubescent, caducous. Leaves 2- or 3-foliolate; petioles 2.3–7 cm long; petiolules 0.6–4.3 cm long; leaflets broadly ovate to rounded, acuminate, cordate at base, 2.5–16 cm long, 2–10.7 cm wide, sparsely glandular-lepidote on both surfaces, the upper surface glabrous except on veins, the lower surface tomentulose with dense, short, stellate trichomes and with longer, usually branched trichomes dense on veins and very sparse elsewhere (younger leaves puberulent above). Panicles ± racemose, terminal; flowers 5-parted, aromatic, white, usually variously tinged with maroon, turning deep purple, 3–4 cm long; calyx enveloping one-third of corolla, the margin broad, double, ± lobed and undulate; corolla 3–4 cm long, fleshy, bilabiate, with 3 narrow lobes opposing a single broad lobe; stamens included, the longer pair to 1.8 cm long, the shorter pair to 1.5 cm long, the thecae divaricate, ca 2 mm long; staminodium 3–5 mm long, inserted 4–5 mm from base of corolla; ovary densely tomentulose. Capsules oblong-ellipsoid, blunt to subcordate on either end, 9–16 cm long, to 4.5 cm wide; valves broadly ridged along middle, with an obscure medial groove or ridge, densely glandular-puberulent; seeds chiefly to 1.5(2) cm long and (2.7–3)6.5 cm wide, the wings narrower toward the deeply lacerate lateral margins. *Croat 12499, Foster 1918.*

Uncommon in the forest. The flowers are few and long-lived from May to January. The fruits mature in the middle of the dry season.

Recognized by the unusual double calyx and the hexagonal stems with stellate trichomes.

Mexico to Argentina; West Indies. In Panama, widespread and ecologically variable; known from tropical dry forest in the Canal Zone, Los Santos, Herrera, and Panamá, from tropical moist forest all along the Atlantic slope and on the Pacific slope in the Canal Zone, Veraguas, Herrera, Panamá, and Darién, from premontane wet forest in Colón, Chiriquí, Veraguas, Coclé, and Panamá, from premontane rain forest in Darién, and from lower montane wet forest in Chiriquí.

ANEMOPAEGMA Mart. ex Meisn.

Anemopaegma chrysoleucum (H.B.K.) Sandw., Lilloa 3:459. 1938

A. punctulatum Pitt. & Standl.

Liana; stems glabrous or sparsely puberulent (especially at nodes), ± terete with interpetiolar ridges; tendrils simple, emerging from between leaflets; pseudostipules foliaceous, obovate, 3–13 mm long, to 10 mm wide, 4 per node. Leaves bifoliolate; petioles 0.7–3.4 cm long; petiolules 0.3–2 cm long; petioles and petiolules puberulent especially on upper surface; leaflets elliptic, acute to acuminate, acute at base and ending abruptly, 5–15 cm long, 1.7–7 cm wide, minutely lepidote. Flowers axillary, emerging from between pairs of pseudostipules; pedicels minutely bracteate, to 1.5 cm long; calyx truncate, ca 1 cm long, with sunken plate-shaped glands near apex, fitting loosely around slender base of corolla tube; corolla 5–10 cm long, 3–4 cm wide at apex, pale yellow outside except for white lobes, papillose-puberulent on lobes inside, 2 lobes opposing the other 3, the lower 2 usually recurved, the tube expanded below lobes, glabrous except in basal third below point where filaments are fused to tube; nectar guides bright yellow, alternating with lobes; stamens included, the longer ones 2.3–3.1 cm long, the shorter ones 1.6–2.1 cm long, the thecae divaricate, 3.5–4.5 mm long; staminodium 3–6 mm long; pistil 5.1–5.6 cm long; ovary rounded-cylindric, to 2.5 mm long. Capsules ellipsoid, smooth, 5–15 cm long, 3–5 cm wide, stipitate at base; seeds oblong, to 1.8 cm long and 2.4 cm wide, with broad opaque wings. *Croat 12683.*

Uncommon; collected chiefly on the shoreline of Gigante Bay. Flowers irregularly through the year, especially during the rainy season, with one or two long-lived flowers open at a time. Most seeds are probably released in the dry season.

The species is easily recognized by its yellow flower and foliaceous pseudostipules. When open, the flowers have a very sweet, pleasant aroma. Corolla tubes are often pierced with a hole 3–4 mm diam just above the edge of the calyx by nectar-stealing insects or birds.

The seeds are water dispersed (Gentry, 1973b).

Belize to Venezuela. In Panama, known from swampy areas of tropical moist forest in the Canal Zone (Atlantic slope), Bocas del Toro, Colón, Panamá, and Darién and from tropical wet forest in Colón and Darién.

ARRABIDAEA DC.

Arrabidaea candicans (L. C. Rich.) DC., Prodr. 9:185. 1845

A. pachycalyx Sprague

Liana; outer bark brown, flaky; branches terete with prominent lenticels even on youngest branchlets, the nodes with interpetiolar glandular fields; tendrils simple; pseudostipules lacking. Leaves 2- or 3-foliolate, bicolorous, often drying reddish above; petioles 1.6–6.1 cm long; petiolules 0.8–3.3 cm long; leaflets oblong-ovate to broadly ovate, sharply to bluntly acuminate, cuneate to

KEY TO THE SPECIES OF ARRABIDAEA

Capsule valves lacking noticeably raised medial rib, the margins sharply angled and raised; calyx drying dark with paler or thinner margin; corolla usually more than 3.6 cm long . . . . *A. corallina* (Jacq.) Sandw.

Capsule valves with somewhat raised medial rib, the margins rounded; calyx not drying paler or thinner along the margin; corolla usually less than 3.5 cm long (except *A. patellifera*):

Calyx bilabiate; fruits coarsely and regularly verrucose-tuberculate, the tubercles often to 2 mm or more . . . . *A. verrucosa* (Standl.) A. Gentry

Calyx truncate, usually minutely 5-denticulate; fruits essentially smooth or with irregular coarse but short lenticels, not coarsely and regularly tuberculate:

Corolla pubescent on lobes, the tube glabrous; calyx reflexed away from base of corolla; pseudostipules foliaceous, falcate, to 1 cm long, caducous; leaves simple or 2-foliolate, the leaflets with the midrib and veins densely villous below, the axils tufted . . . . *A. patellifera* (Schlecht.) Sandw.

Corolla densely pubescent throughout; calyx cupular, enclosing base of corolla; pseudostipules usually inconspicuous; leaves with at least some always 3-foliolate, never simple; leaflets with the midrib glabrous or short-pubescent, the axils glabrous or minutely tufted:

Blades densely whitish-pubescent below, the trichomes contiguous; fruits rough with raised bumps . . . . *A. candicans* (L. C. Rich.) DC.

Blades not whitish-pubescent below; fruits not as above:

Corolla less than 1.8 cm long; leaflets with the veins puberulent below, the surface conspicuously lepidote; blades drying green-brown . . . . *A. florida* DC.

Corolla more than 2 cm long; leaflets with the veins glabrate below, the surface minutely, sparsely lepidote; blades and fruits drying reddish . . . . *A. chica* (H. & B.) Verl.

cordate at base, 4.5–17 cm long, 2.5–11 cm wide, minutely pubescent or glabrous above, densely white-tomentulose below, usually with a few sunken glands in basal axils. Inflorescences terminal, showy, pyramidal thyrses; buds lavender; flowers magenta, drying rusty brown; calyx cupular, truncate, 4–6 mm long; corolla 2–3.5 cm long, 2–3 cm broad, densely pubescent outside and on lobes inside, sometimes extending onto 2-lobed side of corolla, the tube white inside; stamens inserted 5–6 mm from base of corolla; filaments retrorse-barbate at point of fusion with tube, the longer pair 1.5–1.6 cm long, the shorter pair 1.1–1.4 cm long, the thecae divaricate, 2–2.5 mm long, shedding pollen in bud; staminodium 4–6 mm long; pistil 1.7–1.8 cm long; ovary linear-oblong, 2–2.5 mm long, finely lepidote; style open and apparently receptive in bud; stigmas recurved, held at level of longest stamens. Capsules 13–34 cm long, ca 1.2 cm wide, acuminate at apex, acute at base, somewhat warty and glandular-lepidote, with a weak medial ridge; seeds to 1 cm long and 3 cm wide. *Croat 5228, 11099.*

Common over the canopy of the forest, extending down to near the lake. Flowers abundantly from November to February (one shore plant in July); individuals flower for about 1 month, with many flowers open at a time. The fruits mature mostly from February to April.

Recognized by its small flowers and densely white-tomentose lower leaf surface.

Southern Mexico to Colombia and Amazonian Brazil. In Panama, commonly known from tropical moist forest in the Canal Zone, Veraguas, Herrera, Panamá, and Darién; sporadically from tropical dry forest in Herrera and Coclé, from premontane moist forest in the Canal Zone, Coclé, and Panamá, from premontane wet forest in Chiriquí, and from tropical wet forest in Colón.

Arrabidaea chica (H. & B.) Verl., Rev. Hort. 40:154. 1868

Liana; stems terete, lenticellate, usually with interpetiolar glandular fields, puberulent when young with short white trichomes, glabrous in age; tendrils simple; pseudostipules small. Leaves 2- or 3-foliolate; petioles to 7 cm long, sparsely white-puberulent; petiolules 0.3–4.6 mm long; leaflets ovate, oblong-ovate to oval, acuminate, obtuse to rounded and sometimes inequilateral at base, 4–11 cm long, 3–5.5 cm wide, glabrate, sparsely lepidote, sometimes with a few glands near midrib below, usually drying reddish. Thyrses large, terminal or upper-axillary, pyramidal; calyx cupular, truncate, 2–5 mm long, densely tomentose, the teeth 5, minute; corolla red-violet, 2.3–3 cm long, densely tomentose outside and on lobes inside, the tube white inside, glabrous except at point of fusion of stamens just above base; stamens included, the longer pair 11–13 mm long, the shorter pair 9–10 mm long, the thecae divaricate, 1.5–2 mm long; staminodium 3–4 mm long, inserted 3–4 mm from base of corolla tube; pistil to 1.3 cm long; ovary narrowly cylindrical, finely lepidote, to 2.5 mm long. Capsules linear, to 23 cm long, ca 1 cm wide, flat, ± verrucose, often drying reddish; valves with a raised medial vein; seeds transversely oblong, to 1 cm long and 2.5 cm wide, the wings hyaline. *Croat 5038.*

Occasional. Flowers sporadically at the end of the dry season and throughout the rainy season, but mostly in August and September; individuals produce abundant flowers for over a month. The fruits mature principally during the dry season.

Dried herbarium specimens are conspicuously reddish. Sterile live plants are difficult to recognize, but they can

usually be distinguished by a combination of mostly trifoliolate leaves, narrow, dark-green leaflets, and interpetiolar glandular fields (Gentry, 1973b).

Mexico to Brazil. In Panama, known sporadically from wetter parts of tropical moist forest on BCI and in Bocas del Toro, Coclé, Panamá, and Darién and from premontane wet forest in Chiriquí and Veraguas.

Arrabidaea corallina (Jacq.) Sandw., Kew Bull. 1953:460. 1954

Tendrils simple; pseudostipules inconspicuous. Leaves glabrous with tufted axils beneath. Corolla lavender to magenta. Capsules linear, 12–47 cm long, 1.6–2 cm wide, conspicuously glandular-pitted. *Stoutamire 2085 (MICH)*.

Rare, if still present; known from a single recently discovered collection made in 1956 at the Rear Lighthouse Clearing. Flowers during the dry season, from February to April.

Mexico to Argentina. Most common in tropical dry forest and premontane moist forest, especially in edaphically dry locations.

Arrabidaea florida DC., Prodr. 9:184. 1845
A. panamensis Sprague

Liana; branchlets terete, lepidote, lenticellate; tendrils simple; pseudostipules small. Leaves 2- or 3-foliolate, minutely glandular-lepidote; petioles 0.9–6.2 cm long; petiolules 0.5–2.7 cm long; petioles and petiolules canaliculate, puberulent; leaflets ovate to elliptic, acuminate, acute to rounded at base, 5–14 cm long, 3–8.5 cm wide, granular-puberulent to glabrous and lepidote with a few round glands near midrib below. Thyrses terminal or upper-axillary; branches, pedicels, and calyces grayish-tomentose; calyx cupular, truncate, 3–4 mm long, the teeth 5, short; corolla abruptly campanulate, 1.4–1.8 cm long, lavender or white, pubescent outside and on lobes inside, the tube flaring above calyx; stamens subexserted; filaments lavender, the longer pair 8–10 mm long, the shorter pair 5–7 mm long, the thecae divaricate, 1.5 mm long; staminodium 3–4 mm long; pistil 8–10 mm long; ovary linear-cylindric, to 1.5 mm long, lepidote. Capsules linear, flattened, 17–23 cm long, ca 1 cm wide; valves densely glandular-lepidote, with a medial groove; seeds ca 8 mm long, 3.5 cm wide, the seminiferous area ± lunate, opaque, the wings transparent, lacerate on lateral margins. *Croat 11798, 12647*.

Occasional, in the forest canopy and at lower levels on the lakeshore. Flowers from late June through October (to November); the flowers are abundant on an individual for about a month, though each flower lasts only one day. The fruits mature mostly in the late dry season.

Sterile plants are easily confused with *Adenocalymma apurense*, but can be distinguished from that species by lacking cartilaginous leaf margins. *Arrabidaea florida* can also be confused with *Tynnanthus croatianus*, but that species is distinguished by having trifid tendrils (Gentry, 1973b).

Belize to Bolivia. In Panama, known from premontane moist and tropical moist forests in the Canal Zone, Panamá, and Darién.

Arrabidaea patellifera (Schlecht.) Sandw., Kew Bull. 22:413. 1968
Petastoma patelliferum (Schlecht.) Miers; *P. breviflorum* Standl.

Liana; trunk light brown; outer bark soft with large corky lenticels; stems terete, lenticellate, with an interpetiolar ridge, glabrate in age; tendrils simple; pseudostipules foliaceous, to 1 cm long, falcate, caducous. Leaves bifoliolate or rarely simple; petioles 1.1–3.5 cm long; petiolules 0.5–2.2 cm long; petioles and petiolules short-pilose; leaflets ovate to obovate, obtuse to caudate-acuminate, obtuse to rounded or subcordate at base, 5–15 cm long, 3–10 cm wide, lepidote on both surfaces, otherwise glabrous above, villous at least on veins below, usually with tufts or cavelike domatia in axils below, entire or rarely dentate on juveniles. Inflorescences large, pyramidal, axillary or terminal panicles; flowers red-violet to violet-purple, whitish-tipped in bud; branches and pedicels lepidote and white-villous; calyx bowl-shaped, 1–4 mm long, 4–8 mm wide, flared, the margin entire, wavy; corolla 2.5–4.7 cm long, the tube glabrous except for gland-tipped trichomes at point of staminal attachment inside, the lobes sparsely pubescent on both sides; stamens included, with the longer ones 1.1–1.3 cm long, the shorter ones 0.8–1 cm long, the thecae divaricate, to 2 mm long; staminodium 3–4 mm long, held on 2-lobed side of corolla opposite nectar guide, this side whitish and spotted inside; pistil 1.5–2.3 cm long; ovary linear, 1.5–2.5 mm long, slightly lepidote. Capsules linear, flattened, 15–39 cm long, 1.1–1.5 cm wide, with medial rib faint or prominent, densely lepidote, often sparsely whitish-lenticellate; seeds ca 1 cm long and 2.5–4 cm wide, the outer margin rounded to truncate, the lateral margin of wings translucent. *Croat 6052, 13161*.

Common along the shore and occasional in the forest. Flowers most abundantly from July to September (sometimes from June to November), rarely in January and February; the flowers are abundant on an individual vine for over a month, but each flower lasts only one day. The fruits mature mainly in the middle of the dry season.

The species can be recognized by its open, bowl-shaped calyces and white-tipped buds. Vegetative characteristics include the dense pubescence on the main veins beneath, the pubescent petioles and petiolules, and the frequency of simple leaves.

Petastoma breviflorum Standl., reported by Standley as a distinct species on the basis of its shorter corolla, appears to be inseparable from this species. Corolla length in the family is variable.

Southern Mexico to Amazonian Brazil. In Panama, known on the Pacific slope, most commonly from premontane moist forest in the Canal Zone and Panamá; known also from tropical moist forest in the Canal Zone, Veraguas, Herrera, Panamá, and Darién, from tropical

dry forest in Herrera, from premontane wet forest in Chiriquí, Panamá, and Darién, and from tropical wet forest in Panamá.

Arrabidaea verrucosa (Standl.) A. Gentry, Selbeyana 2:43. 1977

Scobinaria verrucosa (Standl.) Seib.; *Martinella verrucosa* (Standl.) Standl.

Liana, glabrous except for moderate pubescence sometimes on leaflet veins below and on young stems, petioles, and petiolules; trunk to 10 cm diam; stems terete, lenticellate, with interpetiolar glandular fields; tendrils simple; pseudostipules small or lacking. Leaves bifoliolate; petioles 1.3–5.1 cm long; petiolules 1–4.2 cm long; petioles and petiolules slightly ribbed at apex; leaflets ovate to ovate-lanceolate, acuminate at apex, obtuse to rounded at base, 8–16(17) cm long, 4–7(11.8) cm wide, conspicuously lepidote on both surfaces, with tufts of trichomes in axils below and a few, small, round glands scattered throughout. Panicles axillary and terminal, bearing few flowers; calyx tubular-campanulate, 1.5–2.8 cm long, thin, bilabiate or irregularly 3-lobed, often widely expanded above, sometimes spotted with minute protuberances; corolla with lavender lobes and a white tube, 4.5–6.5(8) cm long, soft-pubescent outside and on lobes inside, glandular-pubescent on base of stamens and on tube at point of staminal attachment; stamens included, the longer ones 1.7–2 cm long, the shorter ones 1.1–1.4 cm long, the thecae divaricate, ca 3 mm long; staminodium 1–3 mm long; pistil 2.5–2.8 cm long; ovary linear-oblong, ± quadrangular, ca 3 mm long, densely lepidote. Capsules linear, flattened, densely sharp-tuberculate, dark-colored, 18–40(46) cm long, 1.7–2.3 cm wide; seeds ca 1.7 cm long and 4.5 cm wide, the wings with a transparent margin and rounded lateral sides. *Croat 13983, 15564.*

Occasional, in the forest and along the shore. Flowers abundantly from June through August (to November), in extravagant, long-lasting bursts. The fruits mature in March and April. The species may not flower every year.

The species is distinguished by its flattened corolla tube, its large, bilabiate, membranaceous calyx, and its linear verrucose-tuberculate fruits resembling wood rasps.

Gentry treated this species as *Scobinaria japurensis* (DC.) Sandw. in the *Flora of Panama* (1973b), but he has subsequently considered that taxon a distinct Amazonian species.

Fruits make an oil spot on pressing paper when dried.

Belize to Venezuela and Bolivia. In Panama, known from tropical moist forest in the Canal Zone, San Blas, Veraguas, Panamá, and Darién and from tropical wet forest in Colón, Panamá, and Darién.

CALLICHLAMYS Miq.

Callichlamys latifolia (L. C. Rich.) K. Schum. in Engler & Prantl, Nat. Pfl. 4(3b):223. 1894

Liana, ± glabrous and minutely glandular-lepidote throughout; branchlets terete, somewhat lenticellate, striate in age; tendrils simple; pseudostipules inconspicuous. Leaves usually 3-foliolate, sometimes with a tendril replacing terminal leaflet; petioles 2.5–9(18) cm long; petiolules 1.1–9.5 cm long; leaflets ovate to elliptic, long-acuminate, acute to rounded at base, 9–16(37) cm long, 4–7(22) cm wide, with tufts of trichomes in axils below, with a few round glands along midrib below. Racemes short, terminal or axillary; flowers bright yellow, few; calyx globular to oblong-ellipsoid in bud, thick and spongy, to ca 4.5(6) cm long, 2(4.2) cm wide, completely enveloping corolla in bud, splitting somewhat irregularly as corolla emerges; corolla 7–9(11) cm long, glandular-puberulent, the limb broadly spreading, the tube with stalked, glandular trichomes at point of staminal attachment; stamens included, the longer pair 2.5–3.8 cm long, the shorter pair 1.5–2.3 cm long, the thecae divaricate, 2–3 mm long; staminodium 0.6–1.1 mm long; pistil 5.2–5.7 cm long; ovary ovate-cylindric, 3–4 mm long, ± glabrous. Capsules compressed-ellipsoid, to 32 cm long and 6–11.5 cm wide; valves ca 5 mm thick, glabrous; seeds 2–4 cm long, 6–10(13) cm wide, tan, somewhat narrowed toward the rounded lateral margins. *Croat 6835, 8058.*

Occasional, in the forest canopy and at lower levels at the margin of the lake. Flowers mostly in October and November, but sporadically throughout the year; on an individual vine a few flowers are open for about a week, followed by a brief burst of flowering. The fruits probably mature in the dry season.

The species can be recognized when sterile by the conspicuous tufts of trichomes in the lower vein axils. Also useful in recognition is the smooth gray bark with whitish lenticels and the red-drying major lateral veins (Gentry, 1973b).

Mexico to Brazil. In Panama, known from tropical moist forest in the Canal Zone, Bocas del Toro, San Blas, Veraguas, Panamá, and Darién and from tropical wet forest in Colón.

See Fig. 490.

CERATOPHYTUM Pitt.

Ceratophytum tetragonolobum (Jacq.) Sprague & Sandw., Bull. Misc. Inform. 1934:222. 1935

C. tobagense (Urban) Sprague & Sandw.

Liana; stems terete, striate, with prominent elongated lenticels and conspicuous interpetiolar glandular fields; tendrils trifid; pseudostipules at first small, ± triangular, acuminate, paired, diverging in age to expose a third pseudostipule to 8 mm long. Leaves 3-foliolate, or 2-foliolate with a terminal tendril; petioles 3–12 cm long; terminal petiolules 2.1–4.6 cm long, the lateral petiolules 0.3–5.7 cm long; leaflets ovate, ovate-elliptic, oblong-ovate, or rarely obovate, usually abruptly acuminate, usually round to subcordate at base, 6–20 cm long, 4–15 cm wide, weakly coriaceous, domatia in axils of lateral veins cavelike, sometimes barbate at mouth. Panicles terminal, subcorymbose; calyx truncate, 10–13 mm long, coriaceous, glandular-lepidote, with fields of sunken glands in distal half; corolla white, becoming yellowish

Fig. 490. *Callichlamys latifolia*

Fig. 491. *Cydista aequinoctalis*

Fig. 492. *Cydista heterophylla*

in age, densely pubescent outside and on lobes inside with moniliform trichomes, the tube yellow and glabrous inside, abruptly narrowed just above base, strongly 2-sulcate on 3-lobed side; stamens included, the longer pair 2.2–2.4 cm long, slightly shorter than style, set among a dense ring of trichomes near base; anther thecae divaricate, 4.5–6 mm long; pistil 3.2–3.3 cm long; ovary narrowly cylindrical, 5 mm long, lepidote. Capsules cylindrical, thick, elongated, acute at apex, blunt at base, 25–39 cm long, 3–4 cm wide, 2–2.5 cm thick, tan at maturity, densely covered with minute rounded or ± elongated pits; seeds to 1.5(1.8) cm long and 4.5(6.3) cm wide. *Croat 14458.*

Occasional, in the forest canopy; few fertile collections have been made on the island. Flowers at least in April, June, and October; individuals produce a few flowers over a long period. The fruits probably mature in the middle of the dry season.

May be recognized in sterile condition by the woody, trilobate pseudostipules, the prominent interpetiolar glands, the trifid tendrils, and the axillary domatia on the leaflets.

Scattered areas from southern Mexico to Venezuela and Surinam; West Indies. In Panama, known from tropical moist forest in the Canal Zone, San Blas, Panamá, and Darién.

CLYTOSTOMA Miers ex Bur.

Clytostoma binatum (Thunb.) Sandw., Recueil Trav. Bot. Néerl. 34:235. 1937
C. ocositense (Donn. Sm.) Seib.

Liana, ± glabrous; trunk ± smooth, the nodes weakly swollen; stems terete, often ± 4-sided as a result of longitudinal ridges from below petiole bases; lenticels ± prominent; tendrils simple; pseudostipules clustered, ca 3 mm long (resembling miniature bromeliads). Leaves bifoliolate; petioles 1–2 cm long; petiolules 1–2 cm long, conspicuously articulate at base; leaflets ± elliptic, acuminate at apex, obtuse to rounded at base, 6.5–19 cm long, 2.5–8 cm wide, lepidote above and below, with a few round glands along midrib below especially near base. Flowers sweetly aromatic, terminal, on short leafy or leafless branches, usually 3 or 4 per cluster; pedicels 1.5–2.5 cm long; pedicels and calyces minutely puberulent; calyx campanulate, 4–5 mm long, truncate, with ± submarginal teeth; corolla thistle-colored to pale lavender, 4.5–7.5(8.5) cm long, tomentulose outside near base and on lobes inside, the tube often white near base outside, white inside, with nectar guides extending down 3-lobed side to near base; stamens included, the longer

pair 1.4–1.9 cm long, the thecae divaricate, 2–3 mm long; pistil to 3 cm long; ovary rounded, short-cylindrical, 2 cm long, 1.5–2 cm wide, 1–1.5 cm thick, densely glandular-pubescent; stigma truncate. Capsules flattened, elliptical to orbicular, ca 5 cm long, the surface echinate; seeds orbicular, ca 2 cm diam with thick, corky wings. *Croat 13963.*

Rare; seen on the north shore of Gigante Bay and on the island north of Burrunga Point. Flowers and fruits throughout the year, with numerous short flowering periods most common in the late rainy season.

The species is recognized in sterile condition by its bromeliad-like pseudostipules (often not present), by its elliptic leaf blades, and by the four darker lines on a subterete to subtetragonal twig (Gentry, 1973b).

The seeds are dispersed by water.

Mexico to Brazil. Gentry (1973b) reported the species to be most common in tropical wet forest, though most of the collections cited were from regions of tropical moist forest, according to the Holdridge Life-Zone map of Panama. In Panama, known from swampy areas of tropical moist forest in the Canal Zone, Bocas del Toro, Colón, Panamá, and Darién and from tropical wet forest in Colón.

CYDISTA Miers

Cydista aequinoctalis (L.) Miers, Proc. Roy. Hort. Soc. London 3:191. 1863

Liana, mostly glabrous, usually glandular-lepidote; stems terete to strongly tetragonal, striate, usually with interpetiolar ridges; tendrils simple, replacing terminal leaflets; pseudostipules small or lacking. Leaves bifoliolate; petioles and petiolules canaliculate, 1.5–4 cm long; leaflets ovate to elliptic, acuminate, obtuse to rounded and often inequilateral at base, 9–16 cm long, 4–11 cm wide, usually with dense clusters of glands in basal axils below. Inflorescences terminal or axillary racemes or panicles of few flowers; calyx campanulate, 6–10 mm long, the teeth 5, minute; corolla usually 5–8 cm long, weakly bilabiate, pale lavender to white, with yellow throat, glandular-lepidote outside, pubescent inside, the trichomes minute, sessile or short-stalked, brownish, globular, the upper lip less deeply lobed than lower lip; nectar guides lavender, on 3-lobed lower lip of corolla, extending to base of tube; stamens included; filaments broadly arched, the thecae divaricate, 3–4 mm long, the longer pair 1.1–1.7 cm long; pistil 1.6–3.2 cm long; ovary narrowly cylindrical, 2–3 mm long, lepidote. Capsules linear, flattened, 30–56 cm long, 2–2.5 cm wide, glandular-lepidote, shiny as if varnished, lacking an obvious medial

---

KEY TO THE SPECIES OF CYDISTA

Flowers precocious; corolla orchid-colored with a white throat; calyx 2- or 3-lobed; capsules with 2 narrow ridges running the length of each valve; leaves often simple ...... *C. heterophylla* Seib.
Flowers not precocious; corolla white or pale lavender with a yellow throat; calyx ± truncate; capsules lacking ridges; leaves bifoliolate .... *C. aequinoctalis* (L.) Miers

ridge; seeds ca 1.7 cm long and 6 cm broad, the lateral margins smooth, rounded. *Croat 5208, 12645.*

Abundant in the forest canopy and along the shore. Flowers and fruits in numerous short bursts throughout the year.

Sterile material is marked by the ± tetragonal twigs, glandular fields in the leaf axils of the lower blade surface, and the general lack of pseudostipules (Gentry, 1973b).

Throughout tropical America. In Panama, known from tropical moist forest in all provinces; known also from tropical dry forest in Coclé and Panamá, from premontane moist forest in the Canal Zone, Los Santos, and Panamá, from premontane wet forest in Coclé and Panamá, and from tropical wet forest in Colón and Darién.

See Fig. 491.

Cydista heterophylla Seib., Publ. Carnegie Inst. Wash. 522:417. 1940

Liana, ± glabrous and glandular-lepidote; stems terete, lenticellate, striate in age; tendrils simple; pseudostipules small. Leaves usually opposite, sometimes subopposite, rarely alternate, simple or 2- or 3-foliolate, deciduous; petioles 1.5–5 cm long; petiolules 0.5–5.1 cm long; leaflets ovate to ovate-elliptic, acuminate, obtuse to rounded at base, 5–11(17.5) cm long, 3–8(11) cm wide, usually with clusters of glands in axils of basal veins below, the veins sometimes sparsely and minutely puberulent; veins at base 3. Flowers in lax axillary racemes on defoliate branches; pedicels 7–15 mm long; calyx usually 2- or 3-lobed, often truncate at base, densely lepidote, 6–8 mm long; corolla funnelform, bilabiate, orchid-colored, 4–7 cm long, glandular-lepidote outside, the lobes softly pubescent on both sides, the tube white, glandular-puberulent on nectar guides; nectar guides darker than lobes, extending from ca midway on lobes to deep in tube on 3-lobed side; stamens included, equaling length of style, the longer pair 1.7–1.8 cm long, the thecae divaricate, 4–5 mm long; staminodium short, sometimes with a reduced anther; pistil 3–3.1 cm long; ovary linear, 4–5 mm long, lepidote. Capsules linear, to 35 cm long and ca 1.5 cm wide, flattened, with 2 narrow, longitudinal ridges; seeds oblong, ca 1 cm long and 2.6–5.3 cm wide, with transparent wings ca 1 cm long. *Croat 8206.*

Rare; collected from the shore. Flowers from April to May (sometimes from February); the flowers appear in a few brief bursts when the plant is leafless. Mature fruits have been reported from adjacent areas of the Canal Zone in June, July, and November.

This species can be recognized when sterile by its tendency to produce simple leaves and by the strongly arcuate pair of basal veins on the blade, each pair with glandular field in its axil (Gentry, 1973b).

Mexico to Colombia. Gentry (1973b) reported this species as most common in tropical dry forest, but perhaps all of the collections cited were made in drier parts of tropical moist forest. In Panama, known from drier parts of tropical moist forest in the Canal Zone, Herrera, Panamá, and Darién and from premontane moist forest in Panamá.

See Fig. 492.

JACARANDA Ant. Juss.

Jacaranda copaia (Aubl.) D. Don, Edinburgh Philos. J. 9:267. 1823
Palo de bura

Tree, to 45 m tall; trunk 30–50 cm dbh, often buttressed, often wrinkled near base; outer bark fissured, very soft, thin, light brown, flaking; inner bark light tan, with minute reddish-brown sap droplets in time after being slashed; petioles, rachises, and midribs below with granular puberulence. Leaves bipinnate, crowded near apex of branches, to 40 cm long (juvenile branches with the leaves to 1.5 m long); petioles and rachises with medial groove; rachis ± angulate; pinnae 5–20 per leaf; leaflets 3–25 per pinna, ± elliptic, inequilateral, 2.5–7 cm long, 1.5–3.5 cm wide, conspicuously punctate on both surfaces, toothed only on juveniles. Flowers precocious, the panicles very large, narrow, borne among leaves near apex of stem, dense but well spaced on rachis; calyx ± truncate, but often split on one side, 4–7 mm long, densely pubescent with short, branched trichomes, lepidote; corolla lavender-blue, 4–5 cm long, densely pubescent with branched trichomes outside and on lobes inside, the tube constricted 7–8 mm at base, then prominently flaring, somewhat curved, white inside, glabrous except for glandular trichomes near juncture of stamens; stamens included, the longer pair 1.1–1.3 cm long; anthers with one theca very reduced; base of filaments and staminodium with glandular trichomes; staminodium to 3 cm long, tufted at apex; pistil 1.5–1.8 cm long; ovary flattened-cylindrical, 2–2.5 mm long and wide, 1.5 mm thick, glabrous. Capsules shortly and broadly oblong, rounded at both ends, compressed, 8–14 cm long, 5–8 cm broad, glabrous, prominently and minutely lenticellate; seeds suborbicular, to 2.5 cm long and 4 cm wide, the seminiferous area 5–6 mm long, the wings transparent except for veins extending out from center, with a distinct sinus at point of attachment. *Croat 6782, 7888.*

Common to locally abundant in the forest. Flowers from February to May; individuals flower in an extravagant burst that lasts over a month. The fruits mature from July to October, with a few fruits falling as late as April (Foster, 1974). Leaves fall in the early dry season and are replaced after flowering.

Belize to the Guianas, Brazil, and Peru. In Panama, a characteristic tree species of tropical moist forest (Holdridge & Budowski, 1956; Tosi, 1971) and common in secondary areas at low elevations (Holdridge, 1970); known from tropical moist forest in the Canal Zone, Bocas del Toro, San Blas, Panamá, and Darién, from premontane wet forest in Panamá, and from tropical wet forest in Colón and Darién.

See Figs. 493 and 494.

MACFADYENA DC.

Macfadyena unguis-cati (L.) A. Gentry, Brittonia 25:236. 1973

Liana; outer bark thin, gray, forming minute straight longitudinal fissures; stems terete, often with interpetio-

Fig. 493. *Jacaranda copaia*

Fig. 494. *Jacaranda copaia*, large pilose staminodium and smaller glabrous style

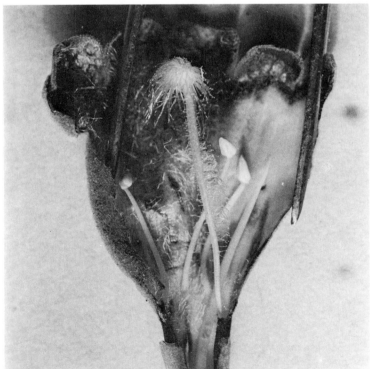

lar glandular fields (very conspicuous on young stems) with a slight ridge above glands, glabrous or minutely pubescent; tendrils terminal, trifid, hooked (rarely rudimentary); pseudostipules ovate, striate, ca 5 mm long, subpersistent. Leaves bifoliolate; petioles 1–5 cm long; petiolules 0.5–2.5 cm long; leaflets ovate-elliptic, acuminate at apex, rounded or obtuse to truncate or subcordate at base, 4–14(16) cm long, 0.7–5(7) cm wide, glabrous or weakly puberulent on midrib or throughout below, with sunken scurfy scales on both surfaces and with a few, round, sunken glands below on either side of midrib, ± entire and somewhat undulate. Inflorescences of 1–3 flowers, mostly in simple dichasia, terminal; flowers with a lemonlike aroma; calyx thin, glabrous, 7–8(18) mm long, the margin truncate or wavy; corolla yellow-orange, often pale yellow on tube, funnel-shaped, 5–9(10) cm long, the lobes ca 2 cm long, minutely ciliate; nectar guides on 3-lobed lip, faint, minutely glandular, with darker yellow-orange lines extending to base of tube; stamens with the upper pair 2.1–2.4 cm long; filaments shortly glandular-puberulent at point of attachment to tube (ca 5 mm from base); staminodium 0.5–1.5 cm long; pistil 2.8–3.8 cm long; ovary minutely papillate; disk doughnut-shaped, conspicuous; style ca 3 cm long; stigmas 2, broadly spatulate. Capsules linear, flattened, 60–108 cm long, 1–2 cm wide, tapered to acute at apex, blunt at base; valves thin, dark-colored, with a faint medial ridge, otherwise ± smooth, lepidote; replum thick, woody; seeds 5.5–6 cm wide, ca 1.5 cm long, opaque except for lateral margins, these deeply lacerate. *Croat 10224, 14062, 14645.*

Occasional but locally common in areas of the older forest. Flowers mainly from March to June; individuals flower in a single brief burst. The peak of seed release occurs in the dry season.

The species may be recognized by its trifid tendrils, which are shaped like a cat's claws. It differs from other species of bignoniaceous vines in its habit of growth: juvenile plants are appressed to the trunks and branches they climb; when the plant becomes established, it branches freely, often dangling, and the leaves are larger and free.

Mexico to Argentina; West Indies. Gentry (1973b) reported the species to be most common in tropical dry forest, but few of the collections he cited were from regions of tropical dry forest. In Panama, known principally from premontane moist forest in the Canal Zone and Panamá, also from drier parts of tropical moist forest in the Canal Zone, Panamá, and Darién, and much less commonly from premontane wet forest in Panamá and Darién.

MARTINELLA Baillon

Martinella obovata (H.B.K.) Bur. & K. Schum. in Mart., Fl. Brasil. 8(2):161, pl. 84. 1896

Liana; branches terete, finely striate, with minute gland-tipped trichomes, the nodes with broad interpetiolar ridges; tendrils trifid, the arms hooked; pseudostipules

replaced by small branches or branch scars. Leaves bifoliolate, glabrous except minutely puberulent on veins below; petioles usually at right angles to branchlets, 2–7 cm long; petiolules 4–6 cm long; leaflets ± ovate, abruptly acuminate to ± falcate-acuminate, cordate at base, 7–21 cm long, 4–15 cm wide, with round glands scattered along midrib below; major veins often darker than surface (lighter on young leaves). Inflorescences usually axillary flexuous racemes, sweetly aromatic, rarely terminal; pedicels slender, ca 1.5 cm long; calyx to 2 cm long, usually 3-lobed at anthesis, minutely glandular-puberulent or glabrous, with a few impressed glands, the lobes short, abruptly acuminate, 5–9 mm long; corolla bilabiate, 5–7 cm long, the tube curved, white at base, pale violet-purple above, sparsely glandular-lepidote outside above, the lobes dark violet-purple, densely glandular-lepidote; nectar guides on lower 3-lobed side of corolla; tube below point of staminal attachment and base of filaments with short glandular pubescence; stamens probably held against 2-lobed side of corolla, the longer pair 1.5–1.7 cm long; anther thecae divaricate, 2.5–3 mm long; staminodium 1–2 mm long; pistil 3.5–3.7 cm long; ovary linear, 4 mm long, sparsely lepidote or puberulent. Capsules brown, linear, 55–80(130) cm long, ca 1.9 cm wide, glandular-lepidote, inconspicuously puberulent, especially along margin; seeds 1–1.7 cm long, 4–6 cm wide, the wings thin. *Croat 14658.*

Rare; seen only in the forest near Armour Trail 500. Flowers sporadically throughout the year, especially in the rainy season, in extravagant bursts of flowering that last about a month. The fruits are released from April to September.

The species may be recognized in sterile condition by the trifid tendrils, the long-petiolate, glossy leaves, the prominent, linear, interpetiolar ridges, and the tendency for the petioles and petiolules to bend or twist (Gentry, 1973b).

Belize to Brazil and Bolivia. In Panama, known from wetter parts of tropical moist forest in the Canal Zone, San Blas, Panamá, and Darién, from premontane wet forest in Chiriquí, Colón, and Panamá, and from tropical wet forest in Panamá.

See Fig. 495.

PACHYPTERA DC. ex Meisn.

Pachyptera kerere (Aubl.) Sandw., Recueil Trav. Bot. Néerl. 34:219. 1937

P. foveolata DC.; *Tanaecium zetekii* Standl.

Liana; stems and petioles ± striate, the young stems sometimes ± square, the nodes with glandular fields and an interpetiolar ridge above glands; tendrils trifid; pseudostipules vertically 3-seriate, acute, to 7 mm long. Leaves 2- or 3-foliolate; petioles 0.9–7.1 cm long, with glandular fields at apex; terminal petiolules 1.5–6.1 cm long, the lateral ones 0.5–3.2 cm long; leaflets ovate to elliptic, acuminate, slightly cordate at base, 10–22 cm long, 4–12 cm wide, glabrous or minutely puberulent below especially on veins, sometimes very sparsely glan-

Fig. 495. *Martinella obovata*

Fig. 496. *Pachyptera kerere*

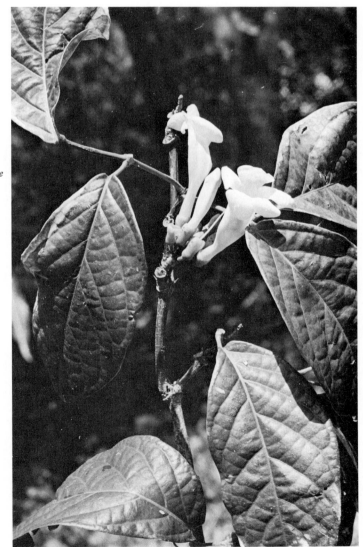

dular throughout. Inflorescences short axillary racemes, densely flowered; calyx 10–16 mm long, often with 2 shallow lobes, one lobe slightly longer and emarginate, short-puberulent, sometimes with conspicuous plate-shaped glands in apical half; corolla 5–10 cm long, white, the tube slender, puberulent outside, glabrous inside except densely glandular-pubescent below point of stami-nal attachment, bent somewhat about midway, the lobes 5, ± obovate, glandular-tomentose, with sunken brown glands on lower side near base, the upper 3 lobes erect, the lower 2 recurved; stamens included; filaments fused to corolla tube more than half their length, the longer pair 1.9–2.1 cm long, the shorter pair 1.4–1.6 cm long, the thecae divaricate, 2–4 mm long, densely white-villous; pistil 4.5–4.8 cm long; ovary flattened-cylindrical, 2–3 mm long, minutely papillose; style with 2 broad stigmas held at level of longer set of stamens. Capsules ± oblong, attenuate at both ends, 10–18(25) cm long, 2–3.8 cm wide, raised along median with a minute medial rib, minutely puberulent and with scattered, sunken, plate-shaped glands; seeds thick, wingless, opaque, 2.5 cm long and ca 3.5 cm wide, thickened below and tapered to knifelike edge. *Croat 8704, 11085.*

Occasional, usually in areas near the shore where the seeds are water dispersed, but sometimes in areas at con-siderable distance from the shore. Flowers sporadically throughout the year, mostly from September to April, with a few long-lived flowers at a time. Mature fruits have been seen from April to July.

The species is recognized by its white flower with pubescent anthers, the three-seriate pseudostipules, and the glandular field on the upper surface of the petioles near the apex.

The interpetiolar glandular areas are sometimes eaten away on the stems. Presumably this is done by insects that feed on the glands.

Belize to Amazonian Brazil. In Panama, known from wetter parts of tropical moist forest in the Canal Zone, Bocas del Toro, Panamá, and Darién, from premontane wet forest in Colón, Darién, and Panamá, and from tropi-cal wet forest in Colón and Darién.

See Fig. 496.

PARAGONIA Bur.

Paragonia pyramidata (L. C. Rich.) Bur., Vidensk. Meddel. 1893:104. 1894

Liana; trunks to 8–9 cm diam; outer bark corky, rough-ened, ± peeling; inner bark light brown, soft; stems terete, with interpetiolar ridges at nodes; tendrils bifid or trifid; pseudostipules lanceolate, sharply pointed, 2–7 mm long. Leaves bifoliolate; petioles 1–4.5 cm long, with large, sunken, plate-shaped glands on upper surface near apex; petiolules 0.7–4.7 cm long; leaflets ovate, obovate, or elliptic, abruptly acuminate and sometimes down-turned at apex, mostly obtuse to rounded at base, 8–24 cm long, 3–13 cm wide, often puberulent on veins below, minutely glandular-lepidote, subcoriaceous, usually drying grayish-green. Inflorescences broad terminal

panicles of many flowers; branches puberulent; flowers imperial purple to dark orchid (drying brown); calyx campanulate, truncate or slightly lobed, 5–9 mm long, somewhat warty and gland-dotted; corolla 2–7 cm long, densely velvety outside with moniliform trichomes except at base, glandular and moniliform-pubescent on lobes and throat inside, the tube glabrous inside except at base, constricted near base, fused to filaments; stamens set among short white bristles, the longer pair 1.6–1.9 cm long, the shorter pair 1.2–1.5 cm long; staminodium short; pistil ca 2 cm long; ovary linear, ca 5 mm long, strongly lepidote. Capsules linear, to 50(61) cm long and 1.5 cm wide, subterete or somewhat flattened with an obscure longitudinal rib, minutely bumpy; seeds 0.8–1.5 cm long, 3.5–4.5 cm wide, rounded and translucent on lateral margins. *Croat 7896, 10161.*

One of the most common bignoniaceous vines on the island. Flowers in extravagant bursts lasting about a month, abundantly in the dry season and erratically throughout the year. Most mature fruits are seen from February to May.

Because of its flower color, it may be confused at a distance with several other species. The high canopy vines can be easily observed along the shore, but, as with other vines of the family, only the fallen flowers are usually seen in the forest.

The species is most easily confused vegetatively with *Ceratophytum tetragonolobum,* but can be distinguished from that species by having simple or bifid tendrils and by lacking intepetiolar glandular fields (Gentry, 1973b). *C. tetragonolobum* has glandular fields and trifid tendrils.

Mexico to Brazil, Bolivia, and Peru. In Panama, wide-spread in tropical moist forest; known also from premon-tane wet forest in the Canal Zone, Coclé, Panamá, and Darién and from tropical wet forest in Colón, Panamá, and Darién.

PHRYGANOCYDIA Mart. ex Bur.

Phryganocydia corymbosa (Vent.) Bur. ex K. Schum. in Engler & Prantl, Nat. Pfl. 4(3b):224, f. 89H. 1894

Liana; trunk usually 5–7 cm diam, conspicuously swollen to ca 9 cm at nodes; outer bark soft, gray-brown; stems terete, striate when dry, with the interpetiolar ridge in-conspicuous; tendrils simple, spoon-shaped, with glandu-lar fields at apex; pseudostipules inconspicuous when young, stalked and with spreading forks in age. Leaves bifoliolate; petioles 0.6–3.3 cm long; petiolules 0.5–3 cm long; leaflets ovate to ovate-elliptic, bluntly acuminate, obtuse to rounded at base, 8–14(22) cm long, 3–10.5(11) cm wide, thick, essentially glabrous, lepidote, with clus-ters of glands at base above, drying gray; midrib some-times conspicuously arched. Panicles short, terminal or upper-axillary, of few flowers; calyx spathaceous, 2.5–4 cm long; corolla tubular-funnelform, 4.5–9 cm long, bilabiate, glabrous except for sparse minute glands especially outside, the lobes 5, lavender, the upper 2 broader than long, united to about middle and often overlapped at margin, the lower 3 free, the tube white-

pubescent only near staminal attachment; nectar guides on lip, dark lavender, fading into tube; stamens included, the longer pair 1.8–2.1 cm long, the shorter pair 1–1.3 cm long, the thecae divaricate, each 3.5–4.5 mm long; staminodium 2–3 mm long. Capsules linear, 35–40(53) cm long, ca 2.5 cm wide, blunt on both ends, densely lepidote, the medial ridge faint; seeds thin, ca 2 cm long, 4.5–5(7) cm wide, the seminiferous area 1 cm wide, the margins scarious. *Croat 6089, 8272.*

An abundant canopy vine in all parts of the forest. Flowers throughout the year in short bursts of few flowers, perhaps more abundantly from June to September. The fruits mature throughout the year, each in about 2 months.

Sterile recognition characters include the weakly winged petiolules and the terete, smooth, light-gray twigs (Gentry, 1973b).

Seed release occurs throughout the year (Gentry, 1973b).

Panama to Brazil. In Panama, known principally from tropical moist forest in the Canal Zone, San Blas, Coclé, Panamá, and Darién; known also from premontane moist forest in the Canal Zone and Panamá, from premontane wet forest in Colón, Darién, and Panamá, and from tropical wet forest in Colón and Darién.

See Fig. 497.

PITHECOCTENIUM Mart. ex Meisn.

Pithecoctenium crucigerum (L.) A. Gentry, Taxon 24:123. 1975

P. echinatum (Jacq.) Baillon

Liana; stems 6–8-angled, becoming ribbed in age, glabrous, whitish-lepidote; tendrils trifid or twice-trifid; pseudostipules 4 at each node, linear-oblong, to 9 mm long. Leaves 2- or 3-foliolate; petioles and petiolules 2–6 cm long, puberulent, lepidote; leaflets ovate, abruptly acuminate, round to cordate at base, 7–14(18) cm long, 6–10(14.7) cm wide, sparsely puberulent to glabrous and glandular-lepidote above and below, with several round glands in vein axils at base of leaf below, usually ciliate; veins at base 5–7. Racemes terminal, bracteate, to 20 cm long; all exposed parts sparsely to densely tomentose; flowers moderately few, with a musky odor; calyx campanulate, truncate, 1–1.5 cm long, the teeth 5, minute, chiefly submarginal; corolla white, yellowish in age, 4–6 cm long, 5-lobed, bilabiate, the lobes longer than broad, reflexed at maturity, the tube prominently curved above base, with minute, moniliform trichomes especially on lobes and tube outside; stamens held against 3-lobed side of corolla, the longer pair just below throat, 1.7–2.1 cm long, the shorter pair 1.2–1.7 cm long; staminodium less than 1 mm long, inserted near base of corolla tube; pistil 3.3–3.5 mm long; ovary short-cylindrical, 5–6 mm long, swollen at middle, with both simple and multicellular trichomes; style with the 2 lobes about as broad as long. Capsules oblong, heavy, densely spiny, to 32 cm long, 6–8 cm wide; replum thick, bearing elongate perforations along margin; seeds ca 4 cm long and

10 cm wide, the tan, ± opaque body radiating into a very broad, transparent wing. *Croat 5804, 7022.*

Occasional in the forest; common locally, often covering entire treetops. Flowers mainly in May and June, with the flowers long-lasting. The fruits mature during the following dry season (February to April).

The seeds of this species are the best fliers I know. To watch the seeds make their slow, irregular trip to the ground is always amazing.

Because of its hexagonal twigs it is confused only with *Amphilophium paniculatum,* but may be distinguished by its lack of pubescence, its much-branched tendrils, its bent, white corolla, and its large echinate fruits.

Mexico to Argentina; West Indies. In Panama, known from tropical moist forest in the Canal Zone, Bocas del Toro, San Blas, Panamá, and Darién, from premontane moist forest in the Canal Zone and Panamá, from premontane wet forest in Chiriquí and Panamá, and from tropical wet forest in Colón and Darién.

See Fig. 498.

PLEONOTOMA Miers

Pleonotoma variabilis (Jacq.) Miers, Proc. Roy. Hort. Soc. London 3:184. 1863

Liana, glabrous except for pubescence on leaflet midribs and veins below and pilose on axils near base of leaflet below; young stems square; tendrils long, trifid near apex; pseudostipules long, forked. Leaves 3-ternate, or 2-ternate with a tendril or tendril scar; petioles 2.1–6.7 cm long; terminal petiolules 1.2–6.5 cm long; leaflets ovate to elliptic, sometimes falcate, acuminate, obtuse to rounded and sometimes inequilateral at base, 5–15(16) cm long, 2.5–8.5(9.5) cm wide, weakly viscid when dried, usually with a few, round, scattered glands. Racemes short, terminal, on short lateral branches; pedicels to 1.5 cm long; calyx campanulate, 6–9 mm long, truncate, minutely toothed; corolla 7–10 cm long, pale yellow throughout or with the lobes white, the tube moderately slender with a few large glands at apex, pubescent inside just below point of staminal attachment, the lobes pubescent and glandular with a few scattered larger glands; stamens included, the longer pair 2.6–2.8 cm long, slightly shorter than style, the shorter pair 1.7–1.9 cm long, the thecae divaricate, to 5 mm long; staminodium 4–5 mm long, inserted ca 2 cm from base of corolla tube. Capsules linear-oblong, acuminate at both ends, to 22(30) cm long and 2.8 cm wide, glabrous, viscid at least on drying; seeds ca 1.2 cm long and 3.7(5) cm wide, the lateral margins ± rounded, the wings ± opaque. *Croat 7876, 7997a.*

Uncommon, in the forest and along the shore. Flowers in the middle to late dry season. The fruits mature a year later in the dry season.

The only bignoniaceous vine with bicompound leaves.

Costa Rica to Venezuela; Trinidad. In Panama, known from tropical moist forest in the Canal Zone, Bocas del Toro, San Blas, Los Santos, Panamá, and Darién and from premontane wet forest and tropical wet forest in Panamá and Darién.

Fig. 497. *Phryganocydia corymbosa*

Fig. 498. *Pithecoctenium crucigerum*

SPATHODEA Beauv.

Spathodea campanulata Beauv., Fl. Oware 1:47. 1806
  African tulip tree

Cultivated tree, 7(25) m tall, pubescent throughout, especially rufous-tomentose on lower leaf surface, inflorescence branches, and calyces; pseudostipules foliaceous, ovate-cordate, 3 cm long, 2 cm wide. Leaves pinnate, 3–19-foliolate, to 50 cm long; petioles 3–7 cm long; petiolules obsolete or to 2 mm long; leaflets ± elliptic, abruptly acuminate, acute to rounded at base, 7–13 cm long, 4–7 cm wide. Racemes terminal, corymbiform; flowers large, showy; calyx spathaceous, to ca 6 cm long, the lobes curved inward; corolla broadly campanulate, asymmetrical, more deeply cleft on one side, to ca 13 cm long, red-orange, the lobes ovate, the margin yellow and undulate; stamens directed inward toward calyx, often exserted between lowermost lobes like the style; anther thecae divaricate, 8 mm long; ovary narrowly oblong, minutely papillate. Capsules ± oblong, flattened, smooth, acuminate at apex, obtuse at base, 6–11(21) cm long, 4–5(6) cm wide, splitting open on one side; replum perforated, loose at base; seeds ca 2 cm long and 3 cm wide, the wings transparent, the margin smooth. *Croat 5177.*

Cultivated in the Laboratory Clearing. Flowers throughout the year.

The species is easily recognized by its large, red-orange flowers and oblong capsules with winged seeds.

Native to tropical West Africa; cultivated throughout the American tropics.

STIZOPHYLLUM Miers

Stizophyllum riparium (H.B.K.) Sandw., Lilloa
  3:462. 1938
  Adenocalymma flos-ardeae Pitt.; *S. flos-ardeae* (Pitt.) Sandw.

Liana; trunk smooth; outer bark soft; stems terete and striate to squarish and hollow in age, lacking interpetiolar glands; stems, petioles, petiolules, and inflorescence branches with short rufous pubescence; tendrils simple or weakly trifid; pseudostipules small. Leaves 2- or 3-foliolate; petioles 2–11 cm long; terminal petiolules 3.2–7.2 cm long, the lateral petiolules 0.7–4 cm long;

leaflets ovate to obovate, acute to acuminate, usually cordate and sometimes inequilateral at base, 3–20 cm long, 2–12 cm wide, sometimes sparsely hispid above, with short dense pubescence on veins above and below and with pellucid dots and conspicuous round glands throughout, dentate to entire. Racemes short, axillary or terminal; calyx campanulate, 1–1.6 cm long, densely rufous-puberulent, sparsely glandular near apex; corolla 3–5 cm long, cream-colored, papillose-puberulent and sparsely glandular-lepidote outside; stamens included, the longer ones 1.5–1.8 cm long, the shorter ones 1.1–1.2 cm long, the thecae divaricate, 2–2.5 mm long; staminodium 3–5 mm long, inserted ca 1 cm from base of corolla; pistil 2.7–3.1 cm long; ovary linear, tetragonal, to 4 mm long, glandular-lepidote. Capsules elongate-linear, somewhat flattened, elliptical in cross section, 30–60 cm long, 7–8 mm broad, somewhat glandular-lepidote and with dense, short, rufous puberulence; seeds 4–5 mm long, 13–24 cm wide, the wings membranaceous, brown, weakly demarcated. *Croat 10200.*

Seedlings common. One fertile collection was made by J. D. Hood (*1040*) in 1933. Flowers mostly from April to November, with few, long-lived flowers at a time. Seed release is aseasonal (Gentry, 1972).

The species has been confused with *Stizophyllum perforatum* (Cham.) Miers (now called *S. longisiligum* (Vell.) Burr. ex Baillon), which grows only in Brazil.

The species is recognized by the pellucid dots and conspicuous round glands on the leaf surfaces.

Mexico to Peru and Bolivia. In Panama, known from tropical moist forest in the Canal Zone, Bocas del Toro, San Blas, Chiriquí, Panamá, and Darién, from premontane moist forest in Panamá, and from premontane wet forest and tropical wet forest in Colón, Panamá, and Darién.

TABEBUIA Gomes ex DC.

Tabebuia guayacan (Seem.) Hemsl., Biol. Centr.-Amer.
  Bot. 2:495. 1882
  Guayacán

Tree, 15–40(50) m tall, to 1.5(2) m dbh, usually lacking prominent buttresses, glabrous except for minute, simple or stellate trichomes in vein axils of leaflets below and on a few flower parts; outer bark thick, deeply fissured

KEY TO THE SPECIES OF TABEBUIA

Flowers pink; leaflets and calyces densely whitish-glandular-lepidote, never with trichomes, with plate-shaped glands at least in lower leaflet axils; fruits with a smooth surface . . . .
. . . . *T. rosea* (Bertol.) DC.
Flowers yellow; leaflets with stellate trichomes below, at least in axils, lacking plate-shaped glands in axils; fruits with a rough surface:
  Leaflets often 7, pubescent only in axils of lateral veins below; calyx pubescent with sparse, thick, stellate trichomes; fruits ± lepidote, with conspicuous irregular ridges and tubercles
  . . . . *T. guayacan* (Seem.) Hemsl.
  Leaflets rarely 7, pubescent throughout below, at least along veins; calyx woolly-pubescent with shorter stellate and longer barbate or simple trichomes; fruits at least sparsely stellate-pubescent, lacking conspicuous ridges and tubercles . . . .
. . . . *T. ochracea* (Cham.) Standl. var. *neochrysantha* A. Gentry

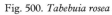
Fig. 499. *Tabebuia guayacan*

Fig. 500. *Tabebuia rosea*

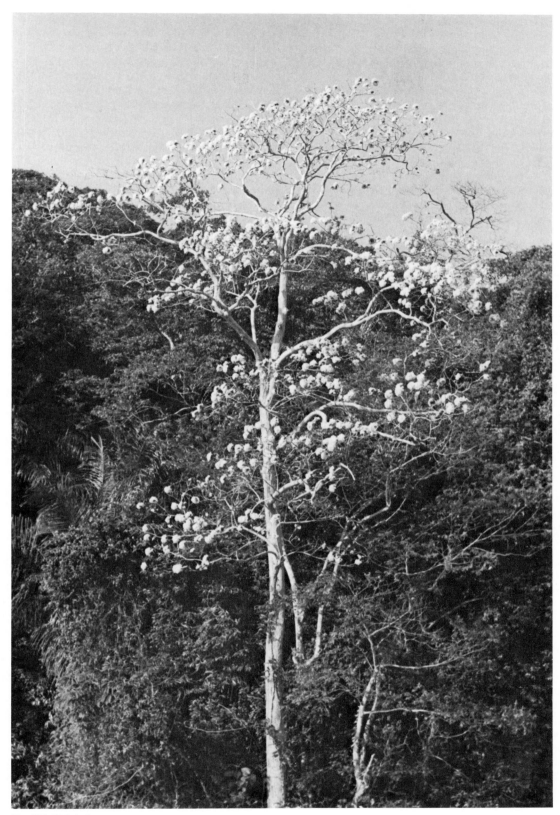

Fig. 501. *Tabebuia rosea*

vertically, the plates between fissures broken into scales, flaky and shaggy on older trees; wood very hard, heavy; branchlets subtetragonal; pseudostipules and interpetiolar glands lacking. Leaves palmately 5- or 7-foliolate, deciduous; petioles 7–10(23) cm long; petiolules 2–4.5 cm long; leaflets ovate to elliptic, acuminate, acute to broadly rounded at base, minutely glandular-dotted, entire, the terminal leaflet largest, 9–30 cm long, 3.5–15.5 cm wide. Flowers precocious, the panicles dense, terminal; calyx campanulate, 1–1.5 cm long, usually shallowly 5-lobed, sparsely stellate-pubescent; corolla yellow, 6–11 cm long, almost as wide, glabrous outside, softly pubescent on nectar guides inside, the tube brownish, the lobes 4–4.5 cm long, obovate; stamens held against glabrous side of tube, the longer pair 1.5–2.5 cm long, the shorter 12–17 mm long; filaments arched, each pair of anthers facing each other, the thecae divaricate, 2–3 mm long; pistil 2.4–3.3 cm long; ovary linear, 3–5 mm long, glabrous to lepidote. Capsules cylindrical, terete with conspicuous irregular ridges and tubercles, 30–50(61) cm long, 1.5–2.5(2.9) cm diam, glandular-lepidote; seeds 9–11 mm long, 3.5–4 cm wide, the wings hyaline-membranaceous, sharply demarcated from body of seed. *Croat 5388, 7940.*

Frequent in the forest. Flowers almost exclusively in one or two brief bursts during the dry season (January to May), but with occasional asynchronous individuals flowering during the rainy season. The fruits mature mostly near the end of the dry season, but some seeds fall in the early rainy season. Leaves fall during the dry season.

The species can be recognized by having leaves often with seven leaflets and stellate trichomes in the lower vein axils (Gentry, 1973b).

Mexico to Colombia. In Panama, known from tropical moist and tropical wet forests in the Canal Zone, Colón, and Darién, from premontane moist forest in Panamá, and from premontane wet forest in Chiriquí.

See Fig. 499.

Tabebuia ochracea (Cham.) Standl. var. **neochrysantha** A. Gentry, Brittonia 22:260. 1970

T. heterotricha DC.; *T. chrysantha* (Jacq.) G. Nicholson sensu Sandw. non Jacq.

Tree, to 25 m tall and to 50 cm dbh; outer bark with shallow vertical furrows separating flat-surfaced ridges; wood hard, heavy; branchlets subtetragonal, stellate-pubescent when young, glabrate in age; glands and pseudostipules lacking. Leaves palmately 5-foliolate, deciduous; petioles 6–18 cm long, stellate-pubescent; petiolules 0.2–5.1 cm long; leaflets mostly oblong-obovate, abruptly acuminate, obtuse to truncate (rarely subcordate) at base, membranaceous, lepidote on both surfaces, the pubescence stellate, sparse above and dense below, the margins entire to serrate, the terminal leaflet largest, 5–22 cm long, 1.8–4.4 cm wide. Inflorescences contracted terminal panicles, branches stellate-pubescent; flowers precocious; calyx campanulate, 5-lobed, 8–15 mm long, woolly-pubescent with shorter stellate and longer (to 7 mm) simple trichomes; corolla yellow, tinged reddish in throat, 4–8.3 cm long, nearly glabrous outside, villous inside, glandular-pubescent at level of stamen insertion, the tube brownish (at least in dried specimens), 3–5.8 cm long; stamens didynamous, the longer ones 1.5–2 cm long, the shorter ones to 1.5 cm long, the thecae divaricate, 1.5–2.5 mm long; staminodium 2–11 mm long; pistil 2–2.7 cm long; ovary linear, to 5 mm long, lepidote to puberulent with simple and stellate trichomes. Capsules cylindrical, terete, tapered at both ends, 13–25 cm long, 1–1.5 cm wide, with golden woolly pubescence of short stellate trichomes and longer, ± simple trichomes; seeds 4–8 mm long, 1.8–2.9 cm wide, the wings hyaline-membranaceous, distinctly demarcated from body of seed. *Croat 6915.*

Known only as a seedling on BCI, but easily occurring as an adult as well. Elsewhere in Panama flowers from January to April with apparently two brief bursts of flowering. The fruits are mature in the middle to late dry season (Gentry, 1972).

Superficially the species looks much like *T. guayacan.* BCI would be about the limit of the species' range toward the wetter Atlantic slope.

The species *T. ochracea* ranges from Honduras to Brazil; var. *neochrysantha* occupies the northern part of the range, extending as far south as Venezuela. In Panama, the variety is known from drier parts of tropical moist forest in the Canal Zone, Veraguas, Panamá, and Darién, from tropical dry forest in Coclé and Herrera, from premontane moist forest in the Canal Zone and Panamá, and from premontane wet forest in Coclé.

Tabebuia rosea (Bertol.) DC., Prodr. 9:215. 1845

T. pentaphylla (L.) Hemsl.

Roble, Roble de sabana

Tree, 5–35 m tall, 1 m dbh; outer bark deeply fissured, very thick; inner bark thick, with annular rings; stems, petioles, and leaflets moderately lepidote throughout. Leaves palmately 3–5-foliolate; petioles 5–25(32) cm long; petiolules 0.2–8 cm long, of varying lengths on one petiole; leaflets broadly elliptic, abruptly acuminate, obtuse to rounded or slightly subcordate at base, 5–22 cm long, 2–11 cm wide, with clusters of round glands in vein axils at base below. Flowers precocious, the panicles terminal, cymose; calyx bilabiate, 1.5–2.5 cm long, densely glandular-lepidote, persisting in fruit; corolla lavender to whitish, 6–11 cm long, glabrous outside, pubescent inside on lobes and especially on nectar guides within tube on 3-lobed side of corolla; stamens with the longer filaments 1.4–2 cm long, the shorter ones 1–1.5 cm long, the thecae divaricate, 2.5–3.5 mm long; staminodium 2–6 mm long, inserted 5–10 mm from base of corolla; pistil 1.9–3.2 cm long; ovary linear, 5–8 mm long, densely lepidote. Capsules cylindrical, elongate, terete or somewhat flattened, 25–35 cm long, ca 1.5 cm wide, densely lepidote and also with a few larger, crateriform glands, lacking projections. *Croat 7771, 8394.*

Frequent throughout the island. Flowers most abundantly in March and April (sometimes in February); individuals produce an extravagant burst of flowers over a long period. The fruits are seen all year; seeds are re-

leased mostly in the early rainy season. Leaves fall possibly twice per year, once in the dry season and again around July.

The species can be recognized when sterile by its leaves, which have five leaflets and are lepidote but otherwise glabrous (Gentry, 1973b).

Mexico to Colombia and Venezuela; West Indies; ranging from near sea level to 1,500 m. In Panama, a characteristic species of tropical dry, premontane moist, and tropical moist forests (Tosi, 1971); known from tropical moist forest in the Canal Zone, Bocas del Toro, San Blas, Veraguas, Herrera, Panamá, and Darién, from tropical dry forest in Coclé, from premontane moist forest in Panamá, from premontane wet forest in Chiriquí, Coclé, Panamá, and Darién, and from tropical wet forest in Colón, Veraguas, and Panamá.

See Figs. 500 and 501.

TYNNANTHUS Miers

Tynnanthus croatianus A. Gentry, Ann. Missouri Bot. Gard. 58:93. 1971

Liana; stems terete, glabrous, the young stems moderately lenticellate, the older stems conspicuously roughened; interpetiolar ridges indistinct; tendrils trifid; pseudostipules minute, acute, the apex caducous, leaving a brown scar on base, the base bulbous, less than 2 mm high. Leaves 3-foliolate, or 2-foliolate with a tendril scar, glabrate to inconspicuously pubescent; petioles longer than petiolules, mostly (1.4)3–7 cm long; terminal petiolule 2.8–3.3 cm long, longer than lateral ones; leaflets ovate-elliptic, abruptly acuminate, obtuse to rounded at base, mostly 5–13 cm long, 2.5–7.5 cm wide, bicolorous, entire; reticulate veins distinct on paler lower surface, the lateral veins in 3–5 pairs. Inflorescences axillary and terminal, paniculate, usually less than 6 cm long, for the most part minutely and densely pubescent; calyx less than 4 mm high, truncate, commonly with 1 obscure tooth, glabrous inside; corolla white, mostly about 1.5 cm long, about as wide, bilabiate, with 2 lobes almost completely fused, glabrous inside, the opposing 3 lobes with nectar guides at base and alternate with them; nectar guides two, yellow; stamens with the longer pair below nectar guides, almost equaling petals; filaments connate near base, the longer ones 1–1.2 cm long, the shorter ones 0.9–1 cm long, the thecae divaricate, ca 1.3 mm long; staminodium ca 5 mm long, inserted 2 mm from base of corolla tube; pistil 16–17 mm long; ovary conical, 1.5–2 mm long, densely pubescent; style pubescent, exceeding anthers; stigmas 2, pointed. Capsules linear, flattened, long-acuminate at apex, to 35 cm long, ca 1 cm wide, 0.5 cm thick, sparsely to densely puberulent (at least when young), the lateral edge with prominently raised margins; seeds 6–7 mm long, to 17 mm wide, the upper edge with rounded corners, the lower edge flat, the corner square. *Croat 11927* (type), *13975*.

Rare, known from the south shore of the island and near the Laboratory Clearing. Flowers mostly in July

and August with extravagant, long-lasting bursts of flowers. The fruits are mature from late November to March.

The species can be recognized in sterile condition by its trifid tendrils, subtetragonal young stems, four-angled pith, and sap with the aroma of cloves (Gentry, 1973b).

Panama and probably Colombia. In Panama, known from tropical moist forest in the Canal Zone, eastern Panamá, and Darién.

XYLOPHRAGMA Sprague

Xylophragma seemannianum (O. Kuntze) Sandw., Kew Bull. 1953:469. 1954
Saldanhaea seemanniana O. Kuntze
Pié de gallo, Pink bell

Liana; stems terete, prominently lenticellate, with interpetiolar glandular fields; most younger parts densely stellate-pubescent; tendrils simple, few, becoming large and woody; pseudostipules inconspicuous or ovate to triangular, to 4 mm long, distichously arranged in dense clusters. Leaves 3-foliolate or with a tendril replacing terminal leaflet, appearing after onset of flowers, deciduous; petioles 7–12 cm long; petiolules 1–7 cm long; leaflets ovate, oblong-ovate, or obovate, abruptly acuminate, obtuse to rounded or subcordate at base, to 13(28) cm long and 7(15) cm wide, stellate-tomentose or glabrate in age except on veins. Panicles dense, in axils of fallen leaves, appearing before new leaves; flowers sweetly aromatic; pedicels, calyces, and corollas (except tube near base) densely pubescent with branched trichomes; calyx narrowly campanulate, 5–7.5(9) mm long, 5-costate, the costae extending beyond margin to form short teeth; corolla lavender-pink, 3.5–5.5(6) cm long, the tube white, glabrous inside; stamens with the longer pair 1.6–2.1 cm long; anther thecae divaricate, 2.3–3.5 mm long; pistil 2.1–2.8 cm long; ovary tapered-cylindrical, ± flattened, 1.5–2.5 mm long, densely lepidote. Capsules oblong, smooth, ± compressed, 9–18(25) cm long, to 5 cm wide; seeds 2–2.5 cm long, to 4.5 cm wide, the body orbicular, tan, cordate at base, to 2 cm wide, the wings transparent. *Croat 13962, Foster 705.*

Moderately uncommon. Flowers in the late dry season (March to May), with extravagant, long-lasting bursts of flowers. The fruits mature in the following dry season. Leaves fall during the dry season and are replaced shortly after flowering.

The species can be recognized when sterile by its relatively large leaves with stellate (dendroid) pubescence and by its interpetiolar glandular fields.

Mexico and Belize to Colombia, Venezuela, and the Guianas; Trinidad. In Panama, known only from the Pacific slope (except in the Canal Zone), from drier areas of tropical moist forest in the Canal Zone, Veraguas, Herrera, Panamá, and Darién, from tropical dry forest in Coclé, from premontane moist forest in the Canal Zone and Panamá, and from premontane wet forest in Chiriquí, Panamá, and Darién.

127. GESNERIACEAE

Epiphytic, hemiepiphytic, or terrestrial shrubs, vines, or perennial herbs, generally succulent; stems usually creeping, loosely rooted. Leaves opposite, petiolate; blades simple, serrate; venation pinnate; stipules lacking. Flowers bisexual, zygomorphic, solitary or in cymose, fasciculate, or racemose inflorescences; peduncles or pedicels bracteate; calyx deeply 5-lobed (may be toothed as well); corolla tubular, bilabiate, 5-lobed, showy; stamens 4, didynamous, epipetalous; anthers connate, 2-celled, dehiscing longitudinally; nectariferous disk ringlike; ovary superior or partly inferior, 1-locular, 2-carpellate; placentation parietal; ovules numerous, anatropous; style 1; stigma 1 and bilobed or scoop-shaped. Fruits loculicidal, 2-valved capsules or berries; seeds many, with abundant endosperm.

Distinguished by being small epiphytic or hemiepiphytic shrubs or vines or soft, juicy-stemmed herbs with opposite, frequently toothed leaves and showy, often somewhat zygomorphic, frequently gibbous flowers with usually four stamens held to one side of the throat.

Columnea billbergiana, *C. purpurata*, *Kohleria tubiflora*, and probably *Besleria laxiflora* are hummingbird pollinated. All other species, with the possible exception of *Drymonia serrulata*, for which bat pollination has been suggested (Vogel, 1958), are probably insect pollinated. Euglossine bees are probably responsible for pollination of *Chrysothemis*, *Codonanthe*, *Drymonia*, and *Nautilocalyx*. It is suspected that *Diastema raciferum* may be pollinated by butterflies (H. Wiehler, pers. comm.).

Most species on BCI probably have bird-dispersed seeds, though *Drymonia serrulata* may be in part bat dispersed and *Codonanthe crassifolia* and possibly *Chrysothemis friedrichsthaliana* and *Nautilocalyx panamensis* may be transported by ants. Seeds of *Diastema raciferum* and *Kohleria tubiflora* are dispersed by the wind with a "salt shaker" action.

Some 140 genera and about 1,800 species; mostly in the subtropics and tropics in Asia and Central and South America.

BESLERIA L.

Besleria laxiflora Benth., London J. Bot. 5:361. 1846

Terrestrial shrub or suffrutescent herb; stems densely strigose when young. Leaves elliptic or narrowly oblong,

---

KEY TO THE SPECIES OF GESNERIACEAE

Flowers white:
  Plants terrestrial:
    Corolla ca 2 cm long; calyx less than 8 mm long; nectary of 5 individual glands; plants with
      scaly rhizomes . . . . *Diastema raciferum* Benth.
    Corolla 4–4.5 cm long; calyx 1.5–2 cm long; nectary of 1 large dorsal gland; plants lacking
      rhizomes but often with tubers . . . . *Nautilocalyx panamensis* (Seem.) Seem.
  Plants epiphytic or hemiepiphytic:
    Calyx leaflike, the lateral lobes more than 4 cm long; corolla ca 6 cm long, the limb more than
      3 cm wide; fruits fleshy, 2-valved, capsular; leaf blades dentate, usually more than 8 cm
      long . . . . *Drymonia serrulata* (Jacq.) Mart.
    Calyx small, the lateral lobes less than 1 cm long; corolla less than 4 cm long, the limb less
      than 2 cm wide; fruits berries; leaf blades entire or serrate, usually less than 6 cm long:
      Calyx winged (with long decurrent keels), the lateral lobes at anthesis less than 3 mm long;
        corolla more than 3 cm long, the limb more than 1.5 cm wide . . *Codonanthe uleana* Fritsch
      Calyx lacking wings or ridges, the lateral lobes at anthesis ca 6 mm long; corolla less than
        2.5 cm long, the limb less than 1 cm wide . . . . *Codonanthe crassifolia* (Focke) Mort.
Flowers red, yellow, or orange:
  Plants usually epiphytic or hemiepiphytic, often high in trees:
    Leaves of a pair strongly unequal, the larger one more than 15 cm long; corolla yellow, tubular, 3 cm long, with a subequal limb; fruits rose-orange, ovoid berries; plants rare or possibly no longer on the island . . . . *Columnea purpurata* Hanst.
    Leaves of a pair subequal, less than 4 cm long; corolla red-orange, strongly bilabiate, 5–7 cm long; fruits white globose berries; plants occasional . . . . *Columnea billbergiana* Beurl.
  Plants terrestrial:
    Calyx lobes connate four-fifths of their length, the calyx 5-angled; corolla 8 mm longer than calyx; nectar guides red lines, on 2 lobes; nectary of 1 large dorsal gland; fruits 2-valved capsules . . . . *Chrysothemis friedrichsthaliana* (Hanst.) H. E. Moore
    Calyx lobes free from one another, the calyx without angles; corolla 15–25 mm longer than calyx; nectar guides lacking; nectary of 5 glands or ring-shaped:
      Corolla densely pubescent, 2.5–3 cm long, constricted in throat, the subequal lobes ca 2 mm long; nectary consisting of 5 glands; fruits 2-valved capsules; leaves conspicuously soft-pubescent below . . . . *Kohleria tubiflora* (Cav.) Hanst.
      Corolla glabrous, ca 2.2 cm long, only slightly constricted in throat, the lobes of the limb ca 3 mm long, flared; nectary ring-shaped; fruits red-orange berries; leaves glabrous or inconspicuously pubescent below . . . . *Besleria laxiflora* Benth.

those of a pair equal or unequal, mostly 10–21 cm long, 4–8.5 cm broad, sparsely strigillose on veins beneath. Flowers in umbellate or cymelike clusters; peduncles to 4.5 cm long; pedicels about as long as peduncles; calyx 7.5–14 mm long, orange or red, one-third to one-half as long as corolla, the lobes ovate to ovate-lanceolate, attenuate to a mucro; corolla red or orange, erect, slightly swollen at base, to 2.2 cm long, glabrous on outside, with a glandular throat inside and a ring of pubescence at the staminal insertion; ovary superior, glabrous; stigma shallowly bilobed, the lobes broadened laterally; nectary ring-shaped. Berries globose, 1–12 mm diam, red or orange; seeds many, small, reddish-brown.

Reported by Kenoyer and Standley (1929) from near the laboratory, but the ecology of the Laboratory Clearing has changed so remarkably since that time that it is unlikely that the species still occurs on the island; no collections of it have been seen. Flowers and fruits elsewhere in Panama throughout the year, mostly in the rainy season. Plants often have both flowers and mature fruits.

Mexico to Colombia, Venezuela, the Guianas, and Brazil; from sea level to 1,400 m. In Panama, ecologically variable, but generally growing in weedier habitats than are now common on BCI; known from tropical moist and premontane wet forests in all provinces and apparently from other life zones at higher elevations as well.

CHRYSOTHEMIS Dec.

Chrysothemis friedrichsthaliana (Hanst.) H. E. Moore, Baileya 2:87. 1954
Tussacia friedrichsthaliana Hanst.

Terrestrial or epiphytic herb; stems obscurely 4-sided, succulent, ca 1 cm thick. Petioles short, flattened; leaf blades broadly elliptic, acute to acuminate at apex, cuneate at base and decurrent onto petiole, 8–40 cm long, 4–16 cm wide, irregularly crenate-serrate, short-pilose above, puberulent below especially on midrib and veins, dark green and shiny above, pale below; veins impressed above, raised below. Inflorescences axillary from upper leaves, usually of 3 pedicellate flowers atop a bracteate peduncle; pedicels 1–4 cm long; calyx green or yellow, inflated, mostly to 2 cm long, irregularly 5-lobed, with 5 broad lateral wings; corolla 5-lobed, slightly asymmetrical, 1.9–2.8 cm long, ± glabrous, yellow or orange at least on the distal part of tube and lobes, the lobes with red lines inside extending somewhat into tube, the tube constricted 3–4 mm above base between the lower 2 lobes; stamens 4, ca 12 mm long; filaments recurved at apex; ovary superior; style mostly glabrous, 5–8 mm long; stigmas bilobed; nectary of 1 large dorsal gland. Capsules 2-valved, fleshy, pubescent, 7–10 mm diam; seeds many, to 0.6 mm long, about twice as long as wide, with a fleshy, enlarged funicle. *Croat 5930, 11708.*

Occasional in the forest and the Laboratory Clearing, often on moist rocks, or rarely on rotting wood. Flowering and fruiting throughout the rainy season (May to December). Individuals flower for a relatively long time,

with some fruits maturing while the plant is still in flower.

Easily recognized by the succulent stems, inflated calyx, and orange flowers with red lines. H. Wiehler (pers. comm.) reports having seen seeds of this species being carried away by ants, with the funicle still attached, both in the field and in the greenhouse. It is not known if this is the sole means of dispersal.

Central America to western Colombia. In Panama, known principally from tropical moist forest in the Canal Zone, Bocas del Toro, Colón, San Blas, Chiriquí, Panamá, and Darién; known also from premontane wet forest in Bocas del Toro (Río Guarumo) and from tropical wet forest in Colón (Portobelo).

CODONANTHE (Mart.) Hanst.

Codonanthe crassifolia (Focke) Mort. in Standl., Publ. Field Mus. Nat. Hist., Bot. Ser. 18:1159. 1938
C. confusa Sandw.

Small epiphytic or hemiepiphytic shrub, usually less than 30 cm long, rooting at nodes, glabrous or with most parts minutely puberulent. Leaves thick; petioles 5–15 mm long; blades variable, ovate to obovate, obtuse to short-acuminate at apex, acute at base, 1.5–6 cm long, 1–2.5 cm wide; lateral veins indistinct. Flowers axillary, 1–4 borne together; pedicels puberulent; calyx puberulent, deeply 5-lobed, ca 5–7 mm long, the lobes slender, the dorsal lobe depressed by corolla, the 4 erect lobes with 3 glands in the sinus, the lateral lobes ca 6 mm long at anthesis; corolla bilabiate, 1.9–2.2 cm long, white, mottled with red inside below the 2-lobed side of lip, minutely puberulent on 3-lobed side, the limb 9–10 mm wide, the lobes rounded, broader than long, the tube gibbous, extending prominently below calyx; stamens 4, included, the longer pair to 2.5 cm long; filaments curved at apex, widened toward base; anthers oblong, to 2 mm long, coherent in pairs or free; staminodium very small; ovary superior, oblong-ovoid, puberulent; style longer than stamens; stigma 1, scoop-shaped, held just above stamens; both stamens and stigmas ± appressed against the 3-lobed side of corolla; nectary of 1 large dorsal gland. Berries subglobose, 10–12 mm long, 7–9 mm wide, red, ± glabrous; seeds reddish, slightly curved, ca 2 mm long and 0.7 mm wide, longitudinally striate. *Croat 5916.*

Frequent in the forest, usually rather high in trees. Flowering and fruiting throughout the year, but most abundantly in the rainy season. Fruit development time is probably about 1 month.

Standley considered *Codonanthe calcarata* Hanst. a synonym of this species, but the name was misapplied according to Leeuwenberg (1958).

Like most other members of the genus, *C. crassifolia* is usually associated with ant nests, though many seedlings are found in cracks in tree bark and are not attended by ants. The ant nests frequently also contain *Aechmea tillandsioides* (22. Bromeliaceae).

The fruit is perhaps most logically adapted for bird

dispersal, but ants may carry the seeds away after the fruits have been pecked open by birds. It is possible that fruits occasionally dehisce. In the greenhouse, H. Wiehler (pers. comm.) has shown that, if plants are not watered for a few days while the fruit is still immature and then watered excessively as the fruit ripens prematurely, the fruit will break open from the excess water.

Guatemala to Colombia, Venezuela, the Guianas, and Peru; Trinidad; from sea level to 1,400 m. In Panama, known principally from tropical moist, premontane moist, and tropical wet forests all along the Atlantic slope and from tropical moist forest in Darién; known also from premontane wet forest in Colón (Santa Rita Ridge) and Panamá (Cerro Jefe).

See Fig. 502.

Codonanthe uleana Fritsch, Bot. Jahrb. Syst. 37:492. 1906

Small epiphytic or hemiepiphytic shrub; very much like *Codonanthe crassifolia* but with the leaf blades oblanceolate to obovate, to 9 cm long and 3.5 cm wide; the calyx 5–6 mm long at anthesis, with the underside of each sepal (especially the lowest one) bearing a long decurrent keel, the lateral lobes 1.3–2.5 mm long at anthesis, to 3.5–4 mm long in fruit; the corolla 3.5–3.8 cm long, the limb 1.5–2 cm wide; the anthers with the pores adorned by 2 minute horns. *Wilson 65.*

Not seen recently. Seasonal behavior uncertain, probably much like that of *C. crassifolia.*

Southern Mexico to Brazil (upper Amazon), Peru, and Bolivia. In Panama, known from tropical moist forest in the Canal Zone and Panamá (San José Island) and from premontane wet forest or tropical wet forest in Colón (Santa Rita Ridge).

COLUMNEA L.

Columnea billbergiana Beurl., Kongl. Vetensk. Acad. Handl. 1854:135. 1854

Small epiphytic or hemiepiphytic shrub, usually rather high in trees and pendent; stems brown, branched, 2–4 mm diam, sparsely strigose when young. Leaves of a pair subequal; petioles 2–6 mm long, strigose; blades ovate-lanceolate, acute, broadly cuneate at base, 2–3 cm long, 8–11 mm wide, thick, glabrous above, pale and strigose below; major lateral veins in 3 pairs. Flowers solitary; peduncles 5–10 mm long; pedicels 5–13 mm long, densely pilose, the trichomes long, white or pink, the bracts linear-lanceolate, 3–6 mm long; calyx 8–12 mm long, red, the lobes 5, ovate-lanceolate, acuminate, 5–8 mm long, white-sericeous outside, glabrous inside, denticulate, the teeth 1–4 on each side, glandular; corolla red, 5–7 cm long, slender, swollen and gibbous at base, sparsely hispid outside, 4-lobed, deeply divided on one side, the upper lobe rounded and apiculate at apex, to 1.5 cm long, the lateral lobes connate to upper lobe, the free parts to ca 8 mm long, acute, the lower lobe 1.2–1.8 cm long, somewhat deflexed; stamens 4, exserted; filaments red

except white at base, nearly as long as and held against upper lobe of corolla; ovary superior, glabrous except at apex; style white, glabrous or sparsely glandular-pilose, equaling or slightly longer than stamens; stigma 1, bilobed; nectary of 1 large dorsal gland. Berries white, globose, ca 1.4 cm diam; seeds yellow, embedded in funicular pulp. *Croat 10088.*

Occasional to frequent in the forest. Flowers sparingly all year, but especially in the rainy season. H. Wiehler (pers. comm.) reports a burst of flowering in October for cultivated plants in Panama.

Distinguished by its epiphytic or hemiepiphytic habit and slender zygomorphic flowers. Ridley (1930) discussed the adaptation to ornithochory obvious in the white berry and contrasting red sepals. H. Wiehler (pers. comm.) suspects that *Columnea arguta* Mort., found in areas surrounding BCI, might also be found on the island.

Known only from Panama, from sea level to 1,000 m, most abundantly from premontane wet forest in Coclé and Panamá, but also from tropical wet forest in Colón (Portobelo) and tropical moist forest in the Canal Zone (vicinity of Gatun Lake).

Columnea purpurata Hanst., Linnaea 34:386. 1865
Dalbergaria sanguinea (Pers.) Steud.

Shrub, 1.2–1.8 m long, usually hemiepiphytic, rarely terrestrial; stems woody, unbranched, 6–10 mm diam, densely yellowish-hirsute. Leaves clustered at apex of stem, those of a pair strongly unequal; larger ones short-petiolate, on petioles 10–15 mm long, densely hirsute, the blades oblanceolate, long-acuminate, cuneate and inequilateral at base, 13–30 cm long, 4–10 cm wide, serrulate, pilose on both sides, with the major lateral veins in 9–11 pairs; smaller leaf of a pair sessile, ovate, oblique, long-acuminate at apex, obtuse at base, to 3 cm long and 1 cm wide, deeply toothed, hirsute. Flowers fasciculate in upper axils, 5-parted; peduncles very short, bracteate, the bracts scarlet, elliptic or lanceolate, ca 3 cm long, pilose outside, strigose inside, spinulose-toothed, the teeth 4 or 5 per side; calyx ca 3 cm long, scarlet, the lobes lanceolate, 5 mm wide near base, long-acuminate, the pubescence similar to that on bracts; corolla tubular, ca 3 cm long, yellow, the tube ca 4 mm diam at base, expanding to 7 mm diam, contracted toward throat, densely brown-sericeous outside, glabrous inside, the limb narrow, regular, about 6 mm wide, the lobes subequal, erect, 4 mm long and 3 mm wide; stamens 4; filaments glabrous at base, short-pilose above; anthers connate, ca 2 mm long and 2 mm wide; ovary superior, long-pilose; style glabrous; nectary of 1 large dorsal gland. Berries oblong-ovoid, rose-orange; seeds light yellow with enlarged funicles. *Standley 31393.*

Collected once on the island by Standley and possibly no longer occurring there. Probably flowering and fruiting all year.

Southern Mexico to Venezuela, the Guianas, and Bolivia; Trinidad; Hispaniola. In Panama, known from tropical moist forest in the Canal Zone, Panamá, and Darién.

DIASTEMA Benth.

Diastema raciferum Benth., Bot. Voy. Sulphur 132. 1845
D. exiguum Mort.

Herb, to 17 cm tall, sparsely pilose; rhizomes scaly; basal nodes of stems often with adventitious roots; stems, petioles, and inflorescences often flushed with maroon. Petioles 4–4.5 (7.5) cm long; leaf blades ovate to elliptic, acute, obtuse to subcordate and ± inequilateral at base, 3.5–10 cm long, 2.5–5.5 cm wide, serrate, with 3 or 4 teeth per cm, sparsely pilose to glabrate above, ± glaucous and glabrous to pubescent at least on the veins below; veins at margins strongly arching upward. Racemes 1.5–8 cm long, long-bracteate, terminal, but secondary racemes often in leaf axils of more mature plants (at first appearing as mere fascicles of leaves); peduncles 10–35 mm long, the bracts foliaceous, elliptic, ca 5 mm long, entire; pedicels 5–15 (25) mm long; calyx turbinate, 5–8 mm long, acutely 5-lobed to two-thirds its length; corolla white with yellow throat (pink in var. *lilacinum* Mort., perhaps to be found on the island), salverform, weakly bilabiate, ca 2 cm long, substrigose outside, glabrous inside, the lobes 5, rounded, 1–2 mm long; stamens 4, didynamous; filaments slender, free, to ca 10 mm long; anthers free, the thecae orbicular; ovary partially inferior, ovoid, minutely puberulent; style slender, to 8.5 mm long, withering from fruit; stigma 1, deeply bilobed, flat, ± winglike, ca 1 mm wide; nectary of 5 individual glands, subulate, to 1.3 mm long. Capsules dry, ca 6 mm long, splitting along only 1 side; seeds brown, very small. *Kenoyer 539a.*

Collected once by Kenoyer; an inconspicuous plant of gulleys and ravines, it is probably still growing on the island, though it has not been seen in recent years. The plant dies back during the dry season in areas having a distinct dry season, and the rhizome is protected from desiccation by a layer of scales.

The pollinator is unknown, but H. Wiehler (pers. comm.) indicates that a very similar species in Venezuela is pollinated by the butterfly *Leucothyris makrena makrena* Hew. (Nymphalidae, Ithomiinae).

Costa Rica and Panama. In Panama, known from tropical moist forest in the Canal Zone, Panamá, and Darién and from premontane wet forest in Coclé.

DRYMONIA Mart.

Drymonia serrulata (Jacq.) Mart., Prodr. 7:543. 1839
D. spectabilis (H.B.K.) Mart. ex G. Don
Calabash vine

Climbing or creeping vine, hemiepiphytic at least when juvenile, often with many aerial roots; older stems with a paperish red bark; younger parts succulent. Petioles 6–30 mm long; blades asymmetrical, acute to shortly acuminate at apex, oblique at base, 5–17 cm long, 3–9 cm broad, slightly fleshy, dark green and scabrous above, paler below (juvenile leaves often purplish below), gla-

brous to slightly pubescent, the margins dentate. Flowers usually solitary in axils; calyx foliaceous, puberulent, 5–6 cm long, the lobes 5, free, cordate, all but the central one very asymmetrical, persistent in fruit; corolla bilabiate, thick, 5–7 cm long, usually white or cream at least on tube and outer surface, the limb 3.5–4.5 cm wide, the lobes 5, dentate, marked with red or violet-purple inside, the upper lip with nectar guides extending into throat, the lobes smaller, the tube asymmetrical, gibbous at base and below the upper lip, the upper concavity with gland-tipped trichomes; stamens 4; filaments broadened at base, twisting in age to draw anthers to bottom of corolla; anthers ca 8 mm long, held together in pairs in throat of corolla, the apex directed downward, the thecae opening at base of anther; ovary superior; style ca 2.5 cm long, densely glandular-pubescent, held between 2 prominent lobes within the tube basal to the lower lip, elongating after opening of anthers, hollow-tubular at apex; stigma 1, with 2 opposing lobes to 4 mm long; nectary of 1 large dorsal gland. Capsules ± globose, 10–20 cm long; valves 2, bright orange to red inside, curving back at maturity to expose the cone-shaped cluster of placental and funicular tissue covered with seeds; seeds many, shiny, black, oblong, tapered at ends, twisted, ca 1 mm long. *Croat 6054, 6121.*

Common along the edge of the lake, occurring occasionally in the forest high in the canopy. Flowers from June to January, principally in the early to middle rainy season. The first fruits mature usually by August, and some are no doubt still present well into the dry season.

The pollinator probably hits the apex of the anther as it enters the corolla, turning the anther over and dumping out the pollen.

Seeds are possibly dispersed in part by birds, which might be attracted to the black cluster of seeds sharply contrasted against the red inner valve surfaces. H. Wiehler (pers. comm.) has indicated that the cluster of seeds sometimes smells like cabbage, indicating possible bat dispersal.

Throughout most of tropical America from Mexico to Brazil (Mato Grosso) and Bolivia; Lesser Antilles. In Panama, known principally from tropical moist forest in all regions (except lowland Chiriquí) where forest still exists; known less frequently from premontane wet forest, tropical wet forest, and premontane rain forest in Chiriquí, the Azuero Peninsula, and central Panamá.

See Fig. 503.

KOHLERIA Reg.

Kohleria tubiflora (Cav.) Hanst., Linnaea 34:442. 1865–66

Coarse perennial herb, often more than 1 m tall, ± villous, the trichomes of varying lengths, longest on stems, petioles, and midribs of lower blade surface; stems often with maroon streaks mottled with smaller, green, elongate spots. Petioles 1–7.5 cm long; blades ovate to oblong, acute, rounded to cordate at base, mostly 4–14 cm long, 2.5–9 cm wide, crenate, bicolorous, often maroon and

Fig. 502. *Codonanthe crassifolia*

Fig. 504. *Kohleria tubiflora*

Fig. 503. *Drymonia serrulata*

soft-pubescent beneath. Flowers axillary, tubular, ca 4.5 cm long, usually 3–5 in each axil, pedicellate; calyx lobes 5, acute, often tipped with maroon, accrescent in fruit; corolla orange-red, 2.5–3 cm long, ca 1 cm wide, constricted near apex, villous outside, glabrous inside except for a few glandular trichomes near apex, the lobes small, rounded, ca 2 mm long, with purple spots on inner surface; stamens 4, ca two-thirds the length of tube; filaments free, curled at apex; anthers nearly as broad as long, at first fused together in a single unit; ovary partly inferior, densely velutinous; style held below constriction of the corolla tube above stamens; stigma bilobed; nectary of 5 distinct glands ca 2.3 mm long, persisting in fruit. Capsules ovoid, ca 1 cm long, reddish, apparently splitting open at maturity by 2 thick valves; seeds reddish, minute, ± oblong, tapered to both ends, loose at fruit maturity (indicating wind dispersal). *Croat 6355.*

Occasional, in clearings, in open areas of the forest, and on the lakeshore, apparently preferring disturbed banks; may be locally abundant. Flowers principally in the rainy season, especially in the last half of the rainy season. Mature-sized fruits may be seen by October, but the actual time of seed dispersal is not known. Since it appears likely that the seeds may be wind dispersed, they may be shed in the early dry season.

Recognized by the brightly colored, tubular flowers, which are somewhat constricted at the apex, and by the semi-inferior ovary.

Central Costa Rica to Colombia and western Venezuela. In Panama, known most commonly from tropical moist, premontane wet, and tropical wet forests along the Atlantic slope in the vicinity of the Canal Zone and also from tropical moist forest in Darién and from premontane wet forest at higher elevations on the Pacific slope; known infrequently from the Pacific slope of the Canal Zone.

See Fig. 504.

NAUTILOCALYX Linden

Nautilocalyx panamensis (Seem.) Seem., Bot. Voy. Herald 250. 1854

Achimenes panamensis (Seem.) Hemsl.; *Episcia panamensis* (Seem.) Mort.

Perennial herb, 10–35 cm tall, lacking rhizomes, often with tubers; stems creeping over soil and weakly rooted; stems and petioles brittle and succulent; all parts except inner surface of corolla moderately to densely villous. Petioles mostly 3–5 cm long, flattened above, often with marginal maroon ribs; blades ovate to oblong, acute or slightly acuminate, acute to rounded at base and ± oblique, mostly 6–15 cm long, 4–8 cm broad, thin, green and shiny above, dull below, the margins markedly crenate. Flowers axillary, usually in clusters of 3, pedicellate; calyx foliaceous, toothed, irregular, the lobes 5, free, to 2 cm long; corolla white, 4–4.5 cm long, the tube narrow, 7–10 mm wide, with linear grooves inside at base of lower

lobes, the lobes 5–7 mm long, to 10 mm broad; stamens 4, equaling or exceeding style or much shorter; filaments ultimately coiled; ovary superior; style 12–16 mm long; nectary of 1 large dorsal gland. Capsules globose, probably to ca 1 cm diam, sparsely pubescent, the trichomes flattened, crisped. *Croat 11814.*

Occasional in the forest, but locally abundant along trails or open areas. Flowers and fruits throughout the rainy season (July to December).

Distinguished by its terrestrial habit and white flowers. When the flower first opens, the styles are two-thirds as long as the anthers. The filaments are held erect, and the anthers are fused in a 2-2 arrangement. Later, the style elongates and the filaments coil up, pulling the anthers well below the style.

The fruits are similar to those of *Chrysothemis freidrichsthaliana* and are probably dispersed in the same manner.

Apparently restricted to Panama, where it is known from tropical moist forest in the Canal Zone and from premontane wet forest in Panamá (8 km southwest of Cerro Brewster).

128. LENTIBULARIACEAE

Glabrous, aquatic, insectivorous herbs; stems stoloniferous, lacking true roots. Leaves inconspicuous, rosulate, or lacking, often replaced by somewhat leaflike, entire or dissected photosynthetic organs which are modified stems; the submerged vegetative parts bearing tubular insect traps. Flowers bisexual, zygomorphic, in bracteolate racemes of few flowers; calyx bilabiate; corolla 5-lobed, bilabiate, the lower lip gibbous; stamens 2, free, epipetalous; anthers 1-celled but constricted, introrse, dehiscing longitudinally; ovary superior, 1-locular, 2-carpellate; placentation free-central; ovules numerous, anatropous; style simple; stigma bilabiate. Fruits 2-valved, loculicidal capsules.

The flowers are very specialized. I have observed visits by skippers and medium-sized bees, but P. Taylor (pers. comm.) believes the flowers to be mostly self-pollinated.

The fruits fall whole and are water dispersed, but the seeds themselves sink (probably a necessary requirement for germination in the soil). Ridley (1930) reported that *Utricularia* is vegetatively dispersed by water currents or by having parts of the plant carried away on the feet of aquatic birds.

Five genera and about 250 species; cosmopolitan.

UTRICULARIA L.

Species of the genus are unusual in that they bear small bladders with trap-door entrances on the submerged parts. Small Crustacea and other small animals are sucked into the bladder by its rapid expansion after coming into contact with a trigger. The decaying animals provide nutrients for the plant (Willis, 1966).

KEY TO THE SPECIES OF UTRICULARIA

Flowers numerous (rarely fewer than 10), ± congested in the upper part of scape; spur of corolla lacking stalked glands at apex; pedicels strongly recurved in fruit; seeds circular, with a very narrow, thin, regular wing; stolons flattened, with 2 separated semicircular series of vascular bundles; foliar organs relatively large, repeatedly pinnately divided into very numerous capillary segments . . . . *U. foliosa* L.

Flowers few (rarely more than 5), if more than 2 then relatively distant; spur of corolla with a few stalked glands at apex; pedicels spreading or only slightly reflexed in fruit; seeds with a relatively thick, very irregular wing; stolons terete, with a single central ring of vascular bundles; foliar organs relatively small, sparsely dichotomously divided 1–3 times only, thus with very few ultimate capillary segments . . . . *U. obtusa* Sw.

This key to *Utricularia* was provided by Peter Taylor. I have modified it slightly.

Utricularia foliosa L., Sp. Pl. 18. 1753

U. mixta Barnh.

Perennial, submerged, freely suspended, insectivorous, aquatic herb; stolons robust, to several m long, distinctly flattened, to 4 mm wide, with fasciculate branches. Photosynthetic organs pinnate many times, ± broadly ovate in outline, to 17 cm long, somewhat dimorphic, some with more traps and fewer segments; insect traps borne laterally near base of penultimate segments. Racemes erect, simple, 7–40 cm long; flowers 3–20, congested at anthesis, later spaced on rachis; scape with 1 or 2 scales below lowest flower; bracts subtending flowers broadly ovate to orbicular; pedicels erect, 4–10 mm long, becoming reflexed in fruit and to 16 mm long; calyx lobes 2, subequal, broadly ovate, the lower lobe sometimes toothed; corolla yellow, sometimes with purple veins, bilabiate, the upper lip ± orbicular, ca twice as long as calyx lobes, the lower lip larger, gibbous, the margin entire or emarginate, the spur conical and ca two-thirds as long as lower lip, its inner surface with stalked glands along central vein; stamens 2; filaments curved; ovary subglobose; ovules few; style short, distinct; stigma bilabiate. Capsules globose, to 8 mm diam, sparsely glandular, indehiscent; seeds 4–12, 2–2.5 mm diam, lenticular, with a narrow wing. *Shattuck 369.*

Collected several times in Gatun Lake, presumably along the shore of the island. This submerged aquatic, easily visible only when in flower, is probably still present. Flowering plants have been seen throughout most of the year, but especially during the early rainy season in July and August. The fruits develop quickly and are often present with flowers and flower buds (P. Taylor, pers. comm.). Vegetative reproduction is probably the prime dispersal tactic, since fruit set is poor (P. Taylor, pers. comm.).

Florida, and Mexico south to Argentina; West Indies, Galápagos Islands; tropical and subtropical Africa. In Panama, known only from the Canal Zone (Gatun Lake).

Utricularia obtusa Sw., Prodr. Veg. Ind. Occ. 14. 1788

Minute, insectivorous, aquatic herb; stolons slender, from base of scape. Photosynthetic organs alternate on stolons, forked 1–3 times, to 2 cm long, the segments hairlike, glabrous, usually inconspicuous; insect traps stalked, ovoid, ca 1 mm long, replacing a leaf segment at a fork. Racemes erect, simple, to 4 cm long, bearing 1–4 flowers; scapes usually with a scale near middle; bracts subtending flowers ovate; pedicels 8–12 mm long; calyx bilabiate, the lobes subequal; corolla yellow, 6–10 mm long, bilabiate, the upper lip erect, ± rounded, the margins curled inward, the lower lip descending, prominently inverted and gibbous with the spur conical, its inner surface puberulent, marked with reddish lines; stamens 2, ca 1.5 mm long; filaments prominently curved and broadened toward apex, extending beyond anthers; anthers adjacent to style; ovary ovoid; style thick, short; stigma thin, bilabiate. Capsules globose, 2–4 mm long; seeds lenticular, ca 1 mm long, the wing relatively thick, very irregular. *Croat 5649.*

Abundance unknown because of its tiny nature, but probably restricted to quiet marshes and floating masses of vegetation on the south side of the island. Flowering specimens have been seen in April, May, and September, but flowering behavior is probably less affected by season than by other factors. Some fruits and flowers may be found on the same inflorescence.

This tiny aquatic is not confused with any other species. Reproduction is probably chiefly vegetative, since fruit set is poor (P. Taylor, pers. comm.).

Mexico to Argentina; West Indies; tropical Africa. In Panama, known only from the Canal Zone (Gatun Lake and the backed-up waters of the Río Chagres above Gamboa).

129. ACANTHACEAE

Trees, shrubs, lianas, vines, or erect or prostrate herbs. Leaves opposite (except *Elytraria*), petiolate, simple; blades entire to serrate; venation pinnate; linear cystoliths often present; stipules lacking. Flowers bisexual, usually zygomorphic (± regular in *Trichanthera*), solitary or on axillary or terminal, bracteate scapes or spikes; calyx deeply lobed (reduced in some); corolla tubular, bilabiate, brightly colored; stamens 4 and didynamous or 2, attached to corolla tube; anthers 2-celled, sometimes widely separated by a connective, dehiscing longitudi-

KEY TO THE SPECIES OF ACANTHACEAE

Plants trees, shrubs, or vines, usually more than 1.5 m tall:
  Plants vines, herbaceous or woody:
    Bracts subtending flowers linear, ca 1 mm wide; flowers solid white; plants usually less than
      4 m tall . . . . *Justicia graciliflora* (Standl.) D. Gibs.
    Bracts subtending flowers ± ovate or oblong, more than 1 cm wide; flowers white marked
      with violet-purple; plants usually vines twining high in forest:
      Bracts conspicuously pubescent with ascending trichomes; corolla less than 3.5 cm long . . .
        . . . . *Mendoncia gracilis* Turr.
      Bracts glabrous or with short, sparse puberulence, not ascending; corolla more than 3.5 cm
        long . . . . *Mendoncia littoralis* Leonard
  Plants trees or shrubs:
    Flowers reddish:
      Inflorescences with many, imbricate, red-orange bracts; blades conspicuously pubescent
        on both sides; corolla tubular and zygomorphic . . . . *Aphelandra sinclairiana* Nees
      Inflorescences lacking large red-orange bracts; blades glabrous except for veins below;
        corolla ± campanulate . . . . *Trichanthera gigantea* (H. & B.) Nees
    Flowers white or purple:
      Flowers white; shrubs vinelike, in the forest . . . . *Justicia graciliflora* (Standl.) D. Gibs.
      Flowers purplish; shrubs cultivated in the Laboratory Clearing . . . .
        . . . . *Thunbergia erecta* (Benth.) T. Anderson
Plants herbs or suffruticose shrubs usually less than 1 m tall:
  Bracts of inflorescence at least in part ovate to obovate, not several times longer than broad:
    Flowers and bracts less than 6 mm long; bracts with dense, woolly trichomes; weed in open
      areas . . . . *Nelsonia brunellodes* (Lam.) O. Kuntze
    Flowers and bracts usually more than 10 mm long; bracts not densely woolly-pubescent:
      Corolla distinctly bilabiate, white with lavender markings medially on inside of lower lip;
        plants rare, in the forest along trails . . . . *Herpetacanthus panamensis* Leonard
      Corolla not distinctly bilabiate (sometimes slightly bilabiate), white to lavender throughout:
      Corolla less than 1.5 cm long; plants low, often sprawling, growing in clearings . . . .
        . . . . *Blechum brownei* Ant. Juss.
      Corolla more than 2 cm long; plants usually erect, often markedly suffruticose, growing
        along forest trails . . . . *Blechum costaricense* Oerst.
  Bracts of inflorescence all slender, several times longer than broad:
    Flowers in loose, open, terminal panicles . . . . *Justicia pectoralis* Jacq.
    Flowers usually closely aggregated in leaf axils or in dense terminal heads or spikes:
      Spikes borne on slender scapes 5–20 cm long . . . . *Elytraria imbricata* (Vahl) Pers.
      Spikes sessile or nearly so:
        Inflorescences terminal on stems and branches:
          Inflorescences often longer than broad; bracts ca 6 mm long; flowers ca 5 mm long,
            white with the lower lip violet . . . . *Teliostachya alopecuroidea* (Vahl) Nees
         Inflorescences as broad as or broader than long; bracts usually more than 15 mm long;
            flowers ca 4 cm long, lavender to violet . . . . *Chaetochlamys panamensis* Lindau*
        Inflorescences axillary, at most nodes, at least in upper part of stem:
          Flowers few at each node, the corolla more than 2 cm long, lavender or lavender tinged
            with white; plants in the forest along trails . . . . *Ruellia metallica* Leonard
         Flowers densely clustered at each node, the corolla 6–8 mm long, white; plants on
            sandbars and at the edge of the lake . . . . *Hygrophila guianensis* Nees

Chaetochlamys panamensis is not included in the flora, but since it is a
weedy plant likely to occur on the island it is included in the key.

nally; ovary superior, 2-carpellate, 2-locular; placentation axile; ovules 1 to several, anatropous or amphitropous; style 1, slender; stigma cupular to 2- or 3-lipped, sometimes with 1 lip reduced. Fruits usually loculicidal, obpandurate capsules (drupes in *Mendoncia*); seeds usually disk-shaped, lacking endosperm, usually forcibly discharged by jaculators (retinacula, hooklike processes on the placenta).

Recognized by their usually densely bracteated inflo-rescences with a two-lipped corolla and by the cysto-liths in the leaves.

Most Acanthaceae flowers are small and are probably pollinated by small to very small bees. *Aphelandra* is probably hummingbird pollinated, and *Mendoncia* and *Trichanthera gigantea* are possibly bat pollinated. The pollination systems of *Justicia graciliflora* and *Thunbergia erecta*, which have larger flowers, are unknown.

The fruits are typically bivalved, often obpandurate

capsules with disk-shaped seeds that are thrown free from the capsules by springlike arms (jaculators or retinacula). Van der Pijl (1968) reported that the seeds of some species of *Hygrophila* have appressed trichomes that become erect and slimy when wet, apparently increasing buoyancy and aiding in establishment of the seed. Van der Pijl also reported that the capsules of some species of *Ruellia* split after a weak spot becomes wet, but this would effect seed dispersal only when soil moisture might be adequate for germination of the seed. *Elytraria, Nelsonia,* and *Mendoncia* lack retinacula. *Mendoncia* fruits are animal dispersed, probably by birds.

Some 250 genera and about 2,500 species; widely distributed in the tropics.

APHELANDRA R. Br.

Aphelandra sinclairiana Nees in Benth., Bot. Voy. Sulphur 146, t. 47. 1846

Shrub, 1.5–3.5(6) m tall; stems densely pubescent with erect trichomes. Leaves thin; petioles and midribs below with pubescence like stem; blades oblanceolate, acuminate, gradually tapered and decurrent onto petiole nearly to base, 10–38 cm long, 3.5–9.5 cm wide, glabrate above except on major veins, densely pubescent below. Inflorescences terminal, densely fine-pubescent, the spikes few to several, densely bracteate, to 13(20) cm long, the bracts ± obovate, acute or abruptly acuminate, 1–2 cm long, red-orange; sepals 5, acute, subtended by 2 similar bracts; corolla tubular, 6–7 cm long, magenta, bilabiate, the upper lip 4-lobed, the lateral lobes very short, the lower lip entire, reflexed; stamens 4, stiffly erect, somewhat enfolded in and almost equaling upper lip; anthers open at anthesis, held together, with densely pubescent lateral margins; filaments fused to tube near base; style slender, equaling and held between fused stamens; stigma cupular, bilabiate, open at anthesis; nectary not obvious but with the nectar accumulating at base of corolla tube. Capsules 2-valved, puberulent, ca 2.5 cm long; seeds disk-shaped, ca 4 mm long, suspended on a long, recurved retinaculum. *Croat 7760.*

Occasional, in the forest. Flowers usually continuously throughout the dry season (December to April); the flowers appear in succession, usually only a single one at a time from each spike, opening in the morning and usually falling off or wilting before the day is out. The fruits develop quickly, and the lower bracts usually contain mature capsules by the middle of the dry season. Though plants continue to flower in the late dry season (even rarely into the rainy season), the spikes are usually heavy with mature capsules.

No doubt hummingbird pollinated.

Costa Rica and Panama. In Panama, known from tropical wet forest in Bocas del Toro (Quebrada Huron), Colón (Río Buenaventura), Coclé (above El Valle), and Panamá (southwest of Cerro Brewster), but perhaps more abundantly from tropical moist forest on both slopes of the Canal Zone.

BLECHUM P. Browne ex Ant. Juss.

Blechum brownei Ant. Juss., Ann. Mus. Natl. Hist. Nat. 9:270. 1807

B. brownei Ant. Juss. forma *puberulum* Leonard; *B. pyramidatum* (Lam.) Urban

Erect or more commonly sprawling herb, to 70 cm long; stems slender, glabrate to crisp-villous, the pubescence often restricted to lines. Petioles short or to 1.5 cm long; blades ovate to ovate-elliptic, acute to acuminate, rounded to attenuate at base, 1.5–5 cm long, 1–3 cm wide, nearly glabrous to densely pubescent usually with a mixture of short, fine trichomes and larger, coarse, often jointed trichomes, the margins usually markedly ciliate; cystoliths very conspicuous, easily confused with trichomes. Flowers in close, terminal, bracteate clusters, the bracts more densely pubescent than leaves, ovate to lanceolate, 5–20 mm long, long-ciliolate; calyx deeply divided, to 3.7 mm long, pubescent, the lobes linear-lanceolate; corolla white or sometimes pale lavender, short-pilose especially on tube, 10–13 mm long, the tube 2.5–3 mm diam, the limb ca 8 mm diam, the short lobes rounded, ciliate; stamens 4, held within tube; filaments fused to one side of tube much of their length, one pair shorter; style held to one side of tube between stamens, one lobe reduced, the other lobe directed into center of throat above anthers. Capsules obovate, 5–6 mm long, densely pubescent, 2-valved, bursting open at maturity to expel several seeds on retinacula; seeds small, dark brown, disk-shaped, ca 1.5 mm diam. *Croat 4604, 6966.*

Abundant in clearings, especially in the Laboratory Clearing. Flowers and fruits principally throughout the dry season (December to May), though flowering also occurs rarely during the rainy season. Plants have often lost many of their leaves by the end of the dry season when the inflorescences contain mostly fruits.

Mexico (east and south) to Guyana, Ecuador, and Peru; West Indies, Galápagos Islands; Oceania. In Panama, an ecologically variable weed; known from tropical moist forest in the Canal Zone, Bocas del Toro, San Blas, Chiriquí, Los Santos, Herrera, Panamá, and Darién, from premontane dry forest in Los Santos, from tropical dry forest in Coclé and Panamá, and from premontane wet forest in Panamá (Cerro Campana).

See Fig. 505.

Blechum costaricense Oerst., Vidensk. Meddel. 168. 1854

Erect or suberect herb, to 1 m tall, often suffruticose; stems pubescent with crisp, apically recurved trichomes mostly in lines above petioles on either side; internodes swollen, often darkened. Petioles 5–25 mm long, often narrowly winged; blades ovate to ovate-elliptic, lanceolate-elliptic, or lanceolate, acuminate, attenuate to rounded and decurrent onto petiole at base, 4–12 cm long, 1–4.5 cm wide, the upper surface often with stout simple trichomes, the midrib usually appressed-pubescent (at least near base), the lower surface glabrate to moderately pubescent with short fine trichomes, the margins glabrous

to ciliate; veins hispidulous to short-villous; cystoliths very conspicuous, easily confused with trichomes. Inflorescences loosely bracteate, terminal; flowers each subtended by 2 linear bracts ca 1 cm long and 1 ovate bract 1.5–2.5 cm long; all bracts puberulent, conspicuously ciliolate; calyx with slender, sharp lobes divided to near base, to ca 5 mm long; corolla ± salverform, 2–3 cm long, hispidulous outside, glabrous inside, lavender-blue above especially on lobes (rarely white), the lobes ca 5 mm long, rounded, the tube 2.5–3 mm diam, whitish, bent outward from below middle; stamens 4; filaments with the basal one-half to two-thirds fused to tube, affixed to one side of tube somewhat below rim, the inner pair held below outer pair; anthers dehiscing introrsely; pollen somewhat tacky, clinging together in small clusters; style slender, held along staminal side of tube; stigmas densely puberulent, 2-lipped, with one lip very reduced, the other extending across mouth of tube, recurved. Capsules ovoid, 5–8 mm long, somewhat flattened, apically beaked, puberulent; seeds disk-shaped, 2–3 mm diam, brown, the outer margin quickly becoming swollen and sticky upon wetting. *Croat 4365* (typical), *7445* (narrow-leaved form).

Frequent along trails in the forest. Flowers mostly throughout the dry season (December to May), rarely flowering and fruiting during the rainy season. The fruits develop quickly.

The small pollinator probably enters the tube to effect pollination, the style acting as a brush to remove pollen picked up from a previous visit. Pollen is probably deposited on the insect's body when it passes the anthers.

Similar to *B. brownei* in most respects but tending to be more robust and erect, to have much larger flowers and leaves, and not to grow in open areas.

Standley reported this species as *B. panamense* Lindau, which, if considered distinct from *B. costaricense*, does not occur on BCI. Leonard has annotated some BCI collections as *B. dariense* Lindau. Both *B. panamense* and *B. dariense* reportedly have corollas less than 1.7 cm long, whereas most of the plants observed in the forest on BCI have corollas 2.5–3 cm long. Perhaps some hybridization takes place between *B. brownei* and *B. costaricense*, since plants in the forest that grow closest to populations of *B. brownei* frequently have the long trichomes on the upper leaf surface associated with that species. Plants farther away in the forest tend to be nearly glabrous on the upper leaf surface.

Costa Rica and Panama. In Panama, known principally from tropical moist forest on the Atlantic slope in the vicinity of the Canal Zone; known also from tropical moist forest in Chiriquí and Darién and from premontane wet forest in Chiriquí (near Concepción) and Panamá.

ELYTRARIA Michx.

Elytraria imbricata (Vahl) Pers., Synops. Pl. 1:23. 1805

Acaulescent to caulescent suffruticose herb, to ca 60 cm tall; stems ± glabrous. Leaves alternate to subopposite; blades mostly ovate to oblanceolate, acute to acuminate, gradually narrowed to slender winged petiole, 3–12 cm

long, 1.5–4 cm wide, glabrate to sparsely villous especially on veins, the margins ± undulate and ciliate; cystoliths apparently lacking. Inflorescences of numerous, upper-axillary or terminal, simple or branched scapes mostly 5–25 cm long (sometimes leafy at apex), closely bracteate, the bracts firm, usually subulate, keeled, with villous margins; spikes 1 to several per scape, to 6 cm long and ca 5 mm wide, the bracts firm, awned-tipped and bearing a pair of thin wings near apex, glabrous outside, densely pubescent inside; bracteoles subulate, ca 3 mm long, the costa ciliate; calyx segments 2, unequal, narrow, the posterior one bidentate; corolla bluish, 5–8 mm long, the tube slender, the limb bilabiate, the lower lip trilobate; stamens 2, barely exserted; anther thecae connate, ca 0.5 mm long; pistillate parts not seen. Capsules narrowly conical, ca 3 mm long and 1 mm diam, glabrous; seeds lacking retinacula. *Croat 6963.*

Rare, on a shady steep slope in the Laboratory Clearing. Flowers in the early dry season. The fruits develop quickly and seeds are dispersed in the dry season as well.

Arizona and Texas through northern and western South America to Ecuador and Brazil (Mato Grosso and Río de Janeíro). In Panama, known from tropical moist forest in the Canal Zone, Colón, Los Santos (Tonosí), and Panamá, from tropical dry forest in Herrera (Pesé), Coclé (Penonomé), and Panamá (Taboga Island), and from premontane wet forest in Chiriquí.

HERPETACANTHUS Nees

Herpetacanthus panamensis Leonard, J. Wash. Acad. Sci. 32:185. 1942

Suffruticose herb, 1 m or more tall, often rooting at basal nodes; stems terete, nodose, brittle, easily breaking at nodes; younger parts with uncinate trichomes. Petioles 3–5 mm long; blades elliptic, acuminate, attenuate and ± decurrent at base, 5–14 cm long, 2.5–7.5 cm wide, essentially entire, glabrous and shiny above with minute linear cystoliths, duller on lower surface; veins strigose, the major lateral veins ± impressed, the reticulate veins conspicuous. Inflorescence bracteate, terminal and axillary, the bracts ovate, acute, pubescent, ciliate; calyx regular, sparsely pubescent, to 5 mm long, deeply lobed, the lobes linear; corolla funnel-shaped, ca 2 cm long, bilabiate, white with lavender markings medially inside on lower lip, glabrous except on lower part of tube inside and on outside of lobes, the tube narrow, the upper lip narrowed to an emarginate tip, the lower lip trilobate; stamens 4, the upper pair with the thecae broadly separated and at different levels on connective; filaments united, the tube ca 1 cm long; anthers 1.2 mm long, longitudinally dehiscent; pistil cylindrical, pubescent, minute; style more than 1 cm long, pubescent; stigmas 2. Fruits capsular, ca 1 cm long, brown; seeds disk-shaped, ca 2 mm diam. *Croat 11900.*

Rare, in open areas of the forest. Seasonal behavior not determined. Collected on BCI in flower in August.

Costa Rica and Panama. In Panama, known from tropical moist forest in the Canal Zone (Atlantic slope)

Fig. 505. *Blechum brownei*

Fig. 506. *Justicia graciliflora*

Fig. 508. *Mendoncia littoralis*

Fig. 507. *Mendoncia gracilis*

and Bocas del Toro and from tropical wet forest in Colón (Santa Rita Ridge).

HYGROPHILA R. Br.

Hygrophila guianensis Nees, London J. Bot. 4:634. 1845

Herb, to 1.2 m tall, sparsely hirtellous to glabrate; stems obtusely quadrangular, branched. Blades lanceolate to lanceolate-linear, acuminate at apex, gradually attenuate at base to an obscure, winged petiole, 5–16 cm long, 0.5–4 cm wide, entire; cystoliths dense but ± obscure. Flowers sessile, in dense axillary clusters subtended by linear-lanceolate bracts; calyx 5–9 mm long, deeply lobed, the lobes ± linear; corolla white, bilabiate, 6–8 mm long, puberulent, the tube ca 1 mm diam, the upper lip bidentate, ca 2.5 mm long, the lower lip trilobate; stamens 4, arranged in 2 unequal pairs and fused to tube at base; pistillate parts not seen. Capsules narrowly oblong, acute at apex, 9–12 mm long, ca 1.5 mm wide, glabrous; seeds 16–18, disk-shaped, less than 1 mm wide, brown, the margin discolorous and becoming sticky on wetting. *Croat 6402.*

Frequent on sandbars in lagoons around the edge of the island, especially on the western side. Flowers from November to March, especially from December to February. Since the fruits develop quickly, the same inflorescence may bear both flowers and fruits.

Mexico to central Argentina; West Indies; in marshy areas and along water courses from sea level to 1,500 m. In Panama, ecologically variable; known from tropical moist forest in the Canal Zone and Panamá, from premontane dry forest in Coclé (Antón), and from tropical wet forest in Bocas del Toro (Santa Catalina).

JUSTICIA L.

Justicia graciliflora (Standl.) D. Gibs., Fieldiana, Bot. 34:69. 1972

Beloperone graciliflora Standl.

Vine or clambering shrub, 1–4 m long; stems moderately pubescent when young, glabrate in age. Petioles 5–8 mm long, often pubescent; blades lanceolate to narrowly elliptic, short-acuminate, acute to obtuse and decurrent at base, glabrous with many prominent cystoliths. Panicles terminal, 4–8 cm long, sessile or short-pedunculate, bearing few flowers, the bracts linear, green, 5–7 mm long, ca 1 mm wide; pedicels ca 1 cm long; sepals lanceolate-linear, 2–2.5 cm long, 3–5 mm wide, pale orange, glabrate with linear cystoliths; corolla white, the tube narrow, 3–5 cm long, the limb bilabiate, to 2 cm long, the lower lip deeply, equally trilobate; stamens 2, nearly equaling upper lip; thecae of anthers large, unequal; pistillate parts not seen. Capsules spatulate to oblong, 1.5–2 cm long, rounded and apiculate to acute at apex, obtuse at base, glabrate; seeds subglobose, ca 3.3 mm diam, brown, glabrous. *Croat 7372, Shattuck 602* (type).

Uncommon, in the old forest; collected on Armour, Drayton, and Wheeler trails. Flowers from late November to late January. Mature fruits have been seen in March. On BCI the plant has not been found in the canopy, but it was found growing at about 25 m in a tree in Veraguas Province.

Known only from Panama, from tropical moist forest on BCI and in Darién and from premontane wet forest in Veraguas (west of Santa Fé) and Panamá (Cerro Jefe).

See Fig. 506.

Justicia pectoralis Jacq., Enum. Pl. Carib. 11. 1760

Erect herb, to 50 (100) cm tall; stems with lines of retrorse pubescence between leaves, often decumbent and rooting at basal nodes. Leaves decussate; petioles 2–12 mm long; blades lanceolate, long-acuminate, acute to obtuse at base, 2–11 cm long, 0.4–2.5 cm wide, ± glabrous except for upper midrib; cystoliths moderately conspicuous above under magnification. Panicles terminal, dichotomous, to 15 cm long, the branches broadly divergent, densely puberulent, with longer glandular trichomes; bracts ca 2 mm long, 1 mm wide; flowers sessile, sparse; calyx ca 2.5 mm long, the pubescence as on inflorescence branches, the lobes long, sharp, the uppermost shorter; corolla bilabiate, lavender except for white base, the upper lip sparsely pubescent outside, narrow, folded, enclosing style in medial groove, the lower lip trilobate, its middle lobe corrugated with prominently raised, white, lateral veins extending into tube on either side of medial groove; stamens 2, affixed laterally, arching to center; anthers horseshoe-shaped, loosely clasping style; style slender, 7–8 mm long, curved downward at apex. Capsules club-shaped, acute at apex, 6–8 mm long, 1.5 mm broad, puberulent; seeds ca 1.5 mm broad, covered with papillae, the papillae minute, immediately becoming sticky upon wetting. *Croat 10253.*

Occasional, in clearings, especially along the path in the Rear Lighthouse Clearing. Flowers from December to May. The fruits develop quickly and are dispersed in the dry season.

Pollination is effected by a small insect, which crawls into the tube (no doubt with the corrugated, somewhat downturned lower lip acting as a landing platform). In crawling beneath the arched stamens, it spreads pollen on its back. At the same time the style is released from its groove and strikes the pollinator's back.

Wide-ranging and weedy throughout the New World tropics, including the West Indies. In Panama, known from tropical moist forest in the Canal Zone (both slopes), Colón, Chiriquí (Davíd), Panamá, and Darién and from premontane wet forest in Panamá (Chimán).

MENDONCIA Vell. ex Vand.

Mendoncia gracilis Turr., Kew Bull. 1919:418. 1919.

Vine; stems, petioles, and blades pubescent, the trichomes sparse to dense, moderately long, ± appressed; stems angulate. Petioles canaliculate, ridged above, 1.5–2.5 cm

long; blades ovate to elliptic-ovate, acuminate, rounded to obtuse at base, 5–12 cm long, 2.5–8 cm wide, the pubescence denser on veins; cystoliths not apparent; veins 3 or 4 on each side. Flowers 1–3, in leaf axils; pedicels 1.4–1.8 cm long, densely appressed-pubescent; bracts ± ovate, keeled, obtuse and minutely cuspidate at apex, truncate at base, 15–17 mm long, 10–13 mm wide, densely pubescent on outer surface with brownish ascending trichomes, glabrous inside; calyx reduced to a minute rim about as high as disk; corolla white, 3–3.5 cm long, bilabiate, glabrous except for stalked, gland-tipped trichomes inside on side of tube below upper lobes, the tube constricted to ca 4 mm wide at middle, expanded at base to ca 5 mm and to ca 7 mm below throat, the lobes to ca 6 mm wide, tinged with violet-purple, especially in throat; stamens 4, in 2, ± equal pairs; filaments and connective with many sessile glands; style ca 2 cm long, slightly protruding over lower edge of throat; stigmas 2, short, unequal, the larger markedly curved. Fruits similar to those of *M. littoralis. Croat 16519.*

Occasional, in the canopy of the forest; probably less abundant than *M. littoralis.* Flowers and fruits principally from December to April.

Panama and Colombia (from the Pacific slope and intermountain valleys in the Departments of Antioquia, Cundinamarca, Valle, and Tolima, usually at 1,000–2,000 m). In Panama, known from tropical moist forest in the Canal Zone and from tropical wet forest in Panamá (Cerro Jefe).

See Fig. 507.

Mendoncia littoralis Leonard, Contr. U.S. Natl. Herb. 31:1. 1951

Herbaceous vine growing into canopy; stems and leaves glabrate to sparsely appressed-pubescent; stems square, the angles ribbed (at least when young). Petioles slender, 2–4 cm long; blades ovate-elliptic, elliptic, or oblong-elliptic, caudate-acuminate, obtuse to rounded at base, 8–16 cm long, 3–7 cm wide, the margins obscurely irregular. Flowers solitary in axils, enclosed in a water-filled pair of bracts before anthesis, emerging from bracts at anthesis, soon falling; pedicels slender, 2.5–3.5 cm long; bracts ovate-elliptic to oblong-elliptic, truncate to weakly bilobed and apiculate at apex, rounded to truncate at base, 2.5–3.3 cm long, ca 2 cm wide, glabrous to sparsely short-puberulent; calyx reduced to a firm annular ring similar to disk (both persistent in fruit); corolla bilabiate, 3.5–5 cm long, glabrous, the tube white, ± curved, to 4 cm long, constricted to ca 5 mm wide above base, to 10 mm wide at base, the throat oblique, the lobes mottled with violet-purple especially near base, to 1.2 cm long and 1.5 cm wide; stamens 4, in 2 unequal pairs, held against one side of tube below upper edge of throat; filaments ± flattened, with sessile glands, fused to tube most of their length; anthers 6–10 mm long, the lower pair sometimes longer, the thecae sometimes unequal; style 1.6 cm long, glabrous; stigmas 2, short, unequal. Drupes ± ovoid, ca 1.5 cm long, purple-black and fleshy at maturity; mesocarp thin; seed 1. *Croat 12413.*

Frequent in the forest, usually in the canopy but often flowering at lower levels, owing to tree falls. Flowers from July to December. The fruits mature from November to March.

Panama and Colombia. In Panama, ecologically variable; known from wetter parts of tropical moist forest in the Canal Zone, San Blas, Chiriquí, Panamá, and Darién, from premontane wet forest in Los Santos, Panamá, and Darién, and from tropical wet forest in San Blas and Chiriquí.

See Fig. 508.

NELSONIA R. Br.

Nelsonia brunellodes (Lam.) O. Kuntze, Rev. Gen. Pl. 2:493. 1891

Prostrate or sprawling herb, densely and softly villous on most parts; stems slender, 15–60 cm long. Petioles 2–20 mm long; blades ovate to ovate-elliptic, acute to obtuse at apex, rounded to attenuate at base, 1.5–7 cm long, 1–3 cm wide. Spikes slender, densely bracteate, terminal and axillary, 1.5–6 cm long and to 6 mm wide; bracts ovate, concave, abruptly acuminate; flowers minute; sepals 4–5 mm long, markedly unequal in width, conspicuously veined, acuminate, villous outside; corolla bilabiate, glabrous, pale violet-purple above middle, the tube white, 4–5 mm long, the limb ca 2 mm broad, the upper lip 2-cleft, the lower lip trilobate, the lobes weakly bilobed; stamens 2, attached near throat; filaments pubescent on free portion; anthers with the thecae moderately separated; style equaling stamens; stigma simple, held beneath anthers. Capsules sessile, ca 4 mm long, narrowly ovate, the valves folding back at maturity; seeds to 10 per capsule, ± globose, subverrucose, lacking retinaculum. *Croat 8221.*

Occasional, in the Laboratory Clearing in open areas. Flowers from February to April. The fruits develop quickly. Late in the dry season plants are often leafless and have principally mature capsules on their spikes.

Mexico to South America as far south as Brazil (Minas Gerais); West Indies; probably introduced from the Old World, where the species is widely distributed in central Africa and South Asia. In Panama, an ecologically variable weed of open areas; known from tropical moist forest in the Canal Zone, Chiriquí, Panamá, and Darién, from premontane dry forest in Los Santos, from premontane wet forest in Panamá, and from tropical wet forest in Coclé (Atlantic slope).

See Fig. 509.

RUELLIA L.

Ruellia metallica Leonard, Publ. Field Mus. Nat. Hist., Bot. Ser. 18:1253. 1938

Suffruticose herb, to ca 50 cm tall; stems weakly quadrangulate, hirtellous or retrorsely strigose on angles, sometimes rooting at basal nodes. Petioles 5–15 mm long; blades elliptic to ovate-elliptic, acuminate at apex, atten-

Fig. 509. *Nelsonia brunellodes*

Fig. 510.
Teliostachya alopecuroidea

uate at base, 6–12 cm long, 2–5 cm wide, ± glabrous except appressed-strigose on veins below, the margins often minutely irregular; cystoliths conspicuous on both surfaces. Flowers borne in small clusters in uppermost axils; bracts irregular, lanceolate to acicular, shorter than calyx; calyx 6–7 mm long, divided to near base, the lobes narrowly pointed; corolla campanulate, to 2.7 cm long, lavender to lavender tinged with white, weakly puberulent in throat, the tube ca 2 cm long, narrow below middle, ca 1 mm wide, flared to ca 6 mm just below throat, the lobes ± rounded, spreading; stamens in 2 unequal pairs held well within throat, the uppermost pair ca 1.5 cm long; anthers 1.3 mm long, the connective extending somewhat above thecae; style with 2 unequal stigma lobes. Capsules clavate, acute at apex, ca 1 cm long, weakly puberulent. *Croat 8642.*

Apparently rare, on the forest trails; collected once on Conrad Trail in the old forest. Flowers in the dry season. The fruits develop quickly and the plant usually bears both flowers and fruits.

Nicaragua, Costa Rica, and Panama. In Panama, known from tropical moist forest in the Canal Zone and Bocas del Toro and from tropical wet forest in Colón, Coclé (Boca de Toabré) and Darién.

TELIOSTACHYA Nees

Teliostachya alopecuroidea (Vahl) Nees in Mart., Fl. Brasil. 9:72. 1847
  Lepidagathis alopecuroides (Vahl) R. Br. ex Griseb.

Erect or decumbent herb, 15–30(50) cm tall, branched many times, often rooting at basal nodes, moderately pubescent especially on stems and veins of lower leaf surface. Petioles to ca 1 cm long; blades mostly narrowly ovate, acute or obtuse at apex, attenuate at base, 3–9 cm long, 2–3.5 cm wide, ± entire. Spikes terminal, densely bracteate, 1–5 cm long and 1 cm wide; bracts similar to calyx lobes, lanceolate to oblong, cuspidate-acuminate, ca 6 mm long, ciliolate with conspicuous veins; flowers numerous; calyx lobes very unequal, 4–6 mm long; corolla to ca 5 mm long, bilabiate, white except for violet lower lip, the tube villous inside near apex, the upper lip entire or emarginate, somewhat enclosing stamens, the lower lip trilobate, recurved at anthesis; stamens 4, fused to tube in basal two-thirds; anthers shedding pollen in bud, the thecae distinct, held at 2 levels; style held next to upper lip and somewhat recurved near apex; stigma simple, held between the 2 pairs of anthers. Capsules sessile, ca 4 mm long, glabrous or pubescent near apex; seeds 4, flattened, ca 1 mm wide, pale. *Croat 7967.*

Occasional, along trails; locally abundant. Flowers and fruits mostly from January to April.

Leonard (1951) reported the species to have stamens free at the base. I have found them to be fused to the basal two-thirds of the tube.

Mexico (Yucatán) to Colombia, Venezuela, Brazil, and Peru; Trinidad, West Indies. In Panama, ecologically wide-ranging; known from tropical moist forest in the

Canal Zone and Veraguas, from tropical wet forest in Veraguas, Coclé (both slopes), and Panamá, and from premontane rain forest in Coclé (Cerro Pilón).

See Fig. 510.

THUNBERGIA Retz.

Thunbergia erecta (Benth.) T. Anderson, J. Linn. Soc., Bot. 7:18. 1864

Shrub, ± glabrous; stems 4-ribbed, branched many times. Blades ovate, acuminate, acute at base and decurrent onto short, canaliculate petiole, the margins entire to obtusely toothed and somewhat undulate. Flowers solitary on axillary pedicels; corolla 7–8 cm long, violet (purple), the tube white outside, yellow inside; stamens 4, in 2 pairs set in a groove on upper side of tube, to 3 cm long, somewhat irregular, with 2 or 3 recurved lobes; filaments with glandular trichomes; anthers bristled around margins of thecae; style set in groove just above upper pair of stamens; stigma scoop-shaped. Fruits not seen (those of the genus capsular, with the seeds ± globose and lacking a retinaculum). *Croat 6918.*

Cultivated at the Laboratory Clearing. Flowers sporadically all year, but especially in the rainy season.

Costa Rica and Panama; Jamaica and Puerto Rico. In Panama, known only from tropical moist forest in the Canal Zone.

TRICHANTHERA Kunth

Trichanthera gigantea (H. & B.) Nees in DC., Prodr. 11:218. 1847
  Palo de agua, Asedera

Small tree, usually to 10(13) m tall; trunks slender, less than 15 cm dbh, often clustered and stilt-rooted, the leaf scars persisting; stems and branches quadrangular. Petioles to 6 cm long, the pairs often unequal; blades ovate to elliptic, acuminate, mostly obtuse to attenuate at base, mostly 11–20(26) cm long, 4–10(14) cm wide, glabrous except on veins of lower surface; cystoliths short, irregularly oriented, linear, obvious only when dry. Inflorescences terminal, compact, ± secund panicles 5–15 cm long; branches to ca 6 cm long, densely tomentose, becoming scurfy in age; bracts triangular, ca 3 mm long; flowers ± regular; calyx 7–10 mm long, deeply lobed, the lobes imbricate, rounded at apex, often tinged with purple; both calyx and corolla densely tomentose outside with both glandular and simple trichomes; corolla 3.5–4 cm long, glabrous and brick red inside, the tube to 3.5 cm long, glabrous near base, the limb 2–3 cm wide; stamens 4, exserted, in 2 pairs, the outermost of each pair slightly shorter, held to one side of tube, to 5 cm long; filaments fused at base in pairs, softly pubescent below with simple and glandular trichomes; anthers ca 6 mm long, introrse, marginally fringed; ovary ovate, sericeous, glabrous and nectariferous at base; style ± equaling stamens, its upper lobe much reduced; nectaries not obvious. Capsules 1.5–2 cm long, 2-valved, obtuse, silky-

pubescent; seeds 1–4 per capsule, flattened, lenticular, 3–4 mm broad. *Croat 4632.*

Rare, usually occurring along streams. Flowers from January to April, especially in February and March. The fruits mature quickly, the first usually by February, and may be found on the same inflorescence with flowers.

The stamens are a little longer than the style at anthesis. Later they are somewhat bent back and rather distant from the style.

Costa Rica to Colombia, Venezuela, and the Guianas. In Panama, known from tropical moist forest in the Canal Zone, Colón, Los Santos, Coclé, Panamá, and Darién.

130. RUBIACEAE

Trees, shrubs, lianas, vines, or herbs, sometimes epiphytic, occasionally with spines. Leaves opposite or ternate, petiolate; blades simple, entire; stipules interpetiolar, sometimes united, forming a sheath. Flowers bisexual or occasionally unisexual (the plant dioecious), actinomorphic or zygomorphic, the cymes variously modified, terminal or axillary, dichasial, sometimes reduced to racemes or dense clusters, the inflorescences often bracteate; bracts sometimes showy; calyx lobes 4 or 5 (sometimes 2, 6, or 8); corolla 4- or 5(6)-lobed,

KEY TO THE TAXA OF RUBIACEAE

- Plants herbs, vines, or lianas:
  - Plants vines or lianas:
    - Flowers both ± sessile *and* clustered in leaf axils, white:
      - Leaves conspicuously scabrous, very asperous above; flowers less than 4 mm long; fruits less than 4.5 mm diam . . . . *Diodia sarmentosa* Sw.
      - Leaves not conspicuously scabrous and asperous above; flowers more than 7 mm long; fruits more than 6 mm diam . . . . . *Sabicea villosa* R. & S. var. *adpressa* (Wernham) Standl.
    - Flowers long-pedicellate *or* not clustered in axils, white, red, or yellow:
      - Leaves with the trichomes sparse, appressed, slender, ca 0.3 mm long and in inconspicuous axillary tufts below; fruits oblong, 1.5–2 cm long, pubescent, with appressed trichomes . . . . *Chomelia psilocarpa* Dwyer & Hayden
      - Leaves densely white-tomentose or glabrate to glabrous, lacking sparse appressed trichomes, lacking axillary tufts; fruits usually less than 1 cm long:
        - Plants armed with large recurved spines; flowers yellow, in globular pedunculate clusters; fruits 6–8(17) mm long . . . . *Uncaria tomentosa* (Willd.) DC.
        - Plants unarmed:
          - Flowers white or yellowish, in racemes or panicles; calyx lobes very short; fruits white at maturity, much flattened . . . . *Chiococca alba* (L.) Hitchc.
          - Flowers red or pink, solitary or in umbels; calyx lobes long and slender; fruits green to black at maturity, not flattened but sulcate medially . . . . *Manettia reclinata* L.
  - Plants herbs, not scandent (sometimes creeping):
    - Flowers and fruits sessile or essentially so, clustered in leaf axils, usually many:
      - Leaves linear or linear-lanceolate, ca 10 times longer than wide; flower clusters 1–2 cm diam . . . . *Borreria densiflora* DC.
      - Leaves otherwise, 2–8 times longer than wide; flower clusters usually less than 1 cm diam:
        - Leaves all less than 3 cm long, the secondary veins conspicuous . . . . *Borreria ocimoides* (Burm.f.) DC.
        - Leaves mostly more than 3.5 cm long or, if less, with secondary veins not obvious:
          - Mature capsules 3.5–4.5 mm long; upper leaf surface asperous, with many short scabrous trichomes . . . . *Diodia sarmentosa* Sw.
          - Mature capsules usually less than 3 mm long:
            - Stems and leaves greatly diminished apically, the distal leaves many times smaller than the leaves lower on the stem; capsules minutely puberulent (the trichomes short and straight), often as broad as or broader than long, often splitting from the bottom; flowers borne in the axils of smaller leaves . . . *Diodia denudata* Standl.
            - Stems and leaves not greatly diminished apically, the distal leaves at most only a few times shorter than the lower ones; capsules either glabrous or pubescent (the trichomes not short and straight), longer than broad, dehiscing largely from apex:
              - Capsules glabrous or sparsely pubescent:
                - Secondary veins on leaves conspicuous below . . . . . . *Borreria laevis* (Lam.) Griseb.
                - Secondary veins on leaves very inconspicuous below . . . . *Spermacoce tenuior* L.
              - Capsules densely pubescent at least apically:
                - Leaves 1.5–3 times longer than broad, their trichomes 0.6–1 mm long, often broad-based; calyx lobes more than 0.5 mm long and sharply triangular; plants drying yellow-green; style bifid . . . *Borreria latifolia* (Aubl.) K. Schum.
                - Leaves 3–4 times longer than broad, scabrid, the trichomes ca 0.2 mm long; calyx lobes less than 0.5 mm long, often blunt or obscure; plants usually drying quite dark . . . . *Diodia ocimifolia* (Willd.) Bremekamp

Flowers and fruits not as above (not sessile, numerous, and crowded in leaf axils):

Leaves slender, less than 5 mm wide; flowers and fruits each less than 4 mm long . . . .
. . . . *Oldenlandia corymbosa* L.

Leaves ovate-cordate, more than 2 cm wide; flowers and fruits more than 4 mm long . . *Geophila*

Plants trees or shrubs:

Ovules 1 per locule:

Flowers in dense heads subtended by large, green, red, or violet-purple bracts . . . . *Cephaelis*

Flowers not in dense heads with large colorful bracts:

Corolla less than 1 cm long:

Plants trees, usually more than 7 m tall; blades and stems densely pubescent with long, appressed, pale trichomes, the corolla lobes pubescent; fruits ellipsoid-oblong, ca 1 cm long, red to black . . . . *Antirrhoea trichantha* (Griseb.) Hemsl.

Plants shrubs or small trees, rarely more than 3 m tall (*Psychotria grandis* sometimes larger), variously pubescent; fruits not as above:

Plants subscandent shrubs, rare; fruits ellipsoid, compressed, white, 4–8 mm diam; seeds 1 or 2; flowers white or yellow, in axillary panicles . . *Chiococca alba* (L.) Hitchc.

Plants not as above; fruits ovoid, not at all or only slightly compressed, red, blue, or black (white by rupturing of the exocarp in *P. deflexa*); seeds mostly 2, rarely 5 . .
. . . . *Psychotria*

Corolla more than 1 cm long:

Flowers 4-parted, white; fruits with 1 seed:

Leaf blades with pitlike axillary domatia below; fruits white at maturity, ellipsoid, compressed, the seed vertical . . . . *Coussarea curvigemmia* Dwyer

Leaf blades lacking axillary domatia; fruits blue or black at maturity, globose or depressed-globose, the seed horizontal . . . . *Faramea*

Flowers 4- or 5-parted, white, red, or yellow; fruits with 2 or more seeds:

Flowers white, the tube sericeous; stems often armed with branch-spines; fruits more than 1 cm diam . . . . *Guettarda foliacea* Standl.

Flowers yellow or red, the tube not sericeous; plants unarmed; fruits less than 1 cm diam:

Flowers red or red-orange; corolla more than 3 cm long; leaves subsessile, often subcordate, less than 7 cm wide; plants cultivated shrubs . . . . *Ixora coccinea* L.

Flowers yellow or orange; corolla less than 2 cm long; leaves petiolate, acute to obtuse at base, more than 7 cm wide; plants not cultivated . . . .
. . . . *Palicourea guianensis* Aubl.

Ovules more than 1 per locule:

Calyx lobes very unequal, some flowers with 1 lobe greatly expanded and leaflike:

Corolla and expanded calyx lobe white; capsules ca 6–10 mm long; seeds winged . . . .
. . . . *Calycophyllum candidissimum* (Vahl) DC.

Corolla red, orange, or yellow, the expanded calyx lobe red; capsules usually less than 7 mm long; seeds not winged:

Corolla less than 8 mm long; stipules more than 10 mm long; flowers in small pedunculate cymes, densely clustered along a long main axis . . . .
. . . . *Warscewiczia coccinea* (Vahl) Klotzsch

Corolla more than 2 cm long; stipules less than 5 mm long; flowers in open panicles . . . .
. . . . *Pogonopus speciosus* (Jacq.) K. Schum.

Calyx lobes ± equal:

Corolla more than 5 cm long:

Corolla funnelform-campanulate, white or yellowish tinged with purple; flowers in dichasial inflorescences; calyx lobes slender, acuminate; fruits compressed-obovoid, the seeds with a thin wing . . . . *Coutarea hexandra* (Jacq.) K. Schum.

Corolla narrowly tubular, whitish or yellowish, not tinged with purple; inflorescences and calyces various; fruits not as above:

Flowers solitary on ends of branches; corolla lobes tapered to slender tip; fruits round, ca 2 cm diam, green with white stripes; calyx with slender teeth, persistent in fruit
. . . . *Randia formosa* (Jacq.) K. Schum.

Flowers clustered on ends of branches; corolla lobes blunt; fruits not as above; calyx lobes short, rounded to acute but not slender:

Corolla yellow; fruits globose to ellipsoid, more than 6 cm diam . . . .
. . . . *Tocoyena pittieri* (Standl.) Standl.

Corolla white; fruits linear-cylindrical or globose, less than 6 cm diam:

Leaves acute to short-acuminate; calyx lobes blunt; style not exserted; anthers exserted; fruits globose, fleshy; seeds probably animal dispersed . . . .
. . . . *Posoqueria latifolia* (Rudge) R. & S.

Leaves rounded at apex; calyx lobes acute; style and anthers exserted; fruits linear-cylindrical, capsular; seeds wind dispersed . . . .
. . . . *Cosmibuena skinneri* (Oerst.) Hemsl.

■ Corolla less than 5 cm long:
  Corolla more than 2 cm long (including lobes):
    Corolla white; fruits globose or ellipsoid, 1.5–3 cm diam, never red:
      Plants armed with short spines; calyx deeply lobed, the lobes broadly triangular; corolla 5-lobed; fruit often longer than broad, usually prominently lenticellate . . . . *Randia armata* (Sw.) DC.
      Plants unarmed; calyx truncate with minute teeth; corolla 4- or 5-lobed; fruit globose or depressed-globose, smooth, topped by the persistent calyx tube . . . . *Alibertia edulis* (A. Rich.) A. Rich.
    Corolla yellow or red (white at anthesis but soon yellowing in *Genipa*); fruits not as above, either less than 1 cm long or more than 3 cm long or red:
      Plants moderately large trees, the trunk usually more than 15 cm dbh; corolla sericeous outside, lobed more than halfway; calyx ± truncate, 7–10 mm wide; fruits more than 5 cm diam; seeds ca 1 cm long . . . . *Genipa americana* L.
      Plants shrubs or small trees, the trunk usually much less than 15 cm dbh; corolla lobed less than halfway; calyx reddish, weakly lobed and less than 4 mm wide, or prominently lobed; fruits less than 2.5 cm diam:
        Leaves 3 at each node . . . . *Hamelia patens* Jacq. var. *glabra* Oerst.
        Leaves merely paired at each node:
          Inflorescences terminal, many-flowered, much-branched; stipules 1–2 cm long . . . . *Isertia haenkeana* DC.
          Inflorescences axillary, short, few-flowered, not much-branched; stipules more than 3.5 cm long . . . . *Pentagonia macrophylla* Benth.
  Corolla less than 2 cm long:
    Inflorescences corymbose cymes, axillary or sometimes subterminal, with at least 25 minute flowers forming hemispherical clusters on peduncles 3–4 cm long; fruits to 2 mm long and ca 1 mm wide . . . . *Chimarrhis parviflora* Standl.
    Inflorescences and fruits not as above:
      Inflorescences axillary; flowers 4-parted; corolla greenish-yellow, lobed beyond middle . . . . *Hoffmannia woodsonii* Standl.
      Inflorescences terminal or subterminal; flowers 5-parted:
        Plants trees; fruits capsular; flowers magenta, in large panicles or white to pale yellow in long slender racemes:
          Inflorescences paniculate; flowers magenta; leaves not clustered at apex of branches . . . . *Macrocnemum glabrescens* (Benth.) Wedd.
          Inflorescences racemose; flowers white to pale yellow; leaves clustered at apex of branchlets, obovate . . . . *Alseis blackiana* Hemsl.
        Plants shrubs or small trees; fruits fleshy; flowers white, yellow, or red:
          Flowers orange to red; leaves verticillate, 3, 4, or 5 per node, sometimes opposite . . . . *Hamelia patens* Jacq. var. *glabra* Oerst.
          Flowers yellow or white; leaves not verticillate:
            Calyx and corolla sericeous; calyx truncate, with minute apiculate teeth; leaves somewhat clustered at ends of branchlets . . . . *Amaioua corymbosa* H.B.K.
            Calyx and corolla not sericeous; calyx regularly lobed; leaves not clustered:
              Flowers yellow, the corolla more than 1 cm long; fruits red to black; inflorescences about as broad as long . . . . *Hamelia axillaris* Sw.
              Flowers white, the corolla less than 8 mm long; fruits purple; inflorescences long and slender, many times longer than broad . . . . *Bertiera guianensis* Aubl.

---

rarely oblique or bilabiate, showy; stamens equal in number to corolla lobes and alternate with them; anthers 2-celled, introrse, dorsifixed, dehiscing longitudinally; pistil 1; ovary inferior, 2-locular, 2-carpellate; placentation axile or seemingly basal; ovules 1 or more per locule; style 1, slender; stigma simple or 2(5)-lobed. Fruits usually berries or sometimes loculicidal capsules; seeds 2 (or 1 by abortion) to many, sometimes winged, the endosperm usually copious, fleshy, rarely none (*Chomelia* and *Guettarda*).

Recognized by the opposite leaves, interpetiolar stipules, and regular, tubular flowers.

A number of morphological features have evolved in flowers of the family to promote outcrossing. Among these are dimorphic heterostyly (e.g., *Psychotria capitata, P. pubescens, Cephaelis ipecacuanha*), protogyny (e.g., *Warscewiczia coccinea, Alseis blackiana*), and marked protandry (e.g., *Posoqueria latifolia, Pogonopus speciosus*). Though basically actinomorphic, flowers of some species have become zygomorphic by modification of the sexual parts, usually by the stamens being clustered on one side of the corolla while the style is held on the opposite side. This zygomorphy is accompanied as well by heterostyly in the case of *Psychotria pubescens*. Probably dimorphic

heterostyly is much more prevalent in *Psychotria* than I have indicated, since it is often not at all apparent from observations of a few individuals and is much more easily observed in the field than from specimens. In *Pogonopus speciosus*, outcrossing is further ensured by a protandry and a marked lateral movement of the stamens after anthesis.

Species with reddish flowers that seem suited for hummingbird pollination include *Hamelia patens*, *Macrocnemum glabrescens* (observed by Chapman, 1931), *Manettia reclinata*, *Ixora coccinea*, *Isertia haenkeana*, *Pogonopus speciosus*, and *Warscewiczia coccinea*. The large red bracts of *W. coccinea* would attract the birds, rather than the flowers themselves. The flowers of *Isertia haenkeana* are yellow, but the calyx and inflorescence branches are reddish. Other species with yellow flowers and bright red bracts, which are possibly also hummingbird pollinated, include *Pentagonia macrophylla* and *Cephaelis tomentosa*. *Cephaelis discolor* with maroon bracts and white flowers may also be hummingbird pollinated, as may be *Hamelia axillaris* with yellow, narrowly funnelform flowers. Hummingbirds have been seen visiting *Psychotria capitata* (pers. obs.) and *Genipa americana* (H. Baker, pers. comm.) and may visit other *Psychotria* species as well.

Butterflies are probably active pollinators of many species but especially *Alseis blackiana*, *Uncaria tomentosa*, *Sabicea villosa*, and possibly *Antirrhoea trichantha*, *Calycophyllum candidissimum*, and many of the suspected moth-pollinated flowers. Butterflies of the genus *Heliconius* (Nymphalidae, Ithomiinae) have been observed visiting the heterostylous flowers of *Psychotria capitata* and species of *Palicouria* (L. Gilbert, pers. comm.). Butterflies may be effective pollinators of many of the other small, white-flowered species such as *Psychotria*, *Borreria*, *Chiococca*, *Bertiera*, *Diodia*, *Geophila*, *Oldenlandia*, and *Spermacoce*. Most of the same species are probably visited by bees as well. In Costa Rica *Heliconius* and pierid butterflies, which frequently visit typical "hummingbird" flowers, visit *Hamelia patens* (Heithaus, 1973).

Flowers with white, mostly slender corollas that generally fall off in the early morning are probably moth pollinated. These include *Faramea occidentalis*, *F. luteovirens*, *Randia armata*, *Guettarda foliacea*, and possibly also *Amaioua corymbosa* and *Alibertia edulis*. Species that seem to be adapted only for hawkmoth pollination include *Posoqueria latifolia*, *Tocoyena pittieri*, *Cosmibuena skinneri*, and *Randia formosa* (see the species description of *Posoqueria latifolia* and Halle, 1967).

Bat pollination is suspected for *Coutarea hexandra* because of floral morphology and time of anthesis.

Despite the fact that few Rubiaceae are definitely adapted to bee pollination, the prevalence of bees as pollinators suggests that many rubiaceous flowers, especially those with shorter tubes, are pollinated by bees. Dressler (1968b) reported that euglossine bees regularly visit flowers of *Sabicea* for nectar. *Genipa americana* in Guanacaste, Costa Rica, is probably bee pollinated (G. Frankie, pers. comm.).

Major dispersal classes in the Rubiaceae are ornithochory and anemochory. Frequently the small black or blue fruits are made more conspicuous by the reddish color of the fruiting inflorescence branches. Species that produce fruits at least partly adapted for dispersal by mammals include *Alibertia edulis*, *Amaioua corymbosa*, *Coussarea curvigemmia*, *Faramea* spp., *Genipa americana*, *Guettarda foliacea*, *Pentagonia macrophylla*, *Posoqueria latifolia*, *Randia* spp., and *Tocoyena pittieri*. Most of these are known to be eaten by white-faced monkeys and a few also by spider monkeys and tamarins (Oppenheimer, 1968; Hladik & Hladik, 1969). Oppenheimer (1968) reported that seeds of *Faramea occidentalis*, *Randia armata*, and *Tocoyena pittieri* all germinate after passing through the gut of white-faced monkeys. Fruits of *Coussarea curvigemmia* are taken by coatis (Kaufmann, 1962).

Fruits of the genus *Genipa* have been reported to be dispersed by lizards in the West Indies (Ridley, 1930). *G. americana* is dispersed in part by fish (van der Pijl, 1968). I have seen whole fruits floating in the water at the lake margin, so they may be water dispersed in part.

Wind-dispersed species include *Cosmibuena skinneri*, *Alseis blackiana*, *Chimarrhis parviflora*, *Coutarea hexandra*, *Calycophyllum candidissimum*, *Macrocnemum glabrescens*, *Manettia reclinata*, *Uncaria tomentosa*, and *Warscewiczia coccinea*.

The seeds of *Oldenlandia corymbosa* and those of the tribe Spermacoceae, including *Borreria*, *Diodia*, and *Spermacoce*, are not clearly adapted for specialized dispersal. Most of these are weedy plants and their seeds are apparently spilled locally. Ridley (1930) reported that seeds of *Borreria*, *Spermacoce*, and *Oldenlandia* are dispersed by rain wash.

Over 450 genera and more than 5,000 species; generally distributed, but most numerous in the tropics.

ALIBERTIA A. Rich. in DC.

Alibertia edulis (A. Rich.) A. Rich. in DC., Prodr. 4:443. 1830

Lagartillo, Madroño, Trompito, Trompo, Wild guava

Glabrous, dioecious shrub or small tree, to 5.5 m tall. Stipules interpetiolar, subpersistent, lanceolate, to 1.5 cm long and 5 mm wide; petioles less than 1 cm long; blades lanceolate-oblong to oblong-elliptic, acuminate, obtuse and decurrent onto petiole at base, 7–20 cm long, 2.5–6 cm wide, with pocketlike domatia in axils below, usually also tufted in axils. Inflorescences very congested, terminal; flowers sessile or subsessile, unisexual, usually few, subtended by one or more pairs of stipules; calyx truncate, 4–6 mm long, with minute, irregular teeth, persisting in fruit; corolla white, 2–3 cm long, densely and inconspicuously tomentose, the tube tomentose outside, sericeous inside except at base, the lobes 4 or 5, lanceolate, narrowly acute, spreading or recurved; stamens 4, sessile, affixed at the middle of the connective; staminate flowers with the anthers to 11 mm long, their apex held near corolla throat, the style elongating to become slightly exserted, the stigmas 2, short and slender, probably not

opening; pistillate flowers with the anthers to 4.7 mm long, deeply included in tube, the style to 1.8 mm long, exserted, the lobes 3–5, spreading and ± appressed to base of petals. Berries round, 2–3 cm diam, yellowish, becoming black and fleshy; seeds many, flattened, irregular, 4–6 mm long, brown, embedded in a grayish, edible matrix. *Croat 5358, 7390.*

Frequent in the forest. Flowers mostly from April to October but especially in the early rainy season (June and July). The fruits mature from July to December, mostly in August and September. Flowering less frequently at the beginning of the dry season, with the fruits maturing by the end of the dry season.

Mexico (Tabasco) to Panama, Colombia, the Guianas, Amazonian Brazil, and Bolivia; Trinidad and Tobago. In Panama, ecologically variable; known principally from tropical moist forest in the Canal Zone and Bocas del Toro, San Blas, Chiriquí, Veraguas, Los Santos, Panamá, and Darién Provinces; known also from tropical dry forest in Coclé, from premontane moist forest in Coclé and Panamá, from premontane wet forest in Chiriquí, Coclé, and Panamá, and from tropical wet forest in Colón, Veraguas, and Darién.

See Fig. 511.

ALSEIS Schott

Alseis blackiana Hemsl., Diag. Pl. Mex. 30. 1879

Tree, to 15(30) m tall; trunk to 1 m dbh, weakly to strongly involuted vertically in age; outer bark light brown with many vertical fissures, soft and loose; inner bark and wood light in color but turning darker soon after being slashed; young stems, petioles, axes of inflorescences, and pedicels minutely puberulent to strigillose. Leaves clustered at apex of branches; stipules slender, acuminate, 8–12 mm long; petioles 1–2.5 cm long; blades oblanceolate, short-acuminate, acute to obtuse at base, 9–20 cm long, 2.5–7 cm wide, almost glabrous above, sparsely appressed-pubescent and densely puberulent on midrib and main veins below. Racemes slender, upper-axillary, 10–23 cm long and ca 2 cm wide; flowers protogynous, sweetly aromatic; calyx lobes minute, triangular, bluntly acute, thick, usually glabrous; corolla white to pale yellow, 3–3.7 mm long, ca as broad as long, glabrous outside, densely pubescent inside, the lobes 5, bluntly acute, erect or curved inward and holding filaments together; filaments curved outward, ca 9 mm long, fused to tube at very base, densely villous except near apex; anthers ca 1.5 mm long, versatile, held perpendicular to axis of flower; style ca 6 mm long, the branches 2, recurved, protruding from flower before it is fully expanded, soon overtopped by stamens and somewhat obscured among trichomes of filaments; nectary green, longer than calyx, with moderate amounts of nectar. Capsules dense along spike, club-shaped, 10–15 mm long, splitting along one side from apex, the valves persisting after seeds are shed; seeds many, 5–10 mm long, very slender, with a narrow subulate wing on either end. *Croat 5389, 14829.*

Common in all parts of the forest. Flowers in April and May. The fruits are of mature size by June, but do not turn brown and shed seeds until late in the following dry season (R. Foster, pers. comm.).

The species is very similar to and probably not separable from *A. yucatanensis* Standl. in Mexico, *A. shippii* Lund. in Belize, and *A. hondurensis* Standl. in Honduras.

Known only from Panama and Colombia, but probably synonymous with plants ranging from Mexico to northern Colombia. In Panama, known from tropical moist forest in the Canal Zone and Darién.

See Fig. 512.

AMAIOUA Aubl.

Amaioua corymbosa H.B.K., Nov. Gen. & Sp. 3:419, pl. 294. 1820
Madroño

Dioecious shrub or tree, to 15 m tall; trunk slender or to 20 cm dbh and deeply invaginated; stems sericeous when very young, glabrate in age, with conspicuous leaf scars. Leaves somewhat clustered at apex of branchlets; stipules caducous, to 1 cm long, sericeous, leaving a conspicuous interpetiolar line of trichomes; petioles 6–14 mm long, sericeous to glabrate; blades ovate-elliptic to obovate-elliptic, abruptly short-acuminate, usually obtuse and ± decurrent onto petiole at base, 7–13(21) cm long, 3.5–8(12) cm wide, glabrous or with sparse long trichomes on major veins below and tufts of trichomes in axils, the margins ciliate and ± revolute. Flowers fragrant, 5–7-parted, sericeous throughout; staminate flowers in terminal corymbs ca 8 cm long; pistillate flowers similar to but slightly smaller than staminate flowers, in capitate or corymbose inflorescences; peduncles to 5 cm long; pedicels 2–4 mm long; calyx tubular, 3–4 mm long, with short apiculate teeth; corolla 10–15 mm long, white, the lobes 5, ca 6 mm long, puberulent, spreading at anthesis; anthers subsessile, to 4.3 mm long; style to 9 mm long, included, longer than stamens. Berries ovoid, 1–1.5 cm long, purplish at maturity, in dense axillary clusters, sparsely sericeous; seeds many, irregular, rounded on one side, flattened, 3–5 mm long. *Croat 16214, Foster 1337.*

Occasional, in the forest and along the shore; common on a small promontory midway between the ends of Zetek and Armour trails (R. Foster, pers. comm.). Flowers mostly in July. The fruits apparently mature over a long period and are seen most months of the year; most probably mature during the rainy season and early dry season.

Mexico to Panama, Colombia, Venezuela, the Guianas, Brazil, and Bolivia; Trinidad and West Indies. In Panama, ecologically variable; known principally from the Pacific slope in tropical moist forest in the Canal Zone, Veraguas, Panamá, and Darién (probably elsewhere as well); known also from tropical dry forest in Panamá (Taboga Island), from premontane moist forest in the Canal Zone, and from tropical wet forest in Colón and Panamá.

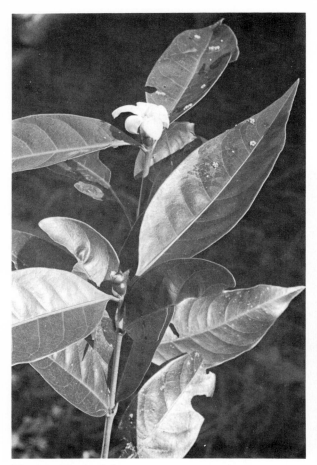

Fig. 511. *Alibertia edulis*

Fig. 512. *Alseis blackiana*

Fig. 513. *Borreria densiflora*

ANTIRRHOEA Commers. ex Ant. Juss.

Antirrhoea trichantha (Griseb.) Hemsl., Biol. Centr.-
Amer. Bot. 2:42. 1881
Pittoniotis trichantha Griseb.

Tree, to 15 m tall and 25 cm dbh; branchlets puberulent
and conspicuously lenticellate; leaf scars prominent.
Leaves deciduous, clustered at ends of branches, emerging at time of flowering; stipules long-attenuate, to 1.5 cm
long, caducous, appressed-pubescent; petioles 1.2–2.8
cm long; blades ovate to elliptic or obovate, acuminate,
acute to obtuse at base, 5–24 cm long, 2–12 cm wide,
thin, glabrate above in age except on veins, with long
appressed trichomes below; most veins raised below, the
smallest ones prominent, parallel, connecting lateral
veins. Inflorescences pubescent to glabrate (in fruit)
panicles, axillary, densely flowered; flowers 4- or 5-parted,
white, with strong sweet aroma, sessile, secund; calyx
lobes acute, ca 0.5 mm long; corolla to 5.3 mm long,
appressed-pubescent outside, villous inside on lobes, the
lobes longer than the tube, reflexed at anthesis; stamens
equal in number and about as long as corolla lobes, ±
erect at anthesis; style ± equaling stamens, divided to
ca middle. Drupes oblong, to 1 cm long, bright red and
shiny at maturity, sweet and fleshy, becoming brownish
when overripe; seed 1, tan, oblong, to 5.3 mm long, ±
flattened, shallowly grooved on each side. *Croat 5214,
11973.*

Rare; known only from the shore on the northern edge
of the island. Flowers from March to May, mostly in
April. The fruits mature from June to October. Plants
apparently lose all their leaves and quickly replace them
just before flowering.

Known only from Panama, from drier parts of tropical
moist forest in the Canal Zone, Panamá, and Darién and
from premontane moist forest in the Canal Zone.

BERTIERA Aubl.

Bertiera guianensis Aubl., Hist. Pl. Guiane Fr. 1:180,
pl. 69. 1775

Slender shrub, to 3 m tall; stems glabrate to appressed-
pubescent. Leaves glabrous except for pubescence on
petioles, veins, and margins, the trichomes sparse, long,
± appressed; stipules long-acuminate, to 5(10) mm long,
sparsely ciliate, subpersistent, the pairs joined toward
base; petioles less than 1 cm long; blades oblong-elliptic,
acute to acuminate, obtuse to acute and decurrent at
base, 10–19 cm long, 4–8 cm wide. Panicles terminal,
to ca 15 cm long, the branches with stiff, long trichomes;
flowers secund; calyx lobes acute, less than 1 mm long;
corolla tubular, white, 5–8 mm long, shallowly 5-lobed,
the lobes narrowly tapered, spreading at anthesis; stamens
affixed to tube near apex, slightly exserted above rim;
anthers ca 2 mm long, acuminate, narrowly sagittate at
base; style to 4.3 mm long, the lobes thick, ca 2 mm long,
exserted at anthesis. Berries ellipsoid, to 8 mm long and
3–4 mm wide, fleshy, 10-costate when dry, blue turning
purple to purple-black at full maturity, pubescent with
long stiff trichomes; seeds several, irregular, usually
obtusely 3-angled, ca 1.5 mm diam, densely muricate,
tan. *Foster 774.*

Occasional, in the forest; collected by Foster on
Wheeler Trail before the junction with Snyder-Molino
Trail. Flowers from March to August, especially in June,
rarely later in the rainy season. The fruits mature in about
1 month.

Mexico to Panama, Colombia, Venezuela, the Guianas,
Brazil, and Bolivia; West Indies. In Panama, known
from tropical moist forest in the Canal Zone, Bocas del
Toro, and Darién and from premontane wet forest in
Panamá (Chimán) and Darién.

BORRERIA G. Meyer

Borreria densiflora DC., Prodr. 4:542. 1830
B. spinosa (L.) Schlecht. & Cham.

Erect, glabrous herb, less than 1 m tall; stems ± square,
the corners ribbed, with occasional blunt short protuberances ca 0.3 mm high. Leaves sessile, clustered at nodes;
stipules with several prominent bristles, fused to base of
leaf; blades linear to linear-lanceolate, acute, gradually
tapered to base, the longer ones 4–7 cm long, less than
8 mm wide. Flowers white, in dense globose clusters
1–2 cm diam, terminal or in upper leaf axils, interspersed
with slender white bristles; calyx lobes 2, slender, ca 1.3
mm long, persistent, pubescent at base; corolla ca 2 mm
long, lobed to near middle, the lobes 4, spreading; stamens with the filaments fused to tube, recurved at apex
in bud; anthers held somewhat below the apparently
receptive stigma, becoming exserted; style shorter than

KEY TO THE SPECIES OF BORRERIA

Leaves linear-lanceolate, ca 10 times longer than broad; flower clusters 10–20 mm diam . . . .
. . . . *B. densiflora* DC.
Leaves 2 or 3 times longer than broad; flower clusters usually less than 7 mm diam:
   Leaves less than 3 cm long . . . . *B. ocimoides* (Burm.f.) DC.
   Leaves mostly more than 3 cm long:
      Stems wing-angled; capsules densely pubescent; seeds not grooved . . . .
. . . . *B. latifolia* (Aubl.) K. Schum.
      Stems angled but not winged; capsules glabrous or sparsely pubescent; seeds with conspicu-
ous transverse grooves . . . . *B. laevis* (Lam.) Griseb.

corolla. Capsules ca 2 mm long and 1 mm wide, pubescent, splitting medially at maturity, the pericarp persistent after seed falls out through the slit opened between calyx lobes; seeds 2, 1 per carpel, oblong, ca 2 mm long and 0.5 mm wide, dark brown. *Croat 12884.*

Occasional, in clearings. Flowers from October to January, mostly in December at the beginning of the dry season. The fruits mature in about 1 month.

Distinguished from other *Borreria* by its habit and the bilobed calyx.

Mexico to Panama, Colombia, and Venezuela; West Indies. In Panama, ecologically variable; known from tropical moist and tropical wet forests at low elevations all along the Atlantic slope and in Chiriquí, Herrera, and Panamá, from tropical dry forest in Panamá (Taboga Island), from tropical wet forest in Veraguas, and from lower montane wet forest in Chiriquí.

See Fig. 513.

Borreria laevis (Lam.) Griseb., Gött. Abh. 7:231. 1857

Herb, erect or ± sprawling, usually less than 60 cm high; stems angulate, scabrid or glabrous except for minutely pubescent lines below stipules. Stipules united, with several bristles, 2–7 mm long; petioles obscure or to 1 cm long, united with stipules at base; blades ovate to lanceolate, obtuse to acuminate at apex, obtuse and attenuate or cuneate and decurrent at base, essentially glabrous or scabrid on veins above, asperous below, bicolorous, the larger ones 3–9 cm long, 1.5–3 cm wide; veins strongly ascending toward apex. Inflorescences axillary, at most upper nodes; flowers many, usually in various stages of development, interspersed with bristles; calyx 4-lobed, the lobes minute, ciliate, persisting in fruit; corolla white, 2.3–2.7 mm long, 4-parted, divided to about middle, the lobes acute, recurved, sparsely pubescent with moniliform trichomes inside; stamens 4, exserted at anthesis, longer than lobes, equaling or exceeding style; filaments fused to tube; anthers bluish, ca 0.3 mm long, longitudinally dehiscent; style exserted, minutely pubescent; stigma globular. Capsules glabrate to sparsely pubescent, oblong, 1–2 mm long, short-pedicellate, dehiscing irregularly, one cell carrying with it 3 of the 4 calyx lobes; seeds 2, oblong, 1–1.5 mm long, less than 1 mm wide, brown, with prominent transverse grooves opposite the longitudinal sulcus. *Croat 4045, 5984.*

A very abundant weedy plant, found mostly in clearings. Flowers to some extent throughout the year, especially at the beginning of the rainy season (from July to September) and at the beginning of the dry season (January). The fruits mature within a month, sometimes while the plant is still flowering.

Florida and Mexico to Panama, Colombia, Venezuela, the Guianas, Brazil, and Bolivia; West Indies. In Panama, ecologically variable; known principally from tropical moist forest all along the Atlantic slope and in the Canal Zone, Los Santos, Coclé, Panamá, and Darién; known also from premontane moist forest in the Canal Zone, from premontane wet forest in Los Santos and Darién, and from premontane rain forest in Chiriquí.

Borreria latifolia (Aubl.) K. Schum. in Mart., Fl. Brasil. 6(6):61. 1888

Erect or decumbent, pubescent herb; stems wing-angled, the wings with scabrous trichomes or blunt protuberances. Leaves sessile or subsessile; stipular bristles ciliate; blades ovate-elliptic, lanceolate, or lanceolate-elliptic, acute at apex, cuneate and decurrent at base, 4–6 cm long, 1.5–2.8 cm wide, often yellowish-green when dried, conspicuously strigose on both surfaces, the trichomes, especially above, often broad-based. Flowers sessile, in dense axillary clusters usually less than 7 mm diam; calyx usually acutely 4-lobed, the lobes to 1 mm long, densely crisp-villous; corolla white, 3–4 mm long, 4-lobed, the lobes ca 1 mm long, pinkish at tip, with a few stiff trichomes outside, strongly recurving; stamens shorter than lobes; filaments united to tube; pollen shed in bud, the open stigma generally completely covered with pollen at anthesis; style prominently bifid at apex. Capsules conspicuously crisp-villous, ca 3 mm long, splitting unequally, with 3 of the 4 calycine lobes remaining with one of the 2 segments; seeds 2, elliptic, ca 1.5 mm long and 1 mm wide, brown, minutely papillate. *Croat 6461.*

Occasional, in clearings. Flowers to some extent throughout the year, but most flowers and fruits are seen in the rainy season (September to December). The fruits develop within a month.

Belize to Panama, Colombia, Venezuela, Guyana, and Brazil; Trinidad and Guadeloupe; West Africa. In Panama, ecologically variable; known from tropical moist forest in the Canal Zone and Panamá, from premontane moist forest in the Canal Zone, from premontane wet forest in Chiriquí, Veraguas, and Panamá, and from tropical wet forest in Colón.

Borreria ocimoides (Burm.f.) DC., Prodr. 4:544. 1830

Weak, mostly glabrous herb; stems often branched many times, decumbent, quadrangular, less than 2 mm diam, the 2 opposing sides with narrow marginal ribs, the ribs glabrous or pubescent especially at nodes. Leaves variable, sessile or subsessile; stipules connate, adnate to petiole base, the sheath cup-shaped, minute, with a few slender teeth to 1 mm long; blades ovate to narrowly elliptic or lanceolate, acute to acuminate at apex, acute at base, mostly 1–3 cm long and 4–13 mm wide, scabridulous above, glabrous and paler below, the margins weakly and minutely scabrous. Flowers white, interspersed with white trichomes, in dense axillary clusters less than 1 cm diam; calyx 2-lobed (sometimes 3- or 4-lobed), the lobes slightly exceeding corolla and persistent in fruit; corolla 0.7–1 mm long, 4-lobed to middle or beyond, the lobes acute, pubescent, with long trichomes near middle; stamens alternate with corolla lobes; filaments fused to tube; anthers oval, minute; style shorter than tube; stigma simple or with 2 obscure lobes. Capsules glabrous or puberulent, dehiscing from base or apex into 2 parts; seeds 2, oblong, ca 1 mm long, brown, markedly reticulate. *Croat 8268.*

Occasional, in clearings on rock outcrops or as an epi-

phyte on tree stumps or floating logs on the lakeshore. Flowers and fruits all year, especially at the beginning of the dry season.

Widespread in the tropics. In Panama, known principally from the Pacific slope, from tropical moist forest in the Canal Zone, Panamá, and Darién, from premontane moist forest in Panamá, and from premontane wet forest in Chiriquí, Coclé, and Panamá.

CALYCOPHYLLUM C. DC.

Calycophyllum candidissimum (Vahl) DC., Prodr. 4:367. 1830

Madroño, Alazano, Harino, Lancewood, Lemonwood, Salamo

Tree, 7–17(25) m tall, to 70 cm dbh; outer bark light, smooth, loose; branches reddish-brown. Stipules ovate, to 1 cm long, caducous; petioles 1–2.5 cm long; blades ± elliptic or ovate, acuminate, abruptly cuneate at base, mostly 5–13 cm long, 2.5–7 cm wide, glabrous above, sparsely pubescent below; midrib sometimes prominently arched. Flowers in terminal, compound dichasia, sweetly fragrant, each cluster of 3 flowers at first enclosed in a deciduous sheath, the lateral flowers of each cluster on pedicels to 2 mm long, the middle flower sessile, often with one, much-enlarged, conspicuous, white, leaflike calycine lobe (sometimes persisting in fruit), to ca 5 cm long and 3.5 cm wide; corolla white, ca 7 mm long, lobed to half its length, the tube short, densely pubescent at rim, the lobes 4, reflexed; stamens held ± erect at anthesis, dehiscing inward, later divergent; style much shorter than stamens; stigmas 2, thick, held mostly together at anthesis, later recurved. Capsules oblong-cylindrical, 6–10 mm long, ca 3 mm wide, widest above middle; seeds several, slender, to 5 mm long, with an acuminate wing on either end. *Croat 8661, 12905.*

Rare on the island, but abundant locally in the Canal Zone. Flowers from November to March, especially in the earliest part of the dry season (December). The fruits mature in February and March.

Conspicuous because of the expanded white calyx lobe.

Mexico (Chiapas) to Colombia and Venezuela; Cuba. In Panama, ecologically variable, though found principally on the Pacific slope and characteristic of tropical dry forest (Holdridge & Budowski, 1956) and tropical moist forest (Tosi, 1971); known also from premontane moist forest in Panamá and from premontane wet and premontane rain forests in Coclé.

See Fig. 514.

CEPHAELIS DC.

Cephaelis discolor Polak., Linnaea 41:572. 1877

Glabrous shrub, less than 60 cm tall; stems usually unbranched, conspicuously marked with old, cupular leaf scars. Leaves decussate, congested at ends of stems; stipules bilobed nearly to base, the lobes maroon, linear, 6–8 mm long; petioles 1.5–4.5 cm long; blades elliptic to oblanceolate-elliptic, abruptly acuminate at apex, cuneate and decurrent at base, 10–20 cm long, 4–10 cm wide, distinctly bicolorous; major lateral veins in 8–13 pairs, arcuate-ascending, raised on both surfaces, the smaller veins reticulate. Inflorescences globose bracteate masses of many tiny flowers, terminal, maroon, congested between upper leaves; basal bracts rounded, becoming narrowed distally, with fine appendages at base; flowers white, sessile, each surrounded by several oblanceolate, violet-purple bracts 7–10 mm long and to 10 mm wide; calyx lobes 5 (rarely 6), alternate with corolla lobes, narrowly triangular, 1–1.7 mm long, divided more than halfway to base, fused with tube (the free part much shorter than corolla lobes), persisting in fruit; stamens 5 (rarely 6); anthers not seen; style slender; stigmas 2, free, ca 0.7 mm long, curved laterally, surrounded by a low, narrow, furfuraceous disk. Capsules round to ± oblong, 4–5 mm long, white turning blue; seeds 2, hemispherical, white, only slightly shorter than fruits. *Croat 11870.*

Rare; seen only in the forest in the vicinity of the junction of Wheeler and Drayton trails. Flowers in July and August, probably sometimes earlier and later as well. The fruits mature from August to December, principally from August to October.

Nicaragua, Costa Rica, and Panama. In Panama, known rarely from tropical moist forest in the Canal Zone and Bocas del Toro; known principally at higher elevations from premontane wet and tropical wet forests in Colón, Veraguas, Panamá, and Darién and possibly from premontane rain forest in Darién.

See Fig. 515.

Cephaelis ipecacuanha (Brot.) A. Rich., Bull. Fac. Med. 4:92. 1818

Raicella

Small shrub, usually less than 50 cm tall, essentially glabrous; internodes moderately long, with ridges on either side of stem below stipules. Leaves sessile or with the petioles to 1.5 cm long; stipules ca 1 cm long, with numerous subulate lobes, subpersistent; blades oblong-obovate or oblong-elliptic, acute to abruptly acuminate

---

KEY TO THE SPECIES OF CEPHAELIS

Bracts red, large; plants densely hirsute; shrubs more than 1.5 m tall; flowers yellow . . . .
. . . . *C. tomentosa* (Aubl.) Vahl
Bracts not red; plants glabrous or sparsely pubescent; shrubs less than 60 cm tall:
    Bracts green; petioles less than 1.5 cm long . . . . *C. ipecacuanha* (Brot.) A. Rich.
    Bracts violet-purple; petioles more than 1.5 cm long . . . . *C. discolor* Polak.

Fig. 514.
Calycophyllum candidissimum

Fig. 515. *Cephaelis discolor*

Fig. 516.
Cephaelis ipecacuanha

at apex, mostly acute to rounded at base, 8–16 cm long, 3.5–10 cm wide, bicolorous, shiny above, often sparsely pubescent below. Inflorescences capitate, becoming globose in fruit; peduncles 2–3.5 cm long, usually oblique to reflexed; bracts 4, green, the outer ones ovate to broadly rounded and acuminate, 5–13 mm long, the inner ones narrower, ± oblanceolate and acuminate, enveloping flower; flowers sessile, 5-parted, heterostylous, numerous, ca 1 cm long; calyx ± truncate or minutely lobed; corolla white, to 1 cm long, glabrous outside, puberulent inside near base of lobes and on most of tube, the lobes recurled; anthers 1.3–1.7 mm long; style lobes 2, short and thick, densely pubescent; short-styled flowers with the style included and the stamens exserted ca 3 mm above rim; long-styled flowers with the style exserted ca 3 mm above rim and the stamens included. Capsules oblong, ca 1 cm long, red becoming blue at maturity; seeds 2, slightly longer than broad, ca 5.3 mm long, flattened on one side, somewhat twisted. *Croat 4104, 15117.*

Common in the forest, especially along trails. Flowers principally from June to August at the beginning of the rainy season, with the fruits developing in about 1 month. Some flowering also occurs at the beginning of the dry season, especially in December.

Distinguished by its small size, its reflexed, bracteate flower heads, and its blue fruits (red when immature).

Central and South America. In Panama, known from tropical moist forest on BCI and in San Blas and Darién and from tropical wet forest in Darién.

See Fig. 516.

Cephaelis tomentosa (Aubl.) Vahl, Eclog. Amer.
1:19. 1796
Psychotria poeppigiana Müll. Arg.
Vasika

Densely hirsute shrub, usually less than 2 m tall. Stipules bilobed nearly to base, linear, 1–2 cm long, persistent; petioles short or to 3 cm long; blades ± elliptic, long-acuminate, obtuse to attenuate at base, 11–24 cm long, 3.5–9 cm wide; major lateral veins arcuate-ascending, often collecting near margin, impressed above, raised below. Inflorescences terminal, capitate, solitary; peduncles 4–9 cm long; flowers 5-parted, surrounded by numerous, narrow, pale bracts and subtended by 2 large bracts, these red, opposite, united at base and mucronate above; calyx minute, lobed midway to base; corolla yellow, rarely tinged with green, ca 12 mm long, the lobes short; filaments fused to corolla tube below middle; anthers 2.5–3.5 mm long, becoming weakly exserted above rim; pollen shed before anthesis; style receptive and held above anthers at anthesis, exserted ca 4 mm above rim; stigma bilobed. Berries bright blue, round to broadly oblong, to 1 cm long; seeds 2, narrow, gradually tapered downward, 7–8 mm long, dark brown, obscurely grooved. *Croat 4362.*

Occasional, in the forest; common elsewhere in the Canal Zone. Flowers and fruits throughout the year, with some flowers and fruits frequently found on the same

inflorescence. The fruits probably mature in 1 or 2 months.

Distinguished by the bright red, opposing bracts, the yellow flowers, and the bright blue fruits.

Mexico to Colombia and the Guianas, south to Amazonian Brazil, Peru, Bolivia, and Argentina; Trinidad. In Panama, ecologically variable, although apparently not occurring in any of the drier areas; known most commonly from wetter parts of tropical moist forest in the Canal Zone, all along the Atlantic slope, and in Los Santos, Panamá, and Darién; known also from premontane wet forest in Colón and Panamá, from tropical wet forest in Colón and Darién, and from premontane rain forest in Coclé and Panamá.

See Fig. 517.

CHIMARRHIS Jacq.

Chimarrhis parviflora Standl., Trop. Woods
11:26. 1927
Yema (llema) de huevo, Platano, Fiddlewood

Probably a large tree, reported on labels to 18 m tall and 25 cm dbh; stems brownish-puberulent when young, often glabrous in age. Stipules caducous, acuminate, 5–11(25) mm long, cinereous; petioles 7–20 mm long; blades obovate to elliptic, acute to acuminate at apex, acute to attenuate at base, 7–11(15) cm long, 1.5–5(6) cm wide, glabrous above, densely sericeous to inconspicuously appressed-pubescent below especially on major veins, glabrate in age. Cymes corymbose in subterminal or axillary, hemispherical clusters; peduncles (2.5)3–4 cm long; branches puberulent; flowers protogynous, the style emerging before petals are reflexed; pedicels very short or to 2 mm long; calyx and hypanthium together to 1 mm long, campanulate, ± glabrous, truncate, with 4(5) blunt, sparsely ciliate teeth; corolla short-funnelform, ca 3 mm long, the tube nearly obsolete, the 4 lobes oblong, rounded at apex, glabrous but with dense white-villous pubescence inside at point of staminal attachment; stamens 4, exserted, to 5 mm long; anthers narrowly ovoid, to 0.7 mm long; style ca 2.5 mm long, the lobes 2, broader than long. Capsules obovoid, ca 2 mm long and 1 mm wide, shallowly longitudinally sulcate with 6 or 8 weak ribs, splitting medially at maturity and releasing seeds; seeds usually 4, ± elliptic, ca 1 mm long and 0.5 mm wide, ± flattened, yellow, covered with wings, the wings many, low, interlocking, scarious, giving the seed a honeycombed appearance. *Croat 11244.*

The single BCI collection, made on Shannon Trail in the middle of July, was in fruit. The branch had fallen from a tree of unknown height. Many of the fruits of this collection had been infected by insects forming brown, fruitlike galls 5 mm long and 4 mm wide. Herbarium material from Panama and Costa Rica shows flowers from April to June, mature fruits in July and August or later.

Costa Rica and Panama. In Panama, known from tropical moist forest on BCI and in Bocas del Toro (the type locality) and from premontane wet forest in Veraguas.

Fig. 517. *Cephaelis tomentosa*

Fig. 518.
Cosmibuena skinneri

Fig. 519.
Coutarea hexandra

CHIOCOCCA P. Browne

Chiococca alba (L.) Hitchc., Annual Rep. Missouri Bot.
Gard. *4*:94. 1893

Lágrimas de María

Glabrous shrub or slender liana, to 3 m or more long,
probably climbing into the canopy, often with pendent,
trailing branches. Stipules ovate, long-acuminate, to 4
mm long, entire, persistent; petioles less than 1 cm long;
blades ovate to elliptic, acuminate, obtuse to rounded at
base, (2)6–10 cm long, (1.3)3–6 cm wide, shiny. Panicles
or racemes axillary, to 10 cm long; peduncles to 3 cm
long; flowers 5-parted; pedicels 2–3 mm long; calyx ca
1.3 mm long, lobed midway, the lobes deltoid, acute;
corolla white, becoming yellowish in age, to 8 mm long,
the lobes 2–3 mm long, spreading to reflexed; stamens
included, ca 5 mm long; filaments ± free from corolla,
villous at base, and united into a short tube; anthers to
3.7 mm long, longer than filaments; style to 8.5 mm
long, exserted, the lobes 2, very short, thick. Berries
compressed-ellipsoid, 4–8 mm diam, fleshy, white at
maturity; seeds 1 or 2, hemispherical, compressed. *Croat
15266* (Madden Forest).

No collections have been made from the island, but
flowers matching this species have been seen in the forest
when the species was in flower elsewhere in the Canal
Zone. Flowers and fruits to some extent all year, but
mostly in August and September. The fruits develop
quickly, probably within a month.

Florida and southern Texas through lowland areas of
Mexico and Central America to Bolivia and Argentina;
West Indies. In Panama, ecologically variable; known
principally from tropical moist forest in the Canal Zone,
Bocas del Toro, San Blas, Veraguas, Los Santos, Herrera,
Panamá, and Darién; known also from tropical dry forest
in Panamá (Taboga Island), from premontane moist
forest in the Canal Zone, Los Santos, and Panamá, from
premontane wet forest in Colón, Chiriquí, Coclé, and
Panamá, and from premontane rain forest in Panamá.

CHOMELIA Jacq.

Chomelia psilocarpa Dwyer & Hayden, Ann. Missouri
Bot. Gard. 54:139. 1967

Arching shrub or liana, widely branched; stems densely
appressed-pubescent, becoming glabrate, terete. Stipules
lanceolate, 6–10 mm long, entire, caducous; petioles
ca 1 cm long, sparsely appressed-pubescent; blades ovate
to elliptic, acuminate, obtuse to attenuate and decurrent at base, 6–11 cm long, 3.5–5.5 cm wide, sparsely
appressed-pubescent, the trichomes denser on veins
below, the axils below inconspicuously tufted, the lower
leaf surface of dry specimens finely wrinkled between
secondary veins. Cymes axillary, on peduncles to 6 cm
long; flowers not seen; sepals 4, oblong, persistent, of
2 lengths, the longer 4–8 mm long, the shorter 2–4 mm
long. Berries sessile or short-pedicellate, oblong, 1.5–2
cm long, 7–10 mm wide, sparsely appressed-pubescent;
seeds oblong, ca 1 cm long. *Croat 7931, Foster 1117.*

Apparently infrequent, but possibly more abundant
than collections and observations indicate; collected as an
arching shrub by R. Foster near Wheeler Trail 1100, as a
vine by Hayden (*143*) along the shore south of the dock,
and high in the canopy (*Croat 7931*). Mature fruits have
been seen from June to August.

Known only from Panama, from tropical moist forest
on BCI and in Darién.

COSMIBUENA R. & P.

Cosmibuena skinneri (Oerst.) Hemsl., Biol. Centr.-
Amer. Bot. 2:12. 1881

C. paludicola Standl.

Glabrous tree, to 10 m tall and 25 cm dbh (usually less),
usually hemiepiphytic. Stipules obovate, 1–2 cm long;
petioles 1–2 cm long; blades oval or obovate, rounded at
apex, acute and somewhat decurrent at base, 7–18 cm
long, 3–9 cm wide, thick, bicolorous, glossy above with
4 or 5 pairs of major lateral veins visible, lighter and dull
below with the veins indistinct. Inflorescences terminal;
flowers 3–6, clustered at apex of short peduncle; calyx
6–8 mm long, acutely lobed ca one-third its length, the
dehiscence circumscissile; corolla white or with the tube
somewhat greenish, the tube 5.5–9 cm long, the lobes
rounded at apex, 2–4.5 cm long, mostly 1–1.7 cm wide;
stamens affixed at apex of tube; filaments fused to corolla
tube; anthers ca 1.5 cm long, attached near base, the cells
free below; ovary cylindrical, ca 3 cm long; style emerging
from tube, with ascending trichomes near apex; stigmas
ca 1 cm long. Capsules linear-cylindrical, 8–12 cm long
at maturity; seeds numerous, thin, narrowly winged,
wind dispersed, 7–13 mm long, less than 1 mm wide, the
seminiferous area ca 1 mm long. *Croat 6326, 8519.*

Occasional, in the forest, often hemiepiphytic in trees
at 30 m or more; also occurring on rocky shores, especially
on the northern edge of Orchid Island and Peña Blanca
Peninsula, where the plant is often attached to large
boulders. Flowers mostly in August and September
(rarely earlier). The fruits mature in the dry season,
mostly from February to April.

Possibly also includes South American material going
by the name *C. grandiflora* (R. & P.) Rusby and *C. macrocarpa* (Benth.) Klotzsch.

Mexico to Colombia, ranging to Brazil if the additional
South American material is included. In Panama, known
from tropical moist forest in the Canal Zone, San Blas,
Panamá, and Darién, from premontane wet forest in
Colón and Coclé, from tropical wet forest in Colón, and
from premontane rain forest in Panamá.

See Fig. 518.

COUSSAREA Aubl.

Coussarea curvigemmia Dwyer, Phytologia
38:215. 1978

Small glabrate tree, usually 3–10 m tall; younger stems
often puberulent. Leaves very short-petiolate; stipules
triangular, 2–3 mm long, caducous; blades elliptic, acumi-

nate, acute to narrowly subcordate at base, 7–17 cm long, 3–9 cm wide, glabrous, the vein axils with pitlike domatia below making bumps above. Inflorescences short congested panicles; branches and flowers minutely puberulent; flowers heterostylous, sweetly aromatic, 4-parted; pedicels 1–2 mm long; calyx ca 1 mm long, with 4 minute teeth, persistent in fruit; corolla tubular, 2–2.5 cm long, slender, white, 4-lobed in apical third; stamens 4, in apical part of tube; anthers pale orange, slender, acuminate at apex, sagittate at base, 5–6 mm long, the lobes extending below free portion of filament; short-styled flowers with the style branched about one-third its length, held between anthers in bud but well below anthers at anthesis (owing to elongation of tube); long-styled flowers with the style 16–18 mm long, lobed ca 1.5 mm, exserted beyond tube opening and beyond anthers ca 3 mm. Drupes ellipsoid and somewhat flattened, to 12 mm long, becoming white, fleshy and sweet at maturity; seed 1(2), ellipsoid, ca 1 cm long. *Croat 5465, 8600.*

Frequent in the forest. Flowers mostly from May to July, sporadically at other times, especially in the early dry season. The fruits develop to mature size by September, but may not ripen until the late rainy or early dry season. J. Oppenheimer (pers. comm.) reports that white-faced monkeys eat the fruits in November and December. According to R. Foster, the species may flower only every other year, alternating with *Faramea occidentalis,* which replaces it in alternate years as a main source of bird food in the forest.

Flowers are apparently moth pollinated and fall in great numbers in the morning.

The species has been confused with *C. impetiolaris* Donn. Sm., which has subquadrangular branches, the corolla tube to 2 mm wide, elongate anthers to 8 mm long, and fruits to 19 mm long (when dry).

Known only from Panama, from tropical moist forest in the Canal Zone, Bocas del Toro, Panamá, and Darién, from premontane wet forest in Colón, and from premontane rain forest in Coclé and Panamá.

COUTAREA Aubl.

Coutarea hexandra (Jacq.) K. Schum. in Mart., Fl. Brasil. 6(6):196. 1889
Niño muerto, Quina

Shrub or tree, to 10 m tall; branches often long and drooping; young stems, petioles, and leaves sparsely pubescent, especially on veins; stems and axes of inflorescences usually lenticellate. Stipules broadly triangular, to 5 mm long, eventually deciduous; petioles to 1 cm long; blades ovate to elliptic, acuminate (sometimes abruptly so), obtuse to rounded at base, 5–13 cm long, 3–7 cm wide, the vein axils often weakly barbate below. Flowers (5)6-parted, with sweet aroma when first open, mostly in 3-flowered clusters, the clusters in terminal or upper-axillary dichasia, subtended by a pair of reduced leaves; peduncles, pedicels, and calyces sparsely pubescent; peduncles to 3 cm long; pedicels 1–1.5 cm long; calyx 7–12 mm long, deeply lobed, the lobes linear; corolla funnelform-campanulate, 7–8.5 cm long, greenish in bud,

white at anthesis or white tinged with violet-purple, sometimes pale lavender outside, pale red-violet inside, glabrous except inside near base, the tube weakly curved, constricted just above base, with faint lavender lines inside below throat, the lobes 2–2.5 cm long, narrowly acute, ± spreading; stamens 6, equaling length of lobes, exserted above throat at anthesis; filaments densely pubescent toward the base; anthers slender, to 1.8 cm long; style about as long as stamens, the lobes 2, very short, blunt. Capsules 2-carpellate, obovoid, compressed, 3–4.5 cm long and ca 2 cm wide, brown, with many conspicuous lenticels, at first splitting medially from apex, followed by spreading of the 2 valves of each carpel; seeds many, 1.5–2 cm long, elongate, thinly flattened, winged, wind dispersed, the seminiferous area medial, round, ca 4 mm wide. *Croat 4937.*

Occasional; known from along the shore between the Laboratory Clearing and Colorado Point and from the forest near Wheeler Trail 800. Flowers from June to October, mostly in July and August. The fruits are full size by October, but are shed mostly in the dry season, from December to March.

Mexico to the Guianas, Argentina, and Peru; West Indies. In Panama, ecologically variable; known from tropical dry forest in Herrera and Panamá (Taboga Island), from premontane moist forest in the Canal Zone and Panamá, from tropical moist forest in the Canal Zone, Colón, Coclé, Panamá, and Darién, and from premontane wet forest in Chiriquí and Panamá.

See Fig. 519.

DIODIA L.

Diodia denudata Standl., J. Wash. Acad. Sci. 15:105. 1925

Erect perennial herb, usually somewhat scandent, often suffruticose, 1–2 m tall; stems usually glabrous (except at leaf nodes), attenuated toward apex, minutely papillate, square, the corners winged, the wings ending just below stipules on each internode, often free at apex. Stipular bristles united at base, prominent, 3–6 mm long, ciliolate; petioles obscure or to 5 mm long; blades ± narrowly elliptic, acute at apex, acute to attenuate and often decurrent at base, greatly diminished toward apex of plant, glabrous to scabrid above, especially near margin, minutely scabrid and puberulent on veins below, the larger ones 4.5–10 cm long, 1–3 cm wide; veins arcuate-ascending, often prominently impressed above. Flowers minute, sessile, interspersed with bristles, in axillary clusters less than 1 cm diam; calyx lobes 4, ca 0.5 mm long, persisting in fruit, 3 of the 4 attached to one fruit valve; corolla greenish-white, ca 1.2 mm long, 4-lobed ca halfway, the tube pubescent at rim inside; stamens included; style held at level of stamens; stigmas opening later than stamens. Capsules ca 1 mm long (excluding the persistent calyx), about as broad as long, puberulent, 2-valved, dehiscing basally; seeds 2, ca 1 mm long. *Croat 4373.*

Apparently rare; known only from the forest near the end of Gross Trail. Elsewhere in central Panama seen at the margins of forest. Seasonal behavior uncertain. Flow-

KEY TO THE SPECIES OF DIODIA

Plants scandent vines, becoming hemiepiphytic; capsules usually more than 4 mm long . . . .
. . . . *D. sarmentosa* Sw.
Plants erect herbs, terrestrial; capsules less than 3 mm long:
  Stems and leaves greatly reduced near apex of plant, the upper leaves many times smaller than
    the lower; corolla and mature fruits ca 1 mm long . . . . *D. denudata* Standl.
  Stems and leaves not greatly reduced near apex of plant, the upper leaves at least one-third as
    long as the lower; corolla ca 6 mm long; fruits ca 3 mm long . . . .
  . . . . *D. ocimifolia* (Willd.) Bremekamp

ers and fruits at least throughout the dry season and in the early rainy season.

Leaf size and shape are variable. The species was reported in the *Flora of the Canal Zone* (Standley, 1928) as *D. nudata* Standl. This was no doubt a typographical error.

Known only from Panama and Colombia (Chocó). In Panama, known from tropical moist forest in the Canal Zone, Panamá, and Darién and from premontane wet forest in Colón (Santa Rita Ridge).

Diodia ocimifolia (Willd.) Bremekamp, Recueil Trav.
  Bot. Néerl. 31:305. 1934
  Hemidiodia ocimifolia (Willd.) K. Schum.
  Hierba (yerba) meona

Erect herb, usually less than 1 m tall; stems pubescent or glabrous except for pairs of pubescent lines extending downward from stipules. Leaves sessile or short-petiolate; stipules oblong, pubescent, with ciliate bristles; blades lanceolate, gradually tapered toward both ends, mostly 3.5–9.5 cm long, 0.7–2.5 cm wide, both surfaces ± scabrous, the trichomes stiff, erect, to ca 0.2 mm long. Flowers sessile, 4-parted, interspersed with long white bristles, in dense axillary clusters less than 1 cm diam; calyx lobes 4, inconspicuous, persisting in fruit; corolla white, sometimes tinged with pink, to 6 mm long, lobed to near middle, glabrous except for throat and base of lobes inside, the lobes spreading; style and stamens subequal, exserted to 2.7 mm above rim; filaments adnate to corolla tube; stigma capitate. Capsules 2-valved, densely pubescent in upper two-thirds, ca 3 mm long; seeds oblong-elliptic, 2.3–2.7 mm long, dark brown, weakly transverse-sulcate. *Croat 12840.*

Common in clearings. Flowers and fruits throughout the year.

This species is in the tribe Spermacoceae, a confusing group including *Borreria*, *Diodia*, and *Spermacoce*. All have minute flowers clustered in the leaf axils and two-celled, capsular, longitudinally sulcate fruits. Several genera in this confusing group are difficult to identify because generic lines drawn largely on the basis of type of fruit dehiscence. It is possible that not all of these genera are valid.

Throughout the American tropics; Malaysia. In Panama, known from tropical moist forest in the Canal Zone, Bocas del Toro, San Blas, Herrera, Coclé, and Panamá, from premontane wet forest in Colón, Coclé, and Panamá, and from tropical wet forest in Colón.

Diodia sarmentosa Sw., Prodr. Veg. Ind. Occ. 30. 1788

Vine, sometimes hemiepiphytic; stems to 1 cm diam, woody at base; younger stems green, herbaceous, square, strigose on the surface, more densely pubescent with longer trichomes on ribbed corners; internodes 12–32 mm long. Stipules pubescent, adnate to petioles, with bristles to 6 mm long; petioles 2–4 mm long; blades narrowly elliptic, acuminate, cuneate at base, 3–7 cm long, 1.2–1.8 cm wide, dark green above, asperous and paler below, sparsely pubescent below especially on veins, the margins ± revolute; lateral veins mostly in 4 pairs, arcuate-ascending and markedly impressed above. Flowers axillary at all upper nodes, usually 6 at each node, white, sessile, soon falling; sepals 5, free, 2.3 mm long, ciliate, ± unequal, persisting in fruit; corolla ca 3.5 mm long, the lobes 5, ca 2 mm long and 1 mm wide, the tube with a ring of trichomes at middle; stamens exserted ca 1 mm; filaments fused with corolla tube; anthers ca 0.7 mm long; style ca 2.7 mm long; stigma capitate and papillose. Capsules 3–4.5 mm long, to 3 mm diam, reddish-brown at maturity, splitting at apex into 2 halves, both halves remaining closed; seeds oblong-elliptic, 2.5–3 mm long, black or dark brown, minutely reticulate, the inner surface deeply invaginate longitudinally. *Croat 5525.*

Occasional; collected only along the shore, sometimes as a hemiepiphyte on logs in the lake. Apparently flowers and fruits all year. Plants usually bear both flowers and fruits simultaneously.

Similar in many ways to *D. ocimifolia*, but that species is a moderately short herb with smaller fruits and much less scabrous leaves.

Mexico to Panama, Venezuela, the Guianas, eastern Brazil; Greater Antilles; Africa. In Panama, known from tropical moist forest in the Canal Zone and Veraguas and from tropical wet forest in Colón (Miguel de la Borda).

FARAMEA Aubl.

Faramea luteovirens Standl., Contr. U.S. Natl. Herb.
  18:138. 1916
  Huesito

Glabrous shrub or small tree, to 4 m tall. Stipule body triangular, ending in a stiff cusp, 3–5 mm long, subpersistent; petioles ca 1 cm long; blades elliptic or obovate, abruptly acuminate, obtuse to acute at base, 7–13 cm long, 3–6 cm wide, moderately coriaceous. Flowers 2 or 3, upper-axillary or terminal, aromatic; pedicels thick,

KEY TO THE SPECIES OF FARAMEA

Flowers 2 or 3 in axils, never in branched inflorescences; pedicels more than 2 cm long . . . .
. . . . *F. luteovirens* Standl.
Flowers usually numerous, in branched terminal dichasia or umbelliform racemes; pedicels less
than 2 cm long . . . . *F. occidentalis* (L.) A. Rich.

2–4 cm long; calyx cupular, truncate, ca 3 mm long, persisting in fruit; corolla white, salverform, 2–3 cm long, 4-lobed; ovary 1-celled. Drupes globose or ellipsoid, 1.5–2 cm long, blue or black at maturity; seed 1, deeply excavate on lower side. *Croat 5217, 6725.*

Probably rare; collected twice along the shore of Miller Peninsula. Flowers mostly in the dry season (February to May). The fruits mature mostly from September to November.

Known only from Panama, from tropical moist forest in the Canal Zone and Darién, from premontane wet forest in Colón, San Blas, Chiriquí, and Darién, and from tropical wet forest in Darién.

Faramea occidentalis (L.) A. Rich., Mém. Soc. Hist.
Nat. Paris 5:176. 1834
F. zetekii Standl.
Benjamín, Bonewood, Huesillo, Huesito, Palo escrito

Tree, to ca 10 m tall and 20 cm dbh, ± glabrous; stems usually divaricately branched. Stipules triangular, with a long acicular apex, 1–1.5 cm long, eventually deciduous; petioles 5–15 mm long; blades oblong-elliptic to elliptic, acuminate and downturned at apex, obtuse to acute at base, 13–18 cm long, 4–8(10) cm wide, ± stiff but not thick. Flowers with strong, sweet aroma, in terminal dichasia or umbelliform racemes to ca 10 cm long; pedicels 4–20 mm long; calyx tubular, truncate, ca 2 mm long at anthesis, persisting in fruit; corolla white, the tube 14–20 mm long, slender, less than 3 mm wide, the lobes 4, narrowly acute, 15–25 mm long, spreading-recurved; stamens 4, attached near apex of tube, included or slightly exserted; anthers to 8 mm long, sharp at apex, blunt at base, attached to filaments near base; style with 2 slender branches more than 4 mm long, at first held below stamens, eventually slightly exserted. Drupes depressed-globose, ca 1 cm diam, bluish-black at maturity; seeds usually 1 by abortion, rounded with a large depression at base. *Croat 6597, 14961.*

Abundant in the forest. Flowers from March to July, mostly in May and June. The fruits mature from April to December, mostly from August to October only every other year.

Flowers are probably moth pollinated; most appear on the ground in late morning. The aroma is intense, much like that of *Quisqualis indica* L. (Combretaceae), a cultivated plant in the Canal Zone, which has flowers opening at dusk.

Scattered from northeastern Mexico to Colombia, Venezuela, Ecuador, and Brazil; West Indies. In Panama, known from tropical moist forest in the Canal Zone, Colón, Veraguas, Los Santos, Panamá, and Darién, from premontane moist forest in Panamá, and from premon-

tane wet forest in Coclé and Darién. Reported from tropical wet and premontane rain forests in Costa Rica (Holdridge et al., 1971).

GENIPA L.

Genipa americana L., Syst. Nat. ed. 10, 931. 1759
G. caruto H.B.K.; *G. americana* var. *caruto* (H.B.K.) K. Schum.
Genipap, Guayatil blanco, Guayatil colorado, Jagua, Jagua amarillo, Jagua blanca, Jagua colorado, Jagua negro, Jagua de montaña

Tree, to 20 m tall and to 30(50) cm dbh; outer bark dark, not fissured but with many raised lenticels; stems stout, with prominent scars. Leaves clustered at ends of branches; stipules triangular, to ca 1.5 cm long, entire, deciduous; petioles 5–10 mm long, thick; blades obovate or elliptic, acute or short-acuminate, narrowed to acute or obtuse base, mostly 15–42 cm long, 5–19 cm wide, glabrous above, almost glabrous to densely pubescent below with short erect trichomes. Cymes stout, terminal or subterminal, peduncles short; flowers 5-parted; pedicels ca 1 cm long; calyx truncate or shallowly lobed, 7–10 mm long; corolla yellow or white becoming yellow, thick, sericeous, 2–3 cm long, the limb ca 4 cm wide, the tube 6–10 mm wide, the lobes divided more than halfway, spreading at anthesis, later recurving; stamens exserted; anthers 8–14 mm long, sessile or with a very short filament, attached well below middle of tube, the anther becoming recurved between corolla lobes; style exserted ca 7 mm above rim, the lobes 2, thick, ca 7 mm long, remaining closed until after shedding of pollen. Berries baccate, globose or ellipsoid, 5–7 cm long, becoming fleshy at maturity, the calyx persistent on fruit and forming a thick-rimmed crater ca 1 cm across; seeds numerous, flattened, yellow, ca 1 cm long. *Croat 6163, 12728.*

Frequent in the forest and occasional along the shore. Flowers to some extent throughout the year, but mostly in the rainy season, usually from May to July. The fruits mature in about a year; they are eaten by tamarins in May and June (Hladik & Hladik, 1969). One tree had mature fruits on it for 2 years, during which time it lost its leaves twice. Leaves are usually replaced during the late dry and early rainy seasons.

The sap of the plant, although clear, quickly turns blue. Bawa and Opler (1975) reported this species to be dioecious, but all trees on BCI that flowered also set fruit.

Mexico to Argentina; West Indies. In Panama, ecologically variable, occurring mostly around the Gulf of Panama on the Pacific slope, extending to the Caribbean coast in the region of the canal, and characteristic of

tropical moist forest (Tosi, 1971); known from premontane dry forest in Los Santos, from tropical dry forest in Los Santos, Coclé, and Panamá, from premontane moist forest in the Canal Zone and Los Santos, from tropical moist forest in the Canal Zone, Bocas del Toro, Herrera, Panamá, and Darién, from premontane wet forest in Chiriquí and Panamá, and from tropical wet forest in Panamá.

See Fig. 520.

GEOPHILA D. Don

Geophila croatii Steyerm., Phytologia 35:401. 1977

Low creeping herb, less than 10 cm high; stems long-trailing, rooting at nodes, minutely papillose, the younger ones purplish. Stipules lanceolate to broadly triangular, subacute to acuminate at apex, 3–5 mm long, curved outward, persistent; petioles 3–8 cm long, with marginal rows of thick trichomes on the upper side, the trichomes sometimes directed toward base; blades ovate, acute at apex, cordate at base, 2.5–6 cm long, 2–4.5 cm wide, minutely papillose; veins often purplish. Inflorescences solitary or few, terminal, on short axillary branches from creeping stems; peduncles 3–20 mm long; heads usually bearing 5–7 flowers; flowers sessile or short-pedicellate, 5-parted; pedicels less than 1 mm long before anthesis, 2–3 mm long in fruit; calyx lobes recurved, linear-oblong or narrowly lanceolate, subcaudate, less than 3 mm long before anthesis, to 8–9 mm long in fruit; corolla white, to 5.5 mm long, the lobes slender, sometimes tipped with maroon; stamens included; filaments fused in basal half, the tube with tufts of trichomes at apex; style equaling stamens. Berries ± globose, 3–4 mm diam, red, often more than 3 per cluster; seeds 2, oblong-elliptic, 3.5–6 mm long, flattened on one side, half twisted, faintly 3-ribbed. *Croat 6647* (type).

Occasional in the forest, but locally abundant along some trails, especially Zetek Trail. Flowering and fruiting throughout the rainy season from June to December, with most flowers seen at the beginning of the rainy season in June and July. The fruits develop in about a month, and some may mature while other plants in the clone are still in flower.

This taxon is most closely related to *G. gracilis* (R. & P.) DC. and has been confused with that species (Williams, 1973). It is distinguished from *G. gracilis* by its glabrous upper leaf surface and longer calyx lobes. It has also been confused with *G. macropoda* (R. & P.) DC., from which it differs by having glabrous, much shorter peduncles, acute to acuminate stipules and leaf apices, glabrous outer surface of floral bracts, glabrous calyx lobes, completely glabrous veins of lower leaf surface,

and abundant cystoliths on the lower surface.

Nicaragua and Panama, probably also in intervening areas. In Panama, known only from BCI.

See Fig. 521.

Geophila repens (L.) I. M. Johnston, Sargentia 8:281. 1949

G. herbacea (Jacq.) K. Schum.

Low creeping herb, less than 10 cm high, much like *G. croatii* in habit and vegetative characteristics. Stipules rounded to broadly ovate, ca 2 mm long; petioles 1–5 cm long; blades ovate, rounded to bluntly acute at apex, cordate at base, mostly less than 4.5 cm long. Inflorescences usually solitary; peduncles to 2.5 cm long; flowers solitary or in groups of 2 or 3, sessile; calyx lobes 2–6 mm long, slender, acute to acuminate, persisting in fruit; corolla white, 1.2–1.5 cm long, glabrous or pubescent, the tube 8–10 mm long, the lobes often tinged with lavender, spreading; stamens 5, included, held just below rim of corolla tube; filaments free near apex; style 6–7 mm long, equaling stamens; style branches 2, puberulent. Capsules 1 or 2 per cluster, globose, ca 6 mm diam, red becoming black; seeds 2, ± elliptical, 4–5 mm long, flattened, twisted, faintly ribbed on both sides. *Croat 10171, 14928.*

Common in the forest and locally abundant, usually on trails or in shady, moist places in open areas. Flowers mostly in the rainy season, from May to October, especially in June and July, and sporadically in the dry season. The fruits mature mostly from August to October, but as late as December.

Mexico to Colombia, the Guianas, Peru, and Bolivia; West Indies; Central and West Africa; Polynesia; the Philippines. In Panama, widespread in tropical moist forest in the Canal Zone, Bocas del Toro, Chiriquí, Panamá, and Darién and known from tropical wet forest in Darién and from premontane rain forest in Los Santos.

See Fig. 522.

GUETTARDA L.

Guettarda foliacea Standl., Contr. U.S. Natl. Herb. 18:139. 1916

Espino, Guayabo

Tree, usually less than 10 m tall, branched from near ground, often with sucker shoots, the branches often vinelike, armed with stout branch-spines (spines terminating branches) to 5 cm long; stems appressed-pubescent when young, glabrate in age. Stipules lanceolate, to 12 mm long, appressed-pubescent, caducous; petioles 1–2 cm long, pilose above, appressed-pubescent below; blades narrowly elliptic to ovate or obovate-elliptic, acuminate,

---

KEY TO THE SPECIES OF GEOPHILA

Blades rounded or blunt at apex; flowers 1–3 per cluster; leaves seldom more than 4.5 cm long . . . .
. . . . *G. repens* (L.) I. M. Johnston
Blades sharply acute at apex; flowers 4–7 per cluster; leaves often more than 4.5 cm long . . . .
. . . . *G. croatii* Steyerm.

Fig. 520.
Genipa americana

Fig. 522. *Geophila repens*

Fig. 521. *Geophila croatii*

Fig. 523.
Guettarda foliacea

obtuse to rounded at base, 6–16 cm long, 2.5–6.5 cm wide, sparsely pubescent (the trichomes denser on veins), sericeous below when young, with tufts of trichomes in lower vein axils below. Inflorescences cymose, axillary, usually spreading, appressed-puberulent, to 7 cm long, bracteate; bracts 5–10 mm long, appressed-pubescent outside, ciliate; peduncles slender, to 3 cm long; flowers sessile; calyx truncate, 2–3 mm long, cupular, appressed-pubescent; corolla narrowly tubular, 2–2.5 cm long, white, shallowly 4-lobed, the lobes spreading, rounded at apex, ca 5 mm long, sericeous outside; anthers 4, sessile, ca 3 mm long, attached just below rim of corolla; pollen golden; style slender, to 2.5 cm long, equaling lobes, exserted above throat at anthesis; stigma globular. Berries globose, 1–3 cm diam, densely short-velutinous, red at maturity, the exocarp thin, the mesocarp sweet and fleshy, 4-loculed; seeds 1 per locule. *Croat 6061, 15160.*

Frequent in the forest. Flowers mostly in late June and early July. The fruits mature mostly by September but often later; they have been observed being eaten during November (J. Oppenheimer, pers. comm.). Sporadic flowering may occur at other times, especially at the beginning of the dry season, with the fruits maturing in the late dry or early rainy seasons.

Animals eat the pericarp and then discard the seed. Elsewhere in the Canal Zone trees have been seen with longer, vinelike branches.

Panama and Colombia. In Panama, known mostly on the Pacific slope, principally from tropical moist forest in the Canal Zone, Veraguas, Panamá, and Darién, but also from premontane wet forest in Chiriquí.

See Fig. 523.

HAMELIA Jacq.

Hamelia axillaris Sw., Prodr. Veg. Ind. Occ. 46. 1788

Panchus chapa, Guayabo negro

Shrub, usually 1.5–3(10) m tall; stems puberulent. Stipules 3.3–3.7 mm long, subpersistent, attenuate, the sides folded together; petioles 1.5–4 cm long, puberulent, caniculate; blades ± elliptic, acuminate, attenuate and decurrent at base, mostly 8–23 cm long, 3.5–7(9) cm wide, glabrous above, puberulent below at least on veins, with short linear cystoliths visible below upon drying; major lateral veins in 6–9 pairs, arcuate-ascending, the axils often with small tufts of trichomes below, the smaller veins reticulate. Inflorescences terminal, short-branched cymes; flowers yellow or orange; pedicels ca 1 mm long; calyx 4–6 mm long, the lobes 5, ca 1 mm long, somewhat accrescent and persisting in fruit; corolla yellow, cylindrical, to 10 mm long in bud, becoming ±

funnelform, to 13 mm long and ca 5 mm wide, the lobes 5, held mostly erect, 1.3–2 mm long; stamens to 8 mm long; filaments ca 2 mm long, continuous with anthers, the connective extended beyond apex and spatulate; pistil 11–13 mm long; style ± equaling petals; stigma oblong, ca 2.7 mm long. Berries oblong, 5–6 mm long, green turning red then black at maturity, fleshy, 5-celled, each cell with many seeds; seeds less than 1 mm long, brown, densely reticulate. *Croat 6299, 12484.*

Frequent in the forest, especially the older forest. Flowers throughout the rainy season, mostly from June to September. The fruits mature from August to December, mostly from August to October.

Distinguished from other shrubs in the family by the funnelform, yellow flowers and many-seeded berries.

Mexico, Belize, Costa Rica, and Panama to Venezuela, Brazil, and Bolivia; West Indies. In Panama, known from tropical moist forest in the Canal Zone, Bocas del Toro, San Blas, Chiriquí, Los Santos, Panamá, and Darién, from premontane wet forest in Panamá and Darién, and from tropical wet forest in Colón and Darién.

Hamelia patens Jacq. var. **glabra** Oerst., Vidensk. Meddel. 1852:42. 1853

H. nodosa Mart. & Gal.

Red berry, Scarletbush, Uvero

Shrub or small tree, usually less than 6 m tall, nearly glabrous. Leaves verticillate, usually 3(5) at a node, or opposite; stipules linear-lanceolate, often curved inward, 3(4) at each node, ca 4 mm long; petioles 1–8 cm long, narrowly winged at least above, often puberulent, sometimes reddish; blades elliptic to oblanceolate, abruptly short-acuminate, obtuse to attenuate and gradually decurrent at base, mostly 9–23 cm long and 4.5–10.5 cm wide, glabrous above, minutely pubescent on veins below, the vein axils barbate, often inconspicuously so. Inflorescences terminal, cymose (the flowers all on upper side of branches), variable in size, to 15 cm long and 20 cm wide; branches often reddish; pedicels 1–5 mm long; calyx lobes short, persisting in fruit; corolla tubular, 15–23 mm long, orange becoming red-orange at anthesis, 5-lobed, prominently ridged below lobes; stamens included; filaments free much of their length; anthers 8–9 mm long, the connective prolonged and acute at apex; pistil 15–24 mm long, equaling corolla tube; stigma 3.5–5 mm long, oblong, swollen. Berries soft, purple or black at maturity, to 13 mm long and 10 mm wide, 5-celled; seeds many, flattened, brown, markedly reticulate. *Croat 4273.*

Occasional, occurring only in clearings, often locally abundant. Flowers and fruits throughout most of the year; the flowering season usually beginning in the late

---

KEY TO THE SPECIES OF HAMELIA

Leaves verticillate, usually 3, 4 or 5 per node, sometimes opposite; flowers orange to red . . . .
. . . . *H. patens* Jacq. var. *glabra* Oerst.
Leaves not verticillate, 2 per node; flowers yellow . . . . *H. axillaris* Sw.

dry and early rainy seasons, with a second flush of flowering activity in the early dry season. The fruits mature in about 1 month.

Leck (1972) reported that four species of birds visit flowers for nectar and nine species take the fruit. Feeding pressures on fruits are so great in the rainy season that even immature fruits are taken. Among the possible pollinators are three hummingbirds, *Thalurania furcata*, *Damophila julie*, and *Amazilia tzacatl*, and one honeycreeper, *Coereba flaveola*. The fruits are taken by a number of birds, including toucans, flycatchers, thrushes, warblers, honeycreepers, and tanagers (Leck, 1972).

Mexico, Nicaragua south to Panama and Venezuela, Amazonian Brazil, Peru, and Bolivia. In Panama, known from tropical moist forest in all provinces and from premontane wet forest in Coclé.

HOFFMANNIA Sw.

Hoffmannia woodsonii Standl., Ann. Missouri Bot. Gard. 28:471. 1941

Weak-stemmed, ± glabrous shrub, to 1.5(3) m tall, sometimes somewhat decumbent. Stipules minute, ca 1 mm long, raised, triangular, deciduous; petioles 2.5–8 cm long; blades oblanceolate to elliptic, acuminate and weakly falcate (the acumen to 1 cm long), attenuate and decurrent onto petiole at base, 6–18 cm long, 3–8 cm wide. Cymules short-pedunculate, axillary, umbelliform clusters ca 3 cm long, bearing 2–6 flowers; pedicels 5–7 mm long; calyx lobes ca 1 mm long, acute, persistent in fruit; corolla 7–8 mm long, greenish-yellow, the lobes 4, ca 4 mm long, acute, recurved at anthesis; stamens exserted; filaments fused to tube near base; anthers ca 4.3 mm long, introrse, held together in a cluster; style to 8.5 mm long, unbranched, thickened at apex, held well above anthers. Berries ellipsoid, 3–4 mm long, glabrous, delicately ribbed, yellow-green or reddish; seeds many, ca 0.5 mm long. *Croat 13511.*

Known only from the creek north of the laboratory. Flowers from April to August, mostly from May to July. The fruits mature in about a month. Elsewhere in Panama the species may flower in the early dry season.

Costa Rica and Panama; mostly at higher elevations. In Panama, ecologically variable; known from premontane moist forest in the Canal Zone, from tropical moist forest in the Canal Zone, Bocas del Toro, Colón, Chiriquí, Los Santos, and Darién, from premontane wet forest in Chiriquí, Coclé, and Panamá, and from premontane rain and lower montane wet forests in Chiriquí.

ISERTIA Schreb.

Isertia haenkeana DC., Prodr. 4:437. 1830
Canelito

Stout shrub, usually 2–3 m tall; stems densely pubescent, usually 4-ribbed. Stipules 1–2 cm long, lobed to base, the lobes separate, oblong-lanceolate, persistent; petioles obscure or to 4 cm long; blades elliptic to obovate, long-acuminate, gradually tapered to a decurrent base, 10–45 cm long, 3.5–16 cm wide, glabrous above, tomentose and ± viscid below especially on veins. Panicles stout, terminal, bracteate, reddish-tomentose, 9–22 cm long, the entire inflorescence turning bright orange sometime before fruit maturation; pedicels to ca 2 mm long; calyx cupular, red-orange, minutely 4-lobed, the lobes ca 1.5 mm long; corolla tubular, to 3.5 cm long, yellow, the lobes 4, ca 7 mm long, bearded inside near tip; filaments united to corolla tube, the free part thick, fleshy, flattened; anthers to 3.7 mm long, attached near their apex, the endothecium dark-spotted; style slender, ca 2.5 cm long; stigmas 5, slender, arcuate. Berries 5-carpellate, 6–8 mm long, red becoming black at maturity, each cartilaginous wedge containing many seeds; seeds irregular, reticulate, ca 2 mm diam. *Croat 5076.*

Occasional along the shore and in clearings. Common and showy along roadsides in the Canal Zone. Flowers throughout the rainy season, but mostly in the early rainy season. The fruits mature from the middle of the rainy season until the early dry season.

Mexico to Panama, Colombia, and Venezuela; Cuba. In Panama, ecologically variable; known from tropical moist forest all along the Atlantic slope and on the Pacific slope in the Canal Zone, Veraguas, Panamá, and Darién, from premontane moist forest in the Canal Zone and Panamá, from premontane wet forest in Panamá, from tropical wet forest in Bocas del Toro, Colón, Veraguas, Coclé, Panamá, and Darién, and from premontane rain forest in Coclé.

IXORA L.

Ixora coccinea L., Sp. Pl. 110. 1753
Jazmin de coral, Cache de tore

Small shrub, to 2 m tall, glabrous nearly throughout. Leaves sessile or subsessile; stipules to 5 mm long, with cusps as long as the broadly flaring base, subpersistent; blades ovate to oblong-elliptic, apiculate at apex, rounded to shallowly cordate at base, 3.5–7 cm long, 2–3 cm wide (some pairs often markedly reduced), moderately thick. Flowers 4-parted, in terminal, short-pedunculate dichasia; branches reddish; calyx 2–3 mm long, the lobes 4, ca 1 mm long, acute, reddish; corolla red or orange-red, 5–7 cm long, the tube ca 1.5 mm wide, the lobes 4, lanceolate, spreading, to 1.5 cm long; stamens attached to corolla rim; filaments with the free part short; anthers appressed to corolla, 1–2 mm long; style exserted ca 3 mm above rim, the branches short, slender. Berries ca 1 cm long, red becoming dark violet at maturity, the mesocarp pulpy, sweet; seeds 2, ovoid, ca 6 mm long. *Croat 7051.*

Cultivated at the Laboratory Clearing. Apparently flowering throughout the year, but not setting fruit.

Apparently native to Asia; cultivated in Central and South America, the West Indies, and Madagascar. In Panama, known from premontane moist forest in Panamá and from tropical moist forest in the Canal Zone, Bocas del Toro, Colón, Coclé, Panamá, and Darién.

MACROCNEMUM P. Browne

Macrocnemum glabrescens (Benth.) Wedd., Ann. Sci. Nat. Bot., sér. 4, 1:76. 1854

Palo cuadrado, Palo blanco, Madroño

Mostly glabrous tree, 8–15(25) m tall; trunk to 30(50) cm dbh, all but the smallest involuted in long vertical sections; outer bark reddish-brown, very thin; inner bark tan, thin; sucker shoots often forming near base of trunk. Stipules ovate, to ca 1.8 cm long and 1 cm wide, caducous; petioles obscure or to 3 cm long, glabrate; blades obovate to elliptic, abruptly acuminate, acute to obtuse at base, 7–21 cm long, 2.5–9.5 cm wide, glabrate above, strigose below especially on veins. Panicles terminal or subterminal, to 25 cm long and 20 cm wide; peduncles to 18 cm long, strigose; flowers sessile or subsessile, many, 5-parted; calyx lobes ca 0.5 mm long, blunt; corolla rotate, 1.5 cm long, magenta, the tube at first green, becoming maroon, nearly 1 cm long, densely glandular on basal half inside, pubescent near staminal tufts, glabrous above, the limb ca 1.5 cm broad, each lobe somewhat creased in the middle; filaments fused to corolla tube in basal half, with stiff trichomes directed both upward and downward near junction with tube; anthers opening in bud, held at one side of tube near its rim at 3 levels, the thecae opening broadly toward opposite side of tube; pollen tacky; ovary 2-celled, with many ovules; style held below anthers at anthesis, with the 2 thick stigmas often closed, later elongating, passing between anthers with the stigmas open; nectar copious. Capsules cylindrical, 10–20 mm long, 2–4 mm wide, splitting longitudinally on 2 sides, exposing seeds; seeds minute, membranaceous, winged, ca 3 mm long. *Croat 4806.*

Frequent in the forest, especially in the younger forest, sometimes locally very abundant. Flowers mostly from December to April, especially in January and February, rarely later or earlier (October to June). The fruits mature from February to June, mostly in April and May.

Costa Rica to Colombia (Chocó). In Panama, known most commonly from tropical moist forest in the Canal Zone, Bocas del Toro, Chiriquí, Veraguas, Panamá, and Darién, but also from premontane wet forest in Veraguas and Coclé and from tropical wet forest in Colón and Darién.

MANETTIA L.

Manettia reclinata L., Mant. Pl. Altera 553. 1771

M. coccinea (Aubl.) Willd.

Churco-guidave

Vine, to 1 m or more long, glabrate except sometimes sparsely pubescent on leaves. Stipules broadly triangular, minute, caducous; petioles to 1 cm long; blades ovate to elliptic, acuminate, obtuse to acute at base, 2.5–7 cm long, 1.5–3 cm wide. Flowers axillary, solitary or in umbels; peduncles 2–3.5 cm long; pedicels 1–3 cm long; calyx lobes 8, linear-lanceolate, irregular, to 9 mm long, puberulent and ciliate, persisting in fruit; corolla red

(fading to pink), 15–25 mm long at anthesis, puberulent outside especially on tube, the tube with a ring of moniliform trichomes at rim and in basal half above the somewhat expanded base, the lobes (3)4, ovate, to ca 5 mm long, obscurely short-acuminate; stamens 4; filaments adnate to corolla tube; anthers tightly affixed near rim; stigmas 2, thick, oblong, papillate, held well above anthers, receptive while pollen is shed. Capsules 2-valved, 7–10 mm long, to ca 8 mm wide, green to black at maturity, somewhat obovate, with a medial groove, the valves persisting after opening; seeds many, ca 2.7 mm wide, flattened, winged, the margin irregular, the wing thin, broad, round. *Croat 12740.*

Occasional in old clearings or at the edge of the forest along the edge of the lake. Flowers and fruits mostly in the early dry season (especially December), rarely also at other times of the year.

Mexico south to Colombia and Venezuela; Cuba and Jamaica. In Panama, ecologically variable; known from premontane moist forest in the Canal Zone and Panamá, from tropical moist forest in the Canal Zone, Bocas del Toro, San Blas, Chiriquí, Veraguas, Los Santos, Panamá, and Darién, from premontane wet forest in Chiriquí, Veraguas, and Panamá, from tropical wet forest in Los Santos, Coclé, and Darién, and from premontane rain forest in Chiriquí and Panamá.

OLDENLANDIA L.

Oldenlandia corymbosa L., Sp. Pl. 119. 1753

Slender herb, often decumbent, ± glabrous except for sparse pubescence on underside of leaves. Leaves ± sessile; stipules minute, persistent, with short flattened trichomes on surface and bristles on upper margin; blades linear-lanceolate or linear, acute at both ends, 1–3 cm long, to 5 mm wide, glaucescent below. Cymes small, axillary, bearing few flowers; peduncles to 12 mm long; flowers 4-parted; pedicels 2–5 mm long; calyx deeply lobed, the lobes 4, narrowly triangular, ca 0.5 mm long, with scabrid margins; corolla white to faintly lavender, to 1.4 mm long, 4-lobed to about middle, the lobes acute to blunt, conspicuously pubescent at base inside; stamens 4, minute, included; anthers introrse, ca 0.3 mm long; style about as long as stigma; stigma held at level of anthers. Capsules 2-carpellate, ca 1.7 mm long and 2.3 mm wide, weakly flattened, with a medial groove, rounded below, ± flat at apex; seeds 1 per carpel, globular, minute (ca 0.5 mm diam), the surface areolate. *Croat 7055.*

Infrequent, in clearings or disturbed areas. Probably flowering and fruiting throughout the year, especially at the beginning of the dry season.

Widespread in the tropics of Western and Eastern hemispheres. In Panama, growing mostly at lower elevations on the Atlantic slope and at higher elevations on the Pacific slope; known from tropical moist forest in the Canal Zone, Bocas del Toro, San Blas, Panamá, and Darién (doubtlessly Colón as well) and from premontane wet forest in Colón and Coclé.

PALICOUREA Aubl.

Palicourea guianensis Aubl., Hist. Pl. Guiane Fr.
1:173, t. 66. 1775

Shrub or small tree, to about 3.5 m tall; stems quadrangular, stout; stems and leaves glabrous to inconspicuously puberulent on veins. Stipules lobed nearly to base, the lobes obtuse, 5–8(15) mm long, long-persisting; petioles to 3 cm long; blades ovate to elliptic, acuminate, acute to obtuse and weakly decurrent at base, mostly 12–25 cm long, 6–13 cm wide, chartaceous; major veins raised on both surfaces, the reticulate veins conspicuous. Inflorescences terminal thyrsiform panicles 12–18 cm long; branchlets minutely puberulent, reddish-orange in flower, becoming red-violet in fruit; peduncles to 10 cm long, quadrangular; pedicels 5–8 mm long; calyx lobes 5, puberulent, spreading at anthesis, to 1 mm long; corolla tubular, ca 1.5 cm long, yellow or orange, weakly inflated near apex, puberulent outside, the lobes 5, minute; stamens included; filaments fused to basal two-thirds of corolla tube, densely pubescent; anthers sagittate, the connective blue above; style bilobed, slightly exserted. Capsules ovoid, 4–8 mm long, purple, apiculate at apex, minutely puberulent; seeds 2, ovate, hemispherical, the inner side flat, the outer side rounded and 3–5-ribbed, roughened. *Croat 11082.*

Frequent at the edges of clearings and along the lakeshore; also in the young forest. Flowers from late April to July, rarely later. The fruits develop in about 1 month, mostly from June to October, but some may be seen throughout the year.

Some collections bearing this name from Darién and Panamá Province (Cerro Jefe) differ from the material described here in having corollas 2–2.5 cm long with the styles not exserted and somewhat shorter than the stamens. They may represent another species or variety of *P. guianensis* or possibly another sexual form. The species is not heterostylous on BCI.

Guatemala to Colombia, Venezuela, the Guianas, Brazil (Pará), and Ecuador; Trinidad and West Indies. In Panama, known from premontane moist forest in the Canal Zone, from tropical moist forest in the Canal Zone, Bocas del Toro, San Blas, Veraguas, Los Santos, Panamá, and Darién, from premontane wet forest in Colón, Panamá, and Darién, and from premontane rain forest in Panamá and Darién.

See Fig. 524.

PENTAGONIA Benth.

Pentagonia macrophylla Benth., Bot. Voy. Sulphur
105, t. 39. 1845

P. pubescens Standl.

Hoja de murciélago, Indian ink, Teta de vaca, Tetilla, Wild grape

Shrub, 1.5–5 m tall, often unbranched; stems usually 4-ribbed near apex, sparsely to densely scurfy. Stipules narrowly triangular, 3.5–7 cm long, acuminate; petioles 6–12 cm long, strigose and often scurfy; blades broadly elliptic to obovate, acute at apex, acute to obtuse at base, 30–65 cm long, 21–37 cm wide, ± glabrate to short-pilose on both surfaces but especially below. Flowers few, sessile or subsessile, in dense axillary clusters subtended by large reddish bracts; calyx lobes 5, 10–20 mm long, rounded to obtuse, reddish, persisting in fruit; corolla thick, yellow, 3.5–4.5 cm long, 6–10 mm diam, puberulent outside, the lobes 5, ovate, acute, spreading; stamens of irregular lengths, attached to one side of corolla tube, the longest to ca 3 cm long; filaments and tube pubescent at point of fusion of stamens ca 1 cm above base; style ca 2 cm long, the lobes thick, held to one side of tube opposite and below anthers; nectary 5-lobed; nectar copious. Fruits baccate, spherical but ± tapering into the persistent calyx, to 2.5 cm diam, red to orange-red, 2-carpellate, indehiscent; seeds many, irregularly oblong, ca 4 mm long, brown, somewhat flattened, embedded in a fleshy white matrix. *Croat 6244, 16542.*

Common in the forest, especially the old forest. Flowers from the late dry to the middle of the rainy season, mostly from April to September. The fruits mature from the late rainy season to the dry season of the following year; they are eaten during September and October by white-faced monkeys (Hladik & Hladik, 1969).

Panama and Colombia (Chocó). In Panama, ecologically variable; known from tropical moist forest in the Canal Zone, Bocas del Toro, San Blas, Panamá, and Darién, from premontane wet forest in Colón and Coclé, from tropical wet forest in Colón, Coclé, Panamá, and Darién, and from premontane rain forest in Darién.

POGONOPUS Klotzsch

Pogonopus speciosus (Jacq.) K. Schum. in Mart., Fl.
Brasil. 6(6):265. 1889

Chorcha de gallo

Tree, 2–8 m tall and to 20(30) cm dbh; stems lenticellate, at least the smallest with appressed trichomes. Stipules triangular, to 3 mm high and at least twice as broad, entire, persistent; petioles short or to 3 cm long; blades elliptic to oblanceolate, abruptly acuminate, gradually tapered to base, mostly 8–25 cm long, 3–8 cm wide, sparsely appressed-puberulent on veins below. Inflorescences of cymes forming terminal panicles; flowers 5-parted; pedicels 3–5 mm long; calyx with one lobe greatly enlarged, leaflike, bright red, 3.5–6.5 cm long, often persisting in fruit, the others slender and short; corolla tubular, 2–3 cm long, dark red outside, densely puberulent, the trichomes longer and thicker inside, the tube white inside, the lobes pink; stamens exserted, ca 3.5 cm long, held erect at first, later recurved to one side; filaments fused to tube ca 7 mm at base, the base of the free part surrounded by dense tufts of fine trichomes; anthers purple; style scarcely emerging from flower, with the stigmas closed, later the style erect, nearly as long as stamens, with the stigmas ± reflexed. Capsules ovoid to obovoid, 5–8 mm long, 4–6 mm wide, with conspicuous whitish lenticels; seeds many, minute, irregularly shaped. *Foster 1411.*

Fig. 524. *Palicourea guianensis*

Fig. 525.
Posoqueria latifolia

Rare, on the shore. Flowers mostly from September to January. The fruits mature in December and January (possibly later).

Mexico to Colombia, Venezuela, and Brazil. In Panama, known from drier parts of tropical moist forest in the Canal Zone, Colón, Panamá, and Darién, from premontane moist forest in the Canal Zone, and from tropical wet forest in Darién.

POSOQUERIA Aubl.

Posoqueria latifolia (Rudge) R. & S., Syst. Veg. 5:227. 1819

Boca vieja, Boca de vieja, Borajo, Fruta de mono, Fruta de murciélago, Huevo de mono, Monkey apple, Mosquitowood, Wild coffee, Borojocito del monte

Glabrous shrub or tree, to 10 m tall; wood red. Stipules acute, 8–12 mm long, caducous; petioles 5–20 mm long; blades elliptic to oblong-elliptic, abruptly short-acuminate, acute to obtuse at base, 7–24 cm long, 3.5–13 cm wide, coriaceous, bicolorous; midrib often arched downward, each side of blade turned somewhat upward along midrib. Corymbs terminal; flowers sweetly aromatic, 5-parted; pedicels obscure or 4–10 mm long; calyx thick and short, ca 5 mm long, the lobes minute and blunt; corolla white, fleshy, the tube 13–16 cm long, glabrous except at rim, the lobes 5, oblong, to 2.5 cm long, blunt, recurved after anthesis; filaments united to corolla tube, exserted above throat after petals recurve; anthers united at anthesis, pushed to one side of tube, bursting apart violently when contacted, 2 pairs remaining united, their filaments reflexing laterally, the fifth filament bending forward across tube, its anther carrying with it almost the entire cobwebby pollen mass; style contained well within tube, ca 9 cm long; stigmas 2, slender, papillate-puberulent, somewhat barbellate at apex. Fruits baccate, 2-locular, round to ovoid, 4–5 cm diam, with a shallow circular crater ca 1 cm diam at apex, mottled green becoming somewhat soft and yellow to orange at maturity; seeds numerous, smooth, irregularly angulate, less than 1.5 cm long, embedded in a sweet orange matrix. *Croat 6057, 15062.*

Occasional to locally common in the forest, especially the younger forest; uncommon along the shore. Probably flowers to some extent all year, but collected from November to June, especially in May and June. Mature fruits are seen from April to December.

The flowers always open in the late afternoon and do not persist through the next day. The anthers are held together until contacted by a flower visitor, presumably by a hawkmoth at night. When the stamens spring apart, pollen is sprinkled on the visitor. Pollen must be deposited on its tongue as well, because the style never reaches the apex of the corolla tube (Percival, 1965). Visited flowers tend to fall sooner than unvisited ones, the corolla sometimes falling free before the style opens. Halle (1967) reported that the flowers are functionally staminate until after the stamens have been sprung and then are functionally pistillate.

Mexico to Colombia, Venezuela, the Guianas, Peru, and Brazil; Trinidad. In Panama, ecologically variable; known mostly from tropical moist forest all along the Atlantic slope and in the Canal Zone, Veraguas, Los Santos, Panamá, and Darién, but also from tropical dry forest in Panamá (Taboga Island), from premontane moist forest in the Canal Zone, Los Santos, and Panamá, from premontane wet forest in Chiriquí, Coclé, and Panamá, and from tropical wet forest in Colón, Veraguas, Panamá, and Darién. Reported from premontane rain forest in Costa Rica (Holdridge et al., 1971).

See Fig. 525.

PSYCHOTRIA L.

Psychotria acuminata Benth., Bot. Voy. Sulphur 107. 1845

Shrub, usually less than 2 m tall; stems puberulent or glabrous, the younger stems, petioles, axes of inflorescences, pedicels, and calyces densely but inconspicuously short-puberulent. Stipules bicuspidate nearly to base, persistent, the lobes widely separated, to 3.5 mm long; petioles 5–10 mm long; blades oblong-ovate or elliptic, cuspidate-acuminate, obtuse or rounded and decurrent onto petiole at base, mostly 9–21 cm long and 4–10 cm wide, shiny above and glabrous except for minute puberulence on midrib, dull below; midrib narrow, raised, the lateral veins raised, the reticulate veins conspicuous. Inflorescences terminal, solitary, much shorter than upper leaves; peduncles 1–3(4.5) cm long, the floriferous part to 4 cm long, 4.5 cm wide; inflorescence branches violet-purple in fruit; flowers short-pedicellate (the pedicel thickening as fruit develops), heterostylous, (3)4- or 5-parted, 4–6 mm long; corolla white, puberulent outside and on lobes inside, the throat yellow, villous at point of staminal attachment; short-styled flowers with the stamens equaling lobes, exserted ca 3 mm above throat, held to one side of tube as in *P. pubescens,* the anthers slender, to 1.2 mm long, the style included, its lobes short; long-styled flowers with the stamens included, ± equaling tube, the styles ca 7 mm long, exserted 2–3

KEY TO THE SPECIES OF PSYCHOTRIA

Inflorescences axillary:
  Stems and petioles essentially glabrous; peduncles more than 3 cm long . . . . *P. uliginosa* Sw.
  Stems and petioles densely pubescent; peduncles less than 3 cm long:
    Mature plants less than 60 cm tall, often unbranched; stipules to ca 4 mm long; fruits blue at maturity . . . . *P. emetica* L.f.
    Mature plants more than 1 m tall, usually branched; stipules to ca 15 mm long; fruits red at maturity . . . . *P. psychotriifolia* (Seem.) Standl.

Inflorescences terminal:
  Leaves and stems pubescent *or* vein axils with pitlike domatia below:
    Leaves with pitlike domatia in vein axils below .... *P. horizontalis* Sw.
    Leaves lacking pitlike domatia in vein axils:
      Leaves densely soft-pubescent below, often more than 20 cm long .... *P. micrantha* H.B.K.
      Leaves not densely pubescent below, at most with sparse, appressed, slender, white tri-
        chomes, usually less than 20 cm long:
        Stipules persistent, green, bilobed, the lobes widely separated; leaves usually pubescent
          at least on midrib above:
          Bracts conspicuous, exceeding flowers .... *P. furcata* DC.
          Bracts inconspicuous or at least shorter than flowers:
            Stipules 10–15 mm long .... *P. racemosa* (Aubl.) Raeuschel
            Stipules less than 9 mm long:
              Inflorescences usually less than 4 cm long, pendent; corolla lobes white through-
                out; mature leaves less than 12 cm long; fruits blue or violet at maturity ....
                .... *P. pittieri* Standl.
              Inflorescences usually more than 4 cm long, erect and ± pyramidal; corolla lobes
                often tinged with violet inside near base; mature leaves more than 12 cm
                long; fruits purple-black at maturity .... *P. pubescens* Sw.
        Stipules caducous from base, brown, usually 1–2 cm long, bilobed near apex, the lobes
          held together; leaves glabrous above:
          Fruits, including persistent calyx, 10–15 mm long; inflorescences terminal, to ca 10 cm
            long (not known with certainty for the island but to be expected) ....
            .... *P. calophylla* Standl.
          Fruits, including persistent calyx, to 8 mm long; inflorescences terminal or axillary,
            usually less than 5 cm long:
            Flowers clustered in small, dense, bracteate heads; bracts conspicuous, often longer
              than flowers .... *P. psychotriifolia* (Seem.) Standl.
            Flowers in small open panicles; bracts much shorter than flowers ....
            .... *P. granadensis* Benth
  Leaves and stems essentially glabrous (some leaves sparsely puberulent along midrib below),
    lacking domatia in vein axils:
    Stipules deeply bilobed, divided to or almost to base:
      Bracts conspicuous, equaling or exceeding flowers:
        Bracts purple; leaves 3.5–9(18) cm long, 1.2–3.6(7.5) cm wide .... *P. furcata* DC.
        Bracts green; leaves 10–16 cm long, 4–6 cm wide .... *P. brachybotrya* Müll. Arg.
      Bracts inconspicuous, shorter than flowers:
        Inflorescences much longer than broad; branchlets glabrous; fruits blue then finally
          white at maturity. .... *P. deflexa* DC.
        Inflorescences not longer than broad; branchlets often puberulent; fruits grayish-blue at
          maturity .... *P. acuminata* Benth.
    Stipules not lobed, or bilobed only halfway:
      Flowers in dense clusters subtended by conspicuous bracts *or* bracts longer than flowers:
        Stipules much shorter than petiole; inflorescences paniculate; floral bracts as broad as
          long; fruits violet-blue at maturity .... *P. brachiata* Sw.
        Stipules nearly as long as or longer than petiole; inflorescences sessile or paniculate with
          slender floral bracts; fruits black or red at maturity:
          Inflorescences minute (less than 2 cm long), sessile; leaves less than 10 cm long; fruits
            red at maturity .... *P. chagrensis* Standl.
          Inflorescences not minute; leaves usually more than 10 cm long; fruits black at
            maturity .... *P. capitata* R. & P.
      Flowers not in dense clusters subtended by conspicuous bracts, the bracts not longer than
        flowers:
        Lower midrib of leaves with triangular flaplike protuberances near axils below ....
        .... *P. marginata* Sw.
        Lower midrib of leaves lacking protuberances:
          Leaf blades on mature stems usually less than 5 cm wide, the petioles less than 1.5 cm
            long; shrubs ca 1.5–2 m tall .... *P. carthagenensis* Jacq.
          Leaf blades on mature stems more than 5 cm wide or, if less than 5 cm wide, the peti-
            oles more than 2 cm long:
            Stipules persistent, cuspidate at apex; peduncles more than 8 cm long ... *P. grandis* Sw.
            Stipules caducous, rounded or cleft at apex; peduncles less than 4 cm long:
              Leaves 6–14 cm wide; petioles 1.5–6 cm long; reticulate veins not visible ....
              .... *P. limonensis* Krause
              Leaves 3.5–7 cm wide; petioles 0.5–4.5 cm long; reticulate veins usually clearly
                visible .... *P. granadensis* Benth.

mm. Berries depressed-globose, 5–7 mm diam, usually bilobed, yellow, turning grayish-blue at maturity; seeds 2, hemispherical, ca 3.7 mm diam, smooth (not ribbed as in most species). *Croat 5877, 15077.*

Common in the forest. Flowers mostly from March to July, especially in May and June, and sporadically during the late rainy and dry seasons as well. The fruits mature in 1 or 2 months, mostly from July to October.

Distinguished by the prominently bilobed fruits, cuspidate-acuminate leaves, and persistent, bicuspidate stipules.

Standley (1930b) considered this taxon *P. cuspidata* Bredem. ex R. & S. in his various treatments of the Rubiaceae for Venezuela, Colombia, Bolivia, and Peru. Steyermark (1972) reported that *P. cuspidata* is a species restricted to the coastal cordillera of northern Venezuela.

Mexico (Chiapas) to Colombia, the Guianas, and Brazil (Bahia); Trinidad. In Panama, ecologically variable; known from premontane moist forest in the Canal Zone, from tropical moist forest in the Canal Zone, San Blas, Veraguas, Los Santos, Herrera, Panamá, and Darién, from premontane wet forest in the Canal Zone, Colón, Coclé, and Darién, and from tropical wet forest in Colón. See Fig. 526.

Psychotria brachiata Sw., Prodr. Veg. Ind. Occ. 45. 1788

Shrub or small tree, to 3 (5) m tall, nearly glabrous; branchlets often canaliculate. Stipules ovate, 4–6 mm long, persistent, weakly cleft at apex; petioles 1–3 cm long; blades elliptic to oblong-elliptic or oblanceolate, acuminate, acute to obtuse and decurrent at base, mostly 10–20 cm long, 4.5–8 cm wide, glabrous or sparsely pubescent along midrib below; major lateral veins raised on both surfaces, forming almost right angles with midrib, widely arcuate-ascending, the reticulate veins visible below. Flowers sessile, 5-parted, in closely aggregated clusters in open terminal panicles to 12 (16) cm long and 10 cm wide in fruit; branches sparsely hispidulous, perpendicular to axis; bracts obtuse, equaling or longer than calyx; calyx ± truncate, to 0.9 mm long, persisting on fruit; corolla campanulate, to 6 mm long, pale yellow to yellow-green, hirtellous outside, glabrous inside, the lobes short, ovate, acute, spreading, bearing a purplish, subapical thickening outside; stamens to 2.7 mm long, included; anthers ca 1 mm long; style to 3.3 mm long, papillate, the branches 2, short, thick, densely papillate-puberulent. Berries closely aggregated, ellipsoid, ca 4 mm long at maturity, violet-blue, sparsely puberulent; seeds 2, hemispherical, ca 4 mm long, the inside face concave, the outside face with 5 ridges. *Croat 12457, 15252a.*

Apparently uncommon in the forest; collected on Shannon and Lathrop trails. Flowers and fruits most of the year, but with most flowers opening from April to August and most fruits maturing in the late rainy season.

Recognized by the yellow flowers, the closely aggregated blue fruits, and the persistent green stipules, which are slightly cleft at the apex. In Panama the species merges almost imperceptibly with the South American *P. caerulea* R. & P., which, if taxonomically distinct, has the flowers even more closely congested into conspicuously bracteate heads.

As defined by Steyermark (1972), the species is known from the West Indies (Greater Antilles), Trinidad, and Central America. *P. caerulea* is known from Panama, Colombia, Venezuela, Ecuador, and Peru. In Panama, *P. brachiata* is ecologically variable; known from tropical moist forest in the Canal Zone, Bocas del Toro, Colón, San Blas, Los Santos, Panamá, and Darién, from premontane wet forest in Veraguas, Panamá, and Darién, and from tropical wet forest in Colón, Coclé (Atlantic slope), and Darién.

Psychotria brachybotrya Müll. Arg. in Mart., Fl. Brasil. 6 (5):327. 1881

Shrub, 1–1.8 m tall; stems and leaves glabrate. Stipules persistent, 3–5 mm long, bifid to near base, the lobes distant, acuminate; petioles 1–1.5 cm long; blades ± elliptic, acute to acuminate, cuneate at base, 10–16 (20) cm long, 4–6 cm wide, drying light green. Racemes short, terminal, usually to 1.5 cm long at anthesis, to 3 cm long in fruit; peduncles lacking or to 1.2 cm long; inflorescence branches, pedicels, and calyces with short, ± erect, sparse trichomes, the lowermost branches subtended by a distinct, oblong-elliptic bract ca 5 mm long; bracts green, ovate, 2–7 mm long, acute, foliate, persistent, subtending groups of 2–4 (7) flowers, exceeding the flowers; flowers white, sessile, ca 3.5 mm long, the tube ca 2 mm long, the lobes oblong, to ca 1.5 mm long; filaments ca 0.5 mm long, affixed above the middle of the tube; anthers ca 1 mm long; style ca 4 mm long; stigmas 0.2 mm long. Berries ± globose, 3–4 mm diam, flattened laterally and somewhat grooved, glabrous or sparsely puberulent at least when immature; seeds 2, flat, with a deep groove on the inside surface, rounded and 4- or 5-grooved on the outside surface. *Croat 11132.*

Apparently rare; found in the younger forest south of Fairchild Point and in the forest north of Barbour Trail. Flowers mostly from June to September, principally in July, rarely also in the dry season. The fruits mature from July to October, principally in August.

In Panama, *Psychotria brachybotrya* is most easily confused with *P. involucrata* Sw., a superfluous name for *P. officinalis* (Aubl.) Sw. (Steyermark, 1972). It differs from that species by having distinct bracts beneath the lower inflorescence branches and leaves that usually dry a lighter green. It usually occurs in drier areas than *P. officinalis,* which occurs principally in premontane wet forest and tropical wet forest. *Psychotria brachybotrya* has also been confused with *P. furcata* DC., but is distinguished from that species by its larger, lighter-colored leaves and green, rather than purple, inflorescence bracts.

Panama to Colombia, Venezuela, the Guianas, Brazil, Peru, and Bolivia; Trinidad. In Panama, ecologically variable; known from tropical moist forest in the Canal Zone, Bocas del Toro, Colón, Herrera, Panamá, and Darién, from tropical dry forest in Panamá, and from premontane wet forest in Colón, Panamá, and Darién.

Fig. 526.
Psychotria acuminata

Fig. 527. *Psychotria capitata*

Fig. 528. *Psychotria carthagenensis*

Psychotria capitata R. & P., Fl. Peruv. 2:59. 1799

Shrub, usually to 3 (4) m tall; stems glabrous or weakly puberulent. Stipules lanceolate, ca 1.5 cm long, bilobed at apex, deciduous, sometimes purplish; petioles 5–20 mm long; blades elliptic to oblong-elliptic, acuminate, tapered to base and somewhat decurrent, mostly 9–20 cm long, 2.5–7 cm wide, stiff when fresh; lateral veins raised on both surfaces, forming nearly right angles with midrib, sharply ascending near margin, the basal pair often pubescent below. Panicles white at anthesis, long-pedunculate, terminal, solitary, usually longer than broad, to 13 cm long and 7 cm wide (usually smaller), branching at right angles; bracts 5–10 mm long, oblanceolate, often purplish in fruit; flowers white, heterostylous, usually exceeded by bracts, these white at anthesis, becoming green; calyx lobes minute; corolla 4-lobed, ca 1 cm long, the lobes spreading, ovate, acute or rounded, the tube pubescent in upper third inside, the throat yellow; short-styled flowers with the stamens attached at apex of tube, exserted 5–7 mm above throat, the style included, held at apex of throat; long-styled flowers with the stamens included and hidden among long trichomes in throat, the style exserted 5–7 mm above throat. Berries subovate to oblong, 4–6 (7) mm long, 3.5–5 mm wide, black; seeds 2, hemispherical, ca 4.7 mm long, 3–4 mm wide, the inside face flat with a slight fissure, the outside face 4- or 5-ridged at maturity. *Croat 12414, 14960.*

Occasional in the forest; common in a few localities, mostly in the old forest; occasional to locally common on Zetek Trail. Flowers and fruits sporadically throughout the year, but mostly in June.

A small black-and-white *Heliconius* butterfly (probably *H. cydno* or possibly *H. sappho*; Nymphalidae, Ithomiinae) visits the flowers at anthesis. The same butterfly visits *Psiguria warscewiczii* (131. Cucurbitaceae). The species is also visited by small hummingbirds.

Belize to Peru and Brazil; Trinidad and Tobago. In Panama, ecologically variable; known from tropical moist forest in the Canal Zone, Colón, San Blas, Panamá, and Darién, from premontane wet forest in Colón and Coclé, and from tropical wet forest in Colón, Panamá, and Darién.

See Fig. 527.

Psychotria carthagenensis Jacq., Enum. Syst. Pl. Ins. Carib. 16. 1760

Shrub, usually 1.5–2 (3) m tall, nearly glabrous; internodes very short. Leaves subsessile or on petioles to 1.5 cm long; stipules ovate-spatulate, 4–6 mm long, abruptly cuspidate, entire, caducous; blades obovate, acute to short-acuminate, attenuate at base, 5–16 cm long, 2.5–5.5 cm wide; veins 6–8 on each side. Panicles solitary, terminal, to 7 cm long; peduncles 2–4 cm long; bracts minute, caducous; flowers sessile; calyx 3–5 mm long, irregularly and usually minutely toothed; corolla white, 3–4 mm long, 5-lobed, the lobes ca 2 mm long; stamens inserted at middle of tube; filaments ca 2 mm long; anthers oblong, ca 1 mm long; style 2–2.5 mm long; stigma bifid, 0.7–1 mm long. Berries oblong to elliptic,

ca 6 mm long and 2.5 mm wide; seeds 2, the inside face flat, the outside face 5- or 6-costate. *Croat 10337.*

Occasional, in the younger forest along Barbour Trail. Seasonal behavior uncertain. Flowers at least throughout the dry season and early in the rainy season, but mostly from April to June. The fruits mature mostly from May to July.

Belize south to Paraguay and southern Brazil; Trinidad and Tobago. In Panama, known from tropical dry forest in Los Santos and Coclé, from premontane moist forest in Veraguas and Los Santos, from tropical moist forest in the Canal Zone, Bocas del Toro, Los Santos, Herrera, Panamá, and Darién, and from premontane wet forest in Coclé.

See Fig. 528.

Psychotria chagrensis Standl., J. Wash. Acad. Sci. 15:105. 1925

Shrub, usually less than 1.2 m tall, glabrous; branchlets slender, mostly ± erect from upper side of branches. Leaves clustered on ends of small branchlets on upper side of branches; stipules with long slender apex, caducous, 8–15 mm long; petioles obscure or to 8 mm long; blades obovate-elliptic to elliptic, abruptly acuminate, gradually tapered to base and decurrent, 2.5–10 cm long, 1.3–3.5 cm wide, bicolorous; major lateral veins drying with minute wrinkles. Inflorescences less than 2 cm long, terminal, the heads sessile, bearing few flowers, surrounded by scarious, brownish bracts to 1 cm long; flowers heterostylous, 5 (6)-parted; calyx thin, enclosed in sheath, irregularly toothed; corolla white, 8–10 mm long, the lobes apiculate inside, recurved at anthesis, the throat densely pubescent above point of staminal attachment; short-styled flowers with the stamens fused to throat near rim, ca 2 mm long, the anthers exserted at anthesis, the style included and ca 2–3 mm long; long-styled flowers with the stamens fused to throat near rim but included, ca 1 mm long, the style ca 10 mm long, exserted ca 5 mm. Berries usually 1–3, narrowly ovate, (5)8–10 mm long, bright red; seeds 2, narrowly ovate, ca 6 mm long, flattened on one side, the other side rounded and with 5 ribs. *Croat 14643.*

Frequent in the forest; collected in the swampy area north of Zetek Trail 300 and on Shannon Trail. Seasonal behavior uncertain. Probably flowers and fruits throughout the year, but most flowers are seen in May and August.

Recognized by the small, obovate-elliptic leaves and minute, sessile flowers in small, brown-bracted heads.

Mexico to Panama. In Panama, known from tropical moist forest in the Canal Zone, Colón, Panamá, and Darién, from premontane wet forest in Panamá, and from tropical wet forest in Colón and Darién.

Psychotria deflexa DC., Prodr. 4:510. 1830

P. patens sensu auct. non Sw.

Garricillo

Glabrous shrub, to 3 m tall; younger stems flattened at nodes. Stipules bilobed nearly to base, the lobes ca 1 cm

long, distant, linear, green, persistent; petioles 2–10 mm long; blades ovate-elliptic, gradually long-acuminate and often somewhat falcate at apex, mostly obtuse to rounded and decurrent at base, 8–18 cm long, 3.5–7 cm wide, entire or minutely crenulate, dark green and glossy above, duller below; midrib narrow, markedly raised, the major lateral veins arcuate-ascending, the reticulate veins conspicuous. Inflorescences pedunculate, terminal, usually 3–10 cm long, narrow, the branches, calyces, and corollas minutely puberulent to glabrous; flowers heterostylous, white, sessile to short-pedicellate; calyx lobes minute, acute; corolla ca 3.3(4) mm long at anthesis, yellow in throat, barbate at point of staminal attachment, the lobes 3–5, recurved; short-styled flowers with the stamens exserted ca 1–2 mm, sometimes unequally so, with 2 ± included, the style included; long-styled flowers with the stamens included, the style exserted ca 1 mm above rim, with 2 slender branches. Berries 2-celled, broader than long, becoming blue (may dry green or black), eventually the exocarp rupturing to expose foamy white mesocarp; seeds 2, globose, 2.5–3 mm diam, the inside face grooved, the outside face subcostate. *Croat 11926, 15575.*

Frequent in the forest, occurring also in clearings. Flowers from May to July. The fruits mature mostly from August to December, especially from August to October.

Distinguished by the slender, terminal inflorescences and persistent, linear, bilobed stipules.

Mexico to Peru and Bolivia; West Indies. In Panama, known from tropical moist forest in the Canal Zone, Bocas del Toro, San Blas, Veraguas, Herrera, Coclé, Panamá, and Darién, from premontane wet forest in Veraguas, Coclé, and Panamá, and from tropical wet forest in Panamá and Darién.

See Fig. 529.

Psychotria emetica L.f., Suppl. Pl. Syst. Veg. 144. 1781
    False ipecac, Rachilla

Shrub, usually less than 60 cm tall; stems strigose, often unbranched. Stipules narrowly lanceolate, ca 4 mm long, strigose, subpersistent; petioles ca 1 cm long, strigose; blades oblong-lanceolate or oblong-obovate, acute to short-acuminate, cuneate to obtuse at base, 8–17 cm long, 2–6.5 cm wide, bicolorous, essentially glabrous above, strigose below especially on veins; major lateral veins often impressed above, arcuate-ascending, forming an obscure collecting vein. Inflorescences axillary, subracemose, bearing few flowers; peduncles to ca 1 cm long; flowers sessile, 4-parted, white, ca 5 mm long, glabrous outside, with very short stiff trichomes inside; corolla lobes acute; stamens 5, attached to corolla tube at about its middle; anthers narrowly oblong, ca 1.5 mm long, pilose. Berries oblong when juvenile, becoming rounded and bright blue, paler and somewhat lavender at maturity, to 1 cm long, the mesocarp foamy; seeds 2, elliptic, ca 5 mm long, brown, flattened on one side, twisted. *Croat 5750.*

Common in the forest. Flowers mostly in the late dry and early rainy seasons, especially in April and May and

again in July and August, with a few flowers seen in June. The fruits mature mostly in the middle rainy season, especially in August and September.

Distinguished by its small size and blue fruits.

Guatemala to Bolivia. In Panama, known from tropical moist forest in the Canal Zone, Bocas del Toro, San Blas, Panamá, and Darién and from tropical wet forest in Colón and Panamá.

See Fig. 530.

Psychotria furcata DC., Prodr. 4:512. 1830
    P. involucrata sensu auct.

Shrub, usually 1–1.5(4) m tall; branches slender, terete to flattened, puberulent. Stipules bilobed to base, persistent, the lobes slender, ca 4 mm long; petioles to 5 mm long; blades broadly elliptic to lanceolate-elliptic, acuminate, cuneate and decurrent at base, 3.5–9(18) cm long, 1.2–3.6(7.5) cm wide, glabrous and shiny above (sparsely white-lineolate when dry), sparsely puberulent below especially on veins, membranaceous; lateral veins impressed above, raised below. Inflorescences terminal, solitary, subcapitate, usually 5–20 mm long, contracted, bearing few flowers; bracts maroon, acute, slightly to much longer than corolla; pedicels very thick and maroon in fruit; corolla white, ca 3 mm long; stamens 4 or 5; filaments short, fused part way to tube; anthers linear-oblong, ca 1 mm long; style to 2 mm long; stigma lobes oblong. Berries globose, (3)4–6 mm diam, maroon becoming purple or black at maturity, puberulent; seeds 2, hemispherical, ca 3.3 mm long, the inside face concave, the outside face rounded with 4 or 6 broad ridges. *Croat 6272.*

Frequent in the forest. Flowers in the late dry season and no doubt in the early rainy season. The fruits mature throughout the rainy season and in the early dry season, especially in the middle of the rainy season.

Recognized by the usually small leaves and tiny inflorescences with prominent, violet-purple bracts.

Costa Rica, Panama, and Colombia (Chocó). In Panama, known principally from tropical moist forest in the Canal Zone, Bocas del Toro, San Blas, Los Santos, Herrera, Coclé, Panamá, and Darién; known also from premontane wet forest in Colón, Chiriquí, Panamá, and Darién.

Psychotria granadensis Benth., Vidensk. Meddel. 12. 1852
    Huesito

Shrub, usually to 2 m tall (shorter along trails probably owing to being cut back), glabrous except for minute puberulence on leaf veins below and on inflorescence branches; stems usually mottled. Stipules 7–15 mm long, bilobed ca one-third the way to base, deciduous, the lobes long-acuminate; petioles 0.5–4.5 cm long; blades oblanceolate to oblong-elliptic, acuminate, attenuate and decurrent at base, 12–18 cm long, 3.5–7 cm wide; midrib and lateral veins often whitish above, the reticulate veins inconspicuous. Inflorescences of usually 3 or 4 panicles,

Fig. 530. *Psychotria emetica*

Fig. 529. *Psychotria deflexa*

Fig. 531. *Psychotria limonensis*

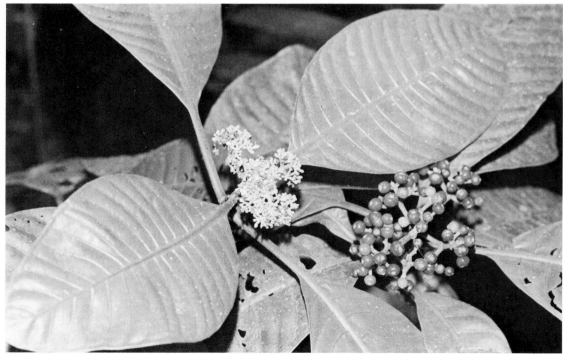

small, terminal, branched, closely compacted, less than 2.5 cm long; flowers sessile, 5-parted; calyx to ca 1 mm long (usually less), the lobes triangular; corolla puberulent outside, glabrous inside except for long trichomes at throat between anthers, the tube campanulate, as long as lobes, the limb to 4 mm diam, the lobes ovate, 1.5–2.7 mm long, spreading to reflexed, acute and swollen at apex, greenish to white; stamens attached at throat and exserted slightly above rim; anthers ca 0.5 mm long, introrse; style 2.5–3 mm long, exserted ca 2 mm; stigmas broader than style, densely papillate-puberulent. Berries ovoid to ellipsoid, ca 4 mm long and wide, becoming red; seeds 2, ca 3 mm long, the inside face flat, with a small longitudinal ridge, the outside face rounded, 4- or 5-ridged. *Croat 7316, 14997.*

Locally common in the old forest above the escarpment, along trails, where it often grows to less than 30 cm tall. This short habit is probably due to clearing of the trails, since elsewhere in the Canal Zone plants reportedly grow much taller. Flowers in June and July. The fruits mature in the late rainy and early dry seasons (October to January).

Guatemala to Panama. In Panama, known from tropical moist forest in the Canal Zone and Veraguas.

Psychotria grandis Sw., Prodr. Veg. Ind. Occ. 43. 1788

Shrub or small tree, to 6 m tall (rarely so tall on BCI); stems glabrous, usually with prominent longitudinal ridges, with prominent leaf scars. Leaves glabrous; stipules broadly ovate, to 12 mm long, 8–13 mm wide, with a cusp 2–3 mm long, cupulate, green, persistent or subpersistent; petioles obscure or to 2 cm long; blades mostly oblong-obovate, short-acuminate, gradually tapered to base, 13–28(50) cm long, 4.5–10(16.5) cm wide; midrib and major lateral veins raised on both sides, the midrib often whitish. Inflorescences panicles, terminal, to about 30 cm long, often about as wide as long, branched many times; peduncles stout, 8–16 cm long; branches and calyces puberulent; flowers heterostylous; calyx lobes broadly triangular; corolla white, ca 4.5 mm long, lobed more than a third of the way to base, the lobes 5(6), spreading, glabrous except for dense beard at point of staminal attachment inside; long-styled flowers with the stamens exserted at anthesis, much shorter than style, the style lobes short, nearly as broad as long; short-styled flowers with the stamens exserted ca 2 mm above throat, the style included. Fruiting inflorescences usually overtopped by new growth; berries globose, to 9 mm diam, orange, the pulp fleshy and sweet; seeds 2, hemispherical, 2.5–3 mm long, the inside face planar, ungrooved, the outside face faintly 6-ribbed. *Croat 14872, Duke 8773.*

Occasional as adult plants; seedlings common in some areas beyond the Tower Clearing. Flowers to some extent throughout the year, especially in the early rainy season in June. The fruits probably mature in 4–6 weeks.

Mexico to Colombia and Venezuela; Greater Antilles. In Panama, known from tropical moist forest in the Canal Zone, Bocas del Toro, San Blas, Chiriquí, Veraguas, Panamá, and Darién and from premontane wet forest in Colón, Coclé, Panamá, and Darién.

Psychotria horizontalis Sw., Prodr. Veg. Ind. Occ. 44. 1788

Shrub, 1.5–3(6) m tall; stems terete or flattened, glabrous. Stipules ovate, to 6 mm long, entire, caducous, often strigose on inside, the trichomes to 1 mm long; petioles 3–11 mm long, slender; blades elliptic to oblanceolate, acuminate, usually ending abruptly and rounded at base, 7–20 cm long, 2.5–7 cm wide, bicolorous, glabrous or sparsely puberulent below, usually with pitlike domatia on bases of secondary veins below, the margins sometimes undulate; veins in 6–11 pairs. Inflorescences terminal, paniculate, pedunculate, usually less than 8 cm long in flower, about as broad as long, somewhat larger in fruit; flowers white, 5-parted, to 7 mm long; calyx lobes to 1 mm long, blunt to lanceolate-linear; corolla lobes acute, strongly recurved at anthesis, the tube bearded inside at point of staminal attachment; anthers in throat held tightly around style; style erect, long-exserted, ca 7 mm long, the lobes 2, enlarged and densely papillate. Berries subglobose, minutely puberulent, red at maturity, 4–5 mm diam; seeds 2, hemispherical, ca 3–5 mm long, the inside face flat, the outside face rounded with 5 broad ribs, the grooves narrow. *Croat 6383.*

Common in the forest. Flowers asynchronously in the late dry and early rainy seasons, from March to August, principally from April to July after the rains begin and only sporadically at other times of the year. The fruits mature principally in August and September, but sometimes as late as December.

Elsewhere the calyx persists, but BCI fruits commonly do not have a persistent calyx. This species and *P. marginata* are the two most abundant species of *Psychotria* on the island.

Mexico to Ecuador; West Indies. In Panama, ecologically variable; known from tropical moist forest in the Canal Zone, San Blas, Veraguas, Los Santos, Herrera, Coclé, Panamá, and Darién, from tropical dry forest in Coclé and Panamá (Taboga Island), from premontane moist forest in the Canal Zone and Panamá, from premontane wet forest in Chiriquí and Coclé, and from tropical wet forest in Darién.

Psychotria limonensis Krause, Bot. Jahrb. Syst. 54, Beibl. 119:43. 1916

Shrub, usually less than 1.5 m tall, glabrate. Leaves opposite-decussate; stipules broad, 5–10 mm long, rounded or sometimes minutely forked at tip, caducous; petioles 1.5–6 cm long; blades broadly elliptic, bluntly acuminate, obtuse to attenuate and decurrent at base, mostly 10–30 cm long and 6–14 cm wide, minutely puberulent on veins, bicolorous; midrib flattened above, the veins forming right angles with midrib, joined by submarginal collecting vein. Inflorescences terminal in flower (sometimes becoming axillary in fruit), 2–8 cm long, somewhat globose, ebracteate; flowers white, congested, to 4 mm long; calyx lobes short, acute; corolla lobed to middle, 5(6)-parted, with tufts of white trichomes alternating with stamens at apex of tube; stamens equaling corolla lobes; anthers shedding pollen before anthesis; style short; stigma bilobed, the lobes blunt, ± open in

bud, held somewhat below anthers. Berries subglobose, 6–7 mm wide, bright red at maturity; seeds 2, hemispherical, 3–4 mm long, 3.5–5 mm wide, the inside face planar, the outside face rounded, usually with 5 ridges. *Croat 8717.*

Frequent in the forest. Flowers and fruits to some extent throughout the year, especially in the dry season. Individuals may flower more than once per year and often bear flowers and fruits simultaneously. Most plants are seen in fruit during the early rainy season.

Distinguished by the short height, the large, broadly elliptic leaves, and the short, almost globular inflorescences.

Southern Mexico to northern Colombia. In Panama, known from tropical moist forest in the Canal Zone, Bocas del Toro, San Blas, and Darién and from tropical wet forest in Colón.

See Fig. 531.

Psychotria marginata Sw., Prodr. Veg. Ind. Occ. 43. 1788

Shrub, 2–3(6) m tall, nearly glabrous. Stipules to 11 mm long, entire, caducous; petioles obscure or to 3 cm long, slender, canaliculate above; blades oblong-oblanceolate to ± elliptic, acuminate, attenuate and decurrent at base, mostly 7–20 cm long and 2.5–6 cm wide, ± bicolorous (drying grayish); major lateral veins impressed above, raised below, the midrib below with flaplike protuberances near axils. Panicles terminal; peduncles 5–8 cm long, often almost as long as floral rachis; flowers white, to 4.7 mm long; pedicels longer than fruit; stamens fused to tube near its pubescent apex, shedding pollen in bud; style held above anthers in bud, exposed by recurved corolla lobes. Berries rounded, 4–8 mm broad, red-orange turning violet-purple at maturity; seeds 2, hemispherical, ca 3.3 mm long, the inside face flat, the outside face rounded, usually with 5 ribs. *Croat 5775.*

Common in the forest; one of the most commonly encountered *Psychotria* on the island. Flowers and fruits throughout the year, mostly in the late rainy and early dry seasons. Individuals usually flower twice per year (R. Foster, pers. comm.). The fruits mature mostly in the early rainy season, from May to July.

The only common *Psychotria* that does not begin flowering after the first rains.

Mexico to Peru and Bolivia; Trinidad, Jamaica, and Cuba. In Panama, known from tropical moist forest in the Canal Zone, Bocas del Toro, Colón, San Blas, Veraguas, Los Santos, Panamá, and Darién, from premontane wet forest in Coclé, Panamá, and Darién, and from tropical wet forest in Colón and Darién.

Psychotria micrantha H.B.K., Nov. Gen. & Sp. 3:363, pl. 284. 1819

P. rufescens sensu auct. non H.B.K., non H. & B. ex R. & S.

Shrub, 1.5–5 m tall, softly short-pilose to hirtellous nearly throughout. Stipules ovate, 1–1.5 cm long, bilobed at apex, caducous, one lobe slightly longer than the other; petioles 5–25 mm long; blades oblong-elliptic or obovate,

acuminate, acute to obtuse at base, 12–30 cm long, 5.5–13 cm wide, short-pilose especially on veins below; midrib often ferruginous. Inflorescences terminal or upper-axillary, solitary, to 15 cm long; peduncles 5–8 cm long; flowers 5-parted, clustered at ends of short branches; calyx short, obscurely lobed; corolla white, ca 4 mm long, lobed somewhat past middle, glabrous or puberulent outside, the tube hispid inside at point of staminal attachment, the lobes recurved; stamens included; filaments adnate to basal half of tube; anthers 0.8–1.3 mm long, rigidly attached to filament below middle; style to 3.2 mm long and exserted above throat, the branches short and thick. Berries globose to ellipsoid, 3–4 mm long, pubescent; seeds 2, the inner face flat, the outer face 5-ribbed. *Croat 6235.*

Uncommon, at least as an adult plant in the forest; juvenile plants are more abundant. Flowers mostly from June to August, rarely earlier or later. The fruits mature mostly from August to October.

Nicaragua to southwestern Colombia, Venezuela, Ecuador, and Peru. In Panama, known from tropical moist forest in the Canal Zone, Bocas del Toro, Colón, San Blas, Coclé, and Darién. Reported from premontane rain forest in Costa Rica (Holdridge et al., 1971).

See Fig. 532.

Psychotria pittieri Standl., Contr. U.S. Natl. Herb. 18:132. 1916

P. dispersa Standl.

Shrub, usually 1–1.6 m tall; stems and inflorescences moderately pubescent with short, mostly erect, stiff trichomes. Stipules bilobed to base, the lobes linear, 5–7 mm long, subpersistent, green, soon becoming brown; petioles mostly 5–10 mm long; blades mostly elliptic to obovate, acuminate, acute at base, 4–12 cm long, 1.5–4.5 cm wide, glabrate above except on midrib, appressed white-pubescent below especially on veins. Panicles terminal, pendent, 2–10 cm long; flowers subtended by a slender bract to ca 3 mm long; calyx lobes ca 0.5 mm long, acute; corolla white, 4–6 mm long, puberulent outside, the throat pale yellow, the tube minutely pubescent at point of staminal attachment at apex of tube, the lobes strongly recurved; stamens regularly arranged and slightly longer than style, or both style and stamens weakly exserted, or the stamens held in a row on one side of corolla and long-exserted above throat, the style included; style densely papillate-puberulent; stigma bifid, ca 1 mm long, thick. Berries subrotund, 5–6 mm diam, minutely pubescent, fleshy, violet becoming blue at maturity, the pericarp inflated; seeds 2, ± obovate, to 2.7 mm long, brown, the inside face flat, the outside face rounded, obscurely 5-ribbed. *Croat 4107.*

Frequent in the forest. Flowers from March to September, but mostly from May to July. The fruits mature mostly from July to September.

Distinguished by its small size, its pubescence, and its small, pendent inflorescence.

Belize, Costa Rica, and Panama. In Panama, known from tropical moist forest on BCI and in Bocas del Toro and San Blas and from premontane wet forest in Coclé.

Psychotria psychotriifolia (Seem.) Standl., Contr.
   U.S. Natl. Herb. 18:133. 1916

Shrub, to 1(2) m tall; young stems, petioles, and parts of inflorescences crisp-pubescent to glabrate. Stipules broad, 7–15 mm long, sharply bilobed at apex, turning brown, caducous; petioles obscure or to 2 cm long; blades oblanceolate, abruptly acuminate, gradually tapered to base and decurrent onto petiole, 9–18 cm long, 3–6 cm wide, glabrous above, puberulent on petiole and major veins below, bicolorous; lateral veins loop-connected near margin. Flower heads dense, short-pedunculate, bracteate, terminal or axillary, the bracts often longer than flowers; flowers sessile, to 4.3 mm long; calyx lobes ± irregular, to ca 1 mm long, acuminate; corolla lobes spreading, the apex cucullate, lobed on outside, the tube very short; stamens attached at apex of tube, interspersed with and exceeded by dense white trichomes; style shortly exserted, bilobed, papillose-puberulent. Berries ovoid to oblong, ca 7 mm long, red, the mesocarp thick, fleshy, sweet; seeds 2, oblong-elliptic, the inside face flattened, the outside face 5(7)-ribbed. *Croat 12883.*

Rare; collected once along the shore south of Fairchild Point. Flowers mostly in the early rainy season, especially in June, but also in the dry season, especially in early dry season (December). The fruits mature throughout the year, but mostly in the middle of the rainy season.

The species is related to *P. granadensis,* but differs in having the prominently lobed calyx, the short corolla with a very reduced, subrotate tube, the capitate, densely flowered heads, and the minutely puberulent veins of the lower leaf surface.

Costa Rica to Colombia, Venezuela, and Ecuador. In Panama, known from tropical moist forest in the Canal Zone, Bocas del Toro, Colón, San Blas, Veraguas, Panamá, and Darién, from premontane wet forest in Chiriquí and Coclé, and from tropical wet forest in Colón.

See Fig. 533.

Psychotria pubescens Sw., Prodr. Veg. Ind. Occ.
   44. 1788
   P. hebeclada DC.

Shrub, usually ca 2 m tall; stems and peduncles hispid. Stipules 3–5 mm long, deeply bifid to near base, the lobes linear, subpersistent; petioles obscure or to 3 cm long; blades mostly elliptic to oblong-elliptic, acuminate, obtuse to acute and long-decurrent at base, 8–21 cm long, 2–9.5 cm wide, glabrate or minutely asperous above (the trichomes often restricted to veins), hispidulous below especially on veins. Inflorescences terminal, erect, paniculate, rounded or elongated, 3–9 cm long; peduncles to 6 cm long; inflorescences and pedicels purplish; flowers heterostylous; calyx lobes acute, ca 1 mm long; corolla white, 4–7 mm long, 5-lobed, the lobes often tinged with violet near base inside, puberulent outside, villous inside about midway; short-styled flowers with the stamens 5–6.5 mm long, held to one side of tube opposite 2 lobes (less commonly regular), the filaments fused to corolla tube in basal half, the anthers exserted at anthesis (owing

to spreading of corolla lobes), the style at first much shorter than stamens, becoming weakly exserted opposite anthers, its 2 short branches densely papillate; long-styled flowers similar to short-styled ones but with the style exserted 4–5 mm, the anthers included and evenly spaced in tube. Berries ± flattened-globose, 4–7 mm diam, minutely pubescent, bilobed, purple or black at maturity; seeds 2, oblong-hemispherical, ca 3.3 mm long, the outside face usually with 3(5) ribs, the inside face with a longitudinal groove. *Croat 10846.*

Occasional, most frequently encountered in clearings or open areas. Flowers mostly in the early rainy season, in June and July, rarely earlier or later in the year. The fruits mature from July to December, mostly by September or October.

Mexico to Colombia and Ecuador; West Indies. In Panama, ecologically variable; known from premontane moist forest in the Canal Zone and Panamá, from tropical moist forest in the Canal Zone, Bocas del Toro, Colón, San Blas, Veraguas, Panamá, and Darién, from premontane wet forest in Chiriquí and Coclé, and from tropical wet forest in Colón.

Psychotria racemosa (Aubl.) Raeuschel, Nomen. Bot.
   ed. 3. 56. 1797

Shrub, usually less than 2 m tall; stems, lower leaf surfaces, and inflorescence branches usually puberulent. Stipules bilobed nearly to base, the lobes widespread, 1–1.5 cm long, linear, persisting, turning brown; petioles obscure or to 2 cm long; blades elliptic or oblong-elliptic, acuminate, attenuate and decurrent at base, mostly 8–19 cm long, 3–7.5 cm wide, bicolorous; lateral veins markedly impressed above, raised below, the reticulate veins very conspicuous. Inflorescences short-pedunculate, mostly 2–5 cm long, ca as long as broad, usually terminal; flowers 4–6 mm long, 4- or 5-parted; corolla white or greenish, papillate-puberulent outside and on lobes inside, the throat yellowish, the tube villous above insertion of stamens, the lobes cucullate, spreading, greenish; stamens included, white or purplish, attached just below throat; style and stigma papillate-puberulent; style becoming exserted at anthesis; stigma 5-lobed. Berries somewhat congested, rounded, 5–6 mm diam, green becoming orange and finally purple or black at maturity; seeds 5, ca 2.7 mm long, shaped like segments of an orange, the outside face slightly costate. *Croat 6189a, 15116.*

Frequent in the forest. Flowers from May to July, mostly in June. The fruits mature from August to December, mainly in September.

Distinguished by the closely congested cluster of usually five-celled fruits, whereas all other species have two hemispherical seeds. The mature black fruits disappear quickly.

Costa Rica to Colombia, Venezuela, the Guianas, Brazil, and Bolivia. In Panama, known from tropical moist forest in the Canal Zone, Bocas del Toro, Colón, San Blas, Chiriquí, Los Santos, Panamá, and Darién, from premontane wet forest in Coclé, Panamá, and

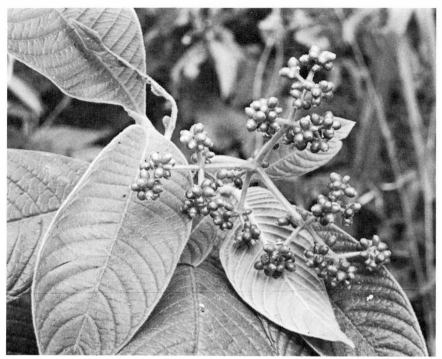

Fig. 532. *Psychotria micrantha*

Fig. 534. *Psychotria racemosa*

Fig. 533. *Psychotria psychotriifolia*

Darién, and possibly from premontane rain forest in Los Santos.

See Fig. 534.

Psychotria uliginosa Sw., Prodr. Veg. Ind. Occ. 43. 1788

Shrub or subshrub, to 2 m tall, ± glabrous; stems usually simple. Stipules obtuse or truncate, 1.5–3.5 mm long, persistent, connate into a horny sheath; petioles 4–7 cm long; blades obovate to oblong-elliptic, short-acuminate, attenuate at base, 15–30 cm long, 7–11 cm wide; major veins 11–17 on each side. Panicles axillary or subterminal, to 7(13) cm long; peduncles 2.5–5(9) cm long; bracts 2–3.5 mm long; flowers sessile or subsessile; calyx cylindrical, ca 1 mm long, ± truncate; corolla white, ca 4 mm long, the lobes pink, reflexed, horned at tip; stamens 5, attached to corolla near mouth of tube; filaments short; anthers oblong, ca 1 mm long; stigmas 2. Berries oblong, capped by the persistent calyx, 6–7 mm long, red at maturity; seeds 2, hemispherical, the inside face flat, occasionally with a median ridge evident on only 1 seed, the outside face 3-ridged, the middle ridge more prominent. *Starry 90.*

Known only from one collection on Wheeler Trail; possibly no longer present on the island. Flowers in the rainy season (July to December). Mature fruits are present much of the year but mostly in the late rainy and early dry seasons.

Mexico to Colombia, Ecuador, and the Guianas; West Indies. In Panama, known mostly from wetter parts of tropical moist forest or at higher elevations; known also from tropical moist forest on BCI and in Bocas del Toro, San Blas, Veraguas, Los Santos, Coclé, and Darién, from premontane wet forest in Colón, Veraguas, Coclé, Panamá, and Darién, and from premontane rain forest in Panamá.

RANDIA L.

Randia armata (Sw.) DC., Prodr. 4:387. 1830

Jagua macho, Miel quema, Rosetillo

Dioecious tree, to 8(10) m tall, nearly glabrous; branches often with stout spines at ends and sometimes at nodes. Leaves sessile or nearly so; stipules lanceolate to vaguely scoop-shaped, 5–10 mm long, caducous; blades oval or obovate, acuminate, attenuate to cuneate at base, 11–22 cm long, 4–10 cm wide, frequently with tufts of trichomes in vein axils below. Flowers white, in clusters of (1)2–4(8) at ends of branches; pedicels slender, 1–2 cm long, sparsely pubescent; calyx 5-lobed, the lobes

spreading and ca 5 mm long; corolla glabrate but with short-villous pubescence in vicinity of anthers, white, the tube greenish, usually 2.5–3 cm long, the limb to ca 3 cm wide, the lobes 5, rounded and oblique at apex; anthers 5, sessile, mounted on rim of tube, ca 5 mm long; style as long as the corolla tube; stigma ca 3 mm long, bilobed, the lobes ca 2 mm wide. Fruits baccate, subglobose to oblong, somewhat longer than broad, 3–4 cm long, 2.5–3 cm wide, densely covered with minute lenticels, becoming pale yellow-orange at maturity; seeds numerous, ca 1 cm long and less than 2 mm thick, in 4 separate longitudinal stacks, ± 3-sided, much flattened. *Croat 7378, 9131.*

Common in the forest. Flowers mostly synchronously between March and June, mostly in April and May; flowering may be completely finished for most individuals in 3 or 4 days. The fruits mature mostly from July to December. Sporadic individuals may flower and fruit at other times of the year. Plants lose their leaves just before flowering, and the flowers usually appear concurrently with new, usually fully formed leaves.

The fruits are eaten by white-faced monkeys (Hladik & Hladik, 1969).

Mexico to Colombia, Venezuela, the Guianas, Brazil, and Bolivia; West Indies. In Panama, known from tropical dry forest in Panamá (Taboga Island), from tropical moist forest in the Canal Zone, Bocas del Toro, Colón, San Blas, Los Santos, Panamá, and Darién, from tropical wet forest in Darién, and from premontane wet forest in Coclé, Panamá, and Darién.

Randia formosa (Jacq.) K. Schum. in Mart., Fl. Brasil. 6(6):342. 1889

Dioecious shrub or small tree, unarmed, widely branched; stems glabrous in age, slender. Leaves clustered at ends of small lateral branches; stipules short, broadly triangular, often ± scoop-shaped, acute, 1.5–3 mm long, brown, persisting below leaf clusters; petioles obscure or to 1 cm long; blades oblanceolate-elliptic, acuminate, attenuate and decurrent at base, mostly 3–11 cm long, 1.5–3 cm wide, sparsely pubescent and dark green above, duller and more densely pubescent below. Flowers single at branch ends, densely sericeous; calyx 13–20 mm long, lobed, persisting in fruit, the lobes 5, linear, 8–10 mm long; corolla tubular, 11–19 cm long, the tube greenish, papillose inside at apex, soon becoming velutinous in upper part, then glabrous, the lobes 5, white, 3–5 cm long, tapered to a slender tip; filaments adnate to tube; anthers 5, ca 5.3 mm long, attached at flared apex of tube; style with ascending trichomes near apex; stigma broadly bilobed, held slightly above anthers. Fruits baccate, ±

KEY TO THE SPECIES OF RANDIA

Flowers solitary at ends of branches; corolla 11–19 cm long; fruits ± spherical, green with whitish bands . . . . *R. formosa* (Jacq.) K. Schum.
Flowers usually in clusters of 2–4 at ends of branches; corolla less than 6 cm long; fruits longer than broad, green turning yellow at maturity . . . . *R. armata* (Sw.) DC.

Fig. 535.
Randia formosa

Fig. 536. *Tocoyena pittieri*

Fig. 537. *Uncaria tomentosa*

spherical, to 25 mm diam, dark green, with several broad white bands extending from apex to middle or beyond, sparsely covered with appressed trichomes; seeds numerous, flattened laterally, stacked in 4 rows, white. *Croat 11296, 11888.*

Occasional; known from the shore near Colorado Point and in the second growth west of the tower. It occurs at the margins of forest elsewhere in the Canal Zone. Flowers in July and August. The fruits mature in August and September.

Panama to Colombia, Venezuela, Peru, and western Amazonian Brazil; Tobago. In Panama, known from tropical dry forest in Panamá (Taboga Island), from premontane moist forest in the Canal Zone, from tropical moist forest in the Canal Zone, Colón, Chiriquí, Panamá, and Darién, and from premontane wet forest in Darién.

See Fig. 535.

SABICEA Aubl.

Sabicea villosa Willd. ex R. & S. var. **adpressa** (Wernham) Standl., Publ. Field Columbian Mus., Bot. Ser. 7:52. 1930

Liana, the younger parts herbaceous; all parts pubescent, the trichomes stiff, ± appressed, sparse on leaf surfaces but conspicuous on veins below. Stipules broadly ovate, acute, 7–10 mm long, persistent; petioles 5–18 mm long; blades ovate to elliptic, acute to acuminate, obtuse to rounded at base, mostly 6–13 cm long and 2.5–7 cm wide, purplish below when young. Flowers sessile, 5(6)-parted, in short, axillary clusters to 3 cm long; calyx lobes 5–7 mm long, recurved, persisting in fruit; corolla white, 8–10 mm long, the tube glabrous inside except near anthers, the lobes short; anthers 5(6), 1.3–2.3 mm long, opening in bud, fused to tube near base of corolla lobes; style variable, to 5(8) mm long, the branches 5, to 3 mm long, held well below anthers within the narrow corolla tube; nectar copious. Berries globose, ca 10 mm diam, sparsely to densely pubescent, red becoming dark purple; seeds many, tan, minute. *Croat 11725.*

Occasional in older clearings, on trails, and at the margin of the forest along the lake; less commonly climbing to the top of the forest canopy and sometimes rooting in water (*Shattuck 704*). Flowers and fruits to some extent throughout the year, but mostly in the dry season, from December to April, and again from June to August. Most fruits mature in the middle to late rainy season. Flowers and mature fruits are common on the same stem.

The genus is so poorly known that the geographic limits are uncertain. The species is found from Belize and Honduras south to Colombia, Venezuela, and Guyana, and the variety *adpressa* ranges from Costa Rica to Colombia. In Panama, the species consists of two intergrading varieties: the typical variety has spreading trichomes and is found in Bocas del Toro; var. *adpressa,* which is more widespread, is known from tropical moist forest in the Canal Zone, Bocas del Toro, San Blas, Panamá, and Darién, from premontane moist forest in the

Canal Zone, and from premontane wet forest in Panamá and Darién. Intermediate forms are found in Bocas del Toro and central Panama on the Atlantic slope.

SPERMACOCE L.

Spermacoce tenuior L., Sp. Pl. 102. 1753

Erect or sprawling herb, glabrous or nearly so, rarely more than 60 cm tall. Leaves short-petiolate; stipules of several slender bristles; blades lanceolate or linear-lanceolate, acuminate, gradually attenuate at base, 3–7 cm long, 5–15 mm wide. Inflorescences 1-sided axillary clusters, bearing many flowers; bracts borne among the sessile flowers, slender but short and inconspicuous; sepals triangular, 0.5–1 mm long (variable throughout the range); corolla white, 1.5–2 mm long, lobed to about middle, conspicuously villous in throat; stamens about as long as tube; pistillate parts not seen. Capsules usually obovoid, ca 3 mm long, glabrous. *Woodworth & Vestal 476.*

Apparently rare; collected only once on the island. Seasonal behavior uncertain. Temperate collections show flowering and fruiting from May to October. A few tropical records show flowering and fruiting mostly in the early rainy season.

The plant is inconspicuous and can be confused with *Diodia ocimifolia,* which can be distinguished by the pubescence on at least the leaves and capsules.

Southwestern Ohio and southern Missouri to Florida, Mexico, Central America, and South America as far south as Argentina and Uruguay; West Indies. In Panama, apparently infrequent; known only from tropical moist forest in the Canal Zone, San Blas, and Panamá, but no doubt growing elsewhere in tropical moist forest.

TOCOYENA Aubl.

Tocoyena pittieri (Standl.) Standl., Contr. Arnold Arbor. 5:151. 1933

Tree, to 15(25) m tall, ± glabrous; stems with prominent leaf scars. Leaves clustered at ends of branchlets; stipules triangular, 5–10 mm long; petioles 1–2 cm long; blades obovate, obtuse to acute at apex, cuneate at base, 18–28 cm long, 7–14 cm wide, sparsely pubescent below with pocketlike domatia in most vein axils. Panicles congested, terminal, nearly sessile or on pedicels to 8 mm long; flowers 5-parted; calyx green, obscurely toothed; corolla yellow, becoming yellow-orange, the tube 8–12 cm long, ca 5 mm diam, the lobes obovate, ca 1.5 cm long, rounded at apex, spreading at anthesis, reflexed in age; anthers sessile, attached at base to rim of corolla; stigma exserted, the lobes 2, very thick, becoming reflexed after anthers have fallen. Fruit baccate, globose to ellipsoid, to 11 cm long and 8(12) cm diam, green, the exocarp to ca 9 mm thick, shiny, yellow inside; seeds many, ovate, ca 19 mm long and 3 mm thick, brown, flattened, irregular, embedded in a fleshy matrix. *Croat 8529, 14877.*

Occasional, in the forest, at least in the vicinity of the Laboratory Clearing. Flowers from March to July, mostly from April to June. The fruits take about a year to mature and are eaten by white-faced and spider monkeys in April and September (Hladik & Hladik, 1969).

Southern Costa Rica to northern Colombia. In Panama, known from tropical moist forest in the Canal Zone, Panamá, and Darién.

See Fig. 536.

UNCARIA Schreb.

Uncaria tomentosa (Willd.) DC., Prodr. 4:349. 1830
Bejuco de agua

Large armed liana; trunk to 25 cm diam near the ground; outer bark thick, deeply fissured, coarse, dark brown; inner bark somewhat orange; younger stems square, ± pubescent, armed at nodes with stout, recurved spines to ca 1.5 cm long and ca 1 cm wide at base. Stipules ± ovate, ca 8 mm long; petioles 1–1.5 cm long; blades ovate to elliptic, bluntly acuminate, rounded to cordate at base, 4–13 cm long, 3–8 cm wide, glabrous above except strigillose on midrib, whitish-strigillose below, the trichomes longer and sparser on veins. Flower heads very dense, globular, pedunculate, axillary or terminal, to 2.5 cm diam, terminating stiff, spinelike peduncles or stiff branches of open panicles ca 10 cm long and 12 cm wide; peduncles 2–4 cm long, flattened, to ca 8 mm wide at base; flowers 5-parted, with sweet aroma, golden-yellow, sessile; calyx cylindrical, ca 2 mm long, truncate, sericeous; corolla tubular, ca 8 mm long, tomentose outside, papillate inside, the lobes to 1.5 mm long, rounded, spreading; stamens fused to rim of corolla tube; anthers exserted, to ca 1.5 mm long; style thickened at apex, unlobed, shorter than anthers in bud, later exserted to ca 3 mm, then often recurving into flower. Capsules ellipsoid, 6–8(17) mm long, 3–4 mm wide, brown-tomentulose, densely clustered in spherical heads ca 1.5 cm diam, irregularly dehiscing along several lines, only a few of the ovaries in the head developing into fruits; seeds many, elliptic to round, flattened, ca 0.5 mm long, bearing a slender scarious wing ca 1.5 mm long on both ends. *Croat 8288, Montgomery 11.*

Common, growing high in the canopy to 30 m or more. Flowers from January to March, mostly in February. The fruits mature quickly, with the seeds dispersed chiefly in the late dry season, but with at least some seeds persisting into the rainy season, sometimes as late as September.

The fruits form a basketlike structure at maturity, similar to those of *Aristolochia* (44. Aristolochiaceae), which permits the seeds to be slowly carried away in the wind.

Central America, Colombia, Venezuela, and the Guianas; Trinidad. In Panama, known from tropical moist forest in the Canal Zone (Atlantic slope), Bocas del Toro, Panamá, and Darién.

See Fig. 537.

WARSCEWICZIA Klotzsch

Warscewiczia coccinea (Vahl) Klotzsch, Monatsb. Akad. Ber. 497. 1853
Crucero, Guna

Tree, to 15 m tall, nearly glabrous. Leaves sometimes minutely pubescent on petioles and veins; stipules ovate, long-acuminate, to 3 cm long, entire, persistent; petioles to 2.5 cm long, stout; blades oblong-elliptic to elliptic, acuminate, obtuse at base, 15–30(50) cm long, 6–15(20) cm wide. Inflorescences terminal and axillary, to 80 cm long, ca 3 cm wide, the cymes short-pedunculate, to 2 cm apart on floral rachis; flowers sessile, 5-parted, protogynous, developing acropetally, densely congested; one flower in each cluster usually bearing a calycine lobe, this red, leaflike, ± elliptic, long-stipitate, to 12 cm long and 4 cm wide; other calyx lobes to 1 mm long, rounded; corolla funnel-shaped, 6–8 mm long, red-orange, lobed to ca one-third its length, the tube weakly pubescent on the outside and at point of staminal attachment inside, the lobes erect, ovate, blunt; stamens attached somewhat below base of lobes; filaments thick; anthers ± oblong, ca 1.5 mm long, exserted, attached near base, often held horizontally; pollen arachnoid, tacky, orange; style 4.5–5.5 mm long, the lobes 2, short and thick, exserted from corolla before it is fully expanded (while anthers unopened). Capsules subglobose, 2-valved, ca 5 mm long, persisting; seeds minute, many, probably wind dispersed. *Croat 6032, 7340.*

Rare; seen only along the shore. Flowers from the late dry season to the middle of the rainy season, especially in July and August, sporadically at other times of the year. The fruits mature quickly and are probably mostly dispersed during the late rainy and dry seasons.

The expanded calycine lobe, though turning brown and persisting, apparently serves no direct dispersal function, since the capsule valves persist after the seeds have been shed. It may cause fluttering of the capsule that dislodges the seeds.

Mexico to Colombia, Venezuela, the Guianas, Brazil, and Bolivia; Trinidad; introduced ornamentally to Africa. In Panama, a characteristic tree of tropical moist forest (Holdridge & Budowski, 1956); known from tropical moist forest in the Canal Zone, Colón, Veraguas, Panamá, and Darién, from premontane wet forest in Chiriquí, Veraguas, Coclé, Panamá, and Darién, and from tropical wet forest in Coclé and Darién.

131. CUCURBITACEAE

Herbaceous tendriled vines (sometimes woody at base), the tendrils simple or branched, spiraled, arising from upper side of petiole base; stems often angled or striate. Leaves alternate, petiolate; blades simple, palmately divided, or merely palmately lobed, entire or sharply serrate, often with apiculate teeth; venation palmate; stipules lacking. Flowers 5-parted, unisexual (plants

KEY TO THE TAXA OF CUCURBITACEAE

Mature leaves divided to the base (each leaflet borne on a distinct petiolule); fruits ± cylindrical:
  Plants with long sparse trichomes; calyx orange, the lobes acuminate, spreading; corolla yellow
    . . . . *Gurania coccinea* Cogn.
  Plants glabrous or nearly so; calyx green, the lobes short, blunt; corolla red or orange . . . . . *Psiguria*
Mature leaves not divided to the base (sometimes deeply lobed but not completely to base); fruits cylindrical or not:
  Calyx orange; stems sparsely pilose; flowers densely aggregated, capituliform or umbelliform
    . . . . *Gurania* (in part)
  Calyx green; stems not pilose; inflorescences not capituliform or umbelliform:
    Most adult leaves with 5 or more lobes:
      Leaves asperous (at least when dry); anthers straight or slightly curved; fruits ovoid, mottled with light and dark green stripes, indehiscent; flower pedicels lacking leafy bracts
        . . . . *Melothria trilobata* Cogn.
      Leaves smooth; anthers conduplicate; fruits turbinate, beaked at apex, orange, fleshy, splitting to expose red seeds; flower pedicels with leafy bract near base . . . .
        . . . . *Momordica charantia* L.
    Most leaves unlobed or to 3-lobed:
      Leaf blades entire, neither toothed nor lobed; corolla less than 2 mm long; plants forest vines; leaves not asperous above . . . . *Sicydium coriaceum* Cogn.
      Leaf blades with the lateral veins ending in sharp or gland-tipped teeth and/or the blades lobed; corolla more than 4 mm long:
        Leaf blades mostly less than 7 cm long, usually not lobed, deeply cordate with a narrow sinus; fruits many-seeded, less than 3 cm long; slender vine . . . . . . *Melothria pendula* L.
        Leaf blades usually more than 8 cm long, lobed or not, not cordate or cordate with a broad sinus; fruits 2–6-seeded or many-seeded and more than 8 cm long:
          Leaf blades mostly obtuse to truncate at base (very slightly cordate in *C. denticulata*); fruits probably always less than 3 cm long . . . . *Cayaponia*
          Leaf blades cordate with a broad sinus; fruits globose or gourdlike, more than 7 cm diam:
            Blades ± entire except for a few glandular teeth; stems glabrous; seeds 6, disk-shaped, more than 4 cm wide . . . . *Fevillea cordifolia* L.
            Blades shallowly or deeply 3(7) lobed; stems densely short-villous; seeds numerous, narrowly obovoid, less than 1 cm diam . . . . *Posadaea sphaerocarpa* Cogn.

dioecious or monoecious), actinomorphic, axillary, solitary or in cymose (often contracted) panicles; calyx united with the base of corolla into a hypanthium; corolla tubular, campanulate or ovate (*Fevillea*); staminate flowers with the stamens usually partially attached to corolla tube, 2 (*Gurania, Psiguria*), or 5 with unilocular anthers longitudinally dehiscent, or 5 variously fused: (a) into 2 pairs and a single stamen (the pairs appearing as simple stamens with 2-locular anthers), (b) into a single column with free or fused anthers variously disposed, or (c) with the anthers fused and the filaments free; pistillode often present; pistillate flowers with the ovary 1, inferior, the carpels basically 5, mostly reduced to 3(4); placentation parietal; ovules usually numerous (solitary in *Sicydium*), anatropous; style 1; style branches or stigmas equaling number of carpels. Fruits pepos (berrylike), usually indehiscent; seeds 1 to many, lacking endosperm.

Members of the family may be recognized by being herbaceous vines with tendrils, by their unisexual, usually tubular to bell-shaped flowers, by their unusually modified staminate flowers, and by their berrylike, many-seeded fruits.

Flowers are often unusually modified, and many no doubt require specialized pollinators. Both *Gurania* and *Psiguria* are pollinated in part by butterflies of the genus

Heliconius, which collect pollen (Ehrlich & Gilbert, 1973). Elsewhere *Gurania* is also pollinated by hummingbirds (L. Gilbert, pers. comm.). Some South American species of *Cayaponia* are bat pollinated (Vogel, 1958).

The fruits of *Melothria, Cayaponia,* and *Momordica* are probably dispersed by birds of clearings, those of *Sicydium* by forest birds, and the larger-fruited forest species, *Fevillea, Gurania,* and *Psiguria,* by mammals. Squirrels and other rodents may be important in the dispersal of seeds of *Gurania* (L. Gilbert, pers. comm.). *Fevillea cordifolia* is dispersed chiefly along waterways, according to Ridley (1930).

About 100–110 genera and 650–850 species; mainly in the tropics and subtropics.

CAYAPONIA Manso

Cayaponia denticulata Killip ex C. Jeffrey, Kew Bull. 25:206. 1971

Vine, climbing into canopy; stems slender, conspicuously ribbed (at least on drying), densely crisp-villous to puberulent at least in grooves. Petioles 2–2.5 cm long, densely villous, some of the trichomes stouter and uncinate; blades ovate, abruptly acuminate, truncate to weakly

KEY TO THE SPECIES OF CAYAPONIA

Leaves divided nearly to base into 3 leaflets .... *C. granatensis* Cogn.
Leaves entire or lobed, not divided to base:
   Lower blades with densely pubescent veins, lacking winglike margins on drying ....
      .... *C. denticulata* C. Jeffrey
   Lower blades with the veins glabrous (except along margins), the blades drying with conspicu-
     ous winglike margins:
      Corolla ca 15 mm long; fruits with 6 seeds .... *C. glandulosa* (Poepp. & Endl.) Cogn.
      Corolla less than 5 mm long; fruits with 2 or 3 seeds .... *C. racemosa* (P. Mill.) Cogn.

cordate at base, 8–15 cm long, 5–9 cm wide, chartaceous, the upper surface glabrate except on major veins, the lower surface densely short-villous, the margins entire except for a regular series of free-ending veins forming minute teeth; veins 3 at base, pinnate above, the basal lateral veins extending ca two-thirds the length of blade, all veins prominently raised on underside. Flowers and fruits not seen. *Montgomery 96.*

Collected once in the canopy near the Laboratory Clearing.

The collection was determined by C. Jeffrey. The identification remains dubious until fertile material is collected.

Cayaponia glandulosa (Poepp. & Endl.) Cogn. in A. DC., Monogr. Phan. 3:755. 1881

   C. poeppigii Cogn.

Monoecious tendriled vine; stems pubescent to glabrate; tendrils 1- or 2-branched below middle. Petioles 2–8 cm long, puberulent; blades cordate to ovate in outline, entire or 3-lobed, 6–19 cm long, 3–18 cm wide, rounded to acute at apex, cordate to cuneate at base, pubescent on both surfaces, the veins usually glabrous except along the margin of the flattened, winglike portion, the upper surface pustulate, the lower surface with sessile glands near the base. Inflorescences axillary, paniculate or race-mose, occasionally 1-flowered; rachis to 20 cm long; pedicels 1–5 mm long; calyx campanulate, 5-lobed, ca 1 cm long, pubescent; corolla greenish-white, 5-lobed to below the middle, the lobes ovate-lanceolate to oblong, ca 1.5 cm long, pubescent; staminate flowers with the stamens 3, ca 1 cm long, the filaments filiform, the anthers folded, forming an irregular mass ca 0.7 mm long; pistillate flowers with the ovary ellipsoid, puberulent to glabrate, 4–6 mm long, the styles ca 1.5 cm long, the staminodia minute. Berries ovoid-ellipsoid to subglobose, ca 2 cm long and 1.5 cm diam, black, the surface smooth, glabrous; seeds 6–15, obovate, ca 6 mm long, ca 4 mm wide, compressed, brown and white mottled, the base somewhat truncate. *Kenoyer 573, Standley 41133.*

Reported by Standley, but probably no longer occur-ring on the island—it is less likely to reoccur than is the more common *C. racemosa.* Seasonal behavior not determined.

Panama and Colombia, Venezuela, Ecuador, Peru, and Bolivia. In Panama, known only from tropical moist forest in the Canal Zone, Bocas del Toro, Colón, Chiri-

quí, and Darién, from tropical wet forest in Colón (Río Guanche), and from either lower montane wet forest or lower montane rain forest in Chiriquí (Las Nubes).

Cayaponia granatensis Cogn. in A. DC., Monogr. Phan. 3:794. 1881

Dioecious vine; most of the vegetative growth ± herba-ceous, but arising from larger woody stems; younger stems and leaves glabrous; tendrils trifid from near base. Petioles 2.5–9 cm long; blades 3-lobed to near base, the lobes oblong to narrowly elliptic, long-acuminate (the acumen sometimes very narrow and to 2.5 cm long), 15–25 cm long, the center lobe 4–9 cm wide, the base truncate to cordate, both surfaces light-punctate, the lower surface also obscurely and minutely papillate; veins at base 3; juvenile leaves entire, ovate-oblong to oblan-ceolate, becoming 2- or 3-lobed. Staminate flowers in axillary fascicles of few flowers, 3–3.5 cm long, white with green lines; calyx campanulate, glabrous, the tube ca 2 cm long, fused to corolla, the lobes 5, slender, 2–3 mm long; corolla lobes 5, ca 1 cm long, 5–7 mm wide, usually 4- or 5-striate, scabridulous outside, crisp-villous inside and on margins; stamens apparently 3 (actually 5 with 2 pairs of fused stamens); pistillate flowers not studied. Berries globose, ca 1.5 cm diam, reddish (drying orange), glabrous at maturity except near the peduncle, the pericarp weakly coriaceous, ca 1 mm thick or less; seeds few, enveloped in a fleshy matrix, flattened, to 11 mm long, 7 mm wide, 2–3 mm thick. *Foster 2208, Croat 11640.*

Occasional, in the old forest, sometimes climbing trunks of trees into the canopy. Flowers at least in March and April and possibly as early as January. The fruits mature in the late dry and early rainy seasons (April and May).

Panama, Colombia, Venezuela, and Peru. In Panama, known from tropical moist forest on BCI and in Darién (Cana), from premontane wet forest in San Blas, and from tropical wet forest in Panamá. It is also known from Cerro Tute in Veraguas (life zone uncertain).

Cayaponia racemosa (P. Mill.) Cogn. in A. DC., Monogr. Phan. 3:768. 1881

Monoecious tendriled vine; stems grooved, glabrate to villous; tendrils 2-branched below the middle. Petioles 1.5–6 cm long; blades entire or lobed, cordate to acute at

base and abruptly decurrent onto petiole, 5–15 cm long, 3–15 cm wide, the lobes 3 or 5 (or 7), obtuse to acuminate, the lower surface and midrib of upper surface puberulent to hispid, the upper surface very asperous, with erect to scabrous pustular-based trichomes, the margins with remote gland-tipped teeth; lower midrib with a broad flattened rib along its margin. Flowers unisexual, solitary or few in axils or in axillary racemes, the pistillate flowers slightly smaller than staminate flowers; calyx broadly campanulate, ca 6 mm long, pubescent, the lobes 5, ovate; corolla greenish-white to yellow, 4–5 mm long, densely tomentose on outside, with long slender trichomes inside; stamens apparently 3 (4 fused in 2 pairs); filaments free; anthers coherent, one of them 1-celled; staminodia minute in pistillate flowers; ovary 3-locular (or 1-locular by abortion); stigmas 3, dilated, reflexed. Berries ellipsoid to ovoid, 1.3–2 cm long, 8–10 mm wide, orange to red at maturity, the surface smooth, glabrous; seeds usually 2 or 3, compressed-ellipsoid, ca 8 mm long, orangish-brown, roughened. *Bailey 577.*

A weed of clearings, possibly once common on the island but probably no longer occurring there. Flowers and fruits throughout the rainy season and in the early dry season (July to January). The fruits mature mostly from November to May.

Standley reported the species to be frequent in thickets (1928).

Florida and Mexico to Panama and Colombia and across northern South America to Venezuela, Guyana, Brazil, and Peru; Trinidad, West Indies. In Panama, known from tropical moist forest in the Canal Zone, Bocas del Toro, Colón, Veraguas, Coclé, Panamá, and Darién and from premontane wet forest in Chiriquí (Boquete).

FEVILLEA L.

Fevillea cordifolia L., Sp. Pl. 1013. 1753

Dioecious tendriled vine, climbing into canopy; stems glabrous, 5-costate; tendrils branched once near apex. Leaves simple; petioles 2–8 cm long; blades ovate-cordate, acuminate, 4–15 cm long, as broad as long, entire or usually with a few blunt glandular teeth, glabrate to weakly short-pubescent especially on veins above, the veins of lower surface drying flattened, with weblike

axillary pockets; venation palmate. Staminate inflorescences paniculate, terminal or axillary, mostly to 25 cm long; axes, branches, and pedicels densely short-puberulent; each branch bearing a fleshy, ovate to rounded bract at base; pedicels filiform, 2–6 mm long; staminate flowers yellow-green, 5-parted, 7–8 mm diam; calyx ca half as long as corolla, lobed to middle, weakly pubescent and ciliate, the lobes ribbed medially in basal half; corolla lobes broadly obovate, rounded at apex, ca 3–3.5 mm long, glabrous on outer surface, weakly pubescent inside and bearing a narrow flap of tissue medially in basal third of petal; stamens 5, free, extrorse, ca 1 mm long; anthers about as broad as long, equaling or longer than filaments; pistillate flowers similar to staminate, the ovary globose, 3–4 mm diam. Fruits globose or nearly so, smooth (drying with very minute bumps), 7–12 cm diam, bearing a circumferential scar one-fourth to one-third of the way down from apex when juvenile, nearer the apex on mature fruits; seeds 6, in 3 carpels in appressed pairs, lenticular, 5–5.5 cm long, 4–5 cm wide, to ca 1 cm thick but the margin very thin. *Croat 11918.*

Apparently uncommon; seen once in the canopy of the old forest south of Zetek Trail. Flowers from May to August. Mature fruits have been found on the ground during October.

The size of the fruits suggests that they are probably dispersed by larger frugivores, perhaps by monkeys. Some fruits with the thick rind broken open have been found on the ground, so they are possibly dispersed by floor foragers also.

Costa Rica to Ecuador and the Amazon basin; Greater Antilles. In Panama, known only from tropical moist forest in the Canal Zone, Chiriquí (Bartolomé), and Los Santos.

See Fig. 538.

GURANIA Cogn.

Gurania coccinea Cogn., Diagn. Cucurb. Nov. fasc. 1. 42. 1876
Ya te vi

Dioecious vine; stems 5–10-sulcate, sparsely white-pilose, tendrils simple. Leaves usually palmately trifoliolate (juvenile leaves sometimes simple and entire or deeply 2- or 3-lobed); petioles 2–6 cm long, sparsely white-

---

KEY TO THE SPECIES OF GURANIA

Mature leaves divided to the base, each leaflet borne on a distinct petiolule . . . . *G. coccinea* Cogn.
Mature leaves not divided to the base (leaves lobed but not to the base):
    Staminate flowers long-pedicellate; calyx puberulent or glabrate, lobed less than one-fourth
        its length, the lobes to 1.2 cm long, much shorter than hypanthium tube; leaves mostly
        3-parted; older stems woody, corky; flowers usually borne on older, leafless stems . . . .
         . . . . *G. megistantha* Donn. Sm.
    Staminate flowers nearly sessile; calyx densely villous, lobed more than half its length, the lobes
        to 4 cm long, equaling or longer than hypanthium tube; many leaves more than 3-parted;
        older stems herbaceous; flowers not borne on older, leafless stems . . . .
         . . . . *G. makoyana* (Lem.) Cogn.

Fig. 538. *Fevillea cordifolia*

Fig. 539.
Gurania makoyana

Fig. 540.
Gurania megistantha

pilose; petiolules to 5 mm long; leaflets acuminate, 12–18 cm long, 5–8 cm wide, sparsely pubescent above, more densely so on veins, pilose below especially on veins, the margins irregular, sparsely ciliate, with apiculate teeth; terminal leaflet elliptic to obovate, equilateral, acute at base; lateral leaflets strongly inequilateral, with a broad cordate lobe on outer side. Inflorescences sparsely pilose; flowers 5-parted; calyx orange; corolla yellow; staminate racemes solitary, umbelliform, pendent; peduncles 15–30 cm long; pedicels 4–15 mm long; flowers 10–15 mm long; calyx bulbous at base, lobed one-third to one-half of the way to base, the outer surface pilose, the inner surface papillose-puberulent, the lobes narrowly triangular, 3–5 mm long; corolla lobes densely papillose-puberulent on both surfaces, somewhat longer than calyx; anthers 2, included, sessile, dorsifixed; pistillate flowers solitary or few in axils or on leafless nodes of the flexuous stem, 5–8 cm apart, the stem then usually pendent for 1 m or more; pedicels less than 1 cm long at anthesis (to 2 cm in fruit); calyx to 1.5 cm long, fused to corolla almost throughout; corolla to 2 cm long, densely papillose-puberulent, the lobes 5, narrowly acute, imbricate, 6–9 mm long; style thick, almost as long as corolla lobes. Fruits oblong-elliptic, to 6.5 cm long, somewhat flattened, to 2.3 cm wide and 1.6 cm thick, pale green with lighter spots, becoming faded in age; seeds many, ovoid, ± flattened, white. *Shattuck 641, Croat 14077.*

Rare, in the older forest, often twining up tree trunks and then hanging pendent from branches. Flowering throughout the dry season (December to April). Mature fruits have been seen during May and June.

The fruits are very bitter before maturity.

Known only from Panama, at elevations to 1,000 m, from tropical moist forest in the Canal Zone (Atlantic slope), Bocas del Toro, and San Blas (Puerto Obaldía), from premontane wet forest in Coclé and Panamá, and from tropical wet forest in Colón, Veraguas, and Coclé.

Gurania makoyana (Lem.) Cogn., Diagn. Cucurb.
  Nov. fasc. 1. 17. 1876
  G. seemanniana Cogn.; *G. donnell-smithii* Cogn.
  Ya te vi

Dioecious vine; stems and leaves sparsely to densely white-villous; tendrils simple. Petioles 4–10 cm long; blades usually deeply (3)5(7)-lobed, 15–30(40) cm long, nearly as wide as long, cordate at base, chartaceous, the lobes acuminate, 4–7 cm wide, with apiculate teeth, the upper surface minutely pustulate; juvenile leaves often entire or shallowly lobed. Inflorescences sparsely to densely pilose; flowers 5-parted; calyx orange; corolla yellow; staminate inflorescences solitary, pendent; peduncles to 15 cm long; staminate flowers sessile, in subglobose clusters; floral rachis 1–2 cm long; calyx ovoid at base, densely villous, the lobes almost filiform, to 4 cm long; corolla fleshy, acute at apex, only slightly exceeding anthers, conspicuously papillose-puberulent, to 15(24) mm long, the tube constricted above anthers; anthers 2, included, dorsifixed, subsessile, narrowly ovate, 5–6 mm

long; pistillate flowers pedicellate, several per axil, on usually leafless nodes at the stem apex; pedicel and ovary together 1.5–2 cm long; pedicels 4–5 cm long in fruit; inner surfaces of calyx and corolla glabrous, otherwise much like staminate flowers. Fruits cylindrical, 4–6 cm long and 2–2.5 cm broad, green at maturity with the old flowers persisting; seeds many, depressed-ellipsoid, 6–9 mm long and 3–5 mm broad, white. *Croat 5674, 9408.*

Occasional, in open areas of the forest and along trails and margins of clearings. Flowering throughout the dry season and in the early rainy season (December to June), perhaps longer. Mature fruits have been seen in the late dry and early rainy seasons.

Guatemala to Panama, perhaps into South America. In Panama, known from tropical moist forest in the Canal Zone (both slopes), Bocas del Toro, San Blas, Chiriquí, Panamá, and Darién, from premontane wet forest in Panamá, and from tropical wet forest in Colón (Santa Rita Ridge), Veraguas, Coclé, and Darién.

See Fig. 539.

Gurania megistantha Donn. Sm., Bot. Gaz. (Crawfordsville) 33:251–52. 1902
  G. suberosa Standl.

Dioecious vine, growing into lower canopy; older stems stout, woody, with a thick, 5–10-sulcate, corky outer bark; younger stems sparsely pilose; tendrils simple. Petioles 3–7(11) cm long, moderately pilose; blades usually deeply trilobate, acuminate, cordate at base, 12–20 cm long, 10–18 cm wide, short, ± erect-pubescent especially on veins. Flowers unisexual, both sexes in pedunculate, capituliform racemes arising from leafless woody stems, rarely branched from short herbaceous branches, arising from the woody stems; peduncles nearly lacking or to 9 cm long, sparsely strigose; rachis 1–3 cm long; pedicels 1–2.5 cm long, puberulent; calyx orange, tubular, with a bulbous base, 1.5–3.5 cm long, 5-lobed to ca one-fourth its length, the lobes linear, 4–12 mm long, spreading at anthesis; corolla yellow, to 4 cm long, equaling or exceeding calyx at anthesis, fused to calyx tube, the lobes 5, narrowly acute, densely papillose-puberulent outside and on tube inside; staminate flowers with the stamens 2, included, to ca 2 cm long, tapered to apex, the thecae lateral, the connective prolonged and granular-puberulent; pistillate flowers with the stigma bilobed. Fruits similar to those of *G. coccinea* but borne in a tight cluster 4–5 cm long and as much as 2 cm wide. *Croat 5782, 12217.*

Occasional, in the forest. Flowering throughout the year; the flowers are seen more easily than the associated vegetation, which may be high in the canopy. The fruits are rarely seen, but have been recorded in May.

Costa Rica and Panama. In Panama, known from tropical moist forest in the Canal Zone (Atlantic slope), Bocas del Toro, and Panamá, from premontane wet forest in Coclé and Panamá, and from tropical wet forest in Colón, Veraguas, and Coclé.

See Fig. 540.

KEY TO THE SPECIES OF MELOTHRIA

Leaves not lobed or weakly lobed; plants common in clearings . . . . *M. pendula* L.
Leaves deeply 5-lobed; plants rare or absent from the island . . . .*M. trilobata* Cogn.

MELOTHRIA L.

Melothria pendula L., Sp. Pl. 35. 1753

M. guadalupensis (Spreng.) Cogn.

Very slender, small, monoecious vine; stems almost glabrous to weakly pubescent; tendrils simple. Petioles 1–4 cm long; blades cordate-ovate, acute to acuminate, mostly 4–7(10) cm long and to 7(10) cm wide, irregularly toothed, sometimes weakly 3–5-lobed, asperous above, the basal sinus open to nearly closed, to 2.5 cm deep. Flowers unisexual, yellow, 5–6 mm long; calyx ca 4 mm long, fused to corolla, the lobes 5, narrow; corolla papillose, the lobes 5, rounded, the limb to 1 cm wide; staminate flowers several, in long-pedunculate racemes, 1–3 cm long; stamens 3; anthers situated in the upper part of corolla tube; filaments fused to corolla in basal half; anthers ca 1.7 mm long, fringed marginally with orange papillae; pistillode globose; pistillate flowers solitary; pedicels to 3 cm long (to 4.5 cm in fruit); calyx and corolla as in staminate flowers; ovary ellipsoid, glabrous, smooth; styles 3, 2–2.7 mm long, connate; stigmas 3, linear. Fruits ovoid, 1.5–2.5 cm long, 1–1.5 cm wide, glabrous, black at maturity, sweet; seeds many, ovoid in outline, flattened, ca 5 mm long, greenish. *Croat 5250, 8277.*

Common in clearings, less so along the shore. Flowering and fruiting throughout the year, especially in the rainy season.

Tropical areas of continental America; Bahamas, West Indies. In Panama, known from tropical moist forest in the Canal Zone (Atlantic slope), Bocas del Toro, Colón, and San Blas, from tropical dry forest in Panamá (Taboga Island), and from tropical wet forest in Bocas del Toro (Santa Catalina).

Melothria trilobata Cogn. in Mart., Fl. Brasil.

4(6):26. 1878

Wild cucumber, Sandillita

Monoecious vine; stems glabrate to sparsely crisped-pilose; tendrils simple. Petioles 1–5 cm long, sparsely pilose and sometimes also with very short, granular trichomes; blades circular in outline, deeply 5–7-parted, cordate at base, 6–10 cm diam, densely short-pubescent on veins above, sparsely to densely pubescent below, bicolorous, asperous, often minutely white-pustulate, the lobes acute to acuminate and apiculate at apex, the lateral lobes with a basal auricle. Flowers small, 5-parted, similar to those of *M. pendula;* calyx campanulate to subcylindrical, ca 3 mm long, the lobes triangular, to 1 mm long; corolla yellow, the lobes ovate-oblong, obtuse, the outer surface villous, the inner surface glabrate; staminate

flowers in axillary, long-pedunculate racemes, the peduncles very slender, the rachis 1–3 cm long, the pedicels very slender, short or to 1 cm long, the stamens 3, free, the filaments short, the anthers oblong, densely ciliate; pistillate flowers solitary, the pedicels ca 1 cm long, the calyx and corolla as in staminate flowers, the ovary ovoid, glabrous, smooth, the styles 3, connate, the stigmas 3, linear, the staminodia 3, minute. Fruits ± ovoid, 4–5 cm long, almost as long as broad, mottled and striped with light and dark green, dark green at apex, light green at base; seeds whitish, flattened, narrowly ovate, ca 7 mm long and 3 mm wide. *Zetek 4996.*

Collected once on the island; possibly no longer occurring there, but to be expected in clearings and at the edge of the forest. Seasonal behavior not determined.

Mexico to Surinam. In Panama, known most commonly from tropical moist forest in the Canal Zone, Bocas del Toro, and Darién; known also from premontane moist forest in the Canal Zone (Ancón Hill) and from tropical wet forest in Colón (Miguel de la Borda).

MOMORDICA L.

Momordica charantia L., Sp. Pl. 1009. 1753

Balsam pear, Basamino

Slender monoecious vine; stems, petioles, and blades (especially on veins) sparsely short-villous; tendrils simple, leaf-opposing. Petioles mostly 1–2 cm long; blades usually deeply 5–7-lobed, subrounded, 3–6(12) cm diam, the margins weakly apiculate-toothed, the lobes mostly obtuse and mucronulate at apex, narrow at base, the outer lobes auriculate. Flowers 5-parted, yellow, unisexual, long-pedunculate; staminate flowers ca 1.5 cm diam; calyx pubescent, to 7 mm long, lobed to about middle, the lobes ovate, acute; petals oblong-obovate, rounded at apex, fused to calyx just below lobes, conspicuously veined and pubescent outside and on upper edge inside, the trichomes in part glandular; staminal cluster orange, 4.5–5 mm long; filaments 3, the basal one-third to one-half fused to corolla; anthers to 3 mm, sigmoid-folded, 1 pair borne on each of 2 filaments, a solitary anther borne on the third filament, the mass somewhat agglutinated; pistillate flowers not studied. Fruits turbinate, 5–12 cm long, pointed on both ends, orange, fleshy, echinate, the pericarp 2–3 mm thick, bursting irregularly to expose several seeds; seeds depressed-ellipsoid, roughened on the ends, ca 9 mm long and 5 mm wide, tan, the covering thin, sweet, tasty, bright red. *Croat 6372.*

Infrequent, in clearings. Flowers and fruits throughout the year.

The bright red seeds, displayed against the orange

interior of the fruit, are no doubt dispersed by birds.

In the tropics and subtropics of the Western and Eastern hemispheres. In Panama, known from tropical moist forest in the Canal Zone, Colón, San Blas, Chiriquí, Veraguas, Los Santos, Panamá, and Darién, from premontane dry forest in Los Santos, from premontane wet forest in Panamá, and from tropical wet forest in Chiriquí (San Bartolo Limite).

See Fig. 541.

POSADAEA Cogn.

Posadaea sphaerocarpa Cogn., Bull. Acad. Roy. Sci. Belgique, sér. 3, 20:477. 1890

Monoecious herbaceous vine, sparsely to densely short-villous on most parts; tendrils simple or bifid, thickened toward base. Petioles 3–12 cm long; blades ovate to orbicular, entire to usually weakly or deeply 3-lobed (to 7-lobed), acuminate, cordate at base, usually to 18 cm long and 19 cm broad, membranaceous, the basal sinus about as broad as long, the margins conspicuously and rather remotely toothed; major lateral veins nearly palmate. Flowers white, ebracteolate or minutely bibracteolate; staminate flowers in short, slender, few-flowered racemes; calyx tube campanulate, 8–9 mm long, the lobes ca 3 mm long; corolla yellow, 5-parted, rotate, puberulent outside, the lobes rounded, notched, 8–9 mm long; stamens 3 (1 unilocular, 2 bilocular), attached to corolla tube; anthers dorsifixed, white, 2 mm long, the thecae recurved at apex; pistillate inflorescences 2–5 cm long; calyx lobes 2 mm long; corolla segments ca 1 cm long, 4–5 mm wide; staminodia 5, obtuse, 1.5–2 mm long; ovary elliptic-ovoid, sparsely pubescent; style 3 mm long; stigmas 3, dilated, obcordate, subreflexed, ca 3 mm wide. Fruits globose, gourdlike, 8–10 cm diam, the pulp white, the cortex ± woody, the surface finely warty; peduncles tapered, swollen; seeds numerous, compressed, narrowly obovate, rounded at apex, attenuate at base, 12–14 mm long, 7–8 mm wide, 2–2.5 mm thick. *Shattuck 90.*

Old collections from in or near the Laboratory Clearing are the only record from the island; probably no longer occurring on the island. Seasonality uncertain. Flowers at least in the late rainy and early dry seasons. Mature-sized fruits have been seen in the dry season and may be accompanied by flowers.

Guatemala to Panama, Colombia, the Guianas, Ecuador, and Brazil, doubtless elsewhere; Trinidad. In Panama, known from tropical moist forest in the Canal Zone and Panamá (Río Tapia).

PSIGURIA Arn. ex Hook.

Psiguria bignoniacea (Poepp. & Endl.) Wunderlin, Phytologia 38:219. 1978
Anguria pachyphylla Donn. Sm.

Dioecious canopy vine, ± glabrous except for sparse pubescence on stems, denser at nodes; stems stout; tendrils simple. Leaves palmately trifoliolate, deeply cordate, thick, coriaceous; petioles 2.5–5 cm long; petiolules 1–2 cm long; terminal leaflet irregularly broadly elliptic, acuminate, acute at base, 12–21 cm long, 8.5–13 cm wide; lateral leaflets asymmetrical, the outer margins irregular, 9–19 cm long, 6–10 cm wide. Flowers red to orange, unisexual; staminate flowers in long-pedunculate spikes; peduncles to 27 cm long; floral rachis 1–2 cm long; calyx tubular, ca 11 mm long, 3–4 mm wide, the lobes 5, short, recurved; corolla orange, extending less than 5 mm beyond calyx, the lobes 5, ± obovate, densely short-pubescent, the tube with a few long trichomes between stamens; anthers 2, sessile, dorsifixed, the connective somewhat produced; pistillate flowers solitary or few in leaf axils, otherwise like staminate flowers. Fruits similar to those of *P. warscewiczii. Croat 8082, 16518.*

Rare, known only from the vicinity of Snyder-Molino Trail 200. Flowering season uncertain. Probably flowers to some extent all year. Single plants certainly flower for several months—the plant over Snyder-Molino Trail flowered from at least February to September 1970, but may not have flowered at all during 1971.

Southern Mexico to Peru; at elevations between 170 and 1,000 m. In Panama, known from tropical moist forest in the Canal Zone, Bocas del Toro, Los Santos, and Darién, from premontane wet forest in Colón and Panamá, and from tropical wet forest in Veraguas and Panamá.

See Fig. 542.

Psiguria warscewiczii (Hook.f.) Wunderlin, Phytologia 38:219. 1978
Anguria warscewiczii Hook.f.

Dioecious vine, ± glabrous; stems slender, usually growing over shrubs and trees in the understory; tendrils simple. Leaves palmately trifoliolate, thin; petioles 3–6 cm long; terminal leaflet ± irregularly elliptic, acuminate, cuneate at base, 11–17 cm long, 5–8 cm wide; lateral leaflets asymmetrical bilobed to near middle, the lobes acuminate or rounded, 11–15 cm long, 5.5–10 cm wide. Flowers unisexual, to 2.5 cm long, the limb 1.5–2.5 cm wide; staminate flowers in axillary racemes; peduncles to 27 cm long; pedicels to ca 5 mm long at anthesis;

KEY TO THE SPECIES OF PSIGURIA

Leaves thick, coriaceous when dry; lateral leaflets often more than 15 cm long, usually only shallowly lobed . . . . *P. bignoniacea* (Poepp. & Endl.) Wunderlin
Leaves very thin, papyraceous when dry; lateral leaflets less than 15 cm long, usually lobed to near middle . . . . *P. warscewiczii* (Hook.f.) Wunderlin

Fig. 541. *Momordica charantia*

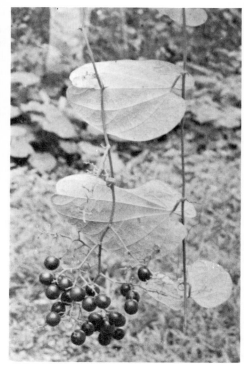

Fig. 544. *Sicydium coriaceum*

Fig. 542. *Psiguria bignoniacea*

Fig. 543. *Psiguria warscewiczii*

calyx tubular, to 2 cm long, slightly bulbous at base, light green, the lobes and warty spots on tube dark green, the lobes 5, short, blunt; petals 5, red-orange, rounded to obovate, 5–10 mm long, densely glandular-tomentose outside and near apex inside, the tube sparsely pubescent inside; anthers 2, sessile, dorsifixed, included, 11–12 mm long, adnate much of their length to corolla tube; pollen grains large, sticky; pistillate flowers solitary or few in axils; styles thick, to 8 mm long, the apex exserted, the 2 lobes bifid, pale orange. Fruits oblong-ellipsoid, 5.5–7 cm long, 1.8–2.5 cm wide, acute at base, beaked at apex, light green with dark green stripes, the flower often persisting; seeds many, stacked in each locule, flattened, irregularly narrow-ovate, light brown, ca 7 mm long. *Croat 4571, 8499.*

Frequent in the forest. Flowers and fruits throughout the year, but especially during the rainy season. Plants may have both flowers and mature fruits at the same time.

Trigona bees have been observed visiting the flowers. The species is pollinated by butterflies of the genus *Heliconius* (Nymphalidae, Ithomiinae) (Ehrlich & Gilbert, 1973).

Southern Mexico south at least as far as Colombia, probably more widespread. In Panama, known from tropical moist forest in the Canal Zone (both slopes), Chiriquí, Veraguas, Los Santos, Panamá, and Darién, from premontane wet forest in Veraguas and Panamá, and from tropical wet forest in Colón and Panamá.

See Fig. 543.

SICYDIUM Schlecht.

Sicydium coriaceum Cogn. in A. DC., Monogr. Phan. 3:906. 1881

Slender dioecious vine; stems striate; tendrils branched once. Petioles 1–5 cm long, inconspicuously puberulent and glandular-puberulent, weakly viscid; blades ovate-cordate, acuminate at apex, 5.5–12 cm long, 3.4–9 cm wide, entire, scabridulous to nearly glabrous above except on veins, densely and inconspicuously puberulent and glandular below, both surfaces with small irregular cystoliths visible when dried; palmate veins 5–7, the basal vein with a branch extending into the basal lobes. Panicles terminal or axillary, to 17 cm long; flowers 5-parted, unisexual, to 2 mm wide, greenish; branches and pedicels densely puberulent and glandular-puberulent; calyx crisp-villous, the lobes obtuse, to ca half as long as petals, ca 0.8 mm long, villous on outer surface, ciliate; petals narrowly triangular to ca 1.5 mm long, papillate-puberulent, erose; staminate flowers with the stamens 3, free, borne on a broad flattened disk, the filaments 3–4 times as long as anthers, the anthers straight, 2 with 2 locules, 1 with 1 locule, the pistillode apparently lacking; pistillate flowers with the calyx and corolla as in staminate flowers, the ovary 1-celled, narrowly ovoid, ca 2 mm long, glandular-pubescent, the ovule 1, pendulous, the styles 3. Fruits globose, ca 1 cm diam, indehiscent, fleshy, black; seed 1, globose, ca 7 mm diam, heavily ornamented

with sharp projections of various lengths. *Croat 9572, 12931.*

Rare, growing into trees; seen fertile only once at about 10 m in a tree. Seasonal behavior uncertain. Mature fruits seen in the middle of April. Peruvian collections have been seen with flowers in November and March and fruits in December.

Standley (1933) reported *S. tamnifolium* (H.B.K.) Cogn. and correctly described that plant as having soft pubescence, but the only collections seen are of *S. coriaceum,* which is not soft-pubescent.

Panama to Peru. In Panama, known only from tropical moist forest in the Canal Zone (BCI and near Madden Dam); doubtlessly more widespread.

See Fig. 544.

132. CAMPANULACEAE

Suffrutescent herbs; sap milky. Leaves alternate, petiolate, simple, subentire to toothed, estipulate. Flowers solitary in the upper axils, pedicellate; pedicels bracteolate; calyx lobes 5; hypanthium present; corolla tubular, 5-lobed, zygomorphic, bilabiate; stamens 5, epipetalous; filaments connate; anthers 2-celled, connate, dehiscing longitudinally, introrse; ovary 1, inferior, the carpels and locules 2; placentas 2, axile; ovules numerous, anatropous; style 1; stigmas broadly bilobed. Fruits leathery berries.

Only *Centropogon* occurs on BCI. It is easily distinguished by its milky sap, bilabiate reddish flowers with a curved tube, and stamens and anthers fused into a tube.

Some 60–70 genera and 2,000 species; in temperate and tropical areas.

CENTROPOGON Presl

Centropogon cornutus (L.) Druce, Bot. Exch. Club Soc. Brit. Isles 3:416. 1914

Suffrutescent herb, 1–3 m tall, often somewhat scandent; sap milky; stems hollow, glabrous at base, with stiff spreading trichomes above. Petioles 0.5–2 cm long; blades mostly oblong to oblong-elliptic, acute or acuminate at apex, rounded or tapering onto petiole at base, 10–15(25) cm long, 3–5 cm wide, subentire to denticulate or shallowly crenate-denticulate, glabrous above, glabrous to minutely scaberulous on veins below. Flowers solitary in upper axils, 6–8 cm long; pedicels erect, puberulent, (3.5)4.5–7(9.5) cm long, the bracteoles paired, ciliate, linear, 4–10(15) mm long, borne at or near base; hypanthium glabrous to bristly-pubescent; calyx lobes 5, ± narrowly lanceolate, 8.5–20 mm long, denticulate, usually wide-spreading at maturity; corolla red or pink (the lobes often white or occasionally the corolla completely white), the tube 3.5–4.5 cm long, curved throughout its length, narrowest at base, gradually widening above, ± glabrous to sparsely short-pubescent outside, glabrous inside, the lobes 5, falcate, triangular, the 2 upper ones 7–9 mm

long, the 3 lower ones slightly shorter with stiff whitish trichomes along margin; filament tube 4.5–6 cm long, white or white and distally pink, sparsely to moderately pilose toward apex, the filaments basally distinct and united to corolla tube; anther tube 7.5–9.5 mm long, densely hirsute, the terminal pubescence of the 2 shorter anthers fused into a triangular, hardened scale about 3 mm long; ovary truncate at apex. Berries depressed-globose, 10–17 mm diam, scarcely inflated, leathery; seeds disk-shaped with rounded edges, ca 0.6–1.8 mm diam, shallowly foveate-reticulate. *Shattuck 424.*

Collected only once on the island, perhaps disappearing as a result of succession, since it is a plant of weedier areas. Specimens have been collected many times in adjoining areas of tropical moist forest in the Canal Zone, but the species is probably more at home in regions of premontane wet forest. Flowers at least from October to February, but also in May and June, possibly all year.

The flowers are well suited to pollination by hummingbirds.

The berries are endozoochorous, possibly taken principally by birds.

Mexico (Oaxaca), Panama, and throughout northern South America, especially in the lowlands, to Bolivia and southern Brazil; Lesser Antilles. In Panama, known from tropical moist forest in the Canal Zone (Atlantic slope), Bocas del Toro, Panamá, and Darién and from premontane wet forest in Colón (Cerro Santa Rita) and Panamá (Cerro Trinidad).

133. COMPOSITAE (ASTERACEAE)

Erect to decumbent herbs or vines, sometimes small trees or shrubs. Leaves cauline and alternate or opposite, or basal, sessile to petiolate; blades simple or pinnately or palmately lobed, entire or serrate; venation pinnate or occasionally palmate at base; stipules lacking. Heads solitary or in various arrays usually surrounded by spirally arranged bracts (phyllaries); flowers bisexual or unisexual (monoecious or dioecious), subtended by bracts (paleae or chaff); calyx presumably represented by the chaffy or setaceous pappus (these most fully developed in fruit) or lacking; corolla of disk flowers tubular and generally 5-lobed, sometimes split on one side, central on head when ray flowers present; corolla of ray flowers prominently extended into a ligule, usually peripheral on head when present, showy; stamens 5 (4 in *Rolandra*), attached to the corolla, syngenesious ($\pm$ connate into a ring), rarely free; anthers 2-celled, introrse, dehiscing longitudinally; ovary inferior, unilocular, 2-carpellate; placentation basal; ovule solitary, anatropous; style 1, 2-branched, usually antrorsely barbed, elongating after anthesis, thus pushing pollen from staminal tube; stigmas slender. Fruits achenes; seeds lacking endosperm.

Members of the family are distinguished on BCI by the bracteate heads, often bearing flowers of two kinds: (1) ray flowers around the periphery, these usually showy and often sterile or staminate, rarely lacking; and (2) disk flowers central on the head, not showy, and usually bisexual or pistillate.

The usually small, tubular flowers are probably pollinated principally by butterflies, but they are visited by a number of other types of small insects, such as short-tongued bees (Percival, 1965). (See the discussion of *Mikania micrantha.*) Pollen grains are small and bees specializing on them have pollen baskets that are densely plumose (Proctor & Yeo, 1973). The style, which is usually roughened or antrorsely barbed, usually pushes the sticky and heavily ornamented pollen out of the tube formed by the syngenesious anthers. This displays the pollen near the apex of the tube.

Achenes are chiefly anemochorous and to a lesser extent epizoochorous; a few are endozoochorous. In wind-dispersed achenes the pappus not only acts as a plume to carry the achene in the air, but may also serve to dislodge the achene by its hygroscopic movements. This usually serves to ensure that the achene is presented to the wind during dry, usually windy weather, providing wider dispersal. The pappus provides the same buoyancy in water as it does in air. Ridley (1930) reported that most Compositae seeds float.

In the case of epizoochorous fruits, the pappus is usually modified with prominent recurved awns, which attach themselves to passing animals. Most species of

KEY TO THE TAXA OF COMPOSITAE

- Leaves alternate or all basal:
  Leaves basal . . . . *Chaptalia nutans* (L.) Polak.
  Leaves borne on stem, not all basal:
    Ray flowers conspicuous; heads bisexual; flowers white; plants often 3 m tall, the leaves usually deeply lobed, narrowed at base and decurrent onto stem . . . . . . *Verbesina gigantea* Jacq.
    Ray flowers lacking or inconspicuous; heads discoid:
      ◆ Individual heads clustered into distinct glomerules or compounded:
        Leaves white-tomentose below (the trichomes in a dense, appressed, cottony mass), not decreasing markedly in size toward apex of plant; pappus of short scales:
          Corolla white; outer heads not enveloped in the base of leaflike bracts; phyllaries mostly 2, coriaceous, the outermost with a straight, spinelike tip . . . .
          . . . . *Rolandra fruticosa* (L.) O. Kuntze
        Corolla violet; outer heads enveloped in the base of leaflike bracts, these with a spinelike tip turned outward; phyllaries ca 6, membranaceous, not armed at apex . . . . .
        . . . . *Spiracantha cornifolia* H.B.K.

Leaves not white-tomentose below, the lowermost largest, markedly decreasing in size toward apex of plant; pappus bristles long and straight or folded:

Compounded heads globose or hemispheroid, subtended by ovoid bracts; leaves densely soft-pubescent below .... *Elephantopus mollis* H.B.K.

Clusters of heads ± cylindroid or ellipsoid; leaves sparsely pubescent below, the surface punctate .... *Pseudoelephantopus spicatus* (Aubl.) C. F. Baker

◆ Individual heads free, not clustered into distinct glomerules or compounded:

Plants weak herbs, usually less than 1 m tall; flowers purplish or greenish; phyllaries in 1 long series (plus a shorter set in *Erechtites*), thin and linear; plants often more than 1 cm high; leaves often deeply lobed along their lateral margins:

Flowers purplish; phyllaries in 1 series; achenes 5-angled; plants abundant .... *Emilia sonchifolia* (L.) DC.

Flowers greenish-yellow; phyllaries in 2 series; achenes 10-angled; plants rare or possibly absent ...... *Erechtites hieracifolia* (L.) Raf. var. *cacalioides* (Spreng.) Griseb.

Plants shrubs or herbs, usually more than 1 m tall (except *Conyza apurensis, Pluchea odorata,* and *Vernonia cinerea*); flower color various; phyllaries in several series, less than 7 mm long; leaves usually entire to toothed but never deeply lobed along their lateral margins (*Neurolaena lobata* sometimes with a pair of lobes near base):

Plants usually less than 1.5 m tall; leaf blades obovate or oblanceolate to spatulate:

Flowers white to yellow; heads with both disk and ray flowers; leaf blades ± sparsely pubescent on both surfaces .... *Conyza*

Flowers violet to pink or purple; heads with or without ray flowers:

Ray flowers present; leaf blades ± granular-puberulent ... *Pluchea odorata* (L.) Cass.

Ray flowers absent; leaf blades densely pubescent below with ± matted trichomes .... *Vernonia cinerea* (L.) Less.

Plants usually more than 1.5 m tall; leaf blades usually not obovate:

Flowers yellow; plants large erect herbs, usually 1.5–2.5 m tall; leaf blades markedly serrate, often lobed to near base .... *Neurolaena lobata* (L.) R. Br.

Flowers white or pinkish to lavender; plants erect or arching shrubs; leaf blades ± entire (not markedly serrate or lobed):

Leaf blades entire, 3-veined; stems and branches prominently ribbed throughout; plants dioecious .... *Baccharis trinervis* Pers.

Leaf blades pinnately veined, often serrate or denticulate; stems not conspicuously ribbed; plants bisexual .... *Vernonia*

● Leaves opposite (the uppermost sometimes alternate):

■ Pappus of numerous bristles or plumose bristle-like scales:

Heads with both ray and disk flowers; plants small white-flowered herbs; pappus of plumose bristle-like scales:

Ray flowers conspicuous; paleae present; pappus of plumose bristles; plants herbaceous ... . *Tridax procumbens* L.

Ray flowers inconspicuous; paleae lacking; pappus lacking; plants shrubs .... *Clibadium*

Heads with only disk flowers (including *Clibadium,* with inconspicuous ray flowers):

Plants vines; phyllaries 4; heads of 4 flowers .... *Mikania*

Plants erect or arching shrubs or herbs; phyllaries more than 5; heads of many flowers:

Leaf blades broadest at about the middle:

Leaves petiolate, the blades not decurrent onto petiole and not extending onto stem .... *Koanophyllon wetmorei* (B. L. Robinson) King & H. Robinson

Leaves sessile, amplexicaul (the leaf tissue extending onto stem) .... *Ayapana elata* (Steetz) King & H. Robinson

Leaf blades broadest below the middle:

Plants small annuals, seldom more than 1 m tall; heads ca 4 mm high (to 6 mm in *F. sinclairii*) .... *Fleischmannia*

Plants much larger, stout herbs or shrubs; heads usually more than 6 mm high:

Leaves glabrous or sparsely pubescent on lower surface, entire or with a few, widely spaced teeth; plants slender arching shrubs:

Heads cylindrical, usually less than 4 mm wide; flowers lavender to white; lower blade surface with round, yellow, raised, glandular dots .... *Chromolaena odorata* (L.) King & H. Robinson

Heads discoid, usually 6 mm or more wide; flowers usually violet; lower blade with glandular dots lacking or at least not yellow, globular, and raised .... *Heterocondylus vitalbis* (DC.) King & H. Robinson

Leaves usually densely pubescent on lower surface, regularly and closely dentate or serrate; plants herbaceous:

Flowers greenish-yellow; blades usually not cordate at base, the margins dentate with mucronate teeth .... *Schistocarpha oppositifolia* (O. Kuntze) Rydb.

Flowers greenish-white; blades cordate at base, the margins crenate .... *Hebeclinium macrophyllum* (L.) DC.

■ Pappus not as above:
  Plants ± woody shrubs (sometimes scandent):
    Flowers white; pappus lacking . . . . *Clibadium surinamense* L.
    Flowers yellow or orange; pappus various:
      Pappus lacking; fruits globose, ± fleshy at maturity; phyllaries acute at apex; ray flowers present; heads with paleae; plants often scandent . . . . . *Wulffia baccata* (L.f.) O. Kuntze
      Pappus of scarious linear scales; fruits narrowly turbinate, not fleshy; phyllaries blunt to rounded at apex; ray flowers lacking; heads lacking paleae; plants erect . . . . *Calea prunifolia* H.B.K.
  Plants herbaceous, sometimes large:
    Leaves pinnately lobed, the lobes several; flowers white; herbs often more than 3 m tall
    . . . . *Verbesina gigantea* Jacq.
    Leaves not pinnately lobed:
      Disk flowers white (sometimes pale yellow):
        Plants nearly glabrous; pappus of 2 or 3 bristles or lacking; ray flowers lacking or few
        . . . . *Spilanthes alba* L'Hér.
        Plants asperous (rough to touch), conspicuously pubescent:
          Leaves oblong-lanceolate to linear, less than 2.5 cm wide; heads with minute ray flowers; anthers whitish . . . . *Eclipta alba* (L.) Hassk.
          Leaves mostly ovate, often with hastate lobes, more than 3 cm wide; heads lacking ray flowers; anthers black . . . . *Melanthera aspera* (Jacq.) Small
      Disk flowers yellow:
        Heads chiefly axillary, sessile or short-pedunculate:
          Plants usually erect; heads very congested in axils, usually sessile; ray flowers present; achenes of 2 types, ± flattened, with spines around the entire margin or with 2 spines at apex . . . . *Synedrella nodiflora* (L.) Gaertn.
          Plants often decumbent, weak-stemmed; heads loose, often somewhat pedunculate; ray flowers absent; achenes ± turbinate, muricate, several-sided; pappus a small crown . . . . *Eleutheranthera ruderalis* (Sw.) Schultz-Bip.
        Heads mostly terminal or at least appearing so, the peduncles 1–10 cm long:
          Petioles obscure, the blades decurrent along petiole and ± ensheathing stem; plants often decumbent or at least rooting at lower nodes; heads usually to 2.5 cm wide (including ray flowers); inner phyllaries not enveloping achenes . . . . *Wedelia trilobata* (L.) Hitchc.
          Petioles obvious, slender; plants erect, not reclining and rooting at nodes; heads less than 2 cm wide:
            Achenes enclosed in subtending phyllaries; pappus lacking; involucre biseriate; blades gradually attenuate at base; plants usually less than 1 m tall . . . . *Melampodium divaricatum* (L. C. Rich.) DC.
            Achenes not enclosed in subtending phyllaries; pappus a crownlike cup; involucre not biseriate; blades truncate to abruptly attenuate at base; plants commonly more than 1 m tall . . . . *Baltimora recta* L.

this nature occur in weedy situations, which are not so common on BCI. Several species occurring there, however, are probably epizoochorous: *Elephantopus mollis, Pseudoelephantopus spicatus, Spiracantha cornifolia,* and *Synedrella nodiflora. Elephantopus* and *Pseudoelephantopus* have a modified pappus capable of becoming entangled in hair. *Synedrella nodiflora* has the pappus or the entire lateral margin of the achene modified with bumpy projections for epizoochorous dispersal. The diaspore of *Spiracantha cornifolia* is probably a glomerule of several one-flowered heads, which acts like a sandbur rather than individual heads. The inflorescence of *Rolandra fruticosa* breaks up at maturity into small, one-seeded heads. These are dispersed perhaps in part by wind or birds (by endozoochory), but may also be epizoochorous by the action of their awnlike bracts. Similarly, the achenes of *Melanthera aspera* may be dispersed by their sharp scalelike chaff. *Eclipta alba* lacks a pappus, but the achene itself is viscid and becomes sticky upon wetting (Ridley, 1930). *Wulffia baccata* is endozoochorous and probably bird dispersed.

Approximately 950 genera and 20,000 species; cosmopolitan.

AYAPANA Spach

Ayapana elata (Steetz) King & H. Robinson, Phytologia 27:235. 1973
Eupatorium elatum Steetz

Herb, 1–2 m tall, branched many times, sparsely short-pubescent. Leaves mostly opposite, sessile, amplexicaul; blades narrowly lanceolate to oblanceolate, acuminate, narrowed to base and sometimes auricled and clasping

stem at base, 8–25 cm long, 1–5 cm wide, with yellowish, globular, viscid glands on lower surface. Panicles large, terminal, open, corymbose; heads ca 5 mm long, short-pedunculate, bearing many flowers; phyllaries 2–6 mm long, in 2 or 3 series; ray flowers lacking; disk flowers bisexual, white or greenish, the anthers syngenesious, the style shortly exserted. Achenes oblong, ca 1.7 mm long, 4- or 5-angled, sparsely hirtellous on angles; pappus of many, long, whitish bristles. *Croat 9554.*

Occasional along the shore. Flowers and fruits in the dry season (February to April).

Costa Rica, Panama, and Peru. In Panama, known from tropical moist forest in the Canal Zone, Bocas del Toro, Chiriquí, and Panamá and from premontane wet forest in Coclé and Panamá.

See Fig. 545.

BACCHARIS L.

Baccharis trinervis Pers., Synops. Pl. 2:423. 1807
Santa María

Dioecious, slender, arching shrub, to 3(4) m tall; branches usually scandent; stems closely ribbed, sparsely pubescent. Leaves alternate; petioles short, canaliculate; blades lanceolate-oblong, acuminate, acute to obtuse at base and decurrent onto petiole, 3–10 cm long, 1.5–3.5 cm wide, usually almost glabrous; veins at base 3. Inflorescences terminal or axillary, of compact rounded corymbs at ends of branches; heads 5–7 mm long, discoid, bearing many flowers; phyllaries in several series, ovate to oblong, the apex scarious and ciliate; receptacle convex, conical, 1–2 mm wide, muricate; ray flowers lacking; disk flowers unisexual, greenish-white, with faint fragrance, the paleae 2–4 mm long, deciduous; staminate flowers with the corolla thick, tubular, ca 3.5 mm long, equaling or exceeding pappus, deeply 5-lobed, reflexed at anthesis, the stamens syngenesious, white, becoming slightly exserted, the style stout, the ovary reduced, sterile; pistillate flowers with the corolla very slender, ca two-thirds to three-fourths as long as paleae, the lobes reduced to glandular trichomes, the style slender, bifurcate near apex, almost equaling pappus. Achenes slender, sparsely to densely pubescent; pappus of many whitish bristles. *Croat 7221.*

Frequent at the edges of clearings and along the shore. Flowers and fruits throughout the year.

The flowers are visited by the bees *Trigona cupira* and *T. tataira.*

Southern Mexico south to Argentina (but not east to Venezuela). In Panama, widespread and ecologically variable; known from tropical moist forest in the Canal Zone, Colón, Veraguas, Los Santos, Herrera, and Darién, from premontane dry forest in Coclé, from tropical dry forest in Los Santos, Herrera, Coclé, and Panamá, from premontane moist forest in the Canal Zone and Panamá, from premontane wet forest in Colón, Chiriquí, Coclé, and Panamá, and from tropical wet forest in Coclé and Panamá.

BALTIMORA L.

Baltimora recta L., Mant. Pl. Altera 228. 1771
Polygamous herb, to 3 m tall, ± strigose on all parts. Leaves opposite; petioles 8–70 mm long; blades ± ovate, acuminate, truncate to abruptly attenuate at base, 2.5–15 cm long, 1.5–12 cm wide. Inflorescences terminal racemes or more often large panicles; peduncles 8–33 mm long; heads 7–22 mm diam, 5–8 mm long; phyllaries 3–6, 3.5–6 mm long, ciliate at apex; paleae 3.5–4.2 mm long; ray flowers yellow, 3–8, pistillate, fertile, the ligule 4–6 mm long, the pappus a small crown; disk flowers 16 or more, bisexual, partly fertile, yellow, ca 2.5 mm long, the anthers with the appendage truncate. Achenes 2.4–3.2 mm long, truncate and puberulent at apex; pappus a crownlike cup. *Kenoyer 593.*

Collected once on Orchid Island; not seen in recent years, but very weedy and likely to occur sporadically in clearings and along the shore. Flowers and fruits throughout the year, especially at the beginning of the dry season (December to January).

Mexico to Panama. In Panama, ecologically variable; known from tropical moist forest in the Canal Zone, Chiriquí, Veraguas, Herrera, Panamá, and Darién, from tropical dry forest in Herrera and Coclé, from premontane moist forest in the Canal Zone, from premontane wet forest in Veraguas, and from tropical wet forest in Colón.

CALEA L.

Calea prunifolia H.B.K., Nov. Gen. & Sp. 4:294, t. 406. 1820
Escobilla

Shrub, to 2.5(5) m tall. Leaves opposite; petioles scabrous, to 1.5 cm long; blades ovate, acute at apex, obtuse to truncate at base and decurrent onto petiole, 4–8(10) cm long, 2.5–5(6.8) cm wide, rugose, scabrous especially on veins, asperous on both surfaces. Inflorescences axillary or terminal, corymbiform or umbellate; heads discoid, (3)8–10 mm long; phyllaries in several series, blunt to rounded at apex, often concave submarginally, the margins scarious, the outer series ovate-oblong to rounded, the innermost oblong, glabrous or scattered-hirtellous, 4.8–6.6 mm long; ray flowers lacking; disk corollas yellow, to 6.5 mm long, 5-lobed, the anthers syngenesious, becoming exserted and often directed to one side of flower, the pollen bright yellow, the style exserted, 2-forked, its branches inconspicuously papillate outside. Achenes narrowly turbinate, subterete to angular, black, pubescent, 1.3–3 mm long, with a distinct carpopodium 0.5 mm long; pappus shorter than corolla, paleaceous; scales narrow, uniseriate. *Croat 12278.*

Occasional, on the shore of Orchid Island and on the shore near Front #8 Lighthouse. Flowers throughout the rainy season and in the early dry season, from June to February. The fruits may be present until June.

Fig. 545. *Ayapana elata*

Fig. 546. *Chromolaena odorata*

Costa Rica to Colombia. In Panama, ecologically variable, but known principally from tropical moist forest in the Canal Zone, Bocas del Toro, Colón, San Blas, Panamá, and Darién; known also from tropical dry forest in Panamá, from premontane moist forest in the Canal Zone, from premontane wet forest in the Canal Zone, Panamá, and Darién, and from tropical wet forest in Colón (Portobelo) and Veraguas.

CHAPTALIA Vent.

Chaptalia nutans (L.) Polak., Linnaea 41:582. 1877

Polygamous erect herb, to 80 cm tall; lower leaf blades and the usually solitary scape densely white-arachnoid-pubescent. Leaves 3–6, all basal; petioles clasping; blades ± lyrate-lobed, much narrowed basally, decurrent onto petiole, 5–10(34) cm long, 1.5–3(10) cm wide, the upper surface glabrous. Scape usually nodding at apex, to 60 cm high, bearing a single head 2.5–3 cm long; ray flowers inconspicuous, pistillate, very slender, red-purple, well exceeding the pappus, the styles ± equaling the pappus; disk flowers bisexual, white, the tube slender, equaling pappus, the anthers syngenesious, not protruding from the tube, the style only slightly emergent from anthers, the stigmas very short. Achenes turbinate, the seminiferous part less than 4 mm long, tapered above into a slender projection; pappus of many bristles, 12–15 mm long, those of the disk flowers longer than those of the ray flowers. *Croat 9205.*

Common in the Laboratory Clearing; seldom seen elsewhere. Flowers and fruits throughout the year.

Southern United States to Argentina; West Indies. In Panama, ecologically variable; known most commonly from tropical moist forest all along the Atlantic slope and in Chiriquí, Los Santos, Herrera, Panamá, and Darién; known also from tropical dry and premontane moist forests in the Canal Zone, from premontane wet forest in Chiriquí, Los Santos, Coclé, and Panamá, and from tropical wet forest in Chiriquí and Panamá.

CHROMOLAENA DC.

Chromolaena odorata (L.) King & H. Robinson, Phytologia 20:204. 1970

Eupatorium odoratum L.

Christmas bush, Hierba de chiva, Paleca

Slender arching shrub, to 2.5(3) m tall; branches decussate; stems nearly glabrous to villous, sometimes reddish-brown. Leaves opposite, petiolate; blades ± rhombic-ovate, acuminate, rounded to cuneate at base, 4–8(10) cm

long, 2–5.5 cm wide, almost glabrous above, sparsely pubescent below with viscid globular glands (especially conspicuous on dry specimens), the margins coarsely toothed. Inflorescences corymbs, terminating the upper branches, bearing many flowers; heads cylindrical, 7–9 mm long, usually less than 4 mm wide; phyllaries in several series, to 7 mm long and 3.5 mm wide, 3-veined, rounded or acute; ray flowers lacking; disk flowers bisexual, lavender to white, fragrant, the corolla ca 5 mm long, only slightly exceeding the pappus, the lobes papillate-puberulent inside, the filaments fused to basal half of corolla tube, the anthers syngenesious, the style lavender, long-exserted after anthesis, bifurcate much of its length, the distal end papillate. Achenes oblong, 5-ribbed and angular, 4–5 mm long, slightly strigose on ribs; pappus of long white bristles. *Croat 7789.*

Abundant in clearings. Flowers and fruits throughout the dry season, especially from December to February. The fruits may persist until July.

Distinguished by the long-exserted style and long cylindrical heads. The styles perhaps serve the same attracting function as ray flowers, which the flowers lack. Some plants investigated appeared to have no pollen in the anthers even while in bud.

Florida and Texas to Paraguay; West Indies; West Africa; Malaya. In Panama, widespread and ecologically variable; known from tropical moist forest in the Canal Zone, Bocas del Toro, San Blas, Chiriquí, Veraguas, Herrera, Panamá, and Darién, from tropical dry forest in Los Santos, Herrera, Coclé, and Panamá, from premontane moist forest in the Canal Zone, Coclé, and Panamá, from premontane wet forest in Bocas del Toro, Colón, Chiriquí, Veraguas, Coclé, and Panamá, and from tropical wet forest in Colón and Panamá.

See Fig. 546.

CLIBADIUM L.

Clibadium asperum (Aubl.) DC., Prodr. 5:506. 1836

Shrub, 1.5–3 m tall; stems densely strigillose. Leaves opposite (or sometimes alternate at branch nodes), asperous; petioles 1.5–7 cm long, strigillose; blades ovate to lanceolate-elliptic, acuminate at apex, attenuate at base, 6–22(28) cm long, 3–10 cm wide, moderately strigillose on both surfaces, denser on the veins, the margins serrate. Inflorescences corymbose panicles; peduncles 1–2 mm long; heads 4–5 mm long; phyllaries 7–10, ovate to obovate, 3–4 mm long, 2–3 mm wide; paleae lacking; ray flowers 5–9, the corollas 3-lobed, 1.5–2.2 mm long; disk flowers 9–22, the corollas 2.5 mm long, pubescent outside, the anthers 1.3 mm long, villous at least near apex.

KEY TO THE SPECIES OF CLIBADIUM

Lower surface of leaves with conspicuous reticulate veins, hispid . . . . *C. surinamense* L.
Lower surface of leaves not with conspicuous reticulate veins, glabrate to strigose . . . .
. . . . *C. asperum* (Aubl.) DC.

Achenes 1.7–2 mm long, ca 1.5 mm diam, puberulent at apex; pappus lacking. *Croat 12283*.

Apparently rare; collected many years ago by Bailey and found recently on the west side of Orchid Island. Flowers and fruits prinicpally from May to September, occasionally during the rest of the year.

Panama to northern South America. In Panama, known from tropical moist forest in the Canal Zone, San Blas, Chiriquí, Panamá, and Darién, from premontane wet forest in Colón, Chiriquí, and Panamá, and from tropical wet forest in Colón and Panamá.

Clibadium surinamense L., Mant. Pl. Altera 294. 1771

Mastranzo de monte

Functionally monoecious shrub, 1–2.5(6) m tall, densely hispidulous to scabrous, the trichomes coarse and erect; stems often mottled with purple. Leaves opposite, asperous; petioles 5–20 mm long; blades ovate-oblong, acute at apex, obtuse to rounded at base, 5–17 cm long, 2.5–8(10) cm wide, scabrous above, ± pilose below, crenate-serrulate; reticulate veins prominently raised. Inflorescences terminal corymbs; heads often secund, sessile, 4–7 mm long; phyllaries 8 or 9, obovate, 3–5 mm long, 2–4.8 mm wide, acute or acuminate, closely imbricate, ciliate; flowers white; ray flowers 3–5, functionally pistillate, lacking stamens, the corolla 3- or 4-lobed, 2–2.7 mm long, the style divided nearly to base, exserted at anthesis while the head appears immature and the disk flowers are still closed; disk flowers 11–14, functionally staminate, the corolla 2.5–3.3 mm long, broadened above, densely pubescent especially on lobes, the anthers syngenesious, the ovary sterile, the style simple, protruding from corolla with anthers. Achenes thick, rounded-obovoid, 2-edged, glabrous except for densely villous apex, black, to 3 mm long; pappus lacking. *Croat 6238, 12584*.

Occasional, on the shore, particularly on the northern edge of the island. Flowers and fruits principally in the rainy season (May to December), but occasionally throughout the rest of the year.

Easily distinguished by the seemingly rayless white heads and by the black phyllaries at maturity.

Stuessy (1975) reported *Croat 12718* as a collection that intergrades with *C. asperum* and suggested that the two taxa may hybridize.

Though the style of the pistillate flower may still be present when staminate flowers open, the pistillate flowers usually have been pollinated and the achene has begun to enlarge. The plants are visited by a variety of insects, including butterflies, small bees, and wasps.

Achenes lack a pappus for dispersal, but the phyllaries become black at maturity. Birds may remove the entire, few-seeded head.

Costa Rica south to Peru, Bolivia, and Brazil; West Indies. In Panama, ecologically variable; known from tropical moist forest in the Canal Zone, Colón, Chiriquí, Veraguas, Herrera, Coclé, Panamá, and Darién, from tropical dry forest in Coclé and Herrera, from premontane moist forest in the Canal Zone and Panamá, and from premontane wet forest in Chiriquí, Veraguas, Coclé, and Panamá.

CONYZA Less.

Conyza apurensis Kunth in H.B.K., Nov. Gen. & Sp. 4:73. 1820

Erigeron spathulatum Vahl

Horseweed

Polygamous weak herb, to 1 m tall (smaller in upland areas); most parts sparsely short-pilose, especially at base of stem; stems often branching, weakly ribbed below each leaf. Leaves alternate; petioles slender and narrowly winged on lower leaves, reduced near apex of plant; blades spatulate, rounded at apex, cuneate toward the base, sparingly to densely pilose on both surfaces, strigose on veins below, thin, crenate in apical half, otherwise entire, ciliate, the lower leaves to 12 cm long, and 5 cm wide, becoming gradually smaller toward apex of plant, the upper leaves to 5 cm long and 1 cm wide. Panicles terminal, open, bearing many flowers, with leafy bracts; peduncles slender, 5–25 mm long; heads ± convex, to 4 mm across; phyllaries linear, imbricate, mostly subequal but with a few forming another series, 3–4 mm long, sharply acute or acuminate, sparsely pilose with long weak trichomes, the margins scarious; ray flowers white, pistillate, many, in about 2 series, filiform, ca 2 mm long, inconspicuous, barely exceeding the pappus, the style long-exserted; disk flowers white, bisexual, ca 3 mm long, shallowly 5-lobed, the anthers ca 0.5 mm long, with narrow deltoid appendages at apex, obtuse, the style branches flattened. Achenes compressed, the margins ± thickened by 2 prominent veins, moderately pubescent; pappus of fine strigillose bristles in 1 series, about as long as disk flowers. *Croat 8676*.

Rare; collected at the Laboratory Clearing. Flowers and fruits principally in the early rainy season (April to July).

Large, lowland specimens of *C. apurensis* may be confused with *C. bonariensis*. However, the larger receptacle of *C. apurensis* distinguishes it.

Native to Asia; wide-ranging through tropical America.

KEY TO THE SPECIES OF CONYZA

Receptacle to 4 mm across; phyllaries 3–4 mm long; leaves crenate in upper half . . . . *C. apurensis* Kunth

Receptacle 2–2.5 mm across; phyllaries 4–6 mm long; leaves entire to sparingly toothed . . . . *C. bonariensis* (L.) Cronq.

In Panama, known from tropical moist forest in the Canal Zone, Chiriquí, Los Santos, and Panamá, from premontane moist forest in Panamá, from premontane wet forest in Chiriquí, Coclé, and Panamá, and from tropical wet forest in Darién.

Conyza bonariensis (L.) Cronq., Bull. Torrey Bot. Club 70:632. 1943

Erigeron bonariensis L.

Tabaquillo

Coarse erect herb, to 1.5 m tall, sparingly branched; most parts sparsely short-pilose; stems striate, green. Leaves alternate, sessile or sometimes with a petiole-like region; blades oblanceolate, rounded to acute, cuneate at base, 6–9 (12) cm long, 1–3 cm wide, largest at base of plant, strigose to glabrate on midrib below, thin, entire to sparingly toothed, ciliate. Panicles terminal, open, bearing many flowers, with leafy bracts; peduncles slender, 5–35 mm long; phyllaries imbricate in ca 2 series, 4–6 mm long, sharply acute or acuminate, sparsely pilose with long weak trichomes, the margins scarious; heads flat or ± convex, 2–2.5 mm diam; ray flowers pistillate, very numerous, white, capillary, inconspicuous, 3–4.5 mm long; disk flowers bisexual, few, yellow, 2–3 mm long, clearly differentiated about halfway up into a slender, cylindrical limb, shallowly and acutely 5-lobed, the anthers ca 0.8 mm long, narrowly appendaged above, basally auriculate, the style branches broadly flattened, with deltoid appendages. Achenes compressed, ca 1 mm long, pubescent especially toward base, the margins ± thickened by 2 prominent veins; pappus of fine, strigillose bristles in 1 series, ca 1.5 mm long. *Bailey 363 (BH)*.

Collected once by Bailey near the laboratory. Apparently flowers and fruits principally in the early rainy season (April to September).

Confused with large plants of *C. apurensis*, which have a receptacle to 4 mm diam.

Nearly cosmopolitan in distribution; Cuatrecasas indicated that it may be native to Argentina (1969). In Panama, widespread and ecologically variable; known from tropical moist forest in the Canal Zone, Bocas del Toro, Chiriquí, Los Santos, Panamá, and Darién, from premontane moist forest in the Canal Zone and Panamá, from premontane wet forest in Chiriquí and Panamá, from premontane rain forest in Coclé (Cerro Pilón), from lower montane wet forest in Chiriquí, and from lower montane rain forest in Chiriquí (Cerro Horqueta).

ECLIPTA L.

Eclipta alba (L.) Hassk., Pl. Jav. Rar. 528. 1848

E. prostrata (L.) L.

Polygamous herb, usually more than 75 cm tall but rarely to 1 m, often epiphytic; stems somewhat succulent, usually weak; all parts except flowers appressed-pubescent with stiff white trichomes. Leaves opposite, sessile; blades lanceolate, acute, tapering to base, 5–10 (16) cm long, 1–2.6 cm wide, subentire to serrulate. Heads 1 to several,

axillary, campanulate and ca 5 mm wide in flower, becoming green, cushion-shaped, and to 1 cm wide in fruit; peduncles 1–4.5 cm long, slender; flowers interspersed with paleae, the paleae slender, ± equaling flowers, bristled near apex; ray flowers pistillate, white, ca 1.5 mm long; disk flowers bisexual, ca 1 mm long, ± campanulate, 4-lobed, pale yellow to whitish, the stamens syngenesious, the filaments free below, the anthers dark, included, the style branches short, slightly exserted. Achenes 2–4-sided, ca 2.5 mm long, white becoming brown, muricate; pappus reduced to an irregular crown. *Croat 9562*.

Occasional, along the shore, often epiphytic on debris or on tree stumps in the lake. Flowers and fruits throughout the year.

Widespread in the tropics and subtropics. In Panama, known from tropical moist forest all along the Atlantic slope and in the Canal Zone, Chiriquí, Los Santos, Panamá, and Darién, from tropical dry forest in Coclé, from premontane moist forest in the Canal Zone and Panamá, from premontane wet forest in Chiriquí, Coclé, and Panamá, and from tropical wet forest in Colón.

ELEPHANTOPUS L.

Elephantopus mollis H.B.K., Nov. Gen. & Sp. 4:26. 1820

Bushy perennial herb, mostly 30–150 cm tall; branches few. Leaves alternate, chiefly cauline but near base of stem; petioles obscure, ciliolate, clasping at base; blades oblong to obovate or oblanceolate, ± acute at apex, gradually tapered to decurrent base, to 25 cm long and 8 cm wide at base, resin-dotted, sparsely appressed-pubescent above, densely soft-puberulent below, crenate. Inflorescence a solitary, terminal, corymbose panicle, branched many times; glomerules ca 1 cm long, 1–2 cm wide, with ca 40 heads, compounded at the ends of branches and subtended by ovate bracts; heads discoid, slender, ca 1 cm long, 4-flowered; phyllaries 8, in 2 series; ray flowers lacking; disk flowers bisexual, the corolla 4–5 mm long, white, 5-lobed, the anthers syngenesious, the style weakly barbed outside, emerging from the tube with anthers but elongating beyond them, the short styles recurving. Achenes narrowly turbinate, 2–3 mm long, 10-ribbed, resin-dotted and strigillose; pappus of 5 barbed bristles. *Croat 13155*.

Uncommon, in clearings. Flowers and fruits principally throughout the dry season (January to April).

Throughout the tropics and subtropics of the world. In Panama, very ecologically variable; known from most life zones and all provinces.

ELEUTHERANTHERA Poit. ex Bosc.

Eleutheranthera ruderalis (Sw.) Schultz-Bip., Bot. Zeitung (Berlin) 24:165. 1866

Small herb, to 50 cm tall, frequently creeping; stems 4–6-angulate, short-pilose with white trichomes. Leaves

opposite; petioles to 1 cm long; blades ovate, acute at apex, abruptly attenuate at base, 2–4(6) cm long, 1–2.5(3.5) cm wide, covered with minute, viscid, globular glands, serrate. Inflorescences terminal aggregates of 2–5 heads; heads discoid, ca 1 cm wide, usually solitary or few in axils, short-pedunculate to nearly sessile; phyllaries leaflike, equaling florets, frequently 5 large and 1 smaller; paleae equaling or exceeding corolla; ray flowers usually absent (if present, small and sterile); disk flowers yellow, mostly bisexual, ca 2 mm long, broadened above the narrow base, the lobes short, ovate, acute, bearing fleshy short trichomes inside, the filaments apparently at first syngenesious, becoming recurved and free, included, the style branches minutely bristled, slightly exserted and strongly recurved. Achenes turbinate, to 2.3(3) mm long, usually 4-sided, coarsely muricate and minutely pubescent; pappus cuplike, pubescent, brown. *Croat 7057*.

Seasonally abundant in clearings. Flowers and fruits primarily from June to December.

Texas south to Peru, Bolivia, and Brazil; West Indies. In Panama, known from tropical moist forest in the Canal Zone, Bocas del Toro, San Blas, Chiriquí, Herrera, Panamá, and Darién, from tropical dry forest in Coclé and Panamá, from premontane moist forest in the Canal Zone and Panamá, and from premontane wet forest in Chiriquí, Coclé, and Panamá.

See Fig. 547.

EMILIA Cass.

Emilia sonchifolia (L.) DC., Prodr. 6:302. 1837
   Tassel flower

Erect, weak-stemmed herb, 10–50 cm tall, glabrous or sparsely pubescent. Leaves alternate, to ca 12 cm long, the lower ones spatulate to lyrate-lobed, the upper ones narrowly triangular, sessile, and with the base sagittate-clasping. Panicles irregular, sparsely branched, terminal, slender-pedunculate, ca 1 cm long; heads few; phyllaries uniseriate, thin, somewhat shorter than the flowers; ray flowers usually lacking; disk flowers bisexual, the corolla tubular, 8–12 mm long, purplish, the stamens syngenesious, apparently remaining within corolla tube, the style very minutely papillate outside (apparently not emerging from anthers). Achenes brown, 5-angled, the angles minutely strigillose; pappus of numerous fine bristles. *Croat 9196*.

Common in clearings, especially at the laboratory. Apparently flowers year-round.

Florida and Mexico to Brazil; West Indies. In Panama, ecologically variable; known most commonly from tropical moist forest in the Canal Zone, Bocas del Toro, and Chiriquí; known also from tropical dry forest in Panamá, from premontane moist forest in Herrera, and from

premontane wet forest in the Canal Zone and Colón (mouth of Río Piedras).

ERECHTITES Raf.

Erechtites hieracifolia (L.) Raf. var. **cacalioides**
   (Fisch. ex Spreng.) Griseb., Fl. Brit. W. Ind. 381. 1861

Polygamous erect herb, to 1(3) m tall, hirsute to glabrate. Leaves alternate, sessile; blades lanceolate to linear-lanceolate, acute to acuminate at apex, auricled and clasping stem at base (especially the lower leaves), 4–12 cm long, 1–4 cm wide, coarsely dentate or incised-lobed. Corymbs pedunculate heads 1–2 cm long; phyllaries linear, ca 1 cm long, uniseriate, sometimes with a few shorter ones at base; ray flowers lacking; disk flowers greenish-yellow, heterogamous, the outer flowers pistillate, very slender, the inner flowers bisexual, tubular, the anthers syngenesious, the style slightly exserted. Achenes linear-oblong, ca 3 mm long, 10-ribbed or -striate, slightly pubescent; pappus of long, soft, white bristles. *Shattuck 757*.

Collected once by Shattuck; not seen in recent years, but to be expected in clearings. Flowering and fruiting to some extent all year, but mostly in the rainy season.

Canada to northern South America; Greater Antilles; central Europe; Hawaii. According to Adams (1972), the typical variety does not occur in Central America. In Panama, var. *cacalioides* is known from tropical moist forest in the Canal Zone, Bocas del Toro, San Blas, Chiriquí, Los Santos, and Darién, from tropical dry forest in Panamá, from premontane moist forest in Panamá, from premontane wet forest in Colón, Chiriquí, Panamá, and Darién, and from tropical wet forest in Panamá.

FLEISCHMANNIA Schultz-Bip.

Fleischmannia microstemon (Cass.) King & H.
   Robinson, Phytologia 19:204. 1970
   Eupatorium microstemon Cass.

Annual herb, 15–100 cm tall; stems and petioles moderately short-villous. Leaves opposite; petioles 5–35 mm long; blades ± ovate, acute to acuminate at apex, abruptly attenuate at base, 1.5–5 cm long, 1–3 cm wide, sparsely strigillose on veins of both surfaces especially below, the surfaces (especially the lower) glandular-dotted, the margins crenate except along lower edge. Inflorescences of terminal or axillary, subcorymbiform panicles; peduncles 2–5 mm long; heads discoid, ca 4 mm long; phyllaries mostly oblong, 1.5–4.5 mm long, 3- or 4-veined, rounded to blunt, glabrous except sometimes weakly pubescent on midrib, usually ciliate and apiculate at

KEY TO THE SPECIES OF FLEISCHMANNIA

Pedicels 2–5 mm long; heads in narrow corymbiform panicles ....
.... F. *microstemon* (Cass.) King & H. Robinson
Pedicels 3–15 mm long; heads in broad open panicles ..... F. *sinclairii* (Benth.) King & H. Robinson

apex; ray flowers lacking; disk flowers bisexual, 5- or 6-lobed, lavender, to 2.5 mm long, slightly longer than pappus, the style branches thick. Achenes prismatic, scarcely constricted near apex, usually blackish with yellowish ribs, the ribs and distal lateral surfaces scabrid; pappus of 25–30 slender bristles. *Standley 31443.*

Collected once in the Laboratory Clearing. Flowers and fruits from June to December, especially in the middle of the rainy season.

Mexico to northern South America; West Indies. In Panama, ecologically variable; known from tropical moist forest in the Canal Zone, Bocas del Toro, San Blas, Panamá, and Darién, from premontane wet forest in Panamá (Cerro Azul), and from tropical wet forest in Bocas del Toro and Panamá (Cerro Jefe).

Fleischmannia sinclairii (Benth. in Oerst.) King & H. Robinson, Phytologia 19:206. 1970

  Eupatorium sinclairii Benth. in Oerst.

Herb, to 1 m tall; stems terete, often branched many times, bearing ± uncinate trichomes. Leaves opposite; petioles 1–3 cm long; blades ± rhombic, mostly acute at apex, acute at base and decurrent onto petiole, 2–5 cm long, 1–3 cm wide, punctate, the upper surface with straight erect trichomes, the lower surface with recurved trichomes on veins, the margins crenate above middle. Panicles lax, open, terminal and upper-axillary; peduncles 3–15 mm long, slender; heads discoid, ca 4–6 mm high; phyllaries in several series, to 5 mm long, the outer acuminate, the inner acute, ciliate near apex; ray flowers lacking; disk flowers bisexual, the corolla ca 2.3 mm long, the tube white, the lobes and style lavender, the stamens syngenesious, included, the style exserted at anthesis. Achenes slender, weakly pubescent, ca 1.3 mm long, 5-striate, bearing a nipple-like projection at base; pappus of numerous awns, ± equaling or shorter than corolla. *Croat 8934.*

Occasional in clearings. Flowers and fruits from December to June, especially throughout the dry season (December to April).

Mexico to Panama. In Panama, ecologically variable; known from tropical moist forest in the Canal Zone, Colón, Veraguas, Herrera, Coclé, Panamá, and Darién (no doubt other provinces as well), from tropical dry forest in Darién (Punta Garachiné), from premontane moist forest in the Canal Zone, from premontane wet forest in Chiriquí, Los Santos (Loma Prieta), and Panamá, and from tropical wet forest in Colón.

HEBECLINIUM DC.

Hebeclinium macrophyllum (L.) DC., Prodr. 5:136. 1836

  Eupatorium macrophyllum L.

Coarse herb, 1–2 m tall; branches striate; young parts densely short-pilose. Leaves opposite; petioles to 5 cm long; blades broadly ovate, acute to acuminate, truncate to cordate at base, 8–13(15) cm long, 6–12(20) cm wide, slightly pubescent above, densely woolly below, crenate. Panicles rounded, compact, corymbose; heads 6–7 mm

long, ca 5 mm wide, bearing many flowers; phyllaries lanceolate, to 5 mm long, in many series, the outer shorter; ray flowers lacking; disk flowers bisexual, the corolla greenish-white, ca 2 mm long, the anthers syngenesious; pistillate parts not seen. Achenes oblong, 5-angled, ca 1.2 mm long; pappus of long white bristles. *Aviles 5.*

Not seen in recent years on the island, but probably occurring periodically in clearings. Flowers and fruits from June to February, especially from the middle of the rainy season to the early dry season.

Mexico to Paraguay; West Indies. In Panama, occurring most commonly at middle elevations in premontane wet and tropical wet forests; known from tropical moist forest in the Canal Zone, Bocas del Toro, Chiriquí, Los Santos, Panamá, and Darién, from premontane wet forest in Panamá and Darién, and from tropical wet forest in Chiriquí, Los Santos, and Panamá.

HETEROCONDYLUS King & H. Robinson

Heterocondylus vitalbis (DC.) King & H. Robinson, Phytologia 24:391. 1972

  Eupatorium vitalbae DC.

Arching shrub or vine, usually 2–3(6) m tall; stems glabrous or sparsely pubescent, glabrous in age. Leaves opposite, short-petiolate; blades oblong-ovate, acuminate, obtuse to rounded at base, mostly 5–10(12) cm long, 3–7 cm wide, glabrous above except on midrib, sparsely pubescent on veins below, entire or the margins remotely toothed in distal two-thirds; basal veins 3–5. Panicles terminal and upper-axillary, bearing few flowers; heads discoid, 1–1.5 cm long, widely spaced; phyllaries glandular, striate, lanceolate-oblong, 7–11 mm long, the outer series much broader than the inner series; ray flowers lacking; disk flowers bisexual, the corolla violet, ca 8 mm long, the anthers syngenesious, remaining in tube, the style long-exserted, white to lavender, minutely papillate outside. Achenes oblong, ca 3.5 mm long, sparsely hispidulous, 5-angulate; pappus of many white bristles ± equaling corolla. *Croat 8776.*

Occasional, at the edges of clearings. Flowers and fruits principally during the dry season (January to April), rarely as late as June.

Honduras south to Peru, Bolivia, and Brazil. In Panama, ecologically variable; known from tropical moist forest in the Canal Zone, Chiriquí, Veraguas, Herrera, and Panamá, from premontane moist forest in Panamá, from premontane wet forest in Chiriquí, Los Santos, and Panamá, from tropical wet forest in Chiriquí, Panamá, and Darién, and from premontane rain and lower montane moist forests in Chiriquí.

KOANOPHYLLON Arruda

Koanophyllon wetmorei (B. L. Robinson) King & H. Robinson, Phytologia 28:67. 1974

  Eupatorium hypomalacum var. *wetmorei* B. L. Robinson

Slender erect shrub, 2–3(5) m tall; stems crisp-pubescent and with viscid, ± globular glands. Leaves opposite; petioles 10–20 cm long; blades broadly to narrowly

Fig. 547.
Eleutheranthera
ruderalis

Fig. 548.
Koanophyllon wetmorei

Fig. 549.
Mikania leiostachya

elliptic, long-acuminate, gradually tapered to base and decurrent onto petiole, 8–20 cm long, 2.6–5 (6.5) cm wide, glabrate or weakly pubescent and densely covered with viscid, ± globular glands on lower surface, sparsely serrate. Inflorescences paniculate, terminal, bearing many flowers; heads discoid, 5–6 mm high; phyllaries in several series, sharp-pointed, 2–5 mm long; ray flowers lacking; disk flowers bisexual, the corolla white, tubular, the lobes small, the anthers syngenesious, remaining in tube, the stigmas papillate outside, long-emergent after anthesis. Achenes narrowly turbinate, sharply angled, very sparsely pubescent, ca 2 mm long; pappus of numerous bristles. *Croat 13273.*

Uncommon along the shore and occasional southwest of Slothia Island on shaded shores. Flowers from the late rainy season through the dry season (November to April).

Panama and adjacent Costa Rica. In Panama, known from tropical moist forest in the Canal Zone (BCI) and in Bocas del Toro and from premontane wet forest in Panamá.

See Fig. 548.

MELAMPODIUM L.

Melampodium divaricatum (L. C. Rich.) DC., Prodr. 5:520. 1836

Polygamous annual herb, usually less than 1 m tall; stems sparsely pubescent, occasionally with purple longitudinal lines. Leaves opposite, petiolate; blades lanceolate to rhombic-ovate, acute to acuminate at apex, cuneate to attenuate and decurrent onto petiole at base, mostly 6.5–15 cm long, 2–7.5 cm wide, scabrous above, sparsely hispid below, the margins usually inconspicuously toothed; basal veins 2, arcuate, ascending, intersecting other lateral veins. Heads solitary, terminal, 10–17 mm wide; peduncles 2–8.5 cm long; flowers yellow; phyllaries in 2 series, the outer 4 or 5, rounded, imbricate, ciliate, the inner not bractlike, enveloping achene; ray flowers pistillate, rounded or emarginate at apex, 4.5–6 mm long, 3–3.5 mm wide, often with green lines on lower surface; disk flowers bisexual (the ovaries abortive), ca 2.7 mm long, each subtended by a fimbriate-margined palea. Achenes 2.8–4 mm long, the lateral surfaces with diagonal striations and enlarged margins; pappus lacking. *Croat 7680.*

Often abundant in the Laboratory Clearing, less frequent in other clearings. Flowers and fruits throughout the year.

According to T. Stuessy (pers. comm.), seeds may fall to the ground and germinate rapidly. Since plants are self-fertile, they are provided with the capability of rapid colonization of an area.

Southern Mexico to Brazil. In Panama, known from tropical moist forest in the Canal Zone, Bocas del Toro, Los Santos, Panamá, and Darién, from premontane moist forest in the Canal Zone, and from premontane wet forest in Coclé and Panamá.

MELANTHERA Rohr

Melanthera aspera (Jacq.) Small, Bull. Torrey Bot. Club 36:164. 1909

Clavellina de monte, Julio, Sirvulaca

Widely branching herb, to ca 1 m tall; stems weakly ribbed, asperous, with appressed trichomes. Leaves opposite, petiolate; blades elliptic to ovate, acuminate, mostly 5–12 cm long, 2.5–6.5 cm wide, asperous, the trichomes ± appressed, the margins crenate, the larger leaves abruptly narrowed and decurrent onto petiole, often hastate-lobed at base; basal veins 3. Heads solitary (or few), terminal or upper-axillary, discoid, long-pedunculate, bearing many flowers; phyllaries ovate, unequal, in 2 or 3 series; ray flowers lacking; disk flowers bisexual, developing centripetally, the outermost falling before the innermost open, ca 4.7 mm long, each subtended by an oblong, acute, keeled scale, the corolla white, ca 6 mm long, narrowed at base, the stamens syngenesious, the filaments free almost to base, the anthers black, shedding pollen in bud, emerging somewhat from tube, later receding as the filaments shrivel, the connective white, the style elongating simultaneously, carrying pollen with it on the bristled outer surface of stigmas, opening and recurving only after emerging from anthers, the nectary prominent, yellow. Achenes slender, angled, ca 3 mm long, glabrous, truncate at apex; pappus of a few, short, barbed awns. *Croat 12886.*

Probably once abundant, now uncommon, occurring sporadically in clearings. Flowers and fruits throughout the dry season (December to April).

The mature heads become loose, and the sharply pointed phyllaries, which often envelop the much shorter achenes at their base, may attach the seeds to passing animals.

Florida and Mexico to northern South America in areas bordering the Caribbean; West Indies. In Panama, widespread and ecologically variable; known from tropical moist forest in the Canal Zone, Bocas del Toro, Colón, San Blas, Chiriquí, Panamá, and Darién, from tropical dry forest in Panamá (Taboga Island), from premontane moist forest in Los Santos and Panamá, from premontane wet forest in Bocas del Toro (Río Guarumo), Colón, Chiriquí, Veraguas, Coclé, and Panamá, and from tropical wet forest in Bocas del Toro (Santa Catalina), Colón, and Coclé (Boca de Toabré).

MIKANIA Willd.

Mikania guaco H. & B., Pl. Aeq. 2:84. 1809

Herbaceous vine or slender liana. Leaves opposite, petiolate; blades ovate, acute or acuminate, acute and decurrent at base, 6–13 (25) cm long, 3.5–10 (15) cm wide, nearly glabrous or scabrous, ± entire. Heads in large paniculate cymes, 4-flowered, ca 1 cm long; phyllaries 4, oblong, 4–6 mm long, rounded at apex, puberulent; ray flowers lacking; disk flowers bisexual, white or greenish-yellow, the anthers syngenesious, the style exserted, forked, its

KEY TO THE SPECIES OF MIKANIA

Heads spicate, evenly disposed on the elongated branches of open panicles . . . . . *M. leiostachya* Benth.
Heads not spicate, conspicuously clustered in globular or flat-topped clusters:
    Blades cordate at base . . . . *M. micrantha* H.B.K.
    Blades not cordate at base:
        Phyllaries usually less than 3 mm high; heads in glomerulate, ± globular corymbs . . . .
        . . . . *M. tonduzii* B. L. Robinson
        Phyllaries 4–5 mm high; heads in flat-topped or ± globular corymbs:
            Heads almost all in groups of 3, often in flat-topped corymbs; style appendages hirsute with
                long papillae; style base papillose . . . . *M. guaco* H. & B.
            Heads not in groups of 3, in glomerulate, ± globular corymbs; style appendages not hirsute,
                with short papillae; style base glabrous . . . . *M. hookeriana* DC.

---

branches long, hirsute. Achenes oblong, 4- or 5-angled, ca 2 (4) mm long, glabrous; pappus tan to reddish, of slender bristles ca 5 mm long. *Wilson 9* (*F*).

Not seen in recent years on the island, but to be expected. The species probably flowers and fruits in the dry season.

Widely distributed in tropical America from Mexico to Peru, Bolivia, and Brazil. In Panama, known from tropical moist forest in the Canal Zone, Bocas del Toro, Panamá, and Darién, from premontane wet forest in Chiriquí and Panamá, and from tropical wet forest in Chiriquí.

Mikania hookeriana DC., Prodr. 5:195. 1836

Slender liana; stems terete, striate, slightly puberulent to sparsely hirsute. Leaves opposite; petioles 1–5 cm long, canaliculate; blades ovate to broadly ovate or ovate-elliptic, rounded to acuminate at apex, rounded to cuneate at base, (3)6–10(15) cm long, (1.5)3–7(15) cm wide, sparsely puberulent to glabrate on upper surface, brownish-puberulent on lower surface; major lateral veins with 2 pairs prominent in the basal fourth of blade, the upper pair of these ascending to apex. Panicles terminal or upper-axillary, the branches short; heads 8–9 mm long, in small clusters at the tips of short branches, sessile; phyllaries 4, ca 4 mm long, ca 1 mm wide, sparsely puberulent and glandular on outside, glabrous and paler on inside; corolla whitish, fragrant, 4.5–5.5 mm long, the tube 1–1.5 mm long, the limb narrowly campanulate, the lobes glandular and short-pubescent near the tip; anther appendages ovate-triangular; style base glabrous, the style appendages short-papillate. Achenes 4- or 5-ribbed, ca 3.5 mm long; pappus bristles 40–50, mostly in one series, sharply angled. *Aviles 500, Bangham 597* (*US*).

Rare or perhaps no longer present. The species flowers and fruits mostly from July to September, but a few collections were made in flower and fruit in January and March. Possibly the species flowers twice per year.

Southern Mexico to Peru and Brazil. In Panama, known mostly from wetter forest life zones at medium elevations; known from tropical moist forest on BCI, from premontane wet forest in Colón, Coclé, and Panamá, from tropical wet forest in Colón and Panamá, and from premontane rain forest in Panamá (Cerro Jefe).

Mikania leiostachya Benth., Pl. Hartweg. 201. 1845

Slender vine; branches nearly perpendicular to axis; stems puberulent and glandular when young, glabrate in age. Leaves opposite; petioles ca 1.5 cm long, moderately puberulent; blades ovate, acuminate, obtuse to rounded (sometimes slightly subcordate) at base, 3.5–12(18) cm long, 2–6(11) cm wide, sparsely pubescent below and on veins above, entire. Panicles axillary, to ca 30 cm long, the branches long, strictly spicate; heads discoid, ca 5 mm long, arranged in spikes and evenly disposed on rachis; phyllaries 4, pubescent, ca 3 mm long; ray flowers lacking; disk flowers bisexual, with a strong sweet aroma, the corolla white, shallowly lobed, the tube bent somewhat outward above the phyllaries, the anthers syngenesious, not becoming exserted, the style well-exserted, the outer edges papillate near apex, the stigmas rolling inward after emerging. Achenes ± oblong, 5-angled, glabrous; pappus bristles many, white, slightly exceeding corolla. *Croat 13245.*

Frequent in the canopy of the forest, occasionally over vegetation near the edge of the lake. Flowers and fruits principally in the dry season (January to April).

The sweet fragrance can be detected at considerable distances.

Honduras to Ecuador. In Panama, known only from tropical moist forest in the Canal Zone and Panamá.

See Fig. 549.

Mikania micrantha H.B.K., Nov. Gen. & Sp. 4:105. ed. fol. 1818

Slender vine; stem weakly pubescent. Leaves opposite; petioles 3–7 cm long; blades ovate-cordate, acuminate, 3.5–13 cm long, 2–8.5(10) cm wide, glabrate or sparsely pubescent, often glandular-dotted below, coarsely serrate; palmate veins at base 5–7. Heads discoid, 4-flowered, 4–5(8) mm long, in long-pedunculate, axillary, sub-corymbiform clusters; phyllaries 4, oblong, acute, 3–6 mm long; ray flowers lacking; disk flowers bisexual, the corolla white above constriction, green below, twisted outward above constriction, the stamens syngenesious, exserted at anthesis, the style long-emergent after anthesis, bending outward away from flower, the stigmas pubescent outside. Achenes oblong, ca 2 mm long, 4- or 5-ribbed, with viscid globular dots (at least on dried specimens); pappus of numerous barbellate bristles,

almost equaling corolla. *Croat 12949.*

Uncommon, usually over low vegetation near the edge of the lake and at the edges of clearings; locally very abundant. Flowers and fruits throughout the year, especially in the late rainy and early dry seasons.

The flowers are visited by two different species of *Trigona* bees.

Mexico to northern South America; West Indies. In Panama, widespread and ecologically variable; known from tropical moist forest in all provinces, from tropical dry forest in Los Santos, Herrera, Coclé, and Panamá, from premontane moist forest in Panamá, from premontane wet forest in Colón, Chiriquí, Veraguas, Coclé, Panamá, and Darién, and from tropical wet forest in Bocas del Toro, Colón, Coclé, Panamá, and Darién.

Mikania tonduzii B. L. Robinson, Proc. Boston Soc. Nat. Hist. 31:256. 1904

Stout herbaceous vine climbing into canopy, glabrous to inconspicuously puberulent on stems, petioles, and lower leaf surfaces; internodes elongate. Leaves opposite; petioles 2–4.5 cm long; blades ovate, acuminate to caudate-acuminate, acute to rounded at base and decurrent onto petiole, 6–15 cm long, 3–10 cm wide, bicolorous, pli-veined, the veins 5. Panicles terminal, 30–50 cm long, branched many times, the branches 8–15 cm long; branches, pedicels, and at least the lower part of phyllaries densely and inconspicuously puberulent and glandular-puberulent; heads 7–8 mm long, sessile or nearly so, in branched semicircular clusters; phyllaries 4, rounded at apex, to 2.5 mm long; ray flowers lacking; disk flowers 4, bisexual, ca 4 mm long, greenish, the lobes 5, narrowly acute, the style exserted, its branches to ca 4 mm long. Achenes narrowly oblong, 1.5–2 mm long, prominently ridged; pappus bristles antrorsely barbed, slightly exceeding corolla. *Croat 7972.*

Apparently rare, in the canopy; collected once in February. Seasonal behavior uncertain. Probably flowers in the dry season.

The corolla of this species distinguishes it from other *Mikania* by being narrow and lacking a distinct tube.

Southern Mexico to Panama. In Panama, known only from BCI.

NEUROLAENA R. Br.

Neurolaena lobata (L.) R. Br., Trans. Linn. Soc. London 12:120. 1817

Contragavilana

Coarse herb, usually 1–3 m tall; stems ribbed, appressed-strigose. Leaves alternate; petioles 5–20 mm long; blades oblong-lanceolate, acute to acuminate, gradually tapered to a decurrent and sometimes inequilateral base, 8–23 cm long, 3–6.5 cm wide, appressed-strigose, the lower surface more densely pubescent, glandular, the glands numerous, viscid, round, the margins remotely and sharply toothed (more so on juveniles), the lower leaves deeply trilobate. Panicles large, terminal, corymbose; heads discoid, ca 7 mm long; phyllaries in several series, blunt, 3-veined;

ray flowers lacking; disk flowers yellow, bisexual, each subtended by a thin blunt bract, the tube constricted somewhat below the syngenesious anthers, the anthers and style emerging simultaneously from tube, the stigmas then emerging from anthers and recurving. Achenes oblong-turbinate, ca 1.5 mm long, black, with a constriction near base; pappus of numerous soft bristles in 1 series, shorter than corolla. *Croat 9112.*

Occasional, in brushy places and clearings; one collection from a tree stump in the lake. Flowers and fruits principally in the dry season, December to April (sometimes to June).

Southern Mexico to Colombia, Ecuador, and Peru; West Indies. In Panama, ecologically variable; known from tropical moist forest in the Canal Zone, Bocas del Toro, Colón, Chiriquí, Los Santos, Panamá, and Darién, from premontane moist forest in the Canal Zone (Ancón), from premontane wet forest in Colón, Chiriquí, Coclé, Panamá, and Darién, and from tropical wet forest in Colón, Chiriquí, Coclé, Panamá, and Darién.

PLUCHEA Cass.

Pluchea odorata (L.) Cass. in Levr., Dict. Sci. Nat. 42:3. 1826

P. purpurascens (Sw.) DC.

Polygamous annual herb, 20–90 (120) cm tall, ± granular-puberulent. Leaves alternate; petioles 2–7 mm long; blades ± oblanceolate-elliptic, acute or acuminate at apex, tapering to base, 4–8 cm long, 1.5–3 cm wide, irregularly and often markedly serrate. Inflorescences terminal, subcorymbiform; heads few, closely aggregated, ca 1 cm wide and 1.5 cm high; phyllaries in several series, 4–7 mm long, weakly pubescent outside, glabrous inside, the margins ± scarious and regularly ciliate, the outermost ± ovate, the inner ones mostly oblong-elliptic; flowers pinkish-purple; ray flowers pistillate, in several series, with the corolla filiform, shorter than the style; disk flowers bisexual but mainly sterile, ca 4 mm long, often with the style undivided, the fertile ones with the anthers exserted above the corolla tube at anthesis. Achenes 4- or 5-angled; pappus bristles many, uniseriate, basally united. *Woodworth & Vestal 611.*

Apparently rare, in marshes along the shore. Seasonal behavior not determined. Flowering plants have been collected on BCI in February and July.

Seeds are probably dispersed by both wind and water.

Atlantic and Gulf states in the United States south to Honduras; Panama and Colombia; West Indies. In Panama, adventive; known only from two collections on BCI.

PSEUDOELEPHANTOPUS Rohr

Pseudoelephantopus spicatus (B. Juss. ex Aubl.) C. F. Baker, Trans. Acad. Sci. St. Louis 12:45, 55. 1902

Escobilla blanca, Chicoria

Coarse herb, to 1 (1.2) m tall; stems dichotomously branched, with loose trichomes borne on the bulbous

base. Leaves alternate; petioles short, sheathing; blades narrowly elliptic to obovate, acute at apex, dilated and clasping at base, mostly 6–13(21) cm long, 1–4(6) cm wide, hispid above, glabrous to sparsely appressed-pubescent and resin-dotted below, sparsely serrate. Heads sessile, 4-flowered, usually in axillary clusters but often forming long terminal spikes; phyllaries ca 1 cm long, in 4 pairs, sharply pointed; ray flowers lacking; disk flowers bisexual, the corolla deeply 5-lobed, the tube white, 5–6 mm long, the lobes violet, the anthers syngenesious, the filaments united to the tube, the style with antrorse bristles, the stigmas curling up after emerging from anthers. Achenes narrowly obovoid, 5–7 mm long, 10-striate, hispidulous principally on ribs; pappus of 5–15 unequal bristles, 2 of them folded and refolded near apex. *Croat 8565.*

Very abundant in flower and fruit in the Laboratory Clearing, especially in the late rainy season and during the dry season.

Continental tropical America; West Indies; introduced in Africa and eastern Asia. In Panama, known principally from tropical moist forest in the Canal Zone, Bocas del Toro, Colón, San Blas, Chiriquí, Los Santos, Panamá, and Darién; known also from tropical dry forest in Panamá, from premontane dry forest in Panamá, from premontane moist forest in the Canal Zone and Panamá, and from premontane wet forest in Colón, Chiriquí, Coclé, and Panamá.

ROLANDRA Rottb.

Rolandra fruticosa (L.) O. Kuntze, Rev. Gen. Pl. 1:360. 1891
Niagurgin

Tall herb, to 1.5 m, mostly unbranched; stems with long, fine, white trichomes. Leaves alternate; petioles short; blades ovate to elliptic, acute at apex, obtuse to acute at base, 4–11 cm long, 1–5 cm wide, white-woolly below, nearly glabrous above in age, entire. Heads very numerous, glomerate in axils, sessile, the glomerules 1–1.5 cm wide, each head consisting of a single, bisexual disk flower subtended by usually 2 phyllaries, these unequal, coriaceous, straight, with spinelike awns; corolla white, ca 3.5 mm long, the lobes 4, sometimes with greenish tips; stamens 4; filaments united to tube below lobes, not elongating after anthesis; anthers syngenesious, ca 1 mm long, exposed after petals open; style barely forked, elongating and emerging after anthesis; stigmas recurved. Achenes oblong-turbinate, ca 1.5 mm long, 4- or 5-costate, resin-dotted; pappus a minute, lacerate crown. *Croat 8589.*

Occasional, locally common in clearings. Flowers and fruits in the dry season, December to March (sometimes November to April).

Honduras to the Guianas, Peru, and Brazil; West Indies. In Panama, ecologically variable, but principally in wetter areas especially on the Atlantic slope; known from tropical moist forest in the Canal Zone, Bocas del Toro, Colón, San Blas, Herrera, Panamá, and Darién,

from tropical dry forest in Coclé (Penonomé), from premontane moist forest in the Canal Zone (Ancón Hill), from premontane wet forest in Colón, Chiriquí, Coclé, and Panamá, from tropical wet forest in Veraguas (Atlantic slope), Colón, and Panamá, and from premontane rain forest in Panamá (Cerro Jefe).

SCHISTOCARPHA Less.

Schistocarpha oppositifolia (O. Kuntze) Rydb., N. Amer. Fl. 34:306. 1927

Hermaphroditic or polygamous, coarse, erect herb, 1–1.5(3) m tall; branches short-hirsute. Leaves opposite; petioles 3–15 cm long, wing-margined; blades broadly ovate, acuminate, abruptly narrowed to cuneate base and decurrent onto petiole, 10–20 cm long, 4–12 cm wide, densely hirsutulous, dentate, with mucronate teeth. Panicles cymose, terminal and upper-axillary; heads numerous, 6–9 mm high, very short-pedunculate; phyllaries in 3 or 4 series, imbricate, obtuse at apex; paleae 6–7 mm long; ray flowers lacking or minute and pistillate; disk flowers bisexual, yellowish or greenish-yellow, ca 5 mm long, the anthers syngenesious, the style forked. Achenes oblong, ca 1.5 mm long, glabrous, black and shiny; pappus of slender soft bristles ca 5 mm long. *Aviles 16.*

Collected once by S. Aviles; not seen in recent years, but to be expected. Flowering throughout the year, especially in the early rainy season (July and August).

Mexico to Bolivia. In Panama, known from tropical moist forest in the Canal Zone, Bocas del Toro, Los Santos, and Panamá, from premontane wet forest in Colón, Los Santos, Coclé, and Panamá, from tropical wet forest in Colón, and from lower montane wet forest in Chiriquí.

SPILANTHES Jacq.

Spilanthes alba L'Hér., Stirp. Nov. 7, t. 4. 1785
S. ocymifolia (Lam.) A. H. Moore f. *radiifera* A. H. Moore

Erect herb, usually to 60 cm tall (rarely less than 10 cm or more than 100 cm); stems spreading-pubescent. Leaves opposite; petioles 5–30 mm long; blades ovate, acute to acuminate, obtuse to rounded and decurrent at base, 1.5–9 cm long, 0.5–5 cm wide, nearly glabrous to sparsely ± appressed-pubescent on both surfaces, dentate to subentire, sparsely appressed-ciliate; lateral veins mostly in 3 pairs, from near base. Inflorescences of 1 or 2 heads emerging terminally, but later in the dichotomies of the stem; peduncles 1.5–10 cm long; heads lenticular to round and finally ovoid, 8–14 mm diam, bearing many flowers; phyllaries lanceolate, ca 8, in 1 or 2 series; paleae ± equaling corolla, yellowish at apex, folded longitudinally to envelop flower; ray flowers yellow, to 1.5 mm long, mostly lacking or few, pistillate; disk flowers numerous, bisexual, the corolla tubular, white, 4–5 mm long, with 5 oblique lobes, minutely papillate-granular inside and out; anthers syngenesious, ca 0.5 mm long, with deltoid appendages; ovary obovate, flattened; style base globose;

style branches yellow or orange. Achenes compressed, oval, black, with a hyaline margin, ca 2 mm long, bearing 2 caducous setae or awns at apex and a dense row of cilia on margin; pappus lacking, or of 2 or 3 weak bristles. *Croat 7782.*

Occasional, in the Lighthouse Clearing. Flowers and fruits from November to May, especially during the dry season.

Mexico to Peru, Bolivia, and Brazil. In Panama, ecologically variable; known from tropical moist forest in the Canal Zone, Veraguas, Los Santos, Herrera, Panamá, and Darién, from premontane dry forest in Los Santos, from tropical dry forest in Panamá, from premontane wet forest in Coclé and Panamá, and from tropical wet forest in Colón.

SPIRACANTHA H.B.K.

Spiracantha cornifolia H.B.K., Nov. Gen. & Sp. 4:28. 1820

Herb, 30–150(200) cm tall; stems often maroon or streaked with maroon, covered with long loose trichomes. Leaves alternate; petioles 5–15 mm long, basally clasping; blades elliptic, acute, with a stiff spine at apex, obtuse to acute at base, 2–11 cm long, 1–4 cm wide, sparsely pubescent above (glabrate in age), densely woolly below with long straight trichomes on veins. Flower clusters dense, capitate, to 1.5 cm diam, mostly at ends of axillary branches, with 20–25 subsessile glomerules, the lowermost glomerules partially enclosed by leaflike bracts; bracts ovate, ca 2 cm long, membranaceous, with a sharp, horizontally directed mucro ca 1.5 mm long; heads 3–11 per glomerule, bearing 1 flower, developing centripetally, never more than 1 or 2 open at a time, soon falling; phyllaries about 6, imbricate, linear, thin and translucent, ca 4 mm long, unarmed at apex, with conspicuous, silky, white pubescence at base; ray flowers lacking; disk flowers bisexual, the corolla violet, to 4 mm long, deeply 4- or 5-lobed, the anthers syngenesious, 0.9 mm long, basally sagittate, the apical appendage ca 0.3 mm long; style branches 0.3 mm long, glabrate. Achenes obovate, ca 2 mm long (possibly immature), smooth, white; pappus with whitish resin dots on upper edge like those on phyllaries; scales narrow, brownish, ciliate, to 1 mm long. *Croat 11998.*

Locally abundant in the Rear #8 Lighthouse Clearing. Flowers and fruits principally during the dry season.

The plant has been confused with *Rolandra fruticosa,* which also has 1-flowered heads arranged in similar spiny clusters. *Rolandra* differs in having white flowers and usually two coriaceous phyllaries, with the outer one produced into a sharp acumen. *Spiracantha cornifolia* has violet flowers and about six membranaceous phyllaries, which are not spinescent. It is confused, however, because the lowermost glomerules of the inflorescences are partially enclosed by even spinier bracts turned outward.

Belize to Venezuela. In Panama, known principally from tropical moist forest in the Canal Zone, Colón, Panamá, and Darién; known also from tropical dry forest

in Herrera and Coclé, from premontane moist forest in the Canal Zone and Panamá, and from premontane wet forest in the Canal Zone and Panamá, and from premontane rain forest in Panamá.

See Fig. 550.

SYNEDRELLA Gaertn.

Synedrella nodiflora (L.) Gaertn., Fruct. & Sem. Pl. 2:456. 1791

Polygamous erect herb, to 75(150) cm tall, with appressed white trichomes. Leaves opposite; petioles winged, ciliate, short; blades ovate, acute to acuminate, abruptly or gradually tapered to base and broadly decurrent onto petiole, mostly 2.5–7.5(10) cm long, 1.5–5 cm wide, rugulose in age, crenate to subentire; basal veins 3. Heads 1 to few, nearly sessile, axillary, to 1 cm long, inconspicuous; phyllaries with the 2 outer pubescent, leaflike, to 1 cm long, exceeding flowers, the inner ones oblong, thin, ciliate; flowers yellow; ray flowers 3–5(9), ca 5 mm long, inconspicuous, pistillate, anthers lacking, the style forked, equaling ligule; disk flowers bisexual, 6–10(13), ca 4 mm long, yellow, the corolla lobes short, round, short-pubescent inside, the filaments fused to tube below middle, the anthers syngenesious, black, emerging somewhat from tube, the stigmas heavily antrorse-barbed outside, emerging from anthers. Achenes dimorphic, those of the ray flowers oblong to obovate, much flattened, the body black, 4–4.5 mm long, the margins pale green, the lateral and upper margins bearing 2 fin-shaped scales to 2 mm long; achenes of the disk flowers ± turbinate, weakly 2-edged, one side with bumpy projections; pappus of 2 or 3 stiff bristled awns to 5 mm long. *Croat 9187.*

Seasonally abundant in clearings, especially at the laboratory. Flowers and fruits principally in the dry season (January to May).

Seeds are probably epizoochorous.

Widespread in the tropics. In Panama, ecologically variable; known from tropical moist forest in the Canal Zone, Bocas del Toro, Colón, San Blas, Chiriquí, Herrera, Panamá, and Darién, from tropical dry forest in Los Santos and Panamá, from premontane moist forest in the Canal Zone and Panamá, and from premontane wet forest in Bocas del Toro, Colón, Coclé, and Panamá.

TRIDAX L.

Tridax procumbens L., Sp. Pl. 900. 1753

Polygamous, prostrate or erect herb, to 40 cm tall, hirsute, branched from the base, sometimes rooting at nodes. Leaves opposite; petioles short; blades ± ovate to ovate-lanceolate, acute or acuminate, cuneate at base, 2–6(12) cm long, 1–4(6) cm wide, incised-dentate (the basal tooth sometimes very large). Heads to 1 cm diam, solitary, on long terminal peduncles; phyllaries with the 2 outer ovate, green, 6–10 mm long; flowers interspersed with paleae, these acute, maroon-tipped; ray flowers pistillate, white, few, the ligule to 7 mm long, 2- or 3-lobed; disk

Fig. 550. *Spiracantha cornifolia*

Fig. 551.
Verbesina gigantea

flowers bisexual, yellow, 6–7 mm long; anthers weakly exserted; style branches slender, flattened, with subulate tips. Achenes hispid, ca 2 mm long; pappus of 18–20 plumose bristle-like scales, on ray-flower achenes to ca 2 mm long, on disk-flower achenes ca 5 mm long. *Croat 7004.*

Very abundant locally in the Laboratory Clearing; occasional elsewhere. Flowers and fruits throughout the year.

Distinguished by the solitary pedunculate heads with few white ray flowers.

Native to Central America; now widespread in the tropics and subtropics. In Panama, ecologically variable, but chiefly from drier areas and mostly on the Pacific slope; known from tropical dry forest in Los Santos, Coclé, and Panamá, from premontane moist forest in the Canal Zone, Los Santos, and Panamá, from tropical moist forest in the Canal Zone, Bocas del Toro, Chiriquí, and Panamá, and from premontane wet forest in Chiriquí, and Coclé, and Panamá.

VERBESINA L.

Verbesina gigantea Jacq., Coll. 1:53. 1787

V. myriocephala Schultz-Bip.

Cerbatana, Lengua de buey

Polygamous, stout, usually unbranched herb, mostly 2.5–3.5 m tall; stems smooth, purplish, glabrous, pruinose when young. Leaves mostly alternate, sessile, the lowermost deciduous; blades deeply pinnatifid, narrowed at base and decurrent onto stem, 20–40 cm long, densely pubescent especially below, the margins sharply and irregularly toothed. Corymbs large, terminal, pubescent; heads ca 7 mm high; phyllaries keeled, pubescent, acuminate, similar to paleae subtending each flower of the head; flowers white, the tube with stiffly ascending trichomes near base; ray flowers 1–3, pistillate, the ligule small, weakly trilobate; disk flowers many, bisexual, the anthers syngenesious, the style emerging from tube simultaneously with anthers but continuing to elongate, the stigmas recurving, their outer surfaces weakly antrorsely barbed. Achenes 2-edged, ciliate; pappus of two barbed bristles, shorter than corolla tube. *Croat 4387.*

Occasional, in clearings. Flowers and fruits principally in the dry season (December to April).

Mexico to Panama. In Panama, known from tropical moist forest in the Canal Zone and Panamá, from premontane moist forest in the Canal Zone, Coclé, and Panamá, and from premontane wet forest in Colón and Panamá.

See Fig. 551.

VERNONIA Schreb.

Vernonia canescens H.B.K., Nov. Gen. & Sp. 4:35. 1820

Hierba de San Juan

Arching shrub, usually 2–3 m long, the stems densely ± appressed-pubescent. Leaves alternate, short-petiolate; blades ovate to ovate-oblong, acuminate, obtuse to rounded at base, 4.5–11(12) cm long, to 5 cm wide, rugulose and short-pubescent above, densely pilose below especially on veins, the margins entire or remotely denticulate. Panicles terminal or upper-axillary; branches usually long, recurving; heads bell-shaped, 6–7 mm high, widely spaced and chiefly on upper side of branches; phyllaries with the outer ones triangular-subulate, spinose-tipped, the inner ones lanceolate, acuminate, scarious; ray flowers lacking; disk flowers bisexual, developing centripetally, the corolla pale lavender (fading white), the lobes long and slender, tipped on outside with glandular droplets, loosely reflexed after anthesis, the filaments not elongating, the anthers syngenesious, the style densely barbed on outside, the stigmas usually not recurving until totally extended. Achenes narrowly turbinate, ca 2 mm long, 10-ribbed, with ascending white trichomes; pappus bristles many, weakly barbed, shorter than corolla. *Croat 8777.*

Common in clearings; occasional on the shore. Flowers and fruits throughout the dry season (January to May).

Mexico to Colombia, Venezuela, and Peru. In Panama, ecologically variable; known from tropical moist forest in the Canal Zone, Veraguas, Los Santos, Herrera, and Panamá, from tropical dry forest in Coclé and Panamá (Taboga Island), from premontane moist forest in the Canal Zone and Panamá, from premontane wet forest in Colón and Panamá, from tropical wet forest in Panamá, and from lower montane wet and rain forests in Chiriquí.

Vernonia cinerea (L.) Less., Linnaea 4:291. 1829

Erect herb, less than 75 cm tall (to 150 cm elsewhere); stems ribbed; pubescence whitish, ± appressed, dense except on upper blade surface. Leaves alternate, often with 1 to several smaller leaves on a very short axillary branch; petioles less than 1 cm long; blades ovate or subrhombic, blunt at apex, obtuse at base and abruptly decurrent nearly to base of petiole, 1.5–5 cm long, 1–3 cm wide, weakly sinuate-dentate. Flowers in terminal panicles; heads discoid, ca 6 mm long, on short slender peduncles; phyllaries in several series, acuminate; ray flowers lacking; disk flowers bisexual, orchid, ca 4 mm long, the corolla lobes long and slender, spreading at anthesis

KEY TO THE SPECIES OF VERNONIA

Plants herbs, usually less than 1.5 m tall . . . . *V. cinerea* (L.) Less.
Plants arching shrubs or small trees, more than 2 m tall:
    Plants erect shrubs or small trees; phyllaries green-tipped, bicolorous; blades mostly appressed-
        pubescent . . . . *V. patens* H.B.K.
    Plants arching shrubs or vines; phyllaries concolorous, their tips not dark green; blades with
        erect trichomes . . . . *V. canescens* H.B.K.

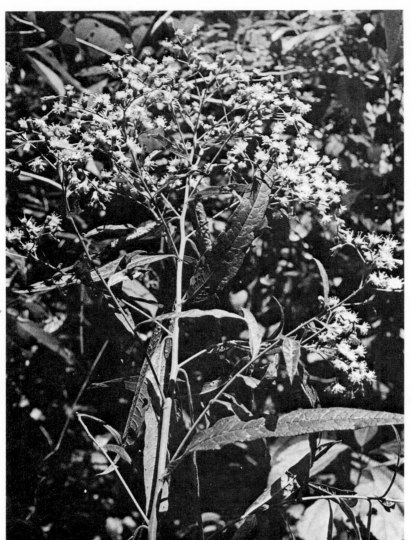

Fig. 552.
Vernonia patens

Fig. 553. *Wulffia baccata*

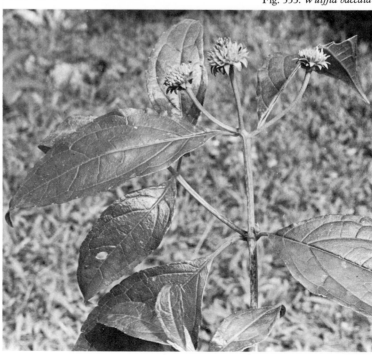

to expose syngenesious anthers, the style exserted, its branches bristled on outside. Achenes terete, ca 1.5 mm long, bristly pubescent; pappus of numerous barbed bristles, slightly shorter than corolla. *Croat 6879.*

Occasional in the Laboratory Clearing. Flowers and fruits throughout the year.

Native to the Old World, but naturalized in the New World and distributed widely in the tropics and subtropics. In Panama, known from tropical moist and premontane moist forests on both slopes of the Canal Zone and from tropical dry forest in Panamá (Taboga Island).

Vernonia patens H.B.K., Nov. Gen. & Sp. 4:41. 1820
  Boton de pegapega, Lengua de buey o vaca, Palo blanco, Salvia, Sanalego, Tuete

Slender shrub or small tree, to 3.5(8) m tall; trunk mostly unbranched to near apex; stems moderately to densely pubescent with ± viscid trichomes, 3-ribbed, the ribs originating below each petiole. Leaves alternate; petioles short; blades narrowly elliptic or lanceolate, acuminate, acute at base, 6–19 cm long, 1–5 cm wide, sparsely appressed-puberulent especially on veins, the margins with small apiculate teeth. Panicles 1-sided, scorpioid; heads sessile, ca 9 mm long; phyllaries in several series, acute, darker green at apex, the margins thin, ciliolate; ray flowers lacking; disk flowers bisexual, white (at least some flowers of the outer ring tinged with lavender), to 5.3 mm long, usually bent outward, the lobes long and slender, often remaining erect, the anthers syngenesious, becoming exserted, the style branches becoming long-exserted, antrorsely barbed outside. Achenes minutely pubescent, ca 1.5 mm long; pappus of many barbed bristles, equaling length of corolla on inner flowers, shorter than length of corolla on outer ring of flowers. *Croat 8778.*

Common in clearings. Flowers and fruits throughout the dry season (January to April).

Mexico to Colombia and Venezuela. In Panama, known from tropical moist forest in the Canal Zone, Bocas del Toro, Colón, Chiriquí, Veraguas, Herrera, Panamá, and Darién, from tropical dry forest in Coclé, from premontane moist forest in Panamá, from premontane wet forest in Colón, Chiriquí, and Panamá, from tropical wet forest in Chiriquí and Panamá, and from lower montane wet forest in Chiriquí.

See Fig. 552.

WEDELIA Jacq.

Wedelia trilobata (L.) Hitchc., Annual Rep. Missouri
  Bot. Gard. 4:99. 1893
  Clavellin de playa

Polygamous low herb, often decumbent, sometimes rooting at lower nodes, to 2 m long; stems sparsely strigose or glabrous, weakly ribbed when dry. Leaves opposite, sessile or short-petiolate or perfoliate; blades obovate-elliptic to subrhombic, acute at apex, conspicuously narrowed in basal third and obtuse to rounded at base, 5–12 (18) cm long, 2.5–4.5 cm wide, sparsely scabrous above, less so below, broadly and irregularly serrate, sometimes shallowly trilobate. Heads ca 2.5 cm diam,

solitary from upper axils; peduncles and phyllaries strigose; peduncles 3–9 cm long; phyllaries in ca 3 series, the outermost foliaceous and longer than the inner ones; flowers yellow; ray flowers pistillate, ca 1 cm long, the style minutely trilobate; disk flowers bisexual, ca 4.7 mm long, the lobes pubescent inside, the anthers black, syngenesious, included, the style yellow, exserted, its lobes 2, recurved. Achenes ca 5 mm long, weakly pubescent, with a few glands; pappus minute, cup-shaped, the margins irregular. *Croat 11350.*

Locally abundant in clearings and on beaches of the lakeshore. Flowers and fruits in the rainy season (April to December).

Florida, Central and South America; West Indies; West Africa. In Panama, found in disturbed, wet places, principally on the Atlantic slope; known from tropical moist forest in the Canal Zone, Bocas del Toro, Colón, San Blas, Los Santos, and Panamá, from tropical dry forest in Panamá, from premontane moist forest in Panamá, from premontane wet forest in Colón and Panamá, and from tropical wet forest in Colón and Veraguas.

WULFFIA Neck. ex Cass.

Wulffia baccata (L.f.) O. Kuntze, Rev. Gen. Pl.
  1:373. 1891

Vine or scrambling shrub, usually 2–4 m long, moderately strigose; branches perpendicular to axis; stems slightly angulate, asperous, mottled with purple. Leaves opposite; petioles 1–2 cm long; blades lanceolate-elliptic to ovate, falcate-acuminate, obtuse to acute and often inequilateral at base, 9–20 cm long, 3.5–9 cm wide, asperous below. Inflorescences terminal or upper-axillary, or 2 or 3 pedunculate heads, often on short lateral branches; peduncles 2–4 cm long; heads large, bearing many flowers; phyllaries in ca 2 series, green to orange, subequal, narrowly acute and recurved; ray flowers 8–15, yellow, sterile, irregular, oblong, weakly bifid at apex, 1–2 cm long; disk flowers bisexual, orange, each subtended by paleae, the paleae orange, to 7 mm long, longer than the disk flowers in bud but considerably shorter at anthesis, the corolla ca 6 mm long, shortly 5-lobed, the tube narrowed just above the base, the stamens syngenesious, the filaments fused to slender part of corolla, the anthers ca 2 mm long, black, the connective acute above, orange, the style branches exserted and spreading, orange, conspicuously antrorse-pubescent. Achenes ± globose, ca 5 mm diam, minutely pubescent at apex, dark green and fleshy at maturity, contrasting sharply with the persistent orange palea; pappus lacking. *Croat 11466.*

Common at margins of the forest or in tree-fall areas within the forest. Flowers and fruits principally in the rainy season (July to December).

Panama south to Bolivia and Brazil; West Indies. In Panama, known from tropical moist forest in the Canal Zone, Bocas del Toro, San Blas, Herrera, Panamá, and Darién, from premontane moist forest in the Canal Zone and Panamá, from premontane wet forest in Colón, Coclé, Panamá, and Darién, and from tropical wet forest in Colón, Panamá, and Darién.

See Fig. 553.

Reference Material

Key to Sterile Woody Plants

Nearly 700 species are included in this key. Excluded are species that are both cultivated and known only from the Laboratory Clearing; species that are usually mostly herbaceous; and some species that are probably no longer on the island. Some of the more common coarse herbaceous vines in the forest *are* included, because they will be more frequently encountered and would otherwise be difficult to identify.

Approximately 60% of the woody plants can be identified with little difficulty, the remainder with varying degrees of difficulty. Species with excessive variation, or with juvenile forms that do not resemble the adult plants, may not key out. A few couplets terminate with two species because no characters could be found that would adequately distinguish the species from each other in sterile condition.

Identifications should be thoroughly checked against the text descriptions and compared with herbarium specimens where possible. Troublesome couplets should be marked and the plant keyed out two or more ways, for even if the process yields several possibilities, it is faster and easier to check these in an herbarium than to attempt identification in most other ways.

Pairs of symbols were added before selected couplets to facilitate progress through the keys.

For further discussion of this key, see p. 53.

KEY TO STERILE WOODY PLANTS

Leaves simple:
    Leaves opposite or whorled . . . . KEY 1, p. 861
    Leaves alternate or spiraled:
        Plants not lianas . . . . KEY 2, p. 877
        Plants lianas . . . . KEY 3, p. 896
Leaves compound:
    Leaves palmately compound with 4 or more leaflets . . . . KEY 4, p. 899
    Leaves pinnately compound or with 3 or fewer leaflets:
        Leaves compound more than once . . . . KEY 5, p. 899
        Leaves pinnate or with 3 or fewer leaflets . . . . KEY 6, p. 901

KEY 1: LEAVES SIMPLE, OPPOSITE OR WHORLED

● Blades with the major veins arising at or near base and ± paralleling midrib for length of leaf, usually with tertiary veins perpendicular to and connecting these major laterals:
  Blades maroon beneath . . . . *Miconia rufostellulata* (107. Melastomataceae)
  Blades not maroon beneath:
    ◆ Blades lacking stellate or densely branched trichomes and lacking fimbriate scales along lower midrib:
      ■ Blades essentially glabrous (possibly pubescent on veins or in axils below):
        Plants lianas; blades entire:
          Stems (at least older ones) armed; blades with white trichomes in small tufts in axils . . . .
          . . . . *Strychnos brachistantha* (115. Loganiaceae)
        Stems not armed; blades lacking tufts of white trichomes in axils:
          Blades with tufts of long, brownish trichomes in axils at base; petioles glabrous or puberulent . . . . . .
          . . . . *Strychnos darienensis* (115. Loganiaceae)
          Blades not tufted in axils at base; petioles with long brownish or white trichomes . . . .
          . . . . *Strychnos panamensis* (Fig. 447; 115. Loganiaceae)
        Plants shrubs or small trees:

Blades usually with small tufts of trichomes in basal vein axils below; larger stems spiny . . . . *Strychnos brachistantha* (115. Loganiaceae)
Blades lacking tufts in axils:
 Blades obscurely crenulate, with sparse stiff setae arising from vein ends near apex; plants shrubs . . . . *Miconia lateriflora* (107. Melastomataceae)
 Blades entire, without marginal setae:
  Plants trees, usually 6–10 m tall . . . . *Miconia hondurensis* (107. Melastomataceae)
  Plants shrubs, usually less than 3.5 m tall (much shorter in parasitic *Oryctanthus*):
   Plants not epiphytic or parasitic; petioles ca 5 mm long . . . . *Conostegia cinnamomea* (Fig. 431; 107. Melastomataceae)
   Plants epiphytic or parasitic; petioles 1–2 cm long:
    Plants large epiphytic shrubs; leaves thin, deciduous in the dry season; tertiary veins conspicuous, perpendicular to major veins . . . . *Topobaea praecox* (107. Melastomataceae)
    Plants small parasitic shrubs; leaves thick, not deciduous; tertiary veins not visible, not perpendicular to major veins:
     Leaf blades obtuse to rounded at base (rarely subcordate) . . . . *Oryctanthus alveolatus* (42. Loranthaceae)
     Leaf blades broadly rounded to subcordate at base . . . *Oryctanthus cordifolius* (42. Loranthaceae)
■ Blades pubescent:
 Petioles and stems glabrous or inconspicuously pubescent in age:
  Blades more than 10 cm wide, entire . . . . *Bellucia grossularioides* (107. Melastomataceae)
  Blades less than 5 cm wide, coarsely toothed:
   Blades asperous above, ovate to elliptic, shallowly toothed . . . . *Calea prunifolia* (133. Compositae)
   Blades smooth above, rhombic-ovate, usually coarsely toothed . . . . *Chromolaena odorata* (Fig. 546; 133. Compositae)
 Petioles and stems conspicuously pubescent:
  Plants lianas:
   Branches not paired and divaricate; blades minutely strigose below; petioles and stems near nodes short-rufous-pubescent . . . . *Adelobotrys adscendens* (107. Melastomataceae)
   Branches always paired and divaricate:
    Pubescence on stems and blades long, conspicuous; blades lacking axillary domatia below . . . . *Strychnos toxifera* (Fig. 448; 115. Loganiaceae)
    Pubescence short; blades with axillary tufts between inner pair of lateral veins below . . . . *Strychnos darienensis* (115. Loganiaceae)
  Plants shrubs or small trees:
   Lateral veins with the inner pair departing midrib 5 mm or more above basal pair:
    Blades ending abruptly at base, often inequilateral; trichomes on stems erect and spreading . . . . *Clidemia dentata* (Fig. 428; 107. Melastomataceae)
    Blades ± decurrent onto petioles; trichomes on stems appressed or acropetal; pairs of leaves often unequal . . . . *Miconia nervosa* (Fig. 436; 107. Melastomataceae)
   Lateral veins all departing midrib at or very near base (less than 5 mm above base):
    Trichomes on stems, petioles, veins below, and margins of blade brown, more than 5 mm long . . . . *Clidemia collina* (107. Melastomataceae)
    Trichomes on stems and petioles less than 5 mm long:
     Larger blades with 7 veins from base; blades broadly rounded to subcordate at base, often with some stellate trichomes on upper surface . . . . *Leandra dichotoma* (Fig. 433; 107. Melastomataceae)
     Blades with 3–5 veins from base; trichomes on upper surface simple or lacking:
      Stems and petioles with purplish trichomes much longer than trichomes on midrib . . . . *Miconia lacera* (Fig. 435; 107. Melastomataceae)
      Stems and petioles with trichomes not purplish, ± the same size as trichomes on midrib:
       Blades conspicuously pubescent above and below, usually acute at base . . . . *Conostegia bracteata* (Fig. 430; 107. Melastomataceae)
       Blades ± glabrous above and minutely strigose below, usually ± rounded at base . . . . *Adelobotrys adscendens* (107. Melastomataceae)
◆ Blades with some stellate or densely branched trichomes or fimbriate scales below:
 ★ Major lateral veins departing midrib more than 5 mm above base:
  ▲ Stellate or densely branched trichomes below very minute (appearing merely granular in hand lens), ± restricted to veins:
   Blades less than 2.5 times longer than broad:
    Blades sparsely pilose above, the margins dentate, with ± irregular, pilose cilia . . . . *Ossaea quinquenervia* (107. Melastomataceae)
    Blades glabrous to granular-puberulent above, the margins ± entire, with sparse, short cilia . . . . *Clidemia septuplinervia* (107. Melastomataceae)
   Blades more than 2.5 times longer than broad:

Petioles more than 1 cm long; blades long-decurrent onto petiole . . . .
. . . . *Miconia lonchophylla* (107. Melastomataceae)
Petioles usually less than 1 cm long; blades not conspicuously decurrent onto petiole:
 Blades usually less than 6 cm wide, often to 8 cm wide; lateral veins spaced mostly 3–5 mm apart,
  the smallest reticulate veins clearly visible, darker than surface . . . .
   . . . . *Conostegia cinnamomea* (Fig. 431; 107. Melastomataceae)
 Blades usually more than 8 cm wide, rarely less than 6 cm wide; lateral veins 5–12 mm apart, the
  smallest reticulate veins not clearly visible, not darker than surface . . . .
   . . . . *Miconia prasina* (Fig. 437; 107. Melastomataceae)
▲ Stellate or densely branched trichomes below very dense and/or conspicuous:
 Blades pubescent above at maturity:
  Blades with 3 veins from base; trichomes above sparse, ± appressed; trichomes below with a single
   strong central axis and with many fine radial arms at base . . . .
    . . . . *Henriettea succosa* (Fig. 432; 107. Melastomataceae)
  Blades with 5–7 veins from base; trichomes above dense, erect; trichomes below of 2 types (simple
   and stellate) . . . . *Conostegia speciosa* (107. Melastomataceae)
 Blades glabrous above at maturity:
  Blades subsessile, subcordate, often more than 10 cm wide . . . .
   . . . . *Miconia impetiolaris* (107. Melastomataceae)
  Blades petiolate:
   Petioles 1–1.5 cm long; blades usually less than 7 cm broad . . . .
    . . . . *Conostegia xalapensis* (107. Melastomataceae)
   Petioles more than 4 cm long; blades more than 10 cm wide . . . .
    . . . . *Miconia serrulata* (107. Melastomataceae)
★ Major lateral veins all departing midrib at or less than 5 mm above base:
 Blades densely pubescent above; plants with simple, glandular, and stellate trichomes:
  Blades cordate at base; trichomes on stem 5 mm long or more; blade surface smooth . . . .
   . . . . *Clidemia octona* (Fig. 429; 107. Melastomataceae)
  Blades obtuse at base; trichomes on stem short; blade surface bullate . . . .
   . . . . *Clidemia capitellata* (Fig. 427; 107. Melastomataceae)
 Blades glabrous or sparsely pubescent above; plants lacking gland-tipped trichomes (except *Miconia elata*):
  Lateral veins obscure or closely spaced (less than 2 mm apart):
   Plants shrubs or small trees, not epiphytic, usually on shore; blades narrowly ovate, long-acuminate;
    lateral veins obscure, not at all raised; stems slender, drying smooth . . . .
    . . . . *Miconia borealis* (107. Melastomataceae)
   Plants shrubs, epiphytic, usually in the forest, rarely on shore; blades ± obovate, not long-acuminate;
    lateral veins distinctly raised; stems thick, drying wrinkled . . . .
    . . . . *Topobaea praecox* (107. Melastomataceae)
  Lateral veins prominent, more than 2 mm apart:
   Trichomes below long, stout, simple (densely branched at base); blades with 3 strong lateral veins . .
   . . . . *Henriettea succosa* (Fig. 432; 107. Melastomataceae)
   Trichomes not as above:
    Lower blade surface completely covered with minute stellate trichomes:
     Lower blade surface brownish, with many, minute, dark-brown, glandular areas scattered among
      the trichomes . . . . *Miconia elata* (107. Melastomataceae)
     Lower blade surface whitish, lacking dark-brown glandular areas . . . .
     . . . . *Miconia argentea* (Fig. 434; 107. Melastomataceae)
    Lower blade surface not completely covered with minute stellate trichomes:
     Lower blade surface often purple, sparsely pubescent . . . .
     . . . . *Miconia rufostellulata* (107. Melastomataceae)
     Lower blade surface green, sparsely to densely pubescent:
      Blades acute at base; stellate trichomes minute, inconspicuous . . . .
      . . . . *Miconia affinis* (107. Melastomataceae)
      Blades cordate at base:
       Trichomes below minute, stellate, restricted to veins . . . .
       . . . . *Clidemia purpureo-violacea* (107. Melastomataceae)
       Trichomes below simple or with long lateral branches, scattered on blade, more dense on
        midrib . . . . *Miconia shattuckii* (107. Melastomataceae)
• Blades with the major veins not arising at or near base and not ± parallel to midrib for length of leaf:
 ⊙ Plants with interpetiolar stipules or their scars continuous between the petioles; milky sap lacking:
  ✿ Stipules not entire, irregularly divided or slightly bifid at top to divided all the way to base and thus appearing
   as 2 separate lobes *or* caducous, leaving a row of interpetiolar setae *or* with the lobes aristate, with tri-
   chomes more than 2 mm long:
   □ Stipules bifid only at tip or divided less than two-thirds their length, otherwise entire (margins may be ciliate
    but not aristate), not represented by an interpetiolar line of stiff setae:

Young stems conspicuously pubescent, the trichomes brown, matted, spreading (not appressed), ca 0.5 mm long:
  Upper blade surface densely short-pubescent, ± scabrous . . . .
  . . . . *Psychotria micrantha* (Fig. 532; 130. Rubiaceae)
  Upper blade surface essentially glabrous, smooth . . . . . . *Psychotria psychotriifolia* (Fig. 533; 130. Rubiaceae)
Young stems glabrous or with very closely appressed trichomes:
  Lateral veins below ± the same thickness and length throughout the blade; reticulate veins inconspicuous, often barely visible on young blades . . . . *Psychotria granadensis* (130. Rubiaceae)
  Every second, or second and third, lateral vein below much thinner and shorter than the others; reticulate veins below conspicuous, connecting the lateral veins:
    Stipules 8–15 mm long, foliaceous, caducous, bifid for less than one-fifth their length . . . .
    . . . . *Psychotria capitata* (Fig. 527; 130. Rubiaceae)
    Stipules 4–10 mm long, mostly persistent, usually bifid more than one-third their length:
      Stipules coriaceous, their margins as durate as the body, the lobes usually flaring away from the stem, some of the stipules usually divided almost to base, the sinus between the lobes usually broadly U-shaped . . . . *Palicourea guianensis* (Fig. 524; 130. Rubiaceae)
      Stipules ± herbaceous, their margins ± thin, usually dissipating before the main body, the lobes not flaring, not divided nearly to base, the sinus between the lobes usually V-shaped . . . .
      . . . . *Psychotria brachiata* (130. Rubiaceae)
◻ Stipules divided completely to base or more than two-thirds their length *or* otherwise not entire *or* with conspicuous setae on margins ca 4 mm long *or* caducous and represented only by an interpetiolar line of stiff setae:
  Young stems glabrous (sometimes with several widely dispersed trichomes):
    Stipule margin with several, long, stiff setae ca 4 mm long; shrubs less than 50 cm tall . . . .
    . . . . *Cephaelis ipecacuanha* (Fig. 516; 130. Rubiaceae)
    Stipule margin lacking long stiff setae:
      Stipules represented by a line of broad-based setae:
        Setae ca 5 mm long . . . . *Mouriri myrtilloides* subsp. *parvifolia* (107. Melastomataceae)
        Setae ca 0.5 mm long . . . . *Psychotria carthagenensis* (Fig. 528; 130. Rubiaceae)
      Stipules not a line of setae:
        Stipule lobes linear:
          Petioles 1.5–4.5 cm long; leaves attenuate at base . . . . . . *Cephaelis discolor* (Fig. 515; 130. Rubiaceae)
          Petioles 0.2–1 cm long; leaves ± rounded at base:
            Stipule lobes ca 10 mm long, their bases 1.5–2 mm wide . . . .
            . . . . *Psychotria deflexa* (Fig. 529; 130. Rubiaceae)
            Stipule lobes ca 3 mm long, their bases ca 1 mm wide . . . .
            . . . . *Psychotria acuminata* (Fig. 526; 130. Rubiaceae)
        Stipule lobes ovate to lanceolate:
          Petioles to 3 cm long; leaves 12–25 cm long; stipule lobes usually ± ovate and ± rounded at apex, sometimes irregularly divided . . . . *Palicourea guianensis* (Fig. 524; 130. Rubiaceae)
          Petioles 1–1.5 cm long; leaves 10–16 cm long; stipule lobes usually ± lanceolate, ± acute or apiculate at apex . . . . *Psychotria brachybotrya* (130. Rubiaceae)
  Young stems variously pubescent, often inconspicuously so:
    Trichomes on young stems densely spreading, ca 1.5 mm long; upper blade surface conspicuously pubescent . . . . *Cephaelis tomentosa* (Fig. 517; 130. Rubiaceae)
    Pubescence not as above:
      Plants lianas; stipules caducous . . . . *Trigonia floribunda* (72. Trigoniaceae)
      Plants shrubs:
        Stipules represented by a line of short bristles ca 0.5 mm long; stems and terminal buds densely sericeous with appressed trichomes; terminal buds conspicuous, ca 8 mm long [if only a line of short bristles are present and the plant does not fit the rest of this description, look for stipules at the very tips of the branches and return to ◔ on the preceding page] . . . .
        . . . . *Amaioua corymbosa* (130. Rubiaceae)
        Stipules not represented by a line of bristles, otherwise not as above:
          Stipules 10–15 mm long or, if less than 10 mm, the lobes separated at base, persistent:
            Stems cinereous, with dense appressed trichomes ca 0.5 mm long; blades 10–45 cm long; stipules ciliate, glabrous to appressed-pubescent . . . . *Isertia haenkeana* (130. Rubiaceae)
            Stems merely puberulent to ± cinereous, with appressed trichomes ca 0.1 mm long; blades 8–19 cm long; stipules not ciliate, glabrous to inconspicuously strigose with minute, black, slender trichomes . . . . *Psychotria racemosa* (Fig. 534; 130. Rubiaceae)
          Stipules usually less than 10 mm long, the lobes not separated at base:
            Blades usually 3.5–12 cm long and 1.2–4.5 cm wide:
              Young stems conspicuously cinereous, with appressed, often matted trichomes ca 1 mm long; lower midrib with dense appressed trichomes ca 0.5 mm long . . . .
              . . . . *Psychotria pittieri* (130. Rubiaceae)

Young stems minutely pubescent, with erect trichomes; lower midrib glabrous or puberulent, with appressed or spreading trichomes ca 0.1 mm long:

Blades usually more than 12 cm long, at least the larger blades more than 3.5 cm wide; major lateral veins usually 10 or more on each side of midrib . . . . *Psychotria acuminata* (Fig. 526; 130. Rubiaceae)

Blades usually less than 10 cm long and 3.5 cm wide; major lateral veins usually 9 or fewer on each side of midrib . . . . *Psychotria furcata* (130. Rubiaceae)

Blades usually 8–21 cm long and 2–10 cm wide:

Younger parts of plant minutely puberulent, appearing almost glabrous, the trichomes scattered, less than 0.1 mm long . . . . *Psychotria acuminata* (Fig. 526; 130. Rubiaceae)

Younger parts of plant more conspicuously puberulent, sometimes appearing cinereous, the trichomes to ca 0.2 mm long:

Stipules 3–5 mm long, triangular, pliable . . . . *Psychotria pubescens* (130. Rubiaceae)

Stipules more than 6 mm long, slender-lanceolate, stiff . . . . *Psychotria racemosa* (Fig. 534; 130. Rubiaceae)

✪ Stipules entire:

Young stems densely pubescent, the trichomes 0.5 mm long or longer:

Stipules glabrous or with at least some part of the outer surface glabrous (may have dense trichomes only on midrib and/or at base):

Youngest blades with the upper surface bearing scattered, broad-based, appressed trichomes to ca 0.4 long:

Blades 3–6(11) cm long and 1.5–3 cm wide; petioles obscure or to 1 cm long; stipules 1.5–3 mm long, triangular, often scoop-shaped and projecting away from stem . . . . *Randia formosa* (Fig. 535; 130. Rubiaceae)

Blades and petioles longer than above; stipules not as above:

Plants arching shrubs or lianas; blades briefly decurrent onto petiole; lower blade surface in dried specimens appearing finely wrinkled between lateral veins . . . . *Chomelia psilocarpa* (130. Rubiaceae)

Plants trees; blades rounded to obtuse at base, not at all decurrent onto petiole; lower blade surface not wrinkled . . . . *Guettarda foliacea* (Fig. 523; 130. Rubiaceae)

Youngest blades with the upper surface glabrous:

Plants trees, to 15(25) m tall; blades blunt to rounded at apex; stipules ovate, to 1.8 cm long; trunk deeply involuted . . . . *Macrocnemum glabrescens* (130. Rubiaceae)

Plants shrubs, to 3 m tall; blades sharply acute at apex; stipules acute, to 0.5(1) cm long; trunk not involuted:

Stipules joining above petiole; trichomes straight, slender . . . . *Bertiera guianensis* (130. Rubiaceae)

Stipules separate above petiole; trichomes ± curly, matted . . . . *Psychotria emetica* (Fig. 530; 130. Rubiaceae)

Stipules with the outer surface completely and evenly pubescent:

Lower blade surface (not veins only) conspicuously and permanently densely pubescent, the trichomes more than 0.5 mm long:

Plants trees, to 15 m tall; stipules long-attenuate; trichomes on upper blade restricted mostly to veins . . . . *Antirrhoea trichantha* (130. Rubiaceae)

Plants shrubs or vines, the younger parts herbaceous; stipules ovate and apiculate at apex; trichomes on upper blade scattered evenly over surface . . . . *Sabicea villosa* var. *adpressa* (130. Rubiaceae)

Lower blade surface ± glabrous to minutely puberulent (may have dense trichomes on veins or widely scattered trichomes less than 0.5 mm long):

Stipules densely brown-tomentose, to 1 cm long; lower stipule scars represented by a line of dark bristles ca 0.5 mm long; trees to 15 m tall . . . . *Amaioua corymbosa* (130. Rubiaceae)

Stipules and habit not as above:

Midrib of upper surface of young blades with dense, appressed trichomes ca 0.5 mm long . . . . *Guettarda foliacea* (Fig. 523; 130. Rubiaceae)

Midrib of upper surface of young blades essentially glabrous:

Plants usually 2–4 m tall; stems and veins of lower blade surface smoothly and densely appressed-pubescent (subsericeous), the trichomes whitish; stipules narrowly triangular, to 5 mm long, pubescent like stems . . . . *Bertiera guianensis* (130. Rubiaceae)

Plants usually less than 1 m tall, often less than 30 cm dbh; stems and veins of lower blade surface coarsely and sparsely strigose and with scattered debris fragments (appearing very dirty), the trichomes white or gray but not conspicuous; stipules inconspicuous, triangular, less than 2 mm long, strigose . . . . *Psychotria emetica* (Fig. 530; 130. Rubiaceae)

Young stems glabrous to densely pubescent, the trichomes less than 0.3 mm long (or longer than 0.3 mm and widely scattered):

○ Reticulate veins not directly visible; tertiary veins completely obscure; plants, if lianas, with persistent stipules:

Blades 2–5(10) cm long and 1.3–4(6) cm wide:

Stipules ca 8–15 mm long, with the cusp ± equaling the main stipular body; plants shrubs, frequent
....*Psychotria chagrensis* (130. Rubiaceae)
Stipules to ca 4 mm long, with the cusp 3 or more times longer than the short, flaring base; plants
shrubs, often climbing and vinelike, rare......................*Chiococca alba* (130. Rubiaceae)
Blades 5–30 cm long and 2.5–14 cm wide:
Blade apex rounded; stipules ± obovate...............*Cosmibuena skinneri* (Fig. 518; 130. Rubiaceae)
Blade apex acuminate to obtuse; stipules not obovate:
Blades 5–10(16) cm long and 2.5–5.5 cm wide; lower stipules represented by a dense line of setae ca
0.7 mm long............................*Psychotria carthagenensis* (Fig. 528; 130. Rubiaceae)
Blades larger than above, lacking setae:
Stipules represented by a low, thick, forward-projecting ridge ca 1.5–3.5 mm high...........
....*Psychotria uliginosa* (130. Rubiaceae)
Regular stipules present, 5–10 mm long..........*Psychotria limonensis* (Fig. 531; 130. Rubiaceae)
○ Reticulate veins directly visible (may be very small); quaternary veins often also visible:
Blades 30–65 cm long and 21–37 cm wide; stipules ca 7 cm long.................................
....*Pentagonia macrophylla* (130. Rubiaceae)
Blades and stipules smaller:
Plants lianas, the stems with thick-based, recurved spines usually less than 10 mm long; blade base
cordate to rounded............................*Uncaria tomentosa* (Fig. 537; 130. Rubiaceae)
Plants with the stems unarmed; blade base not cordate or rounded:
◇ Midrib on lower surface of young blades variously pubescent and/or the blades with tufts of
trichomes in vein axils:
Trichomes on midrib of younger blades below erect to vaguely appressed or less than 0.2 mm
long:
Lower surface of young (to half-mature) blades densely covered with short erect trichomes,
giving the leaf a velvety feel.................*Genipa americana* (Fig. 520; 130. Rubiaceae)
Leaf pubescence not as above:
Branches with the 2–10 cm below the tips mostly glabrous, with conspicuous white lenticels:
Lenticels roundish; vein axils below with conspicuous tufts of trichomes; blades acute to
obtuse......................*Coutarea hexandra* (Fig. 519; 130. Rubiaceae)
Lenticels elongate, slender; vein axils not tufted; blades briefly acuminate at apex......
....*Pogonopus speciosus* (130. Rubiaceae)
Branches not as above:
Lateral veins below usually somewhat enlarged, with pitlike domatia in axils, the trichomes
on veins very short, granular, not densely pilose in axils........................
....*Psychotria horizontalis* (130. Rubiaceae)
Veins and trichomes not as above:
Blades with conspicuous flaplike projections on midrib near vein axils below........
....*Psychotria micrantha* (Fig. 532; 130. Rubiaceae)
Blades lacking axillary flaps:
Stipules glabrous, usually ± scoop-shaped, sharply acute, 5–10 mm long; branches
often spinescent, usually stiff, with the bark peeling into small flakes; leaf scars
often visible, shield-shaped...................*Randia armata* (130. Rubiaceae)
Stipules and branches not as above:
Trichomes on veins below curled, ± matted and appressed, to ca 0.5 mm long;
stems mostly pubescent *and* petioles to ca 1 cm long...................
....*Psychotria emetica* (Fig. 530; 130. Rubiaceae)
Trichomes not as above and/or petioles longer than 1 cm:
Petioles stout, less than 1 cm long; veins below mostly glabrous except for axil-
lary tufts of trichomes; leaves acute at base........................
....*Alibertia edulis* (Fig. 511; 130. Rubiaceae)
Leaves not as above:
Stipules subpersistent, 2–37 mm long, slender-attenuate; midrib below on
young blades glabrous to puberulent; plants not weak-stemmed and
decumbent:
Leaves 3 or 4 at each node...............*Hamelia patens* (130. Rubiaceae)
Leaves 2 at each node.................*Hamelia axillaris* (130. Rubiaceae)
Stipules and leaves not as above:
Underside of veins on half-mature blades densely appressed-pubescent;
plants trees...................*Chimarrhis parviflora* (130. Rubiaceae)
Underside of veins on half-mature blades glabrous or puberulent; plants
weak-stemmed, often decumbent shrubs........................
....*Hoffmannia woodsonii* (130. Rubiaceae)
Trichomes on underside of midrib closely appressed and/or longer than 0.5 mm:

Stipules 1.5–2 cm long, obovate, rounded at apex, not caducous.........................
.... *Macrocnemum glabrescens* (130. Rubiaceae)
Stipules not as above:
  Stipules to ca 15 mm long; leaves 15–42 cm long, the underside covered with short erect
    trichomes, giving the surface a velvety feel ... *Genipa americana* (Fig. 520; 130. Rubiaceae)
  Stipules and leaves not as above:
    Young branches with conspicuous, white, elongate lenticels; stipules persistent, to 2 mm
      long; leaf bases decurrent onto the short, obscure petiole......................
      ....*Pogonopus speciosus* (130. Rubiaceae)
    Branches and leaves not as above:
      Stipules 5 mm long or less:
        Stems usually mostly glabrous; trichomes, when present, usually ca 0.1 mm long ...
        .... *Hoffmannia woodsonii* (130. Rubiaceae)
        Stems usually ± pubescent; trichomes at least 0.3 mm long:
          Stipules joining above petiole; trichomes straight, slender..................
          .... *Bertiera guianensis* (130. Rubiaceae)
          Stipules separate above petiole; trichomes ± curled, matted.................
          ....*Psychotria emetica* (Fig. 530; 130. Rubiaceae)
      Stipules more than 5 mm long:
        Conspicuous lateral veins usually in 5–8 pairs per blade:
          Petioles usually 2–5 mm long; stipules subpersistent, the pairs joined at base ....
          .... *Bertiera guianensis* (130. Rubiaceae)
          Petioles (7)10–25 mm long; stipules caducous, not joined at base:
            Terminal bud area with trichomes to ca 2 mm long ....
            .... *Calycophyllum candidissimum* (Fig. 514; 130. Rubiaceae)
            Terminal bud area lacking such long trichomes, at most densely sericeous:
              Blades with the midrib above densely short-appressed-pubescent; blades not
                long-attenuate at base .... *Guettarda foliacea* (Fig. 523; 130. Rubiaceae)
              Blades with the midrib above glabrous; blades long-attenuate at base ......
              .... *Chimarrhis parviflora* (130. Rubiaceae)
        Lateral veins usually in more than 12 pairs per blade:
          Stipules to 3 cm long, ca 7 mm wide at base, ± persistent...................
          .... *Warscewiczia coccinea* (130. Rubiaceae)
          Stipules to 1.2 cm long, less than 5 mm wide at base, caducous:
            Half-mature blades with the trichomes on veins below less than 0.2 mm long,
              soft, ± curled .... *Chimarrhis parviflora* (130. Rubiaceae)
            Half-mature blades with the trichomes on veins below appressed, straight, more
              than 0.7 mm long.............. *Alseis blackiana* (Fig. 512; 130. Rubiaceae)
◇ Midrib on lower surface of young blades glabrous; blades lacking tufts of trichomes in vein axils
  (the pitlike domatia themselves may be pubescent inside in *Alibertia* and *Tocoyena*):
□ Stipules 7–30 mm long:
  Midrib of blade below bearing conspicuous flaplike projections near each major lateral vein:
    Leaves clustered at branch tips, usually with less than 12 pairs of major lateral veins ......
    .... *Tocoyena pittieri* (Fig. 536; 130. Rubiaceae)
    Leaves not clustered at branch tips, usually with more than 12 pairs of major lateral veins
    .... *Psychotria marginata* (130. Rubiaceae)
  Midrib of blade below lacking flaps:
    Lateral veins at confluence with midrib below usually somewhat enlarged and with pitlike
      domatia, but lacking tufts of trichomes in the pits ....
      .... *Psychotria horizontalis* (130. Rubiaceae)
    Lateral veins lacking pitlike domatia or, if small pits present, the pits usually pubescent
      inside:
    Lower lateral vein axils with pubescent, pitlike axillary domatia:
      Leaf blades drying green or brown, the largest blades less than 17 cm long and 8 cm
        wide .... *Alibertia edulis* (Fig. 511; 130. Rubiaceae)
      Leaf blades drying black, the largest more than 25 cm long and 10 cm wide ....
      .... *Tocoyena pittieri* (Fig. 536; 130. Rubiaceae)
    Lower lateral vein axils not as above:
      Stipules with a cusp ± as long as stipule body ...... *Faramea occidentalis* (130. Rubiaceae)
      Stipules lacking a cusp or the cusp much shorter than stipule body:
        Stipules subpersistent, to 12 mm long, 8–13 mm wide, ovate, with a cusp 2–3 mm
          long................................... *Psychotria grandis* (130. Rubiaceae)
      Stipules not as above:
        △ Stipules caducous, not broadly triangular, usually densely cinereous; blades ± cori-

aceous, usually attenuate at base; plants rare . . . .
. . . . *Chimarrhis parviflora* (130. Rubiaceae)
△ Stipules and leaves not as above:
    Leaves appearing clustered at ends of branches, thin; midrib below often with
        flaps in the axils . . . . *Tocoyena pittieri* (Fig. 536; 130. Rubiaceae)
    Leaves not clustered at ends of branches, ± coriaceous; midrib lacking flaps
        . . . . *Posoqueria latifolia* (Fig. 525; 130. Rubiaceae)
□ Stipules less than 7 mm long or, if longer, lacking conspicuous pits at base of secondary veins:
  Lateral veins with conspicuous pits in the side near axils below . . . .
  . . . . *Psychotria horizontalis* (130. Rubiaceae)
  Lateral veins lacking pits in the side:
    Lower leaf surface with conspicuous pits in the vein axils; petioles ca 5 mm long . . . .
    . . . . *Coussarea curvigemmia* (130. Rubiaceae)
    Lower leaf surface lacking axillary pits; petioles may be much longer than 5 mm:
      Stems with conspicuous white lenticels 2–10 cm from tip, the twigs usually straight, stiff:
        Lenticels round, not more than 2 times longer than wide . . . .
        . . . . *Coutarea hexandra* (Fig. 519; 130. Rubiaceae)
        Lenticels elongate, several times longer than broad . . . *Pogonopus speciosus* (130. Rubiaceae)
      Stems lacking white lenticels, the twigs various:
        Stipules ending in a conspicuous cusp about as long as the broadly flaring base:
          Blades abruptly acuminate, the acumen 7–10 mm long; lateral veins in 7–9 pairs per
            blade; major veins below ± sunken into the leaf matrix; plants in the forest . . .
          . . . . *Faramea luteovirens* (130. Rubiaceae)
          Blades acute or with a slightly projecting acumen; lateral veins in more than 12 pairs
            per leaf; major veins below outstanding on the blade surface; plants cultivated at
            the Laboratory Clearing . . . . *Ixora coccinea* (130. Rubiaceae)
        Stipules lacking a long cusp:
          Plants trees; young branches usually ca 5 mm thick; stipules triangular, subpersistent,
            sharply acute, often somewhat projecting away from stem; blades 18–28 cm long
            and 7–14 cm wide . . . . *Tocoyena pittieri* (Fig. 536; 130. Rubiaceae)
          Plants not as above:
            Petioles to 1.5 cm long; blades 5–16 cm long; all veins arising from midrib of ±
              equal widths and lengths; stipular scars with a minute line of trichomes; stip-
              ules ± blunt at apex . . . . *Psychotria carthagenensis* (Fig. 528; 130. Rubiaceae)
            Petioles and blades longer than above; alternate veins arising from midrib much
              more slender and short than the others; stipule scars lacking a line of trichomes;
              stipules often sharp at apex:
              Stipules ± persistent, linear, ca 4 mm long; leaves often 3 or 4 per node; younger
                parts usually very minutely pubescent . . . .
                . . . . *Hamelia patens* var. *glabra* (130. Rubiaceae)
             Stipules caducous; leaves 2 per node:
              Plants trees, growing near the shore . . . .
                . . . . *Cassipourea elliptica* (104. Rhizophoraceae)
              Plants weak-stemmed shrubs, growing in the forest . . . .
                . . . . *Hoffmannia woodsonii* (130. Rubiaceae)
⊙ Plants lacking interpetiolar stipules (or if stipules or scars present, not continuous between petioles); milky sap
  often present:
  ☆ Leaf pairs markedly unequal (often subopposite):
    ◘ Blades lacking stellate trichomes below:
      Blades drying with prominent cystoliths; stems usually quadrangular in age; blades glabrous above and
        with scattered trichomes on veins below . . . . *Trichanthera gigantea* (129. Acanthaceae)
      Blades not drying with cystoliths:
      Blades with tufts of trichomes in vein axils below; most leaves alternate . . . .
      . . . . *Lycianthes synanthera* (124. Solanaceae)
      Blades lacking axillary tufts (except *Solanum antillarum,* with the leaves mostly opposite):
      Blades ± glabrous below, usually ± equilateral at base:
        Leaves arising opposite one another at nodes; branch tips with the terminal buds naked, reddish-
        brown . . . . *Neea amplifolia* (Fig. 227; 48. Nyctaginaceae)
        Smaller leaves arising from the axils of the larger, the leaves sometimes alternate; branch tips not as
        above:
          Blades with inconspicuous axillary tufts of trichomes . . . . *Solanum antillarum* (124. Solanaceae)
          Blades lacking axillary tufts . . . . *Solanum arboreum* (Fig. 484; 124. Solanaceae)
      Blades pubescent below, usually inequilateral at base:
      Plants armed in leaf axils . . . . *Pisonia aculeata* (Fig. 228; 48. Nyctaginaceae)
      Plants not armed:

Blades pubescent only on midrib below, subcordate on one side at base . . . .
. . . . *Rinorea sylvatica* (Fig. 397; 95. Violaceae)
Blades densely pubescent below:
Blades elliptic-oblong . . . . *Cordia panamensis* (121. Boraginaceae)
Blades ovate-cordate . . . . *Cyphomandra hartwegii* (Fig. 481; 124. Solanaceae)
◻ Blades with stellate trichomes below:
Blades armed on midrib with recurved prickles; prickles on stems recurved:
Plants lianas; blades petiolate . . . . *Solanum lanciifolium* (124. Solanaceae)
Plants shrubs; blades sessile or nearly so (larger blades may be petiolate) . . . .
. . . . *Solanum jamaicense* (Fig. 486; 124. Solanaceae)
Blades unarmed:
Blades glabrous above in age; pubescence below so dense as to obscure surface; stems densely pubescent;
plants often small trees . . . .*Solanum hayesii* (Fig. 485; 124. Solanaceae)
Blades conspicuously pubescent above in age; pubescence below dense but surface visible; stems
densely pubescent when young but sparsely so in age; plants small shrubs . . . .
. . . . *Solanum subinerme* (124. Solanaceae)
☆ Leaf pairs ± equal:
● Blades, when held to light, with punctations or glandular or pellucid dots or lepidote scales below (may be
on veins):
Plants lianas; branch axils sometimes with long recurved spines:
Most parts with stalked or sessile T-shaped trichomes; lower blade surface with a few, large, rounded
glands near base (sunken into tissue, not at all free or peltate) . . . .
. . . . *Tetrapteris macrocarpa* (Fig. 312; 71. Malpighiaceae)
Plants lacking T-shaped trichomes; blades either lacking glands altogether or with glandlike peltate
scales scattered over lower surface:
Blades reddish-brown-pubescent on lower surface, the veins more densely pubescent; stems apparently
lacking recurved axillary spines . . . .
. . . . *Combretum laxum* var. *epiphyticum* (Fig. 420; 105. Combretaceae)
Blades glabrous or nearly so on underside except on veins or in vein axils; stems sometimes with prom-
inent axillary spines:
Stems with axillary spines; underside of blades glabrous on veins, possessing axillary tufts, lacking
conspicuous, pale, peltate scales over the surface . . . .
. . . . *Combretum decandrum* (Figs. 416, 417, and fig. on p. 20; 105. Combretaceae)
Stems lacking axillary spines; underside of blades pubescent on veins, lacking axillary tufts, pos-
sessing conspicuous, pale, peltate scales over the surface . . . .
. . . . *Combretum fruticosum* (Fig. 418; 105. Combretaceae)
Plants not as above:
◆ Blades pubescent at least on veins below:
Blades with at least some stellate trichomes below:
Blades mostly subcordate at base . . . . *Vismia macrophylla* (92. Guttiferae)
Blades usually not subcordate at base:
Blades sparsely pubescent below; sap clear . . . . *Siparuna guianensis* (57. Monimiaceae)
Blade surface below obscured by dense (sometimes minute) pubescence; sap yellow or orange:
Blades usually ovate, broadest near base; petioles more than 1.5 cm long; blades brown or whit-
ish below, conspicuously stellate-pubescent, reddish-pellucid-punctate . . . .
. . . . *Vismia baccifera* (Fig. 394; 92. Guttiferae)
Blades elliptic, broadest near middle; petioles ca 1 cm long; blades usually green below, very
inconspicuously pubescent, black-opaque-punctate . . . .
. . . . *Vismia billbergiana* (Fig. 395; 92. Guttiferae)
Blades lacking stellate trichomes below:
Most blades less than 3 cm wide:
Largest blades less than 6 cm long:
Blade margins distinctly undulate . . . . *Eugenia nesiotica* (Fig. 424; 106. Myrtaceae)
Blade margins not undulate . . . .*Eugenia principium* (106. Myrtaceae)
Largest blades more than 6 cm long:
Punctations opaque when dry . . . . *Adenaria floribunda* (102. Lythraceae)
Punctations pellucid:
Blades caudate-acuminate, the acumen to 1.5 cm long . . . . *Myrcia fosteri* (106. Myrtaceae)
Blades acuminate but not caudate . . . .*Eugenia venezuelensis* (106. Myrtaceae)
Most blades more than 3 cm wide:
■ Plants lianas (*Aegiphila elata* sometimes a climbing shrub); blades ovate or broadly elliptic:
Petioles less than 2 cm long; blades all simple; tendrils lacking . . . .
. . . . *Aegiphila elata* (122. Verbenaceae)
Petioles more than 3 cm long; adult plants with some blades bifoliolate; tendrils simple . . . . .
. . . . *Cydista heterophylla* (Fig. 492; 126. Bignoniaceae)

■ Plants shrubs or trees; blades oblong-elliptic to narrowly elliptic or lanceolate:
    T-shaped trichomes restricted to midrib above and below; blades with a few round glands, especially at tip and base . . . .*Spachea membranacea* (71. Malpighiaceae)
    Trichomes not T-shaped; blades lacking glands:
        Younger parts with conspicuous orange punctations (black and opaque when dry); blades villous mostly on veins below, glabrate above . . . . . . *Adenaria floribunda* (102. Lythraceae)
        Younger parts lacking orange (black or opaque when dry) punctations; blades with appressed, often minute trichomes on surface below:
            Lateral veins in more than 12 pairs, prominently raised below; blades ± thick, obtuse to short-acuminate . . . . *Psidium guajava* (106. Myrtaceae)
            Lateral veins in fewer than 10 pairs; blades not thick, usually acuminate:
                Blades glabrous above, with minute stellate trichomes below; sap yellow or orange . . . .*Vismia billbergiana* (Fig. 395; 92. Guttiferae)
                Blades with trichomes at least on midrib above, with fine, simple, appressed pubescence below; sap clear:
                    Most blades less than 10 cm long and with a long blunt acumen . . . .*Eugenia galalonensis* (106. Myrtaceae)
                    Most blades more than 10 cm long, sometimes mucronate or with a sharp acumen . . . . *Aegiphila panamensis* (Fig. 478; 122. Verbenaceae)
◆ Blades essentially glabrous:
  ○ Petioles less than 5 mm long or lacking:
      Blades sessile, less than 4 cm wide . . . . . . *Mouriri myrtilloides* subsp. *parvifolia* (107. Melastomataceae)
      Blades petiolate:
        Blades less than 7 cm long:
          Young stems quadrangular or with 4 slender winglike ribs, glabrous or inconspicuously appressed-pubescent or minutely and inconspicuously puberulent:
            Blades not noticeably glandular-punctate on surfaces, very sparsely pellucid-punctate, the youngest stems inconspicuously puberulent . . . . *Mouriri myrtilloides* subsp. *parvifolia* (107. Melastomataceae)
            Blades conspicuously glandular-punctate on surfaces, densely pellucid-punctate, the youngest stems glabrous or appressed-pubescent:
               Youngest stems wholly glabrous; tertiary venation of upper blade surface clearly visible, weakly raised, much less conspicuous than major lateral veins . . . . *Psidium friedrichsthalianum* (Fig. 425; 106. Myrtaceae)
               Youngest stems sparsely appressed-pubescent; tertiary venation of upper blade surface obscure but scarcely less conspicuous than major lateral veins . . . . *Calycolpus warscewiczianus* (Fig. 423; 106. Myrtaceae)
         Young stems terete, lacking wings, conspicuously short-puberulent or inconspicuously appressed-pubescent:
          Young stems and petioles inconspicuously appressed-pubescent:
            Leaf blades usually drying greenish (at least on lower surface), the veins all prominulous on upper surface (including the reticulate veins) . . . . *Myrcia gatunensis* (106. Myrtaceae)
            Leaf blades usually drying brown (the upper surface usually dark brown), the veins of upper surface obscure, especially the collecting vein, the reticulate veins not at all visible . . . . *Calycolpus warscewiczianus* (Fig. 423; 106. Myrtaceae)
          Young stems and petioles conspicuously short-puberulent:
            Blades coriaceous, pale brown on drying, seldom more than 5 cm long, the midrib conspicuously puberulent on lower surface, glabrous on upper surface . . . . *Eugenia principium* (106. Myrtaceae)
            Blades not coriaceous, drying green, frequently more than 5 cm long, the midrib glabrous on lower surface, sometimes inconspicuously puberulent along midrib on upper surface . . . . *Eugenia oerstedeana* (106. Myrtaceae)
        Blades more than 7 cm long:
          Blades abruptly acuminate, with a short sharp cusp ca 1 mm long, the punctations below often brown, conspicuous . . . . *Combretum fruticosum* (Fig. 418; 105. Combretaceae)
         Blades not as above:
         ◇ Blades caudate-acuminate:
            Blades more than 12 cm long . . . . *Psidium anglohondurense* (106. Myrtaceae)
            Blades less than 10 cm long:
               Leaf blades usually drying greenish (at least on lower surface), the veins all prominulous on upper surface (including the reticulate veins) . . . . *Myrcia gatunensis* (106. Myrtaceae)
               Leaf blades usually drying brown (the upper surface usually dark brown), the veins of upper surface obscure, especially the collecting vein, the reticulate veins not at all visible . . . . *Calycolpus warscewiczianus* (Fig. 423; 106. Myrtaceae)

◇ Blades abruptly acuminate:
    Blades less than 12 cm long .... *Psidium friedrichsthalianum* (Fig. 425; 106. Myrtaceae)
    Blades more than 12 cm long:
        Blades acute to abruptly and usually bluntly acuminate, drying ± concolorous, usually
            brown; stems tending to be somewhat flattened near nodes ....
            .... *Eugenia coloradensis* (106. Myrtaceae)
        Blades narrowly long and sharply acuminate, drying bicolorous, green, much paler be-
            neath; stems terete; new leaves reddish .... *Psidium anglohondurense* (106. Myrtaceae)
○ Petioles at least sometimes more than 5 mm long:
    Blades more than 3 times longer than broad:
        Blades toothed .... *Koanophyllon wetmorei* (Fig. 548; 133. Compositae)
        Blades entire:
            Plants lianas or small shrubs; blades with lepidote scales below ....
            .... *Combretum fruticosum* (Fig. 418; 105. Combretaceae)
            Plants trees; blades lacking lepidote scales:
                Reticulate veins connecting the major lateral veins very conspicuous, sinuate; blades to 9 cm
                    wide .... *Marila laxiflora* (92. Guttiferae)
                Reticulate veins not conspicuously sinuate; blades to 6 cm wide ....
                .... *Syzygium jambos* (Fig. 426; 106. Myrtaceae)
    Blades less than 3 times longer than broad:
        Sap copious, milky; stems with ridges:
            Plants lianas; reticulate veins conspicuous; punctations restricted to veins, not raised ....
            .... *Odontadenia puncticulosa* (Figs. 458–60; 118. Apocynaceae)
            Plants trees; reticulate veins obscure; punctations raised, not obviously associated with veins
            .... *Tabernaemontana arborea* (Fig. 464; 118. Apocynaceae)
         Sap clear or the stems lacking ridges:
            Basal vein axils with dense clusters of glands below ....
            .... *Cydista heterophylla* (Fig. 492; 126. Bignoniaceae)
            Basal vein axils lacking dense clusters of glands below:
                Reticulate veins connecting the lateral veins very conspicuous, sinuate ....
                .... *Marila laxiflora* (92. Guttiferae)
                Reticulate veins very inconspicuous, not sinuate:
                    Lower blade surface with dense, conspicuous, peltate, usually concave scales ....
                    .... *Combretum fruticosum* (Fig. 418; 105. Combretaceae)
                    Lower blade surface lacking peltate scales, with conspicuous, often raised glands but
                      these never appearing at all free from the leaf tissue:
                        At least the younger stems with 4 ribs, appearing quadrangular ....
                        .... *Psidium friedrichsthalianum* (Fig. 425; 106. Myrtaceae)
                        Stems not ribbed or quadrangular, sometimes flattened laterally so as to appear ±
                        2-edged................................ *Eugenia coloradensis* (106. Myrtaceae)
● Blades lacking punctations, glandular and pellucid dots, and lepidote scales below:
    Leaves often verticillate or congested at tips of branches:
        Leaves not congested at apex of branches but verticillate with 4 or 5 leaves per node; sap milky ....
        .... *Allamanda cathartica* (118. Apocynaceae)
        Leaves clustered at tips of branches; sap not milky:
            Blades less than 10 cm long, with axillary domatia below; stipules lacking ....
            .... *Terminalia amazonica* (Figs. 421, 422; 105. Combretaceae)
            Blades more than 15 cm long, lacking axillary domatia; interpetiolar stipules caducous ....
            .... *Alseis blackiana* (Fig. 512; 130. Rubiaceae)
    Leaves not verticillate, not congested at apex of branches:
▲ Plants parasitic shrubs or lianas, the blades generally coriaceous:
    Stems square, or 2-edged at least near apex:
        Stems square; blades less than 1.5 cm wide .... *Phoradendron quadrangule* (42. Loranthaceae)
        Stems 2-edged at least near apex; blades more than 2 cm wide ....
            .... *Phthirusa pyrifolia* (42. Loranthaceae)
    Stems terete:
        Blades usually ± elliptic, acuminate, ± acute at base ....
        .... *Phoradendron piperoides* (Fig. 212; 42. Loranthaceae)
        Blades ovate, usually rounded at apex, obtuse to subcordate at base:
            Plants lianas, more than 1 m long; blades less than 5 cm long ....
            .... *Struthanthus orbicularis* (42. Loranthaceae)
            Plants shrubby, less than 1 m long; blades more than 5 cm long:
                Blades ± cuneate, the petioles more than 5 mm long ....
                .... *Oryctanthus occidentalis* (Fig. 211; 42. Loranthaceae)
                Blades rounded to subcordate at base, the petioles less than 5 mm long:

Leaf blades broadly rounded to subcordate at base . . . *Oryctanthus cordifolius* (42. Loranthaceae)

Leaf blades obtuse to rounded at base (rarely subcordate) . . . .

. . . . *Oryctanthus alveolatus* (42. Loranthaceae)

▲ Plants not parasitic:

★ Plants lianas:

◉ Blades pubescent below:

Trichomes simple, erect to ± appressed:

Young stems and blades below conspicuously and densely soft-pubescent:

Blades ± rounded or obtuse at base; midrib not arched . . . .

. . . . *Aegiphila cephalophora* (122. Verbenaceae)

Blades often subcordate at base; midrib arched . . . . *Prestonia ipomiifolia* (118. Apocynaceae)

Young stems and blades not conspicuously and densely pubescent:

Blades toothed:

Plants epiphytic lianas, often rooting at nodes; blades inequilateral at base . . . .

. . . . *Drymonia serrulata* (Fig. 503; 127. Gesneriaceae)

Plants arching shrubs or lianas; blades equilateral at base:

Blades rounded to cordate, often with 3–5 palmate veins at base, less than 13 cm long

. . . . *Heterocondylus vitalbis* (133. Compositae)

Blades obtuse to acute at base, not palmately veined, often more than 13 cm long . . . . .

. . . . *Wulffia baccata* (Fig. 553; 133. Compositae)

Blades entire:

Stems with stout axillary spines . . . . *Pisonia aculeata* (Fig. 228; 48. Nyctaginaceae)

Stems lacking spines:

Blades with reticulate venation not conspicuous (tertiary veins often visible on *Mandevilla villosa*); most blades less than 8 cm long:

Blades with major lateral veins obscure, the tertiary veins not visible; canopy lianas

. . . . *Cynanchum recurvum* (119. Asclepiadaceae)

Blades with major lateral and secondary veins conspicuous . . . .

. . . . *Mandevilla villosa* (Fig. 453; 118. Apocynaceae)

Blades with reticulate venation conspicuous without magnification (on dried specimens); at least some blades more than 8 cm long:

Petioles ca 1.5 cm or more long; blades with the lateral veins arising mostly from basal half of midrib . . . . *Mikania leiostachya* (Fig. 549; 133. Compositae)

Petioles less than 1 cm long; blades with the lateral veins distributed evenly along midrib:

Stems usually with interpetiolar ridges; blade midribs and major lateral veins bearing both pilose and puberulent trichomes . . . .

. . . . *Forsteronia myriantha* (118. Apocynaceae)

Stems lacking interpetiolar ridges; blade midribs and major lateral veins glabrous or merely sparsely pilose . . . . *Combretum cacoucia* (105. Combretaceae)

Trichomes T-shaped, often appearing as appressed, 2-ended trichomes but peltately attached (swivels on center):

Blades densely pubescent below:

Trichomes contiguous below, obscuring surface:

Blades ovate, broadly rounded at base, less than twice as long as broad . . . .

. . . . *Stigmaphyllon hypargyreum* (71. Malpighiaceae)

Blades oblong-elliptic, narrowly rounded at base, more than twice as long as broad . . . . .

. . . . *Hiraea faginea* (71. Malpighiaceae)

Trichomes below not contiguous:

Trichomes below stalked:

Blades less than 10 cm wide, long-acuminate; petioles with 2 conspicuous, slender glands at apex; stems sparsely pubescent . . . . *Hiraea quapara* (71. Malpighiaceae)

Blades more than 10 cm wide, cuspidate to short-acuminate; petioles lacking glands at apex; stems densely ferruginous-tomentose . . . .

. . . . *Hiraea grandifolia* (Fig. 308; 71. Malpighiaceae)

Trichomes below not stalked, appressed:

Petioles ca 1 cm long, with small, inconspicuous glands . . . .

. . . . *Banisteriopsis cornifolia* (71. Malpighiaceae)

Petioles more than 1.5 cm long, with a conspicuous pair of glands at apex:

Blades broadly ovate, occasionally lobed, often palmately veined at base, less than twice as long as broad . . . *Stigmaphyllon lindenianum* (Fig. 311; 71. Malpighiaceae)

Blades elliptic or ovate-elliptic, not lobed or palmately veined, more than twice as long as broad . . . . *Stigmaphyllon puberum* (71. Malpighiaceae)

Blades not densely pubescent below:

✿ Blades mostly obovate, the base narrow and subcordate . . . *Hiraea reclinata* (71. Malpighiaceae)

◖ Blades ± ovate or elliptic, the base obtuse to rounded or subcordate:
  Petioles with 2 glands at apex:
    Blades elliptic, not palmately veined at base . . . .
      . . . . *Stigmaphyllon ellipticum* (71. Malpighiaceae)
    Blades ovate, palmately veined at base . . . .
      . . . . *Stigmaphyllon lindenianum* (Fig. 311; 71. Malpighiaceae)
  Petioles lacking glands:
    Stems with sparse to dense, appressed, T-shaped trichomes, soon glabrate to glabrous;
      blades acuminate, obtuse to rounded at base . . . .
      . . . . *Mascagnia nervosa* (Fig. 309; 71. Malpighiaceae)
    Stems tomentose with sessile and stalked T-shaped trichomes, at least the sessile ones
      persisting on stems; blades very short-acuminate or acute . . . .
      . . . . *Tetrapteris macrocarpa* (Fig. 312; 71. Malpighiaceae)
◉ Blades essentially glabrous below (may have cystoliths appearing as trichomes):
  Plants with milky sap:
    Blades mostly less than 5 cm wide:
      Blades with several (1–4) conspicuous glands at base of midrib above:
        Blades lacking axillary pits below, often rounded at base . . . .
          . . . . *Mesechites trifida* (Fig. 454; 118. Apocynaceae)
        Blades with pits in axils below, usually obtuse at base, coriaceous . . . .
          . . . . *Forsteronia viridescens* (118. Apocynaceae)
      Blades lacking conspicuous glands at base of midrib, usually acute at base:
        Blades with small pitlike domatia in axils below . . *Forsteronia peninsularis* (118. Apocynaceae)
        Blades lacking small pitlike domatia in axils:
          Blades markedly bicolorous, the lateral veins inconspicuous, the apex gradually
            acuminate . . . . *Prestonia acutifolia* (118. Apocynaceae)
          Blades not markedly bicolorous, the lateral veins conspicuous, the apex abruptly acu-
            minate or mucronate . . . . *Rhabdadenia biflora* (118. Apocynaceae)
    Blades mostly more than 5 cm wide:
      Blades with pitlike axillary domatia below and 1–4 conic glands at base of midrib above . . .
        . . . . *Forsteronia viridescens* (118. Apocynaceae)
      Blades lacking pitlike axillary domatia and glands:
        Stems corky, with many conspicuous lenticels . . . . *Prestonia obovata* (118. Apocynaceae)
        Stems ± smooth:
          Blades ovate-cordate; petioles ca 6 cm long . . . . . *Cynanchum cubense* (119. Asclepiadaceae)
          Blades not ovate-cordate; petioles less than 6 cm long:
            Blades usually with 10 or more pairs of lateral veins . . . .
              . . . . *Odontadenia macrantha* (Figs. 455–57; 118. Apocynaceae)
            Blades usually with 9 or fewer pairs of lateral veins:
              Reticulate veins conspicuous below, the midrib lacking a raised ridge; veins with
                sessile, opaque, plate-shaped glands; apex of blades not markedly downturned
                . . . . *Odontadenia puncticulosa* (Figs. 458–60; 118. Apocynaceae)
              Reticulate veins inconspicuous and/or the midrib with a raised ridge in center;
                veins lacking glands; apex of blades various:
                Reticulate veins fine, distinct; blades usually obtuse but sometimes rounded at
                  base . . . . *Prestonia portobellensis* (Fig. 461; 118. Apocynaceae)
                Reticulate veins obscure, few; blades mostly rounded at base, becoming sub-
                  cordate . . . . *Marsdenia crassipes* (119. Asclepiadaceae)
  Plants lacking milky sap:
    ◻ Blades broadly rounded at base:
      Blades asperous:
        Blades densely covered with minute lepidote scales especially below; margins often scab-
          ridulous . . . . *Petrea aspera* (Fig. 479; 122. Verbenaceae)
        Blades lacking lepidote scales; margins glabrous . . . .
          . . . . *Prionostemma aspera* (Fig. 334; 78. Hippocrateaceae)
      Blades not asperous:
        Stems with very short, simple trichomes . . . . *Mikania tonduzii* (133. Compositae)
        Stems glabrous or with T-shaped trichomes:
          Petioles with 2 conspicuous glands 1–2 mm long at apex:
            Petioles slender, 1–3 cm long . . . . *Stigmaphyllon ellipticum* (Fig. 310; 71. Malpighiaceae)
            Petioles thick, 0.7–1.5 cm long . . . . *Mascagnia hippocrateoides* (71. Malpighiaceae)
          Petioles lacking 2 conspicuous glands:
            Blades orbicular to very widely elliptic; young stems and buds golden-puberulent . .
              . . . . *Tetrapteris macrocarpa* (Fig. 312; 71. Malpighiaceae)
            Blades elliptic; young stems and buds not golden-puberulent:

Stipule scars lacking . . . . *Hylenaea praecelsa* (78. Hippocrateaceae)
Stipule scars present, sometimes encircling stem:
    Reticulate veins not obvious . . . . . *Tontelea richardii* (Fig. 335; 78. Hippocrateaceae)
    Reticulate and even tertiary veins evident . . . .
      . . . . *Mascagnia nervosa* (Fig. 309; 71. Malpighiaceae)
❑ Blades ± acute at base:
    Blades somewhat toothed:
      Young stems densely puberulent, lacking wings below petiole when dry . . . .
      . . . . *Hippocratea volubilis* (Fig. 333; 78. Hippocrateaceae)
      Young stems glabrous, with minute wings below petiole when dry . . . .
      . . . . *Anthodon panamense* (Figs. 331, 332; 78. Hippocrateaceae)
    Blades entire:
      Blades with 3–5 palmate veins from near base . . . . *Mikania tonduzii* (133. Compositae)
      Blades pinnately veined throughout:
        Blades sometimes broadly ovate, with the lateral veins inconspicuous; plants large,
          much-branched lianas . . . .
          . . . . *Gnetum leyboldii* var. *woodsonianum* (Figs. 41, 42; 13. Gnetaceae)
        Blades ± elliptic to obovate, with the lateral veins distinct:
          Blades asperous and minutely lepidote below, the margins often scabridulous . . . . .
          . . . . *Petrea aspera* (Fig. 479; 122. Verbenaceae)
        Blades not asperous below:
          Blades drying with linear cystoliths; largest blades less than 12 cm long; plants li-
            anas on the forest floor . . . . *Justicia graciliflora* (Fig. 506; 129. Acanthaceae)
          Blades not drying with crystals or cystoliths; largest blades usually more than 12
            cm long:
            Young stems puberulent, the trichomes erect, simple; blades often narrowly sub-
              cordate at base . . . *Combretum laxum* var. *laxum* (Fig. 419; 105. Combretaceae)
            Young stems with appressed or stalked, T-shaped trichomes; blades not nar-
              rowly subcordate at base (usually acute to obtuse or subrounded):
            Blades subcoriaceous, drying brown, the reticulate venation raised, conspicu-
              ous; pubescence reddish-brown, the younger parts tomentose . . . .
              . . . . *Heteropteris laurifolia* (71. Malpighiaceae)
            Blades moderately thin, drying green or brownish, the reticulate venation
              scarcely raised, not conspicuous; pubescence whitish, ± sericeous but
              mostly soon deciduous:
              Blades usually long-acuminate, 2.5–3 times longer than wide; considerable
               pubescence persisting on stems. . . . *Tetrapteris discolor* (71. Malpighiaceae)
             Blades acute to short-acuminate, ca 2(2.5) times longer than wide; stems
              soon essentially glabrous . . . . *Tetrapteris seemannii* (71. Malpighiaceae)
★ Plants shrubs or trees, not lianas:
  ○ Blades pubescent:
    Blades densely and conspicuously pubescent below:
      Blades with 10 or fewer pairs of lateral veins:
        Blades entire, smooth above; veins ± equidistant; stems with prominent leaf scars and stalked
          T-shaped trichomes. . . . . *Byrsonima crassifolia* (71. Malpighiaceae)
        Blades conspicuously toothed, asperous above; veins closer to base than apex; stems lacking
          prominent leaf scars, the trichomes simple:
          Reticulate veins of lower leaf surface conspicuous, densely hispid (the trichomes stiffly
            erect). . . . . *Clibadium surinamense* (133. Compositae)
          Reticulate veins of lower leaf surface not raised, glabrate to strigose (the trichomes ±
            appressed or strongly leaning) . . . . *Clibadium asperum* (133. Compositae)
      Blades with 11 or more pairs of lateral veins:
        Stems ± square; trichomes on blades stellate . . . .
        . . . . *Siparuna pauciflora* (Figs. 246, 247; 57. Monimiaceae)
        Stems terete; trichomes on blades simple:
          Stems and lower blade surfaces brown; blades ending abruptly at base; plants trees, more
            than 6 m tall at maturity . . . . *Vochysia ferruginea* (73. Vochysiaceae)
          Stems and lower blade surfaces not brown; blades decurrent onto petioles; plants usually
            shrubs, less than 3 m tall . . . . *Aphelandra sinclairiana* (129. Acanthaceae)
    Blades not densely pubescent below:
  ◇ Trichomes T-shaped:
    Blades subsessile or with densely ferruginous young stems:
      Blades subsessile; plants shrubs, less than 2 m tall; young stems not densely ferruginous-
        pubescent . . . . *Malpighia romeroana* (71. Malpighiaceae)
      Blades conspicuously petiolate; plants trees, usually more than 10 m tall; young stems
        densely ferruginous-pubescent . . . . *Byrsonima spicata* (71. Malpighiaceae)

Blades petiolate and not with ferruginous pubescence on young stems:

Trichomes on underside of blades restricted to midrib, brownish . . . .
. . . . *Spachea membranacea* (71. Malpighiaceae)

Trichomes on underside of blades sparse to moderate, throughout surface (may be minute), whitish . . . . *Bunchosia cornifolia* (71. Malpighiaceae)

◇ Trichomes not T-shaped:

Petioles mostly more than 1 cm long:

Blades with conspicuous glands at base of midrib and sometimes along margins . . . .
. . . . *Colubrina glandulosa* (81. Rhamnaceae)

Blades lacking glands:

Blades with 3–5 strong basal veins, often remotely toothed . . . .
. . . . *Heterocondylus vitalbis* (133. Compositae)

Blades pinnately veined, entire:

Blades drying without prominent cystoliths; stems terete; blades ± elliptic . . . .
. . . . *Guapira standleyanum* (Fig. 226; 48. Nyctaginaceae)

Blades drying with prominent cystoliths; stems often quadrangular; blades heteromorphic, often ± asymmetrical . . . . *Trichanthera gigantea* (129. Acanthaceae)

Petioles mostly less than 1 cm long:

Blades less than 1.5 cm broad or mostly more than 5.5 cm broad, entire, estipulate:

Plants epiphytic shrubs; blades to 1.5 cm wide . . *Columnea billbergiana* (127. Gesneriaceae)

Plants shrubs or small trees; blades more than 3 cm wide (usually more than 5.5 cm) . .
. . . . *Siparuna guianensis* (57. Monimiaceae)

Blades 1.5–5.5 cm broad, usually toothed, stipulate:

Blades oblique and rounded to slightly subcordate at base . . . .
. . . . *Rinorea sylvatica* (Fig. 397; 95. Violaceae)

Blades usually ± equilateral, acute to obtuse at base:

Blades caudate-acuminate, sparsely hirsute on veins below, less than 13 cm long . . .
. . . . *Rinorea squamata* (95. Violaceae)

Blades acuminate, not caudate, densely puberulent throughout below, to 17 cm long
. . . . *Aegiphila panamensis* (Fig. 478; 122. Verbenaceae)

○ Blades essentially glabrous (some with cystoliths):

□ Petioles mostly more than 1 cm long:

Lateral veins very closely spaced and/or inconspicuous:

Blades oblong, emarginate at apex; lateral veins ± perpendicular to midrib; plants not epiphytic . . . . *Calophyllum longifolium* (92. Guttiferae)

Blades not oblong, not emarginate at apex:

Blades obovate, rounded at apex; lateral veins ascending; plants epiphytic trees:

Blades very thick and leathery, drying opaque . . . *Clusia odorata* (Fig. 387; 92. Guttiferae)

Blades thick but not leathery, drying translucent . . . .
. . . . *Havetiopsis flexilis* (Fig. 388; 92. Guttiferae)

Blades oblong-elliptic, acuminate at apex:

Lateral veins very inconspicuous, ascending:

Blades usually broadest above the middle, coriaceous, the midrib moderately indistinct, smooth, the primary lateral veins numerous and close, the reticulate venation not visible . . . . *Clusia odorata* (Fig. 387; 92. Guttiferae)

Blades usually broadest at about the middle, relatively thin, the midrib obviously raised beneath, bearing close minute projections under magnification, the primary lateral veins mostly in 8–12 pairs, ca 1 cm apart . . . .
. . . . *Tovomitopsis nicaraguensis* (Fig. 393; 92. Guttiferae)

Lateral veins conspicuous, ± perpendicular to midrib:

Blades dull on both surfaces (especially below); veins not raised, the marginal collecting vein not visible; sap in younger parts whitish-yellow; upper surface, when dry, with the reticulate veins not prominulous and the collecting vein obscure
. . . . *Rheedia edulis* (Fig. 389; 92. Guttiferae)

Blades at least weakly shiny; veins weakly raised, the collecting vein clearly visible on both surfaces; sap in younger parts bright yellow; upper surface, when dry, with the reticulate veins prominulous and the collecting vein evident . . . .
. . . . *Rheedia acuminata* (92. Guttiferae)

Lateral veins conspicuous and distant:

△ Larger blades more than 15 cm long:

★ Plants shrubs, less than 2 m tall, or hemiepiphytic shrubs or small trees:

Plants hemiepiphytic shrubs or small trees, usually restricted to higher canopy, rare; blades rounded to acute at apex . . . . *Hydrangea peruviana* (60. Saxifragaceae)

Plants not hemiepiphytic; blades acuminate at apex:

Younger branchlets ferruginous-pubescent, slender, lenticellate; blades lacking a marginal connecting vein . . . . *Neea amplifolia* (Fig. 227; 48. Nyctaginaceae)

Younger branchlets glabrous, thick, not lenticellate; blades with a marginal connecting vein . . . . *Dendropanax stenodontus* (109. Araliaceae)
★ Plants trees, more than 3 m tall:
　Sap yellow:
　　Trees ca 10 m tall with stilt-roots; largest blades usually more than 25 cm long . . . .
　　　. . . . . *Tovomita longifolia* (Fig. 391; 92. Guttiferae)
　　Trees ca 5 m tall, lacking stilt roots; largest blades usually less than 25 cm long
　　　(juveniles sometimes larger):
　　　Petioles less than 1.5 cm long; lateral veins below conspicuously loop-connected,
　　　　with the secondary veins many, closely spaced, parallel, prominulous, connecting the lateral veins . . . . *Marila laxiflora* (92. Guttiferae)
　　　Most petioles more than 2 cm long; lateral veins below at most weakly loop-connected, with the secondary veins reticulate . . . .
　　　　. . . . . *Tovomitopsis nicaraguensis* (Fig. 393; 92. Guttiferae)
　Sap not colored:
　　Petioles to 3 cm long, averaging ca 1.5 cm . . . .
　　　. . . . . *Guapira standleyanum* (Fig. 226; 48. Nyctaginaceae)
　　Petioles less than 1.5 cm long, averaging ca 0.7 cm . . . .
　　　. . . . . *Petrea aspera* (Fig. 479; 122. Verbenaceae)
△ All blades less than 15 cm long:
　Plants hemiepiphytic shrubs; blades usually rounded (sometimes acute) at apex; reticulate veins prominent below when dry . . . . *Hydrangea peruviana* (60. Saxifragaceae)
　Plants not hemiepiphytic; blades acuminate at apex:
　　Blades somewhat palmately veined at base, with 2 glands at base on either side of petiole and with other glands scattered along blade margins . . . .
　　　. . . . . *Colubrina glandulosa* (81. Rhamnaceae)
　　Blades pinnately veined throughout, lacking glands:
　　　Young stems densely ferruginous-pubescent; older stems rough; sap clear; plants usually medium-sized trees; blades drying dark . . . .
　　　　. . . . . *Guapira standleyanum* (Fig. 226; 48. Nyctaginaceae)
　　　Young stems ± glabrous; older stems smooth; sap yellow; plants usually small trees; blades not drying dark . . . . *Tovomita stylosa* (Fig. 392; 92. Guttiferae)
□ Petioles mostly less than 1 cm long:
　Plants with milky or yellow sap:
　　Blades obovate or oblanceolate:
　　　Blades acuminate at apex, the veins inconspicuous; sap milky; plants not hemiepiphytic
　　　　. . . . . *Stemmadenia grandiflora* (Figs. 462, 463; 118. Apocynaceae)
　　　Blades rounded at apex, the veins obscure; sap yellow; plants hemiepiphytic trees:
　　　　Blades very thick, leathery, drying opaque . . . . . . *Clusia odorata* (Fig. 387; 92. Guttiferae)
　　　　Blades thick, not leathery, drying translucent . . . .
　　　　　. . . . . *Havetiopsis flexilis* (Fig. 388; 92. Guttiferae)
　　Blades oblong or elliptic:
　　　Sap milky; lateral veins more than 5 mm apart; stems not wrinkled when dry:
　　　　Blades elliptic or lanceolate-elliptic, long-acuminate, with axillary domatia below . . . .
　　　　　. . . . . *Malouetia guatemalensis* (Fig. 452; 118. Apocynaceae)
　　　　Blades oblong, short-acuminate, lacking axillary domatia . . . .
　　　　　. . . . . *Lacmellea panamensis* (118. Apocynaceae)
　　　Sap yellow; lateral veins either closely spaced or ± obscure; stems with longitudinal wrinkles when dry:
　　　　Blades less than 8 cm long and 3 cm wide; plants very large trees with adventitious roots
　　　　　. . . . . *Symphonia globulifera* (Fig. 390; 92. Guttiferae)
　　　　Blades more than 8 cm long and 3 cm wide; plants small to medium-sized trees lacking adventitious roots:
　　　　　(Fresh) Blades dull on both surfaces, especially below; veins obscure, not at all raised; marginal collecting vein not visible; sap in younger parts whitish-yellow; (Dry) Reticulate veins not prominulous on upper surface; collecting vein obscure to lacking above . . . . *Rheedia edulis* (Fig. 389; 92. Guttiferae)
　　　　　(Fresh) Blades at least weakly shiny; veins weakly raised; collecting vein usually clearly visible on both surfaces; sap in younger parts bright yellow; (Dry) Reticulate veins prominulous on upper surface; collecting vein also prominulous on upper surface . . . . *Rheedia acuminata* (92. Guttiferae)
　Plants lacking yellow or milky sap:
☆ Plants hemiepiphytic:
　Plants small shrubs, less than 1 m tall; blades usually less than 3 cm wide . . . .
　　. *Codonanthe crassifolia* (Fig. 502), *C. uleana* (127. Gesneriaceae)
　Plants trees or large shrubs, more than 1 m tall; blades more than 3 cm wide:

Blades gradually tapered to base, thick, the veins obscure; sap thick and copious; plants common . . . . *Clusia odorata* (Fig. 387; 92. Guttiferae)

Blades ending abruptly at petiole, thin, the veins conspicuous; sap not thick or copious; plants rare. . . . . *Hydrangea peruviana* (60. Saxifragaceae)

☆ Plants not hemiepiphytic:

Mature blades mostly less than 7 cm long; at least the youngest stems winged when dry:

Plant a cultivated shrub at the Laboratory Clearing; most larger stems winged . . . . *Thunbergia erecta* (129. Acanthaceae)

Plants shrubs or trees, not cultivated; older stems not winged:

Blades mostly sessile, sharply acuminate, obtuse to rounded at base, lacking domatia; plants usually shrubs . . . . *Mouriri myrtilloides* subsp. *parvifolia* (107. Melastomataceae)

Blades petiolate, with a rounded acumen, acute at base, bearing a small subterminal cavity; plants medium-sized trees . . . . *Lafoensia punicifolia* (Figs. 410, 411; 102. Lythraceae)

Mature blades mostly more than 7 cm long *or* the younger stems not winged when dry (*Calycolpus warscewiczianus* with leaves to only 8 cm long but the stems not winged):

Largest blades less than 3 cm wide; lateral veins many (more than 15 pairs) and closely spaced (less than 2 mm apart) . . . . *Calycolpus warscewiczianus* (Fig. 423; 106. Myrtaceae)

Largest blades more than 3 cm wide; lateral veins fewer and more widely spaced:

Blades estipulate, drying with prominent linear cystoliths . . . . *Justicia graciliflora* (Fig. 506; 129. Acanthaceae)

Blades stipulate, drying without linear cystoliths:

Stems and petioles essentially glabrous; blades inconspicuously toothed in apical half, the lateral veins loop-connected . . . . *Cassipourea elliptica* (104. Rhizophoraceae)

Stems and petioles with T-shaped trichomes:

Plants shrubs, usually less than 2 m tall; stipules interpetiolar . . . . *Malpighia romeroana* (71. Malpighiaceae)

Plants lianas or trees, usually more than 3 m tall; stipules axillary or lacking:

Plants lianas or scandent shrubs; mature blades glabrous above, coriaceous . . . . *Heteropteris laurifolia* (71. Malpighiaceae)

Plants trees, not scandent; mature blades pubescent on midrib above, not coriaceous:

Most blades with conspicuous glands near midrib at base below and often scattered on surface . . . . *Spachea membranacea* (71. Malpighiaceae)

Most blades eglandular . . . . *Byrsonima spicata* (71. Malpighiaceae)

KEY 2: LEAVES SIMPLE, ALTERNATE OR SPIRALED; PLANTS NOT LIANAS

Petioles with paired, usually stipitate glands near apex; sap milky; leaves very shallowly crenate or entire; plants not stellate-pubescent . . . . *Sapium aucuparium, S. caudatum* (Fig. 325; 75. Euphorbiaceae)

Petioles lacking glands near apex *or* the sap not milky:

● Blades not entire (toothed or sinuate):

◆ Blades essentially glabrous:

■ Petioles mostly more than 1 cm long:

Leaves nearly 1 m long, densely clustered at apices of stout stems; stems with numerous, persistent, imbricate, long stipules . . . . *Cespedezia macrophylla* (89. Ochnaceae)

Leaves and stems not as above:

Blades 5-lobed; petioles more than 20 cm long . . . . *Carica cauliflora* (Fig. 409; 99. Caricaceae)

Blades not lobed; petioles less than 10 cm long:

Petioles with a pair of large glands at apex; blades evenly pinnately veined, lacking axillary domatia below . . . . *Sapium caudatum* (Fig. 325; 75. Euphorbiaceae)

Petioles lacking glands at apex:

Blades with conspicuous glands at base:

Blades with axillary domatia below . . . . *Alchornea latifolia* (75. Euphorbiaceae)

Blades lacking axillary domatia:

Basal glands red on both surfaces when fresh, inconspicuous when dry; twigs tan, not pubescent; basal veins extending less than one-third the length of blade . . . . *Alchornea costaricensis* (Fig. 318; 75. Euphorbiaceae)

Basal glands green above, not visible below, conspicuous when dry; twigs reddish-brown-pubescent when young; basal veins extending more than two-thirds the length of blade . . . . *Hasseltia floribunda* (Fig. 399; 96. Flacourtiaceae)

Blades lacking glands:

Plants shrubs, with slender trunks, usually less than 4 m tall:
　　Blades sparsely toothed, at most 20 teeth on each side of blade . . . .
　　　　. . . . .*Dendropanax stenodontus* (Fig. 443; 109. Araliaceae)
　　Blades with more than 30 teeth on each side of blade:
　　　　Leaves always broadest well above the middle, clustered at apex of stem; plants erect shrubs,
　　　　　　usually more than 2.5 m tall; stems more than 1 cm diam . . . .
　　　　　　. . . . . *Gustavia fosteri* (103. Lecythidaceae)
　　　　Leaves broadest usually at or below the middle; plants shrubs, usually less than 1 m tall or
　　　　　　vinelike; stems usually less than 7 mm diam:
　　　　　　Plants usually less than 1 m tall; blade margins with minute, subulate, regular teeth; peti-
　　　　　　　　oles to 2 cm long . . . .*Ardisia pellucida* (112. Myrsinaceae)
　　　　　　Plants usually vinelike shrubs more than 2 m long; blades irregularly toothed; most peti-
　　　　　　　　oles more than 3 cm long . . . . *Urera eggersii* (Fig. 210; 40. Urticaceae)
Plants trees, with stout trunks:
　　Blades more than 30 cm long, oblanceolate; leaves densely spiraled at apices of branches:
　　　　Lateral veins apparent . . . . *Gustavia superba* (Figs. 414, 415; 103. Lecythidaceae)
　　　　Lateral veins obscure . . . . *Grias fendleri* (Fig. 413; 103. Lecythidaceae)
　　Blades less than 25 cm long, usually broadest near or below the middle; leaves not congested
　　　　at apices of branches:
　　　　Blades ± attenuate to acute at base, decurrent onto petiole; blades ovate . . . .
　　　　　　. . . . . *Roupala montana* (41. Proteaceae)
　　　　Blades obtuse to acute at base, ending abruptly at petiole:
　　　　　　Blades coarsely toothed, mostly with 6 or fewer pairs of lateral veins . . . .
　　　　　　. . . . . *Alchornea costaricensis* (Fig. 318; 75. Euphorbiaceae)
　　　　　　Blades finely toothed, with 7 or more pairs of lateral veins . . . .
　　　　　　. . . . . *Saurauia laevigata* (88. Dilleniaceae)
■ Petioles mostly less than 1 cm long:
　　Blades nearly 1 m long, densely clustered at apices of stout stems; stems with numerous, persistent, imbri-
　　　　cate, long stipules . . . .*Cespedezia macrophylla* (89. Ochnaceae)
　　Blades and stems not as above:
　　　　Blades with pellucid punctations or striations and/or conspicuous axillary domatia:
　　　　　　Blades not pellucid-punctate, rounded to subcordate at base and ± inequilateral; domatia in axils
　　　　　　　　below pocketlike, untufted . . . . *Tetrathylacium johansenii* (Fig. 400; 96. Flacourtiaceae)
　　　　　　Blades with pellucid punctations or striations, not both rounded and inequilateral at base:
　　　　　　　　Branchlets short, with terminal spines; blades less than 5.5 cm long . . . .
　　　　　　　　. . . . . *Casearia aculeata* (96. Flacourtiaceae)
　　　　　　　　Branchlets not short, lacking terminal spines; blades more than 5.5 cm long:
　　　　　　　　　　Petioles 1–4(6) mm long; blades ± equilateral at base, usually with axillary domatia below . . . .
　　　　　　　　　　. . . . .*Casearia corymbosa* (96. Flacourtiaceae)
　　　　　　　　　　Petioles 5–10 mm long; blades often inequilateral at base, lacking axillary domatia:
　　　　　　　　　　　　Blade margins with 2–4 serrations per cm; blades elliptic . . . . *Laetia thamnia* (96. Flacourtiaceae)
　　　　　　　　　　　　Blade margins with 5–8 serrations per cm; blades oblong . . . . . *Laetia procera* (96. Flacourtiaceae)
　　　　Blades lacking pellucid punctations and axillary domatia:
　　　　　　Serrations mostly glandular-tipped; trunk spiny, the spines often branched:
　　　　　　　　Blades usually less than 2.5 times longer than wide, shallowly toothed . . . .
　　　　　　　　. . . . . *Xylosma oligandrum* (96. Flacourtiaceae)
　　　　　　　　Blades usually more than 2.5 times longer than wide, deeply toothed . . . .
　　　　　　　　. . . . . *Xylosma chloranthum* (Fig. 401; 96. Flacourtiaceae)
　　　　　　Serrations not glandular-tipped:
　　　　　　　　Lateral veins loop-connected:
　　　　　　　　　　Blades mostly more than 60 cm long . . . . *Grias fendleri* (Fig. 413; 103. Lecythidaceae)
　　　　　　　　　　Blades less than 25 cm long:
　　　　　　　　　　　　Largest blades less than 14 cm long . . . .*Mabea occidentalis* (75. Euphorbiaceae)
　　　　　　　　　　　　Largest blades more than 15 cm long:
　　　　　　　　　　　　　　Stipules amplexicaul . . . .*Olmedia aspera* (Fig. 205; 39. Moraceae)
　　　　　　　　　　　　　　Stipules lateral, paired:
　　　　　　　　　　　　　　　　Blades asperous . . . . *Trophis racemosa* (39. Moraceae)
　　　　　　　　　　　　　　　　Blades smooth . . . . *Sorocea affinis* (39. Moraceae)
　　　　　　　　Lateral veins not loop-connected:
　　　　　　　　　　Lateral veins not raised below; stipules subpersistent; petioles slightly canaliculate, with sharp
　　　　　　　　　　　　margins . . . . *Ouratea lucens* (89. Ochnaceae)
　　　　　　　　　　Lateral veins raised below:
　　　　　　　　　　　　Petioles canaliculate; blades with inconspicuous, flaplike axillary domatia below, rounded to
　　　　　　　　　　　　　　subcordate at base and inequilateral . . *Tetrathylacium johansenii* (Fig. 400; 96. Flacourtiaceae)
　　　　　　　　　　　　Petioles not canaliculate; blades lacking axillary domatia, acute and equilateral at base . . . . . .
　　　　　　　　　　　　. . . . . *Casearia commersoniana* (Fig. 398; 96. Flacourtiaceae)

◆ Blades pubescent, at least on veins:
  ▲ Blades palmately veined at base:
    Blades with simple trichomes:
      Petioles less than 2 cm long:
        Plants shrubs, often vinelike, armed with recurved spines on stem (inconspicuously on juveniles) . . .
        . . . .*Celtis iguanaeus* (38. Ulmaceae)
        Plants shrubs or small trees, unarmed:
          Blades somewhat cordate at base; stipules caducous:
            Blades 5–15 cm long, 1.5–5 cm wide . . . . *Trema micrantha* (Fig. 193; 38. Ulmaceae)
            Blades (10)15–33 cm long, 6–16 cm wide . . . . *Apeiba tibourbou* (84. Tiliaceae)
          Blades not cordate at base; stipules caducous or persistent, slender:
            Blades rounded onto petiole; tertiary veins prominulous; petioles often more than 1 cm long;
              stipules persistent . . . . *Acalypha diversifolia* (Fig. 315; 75. Euphorbiaceae)
            Blades acute at base; tertiary veins not prominulous; petioles less than 1 cm long; stipules
              caducous . . . . *Casearia guianensis* (96. Flacourtiaceae)
      Petioles more than 2 cm long:
        Petioles ± glabrous; blades lobed or unlobed:
          Blades 3–5-lobed . . . . *Cochlospermum vitifolium* (Fig. 396; 94. Cochlospermaceae)
          Blades not lobed . . . . *Myriocarpa yzabalensis* (40. Urticaceae)
        Petioles conspicuously pubescent with simple trichomes; blades unlobed:
          Blades with conspicuous cystoliths above when dry, acute, obtuse, or rarely subcordate at base:
            Blades often more than 20 cm long . . . .
            . . . . *Myriocarpa yzabalensis* (40. Urticaceae)
            Blades less than 18 cm long . . . . *Urera eggersii* (Fig. 210; 40. Urticaceae)
          Blades lacking cystoliths, cordate or subcordate at base:
            Plants shrubs to 2 m tall; blades less than 10 cm long . . . . *Melochia lupulina* (87. Sterculiaceae)
            Plants trees or shrubs more than 3 m tall; blades more than 10 cm long:
              Blades ovate, 10–20 cm long; stipules less than 13 mm long; petioles more than 4 cm long . . .
              . . . .*Acalypha macrostachya* (Fig. 316; 75. Euphorbiaceae)
              Blades elliptic, more than 20 cm long; stipules more than 15 mm long; petioles less than 3 cm
              long . . . . *Apeiba tibourbou* (84. Tiliaceae)
    Blades with stellate or arachnoid trichomes above or below:
      Blades with the surface below obscured by dense arachnoid pubescence (some with stellate trichomes
      also):
        Petioles more than 2 cm long, ferruginous; sap copious; twigs with large bud scales at apex . . . .
        . . . . *Coussapoa panamensis* (Fig. 198; 39. Moraceae)
        Petioles less than 2 cm long:
          Blades inequilateral at base, cordate on one side, ± acute on the other, usually less than 8 cm long;
            petioles less than 5 mm long . . . . *Muntingia calabura* (83. Elaeocarpaceae)
          Blades ± equilateral at base, more than 10 cm long; petioles more than 8 mm long:
            Lower blade surface with chiefly arachnoid, brownish pubescence . . . .
            . . . . *Luehea seemannii* (84. Tiliaceae)
            Lower blade surface with both arachnoid and stellate pubescence . . . .
            . . . . *Luehea speciosa* (Fig. 359; 84. Tiliaceae)
    Blades lacking arachnoid pubescence:
    ★ Blades cordate or subcordate at base:
      Trichomes on blade below a mixture of simple and stellate:
        Blades lobed, armed with minute prickles . . . . *Hibiscus bifurcatus* (Fig. 363; 85. Malvaceae)
        Blades unlobed, unarmed . . . . *Melochia lupulina* (87. Sterculiaceae)
      All trichomes below stellate:
        Blades with 4 or fewer pairs of lateral veins above the basal pair (usually arising in apical half of
          blade); blades often shallowly lobed:
          Blades deeply cordate, the sinus of the larger leaves usually more than 1.5 cm deep; margins
            crenate (i.e., the teeth blunt) . . . . . *Hibiscus sororius* (Fig. 364 and fig. on p. 12; 85. Malvaceae)
          Blades not so deeply cordate, the sinus usually less than 1 cm deep; margins variously serrate
            with ± sharp teeth:
            Upper blade surface with long, acicular trichomes interspersed with short, stellate, many-
              pointed ones . . . . *Heliocarpus popayanensis* (Fig. 358; 84. Tiliaceae)
            Upper blade surface with trichomes of ± uniform size, either simple or few-pointed:
              Blades 10–20 cm wide, cordate . . . . *Pavonia dasypetala* (85. Malvaceae)
              Blades usually 4–10 cm wide, truncate or only very shallowly cordate . . . .
              . . . . *Pavonia paniculata* (85. Malvaceae)
        Blades with 5 or more pairs of major lateral veins above the basal pair (usually ± equally
          spaced):
          Blades ovate, broadest below the middle; petioles more than 3 cm long, with a row of glands at
            apex . . . . *Croton panamensis* (75. Euphorbiaceae)

Blades usually elliptic or oblong-elliptic; petioles less than 3 cm long, lacking glands:
 Blades softly rufous-pubescent below; stellate trichomes stalked; petioles and young branches long-pilose . . . . *Apeiba tibourbou* (84. Tiliaceae)
 Blades harshly grayish-pubescent below; stellate trichomes sessile:
  Blades usually inequilateral at base; plants trees more than 3 m tall . . . .
   . . . . *Guazuma ulmifolia* (87. Sterculiaceae)
  Blades ± equilateral at base; plants shrubs less than 2.5 m tall . . . .
   . . . . *Waltheria glomerata* (87. Sterculiaceae)
★ Blades not cordate or subcordate at base:
 Lateral veins ± uniformly spaced along midrib, sometimes lacking just above base:
  Petioles more than 1.5 cm long; branchlets essentially glabrous; blades usually with 2 red glands at base . . . . *Alchornea costaricensis* (Fig. 318; 75. Euphorbiaceae)
  Petioles less than 1 cm long; branchlets conspicuously stellate-pubescent; blades lacking glands
   . . . . *Waltheria glomerata* (87. Sterculiaceae)
 Lateral veins clustered in apical half of blade:
  Blades more than twice as long as broad, often with tufts of trichomes in apical axils below, not lobed . . . . *Trichospermum mexicanum* (Fig. 360; 84. Tiliaceae)
  Blades less than twice as long as broad, lacking tufts in axils below, the larger ones sometimes lobed:
   Blades with 3–5 serrations per cm, the serrations ca 2 mm deep . . . .
    . . . . *Pavonia paniculata* (85. Malvaceae)
   Blades with 6 or more serrations per cm, the serrations ca 1 mm deep:
    Plants shrubs less than 2.5 m tall; upper blade surface with only stellate trichomes of various sizes . . . . *Triumfetta lappula* (Fig. 361; 84. Tiliaceae)
    Plants trees more than 6 m tall; upper blade surface with long acicular trichomes interspersed with short stellate ones . . . . *Heliocarpus popayanensis* (Fig. 358; 84. Tiliaceae)
▲ Blades not palmately veined at base:
 Blades with pellucid dots:
  Blades lacking sharp teeth, abruptly acuminate at apex:
   Blades densely pubescent below, rounded to subcordate at base; petioles densely pubescent . . . .
    . . . . *Zuelania guidonia* (Figs. 403, 404; 96. Flacourtiaceae)
   Blades sparsely pubescent below, acute at base; petioles glabrous or sparsely pubescent . . . .
    . . . . *Casearia aculeata* (96. Flacourtiaceae)
  Blades sharply toothed, usually gradually acuminate at apex:
   Blades less than 10 cm long, with conspicuous pellucid dots, the margins finely serrate . . . .
    . . . . *Casearia arborea* (96. Flacourtiaceae)
   Blades more than 10 cm long:
    Pellucid dots minute, red-orange; margins finely serrulate with subulate teeth; blades more than 20 cm long . . . . *Ardisia pellucida* (112. Myrsinaceae)
    Pellucid dots usually clear, conspicuous or inconspicuous; margins coarsely serrate; blades less than 20 cm long:
     Blades broadest above the middle, the pellucid dots conspicuous . . . .
      . . . . *Casearia guianensis* (96. Flacourtiaceae)
     Blades broadest at or below the middle, the pellucid dots inconspicuous . . . .
      . . . . *Casearia arguta* (96. Flacourtiaceae)
 Blades lacking pellucid dots:
  Lower blade surface densely and conspicuously pubescent (sometimes sparsely so in *Banara* but the petioles with glands at apex):
   Blades either more than 10 cm wide and subcordate at base or ± truncate at base with prominent glands at apex of petiole and on tips of serrations:
    Blades with glands at apex of petiole and on serrations, less than 7 cm wide, never subcordate at base . . . . *Banara guianensis* (96. Flacourtiaceae)
    Blades eglandular, more than 10 cm wide, often subcordate at base; teeth consisting of marginal tufts of trichomes . . . . *Castilla elastica* (39. Moraceae)
   Blades less than 10 cm wide, lacking glands:
    Blades mostly acute to attenuate at base; petioles often fused at base with a very short branch . . . . .
     . . . . *Cordia spinescens* (121. Boraginaceae)
    Blades obtuse to subcordate at base; petioles not as above:
     Many leaves opposite or subopposite; blades sharply toothed, sometimes ovate . . . .
      . . . . *Clibadium surinamense* (133. Compositae)
     Leaves all alternate; blades usually obscurely toothed . . . . *Vernonia canescens* (133. Compositae)
  Lower blade surface not densely pubescent:
   ⊙ Petioles more than 2 cm long:
    Blades cordate at base; trunk armed . . . . *Hura crepitans* (75. Euphorbiaceae)
    Blades acute to rounded at base; trunk unarmed:
     Blades less than 12 cm long, toothed only in apical half . . . . *Roupala montana* (41. Proteaceae)

Blades more than 20 cm long, obscurely toothed or sinuate to dentate throughout . . . .
. . . . *Sloanea zuliinsis* (83. Elaeocarpaceae)
⊙ Petioles less than 2 cm long:
  Blades toothed ± throughout, often prominently so:
    Serrations glandular-tipped; stipules subulate, persistent . . . .
    . . . . *Turnera panamensis* (Fig. 405; 97. Turneraceae)
    Serrations not glandular-tipped; stipules ovate and persistent, or caducous, or lacking:
      Stipules to 3 mm long, persistent, ovate, whitish; blades ± elliptic . . . .
      . . . . *Hybanthus prunifolius* (95. Violaceae)
      Stipules small and caducous or lacking:
        Blade margins with 8–12 serrations per cm; petioles 1–2.5 cm long . . . .
        . . . . *Ardisia pellucida* (112. Myrsinaceae)
        Blade margins with 4–6 shallow serrations per cm; petioles less than 0.8 cm long . . . .
        . . . . *Casearia arguta* (96. Flacourtiaceae)
  Blades toothed only near apex or obscurely toothed:
    Blades glabrous above:
      Lateral veins ascending; blades less than 12 cm long . . . . *Lozania pittieri* (37. Lacistemaceae)
      Lateral veins ± perpendicular to midrib; blades more than 12 cm long:
        Trees usually 5–15 m tall; stems greenish when dried, sparsely puberulent, the trichomes ca
          0.5 mm long, the surface smooth, not with brownish structures like sand grains . . . . . .
        . . . . *Olmedia aspera* (Fig. 205; 39. Moraceae)
        Trees 8–25 m tall; stems brown, densely granular-puberulent, the trichomes to ca 0.2 mm
          long, the surface roughened with brown structures like sand grains . . . .
        . . . . *Trophis racemosa* (39. Moraceae)
    Blades pubescent above:
      Blades caudate-acuminate, gradually tapered to base; lateral veins often ± perpendicular to
        midrib; stipules amplexicaul . . . . *Olmedia aspera* (Fig. 205; 39. Moraceae)
      Blades acute to acuminate at apex, ± abrupt at base; lateral veins ascending:
        Stems rufous-tomentose . . . . *Couratari panamensis* (Fig. 412; 103. Lecythidaceae)
        Stems essentially glabrous:
          Plants trees or shrubs, often unbranched except near apex, growing only in clearings,
            estipulate; blades rugose above . . . . *Vernonia patens* (Fig. 552; 133. Compositae)
          Plants trees, usually branched, usually growing in the forest, stipulate; blades not rugose
            above . . . . *Trophis racemosa* (39. Moraceae)
● Blades entire (some lobed but the lobe margins not toothed, not sinuate):
  ❍ Blades lobed:
    Blades deeply lobed, nearly to or beyond middle:
      Blades peltate:
        Blades smooth above, remaining intact and open after falling . . . . *Cecropia insignis* (Fig. 194; 39. Moraceae)
        Blades scabrous or rough above, rolling up after falling:
          Blades lobed more than three-fourths the way to base, flat or slightly folded . . . .
          . . . . *Cecropia obtusifolia* (Fig. 196; 39. Moraceae)
          Blades lobed to or slightly beyond middle, folded (not able to be flattened):
            Basal pulvinus of petiole with uniform velvetlike layer of trichomes; plants common . . . .
            . . . . *Cecropia peltata* (Fig. 197; 39. Moraceae)
            Basal pulvinus of petiole with brown, velvetlike layer of trichomes interspersed with dense, longer,
              white trichomes; plants rare . . . . *Cecropia longipes* (Fig. 195; 39. Moraceae)
      Blades not peltate:
        Blades densely grayish-white-arachnoid below or glabrous and waxy below (waxy lower surface some-
          times appearing pubescent):
          Blades grayish-white-arachnoid below, more than 20 cm wide, asperous above; plants of the forest
          . . . . *Pourouma guianensis* (39. Moraceae)
          Blades glabrous and waxy below, less than 15 cm wide, smooth above; plants cultivated shrubs . . . . .
          . . . . *Manihot esculenta* (75. Euphorbiaceae)
        Blades below not arachnoid-pubescent, not glabrous and waxy:
          Plants suffrutices; spines on stems, petioles, and blades below small, recurved; trichomes on blades
           below of 2 types (forked and simple) . . . . *Hibiscus bifurcatus* (Fig. 363; 85. Malvaceae)
          Plants large trees; spines lacking:
           Trichomes on blades below mostly stellate; trunk without swollen rings . . . .
           . . . . *Sterculia apetala* (Fig. 379; 87. Sterculiaceae)
           Trichomes on blades below mostly simple; trunk with swollen rings . . . .
           . . . . *Cavanillesia platanifolia* (Fig. 369; 86. Bombacaceae)
    Blades shallowly lobed:
    ❑ Blades glabrous or with simple trichomes:
      Blades uniform, almost all lobed; plants cultivated shrubs . . . . *Jatropha curcas* (75. Euphorbiaceae)
      Blades dimorphic, one leaf much smaller, in subopposite pairs, only the younger or lower blades lobed;

plants of open areas in the forest .... *Cyphomandra hartwegii* (Fig. 481; 124. Solanaceae)
❑ Blades with stellate pubescence:
   At least the larger blades cordate or subcordate at base:
     Blades palmately veined, smooth above in age, sparsely puberulent or glabrous above; plants often
       medium-sized trees .... *Ochroma pyramidale* (Fig. 370; 86. Bombacaceae)
     Blades pinnately veined, densely pubescent above; plants shrubs ....
     .... *Solanum ochraceo-ferrugineum* (124. Solanaceae)
   Blades acute to obtuse at base:
     Blades sessile .... *Solanum jamaicense* (Fig. 486; 124. Solanaceae)
     Blades conspicuously petiolate:
       Plants slender shrubs, 1.5–4 m tall; blades to 15 cm long and 9 cm wide, the margins usually weakly
        to strongly sinuate; branchlets armed with stout spines ..... *Solanum subinerme* (124. Solanaceae)
       Plants stout trees, usually more than 5 m tall (on BCI); at least the largest blades 16 cm or more
        long, more than 10 cm wide, the margins usually entire, rarely sinuate; branchlets usually un-
        armed, the larger branches sometimes armed .... *Solanum hayesii* (Fig. 485; 124. Solanaceae)
⊙ Blades not lobed:
   Stems with conspicuous swollen nodes; sap not milky, often with pleasant aroma; plants shrubs or small
     trees, usually sterile only a short part of the year .... *Piper* (36. Piperaceae) [see fertile key]
   Stems and habit not as above:
   ○ Fully formed blades pubescent, perhaps only on veins:
    ◇ Blades above glabrous or essentially so, even on midrib:
     □ Fully grown blades with the pubescence below chiefly restricted to veins:
       Blades conspicuously broadest above or below the middle:
        Blades broadest below the middle:
         Plants epiphytic trees; blades more than 10 cm wide, palmately veined at base; sap red or
          yellowish .... *Coussapoa magnifolia* (39. Moraceae)
         Plants suffruticose shrubs; blades less than 10 cm wide, pinnately veined at base; sap clear
          .... *Witheringia solanacea* (Figs. 488, 489; 124. Solanaceae)
        Blades broadest above the middle:
         Blades less than 10 cm long:
          Blades lacking axillary domatia .... *Ficus colubrinae* (39. Moraceae)
         Blades with axillary domatia:
          Vein axils below with tufts of brownish trichomes; leaves spiral in pseudoverticils; plants
           large trees .... *Terminalia amazonica* (Figs. 421, 422; 105. Combretaceae)
          Vein axils below with flaplike pockets; leaves alternate; plants shrubs or small trees .....
           .... *Annona acuminata* (55. Annonaceae)
        Blades usually more than 12 cm long (to 30 cm):
         Leaves scattered along branchlets, pubescent only on midrib below ....
          .... *Ficus paraensis* (Fig. 203; 39. Moraceae)
         Leaves clustered at apex of branchlets, pubescent on midrib and lateral veins below and
          sometimes on surface:
          Blades with 12–20 pairs of major lateral veins .... *Pouteria fossicola* (113. Sapotaceae)
          Blades with 21–50 pairs of major lateral veins .... *Pouteria sapota* (113. Sapotaceae)
       Blades broadest at or near the middle:
        Plants with stipules:
         Blades lacking axillary domatia:
          Stipules amplexicaul, 5–10 mm long, leaving a distinct scar encircling the stem; major
           lateral veins of blade distinctly and regularly loop-connected, the lower surface drying
           brown, the reticulate veins lighter than the surface... *Perebea xanthochyma* (39. Moraceae)
         Stipules not amplexicaul, 1.5–3 mm long, not leaving a circumferential scar on stem; major
           lateral veins of blade only loosely and indistinctly loop-connected, the lower surface
           drying pale green, the reticulate veins darker than the surface.....................
           .... *Margaritaria nobilis* (75. Euphorbiaceae)
        Blades with axillary domatia:
         Stipules ocreate (encircling stem) .... *Coccoloba acuminata* (46. Polygonaceae)
         Stipules not ocreate (at most half encircling stem) .... *Lozania pittieri* (37. Lacistemaceae)
        Plants lacking stipules:
         Most blades less than 4.5 cm wide:
          Midrib prominently raised above; sap milky; blades lacking axillary domatia; plants trees to
           10 m tall .... *Pouteria stipitata* (113. Sapotaceae)
         Midrib not prominently raised above; sap not milky; blades with axillary domatia; plants
          shrubs or trees usually less than 5 m tall .... *Annona acuminata* (55. Annonaceae)
        Most blades more than 5 cm wide:
        △ Petioles usually more than 1.5 cm long; leaves alternate or subopposite:
         Leaves subopposite .... *Neea amplifolia* (Fig. 227; 48. Nyctaginaceae)
         Leaves strictly alternate:

Leaf blades membranaceous, the tertiary veins not at all raised; plants of clearings and clearing edges .... *Cestrum latifolium* (124. Solanaceae)

Leaf blades coriaceous or subcoriaceous, the tertiary veins prominently raised (on dried specimens); plants of the forest:

Blades ca 7 times longer than petioles; areoles of reticulate veins less than 1 mm diam; plants trees, more than 20 m tall .... *Ocotea pyramidata* (58. Lauraceae)

Blades less than 4 times longer than petioles; areoles of reticulate veins mostly more than 1 mm diam; plants shrubs or trees, less than 15 m tall .... *Garcia nutans* (75. Euphorbiaceae)

△ Petioles less than 1.5 cm long; leaves alternate:

Plants scandent shrubs or vines; young stems glabrous .... *Tournefortia angustiflora* (Fig. 475; 121. Boraginaceae)

Plants small trees; young stems rufous-pubescent:

Blades usually broadest above the middle, the midrib usually raised or flat (rarely slightly sunken), relatively broad, usually densely and conspicuously pubescent, the trichomes long, curved, brownish .... *Annona hayesii* (55. Annonaceae)

Blades usually broadest at the middle, the midrib deeply and narrowly sunken, relatively sparsely and inconspicuously pubescent or glabrous .... *Desmopsis panamensis* (Fig. 236; 55. Annonaceae)

☐ Fully grown blades pubescent on surface below (not just on veins):

Blades stellate-pubescent or fimbrillate-lepidote:

Petioles mostly less than 1 cm long:

Blades with 9 or fewer pairs of lateral veins; plants scandent shrubs..................... .... *Solanum argenteum* (124. Solanaceae)

Blades with 11 or more pairs of lateral veins:

Blades with fewer than 15 pairs of lateral veins, stalked-stellate below; plants medium-sized trees .... *Virola sebifera* (Figs. 242, 243; 56. Myristicaceae)

Blades with 20 or more pairs of lateral veins, sessile, stellate-pubescent below; plants tall trees .... *Virola surinamensis* (Figs. 244, 245; 56. Myristicaceae)

Petioles mostly more than 1 cm long:

Blades narrowly acuminate at apex or conspicuously cordate at base:

Blades attenuate at base; plants shrubs, usually less than 2 m tall; trichomes restricted to a brownish band on either side of midrib on lower surface........................ .... *Parathesis microcalyx* (Fig. 445; 112. Myrsinaceae)

Blades cordate at base; plants small trees, more than 3 m tall; trichomes throughout blade on lower surface .... *Croton billbergianus* (75. Euphorbiaceae)

Blades rounded to obtuse at apex, slightly subcordate at base:

Blades with 11 or more pairs of major lateral veins .... *Virola sebifera* (Figs. 242, 243; 56. Myristicaceae)

Blades with 5–7 pairs of major lateral veins:

Blades inequilateral at base; stems densely stellate-pubescent; plants without a sweet odor .... *Solanum hayesii* (Fig. 485; 124. Solanaceae)

Blades equilateral at base; stems fimbriate-lepidote; plants usually with a sweet odor:

Plants usually trees less than 10 m tall; stems and lower blade surface densely stellate-pubescent .... *Quararibea pterocalyx* (Fig. 376; 86. Bombacaceae)

Plants usually large trees (to 25 m tall); stems and lower blade surface fimbrillate-lepidote .... *Quararibea asterolepis* (Figs. 374, 375; 86. Bombacaceae)

Blades not stellate-pubescent, not fimbrillate-lepidote:

☆ Petioles often more than 1 cm long:

Trichomes on blade below so dense as to be contiguous:

Lower blade surface reddish-brown; trichomes straight, T-shaped; blades broadly elliptic; sap usually not copious .... *Chrysophyllum cainito* (113. Sapotaceae)

Lower blade surface grayish or light green; trichomes arachnoid; blades ovate; sap copious .... *Coussapoa panamensis* (Fig. 198; 39. Moraceae)

Trichomes on blade below not contiguous:

Pubescence of blades below conspicuous, erect, moderately long on veins; blades broadly ovate-elliptic; plants cultivated trees .... *Persea americana* (58. Lauraceae)

Pubescence of blades below inconspicuous and/or appressed, the veins glabrous or appressed-pubescent:

● Axils below with tufts of trichomes:

Blades conspicuously palmately 3-veined ...... *Phoebe mexicana* (Fig. 251; 58. Lauraceae)

Blades not palmately 3-veined:

Largest blades less than 20 cm long, usually broadest above the middle, abruptly short-acuminate .... *Ocotea skutchii* (Fig. 250; 58. Lauraceae)

Largest blades usually more than 20 cm long, usually broadest at middle or below, gradually long-acuminate .... *Nectandra globosa* (58. Lauraceae)

● Axils below lacking tufts of trichomes:

Sap copious, milky, reddish, or turbid:

Petioles more than 5 cm long; blades ovate-oblong, palmately veined; sap reddish or yellowish . . . . *Coussapoa magnifolia* (39. Moraceae)

Petioles less than 4 cm long:

Petioles and stems with long, arachnoid, deciduous trichomes; blades obovate, bluntly acuminate . . . . *Ficus colubrinae* (39. Moraceae)

Petioles and stems not with such pubescence:

Blades caudate-acuminate, asperous below with inconspicuous hirsute pubescence . . . . *Olmedia aspera* (Fig. 205; 39. Moraceae)

Blades not caudate-acuminate:

Petioles and stems with scurfy, brown, peeling outer bark, the petioles scurfy; stipules apical; blades often punctate below with short trichomes, conspicuously 3-veined at base . . . . *Ficus maxima* (Fig. 202; 39. Moraceae)

Petioles and stems smooth; stipules lacking; blades with T-shaped trichomes below, not 3-veined at base . . . . . . *Cynodendron panamense* (113. Sapotaceae)

Sap not colored, copious, or turbid:

Blades 3-veined near base, with the basal pair of lateral veins widely spaced from upper pairs . . . . *Phoebe mexicana* (Fig. 251; 58. Lauraceae)

Blades ± equally veined throughout, acute to narrowly acute at base:

Blades ± elliptic; reticulate veins conspicuous below; pubescence on petioles appressed; blades drying grayish below . . . . *Beilschmiedia pendula* (Fig. 248; 58. Lauraceae)

Blades slightly obovate; reticulate veins obscure below; pubescence on petioles usually ± erect; blades not drying grayish below . . . . *Ocotea oblonga* (58. Lauraceae)

☆ Petioles less than 1 cm long:

Largest blades less than 1.5 cm wide or more than 6 cm wide:

Largest blades less than 1.5 cm wide, lanceolate . . . .*Xylopia frutescens* (Figs. 240, 241; 55. Annonaceae)

Largest blades more than 6 cm wide, elliptic:

Petioles thick, swollen (scarcely so in *Perebea*), usually less than 5 mm long; plants small trees, lacking milky sap:

Plants with conspicuous amplexicaul stipules 5–10 mm long; stems with conspicuous, circumferential stipule scars . . . . *Perebea xanthochyma* (39. Moraceae)

Plants lacking stipules or conspicuous stipule scars:

Larger blades cordate, ± inequilateral at base . . . . *Guatteria amplifolia* (Figs. 237, 238; 55. Annonaceae)

Larger blades not cordate, ± equilateral at base . . . . *Desmopsis panamensis* (Fig. 236; 55. Annonaceae)

Petioles not swollen, usually more than 10 mm long; plants medium-sized trees, with milky sap at least in trunk; blades equilateral at base . . . . *Cynodendron panamense* (113. Sapotaceae)

Largest blades usually 1.5–6 cm wide:

Blades whitish, densely arachnoid below, the trichomes contiguous . . . . *Licania hypoleuca* (61. Chrysobalanaceae)

Blades not whitish, sparsely to densely pubescent below but the trichomes not contiguous:

Blades acute at base, with short trichomes on surface but not on veins; branches smooth, glabrous, divaricately branched, often with a leaf in the branch axil . . . . *Cordia lasiocalyx* (Fig. 474; 121. Boraginaceae)

Blades and branches not as above:

Blades verrucose above . . . . *Guatteria dumetorum* (55. Annonaceae)

Blades not verrucose above:

Blades densely brown-sericeous below . . . . *Chrysophyllum cainito* (113. Sapotaceae)

Blades not densely brown-sericeous below:

Plants with persistent, paired stipules . . . . . . *Margaritaria nobilis* (75. Euphorbiaceae)

Plants lacking stipules . . . . *Xylopia macrantha* (55. Annonaceae)

◇ Blades pubescent above, at least on midrib or veins:

■ Blades pubescent above only on midrib or veins:

▲ Blades below with the pubescence restricted to major veins or glabrous:

★ Stipules ocreate (stipules or scars encircling stems):

Blades more than 2.5 times longer than wide; plants medium-sized trees, the trunk often to 30 cm dbh; stems hollow, inhabited by fiercely stinging ants . . . . *Triplaris cumingiana* (Fig. 223; 46. Polygonaceae)

Blades less than 2 times longer than broad; plants small trees or vines, the trunk slender; stems not hollow:
   Most blades broadly ovate to suborbicular; plants often vinelike . . . .
   . . . . *Coccoloba parimensis* (46. Polygonaceae)
   Most blades obovate; plants trees . . . .
   . . . . *Coccoloba manzanillensis* (Figs. 221, 222; 46. Polygonaceae)
★ Stipules lacking or not ocreate:
  Petioles usually less than 3 mm long:
   Blades ± glabrous below, merely scabrid on midrib above . . . .
   . . . . *Phyllanthus acuminatus* (75. Euphorbiaceae)
   Blades pubescent below, short- or long-pubescent above:
    Leaves usually 8 or more per branchlet; trichomes of ± uniform length . . . .
    . . . . *Hirtella racemosa* (Fig. 252; 61. Chrysobalanaceae)
    Leaves usually 7 or fewer per branchlet; trichomes of 2 types, long acropetal ones interspersed with short erect ones . . . . *Hirtella triandra* (Fig. 253; 61. Chrysobalanaceae)
  Petioles more than 5 mm long:
   Blades ovate, long-acuminate, solitary at the apex of short (ca 1 cm long) branches . . . .
   . . . . *Coccoloba acapulcensis* (Fig. 219; 46. Polygonaceae)
   Blades not as above:
    Blades more than 20 cm long; petioles short, swollen . . . .
    . . . . *Guatteria amplifolia* (Figs. 237, 238; 55. Annonaceae)
    Blades less than 20 cm long; petioles not swollen:
     Blades oblong-elliptic, ca 2.5 times longer than broad; petioles stout:
      Blades usually acute to bluntly acuminate, obtuse to rounded at base; lateral veins in 13 or more pairs . . . . *Couratari panamensis* (Fig. 412; 103. Lecythidaceae)
      Blades gradually acuminate, acute to attenuate at base; lateral veins usually in 9 or fewer pairs . . . . *Ocotea pyramidata* (58. Lauraceae)
     Blades elliptic or obovate, ca 2 times longer than broad:
      Blades rounded or obtuse at apex; petioles densely short-pubescent with appressed, brownish trichomes, slender . . . . *Sloanea terniflora* (83. Elaeocarpaceae)
      Blades abruptly short-acuminate; petioles villous . . . .
      . . . . *Terminalia amazonica* (Figs. 421, 422; 105. Combretaceae)
▲ Blades pubescent throughout surface below:
  Blades stellate-pubescent:
   Blades palmately veined at base . . . . *Croton billbergianus* (75. Euphorbiaceae)
   Blades pinnately veined . . . . *Virola sebifera* (Figs. 242, 243; 56. Myristicaceae)
  Blades with simple pubescence only:
   Plants with stipules; blades with the midrib ± raised above:
    Blades densely to moderately pubescent below . . . . . . *Hirtella americana* (61. Chrysobalanaceae)
    Blades sparsely pubescent below:
     Leaves usually 8 or more per branchlet; trichomes ± uniform in length . . . .
     . . . . *Hirtella racemosa* (Fig. 252; 61. Chrysobalanaceae)
     Leaves usually 7 or fewer per branchlet; trichomes of 2 types, long acropetal ones interspersed with short erect ones . . . . *Hirtella triandra* (Fig. 253; 61. Chrysobalanaceae)
   Plants lacking stipules; blades with the midrib raised above or not:
    Blades with domatia in axils below:
     Domatia pitlike; leaves narrowly elliptic to broadly oblanceolate, well scattered on stem
     . . . . *Ocotea oblonga* (58. Lauraceae)
     Domatia consisting of tufts of trichomes in axils; leaves obovate, clustered at apices of branches . . . . *Terminalia amazonica* (Figs. 421, 422; 105. Combretaceae)
    Blades lacking axillary domatia:
     Lower leaf surface and stems densely rufous-pubescent . . . .
     . . . . *Chrysophyllum cainito* (113. Sapotaceae)
     Lower leaf surface and stems not rufous-pubescent below:
      Pubescence on lower surface uniform, sparse, appressed; blades drying verrucose (bumpy) . . . . *Guatteria dumetorum* (55. Annonaceae)
      Pubescence on lower surface more dense on veins than on surface, not appressed; blades usually not drying verrucose:
       Petioles mostly more than 1 cm long (to 4 cm) . . . . . *Cestrum latifolium* (124. Solanaceae)
       Petioles 3–6 mm long:
        Leaves mostly elliptic, the midrib above not densely pubescent . . . .
        . . . . *Desmopsis panamensis* (Fig. 236; 55. Annonaceae)
        Leaves mostly lanceolate, the midrib above densely pubescent . . . .
        . . . . *Diospyros artanthifolia* (114. Ebenaceae)

■ Blades pubescent above on surface as well as on veins:
  Trichomes stellate on blade below:
    Blades palmately veined at base, obtuse to rounded or cordate, equilateral:
      Petioles less than 4 cm long; tufts of simple trichomes in vein axils below ....
        .... *Apeiba membranacea* (Fig. 375; 84. Tiliaceae)
      Petioles more than 4 cm long; tufts lacking in vein axils below:
        Lateral veins with only 1 or 2 pairs above the basal pair, near apex of leaf ....
          .... *Hampea appendiculata* var. *longicalyx* (Fig. 362; 85. Malvaceae)
        Lateral veins with 3 or more pairs above the basal pair, ± uniformly spaced along midrib:
          Stellate trichomes below reddish-brown; petioles lacking glands ....
            .... *Ochroma pyramidale* (Fig. 370; 86. Bombacaceae)
          Stellate trichomes below whitish; petioles with 2–4 glands at apex:
            Stellate trichomes moderately dense throughout blade above; petioles more than 7 cm
              long.................................. *Croton panamensis* (75. Euphorbiaceae)
            Stellate trichomes very sparse above except on veins; petioles less than 7 cm long ....
             .... *Croton billbergianus* (75. Euphorbiaceae)
    Blades not palmately veined at base, obtuse to attenuate or rounded, inequilateral:
      Blades ovate, less than 2.5 times longer than broad:
        Blades lepidote, not actually stellate-pubescent, the scales appearing like branched
          trichomes .... *Hyeronima laxiflora* (Figs. 322, 323; 75. Euphorbiaceae)
        Blades stellate-pubescent:
          Plants not armed .... *Solanum hayesii* (Fig. 485; 124. Solanaceae)
          Plants armed on stem and/or the petioles with recurved spines:
            Blades densely, softly and conspicuously pubescent, with both sessile and stalked, stout,
              stellate trichomes, the individual trichomes often difficult to distinguish with the
              naked eye, the surface usually not dark, dull ....
              .... *Solanum ochraceo-ferrugineum* (124. Solanaceae)
            Blades sparsely and inconspicuously pubescent with minute, sessile, stellate trichomes,
              the individual trichomes easily distinguishable with the naked eye, the surface
              drying dark, often semiglossy .... *Solanum subinerme* (124. Solanaceae)
      Blades ± oblanceolate or lanceolate, more than 2.5 times longer than broad:
        Trichomes very minute, stellate under magnification; stems becoming sparsely pubescent,
          with conspicuous lenticels; plants trees usually more than 8 m tall; nodes of stems
          prominently enlarged .... *Cordia alliodora* (Fig. 473; 121. Boraginaceae)
        Trichomes conspicuously stellate; stems densely pubescent, lacking lenticels; plants shrubs
         or small trees usually less than 5 m tall; nodes of stems not enlarged:
          Most stellate trichomes on stems stalked, white ....
            .... *Solanum umbellatum* (Fig. 487; 124. Solanaceae)
          Most stellate trichomes on stems ± sessile, generally light brown:
            Leaves mostly less than 5 cm wide .... *Solanum asperum* (124. Solanaceae)
            Leaves mostly more than 5 cm wide .... *Solanum rugosum* (124. Solanaceae)
  Trichomes not stellate on blade below:
◆ Mature blades densely pubescent below:
    Stems conspicuously rufous-pubescent:
      Plants lacking milky sap; stipules paired, persistent, linear ....
        .... *Hirtella americana* (61. Chrysobalanaceae)
      Plants with milky sap; stipules apical, caducous, leaving prominent scars on stem:
        Trichomes on upper midrib whitish, extending onto blade, variable in length, scarcely or
          only slightly longer than those on blade surface .... *Ficus popenoei* (39. Moraceae)
        Trichomes on upper midrib rufous, mostly erect and in a ± narrow line, considerably
          longer than those on blade surface .... *Ficus bullenei* (Fig. 200; 39. Moraceae)
    Stems not conspicuously rufous-pubescent; sap not milky; plants lacking apical, caducous stip-
    ules, lacking prominent stipule scars on stem:
    ⊙ Plants trees, with a large stout trunk, usually more than 6 m tall:
      Pubescence below so dense as to totally obscure blade surface (less dense on juvenile
        leaves)...................................... *Cordia bicolor* (121. Boraginaceae)
      Pubescence below dense but not obscuring surface:
        Blades dimorphic, the larger ± oblong-elliptic, subtended by a smaller ovate or
          rounded blade; young stems with both short and long, stiff, erect trichomes .....
            .... *Cordia panamensis* (121. Boraginaceae)
        Blades ± uniform; young stems sparsely to densely, uniformly crisp-villous or
          appressed-puberulent:
          Lateral veins in 20 or more pairs, densely pubescent ....
            .... *Annona spraguei* (Fig. 235; 55. Annonaceae)
          Lateral veins in 10 or fewer pairs, glabrous ....
            .... *Beilschmiedia pendula* (Fig. 248; 58. Lauraceae)

◉ Plants shrubs or trees, vinelike or slender, without a stout trunk, seldom more than 3 m tall:
    Blades deeply cordate at base, inequilateral, puberulent below and on midrib above; leaves with a large one often subtended by a smaller one, appearing subopposite .... *Cyphomandra hartwegii* (Fig. 481; 124. Solanaceae)
    Blades not as above; leaves strictly alternate:
        Blades ± inequilateral at base; trichomes on smaller stems of 2 sizes:
            Leaves dimorphic, acute to obtuse at base, the larger ones more than 9 cm wide; plants lacking stipules .... *Cordia panamensis* (121. Boraginaceae)
            Leaves uniform, rounded to subcordate at base, less than 5 cm wide; plants with stipules .... *Pouzolzia obliqua* (Fig. 209; 40. Urticaceae)
        Blades ± equilateral at base; trichomes on smaller stems ± uniform in length:
            Blades less than 4.5 cm wide................... *Vernonia canescens* (133. Compositae)
            Blades more than 4.5 cm wide:
                Stems light brown, the trichomes of one type, 1–2 mm long, very dense, often conspicuously curved, the large swollen bases frequently contiguous or nearly so; blade surface usually glossy, lacking a dense layer of short puberulent trichomes .... *Tournefortia hirsutissima* (121. Boraginaceae)
                Stems dark brown, the trichomes of 2 types, the longer 4–6 mm long, moderately sparse, straight, the bases never close to being contiguous; blade surface ± dull, densely covered with a layer of short puberulence .... *Tournefortia cuspidata* (Fig. 477; 121. Boraginaceae)
◆ Mature blades not densely pubescent below:
    Blades conspicuously broadest above the middle; plants large trees; blades with axillary domatia below, obtuse at base and decurrent onto petiole; leaves clustered at tips of branches .... *Terminalia amazonica* (Figs. 421, 422; 105. Combretaceae)
    Blades broadest at or below the middle:
        Blades ovate or ovate-oblong:
            Petioles less than 1 cm long *or* the blades usually less than 5 cm wide:
                Plants with stipules; petioles usually articulate at apex........................... .... *Dalbergia brownei* (63C. Papilionoideae)
                Plants lacking stipules; petioles not articulate at apex........................... .... *Securidaca diversifolia* (Fig. 314; 74. Polygalaceae)
            Petioles more than 1 cm long, the blades more than 5 cm wide; plants soft-wooded shrubs of forests and clearings, lacking stipules:
                Blades not occasionally subopposite with a reduced opposing leaf, short-villous on lower surface, the trichomes jointed .... *Cestrum latifolium* (124. Solanaceae)
                Blades occasionally subopposite with a reduced opposing leaf, puberulent or strigillose, the trichomes not jointed:
                    Blades puberulent throughout below .... *Cyphomandra hartwegii* (Fig. 481; 124. Solanaceae)
                    Blades strigillose only on veins below .... *Witheringia solanacea* (Figs. 488, 489; 124. Solanaceae)
        Blades not ovate:
            Blades caudate-acuminate, with a conspicuous collecting vein:
                Blades elliptic-oblong; stipules 5 mm long, lateral .... *Trophis racemosa* (39. Moraceae)
                Blades obovate; stipules 5–7 mm long, amplexicaul .... *Olmedia aspera* (Fig. 205; 39. Moraceae)
            Blades not caudate-acuminate, lacking a prominent collecting vein; stipules less than 5 mm long or lacking:
                Blades more than 20 cm long, more than 3 times longer than broad (*Annona hayesii* to more than 20 cm long, but less than 3 times longer than broad):
                    Blades rounded to subcordate at base; petioles very stout; some blades more than 30 cm long.................. *Guatteria amplifolia* (Figs. 237, 238; 55. Annonaceae)
                    Blades tapered to an acute, decurrent base; petioles not stout; blades less than 25 cm long..................................... *Nectandra cissiflora* (58. Lauraceae)
                Blades mostly less than 20 cm long:
                  Margins minutely toothed, revolute; blades more than 3 times longer than broad, usually less than 12 cm long .... *Vernonia patens* (Fig. 552; 133. Compositae)
                  Margins entire, not revolute; blades less than 3 times longer than broad:
                    Blades less than 4.5 cm wide...... *Hirtella triandra* (Fig. 253; 61. Chrysobalanaceae)
                    Blades usually more than 5 cm wide:
                        Blades abruptly acuminate, rounded at base ..... *Annona hayesii* (55. Annonaceae)
                        Blades gradually acuminate, tapered to an acute, decurrent base .... *Nectandra cissiflora* (58. Lauraceae)
○ Fully formed blades glabrous even on veins (some with tufts of trichomes in vein axils below):
    ❋ Plants with both amplexicaul stipules and copious, clear, turbid, viscid, yellow-brown, or milky sap:

Lateral veins less than 1 cm apart along midrib of mature blades:
  Stipules more than 2.5 cm long; plants tall trees with regular trunks:
    Blades mostly less than 4.5 cm wide; stipules usually less than 5 cm long; lateral veins 2–4 mm apart . . . . *Ficus yoponensis* (39. Moraceae)
    Blades mostly more than 4.5 cm wide; stipules usually more than 5 cm long; lateral veins 4–10 mm apart . . . . *Ficus insipida* (39. Moraceae)
  Stipules less than 2.5 cm long; plants usually stranglers or with the base ± stilted:
    Outer bark of young stems scurfy; blades ± obovate:
      Leaves (including petioles) less than 11 cm long; petioles less than 1 cm long . . . . *Ficus perforata* (39. Moraceae)
      Leaves (including petioles) more than 11 cm long; petioles mostly more than 1 cm long . . . . *Ficus paraensis* (Fig. 203; 39. Moraceae)
    Outer bark of young stems smooth; blades elliptic to ovate, long-acuminate:
      Lateral veins conspicuously straight and parallel, 1.5–2 mm apart, the reticulate veins very weak . . . . *Ficus yoponensis* (39. Moraceae)
      Lateral veins not conspicuously straight and parallel, more than 2 mm apart, the reticulate veins evident:
        Leaves distichous . . . . *Pseudolmedia spuria* (39. Moraceae)
        Leaves spiral:
          Blade margins not white, at least at apex; blades often narrowly rounded to subcordate at base . . . . *Ficus pertusa* (Fig. 204; 39. Moraceae)
          Blade margins white all around; blades acute at base . . . . *Ficus retusa* (39. Moraceae)
Lateral veins more than 1 cm apart along midrib of mature blades:
  Plants armed . . . . *Poulsenia armata* (Fig. 206; 39. Moraceae)
  Plants unarmed:
    Blades ovate, usually cordate, with 5 palmate veins at base:
      Blades deeply cordate (the sinus narrow, more than 2 cm deep), abruptly short-acuminate at apex, smooth below . . . . *Ficus nymphiifolia* (39. Moraceae)
      Blades slightly and broadly cordate (the sinus broad, less than 1 cm deep), obtuse to rounded at apex, asperous below . . . . *Coussapoa magnifolia* (39. Moraceae)
    Blades not ovate-cordate:
      Blade apex rounded or at most shortly and bluntly acuminate:
        Blades with the lateral veins ± perpendicular to midrib at middle of blade *or* the leaves clustered densely at apex of stem and conspicuously broader beyond the middle; stipules glabrous:
          Blades oblong-obovate, markedly tapered toward base; leaves clustered at apex of stem . . . . *Ficus obtusifolia* (39. Moraceae)
          Blades broadly elliptic; leaves not clustered at apex of stem . . . *Ficus tonduzii* (39. Moraceae)
        Blades and veins not as above; stipules pubescent (except in *Ficus citrifolia*):
          Blades with 3 or fewer pairs of lateral veins . . . . *Ficus colubrinae* (39. Moraceae)
          Blades with 6 or more pairs of lateral veins:
             Submarginal collecting vein and outer ends of lateral veins below weaker than lateral veins near midrib; petioles sometimes more than 5 cm long; stipules glabrous; trees usually less than 12 m tall . . . . *Ficus citrifolia* (39. Moraceae)
            Submarginal collecting vein and lateral veins below ± equally conspicuous throughout; petioles less than 5 cm long; stipules often pubescent; trees often more than 15 m tall:
               Leaf blades generally broadest at the middle; stipules appressed-pubescent (sericeous), the stipular scars on younger stems with long trichomes . . . . *Ficus trigonata* (39. Moraceae)
               Leaf blades often broadest above the middle; stipules patulous-pubescent (appearing almost villous), the stipular scars on younger stems with a few to many long trichomes . . . . *Ficus costaricana* (39. Moraceae)
      Blade apex distinctly acuminate or sharp:
        Petioles mostly less than 1 cm long:
          Blades smooth; stipules lateral to half-amplexicaul . . . . *Brosimum alicastrum* subsp. *bolivarense* (39. Moraceae)
          Blades asperous; stipules fully amplexicaul . . . . *Olmedia aspera* (Fig. 205; 39. Moraceae)
        Petioles mostly more than 1 cm long:
          Blades acute at base, asperous below:
             Blades conspicuously caudate-acuminate at apex; petioles asperous but not scurfy . . . . *Olmedia aspera* (Fig. 205; 39. Moraceae)
             Blades acuminate but not caudate at apex; petioles scurfy . . . . *Ficus maxima* (Fig. 202; 39. Moraceae)

Blades obtuse to rounded at base or subcordate, not asperous below:
  Blades broader toward apex; stems scurfy; petioles generally less than 2.5 cm long. . . .
  . . . . *Ficus paraensis* (Fig. 203; 39. Moraceae)
  Blades generally not broader toward apex; stems smooth; petioles usually more than 2.5
    cm long:
    Trees usually less than 12 m tall; trunk grayish . . . . *Ficus citrifolia* (39. Moraceae)
    Trees often more than 15 m tall; trunk reddish-brown . . . . *Ficus dugandii* (39. Moraceae)
○ Plants lacking amplexicaul stipules and/or the sap not copious, clear, turbid, viscid, yellow-brown, or
    milky:
  Leaves with large, overlapping, persistent stipules . . . . . . *Erythroxylum multiflorum* (66. Erythroxylaceae)
  Leaves lacking such stipules:
    Leaves with a large one subtended by a smaller one, appearing subopposite:
      Lacking tufts of trichomes in vein axils below . . . . *Solanum arboreum* (Fig. 484; 124. Solanaceae)
      Blades with tufts of trichomes in vein axils below . . . . *Lycianthes synanthera* (124. Solanaceae)
    Leaves not appearing subopposite and unequal:
    ● Petioles when fresh conspicuously terete-pulvinate for entire length or at apex (drying wrinkled
        in *Neea amplifolia*) or with a produced wing (sometimes small) at apex:
      Petioles swollen for entire length, either short (usually less than 1 cm long) or with a produced
          wing at apex:
        Blades less than 20 cm long; petioles with a produced wing (sometimes minute) at apex . . .
        . . . . *Swartzia simplex* var. *ochnacea* (Fig. 280; 63B. Caesalpinioideae)
        Blades more than 20 cm long; petioles lacking a wing:
          Leaf blades usually rounded at base, sometimes obtuse, frequently with small lobes onto
              the petiole, these usually turned upward (i.e., at an angle to the remaining blade sur-
              face); petioles 4–6 mm diam; plants shrubs or small trees, (2.5)4–5(6) m tall . . . . . .
          . . . . *Guatteria amplifolia* (Figs. 237, 238; 55. Annonaceae)
          Leaf blades usually acute to obtuse at base, sometimes rounded, but never with small ±
              erect lobes; petioles 1–4 mm diam; plants usually trees more than 7 m tall (on BCI)
              or less than 3 m tall:
            Plants trees, usually more than 7 m tall; major lateral veins often conspicuously loop-
                connected; tertiary veins usually at least in part straight and connecting major
                lateral veins . . . . *Unonopsis pittieri* (Fig. 239; 55. Annonaceae)
            Plants shrubs, usually less than 2.5 m tall; major lateral veins free to the margins; ter-
                tiary veins almost always anastomosing, rarely straight and interconnecting major
                lateral veins . . . . *Neea amplifolia* (Fig. 227; 48. Nyctaginaceae)
      Petioles not swollen for entire length, lacking a produced wing:
        Blades lepidote below; petioles canaliculate . . . .
        . . . . *Hyeronima laxiflora* (Figs. 322, 323; 75. Euphorbiaceae)
        Blades not lepidote; petioles not canaliculate:
          Blades seldom exceeding 8 cm long, rounded to subcordate at base . . . .
          . . . . *Dalbergia brownei* (63C. Papilionoideae)
          Blades mostly (10)15–50 cm long:
            Blades at most acute at apex . . . . *Theobroma cacao* (87. Sterculiaceae)
            Blades mostly acuminate at apex . . . . *Capparis frondosa* (59. Capparidaceae)
    ● Petioles not terete-pulvinate and lacking a produced wing at apex:
      ◻ Most blades widest conspicuously beyond the middle:
        Larger blades more than 50 cm long, clustered at apex of branches; smaller leaves sessile
        . . . . *Grias fendleri* (103. Lecythidaceae)
      Blades less than 50 cm long:
        Blades rounded at apex:
          Plants hemiepiphytic, liana-like shrubs . . . .
          . . . . *Souroubea sympetala* (Fig. 385; 90. Marcgraviaceae)
          Plants trees, not hemiepiphytic:
            Trees less than 10 m tall; blades obovate, 6–15.5 cm long, to twice as long as wide
            . . . . *Anacardium occidentale* (76. Anacardiaceae)
            Trees 10–37 m tall; blades oblong-oblanceolate, 14–31 cm long, more than twice as
                long as wide . . . . *Anacardium excelsum* (Figs. 326, 327; 76. Anacardiaceae)
        Blades acute to acuminate at apex:
          Plants weak-stemmed shrubs, less than 2 m tall, lacking milky sap . . . .
          . . . . *Lycianthes maxonii* (124. Solanaceae)
          Plants shrubs or trees, more than 2 m tall, with or without milky sap:
            Mature blades more than 15 cm long; sap milky . . . .
            . . . . *Thevetia ahouai* (Fig. 465; 118. Apocynaceae)
            Mature blades less than 15 cm long; plants with or without milky sap:

Lateral veins obscure. . . . . *Ternstroemia tepezapote* (Fig. 386; 91. Theaceae)
Lateral veins conspicuous:
  Young stems and blades with T-shaped trichomes ca 0.5 mm long; sap conspic-
    uously milky; trunks not conspicuously buttressed . . . .
    . . . . *Pouteria stipitata* (113. Sapotaceae)
  Young stems and blades minutely rufous-pubescent, soon glabrous, lacking
    T-shaped trichomes; sap not milky; trunk usually with many buttresses
    . . . . *Terminalia chiriquensis* (105. Combretaceae)
◻ Most blades widest at or below the middle:
  ○ Blades with conspicuous pellucid or black punctations or short striations or lepidote scales
    or axillary domatia:
  Blades with axillary domatia (i.e., thickened, pocketlike or flaplike structures, or tufts of
    glands or trichomes in axils of lateral veins below):
  Base of blade with 2–4 glands near midrib, obtuse to rounded . . . .
    . . . . *Alchornea latifolia* (75. Euphorbiaceae)
  Base of blade lacking glands, variable in shape:
    Domatia consisting of a flap in axils of lower surface; lateral veins sometimes prom-
      inulous, flattened before joining midrib; blades glabrous to sparsely pubescent:
    Blades less than 3 cm broad . . . . *Annona acuminata* (55. Annonaceae)
    Blades more than 3 cm broad:
      Blades strictly pinnately veined . . . . *Annona muricata* (55. Annonaceae)
      Blades 3-veined from base . . . . *Celtis schippii* (38. Ulmaceae)
    Domatia consisting of tufts of trichomes in axils of lower surface; lateral veins not
      broadened before joining midrib, usually downturned:
    Base of blade ending abruptly, usually rounded or subcordate:
      Blades ± oblong, the sides often straight and parallel, not conspicuously nar-
        rowed at base; stipules large (17 mm long and 3 mm wide), foliaceous,
        caducous. . . . . *Tetrathylacium johansenii* (Fig. 400; 96. Flacourtiaceae)
      Blades elliptic to obovate; stipules not as above:
        Blade tips minutely mucronate, rounded to subcordate at base . . . .
        . . . . *Adelia triloba* (Fig. 317; 75. Euphorbiaceae)
        Blade tips not mucronate, obtuse at base . . . .
        . . . . *Lycianthes synanthera* (124. Solanaceae)
    Base of blade obtuse to attenuate, continuous with margin of petiole:
      Lowermost pairs of veins much longer than those at the middle of the blade. . .
      . . . . *Nectandra savannarum* (58. Lauraceae)
      Lowermost pairs of veins much shorter than those at the middle of the blade:
        Leaves, at least those near apex of stem, with a smaller subopposite leaf; some
          blades unequal at base . . . . . *Solanum antillarum* (124. Solanaceae)
        Leaves strictly alternate, ± equal in size; blades usually equal at base:
          Reticulate veins conspicuous above . . . *Nectandra purpurascens* (58. Lauraceae)
          Reticulate veins not conspicuous above . . . .
          . . . . *Ocotea skutchii* (Fig. 250; 58. Lauraceae)
  Blades lacking axillary domatia:
    ◇ Lower blade surface lepidote, the scales round, peltate, easily removed, sometimes
      sparse and inconspicuous:
    Scales sparse:
      Blades generally oblong-lanceolate, acute at base, more than 3 times longer than
        broad . . . . *Mangifera indica* (76. Anacardiaceae)
      Blades variously shaped but not oblong-lanceolate, rounded to cordate at base,
        less than 2.5 times longer than broad:
        Petioles more than 3 cm long; stipules borne on stems . . . .
        . . . . *Hyeronima laxiflora* (Figs. 322, 323; 75. Euphorbiaceae)
        Petioles less than 3 cm long:
          Stipules ocreate; scales not fimbrillate . . . .
          . . . . *Coccoloba coronata* (Fig. 220; 46. Polygonaceae)
          Stipules not ocreate; scales fimbrillate . . . .
          . . . . *Quararibea asterolepis* (Figs. 374, 375; 86. Bombacaceae)
    Scales very dense, almost contiguous:
      Blades palmately veined at base, broadest well below middle, brown on lower sur-
        face; scales ± entire; plants small trees . . . . *Bixa orellana* (93. Bixaceae)
      Blades not palmately veined at base, broadest at or above middle, not brown on
        lower surface; scales fimbrillate; plants usually large trees:
        Blades not lepidote on upper surface; plants with sweet aroma (even dried
          specimens) . . . . *Quararibea asterolepis* (Figs. 374, 375; 86. Bombacaceae)

Blades lepidote on upper surface; plants lacking sweet aroma . . . .
. . . . *Hyeronima laxiflora* (Figs. 322, 323; 75. Euphorbiaceae)
◇ Lower blade surface variously punctate, pellucid, or opaque, not lepidote:
At least half the punctations lineate (i.e., many times longer than broad):
Lower blade surface with a brownish band along midrib (composed of minute
stellate trichomes) . . . . *Parathesis microcalyx* (Fig. 445; 112. Myrsinaceae)
Lower blade surface concolorous:
Blades with the visible punctations and lines reddish, darker than the remaining
leaf tissue . . . . *Ardisia bartlettii* (112. Myrsinaceae)
Blades with the visible punctations and lines clear, lighter than the remaining
leaf tissue . . . . *Heisteria* (43. Olacaceae)
Punctations all or chiefly round, not lineate:
Punctations opaque:
Plants moderately large trees with a stout trunk; punctations not obvious when
held to light . . . . *Mangifera indica* (76. Anacardiaceae)
Plants small trees with a slender trunk; punctations visible when held to light,
dark and red . . . . *Ardisia fendleri* (Fig. 444; 112. Myrsinaceae)
Punctations pellucid:
Blades moderately thin or thick, the larger ones much more than 14 cm long;
branches equally leafy throughout; stipules lacking:
Blades not glaucous on lower surface, the major lateral veins in more than 20
pairs, mostly less than 5 mm apart; tertiary veins prominulous; pellucid
punctations reddish . . . . *Stylogyne standleyi* (112. Myrsinaceae)
Blades glaucous on lower surface, the major lateral veins in fewer than 10
pairs, mostly ca 1 cm or more apart; tertiary veins not prominulous;
pellucid punctations ± clear . . . . *Crematosperma* sp. (55. Annonaceae)
Blades moderately thin, less than 14 cm long; branches leafy mostly near apex;
stipules caducous:
One side of blade arising considerably above the other at base . . . .
. . . . *Laetia thamnia* (96. Flacourtiaceae)
Both sides of blade arising at ± the same point at base:
Stipules lanceolate, ocreate, more than 5 mm long at apex of twigs; blades
inconspicuously pellucid-punctate when dry . . . .
. . . . *Ficus pertusa* (Fig. 204; 39. Moraceae)
Stipules not ocreate, small, caducous; blades conspicuously pellucid-
punctate when dry . . . . *Casearia sylvestris* (96. Flacourtiaceae)
○ Blades lacking conspicuous pellucid or black punctations, short striations, lepidote scales,
and axillary domatia (*Unonopsis pittieri* and *Solanum arboreum* with small glandular
white dots when dried; *Dendropanax arboreus* drying with indistinct dark spots):
Leaves ± whorled, clustered at tips of branches, oblanceolate, ca 3 times longer than
broad . . . . *Ternstroemia tepezapote* (Fig. 386; 91. Theaceae)
Leaves not as above:
□ Petioles usually more than 1 cm long:
△ Petioles usually more than 3 cm long:
At least some blades distinctly lobed:
Blades with 5–7 lobes, to 25 cm long and 15 cm wide; plants cultivated in the
Laboratory Clearing . . . . *Jatropha curcas* (75. Euphorbiaceae)
Blades with 3 lobes, to 15 cm long and 7 cm wide; plants growing in the forest
. . . . *Dendropanax arboreus* (Figs. 441, 442; 109. Araliaceae)
Blades never lobed:
Stipules ocreate (leaving scar encircling stem) . . . .
. . . . *Coccoloba manzanillensis* (Figs. 221, 222; 46. Polygonaceae)
Stipules not ocreate or lacking:
Blades less than 15 cm long; petioles usually less than 5(8) cm long . . . .
. . . . *Dendropanax arboreus* (Figs. 441, 442; 109. Araliaceae)
Blades usually more than 15 cm long; petioles usually more than 5 cm long:
Leaves crowded, some subopposite, usually broadly elliptic, tapered and
acute at base . . . . *Oreopanax capitatus* (109. Araliaceae)
Leaves well-spaced, alternate, variously shaped:
Blades ovate . . . . *Roupala montana* (41. Proteaceae)
Blades oblong-elliptic:
Blades with 11 or more pairs of lateral veins . . . .
. . . . *Mangifera indica* (76. Anacardiaceae)
Blades with 10 or fewer pairs of lateral veins . . . .
. . . . *Lindackeria laurina* (96. Flacourtiaceae)

△ Petioles usually less than 3 cm long:

Lateral veins inconspicuous; margins sometimes revolute; sap red (rarely yellow or white); plants large forest trees . . . . *Aspidosperma cruenta* (118. Apocynaceae)

Plants not as above:

Blades 2.5–3 times longer than broad, with a conspicuous collecting vein; sap viscid, becoming milky . . . . *Maquira costaricana* (39. Moraceae)

Blades and sap not as above:

Sap in trunk whitish; blades obtuse to narrowly and gradually attenuate at base, less than 5.5 cm wide:

Blades usually broadest well beyond middle, drying dark green . . . . *Pouteria stipitata* (113. Sapotaceae)

Blades usually broadest near the middle, drying brownish to light green:

Blades with 3–5 palmate veins; stipule rings often visible around stems . . . . *Ficus pertusa* (Fig. 204; 39. Moraceae)

Blades strictly pinnately veined; stipules not leaving ring around stem . . . . *Pouteria unilocularis* (Fig. 446; 113. Sapotaceae)

Sap not whitish; blades mostly obtuse to acute at base, often more than 5.5 cm wide:

Reticulate veins subparallel, ± perpendicular to midrib, prominulous on both surfaces when dry; sap red . . . . *Compsoneura sprucei* (56. Myristicaceae)

Reticulate veins not as above *or* the sap not red:

Lateral veins ± parallel from midrib to a distinct, conspicuous collecting vein; stipule rings inconspicuous . . . . *Maquira costaricana* (39. Moraceae)

Lateral veins not parallel; collecting veins lacking (except in *Coccoloba*); stipule rings, when present, sometimes conspicuous:

Petioles less than 12 mm long; blades caudate-acuminate, ca 2.5–3 times longer than broad, with inconspicuous axillary tufts of trichomes . . . . *Nectandra purpurascens* (58. Lauraceae)

Petioles mostly more than 12 mm long; blades acuminate to rounded, mostly less than 3 times longer than broad:

Twigs herbaceous; twigs and leaves drying black:

Blades membranaceous, usually drying blackish, often appearing punctate beneath with the minute bases of old trichomes . . . . *Tournefortia angustiflora* (Fig. 475; 121. Boraginaceae)

Blades weakly coriaceous, usually drying brownish, not appearing punctate beneath . . . . *Tournefortia bicolor* (Fig. 476; 121. Boraginaceae)

Twigs woody (all but youngest); blades not drying black:

Blades more than 8 cm wide, with conspicuous collecting vein, ± bullate between veins . . . . *Coccoloba manzanillensis* (Figs. 221, 222; 46. Polygonaceae)

Blades less than 8 cm wide (except in *Oreopanax capitatus*):

Lateral veins numerous, mainly restricted to basal half of blade, the blades markedly coriaceous, sinuate to incised around apex . . . . *Roupala montana* (41. Proteaceae)

Lateral veins, if numerous, not restricted to basal half of blade:

Petioles of irregular lengths, some more than 2.5 cm long (to 8 cm):

Blades less than 15 cm long . . . . *Dendropanax arboreus* (Figs. 441, 442; 109. Araliaceae)

Blades to 35 cm long . . . *Oreopanax capitatus* (109. Araliaceae)

Petioles ± uniform in length, 1–3 cm long:

Veins more widely spaced along midrib near base than near apex; blades with weakly developed domatia in basal vein axils below . . . . *Phoebe mexicana* (Fig. 251; 58. Lauraceae)

Veins equally spaced along midrib; blades lacking domatia in vein axils below:

☆ Blades acute at base and decurrent onto petiole:

Margins drying conspicuously wavy; blades often with 5 or fewer pairs of lateral veins; midrib usually raised, the lateral veins not whitish on upper surface . . . . *Ocotea cernua* (58. Lauraceae)

Margins not always drying wavy; blades usually with
6 or more pairs of lateral veins; midrib usually
flat or sunken, the major lateral veins usually
whitish on upper surface . . . .
. . . . *Ocotea pyramidata* (58. Lauraceae)
☆ Blades obtuse to rounded at base, ± ending abruptly at
petiole:
Plants small trees of swamps and marshes . . *Annona
glabra* (Fig. 234 and fig. on p. 12; 55. Annonaceae)
Plants forest trees, often more than 10 m tall:
Blades shiny, coriaceous, the lateral veins not con-
spicuous . . . . . . *Heisteria concinna* (43. Olacaceae)
Blades not shiny, subcoriaceous, the lateral veins
conspicuous:
Most blades elliptic to oblanceolate . . . .
. . . . *Heisteria longipes* (43. Olacaceae)
Most blades lanceolate . . . .
. . . . *Cestrum racemosum* (124. Solanaceae)
□ Petioles usually less than 1 cm long:
● Blades mostly more than 2.5 times longer than broad:
Blades mostly more than 3 times longer than broad:
Shrubs usually less than 1.5 m tall; young stems minutely ribbed . . . .
. . . . *Heisteria costaricensis* (43. Olacaceae)
Shrubs or trees more than 1.5 m tall; stems not ribbed:
Plants with stipules:
Stipules short, persistent, fused to base of petiole; stems and petioles red-
dish; sap red in time after being cut . . . .
. . . . *Licania platypus* (Figs. 254, 255; 61. Chrysobalanaceae)
Stipules not as above; stems, petioles, and sap not reddish:
Blades shiny above:
Blades acute at apex . . . .
. . . . *Erythroxylum panamense* (Fig. 295; 66. Erythroxylaceae)
Blades acuminate at apex . . . . *Mabea occidentalis* (75. Euphorbiaceae)
Blades dull above:
Blades lacking a prominent collecting vein, the acumen long but not
caudate . . . *Lacistema aggregatum* (Figs. 191, 192; 37. Lacistemaceae)
Blades with a prominent collecting vein, the apex abruptly caudate-
acuminate:
Stems granular-puberulent; leaf blades asperous on both surfaces
. . . . *Trophis racemosa* (39. Moraceae)
Stems almost or completely glabrous except perhaps when very
young; leaf blades smooth to touch on both surfaces:
Blades shiny when dried; stems smooth or scurfy but bearing few
if any lenticels; pubescence on youngest parts ± appressed
. . . . *Maquira costaricana* (39. Moraceae)
Blades dull when dried; stems conspicuously and densely lenticel-
late; pubescence on youngest parts puberulent (short and
stiffly erect) . . . . *Sorocea affinis* (39. Moraceae)
Plants lacking stipules:
Trees more than 10 m tall:
Blades drying green, the lower surface glaucous; reticulate veins scarcely
or not at all visible . . . . *Crematosperma* sp. (55. Annonaceae)
Blades drying brown, the lower surface not at all glaucous; reticulate
veins prominulous, close . . . .
. . . . *Pouteria unilocularis* (Fig. 446; 113. Sapotaceae)
Trees or shrubs less than 8 m tall:
Blades oblanceolate, tapered to base; collecting vein obscure or ca 1 mm
from margin . . . . *Cestrum megalophyllum* (Fig. 480; 124. Solanaceae)
Blades narrowly elliptic, acute to obtuse at base; collecting vein usually
conspicuous, 3–6 mm from margin . . . .
. . . . *Anaxagorea panamensis* (Figs. 232, 233; 55. Annonaceae)
Blades mostly 2.5–3 times longer than broad:
◆ Lateral veins ± parallel from midrib to a distinct, conspicuous, loop-connecting
vein near margin (the distance between lateral veins nearly the same at the
connecting vein as at midrib); blades long-acuminate:

Largest blades less than 5 cm wide; margins often undulate . . . .
. . . . *Mabea occidentalis* (75. Euphorbiaceae)
Largest blades more than 5 cm wide; margins not distinctly undulate:
Stems smooth, neither strongly lenticellate nor pubescent; blades smooth
on both sides:
Reticulate veins of blades not raised, the areoles (areas enclosed by the
veinlets) minute, usually less than 0.2 mm diam, usually lighter than
the surface (on dried plants); pubescence of youngest parts minute,
uncinate, usually turned backward somewhat along the axis (but not
appressed) . . . . . . *Brosimum alicastrum* subsp. *bolivarense* (39. Moraceae)
Reticulate veins of blades markedly raised, the areoles more than 1 mm
diam, usually the same color as the surface (on dried plants); pubescence of youngest parts minute, appressed, not uncinate . . . .
. . . . *Maquira costaricana* (39. Moraceae)
Stems not smooth, either densely lenticellate or densely granular-
puberulent; blades smooth or asperous:
Blades smooth to the touch; stems essentially glabrous, conspicuously
lenticellate . . . . *Sorocea affinis* (39. Moraceae)
Blades asperous; stems densely granular-puberulent . . . .
. . . . *Trophis racemosa* (39. Moraceae)
◆ Lateral veins lacking a distinct marginal collecting vein; blade apex various:
Plants with conspicuous milky sap; blades ± obovate or elliptic:
Blades ± elliptic, the major lateral veins in more than 15 pairs, the lower
surface ± glaucous (at least when dried), the midrib sunken on upper
surface . . . . *Aspidosperma megalocarpon* (118. Apocynaceae)
Blades ± oblanceolate, the major lateral veins mostly in 5–7 pairs, the
lower surface not glaucous, the midrib sharply raised on upper surface
. . . . *Pouteria stipitata* (113. Sapotaceae)
Plants lacking conspicuous, milky sap; blades mostly elliptic (obovate in
Cordia lasiocalyx):
Stems divaricately branched, often with a leaf at branch nodes . . . .
. . . . *Cordia lasiocalyx* (Fig. 474; 121. Boraginaceae)
Stems not divaricately branched:
Blades with a discolorous area on both sides of midrib below; plants
shrubs or small trees usually less than 3 m tall; stems often with
persistent inflorescence bases . . . .
. . . . *Erythroxylum panamense* (Fig. 295; 66. Erythroxylaceae)
Blades lacking a discolorous area along midrib below; plants shrubs or
trees usually more than 3 m tall:
Blades ± rounded and ending abruptly at base; blades ± lanceolate
. . . . *Cestrum racemosum* (124. Solanaceae)
Blades acute or at least not ending abruptly at base; blades ± elliptic:
Plants stipulate; midrib of blade not arched:
Stipules persistent, deltoid; young stems reddish-brown; blades
with 7–12 pairs of lateral veins . . . .
. . . . *Margaritaria nobilis* (75. Euphorbiaceae)
Plants not as above:
Stipules deciduous, linear, to 9 mm long, ciliate; blades with 3
or 4 pairs of lateral veins; veins drying conspicuously
wrinkled . . . .
. . . . *Lacistema aggregatum* (Figs. 191, 192; 37. Lacistemaceae)
Stipules caducous, linear-subulate; blades with 5–8 pairs of
lateral veins; veins not drying conspicuously wrinkled . . .
. . . . *Casearia commersoniana* (Fig. 398; 96. Flacourtiaceae)
Plants estipulate; midrib of blade prominently arched (not promi-
nently arched on *Cestrum*):
Blades with domatia in axils below . . . .●. . . . .
. . . . *Nectandra purpurascens* (58. Lauraceae)
Blades lacking axillary domatia:
Blades with the acumen less than 1 cm long; young branchlets
densely rufous-tomentulose . . . .
. . . . *Terminalia chiriquensis* (105. Combretaceae)
Blades with the acumen more than 1 cm long or the blades
gradually acuminate, the acumen not distinct; young
branchlets essentially glabrous:

Leaf blades ± coriaceous, abruptly acuminate, the acumen conspicuous, more than 1 cm long, the reticulate veins clearly visible without magnification; plants occurring in the forest . . . . *Ocotea cernua* (58. Lauraceae)

Leaf blades membranaceous, gradually acuminate, the acumen not conspicuous, the reticulate veins not clearly visible; plants occurring in clearings and open areas . . . . *Cestrum nocturnum* (124. Solanaceae)

- Blades mostly less than 2.5 times longer than broad:

Blades less than 4.5 cm long, inconspicuously scabrid on midrib . . . . *Phyllanthus acuminatus* (75. Euphorbiaceae)

Blades more than 4.5 cm long:

Leaves borne at apex of very short (2–3 mm) branchlets; blades ovate, caudate-acuminate, 6–9 cm long; plants small trees . . . . *Coccoloba acapulcensis* (Fig. 219; 46. Polygonaceae)

Leaves and habit not as above:

☆ Plants with stipules (some caducous):

Plants with the lateral veins ± parallel with prominent collecting vein; blades generally more than 5 cm wide; stipules caducous; petioles scurfy; sap viscid:

Leaves usually less than 8 cm long, frequently less than 6 cm long, the primary lateral veins scarcely more conspicuous than the secondary lateral veins; stipules usually persistent (drying darker than stem) . . . . *Ficus pertusa* (Fig. 204; 39. Moraceae)

Leaves usually more than 10 cm long, frequently more than 12 cm long, the primary lateral veins much more conspicuous than the secondary lateral veins; stipules usually deciduous:

Reticulate veins of blade not raised, the areoles (areas enclosed by the veinlets) minute, usually less than 0.2 mm diam, usually lighter than the surface (on dried plants); pubescence of youngest parts minute, uncinate, usually turned backward somewhat along the axis (but not appressed) . . . . *Brosimum alicastrum* subsp. *bolivarense* (39. Moraceae)

Reticulate veins of blade prominently raised, the areoles more than 1 mm diam, usually the same color as the surface (on dried plants); pubescence of youngest parts minute, appressed, not uncinate . . . . *Maquira costaricana* (39. Moraceae)

Plants not as above:

Petioles forked below base of blade, bearing 2 subulate appendages (vestigial leaflets) . . . . *Swartzia simplex* var. *ochnacea* (Fig. 280; 63B. Caesalpinioideae)

Petioles not bearing subulate appendages:

Blades with lowermost major lateral veins stronger than any of the remaining lateral veins, arch-ascending to apical one-third to one-half of the blade . . . . *Celtis schippii* (38. Ulmaceae)

Blades with lowermost major lateral veins weaker or at least not stronger than the remaining lateral veins, ascending to no more than the basal one-third of the blade:

Blades moderately coriaceous, drying flat, the acumen blunt at tip, the reticulate veins not prominulous (at least not both close and prominulous):

Blades drying grayish, usually long-acuminate at apex; stipules subpersistent . . . . *Maytenus schippii* (77. Celastraceae)

Blades drying brownish, the acumen blunt, emarginate; stipules caducous . . . . *Vantanea occidentalis* (65. Humiriaceae)

Blades mostly membranaceous, the acumen very sharp at tip, the major lateral veins drying wrinkled or the reticulate veins both close and prominulous:

Basal pair of lateral veins reaching margin above the lower third of blade; stipules caducous, 5–9 mm long, marginally ciliate or pubescent . . . . *Lacistema aggregatum* (Figs. 191, 192; 37. Lacistemaceae)

Basal pair of lateral veins reaching margin in the lower fourth of blade; stipules subpersistent, 1.5–3 mm long, the margins glabrous . . . . *Margaritaria nobilis* (75. Euphorbiaceae)

☆ Plants estipulate:
    At least the younger stems usually densely appressed-pubescent:
        Pubescence on younger stems coarse, the trichomes not nearly contiguous and interspersed with many minute glandular globules; sap clear, not viscid; plants slender, erect or climbing shrubs . . . .
        . . . . *Tournefortia bicolor* (Fig. 476; 121. Boraginaceae)
        Pubescence on younger stems fine, the trichomes often so dense as to be contiguous, lacking minute glandular globules; sap milky or clear; plants erect trees:
            Leaf blades usually broadest at the middle, usually drying blackened, the tertiary veins anastomosing and running ± parallel to the major lateral veins . . . . *Cynodendron panamense* (113. Sapotaceae)
            Leaf blades often broadest above the middle, usually drying green, the tertiary veins often running straight between and perpendicular to the major lateral veins . . . . *Annona hayesii* (55. Annonaceae)
    Younger stems glabrous or nearly so (at least not densely appressed-pubescent):
        Stems ± herbaceous; stems and leaves drying blackened . . . .
        . . . . *Tournefortia angustiflora* (Fig. 475; 121. Boraginaceae)
        Stems woody; stems and leaves not drying blackened:
            Leaf margins not revolute and not hyaline; midrib often ± arched:
                Leaves clustered at ends of branchlets in pseudowhorls, minutely verrucose . . . . *Terminalia chiriquensis* (105. Combretaceae)
                Leaves alternate, not clustered at ends of branches, not verrucose:
                    Plants large forest trees, usually more than 20 m tall; leaves thick, the acumen thickened and often discolored at its tip . . . . . .
                    . . . . *Drypetes standleyi* (Fig. 321; 75. Euphorbiaceae)
                    Plants shrubs or small trees, usually in clearings or at the edge of the forest, usually less than 5 m tall; leaves thin, the acumen not at all thickened or discolored . . . .
                    . . . . *Cestrum nocturnum* (124. Solanaceae)
            Leaf margins somewhat revolute and/or hyaline:
                Acumen thickened near its apex and hyaline or discolored, the thickened area extending somewhat down either margin; blade margins weakly hyaline, not noticeably revolute, with a submarginal vein along much of its length . . . .
                . . . . *Drypetes standleyi* (Fig. 321; 75. Euphorbiaceae)
                Acumen not as above; blade margins revolute, lacking a submarginal vein for all or most of its length:
                    Blades coriaceous, shiny, the lateral veins not conspicuous . . . . .
                    . . . . *Heisteria concinna* (43. Olacaceae)
                    Blades subcoriaceous, not shiny, the lateral veins conspicuous below . . . . *Heisteria longipes* (43. Olacaceae)

KEY 3: LEAVES SIMPLE, ALTERNATE OR SPIRALED; PLANTS LIANAS

Blades peltate, ovate:
    Pubescence of sparse, long, spreading trichomes or puberulent . . . . . . *Cissampelos tropaeolifolia* (54. Menispermaceae)
    Pubescence long and sericeous or short and tomentose . . . . *Cissampelos pareira* (54. Menispermaceae)
Blades not peltate:
  ❏ Blades palmately veined at base (the 2–4 pairs of major lateral veins sometimes departing midrib 1–2 mm above base):
    ○ Petioles less than 1 cm long, the plants unarmed, *or* the blades deeply cleft at apex of midrib (juveniles so deeply cleft as to appear bifoliolate):
        Blades deeply cleft at apex:
            Trichomes on petioles and blades below obscure, appressed, less than 0.5 mm long . . . .
            . . . . *Bauhinia guianensis* (Fig. 274; 63B. Caesalpinioideae)
            Trichomes on petioles and blades below ferruginous, erect to reclining, ca 1 mm long . . . .
            . . . . *Bauhinia reflexa* (63B. Caesalpinioideae)
        Blades not cleft at apex:
            Blades glabrous or nearly so:
                Petioles swollen at apex . . . . *Abuta racemosa* (Fig. 230; 54. Menispermaceae)
                Petioles not swollen at apex . . . . *Smilax lanceolata* (26. Smilacaceae)

Blades pubescent:
    Lower blade surface very densely pubescent with erect, crisp trichomes, the upper surface with sparse scabrous pubescence . . . . *Pouzolzia obliqua* (Fig. 209; 40. Urticaceae)
    Lower blade surface sparsely to densely pubescent, the upper surface often glabrate:
        Blades often shallowly cordate at base . . . . *Gouania lupuloides* (81. Rhamnaceae)
        Blades obtuse to rounded at base . . . . *Securidaca diversifolia* (74. Polygalaceae)
○ Petioles more than 1 cm long *or* the plants armed; blades not cleft at apex:
    Mature blades deeply 3–5-lobed:
        Plants glabrous; leaves coriaceous, lacking teeth; tendrils trifid, branched from near base . . . .
        . . . . *Cayaponia granatensis* (131. Cucurbitaceae)
        Plants pubescent; leaves not coriaceous, with ± apiculate teeth; tendrils simple:
            Pubescence dense, minute, ferruginous; blades with several round glands on margin at bases of sinuses between lobes . . . . *Passiflora vitifolia* (98. Passifloraceae)
            Pubescence usually white-pilose or white-villous; blades eglandular:
                Many mature blades 5-lobed, to 30 cm long and wide . . . . *Gurania makoyana* (131. Cucurbitaceae)
                Mature blades mostly 3-lobed, to 20 cm long and wide . . . .
                . . . . *Gurania megistantha* (Fig. 540; 131. Cucurbitaceae)
    Mature blades not lobed:
        Most blades more than 3 times longer than wide, cordate at base, conspicuously punctate above when dry
        . . . . *Aristolochia chapmaniana* (Fig. 215, 216; 44. Aristolochiaceae)
        Blades less than 3 times longer than wide:
        ● Plants armed *or* with conspicuous glands at apex or base of petiole:
            Plants armed:
                Plants armed with recurved spines on stems, petioles, and midribs of blades . . . .
                . . . . *Byttneria aculeata* (Fig. 377; 87. Sterculiaceae)
                Plants armed on stems only; blades toothed . . . . *Celtis iguanaeus* (38. Ulmaceae)
            Plants unarmed:
                Blades rounded at apex; petioles with 2 large glands at apex . . . .
                . . . . *Omphalea diandra* (Fig. 324; 75. Euphorbiaceae)
                Blades long-acuminate; petioles with 2 protruding glands at base . . . .
                . . . . *Smilax lanceolata* (26. Smilacaceae)
        ● Plants unarmed and lacking glands on petiole:
            Lateral veins ending in a sharp tooth on margin:
                Blades densely arachnoid below, the trichomes obscuring the surface . . . .
                . . . . *Vitis tiliifolia* (Figs. 355, 356; 82. Vitaceae)
                Blades sparsely to densely pubescent below, not arachnoid, the surface not obscured:
                    Blades glabrous or with simple, villous pubescence, mostly ovate-oblong, monomorphic, the larger not cordate . . . . *Cissus sicyoides* (82. Vitaceae)
                    Blades pubescent at least on veins below with minute, close puberulence and often with appressed, T-shaped trichomes, ovate, dimorphic, the larger ones cordate . . . .
                    . . . . *Cissus pseudosicyoides* (Fig. 354; 82. Vitaceae)
            Lateral veins not ending in a sharp tooth on margin:
                Blades densely short-white-pubescent below, the trichomes completely obscuring surface (less so on juveniles) . . . . *Chondrodendron tomentosum* (Fig. 231; 54. Menispermaceae)
                Blades not so densely pubescent below:
                Blades with domatia or nectaries in vein axils below at base of blade:
                    Domatia pocketlike . . . . *Fevillea cordifolia* (Fig. 538; 131. Cucurbitaceae)
                    Domatia flat, glandular areas or nectary pores:
                        Blades pubescent at least on veins below . . . .
                        . . . . *Odontocarya tamoides* var. *canescens* (54. Menispermaceae)
                      Blades glabrous . . . . *Odontocarya truncata* (54. Menispermaceae)
                Blades lacking domatia and nectaries in vein axils:
                  Petioles enlarged and usually curved and rugose at apex:
                    Petioles mostly less than 3 cm long . . . . *Abuta racemosa* (Fig. 230; 54. Menispermaceae)
                    Petioles mostly more than 4 cm long . . . . *Abuta panamensis* (54. Menispermaceae)
                Petioles not enlarged or curved at apex:
                  Blade margins usually ± crenate; blades longer than broad, drying with conspicuous cystoliths above . . . . *Urera eggersii* (Fig. 210; 40. Urticaceae)
                  Blade margins entire; blades nearly orbicular, lacking cystoliths . . . .
                  . . . . *Ipomoea phillomega* (Fig. 471; 120. Convolvulaceae)
◘ Blades not palmately veined at base:
◇ Blades pubescent above on surface:
    □ Plants with stellate trichomes and/or armed:
        Plants armed with recurved spines on stems and midribs of blades . . . . *Solanum lanciifolium* (124. Solanaceae)
        Plants unarmed:

Petioles (5)10–20 mm long, the wings flaring upward at base . . . . *Tetracera volubilis* (88. Dilleniaceae)
Petioles 3–10 mm long, the wings obscure or lacking at base . . . .
. . . . *Tetracera portobellensis* (Fig. 383; 88. Dilleniaceae)
□ Plants lacking stellate trichomes and unarmed:
Stems essentially glabrous:
Blades toothed and gradually acuminate (rarely entire); blade surface glabrous or with long, sparse, slender trichomes . . . . *Doliocarpus dentatus* (88. Dilleniaceae)
Blades entire (or obscurely toothed at apex), rounded or bluntly acuminate at apex; blade surface scabridulous except for long, slender, ± appressed trichomes on veins . . . . *Davilla nitida* (88. Dilleniaceae)
Stems conspicuously pubescent:
Stems densely pubescent with T-shaped trichomes . . . . *Heteropteris laurifolia* (71. Malpighiaceae)
Stems lacking T-shaped trichomes:
Stems with 2 types of pubescence: very long (more than 4 mm), spreading trichomes and minute puberulence *or* the trichomes shorter with prominent bulbous bases:
Trichomes on stems of 2 types, very long ones interspersed with short puberulence . . . .
. . . . *Tournefortia cuspidata* (Fig. 477; 121. Boraginaceae)
Trichomes on stems of a single type, bending toward apex with a prominent bulbous base . . . .
. . . . *Tournefortia hirsutissima* (121. Boraginaceae)
Stem pubescence of a single type, lacking bulbous bases:
Stems lacking stipular glands at bases of petioles; blades ciliate, emarginate at apex with tufted apiculum . . . . *Iseia luxurians* (120. Convolvulaceae)
Stems with small, round, stipular glands at bases of petioles; blades ciliate or not:
Pubescence on blades and stems appressed; blades usually acute to obtuse at apex, rounded to cuneate at base . . . . *Securidaca diversifolia* (Fig. 314; 74. Polygalaceae)
Pubescence on blades and stems erect and spreading; blades obtuse to rounded at apex, broadly rounded or truncate at base . . . . *Securidaca tenuifolia* (74. Polygalaceae)
◇ Blades glabrous above or with trichomes on veins only:
Plants parasitic lianas; lateral veins nearly invisible; blades usually less than 5 cm long . . . .
. . . . *Struthanthus orbicularis* (42. Loranthaceae)
Plants not parasitic; lateral veins visible:
Stipules ocreate; blades with very minute peltate scales on both surfaces . . . .
. . . . *Coccoloba parimensis* (46. Polygonaceae)
Stipules lacking:
Plants with long recurved spines in branch axils; vein axils with tufts of trichomes below; midrib raised and pubescent above . . . . *Combretum decandrum* (Figs. 416, 417, and fig. on p. 20; 105. Combretaceae)
Plants lacking spines and axillary tufts:
Midrib conspicuously sunken above:
Midrib arched; lower blade surface smooth; lateral veins scarcely raised below . . . .
. . . . *Maripa panamensis* (120. Convolvulaceae)
Midrib not arched; lower blade surface asperous or not; lateral veins very prominently raised below; petioles narrowly winged or sharply canaliculate below blade:
Leaf blades broadest at or below middle, markedly asperous on lower surface, rounded or obtuse at base, drying grayish-green . . . . *Tetracera hydrophila* (88. Dilleniaceae)
Leaf blades broadest somewhat above middle, smooth on lower surface, acute at base, drying dark brown . . . . *Doliocarpus multiflorus* (88. Dilleniaceae)
Midrib flat or raised above (slightly sunken in *Doliocarpus dentatus* and *D. olivaceus*):
Young stems usually 5-angulate and narrowly winged; blades with conspicuous punctations on both surfaces (inconspicuous on juveniles) . . . . *Doliocarpus major* (Fig. 380; 88. Dilleniaceae)
Stems not markedly ribbed or winged; blades not punctate:
Stems, except oldest parts, herbaceous, green; plants dark brown or black when dry, usually with minute, granular projections on young blades (bases of old trichomes):
Blades membranaceous, usually drying blackish, often appearing punctate beneath with the minute bases of old trichomes . . . . *Tournefortia angustiflora* (Fig. 475; 121. Boraginaceae)
Blades weakly coriaceous, usually drying brownish, not appearing punctate beneath . . . .
. . . . *Tournefortia bicolor* (Fig. 476; 121. Boraginaceae)
Stems woody, not green; plants not blackening and lacking granular projections on blades when dry:
Stems densely pubescent with T-shaped trichomes . . . . *Heteropteris laurifolia* (71. Malpighiaceae)
Plants lacking T-shaped trichomes:
△ Lateral veins conspicuous; blades lacking rows of minute glands near margins, the margins frequently toothed, sometimes pubescent; petioles narrowly winged; plants not epiphytic:
Blades essentially glabrous:
At least some blades conspicuously dentate . . . . *Doliocarpus dentatus* (88. Dilleniaceae)
Blades usually entire to obscurely dentate . . . .
. . . . *Doliocarpus olivaceus* (Figs. 381, 382; 88. Dilleniaceae)

Blades pubescent:

Blades scabridulous over surface, the trichomes on veins longer and ± appressed; margins entire or dentate only at apex . . . . *Davilla nitida* (88. Dilleniaceae)

Blades with long slender trichomes over surface; at least some blades markedly dentate
. . . . *Doliocarpus dentatus* (88. Dilleniaceae)

△ Lateral veins obscure; blades glabrous, with rows of minute glands near margins, the margins entire; petioles not winged; plants hemiepiphytic shrubs:

Blades oblong to oblong-lanceolate, usually rounded or acute and ending abruptly at base, long-acuminate at apex; branches long-pendent . . . .
. . . . *Marcgravia nepenthoides* (Fig. 384; 90. Marcgraviaceae)

Blades not oblong or oblong-lanceolate:

Blades ovate to obovate, acute to rounded at base; leaves irregular on stem, often subopposite . . . . *Markea ulei* (Figs. 482, 483; 124. Solanaceae)

Blades usually obovate, gradually tapered to an acute base; leaves spirally arranged
. . . . *Souroubea sympetala* (Fig. 385; 90. Marcgraviaceae)

KEY 4: LEAVES PALMATELY COMPOUND WITH FOUR OR MORE LEAFLETS

Blades pubescent below:

Blades densely brown-sericeous, the trichomes simple, totally obscuring surface below . . . .
. . . . *Didymopanax morototoni* (109. Araliaceae)

Blades sparsely to densely pubescent below, the trichomes stellate, not obscuring surface:

Leaves alternate; blades pubescent above only on veins; leaflets usually more than 25 cm long . . . .
. . . . *Herrania purpurea* (Fig. 378; 87. Sterculiaceae)

Leaves opposite; blades pubescent above throughout; leaflets less than 15 cm long . . . .
. . . . *Tabebuia ochracea* var. *neochrysantha* (126. Bignoniaceae)

Blades glabrous or lepidote below:

Leaflets conspicuously petiolulate, the petiolules usually more than 1.5 cm long; blades ending abruptly at petiolule; leaves opposite or subopposite (except *Ceiba*):

Blades with plate-shaped glands at least in lower vein axils . . . . . *Tabebuia rosea* (Figs. 500, 501; 126. Bignoniaceae)

Blades lacking plate-shaped glands in vein axils:

Leaflets often more than 5 cm broad, tufted with stellate trichomes in axils below . . . .
. . . . *Tabebuia guayacan* (Fig. 499; 126. Bignoniaceae)

Leaflets less than 4.5 cm broad, not tufted in axils . . . . *Ceiba pentandra* (fig. on p. 10; 86. Bombacaceae)

Leaflets sessile or the petiolules less than 1 cm long; blades tapered at base; leaves alternate:

Older branches and trunk armed:

Leaflets often more than 5 cm wide, usually obovate . . . . *Bombacopsis quinata* (Figs. 366, 367; 86. Bombacaceae)

Leaflets less than 4.5 cm wide, usually oblong-lanceolate to oblong-elliptic:

Leaflets glaucous on lower surface, drying pale bluish-green; reticulate veins not visible . . . .
. . . . *Jacaratia spinosa* (99. Caricaceae)

Leaflets not glaucous on lower surface, drying brownish; reticulate veins prominulous . . . .
. . . . *Ceiba pentandra* (fig. on p. 10; 86. Bombacaceae)

Branches and trunk not armed:

Leaflets rounded to emarginate at apex . . . . *Bombacopsis sessilis* (Fig. 368; 86. Bombacaceae)

Leaflets acute to acuminate or mucronate at apex:

Leaflets articulate; trunk bark lacking green vertical bands . . . . . *Pachira aquatica* (Fig. 371; 86. Bombacaceae)

Leaflets not articulate; trunk bark with prominent, green, vertical bands . . . .
. . . . *Pseudobombax septenatum* (Figs. 372, 373; 86. Bombacaceae)

KEY 5: LEAVES PINNATELY COMPOUND MORE THAN ONCE

☆ Leaves biternate *or* the basal set of leaflets ternate *or* with prominently winged rachis:

● All leaves biternate (i.e., with 3 sets of 3 leaflets each) or with the terminal set replaced by a tendril or tendril scar:

Leaflets densely and softly rufous-pubescent below . . . . *Serjania cornigera* (Fig. 347; 80. Sapindaceae)

Leaflets not rufous-pubescent:

Blades with the midrib glabrous above:

Leaves always biternate; stem unarmed; wood composite, consisting of a central woody core surrounded by 3 smaller peripheral bundles . . . . *Serjania paucidentata* (Fig. 349; 80. Sapindaceae)

Leaves sometimes biternate at apex but always 2- or 3-pinnate lower on stem; stems usually armed with short prickles; wood simple, lacking peripheral bundles . . . *Serjania mexicana* (Fig. 348; 80. Sapindaceae)

Blades with the midrib pubescent above:

Rachis not winged:

Leaflets lacking axillary tufts; stems 5- or 6-costate . . . . *Serjania decapleuria* (80. Sapindaceae)

Leaflets with the axils conspicuously tufted below; stems terete to square:
  Leaflets prominently toothed, the lateral leaflets nearly sessile . . . .
  . . . . *Serjania pluvialiflorens* (80. Sapindaceae)
  Leaflets entire, all distinctly petiolulate . . . . *Pleonotoma variabilis* (126. Bignoniaceae)
Rachis narrowly winged:
  Blades with black lines, especially below . . . . *Serjania atrolineata* (80. Sapindaceae)
  Blades lacking black lines:
    Stems terete; leaflets ± glabrous below except for tufts of trichomes in axils . . . .
    . . . . *Paullinia fuscescens* (80. Sapindaceae)
    Stems ribbed; leaflets pubescent below or lacking tufted axils:
      Leaflets sparsely pubescent below, occasionally with tufts in axils; stems with simple wood . . . . . .
      . . . . *Serjania mexicana* (Fig. 348; 80. Sapindaceae)
      Leaflets densely soft-pubescent below, lacking axillary tufts; stems with composite wood, having 3
        peripheral bundles . . . . *Serjania rhombea* (80. Sapindaceae)
● Blades not strictly biternate, but with the basal set of leaflets ternate, or with more than 3 leaflets on the basal
  pinnae:
  Terminal leaflets less than 4 cm long; leaves partially 3-pinnate; older stems 3-ribbed . . . .
  . . . . *Serjania trachygona* (Fig. 350; 80. Sapindaceae)
  Terminal leaflets more than 4 cm long:
    Stems terete; basal set of leaflets with 3 leaflets . . . . *Paullinia glomerulosa* (Fig. 343; 80. Sapindaceae)
    Stems ribbed; basal set of leaflets with 3–8 leaflets; stems often armed . . . .
    . . . . *Serjania mexicana* (Fig. 348; 80. Sapindaceae)
☆ Leaves not biternate, simply bipinnate, the rachis unwinged:
  Leaflets more than 5 mm wide:
    Petioles or rachises with glands:
      Petioles with a large conspicuous gland at apex (many times larger than other glands on leaf):
        Gland at apex of petiole less than 3 mm high; leaflets 3–5 per pinna, acute at apex, ovate . . . .
        . . . . *Leucaena multicapitula* (Fig. 267; 63A. Mimosoideae)
        Gland at apex of petiole ca 10 mm high; leaflets 7–12 per pinna, often rounded at apex, oblong . . . .
        . . . . *Pithecellobium macradenium* (Figs. 271, 272; 63A. Mimosoideae)
      Petioles lacking a large gland at apex:
        Each pinna with 8 or more pairs of leaflets . . . . *Schizolobium parahybum* (63B. Caesalpinioideae)
        Each pinna with 7 or fewer pairs of leaflets:
          Leaflets rounded at apex, narrowly ovate . . . . *Albizia guachapele* (Fig. 258; 63A. Mimosoideae)
          Leaflets not rounded at apex, not ovate:
            Pinnae with 2 leaflets, with recurved spines at base of leaf . . . .
            . . . . *Pithecellobium hymeneaefolium* (Fig. 270; 63A. Mimosoideae)
            Pinnae with 4–7 pairs of leaflets, lacking spines . . . . *Pithecellobium dinizii* (63A. Mimosoideae)
    Petioles and rachises lacking glands:
      Plants lianas:
        Leaflets subsessile, the sides usually straight, broadest near apex . . . .
        . . . . *Adenopodia polystachya* (63A. Mimosoideae)
        Leaflets petiolulate, the sides curved, broadest at middle . . . *Entada monostachya* (Fig. 259; 63A. Mimosoideae)
      Plants trees:
        Leaflets rounded at both ends, emarginate at apex, equilateral at base with the lateral veins inconspicuous;
          juveniles with resinous stems . . . . *Schizolobium parahybum* (63B. Caesalpinioideae)
        Leaflets acuminate at apex:
          Leaflets inequilateral at base and decurrent onto base of petiole, truly bipinnate; stipules lacking; plants
           large trees, usually more than 20 m tall . . . . *Jacaranda copaia* (Figs. 493, 494; 126. Bignoniaceae)
          Leaflets ± equal at base, the petioles distinct; blades simple, only appearing bipinnate due to deciduous
           branches; stipule at each leaf minute, triangular, persistent; plants shrubs or small trees, less than
           8 m tall . . . . *Phyllanthus acuminatus* (75. Euphorbiaceae)
  Leaflets less than 5 mm wide:
    Stems unarmed:
      Pinnae each with 14 or fewer pairs of leaflets; petioles, rachises, and lower leaflet surfaces tawny-tomentose
      . . . . *Pithecellobium barbourianum* (Fig. 269; 63A. Mimosoideae)
      Pinnae each with 15 or more pairs of leaflets; leaflets not tawny-tomentose:
        Leaflets less than 5 mm long; young stems, petioles, and rachises ferruginous . . . .
        . . . . *Enterolobium schomburgkii* (63A. Mimosoideae)
        Leaflets more than 6 mm long; stems, petioles, and rachises usually glabrate, not ferruginous though
          sometimes ± cinereous:
          Stems, petioles, and rachises usually glabrate; leaflets in 3 or 4 pairs per cm along rachis . . . .
          . . . . *Enterolobium cyclocarpum* (Fig. 260; 63A. Mimosoideae)
          Stems, petioles, and rachises ± cinereous; leaflets in 5–7 pairs per cm along rachis . . . .
          . . . . *Acacia glomerosa* (63A. Mimosoideae)
    Stems armed (often rachises also):

Spines bullhorn-like, black, hollow, more than 3 cm long . . . . *Acacia melanoceras* (63A. Mimosoideae)
Spines not as above:
    Stems and rachises densely rufous-pubescent; rachises armed throughout at each node . . . .
    . . . . *Mimosa pigra* (63A. Mimosoideae)
    Stems and rachises not densely rufous-pubescent; rachises sparsely armed:
        Plants large buttressed trees . . . . *Acacia glomerosa* (63A. Mimosoideae)
        Plants small trees or lianas or shrubs:
            Leaflets 3–7 mm long, glabrous or conspicuously pubescent on lower inner edge . . . .
            . . . . *Acacia riparia* (Fig. 257; 63A. Mimosoideae)
            Leaflets 9–14 mm long, glabrous or pubescent, but the trichomes never restricted to lower inner edge
            nearest rachis:
                Older stems usually deeply 4-sulcate . . . . *Acacia acanthophylla* (63A. Mimosoideae)
                Older stems usually subterete and with 5 prominent ribs but not 5-sulcate . . . .
                . . . . *Acacia hayesii* (63A. Mimosoideae)

KEY 6: LEAVES COMPOUND, PINNATE OR WITH THREE OR FEWER LEAFLETS

★ Leaves at most 2- or 3-foliolate:
    Plants trees or shrubs:
        Leaflets palmately veined at base; trunk spiny:
            Plants large trees; spines on trunk broad, corky; leaflets rounded at apex, less than 8 cm broad, broadest near
                middle . . . . *Erythrina fusca* (Fig. 287; 63C. Papilionoideae)
            Plants small, slender trees; spines short, small, not corky; leaflets acuminate, more than 10 cm broad, broadest
                near base . . . . *Erythrina costaricensis* var. *panamensis* (Figs. 285, 286; 63C. Papilionoideae)
        Leaflets not palmately veined at base; trunk unarmed:
            Leaflets 3, lacking wings on margin of petiole:
                Leaflets sessile; axils of lower blade surface with tufts of pubescence . . . .
                . . . . *Allophylus psilospermus* (80. Sapindaceae)
                Leaflets petiolulate, lacking axillary tufts on lower surface:
                    Leaflets thick, rounded at base, the petiolules with a terete pulvinus, not canaliculate; younger parts
                    reddish-tomentose . . . . *Connarus panamensis* (Fig. 256; 62. Connaraceae)
                    Leaflets moderately thin, cuneate to attenuate at base, the petiolules canaliculate, not at all swollen;
                    younger parts glabrate to puberulent. . . . . *Vitex cooperi* (122. Verbenaceae)
        Leaflets 2 *or* petioles with produced winglike margins at apex:
            Leaflets 3, ± equilateral, often with some simple leaves:
                Leaflets ± elliptic, ending abruptly on petiolule, distinctly petiolulate, obtuse to rounded at base . . . . .
                . . . . *Swartzia simplex* var. *grandiflora* (63B. Caesalpinioideae)
                Leaflets oblanceolate, decurrent onto petiolule or sessile, acute and tapered at base (adults often with 5
                leaflets) . . . . *Quassia amara* (68. Simaroubaceae)
            Leaflets 2, inequilateral:
                Petioles less than 5 mm long; leaflets usually emarginate at apex, sessile at apex of petiole . . . .
                . . . . *Cynometra bauhiniifolia* (63B. Caesalpinioideae)
                Petioles more than 10 mm long; leaflets usually not emarginate at apex:
                    Margins at base of leaflet not meeting petiolule at the same point, separated by 2–3 mm; leaflets not
                    long-acuminate . . . . *Hymenaea courbaril* (63B. Caesalpinioideae)
                  Margins at base of leaflet joining petiolule at ± the same point; leaflets long-acuminate . . . .
                  . . . . *Peltogyne purpurea* (63B. Caesalpinioideae)
    Plants lianas:
    ◆ Leaflets toothed:
        Blades with round, pellucid punctations throughout; stems hollow . . . .
        . . . . *Stizophyllum riparium* (126. Bignoniaceae)
        Blades not pellucid-punctate throughout:
            Tendrils coiled like a watchspring, axillary; leaflets irregularly, broadly toothed, usually crenate or not
            sharply toothed:
                Leaflets lacking tufts of trichomes in axils below . . . . *Serjania circumvallata* (80. Sapindaceae)
                Leaflets with tufts of trichomes in axils below:
                    Leaflets palmately veined at base . . . . *Thinouia myriantha* (80. Sapindaceae)
                    Leaflets not palmately veined at base . . . . *Paullinia turbacensis* (Fig. 346; 80. Sapindaceae)
            Tendrils not coiled like a watchspring; leaflets ± regularly toothed and usually serrate-apiculate or closely
            irregularly crenate (*Cissus erosa*); leaves opposite:
                Terminal leaflet rhombic; plants densely pubescent with glandular and/or eglandular trichomes; stip-
                ules on young branches lanceolate to linear-lanceolate, 3–6 mm long . . . .
                . . . . *Cissus rhombifolia* (82. Vitaceae)
                Terminal leaflet elliptic to obovate; plants usually glabrous or with sparse eglandular trichomes; stipules
                on young branches broadly ovate to subrotund, 2–4 mm long *or* stipules lacking:

Stems tetragonal, often winged, sometimes speckled with maroon at nodes . . . . *Cissus erosa* (82. Vitaceae)
Stems terete to ribbed or subangulate, not winged, not speckled:
    Petioles and petiolules conspicuously but sparsely white-pilose; terminal leaflet usually more than
      12 cm long . . . . *Gurania coccinea* (131. Cucurbitaceae)
    Petioles and petiolules not conspicuously pubescent; terminal leaflet usually less than 12 cm long
      . . . . *Cissus microcarpa* (Fig. 353; 82. Vitaceae)
◆ Leaflets entire:
  Leaflets 2, the tendrils arising from stems or axes of inflorescences:
    Trichomes on petiole and lower blade surface obscure, appressed, less than 0.5 mm long . . . .
      . . . . *Bauhinia guianensis* (Fig. 274; 63B. Caesalpinioideae)
    Trichomes on petiole and lower blade surface erect to reclining, ca 1 mm long . . . .
      . . . . *Bauhinia reflexa* (63B. Caesalpinioideae)
  Leaflets 3 *or* leaflets 2, the terminal (third) leaflet replaced by a tendril or tendril scar:
    Young parts densely rufous-pubescent; lower petioles often more than 10 cm long; leaves trifoliolate, lack-
      ing both tendrils and stipels; plants often arching trees . . . .
      . . . . *Connarus panamensis* (Fig. 256; 62. Connaraceae)
    Young parts not densely rufous-pubescent; all petioles usually less than 10 cm long; leaves trifoliolate,
      with tendrils and/or stipels, the terminal leaflet sometimes replaced by a tendril:
      Plants lacking tendrils (except *Cymbosema*); stipels acicular; pseudostipules lacking:
        Blades sparsely to densely pubescent above and below:
          Trichomes on major veins below mostly more than 1.2 mm long, usually closely appressed . . . . . .
          . . . . *Dioclea reflexa* (Fig. 283; 63C. Papilionoideae)
          Trichomes on major veins below mostly less than 1 mm long, spreading or vaguely appressed:
            Plants often tendriled . . . . *Cymbosema roseum* (63C. Papilionoideae)
            Plants lacking tendrils . . . . *Dioclea guianensis* (63C. Papilionoideae)
        Blades sparsely pubescent to glabrate below, glabrous except on veins above:
          Trichomes on veins below reddish, darker than those on surface . . . .
          . . . . *Dioclea wilsonii* (Fig. 284; 63C. Papilionoideae)
          Trichomes concolorous:
            Leaflets ± elliptic, with more than 10 pairs of lateral veins; stipels lanceolate and persistent, ca 5
              mm long . . . . *Clitoria javitensis* (63C. Papilionoideae)
            Leaflets ovate to ovate-elliptic, with fewer than 10 pairs of lateral veins; stipels acicular, cadu-
              cous:
              Terminal leaflet usually ± rhombic-ovate; leaflets glabrous below, at least the younger ones
                covered on underside with sessile, orange glands . . . .
                . . . . *Rhynchosia pyramidalis* (63C. Papilionoideae)
             Terminal leaflet ovate-elliptic; leaflets lacking orange glands, sometimes with stiff, irritating
                trichomes on underside . . . .
                . . . . *Mucuna mutisiana* (Figs. 291, 292), *M. rostrata* (63C. Papilionoideae)
      Plants with tendrils; acicular stipels lacking; pseudostipules usually present; leaves 2- or 3-foliolate,
        mostly glandular-lepidote; stems usually with either interpetiolar ridges or interpetiolar glandular
        fields (except *Psiguria*, with tendrils opposite trifoliolate leaves):
        Blades densely white- or grayish-pubescent below . . . . *Arrabidaea candicans* (126. Bignoniaceae)
        Blades not whitish-pubescent below:
        ■ Stems (all but the largest) markedly 4-, 6-, or 8-angulate and ribbed:
          Stems 6- or 8- angulate; lateral veins mostly arising from near base of leaf:
            Blades stellate-pubescent in axils below, whitish below; lepidote scales dense, contiguous
              except on veins below; pseudostipules sickle-shaped, to 5 mm long . . . .
             . . . . *Amphilophium paniculatum* (126. Bignoniaceae)
            Blades with plate-shaped glands in axils below, green above and below; lepidote scales sparse
              on both surfaces; pseudostipules spatulate, 7–19 mm long . . . .
             . . . .*Pithecoctenium crucigerum* (126. Bignoniaceae)
          Stems square:
            Leaflets 3, the terminal one never replaced by a tendril; petiolules ± equal on a leaf; blades
              eglandular, not lepidote; plants mostly herbaceous vines lacking pseudostipules . . . .
             . . . . *Psiguria bignoniacea* (Fig. 542; 131. Cucurbitaceae)
            Leaflets sometimes 2, the terminal one replaced by a sometimes deciduous tendril; petiolules
              of terminal leaflets (when present) longer than those of lateral leaflets; blades with few to
              many round glands and scurfy scales; plants lianas, often with pseudostipules:
             Blades pellucid-punctate throughout; stems hollow . . . .
             . . . . *Stizophyllum riparium* (126. Bignoniaceae)
             Blades with few round glands; stems solid:
              Petioles with conspicuous glandular field at apex; pseudostipules vertically 3-seriate . . . .
              . . . . *Pachyptera kerere* (Fig. 496; 126. Bignoniaceae)
             Petioles lacking glands at apex; pseudostipules small or lacking:

Leaflets ± elliptic, with few glands scattered below; pseudostipules small, clustered in leaf axils . . . . *Clytostoma binatum* (126. Bignoniaceae)

Leaflets ± ovate, with dense glands in basal vein axils below; pseudostipules lacking except on youngest stems . . . . *Cydista aequinoctalis* (Fig. 491; 126. Bignoniaceae)

■ Stems terete or subterete, not ribbed (slightly ribbed in *Clytostoma*):

Basal vein axils below with tufts of trichomes, domatia, or glands:

Blades with a few round glands near midrib below:

Stems with interpetiolar glandular fields; most parts drying reddish . . . .

. . . . *Arrabidaea chica* (126. Bignoniaceae)

Stems lacking interpetiolar glands; no parts drying conspicuously reddish:

Axils of major lateral veins of lower blade surface with small, pitlike, pubescent domatia; stems lacking interpetiolar ridges; leaf blades obtuse to broadly rounded at base; tendrils simple . . . . *Callichlamys latifolia* (Fig. 490; 126. Bignoniaceae)

Axils of major lateral veins lacking domatia; stems with interpetiolar ridges; leaf blades truncate to asymmetrically cordate at base; tendrils trifid . . . .

. . . . *Martinella obovata* (Fig. 495; 126. Bignoniaceae)

Blades mostly lacking glands:

Blades often orbicular or ovate, abruptly acuminate, with sunken axillary domatia below

. . . . *Ceratophytum tetragonolobum* (126. Bignoniaceae)

Blades ± elliptic, ± gradually acuminate, with tufted axils below:

Midribs below markedly pubescent (villous); stems lacking interpetiolar glands . . . .

. . . . *Arrabidaea patellifera* (126. Bignoniaceae)

Midribs below inconspicuously pubescent; stems with interpetiolar glands at some nodes

. . . . *Arrabidaea verrucosa* (126. Bignoniaceae)

Basal vein axils below lacking long trichomes, domatia, and glands:

Mature blades conspicuously pubescent, the trichomes at least 0.5 mm long . . . .

. . . . *Cymbosema roseum* (63C. Papilionoideae)

Mature blades glabrous or with trichomes less than 0.5 mm long:

Blades with very conspicuous yellow glands ca 0.1 mm diam on lower surface . . . .

. . . . *Stizophyllum riparium* (126. Bignoniaceae)

Blades lacking such glands:

Stems with conspicuous interpetiolar glandular fields, often with a thin, raised ridge on stem above the field; pseudostipules of several subulate scales 3–4 mm long, ± persistent . . . . *Pachyptera kerere* (Fig. 496; 126. Bignoniaceae)

Stems and pseudostipules not as above:

Pseudostipules ± persistent, foliaceous, not longitudinally striate, 4 at each node, 3–13 mm long, 1.5–10 mm wide . . . . *Anemopaegma chrysoleucum* (126. Bignoniaceae)

Pseudostipules not present or, if present, less than 3 mm long or longitudinally striate:

Pseudostipules ca 5 mm long, longitudinally striate; tendrils trifid and clawlike; phloem rays irregular, many-armed . . . *Macfadyena unguis-cati* (126. Bignoniaceae)

Pseudostipules, tendrils, and phloem rays not as above:

Blades lepidote above and below, yielding a sweet odor when freshly crushed; petioles with a glandular field at apex; pith with 4 phloem arms; pseudostipules of 2 or 3 subulate scales, appressed to twig; young branchlets usually with conspicuous, raised, warty lenticels ca 0.6 mm long . . . .

. . . . *Paragonia pyramidata* (126. Bignoniaceae)

Plants not with the above combination of characters:

Tendrils trifid; pith strongly 4-angular; blades usually broadly ovate, attenuate at apex, bicolorous; fresh vegetative parts with odor of cloves; pseudostipules lacking . . . . *Tynnanthus croatianus* (126. Bignoniaceae)

Plants not with the above combination of characters:

▲ Pith 4-angulate:

Blades drying very reddish, ± ovate, sharply long-attenuate at apex, not subcordate at base . . . . *Arrabidaea chica* (126. Bignoniaceae)

Blades not drying reddish, variously shaped:

Blades usually truncate to subcordate at base, not drying with a conspicuous white margin, lacking branched trichomes on midrib below; petiolules abruptly and conspicuously expanded at apex . . . .

. . . . *Adenocalymma arthropetiolatum* (126. Bignoniaceae)

Blades not as above:

Blades drying with a conspicuous white margin and a dark pewter color; pseudostipules small, pointed, ± ovate . . . .

. . . . *Adenocalymma apurense* (126. Bignoniaceae)

Blades not drying with a white margin, usually drying olive-green to olive-gray; pseudostipules inconspicuous or lacking:

Midribs below with branched trichomes to ca 0.3 mm long . . . . . .
. . . .*Xylophragma seemannianum* (126. Bignoniaceae)
Midribs below lacking such trichomes . . . .
. . . . *Arrabidaea florida* (126. Bignoniaceae)
▲ Pith 8–16-angulate:
Leaves simple to bifoliolate, usually strongly 3-veined from base; petioles
usually less than 3 cm long . . . .
. . . . *Cydista heterophylla* (Fig. 492; 126. Bignoniaceae)
Leaves all bifoliolate; petioles often more than 3 cm long:
Conspicuous glandular area restricted to basal axils on underside of
blade; lateral veins in 3–5 pairs . . . .
. . . . *Cydista aequinoctalis* (Fig. 491; 126. Bignoniaceae)
Dense glandular area in basal axils lacking; lateral veins often in more
than 5 pairs:
Pseudostipules conspicuous, appearing as small bromeliads; blades
elliptic, sharply acuminate, acute at base . . . .
. . . . *Clytostoma binatum* (126. Bignoniaceae)
Pseudostipules lacking or very inconspicuous; blades ± ovate, not
sharply acuminate, rounded to subcordate at base
. . . . *Phryganocydia corymbosa* (Fig. 497; 126. Bignoniaceae)
★ Some leaves with 4 or more leaflets:
⊙ Petioles or rachises bearing conspicuous raised glands:
✪ Rachis winged:
Blades glabrous above except possibly on veins:
Blades pubescent below at least on veins:
Stipules persistent, striate, to 10 mm long; largest leaflets usually more than 5.5 cm wide; rachis winged
throughout or only near apex; young parts ferruginous-pubescent; stems light, not conspicuously
lenticellate . . . . *Inga sapindoides* (63A. Mimosoideae)
Stipules caducous, not striate, 5–6 mm long; largest leaflets less than 5.5 cm wide; rachis winged only at
apex; young parts not ferruginous-pubescent; stems dark brown, conspicuously lenticellate . . . . . .
. . . . *Inga pezizifera* (63A. Mimosoideae)
Blades essentially glabrous below:
Leaflets obovate . . . . *Inga fagifolia* (63A. Mimosoideae)
Leaflets elliptic:
All leaves with 2 pairs of leaflets; rachis wing tapering to base. . . . . . . *Inga marginata* (63A. Mimosoideae)
Some leaves with more than 2 pairs of leaflets:
Leaves with 2 or 3 pairs of leaflets (usually 3); rachis wings becoming obsolete well before base
. . . . *Inga umbellifera* (Fig. 266; 63A. Mimosoideae)
Leaves with 3–5 pairs of leaflets (usually 4); rachis wings narrow . . . .
. . . . *Inga pezizifera* (63A. Mimosoideae)
Blades pubescent above on surface and veins:
Stems conspicuously pubescent with long, dense, reddish-brown trichomes:
Leaflets with gland on upper surface of midrib ca 1–2 cm from base; stipules ovate . . . .
. . . . *Inga goldmanii* (Fig. 261; 63A. Mimosoideae)
Leaflets lacking gland on midrib; stipules apiculate:
Branchlets densely ferruginous-velutinous; terminal leaflets at least 7 cm wide or visibly pubescent
above . . . .*Inga mucuna* (Fig. 263; 63A. Mimosoideae)
Branchlets glabrous in age; terminal leaflets less than 7 cm wide and not visibly pubescent above (may
be densely puberulent) . . . . *Inga pauciflora* (Fig. 264; 63A. Mimosoideae)
Stems not conspicuously pubescent with reddish-brown trichomes, at least in age:
Petioles usually more than 2 cm long:
Lower leaflet surface densely pubescent, the trichomes on veins contiguous; glands on rachis ca 2.5–3
mm diam; stems ribbed between leaves; stipules 2–3 mm long, caducous . . . .
. . . . *Inga minutula* (Fig. 262; 63A. Mimosoideae)
Lower leaflet surface sparsely pubescent, the trichomes on veins not contiguous; glands on rachis 1–2
mm diam; stems not ribbed; stipules to 10 mm long, persistent . . . .
. . . . *Inga sapindoides* (63A. Mimosoideae)
Petioles usually less than 2 cm long:
Apical leaflets usually more than 9 cm wide, glabrous or sparsely pubescent above even when young
(few appressed trichomes); leaves with 2 or 3 pairs of leaflets (usually 2) . . . .
. . . . *Inga spectabilis* (63A. Mimosoideae)
Apical leaflets less than 9 cm wide, moderately to densely pubescent above (at least when young) with
erect trichomes; leaves with 3–6 pairs of leaflets:
Leaves usually with more than 4 pairs of leaflets; rachis glands more than 1 mm diam . . . .
. . . . *Inga vera* (63A. Mimosoideae)
Leaves with 4 or fewer pairs of leaflets; rachis glands less than 1 mm diam:

Blades softly puberulent below .... *Inga pauciflora* (Fig. 264; 63A. Mimosoideae)
Blades glabrate or scabrous below .... *Inga hayesii* (63A. Mimosoideae)
○ Rachis not winged:
● At least some leaves with more than 4 pairs of leaflets:
All leaves with 5 or fewer pairs of leaflets:
Blades densely pubescent below with appressed trichomes, scabrid above with the midrib densely appressed-pubescent .... *Inga thibaudiana* (63A. Mimosoideae)
Blades sparsely pubescent or glabrate below, the trichomes chiefly erect (or at least not markedly acropetal), not scabrid above, the midrib not densely appressed-pubescent:
Glands on rachis more than 1.5 mm wide; blade margins undulate................................
.... *Pithecellobium rufescens* (Fig. 273; 63A. Mimosoideae)
Glands on rachis less than 1.5 mm wide; blade margins not markedly undulate:
Rachis margined to narrowly winged; blades not noticeably puberulent on veins below ....
.... *Inga pezizifera* (63A. Mimosoideae)
Rachis not margined, not winged; blades puberulent on veins below ....
.... *Inga ruiziana* (Fig. 265; 63A. Mimosoideae)
At least some leaves with more than 5 pairs of leaflets:
Some leaves with more than 8 pairs of leaflets; blades sparsely pubescent above except more densely so on midrib, shiny above, pubescent below especially on veins, the trichomes on midrib acropetal; apical leaflets less than 5 cm wide......................... *Inga multijuga* (63A. Mimosoideae)
All leaves with 8 or fewer pairs of leaflets:
Upper surface glabrous or very sparsely puberulent on midrib; leaves predominately narrowly obovate.. *Inga ruiziana* (Fig. 265; 63A. Mimosoideae)
Upper surface villous on midrib; leaves predominately elliptic or lanceolate:
Leaf rachis often with linear appendage ca 4 mm long at apex; basal leaflets usually somewhat asymmetrical .... *Inga thibaudiana* (63A. Mimosoideae)
Leaf rachis without a linear appendage at apex; basal leaflets usually ± symmetrical ....
.... *Inga cocleensis* (63A. Mimosoideae)
● All leaves with 4 or fewer pairs of leaflets:
Petioles longer than rachises; rachis glands subulate to oblong-conic, not at all cupulate; leaflets 4:
Leaflets less than 3.5 cm wide; rachis with glands between both pairs of leaflets ....
.... *Cassia undulata* (Fig. 276; 63B. Caesalpinioideae)
Leaflets more than 3.5 cm wide; rachis with only one gland (between lower pair of leaflets) ....
.... *Cassia fruticosa* (63B. Caesalpinioideae)
Petioles shorter than rachises; rachis glands subcupulate:
Rachis conspicuously pubescent:
Blade surface above softly puberulent with reddish-brown trichomes, softly puberulent to hirtellous below; blades acute or obscurely acuminate at apex; stems densely rufous-pubescent ....
.... *Pithecellobium rufescens* (Fig. 273; 63A. Mimosoideae)
Blade surface not softly puberulent with reddish-brown trichomes; blades short- to long-acuminate:
Young stems essentially glabrous; internodes with ribs extending down from stipule scars ....
.... *Inga punctata* (63A. Mimosoideae)
Young stems moderately pubescent; internodes lacking ribs ..... *Inga quaternata* (63A. Mimosoideae)
Rachis glabrous or inconspicuously pubescent:
Internodes with ribs extending down from stipule scars .... *Inga punctata* (63A. Mimosoideae)
Internodes lacking ribs:
Leaves mostly with 2 pairs of leaflets; leaflets obovate, broadly acuminate or blunt at apex ....
.... *Inga fagifolia* (63A. Mimosoideae)
Leaves with 3 or 4 pairs of leaflets; leaflets ± elliptic or lanceolate, long-acuminate at apex ......
.... *Inga pezizifera* (63A. Mimosoideae)
◉ Petioles and rachises lacking glands:
◻ Rachis winged:
Plants trees or shrubs:
Plants large trees; leaves with 12–14 pairs of leaflets; rachis continuously winged ....
.... *Dipteryx panamensis* (63C. Papilionoideae)
Plants shrubs or trees less than 4 m tall; leaves with 2–5 pairs of leaflets; rachis wing interrupted at leaflets:
Rachis wing sharply toothed at apex; stipules acicular, ca 3 mm long ....
.... *Swartzia simplex,* juvenile (63B. Caesalpinioideae) [see fertile key]
Rachis wing flat or rounded at apex; stipules lacking .... *Quassia amara* (68. Simaroubaceae)
Plants lianas; leaflets 5:
Stems terete; leaflets usually ± entire, with the axils below tufted with trichomes ....
.... *Paullinia fibrigera* (Fig. 342; 80. Sapindaceae)
Stems angulate or ribbed; leaflets usually toothed:
Stems and petioles hispid, the trichomes conspicuous, long, brown, more than 2 mm long; petioles not winged .... *Paullinia baileyi* (80. Sapindaceae)
Stems and petioles not conspicuously hispid:

Axils of veins below shortly tufted; leaflets with few appressed trichomes . . . .
. . . . *Paullinia pinnata* (Fig. 344; 80. Sapindaceae)

Axils of veins not tufted; leaflets with conspicuous erect trichomes below; petioles winged; stipules
more than 4 cm long, persistent, usually brown . . . . . *Paullinia bracteosa* (Fig. 341; 80. Sapindaceae)

□ Rachis not winged:

○ Leaflets opposite (or slightly subopposite on basal pairs of leaflets but no pairs alternate):

Leaves paripinnate:

Petioles pulvinate at base, the pulvinus terete; stipules acicular or foliaceous:

Stipules paired, deciduous, compound, leaflike; stems ribbed; leaves with more than 12 pairs of
leaflets . . . . *Tachigalia versicolor* (63B. Caesalpinioideae)

Stipules not as above:

Blades pellucid-punctate at least on margins (punctations visible when leaf held to light):

Leaflets 2 or 4 (1 or 2 pairs), glabrous; plants often large trees . . . .
. . . . *Prioria copaifera* (Figs. 277, 278; 63B. Caesalpinioideae)

Leaflets 6 or more, glabrous or pubescent; plants small to large trees . . . .
. . . . *Zanthoxylum* [see key to 67. Rutaceae]

Blades not pellucid-punctate; plants shrubs or small trees:

Leaflets densely soft-pubescent below; stems and petioles soft-pubescent . . . .
. . . . *Cassia reticulata* (Fig. 275; 63B. Caesalpinioideae)

Leaflets not densely soft-pubescent; stems and petioles sometimes densely tomentose . . . .
. . . . *Brownea macrophylla* (63B. Caesalpinioideae)

Petioles not pulvinate at base with a terete pulvinus, the pulvinus (when present) usually flat and ±
ribbed on upper side (except *Cedrela odorata*, with petioles often subterete at base and not ribbed);
stipules lacking:

Lateral veins markedly widened at base before joining midrib; plants large forest trees, the bark dark,
very deeply and coarsely fissured on adults; vein axils sometimes with pocketlike domatia . . . . .
. . . . *Cedrela odorata* (Fig. 305; 70. Meliaceae)

Lateral veins not widened before joining midrib, either joining at right angles or turning basally before joining; plants usually small trees, the bark not coarsely fissured:

Leaves large, to 2 m long, often with more than 7 pairs of leaflets; petioles and young stems usually
densely ferruginous-pubescent . . . . *Guarea multiflora* (Figs. 306, 307; 70. Meliaceae)

Leaves less than 0.5 m long, usually with less than 7 pairs of leaflets; petioles and young stems not
densely ferruginous-pubescent:

Leaflets less than 3 cm wide, usually sparsely ciliate and revolute . . . . . *Trichilia hirta* (70. Meliaceae)

Leaflets more than 4 cm wide, not ciliate . . . . *Guarea glabra* (70. Meliaceae)

Leaves imparipinnate:

Leaves opposite; plants glabrous:

Leaflets toothed; stipules not interpetiolar . . . .
. . . . *Turpinia occidentalis* subsp. *breviflora* (Figs. 336, 337; 79. Staphyleaceae)

Leaflets entire; interpetiolar stipules leaving scar . . . . *Platymiscium pinnatum* (63C. Papilionoideae)

Leaves alternate:

Leaflets toothed:

Leaflets more than 5; plants shrubs or trees:

Leaflets pellucid-punctate . . . . *Zanthoxylum* [see key to 67. Rutaceae]

Leaflets not pellucid-punctate . . . . *Astronium graveolens* (76. Anacardiaceae)

Leaflets 5; plants lianas:

Blades with dense simple pubescence on lower surface; stipules large, suborbicular, fimbriate
. . . . *Paullinia rugosa* (Fig. 345; 80. Sapindaceae)

Blades ± glabrous throughout; stipules lanceolate-linear . . . . *Paullinia pterocarpa* (80. Sapindaceae)

Leaflets entire:

◇ Petioles pulvinate at base with a terete pulvinus; petiolules pulvinate and articulate at base; plants
with stipules (these sometimes caducous); sap seldom distinctly aromatic:

Stems with prominent ribs:

Leaflets often more than 5 cm wide; stipules, when present, deltoid, 1–2 mm long . . . .
. . . . *Ormosia coccinea* var. *subsimplex* (Fig. 293; 63C. Papilionoideae)

Leaflets less than 5 cm wide; young stems reddish; stipules deciduous, paired, looking like
reduced compound leaves . . . . *Tachigalia versicolor* (63B. Caesalpinioideae)

Stems lacking ribs:

● Leaves mostly with more than 7 leaflets:

Leaflets each subtended by a stipel, the margins subrevolute, the surfaces glabrous except
on midrib below . . . . *Andira inermis* (63C. Papilionoideae)

Leaflets lacking stipels, the margins not revolute, the lower surface glabrous to densely
puberulent:

Lateral veins on one side of midrib 4–6 per leaflet; leaflets mostly glabrous below . . . . . .
. . . . *Ormosia macrocalyx* (63C. Papilionoideae)

Lateral veins on one side of midrib 8–18 per leaflet; leaflets densely but minutely pubescent below .... *Lonchocarpus pentaphyllus* (63C. Papilionoideae)
- Leaves mostly with 3, 5, or 7 leaflets:
  Reticulate veins all raised and closely anastomosing below:
    Lower surface with the trichomes usually appressed, not raised higher than veinlets, the reticulate veins inconspicuous .... *Lonchocarpus pentaphyllus* (63C. Papilionoideae)
    Lower surface with the trichomes ± erect, raised higher than veinlets, the reticulate veins conspicuous .... *Lonchocarpus velutinus* (Fig. 288; 63C. Papilionoideae)
  Reticulate veins not raised and closely anastomosing below; leaves mostly with 3 or 5 leaflets (*Connarus turczaninowii* with 7):
    Petioles and petiolules densely pubescent; blades densely rufous-pubescent below ......
    .... *Cnestidium rufescens* (62. Connaraceae)
    Petioles glabrous or sparsely pubescent except possibly on pulvinus:
      Plants lianas; blades ± glabrous, usually with fewer than 8 pairs of lateral veins, not caudate-acuminate .... *Connarus turczaninowii* (62. Connaraceae)
      Plants trees; blades not as above:
        Blades moderately golden-pubescent below, with 10–16 pairs of lateral veins ......
        .... *Ormosia panamensis* (63C. Papilionoideae)
        Blades glabrous or with trichomes only on veins, with 9 or fewer pairs of lateral veins, caudate-acuminate ....
        .... *Swartzia panamensis* (Fig. 279; 63B. Caesalpinioideae)
◇ Petioles and petiolules not pulvinate and not articulate at base (or, if pulvinate, the pulvinus not terete); plants lacking stipules; sap usually distinctly aromatic:
  At least some leaves with 6 or more pairs of lateral leaflets, the leaflets less than 3.5 cm wide:
    Leaflets at most bluntly short-acuminate, often acute or even rounded at apex, the midrib on upper surface thick, prominently raised, densely pubescent; reticulate veins obscure ....
    .... *Mosquitoxylum jamaicense* (Fig. 328; 76. Anacardiaceae)
    Leaflets narrowly long-acuminate, the midrib flat or very slender and weakly raised, usually in a sunken trough, glabrous; reticulate veins conspicuously visible ....
    .... *Trichilia hirta* (70. Meliaceae)
  All leaves with fewer than 6 pairs of lateral leaflets *or* the leaflets more than 3.5 cm wide:
    Leaflets pubescent below (trichomes may be restricted to veins):
      Leaflets asperous on upper surface; petioles flat on upper surface, usually more than 7 mm broad near base, the margin prominently raised ....
      .... *Trattinnickia aspera* (Fig. 304; 69. Burseraceae)
      Leaflets not asperous on upper surface; petioles not markedly flattened on upper surface with a raised margin:
        Stems glabrous, the stems and trunk with paper-thin, brown, peeling bark; trichomes on lower blade surface ± erect but not stiff and straight ....
        .... *Bursera simaruba* (69. Burseraceae)
        Stems densely pubescent, the bark not as above; trichomes on lower blade surface stiffly erect, straight .... *Protium costaricense* (69. Burseraceae)
    Leaflets ± glabrous or sparsely pubescent below:
      Petiolules mostly less than 1 cm long:
        Terminal leaflet often more than 8 cm wide; leaflets (3)5–7 per pinna ....
        .... *Trichilia montana* (70. Meliaceae)
        Terminal leaflet usually less than 7 cm wide; leaflets 7–11 per pinna ....
        .... *Tetragastris panamensis* (Fig. 303; 69. Burseraceae)
      Petiolules mostly more than 1 cm long:
        Largest leaflets usually more than 8 cm wide, not papillose; lateral veins not raised below
        .... *Protium panamense* (Fig. 302; 69. Burseraceae)
        Largest leaflets usually less than 8 cm wide:
          Leaflets minutely papillate below; lateral veins conspicuously raised below (a fingernail run down the blade catches on them) ....
          .... *Protium tenuifolium* var. *sessiliflorum* (69. Burseraceae)
          Leaflets not papillate; lateral veins not conspicuously raised below:
            Plants villous to velutinous (the trichomes either not straight or long and very dense); leaves usually drying blackened; blades lacking a submarginal collecting vein
            .... *Spondias radlkoferi* (76. Anacardiaceae)
            Plants glabrous or pubescent, some parts merely puberulent (the trichomes short and straight); leaves not drying blackened; blades with a conspicuous submarginal collecting vein .... *Spondias mombin* (Figs. 329, 330; 76. Anacardiaceae)
○ Leaflets alternate (or subopposite on apical pairs but definitely alternate near base):
□ Plants lianas (or scandent shrubs or trees when young):
△ Leaf nodes on mature plants armed with ± recurved spines:

Most leaves with more than 20 leaflets:
  Largest leaflets more than 5 mm wide . . . . *Machaerium milleflorum* (63C. Papilionoideae)
  Largest leaflets less than 5 mm wide . . . . *Machaerium microphyllum* (63C. Papilionoideae)
Most leaves with fewer than 20 leaflets:
  All leaves with 7 or fewer leaflets . . . . *Machaerium arboreum* (63C. Papilionoideae)
  At least some leaves with 8 or more leaflets:
    Leaflets all less than 3 cm wide:
      Leaflets rounded to retuse at apex . . . . *Machaerium riparium* (63C. Papilionoideae)
      Leaflets acuminate at apex, the acumen blunt . . . . . . *Machaerium seemannii* (63C. Papilionoideae)
    Largest leaflets often more than 3 cm wide:
      Stems (except nearest apex) densely covered with erect spiny trichomes; leaflets ± glabrous
        below (at least in age), long-acuminate . . . .
        . . . .*Machaerium kegelii* (Figs. 289, 290; 63C. Papilionoideae)
      Stems lacking dense spiny trichomes; leaflets densely appressed-pubescent below, obtuse to
        short-acuminate . . . . *Machaerium floribundum* (63C. Papilionoideae)
△ Leaf nodes on mature plants lacking spines:
  Leaves mostly with 4 or 5 leaflets (never more than 5):
    Leaflets ovate; petioles mostly less than 3 cm long; rachis usually flexuous in apical half . . . .
    . . . . *Dalbergia monetaria* (63C. Papilionoideae)
    Leaflets ± elliptic; petioles usually more than 4 cm long; rachis straight
    . . . . *Connarus turczaninowii* (62. Connaraceae)
  Leaves mostly with 3 leaflets *or* some leaves with more than 5 leaflets:
    Leaves mostly with 3 leaflets (rarely 5); blades glabrous above . . . .
    . . . . *Connarus panamensis* (Fig. 256; 62. Connaraceae)
    At least some leaves with more than 5 leaflets; blades pubescent at least on midrib above:
      Leaflets densely rufous-pubescent below . . . . *Cnestidium rufescens* (62. Connaraceae)
      Leaflets inconspicuously puberulent to glabrous below except on veins:
        Leaflets with the reticulate veins somewhat to markedly prominulous on both surfaces . . . . . .
        . . . . *Rourea glabra* (62. Connaraceae)
        Leaflets with the reticulate veins scarcely or not at all visible . . . .
        . . . . *Connarus turczaninowii* (62. Connaraceae)
□ Plants trees or shrubs:
  Leaflets pellucid-punctate, at least on margins:
    Pellucid punctations round . . . . *Zanthoxylum* [see key to 67. Rutaceae]
    Some pellucid punctations elongate . . . . *Myroxylon balsamum* var. *pereirae* (63C. Papilionoideae)
  Leaflets not pellucid-punctate:
  ☆ Leaves usually with 9 or fewer leaflets:
    Leaflets conspicuously pubescent at least on veins below:
      Leaflets toothed, round or acute at apex:
        Leaflets densely reddish-brown-pubescent along midrib above; stems with long reddish-brown
          trichomes; some leaves simple . . . . *Cupania rufescens* (Fig. 339; 80. Sapindaceae)
        Leaflets ± glabrate above or the pubescence not reddish-brown; stems glabrous or with short
          trichomes:
          Blades below whitish, densely tomentose with contiguous trichomes . . . .
          . . . . *Cupania cinerea* (80. Sapindaceae)
          Blades below green, glabrate to sparsely pubescent and papillate; some leaves simple; some
          leaflets subopposite . . . . *Cupania latifolia* (80. Sapindaceae)
      Leaflets entire, acuminate at apex:
        Rachis densely pubescent, the trichomes erect or appressed but not stiff and straight:
          Plants small trees less than 4 m tall; petiolules not equally pulvinate throughout, not articu-
            late at base; stipules lacking . . . . *Cupania sylvatica* (Fig. 340; 80. Sapindaceae)
          Plants large trees to 25 m tall; petiolules equally pulvinate throughout, articulate at base;
            stipules minute . . . . *Pterocarpus rohrii* (63C. Papilionoideae)
        Rachis glabrate to short-puberulent:
          Midrib flat or sunken above when dry . . . . *Trichilia verrucosa* (70. Meliaceae)
          Midrib raised above (markedly in apical half) . . . . *Trichilia cipo* (70. Meliaceae)
    Leaflets essentially glabrous:
      Petiolules pulvinate:
        Leaflets long-acuminate, often falcate; plants lacking stipules . . . .
        . . . . *Picramnia latifolia* (68. Simaroubaceae)
        Leaflets acute or short-acuminate; plants with stipules . . . .
        . . . . *Ormosia coccinea* var. *subsimplex* (Fig. 293; 63C. Papilionoideae)
      Petiolules not pulvinate:
        Terminal leaflets more than 8 cm wide . . . . *Trichilia montana* (70. Meliaceae)
        Most terminal leaflets less than 8 cm wide:

Midrib flat or sunken above (when dry); minute, aborted leaflets often at base; most leaves
with 5 leaflets (not including aborted leaflets) . . . . *Trichilia verrucosa* (70. Meliaceae)
Midrib prominently raised in apical half; minute leaflets lacking, most leaves with 7 leaflets
. . . . *Trichilia cipo* (70. Meliaceae)
☆ At least some leaves with more than 9 leaflets:
  Leaflets mostly more than 5 cm wide:
    Leaves usually with more than 9 pairs of leaflets; all leaflets usually less than 6.5 cm wide, lacking
    a prominent collecting vein; stems often with reduced, leaflike structures at apex . . . .
    . . . . *Talisia princeps* (80. Sapindaceae)
    Leaves usually with 8 or fewer pairs of leaflets; at least the terminal leaflets often more than 6.5
    cm wide; lateral veins with a conspicuous submarginal collecting vein; stems lacking re-
    duced leaflike structures . . . . *Talisia nervosa* (80. Sapindaceae)
  Leaflets mostly less than 5 cm wide:
    Lateral veins mostly obscure; leaflets not acuminate (some acute to retuse at apex) *or* the petioles
    and leaflets pilose below:
      Most leaflets retuse at apex . . . . *Platypodium elegans* (63C. Papilionoideae)
      Leaflets not retuse at apex:
        Lateral veins obscure; leaflets oblong to oblanceolate . . . .
        . . . . *Simarouba amara* (Figs. 300, 301; 68. Simaroubaceae)
        Lateral veins conspicuous, at least below:
          Leaflets widest below the middle; reticulate venation conspicuous . . . .
          . . . . *Dalbergia retusa* (Fig. 282; 63C. Papilionoideae)
          Leaflets widest above the middle; reticulate venation obscure . . . .
          . . . . *Mosquitoxylum jamaicense* (Fig. 328; 76. Anacardiaceae)
    Lateral veins conspicuous; leaflets acuminate:
      Rachis ridged or minutely winged toward apex; lateral veins raised below more than above:
        Plants glabrous or pubescent, parts merely puberulent (the trichomes short and straight);
        leaves not drying blackened; blades with a conspicuous submarginal collecting vein
        . . . . *Spondias mombin* (Figs. 329, 330; 76. Anacardiaceae)
        Plants villous to velutinous (the trichomes either not straight or long and very dense); leaves
        usually drying blackened; blades lacking a submarginal vein . . . .
        . . . . *Spondias radlkoferi* (76. Anacardiaceae)
      Rachis not ridged or winged; lateral veins, if raised below, then about the same below and
      above:
        Petiolules not pulvinate, less than 4 mm long . . . . *Astronium graveolens* (76. Anacardiaceae)
        Petiolules pulvinate, to 4 mm or more long:
          Leaflets acute to obscurely acuminate, short-pilose below, sometimes becoming glabrous
          with age . . . . *Dalbergia retusa* (Fig. 282; 63C. Papilionoideae)
          Leaflets shortly and abruptly acuminate, glabrous below or with minute appressed tri-
          chomes . . . . *Pterocarpus officinalis, Vatairea erythrocarpa* (63C. Papilionoideae)

Species Excluded

A total of 138 names have been excluded from the flora. Most of these are the result of misidentification by Standley at the time of the writing of the *Flora* (1933), but some reflect the different interpretations of later workers. Others are cultivated plants that no longer grow on the island. A few species, such as *Bidens pilosa, Porophyllum ruderale,* and *Lemna cyclostasa,* are species for which no BCI specimens have been found. I believe that Standley reported some of these names merely because he thought they would be found on the island.

The following species and specimens were reported by Standley for BCI unless otherwise indicated.

Acalypha villosa Jacq. (75. Euphorbiaceae): *Shattuck 843* is *A. macrostachya.*

Aeschynomene hispida Sw. (63C. Papilionoideae): *Bailey 763* is *A. ciliata.*

Albizia adinocephala (Donn. Sm.) Britt. & Rose (63A. Mimosoideae): *Bailey 281* is *Leucaena multicapitula.*

Alsophila ternerifrons H. Christ (9. Cyatheaceae): Based on misidentification of *Nephelea cuspidata.*

Annona purpurea Moc. & Sesse ex Dunal (55. Annonaceae): Reported in the *Flora of Panama. Bangham 610 (US)* carries a BCI label that is marked out and replaced with "Government Reservation—Ancón."

Anthurium crassinervium (Jacq.) Schott (21. Araceae): One of several species confused by Standley with *A. tetragonum. A. crassinervium* is a species restricted to South America.

A. maximum (Desf.) Engler: Confused with *A. tetragonum* by Standley. *A. maximum* is a Colombian species of higher elevations.

A. ramonense Engler & Krause: Reported in the *Flora of Panama. Standley 41149* is *A. littorale.*

A. schlechtendalii Kunth: No specimen was cited by Standley. *Shattuck 638,* cited in the *Flora of Panama,* is *A. tetragonum.*

A. turrialbense Engler: Reported in the *Flora of Panama. Standley 31386* and *40887* are probably *A. littorale.*

A. undatum Schott: Based on misidentification of *A. bombacifolium.*

Aphelandra tetragona (Vahl) Nees (129. Acanthaceae): Based on a sterile collection of *Alseis blackiana* (130. Rubiaceae).

Apodanthes flacourtiae Karst. (45. Rafflesiaceae): Based on misidentification of *A. caseariae.*

Ardisia compressa H.B.K. (112. Myrsinaceae): Confused by Standley with *A. bartlettii.*

Bauhinia purpurea L. (63B. Caesalpinioideae): A cultivated species no longer present on the island.

Bellucia costaricensis (L.) Tr. (107. Melastomataceae): A synonym of *B. axinanthera* Tr., which does not occur on the island. The report for the island by Gleason in the *Flora of Panama* (1958) was based on *Shattuck 1119,* which is *B. grossularioides.*

Bidens pilosa L. (133. Compositae): No specimen has been seen and none was cited by Standley.

Blechum panamense Lindau (129. Acanthaceae): Based on misidentification of the narrow-leaved form of *B. costaricense.*

Borreria suaveolens G. Meyer (130. Rubiaceae): *Shattuck 421* is *B. densiflora.*

Bunchosia nitida (Jacq.) DC. (71. Malpighiaceae): Probably based on *Bangham 508,* which is *Malpighia romeroana.*

Byrsonima coriacea (Sw.) Kunth (71. Malpighiaceae): *Starry 129* is *B. spicata.* According to Cuatrecasas (1958), *B. coriacea* is confined to Jamaica.

Calathea altissima Koern. (33. Marantaceae): Confused by Standley and others with *C. inocephala. C. altissima,* which does not occur in Panama, differs in having a smaller head with clavigulate bracteoles and capsule walls that are not fleshy.

C. violacea (Rosc.) Lindl.: Based on misidentification of *C. latifolia. C. violacea* is restricted to Brazil.

Casearia javitensis H.B.K. (96. Flacourtiaceae): Confused by Standley with *C. commersoniana. C. javitensis,* a South American species, does not occur in Panama.

Cassia tora L. (63B. Caesalpinioideae): Confused with *C. obtusifolia. C. tora* is an Old World species.

Celosia argentea L. (47. Amaranthaceae): Reported by Standley as a probable escapee from cultivation. The species no longer occurs on the island.

Centrosema virginianum (L.) Benth. (63C. Papilionoideae): All BCI specimens originally determined as this species are *C. pubescens*.

Chaetochlamys panamensis Lindau (129. Acanthaceae): No specimen has been cited, and the plant has not been seen growing on the island in recent years.

Citrus medica L. (67. Rutaceae): *Shattuck 861*, determined by Standley as this species, is *C. limon*.

Clidemia petiolata (Rich.) DC. (107. Melastomataceae): No specimen has been seen. This species is a narrow endemic (*fide* J. Wurdack, pers. comm.) and definitely does not occur on the island.

Clusia rosea L. (92. Guttiferae): No specimen has been seen. Possibly reported on the basis of *Shattuck 684* (*685?*), which is probably a juvenile collection of *C. odorata*.

Coccoloba changuinola Standl. (46. Polygonaceae): This is a synonym of *C. lehmannii* Lindau, which is not known for BCI. Probably reported on the basis of misidentification of *C. coronata*.

Combretum punctulatum Pitt. (105. Combretaceae): A synonym of *C. spinosum* Bonpl. *Wilson 128* is *C. decandrum*.

Cordia nitida Jacq. (121. Boraginaceae): *Wilson 69* and *Shattuck 813* and *853* are *C. lasiocalyx*.

C. sericicalyx A. DC.: This name was used by Standley for *C. panamensis*. It is similar to *C. panamensis* and ranges from western Colombia to Trinidad and Surinam, but has not been collected in Panama.

Costus spicatus (Jacq.) Sw. (32. Zingiberaceae): No specimen has been seen. *C. spicatus* is a West Indian species.

Croton glandulosus L. (75. Euphorbiaceae): Confused with *C. hirtus*. *C. glandulosus* does not occur in Panama.

Cucurbita pepo L. (131. Cucurbitaceae): Cultivated on the island at one time, but no longer present.

Cuphea wrightii Gray (102. Lythraceae): Apparently based on misidentification of *C. carthagenensis*.

Dalbergia ecastophyllum (L.) Taub. (63C. Papilionoideae): Probably based on *Kenoyer 387* and *629*, which are *D. brownei*.

Davilla kunthii St.-Hil. (88. Dilleniaceae): A synonym of *D. aspera* (Aubl.) Ben., which does not occur on the island. Based on misidentification of *D. nitida* (*Woodworth & Vestal 331*).

D. rugosa Poir.: Based on misidentification of *D. nitida*.

Dichromena pubera Vahl (18. Cyperaceae): *Standley 40831*, reported in the *Flora of Panama*, is *Rhynchospora nervosa*.

Dieffenbachia aurantiaca Engler (21. Araceae): *Bailey 335* is *D. oerstedii*.

Dillenia indica L. (88. Dilleniaceae): Reported by Standley as planted at the laboratory. It no longer occurs there.

Diphysa robinoides Benth. (63C. Papilionoideae): *Shattuck 869* is *Platymiscium pinnatum*.

Epidendrum fragrans Sw. (35. Orchidaceae): *Shattuck 551* is *Encyclia chimborazoensis*.

Epiphyllum pittieri Britt. & Rose (101. Cactaceae): *Shattuck 603* is *E. phyllanthus* var. *columbiense*. *E.*

phyllanthus var. *pittieri* does not occur on the island.

Eriochloa punctata (L.) Desv. (17. Gramineae): *Shattuck 323* is *Brachiaria mutica*.

Eschatogramme furcata (L.) Trev. (10. Polypodiaceae): Based on misidentification of *Dicranoglossum panamense*.

Eugenia sericiflora Benth. (106. Myrtaceae): A synonym of *E. bicolor* L. No specimen was cited and none has been seen. The species is not easily confused with any other, but it is not believed to occur on the island.

Eupatorium billbergianum Beurl. (133. Compositae): Based on misidentification of *Koanophyllon wetmorei*.

Ficus crassiuscula Warb. (39. Moraceae): Based on misidentification of *F. insipida*.

F. glabrata H.B.K.: Based on misidentification of *F. insipida*.

F. hemsleyana Standl.: A synonym of *F. goldmanii* Standl., which does not occur on the island. Many specimens of *F. citrifolia* were annotated by Standley as *F. hemsleyana*.

F. involuta (Liebm.) Miq.: Based on misidentification of *F. obtusifolia*, a distinct species.

F. padifolia H.B.K.: This is a highland species from western Panama and Costa Rica that was confused by Standley with *F. pertusa*.

F. velutina Willd.: A species occurring in Panama only at higher elevations in Chiriquí. *Bailey 408* is a juvenile form of *F. bullenei*.

Gossypium barbadense L. (85. Malvaceae): This is *G. mexicanum* sensu Standl. non Todaro. Probably originally planted near the laboratory, but no longer growing on the island.

Guarea guarea (Jacq.) P. Wils. (70. Meliaceae): *Woodworth & Vestal 747* is *G. multiflora*.

Guatteria dolichopoda Donn. Sm. (55. Annonaceae): *Kenoyer 359* is *G. dumetorum*.

Hampea panamensis Standl. (85. Malvaceae): A synonym of *H. appendiculata* Donn. Sm. var. *appendiculata*, which occurs in western Panama, Costa Rica, and Honduras.

Heisteria macrophylla Oerst. (43. Olacaceae): *Standley 40877*, also listed in the *Flora of Panama*, is *H. costaricensis*.

Heliconia acuminata A. Rich. (31. Musaceae): Based on misidentification of *H. vaginalis*.

H. pendula Wawra: Based on misidentification of *H. irrasa*. *H. pendula* does not occur in Panama.

H. platystachya Baker: Reported by both Standley and the *Flora of Panama*. This South American species has been confused with *H. catheta*.

H. subulata R. & P.: Reported in the *Flora of Panama*, on the basis of misidentification of *H. vaginalis*. *H. subulata* is a Peruvian species.

H. villosa Klotzsch: Restricted to South America. Reported in the *Flora of Panama*, on the basis of misidentification of *H. irrasa*.

Hiraea fagifolia (DC.) Adr. Juss. (71. Malpighiaceae): Probably based on misidentification of *Mascagnia nervosa*. No specimen was cited, but *Shattuck 1057* and *1156* and *Bangham 602* (all *F*), originally deter-

mined by Standley as *H. fagifolia*, are *Mascagnia nervosa*.

Hyeronima alchorneoides Allem. (75. Euphorbiaceae): According to Webster and Burch in the *Flora of Panama* (1967), a species from Brazil that does not occur in Panama. Probably based on misidentification of *H. laxiflora*.

Hymenocallis littoralis (Jacq.) Salisb. (28. Amaryllidaceae): Called *Pancratium littorale* Jacq. in the *Flora of Panama*. Both based on misidentifications of *H. pedalis*.

Ichnanthus nemorosus Doell. (17. Gramineae): Apparently confused by Standley with *I. brevivaginatus*.

Ipomoea triloba L. (120. Convolvulaceae): *Bailey 673* and *Starry 121* are *I. tiliacea*. *I. triloba* does not occur in Panama.

Lasiacis divaricata (L.) Hitchc. (17. Gramineae): *Bailey 285* is *Chusquea simpliciflora*.

L. ruscifolia (H.B.K.) Hitchc.: *Shattuck 528* is *L. sorghoidea*.

Lemna cyclostasa (Ell.) Chev. (Lemnaceae): A synonym for *L. valdiviana* Phil. Reported by Standley on the basis of someone else's report. However, no specimens have been seen on BCI and I have not seen it growing there. It could easily occur in the water of Gatun Lake.

Luffa cylindrica (L.) Roem. (131. Cucurbitaceae): No longer cultivated on the island.

Machaerium darienense Pitt. (63C. Papilionoideae): Reported in the *Flora of Panama* by Dwyer (1965) for BCI on the basis of *Killip 40014* (misspelled *M. darlense* Pitt.). The Killip collection does not differ from other collections of *M. arboreum* and the key characters used to separate the two taxa are quite variable. *M. darienense* is now considered a synonym of *M. striatum* J. R. Johnston.

Malpighia glabra L. (71. Malpighiaceae): *Shattuck 511* is *M. romeroana*.

Monstera pertusa (L.) DeVr. (21. Araceae): Confused with *M. dubia*.

M. pittieri Engler: Based on misidentification of a juvenile specimen of *M. dilacerata*.

Mucuna urens (L.) Medic. (63C. Papilionoideae): Probably based on *Bailey & Bailey 348* and *Shattuck 286*, which are both *M. mutisiana*.

Musa paradisiaca L. (31. Musaceae): No longer cultivated at the Laboratory Clearing.

Myrosma panamensis Standl. (33. Marantaceae): Standley apparently confused this plant with *Calathea panamensis*. *Myrosma panamensis* is a synonym of *Hylaeanthe panamensis* (Standl.) H. Kenn., which does not occur on the island.

Nectandra glabrescens Benth. (58. Lauraceae): A species that occurs in Panama only along the Caribbean coast and most closely resembles *N. globosa*. Confused by Standley with *N. purpurascens*.

Neea psychotrioides sensu Standl. non Donn. Sm. (48. Nyctaginaceae): This is *N. laetevirens* Donn. Sm., which does not occur on the island. *Shattuck 121* is *N. amplifolia*.

Ossaea micrantha (Sw.) Macfad. (107. Melastomataceae): No specimen has been cited, and the plant has not been seen growing on the island.

Ouratea nitida (Sw.) Engler (89. Ochnaceae): *Shattuck 243* is *O. lucens*. The two names may be synonymous.

Paullinia alata Don. (80. Sapindaceae): *Standley 31281* (*US*) is *P. bracteosa*.

Peltophorum sp. (63B. Caesalpinioideae): *Woodworth & Vestal 743* is a fruit of *Tachigalia versicolor*.

Peperomia viridispica Trel. (36. Piperaceae): Reported by Standley on the basis of a single sterile and possibly juvenile specimen (*Shattuck 207*). This specimen does not appear to be any of the other reported species but cannot be assigned to any species with certainty.

Petrea volubilis Jacq. (122. Verbenaceae): Reported by Moldenke for BCI in the *Flora of Panama* (1973). *Aviles 14* and *Shattuck 412* are *P. aspera*.

Pharus glaber H.B.K. (17. Gramineae): Apparently confused by Standley with *P. parvifolius*.

Phaseolus vulgaris L. (63C. Papilionoideae): This garden bean is no longer cultivated on the island.

Philodendron wendlandii Schott (21. Araceae): Based on misidentification of *P. inconcinnum*.

Pilea serpyllacea (H.B.K.) Liebm. (40. Urticaceae): *Shattuck 196* represents a cultivated plant that no longer grows on the island.

Piper breve C. DC. (36. Piperaceae): A synonym for *P. pseudo-fuligineum* C. DC., which does not occur on the island. Possibly Standley confused this with *P. dilatatum*.

P. frostii Trel.: An unpublished name. The specimens cited are *P. aequale*.

P. lucigaudens C. DC.: Reported in the *Flora of Panama*. *Standley 31288* and *31387* are *P. hispidum*.

P. polyneurum C. DC.: A synonym for *P. augustum* var. *cocleanum* (Trel.) Yunck. Though this species has been collected near Frijoles, no specimen has been seen for BCI. *Starry 63* is *P. grande*. Neither *Aviles 36* nor *Starry 95*, also cited by Standley, has been located.

P. pseudo-cativalense Trel.: Confused by Standley with *P. dilatatum*.

Polybotrya caudata Kunze (10. Polypodiaceae): Probably based on misidentification of *P. villosula*. The two species are doubtfully distinct.

P. osmundacea H. & B.: No collection of this species has been seen at lower elevations in Panama.

P. tectum Kaulf.: *Bailey 513* is an aberrant specimen of *P. ciliatum*.

Porophyllum ruderale (Jacq.) Cass. (133. Compositae): No specimen has been seen from the island, and the plant has not been seen growing there.

Pouzolzia occidentalis Wedd. (40. Urticaceae): Based on *Aviles 24*, which has not been located. *Aviles 24b* is *P. obliqua* and was originally determined as *P. occidentalis*.

Prestonia exserta (A. DC.) Standl. (118. Apocynaceae): *Woodworth & Vestal 498* is *P. acutifolia*.

Psychotria calophylla Standl. (130. Rubiaceae): No specimen has been seen for the island. All collections are from areas near the Atlantic coast or from areas of

higher elevation, usually premontane wet or tropical wet forest.

Randia lisiantha Standl. (130. Rubiaceae): *Starry 8* is *R. armata*.

Renealmia occidentalis (Sw.) Sweet (32. Zingiberaceae): This species was treated as a synonym of *R. aromatica* (Aubl.) Griseb. in the *Flora of Panama*, where *Kenoyer 238* was listed for BCI. Presumably this is the same collection on which Standley based his report. *Kenoyer 238* is *R. alpinia*.

Rivea campanulata (L.) House (120. Convolvulaceae): A synonym of *Stictocardia campanulata* (L.) Merr., which Standley reported on the basis of misidentification of *Ipomoea phillomega*.

Rudgea cornifolia (H. & B.) Standl. (130. Rubiaceae): Apparently based on *Standley 41058* (found unlabeled with other *R. cornifolia* specimens). This is *Coussarea curvigemmia*.

Salvinia auriculata Aubl. (11. Salviniaceae): No specimen has been seen from the island. Other specimens identified as *S. auriculata* are *S. radula*.

Schomburgkia sp. (35. Orchidaceae): *Shattuck 500* is *Caularthron bilamellatum*.

Scleria bracteata Cav. (18. Cyperaceae): *Shattuck 144* is *S. secans*.

Securidaca coriacea Bonpl. (74. Polygalaceae): *Woodworth & Vestal 505* is *S. tenuifolia*.

Sicydium tamnifolium (H.B.K.) Cogn. (131. Cucurbitaceae): Apparently confused by Standley with *S. coriaceum*.

Simarouba glauca DC. (68. Simaroubaceae): *Starry 144* has not been located. It is probably *S. amara*.

Solanum bicolor Willd. (124. Solanaceae): The species is probably not found in Panama. The name is applied to several other species, including *S. rugosum*, *S. umbellatum*, *S. asperum*, and *S. erianthum* D. Don. No BCI specimen has been found that originally bore the name.

S. nigrum L.: No specimen has been found. Probably (according to W. G. D'Arcy, pers. comm.) Standley was referring to *S. americana* P. Mill., which is a common garden weed in Panama. Neither *S. americana* nor any of its close relatives occurs on the island.

Spathiphyllum patinii (Hogg) N. E. Brown (21. Araceae): Reported from Colombia and probably confused by Standley with *S. phryniifolium*, though no collection bearing the name *S. patinii* from BCI has been found.

Stachytarpheta cayennensis (L. C. Rich.) Vahl (122. Verbenaceae): *Woodworth & Vestal 739* is *S. jamaicensis*.

Stenorrhynchus sp. (35. Orchidaceae): Reported by Standley on the basis of an undesignated Kenoyer collection. None has been found, but according to R. Dressler, the specimen probably was confused with *Spiranthes lanceolata*.

Stigmaphyllon humboldtianum Adr. Juss. (71. Malpighiaceae): Based on misidentification of *S. lindenianum*.

Stylogyne laevis (Oerst.) Mez (112. Myrsinaceae): Probably based on misidentification of *Ardisia fendleri*.

S. ramiflora (Oerst.) Mez: A synonym of *S. laevis* (Oerst.) Mez, which does not occur on the island. Probably based on misidentification of *S. standleyi*.

Tetracera oblongata DC. (88. Dilleniaceae): Apparently based on misidentification of *T. portobellensis*.

T. sessiliflora Tr. & Planch.: A synonym of *T. portobellensis*, according to the *Flora of Panama*. Probably based on misidentification of *T. hydrophila*.

Tillandsia melanopus E. Morr. (22. Bromeliaceae): *Starry 101* is a juvenile plant that cannot be determined (*fide* L. B. Smith). Since no mature plants of *T. melanopus* have been found, I am assuming that *Starry 101* is a juvenile of another *Tillandsia*.

T. usneoides L.: Reported by Standley on the basis of a sterile fragment (*Shattuck 414*). There is reason to believe that the species could occur on BCI, since it occurs in tropical moist forest in Bocas del Toro and Darién; but *Shattuck 414* has roots and an indumentum different from those of *T. usneoides*, and therefore it cannot be this species. Mason Hale (pers. comm.) has ruled it out as a lichen. It remains undetermined.

Tradescantia cumanensis Kunth (23. Commelinaceae): A synonym of *Tripogandra multiflora*, which does not occur in Panama. Probably confused with *Tripogandra serrulata*.

Triplaris americana L. (46. Polygonaceae): Based on misidentification of *T. cumingiana*.

Tripogandra elongata (G. Meyer) Woods. (23. Commelinaceae): *Standley 41079*, reported in the *Flora of Panama* (Woodson & Schery, 1944), is *Callisia ciliata*.

Vigna repens (L.) O. Kuntze (63C. Papilionoideae): A synonym of *V. luteola* (Jacq.) Benth. *Bailey & Bailey 299* has not been located, but it is probably *Phaseolus trichocarpus*.

Vincetoxicum dubium (Pitt.) Standl. (119. Asclepiadaceae): A synonym of *Gonolobus dubium* Woods. *Wetmore & Abbe 202* is *G. allenii*.

Virola guatemalensis (Hemsl.) Warb. (56. Myristicaceae): *Shattuck 694* and *Wetmore & Abbe 155* are *V. surinamensis*.

Vismia guianensis (Aubl.) Pers. (92. Guttiferae): Based on misidentification of *V. baccifera*.

Xylosma hemsleyana Standl. (96. Flacourtiaceae): A synonym of *X. flexuosa* (H.B.K.) Hemsl. This species apparently occurs only at higher elevations in Chiriquí. Kenoyer (1929) confused this with *Casearia guianensis* and Standley may have done the same.

X. intermedia (Tr. & Planch.) Griseb.: This species is known for certain only from Chiriquí. Confused by Standley with *X. chloranthum*.

Zanthoxylum microcarpum Griseb. (67. Rutaceae): Known only from Panamá Province, but not from BCI. Confused by Standley with *Z. belizense*.

Zeea mays L. (17. Gramineae): No longer cultivated on the island.

Literature Cited

Adams, C. D. 1972. Flowering plants of Jamaica. University of West Indies, Mona, Jamaica: Robert MacLehose & Co., Ltd. 848 pp.

Agharkar, S. P., and I. Benerji. 1930. Studies in the pollination and seed formation of water hyacinth (*Eichhornia speciosa* Kunth). Agric. J. India 35:286–96.

Allan, P. 1963. Pollination of papaws. Farming in South Africa 38(11):13–15.

Allee, W. C. 1926. Measurement of environmental factors in the tropical rain forests of Panama. Ecology 7:273–302.

Allen, C. K. 1945. Studies in Lauraceae VI. Preliminary survey of Mexican and Central American species. J. Arnold Arbor. 26(3):280–434.

Allen, P. H. n. d., unpublished, written between 1936 and 1939. Preliminary index to the timberwoods of Panama. Mimeographed at Library of the Missouri Botanical Garden.

_____. 1949. Orchidaceae (Part 3), in R. E. Woodson, Jr., and R. W. Schery, Flora of Panama. Ann. Missouri Bot. Gard. 36:337–468.

_____. 1956. The rain forests of Golfo Dulce. Gainesville, Fla.: University of Florida Press. 418 pp. Reissued 1977. Stanford, Cal.: Stanford University Press.

Amshoff, G. J. H. 1958. Myrtaceae, in R. E. Woodson, Jr., and R. W. Schery, Flora of Panama. Ann. Missouri Bot. Gard. 45:165–201.

Austin, D. F. 1975. Convolvulaceae, in R. E. Woodson, Jr., and R. W. Schery, Flora of Panama. Ann. Missouri Bot. Gard. 62:157–224.

Badillo, V. M. 1971. Monografia de la Familia Caricaceae. Published by the Association of Professors, Universidad Central de Venezuela, Maracay.

Bailey, L. H. 1933. Certain palms of Panama. Gentes Herb. 3:99–101.

_____. 1943. Palmae, in R. E. Woodson, Jr., and R. W. Schery, Flora of Panama. Ann. Missouri Bot. Gard. 30:327–96.

_____. 1949. Manual of cultivated plants. Rev. ed. New York: Macmillan Co. 1,116 pp.

Baker, H. G. 1963. Evolutionary mechanisms in pollination biology. Science 139:877–83.

_____. 1970. Two cases of bat-pollination in Central America. Revista Biol. Trop. 17:187–97.

_____. 1973. Evolutionary relationships between flowering plants and animals in American and African tropical forests, in B. J. Meggers, E. S. Ayensu, and W. D. Duckworth, eds., Tropical forest ecosystems in Africa and South America: A comparative review. Washington, D.C.: Smithsonian Institution Press.

Baker, H. G., and B. J. Harris. 1959. Bat-pollination of the silk-cotton tree, *Ceiba pentandra* (L.) Gaertn. (*sensu lato*), in Ghana. J. W. African Sci. Assoc. 5:1–9.

Bawa, K. S. 1974. Breeding systems of tree species of a lowland tropical community. Evolution 28:85–92.

Bawa, K. S., and P. Opler. 1975. Dioecism in tropical forest trees. Evolution 29:167–79.

Beard, J. S. 1944. Climax vegetation in tropical America. Ecology 25:127–58.

_____. 1955. Classification of tropical American vegetation types. Ecology 36:89–100.

Bell, C. R. 1971. Breeding systems and floral biology of the Umbelliferae or evidence for specialization in unspecialized flowers, in V. H. Heywood, ed., Biology and chemistry of the Umbelliferae. London: Academic Press, Inc., Ltd. 438 pp.

Bennett, C. F. 1963. A phytophysiognomic reconnaissance of Barro Colorado Island. Smithsonian Misc. Collect. 145(7):1–8.

Bennett, H. H. 1929. Soil reconnaissance of the Panama Canal Zone and contiguous territory. Techn. Bull. U.S.D.A. 94:1–46.

Berg, C. C. 1972. Olmedieae, Brosimeae (Moraceae). Flora Neotropica 7:1–229.

Blohm, H. 1962. Poisonous plants of Venezuela. Cambridge, Mass.: Harvard University Press. 136 pp.

Bonaccorso, F. J. 1975. Foraging and reproductive ecology in a community of bats in Panama. Doctoral dissertation, University of Florida. 119 pp.

Bor, N. L. 1960. Grasses of Burma, Ceylon, India and

Pakistan. New York: Pergamon Press. 767 pp.

Bouriquet, G. 1954. La Vanillier et la vanille dans le Monde. Paris: Editions Paul Lechevalier. 748 pp.

Bunting, G. S. 1960. Revision of *Spathiphyllum* (Araceae). Mem. New York Bot. Gard. 10:1–53.

――――. 1963. A reconsideration of *Philodendron hederaceum*. Baileya 11:62–67.

――――. 1965. Commentary on Mexican Araceae. Gentes Herb. 9:291–382.

――――. 1968. Vegetative anatomy and taxonomy of the *Philodendron scandens* complex. Gentes Herb. 10:136–68.

Burger, W. C. 1971. Piperaceae, in Flora Costaricensis. Fieldiana, Bot. 35:5–218.

――――. 1972. Evolutionary trends in Central American species of *Piper* (Piperaceae). Brittonia 24:356–62.

――――. 1974. Ecological differentiation in some congeneric species of Costa Rican flowering plants. Ann. Missouri Bot. Gard. 61:297–306.

――――. 1975. Moraceae, in Flora Costaricensis. Fieldiana, Bot. 40:94–215.

Carpenter, C. R. 1934. A field study of the behavior and social relations of howling monkeys (*Alouatta palliata*). Comparative Psych. Monogrs. 10(2):1–108.

Chapman, F. M. 1929. My Tropical Air Castle. New York: D. Appleton & Co. 417 pp.

――――. 1931. Seen from a tropical air castle. J. Amer. Mus. Nat. Hist. 31:347–58.

――――. 1938. Life in an air castle. New York: D. Appleton Co., Inc. 250 pp.

Chapman, G. P. 1970. Patterns of change in tropical plants. London: University of London.

Corner, E. J. H. 1952. Wayside trees of Malaya. 2nd ed. Singapore: Government Printing Office.

――――. 1964. The life of plants. Cleveland, Ohio: World Publishing Co. Pp. 1–315.

Croat, T. B. 1969a. Barro Colorado Island, the forest. Missouri Bot. Gard. Bull. 57:16–21.

――――. 1969b. Seasonal flowering behavior in Central Panama. Ann. Missouri Bot. Gard. 56:295–307.

――――. 1973. A new species of Vitaceae for Central and South America. Ann. Missouri Bot. Gard. 60:564–67.

――――. 1974a. A case of selection for delayed fruit maturation in *Spondias* (Anacardiaceae). Biotropica 6:135–37.

――――. 1974b. *Combretum laxum* var. *epiphyticum* (Combretaceae)—a case of selection for water-dispersed fruits. Phytologia 28:188–91.

――――. 1974c. A reconsideration of *Spondias mombin* L. (Anacardiaceae). Ann. Missouri Bot. Gard. 61:483–90.

――――. 1975a. A reconsideration of *Trichilia cipo* (Meliaceae). Ann. Missouri Bot. Gard. 62:491–96.

――――. 1975b. Notes on Flacourtiaceae—*Casearia* and *Xylosma*. Ann. Missouri Bot. Gard. 62:484–90.

――――. 1975c. A new species of *Dracontium* (Araceae) from Panama—a sapromyophilous pollination syndrome. Selbyana 1:168–71.

――――. 1975d. Phenological behavior of habit and habitat classes on Barro Colorado Island (Panama Canal Zone). Biotropica 7:270–77.

――――. 1976. Sapindaceae, in R. E. Woodson, Jr., and R. W. Schery, Flora of Panama. Ann. Missouri Bot. Gard. 63:419–540.

Croat, T. B., and P. Busey. 1975. Geographical affinities of the Barro Colorado Island flora. Brittonia 27:127–35.

Cuatrecasas, J. 1958. Malpighiaceae: Prima Flora Colombiana. Webbia 13(2):343–664.

――――. 1961. A taxonomic revision of the Humiriaceae. Contr. U.S. Natl. Herb. 35:24–214.

――――. 1969. Compositae-Astereae: Prima Flora Colombiana. Webbia 24(1):1–335.

Cutter, E. G. 1957. Studies of morphogenesis in Nymphaeaceae I, Introduction: Some aspects of the morphology of *Nuphar lutea* (L.) Sm. and *Nymphaea alba* L. Phytomorphology 7:45–56.

D'Arcy, W. G., and N. Schanen. 1975. Erythroxylaceae, in R. E. Woodson, Jr., and R. W. Schery, Flora of Panama. Ann. Missouri Bot. Gard. 62:21–33.

Darlington, P. J., Jr. 1957. Zoogeography: The geographical distribution of animals. New York: Wiley Publishing Co. 675 pp.

Davidse, G., and E. Morton. 1973. Bird-mediated fruit dispersal in the tropical grass genus *Lasiacis* (Gramineae: Paniceae). Biotropica 5(3):162–67.

De Dalla Torre, C. G., and H. Harms. 1963. Genera Siphonogamarum ad Systema Englerianum Conscripta. Reprint. Wiesbaden: Verlag für Wissenschaftliche Neudrucke GMBH. 637 pp.

DeWolf, G. 1960. *Ficus* (Moraceae), in R. E. Woodson, Jr., and R. W. Schery, Flora of Panama. Ann. Missouri Bot. Gard. 47:146–65.

Doctors van Leewen, W. M. 1938. Observations about the biology of tropical flowers. Ann. Jard. Bot. Buitenzorg 48:27–68.

Dodson, C. H. 1962. The importance of pollination in the evolution of the orchids of tropical America. Amer. Orchid Soc. Bull. 31:525–35, 641–49, 731–35.

――――. 1965a. Agentes de polinización y su influencia sobre la evolución en la familia orquidacea. Iquitos, Peru: Univ. Nac. Amazonia Peruana.

――――. 1965b. The genus *Coryanthes*. Amer. Orchid Soc. Bull. 34:680–87.

――――. 1966. Ethology of some bees of the tribe Euglossini (Hymenoptera: Apidae). J. Kansas Entomol. Soc. 39:607–29.

――――. 1967a. Relationships between pollinators and orchid flowers, in Atas do Simpósio sôbre a biotica Amazônica, vol. 5. (Zoologia):1–72.

――――. 1967b. Studies in pollination—the genus *Notylia*. Amer. Orchid Soc. Bull. 36:209–14.

Dodson, C. H., and G. P. Frymire. 1961. Natural pollination of orchids. Missouri Bot. Gard. Bull. 49:133–52.

Dodson, C. H., R. L. Dressler, H. G. Hills, R. M. Adams, and N. H. Williams. 1969. Biologically active compounds in orchid fragrances. Science 164:1243–49.

Dressler, R. L. 1966. Some observations on *Gongora* (Orchidaceae). Orchid Digest 30:220–23.

_____. 1968a. Observations on orchids and euglossine bees in Panama and Costa Rica. Revista Biol. Trop. 15(1):143–83.

_____. 1968b. Pollination by euglossine bees. Evolution 22:202–10.

Dugand, A. 1945. Revaluación de *Philodendron hederaceum* Schott (1829, non 1856) como Transfesencia de *Arum hederaceum* Jacq. Caldasia 3:445–52.

Duke, J. A. 1968. Darién ethnobotanical dictionary. Columbus, Ohio: Battelle Memorial Institute. 131 pp.

_____. n.d., unpublished. Plants in certain forest types of Darién, Panama. Written at Columbus, Ohio, Battelle Memorial Institute.

Duke, J. A., and D. M. Porter. 1970. Darién phytosociological dictionary. Columbus, Ohio: Battelle Memorial Institute.

Dunsterville, B. C. K., and L. A. Garay. 1965. Orchids of Venezuela. 5 vols. London: A. Deutsch.

Dwyer, J. D. 1965. Leguminosae subfamily Papilionoideae (in part), in R. E. Woodson, Jr., and R. W. Schery, Flora of Panama. Ann. Missouri Bot. Gard. 52:1–54.

Eames, A. J., and L. MacDaniels. 1947. Introduction to plant anatomy. 2nd ed. New York: McGraw-Hill. 427 pp.

Edwin, G., and D. Hou. 1975. Celastraceae, in R. E. Woodson, Jr., and R. W. Schery, Flora of Panama. Ann. Missouri Bot. Gard. 62:45–56.

Ehrlich, P. R., and L. E. Gilbert. 1973. Population structure and dynamics of the tropical butterfly *Heliconius ethilla*. Biotropica 5:69–82.

Elias, T. S. 1968. Vitaceae, in R. E. Woodson, Jr., and R. W. Schery, Flora of Panama. Ann. Missouri Bot. Gard. 55:81–92.

Emiliana, C., S. Gartner, and B. Lidz. 1972. Neogene sedimentation on the Blake Plateau and the emergence of the Central American isthmus. Paleogeogr. Paleoclimatol. Paleoecol. 6:237–39.

Enders, R. K. 1935. Mammalian life histories from Barro Colorado Island, Panama. Bull. Mus. Comp. Zoo. (Harvard College) 78(4):385–502.

Engler, A. 1905. Araceae, in Das Pflanzenreich IV.23B (Heft 21):53–295.

Engler, A., and K. Krause. 1913. *Philodendron* (Araceae), in Das Pflanzenreich IV.23Db (Heft 60):1–143.

Ernst-Schwarzenbach, M. 1945. Zur Blutenbiologie einiger Hydrocharitaceen. Ber. Schweiz. Bot. Ges. 55:33–69.

Essig, F. B. 1971. Observations of pollination of *Bactris*. Principes 15:20–24.

_____. 1973. Pollination in some New Guinea palms. Principes 17:75–83.

Ewan, J. 1962. Synopsis of the South American species of *Vismia* (Guttiferae). Contr. U.S. Natl. Herb. 35:293–373.

Exell, A. W. 1958. Combretaceae, in R. E. Woodson, Jr., and R. W. Schery, Flora of Panama. Ann. Missouri Bot. Gard. 45:143–64.

Faegri, K., and L. van der Pijl. 1966. The principles of pollination ecology. London: Pergamon Press. 248 pp.

Foster, Robin. 1974. Seasonality of fruit production and seed fall in a tropical forest ecosystem in Panama. Doctoral dissertation, Duke University, Durham, N.C. 155 pp.

Fournier, L. A., and S. Salas. 1966. Algunas observaciónes sobre la dinamica de la floración en el bosque húmedo de Villa Colón. Revista Biol. Trop. 14:75–85.

Fowlie, J. A. 1971. Obscure species: Three distinctive species of *Chysis* from Central America. Orchid Digest 35:85–87.

Frankie, G. W., H. G. Baker, and P. A. Opler. 1974. Comparative phenological studies of trees in tropical lowland wet and dry forest sites of Costa Rica. J. Ecol. 62(3):881–919.

Gentry, A. H. 1972. An eco-evolutionary study of the Bignoniaceae of southern Central America. Doctoral dissertation, Washington University, St. Louis, Missouri.

_____. 1973a. Rafflesiaceae, in R. E. Woodson, Jr., and R. W. Schery, Flora of Panama. Ann. Missouri Bot. Gard. 60(1):17–21.

_____. 1973b. Bignoniaceae, in R. E. Woodson, Jr., and R. W. Schery, Flora of Panama. Ann. Missouri Bot. Gard. 60:781–977.

_____. 1974. Flowering phenology and diversity in tropical Bignoniaceae. Biotropica 6:48–64.

_____. 1975. Humiriaceae, in R. E. Woodson, Jr., and R. W. Schery, Flora of Panama. Ann. Missouri Bot. Gard. 62:35–44.

Gleason, H. A. 1958. Melastomataceae, in R. E. Woodson, Jr., and R. W. Schery, Flora of Panama. Ann. Missouri Bot. Gard. 45:203–304.

Gomez Pampa, A. 1966. Estudio del genero *Aspidosperma* (Apocynaceae) para la flora de Misantla, Ver. Ciencia (Mexico) 24:217–22.

Goodwin, G. G., and A. M. Greenhall. 1961. A review of the bats of Trinidad and Tobago. Bull. Amer. Mus. Nat. Hist. 122(3):191–301.

Gottsberger, G. 1970. Contributions to the biology of the Annonaceae (Angiospermae). Oesterr. Bot. Z. 118:237–79.

Graham, A. 1972. Some aspects of tertiary vegetational history about the Caribbean Basin, in Memorias de symposia, 1 Congreso Latino-Americano de Botanica 97–117.

Guinet, P. 1969. Les Mimosacées, étude de palynologie fondamentale, correlations, évolution. Inst. Franc. Pond. Trav. Sec. Sc. Tecl. 9:1–293.

Guppy, H. B. 1912. Studies in seeds and fruits. London: Williams & Norgate. 528 pp.

Halle, F. 1967. Étude biologique et morphologique de la tribu des Gardeniees (Rubiacées). Paris: Office de la Recherche Scientifique et Technique Outre-Mer. 146 pp.

Harley, R. M. 1971. An explosive pollination in *Eriope crassipes*, a Brazilian Labiatae. Biol. J. Linn. Soc. 3:159–64.

Harling, G. 1958. Monograph of Cyclanthaceae. Lund,

Sweden: Hakan Ohlssons Bok Tryckeri.

Hartog, C. den. 1963. Hydrocharitaceae, in R. E. Woodson, Jr., and R. W. Schery, Flora of Panama. Ann. Missouri Bot. Gard. 60:7–15.

Heithaus, R. 1973. Species diversity and resource partitioning in four neotropical plant-pollinator communities. Doctoral dissertation, Stanford University, Stanford, California.

Heithaus, R., T. H. Fleming, and P. A. Opler. 1975. Foraging patterns and resource utilization of eight species of bats in a seasonal tropical forest. Ecology 56:841–54.

Heithaus, R., P. A. Opler, and H. G. Baker. 1974. Bat activity and pollination of *Bauhinia pauletia*: Plant-pollinator coevolution. Ecology 55:412–19.

Hills, H. G., N. H. Williams, and C. H. Dodson. 1968. Identification of some orchid fragrance components. Amer. Orchid Soc. Bull. 37:967–72.

———. 1972. Floral fragrances and isolating mechanisms in the genus *Catasetum* (Orchidaceae). Biotropica 4:61–76.

Hladik, A. 1970. Contribution a l'étude biologique d'une Araliaceae d'Amérique tropicale: *Didymopanax morototoni*. Adansonia 10:383–407.

Hladik, A., and C. M. Hladik. 1969. Rapports trophiques entre vegetation et primates dans la floret de Barro Colorado (Panama). La Terre et la Vie 1:25–117.

Holdridge, L. R. 1967. Life zone ecology. San José, Costa Rica: Tropical Science Center. 206 pp.

———. 1970. Manual dendrologico para 1000 especies arboreas en la Republica de Panama. Panamá City: Government of Panama.

Holdridge, L. R., and G. Budowski. 1956. Report of an ecological survey of the Republic of Panama. Caribbean Forest. 17:92–110.

———. 1959. Mapa ecologica de Panama (1:1,000,000). Turrialba, Costa Rica: Interamerican Institute of Agricultural Sciences.

Holdridge, L. R., W. C. Grenke, W. H. Hatheway, T. Liang, J. A. Tosi, Jr. 1971. Forest environments in tropical life zones. New York: Pergamon Press.

Hunter, G. E. 1965. Dilleniaceae, in R. E. Woodson, Jr., and R. W. Schery, Flora of Panama. Ann. Missouri Bot. Gard. 52(4):579–98.

Hutchinson, J. 1967. Key to the families of flowering plants of the world. Oxford: Clarendon Press. 117 pp.

Jackson, B. D. 1928. A glossary of botanic terms. 4th ed. London: Duckworth & Co. 481 pp.

Janzen, D. H. 1967a. Interaction of bull's-horn *Acacia* (*Acacia cornigera* L.) with an ant inhabitant (*Pseudomyrmex ferruginea* F. Smith) in eastern Mexico. Univ. Kansas Sci. Bull. 47:315–558.

———. 1967b. Synchronization of sexual reproduction of trees within the dry season in Central America. Evolution 21:620–37.

———. 1968. Reproductive behavior in the Passifloraceae and some of its pollinators in Central America. Behavior 32:33–48.

———. 1969. Allelopathy by myrmecophytes: The ant *Azteca* as an allelopathic agent of *Cecropia*. Ecology 50:147–53.

———. 1971. The fate of *Scheelea rostrata* fruits beneath the parent tree: Predispersal attack by bruchids. Principes 15:89–101.

———. 1972. Escape in space by *Sterculia apetala* seeds from bug *Dysdercus fasciatus* in a Costa Rican deciduous forest. Ecology 53:350–61.

Jimenez S., H. 1970. Los Arboles mas importantes de la serrania de San Lucas. Div. Forestales, Bogota: Instituto de Desarrolo de los Recursos Naturales Renovables. 240 pp.

Johnston, I. M. 1949. The botany of San José Island. Sargentia 8:1–306.

Kaastra, R. C. 1972. Revision of *Chlorophora* (Moraceae) in America. Acta Bot. Neerl. 21(6):657–70.

Karr, J. R. 1976. An association between a grass (*Paspalum virgatum* L.) and moths. Biotropica 8(4):284–85.

Kaufmann, J. H. 1962. Ecology and social behavior of the coati, *Nasua narica*, on Barro Colorado Island, Panama. Univ. Calif. Publ. Zool. 60:95–222.

Kennedy, H. 1973. Notes on Central American Marantaceae I: New species and records from Panama and Costa Rica. Ann. Missouri Bot. Gard. 60:413–26.

Kenoyer, L. A. 1928. Fern ecology of Barro Colorado Island, Panama Canal Zone. Amer. Fern J. 18:6–14.

———. 1929. General and successional ecology of the lower tropical rain-forest at Barro Colorado Island, Panama. Ecology 10(2):201–22.

Kenoyer, L. A., and P. C. Standley. 1929. Supplement to the flora of Barro Colorado Island, Panama. Publ. Field Columbian Mus., Bot. Ser. 4(6):143–58.

Killip, E. P. 1960. Urticaceae, in R. E. Woodson, Jr., and R. W. Schery, Flora of Panama. Ann. Missouri Bot. Gard. 47:179–98.

Kimnach, M. 1964. *Epiphyllum phyllanthus*. Cact. Succ. J. (Los Angeles) 36:105–15.

Knight, D. H. 1970. A field guide to the trees of Barro Colorado Island. Published by the Department of Botany, University of Wyoming, Laramie.

———. 1975a. A phytosociological analysis of species-rich tropical forest: Barro Colorado Island, Panama. Ecol. Monogr. 45:259–84.

———. 1975b. An analysis of late secondary succession in species-rich tropical forest, in F. B. Golley and E. Medina, eds., Tropical ecological systems: Trends in terrestrial and aquatic research. New York: Springer-Verlag.

Krukoff, B. A. 1939. The American species of *Erythrina*. Brittonia 3:205–337.

Kubitski, K. 1971. *Doliocarpus, Davilla* und Verwandte Gattungen (Dilleniaceae). Mitt. Bot. Staatssäml., München 9:1–194.

Kuijt, J. 1969. The biology of parasitic flowering plants. Berkeley and Los Angeles: University of California Press. 246 pp.

———. 1976. Revision of *Oryctanthus* (Loranthaceae). Bot. Jahrb. Syst. 95(4):478–534.

Lamb, F. B. 1953. The forests of Darién, Panama. Caribbean Forest. 14:128–35.

Lang, G. E. 1969. Sampling tree density with quadrats in a species-rich tropical forest. Master's thesis, University of Wyoming, Laramie. 54 pp.

Lang, G. E., D. H. Knight, and D. A. Anderson. 1971. Sampling the density of tree species with quadrats in a species-rich tropical forest. Forest Sci. 17:395–400.

Lawrence, G. H. M. 1955. An introduction to plant taxonomy. New York: Macmillan Co. 179 pp.

———. 1964. Taxonomy of vascular plants. 9th ed. New York: Macmillan Co. 823 pp.

Leck, C. 1972. Seasonal changes in feeding pressures of fruit and nectar eating birds in Panama. The Condor 74:54–60.

Leeuwenberg, A. J. M. 1958. The Gesneriaceae of Guiana. Acta Bot. Neerl. 7:278–81.

Leonard, E. C. 1951. The Acanthaceae of Colombia. Contr. U.S. Natl. Herb. 31:1–781.

Leppik, E. E. 1955. *Dichromena ciliata*, a noteworthy entomophilous plant among Cyperaceae. Amer. J. Bot. 42:455–58.

Lewis, D. 1966. The genetic integration of breeding systems, in J. G. Hawks, Reproductive biology and taxonomy of vascular plants. Report of the Botanical Society of the British Isles. London: Pergamon Press.

Linhart, Y. B. 1973. Ecological and behavioral determinants of pollen dispersal in hummingbird-pollinated *Heliconia*. Amer. Naturalist 107:511–23.

Little, E. L., Jr., and F. H. Wadsworth. 1964. Common trees of Puerto Rico and the Virgin Islands. Agriculture Handbook #249. Washington, D.C.: Forest Service, U.S.D.A.

Lovell, J. H. 1920. The flower and the bee. London: Constable. 286 pp.

Lowden, R. M. 1973. Revision of the genus *Pontederia*. Rhodora 75:426–87.

Lundell, C. L. 1971. Myrsinaceae, in R. E. Woodson, Jr., and R. W. Schery, Flora of Panama. Ann. Missouri Bot. Gard. 58(3):285–353.

Maas, P. J. M. 1972. Flora neotropica monograph no. 8. Costoideae (Zingiberaceae). New York: Hafner Publishing Co.

MacBride, J. F. 1959. Ebenaceae, in Flora of Peru. Field Mus. Nat. Hist. Bot. 13:205–14.

Martius, C. F. P. von. 1824. Historia naturalis palmarum Leipzig (1823–1853). 10 parts, 3 vols. 2:87, t. 69.

McClintock, E. 1950. Saxifragaceae, in R. E. Woodson, Jr., and R. W. Schery, Flora of Panama. Ann. Missouri Bot. Gard. 37:139–45.

———. 1957. A monograph of the genus *Hydrangea*. Proc. Calif. Acad. Sci. 29:147–256.

McDiarmid, R. W., R. E. Ricklefs, and M. S. Foster. 1977. Dispersal of *Stemmadenia donnell-smithii* (Apocynaceae) by birds. Biotropica 9(1):9–25.

McDonald, D. F. 1919. The sedimentary formations of the Panama Canal Zone with special reference to stratigraphic relations of fossiliferous beds. U.S. Natl. Mus. Bull. 103:525–45.

McVaugh, R. 1963. Myrtaceae, in Flora of Guatemala. Fieldiana, Bot. 24(7):283–405.

Meyer, C. W. 1969. The ecological geography of cloud forest in Panama. Amer. Mus. Novit. 2396:1–52.

Mez, C. 1889. Lauraceae Americanae. Jahrb. Königl. Bot. Gart. Berlin 5:377–78.

Moldenke, H. N. 1934. Monograph of *Aegiphila* (Verbenaceae). Brittonia 1:245–477.

———. 1973. Verbenaceae, in R. E. Woodson, Jr., and R. W. Schery, Flora of Panama. Ann. Missouri Bot. Gard. 60:41–148.

Montgomery, G. G., and M. E. Sunquist. 1974. Impact of sloths on neotropical forest energy flow and nutrient cycling, in E. Medina and F. Golley, eds., Trends in tropical ecology. Ecological Studies 4. New York: Springer-Verlag.

Mori, S. 1970. The ecology and uses of species of *Lecythis* in Central America. Turrialba 29(3):344–50.

Mori, S., and J. Kallunki. 1976. Phenology and floral biology of *Gustavia superba* (Lecythidaceae) in Central Panama. Biotropica 8(3):184–92.

Morley, T. 1971. Geographic variation in a widespread neotropical species, *Mouriri myrtilloides* (Melastomataceae). Brittonia 23:413–24.

Morton, C. V. 1945. Dioscoreaceae, in R. E. Woodson, Jr., and R. W. Schery, Flora of Panama. Ann. Missouri Bot. Gard. 32:26–34.

Munz, P. A. 1959. Onagraceae, in R. E. Woodson, Jr., and R. W. Schery, Flora of Panama. Ann. Missouri Bot. Gard. 46(3):305–31.

Nevling, L. I. 1958. Lythraceae, in R. E. Woodson, Jr., and R. W. Schery, Flora of Panama. Ann. Missouri Bot. Gard. 45(2):97–115.

———. 1961. Olacaceae, in R. E. Woodson, Jr., and R. W. Schery, Flora of Panama. Ann. Missouri Bot. Gard. 47:293–302.

Nicolson, D. H. 1960. The occurrence of trichoschlereids in the Monsteroideae (Araceae). Amer. J. Bot. 47:598–602.

Nowicke, J. W. 1968. Palynotaxonomic study of the Phytolaccaceae. Ann. Missouri Bot. Gard. 55:294–363.

———. 1969. Boraginaceae, in R. E. Woodson, Jr., and R. W. Schery, Flora of Panama. Ann. Missouri Bot. Gard. 56:33–69.

———. 1971. Rhamnaceae, in R. E. Woodson, Jr., and R. W. Schery, Flora of Panama. Ann. Missouri Bot. Gard. 58:267–84.

Oppenheimer, J. R. 1968. Behavior and ecology of the white-faced monkey, *Cebus capucinus*, on Barro Colorado Island, Canal Zone. Doctoral dissertation, University of Illinois, Urbana.

Ornduff, R., and T. Mosquin. 1969. Variation in the spectral qualities of flowers in the *Nymphoides indica* complex (Menyanthaceae) and its possible adaptive significance. Canad. J. Bot. 48:603–5.

Ospina, H. M. 1958. Colombian orchids. Bogota: Publicaciones Tecnicas, Ltda. 305 pp.

Parija, P. 1934. Physiological investigations of water hyacinth (*Eichhornia crassipes*) in Orissa with notes on

some other aquatic weeds. Indian J. Sci. 4:399–429.

Peñalosa, J. 1975. Shoot dynamics of tropical lianas. Doctoral dissertation, Harvard University.

Pennington, T. D., and J. Sarukhan. 1968. Arboles tropicales de Mexico. Mexico City: Inst. Nat. de Invest. Forestales. 413 pp.

Percival, M. 1965. Floral biology. London: Pergamon Press. 243 pp.

Petch, T. 1924. Notes on *Aristolochia.* Ann. Roy. Bot. Gard. Peridenya (Ceylon) 8:1–108.

Petersen, O. G. 1890. Musaceae, in C. Martius, Flora Brasiliensis 3(3):1–27.

Pfeifer, H. W. 1966. Revision of North and Central American hexandrous species of *Aristolochia* (Aristolochiaceae). Ann. Missouri Bot. Gard. 53:1–114.

Pijl, L. van der. 1930. Uit het leven van enkele gevoelige tropishe bloemen, speciaal van "Horlogebloemen." Die Tropische Natur 19:161–90.

———. 1968. Principles of dispersal in higher plants. New York: Springer-Verlag. 153 pp.

Porsch, O. 1939. Das Bestaubungsleben der Kakteenblute II, Cactaceae (Jahrb. deutschen Kakt.-Ges.). Pp. 81–142.

Prance, G. T. 1972. Flora neotropica monograph no. 9. Chrysobalanaceae. New York: Hafner Publishing Co. 410 pp.

———. 1976. Pollination and androphore structure of some Amazonian Lecythidaceae. Biotropica 8(4):235–41.

Proctor, M., and P. Yeo. 1973. The pollination of flowers. London: William Collins Sons & Co., Ltd. 418 pp.

Raeder, K. 1961. Phytolaccaceae, in R. E. Woodson, Jr., and R. W. Schery, Flora of Panama. Ann. Missouri Bot. Gard. 48:66–79.

Ramirez B., W. 1970. Taxonomic and biological studies of neotropical fig wasps (Hymenoptera: Agaonidae). Univ. Kansas Sci. Bull. 49(1):1–44.

Ridley, H. N. 1930. The dispersal of plants throughout the world. Ashford, Kent: L. Reeve & Co., Ltd. 774 pp.

Rizzini, C. T. 1960. Loranthaceae, in R. E. Woodson, Jr., and R. W. Schery, Flora of Panama. Ann. Missouri Bot. Gard. 47:263–90.

———. 1971. Arvores e madeiras úteis do Brasil— Manual de dendrologia Brasileira. Sao Paulo: Editora Edgard Blucher Ltda.

Robyns, A. 1968. Flacourtiaceae, in R. E. Woodson, Jr., and R. W. Schery, Flora of Panama. Ann. Missouri Bot. Gard. 55:93–144.

Rovirosa, J. N. 1892. Calendaria botanico de San Juan Bautista y sus Alrededores. La Naturaleza ser. 2, 2:106–26.

Rubinoff, R. W. 1974. Environmental monitoring and baseline data. Washington, D.C.: Smithsonian Institution Press. 466 pp.

Scagel, R. F., R. J. Bandoni, G. E. Rouse, W. B. Schofield, J. Stein, and T. M. C. Taylor. 1965. An evolutionary survey of the plant kingdom. Belmont, Cal.: Wadsworth. 658 pp.

Schlising, R. A. 1972. Sequence and timing of bee foraging in flowers of *Ipomoea* and *Aniseia* (Convolvulaceae). Ecology 51(6):1061–67.

Schmid, R. 1970. Notes on the reproductive biology of *Asterogyne martiana* (Palmae). Principes 14:3–9.

Schultz, A. G. 1942. Las Pontederiaceas de la Argentina. Darwiniana 6:45–82.

Sculthorpe, C. D. 1967. The biology of aquatic vascular plants. London: Edward Arnold Ltd. 610 pp.

Simmonds, N. W. 1950. The Araceae of Trinidad and Tobago, B.W.I. Kew Bull. 1950(5):391–406.

Smith, A. R. 1971. Systematics of the neotropical species of *Thelypteris* section *Cyclosorus.* Univ. Calif. Publ. Bot. 59:1–136.

Smith, C. C. 1975. The coevolution of seeds and seed predators, in L. E. Gilbert and P. H. Raven, eds., Coevolution of animals and plants. Austin: University of Texas Press. Pp. 53–77.

Smith, C. E., Jr. 1965. Meliaceae, in R. E. Woodson, Jr., and R. W. Schery, Flora of Panama. Ann. Missouri Bot. Gard. 52:55–79.

Smith, L. B. 1944. Bromeliaceae, in R. E. Woodson, Jr., and R. W. Schery, Flora of Panama. Ann. Missouri Bot. Gard. 31:73–137.

Smith, R. R. 1968. A taxonomic revision of the genus *Heliconia* in Middle America. Master's thesis, University of Florida, Gainesville.

Smythe, N. 1970. Relationship between fruiting season and seed dispersal methods in a neotropical forest. Amer. Naturalist 104:25–35.

Soderstrom, T. R., and C. E. Calderon. 1971. Insect pollination in tropical forest grasses. Biotropica 3(1):1–16.

Sorsa, P. 1969. Pollen morphological studies on the Mimosaceae. Ann. Bot. Fenn. 6:1–34.

Standley, P. C. 1927. The flora of Barro Colorado Island. Smithsonian Misc. Collect. 78(8):1–32.

———. 1928. Flora of the Panama Canal Zone. Contr. U.S. Natl. Herb. 27:1–416.

———. 1930a. A second supplement to the flora of Barro Colorado Island, Panama. J. Arnold Arbor. 11:119–29.

———. 1930b. The Rubiaceae of Colombia. Publ. Field Columbian Mus., Bot. Ser. 7:1–176.

———. 1933. The flora of Barro Colorado Island. Contr. Arnold Arbor. 5:1–178.

———. 1944. Araceae, in R. E. Woodson, Jr., and R. W. Schery, Flora of Panama. Ann. Missouri Bot. Gard. 31(1):405–64.

Standley, P. C., and J. A. Steyermark. 1946a. Leguminosae, in Flora of Guatemala. Fieldiana, Bot. 24(5):1–368.

———. 1946b. Malpighiaceae, in Flora of Guatemala. Fieldiana, Bot. 24:468–500.

Steyermark, J. A. 1972. Rubiaceae in the botany of the Guayana highlands. Part 9. Mem. New York Bot. Gard. 23:227–832.

Stiles, G. 1975. Ecology, flowering phenology and hummingbird pollination of some Costa Rican *Heliconia* species. Ecology 56:285–301.

Stuessy, T. 1975. Tribe Melampodiinae (Compositae),

in R. E. Woodson, Jr., and R. W. Schery, Flora of Panama. Ann. Missouri Bot. Gard. 62:1062–91.

Styles, B. T. 1972. The flower biology of the Meliaceae and its bearing on tree breeding. Silvae Genet. 21:175–82.

Svenson, H. K. 1943. Cyperaceae, in R. E. Woodson, Jr., and R. W. Schery, Flora of Panama. Ann. Missouri Bot. Gard. 30:281–325.

Swallen, J. R. 1943. Gramineae, in R. E. Woodson, Jr., and R. W. Schery, Flora of Panama. Ann. Missouri Bot. Gard. 30:104–280.

Torre, P. F. 1965. Notas geológicas de la Republica de Panamá, in Comisión de Panama, Atlas de Panama.

Tosi, J. A., Jr. 1971. Zonas de vida, una base ecologica para investigaciones silvícolas y inventariación forestal en la Replíca de Panamá. Rome: Organización de las Naciones Unidas para Agricultura y Alimentación. 123 pp., with multicolor map.

Traub, H. P. 1962. Classification of Amaryllidaceae. Pl. Life 18:55–72.

Traub, H. P., T. R. Robinson, and H. E. Stevens. 1942. Papaya production in the United States. Circ. U.S.D.A. 633:1–36.

Tryon, R. 1964. The ferns of Peru—Polypodiaceae (Dennstaedtieae to Oleandreae). Contr. Gray Herb. 194:1–253.

Uittien, H. 1937. Sapindaceae, in A. Pulle, Flora of Suriname, Kol. Inst. Amsterdam Medel. 30. Vol. 2(1):380.

Underwood, L. M. 1907. American ferns, VIII. A preliminary review of the North American Gleicheniaceae. Bull. Torrey Bot. Club 34:243–62.

U.S. Army Tropic Test Center. 1966. Environmental data base for regional studies in the humid tropics. Semiannual Report #1,2. Fort Clayton, Canal Zone.

———. 1967. Semiannual Report #3.

Vattimo, I. de. 1971. Contribuição ao conhecimento da tribo Apodanthea R. Br. Parte I: Conspecto das especies (Rafflesiaceae). Rodriguésia 26:37–73.

Vogel, S. 1958. Fledermausblümen in Südamerika. Oesterr. Bot. Z. 104:491–530.

Webster, G. L., and D. Burch. 1967. Euphorbiaceae, in R. E. Woodson, Jr., and R. W. Schery, Flora of Panama. Ann. Missouri Bot. Gard. 54:211–350.

Wessels-Boer, J. G. 1965. Palmae, in A. Pulle, Flora of Suriname 5(1):1–172.

———. 1968. The geonomoid palms. Meded. Bot. Mus. Herb. Rijks Univ. Utrecht 282:1–202.

Wheeler, W. M. 1910. Ants. New York: Columbia University Press. 663 pp.

White, F. 1977. Ebenaceae, in R. E. Woodson, Jr., and R. W. Schery, Flora of Panama. Ann. Missouri Bot. Gard. 65:145–54.

Williams, E. C., Jr. 1941. An ecological study of the floor fauna of the Panama rain forest. Chicago Acad. Sci. Bull. 6(4):63–124.

Williams, L. O. 1973. Geophila (Rubiaceae) in North America. Phytologia 26(4):263–64.

Williams, N., and C. H. Dodson. 1972. Selective attraction of male euglossine bees to orchid floral fragrances and its importance in long distance pollen flow. Evolution 26:84–95.

Willis, E. O. 1974. Populations and local extinctions of birds on Barro Colorado Island, Panama. Ecol. Monogr. 44:153–69.

Willis, J. C. 1966. A dictionary of flowering plants and ferns. 7th ed. Cambridge, Mass.: Cambridge University Press. 1,214 pp.

Wilson, D. E. 1971. Food habits of Micronycteris hirsuta (Chiroptera: Phyllostomidae). Extrait de Mammalia 35(1):107–10.

Woodring, W. P. 1958. Geology of Barro Colorado Island. Contr. Smithsonian Misc. Coll. 135(3). 39 pp.

Woods, F. W., and C. M. Gallegos. 1970. Litter accumulation in selected forests of the Republic of Panama. Biotropica 2(1):46–50.

Woodson, R. E., Jr., and R. W. Schery. 1944. Commelinaceae, in Flora of Panama. Ann. Missouri Bot. Gard. 33:138–51.

———. 1945a. Musaceae, in Flora of Panama. Ann. Missouri Bot. Gard. 32:48–57.

———. 1945b. Marantaceae, in Flora of Panama. Ann. Missouri Bot. Gard. 32:81–105.

———. 1948. Capparidaceae, in Flora of Panama. Ann. Missouri Bot. Gard. 35:75–99.

———. 1950. Leguminosae (Mimosoideae), in Flora of Panama. Ann. Missouri Bot. Gard. 37:170–314.

———. 1958a. Lecythidaceae, in Flora of Panama. Ann. Missouri Bot. Gard. 45:115–36.

———. 1958b. Passifloraceae, in Flora of Panama. Ann. Missouri Bot. Gard. 45:1–22.

———. 1960. Moraceae, in Flora of Panama. Ann. Missouri Bot. Gard. 47:114–78.

———. 1961. Nyctaginaceae, in Flora of Panama. Ann. Missouri Bot. Gard. 48:51–65.

Yazquez-Yanes, C., A. Orozco, G. Francois, and L. Trejo. 1975. Observations on seed dispersal by bats in a tropical humid region in Veracruz, Mexico. Biotropica 7(2):73–76.

Yuncker, T. G. 1950. Piperaceae, in R. E. Woodson, Jr., and R. W. Schery, Flora of Panama. Ann. Missouri Bot. Gard. 37:1–120.

Index of Common Names

Index of Botanical Names

Names set in large and small capitals are of taxa at levels above genus. Names set in italic type are treated here as synonyms or are otherwise not recognized. Italic numerals indicate pages that carry photographs. Boldface numerals indicate pages on which principal treatments begin; for each family, the first boldface numeral indicates the page in the Introduction that lists taxonomic placement and numbers of taxa treated. The word *passim* following a given span of pages indicates several separate appearances within that span.